FUSION TECHNOLOGY 1996

FUSION TECHNOLOGY 1996

Proceedings of the 19th Symposium on Fusion Technology,
Lisbon, Portugal, 16-20 September 1996

VOLUME 1

edited by:

C. VARANDAS

and

F. SERRA

Centro de Fusão Nuclear
Instituto Superior Técnico
Lisboa, Portugal

1997

ELSEVIER
AMSTERDAM • LAUSANNE • NEW YORK • OXFORD • SHANNON • TOKYO

ELSEVIER SCIENCE B.V.
Sara Burgerhartstraat 25
P.O. Box 211, 1000 AE Amsterdam, The Netherlands

Pages 1-176 are reprinted from
Fusion Engineering and Design, Volume 36/1 (1997)

ISBN: 0 444 82762 5

©1997 ELSEVIER SCIENCE B.V. All rights reserved.

No part of this publication may be reproduced, stored in a retrieval system or transmitted in any form or by any means, electronic, mechanical, photocopying, recording or otherwise, without the prior written permission of the publisher, Elsevier Science B.V., Copyright & Permissions Department, P.O. Box 521, 1000 AM Amsterdam, The Netherlands.

Special regulations for readers in the U.S.A. - This publication has been registered with the Copyright Clearance Center Inc. (CCC), 222 Rosewood Drive, Danvers, MA 01923. Information can be obtained from the CCC about conditions under which photocopies of parts of this publication may be made in the U.S.A. All other copyright questions, including photocopying outside of the U.S.A., should be referred to the copyright owner, Elsevier Science B.V., unless otherwise specified.

No responsibility is assumed by the publisher for any injury and/or damage to persons or property as a matter of products liability, negligence or otherwise, or from any use or operation of any methods, products, instructions or ideas contained in the material herein.

This book is printed on acid-free paper.

Printed in The Netherlands

PREFACE

The Symposium on Fusion Technology (SOFT) is held every two years with the objective to set the stage for the exchange of information on the design, construction and operation of fusion experiments and on the technology which is being developed for the next step devices and fusion reactors.

The 19th Symposium was held at Culturgest, in Lisbon, on September 16—20, 1996. It was hosted and organized by "Centro de Fusão Nuclear" of "Instituto Superior Técnico" on behalf of the Association EURATOM/IST.

By decision of the International Organizing Committee SOFT-96 has included invited talks, oral contributions and posters in the following topics:

A. First Wall, Divertors and Vacuum Systems
Plasma facing components; wall conditioning; first wall materials.

B. Plasma Heating and Control
Neutral beams, RF systems; current drive; related power supplies.

C. Plasma Engineering and Control
Interaction between plasma and machine components.

D. Experimental Systems
Design and operation of complete systems; diagnostics; data acquisition and control systems; future machines.

E. Magnet and Related Power Supplies

F. Fuel Cycle and Tritium Processing Systems
Fueling and pellet injection, exhaust gas pumping and purification; tritium separation, storage and extraction.

G. Blanket Technology/Materials
Irradiation sources; neutronics; hydraulics; breeder, multiplier and structural materials; shielding.

H. Assembly, Remote Handling and Waste Management and Storage

I. Safety and Environment, Reactor Studies

A total of about 580 papers were submitted for poster presentation. Due to poster space limitations and taking into account their scientific contents, some papers had to be rejected and others had to be rewritten or combined. In the end, 470 papers were accepted, of which 430 were presented at the conference (420 as posters and 10 as oral contributions). Of those, 413 were submitted as author-prepared, camera-ready manuscripts, which are now published in these Proceedings. Table 1 contains the distribution of these papers by the Symposium topics. The editors wish to point out that, even though some editorial improvements were suggested and layout and technical improvements were carried out on some papers, the authors are themselves responsible for the final quality and content of the contributions. The invited papers, reprinted from a special issue of the Journal of Fusion Engineering and Design published after a standard review process, are also included in these Proceedings.

Topic	No.	Topic	No.
A	72	F	43
B	48	G	81
C	22	H	22
D	51	I	41
E	33		

Table 1 — Number of published paper per symposium topic.

Simultaneously with SOFT-96, an exhibition of industrial companies active in fields associated with nuclear fusion (see list of Industrial Exhibitors) was organized.

The International and Local Organizing Committees would like to thank the financial support received from the following Institutions and Companies:

- The Commission of the European Community

- Caixa Geral de Depósitos

- Junta Nacional de Investigação Científica e Tecnológica

- Europa Metalli Spa

- Consortium EFET

- Consortium AGAN

- Air Liquide

Finally, the editors wish success to the organizers of the next Symposium, which will be held in France in the Autumn of 1998.

C. Varandas
F. Serra

COMMITTEE MEMBERS

International Organizing Committee

E. Bertolini	JET, Abingdon, U.K.
H. Conrads	KFA, Julich, Germany
U. Finzi	CEC, Brussels, Belgium
P. Fenici	JRC, Ispra, Italy
M. Gasparotto	ENEA, Frascati, Italy
G. Rostagni	CNR, Padova, Italy
E. Salpietro	NET, Garching, Germany
G. Tonon	CEA, Cadarache, France
C. Varandas (Chairman)	IST, Lisboa, Portugal
G. Vecsey	CRPP, Lausanne, Switzerland
J.E. Vetter	KfK, Karlsruhe, Germany
R. Wilhelm	IPP, Garching, Germany

Local Organizing Committee

J.A.C. Cabral (Chairman)
F. Serra (Scientific Secretary)
C. Vitorino (Administrative Secretary)
M. Fernanda
A. Moreira
A. Soares

INDUSTRIAL EXHIBITORS

- **Dutch Scientific**
  Apeldoorn, The Netherlands

- **Consortium AGAN (Ansaldo, GEC Alsthom, Preussag Noell, ACCEL)**
  Wurzburg, Germany

- **Canadian Fusion Fuels Technology Project**
  Mississauga, Ontario, Canada

- **EFET (Ivo, Belgatom, Citif, Framatome S.A., Ibertef A.I.E., NNC LTD., Siemens AG)**
  Erlangen, Germany

- **JRC-IAM**
  Petten, The Netherlands

- **Europa Matalli Spa**
  Firenze, Italy

- **Air Líquide**
  Sassenage, France

- **SEP**
  Vernon, France

- **IST/OGMA**
  Lisboa, Portugal

TABLE OF CONTENTS

Volume 1

Preface	v
Committee members	vii
Industrial exhibitors	viii

Invited Talks

Nuclear fusion, an energy source
R. Toschi — 1

ITER overview
R. Aymar — 9

The ITER magnet system
M. Huguet — 23

Design of in-vessel components for ITER
R.R. Parker — 33

Management of waste from the International Thermonuclear Experimental Reactor and from future fusion power plants
K. Brodén, M. Lindberg, S. Nisan, P. Rocco, M. Zucchetti, N. Taylor, C. Forty — 49

Status of European breeding blanket technology
L. Giancarli, M. Dalle Donne, W. Dietz — 57

Recent results on vacuum pumping and fuel clean-up
R.-D. Penzhorn, D. Murdoch, R. Haange — 75

Remote handling on fusion experiments
A.C. Rolfe — 91

Possible divertor solutions for a fusion reactor. Part I. Physical aspects based on present day divertor operation
A. Kallenbach, H.-S. Bosch, S. de Peña Hempel, R. Dux, M. Kaufmann, V. Mertens, J. Neuhauser, W. Suttrop, H. Zohm — 101

Possible divertor solutions for a fusion reactor. Part II. Technical aspects of a possible divertor
Ph. Chappuis, F. Escourbiac, M. Lipa, R. Mitteau, J. Schlosser — 109

Engineering experience in JET operations
E. Bertolini — 119

D-T operation on TFTR
M.D. Williams
135

Development of high current negative ion source
T. Kuroda
143

Advances in techniques for diagnosing fusion plasmas
P.E. Stott
157

On-site developed components for control and data acquisition on next generation fusion devices
C.A.F. Varandas, B.B. Carvalho, C. Correia, C. Loureiro, J.L. Malaquias, A.P. Rodrigues, J. Sousa
167

Oral Presentations

Industrial experience in manufacturing the superconducting CIC conductors in EU for the ITER CS model coil
G. Bevilacqua, E. Salpietro, H. Krauth, A. Szulczyk, M. Thoener, S. Angius, P. Gagliardi, A. Laurenti, R. Garré, S. Rossi, M.V. Ricci, M. Spadoni
179

The ITER-QUELL, a QUENCH propagation experiment on long length CICC with central channel
A. Anghel, Y. Takahashi, S. Smith, S. Pourrahimi, M. Zhelamskij, B. Blau, A. Fuchs, B. Heer, K. Hamada, H. Fujisaki, C. Marinucci, G. Vecsey
185

The plasma shield in ITER plasma wall interactions
H. Würz, B. Bazylev, F. Kappler, I. Landman, S. Pestchanyi, G. Piazza
191

In-situ repair concepts for the ITER first wall components
R. Matera, S. Chiocchio, G. Federici, K. Ioki, R. Raffray, A. Cardella, A. Lodato, F. Saint-Antonin, G. Le Marois, R. Castro, M. Akiba, S. Suzuki, D. Karpov, L. Moreschi
197

High performance results with the LHCD system on TORE-Supra and new Launcher design for quasi continuous operation
P. Froissard, P. Bibet, G. Agarici, S. Berio, C. Deck, L. Garampon, M. Goniche, D. Guilhem, P. Hertout, T. Hoang, J.Y. Journeaux, F. Kazarin-Vibert, X. Litaudon, G. Martin, M. Mattioli, Y. Peysson, C. Portafaix, G. Rey, F. Surle, M. Tareb, G. Tonon, J.G. Wegrowe
203

Mathematical model and results of calculations for poloidal magnetic field system and stress analysis for toroidal field system of the TEXTOR 94
I. Baturo, N. Berkhov, H. Bohn, E. Bondarchuk, V. Filatov, B. Giesen, N. Doinikov, M. Korol'kov, B. Kitaev, N. Kozhukhovskaja, N. Maximenkova, B. Mingalev, O. Neubauer, T. Obidenko, A. Panin, A. Simakov
209

Research progress on JT-60U in advanced steady-state operation
O. Naito (for the JT-60 Team)
215

An overview of the DIII-D program — 221
J.L. Luxon for the DIII-D Team

Commissioning tests and enhancements to the JET active gas handling plant — 227
R. Lässer, A. Bell, N. Bainbridge, D. Brennan, B. Grieveson, J.L. Hemmerich, G. Jones, S. Knipe, J. Lupo, J. Mart, A. Perevezentsev, N. Skinner, R. Stagg, K. Walker, R. Warren, J. Yorkshades

Neutronics shield experiment for ITER at the Frascati Neutron Generator FNG — 233
R. Batistoni, M. Angelone, W. Daenner, U. Fischer, L. Petrizzi, M. Pillon, A. Santamarina, K. Seidel

Contributed Papers

Section A: First Wall, Divertors and Vacuum Systems

Experimental validation of the liquid lithium divertor concept — 243
N.V. Antonov, V.B. Petrov, A.S. Pleshakov, A.S. Rupyshev, Y.A. Sokolov, B.I. Khripunov, V.V. Shapkin, A.V. Vertkov, V.A. Evtikhin, I.E. Lublinsky, V.I. Pistunovich, L.G. Golubchikov

Feasibility of liquid gallium cooling for ITER divertor cassette — 247
P.V. Romanov, Yu. Shpanskij, A.V. Klischenko, V.S. Petrov, S.A. Moshkin, A.V. Beznosov

Review of liquid metal divertor concepts for Tokamak reactors — 251
I.R. Kirillov, E.V. Muraviev

Helium retention of B_4C or SiC converted graphite — 255
T. Hino, Y. Yamauchi, Y. Hirohata, K. Mori

ITER outboard Baffle design and small-scale mock-ups test programme — 259
L. Giancarli, B. Bielak, P. Chappuis, C. Cheron, G. Le Marois, Y. Poitevin, J. Quintric-Bossy, J.-F. Salavy, Y. Severi, J. Szczepanski

Comparison between various thermal hydraulic tube concepts for the ITER divertor — 263
I. Smid, J. Schlosser, J. Boscary, F. Escourbiac, G. Vieider

ITER Baffle small-size mock-ups fabrication and testing — 267
G. Le Marois, F. Bernier, D. San Filippo, F. Saint-Antonin, F. Moret

High heat flux performance of divertor modules for ITER with beryllium and carbon armor — 271
J. Linke, G. Breitbach, R. Duwe, A. Gervash, M. Rödig, B. Wiechers

Manufacture and testing of small-scale mock-ups for the ITER divertor — 275
G. Vieider, P. Chappuis, R. Duwe, R. Jakeman, M. Merola, H.D. Pacher, L. Plöchl, F. Rainer, M. Rödig, N. Reheis, I. Smid

Engineering design of the ITER divertor
R. Tivey, M. Akiba, A. Antipenkov, S. Chiocchio, D. Driemeyer, R. Jakeman,
G. Janeschitz, I. Mazul, M. Ulrickson, G. Vieider 279

Manufacturing feasibility study for the ITER divertor vertical target and wing
G. Cai, M. Grattarola, G.B. Ottonello, F. Zacchia 283

Thermal and structural analysis of the ITER divertor outboard channel
B. Riccardi, A. Pizzuto, R.A. Palmieri 287

The outline design of the ITER divertor outboard vertical target dump plate and wing
A. Pizzuto, G. Vieider, M. Baldarelli, G. Brolatti, B. Riccardi, M. Roccella, I. Smid 291

Development and construction of a high efficiency divertor mock-up using a semi-solid thixotropic alloy as thermal bond layer
L.F. Moreschi, E. Visca, S. Storai, M. Sacchetti, F. Salvi 295

Neutron exposure characterisation of ITER divertor materials for source terms evaluation
D.G. Cepraga, G. Cambi, L. Di Pace, M. Vaccari 299

Thermal analysis of sacrificial components
F. Andritsos, A. Inzaghi 303

Development of plasma-facing components for fusion experimental reactors at JAERI
M. Akiba, M. Araki, K. Nakamura, S. Satoh, S. Suzuki, M. Dairaku, K. Yokoyama, Y. Ohara 307

Boiling crisis at heat transfer intensification on divertor surfaces under resistive and e-stream heating condition
V.A. Divavin, S.A. Grigoriev, A.V. Lipko 311

Three dimensional neutronic and activation analysis for the ITER design
J.-Ch. Sublet 315

Creep-fatigue damage in OFHC coolant tubes for plasma facing components
E.E. Reis, R.H. Ryder 319

Radiation-induced desorption of hydrogen and deuterium from candidate plasma-facing materials
R.G. Macaulay-Newcombe, D.A. Thompson 323

Evaluation of an advanced silicon doped CFC for plasma facing material
C.H. Wu, C. Alessandrini, P. Bonal, A. Caso, H. Grote, R. Moormann, A. Perujo, M. Balden, H. Werle, G. Vieider 327

Fracture mechanics investigations of molybdenum alloys
M. Rödig, H. Derz, G. Pott, B. Werner 331

Oxidation of carbon-based plasma facing materials in accident-like temperature transients
H.-K. Hinssen, M. Hofmann, A.-K. Krüssenberg, R. Moormann, C.H. Wu — 335

Thermal fatigue behaviour of first wall mock-ups with defects
G. Hofmann, E. Diegele, M. Kamlah, S. Müller — 339

Development of plasma facing components with ceramic armour
R. Vaßen, A. Yehia, R. Lison, R. Duwe, D. Stöver — 343

Materials aspects of ITER in-vessel components
V. Barabash, J. Dietz, K. Ioki, G. Janeschitz, G. Kalinin, R. Matera, K. Mohri and ITER Home Teams — 347

Analysis of plasma shield formation and graphite target erosion in disruption simulation experiments with high power quasistationary plasma streams
V.V. Chebotarev, I.E. Garkusha, V.V. Garkusha, V.A. Makhlaj, N.I. Mitina, S.E. Pestchanyi, D.G. Solyakov, V.I. Tereshin, S.A. Trubchaninov, A.V. Tsarenko, H. Würz — 351

Tribological behaviour of CFC material
A. Orsini, E. Di Pietro, S. Libera, L. Verdini, E. Visca, G. Vieider — 355

Development of 3D-based CFC with high thermal conductivity for fusion application
M. Araki, Y. Kude, Y. Sohda, K. Nakamura, K. Sato, S. Suzuki, M. Akiba — 359

Evaluation of hydrogen sorption and desorption for Ti-6Al-4V alloy as a vacuum vessel material
Y. Hirohata, Y. Aihara, T. Hino, N. Miki, S. Nakagawa — 363

Crack propagation test under cyclic thermal stress by irradiation of electron beam
H. Kogawa, S. Kimura, T. Teramoto, M. Saito — 367

Creep analysis of first wall in fusion reactor
K. Hatamoto, T. Teramoto, M. Saito — 371

Behaviour of sub-surface defects under thermal fatigue loading in a 316L first wall mock-up
F. Sevini, M. Merola — 375

Erosion damage of nearby plasma-facing components during a disruption on the divertor plate
A. Hassanein, I. Konkashbaev — 379

The study of CVD thick B_4C and SiC coatings on graphites of different types in DIII-D divertor
O.I. Buzhinskij, I.V. Opimach, V.A. Barsuk, P.W. West, N. Brooks, D. Whyte — 383

Mechanical study of JET MKI beryllium tiles after ELMs and deliberate melting
E.B. Deksnis, A. Chankin, F. Hurd, W. Parsons, B. Tubbing — 387

Design, fabrication and testing of helium-cooled vanadium module for fusion applications
C.B. Baxi, E. Chin, B. Laycock, W.R. Johnson, R.J. Junge, E.E. Reis, J.P. Smith — 391

Characterization of copper-stainless steel EXW joints
S. Tähtinen, P. Moilanen, P. Karjalainen-Roikonen — 395

Development and characterization of Be/Cu joint obtained by hot isostatic pressing
F. Saint-Antonin, D. Barberi, G. Le Marois, A. Laillé — 399

Development of mechanically joined divertor plate for large helical device
Y. Kubota, N. Noda, A. Sagara, Y. Kato, T. Morisaki, N. Ohyabu, O. Motojima — 403

Reactivity test between beryllium and dispersion strengthened copper
N. Sakamoto, H. Kawamura, R.R. Solomon — 407

Fabrication of ITER first wall modules from powder by hot isostatic pressing
A. Lind, J. Collén — 411

Development of silver-free bonding techniques and an investigation of the effect of cadium in brazed joints made with silver based brazing alloys
A.T. Peacock, M.R. Harrison, D.M. Jacobson, M. Pick, S.P.S. Sangha, G. Vieider — 415

Thermal fatigue test of a Cu/ss primary wall mock-up
M. Merola, P. Fenici, R. Scholz, B. Weckermann, S. Tähtinen — 419

Technologies of joining between ITER reference grade beryllium and copper alloys by diffusion bonding process
E. Visca, G. Ceccotti, B. Riccardi, G. Mercurio — 423

In-vessel cryo pump for the ASDEX Upgrade divertor II
B. Streibl, S. Deschka, O. Gruber, B. Jüttner, P. Lang, K. Mattes, G. Pautasso, J. Perchermeier, K. Schippl, H. Schneider, U. Seidel, W. Suttrop, G. Teller, M. Weissgerber — 427

The model and experimental basis for the design parameters of the JET divertor cryopump protection system including variations in divertor geometry and first wall materials
P. Ageladarakis, S. Papastergiou, D. Stork, H. van der Beken — 431

Upgrading of the TORE Supra ergodic divertor
L. Doceul, A. Grosman, P. Deschamps, B. Bertrand, J.P. Cocat, J.J. Cordier, L. Garampon, L. Gargiulo, Ph. Ghendrih, M. Lipa, R. Mitteau, H. Viallet — 435

Development and fabrication of improved CFC-brazed components for the inner first wall of TORE Supra
M. Lipa, Ph. Chappuis, G. Chaumat, D. Guilhem, R. Mitteau, L. Ploechl — 439

Non destructive testing of actively cooled plasma facing components by means of thermal transient excitation and infrared imaging
R. Mitteau, S. Berrebi, P. Chappuis, Ph. Darses, A. Dufayet, L. Garampon, D. Guilhem, M. Lipa, V. Martin, H. Roche — 443

Developments for an actively cooled toroidal pump limiter for TORE Supra
J. Schlosser, Ph. Chappuis, L. Doceul, M. Chatelier, J.P. Cocat, L. Garampon, P. Garin, R. Mitteau, A. Moal, L. Plöchl, G. Tonon, E. Tsitrone — 447

The TORE Supra vented limiter: an alternative concept for heat and particle exhaust
E. Tsitrone, T. Loarer, B. Pégourié, H. Roche, P. Chappuis, D. Guilhem — 451

Divertor II plasma facing components for ASDEX Upgrade
S. Deschka, S. Schweizer, B. Streibl, ASDEX Upgrade Team, C. Garcia-Rosales, G. Hofmann, J. Linke — 455

Actively cooled test limiter for TEXTOR
W. Hohenauer, H. Bolt, T. Koppitz, J. Linke, R. Lison, W. Malléner, V. Philipps, M. Sauer, R. Uhlemann, J.H. You, H. Nickel — 459

Divertor engineering for the Stellarator Wendelstein W 7-X
H. Greuner, H. Bitter, R. Holzthüm, O. Jandl, F. Kerl, J. Kisslinger, H. Renner — 463

Design, manufacturing and testing of the W 7-X target element plate prototypes
H. Greuner, T. Huber, J. Kisslinger, E. Parteder, L. Plöchl, H. Renner — 467

Design of a compact W-shaped pumped divertor in JT-60U
S. Sakurai, N. Hosogane, K. Kodama, K. Masaki, T. Sasajima, K. Kishiya, S. Tsurumi, S. Takahashi, M. Saidoh — 471

Structural design of the DIII-D radiative divertor
E.E. Reis, J.P. Smith, C.B. Baxi, A.S. Bozek, E. Chin, M.A. Hollerbach, G.J. Laughon, D.L. Sevier — 475

Fabrication development and usage of vanadium alloys in DIII-D
J.P. Smith, W.R. Johnson, E.E. Reis — 479

The MkII gas box divertor — A new design concept
H. Altmann, E. Deksnis, C. Froger, S. Lawson, C. Lowry, A. Peacock, M. Pick — 483

Advances in porous media heat exchangers for fusion applications
J.H. Rosenfeld, J.E. Lindemuth, M.T. North, R.D. Watson, D.L. Youchison, R.H. Goulding — 487

Evolution of framatone and CEA high thermal flux station for fusion technology experiments needs
M. Diotalevi, M. Febvre, P. Chappuis — 491

Xenon pellet production and acceleration in a pipe-gun
P.V. Reznichenko, I.V. Viniar, B.V. Kuteev — 495

Disruption simulations on tungsten specimens in a plasma accelerator
A. Gervash, E. Wallura, I. Ovchinnikov, A.N. Makhankov, J. Linke, G. Breitbach — 499

Void swelling in proton irradiated Fe-Cr-Ni ternary alloys
Y. Murase, N. Yamamoto, J. Nagakawa, H. Shiraishi — 503

Study of plasma target interactions with plasma streams of power density of 40 MW/cm^2
N.I. Arkhipov, V.P. Bakhtin, S.M. Kurkin, S.E. Pestchanyi, V.M. Safronov, D.A. Toporkov, S.G. Vasenin, H. Würz, A.M. Zhitlukhin — 507

Post-mortem analysis of HIP bonded first wall panel made from SS316 and DS-Cu after high heat flux testing
T. Hatano, K. Fukaya, M. Dairaku, T. Kuroda, H. Takatsu — 511

Ohmic baking system upgrade for wall conditioning of Tokamak-15 discharge chamber
V.N. Garnov, S.V. Kabanovsky, V.A. Khrabrov, P.P. Khvostenko, V.A. Kochin, A.I. Nikonorov, P.N. Orlov, I.A. Posadsky, A.N. Vertiporokh — 515

GlidCop® DSC extruded plus cross-rolled plate tensile properties in the temperature range of 20–500°C
R.R. Solomon, J.D. Troxell, A.V. Nadkarni — 519

Section B: Plasma Heating and Control

The 110 GHz ECR system for TdeV
R. Magne, G.A. Chaudron, R. Cool, Y. Demers, Ph. Cumyn, R. Décoste, A. Dubé, J.M. Guay, D. Larose, A. Robert, C. Trudel, L. Vachon — 525

Low power radiation field pattern measurements of main components of the TORE Supra ECRH transmission line
G. Berger-By, J.P. Crenn, R. Levy, P. Garin, D. Roux, M. Pain — 529

Status of the 118 GHz-Quasi CW gyrotron for the TORE Supra and TCV tokamaks
M. Pain, P. Garin, M.Q. Tran, S. Alberti, M. Thumm, O. Braz, E. Giguet, P. Thouvenin, C. Tran — 533

The 140 GHz-2MW-2 s ECRH system for ASDEX-Upgrade
H. Brinkschulte, A. Fix, W. Förster, G. Gantenbein, W. Kasparek, F. Leuterer, F. Monaco, M. Münich, A. Peeters, F. Ryter, P. Schüller, V. Sigalaev, E. Tai — 537

Conceptual design of the 140 GHz/10MW CW ECRH system for the stellarator W7-X
L. Empacher, W. Förster, G. Gantenbein, W. Kasparek, H. Kumrić, G.A. Müller, P.G. Schüller, K. Schwörer, U. Schumacher, D. Wagner, V. Erckmann, T. Geist, H. Laqua — 541

Advanced high power gyrotrons for ECW application
B. Piosczyk, E. Borie, O. Braz, G. Dammertz, C.T. Iatrou, S. Illy, S. Kern, M. Kuntze,
M.V. Kartikeyan, G. Michel, A. Möbius, M. Thumm
   545

Development of 170GHz long pulse gyrotron with depressed collector
A. Kasugai, K. Sakamoto, M. Tsuneoka, K. Takahashi, S. Maebara, T. Imai, T. Kariya,
K. Hayashi, Y. Mitsunaka, Y. Hirata
   549

Development of an ECRH system for large helical device
T. Shimozuma, M. Sato, Y. Takita, S. Kubo, H. Idei, K. Ohkubo, T. Watari, S. Yasutomi,
Y. Suzuki, F. Saito, S. Sasaki, Y. Saito, T. Okamoto
   553

New developments in the 110 GHz system on the RTP tokamak
O.G. Kruijt, W.A. Bongers, A.B. Sterk, F.J. van Amerongen, A. Bijker, G.G. Denisov,
W. Kooijman, S. Kuyvenhoven, G. Land, A. Montvai, A.A.M. Oomens, R.W. Polman,
F.C. Schüller, A.G.A. Verhoeven, D. Vinogradov and the RTP team
   557

Present status of the 1MW, 130–260 GHz free electron maser
A.B. Sterk, B.S.Q. Elzendoorn, W.A. Bongers, P. Manintveld, P.R. Prins,
A.G.A. Verhoeven, W.H. Urbanus, the VE team and the FEM team
   561

Design and installation of the electron cyclotron wave system for the TCV tokamak
T.P. Goodman, S. Alberti, M.A. Henderson, A. Pochelon, M.Q. Tran
   565

Design and operation of the power installation for the TCV ECR additional heating
D. Fasel, J. Alex, A. Favre, T. Goodman, M. Henderson, P.-F. Isoz, A. Perez,
M-Q. Tran
   569

High-power corrugated waveguide components for mm-wave fusion heating systems
R.A. Olstad, J.L. Doane, C.P. Moeller, R.C. O'Neill, M. DiMartino
   573

ICRF power feedback regulation of the plasma diamagnetic energy content of the
tokamak TEXTOR
C. Königs, F. Durodié, M. Vervier
   577

Generating system for Alfvén waves in TCA/BR
L. Ruchko, R. Galvão, I.C. Nascimento, E. Ozono, F. Degasperi, E. Lerche
   581

Negative ion extraction physics
M. Bacal, F. El Balghiti-Sube, A.A. Ivanov, A.B. Sionov
   585

Effect of cesium seeding in volume negative ion sources
M. Bacal, F. El Balghiti-Sube, L.I. Elizarov, A.J. Tontegode
   589

First results of automatic matching system on Tore Supra ICRH antennas: fast
matching network for ICRH systems
L. Ladurelle, G. Agarici, B. Beaumont, S. Bremond, H. Kuus, G. Lombard
   593

ICRF operation during H-mode with ELMs development status at ASDEX Upgrade
F. Wesner, W. Becker, F. Braun, H. Faugel, R. Fritsch, F. Hofmeister,
J.-M. Noterdaeme, Th. Sperger — 597

An ARC detection system for ICR-heating
F. Braun, Th. Sperger — 601

Matching fast ICRF antenna coupling variations by frequency change
F. Hofmeister and ICRH team — 605

Progress on radio frequency auxiliary heating system designs in ITER
M. Makowski, G. Bosia, F. Elio and the Home Teams — 609

The 1.8 MW-433 MHz RF system for the ion Bernstein waves experiments on FTU
P. Papitto, R. Cesario, A. Cucchiaro, A. Marra, G.L. Ravera, P. Zampelli — 613

Achievment of 1.6MW/5000sec operation of RF oscillator on the large helical device
R. Kumazawa, T. Mutoh, T. Watari, T. Seki, F. Shinbo, G. Nomura, S. Masuda,
T. Ido, T. Kuroda — 617

Operation of high power ICRH with ELMY plasmas at JET
M. Schmid, V. Bhatnagar, C. Gormezano, P.U. Lamalle, A. Sibley, M. Simon,
M. Timms, T. Wade — 621

Improved tuning and matching of ion cyclotron systems
D.W. Swain, R.H. Goulding, F.W. Baity, R.I. Pinsker, J.S. deGrassie, C.C. Petty,
D.J. Hoffman — 625

Development of fast wave systems tolerant of time-varying loading
R.I. Pinsker, C.P. Moeller, C.C. Petty, D.A. Phelps, T. Ogawa, Y. Miura, H. Ikezi,
J.S. deGrassie, R.W. Callis, R.H. Goulding, F.W. Baity — 629

Flexible $N_{//}$ Launcher for lower hybrid wave injection in the first operation phase of ITER
G. Rey, P. Bibet, P. Froissard, J.G. Wegrowe, S. Bério, M. Goniche, G.T. Hoang,
F. Kazarin-Vibert, X. Litaudon, M. Tareb, G. Tonon — 633

Development of a new lower hybrid antenna module using a poloidal power divider
S. Maebara, M. Seki, K. Suganuma, T. Imai, M. Goniche, Ph. Bibet, S. Bério,
J. Brossaud, G. Rey, G. Tonon — 637

Upgrading of the injection power of the W7-AS neutral beam system
W. Ott, F.-P. Penningsfeld, W. Melkus, F. Probst, E. Speth, R. Süss and
W7-AS Team — 641

Development of a novel compact RF-source
B. Heinmann, J.-H. Feist, W. Kraus, F. Probst, R. Riedl, E. Speth, M. Busch,
W. Szcepaniak — 645

Optimization of a large-area RF plasma generator
W. Kraus, J.-H. Feist, E. Speth
649

Modelling of the TEXTOR-94 neutral-injection power supply system in the search of parasitic currents
C.-C. Hering, U. Braunsberger, M. Sauer, W. Schalt
653

Neutral beam injector for steady state superconducting tokamak
A.K. Chakraborty, N. Bisai, M.R. Jana, P.K. Jayakumar, U.K. Baruah, P.J. Patel, K. Rajasekar, S.K. Mattoo
657

Beam properties of the enhanced JET PINIs
D. Ciric, C.D. Challis, H.P.L. de Esch, H.-D. Falter, D. Godden, A.J.T. Holmes, D. Stork, E. Thompson
661

Upgrading of the neutral beam power supplies from 80KV/60A to 140KV/60A
F. Jensen, R. Claesen, H. McBryan, J. Mills, R. Öström, A.P. Vadgama
665

Operations with tritium neutral beams on TFTR
L.R. Grisham, J. Kamperschroer, T. O'Connor, M.E. Oldaker, T.N. Stevenson, A. von Halle and TFTR Group
669

The BEL: a test bed for the positive ions based neutral beam injectors with energy recovery system for TORE Supra
P. Bayetti, F. Bottiglioni, H. Dougnac, F. Imbeaux, J.Y. Journeaux, Ph. Lotte, G. Mayaux
673

Negative D^- ion source relevant for application to ITER neutral beam injectors
C. Jacquot, D. Riz, R. Trainham, K. Miyamoto, Y. Okumura, M.B. Hopkins
677

Extraction system development and stray electron study for high intensity negative ion based NBI
A. Simonin, K. Miyamoto, Y. Fujiwara
681

Results of the Cadarache 1 MeV D^- "SINGAP" experiment
J. Bucalossi, C. Desgranges, M. Fumelli, P. Massmann, J. Paméla, A. Simonin
685

Development of a large RF-driven negative ion source for neutral beam injector
T. Takanashi, Y. Takeiri, O. Kaneko, Y. Oka, K. Tsumori, M. Osakabe, T. Kuroda
689

Initial beam operation of 500KeV negative-ion based NBI system for JT-60U
M. Kuriyama, N. Akino, T. Aoyagi, N. Ebisawa, Y. Fujiwara, N. Isozaki, A. Honda, T. Inoue, T. Itoh, M. Kawai, M. Kazawa, J. Koizumi, K. Miyamoto, N. Miyamoto, K. Mogaki, Y. Ohara, T. Ohga, Y. Okumura, H. Oohara, K. Ohshima, F. Satoh, H. Seki, T. Takenouchi, Y. Toyokawa, K. Usui, K. Watanabe, M. Yamamoto, T. Yamazaki
693

General design of the neutral beam injection system and integration with ITER
A. Krylov, E. Di Pietro, M. Hanada, R. Hemsworth, C. Holloway, S. Stoner, E. Alexandrov,
M. Barinov, E. Dlougach, V. Kulygin, V. Naumov, A. Panasenkov, V. Petrov, Y. Fujiwara,
T. Inoue, K. Miyamoto, N. Miyamoto, Y. Ohara, Y. Okumura, K. Shibata, M. Tanii,
K. Watanabe, J.-H. Feist, B. Heinemann, E. Kussel, P. Lotte, P. Massmann, J. Paméla,
M. Watson   697

Design and R&D of high power negative ion source/accelerator for ITER NBI
T. Inoue, Y. Okumura, Y. Fujiwara, K. Miyamoto, N. Miyamoto, Y. Ohara, K. Watanabe,
B. Heinemann, M. Tanii   701

Experimental results using the JET real time power control system
N.H. Zornig, H.E.O. Brelen, A. Browne, M.L. Browne, C. Gormezano, T. Dobbing,
J.A. How, F.A. Jensen, T.T.C. Jones, F.B. Marcus, Q.A. King, F.G. Rimini, J.A. Romero,
A.G.H. Sibley, F. Söldner, B.J.D. Tubbing   705

E.C.R.H. system for TJ-II experiment
R. Martin, K. Likin, A. Fernández, M. Sorolla, A. Sánchez, C. del Rió Bocio, N. Matveev   709

Sliding contact tests at high RF current under vacuum
G. Agarici, B. Beaumont, L. Ladurelle, G. Lombard, P. Mollard, H. Kuus   713

Section C: Plasma Engineering and Control

Loads on the ITER in-vessel components from electromagnetic transients
S. Chiocchio, K. Ioki, M. Araki, P. Barabaschi, J.B. Bialek, V. Kokotvov, M. Roccella,
R.S. Sayer, J. Wesley, D. Williamson   719

3-D electromagnetic model and electromagnetic analysis of the ITER in-vessel
components
M. Roccella, A. Capriccioli, M. Ferrari, M. Gasparotto, A. Pizzuto, S. Chiocchio,
F. Lucca   723

Modelling and analysis of plasma-first wall contact during vertical instabilities in
Next-Step tokamaks with a 3D eddy current code
S. Fantechi, Y. Crutzen   727

Control of the magnetic configuration in ITER
A. Portone, Y. Gribov, Y. Mitrishkin, P.L. Mondino, J. Wesley, R. Albanese, G. Ambrosino,
D. Ciscato, E. Coccorese, D. Humphreys, S. Jardin, A. Kavin, C. Kessel, J. Lister,
D. Pearlstein, A. Pironti, I. Senda, M. Walker, D. Ward   731

Modelling and engineering aspects of the plasma shape control in ITER
R. Albanese, G. Ambrosino, E. Coccorese, J.B. Lister, A. Pironti, D.J. Ward   735

Enhancement of the JET vacuum vessel supports
A. Miller, J.L. Hemmerich   739

Effects of high frequency disruptions on the JET divertor cryopump, including potential
JET toroidal field upgrades
S. Papastergiou, P. Ageladarakis — 743

Enhancement of JET machine instrumentation and coil protection systems
V. Marchese, T. Businaro, M. Buzio, E. De Marchi, N. Dolgetta, J. Howie, J. Last,
T. Raimondi, L. Scibile, J. van Veen — 747

Present understanding of electromagnetic behaviour during disruptions in JET
P. Noll, P. Andrew, M. Buzio, R. Litunovsky, T. Raimondi, V. Riccardo, M. Verrecchia — 751

Axisymmetric and non-axisymmetric structural effects of disruption-induced
electromechanical forces on the JET tokamak
M. Buzio, P. Noll, T. Raimondi, V. Riccardo, L. Sonnerup — 755

Scientific basis and engineering design to accommodate disruption and halo current
loads for the DIII-D tokamak
P.M. Anderson, A.S. Bozek, M.A. Hollerbach, D.A. Humphreys, J.L. Luxon, E.E. Reis,
M.J. Schaffer — 759

Effects of plasma behaviour on in-vessel components in JT-60 operation
H. Hiratsuka, T. Sasajima, K. Kodama, T. Arai, K. Masaki, Y. Neyatani, J. Yagyu,
A. Kaminaga, M. Saidoh — 763

Numerical and experimental evaluation of halo currents in RFX plasma facing
components
S. Peruzzo, A. Masiello, N. Pomaro, P. Sonato, G. Zollino — 767

Feasibility analysis of an active local field control at the poloidal gap of RFX
F. Bellina, G. Chitarin, P. Fiorentin, E. Gaio, G. Marchiori, P. Sonato, V. Toigo,
P. Zaccaria, G. Zollino — 771

Eddy currents in the RFX shell after short-circuiting one poloidal and one equatorial gap
G. Chitarin, G. Marchiori, A. Masiello, N. Pomaro, G. Zollino — 775

Plasma-material interaction during high transient heat loads in fast probe experiments
at TEXTOR
T. Scholz, A. Hassanein, H. Bolt, K.H. Finken, J. Linke — 779

Minimizing the first wall and blanket loading caused by plasma disruptions by
tuning first wall's resistivity
T. Jordan, D. Schneider — 783

Joining technology and material and shape optimization for the ICRF vacuum transmission
line dielectric window
P. Auerkari, L. Heikinheimo, J.A. Heikkinen, J. Linden, M. Kemppainen, K. Kotikangas,
S. Nuutinen, S. Orivuori, M. Peräniitty, S. Saarelma, M. Sirén, S. Tähtinen,
F. Wasastjerna, G. Bosia, E. Hodgson — 787

Design study on high Tc superconducting plasma stabilizer
T. Uchimoto, K. Miya — 791

Development of a precise long-time digital integrator for magnetic measurements in a tokamak
K. Kurihara, Y. Kawamata — 795

Section D: Experimental Systems

Simulation of the magnetic field penetration in the structure material of the dynamic ergodic divertor coils of TEXTOR-94
D. Styhler, M. Lindmayer, U. Braunsberger, B. Giesen, O. Neubauer — 801

Design of vacuum system and support structure for plasma facing components of SST1 tokamak
M.K. Bhise, P. Chaudhuri, D. Chenna Reddy, R. Gangradey, S. Jacob, S. Khirwadkar, C. Muralidaran, H.A. Pathak, E. Rajendra Kumar, T. Ranga Nath, N. Ravipragash, P. Sinha and SST1 Team — 805

Electromagnetics aspects in ITER: impact on magnet cryogenics, plasma diagnostics and control system
R. Albanese, G. Ambrosino, E. Coccorese, R. Fresa, R. Martone, C. Morabito, A. Pironti, G. Reitano, G. Rubinacci — 809

Plasma engineering in the IGNITOR experiment
B. Coppi, R. Andreani, C. Ferro, M. Gasparotto, C. Rita, A. Pizzuto, M. Roccella, G. Cenacchi, A. Bianchi, G. Galasso, L. Lanzavecchia — 813

Design characteristics of KT-2 tokamak for advanced tokamak operation
B.G. Hong, S.H. Jeong, S.K. Kim — 817

Numerical simulation of plasma equilibrium and shape control in tight tokamak GLOBUS-M
A.A. Kavin, V.A. Belyakov, S.A. Bulgakov, Y.A. Kostsov, E.N. Rumyntsev, S.A. Galkin, L.M. Degtyarev, A.A. Ivanov, Y.Y. Poshekhonov, V.A. Yagnov — 821

Thermal and structural analysis of vacuum and in-vessel components for low aspect ratio tokamak Globus-M
V.A. Divavin, S.A. Grigoriev, A.V. Lipko, V.A. Bykov, V.N. Komarov, E.G. Kuzmin, I.A. Mironov — 825

Globus-M tokamak magnets
A.B. Alekseev, A.F. Arneman, V.A. Belyakov, S.A. Boulgakov, V.A. Bykov, V.A. Divavin, K.E. Egorov, S.A. Grigoriev, V.K. Gusev, A.A. Kavin, V.A. Korotkov, Y.M. Krivchenkov, E.G. Kuzmin, A.A. Malkov, A.G. Panin, N.V. Sakharov, V.F. Soikin — 829

Joint upgrated spherical tokamak (JUST) and support installation — 2 MA spherical tokamak "Selena": concept and status
E.A. Azizov, N.Ya. Dvorkin, O.G. Filatov, G.P. Gardymov, I.S. Garypov, V.E. Golant, V.A. Glukhikh, V.I. Iogansen, V.A. Iagnov, I.A. Kadi-Ogly, R.R. Khayrutdinov, V.V. Korshakov, V.K. Krylov, I.N. Leykin, V.E. Lukash, A.B. Mineev, G.E. Notkin, À.R. Polevoy, K.G. Shakhovets, S.V. Tsaun, E.P. Velikhov, N.I. Vinogradov, G.M. Vorobiev
833

Application of seismic isolation for ITER
D. Dilling, C.E. Ahlfeld, P. Barabaschi, V. Chuyanov, K. Ishimoto, E. Tanaka
837

Calorimetric measurements of energy deposition in TORE Supra
F. Surle, G. Mayaux, J.J. Cordier
841

Application of FIR polarimetry for real-time density control on TEXTOR-94
H.R. Koslowski, B. Giesen, P. Hüttemann, H.T. Lambertz, K.P. Pelzer, W. Schalt, E. Zimmermann
845

Engineering aspects of ITER plasma diagnostic systems
C.I. Walker, T. Ando, A. Costley, L. deKock, K. Ebisawa, G. Janeschitz, L. Johnson, V. Mukhovatov, G. Vayakis, M. Yamada, S. Yamamoto
849

Tracer-encapsulated cryogenic pellet for particle transport diagnostics
S. Sudo, H. Itoh, K. Khlopenkov
853

Experiment of 14MeV neutron induced luminescence on window materials
F. Sato, Y. Oyama, T. Iida, F. Maekawa, J. Datemichi, A. Takahashi, Y. Ikeda
857

Status and characteristics of diagnostics on Korea superconducting tokamak research (KSTAR)
S.G. Lee, S.M. Hwang, H.Y. Chang, G.S. Lee, H.K. Park, J. Kim, D.I. Choi, S.G. Oh, K.K. Choh, J.H. Choi, J.W. Choi, Y.S. Chung, J.H. Han, J. Hong, B.C. Kim, W.C. Kim, Y.J. Kim, H.G. Lee, H.K. Na, Y.K. Oh, H.L. Yang, J.G. Yang, N.S. Yoon
861

High performance drivers for fast sweep microwave reflectometry
L. Cupido, A. Silva, M.E. Manso, F. Serra
865

Engineering aspects of an advanced heavy ion beam diagnostic for the TJ-II stellarator
A. Malaquias, C. Varandas, J.A.C. Cabral, L.I. Krupnik, S.M. Khrebtov, I.S. Nedzelskij, Yu.V. Trofimenko, A. Melnikov, C. Hidalgo, I. Garcia-Cortes
869

Two-frequency correlation reflectometer
J. Fernandes, A. Silva, L. Cupido, M.E. Manso, P. Varela
873

Plasma density profile evaluation in broadband reflectometry using a neural network
F.D. Nunes, J. Santos, M.E. Manso
877

Time resolved energy dispersive X-ray diagnostic for the TCV tokamak
J. Sousa, P. Amaro, P. Amorim, B. Duval, C.A.F. Varandas
881

Experimental tests of an ion source for the diagnostic neutral beam injector of the TEXTOR tokamak
G.F. Abdrashitov, A.A. Ivanov, V.V. Mishagin, A.A. Podyminogin, A.I. Rogozin, I.V. Shikhovtsev
885

Heavy ion beam probing for ITER
A.V. Melnikov, L.G. Zimeleva, K.N. Tarasyan, L.G. Eliseev, L.I. Krupnik, I.S. Nedzelskij, Yu.V. Trofimenko, Y. Hamada, H. Iguchi, K. Connor, T. Crowley, C.A.F. Varandas, A. Malaquias
889

Remote sensing of a capacitance manometer pressure measurement head
A. Browne, D. Cooper, J.F. Davies, L. Hackett, L. Svensson, D. Young
893

Engineering design of the JET edge Thomson scattering system
D.J. Wilson, P. Nielsen, C.W. Gowers, R.J. Eagle
897

A PC based real-time imaging system for the TdeV tokamak
F. Meo, P. de Villers, F. Brunet, G. Ratel
901

Design of an optimal control system for Tore Supra
S. Brémond, J.-M. Ané
905

Evolution of the Tore Supra data acquisition system: towards the steady state
B. Guilherminet, D. Elbèze, J.F. Artaud, S. Balme, Y. Buravand, B. Couturier, L. Ducobu, B. Gagey, M. Leluyer, R. Masset, D. Moulin, B. Rothan, J. Signoret
909

The Tore Supra control computer system: evolutions
E. Chatelier, F. Hennion, M. Hernandez, J.Y. Journeaux, P. Lebourg
913

First results of the new plasma feed-back control for TORE-SUPRA
G. Martin, D. Moulin, D. van Houtte, T. Wijnands
917

A data acquisition, control and visualization system for the upgraded TOSKA facility at FZK
A. Augenstein, H. Barthel, I. Donner, H. Frankrone, P. Gruber, G. Hellmann, P. Klingenstein, U. Padligur, K. Rietzschel, T. Specht, G. Würz
921

Plasma regime guided discharge control at ASDEX Upgrade
T. Zehetbauer, P. Franzen, G. Neu, V. Mertens, G. Raupp, W. Treutterer, D. Zasche and ASDEX Upgrade Team
925

Structure for next generation discharge control systems
G. Raupp, K. Lüddecke, G. Neu, W. Treutterer, D. Zasche, T. Zehetbauer and ASDEX Upgrade Team
929

Plasma shape control design in ASDEX Upgrade
W. Treutterer, J. Gernhard, O. Gruber, P. McCarthy, G. Raupp, U. Seidel and ASDEX Upgrade Team — 933

The use of fuzzy curves for the reconstruction of the plasma shape and the selection of the magnetic sensors
F.C. Morabito, M. Versaci — 937

Measument and control of error field driven magnetic islands in a tokamak reactor by electron cyclotron current drive
E. Lazzaro, S. Cirant, G. D'Antona, S. Nowak, G. Ramponi — 941

Design of central control and man-machine interface systems for large helical device
H. Yamada, K. Yamazaki, K.Y. Watanabe, S. Yamaguchi, K. Nishimura, Y. Taniguchi, H. Ogawa, N. Yamamoto, S. Sakakibara, M. Shoji, O. Motojima — 945

Real time control of the plasma boundary in JET
S. Puppin, M.E. Angoletta, D.J. Campbell, J.J. Ellis, M. Garribba, M. Lennnholm, F. Milani, D. O'Brien, F. Sartori, R. Sartori — 949

Implementation and initial operation of an adaptive plasma density controller at JET
H.E.O. Brelén, T. Budd, J. Ehrenberg, M. Gadeberg, C. Ryle — 953

Real time software for the control and monitoring of DIII-D system interlocks
J.D. Broesch, B.G. Penaflor, R.M. Coon, J.J. Harris, J.T. Scoville — 957

An introduction to FASTCAMAC (60 Megabytes/sec in CAMAC?)
S. Dhawan, C. Hubbard, T. Radway, R. Sumner — 961

A structured architecture for advanced plasma control experiments
B.G. Penaflor, J.R. Ferron, M.L. Walker — 965

Availability analysis of five years of operation of the superconducting tokamak Tore Supra
D. van Houtte, B. de Gentile, C. Grisolia, J.Y. Journeaux, P. Joyer, J.M. Laurens, G. Martin, F. Parlange, T. Wijnands — 969

The Frascati Tokamak Upgrade machine: plant and operation status
S. Ciattaglia, B.M. Angelini, G. Buceti, F. Crisanti, F. Gravanti, G. Mazzitelli, M. Panella, E. Sternini, A. Tuccillo, V. Zanza — 973

Operation of the tokamak ISTTOK in an alternating current regime
H. Fernandes, H. Figueiredo, J. Sousa, C.J. Freitas, J.A.C. Cabral, C.A.F. Varandas — 977

Monitoring and digital control system for the 130 MVA pulse generator system of the Spanish Stellarator TJ-II
L. Kirpitchev, L. Almoguera, M. Blaumoser, P. Mendez, L. Pacios, A. de la Peña, F. Lapayese, R. Carrasco, I. Labrador, A. Pérez, B. Alberdi, J.M. del Río, E. Jauregi — 981

Final assembly and present status of the Spanish Stellarator TJ-II
J. Botija, M. Blaumoser, J. Doncel, J.R. Knaster, J. Alonso and TJ-II Team ... 985

Operational experience with the JET saddle coil system
A. Santagiustina, H. Altmann, M. Buzio, D.J. Campbell, R. Claesen, G. D'Antona,
M. De Benedetti, A. Fasoli, G. Israel, R. Ostrom, S. Peruzzo, T. Raimondi, L. Rossi,
F. Sartori, M. Tabellini, A. Tanga, G. Zullo ... 989

Compatibility analysis of the magnet system of the KT-2 tokamak for long-pulse operation
K.W. Lee, J.M. Han, B.G. Hong, B.J. Yoon, Y.D. Bae, W.S. Song, D.E. Kim, N.S. Shin,
J.E. Milburn ... 993

Volume 2

Section E: Magnet and Related Power Supplies

Switching circuit for generating fast field transients in superconducting coils
M. Darweschsad, P. Komarek, G. Nöther, C. Sihler, A. Ulbricht, W. Weigand,
F. Wüchner ... 999

In-situ shear/compression testing of organic insulators after reactor irradiation at 4 K
H. Gerstenberg, E. Krähling, H. Katheder, R. Maix, M. Söll ... 1003

Determination of compound properties for superconducting magnets by combined
theoretical and experimental method
H. Kronhardt, K.-D. Herrmann, W. Broocks, M. Pillsticker ... 1007

Behaviour of polyimide insulated wire in pressurized hot water
L. Drews, U. Braunsberger, B. Giesen, M. Poier ... 1011

Operation of the Upgraded TOSKA facility and test results of the EU-LCT coil cooled
with forced flow helium II at supercritical pressure
M. Darweschsad, G. Dittrich, A. Grünhagen, S. Fink, G. Friesinger, A. Götz, R. Heller,
W. Herz, A. Hofmann, P. Komarek, O. Langhans, W. Lehmann, W. Maurer, I. Meyer,
G. Nöther, G. Perinic, W. Ratajczak, H.P. Schittenhelm, G. Schleinkofer, K. Schweikert,
Ch. Sihler, E. Specht, H.J. Spiegel, M. Süßer, A. Ulbricht, A. Völker, F. Wüchner,
G. Zahn, V. Zwecker ... 1015

Structural and mechanical design of cryogenic support system for LHD
H. Tamura, S. Imagawa, H. Hayashi, T. Satow, J. Yamamoto, O. Motojima,
T. Takahashi, K. Asano, S. Suzuki and LHD Group ... 1019

Design and components test results of the superconducting current feeder system for
the LHD
S. Yamada, T. Mito, H. Chikaraishi, R. Maekawa, S. Tanahashi, S. Kitagawa,
K. Nishimura, T. Uede, H. Hiue, Y. Yasukawa, I. Itoh, K. Sakaki, T. Satow,
J. Yamamoto, O. Motojima and LHD Group ... 1023

Accuracy of superconducting helical coils for LHD
S. Imagawa, S. Masuzaki, N. Yanagi, S. Yamaguchi, T. Satow, J. Yamamoto,
O. Motojima, LHD Group, K. Nakanishi, K. Uchida, T. Yamagiwa, T. Yamamoto,
K. Asano
1027

Local stress analysis of the W7-X superconducting winding pack
N. Jaksic, J. Simon-Weidner, J. Sapper
1031

On modular coils of a helias reactor
E. Harmeyer, N. Jaksic, J. Simon-Weidner
1035

Test results of the EU subsize conductor joints for ITER
D. Ciazynski, P. Decool, B. Jager, A. Martinez
1039

Simulating the ITER TF magnet with the TF model coil adjacent to the LCT coil
P. Libeyre, P. Decool, B. Turck, B. Dolensky, R. Meyer, S. Raff
1043

Fabrication of prototype conductors for ITER TF model coil
S. Conti, R. Garrè, S. Rossi, A. della Corte, M.V. Ricci, M. Spadoni, G. Bevilacqua,
R. Maix, E. Salpietro
1047

Automatic welding of the ITER-CSMS superconductor sheath
A. Laurenti, E. Bisio, P. Gagliardi, N. Scontrino, C. Aldrighi, G. Nardoni, G. Bevilacqua,
E. Salpietro
1051

Comparison among reduction and compensation methods of reactive power control for
the ITER AC/DC converter system
E. Gaio, R. Piovan, V. Toigo, I. Benfatto
1055

The switching network and discharge circuit of ITER
B. Bareyt, I. Benfatto, P.L. Mondino, A. Roshal, T. Bonicelli, A. Maschio, D. Hrabal,
J.M. Bottereau, S. Bulgakov, V. Kuchinski, N. Mikhailov
1059

Status of the ITER central solenoid and toroidal field model coil program
K. Okuno, R. Vieira, D. Bessette, N. Mitchell, P. Bruzzone, Z. Piec, R.J. Thome,
E. Salpietro, H. Tsuji, S. Egorov, R. Jayakumar
1063

A segmented, alternative, central solenoid design for ITER
D. Bessette, B. Stepanov, V. Vasiliev
1067

ITER tokamak supports: initial sizing and design
R. Gallix, C. Sborchia, P. Barabaschi, G. Johnson
1071

Mechanical structures for the ITER magnet system
C. Sborchia, R. Gallix, N. Mitchell, K. Okuno, B. Stepanov, R.J. Thome, K. Yoshida,
F. Wong, P. Barabaschi, C. Jong, A. Alekseev, A. Malkov, P. Titus
1075

The ITER thermal shields
R. Bourque, A. Alekseev, M. Wykes
1079

Development of Nb_3Al coil for the ITER TF magnet
T. Ando, N. Koizumi, T. Ito, H. Nakajima, Y. Nunoya, M. Sugimoto, M. Nakahira,
H. Tsuji, H. Tsukamoto
1083

The harmonic power filters of the JET poloidal divertor field amplifiers
T. Bonicelli, M. Huart, P. Doyle, M. Rouleau, G. Zullo
1087

The JET high voltage power distribution, analysis of present and future operational requirements
N. Dolgetta, E. Bertolini, M. Huart, G. Murphy
1091

Upgrading the JET toroidal field to exceed 3.45 Tesla
J.R. Last, E. Bertolini, M. Buzio, P. Presle, R. Raimondi, V. Riccardo
1095

Inspection techniques for JT-60 toroidal field coil cooling pipes
T. Arai, M. Honda, T. Koike, M. Saidoh, M. Shimizu
1099

Design and development of sliding joints for the MAST toroidal field coils
G.M. Voss
1103

Control of highly vertically unstable plasmas in TCV with internal coils and fast power supply
A. Favre, J.-M. Moret, R. Chavan, A. Elkjaer, D. Fasel, F. Hofmann, J.B. Lister, J.-M. Mayor, A. Perez
1107

The central solenoid of ignitor: design and manufacturing aspects
G. Galasso, L. Lanzavecchia, J. Rauch, A. Cucchiaro, A. Pizzuto
1111

Selection of jacket materials for Nb_3Sn superconductor
F.M.G. Wong, N.A. Mitchell, R.L. Tobler, M.M. Morra, R.G. Ballinger, H. Nakajima
1115

Simulation of the power supply system for the TJ-II Spanish stellarator
J. Acero, A. Perez, J.M. del Rio, C. Lucia, B. Alberdi
1119

Section F: Fuel Cycle and Tritium Processing Systems

Status of ITER fuelling and wall conditioning system design
H. Nakamura, G. Federici, P. Ladd, G. Janeschitz, J. Andres, A. Antipenkov, K.J. Dietz, P.W. Fisher, M.J. Gouge, H.S. Hurzlmeier, R.A. Marrs, D. Mitin, K.M. Schaubel, M. Sugihara
1125

High repetitive pellet injectors for plasma density control
P.T. Lang, P. Cierpka, P. Kupschus
1129

High speed pellet injection system for large helical device
M. Kanno, S. Sudo, H. Yamada, O. Motojima, T. Baba — 1133

A repetitive pellet injector with screw extruder
I.V. Viniar, S.V. Skoblikov, P.Yu. Koblents, B.V. Kuteev, A.Ya. Lukin — 1137

A tritium compatible pneumatic pellet injector
P. Koblents, I. Viniar, B. Kuteev, S. Skoblikov, A. Shlyahtenko, M. Parshin, G. Saksagansky, V. Skripunov, S. Saksagansky — 1141

A repetitive pipe-gun pellet injector
I.V. Viniar, S. Sudo, B.V. Kuteev, A.P. Umov, V.G. Kapralov, S.V. Skoblikov, P.Yu. Koblents, S.M. Egorov, K.V. Khlopenkov, V.V. Arhipov — 1145

Pellet injector with the liner compression of propellant gas
V.P. Bazilevski, Yu.A. Kareev, A.I. Kolchenko, V.P. Novikov — 1149

Pellet injector development at ORNL
S.K. Combs, S.L. Milora, L.R. Baylor, P.W. Fisher, C.A. Foster, C.R. Foust, M.J. Gouge, T.C. Jernigan, H. Nakamura, B.J. Denny, R.S. Willms, A. Frattolillo, S. Migliori — 1153

Plasma driven superpermeation and its possible fusion applications
M. Bacal, F. El Balghiti-Sube, A.I. Livshits, M.E. Notkin, M.N. Soloviev — 1157

Progress report of a cryomechanical vacuum pump prototype
J.P. Périn, D. Henry, R. Vallcorba, J.J. Cordier, F. Samaille — 1161

Development of cryopanel fast regeneration methods for use in the ITER primary vacuum system
C. Day, H. Hass, A. Mack — 1165

Primary vacuum pump concept, component testing and model pump development for ITER
A. Mack, J.C. Boissin, D.K. Murdoch, D. Röhrig, G. Saksagansky — 1169

The design of the ITER primary pumping system
P. Ladd, H. Hurzlmeier, G. Janeschitz, R. Marrs, K. Schaubel — 1173

Plasma impurity processing with HITEX
L. Rodrigo, J.M. Miller, J. Senohrabek — 1177

Tritium extraction from the coolant of a DEMO solid breeder blanket and from the coolant of an ITER blanket test module
H. Albrecht, E. Hutter — 1181

Separation of tritiated hydrogen species in the TLK isotope separation system
G. Neffe, J. Dehne, E. Hutter, H. Kissel, H. Brunnader — 1185

AMOR facility: regeneration of molecular sieve beds used for the retention of tritium at the Tritium Laboratory Karlsruhe
E. Hutter, H.-D. Adami, U. Besserer, R.-D. Penzhorn — 1189

Tritium tests with a PERMCAT reactor for isotopic swamping
M. Glugla, J. Miller, P. Herrmann, M. Iseli, R.-D. Penzhorn — 1193

Results from tritium operation of the clean-up facility CAPRICE
M. Glugla, R. Kraemer, T.L. Le, K.H. Simon, K. Günther, J. Wendel, R.-D. Penzhorn — 1197

Low tritium retention zeolites for fusion gas processing
F. Toci, C. Malara, T. Mencarelli, I. Ricapito — 1201

Experimental confirmation of the theoretical previsions for the applicability of catalytic membrane reactors for the fusion fuel cycle
V. Violante, S. Tosti, A. Colombini, S. Castelli, M. De Francesco — 1205

Tritium recovery system in helium-cooled ceramic blanket for DEMO
S. Tosti, A. Colombini, V. Violante, C. Rizzello — 1209

An alternative approach to tritium recovery from water-cooled Pb-17Li DEMO blanket
H. Dworschak, C. Malara, I. Ricapito, D. Sarigiannis — 1213

Testing of a water vapour cold trap for atmospheric air detritiation
C. Housiadas, K.H. Schrader — 1217

Development of a fusion fuel processing system at the Japan Atomic Energy Research Institute
S. Konishi, M. Enoeda, T. Yamanishi, K. Okuno — 1221

Breakthrough properties of hydrogen with Zr_9Ni_{11} particle packed bed
K. Tsuchiya, H. Imaizumi, H. Kawamura, T. Kabutomori, Y. Wakisaka, T. Niiho — 1225

Tritium recovery from Li17-Pb83 liquid breeder by permeation window method
T. Terai, A. Suzuki, S. Tanaka — 1229

Experimental closed loop for dynamic modeling of ITER vacuum-tritium system. Main parameters and status of the development
G.L. Saksagansky, A.I. Vedeneev, V.G. Klevtsov, E.L. Koira, V.N. Lobanov, S.A. Pimanikhin, B.N. Tenyaev, Yu.P. Averin — 1233

Effects of tritium and Helium-3 on life time properties of Pd-Ag alloys in ITER tritium purification technology
V. Tebus, G. Arutunova, V. Bulkin, E. Dmitrievsky, V. Filin, Y. Golikov, L. Panteleev, A. Perevezentsev, L. Rivkis — 1237

The mechanism of self radiolysis of tritiated water on molecular sieve: gas phase hydrogen isotopic distribution
R.T. Walters — 1241

Evaluation of the tritium surface activity monitor
W.T. Shmayda, N.P. Kherani, D. Stodilka — 1245

Tritium accountancy in 200 litres drums by calorimetry
D. Devillard, J.-P. Corot, J.Y. Floricourt — 1249

Analysis of hydrogen isotopes by RAMAN spectroscopy and optical fibres
Y. Chaufour, D. Devillard, K. Danger, D. Dall'ava, H. Berger — 1253

Accountancy penalty in case of hidden inventories
R. Avenhaus, G. Spannagel — 1257

Tritium experience at the Tritium Laboratory Karlsruhe
U. Besserer, J. Dehne, L. Doerr, M. Glugla, W. Hellriegel, T. Le, F. Schmitt, K.H. Simon, T. Vollmer, J. Wendel, R.-D. Penzhorn — 1261

Tritium monitoring by gas scintillation
P. Pacenti, F. Campi, C. Mascherpa, S. Terrani — 1265

Hot commissioning of the analytical glovebox system in ETHEL
U. Engelmann, G. Vassallo — 1269

A solid scintillator area monitor for tritiated water vapor in air
F. Campi, R.A.H. Edwards, A. Ossiri, P. Pacenti, S. Terrani — 1273

Tritium technology research and development at the tritium process laboratory of JAERI
K. Okuno, S. Konishi, T. Yamanishi, S. O'Hira, M. Enoeda, T. Hayashi, H. Nakamura, Y. Kawamura, Y. Iwai, K. Kobayashi — 1277

Radiogenic helium thermodesorption from uranium deuterotritide
P.G. Berezhko, A.I. Vedeneev, B.F. Dadonov, V.P. Sorokin — 1281

Specialized mass spectrometers for analysis tritium gas mixes in "on-line" mode in technological systems of TR
I. Milechkine, N. Aryev, L. Gall, B. Mamyrin, N. Riazantseva, G. Saksagansky — 1285

Testing of the cryogenic target handling system for the OMEGA laser
D.T. Goodin, N.B. Alexander, W.A. Baugh, C.T. Beal, G.E. Besenbruch, K.K. Boline, L.C. Brown, W. Egli, J.F. Follin, C.R. Gibson, M.J. Hansink, E.H. Hoffmann, W. Lee, R.A. Mangano, K.R. Schultz, R. Stemke, T.A. Torres — 1289

Section G: Blanket Technology/Materials

DEMO blanket segment fabrication using advanced HIP technology
C. Dellis, G. Chaumat, M. Fütterer, L. Giancarli, G. Le Marois — 1295

Development and design of water-cooled ceramic breeding blanket for ITER
Y. Poitevin, R. Antidormi, B. Bielak, L. Giancarli — 1299

Design and transient thermal analysis of a water-cooled Pb-17Li test blanket for ITER
M.A. Fütterer, B. Bielak, J.-P. Deffain, L. Giancarli, N.B. Morley, J-F. Salavy, J. Szczepanski — 1303

EU water-cooled Pb-17Li DEMO blanket: fabrication issues and future R&D priorities
L. Giancarli, G. Benamati, M.A. Fütterer, C. Nardi, J. Reimann, K. Schleisiek — 1307

Solid HIPed demonstrator of ITER blanket — shield modules
M. Febvre, J.L. Deneuville, P. Lorenzetto, W. Daenner — 1311

Design and analysis of the ITER breeding blanket
Y. Gohar, M. Billone, A. Cardella, I. Danilov, W. Dänner, M. Ferrari, M. Giegerich, K. Ioki, T. Kuroda, D. Lousteau, P. Lorenzetto, S. Majumdar, R. Mattas, K. Mohri, R. Parker, R. Raffray, Y. Strebkov, H. Takatsu, E. Zolti — 1315

HEBLO, a helium blanket test loop for small test sections of helium cooled solid breeder blankets
P. Norajitra, D. Piel, G. Reimann, R. Ruprecht — 1319

Development of fabrication techniques for the European helium cooled pebble bed breeder blanket
T. Heider, G. Reimann, H. Riesch-Oppermann, K. Schleisiek — 1323

The ITER "L-4" blanket project
W. Dänner, A. Cardella, K. Ioki, R. Mattas, Y. Strebkov, H. Takatsu — 1327

ITER blanket system design
K. Ioki, L. Bruno, A. Cardella, W. Dänner, A. Lodato, D. Lousteau, R. Mattas, K. Mohri, R. Parker, R. Raffray, Y. Strebkov, N. Tachikawa, H. Takatsu, D. Williamson, M. Yamada — 1331

He-FUS 3 — European helium cooled blanket test facility for DEMO
G. Dell'Orco, G.C. Bertacci, M. Mazza, L. Borsati, F. Mariotti, R. Penco, P.L. Valente — 1335

Design development of breeding blanket based on pebble bed concept for fusion experimental reactor
H. Miura, K. Kitamura, H. Takatsu, T. Kuroda, T. Kurasawa, S. Sato, K. Furuya, T. Hatano, I. Tokami, Y. Itou, T. Osaki, T. Hashimoto, S. Sato, I. Kawaguchi — 1339

Fabrication of small-scaled shielding blanket module and first wall panel for international thermonuclear experimental reactor
K. Furuya, S. Sato, T. Kuroda, T. Kurasawa, I. Tokami, T. Hatano, H. Miura, H. Takatsu 1343

Development of full-scale sector model for ITER vacuum vessel
K. Koizumi, M. Nakahira, Y. Itou, N. Kanamori, K. Kitamura, E. Tada, G. Jonhson, K. Shimizu, G. Sannazzaro, K. Takahashi, Y. Utin, K. Ioki 1347

Design study of in-pile blanket mockup simulated neutron pulse operation of fusion reactor
M. Nakamichi, H. Sagawa, K. Yamaguchi, T. Ishitsuka, H. Kawamura 1351

Computer simulations of displacement damage in solid breeder blankets
D. Leichtle 1355

High temperature cyclic stress-strain response of structural stainless steels for thermonuclear fusion reactors
A.F. Armas, I. Alvarez-Armas, M. Avalos, C. Peterson, R. Schmitt 1359

Mechanical properties of the martensitic chromium steel F82H Mod., heat 9741
L. Schäfer, M. Schirra, R. Lindau 1363

Mechanical properties of the MANET-II martensitic chromium steel and their optimization
L. Schäfer 1367

Buckling analysis of the W7-X inner vacuum vessel
J. Simon-Weidner, N. Jaksic, J. Sapper 1371

Joining of SiC/SiC$_f$ composites using a SiOC glass
A. Donato, P. Colombo, Th. Dikonimos Makris, R. Giorgi, M.O. Abdirashid, G. Scarinci 1375

SiC/SiC fiber ceramic matrix composites for fusion application: a new manufacturing process
A. Donato, C.A. Nannetti, A. Ortona, S. Botti, G. D'Alessandro, G. Filacchioni, A. Masci 1379

Low cycle fatigue and electrochemical behaviour of F82H martensitic steel in water coolant environments
M.F. Maday 1383

Mechanical and structural properties of Ti-bearing reduced activation martensitic steels
G. Filacchioni, L. Pilloni, F. Attura, E. Casagrande, U. De Angelis, G. De Santis, D. Ferrara 1387

Thermomechanical behaviour of SiC$_f$/SiC composites in helium environments
A.J.F. Rebelo, M. Oksanen, P. Fenici, H. Kolbe 1391

Post-irradiation mechanical properties of low-activation Cr-Mn austenitic steels
R.A.H. Edwards, G.P. Tartaglia, P. Bottelier, P. Fenici — 1395

Effect of neutron irradiation on mechanical properties of Nb-1%Zr/SS304 joints fabricated by friction welding
K. Tsuchiya, H. Kawamura, T. Niiho — 1399

Structural analysis of blanket system and vacuum vessel for International Thermonuclear Experimental Reactor (ITER)
K. Kitamura, K. Koizumi, H. Takatsu, Y. Itou, M. Nakahira, E. Tada, T. Tsunematsu — 1403

Density improvement of Li_2O pebble fabricated by sol-gel method
K. Tsuchiya, H. Kawamura, K. Fuchinoue, S. Yoshimuta, K. Watarumi, T. Niiho — 1407

Model for steady-state tritium analysis in ceramic breeder blanket. Comparsion of inventory results with mistral code predictions
R. Antidormi, N. Roux — 1411

Understanding of experimental results on the radiation enhancement of H-Isotopes' permeability through austenitic/martensitic steels
L.A. Sedano, A. Perujo, B.G. Polosukhin, N.G. Primakov, I.L. Tazhibaeva — 1415

Elaboration of alumino-forming coatings for tritium permeation barriers using "low" temperature CVD
F. Schuster, C. Chabrol, E. Blanquet, C. Bernard, F. Maury, F. Felten, A. Terlain — 1419

The formation of aluminide coatings on MANET stainless steel as tritium permetation barrier by using a new test facility
H. Glasbrenner, J. Konys, G. Reimann, K. Stein, O. Wedemeyer — 1423

Tritium control in the European helium cooled pebble bed blanket
L. Berardinucci, M. Dalle Donne — 1427

Tritium release from neutron irradiated beryllium pebbles
F. Scaffidi-Argentina, H. Werle — 1431

Effect of alloying elements on mechanical and microstructural characteristics of aluminide coatings on MANET steel
G. Benamati, H. Glasbrenner, A. Casagrande, C. Fazio — 1435

TiCN and surface oxidation as hydrogen permeation barrier in F82H: a comparison
E. Serra, A. Perujo, G. Benamati — 1439

The effect of cyclic loads on the hydrogen permeation rate of structural materials
T. Sample, H. Kolbe, A. Perujo, P. Fenici — 1443

Deuterium permeation through TiN and TiN-TiC coatings deposited on F82H steel
A. Sabbioni, N. Laidani, A. Miotello, G. Benamati — 1447

Mechanism of defects production in Li_2O and their influence on tritium release
V. Grishmanov, S. Tanaka, J. Tiliks — 1451

Tritium release from Li_2O single crystals irradiated with fast neutrons
T. Tanifuji, D. Yamaki, K. Noda, O.D. Slagle, F.D. Hobbs, G.W. Hollenberg — 1455

In-situ tritium release behaviour from molten Li_2BeF_4 salt under neutron irradiation
A. Suzuki, T. Terai, S. Tanaka — 1459

Tritium distribution and release from neutron irradiated lithium orthosilicate pebbles
A. Abramenkovs, J. Tiliks, G. Kizane, H. Werle, S. Tanaka — 1463

Neutron irradiated beryllium: fracture toughness
F. Moons, J.L. Puzzolante, A. Rahn, J. Van de Velde — 1467

On the use of zirconate/titanate for the helium-cooled ceramic pebble-bed DEMO blanket
M. Eid, R. Antidormi, B. Bielak, L. Giancarli, V. Mathis, N. Roux, J.-F. Salavy, J. Szczepanski — 1471

Replenishment of lithium lost from Pb-17Li: an assessment of the methods available for addition of lithium to Pb-Li alloys
P. Hubberstey, M.J. Capaldi, F. Barbier — 1475

Behavior of bismuth in lithium-lead mixtures
S. Bucké, H. Feuerstein, L. Hörner, J. Beyer, S. Horn — 1479

Development work for lithium orthosilicate pebbles
M. Dalle Donne, G. Piazza, A. Weisenburger, H. Werle, V. Geiler, B. Speit, D. Sprenger — 1483

Metallographic investigation and mechanical behavior of beryllium pebbles at room temperature
M. Dalle Donne, E. Kaiser, O. Romer, F. Scaffidi-Argentina, P. Weimaar, H. Werle — 1487

Effect of neutron irradiation swelling on the heat transfer behavior of a beryllium pebble bed
F. Scaffidi-Argentina — 1491

Studies on surface hydroxyl group on Li_2O by infrared absorption and ab-initio calculation and their consequence on tritium desorption
M. Taniguchi, S. Tanaka — 1495

Reprocessing technology development for irradiated beryllium
H. Kawamura, K. Tatenuma, Y. Hasegawa, N. Sakamoto, K. Nishida — 1499

Thermal properties of neutron irradiated beryllium
E. Ishitsuka, H. Kawamura, T. Terai, S. Tanaka — 1503

The influence of magnetic field on the radiolysis of lithium orthosilicate ceramics
J. Tiliks, S. Tanaka, G. Kizane, A. Supe, A. Abramenkovs, V. Grishmanovs   1507

Performance of ceramic breeder materials irradiated to high lithium burnups in exotic-7
J.G. van der Laan, M. Stijkel, R. Conrad   1511

Heat transfer and secondary motion in liquid metal flow in horizontal duct under fusion relevant conditions
V.G. Sviridov, N.G. Razuvanov, A.V. Ustinov, Yu.S. Shpanskij   1515

Effect of the presence of ferromagnetic structural material on the DEMO helium cooled pebble bed blanket during plasma disruptions
L.V. Boccaccini, P. Ruatto   1519

Application of a mesh refining procedure to the electromagnetic analysis of the DEMO HCPB blanket concept during disruptions
L.V. Boccaccini, B. Tellini   1523

MHD issues of the European self-cooled and water-cooled 83 Pb-17Li blankets
J. Reimann, G. Benamati, R. Moreau   1527

MHD turbulence generation in bends perpendicular to the magnetic field
J. Reimann, S. Dementjev, A. Flerov, I. Platnieks   1531

Compatibility of insulating ceramic materials with molten lithium metal
T. Yoneoka, T. Mituyama, T. Terai, S. Tanaka   1535

Safety assessment and accident management of a water-cooled Pb-17Li test blanket for ITER
Y. Severi, C. Bertrand, M.A. Fütterer, L. Giancarli, G. Marbach, N.B. Morley   1539

Analysis of loss of flow transients in the first wall cooling system of the water-cooled Pb-17Li blanket concept for the European DEMO fusion reactor
K. Gabel, K. Kleefeldt   1543

Reliability & availability analysis as a decisional meansat early stages of the new machines design
C. Nardi, H. Schnauder, M. Eid   1547

Effects of water micro-leaks in a Pb-17Li DEMO blanket module
L. Rapezzi, M. Guccini, G. Benamati   1551

Neutron flux experiment in the ITER shield mock-up at FNG
A. Santamarina, L. Benmansour, B. Camous, H. Philibert, P. Batistoni, M. Pillon   1555

Three-dimensional activation analysis for DEMO type fusion reactor blankets
H. Tsige-Tamirat   1559

Three-dimensional neutronics analyses of the ITER bulk shield experiment
U. Fischer, P. Batistoni, M. Pillon — 1563

Methods for nuclear heating measurements in an ITER shield-blanket mock-up
H. Freiesleben, W. Hansen, K. Merla, D. Richter, K. Seidel, S. Unholzer — 1567

Measurement and analysis of spectral neutron and photon fluxes in an ITER shield mock-up
H. Freiesleben, W. Hansen, D. Richter, K. Seidel, S. Unholzer, U. Fischer, Y. Wu, M. Angelone, P. Batistone, M. Pillon — 1571

A 14-MeV neutron transmission experiment on vanadium
U. von Möllendorff, B.V. Devkin, U. Fischer, B.I. Fursov, M.G. Kobozev, M.M. Potapenko, S.P. Simakov, V.A. Talalaev, Y. Wu — 1575

ITER nuclear analysis
R.T. Santoro, H. Iida, V. Khripunov, S. Mori, L. Petrizzi, D. Valenza and members of the European Union, Japan, Russian Federation and U.S. Home Teams — 1579

Nuclear analysis of an ITER inboard blanket module
G. Vella, O. Fiorella, D. Valenza, R.T. Santoro — 1583

2-D overall shielding analysis of ITER tokamak machine
S. Sato, H. Takatsu, K. Maki, T. Utsumi, H. Iida, R.T. Santoro — 1587

Characterization of self-powered neutron detector at high temperature under neutron irradiation
M. Nakamichi, C. Yamamura, H. Sagawa, N. Nakazawa, H. Kawamura — 1591

Evaluation and test of high energy neutron cross-section data for the IFMIF intense neutron source
Yu.A. Korovin, A.Yu. Konobeyev, V.P. Lunev, P.E. Pereslavtsev, A.Yu. Stankovsky, U. Fischer, U.v. Möllendorff, M. Soksic-Kostic, P. Wilson, D. Woll — 1595

Analysis and implementation of a Monte Carlo high energy neutrons source for IFMIF
P.P.H. Wilson, U. Fisher — 1599

Fusion materials activation tests with a deuteron-beryllium neutron source
U. von Möllendorff, H. Giese, H. Tsige-Tamirat — 1603

Applicability of the FIMEC indentation test to characterize materials irradiated in the future IFMIF high intensity neutron irradiation source
P. Gondi, R. Montanari, A. Sili, S. Floglietta, A. Donato, G. Filacchioni — 1607

The HRF Petten as a test bed for breeder materials and blanket concepts for ITER and DEMO
R. Conrad, J. van der Laan — 1611

Toward an international fusion materials irradiation facility
F. Cozzani, T. Kondo, T. Shannon, F.W. Wiffen, L. Zavialsky — 1615

Section H: Assembly, Remote Handling and Waste Management and Storage

Measuring sound and vibrations under high radiation for enhanced safety in fusion reactor remote handling
S. Coenen, M. Decréton, I. Baetens — 1621

Motors for the in-vessel handling unit — a radiation hardening achieved up to 80 MGy
M. Decréton, B. Haferkamp, M. Englert, A. Rahn, A. Suppan — 1625

Water hydraulics in ITER divertor refurbishment
M. Siuko, M.J. Vilenius, T. Vivarlo, K.T. Koskinen, E. Mäkinen, E. Luodemaki, A. Timperi — 1629

Design of the ITER divertor remote handling system
E. Martin, T. Burgess, G. Cerdan, C. Damiani, D. Duglué, G. Janeschitz, D. Maisonnier, K. Shibanuma, M. Sironi, E. Tada, A. Tesini, R. Tivey — 1633

International Thermonuclear Experimental Reactor in-vessel viewing system ITER/IVVS
A. Timperi, D. Maisonnier, E.R. Hager, T. Businaro, L. Consano, H. Hannula, S. Kuitunen, V.-P. Lappalainen, M. Lopez, P. Stigell, T. Ylikorpi, M. Aikio, H. Ailisto, V. Heikkinen, M. Lindholm, A. Halme, P. Jakubik, J. Suomela, J. Heimsch, I. Bhandal — 1637

ITER divertor PFC's attachment and replacement
A. Antipenkov, S. Chiocchio, D. Dilling, G. Janeschitz, D. Maisonnier, E. Martin, S. Schleicher, R. Tivey, A. Turner — 1641

Development of a NiTi shape memory alloy vacuum tight flange for JET in-vessel inspection system
S. Besseghini, T. Businaro, S. Ceresara, A. Tuissi — 1645

Blanket cooling pipe maintenance system for fusion experimental reactor
K. Oka, S. Kakudate, M. Nakahira, K. Taguchi, A. Itoh, E. Tada, A. Tesini, K. Shibanuma, R. Haange — 1649

Development of an end-effector for ITER blanket module handling
M. Nakahira, K. Oka, S. Kakudate, S. Fukatsu, K. Taguchi, E. Tada, K. Shibanuma, N. Matsuhira, R. Haange — 1653

Development of locking and mover system for ITER divertor maintenance
S. Fukatsu, N. Takeda, S. Kakudate, E. Martin, T. Burgess, K. Shibanuma, D. Maisonnier, E. Tada — 1657

Remote operations for the repair and replacement of a poloidal field coil in ITER
A. Tesini, D. Dilling, R. Bourque, Y. Nakashima, Z. Piec, M.E.P. Wykes, K. Yoshida, D. Maisonnier, C. Pascual, I. Ibarreche, J. Icaran, A. Mousdell, E. Barratt, J. Gilroy, E. Tada, F. Kimball, F. Vivaldi — 1661

Radiation effects on remote handling system components
R.E. Sharp, S.L. Pater — 1665

The command and control system for JET remote handling equipment
L. Galbiati, P. Carter, B. Haist — 1669

The implementation and operation of a full size mock-up facility in preparation for remote handling of JET divertor modules
R. Cusack, P. Brown, B. Haist, A. Loving, R. Stokes — 1673

Design and development of a new remote handling transporter facility for JET
L.P.D.F. Jones, M. Irving, J. Palmer, D. Hamilton — 1677

Design and development of remote handling tools for the JET divertor exchange
S.F. Mills, A.B. Loving — 1681

The assessment and improvement of JET remote handling equipment availability
A.C. Rolfe, E. Scott, D. Smith — 1685

Tokamak component handling, maintenance and associated service building
D.M. Banks, G. Janeschitz, D. Dilling, M. Pascual — 1689

Steel reprocessing strategy for fusion reactor dismanteling
G. Marbach, L. Boisset, P. Giroux — 1693

French experience in steel detritiation
L. Boisset, C. Lattaud, P. Giroux — 1697

ITER waste management strategies and final disposal
S. Nisan, K. Brodén, M. Lindberg — 1701

Detritiation of graphite and beryllium plasma-facing components
P. Pacenti, R.A.H. Edwards, F. Campi — 1705

The divertor test platform
C. Damiani, D. Cassarini, P.A. Gaggini, P. Scarcella, M. Tarantini, G. Fermani, G. Cerdan, D. Maisonnier, J. Sheppard, J. Millard, J. Blevins, E. Martin, D. Duglué, A. Tesini, E. Tada — 1709

Section I: Safety and Environment, Reactor Studies

Divertor cooling system simulation with the ATHENA code
W. Van Hove, E. Komen, L. Bartsoen, E. Stubbe — 1715

Tritium chronic release estimates for the ITER design
K.M. Kalyanam, A. Natalizio — 1719

Dose monitoring and control at a radioluminescent light manufacturer
W.T. Shmayda, A.B. Antoniazzi, P. Hirst, E. Kettyle — 1723

Hydrogen hazard in a fusion reactor: evaluation and mitigation
V. Chaudron, L. Boisset, A. Laurent, F. Arnould, C. Latge — 1727

Multiple failure sequence in the divertor cooling loop of SEAFP-calculation of an ex-vessel LOCA and the induced in-vessel LOCA using the CATHARE code
P. Sardain, G. Franzoni — 1731

An application of the lines of defence (LOD) method in the studies on the fusion reactor safety
G.L. Fiorini, G. Franzoni — 1735

Effect of PFCs evaporation on plasma shutdown during a loss of flow accident scenario
G. Franzoni, P. Sardain, J. Villar Colomé — 1739

Modeling of thermal hydraulic phenomena for high flux components in fusion reactors during accidental conditions
G. Langlais, C. Girard, G. Marbach — 1743

Safety and dependability for experimental fusion reactor design
D. Soussan, M. Bernard — 1747

Tritium sorption in carbon dust
D. Boyer, Ph. Cétier, L. Dupin, Y. Belot, C.H. Wu — 1751

Analysis of ITER accident sequences
W. Gulden, H.-W. Bartels, D. Holland, B. Kolbasov, D. Petti, S. Piet, A. Poucet, J. Raeder, Y. Seki — 1755

Development of an integrated system of codes for ITER safety analysis, ISAS
S. Nisan, I. Toumi, M-T. Porfiri, T. Boubée de Gramont — 1759

Evaluation of the activated corrosion product source term for ITER heat transfer systems
S. Nisan, L. Di Pace, J.-C. Robin — 1763

Safety assessment of two DEMO blanket concepts: Helium-cooled pebble bed vs. dual coolant
K. Kleefeldt, K. Gabel — 1767

Comparison of site specific probabilistic dose assessments with deterministic generic dose values compiled within the ITER-EDA
W. Raskob, O. Edlund — 1771

Reponse of ITER divertor to loss of coolant and loss of flow accidents
J.-M. Gay, G. Marbach, E. Ebert — 1775

Multiple failure accident sequences for SEAFP reactor
R. Caporali, T. Pinna, M.T. Porfiri … 1779

Evaluation of the enviroment source terms for ITER divertor primary heat transfer system LOCAS
G. Cambi, M.T. Porfiri, D.G. Cepraga … 1783

Envelope analysis of post-accidental thermal transients for ITER
F. Andritsos, D.A. Sarigiannis, L. Daverio … 1787

In-vessel dust removal system using static electricity
I. Aoki, Y. Seki, S. Ueda, R. Kurihara, M. Onozuka, Y. Ueda, Y. Oda … 1791

Accidental overpressure suppression for the ITER vacuum vessel
M. Wykes, Y. Nakashima, L. Topilski, S. Piet, D. Holland, Y. Seki, F. Kasahara, M. Yamauchi … 1795

Design analysis for reducing dose rate in the NBI to realize direct access by workers for a fusion experimental reactor
K. Shibata, T. Inoue, K. Maki, Y. Yamashita … 1799

Tritium containment in the dust and debris of plasma-facing materials produced during operations
I. Konkashbaev, A. Hassanein, Ju. Grebenshikov … 1803

Analysis of possible accidents in ITER cryostat
B.N. Kolbasov, D.K. Kurbatov, D.P. Ivanov, A.Yu. Pashkov … 1807

Tritium permeation through 04X16H11M3T steel in the presence of $Li_{17}Pb_{83}$ molten alloy
V. Tebus, V. Demidov, A. Zyrionov, V. Zakhartsev … 1811

Influence of some structural factors on hydrogen penetration through structural materials during hydrogen ion bombardment
A.G. Zaluzhnyi, V.P. Kopytin, M.V. Tcherednichenko-Alchevskiy … 1815

Evaluation of second confinement concepts for water-cooled and helium-cooled fusion reactors
R. Blomquist, J. Collén, E. Ebert, A. Natalizio, K. Shen, S.K. Sood … 1819

Activation product transport modelling for the liquid metal loop of a fusion power plant using the code TRAP
P.J. Karditsas, C.B.A. Forty … 1823

Monodisperse aerosol modelling for fusion power plant containments and comparison with polydisperse calculations
W.E. Han … 1827

Safety and environment assessment of a variety of blanket concepts and structural materials
N.P. Taylor, C.B.A. Forty, W.E. Han, I. Cook, C. Clair — 1831

Preliminary cooling circuit activation and ORE assessment for ITER
C.B.A. Forty, P.J. Karditsas — 1835

ITER seismic analysis and structural design
P. Barabaschi, V. Chuyanov, D. Dilling, G. Jonhson, R. Gallix, S. Sadakov, G. Sannazzaro, C. Sborchia — 1839

Dose due to mobilization of tungsten activation products in air
K.A. McCarthy, G.R. Smolik, D.L. Hagrman, K. Coates — 1843

Designing a maintainable tokamak power plant
L.M. Waganer, F.R. Cole and the ARIES team — 1847

ITER plasma safety interface models and assessments
N.A. Uckan, H.-W. Bartels, T. Honda, S. Putvinski, T. Amano, D. Boucher, D. Post, J. Wesley — 1851

Engineering overview of Aries-RS tokamak power plant
M.S. Tillack and the ARIES team — 1855

Engineering design of DREAM components
I. Kawaguchi, J. Adachi, S. Yamazaki, S. Ueda, S. Nishio, Y. Seki, R. Kurihara, I. Aoki, T. Kunugi, T. Kuroda, H. Miura — 1859

Blanket energy multiplication factor influence on maximum commercial inertial fusion power plant construction cost
N. Cerullo, S. Lanza, M. Vezzani — 1863

Design of tokamak reactors: a heuristic approach
J.-M. Ané, S. Brémond, X. Garbet, J. Johner, C. Leloup, P. Magaud, M. Pain, B. Pégourié, G. Tonon, B. Turck, J. Villar Colomé, J.G. Wegrowe, J. Weisse — 1867

Operator protection for a future commercial fusion plant
J. Mustoe, S.M. Ali, L. Di Pace, C.B.A. Forty, B.-C. Friedrich, S. Sandri, H.M. Thompson — 1871

Author index — A1

INVITED TALKS

Nuclear fusion, an energy source

R. Toschi

The NET Team, Boltzmannstr. 2, D-85 748 Garching, Germany

Abstract

The final phase of the feasibility demonstration of fusion, namely the construction and operation of ITER, will require a large and prolonged effort and strong determination by all parties involved. The time is therefore appropriate to revisit the motivations in support of fusion development. The supply of energy would become an issue today if due consideration were given not only to the limits 'internal' to the energy systems but also to those 'external', to it. The first ones are not so stringent because reserves in particular of fossil fuels are ample, but the second ones are very stringent because of the limited capability of self-regeneration of the environment. Energy consumption is anticipated to triple in the next 50 years and if the share among the sources remains as of now the risk of a major climate change due to the release of CO_2 from burning of fossil fuels, with catastrophic consequences on the environment, is high. The development of sources with better compatibility with the environment and acceptable to society, such as fusion, as well as of more efficient energy technologies, should be pursued with a great determination. The potential of fusion as an energy source could be demonstrated in all of its main aspects by carrying out the ITER programme. © 1997 Elsevier Science S.A.

1. Introduction

The development of thermonuclear fusion as an energy source has now reached the crucial stage of launching the full feasibility demonstration phase with the construction and operation of an experimental reactor, called ITER (International Thermonuclear Experimental Reactor) [1]. ITER will have the size, the power and most of the technologies of a reactor and it will require a large, prolonged and coordinated effort from the partners involved.

To be successful, such an enterprise, which has no precedent in the history of science, requires on the part of the partners strong determination, which can only stem from the conviction that fusion is an energy source with such a potential of social acceptability that it must be made available to mankind. In the following we shall discuss reasons in support to this conviction, addressing the following questions:

- Is energy an issue today?
- How can future energy demand be met?
- How can fusion contribute to future energy supply?
- How and when is fusion going to demonstrate its potential as an energy source?

2. Is energy an issue today?

2.1. Energy and environment

At this time when the supply of fossil fuel is plentiful, cheap and apparently secure many believe that there is no 'energy issue'. In fact, energy becomes an issue only in conjunction with political instabilities in the oil producing countries. Even a slight turbulence like the one in early September '96 in the Middle East has prompted a few editorials in newspapers reminding us that energy supply should receive more attention and, this time, with some emphasis on environmental compatibility. Once these political and regional instabilities are suppressed then the 'energy issue' rapidly disappears. Only the 1973 energy crisis had somewhat longer lasting effects on the energy supply structure in some countries in favour of nuclear fission and on new energy sources development. In spite of several initiatives, such as the Earth Summit in Rio de Janeiro in 1992, public opinion seems insufficiently aware of other 'instabilities', related to the energy 'quality' rather then 'quantity', which may have far more dramatic consequences and not only on a regional scale but on a planetary scale.

Some energy related issues on a planetary scale were addressed in the early seventies in the MIT study 'The Limits to Growth' [2] where it was argued that due to the depletion of non-renewable resources, population growth and pollution, a material limit to growth would be reached leading to a sudden and uncontrollable economic decline of society. Even this alarm did not last very long because, under the pressure of the 1973 energy crisis, new oil/gas resources were found, nuclear energy was allowed to expand and technical innovations were believed to reduce substantially the 'energy intensity' (i.e. energy required to produce a unit of gross national product). In the seventies the concern was mainly on possible limits 'internal' to the energy system rather than on limits 'external' to it such as the finite absorptive capacity of the environment as a whole. We should instead be aware that the environment simply cannot absorb for much longer at the present rate the 'waste' of human activities in general and of energy production in particular. Since the selection of primary energy sources for electricity production is driven by economic considerations it is now time that energy sources be judged and ranked for the best combination of 'all' costs including the so called 'externalities' [3], namely the costs to the environment and to human health associated with the use of each energy source, costs which are now, in general, passed over to society and to future generations.

2.2. Present energy consumption and sources of energy [4]

The total energy consumption in the world amounts to about 13 TWY[1]. The average consumption pro capita is about 2.2 kWY but vast disparities among different countries exist, e.g. US, 11 kWY; EU 5, kWY; China, 0.8 kWY and India, 03 kWY. The richest 20% of the world population use 55% of the primary energy and their pro capita consumption is almost five times as much as for the rest of the world population.

The global consumption of 13 TWY is shared among different sources as follows: nuclear (6%), hydro (7%), biomass (10%), fossil (77% of which 45% from oil, 30% from coal, 25% from natural gas).

Electricity production is responsible for about 30% (3.7 TWY) of the total energy consumption and the primary sources for it are: fossil (60%), nuclear (20%) and hydro (20%).

These data confirm the dominant role of fossil sources in the supply of energy both globally and for electricity. The role of fossil fuels is justified by the oil price (at a constant dollar today's price is only twice as much as it was before 1973 crisis), by the prospects of abundant reserves which, in spite of increased consumption are today estimated to last longer then 25 years ago (50 vs. 25 years) and, finally by the dramatic increase in the natural gas reserves estimated to be at least as

---

[1] TeraWattYear (TWY) corresponds to the energy produced in 1 year, for instance, by 1000 electrical power stations of 1 GW each.

large as the oil reserves. Furthermore, the present trend is to increase further the role of fossil fuels in electricity generation using natural gas because of its more favourable economic prospects than other sources (e.g. smaller size, shorter construction time, higher efficiency).

2.3. How the energy demand may evolve till mid next century

The world population, according to most studies [4], will increase from the present 5.3 billion to about 10–11 billion in the middle of next century and possibly stabilise thereafter. This increase is expected to occur mostly in the developing countries and to be accompanied by a large concentration in urban areas (from present 50 up to 75%). In the period 1970–1990 the pro capita total energy consumption has increased by 1% per year and electricity consumption by 3% per year. For the future a most prudent scenario assumes a significant decrease of energy intensity (0.5%/year) allowing pro capita primary energy consumption to increase at a lower rate than in the past (0.5%/year). This assumption implies a development of energy efficient technologies which would require important resources not necessarily available at times of abundant fossil fuel supply. According to this scenario by the year 2050, the pro capita primary energy consumption would increase by about 30%. This means that, even if such an increase is concentrated only on developing countries, they would reach in 2050 a pro capita energy consumption still less than half of the one in the EU today.

Under these prudent extrapolations the total energy demand in 2050 would more than double, approaching 30 TWY.

Over a third of the total energy would be used to produce electricity having taken into due account both the likely faster increase rate in the demand and the improvement in the conversion efficiency. Actually, in the last 5 years in the richest 20% of the world population, primary energy demand grew at the same rate as the economy, i.e. energy intensity did not improve. This extrapolation is also prudent considering that today the richest 20% of the population use 75% of all electricity. This share of electricity is very close to the share of the world Gross Domestic Product.

2.4. Can the present energy sources meet the future energy demand?

2.4.1. Energy reserves

The proven recoverable reserves of energy can be summarised as follows (measured in TWY and in years of duration at present consumption rates):

Coal	800 TWY	300 years
Oil	200 TWY	50 years
Gas	200 TWY	80 years
Nuclear (U)	80 TWY	100 years

Nuclear reserves refer to the Uranium cycle as used in most of today's reactors: thorium cycle or fast breeder reactors would increase the reserve by two orders of magnitude or more.

The ultimately recoverable reserves, namely those yet to be identified, and likely at higher cost, are larger by a factor of between three to five than the proven ones.

The 'internal' limits to energy sources, namely the availability of reserves appear not so stringent as to exclude a 'business as usual' approach for the next 50 years. It is therefore appropriate to verify the implications of this approach and in particular its compatibility with 'external' limits such as the finite capacity of the self-regeneration of the environment.

2.4.2. Fossil fuels

The pollution of the atmosphere with CO_2 produced in fossil fuel combustion raises the highest concern. Every day 60 million t of CO_2 are released into the atmosphere, i.e. 730 g kWh^{-1} for coal, 560 g kWh^{-1} for oil and 430 g kWh^{-1} for gas. The 'greenhouse' effect, which controls the earth's climate, is due largely ($\sim 60\%$) to the CO_2 content in the atmosphere which has increased exponentially, due to fossil fuel burning, from 280 to 360 ppm in the last 130 years and the average temperature on earth has increased by 0.6° [5].

If the present rate of fossil fuel consumption continues, by the year 2050, the CO_2 content in the atmosphere will reach 560 ppm and the temperature may further increase by 1° or more. The increase in the CO_2 content corresponds to about half of CO_2 released annually by fossil burning (~ 20 billion t) the other half being absorbed largely by oceans. Although the exchange rate among the large reservoirs of CO_2 (oceans, biosphere and atmosphere) is probably one order of magnitude larger than the CO_2 release rate in the atmosphere, still the delicate equilibrium between these large fluxes may be altered either by saturation phenomena in the ocean absorptive capacity or by approaching an unstable regime. Deep ice drilling in Greenland and in the polar regions indicate [6] that large changes in the climate occurred in a very short time, say a few decades, suggesting that the relatively fast accumulation of CO_2, as occurring now, may lead to an instability in the climate [6]. Although large uncertainties exist in the global warming predictions, it is very dangerous to wait for their experimental validation because the time constant for climate changes may be much shorter than the time constant (at least 100 years) of CO_2 exchange between surface water and deep ocean. It is well known that the consequences of this climate change would be catastrophic and some of them almost irreversible, e.g. sea level increase, increasing climate turbulence and instabilities, desertification, modification of ocean streams, glaciation, etc.

Studies have been made to sequester the CO_2 produced by electrical power stations either in the oceans or in the gas/oil empty well. This approach, in addition to an increase of electricity cost of 40/80%, would not exclude on the long term damage to the environment.

In addition to the environmental damage such a high share of fuel consumption would be likely to induce strain on oil and gas availability, affecting industry and services now based on these fuels (e.g. transport, chemical, pharmaceutical, etc.) and causing political instabilities given the high concentration of these fuels in a few areas. On the other hand the use of fossil fuel is essential in the medium term to the developing countries (for example the coal share in the energy supply of China and India is 75 and 50%, respectively) which will account for most of the increase in energy demand in the next 50 years. A reduction of the overall fossil consumption can only come from economically developed countries by reducing primarily the fossil share in their electricity generation.

2.4.3. Nuclear fission

Uranium cycle reactors, provided that their social acceptability improves, could contribute to lessen the dependence on fossil fuels but, in any case, only for the medium term. Thorium cycle reactors, recently proposed [7], would have the potential for long term energy supply if proven to be feasible and superior to the U-cycle reactors as for safety and environment. Fast breeder reactors would also have the potential for long term energy supply but their social acceptability is seriously in doubt in most countries.

2.4.4. Renewable

- Hydropower: further exploitation of this source is limited and not exempt from a negative impact on the environment and on local community life.
- Biomass: this source is already covering a large fraction ($\sim 10\%$) of the total energy consumption and meets a variety of needs particularly in the developing countries. Any further exploitation for industrial use on large scale would imply an excessive use of land, fertiliser, water (e.g. 15/30% of today's world agricultural land to produce 1 TWY).
- Photovoltaic: this source can be very valuable for specific applications (e.g. limited power, remote locations, high solar radiation) but due to land requirements (e.g. 100 km^2 for 1000 MWe), service discontinuity and cost (e.g. up to ten times conventional sources) photovoltaic sources will not be able to play a quantitatively significant role in future electrical energy supply.

3. How can future energy be met?

The previous brief analysis shows that:
- It is necessary for environmental and socio-economic reasons to reduce the overall share of fossil sources recognising, at the same time, that developing countries will for economic and technical reasons rely heavily upon fossil fuel in the medium term.
- Present non-fossil energy sources may not be able of meeting in the long term the increasing energy demand and, at the same time, to satisfy correct social acceptability criteria. It is therefore a responsibility primarily of industrialised countries to develop energy source options capable to meet such criteria. These criteria can be summarised as follows:
  - fuel: supply abundant and largely available to all countries, i.e. not constrained by political instabilities;
  - safety: all 'internal' accidents caused either by plant failures or by operator mistakes and conceivable 'external' accidents should not disrupt the life of the population (e.g. no evacuation).
- Environment:
  - pollution: human health and eco-system not affected by emission into atmosphere;
  - waste: isolation from environment, if necessary, only for no longer than a few generations;
  - other environmental impacts: minimum use of land and of fresh water.
- Proliferation: no direct linkage to nuclear weaponry.
- Affordability:
  - cost accessibility: energy cost competitive and predictable;
  - technical accessibility: source technology manageable by countries most in need of energy.
- Service suitability: no severe constraints on size and location.

The adoption of these criteria to present energy sources would have an important impact on their economic assessment.

4. How can fusion contribute to future energy supply?

4.1. Basic process

Fusion is the source of energy for the sun and stars. On a laboratory scale the least difficult process is the fusion of two hydrogen isotopes, deuterium (D) and tritium (T), which leads to the release of 17 MeV of energy shared among the two resulting products (a helium nucleus and a neutron) 20 and 80%, respectively. In a fusion reactor about 1 g of D-T mixture, contained in a toroidal vessel of about 2000 m^3 volume, would be brought to the burning temperature of about 100 million degrees, injecting external energy as particles beams or as electromagnetic waves. The hot mixture (called plasma) is thermally isolated from the vessel by magnetic fields. Such insulation should be sufficient to allow the burn to be 'self-sustained' namely by the 'internal' energy carried by the helium nuclei, which, being electrically charged, are trapped by the magnetic fields. As far as radioactivity inventory is concerned, only one element of the fusion reaction, tritium, is radioactive with a decay time constant of only about 12 years and low toxicity. However, the energetic neutrons (~ 14 MeV) can induce radioactivity in the structural material surrounding the plasma to a level which depends on the type of materials used.

4.2. Assessment of fusion against social acceptability criteria

4.2.1. Fuel supply

Fusion fuel supply is available to all countries and is unlimited on the scale of human civilization. Tritium exists in nature only in negligible quantities but it can be produced by bombarding lithium with neutrons. This can be done inside the fusion reactor itself which therefore can breed its own tritium from a layer of lithium surrounding the reactor plasma avoiding fuel transportation outside the plant. Therefore the primary fuels in a fusion reactor are deuterium and lithium. The energy released by a fusion reaction is one million times larger than the energy released by a chemical reaction and 1 g of primary fusion fuel (deuterium and lithium) would deliver as much energy as 1 t of coal. The deuterium and lithium contained in sea water amounts to many billions of tonnes and would cover the full demand of fuel for electricity production for millions of years.

4.2.2. Safety

No population evacuation would be required for all in-plant accidents and conceivable ex-plant accidents, i.e. a fusion reactor would realize the 'walk away' condition because (i) there is no chain reaction and reactivity excursions of the plasma are limited by inherent processes (e.g. plasma instabilities) within seconds, well before the fuel in the vessel would be burnt (less than 1 min); (ii) low energy density and low afterheat. The fusion reactor is known for its low power density (~ 1 MW m^{-3}) which, while having a negative impact on capital cost, favours very much safety. Passive cooling is sufficient to prevent afterheat from causing melting in the structures; (iii) the entire release of mobilisable radioactivity in fusion reactor would not require population evacuation (i.e. dose < 50 mSv) because most of the radioactive inventory would remain immobilised in solid metal structures. No safety issues exist concerning the fuel cycle because, apart from the first start up charge of tritium, no transport of radioactive fuel outside the plant is required.

4.2.3. Environment

During normal operation no pollution of the atmosphere will occur. Leakages of radioactive substances, e.g. tritium, will stay below 10% of natural radioactivity.

After about 100 years the radiotoxicity index for the total activated materials falls to a level comparable with the ashes from coal fire plants. Active cooling may be required for a small portion of the waste but only for a few years.

Use of land and fresh water: similar to a fission reactor of comparable power.

4.2.4. Proliferation

No fissile material is present in a fusion reactor. Production of fissile materials is in principle possible but it would be a very complex operation and easily detectable.

4.2.5. Affordability

4.2.5.1. Cost accessibility. It is obviously very difficult to estimate the cost of energy produced by a process like fusion still under development and which may become commercial in about 50 years from now. However, the degree of uncertainty in extrapolating the cost of already developed sources may not be smaller than in the case of fusion because of the impact of 'externalities' previously discussed. Furthermore, already now, but more and more in the next few years, we will be able to have a very sound data base to ascertain the upper limit of the capital cost of a fusion reactor via the finalisation of the ITER design with world industry. Since the cost of electricity from a fusion reactor will be dominated by the capital cost, for a given availability, the main source of uncertainty will rest with the assumption of the cost reduction going from the 'first of the kind' (i.e. ITER) to the 'tenth of the kind' reactor. A recent study (Hender et al., UKAEA/Euratom Association, personal communication) where the costing code was validated against ITER capital cost established mainly by world industry and assuming a capital cost reduction of 20% for the 'tenth of the kind' reactor, a modest improvement in plasma performance and an availability of 75%, then the cost of electricity would be about 50–80% higher than the electricity cost of a present-day fission reactor. This cost disadvantage could substantially reduce or even disappear if 'externalities' are properly 'internalised' also for other sources of energy and because of savings for fusion resulting from easier siting, licensing and waste disposal and reduced safety requirements.

4.2.5.2. Technical accessibility. Countries having experience with nuclear fission energy would be able to gain the know-how necessary to operate a fusion reactor. Countries expected to share most of the future energy demand such as China, India, Brazil, would have such a know-how.

4.2.6. Service suitability

Fusion reactors will have a large size (e.g. 1 GWe) suitable for base load operation in a powerful network and they will have technical site requirements similar to other power stations of similar size.

5. How and when is fusion going to demonstrate its potential as an energy source?

The world fusion scientific community now agrees that the time is ripe to launch the final phase of fusion scientific and technological feasibility demonstration. Almost 40 years of fusion research will culminate in this demonstration that will be carried out in the experimental reactor ITER, now in the final phase of engineering design.

Since the early eighties the fusion community has focused its R&D activities to provide the database to ensure the successful construction and operation of an experimental reactor. During this time the reactor relevant plasma performance has improved by a factor of 100 and a factor of about five separates the present JET plasma performance from the self-sustained thermonuclear burn (i.e. ignition) required in ITER. A larger extrapolation is still necessary to reach stationary burn conditions which imply an efficient system of disposing of 'ashes' (i.e. helium) and of impurities. On the engineering side, industry, which has full command of the technology of the present machines, will soon, by completing several prototypes of the main components, have gained also the necessary technological know-how to build ITER. The step to ITER is technologically rather large because, among others, it brings fusion research fully into nuclear technology. ITER would pose therefore to the technical fusion community a new great challenge which explains the thoroughness and the length of the design activities and of the supporting R&D which has been carried out since the early eighties. The fusion community's has however recognised the need of a closer partnership with Industry during the design which should continue beyond the construction during the operation phase. The aim is for industry to gain full knowledge of the fusion 'process' and be able to move to the following step of the prototype reactor.

ITER will have the scale and incorporate the key technologies of a reactor because, contrary to fission reactors, in magnetic fusion plasma physics laws impose the 'full' scale if the demonstration really aims at producing a plasma of reactor relevance. Therefore the power produced can reach 2 GW, the wall loading 2 MW m^{-2}, the burn duration is at least 1000 s and possibly steady state and the construction cost about 7 billion ECU. In general, the operating conditions of most components and the requirements on remote handling and safety are as demanding as in a future reactor. For these reasons in many aspects there will be no risky extrapolation necessary from ITER to a reactor with the important exception of the materials of the components exposed to direct plasma interaction in particular to neutron irradiation. They will need further development (to be carried out in parallel to the ITER programme) using an adhoc neutron source to ensure a component lifetime compatible with the availability requirements of a reactor and a residual radioactivity compatible with stringent environmental criteria of limited time waste disposal. ITER construction and operation will also provide a very reliable database to assess the capital cost of a reactor and its suitability to utilities requirements.

In summary the ITER programme to be carried out in the next 25 years will offer the information necessary to make a sound assessment of fusion as an energy source in its scientific, technological, environmental and economical aspects. If this demonstration is successful, commercial fusion could be available by the middle of next century and become gradually the substitute for U-fission reactors, possibly constrained at that time by fuel shortage and severe waste management issues and for fossil fuel large power stations.

6. Conclusions

The conclusions can be summarised by giving concise answers to the questions posed in the introduction.

Is energy an issue today? Yes, because:
- energy demand may triple in the next 50 years;

- present energy share among sources, if maintained, could cause very severe, partly irreversible damage to the environment and affect economic growth.

How can future energy demand be met?

- In the next few decades it is mandatory to develop technologies in order to reduce energy intensity, to improve social acceptability of present sources and to develop fully acceptable new sources.

How can fusion contribute to future energy supply?

- Fusion has the potential to meet the highest standards in social acceptability and therefore to be competitive with other sources if assessment criteria include social and environmental costs.

How and when is fusion going to demonstrate its potential as an energy source?

- The demonstration of fusion potentials as an energy source can only be done by building and operating ITER which has size, plasma parameter, key technologies and operating conditions fully relevant to a reactor.
- The fusion community's ability to carry out this demonstration in partnership with industry is supported by a very successful record of scientific, technological and managerial achievements and by having, since the early eighties, finalised the R&D programme in support of the experimental reactor design.
- The core of the ITER programme can be carried out in the next 25 years offering the possibility, if successful, to have fusion contributing to the electrical energy supply by the middle of the next century.

References

[1] ITER Interim Design Report, IAEA, 1996.
[2] The Limits To Growth, Universe Book, New York, 1972.
[3] Externe, EU 16520, 1995.
[4] Global Energy Prospectives to 2050 and Beyond, IIASA, 1995.
[5] J.J. Haughton et al. (Eds.), IPCC (Intergovernmental Panel on Climate Change) Scientific Assessment, Working Group I, 1995.
[6] W. Dansgaard et al., Nature, 364 (1993) 218.
[7] C. Rubbia et al., CERN/AT/95-44 (ET).

ITER overview

R. Aymar *

11025 North Torrey Pines Road, San Diego, CA 92037, USA

Abstract

The International Thermonuclear Experimental Reactor (ITER) is a joint project of the European Union, Japan, the Russian Federation and the United States with the objective to design, construct and operate a Tokamak burning plasma experiment. The present phase of the project, the 6 year Engineering Design Activity (EDA), is entering its fifth year. The major features of ITER are now well defined. The development of detailed engineering designs for the components, plans for the machine assembly, the support facilities, the site requirements, construction plans, schedule and costs and a safety assessment are progressing to schedule and will be completed by the end of the Engineering Design Activity in July, 1998, in order to provide a basis for a decision by the ITER partners to proceed to the construction phase. © 1997 Elsevier Science S.A.

1. Introduction

Following the successful joint work (1988–90) on a conceptual design for ITER, the ITER Parties—the European Atomic Energy Community (Euratom), and the governments of Japan, the Russian Federation and the US—entered into a 6 year agreement [1] signed in 1992 under IAEA auspices in which they undertook to conduct jointly the Engineering Design Activities (EDA) for ITER. These activities are conducted, under the authority of the ITER Council and the ITER Director, by the ITER joint Central Team, composed of ~170 scientists and engineers from the four ITER partners located at three internationally staffed Joint Work Sites—Garching, (Europe), Naka, (Japan) and San Diego, (USA), and by the Parties' Home Teams, comprising institutes, laboratories, universities and industrial participants spread throughout the four ITER Parties. The JCT has the primary responsibility for developing the ITER design and defining its technology qualifying tests. The Home Teams undertake agreed Design Tasks and Technology R&D Tasks needed to support the design. The Parties also undertake coordinated Physics development in support of ITER.

The overall programmatic objective of ITER, defined in the ITER EDA Agreement is to demonstrate the scientific and technological feasibility of fusion energy for peaceful purposes. ITER will accomplish this by demonstrating controlled ignition and extended burn of deuterium–tritium plasmas in a Tokamak confinement; with steady state as an ultimate goal, by demonstrating technologies essential to a reactor in an integrated

* Tel.: +1 619 6225233.

system, and by performing integrated testing of the high-heat-flux and nuclear components required to utilize fusion energy for practical purposes.

Detailed technical objectives and technical approaches have been set for ITER [2]. In terms of plasma performance, ITER should meet the objective of demonstrating controlled ignition and extended burn, in inductive pulses with a flat-top duration of approximately 1000 s and an average neutron wall loading of about 1 MW m^{-2}, and should also aim to demonstrate steady state operation using non-inductive current drive in reactor relevant conditions. For engineering performance and testing, ITER should demonstrate the availability of technologies essential for a fusion reactor, test components for a reactor and test design concepts of tritium blankets relevant to a reactor. Two operational phases are foreseen for ITER. The Basic Performance Phase is expected to last a decade and to include a few thousand hours of deuterium/tritium operation. It should address the issues of controlled ignition, extended burn, steady-state operation, and the testing of breeding blanket modules. The Enhanced Performance Phase is also expected to last about a decade. Emphasis would be placed on improved overall performance and carrying out higher fluence components and materials testing programmes.

ITER must be designed to operate safely and to demonstrate the safety and environmental potential of fusion. The achievement of these safety-related objectives is integrated into the functional requirements for the design process and related R&D.

The main characteristics and parameters of the ITER design follow from the above programmatic and detailed technical objectives. ITER must provide a full-scale experimental working model of the core of a future fusion reactor in terms of its size, energy and neutron flux, expected activation and active intervention capacity, and overall safety characteristics. At the same time, ITER must also serve as an experimental facility permitting the fusion community to explore a wide range of fusion physics phenomena and operational domains. Flexibility is required in the design to allow access for introducing advanced features and new capabilities and to permit optimization of plasma performance during operation. There must also be capacity for intervention and repair by remote handling, making it necessary to have modularity in major systems where possible.

The technical challenges presented by ITER's objectives are pushing the frontiers of fusion technology into new domains. In particular ITER will embody:

- Unprecedented size of superconducting magnet and structures;
- Remote handling systems for maintenance intervention of an activated Tokamak structure;
- Extremely high heat flux in the Divertor;
- High heat flux and high neutron fluence at the first wall;
- Tritium breeding blanket test modules.

2. Current overall status of the activities

At the mid-point of the Engineering Design Activities, the project prepared an Interim Design Report, Cost Review and Safety Analysis which was approved by the ITER Council at the end of 1995 [3]. This report and related documents provided a basis for the ITER Parties to start discussions towards a possible agreement on ITER construction, operation, exploitation and decommissioning.

The approval of the Interim Design Report in 1995 has made it possible to freeze the main features of the design and has established necessary conditions for undertaking detailed costs estimates, following 'design to cost' principles. A very small number of design options are still open at conceptual level; task forces are working to finalize and recommend choices. Major R&D programmes are in place to validate the main technological issues in the design.

A Detailed Design Report is now being prepared for consideration by the Parties at the end of this year, which will contain a more mature design and cost estimate and the basis for a detailed safety analysis. Other major milestones foreseen in the schedule of the EDA are the Final Design Report (January 1998) and the Comprehensive Report to be delivered at the end of the EDA Phase. These documents and the information which supports them are expected to provide by the end of the EDA the technical information needed by the Parties as a basis for a possible construction decision.

3. Physics and engineering requirements

ITER will meet its objectives by achieving long pulse, ignited operation. Achievement of these objectives requires the attainment of $n_i T \tau_E$ of $\sim 5 \times 10^{21}$ keV s m^{-3} for long pulses [4]. The requirements for achieving this level of energy confinement are based on the results of the past 30 years of Tokamak experiments [5]. The data from these experiments indicates that a plasma current of ~ 20 MA with a toroidal field of ~ 6 T and an aspect ratio of ~ 3 (major radius/minor radius) is required to achieve ignition. The parameters for the ITER design (Table 1) were developed on the basis of a minimum cost engineering design which would provide the required plasma current and meet the other physics requirements, including MHD stability, plasma control, power and particle exhaust and alpha confinement.

The size of ITER is determined by the plasma current required for ignition, the choice of the aspect ratio, the MHD safety factor, q, the maximum plasma elongation for which reliable plasma control is possible with untrapped coaxial coils, the maximum stresses allowed in the toroidal field coils, the thickness of the neutron shield and the required flux swing for the inductively driven current pulse length.

The single null configuration has been chosen since it maximizes the volume and flexibility of the Divertor. The elongation has been set at the moderate figure of about 1.6 to make efficient use of the 'D' shaped TF coils and to have lower VDE loads on passive structures. An intermediate aspect ratio of about 2.9 minimizes the machine size necessary for the required burn time of 1000 s.

One hundred MW of auxiliary heating is provided to ensure that the plasma can be heated to ignition in a high confinement mode and to provide the capability to conduct experiments with significant levels of driven current. This level of auxiliary power ensures that ITER can meet its objective to produce a burning plasma even if there are shortfalls in the projected plasma performance. This level of auxiliary power is needed to meet the objective of studying the potential for steady-state operation. The decision to have 20 TF coils balances the requirement to have enough access space while maintaining the ripple at no more than 2% at the separatrix. The normalized beta toroidal of 2.4% is consistent with results from present Tokamak experiments, which observe a soft limit around this value.

The size of the Tokamak, together with a requirement for the average neutron flux on the

Table 1
Major engineering design parameters for ITER

Parameter	Symbol	Value
Major/minor radius	R/a	8.14 m/2.80 m
Plasma configuration	—	Single-null divertor[a]
Nominal plasma volume and plasma separatrix surface area	V (plasma)	~ 2000 m^3
	A_S	~ 1200 m^2
Nominal plasma current	I	21 MA
Toroidal field	B	5.68 T (at $R = 8.14$ m)
MHD safety factor	q_{95}	~ 3.0 (at $I = 21$ MA)
Fusion power (nominal)	P_{fus}	1.5 GW
Average wall loading	Γ_n	~ 1 MW m^{-2} (at 1.5 GW)
PF flux swing	$\Delta\phi_{PF}$	530 Wb
Flux swing for burn	$\Delta\phi_{burn}$	>80 Wb
Volume-averaged temperature	$\langle T \rangle$	11 keV
Volume-averaged density	$\langle n \rangle$	1.1×10^{20} m^{-3}
Burn duration (ignited)	τ_{burn}	≥ 1000 s
Number of toroidal coils	N	20
TF ripple	δ_{ripple}	$\leq 2\%$ at plasma separatrix
Normalized toroidal beta	β_n	2.4%
Auxiliary heating power	—	up to 100 MW
Plasma magnetic, thermal energy	W_{mag}, W_{th}	1.1, 1.2 GJ

[a] With its channel poloidal extension ≥ 2 m.

wall of ~ 1 MW m^{-2}, determines the magnitude of the fusion power and the stored thermal and magnetic energies. The large fusion power leads to high peak heat loads on the plasma facing components and to the requirement to exhaust the He ash and refuel the plasma if pulses longer than about 50 s are to be achieved. The plasma current and plasma thermal and magnetic energies are much larger than in present Tokamaks, and lead to large energy fluxes and electro-mechanical forces on the Tokamak components during disruptions. Plasma diagnostics are needed to measure the plasma parameters for plasma control and for characterizing the plasma behavior and performance.

Current assessments of likely performance indicate that, with the above parameters, ITER has high probability to achieve the goal of ignition. However there are inevitable uncertainties in the extrapolation from current experience. Driven burn ($Q > 15$) provides a major means for offsetting physics or engineering shortfalls and will still allow ITER to achieve its physics and technology testing objectives (e.g. to provide a flux of 1 MW m^{-2} of 14 MeV neutrons for a long enough time ≥ 1000 s).

4. ITER design

The primary ITER systems are shown in Fig. 1.

The plasma is surrounded by a blanket and shield and divertor. This assembly is located in the vacuum vessel. Access to the Tokamak interior is provided through vertical and horizontal ports in the main chamber and to the divertor through ports at the bottom. The toroidal field magnets encase the vacuum vessel. Plasma control is provided by seven poloidal field coils and a central solenoid.

Detailed designs and requirements are being developed for all of the other support systems and facilities, including the power supply systems, the tritium plant, the heat exchange system, cryogenic plant and the general plant facilities and buildings.

4.1. The blanket and shield

It must protect the other Tokamak components from the neutron power flux of up to ~ 1.2 GW. It will also contribute to the passive stability of the plasma since rapid motions of the plasma produce eddy currents in the shield. The blanket and shield is composed of 720 water cooled individual modules with rough dimensions of $1 \times 2 \times 0.4$ m^3. The modules are attached to a monolithic backplate for support against the force caused by eddy currents due to disruptions and by 'halo' currents from 'vertical displacement events' following the loss of control of the plasma position (Fig. 2). The front surface of the blanket is comprised of castellated Be bonded to a copper alloy heat sink to diffuse the first wall heat loads to the water-cooled stainless steel blanket modules. Toroidally symmetric limiters are built into the blanket and shield for plasma start-up and shut down when a divertor configuration cannot be maintained. The lower section of the blanket and shield also provides baffling to maximize the confinement of neutrals in the divertor chamber. For maintenance, the blanket modules can be individually replaced with a remote manipulator with access through the horizontal ports. Replenishment of eroded portions of the Be cladding can be accomplished with in-situ plasma spraying without removing the blanket modules.

4.2. Poloidal divertor

Alpha power and particle exhaust is provided by a single null poloidal divertor located at the bottom of the vacuum vessel. Most of the plasma heating power is spread out over the first wall in the main chamber by Bremsstrahlung and impurity radiation from the plasma edge and over the divertor chamber walls by impurity radiation in the divertor plasma. Small quantities of recycling impurities such as Ne or Ar will be added to the divertor plasma to enhance the radiation losses. This reduces the peak heat loads on the divertor target plates to the 5–10 MW m^{-2} range.

The ITER divertor comprises 60 cassettes, each 5 m long and 0.5 m to 1 m wide mounted on toroidal rails welded to the bottom of the vacuum

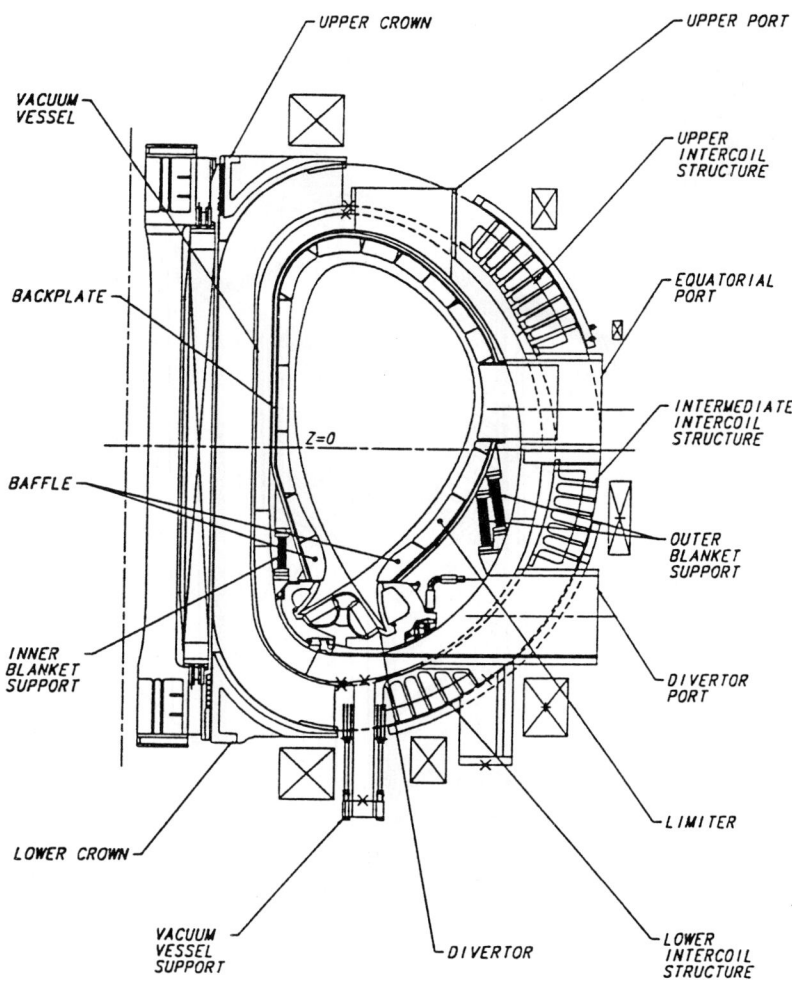

Fig. 1. Cross section view of ITER.

vessel (Fig. 3). The divertor cassette concept has been adopted because cassettes can be rapidly exchanged for repair and easily modified to incorporate new configurations. Primarily because it has the best performance for disruption and other transient heat loads, Carbon Fiber Composite is used as the plasma facing material where the plasma strikes the divertor plate to maximize the erosion lifetime. The rest of the divertor is clad with tungsten to minimize the erosion by low energy charge exchange neutrals. The plasma facing materials are bonded to a water cooled copper alloy heat sink. The divertor is tightly baffled to maximize retention of neutrals in the divertor chamber.

Particle exhaust is accomplished by 16 cryopumps located in the divertor ports. The effective pumping speed of the system is 200 m^3 s^{-1}. Fueling (on average 200 Pa m^3 s^{-1}) will be provided by a gas puffing system in the main chamber and in the divertor, and by two 1.5 km s^{-1} centrifuge pellet injectors. Due to the need to minimize the tritium inventory, the cryopumps will be regenerated during the pulse and a gas purification and separation system will allow recirculation of some of the gas during the pulse to minimize the tritium inventory.

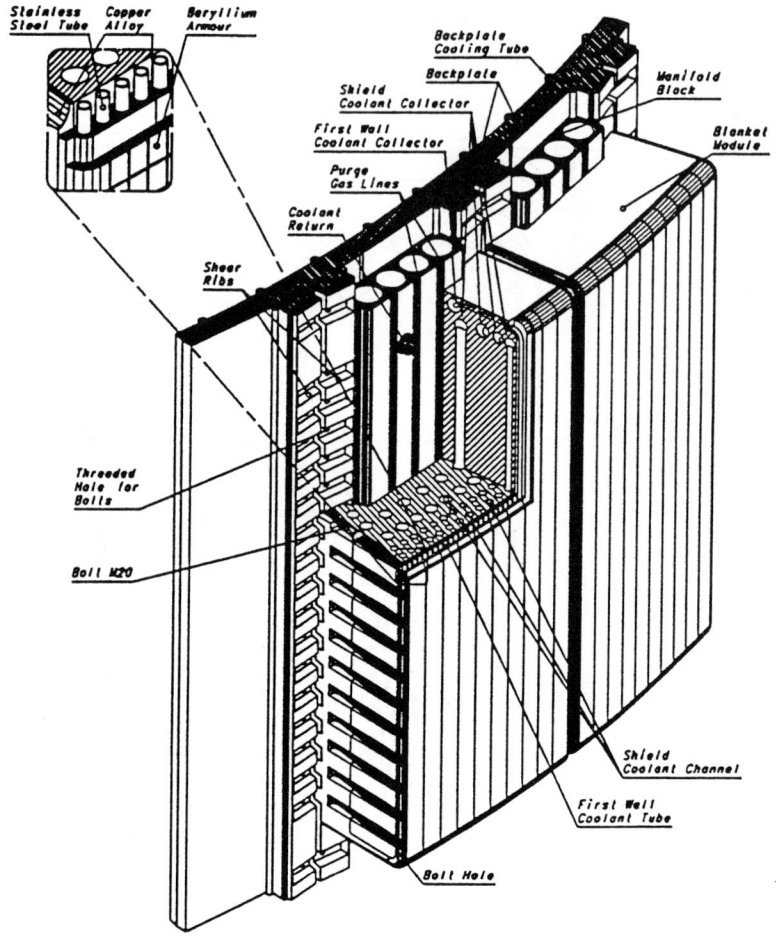

Fig. 2. Cutaway view of the first-wall/shield module located at the inside mid-plane, showing a mechanical attachment to the backplate. Other variants of the method of mechanical attachment, as well as a welded attachment, are under consideration.

4.3. The vacuum vessel

It provides the high vacuum boundary for the plasma and is the primary safety barrier against radioactive releases. It consists of 20 sectors welded together. It is composed of two SS 316 LN ribbed shells with a total thickness of 0.4–0.9 m. Water cooled steel plates between the shells provide neutron shielding. There are 20 sets of ports consisting of an upper, a mid-plane and a divertor port. The upper ports are used for water cooling manifolds for the blanket and the vacuum vessel. The mid-plane ports are used for access for auxiliary heating systems, diagnostics, maintenance systems for the main chamber components and test blanket modules.

4.4. Magnet systems

All magnet systems are superconducting. A very strong toroidal mechanical structure is built with 20 toroidal field coils in stainless steel casings, linked together by upper and lower crowns and bolted intercoil structures. The seven poloidal field coils and the vacuum vessel are directly attached to the TF coil cases. This arrangement of coils and vacuum vessel provides an integrated overall assembly which simplifies the equilibration of electromagnetic loads, relying mainly on the robustness of TF Coil cases. Coil cooling is provided by supercritical helium flow maintained by cryogenic circulation pumps. The toroidal field coils, the central solenoid and the two small diameter poloidal field coils use conductors built of Nb_3Sn strands and housed in Incoloy steel jackets. The larger diameter poloidal field coils conductors are made of NbTi.

The 60 kA TF coil conductor cable is laid in grooved radial plates, forming 'pancakes', which are then enclosed in thick coil cases (Fig. 4). Each toroidal field coil weighs about 750 tons. The magnetic field value reaches 12.5 T on the conductor and the total magnetic energy amounts to 100 GJ.

The central solenoid weighs about 1350 tons and is layer wound in a monolithic structure with the joints between layers in low field regions at the bottom and top. The solenoid supports most of the toroidal field coil centering forces. This mitigates its fatigue stress limits and partially reacts the out of plane forces acting on the TF coils. The large-ring poloidal field coils are supported by the toroidal field coils against vertical forces and are self-supported against radial forces. The five largest poloidal field coils are too large (32 m outside diameter) to be transported and will be manufactured on site. All of the other coils will be manufactured off-site and transported to the site for assembly.

Assessment of an alternative design with a segmented solenoid is underway. Such a variation would increase the flexibility for plasma shape control allowing access to a larger plasma triangularity and to the possibility of replacing the Nb_3Sn in the two smallest poloidal field coils with NbTi as well. One of the major issues that must be resolved for this case is the design of a reliable coil joint located in higher field regions.

4.5. The cryostat

The Tokamak is housed in a large cryostat ($\sim 30\,000$ m^3) which provides thermal insulation for the superconducting magnets and serves as a second safety barrier after the vacuum vessel. The ~ 400 large penetrations make achieving leak-tight conditions challenging.

Four candidate systems to provide 100 MW of auxiliary heating are presently being considered (Table 2). It is presently planned to make use of at least two systems with capability of current drive. In addition a 6 MW ECRH system is planned for start-up assist.

A full set of plasma diagnostics is planned to provide the ability to monitor and control the plasma performance and to measure the plasma properties as completely as possible to maximize the physics information obtained during the operation of ITER. The high heat and disruption loads, the need for extensive shielding and remote handling, the large size of ITER and the radioactive environment all pose challenges to make the required measurements.

4.6. Remote maintenance

Since direct access into the ITER machine will not be possible after the first few DT pulses, it is necessary that all maintenance of in-vessel components be carried out remotely. For those components that are expected to need replacement during the operational period, the main design policies have been to provide modularity (e.g. blanket modules and divertor cassettes), to optimize the combination of remote handling with hands-on assisted options and to minimize shutdown times. The high heat flux components in the divertor are expected to require replacement a number of times during the life of the machine.

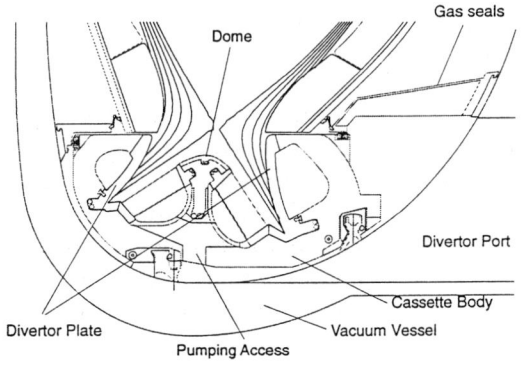

Fig. 3. Outline schematic of the ITER divertor cassette.

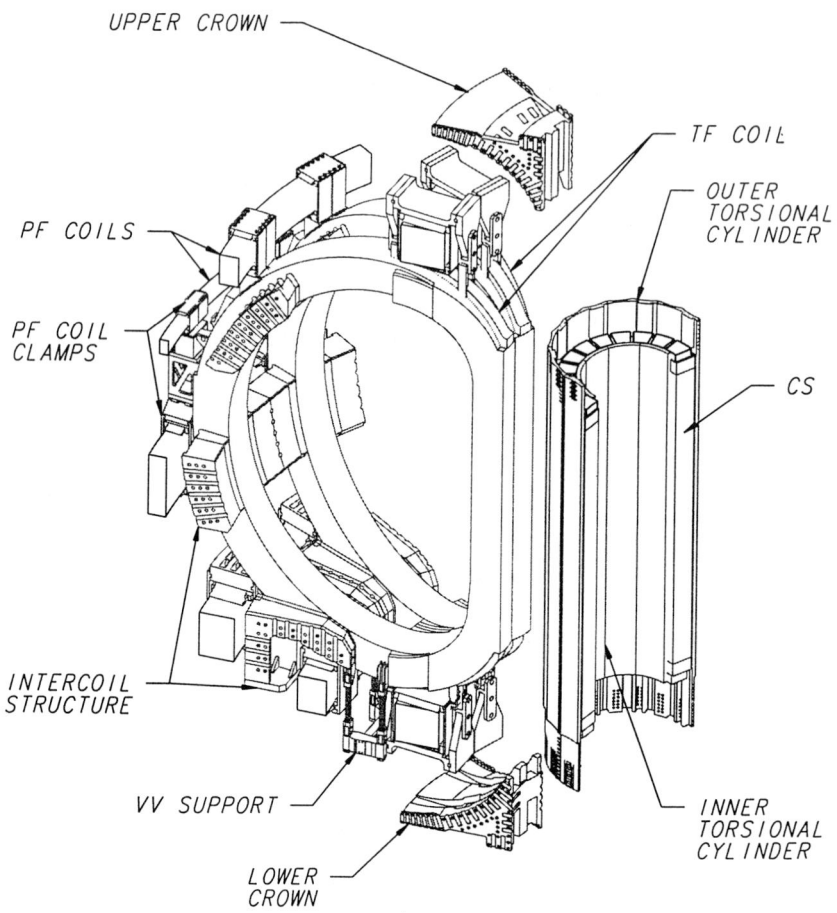

Fig. 4. 3-D view of the magnet systems.

The divertor cassettes are designed to be replaceable through four remote handling divertor ports, with the whole set capable of being replaced within 6 months. The spent cassettes will be transferred to transportation casks through the divertor ports and taken to a hot cell for repair and modification. Similarly the blanket modules will be replaced by an in-vessel manipulator through the equatorial ports. The replacement time for a single module is estimated to be 6–8 weeks, and 2 years for a complete change-out of all of the blanket modules. Ex vessel components have design lifetimes greater than ITER. Design policies for these components are directed at minimizing expected interventions, including the use of component redundancy. At the same time, conceptual designs for replacement options are provided.

Table 2
Candidate heating systems

System	Properties
Neutral Beams	1 MeV
Ion Cyclotron Resonance	40–90 MHz
Electron Cyclotron Resonance	170 GHz
Lower Hybrid	5 GHz

4.7. Site requirements

An initial set of general site requirements and site design assumptions has been developed to enable the ITER partners to begin to assess siting issues. These include specifications for the land area needed, the electrical power requirements, transportation access, water supply requirements and other issues. The technical requirements and assumptions are such that ITER could possibly be built on the territory of any of the Parties.

A plan for Tokamak operations has been developed for the first phase (Basic Performance), which allows for a measured approach to full ignited operation after an initial period of operation with hydrogen plasmas to allow commissioning of the Tokamak systems while 'hands on' maintenance is still possible. The plan calls for about 12 000 pulses with an integrated neutron fluence of approximately 0.3 MW year m^{-2}. This phase will rely on availability of outside procurement of tritium. The second phase, the Enhanced Performance Phase, will place a heavier emphasis on nuclear technology issues in view of a possibly following DEMO reactor. The integrated neutron fluence is then expected to be beyond 1 MW year m^{-2}. Installation of a tritium breeding blanket will be required for this Phase.

The ITER partners are collaborating in ITER to obtain information jointly on the potential of fusion as a practical energy source. It is therefore essential that the information developed during the operation of ITER be transferred from ITER to the institutions in all of the ITER partners. In addition, it is important to minimize the difficulties and costs associated with inter-continental relocation of large numbers of scientists and engineers. To achieve these objectives, plans are being developed to permit remote participation in ITER operations. ITER will have fast computer links to remote control and data collection sites in each of the ITER partners. In such scenarios, a central staff at the ITER site will operate ITER and maintain and service it. Scientists and engineers at remote control rooms will direct the ITER research program, plan and carry out experiments, and collect and analyze data at their home institutions. Present large experiments are already moving toward this type of operation. The expected advances in computer technology and electronic communication will make this type of operation even more feasible than it is now.

4.8. Safety

Safety is a major element of the ITER design development process. From this perspective also, the design has been developed to ensure that ITER can be sited at any of the ITER partners. Emphasis has been placed on maximizing the use of the favorable safety characteristics of fusion. ITER is designed to meet the appropriate dose and release limits. In addition, the design has been developed to minimize the impact and likelihood of accidents, with due consideration to uncertainties in the plasma physics and the performance of the in-vessel components. Detailed safety-related design requirements have been established based on internationally recognized safety criteria and limits. Within the design, major lines of defense against accidental releases have been identified, and their achievable reliability assessed.

A comprehensive non-site specific safety and environmental assessment has recently been completed for the ITER design. It was concluded that the ITER design meets successfully all the safety-related requirements that were established.

However, as a research facility, ITER requires flexibility in operation, acceptance of inevitable uncertainties in plasma behaviour and should be prepared for possible changes of components and scenarios during its lifetime. This means that some safety issues are specific to the design choice of ITER as an experimental machine, e.g. relatively high activation of in-vessel components made of SS316LN, and not necessarily representative of a future fusion reactor. Even so, the safety analysis shows that the design can provide a robust safety envelope, by using well-established concepts of defence in depth and multiple lines of defence against postulated accidents. This achievement relies heavily on the basic favourable safety characteristics of magnetic fusion, mainly intrinsic fail-safe termination of fusion power with off-normal conditions from the auxiliary environment and low decay heat density which cannot melt the structure in case of any loss of coolant. Even more important, mobilizable inventories of radioactive materials in ITER, tritium and activated metallic dust from plasma-wall interaction or from corrosion by water coolant, are such that effluents and emissions during normal operation and envisaged accident sequences are low. Even for the ultimate case of a hypothetical 'worst event' accident, the ground-level emission from 'at risk' radioactive inventories will result in an early dose lower than 50 mSv under average weather conditions: the objective of no off-site evacuation is met.

The EDAs are scheduled to end in July, 1998. The Construction Activity could start at that time if an agreement on construction and a site has been reached. Procurement of many of the long lead time major components such as the magnets could begin then along with site specific design work. Tokamak assembly could begin in 2003, with first plasma at the end of 2008.

5. Fusion technology R&D for ITER

An extensive technology Research and Development program for ITER is being carried out by the Home Teams. The program is defined to test and validate the design solutions developed by the design team. The work is now focused on seven critical areas, each the subject of a large fusion technology project aimed at validating key aspects of the ITER Design, including development and verification of industrial level manufacturing techniques.

5.1. Research and development projects

The so-called 'Seven Large R&D Projects' share certain common features. They are typically multi-stage activities involving multiple Party contributions and cross-dependencies, and high industrial content. Each has a unified management structure and organization in which Project responsibility is shared between the JCT and the Home Teams, with one (or, in one case, two) Home Team(s) designated to take a lead role in overall coordination of the project.

Fig. 5. Internal modules of the Central Solenoid Model Coil being constructed in the US. The Outer Module is being built in Japan.

Two of the Projects are directed towards developing superconducting magnet technology to a level that will allow the various ITER magnets to be built with confidence. The Central Solenoid Model Coil Project and the Toroidal Field Model Coil Project are intended to drive the development of the ITER full-scale conductor including strand, cable, conduit and terminations, and to integrate the supporting R&D programmes on insulators; joints; conductor ac losses and stability; Nb_3Sn conductor wind, react and transfer processes; and quality assurance. In each case the Home Teams concerned are collaborating relevant scale model coils and associated mechanical structures for installation and testing in dedicated facilities—the CS Model Coil in Japan and the TF Model Coil in the EU (Fig. 5).

Three Projects focus on key in vessel components, including development and demonstration of necessary fabrication technologies and initial testing for performance and assembly/integration into the Tokamak system. In the Vacuum Vessel Sector Project, the main objective is to produce a full scale sector of the ITER vacuum vessel, and to undertake initial testing of mechanical and hydraulic performance. The Blanket Module Project is aimed at producing and testing full scale modules of primary wall, limiter, and baffle type, and full scale, partial prototypes of coolant manifolds and backplate, and at demonstrating integration in a model segment. The Divertor

Cassette Project aims to demonstrate that a divertor can be built within tolerances and to withstand the thermal and mechanical loads imposed on it during normal operation and during transients such as ELMs and disruptions. To this end, a full size prototype is being built and tested. Because of the consequences of erosion of the plasma facing materials, the project also includes tasks to understand erosion mechanisms, to develop methods of dust removal and of outgasing tritium codeposited with Be or C, and to demonstrate the feasibility of plasma spray as a possible means for in-vessel repair of armour.

The last two of the large projects focus on ensuring the availability of appropriate remote handling technologies which allow intervention on reasonable time scales so as to provide the flexibility needed for ITER to pursue its scientific and technical goals. The Blanket Module Remote Handling Project is aimed at demonstrating that the ITER Blanket Modules can be replaced remotely. This involves proof of principle and related tests of remote handling transport scenarios including opening and closing of the vacuum vessel and of the use of a transport vehicle on monorail for the installation and removal of blanket modules. In the Divertor Remote Handling Development the main objective is to demonstrate that the ITER Divertor can be maintained and replaced remotely and that divertor cassettes can be remotely refurbished in a Hot Cell. This involves the design and manufacture of full scale prototype remote handling equipment and tools and their testing in a Divertor Test Platform (to simulate a portion of the divertor area of the Tokamak) and a Divertor Refurbishment Platform to simulate the refurbishment facility (Fig. 6).

The total cost of the seven large projects amounts to some $353 M (Jan 1989 value) distributed as shown in Table 3 below. The overall cost of the technology program for ITER was estimated at approximately 750 million dollars (1989 value).

Technical output from the seven large R&D Projects has direct importance in the validation of the ITER Design and in supporting the manufacturing cost estimates for some key cost drivers.

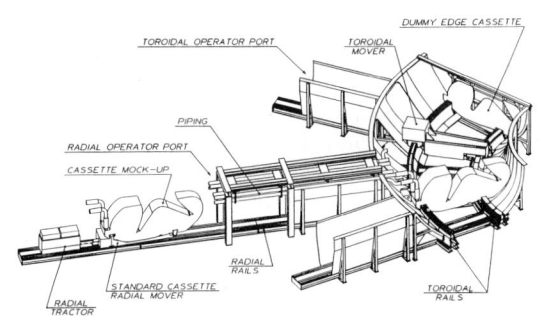

Fig. 6. The Divertor Test Platform.

But the Projects also have a more general importance as exemplars of cross-Party complex ventures and hence as precursors to possible joint construction activities. Already they have provided valuable organizational experience especially in achieving clear project management arrangements in terms of responsibilities, authority and liaison across the JCT, Home Teams and industries involved. Successful performance of the large projects within such an organizational framework will increase confidence in construction proposals.

5.2. Blanket developments

In view of ITER's mission to test DEMO-relevant fusion blanket modules, the Parties are developing their own plans for breeding blanket module design and construction for test in ITER. In addition, it is expected that ITER will itself require a breeding blanket to ensure an adequate

Table 3
ITER seven large technology R&D Projects

Project	Estimated costs $ M (Jan. '89)
Central Solenoid and Toroidal Field Coil	163
Vacuum Vessel	31
Blanket	62
Divertor	50
Blanket Remote Handling	21
Divertor Remote Handling	26
Total	353

Table 4
ITER physics expert groups

Confinement Physics
Confinement Modeling and Database
MHD Stability and Disruptions
Divertor Physics
Divertor Modeling and Database
Auxiliary Heating and Energetic Particles
Diagnostics

supply of tritium fuel during the later (Enhanced Performance) phase of operations. The planning for breeding blanket development and of testing programmes in ITER is proceeding in a coordinated way with the oversight of a Test Blanket Working Group in which the Parties and the JCT are represented.

5.3. ITER physics

A programme of Physics R&D is being carried out by the fusion communities of the four Partners. The programme is coordinated through an ITER Physics Committee supported by seven ITER expert groups composed of senior scientists from the major Tokamak experiments, fusion institutions and universities, and physicists from the Joint Central Team (Table 4). The expert groups are responsible for identifying the key physics issues for ITER design, and for developing and carrying out tasks to resolve those issues. For instance, the ITER Confinement Modeling and Database Expert Group assembles and analyzes energy and particle confinement data from the major Tokamak experiments. Although this program is voluntary in the sense that tasks are considered as part of each Partner's base fusion physics programme, the physics issues for ITER give a focus for Tokamak research on all of the major facilities. ITER also provides an umbrella under which the major facilities can coordinate their research programs and exchange data and information.

6. Conclusion

Each of the ITER Parties has been a full participant from the beginning of the project in 1988. Every major decision has been taken on a basis of consensus of all of the Parties. The ITER EDA represent a truly collaborative joint venture to which the Parties have committed, through assignment to the JCT or participation in the Home Teams and voluntary physics work, the best scientific support. The efficient collaboration between the ITER Joint Central Team (JCT) and the Fusion Community in the four ITER Parties and their associates has proved successful in the areas of technology R&D, detailed design, and manufacturing process and costs. At the same time there has been an impressive development and convergence of the Parties' voluntary contributions in the physics area; the important physics issues have been identified and all the major physics experiments in magnetic controlled fusion are providing high quality, relevant results in a coordinated fashion for consideration through the ITER Physics Expert Groups.

The success of ITER up to this point demonstrates that European, Japanese, Russian and American scientists and engineers can work jointly and effectively on a project at the forefront of science and technology. In this sense, ITER is not only a physics and engineering experiment but is also a very successful experiment in international scientific cooperation: it represents a paradigm for future large international scientific projects.

Recent experiments on TFTR and JET with deuterium and tritium have demonstrated the achievement of significant levels of fusion power. These experiments, together with the progress in Tokamak physics and engineering during the last 20 years gives us a sound basis to define the requirements for building and operating a reactor-relevant burning plasma experiment and confidence that such an experiment would meet its performance objectives. In December 1995, after approving the Interim Design Report Package the ITER Council reaffirmed its position 'that a next step such as ITER is a necessary step in the progress towards fusion energy, that its objectives are valid and timely; that the quadripartite cooperation has shown to be an efficient frame to achieve the ITER objectives, and that the right time for such a step is now'.

The recent technical progress in the ITER EDA reinforces this position. With continued collaborative work planned for the remainder of the EDA, there is every reason to be confident that the potential participants in construction will indeed be in a position to take a positive construction decision with full assurance in the engineering design and parameters and the cost estimate of ITER.

Acknowledgements

The work described in this paper is jointly contributed by members of the ITER Joint Central Team, the ITER Home Teams, and other ITER participants from the ITER Parties. This paper has been prepared as an account of work performed under the Agreement among the European Atomic Energy Community, The Government of Japan, the Government of the Russian Federation, and the Government of the United States of America on cooperation in the Engineering Design Activities for the International Thermonuclear Experimental Reactor ('ITER EDA Agreement') under the auspices of the International Atomic Energy Agency (IAEA).

References

[1] ITER EDA Agreement and Protocol 2, IAEA, Vienna, 1994.
[2] ITER Council Proceedings, 1992, IAEA, Vienna, 1994.
[3] ITER Interim Design Report Package Documents, IAEA, Vienna, 1996.
[4] D. Post, K. Borrass, J.D. Callen, S.A. Cohen, J.G. Cordey et al., ITER Physics, IAEA, Vienna, 1991.
[5] S. Kaye, J. Snipes, R. Granetz, M. Greenwald, F. Ryter, et al., Plasma physics and controlled nuclear fusion research 1994 2 (1995) 525.

The ITER magnet system

M. Huguet

ITER Joint Central Team, Naka Joint Work Site, Naka-machi, Naka-gun, 801-1 Murouyama, Ibaraki-ken 311-01, Japan

Abstract

The main components of the ITER magnet system include 20 toroidal field (TF) coils, a central solenoid (CS), seven poloidal field (PF) coils and structural elements. The TF coils operate at a maximum field of 12.5 T with a total stored energy of about 100 GJ. These coils are enclosed in stainless steel cases that form the major part of an integrated mechanical structure and are bucked on the CS. The CS provides about 140 MA turns at a maximum field of 13 T. In the reference design, the CS is a monolithic, 12 m tall coil. An alternative segmented CS design is being considered to extend the range of operational flexibility and plasma control capability. The PF coil system comprises seven coils, with diameters up to 30 m, for plasma and shape control. The salient activity of the ITER R&D programme is the manufacture of model coils. The conductor production for these coils is well underway and construction of the coils has started. The initiation of the model coil test programme is scheduled for mid 1998. © 1997 Elsevier Science S.A.

1. Introduction

The International Thermonuclear Experimental Reactor (ITER) is a tokamak with a nominal plasma major radius of 8.1 m, plasma minor radius of 2.8 m, elongation of about 1.6 and plasma current of 21 MA. The toroidal magnetic field at the major radius is 5.7 T. The project goals include the demonstration of controlled burn of DT plasmas with a duration of about 1000 s. The average neutron loading at the first wall will be about 1 MW m^{-2} and a total fluence of at least 1 MWy m^{-2} should be achieved for blanket and material tests [1].

The magnet system comprises 20 toroidal field (TF) coils, a central solenoid (CS), seven poloidal field (PF) coils and a mechanical structure. In addition, the PF correction coils provide a small component (about 10^{-4} of the toroidal field) of magnetic field to correct some components of non-axisymmetric field errors. All ITER coils are superconducting and use either Nb$_3$Sn or NbTi superconducting material depending on the field value at the conductor. The magnets and structures are located within a cryostat which provides the vacuum for thermal insulation from the ambient heat load.

2. Description of the mechanical design

Magnet systems are designed for the full machine operation life of 50 000 tokamak pulses without repair or replacement. Since the space available for structures is limited for design efficiency and optimum tokamak performance, the mechanical design is strongly integrated and takes advantage of mutually compensating force systems. Thus, the TF coils are bucked on the CS and the TF coil cases, which form the main structural component of the ITER tokamak, contribute to the containment and equilibration of the electromagnetic and seismic force systems acting on the TF and PF coils and on the vacuum vessel [2]. An elevation view of the magnet systems is shown in Fig. 1.

2.1. Containment of loads acting on the TF coils

The radial plates and cases of the TF coils (Section 3) contain the tensile hoop loads. The centripetal force acting on each of the TF coils is reacted by the CS assembly. This allows the CS to operate under compressive loads, suppressing fatigue crack growth in the conductor conduit. The CS assembly includes the outer cylinder (OC), the CS winding pack and the inner cylinder (IC). The primary function of the OC and IC is to increase the stiffness of the CS as a bucking structure and to keep the compressive hoop stress in the CS winding pack within allowable limits.

A fraction of the centripetal force of the TF coils is also reacted by outer intercoil structures (OIS) linking the outboard legs of the TF coils.

To ensure that the CS operates under compressive stress, a preload structure is provided to compress the CS vertically at assembly and during cool-down of the machine. This is because when the TF coils are energized, the inboard legs of the TF coils stretch and vertical tensile strain can be transferred by friction to the CS assembly.

The tokamak twisting moment due to out-of-plane (azimuthal) forces acting on the TF coils is reacted within the TF system by the OIS linking the outboard legs and by crowns in the upper and lower regions of the TF coils.

The OIS are integral parts of the TF coil cases and are linked in the meridian planes between coils so as to form toroidal shear belts. Shear load transmission at the links is accomplished with insulated bolts and shear keys. The crowns also provide support against shear deformation by means of radial slots which engage rails attached to the TF coil cases. The stiffness of the TF coil cases to out-of-plane bending and torsion is also important for the overall structural rigidity.

The OC of the CS assembly locates the inboard legs of the TF coils by means of fluted grooves on its outer surface that engage the nose of each TF coil, machined as a cylindrical scallop.

2.2. PF coil and vacuum vessel supports

Vertical loads on the PF coil are transferred to and balanced by, the TF coil cases through flexible members which allow relative radial displacements between TF and PF coils, but are rigid in the azimuthal direction. Radial loads on the PF coils are carried by hoop tension in the conductor conduit.

Plasma disruptions and vertical displacement events (VDEs) can generate forces which result from a large fraction of the plasma current flowing in First Wall components. These forces act in the vertical direction (up to 150 MN) and radial direction and are transferred to the vacuum vessel (VV). The forces are not axisymmetric and the radial forces can generate a net horizontal force (up to 50 MN). The VV is attached to the TF cases in the vertical and horizontal directions in order to contain and balance the total vacuum vessel force system within the TF magnet assembly.

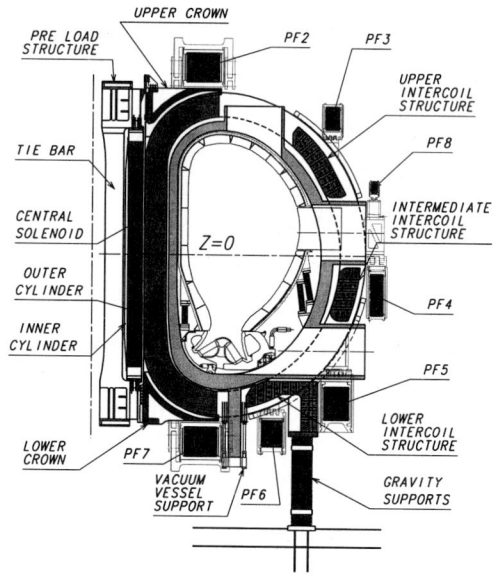

Fig. 1. Elevation of ITER magnet system.

To contain the vertical forces, the VV segments are provided with support legs connected to the TF coil cases by vertical tension rods. The rods allow relative radial displacements between the TF coils and the VV. The horizontal loads acting on the vessel are contained by tension rods that provide a connection in the toroidal direction between the TF coil cases and the equatorial ports of the VV. These tangential rods are also designed to resist the lateral loads generated by seismic events with an acceleration up to 0.2 g and locate the VV with respect to the TF coils [3].

2.3. Gravity supports

The gravity supports (GS) must carry the weight of the tokamak, resist horizontal and vertical loads generated by earthquakes and accommodate the differential displacement due to the thermal contraction of the machine from room temperature down to 4.5 K. These requirements are met by 20 pedestals (one per TF coil) of a laminated construction so as to give radial flexibility while retaining a high azimuthal rigidity [3]. The pedestals are rigidly connected at the top to the lower OIS belt and at the bottom to a stiff ring which is an integral part of the cryostat. Seismic acceleration up to 0.2 g in the vertical and horizontal directions can be resisted without any yielding of the supports. Should the maximum ground acceleration to be considered at the ITER construction site be larger than 0.2 g, it is intended to provide seismic isolation to the building so that the tokamak itself is not subjected to acceleration in excess of 0.2 g.

3. TF coils design description

The TF magnet includes 20 coils. This number was selected as a compromise between the requirement to provide large VV ports to facilitate remote handling of In-Vessel components [4] and the requirement to limit the TF ripple to less than 2.5% at the First Wall contour [5].

The TF coils are pancake wound and enclosed in thick, vacuum tight stainless steel cases (Fig. 2). A specific feature of the TF coil design is the use

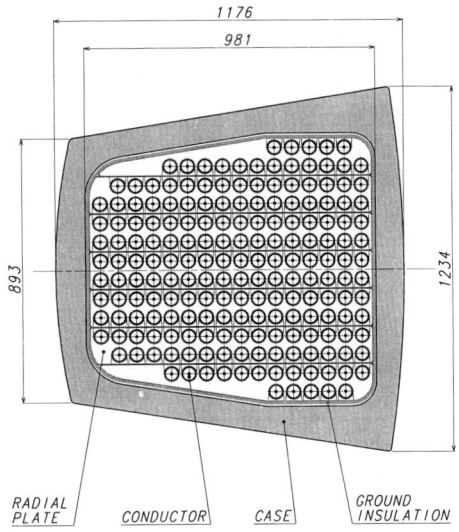

Fig. 2. TF coil inboard leg cross-section.

of radial steel plates with machined grooves that support the insulated conductor. These radial plates transfer the Lorentz forces acting on each conductor directly to the case without accumulation of forces on the conductor and its turn insulation. With this arrangement, the TF coil conductor can use a thin circular Incoloy 908 conduit, the function of which is only to support the local conductor forces and to act as the helium containment.

Some TF coil data are given in Tables 1 and 2.

With the radial plates, the turn insulation is separated from the plate and ground insulations. This is an important factor in the coil reliability. The turn, plate and ground insulations include an electrical barrier material such as a polyimide film. This requires the turn insulation to be applied after the reaction treatment of the Nb_3Sn superconductor.

4. Central solenoid (reference design)

The salient feature of the CS design is that it is layer wound. The layer construction provides an axisymmetrical and vertically uniform bucking structure for the TF coils. The CS uses a conductor with a thick walled Incoloy 908 square jacket.

Table 1
TF coil—geometrical data

Overall height/width	18.7/12 m
Approximate weights	
Cable	64 t
Conduit	19 t
Insul. Miscl.	42 t
Radial plates	145 t
Case + structure	470 t
Total per coil	740 t
Average turn length	44 m

Table 2
TF coil—electromechanical data

No. of turns per coil	192
Current per conductor	60.2 kA
Total stored Energy	101 GJ
Maximum TF at conductor	12.3 T
Maximum poloidal field	2.6 T
Maximum total field	12.5 T
Centering force per coil	762 MN

Table 3
Central solenoid—geometrical data

Height of winding	12.12 m
Inner radius of winding	1.919 m
Outer radius of winding	2.700 m
Maximum conductor length	1020 m
Approximate weights	
Cable	273 t
Conduit	457 t
Insulation	53 t
Buffer zone, etc.	67 t
Outer and inner cylinders	600 t
Total	1450 t
Preload structure weight	710 t

Table 4
Central solenoid—electromagnetic data

No. of turns	3356
Maximum field at conductor	13 T
Current per conductor	39 kA
Flux at outer radius at mid-plane	233 Wb
Total stored energy	14.3 GJ

An advantage of the layer construction is that current leads and interlayer series connections are located about 1 m above and below the winding pack in regions, called buffer zones, well away from mechanical load paths and where the field does not exceed 5.5 T.

Tables 3 and 4 show some CS data.

In the winding pack, layers are wound with four conductors in hand to keep cooling channel lengths within about 1 km. These four in hand conductors are connected in parallel in order to minimize the turn to turn and layer to layer voltages. As a result, the supply current is about 170 kA and research and development (R&D) is required for the circuit breakers that must interrupt this DC current.

Similar to the TF coil insulation, the turn, layer and ground insulations include an electrical barrier material such as a polyimide film. The CS is divided into three modules corresponding to three conductor grades. This allows for modules to be fabricated separately and subjected to factory acceptance tests before the entire solenoid is nested.

5. PF coil system design and maintenance considerations

The PF coil system (PF2–PF8) consists of seven coils built with cable in conduit conductors which use thick walled stainless steel square conduit. Conductors are wound in pancake configuration. PF2 and PF7 which operate at fields up to about 8.5 T will use Nb_3Sn conductors, while all other coils operate at fields not exceeding 5 T and will use NbTi conductors to decrease the cost.

In the event of failure, all PF coils would be difficult to replace because of the need to disconnect many mechanical, electrical and cryo connections. Moreover, all coils, except PF2 and the CS, are either trapped under the machine or by VV ports or cooling pipes for In-Vessel components. As a result, each PF coil consists of four identical modules which are designed to allow full ampere-turn operation with three modules only, but at reduced operating temperature (3.8 K). In the event of a failure in a module, the module would be disabled electrically or mechanically. This redundancy renders the need to replace a coil very unlikely. Should, however, this need arise, schemes have been developed which allow the removal and replacement of each of the PF coils [6]. This scheme is, however, not fully applicable to PF7, which uses Nb_3Sn, without some redesign of this coil.

6. Correction coils

Non axisymmetric field errors can arise due to deviations from theoretical dimensions and locations during manufacture and assembly, non axisymmetric configuration of current in winding packs and stray fields of bus bars. These errors can induce plasma locked modes and disruptions. Avoidance of these locked modes require the $m = 2$, $n = 1$ helical component of the error field normal to the magnetic surface (inside the plasma) defined by $q = 2$ to be limited to about 10^{-5} of the main toroidal field [5]. This requirement is very demanding and, if translated into positional accuracy of the ITER magnets, can only be met if winding packs are typically within 1–3 mm of their theoretical location. Such accuracy is clearly unrealistic and other means must be found to mitigate the effect of error fields. Correction coils will therefore be installed to decrease the amplitude of the most critical helical modes.

The superconducting correction coils will be saddle shaped and located outside of the TF coils as shown in Fig. 3. Four pairs of coils will be required to produce the required helical field component. Estimates indicate a need for a capacity of about 0.2–0.4 MA turns per coil to produce corrections of the order to 10^{-4} of the main toroidal field.

7. Conductor

7.1. Operation conditions and main design features

The CS and TF coils operate at, or close to, 13 T. This, together with the relatively large energy deposition during operation, makes the operating conditions of the conductor very severe. Energy deposition is due to nuclear heat, AC losses and friction forces.

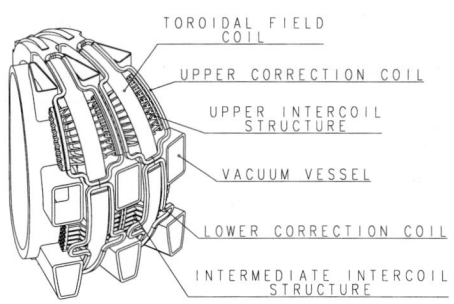

Fig. 3. Illustration of one machine sector with correction coils.

In the case of the TF coils, a significant fraction of the energy deposition is due to nuclear heating, but this heat is essentially deposited in the cases. In these coils, the dominating requirement is for a strand with high critical current. In the CS, the hysteresis losses are about 50% of the total losses and a low loss strand is preferred. For all PF coils, AC losses due to plasma control actions are a significant fraction of the total AC losses. Since these control actions are relatively poorly specified in terms of frequency and amplitude, the PF coils require design margins on the conductor. For the PF2 and PF7 coils, the field is 8.5 T, well above what can be achieved with NbTi at 4.5 K, in view of the high AC losses and these coils require a Nb_3Sn superconductor. NbTi can, however, be used for the outer PF coils where the field is no more than 5 T.

The strand, cabling and jacketing for the conductors of all ITER magnets share many common features. All high field coils (TF, CS, PF2, PF7) use Nb_3Sn superconductor with either a low loss (bronze technique) or high current density (internal tin technique) strand. All conductors are of the cable-in-conduit type and use a circular cable with a central cooling channel. The conductor for the TF coils can use a thin tubular conduit since all mechanical loads are carried by the radial plates. By contrast, the thick CS and PF coil conductor conduits have a major structural function. Because of their high operation field, the CS and TF coils use Incoloy 908 as a conduit material. This material was selected because its thermal contraction coefficient matches that of Nb_3Sn giving a minimum of critical current degradation due to strain induced by thermal contraction.

Table 5
Heat loads (heat loads in kW are averaged over a pulse cycle of 2200 s)

Type of heat load	TF coils (excl cases)	CS	PF	TF cases and structures
AC losses	4.8	3.3	6.3	20.9
Nuclear heat	2.2	0	1.2	6.1
Other heat loads	1.5	1.2	2.8	8.0

A more complete description of the superconducting conductors for ITER magnets can be found in [7].

7.2. Heat loads and cooling

The total cold mass of the magnets and structures is about 25 000 t. Cool-down is expected to take about 1 month. The ITER cryoplant supplies liquid helium to a number of cold boxes which provide forced flow cooling to the coils in a closed loop. The boxes contain helium circulation pumps and a cold bath to recool the circulating helium. Cooling of the conductors is achieved with helium at 4.5 K and 6 bar and flowing at a velocity of about 1.3 m s^{-1} in the central channel. Table 5 shows the typical time-averaged heat loads during operation, the total refrigeration heat load at 4.5 K from the magnets, bus-bars, structures and He pumps is above 95 kW.

The cryoplant also supplies the helium that cools the current lead transitions from 4.5 K to room temperature. The total time averaged requirement for this helium which returns to the cryoplant at room temperature is about 230 g s^{-1} in normal operation. In the event of a fast discharge of the TF coils or a quench in any of the coils, the coil inventory of helium is vented through pressure relief valves to a helium recovery tank.

A thermal shield cooled with 80 K gaseous helium is provided between the TF coils and the VV. This shield is attached by flexible straps to the TF coil cases. The entire cryostat inner surface is also covered by a thermal shield.

8. ITER magnet R&D programme

The ITER EDA magnet R&D programme with an estimated equivalent value of over 200 M$ is well underway with all four ITER participants. This programme aims at developing basic components, production experience and quality assurance (QA) methods for future large scale production and culminates in the construction and testing of model coils that incorporate many design features of the full size ITER coils [8].

8.1. Conductor R&D

The production of conductors able to meet the ITER specifications, in terms of current capability, field and AC loss limitation during operation is a new advance in superconducting magnet technology. The bronze and internal tin types of ITER Nb$_3$Sn strand have already been successfully produced by industries of the four ITER participants. For ITER construction, it is envisaged that multiple strand production facilities will be necessary to produce the nominal 1200 t of Nb$_3$Sn and 650 t of NbTi.

The strand-cable-conductor development provides conductors for the model coils which are identical to the full size conductors, except for the length. An experiment of jacketing (insertion of the cable in the conduit followed by compaction) of a 1 km long TF type conductor is being prepared to complete the demonstration of the full length ITER conductor fabrication.

As already indicated, Incoloy 908 is the selected conduit material for the TF coils and CS conductors. A feature of Incoloy 908, like other high nickel alloys, is a susceptibility to cracking if certain conditions of temperature, stress and oxygen concentration simultaneously occur. This cracking, also called SAGBO for Stress Accelerated Grain Boundary Oxidation, has been investigated in great detail since the reaction treatment for Nb_3Sn requires a temperature of about 700°C, inside the SAGBO range. The study showed that two parameters, namely stress and oxygen concentration, can be independently controlled to avoid SAGBO. For ITER conductors, both parameters are planned to be carefully controlled to give a high safety margin [9].

8.2. Basic component R&D

Joints for Nb_3Sn conductors that combine high current, in the 40–60 kA range, with AC operation are a major area of development. Joints for ITER coils are to be tested up to 12 T in DC conditions at the SULTAN facility in the European Union (EU) and in AC conditions but at lower field (~ 5.5 T) in the Pulse Test facility (PTF) in the United States (US) [10].

Insulation systems have been irradiated using fission reactors in the EU and in the Russian Federation (RF) and tested in cold conditions [11]. An important aspect which requires more investigation is the gas formation in the insulation matrix during irradiation and the behaviour of this gas when coils are warmed up.

The TF coil case requires thick plates, forgings and thick plate welding. Full size portions of the case are planned to be manufactured.

8.3. Central solenoid model coil

The CS model coil assembly is shown in Fig. 4. The model coil is a 640 MJ solenoid of about 1.6 m inner diameter, 1.8 m height and capable of reaching the nominal field of 13 T [12]. The weight of the model solenoid is about 100 t and the total weight of the assembly and mechanical structure is about 150 t. The winding is composed of an inner module built in the US and an outer module built in Japan (JA). The model coil uses full size CS cable, is layer wound and has current leads and interlayer joint designs similar to that of the full size CS.

The model coil will be subjected to a tensile hoop stress when energized, whereas the full scale CS operates in a compressive state. To simulate relevant operating conditions for the conductor, the model coil will be capable of accepting a replaceable single layer insert that can be operated in such a way to simulate the expected hoop compression in the case of the CS conductor, or hoop tension for the TF coil conductor. Inserts will be made in JA (CS type) and RF (TF type). The CS model coil is therefore not only a scaled model for manufacturing development but also a high field test facility for ITER conductors.

Testing of the model coil will be performed in a facility at the Japan Atomic Energy Research Institute, Naka. Testing will include pulsed operation with ramped current and field to demonstrate the ability to continuously operate with the expected level of AC losses in the conductor and surrounding structures.

8.4. TF model coil

The TF model coil is race track shaped with dimensions of about 2.7 × 3.1 m [13]. The coil will be manufactured in the EU. The design and fabrication techniques will closely reflect those of the full scale coils. In particular, the model will use full scale conductor supported in radial plates.

The coil will be tested in conjunction with the EU Large Coil Task (LCT) coil at FZK, Karlsruhe as illustrated in Fig. 5. The test arrangement

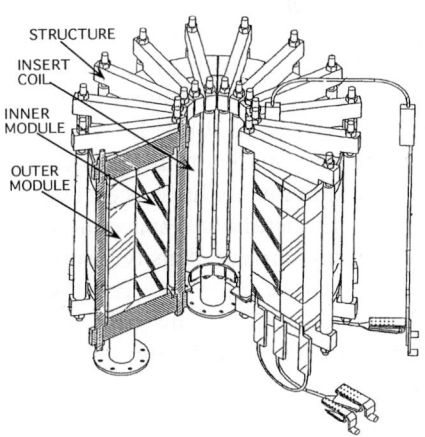

Fig. 4. CS model coil assembly.

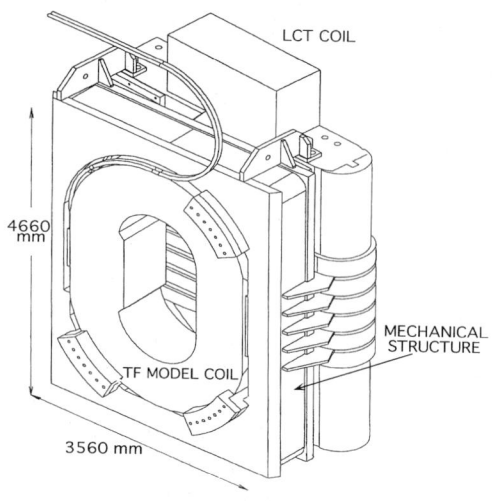

Fig. 5. TF model coil test configuration.

allows for testing under various conditions, including the simulation of some in-plane and out-of-plane loading and for the ability to vary the background field. Even in combination with the EU-LCT coil, the peak field in the model coil will be at most 9.8 T, whereas the ITER TF coil will see a peak field of about 12.5 T. This difference in full field is also the main rationale for testing the TF conductor as an insert in the CS model coil.

9. Summary of ITER plasma control

The ITER PF system is capable of providing the reference plasma scenario at 21 MA in a divertor configuration with an elongation of 1.6 and with an ignited 1000 s burn time. The flux available for burn is 82 Wb [14].

During the current flat top, the maximum PF coil current requirements have been evaluated by considering an operational space defined by a range of plasma current density profiles ($0.7 < l_i < 1.1$) and pressure ($\beta_p < 1.2$). For the reference scenario with $l_i = 0.9$, the safety factor q_{95} varies from 2.8 at start of flat top (SOF) to 3 at end of burn (EOB).

The PF system can also allow a range of alternate plasma scenarios as follows:
- High current ignited at 24 MA.
- High neutron fluence at 17 MA and a burn time of 4000 s.
- Reverse shear steady state at 12 MA.

Plasma position and shape control have been simulated considering different controllers and using linear and non-linear models. These simulations have shown that the PF system provides adequate control in response to typical plasma disturbances such as vertical displacements, ELMs, beta drops and minor disruptions.

10. Alternate central solenoid design

10.1. Rationale for a segmented CS design

The layer-wound reference CS has a vertically uniform current density and plasma shaping must be achieved by the outer PF coils. A more flexible and effective control of the plasma shape and position could, in principle, be achieved with a CS composed of several independently powered segments along the vertical direction.

Another incentive to consider a segmented CS is the possibility to reduce the current-carrying requirements in PF2 and PF7 which could then use the less expensive NbTi instead of Nb_3Sn. Using NbTi, the replacement scheme described in [6] would be fully applicable to PF7.

10.2. Preliminary engineering design and plasma control features of a segmented CS for ITER

The CS is split into five independently powered segments. The central segment, CSO, is 9.5 m tall and subject to the radial compressive load from the TF coils. The upper segments CS1U and CS2U and lower segments CS1L and CS2L are free standing coils with joints in low field regions. CSO is layer wound, similarly to the reference design. To be able to locate the CSO joints in regions where the field does not exceed 7 T (5.5 T for the reference design), a gap of about 1.5 m is provided between CSO and CS1. CS1 and CS2 can be built with the same inner and outer radius as CSO but there is also the option to build CS2 at a radius intermediate between CS1 and PF2 or PF7 to allow the use of NbTi in CS2.

Engineering issues of the segmented design that still need to be addressed include the design of the CSO buffer zone, supports for CS2 and mechanical fatigue in the TF coil cases.

The segmented CS design has the same plasma scenario capability as the reference design but with the following differences:
- The improvement in plasma shaping allows maintaining the safety factor q_{95} above 3 throughout the plasma pulse.
- There is a wider range of available plasma shapes and separatrix to first wall distances that may result in improved performance for the reference as well as the reverse shear steady state scenarios.
- The flux available for burn is slightly reduced to 72 Wb (down from 82 Wb).

For the response to plasma disturbances, the main advantage of the segmented CS design is an improved transient control of the separatrix position in the divertor channels. This capability for a fast transient control is important for high power plasma operation and ensuring a long life of divertor plasma facing materials.

In summary, the segmented design offers some attractive plasma control features but cannot be adopted at this point due to the limited engineering study that has been carried out so far. The overall tokamak design and assembly approach are compatible for both the reference and segmented designs of the CS, so the reference CS will be retained pending more detailed analysis and design effort for the segmented option.

11. Conclusions

The ITER magnets represent a large step in magnet technology because of their size, the use of Nb_3Sn superconductor and the operating conditions. Steady progress has been achieved in design and in advancing the R&D programme. This progress is illustrated by the status of the Model Coil programme where conductor production is well underway and coil winding has started. A large technological data base that is shared by all ITER participants and is essential for full scale construction has already been accumulated. Progress achieved so far shows that international cooperation can be effective, is able to produce a fully integrated design and can manage a large scale R&D programme. This gives great confidence that international collaboration for ITER can proceed successfully to full size construction.

Acknowledgements

This paper reports the work carried out jointly by the ITER Joint Central Team and Home Teams. Within the Joint Central Team, magnet design and analysis is carried out by the Superconducting Coils and Structures Division under the leadership of Dr R.J. Thome and plasma control analysis is carried out by the Plasma and Field Control Division under the leadership of Dr PL. Mondino. Home Teams of the four ITER participants are responsible for R&D activities and have also been key contributors to analysis. The disclaimer contained in ITER Publications Procedures SACPP 1 93-10-12W2 applies to this paper.

References

[1] R. Aymar, ITER Project, A Physics and Technology Experiment, Proc. 16th IAEA Fusion Energy Conf., Montreal, October, 1996, in press.
[2] C. Sborchia et al., The mechanical structure for the ITER magnet system, Proceedings of the 19th Symposium on Fusion Technology, Lisbon, 1996, Elsevier, Amsterdam, 1997.
[3] R. Gallix et al., ITER Tokamak supports: initial sizing and design, Proceedings of the 19th Symposium on Fusion Technology, Lisbon, 1996, Elsevier, Amsterdam, 1997.
[4] R. Haange et al., Maintenance concepts for ITER, Proc. 16th IAEA Fusion Energy Conference, Montreal, October 1996, in press.
[5] ITER Interim Design Report, IAEA publ. ITER EDA/DS/07.
[6] A. Tesini et al., Remote operations for the repair and replacement of a PF coil in ITER, Proceedings of the 19th Symposium on Fusion Technology, Lisbon, 1996, Elsevier, Amsterdam, 1997.
[7] M. Mitchell et al., Conductor design and optimization for ITER, IEEE Trans. Magn. 32(4) (1996) 2997–3000.

[8] K. Okuno et al., Status of the ITER Central Solenoid and Toroidal Field Model Coil programme, Proceedings of the 19th Symposium on Fusion Technology, Lisbon, 1996, Elsevier, Amsterdam, 1997.

[9] F.M. Wong et al., Selection of jacket materials for Nb_3Sn superconductor, Proceedings of the 19th Symposium on Fusion Technology, Lisbon, 1996, Elsevier, Amsterdam, 1997.

[10] P. Bruzzone, N. Mitchell et al., Design and R&D Results of the Joints for the ITER Conductor, Applied Superconductivity Conference, Pittsburg, USA, 1996.

[11] R. Viera et al., ITER coils insulation R and D program, 16th Symp. on Fusion Engineering, Champaign, Ill., 1995.

[12] H. Tsuji, J. Jayakumar, The ITER CS Model Coil Project, Proc. 16th IAEA Fusion Energy Conference, Montreal, October 1996, in press.

[13] E. Salpietro, The Toroidal Field Model Coil Programme for ITER, Proc. 16th IAEA Fusion Energy Conference, Montreal, October 1996, in press.

[14] Y. Gribov et al., The ITER Poloidal Field Configuration and Operation Scenario, Proc. 16th IEEE Symp. Fusion Eng. 2 (1995) 1514–1517.

Design of in-vessel components for ITER

R.R. Parker

ITER Garching Joint Work Site, Boltzmannstraße 2, D-85748 Garching, Germany

Abstract

This paper reviews the present design status of the major in-vessel components of ITER: the blanket, the divertor and the vacuum vessel. Substantial emphasis in the design of all in-vessel systems is given to the maintenance concept. For the blanket, integrating the remote handling approach with a robust design capable of reacting all thermal and mechanical loads is particularly challenging. A modular approach to the blanket design has been selected and both welded and mechanical attachments of the modules to the toroidal backplate are being evaluated. Transient thermal behavior produces deformations and stresses which must be carefully taken into account in the module attachment, in addition to the disruption loads. The divertor design is also modular and consists of 60 cassettes on which the high-heat flux components are mounted. These components can be replaced in hot cells. Transport of the cassettes into and out of the machine is accomplished by rotating them on the mounting rails installed in the vessel and pulling or pushing them radially outwards through dedicated ports. A mix of plasma facing materials has been provisionally identified: CFCs for the high-heat-flux targets, tungsten for areas bombarded by high neutral fluxes and Be for the remainder of the machine. While the vacuum vessel is of conventional double-wall design, it forms the first confinement barrier and all load cases must be carefully considered in the structural analysis. In all cases examined, the stresses are within allowables permitted by the codes when the appropriate load classification is considered. © 1997 Elsevier Science S.A.

1. Introduction

The design of the in-vessel components for ITER requires a significant extrapolation beyond the design of the corresponding systems in today's large tokamaks. In comparison to JET, TFTR and JT-60U, the particle and power loads in ITER are substantially higher, the pulse length is much longer (extending to steady-state) and the loads due to disruptions are far more intense. In addition, the requirements arising from the neutronic and tritium environment and maintenance and safety considerations are much more demanding. Nevertheless, thanks to a concerted effort by the Central Team and Home Teams, which includes implementation of an extensive validating R&D program, robust design concepts which meet the requirements of the major in-vessel components have been developed. While further design detail and improvements continue to be elaborated, there is at this time confidence that the technical basis for the design of in-vessel components will be sufficient at the conclusion of the EDA to justify proceeding to ITER construction.

The major in-vessel components discussed in this paper are the shielding blanket, the divertor and the vacuum vessel systems shown in poloidal cross-section in Fig. 1. Thermal and neutronic requirements for these components can be derived from the main machine parameters and representative full-power operating conditions as given in Table 1 [1]. The shielding blanket (Section 2) consists of shield modules with integrated first wall. The primary function of the shielding blanket is to remove the majority of the thermonuclear generated power produced in the plasma. The heat flux to the blanket associated with the neutron and α-particle production is not difficult to remove, but the competing requirements to implement efficient repair (and eventual replacement with a breeding blanket) by remote handling operations and to assure adequate design margins for all normal and off-normal loading conditions poses a challenging design problem. The divertor (Section 3), like the blanket, is based on a modular or 'cassette' design in which the cassette forms the main structural component while the plasma interaction region is lined with high-heat-flux components in order to remove the remainder of the α-particle heat not radiated to the blanket. Thanks to developments in divertor physics during the last few years, relatively modest heat loads (5–10 MW m^{-2}) can be anticipated on the targets. Nevertheless, higher heat loads must be at least transiently expected as different plasma regimes are explored and the high-heat-flux component design must accommodate such possibilities. The final main in-vessel component described in this paper is the vacuum vessel (Section 4). Although more straightforward in concept than the blanket or divertor, it has a primary safety function and therefore must be designed to be consistent with standard engineering design codes. For all loading conditions investigated including seismic, disruption and overpressure events, the calculated stresses are below allowable limits when appropriate classification of the load category is considered.

2. Blanket

During the Basic Performance Phase (BPP), which is scheduled for the first 10 years of ITER operation and is estimated to produce a fluence up to 0.3 MW·a m^{-2}, it will not be necessary to breed tritium. The function of the blanket then is mainly to provide shielding which, in combination with that provided by the vessel, reduces nuclear heating and lifetime radiation exposure in the coils and surrounding environment to acceptable levels, e.g. 17 kW and 1×10^7 Gy, respectively in the TF coils. The minimum degree of shielding provided by the blanket is determined by the requirement that sectors of the vacuum vessel should be replaceable at any time up to the end of ITER operations. The reweldability requirement implies that the helium production in the vessel should remain below about 1 appm through the Extended Performance Phase (EPP) which follows the BPP and could result in a FW fluence up to 3 MW·a m^{-2}. During the EPP it will be necessary for ITER to breed tritium (TBR ≥ 0.8) as external supplies are not foreseen to be sufficient to supply tritium at the required rate. Thus a complete changeout from the shielding blanket to the breeding blanket after 10 years of operation must be anticipated.

In addition to the shielding function, the first wall (FW) of the blanket must be capable of removing up to 300 MW of power deposited in the plasma, corresponding to an average power

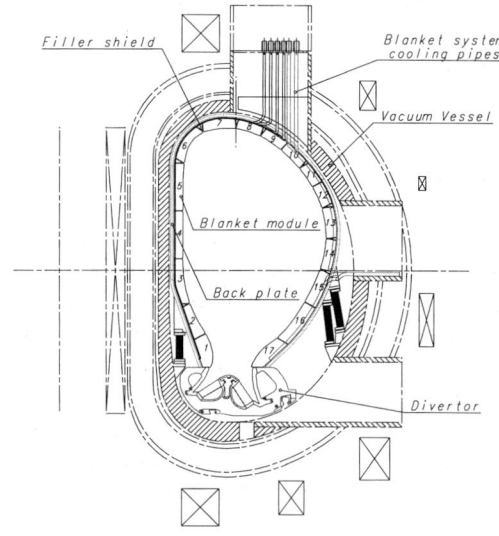

Fig. 1. Poloidal cross-section of the ITER machine indicating main in-vessel components.

Table 1
Parameters and a representative operating point for the ITER design as described in the Interim Design Report

Parameter	Symbol	Value
Major, minor radius	R, a	8.14, 2.8 m
Plasma elongation	κ_{95}, κ_X	~1.6, ~1.75
Plasma triangularity	δ_{95}	~0.24
Nominal plasma current	I	21 MA
Toroidal field	B	5.68 T (at $R =$ 8.14 m)
MHD safety factor	q_{95}	3.05
Volume-averaged temperature	T	10.5 keV
Volume-averaged density	n	1.3×10^{20} m^{-3}
Impurity fractions	f_{Be}, f_{He}	0.02, 0.14
Effective charge	Z_{eff}	1.5
Normalized beta	β_n	2.4
Fusion power (nominal)	P_{fus}	1.5 GW
Average, peak neutron wall loading	Γ_n	~1, 1.2 MW m^{-2}
Burn duration	t_{burn}	1160 s
Plasma magnetic, thermal energy	W_{mag}, W_{th}	1.1, 1.2 GJ

The confinement time used in obtaining the operating point is that derived for ELMy H-Mode by the ITER Confinement Database and Expert Group [2].

loading of 0.25 MW m^{-2}. Except for special areas such as the startup and shutdown limiter, and the baffles located near the X-point, the peak FW power specification is 0.5 MW m^{-2}. This would allow a peaking factor of 2 if the full 300 MW were incident on the FW. However, in normal operation only about half the power should be radiated to the first wall. Radiation from the main plasma is mainly from bremsstrahlung and medium-Z impurities introduced to reduce power flow into the divertor [3] and will not be strongly peaked except possibly near the X-point where MARFES may form. However such radiation would fall largely on the baffle where the design peak heat load is substantially higher (3 MW m^{-2}).

A critical aspect of the blanket system design arises from the need to provide efficient maintenance and replacement of components using only remote handling (RH) tools. Although neither the FW nor the shielding part of the blanket is expected to require regular maintenance during the BPP, it is likely that the FW will occasionally be damaged during disruption events. In-situ repair of the FW armour (Be) using plasma spray deposition appears feasible and may mitigate the need for more substantial intervention [4]. However, in the event that the FW cooling system beneath the armour is also damaged, more complex repairs must be undertaken. The requirement for replacement of such a damaged component is extremely stringent, namely it should not take longer than 8 weeks. After the BPP, the entire shield blanket must be replaced with a breeding blanket, and this is required to be accomplished (via RH tools) within 2 years. Such targets for repair and replacement times are aggressive and require the blanket system design to be highly integrated with the RH approach.

2.1. Shielding blanket design description

Remote handling and manufacturing considerations lead to a modular design in which the basic component is a 316 LN IG (ITER Grade) water-cooled shield block (typical dimensions, $0.8 \times 1.8 \times 0.3$ m^3; weight, ~4 tonnes) [5]. The modules are mounted on a 100 mm thick toroidal shell or backplate, open at the bottom at the interface with the divertor so that its cross-section in the poloidal plane takes the shape of a horseshoe. In the design described here, the modules are oriented with the long dimension in the poloidal direction except in the upper outboard region where the longer dimension runs in the toroidal direction. Manifolds enter the machine from the top and horizontal ports and run along the backplate under the modules. Fig. 2 depicts the arrangement of the modules, the backplate and the manifolds. Fig. 3 shows a cross-section of a typical module as it would appear installed on the backplate. In this case the attachment (see Section 2.4) is by means of a weld running poloidally along the interface between a boss on the module and a corresponding boss on the backplate.

The FW consists of a 20 mm thick Cu alloy (DS copper) mat in which are imbedded SS cooling tubes, 10 (12) mm ID (OD) with a pitch of typically 22 mm [6]. The FW armour material is beryllium and is 10 mm thick. It is fabricated

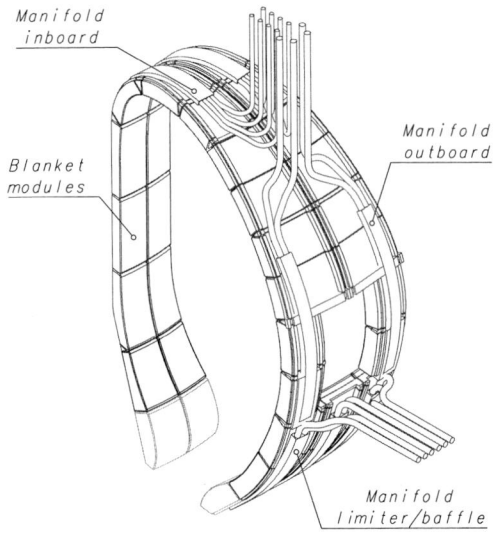

Fig. 2. An 18° sector of the shielding blanket indicating module segmentation and piping.

either from a mat, castellated to reduce thermal stress, or individual tiles. Hot isostatic pressing (HIP) is attractive for bonding the Be armour to the copper mat, and the copper mat to the steel shield block. However, other fabrication methods are under consideration. In this design the FW is permanently bonded to the shield block and the entire module must be replaced in the event of FW damage which cannot be repaired by in-situ methods.

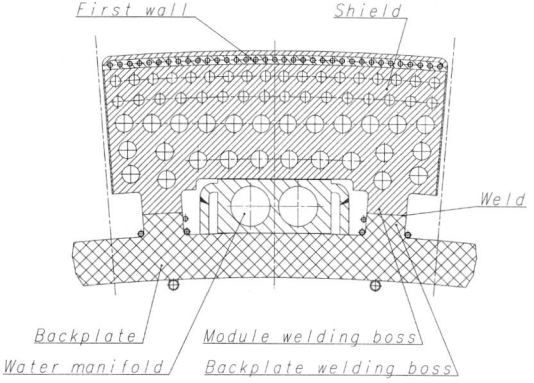

Fig. 3. Cross section of a shielding blanket module installed on the backplate on the inside midplane of torus. A welded attachment is shown.

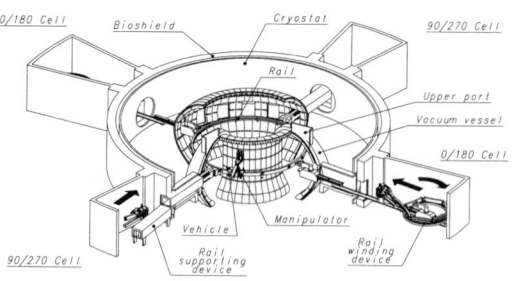

Fig. 4. Schematic of remote handling approach to blanket module replacement. A toroidal rail is deployed in the torus and a manipulator is used to remove and replace modules.

Referring to Fig. 1, the two toroidal belts of modules just below the horizontal ports serve as a limiter for plasma startup and shutdown before a diverted configuration is formed. The two belts of modules just above the divertor on the inside and outside of the machine tangentially intercept the flux surface passing 6 cm outside the separatrix at the midplane and have the main function of baffling the flow of neutrals from the divertor to the main plasma chamber (hence the name 'baffle modules'). Both the limiter and baffle modules have a construction similar to the main FW modules described above, but the 1 mm thick SS tubes are in each case replaced by 0.2 mm SS liners in Cu tubes in order to increase their power handling capability (5 and 3 MW m^{-2}, respectively). In addition, since the neutral flux to the first wall near the X-point is expected to be intense, the Be armour for the lower half of the baffle module FW is replaced by tungsten which is predicted to have a longer lifetime owing to its superior sputtering characteristics.

2.2. Remote handling of blanket modules

The general scheme for replacing blanket modules is illustrated in Fig. 4. Four horizontal ports are dedicated to RH operations, two for storage of a rail system and two for removal of the damaged modules and for the introduction of new ones. In order to replace a module, a complete toroidal rail is inserted into the torus. A vehicle is then guided along the rail to the toroidal position of the damaged module. The module is prepared for removal by cutting the pipe connections to the manifolds and, for a welded attachment, cutting the bosses attaching the module to the backplate. These operations are carried out respectively by pipe cutting tools inserted into the manifolds and by additional cutting tools (plasma torches) which move poloidally on each side of the module through the rectangular RH shaft formed between the backplate and two adjacent modules (Fig. 3). During the cutting of the attachment legs, the module is held in place by a temporary fixture and is then gripped and removed by an articulated arm on the vehicle. After transferring the module into one of the dedicated horizontal ports and transporting it to a hot cell, the new module is introduced, transported to the vacant position, held in place by temporary fixtures and rewelded by TIG welding heads moving in synchronism back and forth in the RH shaft. The time required for a single module replacement is estimated to be 8 weeks. The rail, rail deployment system and vehicle required for the RH of the blanket modules is being developed by the Japanese Home Team as part of one of the large R&D projects.

2.3. Mechanical and thermal loads

The major mechanical loads on the in-vessel systems arise from the electromagnetic forces associated with disruptions. Two main types of disruptions can occur: so-called centered or radial disruptions which occur on a rapid time-scale (10's ms) and Vertical Disruption Events (VDEs) for which the timescale is longer (100's ms). The centered disruption is initiated by a fast (few milliseconds) thermal quench after which the plasma current rapidly decays to zero. As a result of the rapid plasma current decay, a current which is comparable in magnitude to the initial plasma current is inductively coupled to the blanket system.

In the case of a VDE, vertical control is lost (which may also be the result of a thermal quench) and the plasma begins to move mainly in the vertical direction. While undergoing this motion the current slowly decays and the size shrinks until q has reached a sufficiently small value (estimated to be ~ 1.5) that the plasma becomes unstable and a thermal quench followed by a current quench then occurs. The destabilizing force resulting from the interaction of the plasma current with the initial shaping field is balanced by the interaction of: (i) the toroidal plasma current with eddy currents in the vessel; and (ii) a poloidal (halo) plasma current with the toroidal field. The halo current can be substantial (up to 40% of the initial plasma current in existing experiments) and flows through the plasma to the first wall, completing its circuit either through the backplate (upward moving VDE) or through the backplate and divertor cassette (downward moving VDE). The halo current, flowing through the backplate and/or divertor and interacting with the toroidal field, produces a force which, in combination with the force due to the eddy currents, defines the total force acting on these structures. Peak local stresses in the backplate, divertor and vessel depend on the relative strength and distribution of halo and eddy currents, as well as the degree of toroidal peaking. Axisymmetric codes such as MAXFEA and TSC are used to model the VDE dynamics and calculate the resulting loads. The maximum vertical force arising from a downward VDE is calculated to be 150 MN and acts mainly on the divertor and, through the divertor, to the vessel and its supports. In addition, it is known that the halo current can be toroidally asymmetric and this can increase the peak stresses occurring in the in-vessel components, relative to the stresses calculated for an axisymmetrically distributed load. By extrapolating the data base gathered from presently operating tokamaks, it is concluded that the worst-case loading condition for ITER can be bracketed by assuming a halo current equal to 29% of the initial plasma current with a toroidal peaking factor (peak to average) of 2. In addition, non-axisymmetric halo currents can generate horizontal loads. Such loads are not yet well documented or understood and their prescription for ITER has somewhat arbitrarily been set at 50 MN. Work continues on developing models for these loads and understanding the non-axisymmetric currents that generate them.

The bounding load case for the structural behavior of the modules near the (inner) midplane is determined by radial disruptions which, for a 21 MA plasma current disruption, induce 1.4 MA m^{-1} (per meter of poloidal length) of toroidal current in the module, concentrated near the FW [7]. The corresponding average traction on these modules is 1.2 MPa. Modelling has so far not accurately included the effect of the collapse of the diamagnetic component of toroidal flux, which also acts in the radial direction but which peaks much earlier during the thermal quench, nor the slower decaying paramagnetic component which tends to reduce the traction, but these effects are small. As the poloidal area of the inner midplane module is approximately 1.4 m^2, the attachment between module and backplate must have a minimum cross-sectional area of about 130 cm^2 to keep the tensile stress within S_m (the so-called working stress defined as the lessor of 2/3 of the yield or 1/3 of the ultimate strength) for SS.

More difficult than the radial forces to react is the radial moment on the module, which is produced by the interaction of the disruption-induced current along the module sidewalls with the toroidal field (TF). Equal and opposite poloidal forces up to 5.5 MN m^{-1} are exerted on the sides of the modules and as they are balanced the net effect is to produce a radial torque which must be reacted either by the backplate or adjacent modules.

The thermal loads on the blanket system are produced by the heat flux to the first wall and the nuclear heating of the FW, shield and backplate. The heat generated in the steel part of the module ranges from 15 MW m^{-3} just behind the FW to 0.5 MW m^{-3} at the back. The nuclear heating in the 100 mm backplate drops from 0.5 MW m^{-3} on the module side to 0.1 MW m^{-3} on the side facing the vessel. Near the module surface, the cooling passages form a dense grid while toward the rear of the module the passages are larger and have a larger pitch. Consequently the thermal time constant of the module varies from ~10 s near the FW to ~100 s toward the rear of the module and ~1000 s in the separately cooled backplate and transient conditions produce thermal stresses which must be taken into account in the structural analysis of the module and its connection to the backplate.

During a VDE the plasma can contact the first wall and release a substantial fraction of its stored energy to the first wall. The energy density released to the wall is calculated to be in the range 20–60 MJ m^{-2}. The duration of the transient is critical to the survival of the first wall: fast transients melt the Be surface which causes a loss of armour but leaves the cooling system in tact, whereas slow transients occurring over a duration of ~1 s may cause coolant burnout and damage to the first wall which can only be repaired by changeout of the affected module [8]. Modeling indicates that the interaction time should last at most a few hundred milliseconds in which case burnout can be avoided provided the thickness of the Be first wall is ≥ 10 mm. Results from Be plasma spray R&D carried out in the US Home Team are encouraging in regard to the possibility of using this approach to implement in-situ repair of first wall damage due to surface melting of the Be armour.

2.4. Module attachment

Both welded and mechanical connections of the blanket module to the backplate have been considered. In the welded approach, the connection is made by welding a boss running poloidally along each side of the module to a corresponding boss on the backplate (Fig. 3). The deformation of the module-backplate system and stress in the attachment boss for thicknesses of 70, 80 and 90 mm are summarized in Fig. 5 where the loads for a centered disruption described in Section 2.3 have been applied and the inner backplate plus module system has been idealized in the calculation to a cylindrical shape (length ≫ radius). Based on the analysis, a 90 mm thick boss would be required to react the shear load. The technology for the cutting and welding required for module replacement is under development in the Japanese Home Team.

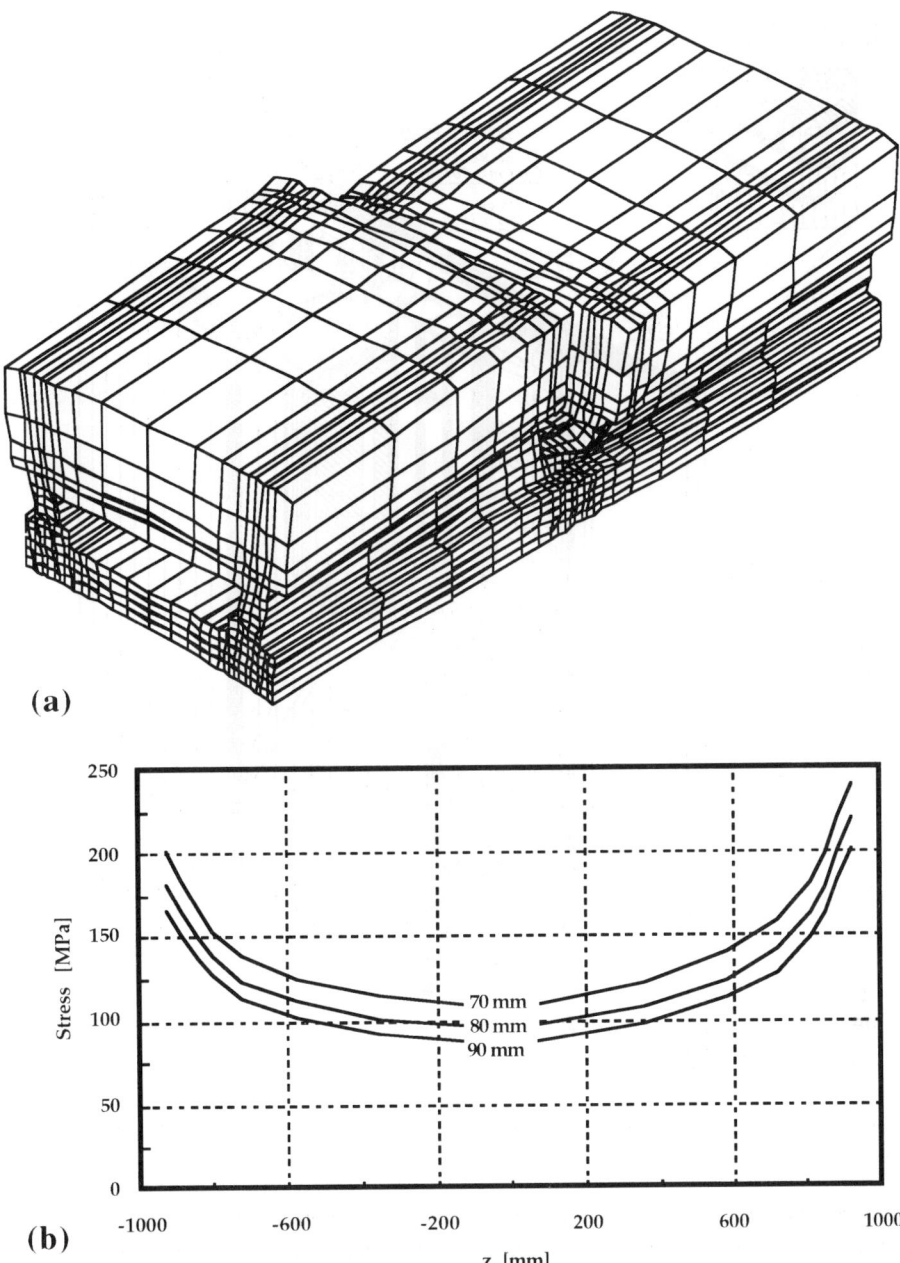

Fig. 5. (a) Distortion of number 4 module (refer to Fig. 1 for location) under the effect of 1.0 MPa traction on first wall and upward (downward) force of 5.5 MN m^{-1} acting on the right (left) side of the module. The maximum displacement is 1.5 mm. (b) Tresca stress in the weld vs. poloidal position under the load conditions of (a) for weld thicknesses of 70, 80 and 90 mm.

A mechanical attachment has also been developed in which the weld joining the bosses on the backplate and module is replaced by a system of shear keys and high-strength bolts or studs (Fig. 6). The latter are accessible from the plasma side through a 20 mm slot between adjacent modules.

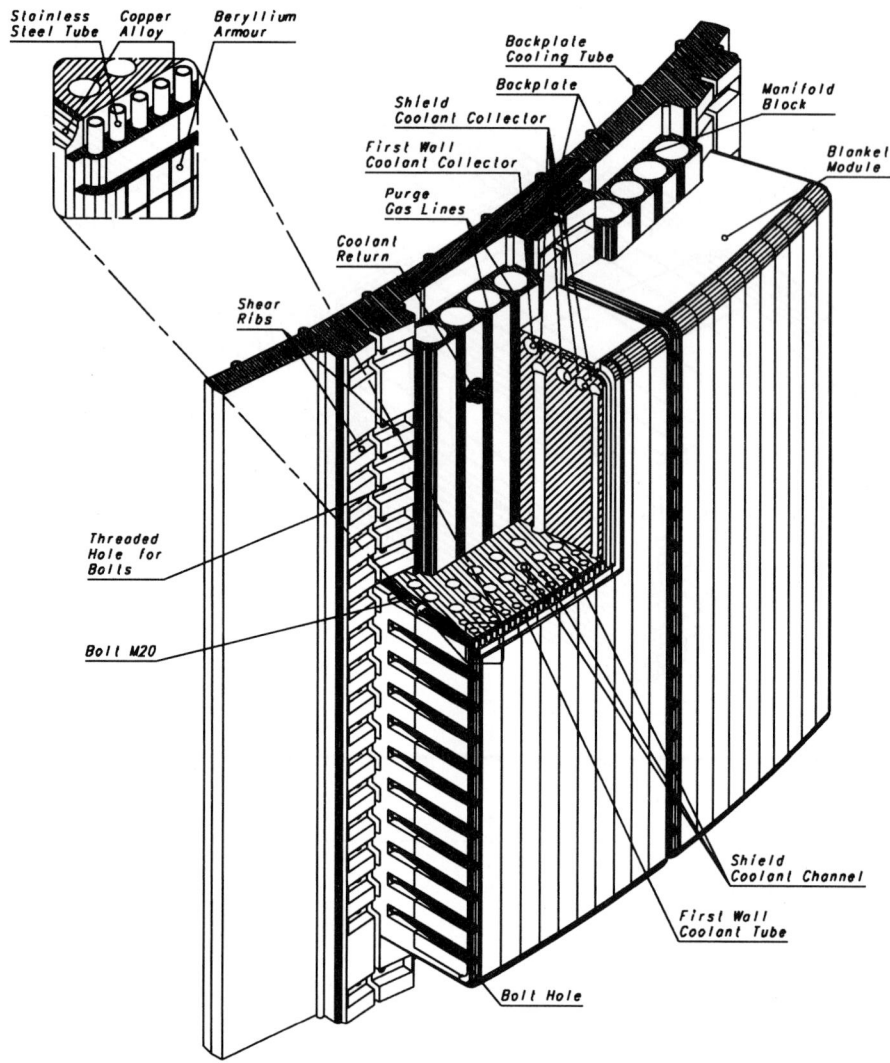

Fig. 6. Cutaway view of a shielding blanket module installed on the backplate on the inside midplane of torus. A mechanical attachment is shown.

Analysis has shown that the shear stress in the attachment arising from the torque exerted on the module during a centered disruption is acceptable and this conclusion has been verified by a laboratory simulation of the shear key system carried out by the EU Home Team [9].

A welded attachment appears to be more robust than a mechanical attachment with regard to reacting the mechanical loads, although additional mechanical and thermal effects discussed below must be more fully analyzed to verify its performance. However, a welded attachment has drawbacks for maintenance due to the need to remotely carry out complex welding and cutting operations within tightly confined spaces. The time required for cleanup of the debris and the residual effect on the plasma quality are additional uncertainties. From this point of view and also with a view toward minimizing assembly time, a mechanical attachment would have advantages and recent design and analysis activities have concentrated on this approach.

Two effects which have important consequences for the module attachment are the deformation of the backplate induced by disruption loads and differential expansions and distortions of the modules relative to the backplate caused by temperature differences. In the case of the mechanical loads, the backplate reacts to the 'pulling' or 'pushing' pressure on the first wall mainly via hoop stress; however, because the backplate is neither a true cylinder nor a closed torus and also because the modules are discrete, distortion of the backplate occurs over the length of the module. This deformation can overload the attachment since the module itself is stiff and cannot easily conform to the backplate deformation. The effect is particularly strong near the outside midplane where the backplate has weak hoop capability since the ports interrupt toroidal continuity. Relatively large deformations can also occur near the bottom of the backplate, where it is open, due to the pressure associated with halo and eddy currents which occur during a VDE.

In the case of thermal effects, the average temperature of the module relative to the backplate during transients such as startup, shutdown and disruptions can be greater than in the steady-state owing to the spatially varying thermal time constant. Calculations indicate a maximum transient ΔT of 80°C can occur between module and backplate after a disruption compared to ~ 40°C steady-state. Not only must the attachment accommodate differential expansions due to such large average ΔT, it must also accommodate the bowing of the module due to the temperature gradient in the module itself. The bowing causes an initially planar module to deform so as to be congruent to a spherical surface with radius $R = d/(\alpha \Delta T)$, where d is the module thickness, α is the coefficient of thermal expansion and ΔT is the difference in temperature across the module. During transient conditions the distortion over the surface of the module is typically 1 mm. A full treatment of these problems and the effect on the module–backplate attachment requires the use of elaborate 3-D models which are now under development.

A new mechanical attachment which recognizes ab initio the above problems of differential expansion and distortion is being developed. The concept is based on a small number of localized attachments (ideally 3) which are compliant with respect to transverse loads, such as those induced by differential thermal expansions of module and backplate, but which are sufficiently stiff in the direction perpendicular to the backplate to react the 'pulling' or 'pushing' loads. In addition, one or more large keys are used, either between the backplate and the modules or from module to module, to react to the torques resulting from interruption of the toroidal current during disruptions and also to keep the module central position. Although this concept is still at an early stage of development, it appears that there is sufficient design 'headroom' for meeting the requirements and detailed design is now being carried out.

2.5. Breeding blanket

The breeding blanket for the EPP of ITER replaces the BPP shielding blanket and will produce a large fraction of the tritium needed for EPP operation [10,11]. The blanket segmentation and water coolant parameters ($T_{in} = 140$°C, $T_{out} = 190$°C, $p_{max} = 4$ MPa) are the same for the two phases. The concept is based on a solid breeder blanket with beryllium neutron multiplier. Lithium zirconate is the reference breeder material but others such as lithium titanate could be used. Highly enriched lithium is required in order to achieve the desired breeding ratio. As for the shielding blanket, the structural material is Type 316LN-IG austenitic steel. A purge gas loop is used for tritium recovery during operation from the multiplier and the breeder. The purge gas is helium with 0.1% hydrogen and ~ 0.1 MPa pressure. The general design requirement of a net tritium breeding ratio ≥ 0.8 and the limited radial blanket thickness lead to the selection of a layered blanket configuration.

The calculated net tritium breeding ratio is 0.87 without breeding in the limiter, baffle and midplane port regions. This breeding capability permits ITER to operate with a fluence goal in the range 1–3 MW a m^{-2}, assuming a yearly tritium supply of 1.5 kg, a breeding blanket installation at the start of the EPP, and less than 3 kg total tritium inventory in the different reactor components (plasma facing materials, fuel cycle, etc.). The thicknesses of the breeding blanket sections are about 24 and 33 cm for the inboard and the outboard sections of the reactor, respectively. The total thermal power of the breeding blanket is about 2% more than the shielding blanket. The shielding performance of the breeding blanket was checked against the design requirements. The results indicate that the breeding blanket can be accommodated within the reference ITER configuration and satisfy the design requirements with adequate safety factors.

3. Divertor

The main requirements for the divertor are to exhaust the fusion and auxiliary power deposited in the plasma and to provide a sufficient partial pressure of helium at the entrance to the pumping ducts to exhaust helium at the rate it is generated (2 Pa m^3 s^{-1} at 1.5 GW fusion power). For high-recycling divertor configurations the power density on the divertor target plates would be near the limit of what is feasible to remove by practical heat removal systems (~ 30 MW m^{-2}). Fortunately, developments in divertor physics in recent years have provided a solution with the discovery of partially and fully detached divertor regimes [12–16]. These regimes are produced by operating at relatively high density and adding trace amounts of medium-Z impurities, which enhances the radiative power loss. Aligning the target plate nearly tangential to the separatrix at the strike point has also been shown to be effective in promoting such divertor behavior while maintaining good confinement performance [16] and this configuration has been adopted as the reference configuration for ITER. Additional requirements arising from lifetime and maintenance considerations are addressed by careful selection of materials for the throat and target plates and use of the modular 'cassette' design.

A key physics issue is whether fully or partially detached operation is compatible with achieving the good confinement required for sustained ignited operation. The issue is compounded by the desire to operate near, or even above, the Greenwald density limit extrapolated to ITER conditions. Resolution of this issue is providing a focus for the confinement research programs of the major tokamaks now in operation.

3.1. Design description

The divertor structure consists of 60 cassettes fabricated from 316 LN SS, each 5 m long and tapering in thickness from 1 to 0.5 m and mechanically attached to toroidal rails welded to the bottom of the vacuum vessel [17,18]. The high-heat-flux (HHF) components are mounted on the cassettes which is then the basic unit that is exchanged via remote operations when repairs are required. The arrangement of the cassettes is shown in Fig. 7 and an exploded view indicating the HHF components is shown in Fig. 8. There are 20 ports at the divertor level: 16 are used for cryopumps, which provide the main vacuum and divertor exhaust, and the remaining four are dedicated to remote handling operations used for the removal and introduction of the cassettes into the machine. Of the 15 cassettes associated with each RH port, the three immediately in front of the RH port are instrumented with diagnostics. The cassette body complements the blanket in providing shielding to the lower part of the vessel and the coils and contains internal cooling channels and manifolds for routing water to the HHF components.

The main interaction of the plasma with the divertor occurs in the W-shaped region formed by the inboard and outboard targets (outer legs of the 'W') and wings (inner legs of the 'W'). These structures are lined with HHF components. The separatrix enters the two openings at the top of the W and runs down into each V-shaped region where it eventually intersects the targets, i.e. the inner and outer legs of the W. These areas must be designed for the highest steady-state heat flux in the machine. Although the peak heat flux under fully detached conditions is expected to be ≤ 5

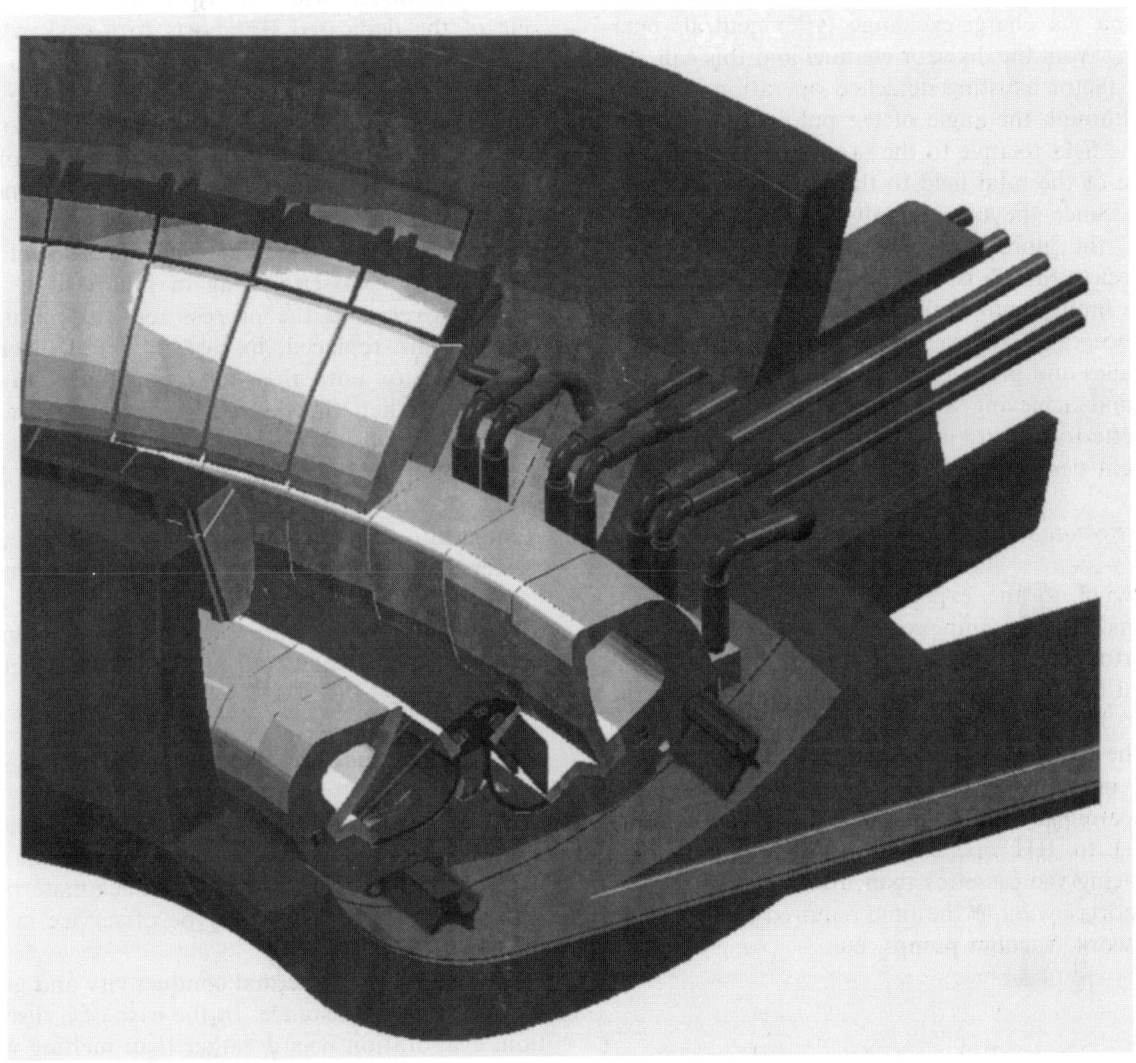

Fig. 7. Divertor cassettes mounted on rails in bottom of vacuum vessel.

MW m^{-2} and ≤ 10 MW m^{-2} for partially detached regimes, transient operation is anticipated to occur which could result in heat fluxes up to 20 MW m^{-2} and the targets are designed to accommodate this power density on a steady-state basis.

The inner legs of the W are each formed by two wings, so named because of the resemblance of these structures to the wings of an aircraft. The wings are mounted parallel to each other but are offset so as to resemble a partially opened Venetian blind. This semi-transparent wall absorbs most of the power radiated on it from the divertor channel (peak power density ≤ 5 MW m^{-2}) but allows gas to flow freely between the private flux region and the divertor channel, thereby simultaneously promoting conditions for detachment while providing the source of gas for pumping. The wings also remove momentum from the plasma via charge-exchange (CX) neutrals emanating from the divertor channel and this can also be a factor assisting detached operation [19].

Although the angle of the poloidal component of the field relative to the target plates is 15°, the angle of the total field to the plates is only ~ 1–1.5°. Since the targets only intercept open field lines, the global alignment of the divertor with respect to the TF is less critical than that required for a limiter and a tolerance of at least + 10 mm is acceptable. However, alignment of adjacent modules and protection of leading edges are critical and achieving a tolerance of 2–4 mm from cassette to cassette in the toroidal direction is the present design goal.

3.2. Remote handling

Owing to the expected finite lifetime of the plasma-facing components in the divertor, the divertor is expected to be replaced on an infrequent but regular basis, e.g. three times during the BPP. Therefore strong emphasis has been placed on the remote handling approach and its integration into the design. Analysis has shown that dedicating a relatively small number of ports (four) to RH operations is more efficient for replacing the cassettes than utilizing all 20 divertor ports owing to the time required for replacing pipework, vacuum pumps, etc.

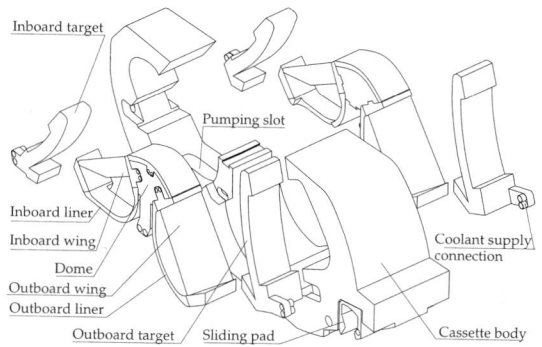

Fig. 8. Exploded view of divertor cassette showing detachable high-heat-flux components.

A cassette is removed by mounting it on a skid or 'radial mover' which moves it radially through one of the dedicated RH ports to a cask which provides containment as it is transported to a hot cell for repairs [20]. Once a cassette is removed, another cassette is prepared for removal by rotating it on its toroidal mounting rails to a toroidal position immediately in front of the RH port. This operation is performed by means of a second RH tool or 'toroidal mover'. Once in front of the RH port the cassette is again removed by the radial mover, and the process repeated until all cassettes are replaced, in case of a full divertor changeout, or until the faulty cassette(s) is (are) removed. When the desired number of cassettes has been removed, new or refurbished cassettes are brought from the hot cell to the divertor RH ports, inserted radially by the radial mover and rotated into their toroidal position by the toroidal mover. The RH goal is to be able to replace all cassettes within 6 months and to replace any one cassette within 2 months. The scheme outlined above has been shown to be capable of meeting this requirement.

3.3. Plasma facing materials selection

Three plasma facing materials are being considered for use in ITER: beryllium, tungsten and carbon (carbon fiber composites.) Each material is best suited for a particular type of service in the machine.

CFCs have high thermal conductivity and good thermal shock resistance. In the case of a disruption, evaporation occurs rather than melting with subsequent loss of melt layer, as in the case for a metal. However carbon has two principal disadvantages, namely chemical sputtering, which lowers the lifetime estimate below that predicted only on the basis of physical sputtering and codeposition, a process in which carbon forms hydrocarbons with hydrogenic species leading to substantial levels of in-vessel tritium retention. These disadvantages can be mitigated by doping with Si and a high thermal conductivity Si-doped 3-D CFC under development by the EU Home Team [21,22] is being considered for the reference divertor target material.

Important issues for the divertor targets are the ability of the armour and its joint to the heat sink to withstand the extremely high heat fluxes expected for ITER operation and for the thermal-hydraulic system to have adequate margin against burnout. Fortunately, the R&D program has recently produced encouraging results in both of these areas. In initial high-heat flux tests, up to 18 MW m^{-2} heat flux has been successfully sustained for 1000 cycles on a small-scale mockup consisting of Si-doped CFC tiles bonded to a Cu substrate. An active metal casting method was used to join the tile to the substrate. In separate tests on the heat sink itself, heat fluxes as high as 27 MW m^{-2} have been stably achieved using swirl tube technology while peak heat fluxes as high as 45 MW m^{-2} have been successfully reached with axially peaked profiles typical of a divertor operation. These results indicate that the basic design approach adopted for ITER divertor targets can meet the stringent requirements. Testing will continue on mid-scale and near full-scale mockups in the near future.

Taking into account chemical and physical sputtering, evaporation during detached (5 MW m^{-2}) and high transient power (20 MW m^{-2}) operation, as well as disruptions (10% frequency), a lifetime of approximately 7000 full power shots for the lower CFC (undoped) vertical targets has been estimated [23]. Thinner targets will be required for a nominal heat flux of 10 MW m^{-2} and the lifetime will be correspondingly shorter. The lifetime of Be targets for the same normal and off-normal conditions is predicted to be unacceptably low.

The principal advantage for tungsten is its high sputtering threshold, which is the basis for its selection for the upper divertor target and lower part of the baffle. Tungsten also has low tritium retention characteristics; however, because of the potential loss of the melt layer during a disruption, its lifetime as the PFM on the lower divertor target is predicted to be less (factor of 2(6) with 10(50)% melt layer loss) than that of C. An additional issue for tungsten is the potentially deleterious effect on plasma performance. Owing to this concern the use of tungsten is limited to those regions of the first wall where CX fluxes are high and the use of beryllium or carbon would result in significantly lower lifetime and where in addition frequent melting due to disruptions is not expected to occur.

Beryllium is chosen as the PFM for the remainder of the first wall, except possibly for the limiter where C is retained as a backup. The principal reasons for the choice of Be over C is its better compatibility with plasma performance, including consideration of issues such as first wall conditioning (particularly after disruptions), density control and lower Z, and its longer expected lifetime in areas of moderate heat flux due to the absence of chemical sputtering. Although Be codeposition with hydrogen takes place in much the same way as for C, there are indications that such co-deposited layers are inhibited from forming at temperatures typical of those found in the divertor where most of the codeposited layers would be formed [24] (R. Causey, personal communication).

4. Vacuum vessel

The vacuum vessel provides the first boundary for tritium confinement as well as the high-quality vacuum necessary for plasma operations. It is fabricated from 20 sectors, each spanning 18°, one of which is shown if Fig. 9 [25,26]. The design is based on a double-wall fabrication, where the inner and outer skins are each 40–60 mm thick and joined by welded stiffening ribs. The vessel is curved poloidally, faceted in the toroidal direction and the overall thickness varies from 0.45 to 0.82 m around the poloidal circumference. In addition to the ribs, plates and/or blocks are fastened in the space between inner and outer skins, in order to provide the requisite shielding. The attachment of the shield material maximizes the loop resistance which for the vessel alone is 10.4 μΩ. The structural material is 316 LN IG SS. The vessel is designed for an internal pressure of 0.5 MPa which is the limiting pressure reached with the aid of a pressure suppression tank in the event of an in-vessel breach of a coolant loop.

The vessel design has stress margins for the loads corresponding to weight, and normal and

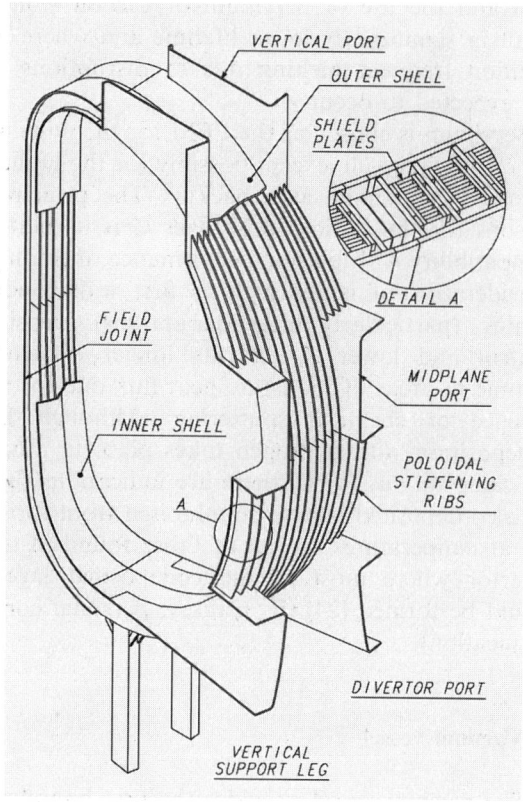

Fig. 9. An 18° sector of the vacuum vessel. Plate shielding is shown in the insert.

off-normal pressure conditions. For example during normal operation, vessel deformation is about 6 mm and stress levels are about 75 MPa (50 MPa membrane). Additional load conditions arise from VDEs, as the VDE force is transmitted through the blanket and divertor supports to the vessel and then through the vessel supports to the magnet cases. Estimates of the net force from VDE modelling are 150 MN vertical and 50 MN horizontal. These values also bound the vessel forces calculated from seismic events up to 0.2 g. In the case of an axisymmetric halo current with $I_{halo}/I_{plasma} = 0.35$ the maximum deformation is 10–15 mm and the maximum stress level is ~ 104 MPa (85 MPa membrane), including the stresses due to gravity and coolant pressure. For the case of an non-axisymmetric halo current with $I_{halo}/I_{plasma} = 0.29$ and peaking factor of 2, the peak VDE vertical load increases by the factor 1.7 and this load condition increases the maximum stress to 140 MPa (116 MPa membrane), about 10% under the allowable stress. To react the assumed 50 MN horizontal load component associated with an asymmetric VDE, links between the vessel midplane ports and the TF cases have been added and the area of the vessel near the port has been reinforced. The calculated stresses in the vessel due to this horizontal load are at least a factor of 1.7 below allowables. The limiting stress case for the vessel is in fact not for a VDE, but for an event leading to an overpressure of 0.5 MPa. However, for such extremely rare events the allowable stresses are somewhat higher and the calculated safety margins in this case are even larger. Thus for normal operation including worst-case disruptions, as well as for extremely rare off-normal events, the stresses in the vessel have adequate margin relative to typical code requirements.

5. Conclusions

The basic design concepts adopted for the main in-vessel components for ITER form a good basis for the design optimization now in progress. The ability to rapidly repair or replace in-vessel components will be a key factor in achieving the goals of ITER, and therefore an integrated RH approach is essential to a successful design.

The modular blanket approach is compatible with these goals. Welding the blanket modules to the backplate results in a robust design which meets the demanding thermal and mechanical requirements; however, a mechanical attachment design is being developed with the aim of reducing the assembly time and simplifying the RH approach. The present design emphasis is on development of a semi-flexible attachment which can accommodate relative deformations between backplate and module, but which can nevertheless react to the pushing and pulling forces as well as the torques exerted on the modules as a result of disruption-induced currents.

The divertor design is based on 60 modules or cassettes mechanically attached to the vessel. The modular cassette concept is somewhat more straightforward to maintain than the blanket modules, which is appropriate since regular maintenance for the divertor is expected to be required. The vertical target design takes advantage of recent developments in divertor physics in which semi- and fully detached regimes are reliably produced with modest heat loads. However, the design can also accommodate attached regimes, albeit for limited durations and shortened lifetimes. A mix of materials has been selected for the PFMs, namely CFCs for the high-heat-flux area of the divertor target, W for the upper part of the target and lower part of the baffle and Be for the rest of the machine.

Although the vacuum vessel is of conventional double-wall design, it represents the first tritium confinement barrier and must be designed as far as possible to meet code specifications to safely react all loads that can occur during normal and off-normal conditions. The limiting loads under 'normal' operating conditions are expected for VDEs. Under all normal and off-normal loading conditions so far examined, including the effects of a toroidally peaked halo current ($I_{halo} = 0.29 I_{plasma}$, peak-to-average = 2) and an 0.5 MPa overpressure, the vessel stresses are calculated to be within allowables when appropriate load classifications are considered.

Acknowledgements

This paper draws on the work of many individuals from the ITER Joint Central Team and Home Teams. In particular, the author would like to acknowledge the contributions of P. Barabaschi, L. Bruno, Y. Gohar, K. Ioki, G. Janeschitz, G. Johnson, E. Martin, G. Sannazzaro and R. Tivey. This paper was presented as an account of work performed under the Agreement among the European Atomic Energy Community, the Government of Japan, the Government of the Russian Federation, and the Government of the United States of America on Co-operation in the Engineering Design Activities for the International Thermonuclear Experimental Reactor ("ITER EDA Agreement") under the auspices of the International Atomic Energy Agency (IAEA).

References

[1] ITER Interim Design Report, ITER San Diego Joint Work Site, San Diego, 1995.
[2] S. Kaye et al., in: S. Kaye et al. (Eds.), Plasma Physics and Controlled Nuclear Fusion Research 1994, Vol. 2, IAEA, Vienna, 1995, pp. 525–534.
[3] D.E. Post, B. Braams and J. Mandrekas et al., in: Proceedings of the 12th International Conference on Plasma Surface Interactions in Controlled Fusion Devices, St. Raphael, 20–24 May, 1996, J. Nucl. Mater., in press.
[4] R. Matera, S. Chiocchio and G. Federici et al., in: C. Varandas et al. (Eds.), Fusion Technology 1996, Proceedings of the 19th Symposium on Fusion Technology, Lisbon, 16–20 September, 1996, North-Holland, Amsterdam, in press.
[5] R.R. Parker, W.B Gauster, et al., Fus. Eng. Design 30 (1995) 119–131.
[6] K. Ioki, A. Cardella, F. Elio et al., SOFE 95, Proceedings of the 16th IEEE/NPSS Symposium on Fusion Engineering 1995, IEEE, Piscataway, NJ, 1996, pp. 150–155.
[7] S. Chiocchio, K. Ioki, M. Araki et al., in: C. Varandas et al. (Eds.), Fusion Technology 1996, Proceedings of the 19th Symposium on Fusion Technology, Lisbon, 16–20 September 1996, North-Holland, Amsterdam, in press.
[8] A.R. Raffray, G. Federici, J. Nucl. Mat. 244 (1977) 101–130.
[9] W. Dänner, P. Batistoni, A. Cardella et al., in: Plasma Physics and Controlled Nuclear Fusion Research 1996, Proceedings of the 16th IAEA Fusion Energy Conference, Montreal, 7–11 October, 1996, IAEA, Vienna, in press.
[10] Y. Gohar, M. Billone, A. Cardella et al., in: C. Varandas et al. (Eds.), Fusion Technology 1996, Proceedings of the 19th Symposium on Fusion Technology, Lisbon, 16–20 September, 1996, North-Holland, Amsterdam, in press.
[11] Y. Gohar, M. Abdou and R. Aymar et al., in: Plasma Physics and Controlled Nuclear Fusion Research 1996, Proceedings of the 16th IAEA Fusion Energy Conference, Montreal, 7–11 October, 1996, IAEA, Vienna, in press.
[12] G.F. Matthews, J. Nucl. Mat. 220-222 (1995) 104–116.
[13] S. Allen, N. Brooks, R. Campbell, et al., J. Nucl. Mat. 220-222 (1995) 336–341.
[14] O. Gruber, A. Kallenbach, K. Kaufmann, et al., Phys. Rev. Lett. 74 (21) (1995) 4217–4220.
[15] M. Nagami, JT-60 Team, J. Nucl. Mat. 220–222 (1994) 1–12.
[16] B. Lipschultz, J. Goetz, B. LaBombard, et al., J. Nucl. Mat. 220-222 (1995) 50–61.
[17] R. Tivey, M. Akiba, A. Antipenkov et al., in: C. Varandas et al. (Eds.), Fusion Technology 1996, Proceedings of the 19th Symposium on Fusion Technology, Lisbon, 16–20 September, 1996, North-Holland, Amsterdam, in press.
[18] G. Janeschitz, H. Pacher, G. Federici et al., in: Plasma Physics and Controlled Nuclear Fusion Research 1996, Proceedings of the 16th IAEA Fusion Energy Conference, Montreal, 7–11 October, 1996, IAEA, Vienna, in press.

[19] G. Janeschitz, K. Borrass, G. Federici, et al., J. Nucl. Mat. 220-222 (1995) 73–88.
[20] E. Martin, T. Burgess, G. Cerdan et al., Proceedings of the 19th Symposium on Fusion Technology, Lisbon, in press.
[21] J. Roth, E. Vietzke, A. Haasz, Atomic and Plasma-Material Interaction Data for Fusion, Nucl. Fus. 1 (1991) 63.
[22] C.H. Wu, C. Alessandrini, P. Bonal et al., in: C. Varandas et al. (Eds.), Fusion Technology 1996, Proceedings of the 19th Symposium on Fusion Technology, Lisbon, 16–20 September, 1996, North-Holland, Amsterdam, in press.
[23] H.D. Pacher, I. Smid, G. Federici et al., Proceedings of the 12th International Conference on Plasma Surface Interactions in Controlled Fusion Devices, St. Raphael, J. Nucl. Mat., in press.
[24] M. Mayer, R. Behrisch, H. Plank, et al., J. Nucl. Mat. 230 (1996) 67–73.
[25] K. Ioki, et al., Fus. Eng. Des. 27 (1995) 39.
[26] K. Ioki, A. Cardella and W. Dänner et al., in: Plasma Physics and Controlled Nuclear Fusion Research 1996, Proceedings of the 16th IAEA Fusion Energy Conference, Montreal, 7–11 October, 1996, IAEA, Vienna, in press.

Management of waste from the International Thermonuclear Experimental Reactor and from future fusion power plants

Karin Brodén [a,*], Maria Lindberg [a], Simon Nisan [1,b], Paolo Rocco [c], Massimo Zucchetti [d], Neill Taylor [e], Cleve Forty [e]

[a] *Association EURATOM, Studsvik RadWaste AB, S-611 82 Nyköping, Sweden*
[b] *The NET Team, Boltzmannstraße 2, D-85 748, Garching bei München, Germany*
[c] *European Commission, Institute for Advanced Materials, Joint Research Centre, I-21 020 Ispra (VA), Italy*
[d] *Energetics Department, Polytechnic of Turin, Corso Duca degli Abruzzi 24, I-10 129 Torino, Italy*
[e] *Association EURATOM-UKAEA, UKAEA Fusion, Culham, Abingdon, Oxfordshire, OX14 3DB, UK*

Abstract

An important inherent advantage of fusion would be the total absence of high-level radioactive spent fuel as produced in fission reactors. Fusion will, however, produce activated material containing both activation products and tritium. Part of the material may also contain chemically toxic substances. This paper describes methods that could be used to manage these materials and also methods to reduce or entirely eliminate the waste quantities. The results are based on studies for the International Thermonuclear Experimental Reactor (ITER) and also for future fusion power station designs currently under investigation within the European programme on the Safety and Environmental Assessment of fusion power, Long-term (SEAL). © 1997 Elsevier Science S.A.

1. Introduction

The operation and decommissioning of fusion power plants will generate radioactive material due to neutron activation and tritium contamination from the deuterium/tritium fuel. Part of this material will also be chemically toxic. Management of waste will be an important issue for public opinion when deciding on the realisation and siting of the facilities.

For several years fusion waste studies (for example quantification and qualification studies) and fusion waste related studies (for example material studies) have been performed within the European Fusion Technology Program. The present paper gives a survey of some important waste aspects that have been studied in Sweden, the UK, Italy and the Joint Research Centre of the European Commission. Results for both the International Thermonuclear Experimental Reactor (ITER) and future fusion power station designs investigated within the Safety and Environmental Assessment of fusion power, Long-term (SEAL) programme are presented.

* Corresponding author.
[1] Present address: DER/SIS, CEA/CEN Cadarache, F-13108, Saint Paul-lez-Durance, France.

More detailed recent results on ITER waste management strategies and final disposal are given in a poster presentation at this conference [1].

2. Use of low-activation materials

By appropriate selection of structural materials, it seems possible to reduce the active inventory, particularly that of the first wall, divertor and blanket. The use of specially-developed low-activation materials, which have a composition optimised to reduce neutron activation, is likely to be of benefit. It is important to note that this optimisation, reducing the active inventory on the time scale relevant to waste management, is not the same as optimisation to reduce activation and decay heat on the short time scale relevant to postulated accident scenarios. Furthermore, it is essential to appreciate the contribution that activated impurities in the composition may have on the long time scale.

In order to illustrate these points, Fig. 1 shows some results for four candidate structural materials obtained with the inventory code FISPACT-4 [2] and the EAF4.1 cross section library [3]. The quantity plotted is the ingestion hazard potential, a biological hazard index important in waste management considerations because of the possible transport of radio-nuclides to the biosphere on geological time spans. Plots of other activation hazard indices show similar behaviour to this one.

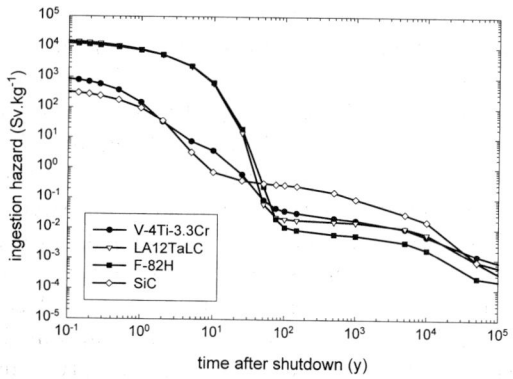

Fig. 1. Ingestion dose hazard potential for four candidate structural materials.

The four materials compared are:
- V-4%Ti–3.3%Cr, a vanadium alloy with nominal composition [4];
- LA12TaLC, a low-activation martensitic steel, considered in the SEAFP study [5];
- F-82H, an alternative low-activation martensitic steel [6];
- silicon carbide composite (SiC/SiC) [4].

All materials contain realistic impurity concentrations. The calculations represent the exposure of each material for 2.5 years in the blanket of a typical fusion power plant with a mean neutron wall loading of 2.0 MW m^{-2}, based on the design developed in the Safety and Environmental Assessment of Fusion Power (SEAFP) and the subsequent decay up to 10^5 years after shutdown.

Fig. 1 shows that the ranking of these four materials in terms of this hazard index is markedly different at the long time scale compared with the short time scale. The better performers after about 50 years tend to be the poorer performers before that time and vice versa, although it is important to note that all four materials are better than stainless steel on all time scales, often substantially so. This conclusion is confirmed when considering other activation indices such as dose rate, decay power and specific activity, which also show the same trend.

A further observation is the critical importance of impurities to the long time response of materials. Impurities present in parts per million concentrations may have a dominating influence on activation response at waste disposal time scales. This is illustrated by a comparison of the two steels, LA12TaLC and F-82H. Fig. 1 shows that within the nuclear data uncertainties, the results for both steels are coincident up to about 70 years. At times longer than this, the F-82H steel is consistently superior to LA12TaLC steel. Examination of the detailed results reveals that this is due to a reduced nitrogen content from 200 to 80 ppm. The important nuclide in this case is carbon-14, produced through the reaction channel [^{14}N](n,p)[^{14}C]. In addition, other impurities become significant when different activation indices are considered. For example, reduction of the niobium impurity content from 50 ppm in LA12TaLC steel to 1 ppm in F-82H, results in a fifty-fold reduction in the gamma dose rate after 100 years in the F-82H steel compared with the LA12TaLC.

3. Detritiation

From the waste management point of view, tritium has to be recovered to reduce outgassing during the interim storage and to comply with the regulatory limits for disposal. Of course, there are also powerful economic reasons for tritium recovery.

Results from detritiation experiments in France indicate that the tritium content in tritium containing metals can be reduced to less than 500 GBq t^{-1} [7]. These results were obtained in a small-scale, one-step industrial melting process for metal parts with initial tritium contents of up to about 30 000 GBq t^{-1}.

The residual tritium content after detritiation will be further decreased during interim storage, due to decay. A specific activity of 500 GBq t^{-1} will be reduced to about 30 GBq t^{-1} after 50 years of interim storage. This figure compares favourably with the tritium limit for waste to the German repository Konrad. The tritium content in metal waste to Konrad must not exceed 190 GBq per package [8]. That means about six tonnes per package.

4. Interim storage

By interim storage the activity in the material is reduced before re-cycling, clearance, or final waste disposal. This can be illustrated by a hypothetical German scenario. Fig. 2 shows that increasing the intermediate storage time allows the waste from the vacuum vessel of ITER to be re-designated from the Gorleben to the Konrad repositories. Gorleben is a deep repository in a salt formation, Konrad is a repository in a former iron mine for non-heat generating waste delivered in packages that can be handled without shielding.

It is clearly seen that if the storage time is prolonged a larger part of the waste can go to a less sophisticated repository. The main reason for

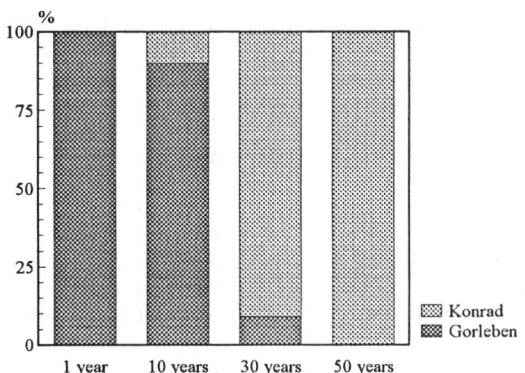

Fig. 2. The effect of intermediate storage on the destination of the waste from decommissioning the vacuum vessel of ITER in a hypothetical German scenario.

requiring waste to be sent to Gorleben is that the surface dose rate of the waste packages exceeds 2 mSv h^{-1}; another is that it is heat generating.

A storage building is sufficient for interim storage of low-level waste before clearance. However, the highly activated waste has to be cooled in a dry or wet storage facility. An interim storage facility like the Swedish CLAB (central storage for spent fuel) could be used.

Waste with high tritium content may require special precautions during storage for retaining the tritium inside the packages, e.g. tight welding, in combination with effective ventilation and air cleaning in the interim store.

5. Recycling and clearance

Assessments on recycling and clearance have been performed in the frame of SEAL [9] for the two power reactor designs, SEAFP Model 1 and 2 (see Table 1).

Most of the detailed recycling and clearance assessments concerned SEAFP Model 2, the configuration adopting more conventional technological solutions. Additional results were produced for SEAFP Model 1, essentially for the sake of comparison.

Activation data were computed with the FISPACT-4 activation code [2]. The irradiation was

Table 1
Material in different components of SEAFP Model 1 and 2

Component	SEAFP Model 1	SEAFP Model 2
Blanket	Li$_2$O	Water cooled Pb-17Li
In-vessel components structure	V-5Ti	LA12TaLC
Divertor	Beryllium-armour	Beryllium-armour
	V-5Ti heat sink	Copper heat sink
	Helium as coolant	Water as coolant
Shield	OPSTAB and water	OPSTAB and water
Vessel	OPSTAB, water, lead and boron carbide	OPSTAB, water, lead and boron carbide
Toroidal field coils	AISI316LN structure	AISI316LN structure
	Nb-Sn superconductor	Nb-Sn superconductor
	Copper conductor	Copper conductor
	Helium as coolant	Helium as coolant
	Glass and epoxy insulator	Glass and epoxy insulator

assumed to be done at the mean neutron wall loading of 2 MW m^{-2}. A fluence of about 10 MWa m^{-2} for the first wall has been considered. Five and 25 years of continuous irradiation were assumed for the in-vessel components and the other zones, respectively. The activation data for the divertor were taken from previous SEAFP analyses. A 15-month irradiation was assumed for the Cu-AISI 316 divertor. All activation levels were computed after 50 years of storage, assumed to be the interim storage period at the plant site.

Concerning recycling (reuse in the fusion industry), surface dose limits of 10 µSv h^{-1} and 10 mSv h^{-1} were assumed respectively for hands-on (HOR) and remote handling (RHR) recycling. These value are modified with respect to those adopted in SEAFP [5], where a surface dose rate limit of 25 µSv h^{-1} was assumed for hand-on recycling and 2 mSv h^{-1} were taken as limits for recycling by remote handling.

The feasibility of clearance (declassification to non-active waste) was evaluated weighting the potential hazard of each radionuclide contributing to the radioactivity concentration of the examined material.

This approach differs from that adopted in previous analyses for SEAFP [5] in SEAFP report where the global limit for clearance was 400 Bq kg^{-1}. This is the limit for Very Low Level Waste currently adopted in Britain and a very low value, compared with the natural radioactivity of many substances, e.g. fertilisers and bricks which may exceed 5000 and 1000 Bq kg^{-1}, respectively.

In the present analyses clearance levels related to each radio-nuclide were taken from an IAEA draft report [10]. In this document, clearance levels for many radionuclides are obtained from a categorisation of various safety analyses of radioactive waste repositories. These analyses assume 10 and 100 µSv year^{-1} as dose limits for the most exposed individual in 'likely' and 'unlikely' accident scenarios happening in the waste repository.

The clearance levels of radionuclides not included in [10] were evaluated with a fitting formula, taking the lowest value (Bq g^{-1}) from the three formulae:

$$\left\{\frac{1}{E_\gamma + 0.1 E_\beta}, \frac{\text{ALI}_{\text{inh}}}{1000}, \frac{\text{ALI}_{\text{ing}}}{100,000}\right\} \quad (1)$$

where E_γ and E_β are the effective energies (MeV) of the gamma and beta emission, ALI$_{\text{inh}}$ and ALI$_{\text{ing}}$ (Bq) are the most restrictive values of the Annual Limits of Intake by inhalation and ingestion.

A 'clearance index' I_c was evaluated for each material, taking into account the contribution of all radionuclides contained. I_c was evaluated from the specific activity A_i and the clearance level L_i of each one of the z radionuclides contained in the material:

$$I_c = \sum_{i=1}^{z} \frac{A_i}{L_i} \quad (2)$$

Table 2
Clearance indices and clearance feasibility for SEAFP Model 2 components

Component	Cooling time (years)	Clearance index	Clearance feasibility
Inboard + outboard shield	50	$\gg 1$	NO
Outboard vessel	10	0.9	YES
Inboard vessel	50	$\gg 1$	NO
Outboard magnets	50	0.02	YES
Inboard magnets	50	$\ll 1$	YES

It is postulated that the material can be cleared if $I_c \leq 1$.

Typical clearance indices and the related clearance feasibility for SEAFP Model 2 components are shown in Table 2. Note that the outboard part of the vessel can be cleared after only 10 years of decay.

Fig. 3 shows fractions of SEAFP Model 1 and 2 activated materials to waste disposal, remote handling recycling, hands-on recycling and non-active waste.

The activated material arising from SEAFP Model 2, is about 70 000 t. This amount includes all structures from the plasma chamber to the magnet zones and takes into account operation, maintenance and decommissioning. Forty-eight percent can be recycled, 39% can be cleared if the I_c are applied, 13% needs to be disposed of as active waste. An extension of the cooling time up to 100 years for the inner components could allow the recycling also of this fraction.

Appropriate detritiation procedures of the in vessel components before interim storage can reduce tritium inventories and tritium outgassing rates to such low levels as to not hinder recycling. Similarly, out-vessel material, subjected to occasional tritium contamination, could be cleared.

The activated material arising from SEAFP Model 1, is about 60 000 t. However, from the radiological point of view, no activated waste needs to be disposed of, 70% of the material could be recycled (41% RHR, 29% HOR) and 30% could be cleared. The material which can be declassified is less than that in SEAFP Model 2, due to the better shielding characteristics of the SEAFP Model 2 Blanket.

6. Volume reduction

The large waste components have to be segmented prior to further handling. For highly activated components segmenting techniques developed for fission reactor internals can be used. For example, the Japan Atomic Energy Research Institute (JAERI) has developed an underwater plasma arc technique for up to 130 mm thick stainless steel components [11]. Other techniques can be used for low level metallic waste. Oxy-acetylene, oxy-lance and plasma torches and also cold cutting with special purpose saws are examples of suitable techniques for this type of waste [12].

Melting can be used for volume reduction of metal waste and also as a treatment method for metals that can be recycled or declassified. By the melting procedure it is possible to get a homogenous distribution of the activity in the metal and to take out representative samples for analysis.

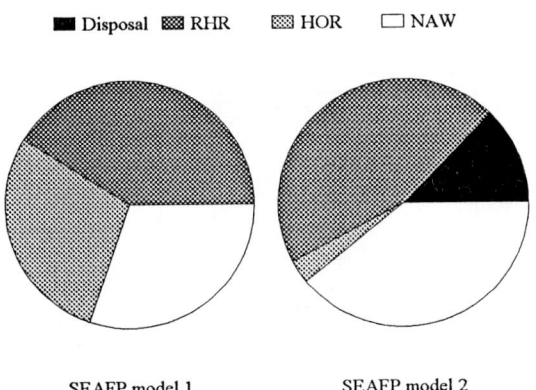

Fig. 3. Fractions of SEAFP Model 1 and Model 2 materials to waste disposal, remote handling recycling, hands on recycling and non-active.

Fig. 4. The melting facility for low-level carbon steel scrap and low-level stainless scrap at Studsvik.

Fig. 4 shows the melting facility at Studsvik RadWaste.

Incineration can be used for volume reduction of burnable dry active waste.

7. Packaging

When putting waste into packagings there are some factors that set the destination and other factors that limit the packages and control the amount of waste. This can be exemplified by a hypothetical German scenario with the deep repositories Konrad and Gorleben. The maximum weight for Konrad is 20 t and for Gorleben 40 t and the maximum surface dose rate for Konrad is 2 mSv h^{-1} and for Gorleben 500 mSv h^{-1}.

When calculating the volume of packaged waste the standard containers used for fission waste in the repository today have been assumed to be used also for the ITER waste. For the present calculation of ITER waste, for Konrad repository in the German scenario, the container which gives the lowest volume is the SBOX-II. This package is a rectangular steel container with a volume of 4.6 m^3. The packing efficiency is generally assumed to be 30% of the available inner volume for all materials.

Fig. 5 shows the repository volume required for packages with permanent steel components of ITER. The blankets are not included. The mixed steel consists of the upper and lower crowns, the gravity support, the pre-load structure and the outer PF-coil support. The central solenoid can be declassificated to non-active waste and is therefore not included.

8. Waste disposal

Potential doses to man from ITER waste in German type repositories have been calculated based on available results from safety assessments for the German repositories Gorleben and Konrad.

Gorleben is a deep geological repository in a salt formation. Two methods for disposal are expected to be used in the repository:
- tunnel disposal for non-heat generating waste at 870 m level and possibly also at 900 m;
- bore-hole disposal at 300–600 m from the 870 m level for heat generating waste.

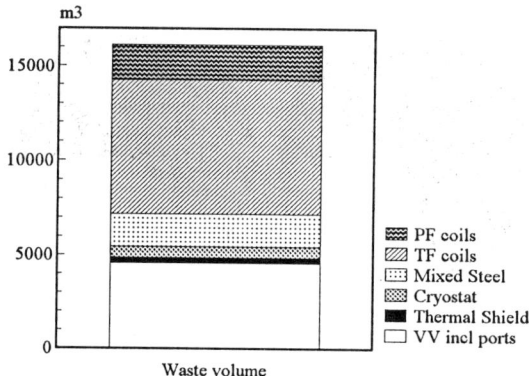

Fig. 5. The repository volumes of the permanent steel components of ITER for a hypothetical German scenario.

A remote handling system will be used for the bore-hole disposal.

Konrad is a deep geological repository (800–1300 m) in a former iron mine. It is a repository for non-heat generating waste delivered in packages that can be handled without shielding.

The safety assessment for Gorleben includes calculated dose rate values for Ni-59, Tc-99 and Nb-94 [13]. It has been assumed that ground-water could intrude to the repository area of Gorleben in the post-operational phase and that there could occur migration of radionuclide with ground-water to surface-water. The results were used to estimate the dose equivalent rate for Ni-59, Tc-99 and Nb-94 in ITER waste. If, for example, the shielding blanket of ITER is placed in Gorleben the maximum dose equivalent rate from Ni-59, Tc-99 and Nb-94 is estimated to be 1.3×10^{-9}, 3.9×10^{-10} and 3.8×10^{-11} Sv year^{-1}, respectively.

Dose rate calculation results are also available for Konrad [14]. The transport times from the repository to the biosphere were extremely long in these calculations. The major parts of the radioactive nuclides will therefore decay before they could reach man. Tc-99 gave a maximum dose equivalent rate after about three million years. If the waste from all permanent steel components inside the biological shield, (i.e. excluding the blanket and free components), is placed in Konrad and if the migration rate of Ni-59 and Nb-94 is estimated to be in the same order of magnitude as Tc-99, only 4.7×10^6 Bq Tc-99, 4 Bq Ni-59 and nothing of Nb-94 will remain after three million years and the dose rate from the nuclides will thus be insignificant.

9. Chemically toxic waste

Depending on the materials used in the different parts of a fusion reactor the waste might not only be activated it may also be toxic. One of the most toxic materials in some designs is beryllium which may be used as armour of the plasma-facing surfaces and as a neutron multiplier in blankets. In Fig. 6 the radiological ALI (Annual Limits on Intake) and the chemical ALI are com-

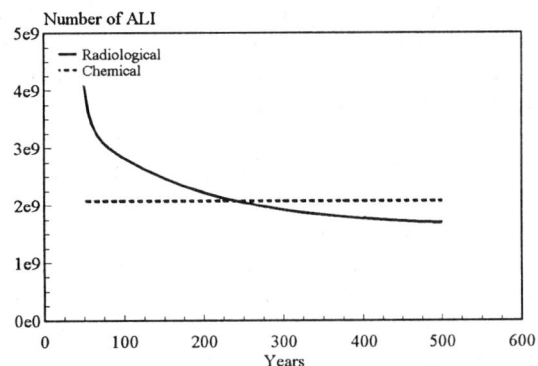

Fig. 6. A comparison between the radiological and the chemical ALI for the shielding blanket of ITER.

pared for the shielding blanket of the ITER machine, assuming beryllium is used as the armour material throughout.

The blanket consists of about 7000 tonnes of SS316LN steel and about 10 t of beryllium. The radiological ALI was in this case calculated as the sum of the number of ALIs of the nuclides, Cr-51, Fe-55, Ni-59, Ni-63 and Co-60 in the shielding blanket while the chemical ALI was derived from the occupational exposure limit values given in units of mg m^{-3} multiplied by the volume of air inhaled per year by one person (which is 2400 m^3).

As seen in Fig. 6 the radiological ALI decreases and 250 years after shut-down is lower than the chemical ALI for Be. The chemical ALI on the other hand does not change with time and in the long run it will dominate the hazard of the ITER waste.

10. Conclusions

Use of low activation structural materials greatly reduces the content of long-lived and hazardous nuclides in the activated material. A vanadium alloy, two different low-activation martensitic steels and silicon carbide have been studied. The hazard potential from all these materials after irradiation will be substantially lower than from stainless steel. One of the low-activation martensitic steels, F-82H, will give the lowest hazard potential after about 100 years of decay.

Tritium is a valuable fusion fuel and should be extracted from the fusion waste. Also for regulatory reasons it is important to reduce the tritium content in the waste. By detritiation followed by 50 years of interim storage it is possible to reduce the tritium concentration below the limit set for the German repository Konrad, for example.

Clearance can be used to reduce the quantity of material finally classified as very low-level waste. In the future it may also be possible to reduce the quantity of waste with higher activity contents by recycling of materials with dose rates up to 10 mSv h^{-1}. This means that almost no material from future power stations may need to be disposed of in a waste repository. However, part of the material from ITER will need to be disposed of in a repository: a less sophisticated one such as the German repository Konrad seems to be adequate.

References

[1] S. Nisan, K. Brodén and M. Lindberg, ITER Waste Management Strategies and Final Disposal, in: Fusion Technology 1996, Proceedings of the 19th Symposium on Fusion Technology, Lisbon, 16–20 September, 1996, North-Holland, Elsevier, Amsterdam, 1997.

[2] R.A. Forrest and J.-Ch. Sublet, FISPACT4 User Manual, UKAEA report UKAEA FUS 287, 1995.

[3] J. Kopecky and D. Nierop, The European Activation File EAF-4, ECN Petten report ECN-C-95-072, 1995.

[4] W. Dietz, private communication with I. Cook/N. Taylor, November 1995.

[5] J. Raeder et al. 'Safety and Environmental Assessment of Fusion Power', European Commission report EURFUBRU XII-217/95, 1995.

[6] N. Yamanouchi, et al., J. Nucl. Mater. 191-194 (1992) 822.

[7] L. Boisset and C. Lattaud, Metallic waste detritiation. Performance of a melting process. CEA Note Technique 95/039, 04/08/95.

[8] P. Brennecke and E. Warnecke, Anforderungen an endzulagernde radioaktive Abfälle (Vorläufige Endlagerungsbedingungen, Stand April 1990 in der Fassung Juli 1991). -Schachtanlage Konrad. Bundesamt für Strahlenschutz ET-3/90-REV-1, 1991.

[9] P. Rocco and M. Zucchetti, Recycling and Clearance of Fusion Waste, final report of SEAL 10.1, JRC Ispra, SEALAG, August 1996. EUR 16453 EN.

[10] Clearance Levels for Radionuclides in Solid Materials: Application of Exemption Principles, IAEA Draft Safety Guide, Safety Series No. 111 G 1–5, Vienna, 1994.

[11] S. Yanagihara, Y. Seiki, H. Nakamura, Dismantling Techniques for Reactor Steel Stuctures, Nucl. Techn. 86 (1989) 148–158.

[12] O. Andersson, Minimising Low Level Waste by Volume Reduction and Recycling. Internationale Zeitschrift für Kernenergie atw 40 Jg (1995) Heft 7–Juli, 461–465.

[13] Entwicklung eines sicherheitsanalytischen Instrumentariums für das geologische Endlager für radioaktive Abfälle in einem Salzstock, Zusammenfassender Abschlußbericht, Kapitel 4, Projekt Sicherheitsstudien Entsorgung, Berlin, 1985.

[14] H. Illi, Safety analyses for the planned Konrad repository, Chemietechnik SI (1987) No. 2.

Status of the European breeding blanket technology

L. Giancarli [a,*], M. Dalle Donne [b], W. Dietz [1,c]

[a] *Euratom-CEA Association, CEA-Saclay, DRN/DMT/SERMA, 91 191 Gif-sur-Yvette, France*
[b] *Euratom-FZK Association, INR, Postfach 3640, 76 021 Karlsruhe, Germany*
[c] *European Commission, DG-XII, 1049 Brussels, Belgium*

Abstract

This paper presents an overview of the activities within European laboratories on the development of breeding blankets for a DEMO reactor which are presently focused on the two selected blanket lines, the Water-Cooled Lithium–Lead (WCLL) blanket and the Helium-Cooled Pebble-Bed (HCPB) blanket. In particular the paper describes the design of the DEMO blanket segments and sub-systems layout, the status of the associated R&D, the preliminary design of the derived test-modules and sub-systems for testing in ITER. The outline of the near-future activities mainly focused on the development of the two EU ITER Test Modules is also discussed. © 1997 Elsevier Science S.A.

1. Introduction

Breeding blanket technology for a tokamak-based D-T Fusion DEMOnstration reactor (DEMO), on which the present paper is focused, has been developed in the European Union (EU) for many years. At present, work is mainly performed by EU associations with limited industrial support. It is recalled that DEMO is assumed to be the only step between ITER and a Fusion Power Reactor (FPR) and has therefore to demonstrate the technology of a FPR. Breeding blankets have to breed the tritium required for D-T reaction and to convert nuclear energy into heat extracted by a coolant under pressure and temperature conditions appropriate for driving an acceptable thermodynamic cycle. These functions have to be ensured in a quasi-continuous way during the expected reactor life-time.

A preliminary assessment, held in 1989, led to the selection of four lines of blankets [1], two of them using the liquid eutectic Pb-17Li as breeder material and the Pb-17Li itself or pressurised water as a coolant [2], the other two using Li-based ceramics as a breeder and helium as a coolant [3]. In 1995 a Blanket Concept Selection Exercise (BCSE) was organised within the EU with the aim of selecting the two best options for further development, in view of manufacturing test modules representative of the selected blanket lines to be tested in ITER. The two selected options are the Water-Cooled Lithium–Lead

* Corresponding author.
[1] Consultant, work performed under contract NET/95-886 to IKL, D-51503 Roesrath, Germany.

(WCLL) blanket [4] and the Helium-Cooled Pebble-Bed (HCPB) blanket [5], both using martensitic steel as structural material.

At the beginning of 1996, the corresponding activities have been integrated in the European Blanket Project (EBP) [24] which includes design and R&D activities associated with the two blanket lines with strong emphasis on the development of a suitable reduced-activation martensitic steel. In particular, they focus on the design and development of two ITER Test Modules (ITM) and associated systems, representative of the two selected breeding blanket lines, to be fabricated by the beginning of ITER operation and tested in the appropriate ITER test ports.

2. Description of the breeder blanket concepts for DEMO

2.1. Assumed DEMO specifications used for design activities

The assumed DEMO specifications, reported in [1], are characterised by a fusion power of 2200 MW, a neutron wall loading of 2.2 MW m^{-2}, an average heat-flux on First Wall (FW) of 0.4 MW m^{-2} and a blanket lifetime of 20 000 h.

As far as the structural material is concerned, because of the severe DEMO operating conditions (neutron irradiation up to 70 dpa, 250–500°C operating temperature range), the selected structural material is martensitic steel. Compared to austenitic steel, it has the advantages of an acceptable swelling in high-fluence, high-temperature conditions, a higher thermal conductivity and adequate mechanical properties at service temperature.

The properties of the so-called MANET, a 11CrMoV-martensitic steel, have been used for design purposes, although, because of its significant irradiation-induced shift (> 250°C) in the Ductile–to–Brittle Transition Temperature (DBTT) determined in recent years, this specific grade is not an acceptable candidate for DEMO blanket fabrication. A significant programme is presently performed to verify the overall performance of the reduced activation martensitic steel of type 8–9 CrWVTa (Section 3.1).

2.2. The WCLL blanket

This blanket line [4,6] uses thermal-hydraulic parameters typical of a PWR, i.e. water pressure of 15.5 MPa, maximum water velocity of 7 m s^{-1} and water outlet temperature of 325°C. The additional choice of the water inlet temperature of 265°C for the breeder zone is dictated by the Pb-17Li melting point of 235°C. The reference blanket design (the 'single-box' design) is based on the following design options: (i) single walls allowed between plasma and liquid metal but no single welds; (ii) two walls between cooling water and plasma and between FW coolant and Pb-17Li (both barriers able to withstand the water-coolant pressure); (iii) two walls between Pb-17Li-pool cooling water and the liquid metal, making use of Double-Walled Tubes (DWTs), with both walls able to withstand the water-pressure; the two concentric tubes are brazed, using a brazing material able to ensure thermal contact but ductile enough to stop propagation of cracks possibly starting, e.g. in the inner tube; (iv) each weld is allowed to leak without requiring an immediate shutdown of the machine for segment replacement; (v) malfunction of one cooling circuit, assuming a delayed plasma shutdown, should not induce a temperature rise which could affect further segment operation.

2.2.1. Design description

Outboard segments have an overall thickness of 85 cm (not including the back plate) and inboard ones have a thickness of 55 cm. Each blanket segment is formed by a single-steel box which acts as a liquid metal container (see Fig. 1). Radial and toroidal stiffeners are added in order to withstand the disruption-induced forces and, at the same time, both the Pb-17Li hydrostatic pressure in normal conditions and the full water-pressure under faulted conditions (see Fig. 2). The design point for an outboard segment, obtained through thermal and thermal-hydraulic analyses, is given in Table 1. The segment-box is directly cooled with an independent circuit formed by toroido-radial brazed cooling tubes with inside/outside diameters of 8/10 mm. In order to reduce thermal stresses, the FW has a corrugated shape towards the plasma. The FW-coolant is collected in vertical headers located behind the back plate (Fig. 2) in order to have water inlet and outlet at the top of the segment. All welds are placed in a double-wall system which provides a leak detection path.

The cooling water in the Pb-17Li pool flows downwards in the front part of the segment and upwards in the rear part within U-shaped DWTs. The DWT diameters are 11.0/13.4 mm for the inner tube and 13.6/16.5 mm for the outer one, leaving a 0.1 mm-gap for the brazing. In order to minimise tritium permeation towards the coolant, the Pb-17Li flows downwards in the back part of the segment and upwards in the front part. Top collectors are formed by three concentric walls, two for fluids flow (water and Pb-17Li), one in between for leak detection (see Fig. 1). This design ensures double containment of water until the outside of the vacuum vessel and the liquid state of Pb-17Li during start-up and shutdown

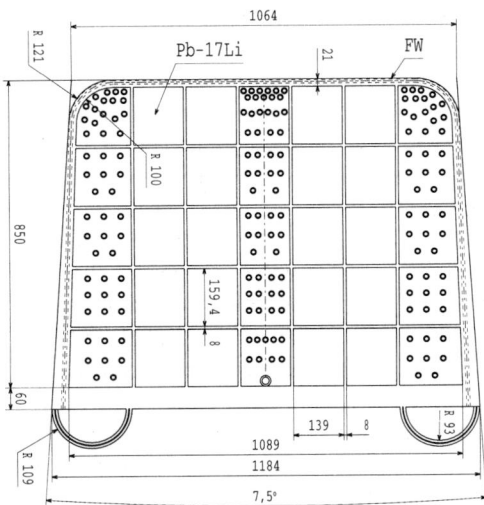

Fig. 2. Horizontal cross-section of an WCLL outboard-segment at the torus mid-plane level.

because of the thermal contact with the water pipes.

2.2.2. Main blanket performances [6]

A 3D-geometry Monte Carlo calculation, taking into account blanket heterogeneity and 3D features including exhaust-pump duct and using a detailed model for the 14 MeV-plasma-neutron source, has indicated a Tritium Breeding Ratio (TBR) of 1.19 (90% enrichment in ^{6}Li, no ports).

Table 1
Design point for an outboard segment in the WCLL DEMO blanket

Total deposited power	33 MW
Pb-17Li pool	25 MW
FW box	8 MW
Pb-17Li-pool/water coolant (15.5 MPa)	
Inlet temperature	265°C
Outlet temperature	325°C
Maximum velocity	5.9 m s^{-1}
nb. DW poloidal U-tubes	205
FW/water coolant (15.5 MPa)	
Inlet temperature	300°C
Outlet temperature	325°C
Average velocity	5.0 m s^{-1}
nb. toroidal tubes	378
Maximum temperature of the Pb-17Li/steel interface	480°C

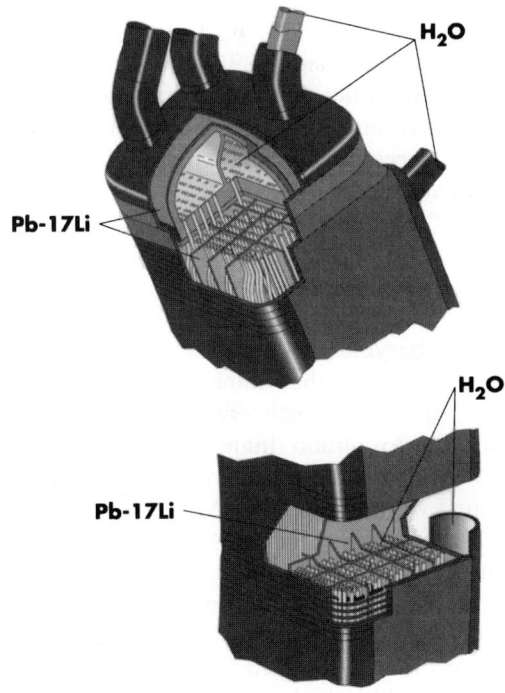

Fig. 1. Isometric view of a poloidal portion of an WCLL outboard-segment at the torus mid-plane level (bottom) and at the top-header level (top).

In the presence of ten horizontal ports the TBR becomes 1.13.

Inboard and outboard blanket segments are cooled by two independent circuits: one for the segment box and one for the Pb-17Li pool. Because these two components are thermally connected, in case of an (out-of-vessel) loss-of-coolant accident (LOCA) in one of the two circuits, the temperature increase due to the afterheat deposition remains within an acceptable level (530°C, 13 s after the accident, in case of a FW LOCA). This conclusion was drawn from 2D-thermal calculations assuming a delayed plasma shutdown (full power for 1 s after the accident, then only surface heat load from the plasma decreasing linearly from maximum to zero in 20 s).

An estimation of the tritium permeation rate from the Pb-17Li through MANET DWTs into the cooling water was performed considering ten recirculations per day and the Tritium extraction efficiency, η, as a parameter. The presence of the brazing material in the DWTs was neglected, assuming a single tube with the two-wall equivalent thickness. The obtained permeation rates range between 32 ($\eta = 1$) and 97 g day^{-1} ($\eta = 0.5$). Assuming a maximum acceptable tritium permeation rate of 1 g day^{-1} and taking into account the effects of the Pb-17Li detritiation, the application of tritium permeation barriers is needed. The required permeation reduction factor (PRF) is, however, moderate. In fact, assuming for instance a reasonable T-extraction efficiency of 0.8, a T-barrier with a PRF slightly below 100 (end-of-life value) would be sufficient. In these conditions, the T-inventory for the whole blanket would be ~ 10 g in the MANET structures and ~ 19 g in the Pb-17Li.

2.2.3. External circuits and components [6]

Water cooling-circuit architecture and most of the component design (e.g. steam generators, pressurizers, pumps, valves, etc.) can heavily rely on the technology developed for PWR cooling circuits. The water detritiation system can exploit the experience gained from the Canadian heavy-water CANDU reactors. The detailed circuits layout includes four primary water-cooling circuits for the breeder regions, each with a flow-rate of approximately 1200 kg s^{-1} and a steam generator (SG) of approximately 450 MW, and two for the segment-boxes, each with a flow-rate of approximately 2100 kg s^{-1} and a SG of approximately 280 MW. Assuming a conventional steam cycle the estimated power conversion efficiency is of approximately 35%.

As far as the Pb-17Li circuit is concerned, the tritium extractor and the facility for Pb-17Li purification are entirely new components on which significant R&D has been performed in the last few years with encouraging results. Other circuit components, such as pumps, valves, Li-content monitor and some instrumentation, can be developed from those already existing for Na-cooled Fast Breeder Reactor technology. There is the same number of circuits as for the breeder-zone cooling circuits: four circuits with a Pb-17Li flow-rate of 140 kg s^{-1} (53 m^3 h^{-1}) each.

2.3. The HCPB blanket

The reference design [5,7] of this blanket line is based on the following design principles:
1. use of ceramics in form of small pebbles to avoid excessive thermal stresses in the breeder material;
2. use of a binary bed of large and small beryllium pebbles as neutron multiplier to achieve an high beryllium density and, at the same time, to leave enough room for irradiation-induced beryllium swelling and to reduce thermal stresses in the material;
3. use of lithium-orthosilicate (Li_4SiO_4) as a breeder to obtain high lithium density (low lithium enrichment, low lithium burn-up), associated with sufficiently fast tritium release (low tritium inventory) and low lithium partial pressure at high operating temperatures (low lithium transport). Lithium-zirconate (Li_2ZrO_3) and lithium-titanate (Li_2TiO_3) are also promising ceramic candidates;
4. use of pebble beds of relatively small dimensions, especially in vertical direction, to avoid thermal ratcheting of the container walls and/or excessive crushing of the pebbles;

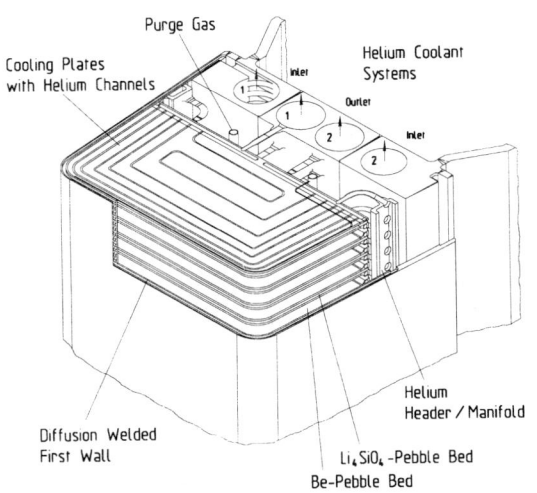

Fig. 3. Isometric view of a poloidal portion of an HCPB outboard-segment at the torus mid-plane level.

5. use of a He-purge flow at atmospheric pressure with a suitable chemistry (e.g. addition of H_2) to reduce the amount and the probability of tritium losses and to reduce the mass flow rate of the gas system and, as a consequence, the need of a permeation barrier between tritium purge flow and helium cooling system;
6. use of He-coolant at 8 MPa with an inlet/outlet temperature of 250°C/450°C. Cooling of the first wall with relatively cold (inlet) helium. The minimum temperature of the breeder is maintained above 300–350°C to keep the tritium inventory low. The beryllium temperature is kept as low as possible to limit irradiation induced swelling;
7. use of radial-toroidal modules to attain a high filling factor for breeder and multiplier pebbles in the blanket region and, furthermore, to allow a subdivision of the modules into the sub-modules. This gives the possibility of making significant out-of-pile tests, and tests in fission reactors and in ITER starting from smaller sub-modules;
8. use of a redundant convective cooling system and of a double containment against tritium losses for safety improvement.

2.3.1. Design description

The main design features can be seen in Figs. 3 and 4 [7]. In particular, Fig. 3 presents an isometric view of a poloidal portion of the outboard blanket segment, where it can be seen the blanket structure made of radial-toroidal cooling plates and, inside the blanket box, the two independent high-pressure He-cooling systems on the back. The Helium purging the tritium produced in the blanket is brought to the front of the blanket by a small tube in each pebble bed layer (beryllium or ceramic) and returns radially to the back of the blanket region from where it is collected.

Fig. 4 shows a horizontal and a vertical cross section of the outboard blanket segment at the mid-plane level of the torus. The flow direction of the He-coolant alternates in opposite directions, both in the first wall and in the blanket, to ensure a better temperature uniformity. Between the cooling plates there are, alternatively, beds (11 mm-thick) of Li_4SiO_4-pebbles ($d = 0.25$–0.63 mm) with few vol.% of TeO_2 and binary beds (40 mm-thick) formed of 2 mm-Be pebbles and 0.1–0.2 mm-Be pebbles. Use of Li-zirconate or Li-titanate would have a minor impact on the design [8].

The arrangement of the cooling channels in the radial-toroidal plates is schematically shown in Fig. 4. The plates are welded to the first wall and to the lateral sides of the blanket box. No high leak tightness is required between the parallel coolant channels. High tightness is required only at the borders of the cooling plates. But even then, operation with small leakages is allowed, as the box can operate at the full He-coolant pressure.

2.3.2. Main blanket performances

The main characteristics and performances of the HCPB blanket are shown in Table 2. The neutronic calculations have been performed in the same way as for the WCLL blanket (Section 2.2.2). The obtained TBR is 1.13 (25% enrichment in ^{6}Li, no ports). The thermal-hydraulic and stress calculations have been performed with the FEM-code ABAQUS.

The ANFIBE-code [9] has been applied to calculate the beryllium swelling and the percentage

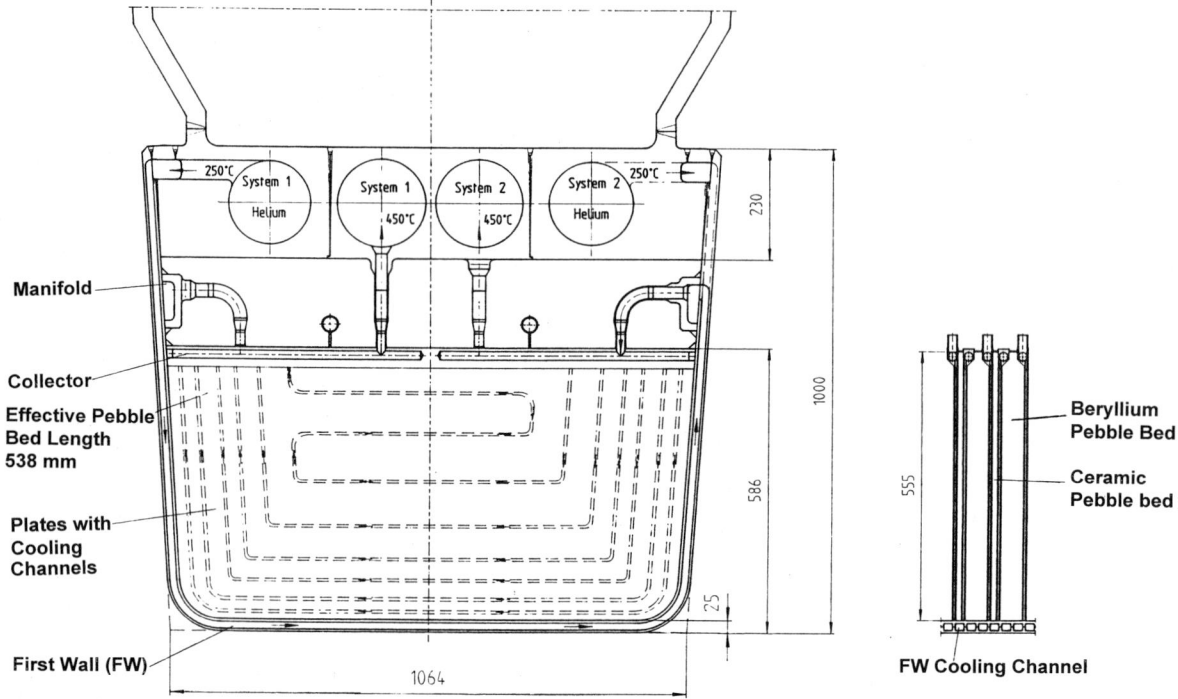

Fig. 4. Horizontal (left) and vertical (right) cross sections of the HCPB outboard-segment at the torus mid-plane level.

tritium inventory remaining in the beryllium. The calculated maximum Be-swelling is 8%, thus the flow of the purge gas is not impeded.

Detailed tritium permeation calculations show that the tritium permeation from the tritium purge flow and from the FW to the main He-cooling system would amount to 0.8 g day^{-1} and to about 12 g day^{-1}, respectively, so that the tritium permeation from the main helium cooling system through the He-heated steam generator can be limited to the acceptable level of less than 20 Ci day^{-1}. This requires a sufficiently high oxygen potential in the main helium coolant system to provide a thin and stable oxide layer on the helium side of the steam generators [10].

Two independent helium loops are used to cool in series First Wall and blanket in adjacent channels, so that, in case of a loss of coolant accident (LOCA) in one of the two loops, the FW temperature increase due to the afterheat deposition amounts to 90°C for a few seconds only, while a loss of flow accident (LOFA) from both blower systems would cause a maximum increase of the FW temperature of 65°C also for a few seconds and stabilise to a level below nominal values by the cooling due to natural convection.

2.3.3. External circuits and components [7]

The external circuits can rely on the technology developed for He-cooled fission reactors (steam generators, blowers, valves, helium purification plant, tritium extraction system). Each of the two independent main Helium-coolant systems, in series between first wall and blanket, is made up of six circuits for the outboard blanket (five in operation, one in reserve) and three circuits for the inboard blanket (two in operation, one in reserve), the helium mass flow rate for each circuit being about 350 kg s^{-1}.

To make full use of the excellent properties of helium (inert and not radioactive) the helium coolant is continuously purified. As in the helium cooled fission reactors, 0.1% of the total helium flow is passed through a Helium purification plant. The main tasks of this system are: (i) to extract hydrogen isotopes as well as solid, liquid or gaseous impurities from the main helium coolant; (ii) to remove condensed water that may be entrained in the helium coolant gas due to leakages or failures of the steam generator tubes; and (iii) to add to the helium returning to the main coolant system the proper amount of H_2O and H_2 to adjust the oxygen potential in the main helium coolant system.

The helium purification is achieved in the following steps: (i) purification from solid particles by means of filters or electrostatic separators; (ii) quantitative conversion of Q_2 to Q_2O (Q = H,D,T) in an oxidizer; (iii) freeze out the water content at $-100°C$ in a cold trap; (iv) send the gas to a molecular sieve bed cooled by liquid nitrogen to adsorb gaseous impurities like N_2 and the rest of the un-oxidized hydrogen isotopes; (v) add to the gas the proper amounts of H_2O and H_2.

The helium from the tritium purge system is directed to the Tritium Extraction System. The tasks of the system are: (i) removal of tritium produced in the blanket test module; (ii) separation and intermediate storage of the two main chemical forms of tritium, i.e. HTO and HT; (iii) purification and conditioning of the purge gas.

Removal of tritium and excess hydrogen from the helium carrier gas is accomplished in two steps: (i) tritiated water (HTO and HO) is frozen out in a cold trap operated at $-100°C$; (ii) molecular hydrogen isotopes (HT, H_2) and gaseous impurities are adsorbed on a molecular sieve bed operated at $-196°C$. The clean helium is then sent through a make-up unit where hydrogen is again added to adjust the H_2 level to 0.1 vol%.

At the end of a cycle, the tritiated water collected in the cold trap is transferred to the Water Detritiation System (WDS) which is part of the installations for the primary plasma fuel cycle. Desorption of the molecular hydrogen isotopes from the molecular sieve bed is carried out in a secondary helium loop containing a circulation pump and a Pd/Ag diffuser. The pure hydrogen isotopes obtained at the secondary side of the diffuser are stored in uranium getter beds and, later on, transferred to the Isotope Separation System (ISS).

Table 2
Main characteristics of the HCPB DEMO solid breeder blanket

Breeding and multiplier material	Separated Beds of Li_4SiO_4- and Be-Pebbles
Total blanket power	2500 MW (+300 MW in divertors)
Coolant helium pressure	
Outboard segments	8 MPa
Inboard segments	8 MPa
Coolant helium pressure drop (FW, blanket, feeding pipes)	
Outboard segments	0.30 MPa
Inboard segments	0.30 MPa
Coolant helium temperature	
Inlet	250°C
Outlet	450°C
Plant thermal efficiency	34.6%
Maximum steel temperature	520°C
Maximum temperature in beryllium	640°C (BOL) 500°C (EOF)
Temperature in breeder material	
Maximum	900°C
Minimum	350°C
3D-geometry TBR (without ports)	1.13 (BOL), 1.11 (EOL)
Accounting for ten ports	1.07 (BOL), 1.05 (EOF)
Peak lithium burn-up	7.25 at.%
Peak He-production in beryllium	16 300 appm of He
Tritium purge-gas system pressure	0.1 MPa
Tritium inventory in breeder material	10 g
Tritium inventory in beryllium	2100 g
Tritium losses to steam/water system	20 Ci day^{-1}

3. Review of the major R&D activities

The R&D activities associated with breeding blankets have been, since several years, mostly devoted to reduce existing uncertainties in the technology required for the proposed blanket designs in terms of system behaviour and material data base. A common R&D to both blanket lines is the development of the structural material, while other R&D is specific to each blanket line.

3.1. Choice of structural material

The selection of a structural material is strongly related to the design requirements, e.g. compatibility with coolant and other blanket materials, anticipated range of temperature, temperatures transients, compatibility aspects with the environment (liquid, solid, gas, impurity effects), the neutron dose level, the thermo-mechanical loading and, finally, the fabricability aspects including non-destructive testing. Moreover, in the long-term, a material with low-activation characteristics is compulsory. An adequate data base must be available for the final selected material to permit a reliable design and validated manufacturing.

The principal materials considered for structural application for the different blanket concepts within international programs include the austenitic and high-Cr martensitic steels and the Vanadium alloys. Scoping investigations have also been made on Ti- and Cr-alloys and fibre-reinforced ceramics such as SiC/SiC, the latter especially under the aspect of radiological behaviour in the short and medium term (accident management timescale).

For ITER, 316 L(N) austenitic stainless steel was selected as the main structural material because this material is already largely qualified by the PWR and FBR efforts for nuclear application in the design codes. For specific ITER conditions some additional work was needed for new product forms, such as HIPping, and for the additional material data at low operational temperature. This work for further qualification is now in progress.

However, because of the requirements of the DEMO relevance for the selection of a material for the ITER Test-blanket Modules (ITMs) in terms of neutron dose (>70 dpa), temperature ($T > 500°C$), environmental aspects, compatibility aspects and also because of programmatic aspects, such as timing and available resources, the reduced-activation martensitic steels appears to be the appropriate choice. The selection of ferritic/ martensitic steels is a continuation of previous work, since this alloy family was for years the prime choice for a design of breeder blankets due to beneficial thermo-mechanical properties and the irradiation behaviour.

3.1.1. EU R&D programme on ferritic/martensitic steels

The EU programme started in the 1980s with a modification of conventional martensitic steel, a 11CrMoVNb-alloy named MANET, for design purposes and with significant efforts in R&D. In the 1992–1994 programme period, experimental results showed that this material had major drawbacks in terms of toughness after low-temperature irradiation and may give design problems for structural integrity verification because of a risk of fast crack propagation [11]. It became also evident that MANET steel had activation levels in the long-term (waste-disposal timescale) not compatible with the goals for reduced activation [12]. It was recommended in 1994 by a Task Force Material (TFM), with representatives of all EU Associations present in the Fusion programme, that future R&D activities in the material field should concentrate on the variants for reduced activation ferritic/martensitic (RAF) steels of type 8-9 CrWV. The activity is included in the international collaborative work for RAFs under the auspices of the International Energy Agency (IEA) [13].

In 1995 a programme for the selection of a primary candidate alloy from eight RAF variants was started in the EU. The principal type of alloy is a ferritic martensitic 8-9Cr-steel with a range (in wt.%) of suitable alloying elements, that is Cr (7.5–9.5), Mn (0.2–0.5), Si (0.02–0.1), Ta (0.02–0.09), Ti (0.02–0.08), V (0.16–0.40), W (0.01–2.0).

The heat treatment of these alloys is in the typical range for ferritic/martensitic (F/M) steels for austenitization and tempering. Post weld heat treatment and preheating for welding will need some investigation to define the appropriate limits. To reduce the activation level, elements such as Nb, Mo and Ni have to be avoided and then, special precaution in the selection of the raw materials is needed [12].

The embrittlement induced by low temperature irradiation, characterized by the DBTT in the impact test behaviour, has been considered for years as the key issue of the F/M steels for application in fusion reactors. The on-going EU structural material programme concentrates on screening the various RAFs to identify those with low DBTT. The programme is aimed at selecting the basic composition of the alloy in late 1998 for further qualification up to ITM fabrication. The selection process for 1998 is based mainly on a series of irradiation experiments in the HFR reactor in Petten up to a fluence level of about 2.5 dpa. This dose level permits the full qualification of selected structural material for the ITM dose level without, however, accounting for possible effects due to the hard fusion spectrum and the corresponding He/dpa ratio. After this step, further irradiation experiments will be needed to prove the validity of the obtained results in the case of specific fabrication routes required for ITMs (e.g. HIP, diffusion welds, or coatings). The qualification of F/M steel for DEMO needs, of course, irradiation experiments with much larger fluence. In this context, the planned facility IFMIF will provide significant information under fusion irradiation conditions.

3.1.2. Status of the ferritic/martensitic steels activities

Only some highlights of the on-going activities are given in this survey. The most significant ones (compatibility aspects, see Section 3.2) are:
1. a 5-ton IEA-heat was produced; it has demonstrated that this new category of steels can be fabricated in an identical scale as other martensitic steels with sufficient homogeneity in the sheet product form [14];
2. previous irradiation experiments have shown that the RAF steels have improved DBTT-behaviour at higher irradiation temperatures (T > 365°C), see [14]. The present information on irradiation behaviour of the EU Low-Activation Ferritic (LAF) steels and the IEA-heats indicates that the basic alloy composition, the 8CrWTaV steel, shows also less irradiation hardening than the MANET

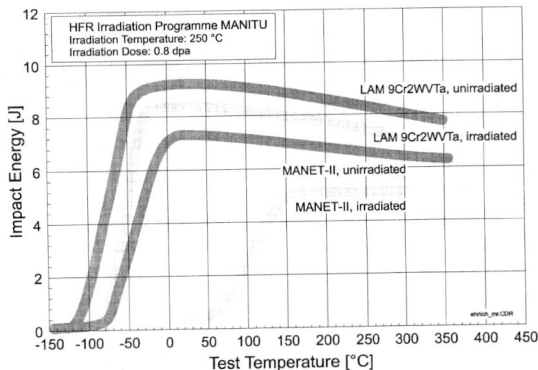

Fig. 5. Variation of DBTT after irradiation for MANET-type steel and for a newly proposed RAF steel (9Cr2WVTa-type).

steel for the lower irradiation temperatures (200–300°C) which is more relevant for the ITMs (see Fig. 5 [15]);
3. creep tests on RAF steels (OPTIFER alloys) showed similar behaviour as MANET/mod.9Cr [16];
4. no indications of radiation-induced stress corrosion cracking of base metal in pure oxygenated water was observed after irradiation at 250°C to approximately 2.3 dpa in 3-point bending tests [17].

3.2. R&D associated with WCLL blanket [6]

3.2.1. Double-wall tubes and segment box fabrication techniques

DWTs are an important feature of the reference WCLL blanket concept because their use leads to a significant reduction of the probability of small water-leaks in the Pb-17Li compared to single-wall tubes. In order to not jeopardize the blanket availability, brazed DWTs are the preferred option (compared to wire-mesh DWTs which would provide a leak detection path). Significant operational experience (> 30 years) for straight brazed-DWTs has been gained in the EBR-II steam generator. Advanced fabrication techniques, based on solid hot isostatic pressure (HIP) technology and on the appropriate choice of the interface material, are under development in order to deal with the DWT bend and irradiation conditions.

For the segment-box fabrication, besides the development for martensitic steel of conventional techniques, such as forging, drilling and laser welding, advanced fabrication techniques, such as powder-HIP technology, are also envisaged. R&D is currently underway. Preliminary results are promising [18].

3.2.2. Tritium permeation barrier development

Activities on the development and testing of Tritium Permeation Barriers (TPB) are going on in various EU laboratories. In the WCLL blanket, TPB can be placed either in contact with Pb-17Li (on the external side of the cooling tube and on the segment-box internal wall) or in contact with water (on the cooling-tube inside wall). Up to now, all activities have focused on TPB on the Pb-17Li-side.

Aluminide coatings with an alumina top layer of few microns yielded the most promising results in terms of permeation reduction factor (PRF) but also of compatibility with Pb-17Li and thermo-mechanical fatigue resistance. PRF of two to three orders of magnitude have been measured. Used substrates are SS-316L and MANET.

Several fabrication techniques have been proposed and tested. Examples are pack and vapour-phase cementation, chemical and physical vapour-deposition, plasma spray, hot dipping, detonation jet. All these techniques are used in non-nuclear industry essentially as anti-corrosion barriers. In general, the processes are not yet adjusted for WCLL blanket application. Compatibility of the different processes with the segment geometry and fabrication sequence and with the required martensitic steel heat treatment have to be further analysed.

The behaviour under high-dose irradiation of such coatings is still unknown. Very low-dose in-pile experiments performed in HFR-reactor in Petten (LIBRETTO-II and III experiments) for aluminide coatings fabricated by pack-cementation on SS-316L-substrate led to an estimated PRF value ranging between 30 and 80 depending on the experimental conditions.

The detrimental effect of cracks in the alumina layer was observed in out-of-pile experiments [19]. Adequate TPB should therefore enable in-service self-healing. No experimental evidence exists yet to support this capability, for any TPB. However, from the thermo-dynamic point of view, aluminide coatings (in contact with both Pb-17Li and water) have this capability. The reaction kinetics, however, are still unknown.

3.2.3. Water/Pb-17Li reactions

Experiments on the water/Pb-17Li interaction simulating a large break of a cooling tube occurring in a WCLL blanket were performed a few years ago in the BLAST facility. They showed that the chemical interaction is self-limiting by the formation of solid LiOH and Li_2O which insulates the melt from the water/steam. Vapour explosions were never observed. The pressurisation of the container did not significantly exceed the actual water-injection pressure and the increase of temperature was limited to 100°C at a Pb-17Li-temperature of 350°C.

Experiments for investigating the evolution of water micro-leaks into Pb-17Li have been performed in the RELA-II facility. After 44 h of testing neither wastage nor leak-blockage was observed. A large quantity of solid products was observed in the test vessel (a tube with $d = 3.2$ cm) but the Pb-17Li flow was never interrupted. This result indicates the importance of assessing the behaviour of the reaction products within the Pb-17Li and, in particular, their redeposition. It must be stressed that the use of DWT should considerably reduce the occurrence of such microleaks.

3.2.4. Tritium extraction techniques

Several techniques have been considered for the on-line tritium extraction from Pb-17Li, such as liquid/gas contactors or getters. The objective is to optimize the design of the extractors in order to reach an extraction efficiency as high as 90%.

Liquid/gas contactors, such as plate columns, bubble columns or packed columns, are currently considered the most promising T-extractors. Their principle is to bring a large Pb-17Li surface in contact with an inert purge-gas which desorbs tritium from Pb-17Li. The purge-gas then has to be treated for the removal of gaseous tritium, e.g. by catalytic oxidation to H_2O and subsequent absorption on molecular sieves.

Several experiments have been performed on liquid/gas contactors in the MELODIE loop, aimed at measuring fundamental quantities such as mass transfer coefficient and liquid/gas contact surface. As far as the T-extraction efficiency is concerned, up to now the best results have been obtained with a packed column. The effects of H-doping of the purge gas has yet to be evaluated.

3.2.5. MHD phenomena

The MHD-related studies are important for the evaluation of the Pb-17Li velocity field within the segments and the connecting pipes which could have an impact on the tritium permeation rate towards the water-coolant and on the temperature distribution. In particular, natural convection phenomena in Pb-17Li have to be better understood. Theoretical estimations and experiments are planned.

3.2.6. Pb-17Li/steel compatibility

Pb-17Li corrosion of martensitic steel (MANET) has been largely investigated under experimental conditions of WCLL blankets [6]. In the WCLL blanket the maximum Pb-17Li/martensitic steel interface temperature has been kept below 480°C, which in association with the low Pb-17Li velocity (~ 5 mm/s), leads to maximum local corrosion rates for bare steel in the blanket below 10 µm year^{-1}. In the feeding pipes, the Pb-17Li temperature is less than 330°C, and, therefore, despite a larger velocity, the Pb-17Li corrosion is negligible. The effects of magnetic fields should not significantly modify this evaluation. Under the same conditions, corrosion data have been produced also for Al-based permeation barriers indicating their good compatibility with Pb-17Li.

A certain number of data is available on the influence of Pb-17Li on the mechanical properties of steels. Specimens of martensitic steel, grade DIN-1.4914, tested under creep rupture and low-cycle fatigue conditions, showed a decrease of the time to rupture when in contact with Pb-17Li but a better low-cycle fatigue behaviour. No effect of stress corrosion cracking has been observed. Martensitic steels present liquid metal embrittlement only in simulated welded conditions, if the post-weld heat treatment is not carried out. Liquid metal embrittlement effects will need future evaluation also for the possible irradiation-hardening expected at the lowest service temperature (265°C).

3.2.7. Water/steel compatibility

The compatibility between water and the materials used in the primary coolant circuit depends on the water chemistry which has still to be adapted to fusion environment. The eventual presence of permeation barriers on the water-side (requiring self-healing) and radiolysis effects have to be taken into account. The materials used in the external cooling circuits are not yet fully defined. R and D is planned using the water chemistry as a parameter. There are indications that the water corrosion rate is affected by irradiation [17]. No information is available for irradiation-assisted corrosion cracking under fatigue loading of martensitic steels hardened by irradiation.

3.2.8. Pb-17Li purification and composition control

Considering the large mass of liquid breeder material involved in a Pb-17Li blanket, the lithium burn-up is relatively small. The only concern could be an increase of the alloy melting point. Such composition changes can be easily compensated by adding lithium or Li-enriched Pb-alloy in the external circuits. Several techniques for on-line Li-concentration monitoring, such as electrical resistivity meters, electrochemical sensors and plugging indicators, are under development and testing. Electrochemical sensors have shown a very high sensitivity to Li-concentration changes in the alloy (0.2 at% Li), although their thermal-shock resistance requires further improvements. Electrical resistivity meters have shown sufficient sensitivity and higher reliability than the sensors although they require a more complex installation.

Purification systems are required to remove impurities and corrosion products in order to minimize the neutron-induced activation of the breeder material. Systems for removing corrosion products, such as cold and magnetic traps, are being developed. Among transmutation products, Hg, Pb and Po radio-isotopes are of major concern. Experiments have shown that a limited amount of ^{210}Po could be acceptable because of the formation of the inter-metallic Pb-Po compound which has a very low vapour pressure. However, it is desirable to develop on-line Bi-removal techniques in order to limit the Bi-level to 1 ppm resulting in approximately 0.1 ppb of ^{210}Po, without needing to remove Po itself.

3.2.9. Minor circuit-component development

Development of minor components for Pb-17Li circuits, such as pumps, valves, and rupture disks, are important for the DEMO reactor, but also for out-of-pile and in-pile experiments requiring Pb-17Li loops. At present, several components derived from Na-technology are used, but limitations (e.g. pumping head and lifetime in electromagnetic pumps) appear already from their use in existing experimental Pb-17Li loops.

3.3. R&D associated with HCPB blanket

3.3.1. Development of the manufacturing process of the blanket segment

The blanket-segment manufacturing requires the development, improvement and/or adaptation of specific fabrication techniques. The current development programme, which is carried out in collaboration with several EU industries and universities, includes the following manufacturing processes: (i) diffusion bonding of large plates with internal cooling channels for the FW and the cooling panels; (ii) bending of the FW plates to obtain the U-shaped blanket box; and (iii) weld connection between the box and the cooling plates.

Encouraging results have been obtained so far for all three steps [20], i.e.:
1. diffusion bonding tests were carried out successfully with specimens of up to 320 mm in diameter. The structures and the mechanical properties of the bonded region are comparable with the base material. The joints are leak-tight and resistant against pressures in the cooling channels of up to 15 MPa;
2. diffusion bonding plates, with a thickness of 25 mm and rectangular cooling channels of 14 mm × 18 mm, were bent at room temperature with a bending radius of 75 mm without failures. The cooling channel geometry was not significantly affected;
3. TIG welding tests for 300 mm-long linear welds were performed with good results at room temperature. No flaws were detected, and the distortions due to shrinkage and thermal stresses were tolerable.

The future activities will concentrate on the manufacturing of medium-scale mock-ups and on the further development of non destructive testing methods.

3.3.2. Development work for the ceramic breeder pebbles

In case of lithium-silicate, the reference pebbles were originally made of Li_4SiO_4 with an excess of SiO_2. The irradiation experiments performed so far indicate that the mechanical integrity of and the tritium release from these pebbles are little affected by lithium burn-ups up to 180% of the DEMO peak value (EXOTIC-7 irradiation [21]). Recently, the chemical composition of the pebbles has been changed by adding to the over-stoichiometric Li_4SiO_4 a small amount of TeO_2. Tellurium oxide has been added because it increases the surface tension of the liquid phase during the fabrication process, thus avoiding the formation of cracks during the cooling of the pebbles. The resulting mechanical properties of the pebbles are considerably improved. Furthermore, recent results of out of pile experiments indicate that the tritium release may be even faster than that of Li_2ZrO_3 and Li_2TiO_3 [22]. Within the EU, the development of Li_2ZrO_3- and Li_2TiO_3-pebbles (which originally were produced in pellet form [23]) had only started at the beginning of 1996 and preliminary results will only be available sometime this year. A new irradiation experiment in the HFR Petten (EXOTIC-8), which started in the fall of 1996, will test the new Li_4SiO_4-pebbles up to high lithium burn ups and also the first Li_2TiO_3-pebbles.

Besides the Li burn-up, the neutron damage can also affect, although to a lower extent, the mechanical integrity of the pebbles. Peak neutron damage expected in DEMO (~ 20 dpa) can be achieved within reasonable times only in FBRs. For this purpose, a joint international irradiation experiment in the Phenix-FBR is under discussion.

3.3.3. Development work for the beryllium pebbles

The data available in the open literature on the behaviour of beryllium under irradiation (swelling, embrittlement and tritium release) are not yet sufficient to cover the peak conditions in the DEMO blanket. The computer code ANFIBE correlates well the data on swelling and tritium release available and is able to extrapolate these data to the DEMO conditions. However, more data are required from high temperature irradiation at high fluence in fast reactors. Capsules containing many beryllium samples irradiated in the U.S. fast reactors FFTF and EBR-II up to the peak DEMO burn up values in DEMO relevant temperature ranges are not being examined due to lack of funds. An effort has been initiated to solve this problem. A joint European beryllium irradiation experiment in the fast reactor Phenix will start as soon as the reactor is in operation (not before the beginning of 1997). However, the burn-up of this experiment will be only 30% of the peak DEMO value. The results of the post irradiation examination (PIE) for the irradiation experiment in the BR2-reactor in Mol will be available by the end of 1996. These data will probably be sufficient to assess the beryllium embrittlement issue, which is expected to be relatively marginal because of the use of small Be-pebbles.

3.3.4. Tritium control

Tritium permeation from the plasma to the FW cooling channels and from the tritium purge system to the plates cooling channels has been calculated on the basis of experimental correlations of the permeation through the structural material. However, these calculations must be validated with further experiments which are already planned within the EU. The major objective is the demonstration of the stability of the thin oxide layer present on the helium-side surface of the steam-generator tubes under temperature transients representative of a DEMO operation [10].

3.3.5. Thermal hydraulic experiments

The flow distribution of the He-coolant in parallel channels, the pressure drops, the behaviour of the blanket under thermal cycle operation and under fast temperature transients due to LOCA and LOFA events are being investigated in the HEBLO loop and will be further investigated with bigger and more relevant mock-ups in the He-FUS3 loop.

Besides these out-of-pile experiments, one experiment with a medium-size mock-up ($10 \times 10 \times 50$ cm^3) should be performed in a fission reactor in order to simulate more correctly the volumetric power distribution in breeder, multiplier and structural material and to assess the combined effects of breeder and beryllium swelling, thermal stresses and neutron irradiation effects on the structural material. Furthermore, the functional behaviour of the tritium purge system will be tested.

These experiments are required to predict with a high degree of confidence the temperature distribution and the pressure drop in the blanket and also to predict a sufficiently low failure probability of the HCPB-ITM.

3.3.6. Helium purification and tritium extraction

A considerable experience has been acquired in the course of the He-cooled fission reactors development on helium purification and detritiation. Furthermore tritium extraction from helium is being investigated for the treatment of the ITER plasma exhaust gases. However, the special requirements of helium purification and tritium extraction from the HCPB blanket need the construction and the operation of small pilot plants, especially in view of constructing these ancillary loops for the ITM.

4. Specific activities related to ITER test blanket modules

The present EU breeding-blanket development programme is focused on the development of an ITM for the two EU blanket lines. The overall time-schedule is based on that of ITER which foresees the first D-T plasma in the year 2009.

Assuming the beginning of the tests for that date, the major ITM milestone becomes the preparation of the detailed design of the ITMs and the associated circuits by the end of 2004, leaving 3–4 years for fabrication and 1–2 years for pre-testing of the components [24].

The development of the EU ITM designs has assumed the boundary conditions imposed by the ITER JCT, both in terms of ITM dimensions and of associated circuits location. In particular, ITMs have to fit in the ITER test port (1.6 m-wide, 2.6 m-high, located in the outboard equatorial plane), they can be poloidally straight and be installed with 5 cm-recess from the ITER shielding blanket FW.

4.1. Test objectives

The ITM system design is very sensitive to the test objectives. In the EU view, the main objectives should be the following: (i) verify and demonstrate the functionality of the integrated system, subsystems and individual components in the fusion environment; (ii) verify and demonstrate the performance of tritium production, extraction and recovery by establishing a complete tritium balance of the system which has to correctly simulate the envisaged DEMO system (e.g. presence of permeation barriers, correct He- and water-chemistry); (iii) verify and demonstrate the performance of high-grade heat production and removal by establishing a complete power balance of the system which has to correctly simulate the envisaged DEMO system (e.g. correct number of independent circuits, presence of DWTs); (iv) calibrate and validate the performed analyses, including the maintenance of the full integrity of the ITM under thermal and electromagnetic loads.

The achievement of these objectives is, of course, related to the possibility of performing the relevant measurements. R&D effort is likely to be required in terms of development of an appropriate instrumentation.

4.2. WCLL test module development

4.2.1. ITM preliminary design and thermal transient analysis

The ITM cross-section has the architecture and dimensions of the DEMO blanket inboard midplane cross-section [25]. It covers only half of the port width. In the remaining half a port, a second WCLL ITM (assumption for the preliminary design) or a module proposed by other ITER partners can be installed. For the preliminary design and analyses, two WCLL-ITMs, MANET properties and the RCC-MR codes have been assumed. The ITM design is shown in Fig. 6. Particular attention had to be paid to the coolant collectors and bottom cap because these areas are located, as opposed to DEMO, in high heat flux areas requiring an actively cooled FW. A 1D neutronic analysis has shown that the amount of produced tritium (using 90% Li enrichment in [^{6}Li]) is about 0.3 g per day (of consecutive com-

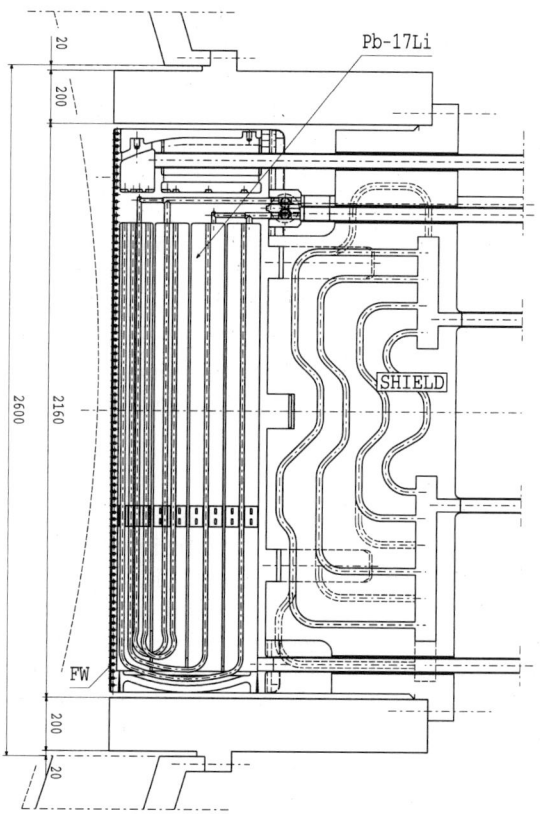

Fig. 6. Vertical cross section of the WCLL-ITM.

plete pulses). The maximum power to be extracted by the ITM is approximately 8.3 MW (including FW heat flux).

The pulsed operating conditions, specific to ITER, requires a thermal transient analysis in order to establish if equilibrium conditions can be reached in ITM during, at least, pulse-to-pulse operation. The results show that equilibrium is reached in less than 200 s in the front part of the ITM, while in the rear part equilibrium is never reached after a full-length pulse (1000 s).

4.2.2. ITM ancillary equipment

A preliminary layout and component design for both cooling and Pb-17Li circuits has been performed in order to evaluate the needs of the WCLL ITM in terms of room required in the ITER pit [25]. Some conclusions could be drawn: (i) the room available in the ITER pit is sufficient to host the required components, including the presence of two independent cooling circuits (for ITM box and for breeder region); (ii) guillotine break of a coolant pipe in the pit (with possible contact with Pb-17Li) is likely to be the dimensioning condition for the pit walls together with shielding requirements during maintenance; (iii) the components requiring the largest volume (18 m^3) are the Pb-17Li extractor, the Pb-17Li purification system, together with the power supply system and, possibly, the safety vessel.

4.3. HCPB test module development

4.3.1. ITM preliminary design and analysis

Fig. 7 shows a cross section of the HCPB test module in a horizontal port of ITER. To facilitate the handling operations, the ITM is bolted to a surrounding water cooled frame fixed to the ITER shield blanket back plate. The HCPB-ITM occupies the lower part of the frame, the upper part being reserved for a He-cooled ITM developed from another ITER partner. For the design of the test module, three dimensional Monte Carlo neutronic calculations and thermo-hydraulic and stress analyses for the operation during the basic performance phase (BPP) and during the extended performance phase (EPP) of ITER have been performed. To achieve higher and thus more DEMO relevant temperatures, an ITM with slightly different geometry features (different routing of the coolant flow in the First Wall, thicker ceramic pebble bed) from the DEMO has also been designed. The behaviour of the test modules during LOCA and LOFA has been investigated.

4.3.2. ITM ancillary equipment

The HCPB-ITM requires three main ancillary loops: (i) the He-coolant system formed by two separate and independent high-pressure He-loops; (ii) the helium purification system to continuously purify the coolant helium; and (iii) the tritium extraction system.

The size of the components of the ancillary loops has been assessed. The tritium extraction system can be placed on one side of the transporter corridor in the pit adjacent to the test module port. The helium coolant and the helium purification systems will be placed in the tritium building located 20 m above the HCPB-ITM, which is advantageous in promoting the natural convection of the coolant in case of a LOFA for both helium cooling loops.

4.4. R&D priorities for ITER test modules

4.4.1. Structural material development

Due to the fact that from the structural material point of view the test of modules in ITER will contribute, together with a suitable neutron source, to the qualification of the DEMO blanket structural material, the same alloy has to be used. Therefore, all activities described in Section 3.1 have to be achieved for ITM fabrication. Related to the structural material, development of ITM box fabrication techniques is also a major item (e.g. HIP, diffusion bonding, welding, etc.). After the selection of the reference material specification, emphasis is on the structural material behaviour under irradiation in the as-fabricated condition and on the generation of a set of data for design activities, including data on irradiation effects and compatibility with coolant. Design code aspects must be addressed both for the mechanical design and the fabrication.

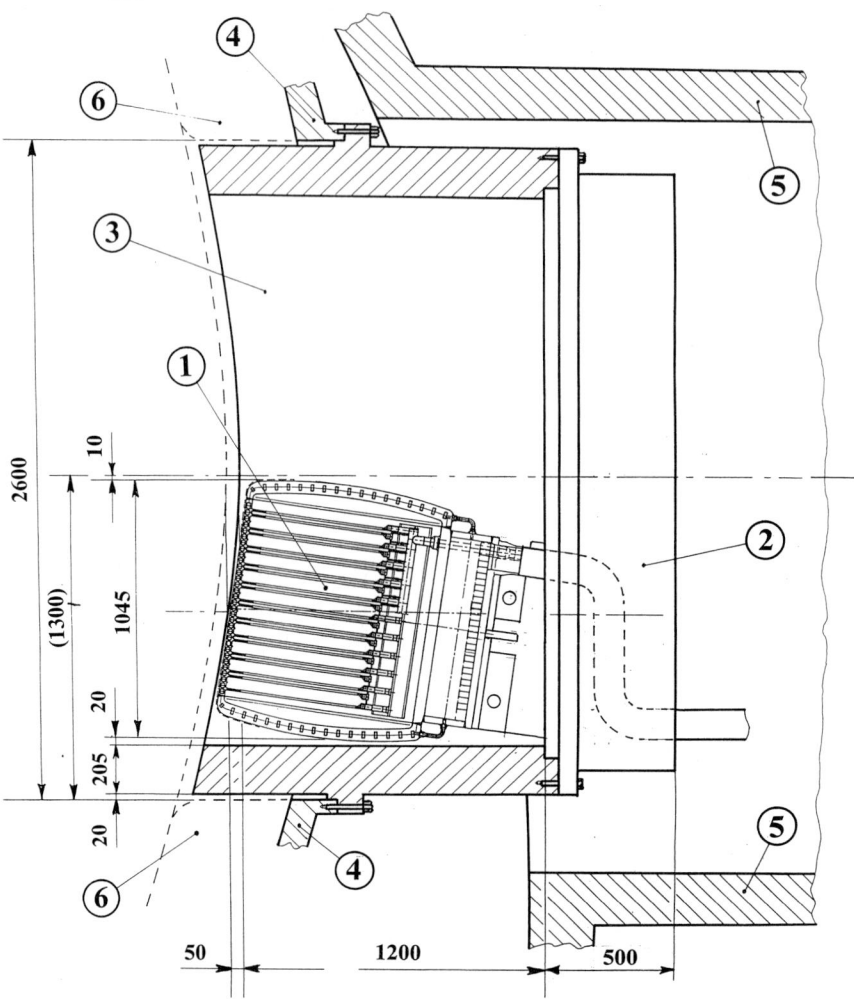

1. Test Module
2. Shield
3. Test Module Frame
4. Back Plate
5. Vacuum Vessel
6. Shield Blanket

Fig. 7. Vertical cross section of the HCPB-ITM in a horizontal port of ITER.

4.4.2. WCLL-ITER test module

Development of DWTs and T-permeation barriers, together with a full understanding of the Pb-17Li/water reaction in case of a large leak are the major items on which considerable effort has still to be made prior to fabrication of the WCLL-ITM.

The first two items are essential for the demonstration of the functionality of the ITM in particular considering its DEMO-relevance. It must be said, however, that if their development is not sufficiently mature by the ITM fabrication time, alternative solutions could be adopted, valid only for the ITM and, thus partly jeopardizing the test objectives. In particular, for the T-permeation barrier, being the T-production rate in the ITM much lower than in DEMO, the use of barrier with a lower PRF could ultimately be envisaged.

The understanding of the Pb-17Li/water reaction and the demonstration that such an accident can be tolerated by the ITM is likely to be required by the safety authorities before licensing. Preliminary accident-management evaluation has shown that this rare event can be handled [26]. This activity has, however, to be developed and completed before ITM fabrication.

4.4.3. HCPB-ITER test module

The main objective of the HCPB-ITM test during the ITER BPP is to test the temperature and helium coolant flow distribution in the modules as well as the operation of the tritium purge system in an environment with the correct neutron spectrum and temperatures and power deposition similar to those in the DEMO blanket. Thus, the out-of-pile (and in-pile) thermal-hydraulic experiments on blanket mock-ups, up to full ITM size, have to be performed before the ITER irradiation.

Clearly, the further development of the manufacturing of the test module (see Section 3.3.1) should have first priority as this is already required for the manufacturing of the biggest mock-ups to be tested in the He-FUS3 loop and of the medium-size mock-up to be tested in a fission reactor [27].

The effect of high neutron fluence of the material behaviour (structure, breeder and beryllium pebbles) is being investigated in fission reactors and will be continued during the ITER EPP under more DEMO relevant conditions (neutron spectrum, size, test module power density distribution). As far as tritium control is concerned, the ITER experiments will allow to demonstrate the performance of the tritium production, extraction and recovery and at the same time, by establishing the correct oxygen potential in the helium coolant system, to verify if the oxide layer on the surface of the blanket cooling plates is stable in the presence of a fusion neutron flux. However, this is not absolutely necessary for the limitation of tritium losses in DEMO to less than 20 Ci day^{-1}. It would be sufficient to show that a thin oxide layer is maintained on the helium facing surface of the DEMO steam generators in the presence of the proper oxygen potential in helium. This can of course be tested out-of-pile (the neutron flux is negligible at the steam generators).

5. Conclusions

The EU activities related to breeding blanket technology have been developed in the last years with a view to define promising DEMO blanket systems. Despite the significant progress, further work has to be done in order to eliminate still existing uncertainties and/or lack of data before a safe and reliable breeding blanket can be retained for DEMO. Most uncertainties depend on or are related to the behaviour of materials and systems submitted to the expected large fusion-neutron fluence which will be even larger for a FPR breeding blanket.

An essential step for the required R&D progress is the test of a blanket Test Module in ITER in order to take advantage of the unique opportunity to test a representative module and the associated systems in a D-T fusion environment before starting the DEMO construction.

In order to fully exploit such an opportunity, the EU breeding blanket technology programme has been focused on the design, fabrication and testing of an ITM for each of the two reference DEMO breeding blanket lines, the WCLL and the HCPB lines, in order to use ITER for a sound large-scale experiment for continuing the way towards a DEMO reactor.

References

[1] E. Proust, L. Anzidei, G. Casini, M. Dalle Donne, L. Giancarli, S. Malang, Breeding Blanket for DEMO, Fus. Eng. Des. 22 (1993) 19–33.

[2] L. Giancarli et al, Overview of the EU Activities on DEMO Liquid Metal Breeder Blankets, Fus. Eng. Des. 27 (1995) 337–352.

[3] M. Dalle Donne et al, Status of EC Solid Breeder-blanket Designs and R and D for DEMO Fusion Reactors, Fus. Eng. Des. 27 (1995) 319–336.

[4] L. Giancarli, Status and further Development of the European Liquid Metal Breeder Blanket, Proceedings of the German Annual Meeting on Nuclear Technology, Mannheim, May 21st–23rd, 1996.
[5] M. Dalle Donne, Status and further Development of the European Ceramic Breeder Blanket, Proceedings of the German Annual Meeting on Nuclear Technology, Mannheim, May 21st–23rd, 1996.
[6] L. Giancarli, M.A. Fütterer, Water-cooled Pb-17Li DEMO Blanket Line-Status Report on the Related EU Activities, CEA Report, DMT/SERMA (November 1995).
[7] M. Dalle Donne et al., European DEMO BOT Solid Breeder Blanket, Report KfK 5429 (November 1994).
[8] M. Eid et al., On the use of Zirconate/Titanate for the Helium-Cooled Pebble-Bed DEMO Blanket, Proceedings of SOFT-19, September 16–20, 1996, Lisbon, Portugal.
[9] F. Scaffidi Argentina, Helium induced swelling and trapping mechanisms in irradiated beryllium, Fus. Eng. Des. 27 (1995) 275.
[10] L. Berardinucci and M. Dalle Donne, Tritium Control in the European Helium Cooled Pebble Bed Blanket, Proceedings of SOFT-19, Sept. 16-20, 1996, Lisbon, Portugal.
[11] ECN-C-96-059, Progress Report 1995 on Fusion Technology Tasks, pp. 43-46.
[12] K. Ehrlich et al., J. Nucl. Mat. 212–215 (1994) 678–683.
[13] IEA Implementing Agreement for a programme of research and development on radiation damage of fusion materials, Task Annex II, IEA Working Group meeting in Baden, September 1995 (ORNL/M-4939).
[14] A. Koyama et al., Proceedings of ICFRM-7, Obninsk, Russia, September 25–29, 1995.
[15] M. Rieth et al., Proceedings of ICFRM-7, Obninsk, Russia, Sept. 25–29, 1995.
[16] M. Schirra, K. Ehrlich, Proceedings of the German Annual Meeting on Nuclear Technology, Mannheim, May 21st–23rd, 1996.
[17] A.C. Nystrand, A. Lind, Fusion Technology, 1994 (SOFT-18), Elsevier, New York, 1995, pp. 1301.
[18] C. Dellis et al., DEMO Blanket Segment Fabrication using Advanced HIP Technology, Proceedings of SOFT-19, Lisbon, Portugal, September 16–20, 1996.
[19] K.S. Forcey, A. Perujo, Tritium Permeation Barriers in contact with Pb-17Li, J. Nucl. Mater. 218 (1995) 224–230.
[20] H. Heider, G. Reimann, H. Riesch-Oppermann, K. Schleisiek, Development of Fabrication Techniques for the European Helium Cooled Pebble Bed Breeder Blanket, Proceedings of SOFT-19, Lisbon, Portugal, September 16–20, 1996.
[21] J.G. van der Laan et al., Performances of Ceramic Breeder Materials Irradiated to High Lithium Burn-up in EXOTIC-7, Proceedings of SOFT-19, Lisbon, Portugal, September 16–20, 1996.
[22] M. Dalle Donne et al., Development Work for Lithium-Orthosilicate Pebbles, Proceedings of SOFT-19, Lisbon, Portugal, September 16–20, 1996.
[23] N. Roux et al., Low-temperature Tritium Releasing Ceramics as Potential Materials for ITER Breeding Blanket, Proceedings of ICFRM-7, Obninsk, Russia, Sept. 25–29, 1995.
[24] S.J. Booth, W. Dietz, and S. Païdassi, The European Breeding Blanket Development Programme, Proceedings of the German Annual Meeting on Nuclear Technology, Mannheim, May 21st–23rd, 1996.
[25] M. Fütterer et al., Design and Transient Thermal Analysis of a Water-cooled Pb-17Li Test Blanket for ITER, Proceedings of SOFT-19, Lisbon, Portugal, September 16–20, 1996.
[26] Y. Severi et al., Safety Assessment and Accident Management of a Water-Cooled Pb-17Li Test Blanket for ITER, Proceedings of SOFT-19, Lisbon, Portugal, September 16–20, 1996.
[27] R. Conrad and J.G. van der Laan, The HFR Petten as a Test Bed for Breeder Materials and Blanket Concepts for ITER and DEMO, Proceedings of SOFT-19, Lisbon, Portugal, September 16–20, 1996.

Recent results on vacuum pumping and fuel clean-up

R.-D. Penzhorn [a,*], D. Murdoch [b], R. Haange [c]

[a] *Forschungszentrum Karlsruhe, HVT, Tritium Labor, Postfach 3640, D-76 021 Karlsruhe, Germany*
[b] *The NET Team/EU-HT, Boltzmann Str. 2, D-85748 Garching, Germany*
[c] *Joint Central Team, ITER, Naka, Japan*

Abstract

Considerable R&D effort has gone into the development of a Torus Exhaust Vacuum Pumping System over the past years. The selected primary pump is of cycling type with discontinuous regeneration, solid cryosorbent for helium pumping and no fuel separation. The pump includes a throttle/regeneration valve to maintain the divertor pressure, minimize the thermal load on the pump and permit regeneration. Modelling of various torus cryopump configurations has been carried out and a research program aimed at developing a primary torus cryopump with rapid regeneration cycles is in progress. Panel tests with tritium will provide additional data required for ITER design. The design, fabrication and demonstration of a model cryopump with integral inlet valve is in progress. Experimental work and modelling to develop a high speed pumping system for the neutral beam injector is under way. Practical experience under ITER relevant conditions and in the presence of tritium has accumulated on the performance of oil-free pumps in various tritium laboratories. Recent world-wide developments in the field of fuel clean-up have been impressive. Several concepts based on palladium/silver diffusors combined with catalytic as well as with electrochemical process steps have the potential to achieve the high overall tritium decontamination factors and low tritium inventory levels required by ITER. Basic research activities aimed at characterizing and optimizing the envisioned essential components for the various ITER fuel clean-up concepts are in progress. Tests with an industrially manufactured facility of technical scale employing realistic concentrations of tritium and using an advanced operation and safety concept have been performed. Optimization of process components with respect to performance, operability and safety criteria is proceeding in parallel with integrated loop tests. Endurance tests to demonstrate complete processes with tritium are in preparation in several laboratories. Basic research aimed at developing a process for the recovery of tritium from the helium glow discharge cleaning exhaust stream is under way. The routine operation with tritium of large civilian research laboratories under conditions specific to the respective sites will contribute significantly to a realistic design of the ITER Tritium Plant. For the operation of the Tritium Plant and tritium accountancy much effort has gone into the improvement of known analytical techniques and the further development of calorimetry. © 1997 Elsevier Science S.A.

* Corresponding author.

0920-3796/97/$17.00 © 1997 Elsevier Science S.A. All rights reserved.
PII S0920-3796(97)00013-6

1. Introduction

The various Pumping Systems and the downstream Fuel Clean-up System are essential infrastructure systems to ensure the routine operation of a fusion reactor.

Important functions of the Primary Pumping System are the roughing of the torus from atmosphere to the crossover pressure of the torus primary pumps, the evacuation of the gas delivery lines of the gas injection system and pumping and purging of the vessel prior to the implementation of remote maintenance activities (see Fig. 1). The Primary Pumping System is also needed for the high vacuum pumping of the torus between pulses (removal of helium, of gaseous impurities evolved from the torus wall and of seeded impurities, e.g. Ne, Ar, Kr introduced to enhance power exhaust from the plasma) following crossover from the torus roughing. Another functional requirement is the pumping of helium and hydrogen isotopes during glow discharge cleaning operations. In addition, the Primary Pumping System will be used for the initial evacuation of the cryostat prior to the cooldown of the magnets and subsequent pumping of minor helium leaks from the magnet cooling circuits as well as for the initial evacuation of the neutral beam injector enclosures prior to cooldown. Other applications of the primary torus pumping system include leak testing, while separate systems are installed for diagnostics, guard vacuum, service vacuum, etc.

The most important functions of the Fuel Clean-up System are to recover the unspent fuel from the plasma exhaust gas and to deliver a highly tritium-decontaminated impurity stream for disposal. The Fuel Clean-up System should operate with an overall decontamination factor such that the waste gas stream could be directly discharged into the atmosphere. However, this stream will be passed through a conventional Tritium Waste Treatment System [1] as an addi-

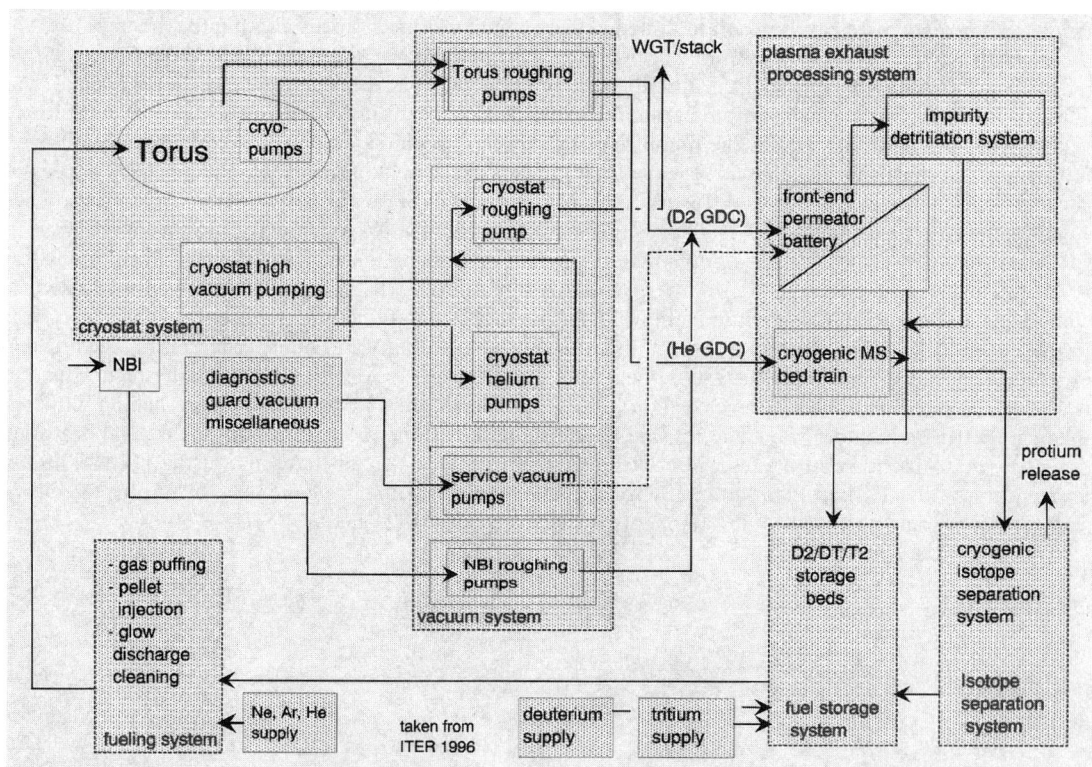

Fig. 1. A simplified fuel cycle block diagram for ITER.

Table 1
Civilian tritium laboratories worldwide[a]

Country	Laboratory/research installation	Licensed inventory	Operation start	Licensing procedure	
Canada	Chalk River Laboratory, CRL, Chalk River	100 g	1982/2 g 1995/100 g	1993	Modification of safety report to raise authorized amount of tritium from 2 to 100 g (for transfer/storage purposes only)
Canada	Ontario Hydro Technologies, OH, Toronto	6 g	1987	1987 1991 1994	License for 0.03 g License for 1 g License for 6 g
European Community, Italy	Joint Research Center, European Tritium Handling Laboratory, ETHEL, Ispra	100 g[b]	Expected in 1997	1984 1988 1993	First laboratory concept as extension of ESSOR Preliminary safety report to the Italian Regulatory Authority (ANPA) Completion of main installation
Germany	Tritium Laboratory Karlsruhe, TLK, Karlsruhe	40 g[c]	1994	1984 1986 1988 1993 1996	First draft concept of tritium laboratory Preliminary safety report to the authorities Formal request for a licence Completion of laboratory, basic license for 10 g License for 40 g
Japan	Tritium Engineering Laboratory, TEL, Tokai-mura	60 g[d]	March 1988	1980 1982/83 1986 1987	First contact with authorities Completion of new building Completion of laboratory License
USA	Tritium Systems Test Assembly, TSTA, Los Alamos	200 g	1984	1978 1982 1984	Start of project Safety report to Department of Energy License

[a] Much tritium experience is also being gained at the large fusion experiments JET (≤ 90 g) and TFTR (≤ 5 g). [b] License pending. [c] Maximum 20 g as free gas. [d] Maximum 25 g per day per experiment.

tional precaution, although no credit for detritiation by this system is given. In addition to the treatment of the exhaust gas pumped from the tokamak, the Fuel Clean-up System must also transfer the purified fuel to the Isotope Separation System and to the Fuel Storage System. The Fuel Clean-up System must, furthermore, handle streams not yet fully specified, e.g. blanket tritium recovery, neutral beam injector, vacuum pump system and possibly pellet propellant.

Whereas a point design has been adopted for the primary pumps and for the mechanical backing pumps of the ITER Vacuum Pumping System, such a decision has not yet been taken for the Fuel Clean-up System.

World-wide there are several civilian tritium laboratories conducting experimental work for ITER (see Table 1). Because of the complexity and high costs of the involved experiments, the work in these laboratories is complementary, either from the point of view of the subject itself or due to the fact that competing alternatives are being investigated. Modern licensing procedures are becoming increasingly complex and the requirements of licensing authorities more demanding. Many stringent demands by authorities (as well as by the ITER Team) have, however, ultimately resulted in significantly improved processes and better overall safety.

2. Plasma exhaust streams

The amount of exhaust gas received from the torus may have substantial variations in composition depending upon the regeneration scenario of the cryopumps and the type of fuelling. The maximum average exhaust rate during burn will be 2.5×10^5 Pa-liter s^{-1}. The typical torus plasma exhaust composition is shown in opposite column [1].

Present ITER divertor designs include the use of gases injected near the separatrix to increase radiative dissipation of plasma energy. These inert impurities become activated by fusion neutrons and need to be contemplated in the processes [2]. Possible impact on existing pumping and plasma exhaust clean-up concepts could be special mea-

Component	Mol%
Protium	4.0
Deuterium	40
Tritium	40
Total H, D, T	84
Waters	0.4
Hydrocarbons	0.8
Ammonia	0.4
Oxygen	0.4
Helium	9
Σ nitrogen, argon, neon	4.0
Carbon oxides	1

sures such as shielding and/or the incorporation of additional processing steps. In addition, all analytical techniques will have to be examined with respect to compatibility with γ-radiating species.

3. Recent experimental and technological developments in the fields of pumping and fuel clean-up

In the following the concepts and present state of development of the interface sequences from the torus to the cryopumps and back-up pumping system via the front-end permeator to the fuel clean-up system will be discussed (see also the simplified block diagram in Fig. 1, taken from ITER).

3.1. Pumping

To deal with the required vacua, throughputs, and type of gas pumped/recirculated in the various ITER systems (primary pumps, the backing or roughing pumps, fuel clean up, neutral beam injection, blanket sweep gas) pumps of different operation principles and sizes are needed. This has triggered world-wide a variety of technological pump developments and numerous performance tests—in some cases with tritium.

The present conceptual design for the primary pumping system of ITER comprises 16 batch regenerating cryopumps located at the divertor level and inside extensions of the vacuum vessel. They are used to evacuate the torus to a high vacuum and to pump the torus during pump down, plasma burn, dwell periods between pulses, bakeout, conditioning and leak testing. The torus roughing system consists of two pumping sets each comprising one first stage roots blower (4200 m^3 h^{-1}), two second stage roots blowers (1200 m^3 h^{-1}) and two four stage piston Toyo pumps (180 m^3 h^{-1}) (one set is on stand by). They are employed to evacuate the torus from atmospheric pressure down to the pressure at which the cryopumps set in; they must also evacuate the cryopumps during regeneration and during pump/purge cycles of the torus prior to maintenance [3].

3.1.1. Cryopump developments

The currently adopted design for the torus primary cryopumps contemplates pumping surfaces or cryopanels covered on one (or both) side(s) with a sorbent material fixed to the metallic surface by an inorganic cement [4]. From a pure thermal point of view the front side of the panel, which faces the pump axis and receives the infrared radiation from the baffle, would advantageously be polished to have a low infrared absorptivity, the back side being coated with the selected sorbent material. In the pumping mode the panels are cooled with supercritical helium to 4–5 K. Gas species whose condensation temperatures lie above panel temperature, e.g. DT, are condensed on the uncovered panel area. Helium is immobilized by cryosorption on the sorbent material. The pump works discontinuously. After loading the panel with gas the pump inlet valve is closed and the panels are regenerated either at 80–90 K (partial regeneration) or at 300 K (total regeneration). The gas evaporating during regeneration is pumped out by an external mechanical vacuum pump. After completion of this step the pump is recooled to the operating temperature.

The high impurity gas fraction (10% Ar, Ne, N$_2$ or Kr) specified by ITER may result in a high level of sorbent pollution during the partial regenerations of the cryopump. These species would remain adsorbed on the sorbent at 80 K. Should that pollution affect the sorbent performance excessively a solution would be to coat also the front side of the panel with sorbent, in order to fix the impurities on that sorbent part and keep the panel back side available for pumping hydrogens and helium. A decision on whether one or both sides of the panels are to be coated with sorbent is not needed urgently and will be addressed once the results of R&D currently in progress are available. The total surface offered by both sides of the panels is approximately 8 m^2 per pump.

The cryopump is designed for operating in the molecular flow regime which exists at pressures lower than 10^{-1} Pa at the pump inlet. Above this pressure transition or viscous flow prevails, which entails an unacceptably high heat conduction and has therefore to be avoided. A throttling valve at the pump inlet maintains the pressure inside the pump within the molecular/transition flow range. The valve is thus used to control the throughput of the pump and to close the pump during regeneration.

To minimize heat loads on the panels by the thermal radiation through the pump, inlet baffles cooled to approximately 80 K are installed. The baffles and panels of the cylindrically shaped cryoumps are assembled into the pump housing. Shields maintained at approximately 80 K are incorporated on the inside of the pump housing to decrease supercritical helium consumption.

The panel technology has been selected for fast temperature cycling. For partial regeneration the panels will be heated to approximately 90 K either by circulating 300 K gaseous helium through the refrigerant channels or by means of electrical resistance heating elements. For total regeneration to recover species such as tritiated water, the pump will be periodically warmed to room temperature, using 300 K gaseous helium. Pump cycle times are based on a tritium inventory of approximately 100 g for the complement of pumps in the pumping mode, corresponding to 140 g for the entire system, including the regenerating pumps. For a full valve opening the pumping speeds of the 1 m inlet port are 76 m^3 s^{-1} for DT and 42.5 m^3 s^{-1} for helium, assuming molecular flow conditions, which prevail at inlet gas pressures lower than 10^{-1} Pa. As the pressure increases above 10^{-1} Pa, the flow conditions enter the transition regime, becoming fully viscous when the pressure reaches 1 Pa.

The R&D program for the development of a primary vacuum pump for ITER at FZK focuses on the design and development of a cryogenic pump with integral inlet valve, together with a program of tests to qualify critical components of the pump. To accommodate new components with additional requirements a facility existing at FZK, i.e. Helitex facility, will be upgraded. The program comprises two steps. First, a reduced scale model pump with linear dimensions of approximately 0.7 scale will be built (corresponding to ~ 50% of the panel area of the ITER pump). Tests are in progress to establish appropriate cryopanel regeneration parameters and to determine panel endurance to temperature cycling. Because experiments on the possible effect of tritium on charcoal panels at 77 K have not yet yielded conclusive results [5] further work on this subject is under way in St. Petersburg [6] and Arzamas, Russia. Much attention must be paid to the throttling/regeneration valve prior to its integration into a follow-up model pump. A full scale prototype pump embodying improvements derived from the model type test results will be built and tested in the follow-up test program after the current Engineering Design Activities phase.

3.1.2. Backing pumps/gas circulators

Oil-free mechanical roots blowers and reciprocating pumps are being contemplated as ITER torus roughing pumps, subject to successful completion of qualification and performance characterization tests. These pumps are preferred for cost and space requirement reasons. A backing pump system based on oil-free scroll pumps is considered an appropriate alternative.

A large oil-free reciprocating pump (Toyo pump) having carbon polyimide composite piston rings and dynamic metal bellows seal has been developed at the TEL, Tokai-mura, Japan [7]. The design pumping speed of a prototype pump was 54 $m^3 h^{-1}$ at 6.7 mbar suction pressure and 1.15 bar discharge pressure. The actual pumping speed determined experimentally with tritium was found to be higher than 120 $m^3 h^{-1}$ at 5.3 mbar suction pressure and 0.66 bar exhaust pressure. The ultimate suction pressure at the same exhaust pressure was determined to be 1.1 mbar. Discontinuous tritium endurance tests covering a period of 350 h were satisfactory. Though tritium was found to be pumped slightly better than the other hydrogen isotopes, the pump can essentially be considered gas-type independent.

Numerous pump performance tests of technical character have been carried out at FZK. Basic data have been obtained systematically for hydrogen isotopes, noble gases, nitrogen and with gas mixtures simulating the composition of the plasma exhaust gas for scroll pumps with nominal throughputs of 15, 18, 60, 150, 600 and 1200 $m^3 h^{-1}$ [8,9]. The investigated parameters include the type of gas, the exhaust pressure, selected pump combination, temperature, etc. The vacua achievable at zero flow with deuterium tritide using a Normetex PV12 scroll pump are better than with protium by more than an order of magnitude, particularly at a low pump discharge pressure of 100 mbar, e.g. DT < 0.001 mbar and H_2 = 0.074 mbar (R.-D. Penzhorn, unpublished data). The observed isotopic effects can be explained by specific back diffusion rates along the helical flow paths and/or across the gaps between adjacent channels. Long-term testing of small scroll pumps with an equimolar deuterium/tritium mixture that also contains plasma exhaust relevant impurities has been under way at TLK since 1995 and is still continuing.

Compressors for Metal Bellows are widely used in tritium facilities. A newly developed double stage compressor manufactured by Siemens KWU, Germany, has been tested at length with tritium without failures. The pump is of positive displacement type and gas-type independent. Any rupture of the metal bellows can be detected by a pressure sensor placed in an isolation vacuum surrounding the bellows. At 1 bar discharge pressure the vacuum attainable with a two stage pump is approximately 120 mbar [10].

Other oil-free pumps that have been used for years in tritium facilities in France and in Canada are manufactured by SRTI, France. For example, the PBT series pumps are fully dry, ceramically sealed, with two coaxial bellows totally separating the gases from the lubricant. Because these bellows are of rolled design, they can withstand pressures of up to 40 bar. In the proposed application this should guarantee high reliability. They are capable of pumping gases down to pressures as low as 1 mbar at high throughput. The models offered include 1–6 compression stages with flow rates ranging from 7 to 20 $m^3 h^{-1}$. The pump is inherently gas-type independent.

Table 2
Qualification of oil-free pumps with tritium

Pump type	Pumping speed ($m^3\ h^{-1}$)	Tested with	Tested at	Test time (h)	Gas type dependent	Remarks
Normetex, France	15	T_2	TEL/TLK/TSTA	>2000	Yes	Reliable pumps in long-term operation, no leaks observed, only occasional maintenance
	18	T_2	TEL/TLK/TSTA		Yes	
	60	H_2/D_2	TLK		Yes	
	150	T_2	TEL/TLK/TSTA	2000	Yes	
	600	H_2/D_2	HIT, FZK		Yes	
	1300	H_2/D_2	HIT, FZK		Yes	
Metal Bellows, MB601, USA	4.2	T_2	TEL/TLK/TSTA	Years	No	Reliable pump, frequent failures of electrical drive (carbon brushes)
SRTI, France	10–15	T_2	Fontenay, Grenoble, Valduc, France	Years	No	Reliable pumps, ceramic/ceramic contact of piston/cylinder, operation in horizontal position only
Tokyo reciprocating pump, Japan	120	T_2	TEL, TSTA	350	No	Reliable pump, carbon polyimide composite piston rings, dust formation from pump possible
Siemens Metal Bellows, Germany	3	T_2	TLK	>2000	No	Reliable pump of smooth performance, double contained head

TEL, Tritium Engineering Laboratory; TLK, Tritium Laboratory Karlsruhe; TSTA, Tritium Systems Test Assembly; HIT/FZK, Hauptabteilung Versuchstechnik/Forschungszentrum Karlsruhe.

Drystar pumps can be used as forepumps to turbomolecular pumps. While they have a rather high noise level and require water cooling, they have been shown to be adequate. Early versions showed severe corrosion by chilled water. This problem has meanwhile been solved.

A few oil-free pumps currently under test with tritium are compared in Table 2.

3.1.3. Turbomolecular pumps

The experience with turbomolecular pumps in a tritium environment is good. Heics et al. [11] investigated systematically the performance of a grease lubricated Leybold Heraeus turbomolecular pump with tritium and found no tritium-related degradation after a period of 12 000 h of operation.

3.1.4. Neutral Beam Injector pumps

The Neutral Beam Injector (NBI) secondary pumping system is used to evacuate the three NBI of ITER and the Diagnostic Neutral Beam (DNB) modules before cool down of the internal cryopanels and for the regeneration of these modules [3]. The system comprises a roughing pump and an external cryopump connected to each module. The roughing pump is used to evacuate each module to the crossover pressure of the cryopump system. The cryopump then evacuates the modules to the required low pressures.

Based on experimental data obtained with hydrogen using a cryosorption pump module of 3.5 $m^3 s^{-1}$ pumping speed provided with charcoal panels, a cryosorption pump of about 300 $m^3 s^{-1}$ pumping speed was designed [12]. This pump was used in NBI experiments at the JIPP-IIU tokamak. An almost constant pumping speed of 330 $m^3 s^{-1}$ was attained with this device within the pressure range 10^{-4}–2×10^{-3} Pa.

A chevron type cryopump of simple design for the NBI of ITER with a pumping speed of 3.8×10^3 $m^3 s^{-1}$ of H_2 has been proposed recently [13]. The pump is designed to operate at 3.6–4.2 K with a liquid He consumption in stand by condition of 11 l h^{-1}. The cool-down of the He panels from 80 to 4.2 K is calculated to take place in 3.5 min and that from 4.2 to 3.6 K in 1 min. Regeneration should be started 30 s after the He flow has been stopped. Pumping of the injector down to 0.1 mbar with a large transfer pump is estimated to occur in about 7.5 min.

3.2. Fuel clean-up for ITER

Campaigns with relevant concentrations of tritium in integrated loop tests as well as in stand-alone experiments have provided evidence that three consecutive and essentially independent process stages ('activity compartmentalization'), i.e. front-end permeator, main impurity detritiation, and a final detritiation step, are needed to achieve the decontamination factors (DF) required by ITER [14], which is currently 10^7. The required DF is derived from the target release of 1 Ci day^{-1} in the detritiated impurity stream separated from the circulating fuel stream which has a flow of up to 10^7 Ci day^{-1} (1 kg tritium day^{-1}).

3.2.1. Front end permeator

The main function of the front end permeator is to recover the bulk of the unspent fuel from the plasma exhaust gas and to isolate non-tritiated and tritiated impurities along with the He product of the fusion reaction for further processing. Several permeator designs of large throughput have been tested extensively with tritium [10,15,16]. The designs include in/out (feed gas is introduced into the palladium/silver permeation fingers (or coils) and the permeate is pumped from the outside) and out/in permeation concepts (feed gas is supplied to the outside and permeate gas is pumped from the inside of the permeation fingers). Experimental results obtained worldwide in several tritium laboratories have provided evidence that permeators can be operated without failure for years in a high tritium concentration environment [10,15–17]. While steam attack by water at the grain boundary level causing failures of the palladium/silver tubes in permeators has been reported, hydride formation during unintentional cooling in the presence of hydrogen (possibility of phase transformation) does not appear to constitute a problem [10,15]. The efficiency of this process step, i.e. achievable decontamination factor (ratio of total tritium in the feed over total tritium in the bleed), is mainly determined by the configuration of the permeator including its total active permeation area, the inlet/permeate pressure ratio of hydrogen isotopes, the concentration of tritium bound to non-permeable impurities and the partial pressure of water. Systematic permeation experiments with H_2, D_2 and DT at permeate pressures in the range 10–300 mbar have been performed at the TLK in Karlsruhe. Relative isotope effects for the permeation of hydrogen isotopes through palladium/silver were determined to be $H_2/D_2 = (1.72 \pm 0.03)$ and $H_2/DT = (2.06 \pm 0.03)$ [17]. The values are consistent with earlier results of Konishi et al. [18] who found $H_2/D_2 = (1.76 \pm 0.09)$ and $H_2/T_2 = (2.37 \pm 0.04)$ and of Fujita et al. [19] who found $H_2/T_2 = (2.12 \pm 0.03)$. Fairly high permeate pressures are being proposed for the ITER front-end permeator. While high permeate pressures necessarily lead to a reduction in the DF, they are acceptable when significant impurity partial pressures of tritiated impurities such as water or hydrocarbons are present in the feed gas, because the latter will limit the attainable DFs. Acceptance of high permeate pressures leads to a considerable simplification of the permeate pumping system, i.e. a complex pumping system comprising several pumps, some of which are large, can be replaced by fewer and smaller ones [20]. This is of great advantage to operation and maintenance in the small available volume of a glove box. In addition, relatively high partial pressures of hydrogen isotopes in the feed zone of a permeator contribute to avoid the condensation of water in the permeation tubes and help to minimize poisoning of the palladium/silver membrane by water [13] or polytritiated hydrocarbons [17]. Decontamination factors for tritium in the range 8–30 will be achievable by the front-end permeator battery depending upon the prevailing operation parameters.

The use of an isolation vacuum for heated components in combination with a getter bed that continuously removes the permeated hydrogen isotopes results in exceedingly low levels of tritium in the glove box, of the order of a few MBq m^{-3} or less even over long periods of time [10]. Without isolation vacuum steady state tritium levels in the glove box orders of magnitude higher can be expected. While components provided with an isolation vacuum are more expensive and of more complex design, in the long run maintenance in less contaminated glove boxes is easier and the risk of contamination of personnel (for instance via rubber gloves) is smaller.

3.2.2. Impurity processing

The bleed gas from the front end permeator contains helium (the 'ash' of the fusion reaction), a residual amount of non-permeated hydrogen isotopes, and non-tritiated (carbon monoxide, carbon dioxide, nitrogen, noble gases) as well as tritiated (saturated and unsaturated hydrocarbons, water, ammonia, etc.) impurities. This gas needs to be processed with the aim of tritium recovery with a decontamination factor of about 10^6 so that the effluent gas stream can be released into the environment via a waste gas treatment system. For this purpose several alternative semi-batch as well as once-through process concepts have been developed and tested experimentally. In the following some of the most important results will be discussed.

An industrially manufactured integral facility, CAPRICE, to demonstrate a process concept with relevant concentrations of tritium, has been in operation at the TLK for about 3 years [13]. The semi-batch closed loop process is based on reactions such as

$$CH_4 = C + 2H_2 \qquad (1)$$

$$CH_4 + H_2O = CO + 3H_2 \qquad (2)$$

$$CH_4 + 2H_2O = CO_2 + 4H_2 \qquad (3)$$

$$H_2O + CO = CO_2 + H_2 \qquad (4)$$

Carbon deposited on the nickel catalyst is regasified by reaction with carbon dioxide ($C + CO_2 = 2CO$) or with water ($C + H_2O = CO + H_2$). Liberated hydrogen is continuously removed from the closed loop via a permeator integrated into the loop. By this procedure the equilibria Eq. (1) to Eq. (4) are shifted towards the products and a very high degree of conversion can be attained. The process concept has been demonstrated at bench scale and technical scale in numerous runs with H_2, D_2 and trace amounts of tritium. More recently technical experiments with DT with a tritium inventory in the loop of up to 7 g have been performed. The results show that decontamination factors of up to 40 000 can be achieved in about 0.5 h of gas recirculation [21]. The tritium inventory in the impurity processing loop, which is determined by that of the injected tritiated impurities, rapidly decreases to very low levels as the chemically bonded tritium is removed with the permeator.

At the Tritium Systems Test Assembly (TSTA), Los Alamos, New Mexico, a related concept, the Palladium Membrane Reactor (PMR), is under investigation [22,23]. The concept is based on the same catalytic reactions described for the CAPRICE facility to recover the bonded hydrogen isotopes from the impurities. Screening tests have indicated that noble metal catalysts have particularly adequate properties to catalyze Eqs. (1)–(4). Thermodynamic equilibrium limitations preclude the simple use of these reactions for the recovery of bonded hydrogen isotopes. In the PMR process concept the catalyst and the permeator are integrated into a single unit. Therefore generated molecular hydrogen is removed in situ from the reactor. Because of this the equilibrium is shifted towards the products and once-through operation becomes possible. The continuous removal of hydrogen isotopes from the reactor leads to very low tritium inventories at all times in this component. For quantitative conversion of bound hydrogen isotopes into molecular hydrogen the oxygen level needs to be measured in real time and continually adjusted. Experiments with tritium have yielded DFs of the order of several hundred in a single first stage unit. In extended tests it was shown that the palladium/silver is not attacked by the catalyst.

A concept using an electrolytic cell has been proposed at TEL [24,25]. As with the previously discussed process concepts, hydrogen isotopes are first separated from the impurities with a front-end permeator battery. The loop comprises a ceramic electrolysis cell, a palladium/silver permeator and a catalytic reactor. The permeator is used to recover molecular hydrogen isotopes from the loop. The ceramic electrolysis cell is made up of a solid oxygen ionic conductor (ZrO) and platinum electrodes. Water vapor is converted to hydrogen gas at the cathode and oxygen migrates to the anode. When oxidizable species are present at the anode they are electrolytically oxidized by the oxygen ions diffusing to the surface. The catalytic reactor may either be designed to promote the steam reforming reaction (reaction Eq. (2)) or the oxidation of hydrocarbons after addition of oxygen. The combination of these three components can be used to recover bonded and molecular tritium from the plasma exhaust gases in a closed loop batch operation.

3.2.3. Final catalytic detritiation step

Following the bulk processing of the impurities a fine polishing step is necessary to achieve the high decontamination factors required by ITER. The alternatives presently under investigation are PERMCAT, HITEX and PMR II.

The PERMCAT reactor developed in Karlsruhe combines a nickel catalyst with palladium/silver permeation fingers [26]. The reactor is based on isotopic swamping in counter-current operation mode. The already highly detritiated gas stream effluent from the CAPRICE facility is passed through an annulus containing catalyst on the outside of palladium/silver fingers while protium is passed in the opposite direction through the inside of the permeation fingers. The catalyst (Ni on Kieselguhr) promotes the isotopic exchange between protium and tritiated impurities. The protium pressure on the permeate side is kept low (~ 50 mbar) to avoid excessive dilution of the impurities. Tritium bonded to the impurities will end up in the effluent permeate stream, and the bleed stream of the PERMCAT reactor will be highly tritium-depleted. An advantage of the counter-current operation is the comparatively low swamping ratio. Using tritium concentrations of 3×10^{14} Bq m^{-3} DFs of several ten thousand have been obtained both in Chalk River as well as in Karlsruhe with the PERMCAT unit.

In Canada the HITEX concept is being investigated with concentrations of tritium of the order of 3.7×10^{12} Bq m^{-3} [27,28]. Basically, tritiated impurities are detritiated catalytically with excess hydrogen in closed loop operation. Tritiated hydrogen is continuously extracted from the loop with a palladium/silver permeator integrated into the loop. The hydrogen removed by permeation is replaced by fresh protium until the desired degree of detritiation is achieved. From systematic tests with various catalysts a hydrophobic platinum/silicalite catalyst was selected. The results indicate that following a fast initial detritiation, yielding DFs in the range 10^2-10^5, a slower detritiation rate, probably limited by the desorption of water from contaminated surfaces, sets in that nevertheless further improves the DF. The detritiation rate and the transition from a fast to a slow detritiation rate decreases with CO and/or CO_2 impurity concentration. This is attributed to methanation reactions, such as

$$CO_2 + 4H_2 = CH_4 + 2H_2O \quad (5)$$

and the inverse reaction Eq. (4), which continuously generate water.

At TSTA, a second stage permeation membrane reactor (PMR II) having noble metal catalyst at the inside of the permeation fingers is being proposed for final detritiation. To achieve high decontamination factors the PMR II is equipped with a turbo molecular pump to remove the permeate. Measured DFs have attained values in excess of 10^4.

3.2.4. Candidate integral process concepts for ITER

To demonstrate fuel clean-up technology for ITER the facility CAPRICE at the TLK is presently being upgraded by incorporating a technical PERMCAT unit to the primary system. The new facility of ITER scale comprising CAPRICE and PERMCAT has been denominated CAPER. The hardware installations will be completed by the end of 1997. An improved catalyst bed combining methane cracking, steam reforming and water gas shift and the carbon gasification reactions into a single unit has been designed, constructed and tested. From data obtained with the single process steps CAPRICE (DF $\geq 10^3$) and PERMCAT (DF $\geq 10^4$) and a DF of 10 for the front-end permeator an integral DF of more than 10^8 can be confidently predicted. An endurance test with realistic tritium concentrations planned for 1997 is aimed at demonstrating the integral concept in once-through operation. A detailed design for ITER based on the CAPER concept including a component definition and description, a process description, the required instrumentation and control and a description of the interfaces has been completed at FZK [29]. The overall DF for CAPER, given by the product of the front-end DF, the DF of the impurity processing loop and the DF of PERMCAT, is expected to be $\geq 10^8$.

A two-stage PMR system designed for 1/5th ITER scale has been constructed and initially tested with tritium at TSTA. The first PMR unit contains 7.1 kg of catalyst and the second less than 100 g. Both PMRs use six palladium/silver tubes with a length of 72 cm and a diameter of 0.635 cm. The retentate pressure of the first PMR is kept at 0.67 mbar and that of the second one at 4×10^{-5} mbar [30]. Coking problems are prevented by adjusting the oxygen content of the gas. The overall decontamination factor achieved during round the clock operation for 31.5 h was greater than 2.2×10^7. Assuming steady state conditions in both PMR stages (saturation of both catalyst beds with tritium) to be attained and including a front-end permeator DF of 10, an overall DF of 2×10^8 was reported from first experiments in which tritium was fed at 24 Ci min^{-1} into the two stage system.

3.3. Glow discharge clean-up

The maximum exhaust rate of the ITER glow discharge cleaning exhaust will be 5×10^4 Pa-liter s^{-1} and its composition either helium or deuterium with 0.3% impurities like methane. A careful systematic experimental study on the adsorption of hydrogen isotopes, including HT, in the presence of methane on cryogenic molecular sieves was made by Enoeda et al. [31]. The best results for the regeneration of a cryomolecular sieve bed were achieved by purging at low pressure and increased temperature [32]. A batch gas chromatographic selective separation method has been tested for the separation of H_2 and He using molecular sieve 5A, and data for the design of an ITER system has been obtained [33]. The regeneration stream from the molecular sieve beds is not released to the environment (see Fig. 1).

3.4. Other recent developments concerning tritium technology

3.4.1. Regeneration of molecular sieves

Molecular sieve (MS) beds are used widely in

Table 3
Comparison of calorimeters at the TLK (in Ci) presented at the Combined Technical Meeting of Tritium Plant and Fuelling and Pumping System, Naka, December (1995)

Vessel no.	FZK calorimeter	JET calorimeter	Antech calorimeter 1	Antech calorimeter 2	Difference (%)
4072	838.4	839.4 ± 0.4	840.8 ± 0.7	—	<0.29
4048	944.7 ± 1.6			948.2 ± 11	0.35
4056	9653.2 ± 5.2			9670 ± 26	0.17
4043	9439.0 ± 6.1			ND	ND
4054	22917.1 ± 6.1			22866.7 ± 15.8	0.22

ND, not determined.

tritium facilities to adsorb tritiated water produced by oxidation of tritiated gas streams, especially in tritium retention systems. At the TLK a new and rather simple technology (AMOR facility) has been developed and demonstrated with molecular sieve beds containing 15 kg of a zeolite 5A/mordenite adsorber mixture [34]. Currently more than 900 kg of zeolite containing up to 2×10^{11} Bq kg^{-1} of tritium in the adsorbed water have been regenerated at temperatures of up to 300°C on a routine basis. The technology is based on the thermal release of adsorbed water with a hot nitrogen carrier gas. Most of the desorbed water is subsequently condensed, collected in a 50 l transport vessel, and then disposed of at FZK. Ultimate drying of the bed is achieved with a second MS bed valved into the loop after most of the adsorption water has been liberated.

3.4.2. Tritium accountancy

The main problem of tritium accountancy is not the lack of accurate tritium detection methods but the fact that tritium is an elusive species difficult to localize. Whereas calorimetry has proven to be highly accurate, the method requires collecting tritium from a facility in a vessel suitable for calorimetry. This is not always possible or practical. Other techniques, such as PVT-c, require appropriate installations (calibrated volumes, calibrated pressure sensors, isothermal conditions and sophisticated tritium and chemical analysis). As with calorimetry, tritium immobilized in structural materials, adsorbers, or catalysts cannot be detected by PVT-c. Recovery of tritium by isotopic swamping should preferably be avoided and only be used when strictly necessary.

3.4.3. Ex-bed calorimetry

Commercially available, user friendly ex-bed calorimeters designed to measure tritium in transportable Amersham getter beds in their containment pot have been developed in recent years [35]. They are provided with an easy to use software. The measurement resolution for routine work is 10^{12} Bq with measurement times of less than 4 h. A new precision calorimeter of unconventional design developed at JET, based on thermal stabilization by inertial feedback control, covers a thermal power range of 1 μW–10 W (10^9–10^{16} Bq) [36]. This calorimeter was compared with two other commercial calorimeters and one built at FZK at the TLK in Karlsruhe, using the same Amersham uranium storage vessels containing tritium amounts in the range 8×10^{11}–2×10^{15} Bq [37] (L. Doerr and H. Hemmerich, unpublished data). The results of these bench mark tests, shown in Table 3, indicate that accuracies much better than 1% can be achieved.

3.4.4. In-bed calorimetry

Gas flow calorimetry has been successfully employed at Savannah River for the routine determination of large quantities of tritium with accuracies of the order of 1%. The measurements require precisely and simultaneously controlled conditions with respect to the inlet temperature of the coolant gas, the mass flow of the coolant gas, the outside temperature of the thermal insulation and the outlet temperature of the coolant gas [38]. At TEL, Japan, first successful tests with up to 4 g tritium have been performed with an isothermal in-bed calorimeter containing ZrCo as getter using He with at a flow rate of 4 l (STP) min^{-1} [39].

In both approaches tritium permeation into the coolant gas needs to be coped with. Another self-assaying storage bed utilizing thermal radiation as the primary mode for heat transfer has been built at TSTA, Los Alamos [40]. A constant emissivity of the radiation surfaces still needs to be demonstrated. At JET Hemmerich [41] has proposed a new thermal bed design for in situ calorimetric measurements with an accuracy of $\pm 0.1\%$. This design uses a microporous thermal insulation operating in a hydrogen atmosphere.

3.4.5. Tritium analytics

While numerous analytical techniques (e.g. mass spectrometry (including omegatron), gas chromatography, laser Raman spectroscopy) are available and have been tested extensively with tritium, work is under way to adapt these techniques to the specific requirements of ITER. Laser Raman was shown to be a useful tool for the on-line analysis of isotopic hydrogens of the TSTA cryogenic isotope separation system [42]. Laser Raman spectroscopy has been used to investigate the radiation-induced reactions of the impurities involved in the fusion fuel cycle [43–45]. A very promising real-time in situ gas analysis technique using a flow through optical cell and optical fibers for the transmission of laser and laser Raman scattering light has been succesfully demonstrated at TEL, Japan [46]. In principle, gas analysis by laser Raman spectroscopy in a glove box or in a remote room becomes possible with this technique.

3.4.6. Hydrogen isotope separation

Cryogenic hydrogen isotope separation is a well established technology [47–49]. Other hydrogen isotope separation technologies like thermodiffusion at TEL [50] or at displacement gas chromatography JET [51] constitute important infrastructure to support tritium experiments. A gas chromatographic system with a throughput suitable for intermediate sized tritium laboratories has recently gone into operation at TLK [52]. Several 100–120 l (STP) samples of hydrogen isotope mixtures from the experiments containing up to 10^{14} Bq of tritium were injected and successfully separated, being the tritium-depleted fractions released via a Waste Treatment System.

4. Conclusions

Many types and capacities of vacuum pumps proposed for diverse applications in the fuel cycle and other ITER systems have been tested under relevant operating conditions, including exposure to tritium. Selected key components are being qualified for critical aspects of their in-service duties (tritium/material compatibility, thermal cycling, etc.) prior to construction of prototype pumps. For the evacuation of the torus and of the Neutral Beam Injector detailed concepts have been developed. Tests in progress or in preparation under relevant conditions will yield the information needed for their engineering design.

The ITER requirements for the Fuel Clean-up System have evolved considerably, in terms of higher flow rates, higher detritiation factors and the range of feed conditions to be handled. This has driven the design towards continuous processes, which offer high capacities with relatively low tritium inventories and the capability to respond rapidly to fluctuating feed conditions. Work performed or underway world-wide in various tritium laboratories with relevant concentrations of tritium indicates that an overall tritium decontamination factor of the plasma exhaust gas of more than the specified 10^7 is possible with several of the process concepts under development. A key issue is the avoidance of cross contamination between consecutive steps. Other criteria such as the effect of neutron activated gas species will have to be considered in the selection of the process for ITER. Endurance tests with tritium will provide the additional information needed to confirm this design selection.

References

[1] J.E. Koonce, O. Kveton, ITER-Design Description Document, Tritium Plant WBS 3.2, June 1995.

[2] G. Saji, H. Iida, Activation of noble gases in tokamak exhaust, ITER-EDA, Interoffice Technical Memorandum, June 1995.

[3] P. Ladd, Primary vacuum pumping system: system overview and plasma exhaust requirements, Fuel Dynamics Meeting, Garching, May 2–3, 1996.

[4] A. Mack, J.C. Boissin, D.K. Murdoch, D. Röhrig, G. Saksagansky, Primary vacuum pump concept, component

testing and model pump development for ITER, Proc. 19th SOFT, Lisbon, Portugal, 1996, Elsevier.
[5] D.W. Sedgley, C. Walthers, E.M. Jenkins, Fusion reactor high vacuum pumping: charcoal cryosorber tritium exposure results, 14th Symposium on Fusion Engineering, IEEE, San Diego, CA, September 30th, 1991.
[6] E.L. Koira, I.N. Moreva, G.L. Saksaganski, A.I. Vedeneev, M.V. Glagolev, N.T. Kazakovsky, D.K. Murdoch, Tritium testing of cryosorption panel coupons for Iter vacuum pumps, Proc. 19th SOFT, Lisbon, Portugal, 1996, Elsevier.
[7] T. Hayashi, M. Yamada, S. Konishi, Y. Matsuda, K. Okuno, J.E. Nasise, R.S. Dahlin, J.L. Anderson, Tritium evacuation performance of a large oil-free reciprocating pump, Fusion Eng. Design 28 (1995) 357.
[8] U. Berndt, E. Kirste, T.L. Le, M. Glugla, R.-D. Penzhorn, Performance characterisics of large scroll pumps, Fusion Eng. Design 18 (1991) 73.
[9] D. Perinic, U. Kirchhof, B. Kammerer, Testing of dry mechanical vacuum pumps for ITER, Proc. 15th IEEE/NPSS Symposium on Fusion Engineering, Hyannis, MA, USA, October 11–15, 1993.
[10] R.-D. Penzhorn, U. Berndt, E. Kirste, W. Hellriegel, W. Jung, R. Pejsa, O. Romer, Long-term permeator experiment PETRA at the Tritium Laboratory Karlsruhe: commissioning tests with tritium, Fusion Technol. 28 (1995) 723.
[11] A.G. Heics, W.T. Shmayda, D. Müller, Tritium effects on the performance of turbomolecular pumps, Proc. 15th IEEE/NPSS Symposium on Fusion Engineering, Hyannis, MA, USA, October 11–15, 1993, p. 73.
[12] Y. Oka, T. Takanashi, R. Akiyama, O. Kaneko, K. Toi, H. Morimoto, M. Terashima, T. Kuroda, Development of cryosorption pump for neutral beam injector, Fusion Eng. Design 31 (1996) 89.
[13] E. Küssel, Cryopump design (Chevron type) for ITER NBI, First Technical Meeting on NBI, 13–16 February, Naka, Japan, 1996.
[14] M. Glugla, R. Kraemer, R.-D. Penzhorn, T.L. Le, K.H. Simon, K. Günther, U. Besserer, P. Schäfer, W. Hellriegel, H. Geißler, Commissioning of the plasma exhaust clean-up facility CAPRICE and first experimental results, Fusion Technol. 28 (1995) 625.
[15] E.A. Clark, D.A. Dauchess, L.K. Heung, R.I. Rabun, T. Motyka, Experience with palladium diffusers in tritium processing, Fusion Technol. 28 (1995) 566.
[16] S. Konishi, H. Yoshida, Y. Naruse, R.V. Carlson, K.E. Binning, J.R. Bartlit, J.L. Anderson, Tritium experiments on components for fusion fuel processing at the Tritium Systems Test Assembly, Fusion Technol. 19 (1991) 1668.
[17] R.-D. Penzhorn, U. Berndt, E. Kirste, J. Chabot, Performance tests of palladium/silver permeators with tritium at the Tritium Laboratory Karlsruhe, Proc. 19th SOFT, Lisbon, Portugal, 1996, Elsevier.
[18] S. Konishi, H. Yoshida, H. Ohno, T. Nagasaki, Y. Naruse, Improvements on some tritium processing components, in: T. Takahashi, S. Tanaka (Eds.), Proceedings of the International Symposium on Fusion Reactor Blanket and Fuel Cycle Technology, Tokai-mura, Ibaraki, October 27–29, 1986.
[19] H. Fujita, S. Okada, K. Fujita, H. Sakamoto, K. Higashi, T. Hyodo, Ratio of permeabilities of tritium to protium through palladium alloy membrane, J. Nucl. Technol. 17 (1980) 436.
[20] J. Koonce, H. Yoshida, O. Kveton, H. Horikiri, R. Haange, Design of ITER plasma exhaust processing systems, Fusion Technol. 28 (1995) 630.
[21] M. Glugla, R. Kraemer, T.L. Le, K.H. Simon, K. Günther, J. Wendel, R.-D. Penzhorn, Results from tritium operation of the clean-up facility CAPRICE, 19th SOFT, Lisbon, Portugal, 1996.
[22] R.S. Willms, S.A. Birdsell, R.C. Wilhelm, Recent palladium membrane reactor development at the Tritium Systems Test Assembly, Fusion Technol. 28 (1995) 772.
[23] R.S. Willms, S.A. Birdsell, Modelling and data analysis of a palladium membrane reactor for tritiated impurities clean-up, Fusion Technol. 28 (1995) 530.
[24] S. Konishi, H. Yoshida, H. Ohno, Y. Naruse, D.O. Coffin, C.R. Walthers, K.E. Binning, Experiments on a ceramic electrolysis cell and a palladium diffuser at the Tritium Systems Test Assembly, Fusion Technol. 8 (1985) 2042.
[25] S. Konishi, M. Hara, K. Okuno, Versatile fuel clean up system based on palladium permeation and vapor electrolysis, Fusion Technol. 28 (1995) 652.
[26] M. Glugla, R.-D. Penzhorn, P. Herrmann, H.J. Ache, Advanced catalytic plasma exhaust clean-up process for ITER-Eda, Fusion Technology, Proceedings of the 18th Symposium on Fusion Technology, Karlsruhe, Germany, 22–26 August 1994, Vol. 2, pp. 1135.
[27] S.K. Sood, C. Fong, K.M. Kalyanam, K.B. Woodall, A. Busigin, HITEX process options for detritiation of impurities in the ITER plasma exhaust, Fusion Technol. 28 (1995) 742.
[28] L. Rodrigo, J.M. Miller, J.A. Senohrabek, Plasma impurity processing with HITEX, Proc. 19th SOFT, Lisbon, Portugal, 1996, Elsevier.
[29] M. Glugla, H. Geißler, Basic Safety Considerations for an integrated catalyst fuel clean-up unit for ITER (CAPER), FZK Report 1996.
[30] S. Birdsell, S. Willms, Palladium membrane reactor ITER Combined Technical Meeting, Naka, Japan, December, 1995.
[31] M. Enoeda, Y. Kawamura, K. Okuno, K. Tanaka, M. Uetake, M. Nishikawa, Recovery of hydrogen isotopes and impurity mixture by cryogenic molecular sieve bed for GDC gas cleanup, Fusion Technol. 28 (1995) 591.
[32] M. Enoeda, Combined Technical Meeting of Tritium Plant and Fuelling and Pumping System, Naka, December, 1995.
[33] T. Abe, S. Hiroki, Y. Murakami, Preliminary trial on H_2/He selective separation at working pressures less than 1 atm, Combined Technical Meeting of Tritium Plant and Fuelling and Pumping System, Naka, December 1995.
[34] E. Hutter, H.-D. Adami, U. Besserer, R.-D. Penzhorn, AMOR facility: regeneration of molecular sieve beds used

for the retention of tritium at the Tritium Laboratory Karlsruhe, Proc. 19th SOFT, Lisbon, Portugal, 1996, Elsevier.

[35] M.I. Thornton, G. Vassallo, J. Miller, J.A. Mason, Design and performance testing of a tritium calorimeter, Nucl. Instr. Methods Phys. Res. A 363 (1995) 598.

[36] J.L. Hemmerich, P. Milverton, G. Newbert, N. Green, A. Miller, Tritium and uranium inventory measurements with the JET AGHS precision calorimeter, Proc. 16th IEEE/NPSS Symposium on Fusion Engineering, Champagne, IL, USA, 30 September–5 October 1995, p. 781.

[37] R. Kraemer, M.I. Thornton, Calorimetric tritium measurements on portable getter beds, Proceedings of the 17th ESARDA Symposium, Aachen, Germany, May 9–11, 1995.

[38] J.E. Klein, M.K. Mallory, A. Nobile Jr, Tritium measurement technique using 'in-bed' calorimetry', Fusion Technol. 21 (1992) 401.

[39] T. Hayashi, Tritium measurement technology by 'in bed' gas flowing calorimeter, 3d Tritium ITER Plant Meeting in Naka, Japan, December 7th, 1995.

[40] J. Nasise, Self-assaying tritium storage bed, ITER Combined Technical Meeting, Naka, Japan, December 5th, 1995.

[41] J.L. Hemmerich, Thermal design of a metal hydride storage bed permitting tritium accountancy to 0.1% resolution and repeatability, Fusion Technol. 28 (1995) 1732.

[42] S. O'Hira, H. Nakamura, S. Konishi, T. Hayashi, K. Okuno, Y. Naruse, R.H. Sherman, D.J. Taylor, M.A. King, J.R. Bartlit, J.L. Anderson, On-line process gas analysis by laser Raman spectroscopy at TSTA, Fusion Technol. 21 (1992) 465.

[43] R.H. Sherman, D.J. Taylor, J.R. Bartlit, J.L. Anderson, S. O'Hira, H. Nakamura, S. Konishi, K. Okuno, Y. Naruse, Radiochemical reaction studies of tritium mixed gases by laser Raman spectroscopy, Fusion Technol. 21 (1992) 457.

[44] U. Engelmann, M. Glugla, R.-D. Penzhorn, H.J. Ache, Laser Raman spectroscopy and omegatron mass spectrometry applied to investigations of the radiochemical reactions between methane and tritium, Fusion Technol. 21 (1992) 430.

[45] T. Uda, K. Okuno, Y. Naruse, Application of laser Raman spectroscopy to analysis of isotopic C2 hydrocarbons in fusion fuel processing, J. Nucl. Sci. Technol. 29 (1992) 94.

[46] S. O'Hira, K. Okuno, Real-time and in situ fuel process gas analysis system by application of laser Raman spectroscopy, ITER Combined Technical Meeting, Naka, Japan, December 5th, 1995.

[47] R.S. Willms, R.H. Sherman, S.P. Cole, J.B. Riggs, K. Okuno, Demonstration of regulatory process controls on the TSTA cryogenic distillation system, Fusion Technol. 28 (1995) 778.

[48] J.C. Buvat, C. Latge, X. Joulia, G.P. Pautrot, Simulation of hydrogen isotope separation by cryogenic distillation, Fusion Technol. 21 (1992) 954.

[49] T.S. Drolet, K.Y. Wong, P.J.C. Dinner, Canadian experience with tritium—The basis of a new fusion project, Nucl. Technol./Fusion 5 (1984) 17.

[50] T. Nakamura, K. Hirata, T. Yamanishi, K. Okuno, Y. Naruse, Experimental study on separation characteristics of thermal diffusion columns using H-D-T system, Fusion Technol. 21 (1992) 942.

[51] R. Lässer, G. Jones, J.L. Hemmerich, R. Stagg, J. Yorkshades, The preparative gas chromatographic system for the JET Gas Active Handling System—inactive commissioning, Fusion Technol. 28 (1995) 681.

[52] G. Neffe, J. Dehne, E. Hutter, H. Kissel, H. Brunnader, Separation of tritiated hydrogen species in the TLK isotope separation system, Proc. 19th SOFT, Lisbon, Portugal, 1996, Elsevier.

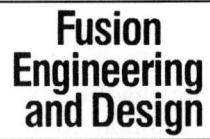

Remote handling on fusion experiments

A.C. Rolfe

JET Joint Undertaking, Abingdon, Oxon, OX14 3EA, UK

1. Introduction

A major part of the maintenance of future fusion reactors will have to be done using remote handling techniques and it is vital that the methodologies and standards to ensure remote handling compatibility are established in advance of reactor concept design. The present programme of experimental fusion machines is the opportunity to define and prove these techniques. In a reactor like machine many interventions will be predictable or routinely required and since availability is a prime consideration these tasks will have to be done quickly and reliably using special purpose remote handling devices with a high degree of automation. The considerations for remote maintenance of fusion experiments is in some ways more stringent than that which will ultimately be required for a reactor. In addition to the general maintenance requirements existing experimental tokamaks have demonstrated that many system designs evolve and require modification during the lifetime of the experiment. In JET this has been accommodated by manual modifications made possible by adherence to a policy of strict neutron economy during physics operations. In future fusion experiments which aim to burn significant amounts of tritium many of the modifications will only be possible using remote handling means [1].

Preparations for the remote maintenance of JET have resulted in a tokamak design and a system of remote handling equipment, personnel and management capable of undertaking a wide range of operations [2,3]. This work has required attention to a number of key issues and constraints which are defined and discussed in this paper.

The requirement for remote maintenance of JET was originally confined to one-off repair and scheduled maintenance type tasks. However, the JET project extension to 1999 includes a major modification to the in-vessel divertor configuration in 1997 using fully remote means. This paper describes how the JET remote handling system has been able to satisfy both the original and the new requirements.

2. Remote handling requirements and constraints

2.1. General

The remote maintenance of a tokamak type fusion experiment requires a system to satisfy unscheduled repair after failure, scheduled replacement of parts due to consumable depletion or erosion and modification of the tokamak in accordance with experimental requirements.

The preparation of remote handling equipment includes the design and manufacture of new equipment, proving of equipment reliability, development and proving of operating procedures, training of operators and setting up an operations support system. The remote handling system is required to be prepared and proven in a time frame dictated by the lead time of the components to be handled. Experience at JET has shown that, in addition to the technical requirements, this presents the remote handling engineers with significant programmatic constraints.

2.2. Unscheduled repair tasks

Repair tasks are characterised as one-off tasks requiring component removal and replacement with a spare or a blank and clean up of surrounding area as required. For these tasks the remote handling system will be required to operate in an incident response type of role and a system adaptable to a wide range of repair conditions will be necessary.

Practical experience at JET during the manual maintenance phase has demonstrated that the failure modes of tokamak components is difficult to predict and the detailed operations required to be undertaken for repair cannot be fully prepared in advance [4].

At JET the preparations for remote repair type tasks is of a generic nature with the primary objective of satisfying task feasibility with little attention to optimisation of task duration. The time available for remote handling preparation consists of the component manufacturing lead time and an undefined proportion of the scheduled component operation period.

2.3. Scheduled maintenance tasks

These tasks are characterised as periodic replacement of components due to wear or depletion whose requirements are prescribed at the component design phase. The tasks may be one-off or multiple operations. The condition of consumables not exposed to the plasma can be assumed to be predictable at the start of operations but the condition of consumables directly exposed to the plasma is less predictable [4].

At JET the preparations for scheduled maintenance type tasks has the objective of satisfying both task feasibility and optimisation of task duration. The time available for remote handling preparation consists of the component manufacturing lead time and a known period of scheduled component operation.

2.4. Task to modify the tokamak configuration

Modification type tasks are characterised as replacement of components due to redesign resulting from changes to the experimental programme.

The manual modification of tokamak configuration as a result of physics experiments has been a major element of the JET programme and its extension beyond the originally planned completion date of 1989. Most of the modifications were not envisaged at the start of the project and have required significant work both in-vessel and ex-vessel with hands-on shutdowns of the order of many months.

The JET project extension to 1999 includes a major modification of the divertor using remote handling. In this case the handling requirements are known during the component design phase. The conditions of the divertor expected to prevail at the time of starting the task are generally predictable with some uncertainty regarding plasma facing components and fixing nuts and bolts.

Remote handling preparations for modification tasks have the objective to satisfy both task feasibility and optimisation of task duration.

The time available for remote handling preparation comprises only the new component design and manufacturing lead time.

2.5. Key issues for remote handling on fusion experiments

The remote handling of tokamak type fusion experiments presents the remote handling community with its most significant challenge. The combination of hostile environment, complex nature and variety of tasks with difficult access constraints presents severe technical challenges. The magnitude of the preparations and remote handling operations also presents a significant project management challenge.

Preparations for the remote maintenance of JET have been dictated by four key issues:
- remote maintenance philosophy;
- tokamak remote handling compatibility;
- remote handling equipment;
- remote handling operations.

2.6. Remote handling philosophy

The remote handling philosophy has a direct influence on the design of remote handling equipment, the design of tokamak components and the remote handling operations.

Since the early 1960s the nuclear remote handling community has developed applications adopting the principles of teleoperation [5]. In parallel, the development of low cost high performance industrial robots has resulted in their use for some remote handling applications [6,7].

The principle of remote handling using robots is characterised by an automation of the handling process with mainly supervisory manual involvement. This philosophy leads to optimisation of the time taken for and safety of the handling task. The robotic approach relies on extensive off-line preparation of the operational strategy and requires detailed preparation of the plant and components to ensure full compatibility with the automation process. It also relies on the predictability of component condition during the handling process.

The principle of component handling using teleoperation is characterised as the man-in-the-loop. The handling task is performed by an operator using a manipulator which provides force, visual and audio feedback of sufficient resolution to create the atmosphere for the operator of actually being in the task environment [5]. This approach provides maximum intelligence and adaptability by direct involvement of the operator. The teleoperation principle requires some preparation of the plant and components to be handled to ensure compatibility with the handling equipment.

A telerobotic approach incorporating the advantages of both robotic and teleoperation philosophies has been adopted at JET.

2.7. Compatibility of tokamak systems and components

The implementation of a telerobotic approach at JET has minimised the constraints that remote handling imposes on the general design of tokamak components. This has been especially important for systems whose probability of requiring remote handling is low and for these implementation of standard components, handling features, alignment methods and access envelopes [8] into the tokamak component designs has been considered sufficient.

The design of JET components which definitely require remote handling are subject to a more rigorous assessment including bench test trials and detailed task feasibility assessment. It is important that the component designers are fully aware of the remote handling capabilities and limitations and to this end the component designers are provided with a definition of remote handling standards and are involved during the remote handling mock-up and bench test operations.

A successful component design can be severely affected by poor quality manufacture or assembly. Experience at JET has shown that even with strict quality assurance procedures mistakes can still be made during detail design and manufacture and with the severe programme time pressures concessions are often sought regarding remote handling compatibility. The JET telerobotic philosophy allows for some flexibility to the handling process and many concessions have been granted for one-off type tasks where operational time is not critical. However, for the repetitive tasks such as divertor module handling it is essential that remote handling compatibility is not compromised.

2.8. Remote handling equipment

The JET remote handling equipment has been specified, designed and developed to satisfy the following requirements:
- functionality and performance;
- reliability and availability;
- recoverability after failure;
- task adaptability.

2.8.1. Functionality and performance

The equipment is specified and designed to satisfy the telerobotic philosophy coupled with computer aided functions to enhance operational safety and performance. The control systems and man-machine interfaces have been designed and implemented to maximise operator efficiency by the use of computer assistance, safety interlocks and simple graphical user interfaces.

The JET remote handling equipment functionality and performance has been reported by many authors [2,3,9–11].

2.8.2. Reliability and availability

The remote handling system is a service to the tokamak experiment and it is essential that the service be efficient and timely. This cannot be achieved if the remote handling system itself is unreliable or has poor availability.

New JET remote handling equipment is specified and designed to maximise its availability by the use of proven sub-components and by adherence to a formal Quality Assurance system for its design, build and acceptance. The equipment is typically procured as a one-off and is therefore of a prototypical nature requiring time for development and reliability proving before being used in active areas. After acceptance testing the equipment is subjected to extensive exercising both formally for reliability proving and implicitly as part of task mock-up operations. Throughout, the equipment is subject to a formal system for fault reporting, location and corrective action which leads to its progressive improvement and development [12]. This process is greatly simplified if there is a standardisation of equipment design and at JET there has been significant effort to rationalise all of the common sub-systems such as actuators, wiring and control system hardware and software.

2.8.3. Recovery after failure

Consideration must be given at the earliest stage of remote handling equipment design to its failure modes and the methods by which it can be repaired and returned to service.

All JET equipment is designed to be 'fail safe' such that any failure will result in automatic cessation of motion of the equipment. A 'fail operational' capability has not been included because it is considered to lead to highly complex and expensive systems which would nevertheless ultimately require recovery in the event of worst case failures.

All JET remote handling equipment has been designed with features to facilitate recovery. For worst case failures the recovery process will require the deployment and use of an emergency remote handling system which for failed in-vessel equipment will be the ex-vessel remote handling system. The recovery system is required to operate within extremely tight space constraints inside the torus concurrently with the failed equipment and to undertake a wide range of disassembly, handling and possibly cutting tasks. Use of the ex-vessel telerobotic system provides the most flexible approach to satisfy these requirements.

The importance of this aspect of the equipment design must not be underestimated. The consequences of an unrecoverable failed remote handling equipment inside a radioactive tokamak will be incalculable both for the experimental programme but by implication for the credibility of fusion reactors.

2.8.4. Task adaptability

The remote handling equipment has to be able to respond to changes of task requirement in both short and long term timeframes.

Short term adaptability is required to be able to handle components which are in some way damaged or not exactly in the condition expected during the design process. Long term adaptability is required for the remote handling equipment to be able to cope with new components designed and implemented as part of the experimental campaign.

The telerobotic approach provides short term adaptability by the direct involvement of the operator in the task who can call upon and use a variety of hand-held general purpose tools as required for the situation. The telerobotic approach facilitates long term adaptability by supplementing a proven remote handling system with specialist tools or end-effectors requiring a minimum lead time to design and qualify operations.

2.9. Remote handling operations

Experience at JET has shown that to ensure efficient manual shutdown work it is essential to comprehensively prepare procedures and train operators using full scale mock-ups. This is even more relevant for remote handling tasks where unexpected changes to a task procedure can cause severe delays. It is essential that the mock-up and the remote handling equipment provide an accurate representation of the real task. At JET this is achieved by deploying the real remote handling equipment for mock-up trials and constructing the mock-up rig using only dimensional information taken from the tokamak as-built photogrammetry surveys.

Selection and training of operators is a key part of the operational preparations and the operator training programme at JET comprises a period of theoretical and practical training for qualification with each remote handling equipment followed by task training to suit the task procedures. Training on equipment operation includes all aspects of normal and fault states up to and including the recovery of equipment from radioactive areas.

Remote handling operations require the support of maintenance engineers who can be called upon in the event of problems with the equipment. The maintenance engineers must be familiar with the equipment and trained to deal systematically with faults and repairs. Experience at JET shows that this training period is many months for a large system. The support function requires the setting up of a library of up to date equipment build documentation with implementation photographs and diagrams, the provision of suitable diagnostic tools for fault location and access to a store of appropriate spares. Also, a programme of periodic inspection and testing must be implemented during operations to ensure that dormant failures have not occurred.

3. A remote handling application from JET

3.1. Requirements for remote handling of the JET divertor

In 1997 it is required to remotely change the existing JET MkIIa divertor to a gas box type.

The JET divertor comprises a water cooled support structure on which are bolted modules of typical weight 35 kg and dimensions $0.5 \times 0.5 \times 0.25$ m [13,14].

The replacement of the MkIIa modules (Fig. 1) with 192 MkIIgasbox divertor modules (Fig. 2) is required. The modules must be removed and replaced with clearances of as little as 1.4 mm between adjacent carbon elements and it is essential that none of the critical component edges or surfaces are damaged during handling. The modules are located in their final position with dowels and are fixed in place with bolt assemblies, there is no requirement for cutting or welding during this task.

With two exceptions the divertor diagnostic systems have all been integrated into the modules. The diagnostic sensors are connected to the measurements systems by means of a remote handling compatible connector integral with the divertor modules. The diagnostic systems which have not been integrated into the modules require the separate handling of small (< 0.5 kg) items.

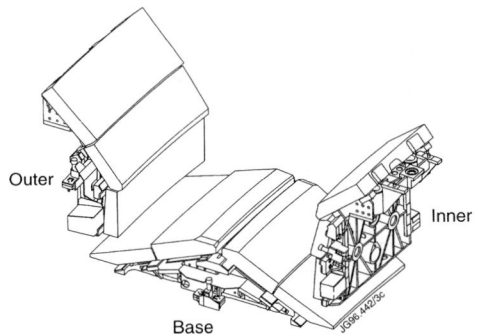

Fig. 1. JET MkIIa divertor modules.

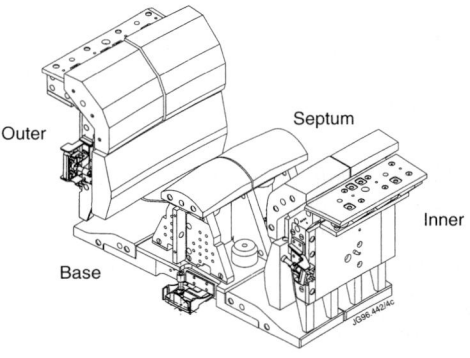

Fig. 2. JET MkII gas box divertor modules.

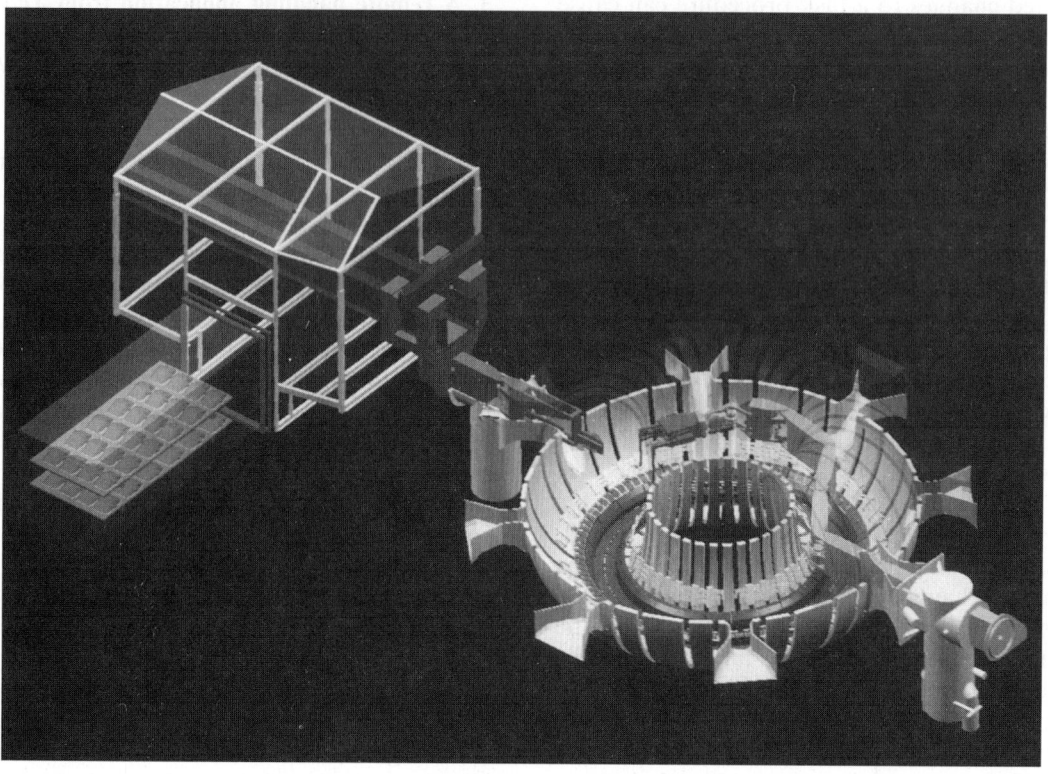

Fig. 3. Overview of JET divertor handling system.

Other tasks include the cleaning of four beryllium evaporator heads, vacuum cleaning of the divertor structure and deployment and use of videogrammetry equipment.

The entire remote handling operation is required to be achieved reliably, safely and within a specified shutdown time frame of 6 months.

The radiation dose rates expected to persist at the start of the shutdown necessitate remote handling inside the torus but all ex-vessel tasks can be done manually.

3.2. Remote handling equipment for the JET divertor

The design and development of remote handling equipment to satisfy the general JET repair and scheduled maintenance task requirements has been reported many times elsewhere [2,3]. The remote handling philosophy adopted by JET has facilitated the utilization of much of the remote handling equipment developed for these tasks with little modification. Some new equipment has been designed and developed to enable parallel handling and storage operations to take place and special divertor module handling tools have been developed (Fig. 3).

The in-vessel operations will be performed making use of the Mascot servomanipulator master-slave system. The Mascot slave is positioned within the torus by means of the Articulated Boom controlled predominantly in teach-repeat mode (Fig. 6). The operations will be supported by 12 cameras located inside the torus.

Special divertor module handling tools have been designed to maximise the speed and safety of the tile carrier manipulations. Stringent physics design constraints have minimised the space for access to the divertor module fixing bolts and special features have been incorporated into the

Fig. 4. JET remote handling control room.

handling tools to ensure that there is no possibility of damage resulting from the handling or bolting operations. Where required, special tools have also been designed and built to facilitate the handling of the other in-vessel components. All of these aspects are described in [10].

The divertor modules are transferred to and from an ex-vessel storage facility using another articulated boom transporter with only one articulated degree of freedom and a special end-effector [9]. This short boom is also able to deploy a Mascot slave unit if required to effect a recovery of the main articulated boom from inside the torus [12].

The remote handling equipment is commanded from the JET remote handling control room using a mixed point-to-point and distributed communications network (Fig. 4)

Considerable attention has been paid to the implementation of the Man-Machine Interface and communications systems and these are de-

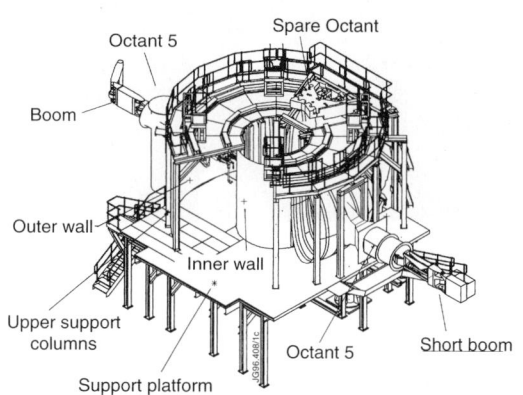

Fig. 5. JET full scale mock-up facility.

Fig. 6. JET Mascot manipulator in service inside the JET torus.

scribed in detail in [11]. In particular, the Man-Machine Interfaces have been implemented on a standard PC hardware platform running under Windows NT with the software written for maximum operator effectiveness and portability.

3.3. Compatibility of JET components

The divertor components have been designed taking into account remote handling requirements. Features have been incorporated to provide accurate attachment of the handling tools, alignment dowels for final positioning, bolt assemblies which ensure bolt control during transport and self alignment during fixing, use of a common bolt head size on all components to minimize wrench requirements, provisions to minimize the likelihood of seizure during operation, pre-drilled bolts and removable nut inserts to facilitate recovery in the event of seizure and visual cues for rough alignment.

The design and build of divertor components has been controlled using the JET QA system and a rigorous process of inspection, measurement and trial fit has been implemented to ensure full compatibility with remote handling.

3.4. Remote handling operational preparations

The remote operations involve the handling of seven types of modules at 144 locations around the torus. In addition there are 12 other in-vessel tasks required to be performed on diagnostic and in-vessel tile systems and a range of possible tasks in the event of remote handling equipment failure in-vessel. Detailed operating procedures are being developed for all of these tasks and proven by demonstration using the real remote handling equipment and a full size mock-up of half the torus [15] (Fig. 5).

To confirm the relevance of mock-up preparations a short period of remote operation inside the JET torus itself was undertaken during the MkII divertor first installation shutdown in 1996.

The preparation of operational methods, task procedures and operating personnel for this 26 week divertor exchange shutdown will have utilised 750 man weeks of effort over a 2 year period by the start of operations in 1997.

4. Conclusions

The remote handling of fusion experiments requires a system which is capable of facilitating scheduled maintenance, unscheduled repair and modification type tasks.

It is not possible to identify every possible failure mode and to prepare a remote handling system to undertake all of the repair type tasks within a plant as complex as an experimental tokamak. The remote handling system must therefore be prepared to undertake repair tasks in an incident response role with equipment and personnel able to handle generic tasks adaptable over a wide range.

The preparation of a remote handling system for scheduled maintenance type tasks, characterised as replacement of consumable components, can be more deterministic. However, the condition of components exposed to the plasma must be assumed to be unpredictable and therefore require an adaptable remote handling capability.

Tokamak modification as a result of experimental demands presents the remote handling system with both a technical and programmatic challenge. A system to remotely modify the tokamak is subject to the technical constraints imposed by a repair or scheduled maintenance type task but in addition is required to be fully prepared and proven within the lead time of the tokamak modification design and manufacturing period.

Experience from preparing a system to satisfy the JET remote handling requirements has identified four key issues: (i) remote handling philosophy; (ii) tokamak compatibility; (iii) remote handling equipment; and (iv) remote handling operations.

The choice of remote handling philosophy fundamentally affects the other three issues. To fully satisfy the requirements a telerobotic approach comprising a teleoperator, or man-in-the-loop, system with extensive computer support has been adopted at JET. This approach achieves maximum adaptability and operational intelligence and offers a number of simplifications for the three remaining key issues.

It is fundamental that the tokamak be constructed for remote handling compatibility. In practise this is achieved by the definition of a set of design standards for remote handling, the education of equipment designers in the capabilities and limitations of the remote handling equipment, the involvement of remote handling engineers in the tokamak design process and finally by strict quality control during design, manufacture and installation of the components.

The remote handling equipment must be designed and built to suit not only the functional and performance requirements but also it is essential to ensure and prove the equipment reliability. The equipment must be designed to fail in a controlled manner avoiding damage to itself or the environment and must be proven to be recoverable after failure. It is considered that designing a system to be fail operational introduces an unjustifiable increase in system and management complexity. The JET approach is to deploy another telerobotic remote handling system to recover failed equipment.

The remote handling operations must be fully prepared in advance. The operational procedures, operator training, equipment support staff training and the support infrastructure must be prepared. Operational procedures for repair type tasks will of necessity be generic in nature but all scheduled maintenance and modification type tasks will require specific operating procedures to be prepared and proven with mock-up trials.

In 1997 immediately after the 3 month D-T operational phase it is intended to fully remotely replace 144 in-vessel divertor modules with 192 new modules of a gas box type divertor configuration. Preparations for this first fully remote handling JET shutdown are well advanced and have included a short period of operation inside the JET torus when 33 divertor modules were successfully installed using fully remote means.

Acknowledgements

The work reported herein is the result of the initiatives and efforts of all of the JET remote handling group personnel. In particular the author would like to acknowledge the technical and inspirational lead provided by Dr T. Raimondi during his many years of involvement with the JET remote handling equipment.

References

[1] R. Haange, Overview of remote-maintenance scenarios for the ITER machine, Fus. Eng. Design 27 (1995) 69–82.
[2] T. Raimondi, The JET experience with remote handling equipment and future prospects, Proceedings of the 15th SOFT, Utrecht, 1988, 197–208.
[3] A. Rolfe, Operational aspects of the JET remote handling system, Proceedings of the ISFNT-1, Tokyo 1988, Part C, pp. 501–507.
[4] E. Bertolini, Engineering analysis of JET operation, 16th SOFE, Illinois, October 1995.
[5] J. Vertut, P. Coiffet, Teleoperation and Robotics, Vol. 3a, Kogan Page, London, 1985.
[6] R. Horn, Gathering of operational experience with robots, EC Teleman/Entorel-RHC-16-1, April 1994.
[7] V. Kiernan, Send in the Robot, New Scientist, 23 September 1995.
[8] JET Remote Handling Manual.
[9] L. Jones, D. Hamilton, M. Irving, J. Palmer, Design and development of a new remote handling transporter facility for JET, SOFT 1996, Lisbon.
[10] S. Mills, A. Loving, Design and development of RH tools for the JET divertor exchange, SOFT 1996, Lisbon.
[11] L. Galbiati, P. Carter, B. Haist, The command and control system for JET remote handling equipment, SOFT 1996, Lisbon.
[12] A. Rolfe, D. Smith, E. Scott, The assessment and improvement of JET remote handling equipment availability, SOFT 1996, Lisbon.
[13] H. Altmann, Design of the JET MkII Divertor with large CFC tiles, SOFT 1994.
[14] H. Altmann, MkIIgas box divertor—a new design concept, SOFT 1996, Lisbon.
[15] R. Cusack, P. Brown, B. Haist, A. Loving, R. Stokes, The implementation of a full size mock-up facility in preparation for remote handling of JET divertor modules, SOFT 1996, Lisbon.

Possible divertor solutions for a fusion reactor. Part I. Physical aspects based on present day divertor operation[1]

A. Kallenbach *, H.-S. Bosch, S. de Peña Hempel, R. Dux, M. Kaufmann, V. Mertens, J. Neuhauser, W. Suttrop, H. Zohm

MPI für Plasmaphysik, EURATOM Association Boltzmannstraße 2, D-85748 Garching b.München, Germany

Abstract

With an anticipated power flux across the separatrix of up to 300 MW of an ITER-like fusion reactor, conventional measures of power spread lead to a peak power load at the target plates in the order of 30 MW m^{-2}, far beyond the technically feasible limit for stationary operation. Radiative cooling by seed impurities appears to be the most promising plasma-physical option to reduce the target power load, but extrapolations of present experiments predict an only marginally tolerable increase of the plasma effective charge Z_{eff}. Key points will be the achievement of very high electron densities, leading to more effective radiative cooling by $\delta P_{rad}/\delta Z_{eff} \propto n_e^2$ while keeping the edge temperature within its optimum range. This range is bounded from below by the H → L mode temperature threshold due to confinement requirements, whereas the upper boundary is given by the ideal ballooning stability limit which is connected to type-I ELM activity which may cause non-tolerable divertor heat loads. The completely detached H-mode (CDH) in ASDEX Upgrade demonstrates radiative H-mode operation within this operational range exhibiting high-frequent type-III ELMs and target power load in the order of 10% of the heating power. At present, open questions on high density reactor operation are related to radiative instabilities as well as edge transport enhancement and H-mode impairment observed in several tokamaks under high density conditions. Measures to overcome these detrimental effects will be investigated with improved divertor concepts in the near future. The possible problems connected to high density reactor operation can be relaxed, if the design of plasma facing components with higher heat flux endurance is successful. © 1997 Elsevier Science S.A.

1. Introduction

The anticipated power flux across the separatrix of a fusion reactor is about 300 MW. With conventional measures for power spread, this will lead to a peak power load in the order of 30 MW m^{-2} [1] perpendicular to the target plates. This value is far above the present technically feasible limit for stationary operation of about 5 MW m^{-2}, based on target lifetime considerations. Consequently, measures have to be taken to reduce the incident power flux and to increase the power resistance of the target plates.

* Corresponding author.
[1] In association with The ASDEX Upgrade-, NI- and ICRH Teams MPI für Plasmaphysik, Garching and Berlin, Germany.

The most promising plasma-physical option to avoid divertor power overload is to increase the radiative losses in the main chamber by impurity seeding. While efficient radiative cooling has been demonstrated in present day experiments, ignited reactor operation imposes a number of critical boundary conditions which have to be necessarily fulfilled. These are, among others, the limitation of the central effective charge Z_{eff}, preservation of good energy confinement and efficient helium ash removal. While good H-mode like confinement is indispensable, accompanying type-I edge localized modes (ELMs) which result in strong, short power bursts on the divertor plates are probably not tolerable. Therefore, the envisaged radial radiation profile must be tailored to prevent type-I ELMs. The most critical aspect of radiative reactor scenarios appears to be the achievement of very high electron densities at the edge. While the anticipated ITER value for an unseeded plasma is critical by itself, adding of impurities is observed to degrade the maximum edge density in present-day experiments.

This paper is organized as follows: after a brief introduction into conventional measures of power spread in a divertor, we illustrate the interplay of the various boundary conditions for radiative discharges utilizing experimental data obtained from the ASDEX Upgrade tokamak. Predictions for the impurity content of a radiative reactor-grade plasma are made using the impurity transport code STRAHL which has been validated against high density radiative experiments. In the subsequent paper, the design of high heat flux components is discussed, the development of which is vital to absorb the excess power possibly not handled by the plasma-physical power spread schemes.

2. Variations of the divertor configuration

Besides providing plasma conditions suitable for power and particle removal with low surface erosion, the geometrical arrangement of a divertor supplies a natural power spread. There are several elements contributing to the dispersal of the high parallel power flux densities occuring around the midplane separatrix until their release at the target plates [2]: midplane field pitch angle (factor ≈ 7), expansion of the flux surfaces in the vicinity of the X-point (≈ 4), tilting of the target plates with respect to the poloidal flux surface (up to ≈ 3), as well as radiation and charge-exchange losses. The geometrical spreading factors lead to a 'wetted' divertor area in ITER of about 10 m^2, which would result in the extreme heat flux of up to 30 MW m^{-2}. At present, several experiments design improved divertor configurations, aiming at better particle retention and power spread capabilities by means of charge exchange neutrals and divertor radiation. As an example, Fig. 1 shows the original divertor I configuration in the ASDEX Upgrade tokamak operated till Summer 1996 and the new divertor II presently under installation [3].

Divertor I is a rather shallow configuration, taking advantage of the large 'natural' spread of the flux surfaces in the vicinity of the X-point caused by the low poloidal field there (flux surface spread translates into power decay lengths and is therefore inversely to peak power loads). Neutrals are reflected from the outer target plate towards the pumped volume, which is beneficial for impurity removal. The deeper Divertor II Lyra configuration (V-shape) compensates for the lower natural power spread by alignment of the target plates leading to near-glancing poloidal incidence of the flux surfaces. The neutrals are reflected towards the energy carrying layer. The extended length of the field lines in the divertor allows higher radiative and charge-exchange losses. Expected improvements of the Divertor II configuration in comparison to Divertor I are the widening of the detached operation window and better neutral retention.

Divertor optimization has to consider various side conditions (low sputtering, effective pumping, disruption endurance) with often opposing constructive demands, e.g. a deeper divertor will improve radiative cooling due to the longer field lines, on the other hand, the flux expansion decreases with separation from the X-point. Therefore, although divertor modelling has undergone significant progress in the past, further experiments are required to decide on the optimum divertor configuration for ITER. However, a reduction of the power flux density under reactor conditions down to 5 MW m^{-2} without additional radiative losses seems improbable at the moment. Seed impurities will be necessary for radiative power removal down to the level the divertor can handle.

3. Optimization criteria for radiative H-mode operation

Ignition and sustaining of an α-particle heated thermonuclear plasma is a formidable task and the chosen reactor size [1] represents a sensitive balance between the conflicting demands of energy confinement considerations on one side and machine cost and tolerable divertor load on the

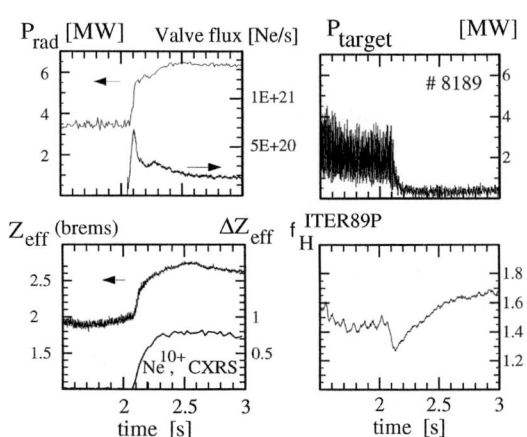

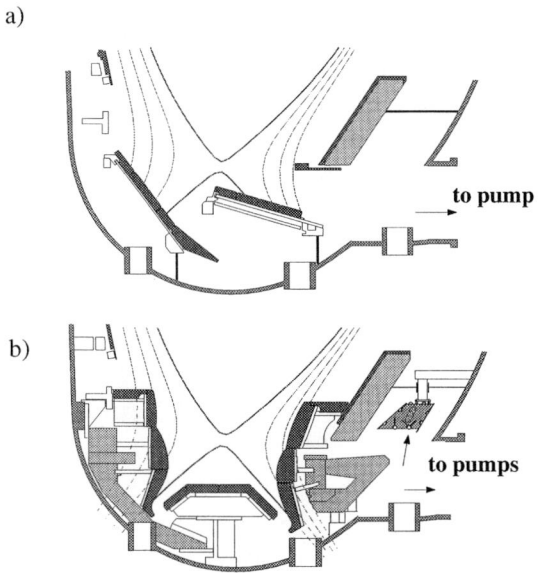

Fig. 1. (a) ASDEX Upgrade Divertor I operated until August 1996. (b) Divertor II presently under installation. The new Divertor II configuration is designed for higher heat flux capability, a widened detached operation window and better neutrals retention. The poloidal field coil system (not shown) represents a reactor-relevant configuration and is situated outside the toroidal field coils.

Fig. 2. Radiated power in the main chamber, feedback-controlled Ne flux, total divertor power load from thermography, central line-averaged Z_{eff} from bremsstrahlung and the contribution of Ne^{10+} at $\rho_{pol} \approx 0.65$ from CXRS and energy confinement normalized to the ITER89P scaling for a CDH-mode discharge. $P_z = 4$ MW, $P_{heat} = 7.5$ MW, $n_0^{div} = 1.5 \times 10^{20}$ D_2 m^{-3}.

other. As a result, the tolerable amount of seed impurities is rather low, mainly due to the narrow confinement margin and the importance of volume bremsstrahlung losses for the low heating power density conditions of a reactor plasma. In the following the various ingredients and boundary conditions for effective and reactor-compatible radiative cooling are discussed.

3.1. Radiative efficiency

Radiative cooling by injection of seed impurities is easily obtained in present-day tokamaks, provided that a dedicated feedback system [4] is available which allows the control of the radiation level with varying plasma and heating conditions. Using low-Z elements, the radiation is typically confined to the region near the separatrix. Surprisingly, improvement of energy confinement is often observed rather than degradation due to the additional losses. Fig. 2 shows typical time traces for a completely detached H-mode (CDH-mode) [5] discharge in ASDEX Upgrade. Immediately after switching on the Ne radiation feedback at $t = 2.05$ s, the CDH-mode is obtained, as indicated by a peak target power load below 1 MW and energy confinement improvement, while the plasma effective charge, Z_{eff}, increases by 0.8. About 55% of the total radiated power is emitted inside the separatrix, while approximately 25% originates from the scrape-off layer above the X-point and only 20% is emitted in the divertor below the X-point. Except for the high Z_{eff} for ASDEX Upgrade conditions, the CDH mode would be very attractive for a reactor due the very low target power load and the effective buffering of the high-frequent type-III ELMs in the divertor.

The key number for the efficiency of radiative cooling is the increase of the central Z_{eff} for a given radiation power, $P_z = \delta P_{\text{rad}}/\delta Z_{\text{eff}}$. A multi-machine study by Matthews et al. [6] has investigated the dependence of the radiative efficiency on plasma parameters and machine size. The relation

$$Z_{\text{eff}} = 1 + 7 \cdot P_{\text{rad}}/(S \cdot \bar{n}_e^2) \text{ (MW, m}^2\text{, }10^{20}\text{ m}^{-3}\text{)} \quad (1)$$

was found to give a good description of the effective charge for given radiated power for various experimental situations, the plasma surface S providing a robust size-scaling. No distinct dependence on transport regime or impurity species is found (note that $\delta Z_{\text{eff}} \propto Z^2$ and therefore a higher Z is connected to a lower impurity concentration at given δZ_{eff} contribution). Notably, P_z increases with the electron density squared.

Impurity profiles observed in the main plasma of ASDEX Upgrade are typically flat. However, if sawtooth activity stops, impurity peaking in the plasma center is observed [7]. This would drastically reduce the fusion rate in a reactor and has to be definitively avoided by current profile control. ITER demands for effective charge values $Z_{\text{eff}} < 1.6$, a contribution δZ_{eff} up to 0.4 of which may stem from the helium ash. For a line-averaged density $\bar{n}_e = 1.3 \cdot 10^{20}$ m^{-3}, Eq. (1) predicts a Z_{eff} increase of 0.4 for additional 100 MW neon (line-) radiation ($P_z = 250$ MW). A somewhat higher radiation level can be expected for a relative separatrix density higher than $0.3 \times \bar{n}_e$ as typical for the experimental conditions Eq. (1) was derived from. Since the total radiated power in the divertor is found to be always low in ASDEX Upgrade radiative H-mode discharges, we focus on main chamber radiation in the following.

3.2. High density operation and energy confinement

The anticipated operational density of ITER is very high [1], about 1.5 times the empirical Greenwald density limit [8] $\bar{n}_e^{GW} = I_p/(\pi \cdot a^2)$ (10^{20} m^{-3}, MA, m^2), which was found to give a good estimate of the maximum achievable density in various experiments for normal conditions. Simultaneously, high energy confinement times have to be achieved expressed by the confinement normalization factor $H^{\text{ITER89p}} > 2$. High electron densities are indispensable for effective radiative cooling due to the n_e^2 dependence of the radiated power at given δZ_{eff} contribution. However, many experiments report strong degradation of the H-mode confinement when high-density H-modes are approached by gas puffing [9–12] and further density buildup is obstructed by the reduced particle confinement. Fig. 3 displays the density operational space for gas-puffed ASDEX Upgrade discharges. The maximum density achievable in L-mode (density limit, DL) exhibits a heating power dependence, which increases with plasma purity [13]. The type-I ELMy H-mode obtains a 'natural' density $\approx 0.6 \times \bar{n}_e^{GW}$, while the radiative CDH-mode exhibits higher line-averaged densities owing to n_e profile peaking. The attempt to increase the H-mode density by strong gas puffing finally leads

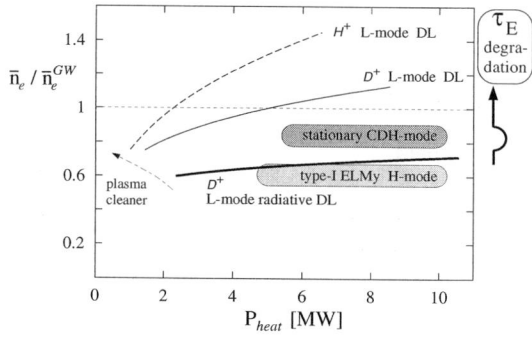

Fig. 3. Density operational diagram for ASDEX Upgrade. The three L-mode density limit curves represent typical parameter variations. The maximum achievable density increases with heating power only for clean plasma conditions. The H-mode exhibits a density resilience ($\bar{n}_e \propto I_p$). Degradation of energy confinement is observed at high densities, except for the CDH-mode with peaked n_e profiles.

to a H→L backtransition in combination with density decrease caused by confinement degradation. Therefore, no direct H-mode density limit exists.

The confinement degradation at high densities is coincident with a high neutral pressure in the main chamber and manifests itself in the erosion of the H-mode transport barrier [14] (it should be noted, that the high neutral pressure and the high edge electron density are observed to be correlated, therefore it is not clear which of these parameters actually causes the confinement degradation). The detrimental effect of the neutral pressure is more pronounced in the JET MkI divertor in comparison to ASDEX Upgrade, which is possibly due to the low neutral compression of the MarkI divertor caused by leaks connecting the divertor volume to the main chamber [11]. The situation is improved considerably when active cryopumping is applied: simultaneous deuterium puffing and cryopumping at JET resulted in higher plasma densities for a given neutral density and confinement degradation ($<n>_e/n_e^{GW} \approx 0.8 \to 1$ at $H^{ITER89p} = 1.6$) [11].

The predicted interplay of high density operation and radiation for ITER is presently not well established when the results of different tokamaks are compared: in ASDEX Upgrade, impurity seeding causes an increase of the line-averaged density due to electron density peaking while the edge is hardly affected. In the type-III ELMy CDH-mode, confinement improvement connected to the density peaking overcompensates the negative effects of radiation and gas puffing. In the circular limiter tokamak TEXTOR, impurity seeding is also coupled to density peaking and increase of $\bar{n}_e$, and confinement improvement up to $H^{ITER89p} = 2$ is observed [15]. In contrast, JET reports confinement degradation and no density peaking with seed impurity radiation. Generally, electron density profiles are flatter in larger devices with elongated plasmas. These observations make it questionable at the moment, whether density peaking and the related confinement improvement could be obtained in ITER.

3.3. Divertor retention and impurity compression

Retention and compression of impurities in the divertor should be maximised for several reasons and is among the divertor design optimization criteria presently pursued on various experiments around the world. ASDEX Upgrade Divertor I shown in Fig. 1(a) exhibits very good impurity and helium compression for high neutral flux conditions even with a high radiation level, allowing for effective helium pumping and active radiation control. The retention of impurities is attributed to internal recycling fluxes [16] which entrain the impurities and sweep them towards the target plate, overcompensating the thermoforces which tend to push the ions towards the midplane. The baffle situated in the outer divertor at the height of the X-point prevents neutrals from escaping into the main plasma. Although much higher divertor impurity densities have been established in relation to the main plasma, total divertor radiation is always small in comparison to main chamber radiation, while the maximum emitted power densities occur around the X-point.

In ASDEX Upgrade divertor I, the retention time τ_{div}^{imp} of both helium and neon increases with the neutral gas density $n_{D_2}^{div}$, while higher absolute values of divertor compression are achieved with neon [17]. Remarkably, the divertor retention is not affected by detachment, as seen in the CDH-mode. The good impurity retention of the divertor I configuration, especially for detached conditions, is attributed to the large distance of the baffle in the (pumped) outer divertor from the target plate. As a consequence, the detachment front, e.g. during the impurity-seeded CDH-mode, is located well below the baffle and the warm plasma around the baffle prevents the neutrals from escaping into the main chamber. As long as the detachment front does not reach the baffle, the divertor retention does not change with the power flux. The difference in the retention of neon and helium can be explained by their different ionization lengths and the presence of deuterium flow patterns: the neon ionization length is shorter in comparison, leading to entrainment of neon in the deuterium flow pattern [17]. Helium has a longer ionization length and may reach regions no longer dominated by strong deuterium flow towards the target plate, where it can be pushed towards the main plasma by thermoforces. For CDH-mode conditions, divertor helium retention is good leading to values of the effective helium confinement time normalized by the energy confinement time, $\tau_{He}^*/\tau_E < 10$, as required for ITER [17].

3.4. Adjustment of separatix power flux to the H-L power threshold

H-mode operation is bound by a lower heating power threshold for the L → H transition with the approximate scaling $P_{thres}^{L \to H} \propto \bar{n}_e B_t$, the H → L backtransition power is typically about a factor of 2 lower for moderate edge densities (H-L hysteresis). Extrapolations of $P_{thres}^{L \to H}$ for ITER reveal, indeed with considerable uncertainties, that a L → H transition at low density may be achieved, but sustainment of the H-mode at high density will be marginal. Local measurements of electron density and temperature near the separatrix identify the temperature as the decisive parameter for the discharge state [18]: H → L as well as L → H transitions occur at the same temperature (measured 2 cm inside the separatrix) with weak dependence on the electron density and the upper bound of the aspired type-III ELMy operational range is also characterized by a fixed T_e value quite independent of other parameters. The L-H hysteresis for the power flux through the separatrix, which is observed without strong gas puffing, is coupled to the density variation between L- and H-mode. The density increase in the H-mode is attributed to the reduced transport in this case. For high density/gas flux levels typical of radiative scenarios, L- and H-mode confinement are found to converge and consequently the power hysteresis is also lost [14]. While the threshold temperature for the type-III ELMy H-mode in ITER is not known so far, the occurence of type-I ELMs can be attributed to the ideal ballooning limit of the pressure gradient [19]. Tailoring of the pressure profile closely inside the separatrix by proper seed radiation should prevent the occurence of type-I ELMs, which are believed to be non-tolerable with regard to target lifetime. A self-consistent

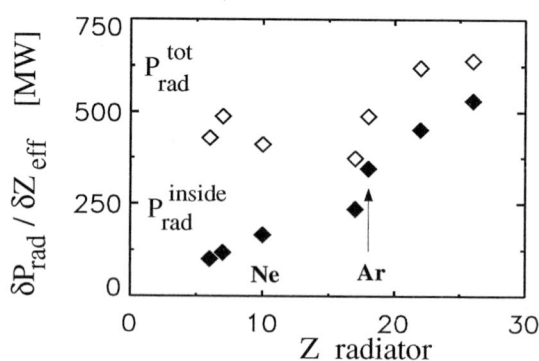

Fig. 4. Total radiated power and radiated power inside the separatrix for ITER parameters and various seed elements for (favourable) high edge density conditions. $n_e(0) = n_e(\text{sep}) = 1.3 \cdot 10^{20}$ m^{-3}, $T_e(0) = 16.4$ keV, $T_e(\text{sep}) = 77$ eV. Hydrogenic bremsstrahlung of $P_{rad} = 47$ MW is not included.

study by Becker [20] of an ignited ITER plasma with high radiative fraction revealed pressure gradients a factor of 2 below the critical ones. Therefore, sustainment of the radiative H-mode at high edge densities appears to be the major problem compared to the avoidance of type-I ELMs.

4. Optimum radiating species for ITER conditions

High density radiative H-mode in ASDEX Upgrade demonstrates a radiative efficiency for neon $\delta P_{rad}/\delta Z_{eff} \approx 4 \cdot \bar{n}_e^2$ (MW, 10^{20}m^{-3}), measured up to ITER-relevant densities of 1.2×10^{20} m^{-3}. The radiation characteristics of these discharges have been successfully modelled with the impurity transport code STRAHL [7], giving confidence into corresponding predictive calculations for ITER parameters. Fig. 4 shows the radiative efficiencies P_z of different radiative species for ITER conditions under the (optimistic) assumption of high electron densities in the scrape-off layer. $T_e(r)$ and $n_e(r)$ profiles are taken from a self-consistent reactor study [20], slightly modified with respect to the actual ITER parameters. Only main chamber radiation is taken into account, the full symbols corresponding to the contribution to P_{rad} emitted inside the separatrix only. While the total radiation per added δZ_{eff} shows only a weak variation with Z, the power fraction radiated inside the separatrix increases considerably with Z. The variation of this fraction for different seed species is a tool to match the bulk radiation to the H → L power threshold and to simultaneously achieve maximum scrape-off-layer (SOL) radiation depending on the actual experimental parameters. The predictions of Fig. 4 are in agreement with Matthews' scaling, Eq. (1), taking into account that $n_e^{sep} = \bar{n}_e$ for the electron density profile used for Fig. 4, resulting in higher SOL radiation than typical of the existing experimental database.

5. Conclusions

The simultaneous achievement of high electron densities and good energy confinement is the key issue for stationary divertor operation in a fusion reactor. High near-separatrix densities are also necessary for effective radiative power removal by seed impurities, as expressed by the relation $\delta P_{rad} \propto n_e^2 \cdot \delta Z_{eff}$. However, the densities envisaged for the ignited burn operation of ITER are in a range where degradation of energy confinement and impairment of the H-mode transport barrier are observed in present day experiments. In addition, a high impurity content degrades the limit for the maximum achievable density below the empirical Greenwald limit, and no electron densities $\bar{n}_e$ beyond $n_e^{Greenwald}$ have been observed in radiative divertor experiments so far. Since the radiative instabilities leading to the density limit are related to the plasma edge, the achievement of high central densities in combination with moderate edge density fulfilling the burn requirements would in principle be conceivable. However, such a scenario seems unlikely in view of the fact that large tokamaks generally exhibit very flat density profiles. In addition, radiative cooling by seed impurities will be highly desirable, and the corresponding increase of Z_{eff} can only be kept on an acceptable level when a high edge electron density is present.

Therefore, the origin of the edge confinement degradation at high densities is a key question. Many experiments relate the enhanced transport to high neutral densities in the main chamber usually connected to high electron density operation. Although improvement of confinement is often achieved by neutral gas reduction, a detrimental effect of the high electron density itself cannot be ruled out owing to the close coupling of electron and neutral densities in the scrape-off-layer. Improved, more closed divertor concepts will have to demonstrate higher energy confinement times in combination with high electron densities. Further assistance for the high density reactor operation may be expected from improved pellet fuelling techniques: Plasma refuelling efficiencies about four times higher with respect to the standard injection scheme have been recently observed in ASDEX Upgrade H-mode discharges using a blower gun to inject pellets from the magnetic high field side [21].

Helium ash removal appears to become a minor problem at least for high neutral flux conditions, since sufficient divertor compression is already achieved [17] and the core transport is fast enough [22]. A further reduction of the helium content in a reactor by improved divertor action would offer more space (in terms of δZ_{eff} contribution) for impurity seeding.

The degree of divertor power load reduction achieved with radiative cooling in present day experiments by itself would fulfill the requirements of a future reactor. Besides the high values of Z_{eff}, the strongest reservations against an impurity-seeded reactor scenario arise from the extrapolation of energy confinement to large machine sizes. Following the line from TEXTOR through ASDEX Upgrade to JET, improvement of energy confinement with radiation diminishes with plasma size, eventually arriving at degraded H-mode confinement in JET radiative H-mode discharges. This finding led to the recent proposal of lightly or non-seeded divertor operation in ITER preserving type-I ELMs [23]. While the prospects of energy confinement are beneficial for such a scenario, a very favourable size scaling, in the sense of low specific divertor power load per ELM is required.

Irrespective of the absolute amount of seed impurity radiation desirable in a fusion reactor, any tailoring of the edge pressure profile, e.g. with respect to the ideal ballooning limit, will be beneficial. Prime candidates as radiating species are the recycling noble gases neon, argon and krypton which offer an extensive range of radiative characteristics.

Finally, it should be noted that some of the critical requirements mentioned above may be relaxed if high heat flux components become available exhibiting higher power handling capabilities. The development status of such components is discussed in the following paper.

References

[1] ITER-JCT and Home Teams (presented by: G. Janeschitz), Plasma Phys. Control Fus. 37 (1995) A19–A35.
[2] C.-S. Pitcher and P.C. Stangeby, Experimental divertor physics, Plasma Phys. Control Fus. 39 in press.
[3] H.-S. Bosch et al., ASDEX Upgrade Divertor II proposal, Report IPP 1/281 (1994).
[4] T. Zehetbauer, P. Franzen, G. Neu, V. Mertens, G. Raupp, W. Treutterer, D. Zasche and ASDEX Upgrade Team, Plasma regime guided discharge control at ASDEX Upgrade, SOFT, Lisbon, September 1996, in press.
[5] O. Gruber, et al., Phys. Rev. Lett. 74 (1995) 4217.
[6] G. Matthews et al., Scaling radiative plasmas to ITER, presented at the PSI 1996 conference, J. Nucl. Mat., in press.
[7] R. Dux, et al., Plasma Phys. Control Fus. 38 (1996) 989.
[8] M. Greenwald, et al., Nucl. Fus. 28 (1988) 2199.
[9] T.W. Petrie, A.G. Kellman, M. Ali Mahdavi, Nucl. Fus. 33 (1993) 929.
[10] A. Kallenbach, et al., Nucl. Fus. 35 (1995) 1231.
[11] G. Saibene et al., Influence of active pumping on density and confinement behaviour of JET plasmas, presented at the PSI 1996 conference, J. Nucl. Mat., in press.
[12] K. Itami, JT-60 Team, Plasma Phys. Control Fus. 37 (1995) A255.
[13] V. Mertens et al., Boundary and divertor physics in ASDEX Upgrade with emphasis on density limit characteristics, Proc. IAEA (1996), in press.
[14] A. Kallenbach, et al., Plasma Phys. Control Fus. 38 (1996) 2097.
[15] J. Ongena, et al., Nucl. Fus. 38 (1996) 279.
[16] H.-S. Bosch, et al., Phys. Rev. Lett. 76 (1996) 2499.
[17] H.-S. Bosch, D. Coster, R. Dux, C. Fuchs, G. Haas, A. Herriman, S. Hirsch, A. Kallenbach, J. Neuhauser, R. Schneider, J. Schweinzer, M. Weinlich, ASDEX Upgrade Team and NBI Team, Particle exhaust in radiative divertor experiments, presented at the PSI 1996 conference, J. Nucl. Mat., in press.
[18] W. Suttrop et al., The role of edge parameters for L–H transitions and ELM behaviour on ASDEX Upgrade, Europhysics Conference Abstracts, 20c, I (1996) 47.
[19] H. Zohm, Plasma Phys. Control Fus. 35 (1995) 869.
[20] G. Becker, Nucl. Fus. 35 (1995) 869.
[21] P.T. Lang, K. Büchl, M. Kaufmann, R.S. Lang, V. Mertens, H.W. Müller, J. Neuhauser, ASDEX Upgrade and NI Teams, High efficiency plasma refuelling by pellet injection from the magnetic high-field side into ASDEX Upgrade, Phys. Rev. Lett., in press.
[22] M.R. Wade, et al., Phys. Plasmas 2 (1995) 2357.
[23] G. Vlases, G. Corrigan, A. Taroni, Lightly- or Non-Seeded, Partially Detached ITER operation, PSI 1996 conference, J. Nucl. Mat., in press.

# Possible divertor solutions for a fusion reactor. Part 2. Technical aspects of a possible divertor

Ph. Chappuis *, F. Escourbiac, M. Lipa, R. Mitteau, J. Schlosser

Association Euratom-CEA, DRFC/STID, CE Cadarache, 13 108 St Paul Lez Durance, France

Abstract

Safe operation of a fusion reactor divertor requires the exhaust of heat fluxes ranging from 5 to 30 MW m^{-2}. The different technical solutions which are proposed rely on various water cooled copper heat sink designs. The protective armour may be Tungsten, Beryllium or Carbon depending on the plasma interaction. Optimisation is achieved by an overall comparison stressing the compliance to industrial defects. Reliable designs have been proposed and tested on large elements with carbon tiles for several thousand cycles, at power levels up to 10 MW m^{-2}. © 1997 Elsevier Science S.A.

1. Introduction

The heat flux management of a reactor divertor involves operation with a good safety margin. The recent progress in Tokamak plasma engineering has shown that a strong reduction (from 30 towards 5 MW m^{-2}) in the convective power can be expected in the divertor but with constraints on the plasma performances. The higher the heat flux a divertor can manage, the easier the plasma may be operated.

One has to suppose that in a fusion reactor the main operating scenario will be steady state operation at constant power and that the off normal events would only occur accidentally. If the gas box divertor design is still the reference solution, then the heat loads should be lower than 10 MW m^{-2} as for the ITER divertor. Removing the power under high heat fluxes in a nuclear environment will remain one of the most critical issues in the design of a fusion reactor. New plasma facing materials could eventually be developed in the future but it seems reasonable to count only on improvements of the ones available. Water remains the best coolant for high heat flux components if no conversion to energy is required for the divertor power. The best solution will be the one having the highest reliability for the lowest cost possible while respecting the design requirements. Designing and testing the ITER divertor components is certainly the best approach to appreciate the reliability of the technological solutions because industrial know-how acquired during this phase will be a predominant factor. The ITER design requirements for the divertor are to remove 400 MW with a peak heat flux of 5 MW m^{-2} on the vertical target, for 10 000 shots with a duration up to 1000 s. Off normal events

* Corresponding author.

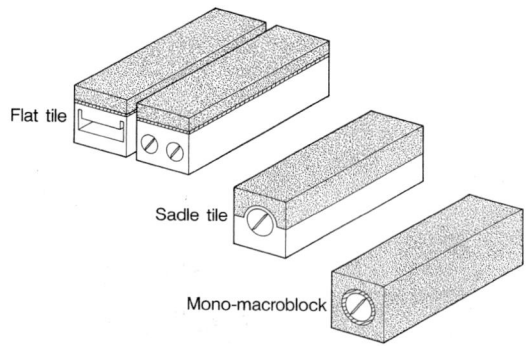

Fig. 1. Possible high heat flux components design options for the divertor cooling.

can increase this value up to 20 MW m^{-2}. Disruptions will dump energy in the range of 100 MJ m^{-2}. The power is removed through a water cooling circuit operating at 4 MPa and 140°C.

For a fusion reactor the design analysis will emphasize time dependent material properties (creep and fatigue) under irradiation.

Several plasma facing components designs are proposed for the ITER divertor. Each of them has specific advantages and extensive testing has shown that they could sustain the requirements.

The three main options depend on the bonding of the plasma facing material to the heat sink. A sketch of each is shown on Fig. 1.

The flat tile design is the most common and certainly the easiest to manufacture. The joining interface is flat and allows an easy fitting with rectangular heat sinks having different types of cooling concepts. Straight tubes with or without turbulence promoters (fins, swirl tapes, etc.) are the simplest cooling structures with reasonable performance. Hypervapotrons are more sophisticated options revealing improved but with higher thermo-hydraulical performance. The saddle tile design is an improvement of the proceeding one with the bonding interface closer to the coolant fluid and an externally finned heat sink which can distribute the power around the cooling channel.

The monoblock or macroblock design takes the most advantage of the carbon composites (CCs) thermomechanical properties (Table 2). The bonding interface is the circular cooling tube which gives the lowest possible temperature for the joint. The high thermal conductivity, the low elastic modulus and the low thermal expansion coefficient reduce drastically the thermomechanical stresses of this concept. This design allows for large massive blocks, having structural properties which reduce local peaking heat fluxes by spreading out the heat flow through the carbon blocks.

As an example (shown on Fig. 2) for a CC (SEP N11) macroblock under a 5 MW m^{-2} heat load, the typical maximum surface temperature would be 550°C and thermomechanical peeked equivalent strain can be as high as 0.8% in this copper tube.

Because of the long and powerful pulse operation of TORE-SUPRA, some of the ITER requirements have already been achieved on the plasma facing components (except neutron irradiation and tritium implantation). The experience acquired on steady state plasma facing component behavior during operation highlights the impor-

Table 1
Comparative properties of the two selected copper materials

	CuCrZr			Glidcop Al 25		
	20°C	400°C	600°C	20°C	400°C	600°C
Yield strength (MPa)	280	170		480	230	140
Ultimate strength (MPa)	400	230		460	250	150
Total elongation (%)	25	20		25	20	20
Heat conductivity (W m^{-1} K)	350	320		360	320	300
Weldability	Good			Medium		
Brasability	Medium (heat treatment)			Good (silver free)		
Hipping	Medium (heat treatment)			Good		

Table 2
Thermal conductivity comparison between different carbon composites values in W m^{-1}°C at room temperature

Manufacturer	Material	Conductivity X	Conductivity Y	Conductivity Z	Total X+Y+Z
Mistsubishi	MFC 1	590	40	40	670
SEP	N11	240	240	150	630
Toyo-tanso	CX 2002	420	330	210	960
SEP	NS 31	300	100	90	490
Dunlop	C2	370	110	90	570

tance of industrial reliability for the manufacturing of large scale elements [11]. Therefore, the design of new high heat flux components for this machine was governed by the choice of materials and technologies with special attention given to eventual defects and an improvement of the components reliability was attained by adding extensive Non Destructive Testing (NDT) at every step of the manufacturing process. As an example of a possible solution for the ITER divertor, the design and testing of the TORE-SUPRA high heat flux elements for the toroidal pumped limiter will be described when analyzing the different parts off the plasma facing component.

2. Heat sink

Various heat sink designs have been evaluated in different test beds under steady state high heat flux loading. The aim is always, for a given surface temperature (the lowest possible), to increase critical heat flux (CHF) margin at moderate water temperature and pressure with a minimum of pumping power. Most experience has been gained with three different solutions (Fig. 3): swirl tapes (thick or thin) inserted in straight tubes, hypervapotrons and externally finned tubes. Comparison of thermohydraulics results has been done by different authors [1–3] and a good agreement has not yet been reached. Schlosser et al. [3] have compared CHF values of smooth tubes with those of swirled tubes and hypervapotron for the same water flow and identical heated surface. Fig. 4 indicates that a constant ratio between the critical heat flux (CHF) and the Tong 75 estimation can be found for different subcooling and various velocities when a coefficient (C_α) depending on the geometry is introduced ([22]). The highest of these coefficients was found for the hypervapotron.

Nevertheless, this data-base shows that if the pumping power allowed for the divertor cooling is high enough to allow velocities in the range of 10 m s^{-1} and the subcooling is sufficient (in the range of 100°C), then any cooling design is satisfactory for incident heat fluxes below 20 MW m^{-2}. Therefore, for TORE-SUPRA (TS), the cooling margin for the High Heat Flux Element (HHFE) of the toroidal pumped limiter was proved sufficient with a straight tube solution (factor >2) thus allowing a simple geometry (peak heat flux = 6 MW m^{-2}, CHF = 15 MW m^{-2}).

The material choice for the heat sink is mainly dictated by the thermomechanical properties and up to now the best results are given by well identified hard copper alloys such as CuCrZr and dispersion strengthened copper (DSCu) as glidcop. Table 1 summarizes some of the main properties of these two materials (no cold work).

Both materials have conductivities close to pure copper but much higher strength at high temperature. The main advantage of DSCu is the conservation of its mechanical properties up to very high temperature. The advantage of CuCrZr is the wide experience in welding and the possibility of heat treatment to restore initial properties. In all cases the neutron irradiation will modify these properties.

The supporting structure and all the coolant connections are made with stainless steel which is one of the most reliable and well known structural materials. Therefore, a specific attachment has to

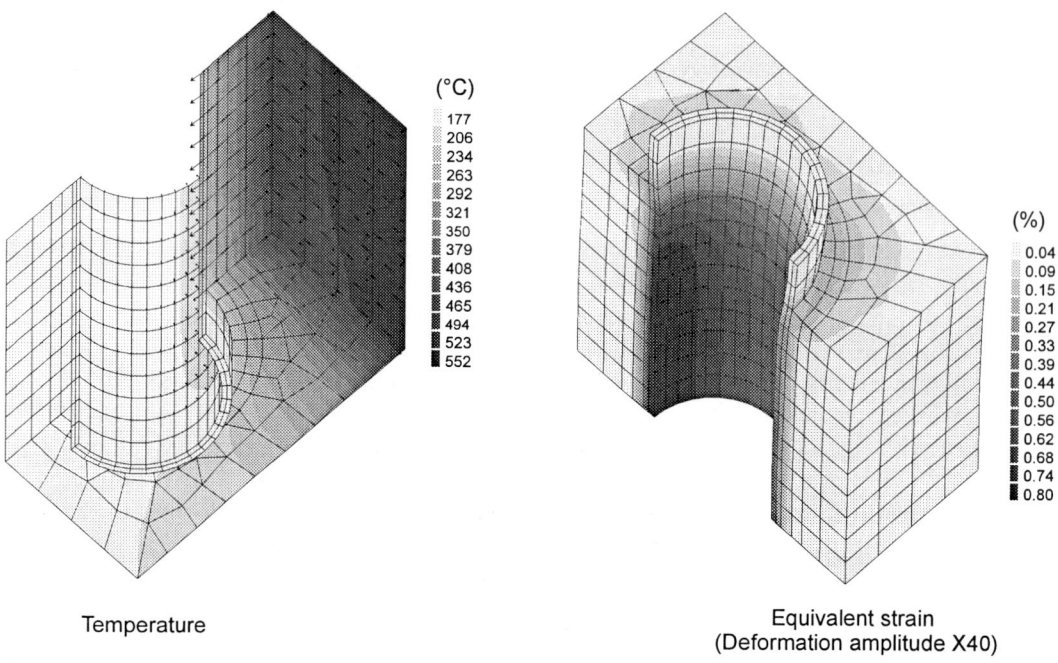

Fig. 2. Thermal and thermomechanical behaviour of a macrobloc design option under 5 MW m^{-2} heat flux.

be carried out between the copper heat sink and the stainless steel structure. The following technologies: hippping, brazing, explosion welding, friction welding and electron beam welding are under assessment. Brazing is the most well known technology but is not well adapted to CuCrZr which requires a specific heat treatment to restore the initial properties. Explosion welding is known to perform well with each material but very little characterization has been done. Friction welding was extensively used for TORE-SUPRA hydraulic connections but has recently shown a brittle behavior on some batches of material. Electron beam welding is well adapted to CuCrZr but not to DSCu because it contains Aluminum oxides and looses its properties after melting. Hipping is a promising technology mainly in the solid state solution at high temperature. With this technology, the whole component is heated to roughly 1000°C and an isostatic 100 MPa pressure is applied. Due to it's insensitiveness to high temperature, hipping is well adapted to DSCu but not for CuCrZr which will undergo a modification of it's properties at this temperature. After trading of the different solutions the TORE-SUPRa choice is justified by the larger diffusion of each metal in the interlayer and by the lowering of the stresses in the whole component. The manufacturing route for these heat sink elements has to prove a high reliability and the most robust design should be preferred. This reliability has to include the attachment of the plasma facing material and the mechanical supporting structure. As a example on TORE-SUPRA, for the toroidal pumped limiter, the choice was oriented towards straight tubes channels drilled in a CuCrZr heat sink mainly for the manufacturing easiness of a pressurized component (Fig. 5). This hardened copper is easy to machine, electron-beam welding is well known and the large ductility of this material enhances security. The dimensions have been optimized to reduce the eddy current forces, increase the pumping efficiency and accommodate the power load. The shape was optimized for the reduction of the heat flux, the flexibility of the plasma position and the simplicity of the element.

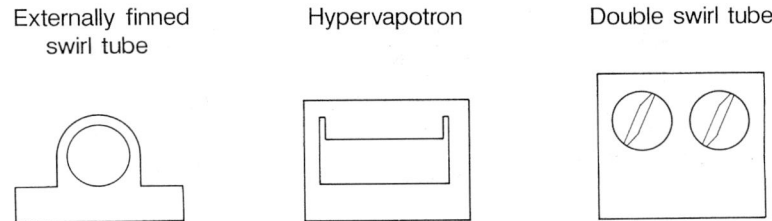

Fig. 3. Geometrical proposals for the divertor heat sinks.

3. Plasma facing materials

CCs and graphite's are the most common plasma facing material in all tokamaks. Results in the past few years from JET [4] and from ASDEX [5] give promising prospects for the use of beryllium and tungsten. These three materials have great potential and are candidates for the ITER divertor.

Carbon is a highly conductive, low Z refractory material which directly sublimates at high temperature and can therefore withstand unexpected off-normal events in a Tokamak. The continuous progress in carbon fibers enables manufacturers to produce materials with increasing conductivity.

These materials can have highly anisotropic properties depending on the fibers arrangements. Therefore, the plasma facing components can be optimized by making use of this anisotropy.

The saddle type tiles (which fit on the externally finned heat sink concept) are well adapted to 1D materials with the heat flux deposited on the tile corners preferentially directed to the foot of the heat sink therefore increasing the power distribution around the coolant tube, (Fig. 6). The monoblock design requires at least a 2D material to allow the heat to flow from the tile corners to the central coolant tube (Fig. 6).

The flat tile solution chosen for TORE-SUPRA high heat flux elements takes the advantage of a close to 3D material to increase the reliability given in case of peaked heat fluxes or when local defects could appear (Fig. 6). In such cases the more isotropic material allows for the reduction of hot spots by diverting the heat flow to the remainder of the tile. The choice of a specified CC grade for TORE-SUPRA was governed by the most isotropic material, this material has also to be available in large quantities with a good industrial reliability and an attractive price. Finally, the SEP N11 was selected.

Nevertheless, in ITER, the anisotropy of conductivity will be diminished by neutron irradiation with a stronger effect on the highest conductive fibers [6].

The thermomechanical properties of these materials are often the limiting factor, mainly for tile attachment to the heat sink. The large mismatch in thermal expansion coefficient (higher in the fibers plane or in the fibers direction) is always a source of high stress at the interface. This stress is partly accommodated by the use of compliant layers (for flat tiles) or by specific designs which reduce singularities (as for monobloc design). All destructive thermomechanical testing has shown that the failure always occurs close to the bond between the tile and the heat sink [7,8,10].

All the different designs have been fatigue tested in different test facilities with very good results: 300 cycles at 30 MW m^{-2} on the hypervapotron concept covered with MFC1 carbon tiles [13], 10 000 cycles at 5 MW m^{-2} for the saddle shape design [14] and 2500 cycles at 10 MW m^{-2} for a flat tiles concept on straight tubes [15].

Carbon is a good trap for hydrogen isotopes. This behavior is due to the large porosity and to the ease of hydrocarbon formation which increases the tritium retention and the material erosion. Recent progress by adding silicon to the material improves the plasma interaction behavior [9]. Therefore, carbon can still be considered as a promising material for the ITER divertor.

Beryllium is a low Z metal with good heat conductivity which showed very good potential in

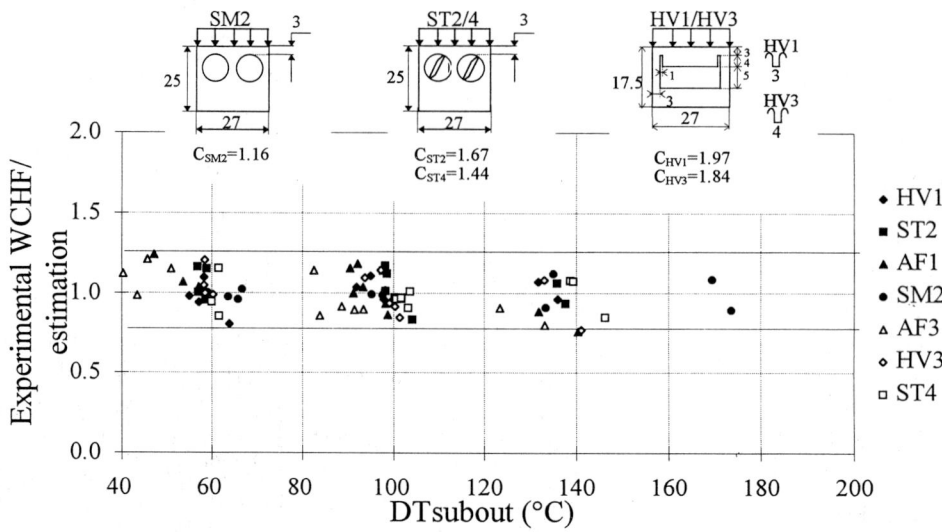

Fig. 4. Critical heat flux measurement on different heat sink geometries. Estimation = Tong 75 calculation × geometrical factor (C_α). AF1 and AF3 are annular flow geometries.

JET [4]. The material erosion by the plasma (for the ITER divertor conditions) is comparable to undoped carbon [21]. The low melting point (1250°C) is a strong limitation mainly when off-normal events (ELMS) produce high heat loads on the material. Joining of beryllium to the metallic heat sink is still a critical point, mainly due to the low ductility of the material and to the trend to produce fragile intermetallic compounds [10,20]. Nevertheless very encouraging results have been obtained with several thousand cycle at 5 MW m^{-2} on diffusion bonded tiles [10] and up to 12 MW m^{-2} for brazed tiles [1,20]. This material might be a serious candidate for areas, close to the main plasma where steady state moderate heat flux are foreseen.

Tungsten is a high Z refractory material with good thermal conductivity. It has a high sputtering threshold and is inert with regards to hydrogen isotopes. This material could be the future best choice for Tokamak plasma facing components if the plasma wall interaction was sufficiently well controlled (low Te) to ensure that no tungsten atoms could migrate in the main plasma. The ductility of this material is very low and the reliability of the bonding to the heat

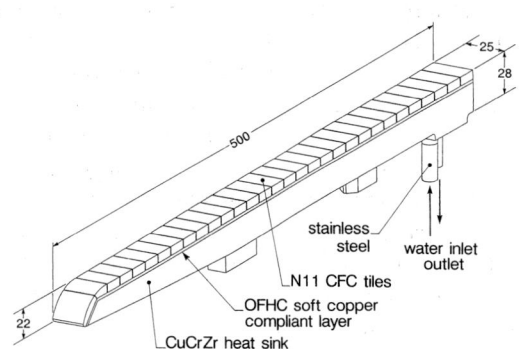

Fig. 5. High heat flux elementary component for the TORE-SUPRA toroidal pumped limiter.

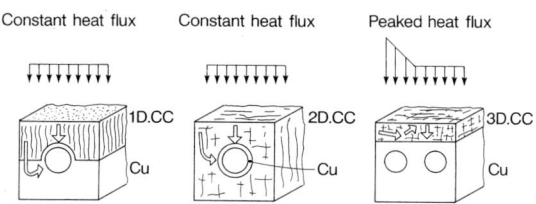

Fig. 6. Effect of carbon structure on the heat flow distribution for various design options.

sinks remains a critical issue. Recently, divertor mock-ups, including tungsten tiles have been fabricated and shall be heat flux tested on the FE200 test facility.

4. Attachment of the plasma facing material to the heat sink

The first aim of the plasma facing component is to remove the heat load in a steady state regime and under off normal events. In this field, a large experience has been gained in TORE-SUPRA [11] and one of the main issues was the reliability of the bonding between the plasma facing material and the heat sink. In the first generation of components the graphite armor material was bonded to stainless steel and copper heat sinks [25]. Little know how was available on this technology and on specific non-destructive testing. The degree of reliability was not achieved by the construction of small scale mock-ups but with the remanufacturing of large scale elements after many years of operation. Increasing local surface temperatures revealed the presence of a crack at the brazed joint, which would brutally propagate in the brittle graphite material until tile detachment. Disruptive heat loads often left a strong impact on the tiles, either by a change of the material coloration or by the tile rupture with no preexisting defect [25].

For the high heat flux elements of the toroidal pumped limiter, the competing technologies were brazing (silver free), Active Metal Casting (AMC) and Hipping. As Copper Chromium Zircon (CuCrZr) was chosen for the heat sink material, the assembling technology had to take into account the heat treatment requirements of this material. After testing, the best compromise was to chose AMC. This process has been developed by PLANSEE ([23]) and allows the direct creation of a compliant layer on the base of the armor tile by casting copper onto a laser treated surface. This technology gives progressive penetration of the copper in the CC which produces a strong hooking and a good transition of the

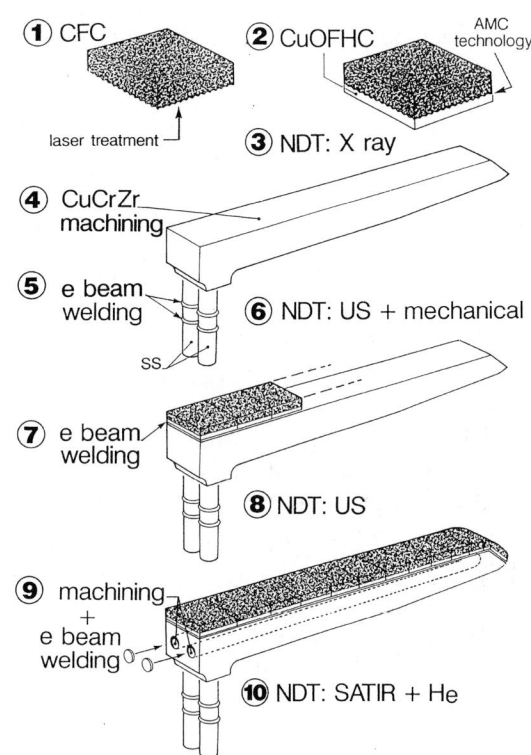

Fig. 7. Manufacturing route for the TORE-SUPRA toroidal pumped limiter high heat flux element.

properties. These elements can be individually tested (X-ray) before they are e-beam welded onto the copper hat sink. To ensure a high reliability every step of the process has to be controlled by an NDT (Fig. 7). An overall final test is done with an infrared imaging test-bed (SATIR) which enables the detection of defects with an equivalent radius larger than 6 mm.

A very aggressive NDT plan has also been applied on the new TORE-SUPRA inner first wall element (manufactured in 1994) where the brazing process has also a high reliability due to the brazed joint penetration in the laser indented armor material which allowed for a good transition to the compliant layer [24].

Hipping in the solid state is actually under assessment for the assembly of metallic materials for the ITER in-vessel components. Preliminary tests with the foreseen plasma facing materials have proven the possibility of bond creation. As no large industrial experience was available on this technology, hipping of carbon material to metallic material was not retained for the TORE-SUPRA HHFE. As this technology has a great potential for assembling other materials, small scale mockups, representative of the ITER divertor, have recently been manufactured with this technology and will be tested.

5. Tore-supra high heat flux experiment

As the first issue for plasma facing components is the reliability of the heat exhaust system, extensive heat flux testing is needed. The interaction with plasma operation for such components can be performed in a tokamak like TORE-SUPRA where long pulses operation is possible (up to 120 s) [12]. The majority of the high heat flux testing was carried out on carbon covered components [16–19]. Often, small scale component mock-ups were very promising while larger ones showed non-reproducible behavior [7], hot spots detected during screening test would quickly lead to a tile detachment or cause a strong local erosion. Critical heat flux was never reached on the correctly attached tiles [13]. Plasma operation was often interrupted by a limiting surface temperature on improperly aligned limiters [15] or by tile detachment [18]. Recently, the new size one high heat flux elements were extensively tested on the FE200 test facility and have proven very satisfactory. As these elements were optimized for a high industrial reliability the results where immediately of good quality on the whole element (500 mm long), no hot spots were detected. The surface temperature distributions were similar for all the HHFE tested. After 1000 cycles at 8 MW m^{-2} no evolution of the surface temperature was detected. Up to 2500 additional cycles were carried out at 10 MW m^{-2} with no tile detachment. A continuous and uniform increase of the tile surface temperature from 1500 to 1750°C was observed and imputed to a slow damaging of the CC material close to the interface. The results are given in Fig. 8.

6. Conclusions

The successful testing of large mock-ups and prototypical industrial high heat flux elements have demonstrated that different solutions exist for the steady state operation of the ITER divertor and therefore could be extrapolated to fusion reactors. Safe operation of such elements at heat flux levels of 10 MW m^{-2} is a first step towards relaxing the stresses on the plasma operating parameters. The experience acquired on TORE-SUPRA has shown that the main source of difficulties is the reliability of the bonding between the plasma facing material and the heat sink. Macroblock or saddle type design have shown a great potential due to the low temperature of the plasma facing material bonding interface and macroblock concepts seem to tolerate defects. The glidcop has very good properties and seems the future material for heat sinks if the assembling technology is assessed. As the plasma loading from a steady state reactor should be less subjected to unexpected off normal events the main source of damage should be the erosion of the plasma facing material, therefore Tungsten has good potential for erosion wear (if TE on the edge < 20 eV).

Nevertheless, the knowledge acquired by industrial companies during the manufacturing of the ITER divertor elements will give the base of reliable solutions for the high heat flux plasma facing components in a fusion reactor.

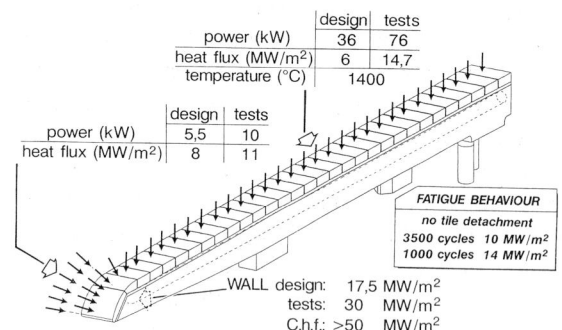

Fig. 8. Heat flux testing results on toroidal pumped limiter element.

References

[1] H.D. Falter, H. Altman, Ch. Baxi, G. Deschamps, R. Hemsworth, D. Martin, P. Massmann, High heat flux engineering in: Thermal test results of the JET Divertor plates, Proc. Soc. Phot Opt. Instrum. Eng. 1739 (1992) 162.

[2] M. Araki, K. Sato, S. Suzuki and M. Akiba, Critical heat flux experiment on the screw tube under one sided heating conditions, Fusion technology, Vol. 29, No. 4, July 1996.

[3] J. Schlosser, F. Escourbiac, J. Boscary, I. Smid and G. Vieder, Comparison between various thermal-hydraulical tube concepts for the ITER divertor, 19th SOFT, Lisbon 1996.

[4] D. Campbell et al., Experimental comparison of carbon and beryllium as divertor target materials in JET, 12th PSI, St Raphaël, 1996.

[5] R. Neu, K. Asmussen, S. Deschka, R. Dux, W. Engelhardt, J. Fuchs, C. Garcia-Rosales, A. Hermann, S. Hirch, F. Mast, A. Thoma, U. Wenzel, The tungstene experiment in ASDEX-upgrade, 12th PSI, St. Raphaël, 1996.

[6] J.P. Bonnal, C.H. Wu, Neutron induced thermal properties changes in carbon fiber composites irradiated from 600–1000°C, J. Nucl. Materials 230 (1996).

[7] Ph. Chappuis, A. Cardella, E. Di Pietro, M. Febvre and H. Viallet, Thermal heat flux testing of prototypical plasma facing element, 18th SOFT, Karlsrube, 1994.

[8] R. Mitteau, Ph. Chappuis, P. Deschamps, M. Febvre, J. Schlosser, H. Viallet and G. Vieder, Development and testing of CFC, copper high heat flux elements, Proc. 18th SOFT, Karlsruhe, 1994.

[9] C. Wu, C. Alessandrini, R. Moormann, M. Ruber, B. Scherzer, Evaluation of silicon doped CFCs for plasma facing material, J. N. M. 220-222 (1995) 860–864.

[10] D. Youchison, R. Guiniatouline, R. Watson, J. McDonald, D. Walsh, V. Beloturov, I. Mazul, A. Zakharov, B. Mills, D. Boehme and V. Savenko, Fusion Technology, Vol. 29, No. 4, July 1996.

[11] Towards long pulse, high performance discharges in TORE-SUPRA, experimental knowledge and technological developments for heat exhaust, Tore-Supra team, Fusion Technology, Vol. 29, No. 4, July 1996.

[12] Ph. Chappuis, Heat exhaust control with active cooling in TORE-SUPRA, towards steady state operation, Tore-Supra PFC group, 12th PSI, St. Raphaël, 1996.

[13] H. Falter, D. Ciric, A. Calentano, C. Ibbot, M. Watson, M. Araki, S. Suzuki and K. Sato, Vaporton as a heat sink for flat high conductivity unidirectional CFC tiles, Fusion Technology, Vol. 29, No. 4, July 1996, pp. 571.

[14] S. Suzuki, M. Araki, K. Sato, K. Nakamura and M. Akiba, High heat flux experiments on a saddle-shaped divertor Mock-up, to be published at IAEA, Montreal 1996.

[15] J. Schlosser, L. Doceul, Ph. Chappuis, M. Chatelier, J.P. Cocat, L. Garampon, P. Garin, R. Mitteau, A. Moal, L. Plöchl, G. Tonon and E. Tsitrone, Developments for an actively cooled toroidal pumped limiter for TORE-SUPRA, 19th SOFT, Lisbon 1996.

[16] V. Barabash, R. Giniatulin, V. Komarov, A. Gervash, Y. Prokofiev, Y. Krivchenkov and E. Privalova, Testing of the divertor Mock-ups under heat flux loading, 17th SOFT 92, Rome, Fusion Technology, 1992, Elsevier, New York, 1993.

[17] M. Araki, S. Suzuki, M. Akiba, M. Dairaku, R. Duwe, J. Linke, E. Wallura, K. Yokohama, I. Smid, M. Ogawa, S. Tanaka, Y. Ohara and K. Nakamura, Development and testing of divertor mock-ups for ITER/FER by JAERI, 17th SOFT, Rome, Fusion Technology, 1992, Elsevier, New York, 1993.

[18] H. Falter, D. Ciric, E. Deksnis, P. Masmann, K. Mellon, A. Peacock, M. Akiba, M. Araki, K. Soto, S. Suzuki and A. Cardela, 18th SOFT, Karlsrube, Fusion Technology, 1994, Elsevier, New York, 1995.

[19] T. Hino, M. Akiba, Summary of Japan-US workshop on high heat flux components and plasma surface interactions for next fusion devices, Fus. Eng. Design 31 (1996) 83–87.

[20] C. Ibbott, H. Falter, P. Meurer, E. Thompson, M. Watson, G. Critchlow, D. Ciric, Further developments in the brazing of Beryllium to CuCrZr, 18th SOFT, Karlsruhe, Fusion Technology, 1994, Elsevier, New York, 1995.

[21] C. Wu, U. Mszanowski, A comparison of lifetimes of beryllium, carbon, molybdenum and tungsten as divertor armour materials, J. Nucl. Mater. 218 (1995) 293–301.

[22] L.S. Tong, A phenomenological study of critical heat flux, ASME paper 75-HT-54, 1975.

[23] H. Huber, L. Plöchl, N. Reheis, J.P. Cocat and J. Schlosser, The manufacturing and testing of the toroidal pumped limiter prototype element for TORE-SUPRA, IEEE/SOFE, Champagne, 1995.

[24] M. Lipa, Ph. Chappuis, G. Chaumat, D. Guilhem, R. Mitteau and L. Plöechl, Development and fabrication of improved CFC-brazed components for the inner first wall of TORE-SUPRA; 19th SOFT, Lisbon, 1996.

[25] M. Lipa, P. Chappuis, P. Deschamps, Brazed graphite for actively cooled plasma-facing components in TORE-SUPRA, description, tests and performances, Fus. Technol. 19 (1991) 2041.

Engineering experience in JET operations

Enzo Bertolini

JET Joint Undertaking, Abingdon, Oxon, OX14 3EA, UK

Abstract

The inherent flexibility of JET's original concept has permitted several engineering upgradings and modifications, to address a large variety of plasma and fusion physics issues. The most recent major modification has been the installation of an axisymmetric single-null pumped divertor (Mark I), successfully operated in the experimental period 1994–1995. Following the divertor optimization programme a new, more closed, divertor configuration has now been installed (Mark II), which has shown a better power handling capability and substantially improved neutral particle retention. A key feature of the new design is the possibility to replace the divertor target plate structure using full remote handling techniques following extended D-T operations. Toroidal asymmetries of vessel forces due to Vertical Displacement Events (VDE) and halo currents were experienced since 1994, leading in some cases to sideways movements of the vessel of 7 mm. This has required modification and upgrading of the vacuum vessel support system. Gap control of plasma position and shape and machine protection systems have been developed further, leading to increased experimental availability. Future development foresees the installation of a Mark II Gas Box divertor structure, while studies are underway to increase the toroidal field capability from 3.45 to 4 T and the additional heating power by increasing the NB injector output from 80 kV, 60 A to 120 kV, 60 A and by using wide band matching for ICRF. © 1997 Elsevier Science S.A.

1. Introduction

1.1. JET objectives and design concept

The global objective of JET is to obtain and study a plasma in dimensions and physics parameters approaching those needed in a thermonuclear reactor, thus providing necessary physics and engineering data for the design and for the operation of a next step device (such as ITER), eventually using D-T gas mixtures.

Since its first operation in June 1983, the configuration of the JET plasma has been modified to allow new physics issues to be studied and its performance has been steadily increased to well above the original technical specifications, culminating in the first ever D-T experiment (PTE, Preliminary Tritium Experiment) in November 1991, with the delivery of 1.7 MW of fusion power and 2 MJ of fusion energy.

During the years, the machine went through a number of modifications and upgradings, suggested by physics results and considered necessary for further progress to be achieved. This engineering development could be implemented without replacing any of the major tokamak components (vacuum vessel, toroidal field coils, poloidal field coils, mechanical structure), thanks to the original design philosophy of JET: large plasma volume (~ 150 m^3) and current (4.8 MA), relatively mod-

0920-3796/97/$17.00 © 1997 Elsevier Science S.A. All rights reserved.
PII S0920-3796(97)00017-3

est toroidal magnetic field (3.45 T), D-shaped plasma, vacuum vessel and toroidal coils, low aspect ratio, achieved by transferring the inward force of the toroidal magnet through the central solenoid to the inner supporting ring. Consequences of these choices are long operating pulses (up to 20 s at full performance), limited stresses in electro-mechanical components and remote handling compatibility. Moreover the main diagnostics are installed beyond the shielding walls and most power and energy is taken directly from the HV grid [1,2].

1.2. Machine development and results

Concerning the upgrading in performance (fusion triple product $n_D \tau_E T_D$, equivalent energy gain Q_{DT} and dilution factor deuteron density/electron density n_D/n_e), there were two main limiting factors: the decay of the energy confinement time τ_E with plasma heating (Fig. 1) [3,4] and the influx of impurities from the vacuum vessel walls. The first problem was counteracted by increasing the plasma current above the design value (from 4.8 to 7.0 MA) and by modifying the poloidal coil configuration to achieve X-point plasmas up to and above 5 MA. This major upgrading of the electro-mechanical system has required JET to modify and enhance the capability of the poloidal coil power supplies, including plasma control and to re-assess the ultimate electro-mechanical capability of the tokamak system, by finite-element modelling of the vacuum vessel, toroidal and poloidal coils and mechanical structure and by load testing of the prototype toroidal coil well above the design values (by a factor of ~ 2). The second limiting factor was first dealt with by using low 'Z' materials as plasma facing components, i.e. carbon and/or beryllium, moving from inconel vessel walls and eight (small) discrete graphite limiters to carbon toroidal belt limiters and a 50% covering of the inconel wall with graphite tiles and finally by using beryllium tiles for one of the belt limiter and for the X-point target plates. This wall structure was supplemented by wall carbonisation first and by beryllium evaporation later (performed in between daily operating sessions).

By progressively implementing all the above modifications, the triple fusion product $n_D \tau_E T_D$ went from 0.12×10^{21} m^{-3} skeV (1983) to 0.9×10^{21} m^{-3} skeV (1991) and the energy gain Q_{DT} equivalent increased from 0.01 to 1.07. Other significant performance achievements were a dilution factor $n_D/n_e \geq 0.9$ and $Z_{eff} \leq 2$ and, in separate pulses, an ion density $n_D = 4 \times 10^{20}$, an ion temperature $T_D = 30$ keV and an energy confinement time of $\tau_E = 1.8$ s. This level of global performance was considered sufficiently attractive to perform the first ever controlled thermonuclear fusion D-T experiment, using a mixture of 11% T and 89% D, producing a peak fusion power of 1.7 MW (more than 50% by thermalized reactions) and approximately 2 MJ of fusion energy.

These results provided a clear indication that a D-shaped machine, with high plasma current and large plasma volume, elongated plasma cross section with X-point configuration and low 'Z' materials for plasma facing components is, at present, the most promising concept for a fusion reactor (see ITER) [5].

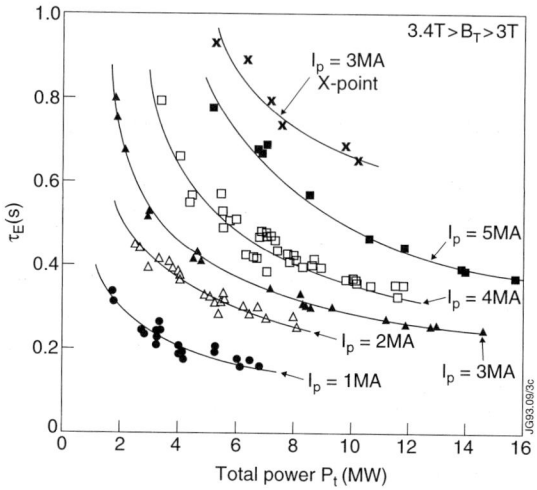

Fig. 1. Behaviour of the energy confinement time τ_E with total heating power P_t in limiter (L-mode) and in X-point (H-mode) configurations.

2. The divertor programme

However, a passive control of impurities is not sufficient to maintain high performance for more than 1 s, due to the combination of MHD instabilities and excessive accumulation of impurities in the X-point region. Therefore, a tokamak reactor requires an active control of the impurity influx into the plasma, because they increase the plasma radiation losses and dilute the fuel. The only known way to implement an active impurity control in a tokamak, is to equip the machine with a pumped divertor, which controls impurity levels, particle and energy exhaust and enhances plasma energy confinement ('H-mode').

The extension of the JET Joint Undertaking from 1992 to 1996, has been specifically granted to study the behaviour of a tokamak plasma with an axisymmetric pumped divertor (Mark I). The recent extension to 1999 has been mainly, but not solely, motivated by the need to tailor the divertor configuration to the ITER requirements. Therefore the divertor programme (Mark I, Mark II, Mark II Gas Box) became the focus of JET physics and engineering activities (Fig. 2). Addressing the divertor issue implied a further major upgrading of the JET machine, involving installation of new poloidal coils, power supplies, ICRF antennae, diagnostics and a completely new configuration of the vacuum vessel first wall [6,7].

3. Mark I divertor

3.1. Design, manufacture and installation

The main components of the divertor are (see Fig. 2):

(a) Four copper poloidal coils, Freon cooled, installed at the bottom inside the vacuum vessel allow magnetic configurations with X-point at a suitable distance from the target plates to be created and are supplied by four AC/DC converters (PDFA), with X-point sweeping capability.

(b) The target plates, arranged in a W-shaped contour, support the CFC tiles and collect the power released from the plasma.

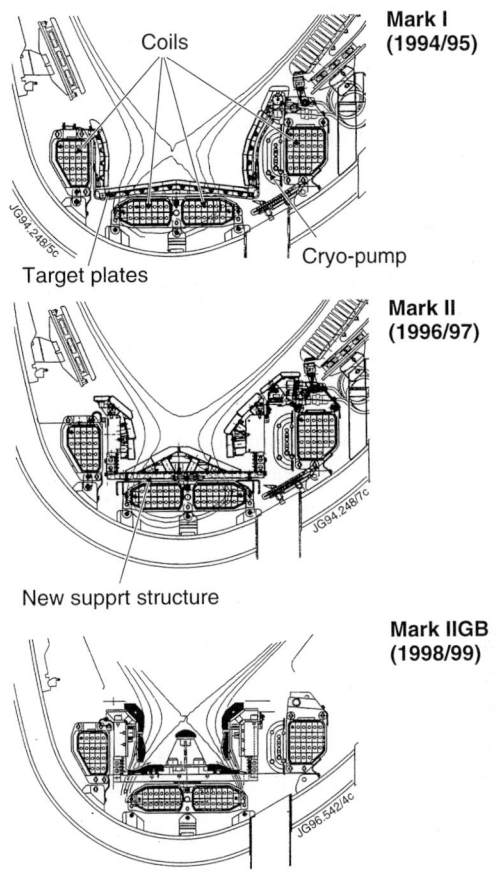

Fig. 2. Sequence of JET divertor configurations with progressively more closed divertors.

(c) The toroidal cryopump, anchored to the outer divertor coil, permits the plasma density in the divertor region to be controlled, when cold gas is injected to minimise impurity release and influx.

The new first wall assembly includes four pairs of ion cyclotron resonance frequency (ICRF) antennae designed to match the divertor plasma shape, a re-shaped and re-positioned lower hybrid current drive (LHCD) launcher, 12 discrete poloidal limiters on the outer wall and 16 inner wall guard limiters, all covered with graphite tiles. In addition two sets of four saddle coils installed at the top and at the bottom of the vessel should allow the control of $m = 2$, $n = 1$ MHD modes to be studied.

The refurbishing of the first wall has also required JET to modify existing diagnostics and to install new ones to measure divertor plasma parameters.

The divertor plasmas are more vertically unstable (open loop instability growth-rate $\gamma \sim 600$–1000 s^{-1}), due to the increase of the average plasma to wall distance, which reduces the passive stabilization effect of the vessel walls and to the higher quadrupolar field associated with the divertor plasma magnetic configuration. A new set of fast radial field amplifiers (FRFA), 5 kV, 5 kA with a 2 ms of response time over the full voltage range (-10 kV, $+10$ kV) was then provided.

The strategy of plasma control had to be completely re-assessed, leading to a new plasma position and current control system (PPCC), which allows control of the plasma boundary in real time, using 'intelligent' software for a selective control of plasma gaps and/or poloidal coil currents [9].

Finally the new machine configuration with greatly enhanced electromagnetic asymmetry relative to the equatorial plane (due to the bottom single null divertor), required a new coil protection system (CPS), which, using the same technology as PPCC, prevents the coils from been operated outside allowed limits (current, voltage, thermal and mechanical stresses) and detects electrical faults.

The frequent modifications and upgrading of the JET machine through the years have resulted in an adjustment of the basic machine parameters [6]; in particular the installation of the divertor structure inside the vessel led to a substantial reduction of plasma volume (by $\sim 20\%$).

3.2. Basic physics results [8]

The care taken in the design and in the installation of the CFC divertor tiles, the successful use of the cryopump and X-point sweeping have been instrumental in achieving longer, cleaner and more stationary H-modes (~ 20 s with $Z_{eff} \sim 1$). With 32 MW of combined NB and RF heating, greater than 180 MJ were injected without any sign of 'carbon bloom', while without the divertor only 15 MJ of injected energy would lead to a 'carbon bloom'.

It is worth noting that, in spite of the reduction of the plasma volume of approximately 20%, global performance almost identical to the one obtained without the divertor (and with a larger plasma volume) was obtained: QDT equivalent ~ 1, $n_D \tau_E T_D \sim 1.0 \times 10^{21}$ m^{-3} skeV and the record value of the reaction parameter $R_{DD} = 9.4 \times 10^{16}$ reactions s^{-1}. Reliable plasma operation up to 6 MA was achieved with the record plasma stored energy of 13 MJ.

An important function of a divertor for a fusion reactor is to limit the energy deposited in the target plate, which can be achieved by spreading the energy over a larger area by enhancing radiation. By injecting a mixture of deuterium and nitrogen (or neon) in JET it was shown that radiation accounts for up to 80% of the energy released by the plasma (Fig. 3).

Finally, two specific experiments requested by ITER were performed: the first to assess the effect of toroidal field ripple on plasma behaviour and performance, proving that the ripple produced in ITER with 20 toroidal coils should be acceptable and the second to compare beryllium and CFC as divertor target plate material, showing very similar plasma performance, although beryllium melting could not be avoided.

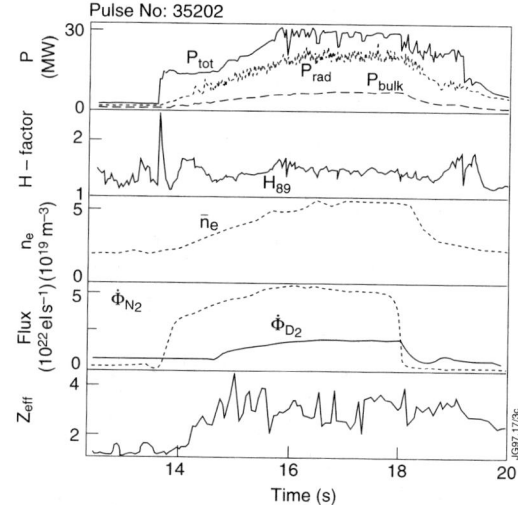

Fig. 3. Radiative divertor H-mode plasmas with high power combined heating, showing up to 80% of radiated power.

Fig. 16. Cuts of TF coil 3.1 for visual inspection and for extracting samples for insulation shear stress testing.

provements implemented in the intervening period, such as magnetic shear optimization, upgrading of the toroidal magnetic field and of both NB and ICRF systems and of the experience gained during DTE1. The study of α-particle heating will then be a major objective of the experiment.

7.2. Machine upgrading

The most straightforward way to increase global fusion performance is to upgrade machine parameters and additional heating power. In the late eighties JET has increased plasma current capability from the design value of 4.8 to 7 MA in limiter mode and more recently, to 6.0 MA in H-mode divertor plasmas. Extensive studies are now being performed to assess the possibility at increasing the toroidal magnetic field from 3.45 to 4.0 T [22]. These studies are conducted on three levels:

(a) FE stress analysis has shown that, for a variety of high performance scenarios, the shear stresses on the interturn insulation and tension on the copper brazed joint are well within the design capability of the coils;

(b) Evaluation of the manufacture documentation, including tests on insulation and on brazed joint samples and on the prototype coil, indicate an acceptable margin of safety (about a factor 1.5) at 4.0 T;

(c) Between 1989 and 1991, three TF coils developed interturn electric short-circuits due to water leaks. The water coolant was replaced by Freon and no more faults became apparent. It has been decided to use these coils to perform further tests. One of the coils has been cut for inner visual inspection and to extract samples, now under testing for shear stress in the interturn insulation (Fig. 16). Deflection tests will also be performed on a spare coil and on one of the (electrically) faulty coils.

When all these tests will be completed the JET governing bodies will consider allowing operations above 3.45 T. To avoid any delay, the order for the upgrading of the TF power supplies has been placed already.

Experiments conducted in the Neutral Beam test bed using a prototype modified accelerator structure have demonstrated the capability to increase the ion beam current from the present 30 A to 60 A at 140 kV [23]. While minor modification would be required by the NB injectors, additional power supplies (PS) are required, since the present PS are for 60 A at 80 kV (or for 30 A at 160 kV). Taking into account cost considerations it has been decided to consider an upgrading of the power supplies to 60 A at 120 kV for one box (with an increase of ~7 MW of beam delivered power). The design is based on AC/DC converters and on IGBTs as switching and regulating elements instead of tetrodes, used in the existing power supplies [24].

Studies have also been initiated on a power upgrade of the ICRF heating system, by using wide band matching [25].

8. Summary and conclusions

- The key results of the first 10 years of JET operation (without divertor) have shown, that large plasma volumes, high plasma currents and X-point magnetic configurations have been essential features for a fusion reactor, such as ITER.
- This has been possible, because the inherent flexibility and design margins built into the original JET design have permitted extension of machine engineering performance well beyond the design parameters.
- The PTE1 fusion experiment, performed in 1991, led to the first production of controlled thermonuclear power up to 1.7 MW for about 1 s, showing however that passive control of impurities with carbon and/or beryllium as first wall material is not sufficient to control the impurity influx into the plasma.
- With the extension of the JET Joint Undertaking to the end of 1999, the JET programme can now address in a comprehensive way the active control of impurity issue, testing different divertor configurations (Mark I, Mark II and Mark IIGB).
- The experiments conducted so far with Mark II have shown the expected performance improvement compared with Mark I, namely a much improved power handling capability and a better retention of neutrals in the divertor region.
- The key physics objective during the next few months is to optimize plasma configuration leading to high plasma performance, so that DTE1 can meet its goal of high Q_{DT} and high neutron yield.
- The technology of remote handling for in-vessel installation of the Mark IIGB structure under active conditions has been fully proved in the IVTF and in the installation of the Mark II divertor target plates.
- The AGHS is now operational and final commissioning with 3 g of tritium is underway, while the infrastructure for an extended D-T operation is in place.
- New physics scenarios, such as magnetic shear optimization, and engineering upgrading such as the increase of the toroidal magnetic field to 4.0 T and the increase of additional heating power of both neutral beam and radio-frequency by several MW, will allow JET to enter in depth into the α-particle physics of a burning plasma with DTE2, planned for the second part of 1999.
- The damaging or potentially damaging effects of plasma behaviour on the structural components of the machine following VDEs, are to be considered a major problem to be addressed in present divertor tokamaks and in ITER.

Acknowledgements

This paper briefly summarises the work recently performed by the whole JET Team. However I wish to specially thank those colleagues, who have provided me with written material and have discussed with me many technical and scientific issues, namely: H. Altmann, Ph. Andrew, T.

Bonicelli, M. Buzio, R. Claesen, K. Fullard, R. Giannella, C. Gormezano, H. Hemmerich, L. Horton, J. Jacquinot, T. Jones, A. Kaye, B. Keen, P. Kupschus, R. Laesser, J. Last, C. Lowry, P. Lomas, V. Marchese, F. Marcus, F. Milani, P. Miele, P. Noll, M. Pick, T. Raimondi, V. Riccardo, A. Rolfe, A. Santagiustina, F. Soeldner, B. Tubbing, T. Wade.

References

[1] The JET Project-Design Proposal, Report of the Commission of the European Communities, EUR-5516-e (1976).
[2] M. Huguet et al, The JET machine: design, construction and operation of major subsystems, Fus. Tech. 11 (1987) 43.
[3] A. Tanga et al, Experimental studies in JET with magnetic separatix configuration, Nucl. Fus. 1 (1987) 65.
[4] M. Keilhacker and the JET Team, The JET H-mode at high current and power levels, Nucl. Fus. 1 (1989) 159.
[5] E. Bertolini (for the JET Team), Impact of JET experimental results and engineering development on the definition of the ITER design concept, Fus. Eng. Design 27 (1995) 27–38.
[6] E. Bertolini (for the JET Team), JET with a pumped divertor: design, construction, commissioning and first operation, Fus. Eng. Design 30 (1995) 53–66.
[7] M. Pick (for the JET Team), JET's latest technical and scientific results and engineering development, Proceedings of the 12th Topical meeting on Technology of Fusion Energy, Reno, 16–20 June 1996, Fusion Technology, in press.
[8] M. Keilhacker (for the JET Team), JET results with the new pumped divertor and implications for ITER Proceedings of the 22nd EPS, Bournemouth, UK, 3–7 July 1995, European Physical Society.
[9] J. Jacquinot (for the JET Team), Features of JET plasma behaviour in two different divertor configurations, presented to the 16th IAEA Fusion Energy Conference, Montreal (Canada) 7–11 October 1996.
[10] A. Rolfe, Remote handling on fusion experiments, Proceedings of the 19th SOFT, Lisbon, 16–20 September 1996, Elsevier Science, 1997.
[11] R. Laesser et al., Commissioning tests and enhancements to the JET active gas handling system, Proceedings of the 19th SOFT, Lisbon, 16–20 September 1996, Elsevier Science, 1997.
[12] S. Puppin et al., Real time control of plasma boundary in JET, Proceedings of the 19th SOFT, Lisbon, 16–20 September 1996, Elsevier Science, 1997.
[13] T. Bonicelli et al., The harmonic power filter of the JET poloidal divertor field amplifiers, Proceedings of the 19th SOFT, Lisbon, 16–20 September 1996, Elsevier Science, 1997.
[14] V. Marchese et al., Enhancement of JET machine instrumentation and coil protection system, Proceedings of the 19th SOFT, Lisbon, 16–20 September 1996, Elsevier Science, 1997.
[15] N.H. Zornig et al., Experimental results using the JET real time power control system, Proceedings of the 19th SOFT, Lisbon, 16–20 September 1996, Elsevier Science, 1997.
[16] P. Stott, New diagnostics techniques, Proceedings of the 19th SOFT, Lisbon, 16–20 September 1996, Elsevier Science, 1997.
[17] D. Summers et al., Edge density profiles in high-performance JET plasmas, Proceedings of the 12th International Conference on Plasma Surface Interactions in Controlled Fusion Devices, St. Raphael, France, May 1996.
[18] M. Buzio et al., Axisymmetric and non-axisymmetric structural effects of disruption-induced electro-mechanical forces on the JET tokamak, Proceedings of the 19th SOFT, Lisbon, 16–20 September 1996, Elsevier Science, 1997.
[19] P. Noll et al., Present understanding of electromagnetic behaviour during disruptions in JET, Proceedings of the 19th SOFT, Lisbon, 16–20 September 1996, Elsevier Science, 1997.
[20] A. Miller and J.L. Hemmerich, Enhancement of the JET vacuum vessels supports, Proceedings of the 19th SOFT, Lisbon, 16–20 September 1996, Elsevier Science, 1997.
[21] H. Altman et al., The Mark II gas box divertor-A new design concept, Proceedings of the 19th SOFT, Lisbon, 16–20 September 1996, Elsevier Science, 1997.
[22] J.R. Last et al., Upgrading the toroidal field to exceed 3.45 Tesla, Proceedings of the 19th SOFT, Lisbon, 16–20 September 1996, Elsevier Science, 1997.
[23] D. Ciric et al., Beam properties of the enhanced JET PINIs, Proceedings of the 19th SOFT, Lisbon, 16–20 September 1996, Elsevier Science, 1997.
[24] F. Jensen et al., Upgrading of the neutral beam power supplies from 80 kV/60 A to 140 kV/60 A, Proceedings of the 19th SOFT, Lisbon, 16–20 September 1996, Elsevier Science, 1997.
[25] M. Schmid et al., Operation of high power ICRH with ELMy plasmas at JET, Proceedings of the 19th SOFT, Lisbon, 16–20 September 1996, Elsevier Science, 1997.

D-T operation on TFTR

Michael D. Williams

Princeton Plasma Physics Laboratory, P.O. Box 451, Princeton, NJ 08 543, USA

Abstract

The Tokamak Fusion Test Reactor (TFTR) has operated safely and routinely with tritium fuel since November 1993. Experiments conducted during this time have demonstrated improved plasma confinement properties, the first measurements of alpha heating and transport and the ICRF heating and current drive of D-T plasmas. Reactor level fusion power densities up to 2.8 MW m^{-3} have been attained as well as more than 10 MW of fusion power and 6.5 MJ of fusion energy per pulse. In support of these accomplishments, the TFTR equipment has met or exceeded the original design values. The toroidal magnetic field has been upgraded from 5.2 to 6.0 Tesla (on-axis), the neutral beams have injected up to 40 MW with a combination of deuterium and tritium fueling and the ICRF systems have been operated at various frequencies, from 30 to 64 MHz, to support an assortment of fast wave experiments. From November 1993 through July 1996, there have been 19 725 plasma shots. These include 841 deuterium-tritium pulses which have produced a total of 4.8×10^{20} D-T neutrons and 1.4 GJ of fusion energy. More than 864 kCi (almost 90 g) of tritium have been processed through the TFTR facility while constrained to a site limit ('in-process') of 50 kCi. A low tritium inventory cryogenic distillation system has been recently installed and tested to provide on-site repurification of the plasma exhaust. Maintenance and repair activities have become routine in this facility. Hundreds of calibrations, repairs and installations requiring work on tritium contaminated and activated systems have been performed. Utilizing As Low As Reasonably Achievable (ALARA) principles, the radiation exposure to workers has been comparable to that experienced during the previous deuterium-only operating phase. The total annual site boundary dose, due to all pathways (prompt neutron radiation, activated air and controlled tritium release), has been less than 3 μSv, or 3% of the design limit of 100 μSv and 0.3% of the regulatory limit of 1 mSv. © 1997 Elsevier Science S.A.

1. Introduction

In 1976, the TFTR Technical Requirements Document described the purposes of the TFTR Project as: (1) to demonstrate fusion energy production from the burning, on a pulsed basis, of deuterium and tritium in a magnetically confined toroidal plasma system; (2) to study the plasma physics of large tokamaks; and (3) to gain experience in the solution of engineering problems associated with large fusion systems that approach the size of planned experimental reactors. These purposes can be satisfied by the production of 1–10 MJ of thermonuclear energy (per pulse) in a deuterium-tritium (D-T) tokamak with neutral beam injection under plasma conditions approximating those of an experimental fusion power reactor [1].

First operation of TFTR began in December 1982. In December 1993, experiments employing a

50/50 mixture of deuterium and tritium were conducted in a tokamak for the first time. As this paper describes, the results of these and subsequent experiments have met or exceeded the original TFTR design objectives.

2. D-T program

The TFTR D-T Program was structured to address four principal elements: (1) heating and confinement of D-T plasmas; (2) effects of alpha particles; (3) D-T technical capability and (4) D-T power production (see Fig. 1).

The primary technology objectives include tritium handling techniques and retention studies, operation and maintenance of a tritium contaminated and activated machine, operation of sensitive diagnostics in the D-T neutron environment and fusion power production greater than 10 MW.

The physics objectives include transport studies in D-T plasmas, study of isotope effects, ICRF heating of D-T plasmas, study of single alpha particle effects, alpha driven instabilities, helium ash transport and indications of alpha heating.

3. D-T facility

With the introduction of tritium, TFTR has been classified by the Department Of Energy (DOE) as a Category 3 (Low Hazard) Nuclear Facility. This classification brought the rigors of Nuclear Conduct of Operations, extensive review and oversight by the DOE and required more than 2 years of D-T preparation activities. Appropriate use of strict management systems and detailed procedures have ultimately contributed to safe operations and compliance with all applicable regulations.

Since the start of D-T operations in December 1993, 864 kCi of tritium have been safely processed through the TFTR facility. Tritium can be injected into TFTR through several gas injectors located around the tokamak or by any of the twelve neutral beam ion sources. (Plans for a tritium pellet injector, which were abandoned for

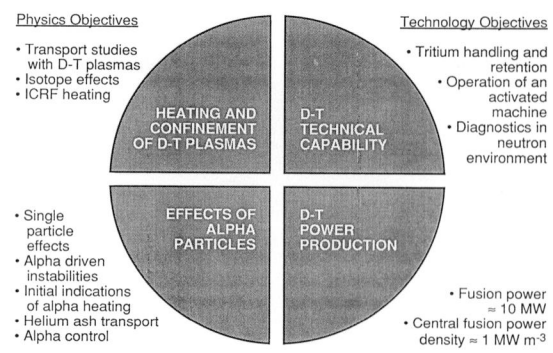

Fig. 1. TFTR D-T program objectives.

financial reasons during the D-T preparation phase, are presently under reconsideration.) Careful accounting of the tritium usage indicates that approximately 2% of the tritium used by the neutral beam and gas puffing systems is held up in the limiters, vacuum vessel and associated piping. The total quantity of tritium allowed in the TFTR vacuum vessel is restricted to a maximum of 20 kCi. This restriction limits the release to the environment in the event of a major vacuum leak and simultaneous failure of the tritium containment systems. The accumulated tritium retained in the TFTR vacuum vessel has been shown to be reduced from 16 to 8 kCi by glow discharge cleaning with deuterium and He–O$_2$ mixtures followed by a moist air purge [2,3].

Operations are conducted with an in-process tritium limit of 50 kCi of which only 25 kCi is 'at-risk' due to any credible failure mechanism. This inventory limit serves to reduce the accident

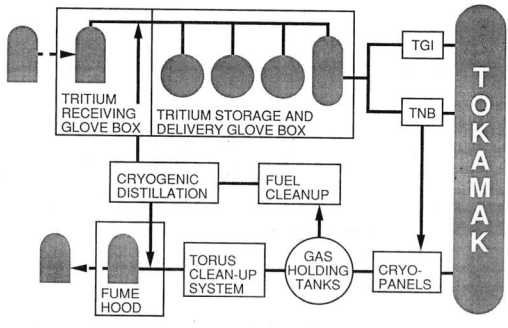

Fig. 2. TFTR tritium process flow.

Table 1
Principal TFTR engineering parameters

	Design	Achieved
Major radius (R (m))	2.45	2.6
Minor radius (a (m))	0.85	0.9
Toroidal field (B_t (T))	5.2	6.0
Plasma current (I_p (MA))	3	3
Heating power (P_{nb} (MW))	33	40
Heating power (P_{rf} (MW))	12	11
Heating pulse length (s)	1.5	2.0

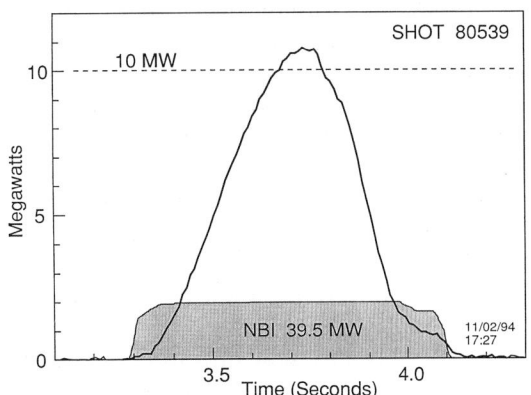

Fig. 4. TFTR 10.7 MW plasma shot.

potential and exposure to on-site personnel as well as the general public. Tritium is received from the Savannah River Site (SRS) in approved shipping containers in typical quantities of 5–20 kCi. The tritium gas is then locally stored on cold uranium beds. These beds are heated when required to transfer the gas to the neutral beams or tokamak gas injectors. The tritium from the plasma exhaust is oxidized, stored on disposable molecular sieve beds and shipped off-site in Type A containers (< 1000 Ci) for burial at Hanford, Washington or Type B containers (typically 5–20 kCi) for reprocessing at Savannah River [4]. (See Fig. 2 for a schematic representation of the TFTR tritium process flow.)

In order to minimize shipments to and from PPPL, an on-site tritium purification system (TPS) has been commissioned. This system, designed and built by the Canadian Fusion Fuels Technology Project (CFFTP), employs a low-inventory cryogenic distillation process [5]. Recently, 2500 Ci of tritium have been processed for the first time from the tokamak plasma exhaust and delivered to the uranium storage beds for use in future TFTR experiments.

There have been hundreds of maintenance activities and repairs requiring work on tritium contaminated and/or activated systems. For example, all 12 neutral beam ion sources have been removed, refurbished and reinstalled. Utilizing As Low As Reasonably Achievable (ALARA) practices, the average worker radiological dose has been maintained at levels not exceeding those experienced during deuterium operations. 'Linebreak' procedures and work permits which control worker stay time and steps to control tritium releases are used extensively when conducting work in critical areas.

The annual site boundary radiological dose has been limited to less than 3 μSv. (Portions of the site boundary are approximately 100 m from the TFTR Test Cell). This level is only 3% of the design limit of 100 μSv and 0.3% of the regulatory limit of 1 mSv. This better than expected performance is because the number and magnitude of planned and unplanned tritium releases have been significantly less than anticipated and the performance of the neutron shielding was more effective than theoretical calculations predicted [6].

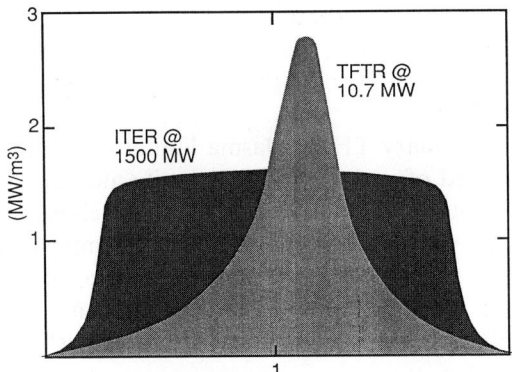

Fig. 3. TFTR fusion power density.

All of the TFTR equipment has been operating with high availability and reliability. Due to strategic relocations, improved local shielding and modifications for tritium compatibility, the plasma diagnostics have operated successfully in the intense 14 MeV neutron environment. The D-T experiments have also provided an opportunity to study the effects of neutrons on fiber optics [7].

Several systems are routinely operated at levels above the original design specifications (see Table 1). The neutral beams, now operating with tritium as well as deuterium fueling, have delivered up to 40 MW total power to the TFTR plasma [8]. This level significantly exceeds the original design expectation of 33 MW. The toroidal field (TF) system has been upgraded and operated at 6 Tesla (T) (on axis) as compared to the original 5.2 T design. This was accomplished primarily by reconfiguring the field coil power supplies and increasing the power drawn from the motor generators from 950 to 1200 MVA. The supporting engineering analysis of the magnet design has increased our confidence that the coils are capable of operating at increased magnetic field [9].

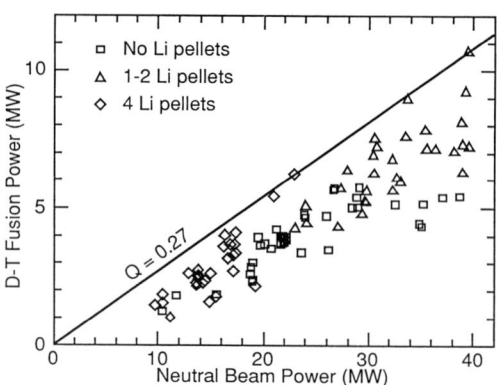

Fig. 5. Effects of lithium conditioning on fusion power.

4. Fusion power

Since December 1993, there have been 19 725 plasma shots including over 841 deuterium-tritium pulses. Plasma conditions for particular experiments are generally set-up in deuterium prior to experimentation with tritium.

The D-T plasmas have so far produced a total of 4.8×10^{20} D-T neutrons and alpha particles and approximately 1.4 GJ of fusion energy. Peak fusion powers up to 10.7 MW, corresponding to $Q \approx 0.27$, and total fusion energy per pulse of 6.5 MJ have been attained. (Q is defined here as the total fusion power divided by the total injected beam power. Shine-through, first-orbit loss, dW/dt terms, etc. are not subtracted from the total injected NB power.) TFTR has produced reactor level fusion power densities of 2.8 MW m^{-3}, significantly exceeding the original design objective of 1 MW m^{-3} (see Fig. 3) [10]. The fusion power flux to the TFTR wall is greater than 0.1 MW m^{-2}.

The primary TFTR plasma limiter (toroidal) is composed of carbon fiber composite and graphite tiles mounted on water-cooled inconnel backing plates. Additional poloidal limiters composed of carbon fiber composite tiles only are used to protect the ICRF antennas. The limiters are capable of handling total heat fluxes up to 30 MW for about 1 s.

Neutral beam heating and fueling have been the most effective approach for producing high reac-

Table 2
TFTR 10.7 MW plasma parameters

Global parameters	
Peak fusion power (MW)	10.7
Neutral beam power (MW)	39.5
Plasma current (MA)	2.7
Toroidal field (T)	5.5
Plasma stored energy (MJ)	6.9
Maximum confinement time (s)	0.21
Plasma major radius (m)	2.52
Plasma minor radius (m)	0.87
q_a ⟨Shafranov⟩	3.8
Normalized beta (%)	1.82
Confinement enhancement (relative to ITER89-P)	2.0
Central parameters	
Fusion power density (MW m^{-3})	2.8
Ion temperature (keV)	32
Electron temperature (keV)	13.5
Electron density (m^{-3})	1.0×10^{20}
$n_e(0)/\langle n_e \rangle$	2.7

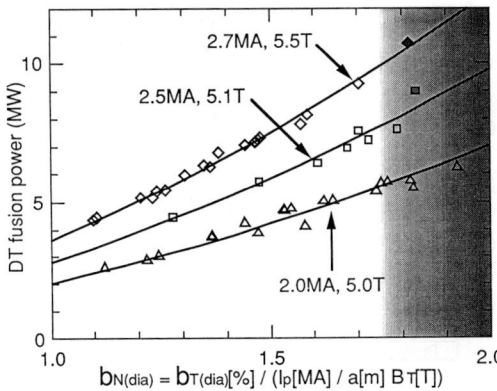

Fig. 6. Effects of increased magnetic field on fusion power.

tivity plasmas in TFTR. The highest performance plasmas (using fusion power as the metric) have been obtained in the 'supershot' mode of operation. These plasmas are typically characterized by ion temperatures, T_i, up to 45 keV, electron temperatures, T_e, up to 14 keV and peaked density profiles with $n_e(0)$ about 10^{20} m^{-3} and $n_e(0)/\langle n_e \rangle$ in the range of 2–4.5 (see Fig. 4 and Table 2) [11].

5. D-T confinement

The improved confinement associated with the best supershot plasmas is typically correlated with low edge recycling which allows the core fueling to be dominated by the neutral beams. Various

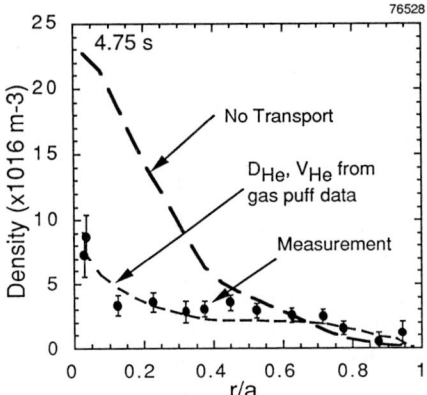

Fig. 7. Transport of helium ash from the plasma core to the edge during the heating phase.

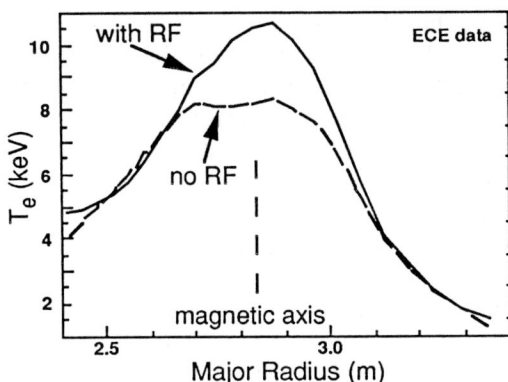

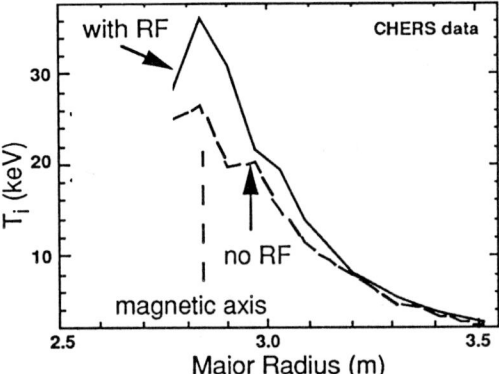

Fig. 8. ICRF heating of ions and electrons in D-T plasmas.

conditioning techniques have been developed to enhance this effect. Limiter conditioning using lithium (Li) pellets has been most successful (see Fig. 5). The lithium pellets not only improve the confinement of the initial plasma they are injected into, but also affect the performance of subsequent discharges. This has led to the investigation of optimal techniques to 'pre-condition' the limiter with a series of plasma discharges varying the number and timing of pellet injections [12].

The fusion triple product for an equivalent steady-state plasma, $n_e \cdot \tau_e^* \cdot T_i$ (where $\tau_e^* = W_p/P_{heat}$) has increased dramatically with lithium conditioning. Previous supershots without lithium wall conditioning have achieved triple product confinement enhancements of up to 13 times typical L-mode plasmas. Lithium conditioning has increased this factor up to 64.

Increasing the toroidal magnetic field at constant β_N has been observed to increase fusion power (see Fig. 6). (Shading represents an increasing probability of disruption in the supershot regime.) The maximum central fusion power density was increased from 2.2 to 2.8 MW m^{-3} as B_T on-axis was increased from 5.1 to 5.5 T.

An isotopic effect has also been observed to improve the confinement of TFTR plasmas. In 1992, the best confinement time, τ_e^*, for a high temperature, beam heated D-D plasma was 0.19 s. In 1994, τ_e^* was increased to 0.24 s in a nominally 50:50 D-T plasma. In 1995, a τ_e^* up to 0.33 s was obtained in a mostly tritium plasma [10].

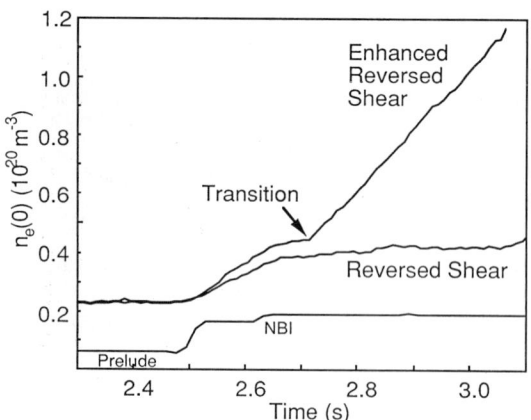

Fig. 10. Central density plot showing transition to enhanced reverse shear (ERS) regime.

6. Alpha physics

The first observations of alpha particles heating the plasma have been made on TFTR. A comparison of electron temperature profiles for D-D and D-T plasmas indicates a temperature increase in excess of that expected due to isotopic effects alone (with parameters such as confinement time and heating power held constant) [19]. Computer analysis of these plasmas (utilizing the TRANSP code) indicates that the increased heating is consistent with that expected from alpha heating.

Measurements of escaping alpha particles suggest that the alpha loss fraction does not increase with increasing fusion power (for stable plasmas). In general, these losses are consistent with expected single particle, first orbit losses. Alpha loss fractions are seen to increase with the onset of MHD activity [13].

The transport of helium ash from the plasma core (for supershot plasmas) is found to be rapid and consistent with modeling based on helium transport predicted from gas puff experiments (see Fig. 7). These measurements suggest that the helium ash transport from the plasma core are in the acceptable range necessary for a fusion reactor for supershot-like operating conditions [11].

In plasmas where the shear has been reduced by raising q(0), an α-driven toroidal Alfven eigenmode (TAE) has been observed that agrees with theoretical predictions. At the observed amplitudes, alpha particles have not been anomolously ejected from the plasma.

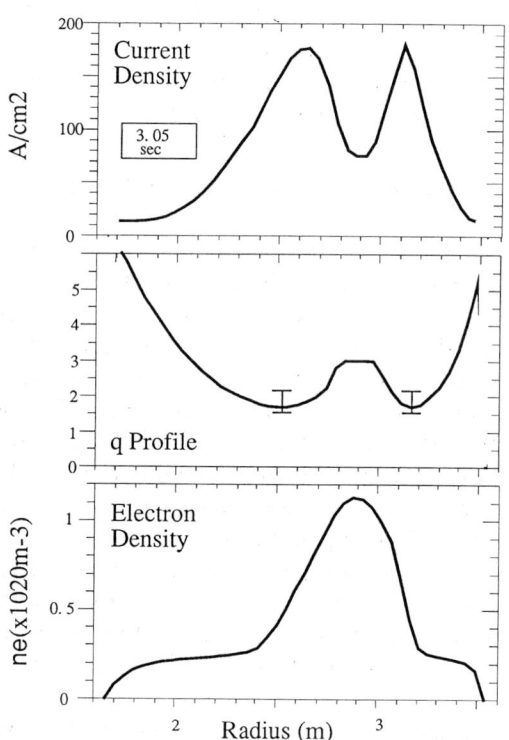

Fig. 9. Reversed shear profiles.

7. ICRF heating and current drive

The ICRF system consists of four dual-strap, inductively coupled antennas with Faraday shields. Six transmitters are utilized in various configurations to provide up to 12 MW of source power at frequencies ranging from 30 to 64 MHz.

ITER relevant ICRF heating at the second harmonic of tritium, $2\Omega_T$, at power levels up to 6 MW have been investigated (see Fig. 8). With a small amount (about 2%) of [^{3}He] added to the plasma to increase the single pass absorption, the central ion temperature, $T_i(0)$, increased from 26 to 36 keV and the central electron temperature, $T_e(0)$, increased from 8 to 10.5 keV.

Other experiments include direct electron heating and current drive by ion Bernstein wave (IBW) excitation. The mode-converted IBW has been shown to heat electrons and drive current both on and off axis as well as exhibiting strong coupling to fusion products (tritons and alphas). Additional experiments to demonstrate direct fast wave current drive have been conducted principally utilizing [^{4}He] target plasmas [14].

8. Enhanced confinement regimes

The TFTR peaked pressure supershot regime has produced 10.7 MW of fusion power at a toroidal field, B_T, of 5.6 T and $\beta_N \approx 1.7$. TFTR is investigating additional enhanced confinement regimes as described below.

Modifying the plasma current density profile, $j(r)$, by ramping up the plasma current and creating a 'hollow' current profile, coupled with a staged neutral beam injection power profile, have produced reversed magnetic shear configurations typical of those proposed for advanced tokamaks such as the Tokamak Physics Experiment (TPX) (see Fig. 9) [15]. The resulting confinement characteristics of these plasmas resemble those of supershots with the same engineering parameters.

However, at neutral beam heating powers exceeding about 16 MW, the core transport changes abruptly (Fig. 10). The electron particle diffusivity drops precipitously by a factor of 50, approaching neoclassical levels. The electron and ion thermal diffusivities also drop by a factor of 2–3. Corresponding to this large increase in central density, the pressure gradient is much larger, by a factor of 3–5, than typical supershot plasmas operating at similar conditions. These TFTR plasmas have been labeled Enhanced-Reverse-Shear (ERS) [11]. Plasmas produced in this regime, which have demonstrated neoclassical confinement in the core and $\beta_N \approx 2$, project to fusion power levels of greater than 20 MW if the available beam power and magnetic field are utilized.

Another path to enhanced confinement involves peaking the current density profile by either ramping down the plasma current or controlling the plasma growth resulting in a transient increase of the plasma internal inductance. Using these techniques, the energy confinement and plasma stability is increased relative to a discharge with a relaxed current profile at the same plasma current level. This mode of operation on TFTR is called the high-l_i (or high-β_p) regime [16–18].

The maximum fusion power produced in this regime is 8.7 MW with $\beta_N \approx 2.3$ at $B_T = 4.8$ T and a corresponding plasma stored energy of 6 MJ. This is comparable to that achieved in Supershots with similar neutral beam power (Fig. 11). Operation at $B_T = 5.6$ T is projected to yield fusion power close to 15 MW.

The potential implications of these enhanced confinement regimes are promising. If the benefits of the predicted enhanced MHD stability can be

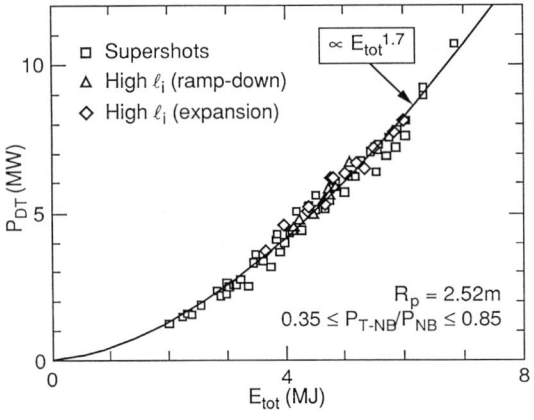

Fig. 11. High l_i plasma performance.

realized along with the improved central confinement, then forthcoming TFTR experiments may demonstrate $P_{\text{fusion}} > 20$ MW. If successful, these regimes will demonstrate some of the features required of a smaller, more economical fusion reactor.

9. Summary

The TFTR D-T phase of operations has proven to be as exciting as it is successful. The facility continues to operate safely and routinely utilizing reactor-relevant D-T fuel and fully satisfying all regulatory requirements. Technological performance and experimental results have met or exceeded all of the original TFTR design criteria and programmatic objectives. New operating regimes demonstrate some of the features required of a more economical fusion reactor.

Acknowledgements

The author wishes to acknowledge all of the PPPL technical and supporting staffs whose contributions have made the TFTR program such a success. This work has been supported by US Department of Energy Contract No. DE-AC02-76-CHO-3073.

References

[1] TFTR Technical Requirements Document, Princeton Plasma Physics Laboratory, 1976.
[2] C.H. Skinner et al., PPPL Report 3172, 1996.
[3] D. Mueller et al, Disruption Avoidance on TFTR, Fus. Technol. 30 (1996) 251.
[4] J.L. Anderson et al., Proceedings of the 15th IEEE/NPSS Symposium On Fusion Engineering, Hyannis, October 1993, IEEE, Piscataway, NJ.
[5] S. Raftopoulos et al., Integration of the Tritium Purification System (TPS) into TFTR Operations, Proceedings of the 16th IEEE/NPSS Symposium on Fusion Engineering, October 1995, IEEE, Piscataway, NJ.
[6] L.P. Ku, S.L. Liew, Proceedings of the 8th International Conference on Radiation Shielding, 1994, ANS, La Grange Park, IL, pp. 1062.
[7] A.T. Ramsey et al., Reduced Optical transmission of SiQ_2 Fibers Used in Controlled Fusion Diagnostics, PPPL Report 2867, Sec. 2.1.4, 1993.
[8] T. Stevenson et al., The TFTR 40 MW Neutral Beam Injection System and D-T Operations, Proceedings of the 16th IEEE/NPSS Symposium on Fusion Engineering, October 1995, IEEE, Piscataway, NJ.
[9] R. Woolley et al., Extension of TFTR Operations to Higher Toroidal Field Levels, Proceedings of the 16th IEEE/NPSS Symposium on Fusion Engineering, October 1995, IEEE, Piscataway, NJ.
[10] K. McGuire et al, Review of D-T Results from TFTR, Phys. Plasmas 2 (1995) 2176.
[11] D. Johnson et al., Recent D-T Results on TFTR, 22nd European Physical Society Conference, Bournemouth, UK, July 1995, European Physics Society, Petit-Lancy, Switzerland.
[12] D. Mansfield et al., Improved Performance of DT Supershots in TFTR Using New Techniques of Limiter Conditioning with Lithium, 22nd European Physical Society Conference, Bournemouth, UK, July 1995, European Physics Society, Petit-Lancy, Switzerland.
[13] R.J. Hawryluk et al., Review of Recent D-T Experiments from TFTR, 15th International Conference on Plasma Physics and Controlled Nuclear Fusion Research, Seville, September 1994, IAEA, Vienna.
[14] R. Majeski et al., ICRF Heating and Current Drive Experiments in TFTR, 22nd European Physical Society Conference, Bournemouth, UK, July 1995, European Physics Society, Petit-Lancy, Switzerland.
[15] C. Kessel et al, Phys. Rev. Lett. 72 (1994) 1212.
[16] M.C. Zarnstorff et al., Controlled Fusion and Plasma Physics, Proceedings of the Sixteenth European Conference, Vol. 13B, Part 1, Venice, 1989, European Physics Society, Petit-Lancy, Switzerland, p. 35.
[17] G. Navritil et al., IAEA Meeting, 1991.
[18] S.A. Sabbagh et al., Workshop on Tokamak Improvement, 1994 School of Plasma Physics Sec. 4.4, Varenna, Italy, 1994.
[19] G. Taylor, et al., Fusion Heating in a Deuterium–Tritium Tokamak Plasma, Phys. Rev. Lett. 76 (1996) 2722.

Development of high current negative ion source

T. Kuroda

National Institute for Fusion Science, Furo-cho, Nagoya 464-01, Japan

Abstract

The development of high current negative ion sources has been intensively continued for application to fusion research, especially neutral beam injection. Recently, the research and development of hydrogen negative ion sources has made rapid progress. Negative ion beam current has been remarkably enhanced by seeding cesium in a magnetically filtered multicusp plasma source. As a result, a negative ion beam current of more than 16 A is obtained at an energy of 120 keV. According to studies of negative ion sources during the last decade a design principle for volume production multi-cusp negative sources has been mostly approved. The present status of the development of high current negative ion sources is briefly reviewed. The performance and the characteristics of a negative ion source are discussed. Emphasis is placed on the important issues and optimized condition: low operating pressure, the relation between the plasma parameters and the extracted beam current, the optimum Cs seeding conditions, the beam extraction characteristics of the negative ion source and the suppression of electron current extracted with the negative ions. The optimum condition is described by considering the physical principle of the negative ion sources and the physical and technological consideration of the design of the negative ion sources are given. © 1997 Elsevier Science S.A.

1. Introduction

Ion beams of various kinds are widely used in industrial applications and scientific research. Many types of positive ion sources have been developed to meet the requirements of their proposed application. Many ion sources are designed for various performance levels and ion species, which are required by different applications. In small current ion sources, which are used for ion assisted technology, high quality of the beam, low emittance and high brightness are important performance features.

The requirements for ion sources for neutral beam injection are somewhat specific, that is, high current compared with the other field applications, high beam power and relatively low beam energy. Especially in a future advanced large scale fusion machines like ITER, a negative ion based neutral beam injector is needed because negative ion beams have a much more effective neutralization efficiency at energies higher than 100 keV. Negative ion source science and technology has been developed in pace with the use of it in NBI. With the progress of studies of negative ion beams, their usefulness attracts attention due to better quality such as small divergence and single charge ion production. Negative ion beams have been available for various species of ions in many fields. Several types of negative ion sources have been studied for volume plasma production, surface plasma production and charge exchange of positive ions. Since it was found that negative ion production is enhanced by the dissociative electron attachment to vibrationally excited molecules in the plasma [1–3], negative ion source development, especially for a large scale high current ion source, has focused on the tandem multicusp source which is a type of volume plasma production source. The physical design principle is proposed on the basis of the negative ion production mechanism due to the dissociative electron attachment [4,5]. The effectiveness of the tandem multicusp source was experimentally demonstrated by Leung et al. [6] and Nicolopulos et al. [7]. A negative ion current of several amperes was obtained by using a multi-cusp tandem source [8]. Recently studies of negative ion sources have made great progress. A hydrogen negative ion beam of more than 10 A can be obtained with cesium seeding of the plasma source [9–11]. The source is optimized to attain a maximum extracted beam current by changing the parameters of the plasma source.

In this report, we describe the concept outline of a negative hydrogen ion source and the key requirements of negative ion sources are discussed from the point of view of physical understanding. The design principle of the ion source is described on the basis of experimental results. We describe engineering problems concerning the design of a large scale and high current negative ion source, technical problems of the beam acceleration and the beam transport. Finally, the present status of hydrogen negative ion sources for fusion research is reviewed briefly.

2. Design considerations of negative ion source

2.1. Concept of the negative ion source

In the tandem model concept, which is the most promising method, the negative ions are generated due to two fundamental processes: (1) producing vibrational excited hydrogen molecules by collisions between hydrogen molecules and energetic electrons of more than 10 eV; and (2) generating negative hydrogen ions due to the dissociative attachment of low temperature electrons to the vibrational hydrogen molecules.

A conceptual design of a tandem negative ion source is schematically shown in Fig. 1. In order to produce negative hydrogen ions effectively by this concept, the tandem multicusp plasma source is divided into two regions by a transverse magnetic field (magnetic filter), i.e., the driver plasma region with high energy electrons of more than 10 eV and the extraction region with low temperature electrons of less than 1 eV. Generally in a d.c. discharge plasma source the plasma is produced due to ionization by the primary electrons emitted from the filaments and these high energy primary electrons also contribute to the production of vibrationally excited hydrogen molecules. Energetic electrons in the driver region are prevented from flowing into the extraction region by the magnetic field of the filter, while the thermal electrons in the driver region diffuse into the extraction region and vibrationally excited hydrogen molecules and low temperature plasmas of less than 1 eV are provided not only to effectively produce the negative ions by the dissociative attachment but also to reduce the destruction of the negative ions.

2.2. Consideration of a high power, multi-cusp, tandem plasma source

Since the plasma source is the most important component in a plasma based ion source, it must be designed with a good physical understanding of the concept. The requirements for a plasma source of the high current negative ion source type are: (1) to produce a more high plasma density plasma with high electron temperature in the driver region; (2) to produce a low temperature plasma of less than 1 eV in the extraction region. The key issues for the plasma source design to meet these requirements are: (1) better confinement of the plasma; (2) effective magnetic filtering; (3) plasma uniformity.

In the multi-cusp plasma source discharge, the primary electrons emitted from the filament mainly leak to the anode through the multi-cusp lines. The leak of the plasma from the magnetic cusp lines is proportional to the width of the cusp lines and this gives an effective anode area, through which the discharge current mainly flows

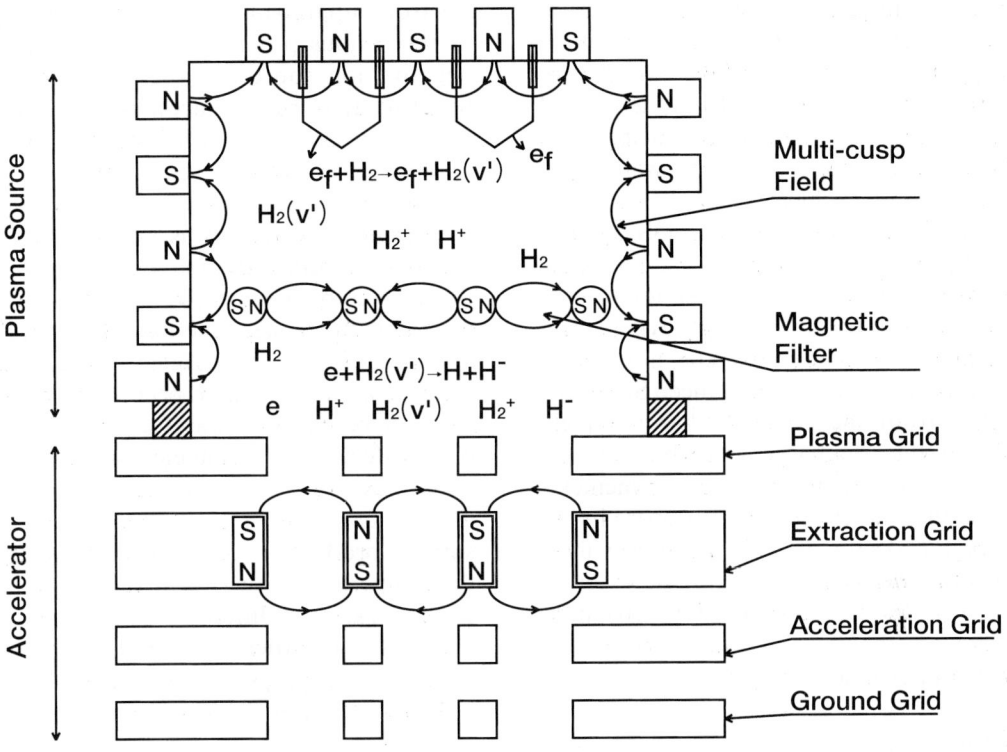

Fig. 1. Conceptual design of a tandem negative ion source.

to the anode. The area is reduced by increasing the magnetic field strength of the cusp and the primary electrons are confined better. Then, in order to increase the plasma density, it is required to operate the multi-cusp plasma source with the small effective anode area.

Nevertheless, it is reported that the multi-cusp plasma source discharge has two discharge modes [12–14] which are classified as high and low impedance discharge. Because the plasma density is higher in the low impedance than in the high impedance discharge, the plasma source must be operated in the low impedance mode. The low impedance mode transits to the high impedance mode when the ratio of the anode area to the cathode area is less than the critical value. The effective anode area must be as small as possible to produce high density plasma with energetic electron but large enough to avoid the mode change during operation of the plasma source. This condition is contrary to the good confinement condition because the stronger magnetic cusp is needed for good particle confinement.

Uniformity of the plasma density is an important issue in a large scale source. The uniformity depends on the condition of the filaments, the distribution of the filaments, the primary electron motion and the operating pressure. The filaments should be located in a week field region to avoid the influence of the magnetic field on the electron motion and interference of the magnet field due to filament current with the plasma density profile. It is reported that strong non-uniformity was observed at high operating pressure and was improved by reducing the pressure [15]. Better density uniformity is obtained during space charge limited operation of the plasma source discharge compared to the temperature limited operation.

There are three types of filters: (1) a rod filter which produces a magnetic dipole field by several pairs of rods having magnets located in front of the plasma grid; (2) an external filter which provides the transverse magnetic field by a pair of permanent magnet array embedded facing each other outside the discharge chamber wall; (3) a plasma grid (PG) filter which generates the transverse magnetic field surrounding the PG by supplying current to the PG directly. The rod filter has the disadvantage, which is to mask the plasma grid, but it does not disturb strongly the plasma confining magnetic field because the filter magnetic field decays in a short distance. Since the magnetic field from the external filter is widely distributed in the discharge chamber of the plasma source, it affects the plasma confinement. It is observed that the electron temperature in the extraction region decreases with increasing filter strength but the electron density in the external region also decays with increasing filter strength. The efficiency of the filter is evaluated by the line-integrated filter field strength.

2.3. Accelerators

Three approaches to accelerate negative ions, i.e. the electrostatic accelerator, Radio frequency quadrupole (RFQ) accelerator [16] and the electrostatic quadrupole accelerator [17] have been studied for a beam energy of around 1 MeV. The electrostatic acceleration presently gives successful results by the multi-stage acceleration method. In most versions a large number of identical parallel beamlets are accelerated by grids with precision-machined and carefully aligned circular holes. Beam extraction characteristics of negative ion sources differ somewhat from those of positive ion sources. A big difference and a serious problem are the extraction of electrons with the negative ions in the negative ion source. It results in a big load on the power supply as well as a heat load on the extraction electrodes. A magnetic field is applied to eliminate the extracted electrons, the stripping electrons and the secondary electrons in the accelerator. Grids need to be water-cooled to reduce the temperature rise due to the extracted electron current and to prevent heating and warping in long-pulse or steady-state operations. An electron suppression grid is sometimes used. One of the serious problems of the accelerator in the negative ion sources is the stripping loss of the negative ions in the accelerator. The stripping causes not only the loss of negative ions but also an increase in the electron current.

The plasma and the extraction grid are vulnerable because the former is exposed to the source plasma and full radiation from the filament and the later is bombarded by deflected electrons extracted with the negative ions. The required cooling channels and the magnet arrays tend to limit the transparency. Their beam-forming structures have to be designed carefully because they seriously effect the performance of the ion source.

Negative ions are produced on the surface of the PG as well as in the plasma volume. Since the surface production of negative ions depends on the temperature of the grid surface as well as on the coverage condition of the cesium, the PG is not cooled in order to raise the temperature due to the discharge power or due to the heater attached to the grid for Cs seeding operation.

3. Characteristics of high current negative ion source and technology

An example of a large high current negative hydrogen ion source is schematically shown in Fig. 2, which is a 1/3 scaled version of the prototype ion source for LHD and delivered a total H$^-$ current of 16 A at a beam energy of 120 keV at an operating pressure of 0.9–0.45 Pa with Cs seeded operation [11]. The plasma chamber is a multicusp-type source and the size of the chamber is 37 cm in width, 62.5 cm in height and 18.5 cm in depth. The plasma source is divided into a driver region and a beam extraction region by a water-cooled rod-type magnetic filter. A thin molybdenum liner is attached to the inner wall to avoid Cs condensation. Up to 48 filaments can be installed as the cathode in the arc chamber, but usually a lower number of filament is used for a discharge of up to 2000 A at a discharge voltage of 150 V. Hairpin tungsten filaments of 1.5 mm diameter are usually used.

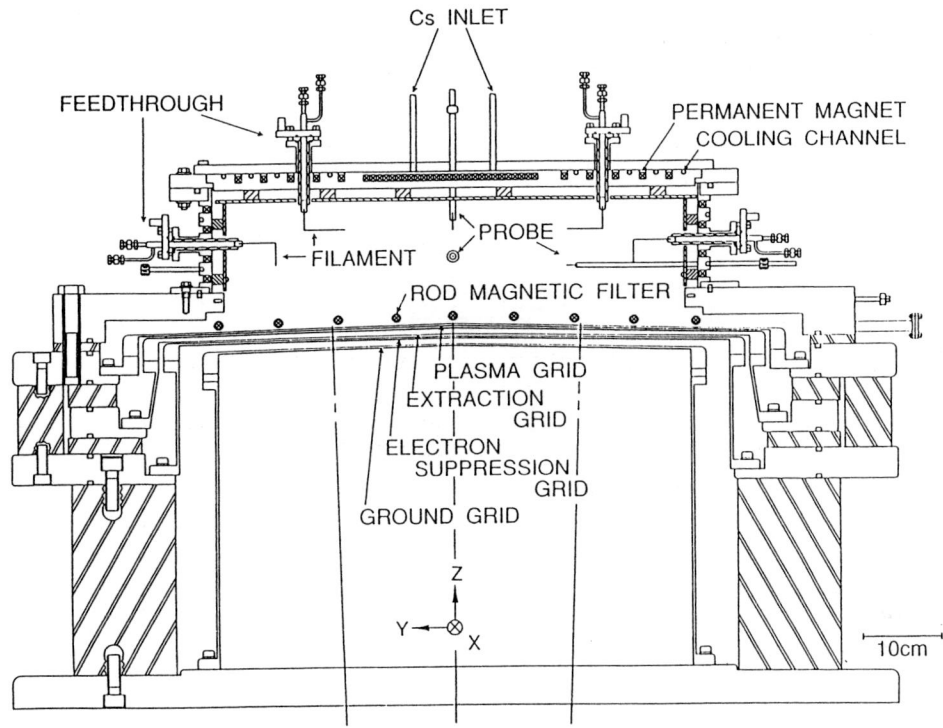

Fig. 2. Schematic drawing of a high current negative hydrogen ion source.

The requirements of long pulse operation influence the design of many parts of the source as filaments, filaments feed throughs and plasma source. All parts of these components require more effective cooling. The accelerator is composed of four grids: a PG, an extracting grid, an electron suppressing grid and a ground grid. Each of these grids has 560 extraction holes of 9 mm in diameter in an area of 25×44 cm^2. The PG is made of molybdenum without water cooling to keep it at a high temperature during Cs operation. The others are made of oxygen-free copper and are partially water cooled. Each grid is divided into two subgrids both of which face towards the focal point 6 m downstream from the ion source. The extracting grid of 10 mm in thickness has electronbending magnets to eliminate electrons extracted with negative ions on to cooling channels.

The magnetic filter is composed of several rods of 1 cm diameter having sector magnets inside. The line strength of the filter field is about 250 G cm. The gap between the rods and the PG is typically about 1 cm. Cs is seeded into the discharge chamber through valves from Cs ovens. The amount of Cs is varied by the oven temperature and the opening duration of the valve. A small amount of Cs vapor is seeded once before starting operation and no more Cs is supplied afterwards.

Fig. 3 shows an example of technical drawing for the grids in the case with water cooling. Each grid in this case has 400 apertures of 9 mm diameter arranged in an area of 25×25 cm. The transparency is about 40%. The grid is divided into modules, each of which is inclined to focalize the negative ion beam at a distance downstream from the source. In the extraction grids, the cooling channels are arranged perpendicular to the permanent magnet rows so that the deflected electrons are incident on the vicinity of the cooling channels. The permanent magnet rows are packed

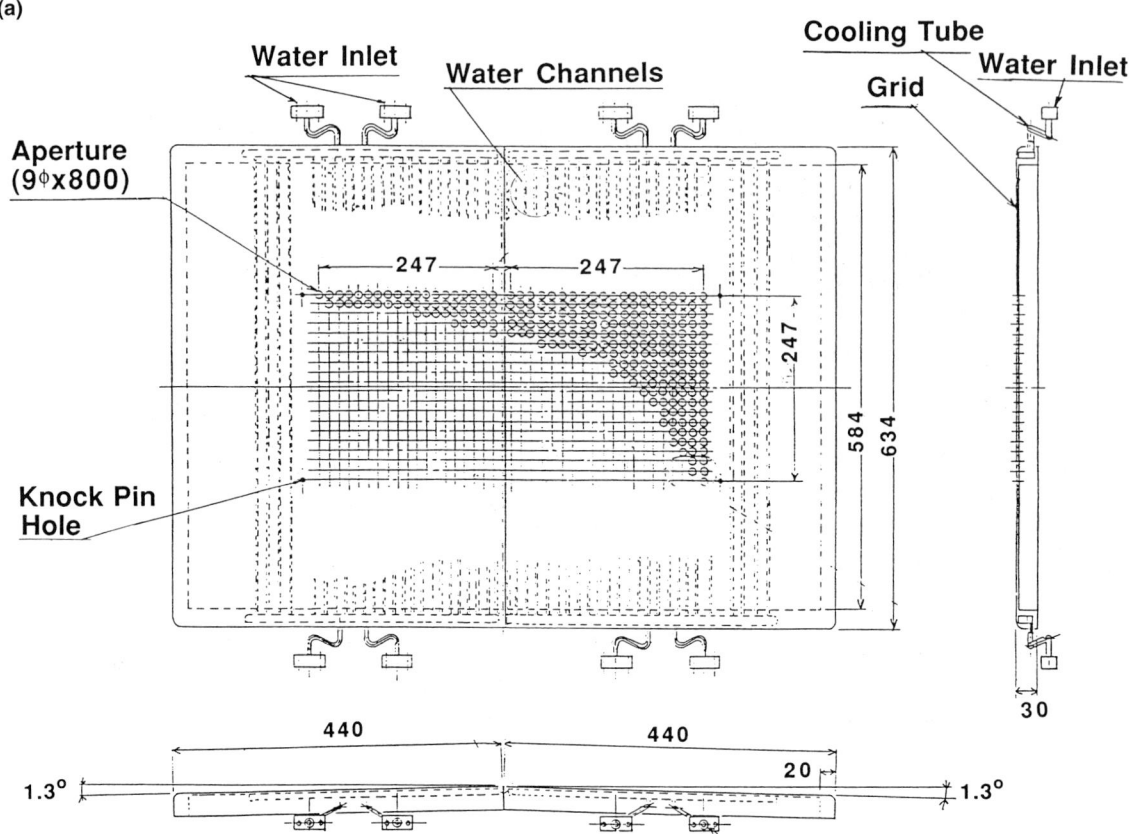

Fig. 3. Schematic drawing of the plasma grids (a), the extraction grid (b), and detailed structures around the aperture of the plasma grid (left) and the extraction grid (right) (c).

in thin stainless steel cans and are inserted in the channels. The extraction electrode system is electrically insulated against an accelerating voltage (120 kV) with the insulator made of synthetic resin.

The results of the beam extraction are shown in Fig. 4 as a function of the arc power, P_{arc}, for different size and different filter type ion sources, respectively. The total H$^-$ current has been obtained by integrating calorimetrically measured beam profile. The H$^-$ current increases nearly with the arc power in the case of the operation with Cs. A maximum H$^-$ current of 16 A is obtained at the arc power of 265 kW the extraction voltage of 7.5 kV and an operating gas pressure of 0.9 Pa. The corresponding current density is 45 mA cm^{-2}, in which the temperature of the grid attains to about 300°C during opera-

tion. The H$^-$ ion current is enhanced three time larger in Cs seeding operation. Simultaneously, the electron extraction current and the operating gas pressure are reduced by Cs seeding. The H$^-$ current of around 16 A is attained in the both ion sources but the arc efficiency is higher in the external filter type source than in the rod filter type source. Since the plasma density increases with the arc power, the total H$^-$ current increases with the plasma density. The electron current can be reduced by changing the plasma source operating condition and the plasma condition in the vicinity of the extraction holes.

Fig. 5 shows a typical example of the extraction current, the hydrogen negative ion current and the electron current as a function of the biasing voltage to the PG. The drain current and the electron extraction current can be reduced at a

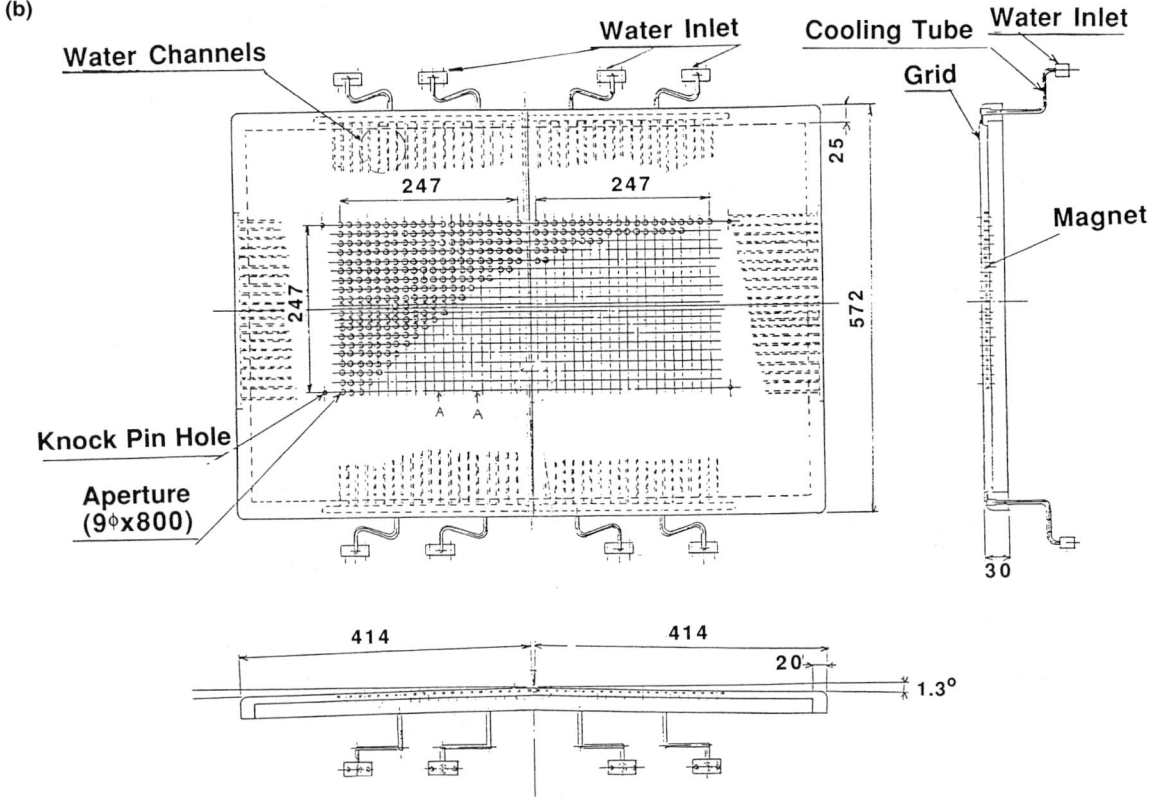

Fig. 3. Continued.

biasing voltage slightly below the plasma potential with a small reduction of the negative ion extraction current. The ratio of the electron current to the negative ion current is much smaller in the ion source with Cs seeding operation than without Cs operation. This shows that in the Cs seeding operation the negative ions are produced at the PG surface covered with Cs as well as in the volume of the plasma at the extraction region.

Fig. 6 shows the H^- current as a function of the operating pressure, p. In a pure hydrogen discharge, the H^- current increases with p. The optimum operating pressure is reduced to 0.8 Pa.

Fig. 7 shows an example of the H^- current as a function of the extraction voltage (a) and beam energy (b). The H^- current increases according to the space charge limited characteristics and shows the 3/2 power dependency on the extraction voltage (beam energy) as similar in the positive ion source. This shows that the ion current density in an optimized space-charge flow is uniquely determined by the accelerating structure (Child–Langmuir relation). However the perveance of the extraction is different from that for a positive ion source. The single stage and two-stage acceleration are examined for the optimization of beam divergence. In both single and two stage acceleration, the beam divergence shows similar dependence on the ratio of the electric field in the acceleration gap to the electric field in the extraction gap, E_{acc}/E_{ext}. A minimum divergence of the beam is obtained by adjusting the ratio E_{accel}/E_{ext} for the single and the multi-hole source, respectively. Beam divergence is less sensitive for the two stage acceleration than for the single-stage acceleration.

Fig. 8 shows the dependence of the beam divergence angle on the ratio E_{acc}/E_{ext} in the case of

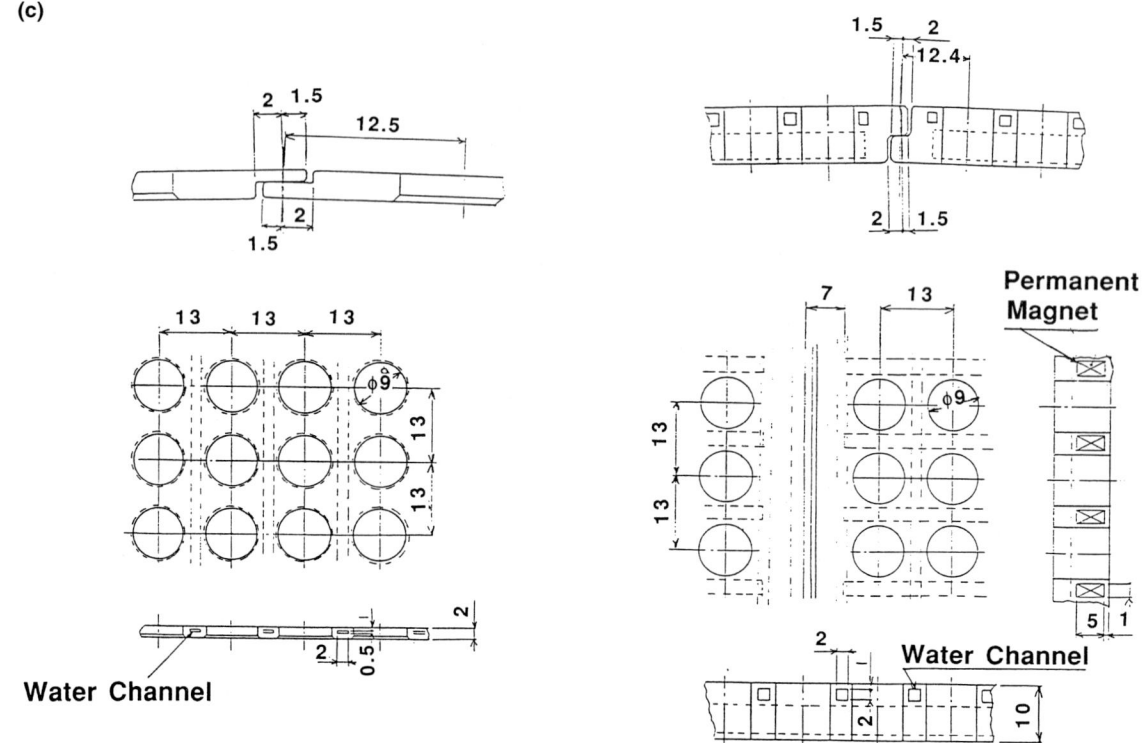

Fig. 3. Continued.

single hole and single stage acceleration and in the case of multi-hole and single stage acceleration. In both cases, minimum beam divergence can be obtained by adjusting the ratio of acceleration electric field to the extraction electric field.

Minimum beam divergences between 5 and 9 mrad are obtained by adjusting the ratio of the acceleration electric field to the extraction electric field E_{acc}/E_{ext} for the single and the multi-hole source, respectively. A normalized emittance of 0.59 mmrad is obtained for a single hydrogen negative beam of 6 mA with an emission current density of about 10 mA cm^{-2} in the simple emittance measurement experiment using capton foil technology [18].

In order to compensate for the deflection of the negative ion beam due to the magnetic field for electron suppression, the effectiveness of the technology of beam deflection due to displacing the apertures of the grid, which is used in the positive ion sources, has been studied for multi-aperture extraction. Fig. 9 shows the results of an aperture displacement experiment [19]. The horizontal beam profile is broad and has two peaks in the case of a non-displacement accelerator. When the amount of the aperture displacement of the ground grid corresponding to the horizontal steering angle of ±8 mrad is added to that for the multi-beamlet focusing in the three-stage accelerator, the beam is well focused at one point.

Fabrication technique of the large area extraction grid, alignment technique of multi-hole grids, the large scale insulator manufacture technology and the aging technology of ion sources have been advanced with the high current positive ion source and the positive ion based NBI became the most promising and reliable heating technology. For the large high current negative ion sources, these engineering developments have been utilized. However, the performance specification of the

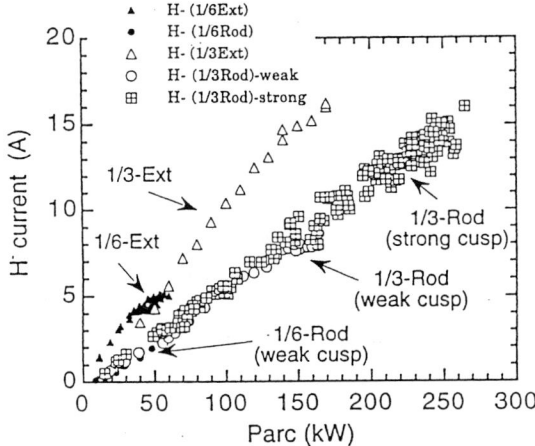

Fig. 4. Beam current as a function of discharge power.

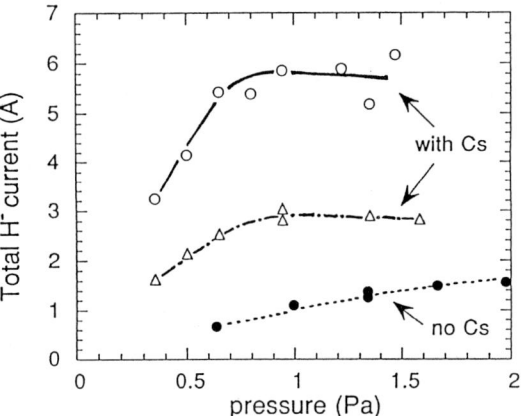

Fig. 6. Negative hydrogen ion current as a function of operating pressure. Extraction voltage, $V_{ext} = 6.5$ kV; acceleration voltage, $V_{acc} = 25$ kV; $P_{arc} = 50$ kW ($\triangle$) and 100 kW ($\bigcirc$). Closed circles show data obtained in pure hydrogen discharge.

negative ion source is upgraded for a longer operating duration of the source, the acceleration voltage and the beam power density. Then peculiar technologies have to be developed for the high current negative ion.

The plasma source conditioning including the Cs seeding is a more important criteria in the negative ion source. There are two methods of Cs feeding to the plasma source. One is to introduce Cs vapor by evaporating pure Cs metal in a Cs oven and the other method is to use special pellets containing a mixture of Cs chromat (30%) and titanium (70%) instead of pure metal Cs [20]. The oven must be carefully monitored during operation.

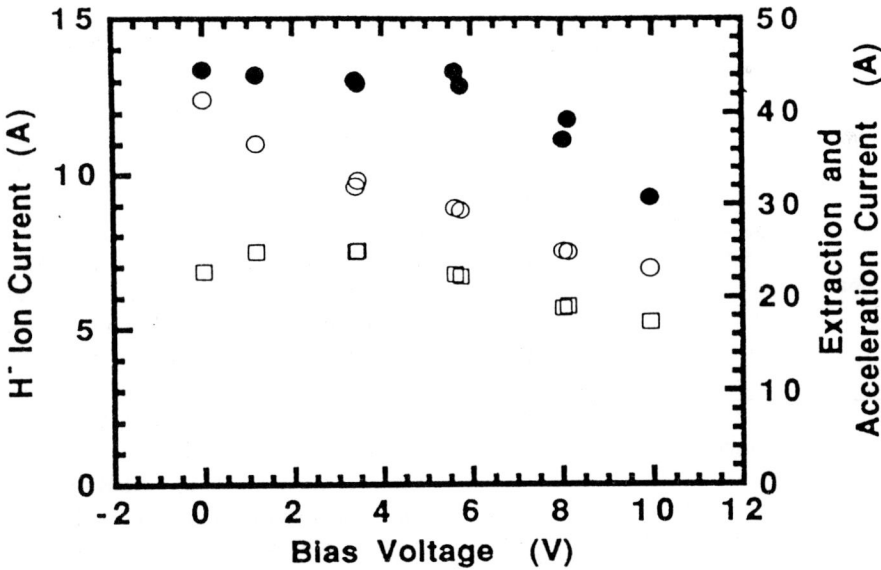

Fig. 5. Extraction current ($\bigcirc$), negative hydrogen ion current ($\bullet$), and the acceleration current ($\square$) as a function of the biasing voltage at the arc power of 130 kW. The operating pressure is 3.8 mTorr.

4. Present development status

Table 1 summarizes the performance and the specification for high current negative ion sources. Three types of magnetic filter are applied. Except for the NIFS negative ion source which requires a higher current beam with a rather low beam energy, the beam energy is more than several hundreds keV. Since current drive and heating with a negative ion based neutral beam injector is being conducted in JT-60 at JAERI and in the Large Helical Device in NIFS, respectively, high current negative hydrogen ion source developments have been carried out intensively both at

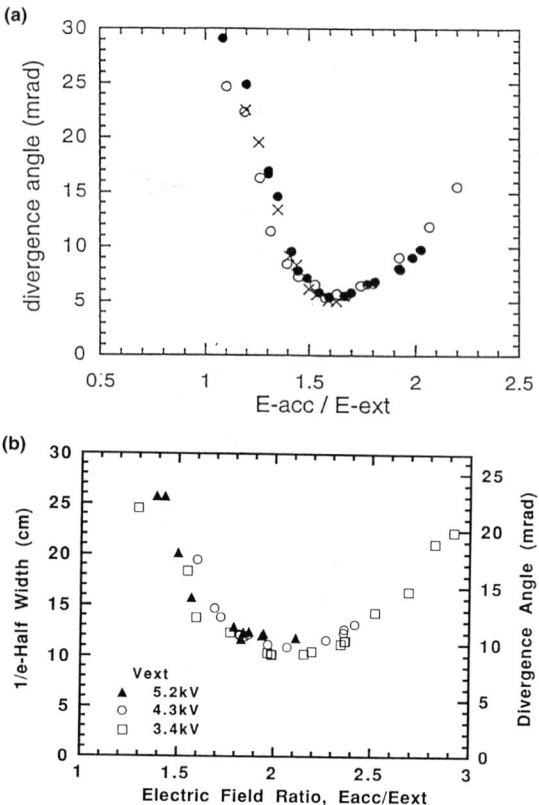

Fig. 8. Dependency of beam divergence on the ratio of the acceleration to the extraction electric field in the case of single stage acceleration (a) for single hole and (b) for multihole.

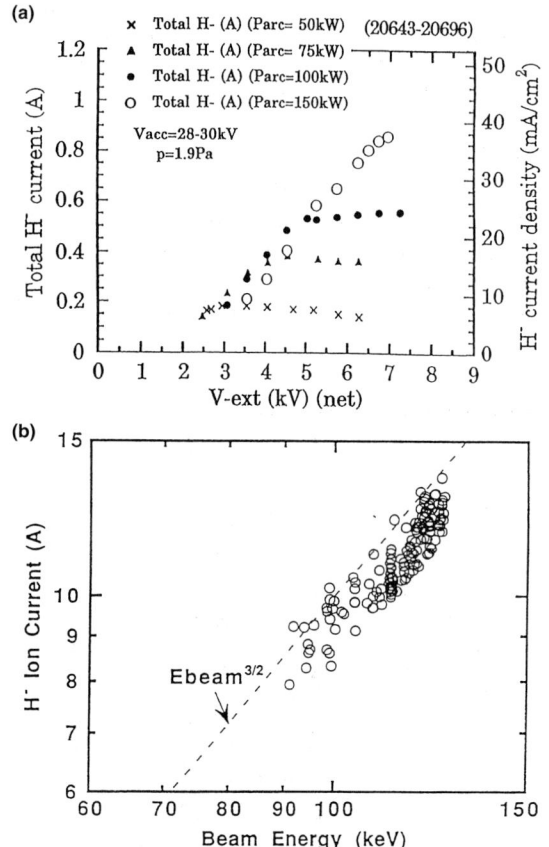

Fig. 7. An example of H$^-$ current as a function of extraction voltage for 1/6 scaled ion source with 36 apertures $V_{acc} = 28$–30 kV and $P = 1.9$ Pa (a) and as a function of beam energy for 1/3 scaled version with 522 apertures $P = 0.45$ Pa and $P_{arc} = 160$ kW (b). A dashed line indicates the 3/2-power dependence on the beam energy.

JAERI and NIFS. So far, it was demonstrated by the JAERI source that Cs seeding in the multi-cusp volume production ion source enhanced the negative ion production. A record negative ion current of 16 A was obtained by the NIFS source for the LHD neutral beam injector with Cs seeding operation. The highest beam energy attained is 400 keV in the JAERI source of a neutral beam injector for current drive experiments in JT-60. A beam with a current of 13 A is successfully accelerated up to 400 keV with electrostatic multi-stage acceleration. The minimum operating pressure of the source is 0.2 Pa at JAERI source. The fraction of the extraction current to the negative ion current is reduced to 2 with Cs operation by optimization of operating pressure. A beam divergence of less than 5 mrad is obtained for the high current negative ion source. It has been

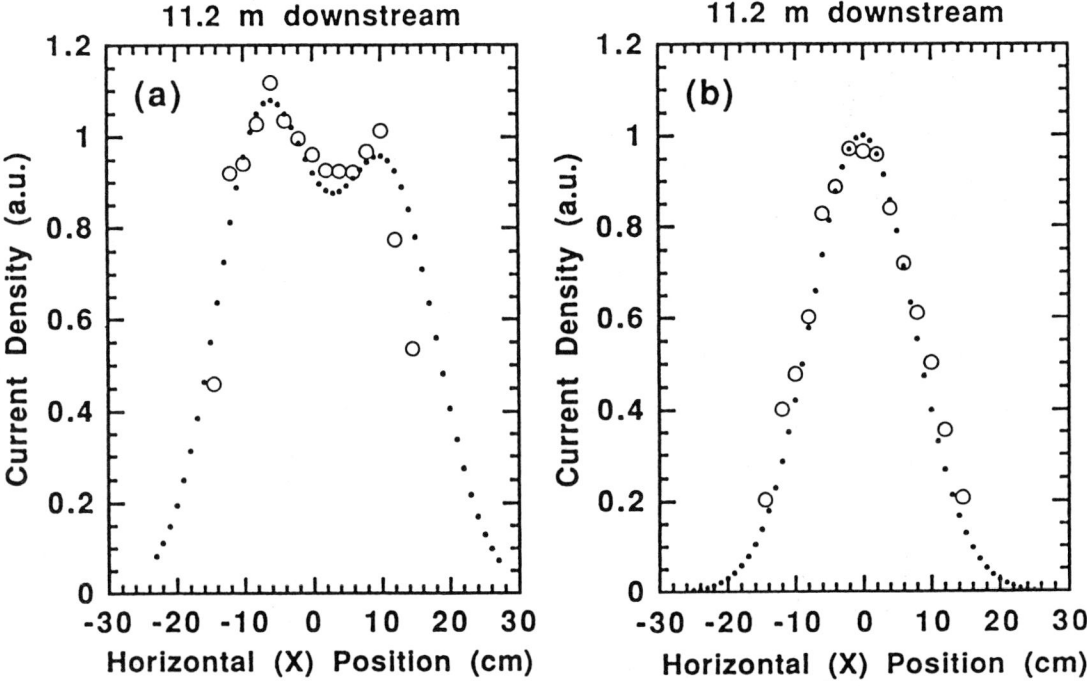

Fig. 9. Results of the aperture displacement experiment. Horizontal beam profile 11.2 m down stream from the grid in the case (a) the aperture displacement for only the multi-beamlet focusing and (b) the aperture displacement for both the multi-beamlet focusing and the compensation of alternate horizontal beamlet deflection by the magnet of the extraction grid.

confirmed that negative ion beams have a smaller divergence due to the lower temperature of negative ions compared to the temperature of positive ions both by calorimetric measurement of the beam profile and emittance measurement of the beam. These results, the beam current density, the operating pressure, the beam energy and the beam divergence, all satisfy the requirement of the negative ion source for a neutral injector.

An engineering design of a prototype ion source for LHD NBI is shown in Fig. 10, which is designed on the basis of the results of research and development so far. An RF negative ion source has been developed for long pulse operation in future but it is still in the fundamental research stage.

5. Conclusion

Large high-current hydrogen negative ion sources have been developed for a neutral beam injector. A multi-cusp tandem source seems to be the most promising method. The maximum negative ion beam current, beam energy and the duration of operation attained are more than 16 A in the 1/3 scale version ion source of LHD, 400 keV in the Negative ion based NBI in JAERI, and approximately 10 s in the 1/3 scale version ion source emphasized the cooling of the grids, respectively. On the basis of the fundamental research into the production mechanism and the technological research with respect to high power, a negative ion source capable of delivering a beam current of 22 A at an energy of 500 keV has already been manufactured and is being conditioned for current drive experiments on JT-60 at JAERI. A negative hydrogen ion source with an ability to deliver 40 A at an energy of 180 keV is being fabricated now for NBI of LHD at NIFS. The design study of a high current negative ion source at an energy of more than 1 MeV has been carried out for NBI in ITER. At this point, the

Table 1
High current negative ion sources

	Present status						Construction and future plane	
	NIFS [11,19]		JAERI [8,21]		Cadarache [22,23]		NIFS	ITER [21]
Ion source	1/3 scaled source		Multi-amper volume source	Kamaboko for JT-60	Pagoda	MANTIS	LHD	Kamaboko
Source type	Tandem volume with Cs		Tandem volume with Cs		Tandem volume with Cs		Tandem volume with Cs	Tandem volume with Cs
Species	H^-	H^-	H^-	D^-	D^-/H^-	D^-/H^-	$H^-/(D^-)$	D^-
Beam current (A)	16	16.2	10	13.5 (22) 2.4	1	1.6	40	15
Beam energy (kV)	120	125	50	400 (500) 460	100	25	180	1300
Current density (mA cm^{-2})	54	30	37	8 (13)	15	4.2	30	25
Operating pressure (Pa)	0.5	0.5	1.2	0.22 (<0.3)	0.65	0.75	0.5	<0.3
Beam divergence (mrad)	5	5	14	5			5	5
Pulse duration (s)	0.3	0.3 10 at 3.6 A, 9.9 keV	0.1	0.12 (10)			10	2 weeks
Plasma source Dimension (cm^3)	$25 \times 62.5 \times 18.5$	$30 \times 62 \times 20.6$	$12 \times 26 \times 15$	$64\phi \times 122$		$79 \times 49 \times 22$	$35 \times 129 \times 25$	$90\phi \times 178$
No. of filament	24	12	24		24	24		
Filter type	Rod	External	Rod	PG	External	External test	External	PG
Accelerator grid area (cm^2)	25×44	25×50	15×40	45×110	240	378	25×110	60×164
Aperture diameter (mm)	9ϕ	11ϕ	9ϕ	114		13		14
No. of Aperture	560 (800)	522 (800)	209	1080		228		1300
Transparency						~38		

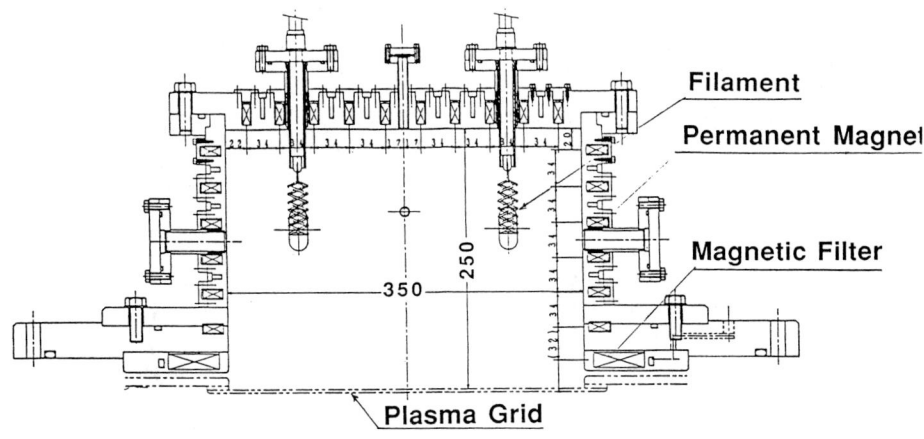

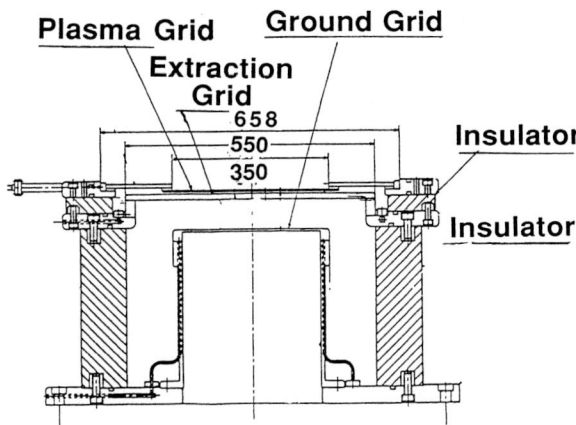

Fig. 10. Engineering design of a prototype negative ion source for LHD NBI.

results of the development seem to be acceptable for the actual large high current negative ion source but the ion source is still in the development stage. Its performance must be highly reliable and reproducible. The applicability of negative ion sources will be demonstrated by the negative ion sources of JT-60 and LHD in Japan in the near future.

References

[1] M. Allan, S.F. Wong, Effect of Vibrational and Rotational Excitation on Dissociative Attachment in Hydrogen, Phys. Rev. Lett. 41 (1978) 1791–1794.

[2] J.M. Wadehra, J.N. Bardsley, Vibrational and Rotational State Dependence of Dissociative Attachment in e-H_2 Collisions, Phys. Rev. Lett. 41 (1978) 1795–1798.

[3] M. Bacal, G.W. Hamilton, H^- and D-Production in Plasma, Phys. Rev. Lett. 42 (1979) 1538–1539.

[4] M. Bacal, A.M. Bruneteau, W.G. Graham, G.W. Hamilton, M. Nachman, Pressure and Electron Temperature Dependence of H^- Density in a Hydrogen Plasma, J. Appl. Phys. 52 (1981) 1247–1254.

[5] J.R. Hiskes, A.M. Karo, Generation of Negative Ions in Tandem High Density Hydrogen Discharges, J. Appl. Phys. 56 (1984) 1927–1938.

[6] K.N. Leung, K.W. Ehlers, M. Bacal, Extraction of Volume Produced H^- Ions from a Multicusp Source, Rev. Sci. Instr. 54 (1983) 56–61.

[7] E. Nicolopoulos, M. Bacal, H.J. Doucet, Equilibrium Density of H^- in a Low Pressure Hydrogen Plasma, J. Phys. (Paris) 38 (1977) 1399–1404.

[8] Y. Okumura, H. Horilke, T. Inoue, T. Kurashima, S. Matsuda, Y. Ohara, S. Tanaka, A High Current Volume H⁻ Ion Source with Multi-Aperture Extractor, 4th International Symposium on the Production and Neutralization of Negative Ions and Beams, Brookhaven National Lab. Oct. 27–31, 1986, American Institute of Physics, NY, pp. 316–325.

[9] Y. Ohara, M. Akiba, M. Araki, M. Hamabe, T. Inoue, H. Kojima, M. Kuriyama, M. Matsuoka, M. Mizuno, Y. Matsuda, K. Okumura, M. Seki, S. Tanaka, K. Watanabe, Recent Activities on Nagoya Ion Beams at JAERI, Proceedings of the 13th Symposium on Fusion Engineering, Knoxville, 1989, IEEE, NY, 284–287.

[10] Y. Okumura, M. Hamabe, T. Inoue, H. Kojima, Y. Matsuda, Y. Ohara, M. Seki and K. Watanabe, Cesium Mixing in the MultiAmpere Volume H⁻ Ion Source, Proceedings of the 5th International Symposium on Production and Neutralization of Negative Ions and Beams, Brookhaven, 1989, American Institute of Physics, NY, 169.

[11] A. Ando, K. Tsumori, Y. Oka, O. Kaneko, Y. Takeiri, E. Asano, T. Kawamoto, R. Aklyama, T. Kuroda, Large Current Negative Hydrogen Ion Beam Production, Phys. Plasma 1 (1994) 2813–2815.

[12] Y. Oka, T. Kuroda, The Effect of a Multipole Magnetic Field on the Characteristics of a Multifllament Plasma Source, Appl. Phys. Lett. 34 (1979) 134–136.

[13] Y. Oka, T. Kuroda, The Effect of the Magnetic Field in the Behavior of Magnetic Multipole Discharge, Jpn. J. Appl. Phys. 15 (1976) 913–994.

[14] D.M. Goebel, Ion Source Discharge Performance and Stability, Phys. Fluids 25 (1982) 1093–1102.

[15] Y. Oka, S.K. Guharay, K. Sakurai, O. Kaneko, T. Kuroda, Development of a Large Bucket Plasma Source, Jpn. J. Appl. Phys. 26 (1987) 1125–1131.

[16] R. Thomae, H. Deitinghoff, H. Hopman, H. Klein, A. Schempp, Fusion Applications of RF Accelerators, Proceedings of the 5th International Symposium on Production and Neutralization of Negative Ions and Beams, Brookhaven, 1989, American Institute of Physics, NY, pp. 661.

[17] O.A. Anderson, I. Soroka, J.W. Kwan, R.P. Wells, Injector for RFQ Using Electrostatically Focused Transport and Matching, Proceedings of the 5th International Symposium on Production and Neutralization of Negative Ions and Beams, Brookhaven, 1989, American Institute of Physics, NY, pp. 676.

[18] S.K. Guharay, K. Tsumori, M. Hamabe, Y. Takeiri, O. Kaneko, T. Kuroda, Simple emittance measurement of H⁻ beams for neutral beam injectors in magnetic fusion, Rev. Sci. Instrum. 67 (1996) 2534–2537.

[19] Y. Takeiri, O. Kaneko, Y. Oka, K. Tsumori, E. Asano, R. Akiyama, T. Kawamoto, T. Kuroda, A. Ando, High-Energy and High-Current Hydrogen Negative-Ion Beam Production with an External-Filter-Type Large Negative-Ion Source, Rev. Sci. Instrum. 67 (1996) 1021–1023.

[20] Y. Belchenko, C. Jacquot, J. Pamela, D. Riz, Directed Cesium Deposition into a Large Volume Negative-Ion Source, Rev. Sci. Instrum. 67 (1996) 1033–1035.

[21] Y. Okumura, Y. Fujiwara, T. Inoue, K. Miyamoto, A. Nagase, Y. Ohara, K. Watanabe, High Power Negative Ion Sources for Fusion at the Japan Atomic Energy Research Institute, Rev. Sci. Instrum. 67 (1996) 1097–1692.

[22] C. Jacquot, J. Pamela, D. Riz, Negative Ion Production in Large Volume Source with Small Deposition of Cesium, Rev. Sci. Instrum. 67 (1996) 1036–1037.

[23] A. Simonin, J. Bucalossi, C. Desgranges, M. Fumelli, C. Jacquot, P. Massman, J. Pamela, D. Riz, R. Trainham, Y. Belchenko, Negative Ion Beam Development at Cadarache, Rev. Sci. Instrum. 67 (1996) 1102–1107.

Advances in techniques for diagnosing fusion plasmas

Peter E. Stott

JET Joint Undertaking, Abingdon, Oxfordshire OX14 3EA, UK

Abstract

Fusion research relies heavily on plasma diagnostics in order to operate safely, to explore and optimise the performance of a fusion experiment and to study in detail the plasma physics. In recent years, there have been many advances in plasma diagnostics, not only as a result of the development of new and innovative techniques, but also by the improvement and careful application of well-established methods. A magnetically-confined plasma has many degrees of freedom and is influenced by many boundary conditions. Consequently, to fully characterise a fusion plasma the measurement of a correspondingly large number of plasma parameters is required. Some parameters, such as the magnetic field, current, temperature and density, are fairly obvious requirements, but some are more subtle or complex. Many of these parameters cover very wide dynamic ranges, either in different regions of the plasma at the same time or at different times during the evolution of the plasma. Obvious examples are plasma temperatures, which range from a few electron volts (a mere ten thousand degrees Celsius) in the divertor to tens of kilo-electron volts (a few hundred million degrees) in the plasma core and neutron yields, which range over many orders of magnitude. Usually, such wide ranges cannot be measured with a single instrument or measurement method. Consequently, many parameters require several different diagnostic instruments to cover their whole dynamic range. Moreover, independent methods of measuring the most important parameters, particularly those used for feed back control, are desirable for reliability and to resolve uncertainties in interpretation or interference from other effects. JET commenced operation in 1983 with about 20 diagnostic systems and now has about 70. Even so, some important parameters are not completely diagnosed.

Measurements in fusion plasmas utilise a wide variety of techniques. Some methods are passive and use the detection of particles or radiation emitted spontaneously from the plasma, other methods use active probing of the plasma with beams of particles or electromagnetic radiation from an external source. Diagnostics based on electromagnetic radiation are particularly important; fusion plasmas emit electromagnetic radiation over a very wide spectral range, from high energy gamma rays through hard and soft X-rays, ultra-violet, visible, infra-red to microwave radiation. Each wavelength range conveys information about specific aspects of the plasma and coverage of the whole range is necessary.

It is impossible in the space of a short paper to cover the wide range of recent diagnostic development. This paper will discuss some of the general principles and requirements. A few examples will be used to illustrate the present state of the art. We will briefly consider the future requirements and challenges of designing diagnostics for ITER and fusion reactors. © 1997 Elsevier Science S.A.

1. Introduction

In the Oxford Concise Dictionary, the word diagnostic has two definitions: 'a symptom' and 'of or assisting diagnosis'. Turning to the word diagnosis reveals two definitions: 'identification of disease by means of a patient's symptoms', which sounds rather depressing and inappropriate to our subject and 'ascertaining the cause of a mechanical fault', which clearly has some connection with plasma diagnostics. Whenever a mechanical fault occurs in a fusion experiment, it is usually assumed initially (unfairly in most cases) that one of the diagnostics must have been the cause. In fact the word diagnosis is derived from the Greek words $\Delta\iota\alpha$ which means 'through', or in this context 'thorough' or 'deep', and $\Gamma\nu\omega\sigma\iota s$—'knowledge'. Hence diagnosis means literally 'a deep knowledge', a definition which is certainly very appropriate for fusion research.

1.1. Why do we need diagnostics?

In fusion research we need good diagnostic measurements for three main applications:
1. for real-time feedback control of the plasma. In most present-day fusion experiments, real-time feedback control is essential for plasma shape and position. These quantities are determined usually from magnetic measurements using pick-up coils mounted on the vacuum vessel walls, although sometimes other parameters have been used. Feedback control of plasma current and plasma density are also fairly common in present-day fusion machines, but the control of almost all other tokamak and plasma parameters involves a human, with considerable experience of tokamak operation and folk lore, in the feedback loop. The next generation of fusion experiments will require much more complex control systems to replace the human link; already at least 20 plasma parameters have been identified as necessary for the control of ITER;
2. to guide the optimisation of performance. It is very difficult to bring a fusion device to the peak of its potential and it would be impossible without good diagnostics. A particular difficulty is that operating modes which have been evolved and optimised in one device usually need additional optimisation when transferred to a new experiment. Diagnostics play a key role in assisting the tokamak physicists to monitor what is happening in the plasma. One of the most significant factors has been the rapid advance in data acquisition technology, in particular the development of real-time data analysis systems which can provide displays of key plasma parameters in the control room;
3. to measure the plasma parameters needed for understanding the physics of fusion plasmas. A wide range of plasma diagnostics is required to provide the data for the detailed understanding of the physics, including such topics as energy and particle confinement, stability, impurity production and transport, H modes, divertor and edge physics, fusion products, heating etc.

JET started operation in 1983 with a basic set of about 20 diagnostic systems and now has about 70 systems. The increase in the number of diagnostics has been driven partly by the steady advances in diagnostic techniques and partly in response to increasing demands from the physicists involved in tokamak operation and interpretation. The number of diagnostics has been reviewed periodically, but each review has concluded that all of these systems are required to support the experimental programme.

1.2. Why do we need so many diagnostics?

The question 'why do we need so many diagnostics' is asked quite frequently. There are several reasons:
1. a magnetically confined fusion plasma has many degrees of freedom and its state is influenced and determined by many boundary conditions, consequently we need to measure many plasma parameters in order to quantify the plasma;
2. each parameter covers a very wide range of values at different times during the evolution of the plasma and also in different regions in the plasma at the same instant in time;
3. it is important to make measurements with independent instruments and techniques in order to ensure reliability, to provide redun-

Table 1
Summary of the main diagnostic methods

Probe method	Emission	Transmission	Scattering
Ions	Lost-alpha detectors		Heavy ion beams
Neutral atoms	Passive neutral particles		Active beam neutral particles
Photons	Electron cyclotron emission	Reflectometry	Thomson scattering
	passive spectrometry	Interferometry	Collective scattering
	Soft X-ray cameras	ECA	Beam emission spectrometry
Neutrons	Neutron flux		
	Neutron spectrometry		
Miscellaneous	Electromagnetic sensors		
	Langmuir probes		
	Surface diagnostics		

dancy, to allow cross-calibration and to eliminate possible ambiguities in the interpretation. In consequence, most plasma parameters require several different diagnostic systems but fortunately many diagnostic methods have the capability to measure several different parameters. Usually these factors roughly cancel each other so that there is a rough numerical equivalence between the number of diagnostics installed on a fusion experiment and the number of parameters that are measured.

2. Diagnostic methods (Table 1)

Some years ago during a debate on fusion in the House of Lords (the upper chamber of the British Parliament) the question 'how do they measure a temperature of 200 million degrees' prompted the remark 'I expect that they use a very long thermometer'. In fact, extrapolating the scale on a conventional mercury-in-glass thermometer (typically 20 cm corresponds to 100°C) to fusion temperatures would require an instrument approximately 400 km long. If such an instrument were practicable, it would allow the temperature in the core of JET to be read in Luxembourg. In fact, although material sensors and probes (Langmuir probes are a typical example) have important applications in the edge and divertor regions of fusion plasmas, where plasma temperatures are less than 100 eV (a mere one million degrees Celsius or about 2 km on the mercury-in-glass scale) measurements in the hotter core regions must use remote sensing techniques.

It is convenient to divide these remote sensing methods into three groups:
1. emission methods which detect and measure particles (usually neutral atoms or neutrons since, with the exception of very energetic alpha particles, charged particles cannot escape across the confining magnetic fields) or electromagnetic radiation (photons) which are emitted spontaneously by the plasma. In most cases the measured signal is an integral of the emission along the line of sight of the detector and so emission methods usually lack inherent spatial resolution, except in certain special cases, such as electron cyclotron emission (where there is a direct correspondence between the frequency of the emitted radiation and the radial position) and some neutrons and soft X-ray diagnostics where the emission is predominantly from the plasma core region;
2. transmission methods which involve the passage of a probing beam, usually photons, through the plasma. The properties of the plasma along the path of the beam are determined by the way that the beam has been modified. As with the emission methods, this generally gives an integral measurement of the relevant plasma parameter(s) along the line of sight of the diagnostic so that measurements along many different lines of sight followed by

an inversion technique, such as Abel inversion, are required to determine spatially localised values. However there are some transmission methods, for example electron cyclotron absorption (ECA) which inherently is sensitive only to a particular point along the line of sight;

3. scattering methods which utilise a beam of photons or particles to penetrate the plasma and measure the particles or photons that are scattered by the plasma into a different direction. Techniques such as charge-exchange spectroscopy, based on neutral atom beams, where the probing beam induces the plasma to emit a neutral atom or photon have a similar topology to scattering methods. Inherently these methods give a localised measurement at the point of intersection of the probing and scattered beams.

Within the constraints of this paper, it is clearly impossible to do justice to the many recent advances in diagnostics. To illustrate the state of the art, I have selected just three examples out of a least ten times that number of candidates.

3. Microwave reflectometry

Although microwave reflectometry is a well-known technique in atmospheric physics, where it is used to probe the density of the ionosphere and it was applied in several early tokamaks, it did not develop as a front line fusion diagnostic until quite recently. The past decade has seen substantial progress, due in part to substantial advances in microwave technology in military and commercial communications fields. Microwave reflectometry has two important applications in fusion plasmas: (i) the measurement of the electron density profile at the plasma edge where density gradients are very steep and good spatial resolution is required; (ii) to study density fluctuations. An electromagnetic wave, launched from an antenna outside the plasma edge, propagates into the plasma until it is reflected at the point where the plasma refractive index becomes zero. In the simplest case, for a wave in the ordinary polarisation, where the electric vector of the wave is parallel to the plasma magnetic field, there is a direct correspondence between the density at the point of reflection and the frequency of the probing wave ($n_e = 12.4 \times 10^{-3} f^2$ where n_e is the local electron density in units of m^{-3} and f is the frequency in GHz). Knowing the frequency and hence the density, the problem, in essence, is to determine the distance from the antenna to the point of reflection. In the extraordinary polarisation, where the electric vector of the wave is perpendicular to the plasma magnetic field, the point of reflection depends on the local magnetic field as well as the density. Problems are caused if there are strong fluctuations in the plasma edge which can confuse the density profile measurement. Several different techniques of reflectometry have been developed [1] including measurements of the phase difference between the incident and reflected beams (similar to interferometry) and time-of-flight methods (similar in principle to radar). There is no overall 'best' method, each has certain merits and certain limitations, and the final choice depends on the particular application. An example of an application of reflectometry to measure edge density profiles in ASDEX-U tokamak is shown in Fig. 1. This system utilises antennas on both the high- and low-field sides of the tokamak [2]. The signals are generated by stable microwave oscillators and active frequency multipliers which provide very reproducible, ultra-fast sweeps with a scan time of ~ 10 μs which effectively 'freezes' the plasma during the scan time and greatly reduces the effects of temporal density fluctuations. Reflectometry is particularly well-suited to measuring the steep density gradients at the plasma edge (Fig. 2).

4. Optical imaging

There is an adage in journalism that a picture is worth a thousand words. Compact colour TV cameras covering the visible and infra-red spectra, which have developed rapidly during the past few years, are finding many applications in fusion including the provision of real-time images of the plasma as a control room monitor, 2-d spectroscopy of impurity fluxes, surface temperature

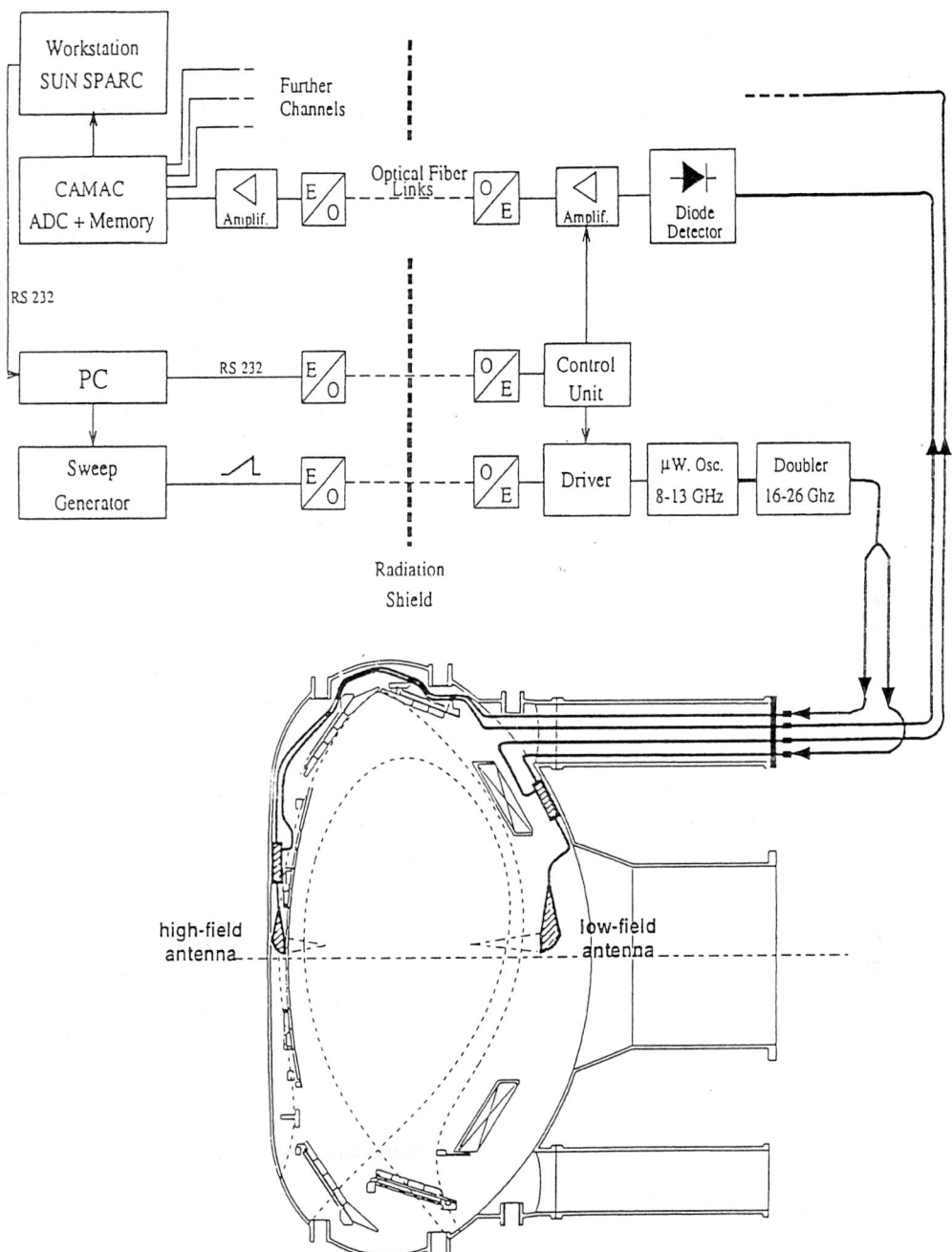

Fig. 1. Schematic of the microwave reflectometer on ASDEX-U showing the high and low-field antennas (reproduced from Ref. [1]).

and power flux measurements. JET has recently developed a new optical relay system, with three separate lens systems housed in a single radial port, which allows the cameras to be placed outside the machine structure about 2 m back from the first wall. One lens system gives a wide angle view looking tangentially into the torus and is used with a colour camera as a control room monitor. This has proved indispensable to the tokamak operators and gives prompt indication of errors in plasma positioning, excessive interaction with the first wall and divertor, etc. Fig. 3 shows a typical view of a divertor plasma in JET. The second lens system gives a telescopic view of the divertor tiles and is used with an interference filter and TV camera to measure the impurity light emission from a region of the divertor. The impurity flux from the divertor can be derived from the light intensity. The third lens system is not yet fully commissioned but will be coupled to an infra red camera to measure the surface temperature and power flux onto the divertor tiles.

5. Neutrons

Fusion reactions produce neutrons, with energy 2.45 MeV from the d–d reaction and 14 MeV from the d–t reaction, which have important applications as diagnostics. The neutron yields cover a wide dynamic range: from $< 10^{10}$ s^{-1} in small tokamaks, to $\sim 10^{19}$ s^{-1} when JET and TFTR operate with tritium, yields will extend to $\sim 10^{21}$ s^{-1} in ITER. No single instrument or measurement technique can cover this very wide dynamic range.

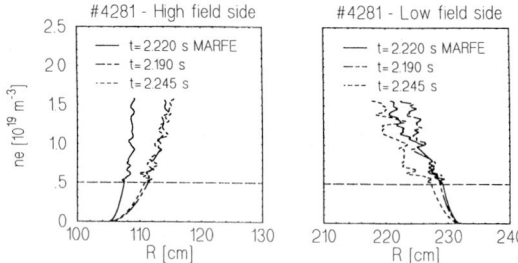

Fig. 2. An example of steep density gradients at the plasma edge in ASDEX-U measured with reflectometry (reproduced from Ref. [1]).

Neutrons are uncharged and therefore cannot be measured directly, so detection techniques are based either on nuclear transformation reactions or on nuclear scattering. There is a wide range of instrumentation, much of which has been derived from nuclear physics instrumentation but there are several factors relevant to fusion which require special attention. These include the need for detectors to be insensitive to (or shielded from) magnetic fields, to have low response to gamma radiation, to be tolerant of high radiation doses, to be insensitive to electromagnetic interference and to be capable of operating at high repetition rates. In most cases the neutron detector needs to view the plasma through a collimator to provide spatial resolution and to be placed inside a massive shield in order to remove the background of neutrons and gammas that have scattered off the machine and building structures. The collimator and shield usually have dimensions of several meters and weights of the order of a hundred tons so that neutron diagnostics are large and massive.

JET has a 'neutron camera' with two large collimator arrays, one viewing horizontally with ten sight lines, the other vertically with nine sight lines, to provide spatially-resolved neutron emission rates (Fig. 4). The multiple sight lines allow a tomographic inversion on a coarse spatial mesh. The neutron emission profiles shown in Fig. 5 were measured during one of the JET Preliminary Tritium Experiments in 1991. These were the first measurements of substantial yields of 14 MeV fusion neutrons (the peak yield in the time frame at 12.85 s corresponds to approximately 1.7 MW of fusion power). The peaking of the neutron yield in the core and the subsequent rapid collapse due to internal instabilities are clearly seen.

6. Summary and conclusions

Plasma diagnostic techniques have advanced substantially during the past decade. The diagnostic systems now in routine use on the present generation of large fusion experiments operate to standards of reliability and accuracy that would have been considered impossible in previous generations of fusion experiments. The three examples of advanced diagnostics that have been selected to

Fig. 3. Wide angle view of a JET plasma. The two strike points in the divertor can be seen at the bottom of the frame. The bright spots in the mid-plane are interactions with the poloidal limiter and the impurities which are released can be seen to steam along and illuminate a magnetic flux tube at the plasma edge.

illustrate this paper show clearly the contributions made by advances in diagnostics to control, optimisation and understanding of fusion plasmas. Microwave reflectometry has proved to be a very valuable technique for measuring the electron density gradients at the plasma edge. Reflectometry is also an important diagnostic for studying density fluctuations. Optical imaging systems have proved indispensable to machine operation, but also contain a vast amount of quantitative data on impurity sources, power fluxes and the temperatures of limiter and divertor surfaces. Neutron diagnostics have become increasingly important as fusion operating conditions have advanced towards reactor conditions and will become even more important in tritium-burning fusion experiments.

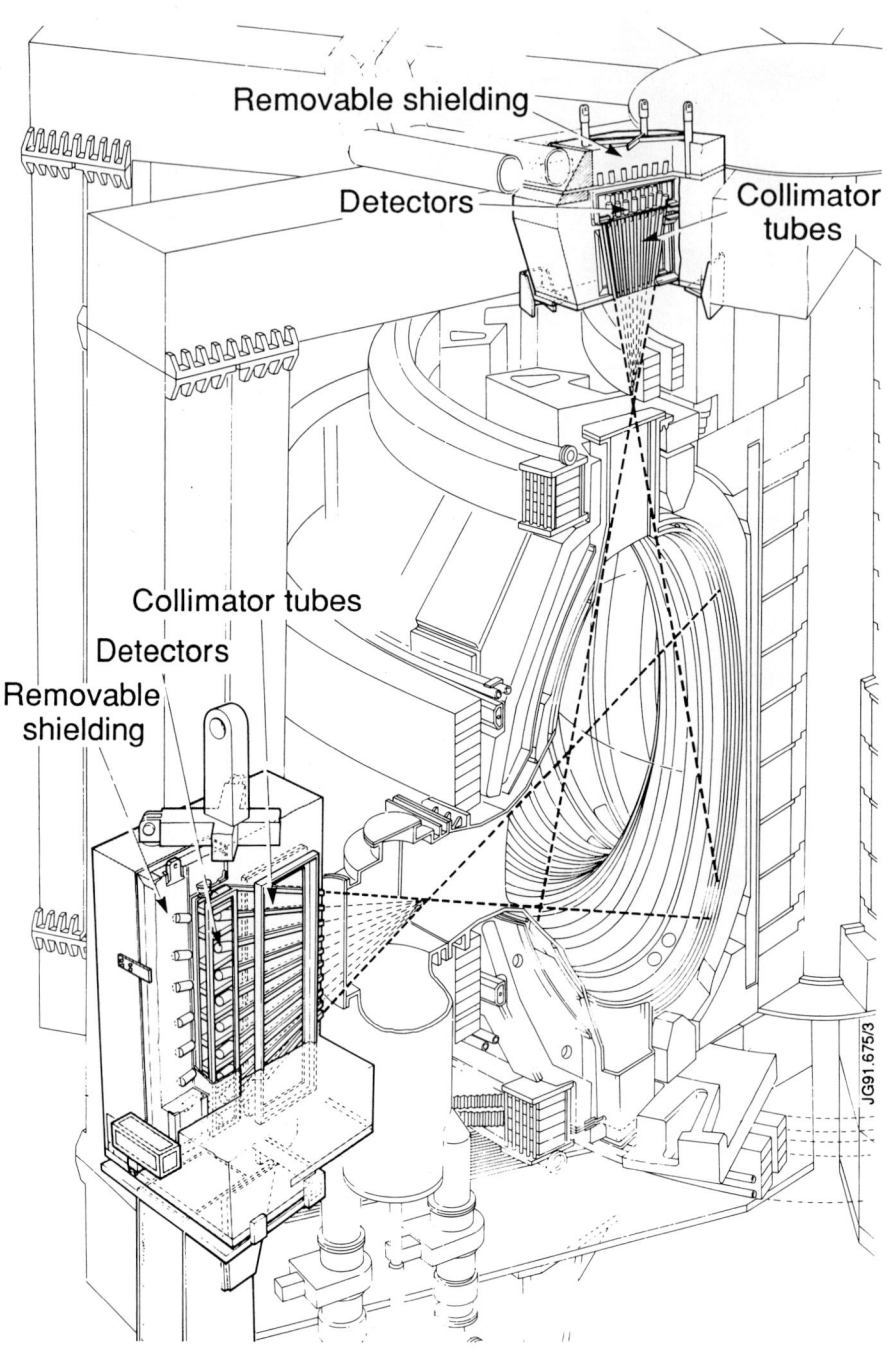

Fig. 4. Schematic of the radial and vertical neutron cameras on JET.

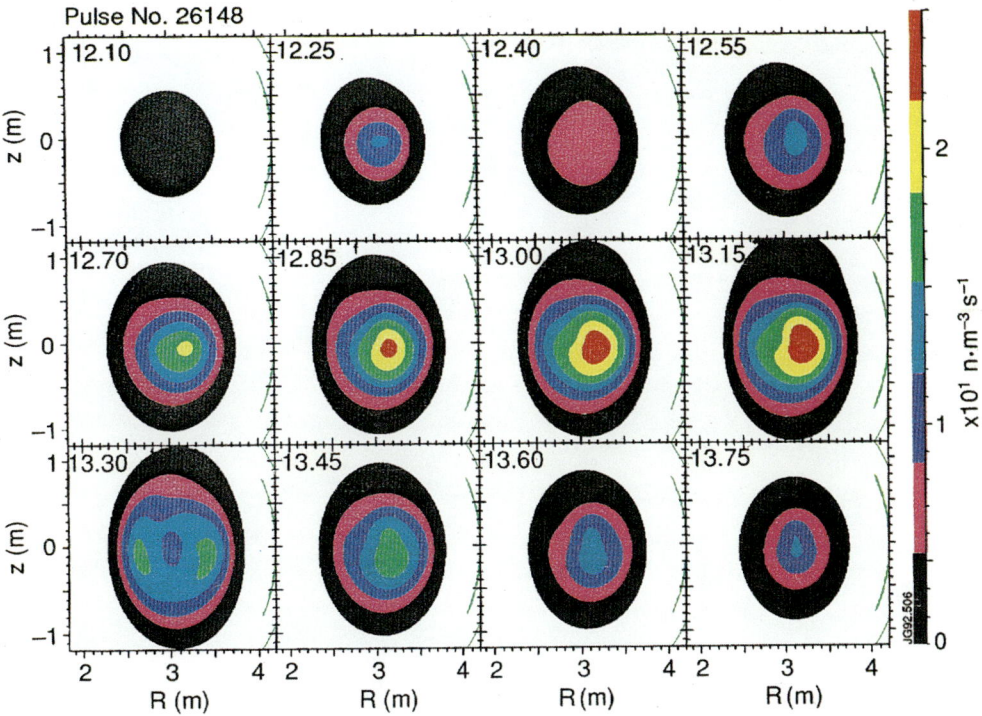

Fig. 5. Neutron emission profiles during one of the JET Preliminary Tritium Experiments in 1991. The peak yield in the time frame at 12.85 s corresponds to approximately 1.7 MW of fusion power. The peaking of the neutron yield in the core and the subsequent rapid collapse due to internal instabilities are clearly seen.

ITER will require an extensive set of plasma diagnostics in order to achieve its operational and programmatic goals. The number systems and the diagnostic methods employed will be broadly similar to those existing on today's large tokamaks such as JET, TFTR and JT60U. However, the demands on the diagnostic systems in ITER will be enhanced. In particular, more measurements will be used in real-time control of the plasma and so will have to be made at very high levels of reliability and availability. Moreover, the implementation of the diagnostics will be especially demanding because of the high levels of radiation, the limited access and the stringent requirements for vacuum integrity and tritium containment. Together these requirements constitute a substantial design challenge to diagnostic physicists and engineers. The overall specifications of the ITER diagnostic system have been determined, and the role which the plasma measurements will play in the ITER physics programme has been summarised. Individual diagnostic systems are now being designed by a combined effort of the ITER Joint Central Team and the Home Teams [3].

References

[1] C. Laviron, A.J. Donné, M.E. Manso, J. Sanchez, Plasma Phys. Control. Fus. 38 (1996) 905.
[2] A. Silva et al., Fusion Technology—Proceedings of the 17th Symposium (Rome 1992) vol. 1, Amsterdam North Holland, 1992, pp. 747.
[3] P.E. Stott, G. Gorini, E. Sindoni, Diagnostics for Experimental Thermonuclear Fusion Reactors, Plenum, New York.

On-site developed components for control and data acquisition on next generation fusion devices

C.A.F. Varandas [a,*], B.B. Carvalho [a], C. Correia [b], C. Loureiro [b], J.L. Malaquias [a], A.P. Rodrigues [a], J. Sousa [a]

[a] *Associação EURATOM/IST, Centro de Fusão Nuclear, Instituto Superior Técnico, 1096 Lisboa Codex, Portugal*
[b] *Associação EURATOM/IST, Departamento de Física, Universidade de Coimbra, P-3000 Coimbra, Portugal*

1. Introduction

The architecture and complexity of the COntrol and Data Acquisition Systems (CODASs) of magnetic confinement nuclear fusion plasma experiments have followed the evolution of the technologies of computer science and information theory as well as of the dimensions of the devices and the specialization of their scientific programmes.

The CODAS of the first generation fusion machines were designed in a centralized philosophy based on CAMAC instrumentation, stand-alone controllers and analog timing units. The systems of the present day fusion experiments have been mainly conceived in a distributed approach, preferentially based on CAMAC and VME instrumentation, coupled multi-variable controllers, digital timing units and feedback control systems for optimization of the plasma performance [1–4]. Recently, the CODAS of some large size machines have been progressively redesigned and upgraded aiming at fulfilling the operation requirements of new experimental components, feedback control and long pulse operation as well as the integration of new technologies [5–7].

The operation of next generation fusion devices will imply special requirements on control and data acquisition due to the complexity of the experiments, the long duration of the discharges, the D-T operation, the specialization of the research programmes and the international collaboration on the design, construction and operation of these machines.

This paper summarizes the main requirements, principles and technical issues to be taken into account in the conceptual design of the CODAS of the new fusion experiments and presents the work that the portuguese EURATOM Association is carrying out aiming at contributing to the development of their control and data acquisition systems.

2. Conceptual design

The control and data acquisition system of a new generation fusion device [8,9] should be conceived in a distributed, integrated, hierarchical, manufacturing independent and user-friendly approach, taking advantage of the most recent improvements in networking, front-end processing and database management.

* Corresponding author.

Distributed means that control and data acquisition are performed by several dedicated subsystems designed in a function oriented structure and linked by fast networks. This architecture allows to develop independently the subsystems, to minimize the data traffic in the network and to increase the system safety. It also provides upgrading capacity to meet new experimental requirements during the life of the experiment and flexibility to include new technologies. Integrated means that there is a central system that supervises the operation of all subsystems and provides the operator-machine interfaces, data storage and on-site and remote communications. Hierarchical means that CODAS can be divided as usual [1,4] into three levels: the supervisor system, the dedicated control of each subsystem and the local instrumentation that guarantees the connection to the machine components.

The utilization of industrial standards for instrumentation hardware, data bus, networking protocol, operating system and data management software reduces the development time and costs and increases the system life-time. However, the design of the hardware and software components should be made in a manufacturing independent approach in order to be able to integrate important technological changes in the commercially available products. Section 3 describes an on-site developed universal hardware bus interface, which allows the simultaneous utilization of instrumentation of any current and future bus in the same crate.

CODAS must perform the typical control tasks of a fusion experiment: coordination of the operation of a large number of plant components, machine and human safety protection, monitoring, timing and real time control of the parameters and processes that influence the plasma equilibrium and stability as well as the discharge performance. Different discharge phases, such as technical preparation and current ramp-up, flat-top or ramp-down, require specific control scenarios.

Safety will be an important issue in the new fusion experiments since their operation close to unstable regimes and on not yet explored parameter ranges for performance optimization and improved confinement scenarios studies will increase the likelihood of hazard. The traditional machine protection based on pre-defined limits of some parameters should be changed by an active real time system that detects deviations between a time dependent set of supervised signals and limits and performs appropriate countermeasures [10]. Moreover D-T operation originates new safety problems related with the presence of nuclear products. The operation of a fusion experiment and of its diagnostics from control rooms located very far from the device site implies new machine and database safety procedures as well as changes in the monitoring processes of the experiment operation.

The central timing system of a fusion device is a key component of the control system since it provides synchronization, clock and triggering for the experimental subsystems. High resolution signals and optimum synchronization are needed to guarantee adequate machine control, efficient operation of some scientific diagnostics as well as the comparability of the measurements of different plasma parameters. Section 4 presents the architecture of a locally developed intelligent timing system, designed in a distributed approach based on advanced triggering and event processing techniques aiming at active control, accurate data acquisition and reduction of the collected data [11].

Real time control is performed by systems utilizing controllers of different technologies: conventional digital based on transputers [12] or Digital Signal Processors (DSPs) [13,14], fuzzy logic [15] and neural networks [16,17]. Section 5 describes a home developed intelligent control system based on an array of four DSPs and provided with two Analog-to-Digital Converter (ADC) channels, two Digital-to-Analog converter (DAC) channels and 16 digital Inputs and Outputs (I/O).

The rise of the discharge duration of the fusion experiments and the refinement of the experimental techniques imply the enhancement of the digitizer local memory, leading to space problems in the design of these modules and to an increase of their costs. Alternative solutions have been proposed based on the substitution of the digitizer memory by a personnel computer [18] and on cyclic processing [19]. Otherwise the high repetition rate of the shots does not allow data transfer, processing and display between two consecutive discharges. Due to these two facts the long duration of the discharges of the new fusion experiments (for instance 1000 s on ITER) implies the utilization of intelligent digitizer and video based modules, providing real time data validation, processing and reduction. These modules acquire data during a short time when a special event is detected by a real time control system or by an intelligent timing system. Their pre-trigger operation guarantees that there is no data lost during the delay between the occurrence of the event and the trigger of the digitizers. Sections 6 and 7 contain brief descriptions of two locally developed real time digitizer modules based on transputers and DSPs. The video based modules will be particularly adequate for data acquisition of applied optics diagnostics as well as for divertor and plasma wall loading studies [20].

The connection of sensors, detectors and diagnostics to computers should be performed by industrial buses allowing multi-processing and fast data transfer. Possible candidates could be VME, PCI and FAST CAMAC [21].

The CODAS software should be conceived aiming at: (i) the easy integration of new subsystems and technologies; (ii) the compatibility of different operating systems; (iii) the remote access for operation and data analysis; (iv) a smooth flow of information from the physical sensors to the final workstations where data will be analysed by the intervening physicists; (v) to keep the amount of information at a manageable level in order not to choke the general data storage system and the valuable processing power in the rear end of the system; (vi) user-friendly operator-machine interfaces and image oriented forms of data acquisition.

An operating system should be found or developed from scratch. It must be flexible enough to run in an heterogeneous environment, easily ported to new hardware platforms as they become available and fully or partially upgradable with minimum pain for users and system administrators. An object oriented operating system might be the closest thing to meeting these goals, but the field is very dynamic nowadays and new approaches, for instance based on the Internet and the Web, might be available shortly.

3. Universal hardware bus interface

The main goal of the Universal Hardware Bus Interface (UHBI) is the connection of any type of host control computer platform with any type of current and future instrumentation buses (CAMAC, VME, VXI, GP-IB, FASTBUS, PCI,...) over standard network components. The advantage of this structure is the total isolation, connection over long distances and mutual independence of the control computers and the desired hardware components. The UHBI has been conceived aiming to facilitate the development and portability of applications on different programming languages and operating systems.

The UHBI has been implemented on a VME crate equipped with M680x0 ($x = 3, 4, 6$) CPU processor boards running the OS-9 operating system. This crate can also be loaded with acquisition and control boards, acting itself as a front-end instrumentation stage. For each subsystem bus the UHBI has a hardware adapter for the VME bus. This device will run a local server program to which remote client application programs will make a network connection to access instruments as well as control and data acquisition modules. Anyone writing client applications will not worry about hardware drivers or subroutine details for a specific hardware bus but will use instead a common application programming interface available on host systems, if possible having a graphical user interface.

Fig. 1 presents the block diagram of a control and data acquisition system with several bus systems and host computers. Communications to the host computers are done via Ethernet or FDDI LANSs using standard TCP/IP, RPC and NFS protocols. The global maximum data rate of 10 and 100 Mbit s^{-1} respectively, can be later extended by the use of ATM, Fibre Channel or SCI technology without major modifications in the application programs.

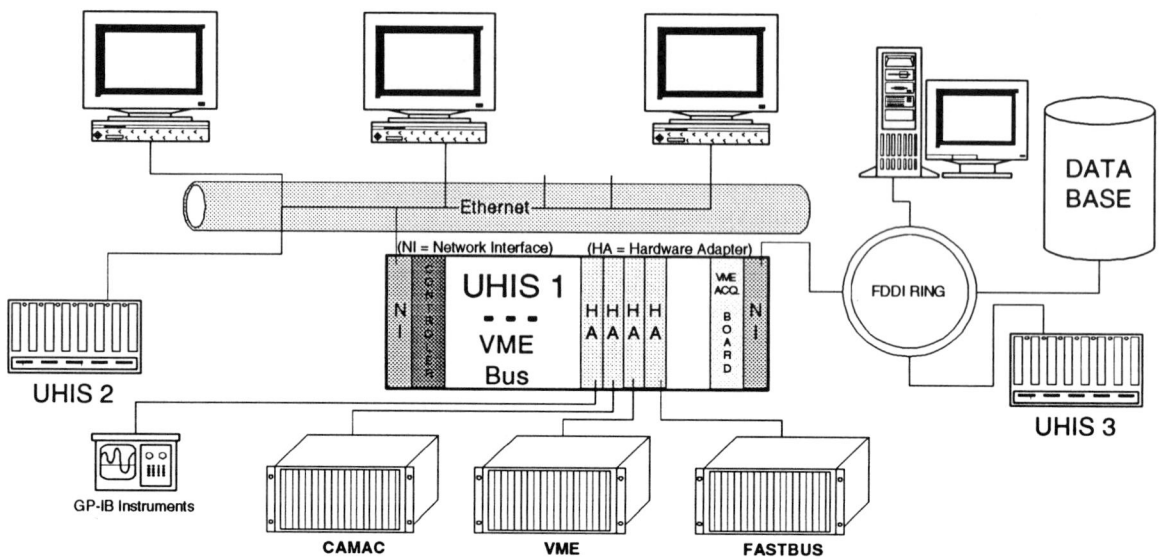

Fig. 1. A CODAS using an universal hardware bus interface.

A set of software packages will be developed in both the UHBI and host systems to create and maintain the necessary TCP/IP socket connections for data transfer, control commands and error/exception events. All the data acquired in different buses is transmitted to the host systems in the same general format. As the ethernet network cannot ensure a minimum latency time when transferring exception events the UHBI must select those with priority and service them locally.

The design of a CODAS using this interface has the following advantages: (i) easy upgrading of older control and data acquisition systems, integrating existing CAMAC modules, with up-to-date VME based equipment; (ii) access to experimental data and remote control over the network for any host enabling distributed processing; (iii) use of local real time processing capabilities for data reduction, re-organisation and synchronization; (iv) improved scalability for the CODAS architecture; (v) easy adaptation to future network communication technologies; (vi) fault tolerant (on software and hardware) data acquisition; and (vii) self-testing and auto-configuration of diagnostic subsystems.

4. Event and timing distribution system

An intelligent event and timing distribution system has been developed, conceived in a decentralized approach based on the existence of a master node, intermediate nodes acting as event commuters and slave nodes, linked by optical fibers (Fig. 2). The master node can synchronize all nodes by sending an event which propagates with the same delay to all nodes. The intermediate nodes allow to minimize the length of the connections, without enlarging significantly the propagation time. Assuming 32 bit transportation cells, 8 bit addresses and reserving one address for broadcasting, this topology allows 210 nodes.

An intermediate node is composed by 16 bi-directional communication units (Fig. 3 shows only eight, designated by COMM), which adapt data to the transportation media (optical signals). The serial FIFOs coupled to the COMM interfaces guarantee that data is not lost in the reception (when several events must be processed) and transmission (if more than one event is sent to the same interface). Data is processed by a central commuting unit, which checks the Code Redundancy Check (CRC) and is sent to the interface that allows routing to the destination address.

The slave node (Fig. 4) has three interfaces that perform the connections to the event distribution network, host computer bus and local instrumentation. Functionally, it contains a high precision programmable complex sequence generator and an event processor. The last receives events from each of the three interfaces and takes actions accordingly. If an event is received from the network, the event processor can, for instance, generate a sequence locally or an interrupt to wake-up the host computer. The COMM interface provides a stable clock (synchronized by the master node) to the sequence generator, which contains a programmable counter that changes state for each clock pulse. The counting value is an absolute time equal in all nodes, because all of them receive the same clock and are synchronized by the same broadcasting event. For each change of counter state, the absolute time is compared to a value stored in an event register. If the two values are equal a sequence (their parameters are stored in the register) will be generated in a local channel. If the counter overflows, a second block of event registers begins being used and an interrupt is sent to the host computer that starts loading another set of events in the first block and so on.

In this architecture the master node is not substantially different from a slave node. However, the master node generates the clock shared by all nodes and commuters. For reliability, all devices should have a local clock permanently running and synchronized from the master node (PLL based).

The maximum number of nodes (210 in the present topology) could be increased by using: (i) a network with two layers of intermediate nodes allowing a maximum of 3360 nodes; or (ii) several masters (the goal is up to eight) inserted in the same VME crate, each one with its own network, allowing a maximum of 1680 nodes. The first solution is disadvantageous in significantly increasing the event propagation time due to, not only the additional delay in the intermediate nodes, but also the need of more address bits in each packet, which will consume more transmission time. The second solution maintains the event propagation time and permits a maximum of 26880 output signals. However it only allows event propagation between networks through the VME bus at a slower rate.

5. DSP based real time control system

This intelligent system has been designed to run signal processing algorithms for real time control, noise suppression in some control signals, detection of events and generation of trigger signals for the fast data acquisition modules.

The architecture of the module (Fig. 5), presently supported on the PCI bus, consists of an array of four DSPs (TMS320C44) with a cycle time of 40 ns and the following 12 bit resolution, independent channels: two ADC inputs with a maximum sampling rate of 40 mega samples per second (MSPS) and two DAC outputs with a maximum update rate of 100 MSPS.

Each ADC channel has independent timing control and can acquire data in a free running mode to a 16 kWord dual port memory. All DSPs

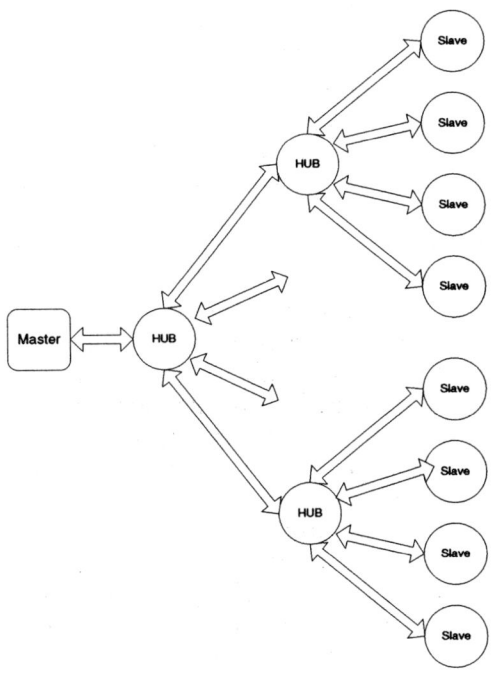

Fig. 2. Architecture of the event and timing distribution system.

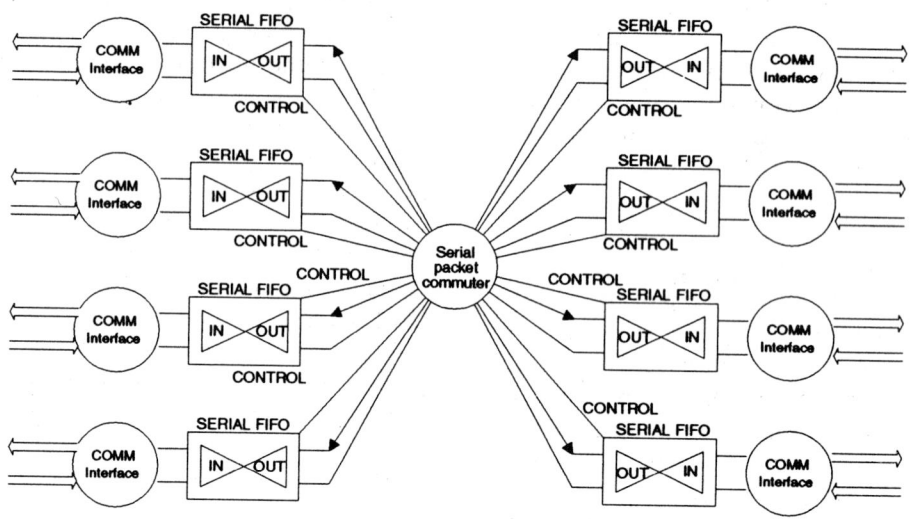

Fig. 3. Block diagram of an intermediate node of the event and timing distribution system.

can read data from any ADC channel and process it performing, for instance, FFT or cross-correlation between two acquired signals. One DAC channel is buffered with a 32 kWord FIFO memory and has an independent timing control that provides the signal generation capability of the channel. Each DAC channel has one DSP connected directly to it, allowing to run fast feedback control algorithms. Eight digital inputs and eight digital outputs are available for multi-purpose use. Some of the digital inputs generate interrupts to the DSPs.

One of the DSPs controls all the internal configurations as well as the PCI bus interface, through which the host computer downloads the board configuration settings and the algorithms that run in the DSPs.

All the logic needed to control the timing and selection of the internal devices is guaranteed by a gate array and several VLSI programmable logic devices.

The software will provide an operator interface based on Labview for Windows, the integration in a local area network and the operation by a central operating team.

A VME version of this module and drivers for the OS-9 operating system is expected to be developed in the near future.

6. 250 MSPS transputer based data acquisition module

The architecture of this module, described in detail elsewhere [22], was conceived to maximize the transputer availability for data processing and control while special care was taken to guarantee a good performance of the analog front-end. The module contains two complete 250 MSPS 8 bit data acquisition channels, each one with 16 kb of FIFO memory and a 25 MHz T800 transputer with 512 kb memory as well as trigger and control hardware (Fig. 6). The different and autonomous trigger sources, along with the pre-programmation of the acquisition channel by writing to a set of registers and the memory acquisition scheme, allow that data acquisition proceeds in parallel with data processing, maximizing the availability of the transputer for data processing and reduction activities.

Several transputer-based modules are now operating routinely in the microwave reflectometry diagnostic of the ASDEX-Upgrade tokamak [23]. A multiprocessor system of eight transputers in hypercube topology was associated with each acquisition channel, providing a local computer power of about 240 Mips and 32 Mb of RAM used to process and store data.

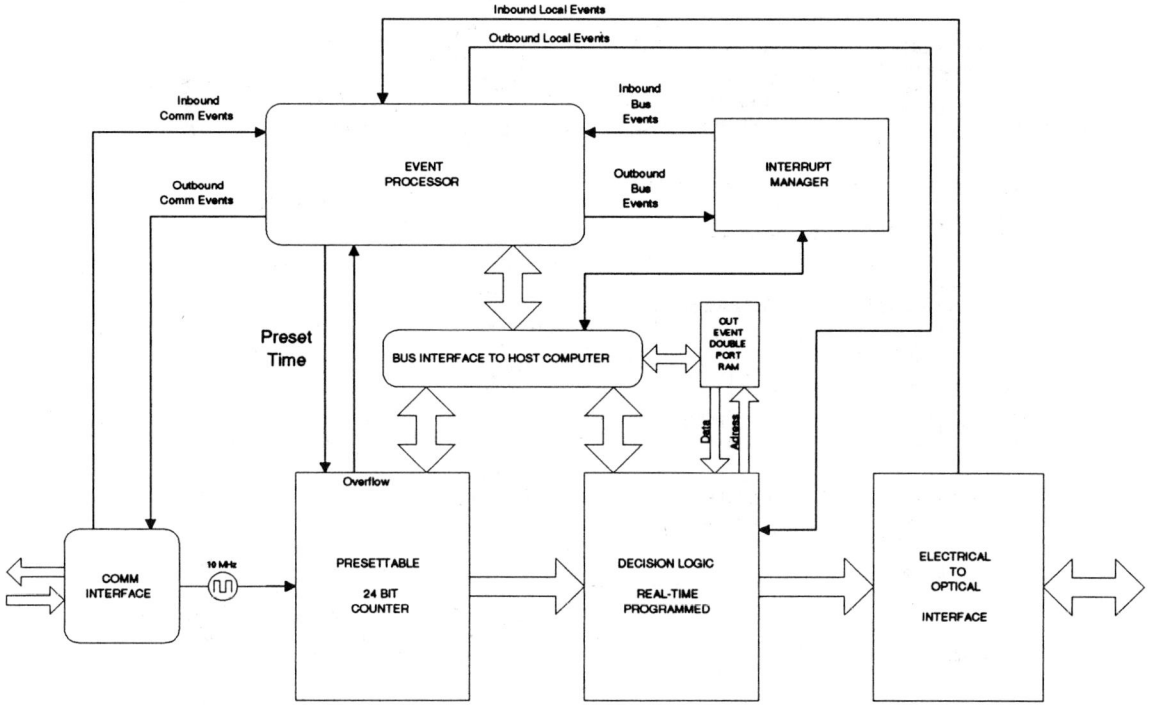

Fig. 4. Block diagram of a slave node of the event and timing distribution system.

7. 1 GSPS data acquisition module

The experience acquired with the transputer based data acquisition module described in the previous section has shown that, in many circumstances, an undesired dead time was perceived due to limitations in the rate at which data could be sent through the transputer links of the acquisition channels. So a new architecture was proposed for the faster acquisition modules, again considering the benefits of a larger data memory buffer under the control of the data acquisition block, but without neglecting the need of a local and powerful processing and control processor.

The design of this VME module (Fig. 7) is especially suited to systems needing multiple similar data acquisition channels operating in parallel and local data processing and reduction capabilities [24]. It includes a local DSP (TMS320C31 at 50 MHz with 2 Mb of 0 wait states SRAM) which can entirely control the module, read data (maybe using DMA) and process it in parallel with the data acquisition. To speed up communication with the VME bus host processor an output FIFO scheme was implemented allowing the host to read the processed data without disturbing the local DSP, improving significantly the availability of the DSP for data processing. Due to the asynchronous characteristics of data acquisition a bi-directional interrupt scheme was also implemented allowing for local DSP and host VME bus interrupts. If desired, the host can also known the status of the module by reading a local port.

The architecture shown allows up to four similar data acquisition channels (Fig. 8), capable of acquiring continuously up to 3 Mb of data at a 250 MHz sampling rate, to be housed in a single width double height VME bus module. Provision is made to operate the channels in an interleaved way, achieving higher sampling rates. The entire control of the data acquisition channels is made

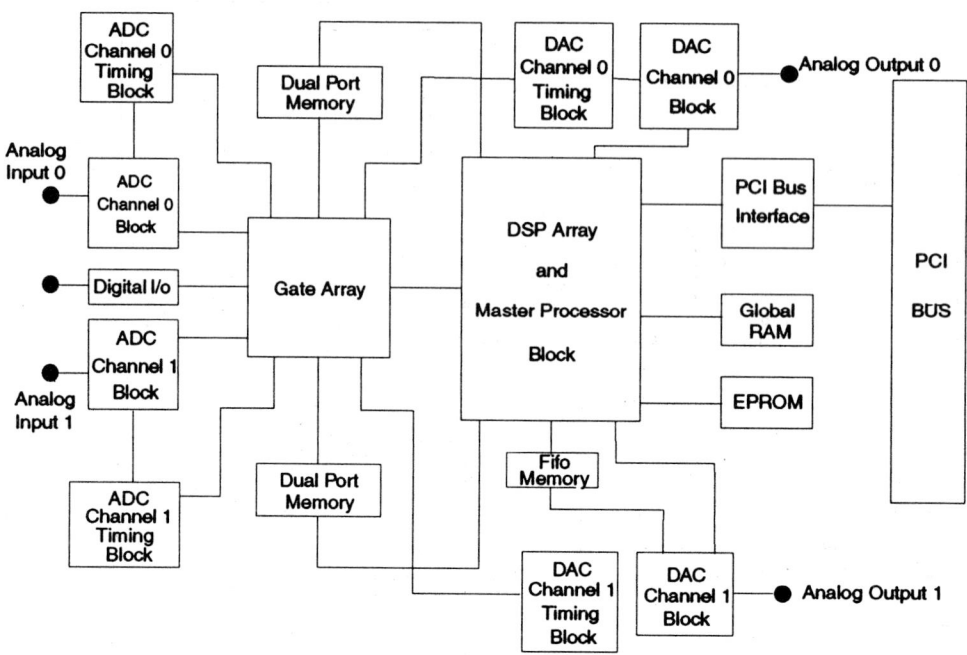

Fig. 5. Block diagram of the hardware of the DSP based real time control system.

throughout a high density gate array, allowing the rather complex control logic to be executed without the need of external intervention other than the initial programmation of some configuration ports. The local gate array can then control the data acquisition process. Several different acquisition strategies are supported, including repeated acquisition cycles of pre-programmed data lengths. The gate array is also responsible for transferring data to the data acquisition channel output FIFOs, from where it can be read and processed in parallel with the data acquisition.

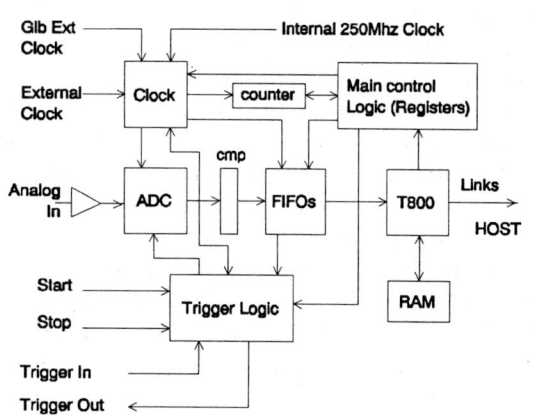

Fig. 6. Block diagram of an acquisition channel of the 250 MSPS transputer based digitizer module.

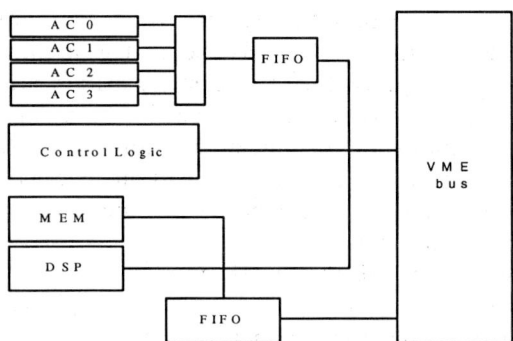

Fig. 7. Block diagram of the 1 GSPS DSP based digitizer module.

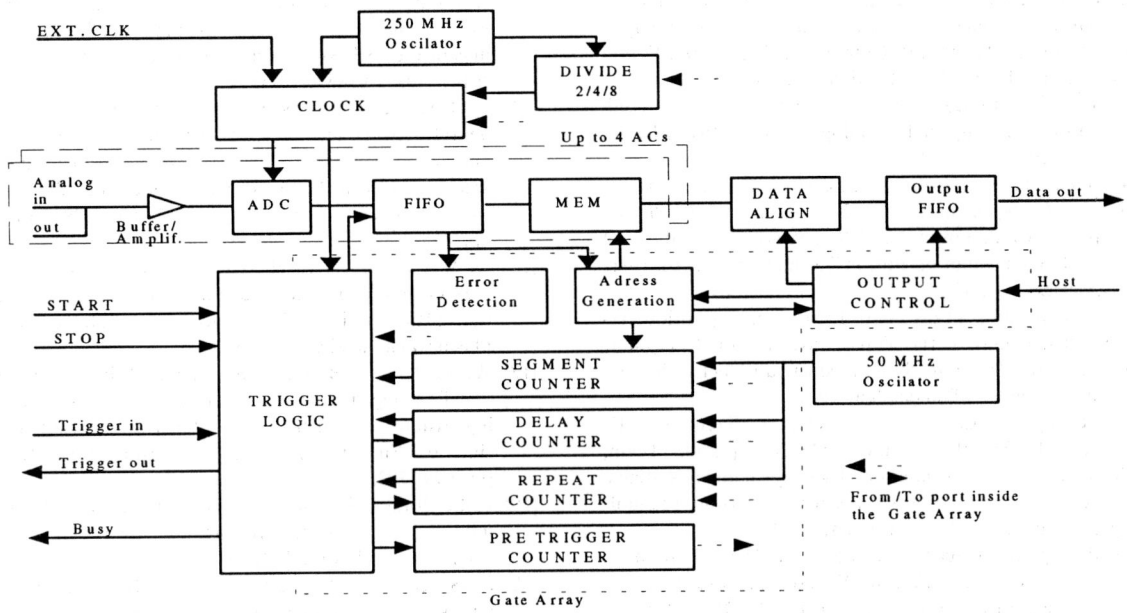

Fig. 8. Block diagram of a data acquisition channel of the 1 GSPS DSP based digitizer module.

8. Conclusions

The control and data acquisition system will play an important role in the safe, reliable and high performance operation of a continuously operated, highly sophisticated, magnetically confined fusion plasma experiment.

CODAS should be conceived following a strict data production policy that will fight all waste in information processing. It should also incorporate as much local intelligence as possible in the information network aiming at reducing the amount of data acquired and stored in the central database.

The portuguese EURATOM Association is developing components that might contribute for the implementation of the control and data acquisition systems of the new fusion experiments.

Acknowledgements

This work has been supported by 'Junta Nacional de Investigação Científica e Tecnológica' and the European Atomic Energy Community. The authors are grateful for helpful discussions with K. Engelhardt, H. Kroiss and F. Schneider from the Max-Planck-Institut für PlasmaPhysik.

References

[1] H. van der Beken, C.H. Best, K. Fullard, R.F. Herzog, E.M. Jones, C.A. Steed, CODAS, the JET control and data acquisition system, Fus. Technol. 11 (1987) 120.
[2] D.N. Butner, Marena Drlik, William H. Meyer, Jeffrey M. Moller, George G. Preckshot, Integrated, multivendor distributed data acquisition system, Rev. Sci. Instrum. 59 (8) (1988) 1786.
[3] P. de Villers, J.M. Larsen, B. Pronovost, J. Caumartin, The data acquisition and management system for the TdeV tokamak, Proceedings of the 17th Symposium on Fusion Technology, Rome, 1994, Elsevier, pp. 1032–1035.
[4] C.A.F. Varandas, B.B. Carvalho, C. Correia, H. Fernandes, C. Freitas, J. Pires, J. Sousa, J.A.C. Cabral, A fully computerized and distributed VME system for control and data acquisition on the tokamak ISTTOK, Nuclear Instr. Meth. Phys. Res. A 349 (1994) 547.
[5] S. Puppin, M.E. Angoletta, D.J. Campbell, J.J. Ellis, M. Garribba, M. Lennholm, F. Milani, D. O'Brien, F. Sartori, Real time control of the plasma boundary at JET, Proceedings of the 19th Symposium on Fusion Technology, Lisbon 1996, Elsevier Science, 1997.

[6] B. Guillerminet, J.F. Artaud, S. Balme, Y. Buravand, B. Couturier, L. Ducobu, D. Elbeze, B. Gagey, M. Leluyer, R. Masset, D. Moulin, B. Rothan and J. Signoret, Evolution of the TORE SUPRA data acquisition system: towards the steady-state, Proceedings of the 19th Symposium on Fusion Technology, Lisbon 1996, Elsevier Science, 1997.

[7] W. Treutterer, J. Gernhardt, O. Gruber, P. McCarthy, G. Raupp, U. Seidel and ASDEX Upgrade Team, Plasma shape control design in ADSEX Upgrade, Proceedings of the 19th Symposium on Fusion Technology, Lisbon 1996, Elsevier Science, 1997.

[8] Izuru Yonekawa, ITER control and data acquisition system, Proceedings of the Topical Meeting on Computer-Based Human Support System, American Nuclear Society Annual Meeting, Philadelphia, 1995.

[9] G. Raupp, K. Lüddecke, G. Neu, W. Treutterer, D. Zasche, T. Zehetbauer and ASDEX Upgrade Team, Structure of next generation discharge control systems, Proceedings of the 19th Symposium on Fusion Technology, Lisbon 1996, Elsevier Science, 1997.

[10] G. Raupp, O. Gruber, V. Mertens, G. Neu, H. Richter, B. Streibl, W. Treutterer, W. Woyke, D. Zasche and T. Zehetbauer, Protection strategy in the ASDEX Upgrade control system, Proceedings of the 18th Symposium on Fusion Technology, Karlsruhe, 1994, Elsevier, pp. 679–682.

[11] K. Blackler, A. W. Edwards, JET Report JET-P (93) 49.

[12] J. Ellis, E. van der Goot and D.P. O'Brien, A real time plasma boundary determination and display system using transputers, Proceedings of the 18th Symposium on Fusion Technology, Karlsruhe, 1994, Elsevier, pp. 743–746.

[13] M. Garibba, M.L. Browne, D.J. Campbell, Z. Hudson, R. Litunovsky, V. Marchese, F. Milani, J. Mills, P. Noll, S. Puppin, F. Sartori, L. Scibile, A. Tanga and I. Young, First operational experience with the new plasma position and current control system of JET, Proceedings of the 18th Symposium on Fusion Technology, Karlsruhe, 1994, Elsevier, pp. 747–750.

[14] T. Kimura, Y. Kawamata and K. Akiba, DSP application to fast parallel processing in the JT-60V plasma control, Proceedings of the 18th Symposium on Fusion Technology, Karlsruhe, 1994, Elsevier, pp. 691–694.

[15] J.E. Lawson, M.G. Bell, R.J. Marsala and D. Mueller, Beta normal control of TFTR using fuzzy logic, Proceedings of the 18th Symposium on Fusion Technology, Karlsruhe, 1994, Elsevier, pp. 739–742.

[16] E. Coccorese, P.J. McCarthy and F.C. Morabito, A neural network technique for the selection of the number and location of the magnetic sensors for plasma shape reconstruction in ITER, Proceedings of the 18th Symposium on Fusion Technology, Karlsruhe, 1994, Elsevier, pp. 795–798.

[17] D. Wroblewski, Neural network determination of tokamak current profile for real time control, Proceedings of the 11th Topical Conference on High-Temperature Plasma Diagnostics, Monterey, 1996.

[18] W. Tenten, G. Bertschunger, K.D. Muller, P. Reinhart and F. Rongen, Introduction of the EISA-PC into existing fusion experiments, Proceedings of the 18th Symposium on Fusion Technology, Karlsruhe, 1994, Elsevier, pp. 699–702.

[19] Eriko Jotaki, Satoshi Itoh, Continuous monitoring and data acquisition system for steady-state tokamak operation, Fus. Technol. 27 (1995) 171.

[20] S. Nordhauser, J.A. Beckstead, Real-time compression and storage of multichannel video, Proceedings of the 11th Topical Conference on High-Temperature Plasma Diagnostics, Monterey, 1996.

[21] S. Dhawan, T. Radway, C. Hubbard and R. Summer, An introduction to FAST CAMAC 60 Megabytes/s in CAMAC, Proceedings of the 19th Symposium on Fusion Technology, Lisbon 1996, Elsevier Science, 1997.

[22] C.F.M. Loureiro, J.M.G.B. Santos, J.B. Simões, C.M.B.A. Correia, M. Zilker, A high-speed transputer-based data acquisition system, Measurement Sci. Technol. 7 (1996) 21.

[23] C. Loureiro, J. Santos, A. Silva, P. Varela, L. Cupido, C. Correia, F. Serra, M.E. Manso, P. Heimann, G. Schramm, M. Zilker, The control and data acquisition system for reflectometry on ASDEX-Upgrade, Proceedings of the 18th Symposium on Fusion Technology, Karlsruhe, 1994, Elsevier, pp. 839–842.

[24] C.F.M. Loureiro, A.M.C.F. Combo, C.M.B.A. Correia, P. Varela, M.E. Manso, C. Varandas, F. Serra, A 1 GSPS VME data acquisition module, IEEE Trans. Nuclear Sci. 43(1) (1996) 184.

ORAL PRESENTATIONS

INDUSTRIAL EXPERIENCE IN MANUFACTURING THE SUPERCONDUCTING CIC CONDUCTORS IN EU FOR THE ITER CS MODEL COIL

G. Bevilacqua[a], E. Salpietro[a], H. Krauth[b], A. Szulczyk[b], M. Thoener[b], S. Angius[c], P. Gagliardi [c], A. Laurenti [c], R. Garré[d], S. Rossi[d], M.V. Ricci, M. Spadoni.

Associazione EURATOM-ENEA sulla Fusione, C.R. Frascati, C.P. 65, I-00044 Frascati, Rome, Italy

[a] The Net Team, Boltzmannstrasse 2, D-85748 Garching bei Muenchen, Germany
[b] Vacuumschmelze Gmbh, Gruener Weg 37, D-63450 Hanau, Germany
[c] Ansaldo Energia, Via N. Lorenzi 8, I-16152 Genova, Italy
[d] Europa Metalli, Centro Ricerche, P. L. Orlando, I-55052 Fornaci di Barga, Italy

In the framework of the activities for the CS Model Coil of ITER, the European Union (EU) is in charge of providing 6.5t of bronze Nb_3Sn strand, cabling for 798m of MC1 and 396m of MC2 cables and final conductor manufacture (jacketing and compaction) for all the conductor lengths of the CS Model Coil, for a total of 5787m. The related industrial activities are in progress. This paper describes in some details the manufacturing experience gained in EU and the present status.

1. INTRODUCTION

The manufacture and testing of two model coils, one representative of the Central Solenoid (CSMC), the second of a TF magnet (TFMC) is one of the qualifying elements of the ITER EDA phase. In both cases, full scale cable-in-conduit (CIC) conductors are used. The CSMC consists of two concentric solenoids, CS1 and CS2, to be respectively manufactured by US and Japan. A high field insert for conductor testing will be also contributed by Japan. The most critical region of the coil is in the inner layers, where a maximum field of 13T will be reached. There, heating by ac losses has to be minimized. In that region it is then required to use a strand which has low hysteresis losses and a significant critical current at high field. The ITER specifications for this strand type, referred as High Performance 2 (HP2), define a non Cu critical current density $Jc > 550 A/mm^2$ at 12T, 4.2K and hysteresis losses $Qh\ (\pm 3T) \leq 200 kJ/m^3$. A bronze Nb_3Sn is the reference material to achieve HP2 performances. European HP2 strand is used in all the conductors for the CS1 solenoid. As far as cabling is concerned, the MC1 cable for CS1 is fully superconducting, with 1152 strands, while the MC2 cable for CS2, made by 1080 strands, contains 33% of Cu wires. A 10 mm central channel is envisaged to keep the pressure drop in the ITER coils within tolerable values. To reduce the coupling losses in the final conductor the individual strands are Cr coated (2μm thickness) and the last-but-one cable is wrapped by a resistive Inconel strip, with 90% surface coverage. Full wrapping of the final cable is foreseen for protection purposes and to make pull-through easier. The jacket material for CSMC is Incoloy 908. The thick, round-in-square jacket is nominally 1.7mm oversize on the nominal cable diameter, but this gap may eventually decrease down to 1.25mm if the tolerances of jacket and cable are taken into account. The final conductor is made by pulling the cable inside the jacket and then by compacting to final dimensions. The conductor layout for CS1 and CS2 is similar, with some difference of dimensions and cabling pattern. Fig. 1 shows a schematic cross section of CS1 conductor before and after compaction. The main parameters of the two conductors are given in Ref. 1. For the CSMC, conductor unit lengths in the range 100-200m are required. The European Union (EU) has to provide 6.5t of HP2 Nb_3Sn strand, the cabling of MC1 cable type for the CS1 solenoid (798m) and of MC2 cable for layer CS2.13 (396m) and the final

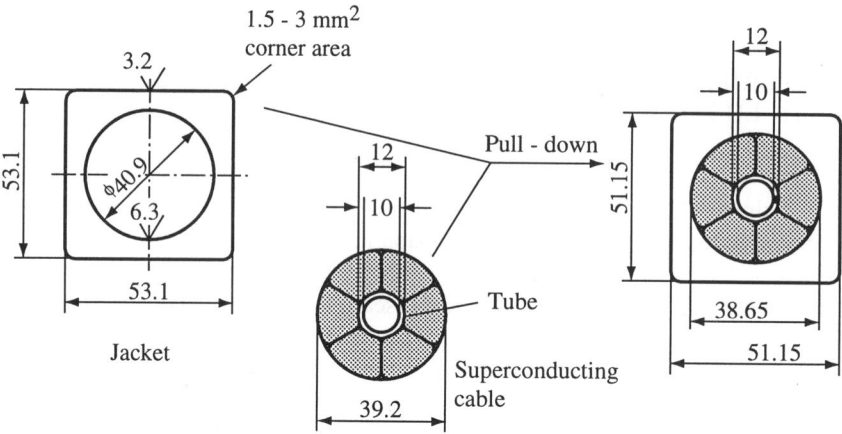

Figure 1. CS1 conductor before and after compaction.

conductor manufacture (jacketing + compaction) for all the conductor for the CSMC, namely 5787m.

This paper is dealing with the status of the industrial production of the CSMC conductor, focusing on the most critical features encountered up to now.

2. THE EU HP2 STRAND FOR THE CSMC CONDUCTOR

Main objectives of the strand production for ITER Model Coils are to assess the stability of the manufacturing process, the yield, the full control of the strand performances and the manufacturing capability. It is worth noting that the overall strand amount for the ITER Model Coils (24t for CSMC +4t for TFMC) represents an increase by one order of magnitude of the world Nb_3Sn advanced strand production capability per year. The HP2 strand specifications have been worked out by ITER Joint Central Team in 1993 with a reasonable extrapolation of the available performances of advanced bronze Nb_3Sn strand. The European company Vacuumschmelze (VAC) is in charge of the HP2 strand production in EU. VAC has demonstrated in 1994, with the manufacture of 500 kg of bronze Nb_3Sn, to have a strand layout fully achieving the ITER HP2 specifications. The standard strand design is 60% Cu-OFHC, 6% Ta diffusion barrier, while 34% is a filamentary region with bronze. The nominal Sn content in bronze is 13.5wt%, the filaments are NbTa with 7.5wt% Ta nominal. The number of filaments is 4675, in groups of 85. Figure 2 shows a cross section of the

Figure 2. The VAC HP2 strand for ITER CSMC.

EU HP2 strand. While manufacturing the strand, VAC has carried out additional R&D with use of different materials inside the billets, to explore potential improvements of the performances. Some problems related mainly to insufficient bonding have been solved during the manufacture. After optimization of the extrusion parameters, a very stable production could be achieved. The strand yield for 120kg billets was stabilized at 100-110 kg, close to the theoretical value. Several billets were drawn to final diameter in one length (~ 22km ≅ 105kg). In some cases the delivered lengths were shorter or not integer multiples of the required unit lengths. This is not only due to the drawing efficiency, but also to cuts, e.g. to accommodate the requirements of the chrome coating line. Maximum production rate is about 300kg per week, limited only by the installed capacity of the chrome coating line at Duralloy AG, CH. Scaling up of the production could be easily achieved by Duralloy, if required. The amount of manufactured strand (6.5t) and the R&D carried out in parallel by VAC have produced a set of valuable information. The optimization of the extrusion parameters has lead to a quite significant improvement of the filament quality. This is indicated by the increase of the quality index n from about 20 to 30-35 and by the parallel increase of non-Cu Jc from about 600 A/mm^2 to nearly 700A/mm^2 for strands with NbTa filaments from one NbTa vendor. However, two billets using NbTa filaments from a second source, although confirming n values above 30, show Jc in the range of 600A/mm^2. Whether the observed difference in Jc is related to the NbTa source has still to be investigated. It is an important point to be clarified, in order to check that the specifications for the raw materials are properly defined and to be sure that all the parameters leading to optimum Jc are under control. Pure Nb filaments, as expected, lead to higher Jc at lower fields. Cross over of the two curves occurs at about 11T. At the reference field of 12T Jc is reduced by about 10%. Hysteresis losses in conductors made from standard bronze and Ta diffusion barrier are below 100 kJ/m^3 for a Jc value of 600A/mm^2. The use of a NbV diffusion barrier reduces Jc by about 10% (Sn is depleted by the formation of VSn intermetallic compounds) but increases the losses to about 200kJ/m^3 due to the low field contribution ($|B|<1T$) of the superconducting Nb ring. The use of Ti added bronze results in high current density, Jc $\cong$ 650A/mm^2 at 12T, 4.2K and high n values ($\cong$ 35) and reduces the losses further to about 80kJ/m^3 at Jc $\cong$ 600A/mm^2. RRR values are above 100, the specified minimum value. However, the production shows a relatively large scatter, from 100 to 259. Although the RRR of OFHC Cu is extremely sensitive to impurities, it would be desirable to understand the origin of the scatter, most probably linked to variations of impurities in different Cu melt ingots or to the Cr coating process. It is worth noting that testing the samples represents a very significant engagement for the manufacturing companies and some simplification will have to be agreed for the ITER magnets mass production.

The EU experience has shown that a low hysteresis loss strand can be produced with performances significantly better than the HP2 ITER strand specifications. Quality Control is being improved on the basis of the manufacturing experience, but it appears basically appropriate to guarantee the final product.

3. MANUFACTURE OF THE CABLES

The EU cabling line for CS and TF Model Coils is located at Europa Metalli, Fornaci di Barga, Italy. The EU has to carry out cabling for 798m of MC1 cable (the four inner layers of the CSMC) and 396m of MC2 cable for the layer 2.13. Three sizes of cabling machines are used. Triplets are cabled by two planetary machines, while a larger cabling system is used for all the intermediate steps, up to the last-but-one (LBO) cable. Manufacture of the final cable is accomplished by a specially designed strander, able to carry out continuous cabling up to a maximum length of 1.1km (6t). The present average manufacturing rate is 400m/month of final cable, limited by the triplets cabling rate. A sketch of the cabling line is shown in fig. 3. A tube or spring or spiral can be introduced from behind the cabling machine to form the central channel of the cable. The two cylinders turk's head is situated next to the cabling machine to compact and shape the

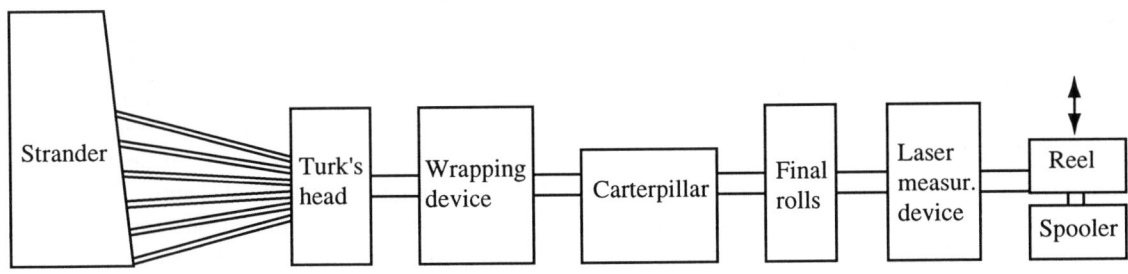

Figure 3. EU cabling line at Europa Metalli

cable. After wrapping, the cable is driven by a Caterpillar, able to generate a maximum pulling force of 500kg. The four shaped rolls guide is then used for the final adjustment of the cable dimensions, continuously measured by a non-contact laser device in two orthogonal directions and recorded by a data acquisition system. The cable collecting spool has a drum diameter of ≥2.4m. The very first trials were performed with non wrapped last-but-one dummy cable, finalized at verifying the correct functioning of the various components which make up the line. Non Cr coated Cu wires were used for the dummy cables. Various lenghts of cable were produced with different cabling parameters to investigate their influence on the dimensional characteristics of the cable itself.

A perforated 12 x 10mm Cu tube, initially used as central element of the cable, was discarded since breakages of the tube with damage to the cable occurred quite often during cable spooling. An AISI 304 stainless steel spring was then successfully used to manufacture dummy cables. However, it was noticed that it was quite difficult to control the pitch of the helical spring. In addition, some wires in contact with the spring showed some deformation [2]. The final choice, used for the manufacture of the superconducting cables, is an AISI 304 stainless steel spiral tape. Wrapping of the LBO subcables is done by a 0.1mm thick, 15mm wide Inconel 600 tape, with a 90% surface coverage. On a dummy cable, which was jacketed into an Incoloy conduit and compacted to final conductor dimensions, some light marks on the outer strands of the LBO, due to the edges of the wrapping tape, were detected [2]. This effect should be reduced inside a superconducting cable made by Cr coated strands. Wrapping with half overlapped 40mm wide tape has been done either with 0.1mm thick Inconel 600 or 0.12mm AISI 304 tapes. Best results are obtained with Inconel. The wrapped cable after the final calibration has a very regular and smooth appearance.

In order to be usable in the EM cabling machines, the VAC strand had to be respooled on suitable reels. To minimise the amount of lost strand, on each reel an integer multiple of the length required for manufacturing the cable was placed. Clearly visible marks were put at the end of each length. The typical yield is 94%.

The prescribed dimensional tolerances have been achieved. The control of the cable dimensions allows to set the average diameter at the desired value inside the tolerance range. For the CS cables, it was agreed to manufacture cables close to the minimum specified diameter, since the jacket bore was also often near to the minimum diameter. After manufacturing 100m MC1, 144m MC2 dummy cables and 20m MC1 superconducting cable for the European Full Size Joint Sample, the cables for CS1.1 and CS1.2 have been completed. Cabling for CS1.3 is expected to be concluded in September 1996, for CS1.4 in November and for CS2.13 in December.

The cabling line at EM is also being used for the MC2 dummy and superconducting cables for the TF Model Coil [3].

As far as fabrication of the superconducting cables for the ITER coils is concerned, the final cabling equipment is already dimensioned to produce the required unit lengths. Increase of the cabling rate can be accomplished by setting up an appropriate number of machines for subcables, specially for triplets.

4. THE CONDUCTOR MANUFACTURING LINE.

The conductor for the CS Model Coil is of the round-in-square type. A thick Incoloy 908 conduit, provided by the US Home Team, is used. The dimensional tolerances initially required for the diameter of the "hole", its eccentricity and the wall thickness, have been found to be too stringent for the industrial production. Present agreed dimensions and tolerances for the CS1 and CS2 conduits before compaction as well as for the final conductors are given in Ref. 1. The EU manufacturing line for the CS conductor is located at Ansaldo, Genoa, Italy.

The process to get the final conductor may be divided into four main steps. First the conduit pieces, received as straight 5-11m unit lengths, are butt-welded to achieve the required jacket length, then the cable is pulled inside the conduit, a compaction by a turk's head is applied to obtain the required conductor dimensions and finally the conductor is bent by a calander to a diameter of 4m to be wound onto a collecting spool. A sketch of the line is shown in fig. 4.

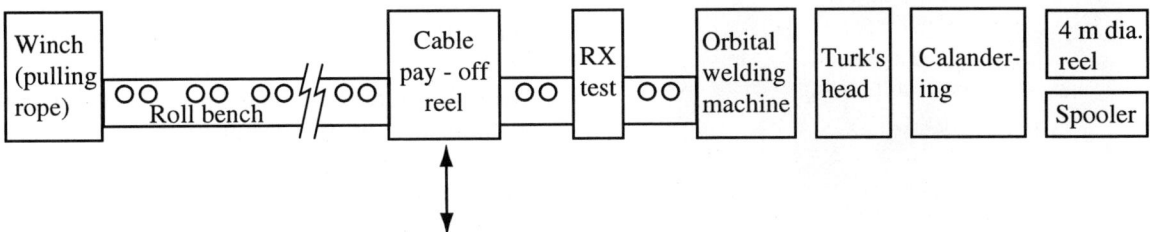

Figure 4. Main components of the Jacketing line at Ansaldo

a) Jacket preparation

Butt-welding and quality control of the welds of the Incoloy conduit have a large impact on the manufacturing rate. High purity gas protection is necessary to avoid oxidation of Incoloy 908 during welding. Two orbital TIG welding machines, designed to the purpose, are used to carry out the Gas Tungsten Arc Welding process. The five controlled axes of the welding head make a full automatic welding cycle possible. A semiautomatic welding process (automatic circular passes + hand corners filling) has also been qualified. A number of destructive tests have been successfully carried out on welded specimens, after compaction, namely: full size tensile; face and root bend; hardness; micrography of base metal, weld and heat affected zone. Non destructive testing is carried out on line, during conductor manufacture. The tests include inner and outer visual inspection, local helium leak test, X-ray test procedure and global helium leak test on the spooled conductor. As the time required for X-ray testing is long, an ultrasonic test procedure (USTP) has been studied, aimed to improve the production rate. Further R&D will be carried out to allow automatic control at the required sensitivity.

Details of the welding process are given in a specific paper presented at this Conference[4]. The use of two welding machines, one on line, the other off-line, allows to achieve a production rate of about 600m/month.

b) Cable Insertion.

Pull-through of the cable inside the conduit is accomplished by a motorized winch, capable of 30000N. The cable spool is installed on another motorized winch, with the same pulling capability, allowing to re-extract the cable, if necessary. A stainless steel rope is used to connect the cable end to the pulling winch. A brazed junction between rope and cable, tested to withstand $\geq$ 30000N is then manufactured. After pull-through, one meter of cable, including the termination, is cut to remove the length damaged by brazing. The experience shows that, if the conductor-cable gap is about 1.5mm, the pull-through operation goes smoothly. If the gap is smaller, or in case of loose or damaged cable wrapping, the pulling force and the risk of stucking the cable increase considerably. Fig. 5 shows the pulling force versus inserted cable length for a case in which the cable got stuck and for a case in which smooth insertion was carried out. In the first case the conduit-cable gap ranged from 1.6mm to 1.1mm. In the second case the gap was around 2mm. Considering all the cables that, up to now, could be inserted smoothly, a friction coefficient can be estimated. A value of 0.7-0.8 is coming from dummy cables and 0.5-0.6 from the superconducting cables. The difference is probably

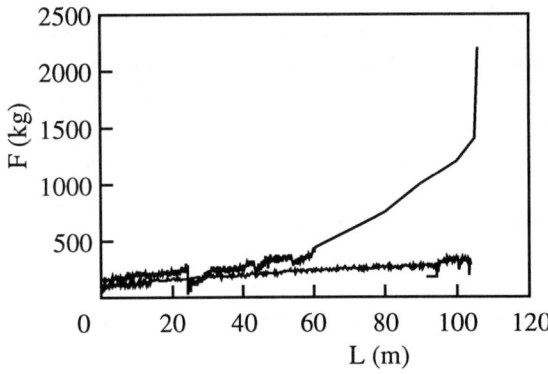

Figure 5. Pulling force versus cable length for two conductors manufactured at Ansaldo.

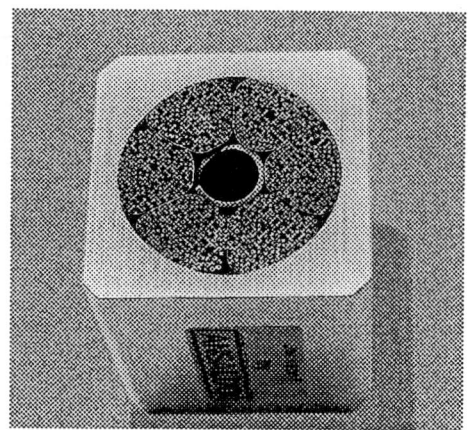

Figure 6. Picture of a CS2 dummy conductor.

due to a better matching of the conduit and cables produced later. Therefore, a maximum pulling force of 30000-35000N can be expected for the pull-through of 1km conductor unit length. The present roughness of the Incoloy conduit seems then to be appropriate.

c) Conductor compaction

Compaction of the jacketed cable is done in one pass by a turk's head. The external cross section is reduced by about 7%, while the observed elongation is about 4%. This means that the cable space is reduced by about 11%, a bit more than the < 9% design values. The radius of curvature at the corners of the final conductor is around 0.5mm. The starting radius of 1.5mm in the as delivered jacket is therefore appropriate to avoid sharp corners. Figure 6 shows a picture of a dummy conductor.

d) Conductor spooling

Calandering is necessary to collect the conductor. However, this procedure leads to some deformation. In particular it is observed that the maximum keystoning is 0.3mm, the wider face being at the inner diameter of the helix, while the face at the outer diameter keeps the nominal dimension. The other two faces become slightly bumped. The conductor dimensional variations, if all the above mentioned effects are taken into account, are ± 0.5mm around the nominal value.

In addition to dummy conductors, 6 CS conductor lengths have been manufactured up to now. Conclusion of conductor fabrication at Ansaldo is foreseen within April 97. As far as the manufacture of the conductors for the ITER magnets is concerned, heavy jacket round-in-square conductors are likely to be used also for the PF coils. The substantial increase of the conductor total and unit length to be manufactured would require the use of many jacketing lines working in parallel, designed to optimise the production rate.

5. CONCLUSIONS.

The industrial experience accumulated up to now in the manufacture of the conductor for the CS Model Coil leads to conclude that basically the specifications worked out for strand, cabling and jacketing can be achieved. The EU production of HP2 bronze Nb_3Sn has shown that a margin for improvement of critical current density and hysteresis losses is available.

A careful tuning of the cabling process allows to manufacture MC1 and MC2 cables within the prescribed mechanical tolerances. The jacketing and compaction can be performed as designed, provided that the cables and conduit tubes are within the specified dimensions.

REFERENCES

1. ITER Joint Central Team: "Conductor Layout for the CS Model Coil with Strand and Cable Allocations", Appendix A Rev. 6, 10 July 1996, to Task Agreement N11TT2494-06-10-FE
2. H. Cloez, P. Decool, J.C. Duchateau "Main characteristics of the ITER CS1 and CS2 dummy conductors".
   NOTE P/EM/96.25 6/06/1996 Association EURATOM-CEA Cadarache (F)
3. G. Bevilacqua et a., "Fabrication of prototype conductors for ITER Model Coil" Paper presented at this Conference.
4. A. Laurenti et al., "Automatic welding of the ITER CSMC superconducting sheating" Paper presented at this Conference.

The ITER-QUELL, a quench propagation experiment on long length CICC with central channel

A. Anghel[a], Y. Takahashi[b], S. Smith[c], S. Pourrahimi[c], M. Zhelamskij[d], B. Blau[a], A. Fuchs[a], B. Heer[a], K. Hamada[b], H. Fujisaki[b], C. Marinucci[a] and G. Vecsey[a]

[a]EPFL-CRPP, Fusion Technology Division, 5232 Villigen, Switzerland
[b]JAERI, Superconducting Magnet Laboratory, Naka, Japan
[c]MIT, Plasma Fusion Center, Cambridge, MA, USA
[d]Scientific Center „SINTEZ", St. Petersburg, Russia

We report on the QUELL experimental results concerning critical current, current sharing temperature, quench propagation and thermal-hydraulic quenchback. We present first a short description of the experiment followed by a detailed analysis of the quench propagation experiments. An important correlation for the temperature margin, operating current and time dependence of the normal zone length have been found.

1. INTRODUCTION

On behalf of ITER a medium scale experiment named QUELL (QUench Experiment on Long Length) was started in January 1996 in the ITER high field test facility SULTAN at CRPP as an international collaboration between EU, JA, US and RF. The goals of QUELL were: 1) Investigation of quench propagation in a typical ITER-relevant CICC, 2) Development, test and qualification of new quench-detector systems, 3) Validation of numerical codes on quench propagation in CICC. The QUELL sample, a double layer, low inductive coil, manufactured using 90m of CICC with central channel was installed in the bore of the SULTAN high field magnet and tested under variable operating conditions (field, current, mass-flow, pressure and temperature). It was equipped with voltage taps, temperature and cold pressure sensors distributed along the whole length in order to monitor the quench propagation. Non-conventional quench sensors such as cowound optical fiber thermometers, cowound voltage taps, Venturi flowmeters, density, magnetization and super high frequency quench detectors were also installed for test purpose. In this paper we present the first results of the evaluation of the experimental data with focus on quench propagation.

2. EXPERIMENT DESCRIPTION

The QUELL sample which is presented schematically in Fig. 1 is installed in the horizontal bore of the SULTAN high field facility and connected to two power supplies of 8kA and 12kA by newly developed 20kA forced cooled current leads. The two power supplies are connected in parallel during the critical current and current sharing temperature measurements when a current up to 20kA is needed. For the quench propagation experiments only one of the two power supplies was used depending on the nominal current needed in the experiment (6-12kA). The sample is cooled with supercritical helium at 4.5-12K provided by a special cryostat [1-2]. Helium pressure, temperature and massflow could be varied.

The EU data acquisition system, based on Visual Designer, had a capacity of 80 channels at a sampling rate of 1kHz and was used to monitor the conventional sensors. The US and RF teams had their own data acquisition systems which were used to monitor the new quench detection sensors.

3. EXPERIMENTAL RESULTS

The QUELL experiment was successfully finished at the begin of May 1996 producing a large quantity of data for the cable qualification (RRR, critical current I_c, current sharing temperature T_{cs}), pressure drop (friction factor), heat slug propagation and quench propagation. Over 100 quench runs were performed with a total of 3GB of data.

3.1 Cable Qualification

At the begin of the experiment the conductor and the sample were qualified by performing RRR,

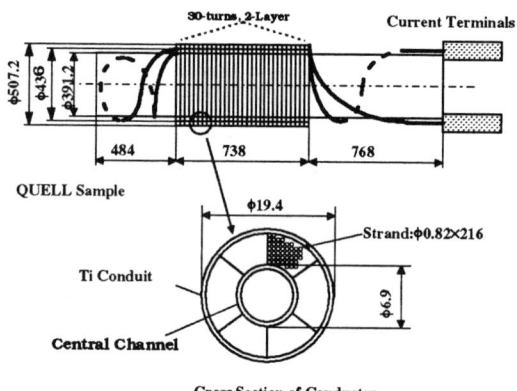

Figure 1. Layout of QUELL sample and conductor cross section

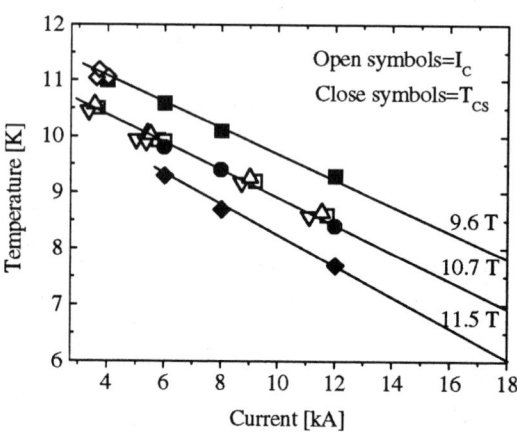

Figure 2. Measured critical current and current sharing temperature of QUELL sample

critical current and current sharing temperature measurements. Based on measurements performed during the cooldown of the sample a value of 260 was found for the overall RRR of the conductor. Excellent agreement was found between the critical current and the current sharing temperature results as can be seen in Fig.2. Work is now in progress to obtain the best fit of measured critical current using the Summers correlation.

3.2 Quench Propagation Measurements

We present here two typical results obtained by measuring the propagation of the normal zone in the QUELL sample. In a typical quench propagation run the sample is first cooled at a given temperature, then charged to the nominal current, and finally an initial normal zone (INZ) is generated using one of the heaters placed in the middle of the sample. The progress of the normal zone is monitored by: 28 voltage taps (V1-V28), 9 temperature sensors (CGR) and 6 cold pressure sensors distributed along the sample. The operating current is hold constant until the quench detection system triggers a ramp down of the power supply. The duration of the experiment depends on the quench threshold set at the begin of the run. For each condition i.e. background field, He mass flow, pressure and operating temperature the quench threshold is increased progressively until the whole cable quenches or the maximum temperature has gone above 100K. The results presented in this paper are from runs with the maximum possible quench threshold, typically around 2 V. The INZ was generated with the 2.3m long resistive heater located half way between inlet and outlet of the sample. Using different pulse energies we could produce both short and long INZ. The short INZ is typically 0.2m long while the long INZ could be as much as 1.2m as measured by the voltage taps near the heater.

The voltage signals measured in a typical quench propagation case (run E0410001) are shown in Fig.3 for the downstream propagation and in Fig.4 for the upstream propagation. The heat pulse is generated at t=0 and it takes about 2s to quench

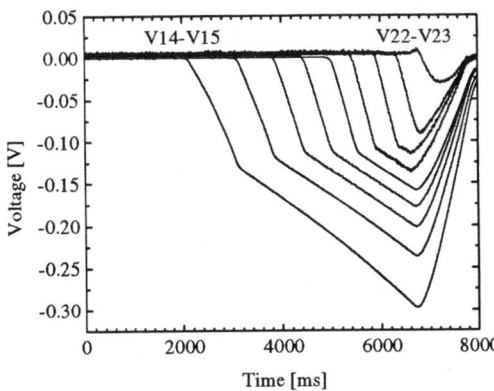

Figure 3. Down stream voltage tap signals for a typical quench propagation run (no quenchback)

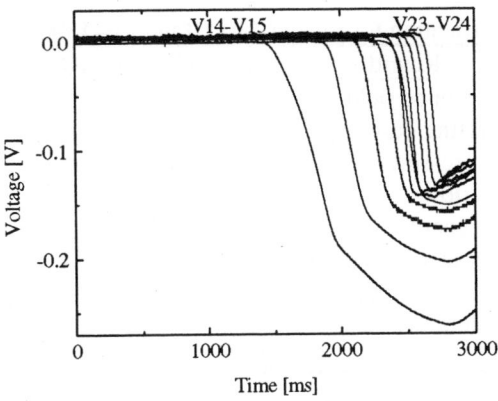

Figure 5. One turn voltage tap signals for a quench back run

the cable for the first time as seen by the runaway of the voltage on the voltage tap pair V14-V15. Then the normal zone propagates downstream reaching the voltage tap V22 which is 12m away from V14. Instead of a generally expected constant quench velocity, a remarkable acceleration was observed. With a distance of 1.5m between adjacent voltage taps we measured an initial velocity of 1.73m/s which increases with time up to 3.37m/s. Similarly in the upstream case the initial propagation velocity is 0.96m/s and ends up with a final value of 1.94m/s. The length of the downstream normal zone is 6.7m and the total normal zone reached in this

run is 21m including the heater length of 2.3m.
No thermal-hydraulic quenchback (THQB) is observed in this run. By reducing the temperature margin to a value lower than 0.5K a quenchback effect could be observed with quench velocity as high as 29m/s. This case, (run E0415003) is presented in Fig.5 and Fig.6 where only the downstream propagation is illustrated. The first graph shows the voltage signals of the one turn voltage taps (1.5m distance). The quench starts also with an acceleration (5.1-13m/s) and develops suddenly a clear cut onset of quenchback. After the quenchback has started, the quench velocity is as

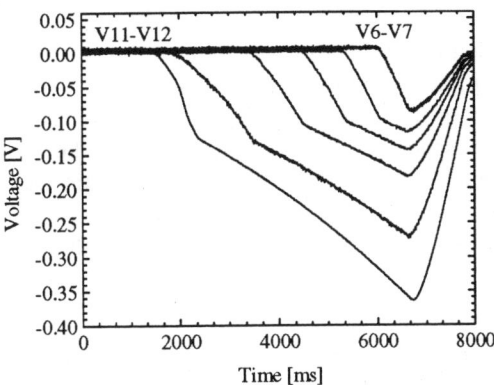

Figure 4. Upstream voltage tap signals for a typical quench propagation run (no quenchback)

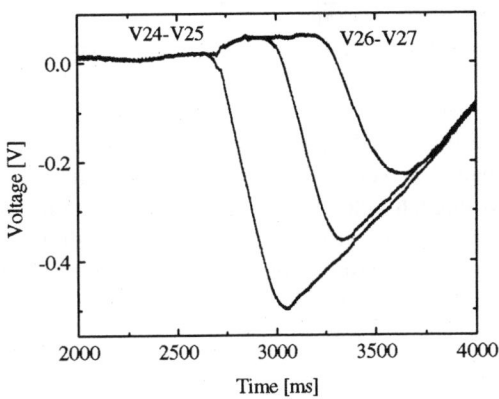

Figure 6. Five turn voltage tap signals for a quench back run

Table 1.
Specification for runs E0410001 and E0415003

Run id	E0410001	E0415003
Background field	11T	11T
Operating current	8kA	8kA
Inlet pressure	6bar	6bar
Mass flow	4g/s	4g/s
Pulse energy	1354J	1654J
Pulse time	0.3s	0.3s
Operating temperature	6.9K	7.3K
Temperature margin	1.85K	0.45K
THQB	no	yes

high as 29m/s as can be measured from the voltage signals of the five turn voltage taps shown in Fig.6.

An interesting effect which can be seen in Fig.6 is that the quenchback is not affected by the current ramp down. Indeed, in this run the current is ramp down at t=2.7s as indicated by the small inductive voltage on V25-V26 and V26-V27 before the resistive runaway. Despite the fast decrease of the sample current (6kA/s) the normal zone propagates further and reaches the highest speed of 29m/s. This indicates that friction and compression are dominant in the THQB case while the joule heating plays a secondary role.

3.3. Quench propagation correlations

Encouraged by these results we decided to make a first attempt to interpret the experimental data and try to extract some correlation for the quench propagation in CICC. We have analyzed the runs with 6, 8, 10 and 12kA, 4g/s, 6bar, 11T, long INZ and variable temperature margin $\Delta T=T_{cs}-T_{op}$ in the range [0.35-2.85K] for the initial and final quench propagation velocity. In case of quenchback, the final quench velocity was chosen to have the value just before the THQB starts. As Fig. 7 shows, the experimental results correlate well if the following expression is assumed:

$$\frac{V_q}{I} = a\left[\left(T_{cs} - T_{op}\right)/T_{cs}\right]^{-b} \quad (1)$$

where a and b are fit parameters. The continuous lines in Fig. 7 are the best fit with this equation and the result is presented in Table 2. The correlation in

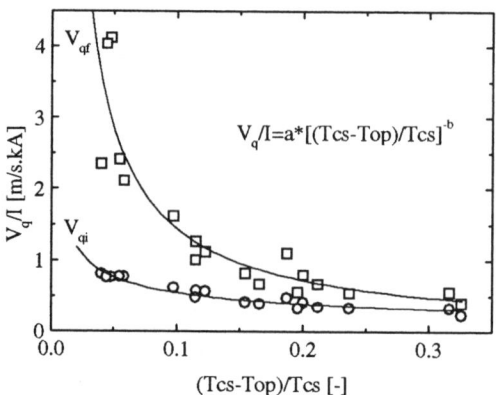

Figure 7. Initial and final values of the quench propagation velocity versus reduced temperature

Eq. 1 is a simple extension of the well known adiabatic propagation velocity [3] which we write here for convenience as

$$\frac{V_{ad}}{I} = \frac{\sqrt{L_0}}{\gamma CA}\left[\left(T_{cs} - T_{op}\right)/T_{cs}\right]^{-0.5} \quad (2)$$

where $L_0 = 2{,}45 \times 10^{-8}$ is the Lorenz number, C=1.14J/kg.K the copper specific heat at T_{cs}, γ=8940kg/m^3 the copper density and A=68.44mm^2 the copper cross section of QUELL conductor.

Looking at Table 2 one can see that the initial value of the quench velocity is well described by the adiabatic formula since b is close to 0.5 and the calculated value of the parameter a, using Eq. 2, is also very close to the fit value. We conclude therefore that the quench propagation starts adiabatic and this can only happen if the cable space is completely depleted of helium. Indeed in the early phase of the quench propagation the helium in the cable space is easily expelled into the central channel through the spiral gaps. In this case we expect a very poor heat exchange between the strands and helium. At a later stage of the quench,

Table 2
Fit results for the quench propagation velocity

Quench velocity	a [m/s.kA]	b [-]	a, calculated [m/s.kA]
initial, V_{qi}	0.18	0.48	0.22
final, V_{qf}	0.15	0.99	-

pressure-driven propagation sets on, as is indicated by the acceleration effect observed in the experiment. The final (asymptotic) value of the quench velocity correlates also with Eq.1 but with a changed value for the exponent, b=1 instead of b=0.5 indicating that we are in a completely new regime.

In order to investigate the pressure-driven regime the time dependence of the quench propagation should be analyzed in detail. Using the same set of experimental data we have extracted the time dependence of the normal zone length for each run, for both the down- and upstream propagation. We found that the data correlate well with a expression of the following general form

$$L_n = A \left[\frac{(T_{cs} - T_{op})}{T_{cs}} \right]^{-1} I^2 t^B \qquad (3)$$

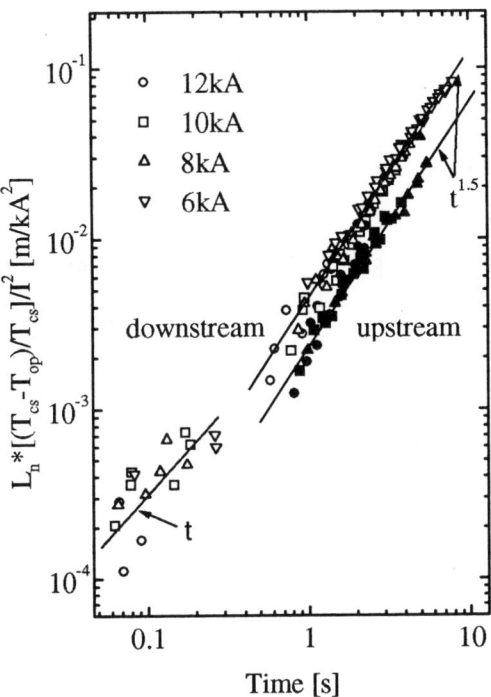

Figure 8. The time dependence of the normal zone length for all non-THQB runs

with the coefficient A and the exponent B having different values for down- and upstream propagation and for non-THQB and THQB runs. In Eq.3, L_n represents the normal zone length, I the transport current and t is the time. In Fig. 8 all the experimental data for the non-THQB runs, $\Delta T = T_{cs} - T_{op} > 0.5K$ are presented in a log-log scale to stress the power law dependence on time. For times greater than ~0.4s the data correlate well while for earlier times the deviation is obvious. This is expected since, as already remarked before, the quench starts in adiabatic (dry) conditions followed by a crossover to the hydrodynamic regime. Accordingly in this time range no power law dependence could be fitted. Still, but with less precision, the data in this range seem to correlate linearly with the time which correspond to the classical result of a time independent (adiabatic) quench propagation velocity $L_n(t) = V_{ad} t$. The THQB data are presented in Fig. 9. Here again we used Eq.3 to fit the experimental data. For the THQB runs we get a quadratic time dependence for both up- and downstream propagating normal zones in the same range t>0.4s. The results of the fit made with Eq.3 are synthesized in Table 3. First of all we note that the coefficient A is different for the non-THQB and THQB cases. They have different units due to the different power of the time dependence. Second, the same coefficient is different for up- and downstream propagation. This is expected since the quench propagation experiments were done under steady flow conditions i.e. helium velocity in the cable was not zero. We expect therefore some sort of influence of this background velocity on the propagation of the normal zone. Indeed we observe that the coefficient A for the downstream propagation is about twice that for the upstream propagation i.e. $A_{down}:A_{up}=2:1$. Interesting enough, the time dependence is not affected.

At this stage of the analysis it will be too speculative to try to infer on the velocity dependence of the coefficient A in Eq.3. More could be said after we finished the analysis of the quench runs with variable helium massflow and after we will analyze the quench runs with other operating pressures which were not included here. The results will be presented elsewhere together with the

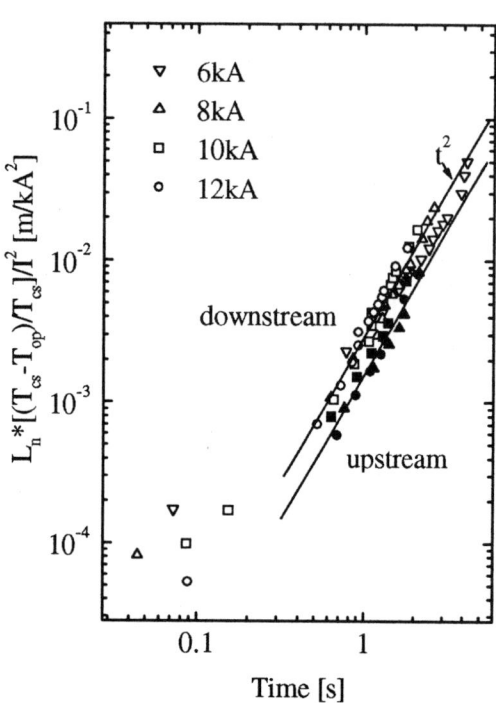

Figure 9. The time dependence of the normal zone length for the THQB runs.

Table 3.
Fit results for the normal zone length

runs	A [m/s^{B}kA^2]		B [-]
	down	up	
non-THQB	4.49×10^{-3}	2.26×10^{-3}	1.5
THQB	2.82×10^{-3}	1.55×10^{-3}	2

correlations for the maximum quench pressure and temperature in the cable.

4. CONCLUSIONS

The QUELL experiment was successfully completed and a large amount of experimental data was produced. The interpretation of the experimental results is still underway.
We concentrated here on the most important aspects i.e. sample characterization and quench propagation. As to date, the main experimental results of the quench propagation in long length CICC with central channel are:
-Shortly after the initiation, the normal zone propagates with the adiabatic propagation velocity due to the expulsion of helium from the cable space
-After about 0.4s the propagation enters in a hydrodynamic regime, characterized by the acceleration effect.
-The time dependence of the normal zone length could be described by $t^{1.5}$ for the non-THQB runs and by t^2 for the THQB runs.
-The temperature margin effect could be described by the inverse of $[(T_{cs}-T_{op})/T_{cs}]$ for both non- and THQB runs
-The normal zone length depends on operating current as I^2 for both non- and THQB runs in the hydrodynamic regime
-A cross-over from the adiabatic propagating zone regime to the hydrodynamic regime was observed. This is a new effect and it is a challenge to find out the mechanism which drives this transition
-The down- and upstream propagation differ only in the proportionality constant A in Eq. 3. To our understanding this is due to the steady state helium flow in the cable. The finite helium velocity in the cable breaks the up- and downstream symmetry of the normal zone propagation problem.

The first experimental results on the performance of the US quench detection systems [4] and code validation[5] are very promising. Work is now in progress to analyze the pressure drop and heat slug propagation experiments.

REFERENCES

1. A.Anghel, C. Marinucci, G. Vecsey, Y. Takahashy and J. Schultz, Proceedings of 18th Symposium on Fusion Technology, Karlsruhe, Germany (1994) 881
2. A.Anghel, J. of Fusion Eng. 14 (1995) 129
3. M. N. Wilson, „Superconducting Magnets", Oxford University Press, New York (1989)
4. S. Pourrahimmi et al., Applied Superconductivity Conference, Pittsburgh (1996), to be published
5. R.Zanino et al., Applied Superconductivity Conference, Pittsburgh (1996), to be published

The plasma shield in ITER plasma wall interactions

H. Würz[1], B. Bazylev[2], F. Kappler[1], I. Landman[3], S. Pestchanyi[3], G. Piazza[1]

1. Forschungszentrum Kalsruhe, INR, Postfach 3640, D-76021 Karlsruhe, Germany

2. Lykov Institute of Heat and Mass Transfer, Minsk, Belarus

3. Troitsk Institute for Innovation and Fusion Research, 142092 Troitsk, Russia

In evaluating the lifetime of ITER plasma facing components (PFC) against not normal heat loads credit is taken from the existence of a plasma shield which effectively protects the wall from excessive evaporation. The plasma shield formed from vaporized PFC material though beneficial for the PFC could become a potential threat for the tokamak because of possible penetration of the impurity ions into the central plasma. This paper discusses various aspects of plasma shields for a rather wide range of heat load conditions. Evaporation and melting of PFCs, physical properties of the plasma shields, plasma shield instabilities and shield efficiencies for tilted targets and estimations on impurity transport from the plasma shield into the SOL of the main plasma are addressed. The numerical results for power densities 0,5 - 10 MW/cm^2 are obtained with the newly developed 2D radiation magnetohydrodynamics (R-MHD) code FOREV-2. For lower power densities 2D heat conductivity calculations with an improved vaporization model are performed.

1. INTRODUCTION

Recently the physical properties of essentially one-dimensional non-LTE carbon plasma shields formed in disruption simulation experiments were studied experimentally and theoretically [1]. The calculated plasma shield parameters such as time dependent plasma temperature and electron density distributions, conversion efficiency of deposited energy into radiation in the plasma shield, total and soft x-ray (SXR) radiation leakage fluxes from and energy balance in the plasma shield were in quite good agreement with the experimental values [2] thus demonstrating that a realistic modelling of ITER plasma wall interactions at heat loads levels of MW/cm^2 is possible.

The validated models then were used in detailed one-dimensional radiation-magnetohydrodynamics (R-MHD) calculations for ITER hard disruptions [3]. Lateral losses of plasma mass by across magnetic field diffusion and radiation from the plasma shield due to the finite width of the incoming hot plasma were taken into account in these 1D calculations by using simplified models [4]. Despite the fact that the inclined magnetic field decreases the expansion of the plasma shield perpendicular to the target, its shielding efficiency increased only slightly in comparison with zero magnetic field [5]. The plasma shields formed in powerful tokamak hot plasma wall interactions are two temperature plasmas with a rather cold, dense plasma close to the wall (atom densities up to 10^{19} cm^{-3}) and a low dense plasma corona (atom densities typically up to 10^{16} cm^{-3}) with temperatures up to a few hunded eV.

Typical results on calculated disruptive erosion are listed in Table 1, including a comparison of calculated and measured results from simulation experiments demonstrating a quite good agreement. For ITER it was assumed that the impact energy of the hot plasma is 10 keV and that ions and Maxwellian distributed electrons contribute equally to the deposited energy. Lateral radiation losses were taken into account. The effective lateral width of the plasma shield was assumed to be 10 cm. 2D radiation transport calculations were also performed to quantify damage of side walls from the intensely radiating plasma shield. The laterally radiated intensity Q_{rad} amounts up to $Q_{rad}/Q_o = 0.04$ with Q_o the power density of the hot plasma. For incoming Q_o above 0.5 MW/cm^2 this radiative heat flux is sufficient for melting and evaporation of side walls and internal structures[6]. From the rather small 1D erosion values as given in Table 1, it can't be concluded that erosion of PFCs during not normal operating conditions is tolerable because firstly an analysis for low power densities but longer deposition times has to be included, secondly the situation with tilted targets has to be analyzed, thirdly the long term stability of the plasma shield under tokamak conditions has to be demonstrated and fourthly the behaviour of the impurity ions has to be investigated.

Table 1 **Erosion for tokamak hard disruptions and ELMs from 1D calculations and comparison with results from disruption simulation experiments**

Simulation facility	carbon erosion (μm)	
	measured	calculated
MK-200 UG[1] 14 MJ/m^2	0.3[3]	0,7
mean pulse duration 25 μs	1.6[4]	
Kh-50[2] 50 MJ/m^2	2.0[3]	1.6[5]
mean pulse duration 35 μs		
tokamak conditions[5]	C	Be W (μm)
10 MJ/m^2 0.1 ms	2.6	3 1
100 MJ/m^2 1 ms	6.5	6
melt depth (μm) 0.1 ms		27 35
1 ms		75 115
erosion 1 MJ/m^2 0.1 ms	0.5	melting at side walls
10 MJ/m^2 1 ms	3.2	starts after 0.5 ms
melt depth (μm) 1 ms		75

[1] directed ion energy 1.5 keV, guiding magnetic field 2T
[2] directed ion energy 0.8 eV, no guiding magnetic field
[3] perpendicular target
[4] tilted target $\alpha = 22°$
[5] lateral radiation losses taken into account

In this study now the above mentioned points are addressed. The range of power densities of the incoming hot plasma stream was extended from 10 MW/cm^2 down to 0.01 MW/cm^2 thus covering hard and soft disruptions, ELMs and other not-normal operational conditions such as VDEs. Results on plasma shield formation, PFC erosion and melting and for the first time results on the MHD behaviour of the plasma shield based on a 2D analysis are presented. Especially addressed are newly found instabilities evolving in the plasma shield for inclined impact of the hot plasma and depletion of shielding efficiency by pushing of plasma along the surface of target tilted in the ploidal plane. Both these effects result in increased erosion. Moreover results from a recently performed analysis of electric potentials in the plasma shield and its consequences on the distribution of the energy deposited by the hot plasma into the plasma shield are discussed.

The 2D analysis is performed with FOREV-2 a 2D R-MHD code with a 2½D MHD model which takes into account that the main component of the magnetic field is in the z (toroidal) direction. The main features of FOREV-2 are shortly described in [3,4].

An electric potential is formed in the plasma shield [7]. Its distribution was calculated recently for perpendicular impact of the hot plasma [8]. According to these results the main part of energy deposition of hot electrons is close to the Langmuir sheath. Only a minor part is deposited into the main volume of the plasma shield. Hot ions carrying a large part of the momentum give only a small contribution to the deposited energy. The assumption of equal fraction of energy deposition by hot ions and electrons used in all previous R-MHD calculations of plasma wall interactions thus describes the actual situation of energy deposition of the hot plama into the plasma shield quite realistically.

2. LOW POWER DENSITY

2.1 Improved vaporization model

For power densities of the impinging hot plasma well below 1 MW/cm^2 an improved vaporization model was developed. Heating of the bulk target, vaporization and melt front propagation are calculated from the 2D heat conductivity equation.

$$\rho c \frac{\partial T}{\partial t} = \frac{\partial}{\partial z}\left(\kappa \frac{\partial T}{\partial z}\right) + \frac{\partial}{\partial y}\left(\kappa \frac{\partial T}{\partial y}\right) + u_v \rho c \frac{\partial T}{\partial z} + Q \quad (1)$$

with $x = z + \int_0^t u_v dt'$

and the boundary conditions

$$\begin{aligned} z = z_\infty \quad & T = T_0 \\ z = 0 \quad & -\kappa \frac{\partial T}{\partial z} = S_w - \rho_s(T_s)u_v H_{vap} \end{aligned} \quad (2)$$

$y = 0, y = y_0 \quad T = T_y$

with H_{vap} - heat of vaporization
 T, ρ - temperature and density of bulk material
 u_v - velocity of vaporization front
 $\rho_s(T_s)$ - density of saturated vapor
 T_s - surface temperature
 c, κ - specific heat and coefficient of heat conductivity
 S_w - flux of incident hot plasma and radiation to the surface
 Q - volumetric source term

The velocity of erosion u_v is determined from the following expression

$$u_v \rho_s = u_g \rho_g \qquad (3)$$

with $\rho_s(T_s)$ the density of saturated vapor.

The expansion velocity u_g of the vapor is calculated from the hydrodynamic equation of motion with the condition that u_g remains below sound velocity. Temperature T_g and density ρ_g of the vapor beyond the Knudsen layer were obtained from a solution of the hydrodynamics problem of motion of a monoatomic vapor inside the Knudsen layer. For evaporation into vacuum temperature T_g and density ρ_g respectively are 0.67 T_s and 0.31 ρ_s with T_s the surface temperature [9]. Temperature dependent data for c, κ and ρ_s are taken from Gmelin's Handbooks for beryllium [10] and carbon [11].

2.2 First results

The importance of the plasma shield for power densities of the hot plasma ranging from 0.01 up to 10 MW/cm^2 was evaluated using the models described above. For the surface flux S_w of incident hot plasma and radiation simplified models derived from detailed 1D R-MHD calculations with FOREV-1 were used. Fig. 1 shows a comparison of calculated and measured values of absorbed energy versus delivered energy for pyrolytic graphite. The experimental results were obtained from calorimetry measurements performed at the 2MK-200 CUSP facility at TRINITI Troitsk [2]. The calculations were performed for κ = 0.2 and 1.2 W/cmK. For delivered energies above 25 J/cm^2 (mean power density 2 MW/cm^2, deposition time 15 µs) there is agreement using κ = 0.2 W/cmK. For delivered energies below 7 J/cm^2 the calculation does not yield any shielding whereas the experimentally determined absorbed energies are less than the delivered energies. This is not due to vaporization of carbon. Desorption of gas, reflection of hot plasma ions, sputtering and/or incomplete confinement of the impacting hot plasma might be responsible for this.

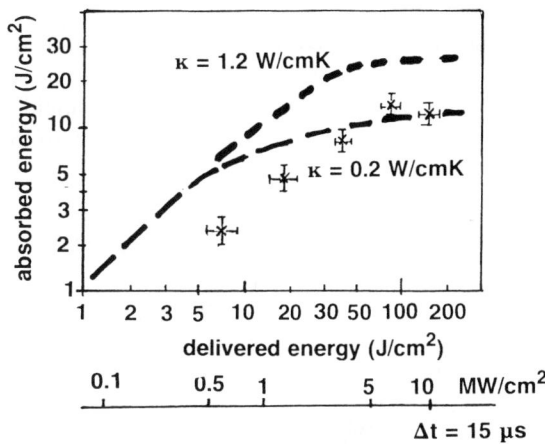

Fig. 1 Comparison of calculated and measured absorbed energy for pyrolytic graphite
 x experimental results from plasma gun
 -- 2D heat conductivity with improved vaporization model

For high power densities of the incoming hot plasma (MW/cm^2 range) the plasma shield reduces the vaporized mass at least by a factor of 40. The melt layer thickness for metal targets remains rather comparable to the case without plasma shield.

At medium power density levels (around 0.1 MW/cm^2) and for deposition times of 0.1 s the vapor shield if it would exist and if it would be stable would reduce the mass of vaporized material by a factor of 500. Again the thickness of the melt layer would be influenced only weakly by the existence of a plasma shield.

At low power densities ($\leq$ 0.01 MW/cm^2) and time duration up to 0.2 s no plasma shield is formed for carbon. Plasma shield formation for beryllium could occur for deposition times larger than 0.2 s. Melt layer formation is dominating. With plasma shield the vaporized mass remains below 0.3 mg/cm^2 and the thickness of the melt layer would be reduced at least by a factor of 4.

At low and medium power densities the properties of plasma shields as well as the stability of the melt layer up to now still are open questions, requiring further detailed analysis.

3. RESULTS FROM 2D MODELLING

3.1 Tilted targets

FOREV-2 was used to analyse the situation with targets tilted in the poloidal plane. Calculated density and velocity evolutions in a carbon plasma shield are shown in Fig. 2a for a constant power density of the

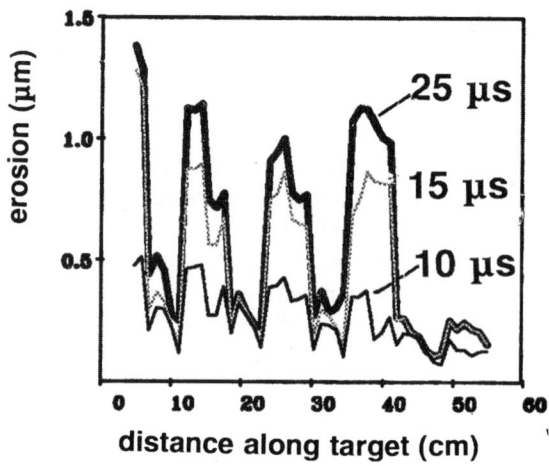

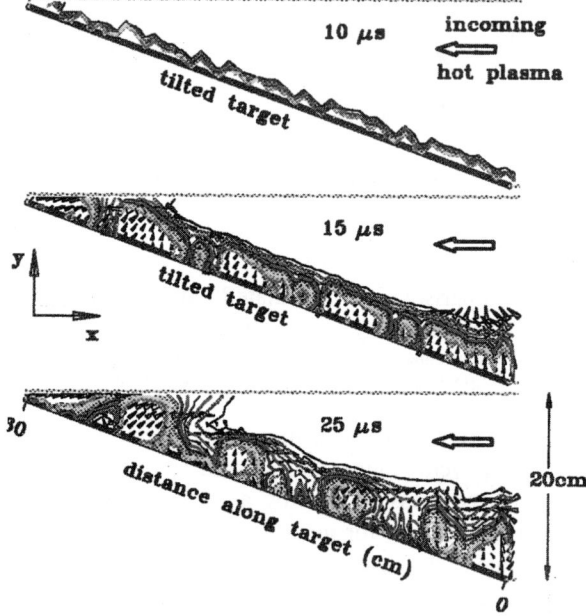

Fig. 2a Time evolution of plasma density and velocity in the plasma shield for inclined target.

hot plasma of 5 MW/cm^2 impacting along $B_x = 0.5$ T, $B_z = 5$ T. Clearly to be seen is an evolving instability in the plasma shield which is triggered by an incidental density fluctuation and which shows a periodic structure along the target. The periodicity is to be seen from the erosion profiles shown in Fig. 2b. The erosion ratio maximum to minimum is linearly growing with time and after 25 μs reaches a value of 3.7. This instability and consequently modulation of erosion also is occuring in the toroidal direction of tokamaks because of hot plasma impact along inclined magnetic field lines. For the same hot

Fig. 2b Time evolution of erosion pattern due to plasma shield instability

plasma the erosion at a target perpendicular in the poloidal plane after 25 μs is 1.6 μm, in case of tilted target the maximum erosion is comparable to the perpendicular target.

3.2 Plasma shield instabilities

A simple analytical model schematically shown in Fig. 3 was developed to confirm the numerical results on instability of the plasma shield in case of inclined impact of the hot plasma. Assuming inverse

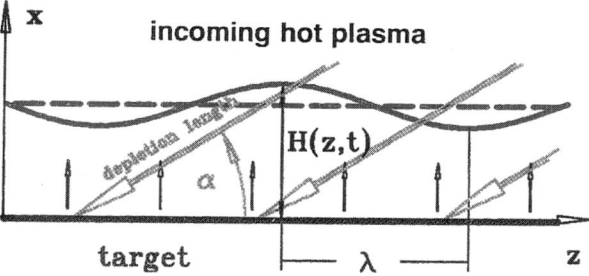

Fig. 3 Scheme forr analytical model of plasma shield instability

proportionality between the heat load S_w deposited to the wall and the thickness H_o of the plasma shield of constant temperature and density S_w is given as

$$S_w = S_o - kH_o \operatorname{ctg}\alpha \qquad (4)$$

with S_o power density of the incoming hot plasma

$k = dS_o/dl$ power density depletion in the plasma shield

α   inclination angle.

cycles at 5 MW/m^2 for 10 s pulses before cracking was observed in the coating, which lead to the formation of hot spots on the surface. The beryllium coating that was applied to a badly melted tile survived 500 cycles (same conditions) before cracking was observed and a total of 680 cycles before the formation of hot spots on the coating surface terminated the test. It is worth mentioning that the beryllium spray process was not completely optimised for this demonstration. One test was carried out on a PVD coating 1.8 mm thick, which survived only several cycles at ~2 MW/m^2 before the coated layer delaminated. The poor performance of PVD coating is attributed to the lack of adequate surface preparation. The widely different HHF results obtained for the three surface conditions prior to LPPS and PVD demonstrate the importance of this parameter.

3. THERMAL BOND LAYER

The TBL concept is based on the presence of a rebrazeable soft compliant layer between the armour and the heat sink. During operation, the TBL has to efficiently transfer the heat without constraining the relative displacement due to the thermal expansion mismatch of the two different materials. Some necessary conditions have to be fulfilled: adequate bond strength, higher than the flow stress of the TBL material, large ductility, metallurgical compatibility with both armour and heat sink materials, stability in any operational condition, reversibility of the bonding process.

The TBL investigated by the four Home Teams were liquid Ga or Pb [6,9], infiltrated felt [10], rheocast Cu-Pb monotectic [11]. Only the results of the TBL's which reached the "proof-of-principle" stage will be discussed here, i.e. the rheocast Al-Ge eutectic [12] and the Pb solder [9] TBL's.

The rheocast processing technology [13] consists in vigorously stirring a two phases alloy featuring a large solidification interval at a temperature corresponding to an approximately equal proportion of the solid and liquid phases. The process leads to a fine dispersion of a globular solid phase into a liquid matrix. The rheocast alloy has a number of features which makes it potentially interesting as TBL. When heated above the solidus temperature, the rheocast alloy transforms into a semi-solid having a viscosity close to that of grease. This property allows the in-situ replacement even on a vertical surface. Moreover it allows the fabrication of thick joints (0.5-1 mm) with obvious advantages from the point of view of thermal stresses and geometrical tolerances. This property is reversible with temperature, the globular structure and its rheological properties being retained after multiple solidification and partial re-melting cycles, provided the rheocast treatment temperature is not exceeded.

The Al-Ge system was chosen by the CEA on the basis of operating temperature range, wettability, compatibility and ductility considerations. The rheocast Al-27.6wt%Ge alloy was fabricated in form of sheets and thoroughly characterised [14] (tensile and creep tests in the solid state, compression tests in the semi-solid state, metallography of the interface with copper and its evolution with temperature, mechanical resistance of the joint, removal and replacement of copper plate). Re-brazing experiments were carried out at 475°C, ~50°C above the solidus temperature (424°C). During this experiments, the brazing alloy reached the semi-solid viscous state without flowing down. The joint between the new Al-Ge sheet and the debrazed part could not be detected by metallographic analysis. The UTS is of ~190 MPa at RT and decreases to ~13 MPa at 400°C. In the same temperature interval the fracture strain increases from 5 to over 90%. The creep rate is ~10^{-6} at 300 °C under a stress of 10 MPa. Shear tests at room temperature on several Cu/Cu sandwiches with various brazing thicknesses (0.15-0.74 mm) showed an average ultimate shear strength of ~25 MPa. The formation of a continuous Ge-rich layer near the copper interface, due to the local Al-Cu reaction and Al depletion was responsible for this low value. In order to get rid of the continuous Ge layer, the Ge content was reduced to 21.8 wt%, with an improvement of the shear strength from 25 to 75 MPa.

The measure of the thermal conductivity and specific heat of Al-Ge alloys was performed by ENEA [15]. The thermal conductivity and specific heat of Al27.6wt%Ge are ~50% and ~75% of those of pure aluminium, respectively. The reduction of Ge content to 21.8 wt% improves also the thermal conductivity of over 25% in the expected temperature operating range. Drawing tests were carried out by the same laboratory to form the alloy sheet in the required shape. The sheet can be easily formed at 430 °C under the weight of a stainless steel punch onto a copper die in form of hemispherical dome of 30 mm diameter, 0.5 mm thickness.

A thermal fatigue test on a Cu/Cu mock-up, brazed with the non-optimised alloy, was performed by CEA with the following results: a) calibration with a heat flux increasing up to 7.5 MW/m^2; water inlet temperature 100°C; b) thermal fatigue at 5.6

MW/m² for 26 cycles. The test was terminated due to damage near one edge, at the cold Cu interface [14]. It is worth noting that these results were obtained with the first rheocast alloy, having a bond strength of only 20 MPa, as compared to the 75 MPa of the optimised braze. Extrapolating these results to the real geometry and heat loads, it is reasonable to assume that the optimised joint can withstand the design number of cycles and the loads of the primary wall (0.5 MW/m²).

4. DESIGN

The application of the TBL concept to two plasma facing components, the Primary Wall and the divertor Dump Plate, is at present being evaluated, both analytically and experimentally.
The most critical design condition for the PW is the heat load during a VDE (20-60 MJ/m², 0.3-1 s). With the progressive reduction of the beryllium tile thickness, as a consequence of the various erosion mechanisms, higher and higher temperatures may be attained in the copper heat sink and at the armour joint. For an armour residual thickness of 5 mm, the temperature at the joint can exceed 700°C, with an increasing failure probability, either of the Be/Cu joint or of the heat sink. In the present design, the only line of defence is the timely refurbishing of the armour by plasma spray. It is worth exploring the possibility of having a second line of defence against VDE, to avoid the replacement of the entire shielding blanket module in case of severe damage.

A representative mock-up of the separable wall shown in Fig. 1 is being fabricated with the rheocast

Two alternative designs of the replaceable sub-component are possible, one (D1) made out entirely of beryllium, the other (D2) with a beryllium tile bonded on the top of a copper T-shaped element. The former design is much simpler, avoiding the additional Be/Cu joint. However, it has to be assessed whether the replacement operation of the sub-component is compatible with the beryllium embrittlement after neutron irradiation at the expected operating temperature.

Since the liquidus and the brazing temperature of Al-21.8wt%Ge are relatively high, 427 and ~525°C, respectively, the problem of heating the old subcomponent for in-situ removal and the new one for rebrazing in not trivial.

One possible method is the use of High-Energy Electron-Beam (HEEB) processing [16] with energies in the 1-10 MeV, in pulses of $10-10^4$ ns duration in beams of centimetre diameters. HEEB can be operated in air or inert gas, because of the reduced beam spreading by the atmosphere. Since the energy deposition of the HEEB is proportional to the atomic number of the elements, the replaceable beryllium (Z=4) sub-component would be rather transparent to the beam, with maximum absorption of energy in the rheocast layer (effective Z=17) and in copper (Z=29). The energy distribution should be sufficient to heat the braze and the substrate without overheating the beryllium tile.

For the divertor dump plate, the most severe design condition is the slow, high-power transient from 5 to 20 MW/m² in 10s. Here the issue is not so much the in-situ replacement, because the whole divertor is designed for a rapid replacement of all HHF components outside the plasma chamber, but the protection against burn-out phenomena and against damage propagation from the armour to the heat sink during off-normal events, the ease of tile replacement, the reduction of rad-waste.

A mock-up (Fig. 2) with a pyramidal CfC tile joined to a water cooled copper heat sink by a 0.5

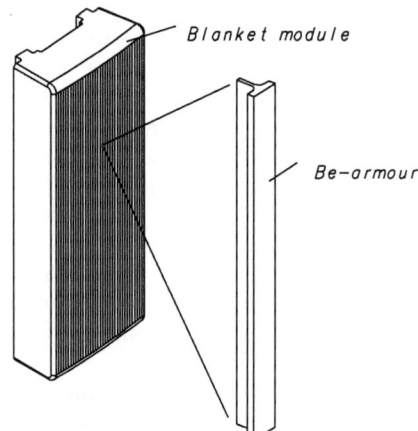

Fig. 1. Separable Primary Wall technology. It will be tested under normal and off-normal heat loads.

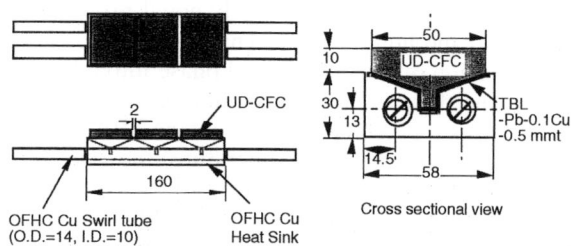

Fig. 2. Mock-up of Dump Plate

mm thick Pb0.1%Cu braze, was high heat flux tested with the following results: 50 cycles at 5 MW/m^2, water inlet temperature 25°C, burn duration 30 s, 30 additional cycles at 10 MW/m^2 without any sign of degradation [17]. Since the Pb braze partially melts at the higher heat flux, the use of this solution is limited to a horizontal target such as the inboard short dump target of ITER.

After having ascertained the thermo-mechanical performance of the Pb TBL, the problem of its stability above the Pb melting point will be addressed. The envisaged solution of the melted braze stability relies on the use of a rheocast Pb-Cu alloy [11] or of a Ni felt infiltrated with liquid metal [10].

5. ANALYSES

A 2D thermomechanical analysis of a separable PW mock-up (D2 option) was performed to evaluate the temperature and the elastic stress distribution when a heat flux of 0.5 MW/m^2 is applied in steady state test condition. Generalised plane strain (with free rotation) boundary condition has been applied out of plane. Other input data were:
a) water inlet temperature 140 °C;
b) heat transfer coefficient 16000 W/m^2 °C;
c) water pressure 4 MPa;
d) stress free temperature 20 °C.

The maximum temperature in the TBL is 193 °C. The presence of the rheocast alloy has a small influence on the temperature distribution in the structure, because its relatively large thermal conductivity and small thickness.

Thermal stresses arise in the rheocast alloy region already at the coolant temperature with no heat flux due to the mismatch in the thermal coefficient of expansion. These are mainly in longitudinal direction (max. -38 MPa). The maximum Von Mises elastic stress when the heat flux is applied is 71 MPa, as shown in Fig. 3.

The stresses due to the coolant pressure are negligible. In the real structure a decrease of stresses is expected because of the castellation in longitudinal direction

To judge whether this stress can be considered acceptable with good engineering margins, a plastic analysis would be needed. However, the very function of the TBL is to accommodate the differential expansion by a small plastic deformation (to be compared with the ductility of the material, as high as 40% at 200 °C) to relax the thermal stresses.

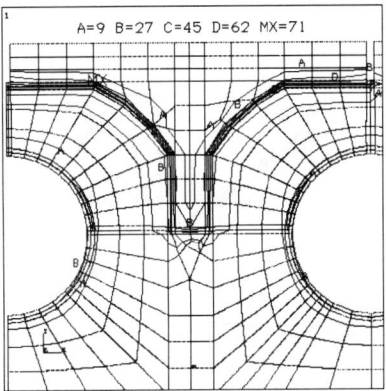

Fig. 3. Von Mises stress for thermal loads (MPa)

By comparing the present FEM results with those of the reference PW design, it can be inferred that he TBL does not change appreciably the stresses level in copper and stainless steel.

A second analysis was conducted to evaluate the thermal response of the separable first wall against the thermal transient expected to take place on the primary first-wall, during accidental plasma contact during burn, e.g., resulting from VDE ($\leq$ 60 MJ/m^2 in times of 100 ms -1s) The model RACLETTE [18,19] was used to perform the calculations. It provides for the solution of the heat conduction problem across a duplex structure and includes evaporation, radiation, melting of the armour material and convective heat removal at the coolant side. The model is based on a simple 1-D geometry, corrected to account for two-dimensional effects in the case of a circular cooling channel. Two cases were analysed, first, referring to option D1, second, referring to option D2. The results of the D1 analysis are shown in Fig. 4; D2 gives similar results. The temperature of the two characteristic points of the TBL (indicated as T1 and T2 in the figure) is plotted considering the armour thickness of 10 mm and 8 mm, respectively. During the transient, the rheocast braze is overheated above the solidus for 2 s at BOL and for increasing time as the armour erodes. During this short permanence in the semi-solid state, the attachment of the subcomponent to the heat sink should be provided by the vertical leg sitting between the cooling channels, which remains in any circumstance solid.

Preliminary calculations were performed to check the capability of the proposed separable PW to withstand the electromagnetic loads during off-

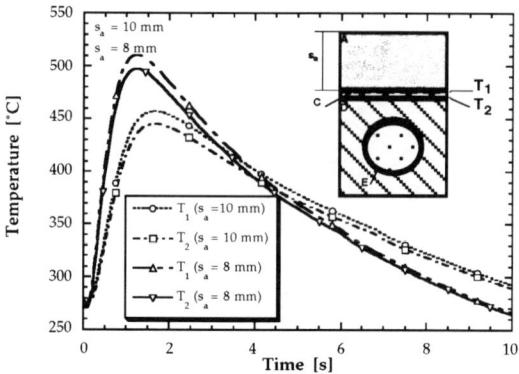

Figure 4. Time history of T_1 and T_2 for the option D1 of the primary first wall (see text). A is beryllium, C is rheocast Al-Ge, D is Cu-alloy, E is stainless steel.

normal events. During plasma abnormal transient, e.g. disruptions or Vertical Displacement Events, we expect an electromagnetic pressure on the PW up to 1.2 MPa (directed towards the plasma) and up to 3 MPa (directed towards the blanket module); the contemporary presence of mechanical and thermal loads due to the large energy deposition has to be considered. In this conditions, as shown by the VDE analysis, the rheocast TBL is above the solidus temperature along the surface of the model; therefore only the joint along the radial rib of the structure can have structural functions. Even in this case, no additional mechanical attachment is needed, if the shear strength of the bond is above 5 MPa, value to be compared with the measured shear strength of the Al-Ge TBL of 75 MPa at RT.

6. SUMMARY

Different technologies are under development to allow the in-situ maintenance of PFC's, according to the extent of damage caused by their interaction with the plasma:
- coating methods for in-situ repair of the erosion or local damage of the tiles;
- in-situ rebrazing of separable sub-components to avoid replacement of a whole FW/Shield module in case of loss of tiles or damage to the underlying structure.

As far as the first operation is concerned, LPPS is by large the most suitable coating technique for in-situ repair. It fulfils all in-situ repair requirements but one, the substrate temperature during the deposition process, that is still to high. VAD presents some advantages in term of substrate temperature and deposition efficiency. It could be considered as a back-up solution, pending an experimental proof to produce coatings of relevant thickness and adequate thermal fatigue lifetime.

For the in-situ rebrazing, the encouraging results obtained in the development of the TBL concept allow to consider alternative designs of the PFC, under experimental validation, which will shorten the replacement time and reduce the rad-waste.

ACKNOWLEDGEMENT

This report was prepared as an account of work performed under the Agreement among the European Atomic Energy Community, the Government of Japan, the Government of the Russian Federation, and the Government of the United States of America on Co-operation in the Engineering Design Activities for the International Thermonuclear Experimental Reactor ("ITER EDA Agreement") under the auspices of the International Atomic Energy Agency (IAEA).

REFERENCES

[1] R. Matera et al., J. Nucl. Mat. (1996) in press.
[2] G. Federici et al., Fus. Eng. Des. 28 (1995) 34-43.
[3] R.G. Castro et al., Proc. 16th IEEE/NPSS SoFE, Vol. 2. October 1995, 381-384.
[4] M. Akiba, private communication.
[5] D.A. Karpov et al., Surface and Coatings Techn. (1996) in press.
[6] I. Mazul, private communication.
[7] Advanced Materials & Processes 6 (1995) 22-23.
[8] R.G. Castro et al., Phys. Scr. T64 (1996) 77-83.
[9] S. Suzuki et al., Proc. ANS 12th Topical Meeting on the Technology of Fusion Energy, Reno, Nevada, USA, June 16-20, 1996.
[10] D. Driemeyer et al., Fus. Techn. 26, No.3, Part 2 (1994) 603-610
[11] M. Salvo et al., J. Nucl. Mat. 226 (1995) 67-71.
[12] F.Saint-Antonin et al., J. Nucl. Mat, (1996) in press.
[13] M. C. Flemings, Met. Trans. 22A (1991)957-981.
[14] F. Saint-Antonin, et al., CEA Report, D.E.M., 23/96, June 1996.
[15] L. Moreschi et al., this conference, paper PA15.
[16] F.H. Froes, JOM Vol. 45 No. 6 (1993) 59-60.
[17] M. Akiba et al., this conference PA18
[18] A.R. Raffray et al., J. Nucl. Mat., in press.
[19] G. Federici et al., J. Nucl. Mat., in press.

High performance results with the LHCD system on Tore-Supra and new Launcher design for quasi continuous operation

P. Froissard, P. Bibet, G. Agarici, S. Berio, C. Deck, L. Garampon, M. Goniche, D. Guilhem, P. Hertout, T. Hoang, J.Y. Journeaux, F. Kazarian-Vibert, X. Litaudon, G. Martin, M. Mattioli, Y. Peysson, C. Portafaix, G. Rey, F. Surle, M. Tareb, G. Tonon, J.G. Wegrowe

Association EURATOM-CEA, Département de Recherches sur la Fusion Contrôlée
Centre de Cadarache, 13108 Saint Paul-lez-Durance Cedex, France

High power and energy performance have been achieved by the Tore-Supra LH system during the last experimental campaign. Far distance coupling as well as plasma pulses up to 120 s in steady state conditions with a power density of 24 MW/m^2 were obtained. A new launcher, made with RF components such as mode converters has been designed in order to extend the present TS performance towards quasi continuous operations.

1. INTRODUCTION

The Lower Hybrid Current Drive system on Tore-Supra is made of 2 identical launchers with a nominal plant power of 8 MW at 3.7 GHz [1]. Hardware improvements have increased the reliability and power handling capability of the system so that record injected energies as well as advanced coupling scenarios were achieved during the 95-96 campaign.

In order to extend the Tore-Supra performance a new launcher has been designed, which includes ITER relevant components such as mode converters and active passive multijunctions.

2. LH SYSTEM IMPROVEMENTS

The Tore-Supra Lower Hybrid system has been successfully operating with a new position control and launcher splitting network, designed for steady state operation.

The launcher position is driven by a Programmable Logic Controller (PLC), which enables real time feedback of the position. The splitting network has been designed to allow the launchers positioning within a 200 mm maximum span using the waveguide flexibility alone as shown figure 1. A maximum acceleration of g/10 was chosen in order to minimise the stress on the launcher back plates. A maximum speed of 20 mm/s and a precision better than 1 mm are obtained under

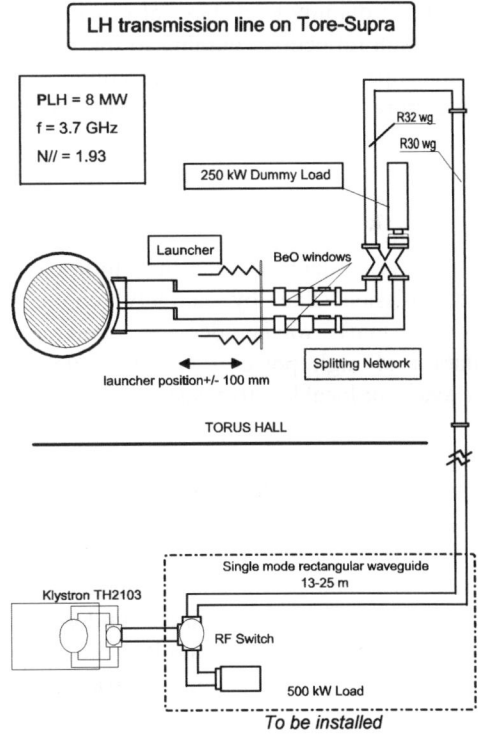

Figure 1: LH system layout on Tore-Supra

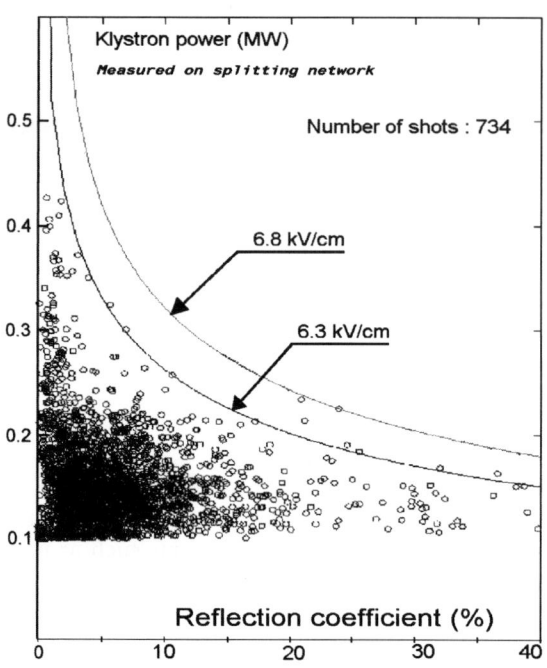

Figure 2: Power limit with reflection coefficient

these conditions.

High power dummy loads have also been installed to permit high reflection coupling to the plasma for continuous operation and HF conditioning.

Furthermore the LH data acquisition system has been modified under VME to enable real time feedback of the LH power, phase and launcher position with plasma parameters and grill mouth heat flux [2].

Future developments include the modification of the transmission line by actively cooled R30 (38.6x77.2 mm) waveguides as well as the installation of high power CW RF switches and power loads for local klystron tests.

3. POWER HANDLING CAPABILITIES

Up to 6 MW corresponding to 85% of the maximum launched power and a power density of 36 MW/m^2 have been obtained under various plasma conditions. The maximum launched power is consistent with a maximum electric field of 6.5 kV/cm (figure 2) inside the narrowest waveguides section of the multijunctions (up to 1.7 times higher than in the input section for a given reflection coefficient).

For reflection below 4% and for pulse time lower than 10s, the injected power is limited by the plant power as shown figure 3. The lower boundary of this figure is given by the LH efficiency. A case corresponding to $\eta cd = 0.8 \cdot 10^{19}$ A/Wm^{-2} for $n_e = 2 \cdot 10^{19}$ m^{-2} and Ip=0.8 MA is given as an indication of this limit. The upper boundary has not yet been explored systematically. The pulse termination, when it is not programmed, does not appear to be directly linked with the total injected energy but rather with the in-vessel components behaviour (outgassing, thermal loads...) as seen on figure 6.

Figure 4 gives the power and field distribution function in time for the 1995-1996 campaign where up to 986 shots have been analysed. An average of 15 seconds of injected LH power per shot was achieved (including commissioning shots) with a total injected energy of 20 GJ. During this single campaign, the LH system has been operated for more than 4000s at an electric field inside the multijunctions between 3.8 and 5.4 kV/cm. This is

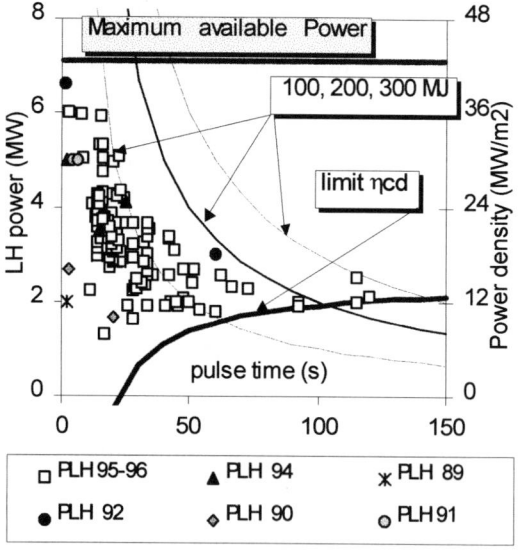

Figure 3: LH Power and duration performance on TS with milestones since 1989

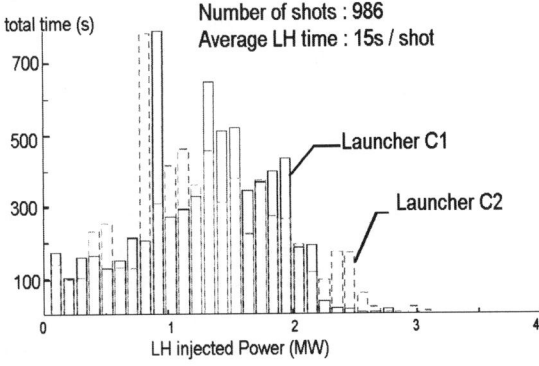

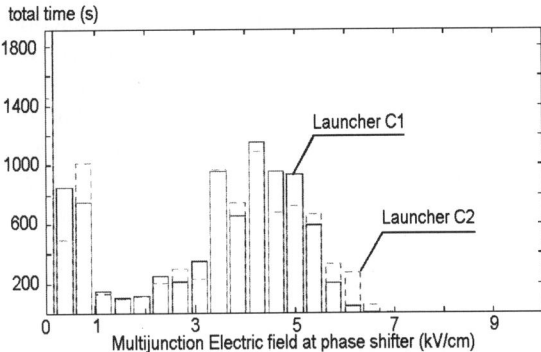

Figure 4: LH power and field distribution for experimental campaign 95-96

consistent with a power density in the grill aperture between 12 and 24 MW/m^2 for a reflection of 5%. The mean electric field value deduced from this distribution function is around 4.1 kV/cm for a standard deviation of 0.8 kV/cm.

4. FAR DISTANCE COUPLING

Far distance coupling has been studied for various plasma parameters, where up to 4 MW have been launched with a reflection coefficient lower than 8% (figure 5) while the grills were positioned at 150 mm from the plasma last closed flux surface (LCFS). Similar experiments were done while the plasma was moved away from the launcher.

The overall LH efficiency is not degraded during the plasma or launcher movement, and the thermal load on the grill protection is much reduced. The LH power is observed to create a sufficient density at the grill mouth for efficient coupling. This has been

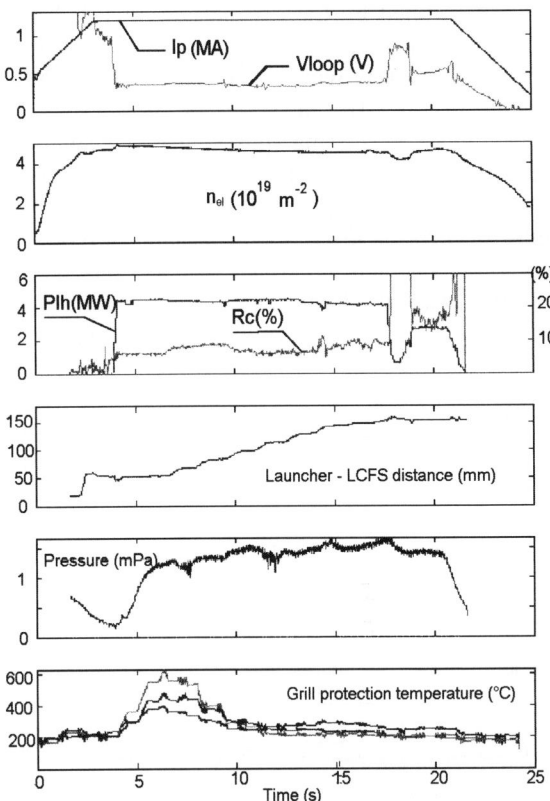

Figure 5: Far distance experiment #18642

found to be independent of the grill position, the connection length in the Scrape of Layer (SOL) or the edge density prevailing before RF, even when the latter is below the cut-off density. The rate of density increase measured using Langmuir probes is about $1.5 \pm 0.7 \; 10^{17}$ m^{-3}/MW. The amount of power needed to create the edge density is estimated with our conditions (density increase of $2 \; 10^{17}$ m^{-3} and edge temperature of 20 eV) to be around 5 kW.

Gas puffing at the grill mouth was applied during some of these shots without noticeable effects on the coupling.

5. LONG PULSE PERFORMANCE

Very long non inductively driven discharges have been achieved using a combination of active feedback loops [2]. For instance, one minute fully non-inductive plasma pulses have been obtained with the loop voltage pre-set to zero and the LH power controlled by the plasma current [3]. Up to 200 MJ were injected in the plasma with an average LH power of 3 MW with full current drive (0.7 MA) at a density of $2.5 \ 10^{19} \ m^{-3}$.

Two minutes plasma pulses have also been obtained with LHCD alone leading to a new record of total injected energy of 280 MJ as indicated on figure 6. 90% of the 0.8 MA plasma current was driven by LH and the total primary flux swing (15 Wb) was used during the discharge. The launcher mouth heat flux, monitored by IR cameras, as well as the plasma impurity flux have been analysed for these steady state conditions. The temperature evolution on the grill protections indicates that a steady state regime is achieved on the launcher. However a density increase is observed starting 60s, as well as an influx of Silver and Oxygen in the plasma. The measured Zeff increases slightly from 2 to 2.2 while the ougassing rate inside the launcher is kept constant at $1.5 \ 10^{-2} \ Pa.m^3.s^{-1}$.

A similar discharge has been achieved using only one launcher with an average power density of 24 MW/m² for 75s with a reflection of 5% and a launcher located 100 mm from the plasma LCFS. The deduced electric field inside the multijunctions is around 5.4 kV/cm. No impurity influx was observed and complete steady state regime at the grill was established, thus demonstrating that this power density is viable for a new launcher design.

For these experiments, programmed launcher position waveforms were used to optimise the thermal load on the grill protections.

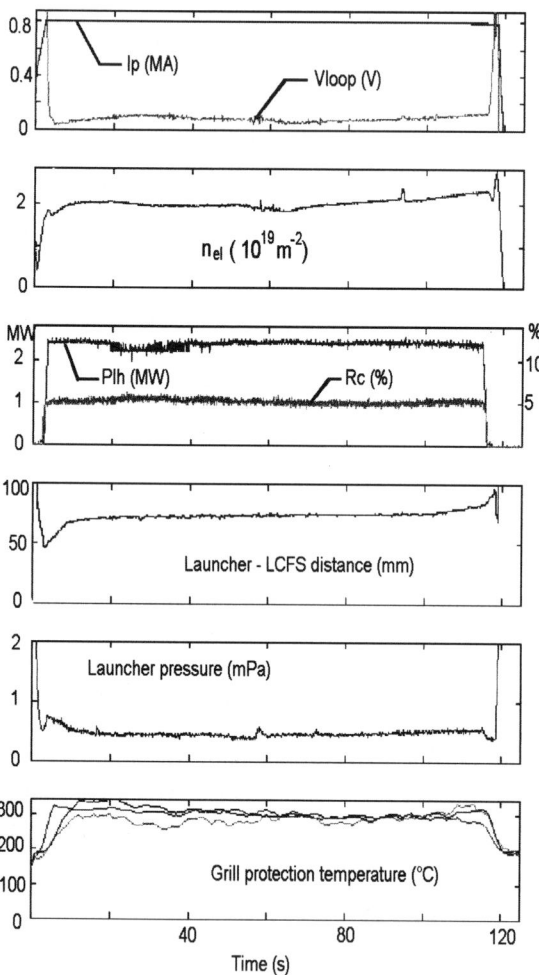

Figure 6: Record TS long pulse

6. NEW LAUNCHER DESIGN

In order to extend the Tore Supra performance towards quasi continuous operation, full current drive profile control is one of the main objective. More than 12 MW of injected power is necessary to drive the main part of a current of 1.7 MA at an average electron density of $6 \ 10^{19} \ m^{-3}$. Taking into account a present routine power density of 24 MW/m², it has been decided to design new antennae, with a radiating surface two times larger in order to inject 4 MW on each launcher with a good reliability at an $N_{//}$ peak value of 2.

The design requirements are:
- keep existing splitting network i.e. 1 klystron for 2 BeO windows.
- fill up the total port section i.e. 590x690 mm².
- keep existing launcher position control (± 10 cm radial movement).
- design grill protection limiter as first wall components, assembled inside the vessel.
- design a removable grill mouth to change by active passive multijunction modules (PAM).

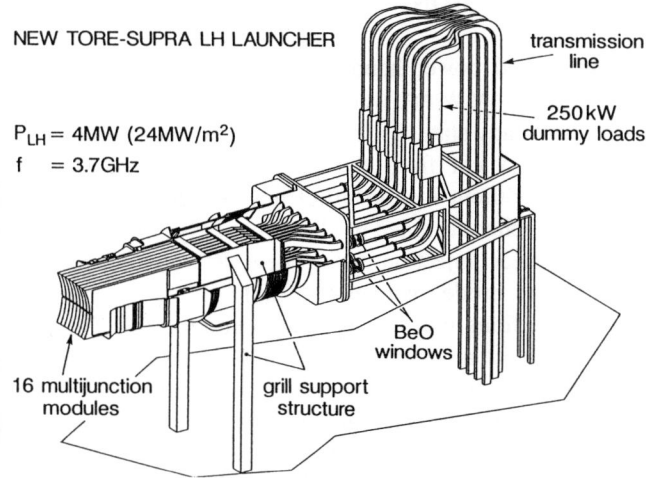

Figure 7: New launcher layout

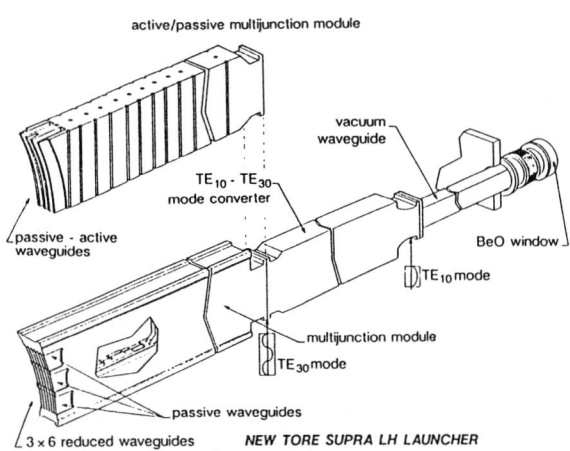

Figure 8: Multijunction module layout

- weld the modules together at the grill mouth in order to reduce over voltage and arcing between waveguides [4].

These new antennae are made of 16 modules composed of 3 poloidal rows of 6 waveguides each fed by mode converters. The result is a total waveguide array of 6 poloidal rows of 48 toroidal active waveguides and 9 passive waveguides on each toroidal row (figure 7). The launched wave spectrum is peaked at $N_{//}=2.03$, and can be changed from 1.68 to 2.37 by varying the phase between modules from $-\pi/2$ to $\pi/2$.

The mouth, made first of standard E plane multijunction (figure 8), can be replaced by an active-passive waveguide array, designed to radiate the power at the same $N_{//}$ peak value. Its main advantage is a cooling circuit closer to the grill mouth with a directivity and spectrum characteristics comparable with the usual E plane multijunction. This new concept must therefore be tested in order to validate it for next step Tokamak such as ITER [5].

The support grill structure has been designed to take the mechanical stress outside the vessel. Therefore the outer structure of the modules is made of a 316L stainless steel and the modules are welded together at the grill mouth. In order to keep a good thermal and electrical conductivity, the inside structure of the multijunction is made of dispersion strengthened copper (DSC for $CuAL_2O_3$) for the inner walls and OFHC copper exploded on 316L plates for the outer walls. Each module is cooled top and bottom by pressurised water at 200°C and 40 bars. The cooling circuit goes as close as 40 mm from the mouth, compared to 115 mm with the actual launcher.

Mechanical and thermal stress analysis were performed using the CASTEM 2000 code with the following assumptions; a radiated flux at the mouth

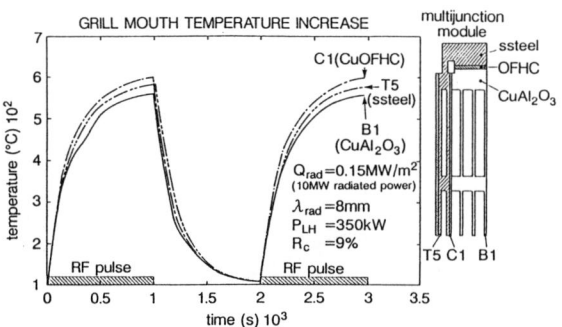

Figure 9: Temperature increase at grill mouth

of 0.15 MW/m^2 corresponding to 10 MW of radiated power, a radiated decay length of 8 mm inside the grill and nominal LH operation at 350 kW per module for 1000s with an electric field value of 4.6 kV/cm. Numerical results indicate that the temperature increase inside the multijunction goes beyond 600°C for 1000s plasma pulses, thus demonstrating the need for DSC (figure 9). As expected a small fraction of the thermal stress comes from the injected RF power, but depends strongly on the radiated power as well as the radiated decay length.

During disruption, the maximum calculated torque is 8.6 10^4 N.m corresponding to a maximum stress in the stainless steel and the DSC within tolerable limits. For nominal plasma parameters i.e. 10 MW radiated power from a 2 MA plasma for 1000s, the maximum allowed number of cycles is 500 and increases to 9000 while working at half performance.

Low power tests of the mode converter have shown that up to 99% transmission efficiency could be achieved in good agreement with theory [6]. High power tests were performed at 300 kW for 1000s as shown on figure 10 and have successfully validated this new component. Further tests will be performed on the remaining components and one complete launcher is foreseen to be installed on Tore-Supra by 1998.

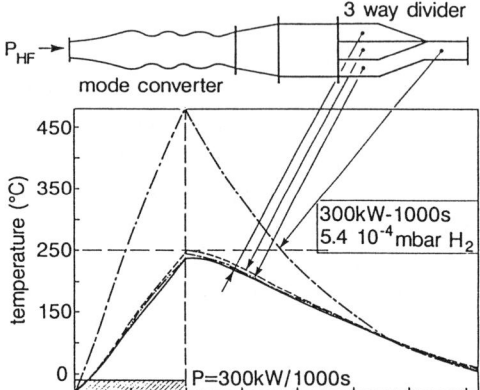

Figure 10: Mode converter high power test

7. CONCLUSION

Improved power handling capabilities leading to routine 24 MW/m^2 were achieved on Tore-Supra during steady state discharges. Far distance coupling beyond 15 cm from the plasma last closed flux surface without loss in efficiency was achieved. A new launcher designed to allow routinely continuous high power injection using new RF components such as mode converters and active passive multijunctions will be installed starting 1998. This, coupled with an upgrade of the LH power and of the Tore-Supra first wall components, should lead to improved performance at high plasma current and density for quasi continuous operations.

ACKNOWLEDGEMENTS

The authors wish to thank the Machine and Operation Groups for their invaluable support during the last experimental campaign.

REFERENCES

[1] G. Rey et al., "High power and long pulse capability of the 3.7 GHz LHCD system on Tore-Supra", 17th SOFT, Rome, 1992

[2] T. Wijnands et al., "Feedback Control of the Current Profile on Tore Supra", 23th EPS, Kiev, 1996

[3] X. Litaudon et al., "Stationary regimes of improved confinement in Tore Supra", 23th EPS, Kiev, 1996

[4] M. Goniche et al., "Acceleration of electrons in the near field of Lower Hybrid frequency grills", 23th EPS, Kiev, 1996

[5] G. Rey et al., "Flexible N// Launcher for Lower Hybrid Wave injection in the first H operation phase of ITER", 19th SOFT, Lisbon, 1996

[6] P. Bibet et al., " Experimental and Theoretical results concerning the development of the main RF components for next Tore Supra LHCD antennae", 18th SOFT, Karlsruhe, 1994

Mathematical model and results of calculations for poloidal magnetic field system and stress analysis for toroidal field system of the TEXTOR 94

I.Baturo[a], N.Berkhov[a], H.Bohn[b], E.Bondarchuk[a], V.Filatov[a], B.Giesen[b], N.Doinikov[a], M.Korol'kov[a], B.Kitaev[a], N.Kozhukhovskaja[a], N.Maximenkova[a], B.Mingalev[a], O.Neubauer[b], T.Obidenko[a], A.Panin[a], A.Simakov[a]

[a]D.V.Efremov Scientific Research Institute, NTC "SINTEZ", 189631 St.-Petersburg, Russia

[b]IPP, Juelich, Forschungszentrum Juelich GmbH, Association EURATOM-KFA, D-52425, Juelich, Germany

The aim of the numerical modelling is to study different regimes of operation of the TEXTOR 94 poloidal magnetic field system. The model allows to take into account the gaps between the parts of joke, nonlinear magnetic characteristics of ferromagnetic, discharge parameters as well as to simulate the operation regime providing scenarios for discharges with equilibrium conditions. The model includes vacuum vessel and liner and makes it possible to study the start up phase. Inductance matrix and efficiencies of the poloidal coils are determined for different levels of the iron magnetization. The results of the coils magnetic field measurements have been compared with the calculated ones. The ways of the stray fields compensation in this region for the start up phase have been proposed. On the basis of the predefined currents in the poloidal field system the possibility is shown to calculate the plasma parameters (current, inner inductance, elongation). The calculated parameters are in a good agreement with the experimental ones.

The TEXTOR 94 magnet system incorporates 16 toroidal field coils (TFC). The coil system forms a vault inside to support centripetal forces. At the outside the coils are connected via intercoil structures taking the tilting moments. 2D and 3D finite element stress analyses of the TFC under electromagnetic and thermal loading have been performed for the upgrade system parameters. Contact interaction between the magnet system components has been taken into account.

1. TEXTOR-94 POLOIDAL MAGNET FIELD SYSTEM AND ITS 2D MODEL

In TEXTOR-94, like in the known facilities of this type (JET, T-15), the conditions are provided for a deep saturation of the iron core (see [1,2]). Its main parameters are shown in Table 1. It comprises a closed iron core, consisting of a central core and a six-sectional external yoke, and a number of circular coils (Fig.1). The correcting coils are not shown in Fig. 1.

The iron core is made of two iron grades. The remaining intact outer part of the yoke is described by the function g_y. The iron core manufactured anew has the characteristic g_c ($g_{Y,C} \equiv \dfrac{\mu_0}{\mu_{Y,C}}$). The specific magnetic characteristics $q_{y,c}$ differ essentially at low flux densities (B<2 T) $g_Y \approx 3\,10^{-5}$, $g_C \approx 5\,10^{-4}$ and are practically the same at large flux densities (B>2 T). The continuity of the iron core is interrupted by 9 structural gaps 1 mm each. Based on the measuring data the filling factor is k_s=0.935-0.97.

The problem of producing and maintaining current in the plasma at a specified level is solved mainly by the BM inductor- coil and that of the plasma column equilibrium control by the BF and BV coils, the first producing in the plasma area mainly the quadrupole field (Fig. 2a) affecting the shape, and the second the

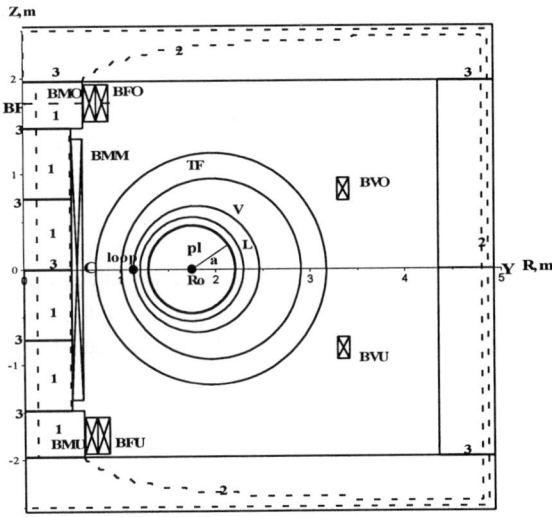

Fig. 1. Meridional section of the installation.
1-core, 2-yoke, 3-air gap, C, BF, Y-core and joke sections, dashed lines - 2D model boundaries, solid lines - actual geometry boundaries.

Table 1. Main parameters of the TEXTOR 94

Major radius	1.75 m
Minor radius	0.46 m
Toroidal magnetic field	3 T
Plasma current	0.8 MA
Long pulse capability	10 s ($\Phi < 9$ v s)

dipole field (Fig. 2b) providing the specified radial position of the column. The inductor stray fields are given in Fig. 2c. The parameters of the main PF coils are presented in Table 2.

Division of coils by a functional feature in this case is not strict, as all of them create to one extent or another a time-varying flux through the core and a nonuniform field in the plasma area.

Finally, the PF system comprises the plasma, vessel, liner and structural elements.

In the investigations of the tokamak PF system wide use is made of numerical results based on the 2D model, into which the central core, coils and plasma easily fit. The following expression can be used to describe the distribution of the longitudinal current in the plasma:

$$j_\varphi(r,z,\psi(r,z)) = j_0 \left[1 - \left(\frac{\psi - \psi_m}{\psi_G - \psi_m} \right)^\alpha \right] \left[\beta_J \frac{r}{R_0} - (1-\beta_J) \frac{R_0}{r} \right] \quad (1)$$

where r, φ, z - cylindrical coordinates; $\psi = 2\pi r A_\varphi$ - poloidal flux; A_φ - vectorial potential of the magnetic field; ψ_G, ψ_m - boundary and extreme values of the flux function; α - parameter characterising the current peak formation; β_J - relation of gas-kinetic pressure to pressure of the magnetic field of the plasma current; R_0 - radius of the geometric plasma centre; j_0 - scale factor.

The plasma model assumes that either ohmic resistance $R_{pl}(t)$ or active voltage $U_{pl}(t) = R_{pl}I_{pl}$ or voltage $U_{loop}(t)$ on the control turn located nearby the plasma boundary are specified. The relation between U_{pl} and U_{loop} can be found [3]:

$$\Delta\psi_{pl} = \Delta\psi_{loop} + \mu_0 \Delta I_{pl} R_0 \left(\frac{l_{in}}{2} + k_{coef} \ln \frac{2a_{loop}}{a+b} \right)$$

$$- \mu_0 I_{pl} R_0 \frac{\Delta a + \Delta b}{a+b} k_{coef} + \mu_0 I_{pl} R_0 \frac{\Delta l_{in}}{2} \quad (2)$$

where I_{pl} and ΔI_{pl} - current in the plasma and its

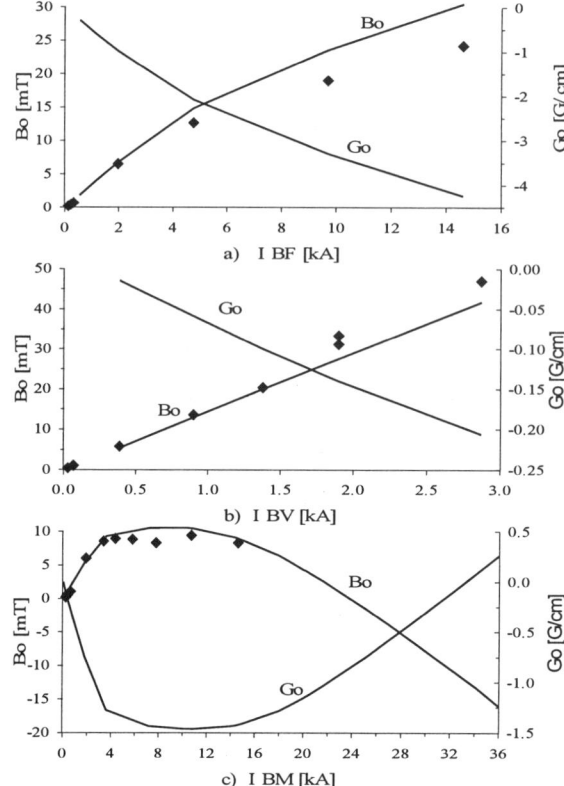

Fig. 2. — calculated data, ♦ experimental data.

change; l_{in} and Δl_{in} - inner plasma inductance per unit length and its change; $a, b, \Delta a, \Delta b$ cross-sectional sizes of the plasma and their change; $k_{coef} = 0.74$.

The geometrically continuous objects, like the vessel and liner, are replaced with a set of circular turns with specified resistance. These resistances can be selected by analyzing the spatial maps of current spread-out. Later on, they can be defined more exactly on the basis of the results of comparison between the experimental and calculation data.

The magnetic field is described by a differential equation of the type [4]:

$$\frac{\partial}{\partial r}\left(\frac{1}{\mu r}\frac{\partial \psi}{\partial r}\right) + \frac{\partial}{\partial z}\left(\frac{1}{\mu r}\frac{\partial \psi}{\partial z}\right) = 2\pi\mu_0 j \quad (3)$$

with the boundary conditions:

$$\psi\big|_{r\to 0} \approx kr^2 \quad (4)$$

$$\left(\frac{\partial \psi}{\partial \rho}\rho + \psi\right)\bigg|_{\rho=\sqrt{r^2+z^2}\to\infty} \to 0 \quad (5)$$

The latter was obtained by analogy with [5].

The system (3)-(5) is to be supplemented with the plasma column equilibrium condition:

$$\psi_G(r,z) = const, \quad (6)$$

equations of voltage balance on all conducting elements:

$$\frac{\partial \psi_\Sigma^{(i)}}{\partial t} + R_i I_i = u_i, \quad (7)$$

where

$$\psi_\Sigma^{(i)} = \int_{(S_i)} \psi j dS / I_i. \quad (8)$$

S_i - the cross-sectional area of the i-th conductor, u_i - external voltage on the i-th conductor. For the plasma turn and turns replacing the vessel and liner $u_i \equiv 0$.

To solve the equation system (3)-(8) the MRUR code [6] was developed based on the integral-differential method of difference equation production and the method of nonlinear successive overrelaxation for their solution.

The time dependencies of the plasma parameters and the initial current values in the coils can serve as the initial data. In this case found should be $I_i(t)$ and $u_i(t)$ in the coils and $I_i(t)$ in structural elements. On the other hand, $I_i(t)$ or $u_i(t)$ in the coils and I_{pl} can be specified, as a result defined are the time dependencies of plasma parameters and currents in the structural elements. Note that the latter are found by the special procedure.

2. POLOIDAL COILS FIELD

The poloidal field (PF) coils have an inductive coupling, the character of which depends on the core saturation degree.

Table 2. Poloidal field-coil parameters

Coil	W	I_{max}	$U_{max}^{(+)}$	$U_{max}^{(-)}$
		kA	kV	kV
BM	294	36	3.3	-2.0
BMM	218			
BMO/U	38			
BF	64	40	1.7	-1.2
BFO/U	32			
BV	48	20	1.7	-1.2
BVO/U	24			

Fig.1 shows three cross-sections of the iron core with corresponding induction values: B_c, B_{BF} and B_y. B_y is always <2T. At the start of the discharge $B_c \cong -5T$ and $|B_{BF}| < |B_c|$; later on the core ends (B_{BF}) and its middle part (B_c) turn out to be counter-magnetized. The interval Δt, when the iron core is not saturated, is extremely short, but Δt increases with plasma current value decreased on the discharge plateau.

The scale of changes in the inductance of the coils, plasma and system made by the vessel and liner amounts to ~500 when inductance B_C changes from ~5T to ~1 T.

The field of the main coils in the area occupied by the plasma was calculated and the calculation results were compared with the measurement data (Fig. 2). They may be said to correlate satisfactorily at a measuring accuracy of ~10G. Measurements were performed with an electronic beam.

Table 3. Starting points for different I_{BM} ($t=0$, $I_{pl}=0$)

N	I_{BM}	I_{BF}	I_{BV}	I_{KV}	before	comp.*	after	comp.
					Bo	Go	Bo	Go
	kA	kA	kA	kA	G	G/cm	G	G/cm
1	-0.428	0.308	0.	0.	-8.77	0.144	-0.12	-0.021
2	-3.028	2.924	0.	0.	-53.7	0.744	-0.37	-0.007
3	-10.0	7.413	0.	0.	-105.4	1.459	-0.013	-0.021
4	-15.0	7.413	0.	0.	-86.5	1.410	-0.16	0.033
5	-19.0	6.767	0.	-0.188	-53.5	1.221	0.19	0.109
6	-20.0	7.69	0.	-0.51	-43.3	1.155	-0.11	-0.09
7	-27.0	3.229	0.	-1.28	38.7	0.579	0.14	0.056

*) $I_{BF} = I_{BV} = I_{KV} = 0$

The inductor stray field (coil BM) in the centre of the vacuum vessel varies within B_0=-160 G - 100 G (Fig. 2c). For its compensation at the discharge start use can be maid of both the main BF and BV coils and special correcting coils. The variants of stray field compensation are presented in Table 3. Note that the current direction in the BF coil is identical to that required at the next discharge stage. As a rule, the field level B<5 G is achievable in the vessel in a wide range of I_{BM} changes. With I_{BM} current increase, use should be made of the KV coil for compensation (the correcting coil installed nearby the BV coil)

3. POLOIDAL FIELD SYSTEM OPERATION

The MRUR code was used to investigate a) the initial discharge stage; b) the stage of I_{pl} growth and sustainment at a specified level; c) the stage of plasma current disruption at a different core saturation. The aim of this calculation was 1) to determine compensation currents; 2) to determine the time laws of current changes in the coils; 3) to determine the mechanical loads and stresses on the facility construction and voltages in the power supply system.

The calculation results of the start phase of one of the discharges are shown in Fig. 3. Currents in the main coils, i.e. I_{BM}, I_{BF}, I_{BV} are used as the initial values. Account was taken of eddy currents in the vessel and liner. The calculation and experimental data relative to the values U_{loop} and I_{V+L+pl} correlate fairly well.

The power supply regimes were analyzed with consideration for available limitations on currents and voltages (Table 2). Considered were 4 discharge scenarios with I_{pl} =0.35, 0.5. 0.65 and 0.8 MA.

The time dependencies of currents in the PF coils and flux density B_c resulting from the numerical simulation for the discharge with current I_{pl}=-0.8 MA are presented in Fig. 4. It follows that the discharge duration is found by the ultimate current in the BF coil . But for scenarios with less current I_{pl} the duration is determined by ultimate current in the BM coil.

The calculations show that for the maximum plasma current the duration of the discharge plateau is about τ=8.5 s.

Of interest is the comparison between the calculation scenario and the experiment (shot № 58163). The agreement between k and k_{exp} is good, their difference doesn't exceed 2 % during the whole pulse. The radial distributions of the plasma current density for various times are obtained as well.

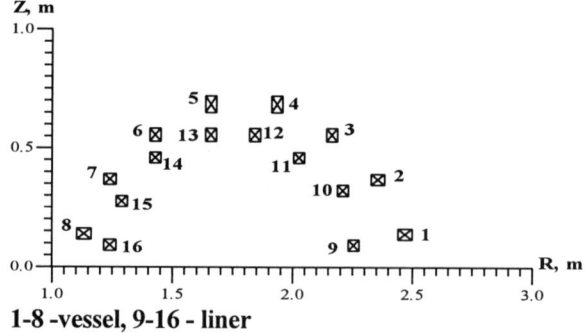

1-8 -vessel, 9-16 - liner

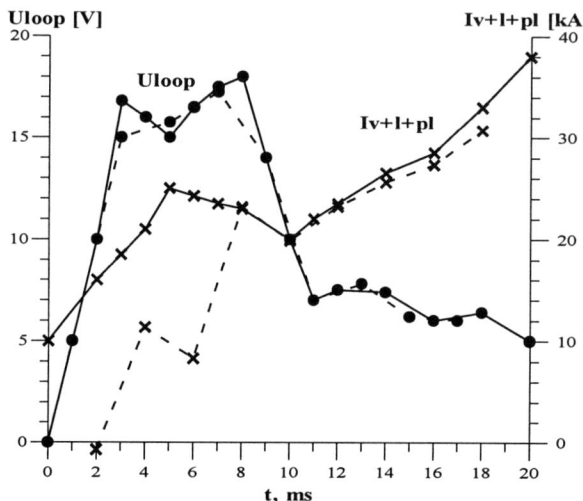

Fig. 3. Comparison of experiment and calculations comparisons (shot №58832, — experimental data, - - - calculated data)

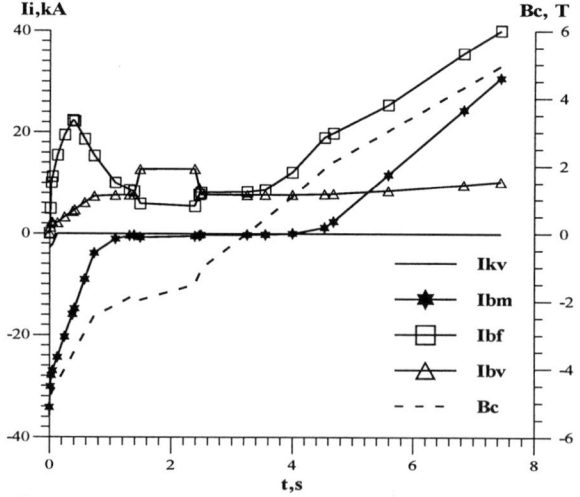

Fig. 4. Operation regime for poloidal system I_{BM}, I_{BF}, I_{BV}, I_{KV}, B_C (Ipl_{max}=-0.8 MA).

These and other comparisons between the results of calculation and experiment testify to a certain reliability of the proposed mathematical model and assurance of the results obtained. This model was used to create the calculation data base, on the basis of which the so-called "fast code" was developed allowing for an effective consideration of the discharge scenarios with sufficiently arbitrary initial data.

4. TFC SYSTEM DESIGN

The TEXTOR magnet system [7] comprises 16 toroidal field coils (TFC). Each coil incorporates a winding pack (WP) embraced with a steel ring for taking the thermal loads and a steel case supporting the electromagnetic loading. To exclude the transmission of the thermal loads to the case some air gap is provided between the case and the steel ring. A low friction material is used at their interface.

The inner portions of the coil cases are wedge-shaped and form a vault for supporting the centripetal electromagnetic forces. Each case has two pairs of clamps (at its top and bottom) to transmit the out-of-plane loads from the winding to the case. The moments due to these forces are taken partly by the shear keys installed between the coils in the vault region and partly by the outer intercoil mechanical structures.

The integral characteristics of the calculated electromagnetic forces as well as the temperature in the winding are given in Table 4 for the case of the plasma disruption (the loads refer to the coil half).

5. FINITE ELEMENT MODELS OF THE TFC

The stress analysis has been performed at two stages. At the first one the 2D finite element (FE) model of the TF coil is employed to evaluate the impact of the in-plane electromagnetic forces coupled with the winding warming on the stress-state in the coil system. At the second stage the 3D FE model is used to carry out the stress analysis of the system under the full set of the loads.

Table 4.

Max. winding temperature rise	K	49
Centripetal load	MN	-2.14
Out-of-plane load	MN	0.061
Vertical load	MN	5.21
Moment around radial axis	kN m	528
Moment around torus axis	kN m	1115

The 2D model of the winding pack crossection representative element is used to calculate the orthotropic thermo-mechanical properties of the winding pack according to [8]. The 2D global model of the coil uses the coil and the in-plane electromagnetic loads symmetry in reference to the coil midplane. The case vault region and outer intercoil structures are modeled with the axisymmetric elements. The rest is modelled with the plane-stress elements.

The 3D FE model of the TFC system is shown in Fig. 5. This model uses the coil symmetry in reference to its horizontal midplane. Symmetry of the in-plane loads and anti-symmetry of the out-of-plane ones in reference to this midplane are used as well.

The model includes the winding pack (WP), steel ring embracing the winding and the coil case. The WP and the steel ring are surrounded with the steel case. The gap between the steel ring and case allows for their relative radial movement. The case begins to bear the in-plane loads while the case/winding gap is closing. The appropriate boundary conditions on the lateral surfaces of the case vault region enable it to work as a wedge and to support the centripetal electromagnetic load.

The clamps are used to transmit the out-of-plane loads from the steel ring to the coil case. For simplicity, these clamps (saddles) are not modeled. In this model the FE nodes at the locations of the screws attaching the saddles to the case are coupled with the appropriate nodes of the ring in proper direction.

The tilting moments acting on the coils are taken with the shear keys installed between the neighbouring coils in the vault region and with the intercoil mechanical structures at the outside.

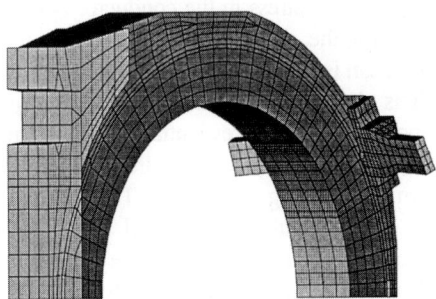

Fig. 5. 3D FE model of the coil.

The shear keys are not modeled. Neglecting their compliance, the equal vertical displacements are prescribed at the opposite edges of the key slot.

The intercoil structures are attached to the case by means of screws and keys. This is modeled by coupling the corresponding case/structure nodes in proper directions. The cyclic symmetry conditions on the lateral surfaces of the intercoil structures enable the structures to work in compression and to support the moments due to the out-of-plane loading.

6. RESULTS OF THE TFC STRESS ANALYSIS

Due to the design features the problem is very sensitive to the value of the case/winding air gap and to the winding temperature which has a strong impact on the gap value.

As expected, the case of the plasma disruption gives higher stresses in the coil components. Stresses in the slim part of the coil case (coil nose, midplane of the coil inner leg), intercoil mechanical structures and steel ring satisfy static requirements for the structural materials [9]. The maximum Tresca stress is 177 MPa, 155 MPa and 209 MPa respectively (Fig. 6). The low cyclic fatigue requirements [9] for designed 80 10^3 cycles seems to be satisfied as well.

It should be noted that compared with the 3D analysis the 2D one shows less tension in the coil nose under the in-plane loading accompanied with the winding warming. On the one hand, this is due to the WP distortion in the through thickness direction under the thermal load (the WP contact surface becomes curved) as it is shown by 3D analysis. This decreases the winding/case gap and, as a result, more vertical load is transmitted to the case from the winding. On the other hand, the 3D distribution of the hoop (wedge) stress in the case nose differs from that obtained in axisymmetry approach.

The Tresca stress in the conductor is below allowable value for the copper. The shear stress in the winding insulation is rather moderate (under 5-7 MPa).

It is shown that the most critical system component appears to be the screws attaching the clamps (saddles) to the coil case. These crews transmit the tilting moment from the winding to the case. The maximum tension in the screws (with the required screw prestress being taking into account) is quite close to the allowable value.

On the basis of the above analysis it is recommended to increase the initial prestress in the bolts clamping the intercoil structures of the neighbouring coils.

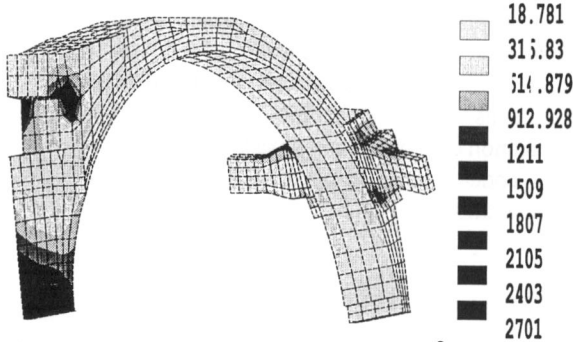

Fig. 6. Tresca stress in the coil case (kg/cm^2)

Hand estimations show that the high peak compression of about 227 MPa can occur in the insulation of the vault shear key. It is recommended to perform detailed local analysis for this region (for instance, similar analysis was done in the work [10]).

REFERENCES

1. Sheffield J. "Status of the tokamak program." Proc. IEEE, 1981, Vol. 69,No. 8, pp. 885-917.
2. Bondarchuk E. et al. "Tokamak-15 Electromagnetic System. Design and test results." Plasma Devices and Operations, 1992, Vol. 2, pp. 1-25.
3. Mirnov S. Plasma Physic Processes in Tokamak. Energoatomizdat, Moscow, 1983, p. 185(Russian).
4. Dnestrovskij Yu., Kostomarov D. Mathematical Plasma Modelling. Nauka, Moscow, 1982, p. 320 .
5. Parzen G. and Jellet K.. Part. Accel., 2, 1971, pp. 169-175.
6. Bondarchuk E.,Doinikov N.,Mingalev B. Zh. Tekh. Fiz., 1977, Vol. 47, No. 3, pp. 521-526 (Russian).
7. Giesen B. et al. "The toroidal field magnet of TEXTOR," Proceedings 10th Symposium on Fusion Technology, Padova, Italia, Sept. 1978, Paper F-1.
8. A.Borovkov et al."FE analysis of effective mechanical characteristics of micro-heterogeneous superconducting toroidal field coils," IEEE Trans. on Magn. ,V.28, No.1, pp.927-930, Jan. 1992.
9. Strength analysis code for the equipment and piping of nuclear power plants (PNAE G-7-002-86), Moscow, ENERGOATOMIZDAT, 1989 (Russian).
10. E. Bondarchuk et al. "Mechanical Modeling of the ITER Toroidal Field Coils Shear Keys Behavior. Design and Choice of Key Mock-Up for Electrical Insulation Testing on the Basis of Numerical Models," IEEE Trans. on Magn., Vol. 32, No.4, Part I, pp. 3024-3027, July 1996.

Research Progress on JT-60U in Advanced Steady-State Operation

O. Naito (for the JT-60 Team) [a]

[a] Japan Atomic Energy Research Institute, Naka-machi, Naka-gun, Ibaraki 311-01, Japan

Recent research progress on JT-60U in advanced steady-state operation is presented. High fusion performance of $Q_{DT} > 0.6$ has been obtained in negative magnetic shear plasmas as well as in high-β_p H-mode. A negative shear discharge is shown to have enhanced core confinement. Its sustainment and compatibility with radiative divertor are also explored. Stability and confinement are improved in high triangularity operation and a non-inductive discharge with high integrated performance has been achieved. The installation of negative-ion-based neutral beam injection system on JT-60U has been completed and beam injection into plasmas has been started. A preparation for the modification of the present JT-60U's open divertor to a W-shaped semi-closed divertor, and a conceptual design and R&D for JT-60 Super Upgrade are in progress.

1. Introduction

The JT-60U project has been devoted to an establishment of high performance steady-state tokamak operation, as well as to the physics R&D of ITER. For this purpose, confinement improvement, non-inductive operation, radiative divertor, high energy particle physics, and operation techniques are investigated at JT-60U. The achievement of high fusion performance in the high-β_p H-mode plasma [1] and the demonstration of high integrated performance discharges with high bootstrap current fraction [2] had shown a prospect of compact economical reactors such as SSTR (Steady-State Tokamak Reactor) [3].

More recently, researches on negative magnetic shear plasmas for an establishment of operational scenario in SSTR have revealed an improved core-confinement for electrons at relatively high densities [5]. At the same time, negative shear plasmas have shown a high fusion performance at moderate ion temperatures ($T_i(0) \sim 15\text{-}20$ keV). In parallel, high triangularity (δ) discharges are explored to improve the stability and confinement of edge plasma and to achieve high integrated performance compatible with radiative divertor. As for a heating and current drive system for reactors such as ITER, the installation of negative-ion-based neutral beam injection system (N-NBI) has completed and beam injection into JT-60U's plasmas has started. For further improvement of divertor function, preparation for the modification of present JT-60U open divertor to a W-shaped semi-closed divertor is in progress.

This paper describes the above research progress in advanced steady-state operation and the status of the conceptual design and R&D of JT-60 Super Upgrade.

2. Progress in Fusion Performance

Further progress in fusion performance has been achieved both in high T_i regime (> 40 keV, high-β_p H-mode) and moderate T_i regime (< 20 keV, negative shear plasma). In the recent campaign, the neutral beam injection power P_{NB} and the beam fueling rate were increased by reducing the gaps of the acceleration electrodes. As a result, 41 MW of power can now be injected at a beam energy of 95 keV.

In high-β_p H-mode, operations at higher plasma current up to $I_p = 2.7$ MA were explored. The maximum fusion triple product reached $n_D(0)\tau_E T_i(0) = 1.5 \times 10^{21}\text{m}^{-3}$ s keV at a high central ion temperature of $T_i(0) = 45$ keV, with central ion density $n_D(0) = 4.6 \times 10^{19}\text{m}^{-3}$, energy confinement time $\tau_E = 0.75$ s, neutron production rate $S_n = 5.2 \times 10^{16}\text{s}^{-1}$, $I_p = 2.4$ MA and $P_{NB} = 33$ MW. The confinement improvement factor defined by $H = \tau_E/\tau_E^{ITER89P}$ reached 3.3, where $\tau_E^{ITER89P}$ is the ITER L-mode scaling, and a normalized beta was $\beta_N \sim 2$

%mT/MA.

On the other hand, JT-60U's highest value of equivalent fusion amplification factor $Q_{DT} = 0.63$, exceeding the previous record of $Q_{DT} = 0.6$ obtained in high-β_p H-mode, has been achieved in a negative shear plasma. This new record was obtained at a relatively high density $n_e(0) = 9.5 \times 10^{19}$m^{-3} and a moderate ion temperature $T_i(0) = 16$ keV, with $\tau_E = 0.9$ s, $H = 3.2$, $S_n = 3 \times 10^{16}$s^{-1}, $W = 9.4$ MJ, $I_p = 2.5$ MA and $P_{NB} = 13$ MW.

3. Negative Magnetic Shear Plasmas

Negative magnetic shear discharges have been proposed as a possible operation scenario in SSTR [4]. The recent research on the negative shear plasma in JT-60U has shown its outstanding features on the improvement of core electron confinement as well as its high fusion performance. A negative shear plasma is produced by applying NBI heating during current ramp-up.

Typical temperature and density profiles in a negative shear plasma are shown in Fig. 1 along with a safety factor profile. Steep gradients in temperature and density are observed just inside the minimum q position, indicating a formation of internal transport barrier (ITB) at $\rho = 0.5$-0.6 (ρ is the normalized flux radius). Inside ITB, the temperature and density profiles are fairly flat and the ratio of ion temperature to electron temperature is relatively small ($T_i/T_e = 1.4$-3). This ratio decreases with increasing target plasma density. From a transport analysis [5], effective thermal diffusivity of electrons χ_e^{eff} drops sharply by a factor of 20 at $\rho = 0.6$. and effective thermal diffusivity of ions χ_i^{eff} is less than 1/5 of the neoclassical value at $\rho = 0.5$.

Figure 2 shows the confinement improvement H versus density normalized to the Greenwald limit for negative shear plasmas and the usual ELMy H-mode plasmas. For ELMy H-mode plasmas, H decreases with increasing density, mainly because of unfavorable effects of gas puff on the energy confinement. On the contrary, higher H is obtained at higher density in negative shear plasmas. Even at $\sim 70\%$ of the Greenwald's density limit, a high confinement of $H = 2.3$ is achieved.

A compatibility of negative shear plasma with radiative divertor was investigated by applying neon and hydrogen gas puff after a formation of

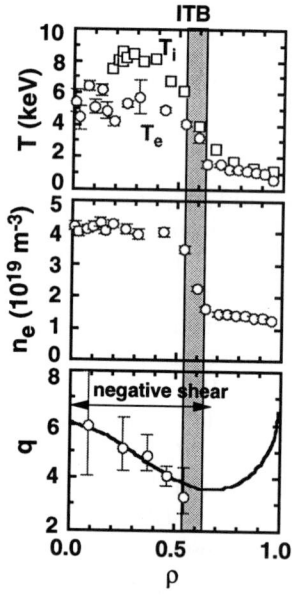

Figure 1. Radial profiles of ion and electron temperatures T_i, T_e and safety factor q in a typical negative shear discharge.

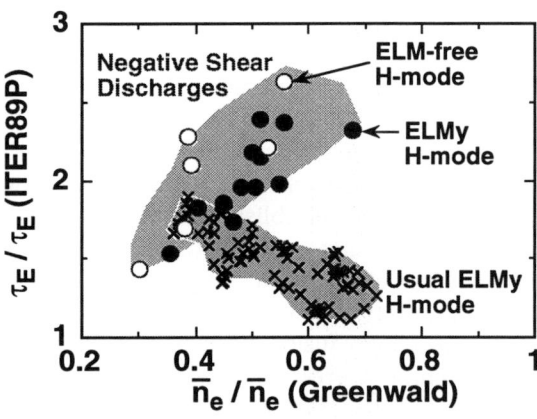

Figure 2. Confinement improvement H versus density normalized to the Greenwald limit $\bar{n}_e/\bar{n}_e^{Greenwald}$ for negative shear plasmas and ELMy H-mode plasmas.

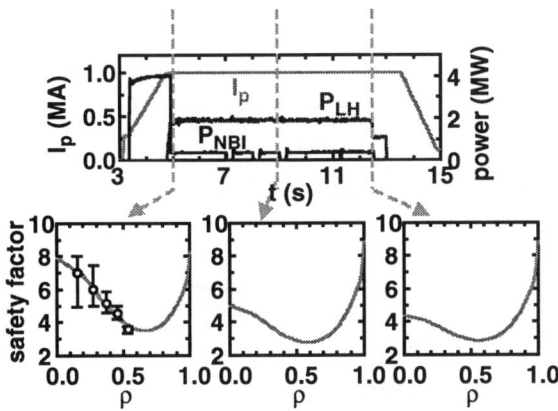

Figure 3. Safety factor profiles at 3 time slices in LH-sustained negative shear discharge. Waveforms of plasma current I_p, NBI power P_{NB}, LH power P_{LH} are also shown.

ITB. A highly radiative divertor was formed and sustained for 1 s without degradation of core confinement. By sequential injection of neon and hydrogen gas into a discharge with $P_{NB} = 13.5$ MW, a significant increase in the radiation loss was observed in the divertor (from 1.4 MW to 7.6 MW) whereas only a slight increase (from 0.5 MW to 1 MW) was seen in the main plasma. The total heat load to the divertor target plates was reduced to 20% of the absorbed NBI power.

The negative shear plasmas thus obtained are inherently transient because the hollow current profiles survive only for the current diffusion time. In JT-60U, lower hybrid current drive (LHCD) was exploited to sustain the hollow current profile and control a magnetic shear. Figure 3 shows q profiles measured at 3 time slices along with the waveforms of I_p and injected powers. The negative magnetic shear is sustained by LHCD for 7.5 s. The slow evolution of q profile is due to the diffusion of residual ohmic current (30-40%). Also, a strong negative magnetic shear, deeper than that formed by NBI heating during current ramp, is obtained only with LH wave injection.

Finally, magnetic shear can be controlled by changing the injected LH wave spectra. When LH waves with current-broadening spectra are injected, clear negative magnetic shear is observed. On the contrary, when LH waves with current-peaking spectra are injected, q becomes lower in the central region. Such non-inductive control method is beneficial to the sustainment of stability and performance of high bootstrap current discharges.

4. High Triangularity Operation

Steady-state reactors such as SSTR require a simultaneous achievement of high fusion performance, non-inductive operation with high bootstrap current fraction, high fusion power density (i.e. high β_N), and efficient divertor function. To fulfill the above requirements, improvement of stability at high plasma pressure is necessary. For this purpose, high triangularity (δ up to 0.5 in NB heated plasmas) operations were exploited by modifying the poloidal field coil power supply.

In JT-60U, the energy confinement in H-mode at high density is limited by an onset of giant edge

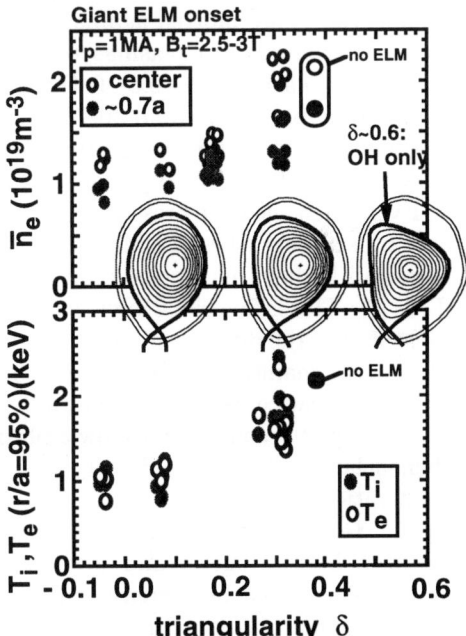

Figure 4. Electron density $\bar{n}_e$ and edge ion and electron temperatures T_i, T_e at the onset of giant ELMs as a function of triangularity δ.

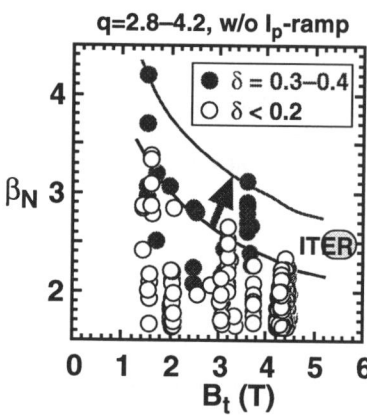

Figure 5. Normalized beta β_N versus toroidal magnetic field B_t, for high δ (closed) and low δ (open) plasmas.

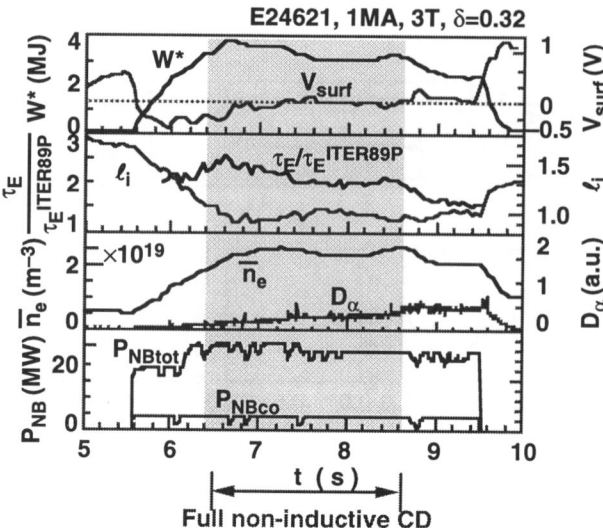

Figure 6. Waveforms of a full non-inductive current driven discharge obtained at $\delta = 0.32$ ($I_p = 1$ MA, $B_t = 3$ T).

localized mode (ELM). In Fig. 4, the electron density, edge ion and electron temperatures at the onset of giant ELMs are plotted against δ. Compared to lower triangularity plasmas ($\delta \sim 0.1$), high triangularity plasmas ($\delta \sim 0.3-0.4$) have 30-40% higher attainable density and 30-70% higher attainable temperature. The increase in δ also extends the duration of ELM-free period by factor of 2~10. Even when the edge pressure reaches the limit and ELMs appear, these ELMs have smaller amplitudes and higher frequencies than those of giant ELMs [7]. This is favorable in view of reducing the heat flux to the divertor plates. Moreover, high δ plasma can maintain the improved confinement ($H > 2$) at twice as high normalized edge density $n_e/n_e^{Greenwald}$ compared to low δ plasma. This is beneficial to a high-edge-density radiative divertor operation.

In Fig. 5, β_N is plotted against the toroidal magnetic field B_t for high δ and low δ cases. Attainable β_N increased by increasing δ. For $\delta = 0.34$, a high value of $\beta_N = 3.1$ %mT/MA was obtained at relatively high B_t of 3.6 T. This result suggests the feasibility of achieving required condition in ITER.

A high integrated performance of high δ operation is demonstrated in a discharge shown in Fig. 6. In this shot, full non-inductive driven state with high bootstrap current fraction (bootstrap current $\sim 60\%$, beam driven current $\sim 40\%$) was sustained for 2 s without giant ELMs. In addition, high confinement ($H = 2-2.5$) and normalized beta ($\beta_N = 2.6-3.1$ %mT/MA) were maintained.

5. Status of Negative-Ion-Based Neutral Beam Injection System

The negative-ion-based neutral beam injection system has potential of producing high energy beams in MeV range and is one of the promising candidates for heating and current drive system in ITER. The installation of tangential N-NBI system on JT-60U was completed and the beam injection into the JT-60U plasma started in March 1996. This system is designed to deliver up to 10 MW of power for 10 s at beam energy of 500 keV.

Up to now, ~ 2.5 MW of deuterium beams were injected for ~ 0.5 s at ~ 350 keV. Figure 7 shows the waveforms of stored energy W, electron density n_e, $\beta_p + \ell_i/2$, $\beta_p^\perp$, neutron production rate S_n, loop voltage V_ℓ, central electron temperature $T_e(0)$ and N-NBI power P_{NNBI}. Here β_p and $\beta_p^\perp$ are respectively average and perpendicu-

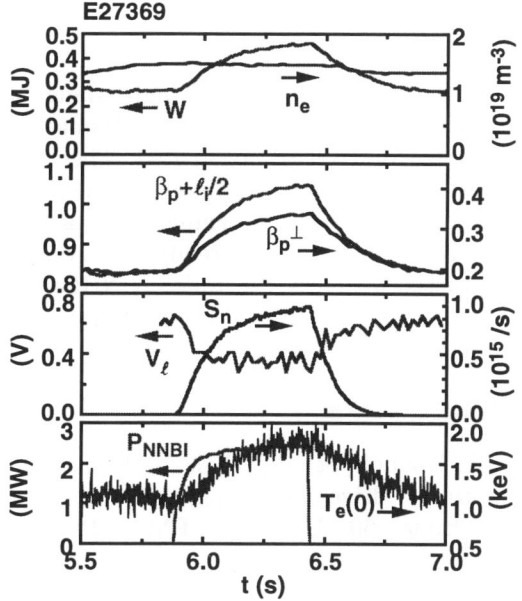

Figure 7. Waveforms of stored energy W, electron density n_e, $\beta_p + \ell_i/2$, $\beta_p^\perp$, neutron production rate S_n, loop voltage V_ℓ, central electron temperature $T_e(0)$ and N-NBI power P_{NNBI}.

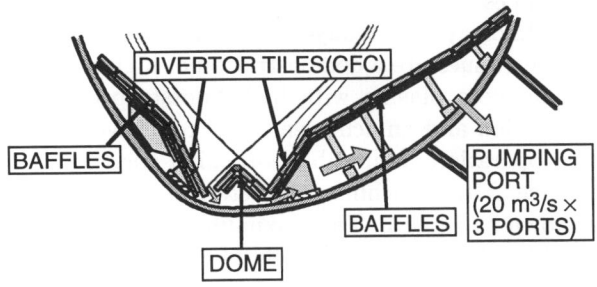

Figure 8. W-shaped semi-closed divertor.

lar poloidal betas, and ℓ_i is internal inductance. Clear increases in W, S_n, $T_e(0)$ and a decrease in V_ℓ were observed. Assuming that the change in ℓ_i is small during N-NBI pulse, the difference in the increments of $\beta_p + \ell_i/2$ and $\beta_p^\perp$ indicates that increase in the plasma pressure is about twice as high in parallel direction compared to perpendicular direction.

By combination of conventional NBI, full current drive of more than 2 MA is expected. Further optimization of the N-NBI system is in progress.

6. W-Shaped Semi-Closed Divertor

In order to establish an effective radiative divertor compatible with a high main plasma performance, the present open divertor in JT-60U is planned to be modified to a W-shaped semi-closed divertor with pumps in February-May, 1997. The new divertor consists of inclined divertor plates, a dome in the private flux region, and a baffle structure, as shown in Fig. 8. Carbon Fiber Composite (CFC) tiles are used where high heat flux of $\sim 10 MW/m^2$ is anticipated (e.g. divertor plates, top tiles of the dome), otherwise graphite tiles are used. The tiles are cooled inertially.

Three cryopumps of the present NBI system are converted to divertor pumps, whose pumping speed is $35 \sim 70$ m^3/s at the exhaust throat, sufficient for exhausting particles fueled by 90 keV, 20 MW of NBI. To obtain a stable operation of radiative divertor, following active control methods are adopted. The fast shutter valves of NBI ports are used to change the pumping speed during a shot. In addition to rearrangement of gas fueling system, the present pneumatic pellet injector is modified to a centrifuge type which can inject pellets of 2mm$^\phi$ × 2mm at a injection speed of 0.5-1 km/s and a repetition rate of 30 Hz.

A simulation study using a two dimensional neutral transport code (NEUT2D) [8] shows that the neutral flux into the main plasma is reduced to a half of that in the open divertor and that the dome structure raises the neutral pressure at the exhaust region by an order of magnitude, favorable for efficient exhaustion. Also a simulation using a two dimensional impurity transport code (IMPMC) [8] shows that the influx of methane gas generated by chemical sputtering in the private flux region (a cause of X-point MARFE) is suppressed by the existence of the dome.

7. JT-60SU (Super Upgrade)

JT-60SU will be a major upgrade of the present JT-60U to a superconducting tokamak. The mission of JT-60SU is to complement the ITER program and establish an integrated basis for physics and technology for steady-state tokamak operation. Key issues are a demonstration of high performance core plasma with high bootstrap current fraction and an achievement of radiative divertor operation compatible with high core performance. Main machine parameters of JT-60SU are shown in Table 1. The maximum plasma current of 10 MA and toroidal field of 6.25 T enable a research on fusion relevant plasmas. A steady-state operation at $I_p = 5$ MA will be demonstrated with ~ 40 MW of N-NBI and ~ 20 MW of electron cyclotron heating (ECH). In order to explore optimum plasma shape, elongation and triangularity can be varied as in Table 1, in addition to the selection of single or double null. The vacuum vessel is made of low cobalt concentration stainless steel (316SS) to reduce the radiation dose rate inside the vessel. For toroidal field (TF) coils, Nb_3Al conductors will be used which have better mechanical and critical current properties than $(NbTi)_3Sn$ conductors. Adjacent pair of TF coils are connected for increasing the rigidity and reducing the weight.

8. Summary

The performance of JT-60U has been improved by adoption of negative magnetic shear and high triangularity operations. High core performance of $Q_{DT} > 0.6$ has been obtained in negative shear plasmas, and high stability and confinement at edge have been realized with high triangularity. A compatibility of negative shear plasmas with a radiative divertor is demonstrated. These results suggest the feasibility of operation scenario in a high bootstrap current tokamak such as SSTR. Further research on the steady-state operation will proceed with newly installed N-NBI and with the introduction of W-shaped semi-closed divertor, and for the future with JT-60SU.

REFERENCES

1. T. Kondoh et al., High performance and current drive experiments in the JAERI Tokamak-60 Upgrade, Phys. Plasmas 1 (1994) 1489-1496.
2. Y. Kamada et al., Non-inductively current driven H mode with high β_N and high β_p values in JT-60U, Nucl. Fusion 34 (1994) 1605-1618.
3. Y. Seki et al., The Steady State Tokamak Reactor, Proc. 13th Int. Conf. on Plasma Physics and Controlled Nuclear Fusion Research, Washington, DC, 1990, Vol. 3, IAEA, Vienna, 1991, pp. 473-485.
4. T. Ozeki et al., Profile control for a stable high β_p tokamak with a large bootstrap current, Proc. 14th Int. Conf. on Plasma Physics and Controlled Nuclear Fusion Research, Würzburg, 1992, Vol. 2, IAEA, Vienna, 1993, pp. 187-194.
5. T. Fujita et al., Internal Transport Barrier for Electrons in JT-60U Reversed Shear Discharges, submitted to Phys. Rev. Lett.
6. S. Ide et al., Sustainment and Modification of Reversed Magnetic Shear by LHCD on JT-60U, to appear in Plasma Phys. Control. Fusion.
7. Y. Kamada et al., Onset condition of ELMs in JT-60U, to appear in Plasma Phys. Control. Fusion.
8. K. Shimizu et al., Simulation of Neutral Particle and Impurity Behavior in JT-60U W-Shaped Divertor, J. Plasma and Fusion Research 71 (1995) 1227-1237.

Table 1
Main machine parameters of JT-60SU

	Inductive	Non-inductive
Plasma current	10 MA	5 MA
Toroidal field	6.25 T	6.25 T
Major radius	4.8-5.2 m	4.8 m
Minor radius	1.3-1.5 m	1.3 m
Elongation	1.6-1.8	1.8
Triangularity	0.4-0.8	0.4
Flat top	200 s	1000 s - 1 hr
N-NBI (750 keV)	~ 40 MW	←
ECH (220 GHz)	~ 20 MW	←

AN OVERVIEW OF THE DIII–D PROGRAM

J.L. Luxon for the DIII–D team

ABSTRACT AND SUMMARY

The DIII–D program focuses on developing fusion physics in an integrated program of tokamak concept improvement. The intent is both to support the present ITER physics R&D and to develop more efficient concepts for the later phases of ITER and eventual power plants. Progress in this effort can be best summarized by recent results for a diverted deuterium discharge with negative central shear which reached a performance level of $Q_{DT} = 0.32$. The ongoing development of the tools needed to carry out this program of understanding and optimization continues to be crucial to its success.

Control of the plasma cross-sectional shape and the internal distributions of plasma current, density, and rotation has been essential to optimizing plasma performance. Recent measurements of the current profile have resulted in improved control of the profile and thus significant progress in optimizing the confinement and stability of discharges. Pellet injection along with strong pumping of the plasma edge has provided operation at high plasma density with good confinement. The addition of FWCD and ECH systems is providing the tools needed to develop steady state control of the current and temperature profiles.

Advanced divertor concepts provide edge power and particle control for future devices such as ITER and provide techniques to help manage the edge power and particle flows for advanced tokamak concepts. New divertor diagnostics and improved modeling are developing excellent divertor understanding. Divertor pumping has proven successful at controlling the density and impurity content of plasmas, and the installation of the radiative divertor configuration will allow pumping of the highly shaped configurations which are optimal for high performance. The use of vanadium in parts of the divertor structure will introduce this low activation material into the tokamak environment for the first time.

Many of the plasma physics issues being posed by ITER are being addressed. Scrapeoff layer power flow is being characterized to provide an accurate basis for the design of reactor devices. Ongoing studies of the density limit focus on identifying ways in which ITER can achieve the required densities in excess of the Greenwald limit. Better understanding of disruptions is crucial to the design of future reactors. Measurements of the halo currents have better quantified the spatial and temporal distribution of the currents. Predictive models are being developed to describe the disruption evolution and associated current distributions. The injection of impurity pellets into the edge of a plasma at the onset of a disruption is found to ameliorate the halo currents and heat loads, and a neural network has had success in identifying the onset of disruptions.

1. INTRODUCTION

The DIII–D tokamak facility is well-equiped with a diverse set of capabilities and diagnostics to provide germane research both for the development of plasma science and in support of engineering and technology, especially ITER. In recent years, the DIII–D program has focussed on 1) the understanding and efficient use of the divertor configuration, and 2) developing the plasma configuration and internal profiles to optimize tokamak performance including power and particle control.

DIII–D is a non-circular cross-section tokamak with major radius $R_0=1.66$ m, minor radius $a=0.66$ m, and plasma chamber elongation of 2.1 with toroidal magnetic field B of up to 2.2 T at R_0 [1,2]. The tokamak operates in a wide range of configurations including single null and double null divertors. The maximum plasma current I_p is presently 2.5 MA. 20 MW of neutral beam heating and 6 MW (source) of fast wave power are routinely available, and 1.0 MW (source) of 110 GHz ECH power has been recently commissioned. The device operates largely in deuterium. DIII–D operation is characterized by highly reproducible discharges due, in part, to the capabilities discussed in this section.

Improvements in the reduction of error fields present at plasma initiation and attention to cleanliness of the vessel walls has resulted in routinely achieving initiation and startup with remarkably low voltages of 3.0 V or less without auxiliary power. This has been aided by the use of helium glow discharge cleaning between tokamak discharges and the occasional boronization of the carbon wall facing

---

*Work supported by U.S. Department of Energy under Contract No. DE-AC03-89ER51114.

surfaces. Low voltage startup is significant for next generation devices where a voltage reduction reduces power supply costs.

Development of the digital plasma control system has been effective in providing enhanced capabilities to manage plasma configurations and profiles [3,4]. This system allows superb flexibility in implementing new control algorithms and developing new operating configurations. Recently, the plasma configuration is controlled to match a desired equilibrium configuration using real time MHD equilibrium calculations. This capability has allowed us to implement new plasma configurations and control algorithms including new crescent shaped plasma cross-sections and precise control of the divertor X–point location over a wide range of plasma parameters. A neural network based algorithm to detect conditions leading to disruptions has been developed [5] and installed into the plasma control computer. This system is capable of predicting disruptions due to exceeding the beta limit better than conventional standards such as the Troyon limit.

2. HIGH PERFORMANCE DISCHARGES

The achievement of high performance tokamak discharges has been an ongoing commitment of the DIII–D program. Continuous improvement of the performance of tokamak discharges as measured by, i.e., the product of the plasma pressure normalized to the magnetic field and the energy confinement time, $\beta\tau_E$, has been demonstrated by exploiting the roles of elongation, triangularity, and shaping of the pressure and current profiles. These improvements, when obtained simultaneously and in steady-state, lead to more cost effective tokamak power plants.

The development of discharges with current profiles modified to be hollow so that the magnetic shear in the central region of the plasma discharge is negative has resulted in discharges with thermonuclear performance well beyond that previously achieved on DIII–D and comparable to that achieved on devices with larger size and higher magnetic fields [6]. The plasma configuration had an elongated triangular cross-section identified as having superior confinement and stability properties. By shaping the current profile transiently using the expeditious application of neutral beam heating early in the discharge, the required hollow current profile was obtained and a discharge with very high central confinement resulted. In the best discharge, $\tau_E = 0.4$ s (4.5 $\tau_{ITER89P}$) and $\beta=6.7\%$ ($\beta_N=\beta aB/I_p=4.0$), resulting in values of the fusion reactivity $Q_{DD}=0.0015$. If this reaction rate is projected to DT fuel using a detailed energy transport analysis code and making no correction for isotopic dependence of confinement, the result is $Q_{DT}=0.32$. This is three times the reactivity achieved previously in DIII–D and remarkable compared to the results of larger machines when normalized to the major radius and toroidal field [6].

High performance discharges have been achieved in single-null divertor configurations with lower triangularity ($\delta=0.3$) and low safety factor ($q\cong3$) prototypical of high performance JET discharges and of ITER by modifying the plasma internal profiles to produce areas of weak negative central shear and a broad pressure profile [7]. Values of $\beta_N=4$ were achieved, comparable to the best previous discharges. The H–mode edge transport and an internal transport barrier presence lead to global energy confinement exceeding ITER-89P by a factor of 4 and fusion reactivity of $Q_{DD}\sim 1 \times 10^{-3}$. The performance of the single-null discharges normalized to I_p is comparable to that achieved in double-null discharges except that higher $_p$ can be achieved with double-null for the same safety factor.

These highest performance single and double-null results are obtained for short durations (0.1 s) owing to the lack of tools for long pulse control of the current and pressure profiles. In single-null discharges with longer duration (1.5 s), performance parameters of H=2.5 and $\beta_N=3.0$ have been achieved. The aim of future experiments will be to use current drive to sustain the optimum profiles.

The presence of negative magnetic shear alone is not sufficient to explain the confinement improvement seen in the discharges described above [8]. The ion energy transport is sharply reduced in the core region to below the neoclassical value, while the electron transport is also reduced to a lesser extent [Fig. 1(a)]. Other characteristics are seen experimentally including the presence of a power threshold and peaking in the ion and electron temperature profiles, and a substantial increase in the core plasma rotation. A likely mechanism for this is the reduction of turbulence driven transport by E×B shear decorrelation of turbulence wherein the the shear dramatically reduces the turbulent eddies causing the transport [9–11]. Experimental data shows that stabilization by the E×B shear increases dramatically as the plasma makes the transition into enhanced confinement, so that it far exceeds the calculated growth rate for the dominant mode in the center of the plasma, in this case the trapped electron η_i mode [Fig. 1(b)] [12]. The incremental radial electric field

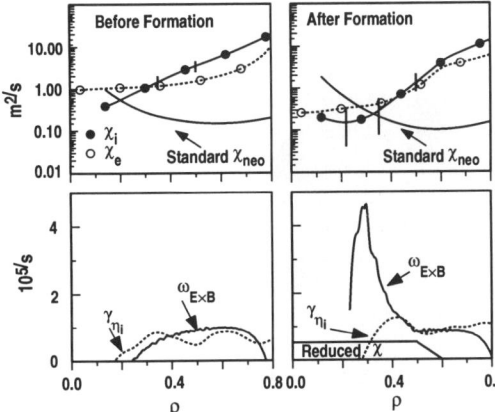

Fig. 1. Transport before and after the transition to negative central magnetic shear. (Top): The ion and electron diffusivity. (Bottom): The calculated growth rate for the (dominant) η_i mode and the damping from E×B shear shown on the same scale.

which provides this stabilization comes largely from an increase in the rotation in the plasma core following the stabilization of the MHD modes by the increased negative magnetic shear. Further evidence in support of this comes from the virtual disappearance of core turbulence following the transition [8].

Fast wave current drive (FWCD) offers a means of providing the steady state current profiles needed for these enhanced performance discharges. Experiments using 2.4 MW of fast wave power have successfully demonstrated that fast wave power can be used to broaden the region of negative shear and make the shear more negative Fig. 2 [13]. The previously observed current drive efficiency [$\eta = n(10^{20})$ I(MA) R(m)/P(MW)] of FWCD scales with temperature to higher temperature. Analytical models using these results indicate that by using up to 6 MW (source) of 110 GHz ECH heating to heat the electrons in DIII–D along with existing FWCD and neutral beam heating capabilities, that the profiles needed to maintain high performance discharges steady state (10–20 s in DIII–D) could be sustained.

Historically, the densities at which the tokamak could operate appeared limited by physical phenomena collectively characterized by the Greenwald limit ($n_{max}^{GW} \sim I_p/\pi a^2$). ITER has identified the advantages of operating at densities above this limit in achieving ignition with successful divertor operation. ITER requirements call for $n \geq 1.5 \, n_{max}^{GW}$, with H–mode confinement (H=2), q_{95}=3, and β_N=2. Experiments on DIII–D have been undertaken to understand the physics of the density limit and devise paths beyond it.

Experiments show there is no fundamental obstacle to achieving densities above the Greenwald limit [14]. Using low neutral-beam-injected power, divertor gas pumping, a progression of regularly spaced deuterium pellets, and stopping pellet injection above a pre-programmed density, we achieved discharges (Fig. 3) with quasi-steady ($\geq$0.5 s)

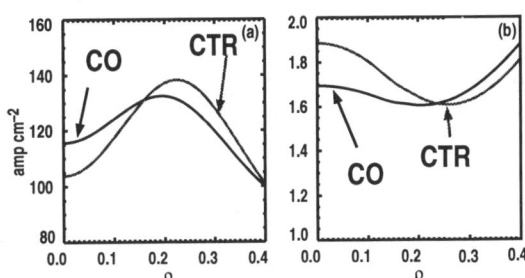

Fig. 2. Comparison of co- and counter-FWCD applied to similar discharges. (a) the current density as a function of the normalized radius in the central region. (b) the resultant profile of the safety factor q. RF power of 2.4 MW was applied and the data waste taken 400 ms into the rf pulse.

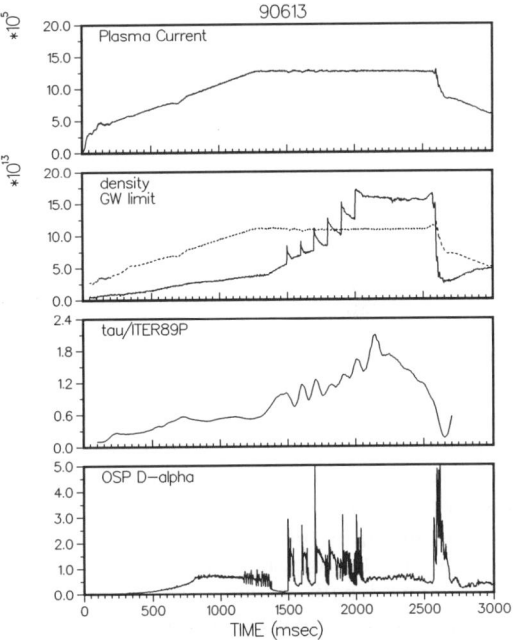

Fig. 3. Pellet injection into an H–mode discharge resulting in density exceeding the density limit. (a) the plasma current, (b) the plasma density and density limit, (c) the confinement time normalize to the ITER-89P confinement scaling, (d) D_α light showing ELM behavior at each pellet and a quiescent period during the high density flat top.

line-average density 40%–60% above n_{max}^{GW}, with global energy confinement times of 1.5–1.8, normalized to ITER89-P scaling. The discharge was ELM-free between pellets, and each pellet triggered an ELM. As the radiated power increased during the ELM-free phase 2.0–2.6 s, the energy confinement time gradually degraded. A H–L transition was observed at t=2570 ms, and a locked mode rapidly ejected particles at t~2600 ms. During the experiments, restrictions on the accessible density window resulted from pellet fueling limits, H–L power transition limit, and MHD activity onset.

3. DIVERTOR DEVELOPMENT

Understanding of the plasma wall interaction and, particularly, the divertor is essential to the next-generation device designs and meaningful upgrades of large tokamaks. Two major issues can be identified; control of the particle flow from the plasma edge including impurity and ash removal, and management of the heat flux and erosion at the divertor surface. Progress has been made in both areas.

To understand the physics of the divertor region and plasma surface interactions, it is essential to characterize the plasma properties in this region, along with those of the core plasma and the edge region well away from the divertor where the local properties are determined by the transport from the core. DIII–D has had an outstanding set of core and edge plasma diagnostics. A comprehensive set of divertor diagnostics has been developed to complement these [15], including Thomson scattering, spectroscopy, and probe measurements. Simultaneously, computer models have been developed to relate the measurements to physical understanding [16]. The combination of capabilities is leading to a better understanding of divertor plasma and to improved divertor configurations.

Long-legged divertor configurations have been developed in which much of the plasma energy is radiated away before it reaches the wall and the plasma interacting with the wall is quite cold (Fig. 4). The divertor legs, normalized to the device size are the same length or longer (ra/=0.9) than the divertor legs of the present ITER design (r/a=0.67). These discharges represent perhaps the closest yet achieved to the proposed ITER design. These partially detached divertors were formed by flowing deuterium gas into the mid-plane region and pumping near the outer divertor leg [17]. These discharges radiate along the entire outer divertor leg, and the heat flux at the divertor surface is reduced a factor of

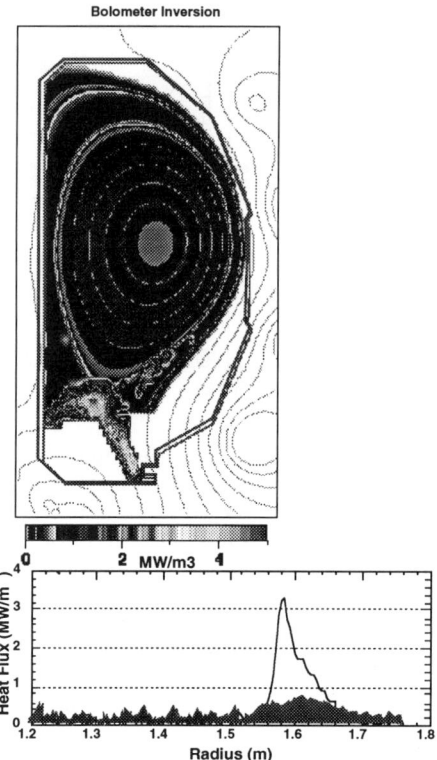

Fig. 4. A reconstruction of the radiated power from a long-legged divertor discharge showing a high level of radiation along the divertor leg following the onset of deuterium injection. Below is the heat flux to the divertor before (line) and after (solid) the radiating zone was established.

3 to 5 while the radiated power increases proportionally. The temperature at the divertor plates is seen to drop to about 1 eV. These low temperatures would substantially reduce the erosion expected in a large steady state fusion device.

The analysis of these discharges is producing reasonable agreement with models [16]. The observed low temperatures at the divertor plate has led to changes in the physics models used to describe the divertor region. In particular, more atomic processes, such as volumetric recombination, were included. The model correctly shows the long cool region of high radiation along the divertor leg and the low temperatures at the divertor plate.

4. DISRUPTIONS

The understanding of disruptions is crucial to the design of next-generation tokamaks. The DIII–D program has taken a strong role in understanding the disruption process, characterizing and modeling disruptions and developing techniques to mitigate disruptions [18].

The heat pulse and the halo currents from the disruption have been characterized. Heat flux measurements during high beta and impurity-induced disruptions have shown that 70%–90% of the energy lost is incident on the divertor region. The peak of the energy distribution is often far from either of the divertor strike points and can move substantially during the disruption. The width of the heat pulse distribution on the divertor plates can range from as little as 3 cm to as much as 10 cm (FWHM).

Halo currents are poloidal currents flowing from the plasma mantle through the vessel wall components and back into the mantle during the transient conditions of a disruption. Understanding of the halo currents has evolved considerably in the last year with the availability of improved diagnostics. Under combined worst case conditions, halo currents of as much as 30% of the total plasma current are observed in DIII–D and the toroidal asymmetries in these currents can be as large as a factor of two at the peak currents [18]. Higher asymmetries have been seen, but only at proportionally less total halo current. On time scales of less than a millisecond, yet higher transient values of the current (40% of I_p) and peaking factor (5) have been seen, but the time scale is so fast they are of little significance to the engineering of DIII–D.

The injection of impurity pellets at the onset of a disruption has been found to significantly mitigate the consequences of disruptions [19]. Neon pellets with diameters of 1.7 and 2.8 mm (up to 20 percent of the plasma particles) were injected into both discharges with triggered vertical displacement events and into normal stable discharges (Fig. 5). The pellets caused 70% of the thermal energy to be lost by the end of the pellet ablation. The current decay begins within 1 ms of the end of the ablation and the halo currents were found to decrease by a factor of two compared to disruptions without pellets injected. There was also very little asymmetry in the halo currents observed under these conditions. Small amounts of runaway electrons were seen following the pellet injection which is significant because runaway electrons were not seen in disruptions of similar discharges without pellet injection and are not normally seen in disruptions of DIII–D discharges. The presence of runaways is consistant with the Dreicer condition for electron avalanche.

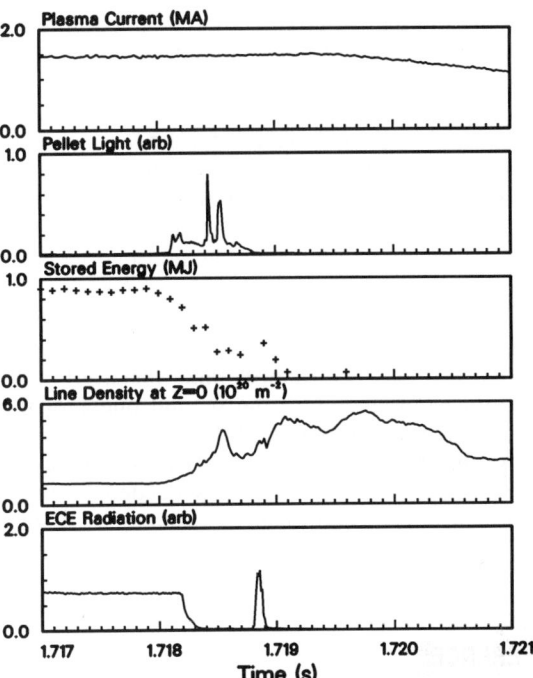

Fig 5. Data from neon pellet injection into a discharge undergoing a vertical displacement event. The discharge began at ~0 on the time scale. (a) plasma current, (b) pellet light indicating ablation takes place in ~0–0.75 ms, (c) plasma stored energy, (d) plasma density, (e) electron cyclotron emission indicating the presence of a runaway tail at 1718.8 s. The emission is non-thermal in character.

5. FUTURE PLANS

Better understanding of divertor configuration has led to the need for a more closed divertor to inhibit recycling of divertor neutrals back to the edge of the main plasma. Understanding of the role of the plasma configuration in the accessibility of high performance regimes has indicated the need for a more triangular crossection for the diverted plasmas. This has led us to develop the Radiative Divertor configuration, the first phase of which is being installed in the DIII–D vessel in the fall of 1996 [20,21]. This configuration will allow us to directly compare triangular divertors formed using the top divertor with the presently achievable less triangular shapes using the lower divertor while allowing us to maintain the excellent diagnostic capabilities of the lower divertor. Future activities, if supported by experimental results, will be to install a second divertor for the lower null of the triangular configuration and re-establish the detailed divertor diagnostic for this new configuration.

One of the major problems facing development of fusion power plants is the available of low activation materials especially metals for the fabrication of structural components. One of the best candidate metals is vanadium which when activated by neutron bombardment substantially decays in 10 years. This metal is abundant, but suitable fabrication techniques are not well developed. General Atomics and the DIII–D program has begun a program of developing vanadium fabrication techniques which will culminate in the fabrication and installation of vanadium components in the second phase of the Radiative Divertor Program [22].

As discussed, additional ECH power is key to achieve steady-state the profiles needed for high performance advanced tokamak discharges. A total of 3 MW (source) of ECH power is planned next year with an additional 3 MW (source) planned later.

REFERENCES

[1] J.L. Luxon, Fusion Engr. Des. **30** (1995) 39.
[2] J.L. Luxon and L. Davis, Fusion Technol. **8** (1985) 441.
[3] M.L. Walker, et al., Proc. 16th Symp. on Fusion Engineering, Champagne, Illinois (1995) 885.
[4] B.G. Penaflor, et al., "A Structured Architecture for Advanced Plasma Control Experiments," this conference.
[5] D. Wroblewski, G.L. Jahns, and J.A.Leuer, "Tokamak Disruption Alarm Based on a Neural Network Model of the High Beta Limit," submitted to Nucl. Fusion.
[6] E.A. Lazarus, "Higher Fusion Power Gain with Profile Control in DIII–D Tokamak Plasmas," submitted to Nucl. Fusion.
[7] E.J. Strait, et al., "Improved Fusion Performance in Low-q, Low Traingularity Plasmas with Negative Central Shear," Proc. 23rd Euro. Conf. on Contr. Fusion and Plasma Phys., Kiev, Ukaraine, 1996 (European Physical Society, Petit-Lancy, Switzerland) to be published.
[8] L.L. Lao, Phys. Plasmas **2** (1995).
[9] Hahm, Burrell, Phys Plasmas **2** (1995) 1648.
[10] P.H. Diamond, et al., submitted to Phys. Rev. Lett.
[11] F.L. Hinton and G.M. Staebler, Phys. Fluids B **5** (1993) 1281.
[12] E.J. Strait, Phys. Rev. Lett. **75** (1995) 4421.
[13] R. Prater, et al., "Fast Wave Discharges and Current Drive in DIII–D in Discharges with Negative Central Shear," Proc. 16th International Conf. Plasma Phys. and Contr. Nucl. Fusion Research, Montreal, Canada, 1996 (Inter-national Atomic Energy Agency, Vienna, Austria) to be published.
[14] R. Maingi, et al., "Investigation of the Physical Processes Limiting Plasma Density in DIII–D," to be published in Phys. Plasmas.
[15] S.L. Allen, "Recent Results from Tokamak Divertor Plasma Measurements," Proc. 11th Topical Conf. High Temperature Plasma Diagnostics, Monterey, California, (1996).
[16] G.D. Porter, et al., "Divertor Characterization Experimentsand Modeling in DIII–D," Proc. 23rd Euro. Conf. on Contr. Fusion and Plasma Phys., Kiev, Ukaraine, 1996 (European Physical Society, Petit-Lancy, Switzerland) to be published.
[17] A.W. Leonard, et al., "Divertor Heat and Particle Flux Due to ELMs in DIII–D and ASDEX-Upgrade," Proc. 12th International Conf. on Plasma Surface Interactions in Contr. Fusion Devices, Saint Raphael, France (1996) to be published
[18] A.G. Kellman, et al., "Disruptions in DIII–D," Proc. 16th International Conf. Plasma Phys. and Contr. Nucl. Fusion Research, Montreal, Canada, 1996 (International Atomic Energy Agency, Vienna, Austria) to be published.
[19] T.E. Evans, et al., "Measurements of Non-axisymmetric Halo Currents With and Without "Killer" Pellets During Disruption in the DIII–D Tokamak," Proc. 12th Internationa Conf. on Plasma Surface Interactions in Contr. Fusion Devices, Saint Raphael, France (1996) to be published.
[20] S.L. Allen, et al., "First Measurements of Electron Temperature and Density with Divertor Thomson Scattering in Radiative Divertor Discharges on DIII–D," Proc. 12th International Conf. on Plasma Surface Interac-tions in Contr. Fusion Devices, Saint Raphael, France (1996) to be published.
[21] E.E. Reis, et al., "Structural Design of the DIII–D Radiative Divertor," this conference.
[22] J.P. Smith, et al., "Fabrication Development and Usage of Vanadium Alloys in DIII–D," this conference.

Commissioning Tests and Enhancements to the JET Active Gas Handling Plant

R Lässer, A Bell, N Bainbridge, D Brennan, B Grieveson, J L Hemmerich, G Jones, S Knipe, J Lupo, J Mart, A Perevezentsev, N Skinner, R Stagg, K Walker, R Warren, J Yorkshades

JET Joint Undertaking, Abingdon, OX14 3EA, United Kingdom

The JET Active Gas Handling (AGH) plant went through an extensive phase of inactive commissioning which involved end-to-end testing of signals, checking of the Distributed Control System, operation of the plant with gas mixtures simulating torus exhaust gas and finally pumping of gas from the torus. At present tritium commissioning is being performed. Trace tritium commissioning with 0.08g of tritium gas collected during the Preliminary Tritium Experiment at JET has been successfully completed. Results of the tritium commissioning as well as enhancements to the AGH plant are presented.

1 INTRODUCTION

The JET Active Gas Handling System (AGHS) was designed for a maximum daily throughput of up to 5 moles of tritium, 15 moles deuterium and 150 moles of protium. During the JET tritium operations phase AGHS will pump the exhaust gases from the JET torus and separate them into helium, hydrogen and impurities. Almost pure tritium and deuterium will be produced in two isotope separation systems for re-injection into the torus. Tritium-containing impurities will be detritiated and the tritium reclaimed.

Process pipework of the AGHS is surrounded by secondary containments, in a few cases even tertiary containments. Detritiated process gas and gas from secondary containments may be discharged through the Exhaust Detritiation (ED) system which achieves a detritiation factor of about 1000.

In this way the JET torus and JET AGHS form an almost closed gas loop, unburned tritium and deuterium will be re-used and risks to the environment are minimised.

Fig 1 shows a schematic flow diagram of the AGHS and its connections with the users.

2 INACTIVE COMMISSIONING

All sub-systems of the AGHS were built in compliance with a strict quality assurance programme. Process pipework and secondary containments were subjected to stringent leak tightness tests. End-to-end commissioning of signals and control loops between the various sub-systems and the Distributed Control System has been performed. Results of the inactive commissioning of the various sub-systems of the AGHS were described previously[1-6].

Pumping tests of the torus exhaust gases during normal plasma operation, during regeneration of the divertor and Neutral Injection Box (NIB) cryopumps, and during helium glow discharge were performed with the Cryogenic Forevacuum (CF) and the Mechanical Forevacuum (MF) systems.

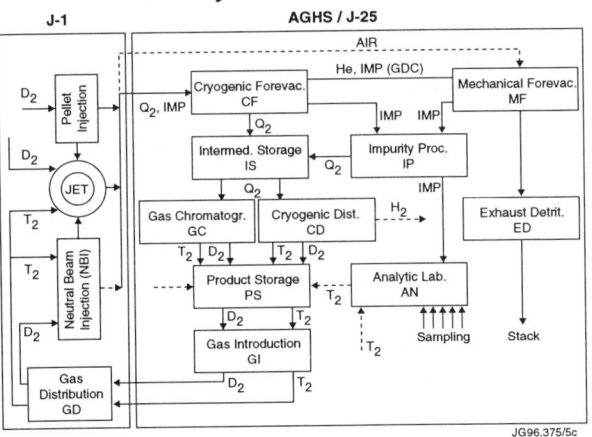

Figure 1 Schematic flow diagram of the AGHS and connected users

3 TRITIUM COMMISSIONING OF AGHS

The tritium commissioning of the AGHS is performed in two steps:

i) trace tritium commissioning with about 30TBq (800Ci or 0.08g); and
ii) full tritium commissioning with 3g of tritium.

Local Rules for the AGHS specify many aspects of management of a tritium plant such as management structure, operating directives, design, installation, commissioning, operation, maintenance, modification, radioactive material control, training, etc.

Operating Instructions were raised for many actions which occur on a regular basis.

Tritium commissioning procedures were written for the various sub-systems, approved and safety assessed.

The Authority to Operate (ATO) for the AGHS was issued by the Director of JET. This document gives the final permission for tritium operation and defines the conditions under which the AGHS can be operated.

4 TRACE TRITIUM COMMISSIONING OF AGHS

Before the commencement of trace tritium commissioning all software interlocks were tested. Hardwired interlocks are checked every six months.

During trace tritium commissioning the Radiation Protection Instrumentation (RPI) system was checked on a weekly basis with a small calibrated radioactive source. Health physics area surveys were also carried out regularly.

4.1 Trace tritium gas mixture

Trace tritium commissioning was performed with the tritium-hydrogen gas mixture stored in the gas collection trolley[7] used during the Preliminary Tritium Experiment (PTE) at JET. The gas collection trolley with its four uranium beds (U-beds) was moved into the AGH building and connected to the make-up pipework of the analytical glove box (AN-GB).

Almost all the tritium was present in one of the U-beds resulting in an initial tritium concentration of about 0.05%.

From the gas collection system the mixture was moved into Product Storage (PS), then from PS to other systems of the AGH plant: Intermediate Storage (IS), Impurity Processing (IP), Gas Chromatography (GC) systems, Analytical Laboratory (AN) and Cryogenic Distillation (CD) system.

4.2 Criteria for AGH sub-systems to be tritium commissioned

The purpose of the trace tritium commissioning was to check tritium-specific equipment, such as the various ionisation chambers, to confirm the overall leak tightness of the whole plant, to check the functioning of various processes by means of the more sensitive tritium-specific instrumentation and to train staff.

4.3 Sub-systems not to be tritium commissioned

Systems without any tritium-specific equipment in the process lines such as CF or MF were not tritium commissioned to avoid contamination of the torus. Functional tests with tritium for these two systems are not necessary because tritium is pumped more easily than deuterium and protium and effects of helium-3 and of the additional decay heat are negligible.

The Gas Introduction (GI) and Gas Distribution (GD) systems will be tritium commissioned after extended periods of supply of deuterium to the torus via these systems.

4.4 Isolation of AGH sub-systems

CF, GI, GD and MF were isolated from the other AGH sub-systems by closing manual valves, by removing the compressed air supply to automatic valves and by the addition of argon plugs into interconnecting lines. Part of MF was occasionally connected for special purposes, eg evacuation of $2m^3$ reservoir in IP.

5 RESULTS OF TRACE TRITIUM COMMISSIONING

In the following a few results of trace tritium commissioning for various AGH sub-systems are presented.

5.1 Product Storage (PS)[2]

The control sequences for the 8 PS U-beds were extensively tested. Blanketing effects of U-beds in the low pressure range were observed after absorption of large gas amounts, but could always be removed by circulation of the gas by means of a $15m^3/h$ Normetex pump.

In the JET U-beds hydrogen permeation can occur through the process walls surrounding the heater into the Hydrogen Isotope Storage Assembly (HISA). The HISA acts as the secondary containment for 4 U-beds. Permeated gas is absorbed by an ST-707 getter cartridge which in combination with a sputter pump creates high vacuum inside the HISA. Hydrogen absorbed in the getter can be regained by heating.

5.2 Intermediate Storage (IS)[2]

Blanketing effects in IS U-beds were observed with trace tritium gas. The blanketing gas was removed with the $15m^3/h$ Normetex pump in GC. A modification (see below) was introduced to allow removal of the blanketing gas without using another sub-system of the AGH plant.

IS was thoroughly trace tritium commissioned during its gas feeding tasks for CD.

5.3 Impurity Processing (IP)[3]

Trace tritium gas mixtures with methane (to simulate impurities) and helium as a carrier gas were oxidised to water and CO_2 and the water cryocondensed on a cold trap. Tritium levels in the helium could be easily reduced to allow discharge of the helium via ED into the environment. The water was cracked in hot iron beds producing hydrogen and iron-oxides. The partly oxidised iron beds were regenerated with protium, but the water produced contained a high fraction of tritium so that discharge via ED was not permissible. The water was finally cracked on heated U-beds producing irreversible uranium-oxide. The cause of the high contamination of the water is not fully understood, because the iron beds were purged thoroughly with helium scrubbed by U-beds, but could be due to absorption on the iron powder used (particle size <450μm).

This means that reduction in uranium consumption from the use of iron beds which was expected from prototype and inactive commissioning could not be achieved with tritium.

During the various IP runs the trace tritium gas was slightly diluted by the generation of hydrogen from the added methane.

5.4 Cryogenic Distillation (CD)[4]

The JET CD system contains 3 columns (COL), 1 feed each to COL1 and COL2, 3 product extractions, 2 recirculation loops each with a catalyst and 3 large expansion volumes (total volume of CD is 900ℓ).

After a few days of operating CD in full recycling mode the following tritium concentrations were observed: 4ppb in the H_2 product, 18ppb in the D_2 product and about 4% in the T_2 product. Therefore, H_2, and even D_2 product, can be discharged into the environment. The enrichment to 4% with only a tritium inventory of 0.08g is reasonable considering that CD was designed for a tritium inventory of 30g.

The tritium in the bottom of COL3 was moved to a GC U-bed. Finally the whole content left in the CD system with the exception of the last 80barℓ was discharged into the environment within the daily discharge limits.

A further by-product of the CD trace tritium commissioning was that the 4 U-beds of the gas collection system were emptied. They can now be used as a further storage capacity.

Checking of the 1.8bar(a) helium atmosphere in the unpurged instrument box of CD revealed a very small tritium concentration of a tenth of a DAC. Tritium was considered to have leaked during the filling time of the CD system when the process pressure was larger than the helium pressure in the secondary containment. This was consistent with the leak rate measured previously.

5.5 Gas Chromatography (GC)[5]

The principle of the JET GC system is displacement gas chromatography. The gas mixture to be separated is injected into a column filled with helium and Pd-packing material and displaced with protium gas. At the column exit first the helium is seen, then tritium, deuterium and protium.

Three diagnostics are located in the column exit: a thermopile to detect the absorption heat of hydrogen in Pd, a katharometer to measure the various hydrogen molecules via their slightly different thermal conductivity and a small ionisation chamber to measure the tritium concentration.

Fig 2 shows the signals of the various diagnostics as a function of time for an injection of 25ℓ of trace tritium gas, CD enriched to 98.8% deuterium and 1.2% tritium, and displacement by protium. The katharometer detected clearly helium, the enriched tritium, deuterium and protium. Peak tritium concentration was calculated to 91% assuming a linear response of the ionisation chamber from 0.05% up to 100% tritium. To obtain a 100% tritium product, the tritium content of the gas mixture injected into the JET GC must be higher than 30TBq (800Ci).

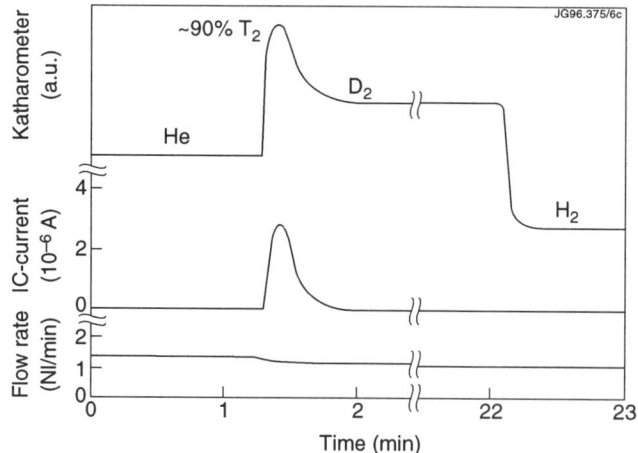

Figure 2 Signals of GC diagnostics for a column injected with 25barℓ of 1.2% tritium and 98.8% deuterium

5.6 Analytical Laboratory (AN)[6]

The trace tritium gas was analysed with the analytical equipment in AN: omegatron, residual gas analyser, large and small ionisation chambers, gas chromatograph. The trace tritium was clearly seen by the large (560cm^3) and small (15cm^3) ionisation chambers, the flow proportional counters and the newly installed, very small ionisation chambers in the analytical GC system.

During the whole commissioning AN made important contributions to the determination of gas mixtures processed in other sub-systems of the AGHS.

5.7 Ionisation chambers (ICs) and pressure gauges

A few electrical wiring faults were found in connection with ICs. The ICs were tested in the inactive phase with strong gamma sources, but a few ICs (eg in the GC columns and in analytical gas chromatograph (AN-GC)) were too small to give any response.

The common pressure instruments in the AGHS are of the strain gauge type, only in very few cases are special capacitance heads used. The strain gauge pressure instruments give unreliable pressure indications at low pressures which makes the operation of the plant more difficult.

5.8 Exhaust Detritiation (ED) and Over/Under Pressure Protection (OUPP)

Both systems were in full use during the trace tritium commissioning and worked very reliably. No tritium was detected by the IC or the passive sampler located at the outlet of ED, whereas the inlet IC - although one decade less sensitive - reacted to the introduction of "detritiated" gas mixtures.

In addition, the pumping efficiency of ED was determined by simulating an air ingress accident in the torus by opening a valve next to the torus, and found to be within the specification of 500m^3/h.

5.9 Maintenance

Many breaches of secondary containment and of process pipework were performed. Well established operating instructions are in use.

The interior of any process or secondary containment must be thoroughly purged before any breach of boundary.

Before a process line breach the stainless steel doors of the JET valve box are replaced by polycarbonate windows with gloves. A high air exchange rate inside the valve box is achieved by the connection of an elephant trunk between the valve box and the ED system.

No tritium contamination inside any secondary containment, with the exception of the unpurged instrument box in CD, has been detected up to now.

6 FULL TRITIUM COMMISSIONING

Full tritium commissioning will start in September 1996. 3g of tritium will be transferred from an Amersham U-bed to PS. The pure tritium will be used to determine the achievable accuracy of the $PS-T_2$ U-beds when used as calorimeters, to calibrate the various ICs in the AGHS and to check the diagnostics of AN. In addition, tritium will be injected into the GC columns to obtain a large enough tritium peak to commission the automatic switching of the valves in the presence of tritium, deuterium and protium.

7 COMMISSIONING OF JET TORUS AND AGHS

Before the start of the DTE1 phase, the AGHS will supply inactive gases to the torus, pump the exhaust gases from the torus and treat the gases within the AGHS as if they were deuterium-tritium gas mixtures.

The tritium operations phase at JET will be started with the introduction of a 1% tritium-99% deuterium gas mixture.

8 ENHANCEMENTS TO THE AGHS

During the inactive and trace tritium commissioning necessary enhancements were identified and modifications raised. Any change to the AGH plant requires the creation of a modification so that changes are detailed, approved, performed in accordance with quality assurance programme and safety assessed. Some of these modifications were installed during the trace tritium commissioning phase. A few of these modifications are discussed below.

8.1 Installation of a further pump in MF

MF comprises 3 Normetex pumps: 1 off $600m^3/h$ and 2 off $150m^3/h$.

In earlier times glow discharge cleaning in the torus was done with deuterium gas, but nowadays mainly helium gas is used. The CF accumulation cryotransfer pumps can cryosorb only about $28bar\ell$ helium per pump due to the limited amount of charcoal.

Pumping tests during helium glow discharge revealed that the 3 mechanical pumps were not capable of supplying the necessary backing pressure for the torus turbopumps.

A roots pump with a pumping speed of $1000m^3/h$ was added to MF and is backed up with a $150m^3/h$ Normetex pump. Tests with the 4 pumps revealed a pressure of less than 0.2mbar for a helium mass flow of about $30mbar\ell/s$. The helium glow discharge gas was compressed into the $2m^3$ reservoir of IP.

8.2 Connection of a metal getter bed to the intermediate volume of JET U-beds

Each of the sub-systems IS, $PS-D_2$, $PS-T_2$ and GC has a HISA which houses 4 JET U-beds (maximum capacity 27 moles for D_2 and 7.5 moles for T_2), acts as secondary containment and can be kept under high vacuum.

The outer process walls of the JET U-beds are surrounded by an intermediate volume to collect permeated hydrogen. A further getter bed (getter material: ST-707 from SAES) was connected inside HISA to the intermediate volumes of the 4 U-beds to absorb permeated hydrogen to minimise further permeation into the HISA module and to reduce heat transfer to the outer wall of the intermediate volume.

8.3 Installation of new pipework for removal of blanketing gas from IS U-beds

IS comprises one HISA with 4 U-beds. The inlet valves and outlet valves of the 4 U-beds are connected via common inlet and outlet manifolds, respectively. Blanketing of these U-beds was observed after transfer of imperfectly distilled hydrogen from CF.

New pipework was installed for the evacuation of the blanketed U-bed via the outlet manifold with a 15m^3/h Normetex pump in the IP leak testing box and for the compression of the gas in a newly installed 17ℓ reservoir at the outlet of the pump. In addition, this volume was connected to the pumping side of the 150m^3/h Normetex pumps in IP.

Any blanketing gas in an IS U-bed can now be pumped off and temporarily stored in the 17ℓ reservoir without any interference with another sub-system of the AGHS. Finally the gas can be moved to IP for further treatment at a suitable time.

8.4 Installation of an additional valve in IP

The IP system was built with the 2m^3 reservoir at the outlet of the two 150m^3/h Normetex pumps. The chemical modules (recombiner, cold trap, 2 off iron beds and 4 off U-beds) are located between the reservoir and the pump inlet side. It was not possible to circulate gas with the pumps through the chemical modules by by-passing the reservoir or to compress gas from the reservoir into the chemical modules.

The installation of one further automatic valve allowed all these tasks to be performed. One of the main advantages of this new valve is that, eg hydrogen gas in the 2m^3 reservoir can be compressed by means of the Normetex pumps into an absorbing U-bed which will speed up the absorption and full evacuation of the reservoir enormously. The pumps and the equipment involved are protected by software and hardwired interlocks from the occurrence of high pressures at the outlet of these pumps.

9 SUMMARY

The JET AGHS has been successfully trace tritium commissioned. All sub-systems performed in the presence of trace tritium gas as expected, with the exception of IP where the tritium concentration in the water collected from the regeneration of the iron was too high. No radiological incident and no exposure of any radiological concern to staff or to members of the public occurred.

The operation team was able to run the daily affairs, gain experience in tritium operation of the AGHS and maintain the plant.

Full tritium commissioning of the AGHS with 3g of tritium can therefore be approached with confidence.

REFERENCES

[1] A C Bell et al, "Status of the JET Active Gas Handling System and Plans for Tritium Operation", Fusion Technology 28 (1995) 1301-1306

[2] R Stagg et al, "The Intermediate and Product Storage Systems for the JET Active Gas Handling System - Inactive Commissioning", Fusion Technology 28 (1995) 1425-1430

[3] J Lupo et al, "The Impurity Processing System for the JET Active Gas Handling System - Inactive Commissioning", Fusion Technology 28 (1995) 1347-1352

[4] P Boucquey et al, "Progress on the Inactive Commissioning and Upgrade of the JET Cryogenic Distillation System", Fusion Technology 1994, Proceedings of the 18th SOFT 1994 in Karlsruhe, Vol 2, 1055-1057

[5] R Lässer et al, "The Preparative Gas Chromatographic System for the JET Active Gas Handling System - Inactive Commissioning", Fusion Technology 28 (1995) 681-686

[6] R Lässer et al "The Analytical Laboratory for the JET Active Gas Handling System - Inactive Commissioning", Fusion Technology 28 (1995) 1033-1038

[7] J L Hemmerich et al, "Gas Recovery System for the First JET Tritium Experiment", FEDEEE 19 (1992) 161-167

Neutronics Shield Experiment For ITER At The Frascati Neutron Generator FNG

P. Batistoni [a], M. Angelone [a], W. Daenner [b], U. Fischer [c], L. Petrizzi [a], M. Pillon [a], A. Santamarina [d], K. Seidel [e]

[a] Associazione EURATOM-ENEA sulla Fusione, Centro Ricerche di Frascati, C.P. 65 -00044 Frascati, Rome (Italy)

[b] The NET Team, D-85748 Garching,

[c] FZK D-76021 Karlsruhe,

[d] CEA Cadarache, F-13108 St.Paul les Durance,

[e] Technical University, D-01062 Dresden

1. INTRODUCTION

Radiation loads are critical issues for the ITER shielding design. Limits on radiation induced displacement damage, on insulator dose, and on He production are imposed for all components assumed to be permanent or semi-permanent, i.e. the toroidal field coil (TFC), the vacuum vessel, the blanket backplate and the coolant manifold. These components are to be designed for at least 1.3 MWa/m^2 of total fluence (0.3 in the Basic Performance Phase and 1.0 in the Extended Performance Phase). Limits on the nuclear heating of the toroidal field magnet are also imposed. Radiation loads to ITER components are routinely calculated by using current state-of-the-art neutron and photon transport codes and nuclear data, e.g. the MCNP-code [1] and the FENDL (Fusion Evaluated Nuclear Data Library [2]) cross-section data file. The calculated values need however to be validated and the associated uncertainties, mainly due to uncertainties in nuclear data, need to be quantified. For that purpose, a Neutronics Bulk Shield Experiment has been initiated by ITER with the main objective to validate the shielding performance predicted by calculations. A mock-up of the ITER shield has been realized and assembled at the Frascati 14-MeV Neutron Generator (FNG) of ENEA, Frascati, that has been expressly designed and characterized to perform experiments in support of the fusion technology program. The Bulk Shield Experiment is being performed in collaboration with CEA Cadarache, TU Dresden, FZK Karlsruhe, ENEA Frascati and with groups of the Russian Federation (RF). The mock-up is made of layers of copper, stainless steel (AISI 316) and water equivalent material and replicates, as closely as practical, the ITER inboard shield (where generally the critical loads are found), including first wall, shielding blanket and vacuum vessel. The experimental assembly includes also a mock-up of the TFC, made of copper and stainless steel (SS), attached behind the shield. The neutron and γ-ray leakage spectra have been measured behind the shielding blanket and the vacuum vessel; the nuclear heating has been measured over a 10^6 variation range from the first wall to the the inner layers of the TF-coil mock-up, and neutron reaction rates have been obtained as a function of the penetration depth in the shield irradiated by 14-MeV neutrons. The measurement of other relevant nuclear responses such as the neutron induced activation in the ss are also planned. The analysis of the experimental data base provides the validation of the calculational predictions for the radiation loads to the inboard shield and magnet, and the assessment of the associated uncertainties. In addition, it contributes to the validation of the FENDL nuclear data library, which is the reference library for ITER design calculations.

2. THE FRASCATI NEUTRON GENERATOR FNG

In FNG [3] a deuterium beam (1 mA) is accelerated (up to 300 keV) and focused on a tritiated titanium target to react with tritium thus producing a nearly isotropic 14-MeV neutron source (1×10^{11} n/s). FNG has been designed and built at the ENEA Frascati Research Centre to perform neutronics experiments in the framework of the research activity on controlled thermonuclear fusion [4,5]. The facility is housed in a large shielded hall (11.5×12 m^2 and 9 m high) and the target (fixed) is located at a minimum distance of 4 m from any walls. This reduces the backscattered flux close to the target and allows for a good definition of the neutron field, particularly important in neutronics benchmark experiments.

The absolute neutron intensity is accurately determined by counting the alpha particles associated with the neutron emission, by means of a small silicon surface barrier detector (SSD) incorporated in the beam line. This method has been cross-calibrated with the absolute emission measured by means of fission chambers, activation technique, and calculation methods based on the use of the Monte Carlo code MCNP [6]. This intercomparison allowed to conclude that the FNG absolute neutron emission is measured at ±2%, at 1 σ level. The time dependent neutron emission is also monitored by using a U-Nat fission chamber, a BF3 counter and a NE-213 scintillator. These counters were also calibrated against the SSD counter.

3. THE EXPERIMENT SET-UP

The configuration and material composition of the experimental mock-up replicate those of the ITER shielding system at the inboard side, accounting for the neutronics relevant features to reproduce the flux attenuation properties. The mock-up, shown in Fig. 1, consists of a 1 cm thick layer of copper simulating the first wall (FW), a shielding block made of SS316 plates and perspex (water equivalent material), and of a block made of alternate plates of SS316 and Cu (both with

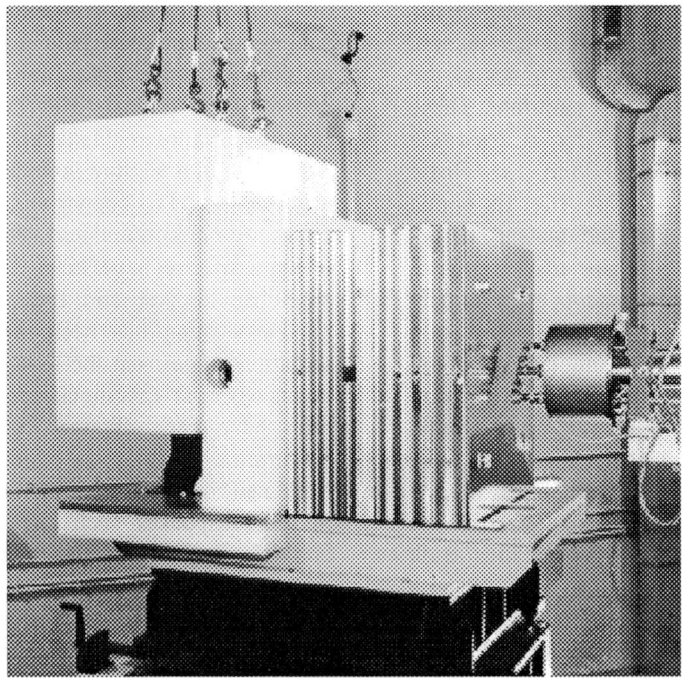

Figure 1. The ITER Bulk Shield Mock-up set up in front of FNG 14-MeV neutron source.

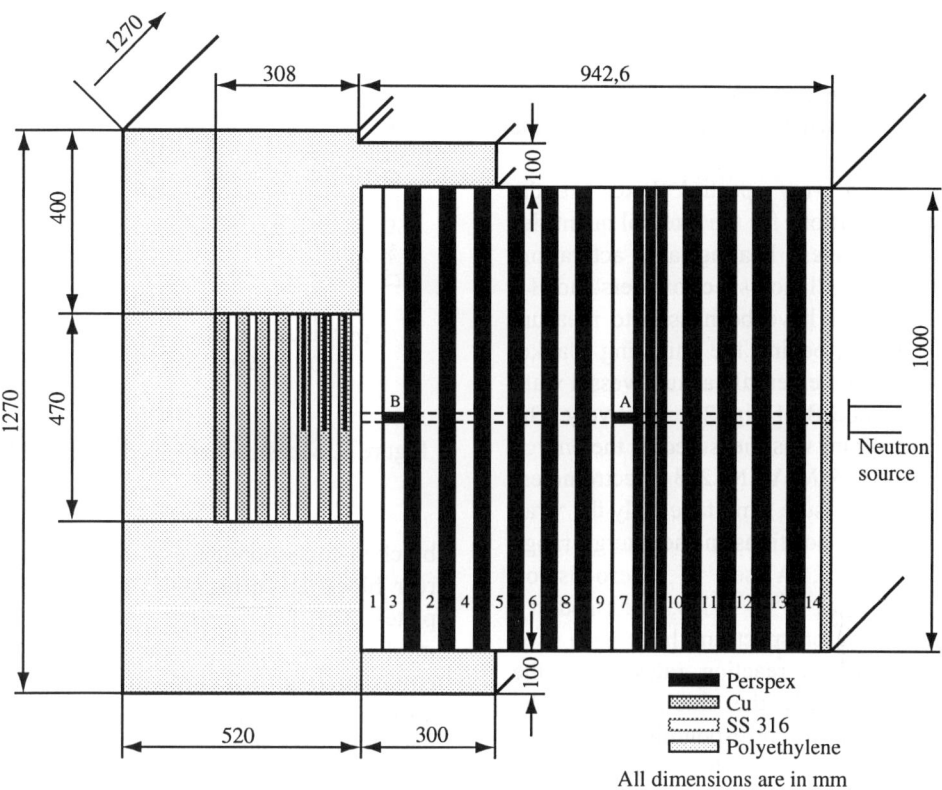

Figure 2. The mock-up configuration.

thickness of 2.2 cm), attached behind the shielding block, and simulating the ITER toroidal field coils (TFC). The plates in the shield block are arranged in such a way as to reproduce the radial features of the shielding blanket (SB) and vacuum vessel (VV) segment, including the manifold region and the backplate (Fig. 2). The total mock-up thickness is 94.26 cm and, apart from the copper first wall, the average material composition is 71.6 % SS316 and 28.4% water equivalent material. The mock-up is shielded in the rear side by a polyethylene shield to reduce to negligible levels (<10% in the last measuring position) the radiation backscattered from the bunker walls with respect to the direct penetration flux. The representativity of the mock-up for ITER has been calculated using the MCNP code and is described in detail in [7]. The conclusion of the analysis is that, apart from local deviations and the difficulty in representing the real volume source in ITER by a point source in the experiment, the mock-up configuration replicates the features of the ITER shielding system at the inboard side. As required, the shielding performance can be reproduced to a total attenuation > 10^5.

4. THE MEASUREMENTS

A number of measurement has been planned in the experiment, using different detecting techniques. The shielding block is provided with a central horizontal channel with diameter equal to 30 mm (dashed lines in Fig. 2). This central channel is used to introduce various detectors (activation foils, micro-fission chambers, ionization chambers, thermo luminescent dosimeters (TLD's) at different penetration depths. This is done by means of a unique ss tube of inner diameter 27 mm and external diameter 30 mm. The tube, between detectors, is filled with perspex and SS316 plugs, arranged in the same order as the plate order in the mock-up. Moreover, two lateral penetrations (50 mm diam.) are provided in two ss layers to locate spectrometers in positions A and B of Fig. 2. Vertical channels are used to introduce TLD's in the TFC mock-up.

The neutron and γ-ray spectral fluxes are the basic weighting functions for the integral quantities such as reaction rates, heating and activation. NE213, stilbene scintillation spectrometers and H_2-proportional counters have been used to measure neutron flux spectra behind the shielding blanket (backplate) and in the external vacuum vessel wall, i.e. in position A and B of Fig. 2 respectively. The neutron flux spectrum was measured in the energy range 30 keV<E<15 MeV. NE213 spectrometers were also used to measure simultaneously the γ-ray flux spectra at these positions in the energy range 200 keV<E<8 MeV. A set of micro-fission chambers, ^{239}Pu, ^{235}U, ^{238}U, ^{237}Np, calibrated in a reactor standard neutron spectrum, have been used to measure the fission reaction rates along the penetration depth inside the shield block. The results obtained and the corresponding analysis is described in detail in [8] while the measured neutron and γ-ray spectra and the relative analysis are described in detail in [9]. The measurement of reaction rates using the activation foil technique [^{197}Au(n,g), ^{55}Mn(n,g), ^{115}In(n,n'), ^{58}Ni(n,p), ^{56}Fe(n,p), ^{27}Al(n,a), ^{93}Nb(n,2n), ^{35}Cl(n,a), ^{32}S(n,p)] will provide further information on the neutron spectrum integrated over selected energy ranges. The measurement of neutron induced activation in ss is also in progress. Here we describe in detail the measurement of nuclear heating in the shield block and in the TFC mock-up.

5. THE NUCLEAR HEATING MEASUREMENT AND ANALYSIS

The nuclear heating was measured using thermoluminescent dosimeters, in particular TLD-300 (CaF_2:Tm) were chosen mainly for their high sensitivity, necessary to measure very low doses in deep locations. These dosimeters were calibrated, studied (sensitivity limit, linearity) and tested in a previous experiment, as described in detail in [10,4]. The TLD's were located in the shielding

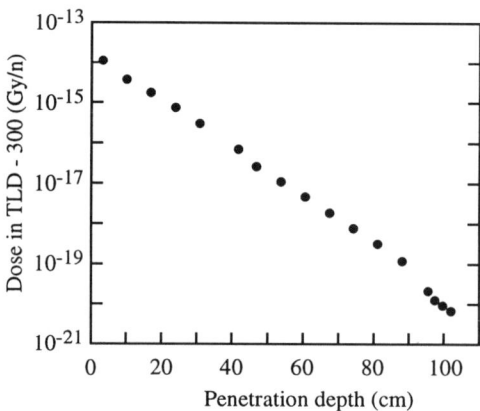

Figure 3. Measured absobed dose in TLD

block in correspondence of the ss plates, and in the rear block in correspondence of both ss and Cu plates. The TLD's were annealed just before the irradiation to minimise the absorbed dose due to environmental background radiation. A set of dosimeters was used to monitor the background dose absorbed during the experimental time. The doses measured in TLD-300 are shown in Fig. 3. The total errors on the measured dose is < ±13% in the shielding block, and < ±30% in the TFC block, and include the uncertainty on the absolute calibration of TLD (±3-5%), on neutron source intensity (±2%), and the standard deviation for each set of dosimeters in each location.

The analysis of the experiment was carried out by means of MCNP.4A Monte Carlo code. The 3-D Monte Carlo simulation allows for an adequate representation of the experimental features and an accurate treatment of the neutron transport with continuous cross sections so that the comparison between the calculated (C) and measured (E) data can be referred essentially to the basic nuclear data. The actual energy and angular distribution of the neutron source was calculated by a subroutine in MCNP taking into account the reaction kinematics and the beam deuteron energy loss in the tritium-titanium target [6].

A very stressed variance reduction was required to improve the statistical accuracy for the deep penetration and the slowing down effect. The weight-window technique was adopted by using an importance distribution which depends on space, energy and detector response. The bunker and walls were included in the model, to properly take into account the room background contribution to the measured doses. The neutron and photon absorbed doses in ss and copper were calculated along the central axis of the assemblies, were the detectors were located, by track length estimator multiplied by KERMA factors (tally F6 of MCNP), using FENDL-1 cross-section library. The calculated absorbed doses in material (ss or Cu) due to photons and to neutrons are shown in Fig. 4. The background contribution to the absorbed dose in the TFC mock-up was obtained by comparing two calculations, with and without the inclusion of the bunker walls and in the last measuring position was found to be comparable with the combined statistical computational uncertainty ($<\pm5\%$ in each calculation), indicating that it is limited to <10%, as expected from the experiment pre-analysis.

The comparison with TLD measurements requires, however, the calculation of the absorbed dose in TLD. This has been calculated using a local MCNP model describing the TLD's as in previous analyses [4], taking into account the effects of the electron transport at the material/TLD interface. The resulting values of the ratios of calculated over measured (C/E) absorbed doses in TLD are given in Fig. 5. The error bars in Fig. 5 represent the total (experimental and computational) uncertainty on C/E values. The comparison shows that the

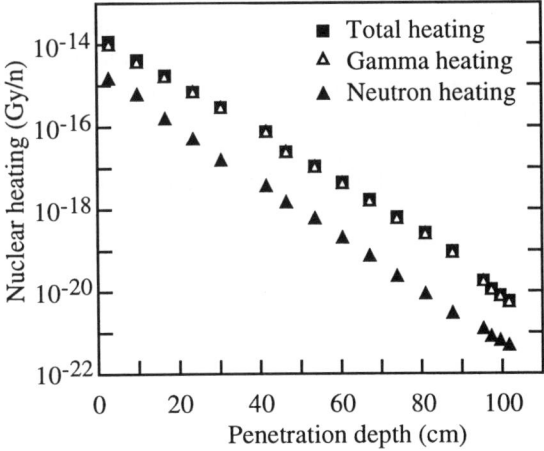

Figure 4. Calculated absorbed dose in the mock-up (stainless steel layers).

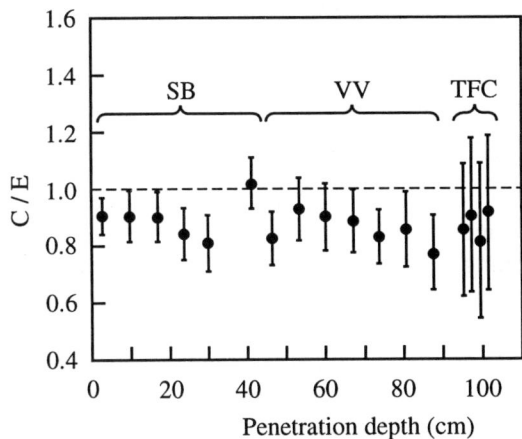

Figure 5. Ratios of calculated (C) and measured (E) dose as a function of penetration depth, using MCNP code and FENDL-1 library.

FENDL-1 calculation constantly underestimates (by about 10-20%) the measurements, as in previous experiments [4] and that there is no trend with the increasing depth. Some of C/E values in the shield and all those in TFC mock-up are compatible with unity within the total error bars. Since most of the heating is due to photons, and since no uncertainties are expected from photon KERMA data, the found underestimation of about 10-20% could be attributed either to the calculation of the neutron spectrum or to the photon production reaction cross sections in the FENDL-1 library. These results are also confirmed by the analysis of the measured γ-ray spectral fluxes in positions A and B given in [9].

Since the total nuclear heating of the TFC is calculated to be about 4 kW, and since the design limit is 17 kW [11], it can be concluded that the underestimation found in the experiment is tolerable as far as the TFC is concerned. However, the ~10% underestimation of the nuclear heating in the first 30 cm (FW and SB), where most of the power is deposited, would result, when extrapolated to ITER, in a non negligible underestimation of the total power generation of about 150 MW at a fusion power level of 1500 MW.

6. CONCLUSIONS

The results of the Neutronics Bulk Shield Experiment at FNG provide experimental data on the relevant nuclear quantities on a mock-up of the ITER inboard shield where, generally, the critical loads are found. The measured quantities are the radiation loads of interest, as in the case of nuclear heating, or closely related to them, as in the case of neutron and γ-ray spectral fluxes and reaction rates. These data allow to verify the expected radiation loads on the inboard shield components and TFC, as well as the corresponding calculated margins with respect to the limits given by the Project.

The analysis of the measurements already available show that FENDL-1 library can satisfactorily describe the features of the neutron and gamma flux attenuation in the large ss/water shield mock-up to about 1 m of penetration depth. In particular, the nuclear heating can be satisfactorily predicted by FENDL-1, although with a constant underestimation (10-20%) all along the shield up to 1 m of depth, and in the first, most exposed layers of the superconducting magnet.

AKNOWLEDGEMENTS

The present work has been performed under ITER Task T-218.

REFERENCES

1. J. Briesmeister (ed), "MCNP - A general Monte Carlo code for neutron and photon transport, Los Alamos National Laboratory Report LA-7396-M, Rev. 2, 1986.
2. S. Ganesan, P.K. McLaughlin, Documentation Series of the IAEA Nuclear Data Section, IAEA Report IAEA-NDS-128 Rev.1, (June) 1995) Vienna
3. M. Martone, M. Angelone, M. Pillon, J. Nucl. Mater. 212-215 (1994) 1661.
4. P. Batistoni, M. Angelone, M. Pillon, V. Rado, "Nuclear Heating Experiment for the validation of the fusion reactor shielding performance", to be published in Fusion Eng.&Des.
5. P. Batistoni, M. Angelone, M. Martone, L. Petrizzi, M. Pillon, V. Rado et al., J. Nuclear Mater. **212-215**, 1724 (1994).
6. M. Angelone, M. Pillon, P. Batistoni, M. Martini, M. Martone, V. Rado, "Absolute experimental and numerical calibration of the 14 MeV neutron source at the Frascati Neutron Generator (FNG)", Rev. Sci. Instrum. 67 (6) (1996) 2189.
7. U. Fischer, P. Batistoni, M. Pillon, "Three-dimensional neutronics analyses of the ITER bulk shield mock-up experiment", this Symposium.
8. A. Santamarina, L. Benmansour, B. Camous, H.Philibert, P .Batistoni, M. Angelone, M. Pillon, "Neutron flux measurements in thITER bulk mockup at FNG and analysis", this Symposium.
9. H. Freiesleben, W. Hansen, D. Richter, K. Seidel, S. Unholzer, U. Fisher, M. Angelone, P. Batistoni, M. Pillon, "Measurement and analysis of specral neutron and photon fluxes in an ITER shield mock-up", this Symposium.
10. M. Angelone, P. Batistoni,A. Esposito, M. Pillon, V. Rado, "Gamma and neutron dosimetry using CaF_2:Tm thermoluminescent dosimeters (TDL-300) for fusion reactor shielding experiments" submitted to Nucl. Sci. and Eng.
11. W. Daenner, Report on Shielding, presented at TAC-JCT Informal Technical Review on In-Vessel Systems and Related Physics, ITER Joint Work Site, Garching, 1-5 May 1995.

CONTRIBUTED PAPERS

Section A

First Wall, Divertors and Vacuum Systems

FEASIBILITY OF LIQUID GALLIUM COOLING FOR ITER DIVERTOR CASSETTE

P.V.Romanov, Yu.Shpanskij, A.V.Klischenko, V.S.Petrov, S.A.Moshkin[a], A.V.Beznosov[b]

[a] Russia Research Centre "Kurchatov Institute"

[b] Nizhnij Novgorod Technical University

This paper involves the results of some preliminary analyses of the conceptual design of an advanced divertor plates system, alternative to the current high heat flux loaded divertor plasma facing components for ITER. In proposed concept the divertor targets are cooled by liquid metal (LM) which is compatible with the main ITER divertor water coolant. At the same time LM design option is compatible with the space envelope, thermal and mechanical interfaces as for the base ITER water cooled divertor option.

1. INTRODUCTION

The objective of the proposed study was to provide a robust divertor plates system for ITER capable to insure the machine operation under highly uncertain working conditions which will remain until the actual tokamak work is start.

During the first stage of this study it was required to perform several material compatibility and design assessment analyses including thermal hydraulics, MHD, thermal mechanical, structural, neutronics, tritium inventory and permeation analyses, to evaluate the feasibility of this concept with LM cooled divertor targets, alternative to the present baseline ITER divertor design with water cooling [1].

2. AVAILABLE DESIGN STUDY

The basic idea of the proposed concept has been considered in a conceptual design study for ARIES project [2] and was already suggested for ITER project in 1992 [3]. At that time the final option was made in favor of the conventional water cooled divertor targets. However, the necessity to improve the existing divertor plates design (for 5 MW/m^2 nominal heat loading) turned back designer views to the more reliable solutions.

Fast thin LM film flows (of Ga-alloy) with almost the same geometry and magnetic field were obtained by V.V.Yakovlev in experiments including regimes of film flow on the ceiling. In other experiments carried by O.Lielausis the feasibility of film formation from jets in strong magnetic fields was demonstrated. It was shown that the heat transfer coefficients above 100 kW/m^2·K are feasible and thus plates could withstand the heat flux at 15-20 MW/m^2 level.

3. INPUT SUGGESTIONS FOR ITER DVERTOR

The reference value of the ITER total fusion power is 1500 MW. Power excursion of ±20% lasting for 10's seconds must be anticipated and therefore accommodated to divertor design [4].

ITER divertor should be capable to conduct up to 315 MW from the main plasma.

Distribution of the heat loads should be very localized (Fig.1) and depends on the type of the divertor target (see Table 1) and divertor operation regime.

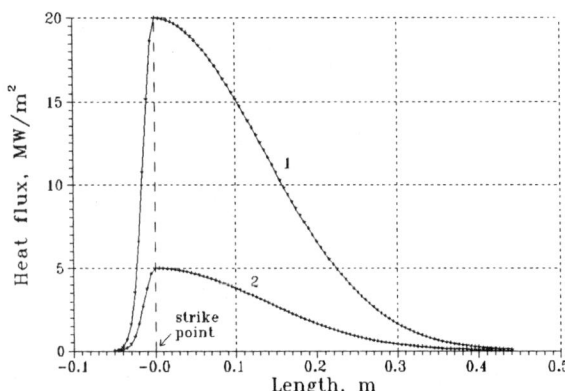

Fig.1. Power deposition onto the surface of the vertical divertor target.
1- for HHF plates; 2- for LHF plates

Significant transient particle and thermal loads on the in-vessel surfaces are expected during plasma configuration control transients which result in plasma-surface contact.

A single null divertor shall be installed in the bottom part of the vacuum chamber. It should be built from modules/cassettes which are remotely replaceable.

The divertor targets should be designed to limit the average continuous surface heat load up to 5 MW/m^2 and to provide capability of accepting 10-15 MW/m^2 at a special plasma striking points for 10-20 s during each 1000 s pulse at nominal parameters.

Divertor as other in-vessel components will be usually baked by heating the coolant to the maximum temperature capable with keeping subcoolant conditions. It is assumed that coolant may be heated by external means to the maximum temperature of 150°C.

The loads onto the divertor targets were chosen according to the assumptions summarized in [5, 6].

4. CONCEPT APPROACH TO THE LM COOLED DIVERTOR

4.1. Concept key features

The basic idea of the proposed concept is depressurized cooling of the most loaded divertor surfaces due to liquid metal cooling compatible with the vacuum chamber environment and application of leak-tolerable cooling system. These basic features provide the following advantages of proposed concept in comparison with water cooling option:

- elimination of the possibility of a whole type of accidents concerned with the internal loss of vacuum (LOVA) that may have the strongest effect on the reactor reliability and availability;
- elimination of the possibility of another type of accidents concerned with the vacuum chamber pressurizing due to internal cooling tract rupture;
- easy replacement of the divertor plates without welding/brazing operations, separately from the piping system, without disturbing of the whole divertor cassette module;
- no coolant interaction with hot plasma facing materials in the case of coolant ingress and no hydrogen production;
- more longer operating life of the pressurized feeding pipes and no need for the cooling system protection from the run-away electrons due to liquid metal ability to absorb large amounts of energy without producing high pressure.

4.2. LM coolant selection

The listed advantages are feasible to a great extend due to unique properties of the proposed LM coolant which is gallium.

It seems that gallium can be fit for divertor application in particular for contact devices with open surface into the vacuum chamber. But gallium has a charge number of 31 and is considered as a dangerous impurity for reactor plasma. At the same time gallium has enough high sputtering coefficient. These properties were the major obstacles for reconsidering very attractive idea of Ga cooled divertor targets with open LM surface directly looking into the vacuum chamber and absorbing the conducted into divertor heat and particle fluxes in favor of proposed concept.

In proposed concept Ga influx into the main plasma should be limited due to the divertor plates cooling from the back side relatively to the incident heat flux

Taking into account all above the use of gallium as a LM coolant is favored due to its low melting temperature (30°C), extremely low vapor pressure at normal working temperatures lower 300°C $p<10^{-13}$ Torr which make it possible to have this coolant with open surface directly in vacuum chamber.

An important Ga feature is that it has not chemical reactions neither with water nor air and fully compatible with water coolant for other ITER components. But Ga is a highly corrosive LM at temperatures above 300-350°C for steels, though plasma facing materials such as beryllium, graphite and tungsten are rather stable and compatible with gallium.

The last means that Ga cooling could be applied directly to the plates made of these materials and recommended for ITER application, without using various intermediate heat sink materials and corresponding interface.

4.3. CONCEPT DESCRIPTION

As in the base water cooled ITER divertor design the complete divertor ring comprises 60 divertor cassettes. At this stage we assumed

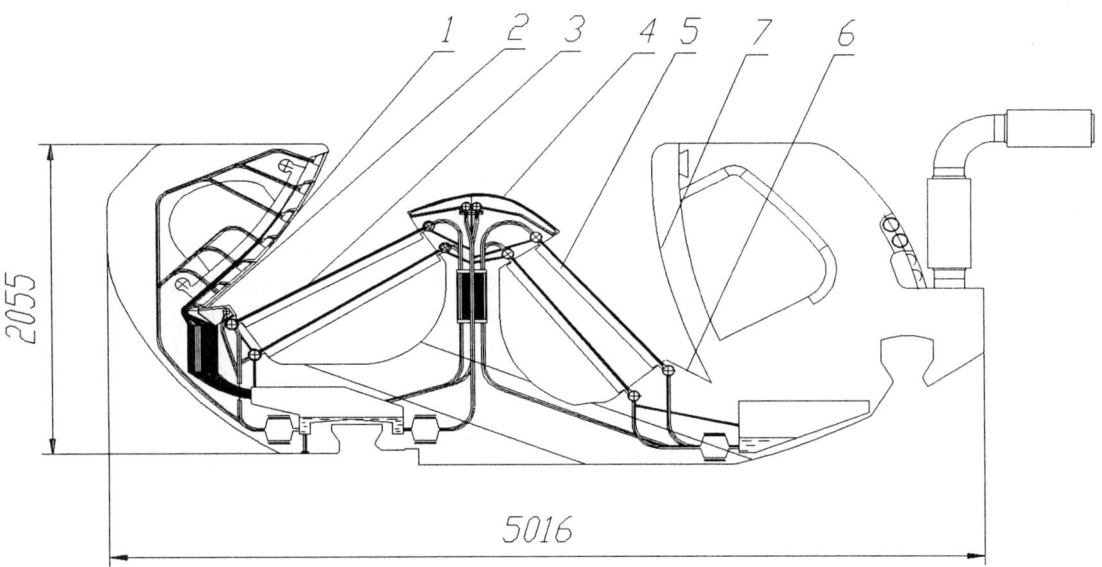

Fig.2. General view of ITER divertor cassette. cooled with liquid metal.
1 - inner vertical target; 2- inner dump target; 3 - inner wings; 4 - dome;
5 - outer wings; 6 - outer dump target; 7 - outer vertical target.

that there are no differences between cassettes. Each divertor cassette includs 8 targets and structurally consists of the two equal parts. However the whole structure of cassette is the same as for vertical target option design for ITER.

Each part includes divertor targets and auxiliary elements making LM loop.

Divertor targets are cooled by three autonomous LM cooling loops. The first loop is for the cooling of the inner vertical and bottom targets, the second - for inner and outer wings and central dome structure, the third - for outer vertical and bottom targets.

The vertical divertor targets (Fig.2) are divided on the set of circles/layers in poloidal direction. In toroidal direction the circle includes a set of replaceable cells/plates. The plates are formed as an inclined troughs fixed in the top and bottom. The header has a set of orifices from which LM jets are ejected on the back side of the divertor plate (Fig.3). LM jets after hitting would form a thin film that flows down cooling the plate.

According to the power distribution divertor plates may be high heat flux (HHF) loaded and low heat flux (LHF) loaded. LHF plates are capable to withstand the heat flux with peak value of 5 MW/m^2 and HHF plates - with peak value of 20 MW/m^2.

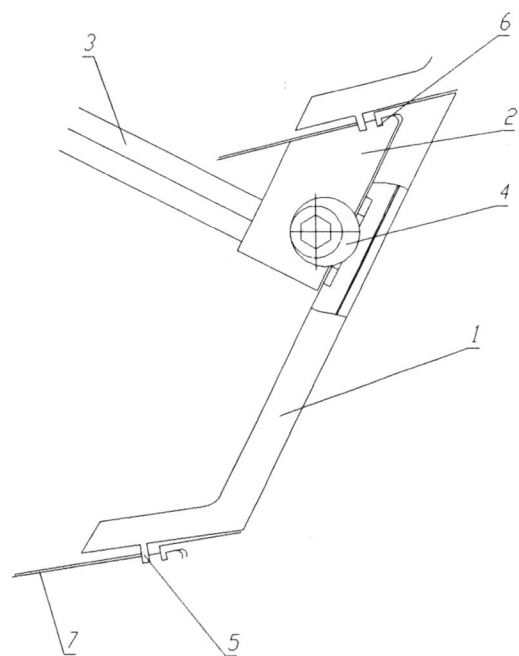

Fig.3. Tile of vertical divertor target.
1- tile; 2- feeding manifold; 3- feeding pipe; 4- lock; 5- ledge; 6- groove; 7- draining plate.

4.4. Inner vertical and most divertor targets

The inner vertical divertor target consists of 6 circles of divertor plates: the 3 upper circles include LHF plates and the 3 lower - HHF plates as the inner most divertor target consisting of one ring of HHF plates.

There are 8 divertor plates in toroidal direction in each circle. Thus 48 divertor plates form the plasma facing surface of the inner vertical divertor target in cassette.

As the position of the separatrix strike point with divertor target was not given and moreover it is assumed that this point can scatter in vertical direction the divertor plates are designed for maximum possible heat loading in according with the heat flux distribution along the divertor plates. Parameters for thermal and thermal stress analyses of the divertor plates are summarized in Table 1.

It was obtained that
- each circle consisting of HHF divertor plates is loaded with the maximum possible thermal power according to the given power distribution and capable to conduct the thermal load ~1.7 MW. That value was obtained by integrating of the thermal power distribution along the plate of 200 mm length;
- for average LM coolant heating of 140°C the maximum required gallium flow rate through the one circle consisting of HHF divertor plates is ~6 l/s;
- HHF divertor plates are cooled by LM film with initial film thickness 2 mm and inlet LM velocity 5 m/s;
- each circle consisting of LHF divertor targets has to conduct 0.54 MW taking into account the maximum possible thermal power and is cooled by gallium flow rate of 1.87 l/s;
- total gallium flow rate through the inner vertical divertor target is 23.6 l/s. It could be decreased if the less forced cooled divertor targets would be used.

ACKNOWLEDGMENTS

The authors wish to thank E.V.Muraviev from ITER JWS (San Diego) for his attention, help and useful discussions during this concept analysis and V.I.Khripunov from Garching JCT who performed gallium activation analysis for the concept under consideration.

REFERENCES

1. P.V.Romanov et al. Gallium Cooled ITER Divertor Target. Final Report. TA No.16-95-01-06 RF.
2. E.V.Muraviev et al. Liquid Metal Cooled Divertor for ARIES. Technical Report of General Atomics. GA-A21755, 1994.
3. Chuyanov et al. New Concept of ITER Divertor. Report presented to ITER Topical Meeting, June 1992, Garching.
4. General Design Requirements Document (GDRD). ITER EDA-JCT. S 10 GDRD 1 95-02-10 W1.2.
5. Design Description Document for the ITER Plasma. Draft. 9 January 1995.
6. S.Chiocchio, R.Tivey. Thermal Loads in the ITER Divertor. Draft. G17 RI 12 95-03-02 W 1.1. 2 March 1995.

Table 1
Parameters for HHF and LHF divertor plates

Parameter	HHF plates	LHF plates
Maximum thermal power, MW:- through one circle	1.72	0.54
- through one divertor plate	0.215	0.067
Peak heat flux, MW/m^2	20	5
Temperature, °C: - inlet	100	100
- outlet	240	240 (190)
Coolant temperature rise, °C	140	140 (90)
Coolant flow, kg/s: - through the most loaded circle	35.96	11.23
- through divertor plate	4.5	1.4
Divertor plates sizes, mm:	200 x 80	200 x 80
Film parameters: - initial thickness, mm	2	0.64 (1)
- initial velocity, m/s	5	5

Review of liquid metal divertor concepts for Tokamak reactors

Igor R. Kirillov[a] and Evgeni V. Muraviev[b]

[a] D.V.Efremov Scientific Research Institute of Electrophysical Apparatus,
189631 St. Petersburg, Russia

[b] Russian Research Center Kurchatov Institute/ITER San Diego Joint Work Site,
11025 N. Torrey Pines Rd., La Jolla, CA 92037, USA

Different types of liquid metal (LM) divertor targets are characterized and compared. Conceptual designs of these targets for a number of Tokamak projects are considered. The conducted R&D and design studies confirm the potential attractiveness of LM divertor targets and require further efforts, including testing in present day Tokamaks.

1. INTRODUCTION

Divertor target is one of the most challenging engineering device of Tokamak reactors. Water cooled solid plate divertor targets can not satisfy all demands for DEMO or power fusion reactor divertors, such as long life time under high heat and particles fluxes, safety etc. So different types of alternative divertor target concepts were proposed in the past 10-15 years and among them liquid metal (LM) divertor targets were mostly studied.

They have a number of advantages over water cooled divertor targets and some concerns to be taken into account.

Among the advantages one can name the following:
- absence of heat transfer crisis due to coolant boiling and, evidently, no channel distruction due to that;
- no need for cooling system protection from runaway electrons (due to LMs ability to absorb large amounts of energy without producing high pressure);
- much lower consequences of large coolant ingress into vacuum chamber during coolant tract rupture (absence of considerable pressure increase due to vaporization; no hydrogen production at interaction with plasma facing materials);
- possibility to keep high temperatures of contact surfaces between the pulses without applying high static pressure in the coolant;
- possibility of using non-vacuum-tight coolant schemes enabling to develop new types of divertor targets - with mechanical motion or with easy changeable contact surfaces.

In addition to that LM divertor targets with LM facing the plasma provide for continuous renewal of surfaces subjected to plasma particles erosion.

The problems, associated with the use of LMs in divertor targets, are the following: LM migration to plasma in plasma facing or non-vacuum-tight LM devices; corrosion; chemical activity; MHD pressure drop in high magnetic fields of Tokamak reactors.

2. LM DIVERTOR OPTIONS DESCRIPTION AND CHARACTERIZATION

A number of liquid metal divertor targets were proposed, studied theoretically and to some extent experimentally. They differ in heat accumulating capabilities, in direct/indirect plasma exposure and in presence or absence of moving mechanical parts (apart from LM).

In LM divertor targets without direct exposure of LM to plasma, it serves as a coolant only, with solid material facing the plasma. The erosion problems are still here and the advantages over water cooled targets are due to the use of LM as coolants or as heat sink layer. Divertor target cooled with LM channel flow (Fig.1,a, [1]) is vacuum tight and LM may enter plasma during emergency events only, for instance at burn-out of divertor target. Estimation show that it may accumulate surface heat fluxes q<10 MW/m^2, and is applicable for single null (SN) and double null (DN) divertor configuration. To prevent large MHD pressure drop it is necessary to develop self-healing electroinsulating coatings on the interface of LM and structure materials.

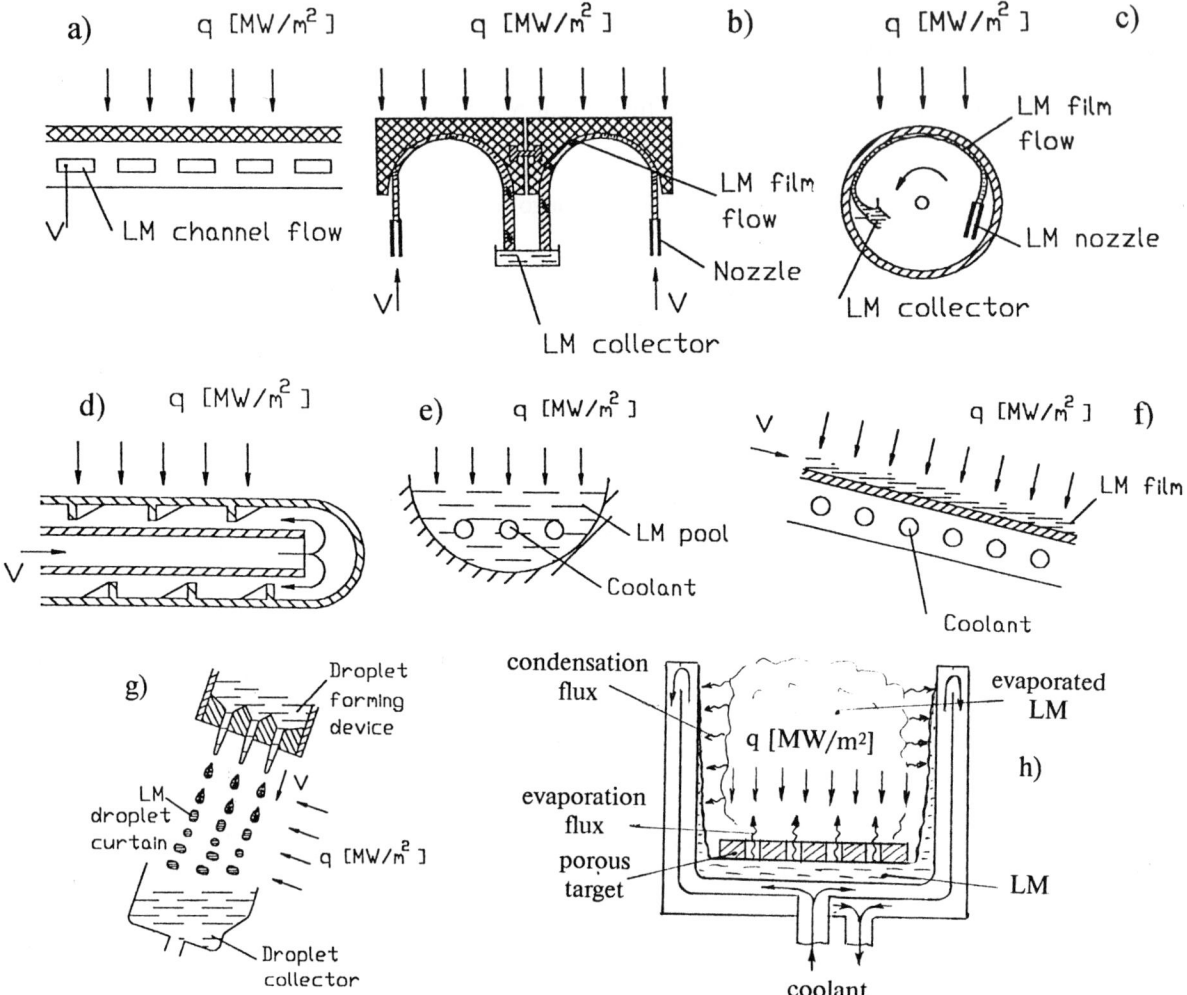

Figure 1. LM divertor targets schemes: a- Solid target cooled by LM channel flow; b- LM film without direct contact with plasma; c- Rotating drum with LM film flow on inner surface; d- Rotating tube with LM channel flow; e- LM pool; f- Slow or fast moving LM film; g- Jet-droplet LM flow; h- Solid target with capillary structure

Device shown in Fig.1,b, [2-4], where solid target is cooled with LM film flowing over it's inner surface, do not provide for LM vacuum tight enclosure and some LM vapor ingress into plasma during normal operation takes place. Its advantages over the device of Fig.1,a are easier tiles replacement, less mechanical stresses and higher heat fluxes (q<15-20 MW/m^2) due to absence of heat sink layer and larger allowed LM velocities. Application of this device for SN configuration does not cause any problems while for upper divertor in DN configuration it is more difficult.

To increase the heat flux removing capability in devices of this type, the mechanical movement (rotation with low frequency) of the target may be used (Fig.1,c,d), [2,5]. Due to spreading of the heat flux over larger surface, its mean value is decreased which allows to accumulate larger maximum heat fluxes (up to some hundreds MW/m^2) than in stationary target. The target rotation may be achieved by electromagnetic force (due to interaction of tokamak reactor field with supplied electric current) or by using the kinetic energy of LM jets (Fig.1,c) or LM channel flow (Fig.1,d). Again they are easily applicable only for SN divertor. Additional concerns here are associated with rotation (strong magnetic field, LM bearings, forces at electromagnetic transients).

Heat pipes technology for transporting heat from tiles to heat sink was proposed by some authors for stationary and rotating targets of Figs.1,a,c. Advantages may arise from the possibility of separation of heat intake and heat sink. In some design schemes easier replacement of tiles than in Fig.1,a may be realized with no LM ingress to plasma during normal operation. On the other side restrictions on facing plasma tiles surface temperature and comparatively large temperatures of heat pipes evaporation zone with LM lead to less allowable surface heat fluxes than in previous schemes.

LM divertor targets with direct plasma exposure provide for continually renewing working surface. Thus the erosion losses will be constantly replenished and erosion process will no longer limit the lifetime of the divertor surface. On the other side LM may enter plasma due to evaporation and sputtering at normal operation and due to splashing during electromagnetic transients. Eroded material may have a detrimental effect on the plasma and this is a main feasibility issue of these devices.

Some of them, like pools or slow moving films (Figs.1,e,f), do not provide for heat removal and serve the only purpose to solve the erosion problem [6,7]. They absorb the impact of plasma particles, but require additional cooling. Their heat accumulating capability is not large ($q<10$ MW/m^2) and they are applicable to SN divertor only. Positive moment is that they require small LM flow rate and no MHD problems appear due to low LM velocity.

Others, like fast moving films, jets or droplets (Figs.1,f,g) provide both - renewal of plasma facing surfaces and heat removal without additional coolants [8-11]. Allowable surface heat fluxes are rather high (up to 100 MW/m^2). They are applicable, in principle, to DN divertors, though the problem of LM ingress in plasma during abnormal conditions is much severe for upper divertor target than for low one.

The idea was also put forward in different times [12,13] to use as a plasma facing component some porous material with liquid metal seeping through the pores. The evaporation of liquid metal keeps the structural material temperature below desired limit. However to use this scheme for removing of heat loads at stationary divertor target conditions is impractical, to our opinion, since even for Li, having very high specific evaporation capacity and low Z, the influx of Li for 10-30 MW/m^2 heat load is some kilograms per second from 1 m^2 surface. If porosity may be organized in such a way that during normal operation LM just fills in the pores not flowing through them, this scheme may be used to protect the solid target from disruption events [12], say for option of Fig.1,a. Authors of [13] claim that proposed by them scheme of Fig.1,h may be applicable to divertors with dynamic gas target and q up to 40-60 MW/m^2. The use of porous substrate for partial feeding of fast film option (Fig.1,f) was also reported in [11].

3. LM DIVERTOR CONCEPTUAL DESIGNS

Some LM divertors were analyzed for a number of recent Tokamak designs (see Table 1). Conceptual drawings and supporting calculations indicated the possibility of such devices incorporation in Tokamak projects with necessary characteristics. Quite extensive experimental back-up was performed for a number of options, though no LM divertor was so far working in any experimental Tokamaks.

Table 1
LM divertor target designs parameters

	ITER CDA [11]		ITER EDA [4,1]		ARIES [3]	Russian DEMO [14]
Type of LM divertor	Fast film flow (Fig.1,f)	Jet-droplet flow (Fig.1,g)	Film flow (Fig.1,b)	Channel flow (Fig.1,a)	Film flow (Fig.1,b)	Rotating drums (Fig.1,c)
Peak surface heat flux, MW/m^2	36.6	62.5	20	5	15	~300[5]
Divertor type	DN	DN	SN	SN	DN	SN
LM	Ga	Ga	Ga	NaK	Ga	Ga
LM film thickness, mm	4	3.6[1]	2	4×16[4]	~1	some mm (0.8×1m)[6]
LM velocity, m/s	~2	10.8	5	5-8.4	1.5-5	5-6.5
LM temperature, °C: inlet	~80	~80	100	100	80	80
outlet	190[2]	280[3]	240	250	230	230
maximum surface	~730	~1000	350	–	–	–

Notes: [1] - droplet diameter; [2,3] -inner and outer targets are switched hydraulically in parallel[2] and in succession[3]; [4] -channel cross-section dimensions; [5] - normal to separatrix; [6] - drum's length and diameter.

4. CONCLUSION

Conducted R&D and design studies confirm the potential attractiveness of LM divertor targets and require further efforts, including testing in present day Tokamaks.

5. REFERENCES

1. I.R.Kirillov, S.I.Sidorenkov, D.L.Smith and T.Q.Hua, Development of ITER Li/V advanced blanket option. Report ITER/RF-US/96/BL.
2. V.Chuyanov, A.Klishenko, E.Murav'ev and V.Petrov, Liquid metal cooled divertor target concept. US-Russia Workshop, (July-August 1992).
3. E.V.Murav'ev, V.S.Petrov, P.V.Romanov et al., Liquid metal cooled divertor for ARIES. Report of General Atomics, GA-A21755, (1994).
4. I.V.Altovskij, A.V.Beznosov, V.I.Khripunov et al., Feasibility of liquid gallium cooling for ITER divertor cassette. Report RRC "Kurchatov Institute", ITER-RF-95-RRCKI-NFI-40/95-117.
5. E.A.Azizov, I.A.Kovan, A.I.Koltchenko et al., The compact volumetric neutron source on the tokamak basis. IEA Technical Workshop on Neutron Sources for Fusion, (July 12-16, 1993), Moscow, Russia.
6. P. Schiller, In: Tokamak Concept Innovations. IAEA, Vienna, (1986) 97.
7. P.I.H.Cooke and A.Bond, Ibid., 93.
8. T.N.Aitov, A.B.Ivanov, E.M. Kirillina et al., Ibid., 488.
9. K.Maki, Ibid., 87.
10. V.O. Vodyanuk, B.G. Karasev, A.F. Kolesnichenko et al., Ibid., 480.
11. I.R.Kirillov, I.V.Mazul and E.V.Murav'ev, Alternative concepts of the divertor targets. Working Material to the ITER Meeting on Advanced Divertor Concepts, (October 1990). ITER-IL-PC-9-0-18.
12. V.A.Divavin, S.P.Gurin and V.N.Odintsov, In: Tokamak Concept Innovations. IAEA, Vienna, (1986) 491.
13. L.G.Golubchikov, V.A.Evtikhin, I.E.Lyublinski et al., In.: Seventh International Conference on Fusion Reactor Materials. Abstracts. Obninsk, Russia, (1995) 76.
14. Yu.A.Sokolov, Fusion Engng. and Design, 29 (1995) 18.

Helium retention of B₄C or SiC converted graphite

T. Hino[a,b], Y. Yamauchi[a], Y. Hirohata[a] and K. Mori[c]

[a] Department of Nuclear Engineering, Hokkaido University, Kita-13, Nishi-8, Kita-ku, Sapporo, 060 Japan
[b] National Institute for Fusion Science, Furo-cho, Chikusa-ku, Nagoya, 464-01 Japan
[c] Gunma Kosen, Maebashi, 371 Japan

The effect of the helium retained in the wall on the helium ash level in a core plasma was analyzed based on the density balance. It is shown that the helium desorption from the wall may increase the level of helium ash.

The amount of the retained helium was obtained by the helium ion irradiation for graphite, B₄C or SiC converted graphite. The retained amount was considerably large, less than or comparable with the retained amount of fuel hydrogen. The major desorption temperature was low, 300 °C for the graphite, but high, 900 °C for B₄C or SiC converted graphite. In the burning plasma, the helium desorption from the wall has to be sufficiently suppressed.

1. INTRODUCTION

In a fusion reactor, the reduction of helium ash concentration is one of most important issues to achieve a burning plasma state with a long time period [1]. Since the helium ash causes the fuel dilution, α-heating power becomes smaller than the energy conduction loss power and then the ignition condition can not be sustained. If the helium ash concentration is not sufficiently reduced, the energy conduction loss has to be decreased, e.g. the energy confinement time be lengthened. Since the energy confinement time, τ_E, is roughly proportional to the plasma radius, a, the plasma size has to be enlarged to obtain the ignition condition [2,3]. Thus, the reduction of the helium ash is a critical issue to achieve the burning plasma.

In the case of the burning plasma, the plasma facing walls may largely retain the heliums produced by the fusion reactions. The heliums may emit into the core plasma and then the helium ash level might be enhanced. This process has not been taken into account so far. The increase of the helium ash level due to this process was discussed based on the density balance of the helium. In order to evaluate the amount of helium retained in the plasma facing material, the helium ion irradiation for graphite, SiC or B₄C converted graphite was conducted by using an ECR ion source. The retained amount of helium was measured by a thermal desorption spectroscopy. The desorption amount and temperature were discussed from the obtained data.

2. EFFECT OF HELIUM RETAINED IN WALL

The plasma facing walls can trap the heliums produced by the fusion reactions. During the discharge shot, these heliums may be desorbed both by ion/neutral impact desorption and the thermal desorption. In this section, these processes are taken into account in the density balance of the helium. If the plasma is sufficiently detached from the first wall, a major process of the helium desorption is due to the charge exchanged particles produced by the reaction between the neutral atoms and the plasma ions in the scrape-off layer. Thus, the density balance for the helium of the core plasma is given by

$$\frac{dn_c}{dt} = -\frac{1-R_c}{\tau_c}n_c + \frac{n_{DT}^2}{4}\langle\sigma v\rangle_f + f_c\Gamma_\alpha \quad , \quad (1)$$

where τ_c is the helium confinement time, R_c the ratio of helium returning to the core plasma from the divertor, n_{DT} the fuel deuterium and tritium ion density,

$<\sigma v>_f$ the fusion reaction rate, $\Gamma_{cx} \sim n_{DT}^{SOL} n_0 <\sigma v>_{cx} g_{cx}$ $(2\Delta/a)$ the helium flow due to the charge exchanged particles, $f_c = \exp(-\Delta n_e <\sigma v_e>_I / v_{He})$ the fraction penetrating into the core plasma, n_{DT}^{SOL} the plasma density of the scrape-off layer, $n_0 = 3.5 \times 10^{22} P_0(\text{Torr})\,\text{m}^{-3}$ the neutral atom density, $<\sigma v>_{cx}$ the charge exchange rate, g_{cx} the ratio of impact desorption, Δ the thickness of the scrape-off layer, n_e the electron density of the scrape-off layer, $<\sigma v_e>_I$ the ionization rate and v_{He} the velocity of the emitted helium. Figure 1 shows the process of helium flow into the core plasma.

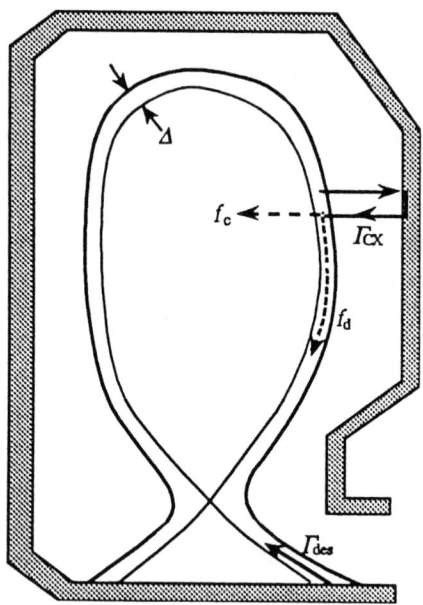

Fig. 1 Emissions of heliums retained in the first wall and the divertor.

The helium density of the divertor may be expressed as

$$\frac{dn_d}{dt} = \frac{R_c}{\tau_c}(1 - R_c - R_p) + f_d \Gamma_{cx} + \Gamma_{des} \quad , \quad (2)$$

where $f_d = 1 - f_c$, $\Gamma_{des} = \Gamma_W + \Gamma_{ion}$ the helium flow emitted from the divertor, Γ_W the helium flow due to the thermal desorption, $\Gamma_{ion} = n_{DT} V_c g_{ion} / V_d \tau_{DT}$ the helium flow due to the ion impact desorption, V_c and V_d the volumes of core plasma and divertor respectively, g_{ion} the ratio of the impact desorption and τ_{DT} the ion confinement time. If the divertor wall temperature is not elevated, it is regarded that $\Gamma_W \sim 0$.

In a case of steady state, Eqs. (1) and (2) become

$$n_c = \frac{\tau_c}{1 - R_c}\left(\frac{n_{DT}^2}{4}<\sigma v>_f + \underline{f_c \Gamma_{cx}}\right) \quad , \quad (3)$$

$$R_c = 1 - R_p + \frac{\tau_c}{n_c}\underline{(f_d \Gamma_{cx} + \Gamma_{ion})} \quad . \quad (4)$$

The terms with the underlines are due to the heliums emitted from the wall. These terms have to be much smaller than those in the case without the helium emission. When $f_c \sim 0.1$, $n_{DT}^{SOL}/n_{DT} = 1/4$, $<\sigma v>_f = 10^{-22}\,\text{m}^3/\text{s}$, $<\sigma v>_{cx} \sim 10^{-13}\,\text{m}^3/\text{s}$, $\tau_{DT} = 2$ s and $V_c/V_d = 10$, these conditions become

$$g_{cx} P_0 \ll 10^{-9}\,(\text{Torr}) \; ,$$

$$g_{ion} \ll 5 \times 10^{-4} \; .$$

If the neutral gas pressure, P_0, is 10^{-6} Torr, g_{cx} or g_{ion} has to be much less than $\sim 10^{-3}$. This example shows that the desorption ratio due to ion or charge exchanged particle should be kept extremely small. Then, the amount of the retained heliums in the wall has to be very largely reduced. In the case of ITER, the divertor wall temperature in the scheme of gas target divertor is estimated to be approximately 300 °C. Even if the heliums are trapped in the wall, there may be no problem only when the heliums are thermally desorbed at the temperature less than 300 °C. In order to avoid the helium emission from the wall, the material in which the helium can be thermally desorbed at the temperature less than the operation wall temperature, is desirable.

3. HELIUM RETENTION OF PLASMA FACING MATERIAL

The helium ion irradiation was carried out by using the ECR ion source (Fig. 2). The energy and the fluence of the helium ion were 5 keV and 3×10^{18}

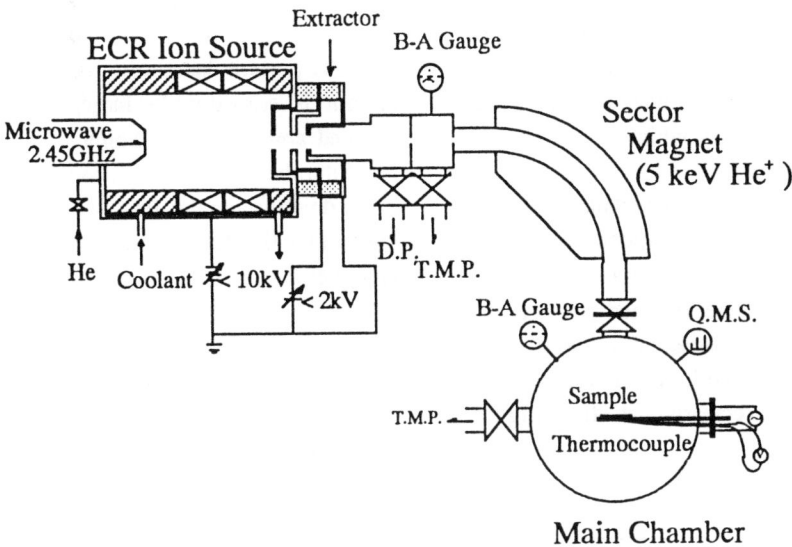

Fig. 2 ECR ion irradiation apparatus.

He/cm², respectively. The irradiation temperature was RT. This fluence was sufficiently enough for the retained amount to saturate for the isotropic graphite, B₄C converted graphite and SiC converted graphite. After the helium ion irradiation, the sample was heated with the rate of 50 °C/min in the same chamber and the desorbed amount was measured by the attached quadrupole mass spectrometer, QMS.

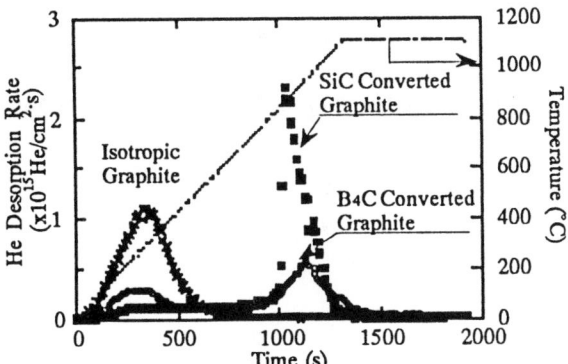

Fig. 3 Helium desorption spectra of isotropic graphite, B₄C converted graphite and SiC converted graphite.

Figure 3 shows the helium thermal desorption spectra of the isotropic graphite, the B₄C converted graphite and the SiC converted graphite. For the case of the isotropic graphite, the spectrum had a single desorption peak at 300 °C. The activation energy of the desorption was 1.6 eV. The desorption amount was 3.4×10^{17} He/cm². In the similar way, the hydrogen ion irradiation was conducted and the retained hydrogen amount was measured. The hydrogen fluence was taken sufficiently large for the retained amount of hydrogen to saturate. The hydrogen energy and the irradiation temperature were 1.7 keV and RT, respectively. Compared with the hydrogen amount, the retained amount of the helium was approximately a half. In the B₄C converted graphite, two desorption peaks appeared at 300 °C and 950 °C. The activation energies of the low and the high temperature peaks were 1.6 eV and 3.5 eV, respectively. The low temperature desorption corresponds to the helium desorption due to the carbon content. The retained amount of helium was 2.7×10^{17} He/cm², which was about 1/4 of the retained amount of hydrogen. The SiC converted graphite had a sharp desorption peak at 850 °C. It was also observed that a small amount of the helium desorbed at the low temperature. The low temperature desorption may be due to the carbon content. The activation energies of the low and the high temperature desorptions were 1.6 eV and 3.3 eV, respectively. The retained amount of the helium was 4.5×10^{17} He/cm^2, which was 3/4 of the retained amount of hydrogen.

Figure 4 shows the amount of the retained amount of helium versus the annealing temperature for the isotropic graphite, the B$_4$C converted graphite and the SiC converted graphite. The retained amount of helium in the graphite largely decreased around at ~400 °C. The retained amount of helium of the B$_4$C or the SiC converted graphite decreased at the higher temperature, ~900 °C.

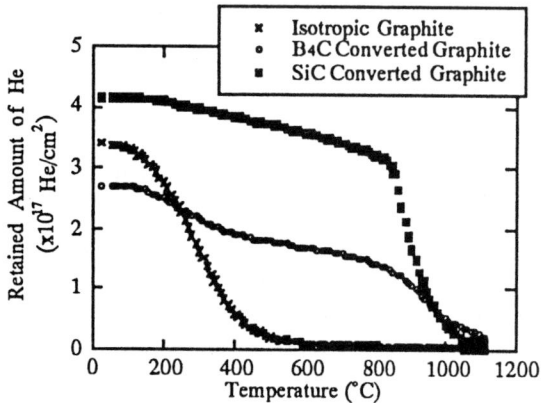

Fig. 4 Annealing temperature dependences of retained helium for isotropic graphite, B$_4$C converted graphite and SiC converted graphite.

4. DISCUSSIONS

The present data show that the retained amount of helium is considerably large, e.g. (1/2~3/4) of the retained amount of hydrogen. In the case of the graphite, the helium desorption temperature is 300 °C. If the divertor and the first wall temperatures are kept higher than this temperature, or the baking temperature is higher than 300 °C, the contamination of the helium from the wall is very small. In the case of the B$_4$C or the SiC converted graphite, the helium desorption temperature is about 900 °C. So the large amount of the helium shall be trapped in the wall during the discharge shot. The helium desorption may take place during the discharge due to the thermal desorption and/or the ion impact desorption from the plasma facing materials. Before the discharge shot, the conditioning such by a high temperature baking is necessary for the reduction of the retained helium.

The present study indicates that the helium retention properties of the candidate plasma facing materials have to be more systematically investigated for the reduction of the helium ash level in the burning plasma.

REFERENCES

1. IAEA, ITER EDA Documentation Series No.7, Technical Basis for the ITER Interim Design Report, Cost Review and Safety Analysis, IAEA, Vienna, 1996.
2. T. Hino, H. Yanagihara and T. Yamashina, Fusion Eng. and Design, 24(1994) 437.
3. T. Hino, Fusion Eng. and Design, 30(1995) 299.

ITER Outboard Baffle Design and Small-Scale Mock-ups Test Programme

L. Giancarli[a,*], B. Bielak[a,*], P. Chappuis[b], C. Cheron[c], G. Le Marois[d], Y. Poitevin[a], J. Quintric-Bossy[e], J.-F. Salavy[a,**], Y. Severi[e], J. Szczepanski[c,***]

Commissariat à l'Energie Atomique
[a] CEA/Saclay, DRN/DMT/SERMA, F-91191 Gif-sur-Yvette Cedex, France
[b] CEA/Cadarache, DSM/DRFC/STID, F-13108 St. Paul-lez-Durance Cedex, France
[c] CEA/Saclay, DRN/DMT/BCCR, F-91191 Gif-sur-Yvette Cedex, France
[d] CEA/Grenoble, CEREM/DEM/SGM, F-38054 Grenoble Cedex 9, France
[e] CEA/Cadarache, DRN/DER/STML, F-13108 St. Paul-lez-Durance Cedex, France

A. Cardella, ITER JWS Garching, Boltzmannstr. 2, D-85748 Garching, Germany

G. Vieider, The NET Team, Boltzmannstr. 2, D-85748 Garching, Germany

The present paper describes the ITER outboard baffle design including the choice of armor and heat sink first wall materials. Two versions of first wall attachment are presented which differ from the applied fabrication techniques. The choice of the best fabrication techniques will be based on the results of the implemented experimental programme which starts with fabrication and testing of small-scale FW-mock-ups.

1. BAFFLE SPECIFICATIONS

The baffle modules main function is to avoid particle back-flow from the divertor chamber. As a consequence, the baffle-FW is submitted to an high thermal flux and its design needs the use of high-heat-flux component technology.

It as been decided to focus the activities on the outboard baffles because of the more severe heat-load conditions and the more complex overall geometry.

The baffle design takes into account the requirements coming from ITER operations and from the integration of other in-vessel sub-systems and components (e.g., maintenance and remote handling, divertor cassette, back plate, shielding blanket and poloidal manifolds, etc.).

1.1. General assumed guidelines

The assumed main guidelines for the baffles are the following:
- integrated First Wall (FW);
- lifetime for the whole BPP (10,000 cycles up to 1,000 s duration);
- nominal heat load: average ~1 MW/m^2,
                          peak   ~3 MW/m^2;
- neutron wall loading: 0.6 MW/m^2;
- transient head loads: 20 - 60 MJ/m^2 in 0.3 s;
- one cooling circuit per module, water coolant: pressure 4MPa, inlet/outlet temperature 150/200°C.

1.2. Baffle geometry

The baffle coincides with the last row of blanket modules at the bottom of the inboard and outboard shielding blanket. There are 60 outboard and 40 inboard baffle modules. The typical dimensions of an outboard baffle are ~1.5 m in the poloidal direction, ~0.6 m in the radial and ~0.9 m in the toroidal ones. Figure 1 gives a 3D-view of an outboard baffle module, where it can be seen that this type of module is the only one whose top and bottom walls are not parallel, with the further requirement of having both cooling circuits inlet and outlet at the top of the module.

1.3. First wall materials

The armor material (facing the plasma) will be finally selected only after having performed further R&D. The possible alternatives actually retained are the use of Beryllium or CFC (a graphite composite). In the bottom corner of the baffle, because of the expected low-energy of the incident particles and the high heat flux, the use of Tungsten is envisaged.

---

[*] Framatome, [**] ALTEN, [***] Concept 21

Copper alloy is required for heat sink material in order to have a more uniform distribution of the heat flux towards the water cooling tubes, themselves made up with the same material which will be DS Copper (Glidcop®) or CuCrZr.

The stainless steel 316L(N) has been selected as the reference shield-block and structural material.

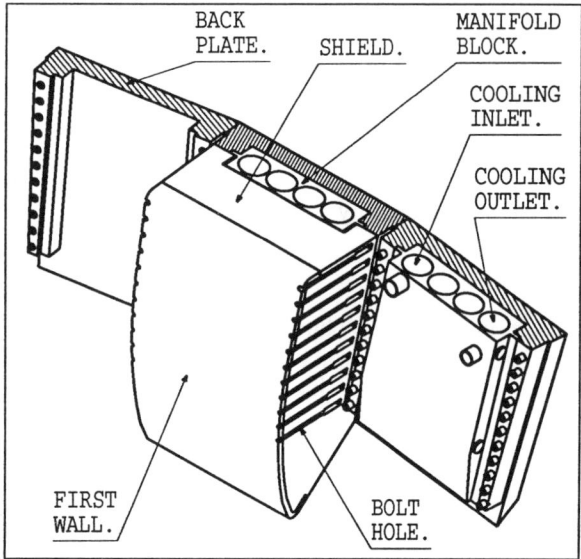

Figure 1: 3D-view of an outboard baffle module

2. DESIGN & MANUFACTURING SEQUENCE

Because of the specific design of the baffle which requires a strong link between the FW and the relatively-cold Shield Block (SB) of about 5 tons, it was recognized that, for all the three armor-material candidates, the use of tiles attached to a water-cooled Cu-alloy plate had to be preferred to other existing designs. As far as this FW/SB link is concerned, two principal options have been retained and developed [1] which differ basically for the adopted fabrication technique.

2.1. Fully integrated-FW (reference option)

As it is shown on figure 2, the fully integrated-FW design foresees a direct attachment of the FW to the SB by solid HIP technique. FW collectors are integrated in the SB.

The baffle design has been established after many iterations between feasibility assessment, thermal-hydraulic and thermo-mechanical analyses, and specifications and geometry constraints. In particular, calculations have lead to the choice of the serial connection option for the single baffle cooling circuit. Analyses have been used to define geometry and position of the four rows of cooling channels within the massive steel SB, the tile thicknesses for each of the three tile-material candidates (10 mm for Be, 30 mm for high conductivity CFC, and 10 mm for W, see figure 2) and a preliminary choice of the castellations lengths.

2.2. Welded-FW (alternative option)

The welded-FW design [2] has the advantage of permitting to fabricate separately FW and SB, and then attach them through welding at the last baffle fabrication step (see Figure 3).

An additional advantage is to permit a wider choice of material (e.g., the use of CuCrZr alloy) and of joining techniques (e.g., the AMC/EB joint between W/CFC-tiles and Cu-alloys). Thermo-mechanical calculations have shown that no castellations are required in the Cu-plate. The major drawback is the large number of welds in the FW-collectors.

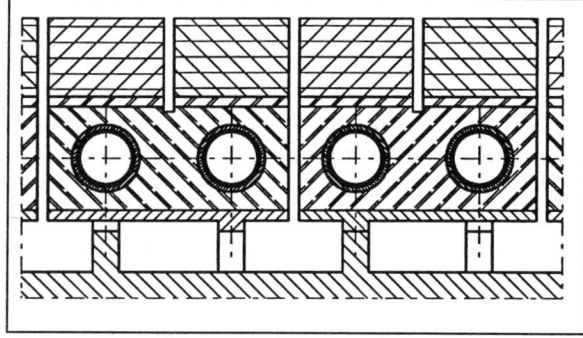

Figure 3: Detail of the welded-FW design

2.3. Manufacturing sequence

For both FW-attachment options, manufacturing sequence has been established and the controllability of the joints performed in the various phases has been verified [1,2]. Reference fabrication is based on solid HIP technology. Solid HIP cycles temperature governs the fabrication route [1,3] which requires to perform at first SS/SS joints (>1050°C), then SS/Cu-alloy joints (900-1000°C), and then Cu-alloy/Cu-alloy joints (800-900°C). Joint with the armor material has to be the last step in the integrated-FW option. Welding of the FW-panels is the last step for the welded-FW option.

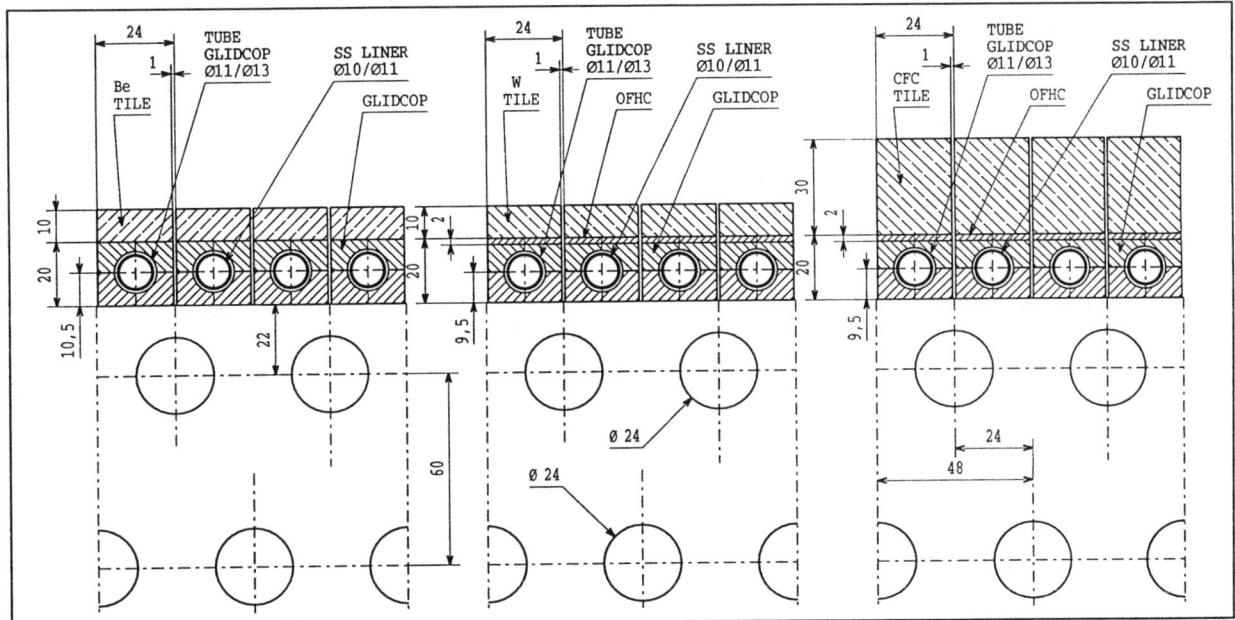

Figure 2. ITER outboard baffle fully integrated-FW design - First Wall detail: Be, W and CFC options

3. BAFFLE SMALL-SCALE MOCK-UPS

The proposed fabrication routes have allowed to identify uncertainties requiring further specific R&D. An R&D programme and planning have then been defined with the overall objective to demonstrate the baffle feasibility and specific performances by mid-1998 [1].

3.1. Test program objectives

The objective is *the test of a relevant-size outboard baffle mock-up which should be fabricated using the same techniques proposed for the fabrication of all the ITER baffle modules*. To reach this objective, R&D has been divided in three stages: i) the small scale mock-ups stage, which should be used to define the best fabrication procedure, including the selection of the joint techniques to be used; ii) the medium-scale mock-ups which should make use of the results of the previous stage and should use fully industrial fabrication techniques and control; iii) the relevant-size mock-ups.

3.2. Design of the small-scale mock-ups

Starting from the baffle module reference design and the general definition of the small-scale mock-ups, the detailed design of the five small-scale mock-ups has been performed taking into account the characteristics of the proposed testing facilities (i.e., the EB in Julich Laboratory (D): JUDITH, and the EB in Le Creusot (F): EB 200). These five mock-ups, fully described in [4], are the following:

• Mock-up 1: Straight three-tubes mock-up formed by a directly cooled Glidcop® plate attached on a 50 mm steel support and holding 4 mm thick Be-tiles. Fabricated by CEA [5].

• Mock-up 2A: Straight three-tubes mock-up formed by a directly cooled Glidcop® plate attached on a 50 mm steel support and holding 10 mm thick CFC-tiles. Each tube has a different CFC grade: SEP NS31, SEP NB31 and Dunlop Concept 2. Fabricated by CEA [5].

• Mock-up 2B: Straight single tube mock-up formed by a directly cooled Glidcop® plate attached on a 50 mm steel support and holding 10 mm thick tiles made of three different materials: W-1%La$_2$O, CFC SEP NS31 and CFC Dunlop Concept 2. Fabricated by Plansee with AMC/EB technique [6].

• Mock-up 3A: Straight three-tubes mock-up formed by a directly cooled Glidcop® plate attached on a 50 mm steel support and holding 10 mm thick W-1%La$_2$O -tiles. Fabricated by CEA [5]

• Mock-up 3B: As mock-up 2B but with CuCrZr instead of Glidcop®. Fabricated by Plansee with AMC/EB technique [6].

3.3. Fatigue testing strategy

During the ITER BPP, baffle components are expected to be submitted to approx. 13000 cycles at 3 MW/m^2 of incident flux. Making testing of all mock-ups at the right heat flux for this full number of cycles will lead to a too long testing time requirement which would lead to excessive cost and possible testing for a very low number of mock-ups. This problem was overcome correlating heat flux and number of cycles based on previous measurement of the Cu/stainless steel joint system. This correlation indicates that an increase of a factor 2 on the heat flux is equivalent to an increase of about a factor 10 on the number of cycles. The use of this rough correlation leads to the following proposal for all mock-ups:

- initial 100 cycles at 5 MW/m^2 for detecting any significant fabrication defects,
- run of 1000 cycles at ≈ 9 MW/m^2 which is the reference fatigue test for simulating the baffle specification,
- run of 1000 cycles at the maximal heat flux corresponding to the maximum acceptable tile-material temperature, in order to get fatigue information on the joint, which are essential for divertor application,
- final simulations of disruption/VDE events (10 times, if possible).

The tests are planned for the autumn 96.

Mock-up	Tile-material	Total testing time	Maximum heat flux
1	Be	10.3 h	11 MW/m^2
2A	NS31	12.7 h	19 MW/m^2
3A	W	34.1 h	14 MW/m^2
2B	NS31	14.9 h	19 MW/m^2
	W	12.7 h	16.5 MW/m^2
3B	NS31	14.3 h	19 MW/m^2
	W	12.9 h	17 MW/m^2

Table 1. Thermal calculations main results

3.4. Thermal-hydraulic calculations

Thermal and thermal-hydraulic calculations have been performed on the basis of the foreseen baffle small-scale mock-ups testing program. They have been used as a guideline to define the testing sequence for each mock-up. Results of the steady-state and transient analyses performed are fully described in [3]. The major boundary condition is the acceptable maximum tile-surface temperature in the mock-ups, that is: Be 800°C, CFC 2200°C, W 2500°C. Table 1 summarizes, for each mock-up and tile-material, the foreseen durations of the fatigue tests based on the estimation of the thermal response (transient) and the maximal heat flux based on the maximal acceptable tile-material temperature (steady state).

4. CONCLUSION

Starting from ITER specifications, baffle module design has been developed and dimensioned from points of view hydraulics, thermal and thermo-mechanical. Manufacturing sequences for the two FW-attachment options have been established. Test programme of mock-ups has been defined in order to validate the FW-design.

Future activities will include fabrication and test of medium- and near-full-scale mock-ups. The selection of materials, fabrication techniques and FW-attachment option will be mainly based on the small-scale mock-up test results.

REFERENCES

1. L. Giancarli et al., ITER Baffle Design and Associated Test-program, CEA/DRN/DMT Report (1995).
2. L. Giancarli et al., Welded-FW Option for ITER Baffle Design, CEA/DRN/DMT Report (1996).
3. G. Le Marois et al., Baffle Fabrication - Integrated Concept, CEA/CEREM/SGN Report (1996).
4. Y. Poitevin et al., Thermal and thermal-hydraulic calculations on the ITER Baffle Small-scale Mock-ups, CEA/DRN/DMT Report (1996).
5. G. Le Marois et al., ITER Baffle Small-size Mock-ups Fabrication and Testing, this conference.
6. G. Vieider et al., Manufacture and testing of small-scale mock-ups for the ITER divertor, this conference.

FUSION TECHNOLOGY 1996
C. Varandas and F. Serra (editors)
1997 Elsevier Science B.V.

Comparison between various thermal hydraulic tube concepts for the ITER divertor

I.Smid[b], J.Schlosser[a], J.Boscary[a], F.Escourbiac[a], G.Vieider[b]

[a] Association EURATOM-CEA, Département de Recherche sur la Fusion Contrôlée, Centre de Cadarache
13108 Saint Paul lez Durance Cedex, France

[b] The NET Team c/o Max Planck Institut für Plasmaphysik, Boltzmannstr. 2
85748 Garching, Germany

High heat flux tests on CuCrZr actively water cooled elements were performed with geometric and thermal hydraulic parameters relevant to ITER (International Thermonuclear Experimental Reactor) divertor conditions. Different types of mock-ups with the same width were tested and compared : double smooth tubes (SM2), swirl tubes (ST2; ST4), annular flow tubes (AF1; AF3) and hypervapotron tubes (HV1; HV3). Analyses of tests were done using the CEA method [1;2] first developed by Sandia Laboratory [3]. Finite Element calculations were used with a set of correlations in order to express the wall heat flux as a function of wall temperature in the convective regime as well as in the subcooled boiling regime (this set is now available in the EUPITER code). Maximum wall heat flux was compared with modified TONG-75 correlation. In terms of ICHF (Incident Critical Heat Flux) and for the same thermal hydraulic conditions, results gave this decreasing order : HV1, HV3, ST2, AF1, ST4, AF3, SM2. Versus lineic pumping power, the previous order was slightly changed : HV1, HV3, ST2, ST4, AF1, SM2, AF3. A typical HV1 result is a 38 MW/m² ICHF for 135°C local subcooling, 10 m/s water velocity, 3.5 MPa local pressure, 0.3 MPa/m lineic pressure drop and 380 W/m lineic pumping power.

1. INTRODUCTION

For several years, different laboratories in the world have been working and collaborating on CHF for one-side heated tubes. These researches aim to get comfortable safety margins in water cooled high heat flux (HHF) plasma facing components (PFC). In Tore Supra, this is crucial for PFCs which are working at steady state (maximum plasma duration 120s up to now, injected energy 240 MJ), but also for the future machine ITER. Historically, in Europe, JET developed a technology based on hypervapotron tubes [4;5] whereas Tore Supra and NET studied swirl tubes [1;2;3]. A comparison between the two concepts was punctually presented by JET [6]. For CHF tests on simple swirl tubes (ID = 10, 14 and 18 mm; twist ratio 2), CEA used intensively the electron beam facility FE200 [7]. This facility is equipped with a pressurized cooling water loop (flow rate up to 6 kg/s; inlet pressure up to 4 MPa; inlet temperature from 50 to 240°C; pressure drop up to 0.7 MPa) : about 50 CHF values with a flat 100 mm long incident heat flux profile were obtained for different velocities (5; 10; 15 m/s), outlet pressure (1 ,2 ,3.5 MPa) and inlet temperatures (50, 100, 150°C) when possible. In the frame of ITER-EDA, it was decided to test different concepts of tubes with the same width in order to establish a direct comparison in terms of incident critical heat flux (ICHF).

The different test sections were installed 2 by 2 on mock-ups into FE200 facility to save mounting and dismounting time. Generally, one mock-up required two weeks of test campaign (one week for each test section). This allows 103 ICHF tests to be performed in 9 months.

2. TEST SECTIONS

Compared with the reference tests sections, the double channel swirl tubes were equipped with thicker twisted tapes made of OFHC copper. The twist annular flow tube is a new concept built with a swirl rod insert made of a stainless steel cylinder on which a helical wire is welded by spots. The head of the insert is machined in spin in order to reduce pressure drop. The Hypervapotron concept is typically the JET concept, originally developed by [4]. Reference and new test sections are presented in Fig 1 and 2.

3. PRESSURE DROP MEASUREMENTS AND CORRELATIONS

In order to compare the different mock-ups, pressure drop of the prototypes were carefully measured at ambient temperature (Fig.3).

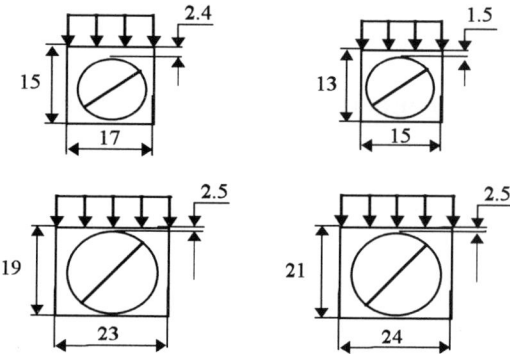

Fig.1 : Reference test sections (1992/1994)

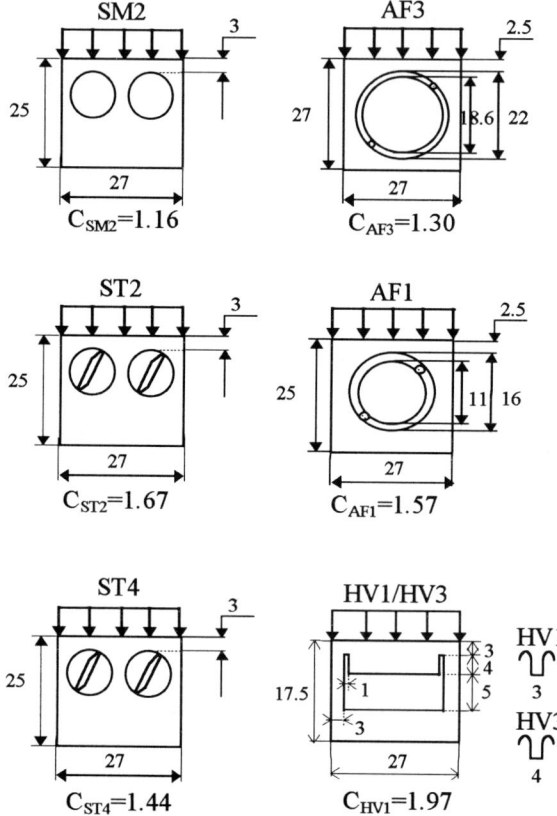

Fig.2 : New test sections (1995/1996)

CEA has developed 2 correlations giving the pressure drop coefficient λ (Fig.4), relations only valid for swirl tubes.

Pressure drop is then calculated from :

$$\Delta P = \lambda \frac{L_{sw}}{D_H} \frac{1}{2} \rho_l V_{sw}^2$$

- For Stainless Steel tubes or poor roughness :

$$\lambda = 0.316 \, D_H^{0.91} (v_l)^{0.09} Re_{Hsw}^{-0.16} \quad (I)$$

- For Rough tubes :

$$\lambda = \left(\frac{\varepsilon}{D_H} + \frac{15}{Re_{Hsw}} \right)^{0.5} \quad (III)$$

L_{sw}, V_{sw} : swirled length and swirled velocity [1;2]

D_H : hydraulic diameter of the tube

ρ_l : density of the liquid

v_l : kinematic viscosity of the liquid

$Re_{Hsw} = \dfrac{V_{sw} D_H}{v_l}$: swirled Reynolds number

ε : roughness of the tube wall

For hypervapotrons and annular flow tubes, experimental data are used.

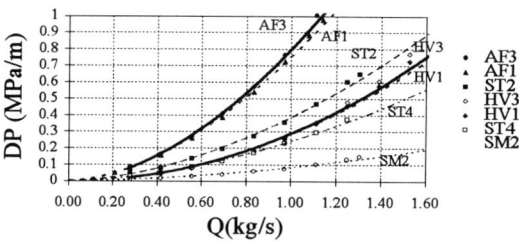

Fig.3 : Lineic pressure drop vs mass flow rate at ambient temperature.

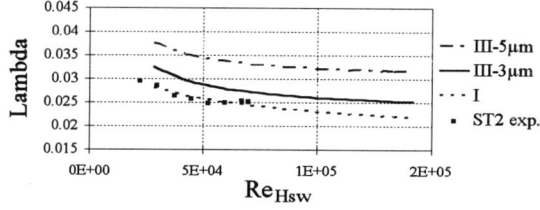

Fig.4 : Pressure drop coefficient vs swirled Reynolds number (for 10mm swirl tube, tape thickness 0.8mm, twist ratio 2).

4. INCIDENT CRITICAL HEAT FLUX TEST-RESULTS AND COMPARISON BETWEEN VARIOUS TUBES

For the 7 new test sections, during each performed test, the power was increased step by step up to reach the ICHF detected by sensible excursion on the temperature measurements or hot spot developing on the IR image. CuCrZr thickness between the heated surface and the cooling channel (2 to 3 mm) induces a high surface temperature, a structural degradation of CuCrZr and a loss of thermal conductivity. This leads to underestimate ICHF (for high heat fluxes, CuCrZr melting point (1050°C) is reached before CHF).

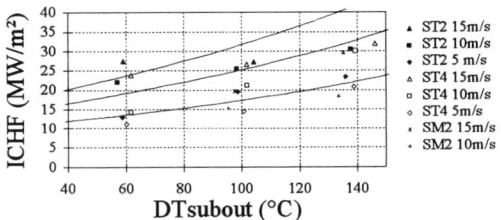

Fig.5 : Incident Critical Heat Flux for ST2, ST4 and SM2 (3.5 MPa, Lh = 100 mm)

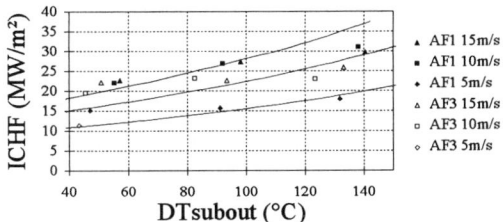

Fig.6 : Incident Critical Heat Flux for AF1 and AF3 (3.5 MPa, Lh = 100 mm)

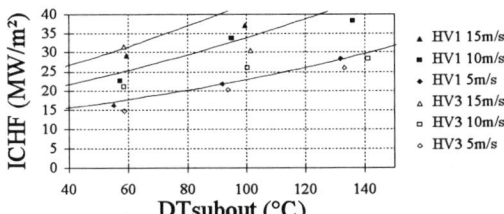

Fig.7 : Incident Critical Heat Flux for HV1 and HV3 (3.5 MPa, Lh = 100 mm)

Fig.8 presents a comparison between different mock-ups in terms of ICHF versus lineic pumping power (defined as volumic flow rate multiplied by pressure drop).

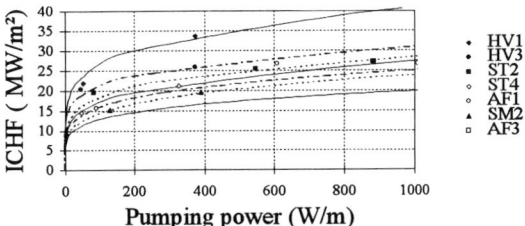

Fig.8 : Incident Critical Heat Flux vs lineic pumping power (100°C outlet subcooling, 3.5MPa)

5. TEST ANALYSIS METHOD AND EUPITER CODE

Special developments were included in the CASTEM 2000 code to perform Finite Element (F.E.) calculations taking into account the heat transfer evolution at the CuCrZr/water wall in the convective and the subcooled boiling regimes. Temperature measurements given by the thermocouples installed into test sections were compared with F.E. calculations. The best agreement was obtained with a Thom modified correlation for the subcooled boiling regime. The Bergles and Roshenow method was used (quadratic sum) for the connection between the convective regime and the subcooled boiling regime [2]. JAERI also proposed a modification of Thom correlation [6] in good agreement with CEA data base. As CHF is a local phenomenon, Wall Critical Heat Flux (WCHF) is defined as the maximum heat flux at the CuCrZr/water wall. In this paper, the WCHF is obtained by F.E. calculations, the peaking factor (Pf) is defined as the ratio of WCHF and ICHF (Table 1).

The EUPITER4.2 (EUropean thermal hydraulic Package for ITER EDA) code was developed by The NET Team to help design of HHF elements. Pressure drop, heat transfer coefficient both in convective and subcooled boiling regime and critical heat flux for one-sided heating conditions are included, taking into account results from European/CEA data base.

6. CRITICAL HEAT FLUX CORRELATION

The TONG75 correlation was extensively used to estimate WCHF. For absolute prediction, it was found necessary to use the hydraulic diameter and

corrective factor C_f depending on the geometry :
$$\Phi_{CHF} = TONG75_H \times C_f$$
$TONG75_H$ is TONG75 correlation with Re_H.
C_f is the corrective factor.

For a design approach, the corrective factor is given for each mock-ups (Table 1). This corrective factor, coming from a mean of experimental data, gives a WCHF prediction in a roughly ± 20% error margin in the range : inlet pressure from 2 MPa to 4 MPa, mass flow rate from 0.5 kg/s to 2 kg/s and inlet temperature from 60°C to 180°C. Reference values and the new CEA data base are compared with $TONG75_H$ on Fig.9 and Fig. 10.

	HV1	HV3	ST2	AF1	ST4	AF3	SM2
Cf	1.97	1.84	1.67	1.57	1.44	1.30	1.16
Pf	1.29	1.29	1.39	1.39	1.39	1.25	1.39

Table 1 : Corrective and peaking factors

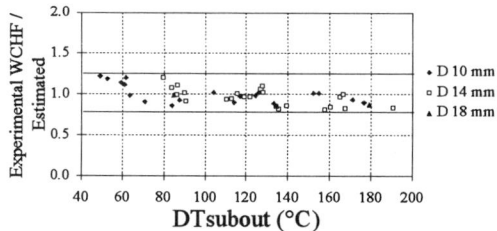

Fig.9 : Comparison between experiments and estimation (reference test sections 1992/1994, 49 points)

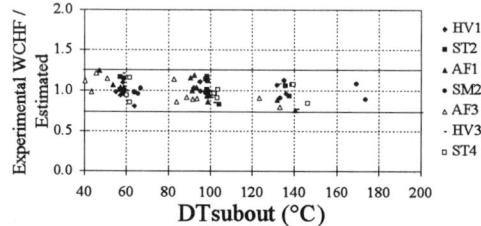

Fig.10 : Comparison between experiments and estimation (new test sections 1995/1996, 83 points)

Table 1 quite agree with results extracted from reference data base where we had :
C_f = 1.25   for one-side heated smooth tube
C_f = 1.67   for one-side heated swirl tube (Tr = 2, tape thickness 0.8mm)

As a consequence of the CuCrZr structural degradation, the modified TONG75 correlation tends to overestimate experimental results at high velocities (i.e. high pumping power) and high subcooling. Despite that, a good ICHF correlation is found for SM2 and ST2 mock-ups up to 1 kW/m lineic pumping power.

For SM2, a 1.16 corrective factor instead of 1.25 for the simple smooth tube could be explained by an interaction between the two channels inducing a loss of cooling efficiency. Nevertheless, ST2 is not affected by this problem.

7. CONCLUSION

As for the reference swirl tubes data base, it has been possible to adapt TONG-75 correlation in order to make a comparison in terms of wall critical heat flux. An experimental corrective coefficient depending on the geometry has been calculated and has allowed mock-ups to be compared directly in terms of ICHF.

The thermo-hydraulic experiments showed the best performance for the HV1 with 30% higher ICHF than for the swirl tube ST2. However, thermo-mechanical analysis indicates that this high ICHF can not be utilized with the flat HV1 geometry above 20 MW/m2 due to limitations by temperatures and stresses in the Cu-alloy heatsink. These limitations are less severe with tubular heat sinks such as ST2 in combination with mono-block type CFC armor.

Definition of the best concept could be a compromise combining hypervapotron thermal hydraulic performance with swirl tube thermo mechanical performance : it could be a circular channel with helical fins.

REFERENCES

[1] J.Schlosser, J.Boscary NURETH 6, Grenoble 93, 815-824.
[2] J.Schlosser, J.Boscary 18th SOFT, Kalsruhe 94, 295-298.
[3] J.A.Koski, A.G.Beattie et al., NHTC, Pittsburgh 87, ASME 87HT-45.
[4] C.A.Beurtheret, 4th Int. Heat Transfer Conf., Paris 70.
[5] C.B.Baxi, H.Falter, 17th SOFT, Rome 92, 186-190.
[6] H.D.Falter, H.Altmann et al., SPIE-HHFS, San Diego 92.
[7] G.Mayaux et al., 17th SOFT, Rome 92, 317-321.
[8] M.Araki, M.Ogawa et al., Int. J. Heat. Mass. Transfer. 1996, Vol.39, N°14, 3045-3055.

ITER Baffle Small-Size Mock-ups Fabrication and Testing

G. Le Marois, F. Bernier, D. San Filippo, F. Saint-Antonin, F. Moret

CEA Grenoble, CEREM/DEM/SGM, 38054 Grenoble, France

Abstract : Solid HIP has been proposed to manufacture ITER Baffle module. To qualify the process, mock-ups with increasing scales are currently tested. This paper describes the proposed mock-up fabrication process and the test protocol defined to simulate Baffles thermal fatigue conditions during the BPP phase.

1- INTRODUCTION

The ITER (International Toroidal Experimental Reactor) tokamak reactor is designed to demonstrate the technical and scientific feasibility of controlled fusion at industrial scale. Its fabrication is based on existing technologies. However for complex and heavily loaded parts of the machine as internal components, quite new technologies have to be considered.

For licensing these advanced technologies and to provide input for the manufacturing of full-scale components, a phased approach has been adopted with the fabrication of mock-ups of increasing scales. Previously, small-scale mock-ups have been designed aiming at demonstrating the feasibility of materials joining by HIP diffusion bonding (solid HIP) and to propose a fabrication route applied to full-scale Baffles.

These mock-ups with different armour materials have been manufactured and are currently tested under high heat flux.

This paper describes the proposed mock-up fabrication process and the test protocol defined to simulate thermal fatigue conditions of baffles during the ITER BPP phase.

2-BAFFLE FABRICATION

2.1 Baffle design

Baffle modules are located at the bottom of the Shielding Blanket. Functions of these components are to drive the plasma boundary layer particles to the divertor chamber and to avoid their return to the main plasma.

Each module incorporates a multi-layered structure consisting of a armour material bonded to a heat sink material supported by a structural material.

For the armour the possible alternatives are the use of Be or CFC (a C composite). The use of Tungsten is envisaged. The maximum acceptable temperature is assumed to be 700°C for Be, 1500°C for CFC and 2000°C for W.

DS (dispersion strengthened) or PH (precipitation hardened) Cu are the 2 heat sink candidates materials. As DS-Cu is the most suitable for solid HIP joining application, only this alloy is considered in the present paper. Its maximum acceptable temperature is 1000°C.

The supporting structure, made of 316LN stainless steel (SS) is designed to work at temperature below 400°C.

These components present a double curvature with an internal cooling. Moreover, in order to guarantee a low leak level, it is required to reduce the number of welds and to perform a double containment of the cooling. These design constraints don't allow to realize these blocks by conventional techniques.

2.2 Manufacturing process

For joining the different parts forming ITER internal components, solid HIP has been proposed as it allows to achieve complex shapes with internal cooling, from different materials. It is now considered as the reference manufacturing technique even if some qualification is still needed.

A fabrication route of these modules has been proposed [1] by HIP'ing successively curved plates and tubes from a thick and tight SS part : grooves are machined on corresponding sides of each machined and folded plates. Tubes are inserted in the grooves, then plates and tubes are HIP'ed.

To achieve the cooling circuit, connections between tubes and manifolds are performed by milling collectors at upper and lower part of the block. A double containment of the circuit is performed.

At each stage of the process, the joint quality is controlled by ultrasonic testing (US). The origin and the type of defects being able to appear according to the HIP process, are previously analysed. Calculation of critical defects sizes based on a mechanical analysis of the joint quality and development of the US technique to define conditions of detection of these critical and process defects are currently studied.

Today large HIP'ing facilities are available for the fabrication of these modules.

3-MOCK-UPS MANUFACTURING

3 small-scale mock-ups, with 3 different armour materials, Be, W or CFC have been manufactured.

3-1 Mock-ups design

These mock-ups are representative of a part of the Baffle first wall [2]. They are straight and made of 3 material layers :
- a armour material, Be, W or CFC in tiles form.
- a heat sink plate made of Glidcop DS-Cu, 20mm thick with 3 cooling tubes.
- a stainless steel structural part, 50mm thick with 2 holes representative of cooling circuits.

Details on design and castellations are shown on Fig 1

Specificities of the mock-ups are directed by the properties of the armour material and the corresponding maximum acceptable temperature.

Be mock-up

10X23X4 mm thick Be tiles are attached directly on the Glidcop plate. They cover a surface of 7X7,2 cm^2.

W and CFC mock-ups

19X23X10 mm thick tiles are bonded. The overall surface covered is 7,1X30 cm^2.

3-2 Materials

As forged 316LN stainless steel (SS) plates and 316L SS tubes have been used for the manufacturing of the supporting structure.

Due to its stability under irradiation and temperature and to its improved mechanical properties DS-Cu Glidcop IG0 has been preferred as heat sink material.

Based on mechanical properties and behavior under irradiation, Be S65C from JET is the reference material for the Be armour and has been selected for the Be mock-up.

Due to availability, NB31, NS31 and Dunlop II CFC grade have been selected for the CFC small-scale mock-up. Each of them covers one cooling tube.

W-1%La$_2$O$_3$ is recommanded by EU HT even the O sputtering that would occur is a concern. Anisotropic tiles of this grade has been used for the W mock-up.

3-3 Manufacturing route

The fabrication sequence by solid HIP is similar to those described for the full-scale baffle.

For the 3 mock-ups with armour, it includes the following steps :

phase I : SS / SS joints (plates and tubes),
phase II : Cu alloy / SS joints (plates, tubes and liners),
phase III : Armour (Be, CFC or W tiles) / supporting plate joint
phase IV : Supporting plate / Cu alloy joint.
NDE of the joints are provided at each step of the fabrication.

Solid HIP is used for the joining of the SS, Cu, Be and W components. Brazing is used for CFC. Mock-ups manufacturing includes material procurement and control, design of fabrication, machining and specific operation related to HIP technology (material and surface treatment, canning and HIP treatment).

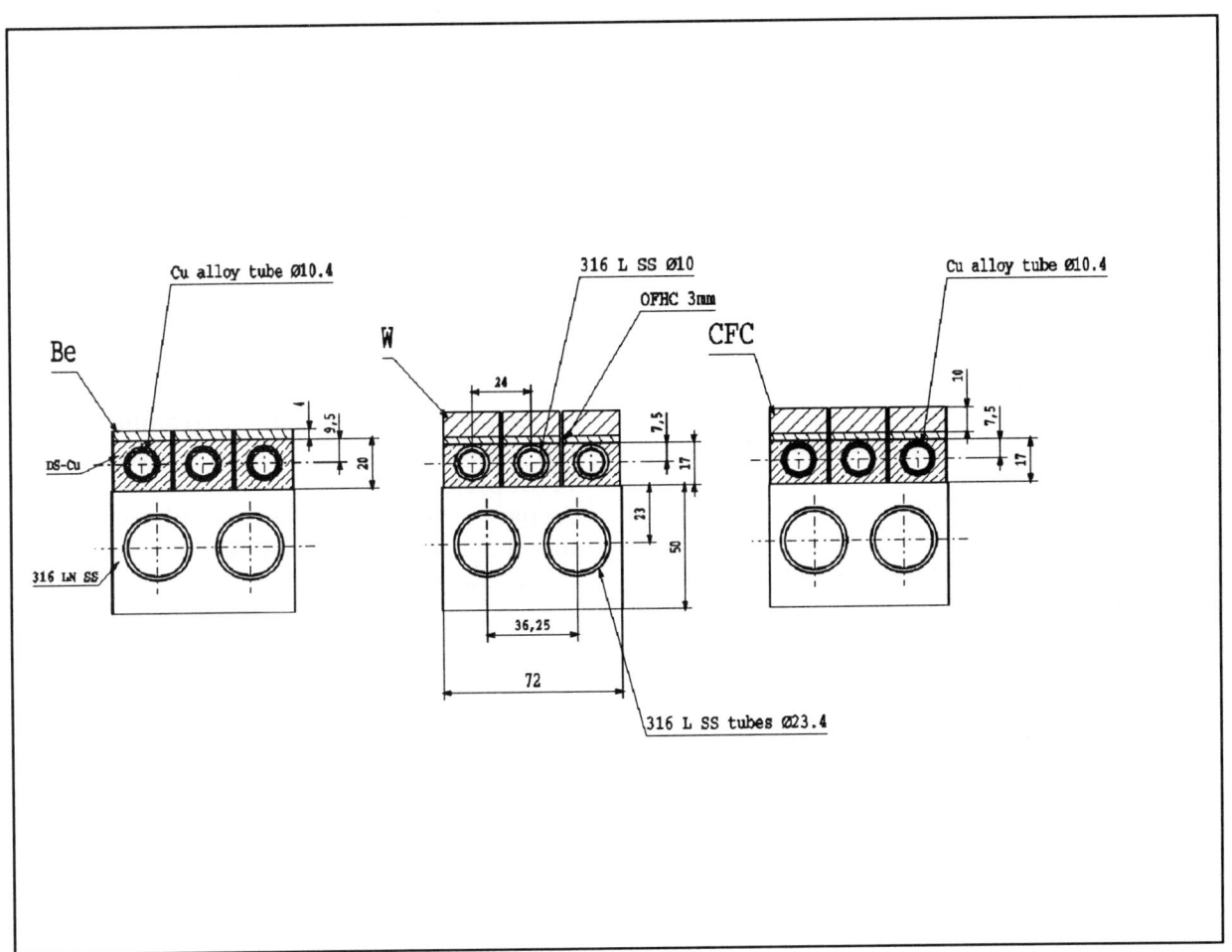

Figure 1 : ITER Baffle Small-Scale Mock-up - Be, W and CFC options

Mock-up with Be armour :

I/ the shield is made of 2 plates and 2 tubes : grooves are machined on corresponding sides of each plate, tubes are inserted in the grooves. Then tubes and plates are joined in a first HIP cycle by a self canning technique. Controls of the joints are achieved by US testing both from external side and inside the tubes.

II/ the heat sink is made of 2 plates and 3 Cu alloy tubes with a SS liner inside : grooves are machined on corresponding sides of each plate. The internal Cu plate is joined by HIP onto the SS after canning : so it allows to control the Cu/SS joint by US testing. The tubes with the liner are inserted in the grooves. The external copper plate is placed and the whole is joined by HIP. Control of the Cu/Cu and Cu/ SS liners joints are achieved by US testing from external Cu side and inside the tubes.

III/ For the joining of Be tiles, a 2 mm thick Glidcop plate is used as supporting plate. Interlayers, 1 mm thick are placed in grooves. It allows to mark the castellation position, to place the tiles and to avoid a collapse of the castellation during the HIP cycle. After canning, the joint is performed by HIP with the help of a reactive layer [3]. US testing of the joint is achieved from the plate side.

IV/ The supporting plate is joined to the heat sink in a last HIP cycle : self canning or canning system will be used according to the industrial availabilities. US control of the joint is achieved. At least the interlayer is removed and the castellation of the copper is achieved by machining.

Mock-up with W armour :
Phase I and II are similar.
In phase III, the supporting plate is made of 3 mm thick OFHC Cu and 2 mm thick Glidcop plate. The W onto Cu joint is also performed by HIP with the help of a reactive layer.
Phase IV is similar. A self canning technique is used.

Mock-up with CFC armour :
Phase I, II are similar.
A same supporting plate, as for W, is used. Interlayers and grooves allow to mark and place the tiles. Cu-Mn filler metal is used to bond CFC onto copper. The joining is performed by HIP assisted brazing.
Phase IV is similar to those described for W.

4- MOCK-UPS HHF TESTING

High Heat Flux (HHF) tests protocol on mock-ups have been defined.
The objectives are to submit the mock-ups to thermal and fatigue conditions representative of those applied to the Baffles during the BPP phase. In addition these mock-ups will be tested at the maximum acceptable conditions in order to get informations on the joint fatigue resistance. The conditions of the tests are as following :
* 1000 cycles at 5 MW/m^2 for detecting any fabrication defect,
* 1000 cycles at 9 MW/m^2 to simulate the fatigue conditions of the baffle,
* 1000 cycles at flux corresponding to the maximum acceptable armour material temperature,
* final simulations of disruption/VDE events (10 times if possible).

These tests are planned for the autumn 96.

5- CONCLUSION

A complete qualification programme of the HIP technology is currently undertaken by ITER project.
As part of the programme, small sizes mock-ups have been manufactured following a representative fabrication route of the baffles demonstrating the feasibility of the manufacturing process.
For baffles, the selection of armour joining techniques and design option will be based on the small-sale mock-up test results.
Fabrication and testing of mock-ups with increasing scales, development of NDE techniques and R&D on joints quality will complete this programme.

REFERENCES

1. G. Le Marois and P. Revirand, Baffle fabrication integrated concept, NT CEA/DEM n°14/96, 1996.
2. L. Giancarli et al, 'ITER Outboard Baffle Design and Small Size Mock-ups Test Programme', paper presented to the SOFT 1996.
3. F. Saint-Antonin, 'Development and characterization of Be/Cu alloy HIP joints', NT CEA/DEM n) 03/96, 1996

High heat flux performance of divertor modules for ITER with beryllium and carbon armor

J. Linke[a], G. Breitbach[a], R. Duwe[a], A. Gervash[b], M. Rödig[a], B. Wiechers[a]

[a] Forschungszentrum Jülich GmbH, Association KFA-EURATOM, D-52425 Jülich, Germany

[b] D.V. Efremov Institute, St. Petersburg 189 631, Russia

To evaluate the performance of divertor modules for ITER under repeated quasi-stationary heat loads of typically 5 MWm^{-2}, high heat flux tests have been performed in the electron beam test facility JUDITH. Main objective of these experiments was the quantification of the heat removal efficiency and the thermal fatigue behavior under cyclic heat loads of modules with a selection of different plasma facing materials which are considered as prime candidates for HHF components in ITER.

Miniaturized divertor modules with geometries also suitable for neutron irradiation experiments have been investigated in high heat flux electron beam experiments. The plasma facing materials investigated so far are different beryllium grades and carbon fiber composites (SEPCARB N31 and DUNLOP Concept 1). The beryllium modules are of the flat tile type with a copper heat sink (CuCrZr). Different joining techniques using INCUSIL and silver free brazes have been included in the test matrix. The CFC modules are primarily of the monoblock type with coolant tubes made of CuCrZr or DS-copper.

1. INTRODUCTION

To simulate the thermomechanical loading scenarios of the divertor in future thermonuclear fusion devices small test modules with ITER relevant cross sections have been exposed to steady state and cyclic heat loads in an electron beam test facility. The high heat flux (HHF) experiments are part of a comprehensive neutron irradiation experiment on plasma facing components [1]. In the frame of this program miniaturized test modules have been irradiated in the High Flux Reactor Petten at 350 and 700°C and a fluence of 0.5 dpa. For HHF tests with neutron activated components the electron beam test facility (JUDITH) is installed inside a hot cell; loading and unloading of the test chamber can be realized by remote handling techniques.

Experiments described in this paper represent a set of pre-irradiation tests which have been performed to evaluate the zero-fluence high heat flux performance of the individual design variants with beryllium and CFC armor. In addition the thermal fatigue behavior of selected test samples has been investigated. The post-irradiation HHF experiments are scheduled to begin by the end of 1996.

2. EXPERIMENTAL PROCEDURE

2.1 Divertor modules

To save volume in the irradiation rig all test modules have been miniaturized; module geometries vary from 15 x 25 x 23 mm^3 to 22 x 50 x 30 mm^3. These test specimens are connected to an instrumented coolant loop by a special clamping mechanism. The modules (flat tile or monoblock design) have been manufactured with different plasma facing and heat sink materials. In general Be-tiles have been brazed using Incusil; in addition one type of test modules has been manufactured with a silver-free braze material (Cu-Mn-Sn-Ce). The monoblock CFC modules have been produced by active metal casting (AMC) [2] and subsequent brazing using a Ti braze.

Table 1
Miniaturized divertor modules for HHF tests in JUDITH (selection)

module type	plasma facing material	braze material	heat sink material	manufacturer
flat tile (3 mm Be)	S65C	Incusil	CuCrZr	ACCEL
flat tile (8 mm Be)	S65C	Incusil	CuCrZr	ACCEL
flat tile (8 mm Be)	S65C	Incusil	CuCrZr	GEC
flat tile (8 mm Be)	S65C	Cu-Mn-Sn-Ce	CuCrZr	GEC
CFC monoblock	Dunlop Concept 1	AMC + Ti-braze	CuCrZr	Plansee AG
CFC monoblock	Dunlop Concept 1	AMC + Ti-braze	DS-Cu	Plansee AG
CFC monoblock	SEPCARB N31	AMC + Ti-braze	DS-Cu	Plansee AG

All Be-tiles have been castellated (spacing 6 - 7 mm). Details concerning the selected materials and the joining process are listed in table 1.

2.2 Electron beam testfacility

High heat flux testing of the modules has been performed in the Juelich Divertor Testfacility in Hot Cells (JUDITH) under quasi-stationary and cyclic heat loads. Safety against leakages is guaranteed by a spring loaded sealing adapter in the high pressure loop [3, 4]. To prevent any damage of the braze joints during the pre-irradiation HHF testing and to avoid severe contaminations originating from evaporated beryllium, the temperature of the Be-surfaces has been limited to approx. 500°C. For additional safety all hypothetical erosion processes on beryllium (or radioactive species in later experiments) are restricted to a special inner containment. Temperature measurement has been performed by thermocouples inside the CFC armor; surface temperatures have been determined by IR pyrometers and a high resolution IR scanner.

2.3 Emittance measurement

The determination of surface temperatures by optical methods is rather sensitive to the emittance ε of the prevailing surface of the plasma facing material (PFC). CFC surfaces show typical ε-values in the order of 0.7 to 0.8 which are rather independent from temperature and the surface morphology. However, beryllium is rather sensitive to oxidation processes [5]. Thin oxide films and surface modifications due to the machining process have significant influence on ε. Thus detailed emittance measurements have been performed in a vacuum chamber on all Be-modules. Typical results are plotted in fig. 1.

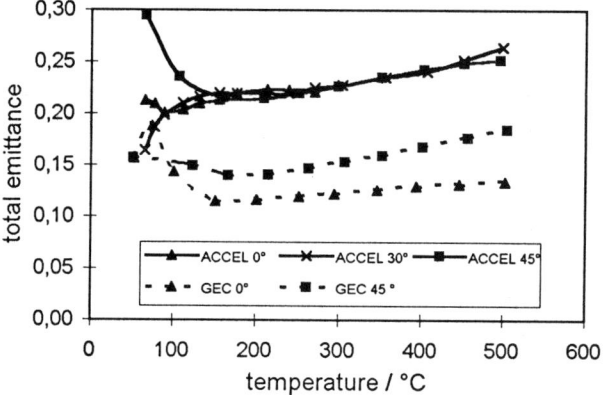

Figure 1. Total emittance for Be/Cu-modules vs. temperature for different angles.

3. RESULTS

3.1 Beryllium modules

During electron beam loading part of the incident electrons (120 keV) is reflected and does not contribute to surface heating. According to calibration experiments which allow to measure the ratio of the absorbed to the incident current [6] the absorbed fraction is 93% for Be surfaces and 90% for carbon surfaces. The resulting surface temperatures (median temperature T_{med}) for HHF screening tests on the Be/Cu-modules with quasi-stationary heat loads which have been increased stepwise, are plotted in fig. 2. These data have been calibrated with the emittance data obtained on each individual divertor module. Test specimens with thin Be armor (3 mm) but otherwise identical geometries show a rather good heat removal efficiency with a resulting surface temperature increase of approx. 55 K per MWm^{-2} absorbed power. Components with thicker Be tiles (8 mm) are characterized by a rather uniform HHF performance; all data lie within a rather narrow scatter band. The silver-free alternative braze (Cu-Mn-Sn-Ce) shows a slightly higher surface temperature ($\leq$ 10%). The temperature distribution of those specimens brazed with Incusil is a little bit more homogeneous than for the silver-free braze. Additional results (including experiments with other geometries and Be-grades) are published elsewhere [7, 8].

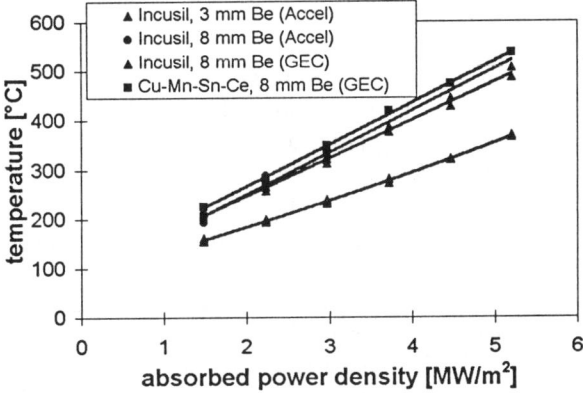

Figure 2. Surface temperature of different Be-Cu-modules vs. absorbed power density (calibrated with sample specific emittance data).

3.2 CFC modules

HHF tests on the CFC monoblock divertor modules have been performed similar to the Be/Cu specimens under quasi-stationary heat loads. Here maximum absorbed heat fluxes of 20 MWm^{-2} have been applied. All tested modules show a rather good and uniform heat removal efficiency. For surface temperatures below approx. 1000°C the electron beam induced temperature rise remains at 80 K / MWm^{-2}; above 1000°C this value increases to 100 or above. This is mainly due to the reduced thermal conductivity of CFCs at elevated temperatures. The differences for all three module types are small; the agreement with results obtained from FEM calculations is good (cf. fig. 3).

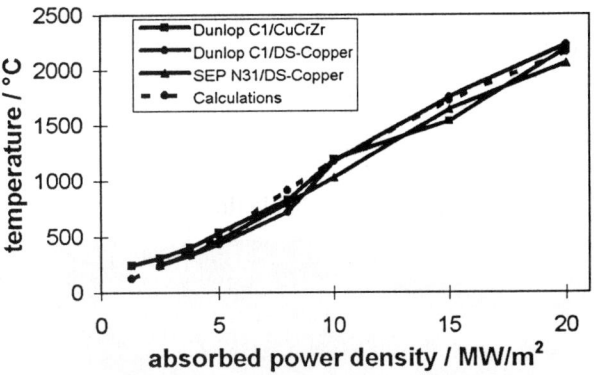

Figure 3. Surface temperature of different CFC monoblock modules vs. absorbed power density.

Up to heat loads of 20 MWm^{-2} all CFC modules show a rather uniform HHF performance. Obviously the heat removal efficiency is not directly correlated with the type of CFC material; results from similar divertor modules with Mo5Re tubes at low power densities [7] show lowest temperatures for the Dunlop Concept 1 material. However, significant inhomogeneities within one batch of CFC material (and even in a single divertor module) have been observed which result in surface temperature variations far beyond the data shown in fig. 3.

Thermal fatigue tests have been performed on all three types of divertor modules; a total of 1000 cycles has been applied at absorbed heat fluxes of 20 MWm^{-2} (10 s pulse duration, 10 s cooling down). In these test the modules have been loaded from the rear side, i.e. with a CFC armor thickness of 6 mm (quasi stationary heat loads for heat removal measurements have been performed on the front face with a CFC armor thickness of 12 mm). During thermal cycling no degradation of the heat removal has been observed (surface temperature and thermocouple signals in the CFC block remained constant throughout the experiment.

Metallographic examination (perpendicular and parallel to the coolant tube) shows an almost intact bond layer, fig. 4). The laser structured interface

CFC / AMC-copper does not show any degradation or morphological changes. However, several larger pores (1,5 mm in diameter) have been detected in the Ti braze layer, i.e. between the AMC-copper and the DS-copper coolant tube. Since the high resolution IR images did not indicate any changes during thermal cycling, it is assumed that these defects did not develop during electron beam loading. Obviously the high thermal conductivity coolant tube represents a reliable thermal bridge.

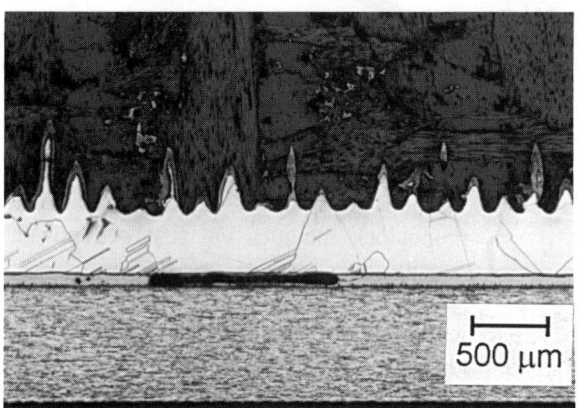

Figure 4. Metallographic examination of the braze zone (Sepcarb N31 / DS-copper) after 1000 cycles at 20 MWm^{-2}.

4. SUMMARY

To evaluate the high heat flux performance of divertor components small scale test samples have been developed and tested under ITER relevant heat loads. Miniaturization (i.e. reduction of the component's length and omission of connector tubes) has significant advantages such as reduced manufacturing cost, easy sample exchange (clamped coolant connectors), and minimum volume requirements in the irradiation rig.

Surface temperature measurements by optical methods depend strongly on the emittance ε of the plasma facing material. Careful ε-measurements are necessary (in particular for Be) to quantify the temperature dependence and the effect of surface structure.

HHF tests on beryllium divertor modules have been limited to surface temperatures $T_{surf} \approx 500°C$.

Under these conditions all tested modules show a rather uniform heat removal efficiency (mean increase of the surface temperature $\Delta T_{surf} \approx 55$ to 85 K / MWm^{-2}). Differences are mainly design-specific; the manufacturing process and the type of the braze material (INCUSIL or Cu-Mn-Sn-Ce) has no significant influence.

Monoblock divertor modules with CFC armor are characterized by a heat removal efficiency with a resulting mean $\Delta T_{surf} \approx 70$ to 110 K / MWm^{-2}. However, variations within a batch of identical modules (and even in a single CFC tile) can be significant. The thermal fatigue behavior for cycles numbers up to 1000 and incident heat fluxes of 20 MWm^{-2} is adequate.

REFERENCES

1. G. Janeschitz, K. Borrass, G. Federici, Y. Igitkhanov, A. Kukushkin, H.D. Pacher, G.W. Pacher, M. Sugihara, J. Nucl. Mater 220-222 (1995) 73-88
2. T. Huber, L. Plöchl, N. Reheis, J.P. Cocat, J. Schlosser, Proc. 16th Symposium on Fusion Engineering, 1996, to be published
3. J. Linke, M. Akiba, M. Araki, H. Bolt, G. Breitbach, R. Duwe, K. Nakamura, J.H. You Fusion Engineering and Design 28, 1995, 72-80
4. R. Duwe, W. Kühnlein, J. Linke, M. Rödig, Bericht des Forschungszentrum Jülich, JÜL-3183, 1996
5. H.D. Falter, D. Ciric, D.J. Godden, C. Ibbot, A. Celetano, JET Report JET-R (96) 02, 1996
6. J. Linke, M. Akiba, H. Bolt, G. Breitbach, R. Duwe, A. Makhankov, I. Ovchinnikov, M. Rödig, E. Wallura, Proc. 12th Int. Conf. on Plasma Surface Interactions in Controlled Fusion Devices, St. Raphaël, 20.-24.05.96
7. M. Rödig, R. Duwe, A. Gervash,. J. Linke, A. Schuster, Physica Scripta Vol. T64, 60-66, 1996
8. J. Linke., R. Duwe, A. Gervash, W. Kühnlein, K. Nakamura, A. Peacock, M. Rödig, Proc. 2nd IEA International Workshop on Beryllium Technology for Fusion, INEL Report CONF-9509218, 1995, 122-130

MANUFACTURE AND TESTING OF SMALL-SCALE MOCK-UPS FOR THE ITER DIVERTOR

G. Vieider[1], P. Chappuis[2], R. Duwe[3], R. Jakeman[4], M. Merola[1], H.D. Pacher[1], L. Plöchl[5], F. Rainer[5], M. Rödig[3], N. Reheis[5], I. Smid[1]

[1] The NET Team, D - 85748 Garching
[2] CEA Cadarache, F-13108 Saint Paul Lez Durance
[3] Forschungszentrum Jülich, D - 52425 Jülich
[4] ITER Joint Central Team, D - 85748 Garching
[5] Plansee AG, A - 6600 Reutte

This task within the EU R&D for ITER is primarily aimed at the development of solutions for the divertor target which has to be designed for up to 1000 off-normal transients at 20 MW/m^2. Representative small scale mock-ups with carbon and tungsten armour have been manufactured using a similar technology. First tests support the analysis which indicates the best high heat flux capability for carbon mono-blocks.

1. INTRODUCTION

The ITER divertor target plates for the initial Basic Performance Phase have to be developed with the following **main requirements** [1]:
- 10^4 nominal pulses at 5 MW/m^2 for up to 10^3s and 10^3 off-normal pulses at 20 MW/m^2 for 10s.
- A total associated neutron damage of 2 dpa.
- Plasma facing armour materials Carbon Fibre Composites (CFC) and tungsten (W) selected for maximum life time [2].
- High temperature Cu-alloys for the water cooled heat sink chosen in view of prospects for acceptable performance at high heat flux under neutron irradiation.
- Silver to be avoided in the joints due to the transmutation Ag → Cd which is undesirable in high vacuum systems.

The **objective** of this EU R&D task is primarily the development of such ITER divertor components which includes the following steps:
- Demonstration of the manufacturing feasibility of representative small scale mock-ups.
- Testing of more than 20 mock-ups with different materials and joining techniques for at least 10^3 pulses at up to 20 MW/m^2 as basis for selection of the prototype designs by October 1996.
- Similar testing in 1997 of mock-ups irradiated to 0.5 dpa.

These divertor solutions can also be utilised for other high heat flux components with somewhat lower heat fluxes, such as the baffle or limiter.

The mock-ups described here were all manufactured with a similar joining technique utilizing "Active Metal Casting" (AMC®) of a ductile Cu interlayer onto the armour tile as interface to the Cu-alloy heat sink [3].

2. DESIGN OF DIVERTOR TARGET AND MOCK-UPS

Fig 1 illustrates the **EU design** of the outboard divertor channel including the vertical target plates with the required high heat flux capability [4]. This target plate consists of plates like modules, each with a single water-cooled swirl tube heat sink in Cu-alloy, to which the armour is joined, i.e.:
- CFC in the lower half of the target due to its resistance to the high off-normal heat fluxes.
- W in the upper half with lower heat fluxes instead of CFC to minimise the sputtering erosion and tritium retention expected with CFC.

The typical **cross sections** of the target modules and the corresponding small scale mock-ups are also shown in Fig 1:
- CFC "mono-block" tiles brazed to the Cu heat sink tube [5].
- Flat tiles in CFC and W electron beam (EB) welded to the Cu heat sink bar, which in turn is EB welded to a stainless steel support.
- For the flat W tiles a 5 mm castellation was foreseen to reduce the thermal stresses.

The armour thickness of the mock-ups is limited to about 10 mm by the maximum temperature of 2000 and 3000°C for CFC and W, respectively, which is limited due to the armour evaporation allowed in the EB test facilities.

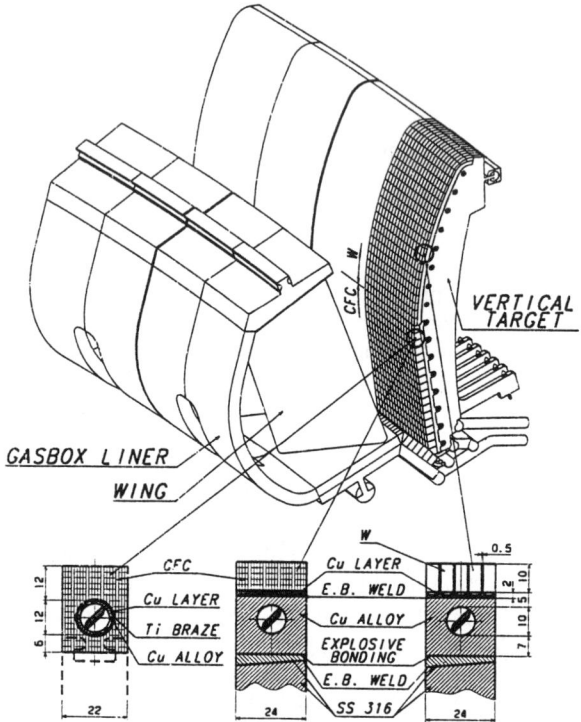

Fig. 1 Outboard divertor target with cross sections of the corresponding mock-ups with mono-block and flat tiles

3. SELECTED MATERIALS

As **CFC armour** for high heat flux components EU industry has developed several advanced materials (SEP N31 & NS31, DUNLOP Concept 1&2) with the following main features:
- High thermal conductivity (> 300 and 150 W/mK at 20 and 800°C, respectively) in one direction using pitch fibres.
- 3-D fibre structure for resistance to thermal shocks, mono-block manufacture and neutron induced swelling.
- One option with 10% Si for reduced chemical sputtering, tritium retention and reactivity with water [6].

As **tungsten armour** W - 1% La_2O_3 was selected because of easier machining, lower cost and better grain size stability compared to pure W. However, as for all W-alloys the fracture toughness is low and depends strongly on the grain direction.

For the **heat sink Cu-alloy** the prime choice was dispersion strengthened (DS) Cu "Glidcop Al25"® because it has the best resistance among the Cu-alloys to temperatures < 900°C and to neutron damage (except for fracture toughness). In addition, precipitation hardened (PH) CuCrZr (~ 0,7% Cr, 0,1% Zr) was considered as an alternative due to better weldability and ductility. However, this PH-Cu suffers from loss of yield strength and conductivity when T > 500°C for more than one hour.

4. MANUFACTURE

The **mono-block** mock-ups each with two 20 mm long CFC tiles surrounding the 12 mm diameter heat sink tube were manufactured with the following main steps - see Fig 1 and 2:
- Active Metal Casting (AMC®) of a 0,5 mm thick ductile OFHC Cu interlayer onto the CFC bore interface, which was strengthened by micro craters created by laser treatment.
- Eutective Ti-brazing (5 minutes at 900°C in a high vacuum induction furnace) of the Cu interlayer of the tile onto the Cu-alloy heat sink; in order to avoid cracks in the CFC-Cu interface due to the strong expansion mismatch, both Cu-alloys had to be softened by heat treatment.
- Testing of the resulting joint by X-ray and shear showing a shear strength of 26 - 32 MPa.

The **flat tile** mock-ups, each having both 150 mm long CFC and W armour (see Fig 1, 3 & 4), were manufactured with a basic technique that had already been successfully applied to the pumped limiter elements for TORE SUPRA, using CFC tiles [3]:
- Production of a 2 mm thick ductile Cu-interlayer via AMC on the laser structured CFC tiles and X-ray inspection of the bond.
- EB welding of the tile Cu-interlayer to the Cu-alloy heat sink and ultrasonic inspection of the weld, which showed acceptable results also for DS-Cu.
- EB welding of the heat sink with a thin explosion bonded stainless steel strip onto the massive steel support plate.
- EB welding of a Ni-adapter between the Cu-alloy heat sink and the stainless steel water pipe, which showed better results than direct Cu-steel joints by e.g. friction welding.

For the W tiles a similar process was used. However, the lack of fracture toughness in W transverse to the grain boundaries seemed to make castellation by cutting difficult. Instead, castellated tiles were produced via AMC from an array of 4.5 x 4.5 mm² rods, which in addition have better mechanical properties than larger blocks and allow reduction of W-Cu interface stresses by e.g. a chamfer on the foot. Such W tile arrays survived 20 cycles of water quenching from 600°C without delamination of the W-Cu interface.

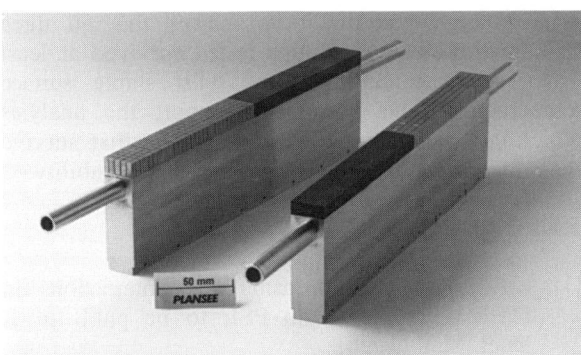

Fig. 4 Detail of the CFC and W flat tile mock-ups

5. NUMERICAL ANALYSIS

5.1. Introduction

Numerical analyses were carried out on the CFC monoblock, CFC flat tile and W flat tile mock-ups by means of the ANSYS finite element code. A 3D model was generated for each component. Thermal transient analyses were performed to simulate the thermal fatigue tests described below in section 6. The heat transfer coefficients were computed by means of the EUPITER code, a thermohydraulic software package developed by the EU Home Team.

For the thermal stress analysis, elastic calculations were carried out for conditions at the end of the heating phase. Particular attention was paid to the choice of the thermal strain reference temperature (TREF), the thermal - stress - free temperature. In the present study a sensitivity analysis was performed with TREF ranging from RT up to 650 °C.

5.2. Results

At an absorbed heat flux of 20 MW/m² the highest computed temperatures on the heated surface were 3050, 2454 and 1833 °C for the W flat tile, the CFC monoblock and the CFC flat tile, respectively.

Fig. 5 shows the maximum temperatures reached in the heat sink. It is to be noted that 500°C, the threshold for overaging to become significant is reached for the flat tile geometries for heat fluxes in excess of ≈14 MW/m².

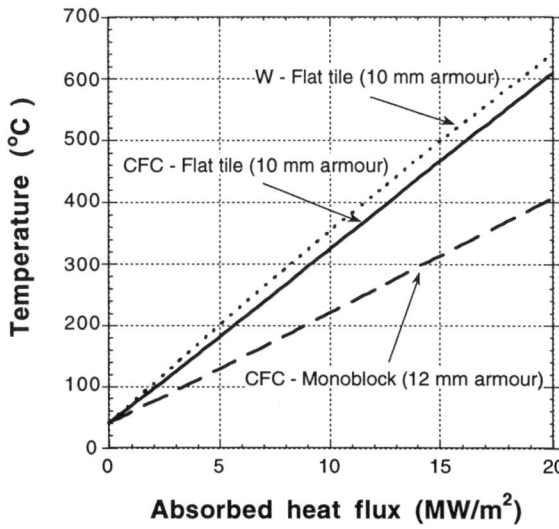

Fig. 5 - Maximum temperature in the heat sink

The thermal stress analysis was mainly aimed at evaluating the stress and strain field in the heat sink.

In fact a failure in this part of the component would be the most critical from a safety standpoint since it would result in water leakage. Table 1 gives the lifetime estimates for the DS-Cu heat sink based on the thermal strain range at 20 MW/m^2 and the DS-Cu fatigue data [7]. Table 1 also gives the manufacturing strain due to armour-heat sink joining.

Table 1
Lifetime estimates for the heat sink at 20 MW/m^2

	Manufact. strain (%)	Strain range (%) at 20MW/m^2	Number of cycles at 20 MW/m^2
CFC Monoblock	1.46	0.57	≈ 8 x 10^3
CFC Flat tile	0.14	1.67	≈ 150
W Flat tile	0.10	1.22	≈ 600

The above predictions are subject to some uncertainty for the following reasons: (1) the strain range is approximate since it was elastically computed; (2) the most highly stressed location was computed in the Cu interlayer / heat sink interface; since the joint is generally weaker than the base material one should have used the fatigue data of the joint, which is not available; (3) the annealing effect of the final joining could not be taken into account in the numerical analysis, whereas it is likely to play a role just in the Cu interlayer / heat sink interface. The above lifetime estimates should be used for comparative purpose only. They indicate that the monoblock solution is the most robust among the geometries under investigation provided the high manufacturing strains do not result in failures at manufacture.

6. HIGH HEAT FLUX (HHF) TESTS

Three **CFC mono-block** mock-ups (see Fig 2) have been HHF tested at the 60 KW EB facility JUDITH at Jülich [8] in accordance with the common test procedure agreed for all small scale mock-ups of HHF components i.e.:
- 100 cycles at 5 MW/m^2 plus 1000 cycles at 8 MW/m^2 plus 1000 cycles at 20 MW/m^2.
- 10 s heating and 10 s dwell time.
- cooling water at ~20°C, ~3.5 MPa and ~15 m/s.

All three CFC mono-block mock-ups (2 with DS-Cu, 1 with PH-Cu) survived the HHF test with stable CFC surface temperatures up to the last cycle with average values well in line with predictions.

It was noted, however, that the CFC temperatures varied across the surface by up to 500°C which could be explained by variations of the CFC thermal conductivity and by local flaws in the brazed joint. Such minor flaws were found in the destructive examination of one of the tested CFC mono-blocks [8].

Two **CFC&W flat tile mock-ups** (see Fig 3) are being HHF tested at the 200 KW EB facility FE 200 at Le Creusot. Similar CFC-CuCrZr mock-ups for TORE SUPRA sustained 1000 cycles at 15 MW/m^2 but started to fail after a few more cycles at 18 MW/m^2 [3].

7. CONCLUSIONS

Manufacturing feasibility has been demonstrated for small scale mock-ups of advanced high heat flux components having CFC and W armour bonded to Glidcop and CuCrZr heat sinks.

First high heat flux tests showed that all three CFC mono-block mock-ups tested survived at least 1000 cycles at 20 MW/m^2 with stable surface temperatures. This seems to support the analysis which indicates that CFC mono-blocks that survive manufacture have the best high heat flux capability.

REFERENCES

[1] R. Parker et al, "Plasma-Wall Interactions in ITER", Proc. 12th on PSI, to be publ. in J. Nucl. Mat. (1996)

[2] H.D. Pacher et al, "Erosion lifetime of ITER divertor plates", ibid

[3] J. Schlosser et al, "Developments for an Actively Cooled Toroidal Pump Limiter for Tore Supra", this conference

[4] A. Pizzuto et al, "The Outline Design of the ITER Divertor Outboard Vertical Target Dump Plate and Wing", this Conference

[5] G. Vieider et al, "ITER plasma facing components - design and development", Proc. of 2. ISFNT, Karlsruhe, 1991, Fusion Eng. Design (1991)

[6] C.H. Wu et al, "Evaluation of an advanced silicon doped CFC", this conference

[7] J. Stubbins et al., "High temperature fatigue testing of CuNiBe, CuCrZr and CuAl25 at Univ. of Illinois", Private Communication (1996)

[8] J. Linke et al, "High heat flux performance of divertor modules for ITER", this conference

ENGINEERING DESIGN OF THE ITER DIVERTOR

R.Tivey[1], M.Akiba[2], A.Antipenkov[1], S.Chiocchio[1], D.Driemeyer[3],
R.Jakeman[1], G.Janeschitz[1], I.Mazul[4], M.Ulrickson[5], G.Vieider[6]

[1]ITER Joint Work Site, Boltzmannstr..2, 85748 Garching, Germany; [2] Japan Atomic Energy Research Institute, Naka-machi,Ibaraki-ken 311-01 Japan, [3] McDonnell Douglas Aerospace, PO Box 516, St Louis, MO 63166-0516, USA; [4]D.V. Efremov Research Institute, St. Petersburg, Russia; [5] Sandia Nat Lab, Albuquerque, USA; [6] NET, Boltzmannstr. 2, 85748 Garching, Germany.

This paper reports on the status of the mechanical and thermo-hydraulic design of the divertor. It describes the current design of the divertor plasma facing components (PFCs), and the steel structure (cassette body) that supports these PFCs.

1 DIVERTOR ASSEMBLY

A divertor assembly comprises 60 cassettes (each 5 m long, ~2 m high and 0.5 -1.0 m wide, and weight 25 tons) installed on toroidal rails [1]. Each cassette assembly consists of a cassette body and of the following components demountable in the Hot Cell :
i. a vertical target (VT) (intercepting the plasma outboard of the separatrix) with a water cooled > 100mm thick stainless steel body on which a water cooled copper heat sink is mounted. Cladding is tungsten (top part) and CFC (lower part).
ii. a tungsten clad water cooled copper louver or wing and a gas box liner (in the private region) to protect the SS cassette body.
iii. a tungsten or beryllium clad dome (located below the X-point) provides a baffle to prevent neutrals from entering the main plasma at the X-point and protects the wings in the private region from direct interaction with the plasma.

An important feature of the design is the modular nature of the components, allowing the exchange of fully tested armoured PFCs on a cassette body

1.1 The Cassette Body

The cassette body is a rigid stainless steel structure cooled by water. Internal cooling channels in the cassette body are optimised to reduce temperature rise and thermal stress due to nuclear heating, to keep mechanical stresses within allowable limits for internal pressure and electromagnetic loads (in particular those transmitted imposed by the PFCs [2]) to distribute required water flow, to drain the cooling water as quickly as possible and to eliminate local air pockets. The body will be constructed using either cast/HIP or powder/HIP technology.

1.2 Plasma Facing Components

The inner and outer channels each consist of a VT, a wing and a liner protecting the cassette body in the private region from high heat fluxes. The PFCs in each channel are fabricated, tested and installed on the cassette body as two separate sub-assemblies, these are: a VT and short dump target; and a wing and gas box liner.

The divertor is designed to exhaust a maximum of 300 MW of heat flux conducted to the surface of the PFCs plus a further 100 MW of neutron bulk heating. Uncertainties in the distribution of the 300 MW mean the inner and the outer channels of the divertor are designed for maximum loads of 150MW and 200MW respectively. In general the PFCs are designed for < 5 MW/m^2, however the lower part of the vertical targets at the ends of the

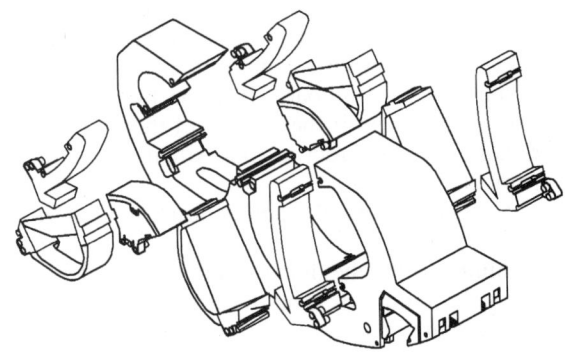

Fig.1   Exploded View of Cassette and PFCs

two divertor channels have to accept almost all the power conducted into the SOL when the formation of a gas target fails or is delayed, (up to 1000 events of 20 MW/m^2 for typically 10 sec are anticipated for the Basic Performance Phase) and the energy loads during disruptions (~ 100 MJ/m^2 in 0.1 to 3 ms) and ELMs (~10 MJ/m^2). In all these abnormal operation conditions, the CFC cladding material will evaporate. A CFC monoblock or saddleblock on a Cu tube is preferred for lower end of the VT and W clad Cu monoblock for the upper end [3,4,5].

The dome consists of a W or Be clad PFC of the hypervapotron design mounted on a stainless steel dome block, and which in turn is attached to the cassette body. This design allows the separate fabrication and testing of the PFC and for initial operation, with low neutron fluence, allows for the exchange of the PFC and reuse of the dome block.

There is a large R&D effort worldwide aimed at developing and assessing the reliability of the ~10^6 armour to heat sink joints in a divertor [6].

1.3 Alignment Of PFCs

The SOL strikes the targets at a glancing angle (as low as 1°). Targets on adjacent cassettes must be aligned accurately with respect to one another and angled slightly so as to shield the leading edges of the adjacent components from direct incidence of the field lines in the SOL. It is necessary to keep the tilting of the target in this way small since it contributes to the increase in thermal peaking factor on the already highly loaded dump target and dome. In computing the peaking factor account is taken of the flux angle, the toroidal gap between adjacent dump or dome targets, poloidal misalignment between adjacent cassettes (manufacturing and assembly tolerance), and allowance is made for thermal distortions. The contribution of field ripple to the peaking factor is considered to be negligable on the inboard vertical target and about 1.05 on the outboard target. An installed maximum step in the poloidal direction between adjacent targets of 4 mm is reasonable giving peaking factors of 1.35 outboard and 1.4 inboard. Peaking factors become unacceptably high if the surface of the target plates are not faceted in order to approximate the radius of the divertor.

In general, normal manufacturing tolerances are sufficient for the cassette body, but precision machining of the the dome and dump target support points is required. The tolerances of adjacent cassettes are met by individually machining the support shoes (which interface with the toroidal rails) according to as-built survey of the rails.

1.4 PFC Attachment On Cassette

PFCs on each toroidal half of a cassette are combined into five units; the dome: the inboard vertical targets, wing and gas box liner; the outboard vertical targets, wing and gas box liner (Fig.1). The units from each toroidal half are mounted onto the cassette body toroidally from opposite sides of the cassette. Tight positioning and alignment of the unit is achieved by measurement and machining accuracy of the registers. Several options are foreseen to be tested for the attachments.

Two types of keys are foreseen: one that will pull a reference surface on the PFC hard against reference surface on the cassette body (restraining the PFC against rotation as well as translation radially and vertically); and a second type in the form of a "dumbbell" that allows rotation in a poloidal plane and a limited linear expansion of the PFCs. To transmit the large EM induced tortional forces from the PFCs to the cassette (of particular concern for the dome block to body attachment) rectangular tongues on the PFCs are inserted in recesses in the cassette body and located with separate keys. The contact through the keys or the surface contact between reference surfaces should be adequate to carry the halo current without using separate earth straps.

The cooling pipes of the PFCs are joined to the cassette body in a region of relatively low neutron fluence where the He production will be less than 0.5 appm allowing the cassette body to be re-used for the entire life of ITER. To ensure that conventional welding tools and inspection methods can be used and to minimise voids in the body the pipes are welded from inside through a hole in the water supply appendix of the PFC, the hole for the tool's entrance then has to be closed with a welded lid. Alignment of the butts to avoid mismatch is achieved by careful survey of the stub remaining on the cassette body and machining an over-length and over-thick stub on the new replacement PFC to suit.

2 THERMALHYDRAULICS

The overall coolant parameters adopted for the Heat Transfer System (HTS) dedicated to the divertor are defined as: 140 °C inlet temperature, 1 m^3/s flow rate in each of four separate circuits, 4 MPa inlet pressure, and 1.5 MPa pressure drop in-vessel through the divertor. The CHF limits of the candidate PFC heat sink designs are in line with these parameters and have been reported [7] for both swirl tapes, and hypervapotrons as > 25MW/m^2 for a flat power profile and $\geq$ 35 MW/m^2. for the highly peaked flux profiles that will be incident on the vertical targets. The mechanical properties of copper alloys under neutron irradiation are very sensitive to the irradiation temperature [8] and limited operating temperature windows are used for both candidate copper alloys for the heat sink designs. In the case of Dispersion Strengthened Cu 150-350 °C and for CuCrZr 150-250 °C (with excursions up to 350 °C during transients). On this basis analysis shows that a flat tile design, with Cu monoblock or hypervapotron, would be one of the suitable options for the regions subjected to the < 5MW/m^2, however the CFC monoblock or saddleblock designs employing swirl tapes are suitable for areas receiving up to 20 MW/m^2 at the SOL strike point.

Since there are large uncertainties in the power distribution in the divertor, several plasma facing components are cooled in series before the coolant is routed to the cassette body to remove the nuclear heat. This maximises the temperature rise in the coolant and minimises the mass flow (and minimises the cost of the PHTS).

The following circuit cools the PFCs (fig2):
• Coolant is fed to each cassette in parallel. Fifteen cassettes are attached to each PHTS loop. The coolant is routed along radial straight feed and return lines. Three cassettes are fed through each port, 6 pipes per port (3 feed and 3 return).

On entering the cassette, coolant is split in two, half to feed each toroidal half of the cassette.
• Coolant is routed through the cassette body to cool in parallel the inner and outer vertical target assemblies. The vertical target and the dump target are fed in series (~19 l/s for each outer channel and ~14 l/s for each inner channel target): firstly to the regions of the vertical targets intercepting the highest heat flux so as to have the maximum sub-cooling in these areas: secondly to the short dump target which only receives this high flux during transient excursions of β and l_i, when the SOL is swept rapidly across its surface and the thermal load is absorbed inertially by the CFC and Cu mitigating the thermal-hydraulic requirements of the short dump.
• The coolant is then fed via the cassette body to cool three sub-circuits in parallel. These are: the wing and liner assemblies of the inner and outer channels, and the dome assembly. Coolant enters the wing foot (incident flux 3 MW/m^2) and then passes along the length of the wing to the upper foot before returning via the gas box liner to the lower foot and then to the cassette body. The wing cooling is using a series of 6 mm diameter holes. A separate annular flow along the wing nose is under consideration as an alternative. Coolant to the dome is used to cool the dome block which, because of the high neutron heating in this region, has closely packed coolant channels, and to cool the dome PFC, a hypervaptron structure.
• The coolant is then recombined in the cassette body before leaving the vessel.

3 GAS SEALS

Electrically insulated gas seals are located between the outermost/innermost upper points of a divertor cassette and the vacuum-vessel wall. They are designed to provide a high impedance barrier between the divertor pumping area and the plasma main chamber: leakage < 5 Pa·m^3·s^{-1} or 1.35·10^{21} particles/s (equivalent to <10% of the fueling rate and ≤10% of pumping speed), with a factor 10^4 divertor and main chamber differential pressure and a pressure of 1 Pa in the divertor area.

The sealing consists of two components: a permanent membrane ("shelf"), cantilevered from the vacuum vessel, and a detachable seal connection attached to the cassette.

A rigid cantilever is necessary: to precisely position (< ±2 mm) the mating sealing surfaces;

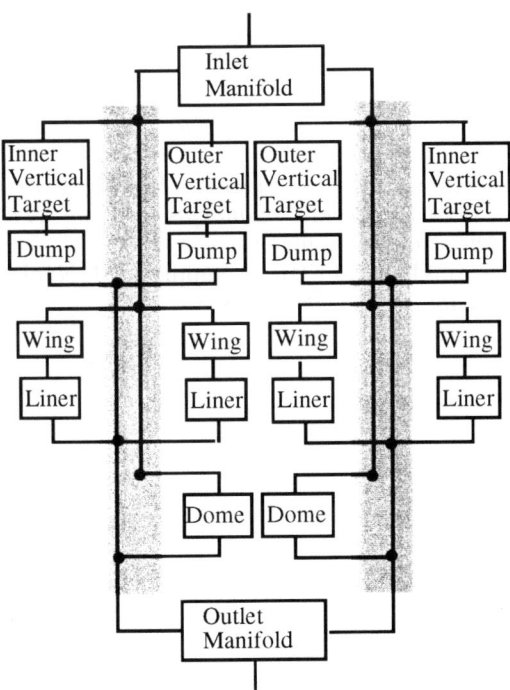

Fig.2   Cassette Coolant Routing Schematic

withstand the JxB forces should there be an electrical breakdown allowing a radial current of up to 60 kA to flow through the shelf; sustain the weight of water should it fill all the space between the blanket back plate and the vacuum vessel. The inner seal uses the blanket support, so only a detachable connection is provided on the cassette.

The detachable sealing comprises a copper ∏-profile bar with ceramic coating to prevent electrical contact. The bar slides inside an inverted bar of a similar profile and makes contact to it using copper spring plates welded to the bars. The spring presses the bar to the cantilever. The assembly is inserted into the cassette body and welded along it's edges.

Similar sealing can be incorporated between the cassettes, closing the 10±5 mm installation gaps in three regions (outboard, inboard and the dome). The bars are pressed against the side surface of the adjacent cassette. The dome seal may require the seal bar to be made of refractory alloy (W or Mo-Re) because of higher nuclear heating and additional thermal load through the gap.

4 DIVERTOR LEAK LOCALISATION

It is proposed to use chemical markers in the 4 divertor PHTS loops so that the source of a water leak in the divertor can be isolated to a quadrant. The following procedure is considered which allows the in-vessel localisation of a leak to a cassette(s) and which minimises the number of valves in the PHTS:

i. All the cassettes in a quadrant will be drained simultaneously .
ii. A branch connection ex-vessel on the 150 mm radial feed pipe to a cassette will be cut
iii. A temporary connection will be made between a gas feed line and the cassette feed. One leg of the gas feed system will be available adjacent to the coolant pipes at each port. The connection shall be made inside the pipe such that the gas fed into the cassette passes through the cassette purging the PFCs and driving any water remaining after draining into the sump at the base of the cassette body. The gas circuit is sized for draining and drying one cassette at a time (25 kW pump and 50 KW heater).
The return gas flow will be redirected via the water return line to the gas condenser/driers/ separators of the gas circuit by closing the valves which allow the HTS HX to be by-passed.
iv. When the cassette is fully dried, then the radial return line will be cut at the branch. Plugs will be inserted inside the pipes employing elastomer 'O' ring seals or gas balloons in order to seal the pipes prior to evacuating the cassette or pressurising it with helium. Gas and vacuum lines at each port allow internal or external leak tests to be performed.
v. After leak testing, or possibly soon as having dried a cassette, a second cassette will have its' feed line opened and the procedure repeated.

5 CONCLUSION

The ITER divertor programme is on schedule to demonstrate technical feasibility by the end of the EDA and to provide design input for the construction of ITER.

REFERENCES

1. E.Martin et al., *Design of the ITER Remote Handling System* This conference.
2. S.Chiocchio et al *Loads on the ITER In-Vessel Components from Electromagnetic Transients* This conference
3. A.Pizzuto et al, *The Outline Design of the ITER Divertor Outboard Vertical Target Dump Plate and Wing* This conference
4. G.Vieider et al, *Manufacture and Testing of the Small Scale Mock-ups for the ITER Divertor* This conference..
5. S. Suzuki et al., *High Heat Flux Experiments on a Saddle-shaped Divertor Mock-up*, IAEA, Montreal, Canada (1996).
6. M.Ulrickson et al *Development of a Full-Size Divertor Prototype for ITER* IAEA Montreal (1996)
7. J.Schlosser et al. *Comparison between Various Thermalhydraulic Concepts for the ITER Divertor* This conference.
8. S.Zinkle,S.Fabritsiev, Suppl. to J Nuclear Fusion, vol5, (1994) 163.

Manufacturing Feasibility Study for the ITER Divertor Vertical Target and Wing

G. Cai, M. Grattarola, G.B. Ottonello and F. Zacchia

Ansaldo Ricerche S.r.L, Corso Perrone 25, 16161 Genova, Italy

In the frame of the stage 1 of the ITER Task T232.6 on Engineering Tests, Divertor Baffle and Cassette / Vertical Target and Wings Ansaldo Ricerche, as EFET member, has collaborated with ENEA to the development of the Divertor outline design. The objective of the activity has been the definition of the concept feasibility through the analysis of the manufacturing process and the production of the design and the manufacturing specifications for the mock-ups of the outboard channel high flux components.

1. INTRODUCTION

In the frame of the stage 1 of the ITER Task T232.6 on Engineering Tests, Divertor Baffle and Cassette / Vertical Target and Wings Ansaldo Ricerche, as EFET member, has collaborated with ENEA to the development of the Divertor outline design. The present reference concept for the ITER Divertor is the so called "Vertical Target Option" composed of the Vertical Target on the outside of the Divertor channels, the Dump Target on the bottom and the Wings (transparent walls) on the inside. The objective of the activity has been the definition of the concept feasibility through the analysis of the manufacturing process in terms of suitable technologies and possibility to implement it on a sound and economical industrial basis.

First, Ansaldo Ricerche provided, for the developed design options, the information needed to demonstrate the feasibility of the proposed design within reasonable time and cost constrains. A particular attention has been paid to relevant crucial aspects including:
- the use of CuCrZr or DS Copper heat sink alloys;
- the selection of the most appropriate technologies for joining Carbon Fibre Composites (CFC) and Tungsten armour tiles to the copper alloy heat sink;
- the application of Tungsten coating by plasma spray to the wing body;
- the application and qualification of the Non Destructive Examinations (NDE) techniques for the various connections;
- the application and qualification of techniques for joining the copper heat sink to the AISI 316 supporting structure and cooling system.

Based on the above mentioned activity, the detailed design and manufacturing specifications for medium and full-scale mock-ups of the outboard channel high flux components have been produced. One of the main purposes of the small and medium scale mock-ups is to test and qualify all the relevant technological solutions and manufacturing sequences proposed for the full-scale mock-up.

1.1. Reference design and critical aspects

The ITER Divertor reference conceptual design is shown in fig. 1 [1]. It is composed by the Vertical Target, the Dump Target and the Wing. The Vertical target consists of copper alloy armoured cooling channels supported by a strong stainless steel back structure. The armour material is Carbon Fibre in monoblock configuration in the lower zone of the cooling channels while in the upper part is tungsten in flat tile configuration. The Dump target is completely armoured with CFC in monoblock configuration. The Wing is composed by a foot and a body in copper alloy, with machined cooling channels, coated with a 5 mm thick layer of tungsten.

A first review of the design has pointed out a list of critical aspects for which the availability and/or the applicability of suitable existing industrial technologies are not well defined and proved.

The main relevant and crucial technological aspects are summarized as follows:
- the joining of the armour material on copper alloys having mechanical properties strongly dependent on the thermal cycling;
- the presence of different armour materials on the vertical target and consequently the necessity to adopt different joining techniques on the same component;
- the 5 mm thick tungsten coating of the wing by plasma spray;
- the connection between the copper alloys and the stainless steel structures.

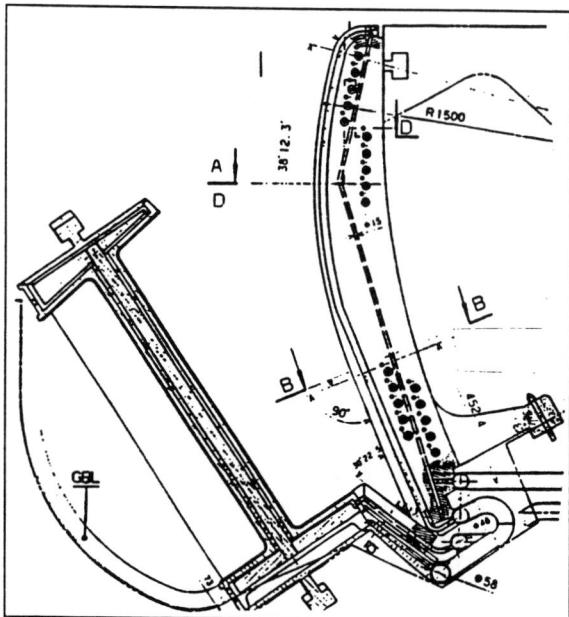

Figure 1. ITER Divertor reference design.

2. BRAZING OF THE ARMOUR MATERIALS

The lower part of the Vertical Target and the whole Dump Target are armoured with CFC in monoblock configuration. In the monoblock version the cooling tube is housed and brazed in the hole drilled in the CFC tiles.

On the upper part of the Vertical Target the tungsten tiles could be bonded in flat geometry to a thick copper block which will be brazed, in monoblock configuration, to the cooling tube.

2.1. Brazing thermal cycle for DS-Cu and CuCrZr

From the thermal point of view Glidcop is fit for brazing due to its excellent resistance to softening even after exposure to temperature close to the melting point of copper. A thick (0.5 ÷ 1.0 mm) soft Cu compliant layer, between Cu and CFC, is recommended to adsorb the differential thermal expansions between Cu and armour materials. The compliant layer should be an interlayer ductile and soft enough to stop crack propagation and to withstand the thermal deformations without building up high stresses during the brazing process.

CuCrZr reaches the optimum in strength after a complex thermal-mechanical treatment involving a solution annealing at high temperature (over 920 °C), quenching at room temperature, cold working and finally ageing at intermediate temperature (450 °C). The material is very sensitive to the thermal cycling and the brazing process strongly softens the alloy. Consequently the brazing thermal cycle for CuCrZr should foresees the brazing at the solution annealing temperature, a cooling at a rate higher than 30 °C / minute followed by ageing at 450 °C for 5÷6 hours. The thermal treatment can restore only the 50 ÷ 70 % of the original cold worked and aged alloy. Also in this case a soft Cu compliant layer could be necessary to accommodate the differential thermal expansions during the cooling after the final ageing.

2.2. Brazing procedure of the High Flux Components

Brazeability of carbons/graphite depends on the porosity and size of pores. The wetting characteristics of all the carbons and graphite are strongly influenced by the surface morphology and impurities such oxygen or water, that are either adsorbed on the surface or in the bulk material. Outgassing prior to brazing is highly helpful. In order to overcame the wetting difficulties of graphite, it could be coated with a more readily wettable layer. When brazing CFC to a Cu alloy, the soft Cu compliant layer can resolve the wetting problems too.

In the monoblock version the cooling tube is brazed in the hole drilled in the CFC tiles. The brazing procedure can be divided in two steps. First the compliant layer can be brazed into the holes of the tiles in form of Cu rings and then the tile can be bonded to the cooling tube by means of a Cu/Cu brazing.

The tungsten tiles are bonded in flat geometry to a thick copper block which will be brazed, in monoblock configuration, to the cooling tube. The main problem in development of a reliable W/Cu junction is the large difference in coefficients thermal expansion. The reliability of the joints depends on the geometry and thickness of the armour material. Also in this case a compliant layer is necessary to accommodate the residual thermal stress. A promising solution seems to be the introduction of a compliant layer, deposed by plasma spray, having a variable composition from 100% Cu to 100% W.

The CFC tiles, equipped with the internal compliant layer, and the W tiles, bonded to the Cu blocks, can be joined to the cooling tubes by means of a Cu/Cu brazing in a single thermal cycle. In case of CuCrZr tubes a very quick cooling from the brazing temperature to the ageing temperature (400 ÷ 450 °C) is required to avoid excessive softening of the material. A cooling circuit inside the furnace is foreseen to allow a forced cooling by means of inert gas or molten salts. The cooling down must be compatible with the capability of the structure to withstand thermal shocks. A final ultra sound examination will inspects the copper/copper joints from inside the tubes.

3. TUNGSTEN COATING OF THE WING

Presently an R&D programme is in progress in Ansaldo Ricerche to verify the possibility to obtain a reliable 5 mm thick coating by plasma spray. Due to the large thermal expansion coefficient mismatch between copper and tungsten, the interlayer between the two materials and the deposition temperature have a crucial importance in the process. Also in this case a compliant layer having a composition variable from 100% Cu to 100% W can be introduced. At the moment the possibility optimize the deposition temperature in order to minimize the thermal stress both at room and operation temperature is under study.

4. CONNECTIONS BETWEEN COPPER ALLOYS AND AISI 316 SS

All the cooling circuits of the divertor high heat flux components have to be connected to the main stainless steel cooling loop. In particular, on the upper section of the vertical target, the Cu armoured tubes are connected to the corresponding cooling channels drilled in the back supporting structure. The lower ends of the armoured tubes are connected to the rear vertical target manifold. In both cases there is a transition copper-stainless steel where brazed joints in direct contact with the cooling water must be avoided. The technological solution for the heterogeneous junction can be different according to the particular type of copper alloy.

4.1. CuCrZr/SS connections.

The two main problems to be taken into consideration in such a particular type of heterogeneous junction are the metallurgical compatibility between the two materials and the softening of the CuCrZr after welding or exposure to temperature higher than 450 °C. An R&D activity on welding by means of TIG and laser processes is in progress.

High heat concentration welding processes like laser thermally affects a very narrow copper zone and consequently are suitable to minimize copper softening. In the case of TIG welding, very good joints have been produced but a larger softened area has been obtained in comparison with the laser process. Regarding the NDT methods, X-rays have been successfully utilized on welded samples obtaining a 1% defect sensitivity on 10 mm thick specimens.

The welding can be fully automated from outside (orbital welding) or inside (internal bore welding) by means of commercial welding machines.

An alternative to laser and TIG welding could be the friction welding process. The friction welding between stainless steel and copper is a standard process but additional R&D should be performed on the particular material CuCrZr. The main advantage of such a method is the very narrow heat affected zone. The process requires a relative rotation and high contact pressure between the two surfaces to be welded so it can be applied only to structure easy to handle and grip. Friction welding can be utilized only for sub components and not for the final assembly of the divertor.

4.2. Glidcop/SS connections.

When conventional welding methods are used, the aluminium oxide separates from the molten copper matrix and segregates at the boundary of the resolidified grains. This creates a brittle joint which has also a lower mechanical resistance due to the aluminium oxide depleted solidification zone. For this reason friction welding seems to be the best welding process for joining DS-Copper; in facts it is a completely mechanical solid-phase process in which the melting of the material does not occur. Literature reports that tests performed on this kind of welding have shown a good compatibility between Glidcop and AISI 316 [2].

5. THE DIVERTOR MOCK-UPS

Based on the industrial feasibility analysis, the design and the relative manufacturing processes for medium and full-scale mock-ups of the outboard channel high flux components have been proposed.

The main purpose of the small and medium scale mock-ups is to test and qualify all the relevant technological solutions described in this paper and the manufacturing sequences proposed for the full-scale component. Fig. 2 shows the Vertical Target medium scale mock-up as an example.

In particular the proposed manufacturing process includes: the technological solutions, the applicable standards for the construction and non destructive examination of the component, the R&D phases required to set up and qualify special manufacturing processes.

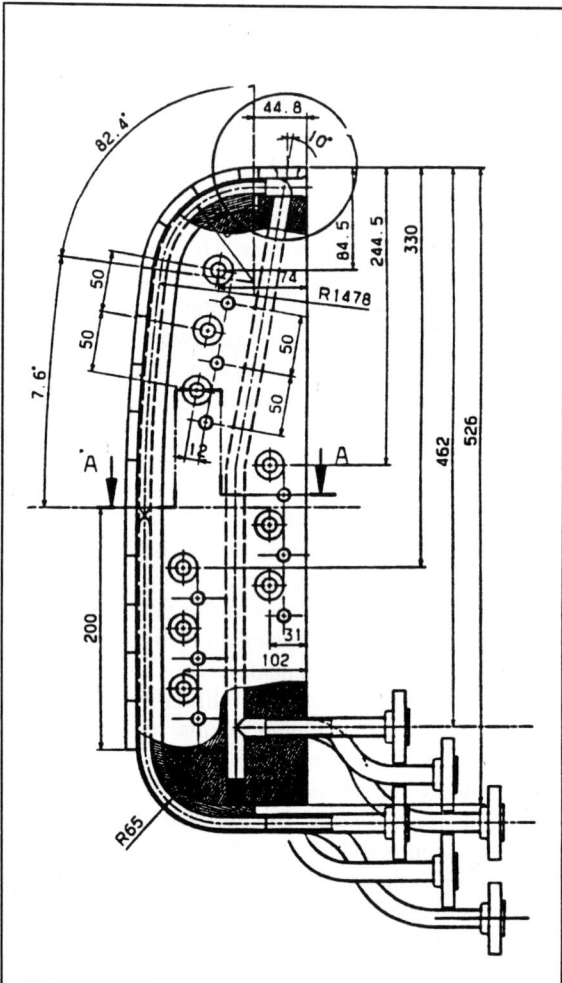

Figure 2. Vertical Target medium scale mock-up

6. ACKNOWLEDGEMENTS

The authors wish to acknowledge the following firms for the their helpful collaboration: LMI - Europa Metalli, the Italian Institute of Welding, RTM, Flametal Coating Service, Ansaldo Energia, CSC, ENEA Faenza and CSM.

REFERENCES

[1] M. Baldelli, G. Brolatti, A. Capriccioli, E. Di Pietro, M. Gasparotto, F. Lucca, A. Pizzuto, A. Palmieri, B. Riccardi, and M. Roccella, "The Outline Design of the ITER Divertor Outboard Vertical Target Dump Plate and Wing" ENEA Frascati.

[2] K. Tsuchiya, H. Kawamura and M. Saito, Fusion Technology 18 (1994) 447 - 450.

Thermal and Structural Analysis of the ITER Divertor Outboard Channel

B. Riccardi, A. Pizzuto, A. Palmieri.

Associazione EURATOM-ENEA sulla Fusione, CR Frascati, C.P. 65 - 00044 Frascati, Rome, Italy

The thermal structural assessment of the most critical components of the ITER divertor outboard channel has a fundamental importance to validate the proposed design solution which provides the flat tile-monoblock option for the Vertical Target (VT) and a structure composed of an hollow blade with upper and lower feet for the wing. The reference thermal loads taken in to account for the VT are 5 MW/m^2 (Steady State) and 20 MW/m^2 (transient) incident heat flux acting on the tiles. The analyses performed were intended to investigate the stress distribution in Steady State and to provide a preliminary assessment of the fatigue lifetime of the heat sink in transient conditions. The wing component has been fully modelled with an extensive Finite Element (FE) mesh using shell elements. Several load cases have been considered : the water pressure, the eddy current and the halo current. The main results obtained will be illustrated in the paper.

1. INTRODUCTION

The european design [1] proposed for the ITER divertor outboard channel provides for VT two different High Heat Flux Components (HHFC) solutions: Carbon Fiber Composite(CFC) monoblock in the region were the highest thermal loads are expected and W flat tile in the region below the Buffle. Both the design concepts are assembled on a Stainless Steel plate that could be considered self standing with respect thermal loads. The wing component consists of a 60 mm thick, 1 m long Copper Chromium Zirconium (CuCrZr) blade with drilled cooling ducts. This blade, armoured by W, is connected to a lower and an upper plenum called 'feet' for coolant path needs.

The present paper concerns the temperature and stress distribution of the VT HHFC under normal thermal loads, the comparison between monoblock and flat tile concepts with respect thermal fatigue under slow transient loads and finally the structural analysis of the wing.

2. VERTICAL TARGET ANALYSES

2.1 Geometries, materials and loads

The basic geometry of the VT provides, in the CFC-armoured region, an heat sink consisting of 10 mm diam. 1.5 mm thick copper alloy tube : this tube is inserted and brazed in a CFC-monoblock tile which has a minimum thickness against heat flux of 30 mm. To enhance the thermohydraulic efficiency a 2 mm thick swirl tape turbulence promoter is inserted in the tube (fig.1). The W-armoured heat sink consists of a rectangular hollow copper bar with 15 mm thick flat W tiles. The hollow is 10 mm

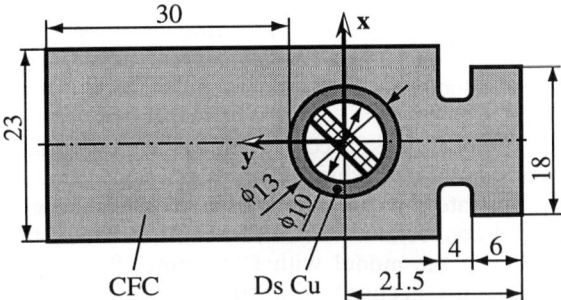

Figure 1. VT Monoblock concept section.

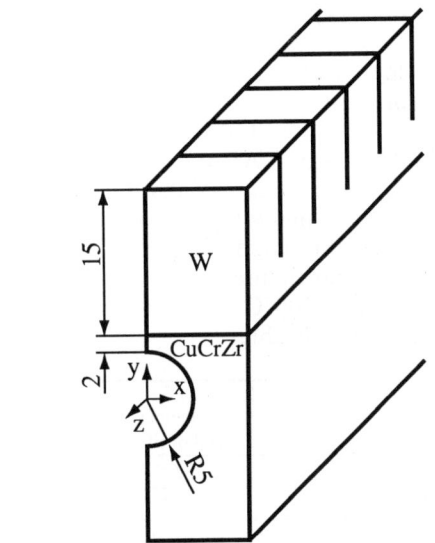

Figure 2. VT flat tile concept isometric view

diameter without swirl tape and lies 2 mm below the interface with the tile (fig.2).

The selected armour materials are: Tungsten with 1% La$_2$O$_3$ upset forged and the 3 dimensional CFC SEP N31 (EU). For the CFC material unirradiated and 0.2 dpa irradiated thermal conductivity values were assumed [2], [3].

The material selected as heat sink are: DS copper 'ITER grade 0' (0.5% Al$_2$O$_3$) for the CFC monoblock and CuCrZr (0.65%Cr-0.08%Zr) from Kabelmetal (EU) for the W flat tile concept.
Steady State thermomechanical analysis has been performed with an incident heat flux of 5 MW/m^2 and a neutron heat generation of 5 MW/m^3 for heat sink and W and 1 MW/m^3 for CFC.
Finally a transient thermomechanical analysis with 20 MW/m^2 incident heat flux acting for 10 s has been carried out.

2.2 Steady State thermal analysis

A 2D model has been used for the thermal Finite Element calculations assuming a water sink temperature of 150°C; water pressure of 4.0 Mpa; water velocity of 12 m/s and a swirl tape twist ratio of 2.

The heat transfer coefficient has been calculated by a subroutine during the FE calculations using the Sieder-Tate correlation being the heat exchange by forced convection. The influence of the swirl tape (only for the model with CFC armour) has been taken into account introducing the hydraulic diameter of the duct between the tape and the tube surface and increasing the water velocity to handle the swirl effect [4]. With this assumption for the swirl tape tube the correlation leads to an heat exchange coefficient ranging from 114 to 122 kW/m^{2}K for 150 and 217 °C of tube wall temperature. The non linear analysis has been carried out by means of ABAQUS code. Two CFC thicknesses have been considered: 30mm unirradiated and 20 mm 0.2 dpa irradiated (material supposed to be at the end of life) [5].
The maximum temperatures (°C), that for the CFC remain under the 1500°C recommended limit, are listed in the following table

	CFC monoblock unirradiated	CFC monoblock end of life	W flat tile
Armour	1228	1265	887
Heat sink	252	252	296

An estimation of the FE heat flux peaking factor on the cooled wall under 5 MW/m^2 has lead to 1.49 with unirradiated CFC and 1.62 for irradiated CFC.

2.3 Stress analysis in normal operations

The basic assumptions of this analysis are to have considered the structure stress free at 150°C, the use of unirradiated properties both for armour and heat sink. Furthermore, no braze layer has been considered between armour and heat sink.

The CFC monoblock configuration have been analysed by means of a generalised plane strain model which allows the thermal expansion along the path of the dove tail groove (fig 1). Orthotropic properties of the CFC (E,ν,G,α) have been considered. The stress results are summarised as follows:

	σ (MPa)	σ$_x$	σ$_y$	σ$_z$	τ$_{xy}$	Mises
DS Copper	max	41	14	97	46	125
	min	-132	-68	-155	-4	
CFC	max	17	49	7	20	58
	min	-14	-20	-7	-22	

It comes out that the copper alloy stresses are under the yield limit (about 285 Mpa at 250 °C) and CFC exhibits a moderate stress localised mainly at the copper interface.

The W flat tile configuration have been analysed by means of a 3 dimensional model in order to take in to account stress concentration at the end of armour tiles due to edge effects. Since originally the configuration analysed provided a rigid connection (welding) to the Stainless Steel backplate; this backplate was also modelled to take into account his influence on the stress distribution. The W tile axial length considered was 50 mm and a 10 mm deep castellation was also taken in to account using orthotropic elastic properties. The results show that for the CuCrZr the maximum Mises stress reaches value close to the yield in the region at the end of the W tile (350 Mpa at 270°C) in the section at the end of tile and reduces to 250 Mpa in a section at the middle of the tile; the Mises in the W are close to the allowable stress at the interface with the heat sink. The max shear at the W-Copper interface is 25 Mpa.

2.4 Thermal fatigue assessment

The main reason of the mixed configuration (flat tile- monoblock) in the current design of the VT is the limited lifetime of the CFC-flat tile concept with respect to the thermal fatigue of the copper heat sink due to the high transient loads. For this reason, an investigation on the thermal fatigue lifetime of CFC monoblock and CFC flat tile solutions was made during the design activities.

Both the concepts were analysed using a generalised plane strain model considering CuCrZr as heat sink material. A transient non linear thermomechanical analysis was carried out imposing 20 MW/m^2 incident heat flux acting for 10 s: the impinging thermal power reduction due to black body radiation and evaporation of the CFC as a function of the surface temperature was also taken into account [5]. The heat exchange coefficient, as function of the wall temperature, has been estimated by Sieder-Tate, Bergles-Rohsenow and Thom relation[4]. The parameters of the CFC flat tile concept remained unchanged (10 mm diam. of the coolant passage, 23 mm pitch, 2 mm average thickness of the swirl tape but 4 mm of minimum copper thickness versus plasma, inlet water temperature equal to 150°C and 12 m/s water velocity, CFC unirradiated with 30 mm thickness). A conservative fatigue lifetime estimation was made based on recent test campaign results [6] and the computed equivalent elastic plastic strain were purged from edge singularity effects.

The following table summarises the results obtained for the CFC flat tile concept:

Absorbed heat flux	Tmax CFC (°C)	Tmax Cu (°C)	Cu equiv strain(%)	Fatigue life(cycles)
20 MW/m²	3882	402	0.53	≈5000
20 MW/m² -radiat-evap	3006	383	0.48	5000

Nevertheless, the estimation of the fatigue life with CFC irradiated and reduced in thickness (end of life) leads to a lifetime of few hundreds of cycles. For the above results the CFC flat tile concepts seems marginal with respect the ITER design requirements [7].

The results of the same analysis performed for the CFC monoblock concept have shown that no plasticity occur and the thermal fatigue lifetime accomplish the design requirements. The following table summarises the results obtained.

Absorbed heat flux	Tmax CFC (°C)	Tmax Cu (°C)	Cu equiv strain(%)	Fatigue life(cycles)
20 MW/m²	3874	304	0.18	>10⁵
20 MW/m²-radiat-evap	3003	291	0.16	>10⁵

3. WING STRUCTURAL ANALYSIS

A linear elastic analysis was performed on a 3D shell elements model that includes the lower foot the wing blade and the upper foot. The mesh that provides also the presence of the fins and plates for sharing the water flow in the foots (fig 3), doesn't take into account the stiffening effect due to the W

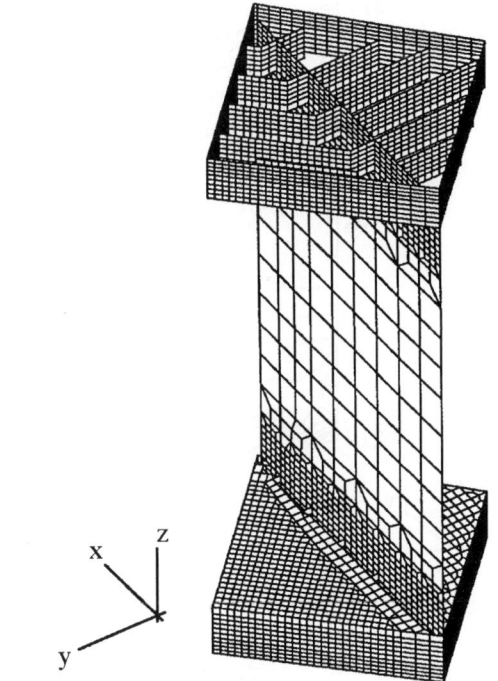

Figure 3. Wing mesh (with top foot open).

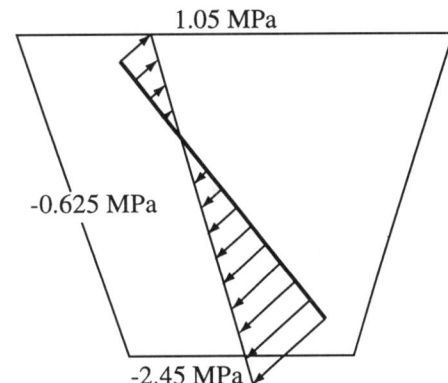

Figure 4. Wing blade VDE loads.

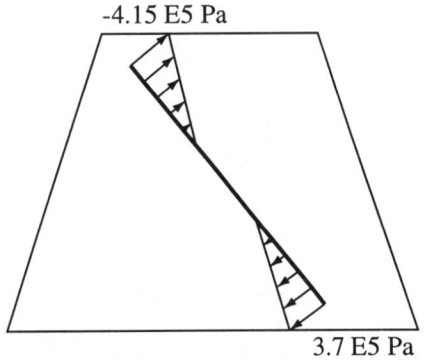

Figure 5. Wing blade centered disruption loads.

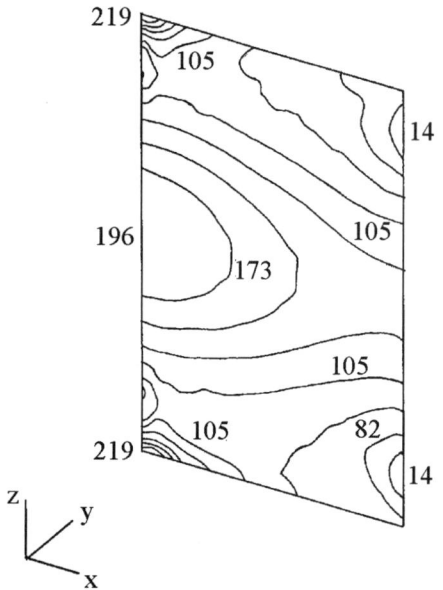

Figure 6. Wing blade max Tresca stress intensity (Mpa) under pressure + VDE+halo currents loads.

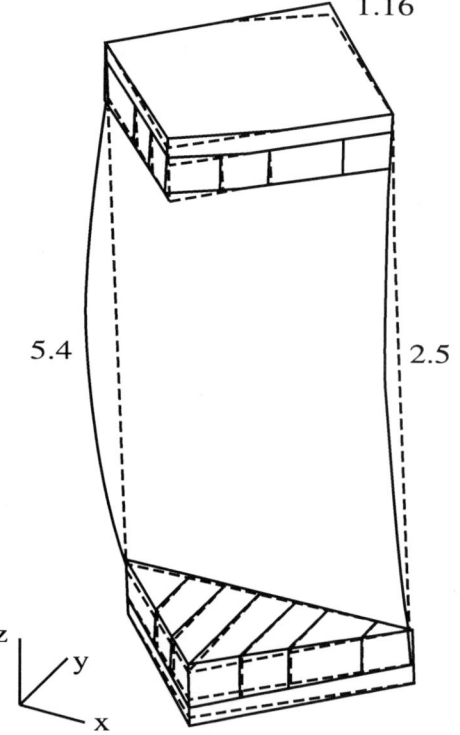

Figure 7. Wing deformed plot under pressure+VDE+ halo current loads (displacements in mm).

Several load case have been considered: 40 bar water pressure in the wing feet ; the eddy current loads[8] for a plasma Vertical Displacement Event (VDE) and Centred Disruption (CD) (fig 4,5); and the halo current loads (HC) assumed as a uniform pressure of 0.76 MPa.

The analyses for each load case and for the possible combinations of them have shown that the most severe conditions occur under pressure+VDE+HC loads. For this case the wing blade shows a maximum Tresca stress intensity of 220 Mpa (fig6). Higher values of Tresca stress intensity in the foot lateral faces region close to the attachments suggest a better mesh definition in that area. The max out of plane deflection of the wing blade is 5.4 mm and the max. displacement of the foot upper plate is about 1 mm (fig 7).

4. CONCLUSIONS

The analysis results show that the design solution envisaged for the VT is able to accomplish the ITER requirements both for normal and transient condition. The wing requires a design and analysis optimisation in order to reduce stress and displacement values also because the effect of thermal stress has not yet been evaluated.

REFERENCES

[1] A.Pizzuto et al. The outline design of the iter divertor outboard vertical target, dump plate and wing . This conference.
[2] J.P.Bonal, D Moulinier. Thermal properties of advanced carbon fiber composites for fusion application. Report CEA DMT/95-495.
[3] J.P.Bonal, C.H.Wu et al. Neutron induced thermal properties changes in carbon fiber composites irradiated from 600 to 1000°C. .J. Nucl. Materials **230** (1996) 271-279.
[4] J.Schlosser, J.Boscary. Thermal hydraulic tests on divertor targets using swirl tubes. CEA Note P/CO -94/004.
[5] I. Smid et al. Lifetime of Be-, CFC- and W-armoured ITER Divertor plates. Presented at ICFRM-7(N 080056), sub. to J.Nucl.Materials.
[6] J. Stubbins. Copper Alloy Performance: Fatigue. Meeting on Materials and Joints for In-Vessel Components. Garghing 1996.
[7] ITER EDA JCT, General Design Requirements Document S10 GDRD 1 95-02-10 W1.2
[8] M.Roccella et al, 3-D Electromagnetic Model and Electromagnetic Analysis of the ITER In-Vessel Components. This concerence

coating in the wing blade. The wing attachments to the divertor cassette body have been preliminary modelled by means of cylindrical hinges positioned at the middle of the external covering plates of the feet .

THE OUTLINE DESIGN OF THE ITER DIVERTOR OUTBOARD VERTICAL TARGET DUMP PLATE AND WING

A. Pizzuto[a], G. Vieider[b], M. Baldarelli[a], G. Brolatti[a], B. Riccardi[a], M. Roccella[a], I. Smid[b].

[a]Associazione ENEA-EURATOM sulla fusione CP65 00044 Frascati Italy

[b]The NET Team-Max Plank Institute fur Plasma Physics-Boltzmannstrasse,2 D-5748 Garching Germany

The ITER divertor outboard channel has to remove about 220 MW out of the 400 MW of the whole divertor. In the normal operation the power is exchanged by radiation leading to average thermal load lower than 1.5 W/m^2 and peak load of 5MW/m^2. During transient phases, the thermal loads are much more severe: up to 20 MW/m^2 for 10s have to be reliably removed. Besides, the eddy current originated during the plasma disruption events generate strong dynamic forces and the divertor components have to resists the neutron irradiation damage. For these reasons, the design of the high heat flux components comes from a trade off made evaluating the thermal hydraulic and thermal mechanical issues of several options. At the end, two solutions are still considered. The reference one is based on a combination of monoblock and flat tile concepts while the full flat tile option is considered as an alternative.

INTRODUCTION

The outboard divertor channel (see fig. 1) consists of the vertical target (VT), the dump plate (DP), the wing and the gas box liner (GBL). The VT and the DP have to remove the heat load coming from plasma via conduction and convection during the normal and transient operation as far as during the plasma disruptions. The Wing is the 'transparent wall' which separate the divertor channel from the 'private region' were the neutrals are pumped out. It is impinged by radiation heat loads only. The tile and coating armour materials are also to be tailored zone by zone because of the different mechanism of the plasma energy release.

1. DESIGN REQUIREMENTS AND CONSTRAINS

The design of the outboard divertor components has been made on the basis of the design requirements fixed by the JCT [1]. The reference armour materials selected are CFC SEP N 31 C for the lower region of VT and for the DP, while the W-La$_2$O$_3$ tungsten alloy has been chosen for the tiles of the VT upper part. The Wing is coated by 5-mm-thick W layer.

Due to the operating temperature and the neutron irradiation resistance requirements, the heat sink material must be high strength copper. Two main candidates have been considered, i.e. CuCrZr and Dispersion Strengthened Copper.

The total flow rate available for the entire (inboard+outboard+dome+cassette) divertor region was fixed in 4 m^3/s. The flow rate available for the outboard channel is about 1.8 m^3/s. The allowable pressure drop is 1.5 MPa . The reference hydraulic scheme assumed here is different from that originally envisaged as reference. The coolant (see fig. 1) enters from the back of the DP, flows along the bottom part of the Wing foot, it is split in two parallel flows entering into the GBL and the Wing rear channel, it is collected in the upper Wing foot, returns through the front channel of the Wing, flows in the DP front and, afterwards, in the VT front. The flow return into the cassette manifolds trough the steel back of the VT.

2. THERMAL HYDRAULICS FEATURES

A very extensive analysis has been carried out in order to find a hydraulic configuration having sufficient critical heat flux (CHF) margin with respect the operating transient thermal loads. To take into account all uncertainties and safety margin, the incident CHF of the selected configuration should be 25 MW/m^2 at least. The heat flux focusing factor has been evaluated by finite element analysis, using a single phase correlation (Sieder Tate).

Having to limit the water flow to 1.8 m^3/s and having to keep the tube pitch within reasonable limit avoiding high flux concentration, 10-mm-dia coolant channels have been chosen for both the VT and DP. To limit at a minimum the pressure drop, the swirl tape has been envisaged in the very high heat flux region such as the bottom part of the VT, for the extension of about 700 mm, and in the entire DP region.

The thermal hydraulics design has been made taken into account the following data:

Thermal power	400 MW/m²
Coolant inlet temperature	140 °C
Total flow rate	3690 kg/s
Overall pressure drop	1.5 MPa
Pumping power	2.2 MW/loop
Coolant inlet pressure	4.0 MPa
Max. incident heat flux	20 MW/m²

The swirl tape is 1.6-mm-thick. The VT and DP steel backs have 15-mm-dia and 14-mm-dia channels respectively.

2.1 Pressure Drop Computations

To evaluate the pressure drops due to concentrated hydraulic losses, standard literature values has been used, while those in the smooth channels have been computed by the D'Arcy formula.

For the swirl tubes the CEA 1 correlation developed by Schlosser has been applied being it extensively validated. It has been derived from data obtained with stainless steel tubes with smaller surface roughness than copper tubes. Nevertheless this correlation leads to a little overestimation of the experimental data.

The results of the hydraulic analysis made for a 3 degree outboard divertor channel module are summarised in the following table n. 1.

Table 1
Pressure drop computation summary

	V_{H2O} (m/s)	D (mm)	N. Tubes	Δp (bar)
DT rear	6.26	14	14	0.26
WING	5	12	11 rear 22 front	1.87
DT front	12	10	18	2.49
VT front	12/9.55	10	18	4.95
VT rear	4.24	15	18	0.24
			tot. Δp	**12.87**

2.2 Critical Heat Flux Results

Assuming a 1.45 of Finite Element focusing factor, the design incident heat flux of 25 MW/m² leads to the reference design wall heat flux of 36.2 MW/m².

The Tong 75 Wall Critical Heat Flux, which has been extensively validated for the swirl tape tubes in the parameter range relevant for the ITER divertor, has been used to evaluate the CHF at the inlet (lowest point) section of the VT for several water velocities.

The results are summarised in tab. 2 (for Twist ratio=2 and tape thickness = 1.6 mm).

Table 2
Computed wall critical heat flux Vs water velocity

Axial speed(m/s)	10	12	14
Module flow rate (l/s)	11.26	13.5	15.76
Temperature(°C)	145	145	145
Local Pressure(bar)	36.67	35.3	33.65
Delta T sat(°C)	100.	97.7	94.85
Calculated WCHF(MW/m²)	31.51	34.47	36.94

The CHF computations have shown that inlet temperature of at last 140 °C, about 14 m/s axial speed in the swirl tube, 40 bar inlet pressure would be needed to accomplish the design CHF of 25 MW/m². Nevertheless, the water velocity of 12 m/s has been chosen as design value on the basis of latest CHF test results [2] which show higher values than those predicted by Tong 95 for 10-mm-dia tubes with 1.6mm thick swirl tape.

3. DESIGN DESCRIPTION.

3.1 Vertical Target.

The VT (see fig. 1) consists of 120 modules. Each module is made of 18 units joined together and fixed to the cassette body. The poloidal profile has been tailored in order to minimise the impinging heat loads taking into account also the 100mm x-point movement. The profile of the lower part, which is the most loaded during the transient and the disruption conditions, is straight for an extension of 700 mm, while in the remaining region it is a curve of constant radius up to the top, where the radius becomes much smaller to realise a suitable transition zone in the region immediately below the Buffle.

Each unit consists of the high heat flux components (HHFC) joined to the 120 mm thick steel back. The HHFC carries the protective tiles. Having the VT to experience different loading conditions along the poloidal profile, the protective tiles are of different materials and thickness. The lower region, which is mainly interested by the disruption and transient loads, is covered by 30 mm thick CFC tiles. The upper region is covered by 15 mm thick W-La_2O_3 tiles. In the CFC-armoured region, the heat sink consists (see fig. 2) of a 10 mm-inner-dia 1.5 mm-thick copper tube. The tube is

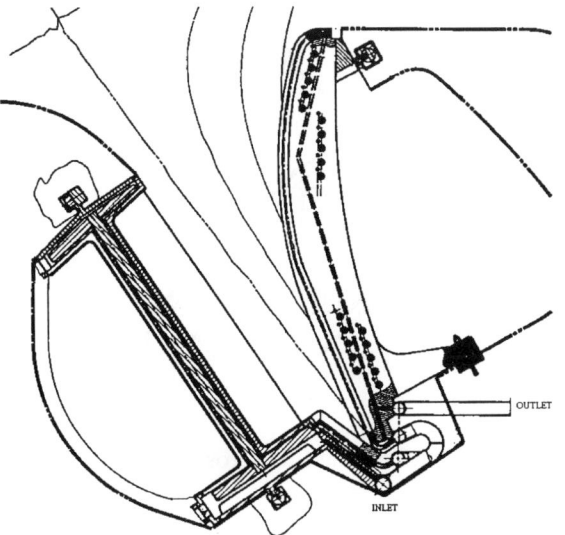

Figure 1. ITER Outboard Channel Assembly

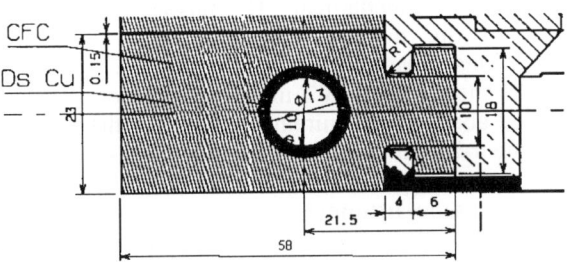

Figure 2. Vertical Target : CFC armoured region cross section

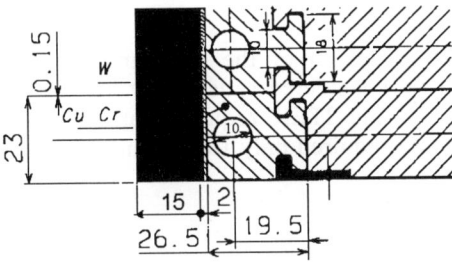

Figure 3. Vertical Target : W armoured region cross section

inserted in the CFC-monoblock tiles. The 1.6 mm-thick swirl tape is inserted in this region to enhance the CHF limit above 25 MW/m^2.

In the W-armoured region, the heat sink consists (see fig. 3) of a rectangular hollow copper bar. The hole is 10-mm-dia, is positioned eccentrically and lies 2 mm below the external surface. The upper end of the bar is machined to get a 2 mm thick annular section, in order to make easier the joining of a sleeve or the direct joint to the steel back.

The critical issue of this solution is the transition between monoblock and flat tile regions. The main option (see fig 4) consists in introducing the DS tube into the hollow bar for an extension to be determined on the basis of a trade off among the strength, sealing and manufacturing factors.

Besides, other two solutions are proposed for the transition zone: in the first the tubular part is machined from the rectangular DS Copper hollow bar; in the second, the CuCrZr copper the rectangular bar is attached by friction welding to the DS copper tube.

The 18 units, per each of the 120 VT modules, are joined together by means of mechanical attachments.

The inlet and outlet manifolds are located at the bottom of the VT. They consist of 60-mm-dia cylindrical boxes (see (fig. 5).

The VT is attached to the cassette body at the two ends. The top joint (see fig. 6) consists of a

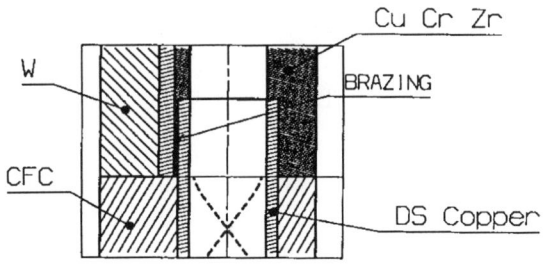

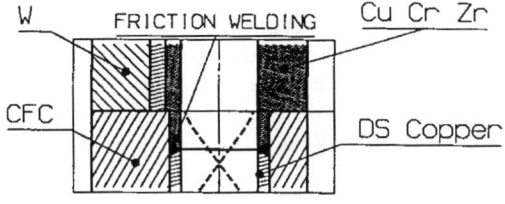

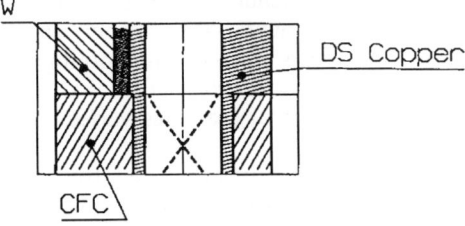

Figure 4. Vertical Target : CFC-W transition zone cross sections

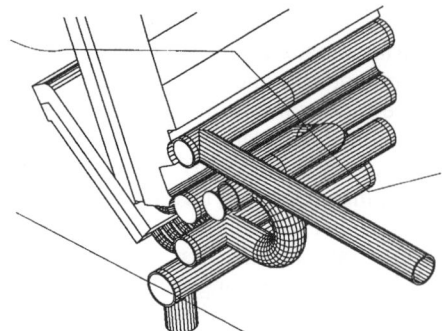

Figure 5. Vertical Target Manifolds

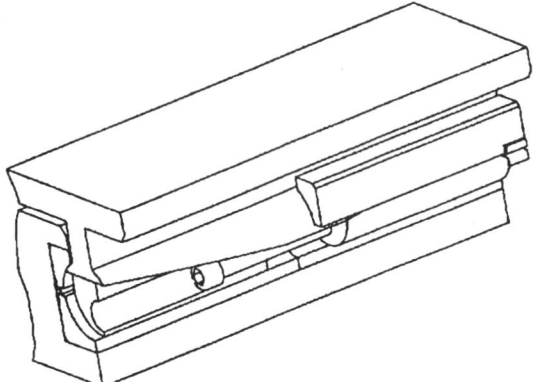

Figure 6. Vertical Target and Wing Attachment : isometric view

hinge which can slide to allow the VT thermal expansion. At the bottom, the VT flanges are bolted to the cassette body.

3.2 Dump Target

The DT has to withstand the same very high heat loads of the VT. This implies the same configuration of the VT lower zone have to be used, even if the geometry is slightly different. The DT modularity is the same of the VT.Wing

3.3 Wing

The wing (see fig. 7) is a 60 mm thick about 1 m long CuCrZr copper blade which goes from the lower to the upper plenum called 'foot'. Due to the weldability requirements, the copper chromium zirconium has been utilized. The cooling channel are realised by drilling 12-mm-dia holes all along the periphery. The pitch of the holes varies depending on the position.

The leading edge is armoured by means of ~10 mm thick W tiles while the remaining surface is coated with 5-mm-thick W layer.

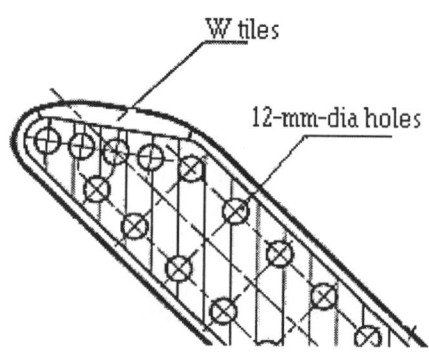

Figure 7. Wing Leading Edge cross section

4. ANALYSES

To verify the lifetime of the outboard divertor channel components, extensive electromagnetic [3], thermal and mechanical [4] analyses taking into account the normal, transient and disruption conditions have been carried out. The analysis results show that the VT monoblock option lifetime is well above the requirements both for the normal and the transient phase conditions while, as far as the transient phase is concerned, the VT full flat tile design option liftime is limited to few thousand cycles. The Wing still requires deeper analyses to assess the local stress and the deflection which, in the shell model used so far, seem to be only marginally acceptable.

5. CONCLUSION

The outline design of the outboard divertor channel can be considered satisfactory because it accomplish all the ITER requirements. Further assessments are required only for few manufacture processes.

REFERENCES

1. ITER EDA JCT, General Design Requirements Document S10 GDRD 1 95-02-10 W1.2
2. J. Schlosser et al, Comparison between various thermohydraulic concepts for the ITER divertor, this conference
3. M.Roccella et al, 3-D Electromagnetic Model and Electromagnetic Analysis of the ITER In Vessel Components- this conference
4. B.Riccardi el al, Thermal and Structural Analysis of the ITER Divertor Outboard Channel, this conference

Development and construction of a high efficiency divertor mock-up using a semi-solid thixotropic alloy as thermal bond layer

L. F. Moreschi[a], E. Visca, S. Storai[a], M. Sacchetti, F. Salvi [a]

Associazione EURATOM-ENEA sulla Fusione, Centro Ricerche Frascati, C.P. 65 - 00044 Frascati, Rome, Italy.
(a) Associazione EURATOM-ENEA sulla Fusione, Centro Ricerche Brasimone, CP 1 - 40032 Camugnano, Bologna, Italy

One of the technological problems that has to be tackled when designing a controlled thermonuclear fusion reactor is the protection of the structural material used for the first wall against damage (usually fusion and partial vaporisation of the most exposed wall) caused by plasma disruption. This problem has been overcome by screening the first wall with a layer of expendable material, with no structural function. Such armour materials (carbon fiber composites or metallic beryllium) after a number of duty cycles should be eroded, and their repair presents several technological problems. This framework forms the basis for the present study on the design of a structure attachment system in which the sacrificial elements of the ITER Divertor are detachable. A possible solution under development is in situ repair, for which a new brazing alloy with semisolid thixotropic behaviour is foreseen. This layer of semisolid thixotropic alloy between the armour material exposed to the plasma, which must be replaced, and the cooled structural part should transfer heat effectively, fill any gaps, compensate for any thermal dilation between the components, stop cracks from spreading, and permit fast and efficient replacement of the armour. Thus, the junction lies between two copper alloy hemispheres bonded with a layer of rheocast alloy. A system has been constructed to demonstrate the possibilities of remote dismantling of a simple attachment system with flat geometry. A 7-part High Efficiency Thermal Shield (HETS) mock-up with a multiple attachment system has been fabricated to perform a feasibility study in one only bonding phase under the expected vacuum and temperature conditions.

1. INTRODUCTION

One of the technological problems which must be faced in planning a controlled thermo-nuclear fusion reactor is that of protecting the structural material of the first wall from damage caused by a break in the plasma, usually caused by fusion and partial vaporisation of the most exposed wall. This problem has been solved by shielding the first wall with a layer of renewable material, which does not have structural functions.

By their very nature, these shields should be replaced on site. This study deals with a system, which can be dismantled, for connecting the so-called sacrificial elements of the Divertor to the structure.

The proposed solution, developed in the frame of the ITER task T6, is that of placing, between the parts exposed to the plasma, which is to be replaced, and the cooled structural part, a layer of semi-solid alloy which is capable of the following:
- transferring heat efficiently
- recovering any gap
- compensating for any thermal expansion between the components
- preventing the propagation of any cracks
- allowing a rapid and efficient replacement of the sacrificial part.

The idea is to keep such rheocast alloy (AlGe, CuPb developed in the CEA-CEREM labs) in the semi-liquid state during application and removal of the CFC armour tiles. These alloys allow a low bonding and detachment temperature (about 500°C) and their intrinsic rheocast structure permits a thickness of about 0.5mm. The junction between the external sacrificial element consists of a brazed graphite tile on a thin copper hemispherical dome and a cooled structural part, which also terminates in a DS copper dome.

2. RHEOCAST ALLOY PROPERTIES

The characteristics which these alloys must have are as follows:
- low vapour pressure

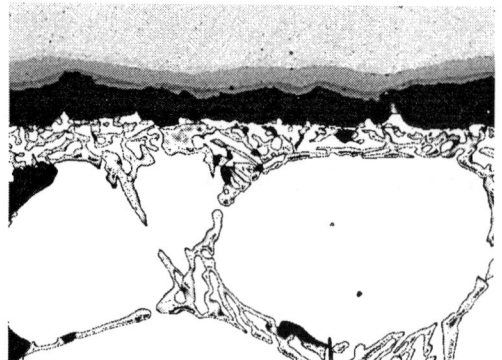

Figure 1. Typical rheocast joint structure Cu/Al27.6Ge

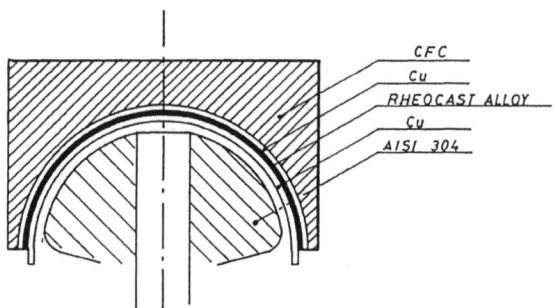

Figure 2. Layout of a HETS single dome

- a high temperature range between the solid and the liquid state
- good thermal and electric conductivity
- sufficient durability under neutronic irradiation.

The main property of these alloys, on which, therefore, the present application is based, is its transition temperature. Below this temperature, the alloys are in the solid state while, above it, and within a certain temperature range, they pass to a soft state, known as semi-solid or semi-liquid. In this state, the alloys can be moulded with low pressure.

This status is obtained by agitating the alloy, while it is slowly solidifying, so that the dendrites tend to form large globules, separated by a mix of small-sized dendrites, around which the globules can slide, within a set temperature range. A typical structure (Al27.6Ge) is shown in Fig. 1, where the large globules can be seen. They have Aluminium as their main component. The small dendrites, around which the globules can move, can also be seen.

The idea is to maintain the alloy in its semi-solid state during the application and removal of the armour material.

Three rheocast alloys were tested, one Lead and Copper based, and two Aluminium and Germanium based.

3. PRELIMINARY TESTS

The aim of this task was to perform a feasibility study on the Copper /Rheocast alloy /Copper jointto be used for the manufacturing of a ITER divertor mock-up adopting the technology developed for High Efficiency Thermal Shield (HETS). This heat exchanger introduces high heat removal capability with low related mechanical stresses. The concept consists in utilising a fluid flow characterised by a strong velocity component perpendicular to the heated wall.

One of the problem to be solved was to mould the rheocast alloy in the hemispherical shape to complain the domes of the HETS heat sink (see fig. 2).

The solution was found after several attempts placing the necessary quantity of alloy in the bottom of a copper alloy hemispherical shaped die.

A stainless steel punch, under the force of gravity, gives the alloy sheet a dome shape on the copper die. The conditions were P= 10^{-5} mbar, T= 430°C.

In this phase, several flat junctions were also carried out in succession to obtain samples for the following:
- tensile stress tests
- thermal conductivity tests
- microstructure analysis.

The brazing procedure consisted of the following steps:
- cleaning the surface of the test items (copper alloy and rheocast alloy) using a suitable cleaning method
- measuring the test items, where necessary
- making junctions in a vacuum furnace by reaching the temperature of the optimum brazing temperature (T=480°C for the Al27.6Ge alloy and T= 520°C for the Al21.8Ge alloy) with a vacuum level of 10^{-5} mbar for approximately 10 minutes.

2.1 Tensile stress tests

As well as the tensile tests carried out on the test sections used for the thermal conductivity measurements, samples were made according to ASTM D2733-70. The test parameters used were the following:

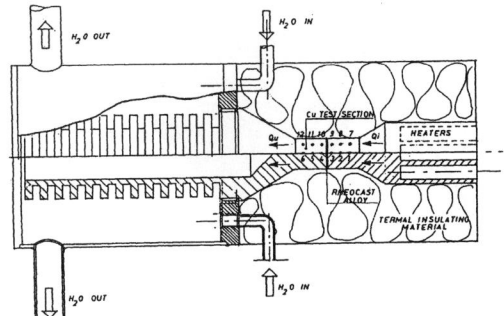

Figure 3. Thermal conductivity test arrangement

- application speed of the load: 50N/s,
- junction area A= a*b = 25.4*12.5 = 317.5 mm^2,
- shear stress s= P/A.

The three samples tested broke, respectively, at 15.7, 21.5 and 22.2 Mpa.

2.2 Thermal conductivity measurements

A small mock-up was also made to measure the thermal conductivity of the junctions (see fig.3).

The thermal flow in the mock-up was created by heating one of its ends with heating plugs with a maximum power of 1000W and cooling the other end, which was suitably shaped, with flowing water. The central part contains the test section, formed by two CuCrZr cylinders, bonded with rheocast alloy. The mock-up was then fitted with thermocouples for measuring the thermal flux at the inlet and the outlet.

The thermal flow was calculated using the indications of the thermocouples with a sensitive element in the centre of the test section and the distances between them as a reference.

Further the thermal conductivity measurements of the junction we have carried out "Laser flash tests" to measure the thermal diffusivity of the rheocast alloys. They allowed us to obtain indirectly the thermal conductivity.

Table 1 shows the difference between the thermal conductivity of the junction (Kj) and the rheocast alloys (Kr) in the same conditions of temperature.

2.3 Microstructure analysis

SEM analysis was also carried out on one of the junctions to highlight the diffusion of the various elements. After measuring the thickness of the two CuCrZr sandwiches with micron accuracy, the junction was made and when the test sample was removed from the vacuum furnace, it was sectioned and subjected to SEM analysis. This analysis (see fig. 1) established the following:

- the structure of the alloy is effectively that of a rheocast alloy, formed by large globules that mainly contain Al, separated by a mix of two dendrite phases of which one practically consists of pure Ge and the other a mix of Ge and Al.
- the rheocast alloy elements diffuse in the CuCrZr alloy, forming phases mixed with the Cu for a thickness of approximately 50 mm.
- the Cu which forms part of the Cu-Cr-Zr alloy also migrates towards the rheocast both inside the globules and forming mixed islands with the Al and the Ge.
- the area which coincides with the border between the Cu-Cr-Zr and the Al-Ge alloys, before the junction is made, shows, after the junction is made, clear initial signs of detachment. This confirms the results of the ultrasonic testing and the mechanical stress tests which always causes the detachment in that area.

Table 1 Rheocast Alloy Thermal Conductivity

T [°C]	Cp [J/gK]	α [cm^2/s]	δ [g/cm^3]	Kj [W/mK]	Kr [W/mK]	RHEOCAST ALLOY
200	0.52	0.56	2.85	82	60	Al/Ge 21.8%
280	0.68	0.63	2.85	122	78	Al/Ge 21.8%
330	0.72	0.62	2.85	123	72	Al/Ge 21.8%
200	0.43	0.56	3.3	70	53	Al/Ge 27.6%
250	0.40	0.59	3.3	70	57	Al/Ge 27.6%
300	0.46	0.62	3.3	83	55	Al/Ge 27.6%
370	0.56	0.55	3.3	88	62	Al/Ge 27.6%

4. DISASSEMBLING-ASSEMBLING TEST

Four samples were manufactured to simulate the assembling and disassembling of a rheocast alloy joint repeated for several times. These tests were performed in a glow box under nitrogen atmosphere and heating one of the ends of the samples till the rheocast interface reaches its transition temperature.

The most relevant results were found by disassembling a sample already joined and then assembled again. The strength of the final joint was found as 10 MPa.

Another similar test was performed by preparing two cylinder with a pre-deposited layer of rheocast alloy in order to simulate a replacement of a damaged tile with a new one. The strength of this joint was found as 13 MPa.

5. MANUFACTURING OF A HETS GEOMETRY MOCK-UP

After checking the feasibility of the junction between the two hemisphere-shaped Cu-Cr-Zr elements, we joined a group of 7 elements of armour material (CFC tiles, already pre-brazed to annealed copper domes with a thickness of 0.5 mm) in a HETS geometry (see fig. 4).

First, the mock-up components were measured. Afterwards, there was a coupling test, carried out by placing a silicon elastomer resin between the two elements and allowing the coupling to take place.

The result was a group of 7 layers of resin from which we measured the volume of the rheocast alloy we need for the joining of the seven tiles.

At this point, before the real junction was made, a thermal mapping was performed in the vacuum furnace under the same test conditions. For this purpose the temperature was measured at the top of four hemisphere-shaped elements and one on the outside. The signals read by the five thermocouples were continuously recorded.
- the stress corresponding to the detachment of the manipulated junctions was 13 and 10 MPa respectively, when the results obtained for similar non-manipulated junctions were between 13 and 25 MPa.
- these tests were carried out manually in a glove-box; therefore the results can be only improved.

It can be stated that, in principle, this junction system can work.

Therefore, the difficulties that arise are:

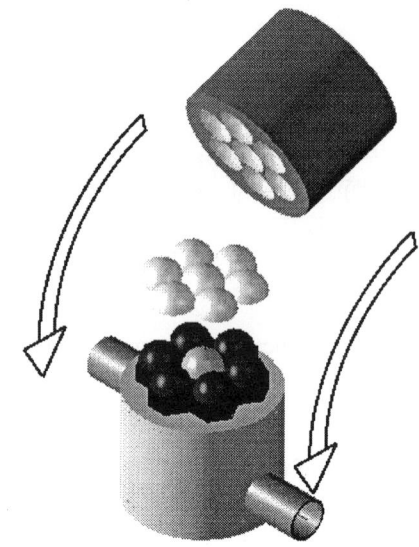

Figure 4. Exploded view of the HETS Mock-up

- the need for a complex preparatory cycle for cleaning of the surfaces
- mechanical problems relating to the positioning of the surfaces to be joined
- the high operating temperatures: T = 420°C is the temperature used for transforming the rheocast to a semi-solid state and T = 520°C is the brazing temperature of the alloy on the CuCrZr.

To complete the activity it will be basic to test the HETS mock-up under high heat flux in order to verify its thermo-mechanical behaviour. These tests are scheduled in the next future. The results of the fatigue tests will give us only the information regarding the thermal conductivity of the rheocast alloy and its capacity to take up the mechanical stress due to the different thermal expansion of the joined materials. In this phase, in fact, it is not required to the rheocast alloy to provide any mechanical strength.

REFERENCES

1. M.C. Flemings, Behavior of Metal Alloys in the Semisolid State, Metallurgical Transaction A, Vol. 22A, (1991)
2. G. Federici et al., Design, materials and R&D issue of innovative thermal contact joints for high heat flux applications, Fusion Eng. and Design 28 (1995) 34-43

Neutron exposure characterisation of ITER divertor materials for source terms evaluation

D.G. Cepraga [a], G. Cambi [b], L. Di Pace [c], M. Vaccari [a]

[a] ENEA, Dipartimento Innovazione, 2 Via Don Fiammelli, 40128 Bologna, Italy

[b] Bologna University, Physics Department, 46 Via Irnerio, 40126 Bologna, Italy

[c] Associazione Euratom-ENEA sulla Fusione, 27 Via E. Fermi, 00044 Frascati, Italy

The paper presents the results of the Sn neutron transport and activation calculation to evaluate the activated corrosion product source terms related to the ITER divertor primary heat transfer system. The integrated radiation transport and activation methodological approach based on the calculation sequence VITAMIN(ENEA)-SCALE-ANITA has been used. The divertor cassette has been subdivided in five regions: inboard and outboard vertical targets, dome, inboard and outboard wings. The heat sink material is Glidcop Al-25 and the support material is AISI 316LN-IG. The divertor cassette body material is AISI 316LN-IG. The impact of various protective layer materials (Beryllium, CFC, Tungsten) on the neutron spectra and on the activation results have been investigated. The reaction rates and the activation data for both the steel and the copper alloy needed by the Pactole code have been produced by means of the RR-PACT module.

1. INTRODUCTION

In order to evaluate the environmental source terms for the various accident sequences to be analysed and included into the Non-site Specific Safety Report-1 (NSSR-1) for ITER, the Activated Corrosion Products (ACP) related to the Divertor Primary Heat Transfer System (DV PHTS) have to be estimated. This is done by using the Pactole [1] code.

So far, to obtain the reaction rates and the activation parameters needed as input data by Pactole, a set of neutron and activation calculation has been performed for the various in-vessel zones of the ITER divertor. The assessment refers to the ITER Basic Performance Phase. The irradiation scenario considered results in a neutron fluence of 0.3 MW-y/m^2 on the Outboard First Wall (neutron wall load of 1 MW/m^2 on the equatorial plane). The neutron wall load for the divertor region zones is based on the values given in [2]. The VITAMIN(ENEA) SCALE ANITA RRPACT computational approach presented in [3] is used. The results of the assessment are fully documented in [4] and [5].

2. METHODOLOGY

The integrated radiation transport and activation methodological approach based on the calculation sequence VITAMIN(ENEA)-SCALE-ANITA has been used to perform neutron and activation calculations. Both the NJOY 91.38 modular code system and the AMPX77 code system have been applied to process the FENDL(ENDF/B-VI) and the EFF-2 data to produce the fine-group cross-section 174-n 38γ group VITAMIN-ENEA library.

The neutron flux spectra have been obtained by the transport sequence BONAMI-S - XSDRNPM-S. The BONAMI-S performs resonance shielding through the application of the Bondarenko shielding factor method. The XSDRNPM-S code is the available module in the SCALE 4.3 Modular Code System [6], for evaluating one-dimensional radiation transport through a shield. It is a highly evolved discrete-ordinates transport program capable to perform neutron and coupled neutron-gamma calculations with the scattering anisotropy represented to any arbitrary order. The finite-differencing is done with the weighted diamond difference model. The shielding analysis sequences produce problem-dependent broad-group cross-sections derived from VITAMIN-ENEA library (100-n group and 18-γ group, named the VIT-ITER n-γ coupled working libraries) and assess the radiation transport through a shield (neutron and gamma flux distributions in different spatial intervals/zones).

The activation characteristics of the various materials have been obtained using the ANITA-3 code [7] (EAF3 activation library).

The RR-PACT module is used to process the neutron flux spectra given by XSDRNPM and the 100-groups EAF3 library in order to evaluate the set of the activation data needed in input by the Pactole code. For each one of the nuclear reaction in the material of the zones interested by the divertor cooling channels, the following data are obtained:

the nuclear density, the reaction rates RR (s^{-1}), the specific reaction rates (RR/cm^3), the averaged spectrum cross sections, the total neutron flux. Only the values of the reaction rates are requested in input by the Pactole code. Anyway, the specific activities of the various materials (both total and detailed for isotope), are used to normalise the results of the Pactole activation model to those provided by the more sophisticated ANITA activation model

3. DIVERTOR CALCULATION MODEL

3.1 Geometry and materials

The divertor cassette has been subdivided in five sub-regions differing in layout, dimensions, materials, and in neutron load: outboard vertical target OVT, inboard vertical target IVT, dome D, outboard wing OW, inboard wing IW (Figure 1).

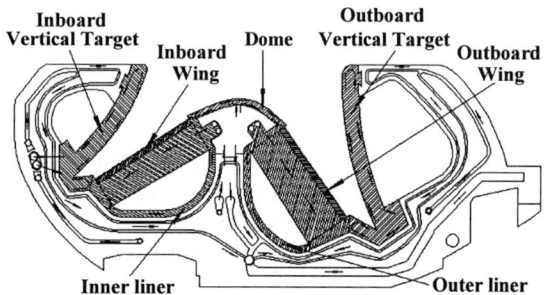

Figure 1. ITER divertor cassette scheme

The assessment described in this paper focus on the outboard vertical target and on the dome.
The OVT is considered poloidally subdivided in three different parts (upper part OVT-UP, lower part OVT-LP, dump plate OVT-DP), while the dome is considered poloidally as a unique part.
A 1D cylindrical geometrical model is considered from the centre of the plasma up to the vacuum vessel. It includes: plasma, scrape off layer, heat sink, support, gap, cassette, and vacuum vessel.
The water coolant flows firstly into Glidcop Al-25 copper alloy cooling channels on the heat sink and then into AISI 316LN-IG cooling channels on the support and on the cassette body. The basic composition of the materials interested by water corrosion/erosion are given in the following. Glidcop Al-25 DS copper alloy: Cu 99.5%, Al 0.25%, O 0.22%, B 0.025%; AISI 316LN-IG: Fe 64.95%, Cr 17.5%, Ni 12.25%, Mo 2.5%, Mn 1.8%, Si 0.50%, Co 0.05%, P 0.025%, N 0.07%.

The detailed material composition, including the impurities considered for the neutronic and activation assessment, are given in [4].
The impact of the protective layer PL materials has been assessed by analysing three cases :
reference case, beryllium is the PL material for all the three OVT parts and for the dome ;
option A, with tungsten for OVT-UP and dome ;
option B, with CFC for OVT-LP and dump plate.

3.2 Divertor irradiation characteristics

The neutron current distribution given in [3] has been used as normalisation factor to obtain the neutron wall loading for the different divertor sub-regions. So far, the neutron wall load values related to outboard vertical target and dome are :
OVT-UP 0.20 MW/m^2; OVT-LP 0.40 MW/m^2; OVT-DP 0.40 MW/m^2; D 0.57 MW/m^2.
A 723 steps irradiation scenario resulting in a total fluence of 0.3 MW-y/m^2 on the outboard first wall is considered:
- fluence build-up operation: continuous operation for 1 month and 3 months down for two cycles;
- transition operation: one month down. Continuous operation for 0.6 month and 1.8 month down;
- pulse build-up operation: one month consisting 1 hour burn and 1 hour down; one month down; one month consisting of 1 hour burn and 1 hour down.

4. ACTIVATION CALCULATION RESULTS

4.1 Material activation characteristics

The total specific activity [Bq/cm^3], the total specific decay heat [W/cm^3], the contact dose rate [Sv/h], and the Annual Limit of Intake [ALI/cm^3] have been obtained for the zones/materials of the three OVT parts and for the dome. The results are obtained with reference to the end of the irradiation scenario (time = 0) and for various cooling times, up to 10^3 years after plasma termination. Tables 1 to 3 show the main activation parameters at t=0, for the reference case and for the two options A and B, respectively. The full set of results, including the isotope detail, is given in [4].
The analysis of the activation results show that the Glidcop Al-25 specific activities for the reference case are higher (18 to 38 %) compared to the corresponding values for the option A, while they are 15 % higher if compared to the option B.
For the support zone, the reference case shows steel activities higher (5 to 13 %) than the corresponding

ones of the option A and 17 % higher with respect to the option B.

Table 1
Divertor OVT and dome activation characteristics (0.3 MW-y/m^2). (Beryllium protective layer)

Zone		Activity [Bq/cm^3]	Decay Heat [W/cm^3]	Contact dose [Sv/h]
OVT-UP	a	1.26E+12	1.80E-01	1.25E+04
	b	8.20E+10	1.99E-02	2.64E+03
	c	8.00E+09	2.05E-03	2.72E+02
OVT-LP	a	2.29E+12	3.34E-01	2.34E+04
	b	1.83E+11	4.57E-02	6.05E+03
	c	1.72E+10	4.47E-03	5.93E+02
OVT-DP	a	2.31E+12	3.35E-01	2.35E+04
	b	1.86E+11	4.65E-02	6.16E+03
	c	1.86E+10	4.83E-03	6.40E+02
D	a	4.15E+12	6.58E-01	4.73E+04
	b	6.34E+11	1.49E-01	1.97E+04
	c	2.49E+10	5.96E-03	7.92E+02

a : heat sink Glidcop cooling channels ; b : support steel cooling channels ; c : cassette body steel cooling channels

Table 2
Divertor OVT and dome activation characteristics (0.3 MW-y/m^2). (Tungsten protective layer)

Zone		Activity [Bq/cm^3]	Decay Heat [W/cm^3]	Contact dose [Sv/h]
OVT-UP	a	1.03E+12	1.52E-01	1.06E+04
	b	7.75E+10	1.90E-02	2.52E+03
	c	7.41E+09	1.90E-03	2.53E+02
D	a	2.59E+12	4.81E-01	3.58E+04
	b	5.50E+11	1.30E-01	1.72E+04
	c	2.26E+10	5.45E-03	7.24E+02

Table 3
Divertor OVT and dome activation characteristics (0.3 MW-y/m^2). (CFC protective layer)

Zone		Activity [Bq/cm^3]	Decay Heat [W/cm^3]	Contact dose [Sv/h]
OVT-LP	a	1.93E+12	2.77E-01	1.93E+04
	b	1.52E+11	3.79E-02	5.02E+03
	c	1.44E+10	3.74E-03	4.96E+02
OVT-DP	a	1.94E+12	2.79E-01	1.94E+04
	b	1.55E+11	3.85E-02	5.11E+03
	c	1.56E+10	4.03E-03	5.35E+02

4.2 Neutron and activation data for Pactole

The RR-PACT module has been used to process the XSDRNPM-S neutron flux spectra and the EAF3 activation library in order to evaluate the set of the activation data needed in input by the Pactole code. For each zone/material and for the relevant activation reaction the nuclear density, the reaction rates RR (s^{-1}), the specific reaction rates (RR/cm^3), the averaged spectrum cross sections, the total neutron flux are obtained.

The "standard fission version" of the Pactole code takes into account five elements: Fe, Ni, Mn, Cr, and Co to perform the activated corrosion product source terms evaluation. The upgrade "fusion version" includes copper as a new element to make possible the corrosion products assessment for the divertor and the baffle-limiter primary coolant systems. Anyway, the number of nuclear reactions that can be managed by the Pactole is still limited to a maximum of ten reactions at a time. The following criteria have been established for the nuclear reaction selection:
1. only the nuclear reactions with father and daughter isotopes of the elements considered by Pactole (Fe, Ni, Mn, Cr, Co, and Cu) are considered;
2. the nuclear reactions producing isotopes with very low half life (e.g. few minutes) are rejected;
3. the nuclear reactions producing isotopes with very low contribute (e.g. < 0.1%) to the activation characteristics (specific activity Bq/cm^3, gamma-ray sources MeV/g$*$s, biological hazard ALI/cm^3) of the material in a specified zone are rejected;
4. the nuclear reactions resulting in low values of the specific reaction rates (for a specified zone/material) are rejected;
5. the nuclear reactions resulting in isotopes of elements that present very limited corrosion parameters for the operating conditions of the divertor primary heat system are rejected.

The relevant activation reaction rates for the glidcop heat sink and the support steel are presented in Tables 4 to 7.
It has to be underlined that the specific activities of the various materials (both total and detailed for isotope) are not directly used by the Pactole code. Anyway, those values constitute an useful parameter to check and/or to adapt the results of the Pactole activation model to those provided by the more sophisticated ANITA activation model.

Table 4
Divertor OVT and dome reaction rates for Glidcop Al-25. (Beryllium protective layer)

Nuclear reaction	Reaction rate RR [s^{-1}]			
	OVT-UP	OVT-LP	OVT-DP	Dome
Cu63(n,α)Co60	1.77E-13	3.36E-13	3.36E-13	7.66E-13
Cu63(n,γ)Cu64	2.23E-11	3.94E-11	3.97E-11	6.40E-11
Cu65(n,2n)Cu64	7.73E-12	1.49E-11	1.49E-11	3.31E-11
Cu65(n,γ)Cu66	9.06E-12	1.58E-11	1.60E-11	2.54E-11
Cu65(n,p)Ni65	2.00E-13	3.81E-13	3.81E-13	8.60E-13

Table 5
Divertor OVT and dome reaction rates for AISI 316LN-IG, zone support (Beryllium PL)

Nuclear reaction	Reaction rate RR [s^{-1}]			
	OVT-UP	OVT-LP	OVT-DP	Dome
Cr52(n,2n)Cr51	7.39E-13	1.39E-12	1.39E-12	6.91E-12
Mn55(n,2n)Mn54	2.10E-12	3.92E-12	3.93E-12	1.95E-11
Mn55(n,γ)Mn56	3.42E-11	8.69E-11	8.89E-11	2.32E-10
Fe54(n, p)Mn54	1.64E-12	2.96E-12	2.97E-12	1.37E-11
Fe56(n,2n)Fe55	1.13E-12	2.12E-12	2.12E-12	1.05E-11
Fe56(n,p)Mn56	3.33E-13	6.11E-13	6.16E-13	3.06E-12
Co59(n,γ)Co60	5.93E-11	1.54E-10	1.57E-10	3.26E-10
Ni58(n,2n)Ni57	6.96E-14	1.31E-13	1.31E-13	6.53E-13
Ni58(n,np)Co57	1.39E-12	2.59E-12	2.59E-12	1.29E-11
Ni58(n, p)Co58	9.83E-13	1.78E-12	1.78E-12	8.29E-12

Table 6
Divertor OVT and dome reaction rates for Glidcop Al-25. (Options A and B)

Nuclear reaction	Reaction rate RR [s^{-1}]			
	Option A		Option B	
	OVT-UP	Dome	OVT-LP	OVT-DP
Cu63(n,α)Co60	1.45E-13	5.98E-13	2.85E-13	2.85E-13
Cu63(n,γ)Cu64	1.76E-11	3.04E-11	3.38E-11	3.40E-11
Cu65(n,2n)Cu64	6.71E-12	2.80E-11	1.21E-11	1.21E-11
Cu65(n,γ)Cu66	7.28E-12	1.22E-11	1.36E-11	1.37E-11
Cu65(n,p)Ni65	1.66E-13	6.84E-13	3.20E-13	3.20E-13

Table 7
Divertor OVT and dome reaction rates for AISI 316LN-IG, zone support. (Options A and B)

Activation reaction	Reaction rate RR [s^{-1}]			
	Option A		Option B	
	OVT-UP	Dome	OVT-LP	OVT-DP
Cr52(n,2n)Cr51	6.74E-13	6.17E-12	1.17E-12	2.11E-13
Mn55(n,2n)Mn54	1.88E-12	1.70E-11	3.32E-12	6.15E-13
Mn55(n,γ)Mn56	3.44E-11	2.09E-10	7.10E-11	9.95E-12
Fe54(n, p)Mn54	1.36E-12	1.07E-11	2.63E-12	5.76E-13
Fe56(n,2n)Fe55	1.02E-12	9.27E-12	1.79E-12	3.26E-13
Fe56(n, p)Mn56	2.88E-13	2.53E-12	5.31E-13	1.06E-13
Co59(n,γ)Co60	6.03E-11	3.01E-10	1.25E-10	1.58E-11
Ni58(n,2n)Ni57	6.39E-14	5.87E-13	1.10E-13	1.97E-14
Ni58(n,np)Co57	1.25E-12	1.13E-13	2.19E-12	4.05E-13
Ni58(n, p)Co58	1.25E-12	6.51E-12	1.58E-12	3.41E-13

5. CONCLUSIONS

To evaluate the activation corrosion products source terms of the ITER heat transfer system, the activation characteristics of the divertor materials have been calculated. The divertor cassette has been subdivided in five sub regions differing in layout, dimensions, materials, and in neutron load. The heat sink material is Glidcop Al-25, the support and the cassette body material is AISI 316LN-IG. Different protective layer materials (Be, W, and CFC) have been considered and their impact on the activation data assessed. The activation data refer to the ITER Basic Performance Phase. The neutron wall loads are based on the distribution provided by ITER-JCT.

As expected, the result analysis indicates that the activation parameters and the reaction rates are higher for the reference case (beryllium protective material) than the corresponding ones of the option A (tungsten protective material) and option B (CFC protective material).

Criteria to chose relevant activation reaction rates for the Pactole corrosion code have been established. The set of reaction rates and neutron/activation data given by the RR-PACT module and the ANITA-3 activation code are presented.

REFERENCES

1. J.-C. Robin et al., PACTOLE: A computer code to predict activation and transport of corrosion products in PWRs, Proc. of the International Symposium on Activity Transport in Water Cooled Nuclear Power Reactors, Ottawa (Canada), 1994, October 24-25.
2. P. Barabaschi, Update of line of sight calculation of the neutron wall loading, Interoffice Memorandum S 73 MD 23 95-08-09 W 1.1, August 9, 1995.
3. D.G. Cepraga et al., Fusion plant material damage due to neutron irradiation: an assessment for ITER in-vessel components, 12th Topical Meeting on the Technology of the Fusion Energy, 1996 ANS annual meeting, June 16-20, Reno, Nevada.
4. G. Cambi et al., ITER IDR-95 divertor regions: neutron transport analyses and activation evaluations, ENEA ERG-FUS/TECN TR 10/96, April 1996.
5. L. Di Pace et al., Evaluation of the activated corrosion products for the ITER heat transfer systems in support of NSSR-1, ENEA ERG-FUS/TECN TR 18/96, Rev. 0, July 1996.
6. O.R.N.L., SCALE 4.3 - Modular Code System for Performing Standardized Computer Analyses for Licensing Evaluation, Martin Marietta Energy Systems Inc. for the U.S. D.O.E., RSIC CCC-545, 1995
7. D.G. Cepraga et al., ANITA-3: A new version of the ANITA (Analysis of Neutron Induced Transmutation and Activation) code, to be published.

Thermal Analysis of Sacrificial Components

F. Andritsos and A. Inzaghi

European Commission - Joint Research Centre, Institute for Systems, Informatics and Safety
T.P. 210, 21020 Ispra (VA), Italy

Sacrificial components are of primary interest to the design of the divertor of ITER where the very high heat flux is likely to cause the requirement for their frequent substitution. Among the alternatives for a viable divertor design is the one where a 1-2 mm thick rheocast alloy layer interfaces the plasma facing sacrificial material (Be, CFC or W) to the heat sink (Copper). Such a solution would relax the thermal stresses due to the inevitable high thermal gradients without penalizing much the overall heat transmission coefficient. Moreover, during maintenance, damaged tiles could be easily detached from their heat sink by heating them above the liquidus temperature. To optimize the experiments and help at the interpretation of the results, a parametric finite element analysis has been undertaken at JRC Ispra during 1995. The work described here refers to the steady state thermal modelling of various specimen geometries, subjected to heat fluxes and cooling conditions such as those that are expected to be faced during ITER operation. Results indicate that the rheocast bonding can indeed be a viable solution for the divertor and, possibly, other sacrificial plasma facing components of ITER.

1. INTRODUCTION

An alternative for a viable divertor design is the one where a 1-2 mm thick rheocast alloy layer interfaces the plasma facing sacrificial material (Be, CFC or W) to the heat sink (Copper). Some rheocast alloys combine a wide temperature area (some hundred degrees) between the solidus and liquidus temperature together a high thermal conductivity and good wetting to materials such as Cu, Be and CFC. In that area, they have a viscous mechanical behavior, so that they are expected to relax the thermal stresses without penalizing much the overall heat transmission coefficient. Moreover, during the divertor maintenance, damaged sacrificial tiles can be detached from their heat sink by heating them close to the liquidus temperature of the rheocast alloy. Experimental work is being performed for the testing and the characterization of such materials. The numerical analysis presented here aims to the interpretation and further optimization of such experiments, performed by ENEA under the European Fusion Program, and contribute to the design of the actual sacrificial components.

To accommodate the high heat flux expected at the divertor region, a high performance cooling method consisting in water impinging (rather than flowing along) on a concave dome shaped surface of the heat sink was adopted for the experimentation, [1]. OFHC Copper was used for the heat sink and liner; Beryllium for the sacrificial tiles. A 50% Copper - Lead rheocast alloy was chosen as interface.

A first issue has to do with the shape of the interface and the relative mass of the heat sink. A second one concerns the optimization of the position and dimensions of the interface. For this reasons, two classes of specimen were considered: one with a dome shaped interface and one with a flat interface. For each class a parametric Finite Element (FE) model was set up. Their respective geometries are shown in figures 1 and 2 while the values of the parameters for the various simulation cases are given in tables 1 and 2.

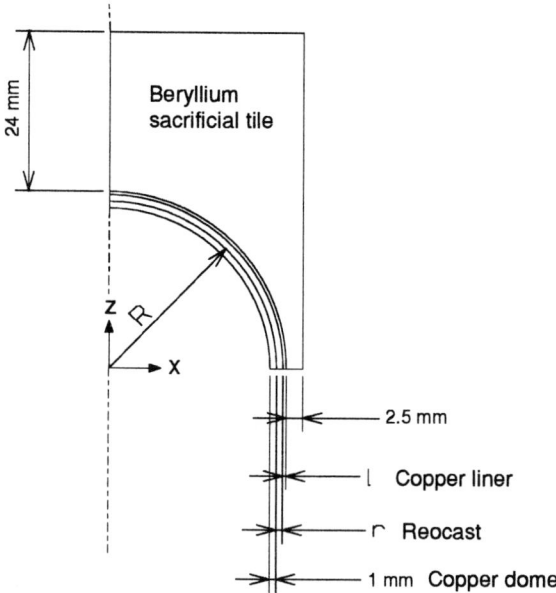

Figure 1. Axial section of the specimen with dome shaped reocast interface

Table 1. Dome shaped interface

case	R	r	l
# D1	15.	0.5	0.0
# D2	15.	1.0	0.0
# D3	15.	1.0	0.5
# D4	15.	0.5	0.5
# D5	25.	0.5	0.0
# D6	25.	1.0	0.0
# D7	25.	1.0	0.5
# D8	25.	0.5	0.5

2. NUMERICAL SIMULATIONS

Two FE models were set up in such a way so as to be possible to get all 12 configurations described in tables 1 and 2 by switching the material flags of certain groups of elements. The tile was intentionally left very thick; the simulation being steady state, its thickness doesn't influence the temperature distribution or the heat fluxes at the rest of the model. The thickness of the actual experimental tile would be it that of the part of the tile which is below the isotherm of the melting temperature of Beryllium (1280 C).

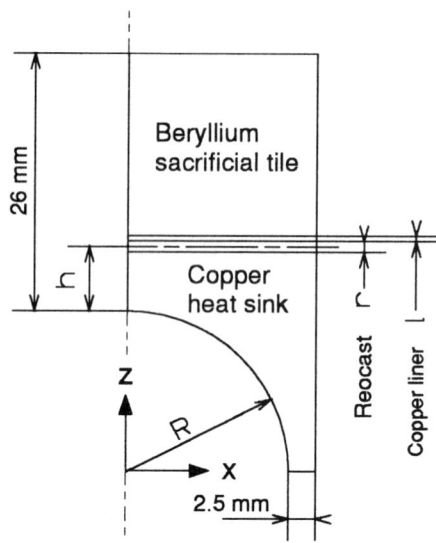

Figure 2. Axial section of the specimen with flat reocast interface

Table 2. Flat interface

case	R	r	l	h
# F1:	14.	1.	0.5	5.5
# F2:	14.	1.	0.5	8.5
# F3:	14.	1.	0.5	11.5
# F4:	14.	1.	0.5	14.5

In all 12 cases, the nominal expected heat flux at the divertor $Q_{nom} = 5$ MW/m^2 was applied to the top surface. Only case #D5 was repeated with a heat flux $Q_{pk} = 20$ MW/m^2. Let this case be called #D5H.

The cooling of the concave dome surface was modeled as follows: At the top third of the dome, where the jet impinges, with a convection coefficient $h_{top} = 180$ kW/(m^2 K), at the rest of surface with $h_{bot} = 180$ kW/(m^2 K) [1,2]. Sink temperature is $T_{sink} = 150$ C everywhere.

Properties for Beryllium and Copper were taken from [3]. As no experimental data was available for the rheocast alloy, its thermal properties were derived from those of its constituents, as described in [4].

The ABAQUS v5.2 finite element package [5] was used for the numerical simulation on a IBM R6000 mod 590 platform.

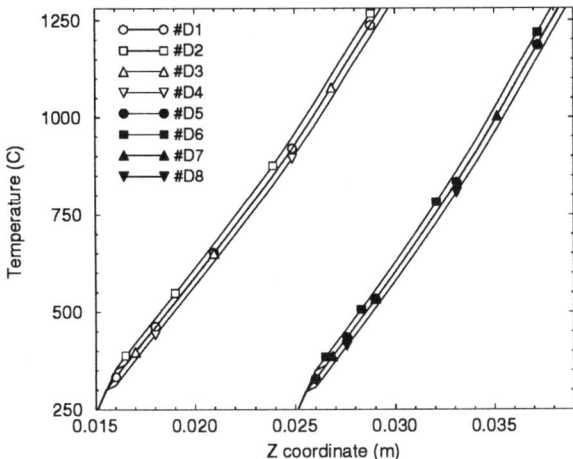

Figure 3. Temperatures along the centerline, dome shaped interface

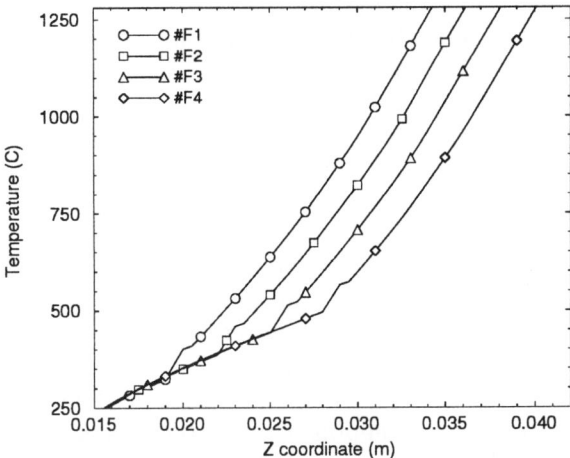

Figure 4. Temperatures along the centerline, flat interface

3. RESULTS

Results are presented in figures 3 and 4 as plots of the temperatures along the specimen centerline (Z), for cases #D1-8 and #F1-4 respectively. Coordinates X, Z start from the center of the dome and are expressed in meters.

The upper bound of the temperature axis of both figures is the melting temperature of Beryllium (1280 C). So, the point where each curve intersect it gives the Z dimension of the respective actual specimen and, consequently, the thickness of the Beryllium tile. For example: case #F4 intersect it at 40 mm. Thus the tile must be of thickness:

$h_t(max) = 40 - R - h - l - r/2 = 10.5$ mm

otherwise its PF surface will melt.

In fig.4 is depicted the effect that the thickness of the heat sink on the efficiency of the flat interface specimen: The maximum allowable tile thickness $h_t(max)$ is an inverse function of h, decreasing from 13.5 to 10.5 mm (about 1 mm for each 3 mm of increment of h). The rheocast layer temperature is distinguished as the high gradient short part at the low left of each curve. It is directly proportional to h, rising from about 400 C (#F1) to 580 C (#F4). This latter is close to the liquidus temperature of the Cu-Pb rheocast alloy.

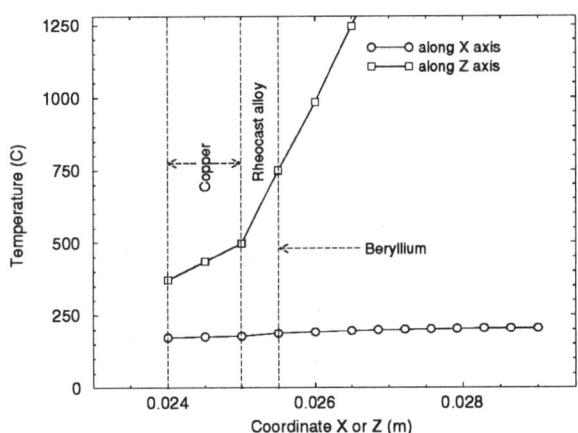

Figure 5. Temperatures along the X and Z axes, case #D5H, $Q_{pk} = 20$ MW/m^2

In fig. 5 is shown the temperature profile for the case #D5H, with 4 times the nominal heat flux: $Q_{pk} = 20$ MW/m^2. We see that such a flux, if applied long enough to cause a steady state profile (few seconds), it will melt most of the Beryllium tile, the melt reaching as close as 1 mm from the rheocast alloy. The rheocast layer, although at the top of the dome is well above its liquidus temperature, at the lower left side it has a temperature of about 200 C, thus providing a good mechanical resistance.

4. CONCLUSIONS

The following conclusions can be deducted from the work summarized in the above sections:

a. The flat interface design has better performance. This is due to the shorter path of the heat flow inside a "low" conductivity material such as Beryllium. Nevertheless, a dome shaped interface presents the advantage of a much better mechanical resistance not only under nominal but also under peak heat flux.

b. There is a marginal gain in thermal performance by keeping r and l (rheocast layer and liner respectively) as low as possible.

c. The flat interface specimen must provide a safety margin relative to the maximum temperature in the interface layer, assuring an operating interface temperature not too close to the alloy liquidus temperature. Otherwise, the isotherms being almost parallel to the interface surface, the rheocast layer is prone to lose at once all mechanical resistance. In practical terms this means the choice of configuration #F2 rather than the #F3 or #F4 even if it implies a significantly smaller thermal sink mass.

As a general conclusion: The use of rheocast alloys as interface between the sacrificial components and their heat sinks in high heat flux areas of fusion reactors seems quite promising. Nevertheless, a lot of work remains to be done for the characterization of these materials as well as for the design of the appropriate engineering solutions that will optimize the expected benefits from their use.

REFERENCES

1. E. Dipietro, personnal communication, ENEA Frascati, (1995).
2. A. Pizzuto, C. Sangiovanni, A High Performance Water-Cooled Thermal Shield Device, proc 15th SOFE (1993) US 855.
3. E. Zolti, Material Data for NET, rep. N/I/3300/5/A, Garching, (1990).
4. M. Merola, On the Use of the Copper-Lead Rheocast Alloy in the ITER Reactor, PT DE 380/IN, Politecnico di Torino (1995).
5. H.K.S. Inc (eds) ABAQUS v. 5.2 Manuals, US (1994).

Development of Plasma-Facing Components for Fusion Experimental Reactors at JAERI

M. Akiba, M. Araki, K. Nakamura, S. Satoh, S. Suzuki, M. Dairaku, K. Yokoyama, Y. Ohara

Japan Atomic Energy Research Institute
801-1 Mukoyama, Naka-machi, Naka-gun, Ibaraki-ken, 311-01 Japan

This paper presents recent experimental results on plasma-facing components for fusion experimental reactors at JAERI. The results include heating tests on a multi-cooling channel divertor element, a 3D-CFC divertor element, a 2D-CFC monoblock element, and divertor elements with a thermal bond layer of Pb-0.1Cu.

1. INTRODUCTION

The divertor plate for fusion experimental reactors such as ITER should withstand high heat loads and particle loads from the fusion plasma. In the design of ITER[1], the divertor plate is assumed to have a steady-state heat load of 5 MW/m^2, and a transient heat load of 15 MW/m^2, 10 s. JAERI has extensively developed divertor elements for fusion experimental reactors, in particular, for ITER. In this paper, recent heating test results on the divertor elements are described.

2. MULTI-COOLING CHANNEL DIVERTOR ELEMENT

JAERI has proposed a saddleblock divertor[2,3] for the ITER divertor plate. It is one of the critical issues to evaluate thermal performance limit of proposed elements. A multi-cooling channel divertor element was fabricated and tested in JAERI. The divertor element has carbon-fiber reinforced-carbon (CFC) armour tiles which are brazed onto a cooling substrate made of OFHC copper as shown in Fig.1. The armour tile is unidirectional CFC (UD-CFC) tiles of 50 mm x 50 mm. The cross section of the armour tile is machined to be a saddle shape in order to decrease residual stress and thermal stress during heating. The dimension of the cooling substrate is 100 mm wide, 400 mm long, and three cooling channels are swirl tubes, whose inner diameter is 15 mm with thickness of 1.5 mm with a twisted tape insert as a heat transfer enhancer.

The divertor element was tested in the JEBIS facility which is an electron beam heating facility in JAERI. Cooling conditions

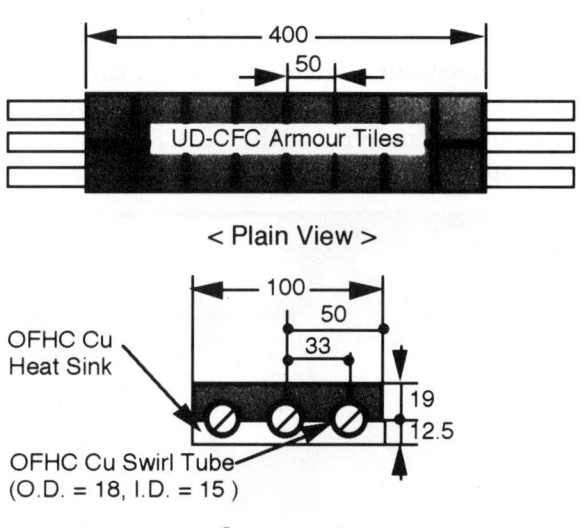

Fig. 1 Schematic of the Multi-Cooling Channel Divertor Element (Unit: mm)

are 2.5 - 10 m/s at a local pressure of 0.25 - 2 MPa, and the inlet temperature of water is about 25 °C. Temperature on the surface and in the copper heat sink were measured by an infrared (IR) camera and thermocouples, respectively. Since the thermocouples located far from the cooling tubes, it was difficult to detect the burnout of the cooling tube. In this experiment, therefore, a rapid temperature rise on the surface is monitored by the IR camera to detect the burnout. The incident critical heat flux of 20 MW/m^2, which is equivalent to the ITER design[1], is successfully achieved with the velocity of 2.5 m/s at the pressure of 0.25 MPa. The subcooling is about 100 K at this condition, which is also relevant to the ITER design. Though the local pressure and the inlet temperature are lower than the ITER requirement, the present result implies that the swirl tube could drastically reduce a flow rate, which would help saving a pumping power. Numerical analyses on the experiments are under way[4].

3. 3D-CFC DIVERTOR ELEMENT

Since the divertor armours should withstand high thermal stresses, the ITER divertor design prefers three directional CFCs (3D-CFC) more than UD-CFCs because of its higher mechanical strength. High thermal conductivity 3D-CFCs have been developed. Thermal conductivity of the 3D-CFC reaches about 500 W/m/K at 25 °C in one direction[5]. A small divertor element was fabricated to evaluate thermal performance of the 3D-CFC armours. A top view of the mock-up is shown in Fig. 2. 3D-CFC armours are brazed onto a cooling structure with a Cu-Mn braze. It should be noted that this braze does not include silver, which is relevant to the ITER design. The heat sink is made of OFHC Cu, and the cooling tube is made of DSCu with a twisted tape insert. For brazing between CFCs and DSCu, an interlayer of 1 mm thick OFHC Cu was used.

Heating tests were carried out in the JEBIS facility up to 15 MW/m^2 for 15 s. Figure 3 shows an IR image and surface temperature evolution monitored by an infrared camera during the electron beam heating. No detachment of the armour tiles were found after several thermal cycles at 15 MW/m^2. As shown in the figure, however, the surface of the armour locally shows rather high temperature, more than 1600 °C. This is caused by the fact that carbon fibers perpendicular to the incident beam are not efficiently cooled, while fibers in parallel to

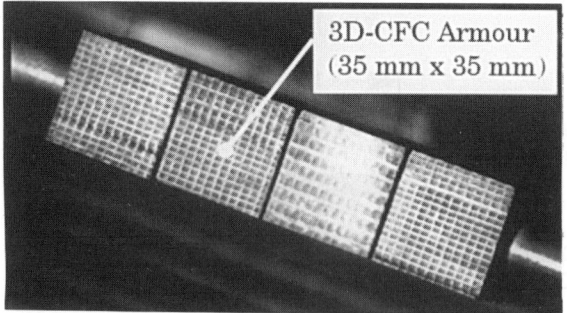

Fig. 2 Top View of the 3D-CFC Divertor Element

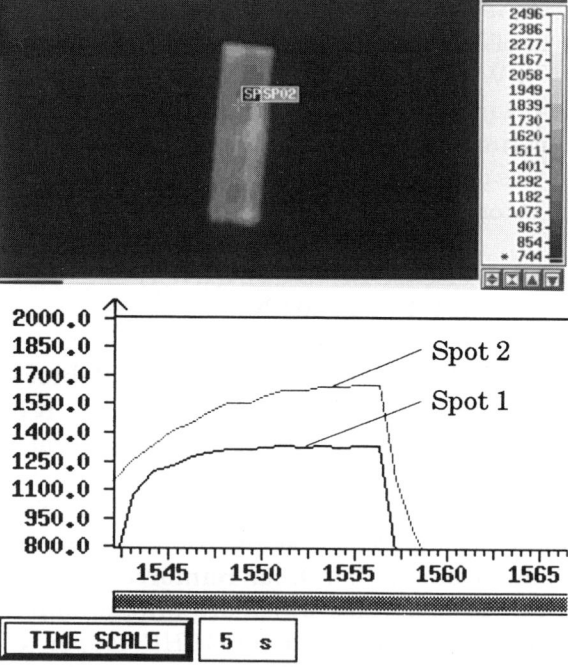

Fig. 3 IR image (upper) and temperature evolution (lower) of the element surface during heating

the beam are sufficiently cooled by the cooling structure with the braze. Based on this results, a new 3D-CFC material has been developed, whose surface region is composed of UD-CFCs[6].

4. 2D-CFC MONOBLOCK DIVERTOR ELEMENT

In the ITER design, divertor components in a louver configuration will be mounted to provide a neutral gas exhausting pass, which is called as a wing. Since the edge of the wing is exposed to the divertor plasma directly, the heat flux reaches up to 5 MW/m^2. JAERI has proposed a monoblock divertor element for the wing, which is composed of tungsten coated CFCs. As a first step, a monoblock divertor element with 2D-CFC armours was fabricated and tested in an ion beam test facility in JAERI. The monoblock configuration has possibility to reduce electro-magnetic forces. For this configuration, 2D-CFCs are preferred much more than UD- or 3D-CFCs, because they can provide rather uniform thermal properties in a cross-sectional plain. A schematic of the element is shown in Fig. 4, which simulates the edge element of the wing. 2D-CFC armours are brazed onto an OFHC Cu swirl tube with Ti-Cu-Ag braze. The inner diameter of the tube is 15 mm, and its thickness is 1.5 mm.

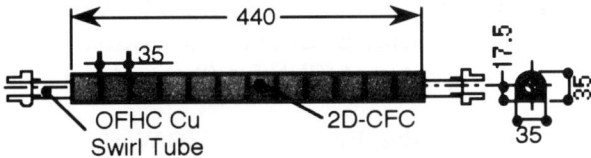

Fig. 4 Schematic of the wing-edge element with 2D-CFC armours

The heating test was performed with emphasis on the thermal fatigue of the CFC-Cu interface. To save machine time, an ion beam of higher heat flux for a shorter pulse is applied for this test as a simulated steady-state heat load. The heating tests started at a heat flux of about 2 MW/m^2, and the heat flux was gradually increased. Finally a heat load of 16 MW/m^2 for 1 s was selected for the thermal fatigue test, which provides almost the same thermal stresses at the CFC-Cu interface under a steady-state heat load of 10 MW/m^2. Neither detachment of armours nor implication of degradation of heat removal performance were found after 2000 thermal cycles. Based on this encouraging results, a bonding technique with a silver-free braze and a tungsten coating onto CFCs are being developed.

5. 1D-CFC DIVERTOR ELEMENT WITH THERMAL BOND LAYER

The bottom plate of the ITER divertor is exposed to not only a high heat load of more than 15 MW/m^2 for 10s, but also a disruption heat load of 100 MJ/m^2 for less than 10 ms. Since the disruption heat load causes severe erosion of armour material, it is preferable for armour tiles in this region to be replaced easily. Further to provide easy replacement of damaged armours of the first wall is critical to save time for maintenance and to reduce radio active wastes. From this point of view, components with a thermal bond layer (TBL) has been proposed for the divertor and the first wall designs in the ITER[1]. The TBL concept is based on the presence of a rebrazeable soft compliant layer between the armour and the heat sink. Several approaches have been proposed for TBL[7]; rheocast material, infiltrated felt, or low melting temperature braze. JAERI started applying pure Pb as TBL,

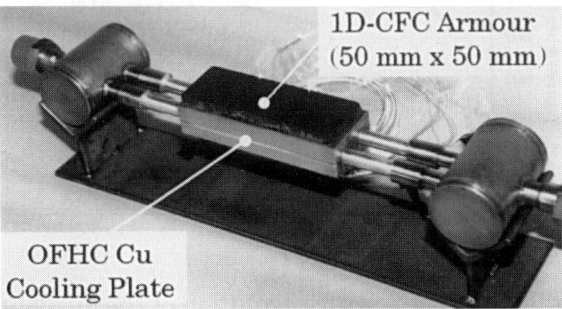

Fig. 5 Divertor element with a Pb-Cu braze

and found that it is critical to control a gap between armour tile and a cooling substrate. As a next step, 1D-CFC divertor element with a Pb-0.1Cu braze, as shown in Fig. 5, was fabricated and tested. 1D-CFC armours are machined in pyramidical, and are bonded onto a OFHC Cu cooling panel with a 0.5 mm thick Pb-Cu braze as shown in Fig. 6.

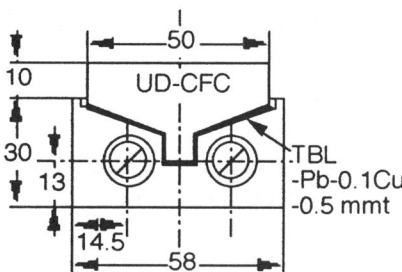

Fig. 6 Cross-section of the element

Heating tests were carried out at a heat load of 10 MW/m^2, 30 s in the JEBIS facility. The temperature evolution of the armour surface monitored by an IR camera and of the heat sink measured by a thermocouple are shown in Fig. 7 with a FEM numerical results. The measured temperature show good agreement with the numerical results. Though the Pb-Cu layer is partially melted at this condition, no indication of degradation of heat removal performance is found for 50 thermal cycles. A thermal fatigue test was also performed at a heat load of 15 MW/m^2, 10 s, which is equivalent to the transient heat load of ITER. The element survived up to 177 thermal cycles. Though maximum surface temperature increased cycle by cycle gradually, no armour detachment was found. Based on this result, further improvement will be continued, e.g., to use higher melting point braze, and to optimize a armour tile shape.

ACKNOWLEDGMENT

The authors would wish to thank the members of NBI Heating Laboratory for their valuable discussions and comments. They would also like to thank Dr. M. Ohta and Dr. S. Shimamoto for their support and encouragement.

REFERENCES

1. ITER, ITER EDA†Documentation Series, No.7, IAEA, Vienna, 1996.
2. S. Suzuki, K. Sato, M. Araki, K. Nakamura, M. Dairaku et al., Proc. ANS 12th Topical Meeting on the Technology of Fusion Energy, Reno, Nevada, US, June 16-20, 1996.
3. S. Suzuki, M. Araki, K. Sato, K. Nakamura, M. Akiba, to be presented at 16 th IAEA Fusion Energy Conference, Montreal, Canada, Oct. 7-11, 1996.
4. P. Sevini, S. Suzuki, M. Akiba, to be presented at ISFNT-4, Tokyo, April 7-11, 1997.
5. M. Akiba, Proc. Japan-US Workshop on High Heat Flux Components and Plasma Surface Interactions for Next Fusion Devices, pp. 196-205, edited T. Hino and M. Ulrickson, Sapporo, Japan, Jan.31-Feb.3, 1995.
6. M. Araki, Y. Kude, Y. Sohda, K. Nakamura, K. Sato et al., at this conference.
7. R. Matera, S. Chiocchio, G. Federici, K. Ioki, R. Raffray et al., at this conference.

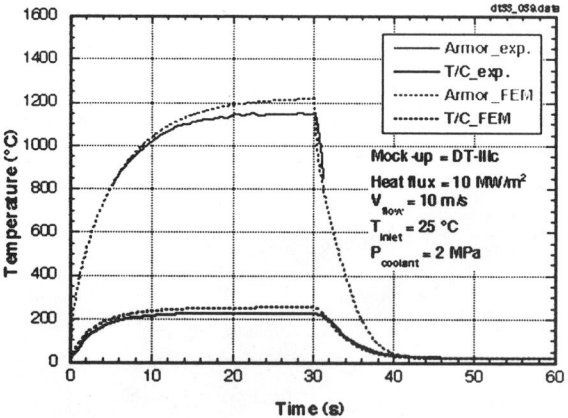

Fig. 7 Temperature evolution of the element

Boiling Crisis at Heat Transfer Intensification on Divertor Surfaces under Resistive and e-Stream Heating Condition

V.A.Divavin, S.A.Grigoriev, A.V.Lipko

Efremov Institute, St-Petersburg, 189631, Russia

Experimental data on determination of temperature state and ultimate heat fluxes for mock-ups with finned inner surface cooled by water (hypervapotron) are reported. Comparison of heat transfer efficiency for hypervapotron and circular tubular channels (with deposited porous coating) is also given.

1. Heat loading condition and geometrical sizes of mock-ups

Experiments on mock-up with finned inner surface cooled by water was carried out on two different facilities with heat loading of absorbed surface by: a) e-stream and b) special ohmic heater.

1.1 Heat loading by e-stream

Finned surface cooled by water was anode of specialy developed facility based on high energy transmitting triode. Scheme of this facility is shown on Fig.1. Cathode is row of parallel tungsten wires. Cathode is surrounded by anode of flattened tube type.

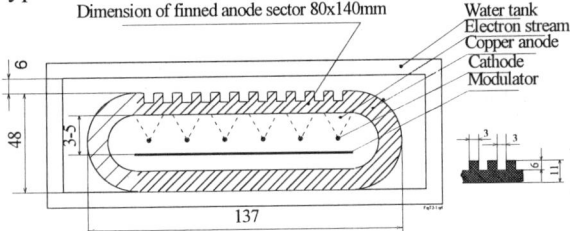

Fig.1. Testing scheme of finned surface at the experimental stand with e-stream.

Control electrode (modulator) is located between row of wires and anode (from non-operating side). The functions of this electrode are following: a) e-current to anode is controlled by variation of its potential and b) saving of non-operating (back) anode surface from e-stream.

Triode is located into the tank with water forced convection. The tank is made as two following variants: a) for water convection along fins and b) for water convection across fins.

Max value of heat flux on this facility is 25 MW/m^2.

1.2 Heat loading of special ohmic heater

The original method of one-sided ohmic heating (which had been used by us earlier at many tests on mock-ups with circular tubular channels) is used on second experimental stand. Carbon-composite heater operating at high temperature is clamped between two identical copper mock-ups (Fig.2). This multi-layer assembly is clamped between two current supplied copper buses. The main voltage drop takes place on flat C-composite heater at current flow across united assembly.

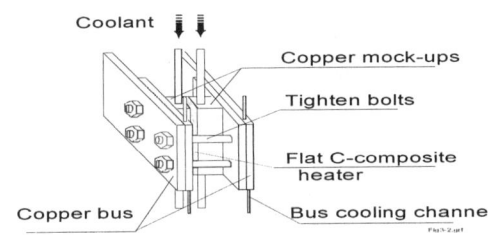

Fig.2. Testing scheme with special ohmic heater.

One-side heating is simulated on two testing mock-ups simultaneously by this ohmic heater operating at high temperature.

Testing of vapotron mock-ups with geometrical dimensions of fins as Falter's [1] ones was carried out on this facility. However, width of our mock-ups is less than Falter's mock-ups width in two times approximately, as there is limit of current value by supply electrical unit. That is why one side-wall of our mock-up was made from stainless steel. The low heat conductivity of the stainless steel allow to simulate here symmetrical-axis of Falter's mock-up. Construction is shown on Fig. 3.

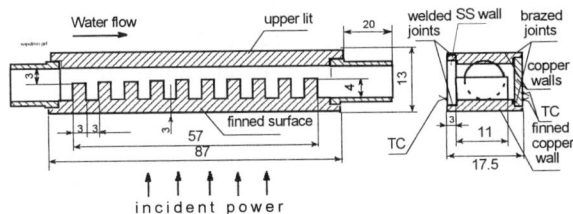

Fig.3. Vapotron mock-up with C-composite heater.

1.3. Test operating parameters for both facilities

Incident heat flux IHF, MW/m²   5 ÷ 20
Inlet pressure, MPa   02÷0.4 (1.1) & 1.8÷2.0 (1.2)
Inlet coolant tempr., C   7÷15 (1.1) & 80÷100 (1.2)
Subcooling Δt_{sub}, C   80 ÷ 100
Flow velocity w, m/s   1 ÷ 6

2. Test results for fined surface

2.1. Tests on facility with e-stream heat load

Ultimate heat flux (UHF) test data as a function of water flow rate is shown on Fig.4.

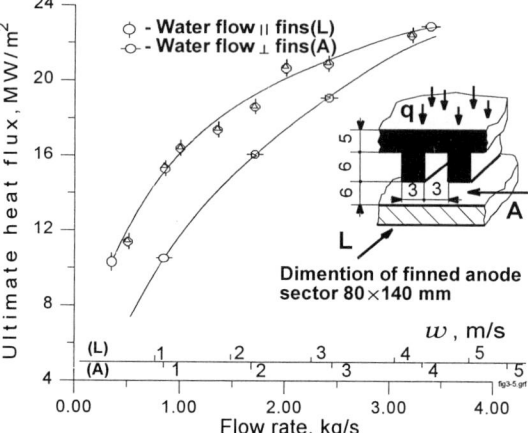

Fig.4. UHF at longitudinal and across flow round.

It is be seen on Fig.4 water flow along fins cause more UHF value than at water flow across fins at the same flow rate. This difference decrease (from 50% at w=1 m/s down to 5% at w = 4 m/s) at flow rate increasing.

2.2. Experiments with C-composite heater

Investigation was carried out for three value of incident heat flux: IHF = 5; 10 and 15 MW/m². Crisis was determined as abrupt increasing of temperature at IHF=idem and appearing of a large pulsation of signal from thermocouples located very closely to heated surface (at fin root).
In Fig.5 test results are given where UHF dependence on flow velocity in gap is shown for hypervapotron with geometrical characteristics the same as Falter's one [1]. The values of UHF obtained by Falter are also shown here. As it seen our test significantly extend investigated velocity range, and both our and Falter's results may be good extrapolated by one common curve UHF = $f(w)$.

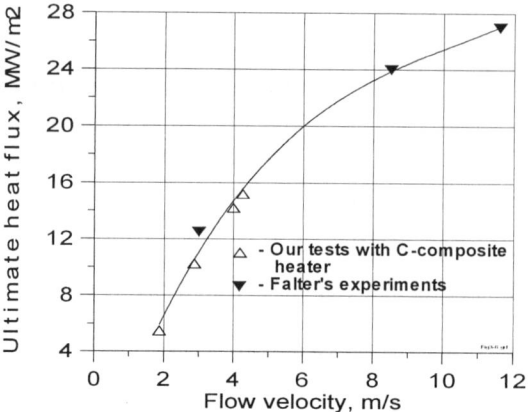

Fig.5. Water flow velocity dependence of UHF.

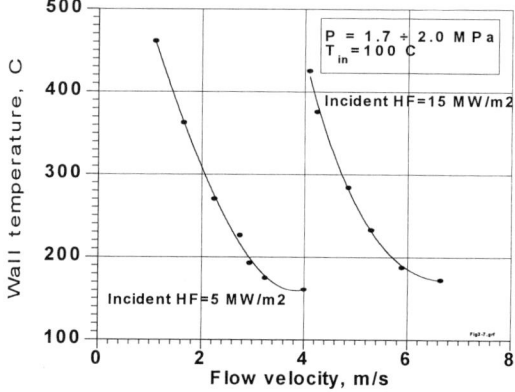

Fig.6 Flow velocity effect on wall temperature state.

In Fig.7 temperature status dependence on IHF, for velocity 4.3 m/s is shown. Analogous dependence was obtained by Falter [1] for w = 8.5 and 11.2 m/s. Here three sectors are also seen clearly: a) single phase convection, b) nucleate boiling, c) transient boiling with deteriorated heat transfer. Let mention, that as in our tests with e-stream, active boiling and acoustic effects appearing at IHF $\geq$ 0.5 UHF.

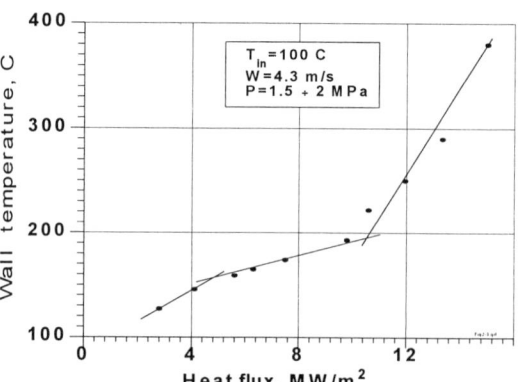

Fig.7. IHF effect on vapotron temper. (at fin root).

3. Testing of circular tubular channels with porous coating

During several years in Thermophysical Laboratory of Efremov Institute experiments under one-side loading of mock-ups with rectangular cross section are carried out. Cooling channels of these mock-ups have circular tubular form. Special attention in these experiments was paid to heat transfer intensification. Porous coating (PC) on inner channels surface was used as heat transfer enhancement. Mock-up with and w/o PC were investigated in a broad range of geometrical and operation parameters. Mock-ups testing was carried out by us on Russian and SNLA (USA) facility. Experiments showed that PC always leads to considerable increasing of UHF, not less than 1.4 - 1.5 times. Results of these investigations were published in materials of 1995 ASME International Mechanical Engineering Congress [2]. We obtained due to generalization of experiments with PC channels dependence $Y=f(X)$ in generalized nondimensional co-ordinates:

$X = A/\pi d$, $Y = UHF/CHF_{Shl}$, where:

UHF - **ultimate** heat flux absorbed by mock-up,
CHF_{Shl} - Critical Heat Flux [MW/m^2] data at uniform loading on circular channels w/o PC calculated by Shlykov correlation [3],

$$CHF_{Shl} = 4.12 \cdot 10^{-2} \sqrt{\rho' w} \; \Delta t_{sub}^{0.33} \left(1 - \frac{\rho''}{\rho'}\right)^{1.8}$$

ρ', ρ'' - liquid and vapour density of coolant, kg/m^3.
UHF dependence on geometrical parameters (Fig.8) has glaring maximum at $A/\pi d = 0.45$. Based on this experimental correlation we can predict UHF for any options of flow rate and geometry.

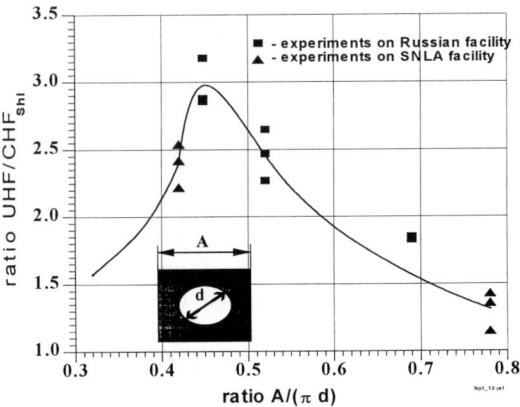

Fig.8. Influence of geometric and operating parameters on UHF at one-side heat loading of mock-ups with PC.

4. Comparison of hypervapotron and circular tubular channels with porous coating

Comparison of the experimental data may be sufficiently carry out for the same cpecific flow rate per unit width of thermally loaded mock-up only.
In Fig.9 Falter's results are compared with our data for circular tubular channel with PC tested on Russian facility. One Falter's channel of 27 mm width corresponds to our 2.5 mock-ups of 10 mm width each of them (with channels of 6 mm) accordingly to geometrical dimensions. In Falter's

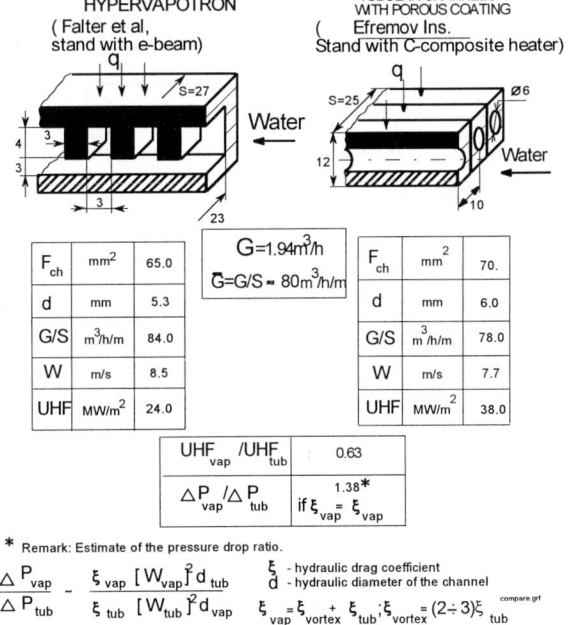

Fig.9. Comparison with hypervapotron (Falter at al) at G=idem.

mock-up at $G=1.94 m^3/h$ crisis came under $UHF=24.0\ MW/m^2$. At the same flow rate trough our 2.5 mock-ups (and at $\overline{G}=G/s=idem$) crisis set in at $UHF=38.0\ MW/m^2$ only, i. e. 1.5 multiple load increasing. In Fig.9 also calculated appraisal of pressure drop ratio between the two type of mock-ups is presented. Even equal drag coefficients (ξ) assumed for both channels, $\Delta P_{vap}/\Delta P_{tub}=1.39$. However it is clear that $\xi_{vap} \gg \xi_{tub}$ as ξ_{vap} has two component: ξ_{tub} - analogous to circular channel and ξ_{vortex} - due to vortex formation in graves between fins. Our estimates for ξ_{vortex} show: $\xi_{vortex}=(2 \div 3)\xi_{tub}$.

So that pressure drop (and pumping power losses) for hyprvapotron under $\overline{G}$ =idem is significantly higher than for circular tubular channel. But at the same time $UHF_{vap} < UHF_{tub}$. In Fig.10 analogous comparison of vapotron with increased fins (our stand with e-stream) and tubular channel had been tested by us on SNLA facility is given. Comparison was carried out under two value of specific flow rate: 48 and 96 m^3/h/m. And here also UHF_{tub} exceeding UHF_{vap} vapotron take place, although somewhat less (due to fin size increasing), Ratio $\Delta P_{vap}/\Delta P_{tub}$ is more large.

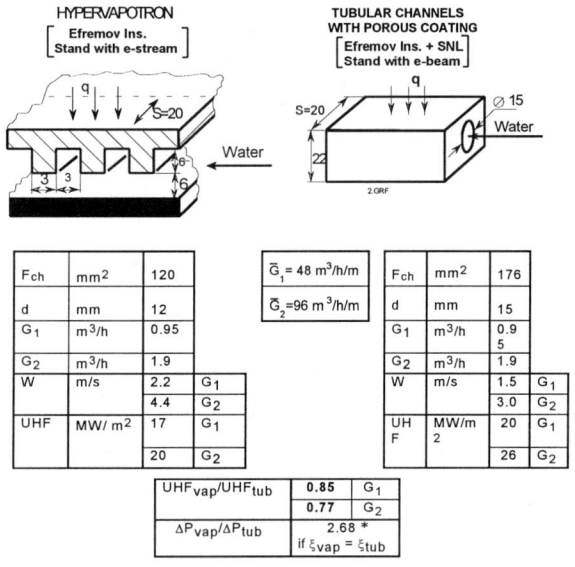

* Remark: The same as Fig.9.

Fig.10. Comparison with hypervapotron (Russian stand) at $\overline{G}$ = idem and S = idem.

This experimental correlation may be used for comparison of hypervapotron with PC tubular channels of the most optimal geometry. Results of such comparison at the same specific flow rate per unit width of loading surface, i.e. $\overline{G}$=G/S=idem are given in Table 1. The data for tubular channel are shown here based on correlation in point of optimal geometric relations (Fig.8), i. e. A/πd=0.45. Curve maximum is located in point $UHF/CHF_{Shl} \approx 3$. However, taking into account abrupt decreasing character of curve, value $UHF/CHF_{Shl} = 2.5$ was used for predicting. Such value of 2.5 was used for formation of Table 1. Because of the data in this Table is a pessimistic estimate.

This Table data shows that (in spite of pessimistic prediction of UHF_{tub}) wall with tubular channels

Table 1. Comparison hypervapotron and PC tubular channel of **optimal geometry**

Specific flow rate per unit width of mock-up G, kg/s/m	UHF, MW/m^2			
	Vapotron		Tubular channels with porous coating	
	h	6	4	d	10	8
	δ	6	3			
	t	3	3	sp	14.13	11.3
	sp	6	6			
20.0	21.0 (W=3.33)	24.0 (8.5)	28.66 (3.6)	32.05 (4.5)
27.16	23.5 (W=4.52)	27.0 (11.5)	33.4 (4.89)	37.35 (6.1)

may be always use to design as a more higher reliable on crisis than hypervapotron (with out increasing of specific flow rate for heat loading surface cooling).

5. Resume

1. Comparison of hypervapotron and circular channels would be carried out under the perfect operating parameters. Specific flow rate per unit of loaded surface width may be determine as the main among others.

2. Comparison shows that using of porous coating for tubular channels is more perspective then hypervapotron using:

$UHF_{tubPC}/UHF_{vap} \approx 1.5 \div 2$, $\Delta P_{tubPC} / \Delta P_{vap} < 1/3$

3. Fined wall temperature at $\overline{G}$=G/S=idem almost does't differ from that for tubular channel with PC.

4. Stiffness characteristics and reliability of tubular channels significantly higher then those for hypervapotron.

References

1. H.D.Falter et al. Thermal Test Results of the JET Divertor Plates. *Proc. SPIE Vol. 1739 High Heat Flux.(1992)*.
2. V.A.Divavin, S.A.Grigoriev, V.N.Tanchuk. High Heat Flux Experiments on Mock-ups with Porous Coating on the Inner Surface of Circular Coolant Channels. *Proc. of the ASME Heat Transfer Division, HTD-Vol.317-1 (1995 ASME International Mechanical Engineering Congress and Exposition)*.
3. Y.P.Shlykov et al. Critical Heat Fluxex in Tubes under Condition of Subcooling Boiling and Low Pressures, *Teploenergetica. 3 (1970)*.

Three Dimensional Neutronic and Activation Analysis for the ITER Design

J-Ch Sublet

UKAEA Fusion, Culham, Abingdon, Oxfordshire OX14 3DB, United Kingdom
(UKAEA/Euratom Fusion Association)[*]

Three-dimensional neutronic modelling of the ITER design out to and including the TF coils has been performed. The Monte Carlo code TRIPOLI was used to obtain the neutron responses required, especially for divertor safety assessment. The divertor cassette was divided into nodes, chosen according to the nuclear properties and materials of the items. FISPACT-FENDL inventory calculations, in each node, solved for the activation response functions and provided residual decay heat, specific activity, dose rates and hazard indices. It has been shown that three-dimensional neutronic and nodal activation and residual decay heat calculations can be performed in a convenient way on this basis.

1. INTRODUCTION

Safety and environmental (S&E) impact issues have acquired increasing importance for the development of fusion power. As part of the ITER programme and in line with the engineering feasibility studies, S&E analyses require a sound and reliable data base for the neutron-induced primary and secondary responses. The activation product inventory and residual decay heat generation depend on the specific design of a tokamak and its components, its geometrical configuration and material choices, as well as the given irradiation conditions: fusion power, operational scenario, and neutron source distribution. It is essential to include in the development activities properly performed activation inventory calculations that are consistent with the overall plant design.

The methodology applied comprises two main steps. Firstly, three dimensional neutron transport calculations to provide the distribution of the neutron flux in space and energy and secondly, activation calculations for each spatial zone taking into account the appropriate irradiation conditions. In principle, when following such a method, no model-related uncertainties are imposed on the calculational steps. This represents a clear advantage over the conventionally performed activation calculations that rely on simplified one or two dimensional models.

2. NEUTRON TRANSPORT

The modelling of neutron transport in a fusion device such as ITER is needed to enable a wide range of assessments to be made. A 3-D model is necessary in order to account accurately for the rather complex torus geometry and particularly the impact it has on the neutronic responses.

2.1 Geometry

An 18° sector model (one twentieth) of the ITER design has been modelled using the Monte Carlo code TRIPOLI [1]. The plan view symmetry of the device allows such a sector model to be an accurate representation of the entire tokamak. It covers a complete sector from the middle of a TF coil to the middle of the next one, assuming 20 TF coils, with reflective boundaries on each side. On the sectional elevation all major items are accurately represented up to the external layer of the TF coils. Each item is modelled as homogeneous within its external limits including: TF coils, TF coils protective layer, vacuum vessel, back plate, blanket element, first wall copper and protective layer, as well as a detailed layout of the divertor (vertical target option). The thickness of most components varies poloidally and this is particularly noticeable between inboard and outboard sections. Such a representation is achieved by first defining a set of surfaces that represents the boundaries of the items or shapes to be modelled and, secondly, building

---

[*] This work was jointly funded by the UK Department of Trade and Industry and Euratom

the volumes as a combination of these surfaces. Each defined volume is then allocated the appropriate material composition to combine geometry and material data. Fig. 1 represents an X-Y section of the model at the Z axis midplane where the grayscale allocation is done on a material basis. The model represented is 17.48 metres in height with minor and major midplane radii of 2.85 and 14.78 metres respectively. The divertor cassette overall dimensions fit in a rectangle of 2.05 by 5.24 metres. The material chemical compositions and densities were mainly taken from ITER SEHD documents.

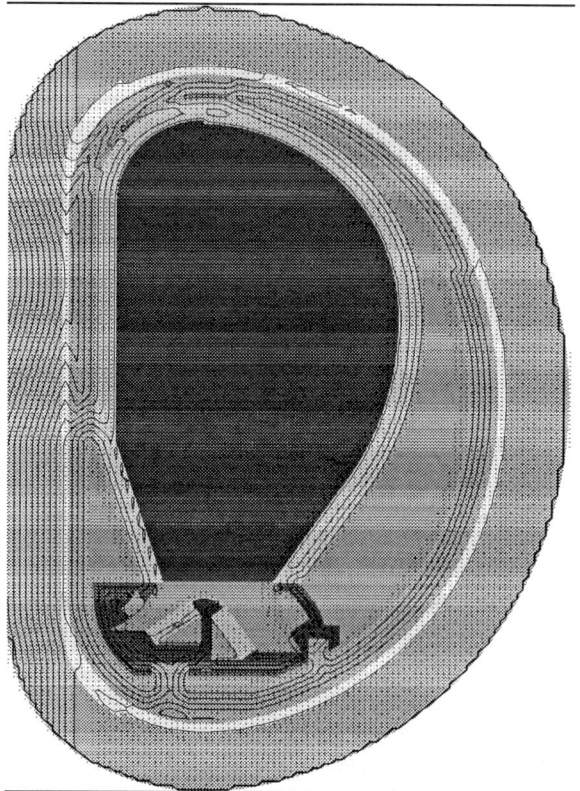

Figure 1 ITER X-Y model and importance map

2.2 Divertor neutron spectra

The spatial distribution of the neutron spectrum and energy integrated neutron flux density were obtained by the three-dimensional TRIPOLI calculations applying appropriate spatial segmentation and importance schemes. In order to give good statistical results for a deep penetration problem in a reasonable amount of time a Monte Carlo calculation needs to be optimised. Such optimisation is done within the module INIPOND of TRIPOLI where an importance map is automatically calculated. However, the determination of the spatial importance of each geometrical point in such a large model cannot be done and spatial discretisation is necessary. A grid in all three axes OX, OY and OZ of 100 x 180 x 50 meshes is superposed on the actual geometry. This leads to mesh sizes of 12, 10 and 10 centimetres that guarantee a proper spatial weighting. Such an importance map, where the lines describe contours of equal importance, has been superposed on a section of the model in Fig. 1 and demonstrates the complex neutronic behaviour encountered in the ITER device. Some additional specific optimisations are still necessary in order to account for each detail encountered in the lower part of the ITER machine. These include an additional weighting in energy to account for the increase in total cross-section for decreasing neutron energy, particularly important when hydrogen is present. A diffusion treatment is employed in the thermal energy region around 3.219 eV. A probability table treatment of the cross-sections has been used for all relevant resonant materials (Fe, W, etc.) in all calculations.

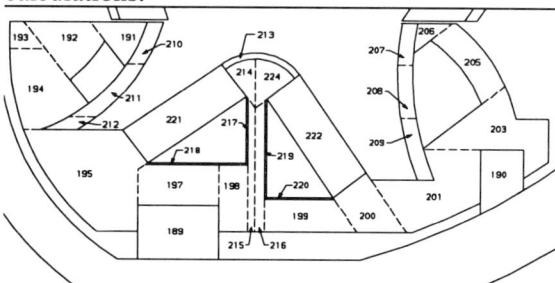

Figure 2 ITER divertor volume description

The zone volumes correspond to an 18 ° toroidal sector that contains 3 divertor modules. The divertor is described by 58 sub-regions or nodes that delineate the boundaries of each internal item. Some of these represent empty space or void for which neutron spectra are of less importance although in such 3D modelling they are accurately represented and accounted for. Each node is allocated a volume number and, as represented in Fig. 2, corresponds to a sub-item for which neutron responses are calculated. The importance map calculated early in the calculational procedure allows the transport calculations to be extremely efficient and leads to satisfactory standard deviations in the approximately 10 metre square that contains the divertor. Even after more than 3 metres biased penetration the standard deviation on the energy integrated flux is always less than 10%, and more often in the one percent range in innermost nodes.

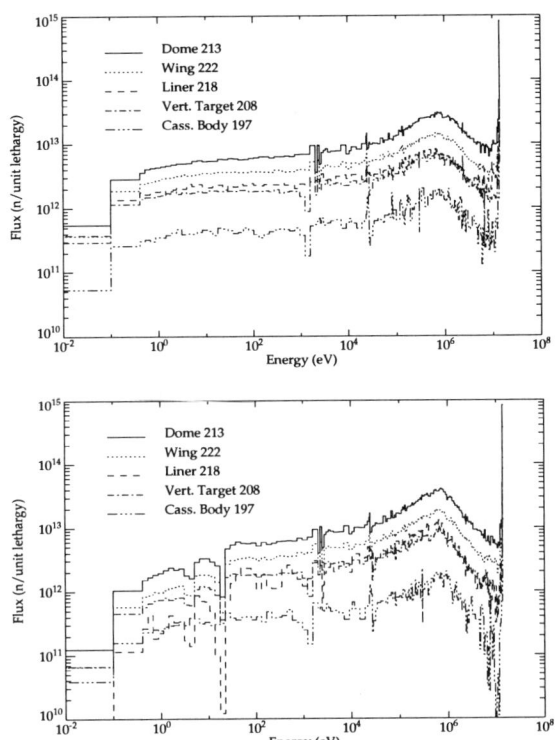

Figure 3 ITER divertor neutron spectra

It can be seen, in Fig. 3, that for neutron transport calculations a change of protective layer materials from beryllium (top) to tungsten (bottom) is significant. Another noticeable aspect is the much better fast flux attenuation experienced when tungsten is used. This is to be expected in view of the material density of the protective layer regions which changes from 1.85 g.cm^{-3} for beryllium, to 2.2 for carbon and 19.1 for tungsten while the fast absorption cross-sections are very different.

3. RESIDUAL DECAY HEAT

FISPACT-4.1 [2] inventory runs have been performed, using the FENDL/A-2.0 cross-sections and FENDL/D-2.0 decay data libraries, independently for all structural materials present in each previously defined ITER divertor node. The pointwise FENDL/A-2.0 library had been previously condensed into its 175 Vitamin-J group format, as FENDL/G-2.0_175, using the SYMPAL-96 processing code. These allow perfect compatibility with the activation library group format and the 175 group node-dependent neutron spectrum provided by the TRIPOLI transport calculation. In each divertor volume the proper node-relevant set of one group activation cross-sections linked with decay data library was then used. This set of inventory calculations was repeated twice in order to account for either a low Z (beryllium and carbon) or high Z (tungsten) protective layer. The nodal activation calculations follows exactly the volume definition and numbering of Fig. 2. More than one material can be present in a given node: if this is the case the inventory runs are not only node dependent but material dependent as well. For material apportionment the neutronic smear composition in each node was used.

3.1 Divertor integral results

Important integral activation results can be calculated, following such a procedure, and the residual decay heat is one of them. In order to calculate such activation responses for the set of 60 divertor modules present in the ITER device it is necessary to account for, in each node, the different material density and volume distribution. Certain activation responses, such as the gamma dose-rate, that are not specific (either per unit weight or volume) cannot be expressed as integral results. For each sub-item (dome, wing, etc.), further divided into nodes as described in Fig. 2, and for each node, the activation responses of all materials present weighted by their density and the node volume are added to form the node integral results. The addition of each node integral response forms a sub-item response. The summation of all sub-item responses forms the divertor integral results. This methodology allows the construction of integral results that account for all effects: 3-D geometrical, spectral shift, material layout, chemical composition etc.

Figure 4 summarises the findings of the activation study. The timescale spans from 1 second to 1 year after shutdown, for convenience of presentation, but ends at 50000 years in the calculations The total residual power integral results are presented for the three cases: beryllium, carbon and tungsten protective layer. It is notable that results with either carbon or beryllium as protective layer are very similar, within 1% at all times. The tungsten case gives rise to higher responses but only for times shorter than 3 months. A 15% difference exists at shutdown, peaking to a factor 2.7 at 1 day while decreasing afterward, reaching a constant few percent after 100 days. This clearly establishes the dominance of the tungsten isotopes, particularly for a time span from hours to months, even if the total amount of materials involved is limited. After a few days the massive amount of steel contained in the cassette bodies generates the bulk of the residual power although at that time the copper is not insignificant. The tungsten case demonstrates very clearly that changing even a minor amount of material (with regard to the total mass) could have a significant impact on the global activation response prediction.

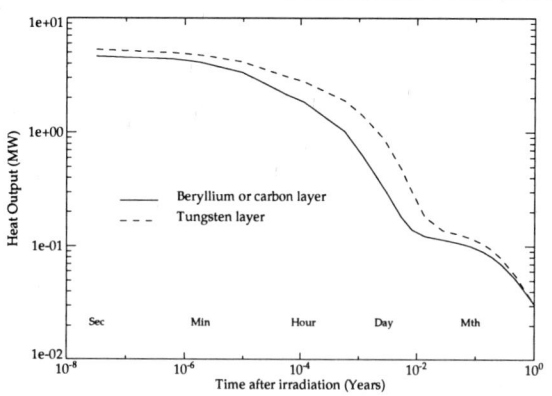

Figure 4 ITER divertor residual power

4. UNCERTAINTIES

Following the three-dimensional route to calculating activation response functions presents some clear advantages. One of them is that, and this is more a feature of the Monte Carlo method, the spectral uncertainties are calculable and usable. The origins of the uncertainties associated with all activation response functions for each material and node can be found in the FENDL-2.0/UN file and the specific handling and pathways routines of the inventory code FISPACT.

The relative importance of the activation results for a given node and/or a material to the global results is time dependent. The standard deviation arising from the Monte Carlo calculations, quoted for each node, is not. The method for the generation of covariance information, as in FENDL-2.0/UN is a significant achievement; merging such data with the statistical standard deviation produced by the Monte Carlo code TRIPOLI cannot be achieved on such a scale. However it would be a safe estimate to systematically add 3% at all times to the integral result uncertainties to account for the spectral uncertainty.

The divertor residual decay heat is in the region of 5 MW initially (total for all 60 divertor cassettes), falling below 1 MW between 5 and 24 hours and to 30 KW at 1 year. The one-sigma uncertainty in the integral divertor residual decay heat is in the region of 6% in the first days, increasing after a week to reach 26% at 1 year.

5. CONCLUSIONS

This integrated TRIPOLI - FISPACT calculational scheme has been applied to the determination of activation and decay heat values in a detailed 3-D model of the ITER divertor The full results of the study, presented in [3], include the activation and decay heat for each material in each geometric node of the divertor model, for both neutronic cases. This data and their associated uncertainties will be directly useful for input to safety and environmental modelling: for example calculations of temperature histories in the period following a postulated loss of coolant accident.

REFERENCES

1. J-C Nimal et al.,TRIPOLI-3 -User Manual, CEA DMT 96/026, 1996.

2. R A Forrest and J-Ch Sublet, FISPACT-4-User Manual, UKAEA FUS 287, 1994.

3. J-Ch Sublet, Three-Dimensional Neutronic, Activation and Residual Decay Heat Analysis for the ITER Design, UKAEA FUS 340, 1996.

Creep-Fatigue Damage in OFHC Coolant Tubes for Plasma Facing Components[*]

E.E. Reis and R.H. Ryder

General Atomics, P.O. Box 85608, San Diego, California 92186-5608, USA

Carbon-carbon (C-C) tiles brazed to water cooled copper tubes provide a concept for heat removal in high flux regions of fusion devices such as ITER. A nearly flawless braze between the C-C and coolant tube can be achieved by the hot isostatic pressing (HIP) process in which high pressure is applied to the surfaces of the components during brazing. The use of annealed oxygen free copper (OFHC) is often considered because its low yield strength not only allows the tube to conform to the C-C interfaces, but also minimizes residual stresses in the C-C that are developed during the brazing process. However, the low creep and fatigue strengths of OFHC cause concern regarding the survivability of the tube through the various operational loading cycles. An elastic-plastic-creep-fatigue analysis was performed for the macroblock design using a simplified axisymmetric structural model. The results show that the strain range induced by 350°C bake cycles exceeds the fatigue criteria after only 160 cycles. Although the creep stress level in the coolant tube is much lower than that for which data is available, use of extrapolated data estimates that high creep damage will occur. Recent high heat flux tests performed by JAERI on C-C saddleblock tile specimens resulted in a water leak due to thermal fatigue stress at the location predicted by the presented creep-fatigue analysis. It is concluded that relatively simple structural models can be used to predict failure in the coolant tubes for high heat flux components. In addition, these results indicate that a copper alloy with greater creep-fatigue properties than OFHC, but still acceptable for the HIP brazing process, is required for the monoblock and similar concepts.

1. INTRODUCTION

The steady-state operation of fusion experiments requires that the plasma facing components (PFCs) be actively cooled. Carbon-carbon (C-C) tiles bonded to water cooled tubes provide the presently preferred concept for heat removal in the high heat flux regions of ITER. The bonded interface for this tile concept is usually accomplished by a brazing process which produces significant residual stresses in the C-C tile and coolant tube. The divertor design favored for the now canceled TPX machine was the macroblock concept. The design utilized a 3-D random weave C-C material bonded to OFHC tubes by a HIP brazing process. In addition to having the capability of maintaining a maximum surface temperature of 1000°C for a steady-state heat flux of 7.5 MW/m^2, the macroblock design provides a robust structure for reacting halo current loads to a divertor support structure. There is concern that the low creep and fatigue strengths of OFHC will cause a coolant leak failure to occur. Therefore, a relatively simple structural model was developed to estimate the creep fatigue damage in the coolant tubes for the TPX macroblock concept.

2. STRUCTURAL MODEL AND METHODS

The elastic-plastic-creep fatigue analysis of a single coolant tube of the macroblock design was performed using the simplified structural model shown in Fig. 1. The analysis was performed with the ANSYS code [1] using eight node axisymmetric elements. This assumption is reasonable in view of the temperature distribution and the large thickness of the C-C block compared with that of the OFHC tube. Generalized plane strain conditions were imposed on the finite elements in the axial direction of the coolant tube.

The temperature dependent elastic-plastic material properties for annealed OFHC presented in Ref. 2 were used for the analysis. The stress-strain curves for OFHC are assumed to be adequately specified by a bilinear relationship. The C-C material was assumed to remain elastic at all times and its creep rate to be negligible.

The steady-state (secondary) creep strain increment $\Delta\varepsilon_{cr}$ as a function of time is computed in the ANSYS code by the following equation:

$$\Delta\varepsilon_{cr} = \dot{\varepsilon}\,\Delta t$$

---
[*]Work supported by IR&D Funds.

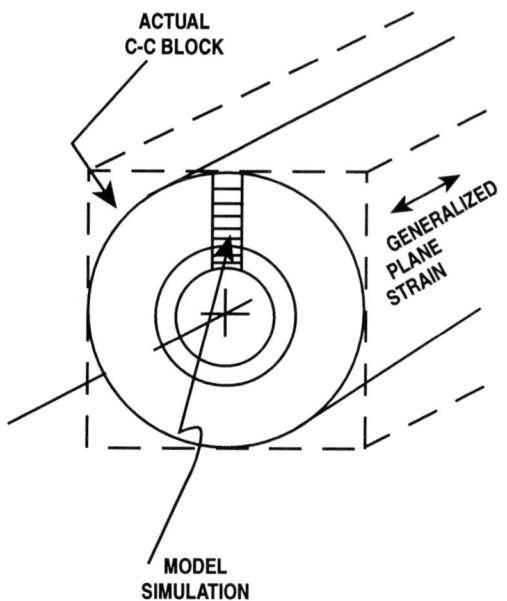

Fig. 1. Plane strain axisymmetric structural model of macroblock.

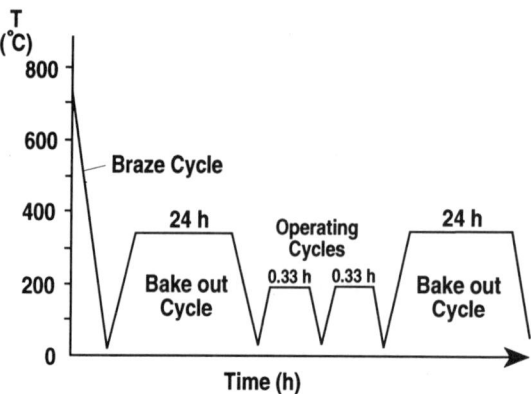

Fig. 2. Load histogram for TPX Macroblock.

where Δt = time increment and the secondary creep strain rate is:

$$\dot{\varepsilon} = A\sigma^B e^{-C/T}$$

and σ = equivalent stress (Von Mises)
T = absolute temperature
e = natural log base

The material creep constants (in SI units) over the temperature range of 260° to 455°C presented in Ref. 3 are:

A = 5.91 x 10^{-4}
B = 3.429
C = 7270

The load cycles for the TPX macroblock assumed for the analysis are shown schematically in Fig. 2. These cycles represent the cooldown from brazing followed by a bakeout and two operating cycles. A final bakeout cycle is then applied to check that residual stresses are not significantly affected by creep relaxation.

3. ANALYTICAL METHODS AND CRITERIA

The fatigue strain ranges over the assumed loading cycles were extracted from the ANSYS results and converted to equivalent strain ranges using the method prescribed in the ASME B&PV Code Case N-47. The number of bakeout cycles specified for TPX was 416 together with 30,000 cycles of normal operation.

Figure 3 presents the relevant data from various authors on low cycle fatigue of OFHC at 300°C. These fatigue tests were conducted in inert environments of vacuum or argon. The dashed curve in Fig. 3 represents the best least square fitted curve for push-pull specimens with test diameters of 6.3 mm. This curve would best fit structural designs with thick copper walls which have been annealed. The solid curve in Fig. 3 represents the design fatigue curve which is based on safety margins of 2 and 20 on stress and number of cycles, respectively, for the data fitted by the dashed curve.

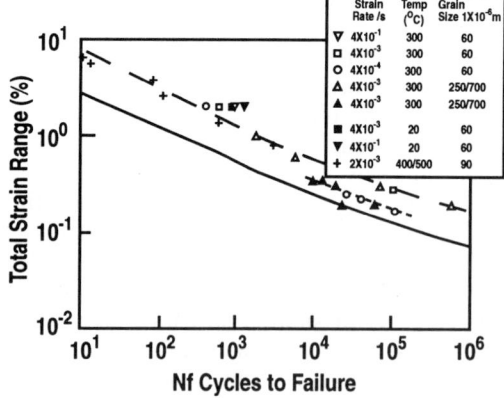

Fig. 3. Fatigue life of annealed copper in inert environment and plastic strain-range.

An estimate of the creep damage in the OFHC coolant tube was made using data from Ref. 3 as shown by Larson-Miller curves in Fig. 4. The curves were extrapolated to provide the time required for creep rupture to occur at the stress and temperature range of the coolant tube. Two possible extrapolations (curves A and B) to the low stress level are presented in Fig. 4.

A significant uncertainty in the evaluation of creep-fatigue damage is whether or not creep strain in compression is damaging in this material. In the majority of materials, it is not. In the calculations summarized, creep in compression has been conservatively assumed to be as damaging as creep in tension. This is particularly significant because, in the configuration investigated, the only elevated temperature hold periods (when creep may occur) are predicted when the stresses are compressive and cracks from creep damage would not be expected to initiate and grow in this constrained situation. The creep effects may indeed be small as indicated by the very small changes in the overall stress history obtained when creep was included in the OFHC material model. At both the bakeout and operating temperatures, the stress levels are low. However, the preliminary extrapolated creep rupture data used, predict very low allowable stresses. No design factor was used on the creep rupture data since the extrapolation was so uncertain.

The combined effects of creep and fatigue were evaluated against the linear creep-fatigue damage rule for stainless steel in the ASME code case N-47. In this case, when the damage contribution from the two phenomena (creep and fatigue) are of similar magnitude the combined usage factor is limited to about 0.3. When the contribution from either is insignificant, the usage factor for the other may approach 1.0.

4. ANALYTICAL RESULTS

Figure 5 shows a typical stress/strain diagram obtained from the analysis. The stress component illustrated in the figure is the hoop stress at a point within the wall of the tube near the braze connection to the carbon. The diagram is typical of all hoop and axial stress/strain components in the tube and is used here for illustration purposes. The actual values of strain used to calculate fatigue damage at the inside and outside surfaces of the tube were obtained by combining component strain ranges to give an equivalent strain range as defined above. The effective stress (Von Mises) was used to compute creep damage.

The average hoop stress in a coolant tube during the 24 hour bakeout at 350°C is 13 MPa. The total duration for the 416 bakeout cycles specified for TPX is approximately 10,000 hours. The creep

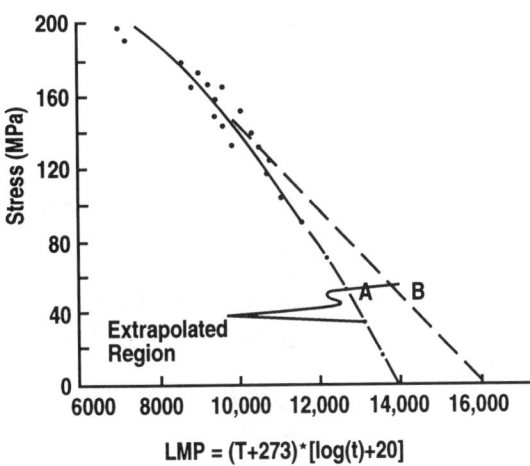

Fig. 4. Creep rupture data for OFHC (ASTM STP 181).

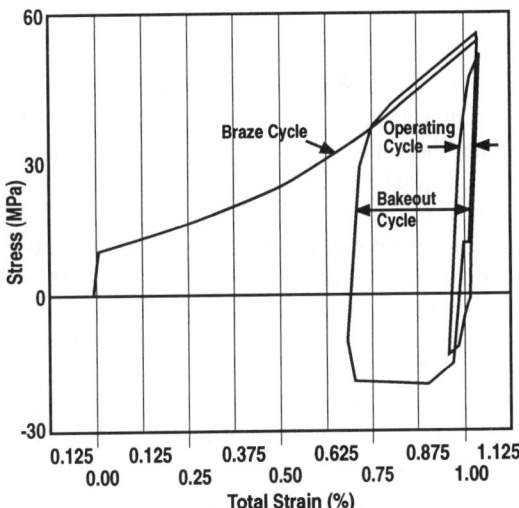

Fig. 5. Stress-strain behavior of OFHC at top of coolant tube.

rupture time corresponding to 13 MPa from extrapolated curve A in Fig. 4 is only 98 hours. Use of its value yields an unacceptable high value of creep damage. The creep rupture time obtained using curve B is 7.58×10^4 hours and results in a creep damage factor of 0.13. Clearly, the estimated creep damage is very sensitive to the method of extrapolation of the Larson-Miller curve and more long term creep data are needed to confirm the acceptability of the bakeout load cycles. The estimated creep damage due to operational cycles is negligible.

The cyclic effective strain ranges at the inside surface and at the interface with the C-C material over a bakeout cycle are 0.86 and 0.67 percent respectively. Using the design fatigue curve (Fig. 4), the allowable number of bakeout cycles is only 160 for the inside of the tube. Based on the specified number of cycles, the fatigue usage factor for this location is therefore 2.60. During operating cycles, the effective strain range at the inside of the coolant tube is only 0.05 percent from the through-the-wall thermal gradient. The effective strain range at the C-C interface is 0.18 percent for which the allowable number of cycles is 35,000 yielding a fatigue damage usage factor of 0.86. The point at the inside of the coolant tube is therefore the location where a crack is predicted to be first initiated. It should be noted that failure of the coolant tube will not occur until the flaw grows to a critical size or progresses right through the tube wall.

To ensure an adequate design margin with a total creep damage factor of 0.13, the summation of the fatigue damage usage factors must be less than 0.77 per the guideline of ASME Code Case N-47 for stainless steel. To satisfy these combined creep-fatigue criteria, the specified lifetime of the TPX brazed macroblock divertor concept would have to be reduced to 120 bakeout cycles and 14,000 cycles of steady-state operation.

High heat flux tests at 25 MW/m^2 were recently performed by JAERI on C-C saddleblock tile specimens (Ref. 4). The temperature of the top of the OFHC coolant tube reached 400°C during the test cycle, similar to bakeout conditions for the TPX macroblock concept. After 1200 thermal cycles, a water leak in the coolant tube was observed. Inspection of failure zone showed that the crack started at the inner top of the coolant tube and propagated toward the C-C interface as well as circumferentially. The vast majority of high heat flux tests on PFCs are terminated after 1000 cycles, far short of the expected lifetimes of the component. Reference 8 and the results presented in this paper support the requirement for extended numbers of cycles in testing high heat flux PFCs.

5. CONCLUSIONS

Although annealed OFHC coolant tubes provide an excellent heat sink for HIP brazed C-C PFC tile concepts, the low creep-fatigue strength of OFHC does not provide adequate lifetimes for divertor designs of present day tokamaks. The results of this analysis show that, for TPX requirements, there was inadequate margin of safety against creep-fatigue for the preferred divertor concept. The use of OFHC for TPX would have required replacement of the divertor modules after 120 bakeout cycles at 350°C and 14,000 cycles of steady-state operation. These results indicate that a copper alloy with greater creep-fatigue properties than OFHC, but still acceptable for HIP brazing to C-C tiles, is required for high heat flux PFCs.

Although the creep stress level in the OFHC coolant tube is much lower than that for which data is available, use of extrapolated data yields high creep damage levels in the TPX macroblock divertor concept. From this analysis it is clear that an adequate design of divertor coolant tube system for a present day tokamak requires careful consideration of creep-fatigue phenomena based on custom-generated materials data.

REFERENCES

1. ANSYS, Ver. 5.2, Swanson Analysis Systems, Inc., Hosuton, PA.
2. E. Chin, et al. "Elastic-Plastic-Creep Analyses of Brazed Carbon-Carbon/OFHC Divertor Tile Concepts for TPX," presented at the 15th IEEE Conference, 1995.
3. C. Upthegrove and H. Burghoff, "Elevated Temperature Properties of Coppers and Copper-Base Alloys," ASTM Special Technical Publication 181, 1956.
4. K. Sato, et al., "Recent R&D Activities on Plasma Facing Components at JAERI," presented at the 15th IEEE conference, 1995.

Radiation-induced desorption of hydrogen and deuterium from candidate plasma-facing materials

R.G. Macaulay-Newcombe[*] and D.A. Thompson

Department of Engineering Physics, McMaster University
1280 Main St. West, Hamilton, ON L8S 4L7, Canada

Abstract

Samples of high purity Be, C, W, BeO and Al_2O_3, either implanted with deuterium or loaded with deuterium by heating in a D_2 atmosphere, were studied using ion-beam analysis. Whether the samples were deuterium-loaded or as-received, there were large concentrations of surface or near-surface hydrogen, ~ 10^{20} at/m^2, and concentrations of > 100 appm at depths > 200 nm. The bulk concentration observed was much higher than expected, particularly in Be and W.

Sequential analyses showed that hydrogen and deuterium were desorbed by the analysis ions, with release cross-sections from 10^{-22} to 10^{-18} m^2. There were usually two desorption cross-sections, one dominant at fluences ≤ 10^{19} ions/m^2, and another, one or two orders of magnitude smaller, dominant at greater fluences.

The desorption cross-sections can be interpreted simply as detrapping cross-sections. Another approach is to model the process and fit an effective radiation-enhanced diffusivity to the data. The effective diffusivities of deuterium in BeO and WESGO-995 alumina during ion-beam analysis were found to be ~ 10^{-18} m^2/s at 300 K.

1. INTRODUCTION

Materials at or near the first wall in a fusion reactor may be exposed to 10^4 Gy/s of ionizing radiation, 10^{-6} dpa/s of elastic (collisional) radiation [1], and high deuterium and tritium fluxes. For safe and reliable reactor operations it is important to be able to predict the resultant tritium inventories and permeation fluxes through these materials. Some of the materials being considered are Be, C, W, BeO and Al_2O_3. While publications dealing with these materials are available, the database is incomplete and the uncertainties are large.

2. EXPERIMENTAL

Three ion-beam analysis techniques were applied: (i) D(^{3}He,^{4}He)p nuclear reaction analysis (NRA) [2], (ii) forward elastic recoil detection (ERD) [3,4], (iii) and Rutherford backscattering (RBS) [5].

In the NRA measurements deuterium was detected using 0.82 MeV ^{3}He ions. NRA is most useful for W and Al_2O_3, where there is very little background noise, enabling the determination of deuterium at concentrations as low as 50 appm. ^{3}He-D NRA is not useful for low levels of deuterium in Be or BeO, due to background from ^{3}He-^{9}Be nuclear reactions.

For the ERD measurements 1.2 - 2.25 MeV ^{4}He$^+$ ions were used, incident at a glancing angle of 75° to the surface normal. Hydrogen and deuterium atoms recoiled in the forward direction were detected and their energy determined by a silicon surface barrier detector. The detector is positioned to select particles emitted at about 30° to the incident ion direction. Forward scattered ^{4}He ions were largely filtered out by positioning a stopping foil of aluminized mylar between the sample and the detector. The ERD sensitivity for hydrogen and deuterium was as good as 5 appm; and the depth resolution was as good as 25 nm.

RBS is used to measure the concentrations of elements at or near the surface. The sensitivity of RBS increases with the square of the atomic number. The depth probed and depth resolution are dependent on the rate of energy loss of the ions in the substrate; for BeO, the the resolution was about 30 nm.

---

[*] Research supported by the Canadian Fusion Fuels Technology Project, and by the Natural Sciences and Engineering Research Council.

3. RESULTS

3.1 Depth distributions

Figure 1 compares the hydrogen depth profiles calculated from ERD analysis of a vacuum-annealed, polished single crystal of beryllium, and a similar sample loaded with deuterium by heating to 900 K in 13 kPa of D_2 for 5 days. The peak at zero depth corresponds to hydrogen adsorbed or trapped at the surface. (The high apparent concentrations of hydrogen at greater depths may include some background counts from the scattered ^{4}He ions.) Energy and angular straggling of the hydrogen signal results in some overlap into the deuterium region. In the case of the loaded sample, RBS indicated that a corrosion layer of nearly pure BeO grew during the deuterium loading process. This layer was much thicker than the region probed by ERD, and its thickness was non-uniform, resulting in increased broadening of the hydrogen peak. What appears to be hydrogen signal at negative depths is primarily due to resolution-broadening of the profile, caused by the finite width of the detector and ion-beam, and by energy losses in the stopping foil. In the D_2 loaded sample shown in figure 1(b), the D signal at greater depths overlaps with the leading hydrogen signal.

Figure 2 shows deuterium depth profiles determined from the same ERD spectra as the hydrogen profiles in figure 1. Here, the primary background signal is from hydrogen recoils. Averaging the apparent deuterium yield in Figure 2(a) up to a depth of 125 nm, it appears that this *unloaded* sample contains about 5 appm: this is an indication of the noise level.

Figure 2(b) is the apparent deuterium profile determined from ERD of the loaded sample. The sharp rise at about 75 nm is background signal due to hydrogen recoils; the estimated H-background is shown in figure 2(c). Figure 2(d) shows the net deuterium profile after subtracting (c) from (b). The net amount of deuterium indicated by figure 2(d) is 1.4×10^{18} at/m^2, or an average of about 117 appm over 85 nm. The uncertainty in the net deuterium profile is large: it depends strongly on the scattering angle and the estimated shape of the hydrogen background. In this case, the uncertainty in the deuterium yield is estimated as $\pm$ 60 appm, and the uncertainty in the location of the surface is $\pm$ 25 nm, due to uncertainty in the angles. In comparison, the solubility of deuterium in BeO would be 28 appm under the above-mentioned loading conditions [6]. The excess deuter-

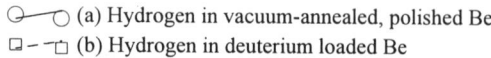
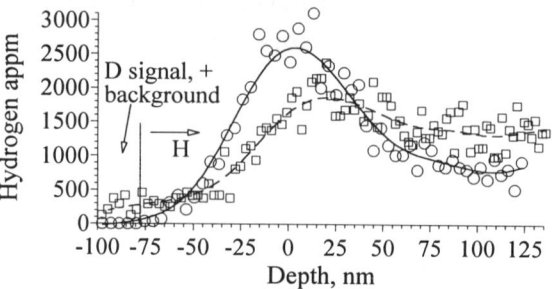

Figure 1: Hydrogen depth profiles in single crystal Be, determined from 1.6 MeV 4-Helium ERD spectra. Lines are arbitrary fits, drawn to guide the eye.

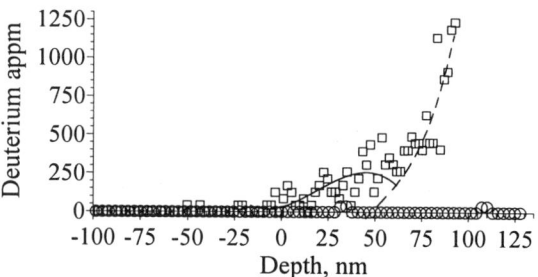

Figure 2: Deuterium depth profiles in single crystal Be, determined from 1.6 MeV 4-Helium ERD spectra. Note that the ERD spectra contain both hydrogen and deuterium recoils, and that there is some overlap.

ium is thought to be trapped, rather than dissolved.

Similar results were obtained for the other materials studied: BeO, EK-98 graphite, W and Al_2O_3. There were large amounts of hydrogen at or near the surface in all samples, with a tail extending well beneath the surface. Deuterium appeared to be trapped near the surface of all D_2-loaded samples.

3.2 Ion-Induced Desorption

During ERD analysis on the same spot, the analysis ions reduced the amount of hydrogen and deuterium present. The damage deposition rates of the various ion beams are: (a) ionization and excitation, $\approx 3 \times 10^6$ Gy/s, and (b) elastic collisions $\approx 2 \times 10^{-6}$ dpa/s. The elastic collision damage rate is close to the ITER first wall damage rate estimated by Zinkle [1], 10^{-6} dpa/s. However, the MeV ion-beam ionization rate is about 300 times higher than the ITER rates estimated by Zinkle, 10^4 Gy/s. Typical analyses lasted 60 s.

Figure 3 shows a hydrogen desorption curve compiled from six sets of spectra collected using 1.95 MeV ^{4}He ERD on a polished single crystal of Be. The yields have been normalised in order to display all of the data sets simultaneously. The initial rate of desorption is about 10 H atoms/^{4}He ion, dropping to 2 H atoms/ion at a fluence of about 10^{19}/m^2. In contrast, no desorption of near-surface hydrogen or deuterium was seen when a deuterium-loaded sample was annealed *in vacuo* for 16 hours at 473 K.

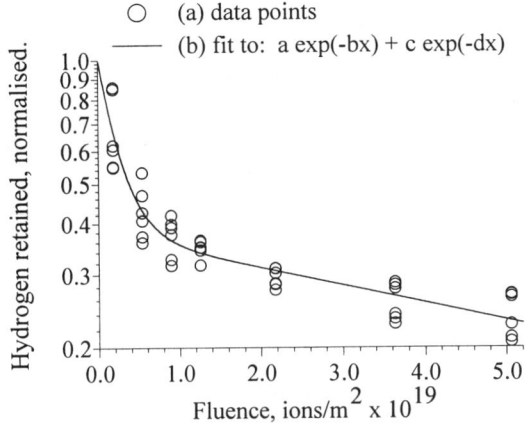

Figure 3: Hydrogen desorption by 1.95 MeV 4-Helium analysis ions during ERD of single crystal Be. Hydrogen yields from 6 sets of spectra have been normalised and plotted together.

The desorption curves determined for BeO were similar to those for Be, except that, since the surface was not roughened by corrosion during the loading, it was possible to determine deuterium desorption curves with more confidence This was also true for EK-98 graphite and Al$_2$O$_3$ samples. In contrast, the tungsten samples showed no measureable desorption of near-surface hydrogen during analysis.

4. CALCULATIONS AND DISCUSSION

4.1 Solubility

Figure 1 and other similar spectra from analysis of beryllium indicated a minimum hydrogen concentration of up to 1000 appm, much greater than the solubility (reported to be of the order of 10 appm, [7]). This might be attributed to trapping at defects during crystal growth. The large hydrogen concentration is consistent with earlier results [6] from thermal desorption of D$_2$-loaded Be: large amounts of HD as well as D$_2$ were desorbed, which could not be accounted for without assuming trapped hydrogen in the bulk of the sample, at about 500 appm.

Distinct from this trapped hydrogen, it appears that the most useful solubility equation for hydrogen in Be is: $S = 2.2 \times 10^{22}$ at/m^3(Pa)$^{1/2}$ exp(-0.17 eV/kT), from Shapovalov and Dukel'ski [7]. However, the intrinsic solubility may be somewhat lower still, since the hydrogen isotope content is dominated by trapping at defects and impurities. Even in well-annealed samples (65 hours in a vacuum at 973 K) the hydrogen content determined by ERD was always >> the deuterium content loaded in following the anneal, indicating deep traps. The presence of this much hydrogen explains the wide range of the literature results of deuterium and tritium solubility measurements, since the hydrogen can be expected to affect the deuterium and tritium absorption and desorption behaviour. Because of this hydrogen, it was necessary to use both thermal desorption and ion-beam analysis to determine deuterium solubilities.

The amount of hydrogen present at or near the surface is 1 - 2 orders of magnitude higher than in the bulk. Most of this hydrogen (and probably some of the deuterium) must be either adsorbed or trapped at the surface. Since neither hydrogen nor deuterium appeared to desorb during the anneal at 473 K, the activation energy for desorption must be ≥ 1.1 eV. This implies trapping rather than adsorption.

The same high concentrations of hydrogen were observed in the bulk and near the surface of BeO. The deuterium concentration after loading by heating in D$_2$ gas was consistent with the solubility previously reported [6], when surface-trapped deuterium is subtracted: $S \approx 10^{18}$ at/m^3(Pa)$^{1/2}$ exp(+0.8 eV/kT). The concentration of deuterium in WESGO-995 Al$_2$O$_3$, after loading in 13 kPa of D$_2$ gas, varied from 200 - 1400 appm, with the larger concentration occuring after a shorter loading time. This leads to the conclusion that the rate of loading is surface-limited, with kinetics controlled by surface impurities such as carbon. If the larger value is taken as representative of the solubility, and if the dissolution is controlled by the O-H bond energy (as appears to be the case for BeO), then $S \approx 5 \times 10^{19}$ at/m^3(Pa)$^{1/2}$ exp(+0.8 eV/kT).

The solubility of deuterium in tungsten appears to be too small to be measured by ion-beam analysis, in agreement with the published value [8].

4.2 Desorption Cross-sections

A simple way of interpreting the observed desorption is to assume that desorption is controlled

Table I: Ion-Beam Induced Desorption of Hydrogen and Deuterium from Be, C, BeO and Al_2O_3

Material	Isotope	Ion Beam	σ_1 (m^2)	σ_2 (m^2)	N_1 (at/m^2)	N_2 (at/m^2)
single crystal Be	H	1.95 MeV ^{4}He	4.3 ($\pm$ 1.4) x 10^{-19}	8.1 ($\pm$ 1.9) x 10^{-21}	5.5 ($\pm$ 1.4) x 10^{20}	3.4 ($\pm$ 0.8) x 10^{20}
EK-98 graphite	H	1.6 MeV ^{4}He	3.1 ($\pm$ 1.0) x 10^{-19}	1.6 ($\pm$ 0.4) x 10^{-20}	2.4 ($\pm$ 0.6) x 10^{20}	4.8 ($\pm$ 1.2) x 10^{20}
polycrystalline BeO	H	1.6 MeV ^{4}He	9.3 ($\pm$ 2.8) x 10^{-20}	9.9 ($\pm$ 3.0) x 10^{-22}	2.9 ($\pm$ 0.7) x 10^{19}	4.8 ($\pm$ 1.2) x 10^{19}
polycrystalline BeO	D*	1.6 MeV ^{4}He	1.4 ($\pm$ 0.7) x 10^{-19}	6.2 ($\pm$ 1.9) x 10^{-21}	6.3 ($\pm$ 1.6) x 10^{17}	2.0 ($\pm$ 0.5) x 10^{18}
polycrystalline BeO	D**	0.8 MeV ^{3}He	7.5 x 10^{-21}	3.9 x 10^{-22}	9.3 x 10^{20}	4.2 x 10^{21}
WESGO-995***	H	1.2 - 2.0 MeV ^{4}He	2.3 ($\pm$ 0.7) x 10^{-19}	9.2 ($\pm$ 2.8) x 10^{-21}	7.6 ($\pm$ 1.9) x 10^{19}	7.4 ($\pm$ 1.8)x 10^{19}
WESGO-995	D*	1.2 - 2.0 MeV ^{4}He	3.2 ($\pm$ 1.6) x 10^{-19}	8.4 ($\pm$ 2.5) x 10^{-21}	5.8 ($\pm$ 1.4) x 10^{18}	10.0 ($\pm$ 2.5) x 10^{18}

*D loaded by heating in D_2 gas **D implanted to saturation at 5 keV, Scherzer et al. [9] ***polycrystalline Al_2O_3

by detrapping of hydrogen and deuterium, with negligible contributions from diffusion and surface recombination.

The desorption data from Figure 3, as well as similar data for C, BeO and WESGO-995 Al_2O_3, were fitted to the desorption equation used by Scherzer et al. [9], where the number of retained atoms is: N = N_1exp(-$\sigma_1\Phi$)+ N_2exp(-$\sigma_2\Phi$). If N_1 and N_2 are taken as the populations of hydrogen (or deuterium) in two kinds of trapping sites, and Φ is the ion fluence, then σ_1 and σ_2 are the de-trapping cross-sections. These cross-sections are listed in Table 1, along with the extrapolated initial population of atoms in the two trap sites. The results of Scherzer et al. [9] on BeO implanted with 5 keV deuterium are included for comparison.

4.3 Radiation Enhanced Diffusion

Effective radiation-enhanced diffusivities of deuterium in BeO and WESGO-995 were estimated using TMAP4 [10]. In these calculations the solubilities used were as mentioned above in section 4.1, the recombination coefficient was taken from Causey and Wilson [11], and trapping was ignored.

The results showed that the effective diffusivities during ion-beam analysis at 300 K were: for polycrystalline BeO, D* $\approx$ 4 x $10^{-17 (\pm 1)}$ m^2/s, and for WESGO-995, D* $\approx$ 2 x $10^{-18 (\pm 1)}$ m^2/s. These values are about 27 orders of magnitude greater than those extrapolated from the thermal diffusivities determined by Fowler et al. [12]. The observed radiation enhanced diffusion is consistent with a reduction of the diffusion activation energy from 2.3 eV to 0.7 eV (for BeO) and from 2.2 eV to 0.8 eV (for the WESGO-995). Such high rates are only observed at temperatures above about 900 K, in the absence of radiation.

5. CONCLUSIONS

(i) Large concentrations of hydrogen were observed both on the sample surfaces and at depths > 200 nm. This may perturb thermal desorption measurements of deuterated and tritiated samples.
(ii) The deuterium solubility in WESGO-995 alumina was estimated as 5 x 10^{19}at/m^3(Pa)$^{1/2}$exp(+0.8 eV/kT).
(iii) Ion-induced desorption cross-sections for near-surface hydrogen and deuterium were in the range of 10^{-22} to 10^{-18} m^2 for elastic damage rates similar to those expected at ITER first walls [1].
(iv) Radiation-enhanced diffusivity appears to be significant for BeO (oxidised surfaces of Be) and Al_2O_3, at temperatures of < 900 K.

REFERENCES

1. S. J. Zinkle, "Ceramic Radiation Effects Issues for ITER", Plasma Devices and Operations **3** (1994) 139.
2. M. Caterini, D.A. Thompson, P.T. Wan and A. Sawicki, Nucl. Instr. and Meth. in Phys. Res. **B15** (1986) 535.
3. F. Besenbacher, I. Stensgaard and P. Vase, Nucl. Instr. and Meth. in Phys. Res. **B15** (1986) 459.
4. J.E.E. Baglin, A.J. Kellock, M.A. Crockett and A.H. Shih, Nucl. Instr. and Meth. in Phys. Res. **B64** (1992) 469.
5. Backscattering Spectrometry, W.-K. Chu, J.W. Mayer and M.-A. Nicolet, Academic Press, New York, 1978.
6. R.G. Macaulay-Newcombe and D.A. Thompson, J. Nucl. Mater. **212 - 215** (1994) 942.
7. V. Shapovalov, Y. Dukel'ski, Russ. Met. **5** (1988) 201.
8. R. Frauenfelder, J. Vac. Sci. Technol. **6** (1969)388.
9. B.M.U. Scherzer, R.S. Blewer, R. Behrisch, R. Schulz, J.Roth, J. Borders and R. Langley, J. Nucl. Mater. **85 & 86** (1979) 1025.
10. Tritium Migration Analysis Program, version 4, developed at I.N.E.L. by J.L. Jones and B.J. Merrill, manual by G.R. Longhurst et al., EGG-FSP-10315 (1992).
11. R.A. Causey and K.L. Wilson, J. Nucl. Mater. **212-215** (1994) 1436.
12. J.D. Fowler, D. Chandra, T.S. Elliman, A.W. Payne and K. Verghese, J. Am. Ceram. Soc. **60** (1977) 155.

Evaluation of an Advanced Silicon Doped CFC for Plasma Facing Material

C.H. Wu [a], C. Alessandrini [b], P. Bonal [c], A. Caso [d], H. Grote [e], R. Moormann [f], A. Perujo [g], M. Balden [e], H. Werle [h], G. Vieider [a]

a NET Team, Max-Planck-Institut für Plasmaphysik, D-85748 Garching, Germany
b Associatione EURATOM-ENEA sulla Fusione, C.R.E. Frascati
P.O. Box 65 - 00044 Frascati, Rome, Italy
d SEP. Division Propulsion A Poudre ET Composite, Les Cinq Chemins - Le Haillan B.P. 37
F 33165 Saint-Médard-en-Jalles Cedex, France
c Centre d'Etudes Nucléaires de Saclay, Laboratoire d'Etudes des Matériaux Absorbants,
91191 Gif-sur-Yvette Cedex, France
e Max-Planck-Institut für Plasmaphysik, D-85748 Garching, Germany
f KFA Jülich-ISR D-52425 Jülich, Germany
g European Commission, Joint Research Centre-Ispra, 21020 Ispra (Va), Italy
h Forschungszentrum Karlsruhe, 76021 Karlsruhe, Germany

To improve the CFCs characteristics of chemical erosion, tritium retention, and $H_s O$/air resistivity, an advanced silicon doped 3D CFC has been developed in the framework of the European Fusion Technology Programme.
This paper presents the manufacturing procedure, the thermal - mechanical properties and the results of experimental investigations on: outgassing behaviours, erosion by plasma, H_2O/air reaction, stability of doped Si-in bulk material under D^+ ion irradiation and D/T retention. The results have been critically analysed and the consequence is discussed.

1. INTRODUCTION

The use of carbon as a first wall and divertor plate protection material in a D-T fusion device is attractive due to its low-Z and excellent thermal shock resistance. It has a low neutron absorption cross section and it retains it strength at high temperature. However, carbon has the following principal limitations [1]:
(a) High specific surface area implying that tritium absorption and, hence, inventory may be a critical problem.
(b) High affinity for hydrogen resulting in a high chemical sputtering by formation of hydro-carbon species.
(c) Enhanced sublimation during ion bombardment (RES) at temperature T>1200K.
(d) High rate of reaction with water/oxygen at elevated temperature (T>1273K).
To improve the properties of carbon materials, the tritium inventory should be reduced, chemical erosion and RES have to be suppressed to increase the resistance to water/oxygen at elevated temperatures. It has been observed experimentally, that by adding small concentrations of impurities (SiC, B, B_4C) [2] to the bulk graphite, resulted in reduction of hydrocarbon formation for energetic hydrogen ions and for atomic hydrogen significantly.
In the framework of the European Fusion Technology programme, an advanced Si doped 3D CFC has been developed, and the properties have been investigated. This report presents the characteristics of this advanced material with respect to thermal properties, H_2O/O_2 reactions, stability of silicon concentration in CFCs at elevated temperatures, D/T retention, and chemical erosion as a function of D^+ energy and target temperature.

2. RESULTS AND DISCUSSIONS

(a) Manufacturing of material

This advanced Si-doped 3D CFC has been developed in co-operation between NET Team and SEP (Société Européenne de Propulsion), France, for

PFC material. The manufacturing procedures are described elsewhere [3]. This 3D CFC contained about 8-10 at. % of silicon. The porosity is around 5%.

(b) Thermal conductivity

Thermal diffusivities have been measured by the Laser Flash Method. The measurements have been carried out from room temperature to 1000°C. The thermal conductivities were then evaluated by equation: $K = D \times \rho \times C_p$.

Table 1 shows the thermal properties as a function of temperature. It is seen, that thermal conductivity is as high as 327 W K^{-1}m^{-1} at 298 K and 154.0 W K^{-1}m^{-1} at 1073 K.

(c) Oxidation behaviour of CFCs and SiC doped CFCs

Oxidation resistance is an important criterion in selection of PFC's due to the fact, that PFC-oxidation may cause severe consequence in loss of vacuum accidents and loss of coolant into vacuum accidents (water/steam ingress into the vacuum vessel). Facilities and first results of oxidation of carbon based doped (Ti, Si) and doped CFCs in steam and air are described elsewhere [4]. Comparing steam and air it was found, that oxidation rates in air are roughly of the same size at 700°C as in steam at about 1000°C; furthermore, sequences of oxidation resistance of different materials in steam are not the same as in air. In general, undoped CFCs have the same oxidation resistance as nuclear carbons V483T and A3. Highly SiC doped INOX A14 (40% vol.SiC) revealed an oxidation resistance in steam about 2-3 orders of magnitude better than undoped materials; doping with Ti, however, does not improve oxidation resistance. From experimental results, it can be seen, reduction of Si contains to 8 - 10 at.% decreases oxidation resistance in comparison with INOX A14 (40% vol. SiC), which, however, remains still better than for undoped carbons by a factor of about 4.

Most probably, the low oxidation rates of Si-doped carbon materials are caused by formation of a SiO_2-layer, which is known to have very low diffusion permeability for steam and oxygen at temperatures <1400°C and, accordingly carbon zones covered with SiO_2 are protected from oxidation. However, as theoretical examinations indicate, SiO_2 is formed only of oxidant pressure is sufficiently high, whereas at low pressures volatile SiO can be formed.

(d) Comparative investigation of tritium retention in C and Si doped CFC

The comparative investigation of tritium retention in undoped carbon and Si doped CFC was carried at ECN (Petten) and FZK (Karlsruhe). At ECN before loading the specimens were conditioned for one hour at 850°C under vacuum to remove air and moisture. Tritium loading was done with a (H2 + 2 ppm T2) gas mixture, at pressure of 800 mbar, temperature of 850°C for 10 hours. The retained tritium is determined

Table 1
Thermal Conductivity of Silicon doped 3D CFC NS31

Temperature (K)	Cp (J/kg·K)	Diffusivity (mm2/s)	Density (g·cm^{-3})	Thermal conductivity (W/m·K)
298	689.7	224.5	2.116	327.6
323	773.4	200.9	2.116	328.7
373	915.7	165.3	2.115	320.1
423	1031.2	139.9	2.115	305.2
473	1126.6	121.1	2.114	288.3
523	1206.6	106.5	2.114	271.6
573	1274.4	95.0	2.113	255.7
623	1332.6	85.6	2.112	241.0
673	1383.1	77.9	2,112	227.5
723	1427.3	71.4	2.111	215.2
773	1466.3	65.9	2.111	204.0
823	1500.9	61.2	2.110	193.8
873	1531.9	57.1	2.110	184.5
923	1559.8	53.5	2.109	176.0
973	1585.0	50.3	2.109	168.2
1023	1608.0	47.5	2.108	161.0
1073	1628.9	45.0	2.108	154.4

with ionization chamber by heating the specimens with 5°C min⁻¹ to 1050°C (purging with He + 0.1% H2) and oxidation at 850°C.

At FZK, Si-doped CFC was loaded with a (H2 + 135 ppm T2) gas mixture, at pressure of 370 mbar, temperature of 850°C for 6 hours. The retained tritium is determined with ionization chamber and proportional counter. For comparison, the tritium retention in various carbon materials is given in table 2. It is seen, that Si-doped (NS31) has the lowest tritium retention. The results also may show the general trend, the T-retention decreases with increasing density of carbon materials.

(e) Investigation of outgassing for SiC doped and undoped carbon

A systematic investigation of outgassing behaviour of Si/SiC doped CFCs and undoped CFCs was carried out in terms of temperature dependence and temperature pre-treatment. The experimental and evaluation procedures have been described in details elsewhere [5]. For comparative investigation, the characteristics of 5 materials; Sep Carb N112 (3D), Dunlop V (2D), SiC 2.5%, SiC 8% doped (CFCs (2D), 3D Novotex (No 3) and NS31 (Si-doped 3D) have been investigated. In order to obtain appropriate comparison, all pre-treated samples (1000°C, 20 hours) were exposed to air for two weeks before the outgassing test. In general the total amount of outgassing of Si-doped CFC is 1-2 order of magnitude lower than that of other carbon based materials tested.

Table 2
Tritium retention in various carbon materials

Materials	Density (g cm^{-3})	T-retention * (ppm)
S1611 (Graphite)	1.75 - 1.78	295
CL5890 (Graphite)	1.81 - 1.83	175
A05 (2 D CFC)	1.81 - 1.86	163
N112 (3 D CFC)	1.94 - 2.0	138
NS31 (Si-doped 3 D CFC)	1.94 - 2.11	71

* Corrected for isotope effects.

Table 3
Thermal Stability of NS31

Sample	Material	Diffusitivity (100°C) before heat treatment (cm^2/s)	Heat treatment	Diffusivity (100°C) after heat treatment (cm^2/s)
1	NS 31	1,4281	1200°C - 1h	1,3039
2	NS 31	1,4477	1600°C - 1h	1,409
3	NS 31	1,4253	1600°C - 20 s	1,4309

The results also indicated, that the release temperature of H_2O of Si-doped CFC is about 100°C lower that another carbon based materials.

(f) Chemical erosion and stability of Si dopant

The measurements of chemical erosion were carried out in two different facilities: a, Plasma Generator Device PSI-1 (flux density 10^{18} cm^{-2} s^{-1}), b, High Current Ion Beam Source (flux density 3×10^{15} D$^+$ cm^{-2} s^{-1}).

At plasma Generator Device PSI-1, the chemical sputtering yield was measured dynamically (during temperature rise) with a calibrated mass spectrometer and the CH-band intensity as well as by the weight loss method. The results of mass spectrometry agree well with the values obtained by weight loss method. It is seen, that the maximum chemical sputtering of Si-doped CFC(NS31) is at least factor 2 lower than that of undoped CFC(DUNLOP 3D).

At High Current Beam Source, the chemical sputtering yield was determined by weight loss method. It is clear that the yield of Si-doped CFC is about factor of 2 lower than that of undoped carbon.

To investigate the stability of Si dopant surface compositon specimens were analysed pre and post-heat treatment. Before the heat treatment, the Si concentration at surface is quite inhomogeneous. After the sample has been heated up to 1800 K for 2 hours, a loss 1-5% of Si has been investigated and enhancement of Si concentration has been observed after heat treatment. The chemical sputtering yield of heat treated sample at 800 K is about a factor of 5 smaller, and the sputtering yield at 40 eV and 300 K is 3-5 times smaller than that of undoped carbon.

(g) Stability of thermal property

To investigate the thermal stability, samples have undergone several temperature treatments. The thermal diffusivity has been measured before and after temperature treatments. The results are given in table 3. 3 samples were used for each set of experiment. The diffusivities presented in the table are the average of 3 measurements. Surprisingly, the decreasing in thermal diffusivity, even heated up to 1600°C, 1 hour, is less than 3%. It implies, that the thermal conductivity of NS31 will sustain through high temperature operation.

CONCLUSION

In the present study, a systematic investigation on the properties of an advanced silicon doped CFC is performed and the following conclusions can be made:
(a) The thermal conductivity of Si-doped 3D CFC is as high as 320 W m^{-1}K^{-1} at 300 K.
(b) It has been demonstrated that doping of Si/SiC can increase the oxidation resistance in steam.
(c) It seems that doping of silicon decreases the total tritium retention.
(d) The outgassing rates of gaseous species, H_2, CH_4, H_2O, CO and CO_2 of Si doped carbon materials are higher than those of undoped carbon materials, The release temperature of H_2 and H_2O is lower for Si doped CFCs.
(e) The chemical erosion yield of the virgin NS31 material decreased at least by a factor 2. Chemical sputtering yields decreased by factor of 5 following the heat treatment at 1800 K, 2 hours.
(f) The thermal conductivity of NS31 remains almost constant undergone high temperature operation to 1600°C.

REFERENCES

[1] C.H. Wu, Ceramic materials for the Next European Torus (NET) thermonuclear fusion reactor, invited paper (B5.2-LO2), presented at 7th CIMTEC World Ceramic Congress, Montecatini Terme, Italy, June 24-30, 1990, Ceramic Today-Tomorrow's Ceramic P3041-3057, P.Vincenzini (Editor), Elsevier Science Publishers B. V., 1991

[2] J. Roth, E. Vietzke and A.A. Haasz, in: Atomic and Plasma-Material Interaction Data for Fusion, Supplement to Nucl. Fusion 1 (1991) 63.

[3] SEP Final Report NS31 3DCFC, 1996

[4] R. Moormann, H.K. Hinssen, A.-K. Krüssenberg, B. Stauch and C.H. Wu, J. Nucl. Mater. 212-215, 1178 (1994)

[5] C.H. Wu, C. Allessandrini, R. Moormann, M. Rubel, B.M.U. Scherzer, J. Nucl. Mat. 220-222, 860 (1995)

Fracture Mechanics Investigations of Molybdenum Alloys

M. Rödig, H. Derz, G. Pott, B. Werner

Forschungszentrum Jülich GmbH, Association KFA-EURATOM,
D-52428 Jülich, Germany

Static fracture toughness K_{Ic} and dynamic fracture toughness K_{Id} have been determined for the molybdenum alloys TZM and Mo5Re. Testing temperatures were RT, 200°C and 450°C.

In a first series of experiments the problem of pre-cracking of samples has been studied. Normal fatigue pre-cracking as described in ASTM-E399 failed. An alternative pre-cracking method with slowly increasing maximum load in the fatigue cycle has been used in the static fracture toughness tests. The results have been compared to samples with machined notches produced by diamond wires. For both materials K_{Ic} is increased by a factor of 1.5 from RT to 450°C.

Results for dynamic fracture toughness surprisingly are 3 to 4 times as high as for the static fracture toughness. Dynamic fracture toughness is independent on temperature up to 450°C.

1. INTRODUCTION

Among the materials considered for high heat flux components in next generation tokamaks are molybdenum alloys. High rate loadings combined with sudden impact forces may be caused either by a loss of coolant accident at sub-cooled boiling conditions or by deposition of energies from runaway electrons in certain parts of the divertor. This may cause unstable crack propagation leading to a spontaneous failure of the component. Cracks may be initiated during the production process or develop during the operation of the component. For the assessment of cracked components, the knowledge of fracture mechanics data are required. In literature only little information is available on the fracture mechanics behaviour of molybdenum alloys. Most of these data has been obtained with specimens of relatively old batches [1-4] which no longer are representative for the grades produced nowadays.

The investigation which are carried out in co-operation with SCK/CEN, Mol (Belgium) cover dynamic and static fracture mechanics (K_{Id}, K_{Ic}) before and after neutron irradiation. For this purpose, several sets of samples have been irradiated in the BR2-reactor at Mol up to 0.2 dpa at 60°C, 200°C and 450°C. This paper reports gives the results of pre-irradiation fracture mechanics tests. Post-irradiation tests are planned for the end of 1996.

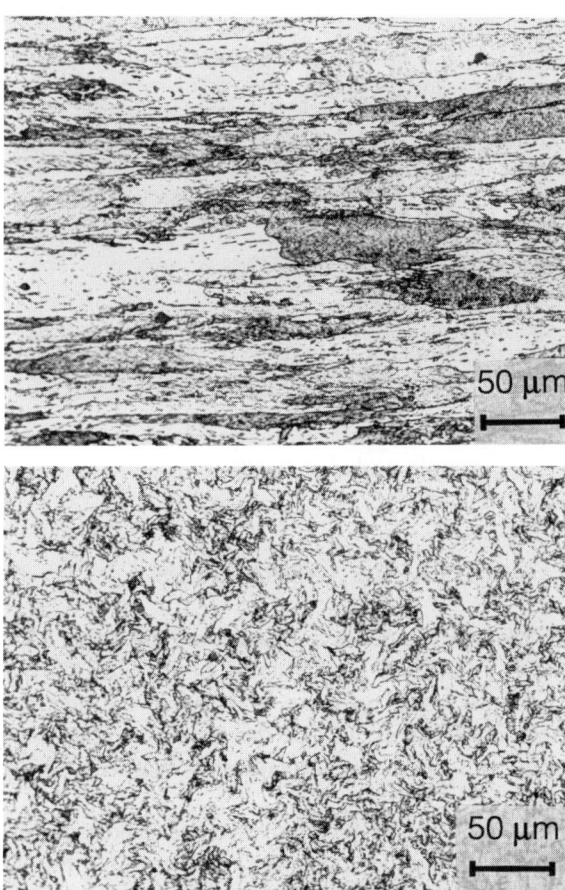

Fig. 1: Metallography for Mo5Re parallel (above) and perpendicular (below) to the axis of rod

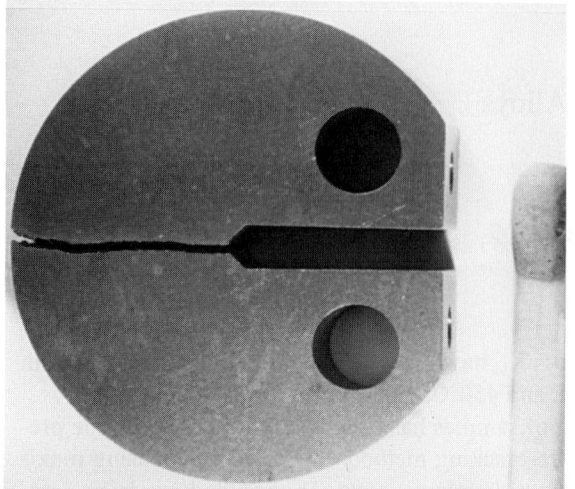

Fig. 2: DCT sample (tested at RT)

Fig. 3: Fracture surface of a pre-cracked Mo5Re sample loaded at 200°C

2. MATERIALS AND SAMPLES

Materials under investigation are the alloys TZM (0.5% Ti, 0.5% Zr) and Mo5Re. In the latter 5% rhenium was added in order to improve the ductility. Both materials as well as the testing samples have been produced by Metallwerk Plansee, Reutte (Austria).

Metallographic pictures have been taken perpendicular and parallel to the axis of the rods (cf. Fig. 1). Both materials show elongated grains with very irregular grain boundaries. The long axis of the grains is more or less parallel to the axis of the rod.

For the determination of static fracture toughness, 8 mm- DCT specimens (cf. Fig. 2) are used. The samples have been taken perpendicular to the axis of the rod (R-C direction according to ASTM-E399 [5]).

Dynamic fracture toughness was measured with impact bending samples corresponding to ASTM-E23 [5] with dimensions $10 \times 10 \times 55$ mm^3. They have been taken parallel to the axis of the rod (L-C or L-R direction according to ASTM-E399).

3. STATIC FRACTURE TOUGHNESS

3.1. Preparation of samples

According to ASTM E-399 [5] the samples shall be pre-cracked by fatigue. The pre-cracking is achieved by a stepwise increase of the upper load in the fatigue cycles. This method failed with the Mo-alloys due to their brittleness. Two alternative methods were tried instead:

1) A slow ramp was superimposed to the fatigue cycles. So the upper load during cycling was increased continuously until a first crack propagation could be monitored by a DC potential drop method. In order to mark the tip of the fatigue crack, heat tinting was used.

2) Experiments on ceramic materials and aluminium alloys have proved that even with machined notches valid K_{Ic}-values can be measured if these notches are sharp enough [6, 7]. It was found that above a critical value the measured fracture toughness is proportional to the square root of the notch tip radius ρ. But below this critical value ρ_c which is in the order of natural inhomogenieties in the material, the fracture toughness is independent of the notch root radius and is considered to be the "real" fracture toughness. In order to measure K_{Ic}-values with machined notches, a set of samples was prepared by means of diamond wires. Several values of notch root radii were produced by wires with diameters between 0.1 and 0.4 mm ($\sqrt{\rho} = 0.22 \ldots 0.45$ mm$^{1/2}$). In addition a notch width of 0.5 mm ($\sqrt{\rho} = 0.5$ mm$^{1/2}$) was produced by spark erosion.

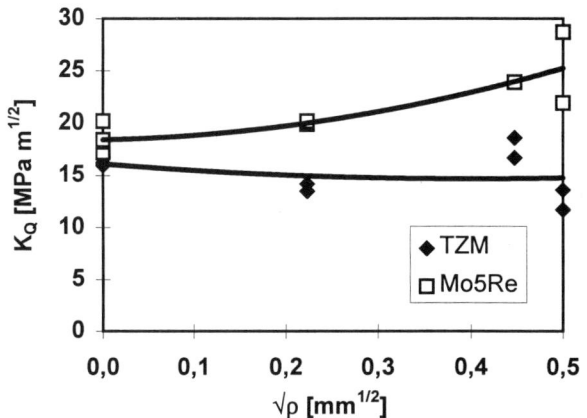

Fig. 4: Influence of notch root radius on fracture toughness

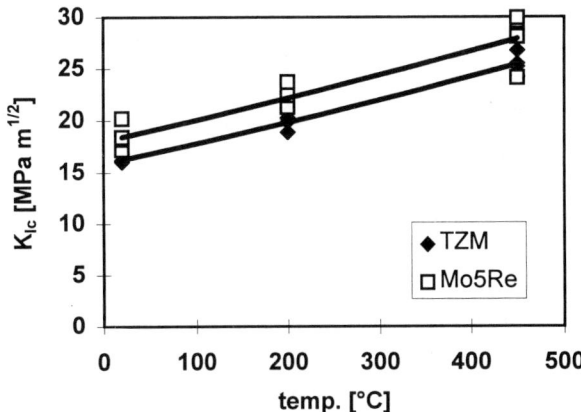

Fig. 5: Influence of temperature on K_{Ic}

3.2. Results

For pre-cracked samples one or more "pop-ins" are found in load displacement curves. They are thought to be a result of large grains with irregular structure surrounded by the fatigue pre-crack, and forming material bridges. During the K_{Ic} tests, these grains are broken (cf. Fig. 3) causing pop-ins. As they are due to local effects, they are not considered as pop-ins in the sense of ASTM E-399, and the maximum load is considered for the calculation of K_{Ic}.

Most of the validity criteria of ASTM-E399 are fulfilled. In particular the plastic zone size is small enough with respect to the dimensions of the DCT samples (up to the highest testing temperatures)

Influence of crack tip

Fig. 4 shows the influence of the notch root radius ρ on fracture toughness. Samples with $\sqrt{\rho} = 0$ are pre-cracked samples. For TZM more or less identical K_{Ic} values are found for pre-cracked samples and for samples with machined notches up to 0.5 mm width. For Mo5Re the dependence of fracture toughness on notch root radius is obvious, although a strict $\sqrt{\sigma}$- dependence cannot be confirmed (probably due to the low number of samples). The large scatter for the samples with a notch root radius of 0.5 is assumed to be due to the creation of fine cracks during the spark erosion process.

So for the determination of K_{Ic}-values in TZM, it should be allowed to use fine machined notches instead of fatigue pre-cracks. For Mo5Re it is preferable to use fatigue pre-cracks instead of machined notches. But for a saw cut of 0.1 mm width the value of the pre-cracked samples is nearly reached.

Influence of Temperature

Only samples with fatigue pre-cracks were used to study the influence of temperature on fracture toughness. For both materials K_{Ic} increases by a factor of 1.5 between room temperature and 450°C (cf. Fig. 5).

4. DYNAMIC FRACTURE TOUGHNESS

4.1. Preparation of samples

Pre-cracking of impact bending samples failed for both materials. From the experience of the static fracture mechanics tests which allow fine saw cuts of 0.1 mm instead of pre-cracked samples (s. above), it was assumed that the same holds for dynamic fracture toughness K_{Id} test. Therefore half of the impact bending samples were cut with a thin diamond wire of 0.08 mm thickness. The depth of these slots was 3 mm approximately. The other half of the samples was produced with ISO-V-notches which were sharpened by a razor blade in addition.

4.2. Results

In general all loading curves and fracture surfaces are similar for both materials, all testing temperatures and both type of notches (ISO-V and saw cut). The evaluation of dynamic fracture toughness K_{Id} followed the recommendations given in [8]. The dependence of K_{Id} on the testing temperature is shown in Fig. 6. Two surprising facts are found from the results:

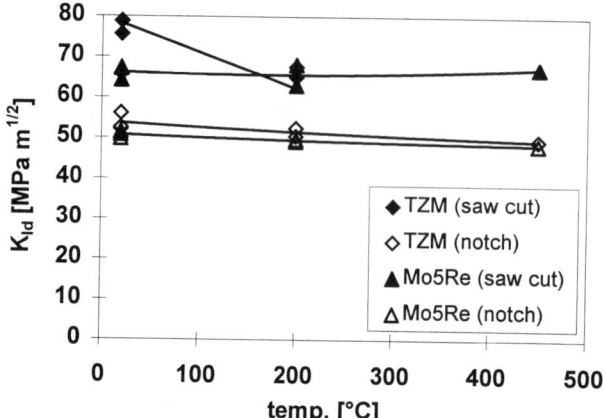

Fig. 6: Influence of temperature on K_{Id}

- The values of K_{Id} for samples with a saw-cut of 0.1 mm are about 30% higher than those for the samples with ISO-V notches.
- The dynamic fracture toughness values are 3 to 4 times as high as the values of the static fracture toughness.

The first fact may be explained by a different kind of stress state under static and dynamic loading conditions. The conservative situation which in the static tests is reached for a saw cut of 0.1 mm width, may no longer be given under dynamic loading conditions.

The second fact surprises, because K_{Id} was expected to be lower than K_{Ic}. This discrepancy may partly be explained from different planes of crack propagation for both types of samples relatively to the grain orientation. The measured values for static fracture toughness fit quite well with results found by SCK/CEN [9]. For TZM dynamic fracture toughness values of 55 MPa m$^{1/2}$ approximately were reported in literature [4]. The reason for the high values of K_{Id} compared to K_{Ic} is not yet understood. To understand this discrepancies, a second set of samples with different crack plane orientation should be investigated

5. SUMMARY AND OUTLOOK

Static fracture toughness

In a first series of experiments the problem of pre-cracking of samples has been studied. Normal fatigue pre-cracking as described in ASTM-E399 failed. An alternative pre-cracking method with slowly increasing maximum load in the fatigue cycle was more successful. The results have been compared to fracture toughness values gained with samples with machined notches. For TZM saw cuts of 0.1 mm width can be used instead of pre-cracking. For Mo5Re the values gained with machined notches of 0.1 mm width are only slightly higher than those gained with pre-cracked samples.

Room temperature values for K_{Ic} were found to be 16 MPa m$^{1/2}$ for TZM and 18 MPa m$^{1/2}$ for Mo5Re. For both materials fracture toughness increases with temperature. This increase is approximately 150% from room temperature to 450°C. The improvement in fracture toughness of Mo5Re compared to TZM is only 10% at all temperatures.

Dynamic fracture toughness

Two sets of samples have been tested. Samples with ISO-V notches show lower values of K_{Id} compared to samples with 0.1 mm saw cuts machined with a diamond wire. The measured values of dynamic fracture toughness are 3 to 4 times as high as the values of the static fracture toughness. Dynamic fracture toughness is insensitive on the testing temperature up to 450°C.

REFERENCES

1. J. Femböck, K. Pfaffinger, Proc. 10th Plansee-Seminar (1981), p. 221-232
2. Aerospace Structural Metals Handbook, AFML-TR-68-115, Vol. 5
3. G. T. Hahn et al., Eng. Fracture Mechanics, 2 (1971), p. 273-286
4. K. Krompholz et al., Zeitschrift für Werkstofftechnik, 15 (1984), p. 117-123
5. Ann. book of ASTM standards, Philadelphia 1994
6. L. Harrod, T. F. Hengstenberg, M. J. Manjoine, J. Mater. Sc. 4(3), 618 (1969)
7. E. Cooper, AWRE Report No. O17/72 (Jan. 1972)
8. Blumenauer, G. Pusch: "Technische Bruchmechanik", Leipzig 1982
9. Moons (SCK/CEN) private communication

Oxidation of Carbon-based Plasma Facing Materials in Accident-like Temperature Transients

H.-K.Hinssen[*], M. Hofmann[*], A.-K.Krüssenberg[*], R.Moormann[*], C.H.Wu[**]

[*]KFA Jülich/ISR, D 52425 Jülich; [**]NET, D 85748 Garching bei München

Accident analyses for steam and air ingress into the vacuum vessel of fusion reactors use kinetic equations for oxidation of the plasma facing materials measured under isothermal conditions and combine them with mass transfer rules. Some sources of errors in this procedure are outlined. In order to prove the applicabilty of this method an experiment is performed with a carbon based plasma facing material (AO5) oxidized by oxygen in an accident like temperature transient. Comparison of measured burn-off data with calculations (executed similar as the accident analyses explained above) shows a relative good agreement. The limits of this validation are discussed and additional experiments to this problem are proposed.

1. INTRODUCTION

Examination of oxidation of plasma facing materials (PFM) is an important task in accident analyses for fusion reactors: In case of an air ingress into the vaccum vessel following loss of vacuum by a vessel leak (LOVA = loss of vacuum accident) oxidation will occur by the oxygen component of ingressing air; rupture of a steam/water containing cooling tube connected to the vaccum vessel (LOCIV = loss of cooling into vacuum) will lead to PFM oxidation by steam. Consequences of these accidents to the environment may be release of radioactive materials (tritium, activated dust), which may even be accelerated by burning of combustible gases, formed during such oxidation processes. Also, severe damage of important components of the reactor may happen in relation to these oxidation, leading to significant economic consequences. Most discussed PFM for ITER are based on carbon, beryllium or tungsten. This paper concentrates on carbon based PFM, although limited work on the other PFM candidates has been performed in our group, too /2/.

Usually, accidents as LOVA or LOCIV are analysed in a manner, that chemical kinetic data of the relevant oxidation reactions are measured under isothermal conditions /1,2,3/ and that these kinetic data in combination with mass transfer rules are used in computer codes for simulation of the accident conditions; mass transfer rules are needed, because in the upper temperature range gas/solid reactions are controlled by mass transfer through the boundary layer /1,6/. Real accidents are accompanied by significant temperature transients and therefore the question has to be answered, whether oxidation kinetics measured under isothermal conditions is applicable on such transients without large error. Also, for accident analyses it has to be shown, that the usual mass transfer rules are applicable even in presence of chemical reactions, i.e. that chemical reactions do not disturb the mass transfer to a significant extent.

A suitable way to answer the questions above is an oxidation experiment on a well known carbon based PFM applying a temperature transient with starting temperatures well in the range of boundary layer mass transfer (regime III), but going down with temperatures through the in-pore diffusion controlled regime (regime II, kinetics is an important factor), until the oxidation rate becomes very small. Comparing the measured weight loss with that of calculations using the computer codes as explained above should give an indication about the errors, made by the beforementioned assumptions. Accordingly, this comparison will also contribute to a validation of our computer models used for accident analyses, in particular, if temperature transients and other experimental conditions are near to real accident conditions. This paper describes such a procedure for PFM oxidation by oxygen.

2. THEORETICAL BACKGROUND

2.1. Regime II oxidation

Under these oxidation conditions both, diffusion within the pore system and chemical reaction, determine the overall rate. For AO5/oxygen-reaction regime II is found approximately between 873 and 1173 K (these limits, however, depend on several influencing factors and are not generally valid). Already under isothermal conditions the overall oxidation process is complicated: There is a concentration and - as a consequence of that - a burn-off gradient within the carbon; however, because both rate determining steps depend on burn-off, these gradients change with time until at high burn-off a quasistationary state is reached. Concerning overall rate that means an increase with burn-off up to a plateau value. This plateau occurs, because the effective gas diffusivity increases continuously with burn-off, whereas the reactivity reaches a maximum at medium burn-off values; this maximum is due to the fact, that burn-off (or porosity) increase only enlarges the inner surface at low burn-off, whereas at high burn-off the pore walls vanish by oxidation, diminishing the reactive surface. In addition, because chemical reaction depends strongly on temperature, but in-pore diffusion does not the beforementioned gradients become steeper at increasing temperatures. Altogether, there is a very complex influence of temperature transients on the overall rate vs. time behaviour. An exact calculation requires effective gas diffusion coefficients and rate constants of the chemical elementary process and the dependence of both on burn-off and temperature. These data are difficult to generate; calculations based on roughly estimated data however indicate, that rates in case of temperature transients may be different by up to some ten percent from that, based on overall rates measured under isothermal conditions. A report on these model calculations is in progress.

2.2. Regime III oxidation

Two effects of carbon oxidation by O_2 may influence the mass transfer process: On the one hand, CO-formation will lead to a volume increase (2 CO molecules are formed from 1 O_2), resulting in a forced flow in opposite direction to O_2-diffusion (it should be noted however, that besides CO also CO_2 may be formed, which does not show a volume increase effect). On the other hand, CO-burning within the boundary layer may generate a significant gas temperature increase /4/. Whereas the first effect decreases slightly the oxygen transfer to the surface, the second one may lead to a significant increase of the mass transfer (factor 3-4), provided, that the Boudouard reaction ($CO_2 + C \rightarrow 2CO$) takes place on the surface, and O_2 is completely consumed by CO-burning within the boundary layer (meaning that CO_2 produced by burning is the oxidizing agent). For HTR-graphites (which show a similar reactivity as most PFM /1/) some experiments have already been performed at 1273 K covering also this problem in our laboratory /5/, indicating, that here Boudouard reaction remains negligible and therefore deviations from mass transfer rules could not be detected. This however may not be true at higher temperatures. It should be noted, that there is no burn-off dependence of rate in regime III and that burn-off remains limited to a small region near to the geometrical surface.

3. EXPERIMENTS

The induction furnace INDEX2, which is an improved version of the INDEX /1/, was used; this furnace allows for fast temperature transients because of its small thermal inertia. It consists of an induction coil containing inside a graphite cylinder with a cylindrical bore hole. The shape of the cylinder was optimized with respect to coupling to the heating middle frequency field. A tube shaped PFC-specimen (length: about 100 mm, inner/outer diameter: 10/20 mm) is brought into the borehole and the oxidizing gas flows through the inside of the tube. Fig. 1 contains a scheme of INDEX2. Herein a tube specimen of the 2D carbon fiber composite AO5 (fabricated by Carbone Lorraine, density 1870 kg·m^{-3}) was oxidized in 5 vol-% oxygen (rest argon) at a flow rate corresponding to Reynolds-numbers within the tube of about 2500, i.e. turbulent flow; before of oxidation, the specimen was treated for 1 h in an ultrasonic/acetone bath. The experiment started at 1473 K, which is certainly in regime III; temperature was decreased with 1.5 - 0.32 K·s^{-1} (typical for LOVA accident conditions /7/) until oxidation rate became very small at 890 K, which is near to the lower limit of the regime II. The temperature transient is shown in figure 2. Weight loss was measured after end of the experiment.

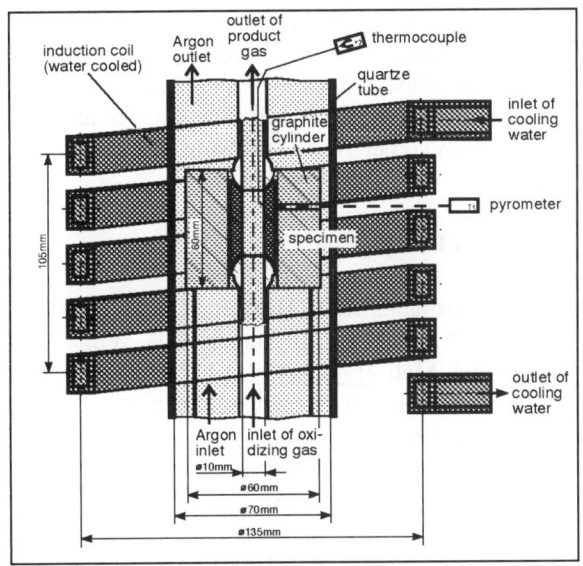

Figure 1. Scheme of the central part of the oxidation facility INDEX 2

4. COMPUTER CALCULATION AND COMPARISON WITH EXPERIMENTAL RESULTS

The computer code REACT/THERMIX /6/, developed originally for graphite oxidation calculations in HTR-accidents, was used. The oxidation model of this code is similar to that of our code COX-RALOC simulating accidents in fusion reactors /7/, which is at present in a testing phase. An isothermal measured (burn-off dependent) kinetic equation for the regime II of AO5 /7/ and a typical mass transfer rule for tube flow /6/ are combined in REACT/THERMIX for estimation of the oxidized amount of carbon. It is assumed, that burn-off caused at temperatures >1273 K does not contribute to the burn-off dependent increase of the regime II rate, whereas at lower temperatures the isothermal measured burn-off dependence of regime II is considered without restriction. This was done, because penetration of burn-off into the carbon material is much smaller in regime III than in regime II, and, accordingly, regime III burn-off influences the rates of regime II to a much smaller extent than isothermal measured rate vs. burn-off relations indicate. It should however be noted, that this procedure is only a rough approximation of the real behaviour.

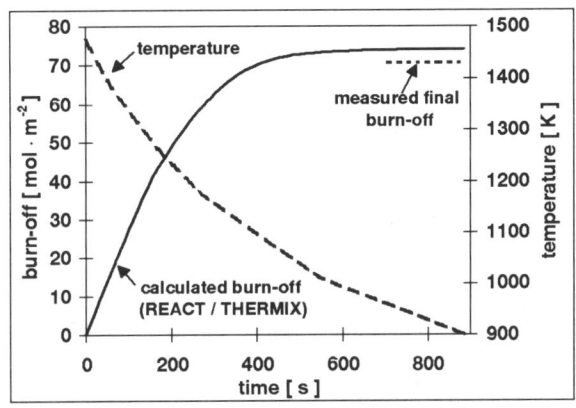

Figure 2. Time dependent measured and calculated burn-off of the 2D CFC AO5 in 5 vol% oxygen during a temperature transient

Figure 2 contains the temperature transient, the measured integral weight loss and the time dependent weight loss calculated with REACT/THERMIX. As can be seen, the difference between measured and calculated burn-off is about 5 % (overestimation by REACT/THERMIX-calculation). If the measured burn-off dependence of rate in regime II is taken into account also for regime III (which is not realistic, but may give a feeling of the errors introduced by the roughly estimated temperature limit of 1273 K) this overestimation increases to about 20 %. Considering, that standard deviations of our measured kinetic equations for AO5 are already in a range of 20 % /7/, the agreement can be quoted as good. In terms of the overall rate the calculated data may be interpreted as follows: A maximum rate of 0.26 mol $\cdot$ m^{-2} $\cdot$ s^{-1} occurs at 1473 K; this rate remains approximately constant until temperature reaches about 1273 K, because the by far dominating boundary layer mass transfer is only weakly temperature dependent. At about 1273K the intermediate zone between regimes III and II begins and here the rate is already 20 % smaller than the upper limit given by mass transfer. The mass transfer influence continuously decreases with decreasing temperatures, which also means, that rates are strongly reduced by falling temperatures. At 1173 K rates are 50 % and at 1073 K 85 % smaller than above mentioned upper limit due to the now dominating in-pore diffusion/chemical reaction. At 973 K the rate is fallen to 3.6 $\cdot$ 10^{-3} mol $\cdot$ m^{-2} $\cdot$ s^{-1} or 1.5 % of upper limit.

5. DISCUSSION

The good agreement between experiment and calculations means probably, that the 3 beforementioned reasons for possible deviations are not significant under conditions selected for the comparison. Therefore, the standard method for calculation of PFM-oxidation in air or steam ingress accidents of fusion reactors is applicable in temperature transients up to at least about 1473 K in case of oxidation by oxygen; however, for PFM significantly more reactive than AO5 used here, deviations cannot be excluded in the upper temperature range, in particular in case of a high reactivity against CO_2. Accordingly, additional experiments should be performed at higher temperatures or with more reactive PFM.

Although oxidation in steam was not studied here experimentally, some conclusions can be drawn for that: Deviations in regime II (which ranges to about 300 K higher temperatures than in case of oxygen due to the lower reactivity of C-based PFM against steam) have not to be expected, because the general reasons for deviations should be the same as for oxygen. In regime III an exothermic secondary reaction as for oxidation by oxygen (CO-burning in the boundary layer etc.) cannot occur; so this reason for acceleration of rate does not exist. However, volume increase by C/H$_2$O reaction may be significantly larger than in case of oxidation by oxygen (because concentration of steam is in most cases larger than that of oxygen, which is always accompanied by nitrogen (sources: air), and because oxidative steam consumption always leads to a volume doubling in contrast to oxygen consumption - see chapter 2.2); this effect however would lead to a rate decrease. An experiment at high temperatures (> 1673 K) with PFM oxidation by steam might be nevertheless reasonable. Such an experiment might also help to answer the question to which extent the good agreement between experiment and calculation found here is induced by the fact, that the possible reasons for deviations cause effects in opposite directions; this question should also be examined by other work studying these reasons separately.

6. LITERATURE

/1/ R. Moormann; H.-K. Hinssen; A.-K. Krüssenberg; B. Stauch; C.H. Wu: J. Nucl. Mat. 212-15 (1994) 1178-82

/2/ H.-K. Hinssen; A.-K. Krüssenberg; R. Moormann; C.H. Wu: Proc. 18th SOFT, Karlsruhe (1994), 1433-36

/3/ C.H. Wu; C. Allessandrini; R. Moormann; M. Rubel; B.M.U. Scherzer: J. Nucl. Mat. 220-222 (1995) 860-4

/4/ E. Specht; R. Jeschar: Ber. Buns. Phys. Chem. 87 (1983) 1099

/5/ M. Ogawa; B. Stauch; R. Moormann; W. Katscher: Jül-Spez-336 (1985)

/6/ R. Moormann; K. Petersen: Jül-1782 (1982)

/7/ A.-K. Krüssenberg: Dissertation Ruhr Universität Bochum (1996), in press

Thermal Fatigue Behaviour of First Wall Mockups with Defects

G. Hofmann, E. Diegele, M. Kamlah, S. Müller

Forschungszentrum Karlsruhe GmbH, Euratom-Association
Postfach 3640, D-76021 Karlsruhe, Germany

First wall mockups with artificial defects were thermal-fatigue-tested. The observed failure is reported and is compared to a prediction of crack growth. Additional cracks that started from intact surfaces and are predictable by standard codes limited the crack growth at some of the defects.

1. Introduction

The first wall (FW) may contain defects in the form of cracks originating from manufacture, overloads (disruptions), or earlier fatigue. Due to the pulsed nature of the present day Tokamak operation the FW area is subjected to cyclic thermal loads and corresponding stresses and strains. In addition stress concentrations caused by defects could shorten the fatigue life time of the component. In this context thermal fatigue experiments under purely thermal load were performed in order to contribute to answers to three questions:

(a) where and how would defective specimens fail, (b) are there indications of crack arrest, and (c) how well predictable by analysis is crack growth under these conditions?

2. Specimens

The specimens were rectangular bars made from stainless steel AISI 316L with a cross section shown in Fig. 1 and 260 mm long. They were designed by the NET-Team and manufactured by Sulzer Innotech in 1993. Each one contained two cooling channels drilled into the bars at dimensions typical for a FW structure; the water supply lines to each channel were connected through the back wall close to both the top and bottom ends. Most of the specimens carried artificial defects in the form of electro-eroded notches in positions at that time expected to be weak points.

There were three specimens selected to be tested simultaneously in the present program: Specimen No. 3, as a reference, did not have artificial defects. Specimen No. 13 carried three notches eroded into the wall of one of its cooling channels at position A of Fig. 1 and oriented normal to the channel axis. Specimen No. 11 carried three notches eroded into the heated surface at position B of Fig. 1. The 0.1 mm wide notches were distributed over the length of the specimens, their shapes were nearly semi-elliptical, and their initial dimensions may be read from Fig. 1. To increase its emissivity the heated surface of the specimens was coated with a 20 μm thick layer of plasma-sprayed Al_2O_3 + 13% TiO_2.

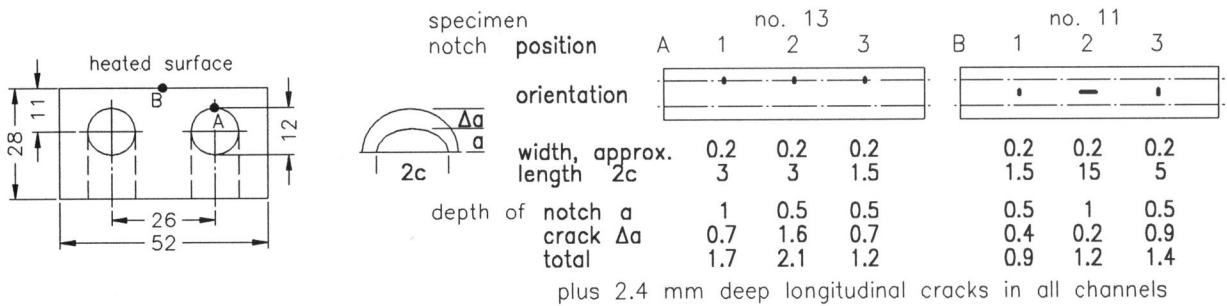

Fig. 1 Shapes and sizes of specimens, notches and cracks

3. Experiment

The three specimens were thermal-fatigue tested in the FIWATKA facility [1] at FZK. They were positioned side by side in a window of the water-cooled heater housing (Fig. 2) and surface-heated cyclically by thermal radiation from a graphite resistance heater in vacuum. For each 180-s-cycle the quick (low-mass) heater was energized for the first 80 seconds and transferred a heat flux of 75 W/cm² to the surface of the specimens. The specimens were cooled continuously with water (8 bar and 30 °C, demineralised and low oxygen); the heat transfer in the cooling channel was 2.2 W/(cm².K) and the water heated up about 4 K when it passed through two cooling channels of a specimen in series. The heat flux received by the surface of the specimen was determined calorimetrically from flow rate and heat-up of the cooling water at the end of a heating phase. The hottest point at the surface of the specimens reached about 450 °C and cooled down during the 100 seconds of the dwell phase to almost room temperature. The specimens were unconstrained since they were fixed in only one point and since the water supply lines were flexible hoses.

As a reference for uniform cycle operation each specimen was equipped at its rear surface with a thermocouple and two strain gauges (longitudinal and transverse) in its center and with three displacement sensors (LVDT) along a 200 mm line to monitor the bending. In Fig.3 the measured quantities are plotted for one cycle indicating that the specimens reached equilibrium at the ends of the heating and cooling phases. The experiment was interrupted every about 10000 cycles when the specimens were taken out for an examination of the crack depth.

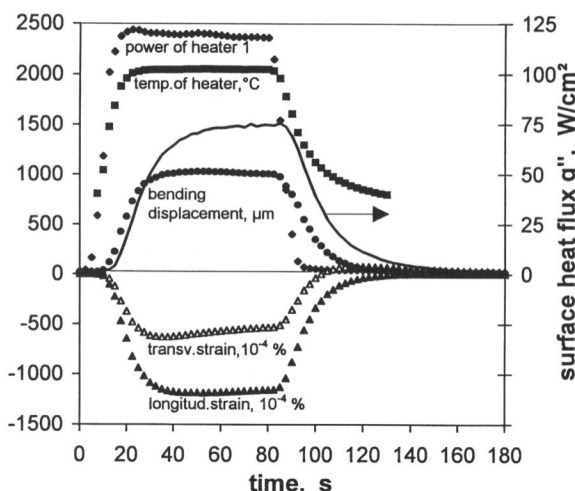

Fig.3 Experimental behaviour during a cycle

4. Observed Failure

4.1 Examination Techniques

Failure was expected to be initiated at the artificial notches and to propagate into the specimens. The eddy current technique was utilized for the non-destructive examination (NDE) of the crack growth since the expected crack propagation is normal to the respective surface, an orientation for which this technique promises to result in high resolution.

A differential sensor (two coils) with 1 mm effective width was operated typically at a frequency of 2 MHz. The sensor was positioned over the center of a notch and moved in steps along its length by rotating it along the inner notch (type A) or by traversing it over the outer ones (type B). All notch areas were destructively examined after the test by breaking them open (mechanical fatigue) which showed the final depths, shapes, and surfaces of the cracks. The sensor signal was interpreted as crack depth by calibrating it with both the signals for the initial notch and the final crack depths.

NDE was applied to the notch areas only; longitudinal cracks away from the notches which were not expected and were discovered only after the test were located with the naked eye and with the dye penetrant method (PT); they also were examined optically after a section was broken open. The crack surface revealed beach-marks that could be correlated with the fatigue test interruptions and hence with the number of cycles.

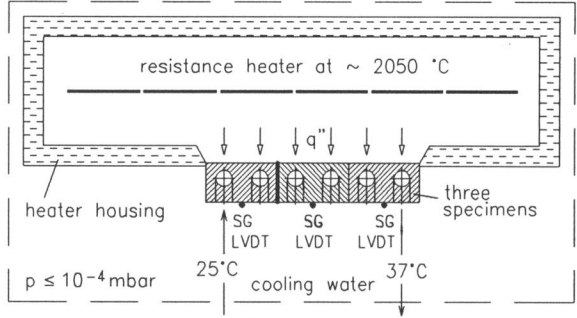

Fig.2 Experimental setup

4.2 Measured Crack Growth

For each of the failures the application of the above techniques yielded the crack growth as a function of the number of cycles:

Crack growth at the inner notches (type A) is plotted in Fig.4 and the final crack depths at the summits are listed in Fig. 1. Cracks were detectable after 10^4 cycles. They grew at a maximum rate around $4 \cdot 10^4$ and stopped growing at about $6.5 \cdot 10^4$ cycles, a reason why the test was terminated; after the longitudinal cracks had been discovered this crack arrest could be explained primarily as a result of a local stress relief by competing and faster growing longitudinal cracks initiated during the test. The maximum growth of a crack was 1.6 mm, i.e. it stopped after it had penetrated 42% of the channel wall. Figure 5 illustrates the elliptical shape of the final crack at A1 and the position of the competing longitudinal cracks.

For cracks at the outer notches (type B) the measurement was disturbed by the blackening coating applied after 10^4 cycles and by some carbon deposition in the notches originating from the heater. Therefore only B3 could be measured and because of data scattering it is not clear whether a crack arrest was reached. Data are not displayed here. The cracks grew up to 0.6 mm from the bottom of the notch.

Growth of the longitudinal cracks is plotted also in Fig.4. These cracks started from the intact surfaces of the cooling channel walls; they appeared uniformly in each of the six cooling channels in positions about 10° left of A in Fig.1 (compare section 5 for the maximum of the strain range in this position). The cracks extend over the whole length of the channels. They obviously were initiated during the first $2 \cdot 10^4$ cycles and grew in depth almost linearly until the end of the test when they had penetrated 2.4 mm or almost half the channel wall. From the data in Fig.4 there is no strong indication of a crack arrest if the test would have been continued. Figure 6 shows the flank of a crack after it was broken open and it indicates the almost straight crack front. Micrographs reveal that the cracks are transgranular.

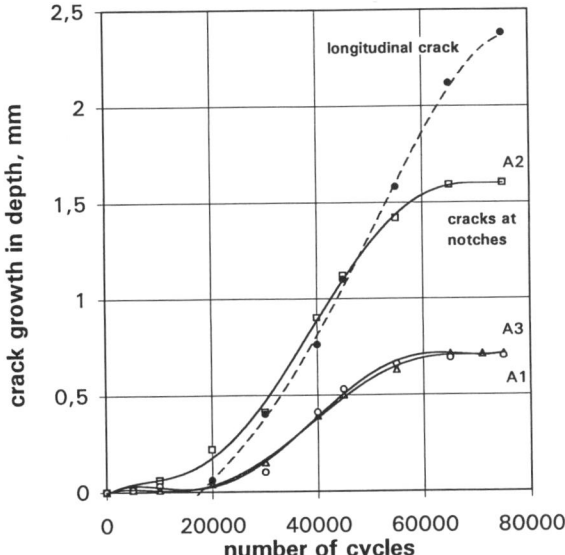

Fig.4 Growth of cracks at channel inside

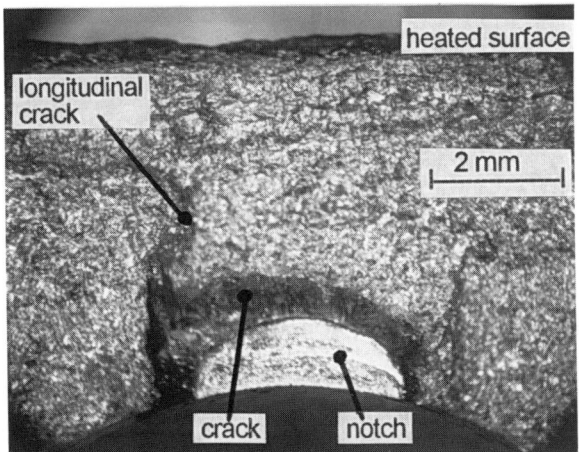

Fig.5 Crack at notch A1 after 75000 cycles

Fig.6 Flank of a longitudinal crack

5. Failure Prediction

Prediction may be focused on two different aspects of failure:

5.1 Crack Initiation from an Intact Surface

As a result of a transient, 2-D, plastic FEM analysis (Fig.7) the maximum mechanical strain range $\Delta\varepsilon$ of 0.52 % is located 10° left of A in Fig.1 where the longitudinal cracks actually appeared. Crack initiation in this study compares reasonably well with other thermal fatigue failure results in Fig.8 and confirms the safety margin of the ASME design curve.

5.2 Crack Growth from a Notched Surface

The stress concentration at the bottom of an artificial notch is assumed to initiate a sharp-edged crack during several thousands of cycles. Growth of such cracks may be predicted on the basis of 'simplified elastic-plastic' fracture mechanics:

The loading parameter controlling fatigue crack growth is the range of the stress intensity factor ΔK which is given as $\Delta K = \Delta\sigma \cdot \sqrt{a} \cdot Y$

A fracture mechanical analysis was carried out
- using the elastically calculated cyclic stress field and assuming elast.-plast. shake-down conditions, i.e. $\Delta\sigma = \Delta\sigma_{elastic}$ and $\sigma_{min}/\sigma_{max} = -1$.
- considering the notch an initially sharp semi-elliptical surface crack with the two axes a (depth) and c.
- applying a geometry function $Y = f(a/c, \text{mock-up geometry, crack shape, and loading conditions})$ which has been developed for the geometry of the benchmark specimen [2].

The relation for incremental crack growth $da/dN = f(\Delta K)$ was taken from [3].

Crack growth was calculated for the conditions of the artificial notches in the test. As an example calculated and measured crack growths are plotted for notch A1 in Fig.9. It seems that the model over-predicts the crack growth but a closer look reveals that in the experiment the crack starts rather slowly (sharp crack forms), reaches the predicted growth rate and comes to an arrest possibly due to the presence and depth of a competing longitudinal crack (not modeled). Yet at other notches in the channel the actual maximum growth rate was under-predicted.

6. Conclusions

All notches inside the channel developed cracks that were arrested, possibly by competing longitudinal cracks that grew faster and continued to grow at the end of the test. Also all notches in the heated surface developed cracks, for which the measured data would not allow a statement on crack arrest.

Initiation of the longitudinal cracks seems predictable from the calculated strain range.

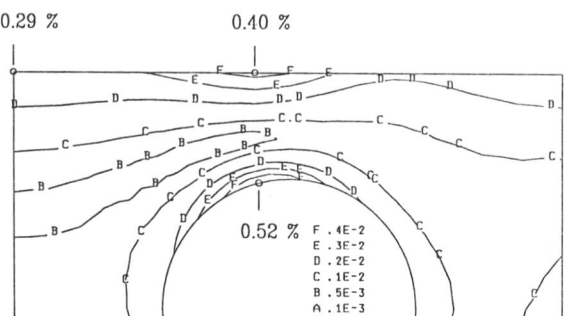

Fig.7 Mech. strain range $\Delta\varepsilon$ from plast.analysis

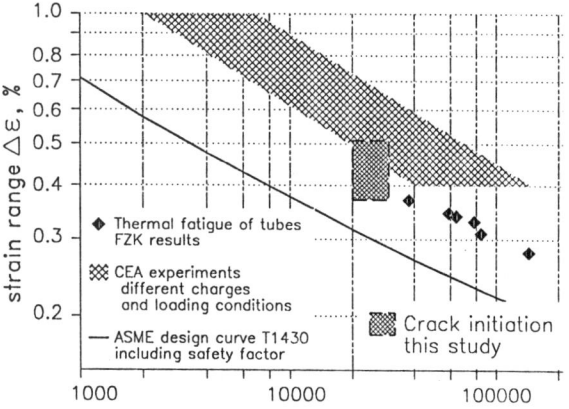

Fig.8 Standard lifetime assessment

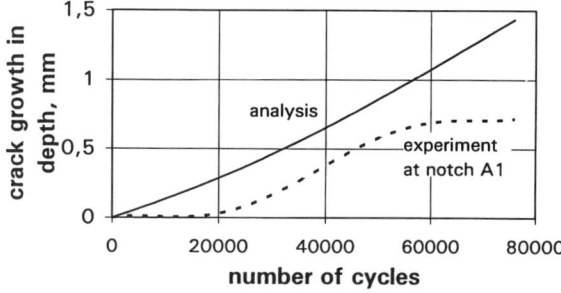

Fig.9 Crack growth at notch A1

References

1. G. Hofmann et al., KfK 5381 (1994)
2. E. Diegele et al., submitted to Int.J.Fract.(1994)
3. R.W.Watson et al., J. Press.Vess.Techn., Vol.105 (1983), pp.144-152.

Development of plasma facing components with ceramic armour

R. Vaßen[a], A. Yehia[a], R. Lison[b], R. Duwe[c], D. Stöver[a]

[a]Institut für Werkstoffe der Energietechnik, [b]Zentralabteilung Technologie, [c]Heiße Zellen
Forschungszentrum Jülich GmbH
D-52425 Jülich

As a result of insufficient thermomechanical properties monolithic SiC, B_4C, and TiC ceramics are not able to withstand the high heat loads during plasma disruptions in a fusion reactor. As demonstrated in previous studies a significant improvement of the thermomechanical properties of SiC can be achieved by the addition of graphite, B_4C and TiC dispersions. The enhanced thermo-mechanical properties result in lower erosion rates and increased tolerable heat loads during electron beam tests in the electron beam facility JUDITH. Optimized samples of the system $SiC/B_4C/C$ (i.e. carbon and boron carbide contents of 20 and 10 or 20 wt%, respectively) were able to withstand 10 pulses (5 ms) with a power density of more than 3 GW/m^2 without fracture.

In order to investigate the behaviour of these materials also under steady state conditions actively cooled modules with molybdenum alloy (TZM) heat sinks were manufactured. Brazing was used as a possible metal/ceramic joining technique. Feasibility studies revealed the possibility of brazing $SiC/B_4C/C$ ceramics with B_4C contents up to 20 wt% to TZM by using a Cu1Cr-brazing agent. Higher boron carbide contents led to an increased formation of brittle phases in the interface and as a result to bad adhesion of the ceramic. Steady state surface and interlayer temperatures have been calculated by a finite element method (FEM) using power densities between 5 and 10 MW/m^2.

In addition a new diffusion bonding process for ceramic(SiC)/TZM bonding was investigated. Highly sinteractive nanocrystalline SiC interlayers have been used to reduce the bonding temperature below 1500°C. Consequently also the thermal stresses which arise during cooling due to differences in the thermal expansion coefficients are reduced.

Bonding experiments have been performed in two steps. First the nanocrystalline layer was joined to a SiC plate in a hot press facility at 1600°C. In the second step a joint of nanocrystalline layer and TZM was formed at temperatures between 1200°C and 1500°C. Results indicate the feasibility of the process at least for small samples (diameter 15 mm) if the bonding conditions are optimised with respect to a minimised reaction zone and reduced defect size in the interlayer.

1. INTRODUCTION

The development of reliable plasma facing materials is a major task in the development of future fusion reactors. For this application low-Z ceramics have been frequently investigated during the past decades [1,2] showing that monolithic ceramics are hardly able to meet the severe demands especially regarding thermomechanical properties. As stated in earlier publications [3,4] the improvement of low-Z ceramic materials for plasma facing applications can be done by producing composite materials via powder metallurgical routes and hot isostatic pressing as consolidation process. Optimised materials reveal significantly improved thermal shock behaviour compared to conventional monolithic ceramics. However an application also implies the joining of these ceramics to a metallic cooling structure. Modules have been produced applying brazing techniques. Simultaneously diffusion bonding experiments have been performed to investigate the potential of this joining technique for the manufacturing of highly temperature resistant bondings.

2. EXPERIMENTAL

2.1 Material processing

SiC, B_4C, and C powders in the size range of 1-10 µm were used for the production of $60SiC/20B_4C/20C$ and $70SiC/10B_4C/20C$

composites. The powders were mixed and cold isostatically pressed. These specimens were degassed and densified by Hot Isostatic Pressing (HIP) using quartz capsules with an argon pressure of 350 MPa applied for 3 hours at 2000°C. Final specimen dimensions were approximatly 50 mm in diameter and 60 mm in length. More details of the sample preparation are given in [4]. From these samples 25*25mm² plates with 4-7 mm height were machined in the ceramic workshop.
The diffusion bonding experiments were performed with monolithic, HIPed SiC qualities.

2.2 Brazing

Before the manufacture of the modules the brazing procedure was optimised using test samples (s. Results). After these preliminary experiments the ceramic plates were brazed to the molybdenum alloy TZM (0.5 Ti, 0.07 Zr) using a 200µm thick CuCr1 foil as brazing alloy and in the same step the ferritic cooling pipes were brazed to the TZM (Fig.1). Samples were heated up to 850°C with 20K/min. After a dwell time of 20 minutes temperature was further increased with 20K/min up to the brazing temperature of 1150°C. After 5 minutes brazing was finished and the samples were cooled to room temperature at a cooling rate of about 5K/min.

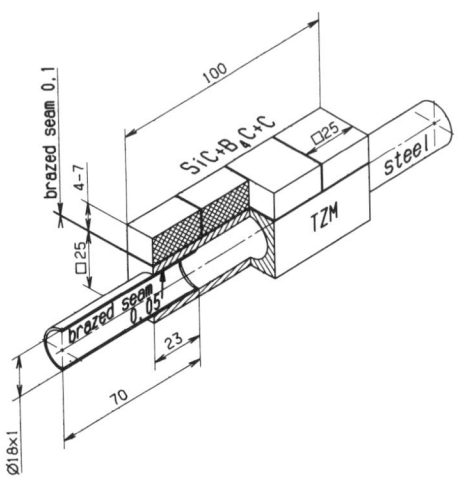

Fig.1 Geometry of the divertor module.

2.3 Diffusion bonding

SiC plates with a height of 5 mm and a diameter of 15mm, similar TZM plates, and about 1 mm thick layers made of fine grained (20nm) SiC powders [5] were used for the diffusion bonding experiments. First the interlayer has been joined to the SiC plate at 1600°C in hot press at a pressure of 113MPa and a dwell time of 3 hours. In a second step the TZM/ceramic bonding was made at lower temperatures (1300-1500°C) and equal dwell times and pressures. The pressure was reduced before cooling to avoid additional stresses at the piston (graphite)/specimen interface.

2.4 JUDITH-tests

Disruption-like heat loads have been simulated at the KFA electron beam test facility JUDITH.

3. Results
3.1 Materials optimisation

SiC/B$_4$C/C-composites have been optimised with respect to thermal shock resistance and erosion loss during disruption simulations. Composites are able to withstand energy densities of more than 20MJ/m² without catastrophic failure. Erosion losses of the composites show an approximately linear dependence on the energy density up to 20 MJ/m² with a treshold energy density between 0.7 and 0.9 MJ/m². The absolute values for a pulse duration of 5ms are in the range of 53 to 65 g/MJ corresponding to 20-24 µm/MJ/m².

3.2 Brazing

Large experience on the brazing of CFC and Si-based materials on TZM are available at KFA [6]. However B$_4$C containing ceramics had not been brazed up to now. Therefore preliminary tests were made. They indicated the formation of brittle phases

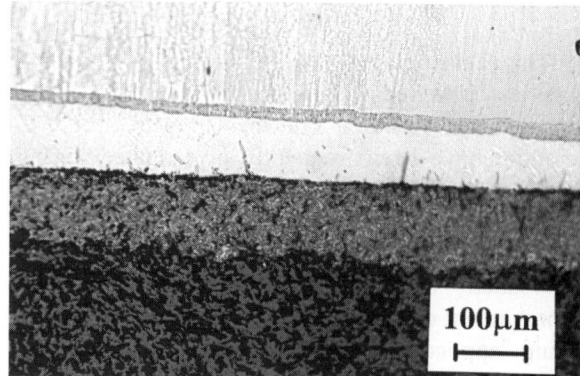

Fig.2 Micrograph of the 60SiC/20B$_4$C/20C//TZM interface brazed with a CuCr1-foil.

at the interphase during brazing of 40SiC/40B$_4$C/20C-composites and relatively bad adhesion. Fortunately the adhesion was sufficiently high for the optimised materials composition with B$_4$C contents of 20 wt% or less. Fig. 2 shows the homogenous structure of the brazed ceramic/TZM interface. After these tests 6 modules have been brazed with different heights of the ceramic composites.

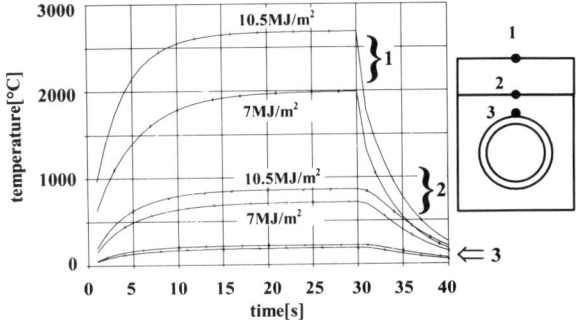

Fig. 3 Results of FEM calculations for the temperature development at three different points of the module (1, 2, 3). 30 s pulses with the given energy densities and a 6 mm thick 60SiC/20B$_4$C/20C -composite have been used.

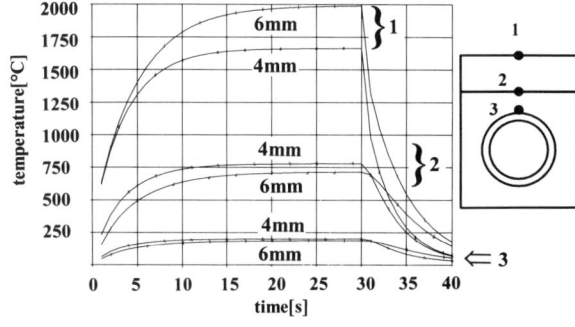

Fig. 4 Temperature development calculated by FEM for different 60SiC/20B$_4$C/20C-composite thicknesses and a constant power density of 7 MJ/m².

Thermal analysis have been performed with a Finite Element Method (FEM) using the ABAQUS code. In the used two dimensional model heat loads of 7 and 10.5 MW/m² were applied on the composite surface. The thermomechanical properties of the ceramic composites have been calculated by a rule of mixtures from the temperature dependant properties of the starting materials.

Fig. 3 and 4 show the temperature development for 30 s pulses at three points (surface, interface composite/TZM, surface cooling pipe) on a line perpendicular to the surface and through the middle of the cooling pipe. The variation of the power density at constant thickness (6 mm, Fig.3) indicates surface temperatures of up to 1990°C at 7 MW/m² and 2680°C at 10.5 MW/m². The second temperature is definitively to high for the used composite. A reduction of the composite thickness from 6 mm to 4 mm at 7 MW/m² (Fig.4) results in a reduction of the surface temperature from 1990°C down to 1660°C while the interface temperature slightly increases from 710°C to 780°C. Disruption lifetime considerations suggest an improvement of the thermal conductivity of the present composites to allow sufficiently thick armour plates. On the other hand uncertainties in the values of the high temperature thermal properties require future thermal response tests in the JUDITH facility to confirm the calculated values.

3.3 Diffusion bonding

Hot pressing of a nanophase SiC pellet to dense conventional SiC at 1600°C resulted in a good diffusion bonding of the two partners and a density of about 90%TD of the nanophase interlayer.
No large defects have been found in the interphase region indicating sufficiently high diffusional transport at these relatively low temperature, which is a result of the high sinteractivity of the used nanophase powder. Fig. 5a and b reveal the microstructure of the nanophase SiC/TZM interphase after hot pressing at 1250°C and 1500°C, respectively. At the lower temperature defects in the interphase regions are present. After hot pressing at 1500°C these defects disappear but on the other hand a broad reaction zone of about 25 µm has been found. Energy dispersive X-ray (EDX) analysis in a scanning electron microscope indicated significant silicon diffusion into the TZM (Fig.6), while at 1250°C only a minor interdiffusion region was found (<1 µm). The large reaction zone leads to a severe embrittlement of the interphase region due to

the formation of brittle intermetallic phases (Mo_5Si_3, $MoSi_2$ [7]). Diffusion bonding experiments

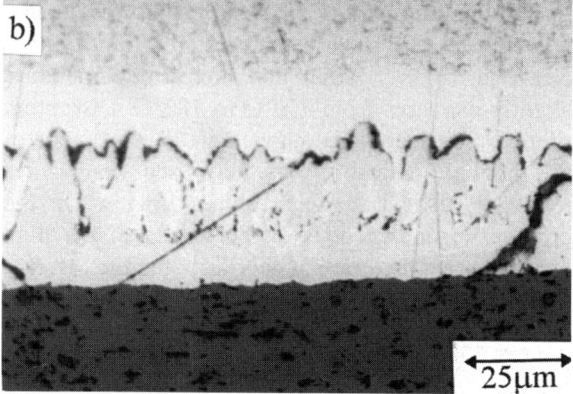

Fig. 5 Microstructure of the TZM (above)/ nanophase SiC (below) interphase after hot pressing at 1250°C (a) and 1500°C (b), respectively.

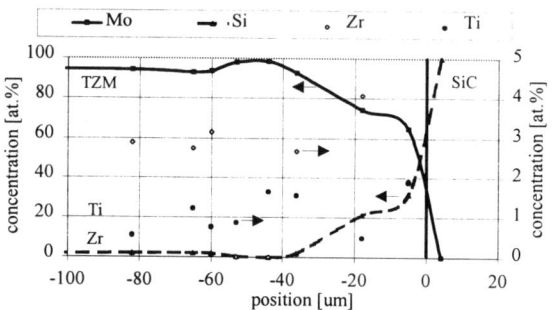

Fig. 6 EDX analysis of the nanophase SiC/TZM interphase hot pressed at 1500°C.

performed at 1400°C reveal a nearly defect free interphase region and a small reaction zone, indicating the principle feasibility of the process at least for small geometries. Future developments are directed towards the improvement of the processing of the nanophase interlayer. A sufficient densification of the nanophase interlayer at temperatures below 1500°C even without high pressures is a future goal.

4. SUMMARY

Ceramic $SiC/B_4C/C$ composites have been optimised with respect to thermal shock resistance and weight loss during high heat load tests in the JUDITH facility. Plates machined from $60SiC/20B_4C/20C$ and $70SiC/10B_4C/20C$ composites have been brazed on TZM cooling structures. Calculation by the FEM code ABAQUS gave surface temperatures of about 2000°C for 7 MW/m² thermal load and 6 mm armour thickness. Diffusion bonding experiments between TZM and SiC with nanophase SiC interlayers at 1300-1500°C show optimistic results for small sample geometries. The applicability of the results to larger geometries has to be demostrated.

REFERENCES

1. H. Bolt, Fusion Engineering and Design 22 (1993) 85.
2. G.P. Pells, J. Nucl. Mat. 122&123 (1984) 1338.
3. R. Vaßen, J. Förster, A. Yehia, K. Hammelmann, H.-P. Buchkremer, H. Bolt, D. Stöver, Proceedings of the 18th SOFT, August 22-26, 1994, Karlsruhe, Germany, in Fusion Technology 1994, K. Herschbach, W. Maurer, J.E. Vetter (editors), 1995 Elsevier Science B.V., p 259-262.
4. A. Yehia, R. Vaßen, R. Duwe, D. Stöver, Proc. of the 7th Int. Conf. on Fusion Reactor Materials (ICFRM7), Obnisnk, Russian Federation, Sept. 25-29, 1995, to be published in the J. Nucl. Mat..
5. R. Vaßen, A. Kaiser, D. Stöver, in s. [4].
6 J. Godziemba-Maliszewski, R. Lison, in the Proc. of the 2th Int. Symp. on Ceramic Materials and Components for Engines, Lübeck-Travemünde, FRG, April 14-17, 1986, W. Brunk, H. Hausner (editors), 1987 Verlag Deutsche Keramische Gesellschaft, p. 425.
7 A.L. Burykina, L.V. Strashinskaya, T.M. Evtushok, Fiziko-Khimicheskaya Mekhanika Materialov 4, 3 (1968) 301-305.

Materials Aspects of ITER In-Vessel Components

V.Barabash, J.Dietz, K.Ioki, G.Janeschitz, G.Kalinin, R.Matera, K.Mohri and ITER Home Teams

ITER Joint Central Team, ITER Garching Joint Work Site,
Boltzmannstraße 2, 85748 Garching, Germany

The status of the materials selection for ITER in-vessel components (primary first wall, limiter, baffle, divertor and shielding blanket) and the R&D program on structural and plasma facing materials are presented in this paper. The joining technologies (armour/heat sink, Cu/SS,) are briefly reviewed in relation to a possible impact on the materials properties.

INTRODUCTION

The materials selection for the ITER in-vessel components is a challenging task because of the complicated operational conditions (ion and high heat fluxes, neutron irradiation) and the safety related requirements (tritium inventory, activation). The selection of materials is also related to the development of the manufacturing technologies, and the technological features of the materials dictate the directions of the design developments. Additionally, the industrial availability of the materials becomes more and more important because at the end of EDA the manufacturing feasibility of the different components should be demonstrated, their cost assessed and the reference materials specified.

1. OPERATION CONDITIONS AND REFERENCE MATERIALS

The detailed operational conditions and the design solutions for the ITER in-vessel components are presented in various papers in this Conference [1-3]. The current status of the materials choice is presented in Table 1. It should be noted that for some components the decision on the materials to be used is already done and only the specific material grades are under study. For others, especially for armour materials, alternative candidates are under consideration for the different components.

The selection of plasma facing materials was done mainly taking into account the plasma compatibility and erosion life time (ion erosion, thermal erosion during disruption and transient events) [4]. Based on these assessments, Be is proposed as a reference materials for FW, limiter, upper part of baffle and for dome. W is proposed to be placed in regions with expected high C-X neutral flux: lower part of baffle and upper part of vertical target. Due to the excellent thermal resistance CFC is proposed for the vertical target and as back-up material for limiter and baffle. Some material operational conditions for the Basic Performance Phase (BPP) are presented in Table 2. It is clear from this table that the fluence range for the BPP phase will be quite moderate, but nevertheless the change of material properties at these conditions could be significant.

Table 1. Materials selection for in-vessel components.

Component	Materials	
	Armour	Heat Sink /Structural
First Wall, Blanket		
Primary First Wall	Be	Cu alloy + SS
Limiter	Be (CFC*)	Cu alloy + SS
Baffle - upper part - lower part	Be (CFC*) W	Cu alloy
Shield Blanket	-	SS
Divertor		
Dome	W or Be	Cu alloy
Vertical Target - upper part - lower part	 W CFC (Be*)	Cu alloy
Wing	W	Cu alloy
Cassette Liner	W	Cu alloy
Cassette Body	-	SS

* - Alternative material.

2. STRUCTURAL MATERIALS

2.1. Stainless Steel

Solution annealed austenitic stainless steel 316L(N)-IG (IG means ITER Grade) has been recommended as a provisional reference grade. The proposed steel is based on EU 316LN-FBR stainless

Table 2. Operational conditions for in-vessel materials, BPP - 0.3 MW*a/m^2.

Materials	Temperature, °C	Fluence, dpa
St.Steel	140 - 320	0.1 - 1.7
Cu alloys	140 - 400 / ~ 700*	0.03 - 3.0
Be	140 - 500 / ~ 700** / MP*	0.1 - 1.2
W	140 - 870 / MP*	0.07 - 0.3
CFC	140-250 / ~ 1200** / ~ 3500*	0.07 - 0.8

* At transient events;
** Limiter start-up and shut-down.
MP - material melting point.

steel with minor modification to meet specific ITER requirements. The main reasons for this choice are the extensive available database (in un- and irradiated conditions), improved mechanical properties, stress corrosion cracking resistance and large industrial experience in nuclear applications [5].

For the ITER applications it was recommended to modify slightly the chemical composition taking into account the following requirements:
- Cobalt content should be reduced to a maximum of 0.1%. The reason for this is to minimise the decay heat.
- Boron content in steel for the reweldable parts of the design should be not more than 10 ppm. In this case the problem of the rewelding of irradiated steel could be significantly reduced.

These requirements will not increase the cost of steel.

The data base on the properties of this material is rather complete. Some data are still missing for specific technological treatments, as an example for steel subjected to high temperature HIP which will be used for the shield block manufacturing, and for cast and HIP'ed steel which is proposed for the divertor cassette body.

2.2. Copper Alloys

As heat sink material the dispersion strengthened (DS) and precipitation hardened (PH) high strength high conductivity copper alloys are under study for first wall and divertor application. The comparison of these two types of alloys have been discussed recently [6]. The major differences of these alloys are:
- thermal stability: DS copper alloys have higher resistance to heating due the microstructural stability of the dispersion particles;
- weldability: PH alloys could be easily fusion welded with partial recovering of properties after additional heat treatment. Fusion welding is not recommended for structural/leak tight welds for DS Cu alloys, but other methods (e.g. friction welding, brazing) could be used;
- neutron irradiation resistance at high temperatures: DS Cu alloys have better properties;

As far as the other properties are concerned, the two alloy types can be considered roughly equivalent.

The following grades are considered as a reference: PH Cu: CuCrZr-IG (Cu-{0.6-0.8}Cr-{0.07-0.15}Zr, heat treatment - 980-1000°C for 1 h/water quench + age at 450-480°C for 1-3 hr); As an additional material CuNiBe (HYCON 3M) is also under consideration by the US Home Team because of its better technological and mechanical properties. However, the existing data on neutron irradiation influence do not permit to recommend it as a reference, and more data are needed.

DS Cu: Glidcop Al25-IG0 (Cu-{0.23-0.27} Al$_2$O$_3$, production: HIP 80 mesh copper powder, high temperature extrusion, cross rolling in transverse direction after extrusion and annealing).

One of the critical issues for both types of Cu alloys is low fracture toughness at elevated temperatures [7]. Fracture toughness of DS Cu alloys (as an example CuAl25-IG0) drops to very low values at 300°C, while the specimen fracture surface shows a ductile deformation mechanism and, at the same temperature, the total elongation is ~30%. Fracture toughness of irradiated DS Cu alloys (e.g. for Glidcop Al15) decreases 2-3 times in comparison with unirradiated material.

The selection of the Cu alloys cannot be made separately from the choice of the manufacturing technologies which include Cu/stainless steel, Cu/armour, Cu/Cu joints and features of the operational conditions for different components. Since there are many different methods under development, the selection of the Cu alloys for different in-vessel components depends on the available results and has to be made latest for the prototype manufacturing.

3. PLASMA FACING MATERIALS

3.1. Beryllium

Some of the issues, such as bonding, melting, and ion erosion, do not depend a priori on the Be grades. Others, such as thermal shock/fatigue resistance, neutron irradiation resistance, and tritium retention, are very sensitive to the impurity levels, grain sizes, methods of production, etc., which usually differ among the Be grades.

For the application as PFC armour, thermal fatigue and thermal shock resistance are the criteria to be taken into account. According to these criteria, S-65C (Brush Wellman, US) and DShG-200 and TShG-56 (RF) were proposed as reference grades.

Additionally, plasma sprayed beryllium is under consideration for the FW and for other easily accessible PFC (for more details see [8]).

As issues for Be application as an armour the following should be mentioned:
- surface damage at transient events (melting, He swelling, etc.) which will lead to a complete change of materials properties;
- embrittlement at neutron irradiation, mainly at low irradiation temperatures.

3.2. Tungsten

The preliminary selection of the W grades has been made taking into account the near term availability, cost, technological features and some advantages in mechanical properties. The following materials have been proposed as a reference for prototype manufacturing: W-1%La_2O_3, PS (EU, US), CVD W (JA), W-0.3%Y_2O_3, W-(5-15%)Mo, W(cast and cold worked) (RF), however the R&D program includes also other grades as W-Re alloys, etc. The criteria for the choice of reference W have to include the thermal fatigue/shock resistance, especially after n-irradiation, cost and joining technologies features. Based on the past experience on unirradiated grades, the ones with fine grain, low impurities content, in cold deformed condition are preferred for their higher thermal fatigue resistance.

For the use of W operation as an armour the following factors are important: change of materials properties at transient events (melting and recrystallisation) and embrittlement due neutron irradiation. Performance of the hard brittle material as an armour should be analysed.

3.3. Carbon Fiber Composites

The following issues were taken into account during the selection of the reference CFCs:
- type of materials: 3-d materials are considered in first priority in comparison with 1 and 2-d materials, because of the more isotropic properties and higher thermal shock resistance;
- Si doping is claimed to be the best dopant to reduce chemical sputtering and improve safety (reduced water interaction rate) [9], but the thermal stability of doped
CFCs at transient events is an open issue;
- neutron irradiation resistance:
    * 3d CFC's show a better dimensional stability under neutron irradiation than 1-2d;
    * CFC's with high initial thermal conductivity keep high conductivity after irradiation.

The following specific CFC grades are under study:
- SEP NS31 (CFC+Si), SEP N31B, Dunlop 3D-P120 - EU Home Team;
- MFC-1, CX-2002U, NIC-01 - JA Home Team.

4. JOINING TECHNOLOGIES

The development of the reliable armour/Cu alloys, Cu/stainless steel and SS/SS joining technologies for the first wall/blanket and divertor components are essential for the ITER design and the on-going R&D program intensively focuses on the solution of this problem.

4.1. Armour/heat sink joining

An extensive work has been performed to develop reliable silver free joining technologies. However, the final selection for manufacturing the divertor and first wall prototypes is subjected to the availability of results of the HHF tests. The following technologies are under development:

- **Be/Cu joining technologies:**

EU	- Brazing, CuMnSnCe alloy;
	- Electroplating + diffusion soldering (Sn);
	- HIP (low temperature) with interlayers;
	- HIP (high temperature);
	- Electroplating + diffusion bonding.
JA	- Brazing (Ag base + coatings)
RF	- Brazing: Cu-Mn, Cu-Sn-In, Ti-Zr alloys;
	- Diffusion Bonding.
US	- Electroplating + low temperature HIP;
	- Brush Wellman "proprietary" (Ag free);
	- HIP (Be to AlBeMet-Expl.Bond To Cu);
	- HIP with Al-13%Si braze filler;
	- Plasma Spray with diffusion barriers layers;
	- Plasma Spray Be/Al FGM on Cu.

For this type of joints the problems to be solved is the minimisation of the brittle intermetallic phase formation which reduces the thermal mechanical performance of the joints.

- **W/Cu joining technologies:**

EU	- Active metal casting + welding;
	- HIP, Ni interlayer;
	- Diffusion bonding;
	- Plasma Spray.
JA	- CVD on OFHC Cu and W-Cu
RF	- Brazing (Cu-Mn);
	- Casting;
US	- HIP/Diffusion bonding + graded W/Cu;
	- Electroplating + low temperature HIP;
	- Brazing (Cu-Mn);
	- Plasma Spray + HIP.

The problem for W/Cu joints is a large mismatch in the coefficient of thermal expansion. Advanced solutions are the use of graded layers with a gradual thick transition between W and Cu or the adaptation of the tile geometry to reduce the thermal stresses (brush or lamella type of W armour).

- **CFC/Cu joining technologies:**

EU	- Active metal casting + welding or brazing;
	- HIP assisted brazing (Cu-Mn, Cu-Ti);
	- Diffusion bonding assisted joining (Cu-Mn);
JA	- Brazing (Ag base);
	- Ag free brazing (Cu-Mn, Cu-Ti, Ti).

This type of joining was formerly widely used mainly with Ag base brazing alloys and new developments are still needed.

4.2. Cu alloys/steel joining

Design of in-vessel components includes the following joints of heat sink and structural materials:
- primary wall, baffle - copper plates with steel cooling tubes and plates;
- baffle and divertor - copper structures with steel plate and/ or copper and steel tubes.

Among the different methods of Cu/steel joining for the primary wall solid HIP is considered as most promising. This method can provide high quality of joints with small distortions. Two ranges of HIP temperatures are under study: high ~ 1000-1050°C, which permit to combine in one treatment the joining of Cu to steel and steel to steel, and moderate ~ 900-980°C, in this case steel to steel joining has to be performed by a separate procedure. The issues for this methods are surface preparation and the use of interlayers to provide high quality joints. For this method the use of DS copper is preferable because the second heat treatment to restore the properties needed for CuCrZr could be eliminated.

As an alternative method powder HIP (with CuCrZr powder) is under consideration. Other joining methods such as explosion bonding seem promising, but they are applicable only for simple geometric shapes of the components. For tube to tube joining the friction welding could be used, there is some experience with promising results.

One problem for all types of joints is the method of joint validation. Specific testing procedures have to be proposed for different joints taking into account the features of the application of these joints in a real design. Neutron irradiation influence is also an important factor to be studied.

5. CONCLUSION

Stainless steel 316LN-IG is selected as reference material for in-vessel components and only additional study of specific material properties and the effect of manufacturing cycle are included in R&D program;

Glidcop Al25-IG0 and CuCrZr-IG are selected as reference, other materials are also under study and the selection will be based on the results of the R&D; selection of the Cu alloys depends on the joining technology and for different components the different Cu alloys could be proposed.

The selection of the armour materials for different components is still open: for some components, a few possible armours are under consideration, and within these material classes different grades are proposed. The final choice of armour, at least for BPP of ITER operation, is planned to be made at the end of EDA.

The armour/heat sink, Cu/stainless steel joining technologies are under development and the selection of the reference is waiting for the results of the R&D program. The characterisation of the joint properties and the predictions of joints behaviour under ITER condition have the highest priority in R&D program.

ACKNOWLEDGEMENT

Thanks to various Home Teams experts for the valuable contributions and advises.

This report was prepared as an account of work performed within the ITER Joint Central Team under the Agreement among the European Atomic Energy Community, the Government of Japan, the Government of the Russian Federation, and the Government of the United States of America on Co-operation in the Engineering Design Activities for the International Thermonuclear Experimental Reactor ("ITER EDA Agreement") under the auspices of the International Atomic Energy Agency (IAEA).

REFERENCES

1. R.R.Parker, this conference.
2. K. Ioki et al., this conference.
3. R. B. Tivey et al., this conference.
4. R.R.Parker, PSI/96, to be published in JNM.
5. G.Kalinin at al., ICFRM-7, 1995, to be published in JNM.
6. S.Zinkle, S.Fabritsiev, Suppl. to J. Nuclear Fusion, vol. 5, (1994) 163.
7. R.Solomon et al., ICFRM-7, 1995, to be published in JNM.
8. R.Matera et al., this conference.
9. C.Wu et al., this conference.

ANALYSIS OF PLASMA SHIELD FORMATION AND GRAPHITE TARGET EROSION IN DISRUPTION SIMULATION EXPERIMENTS WITH HIGH POWER QUASISTATIONARY PLASMA STREAMS

V.V.Chebotarev[1], I.E.Garkusha[1], V.V.Garkusha[1], V.A.Makhlaj[1], N.I.Mitina[1], S.E. Pestchanyi[2], D.G.Solyakov[1], V.I.Tereshin[1], S.A.Trubchaninov[1], A.V.Tsarenko[1], H.Würz[3]

1 Institute of Plasma Physics of the National Science Center
   "Kharkov Institute of Physics & Technology", 310108 Kharkov, Ukraine
2 TRINITI Troitsk, 142092 Troitsk, Russia
3 Forschungszentrum Karlsruhe, INR, Postfach 3640, D-76021, Karlsruhe, Germany

Hydrogen plasma streams with energy densities up to 7 kJ/cm^2, mean proton energies up to 350 eV and pulse duration up to 300 μs were used for disruption simulation experiments. The plasma stream is guided by a longitudinal magnetic field having 0.7 T inside the plasma. Formation of plasma shields with electron densities up to 8×10^{17} cm^{-3} was observed close to the target surface. For incident energy densities in the range 3 to 7 kJ/cm^2 the energy absorbed by the target is constant and amounts up to 50 J/cm2. Average erosion of MPG-7 graphite was around 0.5 μm per 1 kJ/cm^2 of incident energy density.

1. INTRODUCTION

It was shown in our previous experiments [1-3] that the plasma streams generated by the powerful quasi-stationary plasma accelerator QSPA Kh-50 can be used for simulation of plasma wall interactions for ITER disruptions and ELMs. The previous experiments with such plasma streams were carried out without magnetic field. But for adequate simulation one needs to perform these experiments in a magnetic field with magnetic force lines normal or inclined to the target surface. The main goal of the recent experiments was the analysis of the characteristics of the plasma flow in the external axial magnetic field and the analysis of the formation of the transient plasma layers and the target erosion under irradiation by quasistationary plasma streams of high power density.

2. EXPERIMENTAL DEVICE

Experiments were carried out at the quasistationary plasma accelerator QSPA Kh-50 device, described in [1-3]. This device was upgraded [4] by replacing the 2nd and the 3d part of the vacuum chamber (1.5 m in diameter and 1.2 m in length each) by a magnetic system and of smaller in diameter vacuum chamber. An axially-symmetric magnetic field was produced by four large magnetic coils. The first magnetic coil was located at a distance of 120 cm from the accelerator output. Each coil is 32.5 cm in thickness and 120 cm in outer diameter. The inner diameter of the coil, determining the radial size of the vacuum chamber, is 42 cm. The distance between the coils is 15 cm. The inner diameter of the vacuum chamber inserted into the coils was 38 cm. The magnetic coils were supplied by a condenser bank. The electric current was increased from the first coil to the 3d and the 4th in order to produce a magnetic field which slowly increases with length, on a distance of more than 100 cm, and more or less homogeneous magnetic field at the vicinity of the 3d and the 4th coil with a magnetic field strength up to 2 T (<1.2 T in recent experiments). This the magnetic field geometry was chosen to get the best agreement of injected high power plasma with magnetic field. The half period of magnetic field achieved 16 ms. The diagnostic chamber for the experiments was placed between the 3d and the 4th magnetic coil.

Experiments were carried out in different modes of accelerator operation. The ultimate plasma parameters measured in large vacuum chamber with no magnetic field were as follows [2]: maximum plasma power density up to 150 MW/cm^2, plasma energy density up to 2.4 kJ/cm^2 (at a distance of 0.7 m from the accelerator output), maximum energy of protons 0.9 keV, density of plasma stream about 10^{16} cm^{-3} at 1.5 m from the accelerator output, pulse duration - 0.1-0.15 ms.

At the beginning of the recent experiments the influence of the stray magnetic field on the operation of the accelerator was investigated. The plasma parameters radial distribution of plasma energy density, plasma potential and current in the plasma were measured at a distance of 0.3 m from the accelerator output both with and without magnetic field. Switching on the magnetic field didn't influence the accelerator operation and the plasma stream parameters clouse to the accelerator output.

3. PARAMETERS OF THE PLASMA STREAM IN THE MAGNETIC FIELD.

The parameters of the plasma streams were varied by changing the voltage of the main-discharge capacitor bank from 8 kV up to 15 kV and the magnetic field value up to 1,2 T. The value of the magnetic field ΔB_z displaced by the plasma propagating along the magnetic system was measured by local magnetic probes. Radial distributions of magnetic field in plasma B_z^{plasma} given as $(B_{z0}-\Delta B_z)$ and normalized to the value of the vacuum magnetic field B_{z0} are shown in Fig.1 for different distances from the accelerator output. The plasma radius can be estimated from these distributions as distance from the axis where the displaced magnetic field is changing its sign. From this picture it follows that the plasma radius is reduced with propagation along the magnetic system. Plasma stream radius is about 8.5-9 cm at the end of the magnetic system. Simultaneously, the efficiency of the penetration of the magnetic field into the plasma is increasing with the length. The magnetic field inside the plasma achieves 0.7 of B_{z0} at the end of the magnetic system for B_{z0} = 1.2 T. It increases up to 0.85 for B_{z0} = 2 T.

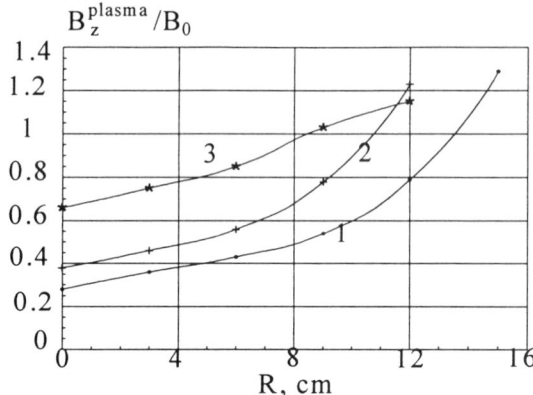

Fig.1. Radial distributions of the normalized magnetic field in presence of plasma
1- distance from the accelerator output z = 155 cm, vacuum magnetic field B_0 = 0.28 T;
2- z = 195 cm, B_0 = 0.41T, 3 - z = 235 cm, B_0 = 0.48T

The time dependent velocity of the plasma stream at the near axis region of the diagnostic chamber and the wave form of the measured magnetic field ΔB_z are shown in Fig.2. One can see from this Figure that the pulse duration of the displaced magnetic field was about 0.25-0.3 ms. The plasma velocity of different plasma stream layers, estimated on the base of magnetic measurements in different axial positions by time of flight, achieved 2.5 x 10^7 cm/s. The time duration of the plasma with velocity exceeding 10^7 cm/s was more than 200 μs and was larger than in the case of plasma propagating in the large diameter vacuum chamber without magnetic field.

The average plasma density (measured by Stark broadening) increased with increasing magnetic field value from (2-3)x10^{15} cm^{-3} for zero magnetic field up to 3 x10^{16} cm^{-3} for B = 1.2 T. The average plasma pressure in the diagnostic chamber, estimated as $n(T_e+T_i)$ = $-\{<B_z^{plasma}>^2 - (B_z^{boundary})^2\}/8\pi$, with $<B_z^{plasma}>$ the average magnetic field inside the plasma stream, achieved (4-6)x10^{17} eV/cm^{-3}. At the near axis region in the diagnostic chamber the plasma pressure achieved a value of $(3 - 4) \times 10^{18}$ eV/cm^{-3}. The total energy of the plasma stream measured behind the magnetic system was about 150-200 kJ. Therefore plasma with energy content of about 40-50 % of the total energy of the plasma stream generated by the accelerator (about 400 kJ) passed through the magnetic system. The plasma energy density at the near axis region of the plasma stream in the magnetic field achieved 5-7 kJ/cm^2.

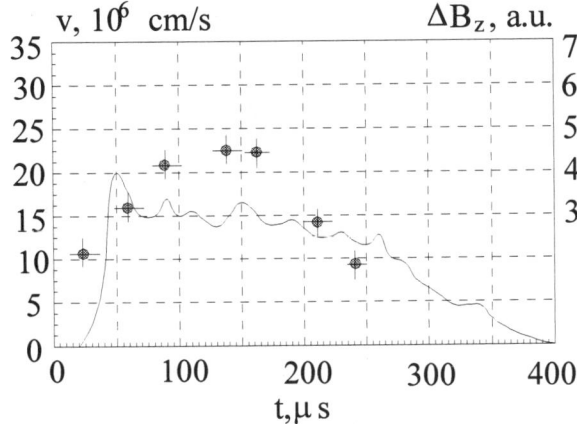

Fig.2 Time dependencies of plasma velocity (+) and magnetic field displaced by plasma

4. DISRUPTION SIMULATION EXPERIMENTS

Plasma streams with energy densities up to 7 kJ/cm^2 and axial magnetic field inside the plasma up to 0.7 T were utilized for first disruption simulation experiments [2, 5-7]. Targets of MPG-7 graphite were used in these experiments. The target diameter was varied from 3 to 8 cm. The target was positioned at the axis of the homogeneous magnetic field (between the 3d and the 4th magnetic coil) and irradiated by plasma stream propagating along the magnetic field perpendicular to the target surface.

4.1. Density behavior in the transient plasma layer.

Plasma density distributions were measured by using optical interferometry with viewing area of 20 cm. A transient plasma layer is formed close to the target surface with plasma densities more than one order of value exceeding the density of the incident hot plasma. A typical electron density distribution along the target surface is shown in Fig.3.

Shock waves are arising during plasma stream target interaction at a distance of 3 - 4 cm from the target surface. The average electron density in the shock-wave estimated from interferometry measurements is 10^{18} cm^{-3}. In comparison with the results from experiments without magnetic field the thickness of the transient layer was increased several times.

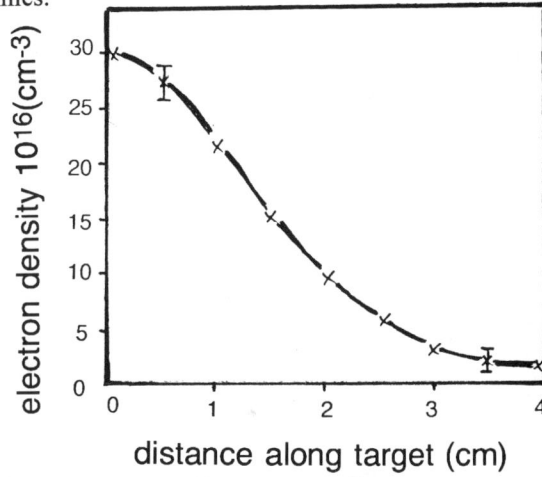

Fig. 3 Typical electron densisty distribution along the target surface at a distance of 0.5 cm from the target.

4.2. Energy characteristics of the transient plasma layer.

These measurements were performed with a copper calorimeter of diameter of 5 mm, placed in the cavity of the target. It was shown that less than 1 % of the incident plasma energy density arrived at the target surface. Maximum absorption of plasma stream energy occured in a layer 0.5 - 1 cm away from the target surface. The value of absored energy increased with decreasing magnetic field and was found to depend weakly on the size of the target. In previous experiments without magnetic field [6] the value of absorbed energy depended on the size of the target. In particular with decreasing size of the target from 8 cm to 3 cm the value of absorbed energy increased 1.5 - 2 times.

4.3. Erosion of the graphite target.

A laser scanning device was used for measurement of erosion of irradiated target. The dependencies of erosion depth on target coordinate are shown in Fig. 4. One can see from this figure that a maximum erosion of the order of 30 µm was obtained for the central region. Full width at half maximum of the erosion profile is about 40 mm. The erosion shows better homogeneity along the surface as compared with irradiated without magnetic field. The average value of graphite erosion is about 0.5 µm per 1 kJ/cm^2 of incident plasma energy density.

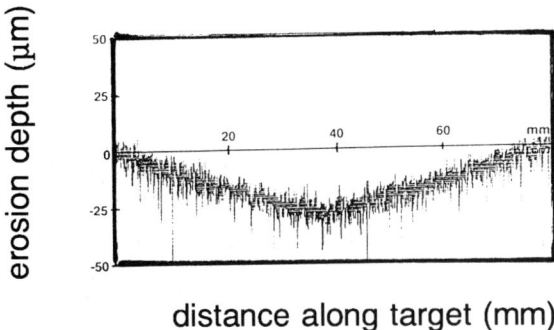

Fig. 4 Erosion profile for MPG graphite after 15 irradiations. Average energy density 5 kJ/cm^2, perpendicular impact of plasma stream.

5. CONCLUSION

Plasma streams with plasma densities up to $6 \cdot 10^{16}$ cm^{-3}, mean proton energy up to 0.35 keV, plasma energy densities up to 7 kJ/cm^2, peak power densities up to 60 MW/cm^2 pulse duration up to 0.2-0.3 ms and value of magnetic field in the plasma up to 0.7 T were used for disruption simulation experiments with graphite target of different size.

High density transient plasma layers with electrons densities up to $(6-8) \cdot 10^{17}$ cm^{-3} were formed close to the target. In comparison with the results obtained in previous experiments without magnetic field [2], the density of the transient layer practically was the same, but its thickness increased several times. The target surface absorbed about 1% of incident plasma energy. The value of absorbed energy weakly depends on the incident energy density and on the size of the target. The region of maximum absorbed energy was 0.5 - 1 cm away from the target surface.

The average erosion coefficient of MPG-7 graphite was 0.4 µm per 1 kJ/cm^2 of incident plasma energy density.

REFERENCES

[1] Tereshin V I 1995 Plasma Physics and Controlled Fusion, 37 A177-A190
[2] Chebotarev V V, Garkusha I E, Garkusha V V et al 1996 J. Nucl. Mater. 11296 C
[3] Tereshin V I et al 1992 Plasma Devices and Operations 2 155-65
[4] Chebotarev V V, Garkusha I E, Garkusha V V et al 1996 Proc. 23d EPS Conf. on Plasma
    Phys. and Controlled Fusion, Kiev, Ukraine
[5] Arkhipov N I et al 1994 Proc. 18th SOFT, Karlsruhe, Germany
[6] Wuerz H et al 1995 J. Nucl. Mater 220-222 1066-70
[7] Arkhipov N I et al J. Nucl. Mater. 11296 C

Tribological Behaviour of CFC Material

A. Orsini[a], E. Di Pietro[a], S. Libera[a], L. Verdini[a], E. Visca[a], G. Vieider[b]

[a]Associazione EURATOM-ENEA sulla Fusione, C.R. di Frascati, C.P. 00044 Frascati, Rome (Italy)

[b]The NET Team, Max Planck Institut für Plasmaphysik, Boltzmannstr. 2, D- Garching (Germany)

One of the monoblock divertor concept for ITER consists of metallic tubes surrounded by an Carbon Fiber Composite (CFC) armour with a sliding support system. In the frame of Subtask NET PDT 2.8 was performed tribological friction tests on Aerolor A05 CFC material coupled with AISI 316 to simulate the displacement due to thermal fatigue of the CFC tiles and their sliding support during the operative conditions of the divertor. For this scope a tribological test facility was developed and manufactured in the ENEA laboratories that is able to perform several type of the friction tests operating to different conditions of temperature (up to 350°C), applied loads and pressure. The mechanical parts, easily modifiables, were assembled inside a suitable vacuum vessel. A PC data acquisition system, developed in the ENEA laboratory using the software Labview, was used for the acquisition and control of the signals directly or through the conditioning units. The main results are reported in this paper in terms of variations of the friction coefficient in function of the process parameters used, simulating the operative life of the component.

1. INTRODUCTION

The aim of this activity was to perform tribological friction tests at different conditions, on Carbon Fiber Composite (CFC) material coupled with AISI 316 stainless steel, to simulate the displacement due to thermal fatigue of the CFC tiles and their sliding supports, according to the current design of the NET/ITER monoblock Divertor concept (Fig. 1), during the operative life [1]. The relevant phases of the test were the following:

- measurement of the friction coefficient during the test, with the goal to verify the variations of values during the operative life and respect to differents status (air, vacuum and hydrogen atmosphere) and temperature;
- visual analysis of the CFC samples at the end of the test.

2. TEST EQUIPMENT AND PROCEDURE

The tests were performed with the tribological test facility shown in Fig. 2, developed and manufactured at the ENEA laboratories that is able to perform several type of the friction tests operating to different conditions of temperature, applied loads and pressure.

Main parameters of the tribological machine are:

- max compression load   30,000 N
- max displacement load   17,500 N
- stroke up/down   0÷10 mm
- max displacement speed   4 mm/s
- max test temperature   350 °C
- pressure   Air, vacuum, gas

The machine has control units and instruments for the measurement and acquisition of all relevant physical parameters. A data acquisition system was

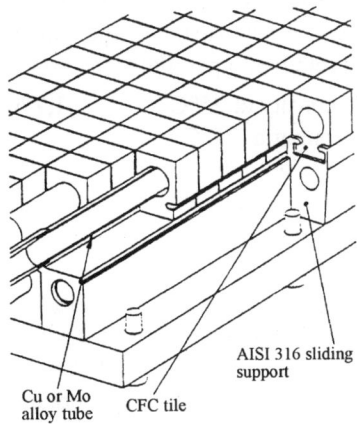

Figure 1. NET/ITER monoblock Divertor concept

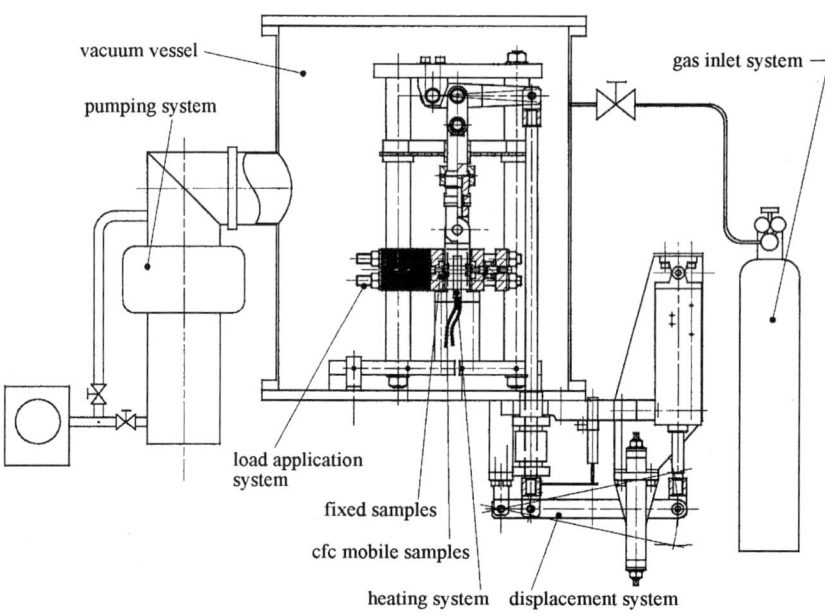

Figure 2. Schematic view of the tribological test facility

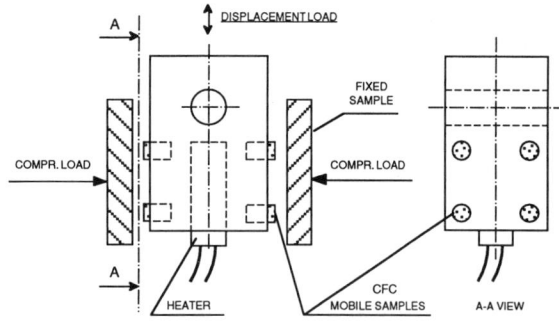

Figure 3. Outline of samples assembly

developed using the "Labview" software, package by NATIONAL INSTRUMENT. This software was used both for the acquisition and the automatic control of the main process parameters.

The fixed metallic samples, the CFC mobile samples and the heating system are shown in Fig. 3.

A procedure has been codified to guarantee the cleaning and drying of the samples, the achievement of the atmosphere (air, vacuum or hydrogen) and temperature conditions, the application of the compression and cyclical displacement loads, the recording of data.

3. TESTED MATERIALS AND PROCESS PARAMETERS

In order to simulate the working conditions of the sliding support and the CFC tiles of the NET monoblock Divertor concept, we decided to use the AISI 316 steel for the fixed sample (sliding) and the CFC "Aerolor A05", a 2D felt type produced by LE CARBONE-LORRAINE - France, for the mobile samples.

The AISI 316 steel was used with sliding surface as milled, roughness R_a=0.20 μm and hardness R_b= 80.

The CFC was tested with the fiber direction normal to contact surface, the main phisical parameters relative to parallel fiber axis were:

- apparent density at 25°C 1,770 kg/m^3
- tensile strenght at 25°C 60 MPa
- compressive strenght at 25°C 140 MPa
- bending strenght at 25°C 100 MPa
- thermal conductivity at 25 °C 200 W/m°C
- contact surface 50x4 mm^2

The process parameters for the tribological tests, to simulate the environment and the displacement due to thermal fatigue of the CFC tiles and their supports choiced from the results obtained from a Finite-Elements calculations of the thermal stress and performed with the ABACUS software [1], were:

- The contact pressure of 15 MPa was used for all tests.
- The displacement, with a conservative value respect to 0.24 mm expected sliding, was between 0.7 and 1.5 mm and was keep fixed during the test. The speed was 0.3 mm/s.
- The number of cycles were 500.

The tests performed in air were taken as refering data to compare the friction coefficient found at the different conditions.

Atmosphere		Temperature °C
Air	1 bar	20-150
Vacuum	$1 \div 10 \times 10^{-6}$ mbar	20-150-300
Hydrogen	$1 \div 10 \times 10^{-4}$ mbar	20-150-300

4. RESULTS

The values and the variations of the friction coefficient, measured during the tests, at the same temperature but in different environment conditions are compared in Figs 4a, 4b and 4c.

Micrographies of the CFC surface before and after the test are shown in Figs 5a and 5b.

<u>Tests performed in air.</u> The values of the friction coefficient measured during the tribological tests are between 0.14 and 0.22 in the range RT÷150 °C.

<u>Tests performed under vacuum.</u> Under vacuum, high friction coefficients were found with a minimun of 0.21 and a maximum of 0.70 in the range RT÷300 °C. The friction coefficient increases of a factor 2.6÷2.7 respect to RT and of a factor 3.2÷3.3 at 150 °C respect to the results obtained in air.

<u>Tests performed in Hydrogen atmosphere.</u> The values measured in this condition ($1 \div 10 \times 10^{-4}$ mbar) change between 0.31 and 0.76. The coefficients increase 2.3÷2.4 respect to RT and 3.1÷3.6 respect to 150 °C in air.

5. SUMMARY

With reference to the objective of the tests, that is the performance of the CFC materials in the NET/ITER monoblock divertor concept, the experimental evidence obtained during the trials has highlighted the following points:

- the tests confirm that the lubriant property of the graphite or CFC is due to absorbed water vapor

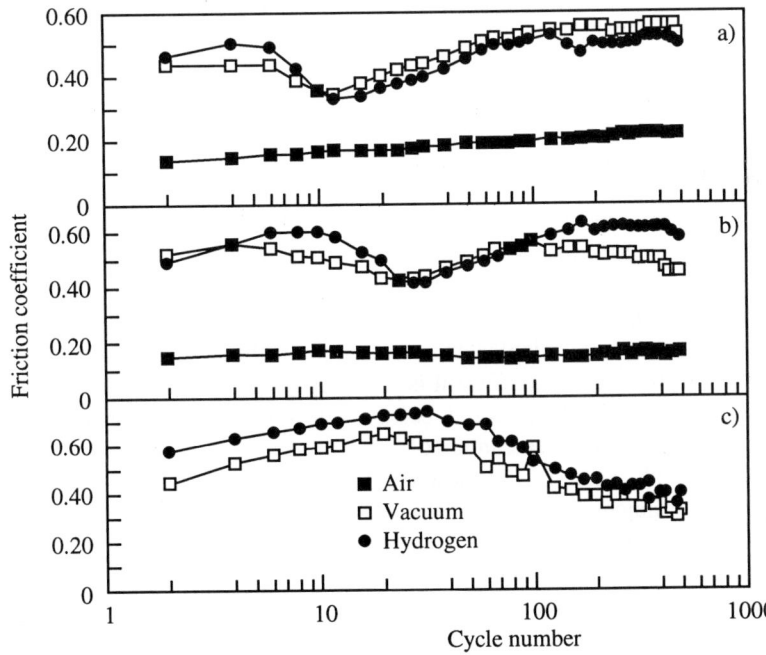

Figure 4. Friction coefficient variation in different status: a) T=20°C,   b) T=150°C,   c) T=300°C

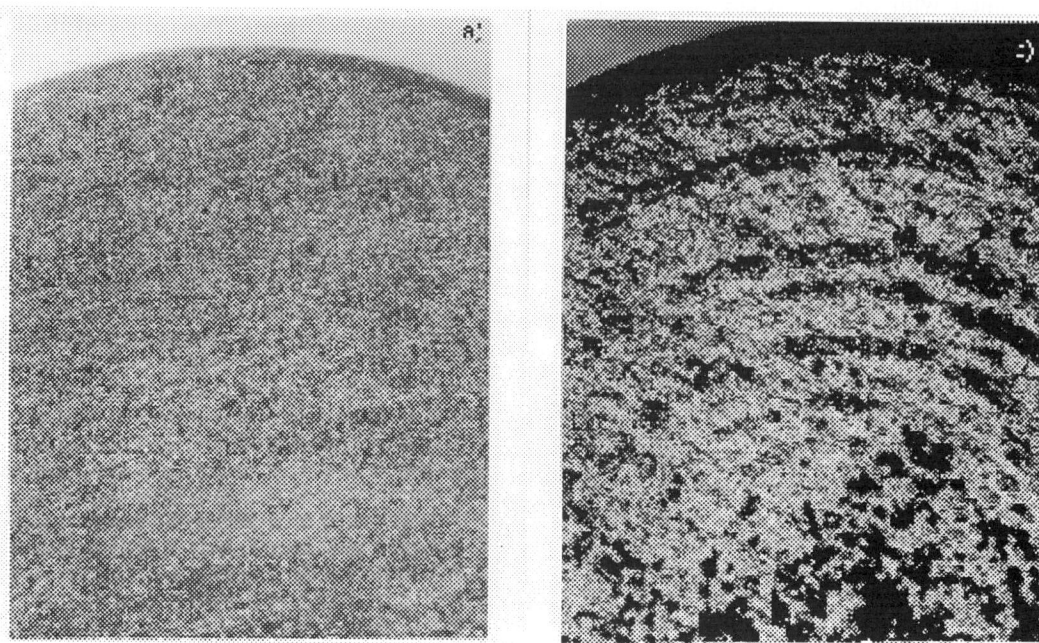

Figure 5.a) 20x micrography of the CFC sample before the test. b) 20x micrography of the CFC sample after 500 cicles

and oxygen. This property is lost in high vacuum or reducer atmosphere. Infact the friction coefficient increase of a factor 2÷4 in vacuum or hydrogen conditions respect the ordinary atmosphere (Figs 4a and 4b)

- the values found under vacuum and in Hydrogen atmosphere seem to be comparable (in hydrogen atmosphere is take as reference the pressure in the V.V. during the shot $10^{-3} \div 10^{-4}$ mbar). Only in Hydrogen atmosphere at 300 °C the friction coefficient increase about 20% respect vacuum condition (Fig. 4c), due to reducing effect of the Hydrogen respect to Oxigen.
- during the 500 cycles of displacement simulation, the trend of the friction coefficient is not constant (see figs. 4a, 4b). This could be justified of the combined action of detachment of the first layers of the graphite that include water and Oxigen (friction coefficient increase) and the filling of the porosities by the CFC powder, that increase the really contact surface (friction coefficient decrease). This phenomena appear to be slow at 300 °C, perhaps due to better mechanical properties of the CFC at high temperature.
- after the tests the CFC samples showed a good wear strength. The loss material measured of the samples was 0.1÷0.2%, equivalent to 10÷20 µm of the thickness. The micrographies taken after the tests show a tipical bright surface on the CFC due to flattening of the graphite powder on the surface. No cracks or damage were seen. The metallic samples do not show any surface damage, only a thin layer of graphite easily removable.

On the basis of the experimental measurements it can be affirm that the "Aerolor A05" CFC coupled with AISI 316 sliding support with no lubricant can stand thermal stress up to 15 MPa in the divertor operative condition and with displacements due to differents temperatures and expansion coefficients.

REFERENCES

1. E. Di Pietro, A. Orsini, M. Sacchetti, S. Libera, A. Cardella, G. Vieider - "Manufacturing and testing of a relevant scale mockup based on monoblock concept" - 15th Symp. on Fusion Engineering, Hyannis, Ma, October 11-15, 1993.

Development of 3D-based CFC with high thermal conductivity for fusion application

M. Araki[a], Y. Kude[b], Y. Sohda[b], K. Nakamura[a], K. Sato[a], S. Suzuki[a], M. Akiba[a]

[a] Naka Fusion Research Establishment, Japan Atomic Energy Research Institute,
801-1, Naka-machi, Naka-gun, Ibaraki-ken, 311-01 Japan.
[b] New Materials R&D Division, Central Technical Research Laboratory, Nippon Oil Company LTD.,
8, Chidori-cho, Naka-ku, Yokohama, 231 Japan.

For divertor high-heat-flux components in next fusion experimental reactors such as ITER, plasma facing material with higher thermal conductivity is essentially required. In addition, it is also required the armor tile itself to have superior mechanical properties at the bonding and fatigue characteristics point of view. To meet their requirements, a new three directional CFC with highly oriented fiber contents in each direction has been basically designed by Japan Atomic Energy Research Institute and fabricated by Nippon Oil Company LTD.. Major idea for improving the thermal conductivity is to use fine-pitch-base fibers with high thermal conductivity and to precise a packing density applying technologies such as HIPing process before graphitatizing. Consequently, the newly developed 3D-CFC has thermal conductivity measured as high as 540 $W \cdot m^{-1} \cdot k^{-1}$ at room temperature. To improve poor compatibility of CTE between 3D-CFC and heat sink made of copper alloy, we also designed and fabricated a 1D/3D functional CFC with high thermal conductivity (~600 $W \cdot m^{-1} \cdot k^{-1}$) and with relatively large coefficient of thermal expansion (~5 x 10^{-6} k^{-1}) based on the full 3D CFC.

1. INTRODUCTION

Plasma facing components for the next generation fusion machines such as the International Thermonuclear Experimental Reactor (ITER) are subjected to not only severe particle fluxes and thermal loads from the main plasma, but also electromagnetic forces during the operations [1]. In particular, divertor plates have the armor tiles bonded onto the actively cooled heat sink to play roles of exhausting helium ash and protecting the structural components such as the divertor cassette body, vacuum vessel and electromagnets from high-enegitc neutrons and heat loads. For developing robust divertor plates, various kinds of R&D efforts, i. e., material developments, bonding technology developments, manufacturing processes and thermal fatigue characterizations, have extensively been conducted in the world [2-7]. Based on their R&D results, ITER divertor plates are designed to have armor tiles made from carbon based materials for high heat flux region, tungsten based materials for high neutron flux region and from beryllium for the other region, respectively. To use thicker armor tiles and decrease the surface temperature of the divertor high-heat-flux components, plasma facing material with higher thermal conductivity is essentially required. In addition, it is also required the armor tile itself to have superior mechanical properties at the bonding and fatigue characteristics point of view. For these purposes, three directional/dimensional carbon fiber reinforced carbon composites (CFC) have been developed. However, their thermal conductivity is seemed to be still low for the fusion application [1, 8]. To meet their requirements, a new three dimensional CFC functioned with highly oriented fiber contents in each direction has been basically designed by Japan Atomic Energy Research Institute and fabricated by Nippon Oil Company LTD..

In the paper, structure and fabrication concept of 3D-CFC describes in section 2, and its thermal and mechanical properties are presented in section 3, respectively. Discussions are presented in section 4.

2. STRUCTURE AND FABRICATION CONCEPT OF 3D BASED CFC

2.1. Full 3D-CFC NIC-01

The existing 3D-CFCs are produced by applying to the textile process. However, their thermal conductivity is limited to be up to 280 $W \cdot m^{-1} \cdot k^{-1}$ due to large porosity around 20 % and low fiber contents [8]. To develop three dimensional (3D) CFC with high thermal conductivity, the conventional textile process was basically considered from the fabrication feasibility and cost relevance points of view. Structural concepts for improving thermal conductivity with high mechanical strength are mainly as follows; one is to reduce porosity of 3D-CFC, results in increase of the density and a packing factor which is defined by a ratio of the actual density to the ideal one. The other is to reasonably optimize the fiber content and orientation in each direction.

In general, a hot isostatic pressing (HIP) process is applied in order to reduce porosity in the bulk material. This process is already well established for metal [9]. Based on this aspect, we fabricated unidirectional CFC by HIPing process, namely U-301, to evaluate whether or not this process can be effective for the carbon fiber composite. HIPing condition was tentatively selected to up to 2000 °C at up to 200 MPa. From the thermal property measurement, it is confirmed that U-301 has a density as high as 2.13 $g \cdot cm^{-3}$ which is close to the ideal density (2.2 $g \cdot cm^{-3}$) and thermal conductivity over 800 $W \cdot m^{-1} \cdot k^{-1}$ at room temperature. Therefore, it is believed that this process be very effective to reduce the porosity and increase the density for 3D-CFC structure.

From the recent textile technology, it can be possible to increase the volumetric fraction of fibers up to 75 vol. % in the case of 1D-CFC [10]. However, in the preliminary study of thermal properties on the fiber density, it revealed that thermal conductivity of CFC has a maximum value at the total fiber content of 55 to 60 vol. % when the fiber content varied from 30 to 75 vol. %. This reason is not clear yet, but possibly due to difficulty of filling pitch derived matrix into the textile with high fiber content efficiently. Therefore, in this development total fiber content of 55 to 66 vol. % is selected for 3D-CFC based on the recent textile technology and the preliminary results. Figure 1 shows a fabrication process of 3D-CFC, namely NIC-01. Fiber contents in each direction are finally provided to have 43 vol. % in a direction which high thermal conductivity is required, 18 vol. % and 6 vol. % in the other two directions.

It is also very important to select pitch-base fibers with high mechanical strength and to provide the graphitization temperature. From the previous heating experiments of 1D- and/or 2D-CFCs [11, 12, 13], it is seen that the inpregnated layers around each fiber are formed with pyrolytic structure and are expected to have higher thermal conductivity than the fiber itself even if the fiber with high thermal conductivity was selected. To create pyrolytic layers around fibers more, high graphitization temperature for long time shall be required. However, it is welknown that the mechanical strength of the 3D-CFC decreases with increasing the graphitization temperature and duration. In this study, commercially available fine-pitch-base fibers, namely Granoc® developed by Nippon Oil Company and manufactured by Nippon Graphite Fiber Corporation, were used. Since Granoc has an advantage of keeping its strength at a temperature around 3000 °C, it is expected that the 3D-CFC can be produced with superior mechanical and thermal properties.

2.2 1D/3D and 1D/3D/1D Functional CFCs

In general, full 3D-CFC has much superior mechanical properties to the other 1D and 2D CFCs. For fusion application, there, however, are further developmental and/or modification items of 3D-CFC to 1) mechanically be matched in metal combinations and 2) improve thermal and mechanical properties on the plasma surface of CFCs. First one is to evaluate the bonding feasibility between 3D-CFC and the heat sink metal due to its poor material compatibility. To solve this problem, based on NIC-01 1D/3D functional CFC, namely NIC-02, is designed and fabricated based on a fact that unidirectional CFCs such as MFC-1 and U-301 have a coefficient of thermal expansion as high as $12 \times 10^{-1} k^{-1}$ which is close to that of the heat sink made of copper alloy. This is one of functional gradient materials without applying any bonding technologies. NIC-02 has structures as follows; in the bonding side of NIC-02, fibers perpendicular to the bond interface with the pitch-base matrix only remain and full three dimensional fiber textile oriented exists from middle region to the surface.

However, large differences of microscopic thermal

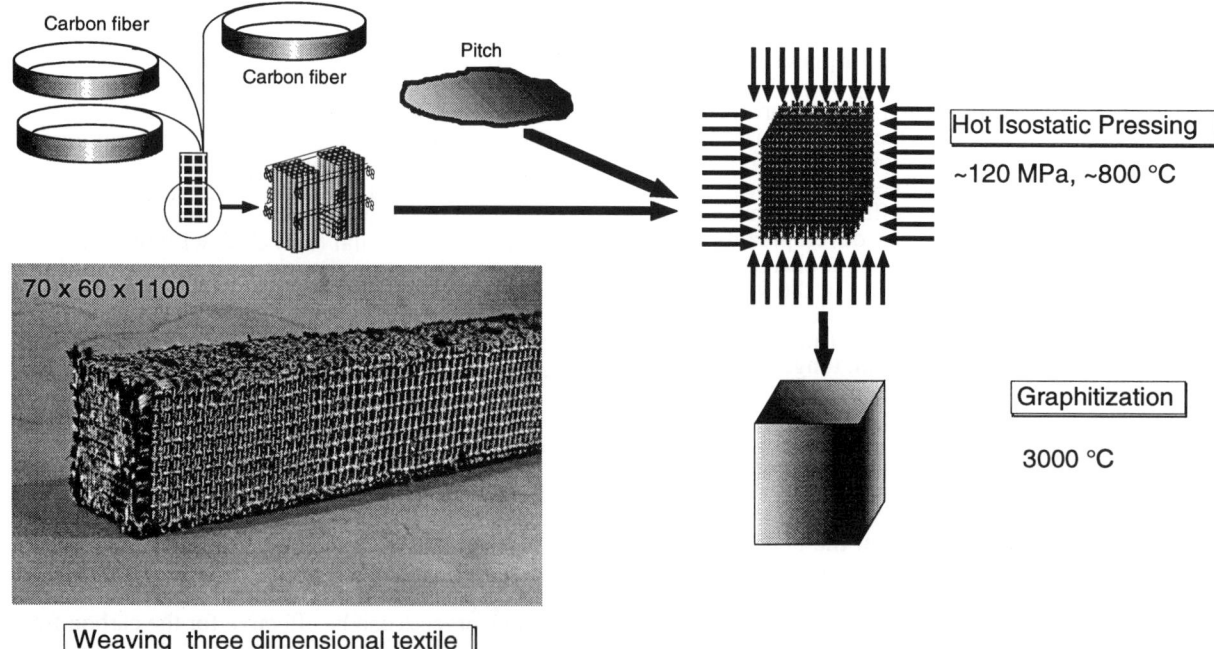

Figure 1. Fabrication Procedure of Newly Developed Three Dimensional Carbon Fiber Reinforced Carbon Composite.

conductivities in each direction of 3D-CFC appear due to the three dimensional structure. This problem was confirmed in preliminary heating test of an actively cooled mock-up with 3D-CFC armor tiles that bright lines were observed during heating, resulting in high temperature along the fiber directions perpendicular to the beam incident angle [14]. Therefore, it might be critical to reduce the high temperature region in order to minimize the erosion and impurities into the plasma. To solve this problem, 1D/3D/1D functional CFC, namely NIC-02A, is designed and fabricated. This functional CFC has one dimensional structures in both of the plasma surface and the interface to the heat sink metal. Quite similar fabrication process of NIC-01 is applied for 1D/3D- and 1D/3D/1D-CFCs, in partial use of special fiber in both of 1D regions, which keeps the proper strength at temperatures up to 500 °C and looses its strength above temperature, during forming the three dimensional textile. After fabricating the 3D textile at room temperature, the HIPing process is applied at a typical condition of up to 800 °C with up to 120 MPa for the proper duration. During HIPing process, special fibers easily loose its strength and becomes brittlement. Also, these fibers are evaporated except for carbon component, so that the pitch base matrix can intrude into their place with collapsing the special fibers. Before the final process, thermal treatment is done at a condition of around 2000 °C at atmospheric pressure to almost exhaust gases in the material. Subsequently, graphitization process is applied at a temperature around 3000 °C for the proper duration. It is also key issue to evaluate the impurity contents. Table 1 shows the measured result. Calcium content was measured relatively high, so that further effort to reduce this impurity is required.

Table 1. Impurity of the 3D-CFC NIC-01.

Material	Al	Ba	Ca	Fe	Zn	others
Contents mass PPM	8	4	160	10	1	undetectable

Qualitative and quantitative analyses were performed using the powder produced from 3D-CFC after burning in air.

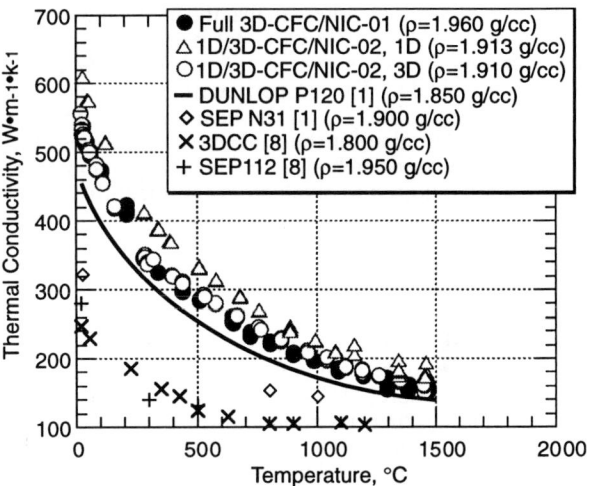

Figure 2. Macroscopic Thermal Conductivity of 3D-CFCs as a Function of Temperature.

3. THERMAL AND MECHNICAL PROPERTIES

3.1 Thermal Conductivity

For measuring their thermal conductivities, laser flashing method was applied for specimen with different sizes which were machined in order to match the equipment. Typical dimension of the specimen was 10 to 11 mm ^{l} x 10 to 11 mm ^{w} x 4.3 to 7.5 mm ^{d} prepared from the large size blocks.

Figure 2 shows typical thermal conductivity of newly developed 3D-base CFCs as a function of temperature. In the figure, thermal conductivity of existing 3D-CFCs are also plotted. It is noted that these values are macroscopic thermal conductivity because of highly oriented composite materials. For fusion application, it is beleived to be more important the armor tile to have higher thermal conductivity perpendicular to the plasma surface than the others due to one-sided heating conditions, so that thermal conductivity in interested direction is measured. From the figure, newly developed 3D-base CFCs appearently have higher thermal conductibity than the existing 3D-CFCs.

3.2 Mechanical Properties

There are many interesting mechanical properties for applying it to the PFM. It is beleived that a mechanical property defined by a ratio of the displacement

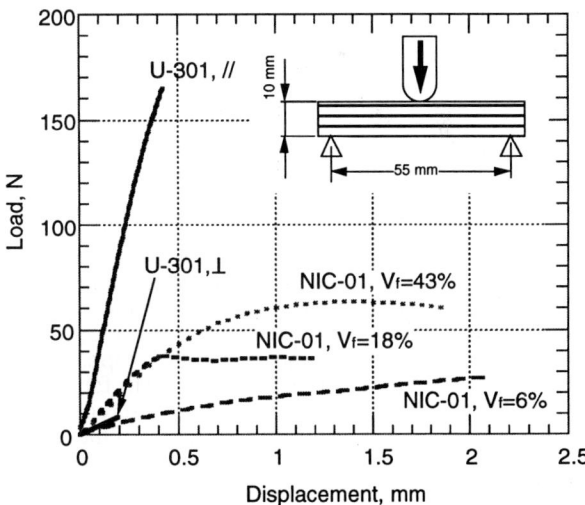

Figure 3. Load History of Bending Test on UD and 3D-CFCs.

The strengths of specimen were measured by short beam method.

to the total length shall be the most important property as well as the strength because the armor tile must be attached to the heat sink and withstand large thermal and elctromagnetic loads. Figure 3 shows a typical displacement of 3D-CFC from the bending test. In the figure, typical displacement of the unidirectional CFC (U-301) are also indicated as comparison. At room temperature, the unidirectional CFC has displacements only up to 0.5 mm for 55 mm total length, in particular around 0.2 mm in Vf = 0 vol. % direction, where is the interface side to the heat sink. On the other hand, NIC-01 has the displacement as large as over 1.2 mm. Based on the bending test results, it is clearly seen that the mechanical property of the developed 3D-CFC defined above (not actual bending strain) is over 2.2 % while 0.4 % in the unidirectional CFC. This property will give much better material compatibility with the heat sink not only for the brazing, but also for thermal loads. However, further measurements such as a tensile strength and shear strength are reuqired to elvaluate its applicability to the plasma facing material.

4. DISCUSSIONS

Newly developed 3D-CFC, NIC-01, has macroscopic thermal conductivity as high as around 540 $W \cdot m^{-1} \cdot k^{-1}$ at room temperature compared with existing 3D-based CFCs. Furthermore, 1D/3D- and 1D/3D/1D functional CFCs with high thermal conductibity were also developed to improve material compatibility with the metal. In particular the measured macroscopic thermal conductivity of 1D/3D functional CFC is over 600 $W \cdot m^{-1} \cdot k^{-1}$ in one dimensional region and around 550 $W \cdot m^{-1} \cdot k^{-1}$ in three dimensional region, respectively. An interested coefficient of thermal expansion at the bond interface of NIC-02 will be expected to be around $5 \times 10^{-6} \cdot k^{-1}$, while CTE of NIC-01 is almost zero. Although development of the bonding technology between the functional CFCs and copper alloy is the ongoing process, we believe that this feature will give better material compatibility with metals.

We also plan to developing 3D-based functional CFCs not only with higher thermal conductibity, but also with much better material compatibility in order to perform more reliable design of the divertor high-heat-flux components. For this purpose, it will be effective to improve the fiber contents in each direction and to make much fine structure. Based on the successful development of the 3D-base CFCs, fiber contents in each direction are rearranged to be 39, 11, and 10 vol.%, respectively. Recently, this 1D/3D/1D functional CFC was fabricated successfully and its properties will be measured soon.

In near future, manufacturing and testing of the newly developed 3D-CFCs with the actively cooled heat sink are also planned in order to confirm its performance and evaluate its applicability for plasma facing materials of the ITER divertor prototype.

ACKNOWLEDGEMENT

The authors would wish to thank Dr. Y. Ohara and the members of NBI Heating Laboratory in JAERI for their valuable discussions and comments. They would also like to thank Dr. T. Baba of National Research Laboratory of Metrogy in Japan and Dr. S. Gotoh of Hitachi Research Institute for measuring the thermal properties within contracts to JAERI. They also would like to acknowledge Drs. S. Matsuda, M. Ohta, M. Seki, T. Tsunematsu, T. Nagashima and Dr. Y. Seki of JAERI for their support and encouragement.

REFERENCES

1. ITER Joint Central Team, "Technical Basis for the ITER Interim Design Report, Cost Review and Safety Analysis," ITER EDA Documentation Series, No. 7, IAEA, Viena, (1996).
2. M. Araki, M. Akiba, M. Dairaku, et al., *J. Nucl. Sci. Technol. 29(9)*, (1992) 901-908; and also see S. Suzuki, M. Akiba, M. Araki, et al., *J. Nucl. Mater. 212-215*, (1994) 1365-1369.
3. I. Smid, C. D. Croessmann, R. D. Watson, et al., *Fusion Technol. 19*, (1991) 2035-2040; and also see A. Cardella, G. Vieider, E. Di Pietro, et al., *Proc. of 18th Symp. on Fusion Technol.*, Karlsruhe, p283-286 (1994).
4. R. D. Watson, F. M. Hosking, M. F. Smith, et al., *Fusion Technol. 19*, (1991) 1794-1798; and also see D. L. Youchison, R. Guiniiatouline, R. D. Watson, et al., *Proc. of 18th Symp. on Fusion Technol.*, Karlsruhe, p287-290 (1994).
5. H. D. Falter, D. Cilic, E. B. Deksnis, et al., *Proc. of 18th Symp. on Fusion Technol.*, Karlsruhe, p291-294 (1994).
6. J. Schlosser, P. Chappuis, P. Deschamps, *Proc. of 17th Symp. on Fusion Technol.*, Rome, p367-371 (1992).
7. A.V. Burdakov, V.V. Filippov, V.S. Koidan, et al., *J. Nucl. Mater. 212-215*, (1994) 1345-1348.
8. I. Smid, M. Akiba, M. Araki, et al., JAERI-M 93-149, *Japan Atomic Energy Research Institute Report* (1993); and also see M. Araki, M. Akiba, R. D. Watson, et al., *Atomic and Plasma-Material Interaction Processes in Controlled Thermonuclear Fusion, Vol. 5*, IAEA, Viena, (1994) 245-265.
9. S. Yamazaki, H. Ise, K. Mohri, et al., *Proc. 17th of Symp. on Fusion Technol.*, Rome, (1992) p.410 and also see Hatano, S. Sato, K. Sato, et al., *presented in the 12th American Nuclear Society Topical Meeting on Technology of Fusion Energy, to be published for Fusion Technology*, (1996).
10. A. Mutoh, Y. Kude, Y. Sohda, et al., *appeared in the extended abstracts of 21th Annual Mtg. Japan Carbon Soc.*, Tokyo, p140 (1994).
11. J. G. van der Laan, *Fusion Technol. 19*, (1991) 2070-2075.
12. M. Araki, M. Akiba, M. Seki, et al., *Fusion Engineering and Design 19*, (1992) 101-109; and also see K. Nakamura, Doctor Thesis, *Hokkaido University* (1995).
13. D. A. Bowers, J. W. Davis, R. B. Dinwiddie, *J. Nucl. Mater. 212-215*, (1994) 1163-1167.
14. M. Akiba, M. Araki, K. Nakamura, et al., *in this series*.

Evaluation of hydrogen sorption and desorption for Ti-6Al-4V alloy as a vacuum vessel material

Y. Hirohata[a], Y. Aihara[a], T. Hino[a,b], N. Miki[c] and S. Nakagawa[c]

[a] Department of Nuclear Engineering, Hokkaido University, Kita-ku, Sapporo, 060 Japan
[b] National Institute for Fusion Science, Furo-cho, Chikusa-ku, Nagoya, 464-01 Japan
[c] Toshiba Corporation, Chiyoda-ku, Tokyo, 100 Japan

The hydrogen absorption and desorption properties of Ti-6Al-4V alloy were evaluated under a condition of a fusion reactor operation, e.g., at low hydrogen pressure and low temperature. When the hydrogen pressure and the temperature were kept 0.3 Pa and 660 K, respectively, for 48 hr, the absorption amount of the polished sample was almost the same as the initial concentration of hydrogen (90 ppm). At a higher pressure, 10 Pa, the absorption amount increased when the temperature exceeded 750 K. In the case of the degassed sample, the absorption amount rapidly increased in the temperature range from 600 K to 650 K when the hydrogen pressure was kept 0.3 Pa. The absorption amount saturated in the absorption time of about 20 hr at 660 K. The saturation level was less than 250 ppm which was lower than the limited concentration for the embrittlement of Ti-6Al-4V alloy (1000 ppm).

1. INTRODUCTION

Since the vacuum vessel is exposed to neutrons in a fusion reactor, the low activation material has to be used for such the component. The dose rate of Ti-6Al-4V alloy after the shutdown at 1 year is estimated as two orders of magnitude smaller than that of 316L SS or Inconel 625 [1-2]. However, the embrittlement due to the absorption of fuel hydrogens may become a problem for the Ti alloy. It is reported that the embrittlement of Ti-6Al-4V alloy occurs when the absorption amount of hydrogen exceeds approximately 1000 ppm [3]. Although the hydrogen absorption properties of pure Ti metal have been well evaluated [4-5], there are little studies for Ti-6Al-4V alloy under the operation condition of a fusion reactor, e.g., such the data at low temperature and low pressure region are required.

In the present study, we investigated the hydrogen absorption and desorption behaviors of mechanically polished and degassed Ti-6Al-4V alloys, under the condition at low hydrogen pressure (0.03-100 Pa) and the low absorption temperature (510-820 K) by using a technique of thermal desorption spectroscopy (TDS). We also examined the hydrogen desorption properties of unpolished and TiN coated samples. The TiN is known as the material with the diffusion barrier for hydrogen [7-8].

2. EXPERIMENTAL

The Ti-6Al-4V alloy with a final treatment of dipping in HNO_3-HF mixed acid was used as the samples. The present Ti-6Al-4V alloy was ($\alpha+\beta$) alloy. Sample geometry was a slab with a size of $15 \times 15 \times 0.4$ mm^3. This sample is called as as-received sample. As-received sample was mechanically polished by both using Emery paper and Al_2O_3 powder, and rinsed by ethanol in an ultrasonic bath and baked at 507 K for 10 min in a vacuum. This sample is called as polished sample. Polished sample was degassed at 923 K for 60 min, and heated at the same temperature for 10 min in hydrogen pressure of 0.3 Pa. The degassing temperature was determined both by the final heat treatment and by the phase diagram of Ti-alloy. This sample is called as degassed sample.

Hydrogen absorption experiments were carried out for polished and degassed samples under the condition of low temperature (T_s= 510-820 K) and low pressure (P_{H_2}= 0.03-100 Pa). The absorption and desorption amounts were measured by using a technique of thermal desorption spectroscopy (TDS).

The sample was heated up to 923 K (degassed sample) or 1053 K (polished sample) with a constant ramp rate of 0.5 K/s, and held for 10 min at such the final heating temperature. We also examined hydrogen desorption behaviors of as-received sample and TiN coated sample. TiN was coated on all faces of the polished sample with a thickness of 0.5 μm by using a reactive sputtering method[6]. The surface morphology, atomic composition and chemical bonding states of samples were analyzed by using techniques of SEM, and ex-situ XPS, respectively.

3. RESULTS AND DISCUSSION
3.1 Hydrogen desorption property

Figure 1 shows typical hydrogen desorption spectra of as-received sample, polished sample and degassed sample. Here, only the TDS spectra of degassed sample was measured after hydrogen absorption at 773 K for 5 hr. As-received sample had one shoulder and one peak in the TDS curve. The temperature of the major peak was about 1020 K. The desorption amount of hydrogen, e.g., initial hydrogen concentration (C_o), was 90 ± 10 ppm. On the contrary, only one peak was observed in a desorption curve of the polished sample at about 900 K. The desorption amount of polished sample was approximately the same as C_o. The chemical binding states and the atomic compositions of as-received and polished samples were analyzed by XPS. The thickness of titanium oxide layer for as-received sample was twice larger than that of polished sample. This results indicates that the desorption peak of hydrogen depends on the oxide layer. It is known that oxide layer of Ti-alloy or stainless steel becomes a diffusion barrier for hydrogen at low temperature region[9-10]. The desorption peak of hydrogen for degassed sample was almost the same as that of polished sample. The degassed sample was again heated up to 1053 K after the TDS measurement, but the hydrogen desorption was not observed. Thus, it is regarded that the hydrogen absorbed in the degassed sample perfectly desorbed during the heating up to 923 K.

The hydrogen desorption spectrum of TiN coated sample is also shown in Fig.1. Although the hydrogen desorption was observed in the temperature range from 900 K to 1050 K, the desorption rate was small. The desorption peak was observed when the temperature was held at 1053 K. This result shows that the TiN coatings can become a diffusion barrier for hydrogen.

3.2 Absorption property of hydrogen

Figure 2 shows the absorption amount of hydrogen as a function of absorption temperature (T_s= 510-820 K) for the polished sample when the absorption time was kept 5 hr. In the case of 0.3 Pa, the absorption amount of the polished sample was almost the same as the initial concentration of hydrogen (C_o= 90 ppm). However, at the higher pressure, 10 Pa, the absorption amount incresed when the temperature

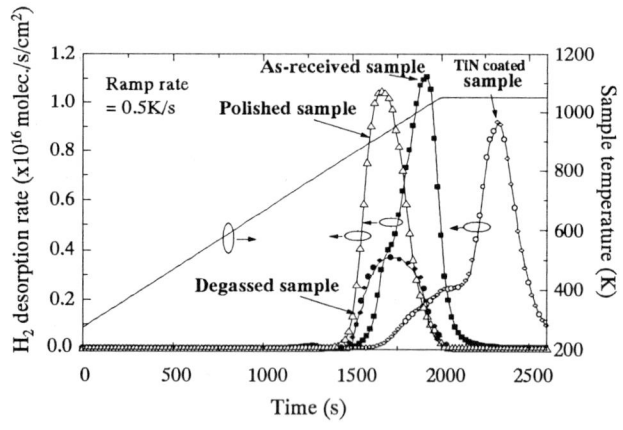

Fig.1 Thermal desorption curves of hydrogen for as-received, polished, degassed and TiN coated samples.

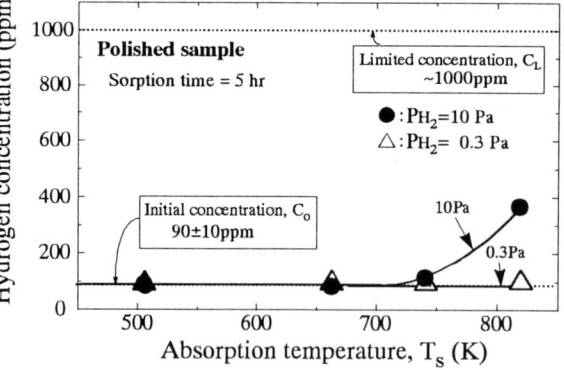

Fig.2 The absorption amount of hydrogen as a function of absorption.

exceeded 750K. When the hydrogen pressure, P_{H_2}, was varied from 0.03 Pa to 100 Pa at 820 K, the absorption amount was observed to be proportional to $P_{H_2}^{1/2}$. This tendency well agrees with the Siverts' law. The Siverts' constant estimated was 170 (ppm/Pa$^{1/2}$), which was between those of α-Ti (140) and β-Ti (340) [4]. At 820 K, the absorption amount saturated before the absorption time reaches to about 20 hr. This saturated level was 490 ppm. The critical pressure, C_L, was estimated 40 Pa, at 820 K. Here, the critical pressure means the absorption amount (1000 ppm) which causes the embrittlement.

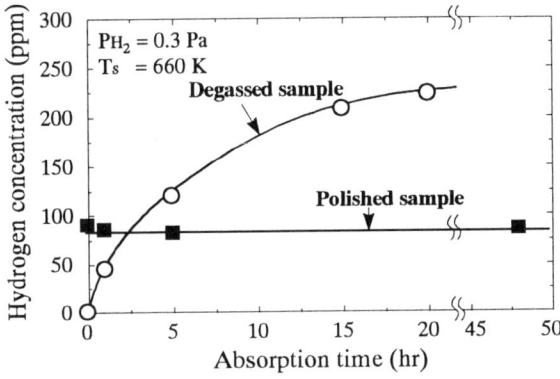

Fig.3 The absorption amount of hydrogen for polished and degassed samples as a function of absorption time.

Figure 3 shows the absorption amount of hydrogen as a function of absorption time for polished and degassed samples when P_{H_2} and T_s were kept 0.3 Pa and 660 K, respectively. In the case of polished sample, the absorption amount was almost the constant, C_o, even if the absorption time was long, 48 hr. However, in the case of degassed sample, the absorption amount increased with the absorption time and saturated at about 20 hr. This saturation level was less than 250 ppm, which is sufficiently smaller than C_L. It was observed by the XPS analysis that the thickness of the oxide layer for polished sample was twice larger than that of degassed sample. In addition, although the oxide layers of Ti and V were the same between these samples, the thickness of aluminium oxide layer for polished sample, was twice larger than that of degassed sample. Thus, the absorption amount of hydrogen largely depends on the oxide layer.

Figure 4 shows the absorption amount of degassed sample as a function of absorption temperature when P_{H_2} was kept 0.3 Pa. Here, the absorption time was changed from 1 hr to 20 hr. The absorption amount observed was very small, less than 30 ppm, when the sorption temperature was lower than 600 K. The absorption amount suddenly increased in the temperature range from 600 K to 650 K, and decreased with the increase of T_s for the temperature higher than 650 K. This tendency was very similar to the case of pure α-Ti, while the temperature region of transition for α-Ti was from 550 K to 590 K[11]. The absorption amount of hydrogen almost saturated at the absorption time of about 20 hr. The saturation level was estimated less than 250 ppm, which is lower than C_L.

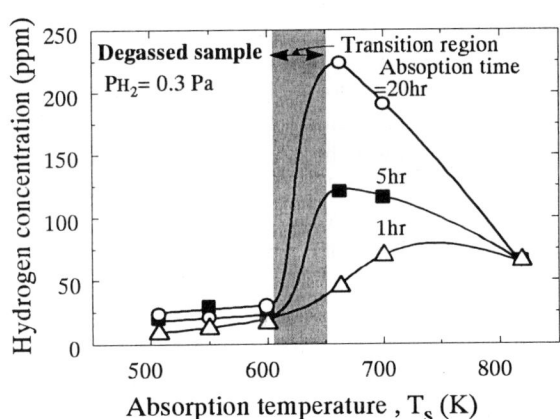

Fig.4 The absorption amount of hydrogen as a function of absorption temperature when the absorption time was changed from 1hr to 20hr.

3.3 Activation energies for diffusion and sorption

Table 1 shows the activation energies for diffusion and sorption of Ti-6Al-4V sample. The activation energies of α-Ti and β-Ti[12] are also shown in the table. In order to obtain the activation energies of the hydrogen diffusion, the desorption curves of as-received samples at constant temperatures (973 K,

Table 1 Activation energies of hydrogen for Ti-6Al-4V alloy.

Activation energy (eV)	Diffusion in titanium			Sorption in titanium		
	Desorption	α-Ti [12]	β-Ti [12]	Absorption	α-Ti [13,14]	β-Ti [13,14]
	0.62-0.73	0.538	0.288	0.15-0.19	0.44	0.50

1023 K, 1073K, 1103K) were measured. The activation energy for diffusion of Ti-6Al-4V alloy was slightly larger than that of α-Ti, and was twice larger than that of β-Ti[12]. Diffusion constant, $D(cm^2/s)$, was obtained as $D=6.6 \times 10^{-3} \exp(-E_a/kT)$, where E_a=0.63-0.73 eV. This value is smaller than that of α-Ti or β-Ti [12]. On the contrary, the activation energy for hydrogen sorption of degassed sample was approximately a half of α-Ti or β-Ti[13-14].

4. Conclusion

The hydrogen absorption and desorption properties for Ti-6Al-4V alloy were evaluated under the operation condition of a fusion reactor. The hydrogen sorption behavior for polished sample was different with that of degassed sample. When P_{H_2} and T_s were kept 0.3 Pa and 660K, respectively, the absorption amount of polished sample remained almost the same as the initial concentration of hydrogen (90 ppm).

In the case of degassed sample, the absorption amount increased with the absorption time, and this absorption behavior was very similar to that of pure α-Ti. The absorption amount suddenly increased in the temperature range from 600 K to 650 K, which was 100 K higher than the case of pure α-Ti. The absorption amount saturated in the absorption time of about 20 hr. The saturation level was less than 250 ppm, e.g., less than the limited concentration for embrittlement (1000 ppm).

Acknowledgment

The work was supported by Grant-in-Aid of the Ministry of Education, Science and Culture of Japan.

References

1. R.W.Conn, J. Nucl. Mater., 76-77 (1978) 103.
2. N.Miya, Y.Kamada, S.Nakagawa, et.al., Fusion Eng. Design, 23 (1993) 351.
3. V.A.Livanov, B.A.Kollachev, Proc. of Ti-Conf. on Hydrogen Embrittlement of Titanium and its alloys, (1968) 561.
4. A.D.McQuillan, Proc. Roy. Soc. London Ser. A204 (1950) 309
5. Yuh Fukai, "The metal-hydrogen system" Spronger-Verlag Berlin Heidelberg, 1993.
6. Y.Aihara, Y.Hirohata, T.Hino and T.Yamashina, Jpn. J. Vac. Soc. 38 (1995) 339.
7. C.Braganza, H.Stussi and S.Veprek, J. Nucl. Mater., 87 (1979) 331.
8. Y.Ikeda, K.Saitoh et.al., Jpn. J. Vac. Soc., 37 (1994) 232.
9. W.A.Swanoiger and R.Bastasz, J. Nucl. Mater., 85-86 (1979) 335.
10. H.Katsuda, and K.Furukawa, Jpn. J. Nucl. Sci. & Technol., 48 (1981) 143.
11. Y.Fukui, J. Less-Common Met., 172-174 (1991) 8.
12. R.T.Wasilewski and G.K.Kehl, Metallugia, 50 (1954) 225., J. Inst.Metals, 83 (1955) 94.
13. M.Nagasaka and T.Yamashina, J. Less-Common Met., 45 (1976) 53.
14. K.Watanabe, J. Nucl. Mater., 136 (1985) 1.

Crack propagation test under cyclic thermal stress by irradiation of electron beam

H. Kogawa, S. Kimura, T. Teramoto and M. Saito

Institute of Engineering Mechanics, University of Tsukuba
1-1-1 Tennodai, Tsukuba 305, Japan

The crack propagation test under cyclic thermal stress is performed in the way that the electron beam is cyclicly radiated to the specimen made of type 316 austenitic stainless steel which has a half ellipic pre-crack. The crack propagation rate decreases with increasing the number of the irradiation cycle. The numerical analysis is performed to evaluate the experimental crack behavior. The crack propagation amount estimated by the analysis agree well the experimental one

1. INTRODUCTION

A first wall of fusion reactor is exposed to high temperature by plasma, and it has unexpected crack by virtue of plasma disruption. While the opposite side of plasma side is cooled for elimination of heat. And since fusion experimental reactors, such as ITER, will be operated in pulse mode, the cyclic thermal stress will take place across the first wall. Hence, it is necessary to predict the crack propagation amount and to clarify the crack propagation behavior for the structural design of first wall.

In this study, to investigate the crack propagation behavior under cyclic thermal stress, the crack propagation test due to cyclic irradiation of electron beam was performed. Using finite element code "MARC", transient heat conduction analysis and elastic plastic stress analysis were performed under this test condition, and the thermal fatigue crack propagation was estimated with non-linear fracture mechanics parameter which is numerically evaluated and crack propagation rule for conventional high temperature fatigue test result.

2. EXPERIMENT AND ANALYSIS

2.1. Electron beam irradiation facility

The experimental facility used in this study consists of vacuum chamber on which electron beam generator is mounted. The shape of the electron beam irradiating the specimen is an ellipse of 12 mm in horizontal direction and 8mm in perpendicular one, and its peak power is about 12 Mw/m^2. As shown in Fig.1(a), the electron beam irradiates the area where the pre-crack is located in the specimen, and it can be moved cyclically to a water cooled copper block by means of deflection coil, so that cyclic thermal stress takes place in the specimen.

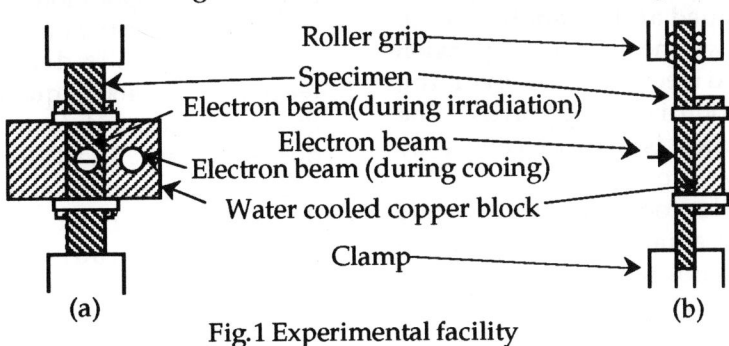

Fig.1 Experimental facility

2.2. Specimen and Analytical model

The specimen tested is made of the type 316 austenitic stainless steel, which is one of the materials proposed for the first wall. The specimen is a 5mm thick plate with gage length of 60 mm and width of 12.5 mm. At the center of specimen front surface, a half elliptic pre-crack with depth of 0.5 mm and length of 4mm is introduced by means of an electro-spark machine and mechanical fatigue load to simulate the unexpected crack, as shown in Fig.2.

The finite element breakdown for the half gage of specimen is shown in Fig.3. Two dimensional analysis is conducted for the section crossing the deepest crack front.

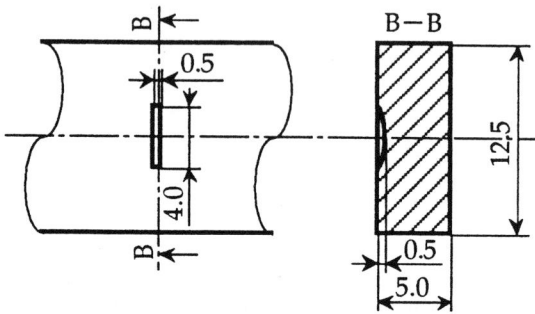

Fig.2 The elliptic pre-crack of specimen

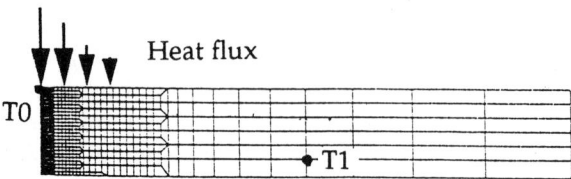

Fig.3 Finite element breakdown

2.3. Experiment

In the irradiation test, as shown in Fig.1(b), the specimen is supported by roller grip at one end and clamped at the another end. This support allows the thermal expansion to longitudinal direction, however the bending deformation is not permitted. The specimen back surface is always cooled by water cooled copper block.

One thermal cycle consists of 30 seconds with 0.5 seconds' irradiation to the specimen and 29.5 seconds' cooling. The specimen is cooled, when the electron beam is moved to irradiate the copper block.

It is too difficult to measure the crack propagation amount during the irradiation test, because this test is carried out under condition that the electron beam is radiated. Therefore, the surface of crack is marked by means of interrupting the irradiation every 500 or 1000 cycles for a while. After finishing the irradiation test, the specimen is ruptured. The ruptured surface is observed by an optical microscope and a scanning electron microscope (SEM). Then the crack propagation length along the specimen thick direction is measured.

2.4. Analysis

Transient heat conduction analysis and elastic plastic stress analysis are performed under the experimental condition, using the finite element code "MARC", in order to investigate the crack propagation characteristics. And the crack propagation is estimated.

A. Transient heat conduction analysis

Thermophysical properties such as thermal conductivity, specific heat and density are given as functions of temperature.

At first, it is performed to compute variations of temperature at T1 denoted in Fig.3 for three different heat transfer coefficients between specimen back and water cooled copper block. The temperature fluctuates due to heat cycle and increases with increasing time. Each temperature fluctuations become stable after several heat cycles, and the temperature fluctuation in the case that heat transfer coefficient is 85 w/m^{2}k seems to agree with experimental one on the whole. This coefficient is used in computation.

The temperature at the center of front surface, where is denoted as T0 in Fig.3, reaches about 460°C quickly due to high heat flux irradiation and decreases in a great degree during cooling. The difference of temperatures between at point T0 and at point T1 is large during irradiation, then it vanishes, that is, the temperature gradient in specimen thick direction becomes small during cooling.

B. Stress analysis

The temperature history concerning 30th heat cycle is used in elastic plastic stress analysis. Mechanical properties such as elastic modulus, Poisson's ratio, and thermal expansion coefficient are what apply at 200 ℃, yield point and strain hardening are given as the functions of temperature. Von mises' criterion and combined hardening are assumed for yield condition.

The stress at the crack tip is compressive one during the irradiation, and when the crack length is 1.0 mm the maximum stress becomes about 400 MPa just before irradiation is stopped. Then stress changes rapidly to tensile one about 400 MPa. This stress fluctuation causes the fatigue crack propagation.

3. RESULT

3.1. Experimental result

The thermal fatigue crack surface observed by SEM is shown in Fig.4. Since transgranular striation is observed, the fatigue crack propagation due to the heat cycle is recognized.

The crack propagation lengths which are measured along the specimen thick direction between pre-crack front and marked crack fronts are plotted with the numbers of irradiation cycles in Fig.5. The crack propagation length increases with increase of the number of the irradiation cycle, however its rate decreases. To investigate in detail, the variation of the crack propagation rate which is

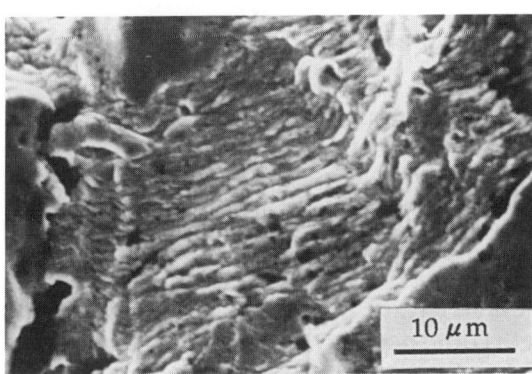

Fig.4 The ruptured surface magnified by SEM

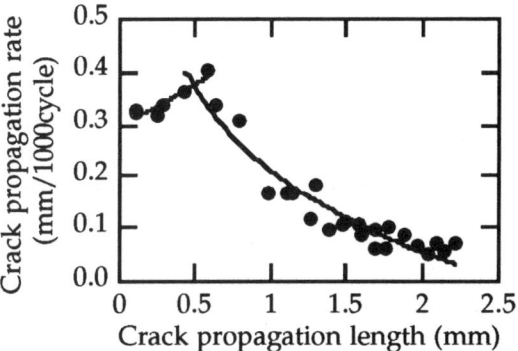

Fig.6 Relationship between crack propagation rate and crack depth

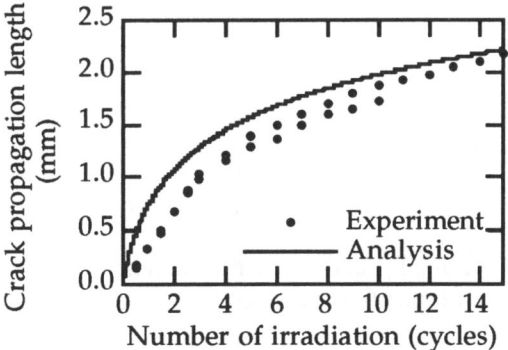

Fig.5 Comparison between experimental crack propagation length and analytical one

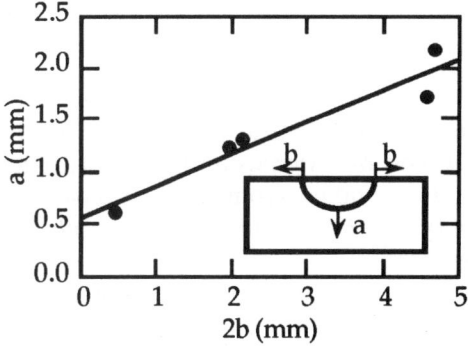

Fig.7 Relationship between the crack propagation length in the specimen thick direction and in its wide direction

crack propagation length per 1000 irradiation cycles is shown in Fig.6 against the crack propagation length along specimen thick direction. The crack propagation rate increases with increase of the crack depth until it propagates about 1.0 mm. When the crack depth reaches this length, the crack seems to begin extending in the specimen wide direction, as shown in Fig.7. Then the crack propagation rate decreases.

3.2. Fracture mechanical analysis

To evaluate the fatigue crack propagation behavior due to heat cycle, non-linear fracture mechanics parameter, J-integral, is introduced, as follow.

$$J = \int_\Gamma (W dy - T_i \frac{\partial u_i}{\partial x}) + \int_A \alpha \sigma_{ii} \frac{\partial \theta}{\partial x} dA \quad (1)$$

where W is elastic strain energy, T_i traction, u_i displacement, Γ integral path, α thermal expansion coefficient, σ_{ii} stress, θ temperature, A area, x direction parallel to the crack in this analysis, y direction perpendicular to the crack.

The computed J-integral value is negative during irradiation since crack tip is closed by compressive stress, and it chugs to positive one after finishing irradiation.

On the other hand, the fatigue crack propagation characteristics is generally represented as follow.

$$\frac{da}{dN} = C(\Delta J)^m \quad (2)$$

where a is crack length, N number of load cycle, C and m material constants. The material constants are taken from conventional fatigue data of type 316 austenitic stainless steel at high temperature[1].

As previous statement in this experiment the crack tip closure may occur. Therefore crack propagation rate, da/dN is rewritten as

$$\frac{da}{dN} = C(\Delta J_{eff})^m \quad (3)$$

$$\Delta J_{eff} = J_{max} - J_{op} \quad (4)$$

where J_{max} is maximum value of J-integral in a heat cycle and J_{op} is J-integral value when the crack tip opening occur in the cycle[2]. The comparison between experimental crack propagation length and analytical one with number of heat cycle is shown in Fig.9. They are different at the early stage of propagation since the analytical J-integral range, ΔJ_{eff}, monotonously decreases, which is different from experimental characteristics that is increases at the early stage as shown in Fig.7. However they agree well each other on the whole.

4. CONCLUSION

(1) Because of rapid change of stress at the crack tip, the crack is extended.
(2) The crack propagation behavior is that the crack length increases with increase of heat cycle.
(3) The crack propagation rate increases until the crack length reaches about 1.0 mm. This length corresponds with beginning of the crack extension to specimen wide direction. Then it decreases.
(4) The crack propagation length estimated in analysis matches the experimental one on the whole.

REFERENCE

[1] S.Taira, R.Ohtani and T.Komatsu, Trans. of ASME, Jour. of Eng. Mat. and Tech. 101 (1979) 162.
[2] S.Kubo, T.Yafuso, M.Nohara and K.Ohji, Trans. of Japan Soc. of Mech. Engin. A-55 (1989) 134.

Creep analysis of first wall in fusion reactor

K. Hatamoto, T. Teramoto and M. Saito

Institute of Engineering Mechanics, University of Tsukuba
Tsukuba, Ibaraki 305, Japan

Inelastic strain of the first wall in steady operation mode is estimated numerically up to 5years. It includes swelling, radiation creep in addition to thermal creep. Structural integrity of the first wall is estimated on the basis of ASME Code Case N-47. First wall may have lifetime more than 5years only if neutron radiation is less than $3MW/m^2$.

1. INTRODUCTION

In future, a fusion reactor is expected to be operated in steady state mode. The deformation due to swelling, radiation creep in addition to thermal creep may be a major factor to decide the structural integrity of the first wall. In long pulse mode operation, it is also necessary to consider the effects of creep deformation in order to estimate the damage accumulation in the first wall due to neutron and high heat flux irradiation. ITER, for example, is designed such that the machine has the capability of pulse operation with burn duration in the range of 1000 second.

In this study, the inelastic strain is estimated numerically by FEM up to 5 years. Neutron wall loading of 3, 4 and 5 MW/m^2 and accompanying surface heat flux of 3/4, 4/4 and 5/4 MW/m^2 are assumed to be radiated to the first wall steadily. The back surface of the wall is assumed to be 100℃ or 320℃ by coolant. The first wall is made of 316SS. Three cases of wall thickness, 3mm, 4mm and 5mm, are examined. As the mechanical boundary condition, the bending constraint is employed, that is, the first wall is constrained not to be permitted bending deformation.

2. NUMERICAL ANALYSIS

2.1 Heat conduction analysis

In two dimensional heat conduction analysis, thermophysical and mechanical properties such as thermal conductivity, specific heat, thermal expansion coefficient and modulus of elasticity are given by functions of temperature. In addition to the high heat flux,

Table 1. Surface temperature

First wall thickness (mm)	Coolant side temperature (℃)	Newtron wall loading (MW/m^2)		
		3	4	5
3	100	232.0 ℃	276.0 ℃	319.9 ℃
	320	432.2 ℃	469.6 ℃	507.0 ℃
4	100	276.0 ℃	334.6 ℃	393.2 ℃
	320	469.6 ℃	519.5 ℃	570.0 ℃
5	100	319.9 ℃	393.2 ℃	466.6 ℃
	320	507.0 ℃	569.4 ℃	631.7 ℃

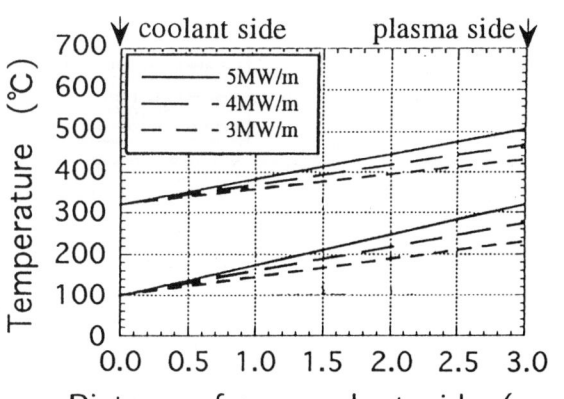

Figure 1. Temperature distribution

nuclear heating of $10W/cm^3$ for neutron irradiation of $1MW/m^2$, is also included in the heat conduction equation.

Table 1 shows surface temperature of the first wall. Figure 1 shows temperature distribution along the wall thickness direction in case of wall thickness of 3mm.

2.2 Inelastic stress analysis

Total strain ε is considered to be a sum of elastic strain ε^e, plastic strain ε^p, thermal strain ε^t, thermal creep strain ε^{ct}, radiation creep strain ε^{cr} and swelling strain ε^s. Stress and inelastic strains are obtained up to 5 years by two dimensional inelastic stress analysis. The constitutive relation for thermal creep is conventional one which is recommended by Japanese Society of Mechanical Engineers. As the constitutive relation for radiation creep, we employ

$$\varepsilon^{cr} = (1.0*10^{-6}*D + 3.0*10^{-5}*S)*9.8*\sigma$$

where D(dpa) is radiation damage accumulated, S (%) is swelling and σ (MPa) is stress. Damage rate 9.47 dpa/year for neutron irradiation of $1MW/m^2$ is applied to austenitic stainless steel 316SS. And as the constitutive relation for the swelling S, we employ

$$S = 1.54*10^{-12} \exp(T/35.5)*D + 5.94*10^{-11} \exp(T/46.3)*D \exp(0.108*D).$$

These relations are the ones proposed previously in JAERI MEMO[1].

Table 2 shows the initial stress on the surface of the first wall. By the bending constraint, the huge compressive stress appears on the high temperature plasma side and tensile stress appears on the coolant side. Figure 2 shows stress distribution along wall thickness direction in case of neutron wall loading of $4MW/m^2$, in which the initial stress is shown by the solid line. The compressive stress is almost reduced during less than one year by virtue of thermal creep when the coolant side temperature is 100℃ and reduced during about half year when the coolant side temperature is 320℃. But as radiation damage due to neutron irradiation is accumulated, the swelling strain increases rapidly in the vicinity of the high temperature plasma side. The positive swelling strain induces compressive stress by the bending constraint. The compressive stress, in turn, causes radiation creep negatively so as to compensate the swelling strain. Stress history and inelastic strain history up to 5 years were calculated in all cases. Even before 5 years if both swelling and radiation creep strains increase over 10%, we broke the calculation because of the lose of reality. Here, three typical cases are shown. Figures 3 (a) and (b) show the stress and inelastic strain history on the surface of the first wall in case of neutron wall loading of $3MW/m^2$. Figures 4 (a) and (b) show these histories in case of neutron wall loading of $4MW/m^2$. Figures 5 (a) and (b) show these histories in case of neutron wall loading of $5MW/m^2$. In these results, wall thickness is 3mm and coolant side temperature is 100℃ in common. It appears that the swelling strain

Table 2. Initial stress

First wall thickness (mm)	Coolant side temperature (℃)	Newtron wall loading (MW/m²)		
		3	4	5
3	100	-157 MPa	-209 MPa	-260 MPa
	320	-125 MPa	-166 MPa	-204 MPa
4	100	-209 MPa	-276 MPa	-339 MPa
	320	-166 MPa	-218 MPa	-270 MPa
5	100	-260 MPa	-339 MPa	-414 MPa
	320	-205 MPa	-270 MPa	-340 MPa

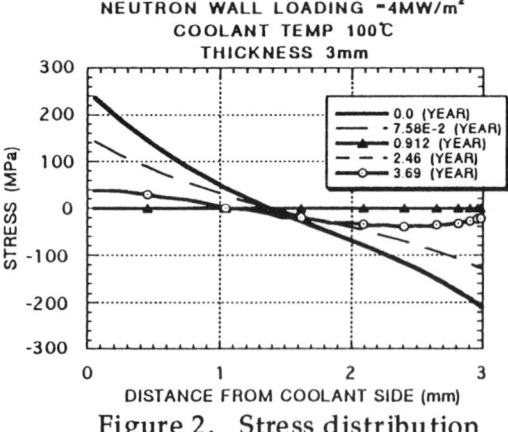

Figure 2. Stress distribution

increases rapidly after 3.5 years by neutron irradiation of 4MW/m² and after 2.5 years by neutron irradiation of 5MW/m². When the wall thickness is 4mm, the rapid increase of the swelling strain is faster by about half year. when the wall thickness is 5mm, the rapid increase is faster by about one year. When coolant side temperature is 320℃, the rapid increase of inelastic strain is much faster.

2.3 Estimation of structural integrity

Structural integrity of the first wall was estimated on the base of the ASME Code Case N-47. This Code is the design criteria for high temperature structural members and gives upper limit to the inelastic strain, that is, the mean strain in the wall must be less than 1%, the equivalent linear strain must be less than 2% and the local strain must be less than 5%.

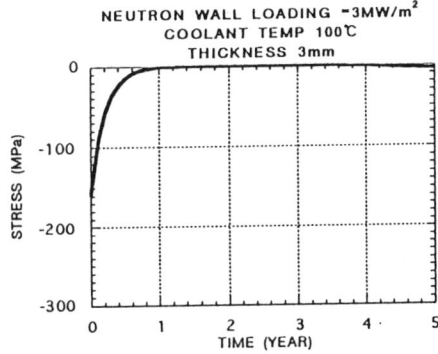

Figure 3. (a) Stress history

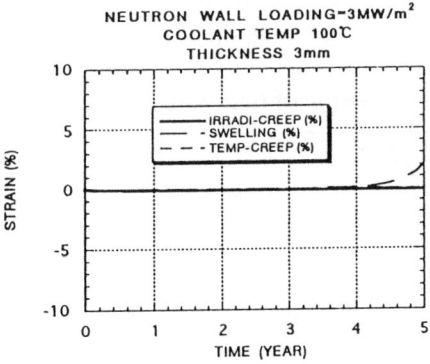

(b) Inelastic strain history

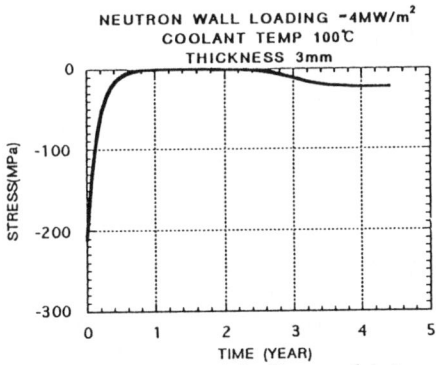

Figure 4. (a) Stress history

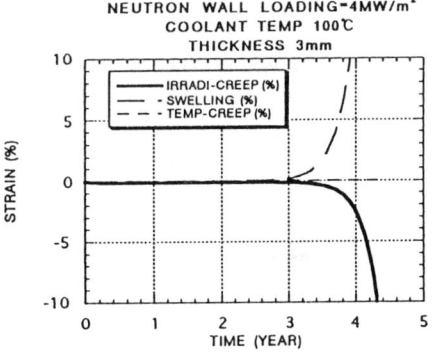

(b) Inelastic strain history

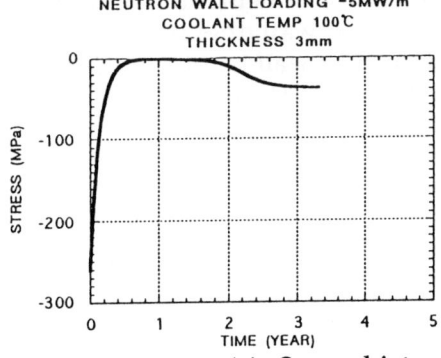

Figure 5. (a) Stress history

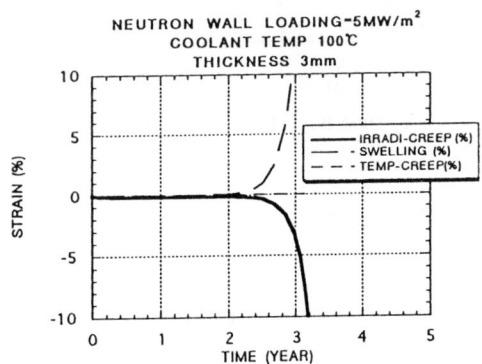

(b) Inelastic strain history

Tables 3 shows the estimated lifetime of the first wall. Among 18 cases examined, only one case, in which wall thickness is 3mm, neutron wall loading is 3MW/m^2 and coolant side temperature is 100°C, gives lifetime more than 5 years. Neutron wall loading is a decisive factor. Wall loading of 4MW/m^2 gives lifetime less than 4 years, wall loading of 5MW/m^2 gives lifetime less than 3 years even in case of wall thickness of 3mm. From consideration of mechanical loading, wall thickness not less than 5mm may be desired. Then neutron wall loading must be less than 3MW/m^2.

Table 3. Lifetime of first wall

First wall thickness (mm)	Coolant side temperature (°C)	Newtron wall loading (MW/m^2)		
		3	4	5
3	100	5.10 years	3.65 years	2.77 years
	320	3.59 years	2.43 years	1.66 years
4	100	4.86 years	3.38 years	2.51 years
	320	3.24 years	1.92 years	1.12 yeras
5	100	4.62 years	3.14 years	2.19 years
	320	2.78 years	1.14 years	0.737 years

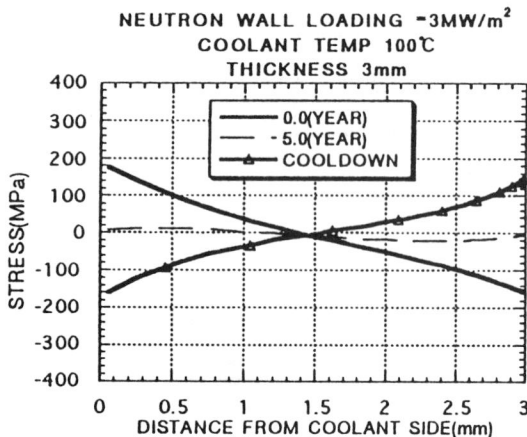

Figure 6. Stress in cooldown

3. DISCUSSION

In the estimation of structural integrity, only inplane component of inelastic strain was considered since effects of the component of inelastic strain perpendicular to the plane does not seem to be recognized clearly. Positive inplane component of swelling is compensated by negative inplane component of radiation creep because of the bending constraint. But this fact does not indicate that the volume expansion due to swelling disappears. The volume expansion appears in swelling strain component perpendicular to the plane. It is about twice of the inplane component of the swelling strain. This swelling strain in the plate-thickness direction may have a bad effect to the structural integrity.

Shutdown of the reactor gives another problem. After steady operation long enough to give relaxation of the initial stress, shutdown induces tensile residual stress in the vicinity of the wall surface. Its magnitude is almost same as the initial stress. Figure 6 shows an example of the shutdown stress. If the first wall may have an unexpected surface crack by radiation damage or others, there is a possibility that the surface crack is propagated by the tensile residual stress in the reactor cooldown.

4. CONCLUSION

(1) High heat flux gives high temperature and huge compressive stress on the surface, but this stress is almost reduced during less than one year by virtue of thermal creep.
(2) On the accumulation of radiation damage, swelling strain increases rapidly, which gives rise to compressive stress by bending constraint. Then the compressive stress, in turn, induces radiation creep negatively in order to compensate the swelling strain.
(3) Except in case of neutron irradiation of 3MW/m^2, the first wall does not have 5years lifetime. If the first wall is thicker than 5mm, neutron radiation must be less than 3MW/m^2.

REFERENCES

1. JAERI, JAERI MEMO 86-176 (1987)

Behaviour of sub-surface defects under thermal fatigue loading in a 316L first wall mock-up

F. Sevini[a], M. Merola[b]

[a] Joint Research Centre, Institute for Advanced Materials, 21020 Ispra, Italy*

[b] Politecnico di Torino, Dipartimento di Energetica, C.so Duca degli Abruzzi 24, 10129 Torino, Italy**

Abstract

The possible reduction of thermal fatigue resistance of 316L first wall mock-ups caused by the presence of sub-surface defects was investigated. The results obtained after 50000 cycles on a mock-up carrying three different electro-eroded flaws, showed that cracks are rather unlikely to originate from such defects, that the shape of the defect itself plays an important role and that in no case was the integrity of the structure affected.

1. INTRODUCTION

The behaviour of defects under thermal fatigue loading is a focus point in the lifetime assessment of plasma facing components. An experimental campaign was carried out on several simplified austenitic 316 L stainless steel (European fusion grade) first wall (FW) mock-ups, manufactured by Sulzer/Innotec, Winterthur, Switzerland on behalf of the NET Team.

Object of the present paper is to present the results obtained on a mock-up containing three artificial notches on the side surface. The test was conducted at the Thermal Fatigue Test Facility of the Joint Research Centre of the European Union, Ispra site, Institute for Advanced Materials [1,2]. The mock-up (Figure 1) was actively cooled by low pressure deionised water, flowing in two cooling channels parallel to the heated surface, obtained by drilling and without insertion of brazed tubes. The overall dimensions are 52×318×28 mm; the heated surface (dimensions of 52×260 mm) is located on the opposite side of the inlet and outlet tubes, and is covered by a black coating, a few microns thick, in order to increase the surface absorptance.

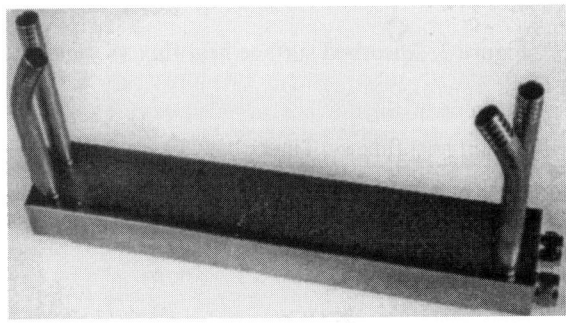

Figure 1. FW mock-up view showing the lateral surface hosting the defects; the heated surface is opposite to the inlet and outlet tubes.

2. EXPERIMENTAL TEST

The three notches, arranged in a row parallel to the heated surface and located at a distance of 2 mm from it, had the following dimensions:
   length: 3.0 mm
   width: 0.1 mm
   depths: 0.5, 1.0, 2.0 mm

---

* Present address: JRC Petten, P.O. Box 2, 1755 ZG Petten, The Netherlands.
**   "   : The NET Team, Max Planck Inst. fuer Plasmaphysik, Boltzmannstr. 2, D-85748 Garching, Germany.

2.1. Applied thermal cycle

Figure 2 shows the surface heat flux applied to the component. Table 1 gives the exact values used in the calculations. The maximum heat flux emitted by the heating lamps was 823 kW/m^2. Taking into account the reduction caused by the geometrical effect, a heat flux of 810 kW/m² effectively reached the component; the peak heat flux absorbed by the component was then estimated equal to 770 kW/m^2, being the heated surface emissivity about 0.95.

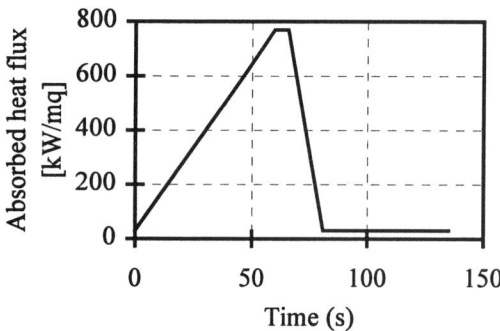

Figure 2. Absorbed surface heat flux vs. time.

Table 1
Absorbed heat flux vs. Time

Time [sec]	Heat flux [kW/m²]
0.	30.
60.	770.
66.	770.
81.	30.
135.	30.

The water flow rate in each cooling channel was 1 m^3/h, at a pressure of 0.15 MPa; an heat transfer coefficient equal to 12 000 W/m²K was calculated by means of Nusselt's correlation, assuming respectively 20 °C and 80 °C as mean values for water sink and wall temperatures.

3. NUMERICAL ANALYSIS

3.1. Introduction

The numerical analysis of the experiment was performed by means of the ABAQUS finite element code (ver. 5.3-1, 1994) using a 2D analytical model. Table 2 reports the material properties introduced in the finite element code. The elastoplastic material data were taken from the RCC-MR design code, Sect. I, Subsect. Z, Technical Appendix A3 as material 1S.

Table 2
316L properties

Temperature	[°C]	20	300	500
Density	[kg/m³]	8000	7870	7780
Thermal conduct.	[W/mK]	14	18	20
Specific heat	[J/kgK]	480	550	580
Thermal exp. (20°C - T°C)	[10^{-6} K^{-1}]	16	17	18
Young's modulus	[GPa]	195	175	155
Poisson's ratio	[-]	0.3	0.3	0.3

Following the thermal one, both elastic and elastoplastic mechanical analyses were performed,. The Oak Ridge National Laboratory (ORNL) hardening modelling was adopted. The input for the calculation requires bilinear representations of the uniaxial stress-strain curve of the virgin material and of the cyclic stress-strain curve of the hardened material, constructed as recommended by the ORNL-TM-3602 (1972), at the maximum and minimum temperatures reached during the cycle. The 316L bilinear curves used for the present and for previous calculations are reported in [3].

3.2. Numerical results

Figure 3 defines the cross-sectional location of the three defects. The temperatures calculated in correspondence to it were 361 °C and 41 °C at the end of the heating phase and at the end of the cycle, respectively. Stresses and strains are reported in Tables 3 and 4 together with the design allowable number of cycles.

3.3 Design allowable number of cycles

Two standard codes were used to evaluate the design allowable number of cycles:
- the American ASME [4], requiring an elastic calculation and adopting a quantity proportional to Tresca equivalent stress range as input for the design curve;
- the French RCC-MR [5], suitable for both elastic and elastoplastic calculations, and adopting Von Mises mechanical equivalent strain range.

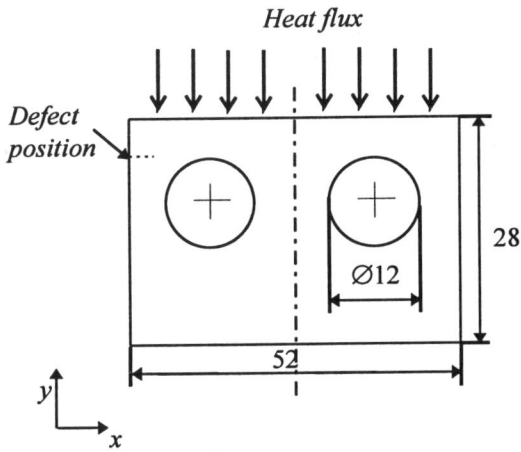

Figure 3. Location of the three defects.

Table 3
Elastic stress components and ASME design cycles*

[MPa]	End of heating	End of cycle
σ_x	-11	-0.5
σ_y	-15	$\cong 0$
σ_z	-319	-21.8
S_{alt}		154
N_{cycles}		∞ (>10^6)

* $S_{alt} = \Delta\sigma_T \cdot K_e \cdot (0.5 \cdot E_{curve} / E_{ref})$;
reference temperature = $(T_{max} + T_{min})/2$.

Table 4
Elastoplastic mechanical strain components and RCC-MR design cycles*

[%]	End of heating	End of cycle
ε_x	0.106	0.057
ε_y	0.085	0.041
ε_z	-0.233	-0.063
$\Delta\varepsilon_{mech, eq}$		0.144
N_{cycles}		671000

* reference temperature = T_{max}

4. EXPERIMENTAL RESULTS

The test was stopped after 50 000 fatigue cycles. After the test, the component was destructively examined. Figure 4 shows the three different section that were selected. Points C1 and C2 were located on the side surface, point D was at the notch maximum depth.

4.1 Metallographic examinations

No crack propagation was observed from 0.5 and 1 mm notches. As for the 2 mm notch, a single crack propagated at point C2 perpendicularly to the heated surface. Figures 5 and 6 show optical images of the crack, observed after chemical etching.

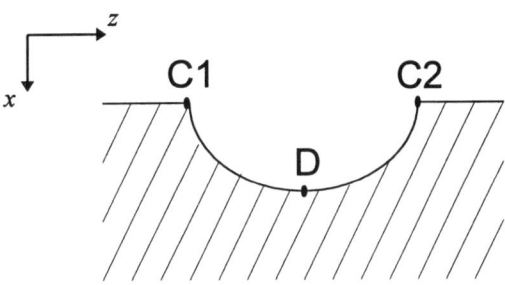

Figure 4. Destructive analysis positions.

4.2 Discussion of the results

According to the elastoplastic analysis results, the σ_y stress (along the component thickness) was quite low; its value ranging from -8 MPa to 9 MPa at the end of heating and at the end of cycle, respectively. This means that the mode I loading (opening) could not be effective in promoting crack nucleation and propagation. This explains why none of the notches developed in a propagating defect. In fact both the ASME code and the RCC-MR code allow practically an infinite number of fatigue cycles at this location.

A single small crack nucleated from one of the two surface tips of the 2.0 mm deep notch. The remarkable aspect was that it did not propagate along the notch plane but perpendicularly to it, opposite to the heated surface. The driving force was therefore the σ_z stress, which ranged from -165 to 149 MPa. This stress acted in the direction of the surface length of the notch and continuously opened and closed the notch, thus generating the observed crack. The nearly rectangular surface cross section of the notch could have played an important role in promoting the crack nucleation (Figure 7). In this respect, it is not sure that a sharp crack-like notch would have generated this crack.

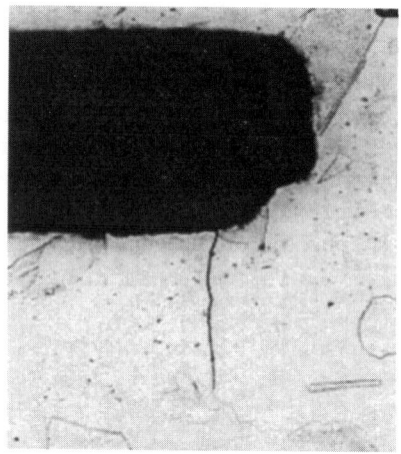

Figure 5. Notch depth 2.0 mm; point C2; 250×.

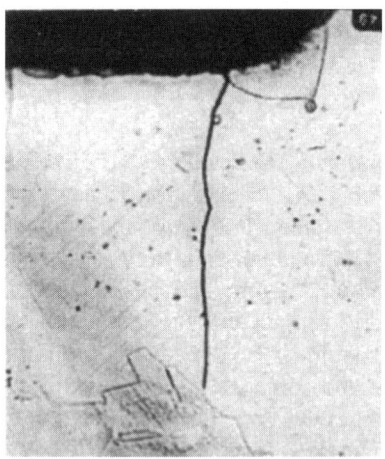

Figure 6. Notch depth 2.0 mm; point C2 crack length 0.16 mm; 500×.

5. CONCLUSIONS

The experimental results showed that subsurface defects, located on a plane parallel to the heated surface, do not seem to have harmful effects on the thermal fatigue behaviour of a first wall component. In fact the specimen could complete the foreseen 50 000 fatigue cycles without any problem. The only detected crack was probably due to the rectangular cross section of the notch and, however, did not impair the thermal fatigue resistance behaviour of the component.

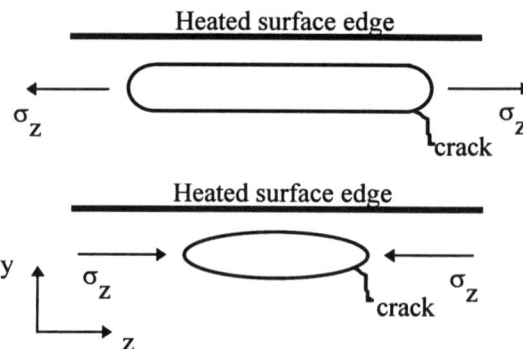

Figure 7. Crack generation by means of the σ_z stress component.

AKNOWLEDGEMENTS

The authors wish to thank R. Matera, ITER JCT Garching, and G. Vieider, The NET Team, for their useful collaboration. The contribution of the Polytechnic of Turin was within the collaboration with the Institute for Advanced Materials of the Joint Research Centre (collaboration No. 10193-94-05 SOED ISP I) and was financially supported by The NET Team (contract NET-94-354).

REFERENCES

1. M. Merola, *Evaluation of First Wall Thermal Fatigue Testing*, Politecnico di Torino, Dip. di Energetica, Italy, Contract No. NET 94-354, Intermediate Report, Part 1, February 1995.
2. M. Merola, R. Matera, F. Sevini, *Experimental results of the thermal fatigue tests for I.A.E.A. benchmark components*, published in Journal of Nucl. Materials (1996), in print.
3. M. Merola, F. Sevini, M. Beghini, L. Bertini, Numerical Analysis of Crack Growth under Thermal Fatigue Loading, Comm. Europ. Commun., Joint Research Centre, Ispra Establishment, EUR Report 16194 EN, 1995.
4. ASME III, *Rules for Construction of Nuclear Power Plant Components*, Division 1, Subsection NB, Class 1 Components, 1992 Edition.
5. RCC-MR, *Design and Construction Rules for Mechanical Components for FBR Nuclear Islands*, Section I, Subsection B: Class 1 Components, AFCEN, June 1985 Edition.

Erosion Damage of Nearby Plasma-Facing Components during a Disruption on the Divertor Plate*

A. Hassanein[a] and I. Konkashbaev[b]

[a]Argonne National Laboratory, 9700 South Cass Avenue, Argonne, IL 60439 USA
[b]Troitsk Institute for Innovation and Fusion Research, Moscow Region, 142092 Russia

Intense energy flow from the disrupting plasma during a thermal quench will cause a sudden vapor cloud to form above the exposed divertor area. The vapor-cloud layer has been proved to significantly reduce the subsequent energy flux of plasma particles to the original disruption location. However, most of the incoming plasma energy is quickly converted to intense photon radiation emitted by heating of the vapor cloud. This radiation energy can cause serious erosion damage of nearby components not directly exposed to the disrupting plasma. The extent of this "secondary damage" will depend on the divertor design, disrupting plasma parameters, and design of nearby components. The secondary erosion damage of these components due to intense radiation can exceed that of the original disruption location.

1. INTRODUCTION

Disruption damage to plasma-facing components (PFCs) remains a major obstacle to a successful tokamak concept. The high energy deposited in short periods on plasma-facing materials (PFMs) can cause severe erosion, plasma contamination, and structural failure of these components. The initial energy flow released at the start of a disruption will cause a sudden vapor cloud to form above the surface of the exposed area. This shielding layer has been proved to significantly reduce the energy flux to the original disruption spot, thus leading to a substantial reduction in erosion rate [1]. Most of the incoming plasma kinetic energy is, however, converted to radiation energy by the expanding vapor-cloud front. Such a large amount of radiation energy can cause significant damage to nearby components not directly exposed to the initial disruption, particularly in a closed divertor configuration such as in the current ITER design [2]. For an open divertor configuration, this problem will be less severe because the radiation will spread over a much larger area.

The models developed in the comprehensive magnetohydrodynamic code A*THERMAL-S have been extended to study the secondary damage of nearby components due to vapor radiation. Originally, three major modeling stages of plasma/material interaction were developed with sufficient detail to accurately simulate disruption effects [3]. Initially, the incident plasma particles from the disrupting plasma will deposit part of their energy on the PFC surface. Models for particle deposition and material thermal evolution that take into account phase change, moving boundaries, and temperature-dependent thermophysical properties, together with kinetic models for surface vaporization, were developed to predict the thermal behavior of PFCs. A shielding vapor cloud will quickly form in front of the incoming plasma particles. Shortly thereafter, the plasma particles will be completely stopped in this vapor cloud. Continuous heating of the vapor cloud will ionize, excite, and generate photon radiation. The kinetic energy of the incoming plasma particles is therefore transformed into radiation energy.

Detailed models for the magnetohydrodynamics and heating of the vapor cloud that shields the original surface and the newly developed secondary vapor cloud were then developed. Finally, models for radiation transport and deposition throughout the vapor cloud were developed to estimate the net heat

---

* Work supported by the U.S. Department of Energy and by the Ministry of Atomic Energy and Industry, Russia.

flux transmitted to the PFMs and to other nearby components. Figure 1 is a schematic illustration of the various interaction zones and processes during the plasma/radiation/material interactions of a thermal quench disruption. The intense radiation from the primary vapor cloud will strike adjacent components in direct line of sight and can cause a secondary vapor cloud of the component's material to form above its surface. The strong primary vapor radiation has already been demonstrated, in laboratory disruption simulation experiments, to cause erosion damage of near-target components [4].

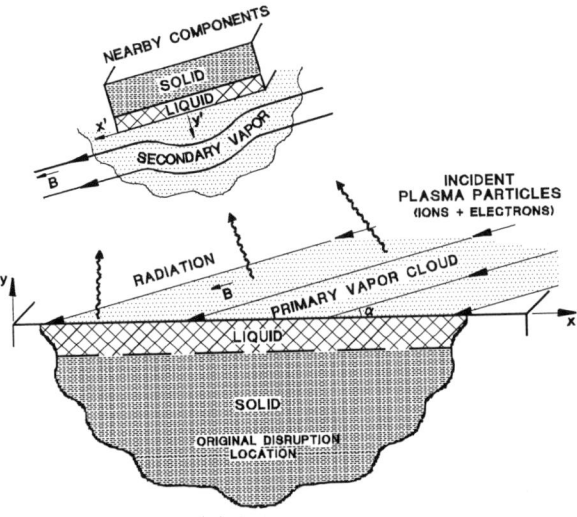

Figure 1. Schematic illustration of interaction phenomena during a disruption.

Because of the importance of radiation transport in the vapor-cloud regions, a self-consistent approach [5] to calculate the actual radiation field has also been developed and implemented in the A*THERMAL-S code. The optical properties of both the "original," primary, and secondary vapor-cloud plasmas are calculated at each time-step during the course of the disruption. The relevant atomic data bases of candidate materials are implemented in the code. The kinetic rate equations are then solved for each ion level population at every time-step. The radiation transport equation is then solved separately for both line- and continuum-generated spectra. The self-consistent model also takes into account the multispecies effect, i.e., mixing between the incoming plasma particles and the primary vaporized material.

To evaluate the extent of the indirect disruption damage to nearby components caused by the primary vapor radiation of the original divertor material, the A*THERMAL-S code was significantly enhanced and new models were developed. Detailed physics of both plasma particles and photon radiation interaction with solid/liquid and vapor materials in a strong magnetic field with various configurations are enhanced and implemented in the self-consistent model. The transport and deposition of the propagating radiation, generated from the primary vapor cloud, in nearby PFCs and in the resulting secondary vapor cloud and its own radiation transport in these components are also calculated in detail. Depending on divertor configuration and design, the energy deposited from the divertor vapor radiation is high enough to cause severe melting and erosion of nearby components. Melt-layer erosion of metallic nearby components can also be significant [6]. The net erosion of these components can, in fact, exceed that of the original disruption location. This can be due to secondary vapor optical properties, vapor diffusion losses, melt layer splashing, and geometrical effects.

2. EVALUATION OF SECONDARY DAMAGE

The amount of energy deposited on nearby components from the primary vapor radiation depends on many parameters, such as distance from the divertor plate, size and orientation of components, and magnetic field structure adjacent to these components. Such parameters are used as input to the A*THERMAL-S code [7].

Figure 2 shows the time dependence of the power density transmitted to the original target and the radiated power density from the developed vapor cloud to other components. Typical plasma disruption parameters are assumed where the incident plasma energy is 10 MJ/m^2 deposited within 100 µs. The kinetic energy of the incident plasma ions is 10 keV. An oblique toroidal magnetic field of 5 T at an angle of 2° is assumed near the original disruption location of the divertor plate [8]. Initially, the power reaching the divertor target is equal to that of the incident disrupting plasma power due to direct deposition of the plasma particles. Shortly after that, the power to the primary target sharply decreases due to shielding and attenuation by the ablated material. After the plasma ions have completely stopped in the target vapor, the primary target heating (<10% of the original value) is mainly from vapor radiation and conduction. The vapor cloud, therefore, significantly shields the originally exposed surface from the disrupting plasma. More than 80% of the incident plasma energy is, however, radiated from the tip of the hot vapor to the other adjacent components, as shown in Fig. 2.

Radiation from the vapor cloud is in the form of low-energy photons. The spectra of this radiation depends on the plasma power deposited and on the target material. Figure 3 shows the calculated photon spectra emitted from the front vapor regions of both C and Be primary target materials. Beryllium vapor emits harder photon spectra with significant line radiation because it has a much higher temperature than C vapor under the plasma conditions shown. Carbon vapor radiation is similar to W vapor radiation and is close to that of a blackbody for the stated conditions [5]. For higher incident plasma energy densities and low-Z target materials, most of the emitted photon radiation is in the form of line radiation. Therefore, a comprehensive treatment of line radiation and its transport is included in the A*THERMAL-S code [8].

The emitted radiation will strike nearby components, deposit its energy, and heat the components, thereby generating a secondary vapor cloud as shown schematically in Fig. 1. To evaluate the potential erosion losses due to such radiation, a nearby Be component such as a part with the divertor cassettes, a small blade, or a fin is analyzed. The secondary vapor cloud evolved above the exposed component surface will also shield its own surface from significant erosion if all the incident radiated energy is deposited at the surface. However, the shielding efficiency in this case is complicated by the expansion and diffusion of the secondary vapor across the magnetic field lines due to both classical and turbulent diffusion and to diffusion along magnetic field lines [7]. This vapor diffusion results in a decrease of the shielding layer away from the incident radiation, therefore allowing more power to reach the surface and cause more material erosion. In addition, as the secondary vapor expands above the surface, it occupies more volume and absorbs more primary radiation power from various sides of the cloud. This further heats the secondary vapor

and its component, also increasing vapor losses and component erosion. More details of the model used in this analysis are published in Ref. 7.

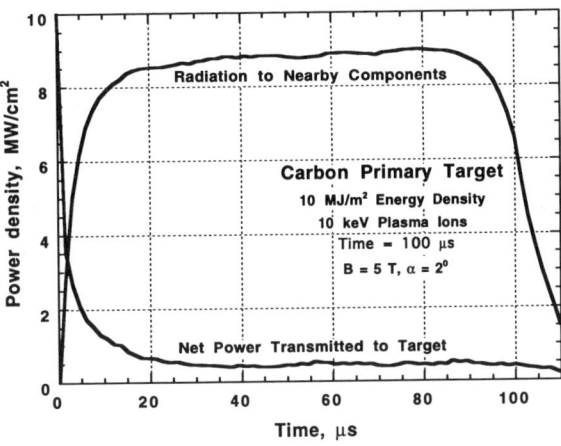

Figure 2. Power density to divertor target and nearby components during a disruption.

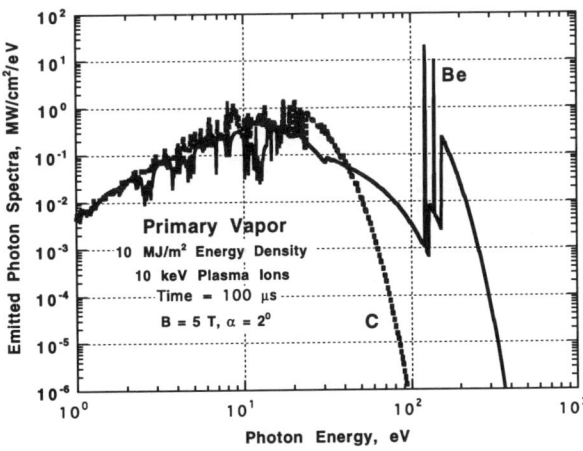

Figure 3. Photon radiation spectra emitted from primary vapor cloud.

Figure 4 compares vapor density and temperature of both primary and secondary beryllium vapor as a function of the normal distance to the exposed surface. The magnetic field lines are assumed to be parallel to the secondary target, as schematically shown in Fig. 1. The secondary vapor expands and accumulates up to only a few millimeters above the surface before it diffuses away from the incoming radiation. The primary vapor, however, expands toroidally for several meters against the incoming plasma and only about 10-50 cm normal to the surface due to diffusion across the field lines [5]. Because of the higher secondary vapor losses away from the incident radiation, the remaining vapor is more optically thin to incoming radiation and its temperature is lower than that of the primary vapor cloud. This results in much high vaporization losses of the nearby component compared than in the original disruption location, as shown in Fig. 5. Additional erosion of the resulting melt layer from splashing due to hydrodynamic instabilities and boiling (the SPLASH code) can further erode nearby metallic components and significantly reduce their lifetimes because of the much thicker melt layer relative to vaporization thickness [6].

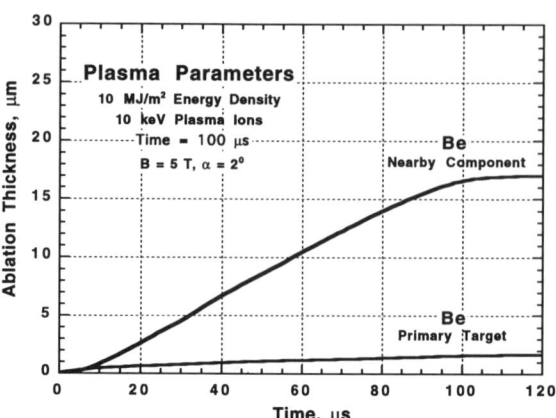

Figure 5. Vaporization losses of a beryllium primary and a beryllium nearby component.

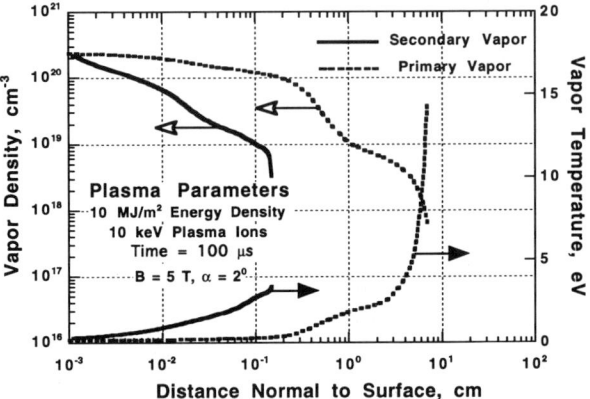

Figure 4. Vapor density and temperature of both primary and secondary vapor above surface.

3. CONCLUSIONS

Preliminary models and calculations to assess the erosion damage of nearby PFCs are presented. The intense radiation emitted from the primary vapor cloud during a thermal quench disruption can cause significant erosion of nearby components, particularly in closed divertor configurations such as the current ITER design. The secondary vapor cloud is not as effective as the primary cloud in protecting adjacent components due to strong vapor diffusion losses, vapor-cloud optical properties, and geometrical effects. More detailed analyses and more relevant experimental data are required to accurately predict lifetimes of plasma-facing and nearby components.

REFERENCES

[1] A. Hassanein and I. Konkashbaev, J. Nucl. Mater. 220-222 (1995) 244.
[2] G. Janeschitz et al., J. Nucl. Mater. 220-222 (1995) 73.
[3] A. Hassanein and I. Konkashbaev, Fusion Engineering and Design 28 (1995) 27.
[4] N. Arkhipov et al., "Study of Structure and Dynamics of Shielding Layer for Inclined Incidence of Plasma Stream at MK-200 Facility," presented at 7th Int. Conf. on Fusion Reactor Materials, Obninsk, Russia, Sept. 25-29, 1995. Accepted for publication in J. Nuclear Material.
[5] A. Hassanein and I. Konkashbaev, "Lifetime Evaluation of Plasma-Facing Materials during a Tokamak Disruption," presented at 7th Int. Conf. on Fusion Reactor Materials, Obninsk, Russia, Sept. 25-29, 1995. Accepted for publication in J. Nuclear Material.
[6] A. Hassanein et al., "Modeling and Simulation of Melt-layer Erosion during a Plasma Disruption," presented at 12th Int. Conf. on Plasma Surface Interactions in Controlled Fusion Devices, St. Raphael, France, May 20-24, 1996. Accepted for publication in J. Nuclear Material.
[7] A. Hassanein et al., to be published.
[8] A. Hassanein, Fusion Technology 26 (1994) 532.

The study of CVD thick B$_4$C and SiC coatings on graphites of different types in DIII-D divertor

O.I. Buzhinskij[a], I.V. Opimach,[a] V.A.Barsuk[a], P.W. West[b], N.Brooks[b], D.Whyte[b],

[a] - TRINITI, Troitsk, Moscow region, Russia;

[b] - General Atomics, San Diego, California, USA;

Thick CVD B$_4$C and SiC coatings with the thickness of 150-300mkm and 200 mkm correspondently on graphites with different thermal conductivity were exposed in DIII-D vessel. in accordance with the DiMES program of testing candidate materials for plasma facing elements in DIII-D divertor. One of B$_4$C samples was implanted by phosphorus to create a special marker in a surface layer at the depth of 250nm for more precise determining erosion rate. It was shown, that B$_4$C coating on graphites substrates possesses higher durability, than SiC, and preserves its phase and chemical composition in conditions of plasma discharges either with stable rejime or with disruptions.

1. INTRODUCTION

The DiMES sample exchange system was developed in the DIII-D tokamak to provide plasma material interaction testbed in the DIII-D divertor region [1]. DiMES allows the incertion of a sample into the divertor strike plate region of DIII-D for a selected set of plasma discharges. The divertor parameters desired for the particular experiment can be established while the sample is incerted. The divertor plasma parameters are measured by an array of diagnostics. In some cases these samples were exposed to ELM free H-mode conditions, while in other cases quasi steady conditions of ELMing H-mode were used.

Graphite materials are widely used in fusion devices, however they possess high chemical sputtering rate bombardment and radiation enhanced sublimation under high temperatures. That's why it is reasonable and interesting to use protection coatings deposited onto surfaces of graphite plasma facing elements of fusion reactors. B$_4$C coating on graphite substrates was tested under electron and plasma beams and in T-10 tokamak and showed a good durability [2,6]. In accordance with the DiMES program, thick CVD B$_4$C (fig.1) and SiC coatings (fig.2) with the thickness of 150-300 mkm and 200 mkm correspondently on MPG-8 fine grain isotropic graphite with the heat conductivity of 130 W/m*K and on RGT recristallized graphite with Ti addition (the heat conductivity is 600-700W/m*K) were exposed in DIII-D vessel. X-ray analysis, scanning electron microscopy (SEM) and Auger electron spectroscopy (AES), and other methods were used to study the surfaces of the samples after the exposure in DIII-D. These experiments were carried out during last five years.

2. EXPERIMENTS AND RESULTS.

The sample of MPG-8 coated with B$_4$C was exposed to low heat fluxes - only 40 W/cm^2, the time of the exposure under outer strike point was 10 sec, plasma current was 0.5 MA, beams power -5 MW. The exposed surface was very similar to the initial surface of the coating, the visual analysis revealed only one damaged spot of 1*2 mm. SEM showed that it is a region of melted and boiled coating material and the level of this spot is higher than the surrounding surface - so the erosion here was negligible. X-ray analysis showed that the exposed and unexposed surface had the same crystalline structure, and only in this damaged spot interplane distances are a little higher (in 1.3 times), that is caused by plasma heating effect and thermal expansion of the B$_4$C lattice.

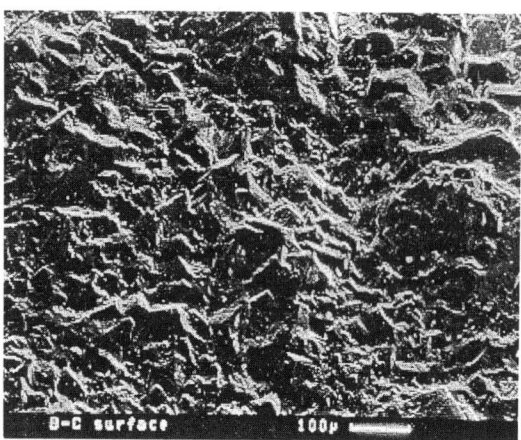

Figure 1. The surface of the B$_4$C coating

To get more detailed information on B$_4$C erosion we have used an implanted marker. Phosphorus was implanted in the surface layer of B$_4$C coating using the ion accelerator HVE-500 (Netherlands) to obtain a reference marker. The P distribution maximum is located at the depth of 2500A. The sample was in the DIII-D vessel during 12 shots, the heat loads varied from 60 MW/cm^2 to 300W/cm^2, some shots were disrupted.

The visual post exposure observation of the sample revealed two regions - dark and light ones. No visible damages are seen. X-ray analysis showed that the phase composition of the coating was not changed and is correspondent to B$_{13}$C$_2$. SEM analysis did not reveal the difference in dark and light regions, the surface was not changed after the exposure in DIII-D. AES study has shown, that the dark surface is enriched by carbon as compared to the ligh region, that can be explained by the deposition of the carbon during the plasma exposure. The thickness of the deposited layer is approximately 150A. The thickness of the eroded layer, obtained using Auger method, was about 150A. The total exposure time of the sample was 6sec., all discharges were with different parameters, so, we can just average the erosion rate for one average shot and the average erosion rate in our experiment is about 20A/sec.

The erosion rate of ATJ graphite was evaluated in experiments carried out also using the DiMES mechanism [5]. The maximum erosion in this experiment was 600 A and the average erosion rate was about 40 A/sec. It is difficult to compare the results of this study and our data, because of different exposure conditions (there were no disruptions and in general, exposure conditions were much softer in experiments with ATJ), however, apparently, we can see the following tendency - the carbon erosion rate is more (at least, in 2 times), that the same for B$_4$C coating, that is in accordance with the result obtained by us in T-10 tokamak [6].

To investigate the influence of the heat conductivity of a tile substrate material, following experiments with B$_4$C and SiC coatings on RGT graphite substrate were carried out. The sample of RGT graphite covered with B$_4$C coating deposited by CVD method, was exposed to the DIII-D divertor strike point plasma fluxes in 14 discharges, in one discharge a locked mode disruption deposited some energy onto the sample surface. The plasma current was about 2.5 MA, B$_t$ was about 1 Tl, neutral beam heating power varied from 5 till 7.5 MW. The accumulated exposure time was about 11 sec. of ELMing H-mode. Te varied from 40 till 80 eV and Ne was about $4*10^{13}$/cm^3. The average heat fluxes were about 200-300 W/cm^2, however, due to imperfect insertion of the sample, a part of its surface was exposed to much higher heat fluxes than average loads and we guess, loads here reached 1000-1200W/cm^2, that corresponds to ITER conditions. The visual post exposure observation showed the appearance of the damaged region in the sample surface , that was exposed to peak heat flux loads. The sample surface was analyzed in addition to previous methods, using scanning Rutherford backscattering spectroscopy (RBS), nuclear reaction analysis(NRA) and neutron elastic recoil detection (NERD) [7] to get data on hydrogen isotopes retention. It was shown that everywhere even in the most damaged spot B$_4$C coating is preserved and the substrate surface did not appear. The thickness of the eroded layer in the damaged spot is about 40 mkm, and the thickness of the remained layer is about 130 mkm. Some melted islands and cracks are observed in the damaged area, however, cracks do not reach the substrate surface. Outside the damaged spot the deuteriom retention R$_D$ obtained by NRA is (2-3)$*10^{17}$D/cm^2, and R$_D$ obtained by NERD -(3-4)$*10^{17}$ D/cm^2 ,; in the damaged spot $12*10^{17}$D/cm^2 and $20*10^{17}$D/cm^2 correspondently. Hydrogen retention R$_H$ is $1.5*10^{18}$ H/cm^2 in undamaged area

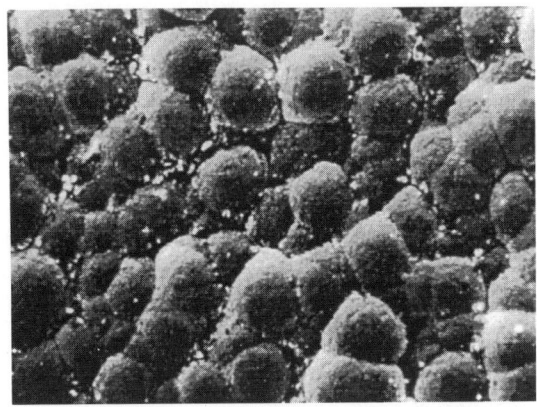

Figure 2. The surface of the SiC coating, *150.

and $5.2*10^{18}$ H/cm^2 - in the damaged region.
The deuterium and hydrogen retention in the damaged spot is higher, than in undamaged area, that is caused apparently by the increase in particle flux and the thickness of the deposited layer and by the development of the surface. Some increase of the width for the main peaks in B$_4$C difractogramm and in lattice spacings are observed for damaged region, that is also due to radiation defects appearance and D retention in B$_4$C surface layer.

The RGT graphite sample coated with SiC coating was exposed to OSP during 10 discharges with plasma current 1 MA and Bt=1.85Tl. During 2 sec portion of these shots when the divertor outer strike point was on the sample or close to it, the heat fluxes varied from 170 till 260 W/cm^2. The visual observation after exposure revealed some defect areas of 1-2 mm diameter on the surface SEM showed 20 mkm of net erosion in these regions. SEM showed also here the change of the surface morphology - namely, the appearance of new anomalous large SiC crystals of about 50mkm in size, possessing cubic structure, corresponding to b-SiC and oriented in such a way, that the crystalline plane (111) was parallel to the sample surface. The neighbouring surface to these large crystals is identical to the unexposed surface and has the same fine-crystalline character .In Auger spectrum peaks of B, C and Si are seen. The source of B and C is plasma (B is in plasma due to boronization of DIII-D and C -due to wide use of graphite tiles in diveror). Si concentration on the surface was very low, however the carbon concentration was very high.

Only after 2-3 hours of ion etching the Si concentration became close to equilibrium level. We think, this phenomena is caused by Si sputtering partly and by the deposition of "tokamakium", containing mainly carbon [8] .

3. CONCLUSION.

1. The complex tests of graphite materials with protection coatings were carried out in divertor area of the DIII-D tokamak.
2. It was shown that graphite materials coated with B$_4$C possess lower erosion comparing to pure graphites and graphites with SiC coating.
3. The use of graphite materials with high heat conductivity as substrate materials allows to increase thermal loads on the surface without essential post exposure damage and even in the disruption regime close to ITER conditions, the coating demonstrates a good durability, preserving its structure and properties. Thus, we can recommend to use B$_4$C coating as a protection on divertor tiles and other plasma facing elements of tokamak vessels.

REFERENCES

1. C. Wong et al., Erosion-redeposition in studies in DIII-D. Gaithburg, July,31-August,1 1990, General Atomics.
2. O.I.Buzhinskij, O.I.Opimach, V.A.Barsuk et al, J. Nucl.Mater., 220/222(1995) 922.
3. V.A.Barsuk, O.I.Buzhinskij, I.V.Opimach et al., ICRFM-7, Obninsk, 1995, Abstracts,N 080001.
4. O.I..Buzhinskij, I.V.Opimach et al., 12 PSI, France, 1996, Book of Abstracts, B48,p.127
5. R.Bastasz, W.R. Wampler, J.W.Cuthbertson et. al. J. Nucl. Mater., 220-222 (1995), 310-314
6. V.A. Barsuk , O.I. Buzhinskij et al., J. Nucl. Mater., 196-198 (1992), 543-548.
7. B.G. Scorodumov et.al 12 PSI, France, 1996, Book of Abstracts, A57,p.62.
8. R.Behrish, M Mayer, ICRFM-7, Obninsk, 1995, Abstracts, N 080052

Mechanical study of JET MKI beryllium tiles after ELMs and deliberate melting

E.B. Deksnis, A. Chankin, F. Hurd, W. Parsons, B. Tubbing

JET Joint Undertaking, Abingdon OX14 3EA, UK

1. INTRODUCTION

The JET MkI divertor configuration was designed for operation using either CFC or Beryllium as target materials. After an initial CFC phase of operations [1], the entire set of tiles, approximately 7300, was exchanged. The beryllium phase of MkI operation, in early 1995 [1], was carefully scheduled to minimise inadvertent melting of the Be target tiles.

A single giant ELM that appeared near the ELM-free divertor discharge [2] was found to have melted approximately 0.075 m^2 of the Be tiles.

In overloading tiles an experiment done in support of ITER [2], surfaces were subjected to a normal flux density of 20 MW/m^2. IR observations [3] show that the surface temperature rises above melting temperatures over 0.3m^2 of tile surface.

This paper presents some conclusions on the extent of melting due to the single giant ELM and surveys the mechanical state of resolidified areas of deliberately melted parts. No gross mechanical failure was observed; no fatigue cracks are visible as observed by JET [4,5] and others [5] in test-bed experiments where melting at flux densities up to 25 MW/m^2 was deliberately induced.

2. MATERIAL AND GEOMETRY OF MKI TILES

Melting of MkI Be tiles did occur as a result of ELM's, deliberate overloading, and possibly due to disruptions. Figure 1 shows the cumulative damage to one module of 4 tile rails. The MkI target tiles form a set of 384 tile rows, mounted inpairs on 192 rails. One tile is slightly higher to

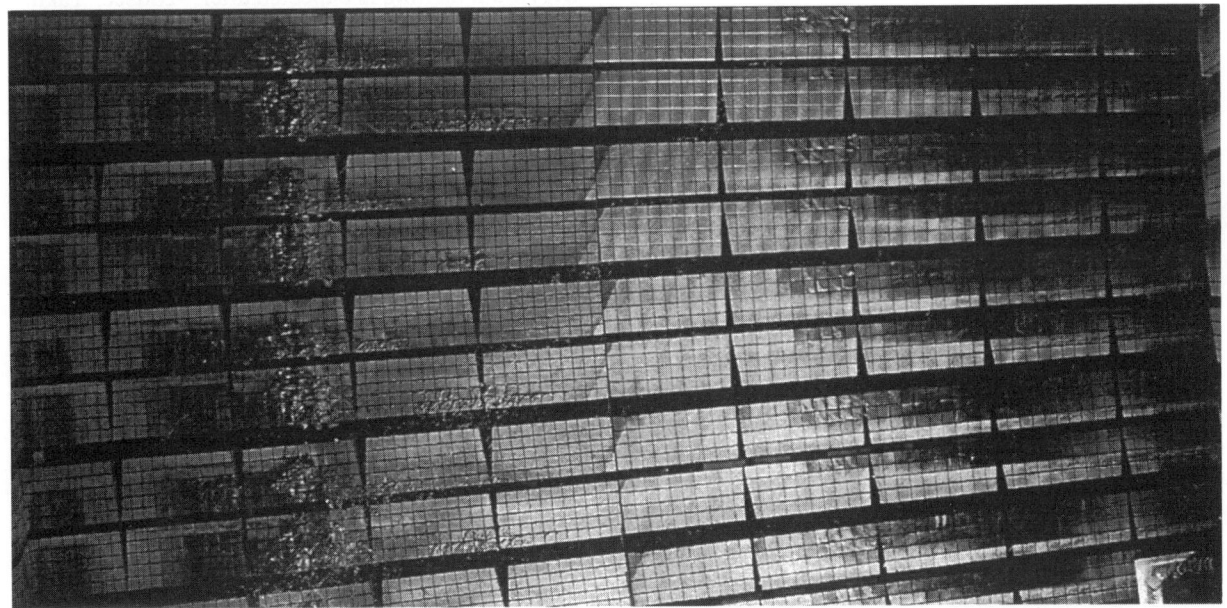

Figure 1. Cumulative damage to MkI Be tiles

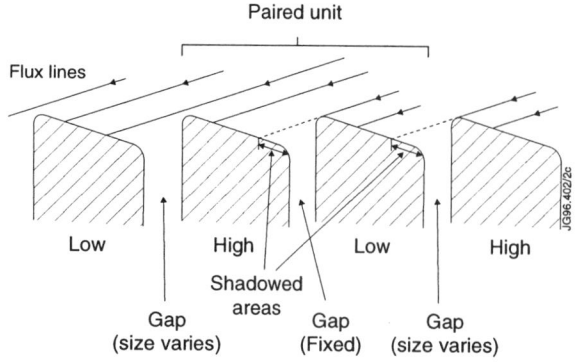

Figure 2. MkI target concept

offset geometry variation, thus tiles have been labelled high and low in figure 2. The high tile is raked at 5.2°, the low tile at 4.2° in the vertical toroidal plane, cf figure 2.

Tile features that determine the melted area for any configuration are (a) relative height of the high and low tile in a tile pair, (b) the tile rail angle in the vertical poloidal plane, (c) the angle of inclination of tiles in the vertical-toridal plane and (d) tile spacing. These values were chosen to handle a range of plasma equilibria. it is noted that at most ~50% of the available area could be melted.

The beryllium tile surface facing plasma flux is castellated in order to improve high-cycle fatigue characteristics. Furthermore consideration of edge exposure produce the raked edge of tiles and the chamfering of the edge of each individual castellation.

Figure 3. Section of base of castellation

Manufacturing and installation have been identified by JET from its experience to influence the global power handling capabilities. Exposed vertical faces imply power handling reduction by orders of magnitude. Localised melting can become a runaway process. MkI design allowed for 1mm tolerance of installation. A detailed survey showed that ≤0.5mm was achieved rail-to-rail.

The tiles were produced by a Brush-Wellman standard manufacturing process out of S65c powder. A near net shape could be achieved on most faces; the faces seating on the support rail and the plasma facing area was machined to better than 0.05mm tolerance.

3. ELM INTERACTION WITH MKI TILES

The divertor configuration of JET and other large tokamaks admits discharges for which short duration bursts of energy (and particles), ELMs, interact strongly with tile material. One set of discharges, a search for the limits to an ELM-free discharge, produced a large ELM whereby approximately 1MJ of stored energy was lost in 100μsec. The footprint of the ELM involved an interaction area of ~0.15m^2. Melting patterns on tile pairs adjacent to and outboard of the centre line of the base cf. figure 1 show that energy was deposited at a glancing angle of incidence, typically ~0.5°. Assuming 50% radiation this implies a peak flux density of approximately 35×10^3MW/m^2. Beryllium melting is expected after ~20μsec. Taking into account melting and evaporation, this implies a layer of no more than 0.5mm resolidified material.

The mechanical properties of beryllium are subject to strain-rate effects. Calculations show that strain-rates of $\sim 5 \times 10^2$/second at the surface are achieved. However, these same calculations show that large strain-rate effects on such a short time scale are confined to near-surface regions which are melted and later solidified. The flux densities incident in an ELM may locally, e.g. on exposed shoulders of a castellation, be high enough to produce vaporisation with little trace of the melted material. Such regions in the MkI design are covered by resolidification of neighbouring material as the affected areas are extremely small.

In the design of an actively - cooled beryllium component the principal problems are of lifetime assessment (loss of material through evaporation or mechanical displacement) due to ELMs. There is a serious problem of residual stresses in the bonded region due to resolidfication of molten material. As the rate of resolidification in a vacuum environment is not determined by incident flux density during a ELM event the JET experience of no visible mechanical damage to the roots of a castellation due to ELM events is encouraging.

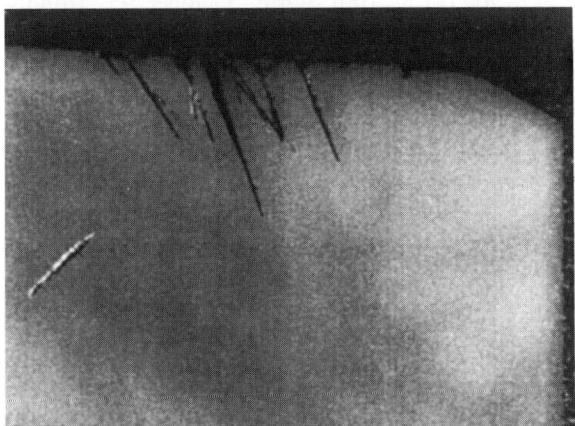

Figure 4. Section of resolidified zone due to single giant ELM crash

4. METALLURGICAL INVESTIGATION

The material melted and resolidified in the ITER-dedicated experiment has been sectionned. Near surface regions are seen to be cracked with long columnar crystals as expected. The roots of castellations were also examined and a crack was seen to have developed at the base, cf. figure 3. Only 1 out of 3 positions examined was shown to have cracked. A more detailed investigation is to be done shortly of this process.

The material melted and resolidified as a result of the single giant ELM crash has also been examined. One section showing long curving fatigue cracks is shown in figure 4. These cracks appear to propagate into the virgin material, i.e. to extend below the resolidified zone. More detailed investigations are in hand.

5. NUMERICAL ANALYSES

Thermo-mechanical assessment of the MkI tiles was carried out largely based on a 2-d picture for thermal calculations with a full 3-d model for the footprint onto a tile. During design a 2-d plane - stress assessment was made. Assessment of actual equilibria required 3-d thermal and mechanical

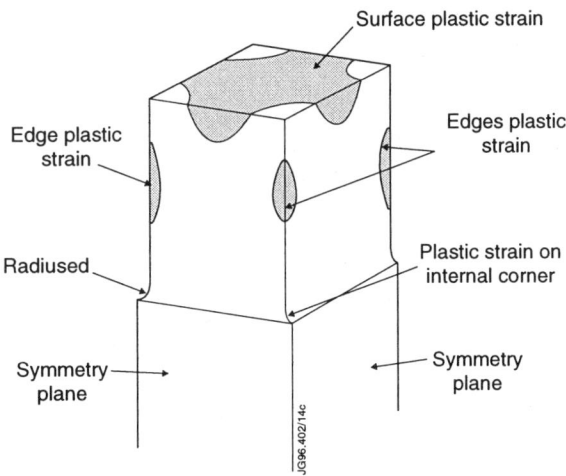

Figure 5. Plastic zones of typical castellation prior to melting

analyses. Plastic zones, ~0.5% equivalent strain develop over most of the irradiated face, also along edges. Only 3-d analyses obtain this last result. Fatigue cracks parallel to the irradiated face have been seen on some MkI melted tiles.

6. CONCLUSIONS

Metallographic studies of the resolidified beryllium region shows characteristics previously observed, microcracks along columnar boundaries, local cracking of the virgin material. Some evidence is seen for lateral cracking at the edges of strongly heated/resolidified castellations. Evidence is seen for crack initiation at the base of a castellation subjected to deliberate melting on several shots.

Attempts to analyse for the consequences of melting on crucial areas of the tile, i.e. roots of castellations show that a full 3d analysis appears to be mandatory though a failure of either plane-stress or plane-strain analyses to take proper account of the diffusion of thermal stress into a typical castellation. As yet there is no evidence to support numerical predictions of large plasticity zones at roots of castellations.

7. BIBLIOGRAPHY

1. The JET Team (presented by M. Keilhacker) Plasma Physics and Controlled Fusion 37, A3, (1995)

2. H. Lingertat et al, Studies of Giant ELM Interaction with Divertor Target in JET, 12th PSI, St. Raphael, France, May 1996.

3. B. Tubbing et al, First Results with JET Melting Experiment. 22nd EPS Conference on Controlled Fusion and Plasmaphysics, Paper 19C, III-453

4. E.B. Deksnis et al, Thermal fatigue of beryllium, 2nd IEA Workshop on Beryllium Technology, Idaho Nat. Eng. Lab. report CONF-9509218, September 1995

5. C. Ibbott et al, JET programme on development of Beryllium Clad components for ITER, 16th SOFE, October 95, Illinois, USA

6. R.D. Watson et al, Low cycle thermal fatigue testing of beryllium grades for ITER PFC, 2nd IEA Workshop, ibid 4.

7. D. Campbell and JET Team, Experimental Comparison of Carbon and Beryllium as Divertor Target Materials for JET, 12th PSI, St. Raphael, France, May 1996

8. D. Dombrowski, E. Deksnis, M. Pick, Thermomechanical properties of Beryllium, Atomic and Plasma - Material Interaction data for Fusion, vol 5, IAEA, Vienna 1994

Design, Fabrication and Testing of Helium-Cooled Vanadium Module for Fusion Applications[*]

C.B. Baxi, E. Chin, B. Laycock, W.R. Johnson, R.J. Junge, E.E. Reis, J.P. Smith

General Atomics, P.O. Box 85608, San Diego, California 92186-5608, USA

Vanadium alloys are attractive materials for fusion applications due to their low neutron activation and rapid decay of radioactivity with time. Design of high heat flux components with vanadium as the structural material is difficult due to its low thermal conductivity relative to copper and the lack of practical experience with fabrication of vanadium components. Similarly, helium is an attractive coolant for fusion power plants due to its chemical inertness, its transparency to neutrons, and stable heat transfer. However, there is a perceived difficulty that the use of helium as a coolant will limit the maximum heat flux on components. Reference 1 discusses the principle that heat transfer enhancement techniques reduce the pumping power for helium cooling, making it practical for cooling plasma facing components and General Atomics (GA) has demonstrated cooling of high heat flux components with helium coolant. A copper module designed by GA was successfully tested to a steady state heat flux level of 3200 W/cm^2 over small area and 1000 W/cm^2 over the entire 20 cm^2 area. As a continued effort to demonstrate practical application of fusion science, GA undertook the present effort to fabricate a vanadium module cooled with helium. Due to lower thermal conductivity of vanadium (6% of copper), this module will withstand about 300 W/cm^2 heat flux over the entire length. The module was fabricated from V-4Cr-4Ti alloy and is 228 mm long and 22.1 mm in diameter. The thickness of the vanadium tube is 1.76 mm. The internal flow path has been designed to enhance the heat transfer coefficient to a value of about 1 W/cm^2-°C at a helium flow rate of 20 g/s. A thermal stress analysis of the design was performed to ensure that the stresses are within limits at a heat flux level of 300 W/cm^2 and a helium pressure of 4 MPa. The test module has been hydrostatically tested to 7 MPa pressure and helium leak checked. The module is ready to be tested at the helium loop (4 MPa pressure 20 g/s flow) at Sandia National Laboratory, Albuquerque. Future high heat flux testing is planned.

1. DESIGN AND ANALYSIS

Thermal conductivity of vanadium is about 6% of copper. Hence, the methods previously used for designing the helium cooled copper module [1] are not attractive for this material. For example, the fin efficiency (ratio of actual heat transfer to heat transfer if entire fin was at the root temperature) for the GA copper module is about 0.5. For vanadium the fin efficiency will be less than 0.2. If the heat transfer area is increased by a factor of 10 using extended surfaces, no enhancement will be obtained.

First, an analysis was performed to find the required effective heat transfer coefficient (HTC). Then, we looked at methods which do not depend on thermal conductivity of the material to increase the heat transfer coefficient.

A finite element (FE) analysis was performed for a vanadium tube of 22.1 mm o.d. and 1.76 mm wall thickness, subjected to heat flux on one side using ANSYS [2]. The results of many analyses are summarized in Fig. 1. With a heat transfer coefficient of 1 W/cm^2-°C for a heat flux of 300 W/cm^2, the peak surface temperature is about 600°C.

We wanted to design our module such that it could be tested at Sandia National Laboratory, Albuquerque (SNLA). The helium loop and beam parameters at SNLA are: e-beam power of 30 kW, flow of 20 g/s, loop pressure 4 MPa, and maximum pressure drop 0.1 MPa.

The design shown in Fig. 2, using the full capacity of SNLA loop, will withstand a steady state heat flux of more than 300 W/cm^2.

The design consists of a vanadium tube of 22.1 mm o.d. and 18.3 mm i.d. with an insert of 18.1 mm o.d. The insert was made from stainless steel. The inside wall of the vanadium tube was

---
[*]Work supported by General Atomics IR&D funds.

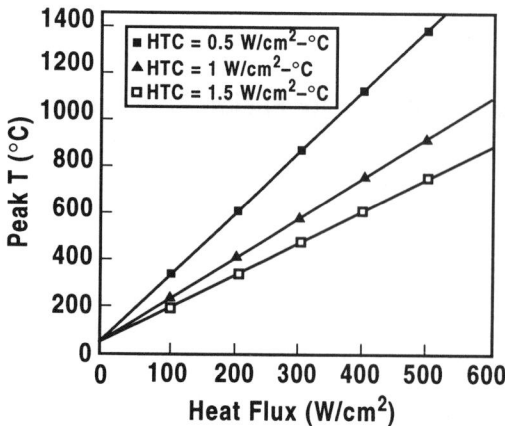

Fig. 1. Peak surface temperature for a vanadium tube of Do = 22.1 mm, Di = 18.3 mm and coolant temperature of 50°C.

roughened with axial ribs shown in Fig. 3 to increase the heat transfer coefficient. The gap between the tube and the insert is maintained by a spiral fin of 1 mm height and 100 mm pitch on the stainless steel rod (Fig. 4). This geometry enhances the heat trasfer coefficient by 1) increasing the flow velocity, 2) reducing the hydraulic diameter, 3) increasing top to bottom flow mixing and 4) breaking up the laminar boundary layer at the wall with the ribs.

Calculations were made using a flow of 20 g/s, an inlet pressure of 4 MPa, an inlet temperature of 20°C, and a heat flux of 300 W/cm^2 on the vanadium module. The results showed a maximum surface temperature of 600°C, a pressure drop of 0.6 bar (9 psi), and a pumping power of 200 W, about 3% of the power removed.

This design has potential of operating near 500 W/cm^2 heat flux by:

1. Reducing the wall thickness to 1 mm.
2. Allowing higher wall temperatures up to 750°C.
3. Using higher flow rates and pressures than currently available at SNLA.

The stresses in the vanadium tube at the maximum pressure (7 MPa) were acceptable (41.3 MPa) for the material (allowable = 124 MPa). This is the pressure at which the module was tested hydrostatically at room temperature to fulfill SNLA requirements. Additionally, the deflections of the tube due to bowing, caused by the temperature gradient of 300°C, were calculated to be small (0.43 mm max). Also, the calculated thermal stress was less than 50% of the allowable.

2. TEMPERATURE LIMITS ON VANADIUM

An assessment was made of the maximum temperature limits for the V-4Cr-4Ti alloy components of the module using Argonne National Laboratory (ANL) and Oak Ridge National laboratory oxidation data for vanadium alloys in air and low partial pressure oxygen environments. Calculations/extrapolations based on tensile properties (ductility) of V-alloys exposed in air at

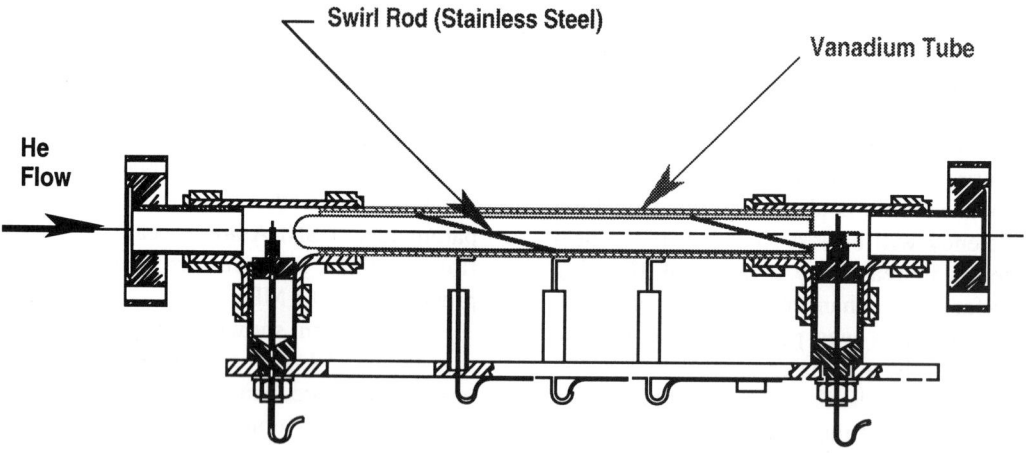

Fig. 2. He cooled Vanadium Module.

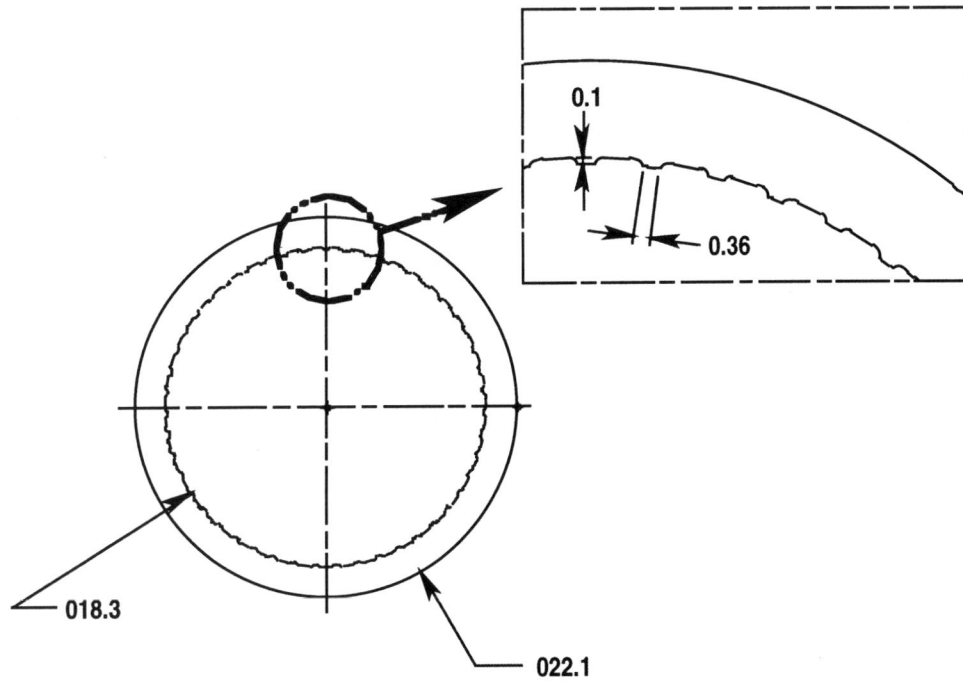

Fig. 3. Vanadium tube cross section and magnification of axial grooves (dimensions in mm).

temperatures of 400 and 500°C yielded a conservative temperature limit of ~750°C for a 1 hour exposure during the high heat flux testing.

3. FABRICATION

The vanadium tube for this module was some of the first tubing made. It was fabricated starting with 29 mm thick plate material produced by Teledyne Wah Chang in Albany, Oregon for ANL. The vanadium was then cold drawn by Century Tubes of San Diego, California. The drawing was done in a series of cycles separated by annealswith the cold work between anneals limited to 30% reduction with 10%–12% per pass. Electric discharge machining was used to make the axial ribs. A vacuum bake was used to degas the hydrogen from the material.

The stainless steel Spiral Rod Insert (SRI) was made from 304 stainless steel rod stock on a four axis lathe by Qualtech Manufacturing in San Diego, California. The rod is rounded at the inlet end to reduce the pressure drop. It is fixed at the exit end to prevent the rotation of the rod.

The vanadium tube is connected to the inlet and exit flanges by standard Swagelock compression fittings. This allowed us to make the assembly without any brazing or welding of the vanadium tube.

The module was equipped with five k type thermocouples. Two thermocouples measure the inlet and outlet temperatures of the helium. The other three measure the vanadium tube temperature and will be used for calibration of the infra-red (IR) camera and two color pyrometer.

The module was assembled, pressure tested to 7 MPa pressure, and leak checked at GA. The helium leak rate was less than 10^{-7} Pa m^3/s.

4. TESTING PLANS

The testing is planned to occur in September 1996.

The test plan consists of three parts. During the first phase, a relation between flow and pressure drop will be established to insure that sufficient flow can be obtained. During the second phase, an IR camera and a two color pyrometer will be calibrated by heating the module slowly to about 300°C using the leakage current for the e-beam. During the final phase, the heat flux will be increased in steps until

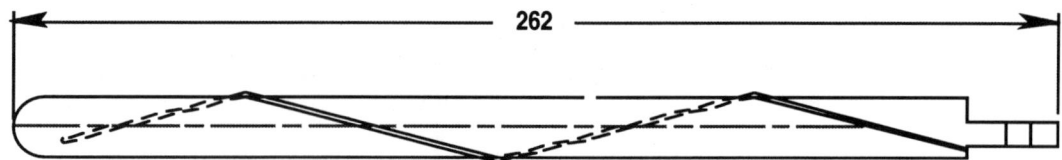

Fig. 4. Sketch of spiral rod insert.

the surface temperature limit of 700°C is reached. The surface temperature will be measured by a two color pyrometer and an IR camera. The heat flux will be calculated from calorimetry. Flow rate, inlet and outlet temperatures, and pressure drop will also be measured during the test.

CONCLUSIONS

We have demonstrated that vanadium components can be fabricated and vanadium can be used as a structural material with helium cooling to remove steady-state fluxes to a level of 300 to 500 W/cm^2.

ACKNOWLEDGMENTS

Authors want to thank K. Redler for help with finite elements analysis and J. Lindgren for material availability evaluation and ANL for supplying the vanadium.

REFERENCES

1. C.B. Baxi, "Evaluation of Helium Cooling for Fusion Divertors," Fusion Engineering and Design, **25** (1994) 263–271.
2. ANSYS Finite Element code, Swanson Analysis System Inc., Houston, PA.

Characterization of copper-stainless steel EXW joints

S.Tähtinen, P. Moilanen and P. Karjalainen-Roikonen

VTT manufacturing Technology, P.O. Box 1704, FIN-02044, Finland

Explosion welding was applied for joining CuCr1Zr copper alloy to 316 LN ITER Grade stainless steel and tensile and fracture resistance properties of joints were determined. Explosion welding results significant hardening and strengthening of both metals. Strengthening is non-uniform after explosion welding showing highest values next to joint interface. Post weld heat treatment recovers hardness and mechanical properties of both copper alloy and stainless steel. The studied joints all show ductile behavior and tensile and fracture resistance properties are consistent with unique dislocation structure induced by explosion welding.

1. INTRODUCTION

The present ITER First Wall design is based on the use of bimetallic structures to manage high heat loads. First candidate structure material is 316 LN ITER grade stainless steel covered with CuAl25 copper alloy and beryllium layers. Hot Isostatic Pressing (HIP) is considered as most promising bonding method for copper alloy and stainless steel. Other options for heat sink materials and bonding methods which have been considered are CuCr1Zr copper alloy bonded to stainless steel by explosion welding (EXW).

The objective of this study is to characterize CuCr1Zr copper alloy to stainless steel EXW joints and to generate experimental data on tensile and fracture resistance properties of these bimetallic joints.

2. MATERIALS AND METHODS

Materials used in this study were AISI 316 LN ITER Grade austenitic stainless steel supplied by Joint Research Centre Ispra and precipitation hardened CuCr1Zr copper alloy supplied by Outokumpu Poricopper Oy. The stainless steel plates were used in hot rolled conditions with a plate thickness of 3.5 mm and the copper alloy was in solution annealed and precipitation heat treated condition e.g. no intermediate cold rolling procedure was applied. Explosion welded joints were prepared by High Speed Tech Oy using a 30 mm thick copper alloy as a base plate for two subsequent 3.5 mm thick stainless steel EXW layers. Final dimensions of explosion welded copper stainless steel compound plates were 1050x140x37 mm^3.

Hardness and tensile properties of base materials as well as fracture resistance properties of copper stainless steel joints were determined both in EXW as received and in post weld heat treated (PWHT) condition. PWHT was performed at 960°C for 180 minutes followed by air cooling and subsequent annealing at 460°C for 120 minutes. Test specimens were machined from compound plates, flat 2.5x7.8 mm^2 tensile specimens with 25 mm gauge length, parallel to joint interface, cross weld round ∅ 5 mm tensile specimens with 25 mm gauge length and cross weld 10x10x55 mm^3 three point bend (3PB) specimens. The cross weld tensile and 3PB specimens were prepared by applying Electron Beam (EB) welding to add extra stainless steel material on specimen blanks.

The notch tip in the 3PB fracture resistance specimens was located in the joint interface along which a crack in the specimens was prefatigued to the a/W-ratio of about 0.5. The specimens were sidegrooved 10 % on both sides. Fracture resistance curves (J-R-curves) were determined using displacement control with constant displacement rate of 3×10^{-4} mm/min. Crack length during the tests was measured using DC-PD method.

3. RESULTS

3.1. Hardness

Hardness across EXW joint interface showed a clear hardness increase in stainless steel induced by explosion welding, Figure 1. Hardness of stainless steel increased from 157 HV_{10} to almost 500 HV_{10} next to joint interface. Hardness of copper alloy increased from 135 HV_{10} to about 150 HV_{10} next to joint interface. Post weld heat treatment recovered the hardness of both component metals. Hardness of stainless steel after PWHT was slightly higher and that of copper alloy slightly lower when compared with hardness values before explosion welding.

3.2. Tensile properties

Hardening due to EXW was clearly seen in tensile properties of stainless steel where yield and tensile strength next to joint interface increased from 325 to 960 MPa and 619 to 1051 MPa, respectively, Table 1. Elongation decreased from original value of 71% to 13% due to explosion welding. Stainless steel plate next to copper joint showed somewhat higher tensile strength and lower elongation values compared to steel plate on top of that due to additional hardening induced by second explosion welding operation.

Yield and tensile strength of copper alloy next to joint interface increased from 248 to 420 MPa and from 371 to 433 MPa. Elongation decreased from original value of 25% to 13% due to explosion welding.

PWHT reduced yield and tensile strength of both stainless steel and copper alloy. Yield strength of stainless steel decreased to 322 MPa and tensile strength to 641 MPa. Yield strength of copper alloy decreased to 190 MPa and tensile strength to 314 MPa next to EXW joint interface. Elongation of stainless steel and copper alloy increased to 66% and 31%, respectively.

Due to EXW the tensile properties of copper alloy varied with the distance from the joint interface, Figure 2. Yield and tensile strength decreased from 420 to 330 MPa and 433 to 370 MPa, respectively, within a distance of about 23 mm from the joint interface.

Table 1. Tensile properties of base materials before and after EXW, specimens were taken next to the joint interface.

Material	$R_{0.2}$ MPa	Rm MPa	A_5 %	n	Note
316 LN IG	325	619	71	-	original
	960	1051	13	0.094	EXW as rec.
	322	641	66	0.281	after PWHT
CuCr1Zr	248	371	25	-	original
	420	433	13	0.05	EXW as rec.
	190	314	31	0.21	after PWHT

n = strain hardening exponent

However, elongation increased from a value 13% to about 20% already with in a distance of about 5 mm from the joint interface. After PWHT no gradients in tensile properties of stainless steel and copper alloy were observed. However, tensile properties of copper alloy next to joint interface showed somewhat lower tensile properties due to large grain size resulted by recrystallization of copper compared with tensile properties measured more than about 5 mm from the EXW joint interface.

Cross weld tensile properties [1] of copper stainless steel tensile specimens are summarized in Table 2. Tensile strength of as received specimens decreased from 412 MPa to 328 MPa due PWHT. At the same time elongation measured across the joint interface increased from 13% to 19%. For as received and PWHT specimens fracture occurred in copper alloy at about 12,5 mm and 7 mm from joint interface, respectively.

Table 2. Tensile properties of cross weld copper stainless steel specimens

$R_{0,2}$ MPa	Rm MPa	A_5 %	Fracture in Cu	Note
364	412	13	12,5mm from joint	EXW as rec.
197	328	19	7mm from joint	PWHT

3.3. Fracture resistance curves

The fracture resistance of the PWHT joint was clearly higher than that of the EXW joints in as received condition, Figure 3. The initiation values, $J_{0.2/BL}$, for ductile crack growth were higher for joints in the PWHT condition, 148 kJ/m², than in the as received condition, 52 kJ/m², Table 3. Higher fracture resistance of PWHT joints compared to that of EXW joints in as received condition is also shown in the higher tearing resistance values, dJ/da, for PWHT joints, Table 3.

Table 3. Fracture resistance properties of copper stainless steel EXW joints.

$J_{0.2/BL}$ (kJ/m²)	dJ/da (kJ/m²/mm)	M	Note
56	84	2.3	EXW as rec
53	80	2.3	EXW as rec
46	52	2.3	EXW as rec
130	133	1.7	after PWHT
167	170	1.7	after PWHT

M = Strength mismatch, $R_{0.2}$(steel)/$R_{0.2}$(copper)

In PWHT condition the fracture took place parallel and close to the joint interface in copper but in as received condition fracture seemed to deviate from interface and propagate into copper. The fracture mode in both conditions was fully ductile.

4. SUMMARY AND CONCLUSIONS

Explosion welding is unique as a result of significant hardening and strengthening which arises from shock wave propagation while the residual strains are small or even negligible. Shock waves are known to generate a high density of dislocations and point defects which are expected to contribute on measured tensile properties [2,3]. High yield strength but also markedly reduced strain hardening ability are expected to result from rearrangement of unstable dislocation substructure induced by shock loading. Strengthening was non-uniform, highest next to joint interface, in copper alloy in as received condition which was clearly indicated by low elongation values within about 5 mm from joint interface. PWHT is expected to recover the shock wave induced dislocation structure and to induce recrystallization where plastic strain or defect structure is developed more than a critical strain. Subsequent tensile properties of copper alloy and stainless steel in PWHT condition were comparable to tensile properties before explosion welding.

The results of the fracture resistance tests are in a good accordance with the results of the tensile tests. The higher elongation, the lower yield strength and the higher strain hardening coefficient of the copper after PWHT near the fusion line compared to the corresponding values of the copper in as received condition indicates higher ductility. This means higher fracture resistance for the joints in PWHT condition as more energy is consumed for plastic work. Strength mismatch is expected to affect the fracture propagation e.g. when strength mismatch was low fracture propagated parallel and close to the joint interface and when strength mismatch was large fracture seemed to deviate from joint interface into the softer material.

Explosion welding is a promising joining method for copper alloy to stainless steel when high interface strength is required. However, more test data is needed especially at elevated temperatures for proper design of copper stainless steel composite structures.

REFERENCES

1. S. Tähtinen, P. Kauppinen, K. Rahka and P. Auerkari, Performance of Copper-Stainless Steel EXW Welds. 2nd International Symposium on Mis-Matching of Welds, April 24-26, 1996, Reinstorf-Lüneburg. In press.
2. B. Crossland, Explosive Welding of Metals and Its Application, Clarendon Press, Oxford, 1982.
3. T.Z. Blazynsky, Explosive Welding, Forming and Compaction, Applied Science Publishers, London, 1983.

ACKNOWLEDGMENTS

This work was performed in the framework of the European Fusion Program by the Association Euratom-TEKES. The work was partly funded by Finnish Fusion Research Program FFUSION and European Fusion Program.

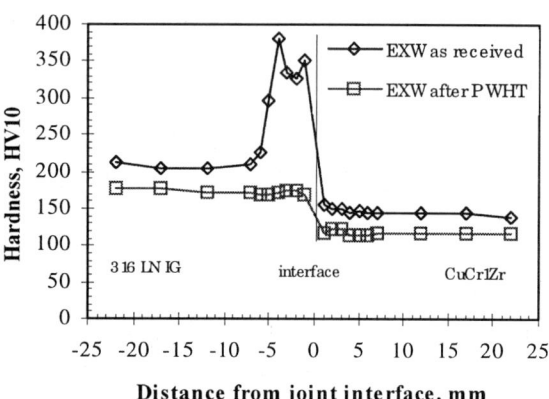

Figure 1. Hardness profile across copper alloy stainless steel EXW joint interface.

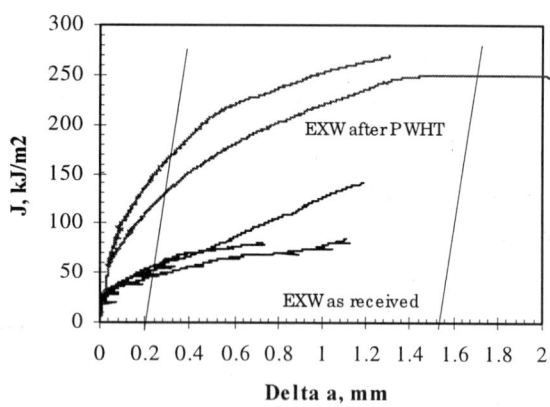

Figure 3. Fracture resistance curves of copper alloy stainless steel EXW joints.

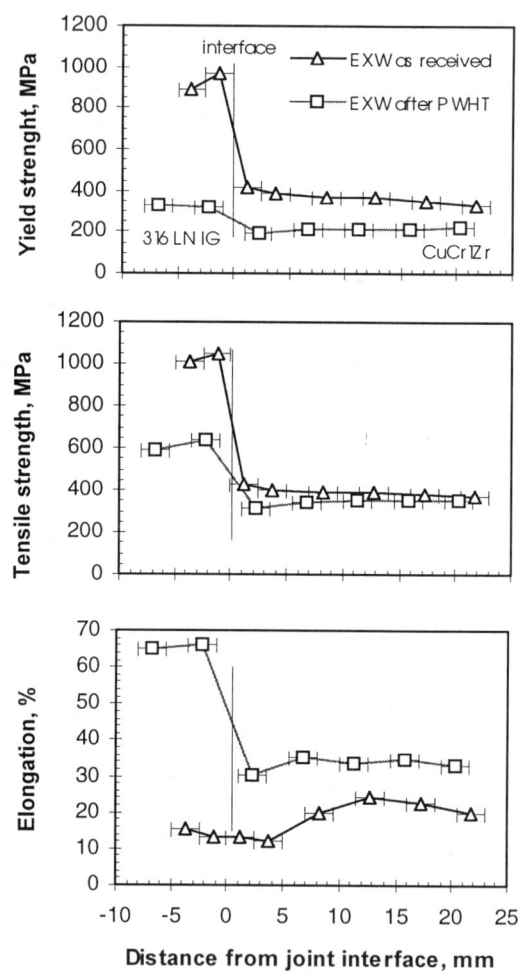

Figure 2. Tensile properties of copper alloy and stainless steel as a function of distance from the EXW joint interface.

Development and characterization of Be/Cu joint obtained by Hot Isostatic Pressing

F. Saint-Antonin[a], D. Barberi[b], G. Le Marois[a], A. Laillé[b]

a CEA/Grenoble, CEREM/DEM/SGM, 38054 Grenoble Cedex 9, France
b CEA/Bruyères-le-Châtel, DRMN/MOS/RDA, BP 12, 91680 Bruyères-le-Châtel, France

Abstract : This paper presents the development and characterization of Be/Cu joints by Hot Isostatic Pressing (Solid HIP). It gives a description of different associated interlayers used as diffusion barriers between copper and beryllium for a low in-service working temperature. The joint structures are described. The best interlayer system was chosen to fabricate a Be/OFHC-Cu tube in tube mock-up as demonstrator for ITER component geometries

1- INTRODUCTION

Beryllium is one of the possible candidate for Plasma Facing Components (PFC) such as divertor or first wall in the ITER project. Due to its high thermal conductivity, copper alloys are used as heat sink material. So, in one way or another, beryllium will have to be fixed onto copper alloys.

In this study, the joining of beryllium onto copper is achieved by Hot Isostatic Pressing (Solid HIP). This joining technique allows an homogeneous bonding. But, as direct bonding between Be and Cu induces intermetallics which are deleterious to the joint (for instance [1]), interlayers are needed to avoid reaction between Be and Cu. This paper gives a description and the role of different associated interlayers used as diffusion barriers between copper and beryllium for a low in-service working temperature

Moreover, a mock-up was fabricated. Shear resistance of the junction was measured from test specimens machined from the mock-up.

2- JUNCTION DESIGN

2.1 General considerations

To avoid reaction and formation of intermetallics between Be and Cu, diffusion barriers is needed. Moreover, the Kirkendall effect (observed in [2]) may be avoided with the use of diffusion barriers

2.2 Diffusion barriers

Diffusion barrier for copper : element such as W, Cr and Mo are immiscible in the solid state, then copper should not diffuse in these elements.
Diffusion barrier for Be : the Al-Be phase diagram is of eutectic type. The solublity of Be in Al below 600°C is negligeable, then no diffusion should occur between Al and Be.

Thus, an interlayer of W, Cr or Mo on the copper side and an interlayer of Al on the beryllium side should prevent diffusion.
The other possibility is to use magnesium which may have the same effect by the formation of an intermetallic.

2.3 Other requirements

Regarding the behavior of these elements under neutron irradiation, W has a high activation potential. Moreover, between Cr and Mo, Mo has a higher thermal conductivity than Be : the Mo interlayer is not a barrier to the heat transfer.

3-EXPERIMENTAL PROCEDURE

3-1 Sample preparations

The different interlayers were pre-placed by a sputtering technique named Physical Vapor Deposition (PVD). This technique allows the deposition of material with a high adhesion with the substrate, a high density and a constant thickness. Previously to the coating, there is a cleaning with ionized argon. The surface roughness of the coating is the same as the substrate. The surface of the OFHC Cu substrate has to be prepared mechanically and chemically. The thickness of the diffusion barrier and the aluminium were chosen to get a high density which is necessary to avoid the diffusion of Cu or Be through the interlayers. The test specimens were made of disk in copper and beryllium with 20 mm diameter and various thickness. Each one of the following chemical elements W, Cr and Mo were sputtered onto different copper disks. Aluminium was deposited on top of these first coatings. The specimens are placed in a container. Vacuum is performed in the container before sealing. The different junction were performed : Cu/Be (reference junction without interlayer), Cu/Al/Be (reference junction without diffusion barrier for copper), Cu/W/Al/Be, Cu/Cr/Al/Be, Cu/Mo/Al/Be, Cu/Mg/Be.

3-2 HIP cycle

The HIP parameters were : 500°C, 100MPa and 1h30. These parameters were chosen essentially to take into account the presence of aluminium or magnesium : the melting point of these two elements are respectively 660°C and 650°C. The temperature 500°C was reached with the rate of 200°/h. After the holding time at 500°C, the furnace power was switched off to cool down the container.

4- RESULTS

From metallographic and X-µprobe analysis, the main results are summarized in the following paragraph :

Cu/Be : the joining did not occur during this HIP bonding cycle.

Cu/Al/Be : A large crack all along the joint is observed at the Al/Be interface. Aluminium has diffused into copper far from the initial Al/Be interface (up to 30µm). An interlayer, near the beryllium side, of about 5µm is composed of Al, Cu and Be.

Cu/W/Al/Be : A thin crack is observed in the tungsten all along the joint, near the aluminium side. There is almost no interdiffusion between tungsten and aluminium. The presence of a small amount of copper in tungsten was observed due to residual porosities in the W layer. Nevertheless, beryllium and copper are not in contact. It should be noted that, in the Al-W phase diagram, the solubility of Al into W is about few percents at about 500°C.

Cu/Cr/Al/Be : No cracks have been observed in the different interfaces. A small interdiffusion layer, of about 6µm, between Cr and Al is present all along the interface, but locally this interdiffusion is larger and forms inclusion in the aluminium layer. In that case, Al, Cr and Be are mixed.

Cu/Mo/Al/Be : From a metallographic point of view, this junction is the more homogeneous, no crack has been observed all along the different interfaces. The profile curves show a small interdiffusion layer between Al and Mo of about 5µm. This could be explained by the presence of pores in the molybdenum layer because from the Al-Mo phase diagram, it could be noticed that the solubility of Al into Mo is negligible at 500°C. The presence of copper was observed in the molybdenum layer due to residual porosities in that layer.

Cu/Mg/Be : Small cracks have been observed in the magnesium interlayer near the beryllium side. The magnesium diffused into copper and beryllium : an interlayer of about 15µm composed of a mixture of Cu, Mg and Be is present. Moreover, it seems that Mg has diffused more on the copper side than on the beryllium one : in the Mg-Cu phase diagram, Mg can be solubilized up to 7at% in Cu.

Conclusion of the metallographic study :

Direct bonding between Be and Cu at 500°C cannot be achieved. The use of diffusion barrier is necessary to avoid 1) contact and large reactivity between Be and Cu and 2) important diffusion of Al in copper. This last effect leads to decohesion between Be and the copper enriched in Al. Among all these junctions, the Mo-Al interlayer has the higher quality : no cracks and homogeneous distribution of the different chemical elements along the joint were observed. This interlayer has been selected for the fabrication of the mock-up.

5- MOCK-UP FABRICATION

A scheme of the mock-up is given in figure 1 : this mock-up is made of a copper tube inserted inside a beryllium tube. The copper tube was cleaned chemically previously to the deposit of Mo and Al by PVD : only the part to be joined was treated. The same HIP cycle as described above in 3.2 was performed. The container was designed in such a way that an internal pressure was applied inside the copper tube.

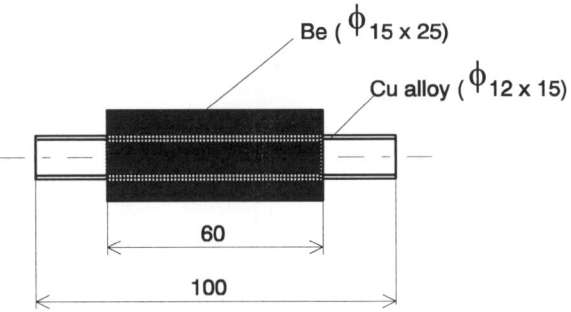

Figure 1 : mock-up Cu/Be joined by HIP

The mechanical resistance of the joint was evaluated with a tensile shear test. Shear specimen were fabricated from the mock-up : slices of 5mm thick perpendicular to the tube axis were cut off. The joint surface was parallel to the tensile axis.

The shear strength of the junction at room temperature is about 55 MPa : at this level, the copper is heavily deformed but the joint did not break. The deformation of copper occured radially.

6- CONCLUSION

The main conclusions are :
- direct Be/Cu bonding was not achieved at 500°C by solid HIP,
- the use of a diffusion barrier is necessary to avoid 1) contact and large reactivity between Be and Cu, and 2) excessive diffusion of aluminium in copper which leads to decohesion at the beryllium interface,
- among the three studied systems with diffusion barrier, Cr-Al, Mo-Al, W-Al, the Mo-Al interlayer has the best metallographic quality (no cracks, homogeneous distribution of the different chemical elements). Moreover, regarding activation behavior under neutron irradiation and the thermal conductivity, Mo is the most suitable solution with Al.

The fabrication of a mock-up Cu-tube into Be-tube was successfully completed with Hot Isostatic Pressing.

The shear strength of the junction reaches at least 55MPa at room temperature : an excessive deformation of the copper occurs without rupture of the junction.

REFERENCES

1. D.L. Youchison, R. Guiniiatouline, R.D. Watson, J.M. McDonald, B.E. Mills and D.R. Boehme, Proceeding of the 18th Symposium on Fusion Technology edited by K. Herschbach, W. Mauer and J.E. Vetter, Karlsruhe, Germany, 22-26 Aug. 1994, p 287.
2. J. Williams and J.W.S. Jones, Powder Metallurgy, n°5 (1960) 45.

Development of Mechanically Joined Divertor Plate for Large Helical Device

Y.Kubota, N.Noda, A.Sagara, Y.Kato, T.Morisaki, N.Ohyabu, and O.Motojima

National Institute for Fusion Science, Furou-cho, Nagoya 464-01, Japan

Mechanically joined(MJ) sample has been developed to realize highly reliable helical divertor plates for large helical device(LHD). To improve the thermal contact of MJ sample, a thin carbon sheet is used at the interface between two materials as a compliant sheet. A set of test results indicates that the real scale MJ using thin carbon sheets and a tight cooling pipe holder can be used even in high heat flux of up to 4 MW/m^2 for 10 sec operation mode and 0.75 MW/m^2 for steady state operation mode, which are nearly maximum heat fluxes to the divertor plate of LHD. The structure and test results of some kinds of MJs are described.

1. INTRODUCTION

Large helical device(LHD)[1,2] with super conductive helical and poloidal coil systems[3] is now under construction at Toki site, the experimental start of which is scheduled in the spring of 1998. The LHD uses a helical divertor configuration[4] together with an innovative divertor concept, local island divertor[5] to control the heat and particle exhaust. A large number of helical divertor plates[6] are used to remove heat due to high power heating whose maximum heat flux in the first phase(B=3T) is estimated to be 5 MW/m^2 for 10 seconds pulse operation mode and 0.75 MW/m^2 for long pulse(1 hour) steady operation mode. At the second phase(B=4T), the heat flux increases to 10 MW/m^2 for 5 sec operation mode. Mechanically joint(MJ) sample for the first phase and brazed joint(BJ) sample will be used as the divertor plate in the LHD. The MJ sample has merits of low cost, simple in construction, easy for replacement. However, the thermal properties of the MJ sample are generally very poor compared those of the BJ sample because of the low thermal conductivity at the mechanical interface between two material. To improve the thermal properties of the MJ sample, some types of compact MJ samples have been fabricated and evaluated using a high heat flux teststand called ACT[7] with a 100kW electron beam source.

Through these experiments[8,9], it was confirmed that a thin carbon sheet (CS) is very effective as a compliant sheet to improve the thermal conductivity at the interface in spite of its low thermal conductivity. Therefore, thin carbon sheet was applied to a real scale MJ sample, to be used as a divertor plate of LHD. In this paper, major results of the development and tests are presented.

2. HIGH HEAT FLUX TEST STAND ACT

The development of MJ samples has been carried out using a high heat flux test stand ACT, which consists of an electron beam source, vacuum chamber, water pumping system for cooling of test sample, and some diagnostics. The beam source is installed at the top of the vacuum chamber with an angle of 20° to avoid contamination of electron gun due to sputtered material from the samples. The main parameters of ACT are shown in Table 1.

Table 1 Main parameters of ACT

Accelerated beam energy	30 keV
Max. beam output(CW)	100 kW
Minimum. beam size	8 mm ϕ
Max.sweeping freq.	200 Hz
Pulse width	1s-CW
Irradiation beam area	$\sim$ 400 cm^2
Max heat flux(for 10 cm^2)	100 MW/m^2
Volume of chamber	400 L
Base pressure	5×10^{-7} Torr

A sample during the heat loading test is actively cooled by water with a flow velocity of up to 8.5m/s.

For the evaluation of the test samples, the temperatures at several points of the test sample, beam current, and vacuum pressure are monitored by a data acquisition system. The heat flux on the test sample can be adjusted using a program controller.

3. TEST USING A COMPACT MJ SAMPLE

To evaluate the effect of compliant sheet on the thermal property of compact MJ sample, a flat plate type was used as a test sample. The MJ sample consists of a graphite tile and a copper heat sink with a copper pipe as shown in Fig.1. The tile is fixed on the copper heat sink with molybdenum bolts.

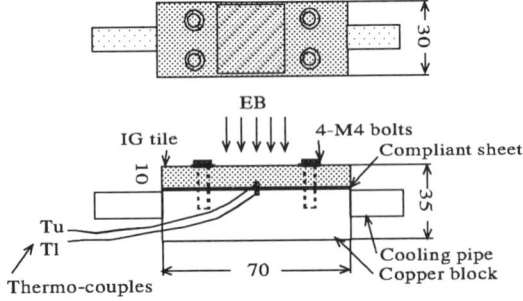

Fig.1 Schematic drawing of a compact MJ sample.

The compliant sheet is inserted between the tile and heat sink. To monitor the temperatures(Tu,Tl) at the tile and the heat sink near the interface, two pairs of thermo-couples are used. The temperature difference ΔT (Tu-Ti) is inversely proportional to the heat transmission efficiency at the interface. Table 2 shows the ΔT(Tu-Ti) as parameter of materials used a compliant sheet under heat fluxes of 4 and 5 MW/m². As shown in the Table, the aluminum sheet has the largest temperature difference, and the carbon sheet has the smallest one although the thermal conductivity of the carbon sheet is very low compared with that of the other materials. This means that the thin carbon sheet is the best material among the tested sheet materials as a compliant sheet. The large temperature difference for aluminum sheet may be due to the thin oxidized layers. The heat transmission efficiency at the interface of compact MJ sample using the carbon sheet increases by about 2-4 times compared with those using the other sheet materials and without sheet.

The reason why the carbon sheet has an effect to

Table 2 Temperature difference ΔT(Tu-Ti) of compact MJ sample as a parameter of materials under heat fluxes of 4 and 5 MW/m².

Material	Difference ΔT(°C)	
	4MW/m²	5MW/m²
50 μm gold sheet	236	328
〃 silver sheet	189	240
〃 aluminum sheet	390	551
〃 copper sheet	357	448
200 μm carbon sheet	82	120
without any sheet	253	551

improve the thermal conductivity at the interface of the compact MJ sample is its softness. Next, to find an optimum thickness of carbon sheet on the thermal property of MJ sample, a heat flux test was carried out using the compact MJ sample under heat fluxes of 3-6 MW/m². The results are shown as a parameter of sheet thickness in Fig.2. The results indicate that temperature difference ΔT(Tu-Ti) has the lowest value when thickness is near 0.1mm. This means that carbon sheet of 0.1mm thickness is effective to improve the thermal property of the compact MJ. Thicker carbon sheet cancels the merit of softness by its low thermal conductivity. On the other hand, a carbon sheet thinner than 0.1mm does not work sufficiently as a compliant sheet.

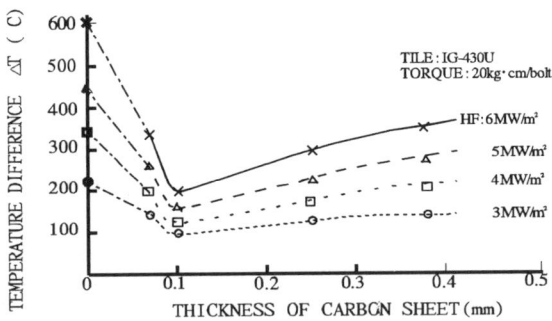

Fig.2 Effect of thickness of carbon sheet

4. REAL SCALE MJ SAMPLE

Some types of real scale MJ samples have been developed to use as an actual divertor plate for LHD. Each real scale MJ sample consists of a graphite tile and a copper backing plate, a stainless steal(SS) backing plate, and a SS water cooling pipe, which are fixed with SS bolts. Thin carbon sheets of 0.1-0.15mm are used at not only between the tile and the copper backing plate, but between the cooling tube and a pair of backing plates to get a good thermal contact between them. As a typical structure of the real scale MJ samples is shown in Fig.3. As the tile material, IG-430U(isotropic graphite made by Toyo Tanso Co. Ltd.) with a thermal conductivity of 139 W/(m·K) is used. Electrolytic tough pitch copper is used as the backing plate materials instead of oxygen free copper(OFC) because of its hardness.

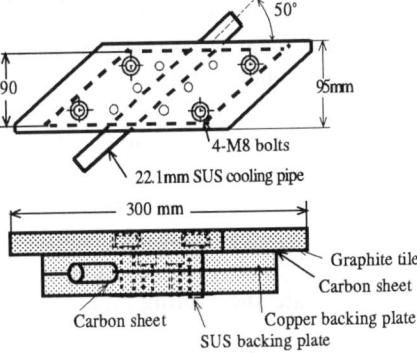

Fig.3 Schematic drawing of typical real scale MJ sample.

5. HEAT FLUX TEST OF REAL SCALE MJ

To evaluate the thermal property of the real scale MJ sample, heat flux test of up to 5 MW/m² for 10 sec pulse and 1.0 MW/m² for long pulse of 1000 sec have been carried out. A beam limiter was used to define beam irradiation area to 50 cm² in rhomboid. The beam area is comparable with that of the divertor trace on divertor plate of LHD.

5.1 Short pulse test

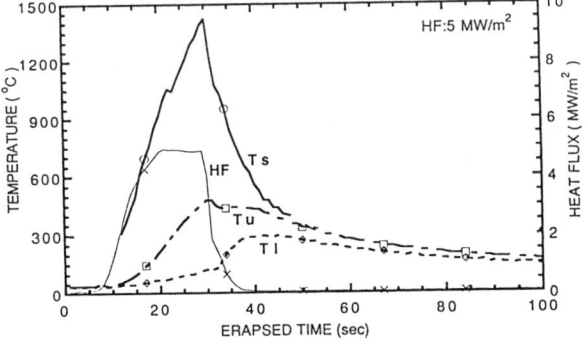

Fig.4 Thermal response curves(Ts,Tu,Tl) of real scale MJ sample under heat flux of 5 MW/m².

Short pulse(10 sec) tests for a real scale MJ sample were carried out under heat fluxes up to 5 MW/m², a cooling water flow velocity of 8.5 m/s, inlet pressure of 0.5 MPa, and inlet water temperature of 25-35 degree in centigrade.

Fig.4 shows typical thermal responses of the real scale MJ sample with time. where, Ts, Tu, and Tl are the temperatures at the surface of graphite tile, bottom of the graphite tile, and top of the copper backing plate, respectively.

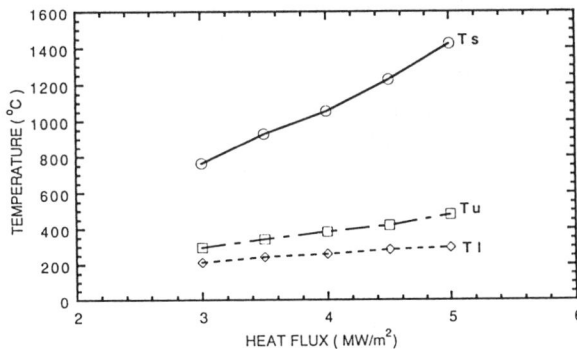

Fig.5 Relationship between the maximum temperatures(Ts,Tu,Tl) and heat flux.

Fig.5 indicates the relationship between the temperatures at the maximum and heat flux(HF). The tile surface temperature Ts reaches 1400 °C, while Tu and Tl are lower than 500 °C and 300 °C under the heat flux of 5 MW/m². The temperature of the tile surface is higher than 1200 °C. Over the temperature, radiation enhanced sublimation (RES) could occur. From this point, this real scale MJ sample should be used with the heat flux lower than 4 MW/m².

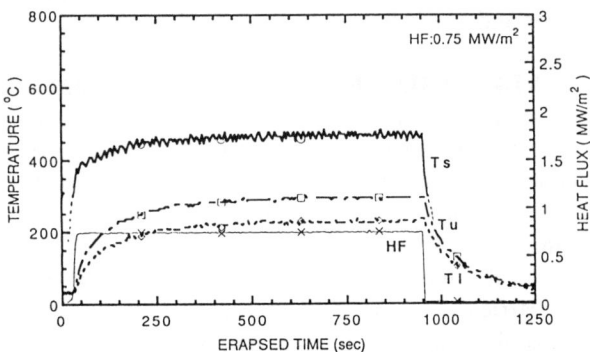

Fig.6 Typical thermal response curves of real scale MJ sample under a steady heat flux of 0.75MW/m².

5.2 Long pulse test of 1000 sec

To evaluate the thermal property of real scale MJ sample under steady load, a test was carried out under heat fluxes 1.0 MW/m² for 1000sec. Typical thermal response curves under a heat flux of 0.75 MW/m² are shown in Fig.6

The temperatures(Ts,Tu,Tl) increase with time. About 200 sec from the start, the temperatures saturate. The saturated temperatures Ts, Tu, and Tl are about 450 ℃, 300 ℃, and 230 ℃. The temperatures(Ts,Tu, and Tl) versus steady state heat flux are indicated in Fig.7. From the figure, the surface temperature(Ts) of the graphite tile is lower than 700℃, which is sufficiently lower than the critical temperature of RES. However, the temperature Tl at copper backing plate reaches about 350℃, which is higher than the temperature at which the mechanical strength of copper becomes week. For mechanical safety, the temperature of copper must be kept lower than 250 ℃. Therefore the real scale MJ sample can be used under lower steady heat flux than about 0.75 MW/m².

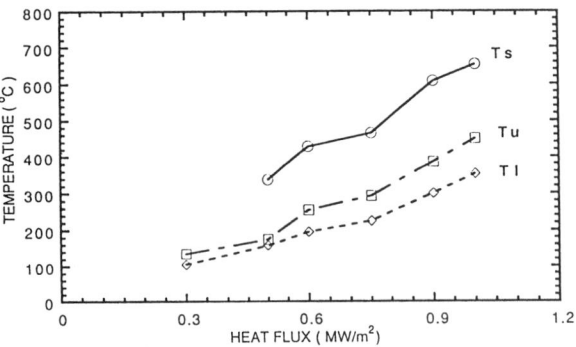

Fig.7 Relationship between the saturated temperatures and steady heat flux.

6. VIBRATION TEST OF REAL SCALE JM

Some kinds of mechanical vibrations due to cooling water flow or induced current of are one of the concerns for the divertor plates used in fusion devices. To solve this problem of the real scale MJ sample assembled with some bolts due to the mechanical vibrations, mechanical safety check must be carried out. Therefore, pulse mechanical vibrations were applied on the real scale MJ sample using a mechanical vibrator, which gives a vibration with an amplitude of 1m/s² and hybrid frequencies of 1000Hz and 160Hz. The vibrations were applied on the sample for 30,000 pulses in the sequence of ON for 10s and OFF for 20s. In spite of the large amplitude of vibration, no slipping of the tile against the backing plate, loosing of bolts used for fix and cracking of the graphite tile has been observed after the mechanical test.

7. CONCLUSIONS

Thermal properties of several kinds of MJ sample developed for divertor plates of LHD have been evaluated using the ACT. The results of heat loading tests for the compact size MJ sample indicate that a thin carbon sheet is very effective as a compliant sheet to improve the thermal contact at the interface. The real scale MJ sample with a copper cooling water pipe and a thin carbon sheet can be used under heat flux up to 0.75 MW/m² for steady state operation and heat flux up to 4 MW/m² for 20 seconds operation.

The most important issue to improve the thermal property of the large real scale MJ sample is to get and maintain a good thermal contact between copper backing plate and cooling pipe.

ACKNOWLEDGMENT

The authors appreciate the encouragement and support given to us by Dr. Junya Yamamoto, a professor of the device engineering division.

REFERENCES

1. A.Iiyoshi, M.Fujiwara, O.Motojima, et al., Fusion Technology, 17(1'990)169.
2. O.Motojima, K.Akaishi, et al., Engineering and Design, 20(1993)3.
3. J.Yamamoto, O.Motojima, T.Satow et al.,IEEE Trans. Magn., 27,2(1991)2220.
4. N.Ohyabu, T.Watanabe, Hantao Ji, et al., Nuclear Fusion, 34,3(1994)387.
5. N.Ohyabu, A.Komori, O.Motojima et al.,J.Nucl. Matr.220-222(1995)298.
6. N.Noda, Y.Kubota, A.Sagara et al., Fusion Technology,1(1992)325.
7. Y.Kubota, N.Noda, A.Sagara et al.,NIFS Internal Report, NIFS-MEMO-13(1994).
8. Y.Kubota, N.Noda, A.Sagara et al, NIFS Internal Report, NIFS-MEMO-16.
9. Y.Kubota, N.Noda, A.Sagara et al., Proceedings of the ASME, HDV-Vol.317-1(1995)159.

Reactivity test between beryllium and dispersion strengthened copper

N. Sakamoto[a], H. Kawamura[a] and R.R. Solomon[b]

[a] Oarai Research Establishment, Japan Atomic Energy Research Institute
Narita-cho, Oarai-machi, Higashi Ibaraki-gun, Ibaraki-ken, 311-13 Japan

[b] SCM METAL PRODUCTS, INC.
2601 Weck Drive, Research Triangle Park, North Carolina, U.S.A.

Beryllium has been expected to use as plasma facing material for the first wall and divertor on ITER. In this application, beryllium and copper alloy will be joined by some kinds of methods (diffusion bonding, brazing and so on). Therefore, we performed the out-of-pile reactivity tests with diffusion couples between beryllium and dispersion strengthened copper which is a first candidate as heat sink material in order to investigate their reaction behavior under joining process and fusion reactor condition. The summary obtained by this test was as follows;
- The reaction layer which consisted of two layers ($Be_2Cu(\delta)$ phase and $BeCu(\gamma)$ phase) was considered above 400°C, and additionally, very thick $Cu+BeCu(\gamma)$ phase which was generated with decomposition of $Be-Cu(\beta)$ phase, was observed at 700°C.
- Large cubical expansion due to $Be-Cu(\beta)$ phase formation was observed by EPMA. And, many cracks were generated in $Cu+BeCu(\gamma)$ phase. Therefore, it is considered that using of joined component between beryllium and dispersion strengthened copper should be avoided at temperature above eutectoid temperature(620°C).

1. INTRODUCTION

Beryllium has been indispensable for the fusion reactor material particularly. For example, beryllium has been considered as plasma facing material in the first wall and divertor (plasma facing components) against plasma. On the other hand, copper alloy has been proposed as heat sink material behind plasma facing components. In those using, they will be joined by some kinds of methods, for example, diffusion bonding, brazing and so on. However, disregarding their bonding methods, their materials will be heated at high temperature under the condition of fusion reactor designing. So in the present work, we noted on the chemical interaction between beryllium and copper alloys under the high temperature condition.

As for the interaction between beryllium and copper, Vickers et al.[1] conducted on the diffusion couple test at the temperature of 600°C for 2000h under vacuum atmosphere. As the results, they reported that a few layers were formed on copper side, and identified BeCu and Be_2Cu as the reaction products. Baird et al.[2] reported that they reacted from the temperature of 500°C, and the reaction layer was 18μm after annealing at 500°C for 29 days. From the results of Monroe et al.[3], in the condition of annealing at 760°C for 4h, beryllium diffused into copper side remarkably, and both of them were bonded. Recently, V.R. Barabash et al.[4] conducted on diffusion bonding test in the condition at the temperature from 800°C to 850°C for about 5 minutes. From the results, they reported that diffusion bonding could be probably used. Then the reaction layers were consisted of a few layers.

From these reports, though it was cleared that the interaction between beryllium and copper took place above 500°C, they were never clarifying about the generation of reaction products and increasing of the reaction layer. Therefore, for clarifying the elementary process of reaction between beryllium and copper, we performed the out-of-pile compatibility tests with diffusion couples of beryllium and dispersion strengthened copper which were inserted in the capsule filled with high purity helium gas (6N). Annealing temperatures were 300, 400, 500, 600 and 700°C, and annealing periods were 100 and 300h.

Table 1
Chemical compositions of beryllium and dispersion strengthened copper (DS-Cu) specimen

contents (wt%)	Be	BeO	C	N	Al	Fe	Mg	Si	Mn	Cr
Be	99.0	1.04	0.083	0.0045	0.053	0.072	0.035	0.0022	0.008	0.008

contents (wt%)	Cu	Al	Fe	Pb	B	Ni	Co
DSCu	99.50	0.26	0.0018	0.0006	0.016	<0.002	<0.0025

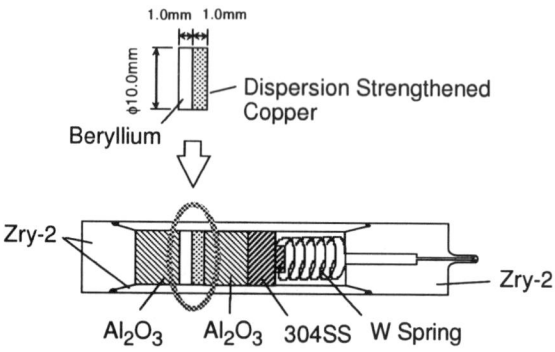

Figure 1. Outline of the capsule.

2. EXPERIMENTS

2.1. Specimens
2.1.1. Beryllium
The beryllium specimens were hot-pressed beryllium disks made by NGK INSULATORS, LTD.. The specimen dimensions were 10.0 mm diameter and 1.0 mm thickness. The purity of the beryllium specimens was almost 99wt.%. The chemical compositions of beryllium specimen are shown in Table 1. The main impurity of the beryllium specimens was beryllium oxide. Hot-pressed beryllium was wrapped with mild steel(SS41), rolled at 900°C, annealed at high temperature, and then polished until shining condition by ϕ1mm diamond on buff cloth. Cleaning of the beryllium specimens was done by ultrasonic washing with acetone.

2.1.2. Dispersion strenghtened copper
The copper specimens were dispersion strengthened copper disks made by SCM METAL PRODUCTS, INC.. The specimen dimensions were 10.0 mm diameter and 1.0 mm thickness. The chemical composition of dispersion strengthened copper specimen are shown in Table 1. The copper specimens were also polished and cleaned by the same method as beryllium one.

2.2. Procedure
The outline of the capsule is shown in Fig.1. A diffusion couple of beryllium and dispersion strengthened copper disk was inserted into the capsule made of Zry-2. Then, after filling the capsule with high purity helium gas(6N purity), the capsule was sealed by TIG welding. The diffusion couple was pressed down by tungsten spring to maintain the constant pressure(32.4 N/contacting area) at the interface. The helium leak test was carried out on the assembled capsule, and the leakage rate was less than $1.0 \times 10^{-9} Pa \cdot m^3/s$. The capsules were annealed in the electric furnace. The annealing conditions were 100 and 300h at the temperatures of 300, 400, 500, 600 and 700°C. After annealing, the interactions between beryllium and dispersion strengthened copper were evaluated.

As for the evaluation, the thickness of reaction layer was investigated by optical microscope and SEM analysis. Then, the reaction products(phase's formation) on the interface between beryllium and dispersion strengthened copper were identified by EPMA(Electron Probe Micro Analyzer). Here, accelerating voltage was 15kV, and electron beam diameter was about ϕ1mm.

3. RESULTS AND DISCUSSION

3.1. Observation of the specimen cross section
It was observed by SEM that reaction layer formed above 400°C by the interaction between beryllium and dispersion strengthened copper. And, the degree of the reaction became remarkable at 700°C. It appeared that observed reaction layers had a tendency to increase with increase of annealing temperature and time.

Authors had performed reactivity tests between beryllium and oxygen free copper[5] / copper-beryllium alloys with diffusion couples, and clarified that above mentioned reaction layers increased with control by mass transfer - namely, diffusion of beryllium and copper. However, in this test, it could not be appeareed that increase of reaction layer was controlled by diffusion.

3.2. Identification of reaction products

At first, it appeared that reaction layer which consisted of two layers (Be$_2$Cu phase and BeCu phase) formed above 400°C, and additionally, very thick Cu+BeCu(γ) phase formed at 700°C with quantitative analysis by EPMA.

SEM photographs of reaction layers formed between beryllium and dispersion strengthened copper after annealing at 700°C for 300h are shown in Fig.2. From these photographs, the reaction area consisted of three layers. Upper thin layer shown in position A was Be$_2$Cu phase (about 30μm), middle layer shown in position A and B was BeCu phase (about 300μm). And, lower layer shown in position C was very thick Cu+BeCu(γ) phase (about 1mm).

Cu-Be phase diagram is shown in Fig.4. Above mentioned Be-Cu(β) phase generates with eutectoid reaction (eutectoid temperature : 620°C). And, on interface of the specimens annealed at 700°C, a large number of cracks were observed in Be-Cu(β) phase. It is considered that they generated due to transformation of Be-Cu(β) phase to BeCu phase and Cu phase with large shrinkage. These cracks will have bad influence upon the joining strength between beryllium and dispersion strengthened copper on plasma facing component. Therefore, it is considered that formation of Be-Cu(β) phase should be avoided concerning diffusion bonding between beryllium and dispersion strengthened copper.

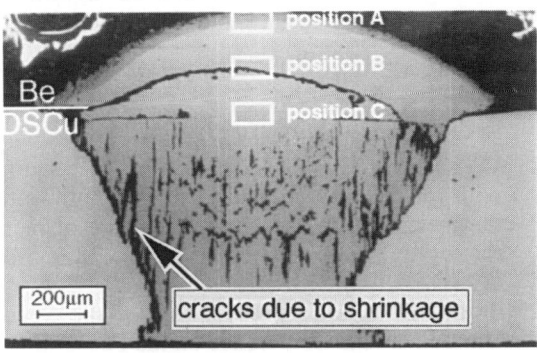

a) Reaction area

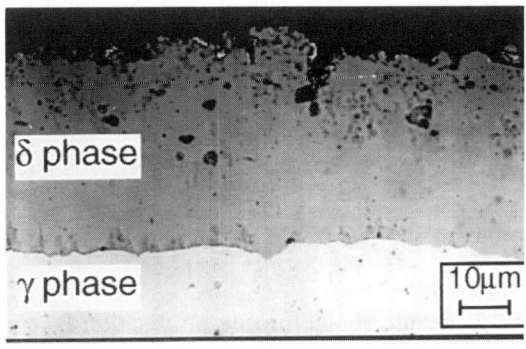

b) position A

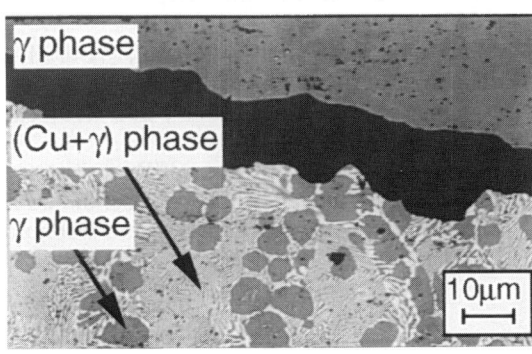

c) position B

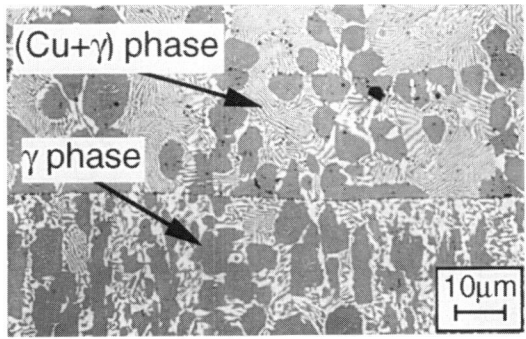

d) position C

Figure 2. SEM photographs of reaction layers formed between beryllium and dispersion strengthened copper after annealing at 700°C for 300h.

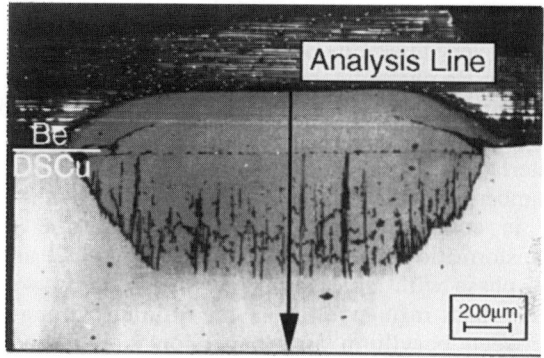

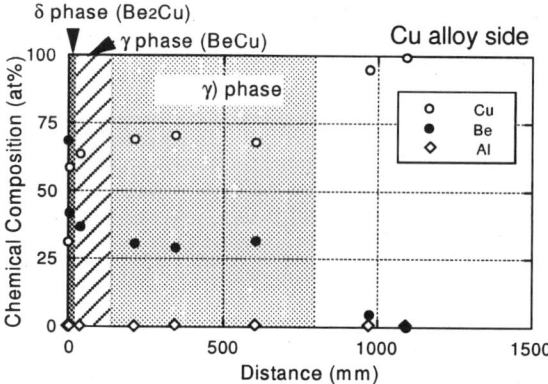

Figure 3. Result of line profile on reaction layers formed between beryllium and dispersion strengthened copper after annealing at 700°C for 100h.

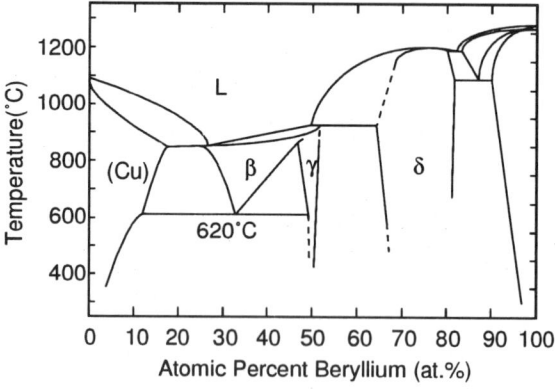

Figure 4. Cu-Be phase diagram.

4. CONCLUSION

We investigated the interaction between beryllium and dispersion strengthened copper, which were promising as fusion reactor material. The summary obtained by this test was as follows;

- The reaction layer which consisted of two layers ($Be_2Cu(\delta)$ phase and $BeCu(\gamma)$ phase) was considered above 400°C, and additionally, very thick Be-Cu(β) phase was observed at 700°C.

- Large cubical expansion due to Be-Cu(β) phase formation was observed by EPMA. And, many cracks were generated in Be-Cu(β) phase. Therefore, it is considered that using of joined component between beryllium and dispersion strengthened copper should be avoided at temperature above eutectoid temperature(620°C).

REFERENCES

1. W.Vickers et al., "The compatibility of beryllium with various reactor materials", Proc.Int.Conf.on the metallurgy of beryllium, institute of metals monogragh and report series, No.28,335-349 (1961).
2. Baird,J.D. et al., "Proceedings of the Second United Nations Conference on the Peaceful Uses of Atomic Energy, Geneva", **24**, (1958).
3. Monroe, R.E., Martin, D.C. and Voldrich,C.B., "U.S. Atom. En. Comm.", Rep. No. BMI-836, (1953).
4. V.R.Barabash, L.S.Gitarsky et al., "Beryllium-metals joints for application in the plasma-facing components", J. Nucl. Mater., **212-215**, 1604-1607 (1994).
5. N.Kawamura and M.Kato, "Reactivity test between beryllium and copper", Proc. of 2nd IEA international workshop on beryllium technology for fusion, 204(1995).

Fabrication of ITER first wall modules from powder by Hot Isostatic Pressing

A. Lind[a] and J. Collén[b]

[a]Studsvik Material AB, S-611 82 Nyköping, Sweden

[b]Studsvik EcoSafe, S-611 82 Nyköping, Sweden

Hot Isostatic Pressing (HIP) technology is presently the primary choice for the fabrication of the ITER first wall and blanket system. In other applications it has been demonstrated that the HIP route starting from powder, offers a good alternative to conventional production routes for the fabrication of complex components with cooling galleries. A comprehensive test program has been conducted in 1994-1996 by Studsvik Material and Studsvik EcoSafe in collaboration with the Swedish Institute of Production Engineering Research (IVF) and Powdermet Sweden. The program comprised the production of steel blocks with cooling channels as well as compound material of steel and different copper alloys.

1. INTRODUCTION

A flexible and cost effective method of producing the first wall and blanket modules for ITER is required and hot isostatic pressing, HIP, has been recognised as one possible alternative. HIP can be used as a joining method of blocks and/or plates of different or identical materials. HIP can also be used to consolidate materials from powder and to fabricate near net shape components of complex design with internal cooling galleries, Ref. 1. Powder consolidation on a block or a plate is another suitable method of joining copper (Cu) to stainless steel (SS).

2. MODULE DESIGN

The present modular design of the ITER first wall and shield blanket structure enables the utilisation of possibilities for flexible design by the HIP powder route.

The modules consist of a blanket part of stainless steel comprising a gallery of cooling channels. The channels are connected to a number of stainless steel tubes introduced as cooling channels inside a copper layer on top of the steel with the main purpose of conducting the heat away from the first wall. On top of the copper a layer of beryllium may be used as a plasma facing material. See Figure 1.

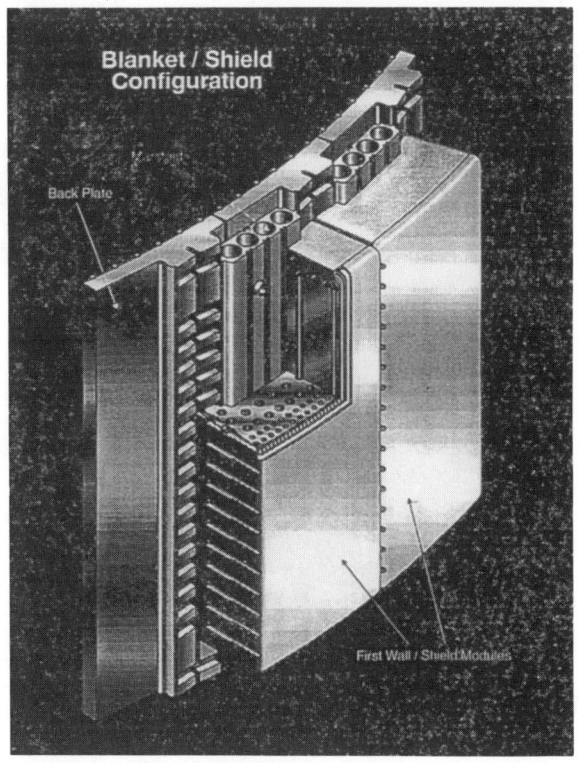

Figure 1. The ITER blanket module.

This design requires a fabrication procedure introducing cooling galleries inside materials as well as between them. Furthermore, the joining of the materials is an important issue. The utilisation of HIP from powder offers a technically interesting and economically attractive route.

3. THE POWDER HIP ROUTE

The main advantages of the powder HIP concept include design flexibility, uniform mechanical properties of the consolidated materials, elimination of leaks in welds, excellent joinability of materials to compound products, and competitive production costs. The main disadvantage is the limited number of test data available to qualify the materials produced by HIP.

The powder route includes the following steps (see also Figure 2).

- capsule design in CAD 3-D to fit module shape
- fabrication of internal tube inserts including helium leak testing and pickling of the produced gallery
- fabrication of the outer capsule designed in CAD 3-D, assembly and tight welding
- helium leak testing
- manufacture and testing of powder
- filling of capsule and a second leak testing, evacuation and sealing
- HIP'ing at 1100-1150°C (steel) 100 MPa, 3-4 h, rapid cooling
- ultrasonic testing of the material and control of tube location
- manufacturing of front surface and first wall tubing connections

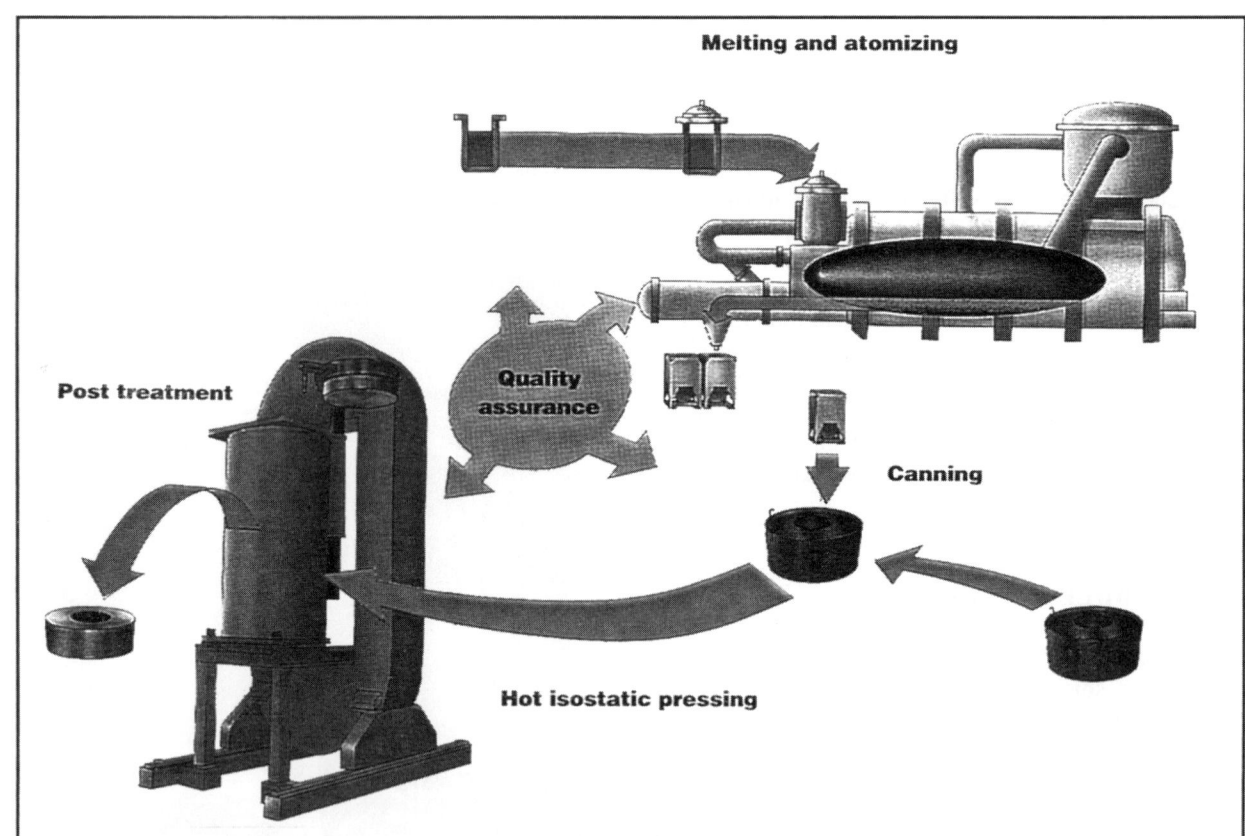

Figure 2. The powder HIP process.

Experiments were performed in 1994-1996 by Studsvik Material and Studsvik EcoSafe in collaboration with Powdermet Sweden and the Swedish Institute of Production Engineering Research (IVF).

4. SMALL SIZE COMPOUND MOCK-UPS

In the middle of 1994 a proposal was presented to fabricate the ITER compound component from powder and consolidate the steel and the copper powder in one HIP step at 1035°C. A demo mock-up was produced this way, see Ref. 2 and 3. After some modification of the copper alloy it was shown that such a process gives excellent bonding between the materials, tensile test data of the steel within the ITER specifications and only limited softening of the Cu alloy, due to grain growth.

For practical reasons the powder HIP route was divided into two HIP steps: one for the steel and one for joining and consolidating the copper alloys. After some initial tensile tests with material HIP'ed from 800°C to 950°C it was concluded that no fractures had occurred in the joint. 850°C was chosen for fabrication of a set of mock-ups. The mock-ups were made from 316LN IG steel produced by a conventional process as well as the powder route joined by HIP to copper powder Cu, 0.7Cr, 0.2Zr and Glidcop® Al-25.

5. MEDIUM SIZE COMPOUND MOCK-UPS

In the beginning of 1995 Powdermet Sweden AB produced two steel blocks (410x210x155 mm) by powder HIP. The blocks contained cooling channels of stainless steel. One block was carefully investigated by ultrasonic testing, tensile testing, metallography, chemical analyses, and direct measurements of the cooling channels, see Ref. 4. On the other block a layer (Cu, 0.7Cr, 0.2Zr), with cooling channels was HIP'ed from powder at 900°C, see Figure 3 and Ref. 5.

6. LARGE SIZE STEEL DEMO BLOCK

In the beginning of 1996 a blanket module with the dimensions 1250 x 650 x 250 mm and a weight of 1500 kg was produced by Powdermet Sweden AB in stainless steel. 316LN ITER grade gas atomised powder was filled into a capsule with internal cooling channels and HIP'ed at 1150°C for three hours at 100 MPa pressure, see Figure 4. A similar block will be produced and subject to a comprehensive measurement and test program.

Figure 3. The medium size Cu/SS mock-up.

Figure 4. The large size demo block.

7. SUMMARY AND CONCLUSIONS

When consolidating powder, shrinkage takes place. This shrinkage is predictable if the powder is well specified and kept within reasonable limits. The extent of the shrinkage depends mainly on the tap density of the powder which is a measure of the density that can be achieved by filling and vibrating the powder according to a special route. The tap density is influenced by the powder particle size and distribution. If an optimised HIP process is desired giving narrow tolerances with good reproducibility, the powder to be used has to be designed specially for HIP.

As in any high temperature process a diffusion process takes place. As a rule the rate of diffusion in solids decreases at low temperature and with large atomic radius. Surface and grain boundary diffusion have a higher rate than that in the bulk.

When joining materials a certain amount of diffusion is desirable in order to achieve good bonding between the materials. If an oxide or other particles or impurities are present at the interface, such diffusion is prevented and the bond will be inferior. One advantage when joining powder is that the total active surface for diffusion is larger compared to the joining of blocks or plates. Furthermore, the powder particles are heavily deformed during the HIP process breaking up surface oxides, if present, to a higher extent than if using blocks/plates. In all bonding techniques it is important to avoid the formation of brittle phases at the interface that will influence the strength of the joint in a negative sense. It is advantageous if the microstructure does not change too much in a region close to the interface which is the case after using melt processes like welding.

In conclusion, joining copper to stainless steel simultaneously consolidating the copper powder offers some advantages provided that the powder particle size and distribution as well as the composition are optimised.

ACKNOWLEDGEMENT

This work has been supported by Powdermed Sweden AB, Studsvik AB, the Swedish National Board for Industrial and Technical Development (NUTEK) and by the European Communities under an association contract between EURATOM and the Swedish Natural Science Research Council (NFR).

REFERENCES

1. A. Lind. Joining of copper alloys to stainless steel by hot isostatic pressing or by explosive welding. A review to the design. Studsvik Material AB, 1994 (STUDSVIK/M-94/52).
2. A. Lind and R. Tegman. First wall and shield components manufactured by hot isostatic pressing. Studsvik Material AB, 1994 (STUDSVIK/M-94/159).
3. A. Lind and R. Tegman. Fabrication, examination of the microstructure and mechanical testing of a Cu(Cr,Zr)-powder alloy joined to 316LN IG and 316LN stainless steels by a one-step hot isostatic pressing (HIP) technique. Studsvik Material AB, 1995 (STUDSVIK/M-95/90).
4. A. Lind. Demo block fabrication by powder HIP route. Studsvik Material AB, 1995 (STUDSVIK/M-95/41).
5. A. Lind. Joining of a first wall copper structure onto a stainless steel shield block by powder HIP. Studsvik Material AB, 1996 (STUDSVIK/M-96/116).

Development of silver-free bonding techniques and an investigation of the effect of cadmium in brazed joints made with silver based brazing alloys

A.T. Peacock, M.R. Harrison[1], D.M. Jacobson[1], M. Pick, S.P.S. Sangha[1], G. Vieider[2]

JET Joint Undertaking, Abingdon, Oxon, OX14 3EA, UK

[1]GEC Marconi Materials Technology Limited, Hirst Division, Borehamwood, Herts WD6 1RX, UK
[2]NET team, Garching bei München, Germany

1. INTRODUCTION

The bonding of Be to CuCrZr has been studied for a number of years at JET (1,2). The best and most developed solution to date is brazing with Incusil ABA, a silver based braze with a Ti active braze addition. High heat flux components brazed using this method have withstood critical heat fluxes of up to 18 MW/m^2, depending upon the thickness of the Be.

For ITER applications the content of the silver in the braze constitutes a dual problem, firstly the transmutation of Ag to Cd presents a radioactive hazard. This question is not addressed here but scenarios could be envisaged for the use of Ag containing braze for periods of low vessel activation. The other problem is the volatilization of Cd from the braze into the plasma. The magnitude of this problem is investigated here.

Many investigators have reported work trying to find alternative bonding techniques which can be used to bond Be to CuCrZr. This paper will detail efforts that have been made by GMMT and JET in this area. These bonding techniques have excluded other elements because of transmutation/volatility concerns. These elements include Au.

Two bonding techniques will be described and results of mechanical testing will be presented. Much discussion has centred upon the acceptability of mechanical testing as a criteria for judging bonding strengths. The test used by JET has been shown to be a very useful screening test and that the only testing that can be used to fully judge the methods of bonding is high heat flux tests. JET is at the present time proceeding with a test programme to high heat flux test the best bonding techniques demonstrated here.

2. Cd STUDIES

2.1 Braze foil preparation

The aim of the program is to study the effect of Cd in braze foils. It was therefore decided to produce foils by ingot casting and strip casting to allow as far as possible the Cd to be distributed throughout the braze foil. The level of Cd was also chosen to be in the region of 5 wt% so that the Cd would not segregate and be unrepresentative of an ITER braze. Two braze foils were produced with the following composition:

Ag 27.4 Cu 12.2Sn 3.32Cd 0.71Ti
Ag 27.0 Cu-12.6Sn 4.57Cd 1.08Ti

2.2 Fabrication and testing of assemblies

To test the braze CuCrZr/CuCrZr assemblies were produced to avoid the complicating effects of the use of Be. Assemblies were produced under identical vacuum and thermal cycle conditions used for brazing samples with Incusil [2].

The testing programme included several investigations.

2.3 Results

2.3.1 Joint microstructure

The joint microstructure is shown in Figure 1 for the 3.3% Cd braze. This is compared to the microstructure of a brazed joint made with Incusil ABA. The microstructure of both samples is similar and void free.

2.3.2 Joint strength

The joint strength for the different braze compositions is shown in table 1.

The joint strengths are much higher than for a CuCrZr/Be joint indicating that the limiting

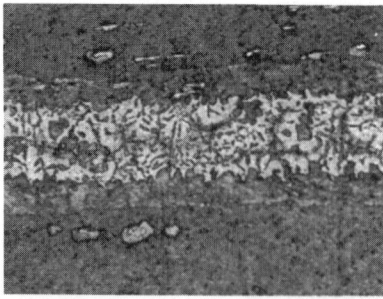
InCuSil ABA joint

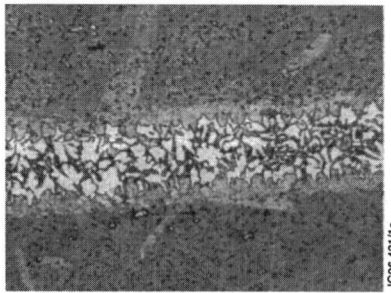
3.3%Cd containing braze

Figure 1. Microstructure of Cd containing braze compared to Incusil ABA

strength in those joints is determined by the Be or Be intermetallic strength. The more inconsistent values for the 4.6% Cd braze is consistent with the finding that these brazes show some porosity perhaps as a result of dross formation or from the volatilisation of the Cd.

Table 1
Joint shear strength for different brazes

Braze	Shear strength (MPa)
Incusil ABA	215,217 Ave 216
3.3% Cd	208,209,212 Ave 210
4.6% Cd	197,203,212 Ave 204

2.3.3 Volatility of Cd

The volatility of Cd from the braze foil was measured by weight loss measurements of braze foils after heating in vacuum and examination of brazed joints by EDAX (energy dispersive X-ray analysis) and atomic emission spectroscopy (AES) after heating in vacuum. Vacuum heating conditions were chosen as 300°C and 400°C, much higher than the predicted joint temperature under ITER conditions. One week exposure times were also chosen.

The % Cd lost from a 100μm thick 3.3% Cd braze foil after vacuum exposure for 1 week was 5% at 300°C and 27% at 400°C.

Attempts were made to collect the volatilized Cd during this period but the vapour pressure of Cd at room temperature (the temperature of the collector plate) is 3.10^{-9} Pa [3] which is sufficient to volatilise and lose all the Cd from the collector over the week of exposure to vacuum.

The braze foil tested in the first series of tests does not represent the geometry of brazed samples in high heat flux components. Therefore, the tests were repeated on brazed joints. These joints were then subsequently sectioned to expose a cross section that had been exposed to vacuum during heating but not during brazing itself.

Figure 2 shows an EDAX map of both Ag and Cd showing a region of Cd depletion near the surface. This particular result is complicated by a large Ti inclusion in the braze. These pictures show that the Cd is associated with the Ag in the braze, as might be expected from the phase diagram.

An attempt was made to quantify the region of depletion of Cd in the braze by performing line scans at different depths. The results of this experiment are shown in figure 3. A depletion depth of 20 μm is observed only for the edge of the sample that was exposed to vacuum during heating.

If it is postulated that Cd can only escape from the free surfaces then this would represent 0.13% of the Cd in the brazed sample with a joint 30mm x 30mm in area and 20% of the Cd in a braze foil. This final figure is quantitatively in line with the weight loss measurements performed upon the braze foils.

3. BONDING TECHNIQUES

Two new silver free bonding techniques for joining Be to CuCrZr have been investigated. These are copper/tin diffusion brazing and brazing with a CuMnSnCe braze.

3.1 Diffusion brazing

This technique is a development of a technique used extensively for bonding with Ag- and Au- based joining systems (Ag-Sn, Ag-In and Au-Sn) [4]. The technique is shown schematically in figure 4. The aim of the investigation was to identify suitable processing conditions for the copper/tin diffusion brazing system that would allow

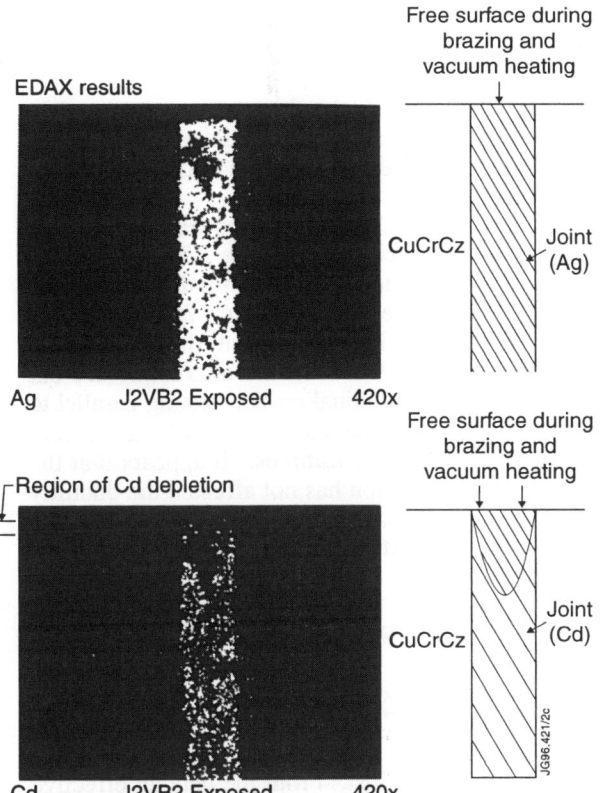

Figure 2. Edax map of Ag and Cd for a braze sample

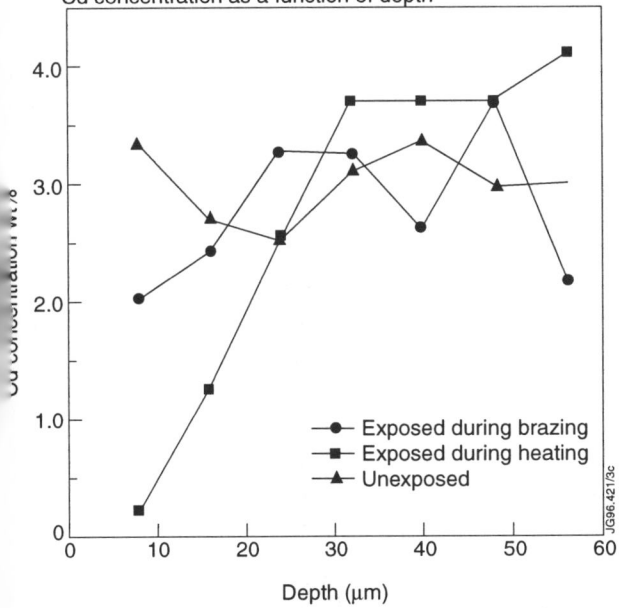

Figure 3. Cd as a function of depth after outgassing

strong joints to be formed. The parameters studied included Be metallization, processing time, processing temperature, Sn thickness and processing pressure.

3.1.1 Be metallization

Both electroplating and Ion plating of Be were assessed to apply a Cu metallization of the Be samples of the order of 20μm thick. Ion plating was found to be by far the better technique from the point of view of adhesion, giving shear test values for joints exceeding 150 MPa. Electroplated Be only achieved a third of this value.

3.1.2 Processing conditions

For the joining technique to work there has to be a minimum thickness of Sn to allow filling of the gaps between samples. If too little Sn is applied voids will be created; with too much it is difficult to diffuse away sufficient Sn to remove embrittling Cu-Sn intermetallic phases.

Therefore processing conditions were optimized to obtain a joint with sufficient Sn diffusion to create a seamless joint.

3.1.3 Testing of joints

Be/CuCrZr joints are assessed by metallography and shear testing. During the development of the technique a range of shear strengths have been measured corresponding to different failure modes. The different range of values all appear to be associated with a different type of failure. This information is shown in table 3.

3.2 CuMnSnCe brazing

This braze was developed as a low cost powder brazing alternative to silver brazing outside the Fusion field. The development undertaken here was to produce this braze as a foil and to use a composition that would allow induction brazing at the same conditions as for the Incusil ABA braze. This would allow the usage of a known Cu-Be

Table 3
Failure mode against typical shear strength

Shear strength	Type of failure
238 MPa	Fracture in the Be component
159 MPa	Failure in Be-Cu intermetallic zone
40 MPa	Failure initiates in residual Cu_3Sn particles

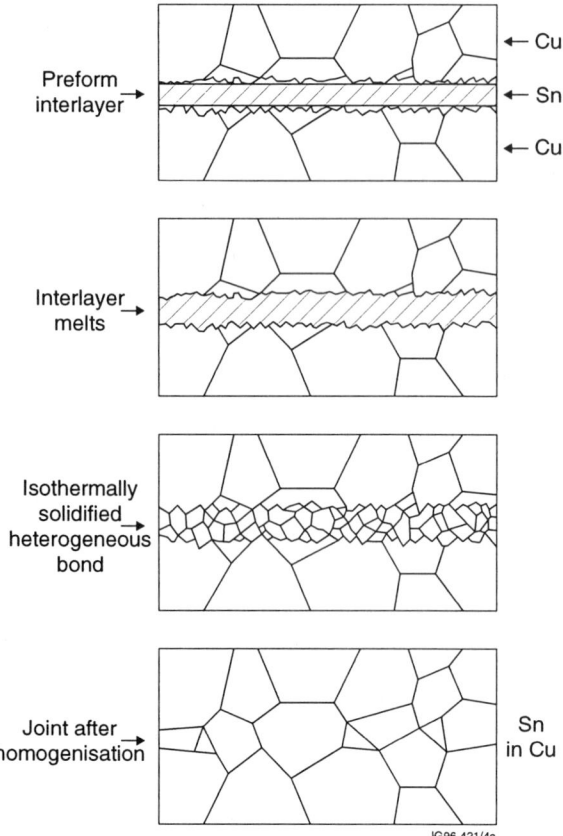

Figure 4. Schematic representation of the diffusion soldering process

intermetallic phase and the same over-aging of the CuCrZr. These temperature requirements gave a very limited range of compositions for the braze. Early tests using the braze gave reasonable strengths, of the order of 50-77 MPa. However, metallography of the joint showed evidence of some shrinkage cracks. The braze foil was very brittle and this gave an indication that brittleness of the braze was a problem. To reduce this problem it was decided to study a range of different compositions to establish the optimum. In particular, the Mn and Ce concentrations were reduced in the expectation that the joints would be less brittle. Also, the substitution of Ce by Ti was tried for the same reason. The original composition was Cu-30Sn-10Mn-1Ce and the new compositions tried were Cu-30Sn-10Mn-0.26, Cu-30SN-9Mn-1Ti and Cu-30Sn-6Mn-1Ti (Wt%). Foils were produced of these three alloy compositions 100μm thick. The melting range of all three brazes were measured to be in the temperature range 630-700°C (±5°C).

3.2.1 Testing of the brazes

The three treatments tried did not improve the room temperature ductility of the braze materials or of the joints.

CuCrZr/CuCrZr joints produced using these brazes demonstrated their viability. The conditions used were the same as for the initial composition. Figure 5 shows the microstructural features of this joint for composition. The joint shows to be an excellent joint, well filled, fine microstructure but with two fine longitudinal cracks running parallel to the entire length of the joint. Similar features were found on all the other samples. It appears that the change in composition has not affected the ductility of the braze and that by going to a CuCrZr/CuCrZr joint the solidification cracking has been exacerbated.

4. CONCLUSIONS

Under the conditions tried in this work the volatilization of Cd in a braze joint is diffusion limited. An upper limit of 0.13% of the Cd in any braze joint will be lost. From the data an effective diffusion coefficient has been estimated.

Cu-30Sn-9Mn-1Ce and the related brazes have been shown to exhibit satisfactory brazing characteristcs, but are limited by their inherent lack of ductility in the composition range studied.

Copper/tin diffusion soldering combined with Ion plated Be has been shown to produce good joints between CuCrZr and Be, as assessed by shear testing and metallographic examination. Further development is underway to produce high heat flux components using this technique. The technique has been shown to work equally well with CuCrZr and D.S. copper substrates.

REFERENCES

[1] H. Altmann et. al. SOFE 1993
[2] C. Ibbott et. al. SOFE 1995
[3] A. Roth, Vacuum Technology, North Holland, 1982
[4] D.J. Jacobson and S.P.S. Sangha, Soldering and Surface Mount Technology, No. 23, June 1996, pp. 12-15.

Thermal Fatigue Test of a Cu/ss Primary Wall Mock-up

M. Merola[a], P. Fenici[b], R. Scholz[b], B. Weckermann[b], S. Tähtinen[c]

[a] Politecnico di Torino, Dipartimento di Energetica, Italy
Present affiliation: The NET Team, c/o Max Planck-Institute für Plasmaphysik, D-85748 Garching, Germany

[b] Joint Research Centre, Institute for Advanced Materials, T.P. 750, I-21020 Ispra, Italy

[c] VTT Manufacturing Technology, Materials and Structural Integrity, P.O. Box 1704, FIN-02044 VTT, Finland

A primary wall mock-up was recently tested in the Thermal Fatigue Test Facility of the Joint Research Centre of the European Union, Ispra site. It was manufactured by High Speed. Tech. LTD Oy in collaboration with VTT Manufacturing Technology, Finland. It consisted of a Cu-Cr-Zr plate in which 6 cooling channels were obtained along the block length. Two steel plates, having a total depth of 5 mm, were explosion welded one after the other onto the surface opposite to the heated one. After that, type 316L steel cooling tubes were tightened into the channels by means of explosive inside tubes. The component was thermal fatigue tested for 13,000 cycles at a maximum heat flux of 570 MW/m^2. The mock-up completed the fatigue test without any problems and without the appearance of any macroscopic defect.

1. INTRODUCTION

During the ITER EDA a new primary wall concept was developed. It is quite different from the ITER CDA design as far as the materials and the manufacturing technologies are concerned. In the CDA design both the heat sink and the cooling channels were made of 316L stainless steel (ss). The tube-to-plate joining technique was brazing. The plasma facing surface was covered by tiles made of carbon fibre reinforced carbon composite.

Presently, the heat sink material is copper mainly because of the much higher thermal conductivity. The alloys envisaged are either dispersion strengthened (DS-Cu), or precipitation hardened (PH-Cu). In the former alloy Cu is reinforced with Al_2O_3 particles. Among the several PH-Cu alloys, one of the main candidate is the Cu-Cr-Zr. Unfortunately, Cu alloys show a drastic reduction in fracture toughness even under very low doses of neutron irradiation. Furthermore the higher erosion rate, if compared with ss, caused by the water flow rate inside the cooling channels, results in an appreciable radioactivity of the coolant. Therefore, at least for the primary wall module, it was decided to made the cooling tubes of ss. The joining of the Cu plate with the ss tubes is a serious problem tackled by the ongoing R&D program. In fact both DS-Cu and PH-Cu reach their optimum strength after cold working but the high temperature employed in the common joining techniques (that is welding or brazing) may change drastically the microstructure of the Cu alloy reducing the strength appreciably. Furthermore in Cu-Cr-Zr alloys, chromium particles, which are the major source of their strength, are likely to dissolve at temperatures in excess of 500 °C, that is in the typical range of the most common joining techniques. As regards DS-Cu, alumina particles may segregate in the melt layer during welding thus destroying the uniform particle spacing which is necessary for high strength. However, due to the high thermal stability of aluminium oxide, if melting is not reached, no significant softening occurs in the base material at usual brazing temperatures (600-950 °C) [1].

The present chapter deals with a first primary wall mock-up manufactured according to the ITER EDA design and tested in the Thermal Fatigue Test Facility of the Joint Research Centre of the European Union, Ispra site.

2. DESCRIPTION OF THE COMPONENT

2.1. Geometry and manufacturing

The primary wall mock-up was manufactured by High Speed Tech. LTD Oy in collaboration with VTT Manufacturing Technology, Finland. It consisted of a Cu-Cr-Zr plate (dimensions 163x100x40 mm) in which 6 cooling channels were obtained along the block length. Two steel plates, having a total depth of 5 mm, were explosion welded one after the other onto the surface opposite to the heated one. After that, type 316L ss cooling tubes were tightened into the 6 cooling channels by means of explosive inside tubes. The tube diameter and thickness was 12 and 1 mm, respectively (fig. 1). The distance between the channel axis and the heated surface was 16 mm. No additional heat treatment was carried out after the manufacturing process therefore high residual stresses can be expected at the joint interface. A few micron depth black-chromium coating was deposited onto the heated surface by means of a galvanic process to increase the absorptance and thus to increase the heat flux effectively absorbed by the component.

Fig. 1 - The VTT component

2.2. Ultrasonic testing

To study the integrity of the explosion welding joints, a C-mode scanning acoustic microscope operating at frequencies up to 100 MHz was used. As far as the interface between the Cu and the ss backplate is concerned, the examination showed some discontinuities at the end of the plate and a rather good bonding in the remaining part where only isolated 1-2 mm in diameter defects were detected.

Two out of six tube-to-copper interfaces were also examined from the ss back-plate. However ultrasonic examinations gave information only along a single line along the joint. There was no bonding between copper and the ss tubes within about 20 mm from both the ends of the tubes. The rest of the tube is mainly bonded even if a certain number of defects could be detected.

3. EXPERIMENTAL CONDITIONS

The component was thermal fatigue tested for 13,000 cycles at a maximum heat flux of 570 MW/m^2. Fig. 2 shows the absorbed heat flux during the test. The full power period of the thermal cycle lasted 24 s. The water coolant velocity inside the cooling channels was 4.4 m/s and its bulk temperature was 12 °C. The computed heat transfer coefficient was about 17 000 W/m^{2}K.

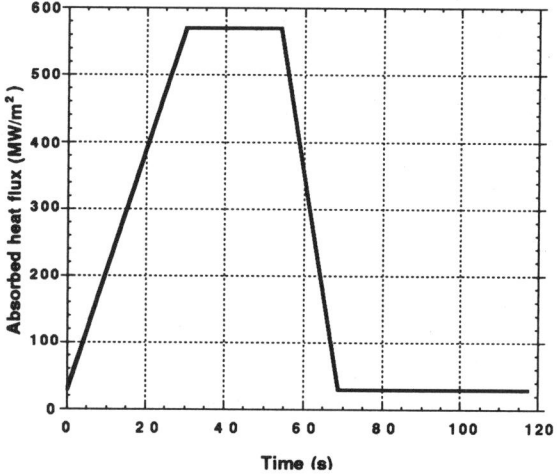

Fig. 2 - Absorbed heat flux during the thermal fatigue cycle

4. NUMERICAL ANALYSIS

4.1. Introduction

The numerical analysis of the experiment was performed by means of the ABAQUS finite element code. A 2D analytical model was adopted. It consisted of 2562 nodes and 804 eight-node elements. Half a model was analysed because of the geometric and experimental symmetries. A linear variation of the material properties vs. temperature was taken into account in the computations between each pair of figures.

Starting from the steady-state temperature at the minimum heat flux, a transient thermal analysis of two cycles was performed to reach the cycle-to-cycle steady-state temperature. The elastic stress analysis was performed in the "steady-state" cycle, at the end of the heating period and at the end of cycle. The generalised plane strain model (bending allowed) was used to compute stress. The thermal strain reference temperature was 12 °C. In other words at this temperature the thermal strain was null. It is worth pointing out that as no thermal treatment was performed after the explosion welding, rather high residual stresses can be expected in the joints. Since their value is unknown, they could not be taken into account in the calculation. Therefore the stress field obtained by the numerical analysis may differ from the real one near the joints. On the other hand, since the yield stress is not exceeded, the computed stress range is still meaningful as well as all the considerations on the fatigue lifetime.

4.2 Results

Fig. 3 shows the maximum surface temperature during a thermal fatigue cycle. It reaches a value of 90 and 20 °C at the end of the full power period and at the end of the dwell phase, respectively.

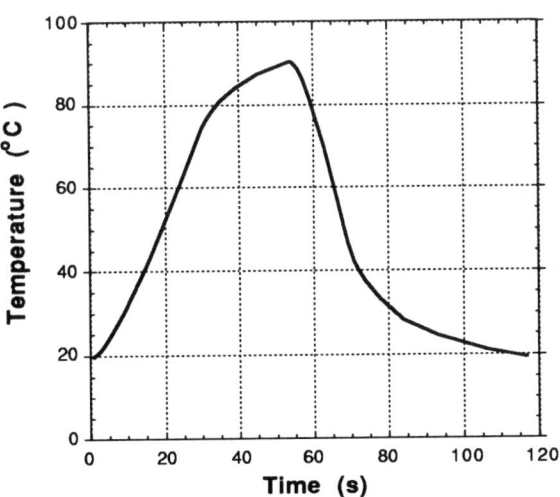

Fig. 3 - Maximum computed temperature vs. time

As far as Von Mises equivalent stress is concerned, a maximum figure of 35 and 158 MPa was computed on the heated surface and on the ss cooling tube, respectively. If one consider the fatigue curve of both CuCrZr [2] and ss [3], an infinite lifetime is expected.

5. EXPERIMENTAL RESULTS

The component was instrumented by means of three thermocouples located along the longitudinal midplane at 5, 10 and 15 mm below the heated surface, respectively. Fig. 4 shows the experimental readings during a thermal fatigue cycle.

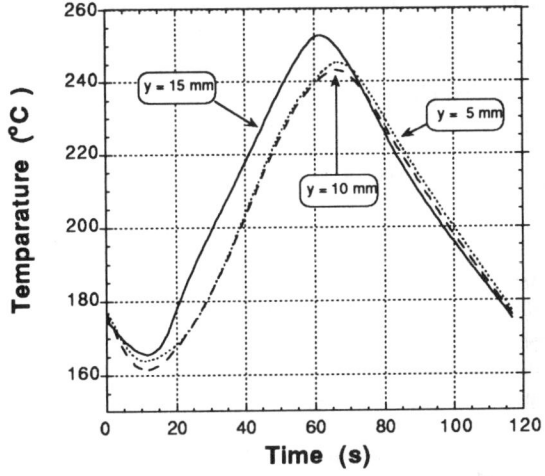

Fig. 4 - Temperature measurements
(y = distance from the heated surface)

Three main considerations can be drawn from the analysis of the temperature readings. At first one can note that the thermal inertia of the component is appreciably higher than that numerically computed. In fact, because of the very high thermal conductivity of the Cu alloy, the temperature field should have followed the heat flux variation according to a much lower time constant. The second point is that all the three thermocouples gave a temperature about 160 °C higher than the maximum computed temperature even if they were located well below the heated surface. The last consideration is the anomalous behaviour of the temperature field that showed a higher temperature in the deepest location from the heated surface.

All these points can be explained by assuming quite large debonded areas in the copper-tube joints. As a matter of fact, this was revealed by the non-destructive examinations and is here confirmed. The high variation of the thermal resistance between the copper plate and the cooling tubes, results in a higher overall thermal inertia of the mock-up, in a higher temperature field and in an unpredictable distortion of the thermal field. Taking this into account some doubts can arise on the relevance of the thermal stress analysis since the computed thermal field is not representative of the real situation.

It is worth pointing out that the temperature field did not show any shift towards higher values during the thermal fatigue test and therefore the debonded regions should not have grown during the test. Furthermore the mock-up could complete the foreseen number of cycles without any problems and without the appearance of any macroscopic defects. Therefore the defect tolerance of this concept proved to be quite high.

6. SUMMARY AND CONCLUSIONS

The results of the thermal fatigue test on a Cu/ss primary wall mock-up were presented. This was the first mock-up manufactured with the explosive bonding joining technique which was therefore not yet fully developed. The ultrasonic examinations carried out before the test showed the presence of some defects in the Cu-to-tube and the Cu-to-ss backplate joints. However the component could withstand the foreseen 13,000 fatigue cycles at 570 MW/m^2 without any problems and the joint defects did not seem to have propagated. In fact the temperature readings have not shown any shift to higher temperatures during the test.

As far as the thermal stress analysis is concerned, it is worth pointing out that a very low stress field was computed and thus an infinite lifetime was predicted. One important conclusion of the test is the appreciable defect tolerance of the concept which makes this design solution very robust from a thermal fatigue stand point.

7. ACKNOWLEDGMENTS

The contribution of the Polytechnic of Turin was supported by the Institute for Advanced Materials of the Joint Research Centre (contract No. 11121-95-07 F1ED ISP I).

REFERENCES

[1] B.N. Singh, *Assessment of physical, mechanical and technological properties of first candidate copper alloys*, ITER R&D Task No. T7 (1994CTA), 1994.
[2] J. Stubbins et al., *High temperature fatigue testing of CuNiBe, CuCrZr and CuAl25 at Univ. of Illinois at Urbana-Champaign*, http://jfs2.ne.uiuc.edu/~kevin/.
[3] R. Matera, *AISI 316 reference book*, Comm. Europ. Commun., JRC Ispra, TN No. I.07.B1.84.62, May 1984.

Technologies of joining between ITER reference grade beryllium and copper alloys by diffusion bonding process

E. Visca[a], G. Ceccotti[b], B. Riccardi,[a] G. Mercurio[b]

[a]Associazione EURATOM-ENEA sulla Fusione, CR Frascati, C.P. 65 - 00044 Frascati (Roma)

[b]ENEA - C.R. SALUGGIA - S.P. per Crescentino - I-13040 Saluggia, Vercelli, Italy

A suitable diffusion bonding process for manufacturing the high heat flux components of ITER have been developed. The process parameters for defining the bonding technology are reported. A shear strength up to 150MPa was obtained by using an interlayer of electrolytic copper deposited on the activated beryllium surface (silver-free joint). After selecting the best process, medium-scale mockups of high heat flux components for testing on a electron beam facility were manufactured. The actively cooled mock-ups had a 50x30x8mm beryllium armour (castellated and non-castellated), with two kinds of heat sink material (Glidcop Al25 and CuCrZr alloy).

1. INTRODUCTION

This activity was carried out in the frame of the ITER task T221-3. The objective of the task was to produce joints between Glidcop and beryllium and copper-chromium-zirconium alloy and beryllium by means of diffusion bonding by replacing silver, used as interlayer in the previous task (1), with a different metal. The choice of silver had been prompted by the fact that this metal can be bonded with itself by diffusion over a wide range of temperatures and hence it makes it possible to select a low enough temperature at which the diffusion of beryllium is virtually nil.

The migration of beryllium, in fact, is extremely harmful, as it gives rise to the formation, from the metals in contact at the interface, of intermetallic compounds, which drastically reduce the mechanical strength of the joints.

2. CHOICE OF THE INTERLAYER

The choice of the interlayer was conditioned by the need to make the joints at temperatures designed to prevent - or at least reduce to the minimum - the diffusion of beryllium. This automatically ruled out all those metals which posses good mechanical properties but have high melting points. For such metals, in fact, the temperature necessary to trigger diffusion processes are high and will also cause the diffusion of beryllium, which as is known, gives rise to the formation of intermetallic compounds in all elements, save for aluminium, silicon and zinc. In metals, in fact, diffusion phenomena become appreciable at a temperature of about half the temperature of the relative melting points. On the basis of the foregoing, the choice of the metal to be used as interlayer in the joints between Glidcop and beryllium and Cu-Cr-Zr alloy and beryllium fell on copper, both on account of the similarity between its mechanical, physical and chemical properties and those of the two alloys to be bonded with beryllium, and because of its relatively low melting temperature (Mp= 1084.87°C). Anodic etching is performed to eliminate the oxide layer on the beryllium surface and at the same time activate the surface and prepare it to receive the electrolytic copper deposit with good adherence.

The mechanical properties of the joints essentially depend on the tenacity with which the copper electrolytic deposit adheres to the beryllium, since the temperatures and duration times of the heat treatment have been selected so as to maximise as much as possible diffusion bonding at the interface between the copper electrolytic deposit and Glidcop or the Cu-Cr-Zr alloy.

3. CHOICE OF OPERATING CONDITIONS

Whilst the literature offers extensive information on the use of silver as interlayer to make joints between different metals, there is no information whatsoever on the use of copper.

t only proved possible to collect some data on the rate of diffusion of copper, in copper, starting from a temperature of 650°C. Although the rate of diffusion at 650°C is quite modest, this temperature is high enough to trigger diffusion phenomena involving the beryllium. Furthermore, in addition to the Glidcop-beryllium joints, the task also included the study of the joints between beryllium and the Cu-Cr-Zr alloy, but the latter cannot be subjected to sustained heat treatments at a temperature of over 460°C. This alloy, in fact, is marketed after under going a heat

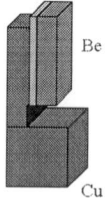

Fig. 1. Shear test sample (Cu alloy/Interlayer/Be tile)

treatment, referred to as ageing treatment, which causes the formation of precipitates in the matrix so as to enhance its mechanical strength characteristics. For the reasons described above, the upper temperature limit for diffusion bonding tests was fixed initially at 450°C and subsequently lowered to 420°C. The effect of pressure is to cause local plastic deformations in the surface and hence to level the irregularities created by machining operations so as to move the atoms making up the surfaces to be joined closer together until the atomic forces intervene enabling the diffusion phenomena to take place.

Since the plasticity of metal is a function of temperature, it follows that, in order to obtain the desired surface levelling, the lower is the temperature, the higher must be the pressure applied. Applied pressures were 98 and 147 MPa, respectively. Some preliminary tests showed that with a heat treatment time of 30', all traces of the original interface between the alloys and the electrolytic copper deposit on the beryllium disappeared. Hence, a heat treatment time of 30' was adopted as a fixed parameters in all tests performed.

In order to determine the optimal thickness to obtain the most satisfactory mechanical properties of the joints, the following thickness ranges were tested:

5-8 μm   10-15 μm   22-26 μm

To assess the shear strength of the joints obtained between beryllium and the alloys, Glidcop and Cu-Cr-Zr by adopting the testing parameters already specified, we used specimens of the shape illustrated in fig. 1.

After the electrolytic coating with copper of the beryllium, the specimen is placed inside a furnace fitted with a hydraulic piston which makes it possible to apply an evenly distributed axial load to the joint to be welded. All operations were performed in a vacuum of over 10^{-4} bar. Stay time at the testing temperature was determined from the instant the pre-determined temperature was reached.

4. PRELIMINARY INSPECTIONS

The measurements performed with the roughness meter produced curves displaying a regular evolution interrupted every now and then by discontinuities arising from the passage of the probe over surface defects. Measurements carried out by means of suitable instruments on beryllium tiles sized 30×50×8 mm revealed marked planarity defects (difference between two opposing tile corners between 200 and 900 μm).

The thickness of copper deposit on beryllium was tested on 100% of the specimens by means of the Micro-Derm apparatus.

5. ULTRASONIC INSPECTION ON THE JOINTS

These checks were performed on 100% of the specimens with joints subjected to shear tests and on the actively cooled elements. From an examination of the C-scan obtained from the Ultrasonic Inspection (U.S.) tests on the joints and through special investigation methods it proved possible to ascertain that, probably due to the cutting operations performed to produce the 10×15×8 tiles from bigger sized pieces, the edges of many specimens had become rounded and hence a certain portion of the surface adjacent to them, even after the copper coating, would not come into contact with the surface of the alloys to which it had to be joined by means of diffusion bonding. Needless to say, the outcome of this is a percentage reduction in the welded surface. Naturally enough, this has resulted in a decrease in the effective values of mechanical strength.

6. RESULTS OF THE SHEAR TESTS

The diagrams in figs. 2,3,4 represent the evolution of the mechanical strength of the joints subjected to shear test (bond strength) as a function of interlayer thickness, temperature and pressure, respectively. Each point of the charts represent the average of the mechanical strength obtained from the shear tests and the number between brackets shows the quantity of shear tests.

Figure 2 shows that the mechanical strength of the joints increases with increasing copper coating thickness from 5 to 12 μ.

Figure 3 illustrates the influence of temperature on the mechanical strength of the joints. It is seen to be highest at the highest temperature in the chosen range (420°C).

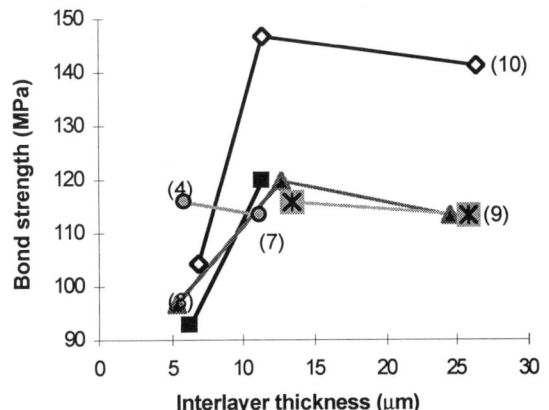

Figure 2. Bond strength of the shear samples vs Interlayer thickness

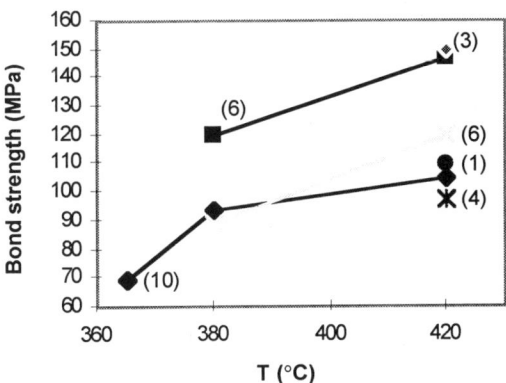

Figure 3. Bond strength of the shear samples vs temperature of joining

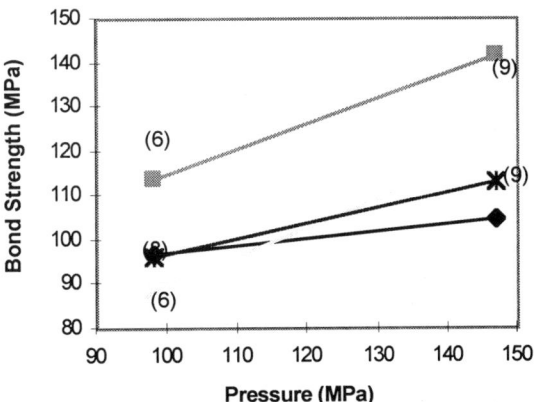

Figure 4. Bond strength of the shear samples vs Pressure applied on the samples

Figure 5. Photo of the actively cooled mock-ups

Figure 4 shows that when the diffusion bonding treatment is performed at higher pressure, the mechanical strength of the joints increases accordingly. Based on the foregoing, this effect is strictly connected with the testing temperature.

7. ACTIVELY COOLED MOCK-UPS

Actively cooled mock-ups were manufactured (see Fig. 5) on the basis of the following process parameters:
Pressure: 98 MPa
Temperature: 420 °C
Time of stay at established temp.:   30'
Interlayer thickness   15 μm

Thermal cycles were always conducted according to the procedure described above.
The choice of a testing pressure of 98 MPa was affected by the consideration that the Glidcop test pieces of the shape envisaged for the manufacture of the dummy elements (dimensions of 50×30×19 mm, area of the surface to be joined to the copper coated beryllium tile by diffusion bonding to be 1500 mm^2), when heated to a temperature of 420°C and tested at the higher pressure level (147 MPa), undergo considerable deformations.

It was decided to use the same pressure of 98 MPa for both the dummy elements made of Glidcop and for those made of Cu-Cr-Zr so as to have a uniform criterion for the interpretation of the results of the heat flux irradiation test to which they will be subjected.

The results of the ultrasonic tests conducted on all the actively cooled mock-ups showed joints with no relevant defects.

8. THERMOMECHANICAL ANALYSIS

Steady State thermo-mechanical analysis was performed with incident heat fluxes of 5 MW/m^2 and of 12 MW/m^2.
A 2D model has been used for the thermal Finite Element calculations assuming a water sink

temperature of 28°C, cooling tube diameter 10mm, water pressure of 4.0 MPa; water velocity of 15 m/s and a swirl tape twist ratio of 2.

The heat transfer coefficient has been estimated by a subroutine during the FE calculations using the Sieder-Tate correlation. In fact for the heat flux level the heat exchange is convective. The influence of the swirl tape has been taken into account introducing the idraulic diameter of the duct between the tape and the tube surface and increasing the water velocity to handle the swirl effect. The non linear analysis was carried out by means of ABAQUS code.

The maximum temperature of the 4mm thick Be tile non-castellated remain under the 350°C for the case of $5MW/m^2$ and reach 960°C for the case of $12MW/m^2$.

The basic assumptions of the stress analysis were to consider the structure stress free at 28°. The Be tile configuration was analyzed by means of a generalized plane strain model which allows the thermal expansion non-castellated Beryllium was considered. The stress results were found to be up to 50MPa for the case of $5MW/m^2$ and about 150MPa for the case of $12MW/m^2$.

The tests that will be performed in the JUDITH facility of KFA at Julich foresee a first phase of screening tests to verify the behavior of the mock-ups in a steady state and a second phase in which they will be tested under thermal fatigue. The results of the analysis show that the mock-ups should stand to the foreseen high heat flux tests in steady state condition.

REFERENCES
1. G. Ceccotti, L. Ingegneri, L. Magnoli, G.Mercurio, S. Sabbioneda, "Solid state bonding of beryllium to DS COPPER", TASK PPM-5.3 Final Report, 1995
2. Mye "Low temperature solubility of Cu in Be, in Be-Ac and Be-Is using ion beams", Metall, Trans A, 8 (1977) pag. 609-616
3. "Phase diagrams of binary beryllium alloys", Edited by H. Okamoto and L. Tanner, Schwartz: Modern Metal Joining Techniques., Wiley Interscience editor
4. "Solid State Welding", Metals Handbook Ninth Edition vol. 6 pag. 672 - 691

IN-VESSEL CRYO PUMP FOR ASDEX UPGRADE DIVERTOR II

B. Streibl, S. Deschka, O. Gruber, B. Jüttner, P. Lang, K. Mattes, G. Pautasso, J. Perchermeier, K. Schippl[a], H. Schneider, U. Seidel, W. Suttrop, G. Teller[a], M. Weissgerber

Max-Planck-Institut für Plasmaphysik, D-85748 Garching, Germany

[a]ALCATEL, Kabelmetal Electro GmbH, Kabelkamp 20, D-30179 Hannover, Germany

This paper summarises major design features and manufacturing aspects of the modular cryo pump now being installed. The cryo feed and the LN2 shielding of the module connections are described in detail. The cryo pump is exposed to considerable magnetic flux changes during major plasma disruptions. Upper limits of the transients are estimated for the thermal energy quench and subsequent current quench. The models developed are based on logarithmic flux functions for circular cross-section and on sheath boundary layer losses.

1 INTRODUCTION

Installation of the 100 m^3/s cryo pump (CP) has been started in September 1996. The CP will condense deuterium on a cryo panel at 4.4 K. The coolant is saturated LHe circulated by an immersed centrifugal pump with magnetic bearings. For pumping H$_2$ the coolant will be sub-cooled to 3.6 K. For gettering He four toroidally equidistant gas inlet valves will supply argon to be frosted on the cryo panel prior to a discharge.

For the pump opening a LN2-cooled chevron was chosen to provide safe shielding of the LHe cryo panel against energetic plasma exhaust particles in the eV range. The CP is positioned between the passively stabilising conductor (PSL) and the outer vessel wall. The PSL, originally a saddle coil, screening only horizontal magnetic flux changes with a time constant of 0.5 s, is meanwhile equipped with two 0.3 mΩ resistors (in parallel) toroidally connecting the current return paths (PSL bridge). Now, as a consequence of the bridge re-sistors, also vertical magnetic flux changes are screened with a time constant of 25 ms. This considerably reduces the induced voltages on the CP.

2 DESIGN OF THE CRYO PUMP.

Module design: Details can be found in a previous paper[1]. Here only major aspects are recalled and new developments commented. The CP cross-section is shown in Figure 1. The chevron, reflector and cryo panel are made of austenitic steel. Originally inconel tubes with steel fittings were envisaged for the cryo panel meander, but cracks on the welded steel-inconel interfaces were detected in metallurgical investigations.

To increase the pump opening, an L-shape was chosen for the chevron and the opening directed towards the vessel wall. The chevron is blackened by browning (iron oxides). The resulting emissivity (ε = 0.65) is rather small since the layer thickness which can be chemically achieved is less than the 300 K radiation wavelength[2]. The reflectors are therefore sandblasted on their inner surface to absorb part of the 300 K leakage radiation.

The chevron and reflector are mechanically joined in a way that allows free toroidal thermal expansion. They are supported via two cantilever beams per module. One of the cantilever beams permits toroidal sliding. The cryo panel meander is suspended on three equidistant stiffeners of its reflector shield via a thin-walled framework. The 6 tubes are clamped by a ceramic block. The toroidal degree of freedom is blocked only in the midplane of a module.

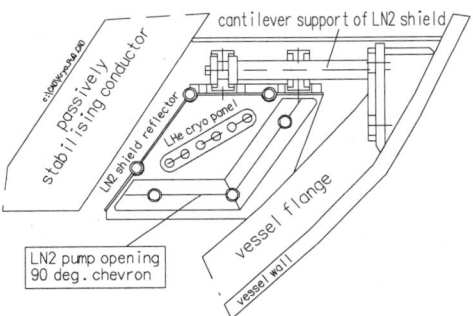

Figure 1. Cross section of cryo pump

Manufacturing tests: Apart from dimensional controls on a template, leak tests and electrical tests were conducted. Leak testing is performed throughout with GHe at a pressure of 13 bar. First all welds were leak tested immediately after completion. Then the assembled modules had to pass a high-voltage test. The breakdown voltage against the reflector for the ceramic cryo panel supports is above 6 kV and for the Vespel-insulated cantilever beam supports above 12 kV. Final leak testing is performed in a vacuum oven, first at 150 C and then at LN2 temperature.

Cryo feed: The cryo feed is shown in Figure 2. Potential separation is realised on the lower flanges of region 8 and on the weld-joined ceramic pieces in section 6. Regions 6 to 8 have their own vacuum system. Regions 5 to 2 belong to the vessel vacuum, where no super insulation was permitted. The He lines are guided there inside a shell onto which the LN2 go and return pipes are welded to the top and bottom. The shell is cut in the midplane to avoid thermal stresses between the cold go and warm return pipes. During cool-down the axial thermal contraction of the cryo feed reaches 10 mm. To keep this movement away from the rather stiff corrugated connections to the first CP module the cryo feed is axially fixed to the vessel and the thermal movement is taken up by bellows in region 6. Vertical displacements of up to 5 mm originate during vessel baking (150 C) in region 3, but do not require special design precautions.

LN2 shielding of module joints: The diagnostic openings require a gap of about 20 cm between adjacent LN2 shields. This gap is toroidally closed by two lids bolted onto the LN2 shield end flanges. Each lid thus has a non-actively cooled and unsupported length of 10 cm. Over this distance the incident radiation heat has to be transported by heat conduction and the eddy current forces taken up by bending. Since thermal and electrical conductivity are closely linked, a low temperature gradient has to be paid for with a high bending stress. As a compromise, the thickness of the connecting steel lids was increased. Compared with the reflectors (2 mm) their plate thickness (4 mm) was doubled. A further problem is the heat transfer over the bolted flange joint. Tests have shown that compared with the raw steel-steel joint the temperature difference can be reduced by a factor of about two with a copper interlayer brazed onto the steel lid. The lids are mounted inside the vacuum vessel, after weld-joining of the module connection pipes.

Cooling circuits and cool-down: All cryogenic cooling circuits are connected in series. The 4 module connections of the LHe cryo panel and LN2 shield (reflector and chevron) encircle the diagnostic paths. The connecting pipe sections are made from corrugated tubes to take up the thermal dilatation during cool-down (4 mm between modules) and to achieve a high electrical resistivity to keep induced currents low. Forces due to disruption induced currents are only a problem for the LN2 tubes. The lateral displacements of these tubes are therefore additionally blocked by supports in the middle of the diagnostic region. The supporting clamps are fixed on the LN2-shield lids.

Four hours of LN2 cool-down time are envisaged. During the first hour the coolant temperature will be linearly reduced from ambient to LN2. Tests have shown that temperature control by gas-liquid mixing is not stable enough and that an electrical heat exchanger is the appropriate solution. After the LN2 cooling front has passed though all chevrons and reflectors the final cool-down to 4.4 K is started by circulating LHe. The expected LHe consumption of the cryo panel is 10 W during stand-by, 20 W during a D_2 discharge, and 50 W during a discharge with a reactor-relevant mix of D_2 and He (3 %). In addition, 10 W of transfer losses is expected.

3 PLASMA DISRUPTION

During plasma disruptions the breakdown voltage on insulated parts can drop to ≈120 V. Upper limit estimates for the induced voltages on the cryo pump are thus required for the events of thermal energy quench (TEQ) and current quench (CQ).

The large total resistivity of the cryo pump (CP) (R_cp = 70 mΩ) permits the induced electrical potential differences (Figure 5) to be distributed favourably over the gap between the end modules (U_gap, Fig. 3) and the insulated parts between the

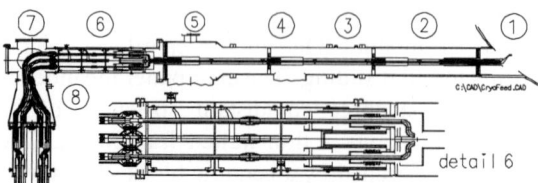

Figure 2. Cryo feed to the vacuum vessel interior

cryo panel/reflector and reflector/vacuum vessel. The means for this are the resistors R_cs bridging the Vespel-insulated cantilever supports (Figure 1). With R_cs ≈ R_cp the gap voltage would reach 50% of the CP loop voltage (Ul_Cp). However, since insulated parts are more sensitive to arcing than air gaps, U_gap ≈ 0.6 Ul_cp will be chosen, which is obtained with R_cs ≈ 2 R_cp. This choice then provides equal maximum potential differences on all insulated components. The larger value of R_cs also entails lower induced currents in the corrugated LN2 module connections (R_ct). A further reduction of these currents is achieved by electrically joining the reflector and chevron at the module extremities.

The reduction of breakdown voltage by dielectric materials was demonstrated in the Berlin plasma simulator ∥ and ⊥ to a magnetic field (B). Measurements with a 16 mm spark gap near the plasma boundary ($n_e \approx 2 \cdot 10^{18}$ m^{-3}) have shown: compared with the empty spark gap ∥ B, the insertion of insulation reduces the ∥ B breakdown voltage by a factor of ≈4. Perpendicular to B the reduction factor is ≈2. The lowest breakdown value found was ≈1000 V.

Thermal energy quench (TEQ): Plasma disruptions start with a sudden loss of thermal energy (TE). Measurements have shown for the electron temperature (Te) decay time values of down to $\tau_{NV} \approx 0.3$ ms. The resulting pressure drop, described by ßp(Te), is the main cause of the inward shift of the plasma column. Additionally, a sudden flattening of the current profile due to MHD instability will occur. The resulting drop of internal inductance per unit length (li) releases magnetic energy, which in turn raises the plasma current(Ip).

For worst-case estimates the time dependence of Te, ßp, and li were approximated by a Gauss function ($\tau_{NV} = 0.3$ ms) of unity amplitude. The electron temperature was varied between Te1, given by ßp1 = 1 and Te2 = 20 eV (ßp2 ≈ 0). For the variation of li three extreme cases were assessed, all starting at ßp1 = 1 with a parabolic current profile li1 = 1: a) only ßp variation: Δßp = -1, b) only li variation: Δli = -0.5, c) superposition of cases a) and b). Experimental experience is only compatible with Ip rises in between cases a) and b). However, for computing an upper limit to the induced voltages the worst case c) will be taken.

Computations were performed with the simplified geometry of Figure 4, which permits all circuit inductances to be expressed in terms of logarithmic flux functions[3]. The model comprises the circuit equations for plasma, PSL, and the lowest two vessel harmonics. The shift of the plasma centre (Rp) is taken into account by the equilibrium condition assuming a negative decay index of the external field (nv ≈ - 0.9 for b/a = 1.6)[4].

Figure 5 shows the loop voltages resulting for the worst case c) on the PSL centre (Ul_Psl), the cryo pump centre (Ul_Cp), and the vessel wall at point A (Ul_Ve). It is obvious that flux conservation by the PSL and vessel wall effectively protects the CP. Without the PSL resistor bridge the resulting voltages would have been a factor of ≈3 larger on the CP and a factor of ≈10 on the PSL.

After ≈0.5 ms the plasma column bumps into the heat shield with a radial velocity of ≈ 200 m/s. Thus on the flux surface ro = ap a voltage of $U_\pi \approx$ 400 V is induced by the toroidal field (B_t = 1.7 T). This voltage results from the integral $\mathbf{E} = \mathbf{v} \times \mathbf{B_t}$ along a field line taken over half of the poloidal

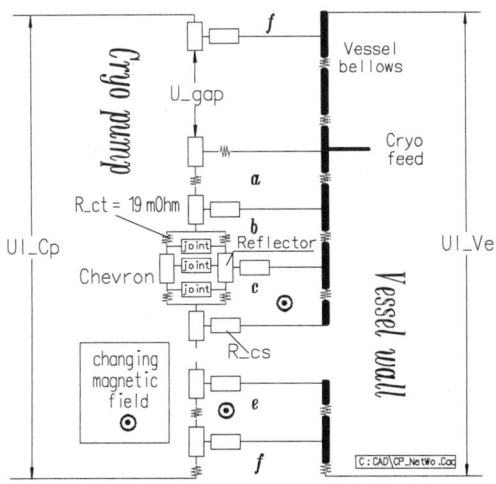

Figure 3. Network for CP potential control

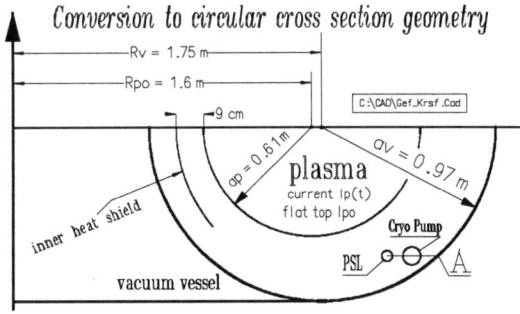

Figure 4. Geometry taken for computations

circumference. It cancels over closed flux surfaces ($U_{2\pi} = 0$), but drives poloidal halo currents Ih_p over open flux surfaces and the vessel structure. The same holds for the consecutive downwards movement resulting from the loss of vertical stability due to the suddenly increased distance between the plasma column and PSL.

Current quench (CQ): The CQ is mainly driven by the Te dependence of the plasma resistivity. Therefore the description of the average Te is based on a physical model for estimating the maximum Ip change rate (Ip'_max). With a sheath model[5] ($P_{sh} \propto ne \cdot t\_b \cdot Te^{3/2}$) sizeable resistive losses ($P_{res} \propto Ip^2 \cdot Te^{-3/2}$) can be transported over the boundary layer (BL) at a typical electron density of $ne = 5 \cdot 10^{19}$ m^{-3}, provided that the BL thickness (t_b) approaches ap. In combination with a BL model Te is thus found via the power balance $P_{res} = P_{sh}$ for temperatures above the radiation level of light impurities (10 to 20 eV).

Circular geometry can be maintained for the BL by decomposing the toroidal current according to Figure 6. The resulting dipole component then feeds Ih_p and produces a horizontal magnetic field B_dp, directed oppositely to the external one. The zero harmonic represents the measurable toroidal current (Ip) and thus defines the q-value inside the BL. With an upper limit of the current density $jpo = Ipo/(ap^2 \pi)$, constant in time and space, all quantities can easily be integrated over t_b and related to the flat-top (Index o) value Ipo.

The maximum values of the halo current and vertical vessel force (Fz) are then found near Ip/Ipo ≈ 0.6. The maximum of the resistive voltage

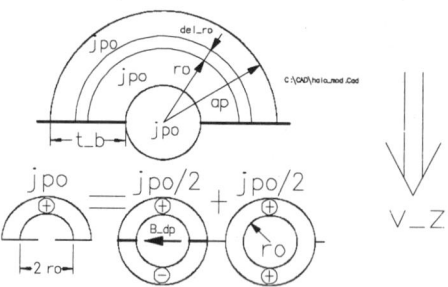

Figure 6. Boundary layer concept for CQ

(U_res) appears near Ip/Ipo = 0.5. For the first two quantities the results are Ih_p_max/Ipo = 0.28 and Fz_max = 0.83 ap Ipo B$_t$/qo (≈ 700 kN for design values and qo = 2). The largest Ip' follows from Ip'_max = U_res_max / Lp_min, where Lp_min ≈ 1 µH is the minimum plasma inductance derived from the flux between the inner vessel wall and PSL. With the sheath model no larger negative Ip' value than Ip'_max ≈ |300 MA/s| can be explained at a reasonable impurity level (3% carbon at flat top, 10% at t_b = ap). The resulting Ip'_max value is almost independent of Ipo since the power balance requires $Te \propto Ip^{2/3}$. Experimentally, typical negative values of Ip'_max ≈ |500 MA/s| are found around Ip/Ipo = 0.7 in combination with intense radiation losses.

According to Figure 5 the CP loop voltage after the thermal energy quench is always between Ul_Psl and Ul_Ve. Since the negative Ip' also reaches a level above |500 MA/s| for the assumptions of Figure 5 it is obvious that during the current quench there is also a sufficient safety margin for the CP against arcing.

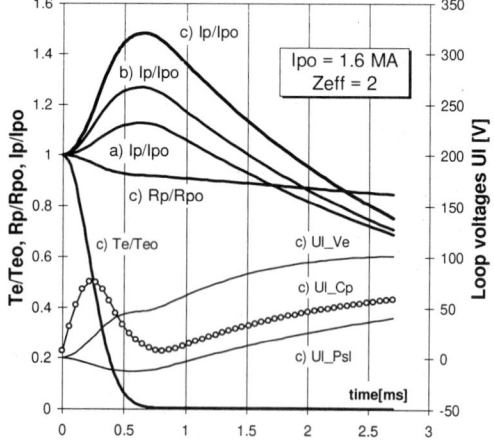

Figure 5. Ip, Rp, Te, loop voltages during TEQ

References

[1] B. Streibl, S. Deschka, G. Hofmann, K. Mattes, J. Perchermeier, H. Schneider, S. Schweizer, M. Weißgerber, ASDEX Upgrade Team, Divertor II for ASDEX Upgrade, 16 SOFE, 1995

[2] T. Ageladarakis, W. Obert, Measurements of the emmisivity by means of a Dornier selectometer, internal JET note, 1996

[3] J. Raeder, IPP 4/174, 1979

[4] V.S. Mukhovatov, V.D. Shafranov, Plasma equilibrium in a tokamak, Nuc. Fusion 11, 1971

[5] M.F.A. Harrison, Impurity control and its impact upon start-up and transformer recharging in NET, Proceedings of the 7th Course of the Intern. School of Fusion Reactor Techn., Erice, 1985

The Model & Experimental Basis for the Design Parameters of the JET Divertor Cryopump Protection System including variations in Divertor Geometry & First Wall Materials

P. Ageladarakis, S. Papastergiou, D. Stork, H. Van der Beken

JET Joint Undertaking, Abingdon, Oxon OX14 3EA, UK

ABSTRACT

A large In-Vessel Cryopump has been operational for several years as part of the JET Pumped Divertor. A model has been developed to analyse the behaviour of the Cryopump under all possible operating or fault conditions, and a suitable Protection System has been defined and implemented. It protects the Cryopump system during baking, "restart", normal operation, glow discharge cleaning, loss of vacuum, loss of cryogens or water flow (in the associated water cooled components). It ensures that no excessive thermal stresses are generated and no freezing or boiling occurs in the neighbouring water cooled components.

The model has been validated through a series of experiments and applied to other Cryopump devices at JET. A freeze-up incident that took place at the early operating stages of the LHCD Cryopump was well simulated.

The In-Vessel Cryopump Protection System has been adapted to respond to changes in the First Wall Material (C, Be) and Divertor Geometry.

1. INTRODUCTION

The large In-Vessel Cryopump system (Figs 1 and 2) has been operational for several years and greatly assisted the JET experimental programme [1], [2]. It demonstrated active density control of the plasma, and contributed in the production and operation of detached plasmas and high fusion performance [3].

The Cryopump system has been highly reliable, while the incorporation of a Cryopump inside the Vacuum Vessel, introduced no significant restrictions in the operations of JET. This was achieved because in the design, manufacture and installation phases, special attention was paid not only to the normal operation requirements, but also to accident scenarios. Under normal operation, the Cryopump has to cope with considerable thermal stresses and eddy currents. In addition, the system should be protected against accidents/abnormal events, like loss of water flow, cryogen flow and/or vacuum. During these events thermal stresses and water freeze-up or boiling (due to the hot Vacuum Vessel) are possible.

The introduction of special design features minimised thermal stresses and removed the need for controlled cool-down or warm-up of the system and for restrictions in the Vacuum Vessel temperatures during baking or normal operation [2].

Despite all these efforts, the Cryopump system is not inherently safe and a Protection System has been designed and implemented. This Protection

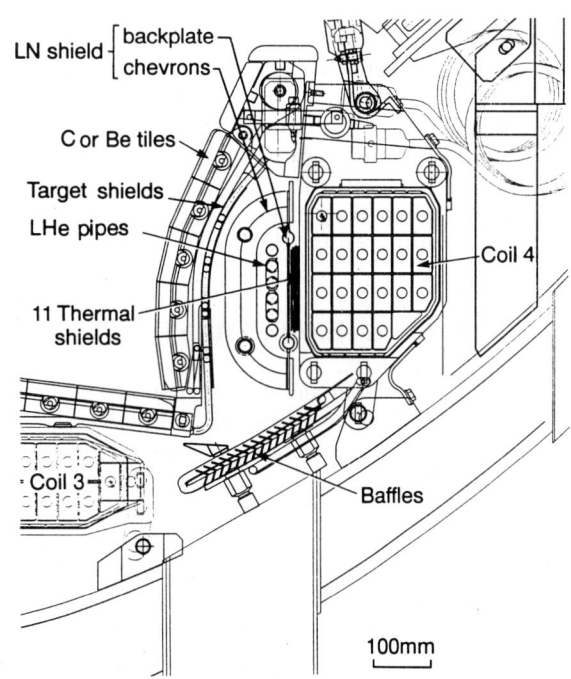

Fig 1. Cross section of the JET MK1 Divertor Cryopump system and its surrounding components

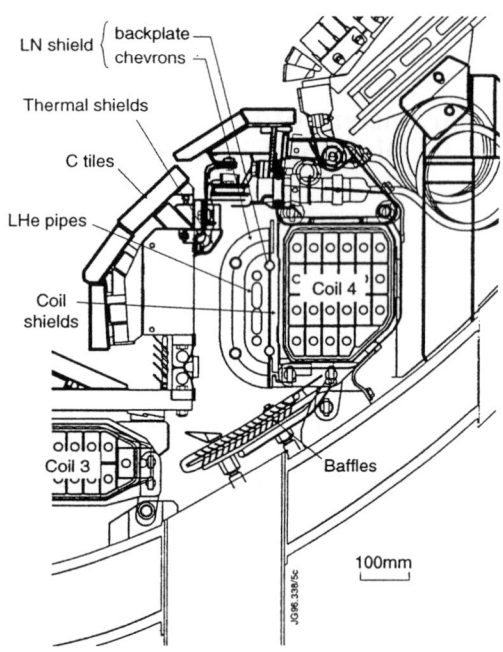

Fig 2. Cross section of the JET MK 2 Divertor Cryopump system and its surrounding components.

System assures that no excessive thermal stresses are generated during abnormal events and controls the draining and refilling of the water circuits within specific time constants, in order to avoid water freeze-up or boiling which may stress the pipework and lead to component failure. The equations that determine the system behaviour during the abnormal events are general, developed from first principles and therefore they can account for variation in the Divertor Geometry, First Wall material and can be applied to other Cryopumping devices at JET (LHCD Cryopump).

2. MODEL & EXPERIMENTAL BASIS

The mathematical model which simulates the system behaviour during the abnormal events together with the experimental validation have been published in Reference 1. The equations are basically heat transfer equations which calculate the bulk transient temperatures of the several system components. They account for all three modes of heat transfer and are solved with a step-by-step integration method. High levels of agreement have been achieved between the model predictions and experiments done either inside the Vacuum Vessel or in a test tank, outside the Machine to allow the safe simulation of abnormal events [1].

Due to the nature of the equations involved, the model can be adjusted to predict the heat transfer behaviour of other Cryopumping (or general heat transfer) devices at JET.

Figure 3 shows the JET LHCD Cryopump. The water circuits of this components are in danger of freeze-up or boiling (from the hot Launcher) should the water flow fail and the system is not drained fast enough. During the early stages of commissioning of this Cryopump a freeze-up incident occurred and the system took a very long time to recover.

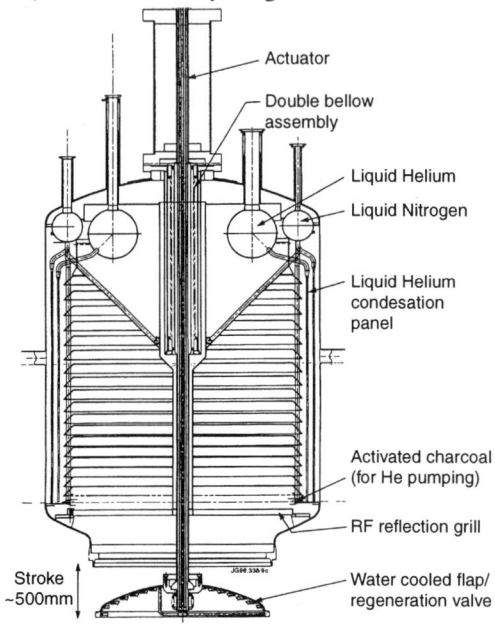

Fig 3. The JET LHCD cryopump.

Figure 4 shows the simulation of this event by the heat transfer model. There is good agreement between the predicted and measured temperatures of the Liquid Nitrogen (LN_2) shield during the long recovery of the system.

Figure 5 indicates in detail the predicted freeze-up time constants during the incident and shows that freeze-up events can be very fast indeed.

3. THE DIVERTOR CRYOPUMP PROTECTION SYSTEM

Following computer simulations, using the heat transfer model, of all possible accident events in the Divertor Cryopump components, a suitable control Protection System has been implemented. This acts as follows:

a) It does not permit operational modes which could potentially result in excessive thermal stresses.

433

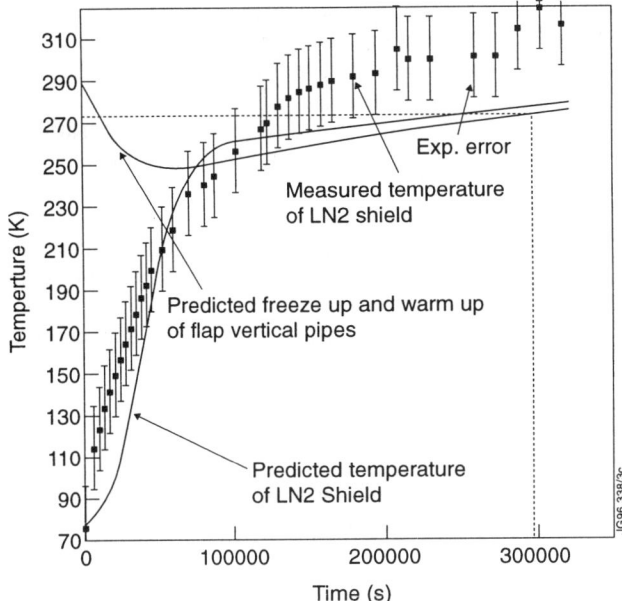

Fig 4. Comparison between model prediction and the LHCD cryopump system behaviour during a freeze up incident.

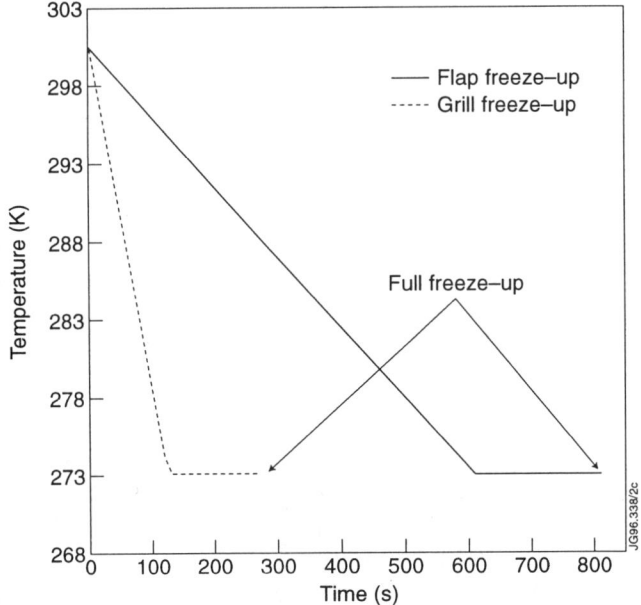

Fig 5. Predicted freeze up of Flap and Grill of LCHD cryopump.

For example, it does not allow the cryopump cooldown, prior to the water introduction inside the Vacuum Vessel, if the Vessel temperature is higher than 100°C.

b) It requests stop of LN_2 flow after a period of time, following a complete loss of water flow inside the Vacuum Vessel, in order to avoid potentially excessive thermal stresses in the LN_2 circuit of the Cryopump

c) It controls draining and refilling of the water circuits within specific time constants to avoid water freeze-up or boiling with loss of water flow.

3.1 Effect of First Wall Material

During the 1994-95 Experimental Campaign, JET operated with two First Wall Materials, C and Be. The emissivity and other properties of Be differ significantly from C. Therefore the mathematical model simulated all abnormal events which are influenced by the different properties of Be. The new First Wall material affected mainly the target shields (Figure 1) behaviour and freeze-up time constants altered. With Be, freeze-up was much faster (within ~200s) while with C considerably slower (~1000s), in the case of loss of water flow and bad vacuum. This was expected since the high emissivity of the hot C delays the onset of freeze-up.

It was thus necessary to recalculate all new accident time constants and make sure that the draining and refilling of the water circuits was compatible with these new restrictions.

3.2 Divertor Geometry

During the 1995-96 shutdown a new Divertor Geometry was installed inside the Vacuum Vessel. This new geometry affected the heat transfer equations of the Protection System. For instance different view factors and different component emissivity in the boundaries affect the exchanged radiated power. In addition it was realised that thermal shields, Fig 2, would significantly improve the system protection against freeze-up incidents. These thermal shields were indeed incorporated and designed to withstand eddy currents and maximise the heat transfer benefit.

With the new In-Vessel Geometry (Fig 2) all abnormal events were re-evaluated and found that:

a) Baking of the systems is not affected, when compared with the MK1 configuration. (Fig 1).

b) The Cryopump and Baffles operational behaviour do not alter significantly i.e. the Cryopump can be baked and operated, following water introduction inside the Vessel, with Vessel temperatures up to 350°C without any restrictions. The Cryopump can withstand any accidents under these conditions.

c) Following loss of water flow in the Louvre, water boiling can occur within ≥6 min depending on the boundary conditions

d) After loss of water flow in the Baffles, freeze-up can occur within ≥20 min (with Vessel at 20°C) or boiling after ≥ 12 min (with Vessel at 320°C) depending again on the boundary conditions.

After the 1996 Experimental Campaign, JET will again alter the Divertor Geometry through a remote handling In-Vessel intervention due to the high radiation levels of the vacuum vessel after the D-T experiments in 1996-1997. The Divertor Cryopump Protection Control System is able again to account for such a change in the Divertor Geometry.

4. DRAINING AND REFILLING OF WATER CIRCUITS

The Divertor Cryopump Protection System requests draining of the water circuits after loss of water flow, to prevent water freezing or boiling. This draining is rather difficult due to the complex In-Vessel Geometry, which incorporates horizontal pipes in parallel. An analysis has therefore been undertaken to determine the minimum percentage of a horizontal pipe cross-section that needs to be drained in order to avoid plastic deformation of the pipe if water freezes.

Figure 6 indicates that practically irrespective of the pipe diameter, thickness and strength of material, more than 10% draining of a horizontal pipe is safe and results in no plastic deformation of the pipe following a freeze-up incident.

Refilling with water is done with the equipment in low temperatures to avoid water boiling and thermal stresses. The hot Vacuum Vessel or the hot LHCD Launcher can raise quickly the temperature of the drained circuits. In the absence or failure of temperature measurements, in these circuits (for instance in the Baffles (Figure 2)), refilling can be permitted only after lowering the Vacuum Vessel temperature. Such an action can result in a major loss of experimental time and undesirable thermal cycling of the Torus. Therefore our computer model, was used to calculate the time constants under which, following a draining action, refilling is permitted. In addition gas cooldown of the circuits to be filled, prior to the water introduction, without significant reduction in the Vacuum Vessel temperature, has been quantified in order to minimise loss of experimental time and to reduce thermal cycling of the Torus.

CONCLUSIONS

The mathematical model developed to determine operational safety of the Divertor Cryopump has been applied successfully to other Cryopumping devices at JET.

This model provided the basis for the JET Divertor Cryopump Protection Control System and accounted for variations in the Divertor Geometry and First Wall material.

This Control System prohibits operational modes which may result to high thermal stresses during abnormal events, like loss of water and/or cryogen flow or loss of vacuum.

In addition it calculates the time constants which control the draining and refilling of the water circuits to prevent water freeze-up and boiling.

The effectiveness of protective actions like draining has been analysed and quantified.

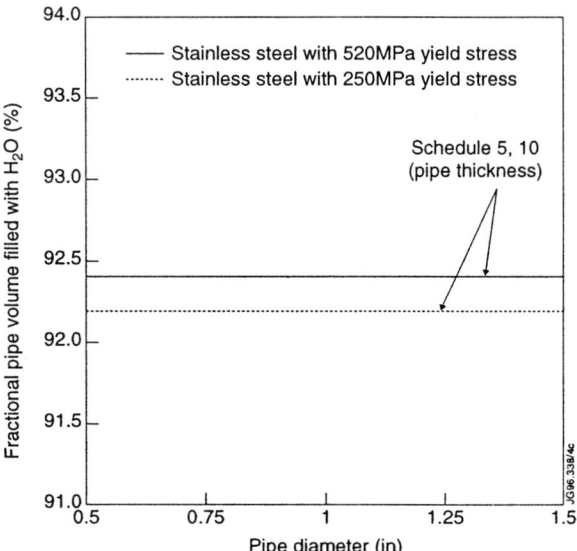

Fig 6. Maximum allowable horizontal pipe cross sections filled with H20 (%) so that plasticity is prevented in the case of a water freeze up incident.

REFERENCES

1. S. Papastergiou, P. Ageladarakis, W. Obert, E.Thompson: Operational Safety of the JET In-Vessel Divertor Cryopump System. Proceedings of Symposium on Fusion Technology, SOFT 18, Karlsruhre, 1994, pp 343-346.
2. S. Papastergiou, W. Obert, E. Thompson: The JET In-Vessel Divertor Cryopump. Design, Manufacture, Assembly, Testing and Operational Safety. Fusion Engineering and Design; to appear.
3. G. Saibene et al: Effect of Active Pumping and Fuelling on Divertor Plasma Discharges in JET. Proceedings of 22nd European Physics Society Conference Bournemouth 1995. Part 2, p121.

Upgrading of the Tore Supra Ergodic Divertor

L. Doceul, A. Grosman, P. Deschamps, B. Bertrand, J.P. Cocat, J.J. Cordier, L. Garampon, L. Gargiulo, Ph. Ghendrih, M. Lipa, R. Mitteau, H. Viallet

Association EURATOM-CEA, Département de recherches sur la Fusion Contrôlée
Centre de Cadarache, 13108 Saint Paul lez Durance Cedex, France

Tore Supra operation in the ergodic divertor configuration has allowed to give evidence of many assets of this plasma-wall interaction control scheme and essentially plasma decontamination and obtention of controlled radiative edge layers. The main aim of the Tore Supra ergodic divertor upgrading is to increase the heat exhaust capability of its six internal modules. The conductive power exhaust capability by the neutralizer plates will be extended from less than 1 to 3 MW, whereas the front face will accommodate up to 6 MW for 30 s. The main interaction zone of the divertor modules with plasma particles is located between the conductors where the particles drift towards the neutraliser plates made of strengthened copper tubes covered with a thin B_4C (150 μm) coating. The radiative and charge exchange power is mainly removed by the front faces of the water cooled module casings which are covered by CFC bolted tiles. The particle control will be allowed by the vented structure of the neutraliser plates.

1. OBJECTIVE

The aim of the Tore Supra ergodic divertor transformation is to increase the power removal capability of the six modules for plasma radiation and particle interactions. The interaction zone of the divertor modules with plasma particles is located between the toroidal bars where the particles drift towards the neutralizer plates made of copper tubes (Fig.1) [1].

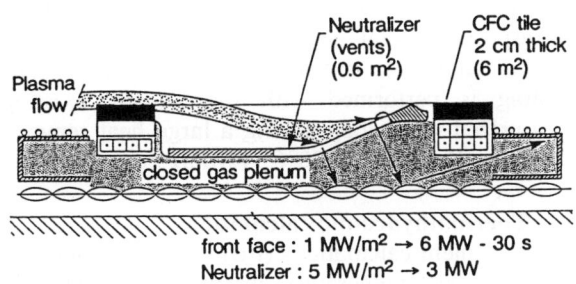

Fig.1: Upgrading description (cross section)

The radiative and charge exchange power is mainly removed by the front face of the module casings. The active pumping is made between the copper tubes in a vented geometry through which the neutralized particles are conducted to a titanium pumping system located on each side of the six casings.

The main goal of the transformation is to increase the maximum extracted power over 30 s : 3 MW exhausted by the neutralizer plate system in a steady state regime and 6 MW by the front face in a semi inertial regime.

2. THE CASING FRONT FACE PROTECTION

The whole front face (6 m^2) of each casing is fully covered with graphite tiles (2 cm thick) made of a three dimensional Carbone Fiber Composite (CFC N_{11} from SEP $\lambda_{//} = 200$ W/mK, $\lambda_\perp = 150$ W/mK). Each tile is bolted with elastic showers to the stainless steel casing (Fig.2).
For a realistic bolting pressure in the range of 0.3 to 0.5 MPa the measured value of the heat transfer coefficient is H = 0.6W/cm^{2}K [2]. The thermal behaviour of the attached CFC tiles will be determined by a semi inertial process: an adiabatic heating of the tiles during the 30 s of plasma heat pulse followed by a cooling phase by conduction to the water cooled casing during the pulse interval. This system increases the exhaust power capability in the range of 6 MW during 30 s by limiting the thermal excursion of the tile surface to 1000°C.

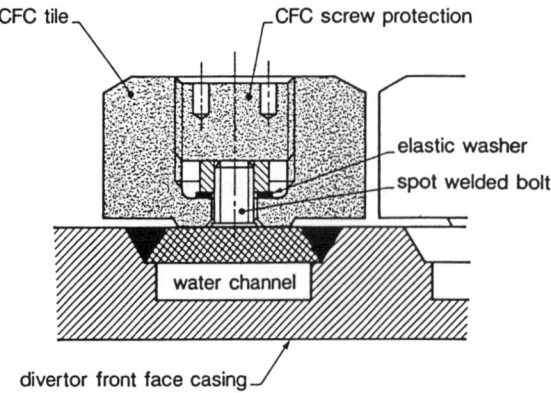

Fig.2: C.F.C. tile attachment

Finite elements calculations taking into account the temperature dependence of the CFC thermal conductivity have been made to determine the tile's temperature during a plasma cycle. The water parameters are: flow velocity 3 m/s, inlet temperature 150°C, $T_{saturation}$ 230°C, heat exchange coefficient between tile and casing 0,5 W/cm²K.

The maximum tile temperature calculated value is 720°C at the end of the 30 s pulse corresponding to an incident heat flux of 1 MW/m². The cooling time constant for the tile is 150 seconds.

A mock-up of the front face has been designed in order to test the thermomechanical resistance of the tile attachment in the electron beam facility FE200. This mock-up has sustained successfully 1000 cycles of 30 s at a nominal value of 1 MW/m² with a maximal surface temperature of 1072°C.

3. THE NEUTRALIZER PLATE SYSTEM (NP)

3.1. Description

Each divertor module will be equipped with new neutralizer plates covering all the available room between the toroidal bars of the casings. Each NP is made of CrZr copper tubes, covered with a B_4C coating (150 μm thick), using a water channel $\varnothing = 12$ mm (four tubes assembly, sites type III, IV, V) and a water channel $\varnothing = 16$ mm (two tube assembly, sites type I, II, VI, VII). Each copper tube is connected in series-parallel with a stainless steel box through a friction welding. Each NP is linked to a water tube collector which is running along the poloidal sides of the casing (see Fig.3).

The vented structure is implemented by a gap between two adjacent bars the width of which varying between 5 and 10 mm with the exception sites VI and VII which do not yield the possibility of a vented structure. Moreover, these two sites are located outside the poloidal extension of the titanium pumps. Only the sites I to V are connected with the titanium pump boxes. Langmuir probes will be introduced between the slots of the vented structure to measure the plasma density and temperature. The whole plate assembly is attached to the divertor casing at each end.

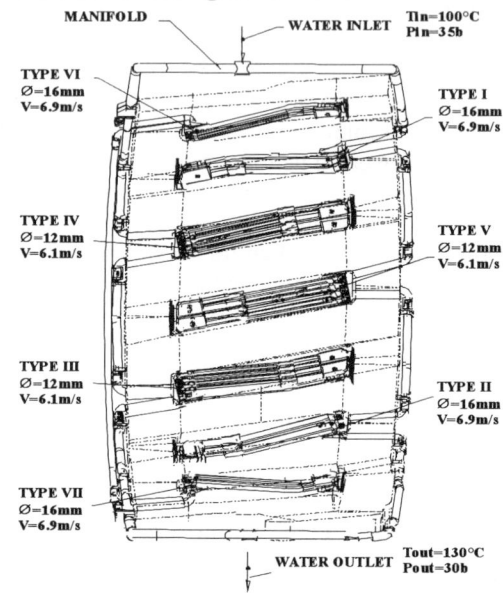

Fig.3: Neutraliser plate system

The water velocity inside the water tube channels lays between 6.1 m/s (sites type III, IV, V) and 6.9 m/s (sites type I, II, VI, VII), the total water flow reaches 86.2 ton/hour.

The temperature increase of the water is 30°C for a corresponding total power of 500 kW per module (i.e. a total heat exhaust $6 \times 0.5 = 3$ MW). The cooling is performed with the already existing pressurized water loop having a large heat removal capability (25 MW at 230°C, 4 MPa).

3.2. Thermalhydraulic calculations

Finite element calculations (FE) have been made to calculate the temperature distribution inside the copper tube section and the corresponding heat flux along the interface between water and copper [3]. Two values of the incident heat flux (on one side of the square section copper tube), 5 and 7.5 MW/m² are taken into account to figure out nominal and off normal operating conditions. The maximum copper temperatures are respectively 320 and 430°C, which remain acceptable values for a chromium zirconium copper material to avoid the softening of the copper alloy during heat pulses (see Fig.5).

The experimental parameters used for the FE calculation are : T_{in} = 130°C, water pressure 3 MPa, water velocity 6.1 and 6.9 m/s, on a total heated length 300 mm.

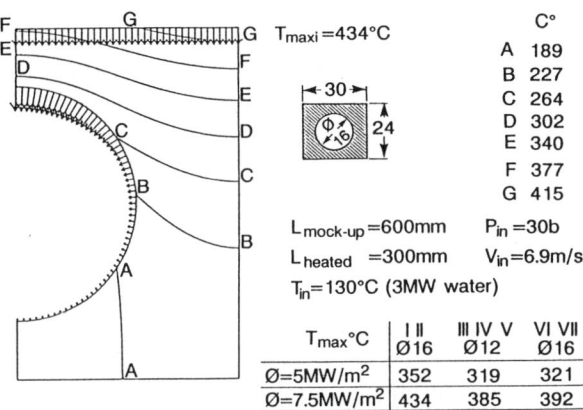

Fig.5: Isotherm curves in a cross section

The safety margin, defined here as the ratio between the critical heat flux (1.3 times the TONG 75 value [4]) and the maximum heat flux at the water interface, lays between 1.71 and 3.25 for incident heat flux values of respectively 7.5 and 5 MW/m². This is considered as acceptable (see Fig.6).

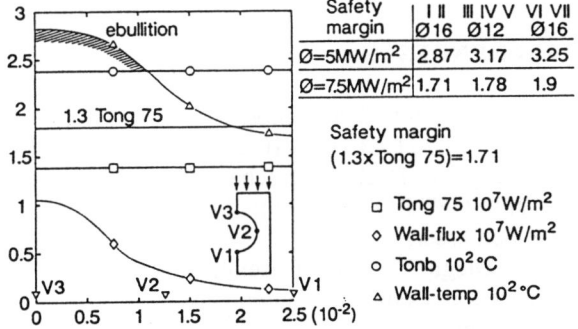

Fig.6: Wall heat flux and temperature

Similar calculations have been performed for a tube with a water channel diameter $\emptyset$ = 12 mm where the B_4C coating (150 µm thick) is meshed. Thermohydraulic calculations have been made with the minimal value of the B_4C thermal conductivity (λ=1 W/mK) with respect to the range of experimental values for λ obtained with plasma sprayed B_4C coatings (1 to 3 W/mK).

The maximum value of the incident heat flux which is acceptable is close to 7.5 MW/m² which yield thus a sufficient safety margin to avoid the B_4C surface erosion ($T_{MAX} B_4C$= 1400°C), a critical heat flux burn out and a softening of the chromium zirconium copper at the surface of the tube.

A mock-up of one neutralizer tube has sustained successfully, in the electron beam facility FE200, 1000 cycles at a nominal value of 5 MW/m² with a maximal surface temperature of 1400°C.

3.3. Protection with plasma sprayed boron carbide

Plasma sprayed boron carbide coating is an alternative technology to the brazing one used for components exposed to a medium plasma heat flux in Tore Supra and having a complex geometry. The thickness of the B_4C layers can reach up to 300 µm without flaking. Plasma sprayed boron carbide coating has been used in Tore Supra for several plasma components, Faraday shields and protective limiters for ICRF antennae [5], protection of the vertical ports against drifting trapped particles, prototype of an ergodic divertor neutralizer.

The coating of the overall neutralizer tube will be achieved by a plasma-spray coating of B_4C (150 µm thick). Such a thickness could sustain 10^4 seconds of full plasma power regarding plasma erosion assuming a maximum value of the sputtering rate for B_4C. That represents at least one year of normal plasma activity in Tore supra at full power on the plates.

This life time could decrease if the surface temperature becomes larger than 1400°C due to repeated abnormal heat flux values.

Moreover, after plasma erosion, deteriorated plates could be renewed by a new plasma spray coating during a shut-down of the Tore Supra tokamak.

3.4. Thermomechanical calculation

Each divertor neutralizer plate made with two or four tubes is linked to a mechanical structure. This structure is attached rigidly at one end and elastically at the other end to the divertor casing (see Fig.7).

Elastic and elastic-plastic thermomechanical calculations have been made for normal and off normal loads (3D finite element code CASTEM). The normal loading is due to the incident particle heat flux over 30 cm length on each tube (5 MW/m^2) and the radiation charge exchange heat flux on the whole surface (1 MW/m^2). The off normal loading corresponds to an increased value of heat flux onto the NP up to 7.5 MW/m^2 to which the forces due to eddy currents during a plasma current disruption are added.

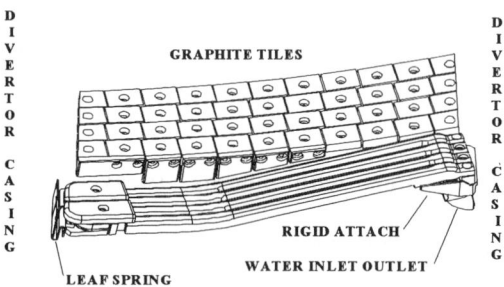

Fig.7: Neutraliser plate system attachment on the divertor casing.

For the normal load, the linear expansion of the tube reaches 2 mm and the maximum Von Mises secondary stresses due to the thermal gradient are comprised between 185 and 206 MPa for the different types of tubes.The acceptable value limit for the secondary stresses is 3 x Sm = 280 MPa (Sm = maximum admissible stress) in agreement with the recommendations of the RCC-MR Code.

For the off-normal conditions, the local efforts induced by the eddy currents due to a disruption have been calculated with poloidal field variation of 130 T/s ($\Delta B = 0.67$ T in 5 ms) and a toroidal magnetic field $B_\varphi = 3.5$ T. The resulting torque reaches 704 Nm for the tube located in site n°V.The maximum Von Mises primary stresses due to the electromagnetic forces are comprised between 25 and 45 MPa for the seven sites. In this case of load the secondary stresses are calculated with the CASTEM 3D elastic plastic code using the kinematic plastification curves of CrZr copper versus temperature. Young modulus and yield stress are also taken as temperature dependent.

The maximum secondary stresses calculated after plastification σ_{ep} are comprised between 188 and 220 MPa which is in the the same order of magnitude as the 3 x Sm value (215 Mpa). The corresponding maximum plastic strain is 0.4 %.

4. NEUTRALIZER PLATE PROTOTYPE EXPERIMENTAL RESULTS

A prototype of the new neutraliser plate was installed in Tore Supra and studied for a whole annual campaign. It was located close to the equatorial plane. The main outcomes of the experimental sessions can be found in ref. [6]. They are summarized as follows :

* the neutraliser plate did not suffer from any mechanical problem. The boron carbide coating was in very good shape and no trace of flaking could be found.

* the heat flux could be extended to very high average values (about 10 MW/m^2 for 1 s). The deposition pattern, deduced from infrared thermographic measurements is somewhat complex but it could be related to the structure of the field lines connections to the plate so that a comparison between experimental results and the outputs of the field line tracing code MASTOC [7] resulted in a good agreement in view of the above mentioned hypothesis. This in particular stressed the importance of the locally important value of the radial component of the toroidal field ripple. Specific shadowing between modules could then be revealed and led to a decrease of the plate inclination to the main field by a factor of about 2 (from 14° to 7°) in the final plate design.

REFERENCES

1. A. Grosman,P. Deschamps,
   PS 95/CEA/1-CCFP64/10.2
2. M. Lipa, C. Deck, P. Deschamps et al.,
   17th SOFT ROME 1992 p.307
3. J. Schlosser,
   San Diego Workshop December 94
   Nureth 6, Conference, Grenoble, 1993
4. F. Escourbiac, J. Schlosser, J. Boscary,
   this conference
5. B.Beaumont , E.Gauthier,
   18th SOFT KARLSRUHE 1994 p.231
6. A. Grosman, Ph. Ghendrih et al.,
   to be published in J. Nucl. Mat.
7. Ph. Ghendrih and A. Grosman,
   to be published in J. Nucl. Mat.

Development and fabrication of improved CFC-brazed components for the inner first wall of Tore Supra

M. Lipa, Ph. Chappuis, G. Chaumat[+], D. Guilhem, R. Mitteau, L. Ploechl[*]

Association EURATOM-CEA, Département de Recherches sur la Fusion Contrôlée
Centre de Cadarache, 13108 Saint Paul lez Durance Cedex, France

[+] CEA-DTA-CEREM, CEN Grenoble

[*] Metallwerk Plansee, A-6600 Reutte, Austria

A new generation of improved carbon fiber reinforced carbon (CFC) brazed inner first wall (IFW) elements for Tore Supra (TS) has been developed and manufactured in order to sustain reliably a continuous heat flux in the 1 MW/m^2 range. The modular stainless steel heat sink structure allows bending to its final polygonal shape after the brazing procedure. The braze joint quality has been assessed systematically by destructive meatallography and non destructive inspection methods such as X-ray radiography and thermography. A toroidal section of the damaged IFW in TS, corresponding to about 1.5 m^2, has been replaced by these new components.

1. INTRODUCTION

The design of the Tore Supra (TS) inner first wall (IFW) was performed during the mid-eighties [1]. A square stainless steel tube with a circular inner channel was chosen for the cooling geometry because of the expected moderate heat flux and lower disruption forces acting on the heat sink structure. A total graphite area (fine grain, high density 5890 PT from Le Carbone Lorraine) of roughly 12 m^2, representing 8700 flat rectangular tiles, faces the plasma on the high field side. The application of non destructive inspection methods to large areas of brazed parts was not proven to be satisfactory at that time [2]. Therefore a certain number of braze voids, depending on size and location, have been accepted or remained undetected. During six years of plasma operation in TS, about 7 % of the brazed IFW tiles were further damaged. A limited number of IFW panels, corresponding to a toroidal section of about 1.5 m^2, were damaged sufficiently to warrant replacement.

2. DESIGN CONCEPT

The design of the new stainless steel IFW heat sink has been oriented towards the construction of modular interchangeable individual elements, allowing replacement of defective components [3]. The heat sink elements are welding structures made of 3 parallel rectangular stainless steel cooling tubes. Cylindrical portions machined on the St-St tubes, located between tile areas allow bending of the elements to their final polygonal shape after brazing. Different elements, comprising one, three and seven tile areas (modules) are assembled as 3°20' (toroidal) sectors and connected together by welding (Fig. 1).

The tile brazing is performed on flat elements in a horizontal position. The length of these elements, with different numbers of tile areas (modules), varies between 20 and 90 cm. Each CFC tile is laser treated in order to increase the surface area and therefore enhance the thermomechanical adherence of the brazing joint [4]. The CFC tiles are brazed by means of a TICUSIL filler metal to a 2 mm thick copper compliant layer in order to compensate the brazing mismatch of the thermal expansion between the CFC and the St-St heat sink.

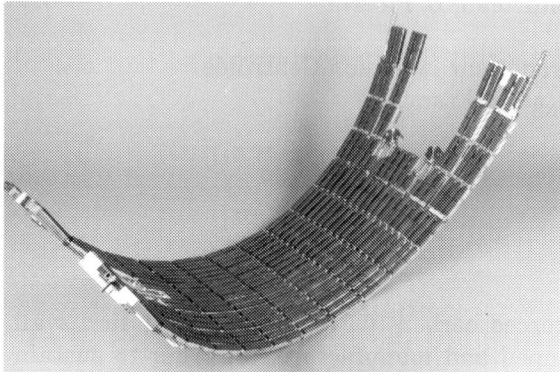

Figure 1. Six 3°20'-toroidal sectors bended to their final shape after brazing.

3. CFC MATERIAL

The fine grain graphite has been replaced by carbon fiber reinforced carbon (CFC), which is generally a material with a higher thermal shock resistance and an enhanced thermal conductivity. The selected material, a PAN fiber CFC type N11 which is close to an isotropic material, has been fabricated by SEP (Societé Europeennes de Propulsion) and supplied in blocks of 200 × 35 × 28 mm.

In order to improve the determination of eventual braze voids detected by means of non-destructive inspection procedures, which have been specified to the brazing company, each tile of size 20 × 65 × 10 mm has been numbered and its average density measured. The densities obtained from about 500 N11 tiles varied between 1.68 and 1.85 g/cm^3. The lowest density is systematically observed on tiles machined from the central part of the CFC blocks. Graphitisation of the base material is performed in a vacuum oven at temperatures above 2000°C. Comparative outgassing properties between N11 and fine grain graphite were measured up to 700°C. No substantial differences were identified (outgassing rate ~ 0.03 Torr.l/g). The thermal conductivity from CFC samples as a function of densities, which represents the full range of measured values, was determined by a laser flash method. The results show that a variation of about 8 % in the CFC density leads to a thermal conductivity change of roughly 25 % at room temperature (Fig. 2).

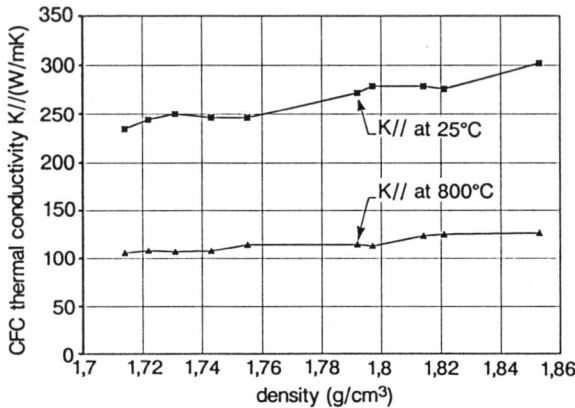

Figure 2. Measured thermal conductivity of CFC N11 as a function of density.

4. BRAZING SPECIFICATION

The main brazing specification has been elaborated in a common agreement with the brazing company. The global braze factor is an indicator for the thermal heat transfer coefficient of the multilayer braze joints. The minimum acceptable value was 70 %. Hereby a minimum brazing factor (100 - {area of braze failures × 100/ theoretical area of brazed surface}) of 90 % between CFC and copper layer and 75 % between copper and stainless steel must be guaranteed. In addition, components with braze flaws larger than 3 mm are rejected. The braze joint quality was assessed by :

- destructive metallography on prototype elements and on reference pieces,
- X-ray radiography of each tile attachment,

- thermographic measurements on each element,
- systematic fracture shear stress measurements on reference pieces

5. DESTRUCTIVE INSPECTION METHODS

Results of fracture shear stress measurements at the CFC/CU interface for reference pieces, accompanying each braze cycle, show average values in the range of 40 to 50 MPa for CFC braze penetrations in the mm range. The rupture spot is systematically located beyond the cone-structure. Shear stresses measured on samples without surface treatment seem to be more sensitive to CFC densities and show also an influence of the braze foil thickness on this phenomenon. Hereby an increasing of a filler metal thickness from 100 µm to 300 µm in Ticusil brazed CFC/CU samples of same densities, leads to an enhancement of the rupture shear-stress from 27 to 43 Mpa (Fig. 3).

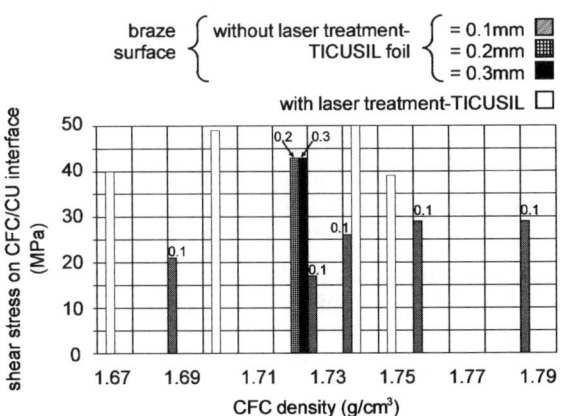

Figure 3. Measured average shear stress values of brazed CFC(N11)/CU samples as a function of CFC density.

In order to test the integrity of the braze joints, thermal quenching of mock-ups has been performed during the prototype development phase. Samples with laser treated tile surfaces survived 20 shocks from 600°C to cold water whereas samples without laser treated tile surfaces were damaged after a lower number of cycles.

6. NON DESTRUCTIVE INSPECTION METHODS

6.1. X-ray radiography

X-ray radiography, perpendicular to the brazed surface, of each tile attachment was performed by the contractor. Following tests on calibrated samples representing the simulation of different braze flaw sizes, a specified value of 3 mm in diameter is considered to be a reasonable guaranteed detection limit of this radiographic inspection procedure. However this failure detection is limited to braze flaws thicker than about 100 µm.

A few defective braze joints of important size (up to 50 % in the CU/ST-ST interface) could only be dedected with transient thermographic measurements at CEA. Following micrographic inspections have confirmed that this X-ray radiography can also be limited for open braze gaps with not well localized boundaries.

6.2. IR inspection by means of hot water

A systematic verification of the thermal continuity of the braze joints on serial elements was performed by transient thermographic measurements at CEA. The method is based on IR measurements of tile surface temperatures during a thermal transient produced by hot water flowing in the heat sink cooling channel. Hereby local surface temperature evolutions of tiles brazed on heat sink elements are compared to well brazed tile temperatures of an equivalent reference element which is hydraulically connected in parallel. The qualification of the reference elements has been performed on three- and seven module prototypes at the e-beam facility at Framatome (Fig. 4). During this qualification test it was verified that the maximum difference in average surface temperature between two CFC tiles was lower than the specified value. The comparison is based on thermohydraulic calculations taking into account the allowable global braze factors and variation in material properties. Experimental results and comparative thermal finite element calculations have shown, that the transient thermographic measurements depends more on braze defect extension than on braze void areas [5] (Fig. 5). However the method has been proven to be complementary to the X-ray radiography.

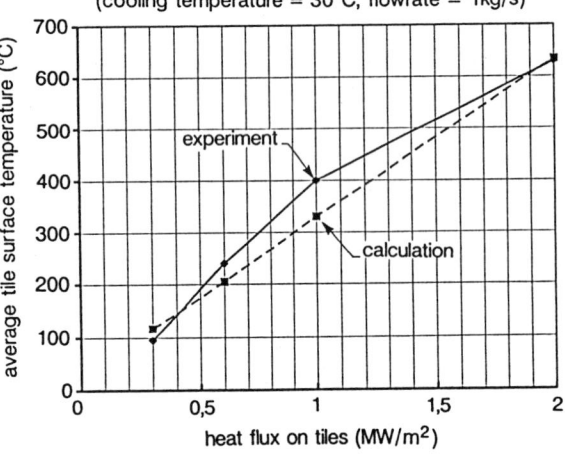

Figure 4. Electron beam test of a reference prototype element PPI-5B.

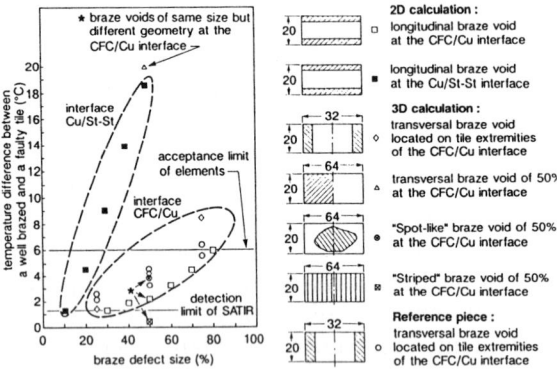

Figure 5. Calculated and measured tile surface temperature differences on a brazed IFW assembly CFC/CU/ST-ST as a function of braze defect size (transient IR measurements).

7. SUMMARY

A new generation of CFC brazed and actively cooled IFW elements for TS has been developed and manufactured in order to sustain reliably a heat flux of 1 MW/m^2 in steady state.

A systematic verification of the thermal continuity of the braze joints on serial elements has been performed at CEA by transient thermographic measurements using hot water. This inspection method has been proven to be complementary to X-ray radiography which is the basic braze failure detection method used by the braze company.

After 1 year of plasma operation in Tore Supra no damage and no hot spots have been observed on the new installed components.

REFERENCES

1. PH. Chappuis et al., "Tore Supra graphite inner first wall", 15th SOFT, Utrecht, Sept. 1988.
2. M. Lipa et al., "Brazed graphite for actively cooled plasma-facing components in Tore Supra", Fusion Technology, Vol. 19, July 1991.
3. M. Lipa et al., "Development and fabrication of a new generation of CFC-brazed plasma facing components for Tore Supra", EUR/CEA-FC-1550, July 1995.
4. N. Reheis et al., "Industrial aspects in the fabrication of prototypes for different divertor concepts", 17th SOFT, Rome, 1992.
5. R. Mitteau et al., "Non destructive testing of actively cooled plasma facing components by means of thermal transient excitation and infrared imaging", presented on this conference.

Non destructive testing of actively cooled plasma facing components by means of thermal transient excitation and infrared imaging

R. Mitteau, S. Berrebi*, P. Chappuis, Ph. Darses*, A. Dufayet, L. Garampon, D. Guilhem, M. Lipa, V. Martin, H. Roche

Association EURATOM-CEA, Département de Recherches sur la Fusion Contrôlée

Centre de Cadarache, 13108 Saint Paul lez Durance Cedex, France

* CEDIP, 19 Bd Bidault, 77183 Croissy Beaubourg, France

SATIR is a new test-bed installed at Tore Supra to perform non destructive examination of actively cooled plasma facing components. Hot and cold water flow successively in the cooling tube of the component and the surface temperature is recorded with an infrared camera. Defects are detected by a slower temperature response above unbrazed areas. The connection between temperature differences and defect sizes is the main difficulty. It is established by tests of standard defects and thermal transient calculations of defective geometries. SATIR has been in use for two years and has proved to be very valuable to test industrial components as well as prototypes.

1. INTRODUCTION

The in situ maintenance of plasma facing components (PFC) is very difficult and a high level of reliability has to be reached. Therefore, non destructive examination (NDE) is systematically applied to test components manufactured by industry. Thermal techniques present the benefit to test what really matters : the thermal transfer of the bonds. Such techniques have already been in use at Tore Supra as well as in other laboratories [1,2]. At Tore Supra, the necessity for a permanent NDE test-bed for PFCs led to the development of SATIR (figure 1). SATIR is an acronym for Station d'Acquisition et de Traitement InfraRouge. This equipment and its use are described in the following pages.

Figure 1 : View of the test-bed

2. TEST - BED DESCRIPTION

The majority of Tore Supra's PFCs are actively cooled. It gives the opportunity to use their cooling channels to heat or cool the components with hot and cold water. An open water circuit is set up (figure 2). Cold water comes from the commercial network and three heating tanks totalling 600 l are installed to deliver hot water. Two elements can be installed in parallel. Various types of connections can be used. This thermal excitation is highly efficient. For a temperature difference of 80 °C between the component and the water, and with an heat transfer coefficient to 17000 W/m²K, the wall heat flux reaches 1.4 MW/m².

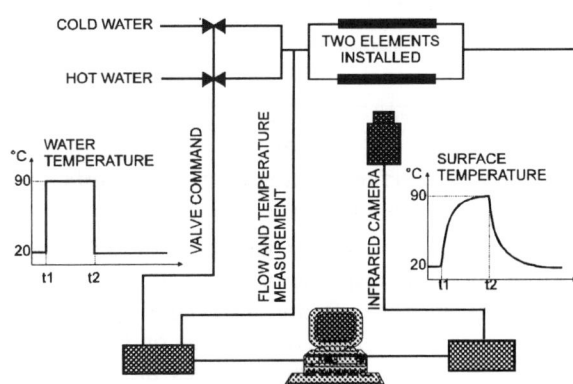

Figure 2 : Test-bed diagram

Depending on the pressure drop of the elements, the hot water flow rate amounts up to 2.5 m³/h and the cold water flow rate up to 4.3 m³/h. These flow rates are sufficient to have a water transit duration through the element (~ 0.1 s) shorter than the thermal time constant of the element (3 to 15 s).

The surface temperature of the elements is measured with an infrared camera (type Inframetrix 600). The video signal is numerised and stored in a PC. Safety copy of the film may be stored with a VCR. The PC works with a PTR-based software (CEDIP) which both remote controls the acquisition sequence as well as it does the thermal analysis. The PC has a 486 processor and 64 Mo RAM which enable to record up to 12.5 images per second. The data are safeguarded on 5 Go cartridges.

3. TEST PROCEDURE

The choice of hot and cold water durations and sampling frequency depends on the element being tested. The inner first wall whose thermal time constant is 15 seconds was tested with 60 seconds fronts and a sampling frequency of an image out of 6 (roughly 2 images per second) [3].

Mock-up aimed at developments are tested individually and get a customised analysis. Elements from large fabrication series are tested simultaneously with a sound reference element, which is chosen after test in a high heat flux test-bed. The subsequent analysis relies on the comparison between the two.

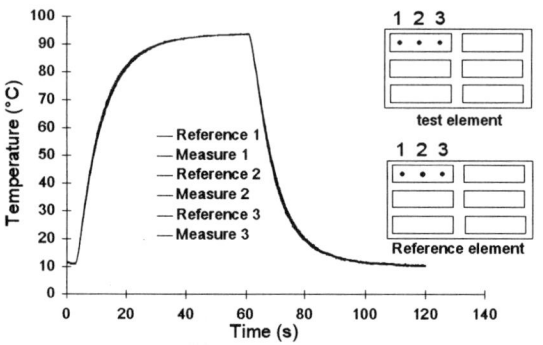

Figure 3 : Surface temperature of 6 points : the 6 lines are merged

The thermal analysis is based on the difference between the time response of each couple of points (test and reference, figure 3). These points are located at the same relative positions on the tiles to allow comparison. The temperatures are extracted from the film, usually from a 3*3 pixel matrix. Figure 3 does not permit to distinguish differences in the surface temperature. One has to display the differences to see the differences (figure 4.a).

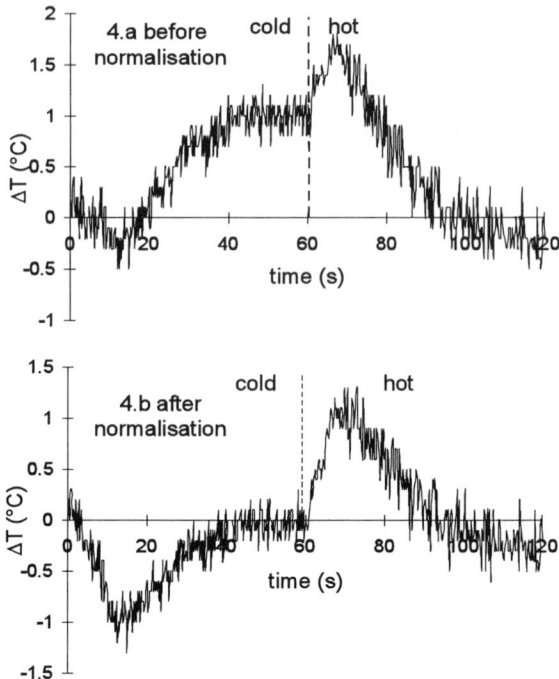

Figure 4 : Temperature difference with and without normalisation

Because of emissivity differences at the surface of the tiles, stable hot and cold temperatures measured by the camera may vary from point to point. In order to calculate the real temperature difference between the points, the temperature curves are normalised linearly (figure 4.b).

Let T_c and T_h be the cold and hot stable temperature as measured by the infrared camera, $\widetilde{T}_c$ and $\widetilde{T}_h$ the cold and hot stable temperatures averaged on all curves. The normalised temperature is given by :

$$T_{normalized} = (T_{measured} - T_c) \cdot \frac{\left(\widetilde{T}_h - \widetilde{T}_c\right)}{\left(T_h - T_c\right)} + \widetilde{T}_c$$

Maximum temperature differences are then deduced from the normalised curves and displayed on the screen (see example table 1). In the case of the inner first wall, the temperature differences were also corrected from the various wall thicknesses measured on the stainless steel heat sink.

A maximal temperature difference is authorized. When an element shows a temperature lag that exceeds the limit, it is more thoroughly investigated and can be rejected.

	ΔT up	ΔT down	defect size
pair N°1	1.3 °C	1.3 °C	2.8 mm
pair N°2	1.1 °C	1.6 °C	3.2 mm
pair N°3	1.3 °C	1.5 °C	3.2 mm

Table 1 : Results of the thermal analysis

Setting the limit is the greatest difficulty. Two methods are employed : test of standard defects and finite element calculations.

The standard defects are either fabricated (e.g. during the brazing cycle by forbidding the braze to wet the armour material using stop-off fluid, figure 5) or created on sound elements (by drilling or grinding the joint with narrow tools). The elements are then tested and the temperature differences plotted against defect sizes.

Figure 5 : Standard defects

Finite elements calculations of faulty geometries are also extensively performed. Both 2D and 3D calculations are made. They are compared to the results of the standard defects. The calculations showed that the temperature difference on the surface of the tile is better correlated to the braze void extension rather than to the braze void area. When no boundary is present, the braze void extension is the radius of the largest circle that can be inscripted in the defect (figure 6).

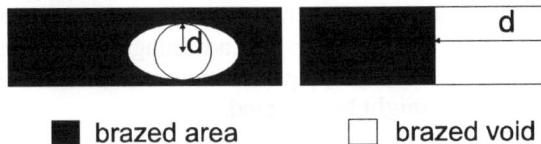

Figure 6 : illustration of braze void extension

However, with boundaries, a more complicated definition has to be used. If we consider the shortest path between a point of the braze void and all points of the brazed area, the braze void extension is the longest of the those paths. This can be mathematically written by the following expression :

$$d = \max_{M \in (\text{braze void})} \left(\min_{N \in (\text{brazed area})} (MN) \right)$$

Overheating of the tile's surface under heat flux is governed by the same parameter, so that setting an acceptance limit on the test bed is equivalent to accept a limited overheating under heat flux. For the inner first wall, those considerations led to a limit of 6°C ([3], figure5).

4. TWO YEARS EXPERIENCE

The experience was gained mainly on the inner first wall, which elements were tested after delivery and after the assembly steps. Testing 1 m² takes approximately 1 month. The test-bed led to the rejection of three elements. One tile had an 11°C lag (figure 7).

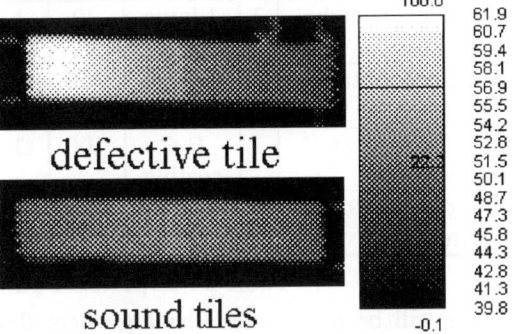

Figure 7 : Defect on a PPI element

X-ray testing of this element had shown no defect, but cuts through the element confirmed the presence of a large defect. This example proved the usefulness of SATIR and its complementarity to X-ray examination.

Many mock-up were also individually tested (fingers, macroblocs, bolted tiles, metallic mock-ups, see table 2). They showed that, as in other NDE techniques, experience is vital and allows to have a better sensibility.

5. FUTURE PROSPECTS

The software is currently being improved to analyse highest and lowest temperatures on surfaces rather than on points. The aim is to detect surface temperature's differences within single tiles. By so doing, the analysis will be independent from temperature lag caused by conductivity or thickness differences between the measured and reference tiles. This will allow to reduce the acceptance limit close to the level of the signal's noise. A comparison to a reference element will however be maintained, to avoid the risk of generalized defects. These

component	Nb	Nb tiles	area (cm²)	tests	results	remarks
Tore Supra inner first wall (IFW)	88	1506	19500	2/3	3 elements rejected	60° toroïdal extension 6 months
Tore Supra IFW standard defects	1	4	24	1	ΔT function of void percentage	correlated to FE calculations
LPT short fingers	8	56	350	2	defaults hardly visible before FE200, easily visible after damage on FE200	studies are under progress to improve the analysis and lower the acceptance limit
LPT long fingers	4	84	500	2	idem	idem
HIP copper - Stainless Steel (ITER)	1	no tile	100	1	non homogenous copper emissivity	task T8
rheocast (ITER)	1	1 copper tile	30	1	very large defect	deteriorated during tests at FE200
macrobloc monotube	3	1 large	30	1	evidence of braze voids	each face has to be tested separately
macrobloc multitube	2	1 large	300	3		
macrobloc	2	1 large	300	1	to be tested	(recessed hole)
ergodic divertor neutraliser	1	B_4C coating	100	3	the thickness of the B_4C coating prevails	no thermographic test for this fabrication
ergodic divertor front face	1	32 bolted CFC	430	1	the loosest tiles lag	semi-inertial element time constant 300 s

Table 2 : Experience gained during two years of operation

techniques will be used to test the high heat flux fingers that are developed for the Tore Supra's Toroidal pump limiter. A 3°C limit is foreseen, which would guaranty a steady state temperature smaller than 1500 °C (in the worst case, figure 8). This value is correlated to a 6 mm defect.

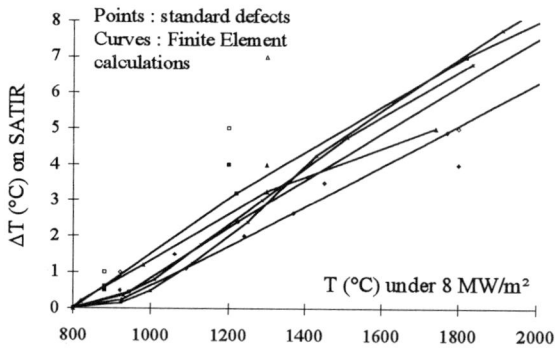

Figure 8 : Setting the limit for the fingers

Beside this improvement, other techniques could be used to increase both the sensibility of the test-bed and its capacity :
- Going towards higher water pressure, velocity and temperature. However, this would require a stronger water loop and stronger connections.
- Cycling the excitation and measure the phase shift.
- Narrowing the temperature range of the camera to increase the precision of the measure. Correcting the emissivity differences on the surface of the tile would require to store a map of the surface's emissivity.

6. CONCLUSION

SATIR has proved to be a very valuable test-bed to test fabrication series as well as prototypes. In comparison to other thermal techniques used in NDE, SATIR presents the advantage of being quantitative. It helps judgement when elements are defective but might be accepted.

REFERENCES

[1] J. Schlosser et al., Fusion Engineering, proceedings of the 14th IEEE/NPSS symposium, San Diego, 1(1991), 350-356

[2] H.D. Falter, Fusion Technology, proceedings of the 18th SOFT, Karlsruhe, 1(1994),467-470

[3] M. Lipa et al., Development and fabrication of improved CFC-brazed components for the inner first wall of Tore Supra, 19th SOFT, Lisbon, 16-20 September 1996, to be published.

Developments for an Actively Cooled Toroidal Pump Limiter for *Tore Supra*

J. Schlosser[a], Ph. Chappuis[a], L. Doceul[a], M. Chatelier[a], J.P. Cocat[a], L. Garampon[a], P. Garin[a], R. Mitteau[a], A. Moal[b], L. Plöchl[c], G. Tonon[a], E. Tsitrone[a]

[a]Association EURATOM-CEA, Département de Recherches sur la Fusion Contrôlée

Centre de Cadarache, 13108 Saint-Paul lez Durance Cedex, France

[b]CISI, Centre de Cadarache 13108 Saint-Paul lez Durance Cedex France

[c]Plansee AG, A-6600 Reute/Tirol, Austria

For power (25 MW) and particle exhaust in *Tore Supra*, a Toroidal Pump Limiter (TPL) structure is under development. The elementary elements in front of the plasma, hereafter called fingers are water cooled and form altogether a flat toroidal ring located in the bottom of the machine with one throat on the low field side to collect the particles. The fingers made of hardened copper (CuCrZr) are covered with CFC (Carbon Fibre Composite) tiles. Many calculations and tests were performed in order to demonstrate the feasibility of such a limiter. Sub-scale and scale one fingers were developed, manufactured and tested under cycling heat load in the *FE200* electron beam facility (up to 3300 cycles at 10 MW/m^2). Insulation of the head of the limiter, pressure drop and water distribution through the different fingers were also tested by means of scale one mock-ups.

1. GENERAL CONCEPT

The TPL is a horizontal annular disk tangent to the plasma in the bottom of the vacuum chamber [1] (Fig. 1). The supporting structure is composed of a lower ring with a radial and elevation adjustment system (allowing in normal operation a 70 mm vertical displacement of the head) and an upper ring which is the support and the water header of the fingers (Fig. 2).

The 12 vertical supports between the two rings are made of a ceramic insulator (allowing a polarisation of the head of the limiter up to 2.5 kV), a stainless steal bellows and two titanium elastic plates. The supports pass through the lower vertical ports of the machine and contain the feeding pipes. The upper ring 30° sectors are connected together through insulating connections (Fig. 3). The head of the limiter is formed by the assembling of 576 fingers water fed at the rear part (external side of the head). The concept of fingers was adopted to limit the Eddy current. Each finger is water cooled (T_{in} = 150°C, P = 3.5 MPa, V = 10 m/s), with a double smooth channel (ID = 8 mm) with a U turn at the leading edge (internal side). The channels are drilled by the rear of the element whereas the U turn is machined in the lower surface of the element. Copper buttons are then electron beam welded to close the double channel.

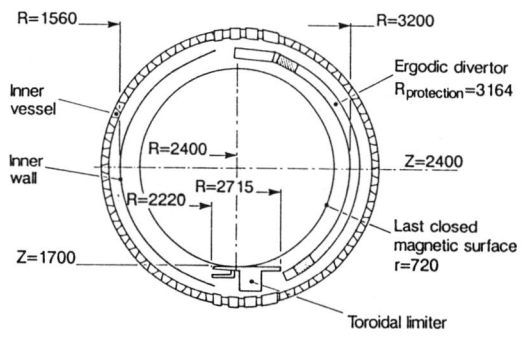

Figure 1 - Position of the TPL in Tore Supra

The head of the limiter is designed to remove a 15 MW convective power at steady state. Under the head of the limiter, 12 neutralisers are positioned. They have a V shape structure as shown in

Figure 2. This shape allows the neutral flux coming from the incident ion flux to be trapped into a "closed" geometry. The configuration is closed by an actively cooled austenitic steel support plate covered with a B_4C coating. This ensures the neutrals to be guided through the aperture of the vertical port to the external pumping system.

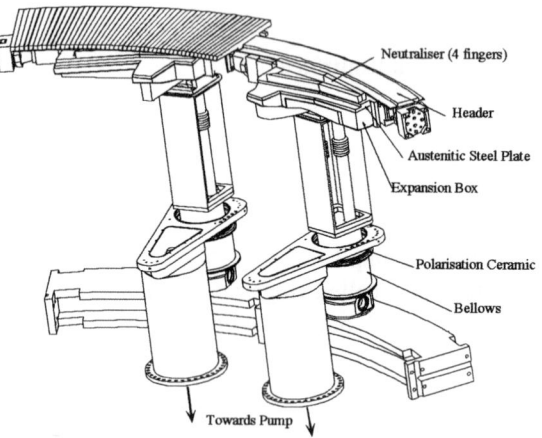

Figure 2 - Concept of the Actively Cooled TPL: two 30° Sectors

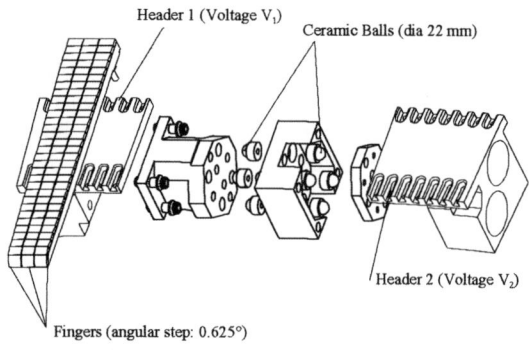

Figure 3 - Concept of the Insulator System between two 30° Sectors

2. DESIGN CALCULATIONS

After the first design calculations already presented [1], the main performed calculations were about power deposition and pumping capability of the limiter.

The power deposition calculation was made in 3D using the CEA *CASTEM 2000* finite element code [2]. Special developments were done to take into account the Shafranov shift and the magnetic surface ripple (due to the 18 toroidal coils). The calculation was made with an exponential decay hypothesis in the Scrape-Off Layer. Several cases of plasma position were studied, a typical one is given Figure 4.

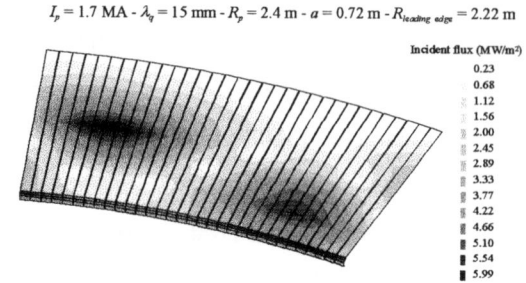

Figure 4 - Incident Heat Flux Isovalues for a TPL 20° Sector

An example of a thermal hydraulic analysis of a finger is given Figure 5. Each finger is 495 mm long and 28 mm thick with an average width of 25.4 mm, and is designed to sustain 8 MW/m^2 on the leading edge and 10 MW/m^2 on the flat part.

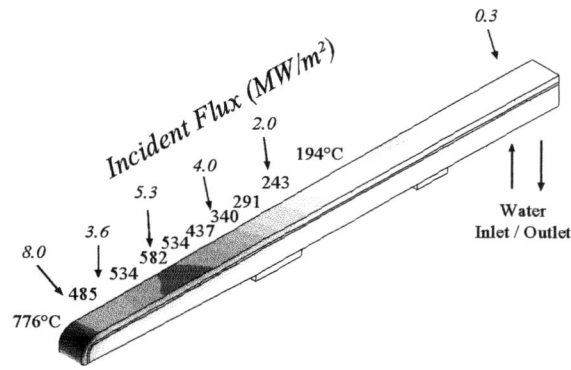

Figure 5 - Isotherms of a Finger

The safety margin for such an element is ranging from 1.4 to 1.8 for a standard plasma, the limits taken into account being 600°C for the CFC/Cu join, 1200°C for the tile surface temperature, critical heat flux on the water wall and 370°C for the CuCrZr structure. The most severe limit is in fact on the leading edge CFC/Cu temperature which can lead in case of overloading to a detachment of the tile. 3D thermomechanical calculations showed that stresses and plasticity strains were acceptable in the copper compliant layer between CFC tiles and CuCrZr heat sink and in the heat sink itself. A life time of 10000 cycles can be predicted. But one cannot predict anything about the CFC/Cu join were the stresses are high especially at the singular points.

The pumping system of the TPL must allow to control efficiently the plasma density by ensuring a pumping efficiency ε (extracted flux / plasma outflux) of the order of 10 %. Ultimately, it should allow to exhaust the largest pellet injection in *Tore Supra*, corresponding to 4 Pa·m^{3}s^{-1}. Some preliminary calculations of the pumping capability of the TPL were performed using the *Eirene* code [3] and a 1D recycling code for typical plasma conditions. The results show that the pumping efficiencies and the corresponding extracted flux are satisfactory, specially at high plasma density (Fig. 6).

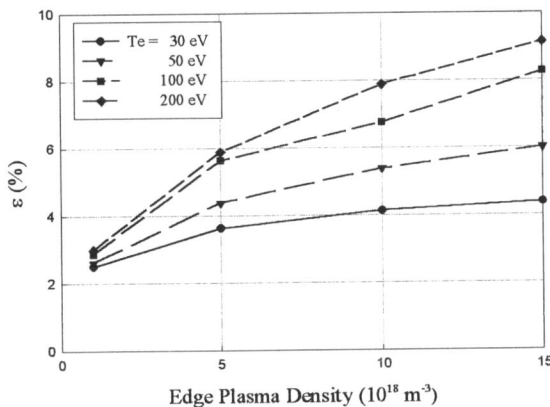

Figure 6 - Pumping Efficiency *vs.* Edge Plasma Density

3. QUALIFICATION TESTS

During the pre-design of the finger, critical heat flux tests were performed on the basis of the critical heat flux campaign performed for swirl tube [4]. From this campaign 1.67 TONG75h was found to be a good correlation for a one side 10 cm flat heat flux profile. Complementary tests showed that 1.25 TONG75 correlates well in the same conditions the smooth tubes. Different finger concepts were tested at critical heat flux (go and return swirl tube, double swirl tube, double smooth tube). Double smooth tube was found to be a good compromise between critical heat flux and pressure drop. From these tests, limit values finally adopted for critical heat flux were 28 MW/m^2 in the U turn and 20 MW/m^2 in the straight part of the channels at a 0.5 kg/s flow (V = 10 m/s).

Tests were then performed about the pressure drop of the fingers. Scale-one mock-ups were manufactured and tested. Improvements were found for the inlet/outlet feedings of the fingers so that at 10 m/s the pressure drop of a finger is less than 0.25 MPa (Fig. 7). This allows two fingers to be put in series.

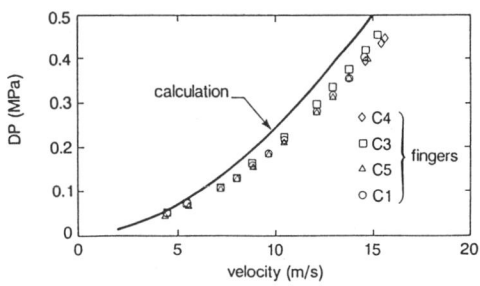

Figure 7 - Finger Pressure Drop: Comparison Calculation/Tests

A manifold of a 10° sector (16 fingers) was also manufactured in order to check the assembly feasibility, to measure the total pressure drop of the system, and to check the good distribution of the flow between the different fingers (Fig. 8). It was shown that the total pressure drop was compatible with the pressurised water cooling loop [1] (ΔP = 0.75 MPa, Q = 208 kg/s).

Figure 8 - Hydraulic 10° Manifold Sector Mock-up

4. MANUFACTURING AND TESTS OF REPRESENTATIVE ELEMENTS

The work was first focused on the bonding of the CFC tiles (N11 from SEP. Bordeaux - France) with the CuCrZr heat sink through a compliant layer of OFHC copper, because it requires a complex temperature cycle for brazing due to the dropping of the CuCrZr strength when exposed to temperatures higher than 500°C. Sub-scale elements were ordered to two manufacturers: *Vide et Traitement* Neuilly en Thelle (France) and *Plansee* Reute (Austria). The emphasis was put on the non destructive tests (NDT) of the two joins: CFC/Cu and Cu/CuCrZr. *Vide et Traitement* adopted a technology of brazing which gave poor results with CuCrZr heat sink (but promising results with Glidcop heat sink) and got difficulties to develop NDT. *Plansee* adopted an original technology: the CFC/Cu join is made by active metal casting (AMC) and controlled by X ray and the tile equipped of its copper layer is then EB (electron beam) welded to CuCrZr [5]. The Cu/CuCrZr join is controlled by UST (Ultra Sonic Testing) which has necessitated developments due to inhomogeneous sonic absorption of the CuCrZr material (difference between image before and after EB welding is made). The sub-scale elements manufactured by *Plansee* were successfully tested under 15 MW/m^2 during 1000 cycles [5].

After this success, 3 scale-one fingers were manufactured by *Plansee*, two of them were tested on the *FE200* test bed. It was shown that the leading edge with the chosen geometry was not able to sustain more than 9 MW/m^2. At this level the two tiles of the leading edge making an angle of 135° were rapidly damaged after a few cycles and the defects gradually increased with cycling. Others parts of this elements sustained without any problem 10 MW/m^2 during 1000 cycles (power removed 40 kW on each element). At 14 MW/m^2 one tile failed after about 500 cycles but no more defects were observed on the elements after 1000 cycles (power removed: 57 kW). A few cycles were performed increasing gradually the incident power (~ 20 cycles at each power). At 18 MW/m^2, other tiles became damaged (power removed: 73 kW).

Considering these tests, it was decided to have a cylindrical shaped tile at the leading edge in order to have a better cooling and no singular angle between tiles. Fortunately this was possible with the *Plansee* technology.

Five new scale-one elements were manufactured by *Plansee* and two of them tested on the *FE200*. They sustained 8 MW/m^2 on the leading edge during 1000 cycles and 10 MW/m^2 on the flat part up to 3300 cycles (power removed: 76 kW).

Other scale-one element mock-ups were manufactured. A polarisation ceramic ring (ID = 270 mm) was successfully brazed on the stainless steel bellows (part of the TPL supports) and sustained 900 mechanical cycles (axial movement with lateral misalignment ranging between 0 and 15 mm). The insulating connections between the 30° sectors were also tested, several materials were tried. The best result was obtained with alumina spheres stainless steel seats and titanium bolts. The mock-up sustained a bending moment of 15·10^3 N·m.

5. CONCLUSION

The design and component development activity for the TPL of *Tore Supra* is almost finished. All the experience of the *Tore Supra* team concerning plasma facing component working at steady state was taken into account for this development. Starting from a concept of flat CFC tiles an interesting technology of bonding was developed: it is silver free and hence *ITER* relevant.

REFERENCES

1. L. Doceul, J. Schlosser, Ph. Chappuis, et al., 18th SOFT, Karlsruhe, 1994, pp.331-334

2. P. Verpaux, et al., Recent Advances in Design Procedures for High Temperature Plant, Risley 1988

3. D. Reiter, J.Nucl.Mater. 196-198 (1992) 80

4. J. Schlosser, J. Boscary, 18th SOFT, Karlsruhe, 1994, pp.295-298

5. T. Huber, L. Plöchl, N. Reheiss, J.P. Cocat, J. Schlosser, IEEE/SOFE, Champaign 1995

The Tore Supra vented limiter : an alternative concept for heat and particle exhaust

E. Tsitrone, T. Loarer, B. Pégourié, H. Roche, P. Chappuis, D. Guilhem

Association Euratom-CEA, Département de Recherches sur la Fusion Contrôlée,
Centre de Cadarache, 13108 Saint-Paul-Lez-Durance Cédex, France

An alternative concept of limiter, the vented pump limiter (VPL), has been tested on Tore Supra. The main advantage of the VPL, which is based upon a different mode of particle collection, is to avoid leading edge close to the last flux surface. The experimental results obtained with a prototype of VPL are presented and compared to the results of an equivalent conventionnal throat pump limiter (TPL) under the same plasma conditions. In a first part, the pumping efficiencies of both limiters are investigated in various configurations : for ohmic discharges with different densities, for highly radiating discharges and for discharges with additionnal power. In a second part, the thermal behaviour of both limiters is explored, with additionnal power up to 6 MW for 3s. The infra-red imaging of the limiter head is also used to deduce the heat flux decay length evolution with the additionnal power and the plasma current.

1. INTRODUCTION

In tokamaks, pump limiters with throats (TPL) have widely demonstrated their ability to control the plasma density [1]. However, modular TPLs are submitted to high heat flux deposition on their leading edge, located just above the throat for particle collection [2]. The current technology limits the acceptable heat flux to ~10 MW/m^2, therefore constraining the position of the throat, which has to be far enough from the last closed flux surface to avoid thermal overload but close enough to the last closed flux surface to pump particles efficiently.

In order to overcome this limitation, an alternative concept of pump limiter has been developped and tested on Tore Supra : the vented pump limiter (VPL) [3]. Instead of collecting directly the ion flux by throats perpendicular to the toroidal direction, the VPL collects the neutrals coming from the recycling of the ion flux on the limiter surface. Indeed, the ion flux impinging on the limiter surface recycles as neutrals, which do not follow the magnetic field lines. A significant part of these neutrals is backscattered to the limiter surface by elastic collisions or atomic physics processes (charge exchange, dissociation) and can then be extracted by slots worked out in the limiter head. These slots are almost parallel to the toroidal direction, allowing to design the limiter in such a way that local thermal overloads are avoided (see Figure 1).

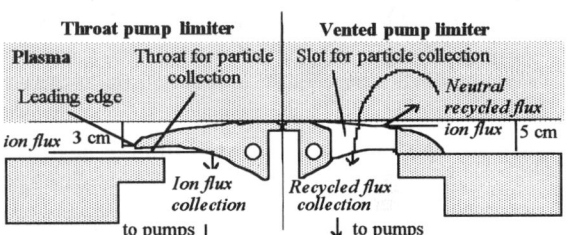

Figure 1 : Comparison between the particle collection mode of a throat and a vented pump limiter

2. EXPERIMENTAL SET UP

Tore Supra is equipped with six vertical modular TPLs located at the bottom of the machine. These limiters are semi inertial components (cooled between shots) made of CFC tiles (Aerolor A05 from Le Carbone Lorraine) assembled with CFC locking pins in order to avoid differential dilatation during long pulses with additionnal power. The only limitation is the CFC surface temperature (a reasonable limit is 1500 °C). Each limiter is then able to sustain 0.7 MW of conducted power for 8 s, corresponding for the 6 limiters to a global power of 5 MW during 5s, or 13 MW during 1s.

One of the vertical TPLs has been removed and replaced by a prototype of VPL. This prototype is also a CFC semi inertial component, in which the throat for particle collection has been removed and replaced by slots between tiles in the toroidal direction (24 slots of 1.2 cm x 9.5 cm), as it is shown on Figure 2. The slots make a small angle with the toroidal direction (0.13 rad), in order to avoid a secondary leading edge on the back of the slots (see Figure 2). The VPL has roughly the same size (40 x 40 cm) and the same back conductance (~4 m^{3}s) than the vertical TPLs of Tore Supra, which allows a direct comparison between the performances of both limiters.

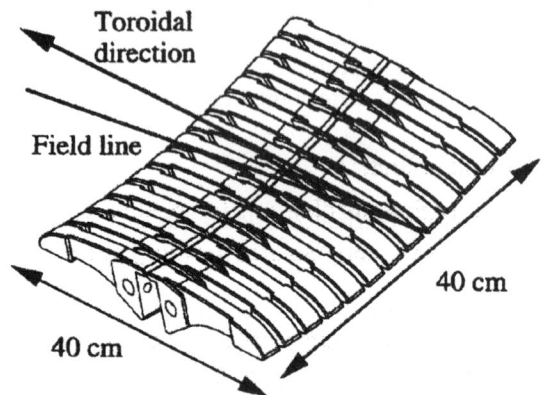

Figure 2 : The Tore Supra vented limiter head

Both limiters are equipped with a baratron pressure gauge and a titanium getter pump, ensuring a pumping speed from 3 to 15 m^{3}s^{-1} for hydrogen or deuterium (helium is not gettered by titanium).

3. EXPERIMENTAL RESULTS

3.1. Particle exhaust

In order to characterize the particle collection by both limiters, experiments have been performed with the plasma lying simultaneously on the VPL and one of TPLs in ohmic discharges for various plasma densities, whithout active pumping.

The neutral pressure in the VPL plenum is found to be lower than the pressure in the TPL plenum by a factor ~3 for deuterium plasmas and ~4 for helium plasmas, as it is shown on Figure 3.

It must be noted that the neutral pressure shows an almost quadratic evolution with the plasma density, for the VPL as well as for the TPL, in helium or deuterium plasmas.

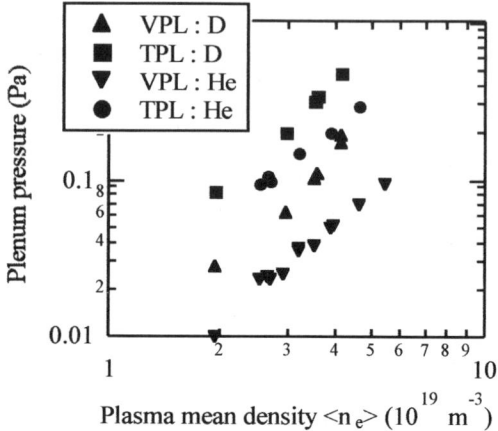

Figure 3 : Pressure evolution as a function of plasma density for the TPL and VPL

Experiments with auxilliary heating have also been performed [4]. For a given plasma density, the plenum pressure increases roughly as the cubic root of the total input power, for the TPL as well as for the VPL.

3.2 Density control

The ability of the VPL to actually control the discharge density has been tested on deuterium shots with the plasma lying on the VPL alone, the titanium getter being activated. It has been shown that the VPL is able to maintain a constant density plateau at a given value for several seconds.

The corresponding pumping efficiency of such a device, which is the ratio of the extracted flux by the plasma outflux, is found to range between 5 to 12 %, depending mainly on the plasma density. These values are relevant for reactor conditions, where a pumping efficiency of the order of 10 % is needed to maintain a sufficiently low helium concentration in the core plasma [5].

Moreover, the VPL has been tested in a discharge with a highly radiating boundary plasma ($P_{rad}/P_{tot} > 80$ %). In this configuration, the parallel ion flux is strongly decreased, leading to a reduced pumping efficiency for TPLs, while the particle exhaust in the VPL is still significant [6]. This might be promising for particle control in highly radiating scenarios, contemplated to reduce the heat load on the plasma facing components for ITER.

However, simulations of the VPL have shown that the pumping efficiency for helium should decrease with decreasing helium concentration [7] (see Figure 4). This leads to an unbalanced pumping for helium and deuterium in typical reactor plasmas with 10 % helium (ratio 3.5 between the pumping efficiencies as shown on Figure 4). This point, linked to a difference in the atomic physics of helium and deuterium, should be further investigated experimentally.

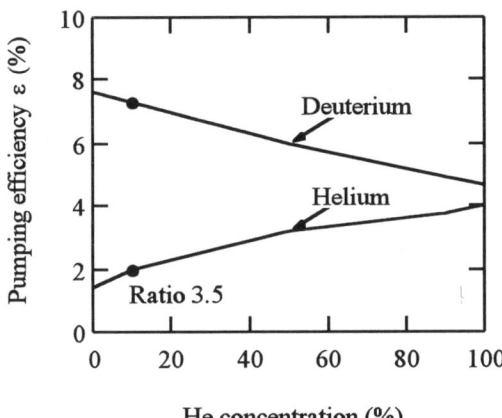

Figure 4 : Simulation of the VPL pumping efficiency for helium and deuterium as a function of helium concentration in the plasma

It has to be noted that this prototype of VPL is not optimised for particle collection. Indeed, the simulations have shown that it is possible to gain a factor 2 on the pumping efficiency by a careful shaping of the slots [7]. However, this optimisation would not reduce significantly the discrepancy between the helium and deuterium pumping.

3.3 Heat exhaust

The surface temperature distribution is presented for both limiters on Figure 5. As is seen from the infra-red imaging, for the VPL, the surface temperature remains quite uniform, while, for the TPL, the surface temperature is significantly higher on the leading edge than on the rest of the tiles. The ratio between the maximum temperature and the average temperature of the tiles is close to 1 for the VPL, while it is around 1.7 for the TPL.

It is important to note that, after a significant experimental use with additionnal power up to 6 MW for 3 s, leading to a surface temperature higher than 2000 °C, no damage has been observed on any of the limiters.

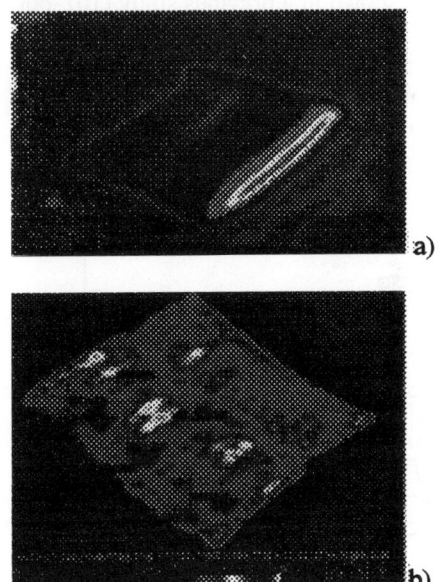

Figure 5 : Infra-red imaging of the TPL (a) and the VPL (b)

3.4 Thermal analysis

A simple model of the scrape off layer [8] shows that the heat flux Q can be written as :

$$Q = Q_0 \exp(-\frac{x}{\lambda_q})$$

where Q_0 is the heat flux at the last closed flux surface, x the distance from the last closed flux surface and λ_q the characteristic decay length.

λ_q is a crucial parameter in the design of the plasma facing components. It is therefore important to estimate its behaviour with the main plasma parameters, such as plasma current or additionnal power.

The infra red imaging of the VPL has been used to study the λ_q behaviour. The heat flux has been estimated from the surface temperature for several radial locations on the tiles (see the crosses on Figure 5), using the CASTEM 2000 code [9]. The characteristic decay length λ_q is then deduced from the spatial evolution of the heat flux.

Discharges with up to 6.5 MW of total input power and plasma current ranging from 1 to 1.6 MA have been analysed. As is seen from Figure 6, λ_q decreases with increasing plasma current, as has been already shown in a previous study [10], but also seems to decrease with increasing input power. It ranges between 1.5 cm (low plasma current, no additionnal power) and 1 cm (high plasma curent, large additionnal power).

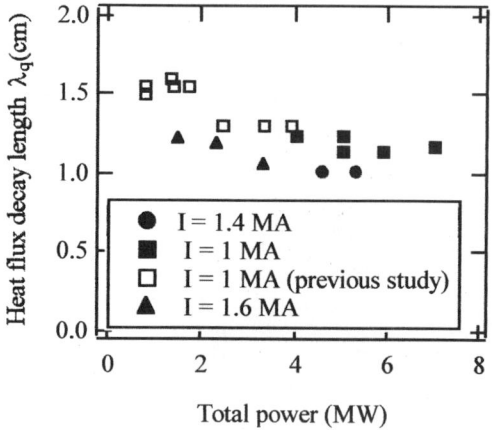

Figure 6 : Evolution of the heat flux decay length as a function of the input power for different plasma currents

4. CONCLUSION

Experiments performed with the Tore Supra VPL have shown that vented structures are a potential candidate to achieve simultaneously heat and particle exhaust.

Indeed, the Tore Supra VPL is able to control the plasma density, even though its pumping efficiency is lower than the pumping efficiency of the TPL. Moreover, the VPL might be a promising concept for pumping in highly radiating scenarios. However, the helium pumping of such a device has to be further investigated, particularly for plasmas with a low helium concentration.

The main advantage of the VPL over the TPL is a uniform heat flux distribution on the limiter head, without any local thermal overload on the leading edge. The heat flux decay length has been estimated from infra-red imaging of the VPL, showing that it decreases with increasing the plasma current and the input power.

In order to further explore the potential of such a concept, vented structures have been installed on the neutralisers of the ergodic divertor for the next experimental campaign [11].

REFERENCES

[1] : P. K. Mioduszewski, "Advanced limiters", in Physics of Plasma-Wall Interaction in Controlled Fusion, (Proc. NATO Advanced Study Institute Val-Morin, 1984), NATO ASI Series, Vol. 131, Plenum Press, New York (1986), 891.

[2] : The Tore Supra team, presented by D. Guilhem, IAEA-CN-60/A2-4-P-17, 15th International Conference on plasma physics and controlled nuclear fusion research, Sevilla, Spain, 1994.

[3] : T. Loarer et al., Nucl. Fus. **36** (1996) 225.

[4] : B. Pegourié et al., « Throat and vented pump limiters particle collection experiments with auxilliary heating in Tore Supra », (Proc. 12th PSI Conf.Saint-Raphaël, 1996), accepted for publication in J. Nucl. Mater.

[5] : M. Chatelier, Vacuum, **47**, 6-8, (1996) 963.

[6] : T. Loarer et al., Plasma Phys. and Contr. Fus. (Proc. 22nd Eur. Conf. Bournemouth, 1995) **37**, 203.

[7] : E. Tsitrone et al., 22nd EPS Conference on Controlled Fusion and Plasma Physics (Proc. 22nd Eur. Conf. Bournemouth, 1995) **Vol 19C part IV** EPS (1995) 301.

[8] : R. W. Conn, J. Nucl. Mater., **128 & 129** (1984) 407.

[9] : P. Verpeaux et al., CASTEM 2000 : A modern approach of computerised structural analysis, Proc. of recent advances in design procedure for high temperature plant, Risley, November 1988.

[10] : D. Guilhem et al., J. Nucl. Mater, **196-198** (1992) 759.

[11] : L. Doceul et al. , « Upgrading of the Tore Supra ergodic divertor », this conference.

Divertor II Plasma Facing Components for ASDEX Upgrade

S. Deschka[1], S. Schweizer[1], B. Streibl[1], ASDEX Upgrade Team[1], C. Garcia-Rosales[2], G. Hofmann[3] and J. Linke[4]

[1] Max-Planck-Institut für Plasmaphysik, EURATOM-Association, D-85748 Garching, Germany
[2] Centro de Estudios e Investigaciones Tecnicas de Guipuzcoa, E-20080 San Sebastian, Spain
[3] Forschungszentrum Karlsruhe, EURATOM- Association, D-76021 Karlsruhe, Germany
[4] Forschungszentrum Jülich, KFA-EURATOM, D-52425 Jülich, Germany

The paper describes in particular the plasma facing components of the LYRE configuration of Divertor II and the results of tests for technological evaluation. The CFC strike point tiles account for heat fluxes up to 15 MW/m². Segmentation reduces thermal stresses and bowing. The baffle tiles are of fine grain graphite. High heat flux tests and heat transfer measurements proved that the strike point tiles provide enough heat conductivity and capacity to allow 3s pulses at full heating power, and the design withstands cyclic loading without degradation. A compliant layer between the carbon tiles and the cooled support plates is mandatory for sufficiently fast cooling.

1. INTRODUCTION

The main configurational modifications of Divertor II, which will be installed in ASDEX Upgrade in late 1996, are the increased length of the divertor fan and the employment of baffles [1]. Three configurations are envisaged: the LYRE (fig. 1) with highly inclined S-shaped strike point tiles, and two GASBAG versions with target tiles intersecting the separatrix at an angle of 45°, and normal to the incident magnetic field lines, respectively. These configurations can be complemented with a roof baffle or a septum in the private flux region [2]. Experimental operation will start in spring 1997 with the LYRE and a roof baffle.

All Divertor II configurations are mounted on the same base support structure. This is obtained by a modular concept enabling exchange of the respective strike point modules and removing the transition modules of the LYRE for installation of the GASBAGs. This can be done during the routine maintenance opening periods of approximately 2.5 months. Cooling loop connections to integrated water manifolds allow short dismantling times.

2. STRIKE POINT MODULES

The plasma facing materials are carbon fibre composites (CFCs) both for LYRE and GASBAG versions. The extremely high heat fluxes during

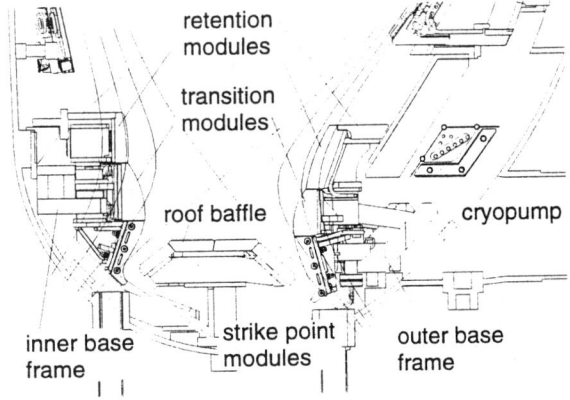

Fig.1: LYRE configuration of Divertor II

start-up of the GASBAGs will demand for CFC materials with even higher thermal conductivity than the one chosen for the LYRE. High-conductivity materials, developed by Industry for NET and ITER, are readily available in the required scales [3]. For the GASBAG versions, boron or silicon doped CFC materials could be an alternative to reduce erosion, however, their main benefit is expected to lie in applications with target temperatures near the maximum of chemical erosion (≈600°C), e.g. baffles [4].

Before installation in the vacuum vessel, the target plates are subjected to a leakage test with helium at 150°C, which is the baking temperature of the vacuum vessel. For inspection of the integrity of the brazings between steel plates and cooling tubes, transient thermography (TT) was performed, where the coolant temperature was changed rapidly from 20°C to 85°C and, after thermally stationary conditions were achieved, to 20°C. From deviations in the infrared images of the plates during the heat-up and the cool-down phases, defects with a length above 3cm at the brazed joints could be detected. In an additional test, the plates were cooled to approximately -30°C, then removed from the freezer and exposed to air to allow the formation of a frost layer on the plates, and subsequently heated with hot water flowing through the cooling tubes. By the pattern of the frosted areas on the plates, defects could be identified with higher resolution than by TT.

2.1. LYRE Target Tiles

The S-shaped strike point tiles are steeply inclined to spread the incident power flux homogeneously over a larger area and concentrate cold particles recycling from there most effectively on the high-heat-flux field lines.

The CFC material of the strike point tiles is 3D SEPCARB N11-012. Its high thermal conductivity in 2 directions (≥ 140 W/mK at 300°C) is required to keep the surface temperatures below 1500°C at heat loads of 15 MW/m^2 for 2s. The CFC is oriented with its lower thermal conductivity (≥ 90 W/mK at 300°C) in toroidal direction, in order to enhance distribution of the peaked temperature profile in poloidal direction and to conduct the heat from the surface into the depth. The tile thickness allows a pulse length of 3s at full power with a resulting surface temperature of ≤ 1800°C.

In toroidal direction, the strike point tiles are subdivided into individual elements to reduce the stresses and misalignment due to warping resulting from thermal loading. The outer tiles each consist of 4 elements with approximately 40 mm width, the inner tiles each of 3 elements. The individual elements of each tile are stacked and connected by 3 tension rods with disk washers. The stack is attached to the cooled support plate by two bolts per element and disk washers to limit the fixation forces and and to ensure mechanically independent movement of the individual element.

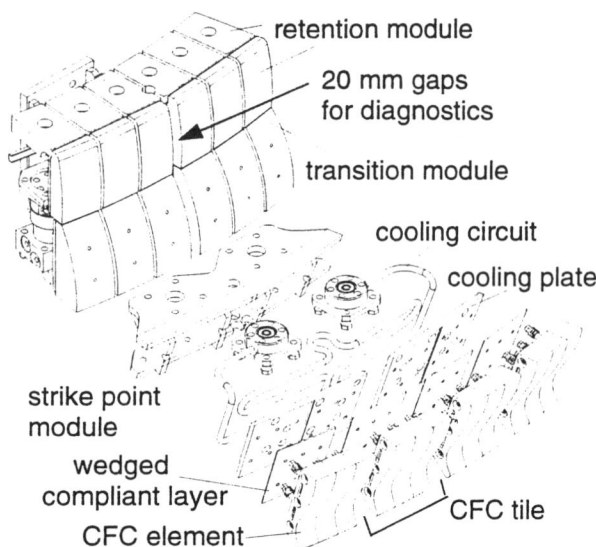

Fig. 2: LYRE strike point modules, transition and retention modules

To avoid hot spots at the edges between neighbouring tiles, the edges are shadowed. The step height is 1.5 mm, leading to a heat load increase by a factor of 1.3 in comparison to non-tilted tiles. Tilting is achieved by a wedged compliant layer of PAPYEX (flexible graphite foil) between CFC tile and support plate. This concept allows to reverse the shadowing direction by exchanging only the compliant layer, without the necessity of machining the CFC tiles. This can be done during routine maintenance. The PAPYEX layer is also mandatory to enhance heat removal and to keep the interval times sufficiently low.

3. BAFFLES

Adjacent to the strike-point modules towards the X-point are the transition modules, which extend, above the X-point, to the retention modules (fig. 2). These component groups assume the function of baffles. After removal of the transition modules, the sideward influx of neutrals to the divertor fans is enhanced by the large gaps, leading to a "flame-like" detachment of the plasma characterizing the GAS-BAGs [1]. The roof baffle in the private flux region improves impurity retention and allows to control the neutral gas pumping efficiency. This is achieved by increasing the gap width between tiles, thus increasing the neutral gas flow towards the plasma. Further, the roof baffle protects the bottom of the vacuum vessel from radiation and energy deposited by plasma disruptions and serves to obtain simultaneous detachment of inner and outer divertor fan.

3.1. Retention and Transition Modules

According to the lower heat fluxes on the retention and transition modules (≤ 5 MW/m²), fine grain graphite is a suitable material for the plasma facing tiles of the baffles. The type chosen is Schunk FP-379. Segmentation of the tiles is not required due to the lower stresses and displacements during thermal loading. Edges are shadowed by tapering only between adjacent modules, as there the gaps are 20 mm for diagnostics, whereas the gap width between neighbouring tiles within the same module is 6 mm. The step height of ≈ 1.5 mm results in a heat flux increase by a factor of 1.8 compared to non-tapered tiles.

The graphite tiles are fixed to the cooled support plates by bolts and disk washers. A PAPYEX compliant layer improves heat conductance to the coolant.

3.2. Roof Baffle

The module design of the roof baffle corresponds to the retention and transition modules, as the thermal loads and mechanical requirements are similar. Fine grain graphite (Ringsdorff R6650 MX2) is used for the plasma interactive tiles.

The edge and top tiles of the roof are a critical zone, as the separatrix may strike these tiles. This transient configuration is expected to occur only during certain "advanced tokamak conditions" with high triangularity. For such discharges, reduction of NBI heating power and plasma current are foreseen, despite a short pulse length, to keep the surface temperatures below 1600°C to avoid excessive carbon erosion and damage of the tiles by thermal shock.

4. TESTS FOR TECHNOLOGICAL EVALUATION

High heat flux tests were performed in the 200 kW electron beam (EB) machine FE-200 at Framatome, Le Creusot [5] and the 6 MW hydrogen beam (HB) test stand MARION at KFA Juelich [6]. Heat transfer measurements were made at the FIWATKA facility at the Research Centre Karlsruhe [7].

4.1. High Heat Flux Tests

Materials tests were carried out in MARION in 1995 by loading CFC materials and fine grain graphites with power densities up to 42 MW/m² for 1-2s. The pulse lengths for reaching a surface temperature of 1500°C and 1800°C, respectively, at 15 MW/m² were determined, and the minimum thermal and mechanical requirements of the target material were deduced from the results. Thus, the average thermal conductivity, determined in the interval between RT and 1000°C, which is proportional to the resulting surface temperature increase, is required to exceed 100 W/mK

Prototypes of the outer strike point modules were subjected to cyclic heat fluxes of 15 MW/m² in the FE-200. The pulse lengths were chosen to reach surface temperatures of 1500°C (level 1) and 1800°C (level 2), respectively. The CFC was Dunlop DMS 704, the only available at that time. A prototype test with SEPCARB N11-012 tiles is scheduled for October 1996 in the MARION facility. In the EB test, the maximum of the peaked heat flux profile simulating the separatrix was set on three positions: nominal strike point, 15 mm and 30 mm towards the lower end of the tile.

During heating, the temperature profile broadened significantly with respect to the applied heat flux profile, demonstrating the high thermal conductivity of the material in the directions normal to the heat flux profile and into the depth. At the peak temperature value of level 1 at the nominal position (pulse length 3.5s), the measured half width of the temperature profile (46 mm) was a factor 2.4 higher than the power half width (19 mm). Due to this broadening, the surface effective for temperature radiation and the heat capacity are increased, thus reducing the surface temperature during the pulse, as predicted by 2D finite element calculations.

The bulk temperatures stayed below 300°C, even at 4.5s pulse duration and 360s interval time, a value which corresponds well with the predictions from the heat transfer measurements.

Under cyclic loading (1000 cycles), no degradation in thermomechanical behaviour occurred. Heat transfer increased slightly with the first 50 pulses at each power level and stayed constant throughout the rest of the test. This indicates slight plastic deformation of the compliant layer under the clamping forces exerted by the tile and better adjustment (nestling) to the CFC and steel surfaces.

4.2. Heat Transfer Measurements

The time required for cooling from a thermally stationary condition with 300°C in the CFC tile (corresponding to loading with the maximum discharge energy assumed for Divertor II) to a temperature of 150°C was measured. From the temperature gradients measured at different positions in the CFC and the steel plate and by using cooling water calorimetry, an average heat transfer coefficient between tile and steel plate, k_{avg}, was calculated. Compliant layers from different manufacturers (PAPYEX by Carbone Lorraine, SIGRAFLEX by SGL) in various thicknesses (1, 2 and 5mm) and configurations, with and without an additional electrically insulating mica layer, were investigated.

The results are shown in fig. 3. With a SIGRAFLEX compliant layer of 1 or 2 mm in thickness, re-cooling of a tile loaded with the maximum discharge energy can reliably be achieved within 6 to 8 minutes. Without the compliant layer the cooling time increased by a factor of 2.8, and even by a factor of 3.6 if a 0.3mm thick mica foil was added to the compliant layer. Therefore mica foil will not be applied for electrical insulation of plasma facing tiles.

Due to the spatial non-uniformity of the heat sink the heat flux through the compliant layer is strongly concentrated at zones near the cooling tubes. This was demonstrated by aligning stripes of the compliant foil on the steel plate only above the cooling tubes, which resulted in almost unchanged cooling times. In case of the 5mm PAPYEX, the time was even shorter, which is probably due to an improved contact by the increased face pressure to the stripes.

For the stripes the heat transfer coefficients k_{str} are much higher, since with the stripe area only the most effective half of the total surface entered the calculation. This is the reason why k_{str} should not be compared directly to values of k_{avg} for a 100% coverage with compliant foil; the cooling time is a more reasonable quantity for a comparison.

For 5mm thick PAPYEX the cooling time is significantly lower than for 2mm SIGRAFLEX, both as stripes: 310s compared to 430s. Though its greater thickness and lower density (0.4 compared to 1.0 g/cm^3) tend to decrease its thermal conductance, this is overcompensated by its better deformability (nestling of surfaces) and hence by lower contact resistance. However, as the 5mm low-density PAPYEX is not available in the required scale, it was not investigated further.

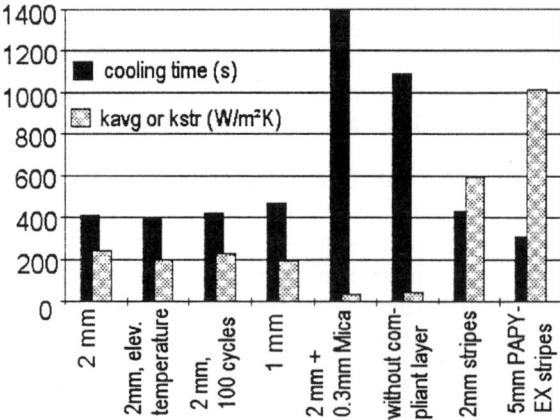

Fig. 3: cooling time (300→150°C), k_{avg} and k_{str}, resp. for different compliant layer configurations

Under cyclic loading (100 cycles), the cooling time did not change significantly. After exposure to elevated temperatures (500°C at thermally stationary condition), the cooling time decreased slightly, but not significantly. This decrease is probably caused by enhanced nestling to the CFC and steel surfaces due to the better deformability at higher temperatures.

REFERENCES

[1] H.S. Bosch, D. Coster, S. Deschka, W. Engelhardt, C. Garcia-Rosales, O. Gruber, M. Kaufmann, W. Köppendörfer, K. Lackner, J. Neuhauser, H. Salzmann, H. Schneider, R. Schneider, S. Schweizer, B. Streibl, M. Troppmann, "Extension of the ASDEX Upgrade Programme: Divertor II and Tungsten Target Plate Experiment", Technical Report 1/281a, IPP, Garching, Germany, 1994.

[2] B. Streibl, S. Deschka, G. Hofmann, K. Mattes, J. Perchermeier, H. Schneider, S. Schweizer, M. Weißgerber, ASDEX Upgrade Team, Proc. 16th SOFE, Champaign, Ill., USA (1995)

[3] C. Wu, NET materials report No.

[4] J. Roth, H. Planck, R. Schwörer, Phys. Scr. T64 (1996), 67

[5] H. Viallet, Note P Co 91/04, CEA Cadarache (1991)

[6] M. Lochter, Fus.Tech. 19, 4 (1991), 2101-2111

[7] G. Hofmann, E. Eggert, The First Wall Test Facility FIWATKA, KfK 5381 (1994)

ACTIVELY COOLED TEST LIMITER FOR TEXTOR

W. Hohenauer, H. Bolt, T. Koppitz, J. Linke, R. Lison, W. Malléner, V. Philipps,
M. Sauer, R. Uhlemann, J.H. You, H. Nickel

Forschungszentrum Jülich GmbH, Association KFA-Euratom, D-52425 Jülich, Germany

To investigate the erosion and redeposition phenomena of fusion related materials under stationary conditions, actively cooled test limiters for TEXTOR (Tokamak Experiment for Technology Orientated Research) were developed. They allow experiments under stationary conditions within the plasma pulse length of 10 s. Heat loads of typically 10 MW/m² are removed by pressurized water: volume flow measures 10 m³/h, pressure 15 bar and the minimum coefficient of heat transfer is nearly 50.000 W/m²K. Prototype limiters were built both as brazed composites of a C/C material (SEPCARB N11) and a TZM substrate as well as modules with thermally sprayed LPPS (Low Pressure Plasma Spraying) - tungsten coatings. The samples were tested successfully in screening tests in the ion beam facility MARION (Material Research Ion Beam Test Facility) with hydrogen beams. Maximum heat loads up to 17 MW/m² were applied without any failure of the cooling system. Steady state of the surface temperature was measured within 5 s. An advanced brazing technique enabled the joining of hemispherically shaped C/C shells with a TZM heat sink without failure. An optimized test limiters was tested in TEXTOR. Analytical and numerical models describing the effects of the heat load distribution, spacial temperatures and stresses were experimentally verified.

1 Introduction

Investigations of plasma wall interactions are one of the main activities in fusion research. Advanced plasma conditions at TEXTOR-94[1] with a maximum plasma pulse length of 10 s allow to study of erosion and redeposition phenomena under stationary conditions. Therefore actively cooled test limiters were designed and tested, which allow different materials to be exposed to fusion relevant particle- and heat fluxes. With respect to the ITER Interim Design report[2] tungsten coatings and C/C materials brazed to a metallic heat sink as plasma facing materials were used.

2 Design

The power load caused by the TEXTOR plasma mainly impacts with its toroidal component. Poloidal and radial components are much smaller and can be neglected. The local power impact $P_{/A}$ is described by equation (1).

$$P_{/A}(\delta,\varphi) = P_{/A}^0 \cdot e^{\frac{z(\delta)}{L}} \cdot \cos\tau(\delta) \cdot \sin\varphi \quad (1)$$

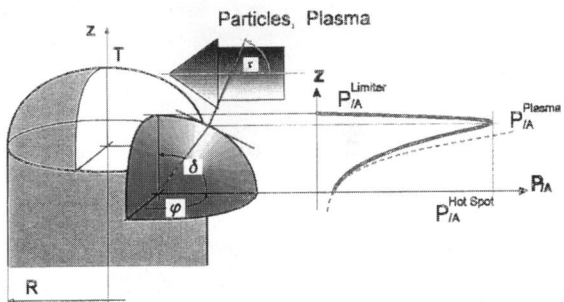

Fig. 1 Geometric details of a rotationally symmetric, elliptically shaped test limiter.

L is a characteristic length of the TEXTOR plasma, typically 8 to 10 mm. Geometric details are shown in fig. 1. $P_{/A}^0$ is the heat load in the origin of the describing coordinate system. The thermal response of the test limiter leads to a hot spot formation, which depends on both, the characteristic length L and its geometry. To egalize shift effects caused by the poloidal and radial flux componets, in general a rotationally symmetical, elliptical shape is used. Hot spot coordinates, which are of interest for the cooling design, can be calculated by the extremal criteria given in equation (2).

$$\frac{\partial P_{/A}}{\partial \varphi} \xrightarrow{extr.} 0 \Rightarrow \varphi^{HotSpot} = 90°$$

$$\frac{\partial P_{/A}(\delta)}{\partial \delta} \xrightarrow{extr.} 0 \Rightarrow \delta_{extr.} = \delta^{HotSpot} \quad (2)$$

The use of the load lock system TEXTOR-Schleuse-III limits the maximum diameter of inserted structures to 120 mm. With respect to stress gradients caused by gradients of the surface distribution of the thermal load, a detailed analysis of the heat distribution on the surface, led to the selection of a hemispherical shape. The diameter of the hemispherical top of the test limiter was set at 80 mm. The heat load distribution is shown in fig. 2 and refers to a hot spot angle of $\delta^{HotSpot} \cong 64°$. It can be calculated analytically by formula (3).

$$\delta^{HotSpot} = \arcsin\left(-\frac{L}{2.R} \pm \sqrt{\left(\frac{L}{2.R}\right)^2 + 1}\right) \quad (3)$$

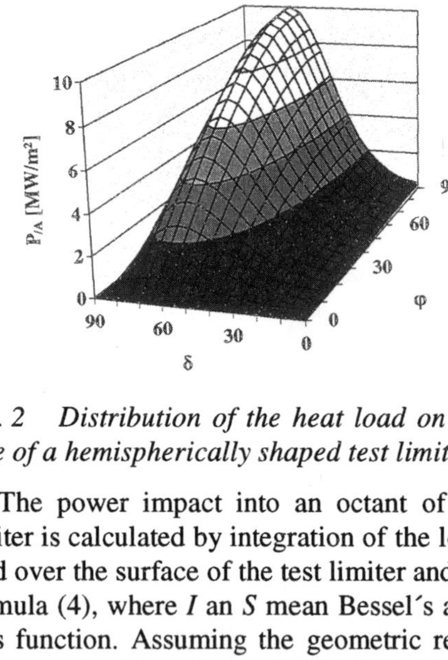

Fig. 2 Distribution of the heat load on the surface of a hemispherically shaped test limiter.

The power impact into an octant of the test limiter is calculated by integration of the local heat load over the surface of the test limiter and leads to formula (4), where I an S mean Bessel´s and Struve´s function. Assuming the geometric restriction and a nominal heat load of 10 MW/m² in the hot spot, it is calculated with 6.25 kW.

$$P_{Oct.} = P_{/A}^0 . R . L . \frac{\pi}{2} . \left[I_1\left(\frac{R}{L}\right) + S_1\left(\frac{R}{L}\right) \right] \quad (4)$$

Due to the hot spot formation, watercooled test limiters are designed as a compound of a plasma facing material and a metallic heat sink, which provides a maximum coefficient of heat transfer in the hot spot area. The principle of design is shown in fig. 3. A central water supply and the coaxial geometry of the cooling channel multiply the coefficient of heat transfer of a linear system, as shown in fig. 4, by a factor of 2 - 3. It can be calculated by equation (5)[3], where $f_{medium}(T)$ describes the temperature dependence of the thermohydraulic properties of the cooling medium, u it´s velocity and l the local hydraulic diameter.

$$\alpha (T) = f_{medium}(T) . \frac{u^{0,8}}{l^{0,2}} \quad (5)$$

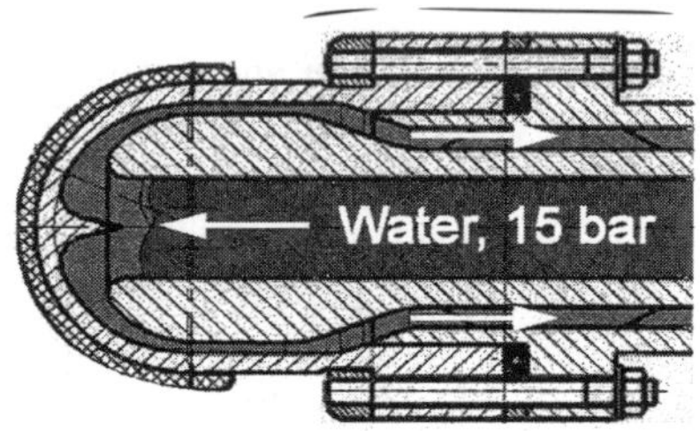

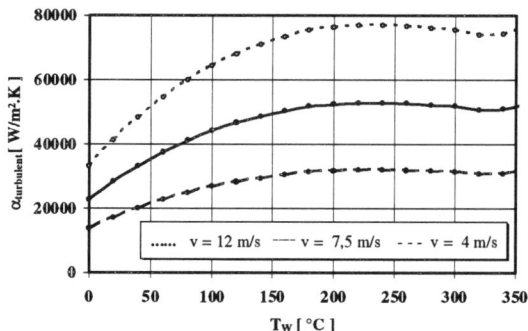

Fig. 3 Design concept of a hemispherically shaped test limiter. Due to the insert in the TEXTOR Schleuse-III the maximum diameter of the half sphere is limited to 80 mm.

Fig. 4 Coefficient of heat transfer, calculated for different velocities of cooling water and specialized at the hydraulic diameter in the hot spot region. The curved geometry in the hot spot region increases these values by a factor of 2-3.

Detailed numerical analysis of the spacial temperature predicts a surface temperarture in the cooling zone below hot spot near 220°C. Pressurized water at 15 bar and a minimum velocity of flow of 4 m/s (standard cooling conditions) ensure subcooling conditions. Therefore the minimum coefficient of heat transfer is 50000 W/m²K. Under standard conditions, the velocity of cooling water is $u \cong 7.5$ m/s and $\alpha \cong 100000$ W/m²K. The critical heat flux[4] can be estimated at 30 MW/m².

For the material combination SEPCARB-N11 / TZM an elasoplastic stress analyses was performed. Due to symmetry, only a quarter of the hemispherical geometry was modelled. The thin, ductile copper braze was neglected, the anisotropy and temperature dependence of material properties were taken into account. The kinematic constraints applied to the finite element model correspond to the actual geometry: the bottom surface of the TZM heat sink in the equator area was assumed to be pinned, fixing all degrees of freedom. Results show that the stress state of C/C armour tiles is not affected by loading. The stress state in the TZM body, however, is markedly relieved during the thermal loading. Stress concentrations occur in the bonding region near the equatorial boundary. In the region below the hot spot, stresses in both materials, C/C and TZM, come up to ≈ 25% of the equatorial maximum values. Due to the fact that thermal stress, caused by a mismatch of the coefficients of thermal expansion, decreases when temperatures in the bonding layer approach to brazing temperatures, in the equatorial region the stress level is reduced significantly during the heat load. The effect is neglegible in the hot spot region. Figure 5 shows the normalized, transient von Mises equivalent stresses generated in the C/C and TZM matertial in both regions of main interest: the hot spot zone and the equatorial area with the maximum stress concentration. Steady state is reached within 5 s and corresponds to the temperature behaviour.

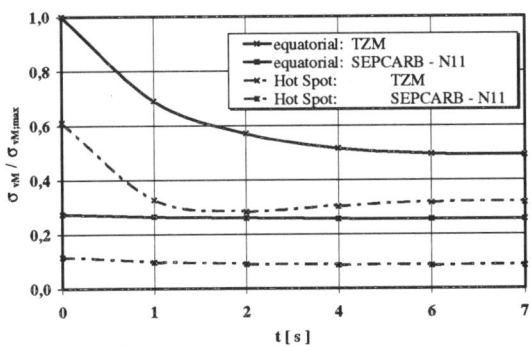

Fig. 5 Transient, normalized von Mises equivalent stress in the equatorial area of maximum stress concentration and in the hot spot zone.

As a consequence of stress analysis, the thickness of the brazing layer was increased towards the equatorial region. Surplus braze depots, provided in the C/C shell, avoid sinkholes during the solidification and form a chamfer between the C/C shell and the TZM body, which minimizes stress singularity effects. Additionally an optimized furnace programme, which ensures an directional solidification of the braze, enabled a defect-free joint. As a ductile braze with good wetting to both joining tiles Cu1Cr is used. In fig. 6 the C/C shell and the brazed test limiter are shown.

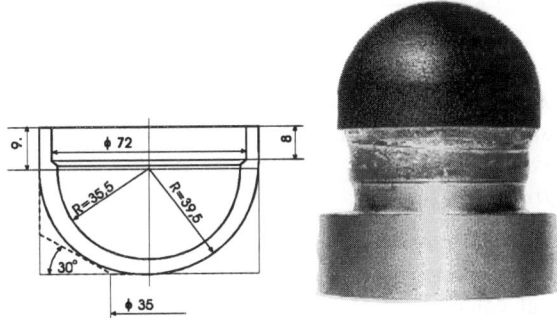

Fig. 6 Drawing of the C/C shell and brazed test limiter; SEPCARB-N11 / Cu1Cr / TZM compound.

Tungsten coated test limiters were also produced. An optimized LPPS (Low Pressure Plasma Spraying) procedure[5] guarantees theoretical coating densities ρ_{th} above 90%, typically 92% and a homogenious distribution of microporosity[6]. Coating thickness were between 500 and 1200 μm.

3 First Experimental Results

Prototype brazed SEPCARB-N11 / Cu1Cr / TZM as well as W-LPPS coated limiters were successfully tested under hydrogen load in KFA´s ion beam test facility. In experiments with W-LPPS coated limiters the numerical predicted transient temperature behaviour could completely be verified. The maximum applied heat load was 17 MW/m², the pulse length was 10 s, steady state was reached within 2 s. In an additional experiment loss of coolant was simulated: the velocity of cooling water was reduced to 0.25 m/s and pressure was decreased to 5 bar. It could be shown, that the nominal heat load of 10 MW/m² could be removed without failure. Under these reduced cooling conditions heat fluxes higher than 10 MW/m² generated boiling regimes. The transient temperature behaviour under standard cooling conditions and under reduced cooling conditions, with a heat load generating transition boiling, is shown in fig. 7.

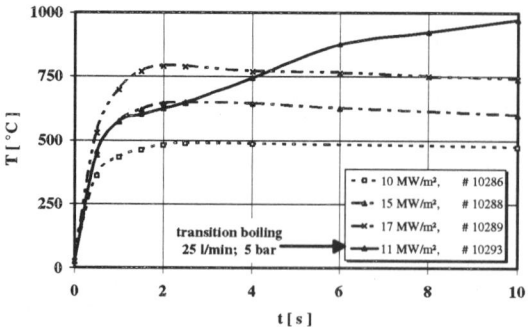

Fig. 7 Transient temperature behaviour under standard cooling conditions and under reduced cooling conditions, with a heat load that generates transition boiling.

The temperature response of prototype brazed compounds was twice that expected from numerical calculations. The reason is the non-otpmized joining technique, which produced local bonding failure. The high surface temperatures led to sublimation and redeposition even on the windows of the test facility. This explains the different transient temperature responses as well as the decreasing measured surface temperatures at consecutive heat loads of 17 MW/m² as shown in fig 8.

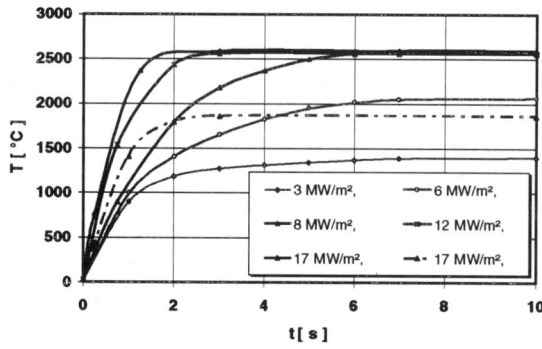

Fig. 8 Transient temperature respons on the surface of prototype brazed testlimiters.

Optimized brazed limiters were inserted in TEXTOR and tested for more than 100 plasma discharges. Heat loads reached 12 MW/m², the avaliiable maximum puls length was ≈ 5 s. The test limiters could nearly reach steady state. The temperature response was in excellent agreement with the numerical predictions. The structural integrity as well as the efficiency of the nominal cooling conditions in TEXTOR could be demonstrated. A comparison between measured and calculated transient temperatures is given in fig. 9. Spacial resolved temperature measurements confirmed the analytically predicted hot spot location.

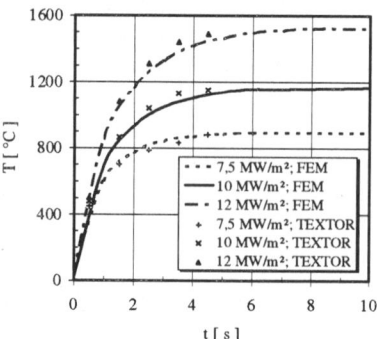

Fig. 9 Comparison of calculated and measured surface temperatures of optimized, brazed test limiters in TEXTOR.

First erosion investigations could be done. Tungsten coated limiters will be tested in TEXTOR in the near future.

4 Conclusions

Water cooled test limiters for TEXTOR were produced as an optimized brazed joint of C/C material SEPCARB-N11 with TZM or W-LPPS coated TZM tiles. They were successfully tested in the ion beam test facility and in TEXTOR. Maximum heat loads reached 17 MW/m². Steady state could be reached within 5,5 s. Numerical models describing transient, spacial temperature response and stresses were confirmed. In the near future actively cooled test limiters will be used in advanced plasma-surface-experiments in TEXTOR.

---

[1] U. Samm; 16th SOFE, Champaign, Illinois; (1995).

[2] G. Janeschitz et al.; J. Nucl. Mater. 220-222; pp. 73; (1995).

[3] W. Wagner; Wärmeaustauscher; 1. Aufl.; Vogel-Verlag, Würzburg; (1993).

[4] J. Schlosser, J. Boscary; Proc. 18th SOFT, Vol.1, pp. 283; (1994).

[5] W. Hohenauer, R. Krismer; Proc. 13th Plansee Seminar, Vol. 3, pp. 397; (1993).

[6] W. Malléner et al.; to be publ. Proc. 9th NTSC; 7.-11. Oct. 1996; Cincinnati, Ohio, USA.

DIVERTOR ENGINEERING FOR THE STELLARATOR WENDELSTEIN W 7-X

H. Greuner, W. Bitter, R. Holzthüm, O. Jandl, F. Kerl, J. Kisslinger, H. Renner
Max-Planck-Institut für Plasmaphysik, D-85748 Garching, GERMANY, EURATOM.-Ass.

For the new HELIAS stellarator W7-X which will become operational on 2004 in Greifswald an "open divertor" concept was developed. The geometry of targets and baffles is optimised in respect of the magnetic structure of the plasma boundary. For the integration inside the vessel a compact and flexible design of the components and interfaces was found. Some critical components, especially the target plates are described.

1. INTRODUCTION

For the new stellarator W 7-X, a divertor system capable of stationary operation with a heating power of 10 MW (ECRF 140 GHz, NBI, ICRF) has been developed in a first step. Related to the five-fold symmetry of the magnetic configuration 10 divertor units (2 units per period) are arranged. A favourable property of the selected magnetic configuration is an inherent divertor with the LCMS (last closed magnetic surface) of the confinement region being either defined by the inner separatrix of islands (intersected by target plates) or by an ergodised boundary with remnants of islands. To achieve effective particle and energy exhaust for a wide range of operating magnetic parameters of W7-X an "open divertor structure" has to be chosen for compromise as a first approach The divertor units combine target plates, baffle plates and a TMP pumping system supplemented by cryogenic panels [1]. The optimisation of the geometry of the divertor (target and baffle plates) was based on field line tracing for the vacuum configurations and simulation of perpendicular transport by "field line diffusion" (Monte-Carlo code) to calculate the power deposition on the targets [2], whereas the pumping efficiency was estimated by means of the EIRENE [3] code taking into account reasonable boundary parameters. The magnetic structures of finite-ß equilibria differ only slightly from the corresponding vacuum field: The edge region ergodises and the width of the macroscopic islands increases with increasing ß, while the O- and X-points of these islands hardly change their positions. Due to this behaviour the island divertor designed for the vacuum field works for finite-ß equilibria also. As in the vacuum case there are no leading edges and the interaction of charged particles leaving the plasma is completely removed from the vessel wall. The intersection angles of the flux bundles on the targets are typically 0 - 3°. Local power densities up to 8 MW/m^2 at the target plates were obtained in a wide range of magnetic parameters for the worst case at low-density and high-temperature plasma. Taking in account neutrals and impurities a significant unloading of the target plates can be achieved already at moderate separatrix densities.[4]. To adjust the magnetic properties of the boundary and influence the plasma outflow, control coils (one saddle coil per unit with dimensions of 0.25 m *1.8 m carrying a relatively small current of up to 20 kA turns) are integrated into the divertor units, as well. Superimposed AC currents with frequencies to the stationary DC currents of the saddle coils significantly enlarge the deposition area for the plasma flow onto the target plates. The target plates are designed for a power load of up to 10 MW/m^2.

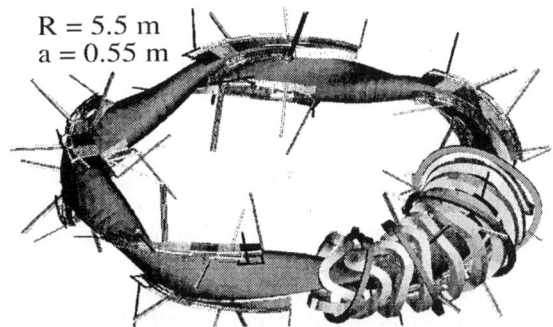

Fig. 1: 3D CAD drawing of W7-X. The plasma column, modular and ancillary coils (one of five periods of the device) are shown. The 10 divertor units, target plates spliced of in modules and the cooling circuits are presented.

2. Divertor Components

An almost complete description and 3D CAD studies of the main divertor components (fig. 1) has been worked out , partly:
- target plates and baffles, including water cooling circuits, feed troughs on the vessel

- control coils
- pumping system, consisting of TM pumps.

Whereas for the additional cryo panels inside the vessel and the wall protection only principal solutions are available, so far. In respect of the geometrical restrictions of the W 7-X vessel (the typical distance between vessel and LCMS is 30-50 cm and between target surface and LCMS is 10 cm) the design of the components and the arrangement must be very compact. To avoid problems with high Z impurities during long-pulse operation for all plasma facing surfaces Carbon was selected. Target plates and baffles will be baked out at 350°C. The wall protection has to be designed for conditioning at 150°C, a limit related to the maximum operational temperature of the inner SS cryostat. Depending on the progress of the physical understanding and control of the boundary during the experiments modifications of the divertor system (vented targets, "closed divertor", change of material etc.) are expected. The design of the components, the supporting and alignment structure of the target and baffle plates, the cooling circuits and the interfaces of the vessel must be flexible for future needs.

2.1. Target Plates

The following criteria are used for the design of the high heat flux components:

- Maximum heat load 10 MW/m^2 during stationary operation
- The typical power deposition area is 3 - 5 m^2. To decouple the plasma from the vessel for all possible magnetic and plasma parameters the wetted area is extended to 22 m^2: per divertor unit 2.2 m^2.
- For easy maintenance and repair, to provide flexibility for the experimental programme and diagnostics the optimised 3D target surfaces are approximated by 2D target elements: Dimensions: width 5 cm, length 27 - 50 cm, partly bended with an angle of 2 - 20°.
- Arrangement of a set of 10 - 15 elements and water manifolds as modules for prefabrication and testing outside the vessel. The target area of one divertor unit being spliced in two parts for effective pumping will be formed by 11 individual modules, finally. 144 elements have to be combined for one divertor unit. By standardisation the number of different types could be reduced to 50, already. The flat elements are mounted on the supporting framework of the modules approximating the calculated 3D surface. Finally, the surface is smoothed by 3D machining to eliminate steps.

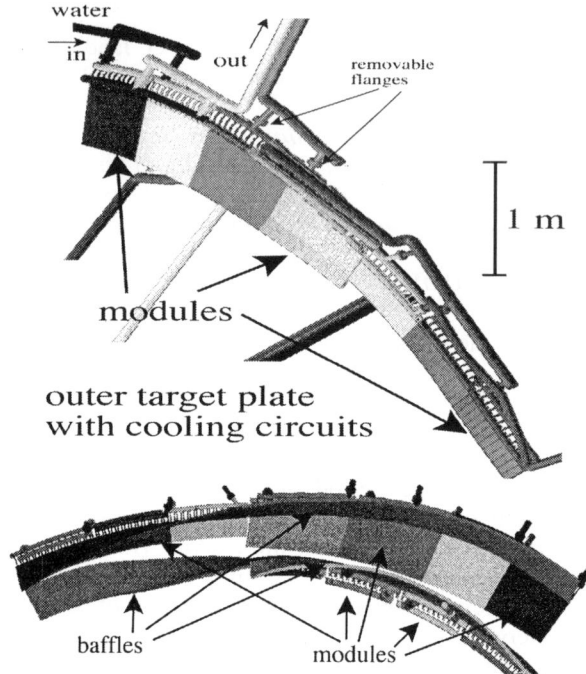

Fig.2 CAD drawing of the outer target plate of one divertor unit: Note, target elements are combined to modules (marked by different grey intensity). The interfaces to the cooling circuits can be identified. The lower part shows the two target plates and baffles of one divertor unit as seen from the magnetic axis. Note the gap for pumping.

The restrictions of the available space for the divertor system inside the vessel W7-X demand a compact design and flexible connections to the in-vessel installed cooling circuits (fig.2).

2.2. Target Elements

A R&D programme has been started for the target element as the most critical component of the divertor. Prototype elements were manufactured by PLANSEE AG and ANSALDO Richerce, already. The manufacturing and testing of the PLANSEE product is described in a special paper at this conference.

The design has to take in account some constraints:
- The surface temperature should not exceed a value of 1200°C at the specified power load

- In respect of the available space a thin and self-supporting element is necessary.
- Water in- and outlet on the side far of the pumping gap.
- Low water flow rates and pressures to minimise the cost of cooling system.

The characteristic data of the elements are summarised in the following table:

Geometry:	
width	5 cm
length	27 - 50 cm
bending angle	0 - 20 °
thickness (complete)	2.2 cm
CFC	0.6 cm
Cooling:	
P/A	10 MW/m^2
av. power	20 kW
max. power flow	90 kW
water:	
temperature	20° in / 70° out
flow velocity	10 m/s
inlet pressure	20 bar

Table 1: Characteristic thermal data of the target element

Several cooling structures have been investigated, such as a smooth tube, swirl tube, hyper vapotron, fin plate. For the prototypes the favourable fin design was selected : A heat transfer up to 10 MW/m^2 can be achieved by forced convection without beginning nucleate boiling safely. The internal fin structure needs the lowest water flow rates and lowest pumping power compared to alternate solutions. The material combination TZM/CFC was chosen to integrate the brazing technique as developed for application of the NET/ITER team. Performing the thermal heat test of the prototype element using the electron beam facility JUDITH of the KFA Jülich (Drs. Bolt, Duwe, Kühnlein) the measured temperature agree very well with results of 3D FEM modelling. The highest temperatures are obtained at the side of the u-bends of the cooling channels (fig.3). To keep the temperature in a tolerable range avoiding degradation of the brazed link of CFC and TZM the power load should be restricted to 8 MW/m^2. Concerning corrosion the location of the brazing link is certainly a critical point of the existing approach. The long brazing line which has to persist over long time without leaks may become problematic.

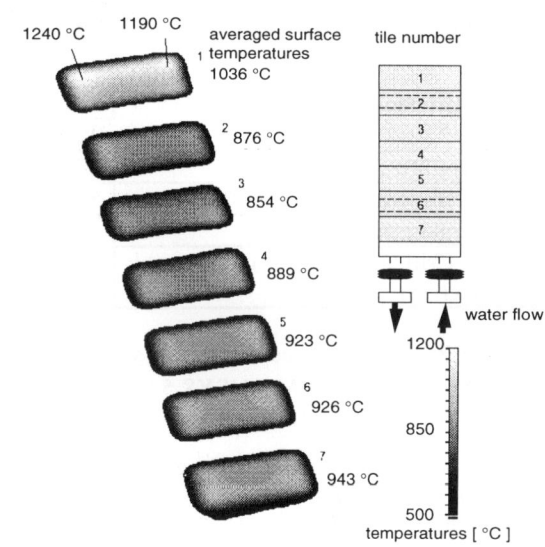

Fig.3 IR images of the prototype target element 2 manufactured for W7-X by PLANSEE AG. The surface temperatures for a power load of 10 MW/m^2 are evaluated (JUDITH / KFA Jülich).

Alternate material and design concepts have to be examined to improve the safety margin for the production. Consequently, a thermal analysis of different solutions for the elements by variation of the design and materials of the cooling structure (Table 2) has been worked out. Both concepts - fin and swirl - using CuCrZr as supporting material lead to lower temperatures at the CFC/metal layer compared to ones based on TZM. The components can be welded by electron beam with a interlayer of OFHC which has to be applied by active metal casting (AMC PLANSEE patent). Recent development by the ToreSupra team has demonstrated the feasibility of this concept [5]. It owns some favourable properties: the brazing is substituted by welding and the lower temperature would allow to increase the thickness of the CFC tiles for longer lifetime of the component becoming exposed to the plasma. Disadvantages concerning material properties and increasing costs have to be carefully ruled out.

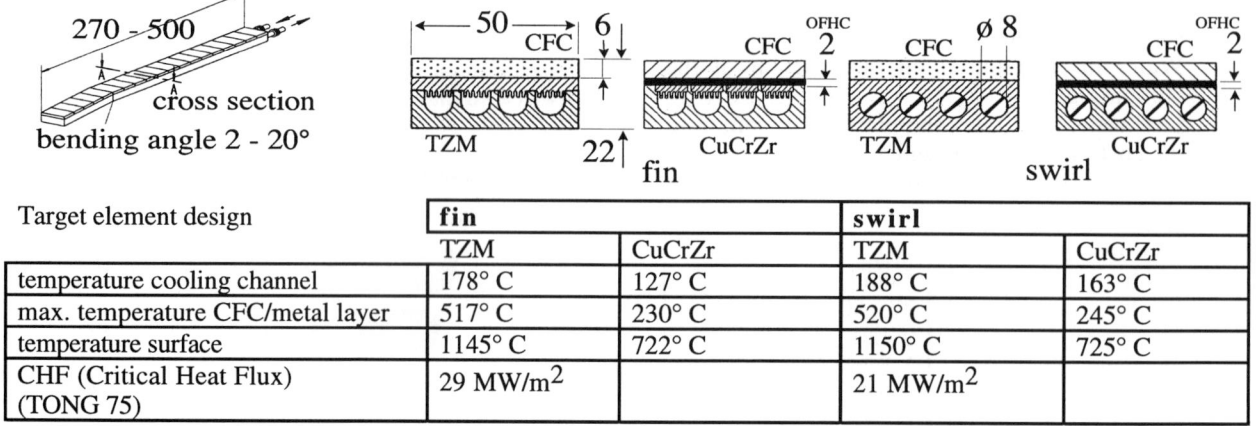

Target element design	fin		swirl	
	TZM	CuCrZr	TZM	CuCrZr
temperature cooling channel	178° C	127° C	188° C	163° C
max. temperature CFC/metal layer	517° C	230° C	520° C	245° C
temperature surface	1145° C	722° C	1150° C	725° C
CHF (Critical Heat Flux) (TONG 75)	29 MW/m^2		21 MW/m^2	

Table 2: Temperatures of W7-X target elements (dimensions in mm) during stationary power deposition of 10 MW/m^2. 2D FEM ANSYS 5.0 results are shown for water as coolant assuming a water velocity of 10m/s, and a coolant temperature of 50° C.

Effective non destructive test methods must be ready for quality control [6] to guarantee save operation of 1500 elements.

2.3. Baffle Plates

To improve the pumping efficiency via the gap between the two target plates neutrals have to be concentrated in the divertor units by means of baffles [7]. The 3D shape of the baffle surface is a compromise for wide parameter operation in W7-X. The power load was calculated in the range of 50 - 200 kW/m^2. A conventional solution on the basis of clamped fine graphite tiles (10*10 cm^2) on a cooling structure was chosen for the first design. Before alternate concepts, for example plasma spraying on 3D cooled surface elements etc. are investigated, a sample was prepared and tested on the plasma generator in Berlin for a power load of 40 - 280 kW/m^2. Using a papyex interlayer of the thickness of 0.5 mm and applying a pressure of 0.1 MPa on the contact to a water cooled Cu support a satisfying heat flow of 2500 W m^{-2}K^{-1} was obtained in stationary operation. Finally, the area spanned by baffle plate is 3 m^2 per divertor unit, 30 m^2 in total. So far similar concepts can be adapted for the shield and protection of the inner cryostat wall.

3. Conclusion

Principal solutions for critical components and some details of a divertor system for the HELIAS configuration W7-X have been worked out. The adaptation on a wide operational range of the experiment requests an "open divertor". Depending on the progress of understanding and modelling of the boundary plasma some changes of the geometry and specifications have to be expected until the start of the experiment planned for 2004. Furthermore, an interchangeable design of interfaces, as adjustable support of the components inside the vessel, positioning and construction of the cooling circuits, flexible manifolds and feed troughs at the cryostat must be provided for later optimisation of the divertor for energy and particle exhaust and effective screening of impurities on the basis of the experimental results.

References:
[1] H. Greuner et al., Proc. 18th SOFT, Karlsr. (1994), **1**, p. 323
[2] E. Strumberger, Nucl. Fusion **36** (1996), p. 891
[3] D. Reiter, Jülich Report Nr. 1947, Jülich (1984)
[4] H. Renner et al. 12th PSI, St. Raphael (1996) to be publ.
[5] ToreSupra Team, Fusion Technology **29** (1996), p. 417
[6] H. Greuner et al., "Design, Manufacturing and testing of W7-X Target Element Prototypes" see this conference
[7] J. Kisslinger et al., EPS Lisboa (1993), 17C, II, p.587

Design, Manufacturing and Testing of the W7-X Target Element Prototypes

H.Greuner[a], T.Huber[b], J.Kisslinger[a], E.Parteder[b], L.Plöchl[b], H.Renner[a]

[a]Max-Planck-Institut für Plasmaphysik, D-85748 Garching, Germany, EURATOM.-Ass.
[b]Plansee AG, A-6600 Reutte/Tirol, Austria

The W7X divertor consists of 10 units with a 3D target area of 22 m^2, which is approximated by flat target elements (dimensions: 50 mm width and 200 to 500 mm length) being capable of a power load of up to 10 MW/m^2. Four prototype target elements were designed, manufactured and partially tested in the JUDITH facility (KFA Jülich), successfully. The design was based on a FEM analysis of the brazing constraints and a J-integral analysis of the braze joint: Even large braze flaws, resulting from a lack of braze alloy or water corrosion, will not critically grow and lead to catastrophic interface failure under the brazing constraints and the water pressure.

1. INTRODUCTION

An important property of the magnetic configuration of the new optimised stellarator Wendelstein W7-X (R=5.5m, a=0.55m, super conducting coils: magnetic field B=3T) is the formation of an inherent divertor structure at the boundary. To concentrate the energy and particle flux and to decouple the vessel wall completely from the plasma in this non-axisymmetric device, a system of 10 divertor units was chosen and the geometry of the interacting surfaces was shaped in order to avoid leading edges and to minimise the local power deposition density [1]. Each unit consists of a set of 3D target plates and baffle plates. For easy fabrication, assembly and maintenance the 3D target area will be approximated by almost flat target elements with a typical width of 50 mm and a length of 200-500 mm. In total, a target area of 22 m^2 combining nearly 1500 elements will be needed [2].

2. ANALYSIS AND MANUFACTURING

2.1. Target Element Design

The W7-X divertor target plates are designed for steady state operation with a local heat load of up to 10 MW/m^2. Because of the geometrical boundary conditions in the W7-X vessel, the target plates have to be very thin, which is why a flat design with SEPCARB® N11 tiles brazed onto a TZM water cooled heat sink was selected. The thickness of the CFC tiles was restricted with 6mm in order to keep a surface temperature limit of 1100° to 1200°C for the highest power load.

In order to increase the thermal transfer coefficient between the heat sink and the cooling water, a fin design is favourable, with five fins per cooling channel. Even for TZM with a relatively low thermal heat conduction the thermal transfer coefficient is increased by a factor 1.3 compared to a flat interface towards the coolant water leading to a homogeneous temperature distribution on the CFC surface. Additionally, the cross section of the coolant channel can be adjusted easily for equal pressure drop of elements with diffrent length enabling parallel operation in the cooling circuit with constant flow velocity.

In fig. 1 a target element cross section is shown. It consists of the CFC flat tiles, the TZM fin plate heat sink and the TZM back plate with four parallel cooling channels.

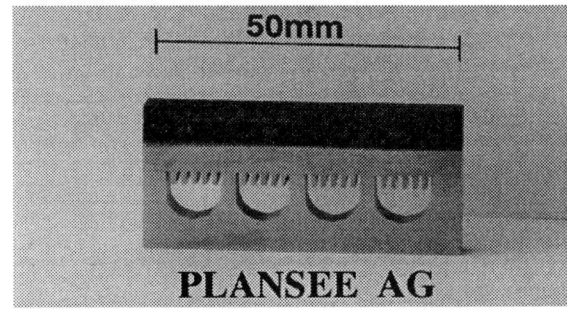

Fig.1: W7X divertor target element cross section

2.2. Manufacturing Route

The manufacturing route of the target element was determined by the functional requirements of its components as they are listed in table 1. The requirement of complete inspection of 100% of all braze interfaces in particular led to a specific feature of the manufacturing route, i.e. the split of the brazing operation in two cycles, the first between the CFC tiles and the TZM fin plate (TiCuSil / 850°C), the second between the TZM fin plate and the TZM back plate (AgCuPd / 810°C). For such a sequence more precision is demanded in the machining and brazing operations, in order to compensate for the warpage induced by the different thermal expansion coefficients

materials	processes
armour material high thermal conductivity high density high vacuum compatibility high erosion resistance high temperature strength low atomic number low thermal expansion	high thermomechanical interface stability between armour and heat sink leak tightness of the interface between fin plate and back plate low thermal constraints
heat sink material high thermal conductivity high temperature strength hot water corrosion resistance low thermal expansion	reliable manufacture in high numbers possibility to inspect 100% of all interface areas

Table 1: Requirements for components of target

of SEPCARB® N11 and TZM (see table 2). A metallographic sample of the target plate cross section is shown in fig.2.

Fig.2 Brazed links of the target element

	SEPCARB®N11	TZM
CTE 0°-1000° x10⁶	// 1.5 ⊥ 2.3	5.7
thermal conductivity at room temperature in W/(mK)	// 240 ⊥ 180	115
thermal conductivity at 1000°C in W/(mK)	// 90 ⊥ 70	109
Young's modulus at room temperature in GPa	// 25 ⊥ 12	300
ultimate tensile strength at room temperature in MPa	// 60 ⊥ 25	
ultimate compressive strength at room temperature in MPa	// 100 ⊥ 100	
interlaminar shear strength at room temperature in MPa	// 20	
yield strength at room temperature in MPa		920

Table 2: Material properties of SEP N11 and TZM

2.3. Stress Analysis

In order to predict the thermal constraints and deformation induced by the brazing process, a 2D FEM analysis (2D generalised plain strain model) was performed, featuring an elasto-plastic material model for the TiCuSil and AgCuPd braze alloys, based on compression tests with cast samples. The strain hardening curves for TiCuSil as example are shown in fig. 3 for different temperatures.

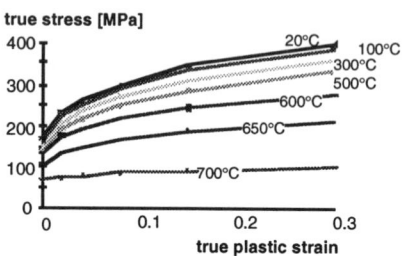

Fig.3: Strain hardening curves for TiCuSil

The results of the calculation can be summarized as follows:
- full plastification of both the TiCuSil as well as the AgCuPd braze occurs, the von Mises stress in the braze alloys is 240 MPa
- the warpage of the fin plate after the TiCuSil brazing cycle is 680μm, the warpage after the AgCuPd brazing cycle is 61μm
- the normal compressive stress in the CFC tiles at the TiCuSil braze interface is 100 MPa, the normal tensile stress in the TZM fin plate is 250 MPa.
- The shear stress maximum in the CFC tiles at the TiCuSil braze interface is 80 MPa, in the TZM fin plate 40 MPa.

2.4. Fracture Mechanical Analysis

The elastic strain field in the TZM being of a non-negligible order of magnitude, especially in the notch-type cooling channel corner which is formed by the AgCuPd braze interface, a fracture mechanical analysis was performed on base of the J-integral, in order to exclude the possibility that initial flaws in the brazing interface at the cooling channel corner, resulting from hot water corrosion or insufficient wetting by the braze alloy, may grow critically and lead to a water leak. The procedure of the J-integral analysis was the following:
- measurement of the fracture toughness K_{IC} on brazed TZM CT specimens with AgCuPd interface
- calculation of the critical load to fracture P_C of the CT-specimen as a function of the crack length a

- calculation of the critical J-integral J_C, as a function of P_C and a, in a FEM model of the CT-specimen
- calculation of the J-integral in a crack in the cooling channel corner in a FEM model of the target plate cross section

The determination of the fracture toughness based on K_{IC} was allowed because the fracture was linear elastic. The fracture of the CT-specimen occured never directly in the braze interface, but the crack path jumped back and forth from one side of the braze interface to the other and propagated in the TZM (see SEM fractograph fig.4).

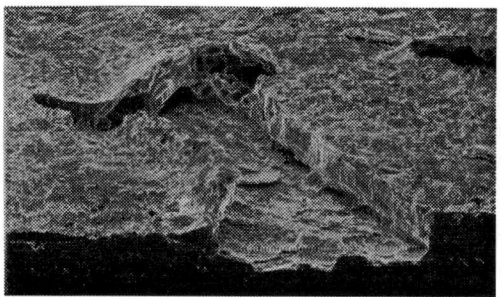

Fig.4: SEM fractograph of brazed TZM CT-specimen

Due to the plain strain state of stress in the braze interface, the inherently ductile AgCuPd braze alloy could not deform plastically. The fracture was brittle, governed by the TZM. A fracture toughness $K_{IC} = 9$ MPa√m was measured. This value corresponds to measurements made on monolithic TZM samples.

Following the ASTM standard, the critical load P_C was determined from the equation for K_{IC} with the constant parameters defined in table 3 and varying the crack length a:

$$K_{IC} = \frac{P_C}{B\sqrt{W}} \cdot \frac{2+\frac{a}{W}}{\left(1-\frac{a}{W}\right)^{\frac{3}{2}}} \left[0.886 + 4.64\left(\frac{a}{W}\right) - 13.32\left(\frac{a}{W}\right)^2 + 14.72\left(\frac{a}{W}\right)^3 - 5.6\left(\frac{a}{W}\right)^4\right]$$

fracture toughness	K_{IC} = 9 MPa√m
sample width	B = 20 mm
sample thickness	W = 40 mm

Table 3: Parameters of the J-integral analysis

The resulting critical loads are shown in table 4. In order to get the „real„ crack length one has to subtract 20 mm from the values in table 4. This is due to the definition of the CT-specimen dimensions in the ASTM standard. In a FEM model the CT-specimen (see fig.5) was loaded up to the calculated value of P_C

a in mm	22	24	26	28	30	32
P_C in kN	3.2	2.6	2.1	1.7	1.2	0.9

Table 4: Critical loads

for a given crack length a. The mesh size along the crack path (equal to the crack length increment Δa) was 50 μm. The analysis was made for an assumed crack path directly in the braze alloy and in the interface between the braze alloy and TZM. The J-integral value was evaluated at the crack tip by the FEM code (ABAQUS). In the latter case of the interface crack path, also the maximum normal stress σ_1 at 50 μm distance from the crack tip in the TZM was evaluated. The results are shown in table 5.

a in mm	22	24	26	28	30	32
crack path directly in the braze alloy						
J in N/m	241	251	253	255	256	257
crack path in the braze interface						
J in N/m	238	248	251	252	254	255
σ_1 MPa	433	460	463	464	465	468

Table 5: Results of J-integral evaluation

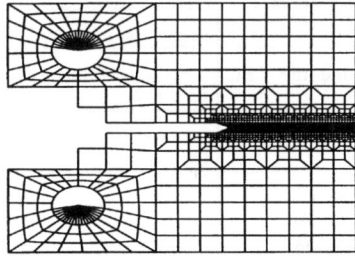

Fig.5: CT-specimen mesh for J-integral analysis

The J-integral is largely independent from the crack length, therefore it is the critical J-integral J_C and a valid fracture mechanical parameter for evaluation of critical crack growth in a brazed TZM part. Its value is around 250 N/m.

In the last step of the J-integral analysis, cracks from 0.1 to 2 mm length, starting from the cooling channel corner, were introduced into the FEM model of the target plate cross section and the corresponding J-integral values calculated at the crack tip. Two load cases were investigated, one including the thermal constraints, the other including the thermal constraints and the cooling water pressure. In fig.6, it can be seen,

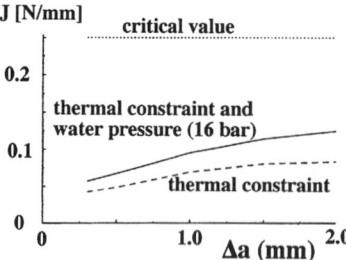

Fig.6 Value of J-integral depending on crack length

that even in the case of 2mm long cracks, the J-integral does not reach the critical value.

2.4. Non-Destructive Inspection (NDI)

There are two major failure scenarios of the divertor target plates in the experiment. One is detachment of armour tiles, the other is a water leak. In order to be able to guarantee the thermo-mechanical stability of the target plates under operation, it is foreseen to install only completely inspected components. Beside providing security of the design by destructive characterisation and fracture mechanical analysis in the development stage, the multi-stage non-destructive inspection of all manufactured parts plays a key roll.

detachment of armour tiles

cause	inspection measure
delamination inside the CFC material	x-ray inspection of blocks by CFC maker
delamination inside the CFC tiles after brazing	transient hot water IR-thermography
TiCuSil braze flaw	x-ray inspection after first brazing cycle (detection limit is appr. 2 mm)

water leak

cause	inspection measure
AgCuPd braze flaw	ultrasonic inspection after second brazing cycle (detection limit is appr. 1.5mm)
	He - leak test
	water pressure test

Table 8: Manufacture and inspection concept

Hence, the manufacturing steps and their sequence were adapted to the technical and physical boundary conditions of the available inspection methods. The integrated manufacturing and inspection concept described in table 8 permits the definition and detection of critical flaws and hence, the elimination of non-conforming parts.

3. Heat Flux Test

Two of the manufactured target elements have been tested using the electron beam device JUDITH of the KFA Jülich (Drs. Bolt, Duwe, Kühnlein) with a stationary power load up to 12 MW/m^2. The area of power deposition was restricted to be 5*3 cm^2 and shifted in direction of the length of the element to simulate the expected pattern on duty. The experimental data of the temperature profiles by means of thermocouples and IR camera agree very well with the 3D finite element thermohydraulic modelling. The surface temperatures as measured by IR technique do not exceed 1200°C (water flow velocity 10 m/s). Fatigue tests with the specified power load of 10 MW/m^2 during periods of 30 s were performed with 1000 cycles. The surface temperature increase of particular tiles is limited to 50° without apparent indications of destruction after the cycles.

4. Conclusion

The manufacturing of prototype target elements combining TZM as the supporting and cooling structure and a brazed CFC layer as the plasma facing material for the W7-X divertor was successful:
An extended study based on a FEM analysis of the brazing constraints and a J-integral analysis of the braze joint demonstrates a stable design. The manufacturing route was adapted to the boundary conditions of the available non-destructive inspection methods. The resulting integrated manufacturing and inspection concept allows the definition and detection of critical flaws and elimination of non-conforming parts. Even large braze flaws resulting from a lack of braze alloy or water corrosion will not critically grow and lead to catastrophic interface failure under the brazing constraints and the water pressure.

Two elements have been tested with a stationary power load of 10 MW/m^2 up to 1000 cycles. This power corresponds to the worst case estimates for operation in W7-X and much lower deposition will be realistic [3]. In respect of the large number of elements which are needed finally a further optimisation of manufacturing and test procedures is envisaged to improve the safety margin. Although the product based on TZM/CFC meets the specified properties of the target element, possible alternate concepts for design and choice of material [4] have to be compared.

References:
[1] E. Strumberger, Nucl. Fusion **36** (1996), p. 891
[2] H. Greuner et al., Proc. 18th SOFT, Karlsr. (1994), **1**, p. 323
[3] H. Renner et al., 12th PSI, St. Raphael (1996) to be publ
[4] ToreSupra Team, Fusion Technology **29** (1996), p. 417

Design of a compact W-shaped pumped divertor in JT-60U

S. Sakurai, N. Hosogane, K. Kodama, K. Masaki, T. Sasajima, K. Kishiya, S. Tsurumi, S. Takahashi and M. Saidoh

Japan Atomic Energy Research Institute, Naka Fusion Research Establishment
Naka-machi, Naka-gun, Ibaraki-ken, 311-01 Japan

M. Inoue, T. Umakoshi, M. Onozuka and M. Morimoto

Mitsubishi Heavy Industries, Ltd.
1-1-1. Wadasaki-cho, Hyogo-ku, Kobe, 652 Japan

In JT-60U, a compact W-shaped pumped divertor, which is optimized for realizing a radiative divertor with high confinement performance in various experiments, will be implemented in February-May, 1997. Pumping and fueling systems are arranged to provide a sufficient capability of particle control. Segmented baffle structure with insulated flexible gas seal is adopted to satisfy experimental and structural requirements.

1. INTRODUCTION

The radiative divertor, which is effective in reduction of heat flux to the target, is considered to be the primary divertor concept in the design of ITER[1]. However, in many radiative divertor experiments[2], the strongly radiating region always moves to the X-point as X-point MARFE or spreads to a main plasma surface as the radiation loss increases up to a certain level. Energy confinement time of H-mode degrades and impurity in the main plasma increases in these states[3]. Optimization of divertor shapes and particle control techniques such as pumping, gas puffing are essential for simultaneously achieving sufficient heat reduction and good confinement of main plasma. Therefore, divertor modification is in progress in the major tokamaks[4,5] to improve the divertor performance and to provide database for the ITER design.

In JT-60U, the present open divertor will be modified to a W-shaped pumped divertor (Fig.1) will be implemented in February-May, 1997, aiming to develop a new compact radiative divertor which realizes radiative divertor plasma for reduction in target heat load and good H-mode confinement simultaneously.

This paper presents the design of divertor modification in JT-60U. Requirements of conceptual design and optimization of divertor configuration, arrangement and capability of particle control system and thermal design of plasma facing components are described in section 2. In section 3, the basic baffle structure and structural analysis are explained and emphasis is laid on the treatment of electromagnetic force due to halo current. Summary is given in section 4.

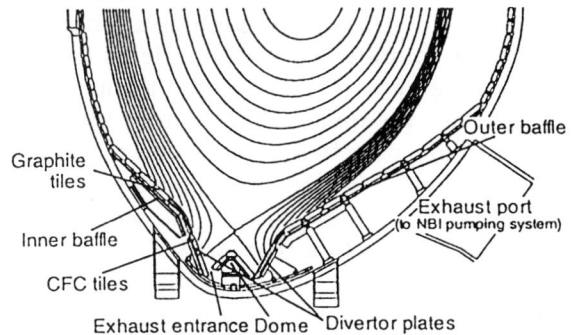

Fig. 1 Design of W-shaped divertor for JT-60U divertor modification.

2. CONCEPTUAL DESIGN

2.1. Design requirements and shape optimization

In order to realize the above-mentioned objectives within the restrictions of present vacuum vessel and poloidal coil system, the shape of divertor and baffle has been optimized, and a W-shaped divertor with inclined divertor plates and a dome in the private region has been designed. Three NBI cryo-pumps will be modified for divertor exhaust.

The designed divertor system fulfills the following requirements:
1) the maximum plasma current of 3 MA, 2) enhancement of neutral pressure at divertor region high enough to obtain a radiative divertor, 3) throughput which balances the particle fueling rate needed to sustain a radiative divertor, 4) reduction in back flow of neutral particles to a main plasma for avoiding degradation of confinement, 5) allowing various plasma configurations: inward-shifted for high βp mode, outward-shifted for RF current driving experiments, etc..

Numerical simulations with the UEDA and NEUT2D codes predict that this W-shaped configuration is effective in enhancing local neutral pressure near the strike-points and in reducing neutral back flow to the main plasma[6,7].

2.2. Pumping and fueling system

Three NBI cryo-pumps, with an exhaust speed of $30 m^3/s$/unit (D_2, 0.1Pa, 300°C) at the port entrance, will be modified to a divertor pumping system. Fast shutter valves, with an aperture changeable during a discharge, will be installed in three ports for active control of particle exhaust. Neutral particles in the divertor region will be exhausted through the space under the dome and the outer divertor in the poloidal direction, and can flow toroidally under the outer baffle. The gap between the dome and the inner divertor will be used as exhaust entrance for the first year experimental campaign, and then gas seals between the dome and the outer divertor will be removed to study effects of pumping location. The net pumping speed and the toroidal distribution of pumping speed at the divertor region were calculated by using an equivalent circuit model which includes the poloidal conductance from the dome to the baffle and the toroidal conductance of baffle ducts estimated from their structure. Obstacles in the baffle duct to the toroidal gas flow were included in the baffle conductance. The net pumping speed in the radiative divertor region, where pressure is possibly larger than 0.2 Pa, is about 40 m^3/s which can balance the particle fueling by NB with absorbed power of 30 MW as shown in Fig. 2.

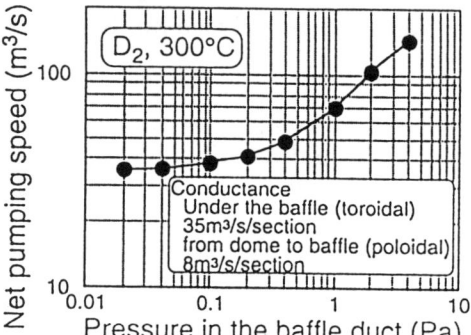

Fig. 2 Puming chracteristics by three cryo-pumps.

The gas puffing system will be re-arranged into 3 systems, fueling at the top of vacuum vessel, wall side of main plasma and divertor region, for experiments such as impurity gas injection to enhance radiation, SOL flow, etc.. Moreover, the present pneumatic pellet injector will be modified to a centrifuge pellet injector to avoid edge cooling by gas puffing.

2.3. Thermal design of plasma facing components

In order to estimate heat loads to plasma facing components, highly radiative steady operation with 70 % radiation loss of the net heating power and low radiative short-pulse operation with 30 % loss were considered for the expected maximum net heating power of 30 MW. CFC tiles are used for the divertor targets, the top of the dome and the baffle tiles at the divertor throat where a heat flux of $5 MW/m^2$ x10 s -10 MW/m^2 x4 s is expected. Thermal analysis shows that the reduction of surface temperature of divertor tiles caused by water cooling of divertor basements is only about 100 °C, because of poor thermal conduction due to long distance and many boundaries between the tile and the basement. Although these tiles are inertially cooled, according to heat analysis, high power operations are possible with intervals of 20 minutes.

These tiles have beveled edges of 0.5 mm in depth to reduce concentration of the heat flux on the edge. Furthermore, the baffles and divertor units will be aligned with accuracies of better than a few millimeters to the plasma surface to avoid heat concentration.

3. STRUCTURAL REQUIREMENTS AND DESIGN

As mentioned in section 2, conductances under the outer divertor and baffle should be large, and careful alignment for plasma facing components is required. Furthermore, the following restriction and requirements are considered in the structural design. 1) size and weight limitation of structural parts, to be admitted through the entrance port of vacuum vessel (500 x 500 mm) and to be handled manually, 2) capability of in-situ adjustment of alignment, accessibility to diagnostics under baffles at maintenance and flexibility for partial modification, 3) robustness against electromagnetic loads and thermal stresses.

3.1. Segmented structure

The structure is toroidally and poloidally segmented to satisfy the requirements. Divertor and private domes are composed of 125 toroidal-segments and are aligned to the plasma surface by using the present cooling basement as the base level of installation. The present cooling basements were already well-aligned to the plasma surface. Inner and outer baffles are toroidally segmented into 72 components at 5 degrees each.

In order to reduce the neutral gas leakage from under the baffle, the gaps between the two adjacent segments are carefully sealed by an inserted-sliding mechanism, which allows thermal expansion during baking and heat loads during a discharge and vibration at disruptions (Fig. 3). Sliding parts are insulated by sprayed ceramic coating to avoid arcing across the gap.

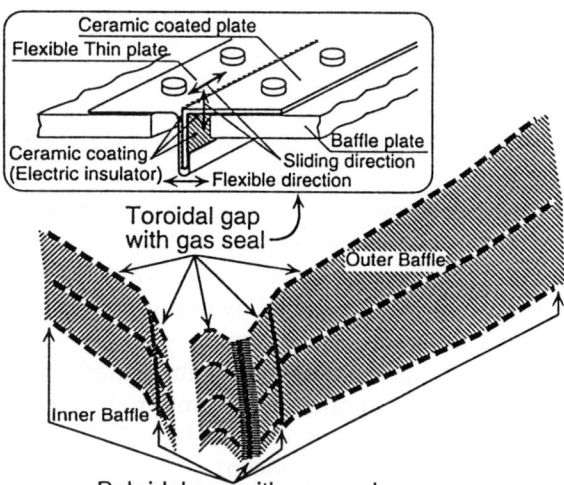

Fig. 3 Schematic of the segmented structure and the gas seal.

3.2. Electromagnetic and structural analysis

Electromagnetic and structural analyses were carried out by Finite Element Methods similarly to the present vacuum vessel[8] to assess robustness of the support structure against 3MA/4T disruptions. Baffles, divertor units, private domes and cooling basements were added to the sector model for computer analysis. Conditions of analysis were revised on the basis of JT-60U experiments.

Two cases were considered to model a disruption. In case 1, a fast current decay time of 5 ms and inward movement of plasma column were assumed. A VDE (Vertical Displacement Event) with halo current on the support structure at current decay time of 10 ms was considered in case 2. Electromagnetic forces due to halo currents through some routes were estimated and added to the electromagnetic loads due to eddy currents.

The total halo current of 26% of plasma current and toroidal peaking factor of 2.5 were assumed on the basis of JT-60U experiments[9]. However, toroidal symmetry was assumed in the analysis except for inner baffles, and some weak points were reinforced to obtain a sufficient safety factor. The reasons are as following: 1) the support structure of outer divertor and baffles must not reduce exhaust conductance, 2) the frequency and its effects of VDE can be reduced by improving the plasma position control, 3) outward-shifted discharges, in which it will be difficult to avoid halo current on the outer baffles due to narrow clearance between the plasma and baffles, will have plasma current limited to 2 MA because of a limit in divertor coil current.

Analysis showed stresses and displacements are within allowable values at all the structural parts. Structural integrity of the welded parts and bolts were estimated by using the results of FEM analysis. Furthermore, the structure and fixing bolts of tiles are confirmed to be strong enough against electromagnetic forces due to halo current on the tile. The gaps between the tiles on adjacent segments will be controlled to allow a relative displacement between segments. A structural analysis for the connected seal was made, in which all the adjacent segments were connected electrically by gas seals, showing integrity of the whole structure, even if all the insulated gaps between adjacent segments were short-circuited by arcing. Toroidal and poloidal arrays of Rogowski coils for measuring halo current on the support structure will be installed to study the distribution and amount of halo current and confirm electromagnetic loads are within the estimation.

4. SUMMARY

The conceptual and structural design of the new compact W-shaped pumped divertor in JT-60U were conducted, aiming to realize radiative divertor plasmas to reduce target heat loads and to achieve good H-mode confinement simultaneously.

The divertor shapes were carefully optimized by using numerical simulations in order to enhance neutral pressure in the divertor region and to reduce neutral back flow to the main plasma. The pumping and fueling systems were designed to provide flexibility in speeds and positions of exhaust and fueling. Sufficient exhaust throughput for steady radiative divertor operation was guaranteed by the large conductances to the exhaust ducts.

Thermal analysis of plasma facing components showed that high power operation up to net heating power of 30 MW will be possible with intervals of 20 minutes.

A segmented baffle structure with insulated flexible gas seals is adopted to satisfy structural requirements. The robustness of whole structure was confirmed by electromagnetic and structural analysis with FEM codes. Electromagnetic forces on the support structure due to halo currents were estimated and taken into account in the structural analysis.

ACKNOWLEDGEMENTS

The authors would like to thank the members of the Japan Atomic Energy Research Institute and Mitsubishi Heavy Industries, Ltd. who have contributed to the JT-60 project.

REFERENCES

1. ITER EDA Documentation series, No.7, "Technical basis for the ITER interim, design report, cost review and safety analysis", Chapter V, IAEA, Vienna, 1996.
2. N. Hosogane, et al., J. Nucl. Mater., 220-222 (1995) 420.
3. N. Asakura et al., in Plasma Physics and Controlled Nuclear Fusion Research 1994 (Proc. 15th Int. Conf. Seville, 1994), Vol. 1, IAEA, Vienna (1996) 515.
4. The JET Team, in Plasma Physics and Controlled Nuclear Fusion Research 1994 (Proc. 15th Int. Conf. Seville, 1994), Vol. 1, IAEA, Vienna (1996).
5. M. A. Mahdavi et al., J. Nucl. Mater., 220-222 (1995) 13.
6. S. Tsuji et al., J. Nucl. Mater. 220-222 (1995) 400.
7. N. Hosogane et al. will presented in 16th. Plasma Physics and Controlled Nuclear Fusion Research 1996, Montreal.
8. H. Kaguchi et al., Proc. 13th Symp. on Fusion Engineering. Knoxville, (1989) pp368-371.
9. Y. Neyatani et al., Fusion Technology 2(1995) 1634.

Structural Design of the DIII–D Radiative Divertor*

E.E. Reis, J.P. Smith, C.B. Baxi, A.S. Bozek, E. Chin, M.A. Hollerbach, G.J. Laughon, D.L. Sevier

General Atomics, P.O. Box 85608, San Diego, California 92186-5608, USA

The divertor of the DIII–D tokamak is being modified to operate as a slot type, disipative divertor. This modification, called the Radiative Divertor Program (RDP) is being carried out in two phases. The design and analysis is complete and hardware is being fabricated for the first phase. This first phase consists of an upper divertor baffle and cryopump to provide some density control for high triangularity, single or double null discharges. Installation of the first phase is scheduled to start in October, 1996. The second phase provides pumping at all four divertor strike points of double null high triangularity discharges and baffling of the neutral particles from transport back to the core plasma. Studies of the effects of varying the slot length and width of the divertor can be easily accomplished with the design of RDP hardware.

Static and dynamic analyses of the baffle structures, new cryopumps, and feedlines were performed during the preliminary and final design phases. Disruption loads and differential thermal displacements must be accommodated in the design of these components. With the full RDP hardware installed, the plasma current in DIII–D will be a maximum of 3.0 MA. Plasma disruptions induce toroidal currents in the cryopump, producing complex dynamic loads. Simultaneously, the vacuum vessel vibrations impose a sinusoidal base excitation to the supports for the cryopump. Static and dynamic analyses of the cryopump demonstrate that the stresses due to disruption and thermal loadings satisfy the stress and deflection criteria.

1. INTRODUCTION

The Advanced Divertor Program (ADP) for the DIII–D tokamak was initiated in 1989 to study enhancement of plasma performance that can be achieved by controlling plasma density and the recycling of impurities to the plasma by divertor pumping and biasing. The program has been extrememely successful. The knowledge gained in the ADP is presently being applied to a divertor modification, the Radiative Divertor Project (RDP), which will provide particle pumping and density control in high triangularity, single or double null tokamak discharges. The RDP also will allow the study of dissipating divertor power by radiation, distributing the power over a larger area while pumping away the neutral particles. Dissipative divertor research is being conducted now in the existing open ADP divertor but gas puffed in the divertor region to enhance radiation sometimes ends up in the plasma core, degrading the quality of confinement. The RDP is designed to limit this core fueling by providing tighter baffles in the divertor region, and the neutral gas is better entrained in the divertor slots.

The RDP hardware consists of inner, private flux, and outer baffle plates in both the top and bottom of the machine (Fig. 1). Three new cryopumps are to be installed to complement the existing ADP pump. The lower baffles and cryopumps are shown in Fig. 1 and are nearly symmetrical about the mid-plane of the tokamak. Experiments and modeling have formed the basis for the new design. Modeling codes, benchmarked with experiments, have helped define the shape of the baffles to entrain the gas in the divertor. The core ionization is anticipated to be reduced by a factor of nine with the addition of the baffle structures. The four divertor pumps installed under the outer and private flux baffles, provide 100 m^3/s of pumping speed to a double null high triangularity discharge, removing neutral gas and particle impurities from the scrape off plasma. The four cryopumps will provide the capability to study inner versus outer strike point pumping for double null plasmas.

Installation of the RDP hardware has been separated into two phases as schematically shown in Fig. 2. The first phase will be completed in December of 1996 and the second phase installation in 1998. The design and analysis for the RDP has

---

*Work supported by U.S. Department of Energy under Contract No. DE-AC03-89ER51114.

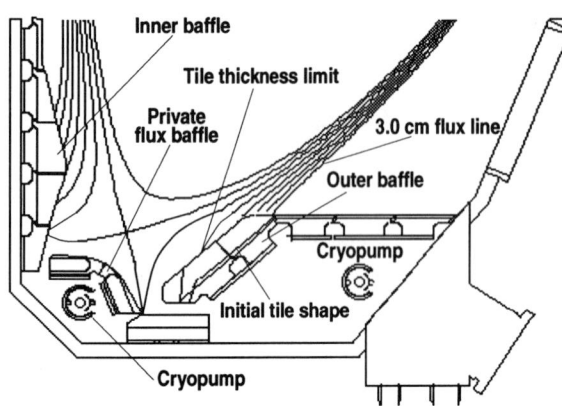

Fig. 1. Cross section of lower RDP structure. Limits of the tile thickness are shown with flux lines.

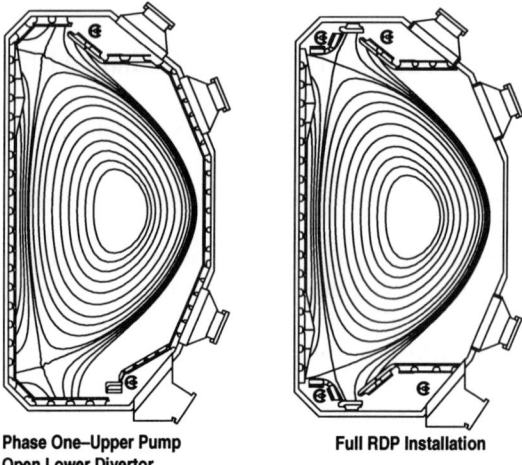

Fig. 2. The Radiative Divertor installation in DIII–D (a) Phase 1 with upper outer baffle and cryopump. (b) The full RDP installation with active pumping and baffling at all four strike points.

been completed and hardware is being fabricated for the first phase.

Flexibility is a key feature of the RDP design. The ability to modify the structure easily as the understanding of the physics evolves has been accounted for in the design. Studies on the effects of varying the slot length and width of the divertor can be easily accomplished with the design of RDP hardware. The slot width is changed by installing a new set of graphite tiles of different size while the slot length can be changed by raising the structural panels and installing longer supports and additional tiles. Slot lengths of 23, 33, and 43 cm have been selected for these studies.

The cryopumps are also toroidally continuous to prevent electrical breakdown in the low density plasma underneath the baffle plates. The design, analysis, and operation of the existing cryopump and its support systems are presented in Refs. [1–2]. The baffles on the inner wall are thick tiles fabricated from graphite mounted to the inner wall of the vacuum vessel.

2. DESIGN DESCRIPTION

The baffles consist of inertially cooled graphite tiles mounted to water cooled support panels. The graphite tiles are of the same design developed for the existing divertor targets that have operated successfully since 1987 [3]. The water cooled panels are toroidally continuous, the design chosen for its hoop strength and reduction of electric potentials during disruptions. The water-cooled baffle rings are attached to the vessel with a set of supports spaced every 15 degrees. Inconel 625 material was selected for the baffle plates, with Inconel 718 alloy required for the supports. More detail on the design can be found in Ref. [4].

The new cryopumps are of the same basic design as the existing pump, a toroidally continuous liquid helium cooled pumping surface surrounded by a liquid nitrogen cooled shield. The small area under the private flux baffle, however, required a more compact support design to limit the vertical deflections and stresses due to disruption loads and provide radial flexibility for thermal contraction of the pump. The cryopump design with its supports and feedlines is shown in Fig. 3. The coolant feed lines utilize a vertical port and require special supports to react dynamic loads on the 2 m long vertical run of concentric tubing.

Upgrades to the cryosystem hardware are required to supply cryogens to the new cryopumps. Two new cryostats will be installed to sub-cool the liquid helium prior to entrance into the pump. A new distribution box at the DIII–D cryoplant will be installed to service the additional coaxial flexible

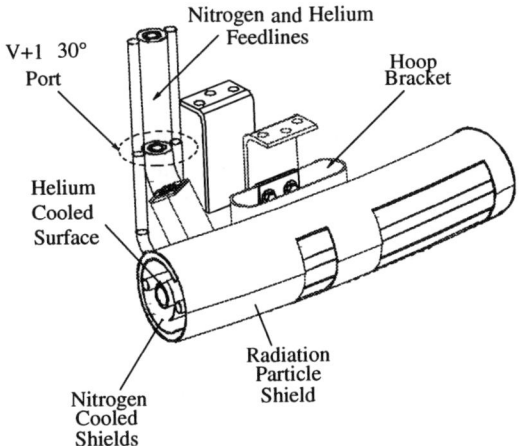

Fig. 3. Cryopump design, with supports and feedlines.

transfer lines. When the full system is installed, additional compressor capacity will be required.

3. STRUCTURAL ANALYSIS

Static and dynamic analyses of the baffle structures, new cryopumps, and feedlines were performed during the preliminary and final design phases. Disruption loads and differential thermal displacements were accommodated in the design of these components. With the full RDP hardware installed, the maximum plasma current in DIII–D is 3.0 MA with disruption induced loads scaled to this plasma current. The maximum differential thermal displacement of the vessel and RDP structural components occurs during baking of the machine to 400°C. While all the components reach nearly the same final temperature, the heating and cooling rates of the elements are different and a temperature difference of 100°C is developed. The differential thermal growth is accomodated by elastic bending of the supports. The cryopump sees a maximum temperature difference while the pump is cooled to liquid nitrogen and helium temperatures, with smaller differences during baking.

The structural design of the baffle plates is governed by the forces induced by halo currents which have been measured on DIII–D [5]. Toroidal currents also flow in the toroidally continuous structural plates due to disruptions, but are significantly smaller than and tend to counteract the halo current forces. The asymmetric halo current induced loads on the divertor structure are based on a 2:1 peak to average factor of a total halo current equal to 20 percent of the 3.0 MA plasma current [6]. The finite element stress analysis of these components therefore required modeling a 180 degree segment of the structure to evaluate the stresses and deflections. Of particular concern was the global offset of the structure and the resultant support stresses caused by the asymmetric loads. The maximum offset of the structure in the radial direction was 0.58 mm and was in the location of the maximum load. The global offset of the model was small, 0.06 mm, measured by the toroidal displacement 90 degrees away from the peak load.

Detailed stress analysis of the baffle plates was performed using a structural model of a 15 degree sector of the water cooled panels and supports. The model of the panels included plate elements simulating the individul sheets of the panels and the structural welds holding the panels together. The peak halo current induced loading was applied in the analysis and the local stresses in the welds and sheets were determined along with the local deflections of the panels. A stiffening ring was added to the structure to limit the vertical deflections of the outermost water cooled ring.

In evaluating the dynamic effects of the loads on the water cooled panels, recent experimental results on DIII–D were considered. The new measurements have shown that the halo current asymmetries remain at a toroidal location for less than 2 ms, while axisymmetric halo currents have a rise-decay time over 15 ms [6]. Since the fundamental frequency of the baffle structure is 104 Hz, a dynamic load factor based on spectra-response curves ranges from 0.6 to 1.0 for the asymmetric and symmetric load cases respectively. Static stress analyses of the baffle structures were performed conservatively using a dynamic load factor of 1.0 for peak loads in all load cases. The stresses and deflections of the baffle structures and their supports are less than the allowable values.

Plasma disruptions also induce toroidal currents in the cryopump, producing complex dynamic loads. Simultaneously, the vacuum vessel vibrations impose a sinusoidal base excitation to the supports for the cryopump.

Dynamic stress analyses of the inner and outer cryopumps and feedlines were performed with beam and spring-mass models using the COSMOS code [7]. The frequency analysis of the spring-mass model showed the helium line and supports have a frequency of 66 Hz and the fundamental frequency of the nitrogen system is 210 Hz. The beam and spring model was used to evaluate the dynamic stresses in the various components of the pump and feedlines due to the disruption currents driven in the pump.

In addition, the vacuum vessel oscillates vertically at 21 Hz due to a vertical disruption event. Although the maximum amplitude of the motion at the center of the floor of the vessel is at most 2.3 mm for a 3.0 MA disruption, the effect of this motion on the cryopump system needed to be determined. The maximum differential displacement calculated between the helium line and nitrogen shield is 7 mm at a support. The helium spring support can displace only about 2.5 mm before bottoming-out. Therefore, there will be some impact forces developed which will change the response of the system. To investigate the structural effects of the support springs impacting during maximum disruption conditions, a spring-mass-gap model was utilized. The results from this analysis showed that a maximum impact force of 818 N acting over 2 ms would occur at a worst case support location. Since there was concern that this impulse may cause cracking of the insulating ceramic ring interfacing with the helium line, an impact test was performed using prototypes. The theoretical impulse was duplicated closely using a calibrated hammer striking the center of the helium support which was held in place by a segment of the helium line. Inspection of the ceramic rings after several strikes showed no damage to the ceramic rings had occurred and the design verified.

4. CONCLUSIONS

Fabrication of hardware for Phase I of the RDP is nearing completion and installation will be completed by the end of 1996. Both the Phase I and II hardware will provide enhanced density control of the plasma. The new baffle structures will control the flow of neutral gas in the divertor region, thereby increasing the power radiated in the divertor slots. Studies of the effects of varying the slot length and width can easily be accomplished with the design of the RDP hardware.

The RDP structure was evaluated using both static and dynamic analyses. Halo current induced loads were critical for the baffle structure, while loads from toroidally induced currents coupled with vessel motion were most significant for the cryopump.

REFERENCES

1. E.E. Reis, et al., "Design and Analysis of the Cryopump for the DIII–D Advanced Divertor," *proc. 17th Symp. on Fusion Technology*, Rome, Italy (1992).
2. J.P. Smith, et al., "Installation and Initial Operation of the DIII–D Advanced Divertor Cryocondensation Pump," *proc. 15th Symp. on Fusion Engineering*, Hyannis, (1993).
3. M.A. Hollerbach, et al., "Upgrade of the DIII–D Vacuum Vessel Protection system," *proc. 15th Symp. on Fusion Engineering*, Hyannis, (1993).
4. J.P. Smith, et al., "The DIII–D Radiative Divertor Project, Status and Plans," *proc. 12th Topical Meeting on Tech. of Fusion Power*, Reno, Nevada, (1996).
5. E.J. Strait, et al., "Observation of Poloidal Current Flow to the Vacuum Vessel During Vertical Instabilities in the DIII–D Tokamak," Nucl. Fusion, **31**, (1991) 527.
6. T.E. Evans, "Measurements of Non-Axisymmetric Halo Currents with and without "killer" Pellets During Disruptive Instabilities," to be published J. Nucl. Materials, 1996.
7. COSMOS/M, version 1.75. Structural Research and Analysis Corp., Los Angeles.

Fabrication Development and Usage of Vanadium Alloys in DIII–D

J.P. Smith, W.R. Johnson, E.E. Reis

General Atomics, P.O. Box 85608, San Diego, California 92186-5608, USA

General Atomics is procuring material, designing components, and developing fabrication techniques for the use of vanadium alloy into the DIII–D divertor as elements of the Radiative Divertor Project modification. This program was developed to assist in the development of low activation alloys for fusion use by demonstrating the fabrication and installation of vanadium alloy components in an operating tokamak. Along with fabrication development, the program includes multiple steps starting with small coupons installed in DIII–D to measure the environmental effects on vanadium. This program is being implemented in collaboration with the Department of Energy (DOE) Fusion Materials Program, particularly efforts at Argonne National Laboratory (ANL) and Oak Ridge National Laboratory (ORNL).

Procurement of the material for this program has been completed. The world's largest heat of vanadium alloy, 1200 kg of V-4Cr-4Ti alloy, has been produced and converted into various product forms. A description of the manufacturing process is presented and results of the chemistry are reported.

As part of the program, research into potential fabrication methods is being performed. Joining of vanadium alloys was identified as the most critical fabrication issue for its use in the Radiative Divertor Program. Successful welding trials have been performed using resistance, friction and electron beam methods. Metallography and mechanical tests have been used to evaluate the welds and results are presented.

1. INTRODUCTION

The environmental benefits of fusion energy can be great if and when low activation materials are utilized. Development of low activation materials has been in progress around the world for many years and many materials have been considered and evaluated for irradiation performance. In the U.S., the primary work on low activation materials has been performed under the auspices of the Fusion Materials Program of the Department of Energy's Office of Fusion Energy Sciences. Vanadium alloys have been identified as the leading fusion structural material based on their high temperature capability and irradiation performance (stability). The alloy V-4Cr-4Ti has been selected by the U.S. program as the primary candidate vanadium alloy. The chrome and titanium additions improve strength and irradiation stability over pure banadium. While vanadium is a fairly common element, it is primarily used as a minor alloying element in steels and not as a base metal for alloys. Outside the fusion program, the uniqueness of vanadium's characteristics have not found other uses, thus the development of the alloys has proceeded only within the limited resources of the fusion program. The production and fabrication of vanadium alloys has been primarily limited to small research heats (<100 kg) to study alloy composition effect on properties and irradiation stability with no development of components. The largest lot of vanadium alloy produced in the U.S. before the present effort was a 500 kg heat of V-4Cr-4Ti [1]. To date, no significant components have been made from vanadium alloys. Welding of vanadium alloys is difficult due to the pick-up of impurities at high temperature and subsequent embrittlement. Welding has been limited to development efforts primarily for looking at the irradiation stability of welds.

2. VANADIUM IN DIII–D

General Atomics, along with Argonne National Laboratory (ANL) and Oak Ridge National Laboratory (ORNL), has developed a plan utilizing vanadium alloys in the DIII–D tokamak for the purpose of enhancing the on-going research on low activation alloys [2]. This plan culminates in the fabrication, installation, and operation of a water cooled vanadium alloy structure as part of the DIII–D Radiative Divertor Program (RDP). The use of vanadium in DIII–D provides a meaningful step towards advancing development of low activation materials for fusion power applications by 1) demonstrating the in-service behavior of a vanadium alloy in a typical tokamak environment, and 2) developing knowledge and experience on design, processing, and fabrication of full-scale vanadium alloy components.

The design, manufacture, and installation of a vanadium private flux baffle structure for the upper divertor of the RDP represents the culmination of the vanadium alloy program (Fig. 1). The structure consists of toroidally-continuous, water-cooled structural panels with mechanically attached graphite tiles making up the plasma facing surface. The panels will be water cooled during tokamak operations, experiencing a maximum temperature of ~60°C, but during baking of the vacuum vessel, hot air replaces the water for baking to 400°C. Due to the lower electrical resistivity of the V-4Cr-4Ti alloy compared to Inconel 625, the toroidal current flow during plasma disruptions will be approximately 4 times larger than in similar Inconel RDP components. The design of the panels and supports account for these larger loads.

Each water cooled ring will be made in six segments, with the segments fabricated by welding two 4.8 mm sheets creating an internal coolant channel. Resistance seam welds will provide the shear strength between the two sheets and a perimeter electron beam weld makes the vacuum seal. Friction welding methods are proposed for attaching studs to the panels to attach the graphite tiles. To facilitate installation, bi-metallic tube joints are planned so that all field welds will be Inconel to Inconel.

3. PRODUCTION OF VANADIUM ALLOY

A heat of vanadium alloy has been produced for the fabrication of the RDP components. In preparation for the production, a detailed material specification for V-4Cr-4Ti alloy was written by General Atomics with assistance from Teledyne Wah Chang Albany, ANL and ORNL. From the starting material to completion of product forms, the specification outlined the production steps along with control and quality tests required. Goals for the levels of impurities were specified, with a focus on keeping the levels of potentially embrittling elements (O, N, C, etc.) low. Consideration was also given to minimizing the levels of long lived neutron activation impurities, Nb, Mo, Ag. In the future, low levels of these elements will be required to achieve low activation properties for the material.

Material processing started in September 1995 with the selection of raw vanadium lots (derbies) to be electron beam melted into pure vanadium ingots. Two ingots were electron beam melted; one (~900 kg) and the other (~1000 kg). Samples were taken from the two ingots for chemical analysis and for rolling by ORNL into sheet and machining into Charpy V-notch specimens by ANL. Impact testing at −196°C was performed to evaluate the fracture behavior of the vanadium. The data agreed reasonably well with test results for previous pure vanadium ingots. The entire 900 kg ingot was chipped and blended with 400 kg of the second ingot. The consolidation of the vanadium along with high purity chromium and double vacuum melted titanium proceeded in February 1996. The ingot was vacuum arc melted twice to form a 1200 kg V-4Cr-4Ti ingot. The ingot outside diameter was machined, cut into two pieces, and vacuum canned in stainless steel for extrusion into sheet bar. The material was extruded into ~11.4 cm × 24.1 cm. sheet bar at 1150°C.

The next step in the processing was the warm rolling of the extrusion into 4.8 mm plate and

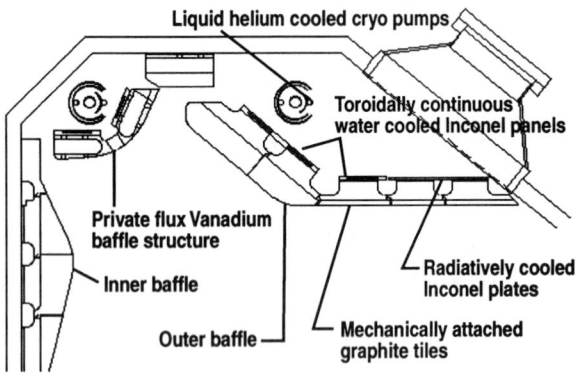

Fig. 1. Location of vanadium structure in radiative divertor.

machining and swaging into 10 mm diameter rod. The sheet bar was longitudinally warm rolled to a thickness of 4.57 cm and then cut into ~59 cm long sections. The pieces were cross rolled to 4.8 mm thick plate, 59 x 170 cm. The warm work of the material was limited to 15% reduction per pass and 50% reduction between vacuum anneals at 1050°C. The rod is presently being manufactured by using the 4.57 cm plate, sawing it into rectangular bars, and machining the bars into rounds. Subsequent swaging will yield 1 cm diameter rod. Again the reduction is being limited to 50% between anneals.

4. WELDING DEVELOPMENT

A key goal of the DIII–D vanadium program is to develop manufacturing techniques. The U.S. vanadium program has been primarily focused upon alloy development and irradiation properties with little work being performed on developing manufacturing techniques or processes. The advent of making components has required research be initiated on fabrication techniques, and in particular, welding. Vanadium, being a refractory metal, is highly susceptible to pick-up and embrittlement by oxygen, nitrogen, carbon, and hydrogen at elevated temperatures, and most welding, therefore, has required a high purity protective environment. Welding studies using gas tungsten arc, e-beam, and laser weld processes are ongoing at ANL and ORNL, but additional processes are required for the DIII–D program. Thus, studies have been initiated on resistance, friction, and electron beam welding of vanadium alloys.

4.1. Resistance Welding

Resistance welding is planned for structurally joining the two halves of vanadium water cooled panels together for the RDP. It will not provide the vacuum-tight weld, however. Resistance welding was chosen as the joining method, placing the weld on the neutral axis of the part, thus potentially minimizing weld distortion. It was also thought that the high pressures necessary to create the weld would help to exclude the embrittling elements. Initial spot welding trials made in air produced a weld nugget with microhardness only 10% greater than the bulk material. This was a very promising result, indicating that a protective atmosphere may not be required. A parameter search of pressure and current for welds of thicker material was started. No weld nugget was initially achieved on 3.8 mm sheet, based on parameters for welding Inconel, although the developed diffusion bond demonstrated considerable strength (up to 135 MPa) in shear tests. Additional trials were made using higher power inputs. These trials were successful, forming weld nuggets approximately 7 mm in diameter. Micro-hardness measurements again showed <10% increase in hardness. Using parameters based on data for welding carbon steel, a material with similar resistivity and strength, spot weld trials were then performed on 4.8 mm thick vanadium material and were also successful. The strength of these spot welds was evaluated in single spot lap shear tests with strengths measured up to 685 MPa. These welds had slight porosity in the center, which further testing has eliminated. Resistance seam welding trials are now planned.

4.2. Friction Welding Studies

Two types of friction welding trials are in progress. Inertia and portable friction welding will be used for joining vanadium alloy to itself, and inertia welding for creating a bi-metallic joint. The first inertia weld trials of vanadium rod to vanadium plate were successful in air, without any protective environment. It is believed that the high pressure used in the process, coupled with the fact that the process is really a hot forging method with the material never reaching a melting point, excludes potentially embrittling elements and minimizes brittle inner metallic formations. Metallographic examination revealed complete bonding with no indications of porosity or cracking. The interface had a fine grain structure, with little or no grain growth. Microhardness measurements showed only slight increases in hardness in the weld and heat affected zones (HAZ). Tensile and torsion tests resulted in failures in the parent material away from the joint and HAZ.

A bi-metallic joint was considered essential to allow for field welding of Inconel components in DIII–D with all vanadium welds carefully characterized in the shop. Inertia welding was considered a good candidate process for a bi-metallic joint because melting of the materials does not occur, minimizing the chance of brittle intermetallic phase formations. After several attempts, trials using different diameter rods for the two materials, to match their forgeabilities, were successful. Metallography showed complete bonding with no porosity or cracking in the joint. Tensile pull tests were performed on three weld trial samples. Two samples failed in the Inconel 625 section well away from the weld area (at ~930 MPa stress) and one sample failed at the approximate weld interface, at a stress level of ~760 MPa. With the joining ability of the two materials demonstrated, the development was shifted to creating a bi-metallic tube joint. The initial trial samples were encouraging but not leak tight and had varying strengths (70–380 MPa). There was significant bonding at the interface, but also a considerable amount of fine porosity. It was determined from metallography, that there was radial displacement of vanadium material at the joint which released some forging pressure. New trials with a different joint configuration have been made and are being evaluated.

Preliminary portable friction welding trials of rod to plate have been performed to develop methods of *in-situ* replacement of studs on the water cooled panels. The initial trials achieved substantial bonding but the hardness of the weld interface increased significantly. In addition, extensive grain growth occurred at the interface and HAZ. It was noted in the trials that the temperature of the interface was significantly higher than in the inertia weld process, and the time to create the weld was longer. It was believed that both of these factors led to the grain growth and increase in hardness. Additional trials are in progress.

4.3. Electron Beam Welding Studies

Preliminary electron beam welding trials have also been initiated at General Atomics to complement the work being performed at ORNL. Initial weld parameters were obtained from ORNL [3] and weld penetration tests were performed using 6.35 mm thick vanadium alloy plate. These trials established specific weld parameters for creating a lap weld of two 3.85 mm thick vanadium alloy sheets. The lap weld, once created, was metalurgical examined. Good weld penetration was obtained with no indications of cracking or porosity. Microhardness measurements showed less than 10% increase in hardness. A weld was made in a single sheet to produce three tensile specimens to develop strength data on the weld. The tensile specimens all failed in the parent material well away from the weld joint and HAZ at values equal to the ultimate strength of the material.

5. CONCLUSIONS

A program for utilizing vanadium alloys in DIII–D has been developed to enhance the development of low activation vanadium alloys for fusion. The production of 1200 kg of V-4Cr-4Ti alloy for this program is nearly complete. Two vanadium ingots have been electron beam melted as base materials for the vanadium alloy ingot. The alloy ingot has been double vacuum arc melted and extruded into sheet bar. Warm rolling the material into 4.8 mm sheet has been completed. Successes have been achieved in making vanadium alloy welds using various methods including resistance, friction, and electron beam welding.

ACKNOWLEDGMENTS

This is a report of work supported by General Atomics internal R&D funding and the U.S. Department of Energy under Contract No. DE-AC03-89ER51114. The authors would like to acknowledge the efforts of Teledyne Wah Chang Albany for their efforts producing the vanadium alloy material, KT Aerofab for resistance welding, Interface Welding for inertia welding work, and to RAM Stud for friction welding work.

REFERENCES

1. H.M. Chung, et al., *Report,* DOE/ER-0313/17, Oak Ridge National Laboratory, Oak Ridge, Tennessee (1994), p. 178.
2. J.P. Smith, et al., "Utilization of Vanadium Alloys in the DIII–D Radiative Divertor Program," to be published in Proc. of the 7th Intnl Conf. *on Fusion Reactor Materials (ICFRM-7),* Obninsk, Russia, September 25–29, 1995.
3. King, J., Oak Ridge National Laboratory, private communication.

The MkII gas box divertor - a new design concept

H. Altmann[a], E. Deksnis[a], C. Froger[a], S. Lawson[b], C. Lowry[a], A. Peacock[a], M. Pick[a]

[a]JET Joint Undertaking, Abingdon OX14 3EA, UK.

[b]Matra Marconi, Hemel Hempstead, UK.

1. INTRODUCTION

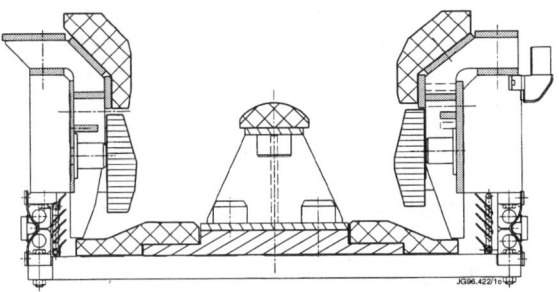

Fig.1: Poloidal X-section of Gas Box Divertor

The Gas Box (GB) Divertor will be the third configuration to be installed in the JET torus since the start of the divertor phase. Details of the MkI [1] and MkII [2] divertors have been reported previously. The design shown in figure 1 is intended to simulate the proposed ITER divertor layout. The main features of the arrangement are:

a) a reduced entrance to minimise the leakage of neutral particles from the divertor region into the main volume and encourage recirculation of impurities and neutrals,
b) a central septum to allow independent control over recirculation of neutral particles along each leg of the divertor;
c) the ability to handle both attached and radiating plasmas.

Detached plasmas that radiate their energy within the divertor volume provide a stringent thermal loading of structural parts.

Divertor physics has had limited success in predicting the effectiveness of the septum or determining a preference for operation on base or side target plates. As a result, maximum flexibility had to be maintained in the design so that additional combinations of components, as shown in figure 2 (a-c), could be tested independently. Further options for the septum were the transparency of the vertical wall and the extent of radial fins to influence neutral particle recirculation. Independent gas feeds to the divertor legs allow fine control of this process.

2. INPUT PARAMETERS

The MkI divertor tested CFC and beryllium targets with the same geometry for both materials. The MkII divertor [2] was optimised for maximum power handling over a wide range of plasma configurations. The GB divertor must accommodate 20MW total conducted power (or 10MW per divertor leg) for:

- attached plasmas with up to $18MW/m^2$ deposited locally on target plates and up to 8 MW/m^2 on structural members (figure 3),

- detached, radiating plasmas with 2 MW/m^2 (max) on to all divertor surfaces including the structure.

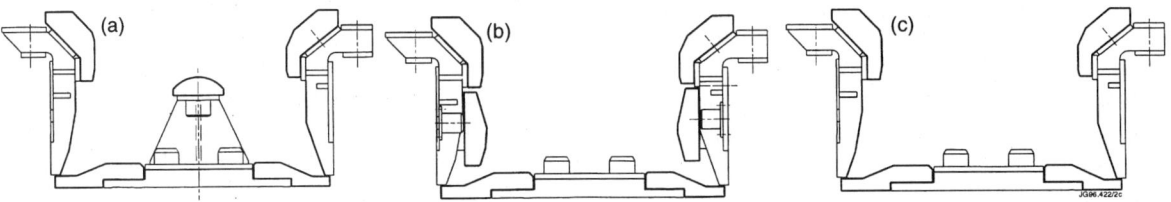

Fig.2(a-c): Options for Divertor Layout.

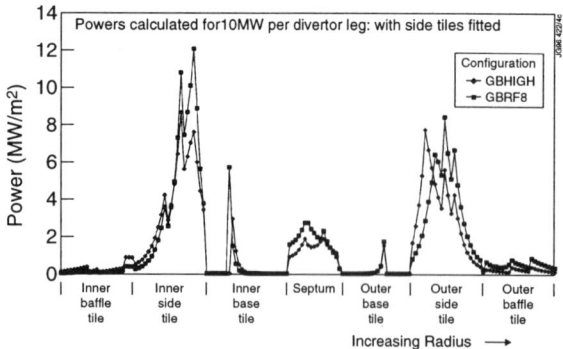

Fig.3: Power Loading over divertor targets.

The principal mechanical loading is due to magnetic forces on the divertor components. These arise from halo currents flowing through the tiles and carriers, then into the support structure (fig. 4). Intrinsic eddy current forces from the rapidly changing magnetic fields appear in the tiles.

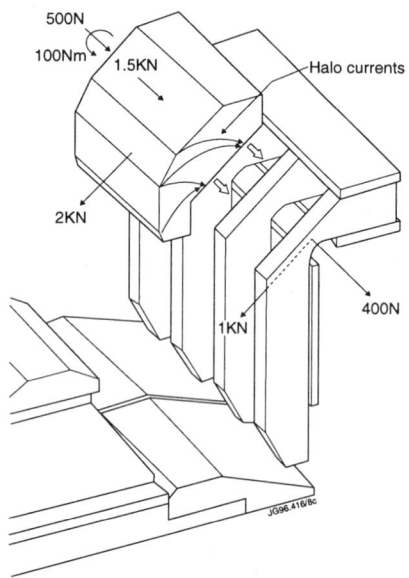

Fig.4: Electro-magnetic forces under halo current.

3. MATERIAL SELECTION

The requirement to accept 2 MW/m^2 anywhere on to the carrier structure led to the decision to use carbon-carbon fibre composite (CFC) plates for all structural members. A structure from metal such as Inconel 600 would distort if 2 MW/m^2 were sustained for 5 seconds. CFC lends itself well to this application, owing to its thermal shock resistance and ability to withstand high temperatures without residual deformation.

Extensive screening tests have been performed during material selection for MkI and MkII divertors. For structural applications a quasi-3D material with nearly isotropic thermal conductivity was chosen. Two grades SEPCARB N11-2 and N11 were selected. The former has high flexural strength but lower thermal conductivity so that use was restricted to areas subject to radiant loads. Structural members that can be exposed to conducted power in some of the options (figure 2) are made from N11 which has the higher isotropic thermal conductivity [3].

Target plates subjected to high heat fluxes on horizontal or vertical plates are similar to those for MkII. A 2D CFC supplied by Dunlop has excellent in-plane conductivity. The transverse direction has low flexural strength so that special fixation methods are used to reduce bending across fibre planes.

4. CARRIER DESIGN

Figure 5 shows a view of the four designs that make up a set. The final assembly has 48 of each design fitted side by side on the divertor support structure.

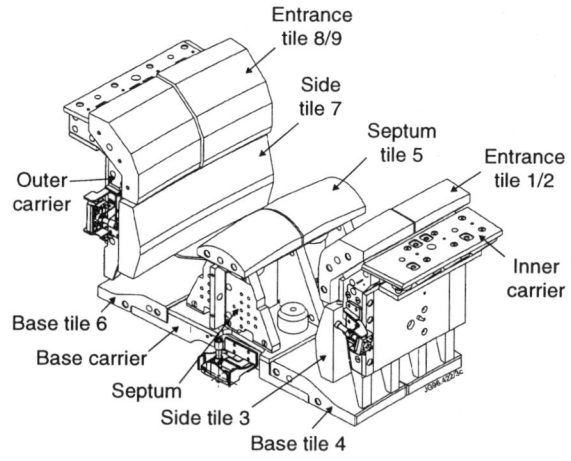

Fig.5: Perspective view of carriers and tiles.

The basic method of joining CFC plates is the barrel nut and bolt shown in figure 6, i.e. an M6 Nimonic 80A countersunk bolt and Inconel 625 barrel nut. Under the assumption of an initial temperature of 200°C and a plasma radiating 10 MW

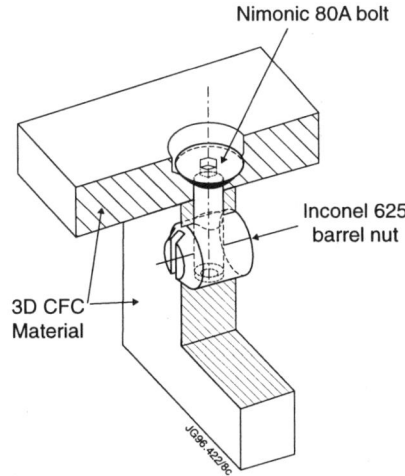

Fig.6: Barrel nut and bolt fixing.

inside each divertor leg for 10 seconds, the worst power loading of 0.7 MW/m^2 on to the ends of an Inconel 625 barrel nut raises the surface temperature to 475°C and a final bulk temperature of 350°C. The surrounding CFC material reaches approximately 450°C. Inconel retains 90% of its yield strength at these temperatures. The problem of seizure between the Nimonic bolts and Inconel nuts is solved by copper plating all bolt threads. Location dowels and anti-vibration nuts by Spiralock are used to ensure alignment and tightness under vibration.

Figure 7 shows the design used for the inner and outer carriers. An arrangement of ribs, webs and plates is used to produce a lightweight, rigid assembly. The target plate tiles are located on to the carriers with alignment dowels and tension bolts loaded with disc springs. Four corner support pads are machined at each tile location to define accurately the tile height with reference to the adjacent tiles for shadowing.

Without the lower side tiles (fig. 2a and c),, radiated power appears directly on the carriers behind the tiles and all exposed metallic components have to be removed or protected with graphite or CFC blocks. The lower ribs on both carriers could be subjected to conducted power up to 8 MW/m^2 while protecting the support structure for certain plasma equilibria. Careful design of the surface slopes and heights is essential.

The manufacturing sequence has to take into account the large number of accurately fitting interfaces that must locate simultaneously to engage all the alignment dowels. A sequence of component and sub-assembly machining has been developed to achieve this task.

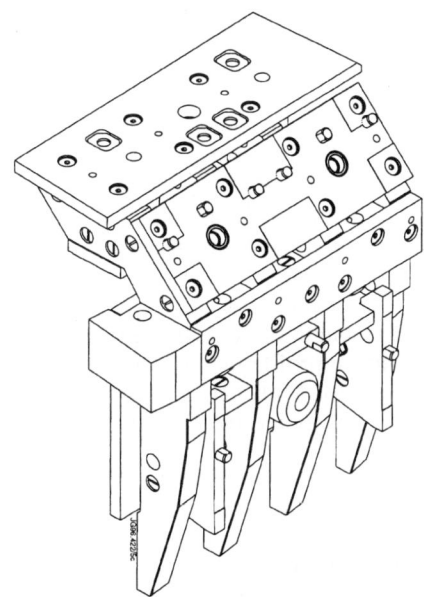

Fig.7: Side Carrier structure.

The base carrier in figure 5 is machined from a single Dunlop 2D tile. The chemical vapour deposition (CVD) process to achieve adequate density in the centre of the 50mm thick tile has proved to be a challenge and multiple oven runs have been required.

The septum in figure 8 is a relatively straightforward structure but one which has to withstand the highest halo current stresses and radiant heat loads.

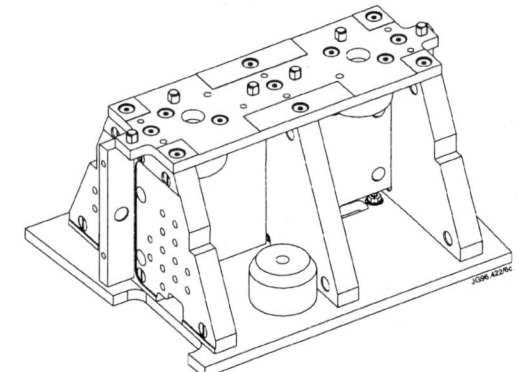

Fig.8: Septum structure.

5. TILE DESIGN

Unlike the tiles in the MkII design [2], shared corner pad support is not used here and each carrier, with its tiles, locates directly on to the support structure. Consequently the build up of tolerances in defining each tile location is larger so that steeper tile angles and shadowing steps are required, resulting in a slightly lower power handling capacity than for MkII.

The orientation of the fibre planes for all except the lower side tiles has been changed from a toroidal to a poloidal plane to improve the spreading of power. A consequence is that all but the side tiles are only half the length of those in MkII.

The surface profile has been designed according to the criteria for MkII [2]. The septum tile surface is bi-directional to receive flux from both toroidal directions.

The method of supporting the tiles mechanically and retaining their integrity against cracking with tie rods is maintained from the MkII. Machining is carried out with ball nosed cutters with vacuum chucks for location.

6. DIAGNOSTICS

The diagnostics may be divided conveniently into three groups - those that measure conditions on the carriers such as thermocouples and Langmuir probes, those that use the carriers as convenient supports such as neutral helium injection and magnetic coils, and those that are mounted independently of the carriers such as bolometers, pressure gauges and µ-wave horns.

Thermocouples are fitted both to tiles and structure with a standard distance of 10mm from plasma facing surfaces. A spring loaded design has been proven in prototype testing.

7. REMOTE HANDLING

Design of the divertor carriers has had to maintain remote handling (RH) capability from the beginning. The side and base carriers have RH lifting points on the top plates while the septum is moved by a special jig locating off the top plate and steps in the ribs. All bolts have been designed for RH operation. Further details are described elsewhere [4].

8. PROTOTYPE TESTING

Thermal and mechanical testing of CFC has been described in [3]. Mechanical testing of a representative septum has been extended to include vibration, pulse and steady state loads. The vibration tests were carried out at 250°C to study the effect upon the copper plated bolts. The top of the septum deflected elastically by 3mm under a 2KN load. After 50 impulses of 1.6KN followed by 80,000 cycles of 900N steady plus 500N cyclic over 10-30Hz, the structure showed no change in compliance. All bolts retained their original tightening torque.

The design of the thermocouple fixing was studied in the neutral beam test bed. Compression springs made from Nimonic 90 do not deteriorate in this environment and the design allows reliable interpretation of the tile surface temperature.

REFERENCES

1. M. Huguet for JET Team, Design of the JET Pumped Divertor, JET report JET-P(91) 51, presented at 14th SOFE, San Diego, USA October 1991.

2. H. Altmann, E. Deksnis, J. Fanthome, C. Froger, C. Lowry, R. Mohanti, M. Nilsen, A. Peacock, M. Pick, D. Spencer, R. Tivey, G. Vlases, Design of the MkII Divertor with Large Carbon-Fibre Composite (CFC) Tiles, 18th SOFT (1994,

3. H. Altmann, E. Deksnis, J. Fanthome, C. Froger, C. Lowry, A. Peacock, M. Pick, The Use of Carbon Fibre Composites in Divertor Target Plate Tiles and Structures, 16th SOFE (1995) 750

4. S.F. Mills, A.B. Loving, Design and development of RH tools for the JET Divertor Exchange, this conference.

ADVANCES IN POROUS MEDIA HEAT EXCHANGERS FOR FUSION APPLICATIONS

J. H. Rosenfeld[a], J. E. Lindemuth[a], M. T. North[a], R. D. Watson[b], D. L. Youchison[b], and R. H. Goulding[c]

[a]Thermacore, Inc.
780 Eden Road, Lancaster, PA 17601 USA
[b]Sandia National Laboratories
Albuquerque, NM 87185 USA
[c]Oak Ridge National Laboratory
Oak Ridge, TN 37831 USA

Advanced technologies are being evaluated for cooling of high heat flux components for fusion applications. One promising class of heat exchanger design uses a porous medium to effect efficient heat removal. High heat flux dissipation has been demonstrated in prototype fusion reactor components using both mechanically-pumped single-phase heat exchanger and capillary-pumped (heat pipe) designs. The state of the art of this technology has been considerably extended in recent years, and an increasing number of applications are being identified.

Water-cooled porous metal heat exchangers are being developed for gyrotron cavities and depressed collectors. Absorbed heat fluxes in excess of 100 MW/m^2 have been demonstrated in prototype testing. Gas-cooled porous metal heat exchangers are being developed for plasma-facing component applications. This approach offers the advantage of no liquid discharge into the tokamak in the event of a component failure. Innovative internal cooling structures based on porous metal cooling are being used to develop helium-cooled Faraday shields and divertors. This approach has demonstrated the capability to dissipate high heat fluxes which are typical for plasma-facing components while minimizing the required helium blower power. Recent tests have demonstrated absorbed heat flux capability in excess of 40 MW/m^2 using this approach. Development is in progress for prototype Faraday shields and divertors using ITER design requirements.

1. INTRODUCTION

Cooling of plasma-facing components provides a significant challenge to Tokamak developers. In addition to removing high heat fluxes, these components must also meet a number of other requirements. These can include exposure to high energetic particle fluxes, exposure to large disruption forces, exposure to high fluence of neutron radiation, remote servicing capability, and compatibility with unusual fluids such as alkali metals. Many of these issues can be addressed through the selection of proper materials. However, materials alone will not solve the heat removal problems often seen in plasma facing components. One major category of heat exchanger that shows the potential to meet the cooling needs of plasma-facing components is one which uses a porous medium to enhance the heat removal. Significant advances have been realized in recent years in development of both mechanically pumped and capillary-pumped (heat pipe) porous media heat exchangers for fusion applications. The purpose of this paper is to summarize this work.

2. HEAT PIPES

One major category of porous media heat exchangers is the heat pipe. A heat pipe is composed of a sealed, evacuated vessel in which the inner surfaces are covered with a porous capillary wick structure (medium). The heat pipe is charged with a working fluid and is sealed. As shown in Figure 1, heat that is incident on the evaporator section of the heat pipe surface causes the liquid in the underlying wick to evaporate and travel as a vapor into cooler regions of the vessel, where the vapor condenses back to a liquid. The capillary wick then transports the condensate back to the heated region. Heat pipes are often categorized by wick structure. Principal categories of wick designs can include sintered powder or fibers, axial grooves, spaced annuli, and wrapped screens. Several requirements for heat pipe operation include containment in a vacuum leak-tight envelope, working fluid compatibility with the containment materials, and operating temperature between the triple and critical points of the working fluid. The reader is referred to general texts on heat pipe operation for specific design techniques [1,2].

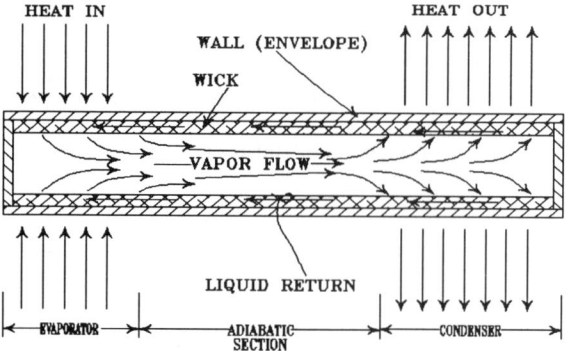

Figure 1: Heat and Working Fluid Flow in a Typical Heat Pipe

Heat pipes can be a preferred method of removing heat for several reasons. Heat pipes can be designed to work against gravity (evaporator above condenser), or to work in the microgravity of space. They are passive, i.e. they require no mechanical pump to function. They also require only a relatively small amount of working fluid, so failure of a heat pipe will only allow a relatively small amount of working fluid to leak out. They can also be used to control heat removal at uniform or constant temperature, or to spread a high heat flux to a larger area to allow heat removal by conventional techniques. It is this last feature that generally makes heat pipes attractive for cooling high heat flux fusion device components.

It is useful to divide discussion of heat pipes into low and high temperature devices. High temperature heat pipes generally employ alkali metals as working fluids. The favorable transport properties of the alkali metals lead to considerably larger heat flux capability of high temperature heat pipes as compared to low temperature heat pipes. A general description of the performance capabilities of both low and high temperature heat pipes is given in [3,4].

2.1 Low Temperature Heat Pipes

For high heat flux applications between ambient temperature and about 300°C, water is usually selected as the working fluid because of its ability to remove high heat fluxes in a heat pipe design in this temperature range. Water heat pipes have recently been developed which use optimized internally extended surface evaporators and bidisperse porous wicks. Tests on these devices have shown their ability to absorb localized heat fluxes in excess of 10 MW/m^2 [5,6]. Plain porous metal layer wicks are typically limited to lower heat fluxes, on the order of 1 to 2 MW/m^2; to date only the latter simpler design has been seriously evaluated for fusion applications.

Water heat pipes are being evaluated for cooling the Faraday shield that protects the radio frequency (RF) antennae used for plasma heating [7,8,9]. The Faraday shield is a nearly horizontal array of tubes located between the current strap and the plasma. Heat input to the Faraday shield arises from two sources: a by-product of RF heating of the reactor plasma by the RF current strap and direct radiation from the plasma. Under DOE Small Business Innovation Research (SBIR) Contract No. DE-FG01-90-ER81058, a Monel K-500/water thermosyphon heat pipe cooled prototype Faraday shield was designed, assembled, and tested. The shield was tested to a maximum input power of 8000 W (or 2000 W per heat pipe) at 586 K (313°C) and a radial heat flux of 0.525 MW/m^2. Testing at higher powers was limited by the heater capabilities. As a result, an operating limit on the heat pipes was never reached.

Testing on the prototype Faraday shield was also performed at Oak Ridge National Laboratory (ORNL) using an RF antenna as a heat source. These tests have shown this design to be fully RF compatible. An equivalent maximum strap current of 1.35 kA and a maximum voltage of 65 kV were successfully applied without failure of the shield.

This work has shown that heat pipes in an appropriate geometry are capable of handling high absorbed power levels and heat fluxes typical of fusion applications. Further uses of this approach for cooling plasma-facing components are being pursued.

2.2 High Temperature Heat Pipes

High temperature heat pipes are capable of removing large absorbed heat fluxes, and could potentially be used as heat flux spreaders for the most intensely heated plasma-facing surfaces. Coupling of such devices to heat sinks could be accomplished by using a design that includes an integral heat exchanger in each panel section. Use of this technology for fusion applications has been proposed but not yet developed.

In a National Aerospace Plane program, absorbed heat fluxes of nearly 1250 MW/m^2 have been demonstrated using tungsten/lithium heat pipes operating at 1923K (1650°C)[4,9].

3. PUMPED SINGLE-PHASE POROUS MEDIA HEAT EXCHANGERS

The pumped single-phase (PSP) porous media heat exchanger (PMHX) design has existed for at least several decades, and has recently demonstrated the capability for removing high absorbed heat fluxes in geometries of interest of fusion component developers. This design of PSP porous media heat exchangers and several test articles recently tested at high heat fluxes are summarized in this section.

3.1 PMHX Design

A pumped single-phase porous media heat exchanger employs a porous layer of a thermally conductive medium beneath the heat exchanger faceplate, which serves as the heat exchanger. The large surface area contained within the porous layer makes it possible to dissipate high heat fluxes from the faceplate with only a modest temperature difference between the faceplate and the bulk coolant temperature; the porous medium acts as an extended surface on the faceplate, improving heat transfer from the heated surface in the same way that fins improve heat transfer from an air-cooled surface, e.g. such as an air conditioner coil. The porous layer is usually bonded between the faceplate and a substrate panel containing coolant injection and removal ports. Either a liquid or a gas can be used as the coolant, with the heat transfer between the coolant and the wick particles occurring by convection.

Recent advances have been made in fundamental understanding of these devices and associated fabrication technologies, which has resulted in a new capability for efficient cooling applications [10,11].

3.2 Water-Cooled PMHX Devices

American and Russian gyrotron developers have produced water-cooled gyrotron cavities using PMHX technology. Absorbed heat fluxes of over 100 MW/m^2 have been removed successfully using this approach in prototype tests conducted at the Plasma under a U.S. Department of Energy SBIR contract at the Plasma Materials Test Facility (PMTF) at Sandia National Laboratories [12].

The PMTF tests on the semi-cylindrical gyrotron cavity sections showed good reliability under fatigue loading typical of gyrotron operation and showed survivability up to absorbed heat fluxes in excess of 100 MW/m^2. These demonstrated heat fluxes are believed to be the highest ever demonstrated for a water-cooled porous metal heat exchanger, easily surpassing the SBIR Phase II design goal of 30 to 50 MW/m^2. The approach is clearly a good candidate for applications where high cooling rates are required, and is achievable at lower pressure drops and pumping power than are possible with channel cooling.

3.3 Helium-Cooled PMHX Devices

Success with the copper/water device has led to evaluation of its use with alternative coolants such as helium or alkali metals. These coolants are being evaluated as candidates for cooling the diverter panels in the International Thermonuclear Experimental Reactor (ITER) Tokamak program and for cooling the blanket and diverter for the Demonstration Power Reactor (DEMO). Helium cooling of Tokamak plasma-facing components is an attractive alternative to water cooling because the helium is easily removed in the event of a leak. Helium is already present in the reactor as a byproduct and therefore does not represent an additional plasma contaminant. Helium is favored over water for safety reasons, especially if liquid metal coolants are used for other in-vessel components. Gas cooling at high heat fluxes challenges the heat exchanger design because of the poor heat transport properties of gases at low to moderate gas pressures. Thus, use of enhanced cooling technologies such as porous metal can be useful for gas cooling applications.

Several helium-cooled porous metal heat exchangers have recently been tested to evaluate the capabilities of this approach for fusion applications [8,9,12-14]. Under DOE Small Business Innovation Research (SBIR) Contract No. DE-FG02-95ER82095, a proof-of-concept helium-cooled Faraday shield test article was fabricated and tested. A cross section of this 12cm long test article is shown in Figure 2. In this application, pressurized (4.1MPa) helium flows semi-circumferentially through a porous metal filled annulus 0.32cm thick and 2.54cm OD. The overall cross section of the test article was 3.2cm square. Absorbed localized heat fluxes of up to 40 MW/m^2 were demonstrated with this test article and effective heat transfer coefficients of up to 9,000 W/m^2·K were inferred with helium flow of 13.3g/s[13]. Further optimization of this structure is expected to yield improved performance. This capability makes gas-cooled PMHX devices a viable choice for cooling of high heat flux plasma-facing components devices for fusion applications.

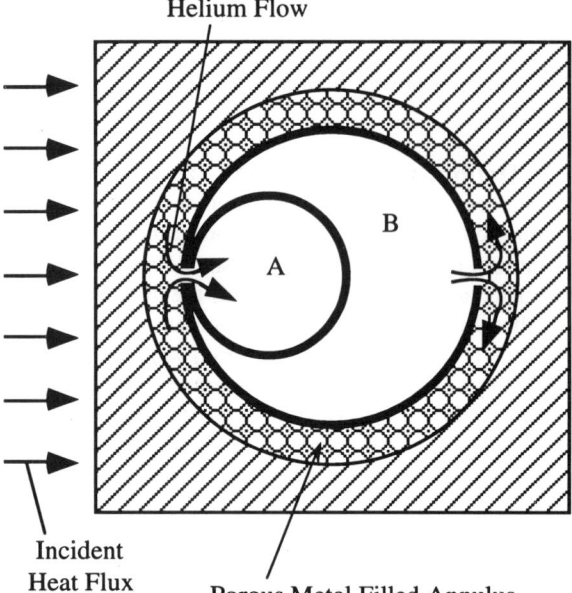

Figure 2: Helium Cooled Faraday Shield Test Article Cross-Section. A = helium exit, B = helium inlet.

4. CONCLUSIONS

A class of heat exchangers known as porous media heat exchangers shows promise for meeting the requirements for several fusion component cooling applications. Several test demonstrations at high heat fluxes have recently been performed. The design eventually selected for each application will ultimately depend on the individual requirements of each application. Advances are being made to extend the capabilities of porous media heat exchangers to meet fusion cooling needs.

5. ACKNOWLEDGMENTS

The authors wish to thank the U.S. Department of Energy for supporting the SBIR development work described herein.

REFERENCES

1. P.D. Dunn and D.A. Reay, *Heat Pipes*, 3rd edition, Pergamon Press, Oxford, UK (1982).
2. S.W. Chi, *Heat Pipe Theory and Practice*, Hemisphere Publishing Corporation, Washington (1976).
3. J.H. Rosenfeld, "Porous Media Heat Exchangers for High Heat Flux Applications," SPIE Vol. 1739 *High Heat Flux Engineering* (1992).
4. J.H. Rosenfeld and M.T. North, "Porous Media Heat Exchangers for Cooling of High-Power Optical Components", *Optical Eng.*, **34**, pp. 335-341 (1995).
5. J.H. Rosenfeld, N.J. Gernert, and M.T. North, "Internally Extended Surface Heat Pipe Evaporators for Microelectronics Cooling," ASME HTD-Vol. 273, S.G. Kandlikar and T.C. Avedisian, ed., pp. 93-100 (1994).
6. M.T. North, R.M. Shaubach, and J.H. Rosenfeld, "Liquid Film Evaporation from Bidisperse Capillary Wicks in Heat Pipe Evaporators," *Proc. 9th Int. Heat Pipe Conf.*, Albuquerque, NM, May 1-5, 1995.
7. J.H. Rosenfeld and J.E. Lindemuth, Final Report entitled, "Heat Pipe Cooling of Faraday Shields," Contract DE-FG01-90-ER81058, 09 July 1993.
8. J.H. Rosenfeld and M.T. North, "Innovative Technologies for Faraday Shield Cooling", *Proc. 16th IEEE INPSS Symposium on Fusion Engineering*, Champaign, IL, October 1995, pp. 972-975.
9. J.H. Rosenfeld, J.E. Lindemuth, M.T. North, R.D. Watson, D.L. Youchison and R.H. Goulding, Fusion Technology, July 1996, pp. 449-458.
10. J.E. Lindemuth, D. M. Johnson, and J.H. Rosenfeld., "Evaluation of Porous Metal Heat Exchangers for High Heat Flux Applications," ASME HTD. Vol. 301, *High Heat Flux Eng.*, A. Khounsary, Ed., pp. 93-98 (1994).
11. J.H. Rosenfeld, J.E. Toth and A.L. Phillips, "Emerging Applications for Porous Media Heat Exchangers", *Proc. Int. Conference on Porous Media and Their Applications in Science, Engineering and Industry*, Kona, Hawaii, June 16-21, 1996, pp. 472-498.
12. "Recent Results of High Heat Flux Testing at the Plasma Materials Test Facility", *Proc. 1996 High Heat Flux Eng. III, SPIE* (in press).
13. M.T. North, J.H. Rosenfeld, and D. L. Youchison, "Test Results from Helium Gas-Cooled Porous Metal Heat Exchanger", *Proc. 1996 High Heat Flux Eng. III, SPIE*, Denver, CO, August 1996 (in press).
14. J.H. Rosenfeld, M.T. North, D.L. Youchison, and R.D. Watson, "Cooling of Plasma-Facing Components Using Helium-Cooled Porous Metal Heat Exchangers", in Fusion Technology 1994: *Proc. 18th Symposium on Fusion Technology*, Karlsruhe, Germany, 22-26 August 1994, pp. 255-258.

EVOLUTION OF FRAMATOME AND CEA HIGH THERMAL FLUX STATION FOR FUSION TECHNOLOGY EXPERIMENTS NEEDS

* Marco Diotalevi, * Max Febvre & ** Philippe Chappuis.

* FRAMATOME Technical Center - High Thermal Flux Station, 3 rue du Guide, 71 200 Le Creusot - France
** CEA - DRFC - 13 115 Cadarache (Saint Paul Lèz Durance) - France

EB200 High Thermal Flux Station evolved with the needs of Fusion Technology experiments. The latests improvements of the station used for testing needs in 1995/96 concern :
- cooling loop flow rate capacity increase to 6 kg/sec
- use of non linear flux shape
- cycling tests on several positions
- use of pulses with variable length

This paper presents these modifications through recent applications on mock-ups.

EB200 High Thermal Flux Station was build in 1991 by FRAMATOME and CEA in frame of NET contract. The station evolved with the needs of Fusion Technology experiments, and can propose a vide randge of experiments for :
- critical flux tests (test of design),
- thermal fatigue tests (test of design, materials and joining techniques),
- disruption tests (test of materials).

The latests improvements of the station used for testing needs in 1995/96 concern :
- cooling loop flow rate capacity increase to 6 kg/sec,
- use of non linear flux shape,
- cycling tests on several positions,
- use of pulses with variable length.

These improvements are requested for a high demand on bigger mock-ups, trials more and more representative of nominal loading conditions, and increased productivity of tests.

1. INCREASE OF FLOW RATE CAPACITY TO 6 kg/s

The increase of flow rate capacity allows the test of bigger mock-ups or several mock-ups at the same time. The modification of the cooling loop has been done in a way that low flow rates are still fully available (figure 1).

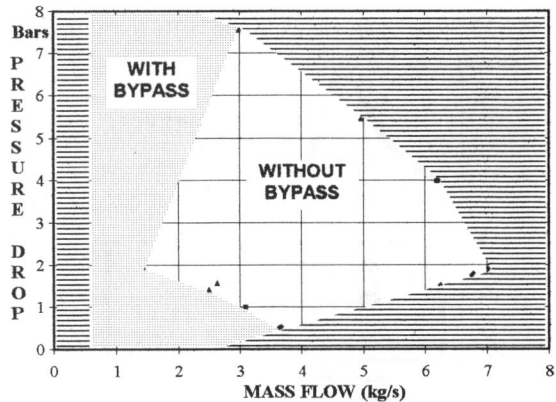

Figure 1 : Mock-up cooling loop capacity

The main characteristics of the mock-up cooling loop are (figure 2) :
- Demineralised water, with control of PH and H_2 levels (AVT system)
- Flow rate : 0.1 to 6 kg/s
- Pressure : 3.3 MPa at the outlet of the mock-up
- Inlet temperature : 50 to 200°C
- Instrumented for calorimetric diagnostic

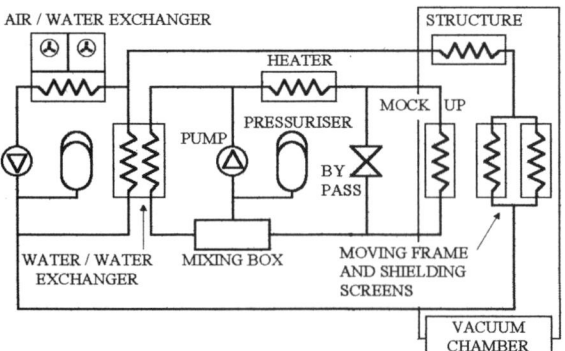

Figure 2 : cooling loops

2. NON LINEAR FLUX SHAPES

The use of a non linear flux is interesting in a way that thermal fatigue tests can be performed with a loading near the real (or theorical) heat deposition profile of the tested component (figure 5).

The heating of the mock-up is obtained through the sweepping of an electron beam (figure 3) by two sets of sweeping coils. The coils can operate at 10 kHz for an angle up to ±10° and 3 kHz for an angle up to ±23° (the distance from the coils to the mock-up is about 1 to 1.5 m).

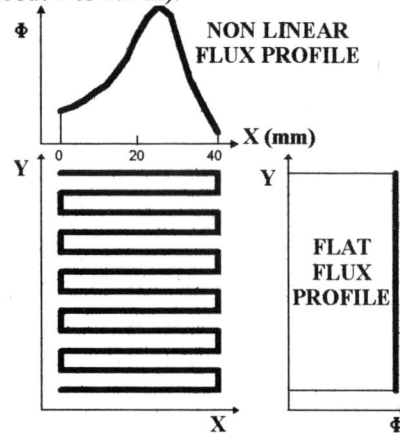

Figure 3 : Electron beam sweeping

These coils are operated by a PC that controls the beam position, the flux pattern (size and shape) and the pulse conditions.

3. AUTOMATIC CYCLING TESTS ON SEVERAL POSITIONS

The initial design of the EB200 station allows preprogrammed cycling tests at one place with constant duration of pulse during the 6000 s of one shot. By adding a wave generator to the system, during cycling we can move the source at several preprogrammed positions. This possibility has been used in 1996 for testing several mock-ups simultaneously or several zones of the same mock-up (figures 4 and 7), but it can also be used to simulate a thermal loading moving on the mock-up.

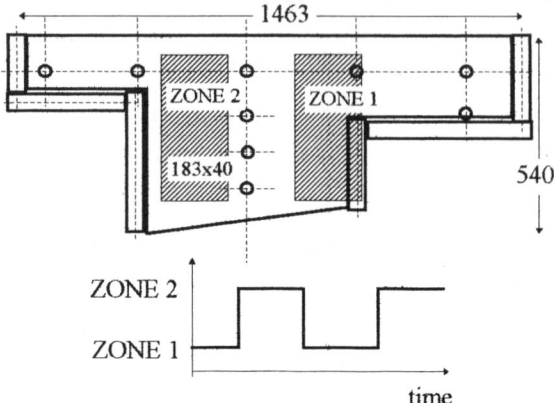

Figure 4 : cycling on two positions on CEA - TORE SUPRA PPE Panel

An upgrading is expected for 1997 in order to modify the acquisition and control system for programming several positions on the 2 axes with possible return to the spare screen.

4. PULSES WITH VARIABLE DURATION

During one shot of 6000 s the pulse period can be changed (it does not have to be preprogrammed). This new system has been used in order to optimise pulse conditions versus obtained temperatures for preparing cycling tests (testing of several pulse conditions without stopping - figure 6).

We use a digital signal (0V- 5V) that controls the 2 shot conditions :
- shot on the mock-up
- shot on the spare screen

Then, all the possibilities of wave generator are available during the shot (modification of the pulse time, triggering) and we can adjust pulse conditions depending on the observed thermal response. This is a big advantage because stopping a shot for modifying the programmed pulse conditions is time consumming.

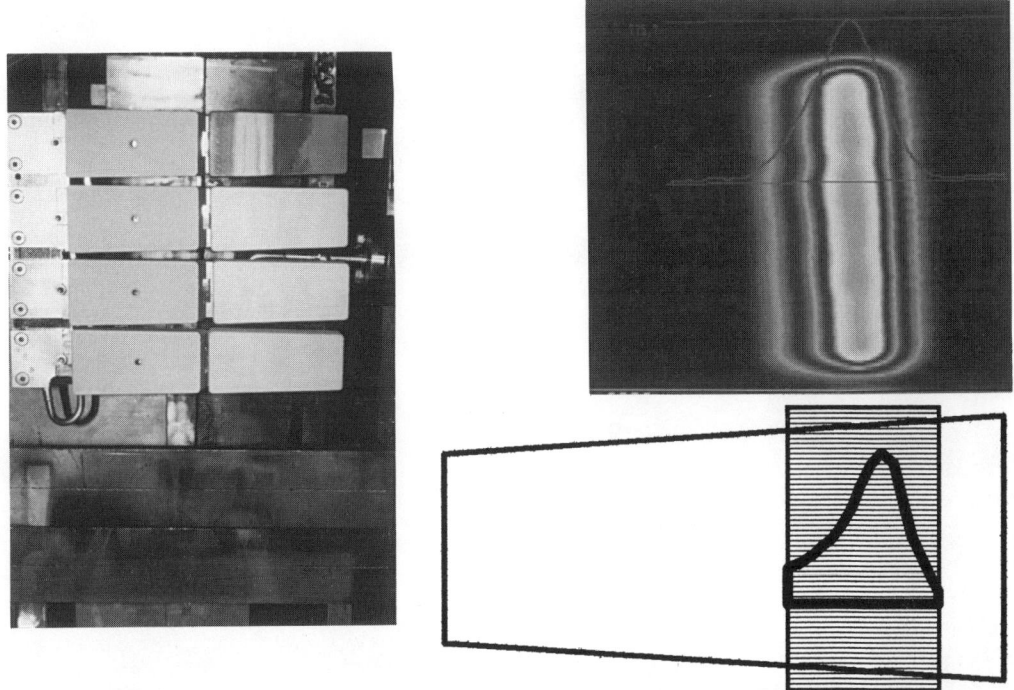

Figure 5 : Non linear flux on IPP - ASDEX Upgrade W-coated Divertor Tiles

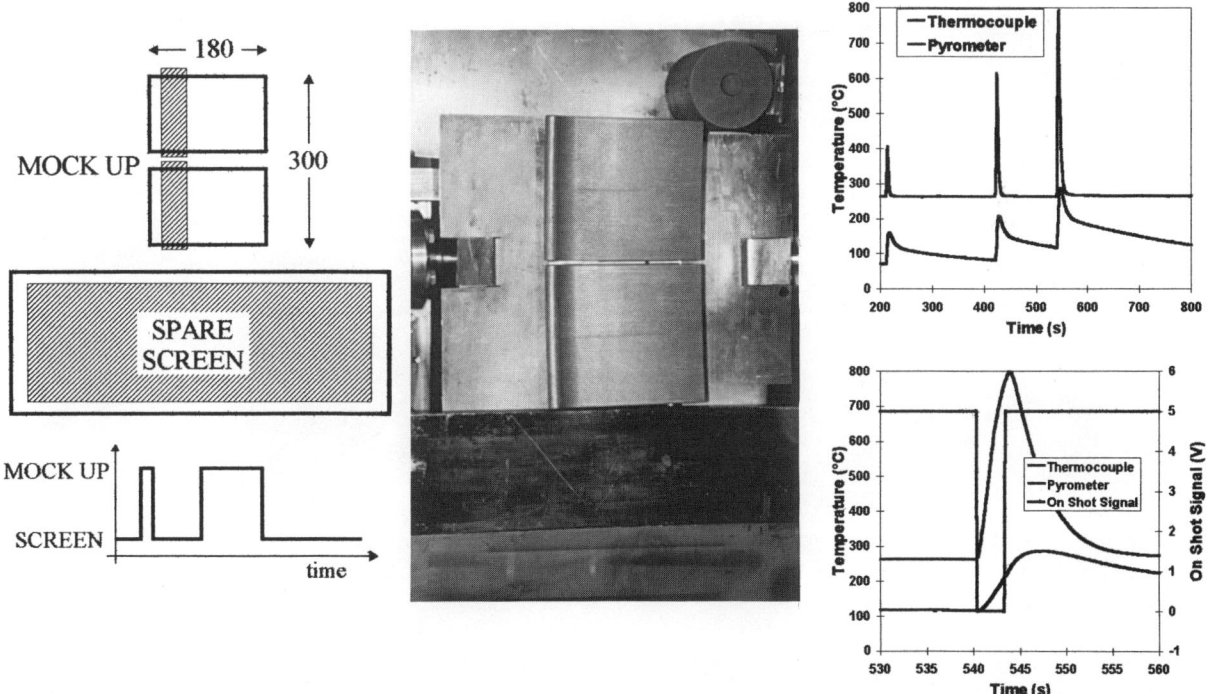

Figure 6 : Pulses with variable length on IPP - Divertor II Strike Point Module Mock-Up

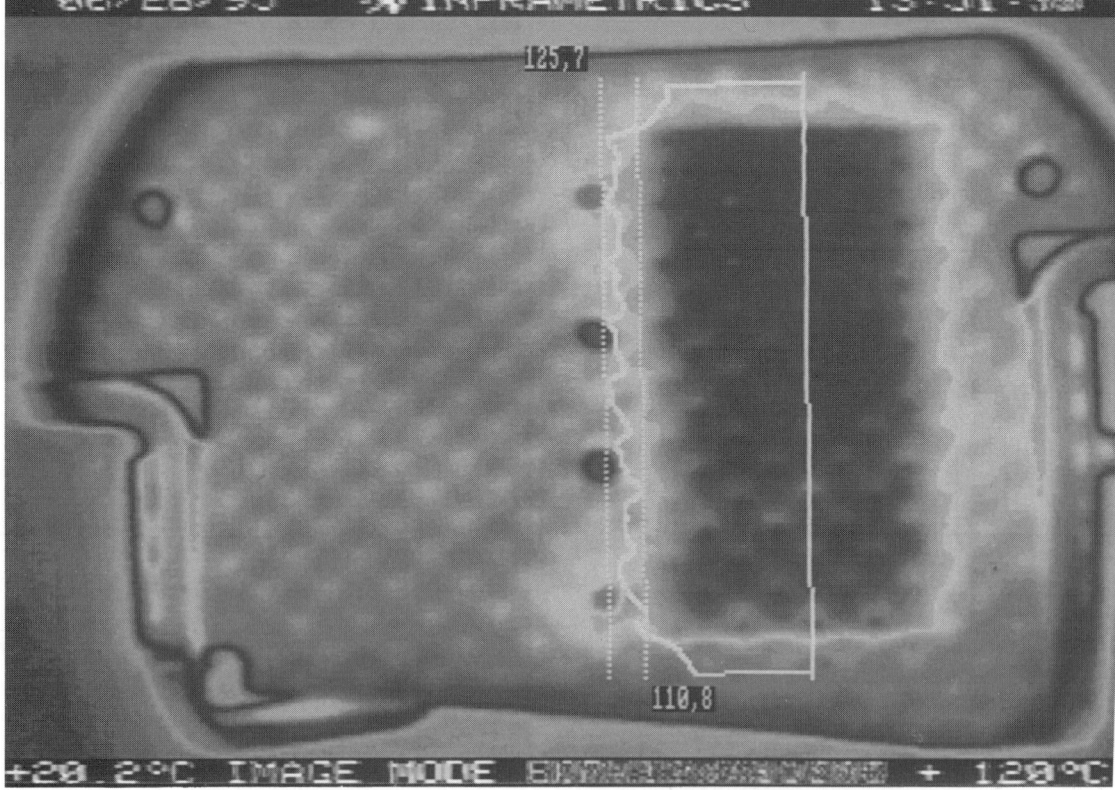

Figure 7 : Multiposition Cycling on CEA - TORE SUPRA PPE Panel

Xenon pellet production and acceleration in a pipe-gun

P.V. Reznichenko, I.V. Viniar and B.V. Kuteev

State Technical University, 29 Politekhnicheskaya, 195251, St. Petersburg, Russia

A new type of double stage gas gun is described. Using of a piston in the pump tube formed from condensed Xe gas is proposed for an improvement of reliability of the double stage gas gun. Technology of the piston formation and results of the fist compression experiments are considered.

1. INTRODUCTION

Central fueling of magnetic confinement devices (tokamaks and stellarators) by pellets requires high pellet velocities in the 10 km/s range [1]. However, the systems, which can provide such pellets, including double stage gas guns, rail guns or beam accelerators [2], demonstrate reliability laying far away from that required for controlled fusion applications. Particularly, the life time of the pistons used in the best double stage gas guns [3] does not exceed 1000 shots, while the desirable operation cycle accounts up to one million shots. As an approach to overcome the troubles related to the short life time of the piston, we offer to use the piston made by condensation or freezing of an appropriate gas or liquid. Such a single shot piston periodically renewed may significantly improve reliability of the injector. Using of the piston produced from a rather plastic material makes it possible as well optimization of acceleration regimes increasing the velocities achieved [4]. This improvement occurs because a piston part penetrates into the barrel at the last stage of compression. In this paper a double stage gas gun with single shot piston is described. The technology of Xe pellet production and acceleration developed can also by applied to ITER discharge termination by killer pellets [5].

2. EXPERIMENTAL SETUP

For proof of principle experiments a relatively small double stage gas gun has been designed and fabricated. A schematic diagram of the injector is shown in Fig. 1. Parameters of the gun are presented in Table 1.

A fast electromagnetic valve accompanied by the storage volume 70 cm^3 was used for supplying propellant pressure to the pump tube. Between the fast valve and the pump tube of the first stage a cooling cell with a heat exchanger was installed. The cooling cell was produced from oxygen free copper. The length of the orifice in the cell where the pistons were condensed was equal to 2.6 mm. The diameter of the orifice was the same as that for the pump tube (4 mm). The cell was connected to a cooling rod, which can be inserted into a dewar vessel with liquid nitrogen. The minimal temperature achieved in the cooling cell was close to 110 K. For compression experiments a piezo transducer was installed near the end of the pump tube. The temporal resolution for the transducer was better than 2 µs. During compression experiments the barrel was disconnected from the pump tube, and the pump tube orifice was closed by a flange. A special hydrogen pellet cell was placed between the barrel and pump tube. The cell allows us to form hydrogen pellet with 1.5 mm in diameter and 2 mm long.. A helium gas flowing cryostat is used for the cell cooling.

A diagnostic chamber installed at the end of the barrel allowed us to make photos of the pistons and pellets in flight and to perform the velocity's measurements. Vacuum and gas supply systems provided necessary parameters of the gases used in the piston or pellet formation and acceleration cycles.

Table 1
Parameters of the double stage gun

Pump tube diameter	4 mm
Pump tube length	1500 mm
Propellant gas pressure	100 bar
Second stage gun diameter	1.5 mm
Second stage gun length	300 mm

3. EXPERIMENTAL RESULTS

In the first series of experiments the parameters of double stage gas gun have been tested using Teflon piston operation. It was shown that with 4x20 mm piston the compression of the gas (N_2) in the first stage up to 250 bars can be achieved. The piston velocity reached was about 200 m/s.

The next step was devoted to searching for appropriate materials for the condensed piston and for the piston formation regimes. Ethanol, carbon dioxide and Xe have been tested as candidates. Condensation was performed in preliminary forepumped cell. The gases or liquids were supplied into the barrel though the vacuum pumping tube. The piston formation was observed through the valve connecting tube while the opposite end of the barrel was highlighted by a lamp. Searching for an appropriate formation regimes was performed in the following ranges of the Ethanol, CO_2, Xe pressure and temperatures shown in Table 2.

For ethanol and CO_2, the pistons were successfully formed, however we did not found regimes which provided good strength and reproducibility The best piston parameters were obtained for Xe condenced near 130 K at the pressure equal to 1.6 bar.

The results of compression experiments with Xe piston are shown in Fig. 2. The initial nitrogen gas pressure in the pump tube for these experiments was equal to 3 bar. The nitrogen gas pressure in the fast valve accelerating the piston was equal to 60 bar.

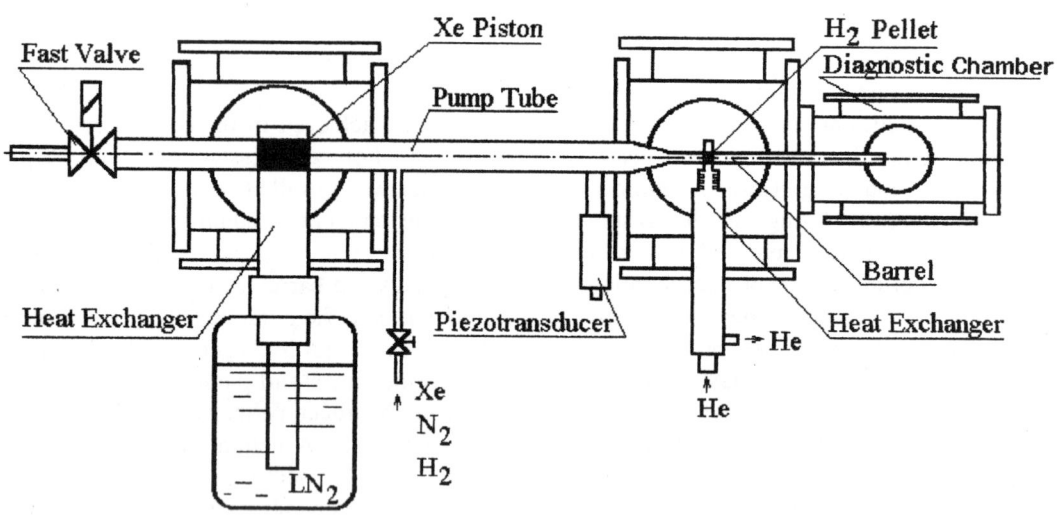

Figure 1. Schematic diagram of the double stage gas gun.

Table 2
Range of the cooling cell temperature and pressure investigated for the piston formation

Piston materials	Ethanol	CO_2	Xe
Cooling cell temperature, K	135 - 150	140 - 200	110 - 140
Pressure, bar	0.025 - 0.06	0.5 - 1	0.5 - 2
Formation time, min	0.5 - 1	1	3-5

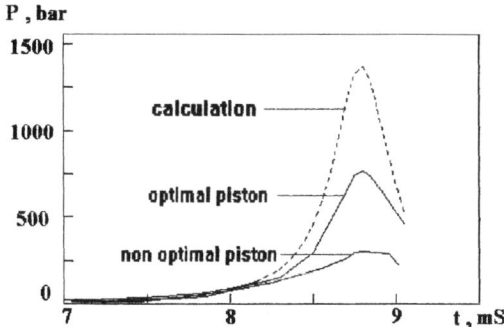

Figure 2. Gas compression curves.

The maximum piston velocity measured was equal to 220-250 m/s while the corresponding calculated value was equal to 280 m/s.

The curve in Fig. 2 marked "optimal piston" corresponds to the best achieved compression parameters with P_{max} = 770 bar. For non optimal pistons the pressure may change as it is demonstrated by the lower curve with P_{max}=300 bar. An ideal compression simulation is illustrated by the curve "calculation". It is seen that condensed Xe piston can provide compression parameters close to the ideal simulation.

After analysis of the piston photos and laser barrier data, the reason of compression deviations from the optimal one has been understood. The low compression corresponds to destruction of the piston into two approximately equal parts. Small defects of the cell manufacturing along with non coaxiality of the cell and the pump tube may be responsible for the piston destruction. A new gesign of the pump tube joint with the cooling cell is in progress now, which will allow us to avoid the piston destruction.

The part dealing with hydrogen pellet production is under tests now and the fist results on hydrogen pellet acceleration by the double stage gas gun will be obtained this year.

4. DISCUSSION

The results obtained confirm principal opportunity to create a double stage gun with cryogenic piston, particularly using Xe gas. Obviously, the operation time for this type of gun will not be restricted be the piston.

The technology developed allows us to produce pistons with size of 4x(20-30) mm and to demonstrate that they can be accelerated and decelerated conserving integrity up to the pressure of 750 bar. Although this pressure value is significantly lower than those in traditional double stage gas guns, there is a hope to increase it in future experiments.

Meanwhile, the size of the piston and the compression pressure reached are consistent with the requirements for ITER killer pellets. Thus our results demonstrate technical possibility to provide Xe killer pellet acceleration.

REFERENCES

1. S.L. Milora et al., Nucl. Fusion 35 (1995) 657.
2. S.K. Combs, Rev. Sci. Instrum. 64 (1993) 1679.
3. S.Sudo et al., Fusion Technol. 20 (1991) 387.
4. "Ballistic apparatuses and their applications", ed. by N.A. Zlatin, Moscow, 1974.
5. B.V. Kuteev, V.Yu. Sergeev, S. Sudo, Nucl. Fusion 35 (1995) 1167.

Disruption simulations on tungsten specimens in plasma accelerator

A. Gervash[b] E. Wallura[a], I. Ovchinnikov[b], A.N. Makhankov[b], J. Linke[a], , G. Breitbach[a]

[a] Forschungszentrum Jülich GmbH, Association KFA-EURATOM, D-52425 Jülich, Germany

[b] D.V. Efremov Institute, St. Petersburg 189 631, Russia

Plasma accelerators are proposed for disruption simulation experiments on plasma facing materials. In this work the VIKA-93 hydrogen plasma gun of the Efremov Institute was used. to provide the plasma disruption scenarios with the pulse length of a shot up to 360 µs and the incident energy during a pulse up to 30 MJ/m^2. Experiments were carried out on tungsten specimens of size 25mmx25mmx10mm. To investigate the influence of ductile-to-brittle transition temperature (DBTT) on tungsten behaviour under disruption the specimens were preheated from RT up to more then 1000 °C. Dependent on the specimen temperature and the loading parameters different phenomena were observed. At the energy of 7.5 MJ/m^2 in case of a preheating temperature of the specimen below DBTT a the number of grain boundary macro cracks was found beside a network of fine surface cracks. For specimen temperatures above the DBTT at the same loading conditions only the fine surface cracks were observed. At the energy of 30 MJ/m^2 remarkable surface melting and a wavy surface morphology was observed for all specimens at temperatures from RT to 1000 °C.

1. INTRODUCTION

The damage of materials in a tokamak based fusion reactor caused by plasma disruptions is one of the main problems to be solved in order to build a tokamak machine like ITER. The expected energy deposition in the thermal quench may reach several tens MJ/m^2 on the divertor dump plates in a short time (0.1-3 ms) [1, 2].

To withstand , among others, short pulse high heat loads induced by plasma disruptions the plasma facing materials (PFM) are being developed. During the plasma disruptions the materials are simultaneously loaded by heat flux, electrical current and accelerated plasma electrons. That is the reason why the application of plasma fluxes generated by plasma guns looks attractive for simulating plasma disruptions.

The aim of the present work was to investigate the tungsten as plasma facing material with different initial temperatures (in the range from ductile-to-brittle transition) under pulsed plasma irradiation. The disruption experiments were carried out on the VIKA-93 hydrogen plasma gun at the Efremov Institute, St. Petersburg. The morphological changes after testing were investigated using metallography and SEM in the Research Centre, Jülich.

2. EXPERIMENTAL PROCEDURE

2.1 Materials

The powder metallurgical (PM) tungsten was sintered in high-purity hydrogen at temperatures of about 2500°C and then heavily worked. The specimens had a thickness of 10 mm and a polished surface of 25mm x 25mm which was exposed to the plasma. The temperatures of the specimens were controlled by thermocouples and by a pyrometer. The time dependent bulk temperatures in the exposed specimens were measured and recorded.

2.2 VIKA-93 plasma accelerator

VIKA-93 is the last version of the VIKA-series plasma accelerators. It is a coaxial plasma gun which can deliver the pure hudrogen plasma (impurity density < 1%) with high enough values of energy densities on the tested specimens.

The power supply system of the VIKA-93 is a 16-sections, 3 regimes forming line with the stored energy of 102 kJ designed for a charging voltage up to 5 kV. This forming line allows to generate current pulses of a quasirectangular shape across the matched load with time duration of 90, 180, 360 μs and amplitude I~120kA. The accelerated plasma is characterized by a strongly directed flow of particles which have a low thermal energy (3-5 eV) and a high velocity (0.5-1.7 x 10^5 m s^{-1}).

The specimen holder is supplied with a heating system which allows to vary the initial temperature of specimens in the range from RT to 1100°C.

The features of VIKA-93 facility are detailed in [3].

2.3 Loading conditions

Assuming the Ductile-Brittle-Transition-Temperature (DBTT) for given powder tungsten around 400°C the initial temperature of specimens was fixed in two temperature intervals:
RT - 300°C (below the DBTT)
>500°C (over the DBTT)
Leaning upon the previous experiments [4] two fixed values of incident energy density per shot were chosen:
7.5 MJ/m^2 - „low energy" which does not lead to surface melting
30 MJ/m^2 - „high energy" which causes remarkable surface melting
The pulse duration was fixed for all specimens as 360 μs. All the specimens were exposed to 10 shots. The summarized test matrix of the experiments is shown in table 1.

Table 1
Test Matrix of disruption experiments

Regime #	Initial temp. of spec. °C	Inc. energy density, MJ/m^2	Pulse duration μs	Number of shots
1	330	7.5	360	10
2	640	7.5	360	10
3	RT	30	360	10
4	800	30	360	10

3. RESULTS AND DISCUSSION

Dependent on the initial specimens temperature and the loading parameters different phenomena were observed.

3.1 Incident energy density of 7.5 MJ/m^2

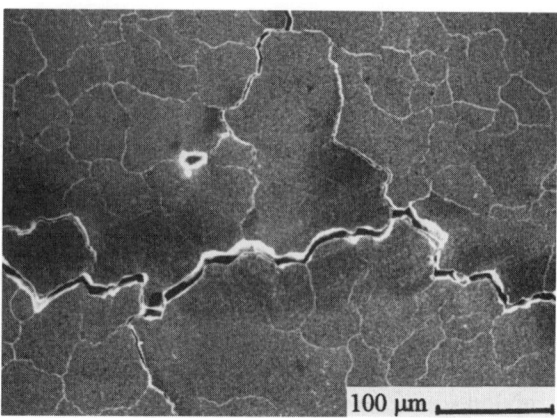

Fig. 1. Surface crack pattern for regime #1

Fig. 2 Surface crack pattern for regime #2

At the low energy deposition (7.5 MJ/m^2) and initial temperatures of the specimens below DBTT (regime #1) it can be clearly seen (see fig. 1) that the surface of the specimen is covered by a net of fine surface cracks. It also can be seen that a number of large macrocracks extend through the specimen surface. Scanning electron microscopy showed that in case of the same parameters of incident energy (7.5 MJ/m^2) but an initial temperature of the specimens over DBTT (regime #2) only the fine surface crack pattern was observed (see fig. 2)

To investigate the depth of the damage zone and crack propagation a metallographic cross-section analysis of the specimens was done. This analysis showed that in case of low initial specimen

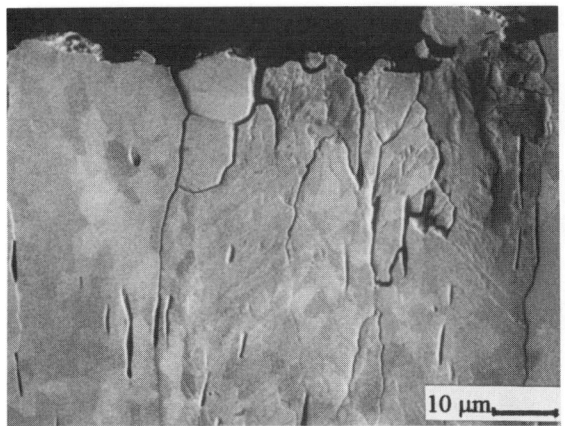

Fig.3. SEM image of the cross-section for regime #1

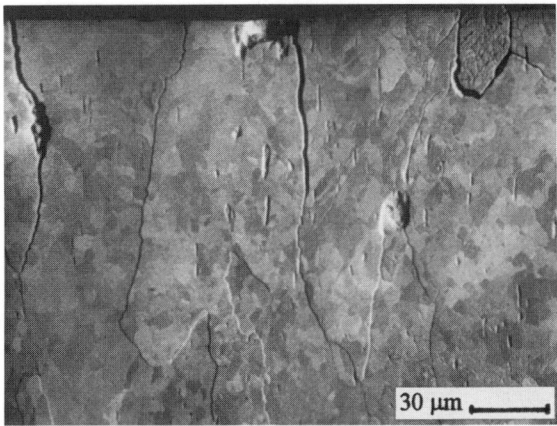

Fig.4. SEM image of the cross-section for regime #2

temperature (regime #1) the most of the grain boundaries are damaged in the layer from the surface up to ~30 μm depth (see fig. 3). The large number of lateral grain boundary cracks (unperpendicular to the loaded surface) were found. It can cause the loss of thermal conductivity in the surface layer and hence serious concern about the durability of this material.

Analyzing the cross-section of the specimens with initial temperature over DBTT (regime #2) only a small number of grain boundary cracks were observed (see fig. 4). These cracks can extend ~50 μm depth but they are mainly perpendicular to the loaded surface and look more attractive with respect to the heat removing from the surface.

It should be also noted that even at low energy regimes the small melted areas with a depth ~1-2 μm were found.

3.2 Incident energy density of 30 MJ/m^2

At the energy deposition of 30 MJ/m^2 remarkable melting occurs at the surfaces for all specimens at the temperatures from RT to 1000°C.
All specimens have characteristic wavy surface structures due to the plasma beam pressure.

In case of the temperature of the specimens below DBTT (regime #3) the number of cooling down surface cracks against the background of the wavy surface can be seen (see fig. 5) but at the same time no surface cracks were found in case of specimen temperature over DBTT (see fig. 6)

The cross-sections of the specimens after

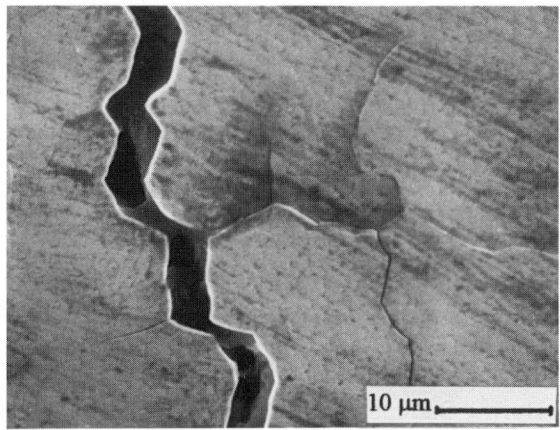

Fig. 5 Surface crack pattern for regime #3

Fig. 6 Surface crack pattern for regime #4

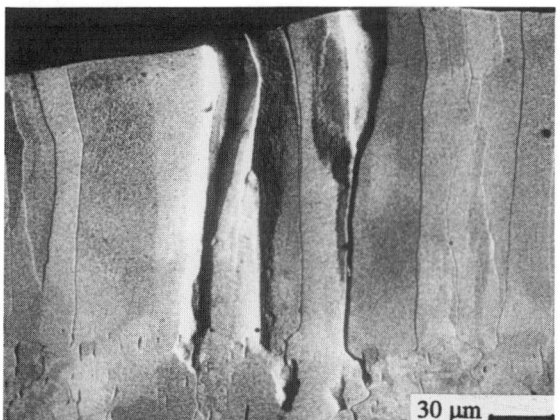

Fig.7. SEM image of the cross-section for regime #3

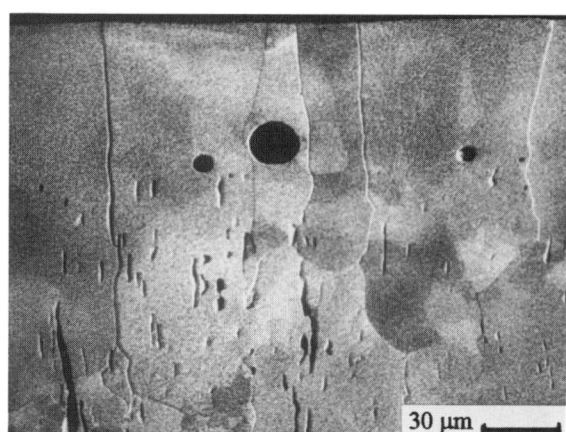

Fig.8. SEM image of the cross-section for regime #4

disruption scenarios in regimes #3 and #4 are shown in fig. 7 and fig. 8. For both regimes of loading the presence of melted layer with typical column structure of resolidified material can be seen. The bottom of the melted zone extends in parallel to the loaded surface but the top has a characteristic wavy shape as a result of the plasma beam pressure on the melted surface. The depth of the melted zone is practically the same for both loading regimes and varies dependent on wavy shape in the range ~ 150÷200 μm.

The loaded specimen with an initial temperature below DBTT (regime #3) has several macrocracks extended from the surface to the bulk material. In comparison with regime #3 the specimen with an initial temperature over DBTT (regime #4) has a more homogeneous structure without detectable macrocracks.

In both regimes a number of voids just under the melted zone can be seen. The voids density seems higher in regime #4.

4. SUMMARY

With the aim to investigate the tungsten behaviour under disruption loads in the temperature interval from ductile-to-brittle transition four different disruption scenarios were realized.

At low energies (without significant surface melting) at the initial temperatures below DBTT a number of large macrocracks against the background of fine surface microcracks can be seen. The grain boundaries of the surface layer are damaged in this regime that can lead to loss of the thermal conductivity. In case of the specimens temperature over DBTT only the fine surface cracks were found.

At the high energy remarkable melting occurs at the surfaces for all specimen temperatures from RT to 1000°C. All specimens have characteristic wavy surface structures due to the plasma beam pressure but the specimen with an initial temperature over DBTT has a more homogeneous structure without detectable macrocracks.

On the whole the specimens preheated over DBTT look more attractive in respect to withstand high heat loads caused by disruption scenarios.

5. REFERENCE

1. T. Kuroda, G. Vieider et al., „ITER plasma facing components", ITER Documentation Series No 30, IAEA, Vienna, 1991
2. G. Janeschitz, K. Borrass et al., „The ITER Divertor Concept", J. Nucl. Mater. 220-222 (1995) 73-88
3. A. Drozdov, V. Kuznetsov et al., „Investigation of Parameters of the Intensive Plasma Flows Generated by High Current Long Pulse Plasma Accelerators", Plasma Devices and Operations, Vol. 4, 1995
4. H. Bolt, V. Barabash et al., „Energy Deposition During Disruption Simulation Experiments in a Plasma Accelerators", Fusion Engineering and Design 30 (1995) 225-232

Void Swelling in Proton Irradiated Fe-Cr-Ni Ternary Alloys

Y. Murase, N. Yamamoto, J. Nagakawa and H. Shiraishi

National Research Institute for Metals
1-2-1 Sengen, Tsukuba, Ibaraki 305, Japan

Void microstructures and swelling were compared between Fe-15Cr-20Ni and Fe-15Cr-25Ni alloys irradiated with 180 keV protons at temperatures from 723 to 873 K. The displacement damage levels were 5, 10 and 20 dpa for the alloys preinjected with 10 appm He at room temperature, and 5 dpa for the alloys without He. In the alloys without He-preinjection, the Ni influence on swelling was pronounced at 823 K where the diminishing H effect was observed. In the He-preinjected alloys, swelling was similar in both alloys at all the temperatures and dose levels examined, and the incubation period, i.e., the major Ni influence on swelling, was suppressed. The major cause of the suppression of Ni influence on swelling would be the highly modified microstructural evolution by gaseous elements such as He and H.

1. INTRODUCTION

The radiation-induced swelling caused by high energy neutrons has been recognized as one of the serious problems for fusion reactor first wall materials. Numerous studies have been focused on the swelling behavior in the candidate materials such as Fe-Cr-Ni austenitic alloys. It is known that microstructure and swelling are strongly dependent on Ni content in Fe-Cr-Ni ternary alloys irradiated with fast neutrons [1-3], electrons [4], and Ni-ions [5]. Void number density and swelling generally decreased with increasing Ni content up to levels of 30-60 wt.% in Fe-15Cr-XNi alloys.

It is well recognized that helium (He) plays a significant role in microstructural evolution and swelling behavior. The He effect was so strong that the Ni influence on swelling was reported to be suppressed in the He-preinjected alloys in case of 1 MeV electron irradiation [6]. The effect of gaseous elements on void microstructure have been considered as an important issue for fusion reactor first wall materials, because the 14 MeV neutrons will produce gaseous products from (n, α) and (n, p) transmutation reaction in materials. Recent studies have reported that hydrogen (H) injected in materials may influence microstructure and swelling in cases of Ni-ion [7] and electron [8] irradiation.

In the present study, the microstructures and swelling were compared between Fe-15Cr-20Ni and Fe-15Cr-25Ni alloys irradiated with 180 keV protons (H-ions) at temperatures from 723 to 873 K. The objective is to investigate the Ni influence on microstructure and swelling under the conditions with and without preinjected 10 appm He.

2. EXPERIMENTAL

The materials used in this study were Fe-15Cr-20Ni and Fe-15Cr-25Ni ternary alloys. The chemical compositions of the alloys are shown in Table 1. Cold-rolled sheets of these alloys with 0.3 mm thickness were annealed for 120 s at 1293 K and spark-cut into 3 mm diameter disks. Surface layer of the disk was removed by mechanical grinding and electrolytic polishing to 120 μm in thickness. Disk specimens were irradiated with 50 to 200 keV He-ions at room temperature so that He level of about 10 appm was distributed uniformly in the range of 200-600 nm. After the He preinjection, the specimens were irradiated with 180 keV protons at temperatures from 723 to 873 K. The dose levels estimated with TRIM-85 code at a depth of 300 nm from the irradiated surface were 5, 10 and 20 dpa for the alloys with He preinjection, and 5 dpa for the alloys without He. The experimental details of He preinjection and proton irradiation have been described elsewhere [9]. The materials were removed by electropolishing from the irradiated surface by 300 nm in thickness, and then the disks were back-thinned to perforation for the transmission electron microscope (TEM) observation. The TEM observation was performed in the foil area of

Table 1 Chemical composition of the Fe-15Cr-XNi alloys (wt.%)

Alloys	Cr	Ni	C	Si	Mn	P	S	N	O	Cu	Fe
20Ni	15.04	20.12	0.0061	0.0020	< 0.001	0.0009	0.0020	0.0017	0.0087	< 0.001	bal.
25Ni	25.09	25.09	0.0049	< 0.005	< 0.001	0.0045	0.0021	0.0012	0.0073	< 0.001	bal.

100 to 200 nm thick. Foil thickness was evaluated by a stereo-graphical technique [10].

3. RESULTS

Voids and dislocation network were observed in both alloys under all the irradiation conditions. Figure 1, 2 and 3 show the temperature dependence of average void diameter, void number density and swelling in both alloys (a) with 10 appm He at 5, 10 and 20 dpa and (b) without He at 5 dpa, respectively. Large voids over 200 nm in diameter were observed not only in He-pre-injected 25Ni alloy at 20 dpa and 873 K but also in both alloys without He at 873 K. In these cases, the thin foil of 200 nm thick necessary for TEM observation was not obtained by using the present electropolishing apparatus (SOUTHBAY MODEL550). Accordingly, quantitative TEM measurements were not carried out in such conditions.

In the He-preinjected alloys, larger voids were generally observed in 25Ni alloy at 10 and 20 dpa as compared with that in 20Ni alloy. However, average void diameter was similar in both alloys at 5 dpa. Void number density in 25Ni alloy was slightly lower than that in 20Ni alloy at 5 dpa. The difference in void number density between the two alloys was more pronounced at 10 and 20 dpa. Swelling was little dependent on Ni content at each dose and temperature.

In the alloys without He preinjection, average void diameter was similar in both alloys at each temperature, and increased rapidly at 823 K. Void number density in 25Ni alloy was lower than that in 20Ni alloy at each temperature, and significant decrease in void number density was observed in both alloys at 823 K. Swelling in 25Ni alloy was lower than that in 20Ni alloy at 823 K, while the difference in swelling was small at 723 and 773 K.

4. DISCUSSION

4-1 He-free alloys

Significant decrease in void number density accom-

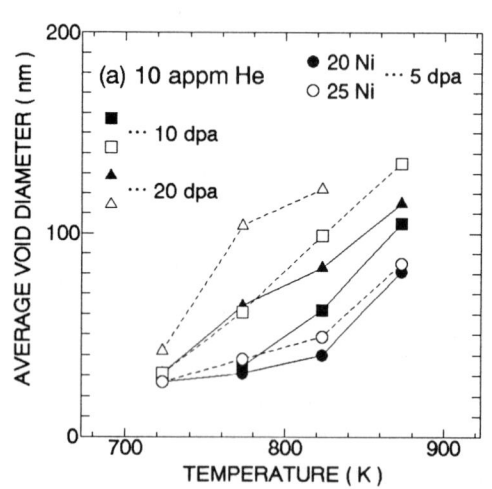
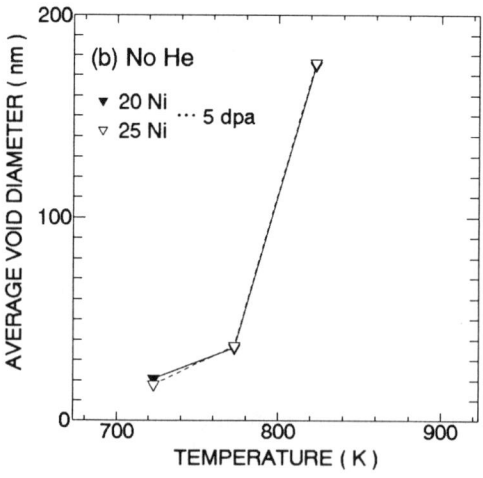

Figure 1. Temperature dependence of average void diameter in both 20Ni and 25Ni alloys (a) with 10 appm He and (b) without He

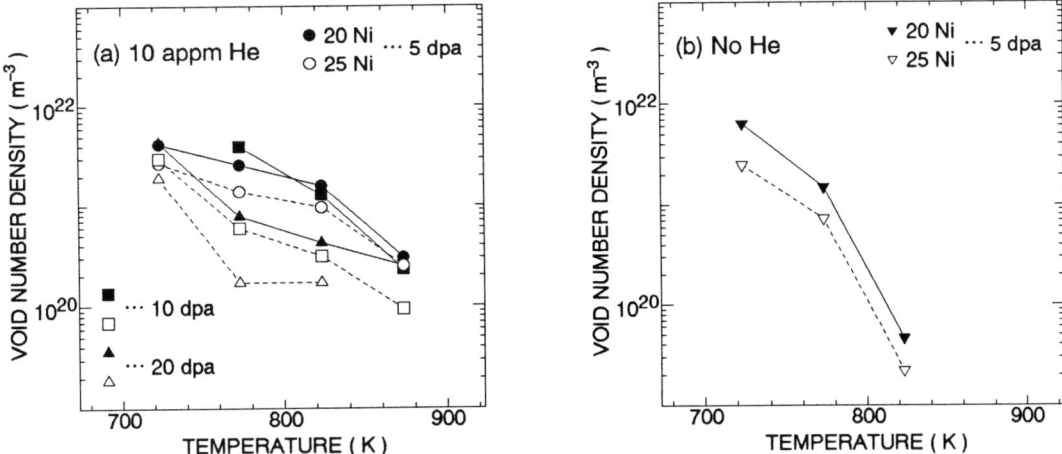

Figure 2. Temperature dependence of void number density in both 20Ni and 25Ni alloys (a) with 10 appm He and (b) without He

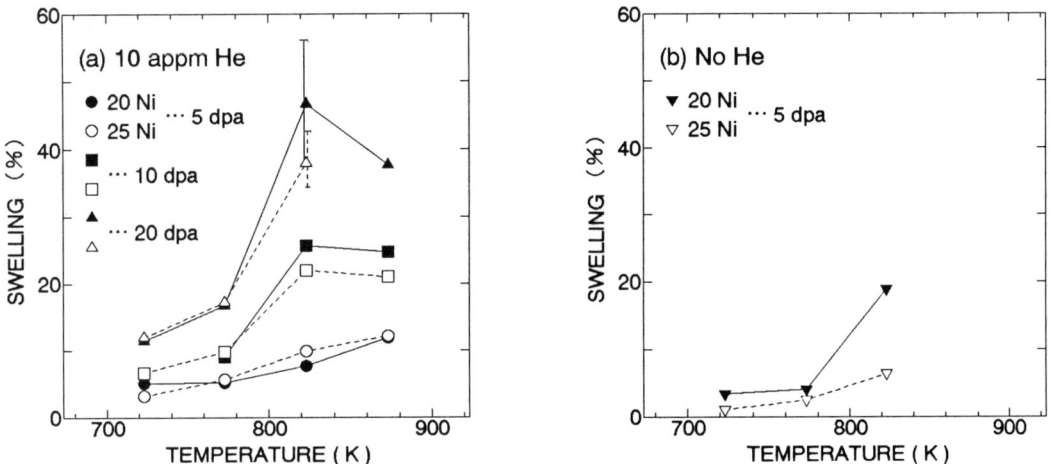

Figure 3. Temperature dependence of swelling in both 20Ni and 25Ni alloys (a) with 10 appm He and (b) without He

panied with rapid void growth was observed in both alloys at 823 K. In the previous report [9], it was concluded that H introduced during proton irradiation seems to play the similar role as He in promoting void nucleation in Fe-15Cr-20Ni alloy at temperatures up to 773 K, while the effect of H appeared to diminish rapidly at 823 K. The drastic changes in the void microstructure at 823 K observed in the present experiment also appear to be associated with the diminishing H effect at high temperatures. The amount of swelling in the lower Ni alloy (20Ni) increased rapidly at 823 K as compared with that in 25Ni alloy. This strongly indicates that the absence of H effect at higher tempera-tures revives the Ni influence on swelling.

4-2 He-preinjected alloys

Relatively higher density and smaller size of voids was persistent in both alloys at 5 dpa and each temperature. This suggests that void microstructures were strongly influenced by He even at 823 and 873 K in contrast with the He-free alloys. Void microstructures were similar in both alloys at 5 dpa, while larger but fewer voids associated with more frequent void coalescence was observed in 25Ni alloy especially at 10 and 20 dpa and higher temperatures above 773 K. Consequently, swelling was little dependent on Ni content. In the fast

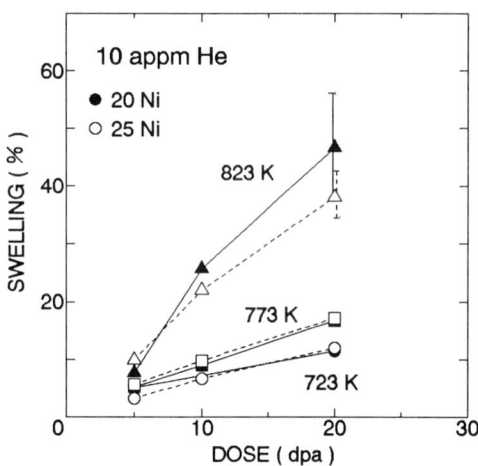

Figure 4. Dose dependence of swelling in the He-preinjected alloys

neutron irradiation conducted by Muroga et al. [2], the major Ni influence on swelling was reported to be an increased incubation dose with increasing Ni content up to 30 %. They have regarded that the enhanced incubation was caused by a weak interstitial bias of the faulted loops around 30-45 %Ni. However in the present proton irradiation, no faulted loops were observed in both alloys even at 5 dpa. Since the plotting of swelling as a function of dpa shown in Figure 4 appears to be running through the origin in both alloys at each temperature, the incubation, namely the major Ni influence on swelling, must be suppressed in the He-preinjected alloys. It is speculated that the presence of gaseous elements in the alloys could promote not only void nucleation but also transformation from faulted loops to dislocation network at an early stage of the irradiation. Consequently, the Ni influence on swelling would be suppressed when the microstructural evolution was highly modified by gaseous elements such as He and H.

5. CONCLUSION

The 180 keV proton irradiation was conducted on solution annealed Fe-15Cr-20Ni and Fe-15Cr-25Ni ternary alloys in the temperature range from 723 to 873 K with and without 10 appm He preinjection. The results of the microstructural measurement can be summarized as follows;

(1) In the alloys without He preinjection, swelling in 20Ni alloy significantly increased at 823 K accompanied by the diminishing H effect on microstructural evolution in both alloys.
(2) In the He-preinjected alloys, swelling was similar in both alloys at each dose level and temperature, and the Ni influence on swelling was suppressed even at 823 and 873 K.
(3) The major cause of the suppression of Ni influence on swelling would be the highly modified microstructural evolution by gaseous elements such as He at 723-873 K and H below 823 K.

ACKNOWLEDGEMENT

The authors wish to thank Dr. I. Shibahara (PNC) for the sample supply and chemical analysis.

REFERENCE

[1]. F.A. Garner and H.R. Brager, ASTM-STP 870 (1985) 187.
[2]. T. Muroga, F.A. Garner, J.M. McCarthy and N. Yoshida, ASTM-STP 1125 (1992) 1015.
[3]. F.A. Garner and A.S. Kumar, ASTM-STP 955 (1987) 289.
[4]. G.P. Walters, J. Nucl. Mater. 136 (1985) 263.
[5]. T. Muroga, N. Yoshida and F.A. Garner, Fusion Reactor Materials Semiannual Progress Report DOE/ER-0313/9 (1990) 87.
[6]. F. Rotman and O. Dimitrov, ASTM-STP 955 (1987) 250.
[7]. D. B. Bullen, G.L. Kulcinski and R.A. Dodd, J. Nucl. Mater. 133&134 (1985) 455.
[8]. S. Ohnuki, H. Takahashi, T. Takeyama and F. Wan, J. Nucl. Mater. 133&134 (1985) 459.
[9]. Y. Murase, A. Hasegawa, N. Yamamoto, J. Nagakawa and H. Shiraishi, J. Nucl. Sci. Technol. 33 (1996) 239.
[10]. P. B. HIRSCH (eds.), Electron Microscopy of Thin Crystals, London, (1965) 419.

Study of plasma target interactions with plasma streams of power density of 40 MW/cm^2

N.I.Arkhipov[1], V.P.Bakhtin[1], S.M.Kurkin[1], S.E.Pestchanyi[1], V.M.Safronov[1], D.A.Toporkov[1], S.G.Vasenin[1], H.Würz[2], A.M.Zhitlukhin[1].

[1]TRINITI, 142092 Troitsk, Russia
[2]Forschungzentrum Karlsruhe, Postfach 3640, 76021 Karlsruhe, Germany

Experimental study of plasma material interaction under simulated disruption conditions was performed at the MK-200 UG facility in the frame of ITER tasks. Different candidate materials for divertor plates were exposed to hydrogen plasma streams with energy densities up to 14 MJ/m^2, directed ion energy of 2-3 keV and pulse duration of 40 μs. Shielding of the target surface by vaporized material plasma was investigated. Formation, structure and properties of such layers was studied in details with interferometry, laser scattering, soft X-ray spectroscopy and profilometry with the aim to produce a database for validation of modelling codes. It was shown that the target plasma for low Z materials expands along the magnetic field lines and effectively protects the surface from the bulk of incoming energy flux. Tilting of the target results in a drift of vaporized material along the inclined sample surface. Erosion increases and efficiency of plasma shield decreases with sample inclination.

1. INTRODUCTION.

Erosion of divertor plates under high heat loads during ITER hard disruptions and ELMs may strongly restrict the lifetime of the divertor plates. Moreover, the erosion products may form a dust layer at the bottom of the divertor and present a real damage for normal ITER operation and environment. As ITER disruptive heat loads (up to 150 MJ/m^2 within a time interval of 0.1-1 ms [1]) are not achievable in existing tokamaks, divertor material erosion is studied in special simulation experiments at the plasma devices [2-4] under conditions as close as possible to the disruption situation.

Disruptive heat loads result in sudden evaporation of a thin surface layer which acts as a plasma shield and protects the divertor from further excessive erosion [5]. The plasma shield dissipates the bulk of the incoming energy flux into radiation [6], reduces the heat load at divertor plates and thus determines the real erosion. The vapor shielding effect is the most important process to be studied in simulation experiments. Experimental results are required for validation of theoretical models presently under development at several laboratories for prediction of divertor plate erosion [7,8].

First series of simulation experiments for perpendicular plasma impact were performed at power densities of 10 MW/cm^2 at the 2MK-200 facility in 1992-94 [9]. Experimental results and first comparison with numerical modelling have been reported in [10].

Facility upgrading was done in 1994-95. MK-200 CUSP and MK-200 UG facilities were created. Power density was increased up to 40 MW/cm^2. This gave the possibility for studying the process of interaction at tilted targets for closer simulation of the real ITER geometry. Plasma-induced erosion and dynamics of target plasma were studied for POCO and RGTi graphite. Laser scattering was used to detected emission of small particles during the erosion process.

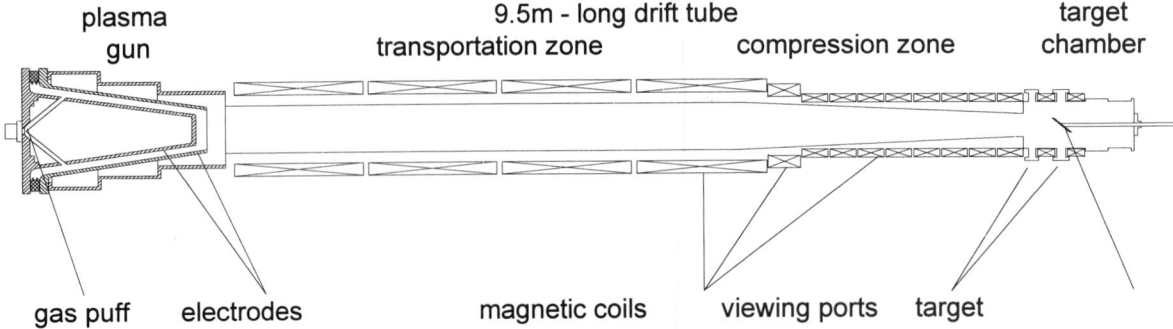

Fig.1 MK-200 UG facility.

2. EXPERIMENTAL SCHEME.

The MK-200 UG facility (Fig.1) consists of a plasma gun, a long drift tube and a diagnostic chamber. The plasma gun generates a hydrogen plasma stream with total energy of 100 kJ, and pulse duration of 10 μs. Pulse duration increases in the drift tube due to plasma velocity dispersion. The drift tube is a cylindrical liner of 30 cm diameter filled with a longitudinal magnetic field and a conical liner of 15 cm diameter at the exit. The magnetic field rises in the cone to conserve magnetic flux and achieves 2 T in the diagnostic chamber. The rising profile of magnetic field is used for radial compression and magnetization of the plasma stream.

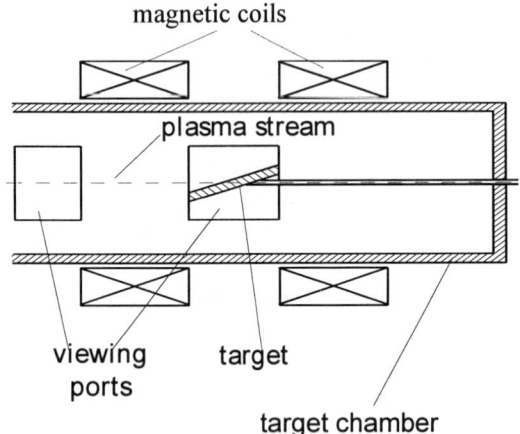

Fig.2 Target position.

The plasma stream parameters at the target position (Fig.2) are as follows: power density W = 35 MW/cm^2, energy density Q = 1.4 kJ/cm^2, stream diameter d = 6 cm, directed ion energy E = 2-3 keV, plasma beta b=0.3, plasma pulse duration t=40 μs.

Optical interferometry, laser scattering technique and soft X-ray spectroscopy have been employed to study the target plasma properties. The interferometry scheme consists of a Mach-Zehnder interferometer, a continuous gas laser (λ= 0.51 mm), a high speed framing camera as a recorder and associated optics. The scheme allows to study the evolution of the electron density in the target plasma with 2-dim. spatial resolution of 1 mm and temporal resolution of 500 ns.

Laser scattering diagnostics was used for local measurements of electron temperature and density in the incoming plasma stream and in the target plasma. A ruby laser beam (λ=0.6943μm) was focused on the plasma axis in the center of the second window (Fig 2). Backward scattered light (at 160°) was analyzed by a grating spectrograph and detected by a set of photomultipliers. The spatial distribution of the target plasma parameters was measured in success shots by replacing the sample along the system axis.

Chemical contamination, ionization state of ions and radiation intensity of the target plasma were studied by means of time-integrated soft X-ray spectroscopy. Absolute spectral measurement was performed with a transmission grating spectrometer (with 5000 l/mm). The recorded spectrum provided 1.5 mm spatial and 2 A° spectral resolution. A Kodak 101-01 film was used as a detector with known sensitivity in the wavelength range of interest of (2-300)A°.

3. EXPERIMENTAL RESULTS.

Interferometric measurements showed that a cloud of dense plasma arises in front of the target within 1-2 μs. Fig.3 shows 2-dim. distribution of the electron density over perpendicular POCO sample at t = 26ms and power density of 35 MW/cm^2. It is clearly seen that a dense plasma cloud expands along the magnetic field lines (vertical direction) towards the incoming plasma stream.

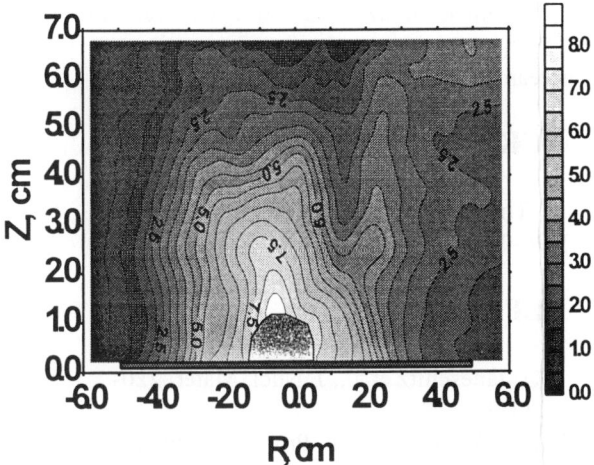

Fig.3 Electron density profile for perpendicular POCO graphite (n_e, 10^{17}cm^{-3})

The spectroscopic measurements showed that the plasma shield consists of highly stripped carbon ions. Fig.4 shows a time-integrated spectrum from the plasma shield at 42mm from the surface. Lines at 33.7A from CVI 1s-2p, and 40.3A from CV 1s^2-1s2p and their second orders are clearly seen.

Laser scattering measurements indicated that electron temperatures of (60-90) eV corresponding to this ionization state occur within the first (25-30) μs of interaction. The temperature later decreases to a few eV.

Although transverse motion of target plasma is limited due to the strong magnetic field the width of shielding layer (D=10 cm, see Fig.4) appears to be higher than the diameter d of the incoming plasma stream (d=6cm).

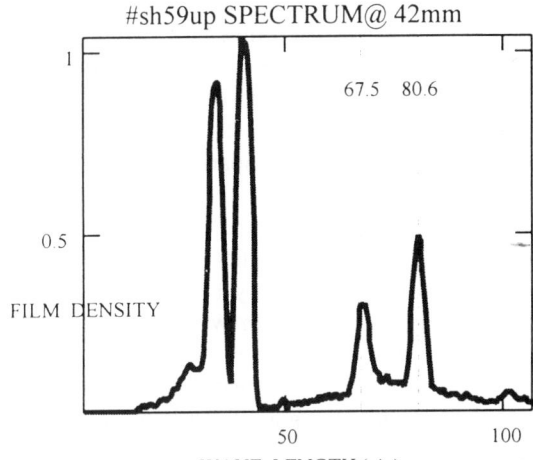

Fig.4. Spectrum of target plasma radiation.

Widening of the target plasma may be caused by lateral mass losses near the surface. Mass transport is more observable for tilted targets in experiments with narrow plasma streams (d=1cm) at the MK-200CUSP facility. Fig.5 demonstrates an increase of the electron density near inclined surface downwards from the point of plasma stream impact (dashed line). Thus a drift of vaporized material along the target surface is occuring.

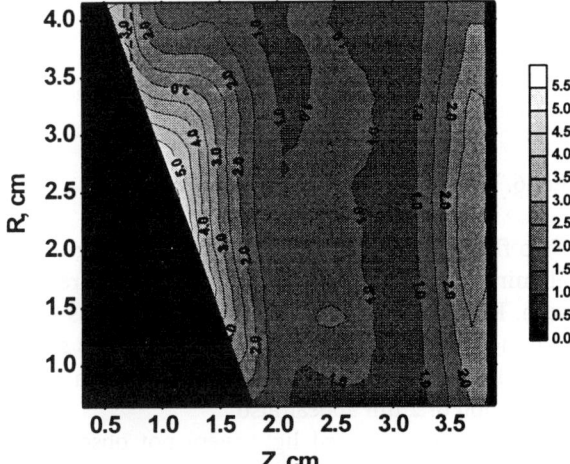

Fig.5 Electron density profile for tilted target at MK-200 CUSP (n_e, 10^{17}cm^{-3})

A consequence of this will be increased erosion for inclined incidence as was demonstrated at the MK-200 UG facility. The erosion rate is measured to be of 0.3 μm per shot under normal strike and 1.6 μm per shot for inclined (22.5°) incidence. Thus the erosion increases 5 times for inclined plasma incidence in spite of the fact that the plasma heat load decreases 3 times. This indicates that the plasma shield effectiveness for tilted target is considerably less in comparison with the normal target position.

Comparison of measured erosion profiles with calculated ones showed that the measured erosion may be explained not only by surface vaporization but also by its destruction with formation of small particles. These particles can't be vaporized completely under high heat loads and a dust of exposed materials may occur after interaction process. First attempts to detect the target particles was carried out by use of a laser scattering technique. Fig.6 demonstrates the scattering of a ruby laser beam (400μs duration, 0.4mm width) at the distance of 4 mm from aluminium target surface at time interval (2.1-2.5) ms after start of interaction.

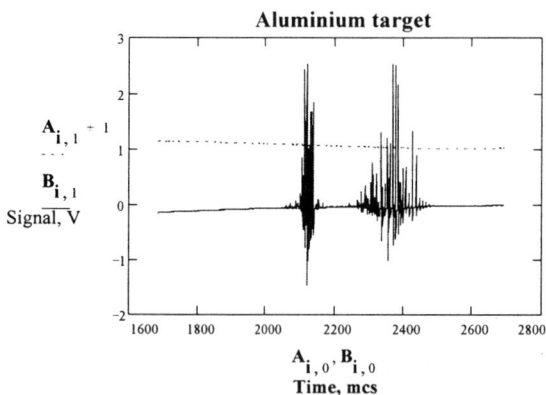

Fig.6. Laser light scattered by Al droplets.

Two flashes of scatterred light corresponding to two aluminium droplets crossing the laser beam are to be seen. Rough estimations give their velocity as 2 m/s and their dimension larger than 3 μm. It was shown that droplets with velocities up to 15 m/s and dimension of 2 μm appear also.

Flashes of scattered light were not observed in experiments with graphite what means absence of carbon particles with dimensions greater than 0.5 μm.

4. CONCLUSIONS

Plasma impact results in evaporation of target material and production of a target plasma which expands mainly along the magnetic field lines. This plasma shield effectively protects the target from the bulk of the incoming energy flux.

Experiments with tilted targets demonstrated drift of vaporized material along the sample surface. This results in increase of erosion. The efficiency of the plasma shield with sample inclination.

Droplets of size of 2-3 μm with velocities up to 15 m/s are detected after interaction of plasma stream with aluminium targets.

ACKNOWLEDGEMENT

This work is partially supported by RFBR, Grant N96-02-19421a.

REFERENCES

1. G. Janeschitz et al., J. Nucl. Mater. 220-222 (1995) 73.
2. N.I. Arkhipov et al., Proc. 12th Int. Conf. Plasma Phys. and Controlled Nucl. Fus. Research, Nice, France, 1988, Vol.2, 683.
3. V.M. Kozhevin et al., Fus. Eng. and Design 28 (1993) 157.
4. J.F. Crawford, J.M. Gahl, J.M. McDonald, J.Nucl.Mater. 203 (1993) 280.
5. A. Sestero, J. Nucl. Fusion 17 (1977) 115.
6. H. Würz et al., J. Nucl. Mater. 212-215 (1994) 1349.
7. I. Landman, H. Würz, Fusion Technology Vol.1 (1994) 335.
8. A. Hassanein, I. Konkashbaev, J. Nucl. Mater. 220-222 (1995) 244.
9. N.I. Arkhipov et al., Fusion Technology Vol.1 (1994) 463.
10. H. Würz et al., J. Nucl. Mater. 220-222 (1995).
11. H. Würz et al., ICFRM-7, Sept. 25-29, 1995, Obninsk, paper 03004 to be published in J. Nucl. Mater.

Post-mortem analysis of HIP bonded first wall panel made of SS316 and DS-Cu after high heat flux testing

Toshihisa Hatano[a], Kiyoshi Fukaya[b], Masayuki Dairaku[a], Toshimasa Kuroda[a] and Hideyuki Takatsu[a]

[a]Naka Fusion Research Establishment, Japan Atomic Energy Research Institute
P.O.Box 311-01, 801-1, Mukouyama, Naka-machi, Naka-gun, Ibaraki-ken, JAPAN

[b]Tokai Research Establishment, Japan Atomic Energy Research Institute
P.O.Box 319-11, 2-4, Shirane, Shirakata, Tokai-mura, Naka-gun, Ibaraki-ken, JAPAN

A HIP bonded DS-Cu/316SS first wall panel of 400 mm(L) x 104 mm(W) x 25 mm(t) with eight built-in cooling channels was thermo-mechanically tested to examine the integrity of the HIP bonded interface, together with the behavior of the DS-Cu layer, under the high heat flux conditions. Test conditions were selected to simulate ITER normal and disruption heat load conditions. After high heat flux testing, this panel was cut and metallurgically observed by means of an optical microscope, SEM and EPMA. There were no cracks and exfoliations observed at the HIP bonded interface, and the integrity of the HIP joint under the normal and disruption conditions was confirmed. Post-mortem analysis by microscope and SEM showed that the melting layer has a dendritic structure. It was also observed that Al_2O_3 particles within DS-Cu were segregated on the surface of DS-Cu, which suggested that ejected particles observed during the disruption high heat flux testing might be the Al_2O_3 particles. There were a number of fine cracks on the surface of the melted DS-Cu. A number of small cavities were also observed within the melt layer close to the surface.

1. INTRODUCTION

According to the operation plan of a fusion experimental reactor such as ITER, the shielding blanket without breeding function will be installed for first ten years, so-called Basic Performance Phase (BPP). The blanket is subjected to huge electromagnetic force, high neutron load and surface heat flux. The first wall (FW) which is integrated with the blanket has been proposed to have built-in cooling channels to remove high thermal loads (surface heat flux of 0.5 MW/m^2 and nuclear heating of 15 MW/m^3). The FW is composed of a beryllium (Be) layer as plasma facing material, dispersion strengthened copper (DS-Cu) layer as heat sink and type 316LN austenitic stainless steel (SS) as structural and shield materials. Hot Isostatic Pressing (HIP) bonding method has been proposed for the joining of DS-Cu/SS and SS/SS. After the investigation of the optimal HIP bonding condition[1], two DS-Cu/SS FW panels with built-in cooling channels were fabricated. SS316L was used for this fabrication and the following testing. One of the panels was thermo-mechanically tested under high heat flux.

2. FW PANEL PREPARATION

FW panel used for the high heat flux testing is shown in Fig.1 and Fig.2. Dimensions of the panel are 400 mm(L) x 104 mm(W) x 25 mm(t) with built-in eight rectangular cooling channels. Internal dimensions of the cooling channel are 10mm x 5mm and thickness 1.5 mm. A 0.5 mm thick SS liner was inserted between DS-Cu and SS channels into gaps to protect DS-Cu incursion between SS cooling channels during the HIP process. A 5mm thick DS-Cu (Gridcop AL-15®) plate, SS channels and SS plate were simultaneously joined by HIP. Six thermocouples were attached to measure DS-Cu and

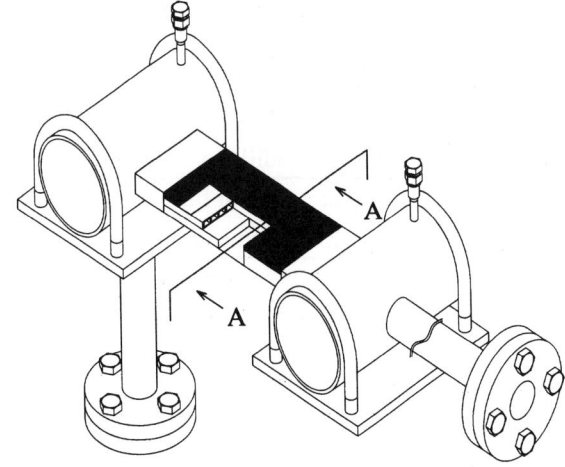

Fig.1 HIP bonded first wall panel

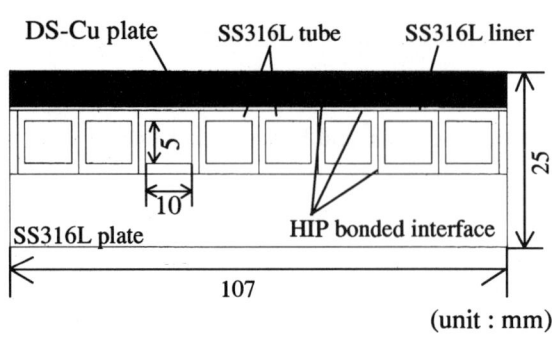

Fig.2 A-A cross-section of FW panel
SS temperatures.

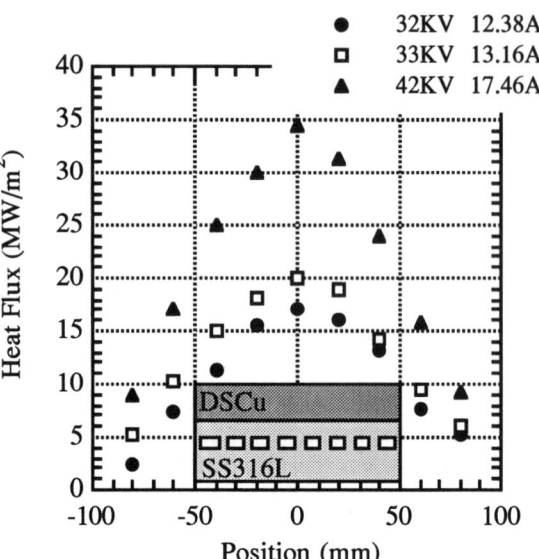

Fig.3 Heat flux distribution of PBEF

3. HIGH HEAT FLUX TEST

3.1. Test conditions

A short heat flux loading time, i.e. 1 sec, was taken in the tests in order to accelerate a thermal fatigue test. Therefore, higher heat fluxes than ITER nominal conditions were imposed on the FW panel so that temperatures and strains at the HIP bonded interface could simulate those under ITER nominal conditions. The heat flux to simulate the HIP bonded DS-Cu/SS interface temperature under ITER normal operation condition is 2.0 MW/m² with 1 sec heat flux loading time[2]. Strains at the HIP bonded DS-Cu/SS interface during ITER normal operation, i.e. 3.6×10^{-3} and 4.2×10^{-3}, can be also simulated with conditions of 1 sec heat flux loading time and heat fluxes of 5.0-6.0 MW/m².

High heat flux tests of the FW panel were performed at the Particle Beam Engineering Facility (PBEF) in JAERI with conditions shown in Table1 in series. The heat flux distribution to the FW panel as a function of accelerated voltage and current of PBEF is shown in Fig. 3.

3.2. Test results

Temperature responses during the thermal fatigue tests (No.1 and No.2) agreed well with the analyses results, and no degradation of thermal performance was observed during these tests[2].

Temperature responses in the first disruption test (No.3) also agreed well with analysis results. Maximum temperature of the DS-Cu surface was about 360°C in this case.

In the last disruption test with a heat flux of 100MW/m²(No.4), ejection of small particles from the heated DS-Cu surface was observed. An appearance of the heated FW panel surface after this disruption test is shown in Fig.4. There are three representative concentric circle patterns on DS-Cu surface along with the beam power distribution, i.e. hoarse, wavy and glossy surface from the center.

Table 1

High heat flux testing conditions

Test No.	Heat flux (MW/m2)	Heat flux loading time (sec)	Number of cycles	
1	2.0-2.5	1.0	1000	simulate the strains during ITER operation
2	5.0-6.0	1.0	1000	simulate the temperature during ITER operation
3	40	0.075*	2	simulate ITER off-normal event (thermal quench)
4	100.0	0.3	1	simulate ITER off-normal event (VDE)

* : In the case of PBEF, the beam has about 50 msec of the strating up time.

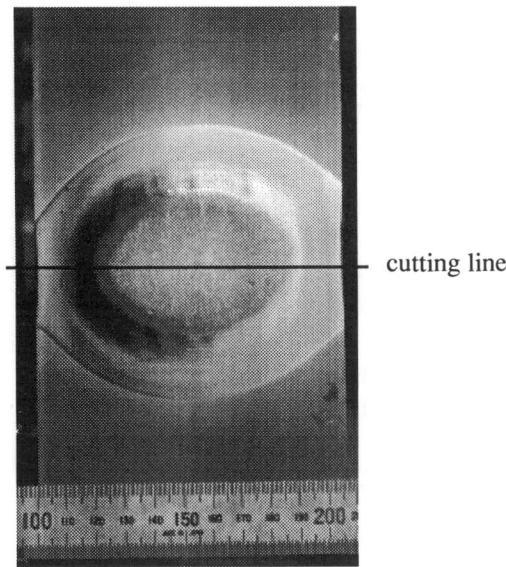

Fig.4 Surface of FW panel after high heat flux test

4. POST-MORTEM ANALYSIS

After the high heat flux tests, the FW panel was cut to observe the HIP bonded interface and the heated surface of DS-Cu. The cutting line is shown in Fig.4. A micrograph of the DS-Cu/SS HIP bonded interface region and the vicinity of the DS-Cu surface is shown in Figs. 5 and 6, respectively. Closer observation was also performed by means of an optical microscope and by SEM. No cracks were observed at the HIP bonded interfaces. Melting zone and heat affected zone (HAZ) up to the depth of 1.5 mm from the DS-Cu surface was observed as shown in Fig.6. There were many fine cracks propagating from the surface into the melting zone. The depth of the cracks was less than 0.1mm. A number of cylindrical voids of about 100 mm in diameter also appeared in the melting zone as shown in Fig.6, which might be due to residual gas release in the DS-Cu. Thickness of melting zone and HAZ, and the main void (more than a diameter of 0.1 mm) locations measured on cross section of FW panel were shown in Fig.7.

Hardness tests with a load of 50gf were performed for the vicinity of the DS-Cu/SS HIP bonded interface. Results of hardness tests of specimens before (without imposing heat loads) and after the high heat flux tests are compared in Fig.8. In the specimen after the high heat flux tests, the hardness of SS peaks near the HIP bonded interface while there is no peak in the specimen without heat loads. This would relate to the residual

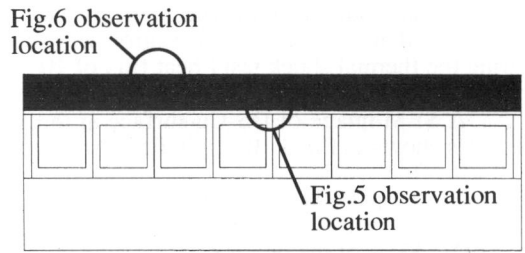

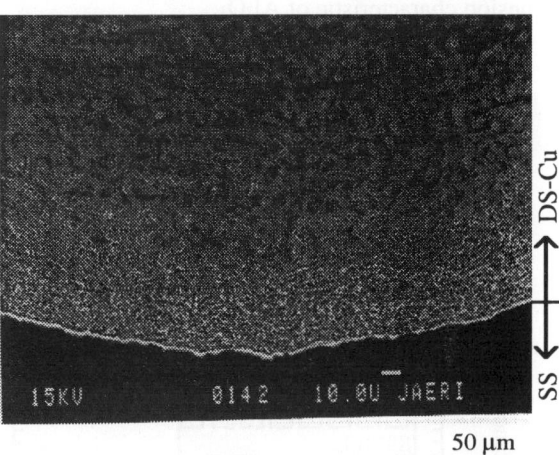

Fig.5 The DS-Cu/SS316 HIP bonded interface

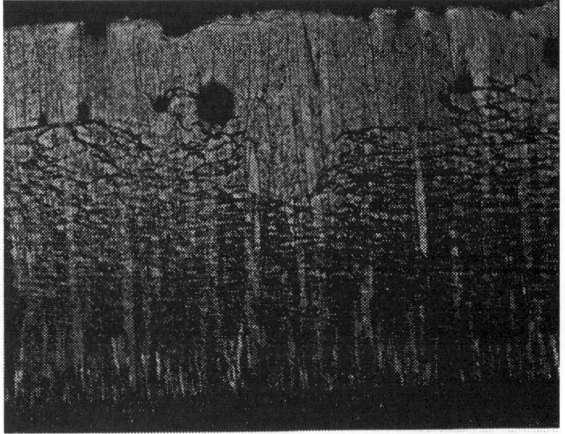

Fig.6 Optical microscope image on cross section in the vicinity of the FW panel surface

stress due to high heat flux loading. DS-Cu hardness near the heated surface of the specimen after the tests was lower than the other. This region was melted during the thermal shock test (heat flux of 100MW/m^2), and the dispersion of Al$_2$O$_3$ was changed greatly.

A SEM image of DS-Cu surface at the beam center is shown in Fig.9. Precipitations in the size of ~10 μm, fine cracks and a dendritic structure are observed. From EPMA analyses for Cu, O and Al, it was concluded that these precipitations were composed of Al$_2$O$_3$. Sizes of the precipitation are much bigger than initially dispersed Al$_2$O$_3$ possibly due to a cohesion characteristic of Al$_2$O$_3$.

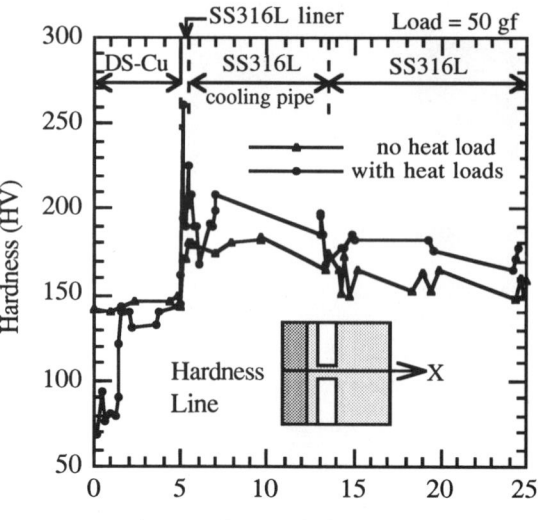

Fig.8 Hardness distribution across the cross-section of FW panel

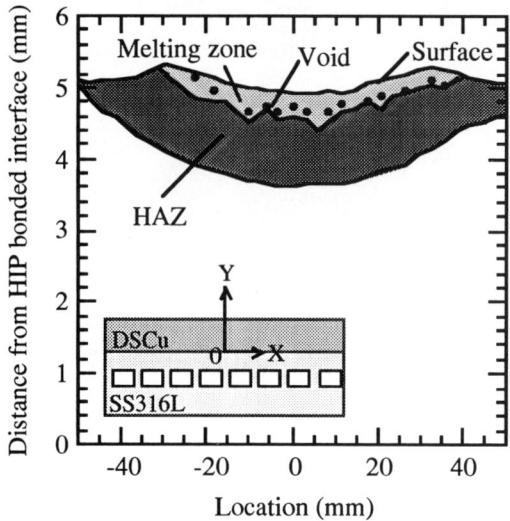

Fig.7 Thicness of melting zone and HAZ on the DS-Cu surface

5. CONCLUSION

High heat flux tests of the HIP bonded DS-Cu/SS316L first wall panel have been performed. Cracks and exfoliations were not observed at the HIP bonded interface after the tests. Melting of the DS-Cu surface occurred under the severe disruption condition with fine cracks and voids formed in the melting zone. Further investigations are needed on the evaluation of residual stress at the HIP bonded interface and the effect of armor material (Be in the ITER design).

ACKNOWLEDGMENT

The authors would like to express their sincere appreciation to Drs. M.Ohta, T.Nagashima and M.Akiba for their continuous support and encouragement.

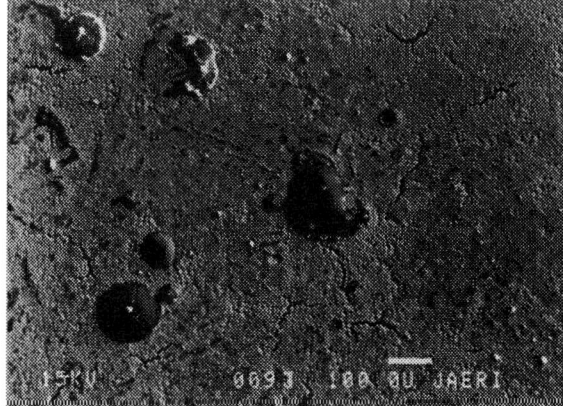

Fig.9 SEM image of the FW panel surface at the beam center

REFERENCES

1. S.Sato et al., "Fabrication of HIPped first wall structure for fusion experimental reactor and preliminary analysis for its thermo-mechanical test", to be published in Proc. 16th IEEE/NPSS Symposium of Fusion Engineering
2. T.Hatano et al, "High Heat Flux Testing of HIP Bonded DS-Cu/316SS First Wall Panel for Fusion Experimental Reactors", to be published in Proc. 12th Topical Meeting on the Technology of Fusion Energy

Ohmic baking system upgrade for wall conditioning of Tokamak-15 discharge chamber

V.N.Garnov, S.V.Kabanovsky, V.A.Khrabrov, P.P.Khvostenko, V.A.Kochin, A.I. Nikonorov, P.N.Orlov, I.A.Posadsky, A.N. Vertiporokh

RRC "Kurchatov Institute", Nuclear Fusion Institute,
Moscow 123182, Kurchatov sq., 1, Russia

For wall conditioning of Tokamak-15 discharge chamber the Ohmic baking system upgrade is used. The results of the designing, tests and operation of the Ohmic baking system upgrade are given in this paper. Locating the heaters inside the discharge chamber and heating to 170°C saves up to 1700 kW•hours and up to 30 tons of liquid nitrogen per 24 hours, compared to the original set-up.

1. INTRODUCTION

Ohmic and inductive baking are used for wall conditioning of Tokamak-15 discharge chamber. The Ohmic technique is used for baking massive rigid support sections and external ports, and inductive baking is used for the bellows sections.

For baking the twelve rigid support sections, 96 Ohmic heaters located between the chamber surface and the inner nitrogen screen were used in the initial design. As the experimental baking regimes have shown, intense chamber surface outgassing occurs when the wall temperature reaches 150-170°C (at a heating power of about 130 kW). Under those regimes, the heating efficiency was approximately 50%.

In the process of continuous Ohmic baking, 22 heaters failed. After termination of the 1990-1991 run, two burn-throughs were detected on the chamber surface between bellows folds. The burn-throughs were caused by breaks in the heaters and by their subsequent contact with the chamber bellows. Open circuits in the heaters was released with insufficient strength in their fastening to the chamber surface at effect of cycling electromagnetic loads (the interaction effect of a.c. current through the heaters with toroidal magnetic field). Further continuous use of the heaters could have resulted in serious chamber damage and in the interruption of the experimental run. Repair of the heaters without disassembly of the whole facility was impossible. The absence of the Ohmic baking system made it impossible to realize a uniform chamber heating up to 150-170°C and to support the chamber at a definite temperature level during the plasma experiments. In this connection, the problem of designing a chamber baking system, which is energy efficient and convenient for assembly and repair, arose.

After considering several proposals, cable-based Ohmic heaters installed on a chamber surface facing the plasma were selected. These heaters employ a single 1.2-mm diam. Ni-Cr alloy wire, housed inside a stainless-steel 6-mm diam. shell, with a magnesium oxide ceramic insulator.

Based on necessary power for the chamber baking (~ 75 kW) a diagram for laying the heaters, consisting of four branches in parallel (7 m long each) and laid down in each of 12 rigid support sections, has been chosen.

2. SHAKEDOWN TESTS OF THE CABLE

Before mounting the heaters inside the discharge chamber, shakedown tests of the cable were performed. The goal of the tests was to study the gas releases from the cable and the processes of gas diffusion through magnesium oxide in case of the cable shell damage, as well as to determine thermophysical characteristics of the cable in the temperature range 440-700°C.

2.1. Study of the processes of gas releases from cable

The cable samples, 80 mm long each, were used for studying the gas releases in the temperature range 440-740°C. The measurements were done at the heated high vacuum stand (Fig.1).

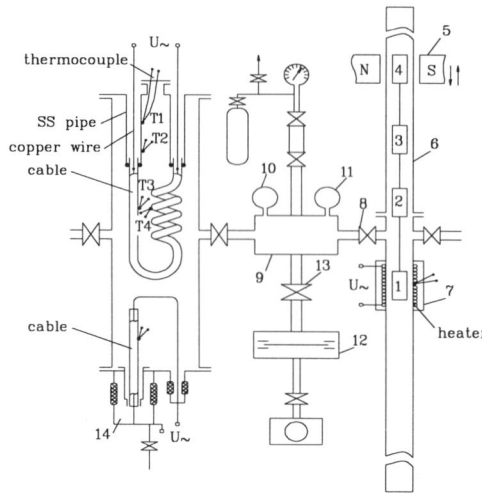

Figure 1. Diagram of a vacuum technological stand.

The measuring part of the stand for determining the gas releases was the pipe (6) joined, through the valve (8) to an analysis chamber (9) with a mass-analyzer sensor (10) and with a high vacuum sensor (11). Evacuation of the analysis chamber was performed with a turbomolecular pump (12) through the valve (13). The evacuation rate was $\cong 40$ l/s.

For outgassing the samples, stand and a high vacuum part of the turbomolecular pump, an electric heater was mounted. The temperature control was performed with thermocouples.

The samples (1,2,3) and an iron core (4) were connected with each other with a thin ($\varnothing$ 0,2 mm) tungsten wire. The distance between the samples was about 200 mm. The displacement of the samples along the vertical axis of the pipe was performed with a ring-like magnet.

The technique of measurements was the following one. The samples were installed (See, Fig.1) in upper position inside the pipe. The stand was evacuated to the limiting vacuum, temperature of the sample heating (7) was set at the preset level, the background pressure (P_b) was fixed, spectrum of a residual gas was registered. Then, the sample under study was introduced into the zone of heating. In the gas release process accompanied by a rise in the total pressure, the time dependencies of pressure in the chamber P(t) and the spectrum of the gas released from the sample were registered. After reduction in the gas release intensity ($P \approx P_b$), the sample was transferred into a cool part of the pipe. The stand was evacuated down to P_b and the next sample was introduced into the zone of heating.

The processes of gas release from the samples was studied at the temperatures 440°C and 740°C. The experimental dependencies P(t), under heating of the samples under mentioned temperatures, are given in Fig.2. The time dependencies of the masses dominant in the spectrum are given in Fig.3. The masses with high numbers were not observed.

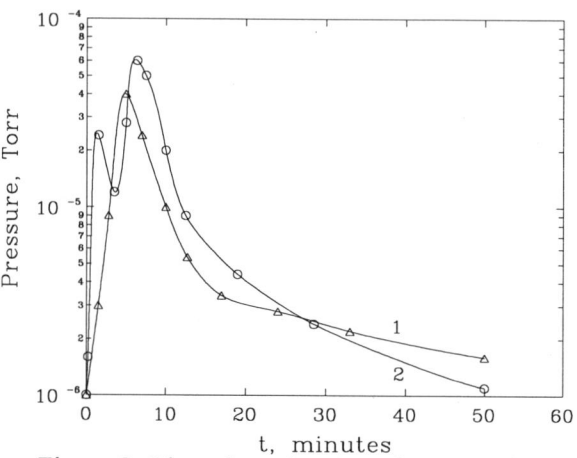

Figure 2. Time dependencies of pressure in the chamber for the samples under their heating up to 440°C (curve 1) and to 740°C (curve 2).

After termination of the samples heating the air was let in into the vacuum volume with the samples. The samples were exposed in the air for 15 hours, then they were heated again up to 440°C and up to 740°C in order to determine the process of gas sorption by the samples. The time dependencies P(t) in the chamber under sample heating up to 440°C and up to 740°C, are shown in Fig.4

2.2. Study of the gas diffusion through the cable

The goal of the experiment is to determine the leakage into the chamber from the atmosphere at the probable violation of the cable shell integrity within the T-15 discharge chamber.

For determining the diffusion of helium throughthe magnesium oxide the samples 10, 20, 100 200 cm long, were studied. One end of the cable was within the vacuum volume; another, in the atmosphere (Fig.1). The Ni-Cr alloy wire was extracted outside through a metalloceramical

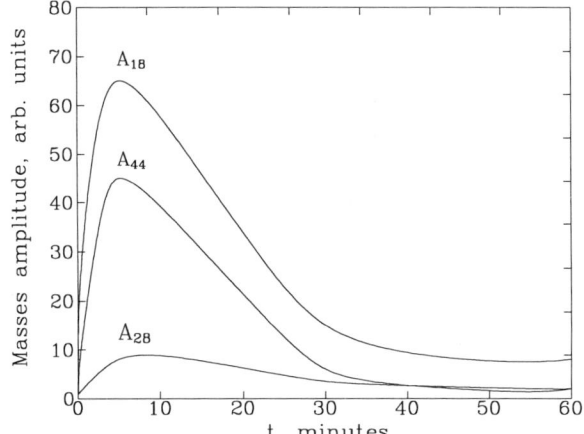

Figure 3. The dependencies of the masses dominant in the spectrum under heating of the sample up to 440°C.

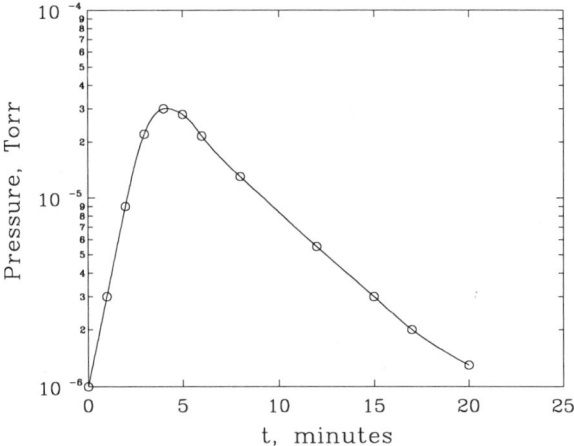

Figure 4. Time dependence of pressure under heating of the sample to 440°C.

insulator. The cable end facing the atmosphere was closed outside with the metallic cap (14) which was evacuated and filled with helium. The control over the helium flux into the vacuum volume through the magnesium oxide was done with a mass-analyzer. After filling the cap with helium, a helium peak in about two hours appeared in the vacuum volume. Helium passed through the cable, 10 cm long, at room temperature. The leakage was equal $\sim 1{,}2 \cdot 10^{-7}$ l·Tor/s. After heating the cable up to $\sim 700°C$ the leakage was doubled. An increase in the leakage after five time thermocycling of the cable was not observed. The helium diffusion coefficient through the magnesium oxide, estimated as $l^2/\Delta t$, was equal $1{,}4 \cdot 10^{-6}$ m²/s. If one assumes that the cable shell is damaged in all the 48 heaters installed in the discharge chamber, the total leakage, at working temperature of the cable of $\sim 600°C$, will be $\sim 1 \cdot 10^{-5}$ l·Tor/s. At the cable shell damage, somewhere in the middle of its length ($\sim 3{,}5$ m), the time of helium appearance in the chamber, in the process of searching the leakage, would be about 240 hours.

2.3. Study of thermophysical cable characteristics

The lead-out of vacuum-tight cable shell is the most preferable approach to the cable lead-out from the vacuum volume. In that case, it is natural that the temperature of the cable shell, facing the atmosphere, should be at the level of 30-50°C. For this purpose, an SS-pipe, 8 mm in diameter, at its wall 1 mm thick, was welded to the cable shell by argon arc, and copper wire, 3 mm in diameter (Fig.1), was welded to a Ni-Cr alloy wire.

The cable shell surface temperature measurement and that of the pipe welded to it were done under pressure in the chamber of about $P \sim 1 \cdot 10^{-6}$ Torr with thermocouples. The temperatures at various points on the cable, dependent on a current passing through it, are given in Fig.5. The cable surface temperature (T1) at the distance of 100 mm from a flange and at the current 30 A was about 50°C. At the working current of 15 A, the T1 temperature was not different from the flange temperature, being equal $\sim 20°C$. The cable surface temperature at the points 3,4 was equal $\sim 560°C$. Thus the chosen heater lead-out design satisfied the above mentioned requirements. The dependence of the reduced blackness degree (ε) for the cable-chamber wall system on the cable temperature (T) is given in Fig.6. The monotonous ε-rise with a rise in T, as well as a reduction in the partial oxygen pressure in the chamber, confirmed the oxidation process realization and 'blackening' of the cable.

3. TESTS AND OPERATION OF THE OHMIC BAKING SYSTEM UPGRADE

During March-October, 1994, 48 branches of heaters were laid within the T-15 discharge chamber.

The first tests of the T-15 Ohmic baking system upgrade were done in December, 1994. At heating power equal to 75 kW, the rigid sections were done in December, 1994. At heating power equal to 75 kW, the rigid sections were heated up to 150-170°C; the bellows shields, to 200°C; and

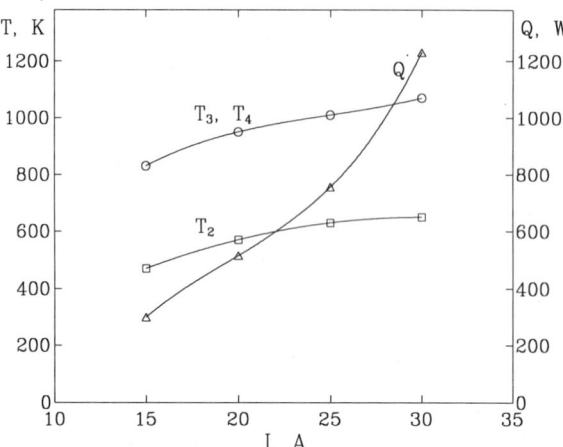

Figure 5. Cable temperature dependence at various points vs. a current passing through it. Q - Ohmic heating power.

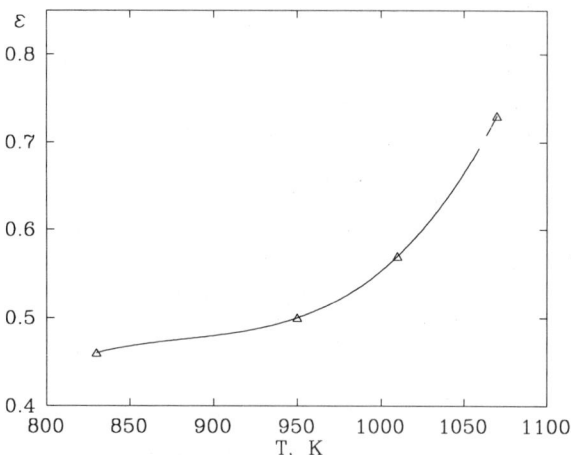

Figure 6. The reduced blackness dependence (ε) in the cable-chamber wall system vs. a cable temperature.

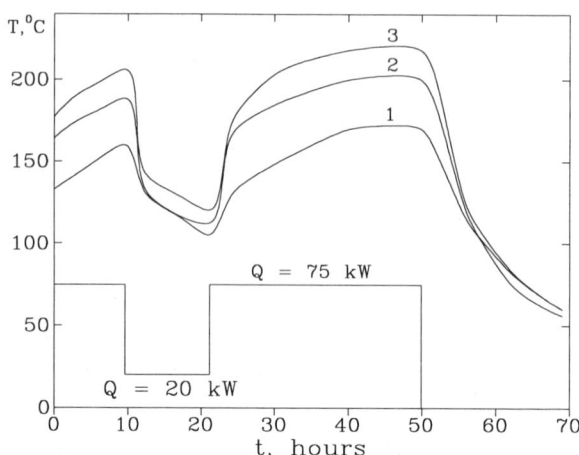

Figure 7. Time dependencies of temperature in the rigid sections (curve 1), bellows shields (curve 2) and moveable graphite limiters (curve 3); Q - baking power.

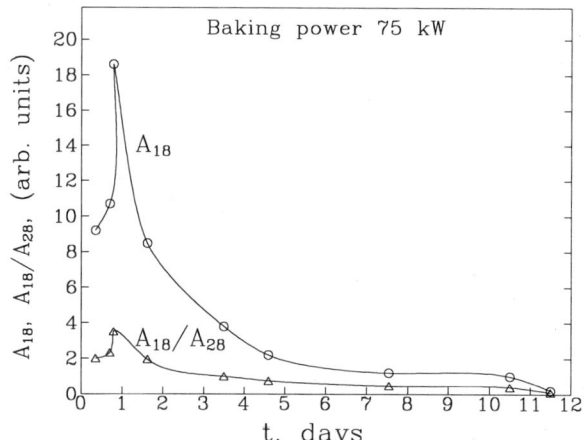

Figure 8. Time dependencies of the 18th mass behavior in the process of baking and those of the ratio between the 18th and the 28th masses.

moveable graphite limiters, to 220°C (Fig.7). The discharge chamber baking during experimental run of 1995 was done for 10 days at the maximal power of baking ~ 75 kW. The chamber wall dehydration process efficiency under baking can be judged by the dynamics of the 18th mass in the process of baking and by a change in the ratio between the 18th and 20th masses (Fig.8). An intense outgassing has allowed one to reduce about 100 times the steam concentration in the chamber after baking and its cooling to room temperatures and to reduce the background pressure to $P \approx 2 \cdot 10^{-7}$·Torr.

4. CONCLUSION

The Ohmic baking system upgrade has demonstrated its serviceability under wall conditioning of the discharge chamber and under operation with plasma.

Locating the heaters inside the discharge chamber and heating to 170°C saves up to 1700 kW•hours and up to 30 tons of liquid nitrogen per 24 hours, compared to the original set-up.

GlidCop® DSC extruded plus cross-rolled plate tensile properties in the temperature range of 20-500°C

Ronald R. Solomon, Jack D. Troxell, Anil V. Nadkarni

SCM Metal Products, Inc., 2601 Weck Drive, Research Triangle Park, NC, 27709, USA

GlidCop® Dispersion Strengthened Copper (DSC with aluminum oxide dispersoids) is manufactured via a proprietary powder metallurgy process whereby Cu-Al alloy powder is internally oxidized to produce $Cu-Al_2O_3$ dispersed powder. The powder must be consolidated to full density for final use as high strength, high conductivity components for use at high temperatures. A previous study examined differences in GlidCop® AL-25 grade plate sections manufactured by processes including Hot Isostatic Pressing (HIP) consolidation, Extrusion (EXT) consolidation, and Extrusion consolidation followed by Cross-Rolling (EXT+XROLL). For plates annealed at 950-1000°C to simulate brazing assembly cycles, the EXT+XROLL plates were found to have superior ductility (total elongation and reduction of area) and fracture toughness (J_{IC}) in the temperature range of 20-350°C compared to plates made by the other two processes. All three processes provided similar UTS, .2% YS, and creep characteristics. This paper presents tensile properties at 20 and 300°C for additional EXT+XROLL plates that were produced with varied percentages of Cross-Roll and Straight-Roll reductions, and presents tensile properties at 500°C for two plate thicknesses (10mm and 20mm) made with maximum Cross-Roll reduction (56%).

1. INTRODUCTION

GlidCop® AL-25 LOX Dispersion Strengthened Copper (DSC) is a premium candidate material for the copper portion of the First Wall, Baffle, Limiter and Divertor sections of current ITER designs. A prior study had indicated that plates made by using extrusion as the powder consolidation method had superior elevated temperature properties (ductility, fracture toughness) compared to plates made by the hot isostatic press (of powder) method. It was further observed that cross-rolling the extruded plates improved the transverse properties, particularly the reduction of area at elevated temperature.

The purpose of this study was to examine the influence of rolling, both cross-rolling and straight-rolling, on properties of extruded plates. Plates used in this study were processed using various cross-roll and straight-roll reduction schedules based on the final size requirements for the plates produced. Only the 10mm and 20mm thick plates were made per the currently defined GlidCop® IG0 process whereby the plates are cross-rolled initially to net a 1016mm wide finished plate, then straight-rolled to final thickness.

Tensile data at temperatures higher than 300-350°C, specifically 500°C, was also deemed necessary for the GlidCop® IG0 plate material based on projected operating and excursion temperatures for the ITER FIRST WALL and other components. Thus, testing at 500°C for the 10mm and 20mm GlidCop® IG0 plates was undertaken.

2. MATERIALS TESTED AND PROCEDURES

All materials tested were made using the GlidCop® AL-25 LOX -80 process. The plates were made by hot extruding powder billets at >500 MPa and >800°C to full density as 63 mm thick X 508 mm wide plates. The plates were then machined to ~51mm thick to remove the copper cladding from the top and bottom surfaces. Individual plates were then cross-rolled to reduce their thickness and increase their width, then straight-rolled to final

Table 1
GlidCop® AL-25 Plates - samples or plates annealed at 980-1000°C Tensiles tested at 20 and 300°C

Size mm	Xroll % red	StrRoll % red	Temp °C	Longitudinal				Transverse			
				UTS	.2%YS	Elong	Red. of Area	UTS	.2%YS	Elong	Red. of Area
10	56	47	20	435	355	26	70	434	362	25	68
12	30	66	20	453	372	24	72	444	365	23	66
20a	56	12	20	419	331	27	69	423	331	27	68
20b	56	12	20	412	324	27	71	415	326	27	68
32	5	34	20	458	371	26	73	440	357	26	63
50	0	0	20	423	330	27	65	431	345	23	57
10	56	47	300	250	219	25	46	232	193	25	36
12	30	66	300	279	250	20	49	253	219	24	35
20a	56	12	300	264	207	28	42	261	207	27	42
20b	56	12	300	238	203	25	40	232	201	22	33
32	5	34	300	284	266	21	39	261	242	27	27
50	0	0	300	265	198	21	36	258	192	15	21

thickness and length requirements for their intended use (these were production plates). All plates were annealed for 1 hour at 950°C to 1000°C to simulate a high temperature brazing cycle.

Tensile tests were performed at 20° and 300° in air (no protective atmosphere) per ASTM E8. Gage diameters and gage lengths were suitably chosen for each plate size per ASTM E8 guidelines. Tensile tests were also performed at 500° for the 10mm thick and one of the 20mm thick plates. All plates were tested in the longitudinal (parallel to extrusion axis) and transverse (perpendicular to the extrusion axis in the width direction) directions. Two tests were done per plate per temperature, and the average of these results are reported. Linear regression analysis was performed on the data generated at 20° and 300° to determine which properties had the strongest correlation with cross-roll and straight-roll reduction percentages.

3. RESULTS

The results of this testing are presented in table and graph form. Table 1 presents the test results at 20° and 300°C for 5 combinations of cross-roll and straight-roll processing conditions (plates 20a and 20b are two different 20mm thick plates produced from different powder batches). Figures 1-3 present best fit linear trends for the properties that showed the best correlation with cross-roll or straight-roll reduction percentage. Note that these were the only three trends noted with an R^2 value of .75 or greater.

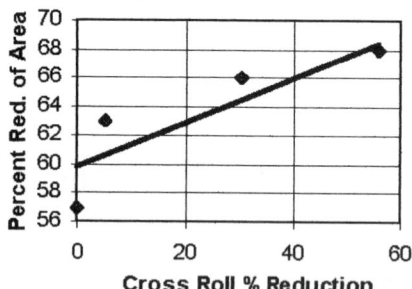

Figure 1. Transverse Reduction of Area at 20°C

Table 2
GlidCop® AL-25 IG0 Plates annealed at 1000°C - Tensiles tested at 20°, 300°, and 500°C

Size mm	Xroll % red	StrRoll % red	Temp °C	Longitudinal				Transverse			
				UTS	.2%YS	Elong	Red. of Area	UTS	.2%YS	Elong	Red. of Area
10	56	47	20	435	355	26	70	434	362	25	68
20	56	12	20	412	324	27	71	415	326	27	68
10	56	47	300	250	219	25	46	232	193	25	36
20	56	12	300	238	203	25	40	232	201	22	33
10	56	47	500	140	123	24	35	106	82	23	35
20	56	12	500	114	96	27	32	111	88	25	30

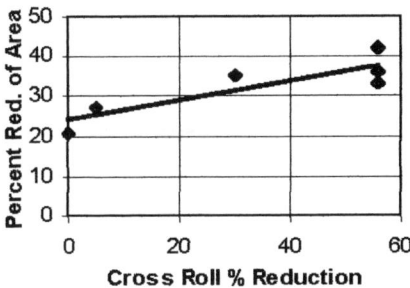

Figure 2. Transverse Reduction of Area at 300°C

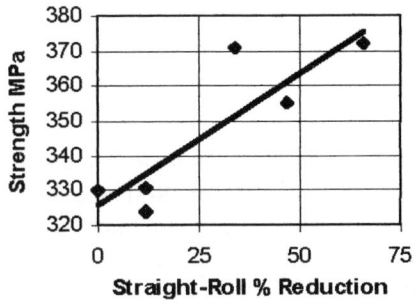

Figure 3. Longitudinal .2% Yield Strength at 20°C.

Table 2 presents the test results in the 20-500°C range for the 10mm and 20mm thick plates. Figures 4-7 present the YS and Reduction of Area data from Table 2 in line graph form.

4. OBSERVATIONS AND CONCLUSIONS

1. All combinations of Cross-Rolling (perpendicular to extrusion axis) and Straight-Rolling (parallel to extrusion axis) resulted in an increase in Reduction of Area in both the Longitudinal and Transverse directions at both 20° and 300°C compared to the EXT (0% rolling) starting plate.

2. In reviewing Table 1, no clear trends were evident in UTS, YS or Elongation as related to Cross-Roll or Straight-Roll reductions (except possibly Longitudinal YS at 20°C related to Straight-Roll reduction). This was also evident in extremely low correlation coefficients for these properties with rolling reductions (not presented in this text).

3. The data suggests that Transverse Reduction of Area at 20° and 300°C increases with the amount of Cross-Rolling, and Longitudinal YS at 20° increases with the amount of Straight-Rolling. Linear best fit lines are presented in Figures 1-3. These properties showed the greatest correlation with rolling reductions. Note that there was inadequate data to conclusively separate the effects of Cross-Rolling from Straight-Rolling.

4. Annealing of the plates at 950-1000°C appears to normalize the effects of rolling on the directionality of strength properties. UTS and YS were essentially the same in both directions, within the scatter of the

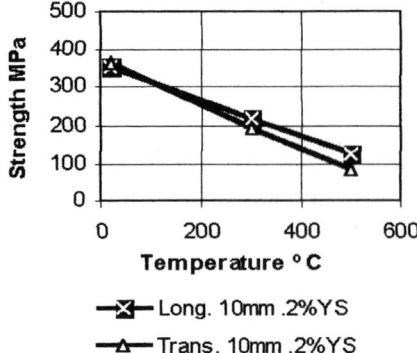

Figure 4. Yield Strength at 20-500°C for GlidCop® AL-25 IG0 10mm thick plates

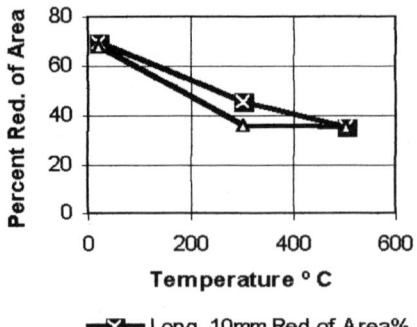

Figure 6. Reduction of Area at 20-500°C for GlidCop® AL-25 IG0 10mm thick plates

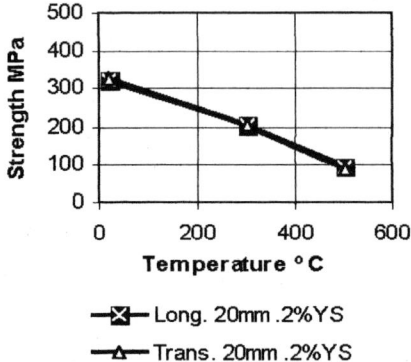

Figure 5. Yield Strength at 20-500°C for GlidCop® AL-25 IG0 20mm thick plates

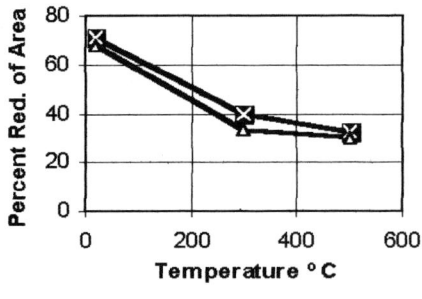

Figure 7. Reduction of Area at 20-500°C for GlidCop® AL-25 IG0 20mm thick plates

data, regardless of the amount of cross-roll or straight-roll reduction percentages

5. Over the test range of 20-500°C, UTS declined almost linearly from 412-435 MPa to 106-140 MPa, and YS dropped from 324-362 MPa to 82-123 MPa for the annealed 10 mm and 20 mm thick AL-25 IG0 plates tested. Figures 4 and 5 graphically show the decline in YS with temperature.

6. At 500°C, the Longitudinal UTS and YS of the 10mm thick plate were higher than that of the 20mm thick plate, while Transverse strengths were similar. This might infer that the much higher Straight Rolling reduction in the Longitudinal direction resulted in higher strength in this direction

7. Over the test range of 20-500°C, Reduction of Area declined non-linearly from 68-71% to 30-35% for the annealed 10 mm and 20 mm thick AL-25 IG0 plates tested. Reduction of Area dipped at 300°C as seen in Figures 6 and 7. Elongation remained essentially constant.

REFERENCES

1. R. Solomon, J. Troxell, A. Nadkarni, GlidCop® DSC properties in the temperature range of 20-350°C, 1995

Section B

Plasma Heating and Control

The 110 GHz ECR System for TdeV*

R.Magne[a], G.A.Chaudron[b], R.Cool[b], Y.Demers[b], Ph.Cumyn[c], R.Décoste[d], A.Dubé[e], J.M.Guay[f], D.Larose[e], A.Robert[d], C.Trudel[b], L.Vachon[g]

[a]Association Euratom-CEA, CE Cadarache, FRANCE [b]MPB Technologies Inc., Dorval, Québec, CANADA [c]Canatom Inc., Montréal, Québec, CANADA [d]Hydro-Québec, Varennes, Québec, CANADA [e]Inspectech 2000 Inc., St-Sulpice, Québec, CANADA [f]INRS, Varennes, Québec, CANADA [g]Instrumentation St-Laurent, St-Laurent, Québec, CANADA.

V.I.Belousov, G.G.Denisov, V.E.Mijasnikov, L.G.Popov

Gycom Ltd, Institute of Applied Physics, Nizhny Novgorod, RUSSIA

A new microwave system operating at the second harmonic of the Electron Cyclotron Resonance (ECR) is currently being installed on the Tokamak de Varennes (TdeV). The generator will be composed of two gyrotrons for a total generated power at 110 GHz over 1.5 MW with pulse length up to 2 s. Two transmission lines made of corrugated waveguides carrying the HE_{11} mode will transmit the rf power from the gyrotrons to the steerable launching structures located inside the tokamak.

1. INTRODUCTION

The main objectives of the ECR system on TdeV (R=0.83 m, a=0.22 m, B_T=1.95 T, I_P=250 kA) are to provide efficient and well localised heating and current drive up to the X2-mode cut-off density (n_e=7.4x10^{19} m^{-3}). Relativistic ray tracing simulations have shown that, for a wide range of parameters, single pass absorption is complete for injection from above the equatorial plane on the low field side. The total injected power of 1.2 MW should be sufficient to study enhanced performance scenarios and synergy with Lower Hybrid Current Drive (LHCD).

2. THE GENERATOR

The generator is built around two gyrotrons GLGF-110/1.0. The tubes are powered by the HV power supply previously installed for the two klystrons of the LHCD system [1]. It has been upgraded to meet the higher working voltage requirements of the gyrotrons. In order to reduce the voltage ripple seen by the cathode (0.5 % maximum required), a modulator regulator using an Eimac 9009 tetrode [2] has been added for each tube. The modulators enable us to set independently, both in time and in magnitude, the waveform of each gyrotron output power. They are also used to switch off very rapidly (T≈1μs) the high voltage on the tube in case of a fault. A third modulator regulator of the same design is required to lower the supply voltage on the two klystrons.

The gyrotrons, the tetrodes and their auxiliaries are cooled with demineralised water using the existing closed loop system of the toroidal field coils of TdeV. The total required flow rate for the whole ECR system is about 5000 l/mn at a pressure of 650 kPa.

Control and protection are done by different subsystems depending on the task to be done and the response time required.

---

* Supported by the Centre canadien de fusion magnétique with funds from the Canadian Government, Hydro-Québec and INRS

Protection of personnel and critical components demands a very short response time and is done by a hardwired failsafe system using the programmable logic device technology. Slow protection, control and general supervision are achieved by programmable logic controllers. The user interface is a Vax working station, VMS system, running a commercial software. Sequencing and data acquisition are done by the relevant central systems of the tokamak.

3. THE GYROTRON

The gyrotron GLGF-110/1.0 is a new model developed by Gycom [3] (fig.1). Its main parameters are given in table 1.

This tube is equipped with an integrated quasi-optical converter which produces a linearly polarized output beam with an 88% gaussian mode content. The rf power at the output window is ≥ 800 kW for pulse length up to 2 s, giving a power ≥ 700 kW after conversion in the HE_{11} mode. For pulse length up to 5 s, the power after conversion is ≥ 320 kW, and for shorter pulse length of 0.5 s, it is expected to reach 1 MW.

The magnetron injection gun is of the diode type, which avoids the use of a modulating anode device. The 133 mm diameter output window is made of a single disk of boron nitride, edge cooled by water circulation. The calorimetric measurement of the power lost within the window (4% absorption) gives a reliable indication of the output power.

Table 1: main parameters of the gyrotron

frequency	110 GHz
oscillation mode	TE 19,5
output power after conversion	700 kW @ 2 s
in HE_{11} mode	320 kW @ 5 s
efficiency	30 %
duty cycle	1/200
cathode voltage	75 kV
beam current	35 A @ 700 kW
	20 A @ 320 kW

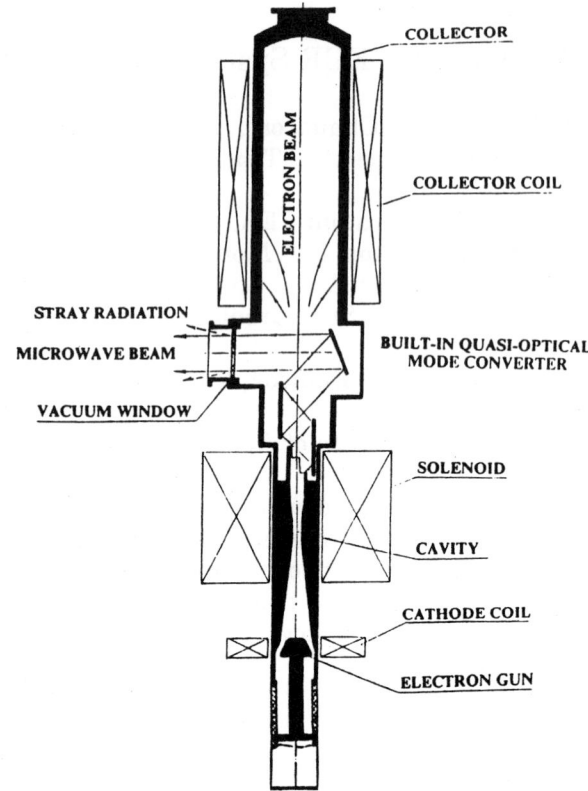

Figure 1. The gyrotron GLGF-110/1.0

The magnetic field of 4.35 T required at the cavity centre is generated by a two-coil superconducting magnet. A conventional magnet located at the cathode level is used to shape adequately the field at this location.

An additional coil, fitted around the collector, is fed with a triangular waveform to sweep the electron beam on the collector wall in order to reduce the thermal load.

The first gyrotron with its superconducting coil assembly will be delivered to TdeV in december 1996, the second set in March 1997.

4. THE TRANSMISSION LINES

The two transmission lines from the outputs of the gyrotrons to the launching structures are about 25 m long. So far, the design of only the first transmission line has been completed. The layout of the second line has not been determined yet, it may use evacuated waveguide to avoid the vacuum barrier window at the tokamak.

The first line could be split in two parts, with two different technologies (fig.2). The control and monitoring components are in quasi-optical (QO) technology in order to simplify their construction and adjustment [4]. The components used to transport the rf power to the launching structure are non-evacuated corrugated waveguides to reduce the space required in the tokamak hall.

The QO components are housed within shielded enclosures and located at the gyrotron output to have a more compact arrangement. They are namely (fig.3):

- the Matching Optics Unit, made of two phase corrector mirrors, conditions the output beam to recover as much power as possible into the TEM_{00} mode,
- the Power Measurement Unit, made of a grated mirror with small-size corrugations acting as a directional coupler, measures the forward and reflected power with a directivity of at least 30 dB,
- the Universal Polarizer, made of two grated mirrors, rotates the wave polarization from 0 to 180° and modify the ellipticity from linear to circular,

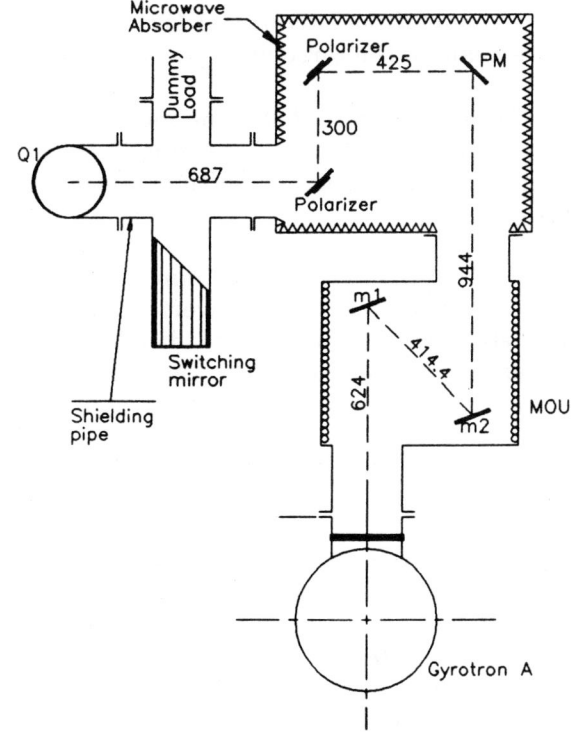

Figure 3. Quasi-optical components

- the rf switch, consisting of a retractable mirror, diverts the beam into a dummy load for test purposes,

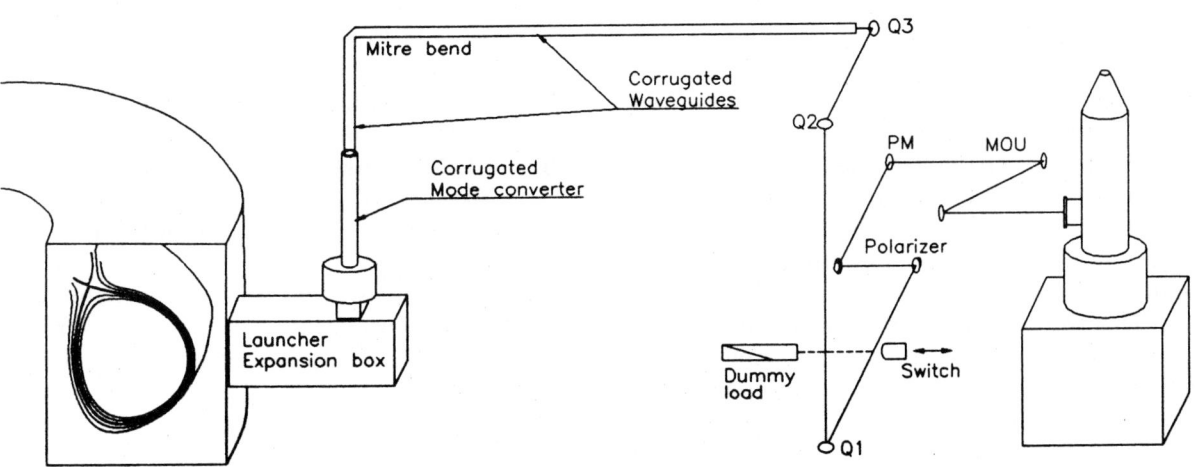

Figure 2. Layout of the first transmission line

- a set of three mirrors (Q1-Q3) couple the wave beam to the corrugated waveguides.

The second part of the line includes:
- a 20 m long horizontal run made of 1 m long 88.9 mm waveguide sections assembled with flanges,
- one mitre bend,
- a 1 m long vertical 88.9 mm waveguide,
- a corrugated mode converter (110 mm diameter and 1.2 m long) to distribute the power over the surface of the vacuum barrier window which follows, with a peaking factor of 2.5,
- a DC break which acts also as a sliding joint.

The power capability of the transmission line is 1 MW, 2 s or 500 kW, 5 s. Its losses from the output of the MOU to the vacuum window are evaluated to be less than 10 %.

All the components of the first transmission line are manufactured by Gycom and will be delivered to TdeV by February 1997.

5. THE LAUNCHING STRUCTURE

The first launching structure is located in one equatorial port, 270x420 mm^2, (fig.4). It consists of three mirrors housed in an expansion box which also supports the vacuum barrier window.

The first two mirrors have special convergent profiles to recover the gaussian beam from the flattened profile at the output of the window with an efficiency of 97 %, and to obtain a $\approx$ 14 mm beam waist at the resonance location. The design and manufacture of these two mirrors are also made by Gycom.

The third mirror is flat and controls the launching angle into the plasma both poloidally (from the axis to the separatrix) and toroidally ($\pm$ 30 °).

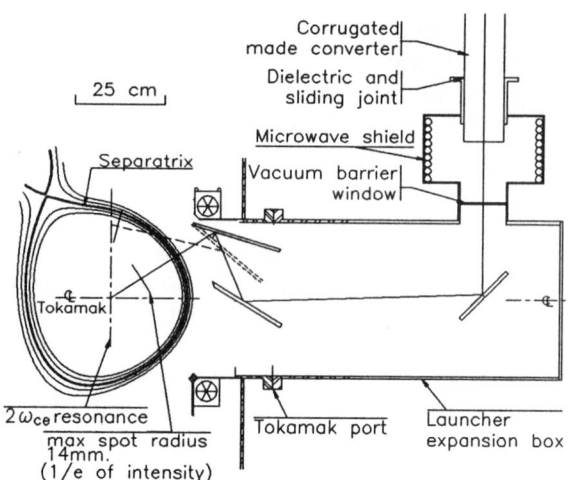

Figure 4. Layout of the first antenna

6. SCHEDULE

The ECR system is planned to be operational in the spring of 1997 with one gyrotron and in the summer of 1997 for the whole system.

REFERENCES

1. Y.Demers et al., Europhysics Conference on Radiofrequency Heating and Current Drive of Fusion Devices, Brussels, 1992.
2. 9009 (X-2062K) Water Cooled Power Tetrode Technical Data, CPI Eimac, San Carlos, California.
3. V.E.Mijasnikov et al., Long-Pulse 110 GHz/1 MW Gyrotron, 12th International Conference on Infrared and Millimetre Waves, Orlando, 1995.
4. V.I.Belousov et al., Auxiliary Elements of High Power Quasi-optical Transmission Lines. Coll.Pap."Gyrotron", IAP, Gorky, 1989.

LOW POWER RADIATION FIELD PATTERN MEASUREMENTS OF MAIN COMPONENTS OF THE TORE SUPRA ECRH TRANSMISSION LINE

G. Berger-By, J.P. Crenn, R. Levy, P. Garin, D. Roux, M. Pain.

Association EURATOM-CEA, Département de Recherches sur la Fusion Contrôlée
Centre de Cadarache, 13108 Saint Paul lez Durance Cedex, France.

1- INTRODUCTION

On Tore Supra tokamak, the building of the ECRH system at 118 GHz is under progress [1]. This 2 MW generator consists of 6 quasi-cw gyrotrons able to deliver about 500 kW each [2] [3]. These gyrotrons have an horizontal output in the HE_{11} mode with quasi-optical mode converter. The 118 GHz antenna consists of 3 assembled modules. It will be connected to generator by six 20 meters long transmission lines.

In order to characterize the different line components in circular oversized corrugated waveguides carrying the HE_{11} mode at 118 GHz, a radiation field pattern measurement have been designed and built.

2- PRINCIPLE OF RADIATION PATTERN MEASUREMENTS

When millimeter waves propagate in oversized waveguides the mechanical defects produce spurious modes. Classical techniques of measurement in transmission with a network analyzer cannot measure and separate ohmic losses and conversion losses into spurious modes. A right method is to feed the device to be tested with HE_{11} pure mode and to measure and analyse the radiated beam. The measured field pattern depends on the modal composition at the guide output.

The boundary region between the near-field and the far field is located at a distance $L=D^2/\lambda$ from the guide output (D= waveguide diameter and λ= beam wave length). In our case, D=63.5 mm and L=1.6m at 118 GHz, for this reason our analysis distance will be around 1.6 m. Theoretical calculations and computer programs are used in order to obtain the mode power content and the modal composition. Some details of our equipment and work method is explained below.

3- TORE SUPRA SYSTEM FOR THE RADIATION PATTERN MEASUREMENTS

The TS system is composed of 5 parts: network analyzers, robot of movement, manual remote control console, master computer driver and mechanical stands of the devices to be tested. See on figure 1 the principle of this system and on figure 2 a photography of our general equipment. This robot have been developed by SEO (Société d'Electronique Occitane) on CEA specifications.

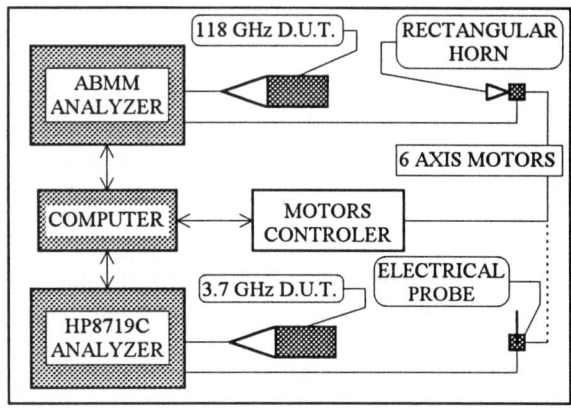

Figure 1: Schematic principle of the full system

- <u>2 network analyzers</u> measure the radiation field pattern near the far field region. An ABmm device is used between 75 to 140 GHz in ECRH activities, while a HP8719C one is used for LHCD activities. This work at 3.7 GHz has been implemented in order to have a complementary test method for the TE_{01}/TE_{03} mode converters in rectangular waveguide [4]. Main microwave characteristics of these vector analysers are summarised in table 1 below. It must be noted that the dynamic range obtained for millimeter waves is very large, which is very useful for a good measurement accuracy.

Figure 2: TS millimeter waves laboratory.

ANALYSER	HP8719C	ABMM	
MEASURE	ELECTRICAL PROBE	RECTANGULAR HORN TH23F	
FREQUENCY RANGE	50 MHz to 13.5 GHz	W	F
FREQUENCY in GHz	3.7	110	118
DYNAMIC in dB — Standard assembly	90	91.3	81.6
DYNAMIC in dB — External Gunn Oscillator		140	132

Table 1: Microwave characteristics.

Converter millimeter ports on 7th harmonic and standard horn are used with the ABmm analyzer. An electrical probe in semi-rigid cable has been developed specially at 3.7 GHz for the centimeter waves.

- <u>A robot</u> moves the 2 measurement sensors with 6 freedom degrees: 3 linear axis and 3 circular axis. 2 circular axis are used in order to obtain a right position of the measurement horn. The measurement transfer function is constant by this way, which make easier the measurement interpretation. The 3rd circular axis turns the horn 90° in order to measure the main field component and the cross one. The 4 different robot movements are configured by the user: circular, plane, spherical and linear.

- <u>A manual remote control console</u> allows to set parameters of 6 axis numerical controls of the motors and to implement simple movements.

- <u>A master computer driver</u> allows to set mechanical and microwave parameters of the automatic measurement method, to display and print out the patterns, to save the measurement files and to do the additional data processings. The whole system works under Labview (Figure 3) which is an industrial software for the measurement equipments.

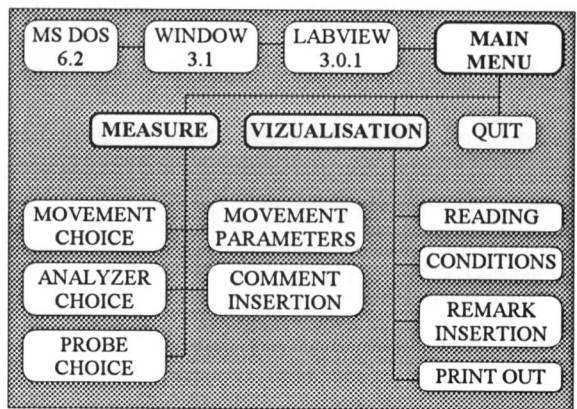

Figure 3: The main jobs of the software.

- <u>Rigid mechanical frames and specific line supports</u> hold up device under test and alignment is achieved with external sights and laser beams. The horn has been put in the robot mechanical center in order to have a maximum measurement area. To get clear conclusions the component to be tested must be aligned very precisely in front of the horn measurement. This alignment is achieved by autocollimatoring with a laser beam on the mirror installed on the measurement horn center.

4- THE HE$_{11}$ MODE CONVERTERS

Well aligned waveguide systems have been used in order to test our different transmission line components in laboratory. For the 2 gyrotrons frequencies 2 different converters have been developed: 110 GHz converter by Thomson TTE and 118 GHz converter by Institute Applied Physics of Russian. They feed the components with a very pure HE$_{11}$ mode

This 118 GHz HE$_{11}$ converter consists of a special horn which radiates into an output corrugated waveguide through a reflectless lens (Figure 4). The standard TE$_{01}$ rectangular mode is converted by this way into a HE$_{11}$ circular mode with compact length. The mode purity is above 97 %. Figure 5 shows a comparison between measurements of the converter followed by a 2 meters long straight waveguide and the HE$_{11}$ theoretical model calculated in near field [5],[6].

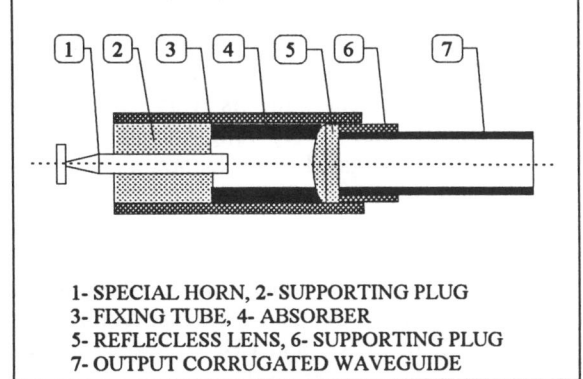

1- SPECIAL HORN, 2- SUPPORTING PLUG
3- FIXING TUBE, 4- ABSORBER
5- REFLECLESS LENS, 6- SUPPORTING PLUG
7- OUTPUT CORRUGATED WAVEGUIDE

Figure 4: 118 GHz HE_{11} converter principle

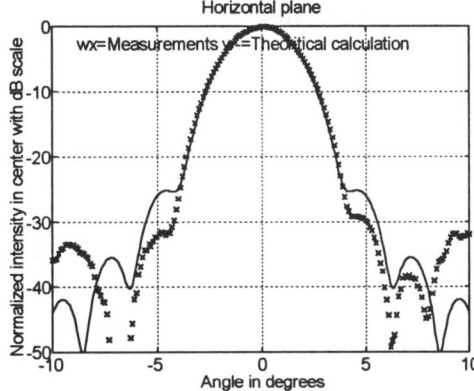

Figure 5: Comparison between measurements and the HE_{11} theoretical model.

The agreement between measurements with a 2 meters long straight waveguide in output and the HE_{11} theoretical model is satisfactory up to 25 dB (Figure 5).

5- TEST OF THE MAIN TRANSMISSION LINE COMPONENTS

After a call for tenders, the contractor of the transmission line components is Spinner who is associated with General Atomics. The TS millimeter wave measurement low power laboratory have been created in 92. In order to reduce the cost of this manufacture, all the HF reception tests have been implemented in Cadarache.

The components are installed on rigid mechanical supports Micro-contrôle and on specific line supports designed by ourselves. They have 5 movement freedom degrees for adjustment and line expansion. The accuracy of this alignment schedule, with laser beam and external sights, is about 1 mm (Figure 6). In these conditions the effects of an alignment defect or a spurious mode mixture can be separated easily.

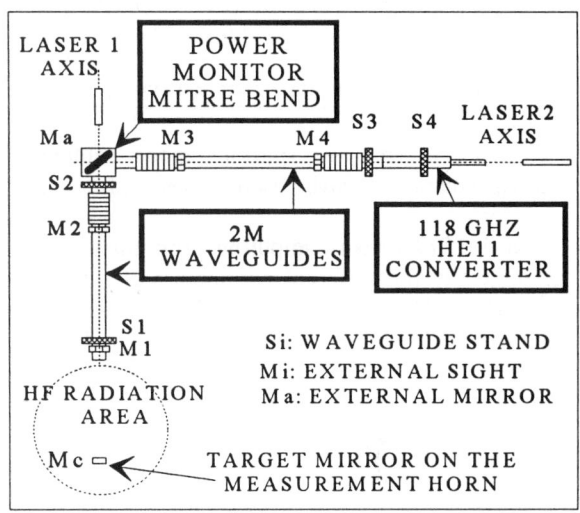

Figure 6: Assembly for the power mitre bend.

The tested components have been: HE_{11} converters, 2 meters aluminium straight corrugated waveguides, pumping section, mitre bends, bellows, DC break, power monitor mitre bend and vacuum valve.

All these components are tested by our automatic radiation measurement system. Main patterns are in 2D and made in horizontal and vertical planes, with main and crossed polarization, with drive in H and E plane. Measurements: in 1D with a distance variation, in 2D with changing plane measure at 1.3, 1.6, 1.9 m and in 3D with 3 analysis distances have been also made depending on the components and the analysis accuracy.

	Measurement area	Nbr. pts	Nbr. pts/°	Recording time	Nbr. pts/s
2D	±20°	401	10	6 mn	1.12
2D	±10°	201	10	3 mn	1.12
2D	±5°	101	10	2 mn	0.84
3D	±20°	70X70	1.75	96 mn	0.85
3D	±5°	50X50	5	44 mn	0.95

Table 2: Measurement time study.

The study of the measurement time (Table 2) clearly shows that a precise analysis including 3x3D patterns and 2 crossed polarizations, requires about 4H 24 mn.

Components more conventional: attenuators, couplers, detectors... of the measurement system put in the power monitor mitre bends output can also easily be calibrated with our equipments.

6- MODE MIXTURE CALCULATIONS

In order to find the spurious mode content 2 different methods have been developed in our laboratory.

- In the first one a computing program searches automatically the different parameters of the main spurious mode mixed with the HE_{11} mode. An arbitrary modes family must be chosen and many iterations are needed, consequently this is an approximate method and the calculation time is very long: several hours. See in figure 7 an example of this method.

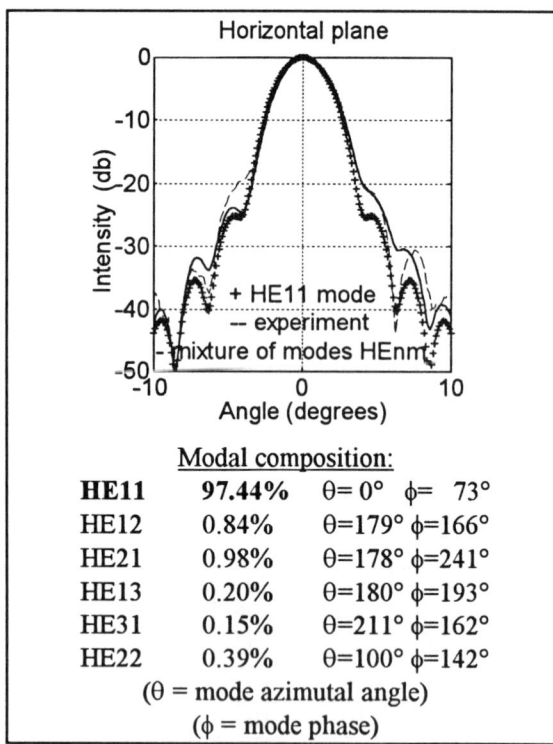

Figure 7: Study of mode mixture patterns for the 118 GHz HE_{11} converter.

- In the second one, the power content of HE_{11} mode is calculated using the measurement results. 2 or 3 x 3D analyse planes have been used here [7] and the measurement time is very long. This method requires not a very long calculation time and it is now investigated in our laboratory.

7- CONCLUSION

The TS low power millimeter waves laboratory is now able to make all the measurements required for TS and ITER in ECRH and LHCD systems. Some theoretical methods to obtain the modal composition or the HE_{11} fundamental mode content at the transmission line output are under investigation and interesting results have already been obtained. Working fields concerned are: vector network analyser in millimeter waves, automated mechanical movable system and calculation codes.

References:

1. Pain, G. Berger-By et al. "Status of the ECRH system of Tore Supra" 18th Soft Karlsruhe 1994 pp 481-484.
2. Whaley et al. "Quasi-CW 0.5 MW-118 GHz gyrotron for ECRH" 18th Soft Karlsruhe 1994 pp 489-492.
3. Pain et al. "Status of the 118 GHz- quasi CW gyrotron for Tore Supra and TCV tokamaks" this conference.
4. Bibet et al. "Experimental and theoretical results concerning the development of main RF components for next Tore Supra LHCD antennae" 18th Soft Karlsruhe 1994 pp 577-580.
5. Rebuffi and J.P. Crenn "Radiation patterns of the HE_{11} mode and gaussian approximations" Int. J. Infrared Millimeter Waves 10 (1989) pp 291-311.
6. Crenn, C. Charollais "Propagation and radiation characteristics of the circular electric, circular magnetic and hybrid waveguide modes" Int. J. Infrared and Millimeter Waves Vol 17 n°9 (Sept. 1996).
7. Chircov, G.G. Denisov, N.L. Aleksandrov "3D wavebeam field reconstruction from intensity measurements in a few cross sections" Optics communications 115, 1995 pp 449-452.

STATUS OF THE 118 GHz - QUASI CW GYROTRON FOR THE TORE SUPRA AND TCV TOKAMAKS

M. Pain, P. Garin
Association CEA-Euratom sur la Fusion Contrôlée
CEN Cadarache, F-13108 St.-Paul-lez-Durance, France.

M. Q. Tran, S. Alberti
Ecole Polytechnique Fédérale de Lausanne, Association Euratom-Confédération Suisse,
PPB-EPFL, CH-1015 Lausanne, Switzerland.

M. Thumm, O. Braz
Forschungszentrum Karlsruhe, Institut für Techniche Physik, Association Euratom-FZK
P.O. Box 3640, D-76021 Karlsruhe, Germany.

E. Giguet, P. Thouvenin, C. Tran
Thomson Tubes Electroniques, 2 Rue Latécoère, F-78140 Vélizy, France.

A 0.5 MW CW gyrotron is being developed to be used in ECRH and current drive experiments on the Tore Supra and TCV Tokamaks. The joint development undertaken by the CEA, the CRPP and Thomson Tubes Electroniques with technical support from FZK is now in it final phase, with the test of a full scale prototype. The main features of this tube are an improved quasi-optical mode converter with three mirrors, and a cryogenic liquid nitrogen cooled window. All tube components are actively cooled in view of CW operation.

A first prototype fitted with the cryogenic window has already produced 5 second pulses at nominal power, and its behaviour showed remarkable agreement with cavity and beam calculations. A second prototype, now under test, is expected to produce up to 210 second pulses once installed at the CEA site in Cadarache, where final tests up to the nominal pulse duration are to take place.

This paper presents the main features of the tube as well as the results of the test campaigns of the first and second prototypes.

1. Introduction

Development of a quasi-CW gyrotron for use in electron cyclotron heating and current drive experiments on the TCV and Tore Supra tokamaks started in 1994 as a joint activity between Thomson Tubes Electroniques (France) as the industrial partner, the Commissariat à l'Energie Atomique (France) and the Ecole Polytechnique Fédérale de Lausanne (Switzerland), with technical support from the Forschungszentrum Karlsruhe (Germany). The specification requirements include a pulse length of at least 210 seconds, a unit power of 500 kW in the HE_{11} output mode and an efficiency above 30%. The frequency of 118 GHz was selected as a compromise between the requirements imposed by the third harmonic heating program of TCV and the first and second harmonic heating on Tore Supra.

2. Technical features

The main characteristics of the tube have been described in some detail in [1]. Its main features are the following (see figures 1a and 1b):

The electron gun is a triode gun, designed for a 22 A beam with a velocity ratio of 1.5 for a total acceleration voltage of 85 kV maximum. The interaction cavity is designed to work on the $TE_{22,6}$ mode, which is converted into a gaussian beam using an improved version of the Vlasov quasi-optical converter, followed by three mirrors (see figure 2). This arrangement converts the $TE_{22,6}$ mode coming from the cavity into a gaussian beam. The RF beam leaves the tube through the window into a matching optics unit (MOU) where the beam is coupled to the HE_{11} carrying transmission line.

Calculations show that a FC75 cooled double window would not work for long pulses at these power levels. Since the CEA and TTE had previously developed a cryogenic window for use on a 110 GHz gyrotron, the design was adapted to the new tube. It consists of an 80mm diameter disk of monocrystaline sapphire, edge cooled with liquid nitrogen at boiling temperature (77 °K). Both the

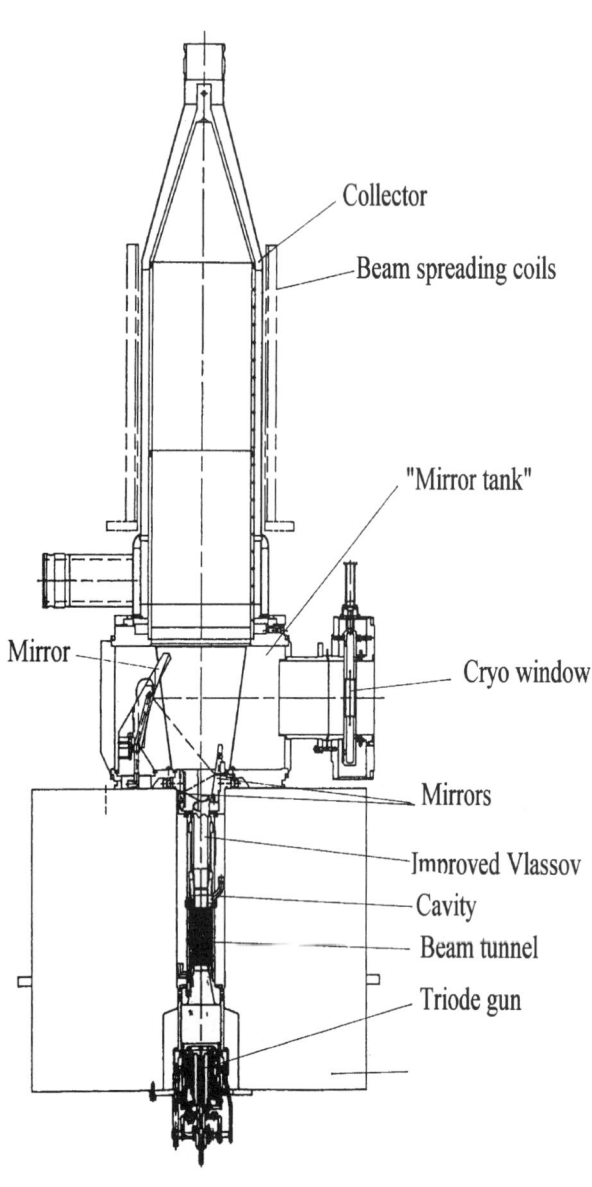

Figure 1a: Gyrotron schematics

Figure 1b: The gyrotron (second prototype)

maquette, the first and the second prototypes were fitted with this window.

The tube is to be connected to a transmission line carrying the HE_{11} mode, but it was thought wiser to output through the window a gaussian beam and to couple it to the waveguide outside the tube, in order to be able to correct any aberration or misalignement of the beam. The coupling from the

Figure 2: Inside the mirror box. The quasi-optical launcher and the three mirrors.

gaussian beam to the HE_{11} mode waveguide takes place in a Matching Optics Unit (MOU) using two flat mirrors to correct the beam position and one ellipsoidal mirror to correct the properties of the beam. The two flat mirrors will also in future be corrugated to change the wave polarisation.
During the development, three tubes have been built: A « Maquette », a first prototype and a second prototype.

3. Maquette results

The maquette was a very simplified version of the tube. It did not include all the cooling systems and was only capable of short pulses (up to a hundred milliseconds). It was used to validate the calculations of the electron gun, interaction cavity and quasi optical coupler. These calculations were run in parallel by the different partners, and cross checked. The measured characteristics were in complete agreement with calculations, as one

example given in figure 3 shows: The figure represents the trajectory of the beam during the gyrotron startup phase in the plane (energy, α) where α is the perpendicular to parallel electron speed

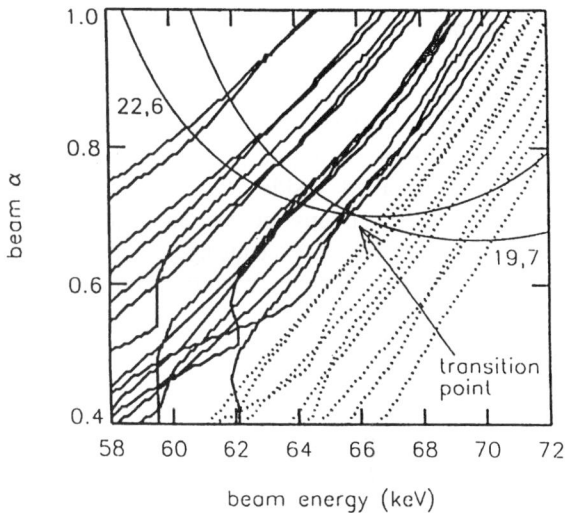

Figure 3: Starting curves of the gyrotron, showing the excellent agreement with theory

ratio. The mode the tube will start in depends of the first starting curve this trajectory crosses. The figure shows the theoretical starting curves for the modes $TE_{22,6}$ and $TE_{19,7}$. The continuos lines indicate the experimental cases with oscillation in the first of those modes, and the dotted lines those corresponding to the latest. As it can be seen, the experimental results are in complete agreement with theoretical prediccions.

The maquette produced pulses of about 100 milliseconds with an output power of 500 kW and an total efficiency of 31% .

4. First prototype results

The first prototype was identical to the maquette, but had been fitted with all the cooling systems designed for CW operation. The mirrors had also been redesigned to correct a mistake made in the maquette, which produced an elliptical output beam instead of a circular one.

The first prototype was in principle capable of achieving the final specification. It was first tested with short pulses, achieving for a pulse length of 1 ms more than 700 kW, with a peak efficiency of 34%. The tube operated in a more stable regime than the maquette, owing probably to the better output beam shape and consequent reduction of parasitic modes.

The window was then cooled to the liquid nitrogen boiling temperature, and the pulse was extended up to 1 second at the nominal power. Then testing had to be stopped due to breakage of the window. After analysis, it appeared that this was caused by fusion of a piece of ceramic material which was installed inside the tube to dampen unwanted modes. Drops of melted material had fallen into the window, causing cracks in the sapphire disk which ultimately led to its breakage.

5. Second Prototype results

Given the quality of the output RF beam found in the first prototype, it was considered that the damping material could be removed, and therefore a second prototype, similar to the first one but without any damping material was built and tested. The testing of this tube is not yet completed, mainly because problems with the auxiliaries and in particular with the load. But it has already achieved very consistent results.

Its behaviour, as far as starting current and mode competition problems are concerned is in all points similar to what was recorded on the first prototype, which shows that the design is now reliable and that the goal of being able to operate three tubes from a single high voltage supply is likely to be achieved.

The tube was first conditioned to 2 second pulses in order to check the parameters of the tube. The pulse was then extended to 5 seconds by steps of 500 ms. The tube went from 2 to 5 seconds in that way without a single pulse failing. This gives us confidence on the possibilities of further extending the pulse to the nominal specification without any major difficulty, given the fact that all internal components are at their equilibrium temperature for a pulse length of 5 seconds.

The tube has now achieved more than thirty pulses of 5 seconds at 480 kW, with an efficiency slightly below 30%.

The mode purity of the tube has also been calculated using radiation pattern measurements done by firing the tube on a ceramic target and measuring the temperature increase by means of an infrared camera. The gaussian content of the radiated spectrum is about 98 %.

6. The next milestones

The formal factory reception tests of the second prototype are scheduled to take place during September 1996. These tests will be performed with the longest pulse compatible with the factory test installation. The tube will then be shipped to the CEA site at Cadarache to be tested up to the maximum pulse length of 210 seconds at nominal power.

7. Conclusion

This R&D project shows that an efficient cooperation can be achieved between the main European laboratories and industry, and that owing to that cooperation Europe has the technical and industrial capability of designing and manufacturing a reliable and industrially sound gyrotron for ECRH applications. Development of this gyrotron has taken about three years, which is reasonable with regard of lead times required by other ECRH projects such as W7X and ITER.

The calculation codes developed by the partners are capable of predicting in a reliable way the behaviour of the tube. From that basis the tube could be scaled for other frequencies economically.

A reliable 500 kW quasi-CW European gyrotron is now available for a frequency of 118 GHz, and the design is easily scaleable up to 140 GHz.

References

[1] D. Whalley et al., « Quasi-CW 0.5 MW - 118 GHz gyrotron for ECRH ». 18th SOFT, Karlsruhe, 1994.
[2] M. Pain et al., « Status of the ECRH system of Tore Supra ». 18th SOFT, Karslruhe, 1994.
[3] D. Whalley et al., « Startup methods for single mode gyrotron operation », Phys. Rev. Letters, 75, 1995.

The 140 GHz - 2MW - 2 s ECRH System for ASDEX-Upgrade

H. Brinkschulte[a], A. Fix[c], W. Förster[b], G. Gantenbein[b], W. Kasparek[b], F. Leuterer[a], F. Monaco[a], M. Münich[a], A. Peeters[a], F. Ryter[a], P. Schüller[b], V. Sigalaev[c], E. Tai[c]

[a]Max Planck Institut für Plasmaphysik, 85740 Garching, Germany

[b]Institut für Plasmaforschung, Universität Stuttgart, 70569 Stuttgart, Germany

[c]Institute of Applied Physics, Nizhny Novgorod, 603600 Russia

A 140 GHz ECRH system is being installed at the ASDEX-Upgrade tokamak. The total power of 2 MW at a pulse length of 2 s - or 2.8 MW at 1 s, respectively - is generated in 4 gyrotrons and transferred to the plasma via 4 transmission lines, partly quasi-optically and partly through HE-11 wave guides. The power can be modulated up to 30 KHz. Rotatable mirrors inside the vacuum chamber facilitate the deposition of the rf power at any desired position within the cross section of the plasma. A first 0.5 MW - 0.5 s unit is operational since summer 1995. Gyrotron, transmission line and in-vessel components worked very reliably. Heating effects and the propagation of heat waves were analysed under various plasma conditions.

1. INTRODUCTION

The ASDEX-Upgrade tokamak - with a major radius of $R = 1.65$ m, minor radius $a = 0.5$ m, elongation $\varepsilon = 1.6$, plasma current $I_p < 1.6$ MA, electron density $n_e < 2 \cdot 10^{20}$ m^{-3}, toroidal field at plasma centre $B_t < 2.8$ T - is designed to investigate the divertor under conditions encountered in future fusion machines. The installed heating power of 10 MW of NBI - to be doubled in the near future - and 8 MW of RF at ion cyclotron resonance between 30 MHz and 120 MHz is sufficient to achieve heat fluxes to the plasma facing components which are similar to those in ITER.

ECRH is a method with very localised deposition of heating power [1]. The absorption layer can easily be shifted anywhere between the centre and the boundary of the plasma either by changing the magnetic field (and thus the radial position of the absorption zone) or by launching the rf beam under an angle with respect to the horizontal (by a rotatable mirror close to the plasma). ECRH is also an excellent tool to study general transport phenomena like electron thermal diffusivity, transition from H - to L -mode, ELMs and MARFEs. In addition, current drive allows the local modification of the current profile which influences plasma transport and which can also be used to influence MHD mode activity and thus the stability of the plasma.

With high power gyrotrons becoming commercially available, it was decided to install 2 MW of ECRH power at ASDEX-Upgrade [2]. Heating is at the 2nd harmonic X-mode which corresponds - at a typical magnetic field of 2.5 T - to a frequency of $f = 140$ GHz. In this case, for a plasma temperature of 1 KeV, absorption is very strong at densities between $1 \cdot 10^{19}$ m^{-3} and the cut-off density of $1.2 \cdot 10^{20}$ m^{-3}.

2. THE RF SYSTEM

4 gyrotrons generate the rf power which is transmitted to the plasma via 4 transmission lines. 1 unit is operational since summer 1995 and the total system will be completed early 1997 [3].

2.1 The Gyrotron

The gyrotrons are manufactured by GYCOM company, Nizhny Novgorod, Russia. The main characteristics of the tubes are as follows

- Frequency 140 ± 0.5 GHz
- Output Power   500 (700) kW
- Pulse Length   2 (1) s

- Duty Factor   0.01
- Beam Voltage   70 (75) kV
- Beam Current   25 (35) A
- Efficiency   33 %
- Output Mode   TEM 00 (>90%)
- Power Mod. Freq.   < 30 kHz

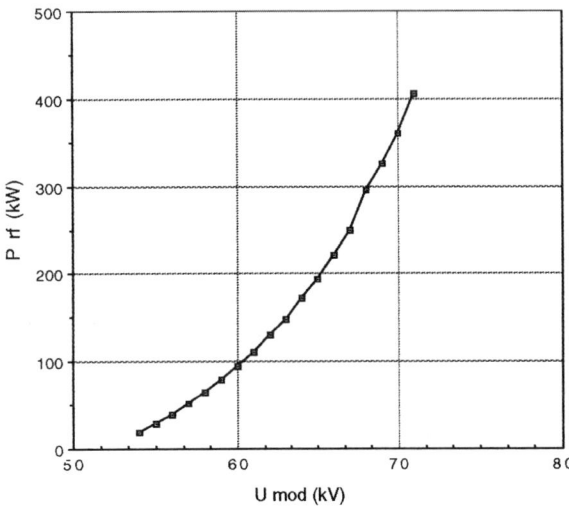

Figure 1. RF power vs. modulator voltage (IAP gyrotron of prototype unit).

Each 2 gyrotrons are fed by a common high voltage power supply unit with a modulator tube in series for switching and modulating the rf power. In figure 1, rf power - measured at the end of the transmission line in front of the vacuum vessel - is plotted versus voltage at the modulator tube. For many plasma experiments, modulated rf power - preferably at frequencies between 10 Hz and 10 kHz - is required. Figure 2 shows beam voltage, beam current and rf power versus time at maximum modulation frequency of 30 kHz. In this case the beam voltage needs to be decreased by only 30 % for 100 % on/off modulation.

2.2 The transmission line

Ca. 90 % of the gyrotron output is in the desired Gaussian mode which is suitable for the transmission over long distances - in our case 35 m between source and plasma. The transmission is partly quasi-optically, partly via HE-11 mode in corrugated wave guides. The entire arrangement with all major components is sketched in figure 3.

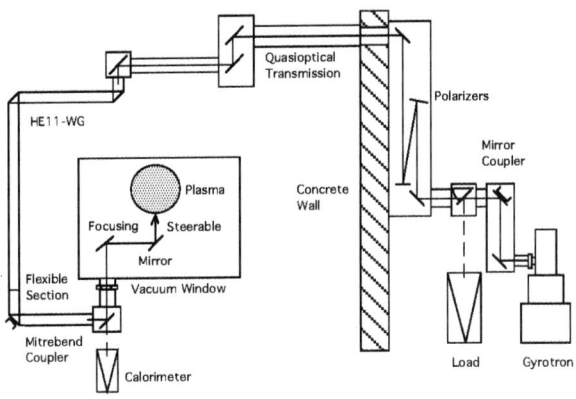

Figure 3. Transmission of rf from source to plasma. Total length: ca. 35 m.

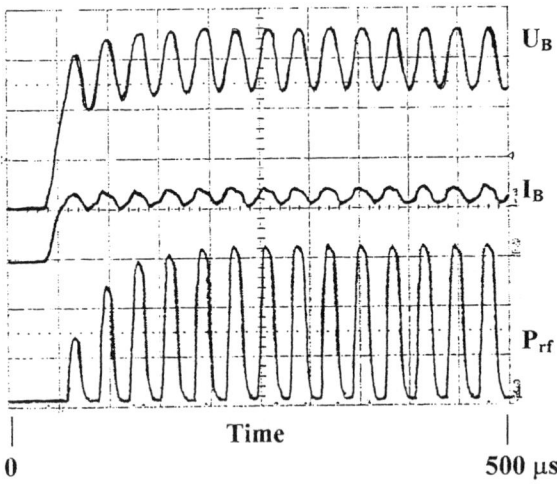

Figure 2. Beam voltage, current and rf power versus time at maximum modulation frequency of 30 kHz.

Spurious power in unwanted modes is dumped in the box right behind the gyrotron. The first 2 mirrors are shaped to compensate the astigmatism of the beam [4], 1 mirror containing a multi-hole directional coupler for measurement of forward and reflected power. The rf diodes provide very good signals and were calibrated against a calorimetric load. During conditioning of the gyrotron, the rf power is diverted into a brick dummy load which can handle maximum energy at maximum pulse repetition rate. A pair of polarisers (corrugated mirrors) enables us to select any plane of polarisation or to excite an elliptically polarised beam, respectively. In the vicinity of the torus - where space becomes a problem - the rf is transferred into a 89 mm diameter wave guide. At the end, its diameter is increased to 123 mm for a broader power profile which is required to reduce thermal stresses in the torus vacuum window [5]. Right in front of the machine, transmission returns to the quasi-optical mode. The last mitre bend contains another (calibrated) directional coupler for measurement of power.

The alignment of the quasi-optical part of the transmission line was straight forward (with a He-Ne laser and by observing the beam patterns from short rf pulses on thermographic screens). However, to achieve good transfer between quasi-optical and wave guide section was quite tedious. For that purpose, the axis's of both parts have to coincide with great accuracy. We found also that slight deviations from a 100% Gaussian beam can lead inside the wave guide to excitation of additional modes which can modify the beam pattern - and thus the power density - substantially along the wave guide.

The loss of the transmission line is ca. 7 %, mainly due to ohmic losses on the 12 mirrors and 3 wave guide mitre bends, respectively.

2.3 In-vessel components.

The rf is launched in the midplane of the vacuum vessel from the low-field side. 2 beams enter the vessel through 2 neighbouring sectors of the machine, toroidally 22.5^0 apart, 2 beams through an opposite sector. Boron nitride resonance windows separate the transmission lines from the torus vacuum. 2 more mirrors per line are needed to transfer the rf to the plasma. One mirror is curved with its focus halfway between centre and inner boundary of the plasma. The waist of the beam is 22.14 mm, calculated without plasma. The second mirror - made from fine-grain graphite and coated with 20 μ copper - is plane and rotatable around the horizontal and vertical axis, respectively. This allows to reach almost the entire cross section of the plasma ($\varphi_{pol} = \pm 35^0$) and also to direct the beam toroidally in both directions ($\varphi_{tor} = \pm 25^0$).

2.4 Shielding of gyrotrons

The gyrotrons are positioned 15 to 18 m away from the centre of ASDEX-Upgrade. At this distance, the magnetic stray field is still up to 30 Gauss and may impede proper operation of the tubes. Therefore, the gyrotrons were mounted with their critical components at the height of the midplane of the plasma where the - particularly harmful - horizontal component of the magnetic field is minimum. To reduce that component, a 1 cm thick plate of soft iron was introduced ca. 2 m above the gyrotrons. A second, 2 cm thick iron plate beneath the tubes shields the magnetic stray field from the machine's high-current bus bars.

3. FIRST EXPERIMENTS

A first unit was set up in close co-operation with the Institute for Applied Physics in Nizhny Novgorod, Russia. The aim was to gain experience with the operation of high power gyrotrons and to perform some exploratory plasma experiments [6]. The prototype unit comprises a IAP gyrotron, a complete transmission line and a fixed and a movable mirror inside the vacuum chamber. We

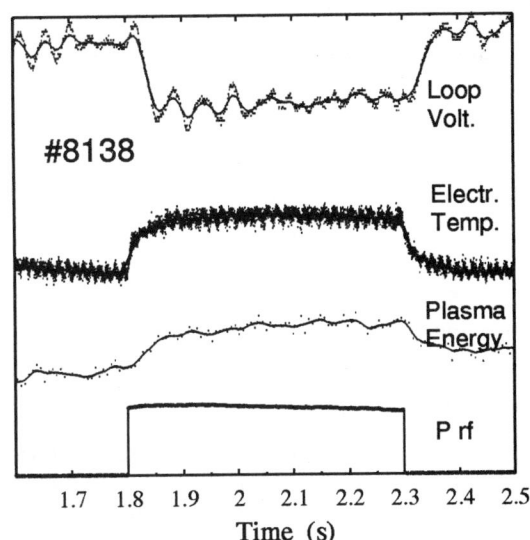

Figure 4. Rf power, plasma energy, electron temperature and loop voltage vs time. $I_p = 1$ MA, $B_t = 2.5$ T, $n_e = 4 \cdot 10^{19}$ m^{-3}, $\varphi_{pol} = + 35^0$.

succeeded to couple 400 kW to the plasma which is

ca. 80 % of the maximum output power of the tube. The pulse length is 0.5 s, or 1 s in modulated mode.

The prototype unit is operational since summer 1995 and during its first year of operation, many plasma pulses with successful additional electron cyclotron resonance heating were achieved. Although the maximum rf power of 0.4 MW is small compared to the ohmic power (P_{ohm} = 1 MW for a typical 1 MA discharge) and far less than the installed ICRH and NBI power, distinct effects upon the plasma were observed. An example for an 1 MA discharge with ECRH on between 1.8 and 2.3 s is depicted in figure 4: plasma energy increases and saturates after 100 ms, electron temperature increases somewhat faster and loop voltage

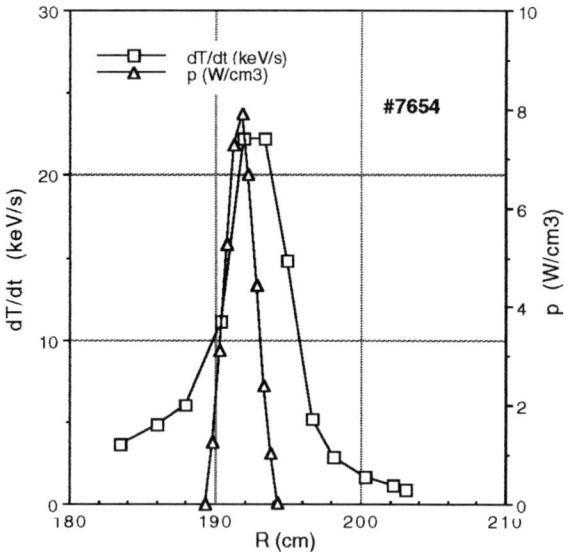

Figure 5. Gradient of electron temperature dT/dt at beginning of rf pulse and calculated power deposition profile. I = 1 MA, B_t = 2.5 T, n_e = 4·10^{19} m^{-3}, φ_{pol} = + 30^0.

decreases due to heating of the (current carrying) electrons. The 45 channel ECE radiometer diagnostic at ASDEX-Upgrade - which allows a complete scan of the electron temperature over the entire cross section of the plasma - was extensively used to study ECRH effects. Figure 5 shows the change of the electron temperature at the beginning of the rf pulse (which is proportional to the absorbed power) and the calculated power deposition profile vs the major radius. The location for maximum heating can readily determined. However, the measured width for dT/dt is broader than the actual profile because the measured values are blurred due to diffusion effects during measuring time.

Since rf power is sufficient for heat pulse studies, most experiments so far were made with modulated rf, typically between 10 Hz and 1 kHz. The electron thermal diffusivity χ_e was measured for a variety of different plasmas [7].

4. DISCUSSION

The prototype unit worked very well and reliably during its first year of operation. The gyrotron needs only little conditioning to reach maximum power and, occasionally, minor re-adjustments are necessary, mainly of the strength of the magnetic field and of the emission current. The transmission line is very stable and requires hardly any re-alignment. However, the inner walls of the wave guide have to be kept clean to avoid rf breakdown.

The first plasma experiments are very encouraging. We found that 80 % of the output power of the gyrotron is coupled to the plasma where it is absorbed in a thin layer. Reflection of rf from the plasma is negligibly small.

REFERENCES

1. U. Gasparino et al., Plasma Physics and Contr. Fusion 30 (1988), 283
2. „A 140 GHz/2 MW ECRH System for ASDEX-Upgrade", MPI für Plasmaphysik, Int. Rep. (Jan.93)
3. H. Brinkschulte et al., Proc. 3rd Inter. Workshop on Strong Microwaves in Plasmas, Russia (1996), in print
4. G. Denisov, 8 th Russian-German Workshop on ECRH and Gyrotrons, Nizhny Novgorod (June 96)
5. L. Empacher et al., Proc. 20th Intern. Conf. on Infrared and MM Waves, 473 (1995)
6. F. Leuterer et al., Proc. 23 rd EPS Conf. on Contr. Fus. and Pl. Phys., Kiev (1996), in print
7. F. Ryter et al., ibid.

Conceptual Design of the 140 GHz/10 MW CW ECRH System for the Stellarator W7-X

L. Empacher, W. Förster, G. Gantenbein, W. Kasparek, H. Kumrić, G.A. Müller, P.G. Schüller, K. Schwörer, U. Schumacher, D. Wagner
Institut für Plasmaforschung, Universität Stuttgart, Pfaffenwaldring 31, D-70569 Stuttgart
V. Erckmann, T. Geist, H. Laqua
Max-Planck Institut für Plasmaphysik, EURATOM-Association, D-85748 Garching

The technical concept of the ECRH system for the planned stellarator W7-X is presented. This system will consist of 10 gyrotrons with 1 MW RF output power each. The microwave transmission from the gyrotrons to the plasma will be performed with purely optical elements including beam matching and polarizing units and a multi-beam waveguide system. The generators and auxiliary supplies are described as well as the elements of the optical transmission system and the microwave diagnostic.

1. INTRODUCTION

ECRH is the main heating system for steady state operation of the planned stellarator W7-X of IPP Garching at Greifswald. Several applications for the ECRH system are foreseen like: plasma start-up, electron heating (X-mode cut-off density: $1.25 \cdot 10^{20}$ m^{-3}, O-mode cut-off density: $2.5 \cdot 10^{20}$ m^{-3}), investigation of improved confinement modes, transport investigations, stimulated heat wave propagation, current drive, combined heating with NBI, density and impurity control by profile shaping, on/off axis power deposition, studies of trapped/passing particle physics by ECCD methods, diagnostics (e.g. collective Thomson scattering and electron cyclotron absorption, ECA). To meet these objectives, a flexible, powerful and reliable design is necessary.

The concept is based on the guideline to simplify the ECRH system making use of the experiences with a similar system running at W7-AS since several years [1] and the most recent advances in source and transmission line technology. In particular the development of gyrotrons in the relevant frequency range of 140 - 160 GHz with a Gaussian optical beam output allowed for a substantial simplification of the overall system. A possible transmission system with obvious advantages is a fully optical one with free space propagation of the power, because no sophisticated waveguide system is required.

In the following the design of the elements and the generators with the auxiliary supplies will be described and its characteristics will be discussed.

2. TECHNICAL CONCEPT OF THE 10 MW, 140 GHz CW ECRH SYSTEM

2.1 GENERATORS AND POWER SUPPLIES

In the ECRH building 10 gyrotrons will be placed in two groups with 5 tubes each (see Fig. 1). These tubes will operate at a frequency of 140 GHz with 1 MW CW output power per tube. The accelerating voltage applied to the electron beam will be in the order of 70-80 kV, the beam current is about 40-45 A. Based on the experience with the gyrotrons at W7-AS a diode type electron gun is favoured because of simplification of the overall system and cost savings in the peripheral equipment. Most recent experiments show that it is possible to increase the (electronic) efficiency of advanced gyrotrons up to 50-60 % (from 30-40 %) by using depressed collector techniques. The availability of such gyrotrons would have a strong influence on the design and costs of the high voltage power supply and other auxiliary systems. Therefore such gyrotrons are favoured.

The output power will be launched through a lateral window in a linearly polarized free space mode. Due to unavoidable microwave absorption in the dielectric window material the pulse length of the existing high power tubes is limited. There is confidence, however, that the present and future developments (diamond window, cryogenic window, distributed window, ...) will reach CW operation at higher power. To reduce the thermomechanical stress and to achieve a homogeneous distribution of the RF power on the window it might be necessary that the output radiation includes higher order modes [2].

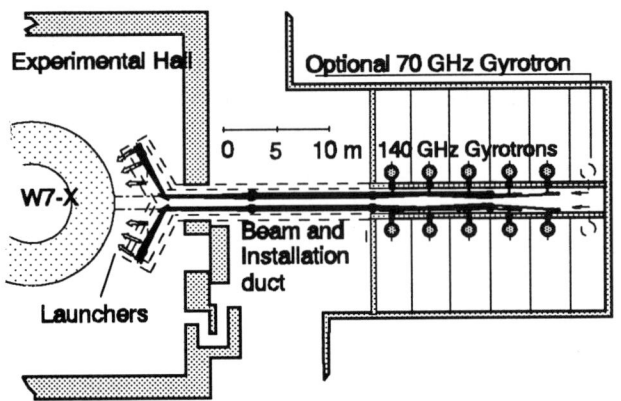

Figure 1: Top view of the ECRH-building.

The superconducting magnets for 140 GHz gyrotrons have resonant fields of about 5.5 T, which is state of the art with conventional superconductors. Due to the diode concept of the gyrotrons the presently running magnets at W7-AS have a simple design (one solenoid, bath cooled at approx. 4 K).

The ECRH-hall houses the gyrotrons, the high voltage power supplies and other auxiliary systems (cooling, LHe and LN-supply). For each gyrotron a power supply for the electron gun and supplies for the magnets (main coil, auxiliary gun coil, collector coil) are required.

There are several possibilities to provide high voltage. For conventional tubes without depressed collector a high voltage supply (for example 80 kV/50 A) with high accuracy and stability will be necessary. These units can be realized by use of series regulators, a well proven and reliable technique. In case of tubes with depressed collectors a solution with a simple power supply and high output current (50 kV/50 A) but low accuracy could be combined with a high voltage supply (80 kV/1 A) with high stability output voltage and low current.

For this purpose a high voltage switched mode power supply (HVSMPS) with 80 kV could be suited for acceleration at depressed collector tubes. Such a test-supply is currently operating at a voltage of 40 kV at 0.75 A and a HVSMPS with 80 kV is in preparation.

Heater supplies in different techniques, sawtooth generators for collector coils were developed and are now in use at W7-AS and will be adapted to the W7-X tubes.

2.2 OPTICAL TRANSMISSION SYSTEM

Based on the excellent experience with a fully optical transmission line which is routinely operating at W7-AS (140 GHz/0.9 MW) [1] an optical transmission line for W7-X was chosen. Such a system is seen to be the most simple, reliable and cost effective solution. In particular, the possibility to use one mirror system for several beams reduces the complexity of the system essentially. The most advantageous characteristics of a beam waveguide are: very high power capability due to low power density, inherent mode filtering properties due to out-scattering (diffraction loss) of unwanted higher-order free-space modes, low Ohmic losses, adjustable bends by using the reflectors, broadband transmission, simple polarizers and power combiners, no need for special waveguide sections for compensation of thermal expansion of the waveguide and movements of the plasma vessel, minimum number of parts in multi-beam waveguide (MBWG), perfect transmission between mirrors (no mode conversion).

In order to produce a stigmatic Gaussian beam (TEM_{00} mode) with appropriate beam waist conversion of the gyrotron output mode mixture is necessary. This can be done with two phase correcting mirrors which will be located in the beam duct close to the gyrotron window. These mirrors are the only elements which have to be adapted to the individual gyrotron output and can be designed and built at a late stage of the installation without affecting the lay-out of the overall system.

For different plasma irradiation (toroidal and poloidal launch angle scan, O- and X- mode launch, current drive), different wave polarizations are needed. Therefore, a polarizer allowing the continuous adjustment of any polarization (linear, elliptical and circular) will be implemented in the transmission line [3].

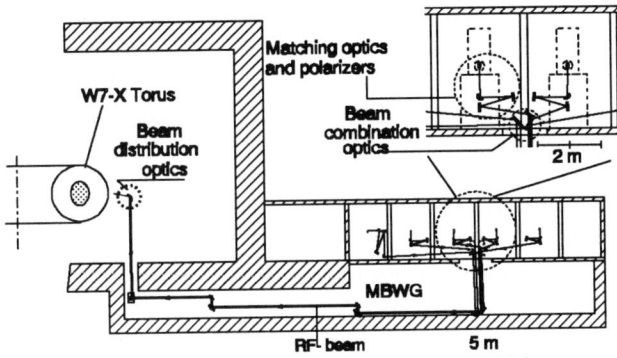

Figure 2: Schematical view of the beam duct showing the optical transmission from the gyrotrons to the torus hall.

Transmission of the microwave power from the ECRH hall to the stellarator port over a distance of ~50 m will be performed in a concrete duct in the basement. A compact solution is a beam waveguide transmitting several beams on a common mirror system. According to the location of the gyrotrons and the symmetric arrangement of the launchers, we use two MBWGs transporting five beams each. In each MBWG, two spare transmission channels are available, one to allow switching to different ports, and one for transmitting a lower frequency beam (e.g. 70 GHz for high β experiments envisaged at a later time). The basic design consists of four common mirrors in a confocal set-up offering a broadband transmission of Gaussian beams and geometrical imaging characteristics. The Gaussian beams are arranged in such a way that their waists are located at the input plane with a separation equal to or larger than the waist diameter (e.g. 4 times the waist radius). At this position, located near the gyrotrons, a plane mirror array is installed which directs the individual beams parallel to the optical axis of the mirror system. At the output plane of the MBWG system the beams are redistributed using similar optics. Due to the imaging characteristics and the low number of parts, the alignment is easy and reliable.

The overall transmission efficiency is determined by several loss-channels. The main contribution of the losses arises from ohmic dissipation in the mirror surfaces, which depends on the electrical conductivity of the material, the frequency, the angle of incidence and the polarization of the beam. A typical value for the maximum ohmic power of each mirror of the MBWG will be 30 kW. Active cooling of the mirrors and a stable frame design are necessary to avoid deformations of the surface. Diffraction losses and cross-polarization are in the order of 0 2.%. If the stability of the adjustment is within 1 mrad, the typical loss due to misalignment does not exceed 0.1 % per 10 m length of the beam waveguide. The atmospheric absorption of radiation at 140 GHz amounts to typically 0.5 % on the 50 m long transmission line (for dry air) which requires forced ventilation of the duct.

Additional losses of the off-axis beams in the MBWG due to coma and astigmatism can be kept negligible for proper arrangement of the mirrors. If the reflection and absorption losses of the torus window are assumed to be 1 %, a total Gaussian beam efficiency of 89 % can be achieved for the line including the antennas. We might have to consider additional stray radiation arising from diffraction losses inside the gyrotron and a small part of non-Gaussian contribution of the gyrotron (5-10 %) which cannot be reconverted to a Gaussian beam.

2.3 OPTICAL LAUNCHERS

The launching ports at W7-X were chosen to provide a design which is fully symmetric with respect to the toroidal and vertical directions. Ten individual launchers will be installed and used at positions where the plasma cross-section and magnetic field configuration minimize the interaction with trapped particles. These ports are equipped with a two mirror launching unit consisting of a fixed, slightly focusing mirror and a movable plane mirror for each beam. The plane mirror can be rotated, allowing a scan of the beam in poloidal and toroidal direction.

The transmission design allows to switch two

microwave beams to two additional ports. The magnetic field at these poloidal planes has a very weak gradient and is close to the toroidal local minimum. At this port the beam is launched to the plasma along the port axis which is tilted by about 10^0 with respect to the magnetic axis. Thus heating and current drive in the presence of a large number of trapped particles can be investigated.

As copper, preferably with a gold coating of the reflecting surface, will be used as launching mirror material the cooling requirements are in the order of 3 kW per mirror for O-mode launch and 1.5 kW for X-mode launch. In order to avoid excessive heat load of the antenna structure and the port duct walls due to stray radiation, active watercooling will be foreseen.

The parameters of the antenna beam are chosen such that on the one hand the power deposition in the plasma is highly localized, on the other hand the distortion of the profile of the electron velocity distribution function is kept as small as possible.

Opposite to the launchers, the graphite tiles forming the first wall of the vessel will reflect the non-absorbed fraction of the beams into a well defined direction to optimize the total power absorption, especially for O2 mode heating, and to avoid direct reflection into the antenna ducts.

2.4 BEAM MONITORING, EHF CONTROL AND DIAGNOSTICS

Monitoring of the high power microwaves in the beam waveguide during gyrotron operation will be performed by waveguide couplers or by grating couplers. In a waveguide coupler small bore holes drilled into the mirror surface are used to match a small amount of the EHF radiation to a monomode waveguide where forward and backward travelling power can be measured. With a grating coupler a small amount (< -50 dB) of the power is reflected into matched horns. For each gyrotron two mirrors are equipped with such a diagnostic, one mirror near to the gyrotron at the beginning of the duct and the one which deflects the beam towards the torus window. These signals will be fed into the protection system and in case of irregularities a fast switch-off of the gyrotrons is affected.

Calibration of these systems will be performed with calorimetric loads. These loads will be installed at different positions in the beam waveguide to determine the output power of each gyrotron individually. Stationary dummy loads will be used for long pulse and CW tests.

Thermal imaging is considered for several applications like window testing, measurement of beam pattern, alignment of the beam, control of heat load on components.

Standard rectangular pick-up waveguides will be integrated into the graphite tiles and used for in-vessel diagnostics of the polarization, the profile of each antenna beam, the calibration and alignment of the antenna geometry as well as for plasma absorption measurements.

3. SUMMARY

The technical design of the ECRH system for the planned stellarator W7-X is discussed. The outline of the system is based on the physical requirements for plasma heating with high availability and reliability. ECRH will operate at a frequency of 140 GHz with 10 MW (1 MW units) output power in CW. The optical transmission line has a modular set-up with independent elements close to the gyrotrons and at the launching ports. A considerable reduction in the number of parts is achieved by using common mirrors for transmitting several beams to the plasma vessel. Due to the thermal loading of the transmission elements a forced cooling is foreseen. Methods for in-situ diagnostics of beam power and beam pattern were described.

References

[1] Kasparek, W., Proc. 8th Joint Workshop on Electron Cyclotron Emission and Electron Cyclotron Heating, Gut Ising, 1992, 439.

[2] Bogdashov, A.A., et al., Int. J. Infrared Millimeter Waves, **16** (1995), 735-744.

[3] Doane, J.L., Int. J. Infrared Millimeter Waves, **13** (1992), 1727-1743.

ADVANCED HIGH POWER GYROTRONS FOR ECW APPLICATION

B. Piosczyk, E. Borie, O. Braz[+], G. Dammertz, C.T. Iatrou, S. Illy, S. Kern[+], M. Kuntze,
M.V. Kartikeyan[++], G. Michel, A. Möbius[+++], M. Thumm[+],

Forschungszentrum Karlsruhe, Association EURATOM-FZK, Institut für Technische Physik,
Postfach 3640, D-76021 Karlsruhe, Germany
[+] also Universität Karlsruhe, Institut für Höchstfrequenztechnik und Elektronik
[++] CEERI, Pilani - 333031, India, [+++] IMT GmbH, Luisenstr. 23, D-76344 Eggenstein, Germany

A 1 MW, 140 GHz gyrotron with a hollow waveguide $TE_{22,6}$ cavity and a 1.5 MW gyrotron with coaxial cavities designed for operation in the $TE_{28,16}$ and $TE_{31,17}$ mode at a frequency of 140 and 165 GHz are under development at FZK. In the $TE_{22,6}$ - cavity gyrotron operated with a depressed collector a maximum output power in a linearly polarized Gaussian rf output beam of 0.94 MW with an efficiency of 31 % has been obtained. A maximum efficiency of 36% was measured for an output power of 0.83 MW. The possibility of slow step frequency tuning has been demonstrated for the frequency range 117.9 to 162.3 GHz. The coaxial cavity gyrotron has been investigated with a set-up with an axial rf-output in short pulse (≤ 0.5 msec) operation. Single mode oscillation has been found over a wide range of operating parameters. A maximum output power of close to 1.2 MW has been measured in the design modes at both frequencies with an efficiency around 28 %. Frequency step tuning has been performed successfully in the frequency range between 115.6 and 164.2 GHz with the $TE_{28,16}$ gyrotron.

1. INTRODUCTION

For ECW application in the next generation of fusion plasma devices, millimeter wave sources with frequencies above 100 GHz operating at long pulses up to cw with an rf output power ≥ MW per unit are required [1]. In particular, for the ITER tokamak 8 MW rf power at variable frequency in the range between 90 and 140 GHz are foreseen for the plasma start-up in addition to about 60 MW rf power for EC Heating and Current Drive (ECH&ECCD) at a single frequency of 170 GHz. For stellarators ECH is essential. For W7-X an rf power of 10 MW in cw operation at 140 GHz is needed.

At FZK Karlsruhe gyrotrons able to fulfill these requirements are under development. In particular a conventional gyrotron with a hollow waveguide cavity operating in the $TE_{22,6}$ mode and two coaxial cavity gyrotrons are under investigation. Conventional gyrotrons with cylindrical waveguide cavities and 1 MW output power at frequencies above 140 GHz are close to their technical limits, mainly because of unacceptably high ohmic wall loading. In coaxial cavities the existence of the inner rod reduces the problem of mode competition and of voltage depression this permits the use of higher order modes than in cylindrical cavities and results in lower ohmic loading. Therefore, coaxial gyrotrons have the potential to generate in cw-operation rf output powers > 1 MW at these frequencies.

2. 1 MW, 140 GHz, $TE_{22,6}$ GYROTRON

The hollow waveguide cavity gyrotron is designed for operation in the $TE_{22,6}$ mode. The rotating high order cavity mode is converted into a linearly polarized Gaussian output beam in a quasi-optical mode converter. An optimization of the efficiency of the quasi-optical output system is in progress. In a first step a non-improved quasi-optical mode converter with internal diffraction losses of approximately 20% has been used. The electron gun is of diode type and uses LaB_6 emitter material. The main design parameters are summarized in Tab. 1. The measured output power and efficiency without the single-stage depressed collector (SDC) vs. the beam current are shown in Fig. 1 together with results of self consistent numerical calculations. A maximum output power of P_{out} = 0.94 MW at 140.1 GHz was obtained with a cathode voltage U_c = 85 kV and I_b = 53 A. This results in an efficiency of 31% with a retarding collector voltage U_{coll} = -27 kV.

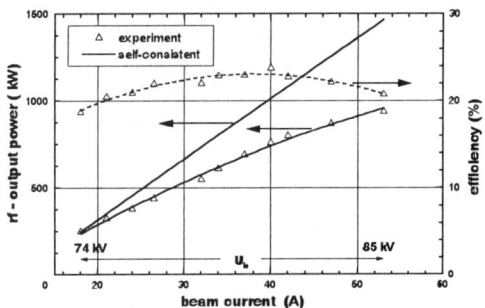

Fig. 1: Output power and efficiency vs. beam current for constant magnetic field. U_b optimized for maximum output power. Self consistent calculations performed with experimental parameters.

Table 1: Design parameters of the investigated gyrotrons at 140 GHz and 165 GHz. The real wall loading in the cavity is about twice the ideal value.

cavity	hollow waveguide	coaxial	
operating mode	$TE_{22,6}$	$TE_{28,16}$	$TE_{31,17}$
frequency f / GHz	140	140	165
rf output power P_{out} / MW	1	1.5	1.5
cathode voltage U_c / kV	80	90	75 - 80
beam current I_b / A	40	50	50
velocity ratio α	1.4	1.35	≈ 1.1
cavity radius R_{cav} / mm	15.57	29.81	27.38
beam radius R_b / mm	7.93	10	9.4
voltage depression ΔU_b / kV	5.8	1.6	1.5
peak wall loading (ideal copper) p_Ω / kWcm^{-2}	1.75	0.63	1.0

A maximum efficiency of 36% was measured for an rf output power of 0.83 MW at U_{cath} = 80 kV, I_b = 42 A and U_{coll} = -28 kV.

2.1. Experiments on slow frequency step tuning

The possibility of slow frequency step tuning over a wide frequency range as required for an advanced ECW start-up scenario has been demonstrated experimentally by variation of the magnetic field and appropriate change of the operating mode. The experimental results for 80kV/40A operation are summarized in Tab.2. The measured power values are corrected for reflections at the window which is transparent to 140 GHz. Values in parenthesis are uncorrected. In the case of large step frequency tuning a triode MIG would be advantageous because of the possibility to adjust the velocity ratio α independently.

Table 2: Frequency step tuning

frequency / GHz	117.9	140.1	162.3
mode	$TE_{-19,5}$	$TE_{-22,6}$	$TE_{-25,7}$
P_{out} / MW	0.75(0.44)	0.83(0.83)	0.97(0.61)
η / % {with SDC}	23 {33}	26 {36}	30 {42}
α	1.2	1.4	1.1

3. COAXIAL CAVITY GYROTRONS

The development is performed in two steps. In a first step, operated at short pulses ($\leq$ 0.5 ms), the gyrotron has an axial waveguide output with 100 mm diameter. The schematic layout is given in Fig.2. The maximum pulse length is limited by the heat load capability of the collector, which is part of the output waveguide. In that version operating problems have been examined and the cavity design has been verified. In a second step a tube design relevant for cw-operation with a radial rf output will be investigated. Especially because of the present power limit of rf output windows the mm-wave power will be split into two beams and coupled out radially through two windows. In addition, a single stage depressed collector will be used in order to enhance the total efficiency and to reduce the power loading at the collector surface. The measurements have been performed with two tubes designed for operation at 140 and 165 GHz in the $TE_{28,16}$ and $TE_{31,17}$ mode, respectively.

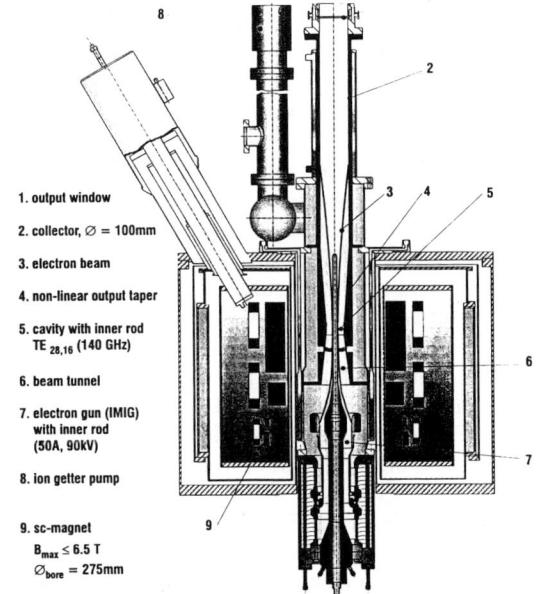

Fig. 2: Schematic layout of the coaxial gyrotron with axial rf-output

1. output window
2. collector, $\varnothing$ = 100mm
3. electron beam
4. non-linear output taper
5. cavity with inner rod $TE_{28,16}$ (140 GHz)
6. beam tunnel
7. electron gun (IMIG) with inner rod (50A, 90kV)
8. ion getter pump
9. sc-magnet $B_{max} \leq 6.5$ T $\varnothing_{bore}$ = 275mm

3.1. 1.5 MW, 140 GHz, $TE_{28,16}$ - coaxial gyrotron

For operation at 140 GHz the $TE_{28,16}$ mode with an eigenvalue χ=87.35 has been chosen [2,3]. The electron gun [4] is of the diode type and has a central rod close to the ground potential surrounded by the cathode (inverse magnetron injection gun, IMIG). The inner rod is electrically isolated but is held close to the ground potential. It is fixed and cooled from the gun side. It reaches up to about 25 cm above the resonator. The length is mainly determined by considerations of voltage depression.

The part of the inner rod within and above the cavity can be replaced and it can be radially adjusted during operation.

For the resonator a cavity with cylindrical outer wall and a radially tapered and longitudinally corrugated inner rod is used (Fig.3). Due to the tapering and impedance corrugation of the inner rod only the azimuthal neighbours with the same radial index such as $TE_{27,16}$ and $TE_{29,16}$ which have almost the same caustic, remain as serious competitors [5]. The impedance corrugation implies no mode conversion because the period of the corrugation is small compared to $\lambda/2$, the free space half wavelength. The ohmic losses on the inner rod are only about 10% of the cavity wall losses and are, therefore, not considered to be a technical problem.

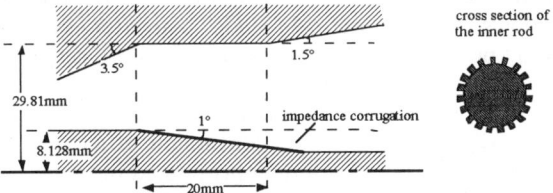

Fig.3: Geometry of the $TE_{28,16}$-coaxial cavity with tapered and longitudinally corrugated inner rod.

Measurements with encouraging results have been performed. The available power supply delivered a maximum voltage of about $U_c \approx 86$ kV at $I_b = 50$ A. The shape of the high voltage pulse was characterized by an overshooting of about 4%. The desired working mode $TE_{28,16}$ was found to work stably over a wide parameter range. This proves the suppression of possible competing modes by the inner rod with impedance corrugations. A maximum rf-output power of 1.17 MW with an efficiency of 27.2% has been achieved in that mode in single mode operation at $U_c = 86$ kV, $I_b = 50$ A and $B_0 = 5.63$ T. The measured frequency of 139.96 GHz is close to the calculated value. The internal losses in the cavity, the output taper and the window are estimated to be 5.5%. Figure 4 shows as an example the measured and the numerically calculated rf output power versus U_c. The calculations have been performed with a multi-mode code using the operating parameters and the geometry of the cavity without any fitting and with an assumed velocity spread $\delta\beta_{\perp rms} = 6\%$. In agreement with the numerical calculations, the azimuthal neighbours $TE_{29,16}$ at 142.02 GHz and $TE_{27,16}$ at 137.86 GHz, which are the remaining competitors, limit the stability region of the working mode in the $U_c - B_0$ parameter space. The experimentally observed regions with single and multi-mode oscillation are indicated at the top of that figure. Single mode oscillation of the $TE_{28,16}$ mode is found within several kV of U_c. The regions of oscillations of the three measured modes agree well with the numerical predictions. The only difference is that the transition region is wider in the experiment than expected. According to calculations the rf power should rise up to $U_c = 87$ kV while the measured values reach a maximum around 84 kV and above about 86 kV the $TE_{27,16}$ mode is oscillating. In the region between $U_c = 84$ to 86 kV there is a small amount of the $TE_{27,16}$ mode present simultaneously with the $TE_{28,16}$ mode. This gives an explanation for the reduced efficiency and output power in that region. However, this multimoding is not predicted by the numerical calculations. The discrepancy is thought to be caused mainly by window reflections which support the competing $TE_{27,16}$ mode, since the rf window is optimized for 140 GHz and has a reflectivity of 10% at the frequency of the competitor. Another reason for the loss of single mode stability in higher voltages is overshooting of about 4% of the accelerating voltage (up to 3 - 4 kV) during a pulse. At the peak of the voltage the competing mode starts to oscillate, and after the voltage dropped again these oscillations may remain.

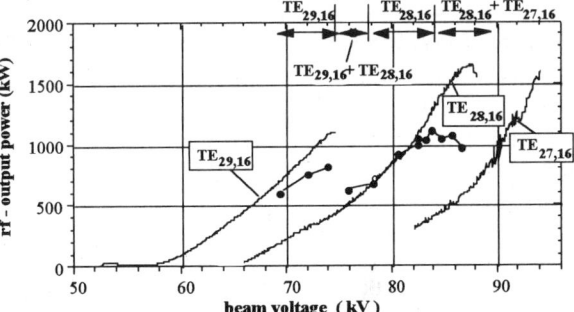

Fig.4 : Rf - output power versus beam voltage with $I_b = 50 - 52$ A and $B_0 = 5.62$ T. Experimental values (•) and calculations (—).

The possibility of step frequency tuning has been proven over a range from 115.6 GHz to 164.2 GHz (Fig.5). Frequency tuning was performed by changing the magnetic field with constant magnetic compression. Output powers around 1 MW with efficiencies above 25% have been achieved near frequencies where the window reflection is minimal (122 GHz, 140 GHz and 158 GHz). At frequencies with high window reflections beam instabilities limited the achievable α to very low values (<1) which resulted in the lower output power. Table 3 gives some examples of excited modes. At frequencies above the design value the maximum achievable velocity ratio is reduced due to the need of higher magnetic field and the diode type of the used gun.

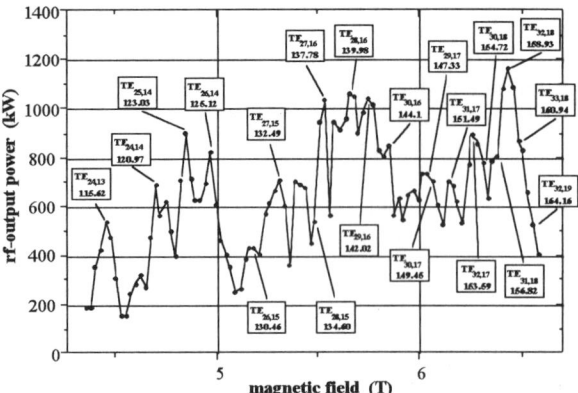

Fig. 5 : Rf-output power versus magnetic field. I_b = 50 - 52 A, R_b = 10 mm. Each point represents stable single mode operation of the indicated mode.

Table 3: Step frequency tuning of $TE_{28,16}$ gyrotron

P_{out} / MW	f / GHz	mode	B_0 / T	η_{tot} / %
0.9	123.03	$TE_{25,14}$	4.84	24.9
0.71	132.49	$TE_{27,15}$	5.31	17.6
1.17	139.98	$TE_{28,16}$	5.63	27.2
1.16	158.93	$TE_{32,18}$	6.43	26.2
0.83	160.94	$TE_{33,18}$	6.50	19.4

3.2. 1.5 MW, 165 GHz, $TE_{31,17}$ - coaxial gyrotron

The two following limitations restricted the selection of the operating mode for the 165 GHz-coaxial gyrotron: (1) The maximum magnetic field ($B_{max} \leq 6.5$ T) of an already existing superconducting (sc) magnet and (2) the use of the electron gun designed originally for the 140 GHz, $TE_{28,16}$ - coaxial cavity gyrotron. Due to the limitations the design had to be performed for reduced beam parameters (U_c = 75 kV; $\alpha \leq 1.1$). The design of the cavity is similar to that of the $TE_{28,16}$ gyrotron (Fig.3). An output window with a transmission adjusted to the 165 GHz has been used. The main design parameters are summarized in Tab. 1.

The maximum delivered output power P_{out} in the $TE_{-31,17}$ mode was 1.17 MW at U_c = 86 kV and I_b = 51 A with B_0 = 6.63 T corresponding to a total efficiency of 26.7%. The maximum output efficiency was 28.2% at P_{out} = 0.9 MW. The corresponding transverse efficiency is above 60% since the operating velocity ratio α was relatively low (<1). The experimental results have been compared with numerical simulations. The principle behaviour is similar as in the $TE_{28,16}$ gyrotron. The tube was also operated at higher frequencies and the maximum power achieved is given in Tab. 4. It is important to mention the megawatt level output power (1.02 MW) at 167.14 GHz frequency with an efficiency of 26.8%.

Table 4: Step frequency tuning of $TE_{31,17}$ gyrotron

mode TE_{mp}	$TE_{31,17}$	$TE_{32,17}$	$TE_{33,17}$	$TE_{34,17}$
f (GHz)	164.98	167.14	169.46	171.80
B_0 (T)	6.63	6.62	6.62	6.62
U_c (kV)	86	76	70	57
I_b (A)	51	50	50	46
P_{out} (MW)	1.17	1.02	0.63	0.35
η_{tot} (%)	26.7	26.8	18.0	13.3

3.3. Coaxial gyrotron with radial output

In order to obtain results relevant for a technical design as a next step a tube with a radial dual rf - beam output and a single - stage depressed collector (SDC) has been designed and is now under construction. For the operation of the two rf - output windows a quasi-optical mode converter system based on a two - step mode conversion scheme, $TE_{-28,16}$ to $TE_{+76,2}$ to TEM_{00} which generates two narrowly directed ($60°$ at the launcher) output wave beams will be used. High conversion efficiency is expected [8]. Operating problems related in the axial version to window reflections are expected to be significantly reduced in the tube with quasi-optical output. The operation may be performed with pulses up to about 20 msec thus reducing the problem of voltage overshooting which is related to short pulse operation.

ACKNOWLEDGEMENTS

The work was supported by the European Fusion Technology Program under the Project Kernfusion of the Forschungszentrum Karlsruhe.

REFERENCES

1. Makowski. M., 21th Int. Conf. on Infrared and Millimeter Waves, Berlin, Germany, 1996, ISBN 3-00-000800-4, paper AW1.
2. M. Thumm, et al., 20th Int. Conf. on Infrared and Millimeter Waves, Lake Buena Vista (Orlando), Florida, USA, 1995, 199-200.
3. Piosczyk, B. et al., ibid, 423-424.
4. Lygin, V.K.et al., 1995, Inter. Journal of Electronics, **79**, 227-235.
5. Iatrou, C.T. et al., 1996, IEEE Transactions on Microwave Theory and Techniques, **41**, 56-64.
6. Kern, S. et al., 20th Int. Conf. on Infrared and Millimeter Waves, Lake Buena Vista (Orlando), Florida, USA, 1995, 429-430.
7. Iatrou, C.T. et al., ibid, 415 - 416.
8. Thumm, M., et al., 21th Int. Conf. on Infrared and Millimeter Waves, Berlin, Germany, 1996, ISBN 3-00-000800-4, paper AT6.

Development of 170GHz long pulse gyrotron with depressed collector

A.Kasugai[a], K.Sakamoto[a], M.Tsuneoka[a], K.Takahashi[a], S.Maebara[a], T.Imai[a], T.Kariya[b], K.Hayashi[c], Y.Mitsunaka[c] and Y.Hirata[c]

[a] Naka Fusion Res. Est., Japan Atomic Energy Research Institute (JAERI), Naka-machi, Ibaraki-ken, 311-01 Japan

[b] Electron Tubes & Device Division, Toshiba Co., Ohtawara-shi, Tochigi-ken, 329-2 Japan

[c] R&D Center, Toshiba Co., Ukishima, Kawasaki-shi 210 Japan

A 170 GHz gyrotron with a single stage depressed collector was developed for high power and long pulse operation. Its oscillation mode is TE31,8 volume mode, and power profile at output window is flat to reduce heat deposition on the output window. At initial experiment, the output of 520 kW-0.6 sec and 230 kW-2.2 sec were obtained, and the maximum of improved efficiency is 38%.

1. INTRODUCTION

Electron cyclotron heating (ECH) and current drive (ECCD) are attractive methods for a fusion reactor since a launching system can be located away from the plasma. In ITER (International Thermonuclear Experimental Reactor), a gyrotron of high efficiency with 170GHz/1MW/CW is required for ECH and ECCD [1]. In JAERI (Japan Atomic Energy Research Institute), the development of 170GHz/1MW/CW gyrotron is being carried out as the R&D Task under ITER-EDA(Engineering Design Activities).

We have already developed 170 GHz short pulse gyrotron prior to the development of a long pulse gyrotron in 1995 [2]. A stable, 1.13 MW single mode oscillation was obtained at 170GHz in the short pulse gyrotron with a very high order mode TE31,8. The maximum efficiency was 30%, and no power degradation due to the mode competition was observed. Based on these results, we fabricated a 170GHz long pulse gyrotron with depressed collector (so called energy recovery system), and started development of a ITER gyrotron with 170GHz/1MW/CW.

In this paper, we report the design and experimental results of the 170 GHz gyrotron with depressed collector.

2. DESIGN

Main design parameters of the 170 GHz gyrotron are summarized in Table 1, and the picture is shown in Fig. 1. The dimension is 3.0 m in height and 650 kg in weight. An oscillation mode is TE31,8 and its cavity is the same one as the short pulse gyrotron demonstrated the maximum power of 1.13 MW. Figure 2 shows the calculated result of the beam current (I_b) dependence of the output power (P_{out}) from the gyrotron, the minimum energy of spent beam (V_{CPD}) and the efficiency with depressed collector (η_{total}) for TE31,8 mode. The calculation was done using a cold beam of $\alpha=1.0$ at V_b (beam voltage) = 80 kV. V_{CPD} is the maximum voltage of depressed collector. The cavity is designed to oscillate at a relatively

Table 1
Main design parameters of the 170 GHz gyrotron.

Frequency	170 GHz
RF output power	1 MW
Cavity power	1.1 MW at 80 kV, 45 A
Pulse duration	CW
Oscillation mode	TE31,8
Beam voltage	70 ~ 90 kV
Beam current	< 30 A (at long pulse)
Pitch factor (α)	1.0
Q-value of cavity	2135
Cavity radius	13 mm
Cavity length	15 mm
Output mode	Flat power profile
Output window	Sapphire (double-disk)
(diameter)	140 mm
Coolant	FC-75
Depressed collector	Single stage ~ 40 kV
Efficiency	Max. 30 % at cavity >50 % with D/C
Collector cooling	Water circulation (with beam sweep)
(Inner diameter)	320 mm
Height	3.0 m
Weight	650 kg

low value of pitch factor, i.e. $\alpha \sim 1.0$. Therefore, the Q-value of the cavity is relatively high, Q~2135. Though the oscillation efficiency is relatively low because of low α, the high level of the total efficiency (>50%) can be obtained by the application of the large voltage of depressed collector.

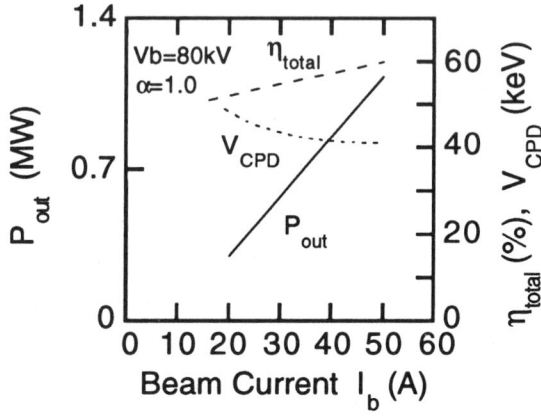

Figure 2. Calculation of an oscillation at cavity.

Oscillated mode TE31,8 is converted to flat profile by an adiabatic radiator (single helix in-waveguide type) and seven reflection mirrors installed inside the gyrotron, which include two phase correlation mirrors. As a result, heat deposition on the output window can be reduced due to decreasing the power density. A calculated result of the mode conversion indicates reduction of the peak power density from 705 MW/m^2 (Gaussian profile) to 155 MW/m^2 (flat profile) and a good form of a flat power profile as shown in Fig. 3. The plane and 3D figures of the power profile in linear scale are shown in (a) and (b), respectively. The contour indicates every 10 % of the peak power density. The 93 % of the oscillation power are included in the flat profile at the output window.

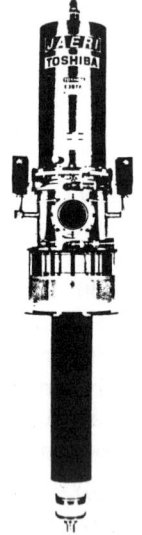

Figure 1. Picture of 170 GHz gyrotron.

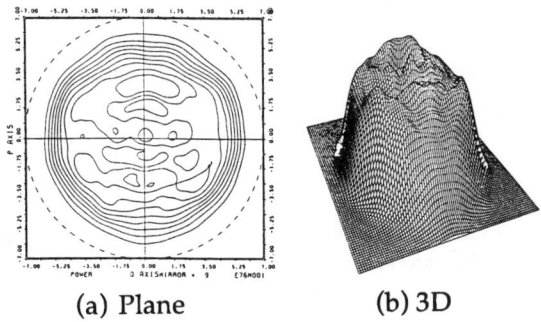

(a) Plane (b) 3D

Figure 3. Calculation of power profile at output window.

An output window is a sapphire double-disk window with face cooling in the gap between two disks by FC-75 coolant. The diameter is 140 mm and the thickness is 3.45 mm. This window is capable for 500kW/several second operation at 170GHz with the flat power profile.

A large ceramic insulator is installed between the body section and the collector. The body section is covered with an insulation jacket made of epoxy regine for application of the voltage of depressed collector up to 50 kV, which improves the gyrotron efficiency.

3. TEST RESULTS

A short pulse experiment was carried out at first. A small dummy load consisting of SiC for a short pulse (< 100 ms) was mounted at the output window directly. RF output power was measured calorimetrically from the temperature increase of the cooling water in the dummy load. Measured frequency by a spectrum analyzer was 170.2 GHz at $V_b = 91$ kV, $I_b = 30$ A and B_c (cavity magnetic field) = 6.90 T. Its frequency indicates an aimed mode TE31,8. Figure 4 shows a beam current dependence of the RF output power, the output efficiency at $V_b \sim 91$ kV, $V_{CPD} \sim 30$ kV and ~50 msec oscillation. The magnetic field at the cavity was optimized for each current. The output power is proportional to the beam current, and the output efficiency is nearly constant at ~19 %.

Oscillation efficiency at the cavity become to ~22 % from considering mode conversion (0.93), window loss (0.95) and ohmic loss (0.985). The oscillation efficiency agrees fairly well with the efficiency (~24 % at $I_b = 30$ A) obtained from the previous experiment of short pulse gyrotron and the calculated efficiency of the oscillation (25 % at $I_b = 30$ A). Figure 5 shows a power profile at distance of ~30 cm from the output window measured by infrared camera. Fairly good flat power profile is obtained.

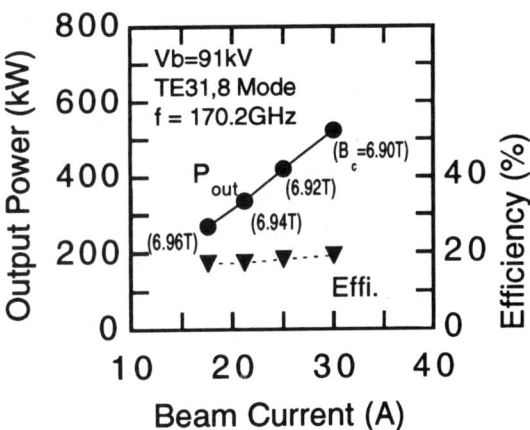

Figure 4. Experimental result of 170 GHz gyrotron. (Output power and efficiency vs. beam current)

After short pulse experiment, the depressed collector voltage and the beam voltage were optimized for improvement of the efficiency. As a result, the maximum efficiency of 38 % was obtained by applying $V_{CPD} = 38$ kV at $V_b = 72$ kV, $P_{out} = ~460$ kW and $I_b = 32$ A.

In a pulse extension (>0.1 sec), RF power is injected into a metal dummy load via a matching section composed of two mirrors, the corrugated waveguide of 88.9 mm in diameter

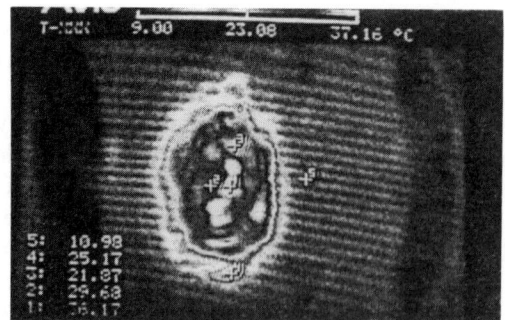

Figure 5. Measurement of power profile.

and two miter bends. A pulse duration was extended up to 0.6 sec. The power is ~520 kW at $I_b = 30$ A and $V_{CPD} = 34$ kV. Dependence of the temperature of the output window and the ceramic insulator on the pulse duration are shown in Fig. 6. Temperature of the window and the ceramic insulator are denoted by closed circles and open circles, respectively.

After 0.6 sec in pulse duration, the window temperature increased from 10 °C to 90 °C. The temperature of the ceramic insulator installed between the body section and the collector for depressed collector increased 10 °C after 0.4 sec. Reinforcement of cooling at the ceramic insulator is needed for further pulse extension.

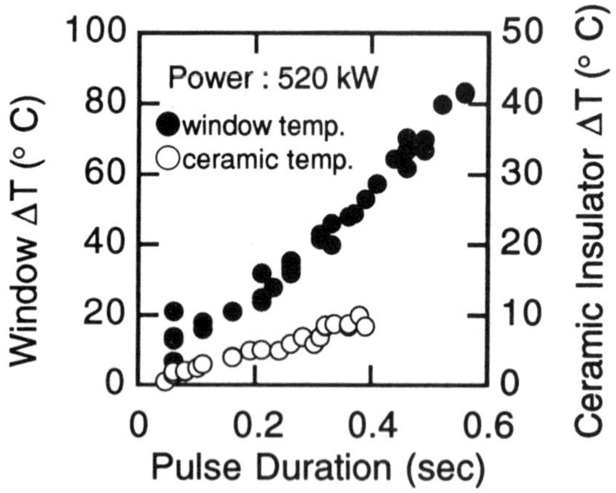

Figure 6. Temperature increase of output window and ceramic insulator.

The maximum pulse duration of 0.6 sec at 520 kW is restricted by the window temperature. The local peaking of the RF power profile was observed, which caused large temperature increase.

After the operation of 520 kW - 0.6 sec, we decreased the output power level to ~230 kW and the pulse duration was also extended up to 2.2 sec. The window temperature limited the pulse extension. The local peaking on the power profile is also recognized. Finally, the 170 GHz gyrotron was overhauled. The deformation due to the thermal stress was observed on the waveguide wall at the downstream of the cavity, which caused a undesired mode conversion. This must be the cause of the local peaking of the power profile in the long pulse operation.

4. SUMMARY

The 170 GHz gyrotron with depressed collector was designed and tested. The maximum power of 520 kW (0.6 sec, 32 %), the maximum efficiency of 38 % (460 kW, 50 msec) and the maximum pulse duration of 2.2 sec (230 kW, 32%) were obtained in the gyrotron tests. In regard to the oscillation at cavity, experimental result agreed fairly well with the calculated one. For the development of ITER gyrotron with 170 GHz/1 MW/CW and high efficiency, improvement of the output window and the mode conversion system will be carried out in JAERI.

REFERENCES

1. M.Makowski, Digest of the 21st International Conference on Infrared and Millimeter waves, Berlin, Germany (1996).
2. K.Sakamoto, A.Kasugai, K.Takahashi, M.Tsuneoka, T.Imai, T.Kariya and K.Hayashi, Journal of Physical Society of Japan, 65(1996) 1888-1890.

Development of an ECRH System for Large Helical Device

T. Shimozuma, M. Sato, Y. Takita, S. Kubo, H. Idei, K. Ohkubo, T. Watari,
S. Yasutomi[a], Y. Suzuki[a], F. Saito[a], S. Sasaki[a], Y. Saito[a] and T. Okamoto[a]

National Institute for Fusion Science, 322-6, Oroshi-cho, Toki-shi, Gifu, 509-52, Japan

[a]Toshiba Corporation, Uchisaiwai-cho, Chiyoda-ku, Tokyo, 100, Japan

84GHz and 168GHz ECRH systems for Large Helical Device (LHD) are described. This system consists of high power gyrotrons, their high voltage power supplies, millimeter-wave transmission lines and antenna system. Long pulse and CW operations of the 84GHz gyrotron were achieved on 500kW-100kW level for steady state plasma production in LHD. A 168GHz system and long transmission lines are being constructed. High power millimeter wave components such as a dummy load, vacuum windows and quasi-optical antenna handling 1MW level power are also designed, fabricated and high-power tested.

1. ECRH SYSTEM FOR LHD

Electron cyclotron resonance heating (ECRH) is one of the most important methods of plasma heating and production for tokamaks and stellarators. LHD is being built at National Institute for Fusion Science (NIFS) in targeting to begin experiments by 1998. It requires 10MW 84GHz and 168GHz ECRH system for initiating, heating and controlling plasma with 10keV and $10^{20}m^{-3}$ electron temperature and density, respectively.

The ECRH system for LHD consists of high power gyrotrons, their high voltage power supplies, millimeter-wave transmission lines, quasi-optical components and antenna system. The details of the system are listed in Table 1. Gyrotrons which we prepare at present are 84GHz gyrotrons with 0.5 - 1MW output for plasma production and steady-state experiments and 0.5MW, 5 seconds 168GHz gyrotrons for high temperature, high density plasma heating.

High power millimeter-waves generated in the gyrotrons are coupled into HE_{11} corrugated waveguides by matching optics unit (MOU) which includes some phase correction mirrors and transmitted over about 100m through some bends, polarizer and waveguide switch. By this waveguide switch, we can change transmission pass to LHD or a dummy load. Through a vacuum barrier window, millimeter-waves are introduced into the LHD vacuum vessel and focused around an equatorial plane by a quasi-optical antenna system.

In the following sections, we describe the designs and obtained test results of each component in detail.

Table 1 High power gyrotrons and power supplies in the ECRH system

Frequency	Power/tube	Duration	Objectives	Cutoff density
84GHz	0.5-1MW	10sec - CW	•Plasma production •Fundamental heating	$8.8 \times 10^{19} m^{-3}$ (O-mode) $4.4 \times 10^{19} m^{-3}$ (X-mode)
168GHz (CPD)	0.5MW	5sec	•Second harmonic heating •High density plasma production	$3.5 \times 10^{20} m^{-3}$ (O-mode) $1.8 \times 10^{20} m^{-3}$ (X-mode)

2. AN 84GHZ HIGH POWER CW GYROTRON

The 84GHz, CW (Continuous Wave) gyrotron consists of a magnetron injection gun, beam tunnel, tapered cavity, nonlinear uptaper, built-in mode converter, collector and output window (FC-75 cooled double-disk window), as shown in Fig. 1. Each component is sufficiently water-cooled and the temperatures at inlets / outlets of the cooling water channels are independently monitored to estimate power losses of an electron beam and electro-magnetic waves at the individual parts.

In this gyrotron, the electromagnetic waves ($TE_{15,3}$ mode), excited in the resonator, are separated from a spent electron beam by a built-in mode converter which consists of a modified Vlasov-type launcher with a visor and beam shaping mirrors. The output mode is converted from excited $TE_{15,3}$ mode in the cavity to a Gaussian-like beam.

The launcher has a visor section on the location of the straight edge in conventional Vlasov one. The visor deforms the beam power profile to the Gaussian on the first mirror, which corrects the beam phase to generate a nearly plane wave. The second mirror, which has an elliptic like shape, focuses the beam to couple to the output window of 4-inch diameter. The converter was designed on the basis of the geometrical optics[1].

Figure 1. Photograph of the 84GHz CW gyrotron.

The radiation patterns on the window was well agreed with the cold test results. The measured diffraction and spill over losses were estimated to be 7.6% in the tube during high power operation[2].

Following test results were obtained.
(1) 500 kW 2 sec.
(2) 400 kW 10.5 sec.
(3) 200 kW 30 sec.
(4) 100 kW CW (30 min.).
(5) 50kW CW (> 1 hr.).

The maximum power at the long pulse operation was limited by window temperature rise. The window peak temperatures during pulses are plotted in Fig. 2. The peak temperature of the window nearly saturated below the boiling point of FC-75 for 400kW and 240kW cases. For 500kW case, however, the peak temperature increased linearly with pulse and we stopped further operation. The gas pressure increase prevented the high duty or CW operations in higher power than 200kW. Bigger ion pumps are required for CW operation.

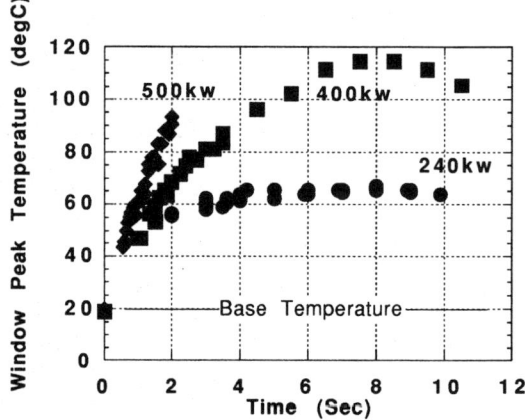

Figure 2. Window peak temperature during pulses.

3. 168GHZ CPD GYROTRONS AND POWER SUPPLIES

The 168GHz gyrotrons have an ability of collector potential depression (CPD), in which spent electron beams impinge on the collector against retarding potential and the total gyrotron efficiency is improved by $V_b/(V_b-V_r)$, where V_b and V_r are the beam voltage and retarding voltage, respectively [3].

The power supplies for 168GHz gyrotrons are all solid-sate. They consist of collector, body and anode power supplies independently. The collector power supply, which has no crowbar circuits, is roughly regulated with GTO switching valves and drives three gyrotrons (60kV, 126A 10sec., 42A CW). Each gyrotron has a body and anode power supplies which are highly regulated with high voltage, low current ability (90kV, 100mA and 50kV 50mA, respectively) and which can be turned off within 10μsec when arcings occur in the gyrotron. This system is working without any damages in gyrotron.

4. HIGH POWER MILLIMETER-WAVE COMPONENTS

4.1. Vacuum windows

Development of millimeter-wave vacuum barrier windows is one of the most important subject to accomplish windows of both high power, CW gyrotrons and ECRH systems of LHD.

Recently, Kyocera Corporation (in Japan) developed new material (silicon nitride composite so called SN-287), which shows low loss tangent at low frequency ranges (7-60GHz). The silicon nitride has higher thermal shock resistance, higher flexural strength and better thermal conductivity than sapphire which is usually used for a gyrotron window.

We measured the loss tangent of this material by observing temperature rises of a disk during high power (80kW level) millimeter-wave transmission from the 84GHz gyrotron. Assuming the rf profile on the disk to be gaussian with the beam waist radius of 18.7mm which is a designed value, we compared observed behaviors of temperature rise with calculated ones and determined the loss tangent. At room temperature, it is estimated to be 1.5×10^{-4}, which is comparable to sapphire. The temperature dependence of the loss tangent is weak and proportional to T[4].

Low power measurements of the loss tangent were performed on another frequency range (140-145GHz) with a sophisticated high Q Fabry-Perot resonator in collaboration with FZK in Germany. The loss tangent in this frequency range was assured to be 2.4×10^{-4}. Figure 3 summarizes the frequency dependence of the loss tangent of the silicon nitride composite. Frequency dependence of tanδ is weak (~ $f^{0.3}$).

Using this material, we fabricated edge-cooled single-disk window with an effective radius of 44.45mm and tested it by 84GHz high power gyrotron. Window temperature increased linearly with average injection power. When 75kW CW power was injected, the peak temperature of the window completely saturated at 230 deg.C. This fact suggests it is promising as a long pulse window material.

Figure 4 shows a fabricated single-disk Si_3N_4 window for a 168GHz gyrotron in which the ceramic disk was brazed in a copper cylinder directly.

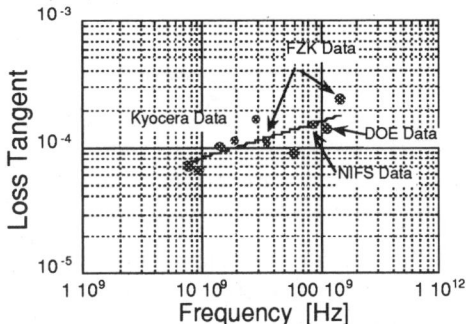

Figure 3. Frequency dependence of Si_3N_4 loss tangent.

Figure 4. Fabricated single-disk Si_3N_4 window for a 168GHz gyrotron.

4.2. High Power CW Dummy Load

During 84GHz high-power long-pulse and CW testings, the output power from the gyrotron was terminated and measured by a water-cooled dummy load that consists of a new silicon nitride (Si_3N_4) composite ceramic of which dielectric loss is lower than alumina as described in previous subsection. The size of this ceramic cylinder is 170 mm in diameter and 215 mm long. We used two sections of this cylinder.

The structure of the dummy load is illustrated in Fig. 5. Millimeter-waves entered into the dummy load are scattered by the end reflector, penetrate the silicon nitride cylinder and absorbed

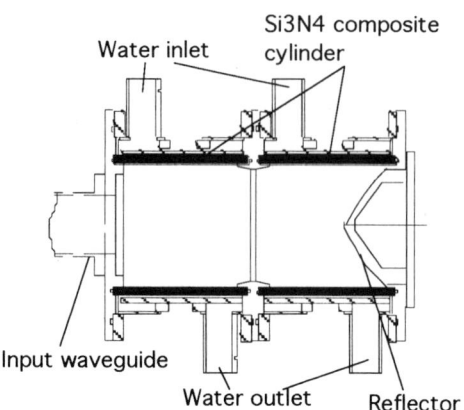

Figure 5. New high power dummy load with Si_3N_4 composite material.

directly by water. Millimeter-wave power is calculated by a temperature difference of flowing water at an inlet and outlet and the water flow rate.

4.3. Long Distance Transmission Lines and Antenna System

Long distance transmission lines consist of straight HE_{11} mode corrugated waveguides with inner diameter of 88.9mm and some bends. The period, width and depth of corrugation are 0.8, 0.6, 0.6mm, respectively. Low power tests of long distance transmission over 62m showed that the attenuation along the corrugated waveguides is lower than measurable limit (2dB/km) [5].

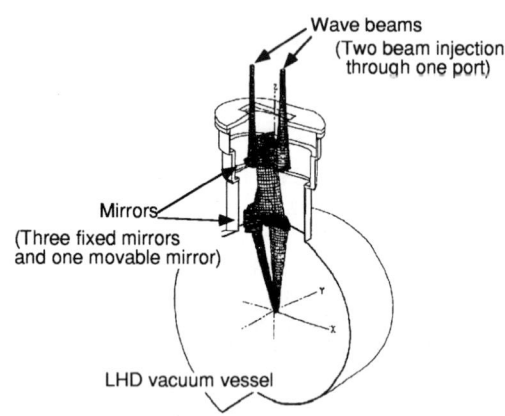

Figure 6. Quasi-optical antenna system.

In-vessel quasi-optical antenna system comprise four mirrors, one of which is movable and the rest are fixed. Launched beams are focused on the equatorial plane of the torus in elliptical shapes with waist sizes of 15 and 50 mm. Injection angle is adjustable by 15 degrees along major radius and 5 degrees along toroidal axis. The configuration of the mirrors and wave beams are shown in Fig. 6. Two wave beams are guided from one vertical port of LHD vacuum vessel.

5. SUMMARY

We have been developing an ECRH system for LHD which consists of high power gyrotrons, their high voltage power supplies, millimeter-wave transmission lines and antenna system. Main power sources are 84GHz CW gyrotron and 168GHz long-pulse gyrotrons. On the 84GHz gyrotron, we obtained 500kW 2sec., 400kW 10.5sec., 200kW 30sec. and 100kW CW output. We also designed and fabricated long low loss corrugated waveguide, a high power dummy load, vacuum windows, some quasi-optical components and antenna system. Testings combining each component will be started successively in this year.

ACKNOWLEDGMENTS

The authors would like to acknowledge collaboration with Drs. M Thumm and R. Heidinger of FZK and Drs. C. M. Loring, Jr., K. Felch, T.S. Chu and P. Borchard of CPI for production of 84GHz gyrotrons and also express appreciation to Professors A. Iiyoshi and M. Fujiwara of NIFS for their continuous encouragement.

REFERENCES

1. T. Shimozuma, M. Sato, Y. Takita, S. Kubo, et al., Proceedings on the 19th International Conference on infrared and millimeter-waves, Sendai, Japan, Oct. 17-21,1994, p65-66.
2. M. Sato, T. Shimozuma, Y. Takita, S. Kubo, et al., Proceedings on the 20th International Conference on infrared and millimeter-waves, Orlando, Florida, Dec. 11-14,1995, p195-196.
3. K. Sakamoto, M. Tsuneoka, A. Kasugai, T. Imai, et al., Phys. Rev. Lett., vol. 73 (1994) p3532-3535.
4. T. Shimozuma, M. Sato, Y. Takita, S. Kubo, et al., Proceedings on the 20th International Conference on infrared and millimeter-waves, Orlando, Florida, Dec. 11-14,1995, p.273-274.
5. K. Ohkubo, S. Kubo, M. Iwase, H. Idei, et al., J. Infrared Millim. Waves, Vol.15, No.9, p.1507-1519 (1994).

New Developments in the 110 GHz System on the RTP Tokamak

O.G. Kruijt, W.A. Bongers, A.B. Sterk, F.J. van Amerongen, A. Bijker, G.G. Denisov[*], W. Kooijman, S. Kuyvenhoven, G. Land, A. Montvai, A.A.M. Oomens, R.W. Polman, F.C. Schüller, A.G.A. Verhoeven, D. Vinogradov[*] and the RTP-team

FOM-Instituut voor Plasmafysica 'Rijnhuizen', Association EURATOM-FOM
P.O. Box 1207, 3430 BE Nieuwegein, The Netherlands

[*] Institute of Applied Physics, Nizhny Novgorod, Russia

Since April 1994 a 110 GHz 500 kW gyrotron is in use at the RTP tokamak. Some of the operating experience is given. The characteristics of a new launcher with the options of sweeping or focusing the deposition region are described, as well as a facility for fast modulation of the gyrotron output power. The microwave beam pattern has been diagnosed with a liquid crystal foil and with an x-y scanner. A comparison of the two methods is given. Finally the results of low-power measurements on a diamond disk, to be used as window material for high power microwave transmission, are discussed.

1. INTRODUCTION.

The research programme of the Rijnhuizen Tokamak Project RTP (R_0 = 0.72 m, a = 0.164 m, boronized vessel, $B_T \leq$ 2.5 T, $I_p \leq$ 150 kA, pulse duration $\leq$ 600 ms) concentrates on two issues: i) physics of transport processes and ii) the interaction of millimetre waves with tokamak plasmas. For this purpose the device is equipped with an comprehensive set of diagnostics, a pellet injector and an ECRH system. The ECRH system consists of two 60 GHz, 200 kW, 0.1 s VARIAN gyrotrons (used for ECRH and ECCD at the first harmonic of the frequency) and one 110 GHz gyrotron (500 kW, 0.2 s) delivered by GYCOM, Nizhny Novgorod, Russia [1]. The latter one is used for heating and current drive studies at the second harmonic frequency. The first results have been reported in [2].

Two years of experience consisting of > 10^4 pulses learned that this gyrotron is a reliable research tool. It can be conditioned quickly and the reproducibility of the gyrotron output characteristics is very good. There was once a failure of the superconducting magnet, which was repaired in a short time. The quasi-optical transmission line needs hardly attention: Only the mirror opposite the tokamak window, at which the field strength is high, needs a polishing cloth now and then.

2. THE LAUNCHER

Until recently the EC beam at RTP was launched radially in the midplane from the outboard side using a fixed mirror in front of the tokamak window. Lately the launching system has been extended with a set of two in-vessel mirrors (Fig. 1).

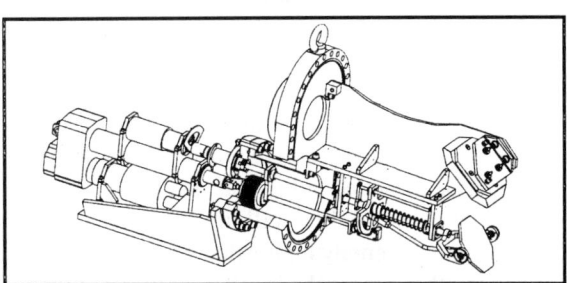

Figure 1. The launcher mounted on the vacuum flange. The RF beam enters from the left, the plasma boundary is 2-3 cm on the right side of the mirrors.

The first is a stationary, concave mirror and the second one is planar and adjustable. The flat mirror allows deposition over the poloidal cross section for poloidal angles in the range -25 to +25 degrees. Toroidally the range of rotation is limited to +/- 20 degrees off-perpendicular by the walls of the diagnostic port, leading to a tangent radius of 0.32 m. Optimum launching in toroidal direction requires elliptical polarisation of the beam.

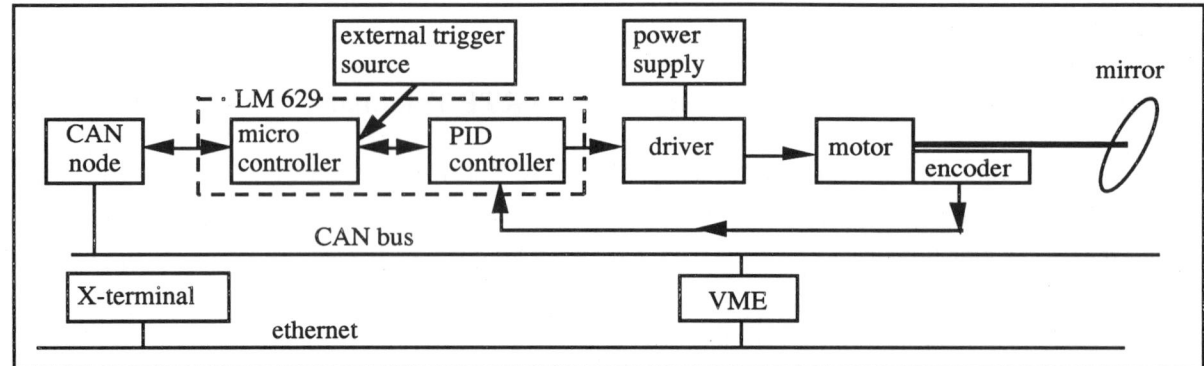

Figure 2. Overview of the mirror control circuit

Therefore an elliptical polarizer, consisting of two grooved, rotatable mirrors, will be added later.

The plane mirror can rotate around two perpendicular axes. by means of two DC servomotors. The motors are placed outside the torus vacuum to facilitate maintenance. The objective is to sweep the mirror, and therewith the deposition of the RF power, over 20 degrees within 0.1 sec. An adequate acceleration is guaranteed by the motors of 400 W peak power and a 24:1 reduction gear, which results in a peak torque of 6.6 Nm, thus controlling the 1000 Nmm^2 moment of inertia. The vacuum force from the bellows feed-through is nullified by a constant force tensator spring.

A second set of mirrors is being prepared to be installed after the campaign with the adjustable launcher. This second set consists of two fixed mirrors, thus constructively allowing a much larger last mirror. With that concave mirror a smaller beam waist can be achieved at the plasma centre: ≤ 5 mm instead of 9 mm. This will increase the power density to an extremely high value which can greatly influence the research in the field of plasma filamentation [3].

2.1 Launcher control system

The control system was developed to meet the following requirements: The mirror should be remotely adjustable with a position accuracy better than one degree and should be able to move, during a plasma pulse, along a pre-programmed path. This movement is started by a trigger derived from the plasma diagnostics as a first step towards active plasma instability control. Fig. 2 shows an overview of the launcher control circuit. In the first stage the operator pre-programs the desired beam path on an X-terminal. A beam path dictated by plasma diagnostics is an option for a later stage. The path is converted to a listing of mirror settings which are downloaded to the buffer of a microcontroller via a VME bus, a CAN bus and a CAN node. After an external trigger, the buffer content is fed sequentially to the PID controller. Each mirror action starts with a position check of the encoders. The motor control hardware is built around an LM629 IC, containing the micro controller and the PID controller as well as the interpreter and memory for the encoder signals. The pre-programmed and the resulting path of the mirror are kept in RAM. The measured sweep speed and position accuracy of the mirror are well within the design objectives.

3. POWER MODULATION

Since the 110 GHz gyrotron is a diode-like source, its RF output power can be modulated only by varying the collector voltage. Measurements have shown that thus a modulation depth of 30% can beobtained by varying the collector voltage from 70 kV to 65 kV. The influence on the electron beam current is negligible. The 60 GHz gyrotrons require a constant 80 kV collector-voltage. To operate the 110 GHz and the 60 GHz gyrotrons simultaneously from a single power supply, a circuit is designed to modulate the 70 kV while the 80 kV remains constant. The principle of the circuit is given in fig. 3a. A fixed resistor of 500Ω is connected in series with the 110 GHz gyrotron to provide a voltage drop of 10 kV at a beam current of 20A. A second resistor of 250Ω can be short-circuited in steps to give

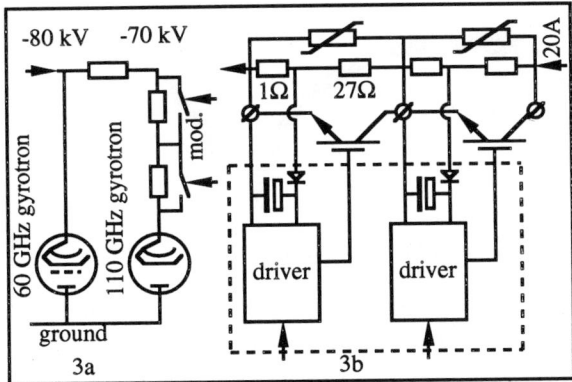

Figure 3. Gyrotron and series resistor (a) and one switch module (b)

additional voltage drops of respectively 1.7 kV, 3.3 kV and 5 kV. The switches across the resistors are optically controlled and consist of IGBT modules of 1200 V/ 75 A each. For reasons of voltage hold-off a number of switches are connected in series and switched on and off simultaneously. Because the switches are located at high-voltage level the driver circuits are self- powered. To limit overvoltages during an arc in the gyrotron the IGBT's are protected with MOV's. The circuit of one switch-module and its corresponding drivers is given in fig. 3b. The switches are optically coupled to a pulse generator and can be individually controlled. In this way the RF power can be modulated by 9, 18 and 27% with a frequency range from DC up to 50 kHz.

4. DIAGNOSIS OF A HIGH POWER MICRO WAVE BEAM PATTERN.

To avoid problems associated with equipment located in a high power microwave beam, the following two methods to investigate the beam pattern are compared: - A temperature sensitive mylar liquid crystal (LC) foil was placed directly in the beam path. A CCD camera, placed outside the beam, looking at the LC-foil delivers its signal to a PC where a frame grabber takes a snapshot after a few high power pulses, when a temperature equilibrium is reached. The foil gives a full colour spectrum in a temperature range of 5°. The colour of the foil is a measure for the local temperature, hence for the absorbed power; - A coupler consisting of a thin polythene wrapping foil under 45° produced an image of the foil in the plane of an x-y translation stage with microwave detection equipment (fig. 4).

Figure 5 gives the two beam patterns with relative power levels. Fits of a Gaussian beam are better for the LC foil measurement than for the x-y scanned profile. This is caused by some large ripples on the scanned profile due to interference with scattered microwave power. The LC-foil is insensitive for this effect because its absorbed power, as calculated from the measurement, is 48 kW. In conclusion: The LC-foil measurement can be used to diagnose a high power beam if the ambient temperature of the foil is under control. A dynamic range of 1: 40 can easily be obtained with a standard video camera.

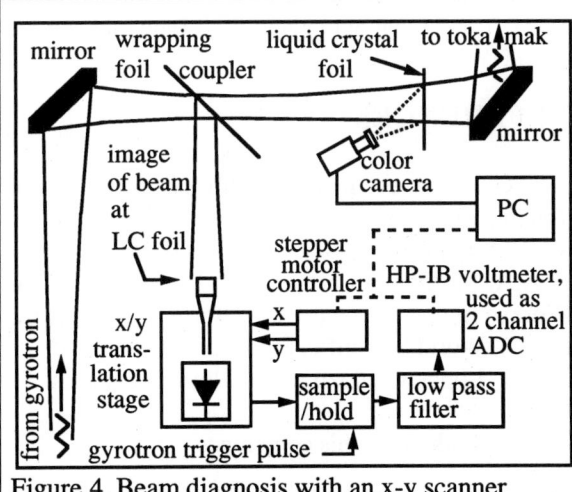

Figure 4. Beam diagnosis with an x-y scanner and with a liquid crystal foil.

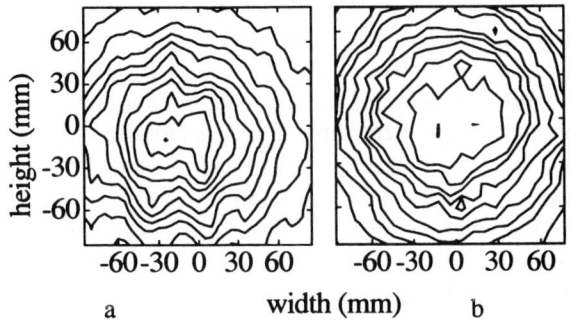

Figure 5. Contour plots of the micro wave beam pattern, measured with an x-y scanner (a) and with a liquid crystal foil (b). The power difference between the levels is 10%.

5. MEASUREMENTS ON A DIAMOND WINDOW FOR HIGH POWER MICROWAVE TRANSMISSION.

For CW broad-band tuneable high power microwave sources such as Free Electron Masers diamond can be an excellent window material because of its high thermal conductivity and its low loss. For the FEM currently being developed at Rijnhuizen a diamond vacuum window can be used in a Brewster angle configuration, by conventional cooling at room temperature [4].

Our low-power measurement set-up has the possibility of antenna pattern and reflection measurements in the range of 95-126 GHz. Variations in the output of the source, a HP-IB controlled BWO, are automatically compensated. Two couplers were installed to measure the forward and reflected power. Those signals were used to calculate the properties of the diamond windows on various locations of its area. Behind the sample was an x-y translation stage with a standard wave-guide antenna and detector to measure the antenna pattern over the frequency range to check for inhomogenities in the material. Patterns in the frequency range from the diamond window were compared with the patterns of the open wave-guide. Because the original pattern of the open wave-guide is very close to a Gaussian one, fits were done to compare radiation patterns with and without the diamond disk.

A model was simulated to interpret the reflection measurements. Thereby it was possible to calculate the refraction index n, the dielectric constant ε_r, the loss angle $\tan(\delta)$ and the loss factor α, of the diamond disk. The result is given in fig 6 for one location on the of the 95 mm diameter disk. The fits on the measurement show that ε_r is close to 5.6 and that $\tan(\delta)$ is lower than 5E-4 and close to 1.5E-4. This means a 0.2% loss factor for a 2 mm thick Brewster-angle window at 200 GHz. To prevent large errors this measurement was done on a location with very little thickness variation (<2 µm). Antenna pattern measurements were done on the wave guide output before and after the disk. Comparison of those patterns show a slight focusing effect, caused by surface roughness and inhomogenity of the diamond. In a Brewster angle application this will cause no negative effect.

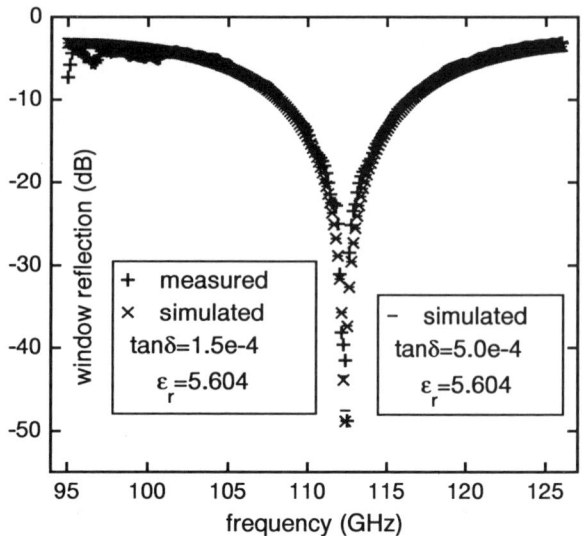

Figure 6. Example of a reflection measurement on the diamond disk. The measured curve (+) and the simulation (x) coincide.

In conclusion: we found diamond to be an excellent material for application in windows for high power transmission. Our findings were affirmed by recent high power measurements on the same diamond disk at the VARIAN, CPI gyrotron test stand in Palo Alto. The tests showed that the disk should be able to hold more than 1 MW CW.

ACKNOWLEDGEMENT

This work was performed under the Euratom-FOM Association agreement with financial support from NWO and FOM.

REFERENCES

1. O.G. Kruijt et al., Proc. 18th Symp. on Fusion Techn., Karlsruhe, Germany (1994), K. Hersbach et al. (editors), Elsevier Science B.V. (1995), p. 493
2. A.A. M. Oomens et al., Proc. of the Ninth Joint Workshop on ECE and ECH, Borrego Springs Ca. (1995), World Scientific, Singapore (1995) p. 85
3. N.J. Lopes Cardozo et al., Phys. Rev. Lett. **73** (1994) 256
4. A.B. Sterk et al., this conference.

Present status of the 1 MW, 130 - 260 GHz Free Electron Maser

A.B. Sterk, B.S.Q. Elzendoorn, W.A. Bongers, P. Manintveld, P.R. Prins, A.G.A. Verhoeven, W.H. Urbanus, the VE team and the FEM team

FOM-Instituut voor Plasmafysica "Rijnhuizen", Association EURATOM-FOM,
P.O. Box 1207, 3430 BE Nieuwegein, The Netherlands

This paper describes the present status of the 1 MW, 130-260 GHz Free Electron Maser for fusion applications (Fusion-FEM), which is being commissioned at the FOM-Institute "Rijnhuizen".

1. INTRODUCTION

The FEM is a new type of millimetre wave source for the next generation of tokamaks. The aim is to generate 1 MW of mm-wave radiation in the range 130 - 260 GHz. The pulse length will be 100 ms and the device will be tuneable over ±3% on ms-scale and over the full frequency range in less than a minute. The FEM is driven by a 12 A, 2 MeV electron beam, which is guided through a periodic magnetic field structure, called the undulator. Electrostatic acceleration is used. In the undulator, part of the electron beam power is converted into mm-wave power. After passing the undulator the beam is decelerated and collected in a multi-stage depressed collector. This concept results in a system efficiency (grid to mm-wave power) of over 50% [1].

2. LAYOUT OF THE SYSTEM

A schematic layout of the Fusion-FEM is shown in fig. 1. Because of the high dc accelerating voltage, the entire system is placed in a vessel filled with SF_6 at 7 bar. The cathode of the electron gun is at earth potential. The undulator and the mm-wave cavity are located inside the high voltage terminal, at 2 MV. To avoid damage during breakdowns of the 2 MV system, the high voltage terminal is connected to the 2 MV power supply, an Insulated Core Transformer (ICT), via a resistive transmission line. The electrodes of the depressed collector are at relatively low voltages, on the order of 10 kV - 250 kV. The advantage of this set-up is that the ICT has to deliver only the electron loss current. The depressed collector power supplies (at ground potential) deliver the 12 A beam current.

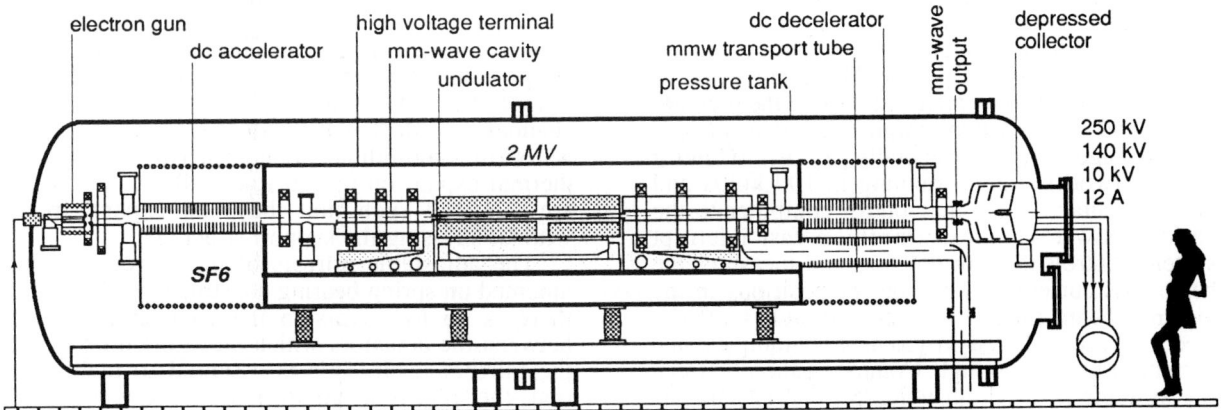

Figure 1. Schematic layout of the Fusion-FEM. The pressure vessel measures 11 m in length and 2.6 m in diameter. The vessel housing the 2 MV power supply is not shown.

3. HIGH VOLTAGE SYSTEMS AND AUXILIARY POWER SUPPLIES

Presently, the FEM is built in a so-called 'inverse' set-up, to give better access to the undulator and the mm-wave system. The electron gun is now mounted inside the high voltage terminal, now at -2 MV, and the undulator and mm-wave system are at ground potential, outside the pressure vessel.

The accelerating voltage of 2 MV is generated by an Insulated Core Transformer. This type of power supply was preferred because of the high current of 25 mA, which for example can not be delivered by commonly used Van de Graaff or Pelletron systems. The high current is needed for two reasons. Firstly, the ICT has to deliver the electron loss current at high voltage. Secondly, part of this current flows through the resistor dividers of the accelerating and decelerating tubes, to set the electrode voltages. Because of the high beam current, and inherent 'high' beam losses, 10 mA flows through the resistor dividers.

The systems inside the high voltage terminal, such as vacuum pumps, high voltage power supplies of the electron gun and computer-control systems, are powered by a 10 kW hydraulic-driven generator. This system consists of a high pressure (175 bar) oil pump outside the pressure vessel, which delivers a constant oil flow to a hydromotor inside the terminal. The hydromotor is coupled to a 10 kW, three-phase generator with built-in voltage stabilization. The oil is transported to the high voltage terminal by two 1.5 m long Delrin tubes, which bridge the 2 MV voltage between ground and terminal. A special oil (Shell Telleus R32) is used because of its insulating properties; many hydraulic oils contain zinc additives, and are therefore unsuitable.

The control system basically consists of three subsystems: a PLC system for controlling 'slow' processes, such as dc voltages, temperatures, coolant flows and vacuum set points. The main PLC, in which the code runs, is mounted outside the pressure vessel and controls two simple 'slaves' inside the high voltage terminal. Secondly, a CAN (Control Area Network) system, running at 100 kbit/s and 10-bit accuracy, links the control computer to the various power supplies. Thirdly, a fast interrupt system controls the gun voltages and measures the beam loss current at a number of positions in the electron beam line. This system switches off the electron beam in less than 400 ns, for example when current losses are too high. Communication between the systems in the high voltage terminal and the systems at ground potential runs via optical links.

4. THE MM-WAVE SYSTEM

4.a. Tolerances of the mm-wave cavities

The mm-wave cavity of the FEM is shown in fig. 2. It consists of the waveguide inside the undulator, where interaction takes place and the power is amplified, a 100% reflector, and a partial reflector and outcoupling system called the splitter/combiner. The reflector and the splitter/combiner are so-called stepped waveguides, which 'split' the mm-wave beam from the undulator waveguide into two off-axis beams. The system is described in detail in ref. [1,2].

The stepped waveguides consist of fixed side walls and moveable top and bottom walls to tune the waveguides to the operating frequency of the FEM. Naturally, there will be a small slit between the moveable parts and the fixed walls.

Low power measurements on the mm-wave system have been performed on a prototype waveguide with adjustable walls and an adjustable reflection mirror, to investigate the frequency behaviour and the sensitivity to tolerances.

The -3 dB power bandwidth of the total cavity (100 % reflector, undulator waveguide and splitter) is 17 GHz, with the side walls adjusted for 200 GHz operation. Further, power losses have been measured of stepped waveguides where openings were made in the top and bottom walls, just after the step. These slots serve for extra vacuum pumping capacity. It was shown that 100 mm long slots result in a power loss of only 0.4% [3]. Since the top and bottom walls of the waveguides are moveable, there must be a small slit between these walls and the fixed side walls. For slits of 0.1 mm over the full length (1.5 m) the extra losses amount to 0.25%.

4.b. Mechanical design

In principle, the splitter/combiner consists of three fixed wave guide walls (vertical) and moveable top and bottom walls for frequency adjustment. Further, two mirrors are tuneable for frequency adjustment and for adjusting the feedback power. Due to the use of corrugated walls, carrying an HE_{11} mode, Ohmic losses are relatively low. To reduce thermal expansion, a number of cooling channels are drilled through the 1.6 m long side walls (in longitudinal direction). This reduces the thermal expansion to 300 µm. Still, the side walls are mounted such that they can expand freely without stressing the support structure.

The adjustable mirrors of the waveguides are mounted on spring bearing constructions. This way, there is no hysteresis in the connecting system between the actuators which move the mirrors and the mirrors itself. This is essential to make sure that the required position is always reached, irrespective of the direction of motion.

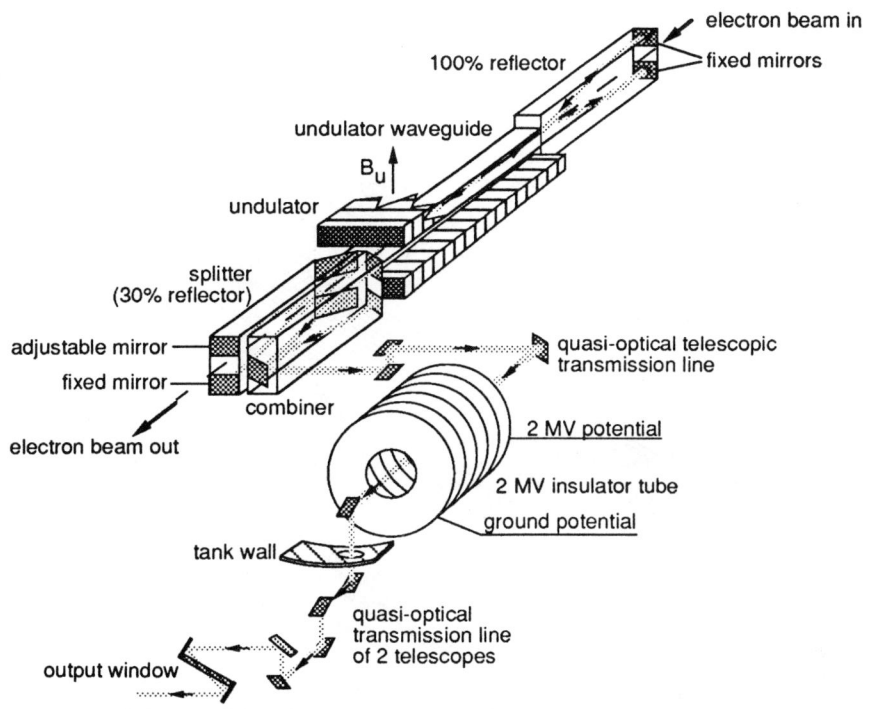

Figure 2. Schematic layout of the mm-wave cavity of the Fusion-FEM.

4.c. The output mirror system

For short-pulse operation a boron-nitride window suffices. A window of 140 mm diameter is used in a Brewster angle configuration to enable broadband, low reflection transmission [3]. To obtain optimal heat transport in the window, to prevent breakdown at the window-air interface and to get low losses, the mm-wave beam is transformed such that the projection on the window is circular. The projected waist measures 27.7 mm, which is less than 30% of the usable diameter. The normal cross section of the beam on the window is elliptical with a waist of 11.2 mm in the direction of the E-field and 27.7 mm in the perpendicular direction. The waist of the beam at the combiner output measures 7.0 mm in the E-field direction and 5.3 mm in the perpendicular direction.

The transformation of the beam dimensions is accomplished by a set of four cylindrical mirrors in an astigmatic telescope configuration, see fig. 4. Such a system covers the full bandwidth of the FEM. As to minimise losses, the mirrors are oriented such that the beam is bend in the plane normal to the electric field.

To calculate the mm-wave beam propagation in this system, the beam is treated as two orthogonal systems; one in the E-field direction and the other in the direction perpendicular to the E-field. In our case, the construction of the Brewster window determines the minimum distance between the window and the mirror next to it. To transform the beam from the combiner to the beam on the window the dimensions were found as given in fig. 4.

The tolerances for mirror positioning are estimated. In first approximation, when the mirror is treated as a thin lens, only parallel displacement of the optical axis changes the beam position on the window [4,5]. Rotation of the mirror surface has to be taken into account since this has the same effect as parallel displacement. Calculations show that the mirrors have to be positioned with an accuracy of 0.1 mm in position and 0.1° in angle.

5. EXPERIMENTAL STATUS

The entire high voltage system, consisting of the ICT, the high voltage terminal and the accelerator column, has been tested up to -2 MV and is now routinely operated at -1.8 MV. Firts-time condition to -1.8 MV takes 9 hours, while during routine operation this takes 2 hours.

During preliminary tests a 100 mA electron beam was accelerated to 1.6 MeV in short pulses (5 μs). Presently, a higher beam current can only be reached at cathode temperatures of 1220°C (nominal temperature 1050°C) due to poor vacuum in the gun. Nevertheless, the tests showed that all beam

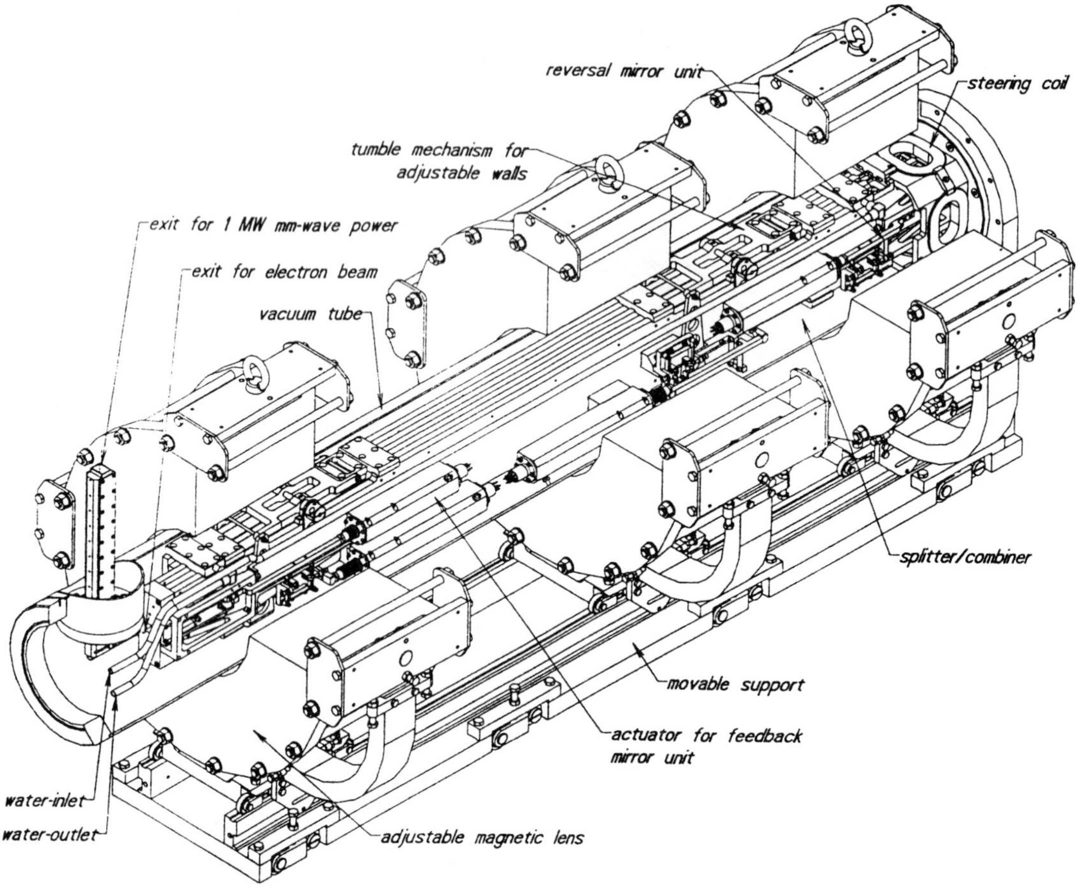

Figure 3. Mechanical construction of the splitter/combiner.

focussing, steering and diagnostic systems, as well as the computer control systems perform as designed. The vacuum system is now being improved.

6. ACKNOWLEDGEMENT

This work was performed under the EURATOM-FOM association agreement, with financial support from NWO and Euratom.

REFERENCES

1. W.H. Urbanus et al., Nucl. Instr. and Meth. **A331** (1993) 235.
2. N. Yu. Peskov, G.G. Denisov, et al., Nucl. Instr. and Meth. **A375** (1996) pp 377-380.
3. G.G. Denisov, M. Yu. Shmelyov and V.L. Bratman, private communication.
4. A.E. Siegman, Lasers, University Science Books, 1986.
5. G. Goubau, Beam waveguides, Advances in Microwaves, vol. 3, pp 110-116.

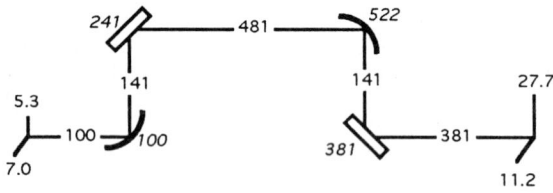

Figure 4 Schematic layout of the transmission line from the splitter/combiner to the output window. *Italic* figures indicate the focal length of the mirrors.

DESIGN AND INSTALLATION OF THE ELECTRON CYCLOTRON WAVE SYSTEM FOR THE TCV TOKAMAK

T.P. Goodman, S. Alberti, M.A. Henderson, A. Pochelon, M.Q. Tran

Association Euratom-Confédération Suisse, Ecole Polytechnique Fédérale de Lausanne, Centre de Recherches en Physique des Plasmas, 1015 Lausanne, Switzerland

The design of the combined 82.7 GHz and 118 GHz, 4.5 MW, 2.0 s electron cyclotron wave (ECW) system for heating and current drive on TCV is described. Low and high power test results of the RF source, transmission line and launching antenna are presented.

1. INTRODUCTION

The Tokamak à Configuration Variable (TCV) will require additional heating and plasma current profiling for the formation and study of strongly shaped plasma configurations. Highly-elongated ($\kappa \leq 3$) plasmas allow third harmonic X-mode (X3) heating to be used since power can be injected quasi-vertically along the resonance resulting in an increase in the optical depth of the plasma and good first pass absorption[1]. X3 enables higher central densities to be reached than with second harmonic heating (X2). Nevertheless, X2 is necessary for higher first pass absorption early in the discharge, when the elongation (X3 path length) is small, in order to shape the current profile and provide stable operation while elongating up to $\kappa=3$. In addition, the elongation evolution can be followed during the pulse.

The following paragraphs describe the design of a flexible, electron cyclotron current-drive/heating system with emphasis placed on the installation and testing of subsystems.

2. SYSTEM OVERVIEW

A total RF power of 4.5 MW will be available during the 2 s flattop of the TCV pulse; provided by nine 0.5 MW gyrotrons grouped into three clusters of three units each. Each cluster is fed by one regulated high-voltage power supply (RHVPS) operating with pulse-step modulation technology[2]. Two of the clusters operate near the second harmonic (82.7 GHz) and one near the third harmonic (118 GHz). The frequencies are determined by the nominal magnetic field of TCV (1.43 T) and the availability of gyrotron sources. Although the frequency ratio of X2/X3≠2/3, simultaneous central heating at both frequencies is still possible[3]. Each gyrotron RF beam is coupled to an evacuated transmission line via a "matching optics unit" (MOU) and propagated to a quasi-optical launching antenna installed on the tokamak.

2.1 Gyrotrons

Gyrotrons are delivered by Gycom (Russia) at 82.7 GHz and Thomson Tube Electronics (France) at 118 GHz. The RF window of the X2 gyrotron has a flattened power density profile on an edge-cooled boron nitride (BN) window while the X3 gyrotron has a gaussian profile on a liquid-nitrogen (LN) edge-cooled sapphire window. Neither window option is desirable at the torus. To avoid the problems associated with either readapting the beam profile or LN cooling at the tokamak, a windowless evacuated transmission line and MOU have been chosen.

2.2 MOU

The gyrotron RF beam contains scattered radiation that needs to be filtered before injection into the waveguide (to prevent arcing). The exact direction and position of the beam at the gyrotron window are not easily controlled during manufacturing within the precision necessary for coupling to the transmission line; therefore, the alignment between the gyrotron and waveguide must be adjustable. Furthermore, the beam waist diameter and location must be modified to provide a proper match to the chosen waveguide diameter. The MOU serves these three functions (filtering, alignment and matching).

To provide optimum coupling of the transmitted RF beam to a given target plasma, the polarisation of the radiation must be adjusted: the MOU is a convenient location to perform this adjustment. To this end, a rotatable two-mirror universal polariser (Al) [4] is included in the MOU of the X2 gyrotron; along with two focusing mirrors (OFHC Cu) used to match the beam to the waveguide (Fig. 1). The four mirrors are aligned before evacuating the MOU and only the rotation of the polariser mirrors is possible under vacuum (between shots). All mirrors are cooled by conduction to the vacuum vessel walls.

The power that is lost in the MOU is absorbed by TiO$_2$-coated OFHC-Cu plates (40% TiO$_2$ / 60% Al$_2$O$_3$, 0.3 mm thick with an additional 0.08 mm thick bonding layer) which are screwed onto rails welded to the vacuum vessel inner wall. The absorption of the plates has been measured at 45° incidence in both the E and H-planes as 34% and 18%, respectively. The rather moderate absorption helps avoid hot spots.

The X3 MOU contains three water-cooled mirrors (one focusing, and a two-mirror polariser) providing X-mode launch after propagation to the torus. The mirrors can be positioned while the MOU is evacuated.

2.3 Transmission lines

Corrugated HE$_{11}$-waveguide transmission lines (Al, 63.5 mm diameter) have been procured from Spinner GmbH (Germany)/General Atomics (USA) and were delivered clean and ready for installation. The lines are bakable to *150 °C*.

A typical line is shown in figure 2. Electrical isolation breaks *(1)* are used to isolate the line from both the gyrotron/MOU and TCV. Miter bends *(2)* act as the fixed points of refer-

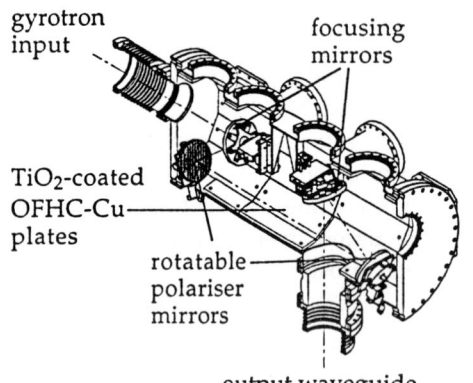

Figure 1. Evacuated MOU box of the X2 gyrotron.

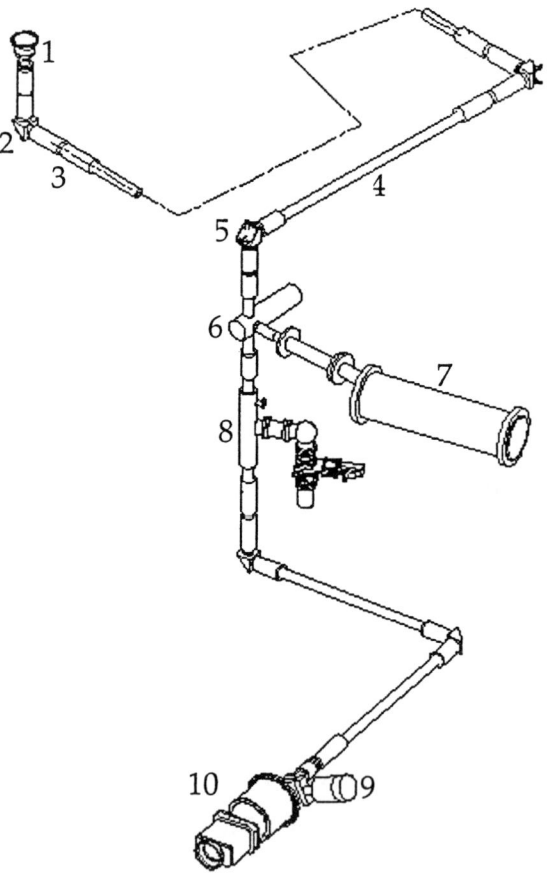

Figure 2. Typical transmission line for X2 gyrotrons. The average length of the lines is approximately *30 m*.

ence for support and alignment. Additional adjustable supports are placed along the lines ~ 3 m apart. They are laser aligned with the miter bends and constrain the line only in the vertical direction to avoid excess gravitational sagging. Each corrugated bellows *(3)* provides *5 mm* extension and *30 mm* compression to compensate for thermal expansion and waveguide length tolerances. Straight sections *(4)* are as long as possible *(2.1 m)* to minimize the total number of joints (potential misalignments) and Helicoflex® vacuum seals (potential leaks). Forward and reflected power signals are provided by a power monitor miter bend with multihole coupling *(5)*, placed as near TCV as possible while still allowing *in situ* calibration against a calorimetric load. A switch *(6)*, with integral gate valves on the outputs, directs the power to either the torus or a stainless steel water-cooled load *(7)* fitted with flow and temperature probes for calorimetric measurements and a small vacuum port for pressure measurements and pumping. To minimize impurity throughput to the plasma, a pumping "tee" *(8)* is located near the torus. An all-metal gate-valve *(9)* separates the launching-antenna vacuum chamber *(10)* (at torus pressure) from the transmission line.

The bellows of the last and next to the last legs of the transmission line compensate for TCV expansion (~7 mm vertically and horizontally) during baking *(300 °C)*. From the torus to the first vertical bellows, the line is kept as simple as possible to ease vertical motion. The last leg can easily be removed to allow access to the launching antenna while maintaining the vacuum from the MOU to the switch; and can be evacuated by the pumping "tee" after reinstallation.

The majority of the line (MOU → load) can be tested under vacuum at high power and long pulse length, independently from TCV.

2.4 Launching antennas

Rectangular boxes *(15×36×36 cm)* extending beyond the TCV shaping coils are welded onto the torus at six entry ports of *15 cm* diameter (four upper lateral and two equatorial). Upper lateral ports are *45.5 cm* above the equatorial (mid-) plane and *28 cm* outward along the major radius. Thus, to be able to aim centrally toward the midplane (downwards), it is necessary to have the final reflection point of the beam located inside the circular port itself. To maintain good localization of the beam, focusing is required between the end of the transmission line and the plasma.

Each X2 launching antenna (Fig. 3) has the capability of aiming in the poloidal and toroidal directions to cover X-mode heating, current profile modification, current drive and breakdown. The last mirror of the four-mirror antenna can be moved over a *24°* range, twice, during a discharge, while the entire antenna can be rotated *0–360°* about the axis of the port between shots: virtually the entire poloidal cross-section can be targeted. The first three mirrors of the antenna are made of OFHC-Cu with the first and third providing focusing. The fourth mirror is made of TZM (a machinable alloy of *Mo: 0.02% C, 0.5% Ti, 0.1% Zr, 99.25% Mo*) due to the mirror's proximity to

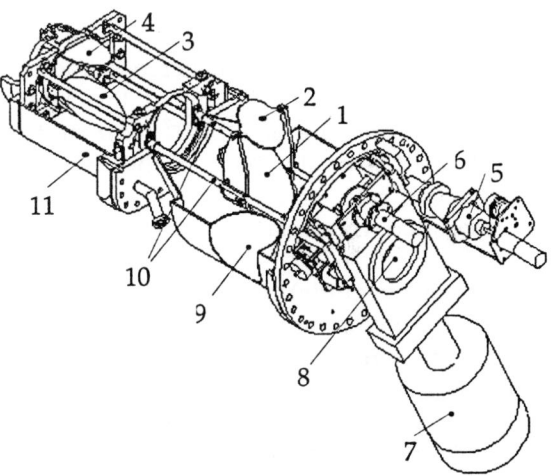

Figure 3. Elements of the X2 4-mirror launching antenna: mirrors *(1-4)*, rotational feedthrough *(5)*, linear actuation system of mirror 4 *(6)*, all-metal gate valve *(7)*, RF input *(8)*, indentation in vacuum vessel to avoid hitting TCV support structure *(9)*, guiding tubes of 4th-mirror actuator rods *(10)*, rectangular box *(11)* welded to the circular TCV entry port.

the plasma. Molybdenum has relatively low sputtering, low thermal expansion, high melting point and reasonably high thermal and electrical conductivity (peak $\Delta T \sim 470°C$, at 0.3% duty cycle: average $\Delta T \sim 110°C$ assuming only radiative cooling).

X3 power will be launched from one port on the top of the machine. All three RF beams are directed onto a single mirror focusing in the poloidal direction only. The beams can be directed, as a group, in the poloidal plane by changing the mirror angle during a shot. The mirror location along the TCV major radius can be varied between shots to optimise first-pass absorption and avoid resonances on the mirror.

3. TEST RESULTS

One X2 gyrotron has been accepted at present, having been tested at *>0.5 MW, 2s, 1% duty cycle* for *4 hours* without missing any shots. Two further gyrotrons have been delivered and are presently under test. The prototype X3 gyrotron is undergoing factory tests [5].

The first three transmission lines have been delivered and most legs of the first cluster have been installed and pumped. One line (*4 miterbends, 20.1m*) has undergone vacuum testing using a *500l/s* turbopump at the MOU; resulting in a pressure of 1.0×10^{-7} *mbar* at the MOU, 3.5×10^{-6} *mbar* at the end of the line and an outgassing rate of 1.1×10^{-11} *mbar·l·s^{-1}·cm^{-3}*. The calculated conductance of the corrugated waveguide is $\sim 0.36\ l \cdot s^{-1}$ (~ 2.4 times lower than for smooth waveguide) yielding a pressure difference of 3.3×10^{-6} *mbar* for the above outgassing rate; in close agreement with the measurements.

The accepted X2 gyrotron was connected to the transmission line (without antenna) and tested at atmospheric pressure. Burn patterns after the MOU and at the output of the line show a well centered beam with some mode impurity. Pulses of *0.1s* and *0.5 MW* - measured at both the end of the transmission line and after the switch – were achieved without arcing. The power monitor miter bend was calibrated against the *100 ms* - calorimetric load used for the acceptance tests of the gyrotron. The coupling is found to be $-87.0 \pm 0.2 dB$. The measured line efficiency (P_{input}/P_{output}) is $93 \pm 5\%$. A flat alignment mirror (used in place of the polariser grating) was rotated to check the alignment properties of the MOU: no systematic power variation was seen.

High power long pulse conditioning tests have also begun. At present, a maximum of *0.6 MJ* (*309kW* for *2.0s* or *405kW* for *1.6s.*) has been transmitted - limited by the large pressure rise seen in the unbaked MOU.

Three X2 launching antennas have been tested (low-power RF; vacuum, mechanical at *150°C*) confirming all the design parameters.

4. CONCLUSION

Initial TCV ECW system tests have begun. Three launchers are now installed and tests with ECW power in TCV are planned for this fall.

ACKNOWLEDGMENTS

The authors would like to thank Dr. J. Doane (General Atomics) for his transmission line calculations and many helpful discussions. The participation of Dr. I. Roy (Kurchatov Institute) during part of the high power tests is gratefully acknowledged. Special thanks go to the drafting, electrical, electronics and vacuum departments of the CRPP for their invaluable support. This work is partially supported by the Swiss National Science Foundation.

REFERENCES

1. Ségui J. L. et al., Nucl. Fusion, 1996, **36**, 237.
2. Fasel D. et al., "Design and operation of the power installation for the TCV ECR additional heating", this conference.
3. Pochelon A. et al., in Proceedings of the 20th EPS Conf. on Controlled Fusion and Plasma Physics, Lisboa, Portugal, 1993, Vol. 17c Part III, 1020.
4. Smits F. M. A. in Proceedings of the 7th Joint Workshop on ECE and ECRH (Heifei, China) 1989.
5. M. Pain et al., this conference.

DESIGN AND OPERATION OF THE POWER INSTALLATION FOR THE TCV ECR ADDITIONAL HEATING

D. Fasel, J. Alex*, A. Favre, T. Goodman, M. Henderson, P-F. Isoz, A. Perez, M-Q. Tran

Centre de Recherches en Physique des Plasmas, Association Euratom - Confédération Suisse
Ecole Polytechnique Fédérale de Lausanne, PPB, CH - 1015 Lausanne, Switzerland

*Thomcast AG, CH-5300 Türgi, Switzerland

Following a brief introduction to the TCV project, this paper concentrates on the Regulated High Voltage Power Supply (RHVPS) system chosen to supply the nine gyrotrons, distributed in three clusters, that will deliver 4.5 MW of Electron Cyclotron Resonance Heating (ECRH) to TCV plasmas. The configuration of these clusters is described in some detail, including the results of site test both with dummy load (80 kV, 85 A, 2 sec) and the gyrotrons themselves (70 kV, 25 A, 2 sec). Some details are also given of gyrotron auxiliaries, interlock circuitry, control and data acquisition, and integration into TCV control environment.

1. INTRODUCTION

The TCV tokamak (Tokamak à Configuration Variable) has been in operation since November 1991. A large number of plasma shapes have already been produced and experiments with fast internal stabilisation coils are presently underway in order to reach the maximum design elongation of $\kappa=3$. **ECRH** has been chosen as additional heating system for TCV and its implementation is progressing in parallel with TCV operation. The system comprises **6 gyrotrons** of the **diode** type at **82 GHz** (2nd harmonic) and **3 gyrotrons** of the **triode** type at **118 GHz** (3rd harmonic). The total power delivered by the gyrotrons to the plasma will be **4.5 MW** for two seconds. The gyrotrons are grouped in three clusters, with each cluster fed by its own power supply. At present, two power supplies have been installed, but only one is completely tested. The third is under study in order to match the specific requirements of the triode gyrotrons.

2. TCV ELECTRICAL NETWORK

Figure 1 illustrates the additional systems designed to feed the power supplies foreseen for the TCV ECRH project together with the internal coil Fast Power Supply (FPS) described in [3]. A description of the main tokamak electrical network may be found in [2]. Electrical power supply to the TCV systems may be derived from either of two sources :

		Alternator	50 Hz network
Pcc	[MVA]	1160	130
f	[Hz]	120 - 96	50
Un	[kV]	0 - 10	20
Sn*	[MVA]	220	7

*Pulsed

2.1 The 50 Hz network

The RHVPS may be directly supplied by the 20 kV grid through a pulsed transformer (named *Test*), also used in other applications (eq TCV rectifier tests, FPS tests).

There are several reasons why the gyrotron clusters should be fed in this way :
- to increase the short circuit impedance
- to ensure lower current (and hence increased installation protection) in case of short circuit
- to work independently with any cluster for conditioning purposes

In order to reach the required 10 kV at the primary side of the RHVPS, an auto-transformer with a voltage ratio of 1÷2 is used. The power rating of both this and the *Test* transformer allows us to work with a complete cluster at the maximum value, of **7 MVA** for **2 sec every 5 min**. This must, however, not be performed simultaneously with a TCV shot in order to avoid too much power being extracted from the grid. The three phase AC cable (1x3x95 mm²) used to feed all the equipment is chained from one isolator unit to another from the *Test* transformer secondary.

2.1 The generator network

Compared to the 50 Hz network, the TCV generator network differs in several respects :
- the working mode : pulsed with a duty cycle of 2 sec every 5 min
- the higher frequency (120 Hz)
- the 10 times higher short circuit power

In addition separate cables supply rectifier groups, FPS and the 3 gyrotron clusters from the generator busbars to the isolator units (see fig. 1).

Using the alternator power source, the RHVPS power modules are active only at relatively low duty cycle, so thermal stresses are lower than at 50 Hz. Since the voltage is ramped up in 2 sec, the mechanical stresses in the RHVPS transformer are also lower.

3. THE CLUSTER STRUCTURE
3.1 General description

Shown in figure 1, each cluster comprises a power supply feeding three gyrotrons in parallel.

Three manually switched copper connections can either be tied to a **M**atching **N**etwork (MN) or to ground. The MN are used to adapt the line impedances to the fast voltage transients which appear, for example, during gyrotron arcing. A separate device places the two RHVPS output polarities to ground in the case of intervention inside the high voltage enclosure. A *triax* cable is used to connect each RHVPS output to the gyrotron cathode filament heating assembly. A *three conductor HV cable* then connects this equipment to the gyrotron.

As illustrated schematically in figure 2, the equipment is distributed physically on three floor levels.

3.2 Heating assembly

This assembly, which is referenced to the cathode potential includes :

• An MN connected to the *triax* cable central conductor
• The power components required to generate the current needed to heat the gyrotron cathode filament
• A control rack including :
- feedback electronics used to attain the current reference and to maintain this value in the range of ± **0.5%** of the demand value.
- fibre optic bus for communication with the main control system
- filament current and voltage measurements
- cathode current measurement
• A high voltage transformer supplying the heating equipment and isolated to 150 kV from ground

In order to improve shielding against fast transients, the electronics are located in a PerAluman (AlMg) enclosure.

3.3 Grounding configuration

The *power* ground is provided by copper bars or wires distributed all around the high voltage equipment in the RHVPS room. This network is referenced to building earth at several points on the grid.

On the AC side, a copper bar follows the power cables up to the *Test* transformer or the generator building in order to ensure a direct (low impedance) connection in case of a fault occur (eq short circuit). All the AC components (feeders, breakers, metallic pieces) are referenced to these bars.

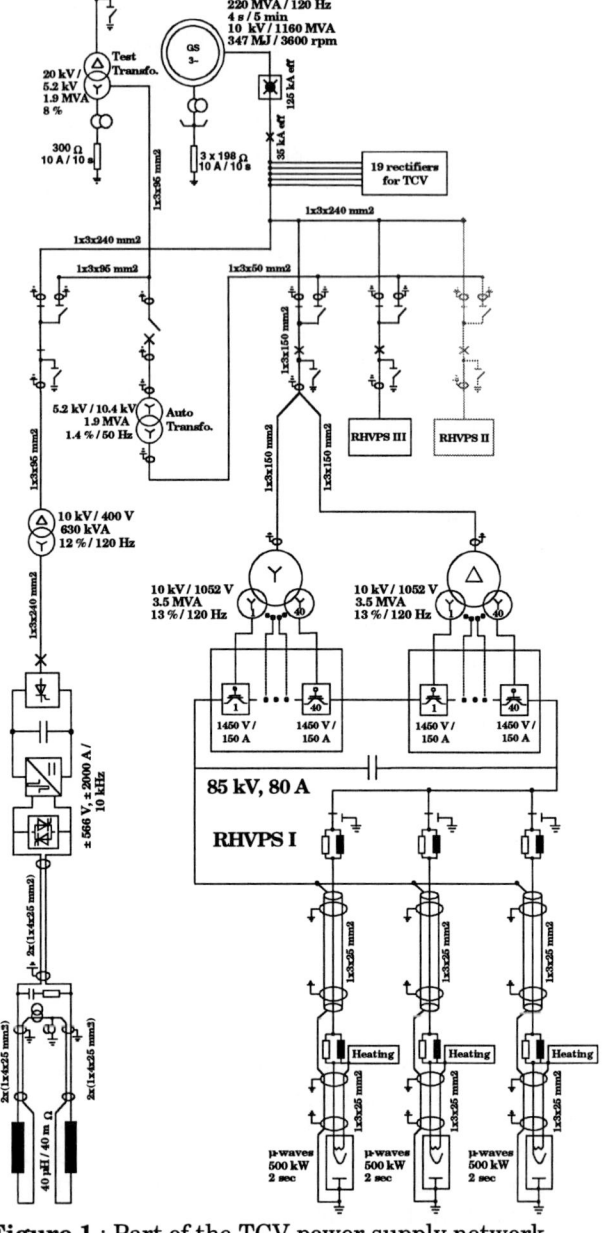

Figure 1 : Part of the TCV power supply network

On the DC side, the *triax* cable shield and the *3 conductor HV cable* are connected through a wire to the *power* ground.

On the level +1, where the gyrotron towers and control frames are located, the total floor surface is covered by a copper band mesh, both to minimise the high frequency impedance of this ground reference and to ensure personnel safety (this is an unrestricted access area) even during fast transients.

Since the gyrotron collector is at the same potential as the tower **and** the "+" RHVPS polarity, there are two grounding possibilities :
- leave the "+" polarity floating and reference the DC part to earth potential via the ground surface described above.
- leave the ground surface floating and reference it to earth via the "+" polarity connected to the *power* ground inside the RHVPS enclosure.

3.4 Triode particularities

The main differences are related to the additional power supply required to feed the gyrotron anode. It has to be placed as close as possible to the gyrotron tower in order to minimise the capacitative leakage. In the existing structure, possible locations are either under the gyrotron tower, or on an additional stage of the heating system (see fig.2)

The HV power supply needed to feed the triode cluster (118 GHz) is presently under study; the requirements are similar to those defined for the 82 GHz gyrotrons, except for the interface to drive the gyrotron anode power supply.

Although the cathode voltage is common to the cluster, the triode will permit independent modulation of the total microwave power using each anode supply.

3.5 Control

All gyrotron auxiliary systems are controlled and driven by distributed CPU nodes (named *slaves*), linked together through the BITBUS fieldbus, described in [4] and used for all TCV plant control.

The different *slave* tasks are:
- for the cooling: control of the pump status and the flow rate in pipes
- for the heating: ON/OFF orders, transfer of I/U cathode filament measurements, current reference, status of the gyrotron heating system
- for the control: centralise warnings and send related orders to the RHVPS and/or to the auxiliaries (pumps, magnetic field etc.), based on programmable logic. Convert fibre optic signals in TTL signals for arc detection on the window, **M**atching **O**ptical **U**nit (MOU) and that part of the waveguide transmission line which is not located in the TCV zone.
- for the pump : control of the waveguide vacuum.

Depending on the necessary tasks, *slaves* are dedicated to individual gyrotrons, a whole cluster or to the complete installation.

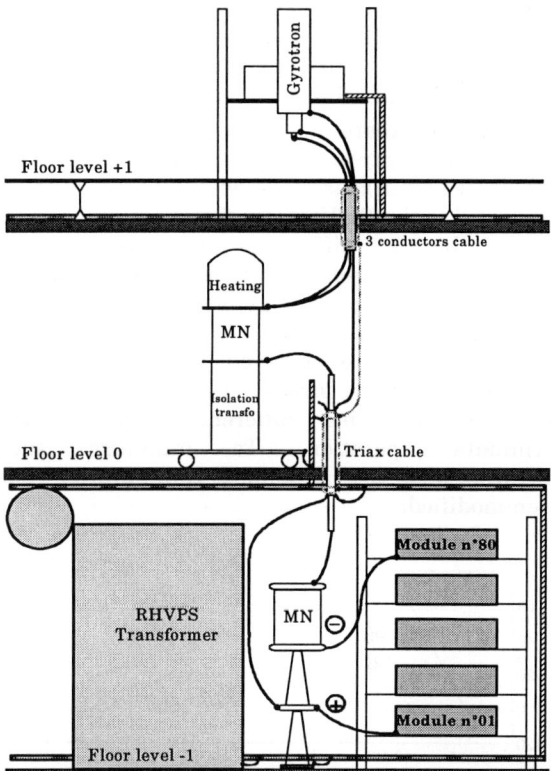

Figure 2 : Physical distribution of the gyrotron and grounding system

Data storage, visualisation, driving orders and parameter settings, have already been integrated into the TCV control system in order to minimise the effort required when microwave power is first injected into TCV plasmas.

4. RESULTS

4.1 Dummy load tests

On site commissioning tests have been performed using a resistive load (1060 Ω, 80A, 100 kV), to simulate the gyrotron environment as closely as possible: same cables, MN and grounding system.

Particular care has been taken with respect to requirements, such as :
- the ripple, which must be < ± **0.5 %** of the nominal HV DC. This requirement is satisfied by the use of an additional capacitor (15 nF) at the power supply output.
- the short circuit energy dissipation, which must be < **10 J**. A *Cu wire* test showed that this requirement is also satisfied.

The ability to work in pulsed mode with the motor generator has also been tested. The main difference here lies in the power-up time of the power module forming the RHVPS : using the generator, the voltage ramp-up time is less than 2 sec. After this time, the voltage supply for the power module must be stabilised and the electronics alive and ready to work.

4.2 Gyrotron tests

To date, gyrotron tests are performed only with the 50 Hz power source and with a single gyrotron connected to the RHVPS (see figure 4).

The biggest difficulties were encountered during the gyrotron conditioning period. When arcs occured either in the gyrotron cavity or in the MOU/transmission line. As a consequence perturbations were observed on the control or heating electronics, strong enough in some cases to interrupt some differential inputs and terminate operation. To minimise such perturbations, the grounding configuration has been modified:

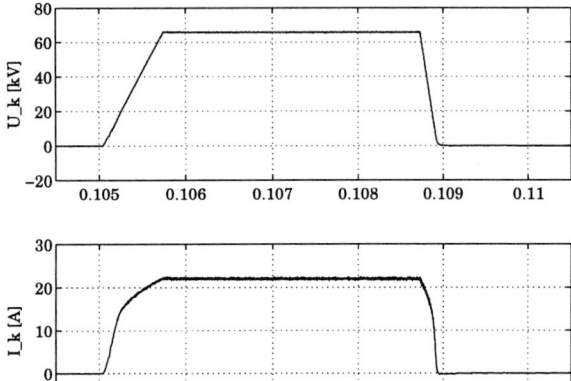

Figure 4: Gyrotron voltage and current

- reference the "+" polarity of the RHVPS to the *power* ground in the HV enclosure.
- reference the copper earth mesh to the "+" polarity through the cable shields
- adapt the gyrotron tower connection to the copper earth mesh, replacing the wire connections by copper band.

Signal cables directly connected to gyrotrons (such as the gun and collector currents and pressure measurement) have also been protected with fast varistors in parallel with a surge voltage limiter. In addition, a copper shield has been installed around the gyrotron coils (gun, collector) to decrease the antenna effect during fast voltage transients.

4.3 RHVPS improvements

Operation with the 50 Hz network has suggested a number of possible improvements to the RHVPS. The most important are:

• the IGBT driver exchange on the power module board. The original driver was being influenced by voltage transients through the galvanic isolation transformers. This problem is now avoided through the use of an optical coupling.

• Due to a new requirement on the permitted energy dissipation in the gyrotron during an arc (**5 J** instead of 10 J **specified**) the inductive stored energy has been decreased by the addition of diodes in parallel with the RL filter in series with each power module.

5 CONCLUSION AND FUTURE PLANS

At present, a single RHVPS is in use, feeding a cluster of 3 gyrotrons of which only one is currently operational. Continuous testing of the unit on the 50 Hz network has demonstrated the ability of this supply to wishtand the gyrotron operation environment. Full power operation for 2 sec at 500 kW gyrotron output power, injected into a waveguide transmission line right up to TCV tokamak, has been obtained

The next step will be to operate the RHVPS from the generator power source, firing microwave power into a dummy load at TCV. Later tests will be performed with the RHVPS supplying a complete cluster in order to investigate the mutual influence of 3 gyrotrons fed by a common power source.

In 1997, tests of the second cluster and the installation of the third RHVPS for supply of the 118 GHz triodes are foreseen . In parallel, delivery and commissioning of the anode power supplies will take place.

REFERENCES

[1] G. Besson et al., Regulated high voltage power supply for gyrotrons based on pulsed step modulator technology, SOFT 1994, vol.1, p.517-520

[2] D. Fasel et al., 19 rectifiers to supply the coils of the TCV tokamak, SOFT 1990, vol.2, p.1492-1496

[3] A. Favre et al., Control of highly vertically unstable Plasmas in TCV ..., this volume

[4] JB. Lister et al., Distributed on the TCV tokamak and modular bitbus nodes, SOFT 1990, vol.2, p. 1268-1272

This project is partly supported by Swiss National Science Foundation

High-Power Corrugated Waveguide Components for mm-Wave Fusion Heating Systems*

R.A. Olstad, J.L. Doane, C.P. Moeller, R.C. O'Neill, M. DiMartino

General Atomics, P.O. Box 85608, San Diego, California 92186-5608, USA

Considerable progress has been made over the last year in the U.S., Japan, Russia, and Europe in developing high power long pulse gyrotrons for fusion plasma heating and current drive. These advanced gyrotrons typically operate at a frequency in the range 82 GHz to 170 GHz at nearly megawatt power levels for pulse lengths up to 5 s. To take advantage of these new microwave sources for fusion research, new and improved transmission line components are needed to reliably transmit microwave power to plasmas with minimal losses.

Over the last year, General Atomics and collaborating companies (Spinner GmbH in Europe and Toshiba Corporation in Japan) have developed a wide variety of new components which meet the demanding power, pulse length, frequency, and vacuum requirements for effective utilization of the new generation of gyrotrons. These components include low-loss straight corrugated waveguides, miter bends, miter bend polarizers, power monitors, waveguide bellows, dc breaks, waveguide switches, dummy loads, and distributed windows. These components have been developed with several different waveguide diameters (32, 64, and 89 mm) and frequency ranges (82 GHz to 170 GHz). This paper describes the design requirements of selected components and their calculated and measured performance characteristics.

1. INTRODUCTION

Electron cyclotron heating (ECH) has become accepted as one of the most efficient and technologically attractive methods for adding energy to fusion plasmas. Until recently, the main limitation of using ECH heating has been the lack of high power mm-wave sources in the 82 GHz to 170 GHz range. Fusion devices with larger magnetic fields and higher plasma densities require higher frequency sources. The gyrotron, a high power mm-wave source, has been under development for a number of years in Russia, Europe, U.S., and Japan; and frequency, power level, and pulse length capabilities are continually improving.

The transmission line system from gyrotron to plasma must be designed to handle the high frequency, power, and pulse length with low losses and low reflection back to the gyrotron. The most demanding component for high power cw transmission is the window. Presently ECH power/pulse length that can be delivered to the plasma is limited by the performance of the window. The development status of ECH windows and other key components is presented below.

2. DIII–D GYROTRONS AND RECENT ECH HEATING RESULTS

The DIII–D program at General Atomics in San Diego is planning to install 10 MW of ECH power to provide the localized heating and current drive needed to execute the Advanced Tokamak program. The first 110 GHz 1 MW gyrotron of a 3 MW initial system is presently being commissioned [1]. This first gyrotron is a state-of-the-art internal-mode-converter gyrotron manufactured by GYCOM, a Russian company. This gyrotron was tested to 960 kW for 2 s into air in Russia. It is designed for cw operation, but the pulse length is limited by the performance of its boron nitride output window. In testing at the DIII–D facility during July 1996, the gyrotron has achieved 0.5 MW for 0.5 s both into a dummy load and into the plasma via transmission through an evacuated 31.75 mm diameter corrugated waveguide system. Increased power and pulse length are planned for late 1996. Two other MW-level 110 GHz gyrotrons manufactured by CPI are scheduled to be installed in late 1996–1997.

Recent experiments at DIII–D dramatically show the effectiveness of mm-waves in heating fu-

---

*Work supported by U.S. Department of Energy under Contract Nos. DE-AC03-89ER51114, DE-AC03-89ER52153 and General Atomics commercial contracts.

sion plasmas. Figure 1 shows that the deposition of 500 kW for 0.5 s into a low density plasma increased the central electron temperature from 3 keV to 10 keV. The power was transmitted to DIII–D using 40 m of small diameter (31.75 mm) evacuated corrugated waveguide with no evidence of breakdown [2]. Further electron temperature increases are expected when the power and pulse length are increased. Future ECH work at DIII–D will concentrate on profile control applications. A key component in obtaining the desired current profile for enhanced confinement regimes is the "off-axis" current driven with the millimeter wave system.

3. COMPONENT DESIGN CONSIDERATIONS

In order to transmit the microwaves from the gyrotron source to the plasma, the transmission line must meet a number of demanding requirements:
1. Low reflected power back to the gyrotron.
2. Low loss in the transmission line, i.e. < 10%.
3. Maintain mode purity so microwaves launch into well-defined locations in the plasma.
4. Ability to control the mm-wave polarization for optimal absorption in the plasma.
5. Ability to monitor the beam power.
6. Ability to operate at the desired frequency for high power operation at long pulse length.
7. Compact transmission because of limited real estate near fusion devices.

The components that need to be designed to meet the above requirements generally include some or all of the following: matching optics unit (MOU), taper, mode converter or mode filter, waveguide switch, dummy (calorimetric) load, straight waveguide, continuous curvature bends, miter bends, power monitor miter bend, polarizer miter bends, pumpout, bellows, d.c. break, window, and launcher.

The use of corrugated waveguide for transmitting mm-waves in the HE_{11} mode results in very low losses. For a 63.5 mm waveguide with corrugation geometry suitable for 100–300 GHz transmission, the losses versus frequency are as shown in Fig. 2(a). These losses were calculated using a space harmonic analysis of the corrugated waveguide to obtain the propagation constants and the electric and magnetic fields. Mode conversion in straight corrugated waveguide propagating HE_{11} mode is negligible. However, mode conversion can result from misalignment of waveguide supports, especially at higher frequencies [Fig. 2(b)]. At 170 GHz, the misalignment must be limited to 1 mm to keep losses down to 0.07 dB (1.6%) per 25 m.

In miter bends, losses are due to ohmic heating, mode conversion in ideal bends, and mode conversion due to mirror misalignment [3]. Results of loss calculations vs. frequency for aluminum mirrors are shown in Fig. 2(c). The dominant loss above 140 GHz is due to mirror misalignment, which was assumed in these calculations to be 0.001 radian. At 170 GHz, the total loss is about 0.05 dB, or 1.2%. These calculations show that low loss transmission can be achieved in a 63.5 mm waveguide system at 100–300 GHz.

4. REPRESENTATIVE RECENTLY-FABRICATED COMPONENTS

GA and its collaborating companies have recently fabricated or are presently fabricating ECH transmission lines for several major devices. In Europe, Spinner GmbH is prime contractor and GA is subcontractor to Spinner in delivering transmission line components to CRPP for the TCV device at Lausanne, Switzerland and to CEA for the Tore Supra device at Cadarache, France. The

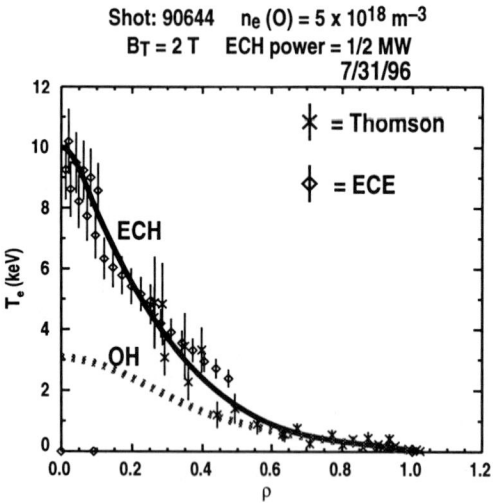

Fig. 1. Plasma electron temperature versus normalized minor radius with ohmic heating alone (OH) and with 0.5 MW/0.5 s ECH.

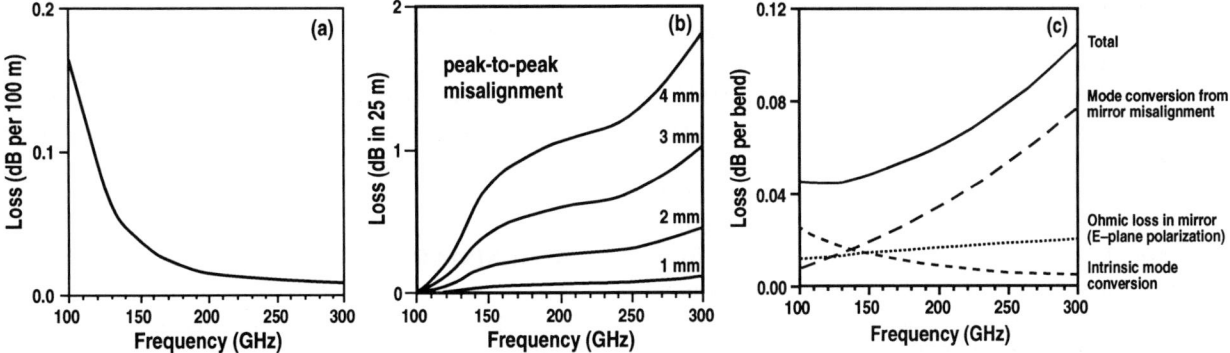

Fig. 2. Calculated losses in 63.5 mm waveguide components for 100–300 GHz HE_{11} transmission: (a) straight corrugated waveguide, (b) waveguide with random misalignment every 2 m, and (c) miter bends.

components for both of these customers were designed by GA, and fabrication efforts are shared between the two companies. In Japan, GA recently supplied a number of specialized components to Toshiba Corporation, which is prime contractor for supplying transmission lines to NIFS for the LHD device at Toki, Japan. In addition to these collaborative efforts, GA fabricates components for DIII–D and other fusion devices worldwide.

Descriptions of typical components are:

4.1. Waveguide Switch for TCV, Lausanne

GA designed and fabricated waveguide switches for Spinner for use on the TCV device at Lausanne. This is a critical component because it enables the gyrotron to stay conditioned by diverting its power to a dummy load between plasma shots. To make the switch, three short 63.5 mm corrugated aluminum waveguides are inserted into the walls of an aluminum housing. Vacuum seals between waveguides and housing are made using metal Helicoflex seals. A pneumatically-controlled linear vacuum feedthrough moves a hardened block inside the housing. In the normal position, the mm-waves pass straight through a corrugated hole in this block from input to output waveguide. In the switched position, a copper mirror diverts the power to the third waveguide at 90° to the others. Pneumatically-controlled vacuum valves can close off either of the output waveguides so they can be at atmospheric pressure while maintaining vacuum in the remaining waveguides. The switches are currently being tested at Lausanne.

4.2. Transmission Line Components for TCV and Tore Supra

Spinner and GA jointly fabricated corrugated waveguides, bellows, miter bends, power monitor miter bends, pumpouts, and d.c. breaks for TCV and Tore Supra. Straight waveguides and miter bends are used to guide the mm-waves from the gyrotron source to the fusion device. Power monitor miter bends are devices for sampling the beam to determine its power level. Bellows are useful to accommodate thermal expansion and contraction of a line during bakeout cycles. All of these components have been successfully tested using 500 kW 1.6 s pulses on one evacuated line at Lausanne.

Spinner and GA also jointly fabricated a dummy load designed for 500 kW 2 s operation for TCV. One of the loads has been tested at Thomson Tubes Electroniques for over twenty 420 kW, 2 s shots using their 118 GHz gyrotron. The loads are now being tested at Lausanne with 500 kW 82.6 GHz 1.6 s pulses. GA also fabricated a similar dummy load for DIII–D designed to handle 500 kW for 10 s. The Lausanne loads absorb power in the stainless steel body; the DIII–D load uses an Inconel liner to handle the longer pulse lengths and consequent higher temperatures.

4.3 Dummy Load for LHD, Toki

This year, GA fabricated four dummy loads (3 at 168 GHz, 1 at 84 GHz) for Toshiba for use on the LHD device. A particularly demanding requirement was that they operate both under vacuum and at 1 atm. This prevented use of materials that would absorb too much energy from the incident beam and cause arcing or plasma discharge in the low density gas in front of the hot surface. Toshiba's specifications called for the load to handle 500 kW for 100 ms with a 1% duty cycle, with <2% reflected power.

To achieve this performance, a carbon-carbon composite material was found with suitable thermal, electrical, and mechanical properties. The loads were designed to have enough bounces so that more than 98% of the power is absorbed by the time the mm-wave beam returns to the input waveguide. Low power reflection measurements on the 84 GHz load showed the reflected power in the HE_{11} mode to be less than 0.5%. High power tests on this load will be performed at Toki later this year.

4.4. Polarizer Miter Bends for LHD, Toki

GA also fabricated four pairs of 88.9 mm miter bend polarizers for Toshiba for the LHD. The mirrors are remotely rotatable to achieve any arbitrary polarization and with ellipticity varying from 0 to at least 30°. The first miter bend acts as a circular polarizer to produce the desired ellipticity. The second mirror acts as a polarization rotator to achieve the desired polarization. The 84 GHz polarizer geometry was confirmed by measurements at GA in which a linear polarized beam was rotated to achieve output polarization varying from −90 to +90 degrees. The desired output polarization was achieved as predicted, with <0.2% of the power in unwanted polarization. The copper mirrors are water-cooled to enable operation at 500 kW for 10 s under vacuum or at 1 atm.

4.5. Distributed Window for DIII–D

GA has been developing distributed windows for use with high-power long-pulse gyrotrons. A 10 cm × 10 cm 110 GHz distributed window consists of 42 sapphire strips separated by water-cooled metal vanes. The geometry of the sapphire and vanes is such that the mm-waves pass through the sapphire with low loss (4%) and low reflection (1%) [4]. A window made last year was tested at CPI at 200 kW with a reduced beam diameter. The tests demonstrated that the window can handle a power density and pulse length equivalent to that in a full size 1.2 MW cw beam with peak-to-average power ratio of 2.7. A new window is presently being fabricated for use on a CPI 110 GHz gyrotron for DIII–D. GA is also developing prototype and full-size 170 GHz windows with improved fabricability, decreased losses and increased bandwidth.

5. SUMMARY AND FUTURE PROSPECTS

The use of mm-waves for plasma heating and current drive/profile control can greatly impact the progress in the development of magnetic fusion as an energy source for the future. A crucial aspect of being able to use the new mm-wave gyrotron sources is having efficient transmission line systems. GA intends to continue to be a leader in designing and making advanced ECH components to meet the demanding requirements of these applications.

6. ACKNOWLEDGMENTS

The authors would like to acknowledge the invaluable contributions of F. Pitschi, W. Loewe, and H. Nickel of Spinner GmbH, and S. Sasaki of Toshiba Corporation.

REFERENCES

1. R.W. Callis et al., "Status of the DIII–D 110 GHz ECH system," (General Atomics Report No. GA-A22378), *Proc. 12th Topical Meeting on Technology of Fusion Energy*, Reno, Nevada, (1996) to be published.
2. R.L. Freeman, "Applications of high power millimeter waves in the DIII–D fusion program," *Proc. International Conf. on Millimeter and Sub-millimeter Waves and Applications III*, (1996) to be published.
3. J.L. Doane and C.P. Moeller, "HE_{11} mitre bends and gaps in a circular corrugated waveguide," Int. J. Electronics, **77**, 489.
4. C.P. Moeller, J.L. Doane, et al., "A vacuum window for a 1 MW cw 110 GHz Gyrotron," 19th Int. Conf. on Infrared and Millimeter Waves, (1994) 279.

ICRF Power Feedback Regulation of the Plasma Diamagnetic Energy Content of the Tokamak TEXTOR.

C. Königs, F. Durodié, M. Vervier

Laboratoire de Physique des Plasmas - Laboratorium voor Plasmafysica
Association "EURATOM-Etat Belge" - Associatie "EURATOM-Belgische Staat"
Ecole Royale Miltaire - B1040 Brussels - Koninklijke Miltaire School

A real time control of the plasma diamagnetic energy content of the tokamak TEXTOR has been realised. The different signals necessary to the computation of the diamagnetic energy are acquired by a 80286 based PC-AT micro computer equipped with a data acquisition board AT-MIO16 of National Instruments. The computed diamagnetic energy content of the plasma is compared to a time-dependent reference signal pre-programmed on the ICRF main control computer and the difference is used to compute the generator output power signal by numerically implemented PID type regulator algorithm. This signal is clipped to a pre-set maximum power level and is output to the generators by the data acquisition board. The acquisition-computation-control cycle is completed in 2 ms. Simulations of the system using a simple energy confinement model showed that a 2 ms cycle time was sufficient to achieve a stable and reliable control of the diamagnetic energy using a continuously control parameter such as the ICRF power. Results so far on TEXTOR show that it is possible to reliably operate in conjunction with radiative edge cooling and to reach operation regimes not to far from the beta- and Greenwald limit in a stable way, with the possible enhancement of some confinement parameters.

1. INTRODUCTION

Radiative edge cooling experiments on TEXTOR, whereby neon is injected in the plasma edge [1,2,3], pose the experimental difficulty that when a confinement transition occurs and the confinement improves, usually the (static) pre-programmed ICRF and NBI power remain too high so that the plasma reaches the ß-limit and the discharge is terminated by a disruption. In order to limit the number of disruptions and to produce more stable and reproducible discharges near the ß-limit a feedback of the power fed to plasma depending on the plasma diamagnetic energy has been realised [4].

Feedback control of the diamagnetic energy content of a tokamak plasma has already been realised on DIII-D by adjusting the duty cycle of the NB injectors [5], as a kind of "bang-bang" control. On TEXTOR the realised system varies the ICRF power in a continuous way.

2. PRINCIPLE

The diamagnetic energy content of the plasma, $E_{dia,m}$, is computed from the plasma current I_p, toroidal magnetic field B_T and the signal of the compensated diamagnetic loop [6] $\Delta\Psi$ by:

$$E_{dia,m} = \frac{3}{8}\mu_0 R_0 I_p^2 - \frac{3\pi R_0 B_T \Delta\Psi}{\mu_0} \quad (1)$$

where R_0 and μ_0 are resp. the tokamak's major radius and the vacuum permeability.

An error, ε, is calculated which is the difference of this measured value and the requested value, $E_{dia,r}$, of the diamagnetic energy. The power requested from the ICRF transmitters, P_{ICRF}, is set proportional to the sum of the error signal itself, the time integral and the time differentiation thereof, weighted with their resp. factors, P, I, D:

$$P_{ICRF} = P\cdot\varepsilon + I\cdot\int\varepsilon\,dt + D\cdot\partial_t\varepsilon \quad (2)$$

The practical implementation using a digital computer introduces a two-fold discretisation : one in time and another in value space. Numerical simulations have been carried out to ascertain under which conditions the thus discretised control loop remained stable. The plasma energy was set to obey the usual equation :

$$\partial_t E_{dia} = P_{tot} - E_{dia}/\tau_E \quad (3)$$

where the energy confinement time, τ_E, depends as in the ITER-L89-P scaling law [7] on the total power, P_{tot}, which is the sum of ohmic, NBI and ICRF power. I_p and B_T were assumed constant while $\Delta\Psi$ was backward calculated from (the continuous) E_{dia} by (1) and then discretised in order to compute the measured $E_{dia,m}$.

With the discretisation in time, equation (2) changes to :

$$P_{ICRF,n+1} = P\cdot\varepsilon_n + I\cdot\sum_{i=1}^{n}\varepsilon_i + D\cdot(\varepsilon_n - \varepsilon_{n-1}) \quad (4)$$

which defines the requested power for the next cycle of the control, $P_{ICRF,n+1}$, through the errors computed from the previous cycles, ε_k.

These simulations have permitted to find practical values for the constants P, I and D and have also shown that in the case of TEXTOR, the control remains stable when the control cycle (acquistion of the signals, computation of the energy and error, and output of requested power) is less than 5ms and furthermore that there is quasi no improvement of the performance of the control loop for cycle times less than 1ms.

3. REALISATION

The plasma current I_p, toroidal magnetic field B_T and the signal of the compensated diamagnetic loop $\Delta\Psi$ as well as the requested value of the diamagnetic energy are acquired by an Intel 80286 based PC-AT micro computer equipped with a AT-MIO16 data acquisition board of National Instruments. By careful programing the cycle time is kept to 2ms and furthermore the diamagnetic loop signal, $\Delta\Psi$, is acquired last in the cycle to minimise the delay between the "measured energy" and the output of the adjusted ICRF power.

The control flow for the diamagnetic feedback system is shown in figure 1. Because the AT-MIO16 data acquisition board has only two analog outputs, extra external hardware was necessary to control two transmitters. One of the two outputs is set to the measured diamagnetic energy $E_{dia,m}$

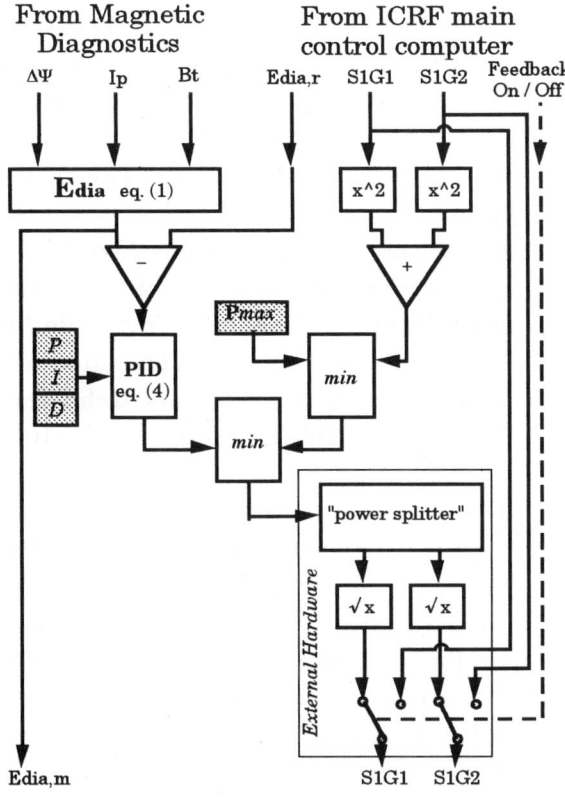

Figure 1 : Control flow of the Energy feedback system : S1G1 and S1G2 are the the control signals for the 2 ICRF transmitters and are proportional to the output voltage rather than power, hence the extra arithmetic (squares and square roots). The constants, P, I, D and P_{max}, can be set from the PC's control display.

while the second output is set to the sum of requested output powers for both transmitters. An external circuit then splits this single signal into a signal for each transmitter whereby the "balance" between the output powers can be choosen. The ouput power is limited by a variable, P_{max}, which, as well as the variables, P, I, D, can be set manualy through a control display on the PC-AT.

4. RESULTS

The energy feedback is used routinely during Radiative I-Mode experiments on

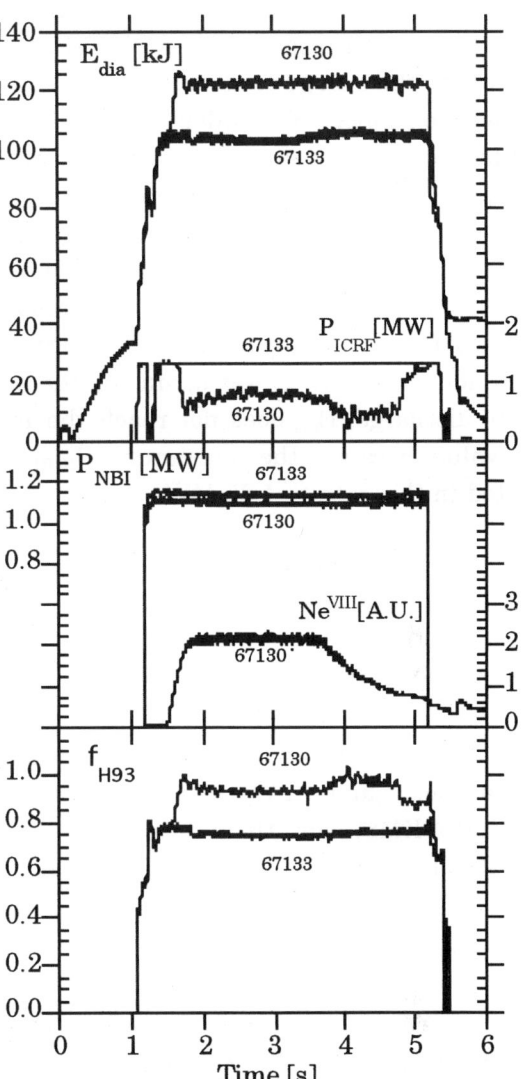

Figure 2 : Evolution of two TEXTOR discharges, 67130 with and 67133 without Neon puff at Ip = 400kA. (Top) Diamagnetic energy and ICRF power. (Middle) NBI power and intensity of Ne-VIII line. (Bottom) the enhancement factor f_{H93} with respect to ITERH93-P.

TEXTOR and has permitted to operate close to ß-limit and up to the Greenwald limit [3,8]. Figure 2 shows two TEXTOR discharges, 67133 without and 67130 with feedback controlled neon injection [9].
In discharge 67130, a clear confinement transition is triggered by the seeding of neon

at t=1.5s as can be seen from the enhancement factor with respect to ITERH93-P, f_{H93}, and where the energy feedback system starts to reduce P_{ICRH} in order to maintain diamagnetic energy, E_{dia}, to the pre-set value of 120 kJ. At t=3.7s the neon valve is closed and as the neon concentration decays, which can be seen from the brilliance of the Ne-VIII line, the improved confinement is gradually lost and the energy feedback has to increase the amount of ICRF power again. At the end of the discharge, E_{dia} does not reach the pre-set value because the maximum P_{ICRH} is limited in this case to 1.35 MW.

For comparison discharge 67133 without neon seeding is shown.

ACKNOWLEDGEMENTS

The authors would like to thank V.Lancellotti and L.Wattiez, members of LPP/ERM-KMS, for their assistance and valuable help.

REFERENCES

1. J.Ongena, A.Messiaen et al., Controlled Fusion and Plasma Physics (Proc. 20th Eur. Conf. Lisboa 1993), vol 17C, Part I, pp. 127.
2. A.M. Messiaen et al.. Nuclear Fusion, Vol. 34, No. 6 (1994), Improved Confinement with Edge Radiative Cooling at High Densities and High Heating Power in TEXTOR.
3. A.M. Messiaen et al., Phys. Rev. Lett., in print.
4. C. Königs, Mesure en Temps Reel et Asservissement du Contenu Energetique d'un Plasma de Tokamak,Travail de fin d'études présenté pour l'obtention du titre d'ingénieur civil polytechnicien. Report LPP-ERM/KMS n° 102. November 1995.
5. DIII-D Team (presented by T.C. Simonen) Plasma Physics and Contr. Nucl. Fusion Research 1992, Recent DIII-D Results.
6. G. Fuchs e al., Interner Bericht KFA-IPP-IB-1/84.
7. A.M. Messiaen, Experimental Transport Analysis and Scaling to Reactors, Transactions on Fusion Technology, Vol. 25.
8. G.Wolf et al., Proc. 16th IAEA Fusion Energy Conference (Montreal 1996), paper IAEA/CN/64O2-5.
9. U. Samm et al., Plasma Phys. Contr. Fusion, 35, B167 (1993).

Generating system for Alfvén waves in TCA/BR

L. Ruchko[*], R. Galvão, I.C. Nascimento, E. Ozono, F. Degasperi, E. Lerche

Instituto de Física, Universidade de São Paulo
Caixa postal 66318, CEP 05389-970, São Paulo, Brasil

The TCA tokamak was transferred from the Centre de Recherches en Physique des Plasmas, Lausanne, to the Institute of Physics of University of São Paulo. The experiments on Alfvén wave heating and current drive initiated at CRPP will be carried on in the new installation, renamed TCA/BR. Although no major modifications are planned for the basic device, the system for exciting Alfvén waves will be substantially different from the one used in Lausanne. The antenna system of TCA/BR will be fed by a four-phase RF generator and will ensure the excitation of travelling waves of different modes with single helicity M=-1, N=-2,-4,-6 and RF power input into the plasma up to 1 MW. The RF system will be based on a four phase generator powered by a pulse forming line. The nominal parameters of the generator are frequency f= 2-8 MHz and pulse duration, τ = 20 msec, that is set by the capacitor bank presently available.

1. ANTENNA DESIGN.

For further progress in the area of Alfvén wave heating and current drive [1-7] improvement of the RF generator and antenna system is essential.

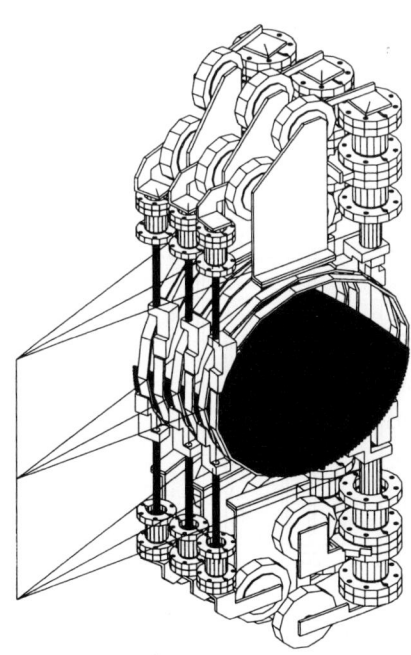

Fig.1 The drawing of TCA/BR antenna unit

Efficient plasma heating and current drive can be achieved only if the possibility to excite monochromatic travelling waves is guaranteed. The location of the RF power deposition zone during Alfvén wave heating depends on the RF frequency and excited wave mode numbers according to the dispersion relation $\omega^2 = k^2_{//}(r)C^2_A(r) (1-\omega^2/\omega^2_{ci})$. An antenna system that launches the waves with double helicity, that is, $N = \pm N_0$, $M = \pm M_0$, is incapable of avoiding RF power deposition close to plasma surface. Most of the previous TCA experiments [3-5] were carried out with excitation of modes N=±2, M=±1 and that was possibly one of the reasons for the central plasma heating to be accompanied by power deposition close to the plasma boundary.

The advanced antenna system of TCA/BR tokamak is formed by four similar units, equally spaced in the toroidal direction, using the same ports for the antenna feeders used in Lausanne. A drawing of a complete unit is shown in the Fig.1. The matching box for each unit will be placed right above and below each unit, as indicated in the drawing. In each unit there are three pairs of complete poloidal loops. Each loop is cut in two equal halves that are fed in parallel and the feeders of the half loops from one pair are displaced 90° in the poloidal direction with respect to each other in order to decrease their mutual coupling. This structure allows the excitation of travelling waves with a main $m=\pm 1$ mode and the

---

[*] *Permanent affiliation*: Applied Physics Institute of National Academy of Sciences of Ukraine, Sumy, 244030, Ukraine

toroidal mode number n can be varied by changing the phase between different loops. The resulting mode spectrum is rather pure, avoiding the excitation of undesirable modes.

The electrical scheme of one loop pair is shown in Fig. 2. The excitation of rotating RF magnetic fields in a definite poloidal direction, that is a choice

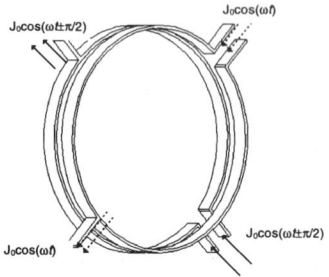

Fig. 2 The electrical scheme of one loop pair

of necessary poloidal wave number sign $M=\pm 1$, is ensured by phasing of feeding currents $\varphi=\pm\pi/2$.

A common Faraday screen will shield each unit and boron nitride plates will protect their sides (Fig. 3).

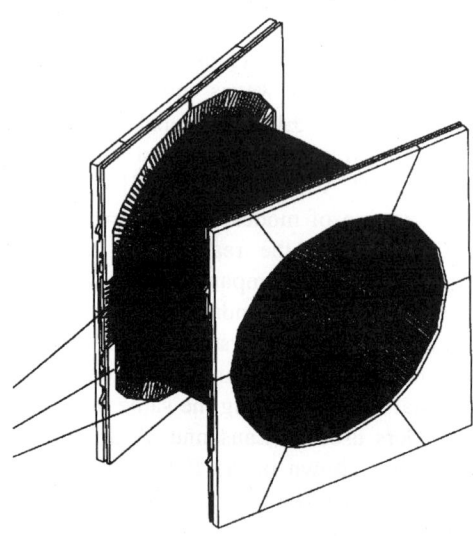

Fig. 3 The Faraday screen with boron nitride (BN) lateral protectors.

2. NUMERICAL SIMULATION OF TCA/BR ANTENNA PERFORMANCE.

For numerical modeling of RF field excitation by the TCA/BR antenna system we used the one dimensional MHD model of current carrying plasma column with a finite ω/ω_{ci} [10]. As it was shown in [8], the antenna radial feeders play significant role in RF field excitation. With the purpose to study the RF field excitation by the antenna system with real geometry, the antennae were modeled by a helical current carrying sheath $I_{\varphi,z}$ $\sim \exp(\iota(M\varphi+Nz-\omega t))$, that was placed at antenna location radius $r_A=0.2$m, and by radial currents in the region $r_A < r < r_W$, where r_W is the conducting wall radius. The radial currents values were chosen in such a way that condition $\vec{\nabla}\vec{J}=0$ was satisfied everywhere in the region $r_A \leq r \leq r_W$. The numerical procedure consisted of finding the numerical solution of the boundary value problem for a reasonable number of field harmonics and then a Fourier analysis was used to determine the loading of the TCA/BR multiloop antenna array and field structure in the vicinity of antenna array. In our analysis we performed the computation for $-24 \leq N \leq 24$, $-6 \leq M \leq 6$. This provided a rather realistic picture of the RF field structure in the region close to the antenna array.

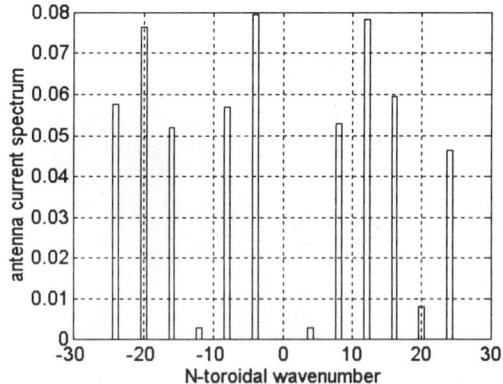

Fig. 4 The spectrum of RF field with poloidal wave number $M=-1$ excited by four module antenna array.

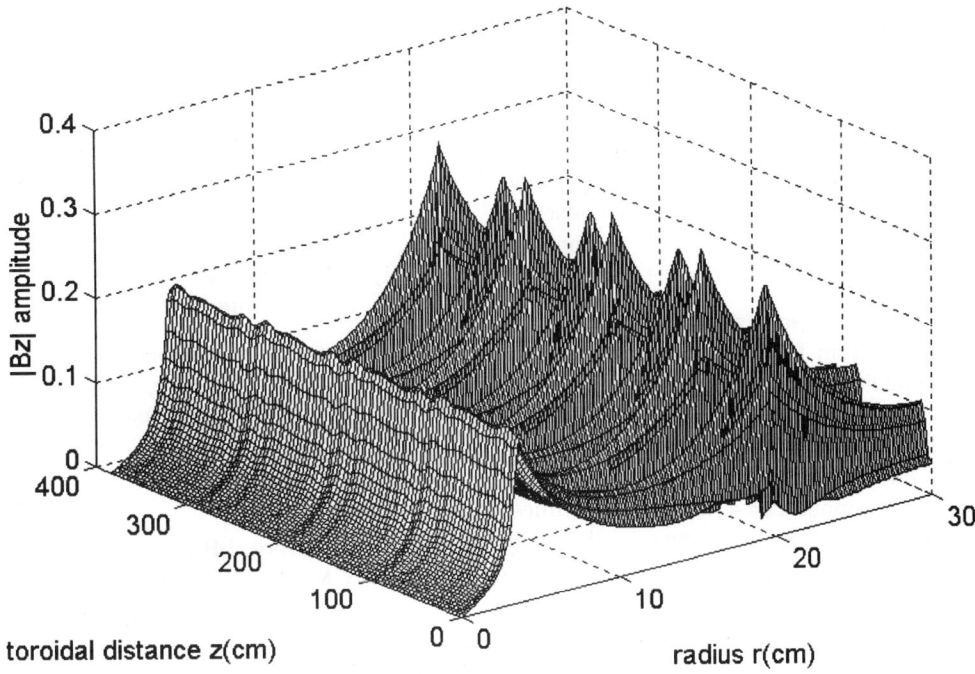

Fig. 5 The structure of RF field excited by four antenna modules with plasma load.

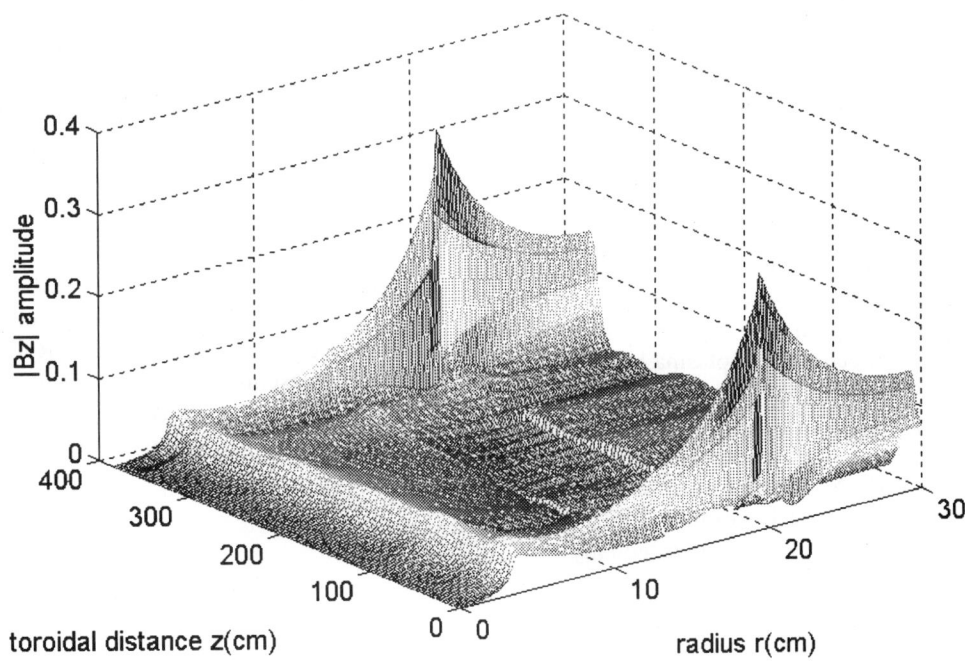

Fig. 6 The structure of RF field excited by one antenna module with plasma load

This analysis permits to study the RF power deposition in the plasma periphery due parasitic mode excitation and to analyze the possibility to decrease the imaginary part of antenna impedance that is connected with additional mode excitation. The following parameters were chosen for RF heating modeling: R=0.61 m, a=0.18 m, toroidal magnetic field $B_0 = 1.0T$, safety factor at boundary q_a =4.4, $q(r)=q_0/(4-6(r/a)^2+4(r/a)^4-(r/a)^6)$, antenna location radius $r_A = 0.2$m, working gas H_2.

The results of the Fourier analysis of the RF antenna current spectrum for antenna array that consists of four modules with phased antenna loop currents is shown in Fig. 4. It is seen that the most dangerous harmonics with low-N mode numbers are effectively damped and optimal conditions for plasma heating by mode N=-4 M=-1 are fulfilled. The corresponding RF field structure is shown in Fig. 5. It is seen that due to discrete nature of antenna array the RF field magnitude at $r = r_A$ has four peaks that correspond to places where antenna modules are located. There is also a sharp increase of the RF field magnitude in the vicinity of Alfvén resonance zone at $r = 0.2_{rA}$. Due to rather good spectrum 'purity', the RF power that is deposited at the plasma boundary because of wrong mode excitation is not large.

The global character of Alfvén wave excitation is clearly seen in the case of plasma heating by antenna array that consists of one module. In that case the RF field in vacuum region close to current carrying antenna elements is large also but at the same time due to resonance properties of plasma load the different Alfvén modes are excited quite efficiently. We see that besides the mode N= -4, that delivers its energy into plasma inner region, the lower-N modes that ensure RF power input close to plasma boundary are excited quite efficiently and it can prevent to efficient plasma heating. .

3. CONCLUSION The RF antenna system of TCA/BR tokamak that consists of four similar units with multiloop poloidal antenna arrays, equally spaced in the toroidal direction, can ensure the excitation of the monochromatic Alfvén modes that are essential for efficient plasma heating and current drive. The developed numerical technique can be used for antenna system design and performance analysis.

Acknowledgments: . The work has been supported by Conselho Nacional de Pesquisa e Desenvolvimento Científico e Tecnológico e Fundação de Amparo à Pesquisa do Estado de São Paulo.

REFERENCES

[1] Weisen, K. Appert, G.G.Borg et al. *Phys. Rev. Lett.* **63**, (1989) 2476 .

[2] T.Intrator, P.Probert, S.Wukitch, et al.; *Phys.Plasmas* **2**, (1995) 2263.

[3] G.A.Collins, F.Hofmann, B.Joye, R.Keller, A.Lietti, J.B.Lister, and A.Pochelon; *Phys.Fluids* **29**, (1986) 2260.

[4] G.G. Borg, and B. Joye; *Nuclear Fusion* **32**, (1992) 801.

[5] G.G.Borg, J.B.Lister, S.Dalla Prazza, Y.Martin; *Nuclear Fusion* **33**, (1993) 841.

[6] R.Majeski, P.H.Probert, T.Tanaka, et al.; *Fusion Eng. and Design* **24**, (1994) 159.

[7] Majeski, P. Probert, P. Moroz et al. *Phys. Fluids* **B 5** (7), (1993) 2506

[8] D.W.Ross, Y.Li, S.M.Mahajan, R.B.Michie; *Nuclear Fusion* **26**, (1986) 139.

[9] Project TCA/BR: a middle size tokamak facility in Brazil. I.C.Nascimento, R.M.O.Galvão, A.G.Tuszel, F.T.Degasperi, L.Ruchko et al. In Proc. of 1994 International Conference on Plasma Physics (ICPP) Foz do Iguaçu-Brazil **v.1**, p.69(1994)

[10] RF system for Alfven wave heating and current drive in the TCA/BR tokamak. L.Ruchko, M.C.R.Andrade, R.M.O.Galvão, I.C.Nascimento. In Proc. of 1994 International Conference on Plasma Physics (ICPP). Foz do Iguaçu-Brazil **v.1**, p.365(1994)

NEGATIVE ION EXTRACTION PHYSICS

M. Bacal, F. El Balghiti-Sube, A.A. Ivanov* and A.B. Sionov*

Laboratoire de Physique des Milieux Ionisés
Laboratoire du C.N.R.S., Ecole Polytechnique
91128 Palaiseau, France

* Permanent adress: Russian Research Center "Kurchatov Institute"

The experimental investigation of the extraction region of a negative ion source by photodetachment is reported. An attempt was done to fit the experimental recovery curves to theoretical ones and to determine thus the negative ion temperature and drift velocity.

1. INTRODUCTION

The negative ion current extracted from a plasma is, obviously, related to the negative ion density in this plasma. Some authors assumed that the extracted negative ion current is determined by the thermal flow of negative ions through the extraction opening, while others believed that the negative ions flow out of the plasma with the ion acoustic velocity. In relation to this problem, a considerable effort has been dedicated to the measurement of the negative ion temperature in the plasma of the negative ion source, but it did not clarify the physics of negative ion extraction, since the measurements have been effected in the central part of the source extraction region. It is the purpose of this work to present the experiments effected in the plasma region next to the extraction opening, a region which is weakly magnetized and limited by a positively biased plasma electrode (PE).

2. EXPERIMENTAL SET-UP AND TECHNIQUE

The investigation has been performed in the hybrid multicusp plasma generator (Camembert III), which has been described in detail elsewhere[1]. The hybrid source contains three distinct regions: i) a driver region, located near the walls, containing the filaments; ii) an extraction region, which extends over all the central, field free region; iii) a weakly magnetized region with high N_i^-/N_e, bounded by the plasma electrode, which contains the extraction opening (N_i^- and N_e are the negative ion and electron densities, respectively). Here the joint action of the positive bias on PE and of the weak magnetic field (20 Gauss at most) results in the redistribution of the plasma components[2,3]. The photodetachment measurements effected at a distance from PE of up to 2 cm show a dramatic increase of the negative ion / electron density ratio, N_i^-/N_e, from 0.5 to 10 (i.e. by a factor of twenty) when the PE bias, V_b, is enhanced from 0 to 2.5 V. In this case N_i^- increases approximately by a factor of two, while N_e drops by a factor of ten (see Figs. 8 and 9 in Ref. 3). Note that the ratio N_i^-/N_e does not exceed 0.1 in the center of the extraction region[1].

The magnetic field in the neighbourhood of the plasma electrode is approximately parallel to this electrode; the electrons are magnetized on a distance of a few centimeters from it. A large fraction of the current into the extraction hole is transported by negative ions. The electrons, as lighter particles, flow to the periphery of the plasma electrode or to the periphery of the plasma chamber along the magnetic field lines.

The H⁻ density, N_i^-, was measured by the photodetachement technique, described in detail in Ref. 3 and 4. In this technique electrons are detached from the H⁻ ions by means of a pulsed laser beam and detected by the cylindrical tungsten probe placed along the axis of the laser beam. The probe is biased positively relative to the plasma and therefore attracts the detached electrons. This results in a probe current pulse whose height is proportional to the H⁻ density. The laser beam is provided by a Nd YAG laser (photon energy 1.2 eV).

The H⁻ negative ion thermal energy was measured using the two-laser-pulse photodetachement technique[4]. This nonresonant optical tagging technique combines two succesive laser pulses with simultaneous Langmuir probe measurements. The first laser pulse destroys all the negative ions in its path by photodetachement. A second laser pulse fired shortly after the first along the same path destroys the negative ions which have originated from outside the laser path. The time evolution of $N_i^-(t)$ is established by making repeated measurements while varying the time delay between the two laser shots. The negative ion temperature T⁻ is determined from the negative ion density recovery curve after the negative ions have been destroyed by photodetachement in a small cylindrical region.

3. RESULTS

Figure 1 presents a typical example of the negative ion recovery observed using the two-laser beam technique.

The measurements were effected on the axis of the source, in the neighbourhood of PE. The four curves shown on Fig. 1 correspond to four values of V_b (0, 1, 2 and 4 V) with respect to the grounded walls.

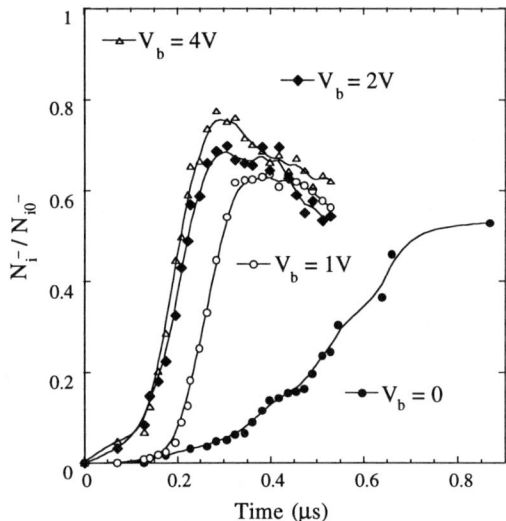

Figure 1. Effect of plasma electrode bias on the negative ion density recovery for V_b = 0, 1 V, 2 V and 4 V. 3 mTorr, 50 V, 30 A discharge, probe bias +50 V, R_L = 0.4 cm.

A small change of V_b leads to a sharp increase of the slope on the time dependence of the negative ion density. The average velocity can be calculated from the relation R_L/t_X, where R_L is the radius of the laser beam (0.4 cm in this case) and t_X is the time required for the increase of the normalized negative ion density to a chosen value N_i^-/N_{io}^- in the center of the laser channel. For example for $N_i^-/N_{io}^- = 0.5$ the average velocity changes from $v \approx 5 \times 10^5$ cm/s at $V_b = 0$, to $v \approx 1.33 \times 10^6$ cm/s at $V_b = 1$ V, and up to $v \approx 2 \times 10^6$ cm/s at $V_b = 4$ V. It is important to note that this velocity increases by approximately a factor of four for any N_i^-/N_{io}^- from 0.1 to 0.5. If, as usual, this velocity is considered to be the thermal velocity of the negative ions, this would correspond to a nonrealistic increase of the temperature by a factor of sixteen. This lead us to assume that there is a flow of negative ions across the laser channel which dominates the negative ion recovery. The velocity of this flow depends on V_b.

Since the voltage applied to electrodes placed at the plasma borders brings about the plasma flow existing in the field-free region (excluding the regions close to electrodes), we assume that the negative ion distribution function before the laser pulse is a shifted Maxwellian, with unknown drift velocity and temperature.

The negative ion flow crosses a channel of radius R_L containing negative ion-free plasma created for a short time by photodetachment. When the negative ion flow velocity exceeds the negative ion thermal velocity, the recovery of the channel plasma is due to the directed flow, rather than to the thermal velocity, so the problem under consideration can be treated in planar geometry and one-dimensional approximation. To simplify we consider that the flow velocity is perpendicular to the laser channel. The point $x = 0$ separates the laser channel from the unperturbed plasma. At the initial moment after the laser pulse the distribution function of the negative ions is supposed to be the shifted maxwellian distribution outside the laser channel, and is zero inside the laser channel. Note that the maxwellian distribution is used as an example and does not decrease the generality of the solution. The influence of the second boundary, at $x = 2R_L$, can be neglected because the flow velocity is large enough.

Since the Larmor radius of electrons is of the order of 0.1 cm, the electrons are magnetized in the vicinity of the plasma electrode. Thus the electrons can move freely along the magnetic lines and the Boltzmann distribution will be valid along these lines only. In the actual situation the magnetic lines cross the plasma electrode, which is an equipotential surface for the electrons. Its potential can be set equal to zero. Therefore we can use the Boltzmann relation across the magnetic field.

When these conditions are satisfied the problem will have the self-similar solution so one could make use of dimensionless variables:

$\xi' = \xi - (V_o/C_s)$, $v' = (V - V_o)/C_s$,

where $\xi = x / t \, C_s$ is the self-similar variable. As a result we have to solve the kinetic equation in form of a nonlinear integro-differential equation, which was solved in Ref. 5 and we will use the solution found there, taking into account the change in argument.

The obtained solutions depend on three parameters N_{io}^-/N_o, V_o/C_s and T_i^-/T_e. The first parameter is determined by laser photodetachment rather precisely[4]. The two other parameters can be found from measurements of the negative ion density recovery at different V_b values by using the self-similar variable $\xi = R_L/(tC_s)$. In the case when the flow is present, this point moves exactly with the flow velocity V_o. Actually the negative ions can reach the probe, placed in the center of the laser channel, from all directions and we should have used the cylindrical geometry.

However, in the case when the flow velocity exceeds the thermal velocity, the growth of the negative ion current to the probe is determined only by the ions arriving from the flow direction. Therefore we can use our planar model when $V_o > V_i^-$.

Finally we plotted all the data of Figure 1 on Figure 2, where N_i^-/N_{io}^- is represented as a function of the self-similar variable ξ.

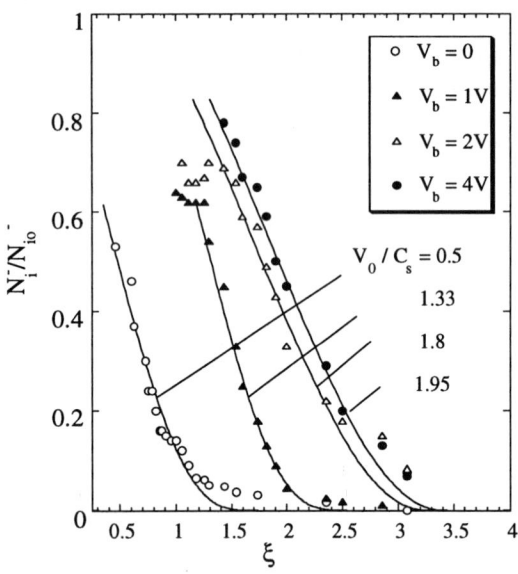

Figure 2. Same as Figure 1, with the variable $\xi = R_L/(tC_s)$. All theoretical curves found by the least square approximation have $T_i^-/T_e \sim 0.2$.

Note that the curves are similar i.e. can be superposed by a translation along the axis ξ. The flow velocity V_o and the thermal velocity, V_i^-, were found for the theoretical curves by the method of least squares. Thus we determine the flow velocity V_o corresponding to each case, as indicated on Figure 2. V_o varies from 0.5 C_s at $V_b = 0$, to 1.95 C_s for $V_b = 4$ V. There is good agreement for $\xi \approx V_o/C_s$.

4. CONCLUSION

These experiments and their theoretical interpretation indicate that the negative ion extracted current is governed by the velocity of the directed negative ion flow to the positively biased PE. In this experiment we found a value as high as 1.95 C_s for $V_b = 4$ V. This explains the physical reason for enhancing the negative ion fraction in the extracted beam when PE is biased positive, which was observed by the experimentalists in this field.

REFERENCES

1. C. Courteille, A.M. Bruneteau and M. Bacal, Rev. Sci. Instrum., 66, 2533 (1995)
2. M. Bacal, J. Bruneteau and P. Devynck, Rev. Sci. Instrum., 59, 2152 (1988)
3. F. El Balghiti-Sube, F.G. Baksht and M. Bacal, Rev. Sci. Instrum., 67, 2221 (1996)
4. M. Bacal, P. Berlemont, A.M. Bruneteau, R. Leroy and R.A. Stern, J. Appl. Phys., 70, 1212 (1991)
5. A.A. Ivanov, L.I. Elizarov, M. Bacal and A.B. Sionov, Phys. Rev. E, 52, 6679 (1995)

EFFECT OF CESIUM SEEDING IN VOLUME NEGATIVE ION SOURCES

M. Bacal, F. El Balghiti-Sube, L.I. Elizarov* and A.J. Tontegode**

Laboratoire de Physique des Milieux Ionisés
Laboratoire du C.N.R.S., Ecole Polytechnique
91128 Palaiseau, France

*Permanent address: Russian Research Center "Kurchatov Institute", Moscow, Russia.
**Permanent address: A.F. Ioffe Physical-Technical Institute, St. Petersburg, Russia.

A systematic study of the effect of cesium vapor pressure on the plasma parameters in a volume negative ion source has been performed. The negative ion relative density is investigated by laser photodetachment in the center of the hybrid source extraction region and in the region near the extraction opening. All the other plasma characteristics and the extracted currents are reported.

1. INTRODUCTION

In hydrogen volume negative ion sources introduction of cesium considerably enhances the yield of negative ions. A review of earlier work can be found in Ref. 1. The mechanism of H$^-$ ion generation in these sources has not been understood yet since the plasma formed in the hydrogen-cesium mixture has a rather complex composition, with four positive ion species, three neutral species, and one negative ion species.

The purpose of this work is to apply the photodetachment technique (2,3) in the study of the characteristics of a multicusp volume H$^-$ source, seeded with cesium. The cesium vapor pressure is determined by a surface ionization gauge. The variation of the extracted electron and ion currents, as well as the resonance radiation of cesium, have also been measured.

2. EXPERIMENTAL SETUP AND TECHNIQUE

In 1995 we have undertaken work with cesium seeding to H$_2$ and D$_2$ plasma in the large hybrid multicusp H$^-$ ion source "Camembert III", of Ecole Polytechnique (4). The first results of this study were reported in Ref. 1. In this source the magnetic filter is represented by the multicusp magnetic field. It separates the driver region, located on the source border, and the extraction region, located in the center of the source. The cesium metal was introduced into the source cesium oven in a sealed glass container, which was broken when heated to a temperature of about 200°C. After this initial introduction of cesium vapor, the cesium reservoir was heated again, to obtain higher cesium pressure. We had no trouble with the electrostatic probes or with operating the filaments and the discharge. We determined the cesium partial pressure with a surface ionization gauge, which has been installed on the source axis in the upper part of the chamber, opposite to the extraction opening. The maximum cesium pressure studied was 1.1×10^{-5} Torr.

The plasma characteristics were measured both in the center of Camembert III, i.e. in the center of its extraction region, and near the extraction opening in the plasma electrode. The negative ion relative density, n^-/n_e, was measured by photodetachment (2,3); the Nd-YAG laser beam diameter was 6 mm; the probe, coaxial with the laser beam, was biased positive at 50 V. In all the experiments the discharge voltage was 50 V and the discharge current was 50 A. The electron density and temperature were measured using the same cylindrical electrostatic probe which was used for the measurement by photodetachment of

n^-/n_e. The principal results for hydrogen will be reported.

We have observed the spectral lines of the neutral cesium. Their intensity is going down when the hydrogen pressure goes up. The intensity of hydrogen lines goes up with this pressure.

3. EFFECT OF CESIUM SEEDING ON THE PLASMA PARAMETERS.

3.1. Study of the center of the extraction region.

The relative negative ion density goes up when the cesium additive is introduced into the hydrogen plasma. The variation of n^-/n_e with hydrogen pressure is illustrated on Figure 1 for pure hydrogen and for two cesium partial pressures.

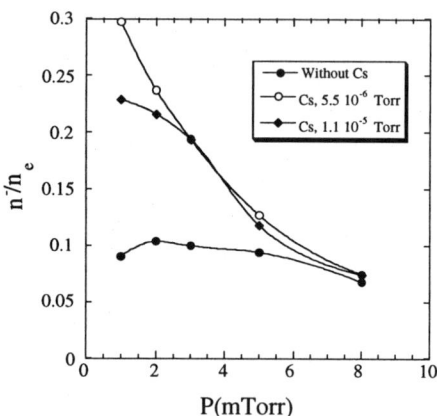

Fig. 1. Dependence of the ratio n^-/n_e on gas pressure in hydrogen plasma for pure hydrogen and two cesium partial pressures.

Note the large enhancement of n^-/n_e, particularly at low pressure (a factor of four in H_2 at 1 mTorr). At 1 mTorr n_-/n_e attains a maximum (0.3) at the intermediate cesium partial pressure, and goes down when the cesium pressure is further enhanced.

Figure 2 presents the variation of the electron density, n_e, electron temperature, T_e, and plasma potential, V_p versus the hydrogen pressure, for the same partial pressures of cesium as in Figure 1.

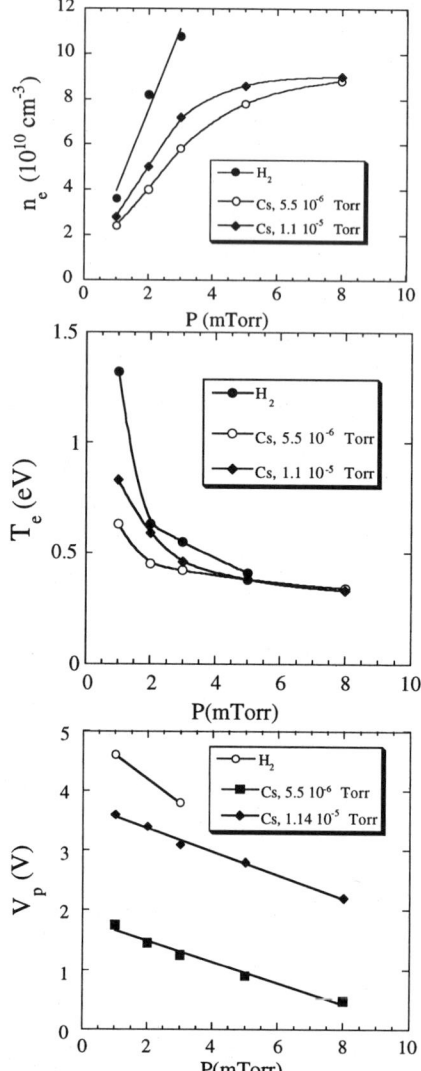

Fig. 2. The variation with the hydrogen pressure of the electron density, temperature and plasma potential for two partial pressures of cesium. For comparison the values measured earlier without cesium are also shown.

The cesium seeding first leads to a decrease of the electron density (as reported in Ref. 1), but the further increase of the cesium partial pressure leads to its small increase; however n_e remains still lower than the value found in pure hydrogen. A similar effect is observed on T_e and V_p. An increase of the plasma potential due to cesium seeding was observed in (1).

Figure 3 shows that the increase of the cesium partial pressure leads to a monotonous enhancement of the negative ion density.

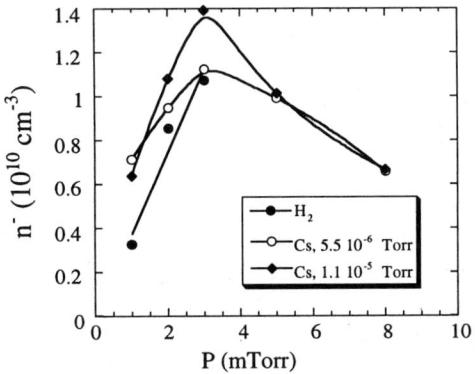

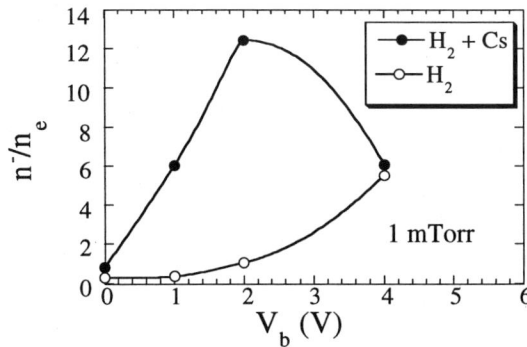

Fig. 3. The variation of the negative ion density with the hydrogen pressure for two partial pressures of cesium. For comparison the values measured earlier without cesium are also shown.

3.2. Study of the plasma near the extraction opening

Figure 4 presents the variation of n^-/n_e versus the plasma electrode bias V_b near the extraction opening, for a partial pressure of cesium of 5.5×10^{-6} Torr. At the lowest hydrogen pressure, the observed enhancement is dramatic, a maximum being identified for $V_b = 2$ V. For the higher hydrogen pressure of 3 mTorr, the enhancement of the maximum value of n^-/n_e is modest, because its value without cesium seeding is very high (approximately 10). One can note a shift of the maximum value towards lower V_b.

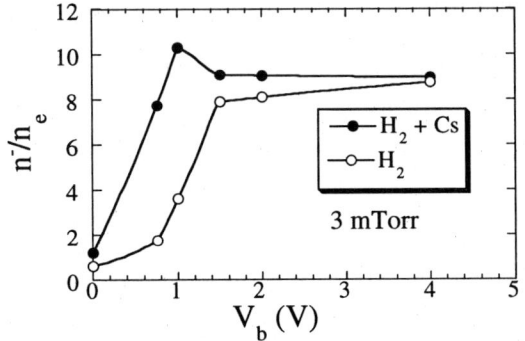

Fig. 4. Dependence of the ratio n^-/n_e on the plasma electrode bias V_b in hydrogen plasma with additive cesium at 3 and 1 mTorr. Partial pressure of cesium: 5.5×10^{-6} Torr.

3.3. Study of the extracted currents

Figure 5 shows the variation with V_b of the extracted electron and negative ion currents for the same discharge conditions and for the hydrogen pressure of 3 mTorr. The maximum partial pressure of cesium (1.1×10^{-5} Torr) was studied here. An enhancement of the negative ion current by a factor 2.5 is observed due to cesium seeding, a maximum being observed for $V_b = 0.75$ V (Fig. 5b).

The reduction of the electron extracted current is larger than the increase of the extracted electron current: a factor of 30 at the optimum V_b.

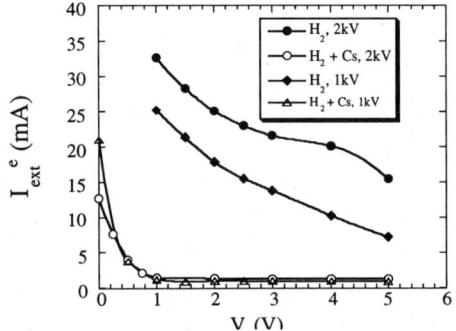

Fig. 5a. Variation of the extracted electron current with the plasma electrode bias at two values of extraction voltage. The partial pressure of cesium was 1.1×10^{-5} Torr. For comparison results without cesium are also shown. Hydrogen pressure was 3 mTorr.

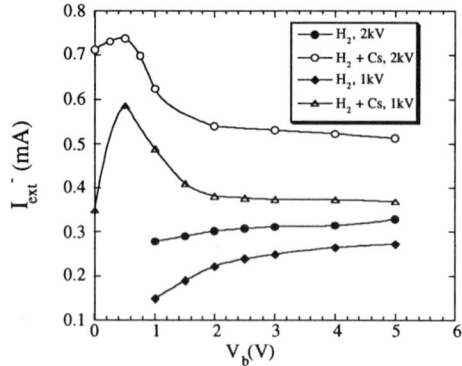

Fig. 5b. Variation of the extracted negative ions current with the plasma electrode bias at two values of extaction voltage. For comparison results without cesium are also shown.

4. DISCUSSION

The reduction of the plasma density is unexpected. It has been suggested that cesium is efficiently ionized in the discharge, which would result in a much higher plasma density for a given discharge power. This was in contradiction with the lower electron current extracted from cesium seeded plasma. We explained (1) the lower plasma density observed in cesium seeded plasma by the enhancement of the positive ion loss due to mutual neutralization, related to the increase of n^-/n_e.

Due to the lower electron density, the negative ion density which is deduced from n^-/n_e is only moderately larger than that measured in the absence of cesium, in spite of a larger effect of cesium on n_-/n_e.

The lower electron temperature observed in the H_2+Cs plasma, compared to that in H_2 was anticipated, since cesium atoms may cool the electrons in the extraction region through inelastic collisions. The lower electron density, associated with the lower electron temperature in the cesium seeded plasma, are the cause of the observed earlier reduction of the extracted electron current.

The variation of the plasma potential after cesium seeding is also unexpected. It was thought earlier that the plasma potential in the extraction region goes down after cesium seeding, as a result of the enhanced secondary electron flow from cesiated walls. It is now clear that the plasma potential passes through a minimum at low cesium partial pressure, and then goes up.

Several causes can be proposed for explaining the increased negative ion density in cesium seeded hydrogen plasma: the enhanced production, due to the increase of the density of vibrationally excited molecules, and the reduced loss, due to absorption (gettering) of atomic hydrogen by cesium.

The present work pointed out (a) a non-monotonous variation of plasma density, electron temperature and plasma potential with the cesium partial pressure; (b) a competition between the excitation of cesium and hydrogen atoms: the radiation of cesium atoms is maximum at the lowest hydrogen pressure, while the radiation of hydrogen atoms goes up with hydrogen pressure.

Acknowledgements. This work was supported by INTAS Contract N° 94-0316.

REFERENCES

1. M. Bacal, C. Michaut, L.I. Elizarov, F. El Balghiti, Rev. Sci. Instrum., 67, 1138 (1996)
2. M. Bacal, Plasma Sources Sci.&Techn., 2, 190 (1993)
3. F. El Balghiti-Sube, F.G. Baksht and M. Bacal, Rev. Sci. Instrum., 67, 2221 (1996)
4. C. Courteille, A.M. Bruneteau and M. Bacal, Rev. Sci. Instrum., 66, 2533 (1995)

First Results of Automatic Matching System on Tore Supra ICRH Antennas ; Fast Matching Network for ICRH Systems

L. Ladurelle, G. Agarici, B. Beaumont, S. Bremond; H. Kuus, G. Lombard,

Association EURATOM-CEA, Departement de Recherches sur la Fusion Controlee
Centre de Cadarache, 13108 Saint Paul lez Durance Cedex, France

Tore Supra ICRH Resonant Double Loop antennas are now equipped with an automatic matching system based on internal capacitors active control. First results on power experiments are exposed.
For future experiments and fast transients, the principle of a fast matching network for ICRH systems has been developed and first tests on a laboratory mock-up have been performed.

1. INTERNAL CAPACITORS FEEDBACK CONTROL

The first approach to automatic matching of Tore Supra RDL antennas consists in tuning the antenna feed point impedance to the feed line characteristic impedance through internal capacitors control. The theoretical method has already been exposed [1], following paragraphs show equipment involved as well as first tests on high power operation.

1.1. Hardware equipment

Error signals derived from line directional voltages and phase measurements are used to drive fast brushless servo motors coupled to the internal matching capacitors C_1 and C_2 of each half antenna (figure 1).

• **Measurements** :
Forward and reflected power are directly measured through a directional coupler and fast diodes supplied by SPINNER[*]. Reflection coefficient phase is given by phase detectors TS35164.

• **Capacitors driving system** :
Each antenna is equipped with a motion coordinator which drives four brushless servo motors. This multi tasking controller is provided with four analog inputs fed with reflection coefficient amplitude and phase to compute error signals. Digital inputs are used to allow capacitors motion according to antenna safety and power thresholds.

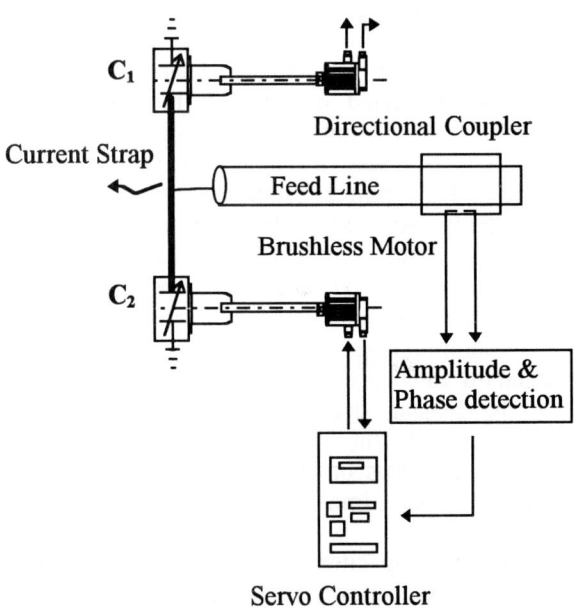

Figure 1 Hardware Equipment

The USASEM05A brushless servo motors with optical encoders have been chosen for their reliability in magnetic environment. Tests up to 0.2 T have been successfully performed.

SPINNER GmbH Erzgiessereistrasse 33 8000 Munich 2 RFA

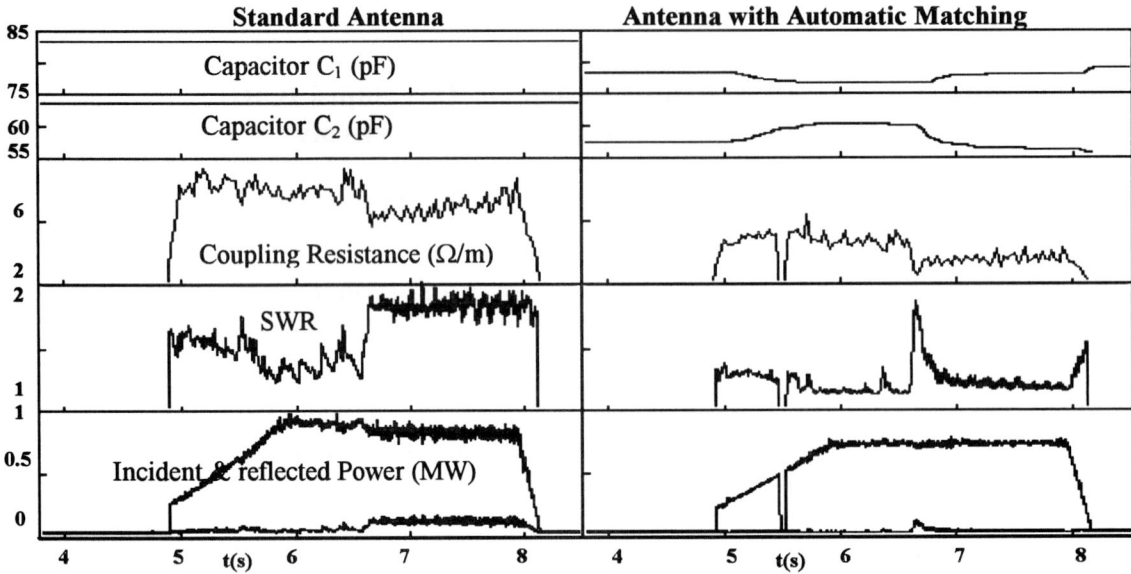

Figure 2 Typical Closed Loop response during shot #19767

1.2. High Power Operation

The automatic matching device has been successfully tested on one of the three TORE SUPRA ICRH antennas during high power experiments at several frequencies [48-57-63Mhz]. During R.F. power ramp up, where coupling conditions are not stable, capacitors vary to minimize error signals and reach a good match. On the shot #19767, after this smooth variation a rapid change in antenna loading occurs and makes the coupling resistance to fall. The SWR of the standard antenna increases, and the power coupled to the plasma decreases, whereas the antenna equipped with the automatic matching device is tuned and keeps on coupling its nominal power after a response time of about 200 ms *(figure 2)*.

As present capacitors life time is dependent on their internal bellows mechanical resistance, speed and acceleration have been limited to minimize stress on this weak part. This system can consequently reach by itself good matching conditions and follow smooth changes in plasma coupling.

The system optimizes also power transmission during macro-phenomena as giant saw teeth *(figure 3)*. It will help to couple ICRF power during long plasma shots with various loading configurations.

For next experimental campaign the whole installation will be equipped with this device. In order to minimize time response, new techniques as RF vacuum sliding contacts are studied to upgrade capacitors [2]. Those could be modified to allow faster movements and lower time response.

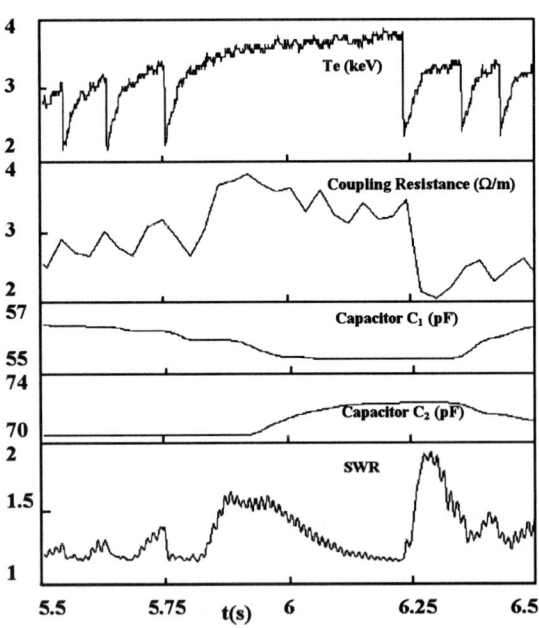

Figure 3 Giant Saw Teeth Perturbation

2. FAST MATCHING NETWORK FOR ICRH SYSTEMS

As no mechanical system can cope with fast transients, such as Elm's or large pellet injection for example, a method using frequency tuning has been developed and will be tested on TORE SUPRA ICRH plant.

2.1. Principle

A double resonant system is created by connecting a fixed stub to a short resonant antenna through a long line *(figure 4-1)*. The short resonant part can consist in short circuited lines in parallel *(figure 4-2)* or in a Tore Supra type antenna with internal matching capacitors *(figure 4-3)*.

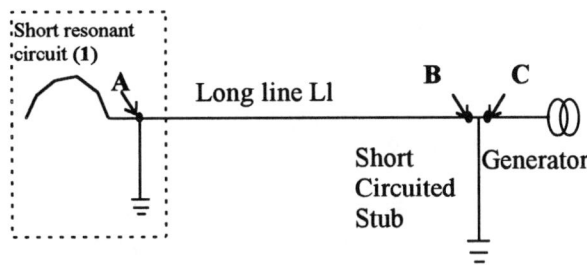

Figure 4-1 Double resonant system

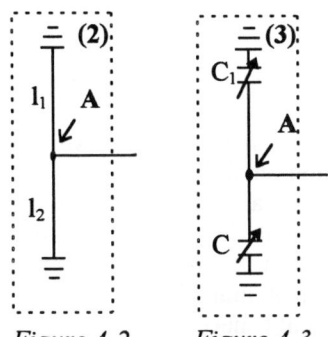

Figure 4-2   Figure 4-3

Line and stub length optimisation allows to keep a low VSWR at the generator on a wide range of coupling resistance by tuning the source's frequency *(figure 5)*. Following simulations are computed with a Tore Supra antenna as short resonator. Reflection coefficient curves are drawn for several coupling resistance with frequency varying from 56.5 to 58 MHz *(figure 6 and 7)*.

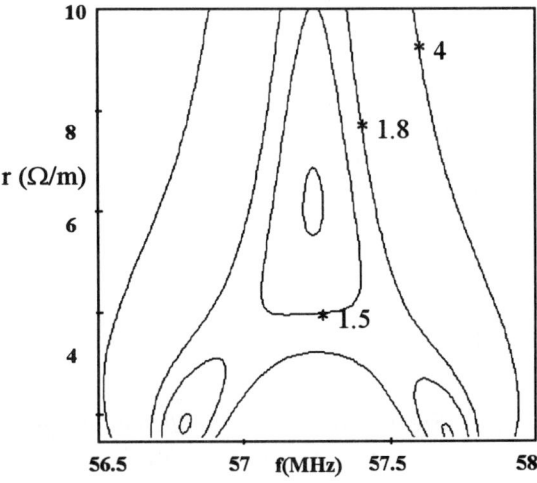

Figure 5 VSWR Contour Map for a TS antenna at point C

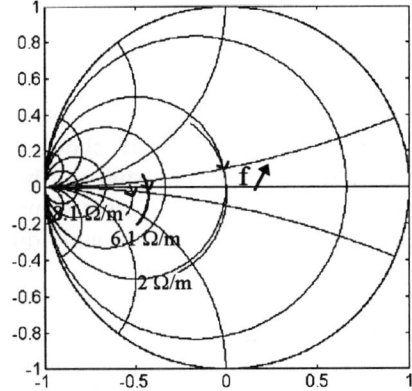

Figure 6 Reflection Coefficient at point A for several coupling resistance : $r=2$, $r=6.1$ and $r=8.1$ Ω/m

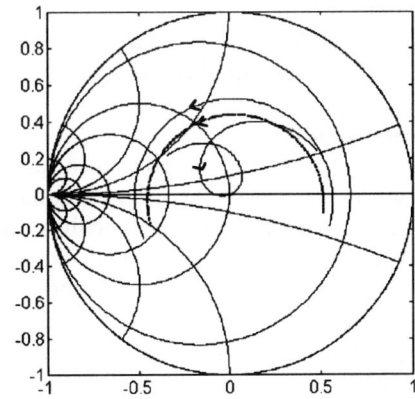

Figure 7 Reflection Coefficient at B

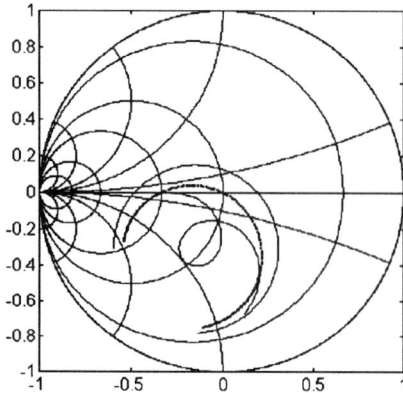

Figure 8 Reflection Coefficient at C for r=2, r=6.1 and r=8.1 Ω/m

Smith charts at point A (figure 6), B (figure 7) and C (figure 8) show reflection coefficient evolution along such a system. After the fixed stub (C), each curve goes near the point of zero reflection.

2.2 Error Signals
Curves shape in Smith chart shows that the reduced admittance after the stub can be used to control frequency value. This error signal can be derived from measurements through directional couplers at the generator output.

2.3 Sensitivity
Sensitivity to the short resonant part parameters is rather high. The accuracy required on Tore Supra antenna capacitors value is +/- 2 pF, which corresponds roughly to +/- 2 cm if capacitors are replaced by short circuited lines.
Sensitivity to transmission line length and to the stub length is rather low. A variation of +/- 10 cm on these length does not have any significant effect on the behaviour of the system.

2.4 Tests Results
This principle has been applied to a laboratory mock up at low power. The short resonant part consists in coaxial lines ended with a tunable load. Frequency feed back is realized through directional line measurements and analog electronic treatments.
Characteristics of the system are :
l1 = 2.758m   l2=3.492 m
Ll= 37.4 m   Lstub = 2.15m

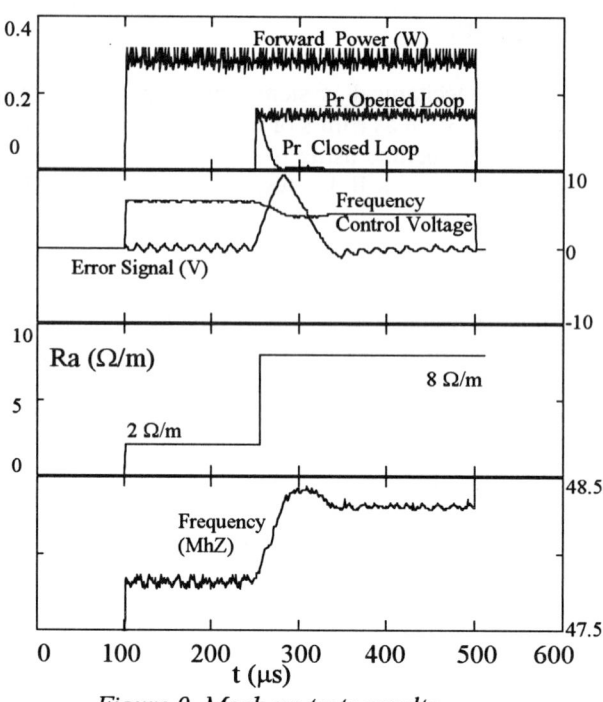

Figure 9 Mock-up tests results

Tuning times as low as 25 µs have been reached to tune the antenna from 2 to 8 Ω/m *(figure 9)*.

3 CONCLUSION
After successful tests on one antenna, the automatic matching system based on capacitors feedback control is being extended to the whole TORE SUPRA ICRH plant to improve plasma long pulses operation reliability.
For fast transients, a method involving no mechanical adjustment is needed, the double resonant circuit design seems to be a promising solution. The effect of mutual coupling between antenna straps should nevertheless be studied. Experiment will be driven on TS in order to check this concept on a full power test.

4 References
[1] L. Ladurelle et al., "Matching of The Tore Supra ICRH Antennas ", in : proc. of 19th SOFT, vol. 1, pp 545

[2] G. Agarici et al., "Sliding Contact Tests at High R.F. Current under Vacuum", in : proc. of 19th SOFT.

ICRF operation during H-mode with ELMs
Development status at ASDEX Upgrade

F. Wesner, W. Becker, F. Braun, H. Faugel, R. Fritsch, F. Hofmeister, J.-M. Noterdaeme, Th. Sperger

Max-Planck-Institut für Plasmaphysik, Euratom Association, D-85748 Garching

Fast plasma boundary variations like ELMs result in heavy antenna coupling changes, strongly affecting the operation and the reliability of the ICRF devices. At ASDEX Upgrade it is tried to overcome these interactions by electronic means and by the use of hybrids. The strategy and status of this development are described.

1. INTRODUCTION

The coupling characteristics of ICRH loop antennas, launching the RF power into the plasma, depend mainly on the plasma boundary and scrape-off layer profiles and on their distance to the antenna. Boundary variations change the antenna impedance, which must be matched to that of the generator by an adequate device. If such variations can not be compensated by feedback means, RF power is reflected back to the generator, which immediately reacts with an output variation. Too large reflections trigger the safety cut-off, normally used as arc protection system.

The generator control is able to correct output deviations slower than typically 1 ms, if load limitations of the power tetrode are not exceeded. These limits depend heavily on mismatching conditions: Only 10% power reflection (VSWR~2) reduces the maximum generator output to about 50%.

Type-1 ELMs are strong boundary variations increasing the antenna coupling resistance by a factor of 2 to 4 within few microseconds, the decay times being within a few ms. The repetition rates, above 3 ms, depend on the heating power. The resulting reflection peaks are comparable to those of arcs, normally interrupting the RF power [1,2,3].

So, reasonable high power ICRF operation above about 1 MW per generator is only possible, if cut-off and a too large reduction of the average power due to the ELMs can be prevented. At ASDEX Upgrade this is being tried with "active" electronic measures, and tests have been made with hybrids, allowing to compensate reflections due to load variations "passively", thus reducing the generator reactions.

2. ELECTRONIC METHODS AGAINST REACTIONS ON ELMs

2.1 Feedback

A matching feedback using fast frequency shifts is the only method which can possibly be made fast enough to cope with the coupling variations at ELMs. At ASDEX Upgrade such a system is being developed, its status and first test results are reported in this conference [4]. Its application would require an increased bandwidth of the ASDEX Upgrade generators. Other groups are also working in this field [5].

2.2 Power reduction

Primarily power cut-off due to the reflection peaks at ELMs must be prevented. At ASDEX Upgrade this can be achieved by an electronic device, triggered by an ELM-signal (H_α or ring voltage). It reduces the generator output to an adjustable value, fast enough to avoid the large peaks. This device, already described in [1,6], was complemented by a feedback with a typical response time of some 10 µs, keeping the generator output at the maximum power allowed by the reflection rate. At smaller reflections the output is kept unchanged. It is reduced if an adjustable reflection rate is exceeded, and immediately increased if it becomes smaller again (Fig. 1). After the triggered power reduction at ELMs it increases the output as fast as allowed by the decaying ELM effect (Fig. 2). This system allows stable operation at arbitrary matching conditions.

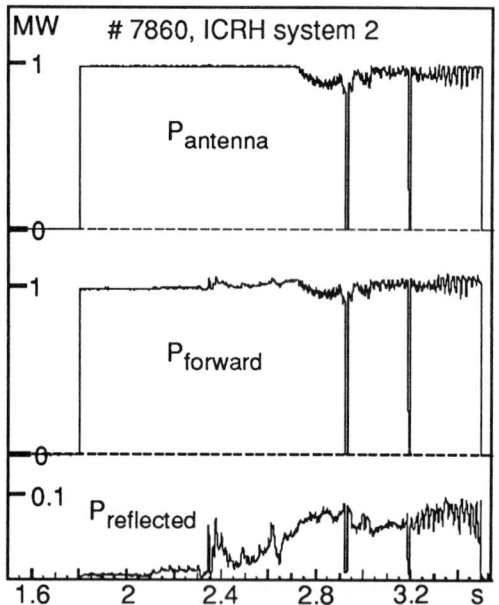

Fig. 1: Typical reaction of the power feedback. Power reflections are compensated by the generator keeping the heating power constant until 2.7 s. Then the power is reduced, keeping the reflection below about 80 kW.

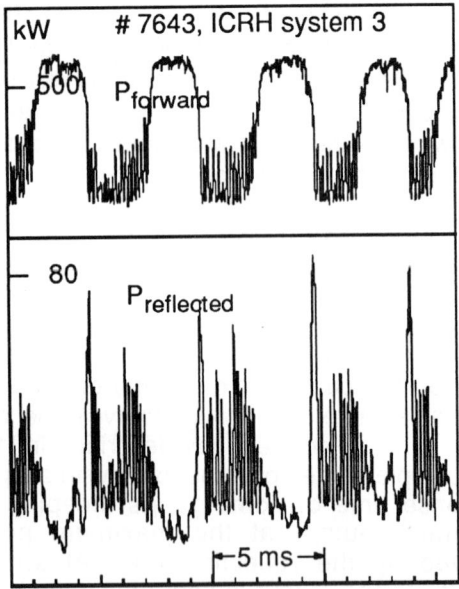

Fig. 2: Power reduction at ELMs with a high repetition rate of about 5 ms. After the power reduction the feedback tries immediately to raise the power again step by step.

2.3 Arc detection

A reliable arc protection method not using the reflected power and insensitive against ELMs and other load variations would allow to tolerate fast reflection peaks and to reduce the average power reduction shown in Fig 2. The tetrode protection, also using the reflection signal, do not need such fast reactions.

Looking for alternatives, two arc sensitive frequency bands besides the nominal frequency have been identified: The second harmonic used at TFTR [7] has proved to be also sensitive to ELMs [8]. At ASDEX Upgrade a system is developed using a frequency band below the generator frequency [9], showing the needed characteristics. If reliability tests, still to be done, confirm its applicability as safety system, it shall routinely be applied, the present reflection based cut-off can then be retarded and used to gain redundancy.

3. HYBRIDS

3 dB Hybrids are 4-port-devices which can be applied as power combiner, adding the power of two generators, or as power splitter feeding two loads. In ICRF systems they have been used so far in circuitries for current drive using the 90° phase difference between the splitter outputs [10]. They can also reduce the effect of plasma variations on the RF device, since reflections from equally mismatched loads are compensated by a power splitter [11].

If, in practice, the impedances of two antennas combined by a hybrid are not equally matched, only the difference of the reflections from both goes back to the generator, the main part being absorbed by a dummy load connected to the hybrid. Coupling variations and the resulting reflection peaks due to ELMs, which can be assumed to be similar at equal antennas, will also be compensated by a hybrid to an extend mainly depending on the equality of the matching. The power lost in the dummy load is much smaller than the output reduction caused by the same reflection without hybrid.

Two of the 4 ICRH systems of ASDEX Upgrade have been equipped with hybrids (which were kindly loaned to us by JET) in a test set up to study the operational characteristics of such circuitries (Fig. 3). The first hybrid is used as power combiner, followed by the other one splitting the power on the two antennas.

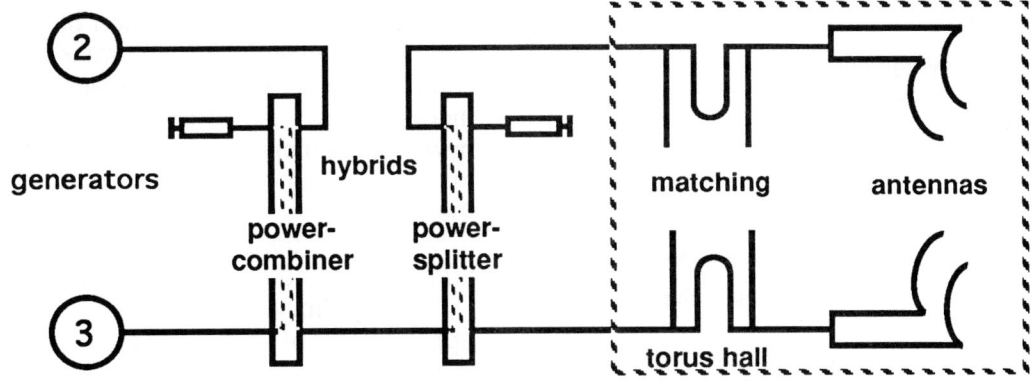

Fig. 3: Hybrid test circuitry combining generators and antennas of the ICRF systems 2 and 3.

This circuitry allows to combine two antennas without losing power.

Different characteristic impedances of the ASDEX Upgrade system (50Ω) and of the JET hybrids (30Ω) were compensated by λ/4 transformers, resulting in a small bandwidth at 30 MHz. The phase between the generators is feedback controlled to 90° to minimise the power lost in the combiner load. The power distribution was determined by measuring the losses in the dummy loads and the forward and reflected power before and after the hybrids.

The results show the expected reduced interaction between generator and plasma due to the heavily reduced power reflection seen by the generator at all operational conditions. A much larger heating power could be achieved more reliably: At a VSWR of about 2 (~10% power reflection) between hybrids and matching elements a generator power of 2x1.8 MW was achieved, about 2x1.6 MW being deposited in the (L- mode) plasma. Without hybrids a generator power of only about 2x1 MW would have been reached at the same VSWR value.

Besides the matching conditions, the power loss in the dummy loads depends also on the phase between the generator outputs (Fig 4). which should be kept within few degrees. These matching and phase conditions are worsened at this test due to the different line and hybrid impedances.

Fig. 5 shows that the reflection peaks seen by the generators at ELMs are heavily reduced, thus allowing a more stable operation at higher power as without hybrids. But it is also seen, that double the power would in spite of the hybrids require an improved matching for antenna 3 to avoid tripping.

The different amplitudes of the remaining reflection peaks at the two generators are due to matching differences. It is not sufficient to achieve equal power reflections from both antennas. When the antenna coupling varies at an ELM the gradients or at least the gradient signs of the reflection must also coincide. This means, that both matching systems must be adjusted to the same sides of the matching optimum. This is a new task for ICRF matching systems not required so far.

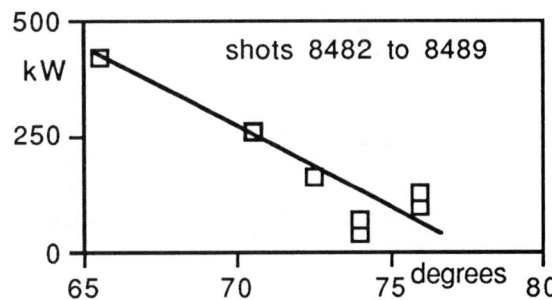

Fig 4: Power loss in the combiner load versus phase at largely equal plasma conditions.

In summary the test proved, that with hybrids the operational characteristics of an ICRF system can be heavily improved and a larger heating power can reliably be achieved. Only at heavy mismatching, e.g. starting with an unknown plasma load, the generators could still be affected, resulting in power variation or even tripping.

#8474, ICRH systems 2 and 3

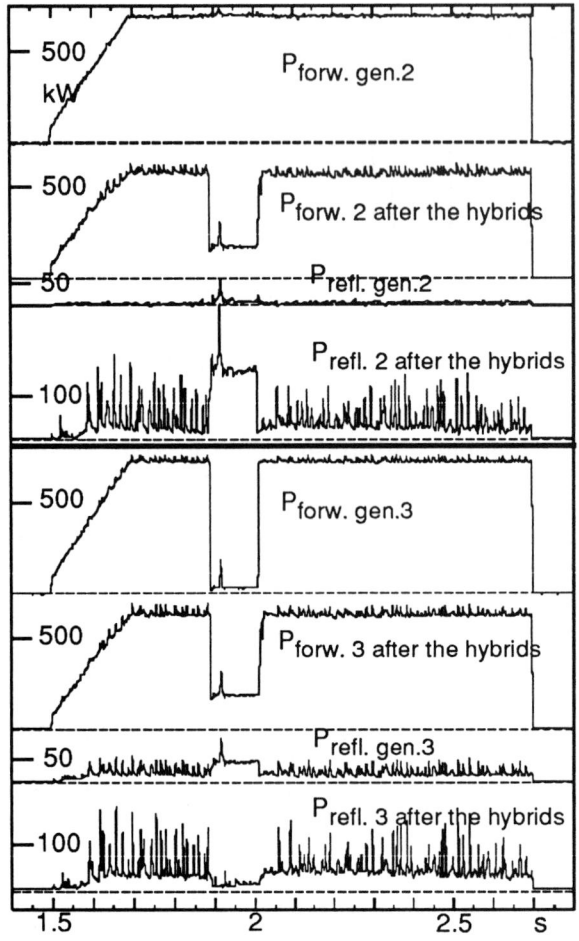

Fig. 5: Typical H-mode shot with hybrids with forward and reflected power before and after the hybrids. Reflection peaks at ELMs are largely compensated, but not equally due to the differently matched coupling impedances.

4. SUMMARY AND CONCLUSION

Electronic devices transiently reducing the heating power in the case of too large power reflection, developed at ASDEX Upgrade, can avoid tripping of the generators and allow stable operation at ELMs. But at high ELM frequencies the average power is reduced up to a factor of 2.

Tests at ASDEX Upgrade have further shown, that by the application of hybrids the interaction between RF generator and plasma can strongly be reduced, thus allowing a much more effective use of the available generator power and a more reliable operation at ELMs. If the advantages of hybrid shall be fully utilised, the impedances of antennas combined by hybrids must be equally matched, calling for further improved matching systems.

Even in circuitries with hybrids the additional application of the described electronic methods remain reasonable to guarantee operation at heavily mismatched conditions, e.g. before the matching elements are properly adjusted. An alternative arc detecting method independent of the power reflection is even more necessary for hybrid circuitries, where the power reflection and the sensitivity of a reflection based arc detection are heavily reduced.

Due to the described results, showing that with hybrids the operational conditions can be much more improved than by electronic measures alone, it is planned to equip the ICRF system of ASDEX Upgrade with hybrids.

References

1. F. Wesner et al., Proc.18th Symp. on Fusion Technology, Karlsruhe (1994), 537.
2. J.R. Wilson, ibid., 533.
3. J.-M. Noterdaeme et al., Proc. 21th Conf. on Controlled Fusion and Plasma Physics, Montpellier (1994), 842.
4. F. Hofmeister, this conference.
5. D.F.H.Start et al., Proc. 11th Top. Conf. on Radio Frequency Power in Plasmas, Palm Springs (1995), 7.
6. F Wesner et al., ibid., 15.
7. J. R. Wilson et al., Proc.18th Symp. on Fusion Technology, Karlsruhe (1994), 533.
8. J. H. Rogers et al., Proc. 16th Symp. on Fusion Engineering, Urbana (1995), 422.
9. F. Braun, this conference.
10. R. W. Callis et al., Proc.18th Symp. on Fusion Technology, Karlsruhe (1994), 573.
11. R. H. Goulding et al., Proc. 11th Top. Conf. on Radio Frequency Power in Plasmas, Palm Springs, (1995), 397.

An ARC Detection System for ICR - Heating

F. Braun, Th. Sperger

Max Planck Institut für Plasmaphysik, D- 85748 Garching, EURATOM Association

Due to resonant matching between generator and load (antenna) in Ion Cyclotron Resonance Heating (ICRH) systems, high Rf-voltages can cause arcing within the coaxial line or antenna. The reflected power at the generator is influenced by ELMs (reflection changes caused by the plasma) and arcs (caused by high Rf-voltage flashovers). The power tube can withstand short-term reflections, but the generator has to shut-off if an arc occurs, to prevent damages within the transmission line system. A system has been designed to distinguish between arcs and ELMs.

1. Introduction

Fast fluctuations in the plasma, result as antenna impedance variations, which can not be compensated by the matching network, lead to reflected power at the generator.   (1, 2)
Too high voltages, caused by the high transformation ratio (in the order of 20) between the antenna and the generator, lead to arcs within the coaxial transmission line or antenna, and the generator has to be switched off. Normally, such a safety system is based on the reflected power.
A system has been developed to distinguish between transients, caused by the plasma and transients, caused by arcs.
For getting more information about the reason of switch-off during a plasma discharge, the characteristics of various arcs are investigated.
Different methods for arc detection are known:

a. Detection by photodiodes
b. Phase measurements
c. Observation of the 2nd harmonic of the generator   (3,4)
d. Frequency independent detection.

This method is discussed in more detail.

2. Frequency independent Detection

If a transient occurs, one gets a frequency spectrum which is the broader, the higher its change. Such a spectrum also can be seen below the operating frequency of the generator. If this method is used as a safety shut-down system, it has to operate independently of frequency and amplitude. The frequency band being observed, only is influenced (but not matched) by the network, therefore such a system is independent of the location of the pickup probe within the system. It also is independent of the type of the probe (directional coupler, voltage / current probe).

Different transients and their spectra have been observed via a filter and a fast digitising storage scope (see Figure 1). In general, transients lead to amplitude modulation of the carrier (generator frequency) and can be written as

$$F = (n \times F_G) \pm (m \times F_{MIX})$$

where n, m are the harmonics of the generator (F_G) and mixing frequency (F_{MIX}).

If an arc occurs, F_G, is off the observed band, therefore the spectrum is independent of the generator frequency and its harmonics.

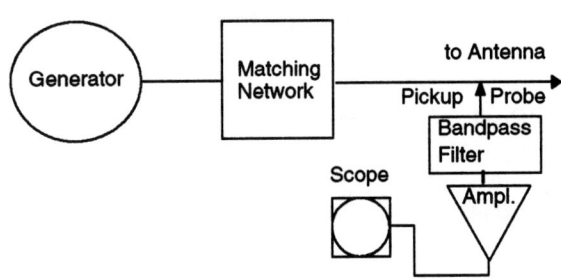

Fig. 1 Test set-up for observing transients

3. Measurements

In a first step, the bandpass filter was removed to verify, that plasma transients only produce a spectrum close to the generator frequency. This has been done with both, ELMs (Fig. 2) and sawteeth. The sawtooth spectrum is more narrow compared to the ELM spectrum.

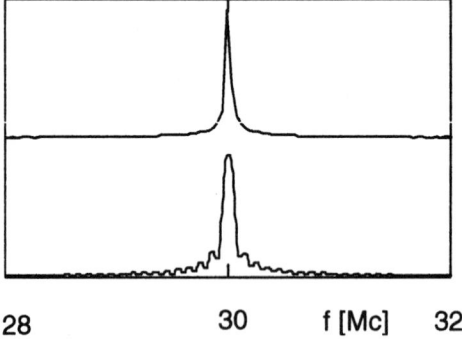

Fig. 2 Upper signal: Spectrum without an ELM
Lower signal: Spectrum during an ELM
Generator frequency: 30 Mc

4. Arc Detection

The fundamental frequency of the generator is now suppressed by the bandpass filter to observe transients, caused by arcs within the Rf-system including the antenna. During the tests, reflected power, the maximum voltage within the unmatched line and the arc signal coming from the above mentioned system, have been observed. A typical measuring diagram is shown in Fig. 3.

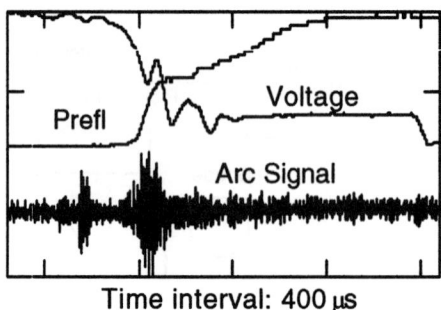

Fig. 3 Observed traces during arcs

The following Fig. 4 shows spectra of different arcs.

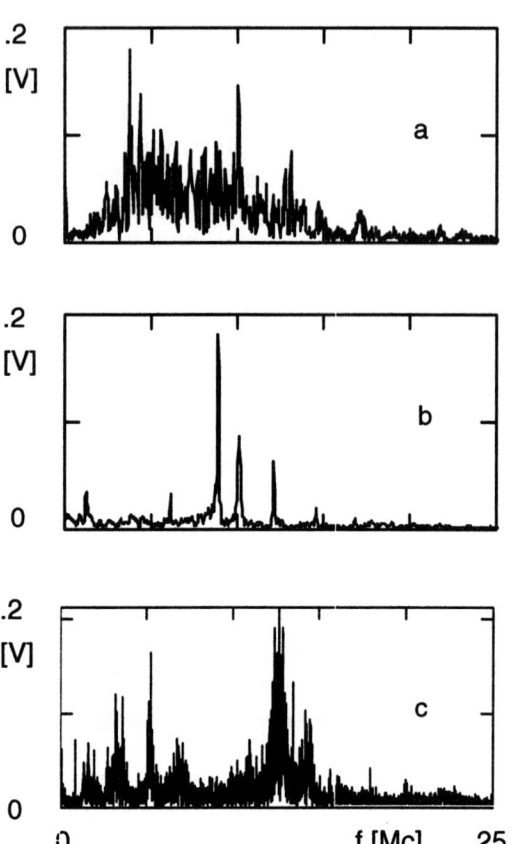

Fig. 4 Frequency spectra of arcs in the
a: Transmission line under atmospheric pressure
b: Transmission line under vacuum conditions
c: antenna during a plasma discharge

With this method also arcs across contacts or DC-breaks (series arc) can be detected (Fig.5). This has been done at a power level, where a 25 μm thin Kapton foil, placed in series of a coaxial line, arcs. This set-up represents a reactance of .2 Ω. Nearly no change of the reflected power has been observed at the generator due this event.

Figures 2 to 5 have been collected with a fast digital storage scope with a sampling rate of 10 ns.

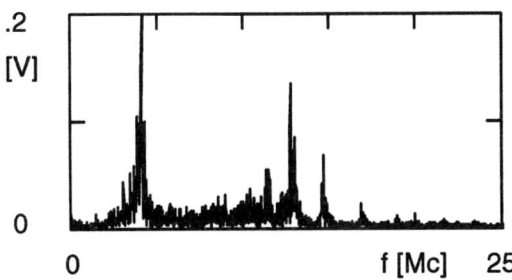

Fig. 5 Spectrum of a series arc
Generator power: 5 kW

The spectrum of different arcs are nearly independent of frequency or amplitude of the generator.

5. Hardware

The most critical part within such an arc detection system is the bandpass filter, which must have a rejection of more than 60 dB for the lowest possible frequency of the generator to avoid error signals, which can be picked up from plasma fluctuations. The limit for the lower cut-off frequency is determined by noise, induced by spikes from thyristor control circuits etc.

Therefore this filter is split into two parts, a lowpass filter with a cut-off frequency of about 20 Mc and a highpass filter with $f_c = 4$ Mc.
The high- and lowpass parts are built as 9 pole Tchebycheff type filter, which attenuates the minimum possible frequency of the ASDEX - Upgrade/ WVII - AS generators by 65 dB. The whole system must be within a shielded enclosure to keep the noise level down.

For a safety system the fast digitising scope is replaced by a 60 dB low noise amplifier and a Rf-rectifier. This signal then is fed through a pulse stretcher to the ICRH data acquisition and to the shut down input of the generator.

6. Conclusions

It has been demonstrated that such a system mentioned above, is able to distinguish between transients caused by the plasma and arcs within the transmission line system independently of the operating frequency. It is also the first demonstration of detecting series arcs (e.g.: in DC-breaks).

Further tests have to be done, to confirm the applicability as a safety system. Its routinely operation is planned, the present reflection based shut down device then acts as a retarded fallback device.

7. References

1. F. Wesner et al., Proc 18th Symp. on Fusion Technology, Karlsruhe (1994), 537
2. F. Wesner et al., this conference.
3. J.R. Wilson et al., Proc 18th Symp. on Fusion Technology, Karlsruhe (1994), 533
4. J. H. Rogers et al.,16th Symp. on Fusion Engineering, Urbana, (1995), 422

Matching Fast ICRF Antenna Coupling Variations by Frequency Change

F. Hofmeister and ICRH Team

Max-Planck-Institut fuer Plasmaphysik,
D 85748 Garching, Germany, EURATOM-Association

At ASDEX Upgrade a test was done to improve matching during an ICRF heating experiment with a slightly modified matching network and an extra generator side stub, thereby being suitable for frequency sensitive matching, by a small frequency offset generated online according to measured antenna coupling variations in a Type I ELM shot.

1. INTRODUCTION

Matching ICRH antennas is still a challenge, when the antenna coupling variations during a shot are so fast that a compromise setting of the tuners taking into account the maximum and minimum coupling is not sufficient. This is for instance the case when ICRH in a divertor tokamak like the ASDEX Upgrade is used with large total heating power, bringing the plasma into the H-mode. Then a regime with large periodic bursts of particles leaving the plasma is reached, the so-called Type I ELMs, which is accompanied by large antenna coupling variations.

2. A FREQUENCY SENSITIVE NETWORK (FSM)

This problem may be solved by tailoring the RF feeder network in such a way that frequency variations have a large effect on the transformation factor of the real part of the antenna impedance to the generator impedance[1]. This method is different from conventional frequency tuning as in JET where the frequency change is used together with an adjustable tuner, functioning as one of two variables needed for matching a fixed impedance to a VSWR of 1.

In contrary the described method intends to bring a range of impedances to a compromize VSWR of 1.5, but using only one variable, namely the frequency change.

2.1 The principle of frequency sensitive matching (FSM)

Figure 1 shows a feeder network with an inductive antenna brought to resonance, i.e. to a mainly real impedance by a stub and to a range of the order of the characteristic line impedance. A varying antenna coupling then transforms approximately to a trajectory y_a near the real axis in the rectangular admittance diagram. This trajectory is now transformed by the antenna-side stub L1 to a VSWR circle of 3 on a relatively long feeder line LL so that the admittance rotates on this circle many times (length is a multiple of $\lambda/2$).

A small frequency change within the bandwidth of the generator then suffices to cover the range A1 to A2 so approximately transforming a range of admittances to unity impedance.

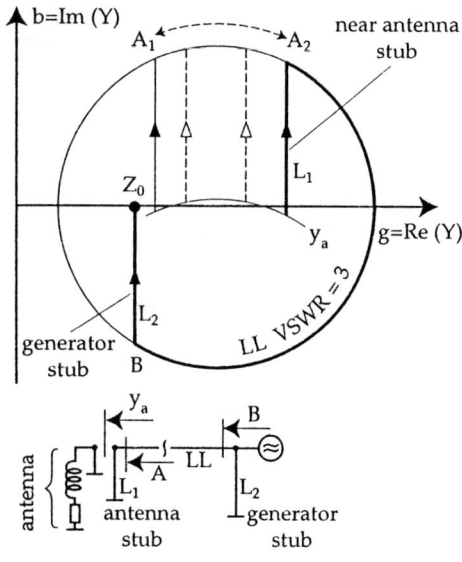

Figure 1. Frequency sensitive matching

A practical network must cope with an antenna resistance range of 1 to 8 Ω and must take into account that the near antenna stub can only be mounted at some distance from the antenna.

2.2 A practical FSM feeder network

The data used in fig. 2 takes care of those restrictions. Here - reading the table from bottom to top - the antenna is brought out of the vessel and toroidal coil space by a 2.5 m long 20 Ω line, followed by a 77 Ω 2.5 m line.

Both lines being $\lambda/4$ for the design frequency 30 MHz serve as RF transformers to bring the impedance to an optimum value of approximately 80 Ω

for a mean antenna coupling at the entrance of the frequency sensitive network following. The FSM-network consists of an antenna side stub of 1.95 m, a 42.95 m long feeder line and a generator stub of 1.45 m.

It is designed for 30 ± 0.5 MHz, a maximum VSWR on the long line of 3 and a maximum VSWR of 1.5 at the generator.

The diagram shows 11 circles for the 11 frequencies 0.1 MHz apart, which are the backwards to the antenna projected VSWR=1.5 generator circles at each frequency. With this generator bandwidth the antenna resistance range covers 1.3 to 9.4 Ω sufficient for L-mode and H-mode ICRH.

3. A FSM-TEST AT ASDEX UPGRADE

While such a FSM scheme was worked out long ago at a time when the ASDEX was still working, it was not brought to use then due to two reasons.

First, the ASDEX half antennas were fed by feeder lines with λ/2 length difference to produce 180 degrees phase shift at the antennas. This is different at ASDEX Upgrade, where the feeder lines are of equal length and the phase shift is produced by the direction of feeding connection into the vessel.

Second, both the ASDEX and ASDEX Upgrade RF generators have a relatively small power bandwidth of < 200 kHz, which is together with the existing long line (LL) of about 40 m not capable to cover the full range of fig. 2.

Yet the ELM problem made it more urgent to find ways of fast tuning and this was the reason to elaborate how useful this scheme of FSM could be despite the existing restrictions. For a test we did not want to change the existing feeder line network essentially, so the first question arising was, whether the existing network when changed only by a minimum could form a FSM network.

We found out that it needed at least a single line length change, the tuner distance length, together with a mandatory generator stub for FSM, which could show a significant effect in ELM matching performance despite the small generator bandwidth. Since all line element lengths are critical in FSM - it practically is a filter circuit - it was necessary to determine all the line lengths with sufficient accuracy.

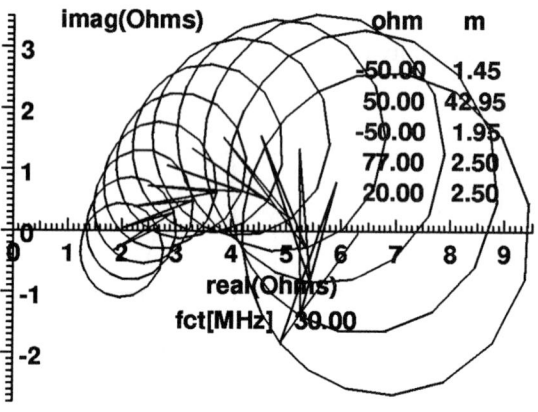

Figure 2. Optimum Network Performance

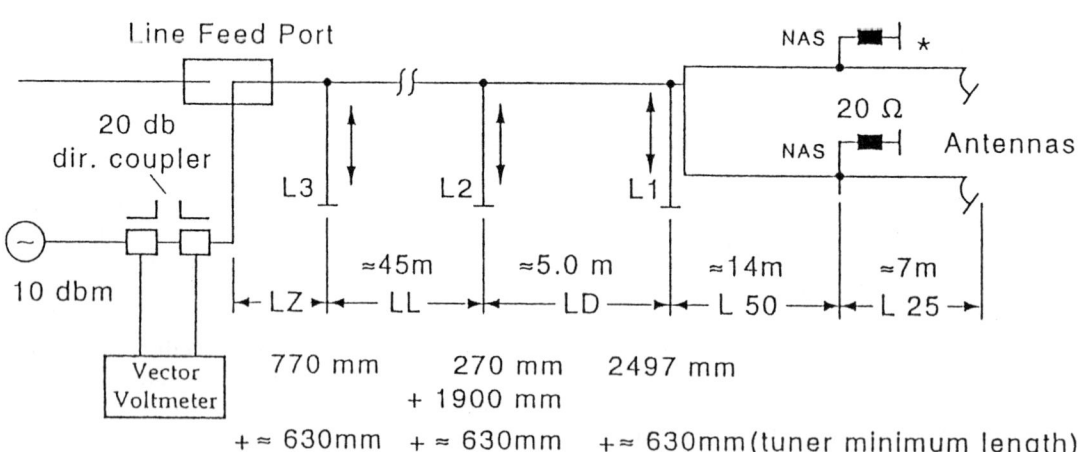

Figure 3. Measurement set-up for determining network element lengths

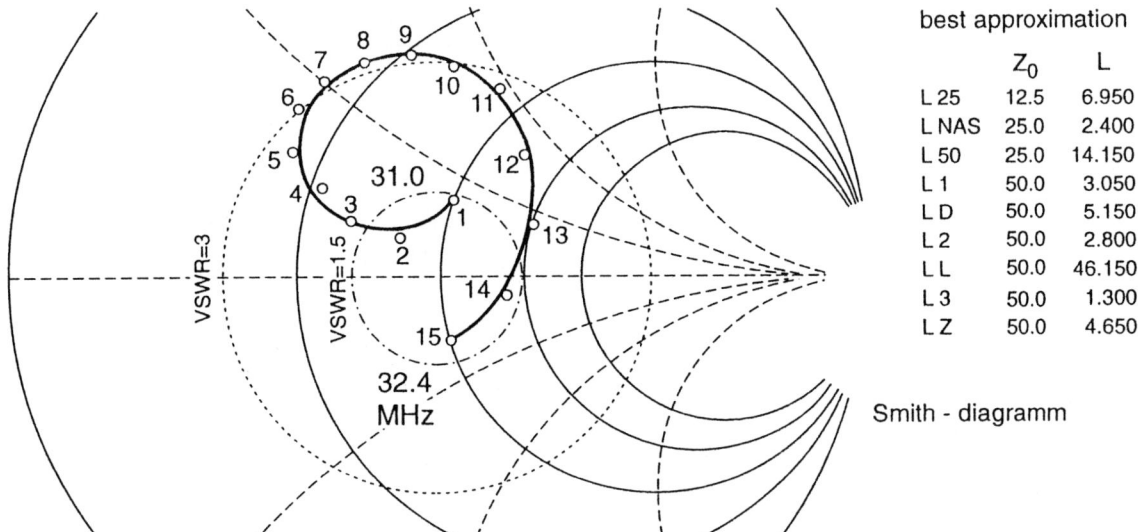

Figure 4. Model lenghts fitting to measured data

This was done by damping the network thus having approximately the losses of an average plasma case and measuring into the network with low power and a vector voltmeter.

Figure 3 shows the feeder line with the variable line elements, i.e. the existing tuner pair L1, L2, the new additional tuner L3, which were set to values for FSM performance and the measurement set-up including the extra damping resistors, which were inserted in exchange to the shortcircuits of the existing T-stubs. The resulting frequency characteristic was then compared with a calculation on a model and the model iterated until the fit of measurement to the model calculation was satisfactory.

Figure 4 shows a measured frequency characteristics with 15 points covering 31.0 to 32.4 MHz and the curve of a model with the line elements given in the included table which fits properly.

Figure 5 shows the line lengths used for the test together with the backwards projected VSWR=1.5 circles giving the performance for the various frequencies 0.1 MHz spaced around a center frequency of 31.2 MHz. Two limitations are obvious in comparison with fig. 2: First, minimum reachable antenna resistance is about 4 Ω. Therefore an antenna coupling with larger distance of plasma to antenna is out of reach. One can expect a reflection percentage only of more than the 4 % corresponding to the VSWR=1.5 circles. Second, having a bandwidth covering only the 3 circles of 31.1, 31.2 and 31.3 MHz - and not the full bandwidth including the slotted circles - the gain at the high side of antenna resistance is limited too.

Therefore taking into account the compromise matching of a VSWR=1.5 situation which covers an antenna resistance ratio of the square of 1.5, i.e. 2.25 we can expect of this FSM test only a limited gain of 10 to 20 % comparing with an optimum FSM network at full bandwidth covering an antenna resistance range of 9.4/1.3 (see fig. 2) corresponding to a gain of 3.2.

4.1 The test set-up with frequency changes

The block diagram of fig. 6 describes the instrumentation used thereby.

A RF probe near the second voltage maximum seen from the antenna, together with a directional coupler in the feeder line, measured the antenna coupling - given approximately as forward power divided by the square of voltage - in the line. A microprocessor reads the associated ADC data and by a table lookup - so avoiding long execution times for multiply and divide - calculates a digital number

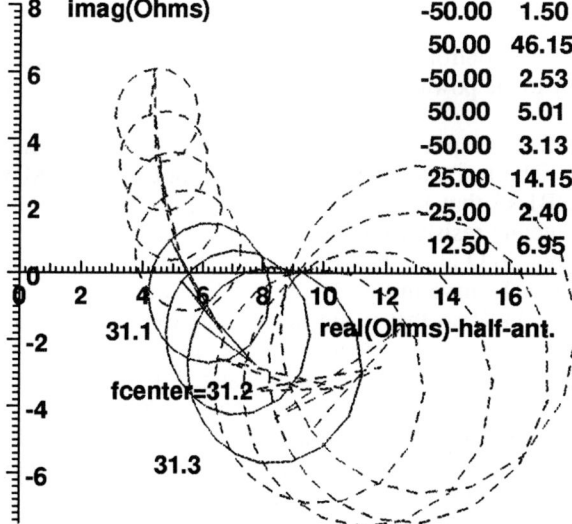

Figure 5. Realizable ASDEX Upgrade Network

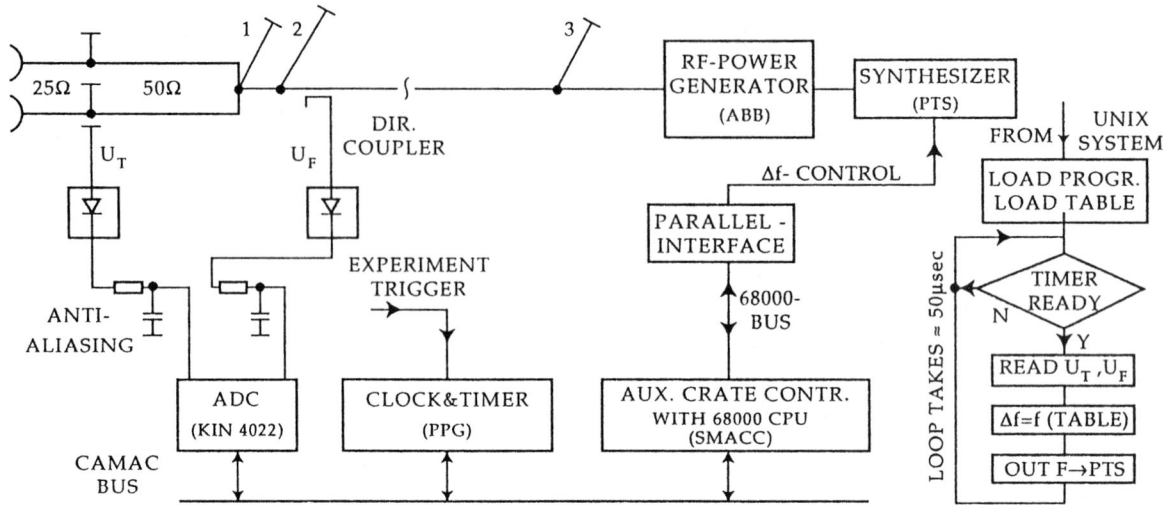

Figure 6. Frequency feedback set-up

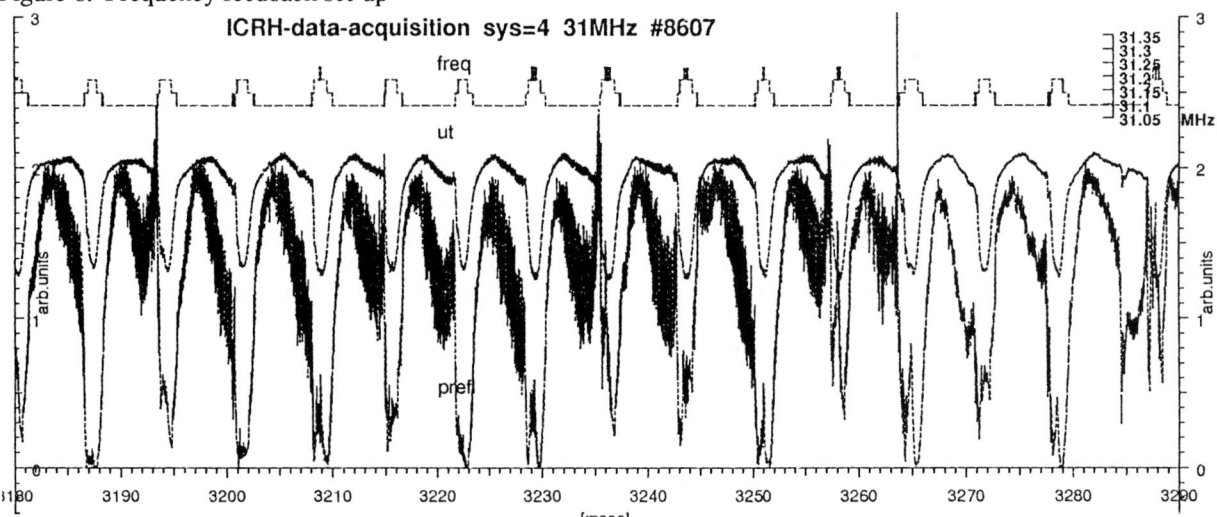

Figure 7. freq, ut, prf = f(t)

which controls a synthesizer for the oscillator frequency to the power generator.

4.2 Test performance in a Type I ELM shot

The test was done in a plasma shot with 7.5 MW NBI so that large Type I ELMs gave rise from an antenna resistance of about 3 Ω in between ELMs to about 6 Ω at the ELM peaks which can be seen from the UT-curve in fig. 7. The ICRF power was limited to 30 kW so that it did not change experiment conditions for other needs. Corresponding to the UT-variation the frequency was changed in 50 kHz steps between 31.15 and 31.3 MHz. The resulting reflection percentage at the generator was about 11 % in the intervals between ELMs showing that the frequency tuning range, limited to 4 Ω did not cover the 3 Ω requested. At the ELM peaks the 6 Ω was well in range, so that the reflection came down to about 2 %.

5. CONCLUSION

Due to the limitation in bandwidth and the compromise network the test did not show the full capabilities of FSM. To cover the full range of 1 Ω to 8 Ω the network would have to be rebuilt incorporating $\lambda/4$ transformers and the generator must be upgraded to 1 MHz bandwidth. Nevertheless it would be worthwhile to evaluate the full capabilities of FSM looking into the ITER future.

REFERENCES

1. F. Hofmeister, ICRF-RF Matching calculations and RF Performance in ICRH on ASDEX, Proc. 14th SOFT, Avignon, 1986, Vol. 1, 801.
2. T. J. Wade et.al, JET contribution, 12th Symp. on Fusion Eng., Monterey, 1987.
3. B. Beaumont et.al, Fast Load Changes and High Power Transmission on ICRH Systems, 11thTopical Conf. on RF Power in Plasmas, Palm Springs, 1995, 376.

Progress on Radio Frequency Auxiliary Heating System Designs in ITER

M. Makowski, G. Bosia, and F. Elio and the Home Teams

ITER Joint Work Site,
Boltzmannstraße 2, D-85748 Garching bei München, Germany

ITER will require over 100 MW of auxiliary power for heating, on- and off-axis current drive, accessing the H-mode, and plasma shut-down. The Electron Cyclotron Range of Frequencies (ECRF) and Ion Cyclotron Range of Frequencies (ICRF) are two forms of Radio Frequency (RF) auxiliary power being developed for these applications. Design concepts for both the ECRF and ICRF systems are presented, key features and critical design issues are discussed, and projected performances outlined.

1. INTRODUCTION

Since the initiation of the EDA phase of ITER both the ECRF and ICRF designs have evolved in parallel with the design of the basic tokamak. The designs have advanced in order to conform to the ever increasing number of design requirements and constraints. The ECRF system has evolved from a completely optical injection system to one using waveguide transmission within the port and optical injection near the blanket. Similarly, the ICRF system design has progressed from an in-blanket to a fully in-port design.

2. ECRF SYSTEM

2.1 ECRF Functional Capabilities

Modeling has shown that heating and current drive with ECRF can be accomplished over a wide range of central magnetic fields, spanning 4.0 - 5.7 T, using fixed frequency, 170 GHz, sources if the toroidal injection angle of the power can be adjusted over a ~30° range. Current profile control is also possible using the same angular steering range.

An on-axis current drive effici-ency of $\eta \sim 0.30 \times 10^{20}$ A/W-m^2 is predicted. Off-axis current drive is also possible with an efficiency in the range $\eta \sim 0.15 - 0.20 \times 10^{20}$ A/W-m^2. Recent calculations have shown that with localized deposition, stabilization of m=2 neo-classical islands is possible.

In addition to the main functions of heating and current drive, the ECRF system will also be used for start-up and wall conditioning.

2.2 ECRF Design Features

A total of 50 MW of 170 GHz power for heating and current drive is delivered to the plasma with an array of 60 waveguides divided equally between two ports. Using the full capacity of both ports an upgrade of the system to a total of 94 MW is possible, 6 MW of which is dedicated to the start-up and wall conditioning (SU&WC) system. All systems are designed for steady-state operation.

Waveguide is used to transmit the millimeter wave power through the equatorial port up to the back plate. Allowance must be made for the relative motion of the vacuum vessel with respect to the cryostat. Radial motion is compensated for through the use of a bellows section between the vacuum vessel and the cryostat. Compensation for the transverse motion is accomplished by allowing an unconstrained length of waveguide to elastically distort between two fixed ends, one rigidly attached to the vacuum vessel and the other rigidly attached to the cryostat. The resulting mode conversion can be minimized through proper choice of the length and diameter of the waveguide.

The waveguide beam illuminates a fixed mirror which deflects the power downward onto a steerable mirror. The vertical axis of rotation allows the beam to be aimed toroidally and injected through a slotted blanket/shield module. Since the elevation of the waveguide and steerable optic differ, shielding can be staggered so that no line-of-sight exists between the first wall and the bioshield (Fig. 1).

The shield/blanket plug forms a separate assembly (Fig. 2). An advantage of this arrangement is that there are no ECRF components attached to the blanket or backplate. Thus no special compensation for the relative motion of the blanket/shield with respect to the vacuum vessel is required and no forces are transmitted from one to the other. The modules forming the shield/blanket plug use a design analogous to the breeder blanket modules in order to provide cooling to the inner surfaces of the slots.

The mirror shield block, containing the injection hardware and shielding for the magnets and pit area, is located within the port. The plasma facing components are located a minimum of 0.8 m from the plasma.

All components are contained within the port; there are no vacuum vessel extensions. The SU&WC systems follow a similar design with the exceptions that the angle of injection is fixed rather than variable.

2.3 Analysis

The plasma-facing steerable optics are mechanically and thermally the most stressed elements of the transmission system. Thermal loads are dominated by Ohmic RF dissipation. Conservative assumptions (Be coating, fully E-plane polarized wave, 1 MW incident, Gaussian distribution) result in a peak heat flux of 3.5 MW/m^2. Thermal modeling shows that this heat load can be accommodated with the available blanket thermo-hydraulic system (140 °C inlet temperature, 5 m/s flow velocity, 4 MPa pressure, 8 mm diameter thin wall SS tube).

An electromagnetic analysis has been performed for a typical optic consisting of a SS body and 5 mm thick copper reflector inclined

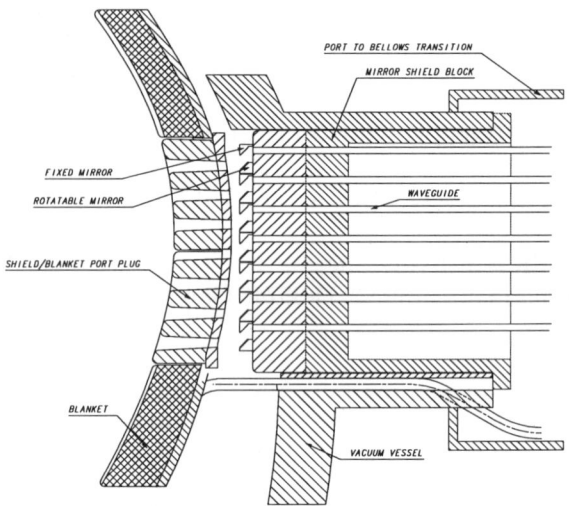

Fig. 1. Elevation view of the ECRF injection system.

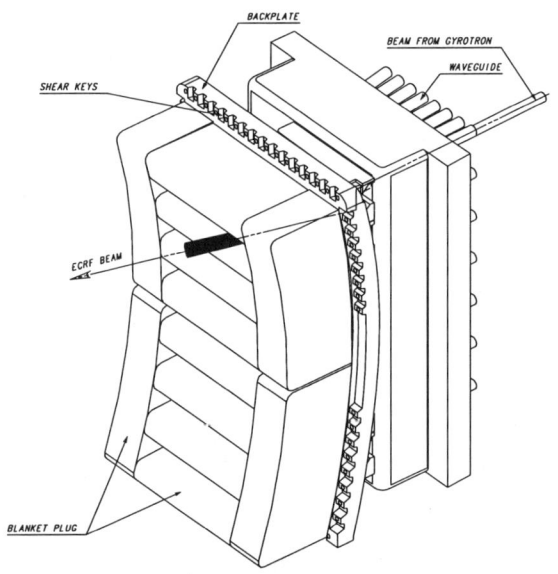

Fig. 2. Shield/blanket plug and mirror shield block for ECRF system.

at 45° to a vertical magnetic field. The peak forces during a centered disruption are found to be negligible. An upper bound on the induced torque is 200 N-m (the analysis did not model the cooling tubes and assumed room temperature copper). At this level, the load is supportable within allowable stresses.

transmission lines was measured to be .01 dB/m so that the total attenuation of each linke is below 1 dB. They transmit power levels as high as 600 kW for 1 sec even if rated for transmitting 40 kW c.w. at 400 MHz. This latter feature has been obtained by pressurizing the transmission lines at 1 bar with dry nitrogen.

A picture showing the final part of one RF linke near a tokamak port is reported in fig. 4. The picture shows some components used in the line, such as the 3 dB power divider, the trombone phase shifter, the pressure conpensated flexible element.

2.3 The antennas

The RF launching system for the experiment is made by waveguide grills installed in the equatorial ports of the tokamak.

Each grill is made of two waveguides each 400 * 29.5 mm wide, joined together along their broad side. This kind of antenna was chosen as it is high and narrow so that it is easily installed between the coils of the tokamak. Moreover, theoretical calculation showed quite a good coupling coefficient between the RF waves and the plasma.

Figs 5 and 6 show pictures of one grill.

The grill was built in stainless steel. The inner surfaces of the waveguides were stochastic gold plated in order to minimize multipactoring effects.

2.4 Special RF components

The construction of waveguide thin vacuum-windows separating the transmission lines from the vacuum vessel turned out to be a major problem.

To solve this problem Duroid pressure windows were built. Duroid is a ceramic fluoroethilene composite that can be easily machined. It is supplied covered by copper layers that allow the construction of the matching iris by chemical etching. RF and vacuum tests performed before the installation of these windows on the antenna gave good results.

This solution however is not fully satisfactory from the vacuum point of view as Duroid requires an in situ lengthy procedure to remove trapped gases.

Figure 7 shows a drawing of a Duroid vacuum window.

Another ceramic pressure windows is also available. It has been built in collaboration with the PPL of Princeton (USA). It is a double waveguide

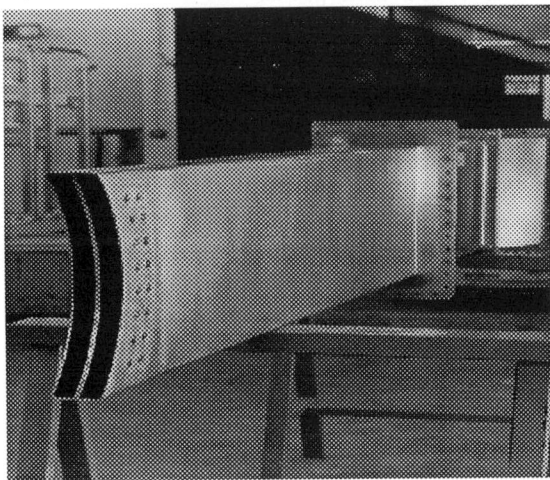

Figure 5. IBW antenna views

Figure 6. IBW antenna views

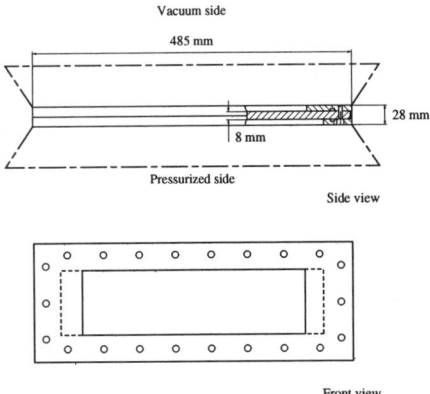

Figure 7. Duroid pressure window

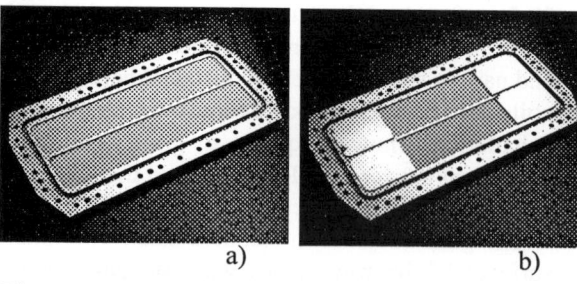

Figure 8. Ceramic pressure window. a) Vacuum side. b) Pressurized side, with adapting iris

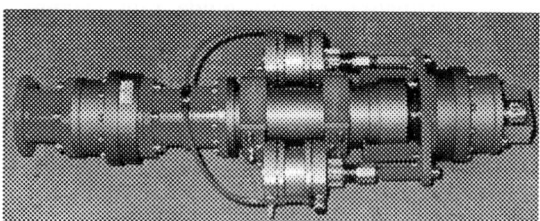

Figure 9. Pressure compensate flexible element

brased on a titanium frame. Vacuum tests on this component have given satisfactory results.

Pictures of this component are shown in fig.8.

Special coaxial pressure compensated flexible elements have been developed by Spinner Gmbh. They are installed in the 3 1/8"section of the transmission lines for allowing the movement of the antennas up to 45 mm inside the vacuum vessel with no stresses. They utilize a pneumatic system for compensating the internal forces due to the pressurization. (Fig. 9)

3. STATUS OF THE PROJECT

Two klystrons have been fully commissioned on a soda-water dummy load with satisfactory results. The available power for the experiments could be 1.2 MW. One antenna was installed in a tokamak port and connected to one generator by the correspondent RF line previosly tested up to 600kW. The experiments started with the aim of delivering to the plasma up to 600 kW in this first phase. A project for connecting two klystrons to the installed antenna is under development in order to push the delivered power to the plasma up to 1.2 MW level in a next phase.

REFERENCES

1. A. Cardinali, R. Cesario, F. De Marco, O. Sauter, Europhysical Topical Conference on Radio-Frequency Heating and Current Drive on Fusion Device, Brussel (1992), Vol. 16, p. 153
2. P. Papitto, R. Cesario, F. De Marco, G.L. Ravera, M. Sassi: Fusion Techn. (1992), **22**, 768

ACHIEVMENT OF 1.6MW/5000sec OPERATION of RF OSCILLATOR on the LARGE HELICAL DEVICE

R.Kumazawa, T.Mutoh, T.Watari, T.Seki, F.Shinbo, G.Nomura, S.Masuda, T.Ido, T.Kuroda
National Institute for Fusion Science, Nagoya 464-01, Japan

The ICRF heating is planned at 3MW with 30min and 12MW with 10sec on the Large Helical Device, which is now being constructed and whose first plasma is scheduled in 1998. For that purpose, we are researching and developing several issues such as high power/long pulse RF oscillator, ceramic feed-through, liquid stub tuner and feedback control for impedance matching etc.. Recent results acquired so far are as follow; (1)1.6MW/5000sec and 1.9MW/10sec operation of RF oscillator, (2)45kV/1200sec and 54kV/10sec operations in liquid stub tuner, (3)34kV/1740sec and 38kV/615sec operation in ICRF heating system and (4)verification of feedback control for impedance matching by frequency modulation and liquid stub tuner.

1. Introduction

The large helical device (LHD) is an l=2, m=10 Heliotron/Torsatron in National Institute for Fusion Science[1,2]. The first plasma is scheduled in 1998. ICRF heating is planned to be applied to the LHD plasma at moderate power, 3MW with steady state, 30min and high power, 12MW with 10sec[3]. The research and development for the ICRF heating has been underway in several issues since 1990.

In this paper, we describe recent results in R&D. Successful achievement of high power and long pulse operation in RF oscillator at low impedance mode is reported in Sec.2. In Sec.3, reliable performance of newly developed liquid stub tuner is described. Test results of RF transmission system including ceramic feed-through, the liquid stub tuner and RF antenna are described in Sec.4. In Sec.5, feedback control for impedance matching is discussed. We summarize in Sec.6.

2. High power and long pulse operation of RF oscillator

We have developed high power RF oscillator capable of long pulse operation, 1.6MW with 5000sec. This operation can be achieved in low impedance mode in tetrode tube, 4CM2500KG. Final amplifier consists of double coaxial cavity, which are terminated by two movable ends. The cavity is 4m long so that it accommodates a wide range of frequency, 25-95MHz. Inner and outer movable ends are referred to T-stub tuner and M-stub tuner, respectively. RF power output transmission line is connected to M-stub tuner. As the length of M-stub tuner becomes shorter, low impedance mode is obtained.

We carried out an experiment to change the impedance of tetrode tube in 1MW with 1sec of 50% duty cycle at 50MHz. Figure 1 shows the dependence of cathode current, I_k, ion current, I_c and RF power conversion efficiency on M-stub tuner position. Here the T-stub tuner position is always selected to acquire high amplifier gain, around 15. When M-stub tuner position is 675, I_k is 70A. Then M-stub tuner becomes shorter, larger I_k is obtained, which shows low impedance mode. Here the ion current, I_c decreases from $0.8\mu A$ to $0.2\mu A$, however RF power conversion efficiency reduces to 46% as shown in Fig.1. The ion current, I_c indicates vacuum pressure inside tetrode tube. When I_c exceeds $3\mu A$, arc discharge prevents a long pulse operation. We choose the M-stub tuner position at 625, where RF power conversion efficiency is 62% with lower I_c.

We executed a high power and long pulse operation as shown in Fig.2. We increased RF power from 1MW to 1.6MW in several steps with 1sec of 50% duty cycle before starting a long pulse operation. 1.6MW with 5000sec operation was achieved without any trouble as shown in Fig.2. The ion current kept a constant value of 0.5µA. Here we describe several important parameters in Table 1. The temperature of inner transmission line increased to 125°C instead of forced air flow with 2m/sec.

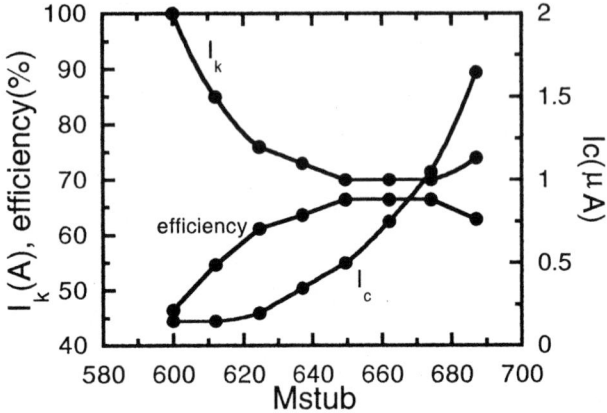

Fig.1 Dependencies of cathode current, ion current and RF power conversion efficiency on M-stub tuner position.

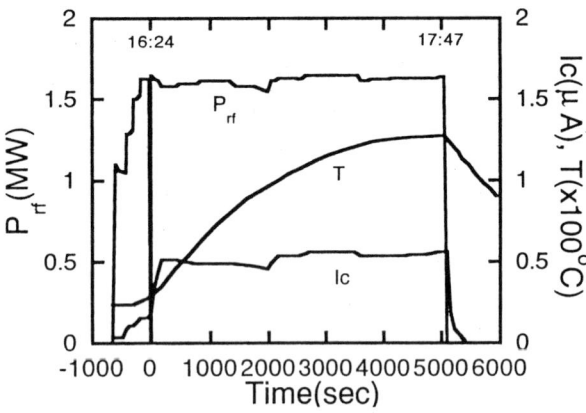

Fig.2 Time evolution of RF power, ion current and inner transmission line temperature in 1.6MW /5000sec operation.

Table 1

Frequency:	50MHz
Plate voltage: 21.5kV, Cathode current:	110A
Control grid voltage/current: -550V /	3A
Screen grid voltage/current: 1450V /	0.8A
Ion current:	0.5µA
Input RF power:	95kW
RF power conversion efficiency:	66%

3. Liquid stub tuner

In ICRF heating experiments, plasma loading resistance changes gradually with density and suddenly in L-H transmission such as JET[4], which results in increase in reflected power from antenna. Liquid stub tuner can be a tool to solve above problem because it is allowable to move its surface, which is equivalent to the movement of terminated end of conventional stub tuner. A difference of RF wave length between in liquid and gas is utilized. We use a diffusion pump oil with relative permittivity, 2.7 because of its low vapor pressure. When the normalized length of the liquid stub tuner is 0.3, it works as 0.5 of normalized length by filling the liquid to its top. We fabricated the liquid stub tuner, whose length is 2600mm with 240mm in diameter. It is equipped with one port for liquid temperature measurement and 8 ports of RF voltage measurement. Its endurance test has been executed with high voltage. The position of highest RF voltage is located near the liquid surface. Figure 3 shows results of RF voltage, kV and pulse length, sec. The steady-state and high voltage operation was achieved at 45kV with 1200sec and at 54kV with 10sec as shown in Fig.3. It will be usable as reliable RF component in RF transmission system in 2MW with 1200sec and 3MW with 10sec, when we assume the LHD plasma loading resistance is 5Ω.

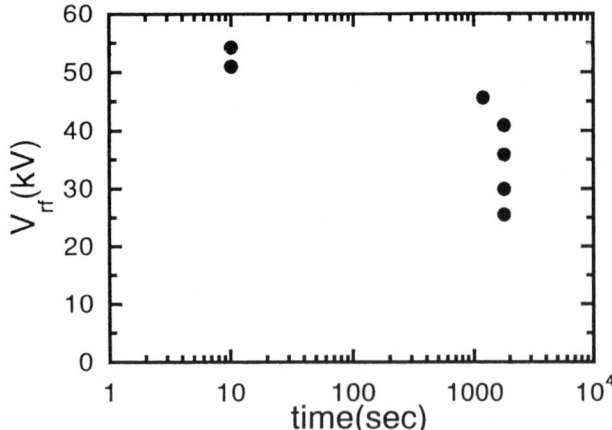

Fig.3 Pulse length of RF voltage in liquid stub tuner.

The liquid temperature increases in long pulse operation. RF power dissipation occurs by ohmic resistance on transmission line and the liquid dielectric loss. The temperature increase is 88°C at 45kV operation. The liquid temperature increase is almost proportional to the square of RF voltage. We found that RF loading resistance is constant within a few percent in the range of the liquid temperature from 22°C to 120°C. Here we calculate RF dielectric loss using $\tan\delta=2\times10^{-4}$ at 50MHz. Thus the RF dielectric loss is estimated 5.4kW at 45kV operation, where RF power input is 66kW. We can manifest that the RF dielectric loss does not increase with liquid temperature increase.

4. Long pulse test of RF power transmission system

In this section, we report results on RF power transmission system for ICRF heating, which consists of RF loop antenna, ceramic feed-through, conventional stub tuner and liquid stub tuner. The whole system is water cooled for long pulse operation. The RF antenna, 430mm wide and 630mm long is installed in a vacuum tank, whose pressure is 2×10^{-7} torr and 7×10^{-8} torr with the aid of Ti-gettering pumping. Frequency feedback control is required to reduce reflected power less than a few % during long pulse operation. The cause may be attributed to thermal extension of transmission line, however, it can not be completely interpreted. The reflected power fraction would have increased to 40% without frequency feedback control. Figure 4 shows a time evolution of a long pulse operation at 34kV with vacuum pressure increase. The pressure keeps less than 1×10^{-6} torr in the beginning, however, it reaches to 1×10^{-5} torr before RF breakdown occurs at 1740sec.

A domain of high RF voltage with long pulse operation achieved so far is shown in Fig.5. Vacuum pressure increase prevents long pulse operation. Temperature of carbon protectors increased up to 300°C, nevertheless it is water cooled. The vacuum pressure increase seems to be by outgassing from the carbon surface. Then we removed them to widen the operation regime plotted by solid circles in Fig.5. The data show that this system will be usable at 1.2MW ICRF heating with 1740sec and 1.45MW with 615sec at the LHD plasma loading resistance of 5Ω. As described in Sec.2, we have already succeeded to achieve 1.6MW with 5000sec, so the R&D for RF transmission system is not enough to make full use of our RF oscillator.

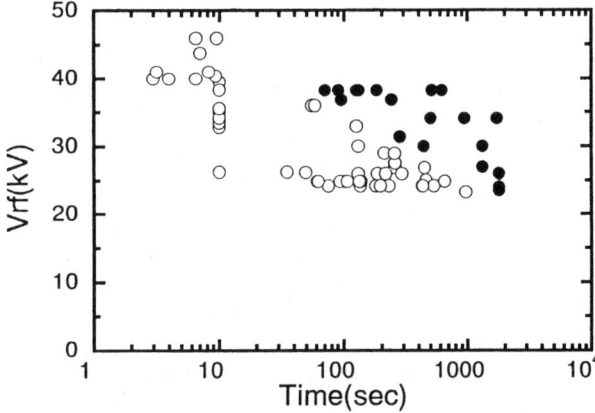

Fig.5 RF voltage and pulse length in RF transmission system.

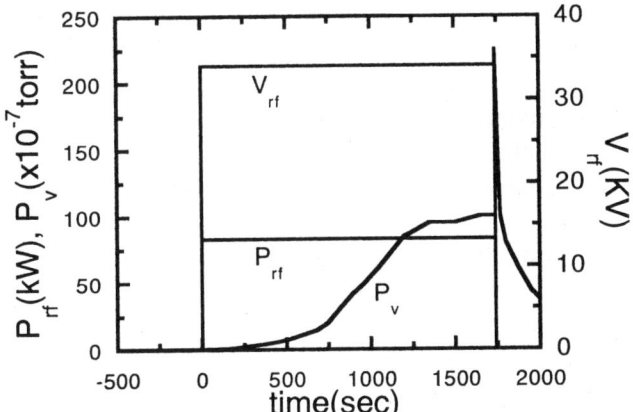

Fig.4 Time evolution of RF power, RF voltage and vacuum pressure in long pulse operation.

5. Feedback control for impedance matching

We reported the feasibility of frequency feedback control using a twin stub[5]. This result predicted that the required frequency modulation should be more than 0.24% in the case of plasma loading change from 2Ω to 6Ω. The reflected power fraction could be reduced to 0.2% by the frequency feedback control with a twin sub tuner. The reflected power fraction would have increased to 11% without frequency feedback control. Then we examined dependence of RF output power on modulated frequency in 1MW level. 95% of RF output power is verified in the range of 0.6% frequency modulation, where the central frequency is 50MHz. The date suggest us to pay attention to increase in control grid current in higher frequency.

Besides frequency feedback control, we propose the liquid stub tuner. During 45kV operation, reflected power fraction gradually increased to 8%, however it can be decreased to negligible amount without RF breakdown as shown in Fig.6 by changing the liquid surface height by 35mm from 1230mm to 1265mm with the aid of frequency feedback control from 40.125MHz to 40.050MHz in 8sec. This response time will be improved, however the twin stub tuner consisting od liquid stub tuner will be a sophisticated stub tuner to alleviate frequency modulation in RF oscillator on the LHD ICRF heating experiment.

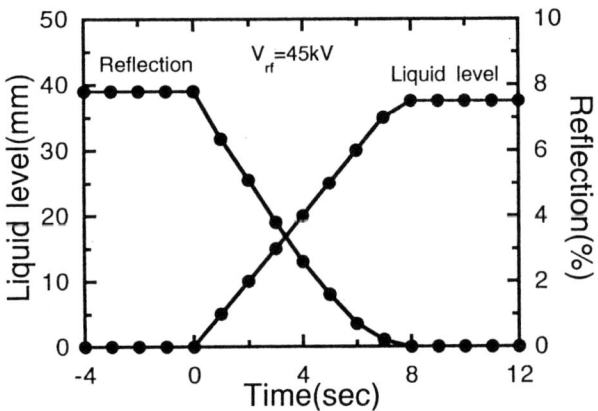

Fig.6 Reduction of reflected power fraction by liquid stub tuner.

6. Summary

We are preparing MW level ICRF heating in steady state experiment on LHD in NIFS. Several developments have been achieved.

(1) Reliable operations of 1.6MW with 5000sec and 1.9MW with 10sec are achieved at a low impedance mode in tetrode tube.
(2) Liquid stub tuner is developed instead of conventional stub tuner. It is proven to be usable in high RF voltage up to 54kV and in steady state operation up to 1800sec. RF power dissipation is deduced several kW in 45kV operation. It can be used as a reliable RF component for steady state and high power ICRF heating on LHD.
(3) The test of RF power transmission line has been executed in high voltage and long pulse, including carried out feed-through, loop antenna and conventional stub tuner and liquid stub tuner. So far 34kV/1740sec and 38kV/615sec operation is achieved, which is equivalent to 1.2MW and 1.45MW for the LHD plasma loading resistance, 5Ω.
(4) We demonstrate feasibility of feedback control for impedance matching, using frequency modulation and liquid stub tuner. 95% of RF power in 1MW can be verified in the range of 0.6% of frequency modulation. The liquid stub tuner is reliable to change virtual length during high voltage operation, whose response time is order of second. More sophisticated system will be fabricated by the combination of liquid twin stub tuner with frequency modulation.

References

[1] A.Iiyoshi, M.Fujiwara et.al., Fusion Technology, 17(1990).
[2] A.Iiyoshi, K.Yamazaki, Physics Plasma, Vol.2, 2449(1995).
[3] T.Mutoh, R.Kumazawa, T.Seki at al., 16th IEEE/NPSS Symposium Fusion Engineering, 1078(1995).
[4] The JET team presented by J.Jacquinot, Plasma Physics and Controlled Fusion Vol.33, 1675(1991).
[5] R.Kumazawa, T.Watari, T.Mutoh et al., Fusion Technology, 554(1992).

OPERATION OF HIGH POWER ICRH WITH ELMY PLASMAS AT JET

M. Schmid, V. Bhatnagar, C. Gormezano, P.U. Lamalle[*], A. Sibley, M. Simon, M.Timms, T Wade

JET Joint Undertaking, Abingdon, Oxon. OX14 3EA, UK

[*)]Laboratoire de Physique des Plasmas, École Royale Militaire, Brussels, Belgium

Abstract:

17 MW coupled power in 0-0-π-π phasing, without trips or breakdowns to the surrounding limiter structures and 15 MW coupled power in 0-π-0-π phasing with only 3 modules out of 4 are the preliminary results of improvements to antennas and generator electronics. Flexible Real Time Power Control is regularly used (see [1]). Investigation into transient effects on the generator output tetrode during arcs and ELMs are presented together with a description of a flexible arc and ELM detection system. Finally a summary is given on the wideband ELM matching schemes which are being studied for implementation at JET. An ELM or Edge Localised Mode, causes rapid density perturbation at plasma edge and therefore rapid, mainly upward changes in antenna coupling resistance.

1. TRANSIENT EFFECTS OF ARCS AND ELMs ON OUTPUT TETRODE

1.1 Arc/Elm test facility

A test-Facility to simulate the transient effects of Arcs and ELM's at high power was set up during the 95 Shutdown. It was possible to electronically trigger an arc at a defined point of the Transmission line for a wide variety of load conditions. It was also possible to simulate ELMs of varying 'severity' by means of an arc in the middle of a $\lambda/2$ stub tuner. Fig.1 shows the schematic layout.

Picture 1 Arc Igniter - used and unused

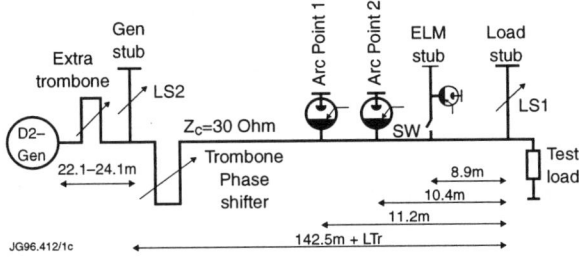

Fig.1 Schematic Diagram of Arc/ELM Test Facility

The **Arc Igniter** (see Pic. 1), consists of a modified spark plug, triggered by an electronic car ignition system (80 kV), positioned 10 mm from the inner conductor of the transmission line. The HV Spark reliably triggers an arc if the RF voltage exceeds 7 kV (in atmospheric dry air). The design protects the Spark Plug Electrode from the RF arc. Residual RF is suppressed by leading the HV Ignition Cable through several Toroidal RF Ferrite Rings.

1.2 Arc Simulation leading to improved crowbar immunity

Fig. 2a and 2b show how the slowing down of the pin diode trip from 1 μsec to 15 μsec avoids a harmful spike on the endstage tetrode anode voltage, which would lead to arcs in the tetrode. The spike is a response of the reactive load to a transient.

This modification has nearly eliminated the incidence of crowbars at JET.

Similar spikes and even oscillations can be created by the onset of the arc itself (see Fig. 3.) While it is possible to control the increase of RF Anode Voltage during the arc "on" time (for example by means of varying the length of the "extra" Trombone shown in Fig.1), no consistent way of suppressing these transitory arc-onset spikes has been found yet.

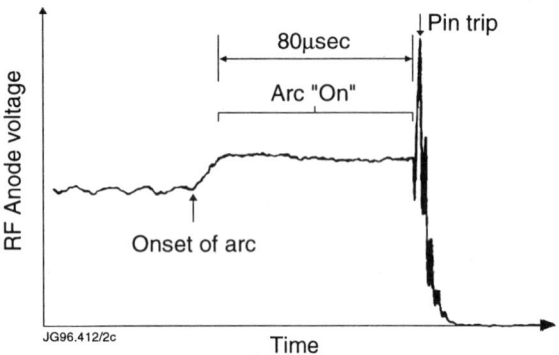

Fig. 2a Anode Voltage Evolution during an Arc terminated with a fast (1μsec) Trip

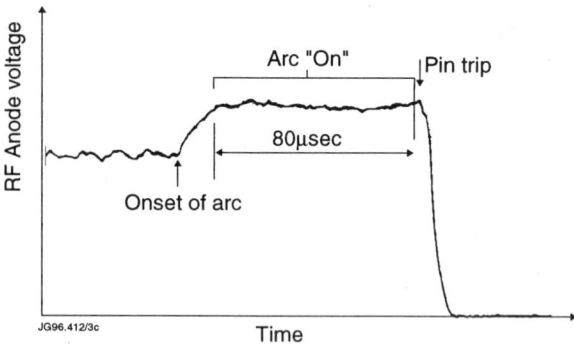

Fig. 2b Anode Voltage Evolution during an Arc terminated with a slow (15 μsec) Trip

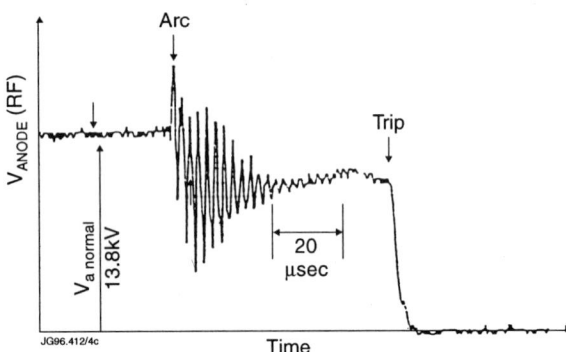

Fig. 3 Anode Voltage Evolution with transitory Oscillation caused by Arc

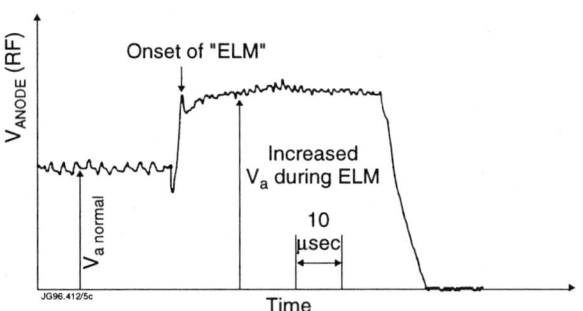

Fig. 4 Anode Voltage Evolution during simulated ELM

1.3 Simulation of ELMs

By triggering an arc at the $\lambda/4$-point of a $\lambda/2$ stub tuner (see ELM Stub in Fig. 1), the impedance of the line is made to step up rapidly to a value adjustable by the load stub. Thus any 'ELM' coupling resistance up to 30 ohms can be simulated. Fig. 4 illustrates the effect of an ELM on the Anode voltage, which is similar to that of an arc. The Anode Voltage can increase to values higher than the DC-Anode Voltage (due to an increase in Anode Impedance with phase angles approaching 90°). Such an increase can be prevented with a trombone at the generator output and the settings for mismatches caused by arcs and ELMs are about the same. The spike seen at the onset of the simulated ELM does not occur with real ELMs, which evolve at a much slower rate (typically 200μsec) compared to less than 1 μsec with this simulator.

2. DETECTION AND DISCRIMINATION OF ARCs AND ELMs

The JET Automatic Matching System delivers error signals for frequency (ferr) and stub length (serr), which are derived from the reflected voltage before the (single) stub tuner and the forward voltage after it (see [2]). These error signals were useful to desensitise the RF Trip System from normally $\rho = 0.5$ to $\rho = 0.8$ in the presence of ELMs, thus avoiding unnecessary trips (ρ = voltage reflection coefficient).

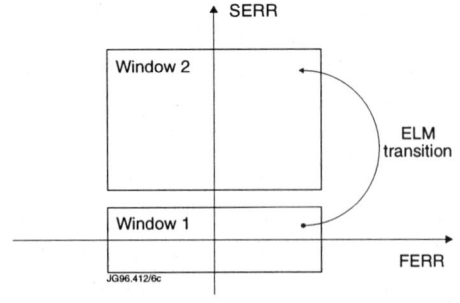

Fig. 5 ELM recognition in Stub- vs. Frequency-Error Domain

A transition from Window 1 to Window 2 (see Fig. 5) in the Stub-versus Frequency error space (and only such a step) is considered as an ELM and the Trip System is then de-sensitised for a maximum period of 20 msec. Arcs cause much larger variations in the frequency error signal and also cause the Stub-error signal to go negative. Arcs and ELMs can thus be clearly distinguished from each other.

Similarly a fast H(or D)-alpha signal is also being used to desensitise the trip system if it fulfils certain ELM criteria.

3. Adaptive RF Trip System and Automatic Power Conditioning Mode

In absence of ELMs if the reflection coefficient ρ for the generator exceeds 0.5 the rf power is tripped for a period of 5 msec and then ramped up again. If the trip-rate exceeds 10/sec then the trip period is lengthened in steps up to 20 msec. Similarly the re-application ramp is slowed down (see Fig. 6).

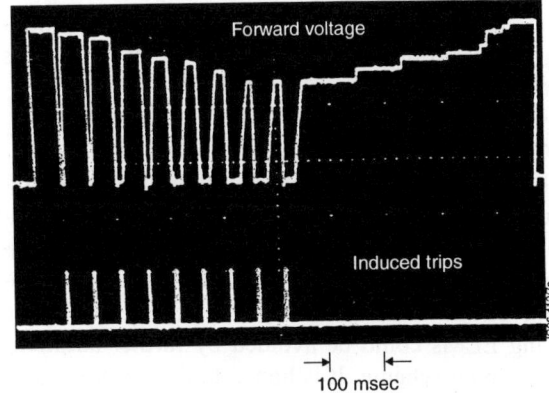

Fig. 6 Flexible Response to Multiple Trips

In the presence of ELMs any rf trips are made to recover with the shortest delay and the fastest re-application ramp (approximately 5 msec each).

An additional feature which can be enabled is called **Power Conditioning** mode. Again if the trip rate exceeds 10/sec the re-application of RF voltage is reduced in steps down to 70% of the demanded waveform (50% in power), and is then stepped up to the original value if no further trips occur. Fig. 7 shows the difference the system can make to a pulse.

4. WIDEBAND ELM MATCHING SYSTEMS

Despite intelligent Elm recognition and trip systems the fact remains that the RF generators can not provide full power in the presence of ELMs, due to the rapid change of coupling resistance on a time-scale with which the present automatic tuning system can not cope. One of the matching parameters which can

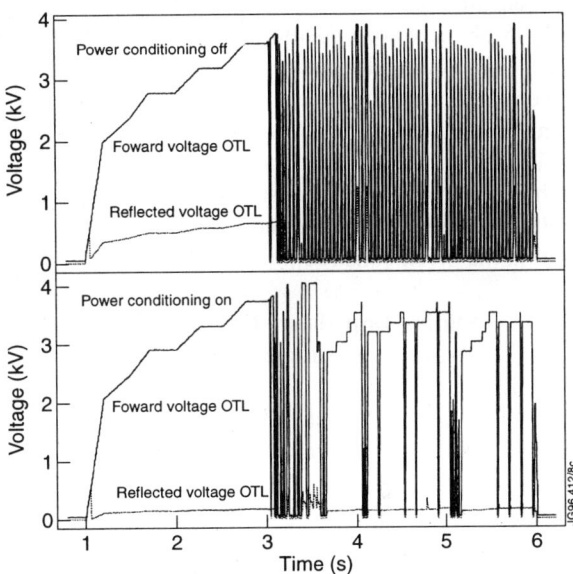

Fig. 7 Effect of Power Conditioning Mode

be changed very rapidly is the generator frequency and several schemes have been studied, which would

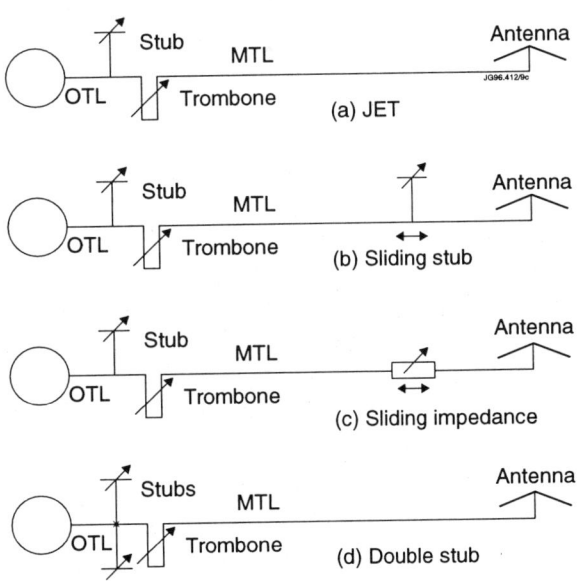

Fig. 8 Wideband ELM Matching Schemes

allow rapid matching of ELMs by means of frequency change alone (see Fig 8 b, c and d).

The scheme which continues to be studied is the sliding impedance (Fig. 8-c). It consists of a section of transmission line with a characteristic impedance of 105 ohms, whose position and length can be varied. This **Sliding Impedance (SLIMP) Tuner** was given preference to scheme (8-d) because it not only achieves wideband matching but also improves the efficiency of the main transmission line (MTL). Scheme (b) was studied first but it proved too difficult to achieve the desired electrical parameters within the given constructional constraints.

4.1 Sliding Impedance (SLIMP) Tuner
(provisional parameters)

Sliding Impedance:	30/105 ohms
Mean Distance to Antenna Short:	18.75 m
Mean Distance to Generator Stub:	74.4 m
Possible Matching:	≈ 28 / 36 / 44 / 52 MHz
Position Adjustment:	± 0.5 m
Length Adjustment:	0.4 to 2.9 m

(Fig. 9 represents the preliminary SLIMP design).

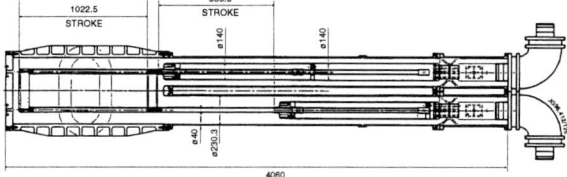

Fig. 9 Sliding Impedance (SLIMP) Tuner

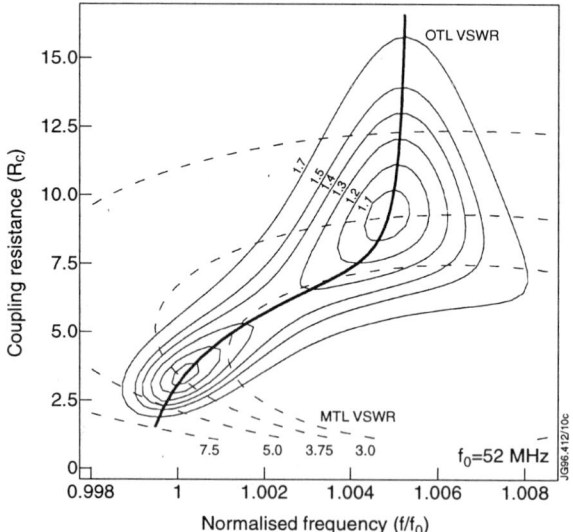

Fig. 10 Contour Plot of OTL and MTL VSWR with SLIMP-Tuner

Fig. 10 illustrates the matching range that can be achieved at 52 MHz with the SLIMP-tuner. A coupling variation from 2.1 to 13.8 ohms can be matched to within a VSWR of 1.5 with a frequency variation of around 300 kHz. The bold trace maps the optimal frequency vs. coupling curve to be followed for minimum mismatch.

The dotted lines are the improved VSWR contour lines on the MTL (Main Transmission Line).

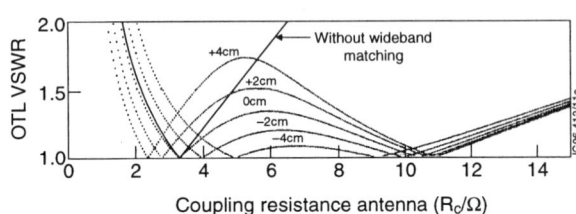

Fig. 11 Sensitivity of OTL VSWR to SLIMP Tuner Position

Figure 11 indicates the sensitivity of positioning the Antenna-end of the SLIMP tuner. Accurate positioning to ± 1 cm has to be achieved with all four straps of a JET Antenna array simultaneously.

5. Conclusions

Imbalance in coupling between straps has been reduced by measures described in [3]. The ICRF Generators now work reliably under all Plasma conditions. Power is limited by Voltage Stand-off / Coupling of the Antennas. Trips or Power Limitation during ELMs could be avoided by further additions to matching scheme. It is hoped to demonstrate such a scheme on JET in the near future.

Acknowledgement

Thanks to Mr. Chris Steele for the construction of the electronics for the arc ignition system.

References:

[1] N. Zornig et al, Experimental Result using the JET Real Time Power Control System, this conference.

[2] G. Bosia, M. Schmid et al., 15th Symposium of Fusion Technology (Utrecht 1988) pp. 459-463.

[3] A. Kaye et al, Proceedings SOFE 1995, Illinois, pp 736-741.

Improved Tuning and Matching of Ion Cyclotron Systems*

D. W. Swain[a], R. H. Goulding[a], F. W. Baity[a], R. I. Pinsker[b], J. S. deGrassie[b], C. C. Petty[b], and D. J. Hoffman[a]

[a] Oak Ridge National Laboratory, Oak Ridge, TN 37831, USA

[b] General Atomics, San Diego, CA, USA

Future fusion devices will require delivery of ion cyclotron heating and current drive power during plasma changes (e.g., L-H transitions, ELMs). The use of a passive circuit ("ELM dump") to protect the rf sources during transients has been demonstrated on DIII-D, and the results are applied to the ITER ion cyclotron system in this analysis. In addition, the use of frequency shifting to compensate for plasma load changes is illustrated for a possible ITER tuning and matching system.

1. INTRODUCTION

Fusion ion cyclotron systems are evolving from earlier configurations that provided a good match (i.e., low reflected power to the transmitters) for only one value of plasma load resistance to systems that can be retuned dynamically during a plasma shot to respond to rapid changes in plasma parameters caused (for example) by an L-H transition or by an ELM. For future fusion devices such as ITER, a large fraction of the ICH power must be coupled to the plasma during these transients. Analyses of a number of tuning and matching (T&M) configurations for ITER have been done [1]. This paper presents results for two cases: one case that relies on purely passive (ELM-dump) means and another case that uses small changes in frequency to rematch during transients.

2. ITER RF SYSTEM CONFIGURATION

Figure 1 shows the antenna configuration that was studied [2]. It consists of eight current straps, each grounded at the center and driven at both ends, in a 4 (toroidal) by 2 (poloidal) array. Overall dimensions are 2.3 m high by 1.3 m wide to fit into an ITER main horizontal port.

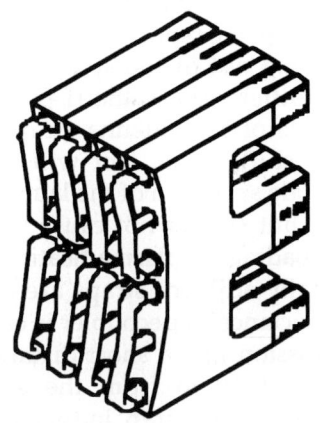

Figure 1. ITER port-mounted antenna (Faraday shield not shown).

Figure 2 shows a schematic of the T&M configurations analyzed here. A 2-MW transmitter drives each double-ended strap.

---

* This research was sponsored by the Office of Fusion Energy, U. S. Department of Energy, under contract DE-AC05-96OR22464 with Oak Ridge National Laboratory managed by Lockheed Martin Energy Research Corp. and under contract DE-AC02-76-CHO3073 with General Atomics.

The passive ELM-dump configuration in this circuit consists of a hybrid power splitter with a dump resistor connected to one port. The power from the transmitter is split and goes to individual T&M circuits for the top and bottom of each strap. The main tuning stub and phase shifter can be located in the main tuning area ($\approx$ 50 m from the machine).

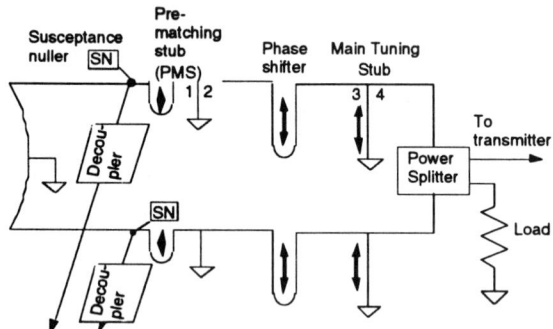

Figure 2. ITER T&M model for coupled center-grounded current straps.

For power reflected from the antenna to wind up in the dump resistor and not be reflected back to the transmitter, the electrical lengths from the antenna top and bottom to the power splitter must be equal. The power coming out of the splitter ports has a $\pi/2$ phase difference, so the rf power driving the two ends of the antenna strap also has a $\pi/2$ phase shift. This fact will change the poloidal mode spectrum of the launched power somewhat, but should not cause a significant change in loading or plasma heating [3].

Toroidally adjacent straps are coupled by mutual inductance. A decoupler is connected to the transmission line a short distance outside the cryostat ($\approx$10 m from the antenna) to compensate for the strap inductive coupling and to allow all of the transmitters to provide the same power to the system [4].

A prematching stub (PMS) can be located a short distance ($\approx$1 m) from the decoupler connection to allow frequency-shift matching. A small phase shifter between the antenna and PMS has been added because of the sensitivity of the circuit operation to PMS locations; this feature is described below.

Wade et al. used a similar configuration (but without the decouplers) for frequency-shift matching on JET [5]. Lamalle recently analyzed it in more detail, again for use on JET, but without the inclusion of the coupling between current straps [6].

3. FREQUENCY-SHIFT MATCHING

With correct placement of the PMS, the circuit mismatch can be kept small by adjusting the frequency of operation as the load changes. Figure 3 shows the calculated frequency shift and the magnitudes of ρ on the generator side of the main tuning stub for three cases at 60 MHz: with no frequency shift (solid curve); with frequency shift but interstrap inductive coupling neglected (dashed curve); with frequency shift and interstrap coupling included (dot-dashed curve). R' values typical of ITER antenna loading have been used [7].

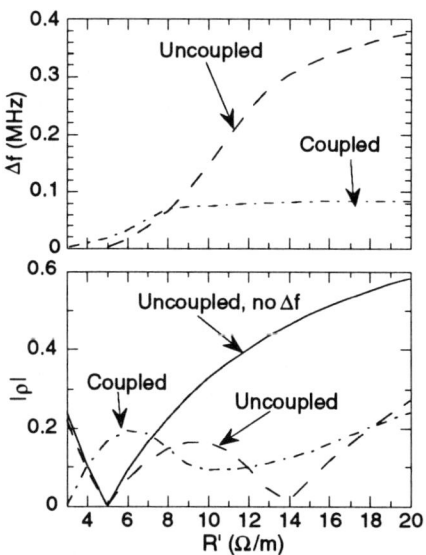

Figure 3. Calculated frequency shift (top), and $|\rho|$ on the generator side of the main tuning stub (bottom) for three cases (see text).

The uncoupled case has all of the PMS locations the same distance from the decouplers. With the frequency shift, $|\rho| \leq 0.2$ (i.e., VSWR $\leq$ 1.5) for R' = 3 – 17 Ω/m (for both the uncoupled and the coupled cases). With no shift (the dashed curves in Fig. 3), $|\rho| \leq 0.2$ only for $R' = 3 - 7.5\,\Omega/m$.

The coupled system is more difficult to evaluate computationally than the uncoupled system, and initial work has only started in optimizing this case using the FDAC code [8]. The dot-dashed curves in Fig. 3 show the frequency shift and the value of $|\rho|$ *averaged* over all four lines as a function of R', obtained using the FDAC code with $\pi/2$ phasing between straps.

Satisfactory behavior of the scheme depends *critically* on the location of the PMS. This is shown in Fig. 4, which plots the VSWR in all four lines for $R' = 20\,\Omega/m$ as a function of PMS location; for this case, the fixed tuning elements are set to provide a match with $R' = 6.3\,\Omega/m$. For each value of PMS position, the frequency shift was changed to minimize the VSWR in the lines.

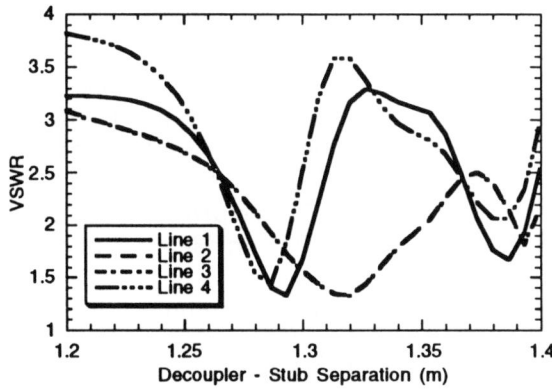

Fig. 4. VSWR at $R'=20\,\Omega/m$ vs. prematching stub position on each of the four lines.

Operation at different phasing is not possible without changing the PMS positions slightly. Interstrap phasing could be changed from $+\pi/2$ to $-\pi/2$ with no changes in PMS position, but operation at 0 or π phasing with the same positions as for $\pi/2$ phasing resulted in very high VSWR in some of the lines.

The results indicate that a frequency shift will maintain the VSWR within an acceptable range over about a factor of four in R'. The disadvantages of this circuit are:

1. the decoupler/PMS must be located near, but outside of, the antenna;
2. the decoupler/PMS separation must be changed to operate effectively at different frequencies; and
3. tuning and operation seem to be sensitive to small changes in component values, particularly for the more realistic case with interstrap coupling included.

4. SIMULATION OF ELMs IN ITER

Characteristic changes in plasma loading have been measured on the DIII-D ICH antennas in an ELMy H-mode discharge. Measurements (Fig. 5a) indicate that R' typically varies between 3 and $20\,\Omega/m$.

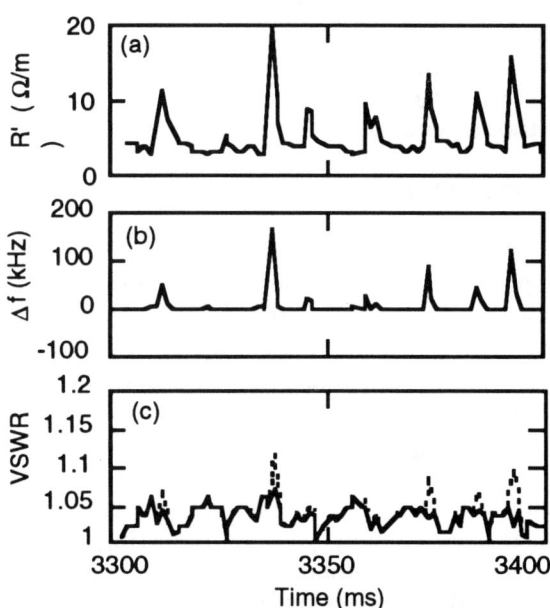

Figure 5. Response of the T&M system to the R' fluctuation
   a Plasma loading during ELMy H-mode (taken from DIII-D measurements).
   b Calculated frequency shift.
   c VSWR on the generator side of the hybrid splitter, with (solid) and without (dashed) frequency shift.

This time-behavior was used as input to the circuit shown in Fig. 2. Results are shown in Fig. 5b-5c. Curve b shows the calculated frequency shift (Δf) vs. time. Curve c shows the VSWR on the generator side of the hybrid splitter.

Components of the T&M system were set to provide a perfect match for R' = 6.3 Ω/m, so the match was not perfect for the lowest values of R'. However, this choice of R' gave good results for the total performance of the system.

The circuit response was modeled both with and without the use of frequency-shift matching. With no frequency shift, the average power delivered to the plasma is about 90% of the input power. The addition of the frequency shift maintains a better match during the very high transients in R', but makes only about a 1% improvement in the *average* power delivered to the plasma, because the high values of R' are rare and of short duration. As can be seen, the VSWR on the generator side of the hybrid splitter is always maintained at a low value with or without frequency shifting.

5. CONCLUSIONS

A simulation of the ITER rf system response to ELMs similar to ones observed on DIII-D indicates that the use of a passive ELM dump *or* frequency matching will maintain a high fraction of power to the plasma during a transient loading excursion.

The use of a hybrid splitter as a passive ELM dump appears promising. Advantages of using the ELM dump to handle transients are:
- it is completely passive, requiring no fast feedback control of the rf system;
- it will function over the entire range of operating frequencies desired for ITER (40 to 90 MHz);
- if used *without* frequency shift matching, the prematch stubs can be eliminated and decouplers can be moved to a location near the generators; VSWR at the generator will remain well below 1.5 for 10:1 changes in plasma load because of ELMs; and
- fast phase shifts between straps can be done under power without re-tuning.

A disadvantage of the ELM dump is that some of the power is diverted to a dummy load, but the diverted power can be made $\leq$ 10% of the input power by properly setting fixed tuning elements for ELMs similar to those observed on DIII-D.

A system that uses fast frequency-shift matching can be used instead of or in addition to the ELM dump to maintain a match during transients. Some advantages of this system relative to the ELM dump are
- fast matching can be done during transients with a high fraction (>95%) of available power delivered to the plasma, and
- $\pi/2$ phase shift is not needed between top and bottom of the current straps.

A major disadvantage of frequency-shift matching is the need for a prematching stub near the antenna, whose position must be changed for operation at different frequencies or different values of interstrap phasing.

REFERENCES

1. D. W. Swain and R. H. Goulding, Design of Fast Tuning Elements for the ITER ICH System, Oak Ridge National Laboratory Report ORNL/TM-13230 (May 1996).
2. ITER Design Description Document, Sec. 5.1, Ion Cyclotron Heating and CD, (June 1995).
3. M. D. Carter, private communication.
4. R. H. Goulding et al., *Proc. 12th Symp. Fusion Tech.*, **1**, 515 (Rome, 1992).
5. T. J. Wade, G. Bosia, M. Schmid, A. Sibley, *Proc. 12th Symp. Fusion Eng.*, p. 1200 (1987). See also T. J. Wade, J. Jacquinot, G. Bosia, A. Sibley, M. Schmid, *Fusion Eng. and Design* **24**, 123 (1994), and references therein.
6. P. Lamalle, private communication.
7. D. W. Swain et al., *Proc. 11th Top. Conf on Radio Frequency Power in Plasmas*, p. 417 (Palm Springs, 1995).
8. R. H. Goulding et al., *Proc. 11th Top. Conf on Radio Frequency Power in Plasmas*, p. 397 (Palm Springs, 1995).

Development of Fast Wave Systems Tolerant of Time-Varying Loading*

R.I. Pinsker,[a] C.P. Moeller,[a] C.C. Petty,[a] D.A. Phelps,[a] T. Ogawa,[b] Y. Miura,[b] H. Ikezi,[a] J.S. deGrassie,[a] R.W. Callis,[a] R.H. Goulding,[c] and F.W. Baity [c]

[a]General Atomics, P.O. Box 85608, San Diego, California 92186-9784

[b]Japan Atomic Energy Research Institute, Tokai, Naka, Ibaraki 319-11, Japan

[c]Oak Ridge National Laboratory, P.O. Box 2009, Bldg 9201-2, Oak Ridge, Tennessee 37831-8072

A new approach to fast wave antenna array design based on the traveling wave antenna has been successfully demonstrated on the JFT-2M tokamak. A traveling wave antenna is powered though a single feed and the power flow from element to element is only via mutual reactive coupling. A *combline* is a particular type of traveling wave antenna, in which only the fed element and the element at the downstream end of the array are connected to vacuum feedthroughs, while the intermediate elements are terminated with reactances inside the vacuum chamber. A twelve element combline for operation at 200 MHz was designed and fabricated at General Atomics, and installed and operated on the JFT-2M tokamak. The full output power of a single transmitter, 0.2 MW, was coupled to tokamak discharges with very little conditioning required. The input impedance of the combline was well matched to the transmission line impedance for all loading conditions, including vacuum (no plasma), Taylor discharge cleaning plasmas, and ohmic, L– and H–mode tokamak discharges with neutral beam heating *without any adjustment of tuning elements.*

1. TRAVELING WAVE ANTENNAS

A single element fast wave antenna can be characterized by a pair of scalar parameters, such as a resistive and reactive impedance. For a loop antenna in the Ion Cyclotron Range of Frequencies (ICRF), typically $R \ll \omega L$. When an array of such loops is operated with a phase difference between the currents in successive elements other than 0 or 180 degrees, a third parameter comes into play, which is the mutual reactance between elements. This reactance is generally inductive, particularly if the elements are equipped with Faraday shields. The usual ordering of these three impedances is $\omega M \ll R \ll \omega L$. This condition can be enforced for arbitrary M and R by connecting an adjustable reactance [1,2] ("decoupler") in parallel with the antenna terminals in such a way that the composite system of antenna and decoupler has an effective mutual reactance $\omega M_{eff} = \omega(M-M_d) \cong 0 \ll R$.

Three such systems with decouplers have been used to power three four-element array antennas on the DIII–D tokamak for some years [2,3]. The decoupling allows the use of a 90 degree hybrid junction in such a configuration that the transmitter output is completely isolated from changes in loading impedance that affect all four elements in the antenna in the same way. This property has enabled recent operation of these systems at the 3 MW level into ELMing H–mode discharges [4].

However, this system still can have a rather high standing wave ratio in the lengths of transmission line between the antenna and the hybrid junction outputs during transients in the antenna loading. The decoupler must be adjusted for any change in the mutual reactance between the antenna elements; changes in the mutual reactance are generally seen to accompany changes in resistive and reactive loading.

A system in which the standing wave ratio remains low in the entire transmission line system for any resistive and reactive antenna loading, requires fewer bulky, expensive, and trouble-prone adjustable tuning elements, and permits full power operation at a wider range of antenna array phasings can be obtained in the following way. By changing the sign and magnitude of M_d so that the effective

---

*Work supported by U.S. Department of Energy under Contract Nos. DE-AC03-89ER51114 and DE-AC05-96OR22464.

mutual coupling between elements is *increased*, so that the decoupler becomes a coupler, and $\omega M_{eff} = \omega(M+M_c)$ with $M_c \gg M$, the power flow from element to element will be dominated by the mutually reactively coupled power. Since the coupling reactance is dominated by a term that is independent of the plasma condition, the input impedance of the element at the "upstream" end will be almost independent of the plasma load to the extent that $\omega M_{eff} \gg R$. This is the basic idea of the traveling wave array configuration, in which the array is fed at the upstream end only. The power flow from element to element is through mutual reactive coupling, and the input impedance of the array is virtually independent of the load condition, as R is now only a small perturbation to a single element in the array.

The traveling wave approach to array design can significantly reduce constraints on the design of fast wave arrays. This is especially true if the design criteria for a given application allow the use of a *combline* [5] traveling wave antenna. A combline can be distinguished from a generic traveling wave antenna by the number of vacuum feedthroughs — a combline is a traveling wave antenna with mutual reactance only on the vacuum side, so only one feedthrough at each end of the array is required. This provides great savings since the feedthroughs are one of the most complex, costly, and least reliable elements in an antenna. Since the input impedance of the structure is practically independent of plasma load, the input impedance can be matched to the characteristic impedance of the transmission line for all loading conditions. The power per feedthrough can be much larger than in a conventional antenna, assuming that the limiting factor is electric field. A large number of antenna elements allows the plasma loading of each element to be much smaller than in an array with a small number of elements, assuming the same total power is coupled to the plasma. This fact permits the array to be far from the plasma surface, leading to a lower heat load on the array. Furthermore, an optically opaque Faraday shield is practical, which can lower the plasma density around the radiating elements, and may permit higher electric fields in the antenna without breakdown than are indicated from the usual design rules. The antenna straps may be closer to the backplane, leading to the possibility of a much thinner radial build, which allows a greater freedom of design.

2. COMBLINE FOR JFT-2M

Design of a new antenna for the JFT-2M tokamak afforded the opportunity to demonstrate some of these advantages of the traveling wave approach. The antenna is designed to couple up to 0.8 MW at 200 MHz (the existing transmitter at JFT-2M consists of four 0.2 MW modules), at $k_{\parallel}$ of 0.21 cm^{-1} ($n_{\parallel}$ =5 at 200 MHz). A pair of existing feedthroughs were to be reused. The combline that was designed is comprised of twelve modules, each of which contains a current strap grounded at one end, open-circuited at the other, a backplane and a three-layer Faraday shield in front and on both sides of the current strap. The shield has two layers in front of the strap and a layer behind the strap, between the strap and the backplane. The purpose of the inner layers of the Faraday shield is to lower the characteristic impedance of the elements and thus minimize the peak electric fields in the structure, and also to provide the capacitative loading necessary to give the combline a non-vanishing bandwidth [6]. The plasma-facing part of the Faraday shield was made of molybdenum coated with a very thin layer of titanium carbide, and was shaped so as to function as a pair of segmented poloidal limiters for each module, to minimize the connection length and hence the stray plasma density at the strap surface. To further reduce the stray plasma density, graphite poloidal limiters were attached to the vacuum vessel wall at each end of the array. The modular antenna design, in which the twelve modules are individually bolted to a pair of toroidal rails inside the tokamak, facilitates installation and maintainance of the array. A photograph of the completed twelve-element array and the pair of associated limiters as installed in the JFT-2M vacuum vessel is shown in Fig. 1.

3. HIGH POWER OPERATION OF THE COMBLINE ON JFT-2M

For this initial set of high power experiments with the combline on JFT-2M, two 0.2 MW, 200 MHz transmitters were connected to the antenna, one at each end. Each transmitter was protected by a circulator. A directional spectrum could be produced by operating one transmitter only, with the power not radiated from the antenna being prevented from reaching the unused transmitter by the circulator. A non-directional spectrum could be produced by feeding both ends of the combline

Fig. 1. The completed twelve-element array and the pair of associated limiters installed in JFT-2M.

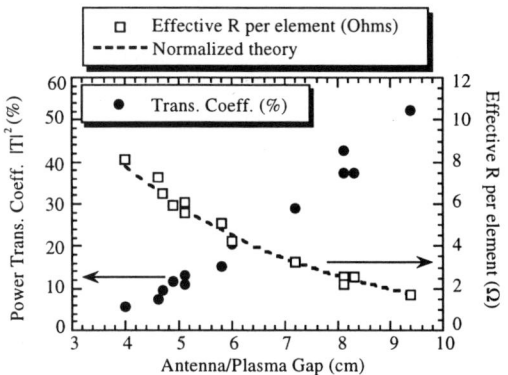

Fig. 2. Data from scan of plasma/combline gap with ohmic discharges, with I_p = 190 kA, B_T = 1.1 T, $\bar{n}_e \approx 1.9 \times 10^{19}$ m^{-3}, lower single-null divertor.

simultaneously, and thus up to 0.4 MW could be applied to the antenna for testing its voltage limits.
In order to determine the $n_\parallel$ that is coupled to the plasma and compare it with theoretical estimates, a scan of the antenna/plasma gap with ohmic tokamak plasmas was performed at a fixed input rf power level of 0.1 MW. Both the magnitude and phase of the transmission coefficient through the combline were measured as the gap was varied from 4 to 9.4 cm. The transmission coefficient magnitude can be related to the resistive loading per element (R) using estimates of the electrical parameters of each element. The decay of the resistive loading as the gap increases can be predicted from a simple model, assuming that the loading is entirely due to the fast wave, using estimates for the edge plasma parameters and normalizing to the measured R at one value of the gap. The experimental data on the magnitude of the transmission coefficient are shown in Fig. 2, along with the corresponding resistive loading per element and the model prediction. The agreement in the decay length between the model and the experiment is excellent.

As expected, the input impedance of the combline was well matched to the transmission line during all conditions studied, including all tokamak plasmas at any gap, in ohmic, neutral beam heated L–mode, and ELMing H–mode plasmas. This is illustrated in Fig. 3, where 1.8 MW of neutral beam heating sustains an ELMing H–mode plasma. The transmission coefficient drops during the short L–mode period from the value observed in the ohmic portion of the discharge, corresponding to increased resistive antenna loading. At the L/H transition, the loading decreases somewhat; at each ELM, the loading transiently rises. Throughout these loading variations, the reflection coefficient from the input of the combline is essentially zero. This was also true for the gap scan, during which the magnitude of the power transmission coefficient varied from 5 to 50% as the gap was increased.

The full output power of one transmitter (0.2 MW) was applied to the antenna without arcing under several different loading conditions. This power level could be sustained even without a plasma load, which corresponds to much higher electric fields in the antenna than are present with a plasma load. In double-ended (symmetric) operation, up to 0.25 MW was coupled to a tokamak plasma; the full two-transmitter output of 0.4 MW was coupled to Taylor discharge cleaning plasmas. These power levels were achieved after only a few half-days of vacuum (mostly multipactor) conditioning. It is important to note that no signs of reaching a power limit, other than the transmitter limits, were observed during this two week campaign. Reliable power handling was facilitated by the fact that since no tuners have to be adjusted at any time, short pulse high voltage vacuum conditioning can be continued up to a few seconds prior to the tokamak shot, and can be resumed immediately after the discharge. This method of maintaining antenna conditioning was applied throughout the high power phase of these experiments. Although the power levels injected into JFT-2M in these experiments were small compared to the multi-MW levels used on large tokamaks, as a result of the small size of the combline, the rf electric fields sustained without breakdown in these experiments actually exceeded the ITER electric field design criteria.

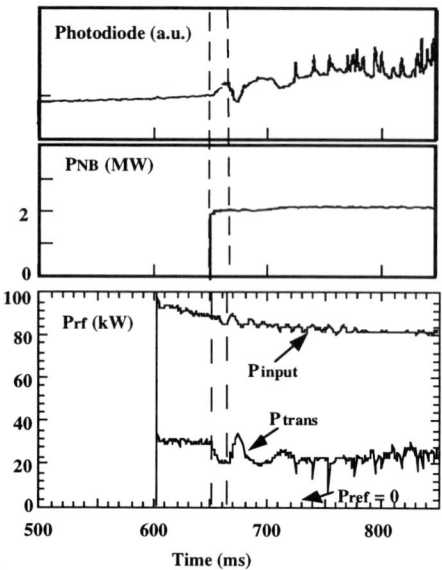

Fig. 3 Power into combline, reflected power from the input (zero), and the power not coupled to the plasma after one pass through the combline, P_{trans}, for a JFT-2M discharge with ohmic (600–650 ms), L–mode (650–665 ms), ELM-free H–mode (665–720 ms), and ELMing H–mode (720–850 ms) conditions.

4. SUMMARY, CONCLUSIONS, AND FUTURE WORK

Low power traveling wave antenna experiments on DIII–D [7] and the high power combline experiments on JFT-2M have demonstrated many of the possibilities of the traveling wave approach to antenna array design. In both cases, the input impedance of the structure was independent of the plasma conditions, and indeed, of whether or not a plasma load was present. The DIII-D experiments used existing antenna arrays, demonstrating the usefulness of this approach with existing conventional antenna arrays, while the JFT-2M case provided an opportunity to design a high power antenna based on the traveling wave concept from the outset. Some of the unique possibilities thus explored were: twelve antenna modules with only two feedthroughs, heavily Faraday shielded elements with light coupling per element and successful operation at antenna to plasma gap of up to 9 cm, up to 0.25 MW coupled to tokamak discharges in a symmetric spectrum using a pair of transmitters and up to the full single transmitter capability of 0.2 MW in a highly directional spectrum with very little antenna conditioning required. No tuning elements were adjusted at any time during the entire experiment.

Future experiments using the JFT-2M combline will use hybrid junctions to combine the four transmitters to work towards the demonstration of 0.8 MW operation of the combline. On DIII–D, several different plans for operating the existing four-strap antenna arrays as traveling wave antennas at high power are being considered. In any case, the relatively small radiation per pass through the antenna structure that is obtained with only four elements necessitates the use of a power recirculation system [5]. Such a resonant ring is characterized by a rather low Q even with only four elements, so that the resonance is not critical. A high power application of this transmission line topology to the existing antenna arrays on DIII–D would demonstrate the usefulness of this approach to existing antenna systems, such as on JET, Alcator C Mod, or to the ITER ICRF system even without any changes to the antenna design.

REFERENCES

1. N. Hershkowitz, R. Majeski, P. Probert, et al., in *Radio Frequency Power in Plasmas (Proc. 9th Top. Conf., Charleston, SC, 1991)*, (AIP, New York, 1992), p. 267.

2. R.I. Pinsker, C.C. Petty, W.P. Cary, et al., Proc. 15th Symp. on Fusion Engineering, IEEE (Hyannis, Massachusetts, 1993) Vol. 2, p. 1077.

3. J.S. deGrassie, R.I. Pinsker, W.P. Cary, et al., Proc. 15th Symp. on Fusion Engineering, IEEE (Hyannis, Massachusetts, 1993) Vol. 2, p. 1073.

4. R. Prater, et al., "Fast Wave Heating and Current Drive in DIII–D Discharges with Negative Central Shear," to be published in Proc. 16th IAEA Fusion Energy Conf. (Montreal, 1996), paper IAEA/CN/64/E-1.

5. C.P. Moeller, S.C. Chiu, and D.A. Phelps, in Proc. Europhysics Top. Conf. on Radiofrequency Heating and Current Drive of Fusion Devices, Brussels, 1992 (European Physical Society, 1992), Vol. 16E, p. 53.

6. C.P. Moeller, R.W. Gould, D.A. Phelps, and R.I. Pinsker, in *Radio Frequency Power in Plasmas (Proc. 10th Top. Conf., Boston, MA, 1993)*, (AIP, New York, 1994), p. 323.

7. H. Ikezi and D.A. Phelps, "Traveling Wave Antenna for Fast Wave Heating and Current Drive in Tokamaks," to be published in Fusion Technology, 1996.

Flexible $N_{//}$ Launcher for Lower Hybrid Wave injection in the first operation phase of ITER

G. Rey, P. Bibet, P. Froissard, J.G. Wégrowe, S. Bério, M. Goniche, GT. Hoang, F. Kazarian-Vibert, X. Litaudon, M. Tareb, G. Tonon,

Association EURATOM-CEA, Département de Recherches sur la Fusion Contrôlée
Centre de Cadarache, 13108 Saint Paul lez Durance Cedex, France

A first conceptual antenna design based on the passive-active multijunction (PAM), was proposed in 93 in the frame of an ITER-TASK [1]. The present work accounts for the dimensional change of the ITER ports and also proposes a new arrangement of the hyperguides feeding the PAM to allow the flexibility of the $N_{//}$ spectrum which is desirable to cover the variety of envisioned scenarios. 35 to 55 MW of LH power could be injected in 1 port of ITER depending on the actual plasma reflection coefficient.

1. MOTIVATIONS

During the first low activation phase of ITER, a large database is required to assess the current drive (CD) efficiency of the heating systems. Lower Hybrid Waves (LHW) are predicted to be the most efficient current driver in any scenario and phase of ITER considered in the IDR [2]. The capability to control the current profile during the early discharge phases and off-axis during the flat-top makes LHW very attractive, especially for the 'advanced Tokamak scenario' which are proposed to achieve steady state operation in ITER.

Progress in physics and technology are clearly illustrated by the achievement in present devices of high performance operations (stationary discharges of up to 2 mn, with 270 MJ of LHW energy injected in Tore Supra, large current driven : 3 MA at JET, 3.6 at JT60), of efficient coupling at large distance of the last closed magnetic surface (up to 15 cm at Tore Supra and JT60). The proposed LHW launcher for ITER is based on the novel concept of the passive-active multijunction (PAM) [3], which associates an efficient coupling of the slow wave to an efficient cooling of the grill mouth.

2. $N_{//}$ FLEXIBILITY REQUIREMENT

To cover the various phases of operation (nominal burn, Advanced Tokamak burn, all ramp-up phases and CD assistance for long pulse in the commissioning phase), a variation (± 0.3) of the main peak $N_{//}0$ of the launched power spectrum is required [2].

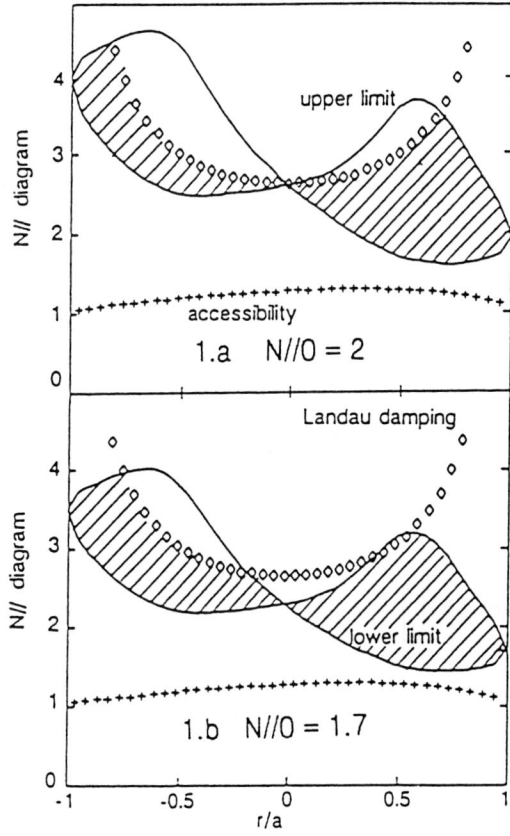

Figure 1a, 1b.   Map of propagating zone : $N_{//}$ vs r/a (shadowed area)

A situation during the ramp-up phase is seen in figure 1 showing the map of authorized $N_{//}$ versus r/a, taking into account the Landau absorption, the accessibility and the toroidal propagation limits. During the ramp-up phase (ne0 = 0.2 e20 m^{-3}, Ip = 7 MA, Te0 = 6 keV), the wave can be absorbed up to the center with a $N_{//}0$ launched at 1.95 (Fig. 1a), while the absorption is marginal at r/a = 0.5 with $N_{//}0$ = 1.7 (Fig. 1b). The later being close to optimal at burn.

3. RF DATABASE FOR LHW DESIGN

3.1 Present devices

The stationary performances, obtained in particular on Tore Supra, are a convenient database for the ITER LHW design.

TE 10-30 mode converters, required for the new launchers of Tore Supra and relevant to ITER [3], have yet been successfully tested at 3.7 GHz, high power and very long pulses (300 kW-1000 s, electric field E = 4.6 kV/cm) [4].

On Tore Supra, long pulses (> 1 mn) were performed in steady-state conditions, at a power density of 24 MW/m² corresponding, with a reflection coefficient of 5 % to an electric field E = 4.6 kV/cm [4]. Figure 2 gives the integrated RF pulse time versus the electric field E at the grill mouth obtained on the 2 launchers during the last experimental campaign of Tore Supra (990 shots- 14 s of average time length) : more than 1200 s have been performed at E > 4.6 kV/cm by each launcher and this value will thus serve as a basic for the ITER design.

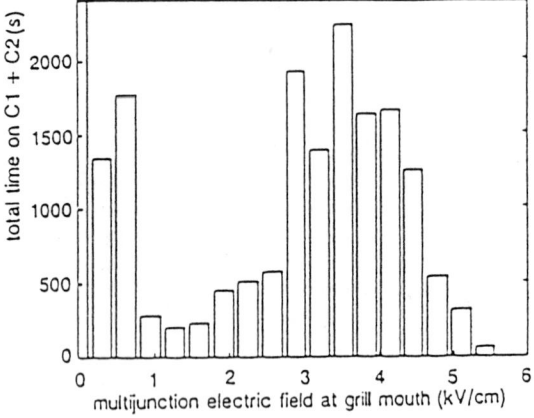

Figure 2. Integrated pulse time vs Electric field during the 95-96 campaign of Tore Supra

3.2 RF power capability for ITER

For ITER at working frequency f = 5 GHz, and taking a conservative extrapolation law (E ~ (v$_f$)) [3], the design value of the maximum electric field is taken equal to E = 5.3 kV/cm and the RF power capability can be deduced from :

$$P = [E/(1+\sqrt{\rho})^2]^2 \, (h \, b/4\eta) \, \sqrt{[1-(n\lambda/2h)^2]} \quad (1)$$

where E is the maximum electric field in the waveguide (wg), h is the free height, b its width, λ the wave length, η the characteristic impedance of the wave, n the order of the working mode TE n0 and ρ the power reflection coefficient at the grill mouth.

With the dimensions defined in table 1, the total power P available in one port of ITER associated to the enhancement factor of the electric field E/E0 (where E0 is the incident field value), are plotted in figure 3 against the power reflection coefficient R or ρ. Due to the multijunction (MJ) effect, R and ρ are related by $R = \rho^2$, where R is referred at the input of the MJ. The optimum working mode is the TE 50, chosen to be low enough to optimize the power capability given by (1) and high enough to be filtered by the height of the phase shifter in the wg.

With working reflection coefficients between R = 1 to 4 % (respectively 10 to 20 % at the grill mouth), 55 to 35 MW could be injected through one port of ITER.

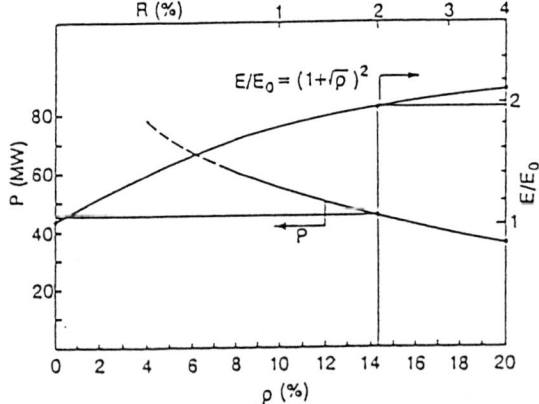

Figure 3. Power capability and electric field enhancement vs reflection ρ and R

3.3 $N_{//}$ flexibility and PAM arrangement

Flexibility of the PAM is illustrated in figure 4, where $N_{//}$ (normalized to $N_{//}$ = 2), is plotted versus the phasing and the arrangement of the passive-active (PA) wg. Though the maximum of flexibility would be obtained using a bijunction, the quadrijunction arrangement (4 active & 4 passive wg) is the good trade-off between $N_{//}$ flexibility ($\Delta N_{//} = \pm 0.3$) and simplicity of the construction.

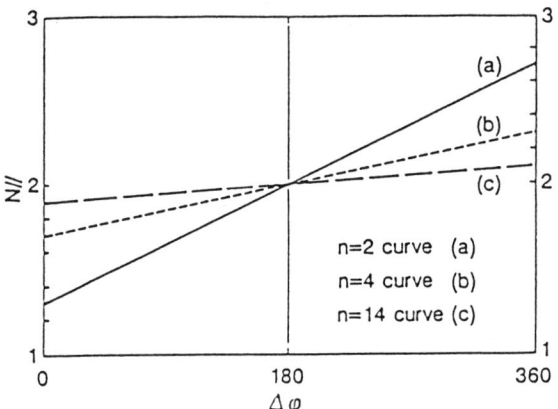

Figure 4. $N_{//}$ flexibility versus the phasing and arrangement of n adjacent active wg

From the Swan code results [5], the electric field can be computed as a function of the density at the grill mouth. On figure 5 is plotted E/E0 and the reflection coefficients ρ & R against the grill mouth density and the phasing between 2 bijunctions made of 2 active and 2 passive wg.

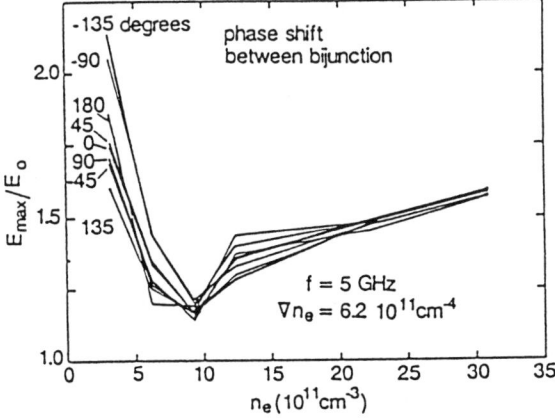

Figure 5. E/E0 vs density at grill mouth and for different phasing between bijunctions

For typical values of 2 to 4 times the cut-off density (beyond this last value, the thermal load on the grill mouth would be prohibitive), R stays below 1 % (at the mouth ρ = 8 %).

The directivity of the $N_{//}$ spectrum (ratio between the power in the $N_{//} > N_{//}0$ to the total $N_{//}$ power) is more than 65 %. For phasing of 180°, almost symmetrical spectrum can be launched which is convenient for heating scenarios.

4. THE $N_{//}$ FLEXIBLE LAUNCHER

Schematic diagrams of the launcher and LHW system are presented in figures 6 & 7 and the main parameters of the launcher are summarized in the table 1.

Table 1

Launcher	
Overall section (mm)	1600 × 2600
Total length (m)	15
Total power (MW)	35 - 55
Total weight (Ton)	65
Number (Nb) of modules	9
Dim. of module (mm)	400 × 750
Free height in module (mm)	2 × 350
Working mode	TE 50
Nb of PA wg per module	2 × 16
PAM arrangement	Quadrijunction
Quadrijunction Nb per mod.	2 × 4
PAM length (m)	1.5
Phasing in adjacent active wg	270°
$N_{//}$ excursion	1.7 - 2.3
Width of active wg (mm)	9.25
Width of passive wg (mm)	13.25
RHG number per module	2 × 4
Size of RHG (mm)	90 × 350
RHG length (m)	4.5
Mode converter (MC)	TE 10 - 50
MC length (m)	3.5
MC number per module	2 × 4
Standard section length (m)	6
Nb of single windows	72

The launcher is made of 9 modules, each composed of 2 stages of 4 quadrijunctons. The power capability of each quadrijunction (4 PA wg of 350 mm high) is 800 kW which is consistent with that of a single window used in the transmission line. The power of each module is 6.4 MW leading to a total power of 58 MW for the 9 modules of the launcher. The dimensioning parameter is the power of the vacuum single window which is limited to 0.8 MW at the level of the launcher. From the window, a standard WR229 waveguide is connected to the input of the TE 10-50 mode converter which feeds a rectangular hyperguide (RHG) connected to a quadrijunction.

The variation of the $N_{//}$ is produced by changing the phase between the circular hyperguides (CHG) feeding the different columns of the quadrijunctions.

5. KLYSTRONS & TRANSMISSION LINES

The working frequency is 5 GHz, frequency high enough to avoid alpha absorption of the wave but low enough to envision the development of 1 MW klystrons and associated windows, to reduce 2F losses.

Circular hyperguides working with the circular mode TE 01 and collecting typically 6 klystrons (Fig. 6) are designed in the transmission ligne.

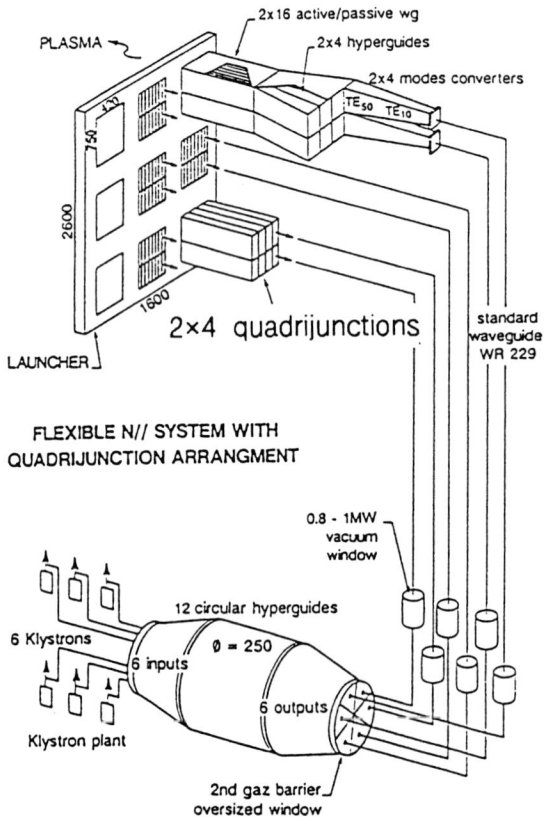

Figure 6. Schematic diagram of the system

A multimegawatt window based on contiguous triangular sectors is proposed to terminate the circular hyperguide and to be used as a second gas barrier of the transmission line.

6. CONCLUSION

A quadrijunction arrangement of the PAM, gives the flexibility of the $\Delta N_{//}$ permitting to cover the various scenarios envisioned during the first phase of operation on ITER. The proposed arrangement preserves the robustness of the ITER LHW system (PAM, Rectangular and Circular Hyperguide) and requires only slight developments to finalize the design. On Tore Supra such a PAM, adapted to the 3.7 GHz klystrons, is planned to be implemented before the end of EDA.

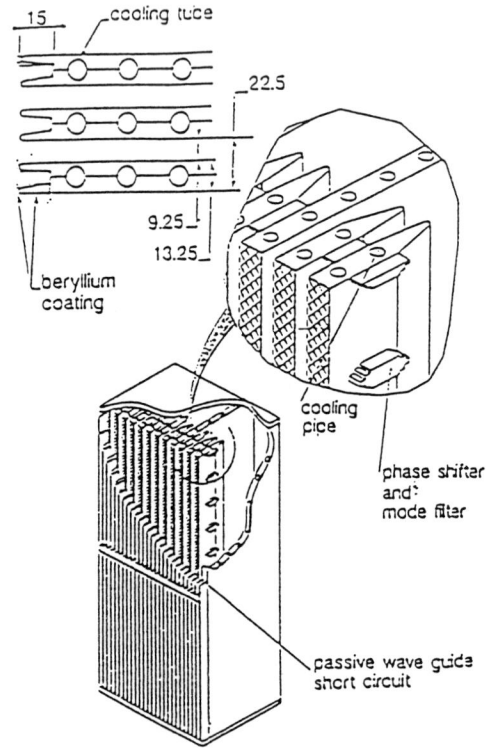

Figure 7. Diagram of one module of the launcher

REFERENCES

1. Report on the ITER Emerg. Task Agreement ETA-EC-IVA-LH, 1993, JET - R(94) 07 or EUR-CEA-FC-1529, Nov. 1994.
2. 94Rt. of the 1st Meeting of the Expert Group on Energetic Particles, H&CD, Moscow, 2 Oct. 1995.
3. Ph. Bibet, X. Litaudon, D. Moreau, Nuclear Fusion 35 (1995) 1213.
4. P. Froissard, P. Bibet et al, 19th SOFT, Lisbon, 1996.
5. D. Moreau, TK. N'Guyen Rep. EUR-CEA-FC1246, 1984.

Development of a new lower hybrid antenna module using a poloidal power divider

S. Maebara, M. Seki, K. Suganuma, T. Imai, M. Goniche[a], Ph. Bibet[a], S. Bério[a], J. Brossaud[a], G. Rey[a], G. Tonon[a]

Naka Fusion Research Establishment, Japan Atomic Energy Research Institute (JAERI), 801-1, Mukoyama, Naka-machi, Naka-gun, Ibaraki-ken, 311-01 JAPAN

[a]Association EURATOM-CEA Cadarache, Departement de Recherches sur la Fusion Controlée, Service Tokamak Ingéniérie et Développement, F-13108 Saint Paul lez Durance Cedex, FRANCE

For simplification of Lower Hybrid Current Drive (LHCD) antenna system and reduction of power density in the antenna, three types of poloidal power divider which split the power in three, and 3 x 6 multi-junction module were developed. A good power dividing ratio of 33 ± 4 % was obtained for each of these dividers. In combination with two dividers connected to the 3 x 6 multi-junction module, a good high power handling capability was assessed : 420 kW-100 sec was transmitted without breakdown. Quasi stationary operation for r.f. injection time of 1000 sec at 300 kW was also demonstrated under water cooling. The outgassing rate at that time is in range of 10^{-7} Pa m^3 s^{-1} m^{-2} with maximum temperature of 100 °C, which requires no active vacuum pumping of the LHCD antenna.

1. INTRODUCTION

Non-inductive current drive is indispensable for a steady-state tokamak reactor in order to provide a steady-state current drive in a tokamak plasma. Lower Hybrid Current Drive (LHCD) is one of the promising method and LHCD experiments have been performed for physical studies and technical assessment on JT-60U, Tore Supra, JET, TdeV tokamaks and others [1-4]. For an antenna design in future LHCD system, evaluation of r.f. properties, outgassing and high power handling capability during a long r.f pulse operation are important issues which are addressed in this paper.

In the LHCD antenna module, the r.f. power has to be split either in the poloidal or in the toroidal direction (or in both directions) in such a way that the port area is used as much as possible to reduce the power density in the waveguides. In the JT-60U LHCD system, wide (12 waveguides) toroidal multi-junction modules are used [5]. $N_{//}$ flexibility (1.5~2.3) is obtained by tuning the klystron frequency (1.74~2.23 GHz). With such conditions, power has been coupled to the plasma up to a power density of 21 MW m^{-2}. Tore Supra antennas use a combination of 3-dB poloidal divider (hybrid junction) and rather narrow (4 waveguides) toroidal multi-junction. In this case, the $N_{//}$ flexibility (1.4 ~ 2.3) is performed by changing the phasing between modules at a fixed frequency (3.7 GHz). With these antennas, 1MA plasma current has been driven for 1 minute with 3 MW of LH power (power density : 18 MW m^{-2}) [6].

For large antennas at high fixed frequency (≥5 GHz), poloidal power dividing systems having at least three outputs will probably be needed in combination with reasonably wide multi-junctions (~ 6 waveguides) to keep a control on the $N_{//}$. For this reason, we designed and developed different types of poloidal power dividers (3 secondary waveguides) and a rather large multi-junction module (3 x 6 waveguides) [7].

2. RF COMPONENTS

Two types of power dividers were considered. The first type divider provides r.f. power into the secondary waveguides by using single eigen-mode in the divider. The second one serves r.f. power by means of multi eigen-modes. JAERI and CEA have agreed to design and test three different poloidal dividers. Schematic drawing of these poloidal dividers is shown in Fig.1. The first type is labeled the E/H plane Multi-junction module (E/H MJ), and two of the second type are labeled the Mode Converter (MC) and the Poloidal Power Divider (PPD), respectively. All these power dividers were designed to be able to feed the 3 x 6 Multi-junction module (3x6 MJ) which consists in three rows of six secondary waveguides. Schematic drawing of the

1) E/H Multi-juction module

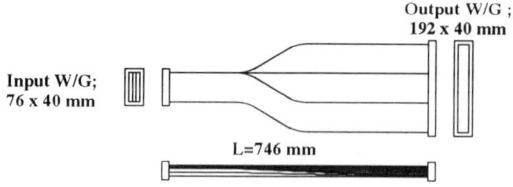

2) Mode converter

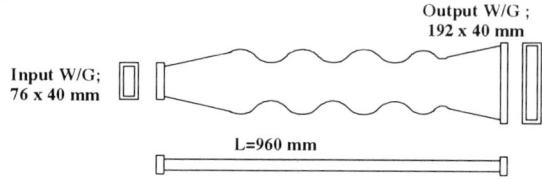

3) Poloidal power divider

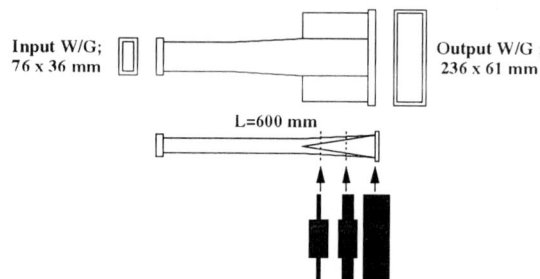

Fig.1 Schematic drawing of Poloidal power dividers

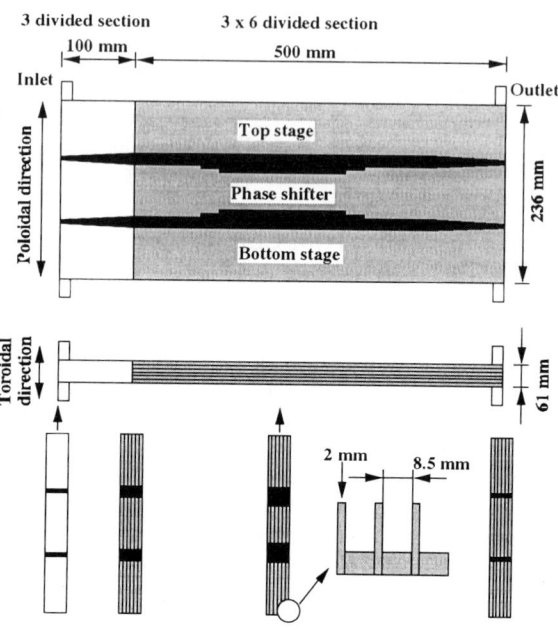

Fig.2 Schematic drawing of 3 x 6 multi-junction module

3 x 6 MJ is indicated in Fig.2.

The E/H MJ has only one eigen-mode (TE_{01}). The E/H MJ takes advantage of the E-plane multi-junction to split the power in three reduced-sections (10 x 76 mm^2). Upper and lower secondary waveguides are twisted in the poloidal direction, and then each of them are widened with a linear taper to lead to three quasi-standard waveguides 40 x 76 mm^2 lined on the same column. A step phase shifter in the central output waveguide insures the good 0 π 0 phasing.

The MC is the combination of a raised cosine taper which enlarges the height from 76 to 192 mm and a quasi periodic TE_{01}-TE_{03} mode converter with a constant beat wave number along the propagation direction [8]. In order to get a quasi perfect mode conversion efficiency, this MC includes 3.5 beat wavelengths.

The PPD has a compact designed, the total length is as short as 600 mm. The PPD is designed using three eigen-modes : TE_{01}, TE_{03}, TE_{05}. R.f power of the TE_{01} mode injected from the inlet, the TE_{01} mode is gradually converted into TE_{03} and TE_{05} modes as transmitting in longitudinal direction in order to fulfill boundary conditions of the PPD. Profile of the r.f. power is governed by combination of these modes. By optimizing the configuration, three equal peaks at the output PPD were obtained.

The 3 x 6 MJ is composed of two sections : three poloidally-divided section and the 3 x 6 MJ section. The 3-divided section exists between PPD and the 3 x 6 MJ, TE_{01} and TE_{11} modes can propagate in that section. The 100 mm length is required to decrease the effect of higher modes at input junction.

3. RF Properties

When tested on a matched load, a good power dividing ratio of 33 ± 4 % was obtained for each of these poloidal dividers, and the reflection coefficient less than 1.5 %. For these dividers, scattering matrix measured in low power tests indicated that reflection coefficient at the front end of the antenna with plasma have to be rather uniform for the different rows in order to have a well-working antenna, and then it is essential to attain a reflection coefficient less than a few % at the rear of the MJ. For the 3 x 6 multi-junction, reflection coefficient is less than 1.3 % and r.f. losses lower than 1.0 % are measured by low power measurements.

Five set-ups were tested at high power : 2 on short-circuit, mainly to check the high electric field withstand of basic components (E/H MJ and PPD in combination with the 3 x 6 MJ) and 3 in transmission on a dummy load to check r.f. properties of the 3 different poloidal dividers in combination with the 3 x 6 MJ. Outgassing evolution of these set-up was also studied in the same time, either by letting the components heating up to 400 °C or with cooling in order to get stationary conditions.

On the short circuit, the components (E/H MJ and PPD+ 3 x 6 MJ) showed good electric withstand : 200 kW and above could be injected corresponding to the maximum electric field of about 7 kV/cm at the input of the E/H MJ and the PPD, and 3.6 kV/cm in the 3 x 6 MJ and at the outputs of the E/H MJ and PPD.

On the dummy load, small resonance ($S_{21} = -0.35$-0.5 dB) occurred in the most set-ups, but high r.f. power transmission capability was not affected in the most cases up to a power density level of 120 MW m^{-2} at the inputs of the poloidal dividers (PPD, E/H MJ, MC) and 31 MW m^{-2} at the 3 x 6 MJ. A good high power handling capability was assessed.

When an output power of 300 kW is injected, variation of the klystron output power, the reflected power, pressure in the vacuum tank, each temperature rise at E/H MJ and MC are shown in Fig.3. A good power dividing ratio is indicated by each temperature rise.

The r.f. losses are deduced from calorimetric measurements for the set-up of PPD+3 x 6 MJ + MC; total losses are 2.5 ± 0.7 % which is slightly below low power measurements (5 ± 1 %), but consistent with calculated losses (2.2 %), assuming losses in each components from the expected propagating mode (TE_{01} or TE_{03}). From this calorimetric loss measurement, it can also been inferred that r.f. loss in each power divider (PPD and MC) is below 1.5 %.

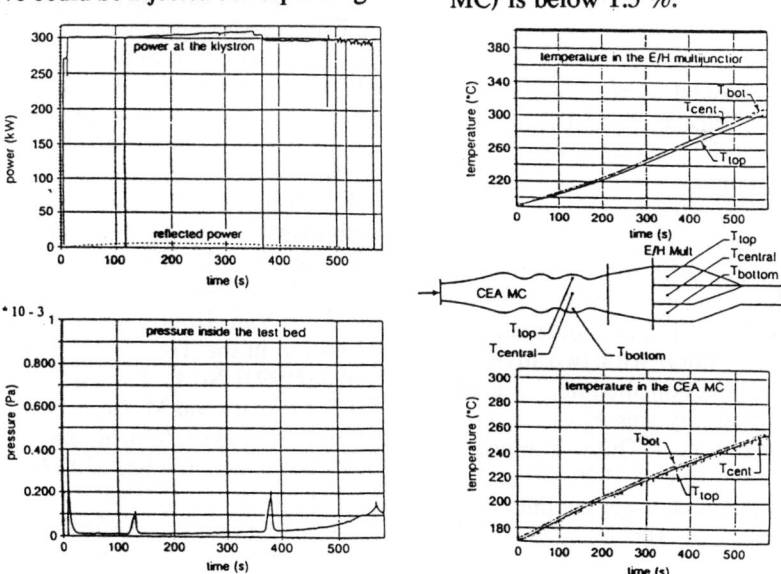

Fig.3 Power dividing ratio

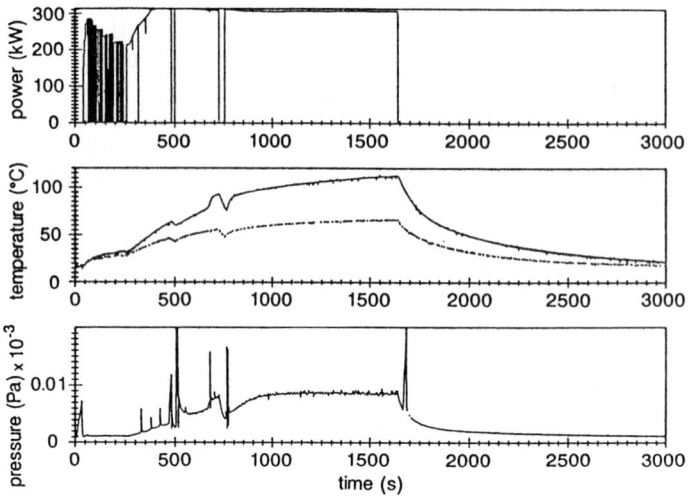

Fig.4 Quasi stationary operation

For the different set-ups, outgassing rate was in the 10^{-6} Pa m^3 s^{-1} m^{-2} range at 300 °C and the 10^{-5} Pa m^3 s^{-1} m^{-2} range at 400 °C. For the set-up of PPD + 3 x 6 MJ + MC, quasi stationary operation for r.f. injection time of 1000 sec at 300 kW was also demonstrated with water cooling as shown in Fig.4. The outgassing rate is in range of 10^{-7} Pa m^3 s^{-1} m^{-2} when the maximum temperature is 100 °C.

4. Conclusion

The antenna using poloidal power divider is an effective method for simplification of LHCD antenna system. This method should allow to reduce the power density in the antenna, while maintaining a good flexibility on the $N_{//}$ spectrum of the wave. For this purpose, three types of poloidal power divider which split the power in three, and 3 x 6 multi-junction module were developed. A good power dividing ratio of 33 ± 4 % was obtained for each dividers. In combination with two dividers connected to the multi-junction module, a good high power handling capability was assessed : 420 kW-100 sec was transmitted. Quasi stationary operation for r.f. injection time of 1000 sec at 300 kW was also demonstrated with water cooling. The outgassing rate is in range of 10^{-7} Pa m^3 s^{-1} m^{-2} when the maximum temperature is 100 °C. Under these test results, we assessed that poloidal power dividing method is feasible for the next LHCD system, and active vacuum pumping system is not needed for the LHCD antenna.

Reference

[1] Y. Ikeda et al., High power Lower Hybrid Current Drive Experiments in JT-60U, IAEA-CN-60/A-3-I-1, Seville, Spain, 26 Sep.-1Oct., 1994.
[2] Tore Supra team, Recent results on Tore Supra, IAEA-CN-60/A-1-II-1, Seville, Spain, 26 Sep.-1Oct., 1994.
[3] The JET team, Lower hybrid current drive in JET and reactor application, IAEA-CN-60/A-3-I-2, Seville, Spain, 26 Sep.-1Oct., 1994.
[4] R. Decoste et al., Biased divertor performance under LH current drive and heating conditions on the TdeV Tokamak, IAEA-CN-60/A-4-II-8, Seville, Spain, 26 Sep.-1Oct., 1994.
[5] Y. Ikeda et al., Simple multi-junction launcher with oversized waveguides for lower hybrid current drive on JT-60U, Fusion Engineering and Design 24(1994)287-298.
[6] D. Van Houtte et al, One minute pulse operation in the Tore Supra Tokamak, Nuc. Fus. Vol.33(1993) No.1, 137-141.
[7] S. Maebara and M. Goniche et al., Development of a new lower hybrid antenna module using poloidal power divider, JAERI-Research 96-036, EUR-CEA-FC 1590 (1996).
[8] Ph. Bibet et al., Experimental and theoretical results concerning the development of the main r.f. components for next Tore Supra LHCD antenna, Proc. 18th Symp. Fusion Technology, Vol.1, Karlsruhe, Germany, 1994, p.577-580.

Upgrading of the Injection Power of the W7-AS Neutral Beam System

W. Ott, F.-P. Penningsfeld, W. Melkus, F. Probst, E. Speth, R. Süss, and W7-AS Team

Max-Planck-Institut für Plasmaphysik, EURATOM Association, D-85748 Garching

Neutral beam injection experiments on W7-AS started at the end of 1988 with four ion sources mounted on two beamlines. The total neutral beam power was 1.5 MW, raising the central beta to $\beta_0 = 2.7\%$ — far from the limit expected to be $\approx 4.5\%$. Aiming at testing β limits in W7-AS, the power has recently been doubled by adding two additional sources to each of the two beamlines. With 3 MW of hydrogen beam in the torus vessel, $\beta_0 = 4\%$ was obtained without any sign of a limit, such as instabilities or enhanced mode activities [1].

One of the two available calorimeters was reinstalled in the W7-AS torus. This way it was demonstrated that the full neutral beam power of 1.5 MW per injector has indeed been injected into the torus. Beam blocking in the port was not observed, but there was a loss of neutral beam power in the front chamber of the injector which was obviously due to insufficient pumping.

It was tried to measure the relative power of the eight different beams by comparing the diamagnetic energy achieved with plasma injection shots. The results, however, were not very promising.

By regapping the extraction grids of the ion sources and raising the extraction voltage from 45 kV to 50 kV for hydrogen, the power is now being increased additionally by $\approx 10\%$.

1. Introduction

The first stage of the injection system for W7-AS has been built with four ion sources mounted on two injectors with beams aiming in opposite directions to balance beam-driven currents as well as momentum input. ASDEX-type injector boxes [2] have been used which were able to carry four ion sources each, but only two sources per box had been installed, yielding 1.5 MW in total.

After several years of successful injection experiments, it became desirable to achieve higher beta values and to approach the equilibrium beta limit of W7-AS. Therefore, two additional ion sources have been mounted on each of the injectors, using as much of the equipment used formerly with the ASDEX injection as possible. In this way, the input power has been doubled with relatively small effort, and injection experiments with 3 MW started in spring 1995.

The physical aspects connected with the upgraded power are described by Penningsfeld et al. [3]. This paper deals with the more technical aspects of the upgraded injectors.

2. Power measurement in the torus with a calorimeter

At the end of the last experimental campaign, one of the two calorimeters was reinstalled in the W7-AS torus. It enabled us to measure the port-through power injected by the "West" beamline. The deflection magnet inside the beamline was active in all cases to remove the unneutralized ions. The gas flux into the beamline was the same in all the cases: When a beam was not intended to run, its ion source was switched on without the ions being accelerated.

The power was measured for three cases: (1) Stellarator field $B = 0$, (2) $B = 1.25$T, and (3) $B = 2.5$T. The gas flux into the beamline was the same, independent of whether only one or more beams were switched on. For each case a number of shots was made. The results scattered by $\approx \pm 5\%$. When the $B = 0$ series was finished, the source W1 had to be switched off because of a water leak. The results of the power measurements are shown in Fig. 1 for extraction currents of 25 – 25.5 A. The first four sets of columns show the results for single-beam operation. The fifth set shows the average power in the case of three beams being fired together. The sources turned out to be of different quality. The worst one (W2) gave about 25% less power than the best one (W4). There were two other surprising results:

- The injected power was reduced by $\approx 10\%$ when the stellarator field was $B = 1.25$T and by $\approx 20\%$ for $B = 2.5$T.
- The average power in multiple-beam operation is not smaller than the average of the single beam powers.

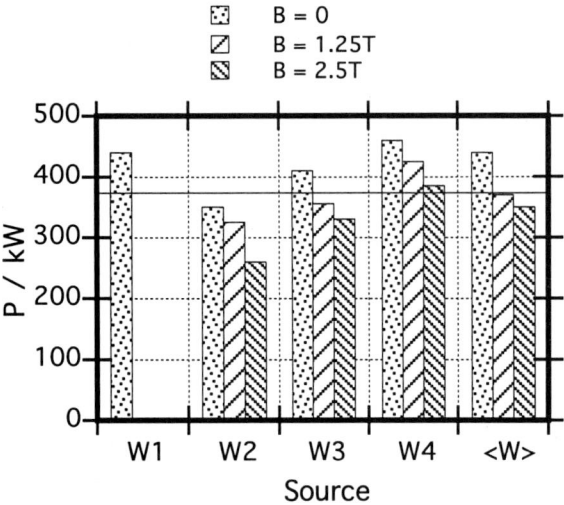

Figure 1: *Power delivered by individual beams into the torus with three different stellarator magnetic field strengths. For the fifth set of columns, the three beams W2, W3 and W4 were shot together and the resulting power divided by 3. In all these cases, the beam powers were kept constant at 45 kV and 25–25.5 A.*

It is concluded from Fig. 1:
- Reionization is caused by rest gas in the beamline. Stronger stellarator magnetic fields cause ions from farther upstream to be deflected and lost.
- Beam blocking due to gas released by a power load in the duct to the torus does not occur. Otherwise the multiple beam power should be less than the sum of the single beam powers.

Reionization should be minimized in future experiments when maximum power is necessary. Care must be taken to guarantee better pumping in the front part of the injector box which due to geometrical restrictions has smaller titanium pumps than the main chamber.

The nominal beam power of 375 kW per source is shown by the horizontal line in Fig. 1. This power was practically achieved on average for $B = 1.25$T, even for the less favourable pumping conditions existing during the time of these measurements.

In a series of shots, it was tested how the neutral power delivered into the torus depends on the accel current when the accel voltage is kept constant at 45 kV and the stellarator magnetic field is $B = 1.25$ T. This was done by increasing the arc voltage of the ion source, starting at low values. Beyond $I_{acc} = 25$ A, it was hardly possible to raise the accel current further by increasing the arc voltage. The torus power, however, continued to rise. Fig. 2 shows that the injection power

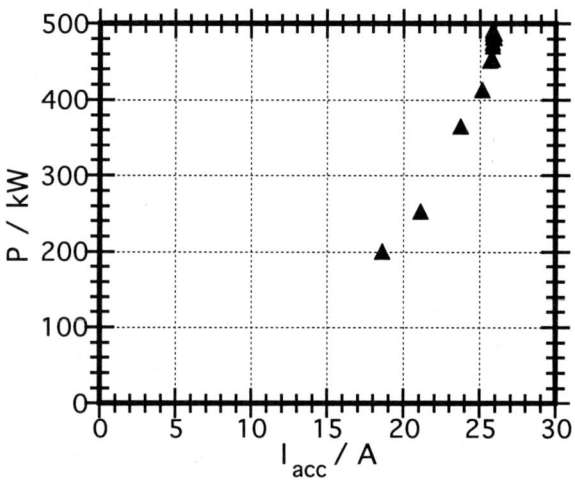

Figure 2: *Power delivered by source W4 into the torus with a stellarator field of 1.25 T when the accel current I_{acc} was varied.*

can be raised by $\approx 10\%$ by raising the accel current beyond 25 A. The steep increase of the torus power with I_{acc} is astonishing and not fully understood because the optimum current with minimum beam divergence was determined earlier [4] to be 23–25 A with 45 kV.

3. Comparison of different sources using injection shots

The measurements as described in section 2 cannot be done during normal operation time of W7-AS because the ports carrying the torus calorimeters are occupied by some diagnostic equipment. It would be desirable to use plasma injection shots to determine the beam power delivered into the torus.

In principle, the time derivatives of the plasma energy after switching from a reference source to a second one could give the difference in heating power of the two sources. There are, however, two difficulties connected with this measurement:
(1) The energy confinement time τ_E of stellarators depends on the heating power [5]. If the power level is switched, the conditions are not well defined for at least one slowing-down time of the injected particles.
(2) The signal W_{dia} of the diamagnetic energy in W7-AS sometimes contains much noise, so that a time derivative on a "short" time scale can hardly be formed.

Therefore a more elementary method is used to check the quality of this comparison: The slowly varying part of W_{dia} after switching from one source to another In this way, plasmas with the same impurity content are compared. Taking

Source	$B = 1.25$ T	$B = 2.5$ T
inner, co	0.892	0.910
outer, co	0.733	0.824
inner, counter	0.594	0.777
outer, counter	0.576	0.760

Table 1: Heating efficiency η for 45 kV beam voltage and a plasma line density of 2.1×10^{19} m^{-2}

Fafner values of the heating efficiency η and the power of the three individual sources as measured with the torus calorimeter (Fig. 1), the plasma energy can be related to the heating power. This is done in Figs. 3 and 4 for a toroidal field strength of 2.5 T and 1.25 T, respectively. The values of η as shown in Table 1 are interpolated for an effective line density of 2.1×10^{19} m^{-2} from the data given in [3].

For Fig. 3 the co-injections could be made with the inner sources only. For Fig. 4 only counter injections were made. The dashed lines in both figures represent a power fit to the measurements. The exponents α of the fit $W \propto (\eta P_{\mathrm{Torus}})^\alpha$ are 0.41 and 0.43, respectively. These values compare very well with the stellarator scaling ISS95 [4] which gives the exponent $\alpha = 0.41$. This scaling would give absolute values of the plasma energy W which agree with the dashed lines in Figs. 3 and 4 within 10 %.

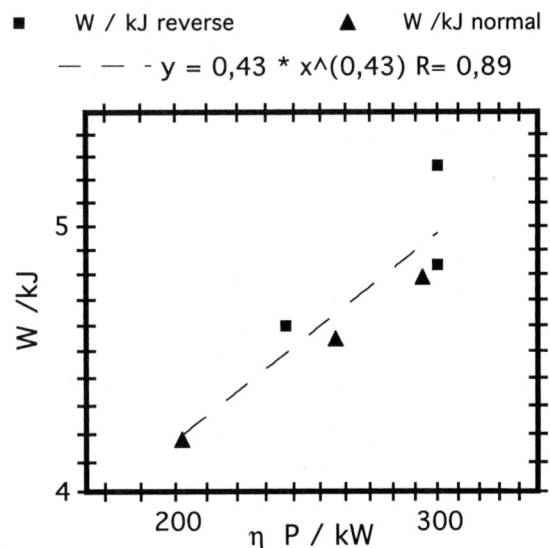

Figure 3: Diamagnetic plasma energy W achieved with different sources into a plasma with a line density of 2.1×10^{19} m^{-2} and a magnetic field of 2.5 T with normal and reversed polarity. The dashed line is a numerical power fit to the data.

Unfortunately, this method is not sufficient to determine the beam powers well enough. The statistical error margin in determining the plasma energy is about 4%. If the power delivered by the individual beams into the torus is to be compared by plasma injections, the error in $\eta \times P_{\mathrm{Torus}}$ becomes 10%. The value of η has sofar only been calculated. Even if the Fafner runs are made for the same line density, profile effects and the unknown charge exchange losses increase the uncertainty by estimated 5–10% to a total error of $\pm 20\%$. Using the torus calorimeters is therefore unavoidable if the individual beam powers are to be determined more accurately.

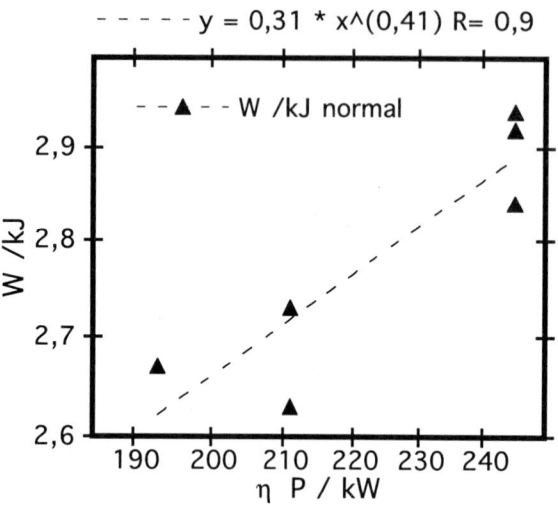

Figure 4: Same as Fig. 3, but with $B = 1.25\,\mathrm{T}$ and normal field polarity only.

4. Regapping the extraction grids

In W7-AS, the plasma density rises in many cases when injection is done with two or more beams. Especially in high-beta experiments, better beam penetration into the plasma is desirable. Therefore the acceleration grids were regapped in order to increase the acceleration voltage. For regapping, the modified perveance law of Uhlemann and Wang [6] was applied. The *effective* gap, calculated as the sum of the metal-to-metal gap between accel and decel electrode (6 mm) + thickness of accel electrode (3.3 mm) + $0.8 \times$ radius of aperture in decel electrode (0.8×2.6 mm), was increased by 1 mm to raise the extraction voltage from 45 to 50 kV.

The regapping has been finished for all eight sources. First experience shows that the sources work as reliably as before and that the beam divergence has improved by $\approx$ 10% from 1.3° to 1.15°. It is, therefore, to be expected that the total beam power is raised additionally by >10 % to ≈ 3.5 MW.

5. Conclusions

Measurements with the torus calorimeter showed that the nominal neutral power of 375 kW per source is indeed delivered into the torus — at least on average. But there are differences between the sources which are not fully understood. Careful titanium gettering of the beam line and especially of the front box is necessary to avoid reionization losses. The use of high accel currents combined with careful source conditioning and beam adjustment is necessary to maximize the beam power. Measuring the diamagnetic energy in plasma injections is a rather unprecise and unreliable method to determine the beam power. It is therefore necessary to repeat the power measurement with the torus calorimeter at least once after the accel voltage has now been raised to 50 kV.

References

[1] R. Jaenicke et al, *High power heating experiments on the W7-AS Stellarator*, Plasma Phys. & Contr. Fus. **37**, A163 (1995)

[2] O. Vollmer et al., *Initial Operation and Performance of the ASDEX Long-Pulse Injection System*, Proc. 15th SOFT, Utrecht, 1988, p. 625 (FUSION TECHNOLOGY 1988, Elsevier Science Publishers, 1989)

[3] F.P. Penningsfeld et al., *3 MW Neutral Injection into the Stellarator W7-AS, Heating Efficiency at High-beta Operation*, 23rd EPS conference on Controlled Fusion and Plasma Physics, Kiev, Ukraine, June 1996.

[4] W. Ott, F.-P. Penningsfeld, *Beam divergence and ion current in multiaperture ion sources*, Laboratory Report IPP 4/252, Max-Planck-Institut für Plasmaphysik, March 1992.

[5] U. Stroth et al, Nuclear Fusion **36**, 1063 (1996)

[6] R. Uhlemann and G. Wang, Rev. Sci. Instrum. **60**, 2879 (1989).

Development of a Novel Compact RF-Source

B. Heinemann, J.-H. Feist, W. Kraus, F. Probst, R. Riedl, E. Speth, M. Busch*, W. Szcepaniak*

Max-Planck-Institut für Plasmaphysik, EURATOM-Association, D-85748 Garching, FRG

*Galvano-T, D-51570 Windeck-Rosbach, FRG

A new design for large rf-sources is under development at IPP, which will solve some of the existing problems and extend the experimental flexibility. It should once substitute the present source design, which has a complicated three-walled plasma containment, namely a vacuum vessel, quartz vessel and Faraday shield (from outside to inside) by a single walled source body, which combines all functions. It is manufactured by electrodeposition of copper and has many longitudinal slits for the permeation of rf-field, which are sealed vacuum tightly either with viton cords or with embedded ceramic stripes.

1. INTRODUCTION

For many years now different types of large area, rf-driven plasma generators have been developed at IPP. They could replace the so far used arc discharge sources, whose tungsten filaments have to be renewed periodically. With this respect the rf-source would be a significant advantage for ITER.

So far the best results were delivered in a configuration, in which the rf-antenna is separated from the plasma by a quartz vessel. A Faraday shield inside the quartz vessel is used to prevent erosion of the quartz and deposition of SiO_2 onto the extraction grids. Furthermore a vacuum vessel outside the quartz vessel is needed as the quartz can not withstand the atmospheric pressure.

This design was successfully tested in the full "PINI"-size (32×61 cm^2, 19 cm long, 33 l volume) at the testbed. A series of 120 pulses was carried out, operating with nominal parameters of 55 kV, 88 A, 5 s (H^+). The needed rf-power was about 110 kW (ref. 1).

Based on these results it was decided to equip the second injector on ASDEX Upgrade with these sources. It will go in operation in autumn 1997.

For future applications (e.g. negative ions, ITER) a new type of rf-source is under development, which simplifies the mechanical construction, erases some of the existing problems and rises the experimental flexibility: it combines all the functions of the different walls (vacuum tank, quartz vessel and Faraday shield) within one single wall. This means the source vessel is now a self supporting vacuum tank with an integrated Faraday screen, permeable to rf-waves but leak tight. The rf-coil as well as the permanent magnets are now outside under air and easily accessible.

The paper first describes the design of the advanced rf-source compared to the present design. Then a description of the status of development, tests and results will be given. Finally the next steps, which are planned, will be outlined.

2. DESCRIPTION OF SOURCE DESIGN

2.1. Present Design

A schematic drawing of the present source design is shown in Fig. 1 (ref. 2).

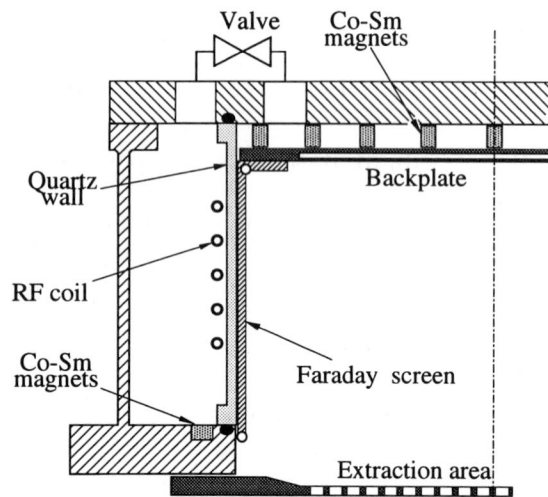

Figure 1: Schematic drawing of the present rf-source

The quartz vessel has almost vacuum tight silicon seals on both ends and separates so the plasma from the vacuum inside the stainless steel tank. The volume outside the quartz vessel is pumped via an external bypass by the source volume itself. The bypass valve is closed during source operation to keep the outside pressure below 10^{-4} mbar and to avoid electrical breakdowns.

The five turn coupling coil around the quartz is insulated with ordinary shrinking tubes and connected via feedthroughs to the rf-generator.

The actively cooled backplate has Co–Sm magnets embedded in a chequerboard pattern and is mounted inside the quartz vessel.

The Faraday shield is made of 3 mm thick copper, with longitudinal slits of 3 mm width every 2 cm. On both ends cooling tubes are brazed to the screen which cool it down sufficiently between the pulses.

During the operation of this source the following disadvantages were hindering and reduced the availability significantly:
- Three different containments and many parts use a lot of space and make the assembly very difficult.
- Bad access for modifications of the coil or magnetic filter causes low flexibility and long maintenance breaks.
- The insulation of coil was unreliable and created some breakdowns.
- The pumping of the outer volume has low conductance and needs reliable valves and control systems.
- Many vacuum feedthroughs are necessary.

2.2. Novel Source Design

The design of the novel rf-source is shown in Fig. 2.

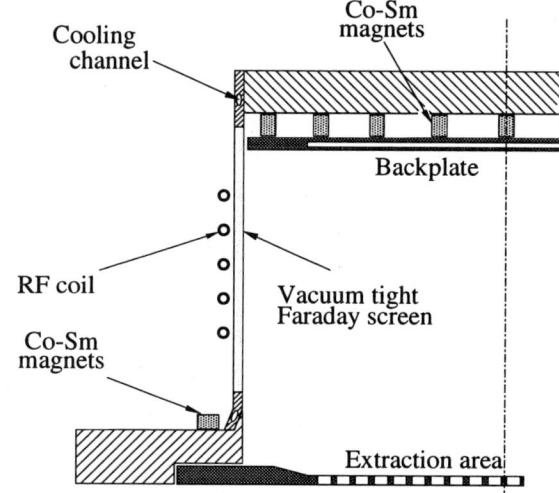

Figure 2: Schematic drawing of the novel rf-source

In this type the Faraday shield is made thicker so that it became a self supporting vacuum tank. The longitudinal slits are vacuum tight sealed by viton cords or by ceramic stripes. Therefore the outer vacuum vessel could be dropped and the coil and magnets are outside under air and easily accessible.

As material for the Faraday shield electrodeposited copper was selected because of its low electrical resistance and relatively high mechanical strength (250 MPa yield strength at low temperatures). For the given geometry a wall thickness of 6 mm is sufficient for the vacuum forces. Furthermore the electrodeposition allows direct connection to the stainless steel base flange and backplate without brazing, gives a lot of flexibility for the cross section of the slits and makes embedding of cooling channels easily possible within the same manufacturing process.

3. STATUS OF DEVELOPMENT

A full scale prototype has been build with 92 slits, sealed with a viton cord. It is mounted now on the negative ion testbed. Plasma has been generated successfully with a rf-power of 80 kW for 3 s. Extraction will start in autumn 1996.

A longer development program was necessary for sealing the slits with ceramic. First a leak tight and mechanically strong bonding between ceramic and electrodeposited copper had to be found. Deposition onto Al_2O_3, which was metallized with the standard MoMn printing process yielded very good results.

Furthermore the ceramic stripes could not be embedded into the thick Faraday shield directly, because their different thermal expansion would crack the ceramic.

As a result of several different approaches a prototype was built now with 150 mm inner diameter and a total length of 235 mm, which is shown in Fig. 3. Again a 6 mm thick Faraday shield forms the source body and is able to stand the atmospheric pressure, but the slits are not vacuum sealed.

Around that is an outer, vacuum tight shell which consists of Al_2O_3-stripes, connected between each other and to the collars on both sides of the Faraday shield by a 0.6 mm thin copper shell. In this way the ceramic stripes are only supported by the F.S., not directly connected to it, but only via the thin outer shell. This thin shell can compensate different thermal expansion up to a certain value.

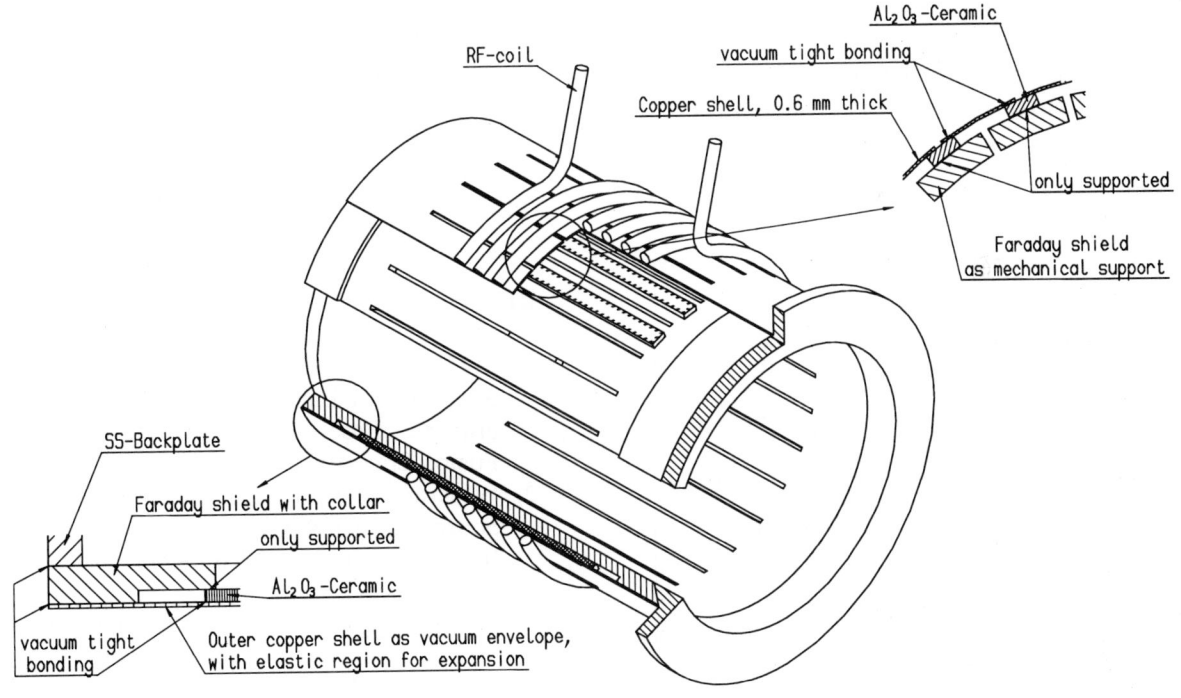

Fig. 3: Prototype of novel rf-source

The slits in the outer shell are displaced relative to those in the FS, so that the ceramic is protected by the FS and will not be eroded by the plasma. Furthermore the space between one ceramic stripe to its neighbours and to the collars of the FS allows flexible deformation for different temperature expansion.

4. EXPERIMENTAL RESULTS

This source was installed at the testbed with an eight turn coil around it. Plasma could be generated without problem with a rf–power of 30 kW. The power density of 20 W/cm² corresponds to about the same value of the present full scale source.

In a first step the source was operated without additional cooling up to 5 second pulses. A series of those pulses every 3.5 minutes increased the temperatures until the ceramic broke.

After installation of water cooling on both ends of the Faraday screen, a series of 9 second pulses with the same parameters could be carried out without damage (Fig. 4). For 11 second pulses cracks occurred again. Therefore for longer pulses or higher power loads an active cooling between the slits is necessary.

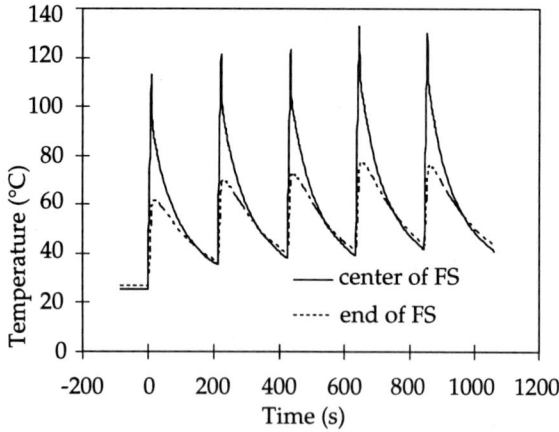

Figure 4: Temperatures of Faraday Shield in the center and at the end

5. FUTURE PLANS

The temperature level for failure of the Al_2O_3–stripes is lower than expected. The cracking of the ceramic can be heard clearly and significantly after finishing the pulse and all cracks occurred at the upper or lower end of the stripes. Therefore it can be assumed that the problem comes from local deformation at both ends which is caused by heat conductance and not directly by the plasma or rf–current. Detailed finite element calculations are in progress at the moment to understand the problem and to modify the design if possible.

Furthermore the outer shell of the prototype will be replaced and active cooling between all slits will be introduced. This type will then be tested up to its limits before a full scale prototype is planned to be manufactured.

6. SUMMARY

A new concept for building big rf–source bodies is under development and has successfully been manufactured. Further tests have to be carried out to complete the ceramic sealing. This type of source promises to be very useful for future experiments and ITER.

REFERENCES

1. J.–H. Feist et al., "Progress in the Development of a Large-Area RF Plasma Generator", Proc. of 16th IEEE/NPSS Symp. Fusion Eng., 1995, p. 976
2. W. Kraus et al., "A High Power RF Plasma Source for ASDEX Upgrade Neutral Beam Injection", Proc. of 18th SOFT, Karlsruhe, 1994, p. 473
3. W. Kraus et al., "A High Power RF Plasma Source for Neutral Beam Injection", Proc. of 17th SOFT, Rome, 1992, p. 549

Optimization of a large-area RF plasma generator

W. Kraus, J.-H. Feist, E. Speth

Max-Planck-Institut für Plasmaphysik, D-85748 Garching, EURATOM-Association

A large-area plasma generator is being developed at IPP for the second injector of ASDEX Upgrade (AUG), scheduled for operation in 1997. The reliability of this source has been demonstrated last year in a demo series of 120 shots at the nominal beam parameters of 88 A, 55 kV, 5 s (1). Further attempts have been made to optimise transmission by improving the uniformity of the plasma density profiles in the extraction area. Variations of the magnetic configuration, geometrical conditions and operational parameters have been carried out. None of those did further improve the transmission so far. The parametric dependence of the transmission on beam energy, beam current/rf power and decel voltage is discussed including experiments up to 100 kV, using a modified extraction system.

Further attempts to reduce power losses and matching problems, caused by eddy currents in the Faraday shield, are under investigation at present.

1. INTRODUCTION

On ASDEX-Upgrade tokamak a second neutral beam injector will start operation mid 1997. It will be based on four positive ion beams at 55 kV in hydrogen or 65 kV in deuterium respectively. In order to have a better deposition of the fast particles in the plasma in the deuterium case, in a second stage it is planned to increase the deuterium energy from 65 keV to 100 keV.

The beamlines will be equipped with powerful RF sources, which are under development at IPP Garching. Compared to conventional arc sources, the fundamental advantages of RF sources are a higher lifetime, a simpler mechanical structure and the high voltage separation by an RF transformer. In addition the costs of the source power supply are much lower.

The source design envisaged for the application on ASDEX Upgrade is based on an external antenna surrounding a quartz vessel (1). In many tests this source has proved to meet the power and reliability requirements in the case of 55 kV. However, although the obtained results meet the design values of the extraction system and the beamline, there are a few features of this source, which make further improvements desirable: (i) the transmission (67% IxU) is slightly below the transmission of the bucket source (72%); (ii) power losses and (iii) matching problems are caused by eddy currents in the Faraday shield.

The subsequent development of the source was hence devoted to improve transmission by optimising the uniformity of the plasma density profiles in the extraction area. To this end variations and modifications of the magnet configuration, the source pressure, the source depth and the Faraday shield have been investigated. However, none of these attempts did further improve the transmission so far. In order to get a clue to this puzzle the parametric dependence of the transmission on beam current/rf power, beam energy and decel current was studied more systematically. In this context also experiments were carried out at voltages up to 100 kV, using a modified extraction system.

Compared to transmission the problem (ii), power losses, i.e. power efficiency is much less of a concern, since only in the case of 95 Amperes hydrogen the required rf power approaches the nominal rf power of the supplies. Nevertheless power efficiency represents a constraint and hence has to be monitored simultaneously with transmission.

The third question (iii) is still under investigation and not part of this paper.

In the following first the present design of the source and the rf power supply is reviewed briefly.

Then the experimental results are described and discussed. The conclusion summarises the findings and further plans.

2. RF SOURCE AND POWER SUPPLY

The six turn RF coil is separated from the plasma by a quartz vessel, which has an area of 32 × 61 cm2 (Fig. 1). In order to avoid implosion side walls, the source is placed inside a metallic vacuum chamber.

The Faraday shield is made of 3 mm thick copper with vertical 3 mm wide slits spaced at 2 cm. It was mounted adjacent to the side walls, covering the entire quartz surface.

The RF generator is a self-excited oscillator with power regulation and a maximum output power of 125 kW at 0.92 MHz. It is connected by a 50 Ω transmission line of 120 m length to the matching unit, which is placed close to the source. For a more detailed description see ref 1.

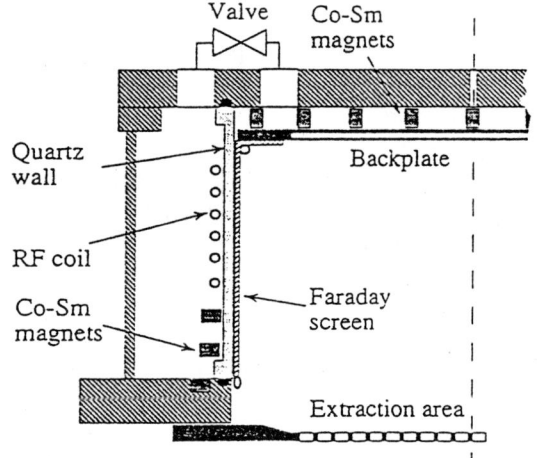

Fig. 1.: Part of the cross section of the RF source with Faraday shield and permanent magnets on the side walls.

3. EXPERIMENTAL RESULTS

The beam transmission is determined as the fraction of the electrical beam power transmitted through a simulated AUG duct onto a calorimeter, which consists of six actively cooled copper panels.

For each configuration a perveance scan at 75 A beam current has been carried out to obtain the maximum beam transmission (see Fig. 2).

3.1 Permanent magnets on the side

From probe measurements it is known, that permanent magnets on the side walls adjacent to the extraction plane significantly influence the plasma density profiles at the edge of the extraction area

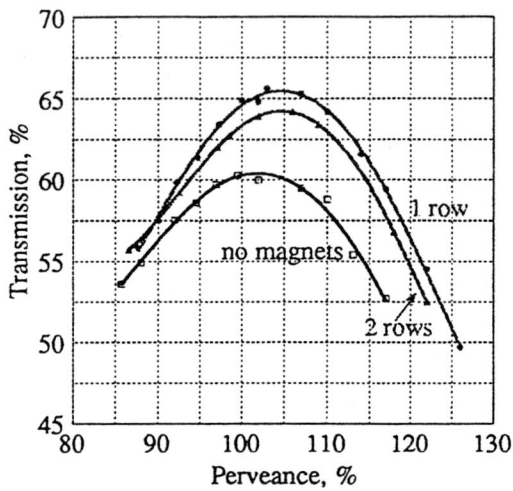

Fig. 2.: Transmission vs. perveance without and with permanent magnets on the side walls at a beam current 75 A and beam energies of 46 - 56 kV

Up to three additional rows of permanent magnets can be mounted in the intermediate space between RF coil and ground plate (Fig. 1). In the used chequer board arrangement, the magnets have alternating polarities within each row and between adjacent rows. Experiments with of one, two and three rows have been carried out. The main results are listed in Tab. 1. It turned out, that with one row the transmission is considerably higher than without magnets at the side. The addition of more rows did not improve transmission, but resulted in reduced power efficiency due to reduced coupling of the RF power to the plasma.

3.2 Variation of the depth of the source

From probe measurements in previous experiments with a slightly narrower source (28 cm width instead of 32 cm) it is known that there is an optimum depth of the source to obtain flat plasma density profiles (2). These experiments have been repeated with beam extraction using the actual source, trying depths of 18, 19 and 20 cm. The highest transmission was obtained with 19 cm, the highest power efficiency with 18 cm, however, the latter not compensating the loss in transmission.

3.3 Variation of the discharge pressure and plasma density

By changing the discharge pressure the profiles can be varied from hollow (high pressure) to peaked (low pressure). The optimum pressure to achieve the highest transmission and power efficiency in the case of the AUG source is around 1 Pa.

There is not such a clear dependence on the RF power: in probe measurements almost no dependence of the plasma uniformity on the plasma density was observed. Because the axial distance between the extraction area and the RF coil (6 cm) is 2 to 3 times the skin depth (2-3 cm), the profile variations caused by different skin depths do not affect the profile at the extraction area very much.

With beam extraction, however, in contrast to probe measurements, a clear improvement of the transmission at higher beam current, i.e. higher RF power is observed (see below).

3.4 Comparison of the source performance with and without Faraday screen

In addition to its function of protecting the wall against sputtering, the Faraday shield shortcircuits the electric field of the coil und thus prevents capacitive coupling to the source plasma. Since it is unclear, if this would have an effect on the beam divergence, the Faraday shield was removed for a comparative test. The measured beam transmission with and without Faraday screen was almost identical. Apparently the electric fields of the coil in the plasma have no impact on the beam divergence.

However, a big difference in power efficiency was observed in the high current range, where the extracted current increases more than linearly with RF power. With the shield about 15 % more power is necessary to achieve a beam current of I = 90 A. An operational problem is the low stability of the working frequency of the RF supply, which makes source matching much more difficult. The power losses and the matching problems, which are assumed to be caused by eddy current, should be reduced, if the ends of the screen are not shortcircuited by the cooling tubes (see Fig. 1). Such a shield, which consists only of insulated straps is presently under construction and will be tested in the near future.

	Efficiency, A/kW	Transm. at 75 A
No magnets at the side walls*	0.75	60.5 %
Magnets at the side walls*		
one row	0.80	66 %
two rows	0.70	64 %
three rows	0.50	-
Depth of the source**		
18 cm	0.83	61 %
20 cm	0.71	65.5 %
Discharge pressure**		
0.7 Pa	0.74	61 %
1.3 Pa	0.78	65 %
Without Faraday shield**	0.92	66 %

*Depth 19 cm, 1 Pa
**Depth 19 cm, 1 Pa, one row of magnets

Tab. 1: Power efficiency (extracted beam current vs. RF power) and maximum beam transmission at 75 A for different source configurations.

4. DISCUSSION OF THE RESULTS

In all experiments the beam transmission did not exceed 66 % at I = 75 A, compared to >70 % obtained with a bucket source. This difference can be simulated by a hypothetical beamlet divergence which is approx. 0.1° higher in the case of the RF source. Since none of the attempts to improve transmission by variation of the plasma density profiles and by omitting capacitive coupling was successful, it cannot be excluded that there may be other reasons for this difference.

If the plasma density profiles are not homogenious, a part of the extraction area is not matched to the perveance optimum. This would result in a locally higher beamlet divergence, and thus in a lower beam transmission.

On the other hand an RF modulation of the plasma density and/or the electron temperature could create fluctuations of the plasma potential and/or the extracted ion curent density. In this case even with a uniform plasma profile there would be a general degradation of the divergence of all beamlets. The role of possible RF fluctuations in the source / beam and its effect on the beam divergence has to be clarified with improved diagnostics. The installation of a single beamlet diagnostic is presently under discussion. First steps are time resolved measurements of the beam current, the plasma density and of the plasma potential.

From the parametric dependence of the transmission on extracted current (i.e. RF power) and extraction voltage there is no clear evidence for the hypothesis of RF-induced degradation of the transmission so far. On the one hand transmission increases with increasing current / RF power (maintaining perveance match, see below) for constant voltage (V = 55 kV). This is hard to get in line with the idea that it is the effect of the RF, which degrades the transmission. On the other hand transmission improves also with increasing extraction voltage (again at perveance match) at constant extracted current (I = 60 A). This could be interpreted as the commen $(T_i/V)^{1/2}$-dependence; any effects of the ion temperature/plasma potential fluctuations on the beam divergence being diminished with increasing extraction voltage. For these experiments it was mandatory to make also use of the 100 kV-PINI with an increased gap setting, in order to maintain perveance match at different combinations of I and V (see above).

If the extracted beam is RF modulated, a dynamic space charge decompensation could occur in the beamline. However, this effect should increase with perveance, which is not supported by the data.

5. 100 kV OPERATION

For perveance match at 100 kV, 65 A (limit of the power supply) in deuterium the first gap has to be increased from 8 mm to 14 mm. For this purpose the grid support structures have been changed in a way that now the distance between the grid holders is the same as between the grids. Therefore it was mandatory to test this arrangement before the release of the manufacturing.

Unfortunately this test could not be carried out in a direct way, because deuterium operation at the testbed is not allowed due to neutron production. Due to restrictions of the power supply (115 kV, 65 A), the titanium pumping system (no pumping of He/Ar) and the power capability of the calorimeter (2.8 MW), testing had to be carried out in steps.
- voltage holding capability: H2, 100 kV, 5 s
- extraction at perveance match: He, 100 kV, 52 A , .15 sec
- long pulse: H2/Ar beam, 100 kV, 30 A, 5 s
- high power, 50 % underperveant: H2, 100 kV, 65 A, a 200 ms on / 200 ms off pulsed , 5 s.

These tests showed a satisfactory performance of the extraction system also under strange conditions, and sbsequently manufacturing was released with the original grid holder dimensions. During these tests, the RF source including the HV transformer worked without any problems.

6. CONCLUSION

The results of the optimised RF source meet the design values of the extraction system and the beamline. Further improvement of the transmission will require better understanding of the limitations using improved diagnostics. The power efficieny and the matching characteristcs in the present design are acceptable; a further improvement may be achieved by a Faraday screen without eddy currents. With respect to the second stage of the neutral injection on ASDEX-Upgrade, the applicability of the modified extraction system for the 100 kV operation has been proved.

ACKNOWLEDGEMENTS
The authors wish to thank P. Pollner, E. Kühn and A. Steinberger and M. Ciric for their assistance.

REFERENCES
(1) J.-H. Feist, W. Kraus, E. Speth et al., Proc. of the 16th IEEE/NPSS Symp. on Fus. Eng., Champaign, IL, 1995, p. 976
(2) W. Kraus, M. Kaufmann, Proc. of the 15th Symp. on Fus. Techn., Utrecht 1988, p. 495

Modelling of the T E X T O R - 9 4 Neutral-Injection Power Supply System in the Search of Parasitic Currents

C.-C. Hering[1], U. Braunsberger[1], M. Sauer[2], W. Schalt[2]

[1]Institut für Hochspannungstechnik und Elektrische Energieanlagen, Technische Universität Braunschweig, PF 3329, D-38106 Braunschweig

[2]Institut für Plasmaphysik, Forschungszentrum Jülich GmbH, Ass. EURATOM-KFA, D-52425 Jülich

In largely distributed power supplies of electrical systems proper potential controlling is necessary. Besides safety requirements for the staff, disturbance problems in measurement systems can occur due to electromagnetic radiation. This is especially relevant in neutral injection power supply systems where frequently fast transients are caused by electrical breakdowns between the accelerator grids.

In order to avoid further problems in the future, a detailed circuit diagram of the complete neutral injection power supply system was established. This diagram took into account, if electrically relevant, the geometric dimensions of the components, as well as those of their connections. These data were transposed to an equivalent circuit to carry out circuit simulations.

In the paper the circuit diagram and the equivalent circuit are delineated. Results of the circuit simulation are reported as well as a comparison with measured waveforms. Possible improvements are discussed.

1. INTRODUCTION

In the course of the increasing complexity of the whole TEXTOR-94-machine including the diagnostics and control problems due to electromagnetic disturbance came up in operating the tokamak. Especially the interaction between the neutral injection systems and the tokamak caused difficulties also if regarding the safety of personnel. In other largely distributed electrical systems like power stations, this problems occurs also and is treated in the literature [1, 2]. Thus is was decided to establish a detailed circuit diagram of the electrical power supply system including not only the main paths of energy flow but also the different grounding points and its connections in the wide spread spatial distribution of all components. On the basis of this diagram an equivalent circuit was drawn for calculating the transient electrical behaviour during breakdown between the main acceleration grids of the injector. With this simulation the elements being responsible for dangerous transient electrical potential elevation should be detected. Then measures for avoiding or diminishing these effect can be found out for appropriate modification of the circuit elements and its connections.

2. THE TEXTOR-94 NI-SYSTEM

Textor-94 has two neutral beam injectors which can be operated independently from each other. The nominal voltage of each of the two power supplies is 72 kV, the current is 100 A. As both systems are nearly identically the study of grounding problems is done on one system only.

2.1. Components of the NI-System

Fig. 1 shows the principal arrangement of the components of the NI-system. The acceleration voltage power supply (1) is situated in a separate building outside the TEXTOR-94 hall. It is connected to the arc- and filament-power supplies (5) via a triaxial cables (3) of 30m and 150 m length and a connecting switch (4). The arc and filament power supplies are located near the TEXTOR-94 bunker and are supplied via an isolating transformer (6) as they are connected to the high voltage potential of the accelerator. From here a special high voltage line (7) leads to the accelerator (8) itself. The arc- and filament connections are conducted in an aluminum tube

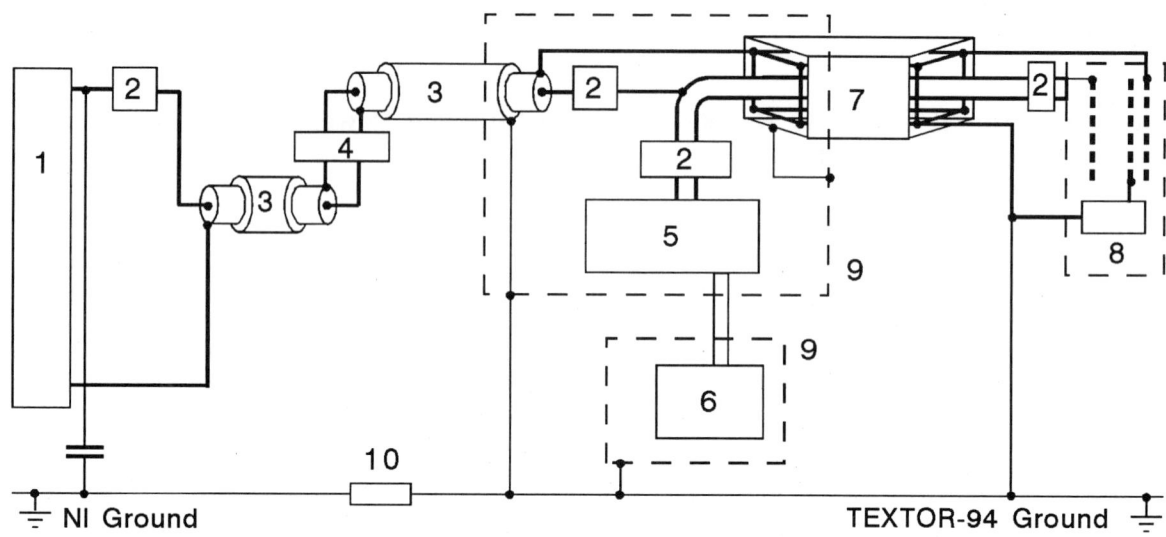

1	High Voltage Power Supply	2	Snubber
3	Triaxial Cables	4	Switch Connection
5	Arc and Filament Power Supply	6	Isolating Transformer
7	Triaxial Line with Accelerator, Arc, and Filament Supply Leads	8	Accelerator
9	Metallic Cubicles	10	Ground Decoupling Resistor

Figure 1: Schematic circuit diagramm of the TEXTOR-94 neutral injection power supply

carrying the acceleration voltage. The current is lead back via four cables surrounding this tube. These leads are situated in a metallic housing for protection purposes.

2.2. The equivalent circuit for simulation

In fig. 2 the components of the NI-system are transferred into the equivalent circuit to be used for transient behaviour simulation during breakdown between the accelerator grids. Most of the components are represented by constant capacitors, inductors, and resistors. Only the voltage of the power supply V101 and the resistor R814 simulating the main load between the accelerator grids G1 and G2 are preset as time dependent circuit elements. All elements and their values are listed in table 1.

The cables (component 2 and 3) and the transmission line (component 7) are described by their values per unit length and combined to the number of sections as given in fig. 2. The other components are drawn with their concentrated elements of L, R, and C, their connections being simulated by L and R, and their stray capacitances (component 5 and 6) between the active parts and the walls of the cubicles.

3. TRANSIENT BEHAVIOUR OF THE CIRCUIT

The current and voltage waveforms of the circuit shown in fig. 2 during breakdown between the accelerator grids were simulated with the PSpice program for transient behaviour calculation. It was the aim to find out whether dangerous transient voltages do occur at circuit elements which are in reach of personnel, i.e. the shielding of the triaxial line (no. 7 in fig. 2, elements 707/8/9) between the arc- and filament power supply and the injector itself. This point of the circuit is marked in fig. 2 with a circle.

3.1. Results of the simulation

During simulation the injector was fed from the voltage source V101 up to its nominal current of 100 A. Then the acceleration gap resistor R814 was switched from the nominal value of 600 Ω to its arcing resistance of a few milliohms.

In fig. 3 the transient voltage on the triaxial line

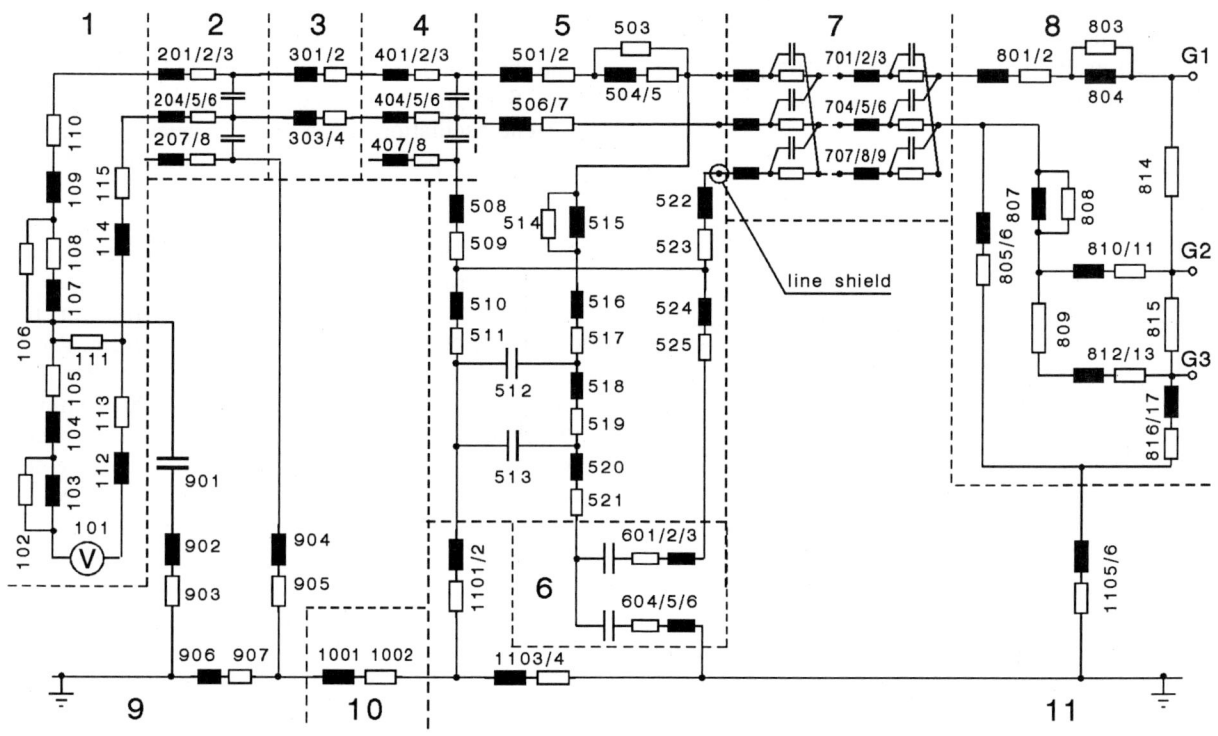

1	High Voltage Power Supply	2	Triaxial Cable 1, 7 m, 7 sections
3	Switch Connection	4	Triaxial Cable 2, 100 m, 10 sections
5	Arc and Filament Power Supply Cubicle	6	Isolating Transformer Cubicle
7	Triaxial Line, 30 m, 6 sections	8	Accelerator
9	NI Power Supply Ground Potential	10	Ground Potential Decoupling Resistor
11	TEXTOR-94 Ground Potential	G1..3	Accelerator Grids

Figure 2: Equivalent circuit of the TEXTOR-94 NI power supply

Table 1: Values of the elements of the equivalent circuit

Element	Value	Element	Value	Element	Value	Element	Value	Element	Value	Element	Value	Element	Value
V101	..60kV	L201	75nH/m	C406	200pF/m	C513	50pF	L603	0.6μH	R803	47Ω	C901	1nF
R102	50Ω	R202	150$\mu\Omega$/m	L407	50nH/m	R514	150Ω	C604	100pF	L804	50μH	L902	10μH
L103	30μH	C203	150pF/m	R408	150$\mu\Omega$/m	L515	10μH	R605	7.4mΩ	R805	7mΩ	R903	5.4mΩ
L104	580μH	L204	75nH/m	L501	1.3μH	L516	1μH	L606	50μH	L806	2.4μH	L904	1μH
R105	75mΩ	R205	150$\mu\Omega$/m	R502	0.4mΩ	R517	10$\mu\Omega$	L701	1.2μH/m	L807	520μH	R905	100kΩ
R106	44Ω	C206	150pF/m	R503	39Ω	L518	1μH	R702	6.6$\mu\Omega$/m	R808	47Ω	L906	3.2μH
L107	4.2mH	L207	100nH/m	L504	413μH	R519	50$\mu\Omega$	C703	42pF/m	R809	25Ω	R907	1.4mΩ
R108	200mΩ	R208	200$\mu\Omega$/m	R505	52mΩ	L520	1.2μH	L704	1.5μH/m	L810	3.4μH	L1001	10H
L109	1μH	L301/3	1μH	L506	2.3μH	R512	10$\mu\Omega$	R705	1.1mΩ/m	R811	3.7mΩ	R1002	2Ω
R110	1.3mΩ	R302/4	0.5mΩ	R507	1.9mΩ	L522	3μH	C706	221pF/m	L812	3.8μH	R1101	7.4$\mu\Omega$
R111	6kΩ	L401	50nH/m	L508	0.2μH	R523	3.4mΩ	L707	1.9μH/m	R813	4.1mΩ	L1102	50μH
L112	39μH	R402	100$\mu\Omega$/m	L509	0.1mΩ	L524	13μH	R708	116$\mu\Omega$/m	R814	600Ω	L1103	0.1μH
R113	13mΩ	C403	100pF/m	L510	1.5μH	R525	1.3mΩ	C709	64pF/m	R815	250Ω	R1104	1$\mu\Omega$
L114	1μH	L404	50nH/m	R511	0.3mΩ	C601	200pF	L801	1μH	R816	1.4mΩ	R1105	4mΩ
R115	1mΩ	R405	100$\mu\Omega$/m	C512	50pF	R602	1.8mΩ	R802	40$\mu\Omega$	L817	1.2μH	L1106	27μH

shield against TEXTOR-94 ground potential is shown. After voltage breakdown in the accelerator a damped oscillating voltage waveform generates a potential difference between the line shield and ground of +17 kV and -11 kV with a few µs duration. The current through the connection between line shield and arc- and filament supply cubicle wall (no. 8 in fig. 2) is drawn in fig. 4. The damped current oscillation has a main frequency of about 250 kHz, superimposed by oscillations with higher frequencies. The peak current is about 210 A.

3.2. Comparison of the measurement and simulation

During operation of the Neutral-Injector the current through the marked connection in fig. 2 was measured when a grid breakdowns occurred. Its waveform is shown in fig. 5 and has a shape similar to calculated current given in fig. 4. But the main frequency is only 200 kHz, a stronger damping can be observed, and the peak current is only 85 A.

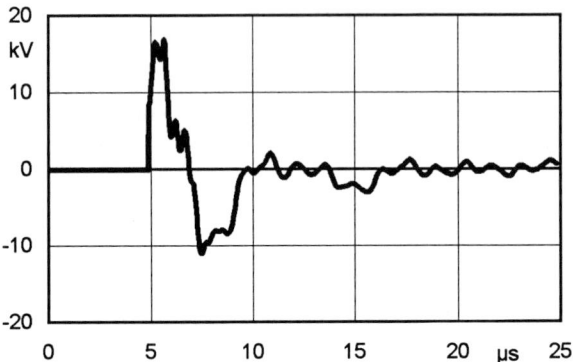

Figure 3: Calculated transient overvoltage of the triaxial line

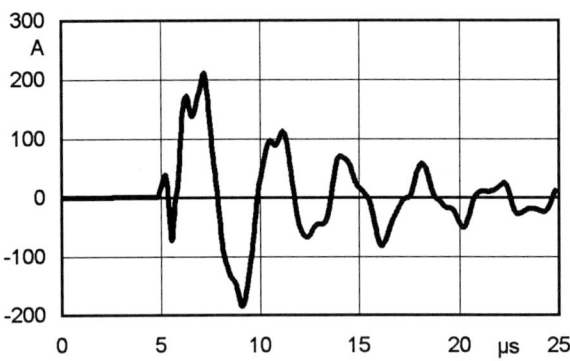

Figure 4: Calculated transient current in the line shield

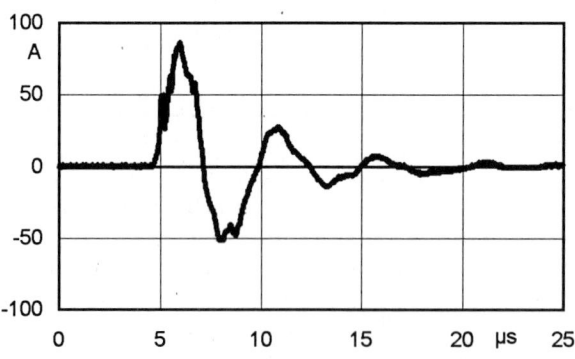

Figure 5: Measured current in the line shield

4. CONCLUSION

The similarity between measured and calculated transient currents shows, that the established equivalent circuit can be used to study the electrical behaviour of the TEXTOR-94 NI power supply system. Some modifications have to be done for matching frequency and amplitude between measurement and simulation.

The calculated high voltage transients show the necessity of measures to reduce the parasitic currents in the system. This can be done by modifying the elements which are mainly responsible for these effects. In future simulations circuit components will be determined where a value modification is technically possible. The improvement by this measures will be proved by simulation before modifications will be carried out.

ACKNOWLEDGEMENT

This work was supported by cooperation contract between Association EURATOM/Forschungszentrum Jülich KFA and Prof. M. Lindmayer, Institut für Hochspannungstechnik und Elektrische Energieanlagen, TU Braunschweig

REFERENCES

1. F. E. Menter, Berechnung elektromagnetischer Ausgleichsvorgänge in ausgedehnten Erdungsanlagen, Dissertation RWTH Aachen, Verlag Shaker, Aachen 1993, Germany.
2. M. Heimbach, EMC Analysis for Grounding Structures in Power Systems, ETEP Vol. 6, No. 3, May/June 1996.

Neutral beam injector for steady state superconducting tokamak

A.K.Chakraborty, N.Bisai, M.R.Jana, P.K.Jayakumar, U.K.Baruah, P.J.Patel, K.Rajasekar and S.K.Mattoo

Institute for Plasma Research, Bhat, Gandhinagar, India -382428

1 INTRODUCTION

Long pulse steady state shaped plasma will be sustained in the SST-1 [1] (R = 1.2 m, a = 0.2 m, κ = 1.7, δ = 0.67, n_e = 2 x 10^{13} cm^{-3}, T_e = 1 keV) through auxiliary power inputs in the form of LH for current drive and ICRH and NB for heating. As the plasma wall equilibration time is ~ a few 100 S, the operational time presently concieved for the SST-1 is 1000 S. Correspondingly the auxiliary heating system including the NB must have pulse lengths of 1000 S.

This paper discusses the design of the NB system for the SST-1 machine. A coupled neutral beam power of ~ 350 kW raises the SST-1 plasma ions to 1 keV [2], at an operational density of 2 x 10^{13} cm^{-3}. In SST-1 maximum beam attenuation is ~ 90 % at 30 keV/amu for an injection tangency radius ~ 98 cm (pivot angle ~ 27°). Shine through is 10%. About 70 % of the power with atomic fraction of 0.4:0.4:0.2 is coupled to the ions and 20% to the electrons. Therefore ~500 KW of power is required for the purpose of ion heating. For an upgraded operation at n_e = 5 x 10^{13} cm^{-3}, a power of ~ 1.5 - 2 MW at 40 - 50 keV/amu (H^o, D^o) is projected. Hence NBI must have a large dynamic range of operation. The port access is 27 cm horz. x 60 cm vert.. The net access available for the duct is 19 cm x 40 cm leaving a space of 4 cm on each side of the duct for mounting of cooling panels. Beam transmission power loss is ~ 150 kW per MW of neutral power launched from the grid, for F_h = 6.5 m, F_v = 8.5 m, & beam divergence = 1°. Losses are reduced for shorter focal lengths and lower beam divergence. The transmission efficiency is 0.36. The design of the NB is based on a single 5 MW source beamline. The physical arrangement of the neutral beam injector system with the SST-1 machine is shown in Fig. 1. The design approach is based on following closely to TEXTOR/JET[3,4] with necessary modifications for 1000S operation. The modifications were made in backplate, first stage neutraliser, operation of cryopumps, cooling panels, the duct protection system, power supplies and data acquisition & control system. Transmission type magnet deflection system has enhanced the background gas density [5] inside it due to the throughput from the dumps.

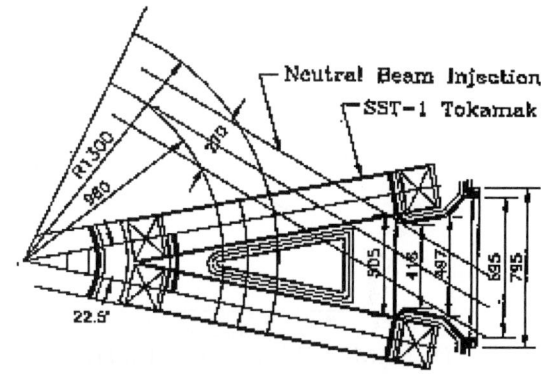

Fig.1: NB Injection in SST-1 Machine

2 ION SOURCE
2.1 Plasma generator

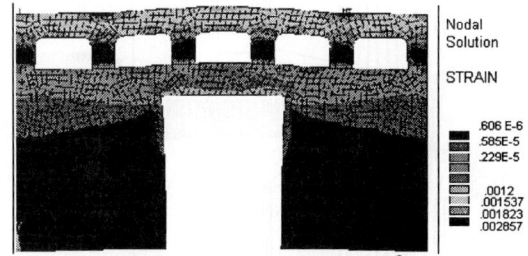

Fig. 2: Strain distribution for 8MW/m² concentrated & 2 MW/m² distributed load.

The plasma generator is a standard multipole bucket type source, which operates in the emission limited discharge regime. Filament temperature of each of the 24 filaments is independently controlled. The focused heat load due to backstreaming electrons increases upto 0.8 kW/cm², while the remaining areas receive upto 0.4 kW/cm². The stress/strain analyses, using ANSYS, shown in Fig.2 predicts that a considerable increase in the fatigue life (~ 10^5 cycles) can be achieved by using a flow rate of 80 g/s per channel at a pressure of 4 bar in the backplate. The arc chamber is water cooled for dissipating 0.2 kW/cm^2 of heat at a flow rate of 240 g/s. Backplate is made of OFHC copper electroform jointed on SS plate. The operating surface temperature of the OFHC chamber is ~150° C. The minimum filament lifetime is ~2x10^5 S, which allows approximately 40 days (4 shots a day) of operation.

2.2 Extractor system

The extractor is a three grid system with extraction area of 1100 cm^2, the beam is extracted from 712 apertures (φ = 8mm, aspect ratio = 0.5).

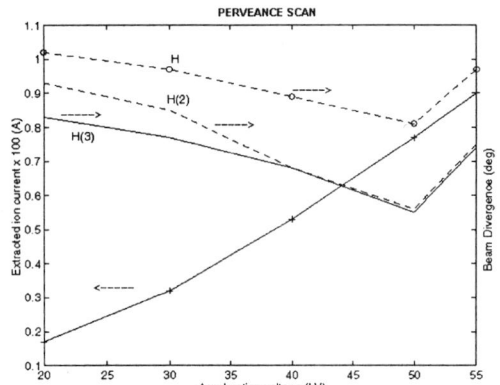

Fig. 3: DYNAMIC RANGE OF EXTRACTOR OPERATION

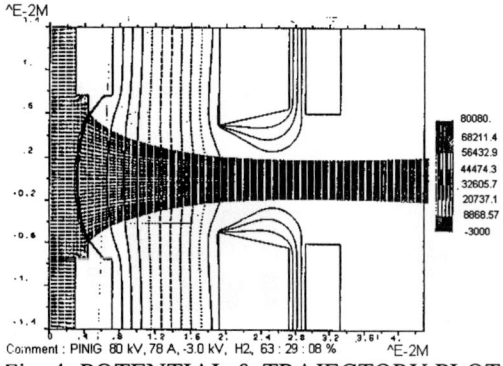

Fig. 4: POTENTIAL & TRAJECTORY PLOT

Focusing (F_h ~6.5 m, F_v ~ 8.5 m) is achieved by electrical and mechanical steering. Improved aperture shapes are being computed to give ion optics with minimised halo and beamlet divergence <1°. Fig. 3 shows the desired dynamic range of operation. Figure.4 shows ion trajectory plot for 80 kV, 78 A for hydrogen [6].

The cooling systems for the 120, 35 and 70 kW heat load on the plasma, deceleration and earth grids respectively are designed under the constraint that the maximum grid temperature rise is <50° C and mechanical integrity is retained within 10 μm. The water flow velocity is 14 m/s at maximum pressure drop ~ 9 bar in the accel grid and 6 bar for the deceleration and earth grids.

3 BEAMLINE (MECHANICAL)

Table I shows the heat load on the various configured beamline components.The cooling channels are electrodeposited in all the elements. Hypervapotrons are conceived for the calorimeter. The length of the calorimeter can be reduced to 70% of its present size and thereby providing tight coupling if superloaded cooling plates can be engineered for 1000 S long operation. Neutralizer (lp ~ 0.3 torr-cm), deflection magnet (LB ~ 0.13 T-m), ion dump (2.0 MW), beam scrapers, liners and duct (0.7 MW), calorimeters (5 MW, ions + neutrals) form the components of the beam line.

Table I- Heat load on beamline elements

ELEMENT	Heat Load (kW)	Flow Rate (m^3/h)
Neutralizer	250	8.6
Dumps	2000	68.8
Liners + ducts scrapers	700	24.1
V-target	5000	172

3.1 Duct

The low conductance of 1.5 x 10^4 l/s for the 1.2 m length results in a blocking current of ~ 100 A. For a desorption factor of unity a current of 50 A can be transported through the duct. The walls of the duct are lined with cooling panels for the removal of 1.0 kW/cm^2 of heat load on it due to beam transmission and reionization.

3.2 Vacuum system

The rough pump(15 l/s), roots pump (100 l/s), and TMP (5000 l/s) forms the external vacuum system for the injector box of 16 m^3 volume and outgassing surface area of ~ 320 m^2. Evacuation time from atmosphere to 10^{-5} torr is ~ 2 hours. The internal vacuum system has 14 m^2 of cryo pumping area; 6 m^2 in the region between neutralizer and magnet,

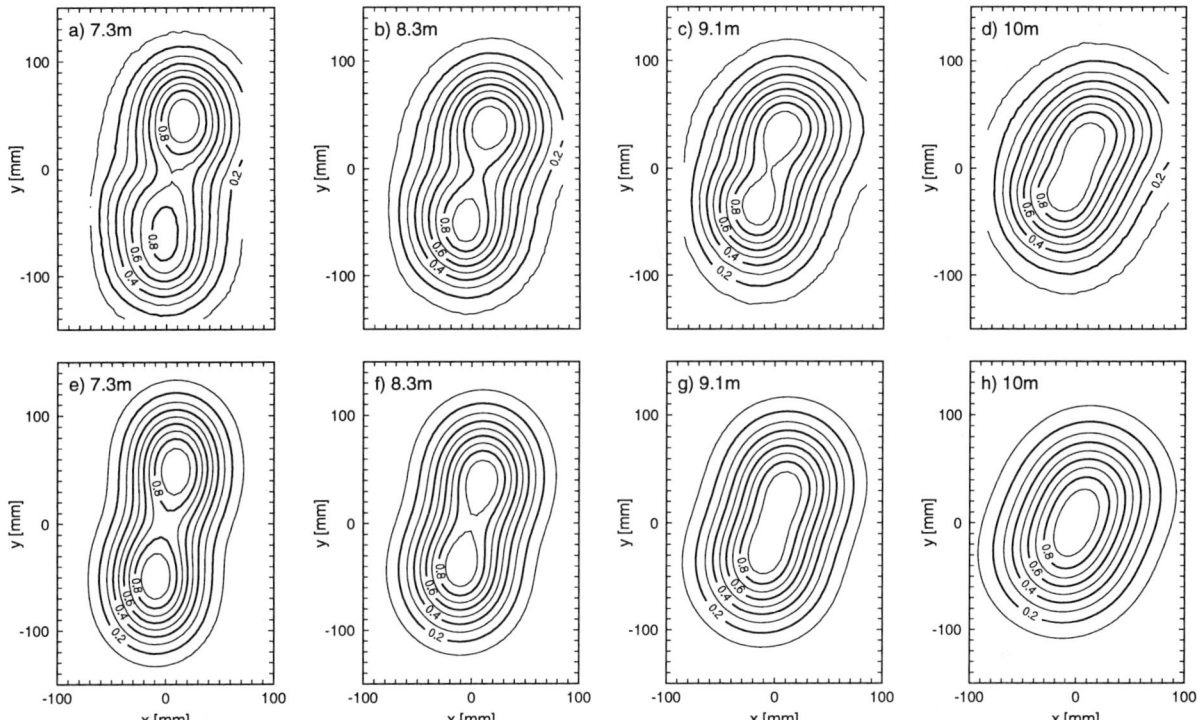

Figure 3. Measured (a-d) and calculated (e-h) normalised power density distributions for the enhanced tetrode PINI (100 kV Helium beam at optimum perveance) at various distances from the source. Best fit to the measured data is obtained for the beam divergence of 0.25° and vertical and horizontal focal lengths of f_V = 12 m and f_H = 8 m.

voltage ($I_{ex} / V_{acc}^{3/2}$) corresponding to the minimum beam divergence, i.e. the minimum beam profile width. Since both enhanced PINIs were fitted with the chequerboard ion source (which has different species composition than the standard JET filter sources) the optimum perveance of the enhanced PINIs fitted with the standard JET filter source was estimated using the effective mass scaling (Figure 2). The effective mass was determined from the measured beam species composition.

Optimum perveance of a Deuterium beam above 50 A is 1.15×10^{-6} A/V$^{3/2}$ for the enhanced triode and 0.95×10^{-6} A/V$^{3/2}$ for the enhanced tetrode. This means that the optimum extracted Deuterium current at 140 kV acceleration for the PINI fitted with a filter source would be 60 A and 50 A for the upgrade triode and upgrade tetrode PINIs respectively.

4.3 Beam optical properties

Beam divergence and steering (vertical and horizontal focal lengths) were determined from the comparison of the measured two-dimensional profiles to the simulated power density distribution (Figure 3). Power density distribution was simulated as a sum of 262 individual beamlets (JET PINIs have 262 apertures in the acceleration grid structures). Beamlet divergence and vertical and horizontal focal lengths were used as free parameters in this procedure and the corresponding values were determined from the best fit to experimental data.

Vertical and horizontal focal lengths and beam divergence are: f_V = 9 m, f_H = 6.5 m, and α = 0.47° for the enhanced triode, and f_V = 12 m, f_H = 8 m, and α = 0.25° for the enhanced tetrode. This means that both accelerators require the modification of the aperture offset between the extraction and deceleration (negative) grids to adjust the focal lengths to the nominal values of 14 and 10 metres respectively.

Power density distributions of the beams from different JET PINIs are compared in Figure 4. As a consequence of very low beam divergence and

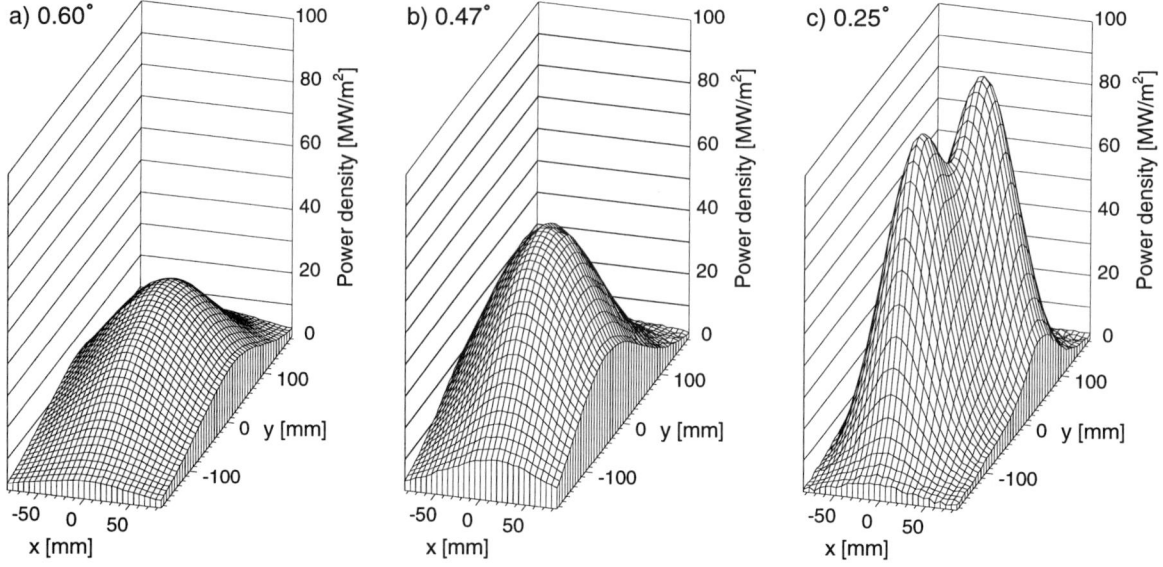

Figure 4: Measured Helium beam power density distribution at 8.3 metres from the source for different JET PINIs fitted with the chequerboard ion source: a) standard JET triode: V_{acc} = 103 kV, I_{ex} = 13 A, p_{max} = 37 MW/m², b) enhanced triode: V_{acc} = 100 kV, I_{ex} = 25.1 A, p_{max} = 74 MW/m², and a) enhanced tetrode: V_{acc} = 100 kV, I_{ex} = 22.6 A, p_{max} = 100 MW/m².

excellent optics, the peak power density of the beam extracted from the upgraded tetrode is extremely high. The two peaks in the distribution correspond to the beams from two grid halves. Note that more than 95% of the beam is contained within an area 150 mm wide and 360 mm high (Figure 4c).

4.4 Power loadings

Power densities of the beams from the enhanced PINIs are extremely high (Figure 4). If those PINIs are installed into JET Neutral Injectors, the power loading on various beamline components (scrapers, dumps and calorimeters) will be considerably higher compared to present situation. Excellent beam properties of the enhanced tetrode PINI practically disqualifies this injector as candidate for neutral injector upgrade whilst keeping the present Ion Dumps. The dump elements could, on the other hand, be subjected to several hundred full power 10s pulses without any danger, if the enhanced triode PINIs were used. Longer lifetime of the beamline components (the most critical being the full energy ion dump) can be accomplished by correcting the beam steering to achieve longer focal lengths, or by reducing the injection energy to ~ 120 kV.

CONCLUSIONS

- Conditioning time is within acceptable limits for both PINI designs (≤ 1500 beam seconds).
- Optimum Deuterium beam current at 140 kV acceleration for PINIs fitted with a filter ion source: 60 A (enhanced triode) and 50 A (enhanced tetrode).
- Beam divergence : 0.47° (enhanced triode) and 0.25° (enhanced tetrode).
- Beam steering: grid modification required for both PINI designs.
- Power density: too high for present beamline components (enhanced tetrode) and just within the design limits (enhanced triode).
- Predicted injected neutral beam power into JET plasma (one injector box fitted with eight enhanced triode PINIs): >15 MW (140kV/60A Deuterium) and >19 MW (140kV/41A Tritium).

REFERENCES

1. G.Düsing et al., Fusion Technology, 11 (1987) 163-202.
2. T. S. Green et al., 10th Int. Conf. on Plasma Phys. and Contr. Nucl. Fusion Research, London, 1984 (IAEA-CN-44/H-I-5).

UPGRADING OF THE NEUTRAL BEAM POWER SUPPLIES FROM 80KV/60A TO 140KV/60A

F Jensen[a], R Claesen, H McBryan, J Mills, R Öström, A P Vadgama

[a]JET Joint Undertaking, Abingdon, OXON OX14 3EA, UK

The present Neutral Beam Power Supplies consist of 16 units, each with a capacity of 80kV/60A. Each of these units can be connected to a 80kV/60A Neutral Injector or alternatively two of these units can be connected in series giving 160kV/60A output capability. In this latter configuration two 160kV/30A Neutral injectors are connected in parallel at the output of the series connected units. In both cases the total DC output power of the Power Supplies is the same and equal to 76.8MW.

In the future JET would like to increase the available neutral injection heating power for the plasma. This can be achieved by either increasing the output current of the existing power supplies or by increasing the output voltage. It is preferred to increase the voltage because this limits the increase in plasma density caused by the injected particles.

The increase in voltage is obtained by connecting a new 60kV/60A power supply in series with the output of the existing unit. This paper explains the main requirements for this design and presents a reference design based on these requirements.

1. Present Neutral Injection Power Supply

The present Neutral Injection Power Supplies are based on a series regulator using a BBC CQK200-4 tetrode. The input voltage to each regulator is supplied from a thyristor controlled power supply. Two regulators are located inside the same cubicle. One regulator is isolated to 240kV DC, allowing the two Power Supplies to be series connected. A series connected unit is rated at 160kV/60A output. An overview of one regulator is shown in figure 1.

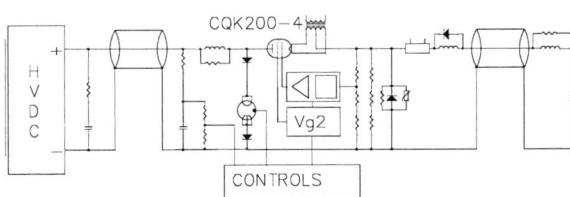

Figure 1. Block diagram of present Neutral injection Power Supply. One unit is shown.

The tetrode acts as a linear regulator and a fast switch. When an over-current is detected, the load is disconnected in less than $10\mu s$. Two separate over-current thresholds are provided. The first threshold is usually set to 10A above the expected load current. The second threshold is set to 30A above nominal load. If the Tetrode fails to block, the second threshold is reached and the input crowbar is fired to protect the Neutral Injector.

A 20mH limiting inductor in series with the output limits the di/dt to $5A/\mu s$ allowing the control electronics to distinguish between the two current thresholds.

The regulated output voltage is supplied to the Neutral Injector via a very low capacitance - approximately 50pF/m - SF6 isolated cable; the Transmission Line. The energy stored in the Transmission Line is dissipated in The Snubber. The Snubber is designed to dissipate 95% of the stored energy, up to a limit of 100J, in the 160kV configuration.

During a breakdown in the Neutral Injector all the energy dissipated inside the injector will be deposited in a very small area. To prevent damage to the Neutral Injector this energy must be

Table 1
Main Design Requirements

Parameter	Definition	Comment
Output Voltage	60kV DC	Stable from No-Load to Full Load.
Output Current	60A	
Voltage Range	0kV–60kV	Adjustable in steps of 1% of nominal output voltage.
Isolation	240kV DC	Continuous DC.
Duty Cycle	20s/600s	On/Off.
Fall-time	max. 2ms	To < 10% of Output Voltage on Neutral Injector Breakdown.
Stored Energy	< 50J	Transferred to the transmission line on turn-off.
Input Voltage	28kV–36kV	50Hz 3-Phase. Varies unpredictably within this range.

limited to less than 5J under all fault conditions.

In the new configuration, both the output voltage and the total capacitance is lower. As a consequence the new power supply can be allowed to transfer up to 50J into the Transmission Line.

2. Booster power supply requirements

The most important requirements for the additional power supply are listed in table 1.

2.1. Design considerations

The Booster Power supply can be inserted either in the ground side of the present power supply or in series with the output. However, Only one unit is isolated to the required voltage level. This would prevent JET from upgrading half the available units.

The stored energy must be minimised to keep the energy dissipation in the Neutral Injector below 5J under all conditions. Because of noise and reliability considerations it is preferred to use passive components to remove the excess energy instead of f.ex. to fire a crowbar on every Neutral Injector breakdown.

A large disadvantage of the present power supply is that the control electronics is located at the output potential. This makes fault-finding very difficult and time-consuming. The Booster Power Supply must have the controls accessible at ground potential.

The size of transformers and filter capacitors decreases with increasing switching frequency, making a high switching frequency attractive. This has to be balanced by the increasing power losses and influence of transformer parasitics at higher frequencies.

Because of the need for maintaining physical separation between the isolation transformer windings, it is difficult to design a transformer with low leakage inductance. Based on literature [3,2] it is believed that the optimum switching frequency for this design will be quite low, typically between 2kHz–6kHz.

It is not practical to design a 3.6MW power converter as one single unit. By splitting the converter stages into a number of identical units, it becomes possible to use commercially available IGBT inverter modules.

3. Design verification

One possible implementation of the proposed design was simulated using ATP version. 2.0 for MSDOS. The principle of this design is given in figure 3.

A matching transformer with a thyristor rectifier creates a stiff DC link voltage of 900 V DC with a maximum current of 4000A for the output stage. The output stage consists of 5 parallel inverters. Each inverter drives a Delta/Star coupled transformer with a 3 phase rectifier. The rectifiers are connected in series to produce the required output voltage. The five inverter modules are described in figure 3.

The present Neutral Injector Power Supply was found to add less than 200mJ to the total stored energy. In is included in the simulation as a DC voltage source of 80kV to reduce the simulation time. This voltage source and the Booster Power

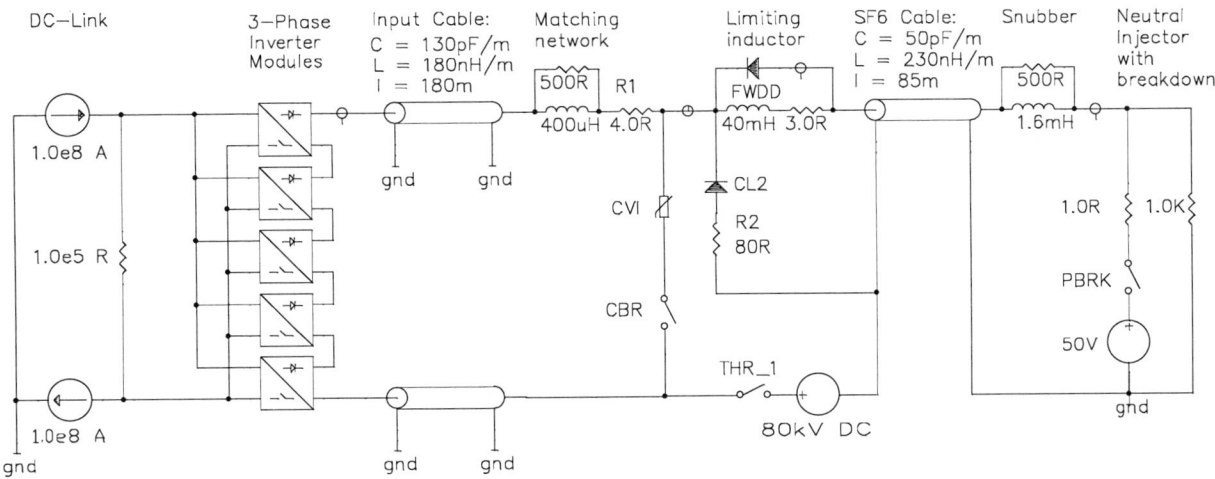

Figure 2. Reference Design Model for computer simulation.

Supply are switched off simultaneously

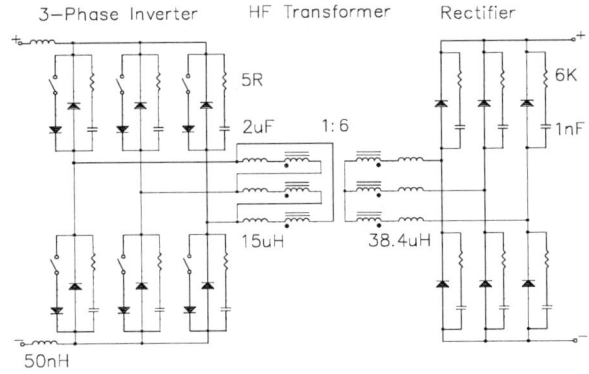

Figure 3. Model of Inverter Modules used in computer simulation.

The switching frequency of the inverters is 2.5 kHz. The five inverters are phase-shifted by 36deg so that the commutation notches of each rectifier does not overlap. The resulting ripple voltage on the output appears at a frequency of 75kHz, allowing the output cable to be used as the filter capacitance.

The output voltage control has been simulated by adopting a basic PWM scheme: The IGBTs of each inverter leg, shifted 120deg, are both turned off for a time proportional to the PWM request. In this way, PWM control of the output voltage with good linearity can be obtained in the range of 60%–100% of nominal output.

Using the DC link voltage to set the nominal output and the PWM to compensate for fast changes, it is possible to stabilise the output within 1% during transients on the 36kV supply.

4. Simulation results

The simulation in figure 4 shows the output voltage and current from the Booster Power Supply when switching on. A Neutral Injector breakdown after 800μs and the current increases to 110A due to the 25μs blocking time of the Inverter Modules.

The energy dissipated in the Neutral Injector following a Neutral Injector Breakdown is calculated by integrating the absolute value of the instantaneous power in the Neutral Injector (see figure 5). The Energy Dissipation only reaches the final value after approximately 15ms. This

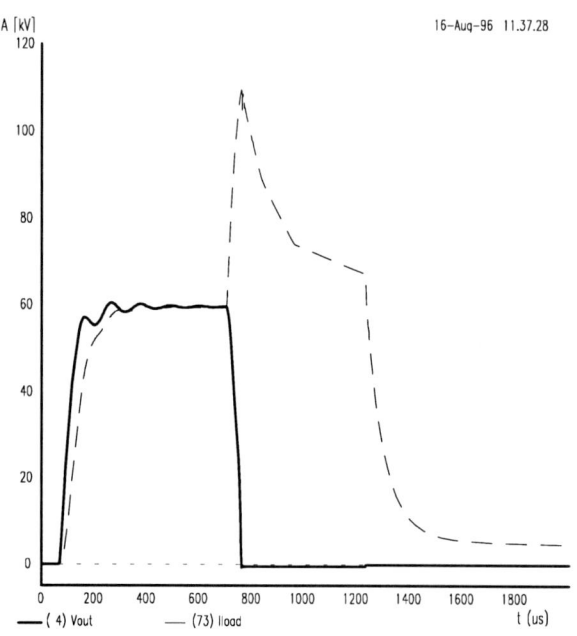

Figure 4. Output rise to 60kV, 60A at 80% duty-cycle. Breakdown occurs at 700μs.

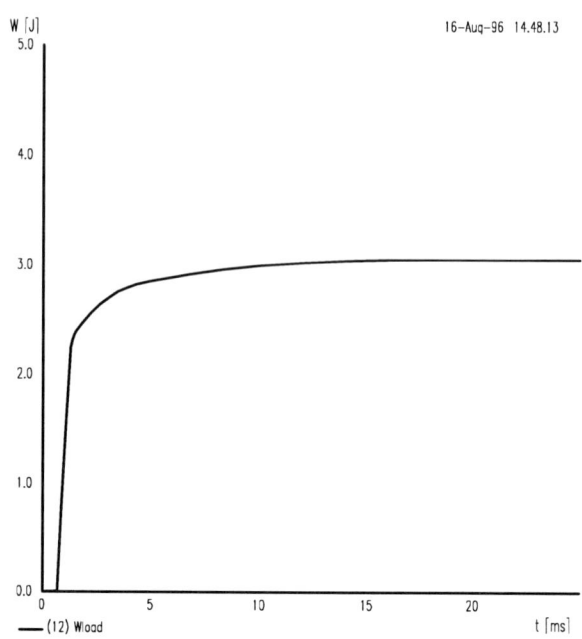

Figure 5. Energy deposited in Neutral Injector on breakdown at 60kV, 60A. Threshold 1 is set to 80A. The time scale differs from figure 4

is caused by the slow commutation of the fault current to the Clipping Diode.

The simulation model shows that the free-wheeling diode across the limiting inductor is essential in reducing the energy deposited in the Neutral Injector: If this diode has a larger forward voltage drop than the rectifiers in the inverters, the current will not comutate to the free-wheeling diode. As a consequence almost all the energy stored in the limiting inductor will be dissipated in the Neutral Injector.

5. Conclusion

The JET reference design demonstrates that a power supply meeting the requirements for operation as a booster supply for the existing Neutral Injector Power Supply can be constructed.

The critical component of the proposed design is the output transformer and the free-wheeling diode: The high isolation voltage of 240kV DC and the requirement of a low leakage inductance are opposing design requirements.

The free-wheeling diode needs to withstand 140kV and have a very low forward voltage drop to ensure that the fault current will comutate to the Free-wheeling diode on turn-off.

REFERENCES

1. W. G. Hurley and D. J. Wilcox, Calculation of Leakage Inductance in Transformer Windings, IEEE Transactions on Power Electronics, Vol. 9, No. 1, Jan 1994.
2. W. H. Flanagan, Handbook of Transformer Design & Applications (2nd ed.), McGraw Hill, 1992, ISBN 0-07-021291-0.
3. A. I. Pressman, Switching Power Supply Design, McGraw Hill, 1991, ISBN 0-07-050806-2.

Operations with Tritium Neutral Beams on TFTR

L. R. Grisham, J. Kamperschroer, T. O'Connor, M. E. Oldaker,
T. N. Stevenson, A. von Halle, and the TFTR Group*

Princeton University Plasma Physics Laboratory
P. O. Box 451, Princeton, New Jersey 08543 USA

In late 1993 the Tokamak Fusion Test Reactor began operating with a deuterium-tritium (DT) fuel mixture instead of the pure deuterium which it had used previously. The major portion of this tritium has initially entered the torus as energetic neutral beam particles. Over 840 deuterium-tritium discharges have now been studied with the aid of more than 2500 tritium ion source shots. The maximum total neutral particle power injected with a mix of deuterium and tritium beams has been 39.6 megawatts, and the maximum injected as tritium neutrals has been 24.3 megawatts. Tritium neutral beam operation has become routine during this time.

1. INTRODUCTION

The principal heating capability for the plasma of the Tokamak Fusion Test Reactor (TFTR) is supplied by the neutral beam injection system, comprised of twelve ion sources in horizontal arrays of three each that are positioned on four beamlines, three of which inject in a direction parallel to the plasma current, and one of which injects counter to the current [1,2].

2. DESIGN CHANGES FOR TRITIUM

This system was originally designed with the idea that it would be used exclusively for the injection of energetic neutral deuterium or hydrogen, with a fast focal plan shutter valve to be located in the beam duct to reduce migration of tritium from the tokamak vacuum vessel into the beamline. However, this valve was never installed, and the discovery that the highest fusion reactivities could be achieved in TFTR in aggressively beam-fueled regimes lead to the realization that, rather than excluding tritium from the beamlines, it would be necessary to introduce large quantities into them for tritium beams in order to maintain the right fuel mix in the strongly reacting plasma core.

A number of experimental studies were carried out [3,4,5] to determine how suitable these beamlines were for tritium beam production. We found that relatively few changes would be required, and that sensitive components such as the epoxy high voltage insulators of the ion source accelerator structures would not degrade in performance due to the tritium exposure they would encounter on TFTR.

The principal changes that were deemed necessary concerned the location of the gas feed, the design of the gas feed system, and the length of the gas pulse relative to the extracted beam pulse. The primary gas feed for each ion source had been located in the rear of its discharge chamber during the pre-tritium phase of TFTR, with a subsidiary feed located in the neutralizer to ensure an adequate line density for efficient beam neutralization. In this configuration, the primary gas feed line had to travel from the source enclosure, which was at ground potential, to the arc chamber, which was at the full acceleration potential of up to 120 kV.

While experiments determined [4] that tritium, with its associated beta decays, would not lead to enhanced breakdown in the gas feed line even with the full 120 kV across the line and with a wide range of tritium pressures, we concluded that it would nonetheless be prudent to change the location of the gas feed. This was because the source feed position required that the gas line be made of a tritium-compatible plastic, and this line, with its couplings at each end, had to cross the 2 bar of SF_6 which is necessary to maintain voltage holdoff within the source enclosure. This presented the possibility that the tritium and SF_6 would become intermingled in the event of leaks in the plastic tubing or its couplings. Such contamination would pose difficulties for TFTR's tritium processing system, since some components are vulnerable to damage or degradation by halogens.

Accordingly, the gas feed point was relocated from the arc chamber to a spot downstream of the exit grid of the ion source. This allowed the gas to be injected through a port at ground potential, so that the entire tritium feed line could be made of stainless steel, and did not cross SF_6 at any point. An added attraction was that, with this location, the ion source could be removed in its enclosure for maintenance or replacement without disturbing the gas feed system. This design was thus quite robust with regards to minimizing the chances of tritium leakage or contamination within the gas injection plumbing. The penalty paid for this added safety is that about 15% more tritium is required for operations than would be the case if all the tritium were introduced within the arc chamber, since the conductance to the low pressure end of the neutralizer is greater for the ground potential gas injection location than for the arc chamber position.

In order to further preclude leakage of tritium to the surrounding air, the tritium gas injection system was entirely double-jacketed in stainless steel, with a secondary vacuum in between the two layers of piping to collect any leakage from the primary containment. In addition, the tritium gas injection system was designed with tritium-compatabile valve seats for its Nupro[tm] pneumatic absolute valves and piezoelectric Maxtek[tm] regulating valves [6]. Unlike the previous gas injection system, the tritium injector was designed to use feedback control of the Maxtek conductance based upon the pressure in a small reservoir in order to make effective use of the small amount of tritium available for experiments. In order to prevent cross-contamination of the tritium and deuterium supplies, independent gas systems are used for the two isotopes, with separate valves and plumbing.

A third change reduced the amount of tritium used in the course of a beam shot. During the pre-tritium phase of TFTR, the beam gas was turned on 1.0 second prior to the initiation of the arc, and the arc began 1.5 seconds before beam extraction commenced. This meant that the gas pulse was 2.5 seconds longer than the beam pulse, which for many cases is about a second. While of little signifance in deuterium operation, this large gas overhead would have severely reduced the number of beam shots we could produce with the allowable tritium inventory. Careful tuning of the power supplies allowed this gas overhead to be reduced from 2.5 seconds to 1.0 second. Currently, the gas begins 0.5 seconds before the arc, and the period allowed for arc stabilization prior to beam extraction is reduced to 0.5 seconds without degrading beam reliability.

3. OPERATIONS WITH TRITIUM BEAMS

The defining restriction which governs the way in which the neutral beam system is operated during the tritium phase of TFTR is the small allowed site inventory of tritium. Only 50 kCi is allowed to be at the TFTR facility at any given time, of which no more than 25 kCi can be in releasable form inside the TFTR vacuum system. Of this, after accounting for tritium occupying plumbing and plenums, roughly 18.7 kCi of tritium is actually available for neutral beam operations from a 25 kCi allotment.

A typical 1 second tritium beam shot from an ion source uses roughly 100 Ci during the second before the acceleration pulse begins the ion beam extraction, and an equal amount during the beam pulse, for a total of roughly 200 Ci. In practice this consumption is often greater due to frequent throughput leaks encountered with the tritium valves and their valve seats. Thus, a high power neutral beam shot with 6 of the 12 ion sources firing tritium uses at least 1200 Ci, and often more with valve throughput leaks. Consequently, if tritium were used for beam conditioning shots, which are many times more numerous than injection shots, all of the availble tritium would be used on the first conditioning shots at the start of each day, with none left over for experiments.

This potential problem has been circumvented by performing all ion source conditioning shots with deuterium. Further, all tokamak plasma conditions are optimized with deuterium beams. Tritium is used only for tokamak injection shots, and these are taken only when the deuterium precursor shots suggest that the desired plasma scenario is obtainable. This switching back and forth between deuterium and tritium is accomplished with the two independent gas systems feeding each neutralizer. Since the perveance for minimized beam divergence is lower for tritium than deuterium, the acceleration voltage is increased for tritium shots and then restored to its lower level for subsequent deuterium conditioning. The acceleration voltage was chosen as the parameter to vary for perveance compensation, for reasons of ease and reproducibility. The acceleration supplies are regulated, but the arc and filaments supplies for the ion sources are not, so adjusting the requested acceleration voltage produces a more predictable response than adjusting the arc control parameters.

Despite the operational complexity associated with this bimodal operation with two isotopes, the reliability of the ion sources and beamlines has been as good with tritium as that with deuterium. Of the problems that have been encountered, none were directly related to the physical characteristics of tritium. There has been no detectable permeation to the outside of the epoxy high voltage insulators in the ion source accelerator structures, as measured by surface swipes taken on the outside of the insulators for beta monitors, and no contamination of the SF_6 in the ion source enclosures has been detected. In addition, whatever neutron activation of this SF_6 which has occured has been too slight to pose a measurable problem for source maintenance.

Experiments conducted with a test sample prior to TFTR tritium operations [4] had shown that the leakage current would increase along the epoxy high voltage insulators during and after tritium exposure. This was not deemed to constitute a problem, especially since increased leakage current usually serves to render insulators more robust against catastrophic breakdown. The hi-potting instrumentation used with the ion sources in place on TFTR is too insensitive to monitor the leakage current. However, as expected, no deterioraton of the voltage holding capabilities has materialized, despite the extended period of tritium operations.

The use of entirely independent gas systems for deuterium and tritium has resulted in beams whose composition is fairly pure in the selected isotope. Some gas does remain adsorbed on ion source and grid surfaces at the end of a beam pulse, and is thus available to contaminate the source plasma on the following shot. An optical multichannel analyzer has measured the doppler-shifted D-alpha and T-alpha light emission to monitor beam cleanup and contamination [7].

For a sequence of cleanup shots with deuterium following a tritium shot, the doppler shift analysis found that two to four beam shots, corresponding to a total of two to four seconds of arc (during half of which beam is being extracted) are sufficient to reduce the contamination of the previously used isotope to negligible levels. These shots each had 0.5 second of gas flow before the arc, followed by 0.5 second of arc, and then 0.5 second of beam extraction. The data also suggest that similar durations of arc and gas exposure without beam extraction may somewhat slow the clean-up process, although more data would be required to be certain of this.

In the normal operating sequence on TFTR, all deuterium beam injection shots into the tokamak plasma following a tritium injection shot have at least seven, and usually far more than seven, conditioning shots coming between them. These intervening conditioning shots each have gas and arc durations at least as long as 0.5 seconds each, although the portion during which beam is extracted is often shortened to 50 msec so it can be fired into the tokamak wall armor (thereby eliminating the need to cycle the beamline calorimeters in and out). Thus, the contamination of deuterium injection shots by tritium should be negligible, and no measurable tritium beam light has been observed on deuterium shots into the plasma.

Such is not the case for tritium beam injection into TFTR. Since no tritium is used for conditioning shots, every tritium injection shot is preceeded by a number, usually a substantial number, of deuterium source pulses. Thus, the only cleanup period available is the half second of tritium gas throughput and the subsequent half second of tritium arc prior to the extraction of beam. Consequently, the deuterium power fraction in tritium beams is typically in the 2% range. Since this is averaged over the beam pulse, it may tend to drop for longer pulse lengths, but most TFTR tritium beam pulses are a second or less. The tritium supplied to the beams has small amounts of protium, deuterium, and sometimes He^3 at the fractional percentage level, and these in turn become minor beam constituents.

4. PROBLEMS ENCOUNTERED

Although no problems unique to tritium have occured over the several years of TFTR tritium operation, the normal failures associated with operating any large system have taken place. Numerous ion sources have been removed for refilamenting and refurbishing following internal water leaks or shorts. Making these changeouts and repairs has become routine, with successive moist air pumps and purges being used to remove most of the tritium contaminating internal surfaces.

The tritium gas injection system for the neutral beams is the only component which has experienced some difficulties [6]. This is partly because it was the one completely new system built for the tritium beam phase of TFTR, and thus there was only limited operational experience to uncover defects prior to full scale usage. Some of these difficulties have arisen from problems with the valve seats chosen for the piezoelectric and pneumatic valves in the tritium injection system. Although the throughput leaks associated with the system do not pose any threat of leaks to air, they do sometimes increase the difficulty of regulating tritium flow to the sources, and they often result in tritium wastage, since some tritium flows into the beamline when it is not needed.

5. CONCLUSION

The TFTR neutral beam system has operated reliably and safely since the start of tritium operations in late 1993. No problems peculiar to tritium have been encountered, while the maximum power injected into TFTR, 39.6 MW, significantly exceeded the previous record of about 34 MW obtained with deuterium only. About 89.8 grams of tritium have passed through the beam systems so far, signifying that tritium beam operations have become routine.

*Work supported by US DOE contract number DE-AC02-76-CH03073.

REFERENCES

1. L. R. Grisham et al, Nucl. Instru. and Meth. B10/11 478 (1985).
2. T. N. Stevenson, T. O'Connor, V. Garzotto et al, 16th IEEE/NPSS Sym. on Fus. Eng., Urbana, Illinois 1 537 (1995).
3. L. R. Grisham, et al, J. Vac. Sci. Technol. A7 944 (1989).
4. L. R. Grisham et al, Rev. Sci. Instrum. 62 376 (1991).
5. L. R. Grisham et al, Rev. Sci. Instrum. 60 3730 (1989).
6. M. E. Oldaker et al, 16th IEEE/NPSS Sym. on Fus. Eng., Urbana, 1 534 (1995).
7. J. H. Kamperschroer et al, Rev. Sci. Instrum. 66 632 (1995).

THE BEL: A TEST BED FOR THE POSITIVE IONS BASED NEUTRAL BEAM INJECTORS WITH ENERGY RECOVERY SYSTEM FOR TORE SUPRA

P. Bayetti, F. Bottiglioni, H. Dougnac, F. Imbeaux, J.Y. Journeaux, Ph. Lotte, G. Mayaux.

Association EURATOM-CEA, Département de Recherches sur la Fusion Contrôlée, Centre de Cadarache
13108 Saint Paul lez Durance Cedex, France

The Neutral Beam Injection system of Tore Supra (six injectors, 100 keV - 40 A - D2 each) is the first one to be based on high power, long pulses injectors using the Electrostatic Energy Recovery concept. So far, the operation of this system has been difficult, since never experimented in full scale. To overcome this difficulty, a test bed, the 'Bâti d'Essais des Lignes' (BEL), has been put into operation at the beginning of 1995. Using the same facilities as those of the Tore Supra Injectors, it is dedicated to the injector qualification at nominal power, the injectors conditioning and the long pulse operation ($\geq$ 30 s). The operation with Helium beams would be also possible by changing the pumping system. This paper describes this new facility and the diagnostics for the beam power inventory and the beam optics. Experimental data, such as beam optics and ion species from profile and Hα ray measurements, are also presented.

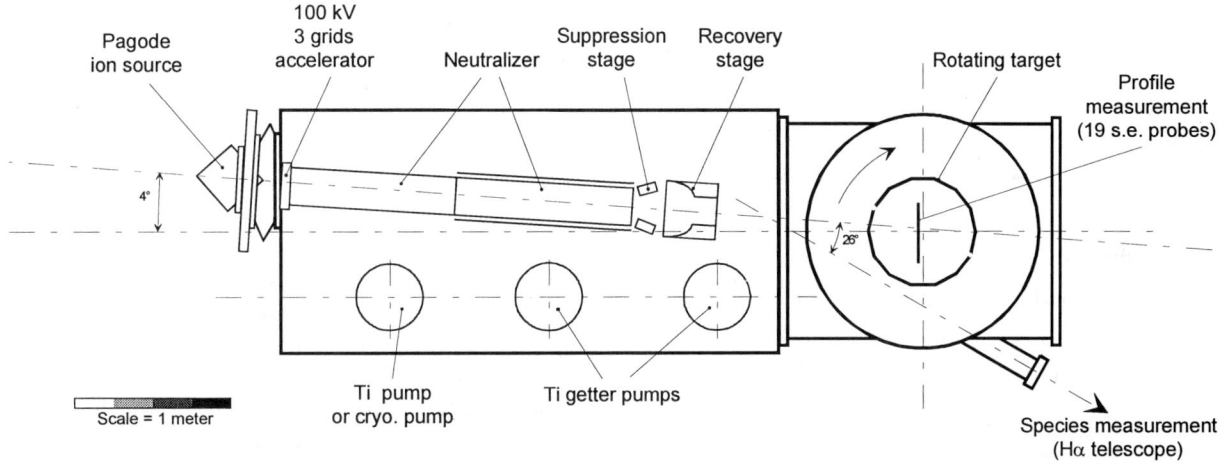

Layout of the BEL test bed facilities

1. INTRODUCTION

The Neutral Beam Injection (NBI) system of Tore Supra has been already described in [1&4]. It is composed of three boxes, each housing two injectors (four in Co, two in Counter-injection) designed to accelerate 40 A of positive ions at 100 keV in D_2, or 80 keV in H_2. The total power at the entrance of the torus is expected to be 5 MW at full energy, and 7.5 MW for all species in D2.

The particularity of this system consists in electrostatically decelerating and recovering the non-neutralised fraction of the beam (Energy Recovery) [2].

Till now, 1.7 MW of D° have been injected into the plasma with three injectors [3], but beams of energy greater than 80 keV are operated with difficulty.

Therefore, a test bed has been put into operation since the beginning of 1995. The main facilities are similar to those of the Tore Supra injectors, namely: the high voltage circuitry and the ion source power supplies [1,4,5] (those of the TS injector N°6) and the pumping system (Ti getters). Based on more recent equipment, the data acquisition and control systems are different. Deuterium operations are not possible because of lack of shielding against neutrons. Diagnostics for beam power inventories and beam optics have been installed.

The first part of this paper is dedicated to the description of the test bed facilities; then the homogeneity of the ion source and the beam characterisation will be presented.

2. FACILITIES

2.1. Mechanical

The vacuum vessel consist in a parallelepiped box, 23 m^3 in volume, which houses one injector and three titanium pumps. This box is connected to a terminal cylindrical box (3 m^3) housing a rotating target [4] to block the neutral beam. The main differences between the BEL box and those used on Tore Supra are the distances between the injector (at high negative voltage) and the box walls (ground), and the fact there is one injector only against two in the TS boxes. As a consequence, the studies of interactions between two injectors into a same box are not possible, and the electrostatic conditions could be somewhat different.

The vessel and other facilities - mainly the high voltage damping self inductance [1,3] and the mechanical pumps are installed into a house (5.4x8.7x5.5x m^3, 25 t) built up with 1 cm thick lead panels which shield against the 140 keV x-ray produced at the suppression electrodes level.

2.2. The rotating target

The rotating target has been designed at Fontenay aux Roses in 1984 to dump the neutral beams for the JET NBI system [6]. It is installed at 4.4 m from the extraction area, 0.70 m downstream the recovery electrode. It is composed of 14 hypervapotron panels vertically screwed on a rotating drum 0.70 m in diameter. Two diametrically opposite vertical slots allow to scan the beam by a fixed diagnostic installed into the cylinder. At nominal power of the neutral beam (1.7 MW for H_2), the surface temperature will not exceed 300 °C for a drum angular velocity of 63 rpm.

2.3. Pumping systems

The conventional pumping system is composed of a high pressure pumping unit and a low pressure pumping unit (65 m^3/h primary pump, plus a 2.2 m^3.s^{-1} turbomolecular pump).

Three titanium pumps (3 m high, 0.4 m in diameter), giving pumping speed up to 280 m^3.s^{-1} for H_2, have first been installed. But the partial pressure of water was pretty high when the target rotates, and the Ti getters are much more efficient for H_2O, O_2 and oxides in general than for H_2. Therefore, we have removed one Ti pump and replaced it by one cryo-pump fed with LN_2 only, as a cold trap, to keep the partial pressures of water and oxides lower than those of H_2. As a consequence, the maximum pumping speed for H_2 decreases, but the pumping capacity increases.

2.4. Data acquisition and control system

The control and acquisition systems are standard industrial devices.

The control system is made up of four Programmable Logic Control modules (PLC) and field bus. Two PLCs are used for the control of the pumping systems, the rotating target, the gas injection and the chronology. The third PLC is devoted to the control of the Low Voltage P.S. The last one is connected to the general TS control network via a jbus optical link to control the High Voltage power supplies.

The data acquisition and processing system is made on two Personal Computers. The A.T.S. software controls the acquisition of 2x32 slow channels (down to 5 ms each) and 3x 16 fast channels (down to 4 µs each). It performs 'on line' calculation facilities, easy signal display and data exportation to standard tools. After shot, data are transferred to the second P.C. for more sophisticated processing.

2.5. Diagnostics

The test bed has been equipped with several diagnostics for the beam qualification:

- **the beam power inventory** is based on 16 electrical measurements for the input power, and 22 calorimetric measurements (water flow and temperature) for the output power.

- **the beam optics** is measured on the rotating target, by scanning horizontal profiles by means of an array of 19 secondary electron emission probes. Each probe is composed of an horizontal wire surrounded by a stainless steel electron collector, positive biased with respect to the wire. The wires are 30 mm vertically spaced The probes are protected by a copper plate with 19 horizontal slots 1 mm width, 280 mm length. So, the beam is scanned in 19 analogic horizontal profiles when the vertical slot of the rotating cylinder passes in front of the plate. Vertical profiles can be inferred.

- **the neutral beam species content** is obtained by H_α Doppler shift measurements, using the Tore Supra Czerny - Turner spectroscope [7]. The line of sight is located in the central horizontal plane of the beam, making an angle of 25.8° with the direction of propagation. The beam is observed 70 cm downstream from the recovery stage, so only the neutrals emitting $H\alpha$ light are seen. The light is collected through a window by a single lens telescope and transported to spectroscope via a 110 m long, 1000 µm in diameter optic fibre.

3. FIRST RESULTS

The extracted power has been increased up to 1.4 MW (21.5 A of H_2 at 65 keV) for shots of 2 s without notable difficulty. Beyond that point, strongly depending on pressure conditions and re-ionisation, breakdowns occur between the suppressor stage, which is at the most negative voltage (-90 kV for -65 kV of accel. voltage), and the recovery stage which is at -5 to -10 kV, close to the grounded source voltage.

The first experiments, dedicated to the qualification of low energy (50 - 60 keV) hydrogen beams, have shown anomalies in the neutral beams profiles. The ion source homogeneity has been measured, and source adjustments have been carried out, so as to improve the source uniformity.

3.1. Beam profiles and source homogeneity

First measurements with secondary electron emission probes have shown dyshomogeneities of the vertical neutral beam profiles, when simulations made with the PROTOS code [8] predict very flat profiles at this distance (fig. 1a). Such neutral beam profile anomalies could not be explained by misalignment or displacement of the extraction grids only.

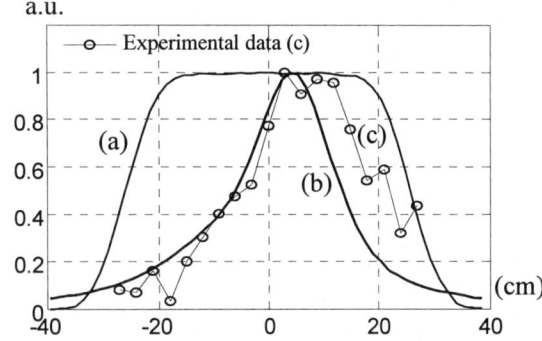

Figure 1. Normalised vertical neutral beam profile at 4.4 m: Calculated (a and b) and measured.

Therefore, the homogeneity of the ion source has been measured (without beam acceleration) by temporarily installing 14 Langmuir probes in the plasma grid plane (two vertical profiles). Experiments have been made for the two possible magnetic source configurations, the Cusp Field regime (high magnetic field, high proton yield) and the Transverse Field regime (low magnetic field, good homogeneity) [9], which show strong plasma drift and non-uniform vertical current density profiles (fig. 2).

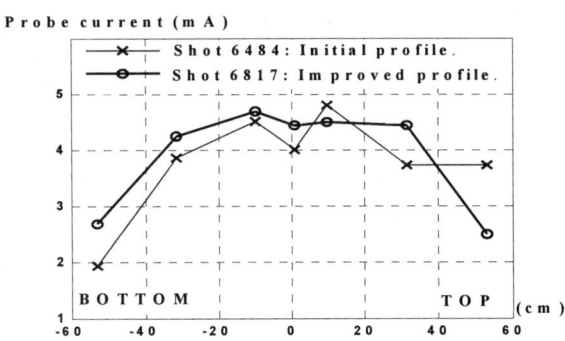

Figure 2. Vertical current density profiles in the ion source (Cusp Field regime 700 A arc discharges).

The PROTOS code has been modified to take into account the source non-uniformity: the positive current density is now fixed for each of the 240 extraction holes, and the divergence of each corresponding beamlet is calculated from the perveance data. By using the source density profiles measurements as input for the PROTOS code, the corresponding neutral beams profiles have been calculated, and show good agreement with the experimental ones (fig. 1b&c).

For both C.F. and T.F. source regimes, the current density vertical profiles have been optimised by adjusting coils and filaments currents (fig. 2), but they are still different from the flat ones measured at Fontenay-aux-Roses [9].

Vertical neutral beam profiles have not yet been measured with the new source adjustments.

3.2. Beam species content

Very preliminary H_α Doppler shift measurements have been made for relative low arc current discharges (fig. 3). Assuming all the molecular are dissociated in the neutraliser, one can estimate the α_i ion species in the source from the β_i ratios of neutrals in the beam:

$$\alpha_i = [\beta_i/i.\eta_i]/[\Sigma(\beta_i/i.\eta_i)]$$

with η_i the neutralisation efficiency for energies E/i. For typical 330 A arc discharges, this leads to ion species α_i: H^+=40%, H^+_2=37% and H^+_3=23%, consistent with the low power level of the discharge.

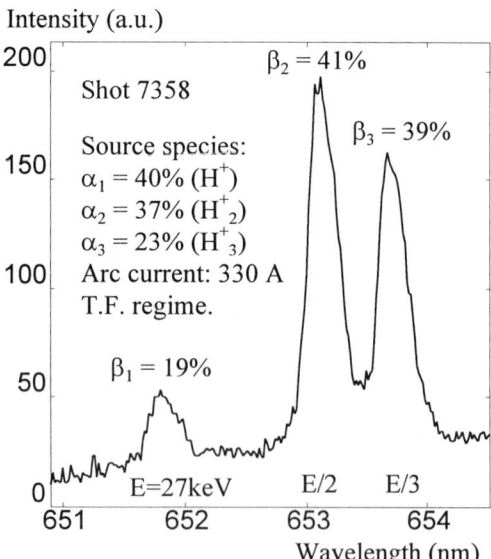

Figure 3. Doppler shifted H_α spectrum.

4. CONCLUSION

The test bed has already permitted to understand certain difficulties we had on the TS injectors, namely:

- at relative low extracted power level (1.4 MW - H_2), breakdowns could occur between the suppressor and the recovery stages, depending on the pressure conditions in this region,

- the vertical neutral beam profiles were not flat since the ion current density of the source was not homogenous enough.

Based on these first observations, design modifications of the suppressor / recovery region are under study; better source regimes have been found, but they still need to be improved.

Ion species determinations and beam power inventories will be performed while the extracted power is increased to its nominal. The next step will consist in preparing the Tore Supra injector long pulse operations (over 30 s).

REFERENCES

1. P. Bayetti et al., Proc. of the 15th SOFT, North-Holland, Utrecht (1988) 604.
2. M. Fumelli et al., Proc. of the 15th SOFT, North-Holland, Utrecht (1988) 610.
3. P. Bayetti et al., Proc. of the 17th SOFT, Italy, Rome (1992) 437.
4. P. Bayetti et al., Proc. of the 16th SOFT, UK, London (1990) 1123.
5. C. Jacquot et al., Proc. of the 16th SOFT, UK, London (1990) 1119.
6. R. Becherer, Nucl. Inst. & Meth. in Phy. Res., A243 (1986) 28.
7. W.R. Hess et al., 17th EPS, Amsterdam (1990) vol. 4, 1580.
8. F. Jéquier, F. Imbeaux, private communication.
9. M. Fumelli et al., Proc. of the 14th SOFT, France, Avignon (1986) 1153.

Acknowledgements: Special thanks to W.R. Hess and E. Chareyre for helping us in the spectroscopic measurements.

Negative D⁻ ion source relevant for application to ITER neutral beam injectors

C. Jacquot[a], D. Riz[a], R. Trainham[a], K. Miyamoto[b], Y. Okumura[b], M B Hopkins[c]

[a]Association EURATOM-CEA, Département de Recherches sur la Fusion Contrôlée Centre de Cadarache, 13108 Saint Paul lez Durance Cedex, France

[b]Japan Atomic Energy Research Institute, Naka Research Establishment NAKA- MACHI, NAKA-GUN, Ibaraki-ken,311-01, Japan

[c]Dublin City University, Glasnevin, DUBLIN 9, Ireland

In a collaborative effort between JAERI and EURATOM- CEA, the KAMABOKO negative ion source, which has produced more than 30 mA/cm^2 of H⁻ in Japan, has been tested for deuterium at Cadarache (France). The ion source is of the same design concept, but at a reduced scale, as the proposed large negative ion source for ITER-NBI. The negative D⁻ ions are produced in a cesium seeded multi-cusp plasma generator of good plasma confinement, which allows efficient operation at low pressure. Routine diagnostics such as Langmuir probes and spectrometer are installed to measure plasma characteristics and atomic population. The transverse magnetic filter of 0.088 T.cm, installed in front of the extraction system, had been optimized for electron suppression in hydrogen operation. Stable arc discharges are observed at 0.35 Pa in deuterium for arc powers of up to 78 kW. For a pressure of 0.35 Pa, the ion source produces 1.4 A of D⁻ with an average current density of 20 mA/cm^2. The extracted electron current, originally 13 A, was prohibitive for a steady state operation of the extraction grid. Stronger transverse magnetic fields (up to 0.18 T.cm) have been installed to reduce the electron flux on the extraction grid. At the maximum field stength, we observe that the production of deuterium negative ions is identical to the original configuration, but that the extracted electron current is strongly reduced. The ratio of the electron current to the D⁻ current is less than one, instead of 10 previously. Under these conditions, 1000 second source operation should be possible, which would satisfy the design nominal values for ITER.

1. INTRODUCTION

One of the most important issues in developing the ITER-NBI negative ion source and accelerator,which is required to produce 1 MeV, 40 A D⁻ ion beams for a duration of more than 1000 s (1) is to reach a uniform current density of 20 mA/cm^2 for D⁻ at the lowest source pressure (< 0.4 Pa) with a multi-holes extraction system. It is also important to prevent electrons from entering.the accelerator. Accelerated electrons cause large power losses and produce parasitic X- rays increasing the possibility of breakdown and damage of the high voltage insulators . Electrons are produced in the accelerator by three processes: leakage of electrons from the source itself, secondary emission on the grid surfaces,and free electrons generated by stripping of D⁻ on the residual gas in the extractor and the accelerator. It is possible to prevent the extracted electrons from entering in the acceleration region by using an external magnetic filter made of permanent magnets installed in front of the plasma grid. Secondary electrons emitted by electrodes are reduced by compressing the beam, thus avoiding interception by the preacceleration grids. The stripped electrons are reduced only by decreasing the source pressure.

2. EXPERIMENTAL APPARATUS

Figure 1 shows a cross-sectional view of the KAMABOKO source (2). The arc discharge chamber, whose dimensions are 340 mm in diameter and 340 mm in length, is covered by 16 line magnetic cusps (10x20 mm SmCo). The magnets inserted in the grid support flange are stronger than the others so as to create a transverse magnetic field, which acts as a magnetic filter. The magnetic filter divides the chamber into two regions and modifies the electron energy distribution and the density so as to enhance the negative ion yield and to reduce the electrons flow

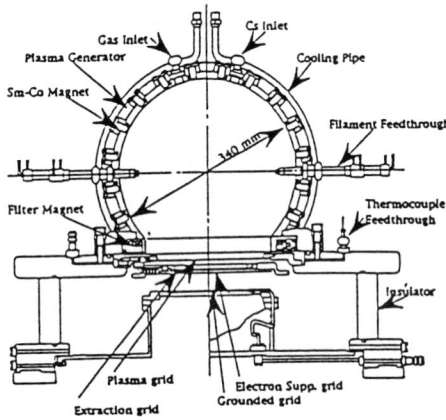

Cross view of the Kamaboko

Figure 1.

Four magnetic filter values have been tested (B.dl=0.088, 0.11, 0.14, 0.18 T.cm). The arc discharge is sustained by electron emission from 12 tungsten filaments (1.5 mm in diameter and 180mm in length) held by molybdenum supports on coaxial filaments feedthroughs. The arc discharge chamber is made of copper and is water cooled for long pulse operation. A cesium oven is mounted on the chamber to enhance the negative ion production. Cesium injection is carefully controlled to a rate of about 40 mg/h. The arc chamber has a rectangular opening of 200 mm x 340 mm for negative ion extraction. The negative ions are extracted from a central 128 x140 mm area where 45 holes of 14 mm diameter are distributed. The total extraction area is 70 cm^2. The extractor consists of 3 grids, called a plasma grid, an extraction grid, and an acceleration grid. The plasma grid is made of molybdenum and thermally insulated from grid frame which is water cooled in order to control its temperature. The optimum temperature is 200-300°C and a possible explanation is that its gives a minimum work function of the plasma grid due to an optimum cesium coverage. The plasma grid (2 mm thickness) is separated from the extraction electrode (18 mm thickness) by a gap of 5 mm. The gap between the extraction grid and the acceleration grid is 8 mm and the final acceleration voltage is 30 kV, limited by biological constraints due to neutrons produced by D°-D° reactions. A multi array calorimeter (72 thermocouples) at the acceleration voltage of 30 kV measures the characteristics of the D$^-$ accelerated beam. Movable Langmuir probes are installed into the arc discharge chamber to measure the plasma characteristics and a spectrometer detects cesium and deuterium spectral lines. Doppler shifts of excited accelerated D° in the beam are also analysed in order to appreciate the effect of the D$^-$ stripping on the residual gas and the quantity of impurities.

3. EXPERIMENTAL RESULTS

3.1. Nominal test with the original magnetic filter

With the original magnetic filter of 0.088 T.cm, and a small quantity of cesium injected (200 mg), the D$^-$ current shows a linear dependance with the arc power (figure 2);

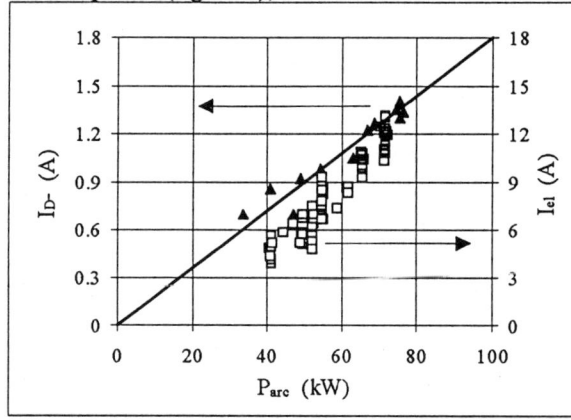

Filter: 0.088T.cm, Pressure:0.35 Pa D2, V bias:+5V

figure 2.

The plasma grid is maintained at high temperature (>180°C) and polarized at +5 V with respect to the source wall. The maximum current density of 20 mA/cm^2 of D- is achieved at 78 kW of the arc discharge or 15 W/cm^2. The electron current drained by the extraction electrode is high (about 13 A) and does not permit a steady state extraction of the beam. The effect of the quantity of cesium has been quantified and shows that small quantity (200 mg) is necessary to produce a surface effect; It is confirmed by the observation of the 8521 Å Cs neutral spectral line (Fig 3) where the switch off of the cesium has the effect to eliminate the Cs from the discharge and does not influence the production of negative ions.

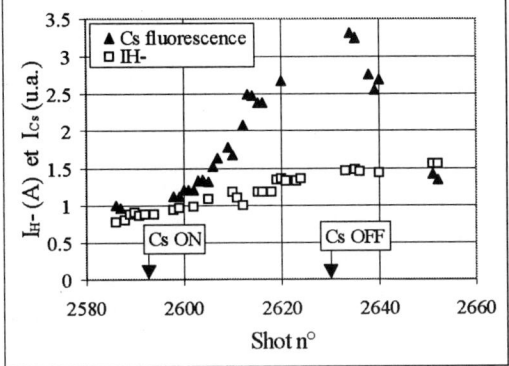

Figure 3

The effect of the bias plasma grid voltage has a small effect on the D⁻ accelerated beam, but decreases strongly the drain electron current (Fig 4).

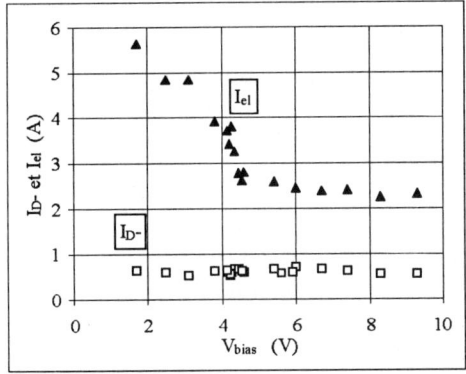

Filt.:0.088 T.cm, 0.35 Pa D2, Parc:36 kW, Vg1:5 V

Figure 4

For a constant acceleration voltage of 30 kV, the extracted current densitry follows a Child-Langmuir space charge law both for H- and D- (Fig 5)

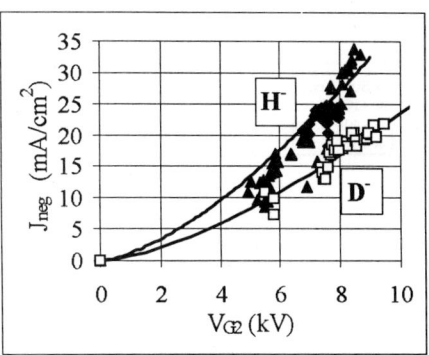

Figure 5

3.2. Tests with stronger magnetic filters

In order to reduce the electron current on the extraction grid to acceptable values for D.C operation, one possible way is to increase the magnitude of the external magnetic filter with additional magnets. The field strength value is changing from 0.088 T.cm to 0.11, 0.14 and 0.18 T.cm. The main results from these tests are shown on figure 6. and 7.

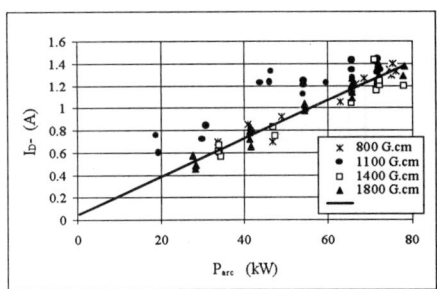

Figure 6

The extracted D⁻ beam is relatively insensitive to the value of the external magnetic filter; That means that the negative ions are probably mainly produced by impinging of D° on the plasma grid. The extracted electron current is strongly reduced when the magnetic field strength is increased (Fig7) For the strongest value (0.18T.cm), the ratio Ie/ID is 0.5 and a steady state operation of the beam is possible with a thermal power deposited theextraction grid less than 200 W/cm².

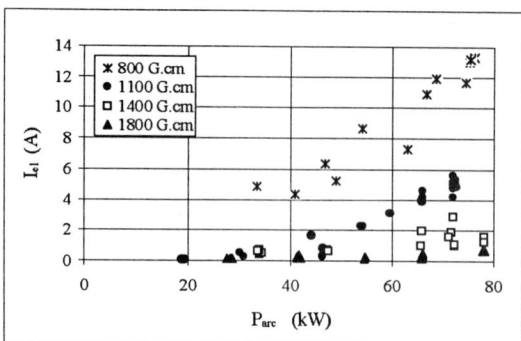

Figure 7

3.3. Source plasma characteristics.

A Langmuir probe immersed in the source discharge is moving from the plasma grid to the middle of the arc chamber. Figure 8 and Figure 9 show the strong effect of the magnetic filter on the density and the temperature profiles of the plasma and the good confinement of the plasma in the center of the discharge. These results are consistent with the reduction of the electron current.

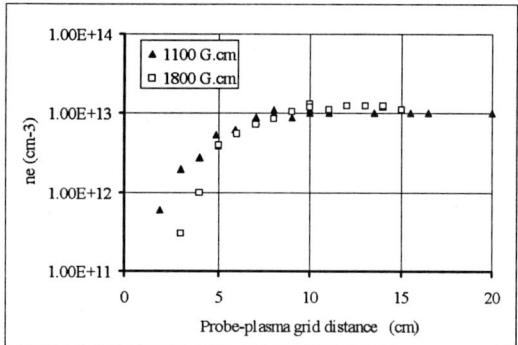

Figure 8

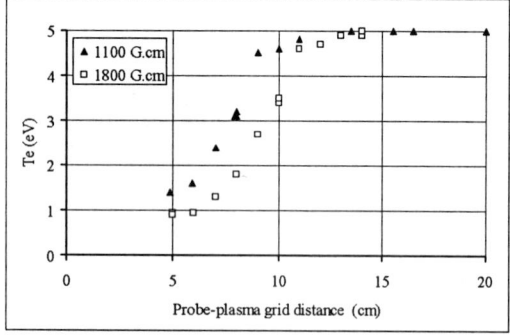

Figure 9

3.4. Stripping of the D⁻ ions in the extraction.

Doppler shifted Balmer alpha of $D^°$ from the beam are presented on Figure 10.

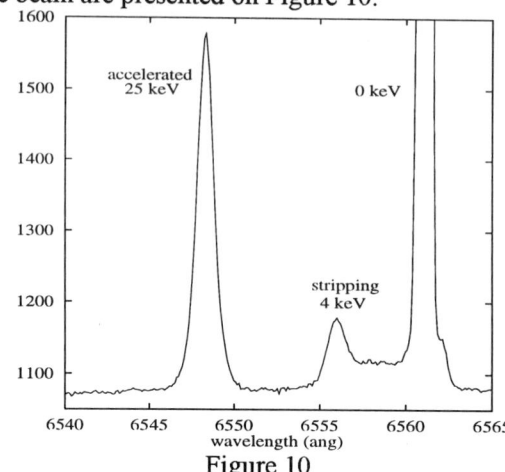

Figure 10

it shows a clear evidence of the stripping of D⁻ on the residual gas D_2 in the first extraction gap and inside the extraction grid. No negative impurities seem to be accelerated.

4. CONCLUSION

To demonstrate the high current density negative D⁻ beam production at a low operating pressure a high confinement surface negative ion source called ITER-NBI Concept source has been tested. D- ion beam of 1.4 A , 30 KeV was produced with an average current density of 20 mA/cm^2 at a pressure of 0.35 Pa. The extracted electron current associated with this D⁻ current is as low as 0.7 A and should permit a possible steady state operation.

REFERENCES

1. ITER Joint Central Team: Design Description Document, Neutral Beam Injection WBS 5.3, 1995.

2. N.Miyamoto, H.Oguri, Y. Okumura, T.Inoue, Y.Fujiwara, K. Miyamoto, A.Nagase, Y.Ohara, K. Watanabe, Experimental Results on ITER-NBI Concept Source, 7 th International Symp.on the Production and Neutralisation of Negative Ion s and Beams and 6 th European Workshop on the production and Application of Light Negative Ions, Brookhaven National Laboratory, Oct. 1995.

EXTRACTION SYSTEM DEVELOPMENT, AND STRAY ELECTRON STUDY FOR HIGH INTENSITY NEGATIVE ION BASED NBI

A. Simonin, K. Miyamoto* and Y. Fujiwara*

Association EURATOM-CEA, Département de Recherches sur la Fusion Contrôlée
Centre de Cadarache, 13108 Saint Paul lez Durance Cedex, FRANCE

*NBI Heating laboratory, JAERI, 801-1 Naka-Machi, Naka-gun, Ibaraki-ken, JAPAN

One of the major issues of Negative Ion Accelerator is to minimize the rate of the stray electrons accelerated simultaneously with Negative Ions (NI) at the high energy (MeV range). During acceleration of NI, particles (electrons, neutrals, ions) are generated mainly by stripping and ionization. Collisions of these particles with metallic accelerating electrodes induce secondary electrons; this process is particularly important in conventional Multi-Aperture Multi Gap (MAMuG) accelerators, as envisaged for ITER NBI. For this purpose, a 3-D Monte-Carlo code has been developed at Cadarache to calculate the 3-D electron trajectories in the accelerating channel, taking into account secondary electron emission from metallic surfaces under particle bombardment. A comparison between the code predictions and experiment on a MAMug-type accelerator (400keV NIAS accelerator JAERI) shows a good agreement.
A collaboration between CEA and JAERI has taking place in April and May 96 at JAERI, to test a new concept of extraction system (named CANIE developed at Cadarache), on a MAMuG-type 400 keV accelerator (NIAS at JAERI). CANIE was developed to allow the acceleration of high density D- current (up to 30 mA/cm^2) with a high efficiency electron suppression, in continuous operation. The experimental results are presented.

1. INTRODUCTION

Neutral Beam Injection (NBI) based on Negative Ions (NI) is one of the candidates for plasma heating and current drive in the new generation of large magnetic fusion devices such as ITER[1]. To achieve ITER NBI requirements, about 50MW of neutral (D_0) beams of 1 MeV energy, considerable development is still necessary, in the production (ion source), as well as in the extraction, acceleration and transport of negative ion beams.

A new concept of extraction system (named CANIE) that should meet the ITER requirements, has been developed at CEA. This development (ITER task) was experimentally tested in the framework of a CEA-JAERI collaboration on the NIAS (400keV, 1A H- acccelerator) test bed in Japan[2]. The experimental results are presented in the first part of this paper.

The acceleration of intense negative ion beams (in the 20-30mA/cm^2 D- range) at high energy (in the MeV range) has not being demonstrated to date. Two different concepts (SINGAP[3] (SINgle GAP, SINGle APerture), MAMuG[4] (Multi-Aperture, Multi-Grid)) have been proposed and are being studied independently (CEA, JAERI) to prove their potentiality and capability to accelerate Negative Ion (NI) beams at the MeV energy. Stripping and ionization, which occur during the acceleration of NI results in the production of particles (electrons, neutrals, positive ions). The bombardment of the accelerating electrodes by these particles can considerably increase the stray electrons rate in the accelerator column. A 3-D electron trajectory code, which takes these secondary phenomena into account has been developed at Cadarache. It simulates the consequences of stripping and ionization on the accelerator parameters, such as the thermal grid loads, the electrical grid currents, etc... The experimental results (database collected on the NIAS accelerator during the collaboration) are consistant with the code predictions.

2. CEA-JAERI COLLABORATION

2.1. The NIAS test bed[2]:
It is composed of a Kamaboko[5]-type ion source, an extraction stage with a molybdenum plasma grid(PG) and an extraction grid (EG), an additionnal Electron Suppression grid (ESG) (only for JAERI systems), and three accelerating electrodes (A1G, A2G, A3G). The beam diagnostic are located downstream from the accelerator.

2.2 Presentation of the Cadarache Negative Ion Extractor (CANIE)[6]

CANIE was developed to test a new concept of extraction system, which could be used in the future high intensity Negative Ion

Accelerators (40A of D⁻ at 1MeV). CANIE should allow the extraction of high density NI currents (20-30mA/cm² of D⁻) with an efficient electron suppression in long shot operation. In addition, its design should allow an increase of the accelerator transparency up to 50%[7]. The transparency of the existing accelerator is limited to less than 40% by the extraction grid, where permanent magnets and water tubes are implemented between each hole for electron suppression purposes. We overcome this limitation by a compression of the beamlets (Figure 1) from the plasma grid (14mm in diameter), to the extraction grid (CANIE) (10mm in diameter).

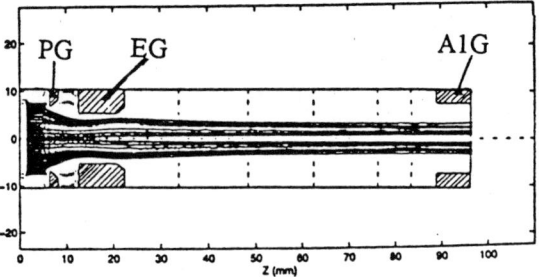

Figure 1. Beam compression from the plasma grid (PG) Φ=14mm, to the extraction grid (EG) Φ=10mm.

The other consequence of smaller hole diameters on CANIE is a reduction of the gas flow and stripping rate in the accelerating column by comparison to other extraction systems (the JAERI extraction grid has hole diameters of 14mm). To overcome a blow up of the compressed beam (strong space charge effect with high density currents) inside the grid (CANIE), the thickness is 10mm (17mm for the JAERI grid[2] (EG + ESG)). The suppression of electrons is more difficult, because it requires a large equipotential area to trap efficiently the electrons in the magnetic field. A 3-D electron trajectory code[8] developed at Cadarache three years ago for electron suppression study allows to optimize the grid design.

2.3 Experimental results

CANIE's first tests have been performed in the JAERI laboratory in Japan, on the NIAS test bed[2]. The latter is a 400keV 1A H⁻ Multi-Aperture MUlti-Grid (MAMuG) type accelerator, with 3 accelerating gaps. This test bed is of particular interest for testing CANIE and comparing its performances with the JAERI grid (extraction system used on JT60U NBI), because of its available means of precise diagnostics for beam analysis. The design of the NIAS accelerator does not allow the extraction of high density current; the maximum value was 10mA/cm² of H⁻. The test at high current density (20-30mA/cm² of D⁻) in continuous mode operation will be performed at Cadarache on the MANTIS test-bed.

a) Electron suppression efficiency:

The electrons extracted from the plasma source must be stopped by the grid, where secondary emission (backscattered electrons) occurs and results in electron leakage[7,8] from the extraction grid. Electron leakage measurement was performed with helium discharges in the source; in these conditions, only electrons are extracted from the plasma (no negative ions). We measure the current flowing on each accelerating grid. The leakage is defined as the ratio of the total accelerated electron current flowing on the accelerating grids and the drain current, by the electrical current on the extraction grid (about 1 to 2 A of extracted electrons). It was observed a linear increase of the leakage from 4 to 7 kV (0% at 4kV and 1.6% at 7 kV); the extrapolation to 10kV corresponding to 30 mA/cm² of D⁻ (ITER relevant NI current density at the source level) gives a leakage of 3%. The same value was obtained by the JAERI team on the JT60U NBI extraction grid.

b) Beam optics analysis:

This experiment was performed on a single beamlet; we displaced a probe (thermo-couple), located at 1.5m downstream from the accelerator, step by step in the transverse direction of the beam axis, to measure its profile (Figure 2).

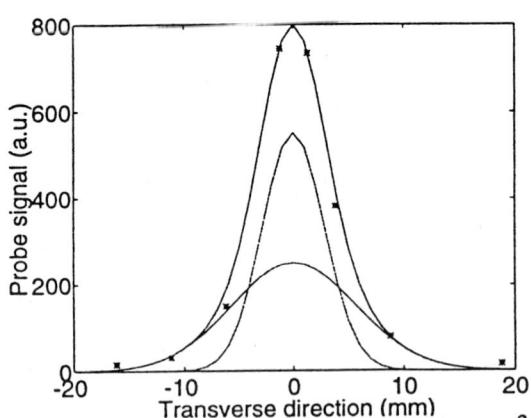

Figure 2. Single beamlet profile (j_{H^-} = 6mA/cm²)

A good fit of the experimental points is obtained by the addition of two Gaussians with different RMS (Root Mean Square) and amplitudes; about 52% of the beamlet has 3 mrad of divergence, and 4.5 mrad for the other part. The H⁻ current density was 6mA/cm².

c) Stripping in the accelerator:

Electrons resulting from stripping propagate in the accelerator; some are intercepted on the accelerating grids where they release their energy, some are accelerated along the accelerating channel with the ion beam.

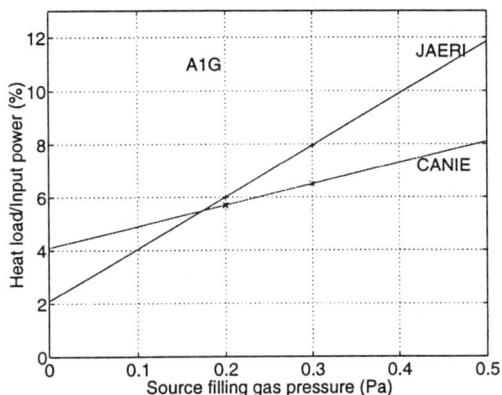

Figure 3. Heat load on A1G for two extraction grids (CANIE and JAERI)

Figures 3 shows the thermal load of the first accelerating grid (A1G) on the NIAS accelerator for two different extraction grids (CANIE and JAERI's extraction grids). We clearly see a linear dependence on the heat loads with the pressure, the slope is lower in the case of CANIE by a factor of 2.3 on AG1 (1.2 on AG2, and 1.5 on AG3), which demonstrates a lower stripping rate with CANIE. The extrapolation of the heat load measurements to zero source pressure shows that in CANIE experiments 4% of the beam was directly intercepted by A1G (see Figure3). The beam optics code (SLAC) shows that the overstepping of the grid support (support of CANIE on the NIAS accelerator), gives rise to a slight deflexion of the beamlets on the first accelerating gap, which should result to beam interception on A1G. Beam halo (resulting from optical aberations) could also contribute.

A Helmoltz coil producing a horizontal B-field located downstream from the accelerator was used to deflected out from the ion beam (270keV j_{H^-}=6mA/cm²) the accelerated electrons. An electrical target was set up (40cm above the calorimetric target) to collect these electrons. A variation of the Helmoltz coil current allows to scan these electrons on the electrical target as function of their energy (see Figure 4: electron spectrum in arbitrary unit). On this figure is also plotted the ratio of the power on the calorimetric target (electrons + ions) by the input power (high voltage power supply).

When all the stray electrons are deflected out from the ion beam, the ratio is about 63%. Using the "electron spectrum" we can identify their emission site in the accelerator. We have found that about 50% of the electrons are emitted at the grid level (A1G : 33%, A2G : 6%, A3G : 4%), certainly due to beam interception (on A1G), and secondary emission under neutral bombardment.

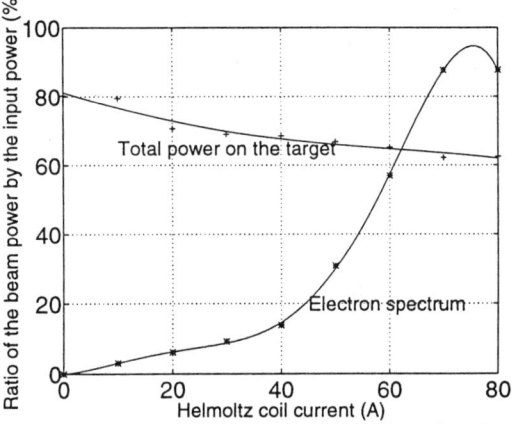

Figure 4. Target power and deflected electron current dependence as function of the Helmoltz coil current.

3. SECONDARY ELECTRONS IN NEGATIVE ION ACCELERATORS

This is a very important problem to be solved; indeed Figure 5 illustrates the dependence of the accelerating grid power loads (NIAS accelerator) as a function of the negative ion current density transmitted on the target (data collected with the CANIE extractor), and 3 mTorr of pressure in the ion source. If we extrapolate these results to the current densities required for ITER, more than 1kW/hole of power has to be removed by active cooling (ITER requires about 1000 holes ==> more than 1MW per grid to remove).

3.2. Numerical simulation principle

We use a code which is the extension of a 3-D Monte Carlo code developed three years ago in the laboratory for the electron suppression study[7]. The principle is based on a Runge Kutta method for the 3-D particle trajectory calculation in the electric and magnetic fields in the accelerating column. Stripping and ionization probability are calculated step by step during the calculation of ion trajectories and generate particles (electrons, neutrals, positive

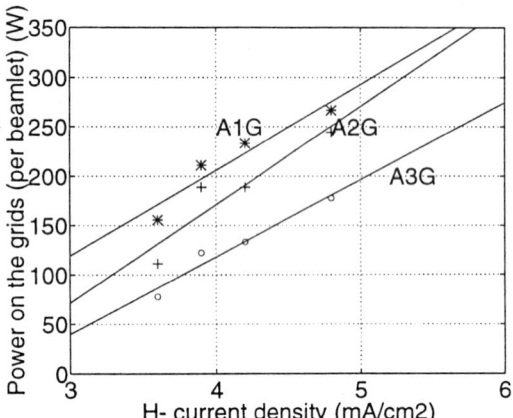

Figure 5. Accelerating grid power load as function of the H⁻ current density on the target on NIAS.

ions) which are also propagated. When collisions of these particles occurs with a metal surface (accelerating grids), a Monte Carlo method is activated to calculate the distribution of the electron secondary emission[9,10]. This emission is particularly intensive under grazing angle bombardment (emission law $\propto 1/\sin(\theta)$). The number of secondary electrons can be as high as 20 electrons per neutral (or ion). The code allows to calculate the calorimetric and electric powers of each accelerating grid, and to perform a complete stray electron study for each accelerator types (MAMuG[11], SINGAP) for ITER.

3.3 Comparison of the code predictions with the experimental results

The data were collected during the CANIE test on the NIAS accelerator (see A1G on Figure 3 with CANIE). Table 1 shows the experimental and simulation values of each accelerating grid power load (ratio of the grid power load by the total electrical power). We observe a quite good agreement between simulation and experiment (we substract the 4% interception fraction;see figure 3).

	Pressure (mTorr)	A1G(%)	A2G(%)	A3G(%)
Exp	2	1.5	2.9	2.4
Sim	2	0.97	2.8	2.43
Exp	3	2.3	4.3	3.4
Sim	3	1.8	3.7	3.5

Table 1. Comparison of simulation(Sim) and experiment(Exp) of the accelerating grid loads (%) for different source pressures.

4. CONCLUSION

The results of the CEA-JAERI collaboration on the extraction grid called CANIE have been presented. Electron suppression efficiency is in the range of 97% in the case of 30mA/cm^2 (on the source level) of D⁻ for ITER. Beam optics with a divergence of less than 5 mrad was achieved. It was also deduced from an accelerated electron spectrum that particle bombardment (ions, and neutrals) of the accelerating grids increases significantly the stray electrons especially in the MAMuG type accelerator columns (about 50% of the electrons on the target are emitted on the grid levels).

The second part of this paper addresses to a study of the stray electrons in the accelerator column. It was demonstrated that if the ITER design follows the NIAS design, the power on the accelerating grids to remove should be higher than 1MW. A 3-D trajectory code based on Monte-Carlo method to simulate secondary emission in the accelerator column has been developed. A good agreement was found between the experimental measurements performed during the collaboration (test of CANIE extractor on NIAS accelerator) and the code prediction. The assessment for the MAMUG and SINGAP type accelerators for ITER will be performed.

ACKNOWLEDGEMENTS:
The authors are indebted to Drs. S. Shimamoto, Y. Ohara, D. Escande and G. Tonon and for having actively supported this collaboration. The authors also wish to thank G. Delogu, P. Van Coillie, and all the JAERI team for their technical support during the collaboration.

REFERENCES:
1. ITER Joint Central Team: Design Description Document, Neutral Beam Injection WBS 5.3,1995
2. K. Miyamoto et al.; 18th Symposium On Fusion Technology(SOFT), Aug 22-26 (1994) Karsruhe, Germany.
3. J. Bucalossi et al.; 19 th SOFT, Sept 16-20 (1996); Lisbonne,Portugal.
4. T. Inoue et al.; 'Design study of Prototype Acceleration and MeV test facility ..'; JAERI-Tech 94-007, Aug 1994.
5. Y. Okumura; 'High Power NI sources for Fusion at JAERI';Rev. Sci. Instru., 67 (1996),p1092
6. A. Simonin; 7th International Symposium on the Production and Neutralisation of NI and beams; Brookhaven, Oct1995.
7. A. Simonin; Negative Ion accelerator development at Cadarache; Rev. Sci. Instru., 67 (1996),p1102
8. A. Simonin; Electron suppression in Negative Ions Accelerators; Euratom-CEA repports; EUR-CEA-FC-1502.
9. E.W. Thomas; 'Particle induced Electron emission'. School of Physics; Georgia Instituteof Technology; Atlanta, USA.
10. H.T. Hopman; 'Estimate of the coefficient of secondary electron emission..'; P/MV 30 July 1992, Internal report: 'Design of a 4 A D-1MV Test bed', R.Hemsworth CE Cadarache, France.
11. DI Petro, E. et al.,F1-CN-64/FP-18; 16th IAEA Fusion Energy Conference; Montreal, Canada, 7-11 Oct 1996.

Results of the Cadarache 1 MeV D⁻ "SINGAP" Experiment

J. Bucalossi, C. Desgranges, M. Fumelli, P. Massmann, J. Paméla, A. Simonin

Association EURATOM - CEA, Département de Recherches sur la Fusion Controlée
Centre de Cadarache, 13108 Saint Paul lez Durance Cedex, France

The objective of the experiment is to demonstrate the acceleration of a long pulse 100 mA D⁻ ion beam in a simplified electrostatic accelerator concept. In contrast to the ITER injector reference design, multiple beamlets are pre-accelerated to about 50 keV, then merged and post-accelerated to 1 MeV into one **SING**le **AP**erture (at high voltage) using one **SIN**gle **GAP** ("SINGAP"). The beam is dumped onto a uni-directional thermal conductivity CFC graphite target allowing infrared-calorimetric beam profile measurements. The experiment is supported by 3d beam trajectory calculations.

Conditioning without beam to 1 MV was achieved after the first 10 days of operation. This required only 35 min of integrated voltage on-time. During voltage application, an intense field-emitted electron current (up to 120 mA of "dark current") is observed. This current can be strongly reduced by conditioning and pressure increase. H⁻ beams of good quality were produced at a level of 910 keV, 40 mA, 1 s pulse.

The only problem encountered was the perforation by breakdowns of one of the 9 insulator rings (epoxy) which constitute the 1 MV bushing. After repairing this insulator, the high voltage operation could be restarted without major difficulties. The damage of the insulator is due to a defect in the manufacturing process and does not impair the qualities of the SINGAP concept.

1. MAJOR OBJECTIVES OF THE SINGAP EXPERIMENT

This experiment is mainly dedicated to beam acceleration and voltage holding studies :
- investigation of the properties of a scaleable high-energy (up to 1 MeV) electrostatic D⁻ accelerator using the simplified concept, SINGAP, proposed by M. Fumelli [1]. Pre-accelerated beamlets (up to 100 keV) are merged into a single beam and post-accelerated at high energy through a large **SING**le **AP**erture, with a minimum number of accelerating electrodes, if possible a **SIN**gle **GAP**.
- study of the voltage-holding capability of a single gap for a large dimensions system in the presence of the beam with a vacuum insulated beamline and the effects of a large stored electrostatic energy released during breakdowns (100-300 Joules range).
- the development of critical HV components : large insulator for bushing, decoupling devices for the external stray capacitances, fast interruption of power supply in case of breakdowns.

All these ITER [2] relevant topics are critical and still not much explored.

2. SINGAP TEST BED

The full description of the apparatus is given in [3] and only relevant details are mentioned here.

The beam is produced in an UHV tank of 40 m³. The source is at the earth potential and the HV electrode is a cylindrical chamber (0.55 m ∅, 2 m length) suspended to the 1 MV vacuum feed-trough insulator (HV bushing) as shown in Fig. 1. The feed-through has to sustain the 4 bar SF6 pressure on one side and vacuum on the other side. It is divided into 9 insulating rings connected in series and surrounded with screens (metal shields) which protect the surface of the insulator from particles bombardment and radiation, and minimize the electric field stress. In the first experiment, 2 of the 9 screens of the inner side of the bushing device had long dimension (in anticipation of future developments). The total applied voltage is shared resistively on the 9 sections by resistors (100 MΩ each). The 1 MV power supply can deliver a maximum d.c. current of 120 mA. The ion beam is directly intercepted (without conversion into neutral) at the end of the HV chamber on an uni-directional carbon fibre composite (CFC) target (2 plates of Mitsubishi MFC-1A material).

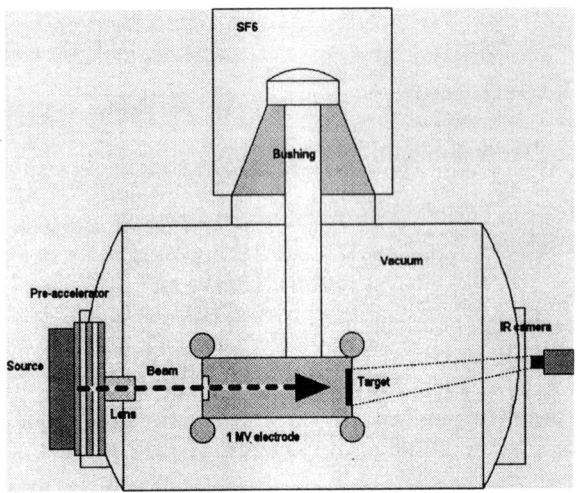

Figure 1. The Cadarache "SINGAP" experiment

3. VOLTAGE HOLDING

3.1. High voltage conditioning

The application of high voltage gives rise to a field-emitted electron current (dark current) with a simultaneous outgassing (typically 5.10^{-6} mbar for 100 mA). Before conditioning the dark current could reach 120 mA (the maximum delivered by the power supply) at only 100 kV.

When accumulating voltage on-time, the dark current decreased and higher voltages could be reached. The Fig 2 displays the current-voltage characteristic of the accelerating gap during the first measurement and after 1200 s of voltage application.

For a given state of conditioning, the dependence of the dark current on the applied voltage has the same functional behavior as the Fowler-Nordheim field emission law:

$$I = AV^2 \exp(-B/V) \qquad (1)$$

But the emission level is larger by orders of magnitude. It is important to note that in this accelerator, the cathode is the whole surface of the vacuum vessel, i.e. 60 m². These unusual dimensions (deliberately chosen to simulate the ITER injector) are probably at the origin of the dark current: the surface has not seen any particular treatment (presence of microtips, dust: lots of emitting points). During conditioning, some sites are destroyed while new sites appear with voltage rising: intense luminescence was observed, in particular at lower pressures.

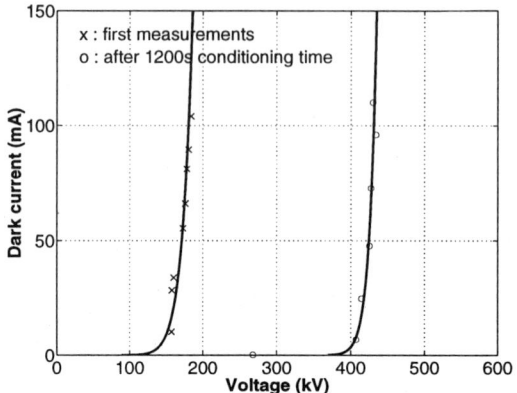

Figure 2. Reduction of dark current by conditioning

Furthermore, it has been observed that there is a strong dependence of the pressure in the vacuum tank on the dark current. Higher pressure reduces the dark current. In order to explain the experimental behavior shown in Fig. 3, coefficient A and B in (1) are necessarily strongly dependant on the pressure.

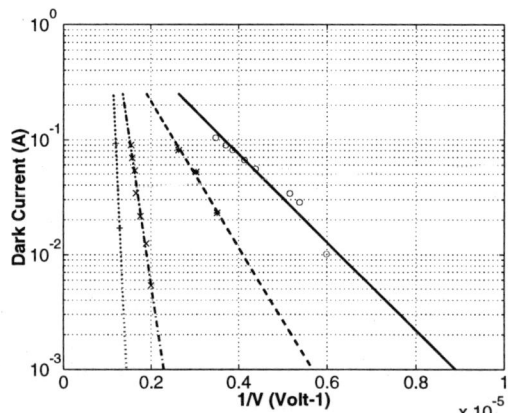

Figure 3. Dark current variation with voltage and pressure (first measurements): [o] 2.10^{-8} mbar, [*] 10^{-6} mbar, [x] 5.10^{-5} mbar, [+] 10^{-4} mbar

Despite of the release of an important electrostatic energy stored in the structure (100-300 J), the breakdowns which occurred during the HV conditioning have not caused any apparent damage on electrodes and do not deteriorate the voltage-holding capability of the system.

3.2. MV insulator issues

A degradation in voltage holding occurred after a few months : the upper bushing insulator rings were damaged and the first one was perforated. This premature ageing was the result of two things : an *overvoltage* due to a non-uniform sharing of the applied potential between the different stages and a *faulty construction* of the ring (important air layer observed between the metal insert and the epoxy ring). The bad distribution of the electrical potential was due to the presence of the two intermediate screens (initially foreseen to support three intermediate accelerating electrodes) in the bushing which intercepted an important fraction of the dark current.

Fortunately, the ring was successfully mended with epoxy glue (tested at 115 kV). After this repair the 2 long screens were replaced by short ones and each stage conditioned up to the appropriate voltage. Since then, the bushing device was operated without problems.

3.3. Voltage holding summary

The reduction of the dark current by increasing the gas pressure (up to the limit of Paschen discharge) was regularly observed in air, He, H_2 gases. The Fig. 4 shows the first conditioning of SINGAP without beam . 1 MV was obtained at 10^{-4} mbar residual gas pressure. During HV conditioning, the dark current intensity slowly decreases and lower operational pressures become possible. In the latest experiment (with the repaired insulator ring), 700 kV at 10^{-8} mbar and 900 kV at 10^{-5} mbar could be obtained.

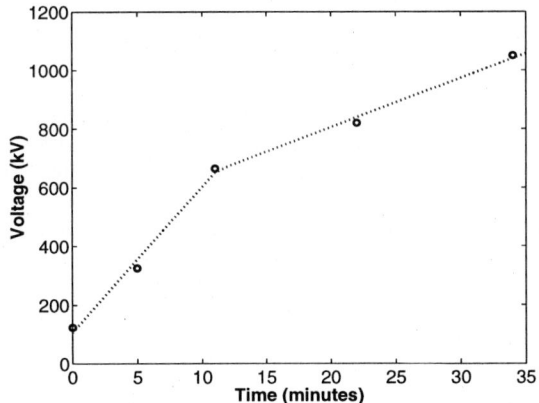

Figure 4. Conditioning of SINGAP without beam (with pressure)

4. BEAM ACCELERATION

In order to operate at a high beam current density (10 mA/cm^2) and given the total current available from the power supply (120 mA), we used 12 preaccelerated beamlets merged through a *single circular aperture* and post-accelerated in *one gap*. This constitute the first experimental test of the SINGAP concept.

4.1. Experimental layout

The D$^-$/H$^-$ beam is extracted from a plasma source trough 12 circular apertures of 1 cm^2 in a quasi-circular array of 65 mm diameter. The 12 beamlets are pre-accelerated by a standard multi-hole triode accelerator system [4] with complete magnetic electron suppression [5]. They are merged and post-accelerated up to 1 MeV through a single circular aperture of 80 mm diameter in a single gap of 630 mm. A cylindrical electrostatic lens of 106 mm inner diameter has been added at the exit of the pre-accelerator to provide a good control of the beam optics. A 3d simulation of the SINGAP accelerator is shown on Fig. 5.

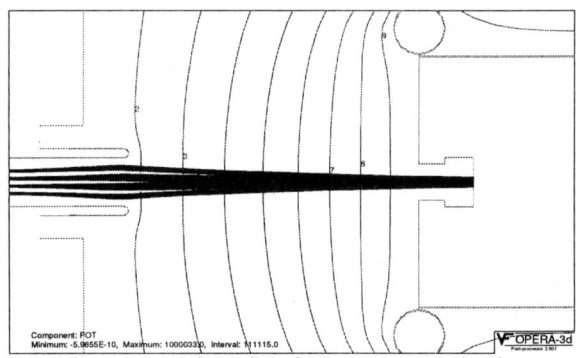

Figure 5. A 3d simulation of SINGAP post-acceleration

4.2. Beam optics

The beam is stopped at the end of the HV accelerating chamber on the CFC target. An infrared camera is viewing the rear face of the target at 2 m distance through a sapphire porthole. Because of the uni-directional heat conductivity of the target along the beam axis, the beam profile can be deduced from temperature rise measurements. Power density profiles can be calculated as well as the total power with an estimated error of ±10% [6].

The Fig. 6 displays the IR images of the beam on the graphite target for 2 different energies (same current). In the first image, the energy is matched to the current density and to the accelerating gap : the 12 beamlets are well focused by the lens. In the second image, the beam is "over-focused" and the reversed image of the 12 beamlets is perfectly resolved. This behavior is exactly that predicted by 3d calculations with a great accuracy.

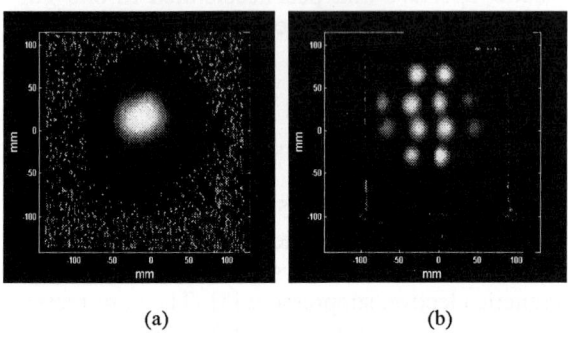

(a) (b)

Figure 6. IR thermography on SINGAP : 25 mA H- beam at 400 keV (a) and 700 keV (b) (1 s pulse)

4.3. Status after one year of operation

Since the beginning of the experiment, the electrical circuit [3] has sustained a large number of breakdowns without any failure.

Negative ion beams have been accelerated at high energy : table 1 shows some significant shots. No particular difficulties encountered with D-/H-.

The beam experimentations have been stopped during 6 months because of the main insulator breakdown.

Table 1

beam energy	beam current	ion	pulse duration	pressure
860 keV	60 mA	H-	1 s	2.10^{-5} mb
910 keV	40 mA	H-	1 s	1.10^{-5} mb
730 keV	35 mA	D-	1 s	4.10^{-8} mb

The beam optics is well mastered : 3d calculations and experimentations are in good agreement. Beams with divergence of less than 5 mrad have been obtained (400 keV, 30 mA H- beam). The size of the gap has to be adapted to the beam energy and the beam current to get the best optics quality. For a 1 MeV, 120 mA D- beam, the simulation predicts an optimum gap of about 900 mm.

5. CONCLUSION AND FUTURE EXPERIMENTAL PROGRAM

This experiment has already fulfilled some of its objectives :
- the SINGAP concept is able to provide an adequate beam quality (ITER relevant),

- the accelerator can be conditioned and operated at very high voltage (1 MV) despite the large stored electrostatic energy (up to 300 J) and the use of a single accelerating gap.

The large size of the electrodes (cathode area of 60 m2) seems to be at the origin of an intense dark electron current.

The production of 1 MeV, 100 mA, D- beams has not yet been accomplished yet because of insulator and D- ion source failures, but should be achieved in the near future : 910 keV, 40 mA, 1 s, H- beam have been accelerated up to now.

The study of a SINGAP beam with a rectangular cross section is planned for the next experiments in order to match ITER needs.

On the HV point of view, the SINGAP experiment presents an interesting opportunity for the study of high voltage phenomena in large dimensions system : dark current, large insulators, X-rays induced effects, etc.

ACKNOWLEDGMENTS
The authors wish to thank G. Delogu, S. Dutheil, P. Heister, P. Van Coillie and R. Yattou for their contribution to the experiment.

REFERENCES
[1] M. Fumelli et al., 6th Int. Symp. Prod. and Neut. of Negative Ions, Brookhaven, NY, 1992.
[2] ITER Joint Central Team : Design Description Document, Neutral Beam Injection, WBS 5.3.1995.
[3] P. Massmann et al., 16th SOFE, Champaign, Il., USA, 1995.
[4] A. Simonin, Negative Ion accelerator development at Cadarache, Rev. Sci. Inst., 67 (1996) 1092.
[5] A. Simonin, Electron suppression in Negative Ion Accelerators, Euratom-CEA repport, EUR-CEA-FC-1502.
[6] D. Ciric et al. 18th SOFT, Karlsruhe, Germany, Sept. 1994.

DEVELOPMENT OF A LARGE RF-DRIVEN NEGATIVE ION SOURCE FOR NEUTRAL BEAM INJECTOR

T.Takanashi, Y.Takeiri, O.Kaneko, Y.Oka, K.Tsumori, M.Osakabe and T.Kuroda

National Institute for Fusion Science, Nagoya 464-01, Japan

A large rf-driven hydrogen negative ion source has been constructed for applying to the neutral beam injector of the next step fusion experimental devices. The negative ion source was optimized for plasma production and negative ion extraction. A high density plasma of 1.5×10^{12} cm^{-3} was generated in the plasma generator with dimensions of $30 \times 30 \times 20$ cm^3 at the rf input power of 18 kW. In the pure-volume operation the hydrogen negative ion current of 5.5 mA, corresponding to 5.0 (mA/cm^2)/(W/cm^3) was extracted at 13.1 mTorr of the gas pressure. Cesium effects which are observed in a filament-arc type negative ion source were also confirmed in the rf-driven negative ion source. The rf-transmission efficiency is improved by using a capacitively coupled DC isolator to the rf ion source.

1. Introduction

In the next step fusion experimental devices such as ITER, long pulse and D-D/D-T burning experiments are proposed. A neutral beam injection (NBI) system is required to be operated for a long period without maintenance, because accessibility to the device is extremely restricted due to its radio activity. The rf-driven ion source has a potential of long lifetime operation compared with filament-arc ion sources, because they have no electrode like a filament which limits a source lifetime by its erosion.

We have constructed a large rf-driven hydrogen negative ion source for the next step NBI application[1], and investigated the operational characteristics.[2] In the filament-arc negative ion sources, cesium injection is utilized for improvement of the negative ion source characteristics. The observed cesium effects are enhancement of the negative ion current, reduction of the extraction current and lowering of the operational gas pressure.[3,4,5] Thus, it is important to investigate the cesium effects in the large rf-driven negative ion source.[6]

In the followings, the plasma production characteristics and the negative ion extraction characteristics in the rf-driven negative ion source are described including the cesium injection results.

2. Setup of Ion Source
2.1 Rf-driven large ion source

A schematic diagram of the large rf driven hydrogen negative ion source is shown in Fig. 1. A rf plasma generator made of stainless steel has dimensions of 30 cm x 30 cm in cross section and 20 cm in depth. Permanent magnets on the outer surface of plasma generator produces plasma confinement line cusp magnetic field. The plasma generator has a one turn rf-induction coil antenna. The antenna which is covered by quartz tube is immersed into the plasma for improvement of the rf coupling. The magnetic filter, which

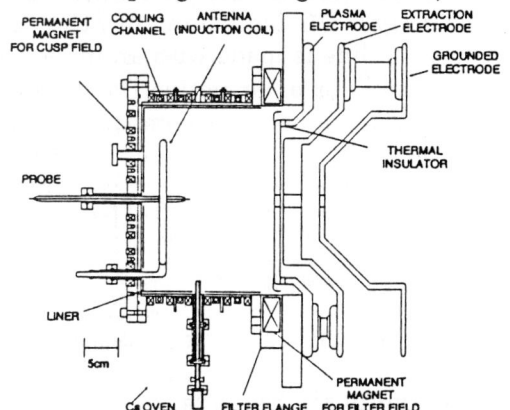

Fig. 1. Schematic diagram of an rf-driven hydrogen negative ion source.

is required for the negative ion source, is generated by a pair of permanent magnet rows outside the plasma generator (external filter). The filer field strength is 100 G at center and the line-integrated filter field strength is 1030 G cm. A cesium oven is attached to the sidewall of rf plasma generator.

A negative ion extraction system consists of a plasma electrode, an extraction electrode and a grounded electrode, as shown in Fig. 1. An extraction aperture is 13 mm in diameter. The extraction electrode has permanent magnet rows for removing the extracted electrons. The extraction electrode and the grounded electrode are electrically connected to the ground potential. The hydrogen negative ion current is measured with a calorimeter located 17 cm downstream from the plasma electrode.

2.2 Rf power supply system

A block diagram of the rf power supply system is shown in Fig. 2. The rf frequency is 2 MHz. The maximum rf power is 30 kW and the pulse duration is 1 sec with a duty factor of 1/30. The output power at the rf amplifier (Pdc) is defined by difference of the forward power and the reflection power which are measured by the directional coupler. The rf power to the antenna ($Pant$) is calculated from the antenna voltage ($Vant$), the current ($Iant$) and their phase shift measured with a high voltage probe and a current transformer, respectively. DC-isolation of the rf-transmission-line is made with an inductive DC isolator[1,2] or a capacitive DC isolator, and these two methods are compared in the next section.

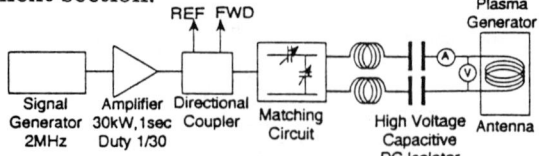

Fig. 2. Block diagram of an rf-transmission line.

3. Experimental Results

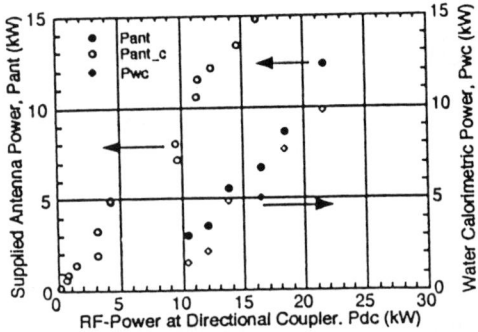

Fig. 3. Supplied rf-powers and water calorimetric power as a function of Pdc.

An rf plasma is successfully generated by matching the rf power with the rf circuit including the rf plasma.

3.1 Rf power flow

The rf power flow into the ion source was estimated by water calorimetry. The total calorimetric power (Pwc), $Pant_L$ and $Pant_c$ are shown in Fig. 3. Pwc is the sum of the thermal powers measured at the antenna, the plasma generator wall and the plasma electrode in the case of the inductive DC-isolator. $Pant_L$ and $Pant_c$ are the rf power to the antenna using the inductive and capacitive DC-isolators in the transmission line, respectively. Since Pwc is comparable to $Pant_L$, it is considered that $Pant_L$ is effectively coupled into the plasma. Pdc is larger than $Pant_L$, and the difference between those powers is an rf power loss corresponding to the rf radiative loss and the resistive loss on the rf

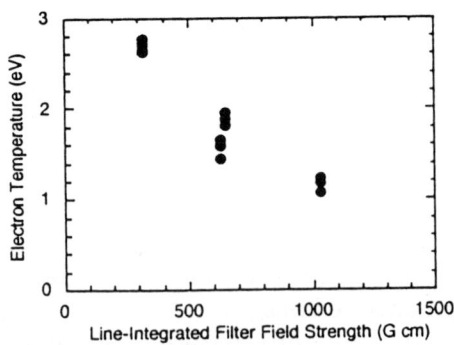

Fig. 4. Electron temperature as a function of the line-integrated filter field strength.

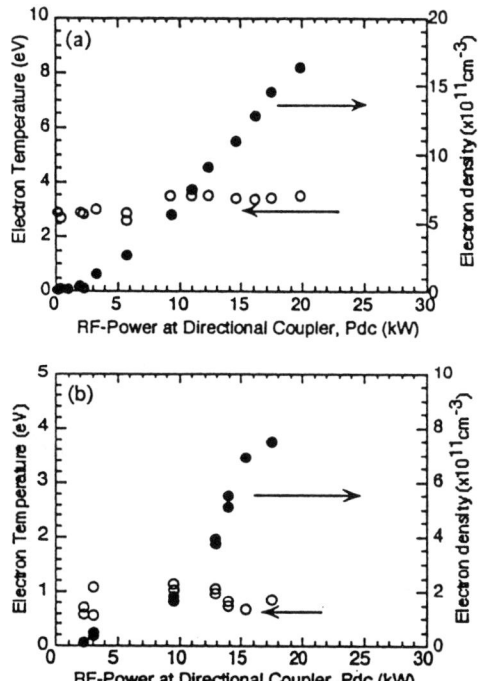

Fig. 5. Electron temperature and electron density in as a function of Pdc. (a). is plotted in the driver region and (b). is in the extraction region.

transmission line. On the other hand, in the case of the capacitive DC-isolator, $Pant\_c$ is roughly the same as Pdc, and the rf-transmission efficiency is high. The rf radiative loss and the resistive loss are suppressed by using the capacitive DC-isolator.

3.2 Rf plasma production

The rf plasma parameters were measured by a Langmuir probe. Figure 4 shows the electron temperature at a position of 1.5 cm from the plasma electrode, which is called the extraction region, as a function of the line-integrated filter field strength. The electron temperature in the extraction region decreases as the filter field strength increases. The electron temperature is about 1 eV at a filter field strength of 1030 G cm. It is found that the stronger magnetic filter is required to reduce the electron temperature in the rf-driven negative ion source compared with the filament-arc negative ion source.

In Fig. 5(a). , the electron temperature and the electron density at a position of 13.5 cm from the plasma electrode, which is in the driver region, are shown as a function of Pdc. Figure 5(b) shows the electron temperature and the electron density in the extraction region, as a function of Pdc. The electron temperature in the driver region is about 3 eV. The electron density is achieved to 1.5×10^{12} cm^{-3} at 18 kW of Pdc, where the ion saturation current density is 300 mA/cm^2. The maximum electron density obtained is 2.5×10^{12} cm^{-3}. A high density plasma is generated in the driver region. In the extraction region, although the plasma parameter dependency on Pdc is nearly the same as that in the driver region, the electron density and the ion saturation current density are about 1/3 of those in the driver region.

3.3 Negative ion extraction

Hydrogen negative ions were extracted from the rf-driven ion source. The plasma electrode is electrically floating against the plasma generator. The total negative ion current and the extraction current are shown in Fig. 6, as a function of Pdc. The negative ion current increases as Pdc increases, and reaches 5.5 mA at Pdc of 15 kW and at an extraction voltage of 8.6 kV. This current corresponds to 4.2 mA/cm^2

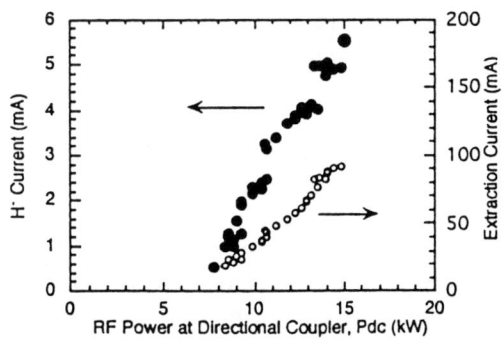

Fig .6. H$^-$ current and extraction current as a function of Pdc.

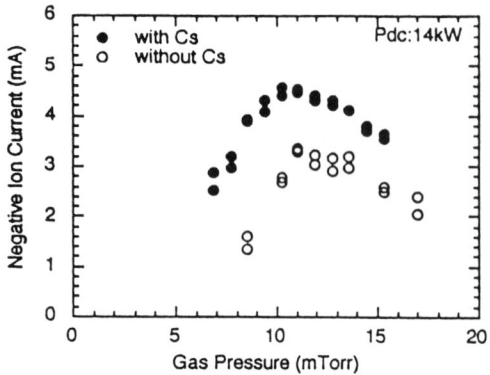

Fig. 7. H⁻ currents with and without cesium injection as a function of the gas pressure.

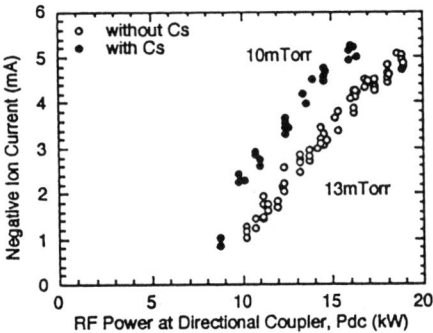

Fig. 8. H⁻ currents with and without cesium injection as a function of P_{dc}.

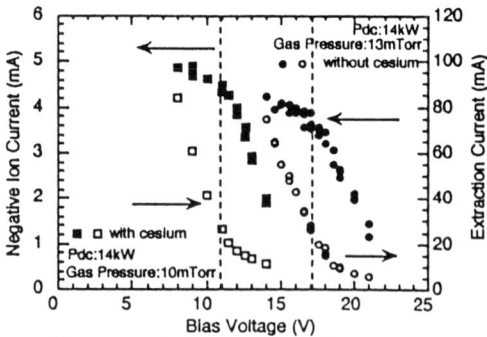

Fig. 9. H⁻ current and extraction current as a function of the bias voltage in the cases of the no cesium injection and the cesium injection.

of the current density.

In order to confirm the cesium effects which are observed in the filament-arc negative ion source, cesium is supplied into the rf plasma generator during the operation. Figure 7 shows the negative hydrogen currents with and without the cesium injection as a function of the gas pressure. The H⁻ current has a peak on the gas pressure in this figure. The optimum gas pressure is lowered from 13 mTorr to 11 mTorr by the cesium injection. The H⁻ currents with and without the cesium injection as a function of the P_{dc} are shown in Fig. 8. The operational gas pressures are 10 mTorr for the cesium injection and 13.1 mTorr for no cesium injection. The power efficiency to the H⁻ current is improved in the case of the cesium injection. Figure 9 shows the H⁻ current and the extraction current as a function of the bias voltage. The optimum bias voltage, where the extraction current is enough suppressed and the H⁻ current is not reduced, is lowered by the cesium injection from 17 V to 11 V.

4. Conclusions

The high density rf plasma was successfully produced under the matching condition. The negative ion current density per rf power density is 5.0 $(mA/cm^2)/(W/cm^3)$ on the pure-volume operation. The cesium-injection effects for the negative ion production are confirmed in the rf-driven ion source.

REFERENCES
1) Y.Takeiri,T.Takanashi,O.Kaneko,Y.Oka, A.Ando,K.Tsumori,T.Kuroda:Fusion Engineering and Design **26** (1995) 501.
2) T.Takanashi, Y.Takeiri, O.Kaneko, Y.Oka, K.Tsumori and T.Kuroda, Jpn. J. Appl. Phys. Part I, **35**, (1996) 2356.
3) K.Leung, C.A.Hauck, W.B.Kunkel, and S.R.Walther, Rev. Sci. Instrum. **60**, 531 (1989).
4) Y.Okumura, etal., Proc. of the 5th Int. Symp. on the Production and Neutralization of Negative Ions and Beams, Brookhaven, 1990, AIP Conf. Proc. No.210,p.169.
5) Y.Takeiri, etal., Rev. Sci. Instrum. **66** (1995) 2541.
6) T.Takanashi, Y.Takeiri, O.Kaneko, Y.Oka, K.Tsumori and T.Kuroda, Rev. Sci. Instrum. **67**, (1996).

Initial Beam Operation of 500keV Negative-Ion Based NBI System for JT-60U

M.Kuriyama, N.Akino, T.Aoyagi, N.Ebisawa, Y.Fujiwara, N.Isozaki, A.Honda, T.Inoue, T.Itoh, M.Kawai, M.Kazawa, J.Koizumi, K.Miyamoto, N.Miyamoto, K.Mogaki, Y.Ohara, T.Ohga, Y.Okumura, H.Oohara, K.Ohshima, F.Satoh, H.Seki, T.Takenouchi, Y.Toyokawa, K.Usui, K.Watanabe, M.Yamamoto and T.Yamazaki

JAPAN ATOMIC ENERGY RESEARCH INSTITUTE, Naka Fusion Research Establishment, Mukouyama, Naka-machi, Naka-gun, Ibaraki-ken, 311-01 Japan

Neutral beam injection with the negative-ion based NBI system for JT-60U started in March 1996. The system is designed to deliver deuterium neutral beams of 10MW at 500keV for 10sec using one beamline with two large-size negative-ion sources. In a preliminary negative ion source test conducted in 1995 ahead of beam injection into JT-60U, a deuterium negative ion beam of 400keV, 13.5A, which is the highest negative ion beam power (5.4MW) in the world, was produced. The first deuterium neutral beam of 180keV, ~0.1MW has been injected into JT-60U successfully in March 1996. The injection power has reached 2.5MW at 350keV as of the beginning of September 1996. The beam energy and power are now being increased further for the plasma heating and NB current drive experiments with high density plasmas in JT-60U.

1. INTRODUCTION

Recent progress of high power negative-ion source and related neutral beam injector technologies has made it possible to construct a high energy/high power negative-ion based NBI(N-NBI) system for a reactor-grade plasma heating[1-2]. In JT-60U, a 500keV neutral beam injection program has progressed for a demonstration of steady-state operation by NB current drive and for plasma core heating experiments in high density plasmas[3]. The N-NBI system has a significant meaning as an intermediate step to a 1MeV NBI system[4] for the ITER in both N-NBI technology and plasma physics standpoints. The system is designed to have a beam injection capability of 10MW at 500keV for 10sec using one beamline with two large-size negative-ion sources. Each negative-ion source accelerates a deuterium negative-ion beam of 22A at 500keV with a low beam divergence angle of 5mrad.

In order to demonstrate an ion source performance prior to the beam injection into JT-60U, a part of the N-NBI system including one negative-ion source and some of the high voltage dc power supply has completed in March 1995. After having confirmed the ion source performance, the rest of the N-NBI component such as the beamline, another ion source and power supply have been installed in November and December 1995. The component tests of the full N-NBI system have been conducted to verify the magnetic field for residual ions bending and magnetic shielding capability of the beamline, high voltage performance of a newly added power supply, following which a beam injection into JT-60U started in March 1996 with two ion sources.

In this report, we describe the outline of the N-NBI system, its performance and the recent progress of neutral beam injection into JT-60U.

Picture 1 N-NBI system installed in JT-60 torus hall

2. OUTLINE OF THE N-NBI SYSTEM

The N-NBI system for JT-60U shown in picture 1 comprises a beamline mounted two ion sources, a set of dc voltage power supply for two ion sources, and subsystems of cooling water, helium refrigeration, auxiliary vacuum pumping and control. The beamline shown in Figure 1 is composed of four components, an ion source tank, two neutralizer cells, an ion dump tank and a beam drift duct. The ion source tank supports two ion sources and houses four modules of cryopump that have a total pumping capability of $1200m^3$/s for deuterium gas. The neutralizer cell is 10m long and maintains a neutralization gas line density of 7.5×10^{15} molecules /cm^2 for 500keV deuterium beams. The ion dump tank houses a pair of ion deflection coils and ion dumps for D^- and D^+ beams, a calorimeter and a set of cryopumps just the same as those of the ion source tank. The drift duct is composed of a large bore (60cm x 110cm) isolation gate valve, a flexible bellows and a duct covered with protection plates made of Mo against a reionized beam.

The ion source[5] comprises three components, a negative ion generator, an extractor and an accelerator. The negative-ion generator is a volume type with cesium seeding, and is used the "KAMABOKO" type[6] arc chamber which yields a high negative-ion current at a low operating pressure due to a good plasma confinement.

The extractor is composed of three grids: a plasma, an extraction and an electron suppression which have an extraction area of $0.45 \times 1.1 \, m^2$. Each grid is composed of five segments. A segment has 216 apertures of 14mm in diameter in an area of $0.45 \times 0.18 \, m^2$. A set of Sm-Co permanent magnets is embedded in the extraction grid to deflect electrons extracted with negative ions.

The accelerator is an electro-static/three-stage acceleration system that consists of the 1st, 2nd acceleration and grounded grids. The gaps spacing in the accelerator are 75mm, 65mm, 55mm in order from the 1st acceleration, thereby gradually making strong a lens effect converged each beamlet. The acceleration area is the same as the extraction grid, though the aperture size is more enlarged the diameter to 16mm.

3. NEGATIVE-ION BEAM GENERATION AND ACCELERATION

3.1 Preliminary test with a full scale ion source

A preliminary test of negative-ion generation and acceleration with a full scale ion source started in the middle of 1995 to verify the ion source performance, using a part of the N-NBI system such as some of the beamline and ion source power supply completed in March 1995. In the test conducted from June to October 1995, a deuterium negative ion beam power has reached 5.4MW (400keV, 13.5A/D^- ; current density: 8mA/cm2) in October as shown in Figure 2. This is the world highest deuterium negative-ion beam power. It has been proved that there is no serious problem in the acceleration of a large current of deuterium beams more than 10 A up to several hundreds keV.

3.2 Negative-ion beam characteristics

After having completed the whole NBI system in the beginning of 1996, a negative-ion generation / acceleration test and a neutral beam injection into JT-

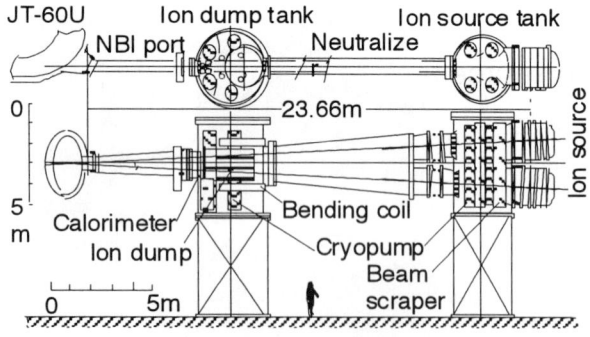

Figure 1 Beamline for JT-60U N-NBI

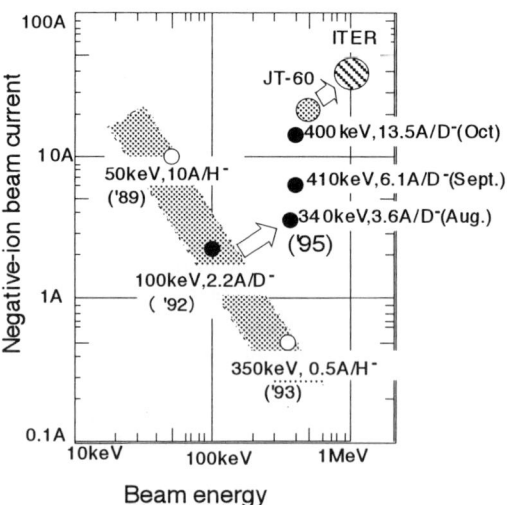

Figure 2 Negative-ion beam progress in the preliminary test

60U with deuterium using two ion sources have been conducted.

The effect of cesium seeding in a negative ion generation is shown in Figure 3. The negative ion generation efficiency, Iacc/Parc (where Iacc: acceleration drain current, Parc: arc discharge power), increased by approximately four times after six hours cesium seeding, and just then the electrons extracted with the negative ions decreased by a factor of 1/4~1/5. This shows clearly the cesium seeding effect in negative ion generation. The negative-ion generation efficiency, Iacc/Parc, has reached approximately 0.17 A/kW in maximum. Figure 4 shows an acceleration drain current and extraction current, I_{ext}, as a function of arc discharge power at a beam energy of 370keV. The Iacc increases linearly with the arc power until an arc power of around 150kW, and thus the power required for obtaining a rated current of 22A/D$^-$ seems to be around 300kW.

4. NEUTRALIZATION EFFICIENCY AND NEUTRON YIELD

The beam component ratios of D$^-$, D$^+$ and D^0 at a beam energy of 370keV are shown in Figure 5 as a function of line density in the neutralizer cell. The line density is controlled by regulating the gas feed into the cell. The highest neutralization efficiency has been confirmed to be approximately 60% as shown in figure 5. The optimum gas line density that gives the highest neutralization efficiency is 8.6×10^{15} molecules/cm^2 at 370keV, which is somewhat higher than a theoretical calculation value of 5.5×10^{15} molecules/cm^2. The reason of the discrepancy is considered be due to a higher gas temperature in the cell than the room temperature. The optimum neutralization efficiency measured with the N-NBI kept a constant value of around 60% in a beam energy range of 250-370keV as predicted with a theoretical calculation.

A neutron yield from a calorimeter and a couple of ion dumps has been measured with ^{235}U detector positioned near the beam drift duct, approximately 3m far away from the calorimeter and the ion dumps.

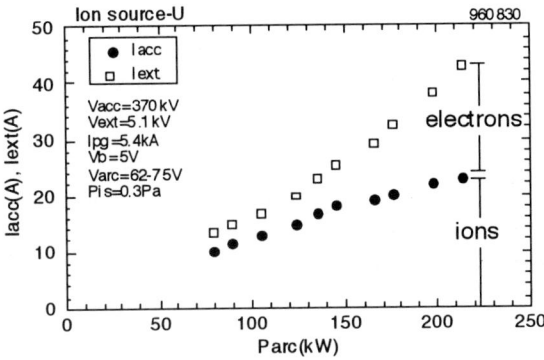

Figure 4 Arc power dependence of extraction and acceleration current

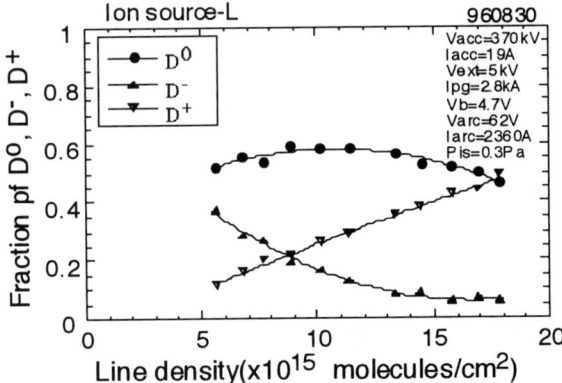

Figure 5 Fraction of D^0, D$^-$, D$^+$ at 370keV

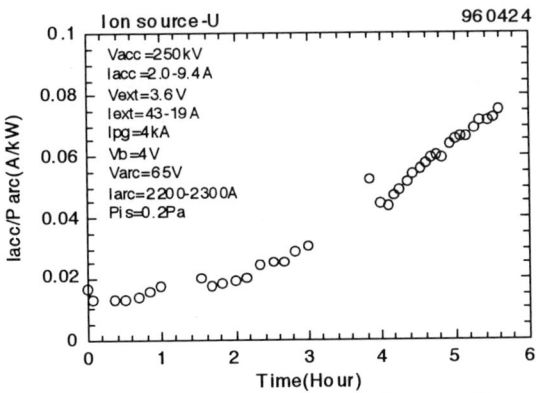

Figure 3 Arc efficiency after the cesium seeding

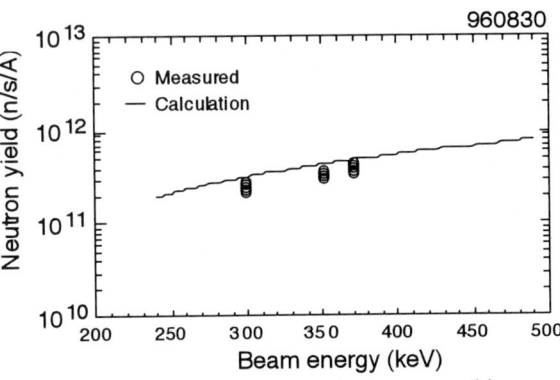

Figure 6 Neutron yield from calorimeter and ion dumps

The detector has been calibrated by inserting a standard neutron source, ^{252}Cf, into the ion dump tank. The neutron yield per deuterium beam current is 2.8-4.5 $\times 10^{11}$ n/sec/A in a beam energy range of 300-370keV, as shown in Figure 6, which agrees with the calculation model reported by J.Kim[7].

5. NEUTRAL BEAM INJECTION INTO JT-60U

The first neutral beam injection of 180keV, ~0.1MW has been conducted successfully in March 1996. The signals of the JT-60U plasma diagnostics such as a neutron yield and an alpha particle analyzer have responded clearly to the beam injection even at a low beam power. The beam energy and power have been increased through an ion source aging. The injection power achieved as of the beginning of September 1996 is 2.5MW at 350keV for 0.7 sec, and the highest beam energy injected is 400keV for 0.35 sec. Figure 7 shows a typical data out of initial beam injections into JT-60U. In the plasma shot[9], a beam power of 2MW at 350keV has injected under injecting a power of ~8MW at 85keV with a positive-ion based NBI. The plasma stored energy and the neutron yield from the plasma increased clearly during a 350keV beam injection. In particular, the neutron yield per injection power with 350keV beams is higher by a factor of four than that of 85keV beams.

The reionization loss in the drift duct during a beam injection has been estimated through the measurement of pressure rise inside the duct. Although the reionization loss was around 50% for several shots from the beginning of the beam injection, it decreased rapidly with shot by shot and has reached a loss of a few %. The reason why the loss decreased rapidly is owing to a helium glow discharge cleaning in the JT-60U vacuum vessel that also has the effect of the NBI drift duct cleaning. The effectiveness has been already proved in positive-ion based NBI experiments[8].

6. SUMMARY

A preliminary test of the negative-ion beam generation and acceleration had been conducted using a part of the JT-60U N-NBI system which included a full scale ion source, some of the beamline and power supply. A negative-ion beam power of 5.4MW(400keV, 13.5A/D-) has been obtained in October 1995. The whole N-NBI system has completed in March 1996, following which a beam injection into JT-60U started. The first beam injection (180keV,~0.1MW) into JT-60U plasma has been conducted successfully at the middle of March 1996. The ion sources aging and debugging of the beamline and power supply have developed, thereby increasing an injection power gradually. An injection power of 2.5MW at 350keV for 0.7 sec has reached as of the beginning of September 1996. The highest neutralization efficiency of negative ion beams has been confirmed to be approximately 60% in a beam energy range of 300 - 370keV.

ACKNOWLEDGMENT

The authors would like to thank the members of JT-60 group. They are also grateful to Drs. A.Funahashi, H.Kishimoto, M.Shimizu, M.Azumi for their continuous encouragement.

REFERENCES

1. T.Inoue, et al., Proc. of the 15th Int.Conf. on Plasma Phys. and Controlled Nucl.Fusion Res., IAEA (1994) 687/IAEA-CN-60/F-10
2. Y.Okumura et al., 6th Int. Conf. on Ion Source, Whistler, B.C, Canada, Sept.10-16 (1995)
3. M.Kuriyama, et al.,Fusion Engineering and Design, 26 (1995)445
4. R.S.Hemsworth et al., Rev. Sci. Instrum. 67-3 (1996) 1120
5. K.Watanabe et al., 7th Int.Symp. on the Production and Neutralization of Negative Ions and Beams, BNL, Oct.23-27(1995)
6. Y.Okumura, et al., Proc 15th Symp. on Fusion Engineering, Hyaniss, Oct.(1993)466
7. J.Kim, Nucl.Instrum. and Meth.145(1977)9
8. M.Kuriyama, et al., Pro.17th Symp. on Fusion Technology, Rome(1992)564
9. K.Ushigusa, to be presented in 16th IAEA Conf.

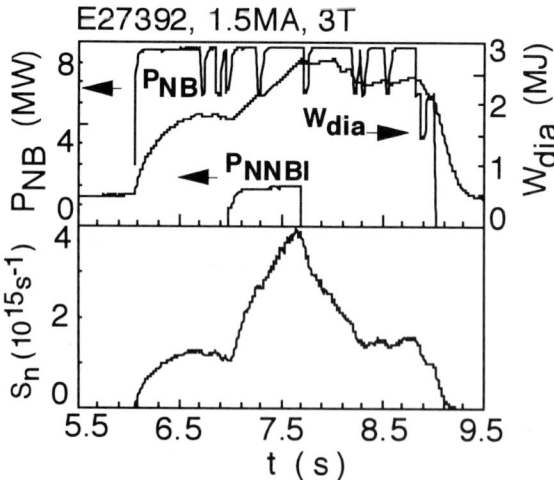

Figure 7 A typical result of N-NBI experiment

General Design of the Neutral Beam Injection System and Integration with ITER

A. Krylov[a], E. Di Pietro[a], M. Hanada[a], R. Hemsworth[a], C. Holloway[a], S. Stoner[a],
E. Alexandrov[b], M. Barinov[b], E.Dlougach[b], V. Kulygin[b], V. Naumov[b], A. Panasenkov[b],
V. Petrov[b], Y. Fujiwara[c], T. Inoue[c], K. Miyamoto[c], N. Miyamoto[c], Y. Ohara[c], Y. Okumura[c], K. Shibata[c], M. Tanii[c], K.Watanabe[c], J-H. Feist[d], B. Heinemann[d], E. Kussel[d], P. Lotte[d],
P. Massmann[d], J. Paméla[d], M. Watson[d]

[a]ITER JCT, Naka; [b]RF Home Team; [c]JA Home Team; [d]EU Home Team

The main technological aspects of the neutral beam (NB) injection system design are described and the constraints and requirements that derive from the integration in the overall reactor design are discussed.

The NB system will deliver 50 MW of 1 MeV D^0 to the ITER plasma with pulse lengths up to 10000 s. Three adjacent quasi-tangential equatorial vacuum vessel ports (0.5m x 1m clear opening) are allocated for NB. The system will consist[1] of three modules, at a power 16.7 MW each. The modules will occupy a common enclosure, the NB cell, located in the ITER pit between the biological shield wall and the inner gallery wall, see Fig. 1. This position, close to the torus, results in relatively high stray magnetic fields and high radiation fields due to neutron and radiation streaming. These, together with restrictions on available space for the arrangement of the system, affect the NB design.

Each of the modules, see Fig. 2, contains a single D^- ion source, connected to a 1 MeV electrostatic accelerator. The accelerated D^- ions pass through a D_2 gas fed neutraliser, then a residual ion dump (RID). The neutraliser is subdivided into five vertical channels to minimise gas flow. The neutraliser is decoupled from the accelerator to reduce the pressure in the accelerator, thus avoiding excessive collisional loss of D^- in the accelerator. Emerging from the neutraliser will be a beam consisting of D^0(60%), D^-(~20%), and D^+(~20%). The RID deflects the beam charged particles onto water-cooled copper surfaces, leaving only the neutral particles to continue to the torus. A water-cooled beam dump, the calorimeter, can be moved into the beam path to allow commissioning of the injector independently of the tokamak. The efficiency of the negative

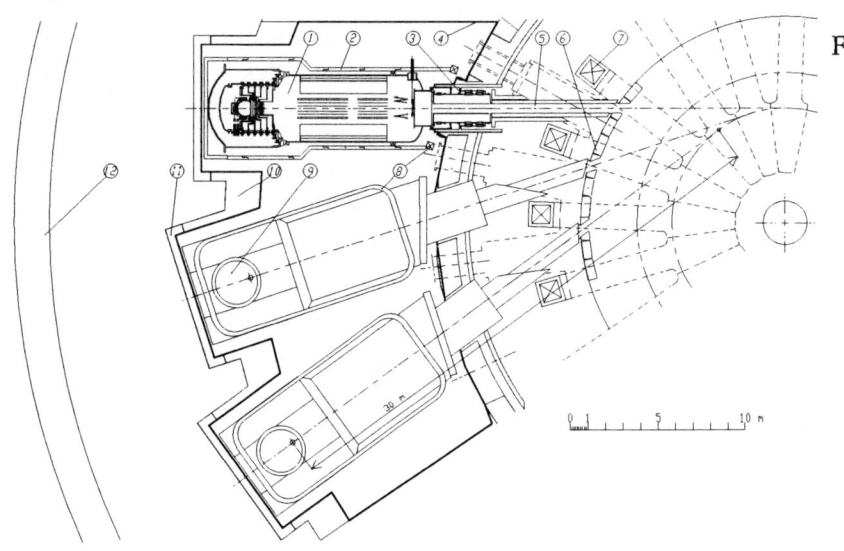

Fig.1 Layout of the NB modules on ITER
1. NB module, 2. Passive shielding, 3. Flexible connection, 4. Second safety barrier (thick line), 5. Duct, 6. Blanket, 7. Tokamak toroidal coils, 8. Active shielding coils, 9. HV transmission line, 10. Wall of the gallery, 11. Door of the NB cell, 12. Wall of the ITER pit.

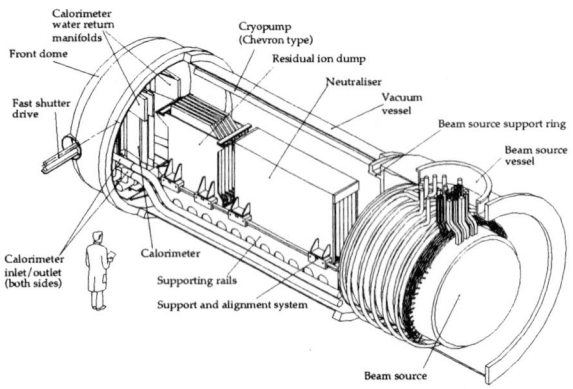

Fig. 2. Isometric cutaway view of an ITER NB module

ion neutralisation primarily defines the injector efficiency. Taking into account other beam power losses, such as stripping during acceleration, direct interception and reionisation the total calculated efficiency is ~35%. Each module also contains a cryopump and a fast shutter. The former removes the gas flowing from the neutraliser and ion source into the module. The latter prevents tritium flowing from the torus to the module during plasma start up and termination, and also allows regeneration of the cryopump without significantly increasing the torus pressure.

The ITER NB system also includes the following systems not described here: power supplies, insulating gas, gas supply, auxiliary pumping, cryoplant, and control and data acquisition.

1. BEAM SOURCE[2]

The D⁻ ions are produced by a filament driven arc discharge within the volume of a caesiated bucket-type source. The ion extraction and acceleration are provided by a 2-grid multi-aperture extraction system, followed by a multi-aperture multi-grid (five-stage) electrostatic accelerator to form a 40 A beam of D⁻. The accelerator grid system is contained within and supported from an insulator column assembly consisting of alternating alumina insulators and metal (Ti) flanges. The vacuum seal between the Ti and alumina is provided by metal O-rings. The ion source is enclosed and supported by a stainless steel dome connected to the insulator column. The ion source dome and the insulator column are surrounded by a pressure vessel connected by the high voltage (HV) transmission line to the HV decks. Both the pressure vessel and the transmission line are filled with high pressure insulating gas.

Large diameter ceramic insulators, consistent with ion source size and radiation environment in source area, have to be developed.

2. HIGH HEAT FLUX (HHF) BEAMLINE COMPONENTS AND WATER COOLING SYSTEM DESIGN

The water cooling for the NB system is split into two subsystems: the cooling for the main beamline components, which all are at or near ground potential, and the cooling for the ion source and accelerator.

HHF components of the beamline include the leading edges of the neutraliser, the RID and the calorimeter. During operation these components, in the worst case, have to remove ~25 MW of power (6.8 MW on the neutraliser and 18 MW on the RID). About 1 MW of power is deposited along the NB duct liner. The calorimeter has to remove 18 MW during commissioning and testing.

The main requirements and constraints defining the design of these components are integrated power, limits in thermal fatigue life time, and beamline geometrical limitations. The design maximum power density attains 3.5 MW/m² on the neutraliser leading edge, 8 MW/m² on the two sides of the RID panels and 15 MW/m² on the calorimeter. The extended pulse length of the NB and the power to be removed make it necessary to take into account dimensions and costs related to the heat exchangers system. Specifically, the water flow for all three injectors and the pressure drop must be minimised, <700 kg/s and <0.7 MPa respectively, and the inlet temperature must be ≈80 °C. Due to the long lifetime required and the activation levels of the components, the water speed has to be limited (< 8.5 m/s in the calorimeter and <5.5 m/s in the RID) to avoid corrosion/erosion phenomena and the problem of mechanical integrity and radioactive inventory in the coolant loops.

The design of RID and calorimeter panels is based on a common, modular, swirl tubes (rectangular hollow section, 20 mm x 25 mm, 16 mm inner diameter) tailored to the various power loads by selection of the swirl tape pitch in the range 56 to 100 mm. This concept guarantees sufficient critical heat flux margins with acceptable velocities and water flow. Different tube elements on RID and calorimeter panels receive different powers. To reduce the total water flow through panels, the tubes on each panel are grouped in pairs connected in series, chosen so that the power to all the pairs is approximately the same. The maximum pressure drop is in the calorimeter, from 2 to 1.3 MPa. During commissioning or testing it is also necessary to dump the D^0 beam power. To limit the additional flow requirement to <130 kg/s the calorimeter cooling system distributes the flow so that one calorimeter can have full flow for full power operation, or the same flow shared among all three calorimeters for fractional power operation.

From the mechanical point of view the only relevant limitation for the HHF elements is fatigue. Thermal cycling results from beam interruptions during normal pulse operation as well as from interruptions during electrical breakdowns in the beam source. During electrical breakdown the tube elements go through (almost) one thermal cycle. The fatigue life of the RID elements and, less importantly, those of the calorimeter, is presently estimated at $5*10^5$ cycles. A total injector operational life of $2.5*10^7$ s at full power implies that the beam source must guarantee an average frequency of less than one breakdown per 50 s. It is part of the ongoing R&D programme to establish the reliability of a 1MeV beam source under these conditions.

The NB HV water cooling is supplied via the HV decks and the transmission line to the ion source and accelerator and is used to cool components at various potentials from -1 MV to -200 kV. The total power to be removed is 4 MW per injector.

The requirements of the HV cooling system are mainly defined by the need to limit the average temperature of the beam source. The reasons are related to the consumption rate of caesium inside the source and thermal expansion limits acceptable for the metal flange to ceramic insulator seals (metal O-rings). On the other hand, due to water velocity and pressure drop limitations in the grid channels, the inlet/outlet temperature increase cannot be reduced below 40 °C. The power loading on the grids does not pose any particular thermal-hydraulic constraints. The HV cooling system is therefore designed for a total flow rate of 70 kg/s to the three injectors, with an inlet temperature of 20 °C and outlet temperature of 60 °C. The specified low inlet temperature necessitates the use of chilled water in the secondary loop. Additionally this must be demineralised/deionised water with a resistivity >20 kΩ*m.

3. PUMPING SYSTEM

Total D_2 gas input from the ion source and the neutraliser is 12 Pa*m^3/s. Gas pumping is needed to maintain gas pressure along the beamline at levels providing acceptable stripping and reionisation loss. Cryocondensation pumps with a speed of $3*10^3$ m^3/s for D_2 are installed. It will fit close to the walls of the injector vacuum vessel from the entrance of the neutraliser to the exit of the RID and cover all but the lower section of the vessel, where the support structure for the other beamline components are located. The present cryopump design is conventional, using a liquid helium filled quilted stainless panel surrounded by a liquid nitrogen structure. The system is designed such that regeneration of the pumps will be performed between ITER pulses, after 4,000 s to 5,000 s of injector opera-tion. The minimum regeneration frequency is dictated by the requirement to maintain deute-rium inventory below the limits of deflagrati-on in case of an air leak accident during regeneration (~2.6 kg of D_2). This limit is reached after ~30,000 s of operation of all three injectors.

4. MAGNETIC FIELD REDUCTION SYSTEM

In the region of the injectors, the stray magnetic field is between $8*10^{-2}$ T and $4*10^{-2}$ T. The principle reasons to reduce it are to ensure proper operation of the ion source (magnetic field $<10^{-3}$ T), and to avoid emittance growth of the D^0 beam due to D^- deflection prior to neutralisation ($<10^{-4}$ T). The RID operation imposes less severe stray field reduction[3]. The proposed system consists of an outer layer of mild steel 200 mm thick around each injector, an inner layer of 10 mm of mu-metal around the ion beam path, and three active compensation coils. Two of these coils (rectangular, 5m x 12 m, ~ 200 kA turns) are located above and below the injector. The third coil (5 m in diameter, ~ 350 kA turns) is located against the ITER biological shielding wall around the injector duct. This combination of passive and active shielding is found to reduce the field within the injectors to the design values with significantly less mild steel than needed for a purely passive system, and it does not produce unacceptable error fields in the tokamak. The passive shield acts also as a part of the neutron and radiation shield.

5. NB MAINTENANCE

The NB system will become activated by neutrons streaming directly into each injector such that the system will need to be remotely maintained. With the exception of the ion source and accelerator, the NB system is designed such that it does not need maintenance during the lifetime of ITER. The ion source will need regular maintenance twice per year for filament replacement since the estimated filament life time is 200 hours. In-situ replacement of the filaments is the reference design option. This will be performed by opening the beam source vessel and ion source dome to provide direct access with a manipulator to "plug in" type filaments. A procedure is also being studied for the removal and replacement of the entire beam source. This procedure will be used for major maintenance of the beam source (e.g. grid replacement and caesium decontamination). Both procedures foresee remote handling operations and full containment of radioactive contamination.

6. SAFETY BARRIERS

The NB system has to comply with the general ITER requirement of providing two confinement barriers for systems containing radioactive contaminants. The injector vacuum vessels (including the beam source accelerator stack and the ion source dome) and ducts (including the flexible bellows that compensate for movements of the NB vacuum vessel ducts) are directly coupled to the tokamak, and they are extensions of the first safety barrier. The water-cooled components inside of the primary vacuum form part of the first safety barrier. The NB cell will form the second safety barrier. The injector vacuum vessel and duct are designed to withstand 0.5 MPa internal pressure due to reference over pressurisation accident of the tokamak vessel, whilst the NB cell can withstand 30 kPa (g). The cell pressure is relieved to the large volume of the heat transfer vaults via bursting disks to avoid over pressurisation.

7. CONCLUSIONS

The present design of the ITER NB system is relying for most of its components on well established technologies or on conservative extrapolations. It is as well integrated within the overall reactor design complying with all the major constraints and interfaces.

REFERENCES

[1] R. S. Hemsworth, et. al., "Neutral beams for ITER", Rev. Sc. Instr., V.67, N3, Part II, 1966, pp. 1120-1125.
[2] T. Inoue et. al., "Design and R&D of high power negative ion source/accelerator for ITER NB", to be published in the proceedings of this symposium.
[3] R. Hemsworth et. al., "ITER neutral beam injector design", to be published in Proc. of the 16th Fus. Energy Conf., Montreal, Canada, October 1996, paper F1-CN-64/FP-18.

The disclaimer contained in ITER Publications Procedures S AC PP 1 93-10-12 W2 applies to this paper.

DESIGN AND R&D OF HIGH POWER NEGATIVE ION SOURCE / ACCELERATOR FOR ITER NBI

T. Inoue, Y. Okumura, Y. Fujiwara, K. Miyamoto, N. Miyamoto,
Y. Ohara, K. Watanabe, B. Heinemann[a], and M. Tanii[b]

Japan Atomic Energy Research Institute
Naka-machi, Naka-gun, Ibaraki-ken 311-01, Japan
[a] IPP Garching, [b] Nissin Electric Co. LTD.

Recent progress in the design and R&D of high power negative ion source/accelerator for ITER NBI are described. In the design of the ion source/accelerator, it was estimated that dose rate in the NB cell is in modest level (~ 10 μSv/hr) after shutdown of ITER. This enables direct access of radiation workers in the NB cell for preparation of remote handling devices by "hands-on". In the R&D of the MeV class accelerator, an H$^-$ ion beam of 805 keV, 0.15 A (power supply drain current) has already been obtained with an ITER prototype accelerator at MeV Test Facility of JAERI. Beam optics study and stray electron consideration are also progressed to accelerate the high current negative ion beam up to MeV energy regime.

1. INTRODUCTION

Negative-ion-based neutral beam injector (N-NBI) is considered as one of the most promising candidates of heating and current drive systems for International Thermonuclear Experimental Reactor (ITER). In the present ITER engineering design activity (EDA), three injector modules are to be used in order to deliver 50 MW, 1 MeV neutral beam in the ITER plasma[1]. Each injector module mounts a single high power negative ion source/accelerator (beam source), which is required to produce a D$^-$ ion beam of 1 MeV, 40 A for > 1000 s. At JAERI, the design and R&D of the ion source/accelerator, as a key component of the ITER NBI system, have been carried out within the framework of the ITER EDA.

In this paper, the design of the ion source/accelerator for the ITER NBI is described, with unique features of neutronics and maintenance aspects. Also recent progress of high power ion source/accelerator R&D, including beam optics study of the MeV prototype accelerator, are reported.

2. DESIGN OF NEGATIVE ION SOURCE/ACCELERATOR

A cross sectional view of the ion source/accelerator designed for the ITER NBI is shown in Fig. 1. The ion source/accelerator is 3 m in diameter and 2.7 m in height, with a total weight of 20 tons.

The negative ion source is required to produce the D$^-$ ions of 40 A at the current density of > 20 mA/cm^2 under the low source filling pressure of < 0.3 Pa. To produce the high density D$^-$ ions efficiently even under the low pressure, small amount of cesium is seeded in the plasma generator to enhance the D$^-$ ion production via both volume and surface processes.

At present, it is believed that the negative ion yield is mainly enhanced via a following surface process:

D^0 -> (cesiated surface at low work function) -> D$^-$.

Hence the plasma generator has to be designed so as to produce high D^0(D$^+$) ratio source plasma. In the present design, the plasma generator has a large volume of 90 cm dia. by 178 cm long, semi-cylindrical (or "KAMABOKO" shape) discharge chamber surrounded with strong magnetic line cusps. The large volume and the magnetic arrangement enable good plasma/primary electron confinement, which results in high dissociation of deuterium molecules to produce D^0. The KAMABOKO shaped positive/negative ion sources developed at JAERI (including the JT-60U N-NBI source) have ever showed a high H$^+$ or D$^+$ ratio in the source plasma.

Produced D$^-$ ions are extracted with a two grid extractor, which has 1300 apertures of φ14 mm over an extraction area of 60 cm x 164 cm. The extracted ions are accelerated up to 1 MeV by an electrostatic accelerator of five-grids, multi-aperture type. The accelerator structure resembles the three-stage one of the JT-60U source, though two more stages are added for the higher energy acceleration. The five accelerator grids are called as A1G, A2G, A3G, A4G and GRG from the extractor to the ground potential. The H⁻ ions are accelerated by 200 keV in each stage, to yield a 1 MeV beam at the GRG exit. The gap lengths are progressively shorten in each downstream stage, so that the electric field is increased in the downstream. This results in the convergent lens to be formed in each aperture to counteract the space charge expansion of the beamlets. Each grid has the apertures of 16 mm in diameter. The accelerator grids are supported with a stack of five large bore alumina ceramics of 99 % high purity, each of which sustains 200 kV DC high voltage.

Figure 2 shows an example of the two dimendional simulation for the 1 MeV D⁻ ion acceleration. By the optimization of the gap length, the extraction voltage, and the current density, it was found that the smallest beamlet divergence of 2.8 mrad was obtained at Vext = 9 kV with the structure shown in the figure, where the perveance (defined as P_{opt} = $I_{D^-} / V_{acc}^{3/2}$ = 24 mA/cm² x 1.53cm² (hole area) / 1 MeV$^{3/2}$) was P_{opt} = 3.65 x 10⁻¹¹ A/V$^{3/2}$ for the D⁻.

To maximize the transmission efficiency in the subdivided beamline components, narrow NB duct and port, following beam steering/focusing techniques are integrated in the accelerator;
1) Geometric focusing: each grid of the extractor/accelerator is divided vertically into five segments, aiming at the magnetic axis of the ITER plasma at Rtan = 6.5 m.
2) Beamlet steering by aperture displacement: To achieve high transmission in the narrow vertical channels of the subdivided beamline components, each beamlet is steered and focused horizontaly by the aperture displacement at the GRG.
3) Electrostatic beam steering: At the accelerator exit, electrostatic beam steering assembly is equipped, which is made up of 5 pairs of parallel metal plates. By applying voltage up to 16 kV to a plate, while the other is connected to the ground, the beams are steered into a proper subdivided channels.

Neutronics environments in the NB cell, where the three injector modules were contained, have been calculated with computation codes of two dimensional neutron and γ ray transport. The neutron flux during the ITER operation is predicted to be < 10¹⁰ n/cm²s at the ion source/accelerator position. As the result, it has been confirmed that functional materials, such as permanent magnets on the source and the accelerator insulator, were durable even under the neutronics environment.

In the present design, only the ion source is fallen on the class 1 component, which requires regular maintenance every 6 months for the filament replacement, cesium cleaning and refilling. From the neutronics consideration, the radiation level in the NB cell is estimated to modest level (~ 10 μSv/hr)[4] after shutdown of ITER. One possible scheme of the maintenance is to prepare the remote handling devices such as manipulator by "hands-on" in the NB cell, outside the module (equipping the shield).Then open up the dome of the beam source to access back of the ion source with the manipulator operation. A "spark plug" type filament holder, which allows remote handling with the conventional manipulator is proposed for the replacement in this maintenance scheme.

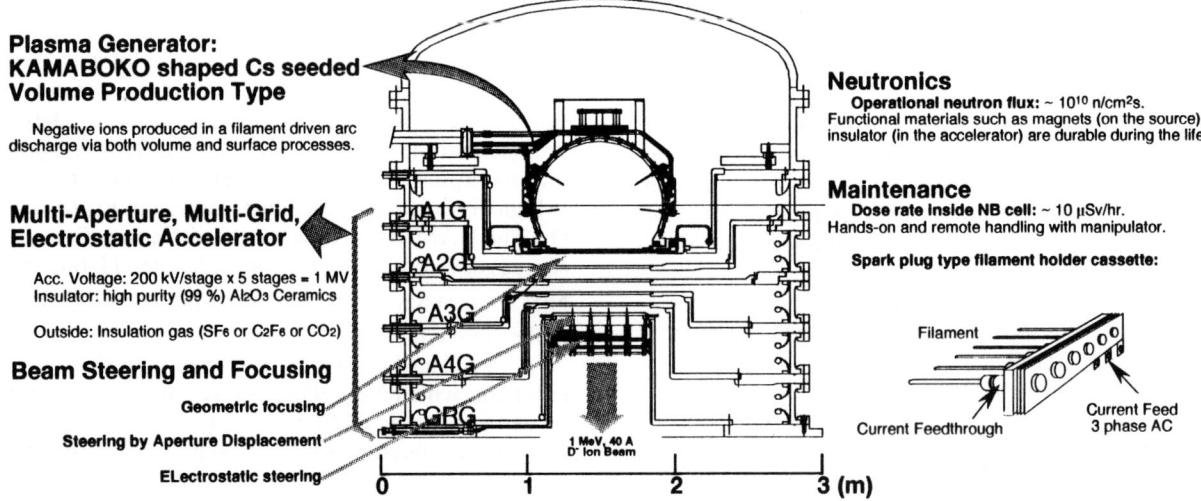

FIg. 1 A cross sectional view of the ion source/accelerator for ITER NBI.

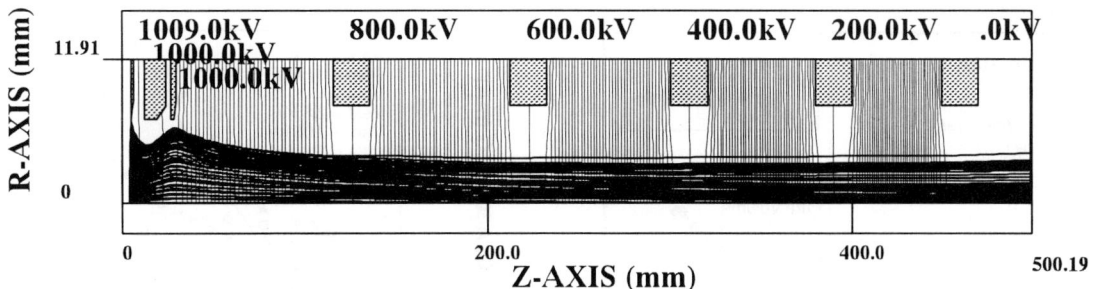

Fig. 2 A two dimendional simulation of the negative ion beam in the accelerator.

3. R&D OF HIGH POWER NEGATIVE ION SOURCE / ACCELERATOR

Two major R&D programs have been proceeded at JAERI to realize the ion source/accelerator for the ITER NBI. They are; 1) Development of high current negative deuterium ion source, and 2) R&D of high energy accelerator.

As for the high current source development, a large negative ion source for the JT-60U N-NBI has been manufactured and tested. Through the ion source test, a D$^-$ ion beam of 13.5 A, 5.4 MW (400 keV) has already been obtained. This is the world record of the high current and high power deuterium negative ion beam. The N-NBI system has been served to injection experiments of JT-60U[2] to explore the new regime of high density, high temperature plasmas with non-inductive current drive.

At present, an urgent and the most important issue of the R&D is a MeV class accelerator which is capable of high current, high energy negative ion beam production. However, any particle beams (positive/negative ions or electrons) of the current higher than 1 A have never been tried to accelerate above 1 MeV. At JAERI, "MeV Test Facility (MTF)" was constructed to develop 1 MeV, 1 A class accelerator technology.

The prototype accelerator being tested at MTF is composed of a plasma generator, an extractor and an accelerator, which are similar to those of the ion source/accelerator in the ITER NBI. The overall dimensions are 2 m in diameter and 1.9 m in height.

The plasma generator is a KAMABOKO shaped cesium seeded volume production type source, of which dimensions are 34 cm in diameter and 34 cm in length. The plasma generator has already demonstrated to produce the H$^-$ ion beam of 30 mA/cm^2 at the pressure of 0.2 Pa[5] and the D$^-$ ion beam of 20.5 mA/cm^2 at 0.35 Pa[6].

The extractor is a three-grids electrostatic one, each of which grid has 49 apertures drilled in an extraction area of 140 mm x 128 mm. In the present experiment, the H$^-$ ions were extracted only from 9 apertures (3 x 3 lattice pattern), the others were masked with a molybdenum plate on the plasma (first) grid.

The accelerator is the same as that of the ITER NBI except for the number of the apertures. The grids are supported with a stack of five voidless FRP (fiber reinforced plastic) insulator columns.

An H$^-$ ion beam of 805 keV, 0.15 A[3] (drain current of the power supply) has already been obtained with an ITER prototype accelerator at MTF. So far, It has been observed that voltage holding capability is not degradated 1) even after several months of air bent, 2) during beam acceleration, 3) with seeding cesium in the source for negative ion enhancement, all of which are encouraging for the MeV class accelerator development.

After the success in the high power H$^-$ ion beam acceleration, beam optics of the prototype accelerator has been studied in detail to optimize the operation window, where the high current beam can get through the multi-grid multi-aperture system. The results of the optimization are summarized in Fig. 3, showing the H$^-$ ion current density at the optimum condition as a function of the acceleration voltage.

In cesium seeded condition, the H$^-$ ion production was enhanced by a factor of two with respect to the pure volume operation, resulting in the arc power of a half gives the optimum. In both pure volume and Cesium seeded operations, the optimum H$^-$ ion current density was well correlated as: $JH^-_{opt} \sim P V^{3/2}$. Through the survey of the arc discharge power, the extraction and the acceleration voltages, the optimized voltage ratio and the perveance was found to be $V_{ext}/V_{acc} = 1.17 \times 10^{-2}$ and $P_{opt} = 3.18 \times 10^{-11}$ A/V$^{3/2}$ for H$^-$. A sharp and fine beamlets with the small divergence were observed in the perveance ranging $-12\% < P_{opt} < +15\%$.

Taking the difference of mass (D/H = 2) into account, the optimum perveance for D$^-$ ion acceleration is estimated to be $P_{D\,opt} = 2.25 \times 10^{-11}$ A/V$^{3/2}$, which agrees with the analytical result of Fig. 2 within 38 %.

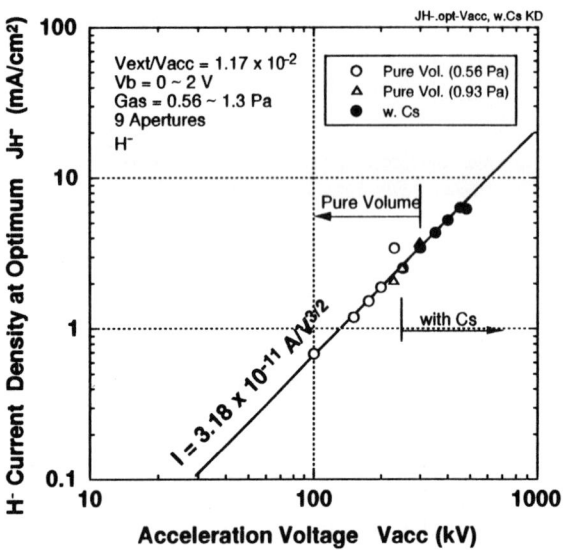

Fig. 3 Optimum current density and acceleration voltage.

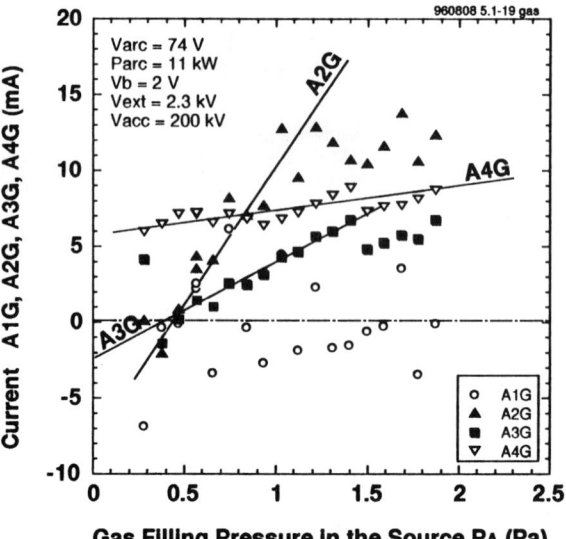

Fig. 4 Grid currents as functions of source pressure.

With increasing the negative ion current by increasing the arc power along the line as indicated in Fig. 3, the high current beam will be accelerated to higher enrgy of the MeV regime with the good optics.

In the optimization of the beam optics, the current flowed into each grid was monitored with shunt resistors inserted between the acceleration power supply and the accelerator. An interesting behavior of the grid currents were observed with varying the gas pressure in the source, i.e. the gas flow rate in the accelerator. Figure 4 shows an experimental result of the grid currents as functions of the pressure. With the pressure raised, the grid currents increased linearly. The slopes of each grid current as linear functions of the gas pressure are A2G > A3G > A4G. According to a brief analysis of the stripping loss of the ions based on a gas conductance calculation in the accelerator, higher fraction of the stripping loss takes place in the upstream gap, where the gas pressure is higher and the beam energy is lower. Thus the behavior of the grid currents dependence on the pressure agrees qualitatively with the analytical results. This might indicate that the stripped electrons were stopped at the nearest grid downstream. This is favorable for a high energy acceleration of the negative ions with high acceleration power efficiency.

4. SUMMARY

The engineering design of the ion source/ accelerator for the ITER NBI is progressing with considerations of neutronics or maintenance point of view, to fit within a constraints of the ITER. In the R&D of the high power negative ion source/ accelerator, high current beam is to be obtained with good optics by raising the energy at the optimum perveance to approach a major milestone of the development, i.e. 1 MeV, 1 A H⁻ ion beam acceleration.

ACKNOWLEDGMENT

The authors would like to thank S. Shimamoto, M. Ohta and T. Nagashima of JAERI Naka for their support and encouragement. They are also grateful to P. L. Mondino, R. S. Hemsworth and ITER Joint Central Team at Naka for their guidance and discussion.

REFERENCES

1 A. Krylov et al., "General Design of the Neutral Beam Injection System and Integration with ITER", in this proceeding.
2 M. Kuriyame et al., " Initial beam operation of 500 keV negative-ion-based NBI system for JT-60U, in this proceeding.
3 K. Watanabe et al., Proc. Int. Symp. on Advanced Nucl. Energy Res., Takasaki (1996).
4 K. Shibata et al., "Design analysis for reducing dose rate in the NBI to realize direct access by workers for a fusion experimental reactor", in this proceeding.
5 Y. Okumura et al., Rev. Sci. Instrum. 67/3, 1092 (1996).
6 C. Jaquot et al., "Negative D⁻ ion source relevant for application to ITER neutral beam injectors", in this proceeding.

Experimental results using the JET Real Time Power Control system.

NH Zornig, HEO Brelen, A Browne, ML Browne, C Gormezano, T Dobbing, JA How, FA Jensen, TTC Jones, FB Marcus, QA King, FG Rimini, JA Romero, AGH Sibley, F Söldner, BJD Tubbing

1. INTRODUCTION.

The control capability of the three JET plasma heating systems (Neutral Beams (NB), Ion Cyclotron Resonance Heating (RF) and Lower Hybrid (LH)) has been significantly enhanced to provide a very flexible alternative to conventional pre-programmed operation. The power can now be varied in real time by a predefined network of software based on algorithmic elements. Using this control system a number of experiments have been carried out in the previous and current JET-campaign, e.g. control of diamagnetic energy (W_{dia},) and ITER scaling experiments. The design of these experiments have been achieved by using a 0-D-plasma model. The model and the experiments will be discussed in this paper as well as the implementation of the new **R**eal **T**ime **P**ower **C**ontrol system (RTPC).

2. THE SIMULATION CODE.

The code used for optimising the controller parameters has been kept simple in order to facilitate quick simulations, with not too many inputs, which can be easily performed on a PC.

2.1. Assumptions and laws used.

The code is a global scaling using the ITER89P scaling law [1] for the energy confinement time. The H-factor (i.e. confinement enhancement) is chosen from experimental data. The density and temperatures profiles are assumed to be:

$$T = T_0(1 - (r/a)^2)^{\gamma_T}$$

where γ_T is the shape parameter. The relative ion and electron heat transport coefficients are taken as inversely proportional to the square root of the temperatures, [2]. Heat exchange between the ions and electrons is purely by a first order collision process [3]. Slowing down of the fast RF minority ions and fast beam ions is also by a first order process following Stix [4]. By using suitable volume averaged quantities, the electron and ion heating rates and loss powers are computed whilst satisfying the global confinement constraint. Fast ion contents are also calculated. Z_{eff} is assured to be inversely proportional to the average electron density. Sawteeth, ELMs and other terminating effects have not been included. The thermal and non-thermal contributions to the DD-reaction rate (R_{DD}) are given by [5] and [6].

2.2. Inputs and outputs.

The H-factor, the NB- and RF-power waveforms and the Ohmic heating waveform are all inputs. The latter is mainly used, together with the profile factors, to tune the initial values of the plasma parameters. Other inputs to the model are the plasma current, I_P, the toroidal field, B_P, the electron density, n_e, and the plasma geometry values. The main outputs are the central ion and electron temperatures, W_{dia} and R_{DD}.

2.3. Results from simulations.

By optimising the code for a reference pulse, Fig 1, the code can be used in a predictive way for similar plasmas, e.g. to optimise values of a PI-controller for closed loop feedback experiments. These values have been used successfully in the experiments.

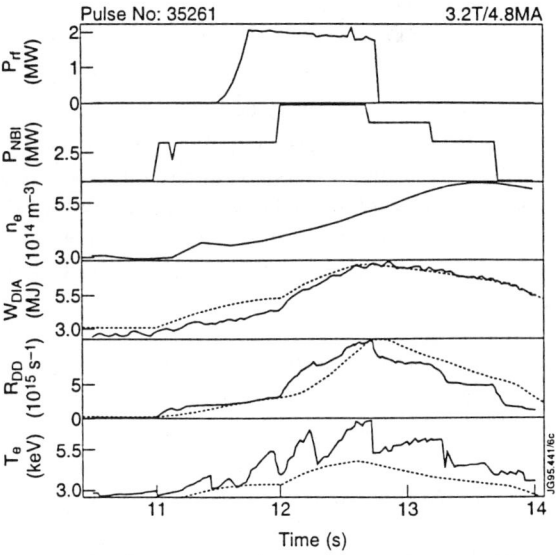

Figure 1. Comparison between simulations (dotted) and a reference pulse (solid) for RF + NB heating.

3. THE REAL TIME POWER CONTROL SYSTEM (RTPC).

The RTPC is a part of the elaborate real time control and protection network at JET. RTPC has three main systems of interest, all communicating with the Real Time General Server (RTGS) and implemented in a VME based system communicating via ethernet, see Fig. 2. The plasma data, e.g. plasma energy, input power and neutron rate, is processed by the Real Time Signal Server (RTSS). This data is received by the Real Time Central Controller (RTCC) where networks can be created resulting in power requests. These requests are being fed into so called Local Managers (LM), which have the task to ensure that the requested power is delivered to the plasma.

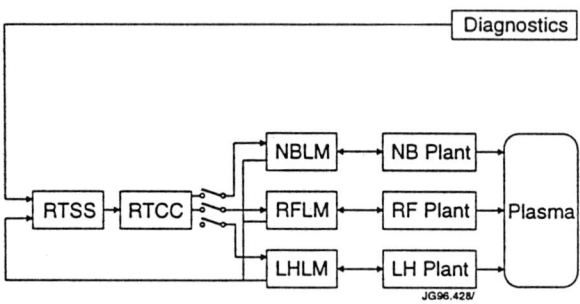

Fig. 2. Overview of the JET RTPC

3.1. The Real Time Central Controller.
The RTCC supports a user defined network. The output from the network can be a combination of different signals, e.g. density and power. In this paper we only discuss the power request. It is possible to pass more than one reference to the same heating system. The RTCC maybe programmed with predefined algorithms, from which all conceivable networks can be created. Examples of predefined algorithms are:

- PID-controller, with optional bumpless transfer and anti windup of the integral term.
- Waveform generator
- Limiter
- Transfer functions.

3.2. The Local Managers (LM).
The three local managers of interest for this paper are the ones for RF, LH and NB. The RFLM and NBLM have local feedback, ensuring that the requested power is delivered. For RF this is achieved by using the measured coupling resistance to calculate the losses in the transmission and adjusting the generator power with a PI-controller. The loss in power by tripped modules is compensated by increasing the power of the others equally. The NBLM has two ways of supplying the requested power (Fig.3). The first one, feedback-off, will ensure that the delivered and requested power match as well as possible, consistent with the discrete power steps of the beams. The other method, feedback-on, minimises the time averaged error between the delivered and requested power, using pulse width modulation. In both modes, there is automatic compensation for interruptions due to HV breakdowns and terminating trips. The LM can also operate without being connected to the RTCC by supplying a local waveform. RFLM has become an integral part of the RF heating plant control system, used for all normal operation. NBLM is an optional feature, used on the majority of injection pulses.

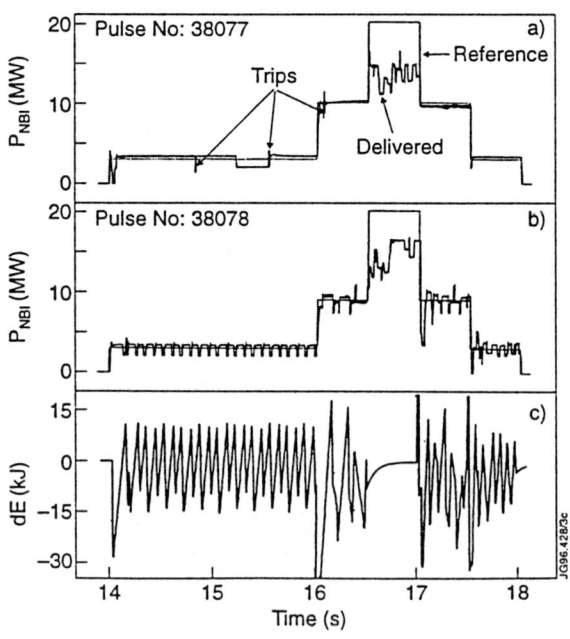

Fig.3. The NB-power following its request. (a) Feedback off. (b) Feedback on. (c) The energy-deficit trace associated with the feedback on mode.

4. EXPERIMENTS WITH RTPC.

The system described above has been used in multiple ways for controlling variables in JET experiments. In the following most of these will be described.

4.1. Neutron economy.
Neutron production in JET is strictly controlled, and it is desirable to limit the reactivity when it becomes clear that a shot is not achieving its goals. This has been successfully used by tripping the beams to a desired level, when they do not deliver the requested power for a specified amount of time or when the neutron rate rise was not as desired. These schemes saved up to 50% of the neutrons which would have been produced if the beams had continued.

4.2. Feedback control of plasma variables.
With the new system it has also been possible to keep the W_{dia}, R_{dd}, the normalised poloidal pressure (β_p) and current profile at a reference level by using the power as an actuator. The values of the controlled variables are received by the RTSS and distributed to the RTCC, where it is subtracted from a reference value. The resulting error is fed into a PI-controller which will calculate a power request to minimise the error between the reference signal and the actual value. The power requests is delivered to the LM. Using this system, the following variables have been controlled:
- W_{dia}, R_{dd} and β_p by RF
- W_{dia} and R_{dd} by NB
- Loop voltage (V_{loop}) by LH

An example of β_p-control by RF is given in Fig 4, of W_{dia} by NB in Fig 5, both kept constant for several energy confinement times, and of V_{loop} by LH in Fig 6. It is shown in Fig 4 that the controller performs well, keeping β_p constant during the density rise. The reason the for the initial high power phase in Fig 5, is to create a fast rise in W_{dia}. In the early part of the feedback phase the controller responds to the initial overshoot in W_{dia} during a hot-ion ELM-free H-mode, and successfully compensates for the fall in confinement after the giant ELM. In Fig 6 it is seen that the control of V_{loop} is not yet optimal, but that V_{loop} can be controlled by LH. The controller succeeds in reducing V_{loop} by 66%, but some oscillation around the set-point is also seen.

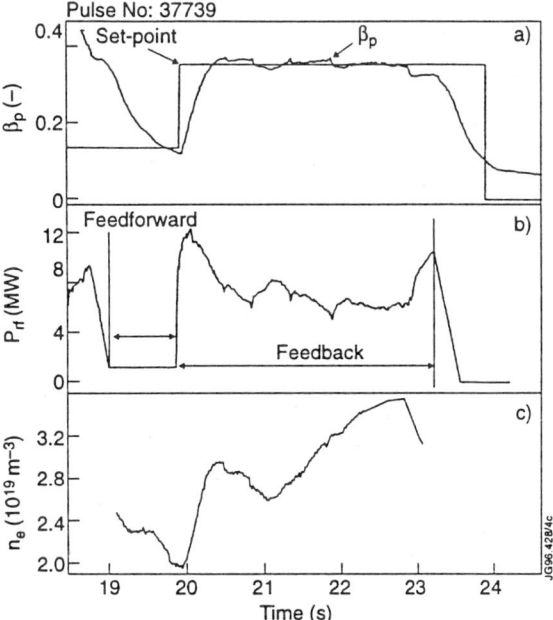

Fig 4. Control of β_p by RF. Local P_{RF} feedback is not used in this case.(a) β_p. (b) Power. (c) Density.

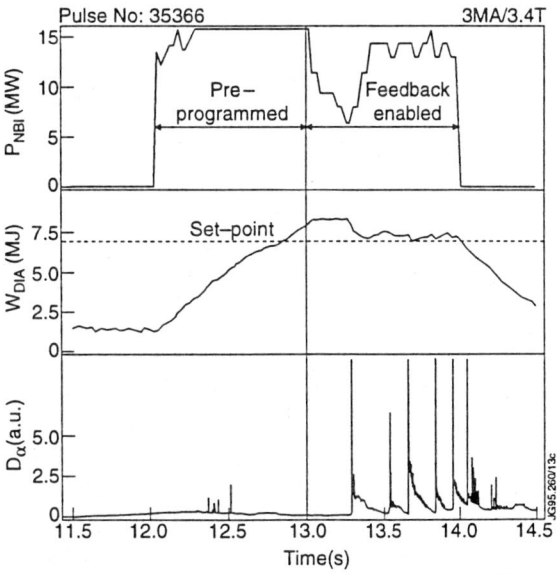

Fig 5. Control of W_{dia} by NB. Local P_{NBI} feedback is not used in this case (a) Power. (b) W_{dia} (c) D_α.

4.3. Simulation experiments of a burning plasma.
As part of studies of ITER-simulation experiments in JET, where an ITER-type plasma is scaled down to a JET-plasma, an initial step is taken in demonstrating a 'thermal runaway' effect in a plasma with significant simulated self heating, see figure 7a. The avoidance

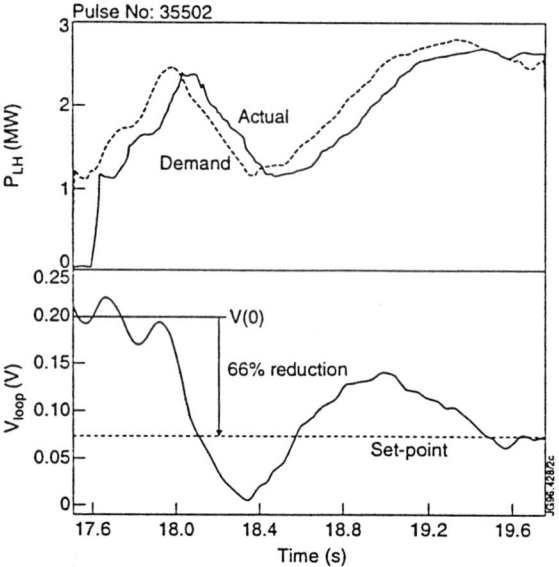

Fig. 6: Control of V_{loop} by LH. (a) Power. (b) V_{loop}.

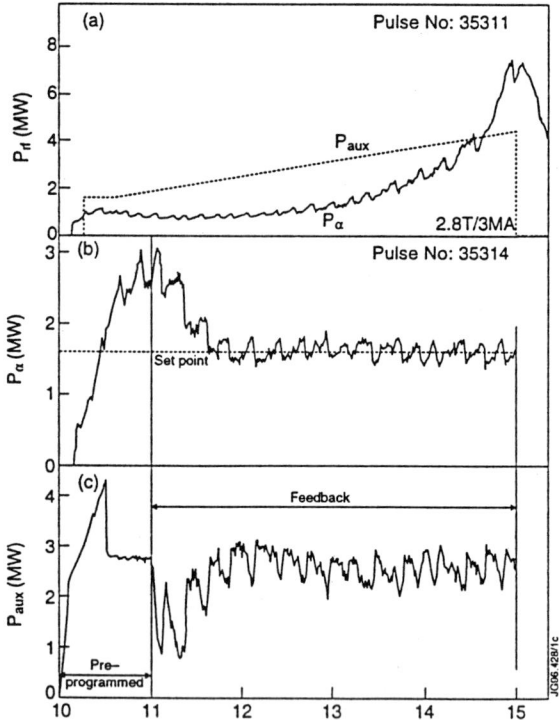

Fig. 7. (a) A 'thermal runaway'. (b/c) Its avoidance.

of this thermal runaway has also been demonstrated, (Fig. 7b/c). In the scheme the 'alpha power', P_α, is made proportional to R_{dd}; this power is given by a part of the RF-plant. An independent RF contribution is used as auxiliary power.

5. OTHER FUTURE SCHEMES USING RTPC.

There are a number of other schemes designed to exploit RTPC. The first one is a scaled down ITER-simulation experiment, in which dimensionless parameters have been preserved, i.e. β_N, q, heat exchange time divided by energy confinement time and the fusion gain factor Q. This results in values for scaling-factors, S, of the density, temperature and reaction rate coefficient, $<\sigma v>$ as a function of the power scaling factor, which depends on the auxiliary power available at ITER and the RF-power available at JET. The alpha power is then calculated using a formula using these scaling factors. The rest of the scheme will be as in section 4.3.

In the second scheme the off-axis current distribution, using a dimensionless quantity derived from the 2nd current moment [7], will be controlled by means of either NB, LH or transformer action. This has been shown to be possible in simulations.

6. CONCLUSIONS AND REMARKS.

The RTPC operates as designed. By using the RFLM and NBLM, to compensate for plant trips, reliability is significantly improved. Both are used extensively in the JET program. Also, closed loop (LM-plasma-RTSS-RTCC-LM) control experiments of W_{dia}, R_{dd}, β_p and loop voltage have proven successful, together with neutron saving. Thirdly, simulated thermal runaway and its avoidance has been demonstrated. A scheme for a more realistic ITER-simulation experiment has been briefly discussed. Finally, an outline experiment to control the off axis current distribution has been given. All feedback experiments have been performed successfully, using values for the PI-controller parameters derived by the 0-D-model described.

REFERENCES:

[1] PN Yushmanov et al., Nuc. Fus., 30, 10 (1990)
[2] D Boucher et al., IAEA TCM, Montreal, 1992
[3] J Wesson, Tokamak, Clarendon Press, 1987
[4] TH Stix, Plasma Physics, 14 (1972)
[5] A Peres, J. Appl. Physics, 50, 9 (1979)
[6] FB Marcus et al, Nuc. Fus., 33, 9 (1993)
[7] VD Shavranov, Plasma Physics., 13 (1971)

E.C.R.H. System for TJ-II Experiment

R. Martin[a], K. Likin[a], A. Fernández[a], M. Sorolla[b], A. Sánchez[b], C. del Río Bocio[b], N. Matveev[c]

[a] Asociación EURATOM-CIEMAT para Fusión
Avda. Complutense 22, 28040 Madrid, Spain

[b] Ingeniería de Telecomunicación, Universidad Pública de Navarra
Campo Arrosadía, 31006 Pamplona, Spain

[c] High Voltage Research Centre of the All-Russian Electrotechnical Institute
Istra, Moscow region, Russia

The special microwave power modulation system PMS was designed to investigate the transport anomalous and low frequency drift turbulence in plasmas. This system allows to modulate output voltage respect to the stable mean level in the range of 30-70 kV with amplitude up to 5 kV. The bandwidth on the 3 dB level is 40 kHz and the frequency response is flat within 0.5 dB over the 30 kHz. That is sufficient to produce the 100% modulation in microwave power according to the gyrotron generation characteristics. Adaptation of this system to TJ-II is under study. For TJ-II ECRH systems we follow the quasi optical tranmission lines way.

The first transmission line has been designed to launch the microwave beam into plasmas with a linear polarization. The design of mirrors which drive the microwaves by means of the free space fundamental mode is based on the method of an iterative beam transformation with proper correction of wave fronts on a mirror. The quasi-optical transmission lines have the following features: simplicity, low ohmic and diffraction losses, high-power capability, inherent mode filtering, simpler bend construction, simpler polarises

1. INTRODUCTION

In addition to the main ECR heating effect there are other physical problems that can be investigated by means of modulation of microwave power. In particular, it is supposed that the drift turbulence, that is observed in the fusion plasmas and may be one of the reason of the anomalous losses, can be supressed by the low frequency plasma waves artificially or feedback excited [1,2]. One way to produce this effect is in the direct changing of the generated microwave power when the electron cyclotron resonance heating is used. According to this considerations a special device to modulate the microwave power in Spanish stellarator TJ-IU was designed and tested. This experiment was done to demonstrate the modulation principle in double electrodes gyrotron microwave system and can be useful for future TJ-II experimental programme also.

2. POWER MODULATION SYSTEM

The ECRH system for TJ-IU consists of a high voltage pulse power supply (PPS), a gyrotron with super-conductive magnet, and a quasi-optical transmission line [3].

The high voltage electron beam valve EBV 50/100 [4] is used in the power supply as a high speed operating switch and regulator at the same time. This tube is remarkable for high rates of hold-off voltage up to 160 KV and plate power dissipation of 800 KW, that provides high transient overload resistence. Due to the flat pentode´s type VA characteristics of this EBV, it is possible to limit the overcurrent under breakdown conditions near the operating level. The switching on/off time of the tube may be less than one microsec, but to reduce the overvoltage during switching the current off this time is chosen of about 10...20 microsecs. It is sufficient to limit the energy dissipation in the gyrotron when breakdown of less than 1 J. In case the gyrotron and regulating tube breakdown occur simultaneously the crowbar is triggering less than in a few microsecs and short circuit the capacitor bank. This multilevel protection allows to prevent the gyrotron damage in breakdown occasions.

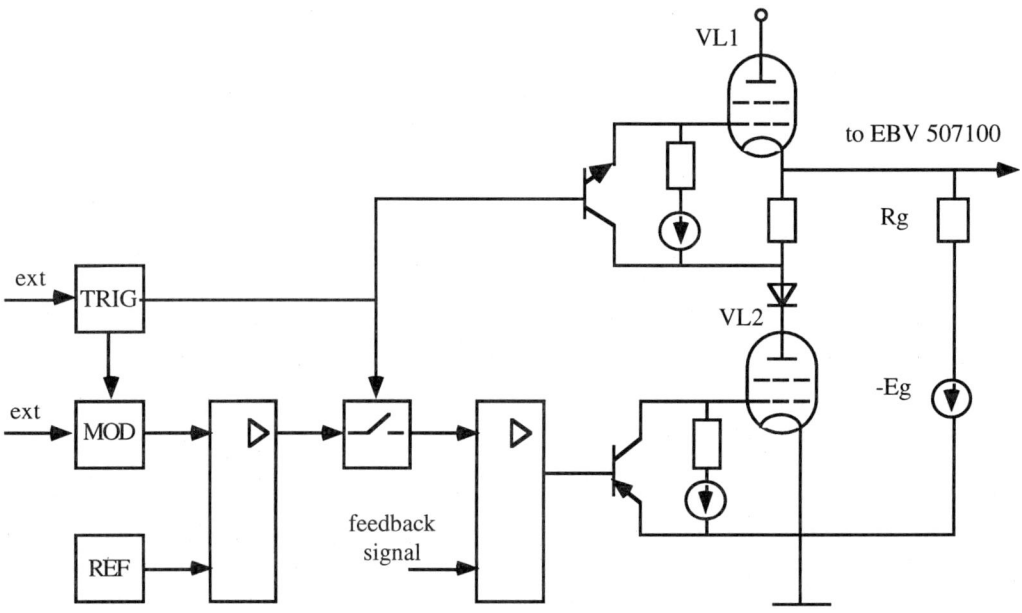

Fig, 1. Pulse Modulation System simplified diagram

The high frequency modulation of the output microwave power in the system based on the double electrodes gyrotron can be provided only by the high voltage control. According to the gyrotron generation unwanted oscillations in the closed regulating system with high factor of the output voltage stabilization at the same time.

The simplified diagram of the electron beam valve driver for power modulation system is presented in Fig. 1. The operation principle is a reference voltage changing by the triggered internal or external modulating signal. Triggering generator TRIG forms the signals to directly start-up the PPS. The special function generator MOD triggering can be delayed for the desirable time. The modulation signal being added with reference set REF pass through the electron switch and feedback amplifier frequency corrected to a booster builded around a two electron tubes VL1,2 to provide the control voltage to EBV. Its value of -1.5 KV to cut off the tube and up to 6 KV under 50 A current. The maximum switching frequency is provided by dividing of the switching and feedback channels of the driver. This circuit allows to get a smooth rising of the gyrotron voltage when it switching on, a high gain in respectively wide bandwidth during the pulse and a high speed of the switching off. To avoid an interference the optical fiber links are widely used.

3. TRANSMISSION LINE DESIGN

The design of quasioptical transmission line where millimeter waves are transmitted as gaussian beams (by means of the TEM_{00} free space fundamental mode) is a well established topic since 1960, but the applications in ECRH are recent.

Gaussian Beams are nice solutions of the paraxial wave equation which approaches very well the output beam of the gyrotron tube. The main propagation parameters that describe the Gaussian Beam are:

- the beam waist w_0 that describes the minimum size of the beam.
- the beam size expansion as the beam propagates.
- the beam curvature radius law as the beam propagates.
- the far divergence angle.

All these parameters can be easily evaluated by means of the classical expressions given in [4,5,6]. This beam gives a good beam focusing and also presents excellent polarization properties. But as the beam propagates, the beam size increases and this has to be avoided in order to reach the inner vessel of the fusion machine with a narrow beam.

The solution consists into transform the properties of the beam by using properly designed focusing mirrors which operate like optical lenses on the beam as it is well known.

The basis of this method is an iterative beam transformation with properly designed mirrors as it is discussed in several papers; see for instance [5,6].

The peak power that can be transported into a metallic mirror beam waveguide is higher than in circular waveguides as it can be deduced by means of simple field theory.

The power density on the mirrors surface is clearly lower than into the beamwaist and this alleviates thermal problems in these components.

Moreover, the quasioptical beamwaveguides present another important property due to the fact of inherent mode filtering capability. Diffraction is the reason of this fact and then, higher order modes are damped strongly.

Waveguide bends are also very easy to realize due to the natural property of the mirror beam waveguide. This helps a lot in order to avoid the obstacles placed into the experiment room. If one thinks about how complicated the circular waveguidebends are, the movement towards the quasioptical solution is clearly explained.

Because its properties, this transmission lines are broadband as it is stated into references [5,6]. The realization of polarizers, powercombiners and dividers is simpler in the quasioptical approach then into the circular waveguides and plays an important role into the design and proper adjustment of the heating system.

Losses have to be taken into account into the transmission line design. From the quasioptical approach, then can be separated into the ohmic contribution, the diffraction losses, atmospheric absorption and modal mismatch at the window.

The diffraction losses come due to the fact of finite size of the mirrors and this can produce that some beam power would be launched out of the waveguide. By proper design of the mirror sizes, near 100% of the power can be focused maintaining the mirrors at reasonable sizes.

At the experiment window, the Gaussian Beam has to be coupled to the modes of circular waveguide, this can produce some mismatch of the power that contributes about 2% to the total losses as proposed in [5,6].

In our case we will take advantage of the use of such quasioptical transmission lines due to the fact of its simplicity, low ohmic and diffraction losses, power capability, inherent mode filtering, simpler bends, simpler polarizers and other interesting properties as it is discussed in [6].

The number of mirrors is another critical aspect of the waveguide. We try to minimize this number in order to reduce losses, alignment problems, cross polarization, mechanical complication and so on.

The proposed solution (see [7]) for the first TJ-II quasioptical transmission line uses 6 cylindrical mirrors coupled in pairs in order to simplify the mechanical work but to obtain the same behaviour as a single double curvature mirror it is compulsive to introduce a second cylindrical mirror (distance in couple of mirrors 400 mm) from the gyrotron side and 2 double curvature mirrors to cover the extra path to the experiment window.

The beamwaist from the output of the gyrotron is 18 mm in X and Y planes and after reflections on the focusing mirrors system we reach a value of 32 mm (for the beamwaist) at the plasma center. The calculations are performed by using classical gaussian optics and taking into accout the proper incidence angle for each mirror.

The results for the first quasioptical transmission line are summarized in Table I.

Mirror	X-Plane Curvature	Y-Plane Curvature	Size in X Plane	Size in Y Plane
1	------	803	326	266
2	------	1387	219	375
3	3443	------	408	162
4	------	1357	181	213
5	------	2132	362	257
6	2813	------	434	230
7	1385	1778	279	449
8	2376	1669	473	317

Table I. Radius of curvature and mirror size in X and Y planes. All the values are given in mm.

REFERENCES

1. C. Hidalgo et al. P.R.L., V 71, p. 3127, 1993.
2. K. Likin et al. Mirowave diagnostic for study of drift turbulence in TJ-IU plasmas. 10th Int. Conf. on Stellarator. CIEMAT, Madrid.
3. R. Martin. The microwave system for ECRH experiments on TJ-IU torsatron. 18th SOFT, Vol. 1, 529.
4. Electron-beam tube Patent, Germany No. 2012681.
5. M. Thumm. Advanced electron cyclotron heating system for next-step fusion experiment. Fusion Engineering and Design 26 (1995), 29-317.
6. M. Thumm. "High-power micronarc transmission systems, external mode converter and antenna technology" C. Edycombe (eds) Gyrotrons Oscillators. Taylor and Francis, London 1993
7. M. Sorolla, R. Martin et al.. "Beam Waveguide for ECRH at TJ-II Experiment " Proc. of Int. Conf. on Infrarred and millimeter waves. December 1990.

Sliding contact tests at high R.F. current under vacuum

G. Agarici, B. Beaumont, L. Ladurelle, G. Lombard, P. Mollard, H. Kuus

Association EURATOM-CEA, Département de Recherches sur la Fusion Contrôlée
Centre de Cadarache, 13108 Saint Paul lez Durance Cedex, France

On each of the Tore Supra ICRH antennas [1], 4 variable vacuum capacitors supplied by COMET* are used to match the antenna feed point impedance to the generator output impedance. On such a system, the whole transmission line is working near matched conditions, and high RF voltages are limited to the front part of the antenna. The quality, performance and reliability of the COMET capacitors, fixed or variable type, have been proved by thousands of units in operation in high power oscillators and amplifier circuits. The variable type (10 to 150 pF) fitted in Tore Supra antennas has allowed remarkable results. A record radiated power density of 1.6 kW/cm^2, has been achieved, and RF peak voltages in excess of 40 kV are commonly used during plasma shots. However, by construction of the capacitor, a water cooled bellows on the RF current path limits the velocity of the capacitance variation.

Reliable RF experiments during transients phenomena like pellet injection, monsters sawteeth or plasma displacements require the matching system to follow rapid variations of loads. Therefore, a new concept of variable vacuum capacitor allowing faster capacitance adjustment, is studied in collaboration with COMET.

In this concept, the RF current flows through a sliding contact strip from PANTECHNIK+ working in the sealed vacuum. The RF current path is therefore dissociated from the vacuum barrier.

Sliding contacts, both in vacuum conditions and handling high RF current are not in common use in the RF field and have to be tested. This paper describes the existing variable capacitor, the new capacitor concept, the RF contact test device as well as the experimental results.

1. VARIABLE VACUUM CAPACITOR

1.1 Existing variable vacuum capacitors

They are composed of two sets of thin concentric cylinders made of oxygen free copper : one is the fixed electrode while the other one, mounted on a shaft, moves for the capacitance adjustment (see figure 1).

On this view, one can see the vacuum sealed envelope which is formed by a vacuumtight ceramic brazed between the fixed and the variable side of the capacitor. The flexible hard copper alloy bellows allows to change the capacitance when distance between electrodes varies. This bellows is also the electrical connection between the adjustable electrode and the body of the capacitor

As in most applications, these capacitors are working in a radio frequency domain, (between 30 and 80 Mhz on Tore Supra), where the skin effect confines the current flow to the surface layers of the conductors. Therefore, on the capacitor, the RF current flows from the connecting flange to the edge of the ceramic, and passes inside the capacitor to the electrodes.

The variable electrode is in the TS antennas on the grounded side of the capacitor, so that the maximum current runs on the outer surface of the bellows. At the maximum operational conditions of the ICRH antennas, capacitors have to handle a current of 1000 Arms and a voltage of 45 kV, both values are near the maximum ratings. The power deposited by skin effect on the outer surface of the

* COMET :   Comet Technik A.G., Berne, Switzerland
+ PANTECHNIK :   Pantechnik S.A., Caen, France

bellows is about 5 kW calculated at 60 MHz frequency. Cooling of the bellows is mandatory for such a power level and is performed by a forced cooling circuit located inside the bellows. The thickness of the bellows is 0.2 mm, so that this critical element supports high velocity water flow on its internal face, and high RF current in high vacuum environment on its outer surface. In addition, this conception is prone to long term corrosion through the bellows

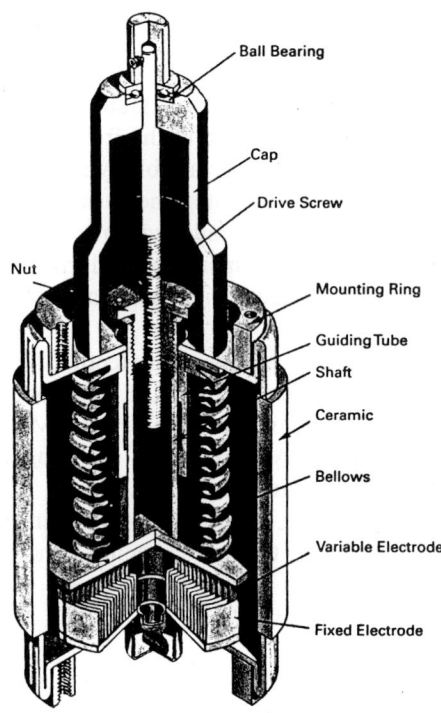

Figure 1. Cut Away View of a COMET Variable Vacuum Capacitor

Flow restrictions are needed in the cooling circuit to obtain high water velocity and therefore good heat exchange. On another hand, fast movements of the electrode leads to pressure peaks that can accelerate the fatigue of the material. These peaks are difficult to be precisely evaluated depending on a lot of parameters as: the static pressure, the water flow, the bellows stroke and the exact geometry.

Therefore, to insure long life of the bellows, the maximum acceleration and velocity are presently limited. On Tore Supra antennas, where the capacitors are working near the maximum operating conditions, we have limited the velocity at 3 cm/s although future experiments [2] would require rapid variations of the capacitance corresponding at velocity between 1.5 and 2 m/s.

1. 2 New variable vacuum capacitor concept

This new concept concerns only the modification of the variable electrode assembly on the existing COMET capacitor. In this design (figure 2) the RF current path always flows from the outside of the capacitor electrode to the inside of the concentric cylinders but now it runs through sliding contacts working under vacuum.

The movement of the shaft and the vacuum-tightness are insured by a bellows now located outside the RF current path and made in edge-welded stainless-steel. This element does need any cooling. The axial movement under vacuum is guided by a nitrided stainless-steel shaft sliding in a Cupro-Aluminium alloy bearing. Friction tests performed under vacuum at 2.10^{-4} Pascal and at temperature of 250°C have been carried out at Cadarache on different material couples. The above couple has allowed to make 50 000 cycles without problem. The water cooling circuit is now built within the shaft, and the internal volume is independent of the electrode movements. Hydraulic parameters can then be adjusted for optimal cooling and are not affected by displacement or acceleration

The contact strip from PANTECHNIK is constituted of small cylindrical fingers in Silver/Carbon compound clinched on a silver platted Copper-Beryllium foil. In our application the contact band, using 59 fingers, is rolled up and fitted by small screws on the moving shaft. Each finger ensures the electrical contact to the fixed part of the variable electrode, where the fingers are in touch, by applying a force between 2.3 and 2.9 Newton.

The contact strip is calculated to allow high density current up to 65 A per cm of length for a frequency range between 30 to 80 MHz. At 60 MHz for a current of 1000A rms., the calculated loss on each finger is of 7.3 W : 2 W for the contact resistance and of 5.3 W for the surface resistance. Total losses in the contact is then around 430 W

If we also take into account the power deposition by skin effect on the outer surface of the shaft and on the fixed part of the electrode, both made in oxygen free high conductivity copper, we find a total power

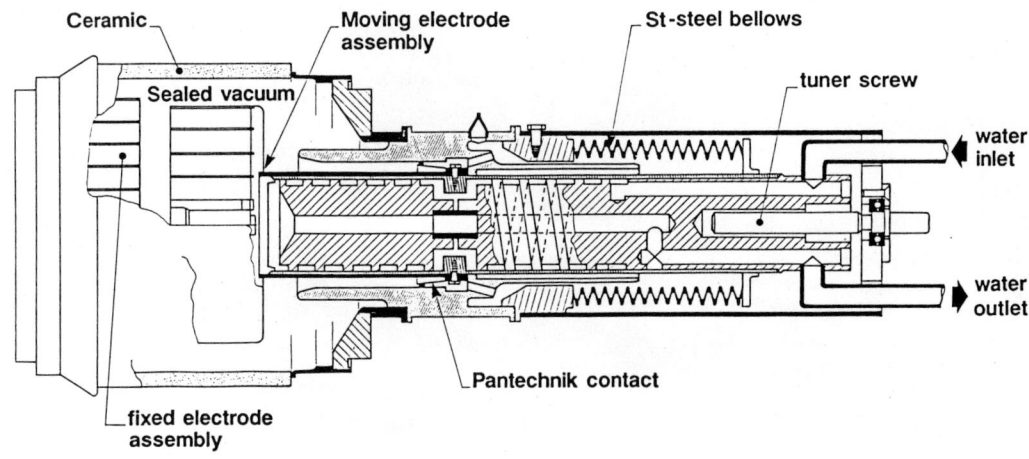

Figure 2. Cut View of the New Variable Vacuum Capacitor Concept

dissipation of 980 W. This total dissipation has to be compare with the 5 kW found above only on the bellows of the existing capacitor. In order to develop this concept, intensive tests on the use of sliding contacts, both in vacuum conditions and conveying high RF current have been carried out at Cadarache.

2. RF SLIDING CONTACT TESTS

2. 1 Tests device.

It is, by the dimensions and the material of its components fully representative of the variable electrode assembly in the new capacitor concept.
It consists (figure 3) of a 30 Ω short-circuited vacuum coaxial line. It is connected to a 400 kW generator, working at 60 MHz, through a 6 inches coaxial line. The matching system consists of 2 variable stubs placed in parallel along the transmission line. A directional coupler incorporated between the stubs and a vacuum window gives the RF current value. The PANTECHNIK RF contact strip is fitted on the front and on the rear part of the sliding shaft in order to keep the electrical length constant during the movement. The water cooled shaft is actuated by an electric motor on 50 mm stroke. The vacuum window is one of those used in the Tore Supra antennas. A stainless steel bellows, outside the RF cavity, allows the movement of the central conductor and a turbomolecular pump evacuates the vacuum line.

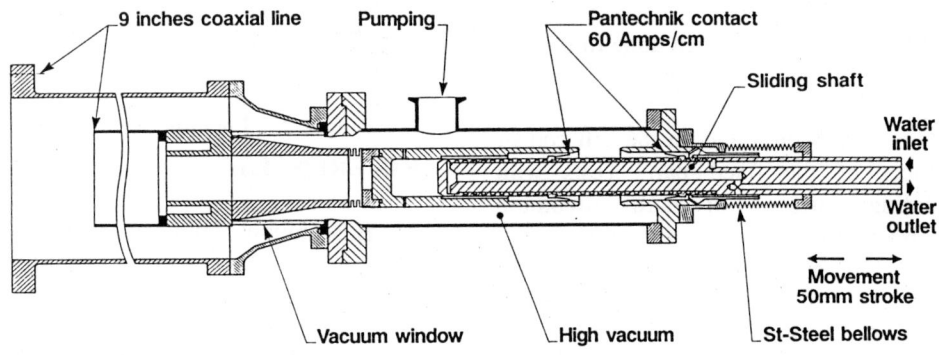

Figure 3. RF Tests Device Cut View

Figure 4. Moving Short Circuit assembly of the Tests Device

A thermocouple situated near the surface in friction with the contact fingers gives the temperature information. Figure 4 shows the moving short-circuit assembly of the tests device with the PANTECHNIK contact fitted on the shaft.

2. 2 Tests and results.

At the time of writing, the full tests allowing to completely qualify the contact are not finished but the first results are very encouraging:

- RF current of 1000 A rms. at 60 MHz has been passed during 60 seconds on the system with the contact sliding under vacuum at velocity of 3 cm/s without any RF problem (figure 5).
- More than 15 minutes of satisfying RF pulses between 30 and 60 seconds with a current ranging from 650 to 1000 A rms. have been done in the test device with the shaft moving at 3 cm/s.

Slight damages have been found on the soft copper sliding surfaces of the fixed part. Tests with different sliding surfaces in hard zirconium chromium copper alloy are in preparation.
Before manufacture of a prototype capacitor, RF tests without movement are also planned as well as the long term qualification of the system by a significant number of RF pulses at 1000 A rms. with and without movement. COMET is also studying for the new capacitor the possibility to water cool the fixed part of the variable electrode assembly for long pulses compatibility.
- For possible future applications, RF tests with the contacts sliding at high velocity, up to 2 m/s are also programmed.

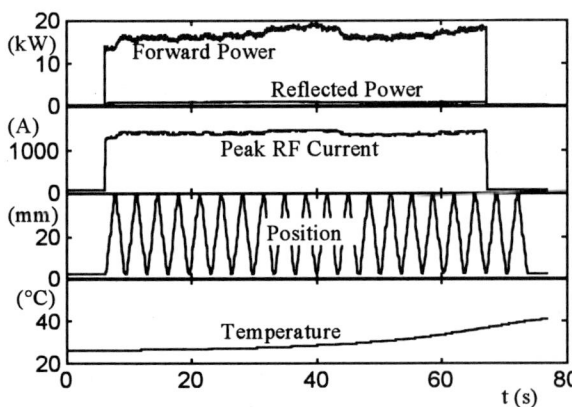

Figure 5. Vacuum Sliding RF contacts tests

REFERENCES

1. B.Beaumont and al., "Tore Supra ICRH Antennae Array", proc. 15th SOFT, Utrecht, September 1992, V1, p503
2. L. Ladurelle and al., "First Results of Automatic Matching System on Tore Supra ICRH Antennas; Fast Matching Network for ICRH Systems", this conference

Section C

Plasma Engineering and Control

Loads on the ITER In-Vessel Components from Electromagnetic Transients

S. Chiocchio[1], K. Ioki[1], M. Araki[2], P. Barabaschi[3], J.B. Bialek[4], V. Kokotkov[5], M. Roccella[6], R.S. Sayer[7], J. Wesley[3], D. Williamson[1]

[1] ITER Joint Central Team, D-85748 Garching bei Muenchen, Germany
[2] Japan Atomic Energy Res. Inst., 801-1, Naka-machi, Naka-gun, Ibaraki-ken 311-01 Japan,
[3] ITER Joint Central Team, 11025 N. Torrey Pines Rd, LaJolla, CA 92037, U.S.A.
[4] Princeton University Plasma Physics Laboratory, P.O. Box 451, Princeton, NJ 08543, USA
[5] D.V. Efremov Research Institute, St. Petersburg, Russia
[6] ENEA, C.R.E Frascati, 00044 Frascati, Italy
[7] Martin Marietta Energy Systems, P. O. Box 2009, Oak Ridge, TN 37931-9068, USA

During the current phase of the design of the ITER plasma facing components, a large effort has been dedicated to the definition of a comprehensive set of worst-case plasma abnormal events and to the evaluation of the associated effects on the other side. The paper summarises the results of these studies and illustrates the response of selected in-vessel components to the electromagnetic excitation.

1. ELECTROMAGNETIC TRANSIENTS IN ITER

The main sources of large electromagnetic (EM) loads in the ITER vacuum vessel (VV) and in-vessel components such as Blanket Modules (BM), First Wall (FW) and Divertor Modules (DM) are related to: **(a)** the fast toroidal flux change during the thermal quench at the beginning of a plasma disruption, **(b)** the poloidal flux change during the plasma current quench phase, and **(c)** the poloidal current flowing in the plasma halo and in the conducting structures (halo currents).

(a) During the thermal quench the magnetic pressure drops to about zero with a typical time scale below 1 ms. The current induced in the structures by the toroidal flux change flows initially in a very narrow depth at the surface of the components, and gives rise to a pressure that pulls the components towards the plasma. For a circular plasma, the toroidal flux change and the induced poloidal current can be estimated analytically. For the ITER paramters (plasma current of about 21 MA, vacuum toroidal field at the plasma axis of 5.7 T and initial β_{pol} of 0.9) the induced current is about 4.5 MA.

(b) The current quench phase has a typical duration of 10-100 ms. This is much shorter than the typical time constant of the toroidally connected structures around the plasma (e.g. FW and VV) but it is comparable to the time constant of segmented structures such as the DM. In the FW and VV, the induced current tends to flow toroidally, whilst in the DM the patterns of the eddy currents are more complicated and depend upon the connection of the different sub-component to the cassette body.

(c) During a Vertical Displacement Event (VDE) the plasma can come in contact with the wall and a fraction of the current flowing in the plasma halo enters the first wall structure, providing a stabilizing vertical forces on the plasma. Experiments performed in tokamaks [1,2,3] show that the current distribution around the torus may not be uniform. Figure 1 shows a survey of the existing data-base from which one can obtain a relation between the magnitude of the halo current (expressed as the ratio of total measured halo current to initial plasma current) and the toroidal peaking factor (TPF, defined as ratio of peak to average current density). Also shown is a set of design curves which indicate the likelihood that an event falls in the region contained below the curve

The time evolution of the halo current is related to the plasma current duration with the peak value reached half way through the current quench phase .

2. EM ANALYSIS OF IN-VESSEL COMPONENTS

2.1 Analytical tools

Different analytical tools and models were used for the prediction of the plasma behaviour and the analysis of the EM effects. Plasma equilibrium codes (TSC [4] and MAXFEA [5]) have been used to solve the magneto-hydrodynamic equation of the plasma with 2-D axisymmetric models.

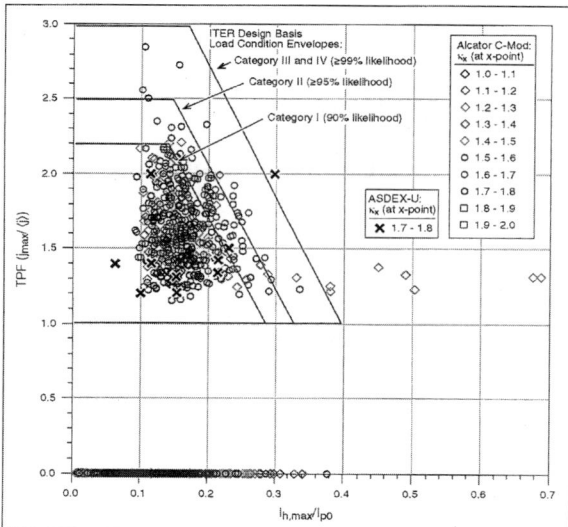

Figure 1. TPF vs. $I_{h,max}/I_{p0}$ data and design curves

Eddy current and EM forces have been computed with the programs EDDYCUFF [6], Typhoon [7] and SPARK [8], which solve the Maxwell equations using an integral formulation for 3-D structure made of thin conductive shells, and EMAS [9], which is based on a differential formulation and allows the study of structure made of solid, shells and beam elements.

2.2 First Wall, Blanket and Divertor

The ITER blanket is comprised of 740 integrated FW/shield modules supported on a structural shell, the Back Plate (BP), interconnected by cooling manifolds [10,11]. The FW is a layered assembly of 10 mm thick Be armour supported by a 20 mm thick copper alloy heat sink and cooled by poloidally oriented stainless steel tubes, 10 mm ID and 1 mm thick[12]. The plasma facing material is beryllium; tungsten is used in the lower baffle region subject to high erosion. The shielding section of the modules is a 316 LN IG stainless steel block, cooled by an array of poloidally oriented circular channels. The toroidal resistance of Blanket system and VV are 7.4 $\mu\Omega$ and 10.4 $\mu\Omega$, respectively; the total toroidal resistance of the assembly is 4.3 $\mu\Omega$.

The divertor consists of 60 steel cassettes, which support replaceable plasma facing components [13]. The Targets are made of steel plates 100 mm thick and copper tubes (10 mm ID) with Carbon Fibre Composite or beryllium monoblocs. The Wings are 60 mm thick tungsten clad copper plates with internal cooling channels. The Dome is a copper structure with rectangular channels and beryllium armour, attached to a massive steel part.

2.3 Electromagnetic analysis of FW/BM

(i) Fast radial disruption

During a disruption the plasma current of 21-MA is assumed to decay linearly to 0 in 10-ms. Electromagnetic analysis has been performed by modelling a module as a box structure (Model SB) or layered plates (Model SP). The plasma current is modelled as multi-filaments derived from a plasma equilibrium at End of Burn. The time evolution of the induced toroidal currents in the FW, SM, BP and VV is shown in Figure 2. The toroidal current in all the modules at 10 ms is 95 % of Ip, ~45 % of which flows in the FW (Fig. 3). The interaction of induced toroidal currents with poloidal field results in a peak radial pressure of 1.1 MPa on the module. The largest EM loads are developed when these currents flow radially along the module sidewall across toroidal field lines. A maximum sidewall force of ~5.5 MN per meter of poloidal length occurs at the inboard midplane, see Fig. 4. Forces due to induced poloidal currents are excluded, that is conservative when $\beta p<1$. EM analysis has been performed also by using solid elements for the modules [14].

(ii) Vertical Displacement Events

Utilising VDE simulation results [5], an EM analysis has been performed [15]. The maximum induced toroidal currents are 5.3 MA in the FW/shield and 10.7 MA in the back plate. The total induced current is 19.8 MA. The maximum magnetic pressure due to toroidal currents is 0.32 MPa on the FW/shield module and 0.69 MPa on the back plate.

The back plate and blanket modules are designed to withstand loads from halo currents which flow poloidally during VDE's. The back plate is a "horse shoe" shaped toroidally continuous stainless steel structure, 100 mm thick. It is sized to react the main loads on the blanket while minimising the loads transmitted to the VV. The magnitude of these currents can be from 10 to 40 % of the initial plasma current and have a corresponding toroidal asymmetry peaking of 3.0 to 1.2. It is assumed that the toroidal distribution of the halo current is given by: $j = j_{ave}(1+(\alpha-1)\cos\phi)$, where ϕ is the toroidal angle, and α is the toroidal asymmetry peaking, see Fig. 5. The structural analysis has been performed on the whole backplate and modules, also shown in Fig. 5. The stresses due to halo currents in the back plate are close to those allowable.

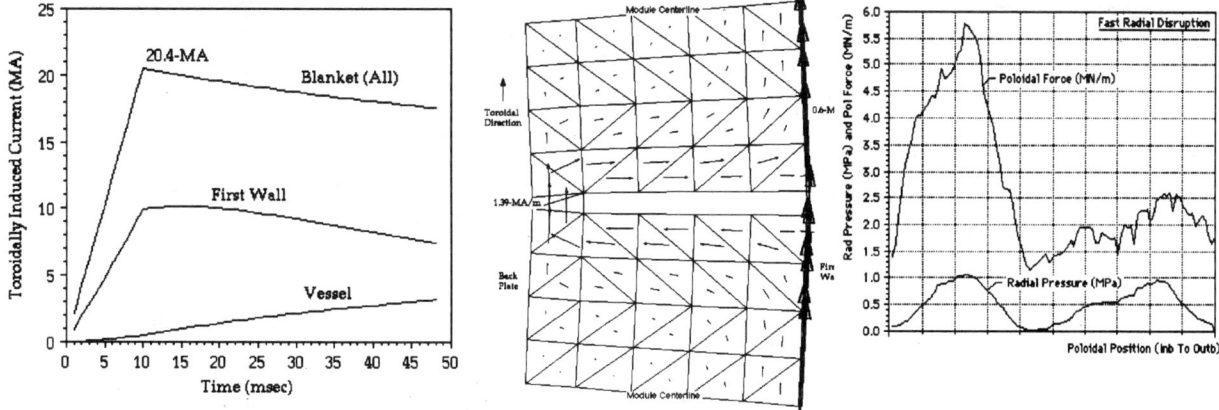

Figure 2 Time evolution of induced toroidal currents due to a disruption (plasma current decay time:10ms)

Figure 3. Eddy current distribution at the horizontal cross section (inboard midplane module, t=10 ms)

Figure 4 Poloidal distribution of force/pressure in blanket modules (time= 10 ms)

2.4 Divertor electromagnetic analysis

(i) Thermal quench

Figure 6 shows the time history of the poloidal current for the fast radial disruption simulated with TSC (initial β_p=0.56). The peak current (3.2 MA) is reached after 0.5 ms; the current then decays during the current quench phase (10 ms). In the DM the current flows entirely in the vertical targets and wings. The average normal pressure is 0.8 MPa at the inner and 0.4 MPa at the outer target, directed toward the plasma. In the inner and outer wings the resultant forces (normal to the poloidal orientation of the wings) are 0.25 and 0.16 MN, respectively.

(ii) Current quench

The plasma current decay induce separate loops of current in each PFC. A typical pattern of the current flow is shown in Figure 7. The largest current density is found in the wings, where the eddy current peaks at the outer edge of the wings. In Table 3 we compare peak current density and pressure in the wing obtained with the programs described above. The maximum deformation of the structure is 34 mm [16]. In the optimised design of the wing, the stresses are below the elastic limits. This was accomplished by reducing the amount of copper in the structure. The eddy current loads in the CB produces an overall moment about the vertical axis of the CB (1.8 MNm) [14]. The maximum toroidal reaction is 0.9 MN (at the inner rail).

(iii) Halo current loads

The halo current was assumed to enter the CB at the outer electrical connection with the Blanket, to flow though CB and PFC, and to return to the BL at the inner connection. Accounting for the peaking factor described previously, the halo current per

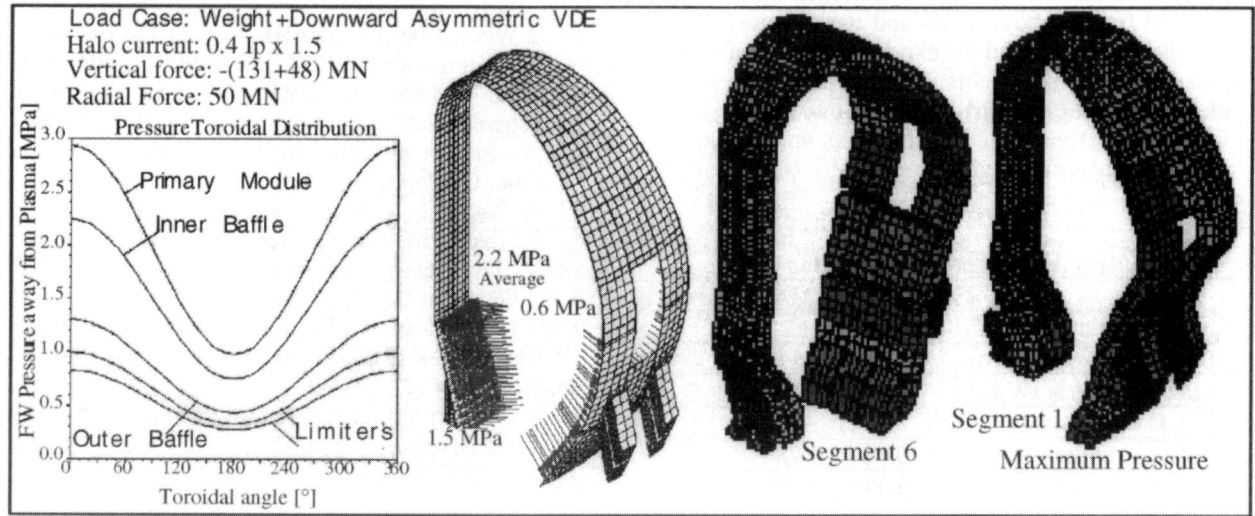

Figure 5. Halo current pressure distribution on back plate and structural analysis results

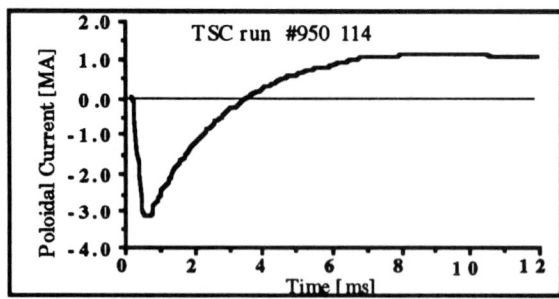

Figure 6. Time-history of the polodial current during a central disruption simulated with TSC

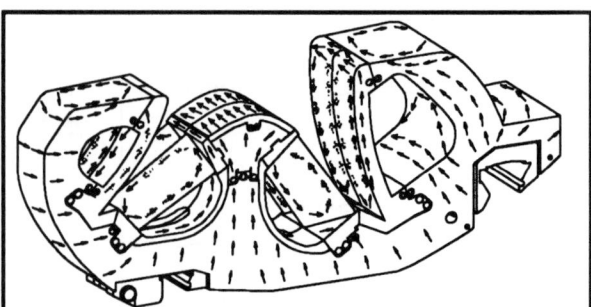

Figure 7. Eddy current distribution in the Divertor

cassette is 0.2 MA. The associated EM loads produce a maximum bending stress in the CB of about 150 MN (below the allowable for 316 LN for bending stress), located near the inner support of the CB.

3. CONCLUSIONS

The response of the in-vessel component to abnormal plasma events has been analysed considering several mechanisms of electromagnetic interaction; where possible we made use of conservative assumptions (e.g. very fast current quench or highly non uniform halo current distribution)..

The most demanding conditions for the Blanket are caused by the vertical forces in the module sidewalls, during a fast plasma current quench, and by the non uniform toroidal distribution of the halo currents during VDE's. In the Divertor, the forces associated to the halo currents produce the largest stresses in the steel structure, whilst the eddy current induced during the plasma current quench produce highly peaked pressure onto the wings.

All ITER in-vessel components have been verified for the above loads and the analyses show that they can withstand the expected number of worst plasma abnormal events. Further analyses, both electromagnetic and structural are under way, with improved models of the components and including considerations of the dynamic effects.

Table 3
Peak current density and pressure in the wings

	EDDYCUFF	EMAS	Typhoon
Radial disruption			
Peak current at IW, A/m	0.37	0.14	
Max pressure at IW, MPa	2.00	0.80	0.60
Downward VDE			
Peak current at IW, A/m	0.65	0.78	
Max pressure at IW, MPa	5.11	5.10	3.38

ACKNOWLEDGEMENT

This report was prepared as an account of work performed under the Agreement among the European Atomic Energy Community, the Government of Japan, the Government of the Russian Federation, and the Government of the United States of America on Co-operation in the Engineering Design Activities for the International Thermonuclear Experimental Reactor ("ITER EDA Agreement") under the auspices of the International Atomic Energy Agency (IAEA).

REFERENCES

1. J. Sorci et al., Halo Current Measurements on Alcator C-Mod, 16th SOFE (1995), Urbana, IL
2. M. Pick et al., Evidence of Halo Current in JET, 14th Symp. on Fus.Eng., San Diego (1991) 87
3. O. Gruber et al, Pl. Ph. & Contr.Fus.35(1993) B191
4. S.C. Jardin, J. Comput. Physics, 66, 481, 1986
5. P. Barabaschi, Fus.Eng.& Des.30 (1995) 1149.
6. A. Kameari, J. Comput. Physics, 42, 124, 1981
7. A.V. Belov et al., Fusion Eng. & Design 31 (1996), 167-180.
8. D. Weissenburger, SPARK Ver.1.1 User Manual, Princeton Univ., NJ (US), rep. PPPL-2494, 1988
9. MSC EMAS User's Manual, The Mac Neal Schwendler Corp., Los Alamos, Ca (USA), 1995
10. K. Ioki et al, Proc. of 16th Symp. on Fusion Eng., Champaign (1995) 150.
11. R. Parker et al., Fus. Eng.&Des.30 (1995) 119.
12. K. Ioki et al., Physica Scripta T64 (1996) 53.
13. R. Tivey et al., Engineering Design of the ITER Divertor, at this conf.
14. M. Roccella et al., 3-D Electr. Model and Electr. Anal. of the ITER In-vessel Comp., this conf.
15. K. Ioki et al., Fus. Eng. & Des.30 (1995) 351.
16. S. Chiocchio et al., Structural Performance of the ITER Divertor during Ab-normal Events, Proc. of 7th ANSYS Conf., Pittsburgh, PA, USA, 1996

3-D Electromagnetic Model and Electromagnetic Analysis of the ITER In-Vessel Components

M. Roccella, A. Capriccioli, M. Ferrari, M. Gasparotto, A. Pizzuto, [a] -S. Chiocchio[b] - F. Lucca [c]

[a] ENEA - Associazione EURATOM-ENEA sulla Fusione, C.R.E. Frascati, C.P. 00044 Frascati, Rome(Italy)
[b] ITER - Joint Central Team - Boltzmann Strasse 2 - D- 85748 Garching (München) -Germany
[c] ENEA consultant - Co ENEA, C.R.E. Frascati, C.P. 00044 Frascati, Rome(Italy)

Using the finite element electromagnetic code EMAS, a single 3-D model, that include all the main in-vessel components of ITER has been developed, with a reasonable number of elements. The electromagnetic loads during plasma fast Centred Disruptions (CD) and during typical Vertical Displacement Event disruptions (VDE) have been evaluated for both the Shielding Blanket Assembly (SBA) and the Divertor Assembly (DA). A 2-D analysis of the ITER Breeding Blanket Assembly (BBA) has been also carried out to account for the very complicate geometry of this structure.

1. INTRODUCTION

The electromagnetic loads due to plasma disruptions represent one of the main engineering concern in the design of the TOKAMAKS aiming to approach ignition conditions. In the present study a 3-D electromagnetic analysis for all the ITER in-vessel components has been carried out using a single model by means of the finite elements electromagnetic code EMAS[1]. This code is capable of mixing elements of different dimensions to overcome the problem of modelling elements of very different aspect ratios in the same mesh and of using non isotropic tensors for the electrical properties (a large use of the non isotropic resistivity tensor has been made to account for the real geometry of the components).

2. THE MODEL

2.1 The Plasma Model

For the CD a ten filaments plasma model has been used. The total plasma current at beginning of CD (the end of the burn configuration) is of 21 MA. All the filament currents decay linearly to zero in 10 ms. The downward VDE followed by a current quench has been simulated using 14 current filaments.

The spatial distribution and the time behaviour of the currents in the filaments have been chosen in such a way to fit the reference plasma model provided by the ITER JCT. The magnetic field at the first wall for the reference plasma model and the model used for the analysis differs less than 1% at the beginning of the CD and within about 10% during the VDE. The current centroid position for the present and the reference model and the total plasma current vs. time are shown in fig. 1.

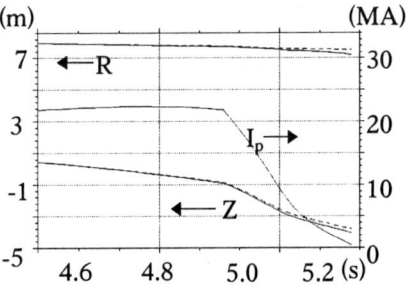

Figure 1. Time history of total plasma current I_p and comparison for the plasma centroid position (radial R and vertical Z); for the reference plasma model (solid line) and the present model (dashed line).

Table 1.
Main material properties. Size is the assumed thickness for the shell elements (in mm) and the assumed cross section for the line elements (in mm^2).

Component	Elem	Size	mat	ρ
Shielding Blanket Assembly				
FW	shell	8	Cu	0.038
SB1*	brick		SS	0.867
SB2*	brick		SS	1.26
SB3*	brick		SS	1.45
VV*	shell	2x 40	SS	0.867
manifold*	Shell	34.5	SS	0.867
BP *	shell	100	SS	0.867
Divertor Assembly				
DC*	brick		SS	1.067
DT*	brick		SS	0.867
DTC	shell	1.4	Cu	0.038
DCP	line	436	Cu	0.038
liner(dome)	shell	1.4 (3)	Cu	0.038
Wings	shell	80	Cu	0.038
PES	shell	DC⇔BP		≅0
PES(rails)	brick	DC⇔VV	SS	0.867
TES	brick	DC⇔DC	SS	0.867

(*)For the Stainless Steel (SS) elements have been used different resistivities ρ (in μΩm in the table) to account for their effective cross sections.

2.2 The 3-D mesh model

The main material properties are given in Table 1.

The full model is shown in fig 2. The model represents a toroidal sector of the full machine. The toroidal periodicity is 60 for the divertor and the outboard blanket while is 40 for the inboard blanket[2], the toroidal angle of the sector used in the analyses is 6° and 9° respectively.

The Vacuum Vessel (VV) has been assumed made of 2 nested surfaces modelled by shell elements.

The SBA mesh includes the First Wall (FW), the Shielding Blanket (SB), the Back Plate (BP) and the manifold.

In fig 3 the detail of DA model is shown. This part includes the Divertor Cassette (DC), the dome, the two inboard and outboard Divertor Targets (DT), the Divertor Target Coverings (DTC), the Target Covering Pipes (TCP), the two wing pairs at the inboard and outboard, the Toroidal Earth Straps (TES) and the Poloidal Earth Straps (PES).

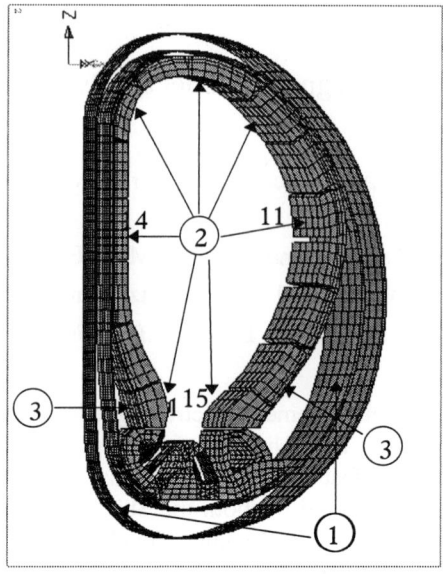

Figure 2. View of the full model. (1) VV; SBA: (2) SB modules (n° 1 to n°15), (3) BP. The FW is superimposed to the plasma facing surface of the SB and modelled in 2-D.

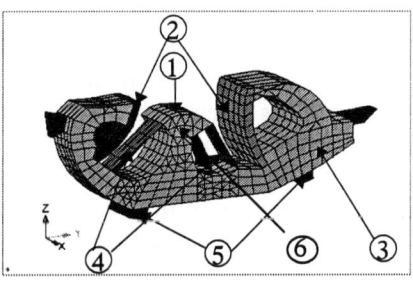

Figure 3. Detail of the mesh showing the DA: (1) dome; (2) inboard and outboard DT; (3) DC; (4) inboard and outboard wing pairs, (5) PES (DC⇔VV), (6) TES, PES (DC⇔BP) are not in view.

3. DIVERTOR RESULTS

3.1 Main loads during plasma disruptions

The most severe load conditions for the DA occur in the wings due to the current loop in each wing pair at the inboard and outboard. In Table 2 the maximum, and the average values of the force per unity area are given per the inner wings at the time

Table 2.
Maximum and average inward and outward force per unity area in MPa at the inner wings for cases a) to e).

Case	a	b	c	d	e
MAX-out	0.7	4.1	3.5	2.6	3.2
MAX-in	-0.75	-4.4	-3.7	-2.7	-3.4
av. -out	0.12	0.9	0.7	0.7	0.08
av.-in	-0.15	-1.1	-0.75	-0.8	-0.08

Table 3.
Toroidal (x), radial (y) and vertical (z) components of the resultant force and torque on the DA.

	Force (MN)	Torque (MNm)
x	-0.27	0.1
y	-0.18	-0.32
z	0.11	1.8

of the maximum loads. The data refer to the following cases: a) CD; b) VDE without TES; c) VDE with TES; d) same as b but doubling the wing resistivities; e) same as b) but doubling the total resistance of the wing loops by increasing the resistance between the wings. In both the cases d) and e) the time constant of the wing loops was about halved; in the last case the main effect is of splitting each current loop in a wing pair in two separate loops within each wing. Since the force are radial and the wings are tilted by 45° the effective pressure on the wings must be scaled by $1/\sqrt{2}$.

In Table 3 we report the maximum values of the components of the resultant force and of the torque about the dome centre, acting on the whole DA.

Table 4.
Electrical field (V/m) and voltage drop (V) between adjacent cassettes in the air gap near the top of the dome (first row) and near the top of the external target (second row).

CD		VDE		VDE + Tor. Con.	
E_ϕ	ΔV	E_ϕ	ΔV	E_ϕ	ΔV
180	4.5	200	5.0	150	3.8
180	5.1	200	5.7	190	5.4

3.2 Voltage drops between adjacent cassettes

In table 4 the maximum induced electrical field E_ϕ and the voltage drop ΔV in the gaps between the cassettes are given, for CD and VDE without and with TES, at the top of the dome and of the external target. The TES at the top of the dome reduces E_ϕ and ΔV in the neighbourhood of about 25%.

3.3 Poloidal current excitation

On the inner wings the forces per unity area due to the halo currents and to the thermal quench are of order of 0.5 MPa (outward) and 0.4 MPa (inward) respectively). The halo current loads have been estimated assuming a toroidal peaking factor of 2 and half of the current flowing in the wings. On the contrary, due to the very short times involved in the thermal quench, all the diamagnetic current has been assumed to flow in the wings.

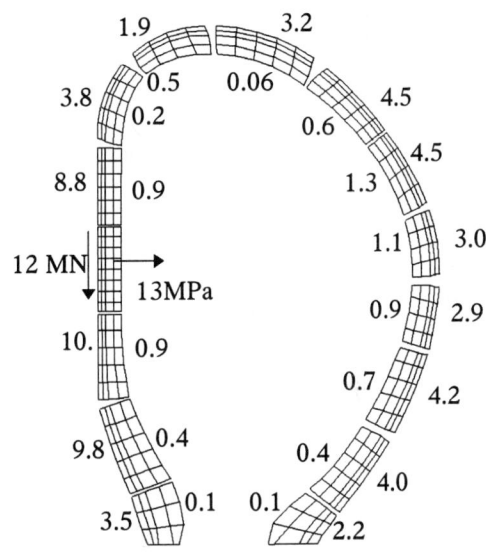

Figure 4. Average pressure (in MPa) due to the total current in each module (Inner values) and total side wall forces (in MN)

4 BLANKET RESULTS

4.1 Shielding blanket results

The most severe load conditions for the SBA occur near the equatorial plane at the end of the CD. The SBA components are in good electrical contact, then the induced currents can flow freely through them according to the induced electrical fields. The SBA is in the inductive limit and the induced current will flow on the outermost blanket surface for a depth equal to the skin depth (about 75% of the total induced current in the FW and the remnant, about 25% in the first SS layer below). The average pressure due to the total current in the SBA and the side wall forces on each blanket module are summarised in the fig.4.

4.2 Breeding blanket preliminary results

With respect to the relatively simple geometrical structure of the SBA, the very complicate one of the BBA produces very meandering current paths and a not easily predictable electromagnetic loads distribution inside the module. However to evaluate the right radial and toroidal current density distribution inside the BBA it can be used a 2-D axis-symmetric model of the equatorial plane of the BBA, that can be detailed enough to account for the geometrical effects. Indeed the spatial distribution of these components are not affected by the poloidal length of the module. In addition, if also the BBA is in the inductive limit, the total current and its poloidal distribution is quite exactly known from the 3-D analysis.

In fig.5 the total current and the currents in the module layers induced by a linearly decreasing excitation current flowing in a solenoid outside the BBA model are shown vs. time. After 10 ms the total induced current per unity length was equal, within some per cent, to the correspondent excitation current variation, thus confirming the validity of the inductive limit hypothesis, for the BBA, during the CD. The current distribution is given from 2-D analysis; the total current per unity length as well as the poloidal field due to the far field sources (the poloidal coils and the far induced currents) from the 3-D analysis. Using these results a preliminary estimate gives a pressure of 0.25 MPa on the FW of the BBA, at the inboard equatorial plane.

CONCLUSIONS

The loads on the divertor wings at the end of VDE increase up to a factor 6 with respect to the CD case, thus becoming the main loads on these components. The presence of TES and the increase in the wing resistivity (case d) does not sensibly reduce these loads. The case e) does not sensibly reduce the maximum or minimum load values but drastically lowers the averages in each wing.

The values of E_Φ and ΔV between the divertor cassettes do not represent serious risk of arching.

From the 3-D analysis the average pressure on the FW of the SBA at the inboard equatorial plane (module n°4) was about 0.75 MPa corresponding to 75% of the total current. The correspondent value for the FW in the BBA was only 0.25 MPa due to the sensible current reduction produced by the particular geometry of the BBA.

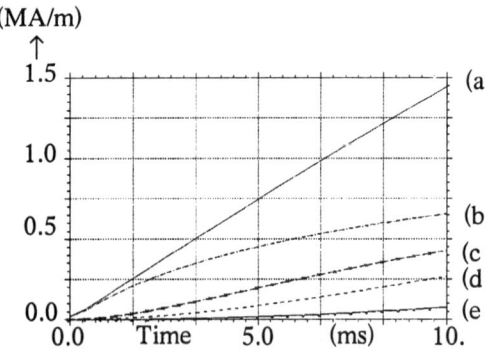

Figure 5. Results from 2-D analysis. Total toroidal current per unity length in the BBA module n°4 at the equatorial plane (curve a). The other curves refers: b) to the FW and first Be layer, c) to the second Be layer, d) to the SS layer behind the second Be layer, e) to the VV.

REFERENCES

1. MSC EMAS User's Manual, Version 3.1, The Mac Neal Schwendler Corporation, Los Alamos, CA (USA), !995.
2. ITER EDA JCT, General Design Requirements Document S10 GDRD 1, 95-02-10 W1.2

Modelling and analysis of plasma-first wall contact during vertical instabilities in Next-Step tokamaks with a 3D eddy current code

Sophia Fantechi, Yves Crutzen

European Commission, Joint Research Centre (JRC)
Institute for Systems, Informatics and Safety (ISIS), T.P. 210, 21020 Ispra (Varese), Italy.

A computer procedure, based on a simplified model of the phenomenon of plasma-first wall (FW) contact, has been developed to take into account the conductive transfer of electric current from the plasma "halo" to the FW during a vertical displacement event (VDE) in a Next-Step tokamak-type fusion device. The model has been introduced in a three-dimensional (3D) eddy current code for electromagnetic (EM) analysis to account for the extra poloidal currents flowing in the reactor wall during the transient vertical displacements of the plasma. These "contact currents" are responsible for very high vertical forces on the reactor FW, which makes them one of the most dramatic occurrences in a tokamak discharge.

1. INTRODUCTION

The complex EM phenomena due to plasma-FW interaction are of primary importance for tokamak technology. Of great concern are those occurring during the so-called *Vertical Displacement Event* (VDE), very common in the current generation of tokamaks because of the elongated plasma cross-section used. The plasma jumps abruptly off its equilibrium position, upwards or downwards, and collides with the surrounding wall. If the core is still able to sustain a large current, the plasma will tend to slow down and reverse its vertical motion remaining very close to the wall for a short time. A disruption will then take place, causing a sudden current quench and the release of the thermal and magnetic energy stored in the plasma. The phenomenon lasts for a time of the order of a few tens of ms.

A VDE thus involves both induced eddy currents *and* electric currents conductively transferred from the narrow plasma *halo* region to the surrounding conductors touched by the plasma during the *contact* phase. These *halo currents* are generated in the scrape-off layer (SOL) when the plasma approaches the conducting walls and are basically characterised by a path which runs partly in the conducting wall and partly in the SOL in contact with it, where the current follows the magnetic field lines. During the *contact* there is a significant flow of *additional poloidal currents* (I_H) between the vessel and the SOL. In a Next-Step fusion reactor, such as ITER, these currents are expected to be in the MA range (up to 30% of the plasma current) and their consequences are of major concern: they interact with the strong toroidal field in the tokamak and generate forces which are so high that they can certainly be classified as the most dangerous source of mechanical loads on a tokamak. Many years of experimental analysis at JET prove that the resulting vertical force on the vessel is about *two to three times higher* than for any other kind of load [1]. Therefore, the investigation of this problem, of the way these currents arise, of the methods to analyse and compute their effects and of the possibilities to prevent them are all key issues for a safe tokamak design and operation.

2. EDDY CURRENT CODE MODEL

The model considered for the EM analysis of ITER is *3D integrated*, i.e. it includes the vacuum vessel (VV), the first wall (FW), the back plate (BP), the poloidal field (PF) coils and the central solenoid. The model represents a 7.5° sector of torus and is used for the EM code CARIDDI [2].

In CARIDDI the plasma is modelled by a toroidal current carrying conductor [3]. The cross section is discretised into elements and the plasma current is carried in each single element by a filament placed in its centre. The properties of this conductor are an input: they are either calculated by an equilibrium code (supposing that the plasma, during its evolution, goes through a series of ideal MHD states), or given by measurements, when available. The plasma represented in this way is the confined plasma: *no SOL*, so *no halo*, is described in the model. Moreover, the plasma thus modelled is an axisymmetric conductor. In this way CARIDDI allows to simulate centered disruptions (current quenches), as well as dynamic disruptions (VDEs ending with a current quench).

3. PLASMA-FIRST WALL CONTACT MODEL

Since the SOL is *not* included in the EM model of the plasma, the transfer of current from the SOL to the FW during the VDE has to be introduced indirectly. This has been done by defining, at each instant, and depending on the plasma centroid position in the vessel, the elements where the SOL intercepts the wall and the total poloidal current circulating between them, closing its path through the wall. In the case of ITER, the halo currents are supposed to be calculated by an equilibrium MHD code, MAXFEA [4]. The plasma motion is therefore 2D, i.e. axisymmetric. For an existing tokamak, this input would be replaced by experimental data. The program also represents a first step of transfer of information between a 2D equilibrium code and a 3D EM code. We suppose that the contact current transferred to the wall has only a poloidal component flows parallel to the surface of the elements and varies quadratically along a curvilinear abscissa (s) joining the centroids of the elements in which the current is calculated:

$$I_C(s,t) = I_H(t) \left\{ 1 - \left[\frac{2(s - s_0(t))}{w(t)} \right]^2 \right\} \quad (1)$$

The plasma-FW contact is described quantitatively by the width $w(t)$ of the contact surface which is assumed to scale with the minor radius a and the total (poloidal) halo current I_H:

$$w(t) = 2 a \frac{I_H(t)}{I_{H_{MAX}}} \quad (2)$$

The simplest case is that of a parabolic distribution $I_c(s,t)$. By taking the maximum value of $I_c(s,t)$ to be $I_H(t)$, we start from the situation where the whole halo intercepts the wall. Considerations on the halo density could bring to slight modifications of (1), but this is not relevant to the calculation and the distribution is reasonably represented by a parabola. With the contact current *density* $\mathbf{J_c}$ in the FW elements centroids, the EM code CARIDDI computes the $\mathbf{J_c} \times \mathbf{B}$ *contact forces* on the FW, in each element centroid, in the same way as it calculates the body forces due to the eddy currents. However, as a first approximation, because of the magnitude of the contact currents, we can disregard the contribution of the toroidal field produced by the contact currents with respect to the tokamak toroidal field (5.7 T for ITER).

4. ELECTROMAGNETIC ANALYSIS

4.1. Approach and inputs used

The simulations have been continued up to the point where all the induced currents in the passive structures have disappeared (8 s). In the modelling procedure the plasma properties have a given time history. Output from simulations performed with the equilibrium code MAXFEA is used as input for CARIDDI. Moreover: the initial equilibrium condition is the reference End of Burn (EOB); for the VDE simulations it was taken $l_i = 0.9$, $\beta_P = 0.7$ and $q_{edge} = 1.5$ as the critical value which triggers the instability; the one turn resistance of the VV is 26 $\mu\Omega$, while for FW and BP 16.1 $\mu\Omega$ and 16 $\mu\Omega$ have been taken respectively; the initial plasma current is 24 MA. The reference upward VDE is a plasma vertical shift of about 3 m in 5.6 s, followed by a slow current quench of about 20 MA in 0.5 s (i.e. $dI_P/dt \approx 40$ MA). Before switching-off, the plasma undergoes an inward radial displacement of 0.8 m, while the vertical drift at that time is +1.95.

The applications concern ITER geometry and PF coils configuration from TAC4 version to the following.

4.2. Eddy current and contact current analysis

The maximum eddy current is obviously reached in the FW: 7.28 MA at 5.6 s, when the plasma switches-off, which is the most violent transient. In the BP, at the same instant, we find 5.59 MA, while in the VV there is a maximum of 1.94 MA, circulating mostly in the inner shell. On the other hand, the poloidal component of the halo current reaches almost 4 MA at 5.3 s, when the width of the contact zone is two times the plasma minor radius. Fig. 1 gives the distribution of the contact currents densities at 5.3 s, when the contact currents reach their maximum value.

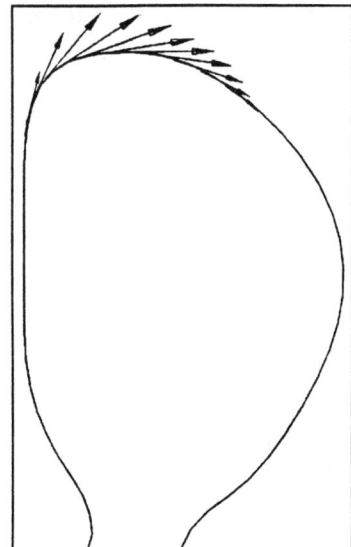

Figure 1. Max. contact currents in the FW (5.3s).

The corresponding forces on the structures, at this same instant, are illustrated in figs. 2 and 3, respectively for the induced currents and for the plasma-FW contact. The contact forces have a vertical resultant directed upwards, while the forces due to the eddy currents are directed towards the plasma region and hence in opposite direction with respect to the contact forces. This is because, for the induction phenomena, the current quench is the dominant effect at this stage, so that the induced currents reverse their direction. Thus, during the current quench, the force due to the eddy currents between the plasma and the passive structures becomes attractive and not repulsive, due to the mutual coupling.

For this reason, the maximum value of the vertical force on the FW due to the contact currents exceeds the total maximum vertical force on the passive structures.

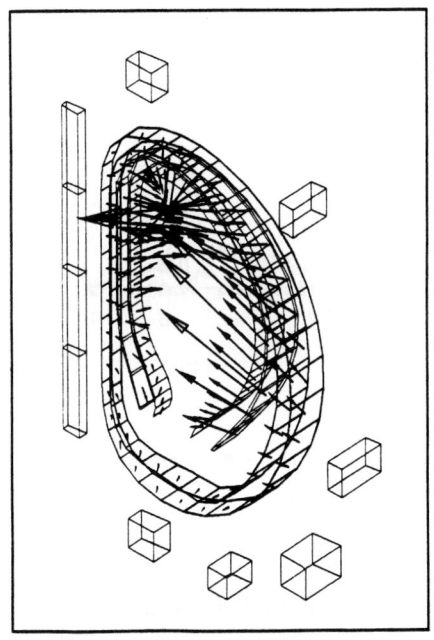

Figure 2. Forces at 5.3 s: eddy.

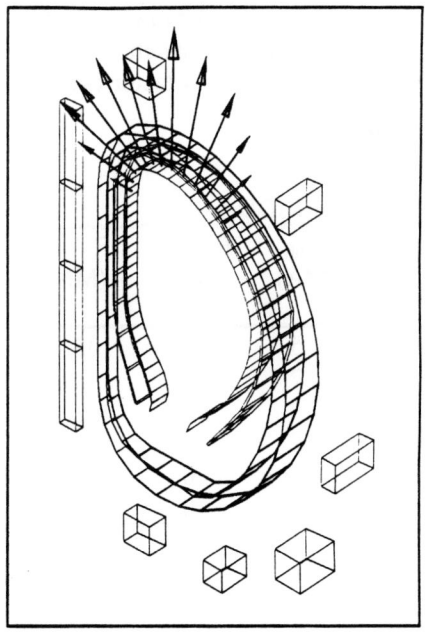

Figure 3. Forces at 5.3 s: contact.

Fig. 4 gives, for comparison, the vertical forces on the FW due to the eddy currents and to the poloidal contact currents separately. The latter attain values that are a *factor two higher* than the former: 85 MN at 5.3 s (when the halo currents are the highest), due to the contact, and 46.5 MN at 5.6 s (when the plasma switches-off), due to the eddy currents.

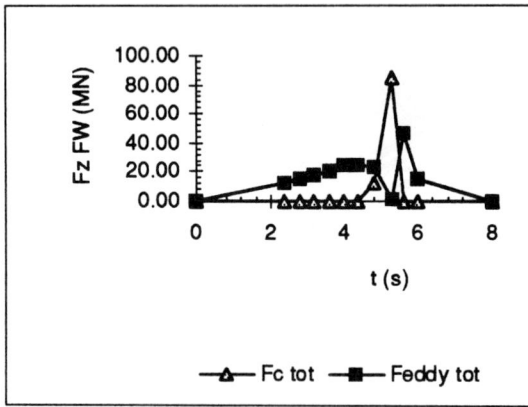

Figure 4. Vertical forces on the FW.

For what concerns the *VDE*, the eddy currents and related forces attain their maximum value during the first phase of the VDE, well before a significant contact of the plasma with the wall. Then, when the plasma motion slows down, the contact currents and related vertical forces reach a maximum value that is of the order of magnitude of those obtained in the first phase of the instability, but about a factor two higher (table 1).

Table 1
Total vertical forces (CARIDDI).

t(s)	Comp.	I_{eddy} (MA)	$F_{z\,contact}$ (MN)	$F_{z\,eddy}$ (MN)
4.8	FW/BP	1.5	11.5	35.4
4.8	VV	0.2	0	0.9
5.3	FW/BP	7.9	84.8	-9.3
5.3	VV	1.2	0	-7.7
5.6	FW/BP	12.9	0	46.2
5.6	VV	1.9	0	-7.0

5. CONCLUSIONS AND COMMENTS

The project has dealt with the full study of a VDE, using a 2D model for the description of the plasma-FW contact introduced into the code CARIDDI. The objective has been the calculation of the contact forces on the structures surrounding the plasma during the vertical instability. These are essential if a correct mechanical analysis has to be performed. The contact model developed can easily be generalised to the contact between the plasma and any in-vessel component. For the study of the EM effects of the VDEs and contact currents, however, a 3D approach of the problem is essential. Owing to the non-uniformity of their distribution, the halo currents can easily reach very high local peaks, which are very hard to estimate with the currently available computational tools; yet they are one of the most dangerous events in a tokamak discharge. Work is in progress in this direction. An asymmetric contact is expected to be caused by an *asymmetric VDE*. On the other hand, induced toroidal currents and related forces are likely to be uniform toroidally.

REFERENCES

1. P. Andrew, P. Noll et al., "Measured Currents in JET Limiters During Disruptions", *16th SOFE*, Urbana, Illinois (1995).
2. R. Albanese, G. Rubinacci, "Integral formulation for 3D eddy current computation using edge elements", *IEEE Proc.* vol. 135, pt. A, no. 7 (1988) 457-462.
3. S.Fantechi, Y. Crutzen et al., "Plasma models for the computation of 3D eddy currents in next tokamaks", in *Proc. of the 2nd Int. Workshop on EM Fields*, Leuven (1994), Belgium.
4. P. Barabaschi, S. Chiocchio, "Disruptions, VDEs and Halo Current: Scenarios and EM Analyses in ITER", in *Proc. of the 3rd Int. Workshop on EM Forces and Related Effects on Blankets and Other Structures surrounding the Plasma Torus*, Capri (1994), Italy.

Acknowledgements - We wish to thank Profs. G. Rubinacci and R. Albanese (CREATE-Naples) for their contribution to this activity and JRC-ISIS for supporting this research.

CONTROL OF THE MAGNETIC CONFIGURATION IN ITER

A. Portone[a], Y. Gribov[a], Y. Mitrishkin[a], P. L. Mondino[a], J. Wesley[b], R. Albanese[c], G. Ambrosino[c], D. Ciscato[d], E. Coccorese[c], D. Humphreys[e], S. Jardin[f], A. Kavin[g], C. Kessel[f], J. Lister[h], D. Pearlstein[i], A. Pironti[c], I. Senda[j], M. Walker[e], D. Ward[h]

[a] ITER Joint Central Team, Naka JWS, 801-1, Mukouyama, Naka-machi, 311-01 Ibaraki, Japan;
[b] ITER Joint Central Team, San Diego JWS, 11025 N. Torrey Pines Rd., La Jolla, CA 92037, US;
[c] EU Home Team, Universita' degli Studi di Napoli, V. Claudio 21, Napoli, 80125 Italy;
[d] EU Home Team, DEI, Universita' degli Studi di Padova, via Gradenigo 6/a, Padova, 35131 Italy;
[e] US Home Team, General Atomics, P. O. Box 85608, San Diego, CA 92186-9874, US;
[f] US Home Team, PPPL, P. O. Box 451, Princeton, NJ 08543, US;
[g] RF Home Team, Institute of Electrophysical Apparatus, D. V. Efremov, S. Petersburg, 189631 Russia;
[h] EU Home Team, CRPP-EPFL, 21 av. des Bains, Lausanne, CH 1007, Switzerland;
[i] US Home Team, LLNL, P. O. Box 808, Livermore, Ca 94550, US;
[j] JA Home Team, JAERI, 801-1, Mukouyama, Naka-machi, Ibaraki, 311-01 Japan.

This paper describes the methodology followed in the design of the feedback system that regulates the plasma current and shape in ITER. The attention is focussed on the design of the feedback controller, which is key to stability. Results obtained by linear and non-linear simulation codes are reported.

1. OVERVIEW

The ITER [1] plasma operation scenario [2] include the following sequence: plasma startup and current rampup, heating to ignition, sustained/controlled burn and controlled fusion power and current shutdown. For each specific step of the operation scenario, scope of plasma control and the corresponding requirements for the Plasma Control System (PCS) comprises: (1) plasma magnetics control (plasma current and shape control), (2) plasma kinetics and divertor control. A Safety Shutdown System is also planned to provide plasma power shutdown following detection of a LOFA or LOCA accident in the coolant circuits for the in-vessel plasma facing and nuclear components.

The control of the ITER magnetic configuration (described in [2]) will be delegated to a sub-system called Poloidal Field Control System (PFCS) (Figure 1). This will be embedded within the higher level PCS which is ultimately responsible to achieve the scenario established before the discharge.

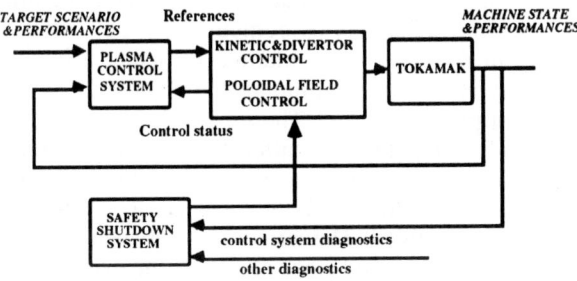

Figure 1. The Plasma Control System.

A Supervisor will receive reference inputs from the PCS and it will issue the reference command signal (y_{ref}) to the lower level controllers (Figure 2). Beside the feedback controller K_{FB}, necessary for stability, the controlled quantities (y) are kept close to y_{ref} also by using a feedforward controller K_{FF}.

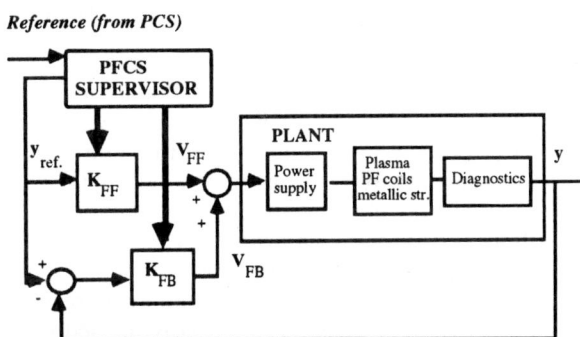

Figure 2. Basic structure of the PF control system.

Updating the controllers gain matrix K_{FF} and K_{FB} is also part of the duty of the PFCS Supervisor. This will choose the controller to be used according to the operation mode, reference inputs and actual plant state and performance. For example, the feedback algorithms designed for the normal operation mode are, at present, restricted to the class of linear controllers to better address the whole multivariable, optimal control problem. In offset conditions, when vertical control is a major concern, nonlinear controllers are envisaged to use at best the power supply capability to control, primarily, the vertical position.

2. FEEDBACK CONTROL DESIGN

2.1. Methodology

The main steps followed in the design and test of the feedback control law are shown in Figure 3 [2].

Firstly, several fiducial equilibria are computed by MHD equilibrium codes at various points of the scenario to be controlled (e.g. the beginning of the diverted configuration, start of burn, etc.).

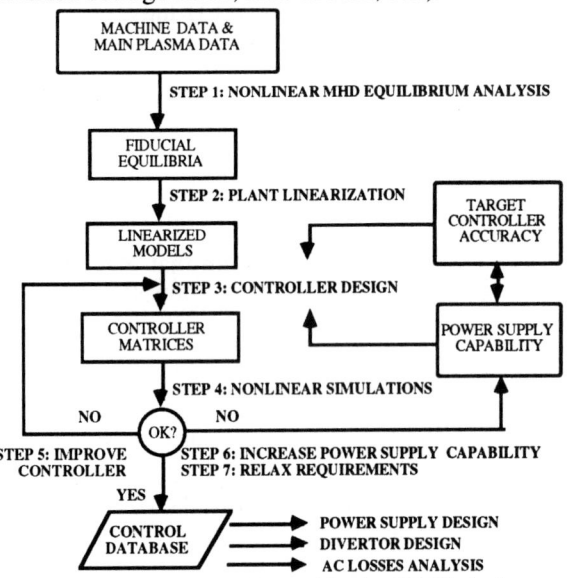

Figure 3. Logic steps followed in the PFCS design.

Linearization is performed about these fiducial equilibria and it aims to derive a linear mapping from the plant input (typically voltage demand to the PF coils power supply) to the controlled variables estimated by the diagnostic system (Figure 2). The reason for deriving linearized models is twofold. Firstly, they can be directly used to design the feedback gain matrix K_{FB} (e.g. proportional and derivative gains for vertical stabilization) addressing features like robust stability. Secondly, closed loop simulations can be quickly run including non-linearities like PF current limitations, voltage saturation as well as diagnostic noise, and the design iterated. Several methodologies can be used to derive a linear model of the tokamak and different nonlinear codes are being used for linearization [2]; having obtained the different linear models, they are compared among themselves and against the nonlinear codes. A parallel R&D activity is underway to benchmark the linear and nonlinear models against experiments [2].

As for the controller design step, several design approaches have proved effective to control the plasma shape in present day experiments (e. g. Decoupling Controllers [3]). However, since the control specifications of the ITER Poloidal Field Control System are very stringent (e.g. required clearance to avoid plasma-wall contact), work is underway to test advanced controllers, based on, for example, optimal LQG and H_∞ methods [2].

At last, the performance of the controllers designed using these different methods has to be tested against the nominal plasma disturbances, not only within the context of the models used to optimize the controller, but also on nonlinear plasma models [2]. These have to include a realistic description of the uncertainties of the measurements being used inside the control loop, of any noise inherent in the system and a model of the power supplies. The results of these simulations can then be compared with the basic requirements for the control system and the design of the controller can be re-iterated in order to meet these requirements better. If it appears that the requirements posed to the controller performances are incompatible with the limits of the power supply system, the design of the latter shall be re-evaluated accordingly.

2.2. Linearization

Here we focus on the linearization of the Input / Output response of the plasma since it is assumed that the power supply and diagnostics introduce no delay nor error (Figure 2).

In air core tokamaks the poloidal flux ψ_i at the point (R_i, Z_i) can be decomposed as:

$$\psi_i = g_{ij} I_j + \psi_i^{(p)}, j = 1,..,N_c \quad (1)$$

where g_{ij} is the Green's function matrix relating the poloidal flux at (R_i, Z_i) to the unit current at (R_j, Z_j); the N_c coils account both for the PF and the eddy currents. The last term in (1) is the poloidal flux contribution from the plasma. By linearizing the mapping fluxes-currents in (1) we get:

$$\delta\psi_i = [g_{ij} + \frac{\partial \psi_i^{(p)}}{\partial I_j}] \, \delta I_j \equiv c_{ij} \, \delta I_j, j = 1,..,N_c. \quad (2)$$

Fig. 4 shows the prediction of the TSC code compared to that of the linear model (2) derived by the perturbation approach coded in PROTEUS [4]. The dashed contour is the plasma separatrix at the reference end of burn equilibrium. The solid line shows the separatrix as computed by TSC after a drop in ℓ_i of 0.1 and obtained by using a hyper-resistivity model to simulate the flattening of the current profile in 1 ms. The separatrix shown in the figure is computed 0.4 s after the sudden flattening. The prediction of the linear model is shown by the dashed-dotted line. In both cases feedback control has been disabled. The agreement between the two models is quite good since after 0.4s the evolution is strongly coupled to the passive structure dynamics, which is well described by the linear approximation.

At shorter times, when the separatrix displacement is primarily influenced by the plasma current profile re-distribution and the inductive current mode in the metallic structures, the comparison (not shown here) is not as good [2]. This indicates that on the time scale of the main resistive modes of the structure, the linear model describes well the plasma boundary dynamics; this is not the case at higher frequencies.

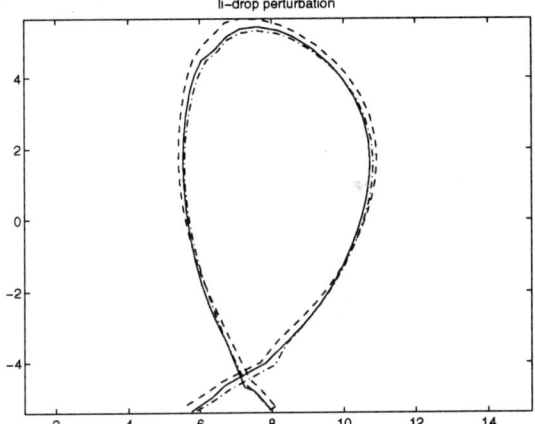

Figure 4. Comparison of a non-linear model (solid) and linearized model (dashed-dotted) prediction.

2.3. Controller design

Two main approaches are used to design control algorithms for ITER [2], differing in the way the last term in (1) is accounted for in the control law that regulates the flux at the plasma boundary[(#)].

Green's function methods. The main feature of this class of methods [5] consists in correcting the flux error $\Delta\psi_i = \psi_i - \psi_{xp}$ at the boundary points by using only the Green's functions plasma-PF coils g_{ij} in (1) (the method can be generalized to include the effect of eddy currents [5]), the plasma contribution being neglected. In this approach the plasma current and current centre position (plasma moments control loop) are controlled separately from the plasma boundary (plasma shape control loop). To minimize the interaction between the two control loops, the shape controller is designed to drive currents which minimize the flux, radial and vertical field linking the plasma [5]. The way in which the shape and the plasma moments control voltages are driven by the flux error signals is part of the specific control algorithm used (e. g. PID, state-space methods, etc.).

[(#)] The plasma separatrix is controlled by regulating the PF currents to achieve $\psi=\psi_{xp}$ at a set of N_p points laying on the desired boundary, ψ_{xp} being the poloidal flux at the desired X-point location where $|\nabla\psi|=0$; the plasma current is controlled by changing ψ_{xp} to account for the inductive and resistive losses. A regularization method is applied to minimize the control currents.

This class of methods has the advantage of avoiding the modeling of the plasma behavior to design the shape control law. The PID gains can be obtained, for example, by direct tuning on accurate nonlinear models [6] or directly on the machine. The disadvantage is that the interaction between the two control loops and the approximation $\Delta\psi_i \approx g_{ij} \Delta I_j$ may lead to several design iterations and, at last, to non-optimal performances as, for example, strong coupling among the controlled variables and/or large power demands.

Jacobian methods. In this approach [4] the coupling between the plasma and the control points is explicitly accounted for by the Jacobian matrix C derived in (2); the matrices of the model derived in the linearization step are used directly to design the controller which can be multivariable, addressing at the same time, vertical stability and shape control issues. Again, the actual technique used to derive the control voltages from the error signal differentiate the control law within the class.

The advantage of this approach consists in the potential accuracy of the resulting control law: in fact, by a careful choice of the C-matrix rows (badly conditioned matrices can be easily obtained by this approach) and by benchmarking the linear model on the experiment, these methods allow excellent performances [3]. The weak point stems from the difficulties in the accurate modeling of the plasma flux variation term [2], that might offset the potential advantages described above.

Application. These two approaches are routinely used to design controllers for ITER [2]. As an example, Figure 5 shows two closed-loop simulations based on a tokamak model linearized about the start of burn condition. By regulating the voltages applied to the set of 8 PF coils [2], the PFCS must control plasma current and shape after a step change $\Delta\beta_p=-0.2$ followed by a recovery in 5 s. The two simulations differ in the feedback controller used. In one case the Green's function approach is used to design two multi input-multi output (MIMO) PIDs to control the plasma current and its centre (first PID) and four points on the boundary (second PID). In the other case, by using the Jacobian matrix, a single (MIMO) PID controls the plasma current and six points on the boundary. Figure 5 shows the plasma major radius variation following the perturbation. In both case the plasma radial position is controlled and the maximum displacement is kept below 0.1m; the Green's function based PID is prompter and it requires more voltage.

It can be concluded that both approaches lead to controllers able to meet the specifications, allowing some freedom in the route to follow to the design.

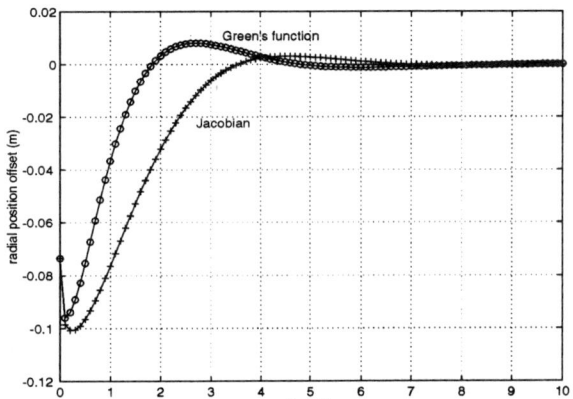

Figure 5. Linear simulation: PID controllers.

3. CONTROL SYSTEM PERFORMANCES

As part of the design and test of the control laws, nonlinear plasma simulations are performed with several specified perturbation and at various points in the discharge [2]. Here we report the closed-loop performances of a H_∞ [2] controller designed on a TEQ-based linear model and simulated on the Corsica code. Two events at the end of burn have been considered: (1) ELMs-like event, simulated as $\Delta\beta_p=-0.2$ in 1ms followed by a recovering occurring on $\tau_E \approx 5$ s; (2) Minor disruption, in which β_p varies as before and ℓ_i drops of 0.1 in 1 ms and recovers on the plasma resistive diffusion time scale $\tau_r >> \tau_E$.

In both events the plasma main displacement is an inward shift that must not exceed 0.15 m to avoid contact with the wall. Figure 6 shows the inward shift after the two events, measured at the innermost plasma point. As shown the controller is able to meet the required performance.

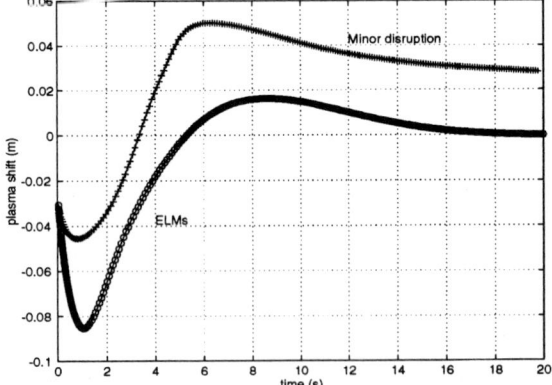

Figure 6. Non-linear simulation of H_∞ controller.

In the early phase, the plasma movement tracks the disturbance since force equilibrium is maintained on the Alfven time-scale. Then the plasma starts drifting on the passive structure field penetration time-scale, as the induced currents from the initial displacement develop. Up to this point the control system has little influence on the plasma evolution. As the simulation exceeds the field penetration timescale (t > 1s), the external fields from the PF coils begin to influence the plasma. The controller performs well for the beta drop case but it lacks an adequate integral action in the minor disruption control event where since the profile flattening persists for a long time (τ_r) the controller must counteract a quasi-steady state disturbance. In both events the control power (not shown here) is within the maximum allowable of 250MW [2].

4. CONCLUSIONS

The multi-step procedure used in the design of the ITER Poloidal Field Control System has been discussed and some of the results obtained shown.

Two different approaches are used in the design of the control law; these are based on the Green's function approximation and linear expansion of the flux at the plasma boundary. Nonlinear simulations have shown the performances of a H_∞ controller designed on the linear expansion method. Although a number of improvements should be made to the design, the controller is able to keep the plasma off the wall after ELM and minor disruption like events.

Additional steps are necessary to complete the design of the PFCS control system; for example, the integration of feedforward and feedback controllers needs to be improved, the goal being the simulation of a whole reference ITER pulse with likely MHD activity and a realistic diagnostic system. Moreover the design of the Supervisor functional logic should be studied in more detail. Under the assumption that the magnetic and kinetic controllers are weakly coupled, the modeling of the kinetic behavior has only been done via prescribed changes to the plasma energy. Further studies must confirm the validity of this assumption.

DISCLAIMER. The disclaimer contained in ITER Publications Procedures S AC PP 1 93-10-12 W2 applies to this paper.

REFERENCES
1. General Design Requirements Document, ITER EDA, S 10 GDRD 1 95 02 03 F1.
2. ITER Design Descr. Doc. 4.7, App.E, June 1996.
3. M. Garribba et al., 15th IEEE / NPSS SOFE, Hyannis, MA, 1993.
4. R. Albanese et. al., Nuclear Fus., 29, no. 6, 1989.
5. F. Hoffmann, S. Jardin, Nuclear Fus., 30, no 10, 1990.
6. C. Kessel, S. Jardin, ITER Task D324, June '96.

Modelling and engineering aspects of the plasma shape control in ITER

R. Albanese[1], G. Ambrosino[1], E. Coccorese[1], J.B. Lister[2], A. Pironti[1], D.J. Ward[2]

[1] Consorzio CREATE, Dipartimento di Ingegneria Elettrica, Universita di Napoli, Italy
[2] CRPP-EPFL, Association EURATOM-Suisse, 1015 Lausanne, Switzerland

As part of the ITER Engineering Design Activity, a number of questions related to plasma control has been addressed, using linearised and non-linear simulation codes to assess the control of the plasma shape given the particular design restrictions of ITER.

1. INTRODUCTION

Due to the size and consequent cost and power limitations of ITER, there will be less margin for plasma control than on current tokamaks and so particular attention is being addressed to the adequacy of the plasma control system. In this paper, we address several issues related to the Poloidal Field Control System, using and comparing two models, the TSC code [1] and a linearisation of PROTEUS [2].

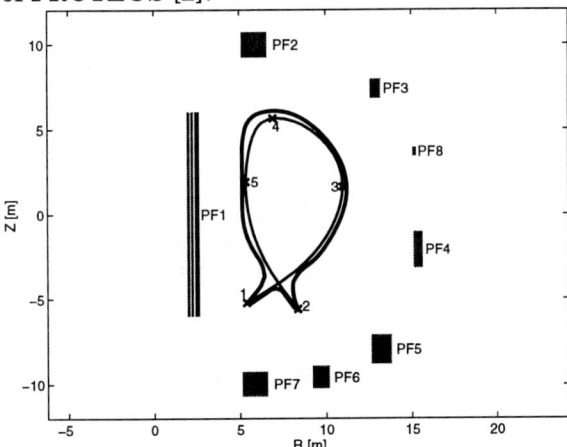

Figure 1. ITER PF coils, plasma facing wall contour, separatrix at End Of Burn and the 6 point separatrix-wall gaps used to define the plasma response.

TSC (Tokamak Simulation Code) is a two-dimensional time-dependent free-boundary MHD simulation code. Perturbed equilibrium calculations from the PROTEUS code are used to determine the non-rigid linearised plasma response to individual variations in conductor currents and to variations of βp and li, given some simple physical constraints on conserved quantities. It is important to verify that the latter model, well suited to the design of controllers and scoping studies of control techniques, agrees with the more physically complete former model. More complete details of this work can be found in [3].

2. MODEL COMPARISON

As a first test, a 1 second square voltage pulse was injected into the PF coils one at a time for the End-Of-Burn equilibrium, in the absence of shape or positional feedback. Figure 2(left) illustrates the case of PF7. The agreement between the two models is good both for the contour displacement seen in the figure and the PF coil currents, not shown.

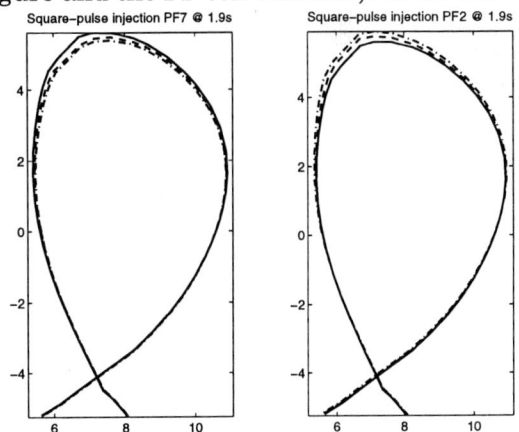

Figure 2. Injection of a square voltage pulse into PF7 (left) and PF2 (right). The TSC evolution of the separatrix-wall gap contour is shown as dashed lines and the linearised simulation is shown as dotted-dashed lines.

Figure 2(right) shows injection into the PF2 coil, with worse agreement, up to a factor of two between the peak contour displacements. The origins of this disagreement are being sought, particularly where it might stem from physical realities missing from the linear model, such as the induction of surface currents.

Secondly, we compare the open loop response of the separatrix to a plasma disturbance, namely $\Delta li = -0.1$ over 1 msec, subsequently evolved for 0.5 sec. Figure 3 compares the separatrix contours of the two models at 0.1 and 0.5 seconds after the disturbance. The initial displacement does not show perfect agreement, but the subsequent evolution agrees well. The PF currents show a significant discrepancy at 0.1 sec, with PF8 showing a sign difference and PF1 showing a significant difference in magnitude. The deformation of the complete contour shows reasonable agreement at the early time. At t = 0.5 sec, the evolution is more dependent on the passive structure dynamics and the agreement is good, for both the contour and the currents.

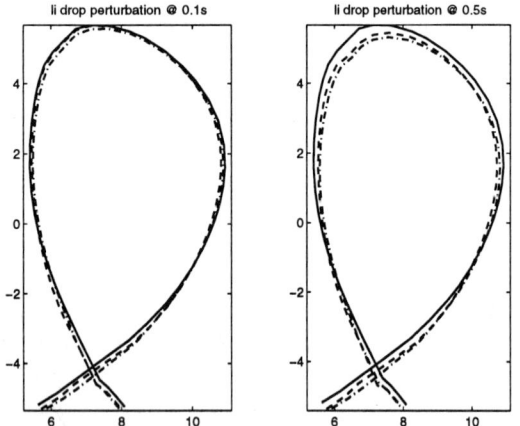

Figure 3. ITER disturbance response to $\Delta li= -0.1$ at t=0.1 sec (left); t=0.5 sec(right).

$\Delta \beta p = -0.2$, modelled in TSC by enhancing the anomalous thermal diffusion coefficient over 4 msec. again gave good agreement for the slow evolution, but the initial displacement is in less good agreement than the case of the li drop. PF8 and PF1 again show anomalous behaviour at the early time. At the later time, the PF currents are not too dissimilar, but the contour shows a significant departure. Combining a βp drop and li drop shows better agreement, the previous disagreements partially cancelling. Again, the reasons for the apparent disagreement are being sought. However, the overall agreement is considered to be excellent, justifying the use of the linearised model in the design of the controllers. Further comparisons will be required to confirm this optimism.

3. PLASMA DISTURBANCE REJECTION

A study was carried out to estimate the hypothetical best possible control of each gap in the presence of a particular disturbance, namely $\Delta \beta p = -0.2$ and $\Delta li = -0.1$. Fully saturated voltages are applied to all PF coils at the time of the disturbance, with each coil voltage chosen to reduce the excursion of that particular gap. This procedure was carried out one gap at a time. Since the system responds better than any real controller, this control is referred to as the "Best Achievable Single Gap Control" (BASGC) and derives a lower limit of the response to the disturbance given the voltage saturation limits. All other gaps are non-optimal and may even have the opposite sign of desired reaction to the disturbance. The task of a full controller is exactly that of finding a suitable compromise for all gaps.

Figure 4 illustrates the time evolution of a typical gap (# 3, Fig. 1) using this BASGC. The solid line is the gap displacement with voltage control and the dashed line is the open loop gap displacement. The control action only has a significant effect after the initial displacement has completed, after which the evolution with no control and with BASGC occur on the same timescale, that of penetration of the flux through the passive structure. Since the control action response grows slowly, the response amplitude is dominated by the disturbance response.

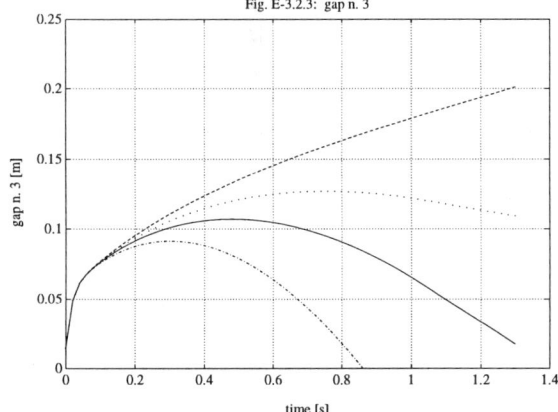

Figure 4. The evolution of gap # 3 following a disturbance given by $\Delta\beta p = -0.2$ and $\Delta li = -0.1$.

Table 1. Performance of the "Best Achievable Single Gap Control" for all 5 gaps with the disturbance $\Delta\beta p = -0.2$ and $\Delta li = -0.1$.

1	Excursion (cm)			Recovery (s)		
%	50	100	200	50	100	200
1	13.1	11.7	10.8	>2	1.2	0.7
2	7.8	7.5	7.2	>2	0.74	0.4
3	12.7	10.7	9.1	>2	1.28	0.8
4	12.0	8.7	7.0	1.8	0.88	0.5
5	11.2	6.3	4.	1.2	0.58	0.3

Halving the voltage limits, the gap displacement is only slightly worsened (dotted curves in Fig. 4). Doubling the voltage limits gives the dotted-dashed curves. The gap responses to this type of disturbance exceed the nominal specifications of the gap control, which should limit the displacement to 10 cm. The results of this analysis are summarised in Table 1, in which we list the gap numbering (Column 1), the maximum excursion (Column 2) and the time for the disturbance response to reduce to 2 cm (Column 3), both measurements assessed for 50%, 100% and 200%. of the nominal PF coil voltage limits. Although the excursion is not reduced much if the voltage is doubled, the recovery time is sensitive to the applied voltages. However, the recovery times with the nominal voltages are already more than adequate.

4. COIL VOLTAGE SATURATION

The most serious non-linearities in the plasma - passive structure - coils - power supply system are the voltage and current limitations of the power supplies themselves. The performance of any optimised controller is no longer guaranteed once the controller response reaches the power supply saturation. The performance of the closed loop will then depend sensitively on the amplitude of any disturbances or reference input changes. It is probable that the coil saturation presents more of a challenge to the controller design than the optimisation of the ideally linear closed loop. The problem of coil voltage saturation is likely to be much more severe in ITER than in present experiments in which it can be avoided by careful scenario programming to retain a sufficient voltage margin for control. The problem of current saturation in ITER will be identical to currently operating tokamaks.

Work has started on an approach in which the design of the controller explicitly considers the presence of coil voltage or current saturation, providing a modification to any A,B,C,D controller. The feedback error is adapted so that the controller outputs remain within the voltage and current saturation limits. Furthermore, the modification to the errors are chosen so that some of them are better respected than others, giving rise to the apellation "Hierarchical Saturation Controller". Any controller which has been designed to be optimal in a small signal sense can be handled, preserving as much of its functionality as possible in the presence of the saturation.

A plasma disturbance given by $\Delta\beta = -0.25$ is illustrated in Figs. 5(a,b). The coil voltage saturation levels were taken as: [10, 16, 14, 15, 17, 10] Volts per turn for PF2-7. For each simulation the following evolutions are shown: a) the 6 gaps; b) the 6 PF coil voltages (PF1 and PF8 were not used); c) the corresponding 6 PF coil currents; d) the total power requirement.

Figure 5(a) illustrates the case of an optimised LQG controller without saturation compensation. The voltage saturation is seen at different times on several coils. The power requirement, shown in the lower figure as a line of crosses, exhibits both positive and negative surges while the voltage saturation occurs. The physical origin of this effect stems from the fact that the energy required to correct a change in beta-poloidal and the corresponding vertical displacement is rather modest. The controller output demand voltage reflects this. However, if the coil voltage saturates, this correction is wrong and the total energy injected during the initial correction is no longer small. Subsequently, the controller has to recover from this energy error, and the power demand is again excessive since this correction itself suffers from a voltage saturation and a further power excursion occurs.

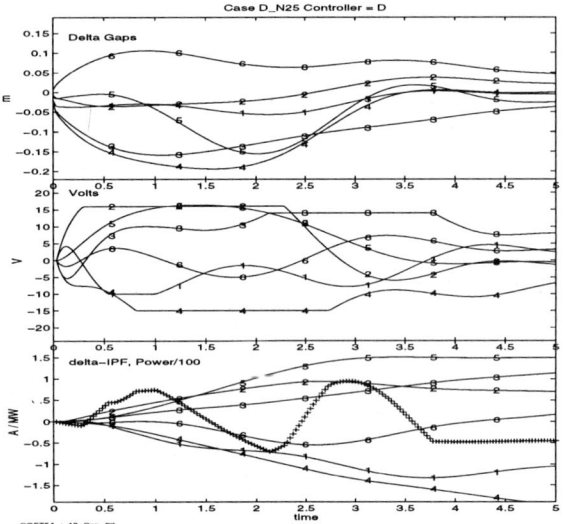

Figure 5(a). Control response without saturation correction.

Figure 5(b) shows the same disturbance, but with the saturation compensating controller in operation. Only a little improvement is found in the evolution of the gaps, and the voltage waveforms show a similar evolution, with trace 1 (PF2) no longer in saturation. Most noticeably, the power requirement has been reduced. Both simulations show asymptotic rejection of the disturbance. The uncompensated controller produced a PF power slew-rate which is outside the nominal specifications of the PF system, whereas the compensated controller had a barely noticeable power and power slew-rate requirement.

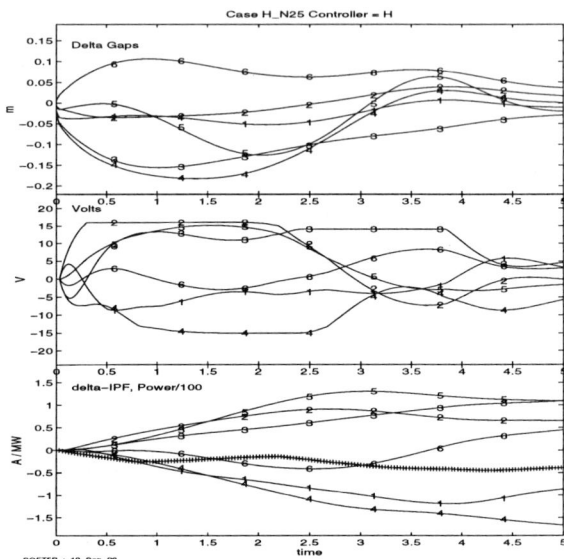

Figure 5(b). Control response with saturation correction.

Acknowledgement: This work was partly funded by NET Contracts. The authors thank to Fabio Villone and Parag Vyas for their help in the preparation of the paper.

REFERENCES
1. S.C. Jardin, N. Pomphrey, J. De Lucia, J. Comput. Phys., 66, 481(1986).
2. R. Albanese et al., Nuclear Fusion 29, 6 (1989)
3. J.B. Lister et al., Lausanne Report LRP 552/96 (1996)

Enhancement of the JET Vacuum Vessel Supports

A Miller and J L Hemmerich
JET Joint Undertaking, Abingdon, Oxon, OX14 3EA, UK

The evolution of the JET vacuum vessel support system is discussed. Originally suspended on flexible straps on its outer periphery in the mid-plane, the vessel is now restrained against vertical and lateral movement by a system of fixed and lockable struts and hydraulic dampers. Recent enhancements of this system are described.

1 INTRODUCTION

The design of the JET vacuum vessel features eight octants, each consisting of five rigid sectors and five bellows sections. The bellows sections have the purpose of increasing the toroidal resistance of the vessel to achieve the high loop voltages deemed necessary for plasma breakdown during the JET design phase. In order to stabilise the shape of this soft structure, two outer restraining rings above and below the mid-plane are rigidised by atmospheric pressure, operating under compressive hoop stress when the vessel is evacuated. Originally, the vessel was suspended by flexible straps and springs around its periphery in the mid-plane, basically like a gravity pendulum on a spring, any forces in addition to gravity leading to movements from its equilibrium position. Such forces, substantially exceeding gravity, can be induced during plasma operation, especially during disruptions. Their origin is subject of extensive studies, see for example references [1, 2].

These forces, particularly high during fast vertical displacement events [VDE] of the plasma centroid made it necessary to restrain the vessel against vertical movements. Vertical restraints were fitted in 1985 with the following features:

- the main horizontal ports were connected to the transformer limbs by fixed vertical struts, articulated to permit thermal expansion of the vacuum vessel;

- the main vertical ports were connected by pairs of angled struts to the transformer limbs. Thermal expansion of the vacuum vessel (the top of each port moves by +12mm from the mid-plane reference when the vessel is heated from room temperature to 320°C) complicated the design: the struts were coupled to the limbs through an inertial locking mechanism, permitting slow movement during vessel expansion but blocking fast movement during plasma disruptions;

- two inner restraining rings were welded inside the vacuum vessel to stiffen it against warping caused by the torque exerted by fixed outboard supports and inboard forces;

- hydraulic dampers were fitted between the mechanical coil support structure and the tips of the main vertical ports to dampen the rocking oscillation of the vessel once excited by vertical acceleration.

Furthermore, recent operational experience has shown that a small fraction (~1 to 2%) of VDEs is toroidally asymmetric, leading to lateral displacement of the vacuum vessel. As the previously installed vertical restraint system was inadequate to counteract lateral vessel movements, such events pose a serious risk to the integrity of vessel attachments. In particular, the rotary high vacuum valves, linking the main horizontal ports on octants 4 and 8 to neutral beam injector boxes (NIB) have shown internal damage (destruction of metal gaskets and imprints on sealing surfaces) directly traceable to lateral vessel displacements. As many operations (eg helium glow discharge, cryopump regeneration and maintenance) require these valves to seal correctly, it was deemed necessary to restrain lateral movements.

In the following, we describe design improvements on vertical restraints, hydraulic dampers and the new lateral restraint system.

2 VERTICAL RESTRAINTS

When first introduced in 1985, the struts connecting the main vertical ports (MVP) to the transformer limbs featured an inertial locking mechanism. Since locking could not be achieved reliably for the forces involved (up to 500kN per strut), these devices were modified several times. In their final form, they were actively locked by an electrically heated expansion rod once the vessel had reached operating temperature. The radii of the mating friction surfaces were shaped to match with close tolerances. This shaping resulted in a free gap of ~1mm between locking positions in opposite directions (up or down movement). Hence these devices had to be preloaded by locking at a temperature ~50°C below the final operating temperature of the vessel, using the thermal expansion of the vessel to lock the devices securely and to prevent sudden jumps or jitter in case of force reversal. This procedure, whilst achieving the required performance, permitted operation only at a pre-determined vessel temperature with the required thermal cycle time reducing available operational time.

This was further exacerbated by the additional weight of the vessel after installation of the in-vessel divertor system: in order to remove this additional load from the original mid-plane spring supports, the restraints on the lower MVPs had to be locked at a precisely calculated temperature below that for the upper MVPs, thus "lifting" the vessel to the operational mid-plane by thermal expansion of the vessel itself. To eliminate this elaborate and time-consuming procedure, the locking mechanism was modified as shown in Fig 1, using most of the parts of the previous system: the drum, excentrically connected to the MVP strut is now freely rotating in its housing on large roller bearings. Locking is achieved by a friction brake, connecting the housing to the drum through a standard spline/lever mechanism by a pneumatic actuator. Since the drum centre is fixed in the housing, there is no possibility for jitter during force direction reversal. Thus, a preload by thermal expansion of the vessel is no longer required. To make the struts on the

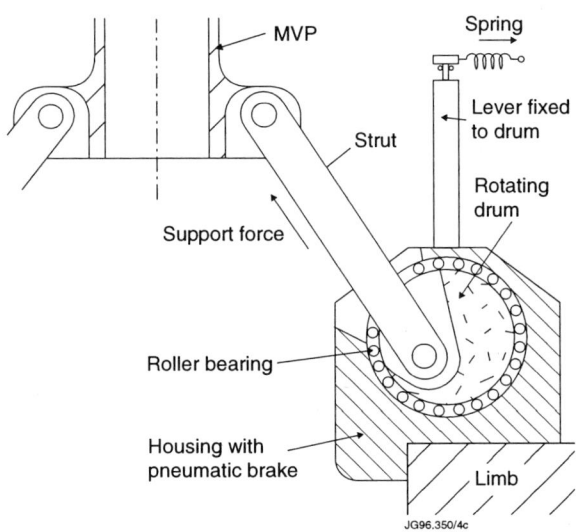

Figure 1 Main vertical port restraint

lower MVPs carry part of the vacuum vessel weight, lever arms loaded by constant force springs are attached to the drums. With this new feature, the lower MVP supports carry up to 80 tonnes of vessel weight independent of vessel temperature with brakes released to permit free thermal expansion. This new system works reliably and has led to substantial savings in operation time and increased flexibility in operating conditions. Operation is now possible at any vessel temperature, the only requirement being locking of brakes during a pulse.

3 HYDRAULIC DAMPERS

Elastic deformation of the restrained vessel due to vertical forces leads to a rocking motion of the vessel around its toroidal axis, with the tips of the MVPs oscillating in radial direction (upper and lower ports in opposite phase). To dampen this oscillation, viscous hydraulic dampers were introduced, connecting the tips of the MVPs to the mechanical structure. To limit dynamic forces/stresses on port attachments, the dampers were specified not to exceed a velocity proportional reaction force of $300 kNm^{-1}s$. During the last shutdown, these dampers were removed and dismantled. Internal damage was found, evidently due to a design flaw: since the piston and both shaft ends had been located in closely toleranced bores (over-defined design), the pistons had partially seized in the cylinders. Correct operation under such conditions cannot be taken for granted. Furthermore, tests performed after repair showed a deviation of the actual vs the intended design performance, with the reaction force being proportional to the square of piston velocity for velocities in excess of $\sim 0.1 ms^{-1}$, once fully turbulent flow in the viscous leakage path had been established.

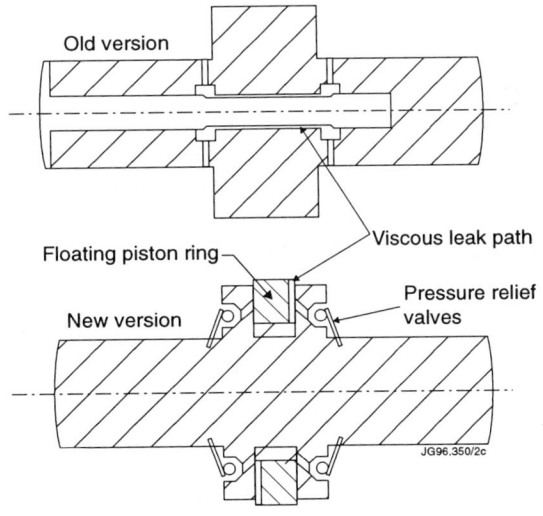

Figure 2 Hydraulic damper pistons

To remedy these deficiencies, the piston/shaft over-definition was removed by a floating piston ring and the reaction force was tailored to the required characteristic by means of internal bi-directional pressure relief valves in addition to viscous flow resistance. The old and new piston designs are shown in Fig 2. The characteristic reaction force vs piston velocity for the old and new designs are shown in Fig 3.

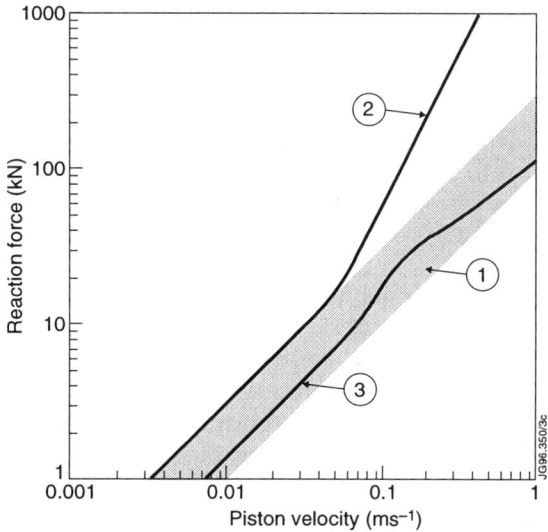

Figure 3 Damper characteristics:
1 design aim, 2 old piston, 3 new piston

4 LATERAL RESTRAINTS

Experience with the original vertical restraint system had shown its capability to cope with vertical vessel accelerations but failed to restrain the vessel horizontally: the MVP strut articulation permits displacements in the direction of a lateral force; at an angle of 90° to a lateral force, the locked angled struts fix the tip of the MVPs but the ports (narrow in toroidal direction) permit movement by their flexibility. It was therefore necessary to use the much stiffer main horizontal ports (MHP) for lateral restraint. The new system links the MHPs (using the MHP attachments for the MHP vertical struts) to mechanically fixed bridges between the mechanical coil support structure and the transformer limbs as shown in Fig 4.

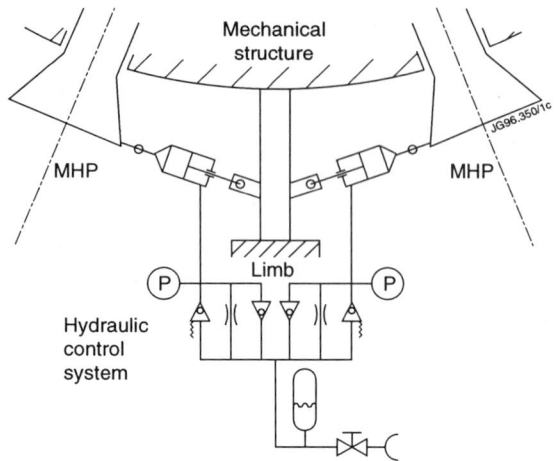

Figure 4 Lateral support schematic

These links use commercially available hydraulic cylinders which are connected to a passive hydraulic control system consisting of pressure relief, non-return and throttle valves, and pressure accumulators. They form a complete belt surrounding the vessel, fixed to the mechanical structure in eight (octant joint) locations. The belt is pretensioned to a force of 50kN by charging the accumulators to 70bar pressure. The throttle valves permit slow movements (thermal expansion, re-centering after displacements). The pressure relief valves are adjusted to limit reaction forces to ~200kN per cylinder during fast displacements, the non-return valves permit fast re-adjustment of the opposite cylinder of each pair during fast displacements. Stress analysis by finite element methods[3] has shown the beneficial effects of this lateral restraint system (for a lateral force of 1500kN):

- reduction of stresses on MVPs from 150MPa to 50MPa;
- acceptable stresses on MHPs of ~50MPa;
- reduction of lateral displacement amplitude from ~9mm to 3mm.

5 CONCLUSIONS

By implementation of a lateral restraint system and improvements on vertical restraint mechanisms and hydraulic dampers, the JET vacuum vessel suspension system was improved to withstand vertical forces of up to 8MN and horizontal forces of up to 2MN of ~10ms duration (typical for disruptions). As this system could only be fitted to MHPs and MVPs penetrating the mechanical coil support structure, sustainable forces are limited by stresses in the root welds where the ports connect to the vessel.

Substantially higher forces, if required, could be counteracted by a rather drastic irreversible solution, presently not considered for implementation: to immobilise the vacuum vessel by filling the interspace between the vessel and the toroidal field coil/mechanical support structure with a sufficiently rigid medium, such as eg polyurethane foam. This would, however, limit the vessel operating temperature to $\leq 100°C$ by inhibiting free thermal expansion, and severely impede access to toroidal field coils and other systems.

The experience at JET has shown, that retrofitting of a vessel support system is a very complex and difficult task and that correct and fully efficient restraint systems should therefore be implemented at an early stage in the design of future large tokamaks such as ITER.

REFERENCES

[1] P Andrew et al, "Measured Currents in JET Limiters During Disruptions", Proc 16th IEEE/NPSS, SOFE 1995, p 770

[2] E Bertolini et al, "Engineering Analysis of JET Operation", ibidem, p 464

[3] M Buzio et al, "Axisymmetric and Non-Axisymmetric Structural Effects of Disruption-Induced Electromechanical Forces on the JET Tokamak", this conference

EFFECTS OF HIGH FREQUENCY DISRUPTIONS ON THE JET DIVERTOR CRYOPUMP, INCLUDING POTENTIAL JET TOROIDAL FIELD UPGRADES

S. Papastergiou and P. Ageladarakis

JET Joint Undertaking, Abingdon, Oxon OX14 3EA, UK

ABSTRACT

The JET Divertor Cryopump System was designed several years ago and the design took account of thermal and eddy current stresses. The design calculations were based on static analyses for disruptions with poloidal magnetic field variation of 110T/s and a toroidal magnetic field (TF) of 3.4T. The cryopump operated during the 1994-95 experimental campaign without any failures. Subsequently detailed endoscope inspections indicated no distortion or damage in any of the cryopump areas.

During this last experimental campaign, a significant portion of plasma disruptions were of high value and relatively high frequency. The eddy current stresses are in principle proportional to both the velocity of the disruption and the TF. The cryopump system was able to withstand these severe disruptions and in order to explain this ability, an analysis has been performed to investigate the dynamic/impact nature of the load, together with the natural frequencies of the structure. In addition, the transient nature of eddy currents, the 'skin' effect and the current decay time have been quantified.

The calculations indicated that, although any damping characteristics of the structure have no significant effect for these high value disruptions, the high stress areas of the pump have a natural frequency much smaller (~4%) than the frequency of the severe disruptions. Therefore, the dynamic effect of these impact loads is approximately 30% of an equivalent static load and resulted in no failure. These analyses confirmed the demonstrated and expected high reliability of the cryopump even for a possible upgrade of the TF to 4T.

1. INTRODUCTION

The JET Divertor Cryopump System (Fig 1) was designed several years ago and the design took account of thermal and eddy current stresses during normal operation and abnormal events like severe disruptions, loss of water, cryogen flow and loss of vacuum [1], [2], [3]. The cryopump was designed to slide on the divertor coil 4 and incorporated expansion devices (bellows, expansion/elongated holes) in order to minimise the thermal stresses. The mechanical, thermal and electrical properties of the materials used [3], [4] were such as to further reduce the thermal gradients and stresses, and in addition to minimise the eddy current stresses. Eddy current stresses are generated during plasma disruptions

2. STATIC ANALYSIS OF DISRUPTIONS

The design calculations of the pump were based on static analyses for disruptions with rates of change up to 110 T/s for the radial and vertical magnetic field and a TF of 3.4T. The radial and vertical magnetic fields before disruption were 1T. Finite Element models simulated the system geometry and the predicted stresses were rather high in certain places, but were accepted because it was felt that the assumed value of 110 T/s for a disruption was conservative.

The maximum stresses in the pump were predicted to be in the large stainless steel backplate and cryoclamps. (Fig 2). In particular, in the backplate ends and last toroidal cryoclamps; large eddy currents (~ 1kA) generate large bending moments. Specially designed cut-outs at the backplate ends and terminal cryoclamps with increased width reduce the eddy current stresses to acceptable levels.

3. IMPACT ANALYSIS OF DISRUPTIONS

During the 1994-95 experimental campaign, a significant portion of disruptions had high values of field rate of change (450 T/s) and high frequency (1 kHz). However the majority of disruptions was still much below 110 T/s and of lower frequency $\leq$ 100 Hz.

Despite these events, the cryopump operated for more than a year without failures. Subsequently detailed endoscope inspections indicated no distortion

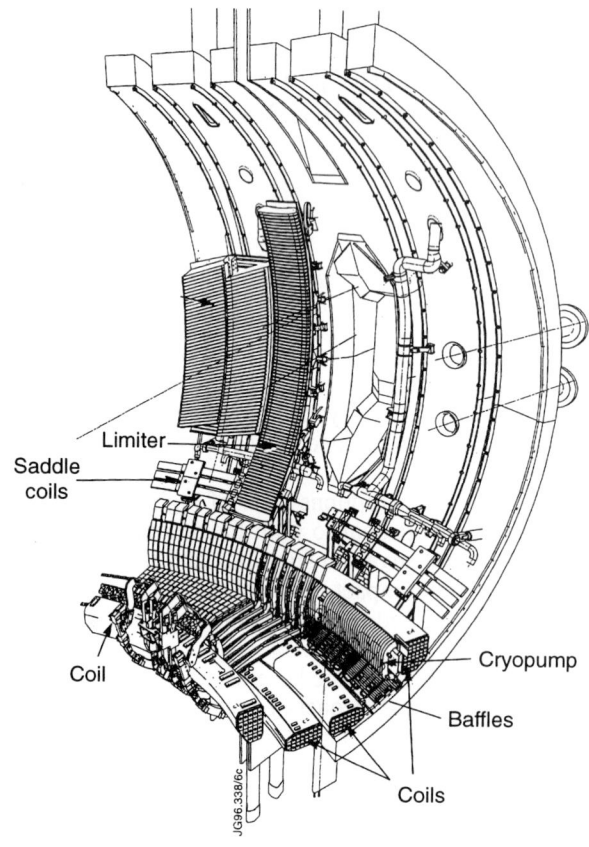

Fig 1. A Three Dimensional View of the JET Torus with a Mk1 Pumped Divertor

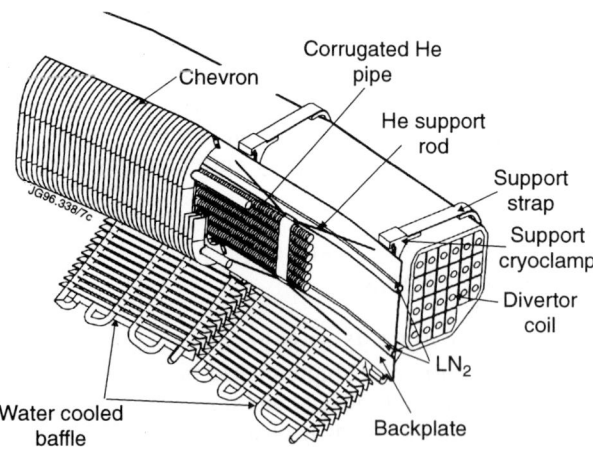

Fig 2. A Three Dimensional view of the JET Divertor Cryopump with associated water cooled baffles and divert coil No 4

or damage in any part of the system. The eddy current stresses are in principle proportional to both the velocity of the disruption and the TF. The cryopump was able to withstand these severe disruptions and in order to explain this ability, an analysis was undertaken to investigate the dynamic/impact nature of the load. In addition, this analysis accounted for a potential TF field upgrade to 4T.

In dynamic/impact phenomena, the higher the ratio of the load period to the natural period of the structure, the more severe the effect. In addition the maximum response to a very fast impulsive load will be reached in a very short time, before any damping forces can absorb much energy [6].

Therefore we investigated the highly stressed areas of the system, with relatively high natural frequencies and simulated these structures under impact/disruption loading with a single degree of freedom undamped models. The end cryoclamps were typical. They were particularly suited, due to their geometry, to one degree of freedom models. Their basic natural frequency is approximately 43Hz. The response of the cryoclamps to an impact load of 1kHz (1 ms) is divided into two phases, the interval, during which the load acts, and the subsequent free vibration stage. The ratio of the natural frequency of the structure to the load frequency is 4.3% and Fig 3 shows the predicted dynamic effect (D) of impact square loads as a function of this frequency ratio [6]. This effect D is 0.27. Due to the very short duration of the loading the maximum response occurs during the free vibration phase of the structure (after 1ms) [6].

Figure 4 shows in addition to the JET MK1 In-Vessel Pumped Divertor Geometry, the direction and measured values of a typical high value, high frequency disruption

Accounting for orientation, the maximum value of the radial magnetic field rate, $\dot{B}_r$, is given by

$$\dot{B}_r = 450 \sin\theta_1 \pm 200 \sin\theta_2 \leq 310 \text{ T/s}$$

For a potential TF increase to 4T, the dynamic analysis above, when compared to the initial static analysis of the design phase, produces a stress magnification factor of

$$0.27 * \frac{4T}{3.4T} * \frac{310 T/s}{110 T/s} < 1$$

It is therefore clear that the static analysis performed during the design phase with 3.4T and 110T/s is more severe than the effect of disruptions with maximum rate of field change of 450T/s, for 1ms, even accounting for a possible increase of the TF to 4T.

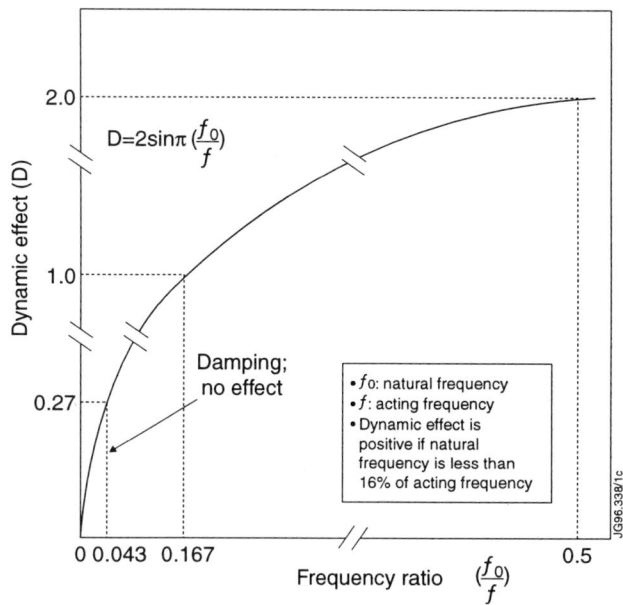

Fig 3. Dynamic Effect (D) of impact Square loads as action of the ratio between the natural and acting frequencies.

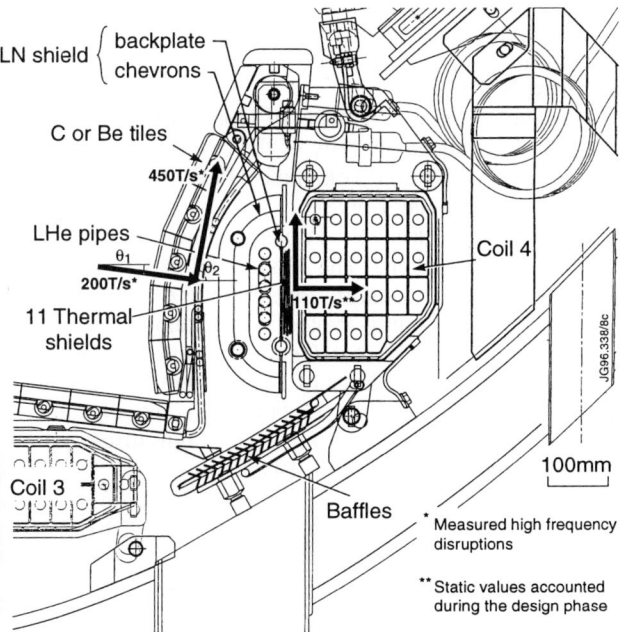

Fig 4. Cross Section of the JET Mk1 Pumped Divertor indicating measurements of high frequency descriptions

4. SECONDARY DYNAMIC EFFECTS

4.1 Skin Effect

The eddy current penetration in a component is proportional to the square root of the ratio of the component's electrical resistivity to the frequency of the disruption [7]. Therefore high frequency disruptions may not penetrate fully components with low electrical resistivity. It can be shown that a 1kHz disruption results to a penetration in Cu of less than 2mm. This phenomena may be present in parts of the Liquid Nitrogen (LN) circuit of the cryopump but the effect in terms of stresses is marginal because of the small thickness (2mm) of the chevrons of this component (Fig 4).

4.2 Eddy Current Decay Time

Eddy currents decay with a time constant which is proportional to the component thickness or skin depth, the electrical conductivity and the square root of the area in which the currents are flowing [7]. For high frequency disruption of ~1kHz and poor conductors (i.e. stainless steel) the current decay time is of the order of only 1-2ms; again not particularly significant to affect the conclusions of the dynamic analysis of the high value, high frequency (1 kHz, 1ms) disruptions on the stainless steel cryoclamp.

5. EFFECTS OF LOW VALUE, LOW FREQUENCY DISRUPTIONS

It can be shown that the dynamic magnification factor D varies as a sine function of the load pulse length ratio to the natural period of the structure and can reach a maximum value of 2 (Fig 3). [6]. This value can be reached only with a rectangular step load, zero damping and for a load pulse ratio of a greater or equal to 0.5.

Therefore for the highly stressed terminal cryoclamps, with a natural frequency of $\simeq$ 43 Hz, disruptions with a frequency of less than 86 Hz (or duration of more than 12ms) and value more than 55 T/s, result in higher stresses than accounted in the static analysis of the design phase. Such disruptions occurred and were really the most severe ones (not the 450 T/s, 1ms events). These could result in stresses very near or just above the component yield point but the detailed endoscope inspection indicated no such deformation.

This analysis ignores structural damping and is therefore conservative. Damping may have a significant mitigating effect in low frequency disruptions. Critical damping (equal to the square root of the stiffness multiplied by the mass of the system) can reduce the parameter D, by a factor of up to two [6]. The cryoclamps are supported on the casing of one of the Divertor coils. Underneath the casing there is the epoxy and copper structure of the coil which would effectively introduce some damping in the structural response.

In addition to structural damping, the effect of structural support position on eddy currents can be very significant indeed. If the system geometry permits, the choice of support points should result to lower stiffness/frequency components. Furthermore careful positioning of supports to minimise the bending moments in the structure can result in a stress reduction of up to a factor of 20 [9]. All these effects result in lower eddy current stresses and explain also the ability of the cryopump to survive the most severe, relatively low frequency and lower absolute value disruptions.

Finally, global sideways displacements of the Torus, up to ~5mm due to disruptions were observed during the 1994-95 experimental campaign. These phenomena are particularly pronounced at Octants 1 and 5, where the cryopump is fed cryogens by the cryofeeds. Such events could damage the concentric thin wall tube cryofeeds and were not predicted during the design phase of the cryopump. However in the design and installation phases adequate gaps between the cryofeed and the vacuum vessel were incorporated to account for baking of the vacuum vessel, to assist installation and to accommodate geometrical inaccuracies of the system. These gaps proved also adequate to protect the cryofeeds against the torus sideways displacements.

CONCLUSIONS

It is the value and relationship of the frequency of the disruption to the component natural frequency and not only the rate of change of field in the disruption that determines the severity of the event. High velocity disruptions may be severe for some components and not for others.

The dynamic analysis performed in the cryopump under the very high value, high frequency disruptions experienced during the JET Experimental Campaign of 1994-95 explained the ability of this component to withstand these events and to cope with a potential increase of the TF to 4T. The high stress areas of the system have still relatively low frequency compared to the high frequency of the applied impact load and therefore the generated dynamic stresses are mitigated and are smaller than equivalent static stresses under similar conditions.

Finally significant global sideways movements of the vessel during disruptions resulted in no major effects on the cryopump feeds.

ACKNOWLEDGEMENTS

The authors would like to acknowledge the assistance of P. Barabarschi and G. Sannazzaro (presently at ITER) for their contribution in the static calculations of the cryopump eddy currents. In addition M. Buzio assisted in providing data for the disruptions and global sideways movement of the Torus.

REFERENCES

1. W. Obert et al: JET Pumped Divertor Cryopump Proceedings of Symposium on Fusion Technology, SOFT 16, London 1990, pp 488-492.
2. S. Papastergiou, W. Obert, E. Thompson: The JET In-Vessel Divertor Cryopump. Design, Manufacture, Assembly, Testing and Operational Safety. Fusion Engineering and Design. To appear.
3. S. Papastergiou, W. Obert, E .Thompson: Material Selection, Qualification and Manufacturing of the In-Vessel Divertor Cryopump for JET. Advances in Cryogenic Engineering, Vol. 40B, pp 1429-1436, Plen. Press NY 1994.
4. P. Ageladarakis: Aspects of Operational Safety and Mechanical Integrity of the cryopump system in the JET Fusion Tokamak. PhD thesis 1996, Imperial College, London.
5. P. Ageladarakis, S. Papastergiou, D. Stork, H. Van der Beken: The Model and experimental basis for the Design Parameters of the JET Divertor Cryopump Protection System including variations in Divertor Geometry and first wall materials. Proceedings of Symposium on Fusion Technology, SOFT 19, Lisbon 1996. To appear.
6. R.W. Clough, J. Penzien; Dynamics of Structures, McGraw Hill, 1975.
7. L.D. Landon et al: Electrodynamics of Continuous Media, Vol. 8, 2nd Edition, Pergamon Press, 1984.
8. E. Bertolini et al: Engineering Analysis of JET Operation. Proceeding of Symposium in Fusion Engineering, Champaign, Illinois, USA, 1995. To appear.
9. S. Papastergiou and P. Ageladarakis: Optimisation of Tile Support Positions to minimise eddy current stresses. JET internal report DN-C(95) 135, 1995.

ENHANCEMENT OF JET MACHINE INSTRUMENTATION AND COIL PROTECTION SYSTEMS

V. Marchese, T. Businaro, M. Buzio, E. De Marchi, N. Dolgetta, J. Howie, J. Last, T. Raimondi, L. Scibile, J. van Veen,

JET Joint Undertaking, Abingdon, Oxfordshire OX14 3EA, UK

INTRODUCTION

A new system for the monitoring of the vacuum vessel motion due to magnetic forces and thermal cycles, called Machine Diagnostic System, has been installed during the 1995-6 shutdown of JET. Due to the very high neutron fluxes expected during operation under active conditions, only passive probes, such as linear variable resistors (LVR) and strain gauges, are used. Furthermore, the planned increase of the toroidal field above 3.45T has required an extension of the protection algorithms of the existing Coil Protection System to enhance the safety of the machine.

MACHINE DIAGNOSTIC SYSTEM (MDS)

The JET vacuum vessel support and restraint system, from its initial design to the present status is shown in Fig. 1.

When the vessel is at room temperature, 66% of its weight (240 tonnes) is taken by the Octant Joint Support (OJS) springs, equipped with 16 strain gauges, and the remaining 34% by the Main Horizontal Port (MHP) springs.

The top and bottom Main Vertical Port (MVP) restraints (lockable brakes), equipped with 64 strain gauges, take most of the vertical force acting on the vacuum vessel during plasma disruptions or Vertical Displacement Events (VDEs). The axial movement of the brakes is measured with 32 LVRs.

Up to 62 LVRs are used to monitor a wide combination of radial, vertical and tangential movements of vertical ports, horizontal ports and inner walls. The radial movement of the MVPs - top and bottom - is used also to detect the rolling

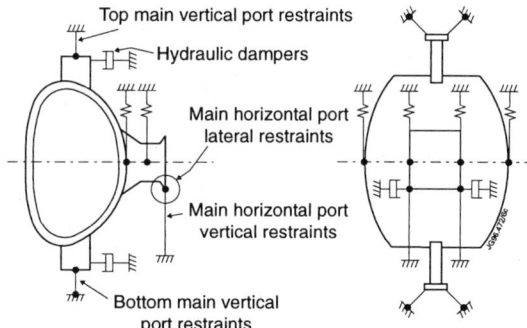

Fig.1 - *JET vacuum vessel supports and restraints*

motion $[(t-b)/2]$ excited by a twisting moment M_t applied to each octant. M_t is generated because the vertical force on the vacuum vessel is not in line with the reaction force of the MVP restraints [1].

The lateral restraints of the MHPs, monitored by 16 pressure gauges, have been introduced very recently to reduce the vessel sideways displacements. Four triaxial accelerometers, in the range 0 to 50g, have been installed at the MHPs of octants 2, 4, 6 and 8

The monitoring of mechanical transients of a tokamak machine is strongly affected by the time-varying magnetic fields. The noise induced by the magnetic field on the sensors is reduced by using a carrier at 5 kHz followed by demodulation.

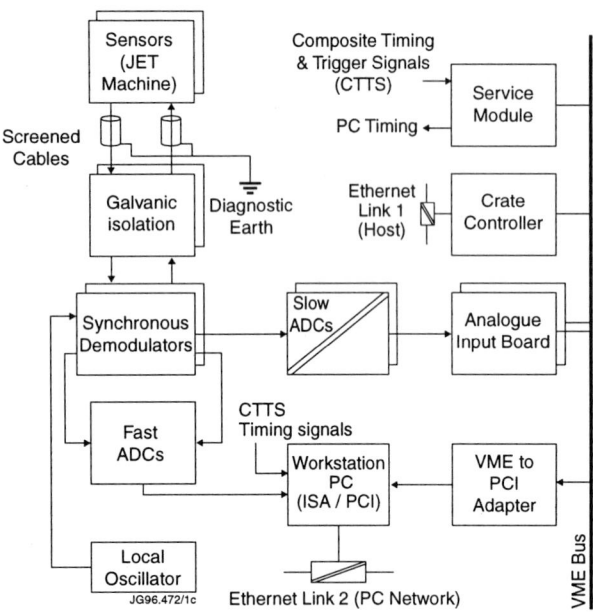

Fig. 2 - MDS hardware configuration

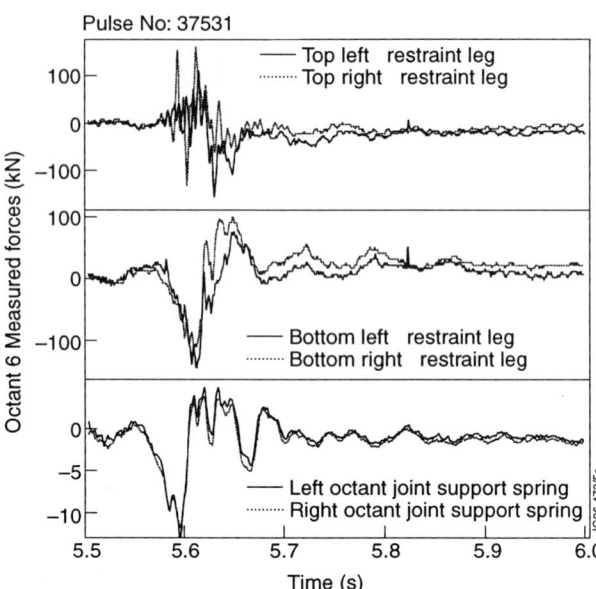

Fig. 3 - Forces on restraint legs and OJS springs

The demodulation is used to separate a sine wave, generally of a few mV, from a noisy signal, such as the output of a strain gauge bridge, and to determine its amplitude. In addition, the demodulators are synchronised to avoid noise generated from beating frequencies.

A new type of accelerometer and displacement sensor, insensitive to electrical noise, based on fibre optics and interferometer, is being tested for future application to JET and next step machines.

The VME based data acquisition system (Slow ADCs) is designed to sample up to 192 channels with a resolution of 16-bit. The data are sampled at 5 kHz and stored at 25 Hz every pulse and at 5 kHz in case of plasma disruption. After the pulse the data are transmitted to the host computer for the post pulse analysis. In addition, a sample of each channel is transmitted to the host continually at time intervals of 4s. The data on VME are accessible in real-time by a standard PC Workstation, via VME-PCI adapter. The PC is used for faster data acquisition (Fast ADCs) of a fewer channels, to compute power spectra and to test some fault detection algorithms, using a standard signal analysis package - LabVIEW - prior to implementation on a final target.

Figure 3 shows the force oscillations (ΔF) at top and bottom of octant 6 MVP restraint legs and OJS springs during an upward VDE with plasma current at 3 MA. The VDE was due to a loss of vertical stabilisation as a consequence of plasma internal perturbations. The net peak force measured on each octant varies from 200 to 300 kN with an estimated total peak force on the vacuum vessel of 2 MN. The net sideways displacement in the direction of octants 1-5 was 2.5 mm.

Figure 4 shows the variation of pressure (ΔP) in the lateral supports of six MHPs, during a plasma disruption at 3.8 MA and maximum sideways displacement of 1.6 mm. The ΔP in each lateral support is about 20 bar which corresponds to a reaction force of 14.3 kN. The peak tangential displacement - in average - is 0.6 mm. The poorly damped oscillation at 15 Hz on octant 2 right restraint is due to the rolling motion of the vessel.

The oscillation at 30 Hz on the tangential displacements has been observed also on the spectra of the radial displacements, not reported here, together with a further poorly damped oscillation at 40 Hz. These oscillations at 15, 30 and 40 Hz coincide with the 2nd, 4th and 6th harmonics estimated with a modal analysis of the vacuum vessel. Harmonics in the band 100-200Hz have been observed on the MHP accelerometer signals.

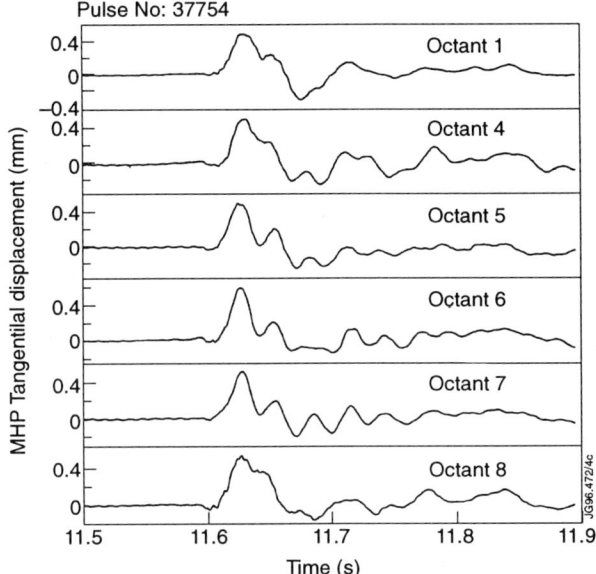

Fig. 4 - *MHP tangential displacements*

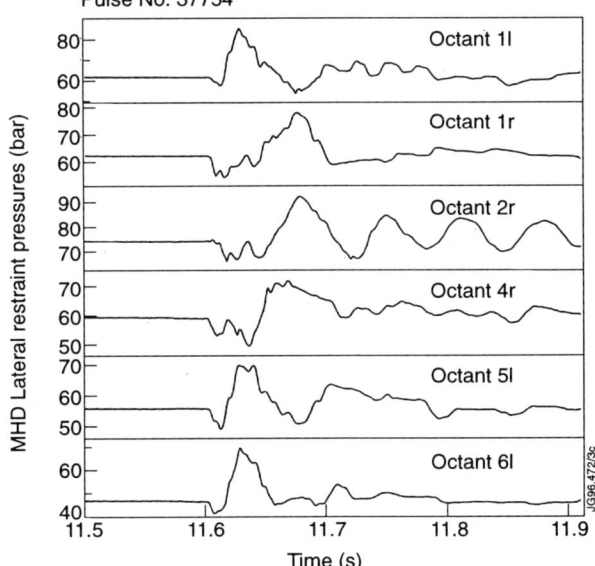

Fig. 5 - *Pressure variations of MHP lateral restraints*

COIL PROTECTION SYSTEM (CPS)

The system became operational in May 1994. The aim of CPS is to detect electrical faults and to protect the coils against mechanical or thermal over-stressing due to operation outside safe limits [2].

The protection implemented are: over voltage and over current for all the circuits; circuit equation simulation and comparison with the measured currents; ampere turn protection; tensile, shear and thermal stresses of poloidal and divertor coils. The protective actions include immediate removal of the voltage from the coils and circuit breaker trip.

The tensile and shear stress of the poloidal coils (P2-4) and divertor coils (D1-4) is computed as a linear combination of the vertical force, radial force and energy dissipated. The radial and vertical force of each coil are computed with flux loops and ampere turn measurements.

The main protection is implemented in VME by means of four high performance DSPs, working in parallel. About 170 electrical signals (e.g., voltages and currents) are sampled at 1 kHz and 30 thermal signals at 10 Hz. During the pulse, samples of the collected and calculated signals are simultaneously stored in 8 Mbytes of shared memory. After each pulse these are collected by the GAP programme on the host and archived. The system is backed-up by a hardware protection, incorporating over current during and outside pulses for all the circuits and over temperatures for the divertor coils.

A comparison between measured ampere turn (AT) and computed ampere turn (nI) for divertor coil D3 and a TF coil is given in Fig. 6. The maximum deviation is 1.33% for D3 and only 0.27% for the TF coil. A further reduction of the error is possible with a refinement of the compensation of the external fields.

FUTURE DEVELOPMENT

TF coil expansion. During pulses the TF coils expand in the radial direction (outward only) and in the vertical directions. This motion can be approximated, on a slow time scale, by a linear combination of the in-plane magnetic force - due to the interaction of the current with the toroidal field - and the dissipated energy.

$$\delta = K_m I_{tf}^2 + K_{th} \int I_{tf}^2 dt \qquad (1)$$

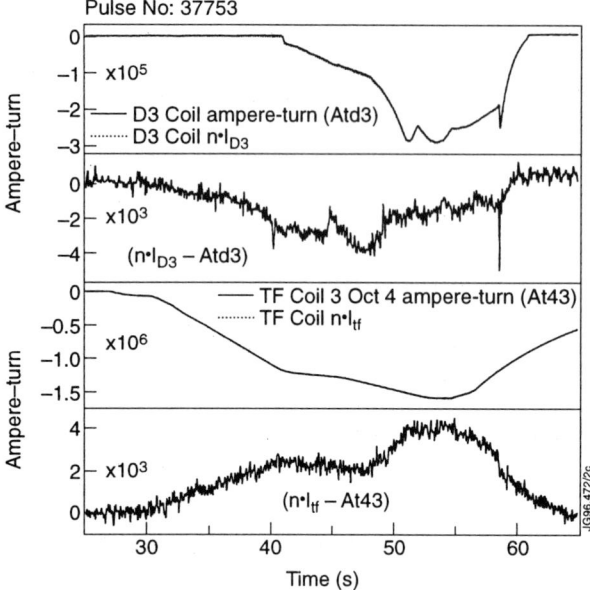

Fig. 6 - Ampere turn of a divertor and TF coil

The parameters K_m and K_{th} have been estimated experimentally for the radial displacement of 12 TF coils using a least square estimation algorithm available in MATLAB: $K_m = 28.1 \cdot 10^{-11} \pm 23\%$ (mmA^{-2}) and $K_{th} = 28.2 \cdot 10^{-12} \pm 5\%$ $(mmA^{-2}s^{-1})$. After the pulses, each coil reaches its rest position as the coil cools down. This simple model can be extended outside pulses introducing the time constant of the cooling. The monitoring in real time of such displacements is quite useful and, since it is not time critical, it has been proposed to implement it on a UNIX workstation connected to the host computer.

TF coil out-of-plane force and stress protection. Algorithms have been established for the calculation in CPS of the out-of-plane forces on the ring and collar teeth of a TF coil on the basis of the measured TF current (I_{tf}) and flux loop measurements (ϕ_{normal}). The maximum shear and tensile stress can occur at several places (at least 5 in the top and 5 in the bottom) within the coil. At each point (i) the mechanical stresses (s) can be expressed as a linear combination of the in-plane force, scaling with $I_{tf}^2 \, r^{-1}$, and the out-of-plane force scaling with I_{tf}:

$$s_i = a_i I_{tf}^2 + I_{tf} \sum_{k=1}^{10} b_{ik} \phi_k \quad (2)$$

Other effects, such as thermal gradients, are left aside. The out-of-plane forces are characterised by an irregular distribution, depending on the magnetic configuration and the real plasma parameters of each pulse, that are difficult to predict a-priori. The fluxes f_k are obtained as averages of the signals from flux loops installed on TF coils D24 and D64. The computed out-of-plane forces and mechanical stresses will be compared with two limits set at 5% and 10% above the allowable levels of stress.

P1 coil stress and controlled decay. During pulses the outward magnetic force on P1 coil is balanced by the inward pressure of the TF coils. Analytical expressions for the hoop, radial, vertical and von Mises stresses, as a function of the P1 current, ΔT and I_{tf} have been specified and are due to be implemented in real-time in CPS. The allowable limit is 60 MPa. A "coherent" decay of P1 and TF currents is required during the termination of the pulse to keep the stress in the safe region.

CONCLUSIONS

The number of sensors for the monitoring of the vacuum vessel motion has been increased substantially during the last shutdown of JET. The positive experience gained at JET with VME suggested to use the same technology, with automated fault detection, for the instrumentation of the machine. Finally, the upgrade of TF to higher fields requires the implementation, in real-time, of additional algorithms for the evaluation of the stress in the magnetising coil (P1) and TF coils.

REFERENCES

1. E. Bertolini et al, "Engineering analysis of JET operation", *Proc. 16th Symposium on Fusion Engineering (SOFE)*, Champaign, Illinois, USA, 1995.
2. V. Marchese et al, "Detailed design, installation and testing of the new coil protection system for JET", *Proc. 18th Symposium of Fusion Technology (SOFT)*, Karlsruhe, D, 1994.

PRESENT UNDERSTANDING OF ELECTROMAGNETIC BEHAVIOUR DURING DISRUPTIONS IN JET

P Noll, P Andrew, M Buzio, R Litunovsky, T Raimondi, V Riccardo, M Verrecchia[*]

JET Joint Undertaking, Abingdon, Oxon, OX12 3EA, U.K.
*ITER JCT, San Diego Work Site, USA

Disruptions in JET cause generally vertical displacement events (VDEs) and vertical forces at the torus. In various disruptions large amplitude locked kink modes were observed which led to lateral displacements of the torus. The toroidal asymmetry of the plasma was investigated on the basis of magnetic measurements.

1. INTRODUCTION

The JET divertor configuration is up/down asymmetric. Large perturbations can cause VDEs due to saturation of the vertical stabilisation. The resulting vertical force can produce significant vessel displacements and stresses [1].

In many disruptions the vessel forces and displacements are toroidally non-uniform. Peaking factors (local/average) of the vertical support forces of up to 1.8 have been reported [2]. Of particular concern is the global sideways displacement of the torus observed in some disruptions. The largest one recorded so far was 5.6 mm in the VDE of pulse 34078 (3.5 MA). Previous damage at vacuum seals of valves between the torus and neutral beam injectors has been ascribed to such sideways movements. The mechanical aspects of asymmetric forces is discussed in [3]. The plasma asymmetry is the main subject of this paper.

2. GLOBAL VDE BEHAVIOUR

In the worst type of a VDE the plasma reaches a large displacement before the start of the current quench, for example about 1m in the pulse 34078. A simulation of this pulse with the MAXFEA code suggests that the main contribution to the vertical force is due to halo currents. PF coil and vessel eddy current forces are not sufficient to provide a force balance at the plasma. The measured poloidal halo current I_H is about 0.6 MA, derived from the top/bottom difference of the toroidal magnetic field. The halo current force is $F_H \approx I_H \cdot B_T \cdot w$, where w is the radial width of halo current recirculation in the torus. With $B_T = 2.7$ T and an estimated width in the range 1 to 1.5 m one expects $F_H \approx 2$ MN while the simulation gives $F_H \approx F_{total} \approx 3$ MN. The reaction force measured at the vertical ports is ≈ 1.8 MN. The consistency among these values is fair considering uncertainties of measurements, of estimates of w, and of simulations.

3. TOROIDAL ASYMMETRY

3.1 Example Pulse 38070

The upward VDE of this pulse (fig.1a) gave a vertical force of 1.5 MN at the vessel supports (fig.1b) with strong toroidal non-uniformity (fig.2) and led to a 5.5 mm sideways displacement of the torus in direction octant 5 $\Rightarrow$ 1 (fig.1c).

The asymmetry of the halo is evident from the difference of the halo current evaluated from magnetic signals at opposite octants (fig.1d), and also from the toroidal distribution of halo currents measured directly from shunts at a subset of the array of mushroom tiles located at the upper outboard part of the torus. The mushroom tile currents exhibit large variations in amplitude and time. This suggests that the halo region is sweeping across the tiles with a non-uniform current density. The net asymmety can be seen from fig.3 which shows a strong peaking of time integrated currents around octant 6. The scaled current intercepted by all mushroom tiles is included in fig.1d. Initially these tiles appear to intercept the whole halo since all three signals shown are then about equal.

The toroidal asymmetry of the plasma is evident from the differences of the vertical and radial plasma current moments at opposite octants shown in fig.4.

These differences indicate a tilt of the plasma about a radial axis through octants 1 and 5 with amplitude

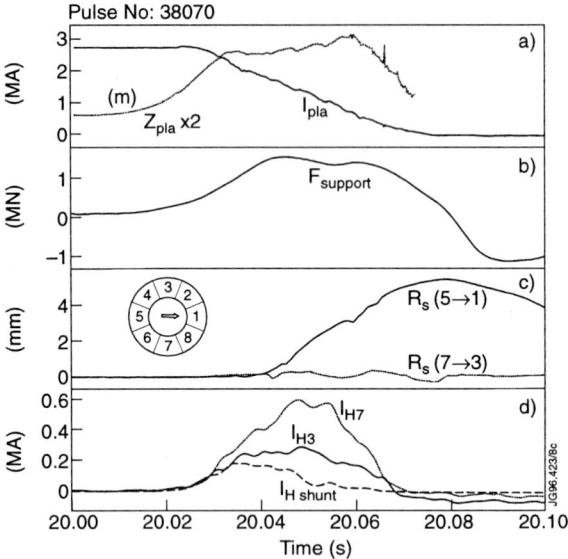

Fig.1: *Experimental data for pulse 38070. $R_s \Rightarrow$ torus sideways displacement in directions octant $5 \Rightarrow 1$ and $7 \Rightarrow 3$. $I_H \Rightarrow$ halo current from octants 3,7. $I_{Hshunt} \Rightarrow$ halo current scaled from mushroom tile currents.*

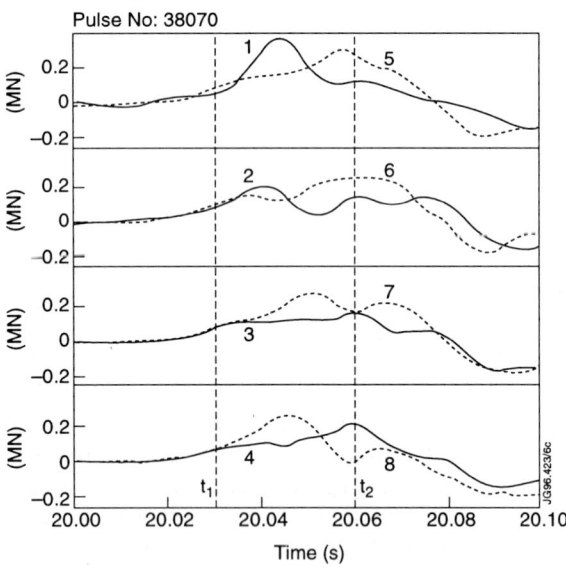

Fig.2: *Increment of vessel vertical support forces at octants 1-8 caused by VDE of pulse 38070. t_1 - t_2 is the interval where the plasma current moments are asymmetric at opposite octants.*

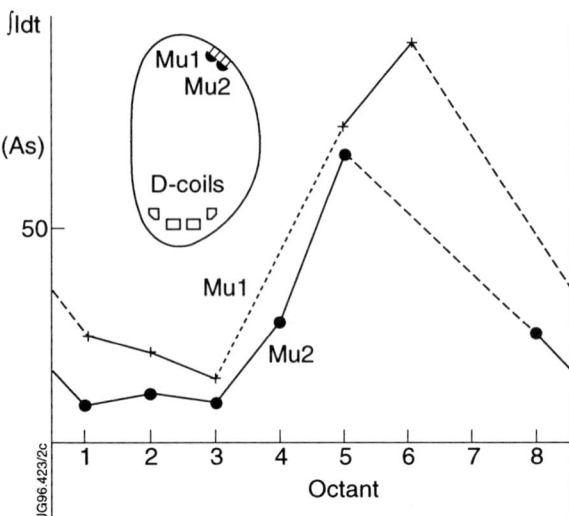

Fig.3: *Toroidal distribution of integrated mushroom tile currents.*

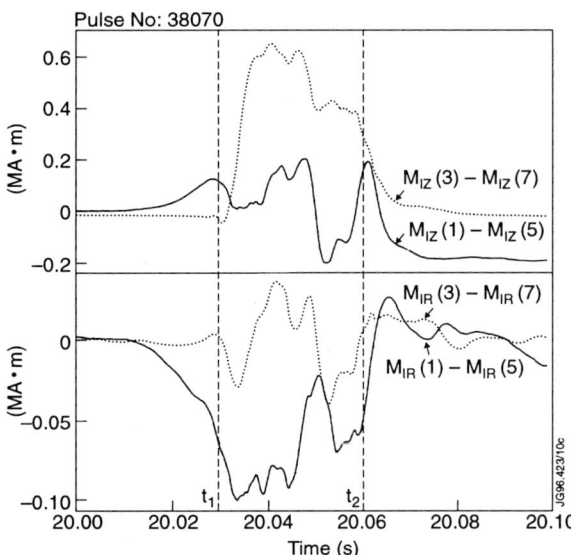

Fig.4: *Differences of the vertical and radial plasma current moments between opposite octants.*

$\delta Z_{max} \approx \pm 0.15$m (at $I_p \approx 2$ MA) and a small sideways shift of about 0.02 m along this axis in direction of octant 1. In octant 7 the plasma is closer to top of the vessel than in octant 3. The toroidal asymmetry of the halo current (fig.1d) and that of local halo currents flowing into the mushroom tiles (fig.3) is as expected from these plasma displacements. The toroidal asymmetry of vertical vessel support forces (fig.2) shows also a consistent correlation: force integrals are larger at octants 6,7,8 than at octants 2,3,4. In this and in a number of other pulses, the plasma mode was locked in one position so that torus displacements could escalate. In other VDEs with asymmetric plasma behaviour the mode appeared to move so that inertia prevented larger sideways torus displacements.

3.2 Simplified Model

The plasma is taken as a current ring tilted about a radial axis by an angle $\Delta Z/R$ and shifted along this axis by ΔR as indicated by current moment measurements, where ΔZ, ΔR are plasma displacement amplitudes from a ring with radius R. The displacement is rigid, the displacement vector projected on a poloidal plane describes an ellipse. A pure m=1/n=1 kink mode would correspond to $\Delta Z = \Delta R$. The destabilising forces acting on the ring may be decomposed into those arising from the toroidal field B_T and from the gradient $\partial B_R/\partial z = \partial B_z/\partial R$ of the equilibrium field. They must be balanced locally and globally by asymmetric repelling eddy and halo current forces between the plasma and the torus. The observed asymmetry of halo currents suggests that it plays a major role in the force balance. The model gives the global force F_x along the tilt axis, the tilt moment M_x and, included here for comparison, the global vertical destabilising force F_z at the plasma:

$$F_x \approx (\pi/2)\cdot[-\Delta M_{IZ}B_T + \Delta M_{IR}R(\partial B_z/\partial R)] \quad (1)$$
$$\approx (+2.4 - 0.03) \text{ MN} \approx +2.4 \text{ MN}$$
$$M_x \approx (\pi R/2)\cdot[-\Delta M_{IR}B_T - \Delta M_{IZ}R(\partial B_R/\partial z)] \quad (2)$$
$$\approx (-1.0 - 0.6) \text{ MN}\cdot\text{m} \approx -1.6 \text{ MN}\cdot\text{m}$$
$$F_z \approx 2\pi R\cdot I_p\cdot(Z_p - Z_{po})\cdot(\partial B_R/\partial z) \approx +3.4 \text{ M} \quad (3)$$

Values, taken from fig.4 and averaged over 30ms, are: $\Delta M_{IZ} \approx 0.5$ MA·m, $\Delta M_{IR} \approx -0.08$ MA·m. Furthermore $B_T = -3T$ at $R = 2.7$ m, and the estimated field gradient required for the original plasma shape is about + 0.1 T/m. The estimated horizontal impulse (2.4MN)·(30ms) is consistent with the dynamic behaviour of the vessel [3], characterised by a measured torus displacement of 5.5 mm, a reaction force at the vessel supports of 0.6 MN, and an observed almost critical damping. The estimated tilt moment acting on the torus is also in rough agreement with the reaction moment implied by the asymmetry of vertical support forces (fig.2).

3.3 Conditions for Torus Sideways Displacement

It can be expected that large displacements are only possible when
(a) the plasma displacement is large before the current quench,
(b) the plasma boundary q-value decreases to about one, to permit kink instability
(c) the mode is locked, not shifting phase or rotating as in Alcator C-mod [4].

The expected relation of sideways displacements with the minimum boundary q-value is shown in fig.5 using $q^* = 5B_T a^2/I_p R_I$ as an approximation. Larger displacements are only seen when $q^* \leq 1.3$. However, low q^* does not always give a large displacement as indicated in fig.5. This may be partly due to changes of the toroidal position of the kink mode giving changes in direction of the force at the torus, and also to the fact that the mode amplitude often remained small.

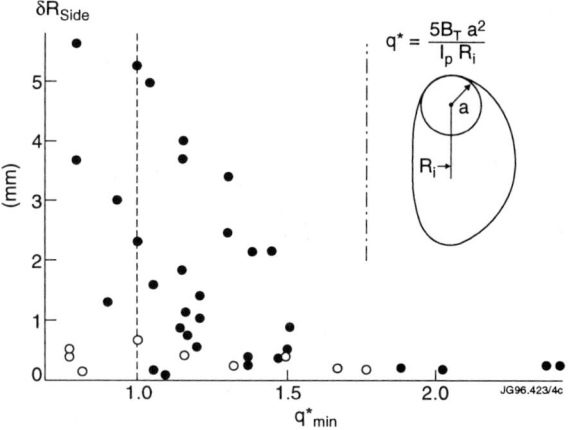

Fig.5: Statistics of sideways torus displacements plotted versus minimum q*. Open circles are downward VDEs.

3.4 Role of Plasma Configuration and In-Vessel Structures

The fig. 6 shows that sideways displacements are small when the initial triangularity is high. An inspection of displaced plasma shapes indicates that large sideways displacements can result when the plasma touches simultaneously structures at some poloidal distance in the upper part of the vessel, such as saddle coil sections, mushroom tiles and the upper dump plate. It appears that the shaping field applied for high triangularity reduces the poloidal extension of the attachment.

It is also suspected that the discreteness of the plasma facing structures in the upper part of the torus encourages a non-uniform development of halo currents intercepting these elements. A toroidal non-uniformity of the halo region could enhance the escalation of the kink mode.

Downward VDEs never produced sideways displacements of the torus exceeding 1 mm, even in cases where the conditions (a) and (b) mentioned above are well satisfied. In all downward VDEs and during the critical phase of high current the plasma/wall contact is poloidally localised at the top of the outer divertor target. This supports the hypothesis that a large poloidal extent of plasma/wall contact is a condition for the generation of a large sideways impulse at the torus.

In the search for causes of sideways displacements it was also noted that upward VDEs with high beta prior to the VDE gave smaller displacements than VDEs with low initial beta, as illustrated in fig.7. With high beta the impurity influx and the speed of current quench are found to be enhanced. The absence of larger sideways displacements may be partly attributed to the relative short duration of the current quench and consequently of large plasma asymmetries.

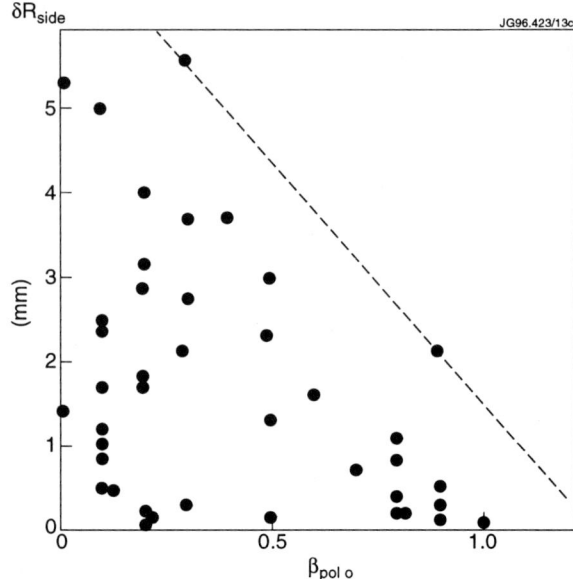

Fig.7: *Sideways torus displacements in upward VDEs plotted versus initial poloidal beta.*

4. CONCLUSIONS

- A large sideways force at the torus in one direction can occur in VDEs with large vertical plasma displacement at large current. It is caused by a locked m=1/n=1 kink mode which can arise when the plasma boundary q-value becomes ≈ 1.
- It is suspected that the choice of a first wall shape giving only a relatively small poloidal extent of plasma contact during VDEs would reduce the danger of creating sideways displacements of the torus.
- Neither are large asymmetric forces expected when the current decreases before the vertical plasma displacement becomes large. In JET this cannot be enforced.

References
1. P Noll et al, Fusion Technology, vol 15, part 2A (1989), p.259.
2. E Bertolini et al, Proceedings SOFE 1995, vol 1, p.464.
3. M Buzio et al, poster at this conference.
4. R S Graenitz et al, Nucl.Fusion, vol 36, no 5 (1996) 545.

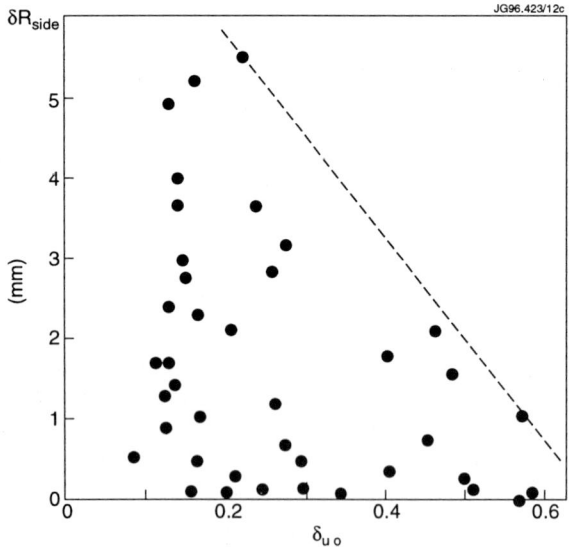

Fig.6: *Sideways torus shifts in upward VDEs plotted versus initial upper triangularity δ_{uo}*

Axisymmetric and non-axisymmetric structural effects of disruption-induced electromechanical forces on the JET Tokamak

M Buzio, P Noll, T Raimondi, V Riccardo, L Sonnerup [*]

JET Joint Undertaking, Abingdon, Oxon., OX14 3EA, UK

[*] Chalmers University, Gothenburg, Sweden

The large-scale mechanical response of the JET Vacuum Vessel and other structural components to disruption-induced loads has been analysed. The observed toroidal variations of the VDE-induced rocking motion can be explained by non-axisymmetric plasma forces. New evidence has been found in support of a plasma kink mode model which could account for the generation of the observed sideways forces.

1. Introduction

Plasma instabilities and, particularly, VDEs (*Vertical Displacement Events*) induce large eddy and halo currents on JET [1] first wall which give rise to deflections of several millimetres and transmit high loads throughout the structure [2]. Research activity is ongoing in order to analyse the structural dynamics of these phenomena from both experimental and computational viewpoints, with the purpose to enable safe machine operation. We shall focus first on the Vacuum Vessel, which is one of the most heavily loaded components, by describing the two main kinds of disruption effects i.e. rocking oscillations and sideways motion.

2. Rocking motion of the vacuum vessel

2.1. Axisymmetric rocking motion

Disruption-induced loads are characterised by radial and vertical components of several MN, with typical time scales ranging from 20 to 50 ms. Since the installation of additional Restraining Rings (1989) the Vessel has become quite rigid with respect to radial axisymmetric forces and, in the present configuration, the most conspicuous mechanical effects are due essentially to the vertical loads. Because of the particular arrangement of the supports, which block the vertical movement of all the ports, vertical forces exert a torque around a rotation centre positioned as in Fig. 1 and set the Vessel in a sort of axisymmetric "rocking" motion at ~14 Hz. This oscillation gets excited practically during all dynamic events and can be easily detected from the difference of the radial displacements δ_{MVP} measured at the top and bottom of the Main Vertical Ports (MVP). The qualitative stress pattern in such an event is also depicted in Fig. 1. The stress level throughout the structure is usually low, except for a concentration at vacuum-critical welds located at the corner at the root of the MVPs. Dynamic simulations using a detailed FE model of half an octant and simplified lumped parameter Simulink models with 2 and 3 d.o.f. have been carried out in order to assess the peak stress in various situations, and their results are in good agreement. The vessel response history for a "typical" case of upwards VDE, the most common and dangerous kind of event, is plotted in Fig. 2. The force history includes: 1) an initial steady-state phase, corresponding to the flat-top of the pulse, due to the attraction between plasma and divertor coils; 2) an upwards force pulse due to eddy and the halo currents induced by the plasma instability; 3) a large downwards force swing ΔF due to the plasma current quench and to the currents induced in the divertor coils; and finally 4) a slow decay of the attractive force between the divertor coils and the iron circuit. The most important feature is the force swing ΔF, which determines the dynamic part of the response and is responsible for the swing of the stress that has to be considered to assess fatigue life. The results of these computations, thoroughly validated against experimental data, show that the ratio between the MVP radial displacement

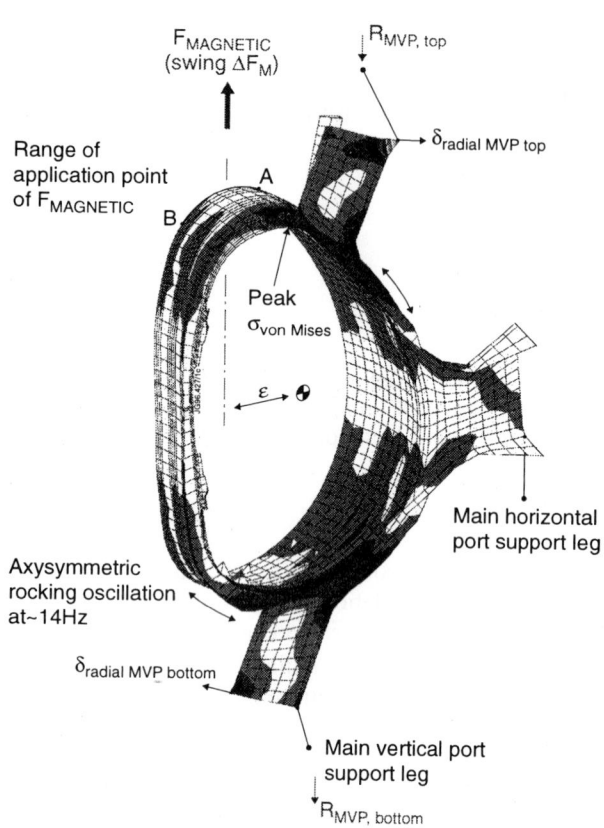

Fig. 1 - FE model showing the rocking motion mode

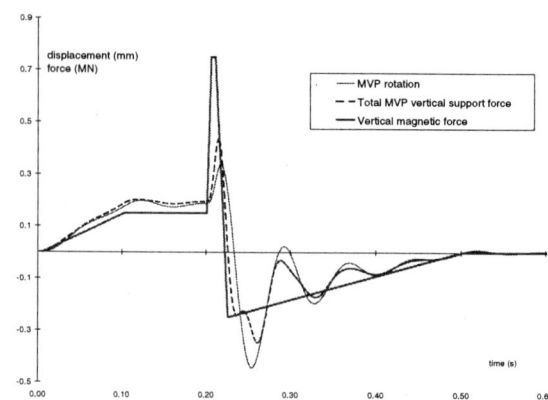

Fig. 2 - Dynamic simulation of vessel rocking motion for a reference case with ΔF=1 MN

swing and the MVP vertical support force swing $\Delta\delta_{MVP}/\Delta R_{MVP}$ is in the range 0.95~1.35 mm/MN, depending on the application point of the resultant ranging from A to B (i.e. going farther from the rotation centre, as shown in Fig. 1). Likewise, the amplification ratio between the magnetic and support force swings $\Delta F/\Delta R_{MVP}$ ranges from 0.8 to about 2.0. This kind of information can be used to deduce the total plasma force for a particular pulse, on the basis of deflection and support force measurements. The stresses at the base of the port have been calculated for many configurations and in the worst expected case, that is a force swing of 11 MN applied in B predicted by MAXFEA for a 6 MA VDE, they would cause a strain not larger than 0.5%. This level of stress could be withstood without danger for 1000+ events.

2.2 Non-axisymmetric effects

Observation of MVP support forces and displacements has led to an estimate of toroidal asymmetries. Deviations from average of strain gauge signals giving support force on each octant up to ±50% were measured, while deviations in the displacement signals are much smaller because of the "equalising" effect of the torsional stiffness of the vessel. Dynamic mechanical simulations indicate that in order to explain these asymmetries a peaking factor[1] of the order of 2 has to be expected in the toroidal distribution of magnetic force. This non-uniformity in the loading has little influence over the rocking motion and the related stresses; however, it may have a localised impact on in-vessel components.

3. Sideways vessel displacements

3.1 Mechanical analysis of sideways events

According to a model proposed by P Noll [3], the sideways force F_{SW} responsible for the observed horizontal movements of the vessel could be assumed proportional to the difference of plasma moment between opposite octants in the *perpendicular* direction, $\Delta(I_{PLA}z_{PLA})_{PERP}$. This difference, in turn, is directly related to the difference in vertical force applied to the Vessel

---

[1] Defined as the ratio (max. peak amplitude)/(average amplitude)

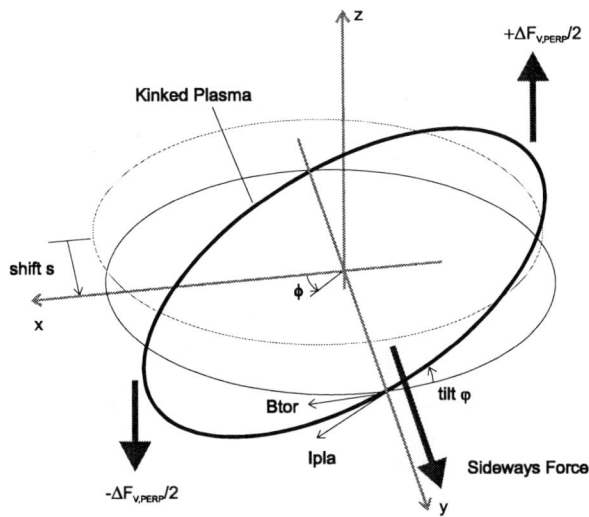

Fig.3 - Kink model for the sideways forces

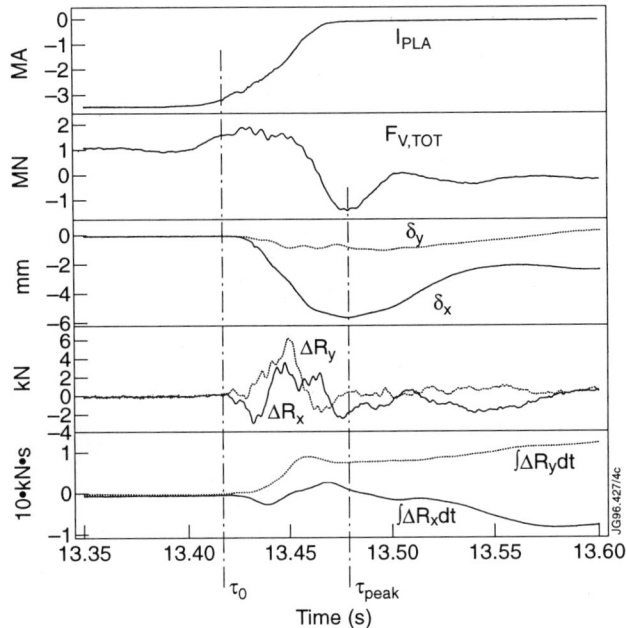

Fig.4 - Experimental data for JET pulse 34078

$\Delta F_{V,PERP}$, which in turn scales with the measured difference in reaction force $\Delta R_{MVP,PERP}$, through a coefficient depending on the vessel's dynamic response. Considering in the first approximation the vessel like a rigid, single d.o.f. system and integrating the sideways impulse equation[2] between the start of the instability τ_0 and the time τ_{peak} when maximum sideways δ_{SW} displacement is reached and the velocity is null, one can see that:

$$\int_{\tau_0}^{\tau_{peak}} (F_{SW} - R_{SW} - D_{SW})dt = \int_{\tau_0}^{\tau_{peak}} mdv = 0 \Rightarrow$$

$$\int_{\tau_0}^{\tau_{peak}} F_{SW}dt \propto \int_{\tau_0}^{\tau_{peak}} \Delta R_{MVP,PERP}dt = \kappa\delta_{SW}$$

where it has been reasonably assumed that both the sideways elastic reaction force of the supports R_{SW} and the damping force D_{SW} increase linearly with the amplitude of the displacement. This kind of approach lends itself to relatively easy order-of-magnitude checks: for example, the F_{SW} time scale for pulse 34078 (worst case so far with δ_{SW}=5.6 mm) as estimated from the history of $\Delta R_{MVP,PERP}$ is about 20~30 ms, corresponding to a dynamic amplification factor of ~0.3. Considering that the sideways support stiffness is $R_{SW}/\delta_{SW} \approx 100$ kN/mm and the natural frequency is about 3 Hz[3], this leads to an estimate F_{SW}=~2 MN (consistent with previous magnetic force calculations [2]). Another interesting correlation comes from the calculation of the proportionality coefficient κ for some JET pulses, which gives consistently $\kappa = \Delta R_{MVP,PERP}/\delta_{SW} \approx 1.5$ kN·s/mm. However, this result is observed only for pulses characterised by sideways displacements > ~4 mm, all of which occur approx. in the direction from Oct.5 to Oct.1. In other cases, where the n=1 mode is probably not locked (or the dynamic response of the vessel to higher modes is not sufficiently attenuated, so that $\Delta R_{MVP} \approx 0$) such correlation is not detectable. The response of the vessel to rotating forces or asymmetric MHP restraints may account for this, but further analysis is needed.

---

[2] Making use of the integrated support force difference signal provides an automatic way of smoothing noisy signals, minimising at the same time the effects of any additional higher frequency modes.

[3] these values have been observed experimentally and confirmed by FE analyses

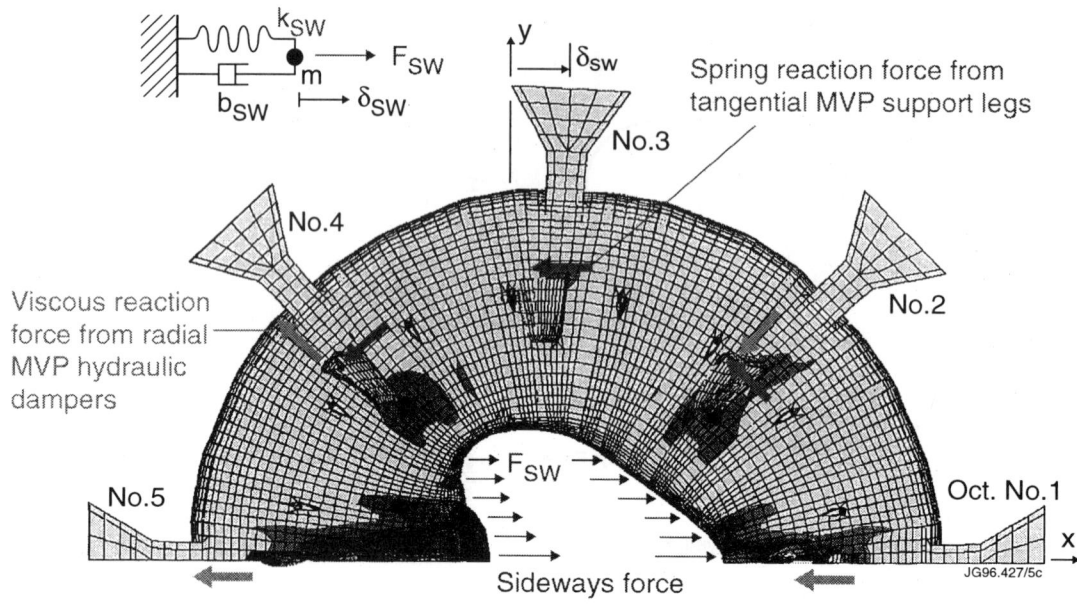

Fig.5 - FE 180° model showing sideways deformation of the Vacuum Vessel

3.2 Implications of sideways motions for other structural components

The presence of sideways forces raises the question of checking their effects not only on the Vessel, but also on all other components that could react such forces. Basically, all axisymmetric loads are reacted by the poloidal coils, while non-axisymmetric ones seem to be taken up by the toroidal system. According to the above mentioned n=1 kink model, the reaction forces applied to the TF coils might have a sinusoidal distribution which should be evaluated by means of a full 3D simulation. The horizontal resultant of such reaction should ultimately be taken by the P1 coil and the iron limbs, and an assessment of their effects is needed.

4. Conclusions

- Mechanical analysis of forces and displacements of JET components (primarily the Vacuum Vessel) can be used "backwards", in order to estimate the characteristics of the magnetic loads that caused them.
- The worst plasma instability that can be reasonably predicted could be sustained by the Vacuum Vessel for ~1000 pulses without any danger of structural damage or vacuum leak. No realistic load can cause failure of the whole Vessel, even if localised force peaks have produced many minor damages in the past and careful analysis is being devoted to in-vessel components [2]. The fatigue life of the MVP root remains the primary concern, especially considering that the machine has already operated considerably above the original design goals and will be probably required to do so for some time to come.
- The effects of non-symmetric reaction forces should certainly be included in the assessment and specifications of any Tokamak machine.

References

1. E Bertolini *et al*, Engineering experience in JET operation, this conference
2. E Bertolini *et al*, Engineering analysis of JET operation, Proceedings of SOFE '95
3. P Noll *et al.*, Present understanding of electromagnetic behaviour during disruption in JET, this conference

Scientific Basis and Engineering Design to Accommodate Disruption and Halo Current Loads for the DIII–D Tokamak*

P.M. Anderson, A.S. Bozek, M.A. Hollerbach, D.A. Humphreys, J.L. Luxon, E.E. Reis, M.J. Schaffer

General Atomics, P.O. Box 85608, San Diego, California 92186-5608, USA

Plasma disruptions and halo current events apply sudden impulsive forces to the interior structures and vacuum vessel walls of tokamaks. These forces arise when induced toroidal currents and attached poloidal halo currents in plasma facing components interact with the poloidal and toroidal magnetic fields respectively. Increasing understanding of plasma disruptions and halo current events has been developed from experiments on DIII–D and other machines. Although the understanding has improved, these events must be planned for in system design because there is no assurance that these events can be eliminated in the operation of tokamaks. Increased understanding has allowed an improved focus of engineering designs.

1. DISRUPTION FORCES

Disruptions drive large electric currents in the vessel and associated components by two identified mechanisms: magnetic induction and contact with halo currents [1,2]. The J×B forces resulting when these currents cross the magnetic field can be very large, possibly damaging in-vessel components or the vessel itself. The global force is reacted magnetically to the external magnetic coils and their support structure. Disruptions also induce electric fields that can break down electrical insulation and allow current to flow in unplanned places. The present DIII–D design philosophy is to avoid all but low voltage standoff in plasma facing components and to ensure that induced currents follow a planned, safe path.

1.1. Induced Current Loads

The DIII–D vessel is all metal (Inconel 625) with no insulating breaks and relatively uniform conductance. Vessel current is magnetically induced in the toroidal direction by the time derivative of the poloidal magnetic flux and is limited by toroidal resistance. Toroidal vessel voltage is measured by 19 toroidal loops attached to the vessel outer surface. The loops also measure poloidal flux for plasma control and diagnostic purposes. Loop voltages at internal components are adjusted for the time derivative of the additional flux between the vessel and that component, calculated with the aid of 31 magnetic pickups on the inner vessel surface. The vessel and internal components are more resistive than inductive on the DIII–D disruption time scale (≥ 3 ms). Therefore, the toroidal current density is approximately $J_T = E_T/h$, where E_T is the toroidal electric field and h the electrical resistivity. The corresponding load is $J_T \times B_P$, where B_P is the poloidal magnetic field.

A review of disruption data confirms that the largest induced toroidal current loads are produced at the top and bottom of the vessel by vertical displacement events (VDE), in which the plasma moves vertically after loss of vertical control, shrinks in cross section as it is limited by top or bottom components, and finally disrupts at low q. The largest loads occur in the vicinity of the disruption (top or bottom), where magnetic coupling to the decaying plasma current loop is greatest. Because VDEs are rare in DIII–D, a semi-empirical scaling law was developed for E_T by combining qualitative theory and available data [3]. This yields: $E_T \sim (B_T I_p)^{1/2}$, $B_P \sim (B_T I_p)^{1/2}$ at the moment of disruption, and force = $J_T \times B_P \sim B_T I_p$, where I_p is the pre-disruption plasma current. This scaling is used to extrapolate from historic VDEs to the anticipated load from a VDE at maximum machine capability. Dynamic loads are calculated using actual VDE waveforms.

Eddy current loads can be important in components that are not toroidally continuous, such as the plates comprising the divertor pump plenums

---

*Work supported by U.S. Department of Energy under Contract No. DE-AC03-89ER51114.

in DIII–D. Eddy currents are induced where changing poloidal magnetic flux penetrates a conducting component. The flux change is derived from actual disruptions, and is extrapolated linearly with I_p to maximum machine capability. Such components are approximated as thin, flat rectangular plates, for which a standard analytical expression yields the eddy current. The largest loads are produced where the current crosses toroidal magnetic field.

Induced current loads dominate in components that do not receive halo currents.

1.2. Halo Current Loads

Halo currents are electric currents flowing in the scrape-off layer (SOL) plasma outside the last closed magnetic surface. Disruptions broaden the SOL, fill it with plasma, induce a large toroidal electric field and drive a large halo current. Halo currents enter and leave the vessel and plasma-facing components (PFC) where they intercept open SOL magnetic lines. Because the halo plasma pressure (β_p) is low, the halo current is almost force-free and flows nearly parallel to B. Only the poloidal component of the halo current flows into the first wall. Poloidal vessel currents originating in the halo and crossed with the toroidal magnetic field produce larger vessel loads than induced toroidal currents [1] and are the principal drivers of vessel motion during VDE's.

Poloidal halo current to the DIII–D vessel is measured by a set of current monitor resistors interposed between selected graphite armour tiles and the vessel wall [4]. A top view of the present tile current monitor (TCM) array in the bottom of the vessel is illustrated in Fig. 1.

TCM data indicate that the peak halo current I_{hPk} is typically greater during VDEs than other disruptive events. Measurements of the vessel vertical displacement, which is an indicator of global VDE vertical impulse $\sim \int I_p B_T \, dt$, show that the worst case impulse increases linearly with I_p [5]. Continued proportionality to $I_{hPk} B_T$ is assumed to extrapolate to the worst case halo current loads using full B_T and $I_{hPk} = 0.2\, I_p$. This empirical scaling can be justified by a model calculation of the force from the quadrupole shaping field acting on a shrunken, off-center plasma, whose size is set by the observation that VDEs always disrupt at about the same safety factor, $q \sim 2$ [6].

Recent data show that the halo current is nonuniformly distributed toroidally, and the nonuniform structure typically rotates at hundreds of Hz [7]. However, there are occasional examples of nonrotating asymmetries. Fig. 2(a) shows an example of another common behavior, where an initially rotating structure later stops at 1.734 s. The nonuniformity is characterized by a toroidal peaking factor (TPF), TPF = (peak local J_h ÷ toroidal average J_h). The TPF is sometimes very large early in the VDE, when I_h is small. However, the TPFs observed to date are ≤ 3 during the time of greatest interest, when the halo current is large. Fig. 2(b) shows the TPF vs. $I_h(t)$.

These DIII–D halo current data are similar to Alcator C-Mod data [C-Mod]. Toroidal peaking, a recently discovered phenomenon, has been included in DIII–D load calculations.

The halo current is believed to be driven by two fundamental effects: decay of the bulk plasma current which induces toroidal current (and thus produces poloidal current) in the force-free halo region, and reduction in the vacuum toroidal flux linked by the halo region as the plasma cross-section shrinks (which produces a poloidal voltage). Continuing theoretical analysis of the disruption-driven axisymmetric halo current has suggested that

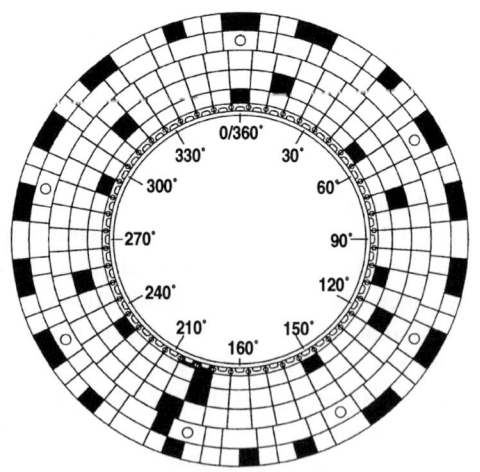

Fig. 1. Array of current monitored tiles in bottom of DIII–D vessel in 1996, viewed from above.

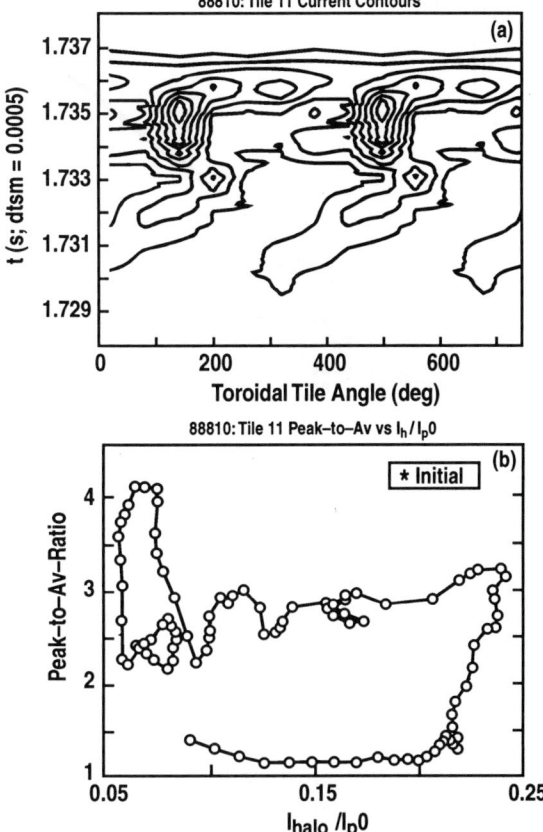

Fig. 2. Toroidal non-uniformity during a VDE. (a) contours of current into tiles versus toroidal angle in second row of tiles from inside vessel. Toroidal angle repeats two periods for viewing ease. Angle from vertical indicates rotation of peak current. (b) evolution of halo current toroidal peaking factor versus halo current normalized to I_p. Points are 0.05 ms apart.

the peak axisymmetric component of the halo current is maximized by a high effective growth rate, high bulk plasma current decay rate and initial plasma current, a low edge safety factor during the current quench phase, and a low halo resistance.

2. VESSEL LOADS AND DEFLECTIONS

The duration of the peak halo currents is about 2 ms, whereas the time for rise and decay of halo currents is about 15 ms. To evaluate the structural effects of halo current forces, a 3D dynamic analysis of the DIII–D vacuum vessel was completed. To encompass all worst case loading conditions, the magnitude of the halo current was taken as 20 percent of a pre-event 3.00 MA plasma with a 15 ms rise-decay time.

A 2:1 toroidal peaking factor is used on the applied loads in the model. The loads were applied on the floor in one case and on the side of the vessel in the second. The loads vary both radially and circumferentially around the vessel. Also, in the radial direction, the loads decrease linearly with increasing radius. The applied load, $P_{applied}$, is equal to $P_{initial}[1 + \cos\phi]$ with $P_{initial}$ varying from 2.86 bar to 1.03 bar in the radial direction on the floor and varying from 1.03 bar to 0.73 bar on the side of the vessel. The loads are applied as a time pulse and rise from zero to peak halo current in 13 ms and then fall to zero in 2 ms.

The dynamic analysis of the DIII–D vacuum vessel was done to find stresses in the support trunnions, the resultant loads placed on the support trunnion bolts, and vacuum vessel stresses and displacements, due to halo current induced loads in the vessel. The vessel is supported by 4 equally spaced horizontal trunnions extending radially at the vessel midplane.

The 3D finite element model of the vessel was subjected to a halo current of 20% of a 3 MA plasma with a peaking factor of 2. Figure 3 shows a vertical displacement plot at the center of the vessel floor at 0 degrees as a function of time. The maximum vertical displacement is 3.5 mm. The loads on the 4 trunnions were nearly equal.

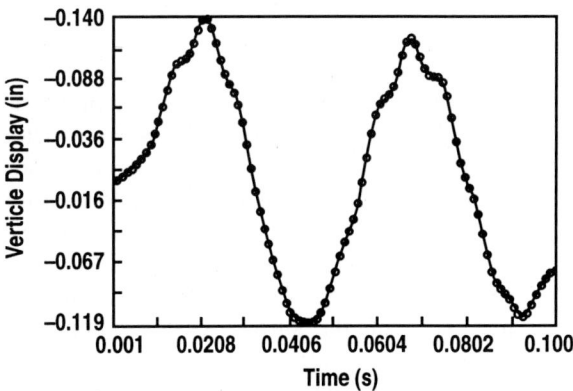

Fig. 3. Model results for halo current induce vessel displacement from a 3 MA plasma.

Measurement of the motions of the vacuum vessel floor have been recorded over the last several years. Plots of the amplitude of the vertical displacements of the vessel versus plasma current are shown in Fig. 4. Since most of the hard disruptions occurred at near maximum toroidal field of 2.2 Tesla, a linear fit through the maximum displacement at each value of plasma current provides the best representation of peak vessel motion.

It is seen that the dynamic analysis results overpredicts the vessel motion. Linear scaling of the measured results indicate a displacement of 0.092 in. vs. the 0.140 in. predicted analytically. This can be partly explained by 2 ms time duration of halo currents with 2:1 peaking factors vs. the 15 ms pulse time used for the dynamic analysis. Based on the conservative structural analysis, it is safe to assume that the DIII–D vacuum vessel can safely react halo current loads resulting from 3 MA plasma operation.

The following procedure is used for the stress analysis of structures in which the design is governed by halo current loads:

1. For preliminary design, apply a static pressure load normal to the plasma facing surface base on 20 percent of a 3 MA plasma, evenly distributed toroidally (no peaking factor). A structural model representing a repetitive sector of the structure is used with a dynamic load factor (DLF) of 1.0 for sizing calculations.
2. Based on the above results, the thickness of the component may be increased or reinforcing ribs and/or gussets added to satisfy the stress allowable for the structure and its supports to the vacuum vessel.
3. Perform a frequency analysis for the reinforced structural model. Using response spectrum curves for impulsively applied loads with a triangular rise-decay time history, determine the DLF for halo currents with a 2:1 peaking factor (2 ms) and symmetric halo current loads (15 ms). Static stress analysis is performed for the sector structural model using the highest pressure loads adjusted for DLF. Various load cases are analyzed for halo current paths that may split on the surface of the component and/or flow through the component supports to the vessel and back to the disrupting plasma.
4. For toroidally continuous structures, the 2:1 peak to average factor for the halo current will produce displacements that are offset globally from the centerline of the vessel. To evaluate these effects, a 180 degrees structural model is required. A frequency analysis is performed to determine the DLF for the 2 ms peaking loads. The equivalent pressure loads, p, are applied statically with p=0 at 0 degrees, p at 90 degrees, and 2p at 180 degrees. A dynamic time-history analysis with a 2 ms impulse load to the structural model is performed if the static analysis does not satisfy the allowable stress values.

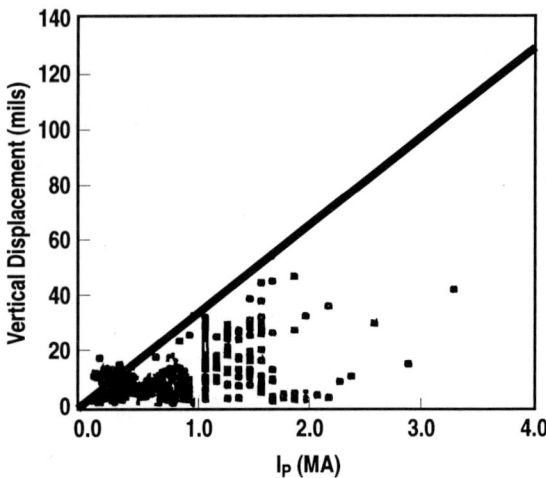

Fig. 4. DIII–D results for vessel vertical deflection versus plasma current.

REFERENCES

1. E.J. Strait, L.L. Lao, J.L. Luxon, E.E. Reis, Nucl. Fusion **31** (1991) 527.
2. A.G. Kellman *et al.*, in Fusion Technology (Proc. 16th Symposium, London, 1990).
3. M.M. Mennon, M.J. Schaffer, A.G. Kellman, Bull. Am. Physical Soc. **37** (1992) 1570.
4. M.J. Schaffer, B.J. Leikind, Nucl. Fusion **31** (1991) 1750.
5. E.E. Reis *et al.*, in Fusion Engineering (Proc. 12th Symposium, Monterey, CA 1987) vol. **1** IEEE, New York (1987) 212.
6. M.J. Schaffer, unpublished.
7. T.E. Evans *et al.*, to be published in J. Nucl. Materials.
8. R.S. Granetz et al., Nucl. Fusion **36** (1996) 545.

Effects of plasma behavior on in-vessel components in JT-60 operation

H. Hiratsuka, T. Sasajima, K. Kodama, T. Arai, K. Masaki, Y. Neyatani, J. Yagyu, A. Kaminaga and M. Saidoh

Naka Fusion Research Establishment, Japan Atomic Energy Research Institute,
Mukouyama, Naka-machi, Naka-gun, Ibaraki-ken, 311-01 Japan

Damage of in-vessel component tiles such as divertor and first wall tiles have been investigated from the view point of the effects of plasma behavior under the JT-60 operation.

1. Introduction

In old JT-60, serious erosion and fractures of the graphite tiles have been observed in the early operation[1,2]. These experiences have been reflected to the design and material selection of the plasma facing components for the upgrade of JT-60 (JT-60U). As a result the damage of the wall tiles has been considerably reduced in the operation of JT-60U[3,4] excluding only some exception. All of the plasma facing surfaces in JT-60U are covered with carbon tiles: as shown in Fig. 1, carbon fiber composite (CFC) is used for the divertor plates and all the other surfaces of the vacuum vessel and the ports are covered with the isotropic graphite tiles and partially with CFC tiles.

The main causes of the in-vessel component damage due to plasma behavior are the followings. First is the halo current which flows from the plasma into the in-vessel components such as the tiles through the contact point. In JT-60U the direction of the plasma movement depends on the plasma position and the plasma current quench speed[5]. This means that the place where the halo current is induced strongly depends on the plasma behavior. The second is arcing which could be induced through scrape-off layer even when the plasma does not touch the wall. The third is the heat flux onto the divertor in the normal operation.

This paper describes recent experiences on the damage of the in-vessel components caused by such events relating to plasma behavior.

2. Damage of tiles by heat flux

Table 1 shows the operation history of JT-60U from July 1991 to July 1996. The heating powers have been increased up to 40 MW for neutral beam (NB), 7 MW for lower hybrid radio frequency and 7 MW for ion cyclotron radio frequency, respectively. The heat flux to the divertor, therefore, has been also increased up to 10 MW/m^2 at the strike point.

Fig. 1 Internal view of the vacuum vessel of JT-60U

Table 1 Operation history of JT-60U

Operation Phases	I	II	III	IV	V	VI
No. of Discharges	873	1880	2335	2766	2568	823
No. of NBI Discharges	556	1085	1305	1653	1612	500
NBI Power (MW)	$\leqq 20$	$\leqq 27$	$\leqq 36$	$\leqq 32$	$\leqq 30$	$\leqq 40$

I : Jul. - Oct. 1991   II : Jan. - Oct. 1992   III : Jan. - Oct. 1993
IV : Jan. - Oct. 1994   V : Jan. - Oct. 1995   VI : Jan. - Jul. 1996

In JT-60U so called carbon bloom due to enhanced sublimation of carbon has not been observed so far. This is mainly due to the precise alignment of the divertor tiles with the toroidal undulation less than ± 2 mm and the successive in-situ taper-shaping of the divertor tiles which eliminates the local heat concentration[4]. In fact only two/three tiles have been so far exchanged annually during regular inspection period due to the damage of the tiles caused by heat load since 1991, though numerous number of carbon tiles (2895 divertor tiles and 9375 first wall tiles) have been used.

3. The fracture of the first wall tiles

The fracture of three CFC first wall tiles located in the lower inner side of the vacuum vessel just adjacent to the divertor tiles was found during density limit disruptions[6]. Three tiles were fractured at the position of the connecting bolts which were not loosened. The main causes of these tile fracture were the followings.

Firstly the plasma touched the wall during plasma current quench resulting in the halo current, which flowed into the CFC tile and induced the vertical force to break the tile as shown in Fig. 2. Secondly these tiles were near the diagnostic port so that the induced current could be concentrated. This means that the induced halo current would be larger than the expected. Thirdly these tiles were overhung to avoid a contact with the cooling pipes of the divertor base plates, which weakened against the bending moment. Halo current required from the vertical force of 9 kN which would fracture the tile was 25 kA, which was in reasonable agreement with the estimated current taking account of the current concentration factor obtained from the ratio of the area of the port to those of the tiles[6,7].

In 1995 the fracture of the CFC armor tile of the Neutral Beam Injection (NBI) port located upper outer side of the vacuum vessel was also found

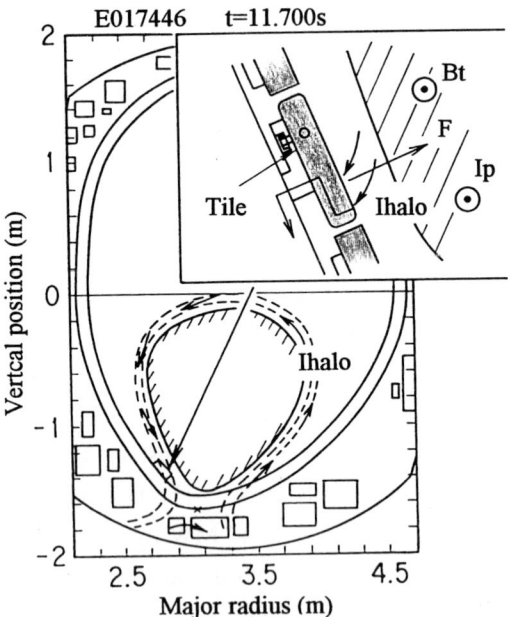

Fig. 2 The halo current induced during disruption flows into the tile and produces the vertical force (F) to break the tile.

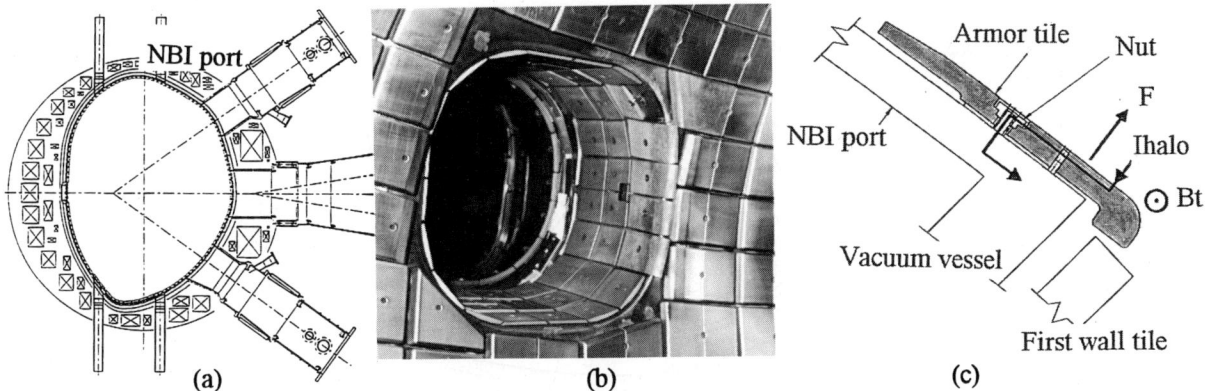

Fig. 3 Location of CFC armor tile of NBI port. (a)cross section of vacuum vessel and NBI port, (b)photograph of broken armor tile of NBI port and (c)layout of armor tile of NBI port and first wall tile, where Ihalo is halo current and F is the induced vertical force, respectively

during disruption. In this typical discharge the plasma position was high and moved upwards. In this case, the tile was also broken at the connection bolt as shown in Fig. 3. Usually this tile is fixed with the connection bolt and the pin which is capable for the tile not to rotate during plasma operation due to the electromagnetic force. When this tile is fixed perfectly, the halo current required to break the tile is estimated to be 5 kA. The main causes of this tile fracture would be the followings. At first the plasma moved upward during disruption which induced halo current or arcing from the plasma into the tiles. Secondly the pin was out of position and raised the tile, which weakened against the bending moment. Thirdly this tile was overhung to protect the vessel. The careful inspection of the fractured tile and the NBI port showed that the backside of the tile was scratched and that the traces of arcing were found at the NBI port just the tile position before and after the fall of the fractured tile.

In both cases the followings were in common. The tiles were overhung and located beside and/or on the port, which easily induced the current concentration. The former will be resolved when we take into account of the structure and the material strength of the tile. In fact three broken first wall tiles were exchanged for the new tiles which were divided into half-size tiles to reduce the bending force. In case of NBI armor tile, the broken tile was exchanged for new tile whose bending strength was about 4 times larger than the previous one from the material selection and the size change of the tile. The latter, however, does depend on the plasma behavior so that the plasma control must be necessary.

4. Melting of the cooling pipe

In 1995 the melting of the divertor cooling pipe were found two times accompanying an abrupt pressure increase up to 0.5 Pa. The location of the melting of the cooling pipe were under the divertor and the adjacent first wall tiles. In this discharge with high triangularity configuration inner separatrix hit point moved across the gap of 3 mm between the divertor tiles and the first wall tiles as shown in Fig.4. Inner separatrix hit point is not usually out of divertor tiles. However, in this case, this hit point shifted inward with the increase in ßp. The high energy particle flux penetrated under the tiles along the magnetic field line, whose angle was between 1 and 2.5 degrees, from a nearby port.

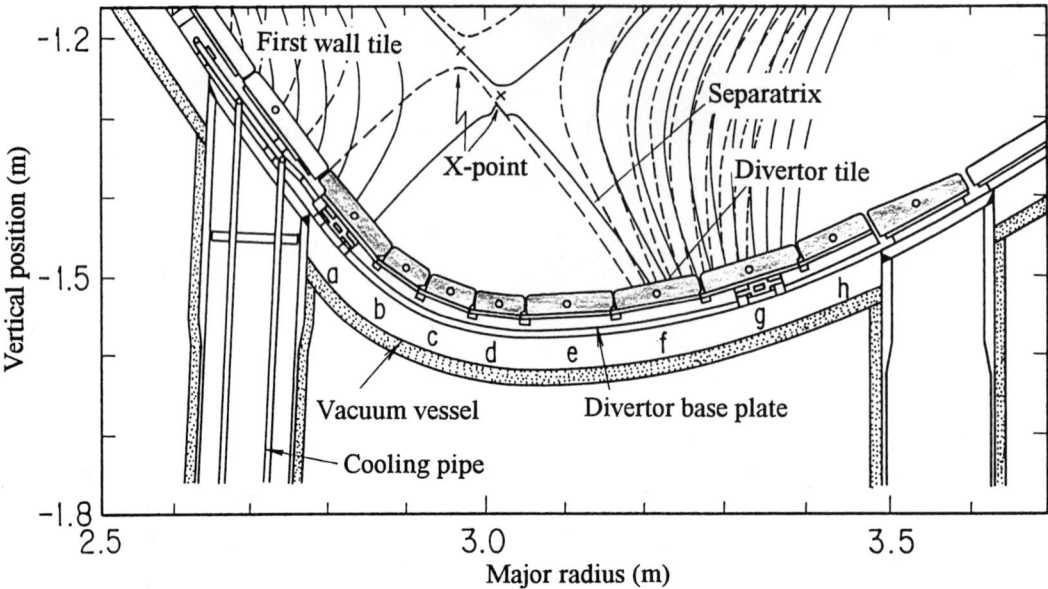

Fig. 4 JT-60U divertor plate and magnetic surfaces in high-ßp (solid) and high triangularity (broken) plasma configurations. In the latter the separatrix was across the gap between divertor and adjacent first-wall tiles.

Heat load was estimated to be 12 MJ/m^2.

In order to avoid the particle penetration, it is necessary to make the tile arrangement to block the magnetic field line. For this purpose every four divertor tiles was made longer in size towards the adjacent first wall tiles to make a jagged structure against the magnetic field line of less than 3 degrees.

5. Summary

The effect of plasma behavior on the in-vessel components in JT-60 operation were presented. The damage of the tiles such as the tile fracture indicates that the electromagnetic interaction should be carefully considered for the design and the arrangement of the plasma facing components.

Acknowledgments

The authors wish to thank the JT-60 Team for the operations and experiments of JT-60U.

References

[1] H. Takatsu et al., Fusion Eng. & Des. 9 (1989) 3.
[2] T. Ando et al., J. Nucl. Mater. 179-181 (1991) 39.
[3] T. Ando et al., J. Nucl. Mater. 191-194 (1992) 1423.
[4] T. Ando et al., Proc. 17th Symp. on Fusion Technol., Rome, Italy, 1992 (Elsevier Science Pub. B. V., 1993) pp.161.
[5] R. Yoshino et al., Nucl. Fusion 36 (1996) 295.
[6] K. Masaki et al., J. Nucl. Mater. 220-222 (1995) 390.
[7] Y. Neyatani et al., Fusion Technol., 28 (1995) 1634.

Numerical and experimental evaluation of halo currents in RFX plasma facing components

S.Peruzzo, A.Masiello, N.Pomaro, P.Sonato, G.Zollino

Consorzio RFX
Corso Stati Uniti 4, I - 35127 Padova, Italy.

The presence of halo currents in the Reversed Field Pinch RFX has been evidenced during a partial replacement of the first wall. The minimum values of halo currents have been evaluated by both an analytical approach and numerical simulations. The estimated value has been experimentally confirmed by tests on a mock-up of the first wall. Finally preliminary results of halo current measurement during RFX pulses are presented.

1. INTRODUCTION

In the last five years in almost all the Tokamak experiments the presence of halo currents, flowing from the plasma open field lines to the plasma facing components and the vacuum vessel during disruptions, has been extensively reported [1,2,3,4,5]. Recently, the halo currents toroidal asymmetries have been evidenced in Alcator C-MOD [3,5] and in JT-60U [4] and further investigations are in progress to identify the asymmetry of the halo current and its relationship with the plasma magnetic structure. Furthermore the mechanical stresses induced into the plasma facing components and the vacuum vessel due to the interaction between the poloidal component of the halo currents and the toroidal field is one of the most critical issue in the design of the next step tokamaks.

For the first time in the Reversed Field Pinch (RFP) experiments the presence of the halo currents has been evidenced in RFX [6]. Halo currents in RFX are strictly correlated with a non axisymmetric perturbation of the plasma column, which is always present in the plasma shots and exhibits a toroidal extension between 30° and 40° [7]. Such a perturbation appears everywhere along the toroidal coordinate, but in 70% of the shots it is locked at the wall in correspondence of one of the two poloidal insulating gaps of the passive stabilising shell that surrounds the vessel [8]. The perturbation has a helical shape dominated by a $m=1$, $n=8$ locked mode, that is internally resonant [9].

In the RFP, at the plasma edge the magnetic field has mainly a poloidal component with a very low reversed toroidal field. The toroidal magnetic flux is sustained by the almost poloidal current at plasma edge. In RFX the typical edge safety factor is $1/q=-20\div-40$.

In the paper the analyses which evidenced halo currents in RFX are reported as well as the evaluation of halo currents amplitude. Finally the preliminari results of a direct measurement are presented and discussed.

2. EVIDENCE OF HALO CURRENTS

RFX has a metallic vacuum vessel made of Inconel 625® [8]. The vessel is composed of 72 cylindrical sectors with an equivalent toroidal amplitude of 5°. At the end of each sector there is a thick poloidal ring (30x30 mm). Twenty-eight graphite tiles are bolted to each poloidal ring, in order to protect the inner vessel surface. The whole first wall includes, in total, 2016 graphite tiles [8].

During the 1994 shut down 112 tiles (four rows) located in the two rings adjacent to each of the two insulating poloidal gaps of the stabilising shell have been replaced [10]. Traces of melted metal have been found on the removed tiles and on the vessel both on the contact surface between the tiles and the vessel. The melted areas appear as a row of spots aligned on one side of the contact area of each tile. The poloidal and toroidal distribution of the melted zones are very similar in the two rings of both the poloidal gaps. The spots are concentrated in the tiles positioned in the lower side, in the external-lower side and in the upper side of each ring (fig. 1). The lines of spots are aligned in the opposite side of the tile contact surface for the tiles positioned in the upper and in the lower side of each ring. The melted area is more extended in the lower side than in the upper side of each poloidal ring.

The understanding of the phenomenon can be based on the poloidal distribution of the observed damage [6], taking into account that before the removal of the 112 tiles all the RFX shots (5000) have been performed without any change of the plasma current direction and of the toroidal field direction. The following considerations can be deduced:

1) a predominant $m=1$ poloidal current flowed in the rings;

2) the damage is in agreement with a flow of electrons that enters into the lower side of the vessel and comes out from the upper side;

3) the exit of the electrons in the upper side is more extended in the toroidal direction with respect to the lower side.

In fig. 1 a schematic view of the possible path of halo currents is shown.

The toroidal distribution is much less clear because only two adjacent rows of tiles have been replaced for each poloidal section. Nevertheless the ring immediately following the two insulating poloidal gaps of the shell, along the toroidal coordinate, are more damaged than the other two rings for both the poloidal sections that have been investigated, thus indicating a strong toroidal asymmetry.

The observed phenomena indicate only the presence of the halo currents that involve a current path between the first wall and the vacuum vessel. Different observations of the plasma wall interaction indicate also the presence of halo currents that enters and exits in the same tile.

3. HALO CURRENTS EVALUATION

Three different methods have been used to estimate the minimum value of the halo currents that could justify the melting of the vessel surface.

3.1. Tilting forces on the tiles

The presence of the tile damages on only one side of the contact surface indicates the presence of a torque that produces a tile detachment on one side and an increase of the contact pressure on the other side as indicated in fig. 2. The electrodynamic torque necessary to produce detachment is originated by the interaction of the halo current, that has a dominant toroidal path in each tile, with the poloidal magnetic field.

The minimum current necessary to tilt the tile is defined from the balance of the electrodynamic torque with the mechanical torque of the tile bolting system. By assuming that the current enters through the front surface exposed to the plasma with a uniform current density, we obtain:

$$I_{Hm} > 48 \cdot \pi \cdot d \cdot F_s \cdot r_w \cdot \frac{1}{l_t^2} \cdot \frac{1}{I_p} \qquad (1)$$

Where:
I_{Hm} = minimum halo current on each tile
d = lever arm of the spring force = 10 mm
F_s = minimum mechanical force = 120 N
r_w = first wall minor radius = 0.457 m
l_t = toroidal length of each tile = 0.2 m
I_p = plasma current

At 850 kA of plasma current, that corresponds to the maximum value in the first 5000 shots, from formula (1) $I_{Hm} > 2$ kA.

3.2. Experimental tests

To obtain a value for the contact resistance between the tile and the poloidal ring and to estimate the minimum current necessary for the melting of the metal surface, experimental tests have been performed on a prototype system including one tile and one portion of the vessel ring.

The current in the tile and in the ring has been simulated by feeding the system at the frequency of 50 Hz with a pulse duration of 100 ms.

As shown in fig. 3 the contact resistance decreases as the test current increases, in agreement with a progressive softening and a consequent

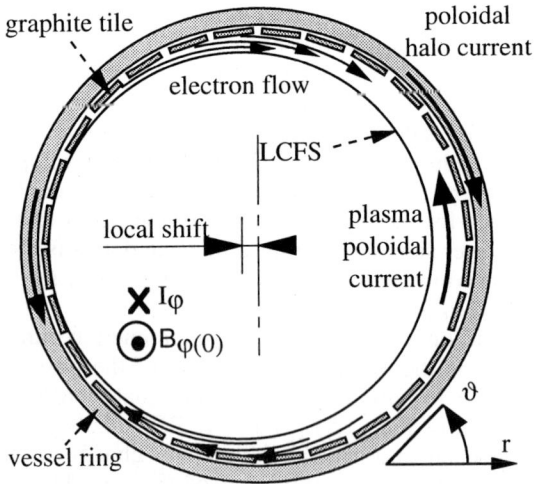

Figure 1. Poloidal path of the Halo Currents in one poloidal ring before the partial replacement of the first wall.

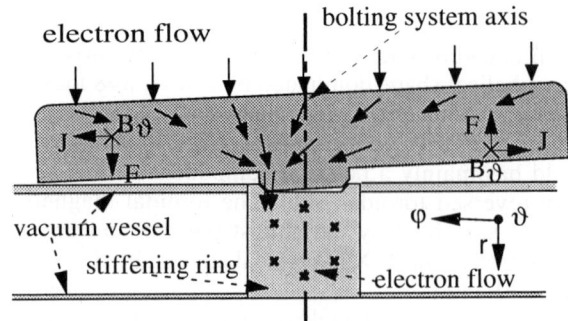

Figure 2. Electrodynamic tilting of each tile.

melting of the materials [11]. Traces of melted metal were observed for a current larger than 2.9 kA, so for the numerical simulation it is possible to consider a 0.5 mΩ contact resistance. This result has been obtained using prototype tiles having the same roughness of tiles installed in the vessel.

Tilting the tile by an adequate force, a little decrease in the contact resistance due to the higher pressure on the contact has been recorded at low current (fig. 3), whereas at high current, when the melting is present, the difference became negligible.

3.3. Numerical simulations

A non-linear transient thermal-electrical analysis has been carried out by means of a finite element code [12] to determine the temperature distribution in the graphite tiles and in the Inconel rings due to joule heating effect. A 2D model has been used, made of first order thermal-electric elements with 4 nodes and two degrees of freedom per node, temperature and voltage.

The mesh is in the parametrical form, so that it is possible to modify the contact area between the tile and the ring. Assuming the measured value of the contact resistance, the actual contact surface has been assessed to be approximately 9 mm^2. Material properties as a function of temperature have been used for both graphite and Inconel. Radiation has been disregarded because the exposed tile surfaces resulted approximately at room temperature, when the contact region reached its maximum temperature.

The mesh has been accurately defined in the contact region, where a very high thermal gradient is present: it includes 824 elements and 942 nodes. A current waveform as reported in fig. 4 has been used, which is consistent with the hypothesis that halo currents rise up during the start up phase and last all over the flat top phase of a typical RFX pulse. The analysis has shown that the Inconel temperature at the contact area reaches 1400 °C approximately, when a current as large as 3.5 kA flows through the tile (fig.4).

4. HALO CURRENTS MEASUREMENT

4.1. Measurement method

A first measurement system to directly investigate the presence of halo currents in RFX has been installed using the pumping ports on the equatorial plane and the upper and lower cooling ports, according to the schematic view of fig. 5. For two poloidal sections (97°30' and 187°30') three couples of insulated wires have been electrically connected to the ports, to measure, respectively, the voltages along the upper external poloidal 90°, in the lower external poloidal 90° and in the external 180°. The third measurement corresponds to the sum of the first two to check the accuracy of the system. As the ports are placed between two poloidal stiffening rings of the vessel, the probes measure an average value between the voltages in the two rings.

Moreover, each wire is installed on the external surface of the shell, so it forms, together with a portion of the vessel, a coil sensitive to the toroidal field. When the toroidal field is not constant probe signals contain an "inductive part", not related with halo currents and proportional to the poloidal loop voltage. This part is almost the same in the two sections, as the wires have been installed in the same way in the two sections, so it is not present when considering the differences between corresponding signals in the two sections.

4.2. Measurement results

If the local perturbation is not in the region where the probes are installed, the signals are in good agreement with the axisymmetric poloidal voltage in the plasma section. When the local

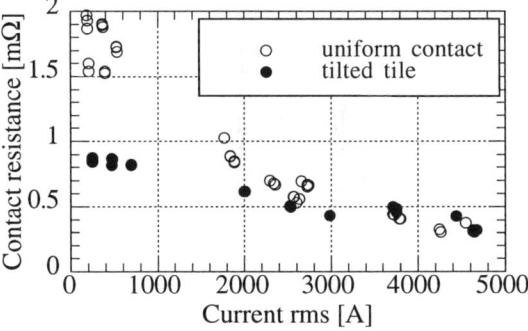

Figure 3. Contact resistance between tile and poloidal ring versus the tile current.

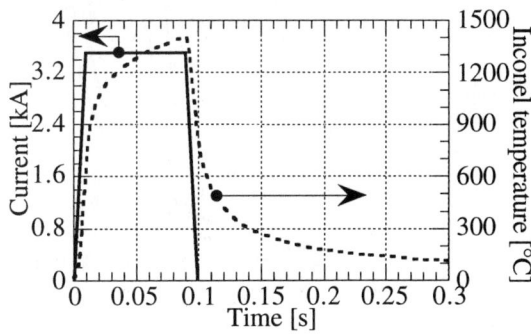

Figure 4. F.E. simulation: imposed current and Inconel temperature in the proximity of the contact.

perturbation is near one of the two monitored sections a significant signal appears (see fig. 6). The voltage difference between the corresponding probes installed in the two poloidal sections is shown in fig. 6 for a shot at 650 kA to evidence the non axisymmetric behaviour. The voltage appears immediately after the formation of the Reversed field Pinch configuration and increases during the shot. The opposite sign of the voltage indicates that the current in the same ring has also an opposite sign as schematically indicated in fig. 5. The poloidal asymmetry of the difference signals is evident. The ELPV (fig. 5) difference signal changes the voltage sign after 70 ms during the plasma current decreasing (fig. 6).

Due to the discretization of the measuring probes and the asymmetry of the two signals it is very difficult to have a real estimation of the halo currents flowing in the two rings. If the current was uniformly distributed in the two rings and in the 90° covered by the probe 1 V of signal would correspond to a current of 1 kA in each ring.

The signals show also a fluctuation with a frequency higher than 5 kHz.

5. CONCLUSION

Experimental and numerical evaluations of the current level necessary to reproduce the melting of the vessel surface, due to halo currents between the first wall and the vessel, are described showing a good agreement of the different approaches.

A first method to measure the halo currents is presently working allowing a first understanding of the halo currents path and intensity. To allow a direct and more accurate measurement, the installation of rogowski coils in the tiles is planned.

REFERENCES

1. E.J.Strait et al., Nuclear Fusion, 31 (1991), No.3, p.527-534.
2. M.A.Pick et al., Proc. 14th Symp. on Fus. Engin., 1991, p.187-190.
3. F.C.Schüller, Plasma Phys. and Control. Fusion, 37 (1995), p.A135-A162.
4. Y.Neyatani et al., Fusion Technology, 28 (1995), p.1634-1643.
5. R.Granetz et al., Nuclear Fusion, 36 (1996), No.5, p.545-556.
6. P.Sonato et al., presented at the 12th Int. Conf. on Plasma Surf. Interact. in Control. Fusion Dev., 1996, to be published J. Nucl. Mater..
7. F.Gnesotto et al., Proc. 21st Conf. on Control. Fus. and Plasma Phys., 1 (1994), p.458-461.
8. F.Gnesotto et al., Fusion Engineering and Design, 25 (1995), p.335-372.
9. E.Martines et al., Proc. 22nd Conf. on Control. Fus. and Plasma Phys., 19C (1995), p.IV181-IV184.
10. W.R.Baker et al, presented at the 12th Top. Meet. on the Techn. of Fus. Energy, 1996, to be published Fus. Techn..
11. R. Holm, Electric Contacts, Springer-Verlag Berlin/Heidelberg/New York, 1981.
12. P.Kohnke, ANSYS User's Manual-Theory rev. 5.1, Swanson Analysis System, 1994.

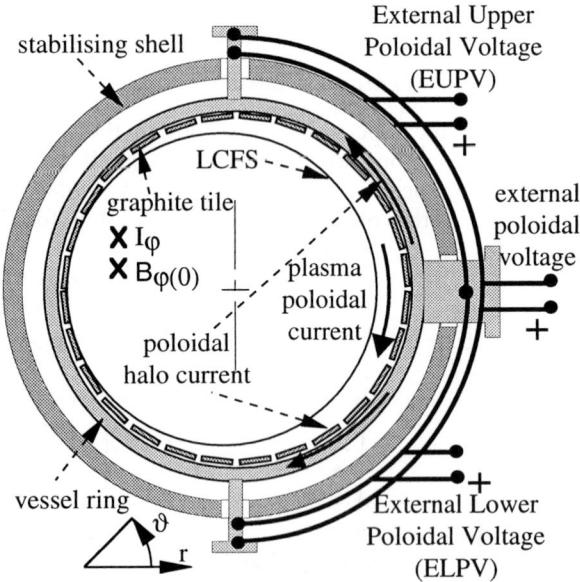

Figure 5. Scheme of the voltage measuring points on one poloidal section.

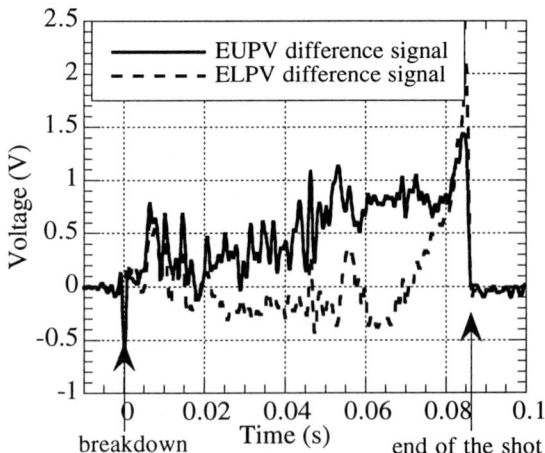

Figure 6. Filtered difference signals (1 kHz) with the local perturbation near the section at 97°30'.

Feasibility analysis of an active local field control at the poloidal gap of RFX

F. Bellina, G. Chitarin, P. Fiorentin, E. Gaio, G. Marchiori, P. Sonato, V. Toigo, P. Zaccaria, G. Zollino

Consorzio RFX, C.so Stati Uniti 4, 35127 Padova, Italy

The paper describes the design of the local active control system of the field error at the poloidal gap of the RFX machine. It consists of eleven coils placed around the poloidal gap, each one fed by a switching converter. The analyses which allowed to define the system specification, and the design of the coils and of a prototype converter are presented.

1. INTRODUCTION

In most of the RFX pulses, the magnetic field configuration presents a locked perturbation which affects the plasma confinement, causing damages to the first wall and the presence of impurities in the plasma. One of the causes of the locking of these perturbations seems to be the field error caused by currents induced in the shell, as it was observed also in MST experiments [1]. In RFX the vacuum vessel is surrounded by a thick aluminium shell which has a global stabilizing effect. However, since the shell has two toroidal and two poloidal insulating gaps, a deformation of the path of the eddy currents is caused, such that a local field error is generated in the gap areas. For this reason some actions have been recently undertaken. First a feedback control system of the field errors at the poloidal gaps has been developed using the existing poloidal Field Shaping (FS) coils: this system controls the poloidal magnetic configuration along the whole machine, therefore it performs an axisymmetric correction [2]. Since the locked perturbation causes different field errors at the two poloidal gaps this system can only compensate for the average error. For this reason one of the two poloidal gaps [3] and the outer equatorial gap have been recently short-circuited and an appreciable reduction of the fast fluctuations of the field error has been obtained in the region of these gaps [4].

The reduction of the field error at the remaining poloidal gap can be achieved by means of both active and passive solutions. A fast response for a short time can be obtained using conductive shields, while an accurate compensation of the field error for a longer time scale can be achieved by means of local coils actively controlled. Implementing both options would allow to cover a wider frequency range of the errors; therefore this type of solution was adopted for RFX.

2. SYSTEM REQUIREMENT ANALYSIS

Analyses were performed on the RFX experimental data to evaluate the spatial and time harmonic content of the field error necessary for the design of the coil and of the power supply systems. In the shots considered the currents in the FS coils were feedback controlled to minimize the deviation from axisymmetry of the first harmonic in $\sin\vartheta$ of the radial field and distributed so as to compensate the harmonics in $\sin(m\vartheta)$ for m = 2,3,4. The analyses showed a spatial uniform radial field (m = 0) and components in $\cos(m\vartheta)$ at low frequency (less than 5Hz), which cannot be suppressed by the currents in the FS coils due to their connection. Moreover high frequency (typically 50-100 Hz) $\sin(m\vartheta)$ components were observed due to the limited bandwidth of the thyristor converters which feed the FS coils.

The amplitudes of the spatial harmonics of the radial magnetic field (B_{rc} and B_{rs} are the $\cos(m\vartheta)$ and $\sin(m\vartheta)$ coefficients, respectively) are shown in table 1 up to the 3th, being the higher order harmonics negligible. These results are referred to a shot characterized by particularly high values of these quantities; therefore this case was assumed as the reference for the estimate of the maximum amplitudes to be compensated.

Table 1
Amplitudes of the spatial harmonics of the radial magnetic field at the poloidal gap.

m = 0		20 [mT]
m	Brc [mT]	Brs [mT]
1	25	30
2	0	5
3	0	10

The observed frequency range of the field errors and its spatial harmonic content suggested to design a system of coils for their compensation. They will be placed on the external surface of the shell, around the poloidal gap. The coils will be independently fed by a power supply and shall be controlled so as to cancel the magnetic flux through measurement windings installed beneath each coil.

The number of coils was chosen according to the results of the numerical analyses and to the following considerations. A system of at least 6 coils is necessary to control up to the third sinusoidal space harmonic, but an higher number could avoid the production of undesirable harmonics (> 3). Moreover the smaller are the coils the easier is the installing procedure, due to several dimensional constraints of the zone near the poloidal gap. On the other hand, costs and complexity of the control claimed for a limitation of the coil number. A trade-off among these issues suggested to examine a solution with 12 coils with the same poloidal extension ($\Delta\theta = 30°$). As a matter of fact, due to the reduced space available it is not possible to insert the coil across the inner equatorial gap, thus a system consisting only of 11 coils was considered (see Fig. 1).

Once defined the coil geometry and position, analytical computation and finite element analyses were performed to obtain the coil system specification on the basis of the residual field error to be vanished. The results suggested to define the value of 5000 for the

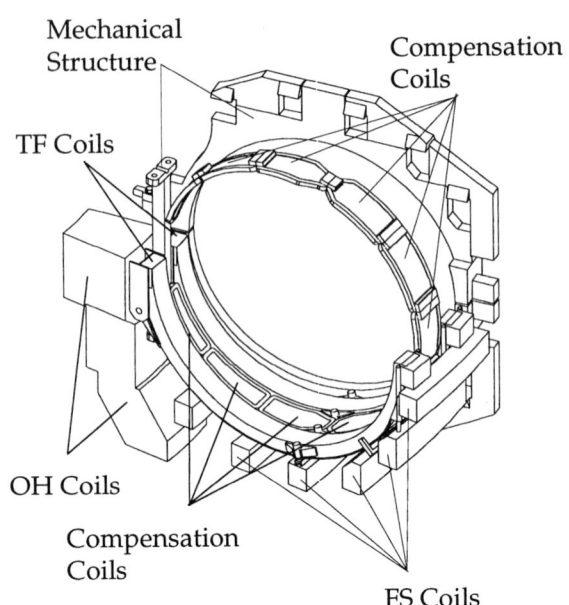

Figure 1. Picture of the gap area where the compensation coils will be installed.

Ampere-turns and of some volts for the one-turn voltage.

3. MAGNETIC ANALYSIS

Finite element analyses were performed to verify the choice of the number of coils and to evaluate their interaction: to this aim both the spatial distribution of the magnetic field and the magnetic coupling between the active and measurement coils were evaluated at several frequencies.

Two models were implemented for 3D eddy current analysis: a cylindrical model and a toroidal one. The former represents 1/4 of the rectified shell, for 1 m length from the poloidal gap in the axial direction, together with the relevant coils. In the latter, 7.5° in the toroidal direction of the whole shell and one couple of coils only are modelled. To assess the effectiveness of the coil system in reducing the field error at the gap, the toroidal and poloidal distribution of the radial field produced by the coils and the distribution due to eddy currents circulating on the shell were compared. At the plasma boundary it was observed a good agreement between the two distributions of the first component in $\sin(\vartheta)$

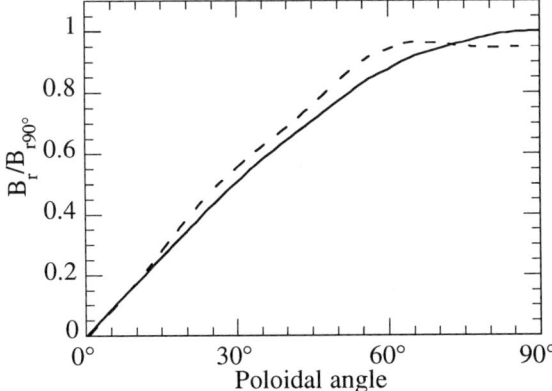

Figure 2. Poloidal distribution of the magnetic field: continuous line - effect of eddy currents on the shell; dashed line - effect of the compensation coil currents.

along the toroidal direction. At the same radius a small difference between the two poloidal distribution is present, as shown in Fig. 2.

Other FE analyses also showed [5] that a passive shield placed in the gap region could allow up to a 30% reduction of the field error for frequencies higher than 100 Hz. This result justifies the choice of including in the design a passive compensation to extend the compensation effectiveness also in the high frequency range.

The computation of self inductances and mutual inductances between each couple of coils and between power and the measurement coils of the control system shows that the maximum coupling coefficient between local coils is approximately 10% at 50 Hz.

4. THE POWER SUPPLY SYSTEM DESIGN

On the basis of the required amperturns and one turn voltage, the definition of the coil turn number and consequently of the maximum voltage and current of the converter derives from a trade-off between the available room for the coils and the opportunity to optimize the performance of the power supply system components. It was chosen a number of turns equal to 80 and the converter rating of 500 V @ 70 A. Considering the harmonic content of the error signal, a conservative approach suggested to require the power supply bandwidth of 0 ÷ 1 kHz to assure a wider margin in the system tracking capability. A switching power supply was chosen to satisfy these specifications with a reasonable cost and volume.

The main devices of each power supply are: one phase isolation transformer, EMC filter, ac/dc converter, dc capacitor bank, dc/ac converter and output filter. The dc/ac converter is based on a H bridge made up of IGBT switches (1200 V @ 200 A). The function of the output filter is to reduce the high frequency harmonics of the output voltage which can induce current oscillation between stray capacitances of the load and inductances of the connection cables. The output filter inductance shall also provide for reducing the load current ripple.

Tests on a converter prototype feeding a dummy load are in progress to optimize the output filter design and to analyze the performance of different modulation techniques; in fact some of these techniques, which control very well the same type of converters in standard applications, do not operate satisfactory at very low voltage reference.

5. THE COIL SYSTEM DESIGN

5.1. Machine geometry

The machine zone where the coils will be installed is a wedge-shaped annular volume, contained between two adjacent toroidal field (TF) coils in the toroidal direction, and between the stabilising shell and the FS coils in the radial direction, as represented in Fig. 1. The compensation coils will be located closely around the shell gap, in order to cover the shell surface between the two TF coils as widely as possible. The compensation coils will be bolted together to form a poloidal belt, which will be clamped on the shell by means of a mechanical structure. The coil cables will lay around the coils in the poloidal direction.

5.2. Design of the compensation coils

The design of the compensation coils has to cope with the tight geometrical constraints

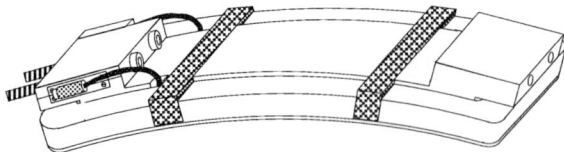

Figure 3. Sketch of a typical compensation coil

imposed by the pre-existing machine components. These have involved a number of problems about the coil leads, the insulation distances and the coil clamping system, whose solution has led to relatively complex coil geometries, as represented in Fig. 3 as an example. To check the design correctness from the geometrical point of view, a number of plastic mock-ups of the coils have been manufactured and successfully assembled on the machine.

The coil main components are the power winding, the measure winding, the shield and the coil bottom tile. An exploded view of a compensation coil is given in Fig. 4. The power winding generates most of the compensation flux and it is made of 80 turns of insulated copper wire, with an outermost insulating layer of glass type. The shield has both an electromagnetic and a mechanical function; the shield withstands the electromagnetic forces acting on the power winding, which is tightly fastened to it. The coil bottom tile has the function of guaranteeing the electrical insulation between the compensation coils and the shell. In case of a fault, a voltage up to 1.1 kV can be present between the coil and each of the two shell halves. The tile is made of a composite structure of stacked glass-epoxy sheets. The measure winding gives the voltage signal necessary to the control of the coil power supplies. The winding is made of 10 turns, obtained in one of the inner tile sheets using the printed board technology. This solution permits to locate the winding very close to the shell gap in a very reliable way.

6. CONCLUSIONS

A compensation system of the field error at the poloidal gap of the RFX shell was

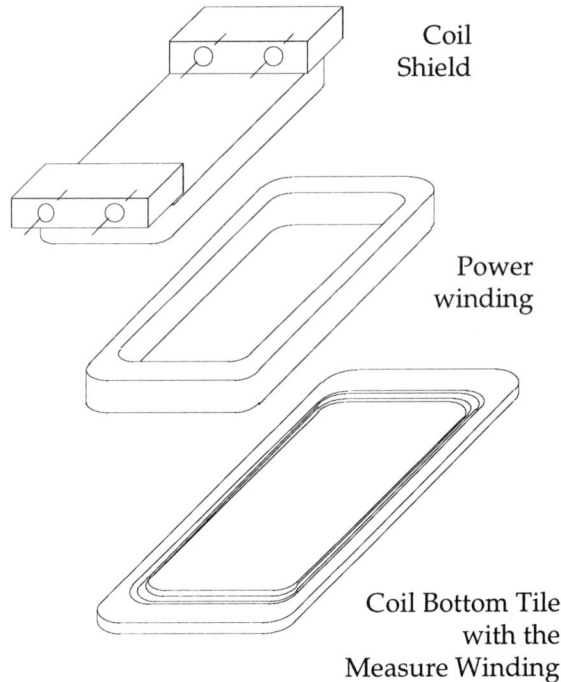

Figure 4. Exploded view of a compensation coil

presented in this paper. The coil design is completed, their construction in progress and they will be installed in the next months together with two prototype converters. Tests will be performed to verify the numerical analyses regarding the mutual coupling among the coils and therefore to define the best solution regarding the control of each converter and of the overall system.

REFERENCES

1. A.F.Almagri et al., Fhys. Fluids B 4 (12), Dec. '92.
2. P.Fiorentin, E.Gaio, G.Marchiori, R. Piovan, V. Toigo, Proc. of the 16th SOFE, Champaign, Il.-USA, Sept. '95.
3. P.Sonato, P.Zaccaria, G.Zollino, Proc. of the 16th SOFE, Champaign, Il.-USA, Sept. '95.
4 T.Bolzonella et al, Proc. of the 23th European Conference on Controlled Fusion and Plasma Physics, Kiev, Ukrain, Jun. '96.
5 G.Zollino, IEEE Transactions on Magnetics, Vol. 31, No. 6, Nov. '95, pp. 3533-3535.

Eddy Currents in the RFX Shell after short-circuiting one poloidal and one equatorial Gap

G.Chitarin, G.Marchiori, A.Masiello, N.Pomaro, G.Zollino

Consorzio RFX, Corso Stati Uniti 4, I - 35020 Padova, Italy.

The RFX machine stabilizing shell was originally provided with two toroidal insulating gaps and with two poloidal gaps. After short-circuiting one equatorial gap and one poloidal gap, the current distribution through the short-circuited gaps and the distribution of the radial magnetic field along the gaps have been measured during plasma pulses. The experimental data, which give a direct measurement of the shell reaction to the plasma magnetic perturbations, have been analysed and compared with the results of the analytical model of the shell.

1. INTRODUCTION

The RFX machine was designed with a stabilising aluminium shell, which was provided with two toroidal (or horizontal) insulating gaps and with two poloidal (or vertical) gaps.

During the recent experimental campaigns, some hints have been found which suggested that the stationary perturbation, which always affects the RFX plasma magnetic configuration at random locations, compromising the confinement and causing severe damage to the first wall, might be related to an excessive field error[1,2,3].

The field errors are produced in the region of the butt-joint gaps by the eddy currents induced in the shell by plasma displacements or by fast magnetic fluctuations.

In order to investigate the importance of a reduction of these field errors on the plasma performances, in November 1995 the outer toroidal gap and one of the poloidal gaps have been short-circuited. The short-circuiting system was designed to produce a substantial reduction of the field error, but avoiding any disassembly or machining of the major components of the machine. For this reason the short-circuited shell is not really equivalent to a continuous shell.

During the subsequent campaign, the distribution of the current through the short-circuiting elements and the field error at the gaps have been measured both during vacuum and plasma pulses.

In order to better understand the effects and the performances of the short circuits, the experimental data have been compared with the results of numerical simulation based on an analytical model previously developed obtaining some interesting information.

2. EQUATORIAL GAP SHORT-CIRCUIT SET-UP

The outer equatorial gap has been short-circuited by means of 96 copper bars bolted on the outer surface of the aluminium shell. The bars are variously shaped to suit the limited space available between the Field Shaping (FS) coils, the Toroidal Field coils and the pumping ports.

Eighteen of the short-circuit bars, all located in a sector of 120° in the toroidal direction, have been equipped with Rogowski coils for current measurement.

The radial component of the magnetic field at the outer equatorial gap has been measured by means of pick-up coils wound around 12 pumping port, which are evenly spaced all around the machine. On the inner equatorial gap it was not possible to install field probes without major disassembly.

3. POLOIDAL GAP SHORT-CIRCUIT SET-UP

The poloidal gap #1 has been short-circuited by means of an articulated chain composed of 22 copper plates with flexible louvre contacts. The chain is placed on the outer surface of the shell across the gap and pressed onto the shell by means of clamping cables. Ten out of the plates have been equipped with Rogowski coils for measuring the current through the short-circuit and also 11 radial field probes have been installed on the chain.

Because of the limited space available and of the unwanted pick-up of the poloidal component of the magnetic field, these current and field probes required special care, such as a robust assembly for precise positioning and, due to the relatively low signal/noise ratio, an accurate conditioning and correction of the signals before acquisition.

The poloidal gap #2 has not been modified and its existing 16 radial field probes have been retained.

Additional information for a reliable comparison of the radial field at the two poloidal gaps was obtained from the existing 2 couples of saddle coil probes which are installed on the inner surface of the shell in correspondence with the two gaps.

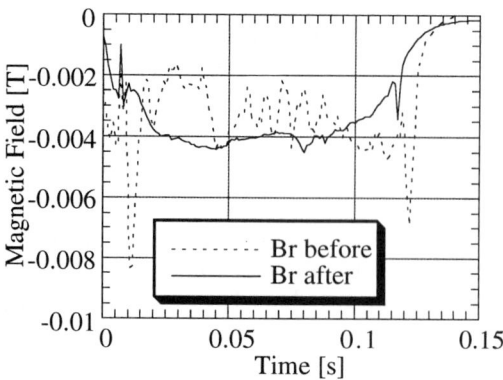

Figure 1. Time evolution of the radial field Br at the equatorial gap before (shot #4234) and after the short-circuit (shot #5504).

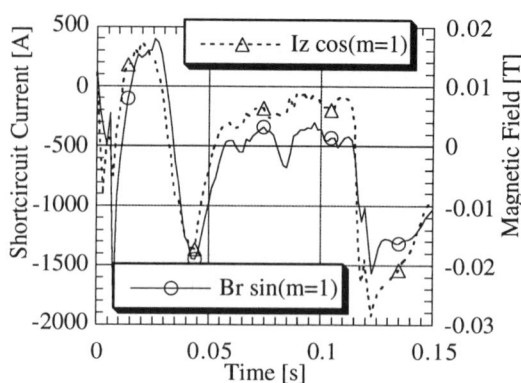

Figure 2. Time behaviour of the $\cos(\vartheta)$ component of the short-circuit current at gap #1 (Iz) and of $\sin(\vartheta)$ component of the radial field at gap #2 (Br), for shot #5504.

4. MEASUREMENT OF THE EFFECT OF THE TOROIDAL GAP SHORT-CIRCUIT

The time evolution of the radial field at the toroidal gap is always similar to that of the plasma current and the toroidal distribution is influenced by the location of the stationary perturbation above mentioned. Thanks to the short-circuit, the radial field measured at the equatorial gap was reduced by a factor of ~3 during the first 10 ms of the pulse.

However, during the flat-top no considerable reduction of the radial field has been obtained, since the major component of the error, which is related to the localized magnetic perturbation is practically stationary. Only the fluctuations of the field faster than 100 Hz have been appreciably reduced (fig. 1).

This result was expected, since the constraints on the design led to a layout of the short-circuiting bars whose estimated equivalent time constant is of the order of 5-7 ms.

Therefore the short-circuit is effective mainly during the start-up phase or in the case of a sudden change of the position of the magnetic perturbation, which sporadically occurred during the pulses.

5. MEASUREMENT OF THE EFFECT OF THE POLOIDAL GAP SHORT-CIRCUIT

In the following discussion, the harmonic decomposition along the poloidal coordinate ϑ is considered for both the radial field at the poloidal gaps and for the toroidal current flowing across the short-circuited poloidal gap.

Since the behaviour of the field error at the gaps is influenced by the position of the localized perturbation, the following two cases must be considered.

5.1. Pulses with magnetic perturbation localised far from both poloidal gaps

Under these conditions, the $\sin(\vartheta)$ component of the field error (i.e. the vertical field) is usually the major component of the field error at both the poloidal gaps, and its value is typically lower than 10 mT during the plasma pulses, thanks to the feedback control of the Field Shaping (FS) winding currents. The horizontal component $\cos(\vartheta)$ and the other components of harmonic order m>3 are usually negligible.

The field measurements show that the short-circuit produced a reduction of the field error at gap #1 by a factor of 3 during the duration of the pulse. This implies that approximately 2/3 of the m=1 current of the shell flow through the short-circuit elements across the gap, while 1/3 of the current does not cross the poloidal gap and flows along it. In fact the measured current distribution across gap #1 is clearly dominated by the $\cos(\vartheta)$ component, and its time evolution is well correlated with the $\sin(\vartheta)$ harmonic of the field error at gap #2 (fig. 2). Correlation of higher order field and current harmonics was not evident. The reduction of the major component of the error field at gap #1 is confirmed by the comparison of the signals from the saddle coils at the two gaps (fig.3).

5.2. Pulses with the magnetic perturbation localised near one of the poloidal gaps

When the localized magnetic perturbations of the plasma is near one of the poloidal gaps, at that gap the horizontal component $\cos(\vartheta)$ of the radial field becomes of the same order as the vertical component $\sin(\vartheta)$ or even larger. For instance if the perturbation is located at the open gap the horizontal field can reach 20 mT.

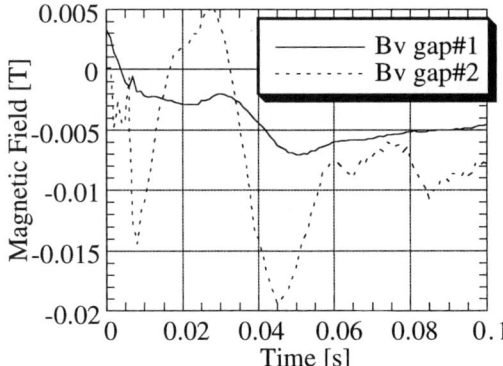

Figure 3. Vertical field measured by saddle coils at both poloidal gaps, with stationary perturbation far from both gaps (shot #5504).

This means that the plasma perturbation produces a localized horizontal field, which however cannot be corrected as long as the currents in the coils are symmetric with respect to the equatorial plane. To this purpose a set of local field control coils will be installed on the open poloidal gap in the next months [4]

Moreover, if the perturbation is located at the short-circuited gap, the short-circuit current distribution is only slightly influenced by the currents induced at the moment of the setting-up of the perturbation. However a more accurate analysis on this matter is in progress.

6. ANALYTICAL MODEL OF THE SHELL WITH THE POLOIDAL GAPS

In order to assess the performances of the short circuit of the poloidal gap in controlled and reproducible conditions a series of tests without plasma was also carried out.

Axisymmetric magnetic fields with a programmed spatial harmonic content were applied by feeding the FS coils with proper current distribution. Due to the topological constraints of the winding, only magnetic field whose radial components is odd functions of ϑ can be created. Each test pulse was characterized by a dominant harmonic component of the radial field (m=1, 2, 3).

An analytical interpolation of the dominant harmonic of the applied field waveform was used as the forcing term in an analytical cylindrical model of the RFX shell [5]. In the model all the electromagnetic quantities are represented by means of their Fourier series along the azimuthal and axial coordinate, corresponding to the poloidal and toroidal coordinate respectively.

The critical issue for the assessment of the short circuit performances is the capability of the model to reproduce the magnetic field at the open gap. Since the calculated field is consistent with the experimental data, the model total current can be taken as a reasonable estimate of the current induced along the real shell far away from both gaps. Then, taking the difference between the estimated total current and the current measured through the short circuit elements, the current which turns poloidally at the short circuited gap can be inferred.

The radial field harmonic components in $\sin(m\vartheta)$ obtained from simulation and from experimental data are compared in figures 4 and 5. The difference is always within 20% and this can be assumed as the confidence interval in the estimate of the current.

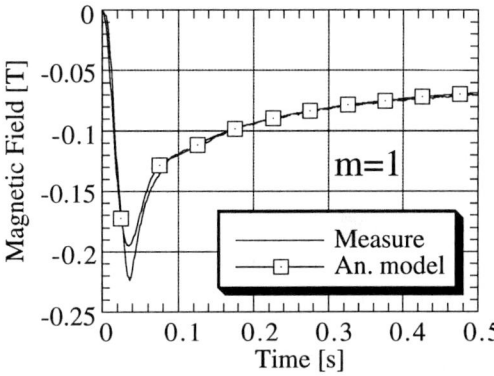

Figure 4. Time evolution of the measured and computed m=1 harmonic component of the radial field Br at gap #2 (no plasma shot).

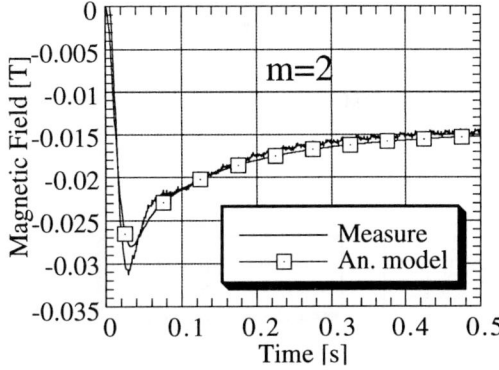

Figure 5. Time evolution of the measured and computed m=2 harmonic components of the radial field at gap #2 (no plasma shot).

In fig. 6 and 7, the evolution of the total axial component of the current linear density and its fraction crossing the gap derived from the measurements are presented. Their ratio is a measure of the effectiveness of the short circuit: in accordance with the design analyses, the maximum effectiveness (up to 60% for m=1) is in the first 50 ms approximately, i.e. the system works fairly well for phenomena whose frequency is higher than 10 Hz. A lower effectiveness was observed in reducing the higher order harmonics (m≥2).

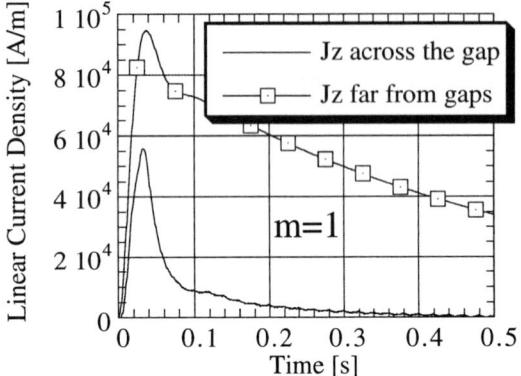

Figure 6. Time evolution of the m=1 harmonic component of the linear current density Jz across gap #1 (measured) and far from the gaps (calculated).

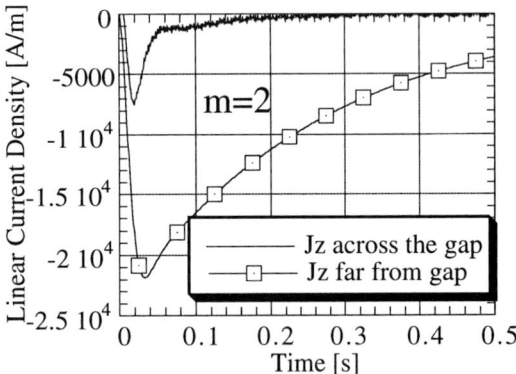

Figure 7. Time evolution of the m=2 harmonic component of the linear current density Jz across gap #1 (measured) and far from the gaps (calculated).

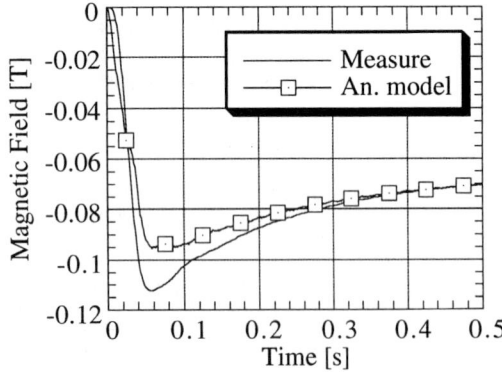

Figure 8. Computed and measured value of the m=1 component of the radial field at the saddle probe on gap #1 (no plasma shot)

In fig. 8 the m=1 average field measured by the saddle probes at the short circuited gap is compared with an estimate of the same quantity provided by the model. The field was calculated considering separately the field due to the current which crosses the gap and the field due to the current which turns poloidally. The sum of the two contributions gave the curve presented in the figure. Again the estimate differs from the measurement by less than 20%.

7. CONCLUSIONS

Thanks to the short-circuit of two gaps, the field errors have been typically reduced by a factor of about 3 at the poloidal gap, and the fast fluctuations of the field at the toroidal gap have also been substantially reduced.

The distribution of the currents measured on the poloidal gap short-circuiting plates in the presence of plasma is in relation with the residual field error at the same gap and at the other poloidal gap and also with the position of the localized perturbation.

The numerical analyses referred to the experimental shots without plasma have also allowed to achieve a better assessment of the poloidal short circuit performances in the case of forcing terms of different poloidal order.

Unfortunately the considerable reduction of the field error did not produce an appreciable reduction of the localized perturbation and of its effects.

REFERENCES

1. V. Antoni et al. "Analysis of Wall Locking Phenomena and their consequences in RFX" Proc. of the 22 EPS conference on Controlled Fusion and Plasma Physics, Bournemouth, UK, jun. 1995
2. A. F. Almagri et al.: "Locked modes and magnetic field errors in the Madison Symmetric Torus, Physics of Fluids B4 (12), dec. 1992
3. A. Buffa et al. "Magnetic configuration and Locked modes in RFX" Proc of the 21 EPS conference on Controlled Fusion and Plasma Physics, Montpellier, jun. 1994
4. F.Bellina, et al. "Feasibility analysis of an active local field control at the poloidal gap of RFX", this conference.
6. P. Fiorentin, G. Marchiori, A. Masiello, G. Zollino: "Computation of field errors due to poloidal cuts on the stabilizing shell of a fusion device", Compel vol. 14 n. 4 dec. 1995.

PLASMA-MATERIAL INTERACTION DURING HIGH TRANSIENT HEAT LOADS IN FAST PROBE EXPERIMENTS AT TEXTOR

T. Scholz, A. Hassanein*, H. Bolt, K. H. Finken, J. Linke

Forschungszentrum Jülich GmbH, Association EURATOM-KFA, D-52425 Jülich, Germany
*Argonne National Laboratory, Fusion Power Program, Argonne, IL 60439, USA

Abstract

In future tokamak fusion devices the plasma facing components are expected to be increasingly subjected to transient high heat fluxes due to various plasma instabilities such as vertical displacement events, edge localized modes (ELMs) and plasma disruption scenarios. The plasma interacting with the solid surface will cause melting and erosion of plasma facing materials. Until now material tests were carried out in various laboratory facilities, using ion and electron beams, plasma accelerators or lasers to simulate the high transient heat fluxes expected. Furthermore simulations with computer codes were made to investigate the interaction of the plasma with the wall and the developed vapor cloud in front of it. The aim of the present work is to complement the modelling and the current experimental efforts by transient high heat flux experiments directly in the plasma of the TEXTOR tokamak.

1. INTRODUCTION

In future tokamak fusion devices plasma facing components such as limiter blades or divertor modules will be frequently subjected to high transient heat fluxes during off-normal events like ELMs, vertical displacement events and plasma disruptions /1/. Due to the high heat flux strong ablation and material damage is to be expected.

Until now materials for PFCs were investigated under pulsed heat fluxes by laser or electron beam irradiation or under exposure with plasma accelerators with varying ablation results depending on the incident species /2-4/.

In the tokamak environment the incident plasma interacts with the ablated material resulting in collisional processes. Due to ionization and excitation subsequently plasma energy is partly lost by radiation and the incident heat flux is moderated by cooling down the incoming plasma.

The local shielding behaviour of carbon fiber composite materials (CFC) being exposed to the edge plasma of the TEXTOR tokamak with a pneumatic fast probe is investigated experimentally. The impact on the Tokamak plasma from its interaction with the eroded material and the investigation of the material damage under incident plasma heatfluxes up to 88 kW/cm² were central aims of this work.

The erosion and the shielding of the probe surface was investigated with the A*THERMAL-S computer code /5,6/ taking the transient plasma cloud conditions within a finite element mesh in front of the surface into account.

2. EXPERIMENTAL

2.1. Fast Probe experiments in TEXTOR

A pneumatic fast probe device of UCSD (University of California at San Diego) was used to insert specimens of various carbon fiber composite materials into the edge plasma of the TEXTOR Tokamak. The time for insertion of the probehead was 50 ms with the head remaining in its top position up to 9 cm inside limiter radius for 80 ms. During insertion and retraction the probe tip travelled through the edge plasma for about 15 ms.

The specimen size was 16 by 16 mm² and 120 mm long. Three types of CFC materials were used in the experiments. To cause a high erosion a two dimensional CFC type Schunk CF260 with a comparatively poor thermal conductivity of 10 W/mK was used initially. In following experiments the three dimensional type SEP N11 and the siliconized SEP NS11 were inserted into the plasma. These materials have a low porosity and good thermal properties (λ = 180 W/mK).

The maximum insertion depth into the plasma was 9 cm as the main limiter was positioned at a plasma radius of r = 46 cm and the maximum tip position of the modified pneumatic fast probe was at r = 37 cm. In this region during NBI heated hydrogen discharges the temperature was around T_e = 600 eV and the density of the undisturbed plasma before insertion of the probe was n_e = 1.2 10^{13} cm^{-3}. With an energy transmission factor through the plasma sheath of f = 6,5 /7/ the incident heat flux would be up to q" = 88 kW/cm^2.

In addition to the usual TEXTOR diagnostics two optical local diagnostics were used. The C I line emission (at λ = 908,8 nm) was observed in an area on the probe surface of 1 cm diameter in a radial position of r = 41 cm, about 4 cm above the maximum position of the probe tip. A high speed CCD - camera system with infrared filter (>950 nm) and a high frame rate of 330 s^{-1} observed the area around the probe shaft just above the tip in end position.

After the repeated exposure to the plasma the specimens were investigated by SEM and also for dimensional changes with a 3D measuring machine and a laser profilometer.

2.2. Electron beam experiments in JUDITH

The electron beam facility JUDITH was used to apply surface heat loads of q" = 30 kW/cm^2 on specimens of CF260 material. This value was chosen according to the experimental parameters of the TEXTOR experiments. The 120 kV beam was directed to an area of 12x16 mm^2 on the side surface of the specimens for a duration of 100 ms.

Finally the erosion values and observed effects were compared with results of calculations using the A*THERMAL-S computer code with the experimental input parameters.

3. RESULTS

3.1. Fast Probe experiments in TEXTOR

Upon insertion of the probe the edge plasma was strongly disturbed as the eroded particles caused a strong radiation of up to 1 MW power. Fig. 1 shows the radiation intensity profiles for a SEP material. The temperature and plasma density in the edge region were strongly reduced. Helium beam measurements showed almost total cooling of the edge plasma in the shadow of the probe. From regions behind the tip no radiation was emitted

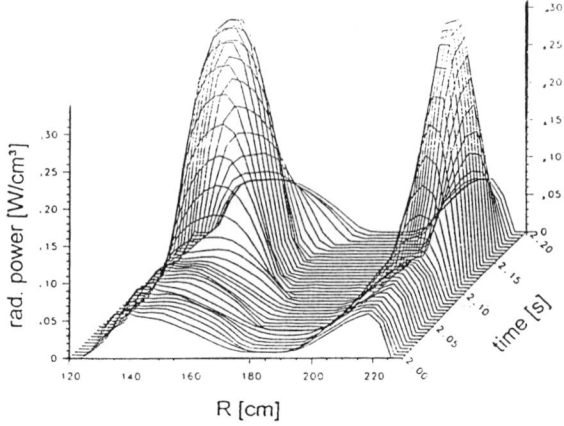

Figure 1. Radiated power upon probe insertion

that could be detected with the CCD camera or the spectrometers which indicates that the probe shaft remained very cold. Fig. 2 shows the radiation of the probe tip upon insertion into the hot plasma.

Figure 2. Radiation around probe tip

3.2. Electron beam experiments in JUDITH

A rapid temperature increase was observed upon irradiation and massive erosion began within less than 10 ms after start. The material was deeply eroded with damaged trunks of reinforcement fibers remaining. The erosion depth per 100 ms pulse was in all cases about δ = 200 μm.

3.3. Simulation with A*THERMAL - S

According to the experimental parameters simulations of the plasma material interaction were made using the A*THERMAL-S computer code. First calculations were carried out to calculate the erosion values for the heat loads of the JUDITH experiments. As shielding does not have to be taken into account only the thermal response was calculated transiently in finite elements. The erosion depth of δ = 238 μm per pulse was well corresponding to the experimental result of δ = 200 μm per pulse.

The simulation of shielding effects due to eroded carbon species was done by calculating stopping powers and radiation emission transiently in the vapor under moving boundary conditions. In this more sophisticated version of the code also the vapor conditions in front of the probe surface were computed in finite elements in a range of up to 10 cm from the surface for each time step.

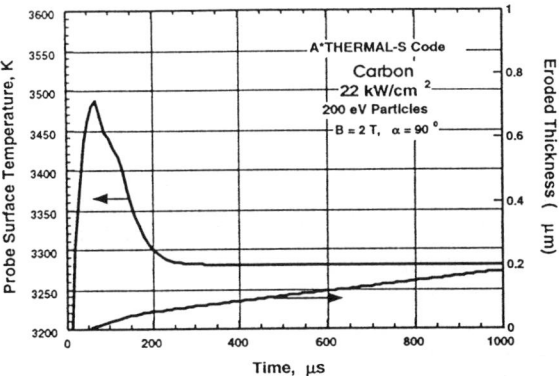

Figure 3. Surface temperature is reduced by shield

Because of the importance of radiation transport in the vapor cloud regions, a self-consistent approach to calculate the actual radiation field is also developed and implemented in the A*THERMAL-S code. The optical properties of vapor-cloud plasma are calculated at each time-step during the course of a disruption or in this case the probe insertion. The relevant atomic data bases of candidate materials are implemented in the code which include all possible transition energy levels, ionization potentials, rate coefficients, photoionization cross sections, statistical probabilities, oscillator strengths, etc. The kinetic rate equations are then solved for each ion level population at every time step. The radiation transport equation is then solved separately for both line- and continuum-generated spectra. The self-consistent model also takes into account the multispecies effect, i.e., mixing between the incoming plasma particles and the emitted vaporized material.

In the calculation the specimen surface was assumed to be a semi-infinite plane in contrast to the actual specimen geometry. However, this assumption does not affect the results as radiation from the surrounding cloud towards the surface is in fact reduced by the geometry.

Fig. 3 shows the rapid rise and following reduction into equilibrium or steady state conditions of the surface temperature, assuming a specimen of 3000K temperature being exposed to the plasma without any vapor cloud around. Within a few microseconds the cloud arises and the plasma particles are completely stopped within the vapor. Energy transport to the probe surface takes place by radiation transport through the cloud only.

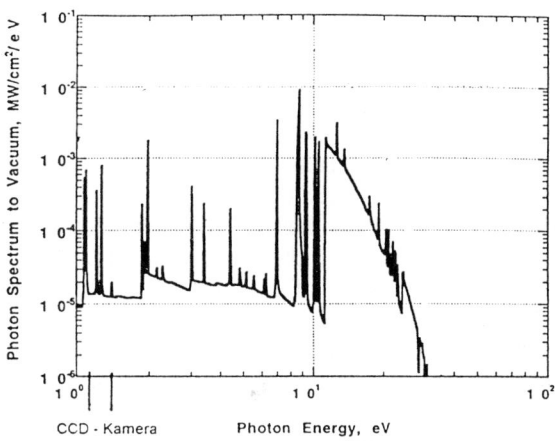

Figure 4. Radiation spectrum to environment

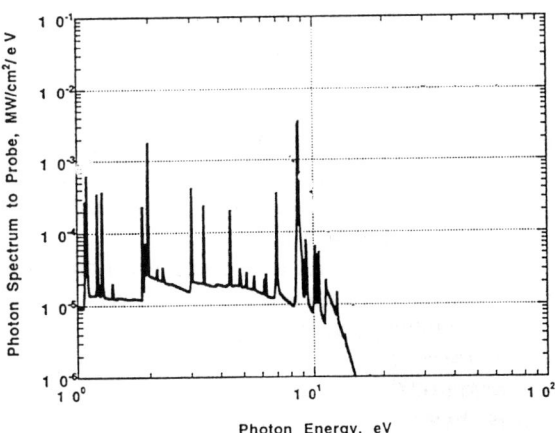

Figure 5. Radiation spectrum to probe surface

Due to opacities in the zone close to the surface radiation and therefore energy transport to the probe are strongly reduced. Figs. 4 and 5 show the radiation spectra to the probe and into the environmental vacuum. The intensity of the radiation into the vacuum is about 10 times higher than towards the probe surface. The sensitivity range of the CCD Camera is marked in fig. 4. The radiation intensity in this range is a bit higher than the blackbody radiation of a hot CFC surface of 3000K, so the camera sees partly line radiation and the thermographic measurement deduced from this is therefore about 3600 K.

4. DISCUSSION

Calculations with A*THERMAL-S code matched the low erosion processes due to shielding of probe surface from the energy influx mostly by the neutral carbon species surrounding the probe tip. The radiation around the tip and the evolution of the radiative cloud was also observed during the experiments in TEXTOR. Fig. 2 shows the area of the probe tip surrounded by a radiating vapor cloud. The radiating cloud has a visible thickness of roughly 10 mm around the specimen. This value is matching the results of the detailed calculations with the A*THERMAL-S code including the local conditions of the cloud for the stationary phase of the probe experiment. Taking plasma zones up to 10 cm from the surface into account it was seen that the main radiative loss is taking place within the area near the surface mostly due to line radiation.

Apart from this the edge plasma was cooled down in wide areas. For the SEP materials a steady state condition with a radiated power of 0.2 MW was reached whereas with the CF260 material the massive impurity influx led to a radiation power peak of 1 MW so that further erosion of the probe was stopped as the edge plasma did not recover within the insertion time. Obviously a steady state or equilibrium condition is reached.

In addition the probe tip partly acts as a limiter but the erosion zone is with less than 2 mm very narrow compared to the decay length of the heat flux in that region. The probe tip seems to be shielded locally by interaction of the eroded particles with the plasma. The local shielding was calculated to be about 90 % of the energy influx and the radiative cloud was observed in a range of 10 mm around the tip.

5. CONCLUSION

Specimens of various types of CFC material were exposed to hot regions of a tokamak plasma. Very low erosion was observed. The simulation of shielding effects due to eroded carbon species was performed using the A*THERMAL-S code by calculating stopping powers and radiation emission spectra in the vapor under moving boundary conditions transiently. The calculations matched the experimental results as the probe surface was strongly shielded from the energy influx mostly by eroded carbon species surrounding the tip. The emitted line radiation could be observed during the experiments as well.

ACKNOWLEDGEMENT

The authors greatly appreciate the support of the UCSD group at TEXTOR providing the experimental facility and support.

REFERENCES

1. R.R. Parker, W.B. Gauster, Fusion Eng. Des. 30 (1995) 119
2. J.G. v.d. Laan, H.Th. Klippel, G.J. Kraaij, R.C.L. v.d. Stad, J. Linke, M. Akiba, J. Nucl. Mater. 196-198 (1992) 612
3. J. Linke, M. Akiba, H. Bolt, J.G. v.d. Laan, H. Nickel, E.v. Osch, S. Suzuki, E. Wallura, J. Nucl. Mater. 196-198 (1992) 607
4. H. Bolt, V. Barabash, A. Gervash, J. Linke, L.P. Lu, I. Ovchinnikov, M. Rödig, Fusion Technology (1994) 383
5. A. Hassanein, I. Konkashbaev in: Atomic and Plasma-Material-Interaction Data for Fusion, Vol.5 (IAEA, Vienna, 1994) pp.193
6. A. Hassanein, D.A. Ehst, J. Gahl, J. Nucl. Mater. 212-215(1994) 1272
7. P.C. Stangeby in: Physics of Plasma Wall Interaction in Controlled Fusion (Plenum Press, New York, 1984) pp. 41

Minimizing the first wall and blanket loading caused by plasma disruptions by tuning first wall's resistivity

T. Jordan[1], D. Schneider[2]

[1]Forschungszentrum Karlsruhe, Institut für Reaktorsicherheit
Postfach 3640, D-76021 Karlsruhe, Germany

[2]Institut für Kerntechnik und Reaktorsicherheit, Universität Karlsruhe (TH)
Postfach 3640, D-76021 Karlsruhe, Germany *

The electrically conducting components next to the plasma of a TOKAMAK fusion reactor will be heavily loaded by plasma disruptions. This paper shows the effects of an electrically conducting first wall on the mechanical loading of the first wall and the blankets. An advanced computational procedure is applied, which no longer prescribes the decay of the plasma current. This current is now part of the solution. The spatial distribution and the time history of the plasma current, furthermore the induced forces and the mechanical stresses in the structure are very sensitive to the electrical design of the first wall. A simple model explains why there is an optimum first wall resistivity which minimizes the stresses in the first wall.

1. INTRODUCTION

During a plasma disruption in a TOKAMAK fusion reactor the electrically conducting structures next to the plasma will be heavily loaded. Especially the blankets which will contain tritium have to withstand these electromagnetic forces. Numerous computations have been conducted [1–3] applying 2d and 3d eddy current codes to determine the volumetric Lorentz forces and the related stress maxima. All these computations introduce considerable simplifications: some, for instance, take only the maximum forces for a static stress analysis, most of them assume nonmagnetic structural materials and prescribe the plasma behaviour in time and space - what is commonly called a "design plasma disruption".

However, the electrical properties of the structural boundary will not only influence, but determine the plasma dynamics. Dropping the assumption of a fixed plasma behaviour will give more realistic results but on the other hand requires some kind of dynamic plasma modelling.

Because it is intended to investigate only the consequences of the bilateral electromagnetic coupling in this paper, the feedback of the induced eddy currents in the structural boundary on the plasma motion will be excluded for the moment - the solution of the plasma momentum equation will be given in a future paper. In contrary to the plasma current the position of the plasma column will still be given as an input.

Experimental observations reveal that disruptions of elongated plasmas are linked to a vertical displacement (VDE) and may be further divided into two chronologically separated phases [4,5]. During the first phase, which is called "thermal quench" or "energy quench", the plasma is cooled down within approximately 1 ms. In the second phase the cold plasma with its high electrical resistance dissipates the magnetic field energy of the plasma current. This inductive-resistive decay, which is called "current quench" takes about 20 ms. Assuming an equilibrium of edge radiation and ohmic dissipation during the current quench the plasma may be treated as isothermal [6]. Furthermore the comparatively short timescaled energy quench is - in contrary to the current quench and the vertical displacement of the plasma - not expected to have any influence on the structure's loading. Therefore modelling of the very quick thermal quench and solving the energy equation is not needed.

---

*This work has been performed in the framework of the Nuclear Fusion Project of the Forschungszentrum Karlsruhe and is supported by the European Community within the European Fusion Program and the DFG

2. COMPUTATIONAL PROCEDURE

The procedure is based on the 3d eddy current code CARIDDI [7]. It was extended by a 2d plasma model which is inductively coupled to the 3d structure. The plasma domain is represented by a set of axissymmetric filaments, all with the same rectangular cross-section. Each filament represents one finite plasma element. The shape function for the electric current in such an element is chosen to be constant. This approach implies that the poloidal components of the plasma and the halo currents are neglected. Introducing the discretization associated with this filamental model and applying Galerkin weighting with the shape function above, the quasistationary Maxwell equations are comprised in a finite element formulation of the plasma current:

$$\boldsymbol{L}_P \frac{d}{dt}\{J_P(t)\} + \left(\boldsymbol{R}_P(t) + \vec{v}\cdot\frac{\partial \boldsymbol{L}_P}{\partial \vec{r}}\right)\{J_P(t)\}$$
$$= \{V_P(t)\} - \oint_{C_i} \boldsymbol{B}^{ext}(\vec{r},t)\cdot\vec{v}\cdot\vec{e}_\Theta \; ds_i$$

Herein $\boldsymbol{L}_P$ is the inductance matrix, the motional inductances $\vec{v}\cdot\frac{\partial \boldsymbol{L}_P}{\partial \vec{r}}$ are added to the resistance matrix $\boldsymbol{R}_P$ and the vector V_P represents the mutual inductance of the 3d structural boundary and the plasma. The last term takes into account the electrical fields induced by the plasma motion through the magnetic fields generated by the poloidal field coils. $\boldsymbol{L}_P$, $\frac{\partial \boldsymbol{L}_P}{\partial \vec{r}}$ and $\boldsymbol{R}_P$ are computed analytically.

The system matrices of CARIDDI were extended with those given above. The system solvers had to be adapted to the non-symmetric properties and the time dependence of the extended $\boldsymbol{R}$ matrix.

Taking into account the free boundary of the plasma column for the plasma material transport a Volume-Of-Fluid (VOF) method [8] was integrated in CARIDDI.

After choosing the appropriate temperature profile and velocity field the currents and the electromagnetic forces in the structures were computed.

3. RESULTS

For all subsequent computations DEMO-relevant geometry and operational data are used. Two different scenarios are assumed for the plasma disruption. The first is the centered or radial disruption which presumes that the position control system is able to stabilise the plasma's vertical position. This scenario is initiated by setting the plasma temperature to 20 eV and keeping the velocities at zero. The observed small radial inward displacement caused by the pressure loss in the plasma during the energy quench is omitted. The second scenario assumes a VDE caused by loss of vertical position control. For the calculations a constant velocity of 100 m/s and a temperature drop to 20 eV are assumed (compare [9]). When the plasma touches the first wall, the layer exterior to the magnetic flux surface, which just contacts the wall, is scraped off. The plasma current of this scraped-off region is coupled inductively into the remaining plasma and the structural boundary. Because the plasma model does not take into account the poloidal component of the plasma current the poloidal halo-currents [9] in the structures can't be modelled.

Three different first wall resistances are investigated. They are supposed to be realized by different electrical resistivities of the gap material between the first wall segments. In option 1 this material is an electrical insulator, in option 3 the conductivity of the gap material is identical to the first wall material and an intermediate value is assigned to option 2.

As expected, the scenario with the vertical displacement of the plasma gives the higher load for each option. This is caused mainly by the contact of the plasma with the structure, where the outer layer of the plasma is scraped off. Subsequently, steep time gradients of the current in the vicinity of the structure induce strong currents and strong forces in this part of the structure. In the DEMO reactor this phenomenon cancels out the effects caused by the motional inductances because the plasma needs only a few centimetres of vertical displacement to contact the wall.

The conductivity of the first wall influences the loading in a complex way. The main effects are:

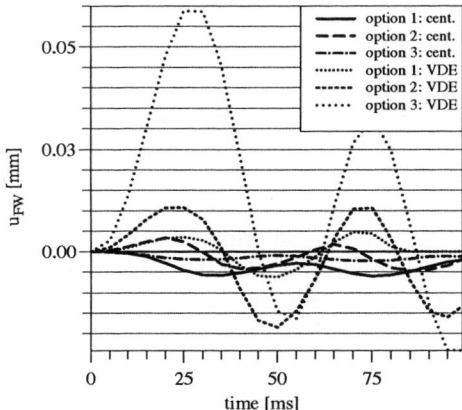

Figure 1. Radial first wall displacements u_{FW} at the equatorial plane

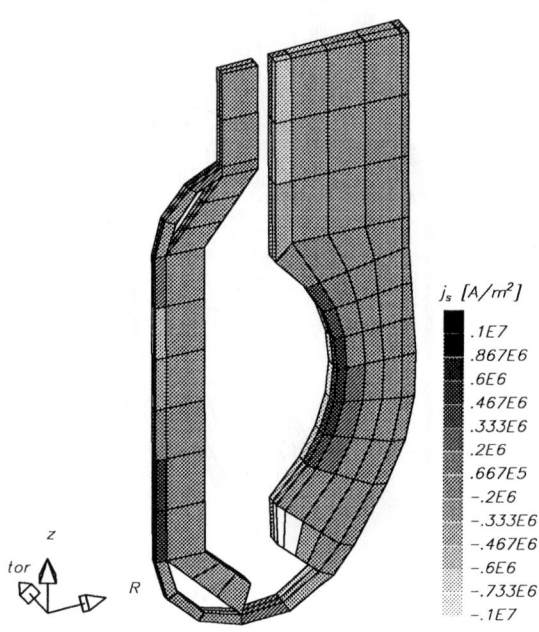

Figure 2. Current density of option 2: VDE at 25 ms

smaller resistance gives a better shielding of the structures behind the first wall from the magnetic field variation (option 1 to 3 with a centered disruption in fig.(1)), while increasing the toroidal current and the forces in the first wall. The rearrangement of the toroidal current in the plasma and the first wall happens very quickly. This implies pulse-like, radial forces, to which the first wall will be quite sensitive (option 3: VDE in fig.(1)).

Option 2 is the best compromise when one takes into account both load patterns. A characteristic current, stress and displacement distribution for this option and a VDE are given in figure (2) and (3). Moreover, there is a minimum load of the blanket's clamping expected for a certain first wall resistance. With a simple model an explanation for this effect will be given.

4. THREE LOOP MODEL

To explain why there is an optimum resistivity to the loading of the complete blanket structure, a simple model consisting of the plasma (index P), the first wall (index F) and the blankets including the back wall (index B) is proposed. These components are represented by a set of concentric current loops. The currents are easily calculated by solving the system of homogeneous ordinary differential equations of first order. The system's matrices are built up with the representative values of the inductances and resistances of all three loops. The actual aim is to minimize the stress in the loops B and F representing the mechanically coupled system which consists of the first wall, the blankets and the back wall, by tuning the first wall resistivity. Assuming a stiff clamping of this system on the vessel allows relating the maximum stress to the sum of the forces F_{FB} acting in loop F and B

$$f_{FB}\left[\frac{N}{m}\right] = \frac{F_{FB}}{2\pi r_{FB}} = i_F(k_{PF}i_P - k_{FB}i_B + B_{pol})$$
$$+ i_B(k_{PB}i_P + k_{FB}i_F + B_{pol})$$

Herein f is the force per length, r_{FB} an average radius of the loop F and B, i denotes the loop current and B_{pol} is the average vertical magnetic induction across the blankets. With the derivative of the mutual coupling $k_{PF} \approx k_{PB} = k$ this reads

$$f_{FB} = (i_F + i_B)(k\,i_P + B_{pol})$$

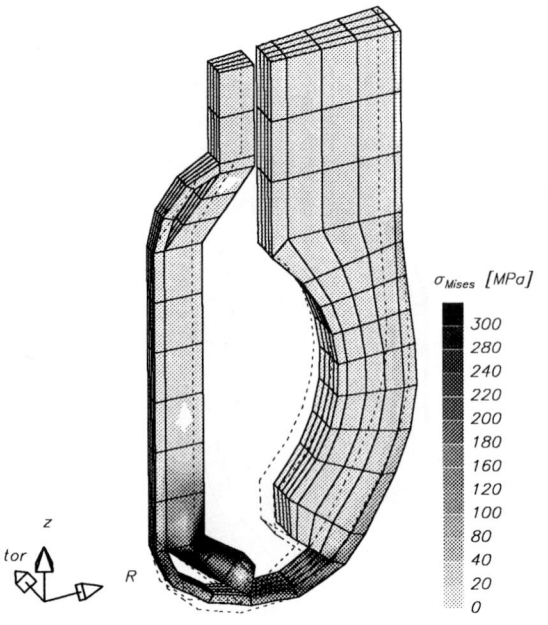

Figure 3. Von-Mises stresses and displacements (×20) of option 2: VDE at 25 ms

With the help of the resistance of the first wall the solution of the differential equations may be fitted such way that $i_F \approx -i_B$ at least during the first oscillation period of the mechanical system.

5. CONCLUSIONS

The results of the detailed computation, as well as the simple model, recommend the selection of an intermediate resistivity for the DEMO first wall. This will keep the stresses in the blanket and in the first wall moderate and minimize the resulting stresses in the blanket's clamping. The actual value of this resistivity depends strongly on the design and has to be determined by iterative computations with a procedure like the one presented above.

Calculations with a further advanced computational procedure are on the way which proof that the feedback of the electromagnetic coupling on the plasma motion will confirm these results.

REFERENCES

1. L. V. Boccaccini. Electromagnetic forces caused by disruptions in the Karlsruhe Demo solid breeder blanket and in the test module for NET/ITER. *Fusion Engineering and Design*, 17:153–159, 1991.
2. P. Chaussecourte, A. Bossavit, J. C. Verite, Y. R. Crutzen. 3d eddy-current distribution in a tokamak first wall during a plasma disruption using TRIFOU. *Fusion Engineering and Design*, 9:101–105, 1989.
3. Y. R. Crutzen, G. Rubinacci. Evaluation of the electromagnetic effects on a tokamak first wall caused by a plasma disruption using a thin shell formulation. *Fusion Engineering and Design*, 11(3):293–303, 1989.
4. P. E. Stott. Density limits and disruptions in plasmas. *Nuclear Fusion*, 28(8):1469–1473, 1988. Report on IAEA Tech. Comm. Meeting held at JET, Jan. 26-28, 1988.
5. J. A. Wesson et al. Disruptions in JET. *Nuclear Fusion*, 29(4):641–666, 1989.
6. TFR Group. Study of post-disruptive plasma in TFR Tokamak. *Nuclear Fusion*, 25:919–930, 1985.
7. R. Albanese, G. Rubinacci. Integral formulation for 3D eddy-current computation using edge elements. *IEEE Proceedings*, 135(7):457–462, 1988.
8. N. Ashgriz, J.Y. Poo. FLAIR: Flux Line-segment model for Advection and Interface Reconstruction. *Journal of Computational Physics*, 93, 449, 1991.
9. O. Gruber, K. Lackner, G. Pautasso, U. Seidel, B. Streibl. Vertical displacement events and halo currents. *Plasma Physics and Controlled Fusion*, 35:B191–B204, 1993.

Joining Technology and Material and Shape Optimization for the ICRF Vacuum Transmission Line Dielectric Window

P. Auerkari[a], L. Heikinheimo[a], J.A. Heikkinen[b], J. Linden[c], M. Kemppainen[c], K. Kotikangas[c], S. Nuutinen[a], S. Orivuori[c], M. Peräniitty[c], S. Saarelma[d], M. Sirén[a], S. Tähtinen[a], F. Wasastjerna[b], G. Bosia[e], E. Hodgson[f]

[a]VTT Manufacturing Technology, P.O. Box 1704, FIN-02044 VTT, Finland
[b]VTT Energy, P.O. Box 1604, FIN-02044 VTT, Finland
[c]IVO Group, Research and Development, FIN-01019 IVO, Finland
[d]Helsinki University of Technology, Department of Engineering Physics and Mathematics, FIN-02150 Espoo, Finland
[e]ITER Joint Central Team, Max-Planck-Institut für Plasmaphysik, D-85748 Garching, Germany
[f]CIEMAT, Avda Complutense 22, E-28040, Madrid, Spain

The choice of dielectric material for the (double) dielectric window of the ITER ICRF Transmission Line is optimized with respect to nuclear, mechanical, and thermal properties. Neutron fluence at the window placed in the vacuum vessel feedthrough is obtained with MCNP-4 to be sufficiently small for allowing beryllia, but not alumina without additional neutron shielding, to be used as dielectrics. For beryllia (10^{-3} dpa) or for unirradiated alumina, the temperature is found to stay between the cooling temperature and 200^0C with maximum hoop stress less than 100 MPa, provided titanium, niobium or materials with similar thermal expansion coefficients are used as support. Based on experimental vacuum tightness and shear strength tests for alumina-titanium joints, three potential brazing filler metals have been identified for bonding.

1. INTRODUCTION

The (double) dielectric window is an essential and the most vulnerable component of the ICRF Vacuum Transmission Line for ITER, as it provides ultimate vacuum and tritium containment. Eight windows are present in each of the four ICRF arrays in the present baseline design. Although several models of dielectric window have been developed and are currently in use [1], the ITER stringent reliability requirements call for special design and prototyping to allow a realistic evaluation of the failure rate. The ITER window design requires specific design and careful selection of the dielectric material because of the long discharge pulse, high electric field strengths, a possible degradation of the dielectric properties due to neutron or gamma irradiation, possible changes in the mechanical, and thermal properties and in gas permeation [2]. Furthermore, the metal-ceramic joints required for the windows and the support structures need to retain reliable vacuum tightness under cyclic operation conditions.

In the ITER VTL, the window assembly will be located on the equatorial duct in the region between the back of the ICRF Array neutron shield and the exit of the cryostat. The earlier calculations of neutron radiation estimates for these positions have been based on simplified homogenized geometries for the shield/blanket, vacuum vessel, cryostat, divertor, midplane port and ICRF array neglecting the neutron streaming. In the present work, detailed neutron radiation calculations are performed with the MCNP-4 code with account for the neutron streaming through the detailed structure of the ICRF Array/Shield/VTL assembly.

BeO and Al_2O_3 (97.5% to 99.99% purity) are considered as dielectric materials. 97.5% Al_2O_3 (polycrystalline) is internationally recognized and stable standard dielectric, with probably the best database available for ICRF conditions. The 5-10 times higher thermal conductivity and the unsensitivity of the loss tangent against modest neutron radiation for BeO, and the good experience with BeO windows at higher RF frequencies, make the latter dielectric interesting for the present study.

The annular windows are assumed to be axisymmetric fitting the characteristic impedance of 30 Ohms and dimensions of the transmission line. The modeling includes the double window structure with a dynamically evacuated intermediate region, 10^{-5}-10^{-2} Pa vacuum conditions on the antenna side, 300 kPa dry air or 100 kPa SF_6 conditions on the pressurized size, cooling on the inner and outer conductor, coronal rings to reduce the electric fields close to the joints, and the joining structure. The design is based on the maximum operating voltage of 50 kV peak RF voltage, with arbitrary amplitude modulation. 2-dimensional finite element codes IVOFEM/IVOHEAT [3] and ANSYS are used for the evaluation of the dielectric losses, related heating of the ceramics window, temperature distribution, created stresses, electric potential distribution, and heat conduction from the ceramics through the joints to the cooling channels. Water cooling with inlet temperature 100 °C and outlet temperature 150 °C with coolant pressure 4 MPa are assumed. The results summarized in this paper are reported in details in the ITER Task 238/2 Final Report.

2. NEUTRON FLUX ANALYSIS

The neutron fluence to which the vacuum window will be exposed during its service life needs to be known, at least to the accuracy needed to determine whether the radiation damage exceeds what is considered permissible. In the calculation, the antenna straps and Faraday shield were omitted, which is a conservative approximation so far as the neutron flux is concerned. The midplane port was considered empty between the array and the cryostat, except when the effect of an additional neutron shield was considered.

According to the present analysis, the fluence on the window placed outside the cryostat remains below 10^{16} n/cm² for 2500 hours radiation. Along the transmission line the fluence may increase up to 10^{19} n/cm² for just behind the neutron shield of the ICRF array, but can remain less than 2×10^{18} n/cm², or even less than 2×10^{16} n/cm² just behind an additional shield with minimum thickness of 55 cm, on the equatorial duct section. By inspection of the material data of irradiated ceramics one can conclude that unirradiated data for the loss tangent, thermal conductivity, and mechanical and electrical strength can be used for alumina and beryllia with fluences below 10^{16} n/cm² [2]. By the fluences up to 10^{18} n/cm² the changes can also be regarded as small (<30%). On the other hand, for alumina a five fold increase of the loss tangent and reduction of the thermal conductivity (depending on the temperature) to one third are expected for the fluence 5×10^{19} n/cm² (5×10^{-2} dpa, in-beam). For that reason the following analysis is made with two sets of parameter data, one for unirradiated case and the other for irradiated case. In the latter, the tangent loss is obtained with the fluence 5×10^{19} n/cm² for the alumina, and with 10^{18} n/cm² (about 10^{-3} dpa; here the data above 10^{18} n/cm² is not available) for beryllia. Conservatively, the saturation level for the reduction of the thermal conductivity by irradiation is used for both ceramics (the reduction is relevant mostly with the fluences above 10^{20} n/cm²).

3. FEM MODELING OF STRESSES, BREAKDOWN, AND COOLING

Due to the long burn time of ITER, 1000 s, stationary condition is met quite quickly from the beginning of the pulse. For unfavourable radiation conditions (5×10^{-2} dpa), alumina is found to heat excessively over 1000 °C with unacceptable stresses. For beryllia (10^{-3} dpa) or for unirradiated alumina (97.5 % purity), the temperature is found to stay between the cooling temperature and 200°C with maximum hoop stress less than 100 MPa, provided niobium, titanium or materials with similar thermal expansion coefficients are used as support. Stresses with steel or aluminium conductor are unacceptable.

An optimized design with properties and constraints
- angle of inclination is 75 degrees
- ceramic thickness varies from 0.01 m to 0.015 m
- radius of inner conductor outer edge is 0.055 m
- radius of outer conductor inner edge is 0.096 m.
- titanium conductor has thickness of 0.006 m
- coolant temperatures are 100 °C and 150 °C at the inner and outer conductor, respectively
- coolant flow velocity is 5 m/s
- stress-free temperature is 20 °C

has been obtained for both irradiated beryllia and unirradiated alumina with titanium or material with similar heat expansion coefficient as a conductor. The tangential and normal components have been evaluated along the ceramic's left surface (against vacuum). The maximum electric fields, E_{tan}, E_{norm}, E_{tot}, in vacuum are: 1.0, 8.1, 8.3 MV/m, and in the ceramic: 1.0, 1.0, 1.6 MV/m. From the temperature histories for BeO and Al$_2$O$_3$ ceramics for the chosen optimized geometry, the maximum temperatures were found to be 157 °C and 246 °C, respectively, in these cases. The total heating power was found to be 1.4 kW for beryllia and 3.0 kW for alumina (the total ceramic volume is 1117 cm^3). For 80 MHz frequency, the maximum temperature for BeO was 166 °C. Table 3.1 gives the found maximum local stresses for both 60 MHz and 80 MHz frequency.

Table 3.1 Maximum local stresses (MPa)

Ceramic	Max. principle stress	Max. shear stress
BeO (60 MHz)	109 (titanium)	34 (titanium)
BeO (60 MHz)	56 (ceramic)	26 (ceramic)
Al$_2$O$_3$ (60 MHz)	204 (titanium)	67 (titanium)
Al$_2$O$_3$ (60 MHz)	121 (ceramic)	51 (ceramic)
BeO (80 MHz)	121 (titanium)	38 (titanium)
BeO (80 MHz)	64 (ceramic)	30 (ceramic)

4. JOINING THE WINDOW TO CONDUCTOR

The objective of the joining experiments was to optimize the joining process for vacuum brazed Al$_2$O$_3$ (96% - 99%) -Ti (Grade 2) joints. The choice of brazing alloys was based on potential commercial active brazing alloys, directly suitable for brazing of the oxide ceramics. The screening was carried out by performing microstructural investigations of the interface areas, vacuum tightness and shear strength tests. The high frequency ultrasonic examination, C-SAM, was used for evaluating the quality of the brazed joint.

The He-leak test samples were made with a tube to plate configuration. Based on these results all the samples except those brazed with the alloy CB6 and the alloy Gold ABA are acceptable showing very high level of leak tightness (the levels of 10^{-8} and 10^{-9} mbar). Prior to the leak test some of the samples were thermally cycled (RT - 200 °C x10). No difference in the leak tightness can be seen between the thermally heat treated and the bonded samples.

Both, bottom and side surfaces, of brazed ceramics were examined together with C-scans. It was possible to see that the brazed joints at the bottom of the ceramics were only partially bonded in all examined samples. The joint quality of the side surfaces of the ceramics showed more variation between the samples studied. Samples brazed with brazing alloys Copper ABA and CB4 showed the fully bonded side surfaces of the joints whereas sample brazed with Incusil ABA was continuous but only partially bonded. The sample brazed with CB6

Table 4.1 The results of the shear strength test, dimensions of the samples were ⌀ 14 mm x 17 mm for Al$_2$O$_3$ and titanium.

Brazing alloy	Maximum force [kN]	Shear strength [MPa]	Fracture path
CB4	7.74	50.3	ceramic
CB4	5.8	37.7	ceramic
CB4	4.9	31.8	ceramic
CB6	9.15	59.4	joint
CB6	-	-	-
CB6	-	-	-
Copper ABA	7.08	46.0	ceramic
Copper ABA	-	-	-
Copper ABA	5.69	37.0	ceramic
Gold ABA	5.26	34.2	ceramic/joint
Gold ABA	9.06	58.9	ceramic/joint
Gold ABA	4.25	27.6	ceramic/joint
Incusil ABA	18.66	121.2	ceramic/joint
Incusil ABA	-	-	-
Incusil ABA	25.05	162.7	ceramic/joint

showed poor bonding quality. Results of the ultrasonic examination were in a good accordance with vacuum tightness test results. Those samples having acceptable vacuum tightness results showed good bonding quality.

The shear strength tests were made for 15 ceramic (A476) to metal (Gr2 Ti) joints and for 15 ceramic to ceramic joints using the same brazing alloys as in vacuum tightness tests. The bonded shear test samples and the measured strength values are shown in Table 4.1.

The alumina to metal joints brazed with the Incusil ABA, CB4 and Copper ABA alloys are the three most potential ones giving the average bond shear strength of 142 ± 21 MPa, 44 ± 9 MPa and 43 MPa respectively.

Based on the experimental work carried out it is possible to recommend the choice of three potential brazing filler metals for further optimization of the oxide ceramic to titanium bonding. The first choice for brazing alloy is evidently the Incusil ABA, a silver-copper based alloy with Cu, In and Ti additions. This alloy presents superior properties in shear strength and is also reliable with respect to the leak tightness. The low bonding temperature of 700 - 750 °C, compared to the other alternatives studied, 900 - 1050 °C, is expected to be favourable in the assembly of the full component, especially with respect to the lower build-up of the thermal residual stresses.

The two other choices are the alloy CB4, an eutectic Ag-Cu alloy with Ti activation, and the Copper ABA alloy, a copper based Ti activated alloy. These brazing alloys will not provide that good ceramic to metal joint strength values than the Incusil ABA alloy does, however, the values are reasonable for many applications where mechanical load is not the basic design criteria. These samples were tested both in an as-bonded and in thermally heat treated (thermal cycling) conditions. No deterioration of the joint could be observed due to this service simulating thermal treatment. As Ag is a high activation material, further work is needed to investigate the compatibility of the corresponding alloys in reactor conditions.

5. SUMMARY

According to the thermal/stress analysis, both alumina (97.5%) (for weak irradiation) and beryllia ($<10^{-3}$ dpa) ceramics can be considered together with both titanium and niobium conductors. The window system is joined to the rest of the transmission line by welding its conductor ends to the transmission line conductors. An alternative modular design, where the window metal supports intrude between the inner and outer conductors can also be considered.

For vacuum brazing experiments alumina and titanium were selected as base materials due to their excellent compatibility of physical and mechanical properties. Vacuum brazing obviously provides sufficiently good conditions for vacuum tightness and shear strength of small scale specimens and also good heat conduction across the interface is expected. But the strength against the relative movements of the outer and inner conductors, thermal expansion, and pressure shocks has to be investigated in full scale models. As the assembly temperature in brazing is high, it is beneficial if the conductor and ceramic have similar thermal expansion coefficients to achieve a sufficient resistance against the loads from manufacturing and service.

Optimization with respect to the breakdown resistance suggests a solution where the angle of inclination of the dielectric is 75 degrees. The tangential component stays below 1.0 MV/m, but the normal component has local peaks which are needed to be decreased by further modifications to appropriate potential rings. Alternatively, an increase in the relative separation of the outer and inner conductors is suggested.

REFERENCES

1. H. Wedler, F. Wesner, W. Becker, R. Fritsch, Vacuum insulated antenna feeding lines for ICRH at ASDEX Upgrade, Fusion Engineering and Design 24 (1994), 75-81
2. Tim Brown, Study of ICW ceramic requirements for NET, Annex K, 15.1.1992.
3. S. Orivuori, Description of IVOFEM program, Imatran Voima Oy, internal report, 1979; S. Orivuori, Int. Journal for Num. Meth. in Engng. 14 (1979) 1461.

Design Study on High Tc Superconducting Plasma Stabilizer

T.Uchimoto[a] and K.Miya[a]

[a]Nuclear Engineering Research Laboratory, Faculty of Engineering
The University of Tokyo, Tokai, Ibaraki 319-11, Japan

This paper presents design study on application of high Tc superconductors (HTSCs) as a stabilizer for improvement of the plasma vertical instability. For this purpose, an numerical simulation of the overall system including the plasma, the structures and the HTSCs was carried out taking an example of the configuration of the ITER tokamak reactor. Through this simulation, arrangements and parameters of the high Tc superconducting stabilizer are optimized with the consideration of various constraints from the technological viewpoint of the fusion reactor.

1. INTRODUCTION

Improvement of the vertical instability is one of the critical issues in the design of fusion reactors because it may make plasma suffer from the vertical disruptions known as "Vertical Displacement Events (VDEs)", which can induce severe forces on vacuum vessel and its components.

High Tc superconductors have the ability to improve the plasma positional instability due to their flux pinning property when they are arranged in plasma periphery. Authors have demonstrated its feasibility in the configuration of ITER [1]: four axisymmetric toroidal coils of Bi-based HTSC in the plasma periphery stabilize the plasma with regard to its vertical position. Here we call the system high Tc superconducting stabilizer. This stabilizing system has some advantages compared with the conventional ones as follows:

1. It requires no power supply unlike feedback control systems.
2. Its stabilizing effect on plasmas is free of decay which is inevitable in that of eddy current induced in structures.

In this paper, parameter survey of the plasma stabilizer is performed in the configuration of ITER/EDA through the numerical simulation of the overall system of the reactor. At first, the relation between the configuration of the stabilizer and its stabilizing effect on the plasma position are investigated. Next, the configuration of the high Tc superconducting stabilizer are optimized with the consideration of various constraints from the technological viewpoint of the fusion reactor.

2. MODEL OF NUMERICAL ANALYSIS

2.1. Modeling of tokamak system

The plasma equilibrium and the eddy currents in the structures were computed applying the modified inductance model [2], which is one of the linearized plasma models and was successfully applied to design study of controller in the ITER tokamak reactor [3]. The system equations based on this model are described after discretization of the continuous problem as follows:

$$d\mathbf{\Phi}/dt + \mathbf{RI} = \mathbf{V} \tag{1}$$
$$\mathbf{\Phi} = \mathbf{\Phi}(\mathbf{I},\mathbf{P}) \tag{2}$$
$$\mathbf{Y} = \mathbf{Y}(\mathbf{I},\mathbf{P}) \tag{3}$$
$$\mathbf{Q} = \mathbf{Q}(\mathbf{I},\mathbf{P}) = \mathbf{Q}(t) \tag{4}$$

where $\mathbf{I}$ is the set of external currents, $\mathbf{\Phi}$ the set of fluxes linked with these currents, $\mathbf{R}$ the resistance matrix, $\mathbf{V}$ the set of applied voltages, $\mathbf{Y}$ a set of output variables to be controlled and $\mathbf{Q}$ a set of variables whose time behavior is supposed to be prescribed. Here, $\mathbf{\Phi}$ was computed by way of the PROTEUS code, which is based on non-rigid, MHD consistent displacement model [4].

2.2. HTSC shielding current analysis

Shielding currents in the HTSCs were estimated based on the flux flow and creep model [5] which properly describes the the quickly changing shielding currents in superconductors like this case. The constitutive relations between the current density, **J**, and the electric filed, **E**, based on the model are as follows:

(1) The creep region ($0 \leq J \leq J_c$)

$$E(J) = 2\rho_c J_c \sinh\left(\frac{U_o}{k\theta}\right) \exp\left(-\frac{U_o}{k\theta}\right), \quad (5)$$

(2) The flow region ($J_c \leq J$)

$$E(J) = E_c + \rho_f J_c (J/J_c - 1), \quad (6)$$

where E_c is the critical electric field, ρ_c the creep resistivity, θ the temperature, U_o the pinning potential, k the Boltzmann constant, ρ_f the flow resistivity and J_c the critical current density. Shielding currents in superconductors can be estimated by applying this constitutive relation to the conventional eddy current analysis [6]. Therefore, in order to introduce the shielding current analysis to the system equations of eqs. (1–4), eq. (1) is changed as follows:

$$d\mathbf{\Phi}/dt + \mathbf{R}(\mathbf{I})\mathbf{I} = \mathbf{V} \quad (7)$$

where,

$$R_i \cdot I_i = \int_c E(J_i) dl. \quad (8)$$

Due to this change, eq. (7) become nonlinear and the Newton-Raphson method was applied to solve eq. (7).

3. RESULTS

3.1. Configuration

The EOB configuration of ITER TAC-8 outline design was considered in this study. The parameters of the plasma and structures are shown in Table 1 and its cross section is shown in Fig. 1.

An example of arrangements of the high Tc superconducting stabilizer is also shown in Fig. 1. It is placed in the form of axisymmetric toroidal coils in the plasma periphery. Silver-sheathed Bi-based HTSC was considered as the material of high Tc superconducting stabilizer. The critical current density, J_c, of Bi-based HTSC at 77 K (liquid nitrogen) drops abruptly in high magnetic field, which is the critical problem of its application. Therefore, the dependence of J_c on the magnetic flux density must be taken into consideration. Here, the following empirical formula of the dependence were approximately applied to the computations [7].

$$J_c(B) = \frac{J_{co}}{\log(B_{\text{self}}/B_{\text{irr}})} \log(B/B_{\text{irr}}), \quad (9)$$

where B_{self} is the self field, B_{irr} the irreversibility magnetic field. The parameters of HTSC in the computation are shown in Table 2.

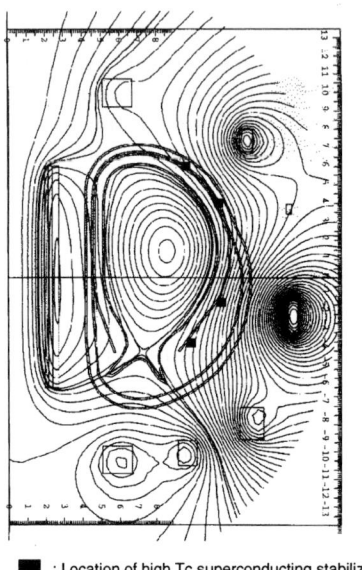

■ : Location of high Tc superconducting stabilizer (Case 2)

Fig. 1: Configuration of ITER TAC-8 (EOB) and high Tc superconducting plasma stabilizer

Table 1
Parameters of ITER/EDA TAC-8 (EOB)

Major radius R (m)	8.171
Minor radius a (m)	2.811
Plasma current I_p (MA)	21
Triangularity δ	0.326
Elongation κ	1.729
Poloidal beta β_p	0.892
Internal inductance li	3.084
Resistances (FW) R_{FW} ($\mu\Omega$)	100
Resistances (BP) R_{BP} ($\mu\Omega$)	8
Resistances (VV) R_{VV} ($\mu\Omega$)	13

Table 2
Parameters of HTSC

Critical current density $J_{co}(A/m^2)$	3.3×10^8
Flow resistivity $\rho_f(\Omega m)$	2.0×10^{-9}
Critical electric field $E_c(\mu V/m)$ ($= \rho_c J_c$)	100
Pinning potential U_o(meV)	96
Temperature (K)	77

3.2. Relation between configuration of stabilizer and its stabilizing effect

At first, poloidal location and size of the cross sections of the stabilizers are examined to optimize the stabilizing effect of this system on the plasma. In this investigation, the followings are assumed for the simplicity; (1) the number of the stabilizing coils is two, (2) stabilizers are located at the first wall and (3) the reference disturbance of the plasma is stepwise change of the poloidal beta ($\Delta \beta_p = -0.2$).

As for the poloidal locations of stabilizers, four cases shown in Table 3 were considered. Here, τ_p is defined as the growth time of the system in which the stabilizing coils are assumed not to be superconductors but perfect conductors. τ_p of each case is also shown in Table 3. Figure 2 shows the displacements of the plasma centroid in each case. From this figure, it is clear that time constants of four cases approximately corresponds to the values of τ_p. Therefore, the poloidal location of the high Tc superconducting stabilizer should be decided taking τ_p into consideration.

Next, size of the cross section of the stabilizing coils was investigated in case A. If the coils of the perfect conductor were placed instead of superconductor in this case, the maximum current, I_{max}, in the coils amounts to 1.16×10^5 A. When the area of the cross section of superconductor coils is larger than I_{max}/J_c, the flux flow phenomenon does not occur in the superconductor, that is, resistivity of shielding current is expected to be quite low. In Case A, $L_m = \sqrt{I_{max}/J_c}$ is 0.07. Figure 3 shows the relation between the displacement of plasma centroid and the size of the cross section of superconducting coils. It indicates that the critical size concerning to the plasma stabilization exists between 0.05 and 0.10 m, which is consistent with the value of L_m and supports for the above explanation.

Table 3
Location of stabilizing coils and τ_p in each case

	Location of stabilizing coils (R, Z)	τ_p (sec)
Case A	(9.83, -2.15) (9.25, 5.22)	-0.787
Case B	(10.37, -1.33) (10.72, 3.49)	-2.171
Case C	(8.31, -3.75) (8.30, 5.82)	-1.044
Case D	(10.98, -0.02) (11.13, 2.49)	3.46

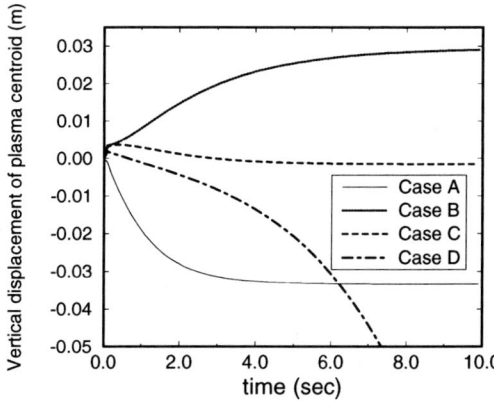

Fig. 2: Relation between poloidal location of high Tc superconducting stabilizer and its stabilizing effect on plasma

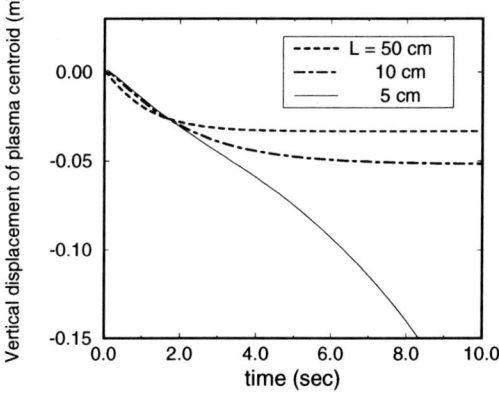

Fig. 3: Relation between size of cross section of high Tc superconducting stabilizer and its stabilizing effect on plasma

3.3. Optimization

Based on the knowledge obtained in the above, optimization of the configuration of the stabilizer was performed taking consideration of technological issues. Three cases were considered; Case 1 in which the coils are placed on the fast wall, Case 2 outside the blanket and Case 3 outside the vacuum vessel. In each case, the number of coils was four and the location and size of the coils were optimized with regard to τ_p and L_m. Details of each cases are shown in Table 4. As for the disturbance of the plasma, we considered instantaneous β_p drop of 0.2 and li drop of 0.1 recovering over 5 sec, which is considered to be the most critical one for the design of ITER. Figure 3 shows the displacement of the plasma centroid in each case. In Case 3, the plasma is not stable because the response of the stabilizer is spoiled due to the presence of the structures between the plasma and the stabilizer. The maximum displacement in Case 1 and Case 2 are 0.055 and 0.111, respectively and they are acceptable. However, in Case 1 where stabilizing coils are arranged on the first wall, there is a problem of the heat and neutron fluxes. On the other hand, the fluxes decay to allowable degree outside the blanket and it is concluded that Case 2 is acceptable from the technological viewpoint of the fusion reactor.

4. CONCLUSION

In this paper, design study on the high Tc high Tc superconducting plasma stabilizer was performed by way of the numerical simulation of overall system of the tokamak reactor including this system. Through the parameter survey of the configuration of the stabilizer, the relation between its configuration and stabilizing effect was made clear and the optimized configuration of the stabilizer was obtained.

ACKNOWLEDGEMENT

The authors wish to thank Prof. Albanese, University of Reggio Calabria, Italy for providing the outputs of the plasma model and for his fruitful advise. This work was partially supported by CREATE, Italy and by a grant-in-aid for scientific research from the Japanese Ministry of Education, Science, Sports and Culture.

REFERENCES

1. T.Uchimoto et al., Stud. Appl. Electromagn. Mech., (1996) in press.
2. R.Albanese et al., Nuclear Fusion, **29**(1989) 1013–1023.
3. R.Albanese et al., Fusion Technology, (1996) in press.
4. R.Albanese et al., 12th Conf. on the Numerical Simulation of Plasmas, S.Francisco, Sept.1987.
5. K. Yamafuji et al., Cryogenics, **32**(1992) 569–577.
6. Y. Yoshida et. al., IEEE Transactions on Magnetics. **30**(1994) 3503–3506.
7. S. Kobayashi et. al., Physica C, **258**(1996) 336–340.

Table 4
Location of stabilizing coils and τ_p in each case

	Location of stabilizing coil (R, Z)	τ_p (sec)
Case 1 (on FW)	(8.31, −3.75) (9.83, −2.15) (10.08, 4.43) (8.30, 5.82)	−0.698
Case 2 (outside BP)	(9.28, 6.04) (11.09, 4.08) (11.25, −1.23) (9.64, −3.43)	−0.916
Case 3 (outside VV)	(9.73, −6.32) (12.06, −4.05) (12.07, 3.31) (9.71, 6.36)	20.12

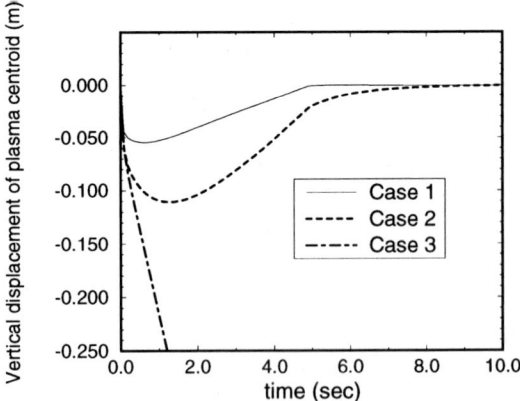

Fig. 4: Relation between location of high Tc superconducting stabilizer and its stabilizing effect on plasma

Development of a Precise Long-time Digital Integrator for Magnetic Measurements in a Tokamak

K. Kurihara and Y. Kawamata

Japan Atomic Energy Research Institute, Naka Fusion Research Establishment
801-1 Mukoyama Naka-machi Naka-gun Ibaraki-ken, 311-01, JAPAN

A new digital integrator based on the method of a VF (voltage-to-frequency) conversion with a UDC (up-down counter) has been developed for a precise long-time integration of magnetic signals in a tokamak. The following steps were taken for the development: (i) Technical causes of signal drift in the VF-UDC system for JT-60 were analyzed especially in an aspect of electronic circuits. (ii) A remodeled VF converter showed that the drift is decreased less than 2 μVs per 1000 s in a stand-alone integration test with zero-voltage input. (iii) In the test application of this new VF-UDC system to JT-60 plasma experiments, the amount of the drift was 3 mVs per 20 s, while the drift without a plasma was 0.15 mVs per 1000 s. The main causes of the drift and their countermeasures in this development are presented.

1. INTRODUCTION

Magnetic measurement is one of the most essential and reliable methods for controls and diagnostics of magnetic confinement systems like a tokamak. The existing methods can be classified into two types; (a) integral of time-derivative of magnetic field/flux using pick-up/Rogowski/saddle coils, and (b) direct measurements of absolute magnetic field using rotating/vibrating coils, Hall-effect-based element sensors, etc.

Long-time D-T burning operation in a tokamak fusion experimental reactor like ITER requires that (1) a sensor must work in an environment of 14-MeV intense neutron field, and that (2) the system must always output precise magnetic field/flux values at the sensor locations during plasma operation of 1000 seconds or longer.

Though the method (a) seems to have more preferable features for the requirement (1) than the method (b), the requirement (2) raises a difficult problem in the method (a), that precise long-time signal time-integral is necessary. Hence, development of a new digital integrator based on the method of a VF (voltage-to-frequency) conversion with a UDC (up-down counter) was motivated. This decision is supported by the fact that JT-60 magnetic measurement system based on the VF-UDC method has been reliably worked since the first plasma.

The VF conversion is divided into two parts: (i) To perform analog time-integral of electric charges proportional to the input voltage by a small capacity condenser, and (ii) To count the repetition number of full charge and discharge of the condenser and add/subtract the series of pulses to/from the basic pulse series. After this conversion, the digitized signal with increased/decreased frequencies depending on integral results and that with the basic frequency are input to the UDC as up-count and down-count, respectively. The counter value in the UDC indicates quantized time-integral result of the voltage.

Analog time-integral in the part (i) is expected to give better approximation to the exact time-integral operation than another digital integrator concept based on the ADC-DSP method, where fast and accurate analog-to-digital converter (ADC) signals are accumulated with interpolations by a digital signal processor (DSP).

The signal drifts, inevitably produced in the integrators have been considered to make it difficult to apply them to the long-time integral operation. It was reported recently, however, that the drift in the ADC-DSP method was successfully reduced to 5 mVs in DIII-D [1], while ITER requires a few mVs for 1000 s [2]. In this report, the result from development of a new VF converter is described.

2. CAUSE ANALYSIS OF THE DRIFT

What produces the drift in the VF conversion process in application to tokamak magnetic measurement should be analyzed at first.

The VF converters equipped in the JT-60 VF-UDC system for plasma equilibrium control show the drift of 1 mVs for a integration period of a pulse discharge (less than 20 s) at its worst case. Now we survey what may cause such drift in the electronic circuit of a VF converter.

The characteristics of raw signal from a magnetic sensor are as follows:
Voltage: ±10 V
Frequency: < 20 kHz (irregularly varying)
The sources of magnetic field are mainly a plasma, the TF (toroidal field) and PF (poloidal field) coils, and they are fluctuated by the asymmetric thyrister ripple (1.2 kHz). In addition, irregular noise signals of up to ±1 V with less than 20 kHz are also picked up by the sensors all the time in the operation. High voltage (>300 kV) breakdowns in the negative-ion-based NBI system make rapid and impulsive voltage fluctuations between the signal lines and the earth.

Figure 1 shows the block diagram of the circuit and a detailed diagram of the VF conversion. The following points were suspected to produce drift in JT-60:

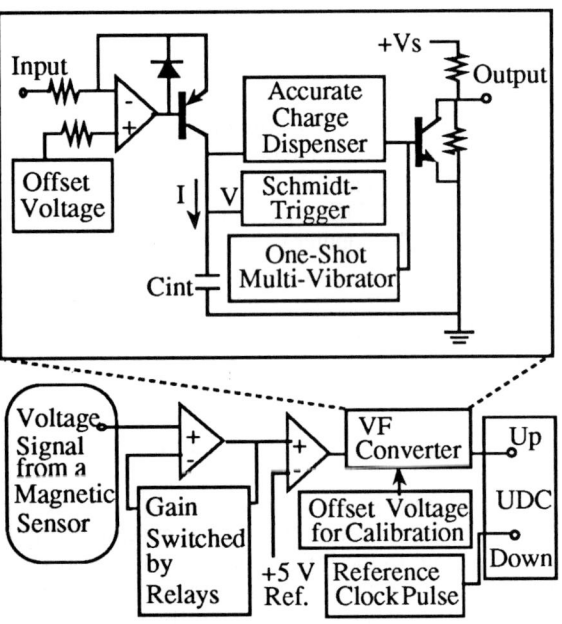

Figure 1. Block diagram of the signal integration process in JT-60U and the VF converter circuit.

(i) Low impedance for voltage input
Since the design philosophy is adopted that the current is induced by the input signal voltage, the input impedance was chosen as less than 1 MΩ. On the other hand, signal amplifying factor (gain) is switched by turning on/off appropriate relays to change the combination electrical resistance. These relays in which the signal current flows may change their internal resistance because of the temperature variation. As even a small amounts of resistance fluctuation directly makes the signal voltage change, this point can be one of the causes to produce drift.

(ii) VF conversion method in the electric circuits
Type-4735 hybrid VF converter [3], made by Teledyne Philbrick Co. Ltd., was chosen for this circuit board. As shown in Fig. 2, discharge from the condenser for integration is controlled by the schmidt-trigger circuit, which monitors the voltage of the condenser. Since the current (charge flow) to be integrated bypasses the condenser during this short period of discharge, charging process is interrupted. This integration error can make drift.

(iii) Non-linearity of VF conversion
As mentioned previously, the voltage signals from the magnetic probes contains the wide variety of amplitudes and frequencies. In particular, during a plasma operation, magnetic field is rapidly fluctuated by a plasma, power supplies, etc. After this signal is conditioned through a filter with the time constant of 1 ms, the resultant signal has a frequency of 1 kHz with amplitude of less than 20 mV.

When such a low signal voltage should be processed in the circuit, we should select an operational (OP) amplifier (Amp) whose offset current is adequately low enough compared with the input signal. The offset current of the OP-Amp in the Type-4735 converter is not available in the manual, but the measured dead-band width for voltage input is about 4 mV. Therefore, this range of the original signal can make an integration error [4], unless the integral results in the concerned plus and minus regions are canceled each other.

(iv) High voltage breakdown noise
The positive/negative-ion-based NBI's are often turned off by breakdowns from high voltage (100/350 kV), and make a large influence on the grounding system. The integration test in the period with breakdowns shows larger amounts of drifts than that without breakdowns.

3. DESIGN OF A TRIAL BOARD

On the basis of the analysis described above, a trial VF converter was designed, where the following countermeasures were taken to avoid or reduce the drift:
(i) High impedance for voltage input
The input impedance for this new board was

chosen as 100 MΩ. (Signal amplifying factor (gain) would be switched by FET circuits without a relay.)

(ii) More precise VF conversion method

Type-AD652BQ monolithic VF converter [5], made by Analog Devices Co. Ltd., was chosen for the new trial board. This is based on the clock-synchronized integration method, shown in Fig. 2: The discharge current from the condenser is exactly regulated to be constant. Hence, the error of integration is minimized.

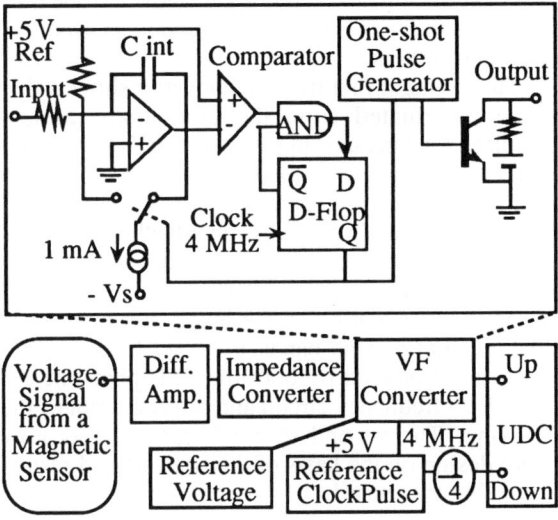

Figure 2. Test system configuration and the new VF converter circuit

(iii) Improvement of VF conversion linearity

The offset current of the OP-Amp in AD652BQ is 70 nA at maximum. The corresponding offset voltage should be measured in the tests.

(iv) Immunity from high voltage breakdown noise

Since a breakdown introduces common-mode noise in the signal lines. Then, the signal is conditioned by a differential amplifier prior to the VF converter.

4. RESULTS FROM THE TESTS

The trial VF converter board was set up to test its drift in the VF-UDC system, as shown Fig. 2.

(a) Stand-alone test (no actual signal input)

The VF-UDC system containing a new trial VF converter was tested without actual signals to measure how much drift is produced from the circuit. As shown in Table 1, the results show that a drift originated solely from the circuit is minimized.

Table 1
Drift of stand-alone test for 1000-s time-integral

	Trial Board	Old Board
Input Short (0V)	1.4 μVs	0.12 Vs
100 kΩ Resistance	2.1 μVs	----------

(b) Measurement test in the JT-60U tokamak

The test with the new trial board was conducted in actual JT-60 plasma experiments. The magnetic probe (pick-up coil) voltage signal was used for this test. The input signal of ±10 V corresponds to the output TTL signal with the frequencies of 1 MHz ± 1 MHz. Figure 3 shows an example of magnetic field measurement in the period excluding and including a pulse of discharge.

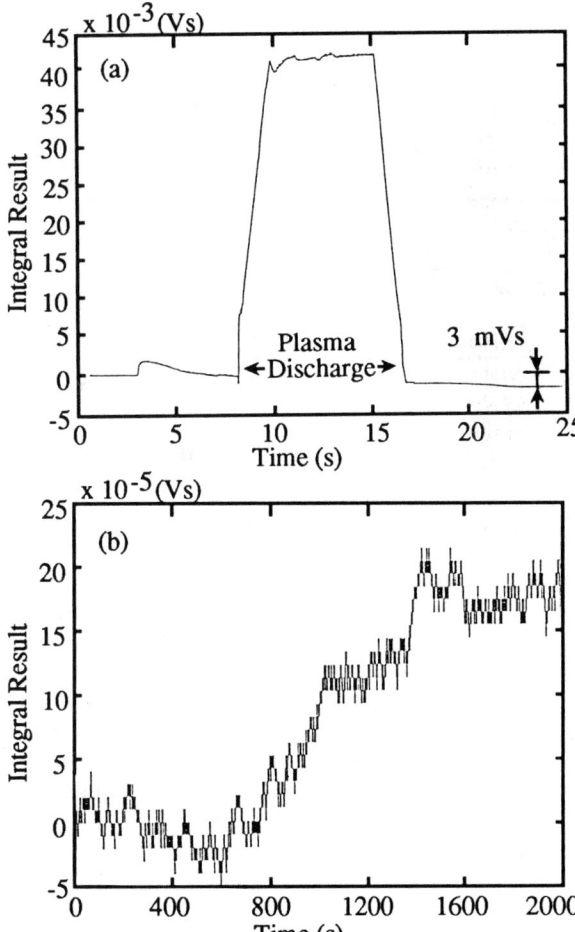

Figure 3. The integration test results. ((a) and (b) are with and without a plasma, respectively.)

In Fig. 3(a) the drift is assumed a difference between the integrated results at the two points of no plasma before and after a pulse of discharge. The amount of the signal drifts with a plasma was not improved as expected, i.e. 3 mVs per 20 s. This drift level is 3 times larger than the old VF-UDC system.

On the other hand, as shown in Fig. 3(b), the drift in integration over a period without a plasma is very small, i.e. 0.15 mVs per 1000 s.

(c) A cause of drift: dead-band measurements

It is known that the VF converter based on the method of clock-synchronized integration must have dead-bands for input voltage due to PLL (Phase Locked Loop) corresponding to the singular frequencies of 1/n of the basic clock. Therefore, this results in imperfect time-integral operation.

For the basic clock of 4 MHz in the trial board, the measured dead-bands at the 1/n frequencies (n=2, 3, 4, ...) are shown in Table 2. This phenomenon possibly causes the drift together with the offset current of the OP-Amp.

Table 2.
Dead-band width at the 1/n frequencies

Frequency	1/n	Input Volt.	Dead-band
1.33 MHz	1/3	+3.33 V	1.0 mV
1.00 MHz	1/4	0.00 V	3.6 mV
0.80 MHz	1/5	-2.00 V	1.7 mV
0.66 MHz	1/6	-3.33 V	2.3 mV
0.57 MHz	1/7	-4.29 V	2.1 mV
0.50 MHz	1/8	-5.00 V	2.5 mV
0.44 MHz	1/9	-5.55 V	1.8 mV
0.40 MHz	1/10	-6.00 V	2.0 mV

5. DISCUSSION

To suppress the drift in the VF-UDC system low enough for a 1000-s integration, the dead-band width must be reduced. A possible cause, the offset current of the OP-Amp, can be solved by choosing a more precise amplifier having the smaller offset values. Another possible cause, the PLL phenomenon, can be avoided by the following method: Since the largest range of frequencies between singular points is the range of 1/2 to 1/3 of the basic frequency. Then, this range must correspond to full span of frequency, e.g. 1 MHz or 2 MHz. Consequently, by adopting the basic frequency of 6 or 12 MHz, the range of 2-3 or 4-6 MHz never include the singular points for PLL. However, under such an extremely high frequency mode, it may be rather difficult to turn on and off the switch to discharge the integration condenser. This seems to require some technological advances in the IC design.

6. CONCLUDING REMARKS

A new digital integrator based on the method of a VF (voltage-to-frequency) conversion with a UDC (up-down counter) has been developed for a precise long-time integration of magnetic signals in a tokamak. Technical causes of signal drift in the VF-UDC system for JT-60 were analyzed in an aspect of electronic circuits.

The resultant first trial VF converter showed that a drift originated from the circuit was almost completely suppressed to less than 2 μVs per 1000 s in a stand-alone integration test with zero-voltage input. In the test application of this new VF-UDC system to JT-60 plasma experiments, the amount of the drift was 3 mVs per 20 s, while the drift without a plasma was 0.15 mVs per 1000 s.

After investigation on the board, it seems that a small amount of dead-bands, less than 4 mV for 0-V input, is introduced by both the PLL phenomenon in the high frequency circuit and offset current of the OP-Amp. The second trial VF converter is now under design on the basis of the analysis and results of the first trial board.

Acknowledgments

The authors thank Dr. T. Kimura (JAERI) for his encouragement and technical advice. They also express their gratitude to Mr. S. Sakamoto (MTT Co. Ltd.) and the members of the JT-60 Control Group in JAERI for helping the authors to test the hardware.

REFERENCES

1. J.D. Broesch, et al., in Proceedings of 16th Symp. on Fusion Engnrg., (Illinois, 1995) 365
2. Documents on Plasma Shape Reconstruction in ITER-EDA
3. Type-4735 IC Manual, Teledyne Philbrick Co. Ltd.
4. M. Okamura, Design of Operational-Amplifier-Based Circuits, CQ Press Co. Ltd., Tokyo (1990) (in Japanese)
5. Type-AD652 IC Manual, Analog Devices Co. Ltd.

Section D

Experimental Systems

Simulation of the magnetic field penetration in the structure material of the Dynamic Ergodic Divertor coils of TEXTOR-94

D. Styhler[a], M. Lindmayer[a], U. Braunsberger[a], B. Giesen[b], O. Neubauer[b]

[a] Institut für Hochspannungstechnik und Elektrische Energieanlagen, TU Braunschweig, Pockelstrasse 4, D-38106 Braunschweig

[b] Institut für Plasmaphysik, KFA Forschungszentrum Jülich GmbH, Postfach 1913, D-52425 Jülich

The **D**ynamic **E**rgodic **D**ivertor DED at TEXTOR-94 will be operated with dc and ac at 50 Hz, 1000 Hz, and 10 kHz [1]. The penetration of the magnetic field in the plasma region especially at 1 kHz and 10 kHz is influenced by the complete structural arrangement near the DED coils. The structure consists of the vessel wall, the liner including coil clamps, and the carbon limiter tiles. For the final design of the coil arrangement and the layout of the power supply the knowledge of the magnetic field strength and its direction is important.

In order to solve the problems mentioned above, magnetic field calculations were performed with the program ANSYS. The paper describes how the complicated geometry of the design was simplified and transformed into the finite element network for simulation. Results of the magnetic field calculation at the different frequencies are discussed.

1 INTRODUCTION

The **D**ynamic **E**rgodic **D**ivertor DED for TEXTOR-94 shall be used to provide a magnetic field perturbation at the plasma periphery [1]. The sixteen helical coils will be fed by a four-phase (rotating) current and are planed to be installed on the inboard side of the liner. This simulation is made for the first draft of the DED. Each coil consists of 42 insulated wires, which are bundled up six times, and is enclosed in a stainless steel tube (figure 1).

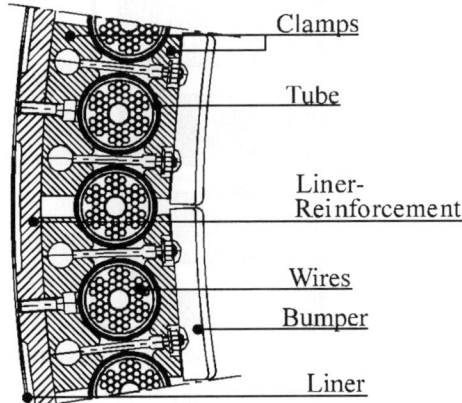

Figure 1 Coil support

The tube is used to cool the wires with water. Because of the electromagnetic forces on the coil the tubes are fixed by stainless steel clamps. Therefore the thin liner has to be reinforced with stainless steel plates. The whole DED is protected by graphite tiles which act as an inner bumper limiter. The current in the wires will generate in all these conducting materials (tube, bumper, clamps, liner and liner reinforcement) eddy currents.

The magnetic field and the eddy currents in the conducting materials are simulated with the ANSYS program.

1.1 Short description of ANSYS

The ANSYS program is a complete, general purpose finite element program which includes, for linear and non-linear structures, static, modal, transient thermal analysis capability; magnetic, fluid, piezoelectric, thermal-electric, and acoustic analysis capability; solid modelling capability; and disign optimisation capability [2].

At the "Institut für Hochspannungstechnik und Elektrische Energieanlagen" of the TU Braunschweig the college Version of the ANSYS program is available. So the number of elements is limited. Therefore the complicated geometry of the coil support has to be minimised.

A second restriction is the wish for quadrilateral element types because some commands are only useable for quadrilateral elements. ANSYS can generate also triangular elements but for symmetric geometry's the results may be unsymmetrical.

2 MODEL SETUP

The support of the Dynamic Ergodic Divertor for TEXTOR-94 has a curvature because it shall be installed on the inboard side of the liner and the liner has a round cross-section.

This symmetric arrangement can be divided into a small part of four coils. The easiest transformation of the coil support (figure 1) into a calculable model is a planar arrangement. Therefore the round cross-section of the liner is neglected. The bolts between the inner part and the outer part of the clamps are neglected because they are very small in relation to the clamps, bumper and the liner reinforcement.

2.1 The two-dimensional model

The arrangement of the coil support has to be transferred into an easy model figure 2.

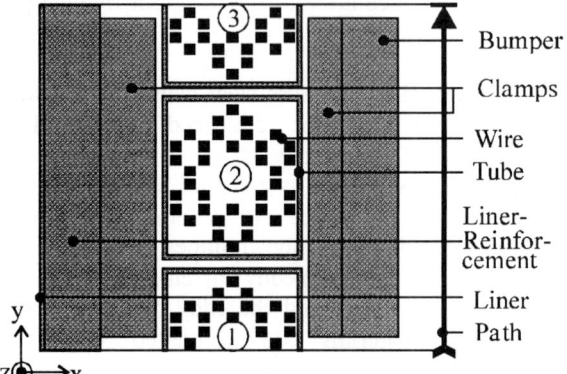
Figure 2 Model: two-dimensional

In order to economise the number of elements the wires and the tubes are build as quadrangular forms. The middle wire of one bundle is neglected and only the half part of the four-phase coil support is considered.

For the calculation of the two-dimensional model ANSYS is able to build an axial symmetric secondary condition. This axis represents the horizontal line centre of TEXTOR-94.

The clamps and the bumper are not closed loops, they have gaps in circumferential direction. The alternating magnetic flux induces voltage in the clamps and bumpers. The result of this voltage are eddy currents. In a greater area the alternating magnetic field would induce higher voltage, so the result would be higher eddy currents. Therefore the calculated eddy currents in the clamps and the bumper would be much too high because of the very great x-z area. With the two-dimensional model only the influence of the tubes, the liner and the liner reinforcement can be determined.

The effects of the eddy currents in the clamps and in the bumper only could be calculated correctly in a three-dimensional model.

2.2 The three-dimensional model

In the circumferential direction of the liner the coils are curvatured, too. This curvature is neglected for the three-dimensional model, too.

The whole arrangement is reduced to four coils and its surrounding materials because it is a symmetric structure. The bolts between the inner part and the outer part of the clamps are neglected because they are very small in relation to the clamps, bumper and the liner reinforcement.

The wires, the liner and the liner reinforcement are at the whole circumference but the bumper and the clamps have gaps in circumferential direction. For a realistic model these gaps have to be taken into consideration. A three-dimensional model has to be set up otherwise in the clamps and the bumper a higher voltage would be induced, which generates higher eddy currents. The front to back size (z-coordinate) of the model (figure 3) has a length of about 12 cm in order to get a symmetric part of the circumference.

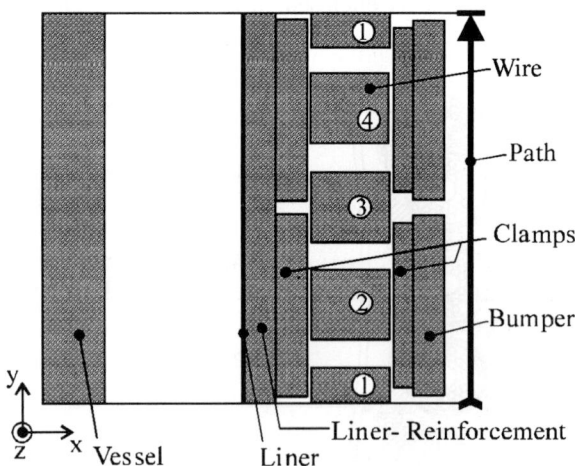
Figure 3 Model: three-dimensional

For a realistic model one coil has to be build up with so many elements that the maximum number of elements is higher than ANSYS would accept. So the 42 single wires of one coil had to be summarised in one single rectangular area.

2.3 Initial conditions

ANSYS needs secondary conditions at each node of the boundary. These are the nodes of the four edges of a rectangle in a two-dimensional model and the six areas of a cuboid in a three-dimensional model.

The boundary conditions of the magnetic field on the inside (left hand side of figure 2 and figure 3) and at the outside (right hand side) of the model are set to far field (which is zero).

The magnetic field of the top and the bottom of the model are coupled. This means that every node on the top has one node with the same coordinates (x-z plane) in the bottom with the same magnetic field.

Additionally in the three-dimensional model the magnetic field of the front and the back of the cuboid has to be defined. Because of the repetition in circumferential direction the magnetic field in the front respectively back is parallel to the plane of projection.

At an appointed time two of the four-phase currents are at zero and the other two phases are at the positive respectively negative maximum. At this point of view the ruling current is a two-phase current.

For a two-phase model the magnetic field penetration results in a image of real components, because the current flows through one coil to the star point and through the next coil but one back to the power supply. So every second coil has a current with an opposite direction.

The simulation was made for two different current loads: the first with 1 kHz / 15 kA and the second with 10 kHz / 7.5 kA. Because of the very big wires in the three-dimensional model, the wires without any current load are not modelled for this simulation.

3 RESULTS OF THE CALCULATION

The component of the magnetic field, which has the direction into the middle of the plasma is responsible for the extra rotation of the plasma. In the model of the arrangement this direction is the x-direction of the magnetic field.

In order to get the influence of the support material the results of the model with the whole material and the results of a model, which includes only the current wires in vacuum, are compared.

The comparison of the magnetic field is made on the q=3 area of the plasma in front of the bumper. This path is from the bottom to the top of the model and is drawn in figure 2 and figure 3. The magnetic field in x-direction is perpendicular and in y-direction is parallel to the path.

3.1 The least influence by the material of the three-dimensional model for 1 kHz

The coils in figure 3 are numbered. The two-phase current can flow through coil number 1 and 3 or through coil number 2 and 4. The magnetic field penetration into the plasma region is different in these two cases because the eddy currents can develop much better if the main field of the x-direction is in the bumper and not in the gap between the bumpers.

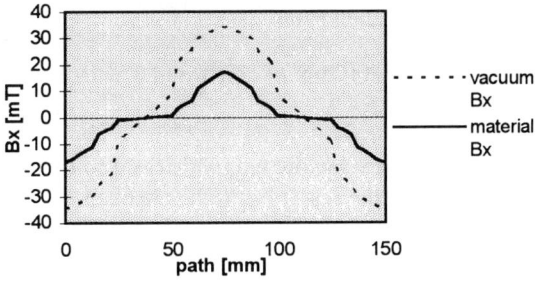

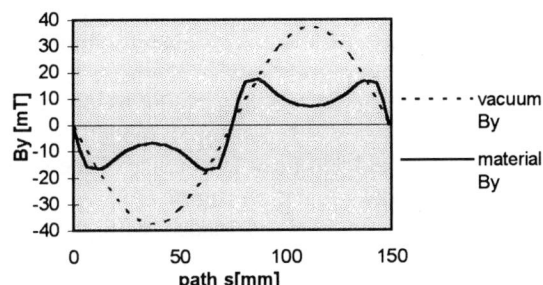

Figure 4 Magnetic flux of coil 2 and 4 with 1 kHz / 15 kA

If the current flows though the coils number 2 and 4 in the middle of the bumper (two phase current) the magnetic field along the path will be decreased to 50 % of the field without material (figure 4). This is the case with the highest field on the path at a model with the whole support material of the arrangement.

3.2 The strongest influence of the material

The lowest magnetic field on the path exists if the current is in coil number 1 and number 3 of figure 3 and if the current has a frequency of 10 kHz. The magnetic field amplitude in vacuum for a frequency of a 10 kHz current is half of the magnetic field amplitude of a 1 kHz current because of half of the current.

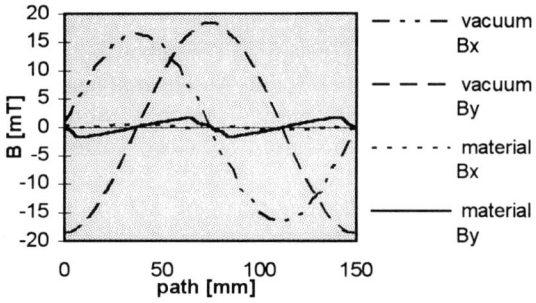

Figure 5 Magnetic flux of coil 1 and 3 with 10 kHz and 7,5 kA

Because of the very low magnetic field in this case (with material) the result has to be checked.

The influence of the very large wires of figure 3 is examined. A comparison between the two-dimensional and the three-dimensional model (coils in vacuum without any other material) has shown that the influence for a two-wire model is immense. For 1 kHz the amplitude of the magnetic field for the three-dimensional model is only 53 % of the amplitude of the two-dimensional model. So the real magnetic field will be at least twice of the result of the value of the simulation (figure 6).

If the model (figure 3) is loaded with the four-phase current one can see that the magnetic field decreases although the number of coils, that are loaded with a current, increases. Figure 6 shows the magnetic flux along the path for the three-dimensional model where only the large area coils were considered. All other components were modelled as vacuum. The parameter is the number of coils and its currents.

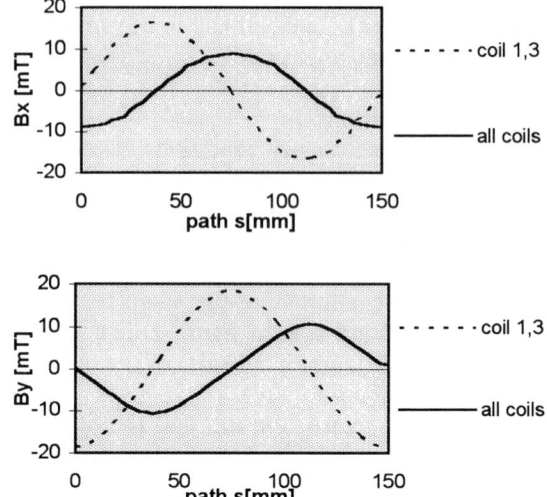

Figure 6 Magnetic flux comparison between two and four coil-current in vacuum 10 kHz

Through the eddy currents in the case when all coils are taken into account the magnetic field decreases by nearly 50 %.

4 CONCLUSION

For a frequency of 1 kHz and 10 kHz the materials near the DED coils reduce the magnetic field between 50 and 95 % at the plasma edge. Due to the effects of the large wire dimension in the three-dimensional model simulation the magnetic field is attenuated. The real magnetic field is probably twice as high as the results of the simulation show.

In future a bigger version of the ANSYS program will be available. So with a finer meshed model the results will get more exact because the coils can be build up with its single wires.

References

1. B.Giesen, 16th Symposium on Fusion Engineering (1995), 0660
2. ANSYS user's manual for Revision 5.1 (September 30,1994)

Design of vacuum system and support structure for plasma facing components of SST1 Tokamak

M. K. Bhise, P. Chaudhuri, D. Chenna Reddy, R. Gangradey, S. Jacob, S. Khirwadkar, C. Muralidaran, H. A. Pathak, E. Rajendra Kumar, T. Ranga Nath, N. Ravipragash, P. Sinha and SST1 Team

Institute for Plasma Research,
Bhat, Gandhinagar : 382 428
INDIA

The design of Steady state Superconducting Tokamak SST1 is in progress at the Institute for Plasma Research, India. The main parameters of SST1 tokamak are, (1) $R_0 = 1.1$ m, (2) $a = 0.20$ m, (3) $I_P = 0.22$ MA, (4) $\kappa = 1.8$, (5) $\delta = 0.66$ and (6) $B_T = 3.0$ T. Poloidal pumped divertor in double null configuration is planned for active particle exhaust. The device will have power handling capacity of 1 MW/m^2 with total power input of 1 MW in the first phase of operation. This paper describes the design of the vacuum system to satisfy the needs for steady state operation. Also, presented is the design of the support structure for various in-vessel components.

1. INTRODUCTION

Over last few years various tokamak experiments with plasma discharge duration varying from few hundred milliseconds to few seconds have generated a large database, useful for the ultimate goal of thermonuclear fusion. However, there are number of processes, e.g. plasma surface interaction, wall pumping and saturation, plasma current profile evolution etc; which has a characteristic time constant much longer than the plasma duration in present tokamak experiments. To generate scientific basis for future reactor grade machines, it is essential to have plasma duration longer than the time required for these processes to be stabilized which is about few 100's of seconds. SST1 is a steady state tokamak being designed at the Institute for Plasma Research, India to investigate these processes with hydrogen plasma. The main objectives of SST1 project is to demonstrate (1) steady state operation (1000 s) of elongated plasma with some triangularity, (2) steady state current drive using LHCD scheme, (3) the heat and particle removal in steady state, (4) the vertical instability control of elongated plasma in high aspect ratio machine and (5) operation of large size superconducting magnetic field coils made of Nb-Ti CIC cooled at 4.2 K. Figure 1 shows the schematic of SST1 tokamak. Vacuum system of SST1 tokamak should satisfy the needs for the

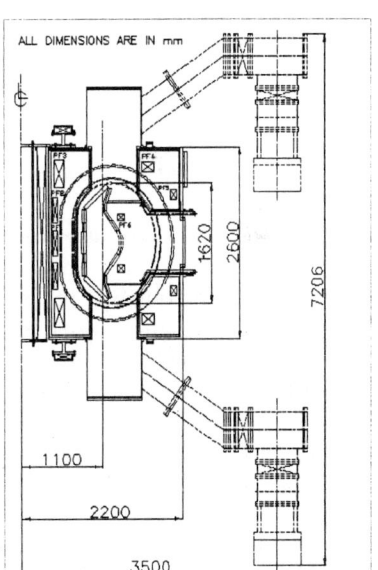

Figure 1. Schematic view of SST1. The machine is symmetric with respect to the major axis

steady state plasma operation. Also, there are in-vessel components, like divertor, passive stabilizers, feedback coils and poloidal field coils, to support the plasma operation. In section 2 the design of vacuum system is presented. Section 3 describes the design of the structure to support various in-vessel components. Section 4 projects the future work to be carried out.

2. VACUUM SYSTEM

This section describes the design of various components of SST1 vacuum system.

2.1 Vacuum vessel

The design and operational requirements are (a) vessel should be UHV compatible, (b) vessel temperature will be 475 K during baking and

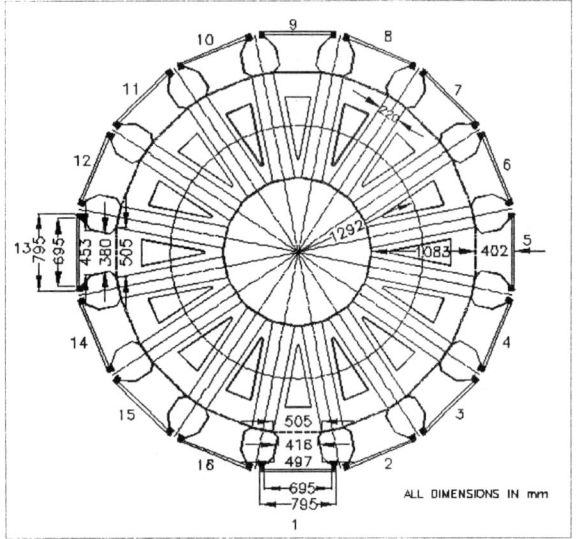

Figure 2. Top view of SST1 vacuum vessel.

425 K during wall conditioning, (c) vessel and radial ports must have enough space for human entry inside the vessel and (d) vessel L/R time constant should be as large as possible to shield the superconducting coils from time varying magnetic field generated due to plasma movement. The estimated L/R time constant of only vacuum vessel is 7×10^{-3} s. The vessel should be modular to remove and insert the vessel module if required for repair of a coil or vessel. This demands the insitu UHV welding of vessel modules. Vessel is made up of sixteen wedge shaped sectors which are joined together by sixteen interconnecting rings to form a complete torus. Each wedge sector contains two vertical ports and one radial port. The material of vessel is SS 304L. The volume of the vessel upto port openings is 16 m^3 and the surface area exposed to UHV is 68 m^2. Figure 2 shows the top view of the vessel. The wall thickness of vessel is 10 mm and that of port is 6 mm. The inlet pressure and temperature of nitrogen gas to maintain the vessel at 475 K will be 5 bar and 550 K respectively with 287 W/m^{2}K heat transfer coefficient between gas and the tube wall. It may not be possible to carry out GDC between plasma discharges due to superconducting toroidal magnetic field coils. For this ECR plasma wall conditioning is planned. The stresses developed on the vessel wall due to internal and external pressure of 850 torr are calculated using pressure vessel design codes [1]. The estimated collapsing pressure is about 10 bar with 10 mm wall thickness. Also, the calculated average maximum stress with 10 mm wall thickness is about 18 times less than the yield strength of SS 304L material.

2.2 Cryostat and LN2 shield

All the superconducting coils and associated supporting structure is known as the cold mass and it will be maintained at 4.2 K. Cryostat is a 16 sided polygon shaped vacuum chamber which encloses the vacuum vessel, cold mass and the thermal radiation shields. The material is SS 304L. The pressure in cryostat chamber will be $\leq 10^{-5}$ torr to reduce the residual gas heat conduction from warm surfaces to the cold mass. To reduce the radiation loss, liquid nitrogen cooled thermal radiation shields at 80 K will be installed between the warm surfaces and the cold mass. The estimated heat load due to radiation on liquid nitrogen is 25 KW at 475 K vessel temperature. The estimated L/R time of the cryostat is $\sim 15 \times 10^{-3}$ s. The volume of cryostat is 39 m^3 and area exposed to HV is 72 m^2.

2.3 Pumping System

The main functions of the vacuum vessel pumping system are (a) to provide clean ultra high vacuum environment for the experiment, (b) to maintain steady state pressure of about 10^{-3} torr for hydrogen in the divertor region for active particle exhaust and density control and (c) to maintain about 10^{-3} torr during wall conditioning. During normal pump down the outgassing from all the in -

vessel components is estimated to be $\leq 10^{-5}$ tl/s. So, the required pumping speed to maintain the base pressure $\leq 10^{-9}$ torr is 10000 l/s. For reference L-mode plasma operation in SST1 ($n_e = 1 \times 10^{19}$ m^{-3}, $\tau_E = 6 \times 10^{-3}$ s) the particle out flux from plasma is

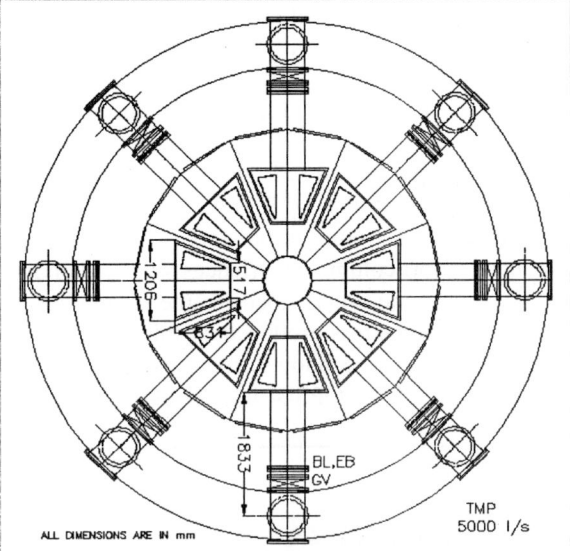

Figure 3. Schematic of SST1 pumping system.

estimated to be 1.38×10^{21} particle /sec which is 22 tl/s. Here the effective particle confinement time is assumed to be $2 \times \tau_E$. The plasma volume for reference discharge parameters is 1.65 m^3. The required pumping speed for hydrogen to maintain the divertor region at 10^{-3} torr is 22000 l/s. The estimated conductance of divertor slot is 50000 l/s, hence the net pumping speed required at torus is about 39000 l/s. The net molecular transmission probability between the vacuum vessel and the pump of one pumping line calculated using HIVAC II code is 0.23 [2]. This gives the net pumping speed of 3950 l/s for hydrogen with one pumping line at the torus. The total pumping speed at the torus for hydrogen is estimated to be 63000 l/s at ambient temperature with 16 turbomolecular pumps (each of 5000 l/s for hydrogen) connected to eight pumping lines on the top of the vessel and eight pumping lines at the bottom of the vessel. Figure 3 shows the schematic of the pumping system. The higher pumping speed will allow us the flexibility in the machine operation for various plasma parameters. We have not considered the option of in-vessel cryo pumping due to the steady state operation of the machine. We have used the Vacuum System Design code VSD-II for pumping calculations for vessel and cryostat.

2.4 Gas Feed System

This system should feed the required gas in required quantity for various machine operations. After wall saturation the recycling coefficient becomes equal to one and hence for the steady state plasma discharge the net quantity of the gas required from external sources is same as the particle exhaust from the plasma. The net particle out flux from plasma is 22 tl/s and the gas load due to 0.8 MW, 30 KeV neutral beam is 4 tl/s. So, the net gas to be fed using external gas feed valves is 18 tl/s. The estimated gas throughput to control the plasma density variation up to 50% of 1×10^{19} m^{-3} within 6×10^{-3} s for L-mode operation is 22 tl/s. For radiative divertor operation it is necessary to inject either fuel gas or impurity gas e.g. neon in the divertor region. The preliminary estimate of the radiated power in the divertor region for 5% neon injection is 0.0095 MW out of 0.354 MW incident on the outer top divertor plate [3]. Table 1 shows the distribution of gas valves for various operations.

Operation	Q	Q / Val (no.)	Location
normal and n_e control	44 44	20 (4) slow 3 (4) fast	midplane 4 pairs radial ports
radiative divertor	10	2 (32)	divertor region,

Table 1. Gas Valve distribution for SST1.

3. SUPPORT STRUCTURES FOR THE PLASMA FACING COMPONENTS OF SST1

The PFC modules and their structure has to be installed in the Vacuum vessel with high accuracy with respect to the toroidal magnetic field.

3.1 Supports and Fixings

The support system of outer divertor(OD) and outer passive stabilizer(OPS) basically consists of three rails running toroidally at the top and bottom of the midplane as in figure 4. The support structure comprises of 16 modules and will be made of SS

304L. The rails are discrete and have toroidal gaps between structure modules. These rails will be

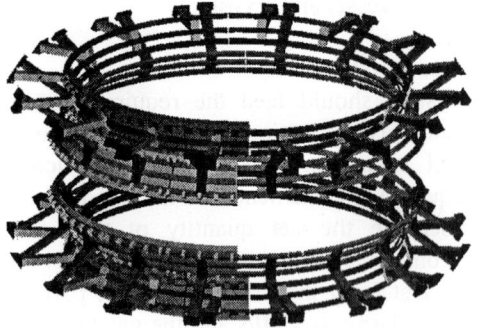

Figure 4. OPS & OD support structure for SST1.

supported by a column and splice plate assembly at 16 toroidal locations. The back plates have five flaps that engage on the rails. A constraint block will be provided for each module that disables the toroidal movement of the OPS/OD modules. The inboard passive stabilizer (IPS) is divided into 8 modules and is supported on the vessel by flexible supports. There is a toroidal gap between any two modules of all PFCs so as to allow for a linear expansion of the module during baking. Toroidal break is present for the OPS and IPS through which the top PS is connected to the bottom PS to form a saddle loop configuration. The structure of inner divertor(ID) consists of rail/channel like structures which extends toroidally ~ 50 mm and there are two such structures per module. Attachment to the structure is by bolts into the tapped holes provided with washers of variable number and thickness to position the ID module correctly.

3.2 Mechanical considerations and analysis

The vacuum vessel will be at 475 K and 325 K during baking and operation respectively while the PFCs baking and operating temperatures are about 650 K and 425 K. As the OPS/IPS itself is a continuous ring except for a break at a single toroidal location, high thermal stresses are expected and in order to avoid them OPS/IPS modules are allowed to expand toroidally by having flexible, least resistance joints between them. Various configurations have been analysed by ANSYS [4], a finite element code, and the predicted maximum stress is found < 32 MPa during baking and 11 MPa during normal operating conditions. However, of particular concern from a mechanical point of view is the case of a disruption. Evidence from other tokamak machines [5] suggests that there will be flow of halo current through the PFCs which would be as high as 40% of the plasma current. Taking 2 as the factor of asymmetry, halo current amounts to 80% of the plasma current. As this will be distributed to several toroidal locations, it is estimated that upto 5% of Plasma current could flow through a single module of OPS or OD. Due to this current, each of the OPS and OD modules will experience a maximum radial force of $\cong$ 33 KN/m and the support column experiences vertical bending force of the same order. Load for the finite element analysis and the stress produced by this can be up to 88 MPa. Figure 5 shows stress intensity on the various parts of the OPS structure estimated using ANSYS.

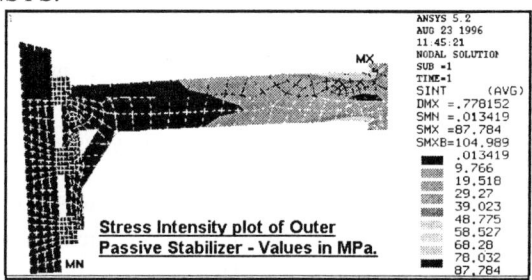

Figure 5. Stress intensity plot of OPS.

4. FUTURE WORK

Stress analysis of various components of vacuum system and PFC structure using F. E. M. is in progress. The development of techniques for CFC brazing with copper for is planned. Also, the calculation of neutral gas pressure in the divertor and pumping duct region using DEGAS code is being carried out.

REFERENCES

1. J. F. Harvey, Theory & Design of pressure vessels, CBS Publisher, 1987.
2. D. N. Ruzic et al, Proc. Int. SOFE, Hyannis, Massachuetts, Oct. 11 - 15, 1993, p. 834.
3. D. E. Post, ADNDT, 20, No. 5, (1977), p. 400.
4. Ansys, Ansys Inc. Houston, PA
5. Pick et al, Proc. Int. SOFE, San Diego, Vol. 1, p. 187

Electromagnetics aspects in ITER: impact on magnet cryogenics, plasma diagnostics and control system

R. Albanese, G. Ambrosino, E. Coccorese, R. Fresa, R. Martone, C. Morabito, A. Pironti, G. Reitano, and G. Rubinacci

Consorzio CREATE, Dipartimento di Ingegneria Elettrica, Università di Napoli "Federico II", Via Claudio 21, I-80125 Napoli, Italy

The paper deals with some electromagnetic aspects of interest for their impact on the ITER project. In particular the following topics are illustrated:
- feedback control of the distance between plasma separatrix and first wall;
- identification of the plasma separatrix from signals detected by magnetic sensors;
- ohmic losses in the cold structures, during normal operation scenario and plasma disruptions.

The numerical results, which mainly refer to the ITER TAC-8 configuration, are discussed in the paper with the aim of contributing to the still open ITER design options.

1. INTRODUCTION

The complexity of the ITER engineering design calls necessarily for a clear subdivision into different specific tasks addressed to the various reactor components. On the other hand, the magnetic field embeds all the torus components, resulting in a strong interaction existing among all design aspects related to the electromagnetic phenomena. The aim of the present paper is to show how the predictions of the electromagnetic analysis influence the input data for the design of magnet cryogenics, plasma diagnostics and control system.

In the framework of the CREATE contribution [1-3] to some activities of the ITER EU Home Team, we have studied in detail the following aspects:
- feedback control of the distance between plasma separatrix and first wall: the controlled variables are 6 gaps located in the most critical zones, including divertor channels; the PF coils provide the field necessary to recovery the nominal equilibrium from disturbances;
- identification of the plasma separatrix from signals detected by magnetic sensors: field and flux measurements are assumed to be available around the torus and in the proximity of the first wall; the above signals are manipulated by a suitable Artificial Neural Network (ANN) to estimate the values of the gaps required by the control system;
- magnetic field change induced losses on PF and TF superconductors, during normal operation scenario and plasma disruptions;
- ohmic losses in the cold structures caused by eddy currents, during normal operation scenario and plasma disruptions: inner and outer cylinder of the CS, TF coil case and intercoil structure, upper and lower crowns are considered in their fully 3D geometry, in the presence of the first wall and vacuum vessel; the knowledge of all currents allows for the calculation of the magnetic field inside the SC coils, which is entered in simplified formulas for the losses.

The ITER TAC-8 configuration [4] has been assumed for most numerical calculations: a selection of relevant results and related comments are presented in the remainder of the paper.

2. CONTROL OF THE PLASMA SEPARATRIX

For the control issue we refer to a linearized open loop plasma response model, derived by a variational formulation based on the balance of the magnetic energy stored in the poloidal field. Ohmic dissipation is taken into account, but exchanges with other forms of energy (i.e. magnetic energy of the toroidal field and kinetic energy) are ignored. The numerical calculations are performed directly on the linearized model, so the calibration of the numerical error is not needed. No skin currents are allowed to appear at the boundary and the plasma is massless. The plasma current profile has three degrees of freedom: I_p, β_p and l_i. The latter are treated as external disturbances, whereas the total plasma current I_p is a result of the calculation.

In this way, we have:

$$dx/dt = Ax + Bu + d[e(w)]/dt \qquad (1)$$

Work performed under EURATOM study Contracts in the framework of the EC ITER Hometeam activities

$$y = Cx + Du + f(w) \quad (2)$$

where x is the set of active and passive currents, y is the output vector (plasma boundary, position of the plasma current centroid and other quantities to be monitored or controlled, including AC losses indicators), $w=[\beta_p, l_i]^T$ is the disturbance vector and A, B, C, D are the plant matrices, directly provided by the linearized model [5]. Obviously, all vectors in (1)-(2) are the deviations from the corresponding nominal values. The effects of the disturbances $e(w)$ and $f(w)$ are determined by applying instantaneous changes of w (at constant poloidal field energy), and computing the corresponding changes of x and y yielding:

$$e(w(0+)) = x(0+) \quad (3)$$
$$f(w(0+)) = y(0+) - Cx(0+) - Du(0+) \quad (4)$$

Actually, we have only computed $e(w)$ and $f(w)$ for $\beta_p=-0.1$ and $l_i=-0.1$ and approximated them as linear functions:

$$dx/dt = Ax + Bu + Edw/dt \quad (5)$$
$$y = Cx + Du + Fw \quad (6)$$

To assess if the linearized model is reliable for the design of the shape controller, we made three different actions. Firstly, we made sure that the linearized model had been correctly derived from the corresponding nonlinear model (no trivial errors in the code). Secondly, we assessed that the expected amplitudes of disturbances and control currents allow us to assume that the system remains in its linearity range without significant approximation errors. Thirdly, we had to assess the basic assumptions of the nonlinear model from which the linearized model is derived. The first two actions showed that the plasma behaviour is nicely linear even for large amplitudes of the disturbances and large current changes, provided that the plasma does not come in contact with the wall. As far as the third point is concerned, we have started a co-operation with CRPP, Lausanne and ASDEX-U, Garching to compare the linearized models with the experimental results on TCV and ASDEX-U. In addition, we are also co-operating with CRPP, Lausanne on the comparison between our results and those of the nonlinear code TSC [6].

Our activity was then addressed to the optimal choice of number and location of the gaps to be controlled for a diverted plasma configuration. This analysis was carried out by considering the effect, at steady state, of an l_i drop of 0.1. We introduced a general and systematic approach aimed at meeting the two clashing requirements of small contour-integrated separatrix excursion and small control current changes. On the basis of our analysis, the ITER JCT decided to select the six gaps illustrated in Fig. 1, for which our performance indexes (about 4.6 cm of rms separatrix displacement with an Euclidean norm of current changes of about 4 MA for an l_i drop of 0.1) are not dramatically worse than the best achievable (1 cm with 26 MA).

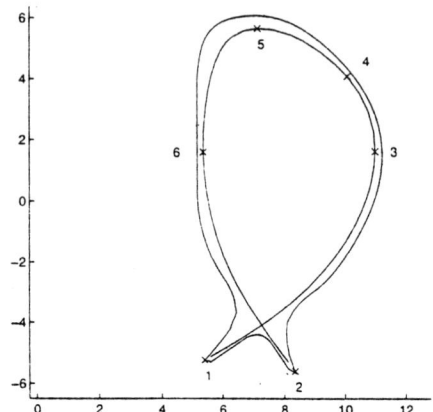

Figure 1. Gap locations selected for ITER TAC-8.

Table 1.
ITER TAC-8 fiducial equilibria and growth rates with a 100 $\mu\Omega$ first wall, 8 $\mu\Omega$ back plate and a 13.2 $\mu\Omega$ vacuum vessel

	XPF	SOF	SOB	EOB
R [m]	8.18	8.15	8.16	8.17
a [m]	2.86	2.84	2.82	2.81
κ	1.68	1.70	1.72	1.73
I_p [MA]	15	21	21	21
β_p	0.10	0.10	0.86	0.89
l_i	0.90	0.91	0.90	0.90
γ_g [s^{-1}]	0.98	1.07	0.85	0.89

We then analyzed the open loop behaviour of the system and the class of disturbances specified by the ITER JCT [4]. We found that the most critical disturbance among them is an l_i drop of 0.1 combined with a β_p drop of 0.2: it yields a displacement of about 3 cm of gap #1 immediately after the appearance of the disturbance. We then assessed the capability of the control system to counteract the disturbance. Focusing on a single output quantity at a time (e.g. gap #1, see Fig. 2) we immediately apply the maximum voltages with the opportune signs to each coil. From this analysis we have an indication of the best achievable performance of the control system. This philosophy also suggested the use of a control strategy based on a variable structure for plasma vertical stabilization in emergency situations [2].

Various multivariable control techniques (i.e., LQG-LTR, H∞, generalized PID) have been applied to meet the requirements specified in [4] for plasma shape control. In particular, the separatrix never touches the wall at the controlled gap locations. The performances are not very far from the optimal ones (see Fig. 2). The separatrix touches the wall for a short time (<2.5 s) in the divertor region because the channels do not fit the fiducial EOB equilibrium. The voltage saturation time is less than 1 s. The problems left open in [5], i.e., large control power demand, divergence of active currents with constant disturbances, nonzero steady state currents with vanishing disturbances and presence of steady state gap offsets with constant disturbances have been overcome. Additional details can be found in [7].

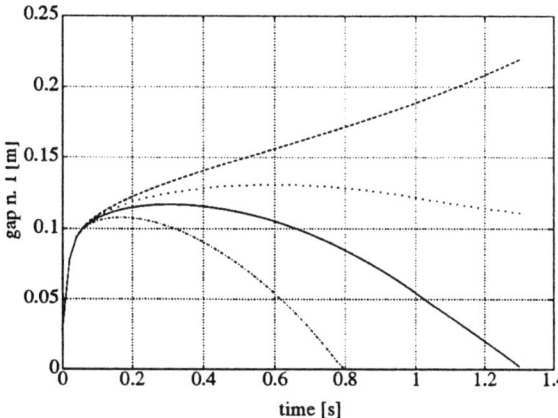

Figure 2. Best achievable performance of the control system. Displacement of gap #1 for an l_i drop of 0.1 and a beta drop of 0.2: no control (dash); nominal saturation voltages (solid); doubling the saturation level (dash-dot); halving the saturation level (dot).

3. IDENTIFICATION OF THE PLASMA SEPARATRIX

For control purposes, a real time plasma reconstruction procedure is needed to identify the value of the gaps from the magnetic measurements.

The nominal values of the gaps are within 10-20 cm range. Therefore, a maximum identification error of 1 cm has been assumed as a requirement for the real time reconstruction algorithm.

The latter is based on the ANN properties of interpolating a dataset of equilibria generated in advance. The ANN is then capable to establish a mapping between the set of magnetic measurements and the set of gaps defining the position of the separatrix in the chamber. Another interesting feature of the ANN method proposed for ITER is the possibility of providing an algorithm for the selection of the optimal number an location of the sensors [8].

The expected identification errors are shown Table 2. Although the results refer to the previous ITER TAC-4 configuration, similar performance are expected for ITER TAC-8. Work is under way for the most recent ITER configuration. In addition, some modifications will be introduced in the ANN topology, that are expected to further reduce the identification errors.

Table 2
Identification error of gaps between plasma boundary and first wall (ITER TAC-4 outline design). A gaussian error of 1% is added to the measurements.

gap	rms error (cm)	max. error (cm)
g1	0.08	0.32
g2	0.35	1.01
g3	0.11	0.20
g4	0.12	0.46
g5	0.23	0.97
g6	0.07	0.38

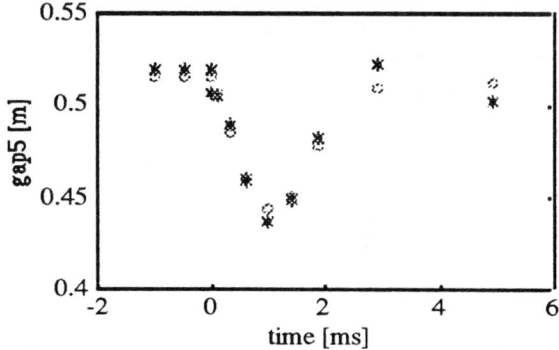

Figure 3. Recovery of gap #5 variation produced by a beta drop of 0.2 in presence of axisymmetric eddy currents flowing in the structure. The values identified by ANN (o) are compared to the ones (*) predicted by running an equilibrium code which includes the eddy currents.

The work in progress is also addressing the integration of the plasma reconstruction within the above described control. A dynamic error adds up to the static error of the algorithm, since the magnetic sensors are shielded by the first wall. It is expected that this effect will keep the signal-to-noise ratio at an acceptable value. Preliminary estimates are encouraging in this respect (Fig. 3), although they are rather optimistic, since the feedback control system for the recovery of the disturbance is driven by ideal gap measurements (i.e. neglecting noise and delays). A degradation is then expected when introducing reconstructed gaps.

4. EDDY CURRENT LOSSES IN THE COLD STRUCTURES

The pattern of the eddy currents for the ITER-TAC8 configuration has been calculated during normal operation scenario (Figure 4) as well as during disruption related to an instantaneous current quench.

The analysis of the inner and outer cylinder structures has highlighted that power losses become negligible after few seconds from the start throughout scenario. The peak power in the inner cylinder is less than in the outer cylinder.

The normal operation scenario has been used to provide the driving field for computing the eddy currents in the remainder of the structure taken into consideration (i.e. TF coil case, TF intercoil structure, TF plates, vacuum vessel, first wall and blanket). The PF coils have been assumed to be current driven. The study has also estimated the shielding effect among the various components: the results confirm the satisfying accuracy of the resistive assumption about the eddy current distribution.

A synthetic view of the total energy dissipated in the various structures throughout both the normal operation scenario (0-2200 s) and a centred instantaneous disruption of a 24 MA plasma is reported in Table 3.

ACKNOWLEDGEMENTS

The authors are grateful to CNR and MURST for financial support and to ITER JCT and EUHT for the technical information. The Authors are indebted in particular with Drs. de Kock, Duchateau, Engelmann, Gruber, Lister, Portone and Salpietro for fruitful cooperation.

REFERENCES

1. Consorzio CREATE, Final Report on Contract ERB5000 CT 940028 NET (NET 94-344), Eddy Current Effects in ITER, June 1996.
2. Consorzio CREATE, Progress Report on Contract ERB5000 CT 950115 NET (NET 95-395), ITER Tasks D255 and D324-1 (Plasma Control Engineering), June 1996.
3. Consorzio CREATE, Final Report on Task D18, Plasma Shape Identification in ITER, Nov. 1995.
4. The ITER Team, ITER TAC-8 Design Document, 1995.
5. R.Albanese et al., "Plasma current, shape and position control in ITER", to be published in the Nov. 1996 issue of Fusion Technology.
6. S.C. Jardin, N. Pomphrey, J. De Lucia, J. Comp. Phys., 50 (1986), p. 481.
7. R. Albanese et al., "Modelling and Engineering Aspects of the Plasma Shape Control in ITER", this conference
8. R. Albanese et al., "Identification of plasma equilibria in ITER from magnetic measurements via Functional parameterization and Artificial Neural Networks", to be published in the Nov. 1996 issue of Fusion Technology.

Table 3
Total Energies dissipated in the various conducting structures of ITER (TAC-8 configuration).

Structure	normal operation	disruption
Inner cylinder	0.5 MJ	–
Outer cylinder	1.3 MJ	–
First wall and blanket	144.0 MJ	1227.6 MJ
Vacuum vessel	90.4 MJ	238.4 MJ
Cryostat	50.8 MJ	2.3 MJ
TF coil case and intercoil structure	12.9 MJ	34.9 MJ
TF coil plates	10.4 MJ	0.5 MJ

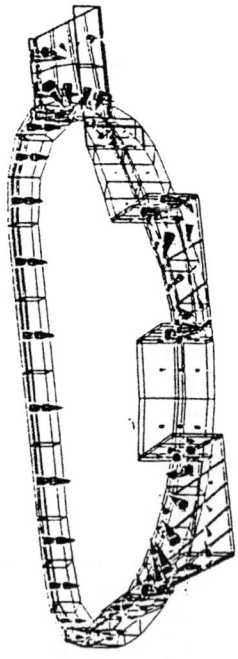

Fig. 4. Eddy current distribution in the vacuum vessel during normal operation scenario.

PLASMA ENGINEERING IN THE IGNITOR EXPERIMENT

B. Coppi[a], R. Andreani[b], C. Ferro[b)], M. Gasparotto[b], C. Rita[b], A. Pizzuto[b], M. Roccella[b], G. Cenacchi[d], A. Bianchi[d], G. Galasso[d], L. Lanzavecchia[d]

[a]M.I.T. Cambridge MA 02139 USA

[b]Associazione EURATOM-ENEA sulla Fusione, C.R. di Frascati, C.P. 00044 Frascati, Rome (Italy)

[c]ENEA - C.R.E. Viale G. B. Ercolani, 8 - 40131 Bologna (Italy)

[d]CITIF, Consorzio Industriale per le Tecnologie e l'Ingegneria della Fusione, Corso Perrone, 25 - 10161 Genova (Italy)

IGNITOR is a compact, high-field tokamak machine. The reference scanarios have been investigated while optimizing the poloidal field system so as to maximize the magnetic flux variation. Two start-up configurations have been verified. A description of the plasma disruption model adopted in the design of the machine is also given.

1. INTRODUCTION

IGNITOR is a compact toroidal device designed to produce sufficiently high plasma currents to reach the temperature and energy confinement time necessary for ignition [1]. An Ion Cyclotron Resonant Frequency (ICRF) heating system (~18MW) is also foreseen to speed up the approach to burning conditions and to control the evolution of the plasma current density profile. The IGNITOR machine is designed to optimize its performance through a complete integration of all its main components. Table 1 shows the main design parameters of the machine.

2. MACHINE LAYOUT

The mechanical structure of the machine (fig. 1) shows a complete integration among the different components:Toroidal Field Coils (TFCs), C-clamps (magnet reinforcing structures), central post, Poloidal Field Coils (PFCs) and vacuum vessel [2].

The structural performance of the machine relies upon an optimized combination of "bucking" between TFCs, central solenoid and central post, and "wedging" in well defined areas of the TFCs and of the C-clamps. The necessary mechanical strength has been obtained designing the copper TFCs in such a way that they can withstand the acting forces, with the co-operation of the C-clamps, preloaded through bracing rings, and of an electromagnetic press.

Table 1
IGNITOR design parameters

Plasma major radius R_o (m)	1.32
Plasma minor radius a (m)	0.47
Aspect ratio R_o/a	2.8
Plasma elongation	1.85
Triangularity	0.43
Vacuum toroidal field at plasma axis (T)	13
Toroidal plasma current (MA)	12
Poloidal plasma current I_θ (MA)	~9
Paramagnetic field (T) produced by I_θ	1.36
Magnetic flux swing (Vs)	37
Flattop (s)	4 at 13 T
Additional radio frequency heating power (MW)	18

3. POLOIDAL FIELD SYSTEM

The IGNITOR PFC system shown in fig. 1, consists of 13 pairs of coils symmetric with respect to the equatorial plane. The PFCs include 2×7 coils for the Central Solenoid (CS) and 2×6 external coils for the plasma equilibrium and shaping.

Bucking by the TFCs allows the 4 coil pairs of

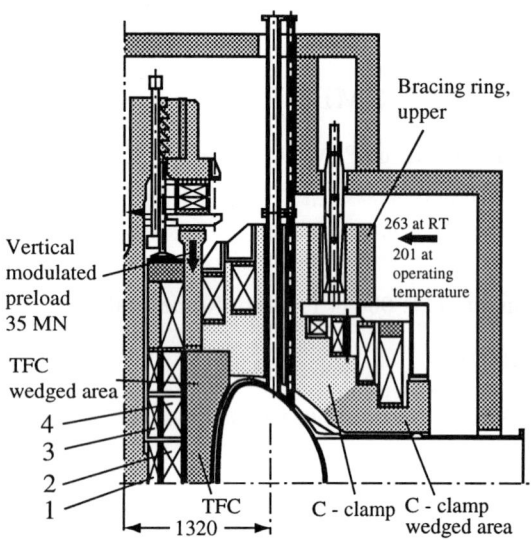

Figure 1. Cross section view of the PFC system.

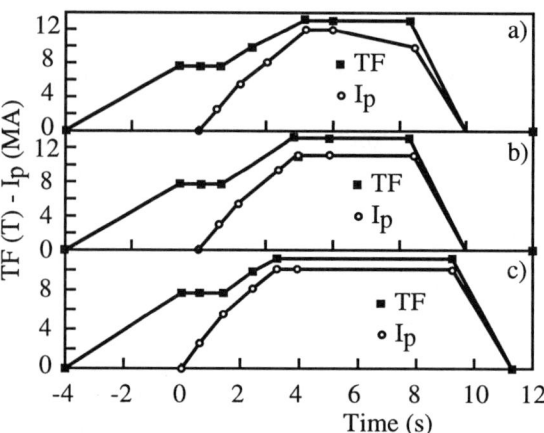

Figure 2. Plasma current I_p and toroidal field TF vs time for the: a) 12 MA plasma scenario, b) 11 MA plasma scenario, c) 10 MA plasma scenario.

the CS nearest to the equatorial plane (coils 1, 2 3 and 4) to resist their own centrifugal force. Independent power supplies for all the 13 coil pairs guarantees a good control of the plasma shape and allows optimization of the flux coupling with the plasma.

At the operating temperature of the PFCs (about 30K at the beginning of the pulse), the magneto resistive effect plays an important role in coil pairs 2 and 4. In particular, the magnetic field gradient can produce a significant temperature gradient. By suitably varying the conductor cross section and the material in the different layers of the coils, the temperatures can be equalized inside the coils, lowering their peak values by some tens of degrees with respect to a homogeneous conductor of constant cross section. Each coil is divided into two parts: in the inner one oxygen free copper conductor has been used, while in the outer, the coil is made of a copper alloy (CuAg) having lower electrical conductivity and slightly lower cross-section. The maximum temperature after a pulse is kept below 280K for the CS coils and below 90 K for the external coils.

4. PLASMA SCENARIOS

The reference plasma scenarios of IGNITOR with maximum plasma current of 12-11 and 10MA and flat-top duration up to 5 s have been investigated. In Fig. 2, plasma current and toroidal field versus time are shown for the three scenarios. It must be noted that to realize the scenarios at 13 T without exceeding the engineering limits in the machine (temperatures and stresses), the co-operating effects of the four coil pairs (no 1, 2, 3 and 4) of the CS must be optimized. In particular, at the end of the toroidal field flat-top, the differences in temperatures and currents among these coils must compensate the different bucking due to the toroidal magnet stiffness variation along the inner part of the coils.

It has been verified [3] that the desired magnetic configurations can be obtained over a wide range of plasma current profiles with only modest changes in the PFC currents. This is quite important due to the uncertainty in predicting the actual evolution of the plasma current density profile in different plasma regimes.

5. PLASMA DISRUPTION

5.1. VDE and Toroidal Current Quench.

A model of plasma hard disruption in IGNITOR has been developed on the basis of previous studies that include the effects of the plasma shape deformation during the vertical displacement event (VDE) on the instability growth rate [4]. This model is used to provide the input data for the electromagnetic and stress analysis in IGNITOR. In the disruption model it is assumed that the plasma vertical displacement occurs at constant plasma current with a time constant of about 10 ms. When the safety factor q at the plasma boundary decreases to a value of about 1.5, a low-q limit disruption occurs and the plasma current quench follows. For this last disruption phase, a maximum current quench rate of 2.5 MA/ms has been assumed.

5.2. Poloidal Currents induced by the thermal and current quench.

The toroidal flux variation due to the thermal quench is about 0.24 Vs. This value has been estimated by an equilibrium code assuming: poloidal beta variation $\Delta\beta_p \cong \beta_p \cong 0.25$; plasma radial dispalacement $\Delta R \cong 0.02$ m, and plasma current increase $\Delta I_p \cong 1$ MA. The toroidal flux decrease during the current quench is given by

$$\Delta\phi_I \cong \Delta\phi_{I0} \times I_p^2(t)/I_p^2(0)$$

where $\Delta\phi_{I0} \cong -0.66$ Vs has been evaluated by equilibrium code assuming $\beta_p \cong 0$; $I_p(0)$ and $I_p(t)$ are the plasma current at the beginning of the current quench and at time t.

5.3. The Halo Currents.

The plasma dynamics during the first phase of the VDE has been evaluated. In this first phase, the plasma is force-free under the effect of the external poloidal field and of the toroidal currents induced on the vacuum vessel. In the second disruption phase, the plasma position is assumed to be at rest. We must then introduce a new force term to ensure that the plasma continues to be force-free. It is assumed that the force term needed to balance all the other forces acting on the plasma is due to the halo currents. Experimental data both from JET and from ALCATOR-C-MOD have shown that halo currents can be non-axis-symmetric; this leads to an increase (up to a factor of two) in the local vertical force acting on the vacuum chamber and to a maximum halo current density in the first wall of about 4.5 MA/m^2.

6. PLASMA START-UP

The start-up magnetic configuration must produce a "suitable" poloidal field null on the equatorial plane. Suitable means that the average poloidal field gradient around the null point, in a region of radius about equal to the minimum distance between the null point itself and the first wall, must be low enough to minimize the flux consumption during the plasma start-up phase.

In a typical magnetic configuration of plasma start-up, the null point is achieved on the equatorial plane at a radial distance R=1.65 m and with an average magnetic field gradient around this point of 1.8×10^{-2} T/m. The minimum required loop voltage (Vloop) in this case is 11 V. The power supplies have been sized to achieve a value up to 30 V. The induced electrical field will penetrate inside the vacuum chamber with a time constant τ_ϕ of ~15 ms. Consequently, the flux losses will be about 3 τ_ϕVloop $\cong$ 0.5 Vs. Start-up configuration, with the plasma near the inboard edge of the plasma chamber has been also investigated. The outer start-up is more advantageous in terms of flux availability and is needed when ICRF heating is applied at the beginning of the discharge.

7. PLASMA EQUILIBRIA

The compatibility of the evolving MHD configurations with the poloidal flux consumption and the thermal loads on the first wall has been verified by avoiding that the temperature rise and the mechanical stresses, in any of the PFCs, exceed the engineering limits.

In each equilibrium case, care has been taken to keep the safety factor q at the plasma boundary well above 3, to reduce the risk of q-limit disruptions. The requirement that the last plasma magnetic surface touches a large fraction of the first wall at very low angles (for reducing the thermal loads), can be satisfied thanks to the versatility of the IGNITOR highly integrated PFCs. A number of different plasma configurations can be produced in IGNITOR:
- partially detached configurations, as shown in fig. 3,
- fully attached plasmas, with the plasma filling the entire plasma chamber cross section,
- double x configurations at somewhat reduced plasma current.

The power load peaking factor [5] on the IGNITOR first wall during normal operation is ~1.5 (corresponding to a value of 0.74 MW/m^2 for a

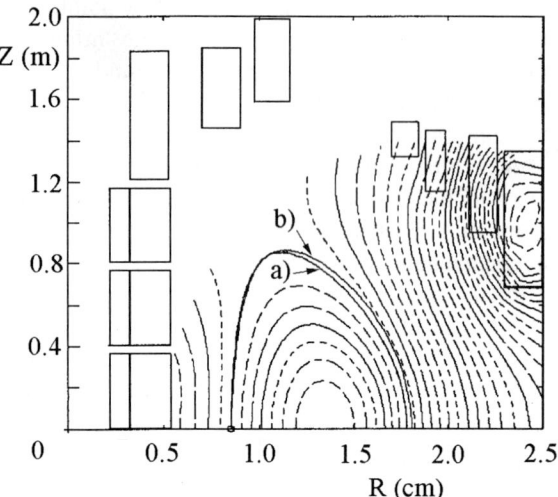

Figure 3. Typical equilibrium configuration for 12 MA plasma current scenario. a) The last closed flux surface, b) first wall.

Table 2.
Flux balance in IGNITOR at various times during 12 MA scenario

Flux balance (Faraday method)					
Time (s)	1.5	2.5	4.0	5.0	8.0
I_p (MA)	5.2	8.0	12.0	12.0	10.0
I_{bs} (MA)	0.3	0.5	0.8	1.35	1.6
$\Delta\Psi^{(av)}$ (Vs)	15.8	22.8	35.0	37.0	33.0
Ψ^{pla} (Vs)	13.5	20.0	31.2	32.6	27.5
Ψ_{ohm} (Vs)	1.9	2.6	3.8	4.2	5.3
Ψ_M (Vs)	0.4	0.2	0.2	0.2	0.2
$\Psi_M^{(bs)}$ (Vs)	0.8	0.85	1.3	2.2	2.8

18 MW discharge) when the last magnetic surface is close to the first wall in the inner region (see fig. 3). A displacement of the plasma column of 1cm in all directions increases, in the worst case, the peaking factor up to 2.8 (which corresponds to a power density of 1.4 MW/m^2 for a 18 MW discharge). The maximum acceptable power density for the envisaged first wall materials (graphite or molybdenum) is 6 MW/m^2.

8. POLOIDAL FLUX REQUIREMENT

Table 2 shows the flux balance referred to the plasma axis at different times during 12 MA plasma scenario evaluated by the Faraday method. In the table, $\Delta\Psi^{(av)}$ is the available flux, Ψ_{pla} is the inductive plasma flux at the plasma axis, and Ψ_{ohm} is the ohm flux consumption. Ψ_{ohm} has been computed by the Ejima scaling law during the plasma ramp-up, while a flux consumption rate of 0.4 Vs/s has been assumed during the flat-top. The flux margin Ψ_M is given with and without the beneficial effect of the bootstrap currents (I_{bs}).

Discharges with additional heating power (18 MW) during the current ramp-up phase require less Vs than the equivalent ohm discharges due to the lower value of the plasma current needed for ignition, the increased bootstrap current fraction, and the higher plasma temperature.

9. CONCLUSION

The IGNITOR PFC system can guarantee "robust" elongated MHD equilibria. The compatibility of the evolving plasma configurations with the poloidal flux consumption and the thermal load on the first wall has been verified. The currents in the PFCs have been optimized avoiding that the coil temperatures and stresses exceed the engineering limits.

In the design of the IGNITOR machine, the elettromagnetic loads due to plasma disruptions (VDE, thermal and current quench and halo currents) have been suitably taken into account, including the recent experimental results obtained in D-shaped tokamak machines.

REFERENCES

1. B. Coppi, M. Nassi L.E. Sugiyama, Physica Scripta 45, (1992) 112.
2. R. Andreani, Proc. 17th Symp. on Fusion Tech., Rome, Italy (1992), 1, (1993) 129.
3. G. Cenacchi, B. Coppi, L. Lanzavecchia, ENEA Report RTI/INN(92)16.
4. C. Rita, M. Roccella, M. Nassi, S. Graziadei, Proc. 18th Symp. on Fus. Tech., Karlsruhe, Germany, (1994) 1, (1995) 707
5. C. Ferro, G. Franzoni, R. Zanino, ENEA Report RT/ERG/FUS(94)14

Design Characteristics of KT-2 Tokamak for Advanced Tokamak Operation *

B.G. Hong, S.H. Jeong and S.K. Kim

Nuclear Fusion Laboratory, Korea Atomic Energy Research Institute,
P.O. Box 105 Yusong, Taejon, Korea 305-600.

Adaptability of the advanced tokamak discharge, i.e., 100 % non-inductive current drive with high(> 70%) bootstrap current fraction, in a large-aspect-ratio (LAR), medium-size, diverted tokamak KT-2 [1] is investigated with the time-dependent transport simulation. Through the dynamic simulation, the stable route to high β_p, high bootstrap plasma with negative shear over a central plasma region is found. And the required auxiliary heating and current drive power and design requirements for the advanced tokamak operation are obtained. The compatibility of the advanced tokamak operation mode with the poloidal field coil system and the divertor system is demonstrated.

1. INTRODUCTION

In advanced tokamak operation, utilizing bootstrap current driven by pressure gradient is advantageous since it reduces power required for the non-inductive current drive. With a high bootstrap current, the current profile is naturally hollow and leads to negative magnetic shear configuration. This configuration has been thought to be promising for the advanced tokamak operation since it produces high β_N due to high-n ballooning stability and enhanced confinement due to suppression of anomalous transport mechanism such as ITG and trapped electron mode [2].

Dynamic simulation is necessary to investigate the stable route to high β_p, high bootstrap current plasma with negative shear over a central plasma region and to calculate the required auxiliary heating and current drive power. For the bootstrap current dominant plasma, the control of pressure profile by the heating and current drive methods for the alignment of the bootstrap current to the desired total current is important since the bootstrap current is strongly dependent on the pressure profile through the plasma transport. The strongly shaped configuration also requires the study of the compatability with the poloidal coil system and the in-vessel components.

We present time-dependent simulation studies of advanced tokamak discharge in a large-aspect-ratio (LAR), medium-size, diverted tokamak KT-2 which was conceptualized [1] with a research goal of 100% non-inductive current drive exploiting high bootstrap current fraction(> 70%). TSC [3] transport simulation code is used to determine dynamic evolution of the profiles of plasma pressure, current and bootstrap current.

2. DYNAMIC EVOLUTION

The reference KT-2 tokamak parameters used in the simulation are; major radius $R_0 = 1.4\,m$, minor radius $a = 0.25\,m$, elongation $\kappa = 1.8$ and triangularity $\delta = 0.6$ with double null plasma.

2.1. Transport Model

TSC solves transport equations with respect to evolving magnetic surface containing a fixed toroidal flux. Also, plasma force balance equations are solved to maintain plasma in near equilibrium during its evolution. For details of TSC model equations, we refer the original paper in Ref. [3]. For the density profile, we take the following parametrized form ;

$$n_e(\hat{\Psi}, t) = n_{0,e}(t)[1 - \hat{\Psi}^{\beta_N(t)}]^{\alpha_N(t)} + n_b(t), \quad (1)$$

where $\hat{\Psi}$ is the normalized poloidal flux and $n_b(t)$ is the density at plasma boundary. The exponents $\alpha_N(t)$ and $\beta_N(t)$ can be adjusted to match the

---
*Research supported by the Ministry of Science and Technology(MOST).

experimantal data, but in this study we assumed $\alpha_N(t) = 0.5$ and $\beta_N(t) = 2.0$. The electron and ion thermal conductivities are modelled as [4]

$$\chi_i = a_{126}\chi_e \qquad (2)$$
$$\chi_e = f_m[(\chi_{TEM})^2 + (\chi_{\eta_i})^2]^{1/2} F(\Phi)$$
$$\times (|\nabla \Phi|)^{-2} \qquad (3)$$

where Φ is toroidal flux, $F(\Phi)$ is a profile factor given in Ref. [5], and

$$\chi_{TEM} = a_{122}(1.25 \times 10^{20})\frac{a}{n(\Phi_b)}[xB_T]^{0.3}$$
$$\times Z^{0.2}\frac{\alpha_N}{4} \qquad (4)$$
$$\chi_{\eta_i} = a_{121}(7.5 \times 10^8)[\frac{P(\Phi_b)}{n(\Phi_b)}]^{0.6}$$
$$\times [RB_T q_{95}]^{-0.8} a^{-0.2} \qquad (5)$$

Here $P(\Phi)$ is the total heating power minus the total radiated power inside the flux surface Φ.

The time averaged effect of sawtooth instability inside $q = 1$ surface is included in Eq.(3). Thus thermal conductivity is enhanced inside the q=1 surface according to the prescription;

$$f_m = 1, \quad \text{for } q > 1 \qquad (6)$$
$$f_m = a_{124}, \quad \text{for } q < 1 \qquad (7)$$

For the calculation of bootstrap current, we use Harris's extension [6] (into Plateau and Pfirsch-Schulter collisionality regime) of Hirshman's collisionless expression [7] which is valid in the neoclassical banana regime. Neoclassical corrections to resistivity are used as in Ref. [8], and the effect of sawtooth instability is taken into account by enhancing resistivity inside $q = 1$ surface. In Eqs. (2)-(7), a_{121}, a_{122}, a_{124} and a_{126} are parameters to control transport properties.

2.2. Scenario I ; Off-axis Heating

The schematic of the scenario is shown in Fig. 1. Throughout simulations, toroidal magnetic field is maintained $2.0 T$ for the entire simulation and plasma current is ramped from $20 kA$ at $t = 0.0$ sec to $300 kA$ at $t = 0.5$ sec. The plasma evolves from an inboard limited circular shape at $t = 0.0$ sec, becomes diverted at $t \sim 0.4$ sec, and reaches an elongation of $\kappa = 1.8$ and a triangularity of $\delta = 0.6$ at $t = 0.5$ sec. The plasma shape and plasma currents are maintained by feedback control on PF coil currents.

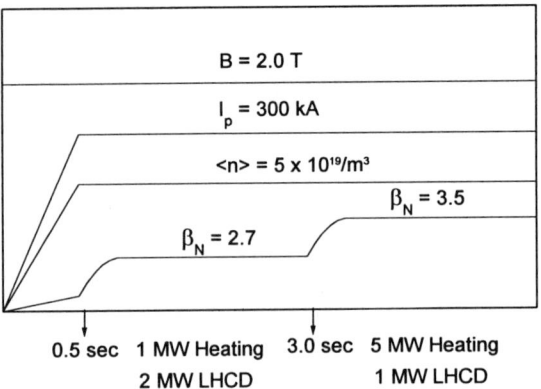

Figure 1. The schematic of AT scenario

The electron density is programmed to increase linearly with time to the current flattop value of $<n> \cong 5.0 \times 10^{19} m^{-3}$ with peak to average value of 0.6. The effective charge Z_{eff} is set to 2.0 and the control parameters, $a_{121} = 0.06$, $a_{122} = 0.4$ in Eqs.(4) and (5) are used to simulate improved confinement. Deposition profiles of heating and current drive are modelled using simplified analytic formula; $S_{Heating,CD}(\hat{\Psi}) = \frac{d^2(\hat{\Psi})^{a_1}(1-(\hat{\Psi})^{a_2})}{(\hat{\Psi}-a)^2+d^2}$, where $\hat{\Psi}$ is the normalized poloidal flux, and a, d, a_1 and a_2 are profile control parameters. At the end of plasma current ramp-up, $2 MW$ lower hybrid current drive with $1 MW$ heating is initiated and as a result, β_N increases to 2.7 with bootstrap current fraction, 57 %. The current profile redistributes from peaked OH profile to hollow profile on a long diffusive time scale. It takes more than 2.5 sec for a full relaxation of $300 kA$ plasma current. After the current profile has broadened sufficiently (non-monotonic q profile develops) to allow second stable regime to high-n ballooning mode, $5 MW$ off-axis heating with $1 MW$ LHCD is initiated and β_N increases to 3.5 with bootstrap current fraction = 73 %. Fig. 2 shows q and current density profiles before and after second phase heating. Due to high bootstrap current fraction(> 70%), the current profile is hollow. Broad pressure profile due to off-axis heating cause the bootstrap current profile to peak on off-axis. A seed current(20 kA) on magnetic axis driven by fast wave is necessary to control q_0 value. Time history of q_0(Fig. 3) in-

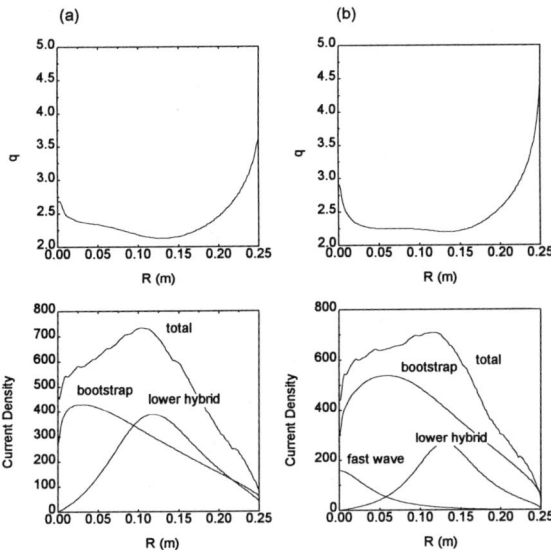

Figure 2. q and current density profile (a) t=3.0 sec and (b) t=7.0 sec

dicates that the current profile is frozen after 4.5 sec.

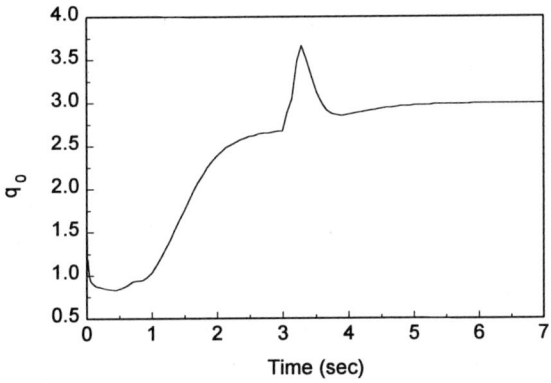

Figure 3. q_0 as a function of time

2.3. Scenario II ; Central Heating

The schematic of the scenario is similar to scenario I except that we initiate heating in initial current ramp-up phase of discharge. Central electron heating by 0.5 MW heating and 1 MW LHCD increases the electrical conductivity and prevent the penetration of the inductively driven OH current by prolonging the current diffusion time. At the end of current ramp-up, 2.5 MW central heating and 1.5 MW LHCD freeze the current profile and raise plasma β_N to 3.6 and the bootstrap current fraction to 70 %. By this way, we can shorten the time which is needed for current profile redistribution and avoid MHD unstable transitions from positive to negative magnetic shear in the region of high pressure. At this state, LHCD deposition profile shifts off-axis with increasing density and controls the location and the value of q_{min} for MHD stability. Due to central heating, peaked pressure profile causes the bootstrap current profile unfavorable. Control of plasma profiles is important since bootstrap current and profile are strongly dependent on plasma transport and resulting pressure profile. To increase $\beta_N > 4$, not only more heating power but strong profile control capability are required. The current profile is frozen after 2.0 sec which is shorter than 4.0 sec in senario I.

2.4. Sensitivity on Transport Models

We repeat the simulation of scenario II with the Hirshman-Sigmar bootstrap current formulation [9] which is valid in all coollisionality regimes. The calculation shows that the magnitude of the bootstrap current is similar to the case with Harris model but the bootstrap current density near the magnetic axis is smaller, so the resulting q profile shows a larger shear in central region. When we use the Kaye-Goldston model for the thermal energy transport, the bootstrap current profile is broader and $q_{min} \sim 2$ compared to $q_{min} \sim 1.6$ with the Coppi-Tang thermal energy transport model. The details of the profiles depend on the transport models used, but the required heating and current drive power to maintain the negative magnetic shear configuration is found to be within the KT-2 design specification.

2.5. MHD Stability

A trajectory of KT-2 equilibria in $l_i - q$ space indicates that the kink mode may be unstable but MHD stability analysis for the equilibria of the two scenarios show that $n = 1$ external kink mode is stable with the presence of a conducting wall at 1.4 times the plasma radius. Profile optimization study [10] for the KT-2 advanced tokamak scenario yielded a marginal ballooning stable equilibrium with $\beta_p = 4.23$ and $\beta = 3.542\%$ corresponding to high Troyon factor, $\beta_N = 5.902$. The infernal mode stability analysis for the optimized equilibrium showed the growth rates are small with the triangularity $\delta = 0.6$ and the mode can be stabilized by shifting the wall towards the plasma surface or lowering the plasma current.

3. COMPATABILITY WITH THE SYSTEM DESIGN

3.1. Poloidal Coil System

KT-2 tokamak poloidal field coil system is designed to provide both shape and plasma current feedback control during the discharges. Poloidal field coil currents waveforms [11] obtained as an output from the simulation for the scenario I show that they provide the proper fields to maintain the high β_p, high bootstrap equilibria within the design specification. The temperature rise during 20 sec discharge is calculated to be $50°C$. The study on the optimization of the magnetization level shows that it is possible to keep the overall temperature rise less than $40°C$ [11].

3.2. Divertor System

Fig. 4 shows the mapping of the scrape off layer in the steady state of the advanced tokamak operation(scenario I) into the divertor plates. The flux surfaces which pass through a point 3 cm from the outer edge of the plamsa and and a point 3 cm inboard from the inner edge of the plasma fit into the divertor plates. Thus the divertor system is compatible with the advanced tokamak operation.

4. SUMMARY

Through the time-dependent transport simulation, MHD stable scenarios to high β_p, high bootstrap, negative magnetic shear configuration are investigated. Two scenarios(off-axis and central heating) are considered. Simulations indicate that the advanced tokamak operation exploiting high bootstrap current in KT-2 tokamak is adaptable with profiles control using heating and non-inductive current drive methods. Also, the magnetic configuration found to be compatible with the KT-2 poloidal field coil and divertor systems.

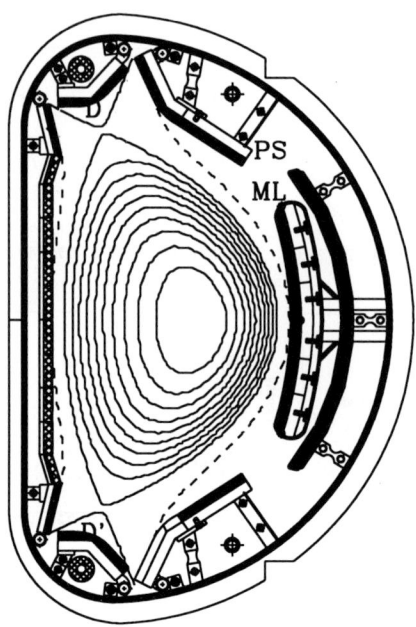

Figure 4. Flux surfaces of SOL

REFERENCES

1. I.S. Chang, et. al., *Concept Definition of KT-2*, KAERI/TR-472/94, (1994).
2. C. Kessel, J. Manickam, G. Rewoldt, and W. Tang, Phys. Rev. Letters **72**, 1212, (1994).
3. S.C. Jardin, N. Pomphrey, and J. Delucia, J. of Comp. Phys. **66**, p.481, (1986).
4. W.M. Tang, Nucl. Fusion **26**, p.1605, (1986).
5. S.C. Jardin, M.G. Bell and N. Pomphrey, Nucl. Fusion **33**, p.371, (1993).
6. G.N. Harris, EUR-CEA- FC 1436, (1991).
7. S.P. Hirshman, Physics of Fluids **31**, p.3150, (1988).
8. S.P. Hirshman, R.J. Hawryliuk, and B. Birge, Nucl. Fusion **17**, p.611, (1977).
9. S.P. Hirshman and D.J. Sigmar, Nucl. Fusion **21**, p.1079, (1981).
10. S. Poedts, A. De Ploey, and J.P. Goedbloed, Rijnhuizen Report 96-227, (1995).
11. K.W. Lee et al., in this symposium.

Numerical simulation of plasma equilibrium and shape control in tight tokamak GLOBUS-M.

A. A.Kavin[a], V.A. Belyakov[a], S.A. Bulgakov[a], Y.A. Kostsov[a], E.N. Rumyntsev[a], S.A. Galkin[b], L.M. Degtyarev[b], A.A. Ivanov[b], Y.Y. Poshekhonov[b], V.A. Yagnov[c]

[a] STC "SINTEZ", D. V. Efremov Research Institute, S. Petersburg, 189631 Russia;
[b] Keldysh Institute of Applied Mathematics, Russian Academy of Sciences, Moscow, Russia;
[c] TRINITI, Troitsk, 142092, Russia;

This paper describes the poloidal field system and plasma control including plasma scenario of small aspect ratio tokamak GLOBUS-M. Plasma configurations and parameters are presented. The main attention is paid to problem of plasma shape control and especially vertical position control. Simulations of control system taken into account real power supply system have been done.

1. POLOIDAL SYSTEM AND PLASMA PARAMETERS

Investigations of small aspect ratio plasmas properties are the main purposes of GLOBUS-M tokamak. Poloidal system has to provide the plasma parameters indicated in Table 1.

Table 1
GLOBUS-M Basic Parameters

Plasma major radius, R_p (m)	0.36
Plasma minor radius, a_p (m)	0.24
Aspect ratio, R_p/a_p	1.5
Plasma vertical elongation, k	1.5-2.15
Plasma triangularity, δ	~0.2
Plasma current, I_p (MA)	0.3
Toroidal magnetic field, B_T (T)	0.5
Pulse duration, (s)	0.3
Required volt-second consumption, (Wb)	~0.31
Weight of tokamak, kg	~5000

Plasma configuration at flat top of plasma current (I_p=0.3MA) is shown in Fig. 1. Symmetrical respectively middle plane poloidal system creates high elongated (k=2.15) plasma with two x-points. Three pairs of PF coils (PF1, PF2, PF3) form the given separatrix and perform the plasma shape control. Central colenoid (CS) wound in two layers consists of 124 turns and produces magnetic flux of ~0.155Wb at maximum coil current 70kA. CS performs the slow plasma current control. Three pairs of compensation field coils (CC1, CC2, CC3) are designed mainly for compensation of stray fields (up to 1G) at the large breakdown region.

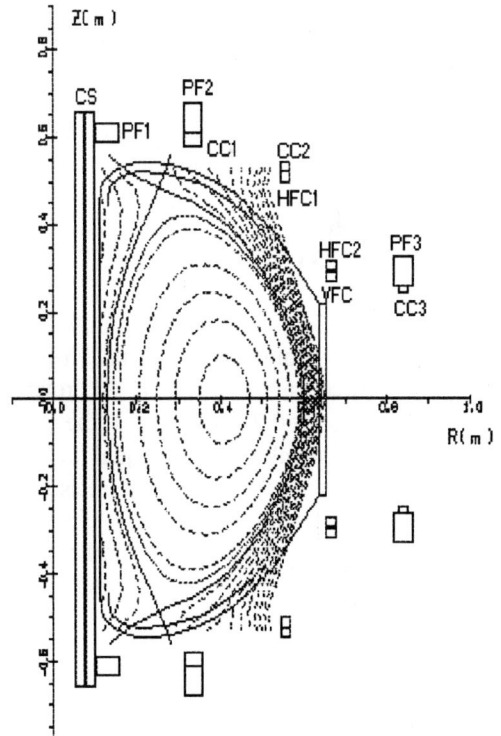

Figure 1. Poloidal system and base plasma configuration.

Also they assist to the plasma shape control at flat top phase. Additional vertical field coil (VF) fed by power supply with low electric power and high frequency is used as corrective coil for the stray field compensation at breakdown and then for the shape control. One of the most important coils is two pairs of horizontal field coils (HFC1, HFC2). The main purpose of this set of coils is to stabilize the vertical plasma displacements for the short time.

2. PLASMA SCENARIO.

As a reference scenario for GLOBUS-M tokamak it was considered ohmic heating regime (I_p=0.3MA) using evolution of plasma parameters according to ITER-P scaling. Plasma discharge starts from inner wall of vacuum vessel. During rump-up phase plasma configurations correspond to natural elongated plasmas with different aspect ratio. Note that homogeneous vertical magnetic field provides the equilibrium of such configurations and, consequently, they are stable respectively vertical displacements. At aspect ratio R_p/a_p=2 elongation of that plasma is equal to k=1.4 and at aspect ratio R_p/a_p=1.5 (minimum aspect ratio of GLOBUS-M tokamak) is equal to k=1.6. This configuration is shown in Figure 2. PF3 coil provides approximately the homogeneous vertical field. Configurations with x-point and elongation up to k=2.15 are obtained by means of PF1, PF2, PF3 coils.

3. VERTICAL INSTABILITY.

The main problem of feedback control is to stabilize strong elongated plasma (k=2.15) respectively vertical displacements. At reference plasma parameters li=0.7 and β_p=0.3 the growth time of vertical instability (τ_γ) is equal to 2.4ms. Such low value is connected with high resistivity of vacuum vessel surrounding plasma (thickness of inner cylinder is 2mm, thickness of middle toroidal part-domes is 3mm, thickness of outer ring is 14mm, material is stainless steel).

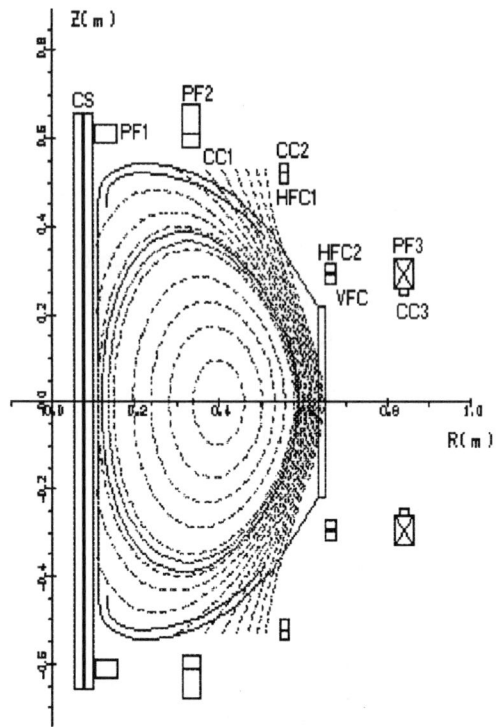

Figure 2. Natural elongated plasma configuration.

4. LINEAR MODEL FOR THE SIMULATION OF VERTICAL AND PLASMA SHAPE CONTROL.

Due to the high increment of instability vertical feedback control is more faster then shape and plasma current control. Besides vertical position control does not depend on the shape one because of symmetry of poloidal system and decoupling of HFC with other poloidal coils. Obviously that only HFC coil is capable to stabilize vertical position.

The other requirements imposed on poloidal system is to control the some selected points of separatrix which are most important to provide stability of plasma operation. For GLOBUS-M tokamak these points are following:
- R_i-innermost point of separatrix;
- R_o-outermost point of separatrix;

The control of these points allows to keep up the needed aspect ratio of tight tokamak and prevents contact of plasma with vacuum wall.
- R_x-radial position of x-point;
- Z_x-vertical position of x-point.

So as x-point is close to divertor plates control of its

position maintains the separatrix branches on divertor plates.

Linear model derived for simulation of GLOBUS-M plasma control system is based on approach [1] developed to the plasma shape control system design of ITER. Realisation of this approach has been made by means of PET-code. The PET code had been developed in Keldysh Institute to compute evolution of free boundary plasma equilibrium including a separatrix and eddy currents in surrounding plasma conductors. A special iterative technique gives fast convergence when the nonlinear system of the 2D Grad-Shafranov equation and the circuit equations is solved even if the conductors are located close to the plasma. Various equilibrium output data including computation of gaps between control points and separatrix allow to introduce plasma shape and position control. The code allows to take into account equilibrium plasma current inside and outside of the separatrix. The code can be adapted to the design of the real machines including description of the geometry of the passive structure and PFC system. The different methods have been used to estimate the growth time with help of simulations of free boundary plasma evolution. All these methods give the close results which are in a good agreement with growth time above for the linear model.

5. LINEAR SIMULATION RESULTS

As most critical for the simulations of control system the following disturbances of plasma parameters were taken:
- initial vertical displacement of the plasma current centre (Z_c) up to 2cm;
- instantaneous β_p drop of 0.1 and l_i drop of 0.05.

At first it was examined the vertical control system with real power supply. HFC is supplied by the current invertor with maximum voltage 500V, current 500A and frequency up to 2kH. PD-controller relatively Z_c and dZ_c/dt as input signal drives up this device. Behaviour of Z_c, I_{HFC}, U_{HFC} for this case is shown in Fig.3. The settling time of Z_c is about 3ms. Note that in this case the separatrix points selected for the shape control are recovered after t>3ms at the initial positions.

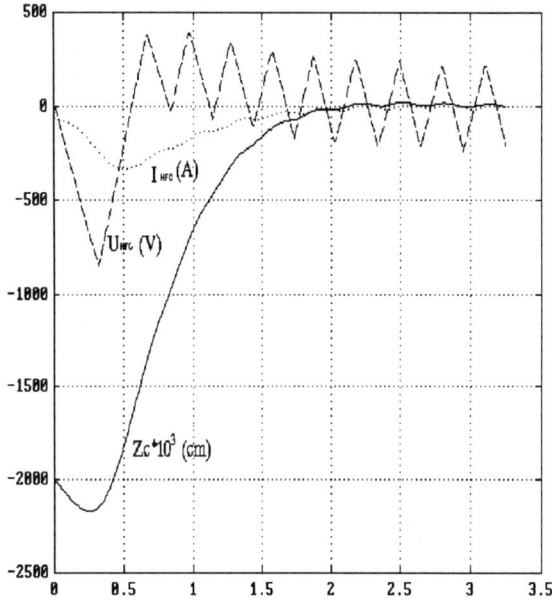

Figure 3. Waveforms of current centre, HFC current and voltage at the initial vertical displacement.

During transient shown in Fig. 3 separatrix has one x-point because of destroying of symmetry respectively middle plane. High frequency oscillations are practically invisible therefore the vertical displacements can be simulated by means of "ideal" PD-controller (with infinity frequency) for the time scale t>3ms. Such time scale fits for control of selected separatrix points. Besides poloidal coils PF1, PF2, PF3, CS, CC are fed from 6-pulse thyristor converters with industrial frequency significantly lower than the current invertor of HFC and VFC. Consequently it is reasonable to use the separatrix points control after stabilization of Z_c. Operation of 6-pulse thyristor converter can be approximated by the first order transform function:

$$U_{out} + T_d \times dU_{out}/dt = U_{in} \qquad (1)$$

where U_{in}-input voltage (controller), U_{out}-output voltage of converter, time constant T_d=5ms. Such model of the converter operation suppresses the control signals with frequency higher than $\omega=1/2\pi T_d$. It should be mentioned that multi input-multi output (MIMO) algorithms are applied for the shape (or separatrix points) control. As an example simulation of combined (vertical and shape) control system at the instantaneous β_p drop of 0.1, l_i drop of 0.05 and initial Z_c displacement by 2cm is shown in Fig. 4.

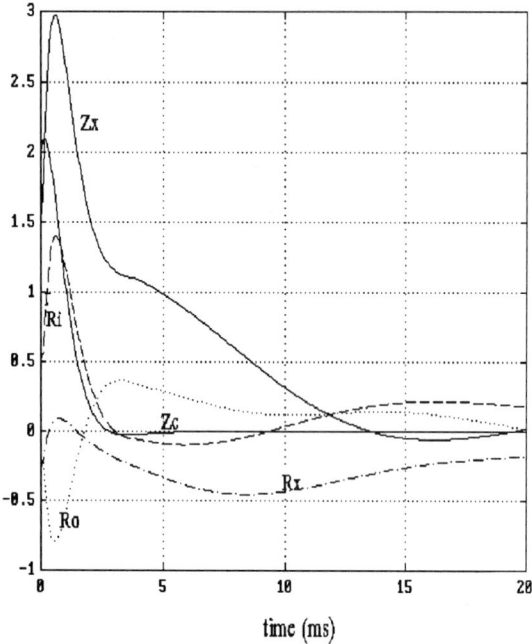

Figure 4. Dynamic of the current centre and the controllable points of separatrix (cm).

Here "ideal" PD-controller is applied to HFC coil and MIMO LQG-controller with transform function (1) begin to operate when Z_c reaches the insensitive zone. As seen from Fig. 4 the time scale of the shape control is about 20ms and the control accuracy is not worse than 3cm (Z_x) that allows to prevent contact of separatrix with divertor plates (acceptable value is 4cm). Maximum sweeping of separatrix branches along divertor plates is less than 7cm at this transient that is acceptable too. The slow waveforms of PF-coil voltages are observed in this case. CS provides the plasma current stabilization on the same time scale.

It can remark one of the application of the vertical and shape feedback control to obtain the single null configuration when the selected points of separatrix save the nominal position. For that purpose it can stabilize the current centre displaced by 2cm respectively middle plane by means of HFC. The other PF coils perform the shape control.

Dynamic of this process is the same as shown in Fig.4 with the exception that settling time of shape control is some shorter. Ultimate single null configuration after damping of transient is shown in Fig. 5.

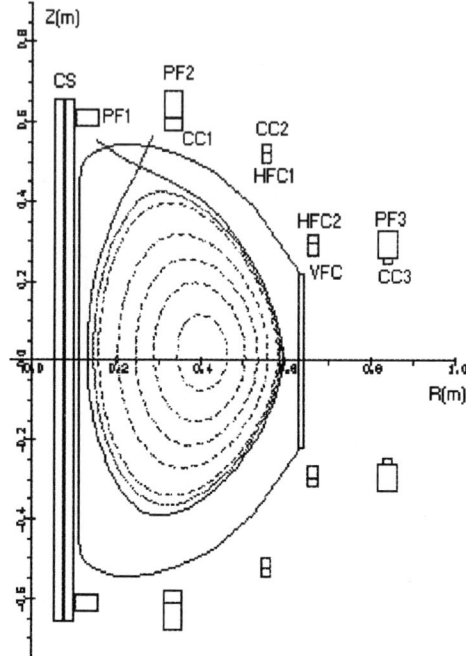

Figure 5. Single null configuration produced by the vertical and shape feedback control.

6. CONCLUSIONS.

Poloidal system of GLOBUS-M provides tokamak operation with limiter (natural elongated as well) and one or two x-points plasmas of high elongation. Plasma vertical control system is more fast than the shape and plasma current one. It is possible to stabilize the controllable parameters above at the considered disturbances. The given power supply system allows to provide the control voltages with needed frequencies and amplitudes.

REFERENCES
1. ITER Design Descr. Doc. 4.7, App.B, June 1995.

Thermal and Structural Analysis of Vacuum Vessel and in-Vessel Components for Low Aspect Ratio Tokamak Globus-M

V.A.Divavin, S.A.Grigoriev, A.V.Lipko, V.A.Bykov, V.N.Komarov, E.G.Kuzmin, I.A.Mironov

Efremov Institute, St-Petersburg, 189631, Russia

Calculation thermal analysis of vacuum vessel and in-vessel elements in backing regime and regime of regular sequence of operating pulses is given. Structural analysis is also presented.

1. Design features of vacuum vessel and in-vessel elements

Vacuum Vessel (VV) of experimental Tokamak GLOBUS-M represents a single-layer thin-shell stainless steel chamber that consists of the central sphero-toroidal central part (dome) with thickness 3 mm, inner cylindrical part with diameter 220 mm and thickness 2 mm and outer cylindrical part (outer ring) with diameter 1270 mm and thickness 14 mm. In-vessel components made from stainless steel are 2 mm limiter plate attached to inner cylinder and two 3 mm divertor plates attached to upper and lower part of vacuum vessel by means of stiffening ribs. The number of vertical ports is 24 and the number of horizontal ports is 14, the range of ports cross-sections changes from the smallest vertical round port with diameter 26 mm to 4 horizontal rectangular ports with dimensions 400 × 200 mm each of them. Vacuum vessel has 4 mechanical supports located symmetrically around torus on the lower walls of rectangular port.

Supports allow the free radial thermal expansion of vessel but they restricted its toroidal and vertical movements. Design of vacuum vessel and in-vessel elements is presented schematically in Fig. 1.

2. Elements temperature state

Two main temperature regimes were considered:
- backing regime (at initial vessel vacuum pumping)
- operating regime (pulses regular sequence).

2.1. Backing regime

It is necessary to provide temperature state of vacuum vessel and all it's components in heating regime at level 200 C during several tens of hours. Forced cooling of vessel in backing regime is not provided (as well as in operating regime). Cooling is due to natural convection in surrounding air.

During heating process temperature difference between any two points of vessel may not exceed value 25-30 C. Given request is caused by necessity to avoid excessive temperature stress (and strain).

It was suggested previously that induction heating by means of using central solenoid is sufficient for backing regime purposes. Indeed, calculations showed that under U=42 V (alternative current I=582 A, frequency f = 50 Hz) value of inducted heat in vessel cylindrical part (near solenoid) sufficient to have in this zone T_{vv} =200 C. However, while using of induction heating only vessel temperature non-uniformity is too large, in spit of glassware insulation using (δ = 25 mm) on dome and outer vessel part. Such temperature non-uniformity is due to abrupt decreasing of inducted heat with radius increasing.

Inducted heating distribution in vacuum vessel body and related temperature distribution is shown

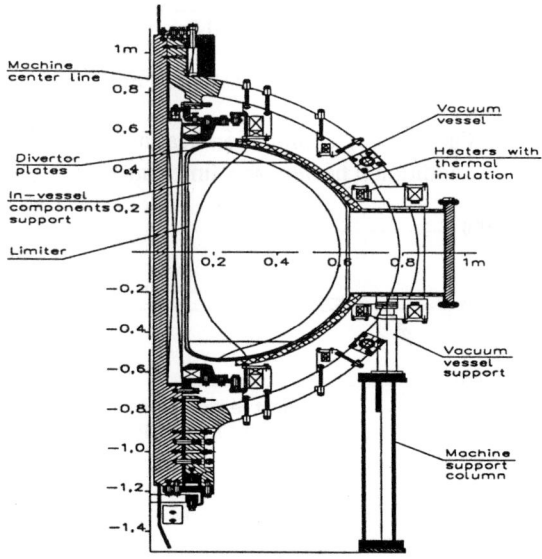

Fig.1. Design of vessel and in-vessel elements.

in Fig.2. Abscissa axis on Fig.2. is evolvent of VV wall (in vertical cross section). The origin of abscissa axis is at the cross point of inner cylinder with horizontal symmetric axis of VV cross section. It is seen that temperature difference between most hot and most cold vessel points is about 160 C in the case of using induction heating only. Such value is not acceptable.

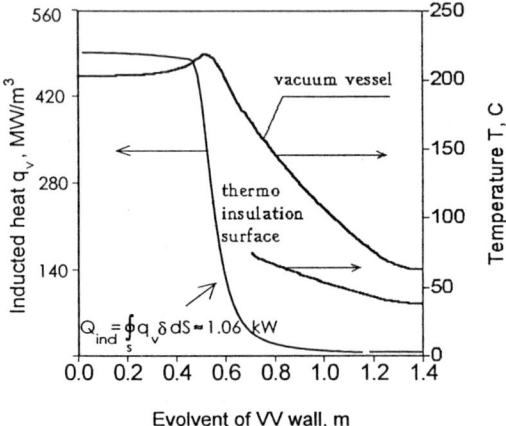

Fig.2. Temperature distribution on vacuum vessel.

It is suggested to use combination of induction heating and additional ohmic heaters located on outer vessel parts (under insulation) to decrease such temperature non-uniformity in backing regime. These additional heaters compensate insufficient induction heating on large radiuses.

Single rode cable is used as an additional heater (rode - nichrome, envelope - stainless steel, insulation - quartz sand). It is sited by two branches of this cable on upper and lower domes, by 4 branches on outer ring, and one branch on each four horizontal ports. While determining operating it was necessary to provide small temperature difference ($T \leq 25\text{-}30$ C) in non-stationary mode of temperature increasing.

Walls of dome, outer ring and ports have different width. That is why supplying voltage of step change character was used on separate cable branches to control an additional heaters power.

Graphic of voltage changing in time for dome, outer ring and ports is shown on Fig.3. Additional heaters power graphic is also given here. Temperature distribution corresponding to this graphic is given in Fig.4. It is seen from figure that combination of induction heating and additional

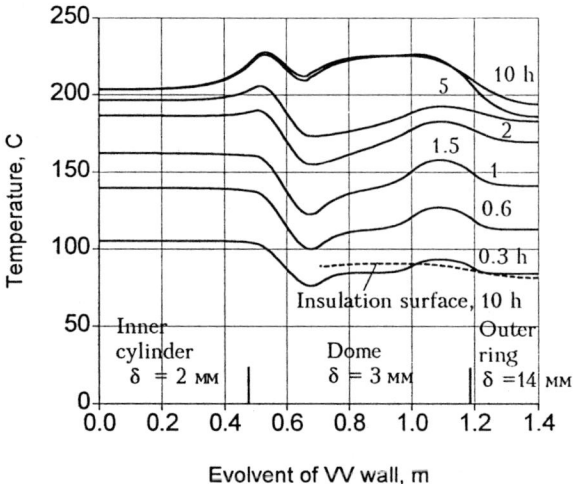

Fig.4. Vacuum vessel induction heating with additional heaters as a time function.

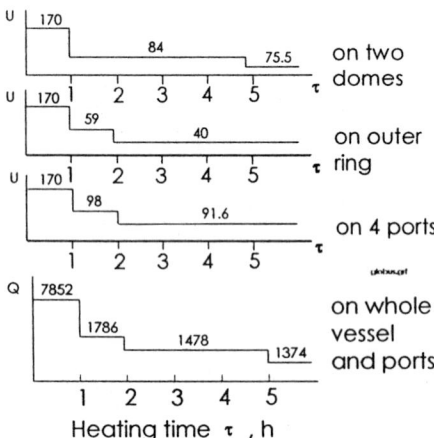

Fig.3. Step controlling of voltage U (V) and power Q (W) for an additional heaters.

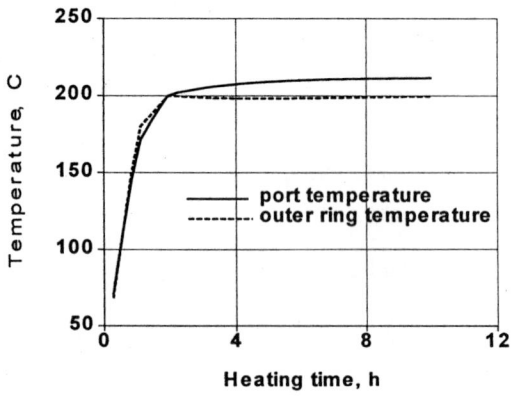

Fig.5. Heating up of connection as time function.

controlled heaters leads to desirable result, i. e. ΔT_{vv} < 25-30 C. Entire heating process (till steady state) is about 5-10 hours. As it is seen, insulation surface temperature does not exceed 90 C. In spit of different width of outer ring and insulation walls in place of VV and ports connection (δ_{ring} = 14 mm, δ_{ins} = 25 mm & δ_{port} = 6 mm, δ_{ins} = 10 mm) this regime of additional power changing provides the equal temp of their heating (see Fig.5).

2.2. Regime of operating pulses regular sequence

Energy release happens on divertor and limiter surfaces while each operating pulse duration. Accumulation of this energy followed by heating of divertor and limiter which are not cooled. Vacuum vessel is also heated by impact of heat radiation from divertor and limiter hot surfaces. Heating process has pulse character. Temperature increasing happens during each pulse, during pause between pulses - partial cooling and partial temperature alignment on surface of these elements.

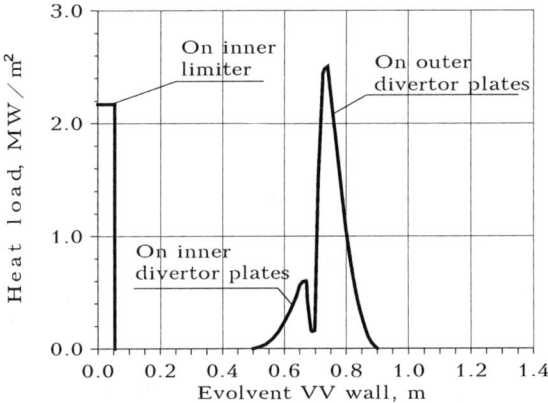

Fig.6 Heat load on in-vessel elements.

Heat load on in-vessel elements during pulse having duration τ_{pulse} = 100 ms is shown on Fig.6. Heat load distribution on co-ordinate corresponds to integral power (by pulse):
on internal divertor plates   (0.08 × 2) MW
on outer divertor plates   (0.32 × 2) MW
on internal limiter
(height L = 100 mm)   0.17 MW
Pause between pulses is suggested τ_{pause} = 600 s. Max pulse number per day - 50.

In Fig.7 temperature state of vessel and in-vessel elements on 50 pulse time moment is given. As it is seen from figure even at natural cooling vessel temperature will not exceed 80 C till the end of operating period. The most hot divertor plates zones will have temperature about 150 C.

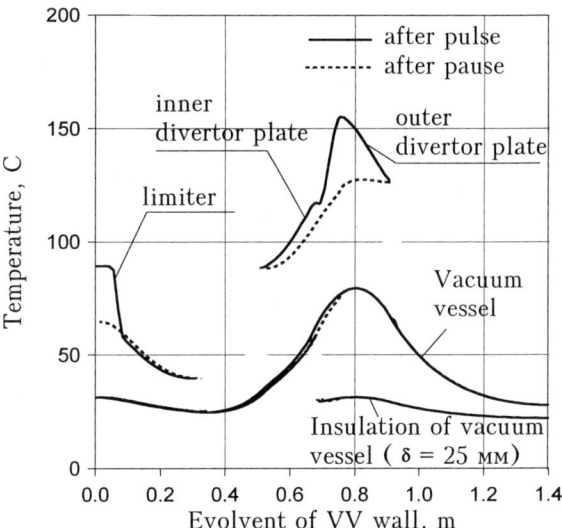

Fig.7 Vessel and in-vessel elements temperature state after 50^{-th} pulse (operating regime).

3. Structural analysis of Globus-M vacuum vessel

Main design loads applied to body of vessel, ports and in-vessel components at normal operating conditions are the following:
- external uniform pressure 0.1 MPa,
- electromagnetic loads (EML) due to plasma disruptions or "halo" currents,
- temperature fields due to overheating at training before plasma operational stage.

The check of leak tightness, strength and buckling is assumed by testing under the external uniform pressure 0.2 MPa at room temperature.

The analysis of electromagnetic loads at central plasma disruption applied on body of vessel and in-vessel components was done for two cases of the in-vessel components (limiter and divertor plates) electrical connection :
- full connection along the torus,
- electrically isolated sections of limiter and divertor plates.

Plasma scenario for determination of electromagnetic loads was the following: total current 0.45 MA decreases linearly with constant speed 2.5 MA/msec till zero. It gives in axially symmetric approach the maximum level of electromagnetic (EM) pressure on outer cylinder 0.03 MPa and on inner cylinder 0.02 MPa that is illustrated on the Fig. 8.

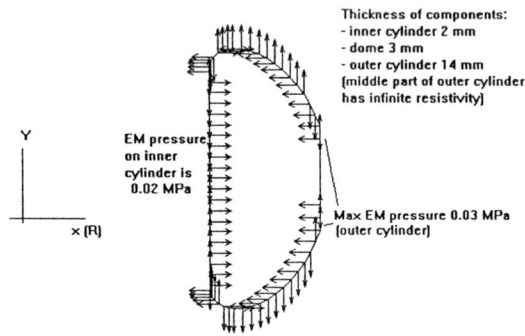

Fig. 8 Distribution of EM loads on Vacuum Vessel of GLOBUS (non-scaled) at time 120 msec after start of disruption (axisymmetric approach)

. Maximum EM pressure on limiter and divertor plates that have electrical insulation along torus is 0.07 and 0.027 MPa respectively in case of 2 electrical breakes of their sections, and 0.03 and 0.016 MPa in case of 8 electrical breakes.

Under these electromagnetic loads, under

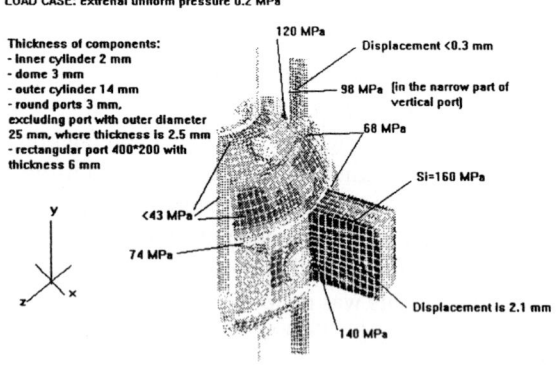

Fig. 9. Stress intensity and displacements in module 1 of GLOBUS Vacuum Vessel at testing by external uniform pressure 0.2 MPa (static elastic approach).

operating pressure 0.1 MPa and under testing external pressure 0.2 MPa the stress analysis and buckling estimations of structural components were carried out with help of 3-D finite elements models.

The results of analyses show that at operating regimes (external pressure 0.1 MPa and EM loads applied simultaneously) the mechanical stresses in vessel body are less in comparison with the case of its testing by external uniform pressure 0.2 MPa at room temperature ; this case is illustrated below on Figs. 9 and 10 for 2 characteristic modules of vacuum vessel body.

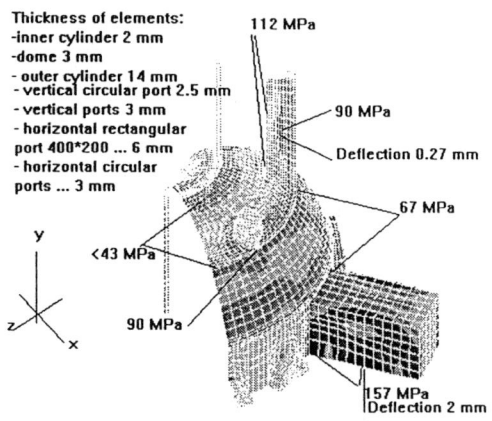

Fig. 10. Stress intensity and displacements in module 2 of GLOBUS Vacuum Vessel at testing by external uniform pressure 0.2 MPa (static elastic approach).

. The level of mechanical stresses in limiter and divertor plates loaded by EM loads strongly depends on their mechanical and electrical insulation along torus, it satisfies to the design criteria (RF Regulations and Norms for atomic energy PNAE-G-7-002-86) in case of 8 electrical breakes or at their full electrical connections.

In principle, the chosen geometric parameters of all construction elements (vessel body, ports, supports, limiter and divertor) mentioned above satisfy to the regulations of static strength in accordance with design criteria, but at the same time the checking of form stability (buckling) of thin sphero-toroidal part of vessel under external uniform pressure 0.1- 0.2 MPa is desirable.

GLOBUS-M Tokamak Magnets

A.B.Alekseev[a], A.F.Arneman[a], V.A.Belyakov[a], S.A.Boulgakov[a], V.A.Bykov[a], V.A.Divavin[a], K.E.Egorov[a], S.A.Grigorev[a], V.K.Gusev[b], A.A.Kavin[a], V.A.Korotkov[a], Y.M.Krivchenkov[a], E.G.Kuzmin[a], A.A.Malkov[a], A.G.Panin[a], N.V.Sakharov[b], V.F.Soikin[a].

[a]STC "SINTEZ" D.V.Efremov Institute, Metallostroy, 189631 St.Petersburg, Russia

[b]A.F.Ioffe Institute, 26, Politekhnicheskaya St., St.Petersburg 194021, Russia

The GLOBUS-M magnet system consists of 16 toroidal field (TF) coils, one central solenoid (CS) coil and 20 poloidal field (PF) coils. The TF coil is a single turn which consists of the bronze straight inner leg and copper outer leg. The inner legs of the central core have the wedged form. The outer turns attached to inner legs with the bolted connections at the bottom and special bandage at the top. The PF coils are fasten to TF coils. The central solenoid (CS) is wound around the inner legs of the toroidal field coils. The magnetic field at the solenoid axis amounts to 8.3 T. The CS incorporates two layers of conductor baked into insulation. The GLOBUS-M vacuum vessel (VV) is designed as all-welded stainless steel construction equipped with ports, manholes and in-vessel components. The basic parameters of the GLOBUS-M magnet system and PF coils and CS and VV are presented. The PF coils fastening and the support and the intercoil structures are described. The circuit of assembly and disassembly of the Globus-M tokamak magnet system is submitted.

1. INTRODUCTION

Globus-M is designed as a spherical tokamak with small aspect ratio plasmas properties [1]. It's construction consists of vacuum vessel, central core, outer parts of toroidal field coils (TF-coils), poloidal field (PF), ohmic heating (OH), compensation (CC) coils, intercoil structures (IS) and supports. The basic parameters of the tokamak are listed in Table1.

Table 1
GLOBUS-M Basic Parameters

Plasma major radius, R (m)	0.36
Plasma minor radius, a (m)	0.24
Aspect ratio, R/a	1.5
Plasma vertical elongation, k	1.5-2.2
Plasma triangularity, δ	~ 0.2
Plasma current, I_p (MA)	0.3
Toroidal magnetic field, B_T (T)	0.5
Pulse length, (sec)	0.3
Required volt-second consumption,(Wb)	~0.3
Weight of tokamak, kg	~ 5000

The vacuum vessel and the magnets will be designed and manufactured during a three year period (April 1995 - March 1998). 3D-view is shown in Fig. 1.

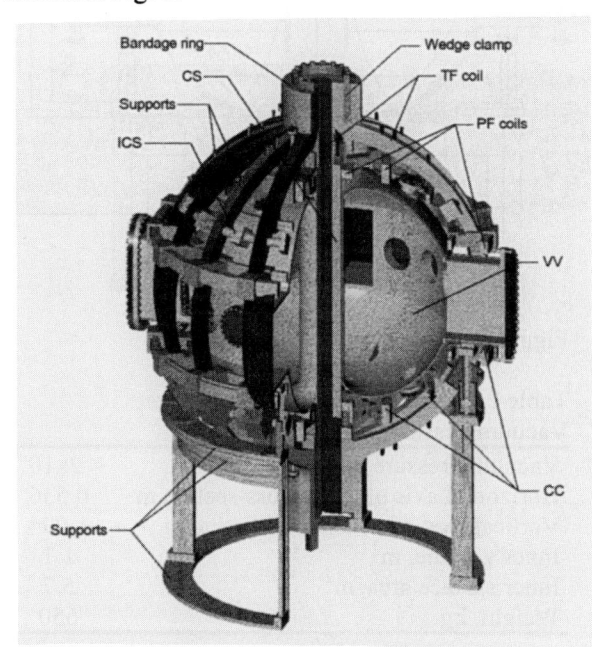

Figure 1. GLOBUS-M 3D-view with cross-section.

2. DESIGN PRINCIPLES

2.1. Vacuum vessel

All welded stainless steel vacuum vessel has a "near-pure compression" shape to increase the vessel stability relatively to atmosphere pressure [2]. Theoretical buckling safety factor is not less than 3. It is supported through the rather thick equatorial plane ring, supplied with ports and manholes for plasma observation, heating and in-vessel components maintenance. 14 mm thickness of the equatorial plane ring provides more favorable conditions for plasma column vertical stability. The other basic parts of the vessel are the inner cylinder (2 mm thick), the upper and lower domes (3 mm thick) and intermediate cups of variable thickness (Fig. 2). Max stress intensity is not greater than 80Mpa under operating loads. The vacuum vessel temperature is 200°C. The vessel parameters are listed in Table 2.

To improve the device performance in future it is possible to provide a 3-fold changing of the inner cylinder with decreasing its diameter every time. This procedure will allow to decrease the tokamak aspect ratio without significant changes in the basic vacuum vessel construction.

2.2. Toroidal field coils

Toroidal field coils (TF) generate a toroidal magnetic field inside plasma. There are 16 one-turn TF-coils. Each coil consists of the straight inner leg constructed of bronze and the outer turn constructed of copper. The coils are connected in series through the cross-overs at the bottom. The number of coils, sixteen, provides the necessary space for the vessel ports disposition. The reverse turn is located on the both sides of the cross-overs for better compensation of stray fields. The main parameters of the TF-coils are listed in Table 3.

Table 3.
TF-coils main parameters

Number of coils	16
Number of turns per coil	1
Axial toroidal field at R=0.36m, T	0.5
Coil current, kA	54.7
Field ripple at R=0.36 m, %	1.6×10^{-4}
Field ripple at R=0.6 m, %	0.785

The inner legs are provided with copper tubes for water cooling and have a wedged form. Inner legs will form after assembly a central core and are pinned from each other. From above on the central core is winded a bandage by thickness 2mm (Fig.3).

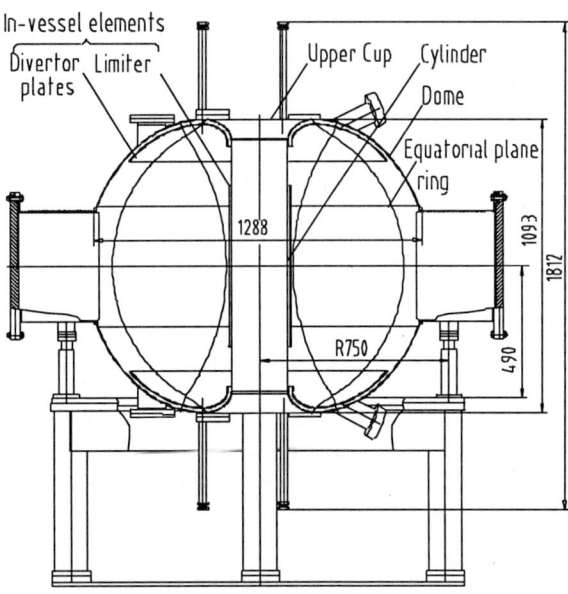

Figure 2. Vacuum vessel basic parts.

Table 2
Vacuum vessel main Parameters

Vacuum pressure, Pa	$< 2 \times 10^{-6}$
Horizontal axis of torus cross-section, m	0.536
Vertical axis of torus cross-section, m	1.094
Inner volume, m^3	1.1
Inner surface area, m^2	5.7
Weight, kg	650

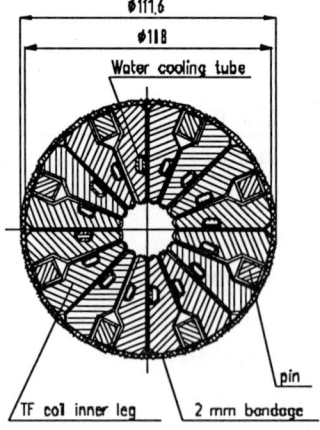

Figure 3. Central core cross-section.

The outer turns attached to inner legs with the special bandage ring at the top (Fig. 4). Electric contacts are provided by horizontal contact straps and by a bandage ring, which press the outer turn to the inner leg.

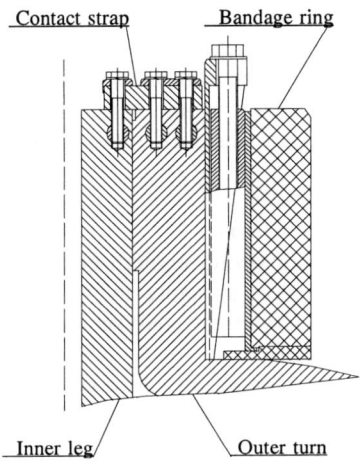

Figure 4. TF coil top contact joint.

At the bottom the outer turns attached to inner legs by the bolted connections (Fig. 5).

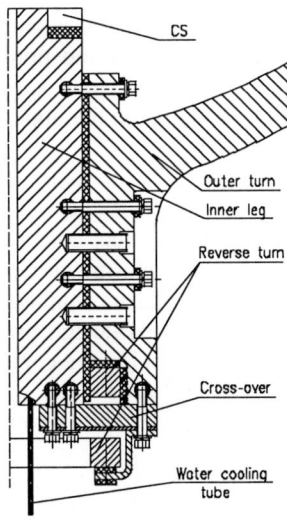

Figure 5. TF coil bottom contact joint.

For the convenience of the machine assembly the outer turn is made of two parts connected below the equatorial plane. To provide the necessary stiffness against TF coils toppling the outer turns are embraced with the hoop closed intercoil structure (IS) below and above an equatorial plane. The outer turns of the TF coils are the part of the tokamak support structure. They are used to support the poloidal field coils and should be stiff enough to withstand in-plane and out-of-plane forces. The overturning loads are taken up by the mutual action of the outer turns and intercoil structures. The part of the overturning moment is resisted by the wedged central core. The global behaviour of the magnet system has been analysed by using the 3D finite element shell-beam model. In this model the outer turn of the TF coil, PF coils and intercoil structure are modelled using beam elements with the equivalent stiffness properties. The in-plane and out-of-plane loads acting on the TF coil and forces acting on the PF coils have been considered for normal and abnormal conditions. The thermal gradients developed under working conditions have been also applied to the model. 2D FE models have been used for detailed analysis of the connections between outer turns and inner legs of the TF coil and other structures. The analysis shows that the maximum out-of-plane displacement of the outer turn is 1.15mm and maximum stress intensity in the turn is 56 Mpa. The maximum tensile stress in the bolt is equal to 96 Mpa with insulation compression being 60 Mpa. The stress intensity in the inner leg is 290Mpa and shear stress in the pins of the connection does not exceed 30 Mpa. The out-of-plane torsion analysis of the central core formed by the inner legs gives the allowable spacing of the pins of 30 mm.

2.3. Central solenoid

The most critical component of GLOBUS-M electromagnetic system is the central solenoid [3]. The basic solenoid parameters are listed in Table 4.

Table 4
Central solenoid main parameters

Number of layers	2
Number of turns per layer	62
Conductor cross section, mm	20×20
Water cooling hole diameter, mm	6
Inner diameter, mm	112
Outer diameter, mm	200
Length of the conductor, m	~ 66
Current in the conductor, kA	± 70
Axial magnetic field, T	8.3
Magnetic flux (double swing), Wb	0.31
Design number of cycles	8×10^4

The central solenoid (CS) coil is located in between 112 mm diameter TF coils central core and the 218 mm diameter bore of the vacuum vessel (Fig. 6).

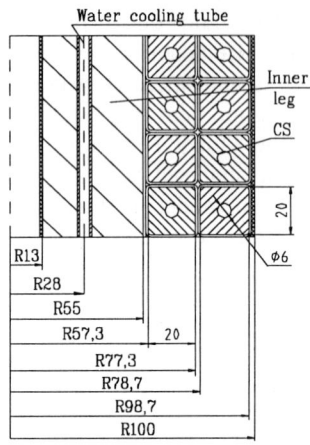

Figure 6. Cross-section of the central solenoid.

A space restriction requires a high accuracy of the CS conductor winding. High stresses arising in the solenoid under cyclic radial and vertical loads (14.6MN and 0.43MN respectively) require high strength continuous conductor of 66 meters total length for the solenoid manufacturing. After careful investigation of candidate materials, accompanied by static and cyclic mechanical tests, the conductor from CuAg alloy of HK01 type manufactured by OUTOKUMPU OY in Pori, Finland was selected (yield strength 270 MPa, ultimate tensile strength 330 MPa). The conductor will be manufactured with initial specially determined trapezoidal cross-section (Fig. 7).

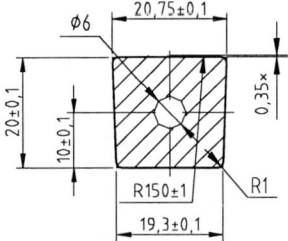

Figure 7. The solenoid conductor cross-section.

2.4. Poloidal field coils.

Three pairs of poloidal field coils (PF1-PF3) are used for plasma shaping and plasma position monitoring. Three pairs of additional coils are used for vertical and horizontal control (VFC, HFC). To simplify plasma breakdown and plasma shaping the compensation coils (CC) together with PF3 coil are used to compensate the central solenoid (CS) stray field. The coils are fixed on the central core and the outer turns of the TF coils. PF1 coils are arranged in 8 stainless steel clamps. The clamps are bolted to the TF coils inner legs and outer turns. Electromagnetic loads acting on the coils are resisted by the clamps. For this reason each joint is supplied with a cylindrical pin, which withstand vertical forces. The other coils are fixed in the stainless steel bearing casings, which are fastened on the TF coils outer turns.

2.5. Support structure.

TF coils are strong enough and do not require any special support structure to withstand in-plane forces. Preliminary stress analysis has shown that to withstand out-of-plane loads TF coils must be provided with intercoil structure (IS). IS fasten the outer turns together above and below the equatorial plane. In the place of IS connection each TF coil is embraced with stainless steel clip. The hollow rectangular rods with flanges on the ends are welded in situ to the clips. The in situ welding allows to compensate inaccuracies of assembly and elements manufacturing. The tokamak magnets total weight is about 5 tons. For devise installation and assembly-disassembly convenience is used a support (Fig. 1). The support consist of 4 legs and upper and lower rings. The outer turns and four supports of the vacuum vessel are fastened on the upper ring.

3. SUMMARY

GLOBUS-M tokamak magnets construction allows to use routine methods of plasma current generation, plasma equilibrium and plasma shaping control. The vacuum vessel and magnets design provide a good access to plasma and vessel inner components. The tokamak construction does not require any sophisticated technology of manufacturing.

REFERENCES

[1] V.A.Belyakov, V.A.Divavin, N.Ya.Dvorkin, et al., Ioffe Inst. Preprint 1629 (1994).
[2] V.E.Golant, et al., in Controlled Fusion and Plasma Physics (Proc.23rd Eur. Conf. Kiev, 1996), a071, European Physical Society, Geneva.
[3] V.K.Gusev, V.I.Nikolaev, K.A.Podushnikova, et al., IEEE/NPSS 16th Symposium Fusion Technology, Vol. 2, pp. 1460-1463 (1996).

Joint Upgrated Spherical Tokamak (JUST) and support Installation - 2 MA Spherical Tokamak "Selena": Concept and Status

E.A.Azizov, N.Ya.Dvorkin[4], O.G.Filatov[2], G.P.Gardymov[4], I.S.Garypov[4], V.E.Golant[1], V.A.Glukhikh[2], V.I.Iogansen[5], V.A.Iagnov, I.A.Kadi-Ogly[5], R.R.Khayrutdinov, V.V.Korshakov[2], V.K.Krylov[2], I.N.Leykin[4], V.E.Lukash, A.B.Mineev[2], G.E.Notkin[3], À.R.Polevoy[3], K.G.Shakhovets[1], S.V.Tsaun[3], E.P.Velikhov[3], N.I.Vinogradov[1], G.M.Vorobiev[2]

Troitsk Institute for Innovation's and Thermonuclear Researches (TRINITI) ,142092 , Troitsk.
  [1] Ioffe Physical-Technical Institute
  [2] Efremov Institute of Electrophysical Apparatus
  [3] RNC "Kurchatov Institute"
  [4] State Enterprise "Leningradsky Severny Zavod"
  [5] Sci. Res. Inst., "Electrosila" Enterprise
  Russia

1. Stages of JUST concept development

The development of spherical tokamak project JUST (fig.1) came through several stages. Each of them allowed to define key points of project, leading to the changes of the parameters. The brief description of this stages is presented below.
Their main parameters are shown in Table 1.
I. On the first stage JUST was in many details by the continuation of START line. The analysis showed that Ip=6-8 MA is the minimum level for burn investigations.
The design was acceptable in the whole, but there was difficulties with plasma current scenario based on central inductor magnetic flux. The flux reserve was obtained with external poloidal field coils assist. This possibilities were very promising and they define much of further development of the JUST concept.
II. The second stage was devoted to the possibilities of flux and plasma current maximization by external poloidal field coils assist. Moreover this stage allow to formulate the main goals and tasks of JUST: burn investigation; ways of the divertor problem solving working out; plasma parameters limits study; some reactor problems (tritium system, blanket modules, current drive systems) investigations.

In the result, the advanced plasma current scenario was developed. It allowed to increase the plasma current from 6-8 up to 15-20 MA in the same geometry, to achieve the project values of Q > 3-5, β_T up to ~ 40 %, average value of the neutron flux on the first wall δ_n ~ 0.4 ÌW/m^2. Part of central inductor to the volt-second balance was ψ ~ 15-20 % ; the rate of the current rise was about 10 MA/s.

Detailed investigation of this concept showed, that plasma scenario development run into difficulties with vertical stability of plasma position and with MHD plasma stability at the early stages of plasma scenario.

This difficulties forced to decrease the part of the external poloidal field coil flux down to 50 % and to pay special attention to the central electrical core. The arising difficulties was decreased by R increasing.

In the whole this stage allow to define the central inductor input ψ and the plasma current operating limits.

III. On the third stage value ψ was taken at the level 50-60 %. To decrease problems with central kern increasing of the major radius and the aspect ratio (R=1.3-->1.7 m, A=1.4-->1.5) and some decreasing of the plasma

Table 1.

	JUST	"Selena"
Major radius, m	1.8	0.72
Minor plasma radius, m	1.2	0.48
Aspect ratio	1.5	1.5
Toroidal magnetic field on axis, T	2.1	1
Plasma current, MA	10-14	2
Plasma elongation, k_{95}	2.3	2 - 3
Auxiliary heating power, MW	20	~ 4
Fusion factor, Q	2.6 (d-t)	0 (d-d)

elongation (k_{95}=3-->2.5-2.6) was considered to be reasonable. Plasma scenario with I_P = 14 MA was worked out.

During this stage the design of the concept have been developed. The usage of newest elaborations and technologies developed by Russian industry. Namely the experience of "Electrosila" Enterprise in 1000 MW generators design and building was used for constructions with current density at the level 80 - 100 A/mm^2.

The questions related to III stage are presented in details in the first preprint on JUST /1/.

For the purposes of plasma behaviour investigations in low aspect tokamaks, ASTRA code /2/ was modified. It was checked on OH regimes simulations for START tokamak and show the good agreement with these experiments /3/ and also with experiments on the modern large tokamaks /4/. So its application to JUST plasma simulations was reasonable. Calculations show the possibility to achieve Q values at the level 0.8-1 within time interval 1-2 s. After that the density profile became more peaked and the plasma energy confinement time degraded. It was detected that Q factor has only a weak dependence on plasma current I_P (for I_P = 8 --> 14 MA, Q = 0.8 --> 1). The possible explanation is in essential role of ion neoclassical transport in the energy balance.

IV. On the fourth stage the main attention was given to the opportunies of the high Q values reaching, on the level 2 - 3 and more. Under the parametrical analysis on 0D models, it was considered to be reasonable not to exceed the limits of enhancement confinement factor H = 2, for the number of widely used energy confinement time extrapolations, namely ITER-P-89 and Kaye-Goldston scalings.

To simplify the problem with JUST physical programme realisation it was considered as reasonable to facilitate several limits. Namely - for the plasma elongation k_{95} < 2.3 (it was earlier 2.5); for q_{95} >4 (was 3-3.3); for normalized plasma beta β_N < 2.5 (earlier the higher values were allowed). Finally for the aim to increase Q it is desirable to rise the level of neutron flux up to the level p_n = 0.3 - 0.4 MW/m^2 (was 0.2). It must lead to the growing of the toroidal magnetic field and to the decreasing of several problems with plasma forming on the earlier stages of the discharge scenario.

2. Results of parametrical analysis

The following set of starting parameters was taked for the parametrical analysis: R = 1.8 m; A = 1.5, a = 1.2 m, k_{95} = 2.3, q_{95} = 4.

For fixed value of neutron flux p_n, toroidal field B_{To} and power consumption P_{sum} increases when normalized beta decreases. But concurrently the necessary value of enhancement confinement factor H decreases. In the result of P_{sum} - H competition leads to the normalized beta reasonable values $\beta_N \approx$ 0.02 - 0.025. So very good normalized beta values are not necessary for given plasma confinement. This situation improve MHD plasma stability for the not very high values of internal plasma inductance (l_i = 0.5-0.7). In this case $H_{(P-89)} \approx$ 2.32 - 2.63; $H_{(KG)} \approx$ 1.48 - 1.69. If the plasma confinement will be better, the higher values of normalized beta will be preferable.

The main results of parametrical analysis are the following:

- too large values of normalised plasma beta are not demanded at used scalings of the plasma confinement H ≤ 2. That is necessary from the point of view of MHD plasma stability. At the same time if the plasma confinement will be better, it becomes profitable to come to the higher (β_N ≥ 2-2.5) values of normalized beta;
- to facilitate the problem with the heat removal from the central core it is desirable to increase major radius up to the R = 1.8 m;
- the achievement of values Q = 2 - 3 is possible at the level of auxiliary plasma heating P_{AUX} = 20-25 MW and at B_{to} ≈ 2 T.

3.1. Electromagnetic system.

The most complex element of JUST device electromagnetic system (EMS) is a central core (CC), consisting from an ohmic heating solenoid (OHS) and a toroidal field coils center parts (TFCP), and distinguished extremely to high level electromagetic and thermal stresses. It should note, that the EMS parameters were incorporated with some reserve concerning those, which requires under the scenario, namely:

- the toroidal field on radius 1.8 m was accepted 2.6 T; (ΣI_{TF} = 23.4 MA-turns). flux from OHS was accepted 9 volt - seconds.

3.2. Ohmic heating solenoid proposal design.

The ohmic heating solenoid should consist from two coaxial coils (internal and external one), divided to cylindrical insulating layer. A power supply feeder is assumed to execute in a zone of the bottom end face of OHS.

Each coil is reeled up from three layers of copper coductors with two ring backlashes formation between them for passage of a cooling water. In backlashes a dielectric distance bars are established.

Input and outpur of a cooling water is executed through a ring cover at the top and bottom end faces of a solenoid. In covers extended ring channels, transient in a "sheaf" of radial oriented pipes are executed, which extends outside through intervals between a toroidal turns external parts. As a bearing structure element for elastic axial covers support, a mentioned dielectric cylinder may be used.

At OHS turn cross-section height of 75 mm, the current in turn will make 250 kA, the OHS power supply voltage - about 1kV, a turn-to-turn voltage - 5V. At insulation between turns thickness of 4.5 mm and specific resistance of distilled water ≥100 kOm.sm, current of outflow between next turns not exceed 2.1 µA.

3.3. Poloidal system element design.

From all poloidal system coils (PF), the coils of group PF-2 are most difficult to realize, because they are located in the limited space near central core, and have maximum current - 6 MA-turns. So expediently a design of these coils and their cooling system to accept such as well as in a ohmic heating solenoid. The main distinctions between them will consist only in following: - the smaller density of a current permit to lower speed of a flow;

- absence of an easy approach to the bottom end face of coils compels to dispose as input as output of a cooling water with the party of its upper end face, that, in turn, results an undesirable, from the point of view of hydraulic, local pressure losses when water flow on 180° turnning.

But negative influence of this local pressure losses not very essential, in view of smaller speed of a water flow.

3.4. Vacuum chamber.

Preliminary calculations of stresses, acting on vacuum chamber as a result of plasma current disruptures, has shown, that they can be equivalent to an excess pressure of the order 0.5 MPa, and the most loaded element of vacuum chamber there will be it central part. It was earlier assumed /1/ to execute it in a kind of a cylinder from a steel sheet, thickness of 5 mm. The new variant of a central part design provides manufacturing of this cylinder in a kind of a double thin-walled steel shell with.

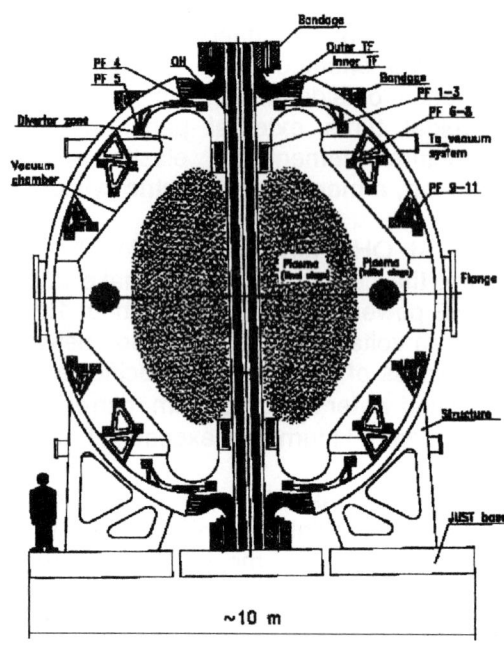

Figure 1. Tokamak JUST.

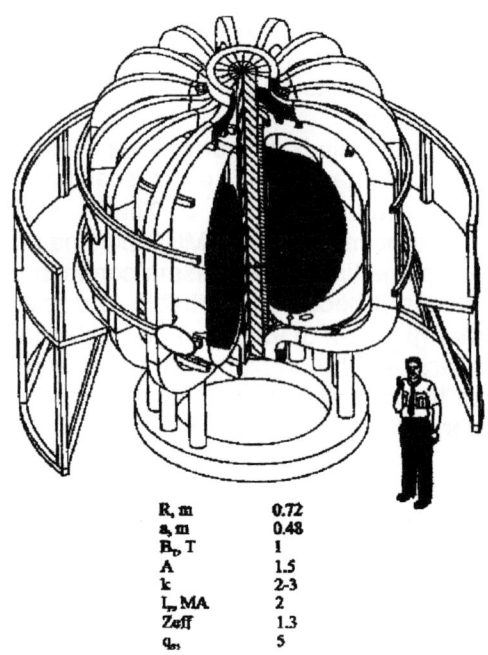

Figure 2. Tokamak SELENA.

R, m	0.72
a, m	0.48
B_t, T	1
A	1.5
k	2-3
I_p, MA	2
Z_{eff}	1.3
q_{95}	5

inner corrugated construction. Such composite shell at general thickness of 10 mm can maintain excess pressure up to 1... 1.2 MPa

Thus the electrical toroidal resistance of vacuum vessel central part will be almost in three times smaller, than in case of 5-mm solid steel wall. It should also note, that the manufacturing of such design can be based on domestic high technologies of combustion chambers and nozzles of large jet engines manufacturing.

4. Concept of 2 MA Spherical tokamak "SELENA"

Many non-traditional decisions and approaches to the discharge scenarios were proposed under JUST project design.

Note some of them:
- ECR assisted breakdown and plasma start-up at high value of stray magnetic field;
- the very high plasma current rate ~ 10 MA/s, usage of plasma flux "lowering" principle;
- DN plasma configuration during the plasma scenario up to plasma elongation $k_{95} \approx 2.3$;
- OH solenoid and the inner part of TF coil with the current density 100 A/mm^2 within 10-15 sec.

It is possible to check these decisions in the installation of sufficiently smaller size. We propose, as the example of such kind tokamak, the 2 MA installation with the aspect ratio $A \approx 1.5$ and the plasma elongation $k \geq 2$. Some parameters of this installation (named "SELENA"-fig.2) are presented in the Table 1.

REFERENCES

/1/ JUST: Joint Upgrated Spherical Tokamak, Joint Team: 1995, Preprint P-0941

/2/ G.V.Pereverzev et al., "ASTRA, An Automatical System for Transport Analysis in a Tokamak", Preprint IAE 5358/6, IPP 5/42, 1991

/3/ J.W.Connor, G.Counsell, M.Gryaznevich et al. "Transport Modelling in Tight Aspect Ratio:", 22 EPS Conf. on Contr. Fusion and Plasma Physics, Bournemouth, 1995, v.19C, Part II, p.205

APPLICATION OF SEISMIC ISOLATION FOR ITER

D. Dilling, C. E. Ahlfeld, P. Barabaschi, V. Chuyanov, K. Ishimoto, E. Tanaka

ITER EDA San Diego JWS
11025 North Torrey Pines Rd
La Jolla CA92037- USA

In this paper we report on trade study results relating to design features needed to accommodate seismic isolation in the ITER design and on selected features of a proposed seismic isolation system. ITER will implement the layout changes identified in this study, but seismic isolation will only be applied if detailed dynamic analysis based on the EDA design and selected site conditions indicate that it is essential to protect the viability of the project.

INTRODUCTION

During 1995, the Special Review Group on ITER Site Requirements and Site Design Assumptions requested a cost sensitivity study of the design, assuming a higher level of seismicity. The JCT performed a study [1] and selected a set of design revisions which would be needed for a site where the peak ground acceleration was 0.4 g, instead of 0.2 g. This study was influenced by preliminary calculations showing that the ITER magnet and vacuum vessel supports could be inadequate to prevent damaging contact between these components during seismic events [2]. Because of the inherent limitations of the support system and the effect that the supports could have on heat transfer to the magnets, it was concluded that the most economic way to protect the tokamak against higher seismic accelerations would be to provide it with horizontal seismic isolation.

SEISMIC ISOLATION CONCEPT

Seismic isolation is accomplished by placing flexing or sliding bearings in the load path of the structure or component to be isolated. A large body of experience exists with the use of this concept in buildings, bridges, and power stations. Many designs for isolation bearings have been developed, but the most widely used consists of bonded, alternating layers of steel plate and synthetic rubber. Bearings of this type (Figure 1) are typically 1 m² in area, and can carry up to 1000 tonnes. They are very stiff vertically, but allow large horizontal movement. In some applications, the entire building or complex has been isolated, while in other applications, delicate parts of the system are isolated and more robust parts remain founded directly on the earth. Where a joint between the isolated and non-isolated parts of the system is located, space must be allowed for relative movement. This space is referred to as the seismic gap. A key aspect of the Trade Study was to identify the optimal location for the seismic gap.

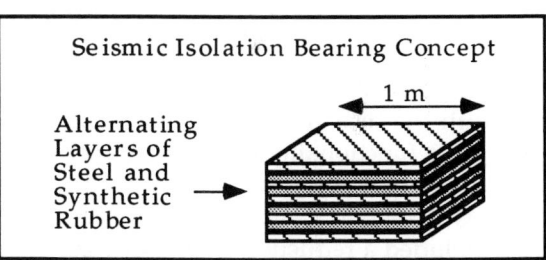

Figure 1

ALTERNATIVES EVALUATED

The starting point for the Trade Study was the tokamak building arrangement reported in the Interim Design Report (IDR). In this design, the tokamak pit diameter was 67 m, tokamak horizontal centerline at -25 m, and heat transfer vaults were located above grade in the tokamak hall. Various alternative locations for the seismic bearings and seismic gap were identified and compared. These alternatives included the following:

Alternate 1: Seismic bearings located inside the cryostat, in the tokamak support legs. This choice was rejected because it required large relative movement between the cryostat and the vacuum vessel ports, at all cooling pipes, and magnet busbar connections.

Alternate 2: Seismic bearings located outside the cryostat, supporting the tokamak and the cryostat. This choice was rejected because it required large relative movement between the cryostat and equipment mounted outside the bioshield, such as neutral beams, pellet injector, and remote handling equipment.

Alternate 3: Seismic bearings located where they would include the bioshield in the isolated structure. This choice was rejected because it still exhibited the need for large motion in all the piping and electrical connections.

Alternate 4: Seismic bearings were located in all the vertical loads for the tokamak, bioshield, and spaces between the bioshield and pit walls. To accomplish this structurally, an intermediate basemat slab was introduced, and a secondary vertical wall was introduced at the edge of the pit. This alternative appeared promising, however the circular shape of the pit did not extend completely beneath the heat transfer vaults. In order for the heat transfer vaults to provide a confinement function, it was essential that they not intersect with the seismic gap. To overcome the problems with the heat transfer vaults, while preserving the functionality of the port access regions in the pit, the general configuration of the pit was revised. The changes which were made included a reduction in the pit diameter, relocation of heat transfer vaults, and introduction of gallery spaces outside of the pit diameter. With these revisions, this alternative was selected.

Additional alternatives were also reviewed, including extending the seismically isolated structure to include the tritium building, electrical termination building, and tokamak hall superstructure; and extending the isolation concept to include all of the tokamak hall, assembly hall, and laydown hall. These alternatives appeared to be technically valid, but significantly more expensive than Alternate 4.

Figure 2 shows a composite view of the isolation location alternatives superimposed on the IDR building design.

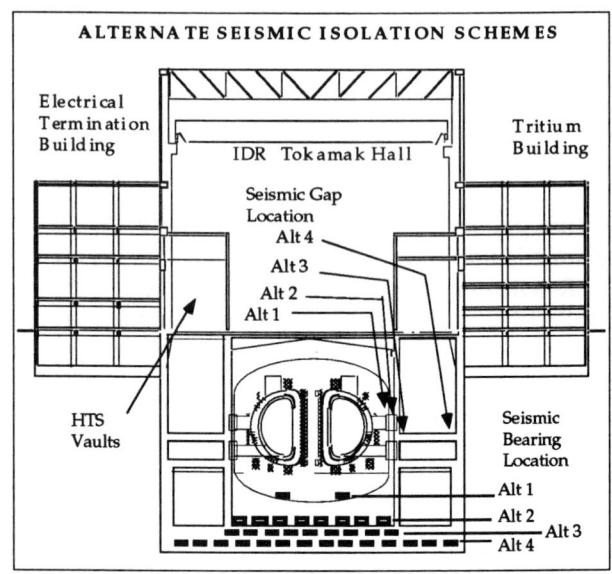

Figure 2

REVISIONS TO THE TOKAMAK PIT

To create a building design which meets all the layout and structural requirements, and which is relatively unaffected by the use of seismic isolation, numerous changes were made. In order to avoid a design which would require primary coolant lines to cross the seismic gap, space was allocated around the top and bottom of the bioshield. Cooling systems which connect to the tokamak through the top ports are re-located to the upper region. Cooling systems connected at the equatorial or divertor ports are moved to the lower region. The two regions, or vaults, are interconnected by ducts so that a single confinement space is created. In the IDR design, the dimension between the bioshield outside diameter and the pit wall inside diameter (13.5 m) was set by the needs of the equipment used to perform remote handling, including clearance to allow it to move azimuthally around the pit. In the revised design, the pit wall inside diameter is reduced to 60 m, providing just 10 m for port access equipment. To provide azimuthal transportation, gallery spaces are provided outside of the pit wall at the equatorial and divertor levels.

The spaces above and below the port access areas are used for the entry of magnet electrical and cryogenic feeds. These areas are made high enough to accommodate the magnet cold terminal boxes and switchgear (about 4 m), and are separated by shielding slabs. To get access to the lower magnet feed area, an additional gallery space is provided below the divertor access level. This lowest gallery also facilitates routing of busbars and cooling pipes to equipment located below the divertor level.

The height of all the spaces and structures in the space between the bioshield and the pit wall control both the elevation of the building basemat, which becomes -44 m, and the center line of the tokamak, which becomes -21 m.

The space which is created below the tokamak

SEISMIC ISOLATION DESIGN

If seismic isolation is required, some additional structural changes will be required. An intermediate basemat slab must be introduced which will support the tokamak, bioshield, and pit wall (inside diameter 60 m). The intermediate slab, in turn, will be supported on the seismic bearings. The thickness of the intermediate slab and the space required to install and maintain the seismic bearings will require lowering the top of the basemat from -44 m to -49 m. The tokamak pit wall will become part of the isolated structure. A new, non-isolated cylindrical wall with an inside diameter of about 66 m. must be introduced. This wall will connect the building basemat to the gallery basemat and gallery slabs, and will support the building superstructure. Figure 4 shows the cross section view of the structure

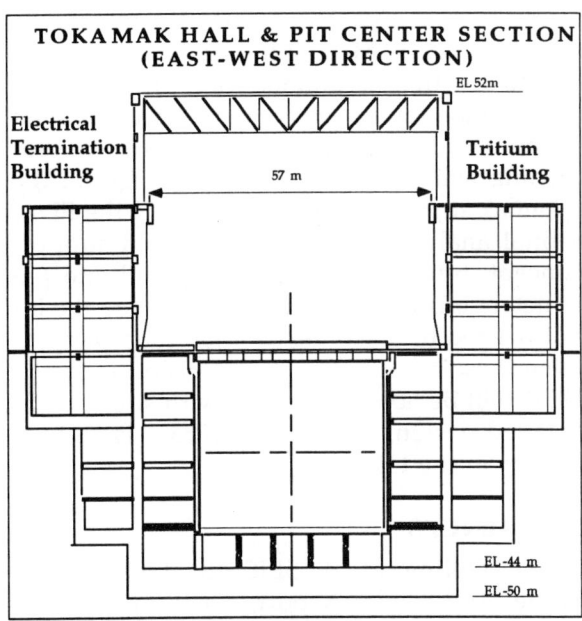

Figure 3

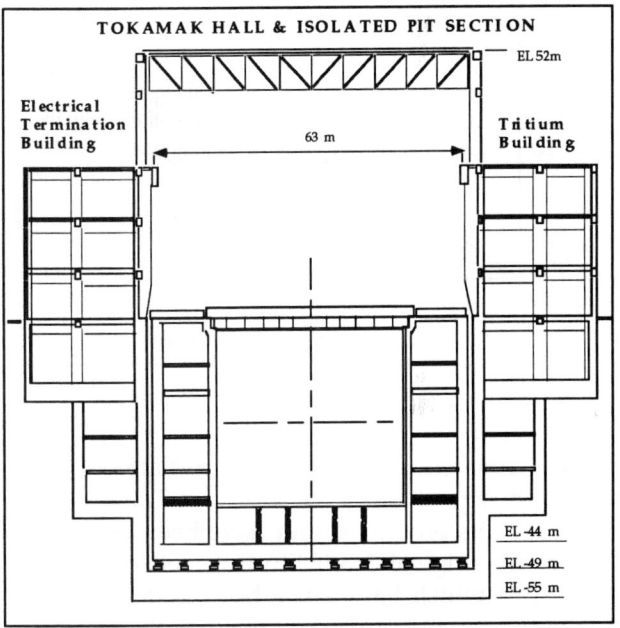

Figure 4

is used for the vacuum vessel pressure suppression tank and the lower heat transfer system drain tanks. It would also be used for in-situ rewinding of replacement coils for poloidal field coils 4, 5, 6, or 7, if needed. Figure 3 shows the cross section of the revised tokamak building design, without seismic isolation.

with seismic isolation.

With this design, the seismic gap will be located outside of the pit areas which contain the tokamak, primary heat transport systems, cold terminal boxes, and remote handling systems. Neutral beams and in-vessel transporters are large enough that they extend part way into the galleries, and these protruding sections must be cantilevered from the isolated section. Piping and service systems which cross the seismic gap must be designed to accommodate the relative motion at the gap, or they will fail. The cost of introducing flexibility must be balanced against the expected cost of repair following a major seismic event.

PERFORMANCE AND COST

The natural frequency of the isolated system and the relative motion at the seismic gap can be influenced by the choice of materials in the seismic bearings, and is virtually independent of the isolated mass. The amount of relative motion is approximately linear with the magnitude of the seismic event. Figure 5 shows the expected movement for the ITER design.

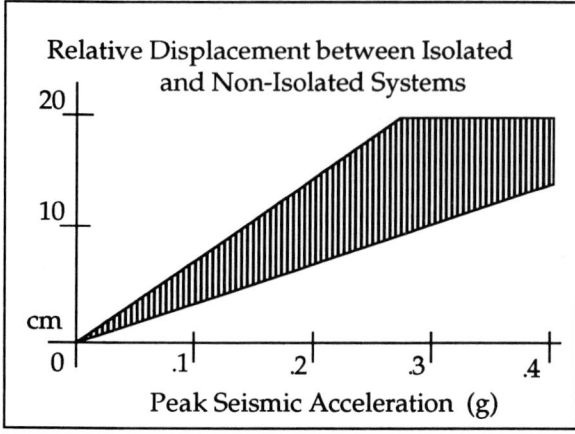

Figure 5

The estimated cost of the structures affected by these design changes are shown in Table 1, below. Units are kIUA, and have been adjusted since the original Trade Study to reflect design features which have since been added. The cost for services which must cross the seismic gap is a conceptual estimate. Not accounted for in this table is the anticipated cost reduction associated with improved layouts (relative to the IDR) of the heat transport system, and value of the reduction in the length of superconducting magnet leads. Such comparisons were outside the scope of the study.

Table 1

Building	IDR Design	Revised Layout	Revised, Isolated
Tok. Hall & Pit	363	386	431
Tritium Building	44	48	52
Elect/Term. Bldg	35	41	49
Subtotal	442	475	532
Gap Crossing	0	0	80
Total	**442**	**475**	**612**

CONCLUSIONS

The ITER design has been revised so that accommodation of a site with seismicity significantly higher than the assumed value can be accommodated. Most of the layout changes required for this accommodation are improvements in the equipment configuration presented in the IDR. In parallel, design efforts are proceeding which will improve the design for the tokamak support systems. Following the EDA, a final analysis using the completed support concept, building configuration, and site parameters must be performed to provide the input needed to make the decision to implement seismic isolation. If such a decision is made, ITER will be able to proceed with site-specific detailed design, without significant layout revisions in the tokamak pit.

REFERENCES

[1] D. Dilling et al, SEISMIC ISOLATION Trade Study Final Report, S 62RI 1 95-12-21 W1.1, December 21, 1995 (ITER JCT)
[2] P. Barabaschi et al, ITER Seismic Analysis and Structural Design, presented at 19th Symposium on Fusion Technology, Lisbon, 1996.

Calorimetric Measurements of Energy Deposition in Tore Supra

F. Surle, G. Mayaux, J.J. Cordier

Association EURATOM-CEA, Département de Recherches sur la Fusion Contrôlée
Centre d'Etudes de Cadarache, 13108 Saint Paul lez Durance Cedex, France

In the framework of the Tore Supra upgrade, one of the final main purpose of the experimental programme is to inject up to 25 MW during up to 1000 seconds and to investigate associated physics and technology issues. Among the latter, development and use of high heat flux components is a central subject, the investigation of which requires a good knowledge of the energy deposition in all parts of the tokamak. Owing to its enclosure into a cryostat, Tore Supra inner vessel and plasma facing components permit precise calorimetric measurements.

1. DESCRIPTION

Tore Supra is a large superconducting tokamak (major radius : R = 2.4 m, minor radius : a < 0.8 m) routinely operated with a toroidal magnetic field of 4 Teslas on axis.

High power Ion Cyclotron Resonant Heating (ICRH), Lower Hybrid Current Drive (LHCD) and Neutral Beam Injection (NBI) have been used throughout the 1992-1996 experimental campaigns with powers up to 8 MW, 6 MW and 0.5 MW respectively (several second pulses) and long pulse discharges have been produced with up to a 120 second flat top at reduced power Lower Hybrid Current Drive.

Almost all the plasma facing components (6 ergodic divertor modules, first inner and outer walls, 6 heating antennae) are actively cooled by pressurised water flow.

Figure 1: The water cooled plasma facing components

This cooling system is designed to be pressurised at 4 MPa allowing routine operation at a water temperature of 373 or 503 K.

Limiters (bottom or outboard) are half-passively cooled. The water flow through the limiter is not enough to extract the total energy received by the walls.

The other plasma facing components (ex. horizontal, vertical windows) and the vacuum vessel are passively cooled.

The design of refrigeration plant is 540 tons/hour which corresponds to a power exhaust capability of 35 MW.

At present, the water flow of 190 tons/hours allows a power exhaust of 13 MW for a temperature increase of 60°.

2. ENERGY BALANCE

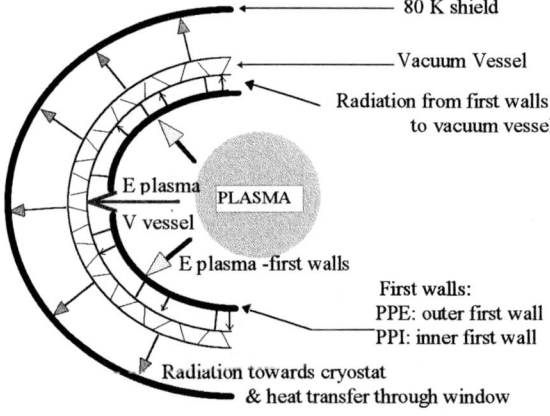

Figure 2 : Heating balance

$$E_{plasma} = E_{plasma \to first\ wall} + E_{plasma \to components} + E_{plasma \to vacuum\ vessel} \quad (1)$$

$E_{plasma \to first\ wall}$: Fraction of the plasma energy transferred to the inner and outer first walls,

$E_{plasma \to components}$: Fraction of the plasma energy transferred to all the components inside the vacuum vessel,

$E_{plasma \to vacuum\ vessel}$: Fraction of the plasma energy transferred to the vacuum vessel.

2.1. Heat transfers with the first walls

$$E_{plasma \to first\ wall} = \Delta H_{first\ wall} + E_{first\ wall \to vacuum\ vessel} + E_{first\ wall \to cooling\ system} \quad (2)$$

$\Delta H_{first\ wall}$: First walls enthalpy variation,
$E_{first\ wall \to vacuum\ vessel}$: Heat transfer from the first walls to the vacuum vessel. The transfers are mainly by radiation because there is ultra vacuum in the vessel,
$E_{first\ wall \to cooling\ system}$: Heat transfer from the first walls to the cooling system by conduction.

Heat transfers between the first walls and the components have been neglected.

2.2. Heat transfers with the plasma facing components

$$E_{plasma \to components} = \Delta H_{components} + E_{components \to cooling\ system} \quad (3)$$

$\Delta H_{components}$: Components (ergodic divertor modules, limiters, heating antennae) enthalpy variation,
$E_{components \to cooling\ system}$: Heat transfer from the components to the cooling system.

Heat transfers between the components and the first walls have been neglected.

2.3. Heat transfers with the vacuum vessel

$$E_{plasma \to v.\ vessel} + E_{first\ wall \to v.\ vessel} = \Delta H_{vessel} + E_{cryogenic\ system} + E_{ports} \quad (4)$$

ΔH_{vessel} : Vacuum vessel enthalpy variation,
$E_{cryogenic\ system}$: Energy radiated towards the cryostat,

E ports : Heat transfer through the ports.

2.4. Global heating balance

It is thought that the whole injected energy is transferred to the plasma.

$E_{injected} = \Delta H_{first\ wall} +$
$\Delta H_{components} +$
$\Delta H_{vessel} +$
$E_{cryogenic\ system} +$
$E_{ports} +$
$E_{cooling\ system}$ (5)

$E_{cooling\ system} = E_{first\ wall\ ->\ cooling\ system} +$
$E_{components\ ->\ cooling\ system}$ (6)

The energy discharged into the cooling system is estimated from the temperature and the flow rate monitoring of the cooling water loops and also from the temperature and the flow measurements of every cooled components including the first walls. The accuracy is around 3 %.

Other energies are calculated from the mean radiated heating surface temperature.

The mid planel or vertical ports are equipped with 21 temperature measurements, the vacuum vessel to 18 and first walls to 13.

Emissivities and heat transfer coefficients are estimated by calibration. Several measurements are obtained with different temperatures and different configurations (with and without cooling system, conserving or not the same temperature on the first walls and on the vacuum vessel....). The accuracy is around 10 %.

3. RESULTS

A study has been carried out on energising shots ($E_{injected}$ > 60 MJ). For all shots studied, the energy source terms are the ohmic and LHCD heating (mainly). Some shots (around 6) are realised with ICRH heating.

3.1. Energy deposition

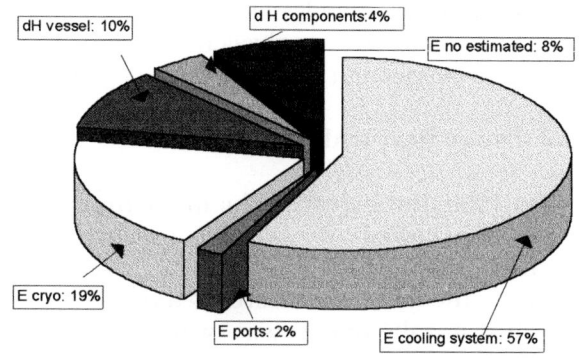

Figure 3 : Energy deposition

Averaging these 35 last energising shots (the configuration of all components is the same for all these shots), we notice that the settled energy represents 92 % of the injected energy with the following deposition : $E_{cooling\ system}$ 57 %, $E_{cryogenic\ system}$ 19 %, E_{ports} 2 %, ΔH_{vessel} 10 %, $\Delta H_{first\ walls} + \Delta H_{components}$ 4 %.

At present, the energy discharged on the components cannot exactly be calculated according to the current measurement.

For shots with other components (the six identical ergodic divertor modules, or the same first wall on the module 2 or ...), the energy discharged into the cooling system is around 60 % of the injected energy.

According to the accuracy, the results obtained for the cooling system are equivalent.

For three energising pulses, the ΔH_{vessel} is estimated from the measurements of the temperature and the pressure of the gas injected into the vacuum vessel double wall. The gas injected is nitrogen, at 293 K, the pressure is 0.150 10^5 Pa.

The stored energy determined from temperature measurements agrees with the one obtained with pressure measurement.

The settled energy into the vacuum vessel can also be assessed by flux measurements and infrared camera measurements. The analysis of the data has not been carried out yet.

3.2. Inner first wall alignment

In 1990, the survey of the inner first wall position showed a peaked excentration located on module number 2 (5.8 mm) towards the magnetic reference axis.

This misalignment was well correlated with the infrared images. Tile damages noticed on the 4 adjacent inner first wall sectors of the module 2 were also well explained by this high excentration value.

After realignment, calorimetric measurements showed that the power deposition on the inner first wall was more homogeneous. The last survey results showed a final inner first wall realignment obtained within an excentration value less than 1 mm.

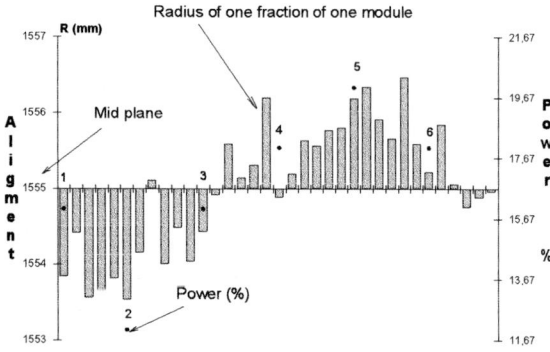

Figure 4 : First wall midplane shape (1996)

In 1996, the survey of the inner first wall showed disparity in the power deposition : 12 % (of power discharged on the inner first wall) on the module 2, 16 % on modules 1 and 3, 18 % on modules 4 and 6 and 20 % on the module 5 the expected average value is 16.6 % on each module).

This power deposition is very well correlated with tile damages noticed on the modules 4, 5 and 6. These results are well explained by the 1 mm excentration towards the plasma for the module 5.

4. CONCLUSION

Averaging, the missed energy corresponds to 8 % of the injected energy, some energies are unable to be calculated according to the current measurements (ex : passive cooling components).

To improve our understanding of the origin of this energy distribution we plan to :

* pursue the estimation of ΔH vessel with the pressure rise of the gas contained in the double walled vacuum vessel ;

* quantify accurately the energy injected into the plasma. At present, the accuracy of the injected power has not been calculated ;

* quantify the energy discharged by the passive cooled components. We plan to increase the radiating surface temperature measurements number on the plasma face components ;

* validate the power deposition with the flux measurements and with infrared camera measurement.

The energy deposition surveyed on the inner first wall is very well correlated with its alignment measured towards the magnetic axis.

REFERENCE

1. J.J. CORDIER, Ph. CHAPPUIS, D. CHATAIN, D. GUILHEM, F. SAMAILLE, Effect of Misalignment and Braze Flaws on the Tore Supra Inner First Wall Behaviour, 17th SOFT, Italy, Rome, 1992.

Application of FIR polarimetry for real-time density control on TEXTOR-94

H. R. Koslowski[a], B. Giesen[a], P. Hüttemann[a], H. T. Lambertz[a], K. P. Pelzer[b], W. Schalt[a], E. Zimmermann[b]

[a]Institut für Plasmaphysik, Forschungszentrum Jülich GmbH, Ass. EURATOM-KFA, D-52425 Jülich, Germany

[b]Zentrallabor für Elektronik, Forschungszentrum Jülich GmbH, Ass. EURATOM-KFA, D-52425 Jülich, Germany

The HCN polarimeter on TEXTOR-94 has been used for feedback control of the electron density and the horizontal plasma position. The poloidal field distribution is assumed to be in accordance with a unique current density profile found in sawtoothing OH discharges [1], thus allowing a determination of the electron density from the measured Faraday rotation angles. The signals are calculated in real-time using a digital signal processor.

1. INTRODUCTION

On the tokamak TEXTOR-94 the electron density and the horizontal plasma position are normally feedback controlled using the nine-channel HCN-interferometer [2]. This approach works for nearly all plasma conditions. A limitation is encountered when the plasma is fuelled by the injection of pellets of frozen hydrogen. The dense ablation cloud can lead to a strong deflection of the laser beams and the rate of change of the line integrated electron density can be too large for the applied real-time phase-detectors [3]. As a result the phase measurement can undergo fringe jumps, yielding a false measurement for the rest of the discharge. The signals from the polarimeter show a disturbance, too, but recover after the perturbation.

The determination of the electron density from the Faraday rotation of a FIR laser beam has been reported for a laboratory plasma [4]. A Faraday rotation measurement along tangential viewing lines is proposed as a density profile measurement on ITER [5], where the required short wavelength makes interferometry very sensitive to vibrations and a robust measurement is required due to the long pulse length [6]. On stellarators, where the magnetic field is usually known, Faraday rotation measurements allow a reconstruction of the electron density profile [7].

2. EXPERIMENT

2.1. Faraday rotation

A FIR laser beam with linear polarization which crosses the plasma column undergoes a rotation of the direction of polarization given by

$$\alpha = c_F \, \lambda^2 \int_L n_e \, \vec{B} \cdot \vec{e_l} \, dl, \qquad (1)$$

where the integral has to be calculated along the line of sight [8] (n_e is the electron density, λ is the laser-wavelength, $\vec{B}$ is the total magnetic field, and $\vec{e_l}$ is the unity vector in direction of the line of sight). If the probing beams are perpendicular to the toroidal field only the poloidal field component contributes and the current density can be measured if the electron density is known. For long wavelength the perpendicular field components leads to a phase shift between the characteristic modes (Cotton-Mouton effect) and allows either by choosing a suitable initial polarization [9] or by an analysis of the beam elliptization the determination of the electron density [10].

2.2. Determination of the electron density

The HCN interferometer/polarimeter [11] and the measuring method of the Faraday rotation angle [12] have been described earlier. The Faraday rotation on TEXTOR-94 is measured along nine vertical chords. In order to utilize this measurement for density control the poloidal field distribution has to be known. Using the assumption

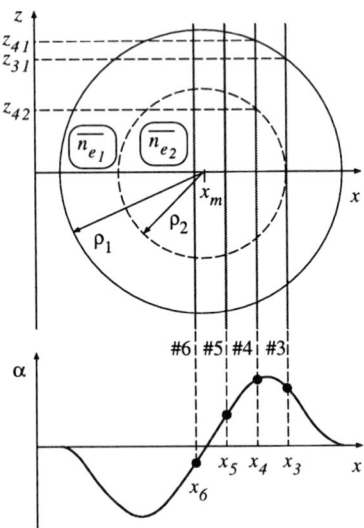

Figure 1. Top: intersection of the probing beams with the plasma column, Bottom: schematic Faraday rotation profile

that the poloidal field is in accordance with a unique current distribution found for sawtoothing quasi-stationary Ohmic discharges [1], and applying a cylindrical approximation for the flux surfaces allows to calculate a signal similar to the line-averaged density from the polarimetric measurements. In a first step the position of the magnetic axis has to approximated. Figure 1 shows the intersection of the probing beams with the plasma column and a schematic Faraday rotation profile. In the cylindrical approximation this position coincides with the zero crossing of the Faraday rotation profile given by

$$x_m = (x_5 - x_6)\frac{\alpha_6}{\alpha_6 - \alpha_5}. \qquad (2)$$

The Faraday rotation angle

$$\alpha(x_i) = 2c_F \lambda^2 \int_0^{z_{max}} n_e(x_i, z) B_{p\parallel}(x_i, z) dz \qquad (3)$$

can be expressed as a summation

$$\alpha(x_i) \approx \sum_{j=1}^{n} \overline{n_{e_j}} w_{i,j} \qquad (4)$$

where the weighting factors are defined as follows

$$w_{ij} = 2c_F \lambda^2 \int_{z_{i,j+1}}^{z_{i,j}} B_{p\parallel} dz. \qquad (5)$$

Taking the probing beams labeled 3 ($x_3 = 30$ cm) and 4 ($x_4 = 20$ cm) for the determination of the line integrated density yields the set of linear equations

$$\begin{aligned} \alpha_3 &= \overline{n_{e_1}} w_{31} \\ \alpha_4 &= \overline{n_{e_2}} w_{42} + \overline{n_{e_1}} w_{41} \end{aligned} \qquad (6)$$

with the solution

$$\begin{aligned} \overline{n_{e_1}} &= w_{31}^{-1} \alpha_3 \\ \overline{n_{e_2}} &= w_{42}^{-1} (\alpha_4 - w_{41} w_{31}^{-1} \alpha_3). \end{aligned} \qquad (7)$$

An electron density signal for feedback purposes can be calculated as a sum of the densities in the ring zones weighted by the intersection lengths

$$\overline{n_e} = l_2 \overline{n_{e_2}} + l_1 \overline{n_{e_1}}, \qquad (8)$$

with the following abbreviations: $l_1 = (a - \varrho_2)/a$, $l_2 = \varrho_2/a$ and $\varrho_2 = x_3 - x_m$.

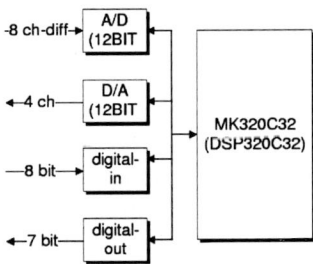

Figure 2. Schematic circuit diagram of the DSP-control board

The sum of the polarimeter signals no. 3 - 5 can be calculated and compared with a lower level

$$S = \begin{cases} 1 & if \quad \alpha_3 + \alpha_4 + \alpha_5 > V_{min} \\ 0 & else \end{cases} \qquad (9)$$

to obtain a signal for the interlock and safety system which indicates the presence of plasma.

2.3. Description of the DSP board

Digital signal processors with floating point unit are well suited to perform the above calculations with a reaction time less than 0.1 ms. A Euro standard based board containing a central DSP module and eight differential A/D–, four D/A–, and several digital I/O–channels has been implemented. The algorithms for calculating the

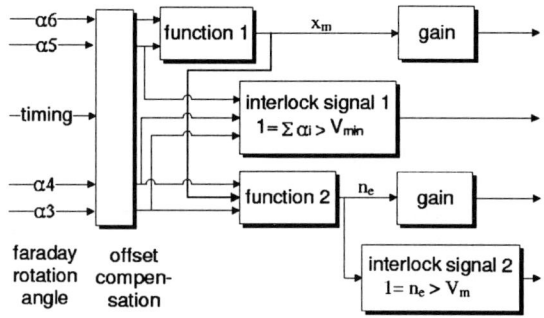

Figure 3. Function diagram of the DSP–program

electron density, the horizontal plasma position, and the interlock signal are programmed in C and down loaded into the DSP memory. The weighting factors (equation 5) were calculated for a fixed plasma current of $I_p = 350$ kA and stored in a look–up table. The offsets of the measured Faraday rotation signals are compensated by software. The memory resident program supports stand–alone and flexible application.

3. RESULTS AND DISCUSSION

3.1. Simulation of the polarimetric density signal

Figure 4 shows the line averaged electron density and the density signal calculated offline from the measured Faraday rotation angles. Four pellets have been injected into this discharge. The interferometric measurement shows a shifted baseline after the end of the discharge indicating some fringe jumps, whereas the signal derived from the polarimeter is robust with respect to those fast density changes (and probably beam losses due to deflection) and returns to the baseline after the discharge. Besides the slightly dif-

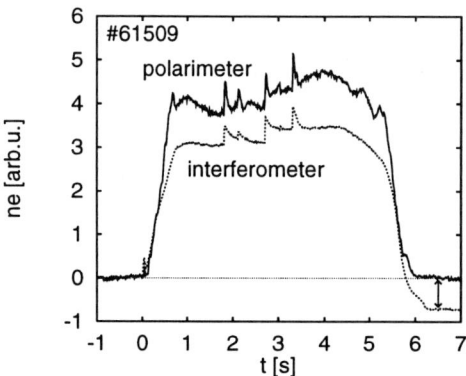

Figure 4. Electron density calculated from the polarimeter signals (full curve) compared to the interferometric measurement at $R_0 = 175$ cm (broken curve).

ferent amplitudes of both signals (what can be corrected by adjusting the amplitude of the polarimetric density signal) the agreement between them is rather good.

3.2. Discharge with polarimetric density control

Figure 5 shows the density traces of a OH standard discharge where the feedback signal for the gas inlet was calculated in real–time from the polarimetric measurements. Both curves show a good agreement. The amplitudes have been adjusted to give the same reading during the flat top phase. Differences are visible during the current

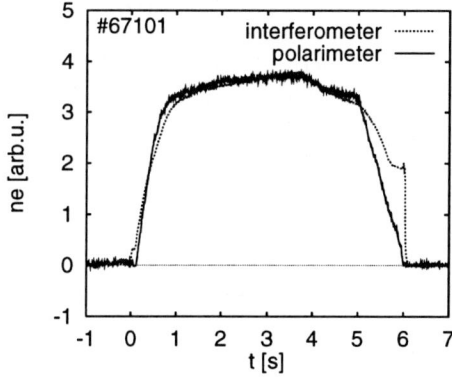

Figure 5. Comparison between the polarimetric and the interferometric electron density signals.

ramp up and current decay phases. One reason for this is that the dependence of the weighting factor (equation 5) on the plasma current is not taken into account in the present version of the DSP-program. This can be corrected by multiplying the weighting factors with $350\ \text{kA}/I_p(t)$, because they depend in first order linear on the plasma current.

3.3. Polarimetric position control

In addition to the polarimetric density control of discharge #67101 we applied the real-time signal of the horizontal plasma position for feedback (figure 6). The plasma position has been shifted

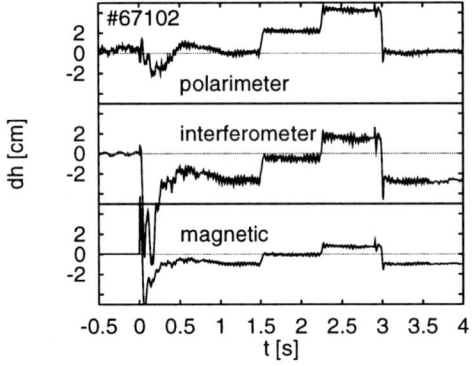

Figure 6. Discharge with polarimetric position feedback and a pre-programmed horizontal displacement.

in two steps by 2 cm and 4 cm, respectively, to the low field side. For comparison the interferometric position signal [2] and the signal derived from the magnetic diagnostic [13] are displayed in the figure. The average values for all three signals differ slightly because the underlying definitions of the plasma position measured by the respective diagnostics are different. The response of the polarimetric position signal (limited by the phase synchronous detection method [12]) is fast enough to allow stable feedback operation.

4. CONCLUSION

We have demonstrated the feedback control of plasma density and horizontal plasma position using the Faraday rotation signals measured with the HCN polarimeter on TEXTOR-94. The vertical arrangement of the probing beams is not ideal for the determination of the electron density and therefore requires the assumption that the poloidal field distribution is in accordance with a standard current profile found for sawtoothing OH discharges. field distribution. The signals obtained in real-time with a specially designed DSP board were successfully applied to the feedback system. Polarimetry offers the possibility of a robust density feedback for long pulse machines like ITER.

REFERENCES

1. H. Soltwisch, in Contributions to High-Temperature Plasma Physics, eds. K. H. Spatschek and J. Uhlenbusch, Akademie-Verlag, Berlin, 1994, p. 471
2. H. Soltwisch, Nucl. Fusion 23 (1983) 1681
3. H. R. Koslowski, Meas. Sci. Technol. 5 (1994) 307
4. A. N. Dellis, W. H. F. Earl, A. Malein and S. Ward, Nature 207 (1965) 56
5. F. C. Jobes and D. K. Mansfield, Rev. Sci. Instrum. 63 (1992) 5154
6. Minutes of the 4th meeting of the ITER physics expert group on diagnostics and technical meeting on diagnostics, Moscow (1996)
7. H. J. Gardner and J. Howard, Plasma Phys. Control. Fusion 36 (1994) 245
8. F. De Marco and S. E. Segre, Plasma Physics 14 (1972) 245
9. A. D. Craig, Plasma Physics 18 (1976) 777
10. S. E. Segre, Plasma Phys. Control. Fusion 38 (1996) 883
11. A. Cosler, E. Kemmereit and H. Soltwisch, Proc. 12th Symp. on Fusion Engineering, Monterey (1987), vol. II, IEEE 87CH2507-2, p. 1488
12. H. Soltwisch, Rev. Sci. Instrum. 57 (1986) 1939
13. B. Kardon, H. Soltwisch and G. Waidmann, KFA report JÜL-2142 (1987)

ENGINEERING ASPECTS OF ITER PLASMA DIAGNOSTIC SYSTEMS

C.I.Walker[1], T.Ando[1], A.Costley[2], L.deKock[1], K.Ebisawa[2], G.Janeschitz[1], L.Johnson[2], V.Mukhovatov[2], G.Vayakis[2], M.Yamada[1], S.Yamamoto[1]

(1) ITER Joint Work Site, Boltzmannstr. 2, 85748 Garching, Germany;
(2) ITER Joint Work Site, 11025 N.Torrey Pines Road, San Diego, Ca 92037, USA

The ITER Diagnostic System will consist of about 40 individual measurement systems to control and evaluate the ignited plasmas[1]. These involve magnetics, neutronics, laser and optical systems, X-Ray to visible spectroscopy, microwaves, probes, pressure gauges, thermocouples etc. Components are located in several different locations: at main ports and in the vacuum vessel, inside and outside the cryostat, and in dedicated areas in nearby buildings. Each area poses its own set of design requirements on the equipment.

1. Diagnostics on ITER

Table 1 lists the proposed diagnostic systems and the principal locations on ITER. Some diagnostics are not required until later in the operational campaign, but it is necessary to construct all of the interfaces and meet the requirements for space and services during the machine construction phase. The set of diagnostics available for first plasma measurements is known as the 'start-up set' and is indicated by an asterix.

Table 1. ITER Proposed Diagnostics and Allocated Locations

ITER WBS 5.5	DIAGNOSTIC SYSTEM	Mid-Port	Vert Port	Divertor Diag Cassette	Divertor Instrum Cassette
A	MAGNETIC DIAGNOSTICS				
A.01	Ex-Blanket Magnetics*	In Back Plate & V.Vessel			
A.02	In-Blanket Magnetics*	In Blanket and Mid-plane Port			
A.03	Divertor Magnetics*			All	
A.04	Rogowski Coils*				
A.05	Diamagnetic Loop*				
B	NEUTRON DIAGNOSTICS				
B.01	Radial Neutron Camera*	16			
B.02	Vertical Neutron Camera*		18		
B.03	In-V Microfission Chambers*				
B.04	Neutron Flux Monitors, ex-v*	many			
B.05	Radial NeutronSpectrometer	16			
B.07	Gamma-Ray Spectrometers	16			
B.08	Activation System, solid sample /fluid	7,19	+5 to assign 3 to assign		
B.09	Lost Alpha Detectors*	15			
B.10	Knock-on Tail Neutr Spectr	Unassigned			
C	OPTICAL/IR SYSTEMS				
C.01	Thomson Scattering, Core*	15			
C.02	Thomson Scattering, Edge	Unassigned			
C.03	Thoms Scattering, X-Point	Unassigned		13	
C.04	Thoms Scattering, Divertor	Unassigned			
C.05	Toroidal Interferometry/ Polarimetry* [‡Retroreflectors	16 ‡19,+4 to assign			
C.06	Polarimetry, for B_p	Unassigned			
C.07	Collective Scattering	Unassigned			
D	BOLOMETRIC SYSTEM				
D.01	Bolometry Array, mainplasma*	19	1,17,19		
D.02	Bolometry, divertor plasma*			3.1	/.2
E	SPECTROSCOPY AND NPA S				
E.01	Active Spectroscopy (DNB)	3			
E.02	H Alpha Spectroscopy*	15	10		
E.03	Impurity Monitors, main pl († Even number Vert Ports)	15	†		
E.04	Impurity Monitor, divertor*			3, 13	
E.05	X-Ray Crystal Spectrometers	7			
E.06	Visible Continuum Array*	15			
E.07	Soft X-Ray Array*	Unassigned			
E.08	Neutral Particle Analyzers	19	4		
E.09	Two Photon Ly-Alpha Fluor	Unassigned			
E.10	Laser Induced Fluorescence	Unassigned			
F	MICROWAVE DIAGNOSTICS				
F.01	ECE Diagnostics, main pl *	19			
F.02	Reflectometers, mainpl*	7	7		
F.03	Reflectometers, pl posit	Unass-	7		
F.04	Reflectometers, divertor pl			18	
F.05	ECA for Divertor Plasma			8	
F.06	Microwave Scattery, main pl				
F.07	Fast Wave Reflectometry	19			
F.08	Microwave Scatter, divertor				
G	PLASMA FACING COMPONENTS AND OPERATIONAL DIAGN				
G.01	IR Cameras (Divertor)* +3 to be assigned	19	8, 13,	8, 18	
G.02	Thermocouples*	All		All	All
G.03	Pressure Gauges*				All
G.04	Residual Gas Analyzers*				
G.05	Hard X-Ray Monitor*	16			
G.06	Visible/IR TV, main pl*	Unass-	2,14,16		
G.07	Langmuir Probes* Tile Shunts*			18, 18.1,	18.2 All
H	DIAGNOSTIC NEUTRAL BEAM	4			

Normal operations will produce fusion burns of more than 1000s, neutron wall loading of 1MW/m^2 and accumulated fluences of 1MWa/m^2. Disruptions with 1.5 GJ of thermal and comparable values of magnetic energy as well as halo currents of up to 8MA may occur. For components inside the vessel, solutions have been developed for the removal of nuclear heat, shielding from radiation damage, protection from large mechanical forces and thermal loads resulting from disruptions. Components on the vacuum boundary must incorporate first safety boundaries and secondary enclosures capable of maintaining integrity under 0.5 MPa over-pressure during an in-vessel loss of coolant accident (LOCA). Remote handling installation is required for all equipment between the plasma and the biological shield. Although these requirements are not always new for Tokamak diagnostics, the engineering aspects have had to be developed for ITER conditions and requirements[2]. Here we demonstrate the solutions under development by considering a few specific systems.

2. Mid-plane Ports

Most of the diagnostic systems are situated in mid-plane ports specifically allocated to diagnostics. An example of an optical diagnostic system here is the Core Thomson Scattering System[3] shown in Figure 1.

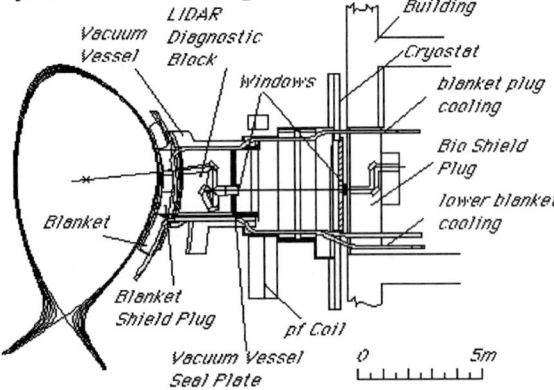

Figure 1. Arrangement of LIDAR System in ITER Mid-Plane Port

The Blanket Shield Plug consists of modified Blanket Modules in a single unit of ~ 10te. that can be pushed into the port. This plug is water cooled by the blanket cooling system with its own pipes routed through the port. Diagnostic viewing holes penetrate this plug.

A water cooled Diagnostic Block incorporates all the equipment, such as shutters and mirrors needed for all the diagnostics at this port. These are installed in sub-plugs that can be maintained remotely in the hot cells. The first active optical element of the LIDAR system is thus the water cooled metallic mirror more than 1.5m from the plasma facing first wall.

The labyrinth shielding arrangement shown here has been demonstrated to attenuate the scattered neutrons adequately when considering neutron heating of the main magnet coils. A current design activity is to assess the effect of integrating all the required diagnostic penetrations at the port.

The Diagnostic Block is rigidly coupled to the Vacuum Vessel Seal Plate to form one, 30te, handleable module closing the primary vacuum. In-vessel equipment is thus removed without venting the Cryostat, reducing maintenance times.

All services to components within the Diagnostic Block are routed to the Vacuum Vessel Seal Plate with remotely handleable connectors on the outside. This typical example of the mid-plane port at sector #15 also contains electrical feedthroughs (for ~200 signals) 7 mechanical and 6 fluid feedthroughs. There are also 37 windows (3 single optical (including the LIDAR), 20 double optical, 12 microwave and two 'windows' for neutron diagnostics).

Windows are common elements in all the access ports. All windows and feedthroughs can withstand 0.5MPa over pressure of a LOCA as well as the normal 0.1 MPa vacuum pressure. The life requirement for these windows is 42,000 hours at operating temperatures between 40°C and 150°C, with 30 baking (to 300°C) and venting cycles. Window sizes range from 25mm to 400mm aperture (120mm for LIDAR); materials include fused silica, sapphire, zinc selenide and possibly diamond. Sealing to stainless steel will be by aluminium diffusion bonding or eutectic brazing. Neutron windows of 1-2mm stainless steel and soft X-ray windows of 0.2mm bonded to thicker aluminium micro-honey-comb will also be used. All windows can withstand the vacuum and accident pressures without detaching. A planned R&D programme aims to establish the viability of the windows and bonds in the irradiation environment.

All connections between the vacuum vessel and the cryostat are sufficiently flexible to accommodate the relative movements of these (up to 75mm) during baking, operation, VDE's, LOCA and earthquake. Diagnostic operation is expected to resume after baking and VDE's, but not after a LOCA. For the LIDAR, the connection is simply optical with a larger window provided on the secondary vacuum seal plate; in some cases self aligning mirror systems are proposed.

Features on the secondary vacuum boundary allow for the docking of the Transport Cask that can contain and remove the secondary vacuum seal plate, the vacuum vessel seal plate (complete with diagnostic block) and the blanket shield plug.

Some diagnostic equipment is located in the pit area; for the LIDAR this comprises gas filled light guides with steerable mirrors. This is either as removable modules (up to 8x3x3 m^3, 80te) or on permanent platforms to the sides and above the access route for the transport cask. In some cases additional shielding, handled in 20teblocks is required in front of the biological shield.

A variation of the general remote maintenance procedure is adopted for the port #7 which is in the NBI vault and obscured by the tangential alignment of the NB injector. Only components unlikely to require maintenance are installed in the port. Components requiring possible access are situated in the NBI enclosure with access from above.

Some diagnostics require an unobstructed straight-through view of the plasma (NPA's, VUV and X-Ray spectroscopy). For these systems the vacuum boundary extends outside the cryostat and biological shield. In this particular position the secondary containment is afforded by the NBI vault. The neutron flux in the NBI vault is considerably higher than elsewhere outside the biological shield. Diagnostic detectors in this region require additional shielding.

3. Vertical Port

Although the equipment in the vertical port varies, a standard hole is provided for the plugs of diagnostic equipment required to be installed in the penetrations.

The 20mm gaps between the blanket modules are used as much as possible to view the plasma.

Designs exist, however, for the modules to be locally customised to give viewing lines of sight (up to 200mm wide) up the port. The blanket back plate is perforated with holes or bridged slots and sometimes reinforced. The blanket cooling pipes and port shielding leave a standard clear 200mm wide aperture along the port centre-line. There are few requirements for complex equipment within the vacuum vessel so this slot is generally filled with shielding. Where diagnostic equipment is required on the boundary (windows) or in the vacuum (mirrors), it is proposed that Diagnostic Plugs are welded to the Port Closure Plate. Conceptual designs exist for the four types of requirement: optical (e.g. Impurity Spectroscopy), neutrons (e.g. Vertical Neutron Camera), straight vacuum extensions (e.g. NPA's) and waveguides (e.g. Reflectometry). The neutron flux from the largest of these diagnostic penetrations (NPA's and Neutron Camera) has been shown to be acceptably low at the PF & TF coils. Blanket coolant pipes use these ports. Water activation radiation has still to be assessed fully but will probably require a 130mm steel boundary to the secondary containment. The flux streaming through the diagnostic penetration is shielded with an

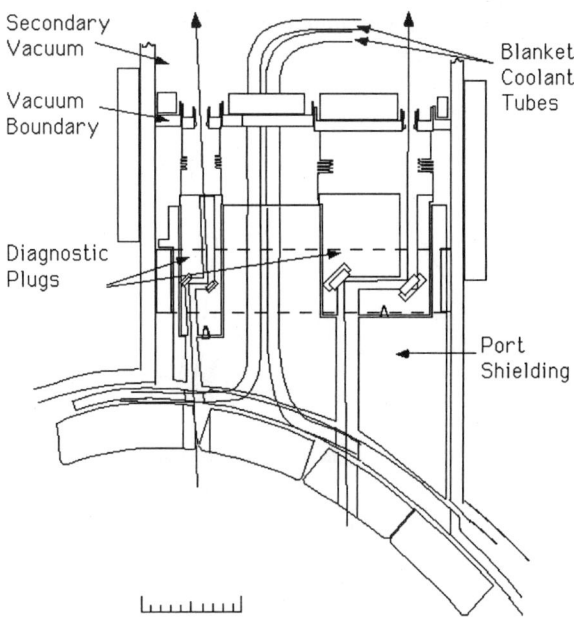

Fig. 2 The functional parts of Plasma Impurity Spectrometer in the Vertical Port.

additional concrete housing around the detectors on top of the biological shield.

Access to the diagnostic plug is not considered to be a routine requirement, although the ability to service the area with standard remote handling tools will be preserved. Activated components, such as the Diagnostic Plugs will be hoisted into contamination control containment flasks which are themselves contained within a shielded Containment Cask. This can be docked on to the Cryostat Flange and transported to the hot cell.

4. Divertor

The central divertor cassettes in the four RH ports are specially adapted for diagnostics. Here maintenance access is simpler and faster. Signal access (optical, microwaveguides, etc.) is straight through a port. These diagnostic cassettes house the non-electrical diagnostics.

Use is made of the gaps, nominally 10mm, between cassettes to avoid penetrating the high heat flux components and target plates. The Divertor Impurity Spectrometer shown here demonstrates the principle for an optical system where the optical channels follow the plane of symmetry through the cassette body giving the mirrors the maximum protection.

A full view of the Target Plates is given by sets of mirrors under the dome and under the body and 30-36mm wide viewing slots. Other systems (e.g.

waveguides) are mounted in grooves in the sides of the cassette avoiding an extra penetration in the target plates. Space is severely limited by the water cooling connections.

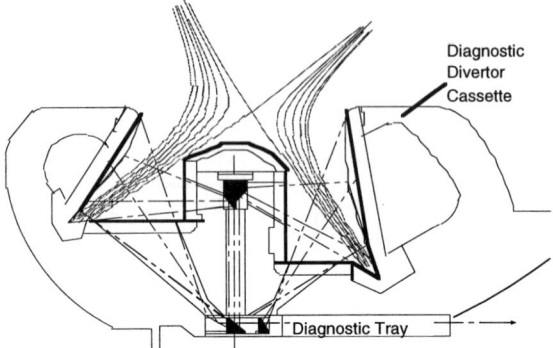

Fig .3 Impurity Spectrometry System sight lines in Diagnostic Divertor Cassette

After installation and removal of the RH tooling[4], a space 200mm high is left under the Diagnostic Cassette. The Diagnostic Tray, with optical relay systems and visible/IR imaging systems, is inserted in this space. Inside the 4 RH ports water cooled, stainless steel Diagnostic Blocks provide a base for the diagnostic equipment. A thermal shield layer, neutron/gamma shielding, and additional cooling inside the Diagnostic Block keep the temperature to an acceptable level for the electronic and optical devices.

The standard Cassettes on either side of the Diagnostic Cassette are used for a wide range of instrumentation and sensors (e.g. bolometers, Langmuir probes). The sensors are integrated in the cable conduiting to facilitate maintenance in the hot cell. All wiring is routed to the plug that automatically connects to a socket on the vacuum vessel when the cassette is installed.

5. Blanket and Back Plate

Diagnostic sensors (e.g. magnetic pick-up coils, bolometer cameras, retroreflectors, antennas) are installed in the Blanket Modules. The modules also require instrumentation (thermocouples and possibly strain gauges). Several modules are modified with holes or enlarged edge gaps for diagnostic purposes, although as much use as possible is made of the 20mm gap which exists between modules for installation purposes. The example shown here is of the High Frequency Magnetic Pick-up Coil in a standard module, fitted with an electrical plug that automatically engages with a socket on the Back Plate when the module is installed. Standard mineral insulated cables running in conduits through to eletrical feedthroughs in the vertical ports complete the circuitry in-vessel. The pick-up coils, with no special cooling circuitry, are attached in open pockets on the side to maximise the frequency response.

Diagnostics such as voltage loops, and equipment such as connectors, on the back plate are permanant features of the vessel and will not be replaced. They are placed in well shielded areas and protected against mechanical damage.

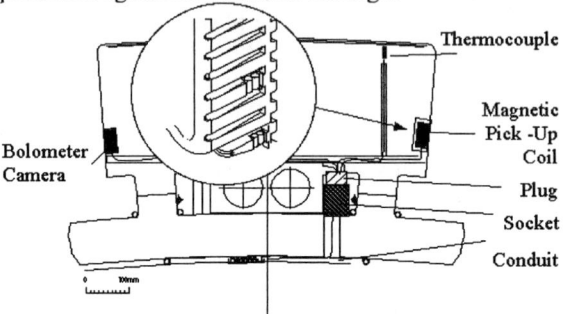

Fig. 4 Blanket Module with Automatic Electrical Connector and Diagnostic Sensors.

6. Conclusion.

In order to install diagnostics components it is necessary to modify some of the main components of the Tokamak. For the interfaces so far considered conceptual solutions have been demonstrated but the detailed designs have yet to be done. The full integration of the diagnostics requires resolution of some difficult design issues which in some cases will require innovative design.

References

(1) A.E.Costley et. al., Requirements for ITER Diagnostics. Proc.of the Int. Workshop on Diagnostics for ITER, (International School of Plasma Physics,Varenna, Italy, Sept 1995) .
(2) C.I.Walker, L.deKock, ITER Plasma Diagnostics Generic Access, ibid..
(3) H.Salzmann et. al.,Thomson Scattering for the Core Plasma of ITER.sProc. of the 11th Topical Conf. on High Temperature Diagnostics, Monterey,USA. May 1996
(4) E.Martin et. al, The ITER Divertor Remote Handling System, this Symposium.

Acknowledgement

The designs presented represent the work of the ITER Joint Central Team and the ITER Home Teams undertaken within the framework of the ITER EDA Agreement under the auspices of the International Atomic Agency (IAEA).

Tracer-encapsulated cryogenic pellet for particle transport diagnostics

S. Sudo[a], H. Itoh[b] and K. Khlopenkov[c]

[a]National Institute for Fusion Science, Furo-cho, Chikusa, Nagoya, 464-01 Japan

[b]Nippon Sanso Corp., Plant & Machinery Div., Tech. Development Dept., 6-2 Kojima-cho, Kawasaki-ku, Kawasaki-city, 210 Japan

[c]Graduate University for Advanced Studies, Shonan Village, Hayama, Kanagawa, 240-01 Japan.

An innovative device for producing a tracer-encapsulated cryogenic pellet is constructed for aiming the demonstration of proof-of-principle of an accurate transport diagnostic system to measure particle transport both in parallel and in perpendicular to the magnetic field lines for magnetic confinement devices. With the device, a tracer-encapsulated cryogenic pellet has been successfully produced.

1. INTRODUCTION

The transport of the plasma of magnetic confinement devices has been still one of most important subjects to be clarified. Main difficulty in analyzing the transport of the plasma is due to lack of the accurate experimental information, and the physics analysis becomes very complicated. Therefore, here we propose the method and device for the accurate diagnostics of the particle transport, and for making interpretation more straight-forward.

The essential point of the diagnostics is based upon the poloidally and toroidally localized particle source as a tracer within a limited small volume with the order of about 1 cm^3 in the plasma [1]. The tracer particles are deposited by a tracer-encapsulated cryogenic pellet which consists of small core as tracer of light atom such as lithium and the major outer layer of hydrogen isotope which is the same species as the bulk plasma ions. The ions which are not contained in the plasma intrinsically, can be used as tracer particles for tracing particle motion.

At the relatively early stage after pellet injection, Li ions will flow approximately along magnetic field lines with being affected E x B drift motion and the other effects.

Then, the annular domain of Li ions will be established. Particles on such surface will move outward due to particle diffusion or convection, or in some case, they will move inward due to pinch effect. We will be able to observe such motion in detail and accurately owing to very local deposition of the tracer particles in the plasma.

In order to deposit the tracer material in the localized area, the typical diameter of the core is from 50 to 200 µm for a plasma temperature in the keV range. In this case, the length of the flight during ablation of the core is around 1 cm. This localization can be adjusted with pellet size and pellet velocity depending on plasma parameters such as electron temperature. The relation between the location of the tracer particle deposition in the plasma, and the pellet size is studied with pellet velocity as a parameter. As an example of an application of a tracer-encapsulated cryogenic pellet, from a calculation of ablation in a superconducting helical device: Large Helical Device (LHD) [2] now under construction at our Institute, the outer pellet diameter ranges from 1 mm to 3 mm and the appropriate diameter of the core is in the range of 50 - 200 μm.

2. DEVICE CONFIGURATION

For technical demonstration of a device producing a tracer-encapsulated cryogenic pellet, we chose the cylindrical form of the pellet: the outer diameter of 3 mm with the length of 3 mm, and the inner core diameter of 240 μm in a sphere form. Schematic of a cryohead part of the device for producing a tracer-encapsulated cryogenic pellet is shown in Figure 1. There are two driving systems in two directions to drive the main and sub disks and the tracer core pushing wire.

The cryohead made of OFC (oxygen free copper) is supported with three Invar rods. Two stepping motor systems with linear scale position sensors are used for driving the pellet carrier disks and also for driving the supplier mechanism to insert a tracer material into the hydrogen pellet with high accuracy of order of 5 μm.

Thermal contraction of the driving systems due to cooling of the cryostat with liquid helium can be compensated with bellows. The thermal contraction of the OFC disk with a length of 153 mm is 450 - 600 μm depending on the temperature profile of the disk. The locations of the pellet production and the core supplier which move after cooling can be also adjusted by monitoring the new position with electric sensors (contact switches) as shown in Figure 1 which are set up in both directions. For the precise construction of the cryohead, a special brazing method in vacuum has been used, and the whole device has been successfully completed.

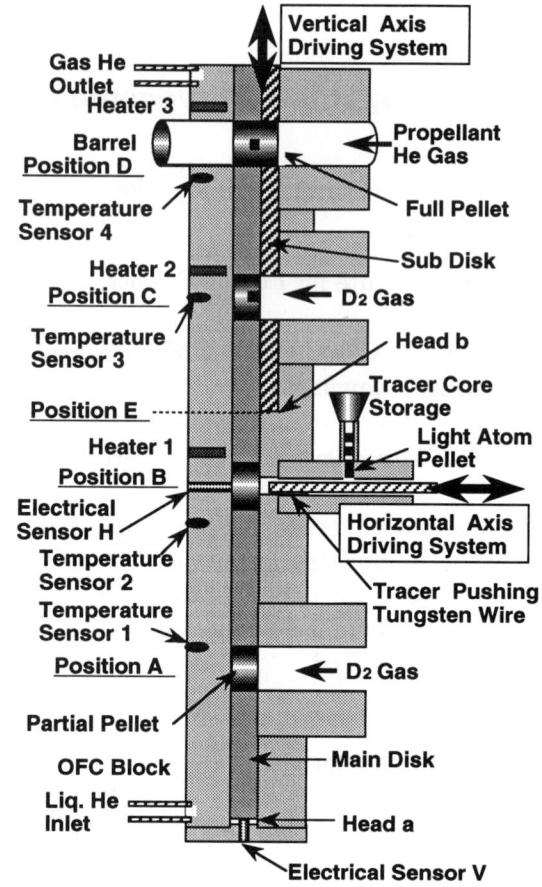

Figure 1. Schematic of cryohead part of the device for producing a tracer-encapsulated cryogenic pellet.

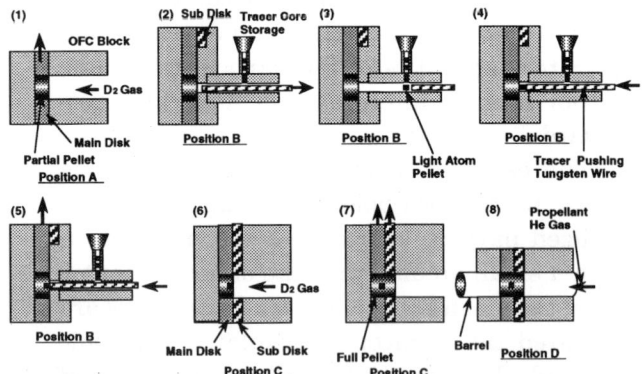

Figure 2. Concept of the procedure of the production method of a tracer-encapsulated cryogenic pellet.

3. OPERATION OF DEVICE

The concept of the procedure of the production method of a tracer-encapsulated cryogenic pellet is shown in Figure. 2. There are four different stopping positions : A, B, C, and D in the vertical direction. The accuracy of the position B should be maintained within 5-10 μm.

For producing a cryogenic pellet, hydrogen isotope gas is first introduced for the partial pellet production with thickness of 1.6 mm in the main disk as shown in (1) of Figure 2. Then, the pellet pushing tungsten wire is pulled at first (2) to the position where a light atom pellet is supplied from the "tracer core storage" (3). This can be done with a pushing spring at the head of the storage. One core tracer pellet is supplied automatically when the tungsten wire is pulled out up to the position indicated in (3) of Figure 2. Once a tracer core is supplied, the tungsten wire pushes it (4), and finally the tracer core is inserted into the partial pellet (5). In our experiment, the carbon core with diameter of 240 μm is fed by the tungsten wire with diameter of 230 μm. The thin tungsten wire can be easily broken, so special care on the points such as wire connection is necessary.

Then, the partial pellet containing the tracer core is transferred to the position C, where the hydrogen isotope gas is introduced again as shown in (6). Here, the additional pellet with thickness of 1.4 mm is put in order to cover with hydrogen isotope the part of the core which remains exposed as shown in (6). Then, the full pellet is completed (7), and it is transferred to the place as shown in (8), where the pellet is accelerated through the barrel by a high pressure gas.

The procedure can be recently controlled by an automatic remote control system with two computers and I/O capacity with 128 ch digital input, 48 ch digital output, 16 ch analog input, and 16 ch analog output. The basic sequence for automatic and remote control is established. The optimum values of the detail temperature and gas feeding quantities are under investigation. The valve operation for gas feeding, and pressure control, and temperature feedback control with five temperature sensors (silicon diodes), heaters at three different locations can be also done by the automatic system. With a mass flow meter, the real time flow rate and the total amount of the supplied gas can be monitored. The testing shows that the stepping motor systems can drive the pellet carriers smoothly in spite of small clearance of the order of several tens μm.

4. EXPERIMENTAL RESULTS

The ejected pellets are photographed by CCD cameras with light sources of a short pulse laser and fast flash lamp during flight of a pellet with a velocity of 400 - 800 m/s with the pellet observing system as shown in Figure 3. The pulse width of the flash lamp is about 180 ns. As for the dye laser, the peak wavelength in case of Rhodamine 640 is 644 nm with spectral width of 80 nm, and the pulse width and power are 3 ns and 20 kW. The dye laser is pumped with a nitrogen laser with the power of 75 kW and pulse width of 4 ns. The use of this dye laser is convenient for obtaining an image of a pellet in flight, because there is no high demand concerning the time response to a CCD camera owing to the short pulse width of the illuminating laser. The spatial resolution should be high as much as possible because of necessity of observing a tiny tracer core. We are using digital CCD cameras having high spatial resolution. These are an Electrim EDC-1000U camera with CCD elements of 1134 x 972 and 8 bit A/D conversion, and Xiliix MI-1400 camera with 1317 x 1035 elements and 12 bit. Digital and analog cameras are set up in the same distance from the gun barrel outlet, and one of those is in the vertical direction, and the other in the horizontal direction.

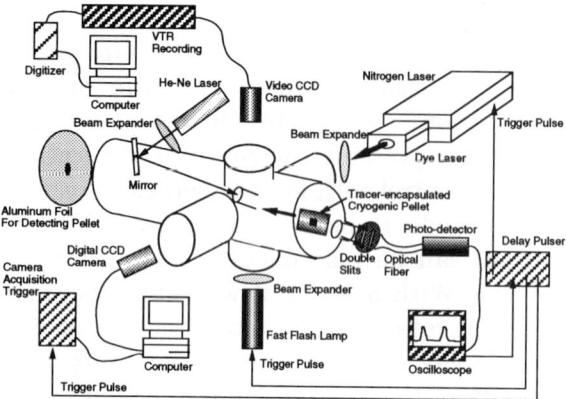

Figure 3. Experimental set-up for observing a tracer-encapsulated cryogenic pellet.

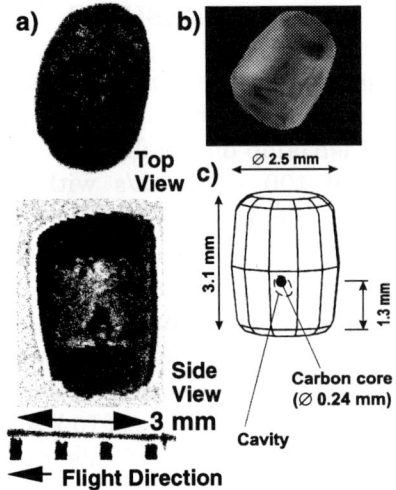

Figure 4. Simultaneous two direction images of a tracer-encapsulated cryogenic pellet.

For synchronizing the fast light sources, a 1 mW He-Ne laser beam is expanded to the diameter of 20 mm so that the beam may catch the pellet. The expanded beam illuminates the mask having two slits which are connected to the photo detectors through an optical fiber bundle. When the pellet passes through the He-Ne laser beam, then the detected light intensity decreases due to shadow of the pellet. The double slits allow to observe the pellet velocity from the separate two pulses. The first pulse triggers the fast light sources and the digital camera acquisition after the appropriate delay with the preset delay circuit.

To avoid the cloud around the pellet which obscures the clear pellet image, the observation position is located about 250 mm remote from the gun barrel, although some gas on the pellet surface can be often still seen.

The tracer-encapsulated cryogenic pellet having a carbon core is shown in the image as shown in Figure 4 a). The images of the ejected pellet have been taken simultaneously in the two directions, that is, vertically and horizontally. These images have clearly proved the sound configuration of the tracer-encapsulated cryogenic pellet. These images are processed on the computer from the digitized images of the pellet as shown in Figure 4 b) and c).

5. CONCLUSION

A device for producing a diagnostic pellet to measure particle transport in the directions both parallel and perpendicular to the magnetic field lines of magnetic confinement devices is constructed and operated. The simultaneous two direction images have clearly proved the double-layer configuration of the pellet.

ACKNOWLEDGMENTS

The authors would like to thank Director-General A. Iiyoshi, Mr. H. Matsuda, Mr. T. Fukano, Mr. S. Sakakibara, and Mr. T. Aoyama for the continuous encouragement to our work. They also thank Mr. S. Kato for helping the overall experiments.

REFERENCES
[1] S. Sudo, J. Plasma and Fusion Research, **69** (1993) 1349.
[2] A. Iiyoshi, M. Fujiwara, O. Motojima, et al., Fusion Technology **17** (1990) 169.

Experiment of 14MeV neutron induced luminescence on window materials

F. Sato[a], Y. Oyama[b], T. Iida[a], F. Maekawa[b], J. Datemichi[a], A. Takahashi[a] and Y. Ikeda[b]

[a] Department of Nuclear Engineering, Faculty of Engineering, Osaka University, Yamada-oka, Suita-shi, Osaka 565, Japan
[b] Japan Atomic Energy Research Institute, Tokai-mura, Ibaraki-ken 319-11, Japan

A photon counting system, composed of a sample holder containing focusing-lenses, radaition-resistant optical fibers, a low-noise photomultiplier tube and others, has been developed to examine the 14 MeV neutron induced luminescence on window materials. *In-situ* measurements of a small amount of the photon emission in the wavelength region from 350 to 650 nm were successfully performed during 14 MeV neutron irradiation for pure and Ge-doped silica, sapphire and other optical samples. Moreover, dose calculations on the neutron irradiation field and similar experiments with a ^{60}Co gamma-ray source were done for the subtraction of the contribution of neutron-induced gamma-rays to the photon emission. The efficiency of the 14 MeV neutron induced luminescence of a typical pure silica sample has been found to be 5 ± 3 photons/MeV, while that of the ^{60}Co gamma-ray induced luminescence to be 135 ± 50 photons/MeV.

1. INTRODUCTION

For reliable application of optical windows to fusion plasma diagnostic systems, it is necessary to investigate not only permanent damage but also transient effects of the 14 MeV neutron irradiation on window materials.[1] The neutron induced luminescence from the window, for instance, may cause a noise problem for a sensitive photon detecting system with the window. Also the transient reduction of transparency of windows caused by a neutron burst should be concerned with the reliability of light measuring systems for fusion diagnostics. Data on the transient optical effects like luminescence of the 14 MeV neutron irradiation are necessary for elaborate design of optical fusion diagnostic systems. The purpose of this paper is to show experimental data on the 14 MeV neutron induced luminescence of some typical window materials. At first, this paper describes the experimental method and apparatus with a DT neutron source, FNS[2] for measuring the luminescence and then shows results obtained for typical window material samples.

2. EXPERIMENT

A photon counting system has been developed to *in-situ* measure the intensity and spectrum of the luminescence on window materials during 14 MeV neutron irradiation at FNS facility. Figure 1 shows a schematic drawing of the experimental arrangement and measuring system. The system was composed of a sample holder, radiation-resistant optical fibers, a photon counting equipment containing a photomultiplier tube and related electronics circuits, a photon polychromator

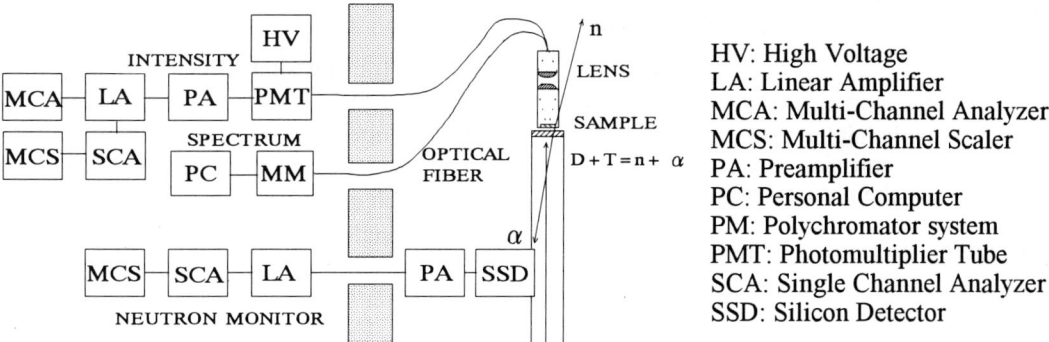

Fig. 1. Schematic drawing of the experimental arrangement and measuring system.

Table 1 Calculated dose rates (Gy/s) for samples.
(DT neutron intensity : 3.0×10^{12} n/sec)

Sample	$D_{neutron}$	$D_{\gamma\text{-ray}}$	D_{total}
Pure silica	1.0	0.07	1.1
Ge doped silica	1.0	0.07	1.1
Sapphire	0.92	0.11	1.0
Calcium fluoride	1.3	0.06	1.4
Plastic scintillator	5.8	0.15	6.0

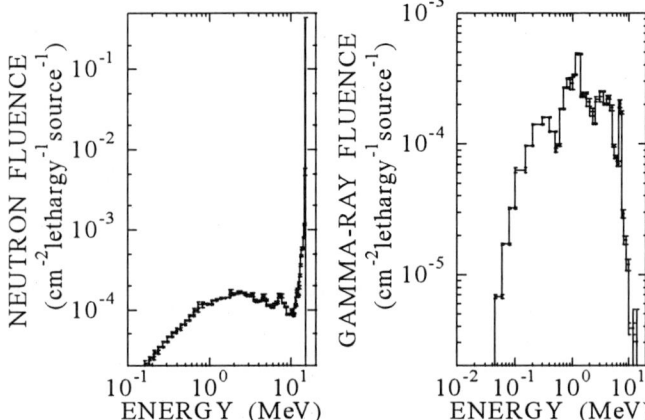

Fig. 2. Calculated energy spectra of neutrons and induced gamma-rays at the position of samples.

system and a personal computer. The sample holder included a window material sample and radiation-resistant focusing lenses and was placed near the tritium target of the deuteron accelerator. The lenses successfully transmitted photons from the sample into the optical fibers of which diameter was 1.2 mm. The photomultiplier tube (R1635P) was the photon counting grade and was moreover cooled to about 260 K with a Peltier-effect device to reduce its thermal noise and sensitively detected photons of 350 to 650 nm in wavelength. The wavelength spectrum of photons was also measured with a calibrated grating polychromator system (M5098) with an image-intensifier. Measured data were all stored in a personal computer. The lenses and optical fibers were away from the sample set near the tritium target, and thus the intensity of the background photons from them was sufficiently lower than that of the photons from the sample. The irradiated samples are listed in Table 1. The size of the samples was 10 mm in diameter and 2.0 mm in thickness. A plastic scintillator (NE102A), of which detailed response to radiations has been examined, was used for the check of the photon detection efficiency of the present system.

The 14 MeV neutron yield at the target was determined from the measurement of alpha particles associated with DT neutrons. A silicon surface barrier detector, set in the beam tube of the deuteron accelerator, detected the alpha particles. And the fluxes and energy spectra of the neutrons and induced gamma-rays at the position of the sample were precisely calculated with the Monte Carlo neutron-gamma transport code MCNP[3] in consideration of the influence of the target assembly and the sample holder. Here, it should be noted that the photon emission is caused by not only neutrons but also the induced gamma-rays in the neutron irradiation experiment. Gamma-rays produce high energy electrons in materials and the electrons induce the luminescence more easily and the Cherenkov radiation as well. Figure 2 shows the calculated energy spectra of the neutrons and induced gamma-rays at the position of the sample. The neutron- and gamma-ray- absorbed doses for the irradiated samples were also calculated using the energy spectra shown in Fig. 2 and the KERMA factors[4] for the samples. Typical examples of the calculated dose rates are summarized in Table 1. The intensity of the 14 MeV neutron induced luminescence on the samples was measured in the neutron flux region below 1×10^{11} n/cm²/sec. Similar experiments to examine the contribution of gamma-rays to the photon emission were carried out using a ^{60}Co gamma-ray source.

The relation between the photon emission efficiency of the sample and the count of the photons measured with the present system is expressed as follows:

$$\text{Count (with sample)} - \text{Count (without sample)} = D\eta_f\eta_g \int \eta_p(\lambda)Y(\lambda)d\lambda, \quad (1)$$

where
D : Dose,
η_f: Photon transmission efficiency of optical fiber,
η_g : Geometrical efficiency between sample and optical fiber, including lens system,
$\eta_p(\lambda)$: Quantum efficiency of photomultiplier tube,
$Y(\lambda)$: Photon emission efficiency of sample,
λ : Wavelength.

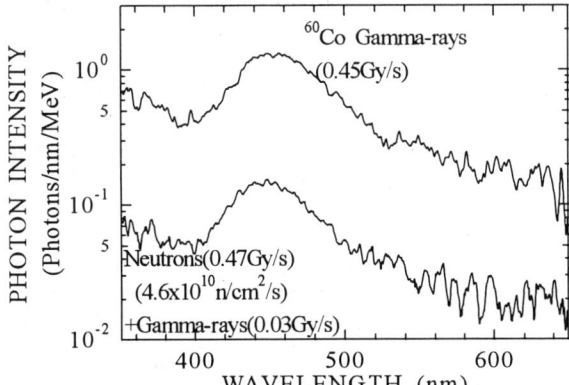

Fig. 3. Wavelength spectra of emitted photons in 14 MeV neutron and ^{60}Co gamma-ray irradiation experiments for the same pure silica sample.

The photon transmission efficiency of the optical fibers was measured with a few types of light emitting diodes and was 0.95 ± 0.02. The geometrical efficiency was calculated on the basis of the lens optics and the numerical aperture (=0.21 ± 0.01) and the diameter of the optical fibers and it became 0.0013 ± 0.0003. The specification of the photomultiplier tube gave the quantum efficiency of about 25% in the wavelength region around 450 nm. In all, the photon detection efficiency of the system was expected to be (3.2 ± 0.7) × 10^{-4} in the wavelength region around 450 nm, and this value agreed within 40% with the efficiency obtained from the calibration experiment with the plastic scintillator and a ^{60}Co gamma-ray source.[5]

3. RESULTS AND DISCUSSION

The photon counting rate was proportional to the 14 MeV neutron flux for all the samples, though the fluence was too low to damage the sample[6]. Table 2 summarizes the photon emission rate, i.e. the number of emitted photons per unit absorbed energy for the samples. Figure 3 also shows the wavelength spectra of emitted photons in 14 MeV neutron and ^{60}Co gamma-ray irradiation experiments for the same pure silica sample. Though the neutron effect is apparently different from the gamma-ray one in the photon emission rate as shown in Fig. 3 and Table 2, both spectra have the same large peak around 450 nm, which means that there is not large difference in the luminescence mechanism between the neutron and gamma-ray irradiations. The same peak has also been observed in the electron and ion beam irradiation experiments on silica glass.[7,8] And the luminescence for the peak is considered to be concerned with self-trapped excitations in oxygen vacancies.[7] It is known that the luminescence intensity of silica depends on the sample itself, in other words, varies with its structure, impurity concentration and so on. It has been found from the present neutron and gamma-ray irradiation experiments with the same silica sample that the luminescence efficiency of 14 MeV neutron is one order and a half smaller than that of the gamma-ray. This result is consistent with previous studies on the response of scintillators to radiations.[9] The scintillation intensity for heavy charged particles is fairly smaller than that for electrons. Most photons are emitted from the region excited by the secondary electrons generated around the track which the energetic primary particle leaves in a scintillator. Energetic primary electrons have lower LET and produce a larger excited region and more secondary electrons than heavy ions. Thus 14 MeV neutrons release high energy charged particles of protons, alphas, recoil atoms and others from almost every material but the intensity of the luminescence by these particles is much lower than that by high energy electrons generated by the gamma-ray reactions. The

Table 2 Number of emitted photons per unit absorbed energy in the wavelength region from 350 to 650 nm for samples.

Sample	Peak wavelength (nm)	14 MeV neutrons+induced γ-rays (photons/MeV)	^{60}Co γ-rays (photons/MeV)
Pure Silica	450	17±6	170±60
Ge doped silica	390	83±30	410±140
Sapphire	410	2500±1000	27000±11000
Calcium fluoride	<350, 550	270±110	1300±500
Plastic Scintillator	422	3000±1200	12000±5000

Table 3 Summary results of the photon emission efficiency (photons/MeV) on the pure silica sample obtained in the 14 MeV neutron and ^{60}Co gamma-ray irradiation experiments.

Irradiation	Photon emission rate	
14 MeV neutron	Luminescence (neutron)	5±3
	Luminescence (γ-ray)	9±3
	Cherenkov (γ-ray)	3±1
	Total	17±6
^{60}Co γ-ray	Luminescence (γ-ray)	135±50
	Cherenkov (γ-ray)	35±10
	Total	170±60

efficiency of the 14 MeV neutron induced luminescence was determined by the subtraction of the portions of the gamma-ray induced luminescence and the Cherenkov photons from the total amount of photons measured in the 14 MeV neutron irradiation experiment. The neutron and gamma-ray doses for the sample were separately calculated as shown in Table 1. The portion of the gamma-ray induced luminescence was estimated from the dose calculations on the sample and the data on the luminescence efficiency per unit absorbed energy which were obtained in the ^{60}Co gamma-ray irradiation experiment, though the difference in energy spectrum between ^{60}Co gamma-rays and neutron induced gamma-rays was neglected. The portion of the Cherenkov photons was calculated on the basis of the Cherenkov radiation formula.[10] The Cherenkov radiation is caused mainly by Compton electrons in gamma-ray field, and the electron flux and energy spectrum in the sample was calculated with a code based on the gamma-ray transport code MCNP. Table 3 shows the summary results of the photon emission efficiency on the pure silica sample obtained in the present 14 MeV neutron and ^{60}Co gamma-ray irradiation experiments.

4. CONCLUSION

The photon counting system, composed of the sample holder containing the lenses to focus photons from the sample, the optical fibers for the photon transmission, the cooled low-noise photomultiplier tube, the grating polychromator system and the personal computer, has been developed to measure the 14 MeV neutron induced luminescence on window materials. The *in-situ* measurements of a small amount of the photon emission in the wavelength region from 350 to 650 nm were successfully performed during 14 MeV neutron irradiation for pure and Ge-doped silica, sapphire and other optical samples. The efficiency of the 14 MeV neutron induced luminescence was determined by the subtraction of the contribution of the induced gamma-ray to the photon emission from the data obtained in the neutron experiments. The efficiency of the 14 MeV neutron induced luminescence on the typical pure silica has been found to be 5 ±3 photons/MeV and one order and a half smaller than that of the gamma-ray induced photon emission. These data should be useful for the estimation of the background level of optical fusion diagnostic systems exposed to 14 McV neutrons. It should be also noted that the results of the present experiments suggest the induced gamma-ray rather than the 14 MeV neutron may be serious for the transient effect like luminescence on optical materials used in a DT fusion reactor.

ACKNOWLEDGMENTS

The authors express their sincere thanks to the staffs of the FNS and ^{60}Co facilities at JAERI for their valuable suggestions in carrying out the irradiation experiments. They are also grateful to Dr. T. Nisitani and Dr. T. Matoba of JAERI for a review of the manuscript.

REFERENCES

1. Reports presented at ITER workshop, St. Petersburg, USSR (1991).
2. T. Nakamura et al., Proc. 3rd Symp. on Accelerator Sci. & Technol., (1980) 55.
3. Judith F.Briesmeister, (editor) RSIC/CCC-200 (1993).
4. K. Maki et al., JAERI-M 91-073 (1991).
5. D. Clark, Nucl. Instr. and Meth.,117 (1974) 295.
6. D.V.Orlinski et al., J.Nucl.Mater 212-215 (1994) 1059.
7. K.Tanimura et al., Phys. Rev. Lett., 51 (1983) 423.
8. P.W.Wang et al., Nucl. Instr. and Meth. B59/60 (1991) 1317.
9. Laura M. I. de Carvalho et al., Nucl. Instr. and Meth., 84 (1981) 563.
10. B. D. Sowerby et al., Nucl. Instr. and Meth.,97 (1971) 145.

Status and characteristics of diagnostics on Korea Superconducting Tokamak Research (KSTAR)

S. G. Lee, S. M. Hwang, H. Y. Chang*, G. S. Lee, H. K. Park, J. Kim, D. I. Choi, S. G. Oh**, K. K. Choh, J. H. Choi, J. W. Choi, Y. S. Chung, J. H. Han, J. Hong, B. C. Kim, W. C. Kim, Y. J. Kim, H. G. Lee, H. K. Na, Y. K. Oh, H. L. Yang, J. G. Yang and N. S. Yoon

Korea Basic Science Institute, Taejeon, Korea

Diagnostic plan for KSTAR device to support the project goal and physics mission is presented in this paper. Merits and weaknesses of some specific diagnostics which could impact the physics mission of KSTAR are discussed in detail.

1. PROJECT GOAL AND PHYSICS MISSION OF KSTAR

Korean National Fusion Program (KNFP) is the national research and development program for a superconducting tokamak capable of steady state operation. Project goal is to bring up the national technology basis to construct and operate a tokamak based on superconducting magnet technology. In addition to the project goal, construction of Korea Superconducting Tokamak Research (KSTAR) device, the physics mission of KSTAR[1] is to explore a new magnetic equilibrium configuration which can extend present stability boundary and improve the performance via a control of local current and pressure profiles. If this mission is successful, the new regime can be tested in a steady state operation. KSTAR can be used as a test bed for ITER program[2] which is also a fully superconducting device. The proposed parameters of KSTAR device are listed in Table 1. The details of engineering and physics parameters are currently under study to meet the project goal and physics mission.

2. STATUS AND CHARACTERISTICS OF DIAGNOSTICS

2.1. Status

Currently, KSTAR is in conceptual design phase and efforts are concentrated to optimize the machine design. In order to support the physics mission based on plasma current and pressure profile control, the KSTAR diagnostics will be focused on an accurate and reliable measurement of plasma parameters closely related with the plasma current and pressure profiles. The diagnostics on KSTAR are divided into three different categories. They are basics, baselines and diagnostics aiming specific goals. Basic diagnostics are tools needed to operate and control the KSTAR device. Baseline items are divided into two section to accommodate funding levels. The specific diagnostic items are aimed to support the mission of physics. In parallel with the KSTAR design, basic concepts and scaled devices are fabricated and tested using HANBIT device[3], magnetic mirror as a test facility. A conceptual arrangement of major diagnostics for KSTAR is illustrated in Figure 1.

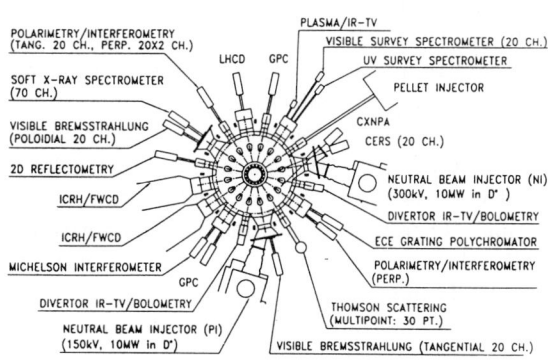

Figure 1. Major diagnostic layout of KSTAR.

2.2. Characteristics

The basic tools such as magnetics will be similar to the conventional methods used in most of tokamaks. However, considering a superconducting tokamak, diagnostic design must consider a steady state operation in the future. As an example, a non-inductive method to measure magnetic properties will be employed. Another obstacle on diagnostic system, will be interfacing optical window that has to be designed considering a high radiation heating during steady state operation. Other constraint is diagnostic access which is often lacking in a superconducting device. For instance, the access required in multi-view diagnostics, such as multichannel interferometer and/or polarimeter system, will be a challenge. Most importantly, diagnostics to monitor divertor regions are critical. Constant monitoring of erosions, melting, and other abnormal behaviors of plasma facing plate is mandatory.

Since the physics mission of KSTAR will be focused on a local control of plasma current and pressure profiles, it is imperative to design a diagnostic tools to measure plasma parameters related to these parameters as accurately as possible. During CDR phase, techniques developed to measure plasma parameters related to current and pressure of the plasma, up to date, are reviewed in depth. The choice of each diagnostics will be based on accuracy and reliability of each instrumentation. Among the listed diagnostics in Table 2, a brief summary of the essential diagnostics is provided.

2.2.1. Plasma current profile measurement

The measurement will be focused on profile information of the plasma current. It is extremely important to address how the variation of plasma current distribution affects the performance and stability of tokamak plasmas. It is well known that most of the MHD phenomena are related with the variation of current distribution in the plasmas. Therefore, it is valuable to quantify the MHD related transport, under the assumption that the frequency and amplitude of the MHD activity are associated with the local variation of plasma current distribution.

The measurement of plasma current distribution has been the most challenging in plasma diagnostics and has a relatively short history. Especially the requirement is a good spatial and temporal resolution of current distribution, the design must emphasize a crucial elements of this type of diagnostics to meet the requirements. There are two techniques commonly practiced in most of tokamaks; they are methods based on Motional Stark Effect (MSE)[4] and Faraday rotation[5], respectively.

Faraday rotation technique is practiced in many devices, since it was tested on TEXTOR. This technique requires a good access to cover a full view of tokamak plasmas, since the measurement requires a double inversion process. Most of uncertainty of this technique will be concentrated near the edge and on axis of the plasma. This difficulty can easily overcome, when the measurement is combined with the information from other diagnostics. Note that intrinsic advantage of this method is an excellent temporal resolution.

On the other hand, MSE technique requires a beam line, either from the heating or diagnostic beam. It is important to note that this measurement is relatively new and a significant improvement in temporal resolution and accuracy are expected in the future. This method is based on pitch angle measurement and has to be translated into a local poloidal magnetic field and current. The analysis requires local magnetic field shape which can only be obtained from equilibrium analysis that, in turn, needs a q profile information. Thus the analysis often goes through many iteration to find a converging solution. The other improvement should be made in temporal resolution which requires a good beam penetration and estimation of background signal contribution. However, this method can measure the current distribution with a good time resolution for an outer half of plasmas where the accuracy is reasonably good.

It is worth to consider a combination of these two methods. The Faraday rotation technique has advantages near the core of the plasma where the machine access may not be a problem, whereas MSE method is certainly valuable for an outer half of plasmas. The details of hybrid techniques for the measurement of current profiles on

KSTAR device are under consideration.

2.2.2. Plasma pressure profile measurement

Plasma pressure consists of many parameters; they are plasma density ($n_{e,i}$), plasma temperature ($T_{e,i}$), impurity (Z_{eff}), and fast beam ions. Among these parameters, ion density (n_i) and fast beam ions can not be measured directly in practice. In principle, the ion distribution including impurity ions can be measured using thermal scattering method which has not been hardened yet. Therefore, the ion density has to be estimated based on electron density (n_e) and absolute impurity measurement (Z_{eff}). Fast ion counts must be estimated from the calculation. Considering that the bulk ion temperature was recently addressed experimentally through impurity ion information, we have had a very little knowledge on the plasma pressures.

The most common diagnostic to measure electron information is Thomson scattering since it was tested on the T-3 tokamak using a single pulse Ruby laser. This method provides local electron density and temperature simultaneously. With the advances in infrared laser and detection technology, a multichannel system has practiced in many different tokamak devices. Note that the temporal resolution is limited by a duty cycle of high power laser source (~50 Hz). To overcome this limitation, often few lasers are stacked temporally and the time resolution can be improved. However, the time scale required for study of the MHD phenomena such as disruption and sawtooth crash, could be up to a sub μ-second level. It is imperative to employ diagnostics that can measure a fast phenomena in a required time scale.

Electron density diagnostics that can fulfill the need are interferometry based on the measurement of phase change of electromagnetic waves in the plasma and densitometry based on Faraday rotation angle measurement arising from toroidal magnetic field which is well known. Since both systems are using CW lasers, the time resolution is inherently limited by a detection speed. One disadvantage is that the information has to be transformed into a local value via Abel inversion, since these methods are fundamentally a chordal measurement. However, if the measurement is made on a horizontal plane, the inversion process can be independent of flux surface.

The theory of cyclotron emission is well understood in tokamak plasmas, when the optical depth is deep enough and the supra-thermal emission is absent. Especially, interpretation of Electron Cyclotron Emission (ECE) become complex during current drive experiment at a low plasma density due to expected large runaway electron population. ECE measurement was used to monitor time dependent electron temperature profiles in a tokamak. There are a number of techniques being used in the present tokamaks. Michelson interferometry, based on the measurement of Doppler broadening of the emission, is rather slow due to the mechanical motion of mirror but absolute calibration is possible. Heterodyne method uses a broad band detection system which can be also absolutely calibrated. Recently this method has made a significant engineering advances. 2D (toroidal and poloidal) measurement based on array technology was applied to map the electron temperature on TEXT-U tokamak. However, this may require a good access to project the plasma image.

All these methods will be reviewed and the most suitable systems will be implemented on KSTAR. Since Thomson scattering system can not be used as a tool for the study of temporal behaviors of electrons, ECE and densitometry or interferometry must be combined to fulfill the physics mission of KSTAR.

2.2.3. Impurity measurements

In order to estimate the ion density in KSTAR device until we have a reliable tool to make a measurement, one has to rely on information of impurities. A common technique is based on bremsstrahlung emission spectrum from plasmas. One has to be extremely cautious to avoid line emission in a band where the measurement is made. In addition, problems arising from a reflection and scattering of the spectrum have to be avoided the spectrum falls in a visible range. It is valuable to have second band to cross check the measurement. Since the emission has a strong dependence on electron density ($\propto n_e^2 T_e^{0.3}$) and linearly proportional to Z_{eff}, an accurate information of density and temperature is equally important. In the presence of multiful ion impurity species, one has to measure absolute level of impurities and their charge states.

2.2.4. Others

In addition to the diagnostics emphasized in previous section, there are many other diagnostics equally important. Bolometry is needed to complete power balance of a discharge. UV and X-ray diagnostics are essential to address high Z impurities. The array of these spectrums can be used to measure plasma parameters such as electron temperature and Z_{eff}. Fluctuation diagnostics such as reflectometry, and Beam Emission Spectroscopy (BES) are needed not only to study transport issues associated with electrostatic fluctuation but also to study localized MHD phenomena. Since the KSTAR is expecting to operate D^+ and auxiliary heating power level is significant, monitoring systems for a fusion products are required. In addition to 2.5 MeV DD neutrons, there will be 14 MeV DT neutrons at about 1% level of DD neutrons. The profile information of DD neutron will be extremely useful to cross-check other performance. These informations will serve as a basis of DT equivalent fusion power estimation and can be used for a reactor projection.

* Permanent Address: Department of Physics, KAIST, Taejeon, Korea.
** Permanent Address: Department of Physics, Ajou University, Suwon, Korea.

REFERENCES

1. D.I. Choi et al., International Conference on Plasma Physics, Japan, 1996.
2. P.H. Rebut, Fusion Eng. Design 27 (1995) 3.
3. G.S. Lee et al., 6th International Toki Conference, Japan, 1994.
4. F.M. Levinton et al., Rev. Sci. Instrum. 61 (1990) 2914.
5. I.H. Hutchinson, Principles of plasma diagnostics, Cambridge University Press, 1990.

Table 1
Proposed key parameters for KSTAR

	Parameters
Major Radius, R_0	1.8 m
Minor Radius, a	0.5 m
Toroidal Field, B_{T0}	4.0 T
Plasma Current, I_P	2.0 MA
Elongation, κ	2.0
Triangularity, δ	0.8
Auxiliary Heating	NBI: 8MW ×2
	ICRH/FWCD: 6 MW
	LHH/LHCD : 1.5 MW
Pulse Length	$t_{pulse,Init} = 20$ sec
	$t_{pulse,CD} = 500$ sec

Table 2
List of diagnostics for KSTAR

	Diagnostics
Basic	Magnetic Coils/Loops
	Fixed Edge Probes
	mm-wave Interferometer
	Hard X-ray Detector
	H_α
Baseline I	Bolometers
	Visible Survey Spectrometer
	Visible Bremsstrahlung
	Thomson Scattering
	ECE Michelson Interferometer
	IR TV
	Neutron Detectors
Baseline II	CHERS
	Motional Stark Effect
	Soft X-ray Arrays
	Multichord Visible Spectrometer
	Multichord Thomson Scattering
	Charge Fusion Products
	Moveable Edge Probes
Mission-Oriented	Interferometer/Polarimeter
	mm-wave Reflectometer
	ECE Grating Polychromator
	Beam Emission Spectroscopy
	Multichannel Neutron Collimator
	Divertor Thomson Scattering
	CHENA

High performance drivers for fast sweep microwave reflectometry

L. Cupido, A. Silva, M.E. Manso and F. Serra

Centro de Fusão Nuclear, EURATOM/IST Association,
Instituto Superior Técnico - Av. Rovisco Pais, 1096 Lisboa Codex, Portugal

Radar measurement techniques are being used to measure the fusion plasma density profile. The FM radar requires fast tunable sources capable of broadband operation In this paper we refer the principles underlying the development of the drivers for both YIG's and VTO's sources, considering the application of reflectometry in next fusion devices.

1 Introduction

With the application of standard measurement techniques to new areas, namely to experimental physics, new developments of the electronics are needed. This is the case of microwave reflectometry for fusion plasmas, where several radar techniques are being used to probe fusion plasmas.

The sweep time required for most of the applications (receivers, spectrum analysers, network analysers, radar's etc.) is long enough to allow simple control electronics and straightforward design. However, one crucial aspect for the high performance of the FM broadband reflectometry diagnostics is to launch signals that can be swept over a full frequency band in time intervals less than the periods of the typical plasma fluctuations ($\leq$ 10ms).

We used for the first time in reflectometry low noise solid-state broadband oscillators and developed high performance drivers for fast sweep operation for both YIG (Yttrium Iron Garnet) and VTO (Varactor Tuned Oscillator). In both cases the measurements show a big increase in accuracy due to the microwave swept sources. Ultra fast probing ($\leq$ 100 μs), crucial to obtain profiles with high temporal and spatial resolution was used in a later stage, based on GaAs-FET VTO's, for the reflectometry diagnostics on IS-TOK and ASDEX-Upgrade. With the developed driver the sweeping times of 10 μs were achieved, about two orders of magnitude faster than the commercially available equipment.

Here we refer the principles underlying the development of the drivers for both the YIG and VTO technologies, which represent the state of the art at the time the diagnostics were designed.

2 Yig tuned devices

A YIG tuned device is controlled by means of a magnetic field which determines the resonant frequency of the YI. The magnetic field is created by an inductor mounted in a ferro-magnetic circuit with a gap in which the YIG is placed. The resonant frequency f_0, of an isotropic YIG sphere in a uniform magnetic field is given by:

$$f_0 = \gamma \cdot H_0 \qquad (1)$$

where f_0 is in MHz, γ is the giromagnetic ratio 2.8 [sec^{-1} · oersted^{-1}] and H_0 is the applied DC magnetic field in Oersteds.

For the reflectometers of ASDEX tokamak, YIG Tuned devices where used in the two frequency ranges 18 - 26.5 GHz and 26.5 - 40 GHz. In order to tune the YIG oscillator from 26.5 to 40 GHz, a DC magnetic field of 9.5 to 14.3 KGauss was required /3/.

The tuning coil was designed by the manufacturer to minimise the power consumption and to allow operation at reasonable supply voltages; tuning speed was not a priority in view of the standard applications. A compromise between the tuning speed and the power dissipated in the coil takes into account the coil resistance, as follows:

$$P = nl \cdot \rho s \cdot I^2 \quad (2)$$

$$H_0 = nI \quad (3)$$

$I=$ the operating current (A)
$n=$ numbers of turns
$\rho=$ coil material resistivity $(\Omega \cdot m^{-3}$)
$s=$ section of the coil wire (m^2)
$l=$ length of an average coil turn (m)
$P=$ dissipated power (W)

Almost all the YIG devices available have a main coil inductance of about 100 mH; this value is too large to allow fast sweeping. In order to develop the electronics for fast sweep operations, attention should be payed to the voltage values that can be developed at the main coil terminals by the rising or falling current. The generated voltage can be calculated by assuming a linear sweeping current $i(t)$:

$$V(t) = R \cdot i(t) + L \frac{di(t)}{dt} \quad (4)$$

$$i(t) = i_0 + kt \quad (5)$$

$$V(t) = R(i_0 + k \cdot t) + k \cdot L \quad (6)$$

$R=$ the total coil resistance (Ω)
$L=$ the total coil inductance (Hy)
$k=$ current rise (or fall) rate in $(A \cdot s^{-1})$
$i_0=$ offset current, YIG at minimum frequency (A)

As we are interested to sweep as fast as possible the term $k \cdot L$ becomes much larger than $R.(i_0+k.t)$ therefore $V(t) \approx kL = L\frac{di}{dt}$. From this expression we can conclude that the sweep time is limited by the maximum voltage in the driver output at which the drivers response is still linear. The driver should also withstand high supply voltages that involve high dissipated power both for mixed DC and swept operation.

From eqs.5 and 6, the minimum sweep time for a maximum voltage V_{tl} (voltage at the end of the sweep), at a given offset current I0 and for a current variation ΔI is:

$$V_{t1} = R \cdot (I_0 + \Delta I) + \frac{\Delta I}{\Delta t} L \quad (7)$$

$$\Delta t = \frac{L \Delta I}{V_{t1} - R(I_0 + \Delta I)} \quad (8)$$

It is desirable to operate at voltages less than 100 Volt in order to use solid state conventional electronics. For V $\leq$ 100V, less than 80 Volt are available for the swing voltage across the main tuning coil (a 10% safety margin on the lower limit and another 10% margin on the upper limit). It should be noted that as on a current fall the voltage developed has the opposite direction so in the case of repetitive sweeps we have in the circuit signals with 200Vpp.

A typical current driver (fig. 1) has several

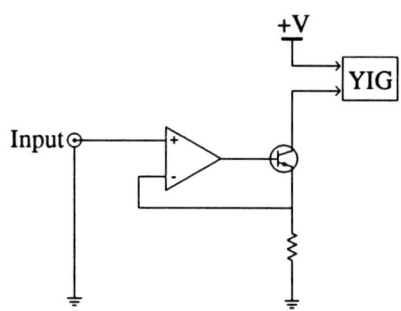

Figure 1 - A typical current driver that has several disadvantages to operate at high voltage and fast sweep times.

disadvantages when operated at high voltage and fast sweeping, namely due to the voltage drop across the resistor R that reduces the useful voltage. Also, as the current pass device (Q1) is above ground making the implementation of protection circuits less efficient. Protection circuits are a critical issue because the inductive nature of the load that can produce high voltage spikes and high frequency spurious oscillations.

For the ASDEX tokamak application, the driver circuits were designed for a 0 to 1 Volt input in order to control the YIG tuned oscillators for both K band (18 to 26.5 GHz) and Ka band (26.5 to 40 GHz). Implementing the driver concept mentioned above enabled swept operation over each full bandwidths at 2ms with very good reliability /5/.

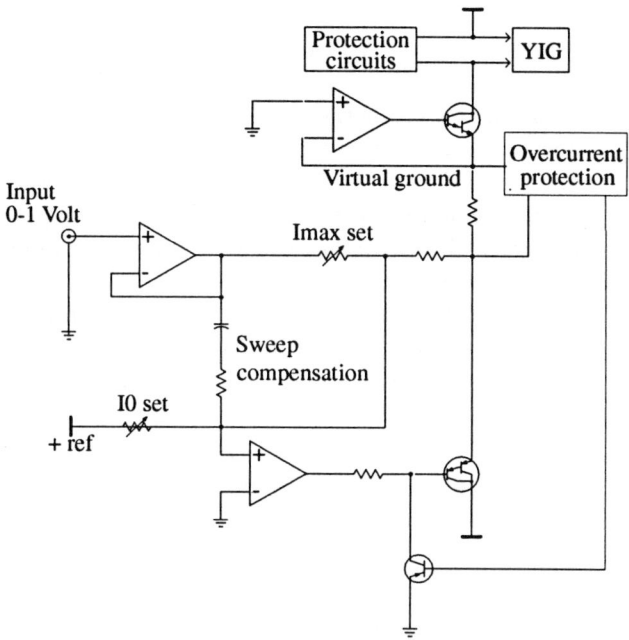

Figure 2 - Current driver with control to the negative rail and load transfered to the positive side by a virtual ground; protection circuits are not detailed for clarity.

3 Varactor tuned devices

The frequency control of a varactor device is done by applying a voltage to the tuning port which presents a high impedance to the driver and is associated with a low impedance capacity. For the same sweep speed driving varactor devices is by far simpler than for YIGs.

Varactor tuned oscillators at frequencies up to 220 GHz are available. However as the frequency increases above 18 GHz the available devices have narrower bandwidths. At millimetre wavelengths typically 10% of the full bandwidth is available. Since for our application we want to explore the full waveguide bandwidth of each channel, we have employed Varactor Tuned Oscillators (VTO) up to 18 GHz with frequency multipliers to achieve the required millimetre bands enabling full band capability.

The use of VTO's are necessary to obtain sweep times less than 500 μs. However, we should accept a non linear tuning curve, less thermal stability, higher pulling and pushing figures, and less clean output spectra. In this respect YIG devices would be much more favourable.

The maximum tuning speed is limited by the driver bandwidth and by the capability to drive the capacitive load due to the decoupling capacitors of the VTO tuning port. There is also another limitation resulting from the oscillator design, that is the intrinsic stabilisation time after a frequency change. This information is provided by the manufacturer in the form of a settling time. The stabilisation times involved should not present any significant degradation for sweep times longer than 10 μs.

Since the capability of driving a capacitive load is, nowadays, an easy task, the main problem consists in having a sufficient large driver bandwidth to allow the full band sweep with a minimum distortion.

It is necessary that the driver has enough bandwidth to accommodate the triangular sweep spectrum (we are referring to a large signal bandwidth). Lets consider a triangular pulse, corresponding to a sweep up and down, and its Fourier transform.

$$w(t) = \begin{cases} A\left(1 - \frac{|t|}{\tau}\right) & |t| < \tau \\ 0 & |t| < \tau \end{cases} \qquad (9)$$

$$W(f) = A\tau sinc^2 2\pi f\tau \qquad (10)$$

For a given distortion (1-D) and a signal amplitude A, the input spectrum to be maintained by the driver can be calculated from:

$$A\tau \int_{-f0}^{+f0} sinc^2 f\tau df \geq DA\tau \int_{-\infty}^{+\infty} sinc^2 f\tau df \qquad (11)$$

$$A\tau \int_{-f0}^{+f0} sinc^2 f\tau df \geq DA \quad (12)$$

Computing equation 12 numerically for a 1% distortion with a sweeping time $\tau=10$ μs, a bandwidth of 1.1 MHz is needed. However, in our application a distortion of less than 1% should be obtained. For a 0.1% distortion a bandwidth of 10 MHz is adequate.

Although a large signal bandwidth is available with operational amplifiers, it seems impossible to find a commercial integrated circuit with the requirements needed for our application: large signal bandwidth > 10 MHz; product gain bandwidth > 300 MHz; output swing 0 - 22V (slew rate > 5 V/μs with a capacitive load); capacitive load drive capability; constant input/output delay.

A discrete amplifier was developed using a classical operational amplifier topology (Fig.3) with

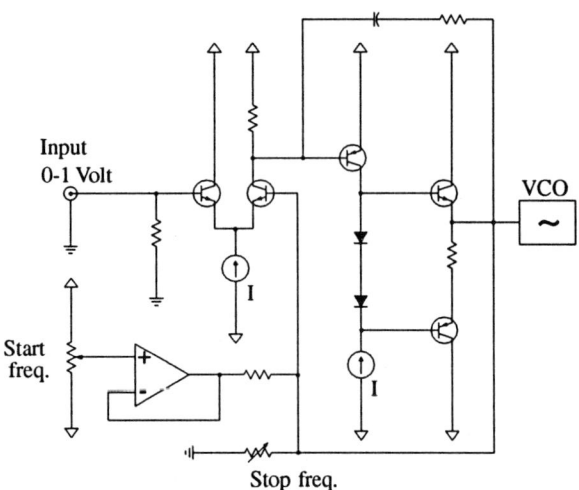

Figure 3 - Very fast amplifier capable of $10\mu s$ sweep of a VTO (Varactor Tuned Oscillator) with less than 0.1% distortion.

differential bipolar input devices and complementary output transistors. Frequency compensation was done to minimise ringing and also to obtain a clean output spectrum of the driven oscillator (Fig 3). An additional speed up compensation was implemented (not shown in this figure) in order to overcome additional delay at the beginning of the sweep.

4 Concluding remarks

Due to their performance, namely the possibility of fast sweeping and insensitivity to magnetic fields, solid state VTO's using hyperabrupt tuning diodes (HTO) were selected for the ASDEX-upgrade reflectometry system. Four bands are now operating from 16 to 74 GHz using two different HTO's (8-12 and 12-18GHz) followed by active multipliers. The specially developed drivers enable ultra fast sweeping (10 μs) of a full waveguide bandwidth with low distortion. Another system using the same type of electronics is installed for the extraordinary mode broadband JET (Joint European Torus) reflectometer /6/. Further improvements of the adopted solutions are being conducted for the application of broadband reflectometry in next fusion machines.

References

[1] M.E. Manso, Plasma Physics in Controlled Fusion, 35, 141, (1993)

[2] M.E. Manso, F. Serra, J. Mata, J. Barroso, J. Comprido et al., Proc. 15th European Conference on Controlled Fusion and Plasma Heating, Dubrovnik, 12 B, III, 1127 (1988).

[3] A. Silva, L. Cupido, M.E. Manso, F. Serra, F.X. Söldner, P. varela et al., Fusion Technology 1992, 747 (1993).

[4] D.P. Zensius, J.L. Hauptman et al., Microwaves and RF (1983).

[5] Avantek catalogue 1986 on signal sources and filters.

[6] L. Cupido, A. Silva, F. Serra, M.E. Manso, P. Varela, R. Prentice and A. Costley, Proc. 21st EPS Conf. on Contr. Fus. and Plas. Phys., Montpellier, 18B, III, 1184 (1994)

Engineering aspects of an advanced heavy ion beam diagnostic for the TJ-II stellarator

A. Malaquias[a], C. Varandas[a], J.A.C. Cabral[a], L.I. Krupnik[b], S.M. Khrebtov[b], I.S. Nedzelskij[b], Yu.V. Trofimenko[b], A. Melnikov[c], C. Hidalgo[d], I. Garcia-Cortes[d]

[a]Centro de Fusão Nuclear, Associação EURATOM/IST, 1096 Lisboa Codex, Portugal
[b]Institute of Plasma Physics, NSC KhIPT, Kharkov, Ukraine
[c]Institute of Nuclear Fusion, RRC "Kurchatov Institute", Moscow, Russia
[d]Associacion EURATOM/CIEMAT, Madrid, Spain

An advanced heavy ion beam diagnostic has been developed for the TJ-II stellarator based on the simultaneous utilisation of two different detection systems for the secondary ions: a multiple cell array detector and a 30° Proca-Green electrostatic energy analyser. This innovative design aims at enlarging the HIBD capabilities in order to allow the instantaneous measurements of electronic density profile and local plasma potential together with their respective fluctuations. In this paper we present the engineering solutions adopted for this diagnostic implementation on this non-conventional approach.

1. THE HEAVY ION BEAM DIAGNOSTIC DESIGN

TJ-II is a medium size stellarator[1] with 4 periods, major radius of 1.5m and 1T typical magnetic field. Its various operation regimes are characterised by strongly different magnetic configurations and large differences on plasma radius. The relation between the poloidal and toroidal components of the magnetic field is high enough to determine sharp 3D trajectories for the probing ions. In addition two different plasma heating scenarios are used: electron cyclotron resonant heating (ECRH) and neutral beam injection (NBI) leading to quite different peak densities, $2 \times 10^{19} m^{-3}$ for ECRH and $1.5 \times 10^{20} m^{-3}$ for NBI. The Heavy Ion Beam Diagnostic (HIBD) design optimisation was based on full 3D trajectories calculations together with the computation of attenuation effects on primary and secondary ions, for all the TJ-II regimes and heating scenarios, using realistic profiles for temperature and density. A Multiple Cell Array Detector (MCAD)[2] is used to collect all the fan of secondary ions. The MCAD position was determined by the minimisation of the overlapping effect of tertiary ions over the secondary ions at the detection plane. The Electrostatic Energy Analyser (EEA) entrance slit position was chosen in order to collect secondary ions with minimum toroidal angular dispersion. In fig.1 we present the HIBD schematic layout obtained after optimisation. The HIBD main components are described in the next sections.

1.1. The ion gun and primary beam line

The ion gun is based on a thermoionic emitter source type, producing at its output a 30 µA Cs^+ beam with variable energy from 80 to 200 KeV. The beam diameter (~ 8mm) and divergence (~ 4mrad) are determined by the focusing properties of the accelerating tube. When energy is changed a proper voltage is applied to a specific electrode in order to maintain 'perveance match' and thus, keep optimum beam parameters. A 200 KV power supply based on diode multiplying scheme is used for beam acceleration. Its current limit of 1mA and ripple level of 5V ensures good energy stability minimising in this way the potential measurements error.

The primary beam line contains four sets of electrostatic plates for beam positioning and injection angle control. The first two pairs, are used to sweep the beam on the poloidal plane (± 8 degrees) through the plasma cross section and to

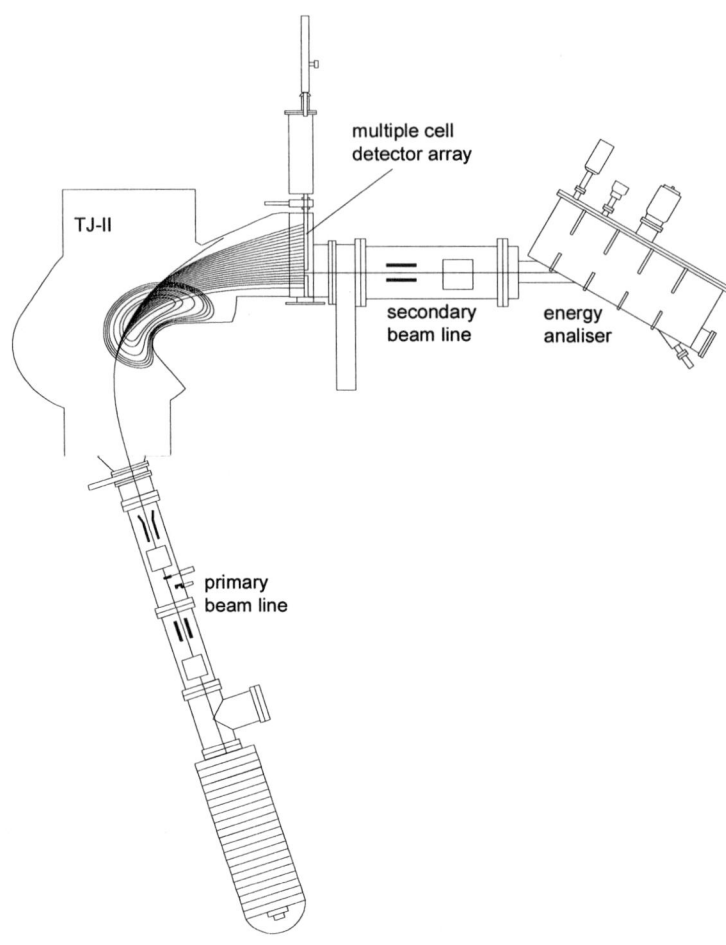

Fig.1 - Global layout of the heavy ion beam diagnostic for the TJ-II sttelarator

steer the beam on the toroidal direction (± 4 degrees). The last two pairs are used to maintain the beam alignment correction. The required angular deflections are provided by a 4 KV power supply with a 100V/µs response. With this power supply, the full plasma cross section can be scanned by the beam in ~ 0.1 ms. Between the two groups of electrostatic plates is placed a mesh detector for beam profile measurements and a faraday cup for beam intensity control.

The ion injector can be operated either continuously or in pulsed mode. The first pair of steering plates, is used to pulse the beam between the faraday cup detector and the injection line. Injecting the beam in this way allow us to control the beam intensity and measure the detectors noise before each injection pulse.

1.2. The multiple cell array detector

The MCAD is used to collect instantaneously all the secondary ions produced along the primary beam trajectory inside the plasma. It is composed by a large number of copper cells forming, in its standard configuration, a rectangular matrix of 30×4 with global dimensions of 35×25 cm² (fig.2).

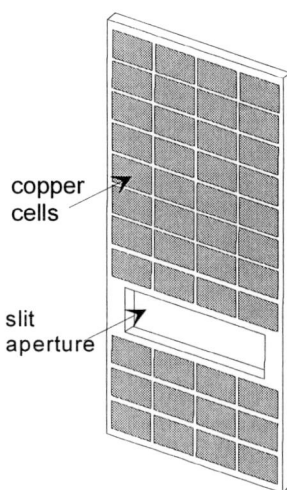

Fig.2 - Schematic of the MCDA. The support material for the copper cells was chosen to be TEFLON due to its good vacuum and thermal properties.

The cell dimensions and the number of rows and columns can be changed giving high flexibility to perform specific measurements. If we are interested in measuring toroidal correlation's, we may increase the number of columns in order to get more detailed information inside the secondary beam toroidal distribution. On the other hand, the rows number can be increased by reducing the vertical length of the cells, improving in this way the radial resolution.

The total cells number is limited by the available number of channels (120) and the minimum cell size is determined by the lower limit of detectable current (~ 2nA). It is clear that the cells size may also vary along the detector matrix if desired. Another advantage of such a detection system, is its flexibility to accept any detector configuration with more than 120 cells and allow us to play with the cell dimensions or with the number of rows and columns by adding the cell signals before the amplifiers.

During normal operation the detector is positioned in front of the output port of TJ-II covering all the port area. A slit (of 5×18 cm²) in the detector matrix allows part of the secondary ions to pass to the EEA. The slit position and dimensions are optimised in order to allow potential measurements in all of the plasma cross section for all the TJ-II regimes. The detector is inserted from the vacuum pre-chamber to its final position by means of a vacuum sealed motion bar (fig.1). All the detector cells are connected to an ending male plug through electrical routes distributed in the back of the detector. The electrical connections between the detector and the atmosphere side are performed by plugging the detector into a special flange containing a vacuum sealed female plug.

1.3. Secondary beam line and energy analyser

The secondary beam line is placed between the MCAD and the EEA. In this interface a pair of steering plates is used to perform active beam control i.e. secondary ions coming to the energy analyser will be toroidaly positioned at is entrance slit by means of these electrostatic plates.

This corrections on the trajectories do not significantly change the ions energy and allows to reduce the toroidal angular dispersion. The EEA is build from the conventional 30° Proca-Green energy analyser. However, due to the high flexibility of TJ-II operation it was impossible to find one common toroidal position for the steering plates positioning. In fact, the differences on the secondary ions toroidal positions and angles were found to be quit large from one TJ-II regime to other. Therefore, instead of a collection point an optimum collection line along the toroidal direction was found. Due to these constraints the steering plates in the secondary beam line and the EEA entrance slit and its internal electrostatic plates were designed to be moveable along the collection line and without vacuum violation. Various calibration proceedings for the EEA are foreseen due to its changeable geometry with TJ-II regimes.

The energy analyser is absent of guard rings and its layout was optimised by means of a computational code restricting the field non-uniformity to values below 5E-4 V/m. The voltage across the EEA plates will be maintained by a 50 KV power supply with 5V ripple and able to deliver currents up to 60 mA.

The secondary ions passing through the EEA slit, will be collected by an array detector of copper cells placed at the end of the EEA electrostatic plates. The detector cells will be distributed in 8 rows with enough toroidal length to ensure the secondary ions detection in all TJ-II regimes. These rows will be divided in to columns to noise reduction proposes as the beam dimensions are much smaller than the toroidal length of the cells.

2. THE CONTROL AND DATA ACQUISITION SYSTEM.

The control and data acquisition system (CDAS) for this experiment is based on VME bus. This standard was chosen due to its characteristics of self running system and high performance in data handling. Also, its flexibility for implementation of multiple purposes boards allow us to easily upgrade the system accompanying the last evolution on data control and acquisition with relative low costs.

The current signals on the MCAD cells and on the EEA array detector will be conditioned by fast amplifiers (band width of 500 KHz) to acceptable values of amplitudes ($\pm$ 5 Volt) for the digitizers. Remotely programmable gain amplifiers controlled by means of multiplexed DAC channels will be used due to the fact that the estimate values for the detector currents can vary a few orders of magnitude with the operational TJ-II regimes.

The total number of ADC channels is around 120 with acquisition rate up to 1.2 MHz. The data will be stored during the acquisition cycle on memory blocks of 512 Kbytes per channel.

The power supplies control will be performed by non multiplexed DAC channels with actualising frequency of 1 MHz allowing us to sweep or steer the beam between two close points in the plasma at high frequencies for radial corrclation's measurements purposes.

3. CONCLUSIONS

A non conventional heavy ion beam diagnostic has been designed for the TJ-II stellarator based in the combination of two different detection systems: a traditional 30° Proca-Green electrostatic energy analyser and a multiple cell array detector. In this new approach it will be possible to simultaneously measure the plasma density and potential profiles together with their fluctuations, enlarging thus the resolution for transport flux determination by means of the Powers relation[3].

REFERENCES:

[1] - *C. Alejaldre, J. Alonso, J. Botija, F. Castejón et al.,* Fusion Technology 17, 131 (1990).

[2] - *J.A.C. Cabral, A. Malaquias, C. varandas, A. Praxedes and W. Toledo* IEEE Transations on Plasma Science V22 N4 AUG1994.

[3] - *T. P. Crowley et al.* IEEE Transations on Plasma Science V22 N4 AUG1994.

Two-Frequency Correlation Reflectometer

J. Fernandes, A. Silva, L. Cupido, M.E. Manso, P. Varela

Associação EURATOM/IST - Centro de Fusão Nuclear, Instituto Superior Técnico,
Av. Rovisco Pais n°1, 1096 Lisboa Codex, Portugal

An O-mode correlation reflectometer, broadband tuneable in the frequency range 18-26.5GHz, is being developed to study density fluctuations on the ISTTOK tokamak namely to determine the fluctuations correlation length. In this paper the basic principles of correlation reflectometry measurements are presented and the developed microwave and control systems for the diagnostic are described.

1. INTRODUCTION

The reflection of electromagnetic waves from a plasma cut off layer enables the study density fluctuations in fusion plasmas [1]. The phase variations of the reflected signal have been interpreted as due to fluctuations located close to the reflecting layer. Assuming a WKB approximation, the total phase change suffered by the incident wave, after the propagation and reflection in the plasma is given by:

$$\phi = \frac{4\pi f}{c} \int_a^{x_c} N(x,f)\,dx - \frac{\pi}{2} \quad (1)$$

where x_c is the location of the cut off layer, a is the position where the plasma density is zero and $N(x,f)$ is the refractive index of the plasma[2]. In O-mode, the electric field of the probing wave is parallel to the magnetic field (E‖B). In this case, the refractive index is:

$$N_o = \sqrt{1 - \frac{f_{pe}^2}{f^2}} \quad (2)$$

where f_{pe} is the plasma frequency.

Each wave in the fluctuation spectrum has a frequency (ω), a wavelength (Λ) and a finite coherence length (l_c). When $\Lambda \gg \lambda$ (λ is the wavelength of the probing wave) the scattering effects are negligible and the reflected signal exhibit phase modulations ($\Delta\phi$) that are mainly due to fluctuations localised in the reflecting region [3].

Probing the plasma with two signals with slightly different frequencies along the same radial chord the reflection occurs at two different radial locations separated by a small distance Δx.

The power that is common to both reflectometers is determined by the crosspower spectrum $G_{12}(\omega)$. As Δx increases, the power decreases according to the value l_c. The coherence function $\gamma_{12}(\omega)$ measures the power common to both signal, normalised to the total power in each signal.

$$\gamma_{12}^2(\omega) = \frac{|G_{12}(\omega)|^2}{G_1(\omega)G_2(\omega)} \quad (3)$$

The correlation length is found by varying the probing frequencies and analysing the variation of γ_{12} as a function of Δx.

2. MICROWAVE SYSTEM

The ISTTOK tokamak is provided with a reflectometry diagnostic to measure density profiles. The system shall be upgraded to measure density fluctuations. The new correlation reflectometer of the ISTTOK tokamak (fig. 1) consists of two tuneable microwave sources, an Hyperabrupt Varactor-Tuned Oscillator (HTO) with an active duplicator and an YIG-Tuned Oscillator (YTO).

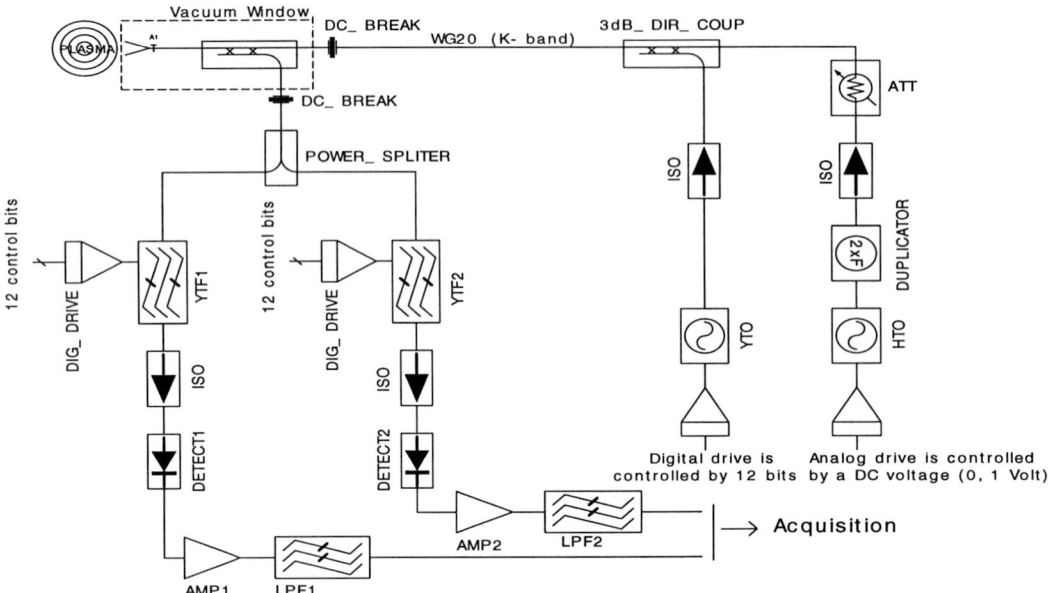

Figure 1. Microwave System

The two signals are combined through a 3 dB directional coupler and are launched into the plasma along the same radial chord. A single pyramidal antenna is used both to emit and receive the probing microwave signal. A pin placed near the antenna is used to obtain a reference signal.

To separate the information produced by the two emitting signals we use two band-pass YIG-Tuned Filters (YTF). The center frequency of the filters can be set digitally in the range [18.0, 26.5GHz]. The detection is homodyne and uses two broadband detectors to mix the plasma and reference signals. The reflected signals, after being detected, are amplified with broadband amplifiers and filtered by low-pass filters, with programmable gain and cut-off frequency. The system uses standard waveguides for the K-band (WG20).

3. CONTROL SYSTEM

The control of the diagnostic (fig. 2) is made through a PC. The distance between the computer and the system is about 15 meters, which permits, to use an interface RS232. This communication is asynchronous and the transmission velocity is 4800 baud. The acquisition and the synchronisation of the diagnostic is controlled by the central system, that communicates also with the PC operator. Parameters that can be controlled are: the frequency of the microwave sources; the central frequency of the microwave filters; the amplifiers gain and the cut-off frequency of the low-pass filters. The control also switches the diagnostic between the two modes of operation (sweeping/fixed frequency). An I/O control board was developed to permit the communication between the computer's operator and the diagnostic.

4. I/O CONTROL BOARD

Figure 3 shows the block diagram of the I/O control board. This board is based on the 8031 microcontroller which is programmable and permits a serial communication (RS232). The velocity of data transmission and the format of the characters are also programmable. The communication can be bi-directional.

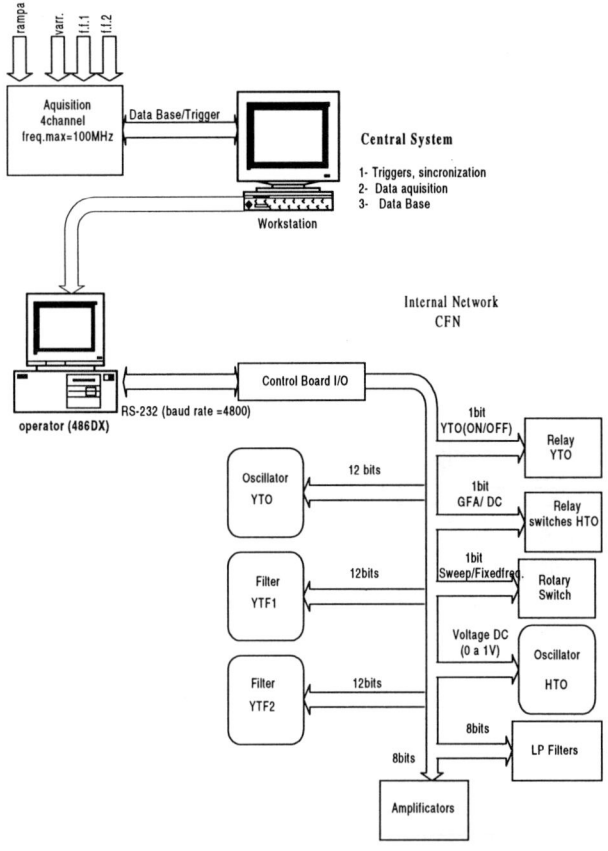

Figure 2 - Control System

The outputs of the board are: a DC voltage (0 to 1V), that controls the HTO frequency, and 72 programmable input/output pins (TTL) to control the other parts of the system.

5. SIGNAL AMPLIFICATION AND FILTERING

Before being acquired, the signals are amplified and filtered. A board with four channels of amplification was projected and implemented based on OPA621BB. These amplifiers have four different levels of programmable gain controlled by a 8 bit word and are wide bandwidth (LB=30MHz) and low noise.

In order to reduce the influence of the high frequency noise and also aliasing effects, low pass filters (Butterwoth) have been projected. These filters have four programmable cut off frequencies (f_c) and the input and output impedance is 50 Ω. The frequencies f_c of all filters are controlled by a 8 bit word.

6. LABORATORIAL TESTS

Table 1 shows the results of laboratorial tests of individual microwave components. It was found that the temperature drift of these components is not significant for the expected maximum temperature variation of the system of 2°C.

7. FUTURE WORK

Future work concerns the characterisation of the system performance at the laboratory. Experiments aim also at the determination of the minimum probing frequency separation to avoid cross-talking between the two probing waves. Finally of the phase shift produced by the periodic movement of a metallic mirror shall be measured.

The tests on the Tokamak ISTTOK will be the final step for the characterisation of the system

Components	frequency linearity	output power	Temp. Drifts
HTO+Act. Duplicator	non linear	20.5 - 22.5 dBm	≈2 MHz/°C
YTO	good	16.5 - 20.5 dBm	≈0.5 MHz/°C
YTF	very good	in. loss ≈ 7 dB	≈0.1 MHz/°C

Table 1 - Laboratorial measurements

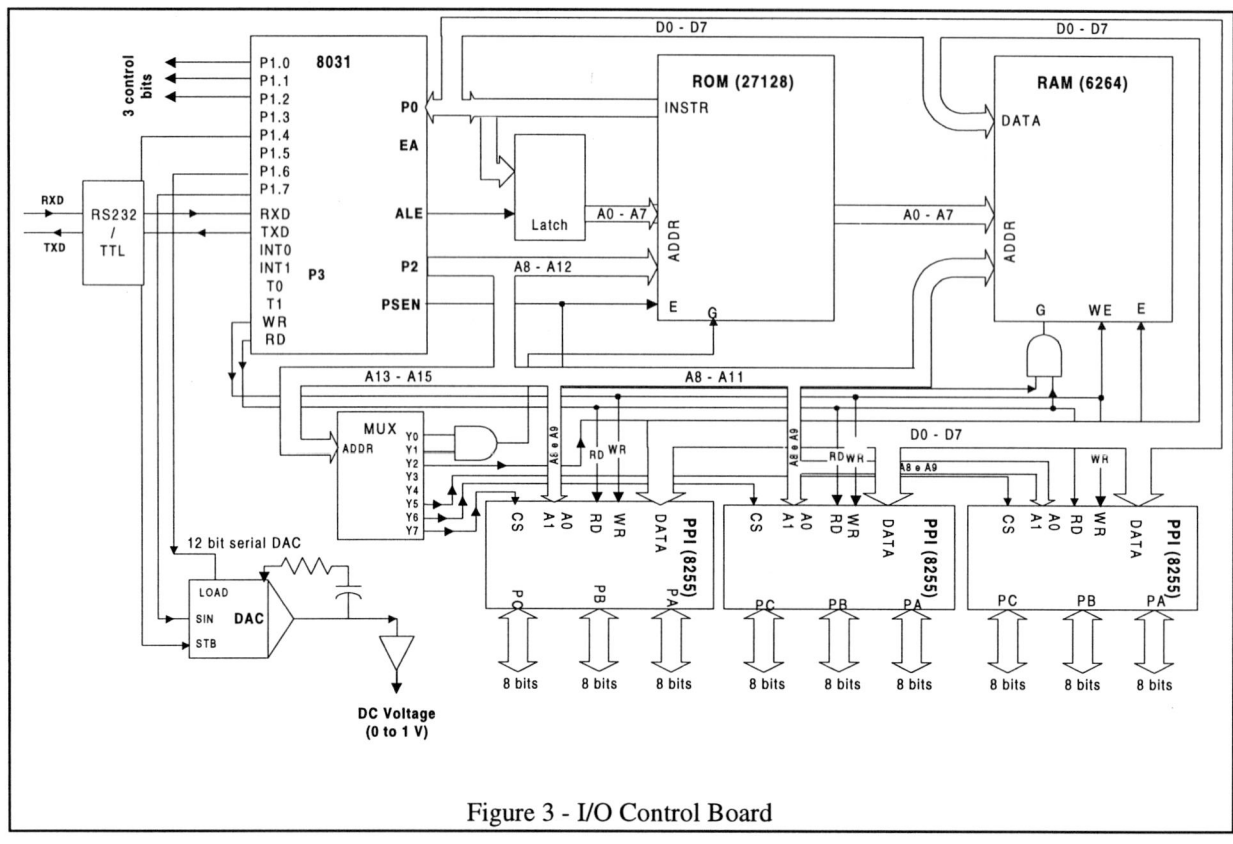

Figure 3 - I/O Control Board

8. CONCLUSIONS

We are developing a two-frequency correlation reflectometer to measure density fluctuations properties in the ISTTOK tokamak. The diagnostic is presently being installed in the tokamak ISTTOK. The system has been designed, constructed and tested in the laboratory. This article presents the microwave circuits and the operator that enables to change by remote control the parameters of the diagnostic and the mode of operation, in between shots. Parameters that can be modified are: the frequencies of operation, the amplification levels; the cut-off frequencies and the sampling rate.

Tests in the laboratory confirm the good performance of the system but showed also that temperature is a critical parameter for the stability of some microwave components (YIG oscillators and the YIG band pass filters).

The diagnostic has the possibility to probe very closely spaced plasma layers and therefore will enable to measure the coherent function of the plasma fluctuations with good spatial resolution.

REFERENCES

1. M.E.Manso et al. A microwave reflectometric system for the ASDEX tokamak. Proceedings on 15th European Conference on Controlled Fusion and Plasma Physics, pages 1127-1130. European Physical Society, June 1988.

2. P. Varela. Sistema de Reflectometria para o Tokamak ISTTOK. Master Thesis, Technical University of Lisbon, I.S.T, Sept. 1993.

3. E. Mazzucato. Density Fluctuations in the A.T.C. Matt-1151, Report of the Plasma Physics Laboratory, Princeton University, Sept 1975

Plasma density profile evaluation in broadband reflectometry using a neural network

F. D. Nunes[a,b], J. Santos[b] and M. E. Manso[b]

[a]Instituto de Telecomunicações,
[b]Centro de Fusão Nuclear, EURATOM/IST Association,
Instituto Superior Técnico - Av. Rovisco Pais, 1096 Lisboa Codex, Portugal

Broadband reflectometry is a diagnostic suited to measure plasma density profiles in present and forthcoming fusion devices like ITER. A major issue is the possibility of using profile measurements from reflectometry at the plasma boundary as a reference for the magnetic diagnostics in long pulse discharges. This requires fast algorithms to process data on-line. Here we present an approach to profile evaluation, based on a neural network, that has great potentialities to meet the speed and accuracy requirements for the position control reflectometer on ITER. The proposed network is a two-layer perceptron which is trained with a set of expected density profiles. We adopted a Fourier series to represent the profile, thus the network output is a set of Fourier coefficients and the entire profile can be conveniently reconstructed with a small number of parameters. Since the Abel inversion integral is not used the algorithm is very fast. In addition, simulations show that the proposed solution is robust to input noise.

1 Introduction

Frequency-modulation broadband reflectometry is a radar technique that uses frequency-swept electromagnetic waves for probing inhomogeneous plasmas in fusion devices (eg, tokamaks). For O-mode waves the spatial distribution of the plasma electron density (density profile) is given by the Abel inversion integral according to [1]

$$X_c(F) = X_0 - \frac{c}{\pi} \int_0^F \frac{\tau(f)\,df}{\sqrt{F^2 - f^2}}, \quad (1)$$

where $X_c(F)$ denotes the location of the reflecting layer for a wave of frequency F. The corresponding electron density is $n_e = (4\pi^2\epsilon_0 m_e/e^2)F^2 \approx (F/8.979)^2$ [mks], where ϵ_0 is the free-space permittivity, e and m_e are, respectively, the charge and mass of the electron, X_0 is the position of the plasma edge ($n_e = 0$); c is the speed of light and $\tau(F)$ is the (round-trip) group delay associated with the reflected wave.

Several methods are currently utilized for determining the group delays from the reflectometric signal which results from mixing the reference and the plasma reflected waves (see, for example, [2] and the references therein).

The evaluation of (1) with good accuracy usually involves a computationally demanding numerical integration over a wide range of frequencies which can take too long if the diagnostic is used for monitoring the plasma in real-time. Thus, we seek for an approach to the profile evaluation that is computationally efficient, accurate and robust to noise.

Assume that the density profile $n_e(x)$ can be well-approximated by the following truncated Fourier series

$$\tilde{n}_e(x) = \sum_{n=1}^{N} \gamma_n C_n(x), \quad 0 \leq x \leq X_0, \quad (2)$$

with

$$C_n(x) \equiv \cos\left[\pi\left(n - \frac{1}{2}\right)\frac{x}{X_0}\right], \quad (3)$$

where, in general, a small number of Fourier coefficients in the vector $\Gamma = (\gamma_1, \gamma_2, \ldots, \gamma_N)$ is sufficient to represent the profile with a good accuracy. The problem of profile evaluation can be stated as follows: given the vector $T = (\tau_1, \tau_2, \ldots, \tau_M)$, with $\tau_m \equiv \tau(m\Delta F)$, where ΔF is a constant frequency increment, determine Γ such that the associated profile, $\tilde{n}_e(x)$, is a good approximation to the original profile, $n_e(x)$.

2 Proposed solution

The problem in hand is twofold consisting basically in (i) finding the nonlinear mapping from T to Γ when the density profile $\tilde{n}_e(x)$ is known, and (ii) computing new profiles from the available data, T, based on that mapping.

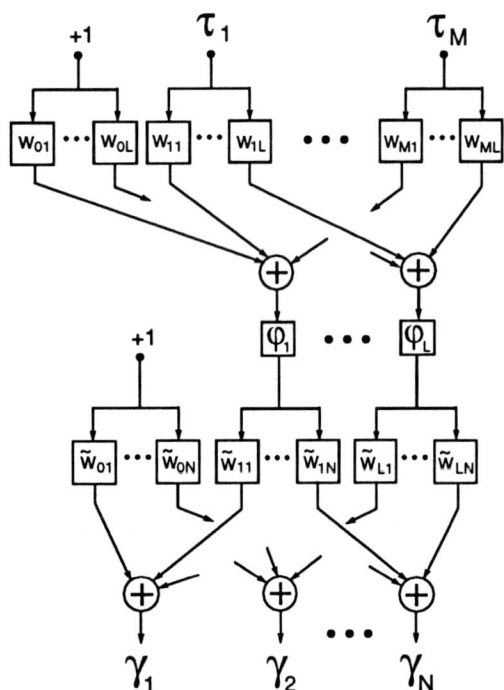

Figure 1. Adopted neural network

The problem formulation suggests the use of a neural network to estimate the density profile from T. In fact, neural networks provide general mechanisms for approximating mappings from data applying nonlinear optimization techniques to select the appropriate models. Besides being widely utilized in image classification, data compression, etc., these structures have also found application in nuclear fusion problems, [4], [5].

The proposed solution is a two-layer perceptron (see Figure 1) which is able to approximate, with arbitrary accuracy, any nonlinear mapping, $\mathcal{N}$, given a training set represented by the set of inputs T and desired (target) outputs $\Gamma = \mathcal{N}(T)$. A sufficient condition is that the activation functions φ_l in the hidden-layer are sigmoidal and the number of neurons is large enough [3]. The perceptron has M inputs, L units in the hidden-layer and N outputs. The adopted sigmoid function is the hyperbolic tangent. The perceptron outputs are given by

$$\gamma_n = \tilde{w}_{0n} + \sum_{l=1}^{L} \tilde{w}_{ln} \varphi_l \left(w_{0l} + \sum_{m=1}^{M} w_{ml} \tau_m \right),$$
$$n = 1, \ldots, N, \qquad (4)$$

where w_{ml} and $\tilde{w}_{ln}$ are (synaptic) weights to be computed during the supervised learning (training phase).

3 Network training

Figure 2 describes the operations performed during the network training. A set of S density profiles with realistic shapes is randomly produced according to the following procedure:

(i) A piecewise linear profile $\overline{n}_e(x)$ with nodes $(x_r, n_e(x_r))$, $r = 0, 1, \ldots, R$ is generated, where x_r and $n_e(x_r)$ are random quantities verifying $0 \leq x_{r+1} < x_r \leq X_0$ and $n_e(x_{r+1}) > n_e(x_r) \geq 0$.

(ii) The profile $\tilde{n}_e(x)$ defined in eq. (2) results from $\overline{n}_e(x)$ by using the least squares approximation criterion (see, for instance, [6], chapter 22). Let

$$(f, g) = \int_0^{X_0} f(x)g(x)\,dx,$$

be the inner product between generic functions $f(x)$ and $g(x)$ existing in $(0, X_0)$ and $\|f\|^2 = (f, f)$. As the functions $C_n(x)$ defined in (3) are orthogonal in the interval $(0, X_0)$, ie $(C_k, C_n) =$

$0, k \neq n$, the approximation to $\overline{n}_e(x)$ that minimizes the quadratic error $\|\overline{n}_e(x) - n_e(x)\|^2$ is given by eq. (2) with $\gamma_n = (\overline{n}_e, C_n)/\|C_n\|^2$.

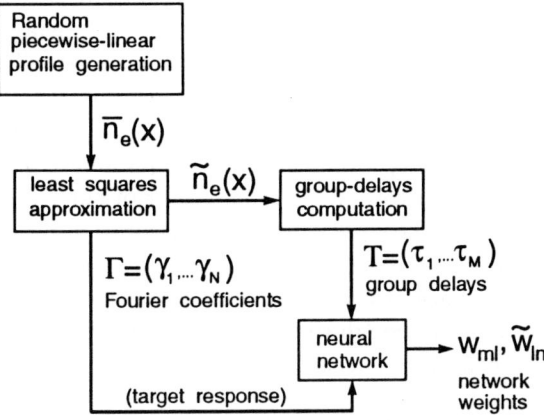

Figure 2. Training phase of the perceptron

The group delays associated to the density profile $\widetilde{n}_e(x)$ are given by [1]

$$\tau(F) = \frac{2}{c} \int_{X_c(F)}^{X_0} \left[1 - \left(\frac{F_p}{F}\right) \right]^{-1/2} dx,$$

where $F_p \approx 8.979\sqrt{\widetilde{n}_e(X_c(F))}$ [mks] is the plasma frequency corresponding to the reflecting layer $X_c(F)$. Since, in general, this integral has no analytic solution each perceptron input, τ_m, is evaluated numerically.

The network is trained in a supervised manner with a set of S density profiles aiming to determine the optimal synaptic weights. We adopted the well-known back-propagation algorithm [3] to accomplish the optimization criterion which consists in minimizing the square error

$$E = \frac{1}{2} \sum_{n=1}^{N} (\gamma_n - \widehat{\gamma}_n)^2,$$

averaged over the S profiles of the training set, where $\widehat{\gamma}_n$ is the actual response of the n.th perceptron output and γ_n the corresponding desired value. To apply the back-propagation process a known input vector T is supplied to the network and an estimate $\widehat{\Gamma}$ of the target vector Γ is produced. The error $\widehat{\Gamma} - \Gamma$ is then propagated backward through the perceptron and the weights are adjusted so as to make the actual response of the network move closer to the desired response. At iteration $m+1$ each weight w_{ij} (or $\widetilde{w}_{ij}$) is updated according to

$$w_{ij}^{(m+1)} = w_{ij}^{(m)} - \eta \frac{\partial E}{\partial w_{ij}^{(m)}},$$

where the learning-rate parameter η controls the amount of change between successive iterations.

4 Simulation results

Our tests were conducted using a perceptron with $M = 50$ inputs, $N = 6$ outputs, a learning-rate parameter $\eta = 0.0025$ and a training set $S = 1000$ profiles. The results of Figure 3 refer to the profile mean square error averaged over 100 profiles not included in the training set. The plots show that better accuracies are achieved by increasing the number of hidden-layer units although the improvements are only marginal for $L > 80$.

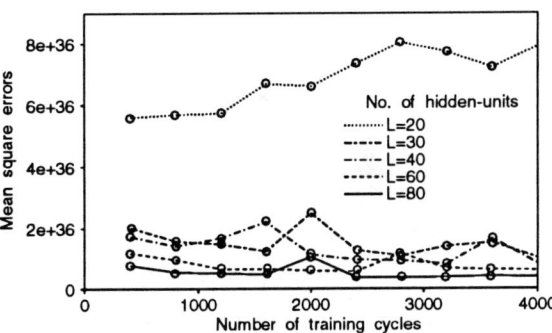

Figure 3. Mean square errors versus the number of training cycles for several values of L

Figure 4 depicts an example of profile reconstruction using a perceptron with $(M, L, N, S) = (50, 80, 6, 1000)$. The solid line corresponds to the exact profile (typical in H-mode plasmas) and the dashed line to the reconstructed one. The network has been previously trained with 4000 training cycles, thus adding up to 4 million iterations.

To test the robustness of this approach to noise (which affects the group delays derived from experimental reflectometric data) we added a zero-mean Gaussian vector to the input, T, characterized by a covariance matrix $10^{-19}I$, where I is the identity matrix. The profile was then evaluated with the neural network (dash-dotted line) and with the Abel inversion integral (dotted line). The results show that the perceptron yields a smoothed profile similar to the noiseless solution (except at the density plateau), whereas the Abel integration produces a slightly erratic profile specially at the plasma edge. Thus, the use of neural networks may be beneficial even when the reflectometric data is noisy (as is often the case).

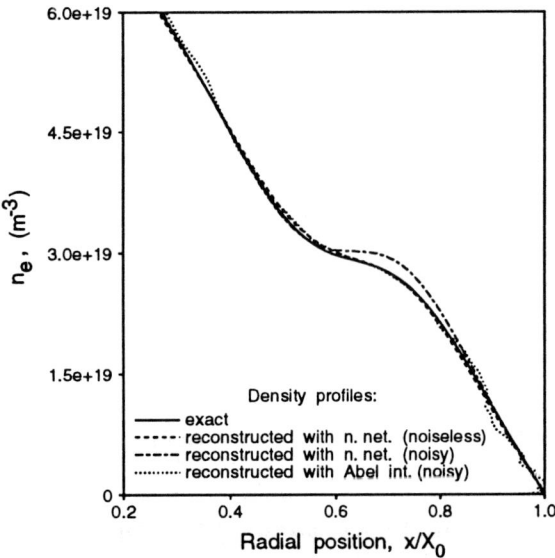

Figure 4. Profile reconstruction using the perceptron with noiseless and noisy inputs

5 Conclusions

We propose a new algorithm to evaluate density profiles from reflectometry which is faster than the standard Abel inversion integral. The approach resorts to a two-layer perceptron whose training is intensive (and, thus, must be performed off-line) but exhibits a small computational burden when used to reconstruct new profiles. In fact, if speed is at a premium, an hidden-layer with $L = 30$ units could be employed with satisfactory results, as shown in Figure 3. Of course, performance is improved by increasing L and by carefully selecting the training set. The selected value of L will depend basically on the specific diagnostic application.

The density profile is defined parametrically, thus the network output is a vector of Fourier coefficients. The parametric representation allows to store and retrieve information in a very efficient way because a few coefficients (typically $N = 6$) are usually enough to approximate the entire profile with a good accuracy. This aspect may be important when dealing with large amounts of data as significant computer memory savings can be obtained.

Future work will concern the application of the new method to experimental data and the study of faster training algorithms.

References

[1] J. Doane, E. Mazzucato and G. Schmidt, Rev. Sci. Instrum. 52(1), Jan. 1981.

[2] C. Laviron, A. Donné, M. E. Manso and J. Sanchez, Plasma Phys. Control. Fusion 38(1996) 905-936.

[3] Neural Networks - A Comprehensive Foundation, S. Haykin, MacMillan, N. Y., 1994.

[4] L. Allen and C. Bishop, Plasma Phys. Control. Fusion, 34(1992), 1291-1302.

[5] E. Luna, V. Zhuravlev, B. Brañas, J. Sanchez, T. Estrada, J. Segovia, J. Oramas, proceedings of the 20.th EPS Conference on Controlled Fusion and Plasma Physics, vol. 17C, part III, pp. 1159-1162, Lisbon, 1993.

[6] Handbook of Mathematical Functions, M. Abramowitz and I. Stegun, editors, Dover Publications, N. Y., 1968.

Time resolved energy dispersive X-ray diagnostic for the TCV tokamak

J. Sousa[1], P. Amaro[1], P. Amorim[1], B. Duval[2] and C.A.F. Varandas[1]

[1] Associação EURATOM/IST, Centro de Fusão Nuclear,
Instituto Superior Técnico, 1096 Lisboa Codex, Portugal.

[2] Centre de Recherches en Physique des Plasmas, EPFL,
Association EURATOM/Confédération Suisse, 1015 Lausanne, Switzerland.

A time resolved energy dispersive X-ray diagnostic is being developed for the TCV tokamak (CRPP - Lausanne) to measure the evolution of the plasma impurities, runaway electrons and electron temperature. A liquid nitrogen cooled Ge diode detects the X-ray photons which are processed by a spectroscopic amplifier and a locally developed interface amplifier and timing generator (IATG) unit. The energy spectrum is obtained using a fast digitiser and a software histogramming algorithm. These electronics components have been optimised to improve the data throughput to match high flux 2 seconds time duration of a TCV plasma pulse. This paper describes the diagnostic hardware with particular emphasis on the IATG unit.

1. INTRODUCTION

Measurement of the X-ray spectrum by means of pulse height analysis is used nowadays by many tokamaks to identify the main high Z plasma impurities, to measure the electron temperature (T_e) and to detect the runaway electrons [1]. This paper describes an energy dispersive X-ray diagnostic that has been developed for the TCV tokamak [2] aiming at performing time resolved measurements in L and H mode regimes [3].

Section 2 contains the general description of the diagnostic, which is based on a liquid Nitrogen cooled solid state Germanium diode. This detector is sensitive to X-rays in the 1-100 keV range, with energy resolution of about 130 eV. Section 3 presents the specially developed electronic unit that is used to post-reconstruct a sequence of groups of X-ray events which arrive during periodic intervals of time. Section 4 describes the installation of the diagnostic on the TCV tokamak and section 5 includes the experimental results.

Finally section 6 contains the conclusions and planned system enhancements.

2. GENERAL DESCRIPTION OF THE DIAGNOSTIC

A block diagram of the hardware is shown in Fig. 2. Each photon event generates a current pulse in a liquid Nitrogen cooled solid state Germanium diode (Canberra GUL0055P) which is pre-amplified (Canberra 2008) and further shaped by a gated integrator (GI) (ORTEC 673). Time coincident pulses that result in overestimated photon energies measurements (pileup) are detected and omitted from the output. The amplitude of the positive unipolar pulses is recorded by a CAMAC digitiser module (INCAA TRCH) which has 12 bits resolution, 1 Mega samples of memory and an 1 µs cycle time. The spectrum is determined by a software histogramming of the acquired pulse heights. The post-reconstruction of the arrival time of a group of events is permitted by a periodic (1-50 ms) negative time pulses of linear amplitude variation, added in the signal path,

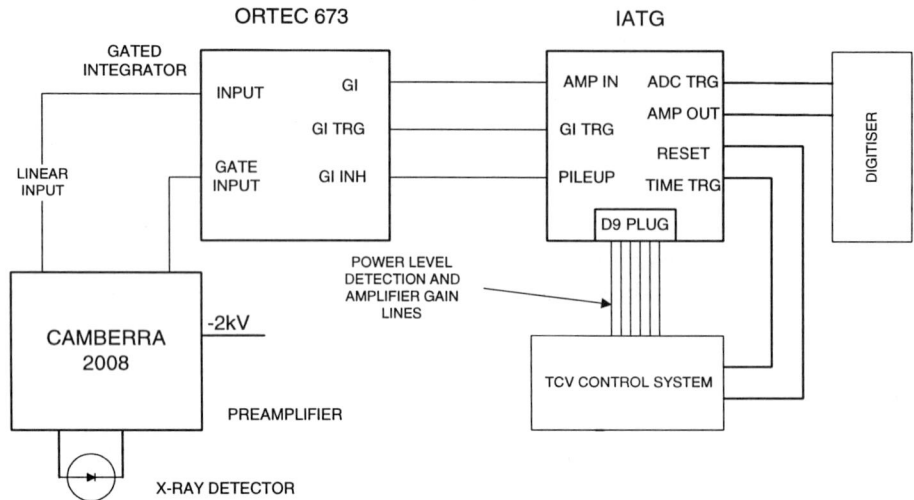

Fig. 2. Block diagram of the hardware setup.

by the dedicated electronic unit described in the next section.

3. INTERFACE AMPLIFIER AND TIMING GENERATOR MODULE

The specially developed interface amplifier and timing generator (IATG) module receives a programmable train of periodic pulses from the TCV timing system for generating the periodic time pulses. When a collision occurs between a signal pulse and a time pulse it discriminates them by delaying the second event for a time of 1 µs.

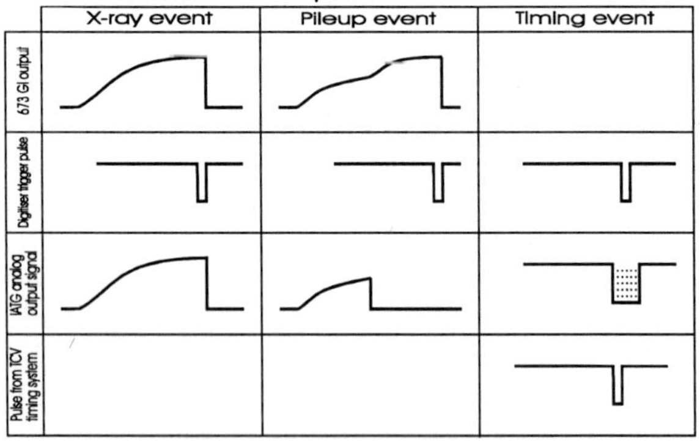

Fig. 1. Timing diagram of IATG.

The main requirements of this module are:
a) It should generate a pulse to trigger the digitiser whenever an event peak occurs. In Fig. 1, under the heading "X-ray event", is depicted the integrated event pulse at the GI output and the generated IATG trigger output.
b) Whenever a pulse pileup is reported by the gated integrator, the signal at the digitiser input should be zero. In Fig. 1, under the heading "Pile-up event", we can see the IATG output analog signal falling to zero immediately after the pileup occurrence.
c) It must output the time information that comes from the TCV timing system to the digitiser, in the form of a negative value which amplitude depends on its time position. This situation is depicted in Fig. 1, under the heading "Timing event".
d) It must prevent collision between time and event pulses by queuing them in two time slots of 1 µs. In this process an error of ±1 event pulse can occur for that spectrum, which is similar to the error of closing the gate in a multichannel analyser.

The IATG unit can be graphically represented as a group of interconnected functional blocks which is presented in Fig. 3.

The programmable gain amplifier (PGA) amplifies the event pulses that come from the GI output and can be set for gains of 1, 2, 4 and 8. Each pulse peak value is stored in an track and hold device (T&H) during a 1 µs (single event) or 2 µs (when a pulse coincidence occurs) interval. The holding time is defined by the GIPULSE, HOLD1 and HOLD2 variables.

An analog multiplexer selects one of these three signals: stored event pulse (AGI), timing value and zero volt. The selection depends on the levels/transitions of the three signal sources: PILEUP, GIPULSE and TPULSE. The multiplexer output is connected to an analog buffer of low output impedance, driving loads down to 50 ohm. The output (ANLOUT) is then connected to the digitiser.

TPULSE comes from the TCV timing system and provides the clock for generating the timing values. Each time a transition occurs in this input, a binary counter (COUNTER) is incremented and a digital to analog converter (DAC) produces a parametric time signal (ATIME).

The coincidence discriminator generates one trigger pulse for a each single event or a sequence of two pulses when a event coincidence happens. The buffered output (DIGOUT) of this signal is connected to the digitiser trigger input. If a coincidence in time of HOLD1 (GIPULSE, 1 µs width) and TP (TPULSE, 1 µs) happens, two trigger pulses within an interval of 1µs are generated, and for each one the multiplexer selects (through ASEL and PILE) the corresponding analog signal: AGI, ATIME or 0V.

The power present detector provides remotely readable control signals indicating the module's operational state.

The mixed analog-digital design and asynchronous nature of this circuit has implied a careful design of the analog block, specially on the printed circuit and power supply decoupling, in order to suppress the digital noise and guaranteeing the specified voltage resolution.

4. IMPLEMENTATION ON THE TCV TOKAMAK

The diode cryostat forced a horizontal mounting on an equatorial TCV port (Fig. 4). In the inherently "noisy" electrical environment of a fusion experiment, the diode, vacuum pumps and tokamak were all separately earthed and the electrical signal and power cables bundled together to minimise the effect of fluctuating magnetic

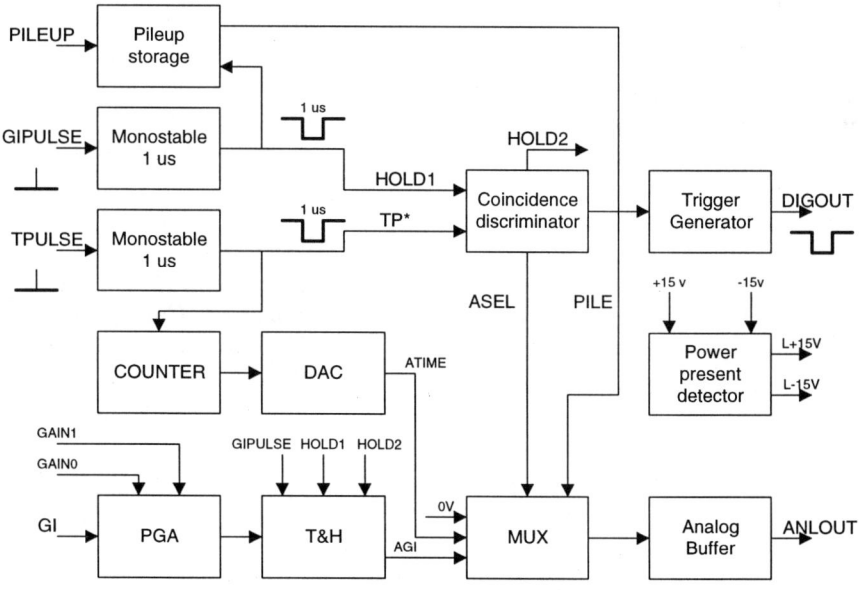

Fig. 3. Block diagram of the IATG unit.

fields. The diode was mechanically coupled via a high vacuum flight line (< 10-7 torr) to a TCV port which holds the possibility of externally change the viewing aperture and the introduction of Be filters into the flight line. For calibration purposes, it is also possible to place a Fe55 5.9 keV radioactive source in one of the filter positions. A pair of apertures with 0.3 mm diameter at each end of the flight tube limits the atendue of the detector to the plasma to 1.2×10^{-6} Str m^{-2} in order to set a maximum observed count rate of 100 kHz in L-mode discharges.

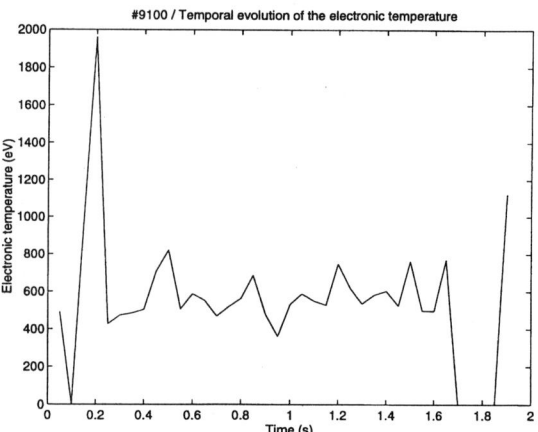

Fig. 5. Evolution of the electron temperature during a TCV discharge.

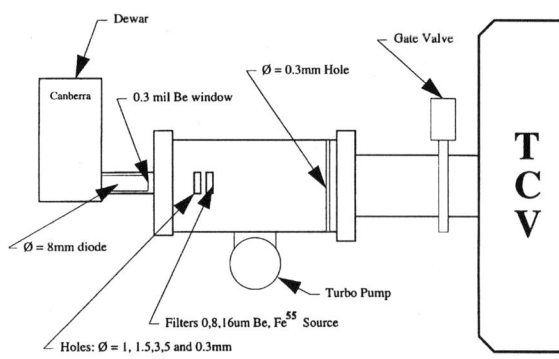

Fig. 4. Schematic view of the diagnostic.

5. RESULTS

The experimental results have shown that energy resolved spectra can be obtained during a TCV discharge. The values of the electron temperature are in good agreement with those determined with other diagnostics.

Fig. 5 shows the evolution of the electron temperature during a TCV discharge.

Another paper [4] presents the detailed studies of the influence of the apertures, Be filters and pulse pileup on the observed count rate and spectral shape. We have concluded that the maximum count rate is currently limited by the analog signal processing electronics.

6. CONCLUSIONS AND FUTURE ENHANCEMENTS

An energy dispersive diagnostic has been successfully mounted on TCV for the analysis of the X-ray emission from a variety of plasma configurations. The interface amplifier and timing generator unit has allowed time resolved measurements of the electron temperature.

Future enhancements are needed in the analog electronic aiming at improving its operation in the electrically noisy TCV environment and to solve the pileup problems that presently limit the operation of this diagnostic.

REFERENCES

1. I. Hutchinson (1987). "Principles of Plasma Diagnostics", Cambridge University Press.
2. F. Hoffman et al (1994). Plasma Physics Controlled Fusion, B36, 277.
3. H. Weisen et al (1996). Plasma Physics Controlled Fusion.
4. B. P. Duval et al. Proceedings of the X-Ray spectroscopy conference, Lisbon 1996.

Experimental tests of an ion source for the diagnostic neutral beam injector of the TEXTOR tokamak.

G.F.Abdrashitov, A.A.Ivanov, V.V.Mishagin, A.A.Podyminogin,

A.I.Rogozin, I.V.Shikhovtsev

The Budker Institute of Nuclear Physics, Novosibirsk, Russia

The beam extraction was experimentally studied using the prototype whose structure is the same as it is developed at the Budker INP for the beam emission spectroscopy in the TEXTOR tokamak. As result of the test the modulated ion beam 30keV, 0.8A, 0.3s with angular divergence below 1° was obtained. The experimental data with results of the numerical simulations are compared.

1. INTRODUCTION.

This work is an extension of our previous research on the development of the neutral beam injector for diagnostics in the TEXTOR tokamak [1,2]. Previously developed design of the ion source as a main part of the injector was realized in the ion source prototype which was used to study the extracted beam parameters.

The main objectives of the work were the following:
1. To demonstrate experimentally at scaleable beam parameters that the proposed RF-driven ion source can produce a low divergence (less then 1°) modulated beam.
2. To study the beam focusing by the multi-holes conceived ion optics grid system.
3. To assess the beam composition.

The ion source prototype tested and the results obtained are considered below.

2. ION SOURCE PROTOTYPE.

Layout of the ion source prototype is shown in Fig.1. Our basic design philosophy was constrained to a four grid type extractor in oder to minimize the beam divergence.
The ion optics system 8 comprises grids which are made of molybdenum. The grid thickness was chosen to be 2mm instead of 4mm assumed for 50 keV ion source. The diameter of the apertures in each electrode is 3.8mm. The apertures are arranged in hexagonal structure with a step of 5mm at the entrance grid, so that the transparency of the grids is about 50%. In order to obtain geometrical focusing of the beam the grids are made as spherical segments with a curvature radius 2m. To achieve maximum power density at a certain point the 151 beam

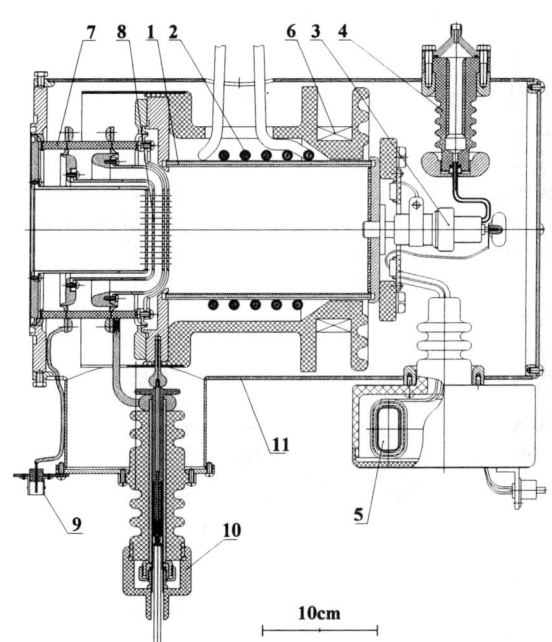

Fig.1.General layout of the ion source prototype. 1-quartz tube. 2-RF-antenna. 3-gas-valve. 4-gas-inlet. 5-transformer with H/V insulation. 6-magnetic field coil. 7-ion optics system. 8-grids. 9,10-feed through. 11-magnetic screen of ion source.

lets of the source were thus aimed at a common point 200cm from the grids. The correct elementary beams focusing requires displacement of the beamlets in the grids with respect to each other which was done by changing of the step of the structure from grid to grid proportionally to the gaps between them.

A hydrogen plasma used as an ion emitter is produced by an inductively excited RF-discharge with a frequency of 6 MHz in a nonhomogenious magnetic field of max.60 G inside quartz tube 1 by the multi-turn antenna 2. Coil 6 generates the magnetic field used to provide the required plasma uniformity [1].

Hydrogen gas was puffed into the discharge chamber using a electro-magnetic gas-valve 3, which is energized through the transformer 5 with high-voltage insulation. The feeds 9,10 are used to apply voltage to the grids.

The used high voltage power supply system was operated in the multi-pulse regime and allowed to produce sequence of the pulses following with a frequency 200Hz and signal pulse duration 1ms. The beam energy was varied from 12 to 30keV. The maximum ion source operating time was equal 0.3s. It was limited by the used test stend pumping capabilities.

3. EXPERIMENTAL RESULTS.

Ion source tests was started from the beam energy of 12 keV in the signal pulse regime to perform the grids conditioning against of breakdowns. Process of the conditioning allows as to increase the beam energy up to 30keV in some hours.

Than the 200Hz modulated beam with an energy up to 30keV and beam current about 0.8A was obtained during 0.3s without any breakdowns into the ion optics system. The maximum beam energy obtained was limited by the existing high voltage power supply capabilities. The hydrogen pressure into the test stand volume by the end of operation time was nearly 10^{-1}Pa.

Note that only the grid voltages were modulated, whereas the RF-discharge was operated continuously.

By this means the operation of the proposed ion source for the modulated beam production was demonstrated. Then the measurements to determine the main beam parameters were performed.

3.1. Beam divergence measurements.

The angular divergence was measured by observing the current density distribution of the beam striking the focal plane at 200cm about the beam axis. At the focal point, the width of the current profile is purely determined by the beam divergence.

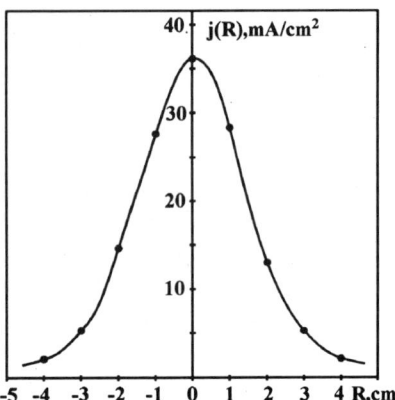

Fig.2 Current beam profile at the focal plane.

The current density profile of the beam measured with the space resolution of ±4mm by the movable calorimeter equipped with a thermocouple is shown in Fig.2. The power distribution was fit to a gaussian to find the 1/e half width divergence angles. The beam divergence (at the beam energy 26keV) calculated from these data was 0.5^0-0.6^0. The total equivalent current of the beam can be found by the integration of the measured beam power profile and dividing by the extracting voltage. The equivalent beam current determined in this way was 0.6A.

To check the results presented above additional local measurements of the beam divergence were performed. On the way of the beam the diaphragm with a narrow slit was installed. A gating slit chops an elementary ribbon beam which then enters a 635 mm drift space. At the end of the drift space, the arriving particles are detected by a small probe moved along the direction perpendicular to the slit. Since the initial size of the elementary beam was chosen to be small enough (1.5mm), the resulted current density profile beyond the slit was determined by the local angular spread of the beam particles at the slit position and the regular deflection angle corresponding to the initial beam focusing.

Measured current profile of the ribbon beam behind the slit installed on the beam axis is shown on the Fig.3.

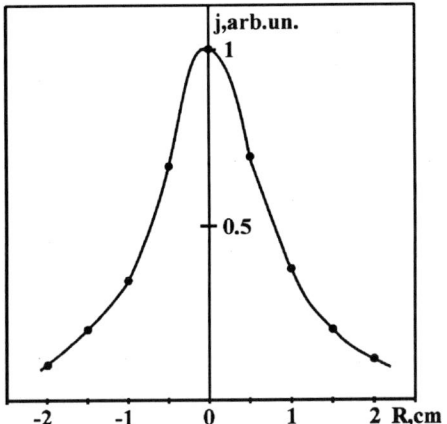

Fig.3 Current profile of the beam passed through the on-axis slit.

The beam divergence determined from width of the profile at e^{-1} level was about 0.8^0 which is close to the value previously obtained taking into account higher energy of the beam.

It is interesting to compare the experimental data with results of the numerical simulation performed for the tested ion optics system. These simulation has been carried out with the 2D-computer code AXCEL [2,3].

Fig.4. shows the real electrode geometry, electrode voltages, and calculated ion trajectories.

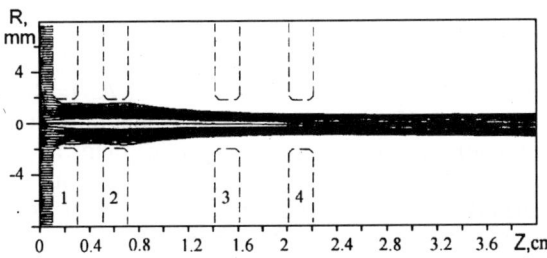

Fig.4 Beam formation in the elementary cell.
U_1= 26 kV; U_2=23.85kV; U_3=0.4kV; I_i=0.8A

Fig.5. shows the diagram of the beam emittance and ellipses embracing 100% of particles for H^+, H_2^+, and H_3^+ fractions. One can seen that the almost all output particles have angular spread $\Delta\Theta \leq 0.7°$.

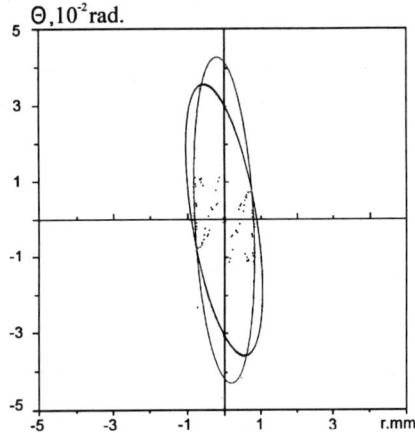

Fig.5 Radial emittance plot

3.2. Beam focusing.

As it was above mentioned the grids of the ion optics system were manufactured with the bend surface to obtain the geometrical focusing of the beam.

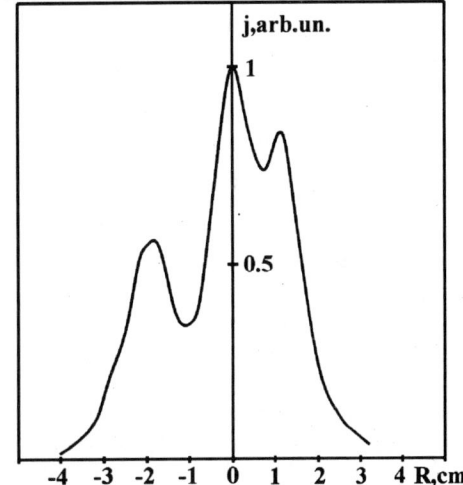

Fig.6 Profile of the beam passed through the three slits.

The focal length of the beam was determined in the case when we use the chopping diaphragm with three entrance slits. The central one was placed on the geometrical axis of the beam whereas the two backside slits were shifted from the axis by 16 and 26mm (the initial beam radius was 32.5mm). Fig.6. shows the measured current profile obtained behind the slits. The middle peak corresponds to the beam from the central slit whereas the two backside satellites correspond to the slits shifted from the axis. Measurements of the regular shift of the slit footprints as a function of their off sets from the beam axis indicated that the focal length is equal to 226 ±16 cm which is in reasonable agreement with the expected value of 2m.

3.3 Assessment of the beam composition.

The beam composition was determined from the measurements of the beam species deflected by magnetic field. In actual practice the beam appeared to be almost completely neutralized by the gas flowing out from the ion source even without use of special gas cell. This hampered the application of above mentioned scheme of the beam composition measurements. To avoid this difficulty, the ion optics system of the source was modified to reduce the puffed gas density. The holes in the plasma grid were covered by a quartz plate thus allowing for the beam extraction just through the central beamlet. As a result, the gas density was reduced so that less then 10% of the beam particles were neutralized.

Extracted elementary beam was deflected in magnetic field produced by a set of the permanent magnets. We measured the current of each beam species by a gridded Faraday cup. The entrance slit of the cup is equipped with the metal strips behind which , in its shades,negatively biased thin wires are placed. This wires suppress the electron emission from the strips as well as from the ion collector. The Faraday cup can be moved to measure the deflected beam fractions. The beam composition data were obtained at the energy of the beam of 26 keV under conditions corresponded to the extracted current of 0.6A if all the 151 beamlets were opened. From these data it can be concluded that the species value are: H^+ - 65%, H_2^+- 5%, H_3^+-25% and the total amount of the ions with a mass in the range of 12-28 is about 5%. Additional experiments show us that the relative content of H_3^+ ions reduced if the RF-power absorbed in the discharge was increased. Thus we can conclude that at the energy of 50 keV required for the diagnostic injector, which is to be operated with the increased RF-power,one can expected larger full energy fraction.

CONCLUSION

The RF-driven ion source prototype has been tested. The modulated beam with an equivalent current 0.8A and energy 30 keV during 0.3 s was extracted.

An optimum beam divergence was evaluated to be 0.6° for the energy 26 keV, which satisfies the design value and corresponds to the numerical simulations.

The beam was focused at the distance of 2m as expected.

The beam species values was estimated. However, further efforts are needed to measure the proton fraction for the relevant RF-power in the discharge.

ACKNOWLEDGMENTS

The authors wish to thank V.I.Davydenko for the useful discussions and R.Uhlemann from IPP,KFA Juelich for the help in the numerical simulations.

REFERENCES

[1] G. F. Abdrashitov, E. D. Bender, V. I. Davydenko, P. P. Deichuli, A. A. Ivanov, A. N. Karpushov, V. A. Feller, A. I. Rogozin, I. V. Shikhovtsev, and B. Schweer, Proc. 18th SOFT, Karlsruhe, V. I, 601.

[2] G. F. Abdrashitov, A. A. Ivanov, A. I. Rogozin, and I. V. Shikhovtsev. Experimental study of RF-discharges in the magnetic field as a plasma emitter for the beam injectors. Proc. of International Conference on Plasma Science, University of Wisconsin-Madison, USA.

[3] AXEL-Code, P. Spaedtke, Ing. Buero fuer Naturwissenshaft und Programmentwicklung, Junkerustr. 99, D-65205 Wiesbaden, Germany.

[4] The physics and Technology of Ion Sources, edited by Ian G. Brown, John Wiley & Sons, New York (1980), chapt. 3.

Heavy Ion Beam Probing for ITER

A.V. Melnikov[a], L.G. Zimeleva[a], K.N. Tarasyan[a], L.G. Eliseev[a], L.I.Krupnik[b], I.S. Nedzelskij[b], Yu. V. Trofimenko[b], Y.Hamada[c], H.Iguchi[c], K. Connor[d], T. Crowley[d], C.A.F. Varandas[e], A. Malaquias[e].

[a]Institute of Nuclear Fusion, Russian Research Centre "Kurchatov Institute",
Moscow 123182, Kurchatov sq., 1, Russia*

[b]Institute of Plasma Physics, National Scientific Centre "KhIPT",
Kharkov, Academicheskaja st., 1, Ukraine

[c]National Institute for Fusion Science,
Nagoya 464-01, Japan

[d]Rensselaer Polytechnic Institute,
Troy, New York, USA

[e]Center of Nuclear Fusion, Association EUROATOM/IST,
1096 Lisbon, Portugal.

The conceptual design of a Heavy Ion Beam Probe (HIBP) diagnostics has been developed for ITER by international HIBP workgroup aiming to contribute for: (i) understanding of the H-mode and Advanced Tokamak mode by the measurement of the plasma potential profile, which allows to monitor and control the electric field at the plasma edge; (ii) the accurate investigation of the plasma edge by simultaneous determination of the plasma potential density and poloidal magnetic field profiles and fluctuations; and (iii) the study of new mechanisms of plasma rotation in ignited plasmas such as losses of alpha particles. In comparison with the traditional scheme this design presents the innovative modification - the electrostatic bending system for primary and secondary beams aiming to transfer them throw the strong B-field area. The estimation of the energy range, beam attenuation and the cost lead to the conclusion that proposed project could make HIBP a realistic diagnostics for ITER.

" ... understanding the role of the electric field in confinement is almost equivalent to understanding the magnetic confinement itself"
(K.Itoh, S.Itoh)

1. INTRODUCTION

1.1. The objectives of the project:

Key objectives of the first ten years of ITER operation are the investigation of the physics of burning plasmas and the demonstration of the long-pulse ignited plasma technologies[1].

This includes studies of the plasma confinement and stability, disruption mitigation and control, steady-state operation under conditions when plasma is heated predominantly by alpha particles. Heavy Ion Beam Probe (HIBP) diagnostics is the necessary tool addressed for this goals, particularly:

1. Understanding the driven mechanisms of the energy and particle transport requires the systematic plasma electric potential measurements. The only direct method to measure the bulk plasma potential is the Heavy Ion Beam Probing.

2. An idea to operate in the H-mode requires the monitoring and control the electric field at the gradient area of plasma. For the flat profiles of ITER it means $\rho = 0.8 -1$.

3. Study of the plasma instabilities needs the plasma potential fluctuation measurements as well as the absolute mean values.

---

* *Russian team was partly supported by Russian Foundation for Fundamental Research. Grant N 96-02-18702.*

4. The plasma edge needs an accurate investigations by various independent methods. HIBP provides the local potential measurements and semi-local density measurements from the plasma core to the scrape-off layer.

5. The ignited plasma needs to investigate the new mechanisms of the plasma rotation such as losses of the alphas.

6. The problem of disruption control, prediction an mitigation require the knowledge of internal data of the magnetic potential fluctuation. HIBP can provide this measurements.

The measurements of the electric potential should be used also for the tokamak operation reasons as an indication of the H-mode proximity and the peripheral rotation level in the "Advanced Tokamak discharges" (inverse shear mode). It is more direct information in comparison with plasma rotation measurements, because the initial mechanism of the phenomenon is the inequality of the particle losses, inducing the E-field that initiate the plasma rotation.

1.2. The principles of HIBP measurements:

A high energy high mass ionic beam is injected perpendicular to the toroidal field. The primary trajectory can penetrate deep into the plasma. Secondary ionized particles are produced by electron impact ionization mainly. The secondary ions exit the plasma and are energy analyzed and detected.

The secondary ion flux I_s is proportional to the local electron density n_e:

$$I_s = n_e I_0 \sigma \lambda R,$$

where I_0 is initial beam intensity, σ is the reaction cross-section, λ is geometric factor (the length of the observation region) and R is the attenuation factor.

The difference between secondary E_s and the primary E_0 ions energies is dependent directly on the local space potential φ_{pl} in the birth point:

$$\varphi_{pl} = E_s - E_0$$

Toroidal deflection of the beam trajectories is the measure of the poloidal magnetic field or plasma current density.

The secondary birth point (region) is the observation point. Its location is variable with changing the initial beam energy and entrance angle into the plasma. The position of the entrance and detection are fixed usually.

The spatial resolution is the size of the observation point. It is determined by the geometry of the trajectories and primary beam diameter. Typically it is from a few mm to a few cm.

The secondary signal is continuous. Temporal resolution is determined by acquisition electronics.

Measurements at two or more adjacent locations using multiple detectors provides fluctuation wavelength resolution.

1.3. Current status of HIBP diagnostics

HIBP was used in a big number of various types of the thermonuclear devices since seventies[2]. The sizes and energy range grown up with the dimensions and toroidal field of the machines. The top achievement was TEXT-U HIBP with 2 MeV Tl beam. The next step - 6 MeV HIBP system with Au beam for Large Helical Device is under construction now and close to be finished. Its total cost is estimated as about M$ 4.

Firstly the HIBP project for ITER was initiated in [3], later discussed in ITER Diagnostic Meeting. Here we present the up to date status of the project elaborated by international HIBP workgroup.

2. THE PROJECT DESCRIPTION

2.1. The probing scheme.

The probing scheme proposed for the outer part of plasma (0.8 < r/a < 1) is shown in Fig 1. We propose to follow the classical probing idea with singly ionized primaries and doubly ionized secondaries. The geometrical restrictions force us to locate the injection and detection points in the horizontal port.

In comparison with the traditional scheme there are some modifications:

(i) Transfer of the primary beam throw the channel with deflecting optics - to let the primaries pass through the long straight diagnostic port (the ions have curvilinear orbits), and to avoid the

3.2 Limiting amplifier

The bandpass filter is followed by a limiting amplifier. A 5 stage design, based on logarithmic function blocks, compresses the exponentially decaying sinusoid reducing amplitude variation while retaining frequency information. The output is a nominally constant 0dBm into 50 ohms for an input signal range of -60dBm to +10dBm.

3.3 Comparator

A comparator circuit converts the limiting amplifier output to a pulse train at logic levels. The comparator output consists of bursts of approximately 100 cycles at the resonant frequency of the head. The repetition rate of the burst signal is 40KHz.

4. DIGITAL SIGNAL PROCESSING

The pulse train from the comparator is processed by a MACH 445 complex programmable logic device (CPLD). The functional requirements of the CPLD are described using Boolean logic statements and programmed into the device.

4.1 Mode of operation

The system uses two clocks; CLK1, a 14.31818MHz quartz oscillator, generates TRIG (see Fig. 4). One clock pulse is gated to the TRIG output every 358 clock cycles giving a pulse repetition rate of 40KHz. This signal feeds a pulse amplifier which in turn drives the head circuit. CLK0 is a 50MHz quartz oscillator and is used for all other timing and control functions including frequency measurement, digital filtering, sample rate decimation, signal scaling and D/A interface. By using separate clocks for generating TRIG and measuring frequency the phase of the input signal WRT CLK0 changes from sample to sample. This causes a one bit jitter on the measured input signal. The jitter on a sequence of readings contains information about quantisation noise. A low pass digital filter removes the quantisation noise, increasing the signal resolution in the process. The signal, which is now oversampled, is decimated producing an output signal which is still Nyquist sampled. A D/A converter produces a 0-10V output covering the measurement range of the head.

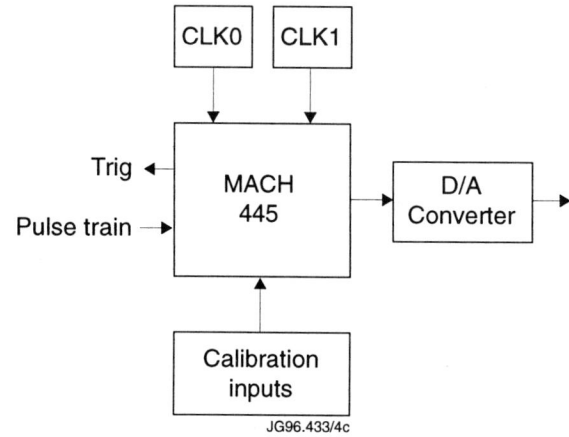

Fig. 4 Digital processing and analog interface.

4.2 Frequency measurement

The frequency of the input signal is measured by counting cycles of CLK0 (50MHz) for a predetermined number of cycles, typically 100, of the input signal. The resulting sample is inversely proportional to the resonant frequency of the head circuit. The resolution of the sample is 6 bits.

4.3 Filtering and sample rate decimation

The readings from a number of samples are accumulated in a binary counter. The high order bits of the counter give a block average of 2,4,8 or more samples. Averaging is a low pass filtering process with a magnitude characteristic which tends to $SIN(x)/x$. Averaging up to 40,000 samples increases the resolution of the signal to 12 bits with a reduction in the sample rate to 1 reading per second. The variation of vacuum pressure during regeneration is relatively slow and is well covered by this sample rate. Other applications, for example vacuum interlock, require higher sample rates which can be achieved but at lower resolution.

4.4 Scaling and calibration.

The signal is scaled to give a 0-10V output over the input pressure range of the head. Inputs to the MACH445 allow the following to be set.

1. The number of cycles of the input signal per sample.
2. The number of samples to accumulate (gain control).
3. The number of bits in the accumulator (resolution and sample rate control).
4. An accumulator pre-load value (zero offset).

5. RESULTS.

Laboratory tests were carried out on the design and the system output plotted with the output from a standard Baratron as a vacuum vessel was vented from <1mbar to atmosphere. Fig. 5 shows a graph of the two signals.

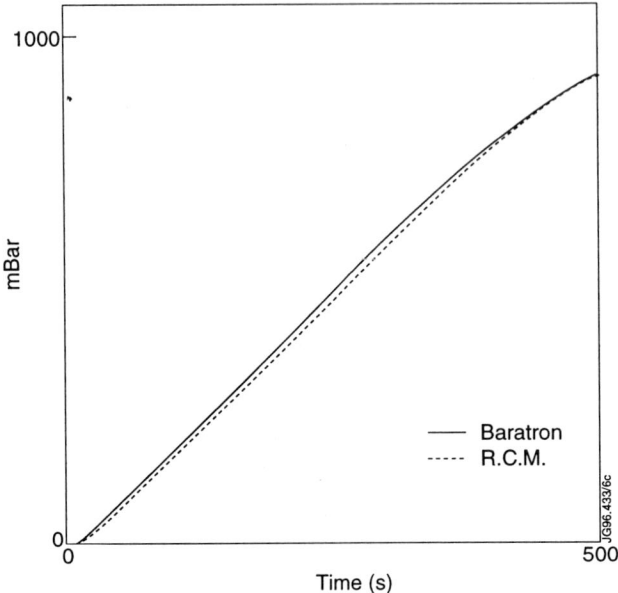

Fig. 5 Comparison between a Baratron and the Remote Capacitance Manometer (R.C.M).

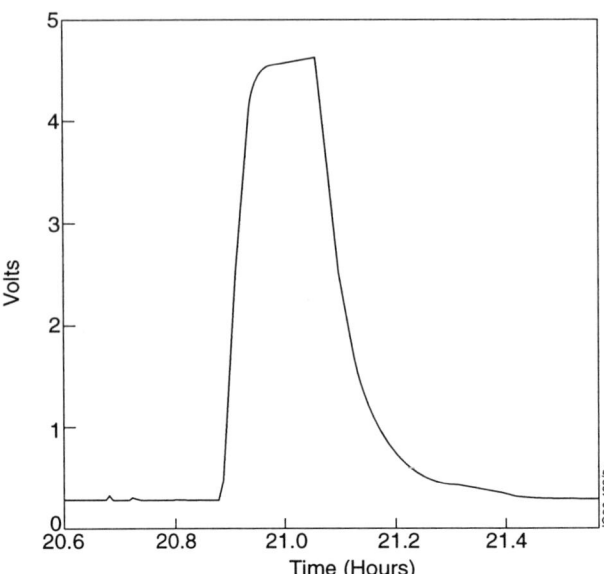

Fig. 6 Remote capacitance manometer output signal during regeneration

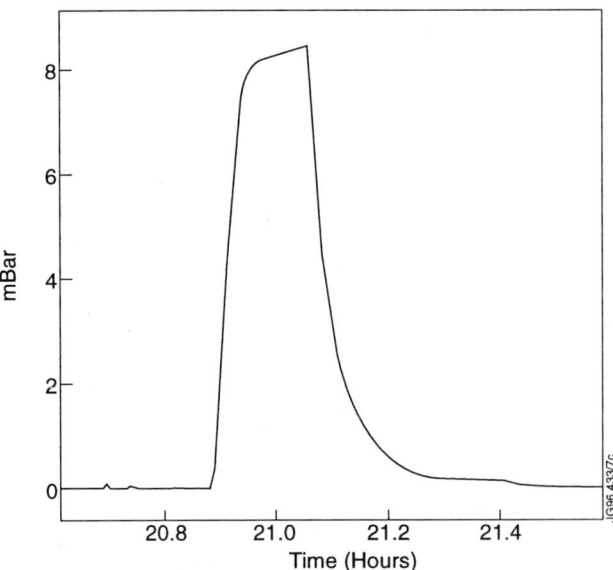

Fig. 7 Pirani gauge output during regeneration

The output is linear with pressure over most of the signal range with some lack of sensitivity at the low end of the scale and too much sensitivity at the top end.

During the active phase of JET it will be necessary to have an absolute pressure measurement in the range 0.1 mbar to 1 bar to perform accurate accounting of tritium released from the Neutral Injection cryopumps. Fig 6 is the output from the Remote Capacitance Manometer during a regeneration of one of the Neutral Injector cryopumps, Fig.7 is the output from a Pirani gauge recorded at the same time. The innacuracy in the reading from the gas dependant Pirani gauge is clearly seen.

6. FUTURE DEVELOPMENTS

The resolution/sample rate trade-off is affected by the cut-off characteristic of the digital decimating filter. Block average filtering is easily achieved with binary counters and acts as a useful but non ideal filter. A dedicated DSP device would allow a better filter to be defined at the cost of a more complex system. The logic device would still be required but a smaller, faster device could be used resulting in an improved measurement of the head resonant frequency. A modification to the system to linearise the D/A output by using a table lookup EPROM between the CPLD and the D/A is currently being pursued.

ENGINEERING DESIGN OF THE JET EDGE THOMSON SCATTERING SYSTEM

D. J. Wilson, P. Nielsen, C. W. Gowers, R. J. Eagle
Jet Joint Undertaking, Abingdon, Oxon, OX14 3EA, UK.

ABSTRACT

A new Thomson Scattering System has been designed to measure electron temperature and density in the Jet plasma, near the last closed flux surface with high spatial resolution. The design overcomes the severe access restrictions around this region.

1. INTRODUCTION

Simultaneous measurements of temperature and density at the plasma boundary are important for understanding of both the Scrape Off Layer and the limiting gradients at the main plasma edge (Ballooning Limits). The spatial resolution required in this region of the plasma is ~1 cm or less perpendicular to the surface. These measurements can be made using reciprocating probes, but these are restricted in the penetration depth and are of a much higher risk. Thomson Scattering can in principle provide these measurements, given enough laser energy and solid angle of collection. Better spatial resolution can be obtained from a conventional scattering geometry, if the measurements are made in a region with some flux expansion and/or if the laser light path is at a small angle to the flux lines.

2 THE THOMSON SCATTERING SYSTEM DESIGN

The system uses two inner vertical ports in the C-Sector of Octant 2 as per Figure 1. The laser passes into the plasma vertically from the outer port, with the collection optics located in the innermost port. In this region the plasma tends to span the last closed flux surface. Twelve scattering locations along the laser path are each imaged onto 2mm nom. diameter quartz fibres. Taking account of the flux expansion in this region the system has an equivalent spatial resolution of 1 to 2 mm at the equatorial midplane.

The light emerging from the fibres is relayed to the ceiling of the torus hall (~ 10 m) by a cassegrain telescope (F/2). In the ceiling the light is relayed through a labyrinth to the spectrometer and detector. The telescope is mechanically linked to the fibre optics periscope which is inserted into the vessel during a pulse.

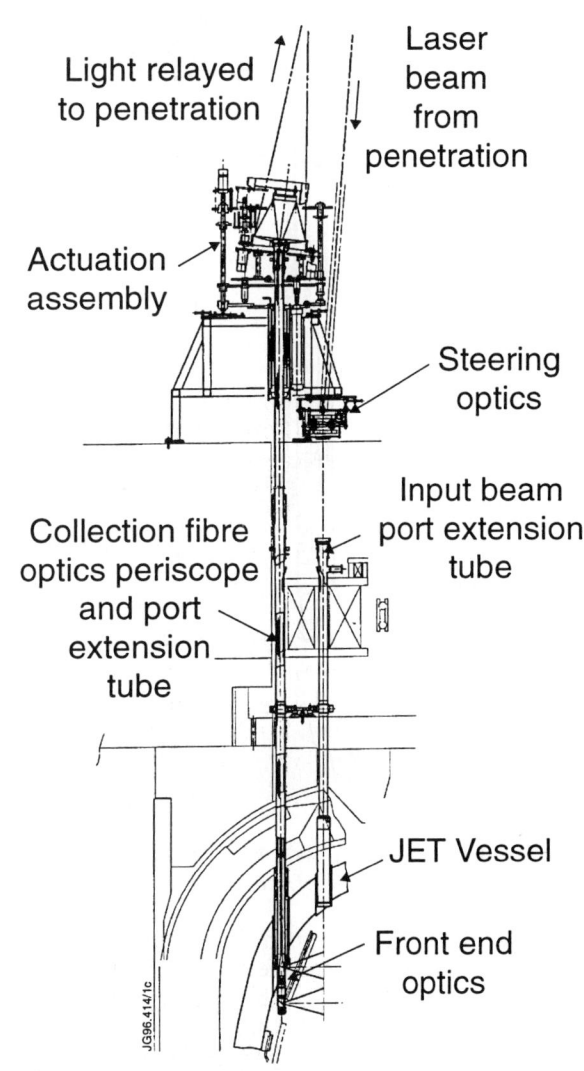

Figure 1 General Layout

An adjustable stroke between 200 and 400 mm with an insertion and withdrawal time of less than 1 second has been adopted. In order to protect the front of the gold coated mirror from plasma deposits, the periscope is pneumatically inserted only during periods of measurement. The length of the downward stroke and hence the selection of scattering positions is controlled by an adjustable stop.

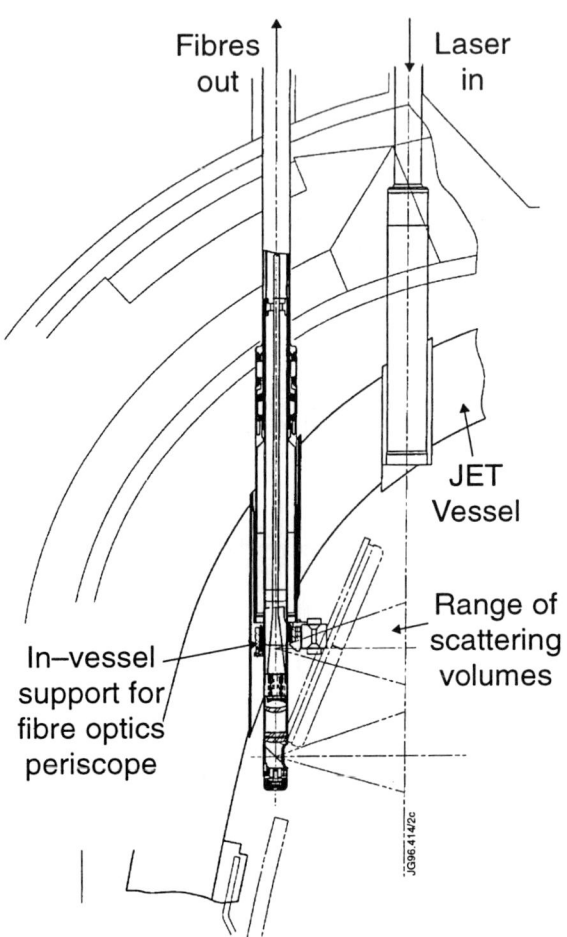

Figure 2 In-Vessel Assembly in fully inserted position, showing range of scattered volumes.

Access restrictions around the Torus ports have made it necessary to extend the vacuum boundary up to the top of the limbs, so that a more substantial structure can be erected. Double bellows have been used as a general policy with regard to maintaining vacuum integrity. The lower bellows are used to provide the flexibility between the heated vacuum vessel and the cold mechanical structure. They also provide the necessary isolation from vessel movements, during plasma disruptions. The upper edge welded bellows provide the necessary stroke for the internally mounted fibre optics collection assembly.

3. IN-VESSEL ASSEMBLY

Figure 2 shows the collection assembly located in-vessel inside a graphite support ring. This support is necessary as high forces are possible during collapsing halo currents. The collection assembly normally rests inside the small circular port, out of view of the plasma. At no point does this assembly

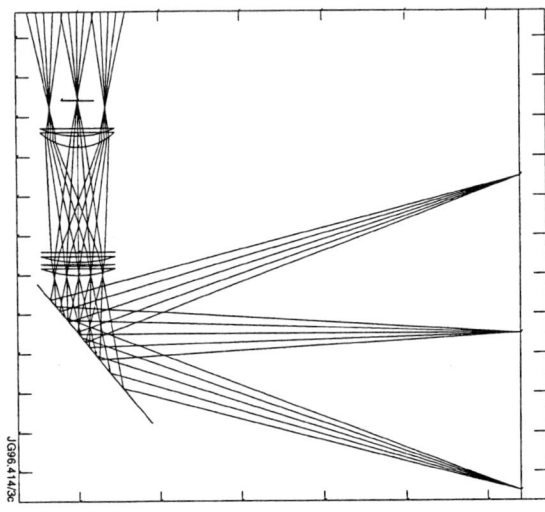

Figures 3a Ray tracing diagram showing performance of telecentric lens

enter the plasma and is protected behind the relevant inner wall tiles.

The collection system uses a fibre optics periscope constructed using sixteen fibres. Four fibres are used for alignment and are mounted on both sides, of the line of 12 collecting fibres. The scattering volumes are imaged onto the fibre ends by a three element telecentric sapphire lens as per Figure 3a. The effective scattering length is 10mm with F/10 collection. The image is deflected through 90 degrees by a gold coated copper mirror at the front of the periscope. All these components are located inside the vacuum vessel and can be heated to 350 °C.

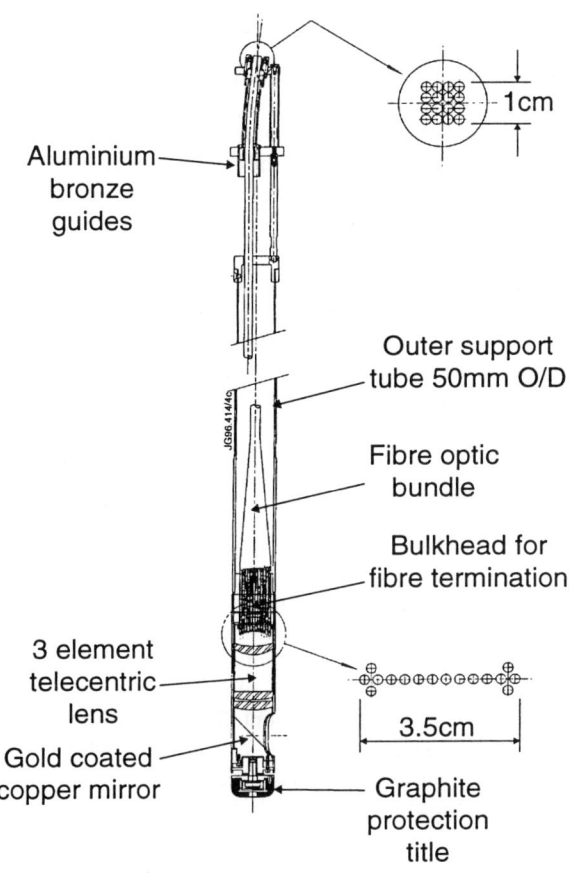

Figure 3b The Fibre Optics Periscope. Inserts show layout of fibres at each end.

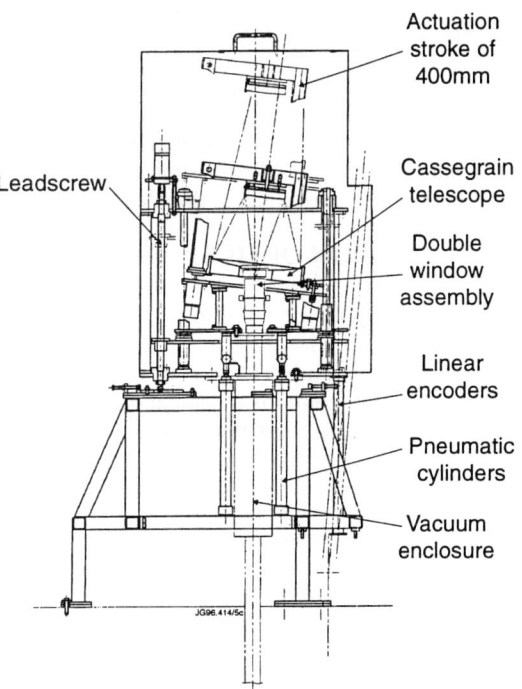

Figure 4 Actuation Assembly with Cassegrain telescope shown in the fully inserted and in the Stand-by positions.

The fibres used are aluminium clad quartz, of approximately 2.6 mm outside diameter. These were selected for vacuum compatibility, maximum radiation hardness and good mechanical strength properties. The ends were cleaved to within 2mm of final length, hand dressed down, then finally polished using a conventional fibre polishing machine. Small ferrules were attached to the lower end so that after final positioning within the bulk head with respect to the lens, the fibres were potted in a Ceramabond 571 ceramic compound. After curing, this joint is vacuum compatible and is a useful technique for bonding in these situations. At the top end the fibres are closely packed in a 4 x 4 matrix. This end is also potted in ceramic compound.

4. THE FIBRE OPTICS PERISCOPE

Figure 3b shows the periscope with its collection optics and the fibre bundle at the top of the vacuum assembly just below the double fused silica window. The periscope is approximately 6 m long and clears both the toroidal field coils and vessel support structures. In this space the transition between the fibre optics to conventional mirror optics is made. Two cam shaped bolts lock the periscope into its correct angular position, using two knurled pressure plates inside the vacuum tube. Thermal expansion of the outer structural tube is downwards at this point. The fibres are attached to a bulk head at the collection optics end of the assembly. Thermal expansion of approximately 5 mm at the window end is achieved with the fibre bundle sliding in a aluminium bronze guide and on three sprung loaded pillars.

5. ACTUATION ASSEMBLY AND CASSEGRAIN TELESCOPE

Figure 4 shows the layout of the cassegrain telescope mounted inside the actuation assembly. Although rather complicated looking its operation is simple. The vacuum tube and hence the fibre optics assembly is pneumatically driven up and down between a fixed top plate and a variable bottom plate. The position of the bottom plate is set using a dc motor driven lead screw, which determines the depth to which the collection assembly reaches. Linear encoders are used to measure the position of the bottom plate as well as the driven plate. Constant force shock absorbers are mounted in pairs in both directions and smooth out the motion, together with flow restriction of the pneumatic cylinders. Access to the double window for remote handling replacement, is achieved by the removal of the cassegrain telescope sub-assembly. The two mirror elements are spaced apart accurately using a three legged frame that gives minimal blockage to the collected light. Mounted off the support frame is the laser steering assembly, consisting of two counter-rotating wedges. Minimising the effects of reflected laser light is implicit in this design.

6. PENETRATION OPTICS

Figure 5 shows the penetration assembly. The spectrometer and detection system are mounted inside of a removable tubular insert. This in turn sits over another insert which contains a conical cavity for the F/10 optics. A 250 mm diameter lens is mounted at the very end of the lower insert and aligns the axes of the penetration with the torus mounted optics. This arrangement was adopted as it minimised the amount of radiation shielding required above the floor level of the roof penetration. All inserts are double skinned so that they can be filled with shielding material. In order to achieve sufficient spatial resolution we have opted for a conventional ruby laser system. The ruby laser provides 2 joules in 1.2 ns pulses at 4 Hz operation.

The square image of the 16 fibres is divided into four spectral channels by dichroic filters. All four channels are imaged onto the same image intensifier - CCD camera detection system. The image intensifier can be gated to 5 nsec. With the short laser pulse it may be possible to gate out laser stray light pulses!

7. SUMMARY

An engineering solution for the new Edge Thomson Scattering System has been made. Operation of these components is required to be both accurate and reliable in an ultra high vacuum and high temperature environment. The spectrometer and detection system for this diagnostic, where the scattered light will be analysed, will be installed in the penetration by the end of 1996.

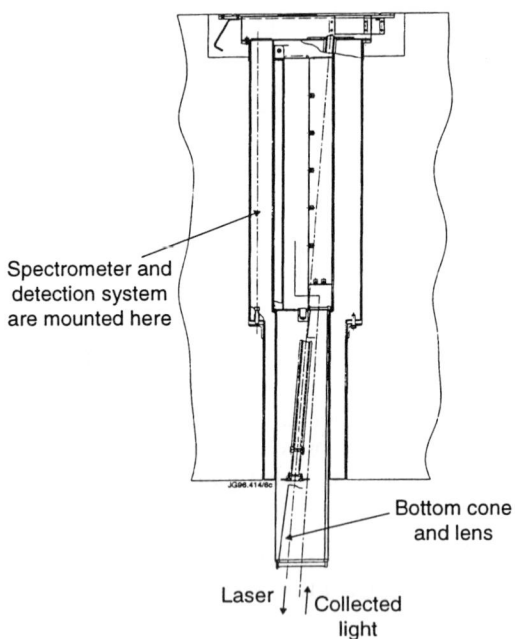

Figure 5 Insert in biological shield above Torus showing the optical layout for the spectrometer/detection system.

A PC based real-time imaging system for the TdeV Tokamak

F. Meo[1], P. de Villers[2], F. Brunet[2], G. Ratel[3]

Centre canadien de fusion magnétique*, 1804 boul. Lionel-Boulet, Varennes, Québec, Canada J3X 1S1
*Supported by the Canadian government, Hydro-Québec and INRS

[1] INRS-Énergie et matériaux, Varennes, Québec, Canada
[2] MPB Technologies Inc., Dorval, Québec, Canada
[3] Hydro-Québec, Varennes, Québec, Canada

CCFM has developed a flexible and generic imaging system to digitize video signals and transfer the data, in real-time, to the host computer. The hardware consists of a Matrox frame grabber board installed in a PC computer. A graphical user interface, developed with the IDL package under MS Windows, controls the acquisition of images, in synchronization with the TdeV experiment. The user interface also includes a virtual VCR to display animated sequences of images. The interface can be configured by users to include parameters describing their own experiment setup. The parameters and compressed images are stored in Hierarchical Data Format (HDF) which is supported by many operating systems such as DOS/Windows, Unix and VMS.

1. INTRODUCTION

Due to the improvement in the technology of frame grabbers, the interest in digital imaging for fusion research is on the increase. At CCFM, we have developed an imaging system (fig.1) whereby two synchronized RS-170 (60 fields/sec) analog camera signals are digitized and transferred, in real-time, to the host computer's memory during the TdeV plasma discharge. The digitized images of both cameras can be viewed in a virtual VCR graphical user interface. The images can be compressed and saved with other parameters of interest, on the computer disk. After a day of experiments, using a network transfer routine, the data files are collected by the TdeV data-acquisition server

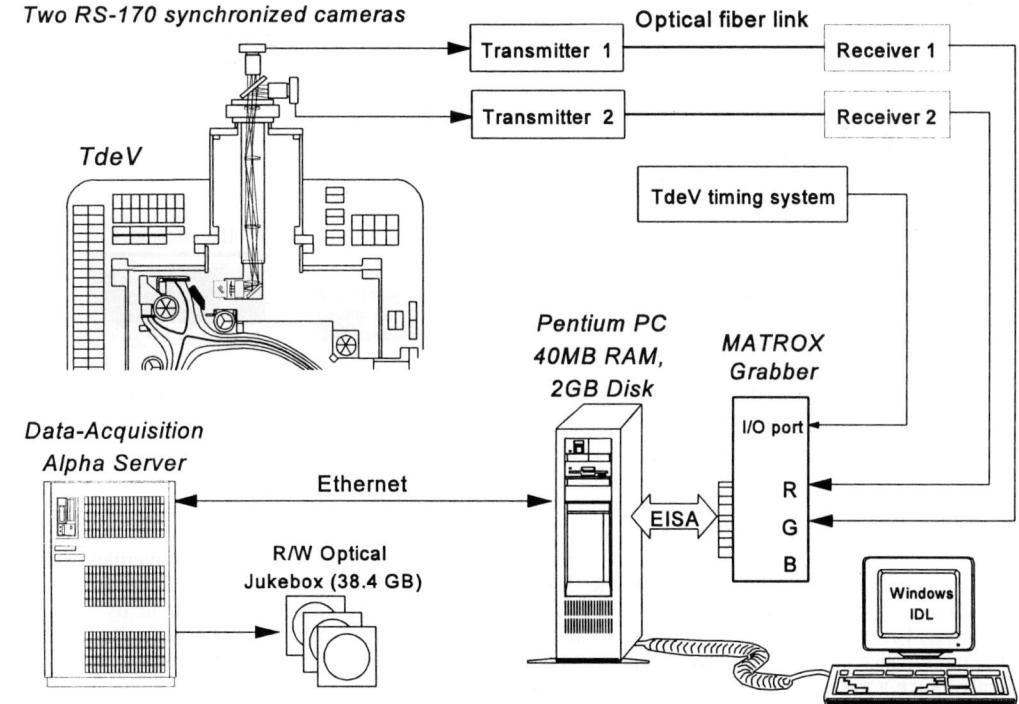

Figure 1. The imaging system's hardware

[1] for permanent archiving on optical disk. The system has been developed and used during the TdeV 1995 experimental campaign for three diagnostics: divertor spectroscopic imaging, infrared camera and the intensified camera for plasma rotation measurements.

2. HARDWARE CONFIGURATION

The two basic components of the imaging system (fig.1) are the frame grabber board and the PC computer. The frame grabber presently used is a MATROX MAGIC/RGB [2] board combining high-speed acquisition and display capabilities with direct memory access (DMA) burst mode. It can digitize images from most cameras, including video standards like RS-170, NTSC, PAL, Y/C. It can support either one 24-bit color signal or up to three 8-bit synchronized monochrome signals each connected to the RGB inputs. Presently, on TdeV, there are two synchronized RS-170 cameras per card. The setup for each type of camera is stored in a digitizer configuration file (DCF) which is loaded when the grabber is initialized.

The grabber also includes a SVGA (1024 x 768) display adapter with a 3 MB on-board frame buffer supporting images up to 1024 x 1024 pixels. The DMA capability allows the card, as it digitizes an image in the frame buffer in real-time, to transfer the images to the PC memory via the EISA bus. Thus, the size of the PC RAM determines the number of images the system can acquire without interruption. The user can specify the frames to be kept and skipped over, to either save memory space or digitize during a longer time interval.

The frame grabber can be triggered by either a software call or by an external signal connected to the digital I/O port of the MAGIC board . The external timing signal (fig. 2) provided by the TdeV timing system [3] , is a TTL signal that indicates when the countdown phase is executing and when the plasma startup is triggered (T = 0.0 sec). The countdown starts 40 to 50 seconds before the plasma startup. At this moment, the grabber enters the PreShot phase where it starts to digitize the video signals in its own frame buffer but with no DMA transfer. At the plasma startup, the card enters the Acquisition phase where the DMA is triggered and starts to transfer the frames of interest to the PC memory for the rest of the plasma pulse.

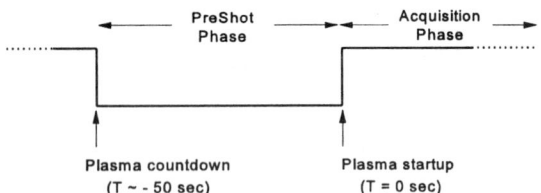

Figure 2. *TdeV timing signal*

The second component of the system is the host computer which consists of a Pentium PC running under MS Windows. It has an EISA bus to accommodate the MAGIC board and is equipped with 40 MB of RAM and a 2 GB disk. Typically, such a system digitizes two cameras at 60 fields/sec during a 1 second pulse producing around 10 MB of raw data per shot.

3. SOFTWARE DESCRIPTION

3.1 Overview

Much of the data processing and image analysis software at CCFM is written in the IDL package [4]. Thus it was suitable to use IDL to write a graphical user interface for the PC imaging system. In addition to supporting many file formats such as HDF, JPEG and TIFF, the IDL language is also portable on different computer platforms (e.g. Windows, NT, Unix, VMS). Thus using IDL makes it possible to acquire, display, and store the images on the PC, then transfer them to the VMS data-acquisition server for archiving. As a result of this portability, image analysis could be done on either the VMS or PC platforms.

The system's software (fig. 3) is composed of three independent components. At the heart of the system's software is the Executive User Interface main program (IMaster) that controls the data flow, manages the images in memory, and sets up the graphical interface that enables the user to have control of the session. The other two components (used by IMaster) is the Grabber Control Routines that manage the grabber board and the Input/Output (I/O) Routines that control the data compression and I/O operations with the disk. The three are linked by a user configuration file therefore making the entire software easily modifiable.

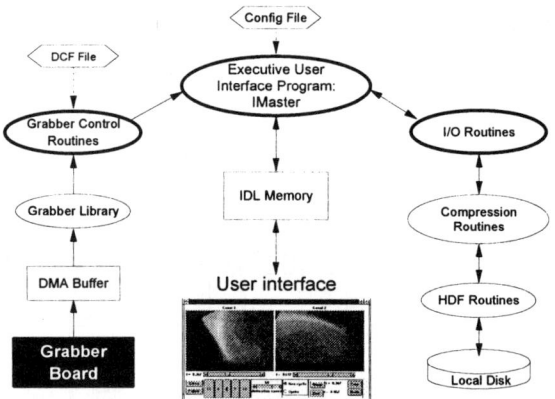

Figure 3. *Software configuration*

3.2 Configuration File

IMaster's characteristics and relation to the Grabber Control and I/O Routines are governed by the configuration file. This is an ASCII file containing information, specific to the user application, to be read by the IMaster program during the software startup. Such information contains the experiment characteristics, interface default settings, and user-defined parameters. The user-defined parameters are auxiliary data information to be saved with the images. As a part of its interface, IMaster creates the appropriate input fields for each parameter whereby the user enters the values to describe an experiment thus distinguishing a series of images related to a tokamak pulse.

In addition, the configuration file contains the names of the Grabber Control, Image Processing and I/O Routines to be used by Imaster. This enables the system to support different types of grabbers, compression schemes, and image processing modules without having to modify the core of the system.

3.3 Executive User Interface Program: IMaster

The IDL package provides a set of high-level graphical objects, named widgets, to construct graphical user interfaces. It includes viewing windows, buttons, sliders, and pull down menus. Using these tools, it is possible to rapidly develop display panels which are portable to the most common windowing systems (e.g. Motif, MS Windows).

The Executive User Interface Program (fig. 4), was created as a generic utility to manage the acquisition, visualization and data manipulation of any image sequence. When IMaster is started, it prompts the user to open one of two working sessions: Acquisition or Analysis session.

3.3.1 Acquisition Session

The acquisition session (fig. 4a) sets up the panel giving the user control of the experiment and acquisition process. Using buttons and pull-down menu, this session displays and controls the grabber's initialization, acquisition, and settings via the Grabber Control routines. It includes the DCF file to use, the active camera channels, the video gain, the number of images to digitize, and the images of interest.

Three modes of synchronization of the acquisition are possible: Internal, Keyboard and External. The Internal mode sends a software trigger to immediately start the card's Acquisition mode. The Keyboard mode rather starts the digitizer's PreShot phase and waits for a keyboard event to start the Acquisition phase. Both modes are useful for monitoring and digitizing images which do not require an accurate time reference such as calibration data. In External mode, the card is triggered by the digital signal provided by

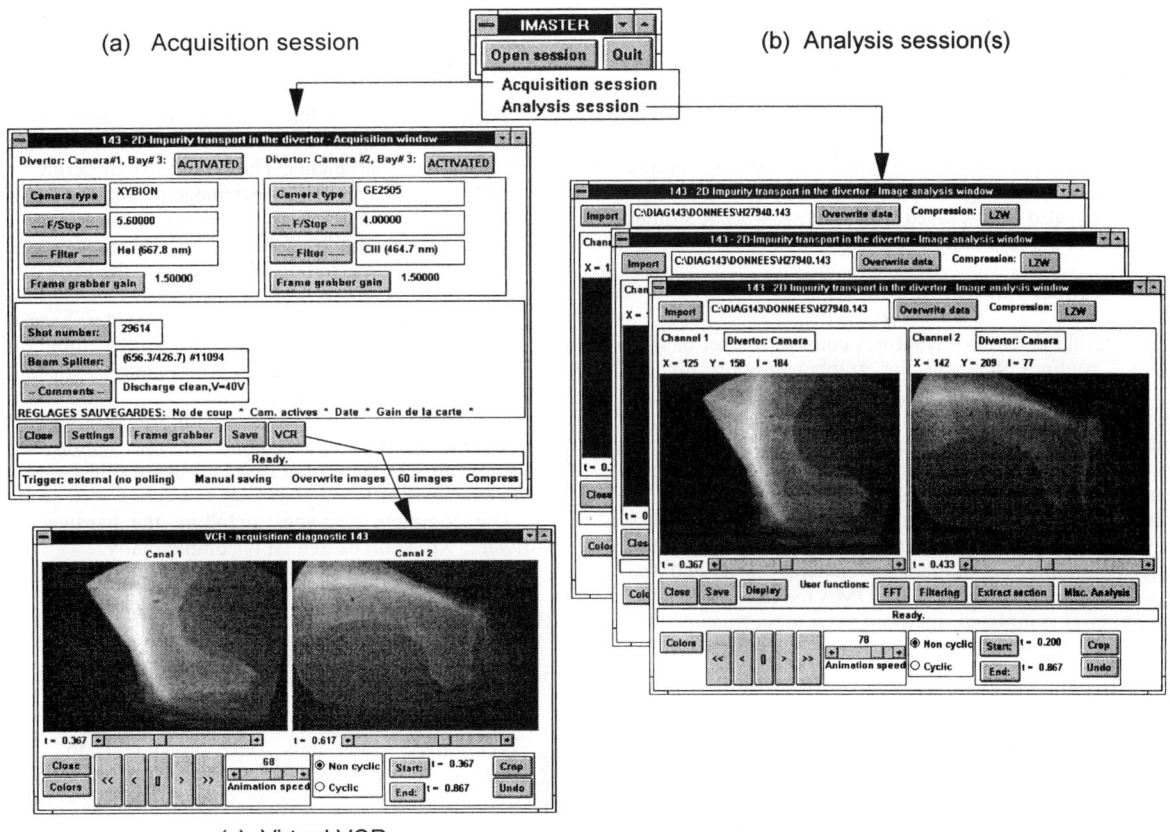

Figure 4. The IMaster's interface. (a) Acquisition session, (b) Analysis sessions, (c) Virtual VCR panel.

the TdeV Timing System. This signal is polled periodically for the start of the countdown phase (fig. 2). When detected, the card's PreShot phase is immediately begun preparing for the Acquisition phase at plasma startup.

After the acquisition is completed, images of interest are transferred from the DMA memory to the IDL environment where they can be viewed with a virtual VCR panel (fig. 4c). It can display specific images using the scroll bar or animate the image sequences, with an adjustable speed, using the "play" and "rewind" buttons..

Shown in the acquisition panel (fig. 4a), is an example of the input fields for the user defined parameters: f/stop, interference filter, camera type, and comments. The user can either input the values directly or choose from the pull down menu to the left of the input field containing suggested values that where previously defined in the configuration file by the user. From the user's command, the images and parameters are saved to the disk under different compression formats specified in the session's Save Settings.

The storage of the data can also be set in Automatic mode where images are saved at the end of each acquisition process. Therefore, it is possible to automate the acquisition process without requiring user intervention.

3.3.2 Analysis Session

This session (fig. 4b) is used mainly for image analysis and can be used simultaneously with the acquisition session or with other analysis sessions. Image sequences can be imported either from the acquisition session or an existent data file. Using the pull down menus, the images can then be analyzed and modified with any image processing routine defined inside the configuration file by the user. The image sequence can also be displayed with the virtual VCR similar to the one in the acquisition session.

3.4 Grabber Control Routines

The Grabber Control Routines consist of a set of IDL and "C" routines that call the grabber support library, provided by the manufacturer, to control the digitizer. The grabber control process is divided in three operations: initialization, acquisition and reading.

The initialization routine is called to set up the digitizer to support the type of camera as specified in a DCF file. The acquisition routine is called to start the digitizing process and to wait for a trigger. In addition, it specifies the triggering mode, the number of cameras used, and the sequence of images to store in the DMA memory. When the acquisition is completed, the reading routine transfers the images from the DMA memory to the user environment.

3.5 I/O Routines

This set of routines is responsible for the image compression and storage/retrieval of the data (image sequences and parameters) to/from the local computer disk.

To save significant space on the disk, a compression algorithm can be applied to the images before storage. Presently, two types of compression are supported: LZW and JPEG. The LZW type is a single-frame lossless compression format mainly used for scientific data with typical compression ratio of approximately 3:2. The JPEG type, a single-frame lossy compression scheme, is optimized for the human eye and is used mainly for qualitative applications. The compression ratio is variable according to the quality of image required. This is specified in the save settings by the user through the "quality index" factor. The system's supported compression formats are defined in the configuration file, therefore other compression routines can be easily incorporated to the software.

The file format used to save the parameters and the compressed images is the Hierarchical Data Format (HDF) developed by the National Center for Super Computing which is supported by IDL on different computer platforms (DOS, Unix, VMS). HDF also has a data index to describe the contents of a file that makes it convenient for sharing data without having to specify how and what data is stored in the file.

4. CONCLUSION

This flexible PC-based imaging system has been used in both scientific diagnostics and for monitoring purposes on TdeV since mid-1995. The use of a frame grabber to digitize video signals in real-time has eliminated all the non-linearities introduced by analog VCR, traditionally used to store and play-back images. It has also drastically sped-up the digitization process of images stored on VCR tapes which were previously digitized frame by frame using a still image frame grabber.

The design of a flexible interface, easily configured, has enabled users to rapidly customize the system according to their specific requirements. The possibility to define user-parameters and to specify its own image processing routines allows the support of many types of applications. The portability of the datafiles and IMaster on different computer platforms gives users the possibility to run the code on the platform of their choice.

Finally, the use of a PC like host computer, keeps the cost of the system as low as possible, and facilitates any upgrade of the system by adding RAM memory, installing a larger disk, or using a faster processor.

Future developments will include the support of the MATROX Magic frame grabber in the Windows NT environment. The support of new PCI-based frame grabbers is also planned.

The complete package, including the hardware and software, will soon be available commercially.

References

1. P. de Villers et al, "The Data-Acquisition and Management System for the TdeV Tokamak", *Proceedings of the 17th Symposium on Fusion Technology*, Rome, Italy, 1992, pp. 1032-1036.
2. MAGIC/RGB, Matrox Electronic Systems, 1055 St. Regis Blvd, Dorval, Québec, CANADA, H9P-2T4
3. P. de Villers et al, "The TdeV timing system", *Proceedings of IEEE 14th Symposium on Fusion Engineering*, San Diego, USA, 1991, pp. 806-809
4. IDL (Interactive Data Language), Research Systems Inc, 777 29th Street, Suite 302, Boulder, CO, USA

Design of an optimal control system for Tore Supra

S. Brémond and J.-M. Ané

Association EURATOM-CEA, Département de Recherches sur la Fusion Contrôlée
Centre de Cadarache, 13108 Saint Paul-lez-Durance Cedex, France

In order to take full advantage of the integrated shape and position control system of the Tore Supra tokamak, we derived a multivariable linear model through perturbed equilibrium calculations performed with the code Cèdres. The model is shown to reproduce quite well experimental time evolutions, and is used as a basis for Linear Quadratic Gaussian design, whose potential optimisation benefits are presented.

1. INTRODUCTION

In most of the present devices, the equilibrium control system was designed on the basis of simplified single input - single output models. Such descriptions remain a precious intuitive background. However, as the demand of shaping accuracy and control stability is increased, and as the installed power is to be minimised, the coupling between control parameters has to be taken into account. In this context, it is essential to have a reliable multivariable plasma response model.

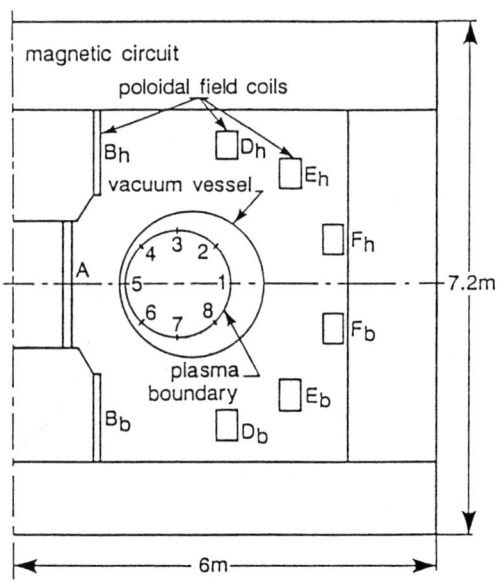

FIG. 1 : Meridian cross-section of Tore Supra Poloidal Field system.

The Tore Supra tokamak has from the origin a fully integrated Poloidal Field coils and power supply system ([1]). FIG. 1 shows the cross-section of Tore Supra. The points 1 to 8 denotes the flux differences control variable locations. Recently, an equivalent plasma-wall gaps control variable set has been introduced ([2]).

The functional coupling of the PF system, which offers the largest number of free parameters for plasma control, is much profitable. But it requires a special care with regard to the control law. In particular, one has to prevent the growth of currents in reverse directions flowing in adjacent PF coils.

In the present work, multivariable models are derived using the equilibrium code Cèdres, and extensively checked against experimental discharges. An optimal control law, in the sense of Linear Quadratic Gaussian method, is then designed. The closed loop system is finally simulated and compared with the current control law.

2. LINEAR PLASMA RESPONSE CALCULATION

2.1 Principle

Provided that the conducting structures around the plasma (vacuum vessel, PF coils) ensure the passive stabilisation of the plasma ([3]), the characteristic instability time is much longer than the Alfvén time. In the case of Tore Supra, the PF coils alone can passively stabilise the positional instability due to the variation of the iron core attraction resulting from the variation of the iron core saturation when a plasma displacement occurs([4], [5]).

Therefore, the plasma can be considered to be in equilibrium at all times, and equilibrium calculations can be used to estimate the linearized plasma response to the conductor currents ([6]). It follows that we can write:

$$\delta g = C \, \delta I \quad (1)$$

where the vector δI consists of the perturbed PF coil, vessel and plasma toroidal currents and the vector δg consists of the resulting variations of position and shape control variables, such as plasma-wall gaps or magnetic flux differences.

Each column in the matrix C is obtained by computing a new equilibrium where a small variation is imposed on one of the currents. By the way, we can get the square matrix D, which relates conductor current variations to changes in magnetic flux through the conductor circuits, put together in the vector $\delta\psi$:

$$\delta \psi = D \, \delta I \quad (2)$$

It is to be noted that the coefficients of the matrix D are not magnetostatic mutual coefficients as they measure not only the flux changes directly resulting from the current variation but also the flux changes resulting from the displacement of the current carrying plasma. In accordance with Ohm's law, we then have the circuit type equations:

$$\dot{\delta I} = -DR^{-1} \, \delta I + D^{-1} \delta V \quad (3)$$

where R is the conductor resistance matrix and the vector δV consists of voltages applied to PF coils.

Equation (1) and (3) form a state space representation of the system. Equation (3) describes the time evolution of the state variables -toroidal currents flowing in the conductors, including the plasma-, and equation (1) describes the dependence of the control quantities on the state variables.

2.2 Derivation with the Cèdres code

The Cèdres code models the evolution of an axisymmetric tokamak plasma at the plasma current diffusion time scale. A free boundary 2D equilibrium solver is coupled with a 1D flux surface averaged resistive diffusion equation solver ([6]).

In this work, we used the equilibrium solver of Cèdres in its second order Finite Elements version. It allows a free positioning of the magnetic axis and avoids spurious effects that might be generated by the compulsory location of the magnetic axis on a mesh node with first order Finite Elements.

Typical Tore Supra plasma geometrical parameters and plasma current profile have been considered in order to compute a basic equilibrium.

The dynamics of the plasma shape and position is influenced by the resistive diffusion of the plasma current density which takes place during the plasma displacement. In the process of determination of the plasma response from equilibrium calculations, the resistive diffusion is implicitly imposed by the constraints that are put on the prescribed plasma current profile. In this work, the functional dependence of the plasma current density on the normalised poloidal flux was kept constant.

On the other hand, since the saturation level of the iron core play an important part on Tore Supra stability characteristics, we actually derived three models, corresponding to the unsaturated, moderately saturated and highly saturated cases.

3. EXPERIMENTAL VALIDATION

Before we give an account of model simulations compared with experimental data, we examine the instability time of the response models. The unstable eigenvalue obtained is all the higher as the iron core is less saturated; we found $\gamma=0.17$ s^{-1} for the unsaturated case, $\gamma=0.15$ s^{-1} for the moderately saturated case and $\gamma=0.08$ s^{-1} for the highly saturated case.

The instability growth rate is of the order of the typical decay rate of the currents in PF coils, in agreement with the analytic predictions obtained with the rigid displacement model on the Tore Supra tokamak ([5]).

3.1 Open-loop behaviour

In the middle of the plasma current flat-top of discharge 15602, the feedback loop was opened and the voltages across the PF coils had been maintained constant for 0.2 seconds. In each of the following discharges (15603 to 15611), similar operation was carried out, except on one coil, each time different, where the voltage was set 100 V higher than the voltage it had on discharge 15602.

We extensively compared the experimental open-loop evolution of the conductor currents and shape variables -obtained by subtracting the data from discharges 15603-15611 to the data from discharge 15602- against our model prediction. The unsaturated case model was used since it corresponded the best to the saturation level in the iron core.

FIG. 2 shows as an example experimental measurement of two flux differences along with estimate from model 1 for a 100V step on coil Bh. Quit a good agreement is found.

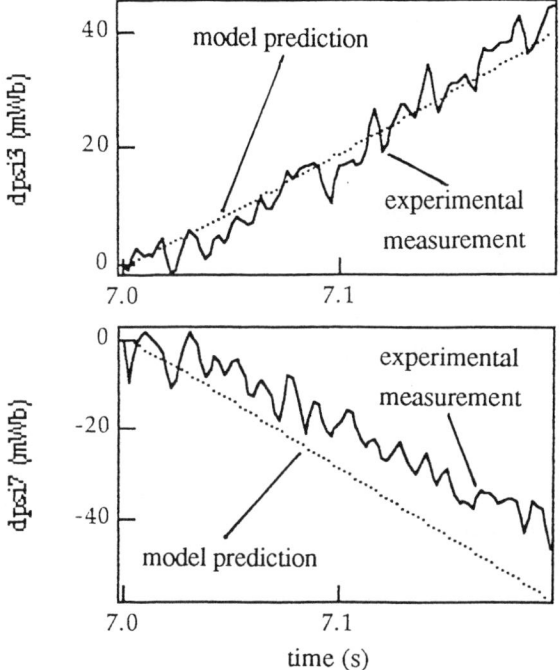

FIG. 2 : Open-loop experimental and calculated evolution of flux difference 3 and 7 for a 100 V step on coil Bh.

3.2 Closed-loop behaviour

We fed back the numerical models with the control law presently implemented on Tore Supra discharges, and compared closed-loop simulations against experimental data. We set out as an example the dynamic response to a plasma current ramp demand during the flattop of discharge 15757. FIG. 3 shows the time evolution of two flux differences.

The sign and the "mean" amplitude of the experimental and calculated time evolution are much comparable. Yet, the model doesn't happen to reproduce some quite large oscillations occurring on the flux difference 3. This discrepancy, which could be due to MHD modes, is not completely understood, but could reveal the limitations of this type of model with respect to internal plasma parameter variations.

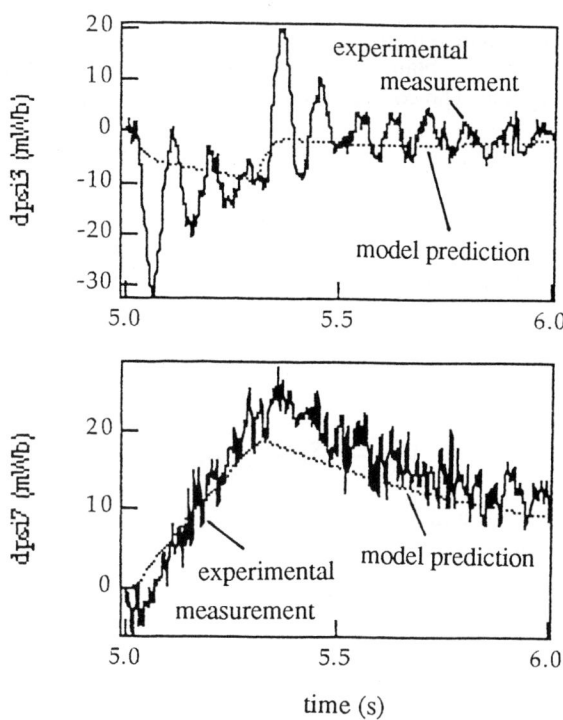

FIG. 3 : Closed-loop experimental and calculated response of flux difference 3 and 7 to a plasma current ramp demand.

4. LQG CONTROL LAW DESIGN AND SIMULATION

The present control law is essentially proportional-integral to shape control variables, but two extra terms are added in order to form the voltages applied to PF coils:

-"Resistive correction" aim to compensate the slow positional drift ([8])

- Feedback is also applied to minimise the difference between adjacent external PF coil currents to prevent the growth of these currents in reverse direction, which lead to overheating of one of them during long pulse operation.

Satisfactory control of the plasma is routinely obtained, but the feedback on PF coil currents is believed to limit the flexibility of the plasma shape control. Moreover, important coupling between the control variables have appeared during some discharges.

LQG methods, which include in the design criteria a balance between tracking errors and energy expenditure was foreseen to be well suited to deal with the multipolar current distribution issue. As plasma current control is not a problem, we consider the system obtained once the plasma current has been fed back. The cost function stands:

$$J = \int \left[\delta g^T Q_1 \delta g + \int \delta g^T Q_2 \int \delta g + \delta I^T Q_3 \delta I \right] dt$$

$$+ \int \delta V^T \delta V \, dt \qquad (4)$$

And the control law which minimises the above integral write:

$$V = -L \delta I - L_i \int \delta g \quad (5)$$

where L and L_i are obtained as solutions of Ricatti equation ([9]).

The choice of the weighting matrix from equation (4), Q_1, Q_2 and Q_3, is done by a "test and try" procedure in relation to the closed loop response characteristics.

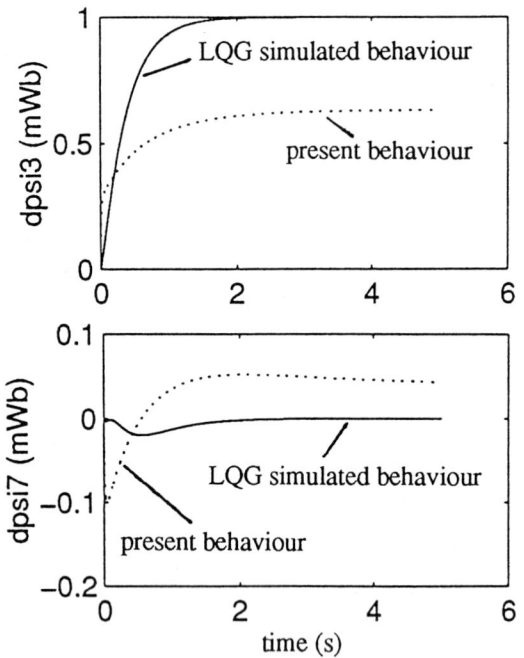

FIG. 4 : Present and LQG simulated flux difference 3 and 7 response to a step on flux difference 3.

We finally simulated the resulting LQG control law against the present one for different scenarios. As an example, we show on FIG. 4 the response to a step demand on one of the flux differences, which corresponds to a local deformation of the plasma.

We can see that, due to the integral action, no more permanent coupling is present between the control variables. It appears finally that currents doublets are not occurring, without being obliged to apply a specific feedback against this problem.

5. CONCLUSION

Plasma shape and stability optimal control require the coupling between equilibrium parameters to be explicitly taken into account. We derived a multivariable linear plasma response model, and, to our knowledge for the first time, extensively checked the model predictions against experimental discharges on Tore Supra. Reasonably good agreement is found, which allows us to rely on the model in order to design an optimal control law, in the sense of LQG methods. As a balance between tracking errors and energy expenditure is included in the design criteria, it could be proved to be a solution to the multipolar PF coils current distribution issue on Tore Supra.

REFERENCES

[1] J.-M. Ané et al., in 15th Symposium on Fusion Technology, Utrecht (1988), p. 657.

[2] E.A. Lazarus et al., Physics of Fluids B3 (1991), p. 2220.

[3] J. Blum et al., in 7th IAEA International Conference on Plasma Physics and Controlled Nuclear Fusion Research, Innsbruck (1978), p. 521.

[4] S. Brémond et al., in 21st EPS Conference on Controlled Fusion and Plasma Physics, Montpellier (1994), p. 688.

[5] R. Albanese et al., Nuclear Fusion 29 (1989), p. 1013.

[6] H. Grad, P.N. Hu, D.C. Stevens, E. Turkel, Plasma Physics and Controlled Nuclear Fusion Research, Berchtesgaden (1976), p. 355.

[7] T. Wijnands, G. Martin, "An advanced plasma control system for Tore Supra", to be published in Fusion Technology.

[8] S. Brémond et al., in 18th Symposium on Fusion Technology, Karlsruhe (1994), p. 711.

[9] H. Kwakernaak, R. Sivan, Linear Optimal Control Systems, New York, Wiley (1972).

Evolution of the Tore Supra data acquisition system : towards the steady state

B. Guillerminet, D. Elbèze, J.F. Artaud, S. Balme, Y. Buravand, B. Couturier, L. Ducobu, B. Gagey, M. Leluyer, R. Masset, D. Moulin, B. Rothan, J. Signoret

Association EURATOM-CEA, Département de Recherches sur la Fusion Contrôlée
Centre de Cadarache, 13108 Saint Paul lez Durance Cedex, France

In operation since 1988, the Tore Supra data acquisition system provides continuous data taking, storage and displays. But several limitations are observed, as the weak errors management, the performances of data transfer and of feedback control. The tied proprietary operating systems and as outcome, the lack of processors evolution prevent from upgrading the original system. So, the data acquisition system has been completely re-engineered aiming long pulse operations. It is based on a commercial package which provides all the basic modules to build a distributed on-line software from the data acquisition level up to the run control and data display. The continuous operation requests a few enhancements we explore in the last part.

1. LIMITATIONS OF THE PRESENT DATA ACQUISITION SYSTEM

For past few years, a few limitations have been encountered in the data acquisition system, as for instance the data flow rate or the number of connected diagnostics, which come from :

- inability to update our hardware architecture : our central computer power is up to 4 Mips for all the operations, namely data taking, run control, data and task monitoring and real-time processing. Moreover the network data rate is less than 120 KB/s on each of the 3 Ethernet lines,
- based on a proprietary real-time operating system. We use iRMX® III in the INTEL™ data acquisition crates (called low level), and SINTRAN® in the central NORSK DATA™ computer, both without progress.

The low processing power does not allow us to use some facilities we have implemented in the original version [1] of the data acquisition system. For instance only the computation of the hybrid power is done in real-time and dispatched to the other consumer tasks. All the processing tasks which are time consuming, are launched on an DEC™ Alphaserver at the end of the pulse, gaining their data from files. It is certainly not the most efficient way, with the drawback of increasing the time delay between the pulses.

During the Tore Supra operation, several problems in the original design of the control software, were met, namely :

- the scheduling and the monitoring tasks are two separate tasks. Operators are then in charge of detecting and correcting any defect, mainly due to a lack of genuine supervision,
- the message-passing is too slow and not enough reliable. For instance, we are able to send only a few messages per second which avoid to monitor the tasks during the pulse,
- at the beginning of each pulse, the parameters are loaded in memory from the users files and dispatched to the data acquisition computers. This behaviour has a severe drawback : indeed it requests a synchronous tasks scheduling, all of them must be in the appropriate state to allow the parameters loading. If one or more fails, the pulse must be stopped and restarted,
- the parameters loading in memory and the user access software are strongly tied to the NORSK DATA™ computer.

2. PRINCIPLES AND PREREQUISITES INVOLVED IN THE EVOLUTION OF THE TORE SUPRA ON-LINE SOFTWARE

A few guidelines have been used to redesign our data acquisition system :

- **Standard operating system.** In our computer architecture, we have two kinds of computers : the so called low level ones which are in charge of the real-time data acquisition, they must have a very short response time to any interrupt, and the high level computers which collect, process and distribute the data, their ability to manage high data rates being predominant. The selected operating systems are UNIX for the high level and LynxOS™, a real-time UNIX-like operating system, for the data acquisition units.
- **Communications as the basis.** In a distributed data acquisition system, the largest part of the software is directly connected with moving the data around. Therefore, the communication system must be reliable and fast enough to drive at

the same time control messages (for instance, the task states or the variable values which have usually a small size) and large binary data messages. The representation of data must be platform-independent: we do not care of the computer type and location, only the task name is of concern. The communication system must also be able to send the same data to many consumers. Of course, producers do not know the consumers: they can subscribe or unsubscribe to some message types at any time. To reduce the inter-process dependencies, the communication system must take care of the asynchronous tasks behaviour. It must be enough open to allow the creation of our own message types and to connect our already existing INTEL™ acquisition units.

- **Expert system.** The growing complexity of the distributed data acquisition system imposes stronger and stronger requirements on the operator who runs the experiment. Nowadays, real-time expert system shells offer a chance to solve this problem, provided that they are connected to the real world and fast. They must accept time-driven rules and their inference mechanism must use forward (data driven) and backward (goal driven) chaining. Such an expert system can then be used for the tasks scheduling and for the errors management.
- **High performance.** In a fully distributed system, it is easier to increase the computing power through new connected computer as far as the network is available and not saturated. To be ready for the future, we based our network over Fast Ethernet. Between two tasks, we achieve up to 6.8 MB/s using Tcp/Ip.
- **Safety and crash recovery.** To reach the continuous operation, the software and the hardware must be fault tolerant, but in our case we request a hardware crash recovery in less than one hour. We ask for removable disks, disks mirroring, multi-licences servers.
- **Connection to the existing data acquisition units.** Around 40 diagnostics based on Multibus I/iRMX® III are already running and data taking. They must be connected to the new on-line software with as less modifications as possible. They behave as data acquisition automata which are fully driven by the tasks scheduler.

3. REALISATION

3.1. Overall presentation

Our hardware architecture is shown in figure 1.

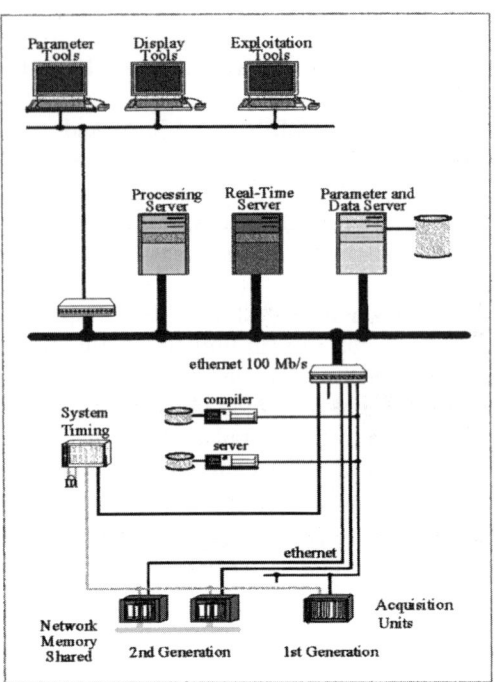

Figure 1. Tore Supra data acquisition architecture

After having studied various ways to improve our on-line software, we decided to implement it on the commercial Rtworks™[2] package. It fulfils completely our requirements and even opens the path towards the continuous operation. Built of many dedicated modules (acquisition, display, expert system,...) in a distributed environment, we were worried about the ability of the Rtserver node, to collect and dispatch in an efficient way our data. In fact, it sustains on a DEC™ 2100 Alphaserver (biprocessors at 275 MHz), 3 MB/s in input and more than 1 MB/s for each consumer with our data type (4 KB/message).

The use of commercial product for data acquisition which was not even possible a few years ago, allows us to access softwares largely debugged, running over many kinds of computers and releases us from the support. One original point which must be emphasized, is the use of the same product for the run control, for the data and process monitoring and for the data acquisition itself.

3.2. Scheduling and control

To schedule the tasks involved in our on-line software, we use the rule-based expert system shell included in Rtworks™[2]. Time-triggered rules give us continuously the tasks states while the backward chaining strategy is used in the inference engine, to find the appropriate rules which must be activated to make a pulse. The use of the same channel for the data and the messages, allows the inference engine to use not only the tasks states but also the data to detect any malfunction. Run control, monitoring and diagnosis are done within the same expert system shell. It eases the defects prevention and their corrections.

Task monitoring and data displays are built with the same tool (Rthci). They are distributed among 5 X-terminals or workstations, each being dedicated to a peculiar function : pulse operation, data display, data acquisition units monitoring, overall monitoring, slow control monitoring and general informations about the previous pulses.

As part of our work, we realized the gateways to the Intel units, one for the message-passing and the other for the data.

For plasma and diagnostic control, specific softwares have been written by the users in FORTRAN (iRMX®) or C (LynxOS®) and integrated in the data acquisition units. To assure a genuine plasma stability, multi-processors VME units are used. So, in a crate equipped with a PowerPC® 604 board, the processing loop in charge of the plasma feedback control [3] is done in less than 2 ms. A shared memory network connects these feedback units in order to achieve a distributed plasma control. The reflective memory SCRAMNET® (SYSTRAN) has been selected and transmits up to 29.5 MB/s.

3.3. Data acquisition

34 first generation units are still used for the data acquisitions. Based on Multibus I crates and iRMX® III operating system, we keep the ones whose performances are sufficient. 7 second generation units (VME crates with LynxOS™) are now in operation, 3 of them are dedicated to the plasma feedback control. Continuous data acquisition is done through specific hardware : double buffering and triggered acquisition boards. The data throughput is up to 100 KB/s for each second generation units.

3.4. Parameters

Previously embedded with the NORSK DATA™ proprietary operating system, the parameters are now stored in a CA-Ingres™ database. In addition to be more reliable, it permits the early detection of wrong parameters and the inter-diagnostics checks. To access the parameters database from any workstation, a user-friendly MOTIF™ interface called TOP with spread-sheet and graph widgets has been designed. A simulation of the plasma position and of its stability is done in TOP. The user may too compare on line the acquired data with some predefined data. The major change concerns the parameters downloading : indeed the acquisition units can fetch and store their current parameters through the Tcp/Ip client-server library (TSLib) at any time. Archived parameters for the older pulses can be read with the same function.

3.5. Real-time processing

Our aim is not only the continuous data taking but processing. Two libraries have been written to access the real-time data frames for the Matlab™ or C programs. Users receive their requested data for a time slice, and once they have collected all of them, they can start their computation. Produced data are then put in Rtworks™ to be available to the other processes. The data flow schedules the tasks, avoiding in this way the burden of the process planning.

3.6. Data storage

As explained in a previous document [4], data are written in files, one per pulse and per diagnostic giving an amount of 30 GB/year. Long time storage is done in an optical jukebox, every previous Tore Supra data being available in less than 30s. Informations about the pulse and pointers to the data files are stored in an CA-Ingres™ database. These data files being available only at the end of the pulse, they are used by the batch processes (launched automatically between the pulses) and by the off-line computing.

To recover previous tasks state or historical data, we use the Rtarchive and Rtplayback processes furnished with Rtworks™. They work as a fast in-memory database with backup and checkpoint done periodically. These processes are used mainly to keep the events needed to analyze malfunctions. Data frames are also stored through Rtarchive to enforce the data safety (disk crash or network failure are always possible), but it could be used in a continuous acquisition mode. Indeed, data are recovered not through pulse number but through time stamp or previous event.

3.7. Safety

A fault tolerant data acquisition system is not requested. Nevertheless, recovery time must be less than one hour in case of hardware failure. At the software level, reliability is assured through

Rtworks™ software. For sensible communications we may even use the guaranteed message delivery mode : messages are saved on disk so that they can be resent after a delivery failure. Moreover a dedicated task is in charge of surveying all the processes involved in the data acquisition system and warns the inference engine if a malfunction happens. Thus we are able to restart a process even during data taking. To be safe from a hardware failure, the central real-time computer has been designed with two processors boards and two fast wide SCSI boards. The sensible disks are mirrored and the application is stored on two removable 4 GB disks. The software development machine serves as a backup computer for our application. In case of hard failure of the real-time computer, operators are able to plug the removable disks even in a running computer and restart the application in less than one hour.

4. EVOLUTION

Reaching the steady state[5] implies a few changes in the forthcoming years :
- remove all the data storage in the acquisition modules. Only data buffering is allowed to take care of the asynchronous behaviour,
- put VME units to manage the diagnostics or heatings involved in plasma control and extend the shared memory network,
- direct links to the operators to drive manually the plasma parameters,
- permanently read and write access to the parameters database. The parameters values must be time-stamped,
- move the batch to real-time processing. We have to change the data access and arrange ourselves to have all the informations as fast as possible. No data should be sent at the end of the pulse if they are used by a computation task. Processes must be modified to adapt this time-slice philosophy,
- increase the processing power. Actually we need 10 times more processing than acquisition power, then we must gain one order of magnitude in processor power. It is not out of range : multiprocessors computer, farm of computers, hardware capability themselves are able to solve this request,
- store and retrieve continuously the data. Only record locking is allowed.

5. CONCLUSION

Although our data acquisition system has been used successfully during many years, it needed to be completely re-engineered when we passed from proprietary to UNIX operating system. The long pulse operation has been always kept in mind, and if there is still some works to be done to achieve it, the continuous data taking is not far to be reached. What we have learnt too, is the availability of commercial software which fulfils all the basic software, namely message-passing, buffer management, inter-processes communication, continuous storage and retrieve, data representation independence, human interface which were the main jobs for the real-time programmers. Then our job is redirected towards our own features, we had great benefits to use real-time expert system shell for the tasks scheduling, errors management and overall monitoring. Indeed it releases the operators from a continuous survey and helps to recover from a malfunction when the experiments become more and more complex. We are going to take data with our new data acquisition system in fall 96 and we will enhance it for the long pulse operation and even for the steady state during the next years.

6. ACKNOWLEDGEMENT

The authors express their sincere thanks to all the members of the computer and electronic groups.

REFERENCES

1. S. Balme & al., The Tore Supra Data Acquisition System, Proceedings of the XVth Symposium On Fusion Technology, Utrecht 1988.
2. T. Barbagallo, Monitoring and Control Software for Real-Time Processes, Real-Time Magazine Vol. 3, 1994.
3. G. Martin & al., First results of the new plasma feedback control of Tore Supra, Contribution in these proceedings.
4. Y. Buravand & al., Tore Supra Data Analysis Facilities, Proceedings of the XVIIth Symposium On Fusion Technology, Rome 1992.
5. B. Gagey & al., Tore Supra Data Acquisition : Implication of the steady-state for the computer system, Workshop on Technical Aspect of Steady-State Device, IPP Garching 1995.

The Tore Supra control computer system : evolutions

E. Chatelier, F. Hennion, M. Hernandez, J.Y. Journeaux, P. Lebourg

Association EURATOM-CEA, Département de Recherches sur la Fusion Contrôlée
Centre de Cadarache, 13108 Saint Paul lez Durance Cedex, France

The Tore Supra Control Computer System (CCS) named Architecture 7 (A7) has been operational since 1985. Its architecture, its modular conception and its user-friendly programmation facility made it highly efficient. Recently, the control command system was shown to require improvements. Opening the system was necessary for the introduction of industrial products providing reliability and maintainability. The overall control computer system has been upgraded : with respect to the human interface, the automation and the communication. A commercial supervisory software : PANORAMA runs on several Personal Computers (PC), connected on the Ethernet network. These PCs are connected to Programmable Logic Controllers (PLC) chosen for their industrial quality and reliability : Telemecanique PLC. A field bus allows the PLC of each Tore Supra subsystem, to communicate among themselves. The new architecture has to exchange data with the old one. Migration is progressive because the Tore Supra control computer system runs continuously so every change must be executed during Tore Supra shut-downs. In the year 2000, about half the units will have migrated from Architecture 7 to Personal Computers and Programmable Logic Controllers.

1. INTRODUCTION

The Tore Supra CCS with A7, has been effective on all the sub-systems : continuous systems such as Cryogeny, Vacuum, Water Cooling and sequential systems such as Additional Heating. After such a long period of time, the need for improvements was recognized to take into account the existence of new industrial products and processes (Ethernet, PCs and PLCs), providing with enhanced reliability and maintainability. A major consideration, motivating this migration, is the increased vulnerability of the CCS, and therefore of Tore Supra as a whole, because of hardware obsolescence and the dispersion or loss of competence with respect to the original system. Modifications started in 1995, at three different levels : the human interface, automation and communication. Due to the ongoing operation of Tore Supra, the migration from the old architecture to the new one must be cautious and progressive, and modifications must take place during shut-down periods.

2. DESCRIPTION OF THE OLD SYSTEM A7

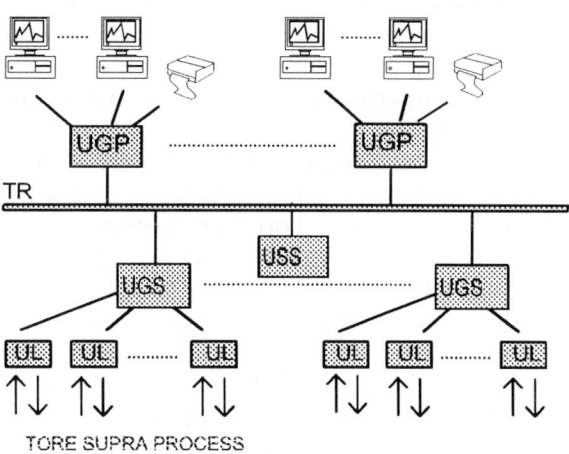

Figure 1. The system Architecture 7

The Control Computer System ARCHITECTURE 7 [1] has been divided into several sub-structures according to the sub-systems of Tore Supra. Some of them are run continuously: Cryogenic, Water Cooling and Vacuum and these sub-systems must be very reliable ; in others, operation is pulsed. The CCS is composed of about 60 units : the human interface units (UGPs) permit conducting and controlling the process, animating with process variables and, in real time, modifying synoptics (Fig. 1). Each sub-system is divided into two levels : local units (ULs) that contain data input/output circuit boards analog and binary (15 000 I/O are processed) and the sub-system units (UGS) that control and supervise the ULs.

Units can communicate on a specific network: Table Ronde (TR). The sub-systems supervisor unit (USS) controls all the units. UGSs, ULs and USS are dedicated to automatisms ; they are programmed in GRAFCET [2] (a norm of automata description) with a keyboard language that enables translating graphs and variables into a computerized language. the users have written their automatisms and defined their man machine (15 000 steps of Grafcet written).

3. REASONS FOR EVOLUTION

Although this system is very efficient (very few stops of Tore Supra are involved by the CCS), evolution was necessary.

Two important conditions must be fulfilled :
* The coexistence of both current and new systems is required during modifications due to Tore Supra operation.
* The migrations are progressive because the Tore Supra CCS runs continuously ; every change must be executed during Tore Supra shut-downs.

3.1. Reliability improvements
Some electronic cards are no longer manufactured, so, maintenance is increasingly difficult.

The original A7 sub-contractor has disappeared, resulting in considerable loss of competence.

Hardware obsolescence and impossible software developments have made the CCS very vulnerable.

3.2. Standardization
The considerable advance in hardware and in software technologies during the last 10 years has permitted choosing industrial solutions with respect to the network, human interface and automata levels.

The A7 technology was too specific. The network could not authorize interaction with other equipment (personal computers, work stations,...). In particular, interaction was necessary to establish communication with the Data Acquisition System [3].

4. NATURE OF THE EVOLUTION

4.1. At the network level
The specific A7 network TR is being progressively replaced by ETHERNET with ISO norm ; its has meant changing the operating system on the UGP and UGS (RMXI became RMXII).

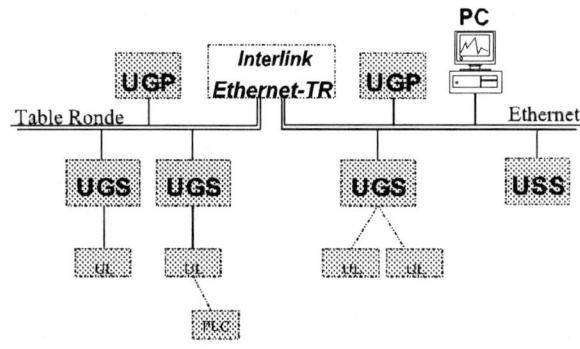

Figure 2. Evolution of A7 towards Ethernet

A PC server was installed for downloading the UGS. It runs with the RMX for Windows operating system.

A temporary bridge was installed between the two networks during Ethernet migration for necessary communication between sub-systems (Fig. 2).

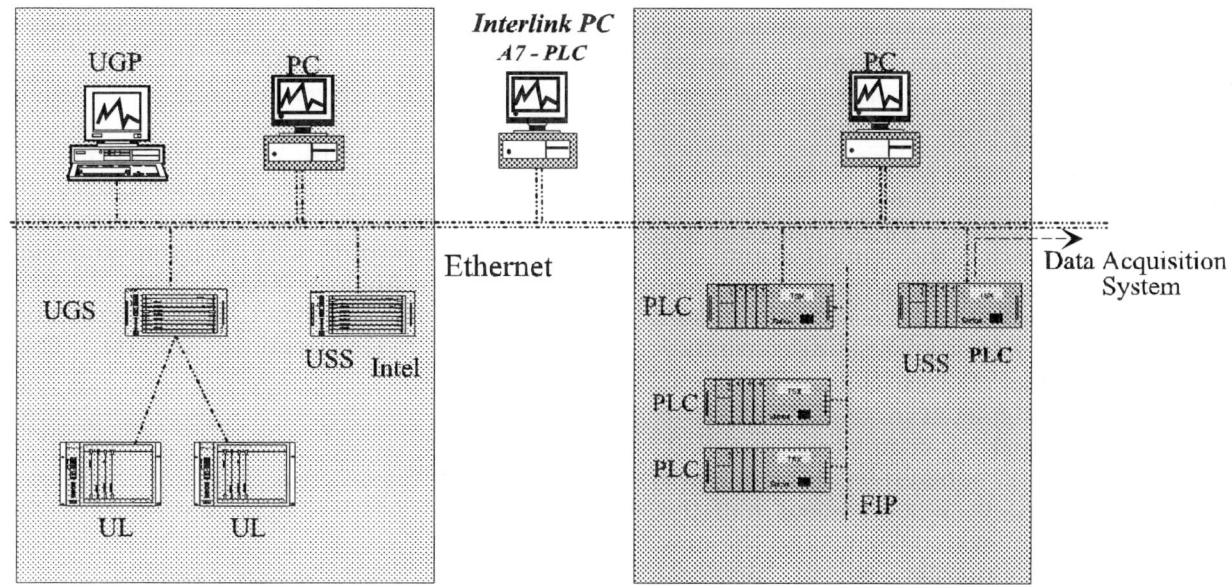

Coexistence between old system A7 and new system PLCs and PCS

4.2. At the human interface level

Several PCs are implemented fitted with PANORAMA, a multi-task supervision software under Windows, including specific modules: a CAD (Computer Aided Design) software permits development of graphic panels, a storage module permits historical analysis and playback Trends, statistics,...). Panorama also includes, an alarm supervisor and simulation. These PCs can supervise all the sub-systems of Tore Supra (A7 subsystems and PLC sub-systems).

An Interlink PC is necessary because of the two different protocols (A7 and Telemecanique world) and permits communication between the UGSs and PLCs.

4.3. At the automata level

Progressively, a part of the UGSs and ULs is beingreplaced by Telemecanique PLCs offering modularity, industrial design and easy maintainability because of their industrialization and their extensive distribution.

These PLC utilize the protocol Telemecanique Uni-Te. They can be connected to Ethernet for communication with PCs which manage the man machine interface and they can be connected to a field bus FIP : FIPWAY for communication between the PLCs belonging to the same sub-system and FIPIO for remote I/Os.

As in the old system, the PLC are programmed by the users themselves using three languages: The GRAFCET language best-suited as the graphic language best suited to representing the sequential part of automated production systems. The LADDER language based on electronic contacts and relay logic, very easy to describe combinational expressions, used world-wide because well-known by electrical engineers and electricians. The LITERAL language provides clear, concise representation by graphically combining basic logic functions, used to represent logic relationships.

The USS PLC, in communication with the USS A7, supervises all the other PLC and permits communication between sub-systems. It controls parameters and authorizes Tore Supra shots. It cans stop the discharge if a sub-system fault occurs. It, also, enables the link with the data acquisition system [3].

5. CONCLUSION

The original Tore Supra Control Computer System, Architecture 7, has been working satisfactorily for more than ten years. This system is in the process of being improved by integrating new industrial products (PCs and Programmable Logic Controllers) with the main aim of overcoming consequences of hardware obsolescence and loss of competence, and thus avoiding any accidental discontinuity in the operation of Tore Supra. The first PLC migrations started in 1995. Today, the Ethernet Network is completely installed. Many A7 and PLC sub-systems are equipped with PCs. In the year 2000, about half the units will have migrated from A7 to the new system. This situation is, at present, considered sufficient to allow satisfactory operation of Tore Supra in the future but more migrations would take place if more obsolescence occurs.

REFERENCES

1. The TORE SUPRA Control Computer System: B. Gagey and Al., Proceedings of the XIVth Symposium on Fusion Technology, Avignon 1986.
2. The TORE SUPRA Control Computer System, programmed and configured by the users: J.Y. Journeaux and Al., Proceedings of the XVIth Symposium on Fusion Technology, London 1990.
3. The TORE SUPRA Data Acquisition System: B. Guillerminet and Al., This conference.

First Results of the New Plasma Feed-back Control for TORE-SUPRA.

G. Martin, D. Moulin, D. van Houtte, T. Wijnands

Association Euratom-CEA, CEN Cadarache, 13108 Saint-Paul-lez-Durance, France

After 6 years of operation, the poloïdal control system of TORE-SUPRA [1] has been upgraded, with two main goals : i) to be able to calculate plasma global parameters in real time to feed-back control on them ; ii) to include several diagnostics in a global "Plasma control" rather than the pure "Poloïdal control" in operation previously.

Based on a 100 MHz processor in a VME crate, this system performed each two millisecond a full control cycle composed of the acquisition of 80 measurements, the determination of the plasma position, the calculation of the generators voltages and several safety checks on the plasma quality. Written in C, it provides much flexibility to implement additional calculations or feed-back loops when they are needed.

Many results have already successfully been obtained with the control of the plasma position, the plasma surface loop voltage during long pulse discharges, the safety factor and the internal inductance, used to characterise the current profile. The plasma current value has been controlled by LHCD, with zero ohmic power, during a steady-state scenario, which has been tested for more than one minute.

1. INTRODUCTION

With its circular shape, TORE-SUPRA is not subject to fast vertical instabilities : the typical time constant of plasma movements is 20 ms. This gives time enough to control its position with a software based system. A first stage, implemented in 1988 for the start of TORE-SUPRA, performed only the minimal calculations needed for plasma positioning. Due to the increase in microcomputer speed, its was decided in 1995 to switch for a more powerful system, to allow a complete determination of the magnetic topology in real time.

Based on a 100 MHz Motorola® MVE-1603 processor in a VME crate, this new feed-back system opens many possibilities, either by the determination of several plasma parameters, like q_ψ, β, l_i, ... or by the possibility to interconnect several other systems through the fast VME bus, e.g. Lower Hybrid or Ion Cyclotron power.

2. PLASMA CHARATERIZATION

The determination of the plasma topology is performed in successive steps :

✤ 33 values of the magnetic field are measured on the vacuum vessel, in the 3 directions of space, using very low drift integrators (< 1% / hour).

✤ The vertical flux Ψ_v and its derivatives are calculated on a reference surface Σ_R, either by fitting the measurements, or by using the Grad-Shafranov equation in vacuum.

✤ The flux function Ψ_v is extrapolated in all the space, with Fourier and Taylor expansions.

✤ The plasma is located along 10 axes, by looking at points with the same values for Ψ_v. One and only one of these touches the limiters : it defines the contact between the plasma and the wall. These 10 positions are compared to desired values (programmed by the pilot), and the differences are used to drive the generators with a gain matrix.

✤ An ellipse [R,Z,a,b] is fitted on these 10 points, to show a global view of the plasma.

✤ The knowledge of the flux map allows to determine several global parameters of the plasma : the plasma current I_p, the safety factor q_ψ, and the stored energy $\beta+li/2$. These two last terms are separated with a diamagnetic loop.

All these values are then used in various feed-back loops, described in the following chapters.

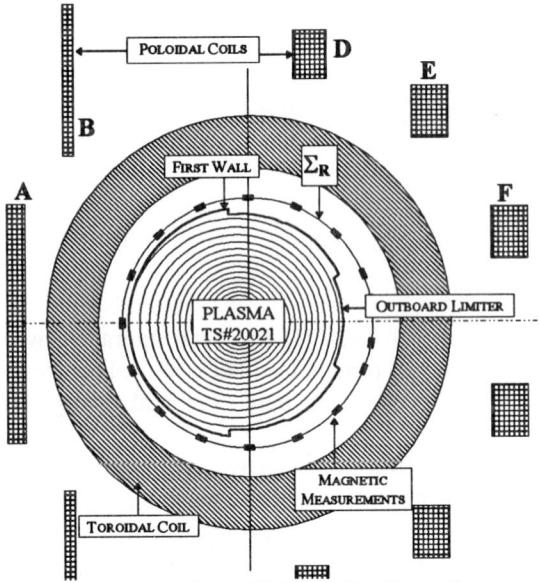

Figure 1 : TORE-SUPRA Configuration.

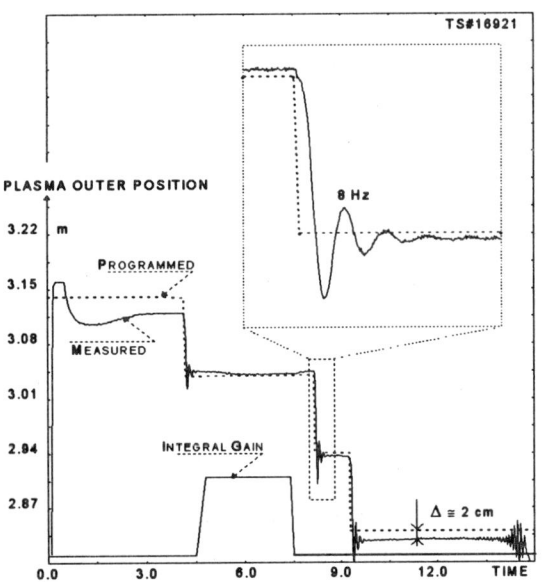

Figure 2 : Control of the Plasma Position.

3. PLASMA POSITION CONTROL

The poloidal field system of TORE-SUPRA fulfils in a single set of coils the ohmic heating and the position control of the plasma. It is constituted of 9 coils (figure 1) all connected in parallel, associated to 9 voltage generators. The main one, G_0, drives the central coil A, mainly used for plasma current control, while the 8 others drive the external coils B to F. Their voltages are adjusted to insure that the plasma lies at the desired position relative to the various limiters : the 10 radial errors are multiplied by a 10x8 matrix, and injected in a full P.I.D. control loop.

This loop was tested during shot TS#16921 (figure 2). With the proportional gain only, the outer position (crucial for wave coupling) is obtained within 2 centimetres, except during the current ramp-up. The switch on of the integral gain between 4 and 7 seconds allows an additional decrease of this gap, down to a few millimetres.

During an abrupt change in position, the plasma exhibit a small oscillation, with its characteristic time of 20 ms ($\tau = 1/2\pi f$). The main origin of this constant lies in the time response of the generators themselves.

4. SAFETY FACTOR CONTROL

On TORE-SUPRA, several scenarii require a good control of the safety factor : current ramp-up at constant q_ψ allows fast ramp-up, useful to save volt-seconds for the flat top ; resonant use of the Ergodic Divertor needs a fine tuning of the q_ψ at the edge. This q_ψ control can either be made at fixed position by acting on the generator G_0 (instead of I_p), or for a given current by changing the large radius R of the plasma.

This second mode was tested on shot TS#17357 (figure 3). The q_ψ value is required to stay as close as possible to 4.8. The radius R of the plasma adjusts itself as I_p is ramp up and down. The small radius a follows, as the plasma keep always its contact with the central column (inboard limiter).

5. CURRENT PROFILE CONTROL

A first round of experiments have been performed to control the current profile [2]. As the polarimetry is not yet available in real time, the internal inductance l_i has been used to characterise this profile. Several results have been obtained with a feed-back on Lower Hybrid waves amplitude and spectrum.

During shot TS#20049 (figure 4), the internal inductance l_i has been controlled by tuning the L.H. waves spectrum, with a change in the relative phases Φ of adjacent klystrons on the launcher.

When the l_i reference is ramp down from 1.7 to 1.5, the phase shift changes from $\approx -.5$ to -1. rd : the mean index of the waves decreases, giving a more peripherical deposition of the power, hence a broader profile.

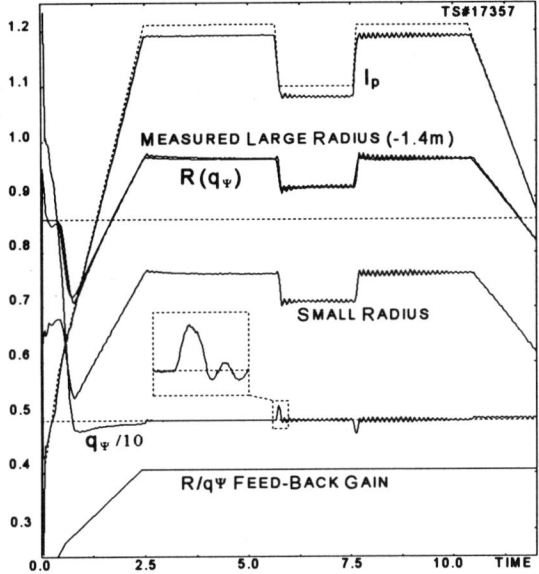

Figure 3 : Constant q_ψ Plasma Operation.

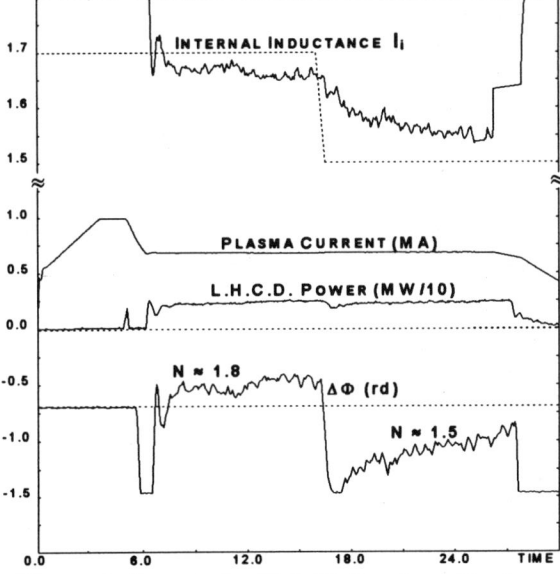

Figure 4 : l_i control with L.H. Spectrum Tuning

To optimise the current ramp-up, one can control the trajectory in the (q_ψ, l_i) plane. In this plane, only a limited area is stable with respect to M.H.D kink modes. Low l_i, obtained with fast ramp-up of the current, must be avoided.

On figure 5, TS#20000 is a « standard » shot, well in the stable area but not optimised in ramp-up velocity (flat top is only reached after 4 seconds). TS#20030 is too fast : reaching the plateau at 2 seconds, it exhibit strong MHD bursts which lead finally to a locked mode and a disruption a few seconds later. During TS#20032, the same I_p was programmed but the (q_ψ, l_i) feed-back slows the dI_P/dt rate to follow the optimised line (heavy dotted line on the figure). This last shot has the fastest ramp available for I_p, without MHD kinks.

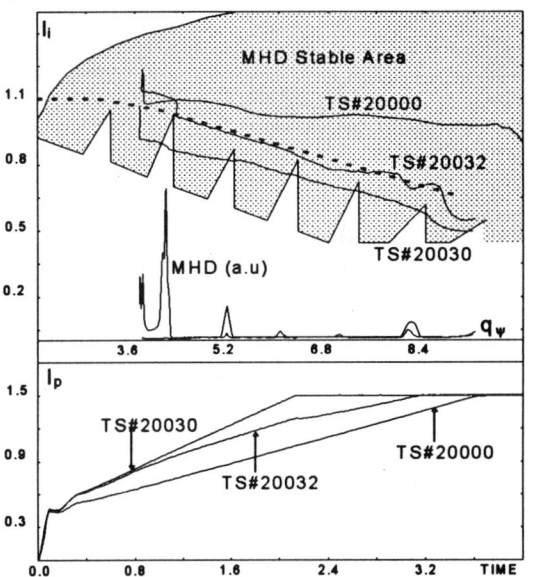

Figure 5 : Current Profile Control during Ramp-up.

6. STEADY-STATE SCENARIO

In the standard mode of operation of a Tokamak, the plasma current I_p is kept constant by a continuous decrease of the vertical field Ψ_v encircled by the plasma ring. Conversely, in a steady-state scenario, Ψ_v must stay constant, as everything. This requires of course a non inductive source of current : Lower Hybrid waves (L.H) are used on TORE-SUPRA. Two new feed-back loops have been implemented to achieve this mode of operation (figure 5) :

i) During shot TS#17018, the standard feed-back loop, linking G_0 voltage to I_p, is replaced between 5 and 10 seconds by the vertical flux control loop. As no more ohmic power flows to the plasma, I_p

decreases with its L/R time of ≈ 3 seconds.

ii) As G_0 is used for Ψ_v, I_p is controlled by modulating the L.H power between 0 and a pre-set maximal value. During shot TS#17970, several changes have been programmed on I_p to test the response of this system.

Both loops were used successfully during TS#19249, one of the first truly steady-state discharge obtained on TORE-SUPRA (figure 8). Only the density rises slowly, as the wall fill itself with D_2, preventing the complete steadiness of the plasma.

7. FUTURE PROSPECTS

The 100 MHz processor performs the full control cycle in around 700 μs (figure 7). The basic clock has been kept at 2048 μs, allowing the possibility to add several new calculations in the loop. Near future plans include more precise position determination (triangularity) and interconnection with several other systems as the gas control unit, infra-red cameras, polarimetry, ICRH, ...

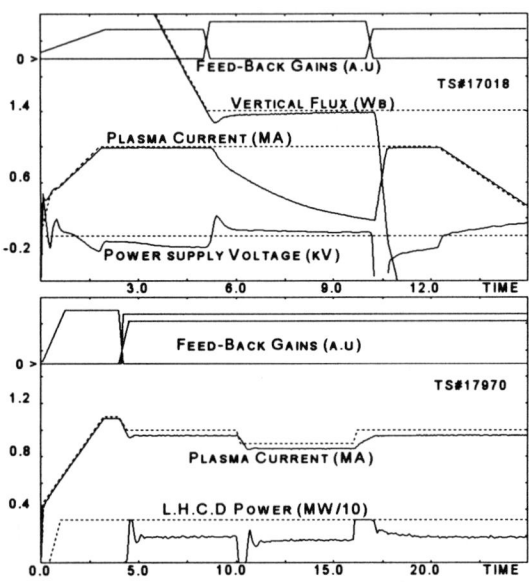

Figure 6 : Steady-state Feed-back loops.

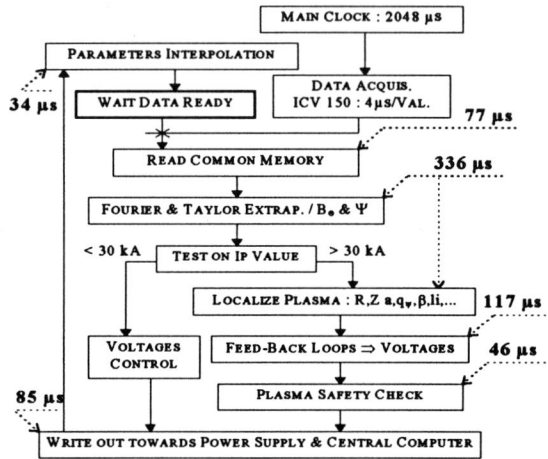

Figure 7 : Program Algorithm with CPU Times

REFERENCES

1. J.M. Ané et al., 15th SOFT, UTRECHT [1988], p.657.
2. T. Wijnands et al., 23rd EPS, KIEV [1996].

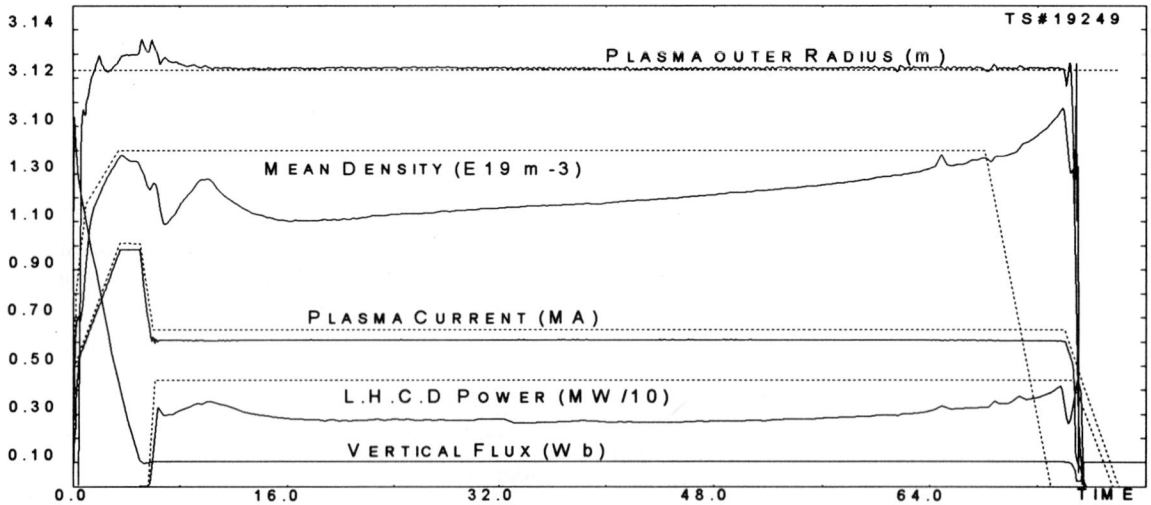

Figure 8 : One minute Steady-state Discharge on TORE-SUPRA

A DATA ACQUISITION, CONTROL AND VISUALIZATION SYSTEM FOR THE UPGRADED TOSKA FACILITY AT FZK

A.Augenstein, H.Barthel, I.Donner, H.Frankrone, P.Gruber, G.Hellmann, P.Klingenstein, U.Padligur, K.Rietzschel, T.Specht, G.Würz

Forschungszentrum Karlsruhe GmbH,
Hauptabteilung Prozeßdatenverarbeitung und Elektronik, Postfach 3640, D76021 Karlsruhe, Germany

The paper describes a distributed data acquisition-, control and monitoring system designed and being used for testing large superconducting prototype coils in the upgraded TOSKA facility needed for magnetic confinement in fusion. The system will be used in 1996 for testing the inforced EURATOM LCT coil at 1.8 K, later in a combination with the W7-X demo coil and finally the ITER TF model coil. The system was designed for monitoring and archiving long time operation of the cryogenic facility over several month as well as single shots in experimental operation. To control the cryogenic facility a number of programmable logic controllers mainly SIMATIC-135U were used as front ends. To keep the system flexible, all parameters and configuration data which describe hardware and software set-ups are kept in a database which is maintained by the operating team using ORACLE FORMS [1] applications on PCs. Periodically sampled and transient recorded data are archived in an ORACLE database residing on an DEC ALPHA Station 250-4/266 starting with an initial storage capacity of 60 GB.

1. INTRODUCTION AND HISTORY

In the past testing superconducting magnetic prototype coils such as LCT and POLO at TOSKA facility DEC hardware platforms and proved operating systems such as RSX11-M+ and VMS were used. All hardware components and developed application software run in stable operations over many months. The accumulated knowledge in our data processing department and by the facility operating team in hardware and software development, maintenance and about operating system VMS lead to the decision for DEC VAXstations and ALPHA platforms. Figure 1 shows the basic components of the new system without the failover node.

The front ends to control the cryogenic facility consists of programmable logic controllers SIMATIC-135U. One of them is dedicated as data concentrator using a CP143 communications processor for data exchange with the control system.

PC hardware is in use as server for data acquisition and control of the subsystems HOTTINGER [2] and DR22 [3] and as interface to the database which keeps the configuration data for the subsystems besides the long term archived data. DEC X-terminals VXT2000+ are used mainly for visualization of process and transient recorded data.

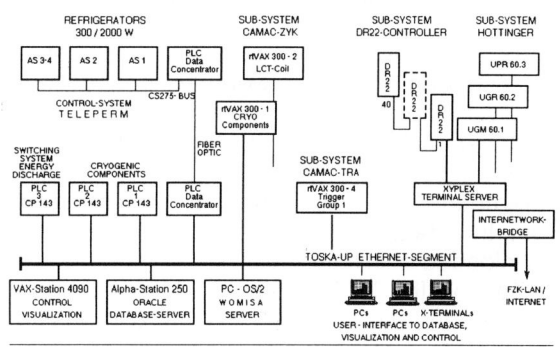

Fig. 1 SYSTEM COMPONENTS

Modified CAMAC modules for temperature measurements down to 1.8K as well as for recording transient data up to 1 MHz have been developed and integrated.

A high voltage resistant electronic unit has been developed to handle low voltage signals obtained from temperature sensors on high voltage potential parts of the test object. 3 CAMAC modules type 3623 from Kinetic Systems Corp. [4] each having 6 16-bit counters take care of these signals.

2. SUBSYSTEMS

2.1. CAMAC Based Systems

In the upgraded facility two CAMAC crates containing groups of improved acquisition modules are installed to sample periodically temperature signals on the test objects and on parts of the cryogenic facility.

A CAMAC module 3968-Z1A from Kinetic Systems with an embedded rtVAX-300 processor in each crate serves as a diskless controller which is directly connected to the Ethernet. The VAXeln real-time kernel image including a communications task and the CAMAC application program can be downloaded by the operator clicking the appropriate buttons in the window provided by the control system VXL (see section 3). Configuration and calculation data will be retrieved from the central database by a server process and downloaded. Calculated temperature value are then inserted in VXLs real-time database for visualization and archiving.

A new CAMAC transient recorder module has been developed to meet the requirement for collecting data during the energy discharge operation. Each module has 4 independent data channels containing a 12-bit analogue-to-digital converter. Data of each channel are stored in local memory of 64k 16-bit words. The user can choose between 10 input voltage ranges. The following sampling rates are available: 1Hz, 10Hz, 100Hz, 1kHz, 10kHz, 50kHz, 100kHz, 500kHz and 1 MHz. Parameters for each channel are retrieved from the database by a server process and downloaded at initialization.

Testing the LCT coil one CAMAC crate is required and equipped with 8 CAMAC transient modules to make up one trigger group of 32 data channels. Calculation of physical values is done in the rtVAX-300 controller and data are transferred as VMS-file to a disk on the database server. Before finally inserting the transient shots into the database previewing is possible using an IDL [5] application program.

2.2 Programmable Logic Controller (P L C)

For data acquisition and process control of the cryogenic facility programmable logic controllers typically SIMATIC-135U are integrated in the system. The TELEPERM system of SIEMENS controlling the two refrigerators is used independently from the VXL control system. Command and data exchange between the main control system VXL and the PLCs CP143 is based on DECnet Phase V (OSI) protocols. The CP143 card firmware implements the complete OSI 7 Layer protocol stack. The CP143 is configured using the COM 143 software package.

VXLs communications driver for the CP143 requires on the VMS operating system the DEComni server to be installed and running. The OMNI Definition Facility (ODF) is used to create and manage the locally stored definitions of remote VMD objects. Using the DEC diagnostic tool OmniView data associated with a remote VMD can be displayed and values can be written into variables.

2.3 The WOMISA Subsystem

The WOMISA (**W**indow **O**riented **M**easurement **I**ntegration **S**ystem **A**rchitecture) subsystem is our middleware to integrate external, distributed, serial devices transparently into the TOSKA data acquisition system. The kernel of the subsystem runs on an IBM PC with the multitasking OS/2 WARP CONNECT operating system. The architecture of WOMISA is strongly client-server-oriented. The communication between the PC and the main system is based on the TCP/IP - RPC (Remote Procedure Calls) technology. The distributed, serial devices are connected to a standalone terminal server that provides 16 asynchronous serial communication ports and an Ethernet interface. The interprocess communication between the WOMISA subsystem and the terminal server is built with the low-level TCP/IP - socket - library. Each external device has a hardware specific driver, that hides the complexity of the devices.

The simplest feature of WOMISA is to scan all channels from all devices periodically in an internal cache. Every measurement value may be manipulated with specific functions for calculation. Several channels can be correlated to produce virtual channels. All practicable manipulation parameters are stored in the database and can be downloaded if necessary.

A more important feature of WOMISA is the bi-directional, asynchronous, event-driven, transparent communication between the visualization system and the external devices. That means, if an input value is changed by an user interaction on the visualization surface, the changed value is transmitted immediately to the WOMISA subsystem and after corresponding, plausible modifications downloaded to the external device.

2.3.1 The Hottinger Subsystem

The Hottinger UGR 60 subsystem consists of a multichannel measurement acquisition system, which has a RS-232-C (V24) interface. This instrument can record, process and issue measured signals up to 60 channels. All strain gauge bridge circuits, thermal elements, resistor thermometers and strain gauge-, inductive- and potentiometer- pick-ups may be attachable with the appropriate units. Our multithreaded device driver enables parallel operations on the 3 Hottinger devices included in the TOSKA data acquisition system.

2.3.2 The DR22 Subsystem

The 40 SIPART DR22 process regulators, used in the TOSKA data acquisition system, control the cryogenic infrastructure. Up to 32 regulators can be connected to one V28 bus co-operating with a bus driver attached to the RS-232-port of the terminal server.

The WOMISA DR22 multithreaded device driver hides the complexity of the SIEMENS - specific communication procedure from and to the process regulator. All configuration-, parameter- and process values are transparently manageable from the superior computer system. The WOMISA subsystem scans periodically the relevant process values from all regulators. In addition several control values can be set asynchronously.

3. CONTROL SYSTEM

The control of operations of the large and complex facility TOSKA-UP requires a flexible control system to meet the need for large-scale control applications in a distributed architecture. It should provide failover capability, data distribution, visualization, alarm- and event handling and Ethernet communication with the CP143 of PLC data concentrator. The control system should also provide full graphics, multi-user privileged access on high resolution X terminals and an interface to exchange data with the subsystems. Most important is to guarantee stable operations over a time period of several month.

VXL [6] was selected as control system running under Open VMS on VAXstations as well as on Alpha platforms.

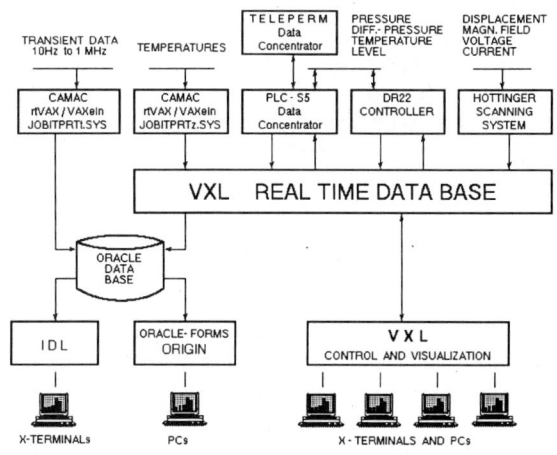

Fig. 2 DATAFLOW

ORACLE RdB, is a database management system specially developed for data retrieval in a real-time environment. The tables of ORACLE RTDB, VXL´s Real-Time Database using RdB contain all the information needed to describe graphical objects of a screen or subwindow and are accessible by the application programmer. Using VXL ACCESS, a library of database access routines, a programmer can write custom application programs to run concurrently to exchange data with subsystems. The data flow into and out of RTDB is shown in figure 2. Except for the transient recorded data all subsystems data are stored in RTDB from which they are accessed and periodically archived.

VXL provides utilities to set up the data blocks and data types used in the PLCs data concentrator. About 22 data blocks are defined each holding 256 16-bit unsigned integer words for digital and analogue input-output-data. The above mentioned data blocks are dedicated in VXL to an external data source. A set of server programs provided in VXL take care of the incoming data to do calculations, visualization, alarm- and event-handling.

VXL has integrated the powerful graphical editor DATA VIEWS which allows for designing the users interface to the process and windows to display and monitor process data. To control and visualizise all the operations in TOSKA facility about 40 high resolution (1280*1024) windows and 25 subwindows are available to the operator.

4. DATABASE

The relational database ORACLE Enterprise Server 7.1 running on a dedicated DEC Alpha is used to store configuration data, long-term archiving data and transient data in one common database. It was selected because of its high stability and its availability on many hard- and software platforms, including most UNIX derivatives and OpenVMS. In contrast, the database client applications (e.g. the graphical end user interface ORACLE Forms) reside on PCs running Microsoft Windows. So the stability of DEC workstations and easy handling of PC software under Microsoft Windows are combined.

4.1 Modelling Configuration Data

The relational database in TOSKA-UP manages more than 1000 sensors, different types of acquisition modules in various subsystems and calculation and set-up parameters for each data channel. As first step, the entity relationship model, a graphical representation of the data model, was designed using ORACLE's integrated CASE environment called Designer 2000. By entering all relevant information into a common dictionary, the further design was simplified by partial automatic generation of database tables and a prototype for the graphical user interface using ORACLE Forms and Browser. After some manual refinement, a first working model of the graphical end user interface could be delivered, which was later refined by the experience of the operating team at TOSKA..

The resulting Forms application allows the operating team to compile the appropriate configuration for the subsystems in use. By pressing a button in one of the configuration forms, a trigger activates a server process dedicated to the subsystem to download new configuration data. For Hottinger and DR22 subsystems it is also possible to upload configuration data from the hardware server into the database.

4.2 Archived Data

The physical data collected by the VXL real time database (RTDB) are archived periodically into the relational ORACLE database by the archive server. After a session archived data may be concentrated to save harddisk space by requesting to delete any archived data of a user-specified time period except e.g. one value per hour.

For visualization of archived data, the scientific data analysis software package ORIGIN [7] is coupled to the graphical database user interface ORACLE Forms via DDE, a standardized interface to transfer data between MS Windows applications.

To reduce transmission time the user may choose a time interval and specify to see e.g. only one value per 20 minutes for a quick look at the data.

In addition to periodically archiving physical data, transient data may also be stored in the ORACLE database. For visualization there is an IDL interface similar to the one described above.

5. CONCLUSIONS

The system is in operation since June 1996 testing the EURATOM LCT coil at 1.8 K. Improvements will be made to speed up the window display by porting the control system VXL and the application software to ALPHA workstations. The results of the test will be presented in the paper by Zahn[8].

REFERENCES

[1] ORACLE Forms, ORACLE Server
Oracle Corporation, Redwood Shores, CA 94065
[2] UGR60, Vielstellen-Meßgerät
Hottinger Baldwin Meßtechnik GmbH, Postfach 4235, Darmstadt
[3] Gerätehandbuch Regler SIPART DR22, SIEMENS AG, Karlsruhe
[4] Kinetic Systems Corporation,
11 Maryknoll Drive, Lockport, Illinois 60441
[5] IDL, Research Systems, Inc., 2995 Wilderness Place, Boulder, CO 80301
[6] VXL, Control Systems International, Inc., Fairway (Kansas City), Kansas 66205
[7] ORIGIN, Microcal Software, Inc., Northampton, MA 01060
[8] G. Zahn et al., Proc. 19th SOFT, Lisbon, Portugal, 16-20 Sept.1996

Plasma Regime Guided Discharge Control At ASDEX Upgrade

T. Zehetbauer, P. Franzen, G. Neu, V. Mertens, G. Raupp, W. Treutterer, D. Zasche
and ASDEX Upgrade Team

Max-Planck-Institut für Plasmaphysik, EURATOM Association, Garching, Germany

In this paper we describe how plasma regime information is used for real-time optimization of discharge control during the flat-top phase of ASDEX Upgrade tokamak discharges. The discharge control system runs a new real-time algorithm that detects and classifies a number of plasma confinement regimes. Supervision control receives the regime information, evaluates it and triggers local or global reaction mechanisms in the controllers.

1. INTRODUCTION

Progress in fusion physics today increasingly depends on the ability to establish adequate control scenarios for specific plasma configurations and regimes.

This usually involves using cooperating feedback control processes to stabilize a number of plasma parameters within narrow operational windows.

To ensure that only valid control configurations can be activated (no two processes may simultaneously access the same actuator) and that transitions between configurations are performed coherently, a recipe concept [1] has been introduced and implemented in ASDEX Upgrade's (AUG) plasma performance controller (R4) [2].

In the first period of operation, switching between recipes was performed according to a predefined schedule or triggered by technical monitoring processes for machine protection.

With growing experimental experience, however, it became clear that considerable benefit would be gained from a dynamic scheduling technique driven by the actual plasma regime.

In fact, from the physicist's point of view, the ideal control system is one that helps to quickly attain a certain regimes, to keep it stable for a determined period of time, to perform desired transitions and promptly react to accidental ones.

A project was set up aimed at devising a robust detection algorithm to recognize regimes with a high identification rate, a supervision mechanism, and reaction strategies for regime guided plasma control.

2. REGIME DETECTION

In a pragmatic definition, plasma regimes can be characterized as (possibly interdependent) operational windows for specific sets of plasma parameters, either directly measured or inferred from online available diagnostic signals.

Examples of such parameters are mean electron density, plasma current, loop voltage, toroidal magnetic field, safety factor, confinement time, internal inductance, stored energy, total radiated power P_{rad}, or complex quantities such as radiation and current profiles.

Criteria which determine the online applicability of a regime detection algorithm are: a high identification rate (including transitions), simplicity (using a few robust signals), and stability of the computed regime information.

To meet these requirements more than 1800 stable plasmas ($\geq$ 200ms) from 530 discharges in various experimental periods of AUG have been identified and their parameters statistically analyzed offline. The results flowed into an identification procedure relying on some 20 input signals and giving a identification rate of better than 95%. A detailed description, with emphasis on the physical background, can be found in [3]. Here, suffice it to say that all involved parameters are either directly available or easily approximated using online input signals.

A somewhat simplified real-time version of this algorithm has been implemented in the R4 and is summarized in Fig 1:

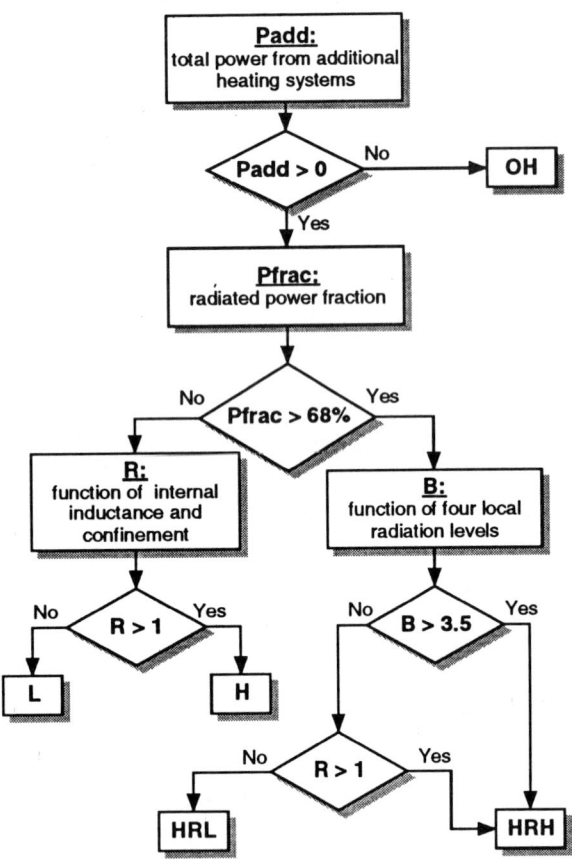

Figure 1: Regime detection

Five different plasma regimes can be detected using equilibrium parameters computed in the the plasma position and shape controller (R2), technical data from heating devices, diagnostic signals, and. Apart from the ohmic (OH) regime which is simply characterized by the absence of additional heating, the algorithm performs a two-level decision cascade to distinguish between high and low radiation and high or low confinement respectively

The first distinction is simple: The radiated power fraction P_{frac} is defined as the quotient between total radiated power P_{rad} and the total heating power P_{heat}. The former is computed from a weighted sum of bolometric signals and the latter is given as the sum of ohmic heating power and of the additional heating systems (signals indicating the emitted power are available online and adjusted by adsorption factors). Transitions from low to high radiation regimes occur when $P_{frac} > 0.68$ and transitions from high to low when $P_{frac} \leq 0.63$.

For low radiation regimes the difference between high (H) and low (L) confinement is established according to a function R where energy confinement time and internal inductance are related to a global scaling law and to a transition threshold respectively.

For regimes with high radiation the distinction between high (HRH) and low (HRL) confinement is slightly more complex. An unambiguous and sufficient condition for HRH can be found from the values of four local radiation levels taken from bolometric measurement. Where this does not hold, the same procedure as in the low radiation case must be applied to separate HRH from HRL.

To complement these threshold governed rules, prescriptions exist which account for the settling times of certain signals in the case of regime transitions. This means that some transitions are "forbidden" (e.g. H -> HRL) and others follow a hysteresis (e.g. L -> HRL -> L)

The algorithm just outlined is implemented on one of the R4´s processors. In each cycle (~2.5 ms) the actual regime is computed and periodically passed on to the supervisor controller (R1) for evaluation.

3. REGIME SUPERVISION

There are various reasons why a detected regime should not always cause immediate reaction: there are phases in a discharge where parameters cannot be reliably measured. It may be desirable to study the behaviour of the plasma maintaining a fixed control configuration or scenario even though a regime transition occurs. The control system itself may be in a state where transition to a new configuration would collide with safety requirements.

Therefore some evaluation of the detected regime is necessary before it is actually used to trigger reaction mechanisms.

In AUG´s control environment two separate evaluation schemes have been introduced. Both involve the discharge supervision controller R1 and the discharge programme (DP) [4].

The DP, which is downloaded into all controllers prior to a discharge, is logically structured into segments that describe distinct phases of a discharge in terms of time-varying set values for the controllers. R1 interprets the segment structure by repeatedly generating system state vectors (representing boolean expressions gained from monitoring of analog and digital input signals) and comparing them to a segment's priority ordered branch condition list. Each condition contains a reference system state, a time window, and a target segment number. When the system state corresponds to a reference state of a condition, and the momentary time lies within the condition's time window, the target segment number is broadcast to all real-time controllers. Upon receiving a new segment number these instantly switch to the new segment's set values.

The first scheme simply extends the existing "branch-on-condition" mechanism. In each supervisor cycle the detected regime information is integrated into the system state vector to form an extended state vector. By setting appropriate time windows and giving conditions with safety but not regime relevant reference states a higher priority this mechanism allows to mask out regime information and ensure that safety relevant branch request are always complied with.

The second mechanism compares the detected regime with a predefined list of permissions, each containing a regime identifier and a time window. When detected regime and regime id of a permission correspond and the momentary time lies in the permission's time window the then validated regime is broadcast to the remaining controllers.

4. REACTION STRATEGIES

For each of the evaluation mechanisms there is a distinct reaction strategy.

The first, enforced by switching to a new discharge segment, attains all control processes which use DP data and, since the recipe can be different in the new segment, may even cause a change in the current control configuration (Fig 2a).

The responsibility for adequacy and correctness of the ensueing control scenario lies with the experimentalist designing the discharge programme.

The second, which is triggered by the arrival of the validated regime information from the discharge supervisor, involves local adaptation of parameters of some elementary control processes, always leaving the control configuration (selection of feedback processes) unaltered (Fig 2b)

Examples of such local and autonomously performed adaptations are: switching to other input signals, modification of the desired value(s) using a predefined internal function, or adjustment of gain factors to adapt a PI-controller's performance to the requirements of a particular regime.

Here, adequacy and correctness of the resulting control scenario depends on the appropriateness and correct implementation of the modification functions in the elementary processes.

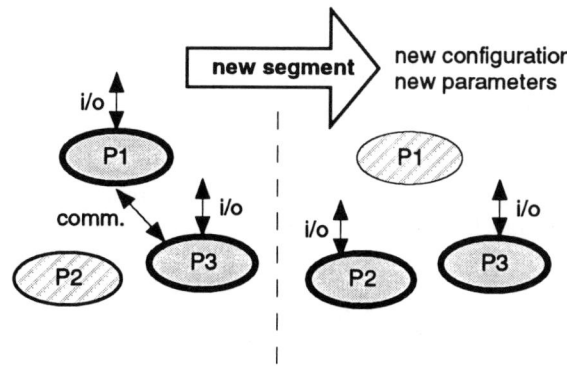

a) enforced segment change

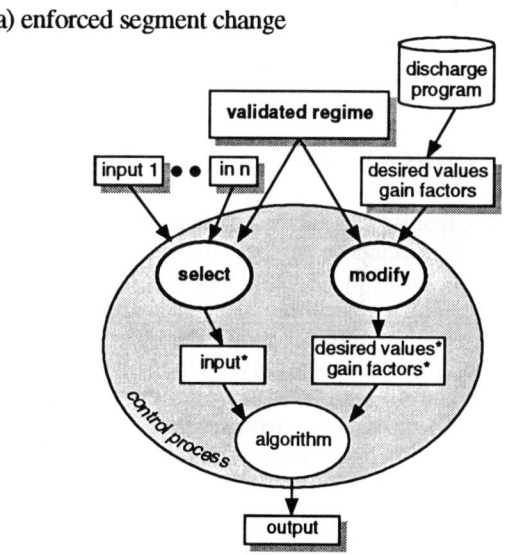

b) automatic local adaptation

Figure 2: Regime reaction

5. APPLICATION EXAMPLE

An application example taken from (3) shows both the correct working of the regime detection and the effect of the global reaction mechanism (Fig 3).

The discharge was aimed at establishing a completely detached high confinement regime (CDH, a special case of HRH, [5]). This regime is characterized by a narrow operational window, in terms of the irradiated power fraction P_{frac}. When the upper limit is exceeded, confinement degradation occurs (at t=2.2s in this discharge), leading to reduction of boundary electron density which in turn diminishes P_{frac}.

If no countermeasures are taken, the P_{frac} feedback controller will attempt to react by output of additional neon, a scenario that inevitably leads to disruption. Here, however, the transition to the low confinement regime HRL is detected in time and used to branch into a new segment which begins with a reduced desired value for P_{frac}. Strictly speaking, branching would not have been necessary in this particular discharge, because due to a MARFE event the apparent radiated power fraction rises above the desired value after transition to HRL. As a reaction the neon valve is temporarily closed. Since the desired value for P_{frac} is gradually ramped up the controller stabilizes the plasma after it regains the intended HRH mode.

6. CONCLUSION

The capability to identify plasma regimes and to react to the occurence of specific regimes have been incorporated into the Asdex Upgrade plasma control system. The algorithm developed for regime classification exhibits good identification rate, reproducibility and stability. Due to the well-designed control system structure, the procedures for regime evaluation and reaction were easy to embed. First application of the regime-guided plasma control shows the viability of the method. We have added a new quality to our control system : Real-time regime information allows putting to work the experimentalist´s knowlege to design discharge programmes for fine-tuned control scenarios.

REFERENCES

[1] T. Zehetbauer et al., "Management of real-time processes for plasma parameter optimization at ASDEX Upgrade", in Proceedings of the 1995 IEEE Conference on Real-Time Computer Applications in Nuclear Particle and Plasma Physics, East Lansing, USA, 1995

[2] Neu, G. et al.: "An Enhanced Plasma Control for ASDEX Upgrade"; Proc. 18th Symposium on Fusion Technology, Karlsruhe (D), 1994, p. 675

[3] P. Franzen, V. Mertens, G. Neu, T. Zehetbauer, G. Raupp and M. Kaufmann, "Online Confinement Regime Identification for the Discharge Control System at ASDEX Upgrade", in Proceedings of the 23rd European Physical Society Conference on Controlled Fusion and Plasma Physics, Kiev, Ukraine, 1996

[4] D. Zasche et al.: "Tokamak Discharge Description at ASDEX Upgrade" ; 8th Conference on Real-Time Computer Applications in Nuclear Particle and Plasma Physics, Vancouver, CDN, 1993

[5] O. Gruber et al.: Phys. Rev. Lett 74, 1995, p. 4217

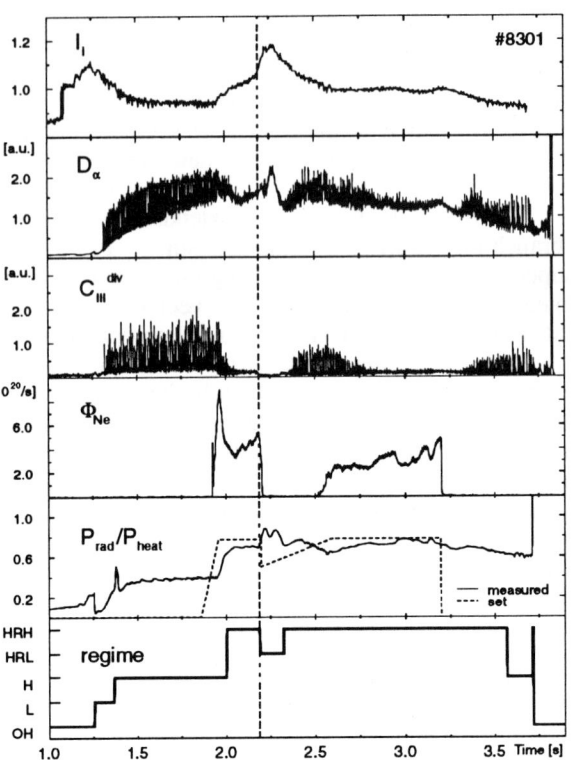

Figure 3: Application example

Structure for Next Generation Discharge Control Systems

G. Raupp, K. Lüddecke*, G. Neu, W. Treutterer, D. Zasche, T. Zehetbauer
and ASDEX Upgrade Team

Max-Planck-Institut für Plasmaphysik, EURATOM Association, Garching, Germany
* Unlimited Computer Systems, Neuried, Germany

This paper outlines a function-oriented structure, applicable for future discharge control systems. The number of infrastructure and application functions is minimized, control tasks are mapped onto separate controller units. The resulting distributed structure does not distinguish I/O from controller units and supports heterogeneous controller hardware. A common communication layer integrates all units. This set-up results in improvements for rapid modification, flexible discharge control including supervision and emergency handling, long pulse operation functions like instant protocol and display or reference value adjustment, and development support with shot replay and test unit mounting.

1. INTRODUCTION

Discharge control and experiment operation of fusion devices rely on digital real-time control systems with their inherent reproducibility, flexibility, extendability and integrability. Consequently, their scope, size, and complexity have evolved continuously: from simple stand-alone controllers to coupled, multi-variable controllers with monitoring and supervision tasks. Beside well-defined technical control tasks new methods are applied to plasma parameter control and performance optimization, where controllers serve as tools for experimental physicists. While safe and reliable operation remains essential, requirements for structural and operational flexibility, performance and migratability have increased. Further requirements come from long-pulse operation.

We describe current scope and structure of controllers for real-time applications (chapter 2). Changing requirements from physical applications, system development and operation procedures are discussed (chapter 3). We propose a new function-oriented structure for future control systems, showing advantages for implementation and operation (chapter 4).

2. HARDWARE-ORIENTED CONTROL

Discharge control requires to execute the sequence of actions defined by the experimentalist, and to feedback control those quantities not accessible to feedforward control with the required accuracy.

In early Tokamak research sequencing of the discharge followed a fixed scheme. A clock was used to generate predefined triggers at predefined times initiating desired actions. Embedded in this scheme was feedback control, limited to few quantities and diagnostic inputs. Analogue control systems were implemented with central trigger to sequence independent PID controllers, for plasma current, density and centre position, each with fixed signal inputs and outputs, and a hardware-based algorithm.

When digital controllers became available, this hardware-oriented set-up was kept, now built around the central processor, with fixed private I/O. Increased computing power allowed software-based control laws and elaborate multi-variable algorithms. Logical structures in software enabled to switch between algorithms to react to plasma events. Operation functions were added, e.g. to download the discharge program into memory, to protocol process data, and to access protection systems. However, as controller behaviour is determined by the fixed input-calculation-output cycle, dynamic collaboration of controllers is difficult, resulting in weakly coupled or stand-alone configurations.

Future plasma control systems will have to process more signals, feedback control more and more complex quantities, perform profile control, and increase accuracy. The question is whether the progress of digital technique with steadily increasing I/O capacity and processing power will be sufficient to comply with future requirements or whether there are reasons to search for a different structure for control systems.

3. FUTURE CONTROL REQUIREMENTS

Requirements arise from physical applications, system development and operation procedures.

The physical application of performing discharges demands improved feedback control and monitoring when operating close to physical stability margins and technical machine limits. This increases not only the number of control tasks, but also adds to the complexity of their configuration. To perform adequate feedback control under different plasma conditions, different sets of input signals, control parameters and control algorithms are required. Control processes have to share information on an increasing number of input or preprocessed quantities. Simultaneous control of plasma quantities has to be performed, demanding collaboration of control processes in a control scenario rather than isolated operation. Different phases of a discharge need different scenarios to better adapt control to actual plasma state. This makes dynamic selection of scenarios in real-time desirable, i.e. supervision control, when transitions in plasma state can not be anticipated accurately.

System development demands for support of rapid modifications because discharge control has evolved from a technical operation tool to an experimental research tool. Flexibility for new control scenarios is essential to participate in physics progress. To make implementation transparent and facilitate extension of control tasks and I/O, a number of measures are required: The bulk of infrastructur tasks such as signal I/O, discharge programme administration, process data protocolling, scheduling, etc. must be separated from the embedded physical core, which is only about 25% of the code. To minimize system size infrastructur tasks must be concentrated rather than being distributed among controllers. Information on how to process signals and execute control tasks must be kept local to reduce knowledge required in other places when accessing signals or managing processes. Simple extendability of I/O at any location is a prerequisite in a system under continuous development. Immediate access to all I/O or processed signals by all controllers is the base to supply dynamically switched control processes with the information needed.

Free choice of control method is another concern. Possible integration of emerging control techniques, e.g. neural networks or fuzzy, must be assured to provide task-optimized choice of components. Use of commercially available hardware components and tools for software development and analysis is required to provide a base for system engineering and a path for performance upgrading where needed.

Operation requirements come from long pulse duration, extending with machine size and type up to quasi-continuous operation. Large amounts of process data must be protocolled and extracted from the control system parallel to execution, and immediate archiving even in case of system failures be guaranteed for reliable fault detection. Long pulse duration demands provisions for real-time display and adjustment of reference values and parameters during discharges. With more knowledge about machine behaviour available in the future, extension of

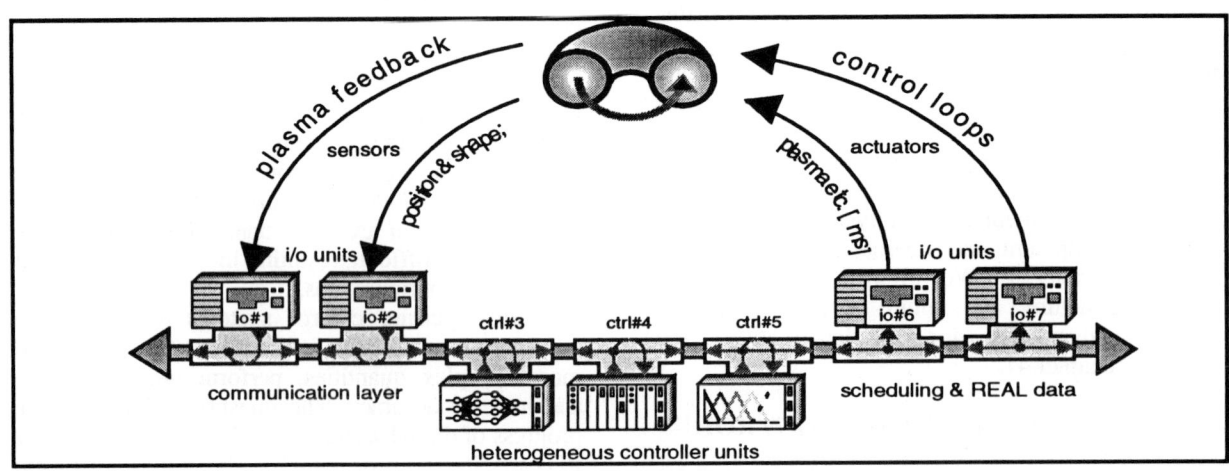

Fig. 1: feedback control loops in a distributed heterogeneous system with common communication layer

human interaction with an automated optimization loop is possible. For safe machine operation embedding of the discharge control in the protection system is mandatory and redundancy desirable.

4. FUNCTION-ORIENTED CONTROL

To find a hardware-independent structure for the control system we decompose the various functions to be performed, map them onto separate units and integrate these for operation.

As a design requirements, the system should be minimal to facilitate modification. Signals shall be produced in just one instance and processes shall exist just once to reduce overhead. We introduce I/O nodes, each responsible to I/O a unique set of quantities, and controller nodes, each responsible to perform specific control tasks. I/O nodes have pre- or post-processing capability to translate between analogue sensor or actuator signals and signals internal to the control system. Controller nodes have algorithms to compute signals for feedback control, monitoring or supervision. The structure between I/O and controller nodes must be symmetric as both produce and use signals. Functions performed by the various I/O and controller nodes can be directly mapped onto separate hardware units (Fig. 1).

Free exchange of internal signals requires a communication layer connecting all units and hiding hardware architectures. When using a common format, most natural as REAL quantities in SI units, immediate understanding of exchanged signal information is possible without further knowledge. Within this framework of a function-oriented distributed and heterogeneous system new signals or processes can be added or removed whenever required.

Equilibrium control will be used as an example to illustrate Fig. 1. I/O unit #1 inputs all PF coil currents. I/O unit #2 produces integrated magnetic fields and magnetic fluxes, preprocessed to correct the integral constants with PF currents measured before ignition by I/O unit #1. From magnetic fields and fluxes, controller unit #3 (e.g. a neural network) derives position and shape. Controller unit #4 uses these to calculate feedback control of position and shape through PF coils, with coil current output by I/O unit #7. Feedback control processes include monitoring of controlled quantities and clipping to actuator limits.

Protocolling is performed with the process data on the communication layer (Fig. 2). Then, the distributed system requires just one instance to take data when updated or when a control cycle is finished. The stream of current data containing signals from I/O units, results from controller units or internal state information required for later analysis must be written instantaneously to a storage medium, with timestamps and tags for identification.

The inverse process is the input of discharge program into the control system. Again only one instance is required to write just-in-time a stream of data containing current reference values, monitoring ranges or mode switches onto the communication layer, where these are accessible by all units processing that information. The time information to correctly fetch reference values from a time-ordered list and protocol process data must be provided by a central real-time clock.

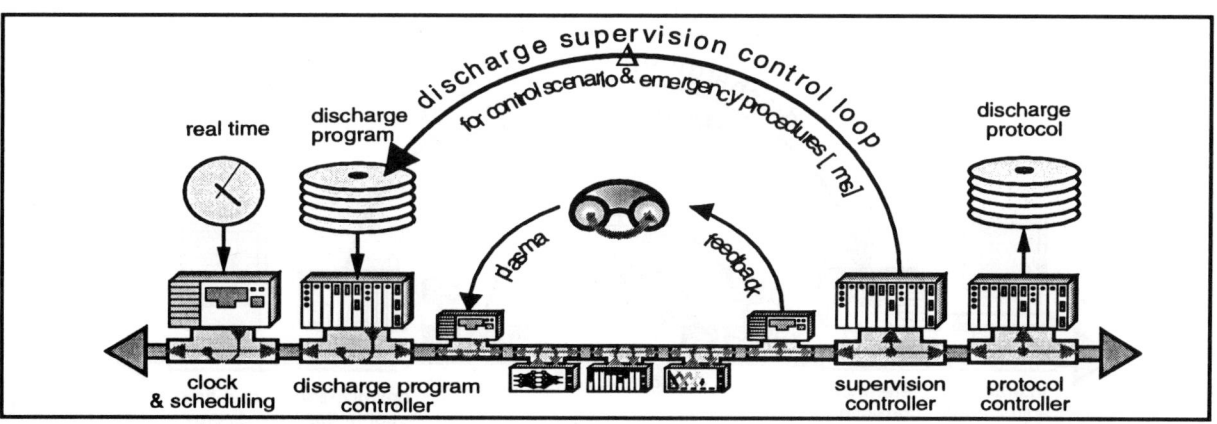

Fig. 2: real-time supervision control loop and handling of discharge program and protocol; the real-time feedback and monitoring loops are embedded

Such a structure relieves controller and I/O units in feedback and monitoring loops from knowing how to provide reference values or protocol data.

The real-time supervision controller (Fig. 2) checks current technical and physical states and decides with execution conditions and transition rules whether to continue, adapt the control scenario, change the discharge's goal, or terminate it. Monitored quantities and controller states required for the decision must be available on the communication layer by the underlying controller units. Adaptation of the control scenario, change of the discharge goal or termination can be performed if the discharge program contains the referenced alternate discharge phase with new desired values. To activate this, the supervisor commands the discharge program controller to run the alternate phase, that will be performed by feeding the new desired values into the communication layer for execution by the controller units.

Display of process data is a real-time function similar to protocolling. The display unit listens to information available on the communication layer and outputs it with analysis algorithms to a graphic display(Fig. 3). Real-time display is a prerequisite for operators and experimentalists to optimize long-pulse discharges under execution through interactive re-adjustment of the pre-programmed desired values in a narrow range. Such a function can also be performed by an automated optimization loop when the gap between display and correction input is closed by an intelligent optimization controller.

Adjustment or correction values are superimposed to desired values from the discharge programme.

The function-oriented structure eases operation of a system under continuous development. The system is open to new signals and control processes through new or enhanced I/O and controller units. This is particularly important for test purposes, e.g. to mount a new controller unit and run it parallel to existing ones with the same input data and full protocol. The structure also allows to real-time switch between the existing controller unit and the new one to evaluate its performance in a time window or under specific conditions.

Another option is discharge replay: if all signals exchanged on the communication layer are protocolled, these can later again be input there together with the respective reference values to analyse control system behaviour, check integrity after modifications or perform step-by-step debugging after system failures.

Safety issues are addressed by facilitated implementation of redundancy and integration into machine protection systems. When redundancy in hardware or software is required, the system is open to add more input signals, more controller units, more algorithms, or an additional communication layer. As resources and knowledge are concentrated, only those parts required have to be doubled.

Embedding of discharge control into machine protection must be done in two ways. Deadman monitoring must guarantee proper shutdown in case of severe component or software failures resulting in a controller cycle violation. Supervision control must initiate emergency procedures when plasma or machine quantities exceed absolute limits or controller states indicate improper function. In both cases the protection system must be alarmed.

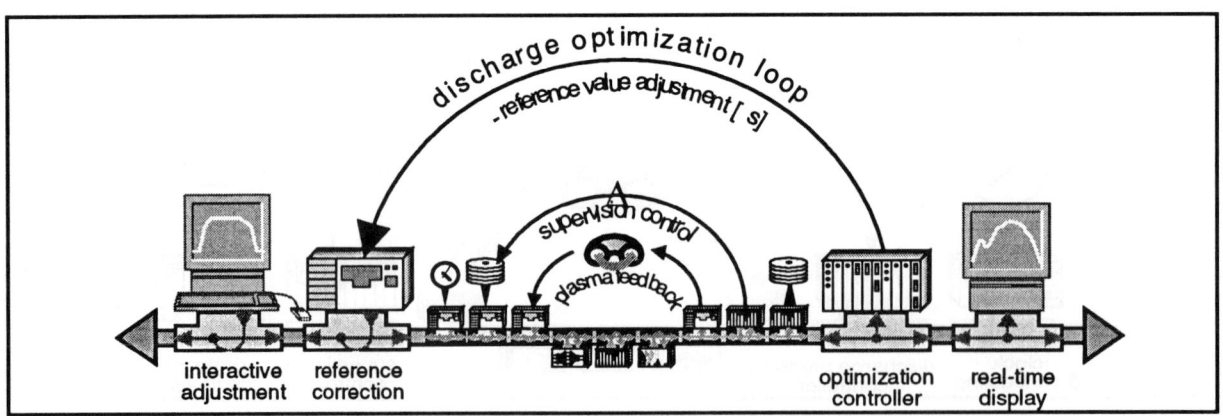

Fig. 3: discharge optimization loop to adjust long-pulse discharges manually or automatically; the real-time supervision and plasma feedback loops are embedded

Plasma Shape Control Design in ASDEX Upgrade

W. Treutterer, J. Gernhard, O. Gruber, P. Mc Carthy[a], G. Raupp, U. Seidel and ASDEX Upgrade Team

Max-Planck Institut für Plasmaphysik, EURATOM Association, Garching, Germany

[a]University of Cork, Ireland

At ASDEX Upgrade shape control has been implemented in addition to the existing position and current control loops to complete the digital feedback control with poloidal field coils. The primary goal was the feedback control of the divertor strike points, though other parameters like gaps and geometrical moments are configurable as well. The approach comprises a multivariable control concept whose core is a matrix PI-controller designed for dynamically decoupled and stationary accurate adjustment and robust operation. The controller gains are computed with a transfer function collocation method. The algorithm has been embedded in ASDEX Upgrade's digital discharge control and has successfully been used in experiment.

1. INTRODUCTION

ASDEX Upgrade is a nuclear fusion experiment of the tokamak type. At a tokamak the performance of a plasma discharge can be significantly improved by optimizing the shape of the plasma column. State of the art are elongated, D-shaped cross-sections for optimal confinement in a divertor configuration.

The task of shape control is to establish, stabilize and maintain a prescribed cross-section. It also provides the possibility to adapt the shape to varying discharge states and to obey technical boundary conditions such as adjusting the strike points on the divertor target plates or optimizing the energy- coupling between plasma and ICRH-antenna via gap control. The examples show that for control purposes the cross-section has to be quantified by a set of shape parameters. These may be coordinates [1] or gaps [2,3] as well as area averaged moments like elongation and triangularity. Shape control uses active poloidal field coils as actuators, while other influences like confinement changes due to external plasma heating or parameter variations act as disturbances. Under reactor relevant conditions with a small number of actuator coils distant from the plasma it is the challenge of shape control to provide high accuracy rejecting disturbances and to offer acceptable tracking behavior despite of strong cross-couplings between coils and between shape parameters.

2. CONTROL ARCHITECTURE

Corresponding to ASDEX Upgrade's poloidal field coil configuration (Fig 1) functional distinction is made : plasma position is controlled separately from plasma shape. This is due to the inherent vertical

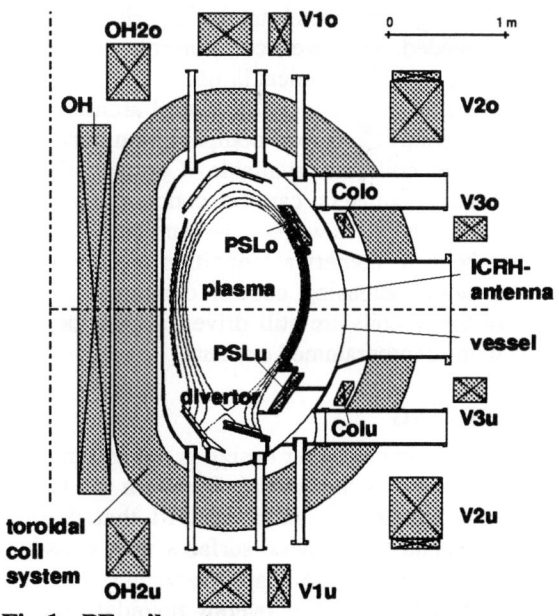

Fig 1 : PF coil system

instability of elongated plasmas which requires a fast separate feedback loop referred to as plasma position control [4]. Plasma position is defined by one radial and one vertical coordinate. These are controlled on a fast timescale of 30 ms using a pair of dedicated control coils (CoIo, CoIu) and a pair of passive saddle loops (PSLo, PSLu). Shape control covers the remainder of the shape parameters. Because plasma is already stabilized by position control, slower response times of about 200 ms are permitted allowing for a power saving, reactor relevant configuration of the main shaping coils (V-coils) distant from plasma.

On the slow time scale, however, position control is linked back to shape control for technical reasons. To obtain always optimal modulation ranges and to avoid thermal overload the stationary currents of the position control coils must be eliminated by load compensation with the V-coils.

A third feedback control of the plasma current accounts for resistive losses. It makes use of the ohmic-heating coil system made up by the OH, OH2o and OH2u coils which are connected in series. From these the OH2u coil is equipped with a parallel power source so that it can concurrently be used by shape control.

Underlying these major control circuits for plasma position, shape and current are current controlled thyristor bridges feed the active coils. Benefits are the reduction of inductive cross-couplings, the simplified dynamic behavior and the fact that no differentiating components in the superior controllers are needed. Moreover, coil currents can easily be confined to their technical limits.

Fig 2 gives an overview of the control architecture. For shape control and load compensation currently seven active coils are available. Correspondingly five shape parameters can be concurrently controlled. In reality, however, the number is often smaller because numerical condition rapidly deteriorates with increasing control dimension. Then some of the V-coils are still driven in feedforward mode with preprogrammed currents.

3. MODELING

The first step in controller design is the assumption of an open-loop system model. Based on the Grad-Shafranov-equation it can be derived that plasma shape is bound to magnetic surfaces with constant poloidal flux. Hence the shape parameters are immediate functions of the overall poloidal current distribution. This current distribution which includes also vessel eddy currents can be approximated with concentrated currents.

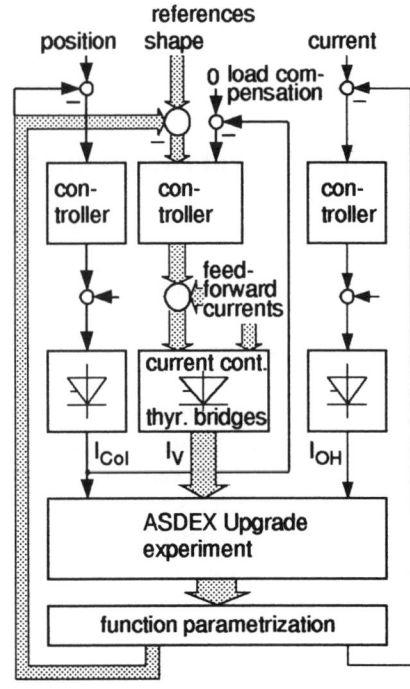

Fig 2 : control architecture

This model can be split into two sections. The static section represents the mapping between the poloidal currents and plasma shape. The dynamic section describes how the currents, coupled by mutual inductances, evolve as a function of induced currents, external voltages and events.

The static section can be described using a function parametrization (FP) approach [5],

$$Y = c + \underline{b}^T \cdot \underline{I} + \underline{I}^T \cdot a \cdot \underline{I}$$

where vector $\underline{I}$ comprises all PF-currents, the poloidal confinement efficiency β_p and the internal plasma inductivity l_i. Y is any shape parameter.

Its coefficients a, b and c are obtained by quadratic regression from a database of numerical solutions of the Grad-Shafranov Equation.

The dynamic model results from the voltage equations of each current loop and from a force balance between Lorenz-force and pressure gradient and depends on the current distribution $\underline{I}$ and the plasma position $\underline{r}$ [6,7].

$$L_I \cdot \underline{\dot{I}} + L_r \cdot \underline{\dot{r}} + R_I \cdot \underline{I} + R_r \cdot \underline{r} = \underline{U}$$
$$F_I \cdot \underline{I} + F_r \cdot \underline{r} = 0$$

For the purpose of controller design the vessel eddy current distribution may be neglected in this model as the decay of the eddy currents with a typical time constant of 5 ms is much faster than the response time of shape control.

After the equations have been linearized and the control laws of coil currents, plasma position and plasma current have been applied the model is written in state space notation. It is still divided into a dynamic part and a static output equation which reflects the linearized FP-coefficients b^T and a.

$$\dot{\underline{i}} = A\underline{i} + B\underline{u}$$
$$\underline{y} = C\underline{i} + D\underline{u}$$

The excitation vector $\underline{u}$ represents the reference values of coil currents, plasma position and plasma current as well as β_p and l_i, whereas $\underline{i}$ denotes the actual currents considered as state variables. The coefficients of the dynamic part in the system matrix A and the input matrix B have been extracted from experimental measurements in order to include also parasitic effects like control and sensor delays. For this we excited the V and OH2u coils with a multidimensional pseudo-random binary signal and used a state-space system identification method to estimate the parameters.

4. SELECTION OF CONTROL COILS

The obtained model comprises the active coils as well as the passive PSLo/u saddle loops and a large number of shape parameters. For a specific control task a set of shape parameters and a corresponding number of coils for control must carefully be selected. A well posed control problem is a prerequisite to avoid performance degradation by technical constraints due to exaggerate current amplitude and rate demands.

In a first approach we use the stationary gains to judge the composition of the chosen sets. The condition number of the gain matrix is a measure for orthogonality. An orthogonal setup on the other hand yields optimal numerical posedness. In addition the gains of the closed loop system have to be checked with respect to changes in β_p and l_i. Supplementary information is provided by the controller gain matrices which are obtained as a result of the design process. They reflect the controllability of the transients.

Further investigations are desirable on this subject. A possible extension is to examine the transfer function gains near the proposed crossover frequency of the closed loop.

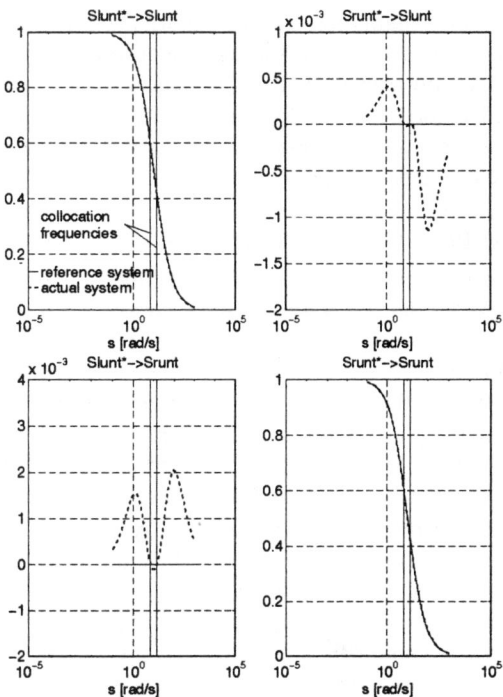

Fig 3 : collocation of transfer functions

5. CONTROLLER DESIGN

Summarizing the properties of the controlled system multiple shape signals must be adjusted with multiple coil commands. Significant cross-couplings arise as each coil influences any shape parameter. Hence, only a multivariable control law is capable to achieve satisfactory decoupled tracking behavior.

In order to guarantee stationary accuracy even in the presence of disturbances and parameter fluctuations the controller is furnished with an integral component. The resulting controller thus has a matrix PI-structure. Under the given circumstances - varying parameters and uncertain system model - this solution is more robust and therefore superior to a state space controller.

To parameterize the controller matrices we used a collocation method which adapts the transfer function of the closed loop system to the transfer function of a reference system at two real Laplacian-frequency points (Fig 3) [8]. Thus the shape parameters behave approximately like the reference model. The fact that this design method does not depend on uncertain assumptions like model order and model structure provides additional robustness. As reference model we chose a diagonal matrix of first-order lags with a time constant of 0.1 s.

The implementation of the multivariable control concept is fully digital and embedded into discharge supervision control [9,10]. This allows for sophisticated algorithms and high flexibility in scheduling shape and gain parameter sets depending on the state of the discharge [11].

6. EXPERIMENTAL RESULTS

Experimental results have been obtained with shape control of the two lower divertor strike points *Slunt* and *Srunt* as well as load compensation of the CoIo, CoIu-coils. Position control involved the vertical position of the plasma center *Zsquad* together with the outer plasma radius *Raus*. The plasma current has been held constant. As an example disturbance suppression is shown in Fig 4. While the confinement parameter β_p is modulated by neutral beam injection in steps of more than 20 % the strike point errors remain less than 4 mm.

A problem left are the currently small operation bands for some coil currents which impair performance for large β_p variations. An extension of the power supply which is currently in progress and a disturbance compensation algorithm using the remaining feedforward controlled coils will eliminate this drawback.

7. CONCLUSIONS

Shape control at ASDEX Upgrade is based on a reactor relevant configuration of the shaping coils. The small number and the wide dispersion of the coils yield strong cross-couplings with the shape parameters. Therefore considerable effort must be made in choosing appropriate control coils for a given set of shape controls, in order to obtain a numerically well posed solution and to avoid power supply limitations. Control is performed using a matrix PI-control law. Robust controller gains are obtained on base of transfer functions identified from experimental data.

Experiments have proved that the design goals - accuracy, decoupling, robustness and reliability - have well been achieved. Future work will deal with an upgrade of power supply, the extension of the shape parameter set and the refinement of synthesis algorithms.

REFERENCES

1. F. Hofmann, S.C. Jardin, "Plasma shape and position control in highly elongated tokamaks", Nuclear Fusion, Vol. 30, No. 10 (1990), p.2013.

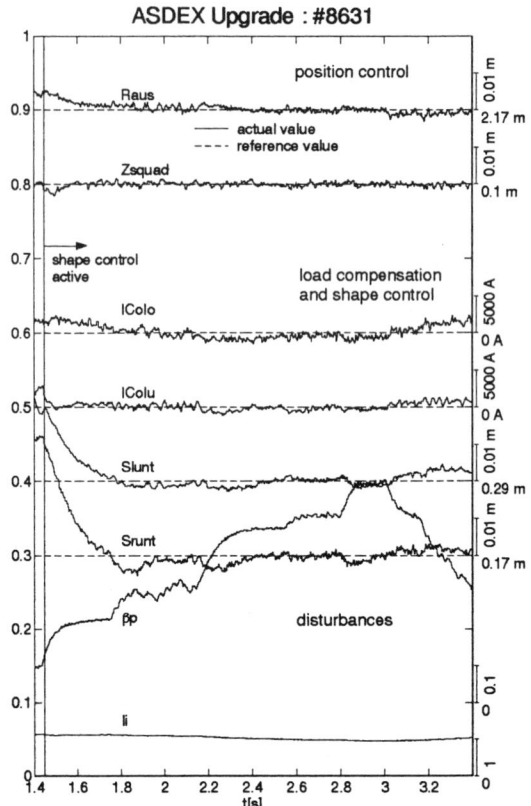

Fig 4 : disturbance suppression in experiment

2. M. Garribba et. al., "First Operational Experience with the New Plasma Position and Control System of JET", Proc. 18th Symp. on Fusion Technology, Karlsruhe (1994), p. 747.
3. D.A. Humphreys et. al., "Design of a Plasma Shape and Stability Control System for Advanced Tokamaks", Proc. 18th Symp. on Fusion Technology, Karlsruhe (1994), p. 747.
4. O. Gruber et. al., "Position and Shape Control on ASDEX Upgrade", Proc. 17th Symp. on Fusion Technology, Rome (1992), p. 1042.
5. P.J. Mc Carthy et. al., "MHD Equilibrium Identification on ASDEX Upgrade", Proc. 19th EPS Conference on Controlled Fusion and Plasma Physics, Innsbruck (1992), p.459.
6. W. Woyke et. al., "Performance of the Equilibrium Control System for ASDEX Upgrade", Proc. 19th EPS Conf. on Controlled Fusion and Plasma Physics, Innbruck (1992), p. 455.
7. U. Seidel et. al., "Plasma Position Control in ASDEX Upgrade", Proc. 7th course of international school of fusion reactors technology, Erice (1985)
8. K. Krüger, "Direct shaping of the reference input response for MIMO systems via output feedback", Proc. 1st IFAC Symp. of Design Methods of Control Systems, Zürich (1991).
9. G. Raupp et. al., "Discharge Supervision Control", Fusion Technology , to be published
10. T. Zehetbauer et. al., "Plasma regime guided discharge control", this conf.
11. W. Treutterer et. al., "Redesign of the ASDEX Upgrade Plasma Position and Shape Controller", Proc. of IEEE Conf. on Real Time Computer Applications in Nuclear Particle and Plasma Physics., East Lansing (1995), p. 287.

The Use of Fuzzy Curves for the Reconstruction of the Plasma Shape and the Selection of the Magnetic Sensors

F.C. Morabito and M.Versaci

DIMET, Università di Reggio Calabria, Via E.Cuzzocrea, 48 I-89127 Reggio Calabria, Italy

In this paper we present a novel tool which can be advantageously used within commonly used plasma shape identification procedures with the aim of carrying out a guided dimensionality reduction of the available pattern of measurements. The tool is referred to as *Fuzzy Curve* being related to *Fuzzy Systems Theory*. The use of fuzzy curves is compared to standard linear correlation analysis and to a ranking technique used within Neural Network approaches. The results of the study show that the fuzzy curves yield relevant insights about the nonlinear relationship between magnetic measurements and plasma parameters which can also be used to build an extremely simplified identification model. While this model is not adequately accurate for copying with actual identification requirements, it can yet be used to have very fast information on the evolution of a discharge as well as to reduce the computational complexity of the training step required by standard identification procedures.

1. INTRODUCTION

In present day tokamaks as well as in future ITER-like reactors, the real time control system of the plasma current, position and shape is driven by a set of signals coming from some manipulations of experimental measurements [1]. The magnetic diagnostics provide the basic pattern of raw data from which the reconstruction algorithms are started.

In this paper, the concept of *fuzzy curve* is adopted for ranking the input variables in order to select the most relevant ones in a plasma identification approach based on the interpolation of a data base of previously generated equilibria. The fuzzy curve tool can also be used to get, by inspection, some insights about the model structure [2]. In this sense, the technique here presented could provide a straightforward extension to neural network based model of reconstruction.

2. THE PROBLEM

The selection of the most relevant reconstruction procedure's inputs is an important step in designing an identification model of a plasma column evolving in a tokamak. Starting from a set of magnetic measurements taken as close as possible to the plasma boundary, we would like to derive an estimation of a set of geometrical plasma shape parameters.

The study can start from a high number of candidate measurements aiming to determine the number and location of the final configuration.

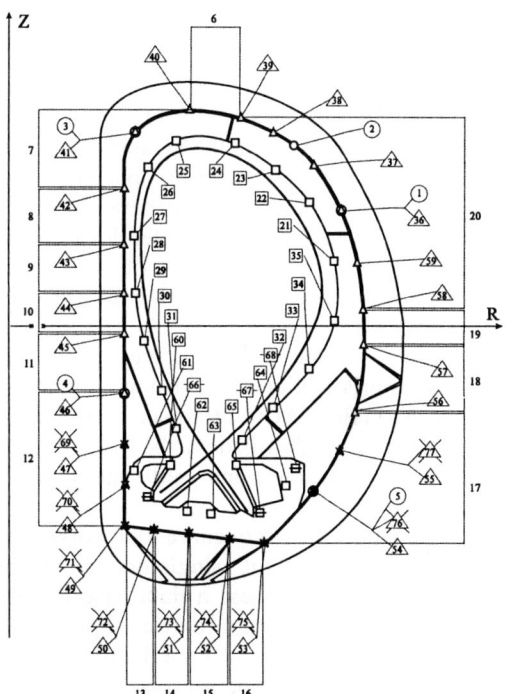

Figure 1. Pictorial representation of the tentative location of the various sensors. Different symbols indicates flux differences, absolute flux and field measurements. Details can be found in [4].

A ranking procedure is then of utmost importance. The most simple approach to selection could be based on the estimation of the linear cross-correlations within inputs and outputs on an available data base of equilibria. However, this technique does not take into account nonlinear effects.

A different technique is proposed in [3,4]: a neural network model of the input-output mapping is generated by suitable training and then the importance of the sensors is assessed by zeroing in turn each one of the N input variables and checking the performance index (average squared error). In such a way a ranking of the sensors is straightforwardly obtained; however, this analysis is carried out after the training procedure. On the other hand, the procedure is a sub-optimal one, in that the correct procedure would be to train a different neural network with (N-1) inputs by excluding the inputs one at a time. This approach would clearly be rather cumbersome when considering a system with a hundred of inputs. In the next Section, we propose a nonlinear approach whose computational complexity is linear with respect to the number of input candidates (N).

3. THE METHOD

The input variables of the procedure are the flux and field measurements from the sensors. The output variables are some geometric plasma shape parameters as well as the gaps between plasma boundary and the first wall at selected locations [4]. The accurate real time control of the plasma boundary will be indeed a critical issue for future ITER-like reactors.

In our case, we use as test data base a set of 976 X-point equilibria obtained by using a numerical code with the ITER TAC-4 reference configuration. The data base generation is described in [4]. Fig. 1 pictorially depicts the analysed measurements configuration.

To explain how the fuzzy curve tool works, let us consider a multiple-input single-output (MISO) system for which we possess a data base of input-output pairs with possible not relevant inputs. In our problem, the input of the model are the flux and field measurements from magnetic sensors, x_i (i=1,..., N), and the considered output, y, is the major radius, Ro, of the plasma column. Likewise, we will consider as output of the procedure the gap between the plasma boundary and a selected location on the first wall. We wish to determine which inputs are the most relevant among N possible candidates.

We assume that m training data are available, thus x_{ik} (k=1,..., m) are the ith co-ordinate of each of the m training patterns. The *fuzzy curve* is defined as:

$$c_i(x_i) = \sum_k \Phi_{ik}(x_i) y_k / \sum_k \Phi_{ik}(x_i), \quad k=1, ..., m,$$

where $\Phi_{ik}(x_i) = exp\left[-((x_{ik} - x_i)/\sigma)^2\right]$ is a Gaussian function (other different local functions could advantageously be introduced). We take σ as a fraction (about 20%) of the range for the corresponding input measurement within the data base.

Each pattern on the database is here treated as a fuzzy rule of the type:

IF x_i is $\Phi_{ik}(x_i)$ THEN y is y_k

for each (x_i,y) input-output pair of variables. Each membership function Φ_{ik} is drawn so that $\Phi_{ik}=1$ coincides with the point (x_{ik}, y_k).

The basic idea behind the method is to assess the flatness of the fuzzy curve, c_i, characterising a given input parameter, since the output is scarcely influenced by the input value if the related fuzzy curve is nearly flat.

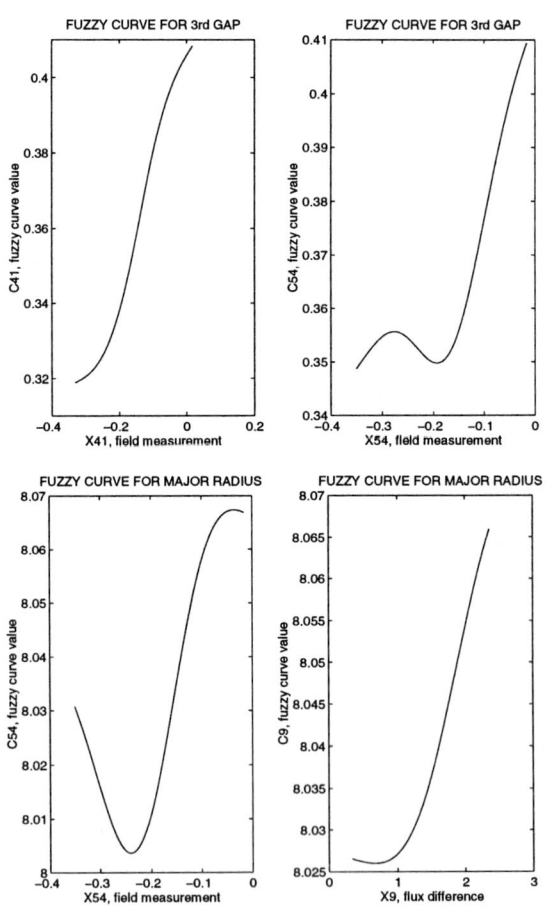

Figure 2. Examples of fuzzy curves for the 3rd gap and Ro.

The importance of the input in affecting the estimation of the output is determined on the basis of a *figure of merit* which is defined as the range of the fuzzy curve, ($c_{i\ max} - c_{i\ min}$). This range is typically a fraction of the range spanned by the corresponding output variable, y, on the whole data set of examples.

If the output variable is independent of x_i, that is $y(x_i)$=cost, the fuzzy curve c_i is also independent of x_i, and then ($c_{i\ max} - c_{i\ min}$)= 0.

The comparison between the different techniques of ranking is based on the linear cross-correlation for the statistical technique, on the rms error degradation in the neural network approach and on the figure of merit for the proposed fuzzy curve approach.

In the case of a multiple-output (MIMO) problem, there is a different fuzzy curve for each input-output variable pair.

4. THE RESULTS

In Figure 2 a subset of the 77 fuzzy curves drawn for Ro and the 3rd gap [4] are drawn. The relevance of a sensor is judged not only on the basis of the figure of merit but also on the shape of the curve. The presence of peaks and valleys is indeed associated to fuzzy rules which will be used in the fuzzy inference procedure to build a rough identification model.

Figures 3 and 4 show the linear cross-correlations between the magnetic measurements and respectively Ro and the gap computed on the available data base.

Figures 5 and 6 plot the figures of merit determined starting from the fuzzy curves vs. the corresponding input measurement for the above mentioned output parameters.

Fig.7 reports the results of the ranking procedure concerning Ro based on the rms reconstruction error achieved by zeroing each input of the trained modell one at a time. The ranking based on different procedures allows to draw some interesting conclusions: while the resulting rankings are basically different, all the approaches recognize the importance of some measurements.

For example, the following measurements are anyway present in the top ten: 11, a flux difference, 30 and 26, two field measurements in the blanket shield, 54, a field measurement in the vessel wall. A basic difference between the approaches is that the NN approach estimates the importance of a sensor in the presence of the remaining 76. The training has been indeed carried out before the ranking procedure. Both the fuzzy curve and the linear approaches estimates the importance of each sensor without taking into account the other sensors. This means that only the NN approach uses an information of self-correlation between the sensors.

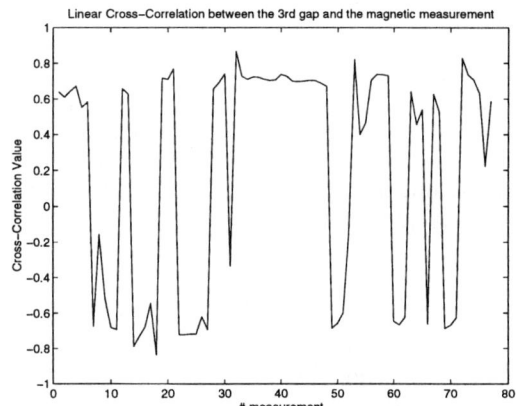

Figure 4. Cross-correlation curve for the 3rd gap.

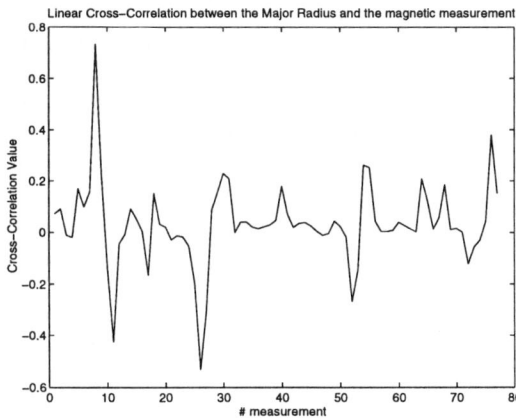

Figure 3. Cross-correlation curve for Ro.

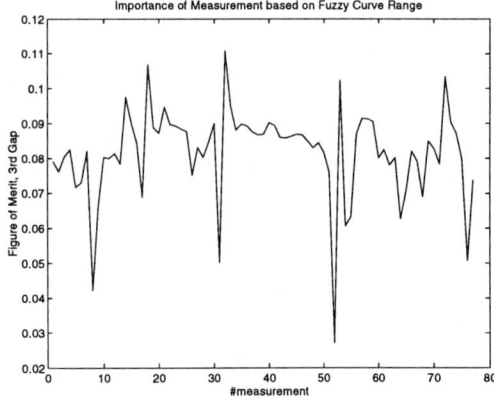

Figure 5. Importance of a sensor based on fuzzy curve figure of merit (gap).

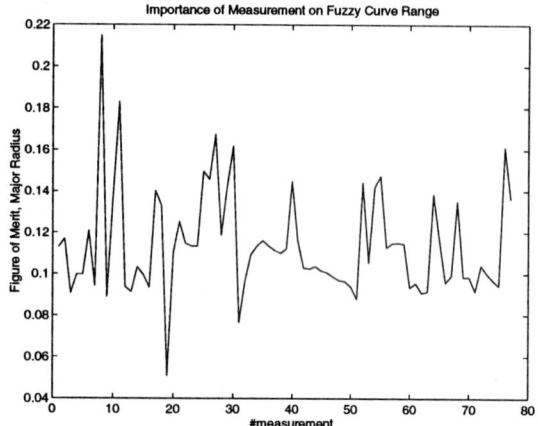

Figure 6. Importance of a sensor based on fuzzy curve figure of merit (Ro).

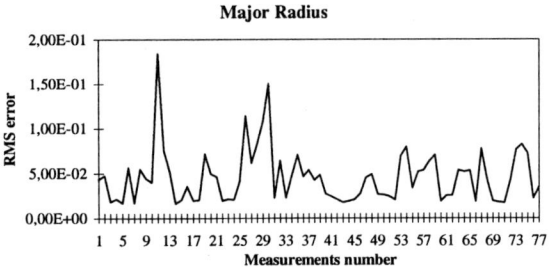

Figure 7. Importance of different measurements for identifying Ro by using the neural network approach.

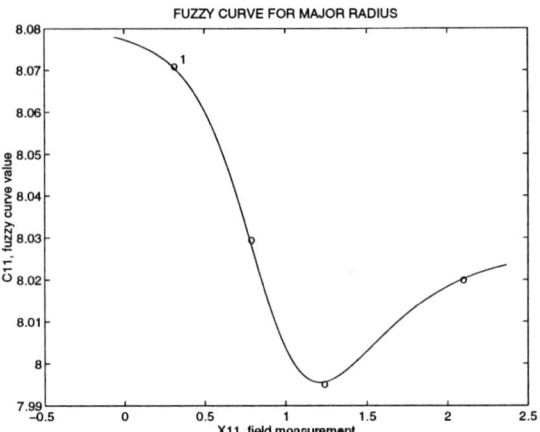

Figure 8. Extraction of fuzzy rules from a fuzzy curve.

5. CONCLUDING REMARKS

The fuzzy curve shape and range are influenced by the choice of the standard deviation parameter, σ. In the limit of $\sigma = 0$, the model is no more fuzzy. The increasing of σ reduces to an excessive weight of the areas covered by most samples. By considering reasonable choices of σ, the shape of the fuzzy curve is not significantly modified. A lot of results on the subject is presented in [5].

REFERENCES

1. L. deKock, Yu Kuznetsov, Magnetic Diagnostics for Fusion Plasmas, Nuclear Fusion, Vol.36, No.3 (March 1996) 387
2. Y.Lin, and G.A.Cunningham III, A New Approach to Fuzzy-Neural System Modeling, IEEE Trans. on Fuzzy Systems, Vol.3, No.2 (May 1995) 190.
3. E.Coccorese, R.Martone, and F.C.Morabito, Identification of Non-Circular Plasma Equilibria Using a Neural Network Approach, Nuclear Fusion, Vol.34, No.10 (1994) 1349.
4. R.Albanese, F.C.Morabito et al, Identification of Plasma Equilibria in ITER from Magnetic Measurements via Functional Parameterization and Neural Networks, paper accepted for publication on Fusion Technology, Special Issue on Plasma Control, Oct 1996.
5. F.C.Morabito and M.Versaci, A Neuro-Fuzzy Approach to Extremely Fast Plasma Shape Identification, in preparation.

Figure 8 shows how some fuzzy inference rules can be extracted by a fuzzy curve. We subdivide the input and the output ranges in 5 overlapping regions and by labelling the corresponding value of the variable with fuzzy values (i.e., VERY SMALL, SMALL, MEDIUM, LARGE, VERY LARGE), we may derive by inspection of the fuzzy curve a set of rules vaguely describing the input-output relationship. In this case, we heuristically decide of selecting 4 fuzzy rules, which are related to the circles in the Figure. For example, the first rule extracted, which is associated to the circle numbered with 1 is:

IF the input measurement is *VERY SMALL*
 THEN the plasma major radius is *VERY LARGE*.

  Likewise, we have:

IF the input measurement is *MEDIUM*
 THEN the plasma major radius is *VERY SMALL*

Measurement and control of error field driven magnetic islands in a tokamak reactor by electron cyclotron current drive

E. Lazzaro, S. Cirant, G. D'Antona[a], S. Nowak, G. Ramponi

Istituto di Fisica del Plasma CNR, Assoc. EURATOM-ENEA-CNR, Via Bassini 15, 20133 Milano, Italy.

[a]Dipartimento di Elettrotecnica, Politecnico di Milano, P.za L. Da Vinci 32, 20133 Milano, Italy

An active feedback system to control magnetic islands is proposed based on the application of Electron Cyclotron Current Drive to a reactor grade tokamak with an ITER like configuration. It includes a digital real time estimator of the frequency and amplitude of the radial component of the magnetic perturbation, and a controller providing an adequate modulation of the Gyrotron power source. The physics and engineering limits of operation for avoidance of the locked mode instabilities are discussed.

1. INTRODUCTION

Magnetic field errors with helical components resonating with closed lines magnetic surfaces, having rational q winding number are a serious threat for reactor grade tokamaks since it is estimated that they can cause the formation of locked magnetic islands, leading to pre-disruption conditions, at amplitudes as small as 10^{-5} times the toroidal field [1]. It can be shown that the threshold for error field driven islands can be made less severe by injection of non-inductive current driven locally around a q= m/n rational surface by Electron Cyclotron Wave collimated beams [2]. With a phase locked modulation of the Gyrotron power source it is possible to control the growth of rotating islands. The feedback loop design and the power requirements for ITER are investigated on the basis of a nonlinear model of island dynamics and Electron Cyclotron Current Drive (ECCD) [3,4].

2. CURRENT DRIVE AND DYNAMICAL MODEL OF MAGNETIC ISLANDS

The ECCD current density parallel to $\underline{B}$, and of appropriate sign can be sharply localized on the "O" point of a magnetic island to counterbalance, through appropriate phasing, the plasma destabilizing current perturbation.
The total current driven by ECW beams of a certain frequency and power P^{EC} is given by

$$I^{EC} = \gamma_{20} \frac{P^{EC}}{R_0 \langle n_e \rangle} \quad (1)$$

where the efficiency γ_{20} (AW^{-1}10^{20}m^{-2}) depends on electron temperature, wave frequency and launching angle β as shown e.g. in Fig.1 for the ITER reference case of Table I.

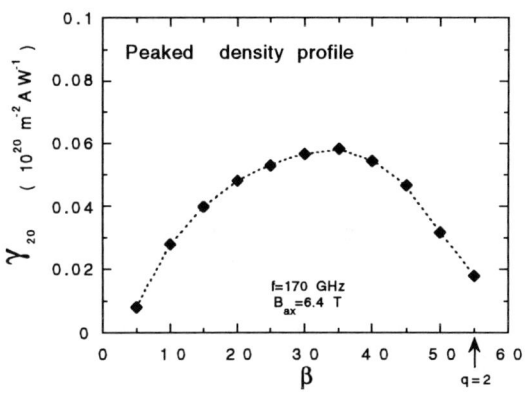

Fig.1 - Current drive efficiency vs injection angle β at f=170 Ghz.

The growth of the *amplitude* $\tilde{B}_{r(m)}$ of the (measurable) radial magnetic field perturbation and the evolution of its *rotation frequency* in presence of ECCD and of external helical "error" or "bias" field [2,3] is ruled by nonlinear state equations describing the superposition of several physical effects in the magnetic reconnection process:

Table I

a (m)	R_0 (m)	k	B_0 (T)	Z	n_{e0} ($10^{20}m^{-3}$)	T_{e0} (KeV)	p^{EC} (MW)
2.6	8.36	1.55	6.4	2	1.27	28	50

$$\frac{d\tilde{B}_{r(m)}}{dt} = c_1 \Delta'_0 \left[1 - \sqrt{\frac{\tilde{B}_{r(m)}}{\tilde{B}_{sat(m)}}}\right] \sqrt{\tilde{B}_{r(m)}}$$
$$+ \left[c_2 \frac{f_V^{1/2} \omega^2 \tau_V^2}{f_V + \omega^2 \tau_V^2} - c_3 \frac{(\omega - \omega_T)^2}{\tilde{B}_{r(m)}^{3/2}}\right] \sqrt{\tilde{B}_{r(m)}}$$
$$+ c_4 \Delta'_{0RF} \sqrt{\tilde{B}_{sat(m)}} + c_5 \frac{\tilde{B}_{r,d}^{ext}}{\sqrt{\tilde{B}_{r(m)}}} \quad (2)$$

$$c_6 \sqrt{\tilde{B}_{r(m)}} \frac{d\omega}{dt} = c_7 [\omega_T - \omega] - c_8 \frac{f_V \omega \tau_V}{f_V + \omega^2 \tau_V^2} \tilde{B}_{r(m)}^2$$
$$+ c_9 \tilde{B}_{r(m)} \tilde{B}_{r,q}^{ext} \quad (3)$$

The coefficients c_k depend on the plasma equilibrium parameters, $\tilde{B}_{sat(m)}$ is the natural (uncontrolled) saturation value of the perturbation, $\tau_V = \frac{\mu_0 \sigma_V \delta_V r_V}{2m}$ is the vessel resistive time constant and $f_V = \left(1 - \left(\frac{r_s}{r_V}\right)^{2m}\right)^{-2}$ is the attenuation factor for the currents induced in the vessel, of minor radius r_V, thickness δ_V and electrical conductivity σ_V, by the plasma perturbation located at $r_s = r_{q=\frac{m}{n}}$.

On the rhs of eq.(2) there is the natural quasilinear growth rate of the perturbation, $\propto \Delta'_0$, the vessel eddy current effect, a stabilizing term due to ion inertia, the RF control term Δ'_{0RF} and the external ("error") field contribution.

The next equation (3) governs angular speed ω of the magnetic island; on the rhs there is an effective viscous torque term, the torque due to eddy currents in the vessel and that due to the external field. The undriven *natural* mode angular speed is ω_T. Here $\tilde{B}_d^{ext}, \tilde{B}_q^{ext}$ are the "direct" and "quadrature" components of the external ("error"or "bias") field in the island reference frame. These equations are supplemented by a Proportional-Integral-Derivative (PID) or a bang-bang feedback relation linking the *state* variable $\tilde{B}_{r(m)}$ to the *control* variable Δ'_{0RF}, proportional to the Gyrotron power [2].

2.1. Data acquisition and control system.

The basic requirement for the control system is the real time amplitude and phase estimation of the radial component $\tilde{B}_{r(m)}$ of the magnetic field perturbation produced by the rotating magnetic island with poloidal and toroidal mode numbers m=2,n=1 respectively.

A comprehensive system for the ITER tokamak can be composed of an analog front end measurement stage and an adaptive digital controller (see Fig.2). The analog front end implements the basic operations for the filtering of the instantaneous amplitude (BrD and BrQ in Fig.2) of the m=2, n=1 magnetic perturbation $\tilde{B}_{r(m)}$ at two different toroidal locations. At set of in-vessel pick-up-coils is available for poloidal field measurements with a bandwidth of the order of 20 kHz. The identification of a rotating mode of poloidal and toroidal numbers m=2, n=1 requires at least 4 coils, placed around the poloidal contour and 2 coils placed toroidally. Furthermore the identification of the position of the mode needs two separate sets of pick-up coils. Thus the minimum number of channels to be acquired is 16. A larger number would be desirable to avoid aliasing effects caused by modes with larger m, n numbers. Furthermore a larger number of coils increases the reliability of the system. The high number of measuring channels to be processed in real time requires an analog front end before the A/D converters of the controller. The phase of the mode (Phase_est in Fig.2) is numerically estimated by a digital PLL implemented using Digital Signal Processors (DSP). On a much longer time scale also the amplitude (Br_est in Fig.2) is numerically estimated. The DSP's are is interfaced to the front end analog system not only for the acquisition of the two signals but also for the real time adaptation of the filter parameters, following different plasma conditions. The synthesis of the control signals is obtained by locking them to the magnetic islands amplitude and phase evolution (Fig.2). The control signals are modulated in amplitude by a bang-bang (or PID) controller on the basis of the internally estimated mode amplitude. The control signal synthetized by the controller is transmitted to the RF Gyrotron tubes Regulated High Voltage Power Supply.

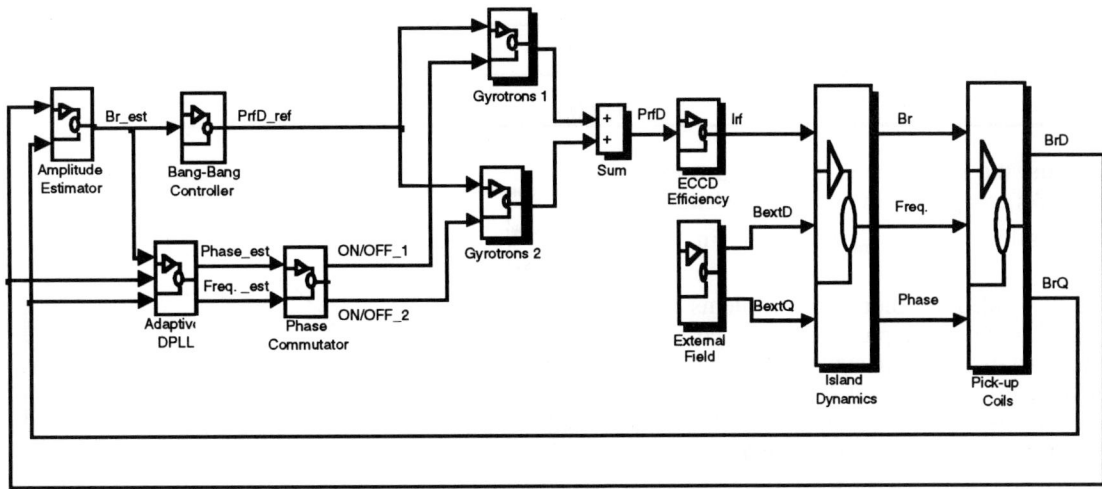

Fig.2 Block diagram of the Digital Data Acquisition and Control System.

3. PERFORMANCE SCENARIOS

The "standard" ITER equilibrium of Table I has the q=2 flux surface located at $\psi \approx 0.7$. With the "standard" choice of Gyrotrons at 170 Ghz this case requires an extremely oblique toroidal launch with $\beta=55°$. At this injection angle the global efficiency γ_{20} drops dramatically to 0.018 as shown in Fig.1.

A much better option is to operate at B=5.77 T with 140 GHz gyrotrons as shown in Table II, and Fig.3.

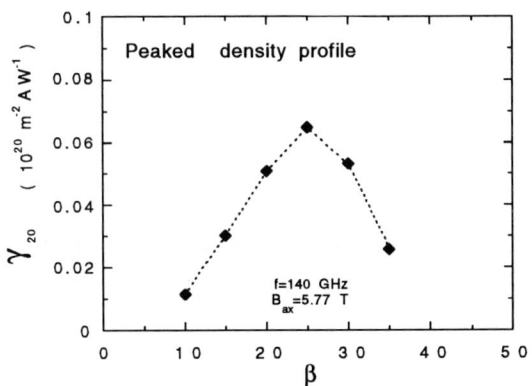

Fig.3-Current drive efficiency vs β at f=140 Ghz.

Launching the wave beam at $\beta=27.5°$ a control current up to 0.7 MA, localized on the q=2 surface over the width of an island can be obtained with ~40 MW of modulated ECW power.

Fig.4 shows the current density profile J^{EC} generated by a Gaussian ECW beam inpinging on the q=2 surface.

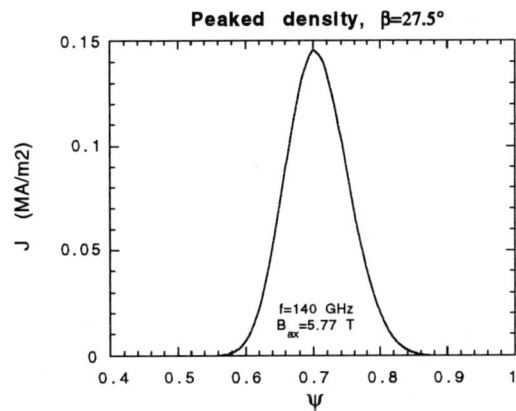

Fig.4- J^{EC} profile vs ψ, f=140 Ghz, $\beta=27.5°$

The simulation of the performance of the island control system is shown in Figs. 5-7. In ITER the time scales of the mode locking (due for instance to interaction with eddy currents in the vessel) is much shorter than that of the mode amplitude growth.

Table II

f (GHz)	β^o	γ_{20}	I^{EC} (MA)
140	27.5°	0.0605	0.696

Typically, using the reference ITER data, a rotating m=2 island locks completely in 5s and the island reaches saturation at $\tilde{B}_{sat(m)}$=10 mT in 100s (Figs.5,6)

In such conditions there are problems for a control system because the phase of a locked island cannot be tracked by magnetic signals. Assuming that some non magnetic diagnostic can locate the "O" point a possible strategy of control can be based on the simple feedback of the *amplitude* of the measured magnetic perturbation $\tilde{B}_{r(m)}$.

With the data of Table II the island growth can be restrained, as shown in Fig.7, and the amplitude reduced to a level $\tilde{B}_{r(m)}$ 0.13 mT using a tandem system of Gyrotrons delivering *each* 40-50 MW at opposite toroidal locations and controlled through an (On/Off) phase commutator (see Fig.2) and a simple bang-bang controller switching on the power whenever Br_est exceeds the amplitude target value.

A reduction of the power required to restrain the island growth is possible by keeping the islands in rotation and using a PID contoller in place of the Bang-Bang one.

An attractive aternative strategy of control consists in providing a magnetic trap in order to lock the mode at a *predetermined* position related to the most efficient launch angle of the ECW beam and measure and control only the mode amplitude.

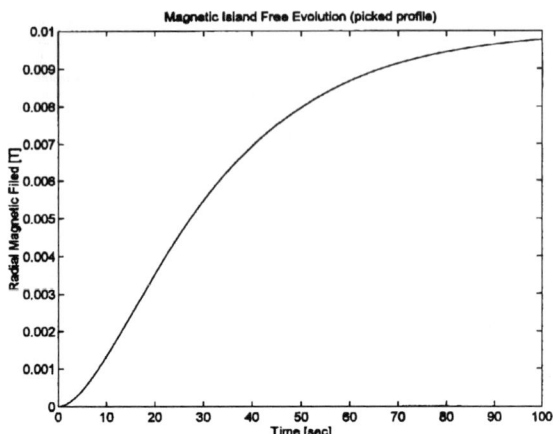

Fig.6-Typical uncontrolled growth of ITER island amplitude $\tilde{B}_{r(m)}$ mT

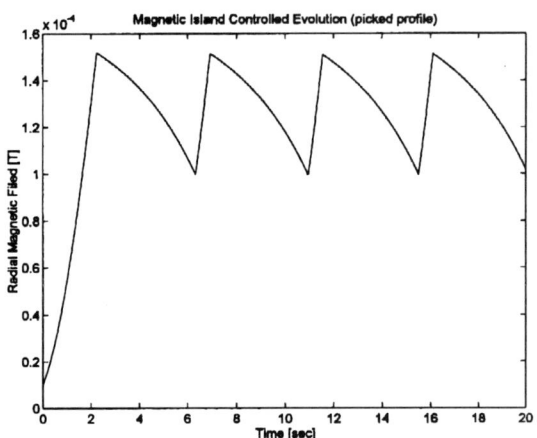

Fig.7-ECCD Bang-Bang control of ITER island amplitude $\tilde{B}_{r(m)}$ mT, with a tandem power supply of 50 MW f=140 Ghz, β=27.5°

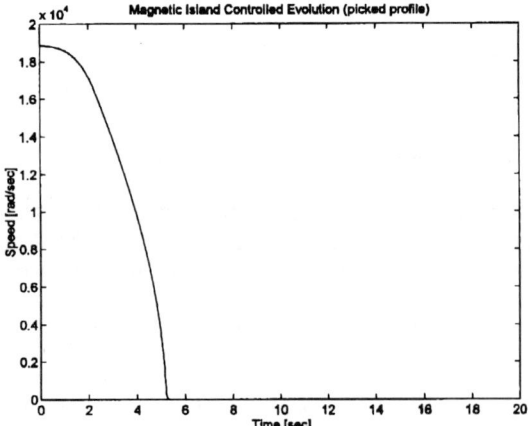

Fig.5-Typical slowing down of ITER island angular speed ω rad/s, due to eddy currents in the vessel.

REFERENCES
1. G.M.Fishpool, P.S.Haynes, Nucl.Fus., 34 (1994) 109
2. E. Lazzaro, G. Ramponi, Phys. of Plasmas, 3 (1996) 978
3. P.Savrukhin,D.J.Campbell,G.D'Antona, A.Santagiustina, IEEE Transactions on Nucl. Science, A1(1996)
4. S.Nowak, E. Lazzaro, G. Ramponi, to appear on Phys. of Plasmas,11 (1996)

Design of central control and man-machine interface systems for large helical device

H.Yamada, K.Yamazaki, K.Y.Watanabe, S.Yamaguchi, K.Nishimura,
Y.Taniguchi, H.Ogawa, N.Yamamoto, S.Sakakibara, M.Shoji, O.Motojima

National Institute for Fusion Science, Oroshi-cho 322-6, Toki-shi, Gifu 509-52, Japan

Design of the central control system for the Large Helical Device (LHD) has been continued. The central control system consists of a central programmable logical controller, a torus supervision monitoring system, a timing system, a protective interlock, and a man-machine interface system. Composition of conservative hard-wired logic control and a client/server system on LAN fulfills requirements of reliability, flexibility and extensibility.

1. DESIGN CONCEPT

The Large Helical Device (LHD) [1] is a superconducting toroidal facility with a mission to steady-state operation of high temperature plasmas. The major radius, the minor radius of plasmas, and the designed toroidal field are 3.9m, 0.65m, and 4T, respectively. While LHD is a plasma physics experimental device, it has a specific feature of a large-scale plant due to steady-state-operation facilities and cryogenic systems. The control system of LHD [2] is required to cooperate a number of plant component devices/facilities (sub-systems) as well as to manage flexible plasma experiments. Also safety protection against all expected occasions and easy extension for system up-grade are key issues. The

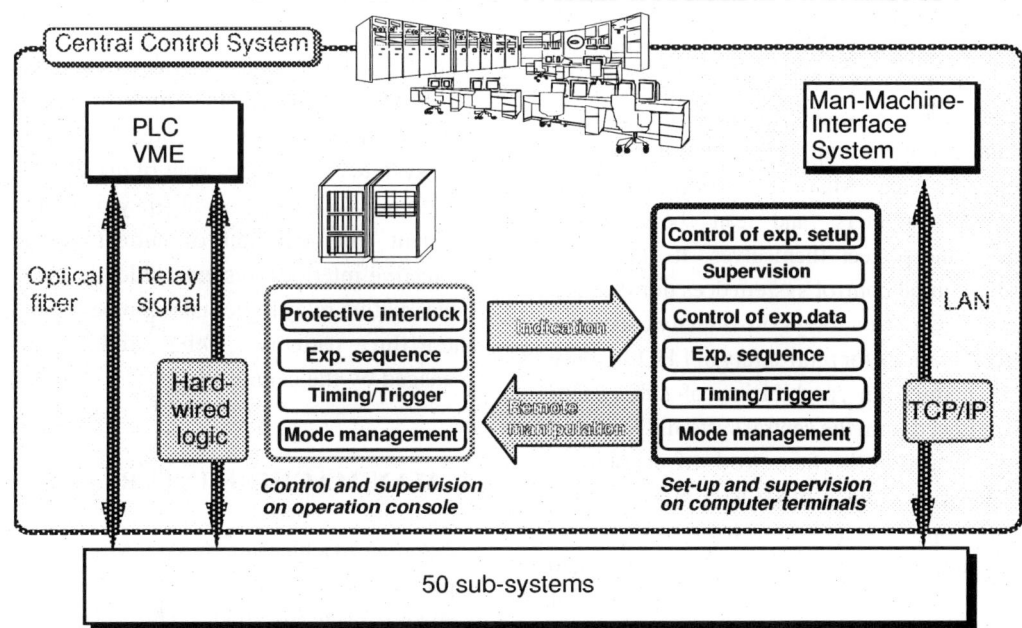

Fig. 1 Architecture of the central control system of LHD

control system of LHD should be, therefore, reliable, flexible and extensible. Composition of conservative hard-wired logic control with a programmable logic controller and the man-machine interface (MMIF) system with a client/server system fulfills these requirements complementarily (see Fig.1). The central control system has been designed to operate LHD safely without help of the MMIF system because hard-wired logic has priority over information transmitted through LAN, however, the MMIF system greatly facilitates procedures of experimental set-up, supervision of facility condition and sequential control. It also contributes to prevention of operational human errors and consequent accidents.

2. COMPOSITION OF CENTRAL CONTROL SYSTEM

The central control system consists of a central programmable logic controller (PLC) for management of sequence and experimental mode transition, a torus supervision monitoring system for more than 1300 instrumentation elements, a timing system, a protective interlock, and the server computer with databases providing man-machine interface environment. VME systems are used for a torus supervision monitoring system and a timing system. The facilities composing LHD are arranged into 50 sub-systems, e.g., a vacuum pumping unit, a cryogenic unit, a power supply for helical coils, an ECH heating unit, etc. Major sub-systems have their own control computers and can be operated independently and protect themselves. It should be noted that the central control system does not dictate a device control layer of a sub-system directly except for the protective interlock for total safety of LHD. Only signals for permission/prohibition of sub-system operation are issued from the central system. Operation of LHD is managed in the framework of operational modes. Transition from a mode to a mode and the process of the operation is controlled by the central PLC which judges consistency of status with hard wired logic. Figure 2 shows the structure of operational modes and an experimental sequence.

The timing system unit controls the process of plasma

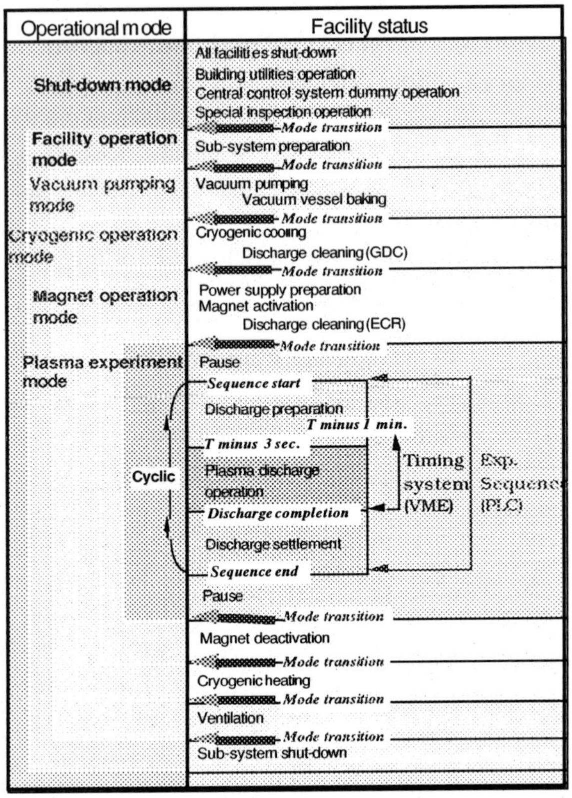

Operational mode	Facility status
Shut-down mode	All facilities shut-down Building utilities operation Central control system dummy operation Special inspection operation *Mode transition*
Facility operation mode	Sub-system preparation *Mode transition*
Vacuum pumping mode	Vacuum pumping Vacuum vessel baking *Mode transition*
Cryogenic operation mode	Cryogenic cooling Discharge cleaning (GDC) *Mode transition*
Magnet operation mode	Power supply preparation Magnet activation Discharge cleaning (ECR) *Mode transition*
Plasma experiment mode	Pause *Sequence start* Discharge preparation — T minus 1 min. *T minus 3 sec.* Plasma discharge operation *Discharge completion* Discharge settlement *Sequence end* Pause *Mode transition* Magnet deactivation *Mode transition* Cryogenic heating *Mode transition* Ventilation *Mode transition* Sub-system shut-down

Fig. 2 Structure of operational modes and experimental sequence.

discharges precisely and accommodate discharges up to 10 hours. Since LHD employs superconducting coils, particular care has been paid to protection of the coil systems [3]. The event of quench is graded three categories, depending on urgency of protection, with different time constant of current shut down. A protective interlock system, which is made up by the conservative relay logic, takes proper steps along with these three patterns. Other slighter interlocks are managed by the central PLC.

3. MAN-MACHINE-INTERFACE SYSTEM

A man-machine-interface (MMIF) system which is a primary component of the central control system for LHD has been specified. The MMIF system, here, involves a variety of intelligent functions needed in the LHD experiments as well as a scheme of graphical user interface (GUI) in a narrow sense. The MMIF system provides a variety of information transmission through LAN.

Major functions of the MMIF system are the following three:

(1) *Manipulation of experimental sequence and mode management*. The duration of discharge, the interval of each pulsed shot and the mode transition are set from the graphical terminal by a permitted operator. Usual users can check the status of LHD experiment on a client computers.

(2) *Conduction of experimental set-up on component devices/facilities*. The number of subsystem and parameters to be controlled are 50 and more than 300, respectively. Here major parameters have physical meanings and substantial set-up of each device is conducted by an individual control computer based on transmitted condition. While usual users can write out experimental condition and store it, registration of data can be done by only a specific operator.

(3) *Supervision of status of component devices/facilities*. This system does not have own data acquisition system but integrate monitoring data from a number of sub-system through data transmission. Although the data are primarily representing ones to watch status of whole LHD system, the number reaches 3000 in total. When an accident is detected by the central PLC, information related to the cause of the accident is accumulated in detail with this function.

The architecture of platform of the MMIF system is shown in Fig.3. This system as well as the computer network is organized on so-called de facto standard so that the system can develop along with rapid progress in this field. The transmission protocol between the server machine and the sub-systems is limited to TCP/IP, since a variety of computers are utilized for the sub-systems. An experimental information LAN of LHD is protected by a firewall system from attacks via the backbone network. All manipulation and indication are conducted in human friendly graphical environment on client terminals (a presentation layer). A server manages information between client terminals and control computers for component devices/facilities with a relational database (a transmission and data house layer). It should be here noted that the MMIF system does not dictate directly a device control layer of each control system for component.

Examination on a prototype has been continued intensively to demonstrate performance capability and usefulness of the planned MMIF system. The network OS, relational database managing system (RDBMS), graphical user interface development environment in a prototype are WindowsNT, ORACLE 7, and VisualBasic. Transmission between GUI terminals (client) and a server is done by SQL*net to promote efficiency. Although the system utilizes a de facto standard, a prototype test is indispensable to clarify performance. because the system has own characteristics. Since compatibility in combination of softwares (OS, GUI, RDBMS, etc.) and hirdwares determines performance and required labor in development, comparative studies with other

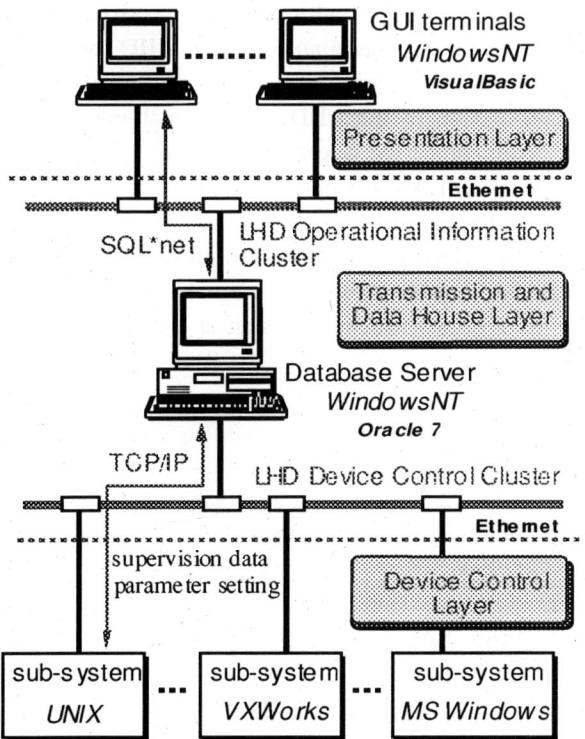

Fig. 3 Architecture of MMIF.

combination on a prototype are of much importance before the start of coding of a real application program. The virtual experiments have been started on an MMIF prototype and a couple of sub systems. Also load tests and behavior of response in over-load condition are examined in terms of transmission speed, time for database access, time for graphical drawings, etc.

4. LHD EXPERIMENTAL LAN

The LHD operation LAN is one component of the LHD experimental LAN, on which the information of LHD experiment is transmitted. The LHD experimental LAN consists of three sub-networks (clusters), i.e., operation, analysis, and plasma diagnostics clusters. The LHD experimental LAN is designed based on a concept that data with different purposes are transmitted in an independent cluster in order to transmit network data reliably, safely and efficiently. This architecture facilitates management of network security and control network traffic. The LHD operation cluster composes an operational information LAN, torus-device control LAN and peripheral-device control LAN. Information exchange for operation and discharge condition, and management of the sequence of experiments and plasma discharges are conducted on these LAN's. In 1995 fiscal year, a backbone part of the LHD operation cluster LAN consisting of FDDI, FDDI-switch, CDDI, Ethernet and Ethernet-switch system has been constructed in the LHD main building and buildings supporting the LHD experiment except for the LHD control building which is under construction. Concerning the protocol routing and packet bridging, routers in torus-device and peripheral-device control LAN's permit only TCP/IP routing and prevent no other protocol packets in order to keep reliability and security of network. On the other hand, in the operational information LAN, various network terminals exist comparing with the devices operation LAN's. Therefore NetBEUI and AppleTalk protocol in addition to TCP/IP are available in the network. In order to keep security of the whole LHD operation LAN clusters, the firewall system which controls security in the level of network application software

carefully is employed on the upper network stream side. The test run of these network system has been started from viewpoint of keeping security and controlling trafic. The entire system of LHD operation LAN will be completed another year later after the completion of the LHD Control Building in 1996 fiscal year.

5. CONCLUSION

Reliablity for superconducting coil protection and flexibility for plasma physics experiments are simultaneously prerequisite to the central control system for LHD. The central control system for LHD is based on composition of conservative hard-wired logic and information transmission by client/server system with LAN. Safety protective interlock, mode transition and experimental sequence are conducted by a programmable logic controller with hard-wired logic. Timing and triggering are provided by a VME system through optical fibers. A man-machine-interface system serves operators for preventing errors with efficient, visual, and comprehensible environment. The LHD operation LAN cluster was designed to guarantee safe and reliable transmission of the LHD experimental information and data of device operation. A part of the LHD operational LAN was completed in the middle of this May. The LAN consists of FDDI switches and Ethernet system. The central control system as well as an MMIF system will be completed by the end of 1997. The LHD experiment will start at the beginning of 1998.

REFERENCES

[1] O.Motojima et al., Fusion Engr. and Design **20** (1993) 3.
[2] K.Yamazaki et al., Nucl. Instr. and Meth. in Phys. Res. A **352** (1994) 43.
[3] O.Kaneko et al., Fusion Engr. and Design **20** (1993) 121.

Real Time Control of Plasma Boundary In JET

S. Puppin, M. E. Angoletta, D. J. Campbell, J. J. Ellis, M. Garribba, M. Lennholm,
F. Milani, D. O'Brien, F. Sartori, R. Sartori

JET Joint Undertaking, Abingdon, Oxfordshire, OX14 3EA

1. Abstract

The Pumped Divertor phase of JET has involved the use of plasma configurations with complex control requirements. Therefore, a new JET Plasma Shape Controller (SC) was developed. This system was brought into operation in February 94 and had been used for all the Mark I campaign (94-95).

The experience gained during that campaign, and the new and more stringent control requirements resulting from the introduction in JET of the Mark II divertor have prompted the development of a new and more powerful version of Shape Controller. This paper describes the new system, and the performance achieved during the Mark II campaign.

2. Shape Controller Main Features

- VME based system utilising 4+1 TMS320C40 floating-point DSPs.
- Real-time plasma boundary reconstruction using the XLOC algorithm [1].
- Multivariable decoupling algorithm with resistive compensation.
- At each moment, it is possible to control 9 control variables, selected from a set of 26 potential control parameters.
- In addition to the plasma current, up to 6 of these control variables can be plasma shape parameters.
- Each pulse can be subdivided in up to 11 control time windows. In each window, a different set of control variables can be selected.
- Transitions between different sets of control variables are handled by a real-time modification of the reference waveforms.
- Circuit current limits are dealt with.

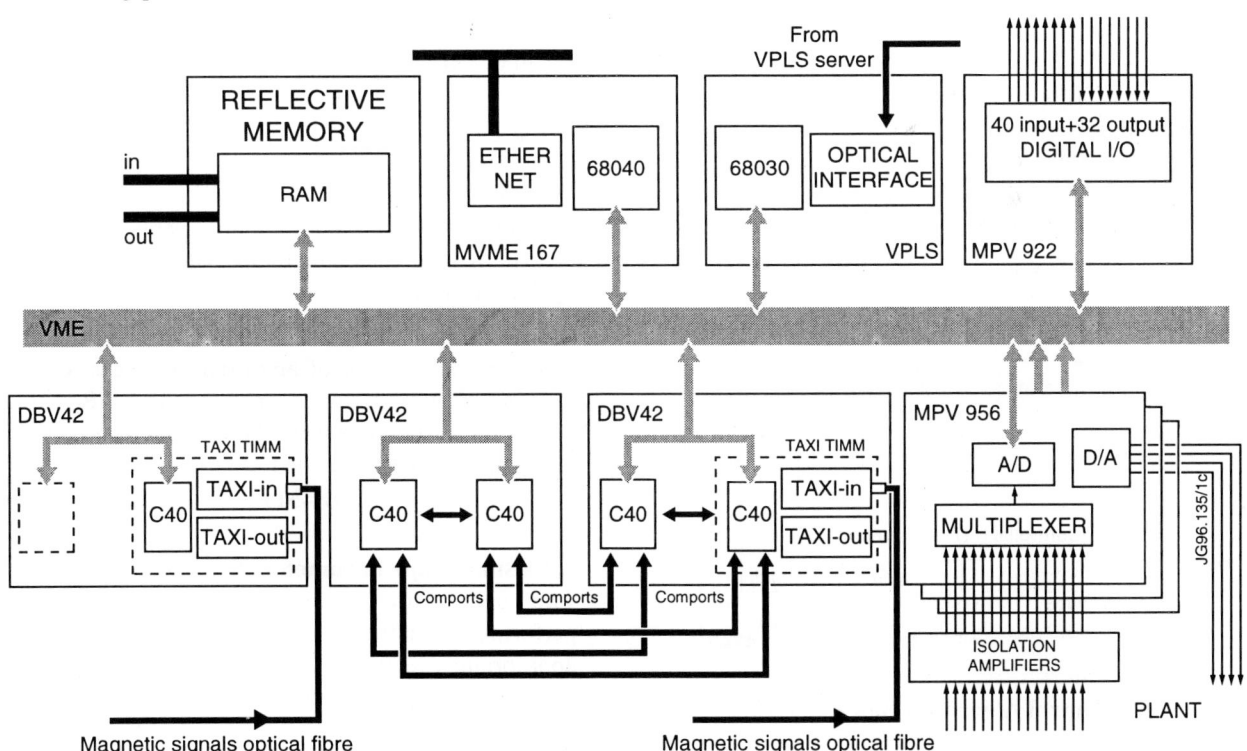

Figure 1 Block diagram of the Shape Controller VME system.

3. Hardware Overview

Shape Controller is a multiple DSPs VME based system. The heart of the system consists of 3 VME master DSP carrier boards, each equipped with 16 Mbyte of DRAM and capable of housing 2 TMS320C40 DSPs. Five DSPs are used in total, two of which are interfaced with optical links (TAXI link, 15.5 Mbyte/s) for high speed communication with the magnetic acquisition system.

In addition to some analogue and digital I/O boards, the system includes a 68040 board running the VxWorks real-time operating system. This board is used for the connection of the system with the UNIX host through Ethernet.

A custom board (VPLS) is used for the connection with the JET Central Timing & Triggering System.

Finally, a reflective memory board is used to provide SC with a very high performance real-time connectivity with other JET real-time systems.

4. Plant Model

The system plasma - PF circuits is a multivariable, time-varying, non-linear system.

In general, the plasma equilibrium can be described by the following equation:

$$\mathbf{X} = \mathbf{X}(\mathbf{I}_C / I_P, \beta_P, l_i, \psi_{iron}) \quad (1)$$

where $\mathbf{I}_C$ is a vector containing the value of all PF circuit currents, and $\mathbf{X}$ is a vector of plasma-shape parameters, β_p e l_i are used to describe the plasma current density profile and ψ_{iron} represents the iron saturation status.

In addition, it is necessary to consider the electrical equation of the system:

$$\frac{d}{dt}\big(\mathbf{M}(\mathbf{X}, \psi_{iron}, \beta_p, l_i)\mathbf{I}\big) + \mathbf{RI} = \mathbf{V} \quad (2)$$

Where $\mathbf{M}$ is the inductance matrix, $\mathbf{R}$ is the

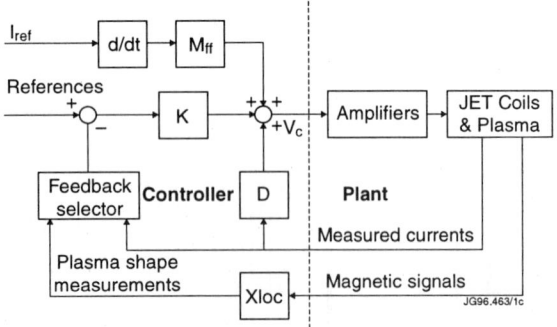

Figure 2 Control Algorithm Block Diagram

resistance matrix, $\mathbf{I} = [I_P\ I_C]$ and $\mathbf{V} = [0\ \mathbf{V}_C]$ ($\mathbf{V}_C$ are the voltages applied to the PF coils). It must be noted that the $\mathbf{M}$ matrix is not constant, but is a function of the plasma shape, plasma current density profile and iron saturation status.

Equations (1) and (2) can be linearised around a plasma equilibrium [2,3,4,5]. The linearised model becomes:

$$\mathbf{M}_S \dot{\mathbf{I}}_C + \mathbf{R}_C \mathbf{I}_C = \mathbf{V}_C + \tilde{\mathbf{f}}(\beta_p, l_i, \psi_{iron}) \quad (3)$$

$$I_P \delta \mathbf{X} = \mathbf{B}_S \delta \mathbf{I}_C + I_P \mathbf{G} \begin{bmatrix} \delta \beta_p \\ \delta l_i \\ \delta \psi_{iron} \end{bmatrix} \quad (4)$$

$$\delta I_P = -\frac{1}{\tilde{I}_P} \tilde{\mathbf{M}}_{PC} \delta \mathbf{I}_C \quad (5)$$

5. Control Algorithm

Since Shape Controller is controlling 9 PF circuits, it is possible to control 9 variables at the same time. These control variables can be a mixture of circuit currents, plasma parameters and plasma-wall distances.

If $\mathbf{Y}$ is the vector of control variable, using the linear model of equations (4) and (5), the transition matrix $\mathbf{T}$ can be defined as:

$$\mathbf{Y} = \begin{bmatrix} I_P \\ I_P \mathbf{X} \\ \mathbf{I} \end{bmatrix} = \begin{bmatrix} -\tilde{\mathbf{M}}_{PC}/\tilde{I}_P \\ \text{rows of } \mathbf{B}_S \\ \text{rows of unity matrix} \end{bmatrix} \mathbf{I}_C = \mathbf{T}\mathbf{I}_C \quad (6)$$

The chosen control law is a multivariable decoupling algorithm with resistive compensation. It is defined by the following equation:

$$\mathbf{V}_c = \mathbf{EC}(\mathbf{Y}_{ref} - \mathbf{Y}) + \mathbf{DI}_c \quad (7)$$

where $\mathbf{V}_C$ is a vector of amplifiers voltages, $\mathbf{C}$ is a diagonal matrix of time constants, $\mathbf{D}$ ($\approx \mathbf{R}_C$) is a matrix used to compensate the resistive drop on the PF circuits, and $\mathbf{E}$ is defined as:

$$\mathbf{E} = \mathbf{MT}^{-1} \quad (8)$$

6. Plasma Shape Parameters Selection

The secret in achieving good and robust performances in the control of the plasma shape is an appropriate selection of the plasma shape parameters to be controlled. The parameters chosen for Shape Controller are shown in Figure 3 and Figure 4. In

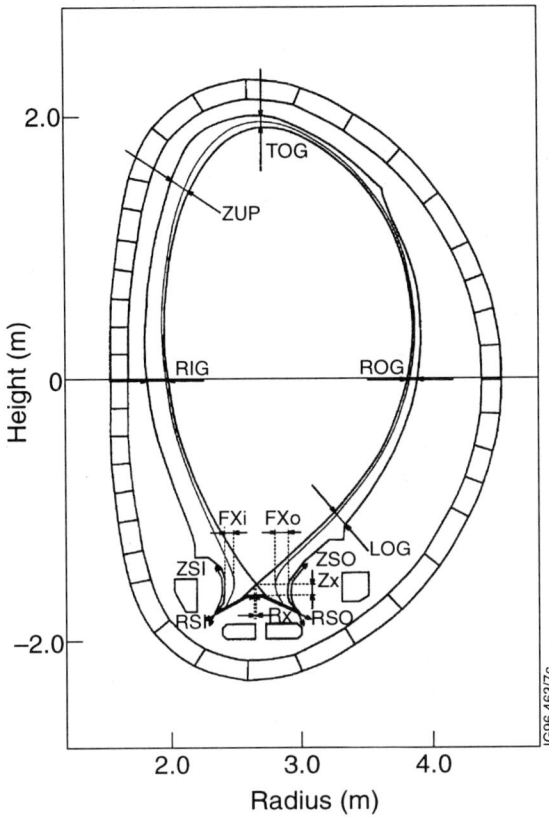

Figure 3 Definition of the plasma shape parameters that are controllable by SC

addition to several plasma wall distances, Shape Controller is able to control the position of the strike points and the aperture of the flux expansion in the divertor region.

The key points in the selection of the plasma shape parameters are:
1. The validity of the linear relation between the plasma shape parameters and the circuit currents (B_s matrix of equation (4)).
2. A sensible selection in the set of control variables that are controlled at each moment. The set of control variables **Y** must be strongly linearly independent with respect to the circuit currents.
3. The requested plasma configuration must be achievable with the PF circuit configuration and the power supplies current limits.

7. Plasma Shape Control Performance

Provided that a reliable measurement of the plasma shape parameters is available, the achievable performance is remarkable. The following points must be noted.
- During the Mark II campaign, in addition to the plasma current, 6 plasma shape parameters have been controlled simultaneously.
- The typical JET pulse uses four plasma shape parameters (Rog, Tog, Rsi and Rso. Figure 5).
- The direct control of the plasma shape has been used in virtually all the JET experimental programs (high currents, reverse shear, etc.), and in all phases of a pulse.
- With the careful selection of the controllable shape parameters, the control algorithm has proved to be extremely robust. The same control coefficients have been used for all plasma configurations.
- With the use of shape control, the magnetic

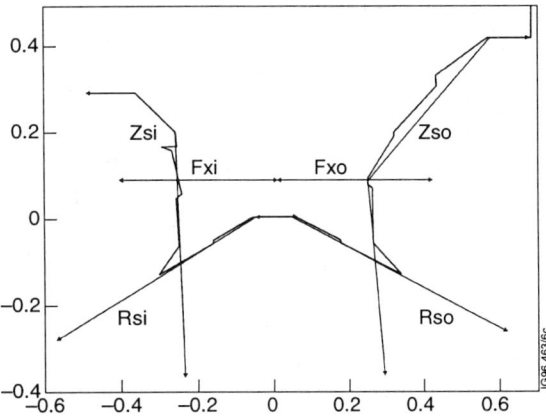

Figure 4 Controllable shape parameters in the divertor region

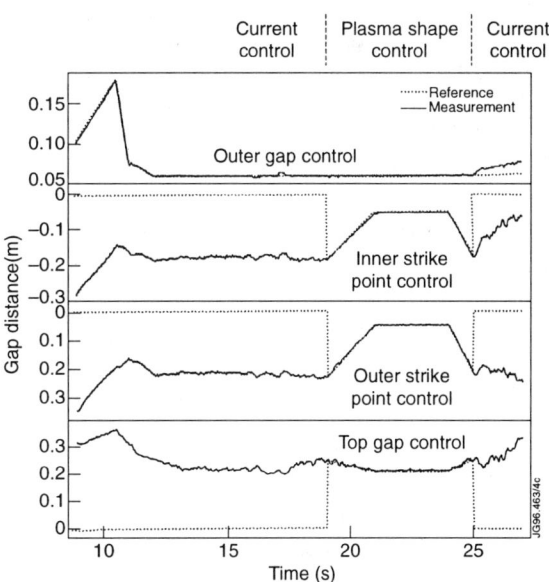

Figure 5 Example of control of shape parameters

geometry in the divertor region is easily controllable even with large beta shifts in the plasma.

8. Plasma Boundary Reconstruction

Shape Controller is using the Xloc algorithm to measure the plasma shape parameters. To reconstruct the plasma boundary, Xloc uses Taylor expansions of the flux function in vacuum in five adjacent regions.

The accuracy of the code is limited by many factors, the most important of which are:
- The limited number of magnetic sensors available.
- The large distance present between many of the sensors and the plasma.
- The high gradients of the magnetic field in the divertor region.

The precision of the plasma boundary reconstruction has proved to be the major limitation in the performance achievable in the control of the plasma shape. Typically the plasma shape parameters are calculated with a precision of a couple of centimetres, but in the most difficult scenarios measurement errors of up to 5 cm can be present.

9. PF Generator Control Loop

One fundamental hypothesis of the decoupling control algorithm is that all power supplies should behave as ideal voltage amplifiers. In JET, with the exception of the P1 coil circuit (the central solenoid), this assumption is well fulfilled.

The central solenoid is powered by the PFGC (Poloidal Field Generator-Converter, synchronous machine with diode rectifier). The input-output characteristic of this assembly is highly non-linear. In the Mark I campaign this non linearity was limiting the plasma current control performance (Figure 6 (a)).

To compensate for this limitation, a new voltage loop (equation (9)) has been added around the PFGC, using as feedback quantity the measurement of the central solenoid voltage.

$$V_{PFgen\_ref} = \frac{255}{N_{PFGC}} \cdot K_p \left(1 + s\tau_d + \frac{1}{s}\tau_i\right) \cdot \Delta V \quad (9)$$

$$\Delta V = \left(V_{p1\_ref} - V_{p1}\right)$$

N_{PFGC} is the PFGC speed. With the new loop, the control of the plasma current can be made much faster than what is required in normal operation (Figure 6 (b)).

10. Conclusions

In its latest revision, the JET Shape Controller has proved to be an extremely useful and reliable tool. The direct control of the plasma shape is now routinely used in most of JETs experimental programs. Even though many new shape parameters are used, the performance and robustness of the control algorithm is very good. The only real limitation of the system is the accuracy of the available boundary reconstruction.

References

[1] D. O'Brien, J. J. Ellis J. Lingertat, "Local Expansion Method for Fast Boundary Identification at JET", Nuclear Fusion 33, 467 (1993).

[2] Garribba M, Litunovsky R, Noll P, Puppin S, "The New Control Schema for the JET Plasma Position and Current Control System", XV IEEE Symposium of Fusion Technology 2993, Hyannis, USA

[3] Garribba et all, "First Operational Experience with the new Plasma Position and Current Control System of JET", SOFT conference, Karlsruhe, August 22-25, 1994.

[4] Puppin S, "sviluppo del controllo della forma del plasma per l'esperimento JET", Phd Thesis, Padova 1995.

[5] Garribba M, Litunovsky R, Noll P, Puppin S, "The Plasma Shape and Current Control System of JET", accepted for publication on Fusion Technology.

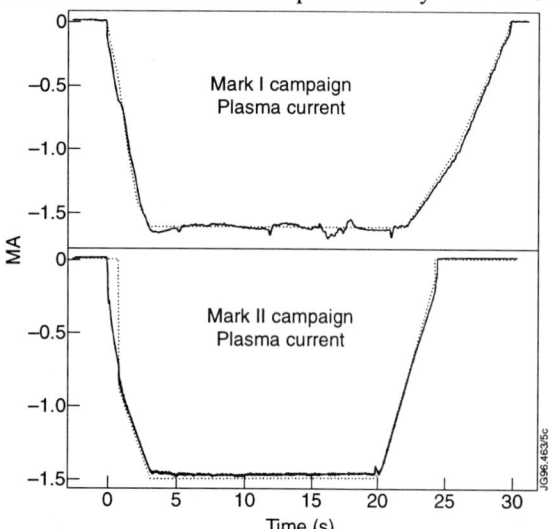

Figure 6 Comparison of the performance of the plasma current control.

Implementation and Initial Operation of an Adaptive Plasma Density Controller at JET.

H. E. O. Brelén, T. Budd, J. Ehrenberg, M. Gadeberg, C. Ryle

JET Joint Undertaking
Abingdon, Oxfordshire OX14 3EA, England

An adaptive controller has been tried in the plasma density feedback system at JET in an effort to improve the control performance as produced by the present conventional controller. Changing plant dynamics calls for retuning of the controller to retain the required performance of the control system. This controller tuning must in the conventional case be done manually. In the adaptive controller, the retuning process is automated and runs in parallel with the control function. The adaptive controller was implemented and commissioned for on-line operation. Experience from the use of the controller and its performance is presented and analysed. It is the first time this controller technology has been applied at JET

1. REQUIREMENTS

The new plasma density controller was designed under the following performance requirements:
- 1.5 - 2 seconds settling time to within 1% control accuracy following a stepwise change in the reference signal.
- maintained stability and recovery within 20 s to above performance figures due to sudden changes over the whole expected plant parameter range.
- maximum 5% overshoot following a stepwise change in the reference signal.
- disturbance rejection factor of at least 20.
- no steady state control deviation.

2. CONTROLLER

A controller with the following control law was introduced:

$$R \cdot u = T \cdot x - S \cdot y \qquad (1)$$

where R, T and S are polynomials of the discrete time shift operator z. x represents the requested plasma density, y the real plasma density and u the actuation signal to the gas introduction. With the plant transfer function B/A, the closed loop transfer function for the control loop is:

$$\frac{y}{x} = \frac{T \cdot B}{A \cdot R + S \cdot B} \qquad (2)$$

3. DESIGNER

Required control performance is translated to required pole locations for the closed loop transfer function, equation 2. The coefficients in the controller polynomials, R and S, are calculated in the designer routine so that for any plant polynomials, B and A, the required pole locations are obtained.

The controlled system has zeros in common with the plant one of which can approach z=1. Its effect is made worse by the fact that, unlike the open loop transfer function, the controlled system does not have an integrator pole to reduce it. Overshoots caused by this zero can be several times larger than the steady state level. By making one of the poles of the controlled system coincide with this zero, that is the zero in B should also be a zero in AR+BS, its effect is cancelled and the intended system performance should be seen.

The situation is complicated by the fact that the plant zero and poles move due to changing plant parameters. In order to maintain the performance of the system, keeping its poles in the required positions including the one cancelling the zero, the calculation of the coefficients of the controller polynomials R and S have to be repeated at each sampling occasion, with the latest estimate of the A and B polynomials.

Finally the polynomial T is used to produce zeros for avoiding unnecessary delays in the controlled system.

4. IDENTIFIER

The identifier produces estimates of the coefficients in the plant polynomials A and B that are necessary for the above controller design. The recursive least square identifier is chosen for this purpose, and the algorithm for this is:

$$\theta_{n+1} = \theta_n + K_{n+1}(y_{n+1} - \varphi^T_{n+1}\theta_n) \qquad (3)$$

where

$$K_{n+1} = \frac{P_n}{\beta}\varphi_{n+1}\left(\frac{1}{1-\beta} + \varphi^T_{n+1}\frac{P_n}{\beta}\varphi_{n+1}\right)^{-1} \qquad (4)$$

$$P_{n+1} = \frac{1}{\beta}(I - K_{n+1}\varphi^T_{n+1})P_n \qquad (5)$$

where y is the plant output, φ contains recent and historic values of the y and u variables from the plant and θ is the identified parameter vector. $\varphi^T\theta$ is thus the identifier's estimate of the plant output. ß determines the length of the identifier's 'forget function'. Figure 1 shows the structure of the adaptive controller. A more detailed description of the controller can be found in [1].

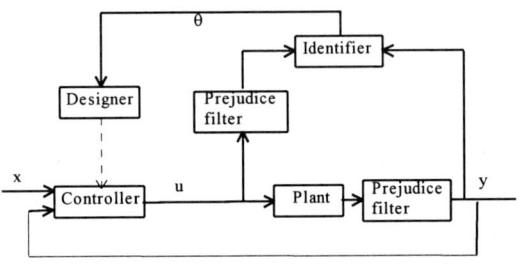

Figure 1: Flow chart of adaptive controller with prejudice filters.

5. ROBUSTNESS

A crucial point in the implementation of the adaptive controller is the robustness of its identifier. To avoid distracting the identifier with uncorrelated data it is only made to operate as long as plasma is being formed with confidence. The controller action is however executed at all times, be it with fixed parameters when the identifier is not running. In order to protect the identifier from the strong uncorrelated influence on the density from neutral beams, radiofrequent heating, X-point formation and plasma touching the limiters, the following modification was made to equation (3).

$$\theta_{n+1} = \theta_n + K_{n+1}\cdot\frac{e}{1+\alpha|e|} \qquad (6)$$

where e stands for the identification error $y_{n+1} - \varphi^T_{n+1}\theta_n$ and α is a tuning parameter.

In order to safeguard the identifier from being distracted by high frequency poles and zeros, also so called prejudice filters were introduced as indicated in Figure 1.

6. RESULTS FROM FIRST ON-LINE TESTS.

The adaptive controller was brought on-line in May 1995 and was operational during 30 successive pulses leading up to the 1995 shutdown. The experience from its operation was very positive.

Results from pulse 35253 during which the adaptive controller was in control of the plant is here used to show the performance of the controller. At about 8 s into the pulse the X-point was formed, see Figure 2, which drains the plasma of particles. In order to avoid an excessive inflow of gas as a result of the anticipated controller reaction to this loss of density, a certain allowance was edited in the reference waveform. The large density peak following at 12 s is due to a short pulse of RF power, altogether a good test of the adaptive controller.

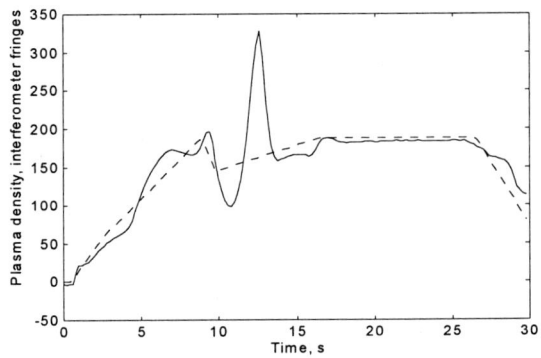

Figure 2: Reference signal, dashed line, and controlled plasma density signal, continuous line..

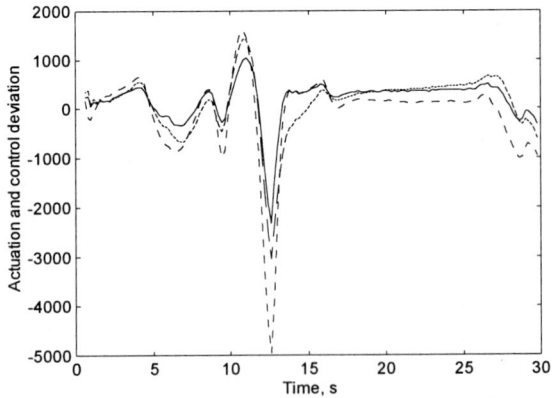

Figure 3: Plant actuation signal, dashed line. Plant actuation with new weighting function, dotted line. Control deviation magnified with a factor 30, continuous line.

Figure 3 shows the actuation in comparison to the plasma density deviation (scaled up a factor of 30 to show on the same plot). It is apparent how the simultaneous redesign of the controller has changed the "proportionality" to the deviation signal. The controller response to the X-point formation was, despite the allowance in reference waveform, a prompt and nearly full opening of the valve. The RF pulse caused an actuation 2.5 times the full range. (Negative actuations are limited to zero before being sent to the plant.)

Figure 4 shows the identification error relative to the real density signal. The large excursions in the beginning of the pulse when the identifier was engaged is partly due to the initial parameter values not being accurate and partly due to the real density signal being quite low. The relative identification error stays mostly within 5 percent, but increases of course during disturbances like the RF pulse. The identification accuracy seems to be sufficient as far as the pole placement is concerned. The estimated gain would if it were obtained more accurately and faster, possibly have led to a more accurate control in the stationary part, between 18 and 26 s, of the pulse.

what is tolerable for the controller as a whole. It is also conceivable that the X-point formation and the RF pulse caused some change in the plant giving rise to some true change in the identified transfer function.

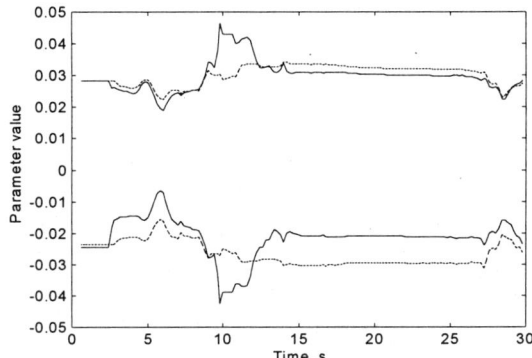

Figure 5: Numerator parameters as identified during the pulse continuous line and simulated with suggested identification error weighting function for the same pulse dashed line.

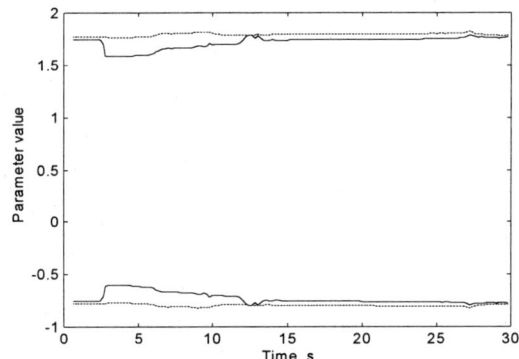

Figure 6: Denominator parameters as identified during the pulse continuous line and simulated with suggested identification error weighting function for the same pulse dashed line.

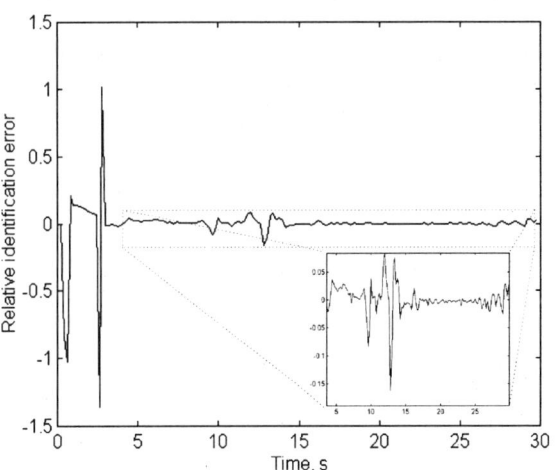

Figure 4: Relative identification error

Time graphs showing how the identified parameters associated with the numerator and the denominator of the transfer function developed during the pulse are given in Figure 5 and Figure 6 respectively. All parameters are of course affected by the X-point and the RF pulse but no worse than

The movements of the identified transfer function's zero and poles during the pulse are shown in Figure 7 and Figure 8 respectively. The pattern of the these zero and pole locations are as expected and are also comparable to what is experienced from all other pulses on which the identifier has been operated.

Finally, step responses from every fifth identified transfer function are presented in Figure 9. A fairly wide range of dominating time constants and gains is displayed. A few very long time constant responses are just distinguishable among the majority of responses which tend to show time constants around 4 s. It is also remarkable to see how the gain of the identified transfer functions varied by a factor two.

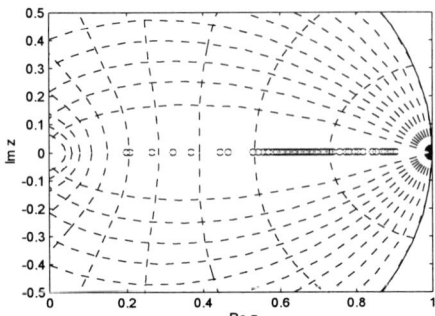

Figure 7: Locations of identified zeros during the pulse.

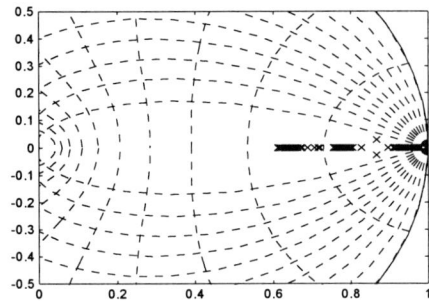

Figure 8: Identified pole locations during the pulse

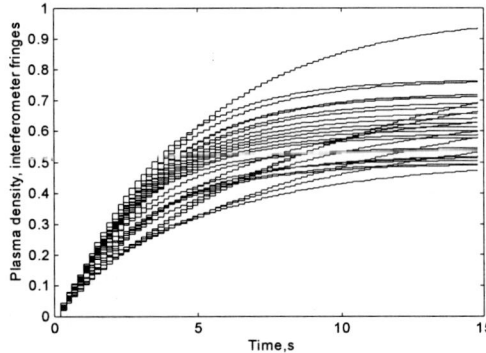

Figure 9: Unit stepresponses from every fifth identified model

7. IMPROVING THE CONTROL ACCURACY.

Even though the plasma density is nearly stationary at times, as between 18 and 26 s in the pulse above see Figure 2, parametric changes are most likely taking place throughout the pulse. Decreasing gas pressures in the reservoirs is experienced as loss of gain in the valves and also the condition of the vessel wall keeps changing. This ongoing drift of the parameters causes the identifier to lag behind the true parameter values. An average identification error of less than half a percent is just visible in Figure 4. Some drift in the identified parameters during stationary plasma density can be distinguished in Figure 5.

All identification errors greater than zero are weighted down in its present implementation, Equation 6, reducing the rate of the identification. It would be useful if this weighting function were lifted up to one for small errors, say a few fringes which seems to be adequate in this context, but still have a strong discrimination of large errors. For this purpose the so far used weighting function is replaced with the following:

$$\frac{1}{1+\delta \cdot e^4} \qquad (7)$$

where δ is a tuning parameter nominally set to 0.01.

The identifier run off-line on the plant signals from the same pulse with the new weighting function makes a significant difference to the identified parameter values as indicated in Figure 5 and in Figure 6. Consequently the controller design was influenced so that its plant actuation signal changed to what is shown as the dotted line in Figure 3. The new actuation suggests that the control deviation would have been reduced. It should be remembered though that this is the result from an off line open loop simulation and that all signals would have been different in a real closed loop case.

8. CONCLUSION

The adaptive controller appears to perform as specified in almost all aspects. It is stable, it does not show any oscillations or excessive overshoots and it settles within the prescribed time. The adaptive process shows robustness to disturbances and speedy recovery. However, the plasma density deviation to a constant reference is not completely eliminated, as can be seen between 18 s and 25 s in Figure 2. The remaining deviation is negligible from a practical operations point of view but is nevertheless of some interest and will be attended to further.

9. REFERENCE

1. H.E.O. BRELÉN, "An Adaptive Plasma Density Controller at Joint European Torus", Fusion Technology, Vol 27, March 1995.

Real Time Software for the Control and Monitoring of DIII–D System Interlocks*

J.D. Broesch, B.G. Penaflor, R.M. Coon, J.J. Harris, J.T. Scoville

General Atomics, P.O. Box 85608, San Diego, California 92186-5608, USA

This paper describes the real time, multi-tasking, multi-user software and communications of the E-Power Supply System Integrated Controller (EPSSIC) for the DIII–D tokamak. EPSSIC performs the DIII–D system wide go/no-go determination for the plasma sequencing. This paper discusses the data module handling, task work load balancing, and communications requirements. Operational experience with the new EPSSIC and recent improvements to this system are also described.

1. BACKGROUND

Diverse control and instrumentation requirements of tokamaks must include robust, reliable, easily maintained, and expandable interlock and protection capability. Such systems must possess efficient and flexible communications capability. These systems are typically distributed through a variety of high EMF environments. They must operate in both cooperative and independent modes.

A variety of subsystems are used to meet these requirements for DIII–D: Programmable Logic Controllers (PLCs); high speed embedded controllers; high performance VME based systems using both real time and non-real time operating systems, and main-frame computers.

EPSSIC's function is to provide shot sequencing, control of the main power switches, and to act as a system wide safety interlock. EPSSIC monitors all key DIII–D sub-systems. If a sub-system abort is detected, EPSSIC generates a system abort and safely shuts down power to the tokamak.

The EPSSIC portion of the control system consists of a VME based multi-processor system employing both a PLC and a 68030 processor. The high EMF environments and vibration found in the E-Power Supply building dictated the use of non-disk based operating systems. Additional requirements included the need to communicate with a variety of network based applications, implement redundant hardware and software interlocks to provide a very high degree of reliability, and to be easily operated remotely by control room personnel.

2. OVERVIEW

Figure 1 shows a high level view of where EPSSIC fits into the overall DIII–D organization. One important category of inputs is the analog voltage commands received from the Plasma Control System (PCS) [1–3]. The PCS commands plasma position and configuration via these analog voltage commands. EPSSIC routes these commands to the appropriate controllers.

Figure 2 shows the details of the EPSSIC block shown in Fig. 1.

There are three main functional areas of EPSSIC: Logic Evaluation/Control; Communications; and hardware interlocking. The logic evaluation and control functions are principally handled by the PLC. Communication, which includes the user interface to the control room, is handled by the CPU. Commands to the high power switches are generated by the PLC based upon inputs from the rest of the DIII–D sub-systems. These commands are validated by the hardware interlock module before being passed on to the switches. Should an error in the program occur that would command the switches to fire at a potentially dangerous point in the cycle or to assume invalid states, the switches will instead be automatically placed in a safe condition and a system abort is generated.

The real time control software is organized around two key concepts. The first is an absolute

---

*Work supported by U.S. Department of Energy under Contract No. DE-AC03-89ER51114.

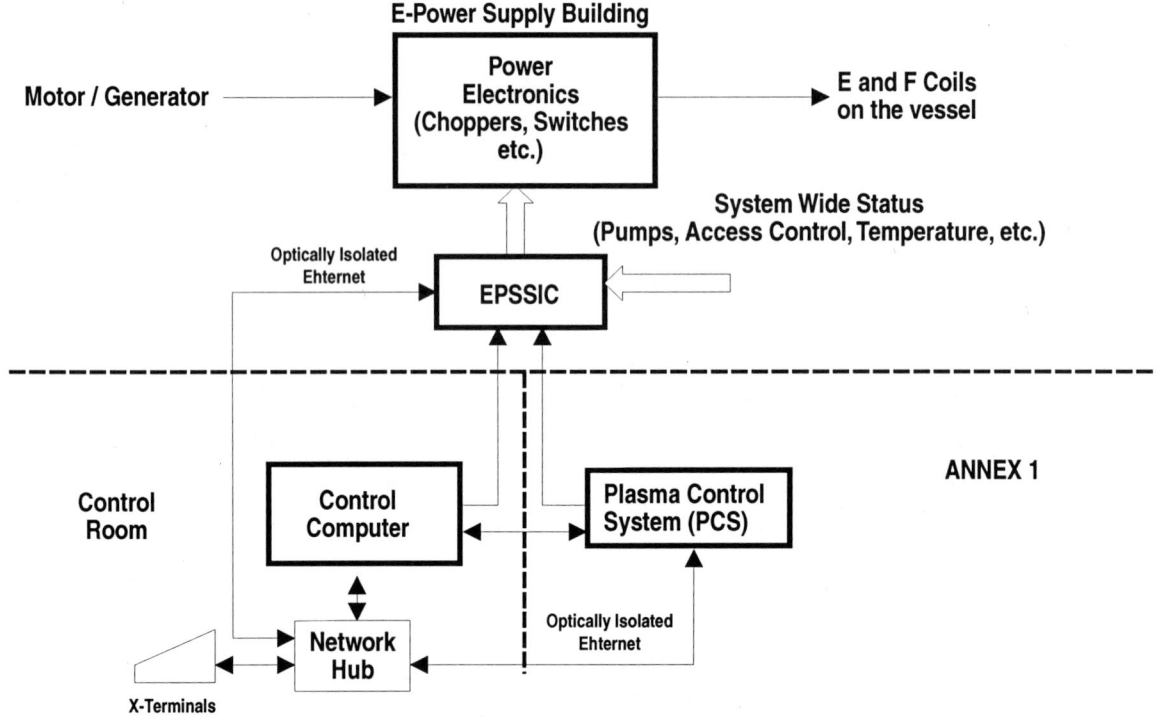

Fig. 1. High level view showing EPSSIC.

addressed memory module. This module is shared by the PLC and the 68030, and is the key communications path for inter-processor communications. The second is a relative memory addressed data module that resides in the 68030 memory space. This second data module provides inter-process communications for the OS-9 based real time control, monitoring, and data communications tasks running on the 68030.

The communications implementation is based on a standard Berkeley Software Distribution (BSD) Sockets interface, with appropriate modifications to support the disk-less operation.

3. SOFTWARE ARCHITECTURE

During operations, the data in the real time module is monitored in the control room and from other diagnostics sites. Further, our plan calls for both user monitoring of the data via simple telnet sessions and for automated monitoring of the real time data via the control computer. These considerations clearly dictated a multi-tasking environment.

OS-9 possesses the standard IPC resources such as signals, pipes, etc. However, it also possesses a more powerful resource for real time data exchange: the data module.

Physically, a data module is a contiguous block of memory that is allocated by the operating system. Logically, it consists of a header, a data area, and a cyclical redundancy check (CRC) based checksum. In most practical applications the CRC is ignored. The header consists of information used by the operating system. The data area is user defined. In their use data modules are similar to memory allocations using malloc. Data modules posses a user defined name that is available system wide. Once a data module has been created, a pointer to its data area can be obtained by making a system call using the data module name as a parameter. In effect, this area of random access memory (RAM) becomes a global memory resource, the pointer to which is globally available to all tasks. While the ability to pass memory objects is seen in other operating systems (for example, the Microsoft Windows 95 ability to lock a memory object and pass a handle), the modular structure of OS-9 makes the use

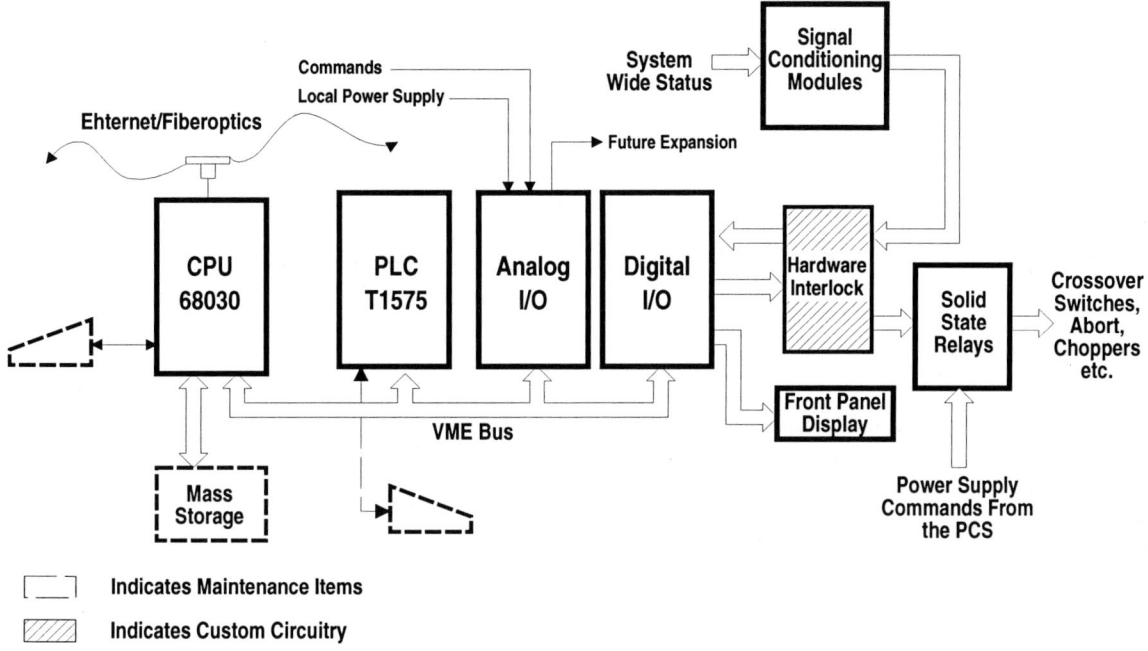

Fig. 2. Conceptual architecture for the updated EPPSIC.

and management of data modules particularly easy to use for real time data interchange.

The one disadvantage of this approach is due to the fact that pointers are used to access memory. No mechanism exists to strongly type check the variables among the various modules. It is the responsibility of the programmer, via the header file, to ensure variable coherency.

Figure 3 shows the high level software architecture. Note that daemons (in this paper, a daemon is defined as a console-less process) are used to perform the data updating between the data modules, I/O ports, and the PLC memory. Since this data is principally viewed by humans during and shortly after a shot, it was determined that a relatively long latency in updating the data module from the I/O and the PLC memory was acceptable. As general rule, 100 ms latency is nearly undetectable to a human operator. Therefore, the daemons are invoked periodically at a 50 ms rate, and the user interface routine is updated at a 50 ms rate. The worst case latency is therefore 100 ms and is sufficiently fast that it presents no significant delay in operator actions. This period is also sufficiently long enough that the system is not heavily loaded. Daemon execution time is approximately 2% of the overall CPU loading. The operator interface accounts for approximately another 10% of the processor bandwidth. This scheduling will be reviewed when the control computer interfaces are implemented; it is not anticipated, however, that a significant increase in the periodic execution rate will be required.

The actual scheduling of the daemons is accomplished by using the signaling capabilities of OS-9. Each daemon is set to respond to a signal. The system is then programmed to generate the signals periodically. At the end of each invocation each daemon voluntarily sleeps.

Due to the efficient inter-process communications, a number of users can be supported while ensuring a highly responsive user interface. The daemons have very little impact on the PLC ladder process accessing the I/O ports. This is important because the low latency allowed between inputs and outputs for certain key control events. Theoretically, the number of users is only limited by the maximum number of processes that the OS-9 system can support. In practice, we find that 68030 CPU can support three user logins executing the user interface routine without suffering any noticeable degradation in performance.

As noted above, approximately 12% of the processor bandwidth is used for user interface and daemons. Current loading analysis indicates that approximately 25% of the available CPU's cycles are being utilized. The other 13% appear to be system overhead, TCP/IP, etc. Empirical testing has indicated that with up to approximately 50% bandwidth utilization no deterioration in responsiveness is detected by the operators.

4. FUTURE WORK

Much of the remaining CPU capacity is targeted for implementing a shadow control algorithm. This algorithm will monitor the control functions of the PLC, and if it detects a failure in the PLC, step in and perform a controlled abort sequence.

An additional design effort is to interface EPSSIC with the tokamak control computer via the TCP/IP protocol. This will allow the operators in the control room to view EPSSIC status from a common and uniform interface screen on the control computer.

5. CONCLUSIONS

The ability to monitor the various system status in the E-Power Supply from the control room has greatly improved the ability to diagnose system problems. This improved capability has meant less down time and a higher number of shots per operating cycle.

REFERENCES

1. G. Campbell, *et al.*, "New DIII–D Tokamak Plasma Control System," *Proc. 17th Symp. on Fusion Tech.*, (1992).
2. J.J. Harris, *et al.*, "A Combined PLC and CPU Approach to Multiprocessor Control," *Proc. Symp. on Fusion Engineering* (1995).
3. B.G. Penaflor, *et al.*, "A Structured Architecture for Advanced Plasma Control Experiments," this conf.

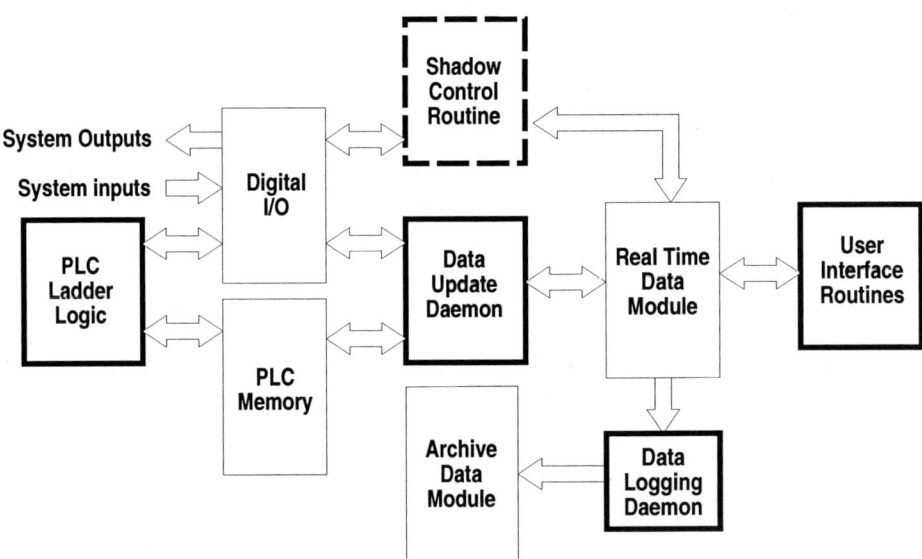

Notes: Thin boxes indicate memory blocks
Thick boxes indicate processes
Dashed boxes indicate future additions

Fig. 3. EPSSIC software architecture.

AN INTRODUCTION TO FASTCAMAC (60 Megabytes/sec in CAMAC?)[*]

Satish Dhawan [a], Charles Hubbard [b], Tim Radway [b], and Richard Sumner [c]

[a] Yale University Physics Department, 260 Whitney Avenue, New Haven, CT 06511
[b] Jorway Corporation, 27 Bond Street, Westbury, NY 11590
[c] LeCroy Corporation, 700 Chestnut Ridge Road, Chestnut Ridge, NY 10977

FASTCAMAC is a high speed block transfer protocol with data rates up to twenty times faster than normal CAMAC [1]. FASTCAMAC is compatible with the CAMAC data acquisition standard. The new protocol is presented as a short tutorial for the data acquisition user. Emphasis is on FASTCAMAC as an extension to CAMAC with discussion of compatibility, performance and system aspects.

1. INTRODUCTION

Despite the slow speed of CAMAC (IEEE STD 583), it is the most widely used physics data acquisition standard. There have been previous proposals[2] to increase the CAMAC transfer rate, but these have usually been proprietary or not compatible with standard CAMAC. None has achieved widespread acceptance by CAMAC users.

By a simple modification of the CAMAC protocol, the transfer rate of essentially all existing CAMAC crates can be increased by 2.5 times, to 7.5 Megabytes per second.

Measurements indicate that even higher rates are possible, up to 30 Megabytes per second with existing dataways, and up to 60 Megabytes per second, with improved dataways. These rates can be achieved while retaining complete compatibility with existing CAMAC. Old CAMAC and new FASTCAMAC modules can be mixed in the same CAMAC crate without interference.

2. STANDARD CAMAC

The CAMAC standard was designed before 1970, and is still widely used in the physics community. There have been few changes since then, the standard has been quite stable for more than 20 years.

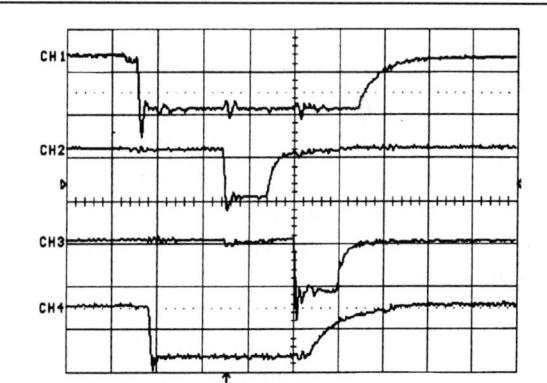

Figure 1. The standard CAMAC dataway timing. Ch1 is CAMAC BUSY, CH2 is S1, CH3 is S2, Ch4 is R1 (one of the read lines). The scales are 5V per division and 200 ns/division.

A CAMAC dataway data cycle involves the assertion by the controller of the N line, function code, and subaddress, followed 400 ns later by S1 and S2 (300 ns after S1), as shown in Figure 1. For a read operation, the CAMAC module must place the 24 bit wide data word on the Dataway before S1 and maintain it until S2, 300 ns later. This long "hold time" is not required by modern integrated circuits. This one microsecond dataway cycle (the state of the art in 1969) is very inefficient by contemporary standards

---

[*] This work is supported by the Department of Energy, Small Business Technology Transfer grant number DE-FG02-94ER86016

3. FASTCAMAC

The basic change to the CAMAC standard, which allows the increases in data transfer rates, is to lengthen the CAMAC dataway cycle, provide additional S1 pulses and use these pulses as additional data transfer strobes.

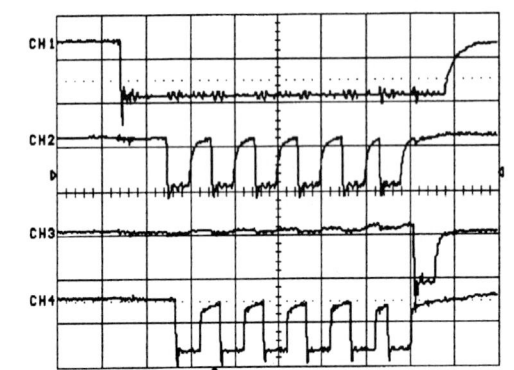

Figure 2. The FASTCAMAC dataway timing. Ch1 is CAMAC BUSY, CH2 is S1, CH3 is S2, Ch4 is R1. Eleven data transfers are shown. Data transfers occur on both edges of each S1 pulse. R1 changes after every transfer. The scales are 5V/division and 400 ns/division.

To provide compatibility with normal CAMAC, the 400 ns at the beginning of the cycle is unchanged, and conforms to the existing CAMAC specification. In a FASTCAMAC read command, the controller stores the data at the leading edge of S1 (or both edges for faster transfer rates). At the same time, the module begins to put the next word on the dataway. This next word is stored in the controller at the leading edge of the next S1 pulse. This process continues until a block of data has been read from the module. Figure 2 illustrates a FASTCAMAC read cycle. At the end of the data transfer portion of the cycle, an S2 strobe along with a end of cycle, as defined in the standard CAMAC timing, is provided for compatibility

Testing of existing equipment has shown that the present CAMAC dataways easily support multiple S1 strobes with 200 ns width and 200 ns spacing (400 ns leading edge to leading edge). Transferring data at the leading edge of each S1 brings the dataway transfer rate from 3 megabytes per second to 7.5 megabytes per second.

The rise time of the open collector dataway drivers (as required by the CAMAC specification) is a limiting factor in data transfer speed. A fully loaded CAMAC crate can have an RC (for the open collector pull up) time constant as large as 250 nanoseconds. Figure 3 shows a FASTCAMAC cycle with open collector drivers. This limitation can be removed if FASTCAMAC modules and controllers use modern tri-state bus drivers (which were unheard of in 1969). These provide nearly symmetric rise and fall times (comparable to the fall time of the open collector drivers).

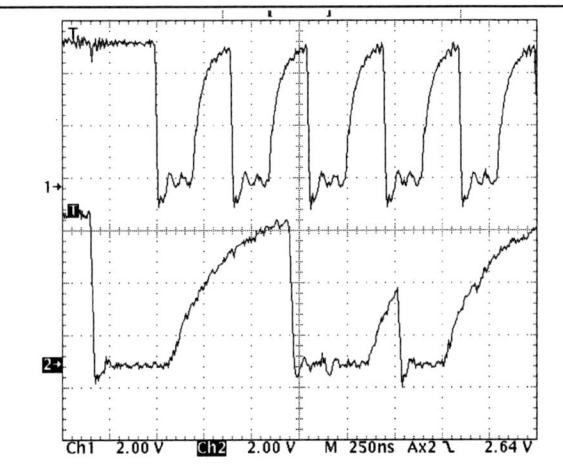

Figure 3. An oscilloscope record of a FASTCAMAC operation with open collector drivers. The data is strobed into the controller on both edges of S1. The data transfer rate is 15 Megabytes per second.

With tri-state drivers (shown in Figure 2, and in Figure 4) both edges of the S1 pulse can be used to produce data strobes. By using both edges of S1, the data transfer rate is increased to 15 Megabytes per second. By reducing, the S1 pulse to 100 ns width and 100 ns spacing, the data transfer rate is increased to 30 Megabytes per second.

The CAMAC dataway provides 24 read lines and 24 write lines, which are used in a mutually exclusive manner. FASTCAMAC treats the read and write busses as one 48 bit wide bi-directional data bus. Modules (and controllers) designed with this 48 bit capability can realize a twofold increase in transfer rates, to as high as 60 megabytes per second.

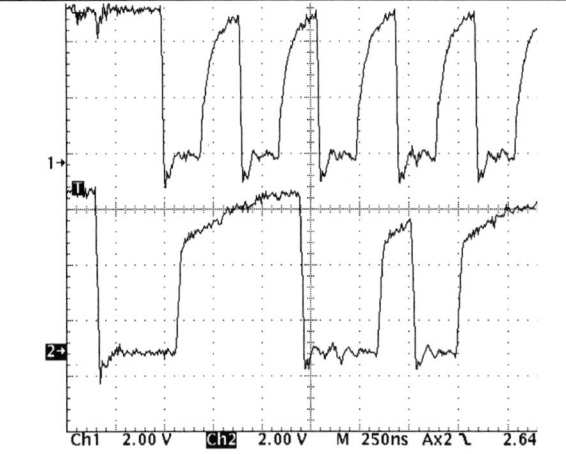

Figure 4. An oscilloscope record of a FASTCAMAC operation with tri-state drivers. The S1 pulses are 200 ns wide, with 200 ns spacing. The data is strobed into the controller on both edges of S1. The data transfer rate is 15 Megabytes per second.

Existing CAMAC dataway design is based on meeting the current performance specification. Not all existing dataway backplanes are suitable for the highest transfer rate, which will require 48 bit transfers and 100 ns S1 pulses. All existing CAMAC crates that we measured (even the very early ones with wire wrapped backplanes) will however operate at the lowest FASTCAMAC rate, single edge operation with 400 ns between data transfers, open collector drivers and 24 bit data width (7.5 megabytes per second).

Use of these techniques described above will enable the CAMAC user to transfer large blocks of data at speeds well above that of a normal CAMAC system. The rates that can be achieved with FASTCAMAC are summarized in Table 1.

FASTCAMAC also defines a multiple module protocol. Many CAMAC controllers have an N register which allows more than one module to be addressed simultaneously. A position register in each FASTCAMAC module defines the order of participation in the multiple module dataway operation.

Table 1
Rates Achievable with FASTCAMAC

output drivers	dataway width bits	# of edges	S1 width ns	S1 spacing ns	transfer rate MBy/s
open collector	24	1	200	-	3
"	24	1	200	200	7.5
tri-state	24	2	200	200	15
"	24	1	100	100	15
"	24	2	100	100	30
"	48	1	200	200	15
"	48	2	200	200	30
"	48	1	100	100	30
"	48	2	100	100	60

As an a example, consider a read operation with a variable (but small) amount of data in 20 modules in the same crate. The FASTCAMAC controller asserts the N lines for all 20 modules and begins the dataway cycle. The position register in each module has been loaded with its position in the cycle, 0 through 19. The module with 0 in its position register responds with data for the first S1, and asserts Q. When its data is exhausted the module asserts Q=0. The controller sends an S2 pulse, then more S1 pulses. Each module counts S2 pulses to determine when it is time to send data. After the first S2, the module with position 1 responds, and so on until the total number of S2s (equal to the number of modules participating) have been sent by the controller, and the CAMAC cycle ends.

4. FASTCAMAC LEVELS

FASTCAMAC has been designed as a series of levels, with increasing speed and capabilities.

Level 1. Define multiple S1 pulses, with 200 ns width and spacing, transfer 24 bit data on the leading edge only, and use open collector drivers on the dataway. The maximum rate is 7.5 megabytes per second. The advantage of this simple entry level is the ease of implementation. Nearly all current CAMAC modules, which support block transfers, could be modified to satisfy this protocol, with minimal effort.

Level 2. Add the tri-state drivers only for the FASTCAMAC mode (open collector drivers will still be used for normal CAMAC operations), allow transfers on both edges of S1, and allow shorter S1 pulses. The addition of new tri-state drivers is probably impractical for most existing modules. It is a simple change to incorporate in new designs, however. The maximum rate becomes 30 Megabytes per second for 100 ns S1 pulses.

Level 3. Add the capability for 48 bit bus widths. This is the most difficult to achieve, and requires some improvement in the dataway for successful operation at the highest speeds. The maximum data transfer rate is 60 Megabytes per second.

The capability for the multiple module block protocols can be added at any FASTCAMAC level, as an optional feature.

To be compatible with FASTCAMAC, a module must implement at least level 1. This provides a substantial performance boost with minimal effort. Level 1, which can be retrofitted easily into many existing module designs, is selected by a unique function code. The higher levels, are selected by a combination of a function code and a control register in the module.

5. AUXILIARY FASTCAMAC CONTROLLERS.

Most existing CAMAC crate controllers have provisions for the auxiliary controller defined in the CAMAC standard. The FASTCAMAC protocol can be implemented in an auxiliary controller, allowing continued use of existing normal CAMAC controllers and interfaces to host computers. The data collected by the auxiliary controller must either be buffered inside the auxiliary controller for later readout, or output on a high speed link to a module in another crate (not necessarily a CAMAC crate). A differential ECL bus such as the LeCroy FERA bus, or an RS-485 bus (as used at Fermilab, for example) are obvious candidates for which destination memory modules already exist.

6. SUMMARY

Improved performance of existing CAMAC systems can be obtained by adding FASTCAMAC components only where the performance is needed. Complete replacement of existing CAMAC systems is not required to obtain the substantial increases in data rates that FASTCAMAC provides. Measurements made with existing crates and dataways show that essentially all existing crates and dataways will be able to support transfer rates 2.5 times normal CAMAC, while still using open collector drivers (Level 1 FASTCAMAC). By using tri-state drivers (Levels 2 & 3), transfer rates up to 20 times faster than normal CAMAC are achievable.

The commercial potential of FASTCAMAC is difficult to predict, but the most often cited reason for not using CAMAC is that it is not fast enough. High speed digitizers with large record sizes are rarely found in CAMAC. However,. sensitive analog circuits, such as high resolution ADCs are easily produced in the CAMAC format, with its multiple power supplies and low noise characteristics. FASTCAMAC will make CAMAC suitable for a wider range of applications, and ensure its future usefulness as an easy to use data acquisition standard.

Further information about FASTCAMAc, including the complete preliminary specification, is available at http://www.yale.edu/fastcamac on the internet.

---

[1] CAMAC, A Modular Instrumentation System for Data Handling, EUR 4100 e, March, 1969
Modular Instrumentation and Interface Standards (CAMAC), ANSI/IEEE Std 583-1982

[2] For Example:
COMPEX, Compatible extended use of the CAMAC Dataway, EUR 8500 EN, 1984
The Smart Crate Controller (SCC) designed at Fermilab, and manufactured by BIRA Electronics

A Structured Architecture for Advanced Plasma Control Experiments*

B.G. Penaflor, J.R. Ferron, M.L. Walker

General Atomics, P.O. Box 85608, San Diego, California 92186-5608, USA

Recent new and improved plasma control regimes have evolved from enhancements to the systems responsible for managing the plasma configuration on the DIII-D tokamak [1]. The collection of hardware and software components designed for this purpose is known at DIII-D as the Plasma Control System or PCS [2]. Several new user requirements have contributed to the rapid growth of the PCS. Experiments involving digital control of the plasma vertical position have resulted in the addition of new high performance processors to operate in real-time. Recent studies in plasma disruptions involving the use of neural network based software have resulted in an increase in the number of input diagnostic signals sampled. Better methods for estimating the plasma shape and position have brought about numerous software changes and the addition of several new code modules. Furthermore, requests for performing multivariable control and feedback on the current profile are continuing to add to the demands being placed on the PCS.

To support all of these demands has required a structured yet flexible hardware and software architecture for maintaining existing capabilities and easily adding new ones. This architecture along with a general overview of the DIII-D Plasma Control System is described. In addition, the latest improvements to the PCS are presented.

1. INTRODUCTION

The architecture designed for the PCS has been demonstrated to provide a reliable framework which has been well suited for the implementation of numerous control schemes. Current capabilities of the PCS include feedback control of various discharge attributes such as plasma shape and position, total plasma current, plasma energy, particle density, magnetic field error correction, loading resistance for the rf antennas, and amount of radiation from the plasma. In order to achieve feedback control in the PCS, a number of tasks are required, including processing user inputs, synchronizing with the DIII-D discharge cycle, sampling data from the tokamak, performing the real-time feedback calculations and sending the necessary control commands to the various tokamak "actuators" or output control devices such as the magnetic coil power supplies, and gas valves (Fig. 1.).

2. RUN TIME SYSTEM

The primary users of the PCS are DIII-D physicists responsible for defining the characteristics of the discharge. From the standpoint of users, the PCS is a single application. In actual operation, the PCS is comprised of several programs or "processes" active at run time. This PCS run time system is organized into three types of processes which include, the user interface, the real-time feedback control and the coordinator processes which synchronize the PCS with the discharge cycle.

2.1. User Interface

For any given discharge there are literally hundreds of parameters required for defining a plasma configuration. To simplify the task of specifying these values, a graphical user interface to the PCS has been developed.

The primary type of input to this interface is a generic construct called a "waveform". Waveforms specify values which are to be used by a set of feedback control routines. Waveforms are entered by users onto a two dimensional display grid showing the discharge time, and desired input as it may evolve over this time period. There are hundreds of possible waveforms which can be modified in the user interface. In a typical discharge, most values remain untouched by the user or are simply loaded in from an archive of a previous discharge. Some examples of waveform data include target values for

---

*Work supported by U.S. Department of Energy under Contract No. DE-AC03-89ER51114.

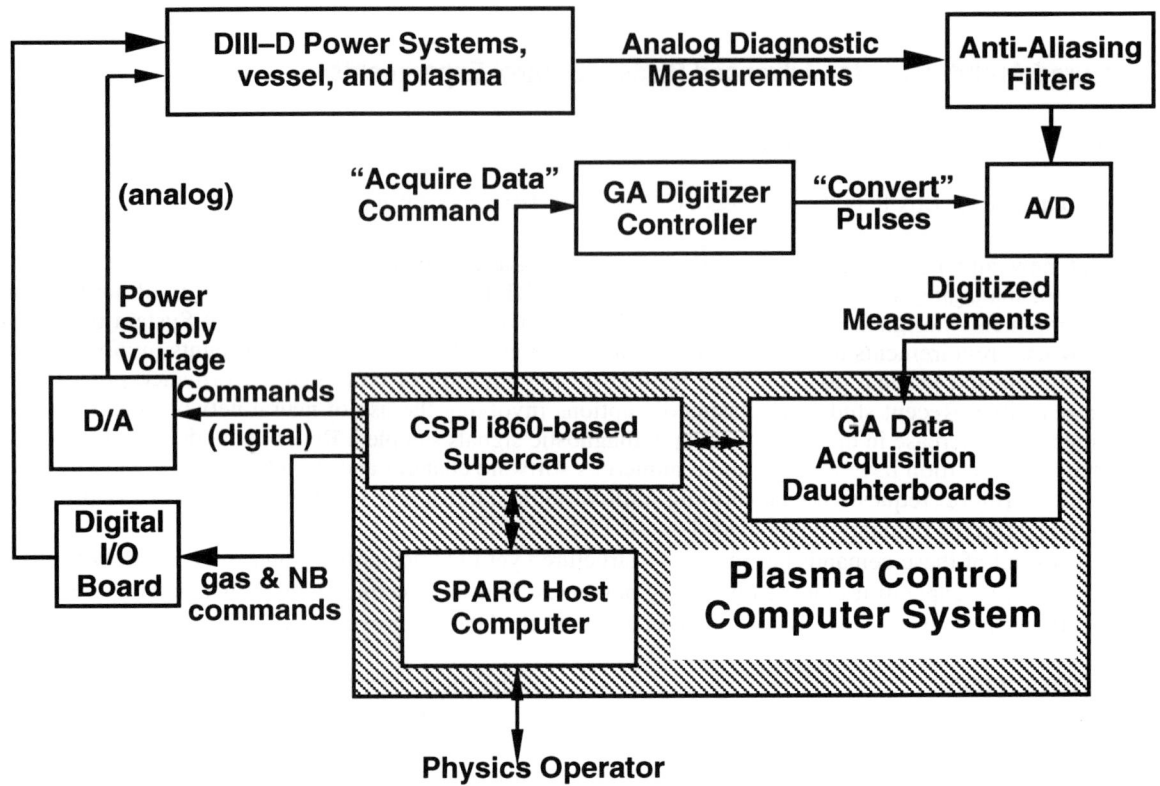

Fig. 1. DIII–D digital plasma control system (PCS) block diagram.

specifying the plasma position (inside gap distance, top gap distance, vertical position), desired density, gas flow rate, and beam modulation.

Waveforms are grouped according to the "algorithms" to which they supply inputs. An algorithm is a collection of one or more routines which execute in real-time to perform a specific function. One basic type of algorithm function is feedback control. There are numerous algorithms implemented in the PCS which serve this function. The most common are algorithms for achieving a desired plasma configuration. Examples are the algorithms for creating single and double null divertor shapes. Other kinds of control include neutral beam modulation and density feedback. Algorithms can also be used to perform calculations in real-time or execute tasks such as collect the base line data for the input signals.

The method for keeping related algorithms together in the PCS is the "category". A category is a grouping typically based on the type of actuator involved. For example, the plasma shape category groups algorithms responsible for performing feedback control for the plasma shaping coils. Another category for gas algorithms includes all routines which control the gas valves. The current implementation of the PCS contains eleven categories which include shape, density, gas, plasma current, power supplies, error field correction coils, rf, neutral beams, equilibrium, alarms and current profile.

2.2. Real-Time Subsystem

The real-time component of the PCS employs several high speed processors running in parallel to perform the calculations required for feedback control. Each real-time processor is responsible for performing one or more specific tasks. In the current PCS setup, there is a processor assigned to the primary shape control, one which serves as the master for triggering data acquisition on all of the processors and also performs vertical position control, another dedicated to running an algorithm for predicting plasma disruptions, and a set of processors which are used in a real-time equilibrium reconstruction calculation.

Each processor is capable of communicating with any other over a single VME bus. The processor for primary shape control sends its calculated commands to the processor for vertical position control on each control cycle. The processors for equilibrium calculations also communicate various pieces of information between themselves and the other processors.

The master processor runs at the highest speed of approximately 60 µs per cycle. The processor running primary shape control, has a cycle rate of about 400 µs. Data sampled from the tokamak for use in feedback calculations is available to each of the processors. The data is sampled at the rate of the fastest processor and written directly into the memory of each processor and made available on each cycle.

During a control cycle each processor performs the following tasks. A processor first acquires its latest set of input data from the tokamak after a trigger to start data acquisition from the master has arrived. The errors for a specific type of control algorithm are computed by comparing user specified target values against actual measured values from the input data. If specified, a routine to perform some transformation of the error values such as a Proportional Integral Derivative Gain filter is executed. The resultant error along with other parameters input by the user such as gains from a matrix, an output offset or cutoff values are used to derive the commands that are sent back to the tokamak for control.

2.3. Coordinator Processes

The PCS is designed to run synchronously with the DIII–D discharge cycle. A separate computer system is responsible for setting triggers which cause the PCS to transition through different discharge cycle states. At each of these states the PCS performs tasks specific to the current state it is in, such as processing inputs, locking out users from making any more changes to the discharge parameters, setting up and running the real-time feedback control, and archiving data collected once a discharge is complete. The tasks associated with each of these states and the transitioning between states is accomplished by a group of coordinator processes.

A waveform server process, or "waveserver" is used to gather and coordinate raw inputs from one or more user interfaces and convert the inputs into data needed by the real-time computers. The waveserver process manages the entire set of discharge parameters and supplies the latest information about the parameters to other requesting processes.

A lockout or "lockserver" process is responsible for coordinating all of the PCS processes to synchronize with the DIII–D discharge cycle and to transition the PCS from one state to the next. The lockserver monitors the triggers which are sent from other DIII–D computer systems.

A set of routines referred to as the "host real-time client" processes, executes shortly before the discharge to load information from the waveserver into the memory of a single real-time processor. These routines are also used to start and monitor the processing on the real-time processors at the start of a discharge. At the end of the discharge, the routines evaluate and report the return status from the real-time computers and archive the results obtained.

3. HARDWARE

Hardware for the PCS consists of the computer needed for the run time system and the interface between the PCS and the tokamak. The user interface requires an X display terminal or workstation. A unix based SparcStation server is used to run the coordinator processes. The computer for the real-time routines is the SuperCard-2 manufactured by CSP inc, a VME format, single board computer based on the Intel i860 RISC-design microprocessor.

The input from the tokamak to the PCS comes from 208 analog signals that originate from various places within the DIII–D vessel. The types of signals sampled include inputs from flux measurements, magnetic probes, measured poloidal field coil currents, chopper voltages, bus voltages, ohmic heating coil currents, soft X–ray signals, and loop voltages. The analog signals are digitized using eight channel DSP Technology TRAQ digitizers. Data from the digitizers is dumped into the memory of the real-time computers using a General Atomics custom designed data acquisition daughter board.

The output ifrom the PCS to the tokamak is through a set of five DATEL Inc., eight channel D/A converters and a VME microsystems digital I/O board. The D/A converter channels include outputs to poloidal field coil power supplies, channels for control of the rf transmitters, channels for the C supplies, a single E supply channel, d.c. power supply channels, gas valve channels and channels reserved for future use. The digital I/O board ports are used to send commands which control neutral beams, gas wave enables, a lithium pellet injector and an oak ridge pellet injector.

4. SOFTWARE

The PCS software consists of a collection of over 500 source files written mostly in C, assembly and a separate high level language for implementing graphical user interfaces. The source is organized into an infrastructure library, installation specific code and application specific code.

An infrastructure library contains code for implementing a generic real-time control system. Contained here are a number of routines which can be used in defining basic user interfaces and server processes. The installation specific source contains the hardware dependent specifications which would vary across different installations of the PCS. Different installation attributes which can be specified include the number of real-time computers used, number and types of input diagnostic channels to be sampled, and characteristics of the outputs such as number of D/A converter channels and purpose for each. The application source includes the code for specifying the types of control categories and algorithms available with the PCS.

The PCS software architecture is built upon a framework consisting of "master" files. A master file contains the definitions and source code for implementing specific functionality. Three kinds of master files are used. These include the category, the algorithm and the cpu masters. Each of these master files contains most of the information necessary for implementing new categories or algorithms or adding new real-time processors to the PCS.

5. RECENT DEVELOPMENTS

The PCS has been upgraded to employ six high speed real-time processors. A number of new diagnostic input channels have been added including signals to provide better information for determining the precise plasma shape and position. Approximately fifty different control algorithms are now available.

Digital vertical position control of the plasma tested in late 1995 has been incorporated and made available for everyday operations use. Steady progress has been made toward true multivariable control of plasma shape and position.

An algorithm known as "real-time EFIT" or rtefit which provides real-time estimates of flux values at chosen poloidal (r,z) locations has been implemented and tested on DIII–D. The rtefit algorithm is based on the EFIT code [3] which has been used for plasma equilibrium reconstruction at GA for several years. Data from flux loops, magnetic probes, and Motional Stark Effect (MSE) channels, are processed to produce the estimated plasma equilibrium.

A control technique known as isoflux control [4] which controls flux at designated poloidal "control points" has also been implemented and used for real-time plasma shape control. This technique exploits the improved accuracy in plasma shape estimation (as defined by flux contours) available from the rtefit algorithm.

6. SUMMARY

A structured system architecture for implementing advanced plasma control experiments has been described. Working within this framework, users at DIII–D have been able to demonstrate a number of new and different types of control capabilities, ranging from digital control of the plasma position to control based on flux at poloidal control points. With more enhancements, and even more control requirements being generated each day, the need for maintaining a structured and well defined system architecture for the DIII–D Plasma Control System increases in importance. A structured architecture for the PCS has resulted in a control system which has proven to be highly reliable despite undergoing numerous changes.

REFERENCES

1. J.L. Luxon, et al., Plasma Phys. and Contr. Fusion **32**, (1990) 869.
2. J.R. Ferron, et al., proc. 16th Symp. on Fusion Engineering, Illinois, (1995).
3. L. Lao, et al., Nuc. Fusion, **25** (1985) 1611.
4. J.R. Ferron, et al., in Bull. of the Amer. Phys. Soc. **40** (1995) 1791.

Availability analysis of five years of operation of the superconducting tokamak Tore Supra

D. van Houtte, B. de Gentile, C. Grisolia, J.Y. Journeaux, P. Joyer, J.M. Laurens, G. Martin, F. Parlange, T. Wijnands

Département de Recherches sur la Fusion Contrôlée / Association Euratom-CEA
CEA/CADARACHE, 13108 Saint-Paul-lez-Durance, France

The superconducting tokamak Tore Supra (TS) is in operation since march 1988. During this period, operation was interrupted for commissioning, new components installation or failure of operating systems. An analysis of the operation calendar of TS is made by comparing the number of actual days with the number of potential operation days. During one day, the actual experimentation time is compared with potential time reduced by various system faults. Then some availability indicators over the last five years of TS are given

1. INTRODUCTION

The main goal of Tore Supra (Fig. 1) is to open the way for controlling long pulse high performance discharges (up to steady state). Experiments were carried out with plasma currents up to 2.1 MA and pulse lengths up to 120 s [1]. This paper presents an overview of operational experience and reports the most relevant events which hampered the experimentation time of Tore Supra. About one down-time per year (8 between 1988 and 1996) is scheduled for implementation of planned new components combined with maintenance. Major technical problems concerned a failure of a toroidal coil (1988-89) and faults of an OH power supply transformer (1988 and 1995). The total number of leaks in vacuum chamber during the tokamak operation was 28 [2].

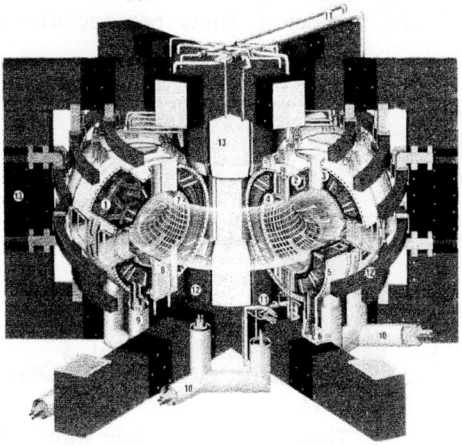

Fig. 1 : the tokamak TORE SUPRA

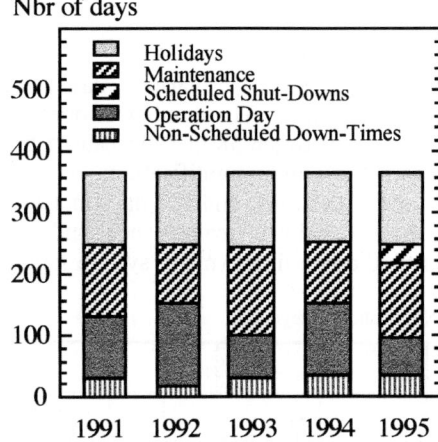

Fig. 2 : distribution of days during a year

About half of them, water leaks mainly (all plasma facing components such as limiters or antenna protections are actively cooled by pressurized water), required immediate shut-down (Fig. 3).

2. MACHINE OPERATION

Tore Supra is in operation 4 days per week, from tuesday to friday. The number of real operation days per year is limited by [3] :
- **Maintenance**,
- Installation of new systems or components during **Scheduled Shut-Downs**,
- **Non-Scheduled Down-Times** due to unexpected system faults,
- **Holidays**.

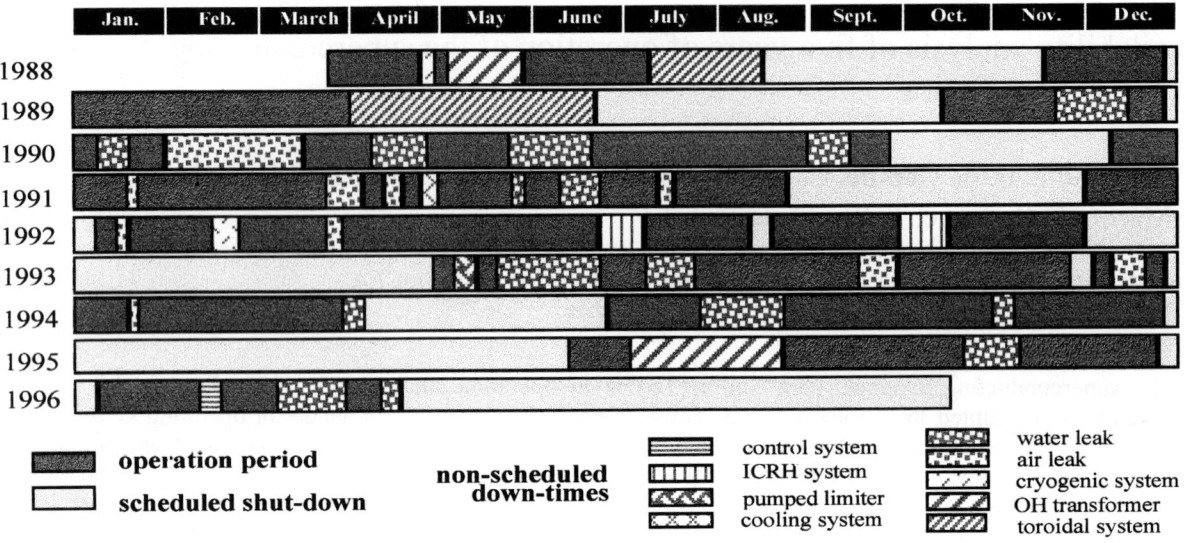

Fig. 3 : Operation calendar of TORE SUPRA

The repartition of these various day patterns is shown in the figure 2. The number of operation days per year varied between 58 in 1995 caused by a long shut-down for major modifications and a severe failure of a transformer, and 137 days in 1992 due to a very short shut-down (maintenance) and only short non planned down-times (major system failure).

The number of discharges carried out per year varied between 2442 (1300 successful plasma shots) in 1995 and 3389 (2397 successful plasma shots) in 1994 (Fig. 4). The operation time per day is about 12 hours. One and half hour is needed before this operation time to decrease the toroidal coil temperature from 2.1 K to 1.7 K and to raise the current in coils. Since 1992, the mean number of successful shots has increased from 17 up to 25 shots per day in 1995. The time interval between two discharges is about 10 minutes (a minimum delay of 4 is required by the poloidal system) to 20 minutes, depending in most cases on the number of diagnostics in operation (Fig. 5). The worst mean delay between two shots is 20 minutes in 1992. Over the years this delay tends to decrease and is about 15 minutes in 1995. The number of shots per operation day is a

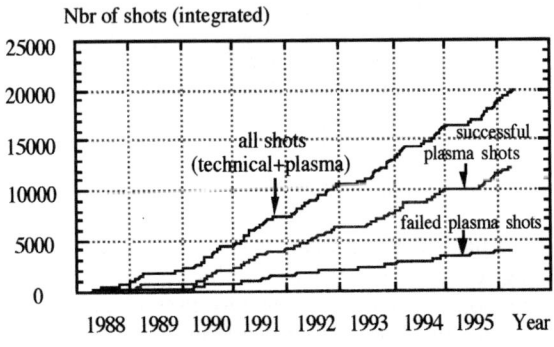

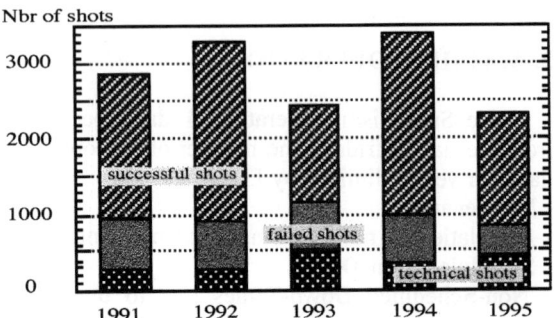

Fig. 4 : number and repartition of shots types

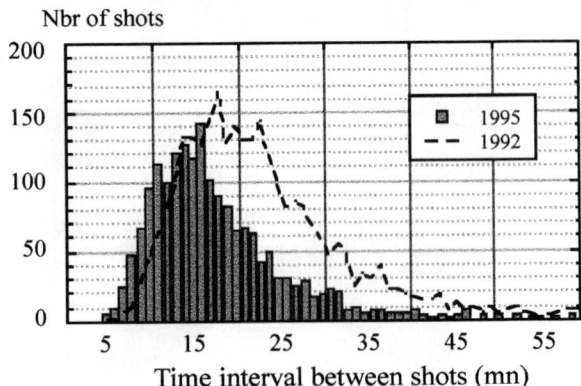

Fig. 5 : time interval between plasma discharges

global criterion of the reliability of a machine : down times during an operation day, hour of the first shot and delays between shots are taken into account at one and the same time. In 1995, 31 plasma shots per day were carried out when only 22 shots was realized in 1992.

3. BASIC SYSTEMS FAULTS

Operation of TORE SUPRA involved many systems [4]. The main systems which are necessary to obtain plasma discharges are :
- *Toroidal* : super-conducting magnet (Nb-Ti cooled by superfluid helium at 1.8 K), producing a magnetic field in the plasma center of $B_T(0) \leq 4.5$ T,
- *Cooling* : first wall actively cooled by pressurized water (220° C, 3 MPa),
- *Poloidal* : nine poloidal field coils with 100 MVA power ($I_p \leq 2.5$ MA),
- *Heating* : 20 MW of lower hybrid (3.7 GHz), ion cyclotron (35 to 80 MHz) and neutral beams (100 keV) powers,
- *Machine & vacuum* : the vacuum vessel is circular with an ergodic divertor. The plasma ring has a major and a minor radius of R=2.4m and a=0.8m.
- *Diagnostics and Acquisition* : real time processing

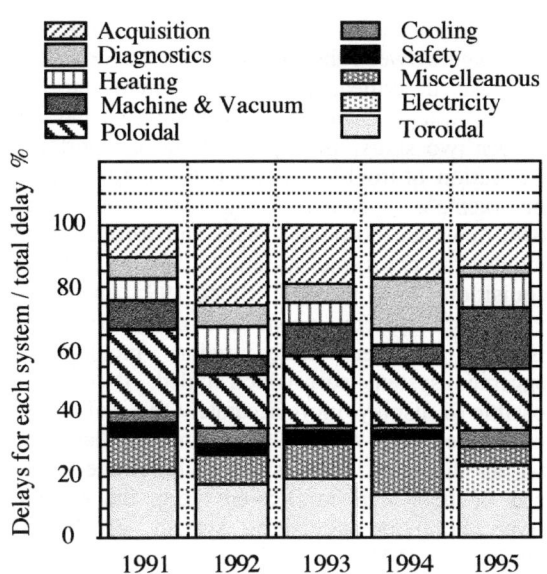

Fig. 6 : repartition of down-times between systems

of more than 30 diagnostics,
- *Electricity*,
- *Safety* : access control to the Tore Supra hall and radiological survey.

During operation days the events which hampered

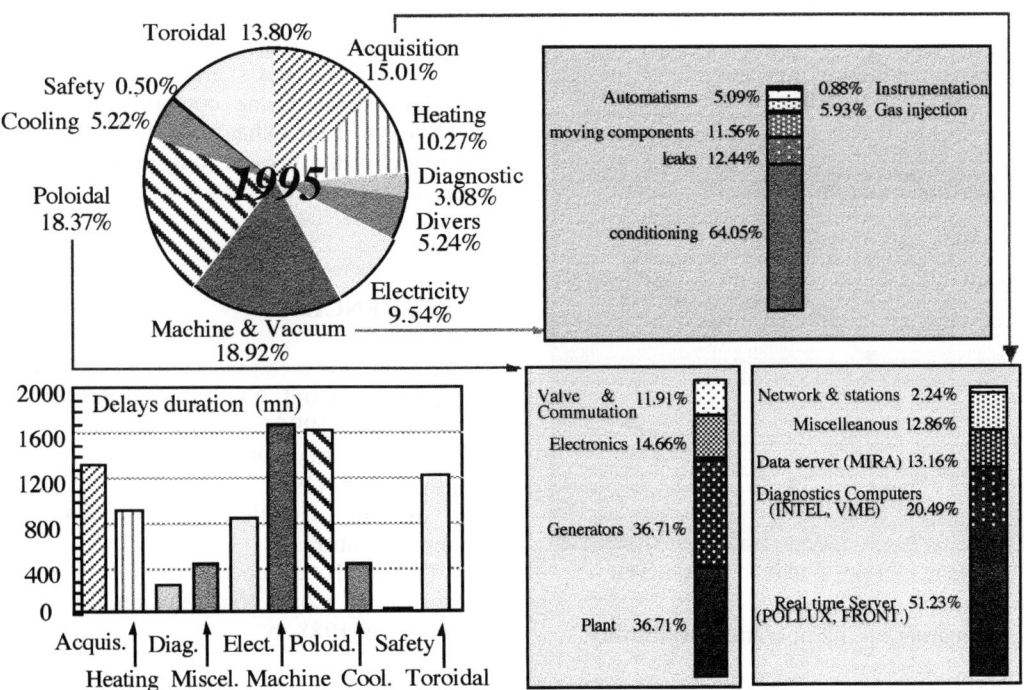

Fig. 7 : repartition of down-times between main systems in 1995 on Tore Supra

the experimentation time of Tore Supra are distributed between the various Tore Supra systems previously described. Technical problems are daily reported when down-times exceed the usual time between two shots. In Fig. 6, we have reported the distribution of troubles between the main systems. For exemple, systems which have caused most down-times are poloidal (26%) and toroidal which includes the cryogenic system (22%) in 1991, and acquisition (25%) in 1992. Since 1995, in addition to involved system, down-time duration and last discharge number, daily report includes the concerned system part which failed. It was thus possible to display, more accurately the real cause of down-times (Fig. 7). In 1995 an analysis has shown that for the main faulty systems only one or two parts of them were implicated : e.g. the real time server (51%) for acquisition system, or the wall conditioning (64%) for machine & vacuum system.

4. AVAILABILITY INDICATORS

An availability analysis (Fig. 8) of the Tore Supra tokamak was made over the last five years of operation.
On the one hand, the number of real operation days during the last five years is compared with the number of potential operation days : the ratio remains near 75% over the years. The worst result is 63% in 1995 due to major problems (OH transformer fault and water leak on a neutraliser of the ergodic divertor) during a short operation period of 6 months only. On the other hand, during one operation day, various systems faults reduce the experimentation time and lead to an actual experimentation time smaller than the full duration of the day : the proportion of down-times decreases from 33% in 1991 down to 22% in 1995, indicating an increase in the reliability of Tore Supra with operational experience.

At last, correlated with the increase of the experimentation time, the number of discharges carried out per day increases over the years. The number of shots per day is a good indicator for the machine operation, but it does not indicate the « quality » of the plasma discharges. However in accordance with the previous analysis, the number of the successful shots increases over the last four years.

5. CONCLUSION

Analysing these datas shows a good machine reliability over the years and an availability increasing from year to year. Except for a short-circuit in the early days of operation of Tore Supra, the super conducting toroidal magnet very satisfactorily works. In fact, the main down-time cause over the years are water leaks on actively cooled plasma facing components, indicating that continuous heat exhaust is the main technological difficulty of the present long pulse high power large tokamaks.

REFERENCES

1. Equipe Tore Supra, 16th Int. Conf. on Plasma Physics and Cont. Nucl. Fusion Research, Montreal, 7-11 october 1996.
2. Tore Supra Team, Fusion technology, Vol. 29, july 1996, p 417.
3. B. de Gentile and F. Parlange, private communication.
4. Equipe Tore Supra, 12th Int. Conf. on Plasma Physics and Cont. Nucl. Fusion Research, Nice, 12-19 october 1988, p 9.

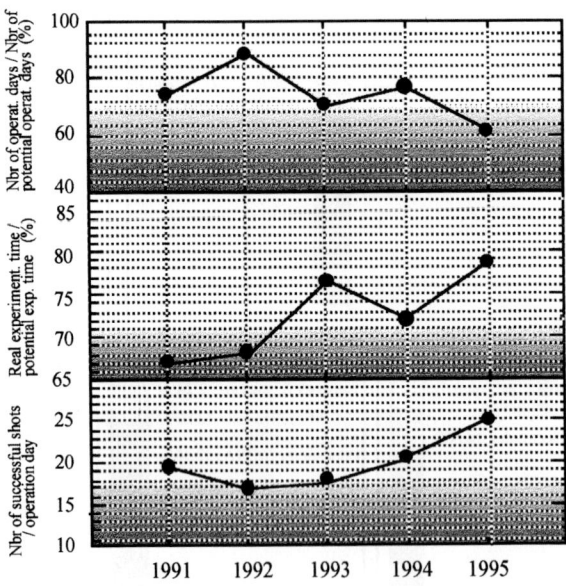

Fig. 8 : indicators of TS operation availibility

The Frascati Tokamak Upgrade machine : plant and operation status

S.Ciattaglia, B.M.Angelini, G.Buceti, F.Crisanti, F.Gravanti, G.Mazzitelli, M.Panella, E.Sternini, A.Tuccillo, V.Zanza

ENEA, Associazione Euratom-ENEA sulla Fusione, C.R. Frascati, CP 65, I-00044, Frascati, Rome, Italy

This paper presents a general overview of the Frascati Tokamak Upgrade (FTU) machine. The status of the major systems and the relevant upgrades are discussed together with a few data on the operation status and the operating experience, and the near term improvements and programme.

1. INTRODUCTION

FTU is a compact medium-high-field tokamak machine (R=0.93m, A=3 $B_T \leq 8T$, $I_p \leq 1.6MA$) cooled by liquid nitrogen (LN) to limit the electrical power and energy requirements and to profit by higher mechanical properties of the structural materials.

FTU operates over a wide line average density range (0.2-3 10^{20} m^{-3}) with $Z_{eff}<2$.

It became fully operational during February 1990.

From then till July 1994, experimental campaigns were carried out with a few short shutdown periods devoted to maintenance, repair and commissioning of new equipment, mainly diagnostic and radiofrequency (RF) equipment.

From July 1994 till May 1995 a major shutdown was performed in order to install a toroidal limiter (TL).

Other two RF structures and new diagnostics were installed and a major general revision, inspection and maintenance of systems and components have been performed.

The main control system was largely redesigned.

Up to the middle of September 1996, 10800 pulses have been done in five year operation: about 100 days a year dedicated to the experimental campaign, 10 h/day with about 20 minutes between pulses.

2. THE FTU SYSTEMS: STATUS AND MAIN IMPROVEMENTS

2.1 In-vessel components and vacuum systems

The vacuum vessel (VV) is a compact, fully welded, stainless steel (AISI 304 LN) structure. The inboard wall is completely shielded by the TL made of TZM tiles [1] introduced in the 94-95 shutdown. The tiles are bolted on 12 equal sectors of a stainless steel structure fixed to the VV. Three sectors are equipped with termoucouples and rogowski coils for temperature, power load and halo current measurements. A metal poloidal limiter is used to protect the outboard wall of the VV. It consists of one halve supporting a set of mushroom-shaped pieces of different material according to the experimental programme.

The total volume of the plasma chamber and ports is 2.5 m^3. The outgassing surface is around 95 m^2 from VV walls and in-vessel components, and the extensions of VV (ports, waweguides) save diagnostics. The VV is evacuated by six turbomolecular pumps with a pumping speed of 2200 l/s for N_2.

Up to now the reliability of the vacuum system has been pretty good and the working pressure, at LN temperature, is 4-5 10^{-8} mbar.

2.2 Fuelling system

The gas introduction is performed by the gas puffing method and, where required from the experimental programme, by pellet injector. In the gas puffing six piezoelectric fast valves, controlled by the automatic sequence of the pulse, inject the request fuel amount into the VV. A storage system may supply purified fuel as D_2 or H_2, and He. A high speed (up to 2.4 km/s) D_2 single-pellet injector has been in operation in 1992 during the ohmic plasma campaign [2]. A single stage multipellet, able to inject D_2 pellets up to 1.3 km/s, has been installed and tested in 1993-94.

2.4 Electrical power supply systems

Two flywheel generators are used to supply the toroidal and poloidal field coils (TF and PF) respectively [3] as the grid cannot withstand the relevant pulse loads. The TF magnet is fed by a twelve-pulse diode bridge, while the PF coils are fed by bidirectional thyristor convertor bridges. Commutation system provides pulses up to 30 KV on the transformer windings to produce the plasma start-up. Such pulses are obtained commutating the transformer current through resistors in parallel with a two pole vacuum breaker, equipped with saturable inductances and capacitor banks to produce an artificial current zero. After the initial operation period, the reliability of the system has been enough acceptable: the major problems have been encountered with the rotating machines.

2.5 Control systems

Due to the reliability diminution along the time and to the further tasks requested to the main FTU control system, it was decided to redesign it completely [4] adopting a widely used software package for the real time database and the man-machine interface. The new system was installed during the 1994-95 shutdown period, the database configured, as required by the existing machine subsystems, and new control panels (mimics) produced. The new system is based on several industrial PLCs, for front-end data acquisition, and three VME stations devoted to PLC data preprocessing. The new design emphasises the man-machine interaction in order to facilitate the operator tasks during the experiment. It includes a number of services, such as dynamic display of the experimental plant on X-terminals, easy mouse-controlled interfacing, access to the local operating systems and sophisticated data plotting facilities.

The plasma current and position control has been improved allowing a fine tuning ($\leq$1cm along r and z axes) of the last magnetic surface. This is quite useful to achieve a good plasma-RF coupling. After an enough long debugging period of a few months the system is working pretty well, improving the reliability of the tokamak operation.

2.6 Radiofrequency systems

The lower hybrid (LH, 8GHz)) system is the most important heating and current drive experiment planned for FTU [5]. The main experimental activity started in 1996 with four gyrotrons operative at a total power of 3.5 MW. A fifth gyrotron (1MW) is going to start operations and the last one (of this phase) will be operative by the end of the year.

The commissioning of the four electron cyclotron resonance heating (ECRH) generators (140 GHz, 0.45 MW per gyratron), together with the relevant transmission lines and coupling structures, are under way to be operative by the end of 1997.

The operation with ion Bernstein waves (IBW, 433 MHz, 0.6 MW per Klystrons) system is being commissioned.

2.7 Diagnostic systems

A full set of diagnostics for the measurements of the most important plasma parameters has been implemented and operated during the experimental campaigns : the five channel DCN laser interferometer, equilibrium magnetic measurements, 19 channel multipulse, Thomson scattering system, UV and visible spectroscopy, 16 channel bolometer, ECE interferometer and polychromator, neutral particle analyser, Langmuir probes, etc.

In view of the full experiments with RF systems, new diagnostic was added in the 1994-95 shutdown: hard x-ray and neutron multicollimator, the termography to measure the toroidal limiter power load, a first wall sample introduction system [6].

2.8 Radioprotection and safety systems

Radiological hazards are limited to the area inside the machine hall, due to the prompt fusion neutrons of 2.45 MeV and to the delayed radiations produced by the neutrons.

An access control system monitors continuously the presence of personnel in the machine hall by a gate counter system.

3. OPERATION STATUS OF FTU

The objectives of FTU are to reach a range of plasma parameters of thermonuclear interest:
- to study plasma transport at medium-high plasma density with strong additional RF heating
- to test plasma heating and current drive by RF systems at high density
- to study plasma-wall interaction at high power loading (~10 kW/cm^2) with metallic limiters
- to study the influence of plasma profile control on plasma performance by pellet injection and different RF systems.

Until the 1994-95 shutdown, the above objectives were pursued mainly in the ohmic regimes [1], [7]. Preliminary tests with LH and ECRH prototype gyratrons were performed [5]. In 1996 the experimental campaign is mainly based on RF with increasing generator power up to 6 MW available by the end of the year. The preliminary LH experiments show promising results with high density plasma ($\geq 10^{14}$/cm^3). Furthermore stable operations with TZM toroidal limited plasma, up to 1.5 MW of coupled LH power are routinely obtained without plasma contamination by high Z impurities.

4. OPERATING EXPERIENCE AND PROCEDURES

Each event on component and systems is recorded and analysed in order to take under control the reliability of the system. The most critical parts, from this point of view, are the control system and the electrical power supply. The first one owing to so many interfaces and complex sequences to be performed for each pulse: the average unavailability is of the order of a few tens of minutes, but the frequency has been relatively high until the replacement of the system in the last shutdown. The electrical power system unavailability is dominated by the lack of redundancy of important components of the flywheel generators: the typical unavailability time has been enough large (several hours) even if the frequency is low.

4.1 Conditioning of the Vacuum Chamber

An important factor for the reliability of the experiments is the vacuum conditions and cleanliness. Rigid procedures are needed, in all phases, for any system/component interfacing with the vacuum system.

After each vent to the atmosphere, backing of VV and ports at T~130°C for 3 days, as minimum, removes water from the VV walls. Baking is normally followed by glow discharge cleaning (GDC) in hydrogen in order to break the oxides on the first VV layer, releasing O_2 and CO. Taylor discharge cleaning is used sometime for conditioning at cryogenic temperature. During the baking phase, fixed limits of ΔT between VV and ports, as well as ΔT between stainless steel and copper of TF coils cannot be exceeded. Fig. 1 shows the partial pressure of the most important impurities versus time during a GDC phase.

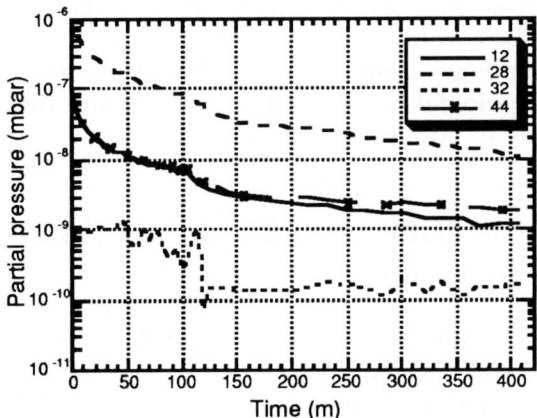

Figure 1. Impurities during the GDC phase.

No other conditioning phase is necessary during the experimental campaign: the machine remains clean even after major plasma disruptions.

4.2 Operational limits

At 8 Tesla pulses, ΔT between stainless steel and copper of TF coils has to be lower than 8°C and the initial temperature of the Cu has to be lower than 88 K. The maximum measured stress (in the internal leg of the TF coil casing) at 6 T is 230 MPa (σ_{VM}). The maximum evaluated value at 8 T is 560 MPa.

TF and PF coils are protected by dedicated Imax and I^2t protection. Electrical resistances are inserted at the end of each pulse, through redundant circuit breakers, to dump the magnetic energy. Major disruption of 1.6 MA plasma current quench in 3 ms, have been considered, together with an inward plasma radial displacement of about 12 cm, to design and verify all the in-vessel components [8].

Any new kind of discharge has to pass the validation pulse programme before to be charged in the control system.

4.3 Safety and radioprotection system

The neutron production rates in the shots already done in FTU is of the order of a few 10^{12} N/pulse. The activation of the FW material is practically negligible. All the in-vessel components, before their removal, must be checked by the health physic staff. Some activation, at low level (a few tens of μSv/hr) as dose contact rate) has been found in a few components, especially in the mushrooms of the poloidal limiter, due mainly to the runaway electrons.

5. FURTHER IMPROVEMENTS AND PROGRAMME

The main program up to 1998 will be largely devoted to experiments with the three RF systems (LW, IBW and ECRH) at 8 MW as total power at generators. Improvements are under analyses to be able to produce high shaped and triangular plasmas (K~1,7, ∂~ 0.7) at low plasma current (200÷300 KA).

In the next shutdown, scheduled at the end of 1996, a miniaturised in-vessel inspection system will be installed : three cameras, located in three different ports, will give a general view of the vacuum chamber. Such tool, working under vacuum, is particularly useful for a cryogenic machine as FTU; its use as plasma diagnostic is foreseen as well.

Furthermore the collective Thomson scattering will be installed to measure the velocity distribution of plasma ions, as well as the CO_2 interferometer to get reliable density measurement at higher densities.

A double stage multipellet injector (2.7 Km/s) that will produce up to eight D_2 pellets of different sizes, is under development as well as an articulated robotic boom for the in-vessel FW component maintenance and inspection.

6. ACKNOWLEDGEMENT

We are very grateful to the FTU experimental and operation group.

REFERENCES

1. M.Gasparotto et al. The FTU after 4 years of operation. SOFE, 1995.
2. F.Alladio et al. MHD and conf. during PI on FTU. Plasma Phys. Contr. Fusion 35, 1993.
3. A.Coletti et al. Sistemi di alimentazione elettrica pulsata. AEI 1993.
4. M. Panella et al. The New FTU Control System. SOFE,1995.
5. ENEA Fusion Division. 1994-95 Progress report.
6. A.Tuccillo et al. 2° Workshop "Strong Microwaves in Plasmas". Novgorod, 1993
7. M.L. Apicella et al, Plasma char. in FTU with different limiter materials. (sub. to Nuclear Fusion)
8. M.Gasparotto et al. Electrom. analysis using MSC/EMAS on FTU in-vessel components. MSC/EMAS users conf. Munich, 1995.

Operation of the Tokamak ISTTOK in an Alternating Current Regime

H. Fernandes, H. Figueiredo, J. Sousa, C. J. Freitas, J. A. C. Cabral and C.A.F. Varandas

Associação EURATOM/IST, Centro de Fusão Nuclear
Instituto Superior Técnico, 1096 Lisboa Codex, Portugal

This paper reports the AC operation of the tokamak ISTTOK performed using, besides the pre-discharge capacitor bank, one single electrolytic bank and a specially designed electronic unit, based on four Insulated Gate Bipolar Transistors (IGBTs), that allows multiple-reversing of the voltage applied to the transformer's primary. Specific setup of the external horizontal and vertical fields allowed the achievement of one cycle with finite plasma density (no dwell time) at current reversal, as well as multiple forward discharges with growing plasma parameters. Both kind of discharges were obtained with fair reproducibility.

1. Introduction

Present day fusion research programmes are oriented towards the construction of a demonstration fusion reactor, which must prove to be easily integrated in a electric power network. In this sense, a continuous fusion burn is preferable, but it is essential for the reactor to provide a steady electric power output with a minimum thermal energy storage.

The tokamak plasma current, which generates the poloidal magnetic field required for confinement and stability of the plasma column, may be externally driven by inductive or non-inductive methods. Steady state operation can be achieved by means of non inductive current drive (NICD), by injection of a neutral beam or radio-frequency waves [1]. The main disadvantages of this processes are the low efficiency provided by the available systems, which leads to high levels of power recirculation in a reactor scenario, and the restrictions imposed on plasma parameters. In the case of inductive current drive (ICD), the plasma current is driven by transformer action. In spite of the higher efficiency, limited flux capability of the transformer's iron core implies that an ICD tokamak is a pulsed device, which is a disadvantage in a power generating reactor [2]. Two techniques can be considered for ICD semi-continuous operation, namely, operation with uni-directional current, which is the conventional way to operate a tokamak, and operation with alternating current. In conventional tokamak operation, recharging of the central solenoid is required between plasmas, being the main contribution to the down time, where no burn occurs. In AC operation the down time is determined by the sum of the plasma ramp-down and ramp-up times, thus minimising the period without burn.

Feasibility of tokamak AC operation was first demonstrated in STOR-1M [3, 4] and STOR-M [5]. In these previous experiments, modifications of the tokamak's power systems were performed aiming at producing one single cycle of alternating plasma current, by means of an LRC circuit to generate a sinusoidal current in STOR-1M, and by using two capacitor banks in STOR-M. In both STOR tokamaks it was found that the plasma density remains finite during the current reversal phase, eliminating the need of pre-ionisation at the beginning of the second half-cycle. AC tokamak with reactor relevant current (2 MA) has subsequently been demonstrated in JET [6, 7]. However, in JET experiments, AC discharges were obtained with a finite dwell time (50 ms to 6 s) between the first and second plasmas, during which ionisation was lost, thus making the second plasma formation equivalent to a normal JET breakdown.

In ISTTOK [8], we built an electronic unit based on four IGBT's mounted in an H-bridge configuration which, connected to one single electrolytic bank, allows multiple AC discharges.

This paper is organised as follows: Section II

describes the experimental set up. Section III presents the experimental results on multiple forward discharges and on AC operation, with and without dwell time. Section IV contains the conclusions.

2. Experimental set-up

Experiments on AC operation have been performed in the tokamak ISTTOK (major radius R = 46 cm, minor radius a = 8 cm and toroidal field B_t = 0.45 T). A new electrolitic bank (ELCO) was installed, providing a higher energy for multiple discharges. It consists of 10x20x8 capacitors of 2200 µF/350 V each, resulting in a total capacity of 3.52 F. The bank is charged by a programmed, phase angle controlled thyristors power supply, driving currents and voltages up to 200 A and 400 V respectively.

The ISTTOK power supply system is integrated in the VME control system [9, 10] that drives the ISTTOK experiment (Fig. 1). The interface between the power supply and the VME controller consists of three optical signals (start, stop and clock), which drives respectively, the beginning of ELCO's discharge over the transformer's primary, the end of the discharge and the alternating period.

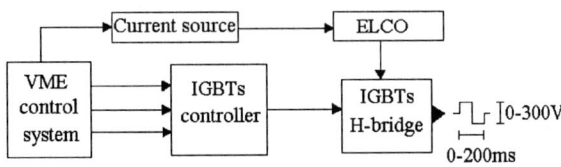

Fig. 1 - Block diagram of the ISTTOK power supply.

The voltage reversible system relies on an H-bridge of four IGBT's, where each pair drives the current in one direction. The IGBT's (MBN300A from Hitachi) can operate till 600 V and can drive currents up to 300 A. The maximum values for rise, fall, turn-on and turn-off times are less than 1 µs. The IGBTs controller circuit consists of a digital interpreter for the VME signals and four drivers (HTK0030) for the IGBTs. Special care was taken to avoid simultaneous conduction of both pairs of IGBTs at reversal.

The pre-discharge was performed as usual with the 1.5 kV capacitor bank (PRECO), to which follows the discharge of ELCO through the IGBTs circuit (Fig. 2). The vertical field is also fed by the same power supply, as it has to be reversed to allow negative cycles. On the other hand, the horizontal field, usually in a series with the vertical one, had to be fed by an external power supply, as it must remain unidirectional, although assuming different values on each semi-cycle

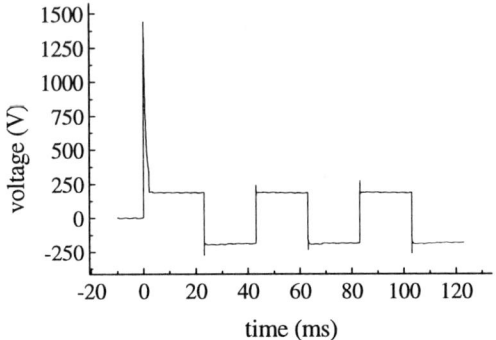

Fig. 2 - Output of the ISTTOK power supply.

Fig. 3 shows the ISTTOK power supply. The IGBTs bridge is isolated from the PRECO by a serie of three diodes in parallel with a serie of two thyristors. The latest, prevents free-wheeling of the IGBT's under PRECO's discharge. The IGBTs gates (G1..G4) are connected to the IGBTs controller unit. The safety operation of the high power devices is guaranteed by snubers composed by associations of resistors, capacitors and voltage dependent resistors (VDRs).

Laboratory tests have shown that this unit can drive fast switching (less than 60 µs) rectangular alternate currents of about ±600 A with cycle duration up to 200 ms.

3. Experimental results

In fig. 4 we present shot n°3274. The plasma is formed under PRECO's discharge, but the horizontal field adjusted in such a way, it favours the plasma vertical position when negative pulses are applied, positive semi-cycles could not be obtained.

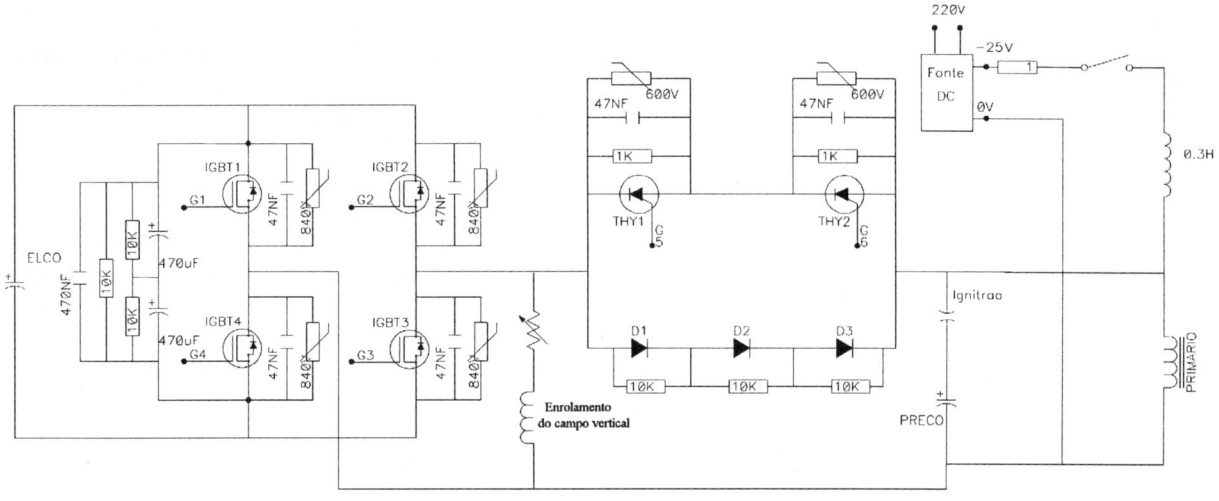

Fig. 3 - Schematic of the ISTTOK power supply.

However, plasma current and density improves in each negative pulse. An interesting question arises when one notes that a dwell time of approximately 10 ms always occurs between voltage inversion and current threshold. Gas puffing was not used and a continuous gas fill was applied.

If the horizontal field was programmed with a compromise value between the correct values for positive and negative plasma current, a full AC cycle, with a dwell time from 2 to 5 ms, could be obtained. Fig. 5 shows a typical discharge with a 5 ms dwell time.

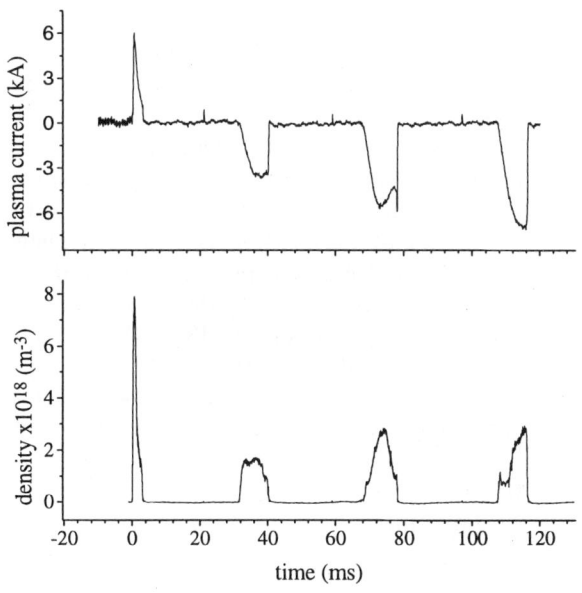

Fig. 4 - Temporal variation of the plasma current and density (shot # 3274).

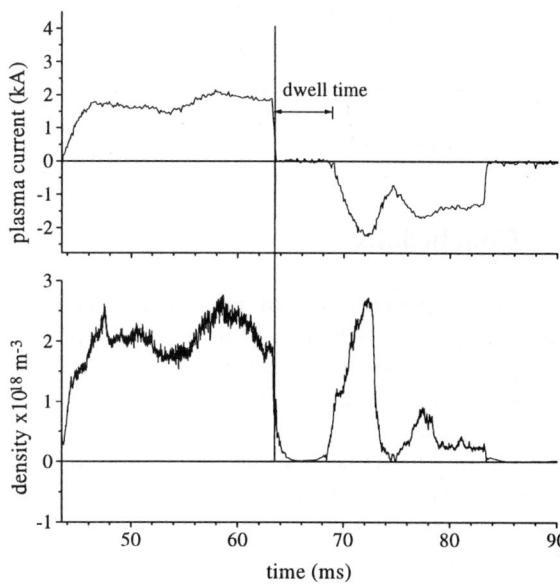

Fig. 5 - Temporal variation of the plasma current and density (shot # 4053).

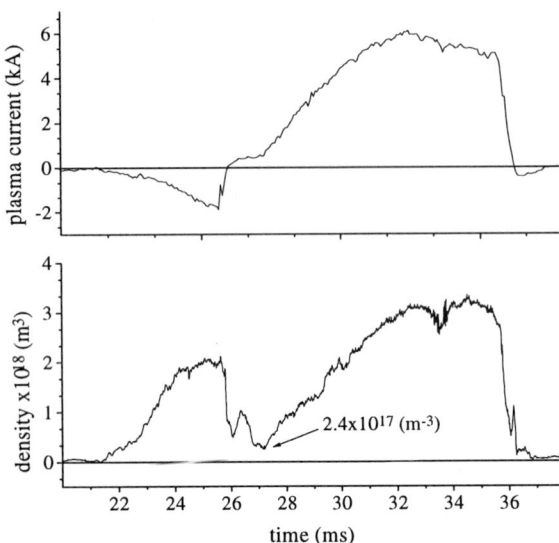

Fig. 6 - Evolution of the plasma current and density (shot # 4818).

A slight change in the horizontal field from the compromise value, favouring the equilibrium of negative current plasmas, could lead to an improvement on far positive semi-cycles, enabling a full AC cycle without dwelling (fig. 6). A curious situation because it was expected the opposite. This could be related with other situations, namely the cooling of the plasma or the peaking of the plasma profile. We note a finite residual density of 2.4×10^{17} m^{-3} during current reversal, in spite of the absence of rotational transform.

4. Conclusions

The next generation of magnetically confined fusion plasma machines will be either AC devices or quasi-DC operational machines, using NICD for additional heating and profile control. ISTTOK is being upgraded in order to operate with many cycles of alternating current, so it will provide information on cumulative effects. Now, our efforts are concentrated in programmed, feedback controlled power supplies ($\tau<1$ms), which will allow the driving of many AC cycles. The main difficulties are due to asymmetric magnetic field position coils caused by hardware constrains and neutral pressure control.

We have demonstrated that soft landing of plasma current is not a necessary condition for current reversal without dwelling, as a finite residual density of 2.4×10^{17} m^{-3} was obtained during reversal. Plasma parameters were improved within consecutive uni-directional cycles, most probably due to the gas exchange between the vessel walls and the plasma, which alters the neutral pressure.

References

1. Fisch, N. J., Rev. Mod. Physics, **1**, 175.
2. Ehst D., et al., "A Comparison of tokamak burn cycle options", in Tokamak Start-up, edited by H. knoepfel, Plenum Press, New York
3. Mitarai O., et al, "Stable AC Tokamak Discharges in the STOR-1M Device", Nuclear Fusion, **27** (1987) 604.
4. Mitarai O., et al, "Plasma Density at the Current Reversal in the STOR-1M Tokamak with AC Operation", Nuclear Fusion, **32**, No. 10, (1992) 1801.
5. Mitarai O., et al, "Alternating Current Plasma Operation in the STOR-M Tokamak"
6. Tubbing B. J. D., et al, "AC Plasma Current Operation in the JET Tokamak", Nuclear Fusion, **32**, (1992), 967.
7. Huart M., et al, "AC Operation of JET Tokamak: Modification of the JET Poloidal Field System", in Proceedings of the 14th IEEE Synposium of Fusion Engineering, San Diego 1991, Vol. 1 (1992), 181 published by the IEEE.
8. Varandas C. A. F. et al, 1994, "Engineering Aspects of the Tokamak ISTTOK", Fusion Technology, VOL. 29, 1996.
9. Varandas C. A. F. et al, 1994, Nuclear Instruments & Methods in Physics Research, A349 547-553.
10. Varandas C. A. F. et al, 1995, Review of Scientific Instruments, 66, **5**, pp3382.

MONITORING AND DIGITAL CONTROL SYSTEM FOR THE 130MVA PULSE GENERATOR SYSTEM OF THE SPANISH STELLARATOR TJ-II

L. Kirpitchev, L. Almoguera, M. Blaumoser, P. Mendez, L. Pacios,
A. de la Peña, F. Lapayese, Ricardo Carrasco, Ignacio Labrador,
Asociación EURATOM-CIEMAT para Fusión
Avda. Complutense 22, 28040 Madrid, Spain

A. Pérez, B. Alberdi, J.M. del Río, E. Jauregi
JEMA S.A., 20160 Lasarte-Oria, Spain

The power supply of the Spanish Stellarator TJ-II will work in a wide range of the currents and voltages in order to provide different magnetic field configurations. The DC currents in the coils of the stellarator are variable between 5% and 100% of their maximum values. A digital control system will allow to drive the coil current with high precision at the required values. The control system must provide stability of entire power supply which is a very important aspect for the reliable operation. The paper describes the monitoring and control system of the entire pulse power supply of TJ-II and concentrates on the digital DC current control.

INTRODUCTION

The Spanish Stellarator TJ-II is a highly flexible, medium-size fusion device of the heliac-type under construction in Madrid. This device allows to do experiments in a large variety of magnetic field configurations. The rotational transform can be varied from 0,9 to 2,5. Shear variation is possible up to 10%. Theory predicts maximum beta values as high as 6%. The average major plasma radius is 1,5 m. The minor dimensions of the bean-shaped plasma are approximately 0,4 m by 0,2 m. Toroidal field on the axis: 1T. Pulse length: 0,2 s to 1 s. Pulse break: 5min. Auxiliary heating: 400 kW ECRH at 53,2 GHz in stage I; 4 MW NBI will be added in stage II. The project is in its final assembly phase and is expected to start operation early in 1997. [1]

Seven independent coil systems allow to create a wide range of magnetic field configurations. Each coil systems is fed from a digitally controlled twelve pulse thyristor converter. The output DC currents range from 5 kA to about 35 kA and are continuously controllable between 5 % and their maximum rated values. The DC output voltages of the converters vary from 150 V to 1050 V. Each converter consist of two 6-pulse bridges working in parallel through an iron core interphase reactor. The converters are supplied by a salient-pole synchronous generator which has a nominal output pulse power of 130 MVA for 3 s and an extractable energy of 100 MJ per pulse. The operational frequency of the generator changes during the pulse from 100 Hz to 80 Hz. The generator is driven by a DC pony-motor with a rated power of 1,5 MW and a nominal speed of 1500 rpm. [2]

CONTROL SYSTEM HARDWARE

The monitoring and control system comprises one central computer HP/E25 and two satellite computers HP rt-743. The central computer characteristics are: 32-bit RISC-PA architecture, 48 Mbytes of RAM, SCSI disk of 4 Gbytes, digital tape of 2 Gbytes. This computer is connected to the two satellites by a fibre optic Ethernet LAN 8023. One of the satellite computers with a RAM of 16 Mbyte does the input/output control and data acquisition for the generator and the other for the coil converters. Three UNIX work stations HP-715 and a synoptic panel containing an unifilar diagram will be used as operator interfaces for the local control. Another UNIX work station of the same type, located at the central control room of TJ-II, is used for the remote control. A fibre optic LAN with the TCP/IP protocol connects the local and the remote control units. Another fibre optic link to the general control of the experiment will allow the TJ-II general control to govern directly the coil converters. [3] Figure 1 shows the monitoring and control system for the power supply of TJ-II.

GENERAL FUNCTIONS OF THE SYSTEM

In order to obtain the required coil current profiles, the control system must generate the suitable references for rectifier regulators. The current references that define the pulse can be memorised in the power supply control system or transmitted in real time from the TJ-II general control (Fast-Control). Synchronization of the memorized references can be defined in the control system or sent from the Fast-Control. As a result, three modes of operation are possible: memorized references and synchronization times, memorized references with synchronization from Fast-Control and real time references from Fast-Control.

The control of the equipment is carried out by logics organized in transitions and stable states. The system may stay permanently in a state, waiting for an operator order which will activate a transition. This transition will bring the equipment to another state. This block diagram structure facilitates operator´s job and avoids forbidden orders, as for example any order during transitions, except stop order. The pulse has been considered as a transition whose time is limited. The figure 2 shows the process state diagram. The authorized operator will be able to activate the different transitions.

Each component of the supply system protects itself against faults. The control program also incorporates protection logics which will coordinate the actions of all the components in case of faults. In addition it also carries out protection tasks that are redundant with the subsystem ones. According to the importance of the fault it will generate an alarm, avoid a transition or stop the whole system.

The acquisition of events, alarms and analog signals (currents, voltages, temperatures,...) will take place continously. These data will be processed and, under operator request, will be displayed on the work stations. During the pulse, the more significant data such as currents, voltages, etc. are stored on a hard disk in form of an experiment register (number, date,...).

The man-machine interface consists of pictures that show real time states of the system and parameter

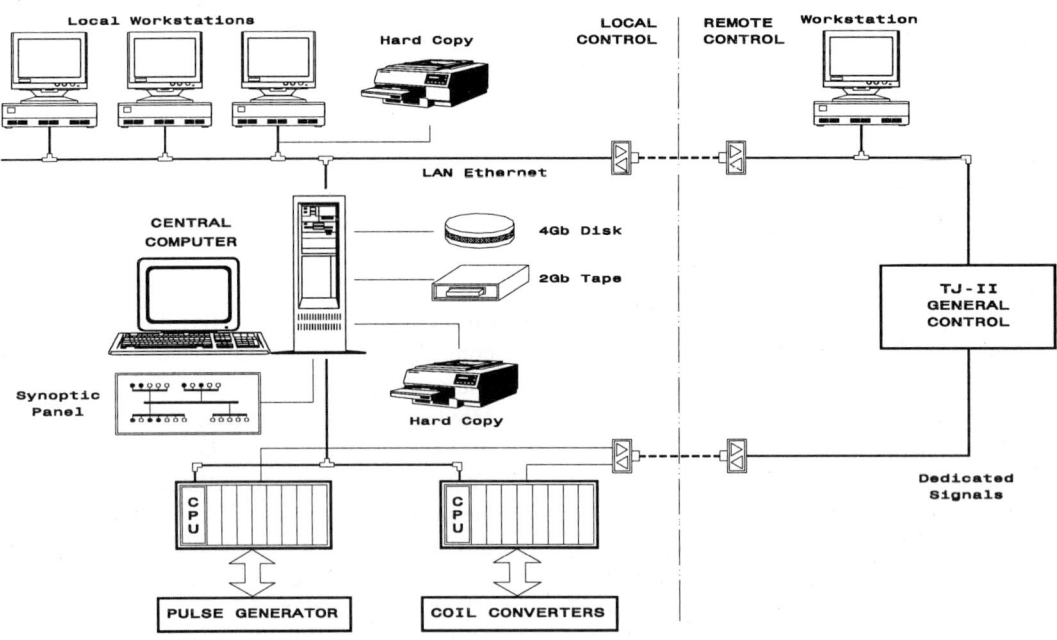

Figure 1 - Monitoring and Control System Layout

lists which can be modified or validated by the operator in order to prepare an experiment. The control system provides the posibility of loading existing parameters (old experiment) to generate a pulse. Modification of a parameter list is only possible for authorized operators and only at one work station simultaneously.

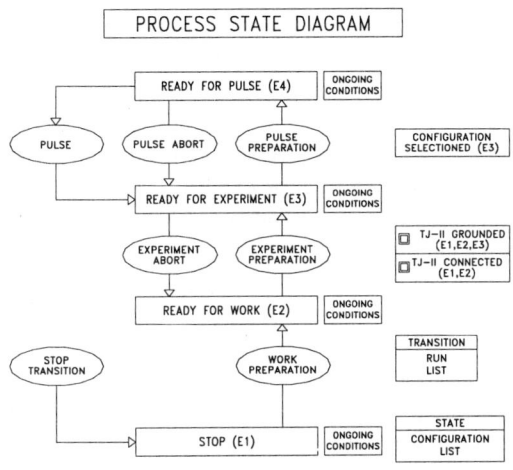

Figure 2. Process state diagram

DISTRIBUTION OF THE CONTROL TASKS

The central computer HP/E25 contains the whole information related to the state of the entire power supply system. It takes in charge the management of the man-machine interface in connection with local and remote work stations. Less critical control of the equipment is processed from HP/E25 computer to prevent overload of the two satellites. All the experiment registers are backed-up on the tape system. When a given experiment is required, it is restored in the disk to be processed by the central computer.

The two satellites interchange up to more than 900 signals with the devices. The real time CPU incorporated in the satellite manage the data acquisition and the more critical logics. So, protection functions as well as stop transition for example, are straightly processed by the satellite respecting the time accuracy requirements and without interactions neither with the other satellite nor with the central computer.

The computer satellite that controls the rectifiers generates current references. During the pulse, a current reference must be generated for each of the seven rectifiers with a 1 msec. time accuracy. Simultaneously, DC voltage and current feedbacks must be acquired with a sample frequency of 1 kHz. All these values are stored in the satellite and when the pulse finishes they are transfered to the central computer for processing. The other satellite controls the generator operation and its auxiliary systems as e.g. lubrication, refrigeration, etc.

CONTROL OF THYRISTOR CONVERTERS

Very special attention was given to the development and construction of the digital control circuits of the thyristor converters because of the strong mutual coupling of some of the coils, the required accuracy of the DC-currents of 0,1% and the danger of power oscillations between the mutually coupled circuits via the 15 kV common bus. The current ripple allowed on the DC side is as low as 0,15%. [4]

The rectifier's control cabinet, are physically separated from the thyristor bridges. The firing pulse signals are transmitted through fibre optics.

The regulation, the PLL and the control logic are implemented on a TI (TM3200C51) DSP. A 80C186 microprocessor deals with the data base, events storage, parameter display and serial port communication tasks. A dual port RAM is shared by the DSP and the µP, to improve the communication speed between both devices. Four FPGA generate the firing pulses, control the DMAs and the system synchronisation.

Normal zero-crossing detection PLLs are too slow for this system (100 Hz to 80 Hz drop in one second). An algorithm based on the product of input Vac waves and sinusoidal references provides a much faster response. Using this variable frecuency for the DSP timing, it will be possible to implement very efficient comb filters (which allow optimal filtering of the rectifier harmonics).

To comply with the accuracy requirements (0.1% of the output current), a two channel, 16 bit ADC is used for the current reference and feedback inputs. A high precision DCCT (20 kHz, <0.05%) provides an accurate measurement.

The in-built state machine controls the rectifier state sequencing, depending on the previous state, the input commands and the faults detection. This way, the controler saves time and memory space by executing only the specific routines to each state. In case of error detection, two possible stop sequences (normal stop, and freewheeling) are considered. The normal stop shifts the firing angle to its maximum value. The freewheeling fires all the thyristors at a time if the AC breaker opens. By this way, the system prevents overvoltage and limits the risk of having the all coil current flowing through 2 thyristors legs.

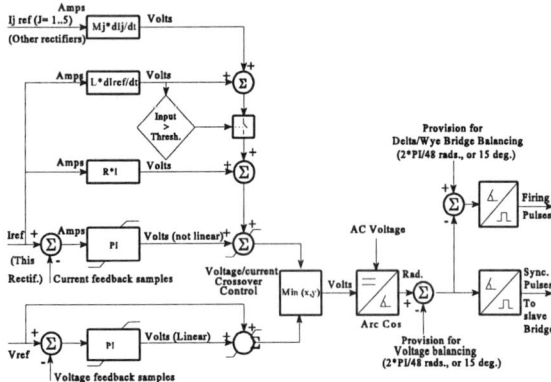

Figure 3. TJ-II Precision Rectifier Control System

Fig. 3 shows the flow diagram of the regulator. The rectifier is controlled in current. Depending on the current reference, the $L*dIef/dt + R*Iref$ provides the value of the required output voltage. To compensate the coil resistance change due to temperature elevations, a PI loop corrects the drift between feedback and reference. The coupling between coils is compensated by current reference feedforward inputs (the effect of the R coil is negligible). $Mj*dIj/dt$ defines the required voltage correction. Both derivative terms $Mj*dIj/dt$ and $L*dIref/dt$ improve the response time. However, derivative terms are very sensible to input noise, which could lead to a relative loss of accuracy. Bearing in mind that the maximum accuracy is required when the current references are constant, the derivative loop is disabled when its value does not exceed an adjustable threshold.

Vref, which is a constant adjustable by the operator, represents the maximum allowed voltage. This loop will, for protection purposes, impede the rectifier to apply a voltage higher than allowed.

As usual, the DC voltage required is converted in a firing angle by means of an Arc Cos table. Nevertheless, in order to compensate AC voltage variations, the applied Arc Cos function is related to the AC voltage amplitude.

The provision for voltage balancing is only used under the special configuration where two rectifiers are working in series. A pulse stream delayed proportionally to the unbalance will affect, with opposite signs, the position of the thyristor firing pulses of both rectifiers.

A rectifier scale model with high precision DCCT and transformer has been constructed to test the rectifier electronics in detail. The rated values are 20 A, 50 V. Freewheeling operation and system response to voltage and frequency drop (which will occur during the stellarator experiments) have been tested. All the manoeuvre and emergency stops have been checked as well.

CONCLUSIONS

The design of the monitoring and control system is based on standard and currently used electronic equipments what should provide a good reliability. Nevertheless, the special technical solutions are necessary in order to meet specified requirements for the accuracy of the current and the stability of entire power supply system. The test which have been carried out by JEMA confirm the validity of the selected design and the required characteristics are meet.

The full scale tests of all monitoring and control system are planned in 1997.

REFERENCES

[1] Luis de Almoguera et al., *Power Supply for the Spanish Stellarator TJ-II*, Fusion Technology, vol 2, 1009.

[2] A. Pérez et al., *Power Supply for the Spanish Stellarator TJ-II, design, construction and tests*, Symposium on Fusion Engineering, 1995, 1066.

[3] L. Pacios et al., *A versatile timing system based on OS9 for the Spanish Stellarator TJ-II*, Symposium on Fusion Engineering, 1995, 1047.

[4] A. Pérez et al., *Current ripple in the coils of the TJ-II Spanish Stellarator*, Symposium on Fusion Engineering, 1995, 1070.

FINAL ASSEMBLY AND PRESENT STATUS OF THE SPANISH STELLARATOR TJ-II

J. Botija, M. Blaumoser, J. Doncel, J.R. Knaster, J. Alonso and TJ-II Team
Asociación EURATOM-CIEMAT para Fusión.
Avda. Complutense 22, 28040 Madrid, Spain

D. Rigadello[1], A. Maistro[1], R. Carretta[1], S. Lorca[1], H. Lüscher[2], E. Kaplan[3]
1. Sulzer/De Pretto-Escher Wyss, Via Daniele Manin, 36015 Schio, Italy
2. Sulzer Innotec, Postfach 65, CH-8404 Winterthur, Switzerland
3. Max Planck Institut fuer Plasmaphysik, Boltzmannstr. 2, 85748 Garching, Germany

1. INTRODUCTION

The Spanish Stellarator TJ-II [1] is a highly flexible, medium-size fusion device of the heliac-type. It is under construction in Madrid, Spain. This device allows experiments in a large variety of magnetic field configurations to be carried out. The rotational transform can be varied from 0.9 to 2.5. Shear variation is possible up to 10 %. Theory predicts maximum beta values as high as 6 %. The average major plasma radius is 1.5m. The minor dimensions of the bean-shaped plasma are approximately 0.4m by 0.2 m. Toroidal field on the axis: 1T. Pulse length: 0.2s to 1s. Pulse break: 5 min. Auxiliary heating: 400 kW ECRH at 53.2 GHz in stage I; 4 MW NBI will be added in stage II. The project is in its final assembly phase. Magnetic field mapping is scheduled for February 1997. The commissioning of the entire device will follow.

The plasma is contained in a vacuum vessel [2] of helical geometry with 96 ports and a wall thickness of 10 mm. The central coil system [3] consists of a solenoid with a diameter of 3 m and two helical coils which spiral around this solenoid. The 32 toroidal magnetic field coils [3] are splitable into halves for assembly reasons. Four sets of poloidal field coils [3], embedded in the general support structure, provide the required vertical, radial and OH fields. A heavy support structure supports each of the main components individually.

2. GENERAL ASSEMBLY ASPECTS

The TJ-II assembly is characterized by its complicated geometry, its narrow tolerances, the small clearances between components and the high geometrical precision required. In order to minimize the assembly risks computer aided design studies have been performed and 1-scale mock-ups of the components and of a sector of the entire device were built. These mock-ups allowed to study the feasibility of the complicated assembly activities and lead to substantial and cost-saving simplifications.

The fundamental assembly strategy was to assemble or pre-assemble the main components in the factories in order to facilitate corrections if necessary and to allow assembly on site with conventional and standardized tools. Robots were considered to be too expensive and complicated. For this reason the design of the vacuum chamber was changed in order to allow the entrance of persons into it for all the in-vessel operations. To transport parts and components and to do their assembly, a crane was installed with low speed capabilities. Its minimum velocity is 0.16 mm per second.

For the position control in the factories and on site a computerized theodolite system [4] was employed with a precision of several tenths of a millimeter.

3. ASSEMBLY ACTIVITIES AT FACTORIES

The general support structure, with a total weight of 25 tons, was entirely assembled at the manufacturer. The dimensions were checked by a computerized theodolite system. In order to meet the tolerance requirement shimming has been done on some of the vertical columns. Afterwards reference pins have been put in place to facilitate the assembly on site. The central coil system consisting of the central solenoid, the coil casing and the two helical coils was totally assembled at factory and delivered as ready for assembly on site. After a very careful dimensional control of each octant of the vacuum vessel with a computerised control system, each of the octants was pre-assembled at factory on a rigid assembly table. On this table the octants were precisely positioned and the interfaces between octants and other components were checked. Casings of the TF coils were transported to the manufacturer and assembled onto the octants of the vacuum vessel in order to verify the clearances between vacuum vessel and toroidal field coils. All the toroidal field coils were assembled at the manufacturer though later they had to be dismounted for the final assembly on site. This assembly was necessary for the dimensional control and the electrical and hydraulical tests.

4. ASSEMBLY SEQUENCE ON SITE

As a first step the four sets of poloidal field coils were assembled onto the four rings of the general support structure. Then the lower part of the support structure and the inner lower ring were assembled. The lower outer ring was put on the floor underneath its final position for later assembly. Then, the assembly platform of the vacuum vessel was mounted. The central coil system was assembled afterwards together with the four lower octants of the vacuum vessel in one assembly operation. Fig. 1 shows this assembly step. The assembly and welding of the four upper octants of the vacuum vessel and the closing rings followed. After the assembly of the inner upper ring and the radial beams of the support structure the 32 splitable toroidal field coils were assembled around the vacuum vessel. The outer rings of the support structure will be put in their final position after the dismantling of the assembly platform. The assembly of the stellarator will be finished with the final positioning of the toroidal field coils. All the assembly steps of TJ-II are considered to be standard except for the vacuum vessel on which we concentrate in the following section.

5. ASSEMBLY OF THE VACUUM VESSEL

The assembly and welding of the octants of the vacuum vessel was the most challenging activity. In order to meet the very narrow tolerances of about one millimeter with respect to the 1.5 m radius, a sophisticated welding procedure was developed in order to minimize the welding shrink and to obtain a homogeneous welding deformation.

Although the inner diameter of the vessel is only 0.74 m, this welding was manually done from inside. For the welders access into the vessel, ports of

Fig1. Assembly of the Central Coil System and 4 lower octants of the Vacuum Vessel

only 21.7 cm x 53.5 cm each were available. The original idea to weld with a robot and to assembly the in-vessel components by special tools was discarded for feasibility and cost reasons.

Safety aspects played an important role for this assembly operation. Only non-inflammable materials and special welder suits have been used. Fresh air supply and smoke exhaust were achieved by independent ventilation systems. For test purposes a person was rescued from the inside of the vessel.

A rigid assembly platform had been installed to provide a stiff base for the support of the octants in order to keep them precisely in position and thus to guarantee reproducible and precise position measurements.

Between the octants closing rings having a width of about 15cm are located. These closing rings were inserted after the assembly of the adjacent octants and were welded on both sides to the octants. Precisely machined flanges at both sides of the closing rings match the corresponding flanges of the octants. The final machining of the closing rings was done after the precise measurement of the distance between the corresponding flanges of the octants and thus compensating manufacturing and assembly tolerances. For each of the two weldings of closing ring and octant, 1 mm was added to the width of the closing ring to account for the welding shrink.

Firstly one of the 4 pre-assembled lower octants was precisely positioned and fixed to the rigid assembly platform. Then the adjacent upper octant was brought into position from above by crane and then supported by the assembly platform. In the supports area fiber-slip sheets were located. This allowed the octant to move during the welding process and thus to reduce the welding stresses. The corresponding closing ring was assembled after being machined to its final width as mentioned above. The welding sequence was chosen in a way such, that the octant set on fiber-slip was in its theoretical position after the termination of the weldings. In this way two half-rings of the vacuum vessel were created and welded together. One half-ring was fixed to the rigid platform, the other sliding entirely on the fiber slip.

Due to the non-uniform geometry of the welds and the reduced thickness of the vacuum vessel walls in the groove region, together with the absence of a flange in this part, time consuming and complicated test weldings had to be done in order to obtain the required accuracy of several tenths of a millimeter for the position of the freely sliding octant.

The welding sequence begins with an internal tack-welding to fix the starting position of the free octant. After that the flanges of the closing ring and the octants were welded outside with intermittent welds. These welds contribute substantially to the mechanical strength of the welding connection. Afterwards the internal welding which connects octants and closing rings was done. These welds serve as mechanical connection of both parts and also as an ultra high vacuum barrier. The V-type welding has a thickness of about 5 mm.

The welding sequence of the internal welds is shown in Fig. 2. Three runs have been done manually in the flange area and two in the groove region. TIG welding had been chosen with the filler material 25.22.2 L Mn (rods Sandvik). The welding zone had been continuously cooled by compressed air from outside.

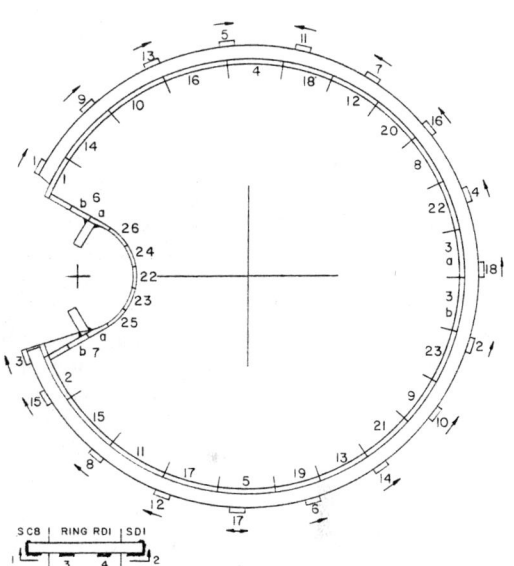

Fig. 2. Welding sequence of external /internal welds

A continuous dimensional control was done on the freely sliding octant with theodolites and mechanical measurement instruments. The two opposite ends of both octants had been specially controlled to verify the immobility of the fixed octant and to see the displacements of the free one during the welding process.

Forces were applied in the welding zones by means of external jigs in order to press the flanges of the closing rings against the octant and thus to reduce the welding deformations. High forces were chosen in the groove area where the expected welding shrink was highest. The forces were varied during the welding process according to the measured deformations.

The central coil system which is very near to the welding area was protected with ceramic insulation and copper plates. The temperature on the ground insulation of the external helical coil was controlled with several thermocouples.

6. DIMENSIONAL CONTROL

The measurement of the position of all the components during the assembly was done by a computer-based theodolite system. Three digital high-precision theodolites have been used. A total of 64 reference marks were put on both the walls of the torus hall and the assembly platform. These marks allowed the determination of the exact position of each theodolite in any possible position within the torus hall and hence the precise measurements of the position of the stellarator components.

The control of the toroidal position of the vacuum vessel was done by means of two theodolites. One was located on the central vertical axis of the vacuum vessel and the other one outside of the assembly platform. For radial position control, standard precision gauges were used. The vertical position of the vacuum vessel was determined by an optical precision level.

Measurements inside of the vacuum vessel were done after the partial and complete welding of each closing ring. Several precise reference points were marked on both sides of each weld. The local welding shrinkage of each weld and the reduction of the closing ring diameter was measured with respect to these reference marks. The reduction in diameter was practically zero. The total local welding shrinkage was about 1mm and depended slightly on the local coupling of closing ring and octants.

Measurements of clearances between all the main components have proven that the overall precision of the geometry of the entire Stellarator is very good. The geometrical deviations including all the manufacturing and assembly tolerances are in the order of several millimeters as expected.

7. PRESENT STATUS OF TJ-II

The intense phase of the assembly of TJ-II was commenced in August 1995 with the assembly of the central coil system and the four lower octants of the vacuum vessel.

First welding activities on the vacuum vessel were started in September 1995. The final welding of the vacuum vessel began in January 1996 and was finished in April. The first vacuum tests have been done in May 1996. No problems have been encountered during the vacuum tests. A global leak rate as low as 1×10^{-9} mbar l/s was measured.

In July the first toroidal field coil was assembled. The last of the 32 coils was assembled in the second week of September 1996. A number of hydraulical and electrical coil tests will be done in order to assure safe operation of this system.

As the next and last step, the outer rings of the general support structure will be assembled still in September. Then the assembly of the stellarator will be finished. The assembly of the diagnostic platform and the installation of the peripheral systems inside the torus hall will follow.

Magnetic field mapping is scheduled for February 1997. This will be done with power supplies connected to the 50 Hz net. The commissioning of TJ-II at high currents will be done afterwards. The start of the experimental operation of TJ-II is scheduled for the middle of 1997 after which some of the diagnostic equipment will be installed.

8. CONCLUSIONS

A total time of about one year and a half is required for the assembly of TJ-II. This is well within the expected period of time. The envisaged tolerances, though very tight, could be met. The computer-based theodolite system has successfully been used for the dimensional control of TJ-II. The assembly strategy with numerous assembly tests and extended pre-assemblies at factories and on site have shown to be useful and necessary.

REFERENCES

1. C. Alejaldre et al.: "TJ-II Project: A Flexible Heliac Stellarator". Fusion Technology 1990.

2. J. Botija et al.: "Fabrication of the Vacuum Vessel of the Spanish Stellarator TJ-II". Fusion Technology 1994.

3. J.Alonso et al.: "TJ-II Magnetic Field Coils". Fusion Technology 1992.

4. A. Martinez et al.: "Assembly and dimensional control of the Spanish Stellarator TJ-II". Fusion Technology 1994.

OPERATIONAL EXPERIENCE WITH THE JET SADDLE COIL SYSTEM

A. Santagiustina, H. Altmann, M. Buzio, D. J. Campbell, R. Claesen, G. D`Antona[1], M. De Benedetti, A. Fasoli[2], G. Israel, R. Ostrom, S. Peruzzo, T. Raimondi, L. Rossi, F. Sartori, M. Tabellini, A. Tanga, G Zullo

JET Joint Undertaking, Abingdon, Oxon, OX14 1XA, UK
[1] Euratom-ENEA-CNR, 20123, Milan, Italy
[2] CRPP-EPFL, Ass. Euratom-Swiss Confederation, 21 av. des Bains, CH-1007 Lausanne

JET is equipped with a set of in-vessel Saddle Coils which have been used for experiments on the stabilisation of tearing modes by magnetic feedback, for studies on the error fields modes, and for the excitation of Alfven Eigenmodes. The system comprising the coils, the power supplies and the respective controllers, has been extensively operated since 1994. The strong interactions with plasma and the arcing during plasma vertically displaced disruptions have produced currents through the upper coils in excess of their mechanical capabilities.

1. INTRODUCTION

Active stabilisation of vertical position has extended the operative range of tokamaks, allowing the exploitation of elongated configurations by the addition of a low power magnetic feedback system.

With an " intelligent magnetic bottle " some of the plasma resistive instabilities might also be actively stabilised at small level by magnetic feedback, increasing the operating space of tokamaks. The feasibility of this idea has been supported by experiments [1].

The rapid development of power and digital electronics now allows the construction of affordable audio bandwidth amplifiers on the 20-40 MVA scale and digital controllers capable of computing hundreds of MFlops per second. Moreover the time scale for tearing modes growth and ohmic plasma rotation speed (now in the 10 ms and 1 kHz range), that dictate the speed of such a control system, are such that active feedback control is feasible in large tokamaks such as JET.

The coils, which launch the high frequency stabilising fields, must withstand the difficult in-vessel environment and are particularly critical.

2. JET SADDLE COIL (SC) SYSTEM

The main elements of the SC System, are:

2.1 Saddle Coils

Eight Saddle Coils (fig.1) [2] have been installed inside the vacuum vessel. Each coil consists of three turns made of bare Inconel bars of 9-18 cm^2 of section. Bars are isolated from their supports by hundreds of alumina balls (diameter~15 mm). Each coil has an impedance at 1kHz of 90mΩ and 32μH.

Fig. 1: *The eight saddle coils installed in 1994. From September 1994 only the four bottom saddle coils have been used in operation.*

2.2 Saddle Coils Circuits (No. 1, 2, 3 and 4)

Two Saddle Coils in opposite octants are connected in series to each DFAS amplifier to generate an n=1 field (fig. 2). Alternatively the TAEE generator can be connected to each Coil.

2.3 DFAS: Disruption Feedback Amplifier System

The DFAS system [3] is composed of four identical audio amplifiers with the following ratings: peak AC Voltage 1.5 kV; peak AC current I=3 kA for frequencies f = 0-1 (kHz); I = 3 kA/f for f=1-10 (kHz). The DFAS amplifiers are four quadrants, 5 levels, chopper amplifiers, made by IGBTs (Insulated Gate Bipolar Transistors).

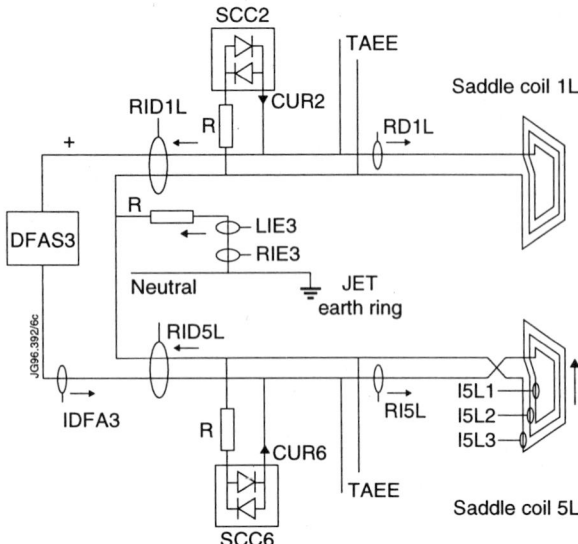

Fig. 2: Lower Saddle Coil Circuit No. 3 including the coils, the DFAS amplifier, the SCC crowbars and the transducers used to monitor the currents at disruption.

2.4 TAEE: Toroidal Alfven Eigenmodes Exciter

TAEE is a 3 kW, 30-500 kHz generator used to excite Alfven Eigenmodes [4]. Associated diagnostics reveal amplitude and phase of the plasma magnetic response. The system has allowed to discover new areas in the field of MHD spectroscopy. It is in particular possible to measure precisely the damping rates of the Eigenmodes in all plasma conditions. New modes have been discovered recently.

2.5 Saddle Coil Protections and Instrumentation

The Saddle Coil Protections (fig. 3) switch off the amplifiers, in 20 µs, when an earth fault or a short circuit is detected on the Coils, limiting the fault energy to 20 J, avoiding metal sputtering on the isolators.

Vertical disruptions, locked modes or overvoltages across the coils are also detected and the SC crowbars are closed to short circuit the coils.

The Coil Instrumentation (fig. 2) includes current measurements taken during faults and disruptions from: six in-vessel Rogowsky Coils and several ex-vessel measurements including one differential (earth) current measurement in each coil.

2.6 SCCs: Saddle Coil Crowbars

A bipolar thyristor crowbar is connected across each Saddle Coil to reduce the Coil terminal voltage by keeping them short-circuited, thereby reducing or extinguishing arc currents during disruptions.

2.7 DFC: Disruption Feedback Controller

The DFC closes the feedback loop for tearing mode stabilisation.

Four fast magnetic pick-up coils, installed in-vessel, measure the tearing mode phase and amplitude. Two measurements of the coil current are used to compensate the spurious signals from the pickup coils due to the Saddle Coil field in vacuum and the fast MHD plasma response to this field.

The digital controller consists of six 16 bits ADCs with insulated input and 200 kHz sampling rate and a cluster of six 40 MHz Digital Signal Processors (TMS320 C40) interconnected through the C40 communication ports. DFC functions and performances are described in references [5] and [7].

3. OPERATION WITH THE SADDLE COILS

3.1 Intermittent earth faults

Lower coils have been affected by several low impedance earth faults generated by small metallic debris, produced during in-vessel installations, bridging the small gap of 5-8 mm between the crossover bars and the vessel. Fortunately the faults disappeared after some weeks of operation because the small objects causing them have been melted or displaced by disruption-induced currents.

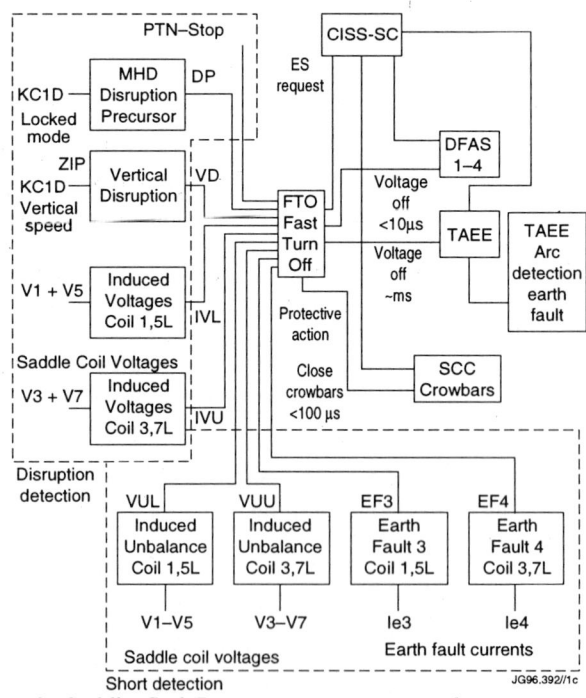

Fig. 3: Saddle Coil Protections comprising the systems detecting disruptions, short-circuits and earth faults.

3.2 Metallization of the alumina balls

Operation of Glow Discharge Cleaning sputters a metallic layer (Inconel) onto the insulating alumina balls. The layer resistance is on the order of 1-10 kΩ. As result the voltage withstood by the coils' insulation in high voltage tests has been reduced from 5 kV to 2 kV DC. The operational experience shows that coils can still withstand the power supplies voltages (~750 V) with only few sporadic arcs.

3.3 Damage to the Upper Saddle Coils

In September 1994 following significant disruptions, the inner wall bars at the ends of the upper coils were found to have been bent by 1-4 cm. Both ends of all 12 inner wall bars and the associated crossover bars, connecting the inner wall bars to the outer wall bars, were bent in a direction consistent with the force produced by the toroidal field and a clockwise induced current (current in the inner wall bars having the same direction of plasma current). A current of 14 kA, well above the design estimates, must have flown in the bars to have produced the observed damage.

The bent elements of the upper coils were removed and operation continued with the lower coils. During the Mark 2 shutdown the upper inner coils were removed, to make more space for highly triangular plasma configurations. The weakest accessible elements of the lower coils were strengthened: this improvement was not possible with the inner wall vertical links which are inaccessible below the divertor D1 coil.

4. SADDLE COIL MECHANICAL ANALYSIS

A detailed mechanical analysis has been performed to identify the weakest points of the coils, their dynamic response and their stresses in the elastic regime. For the most critical element, the inner wall vertical links, a plastic deformation assessment and a dynamic stress calculation, using an ABAQUS FE model, has also been done.

The analysis shows that, despite the strengthening of the coils, the mechanical resonance of some elements limits the maximum AC coil current capability between 25 and 1100 Hz to one third of the design value (i.e. to 1 kA).

The critical element has a resonance of about 470 Hz and the short current spikes of 1-2 ms observed during disruption produce stresses dynamically amplified by a factor ~1.7. A 1-2 ms current spike of ~5.5 kA brings this element to its elastic limits. A continuous ratcheting at 10 kA would produce a plastic deformation of the bar of less then 2 mm which is considered to be acceptable.

5. LOWER SADDLE COIL CURRENTS DURING DISRUPTIONS.

In April 1995 a 100 mΩ resistor was connected in series to each SC Crowbar to reduce the current induced by disruptions in the external circuit by a factor of 2. In this new configuration the coil terminal voltage at disruption is reduced by the SCC crowbars only by a factor 2-3.

Six in-vessel Rogowsky Coils (fig. 5) have been installed across each turn of coils 5L and 7L during the Mark 2 shutdown to detect the presence of internal arcs, halo currents and over-currents.

The measurements on the lower Saddle Coils currents in 86 disruptions have been organised in a data base that indicates that:
- Arcs in the coils can be detected as sporadic differences between the three in-vessel current measurements, the ex-vessel measurements, or as earth currents measured by the differential Rogowsky coils (fig. 4)
- Arcs of 1-3 kA have been observed when the coil terminal voltage exceeds 50 V.
- Arcs appear more frequently at the energy quench at high plasma currents (>2 MA) or during downwards VDE.
- The linear scaling of the observed coils' currents to a 6 MA disruption predicts a worst case current of 9 kA (fig. 5). Such a current can be produced in a downward VDE with a short or multiple arcing in the coils.

Different mechanisms contribute to the generation of the ~1 kV voltages that drives these current spikes:
- Flattening of plasma current profile following the energy quench produces spikes of 0.7-1.5 ms.
- Non axisymmetric structures (n=1 modes) have been seen after the disruption rotating at 500 Hz once around the machine, generating voltages of ~350 V and currents of ~1.8 kA in the coils and up to 3.8 kA in one shorted turn.

Halo currents can justify the very irregular behaviour of the SC currents in downwards VDEs.

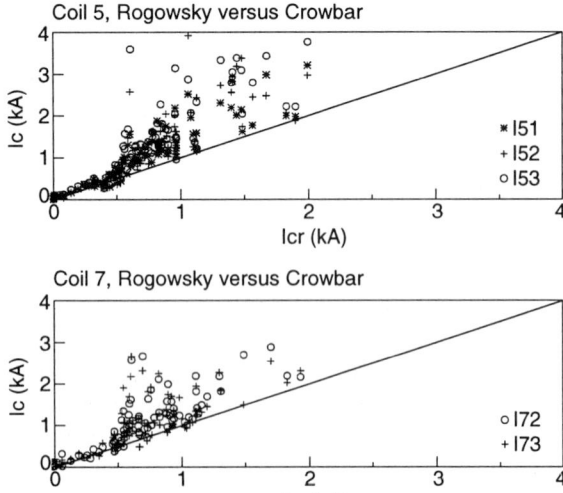

Fig. 4: Internal Rogowsky coil peak current at disruption plotted versus crowbar current. The difference between the two currents implies internal arcing of the coils, which occurs when the current of the crowbar is larger then 500A and the coil terminal voltage is larger then 50 V.

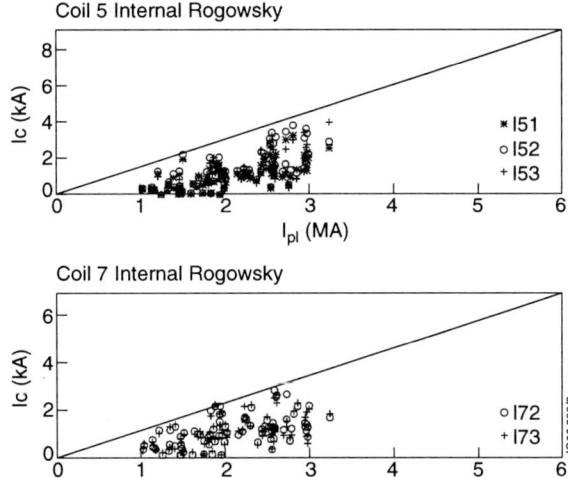

Fig. 5: Internal Rogowsky coil peak current at disruption versus plasma current. A linear scaling gives 9 kA peak current for a 6 MA disruption.

6. MAIN TECHNICAL AND SCIENTIFIC RESULTS

All eight Saddle Coils and their power systems have been commissioned launching n=1 fields of ~750 µT in DC, and ~200 µT at 1 kHz, up to the system maximum capabilities.

Experiments on the DC error field penetration have shown that n=1 field errors as small as 120 µT have been sufficient to generate large tearing modes, confirming the stringent limits on error fields in large tokamaks (in JET: $B_{r21}/B_{tor} < 5 \times 10^{-5}$ for $n_e \sim 1.5 \; 10^{19}$ m^{-3}). Rotating tearing modes have been generated by the application of n=1 fields rotating at the plasma speed.

Plasma MHD fast response to the external AC fields has been measured and corrected in the DFC controller.

The tearing mode's stabilisation feedback loop has been closed, at low gain, through the DFC controller, on tearing modes with amplitude of ~100 µT, showing changes of modes growth rates at the application of the external current [7].

Stable TAE and EAE modes have been driven by the coils and damping rates have been measured.

CONCLUSIONS

The Saddle Coils have been successfully operated, together with their high frequency power supplies and control systems to investigate several aspects of plasma MHD stability, tearing mode feedback control, and Alfven Eigenmode characteristics.

The currents induced in the coils, extrapolated for a 6 MA disruption, are just within the coils' mechanical capabilities. A very careful analysis is in progress to verify the safety margins on the predictions of these currents, in view of the JET DTE-1 experimental campaign.

REFERENCES

1. A. W. Morris et al., "Feedback Stabilisation of Disruption Precursors Oscillations in a tokamak", Phys. Rev. Lett., Vol. 64, No. 11 (1990) 1254
2. M. Pick et al., "Integrated engineering design of new in-vessel components in JET", Proc.15th SOFT, (1988).
3. P.L. Mondino et al., "The High Power, Wide Bandwidth, Disruption Feedback Amplifiers for JET", Proc. 14th SOFT, Vol. 2 (1990) 193.
4. A. Fasoli et al., "Direct Measurements of the Damping of Toroidicity-Induced Alfven Eigenmodes" Phys. Rev. Lett., Vol. 75, No. 4 (1995) 645.
5. A. Santagiustina et al., " Design of the M=2, N=1 Tearing Mode Control System for JET", Proc.15th SOFE, Supplement (1993) 58.
6. A. Santagiusina et al., "Studies of Tearing Mode Control in JET", Proc. 22nd EPS, Vol.4 (1995) 461.
7. M. De Benedetti et al., "Identification of the Physical Mechanism of Low-M, N=1 MHD Mode Control in JET", Proc. 23rd EPS Conf., (1996).

Compatibility Analysis of the Magnet System of the KT-2 Tokamak for Long-Pulse Operation

K.W. Lee[a], J.M. Han[a], B.G. Hong[a], B.J. Yoon[a], Y.D. Bae[a], W.S. Song[a], D.E. Kim[b], N.S. Shin[b], J.E. Milburn[b]

[a]Korea Atomic Energy Research Institute(KAERI), P.O.Box 105 Yusong, 305-600 Taejon, Korea
[b]PAL, Pohang Institute of Science and Technology, P.O.Box 125 Pohang, 790-600 Pohang, Korea

The KT-2 tokamak aims investigations of the advanced tokamak physics issues, thus necessitating steady state operation capabilities with superconducting magnets. In KT-2, where normal copper magnets with active cooling are employed, this need is circumvented by incorporating long-pulse operation capabilities with minimum 4 to over 20 seconds current/field flattop for OH and high bootstrap plasmas respectively. In this paper, compatibility of the magnetic system of KT-2 tokamak with the long pulse operation necessity is investigated in terms of thermal, mechanical and VDE controllability.

1. INTRODUCTION

Design study of a medium-sized, large-aspect-ratio divertor tokamak KT-2 [1] has been recently completed. Plasma and machine parameters are: R/a= 1.4/0.25m, B_t=3T, I_p=500kA, with auxiliary heating power of 7MW, resulting in relatively high heating power density and β_N for the machine size. The machine concept facilitates long-pulse operation of over 20s with ordinary copper magnets, with high bootstrap current fraction relevant to advanced tokamak studies. A major feature on design basis is introduced in table 1 and the magnet design directions are:
- Toroidal field of 3T at major radius of R=1.4m to maintain a plasma current of 500kA with minor radius a=0.25m (Aspect ratio=5.6).
- Long pulse operation capability with normal conductor for the advanced tokamak relevance regarding current diffusion time scale for KT-2 plasmas, which is estimated to be 2~3 s from numerical discharge simulations [2].
- A hybrid controlled poloidal field system to reduce the number of coils and the power.
- Most PF coils located outside the TF magnet for the simplicity in construction and the maintenance of the device.
- An active control coil set located between TF coil and vacuum vessel for the vertical plasma displacement control.
- In-vessel passive conductors for the controllability of highly elongated plasma(κ=1.8, δ=0.6)
- Optimum radial build-up to have maximum toroidal B_t/P and poloidal $\Delta\psi$/P (P: Electric power consumed in magnets, ψ: Magnetic flux)

Table 1. Main KT-2 tokamak parameters

Major/minor radius	1.4m / 0.25m
Aspect ratio R/a (A)	5.6
Elongation κ	1.8
Triangularity δ	0.6
Max. Toroidal field B_t	3 Tesla
Plasma current I_p	500+ kA
Max. density $<n_e>$	$\cong 5.0 \times 10^{19}$ m^{-3}
Electron temp. $<T_e>$	0.5/2 keV (OH/Heating)
ICR/LH/ (ECR/NBI for backup)	5.0/2.0/(1.0/1.0) MW

2. KT-2 MAGNETS

The magnetic system design is developed from an axisymmetric, free-boundary, ideal-MHD equilibrium model and results in figure 1.

2.1 Toroidal Field Magnet

To maintain 3T at R=1.4m, 21MAT of current is required in 16 D-shaped TF coils, resulting ripple of 0.83% at the outboard plasma boundary. The epoxy-impregnated coil is covered with a stainless steel case to facilitate installation and to enhance mechanical strength. The centering forces of 270 tons on the TF coils is countered by the wedge contact face. Keys are designed at the wedging surface to prevent slip, which also assists in resisting torsional

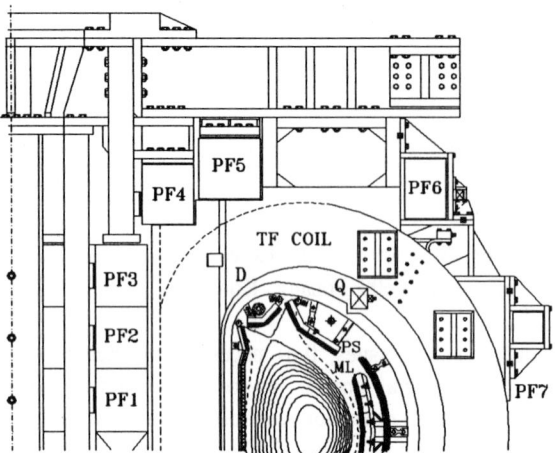

Figure 1. The magnet layout(PF#: PF coil operated in hybrid control scheme, Q: active control coil, PS: passive conductor, D: auxiliary single turn divertor coil, ML: movable limiter)

forces. Anti-torque beams are installed between TF coils to resist the out of plane forces. The conductor material is silver-bearing oxygen free copper which has a 7mm dia. cooling water passage. The conductor width is 15.7mm and the height is 150 mm (123mm, at wedge). One coil is composed of two layers, with each layer having 15 turns, for 30 turns per coil. The cooling system is designed to have one cooling circuit for every 3 turns, to cool the coil within 10 minutes after a full power pulse. All TF coils are connected in series, and the return bus is splitted into two, one runs inside of TF coil the other outside, parallel to the feeding bus. By adjusting the current ratio of these two conductors the error field from bus work can be reduced quite significantly. Through detailed field analysis the current ratio is determined to be 7/3, where the out side return conductor carries a bigger portion of the return current. With this design, the maximum error fields in the plasma region due to the bus structure is estimated to be about 20G.

2.2 Poloidal Field System

Three sets of "design basis" DN equilibrium have been defined as in table 2 [1], and its' stability checked in the framework of ideal and resistive linearized MHD [3]. OH- and Heating- baseline modes are for the conventional q profile and HiBS(High Bootstrap) mode is for the negative shear configuration[2].

A hybrid control scheme is employed for the PF coil system, where the currents for ohmic heating and shape/position control flows in the same coils. Thus each coil is feedback-controlled in order to assure stable operation in the advanced operation modes and plasma configurations, expressed primarily in the high elongation and triangularity of 1.8 and 0.6, respectively. The plasma geometry parameters and plasma current are controlled independently via the use of a set of decoupling vectors and a standard PID feedback system[4]. The plasma parameters to be controlled are deduced from the data provided by the diagnostic coil set, by operating on the vector of data values using a mapping matrix.

The PF magnet consists of 9-coil sets connected in up-down symmetry. There are two pairs of outboard 'ring coils' [PF6,7] mainly for the vertical field, two pairs of 'divertor coils' [PF4,5] assisted by in-vessel D,D' coils, one pair of 'control coils' [Q,Q'], and the central solenoid is divided into 3 pairs of modules [PF1,2,3] (see Figure 1). Q and Q' windings are driven by its own power supply to maintain average bias current level of -20kA for the equilibrium in addition to the control current I_c, i.e., $I_{Q,Q'} = -20kA \pm I_c$. D and D' are to modulate the X-points as well as to reduce the PF4,5 coils' duty.

Table2. KT-2 Operating Scenarios for the PF Design Basis

	OH baseline			Heating baseline				HiBS			
	EOM	SOF	EOF	EOM	SOF	SOH	EOF	EOM	SOF	SOH	EOF
I_p	0	500	500	0	500	500	500	0	300	300	300
$B_t(T)$	3	3	3	3	3	3	3	2	2	2	2
$\beta(\%)$	-	0.53	0.53	-	0.53	1.89	1.88	-	0.47	2.09	2.09
β_p	-	0.47	0.47	-	0.47	1.68	1.68	-	0.51	2.23	2.23
q_0	-	1.0	1.0	-	1.0	1.0	1.0	-	1.06	3.20	3.20
q_{95}	-	2.82	2.82	-	2.82	2.92	2.93	-	3.11	3.47	3.47
l_i	-	0.84	0.84	-	0.84	0.82	0.82	-	0.88	0.50	0.50
V_{loop}	6	1.8	6.2	6	1.8	0.25	3.81	4	0.25	0.0	0.67

EOM: End Of Magnetization   SOF: Start Of Flat-top
SOH: Start Of Heating   EOF: End Of Flat-top

3. COMPATIBILITY WITH THE OPERATION SCENARIO

3.1 Thermal

The discharge duration of a normal conductor to-

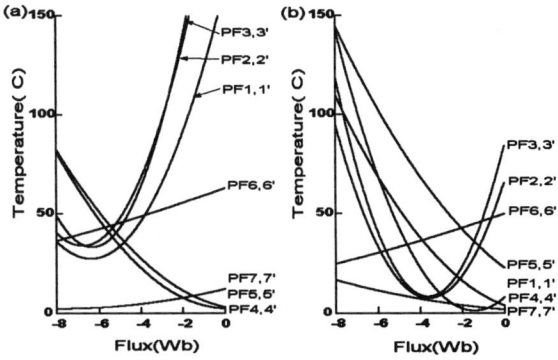

Figure 2. Temperature rises of PF coils of KT-2 tokamak vs. initial magnetization level in 20 seconds of flat top discharge for (a)Heating Baseline and (b)HiBS.

kamak with intensive auxiliary plasma heating and/or high bootstrap current fraction is subject to restriction mainly due to the temperature rise of magnets rather than the limited capacity of poloidal flux swing. If tokamak starts discharge in proper magnetization level so as to minimize the temperature rise of coils for a given plasma scenario, the pulse duration can be extended [5]. Figure 2 shows the temperature rise of each PF coil of KT-2 tokamak vs. magnetization level for two different plasma scenarios, i.e., the heating baseline (figure 2-a) and the HiBS(figure 2-b), with the same pulse length of 20s, which indicates that the optimal magnetization level of -5.4Wb for the heating baseline mode and -2.1Wb for the HiBS mode should be chosen to minimize the overall temperature rise of PF coils less than 45℃. With this scheme, OH baseline scenario is also adjusted to have maximum pulse length under the same cooling guide. From these results, PF coil current waveforms are reconstructed to Figure 3 so that a PF system design is obtained for a 4s OH baseline mode at maximum toroidal field and plasma current, extendable to 20s with a reduced loop voltage when heated as in heating Baseline and HiBS modes. This facilitates virtually steady-state operation where discharge time is much longer than current diffusion time scales, typically 2~3s for KT-2 parameters.

Temperature rise of TF magnet is 50℃ after full power pulse lasting 17s, and the 3turn cooling length is short enough to cool the conductor completely before the next pulse, 10 min. later. Cooling of PF

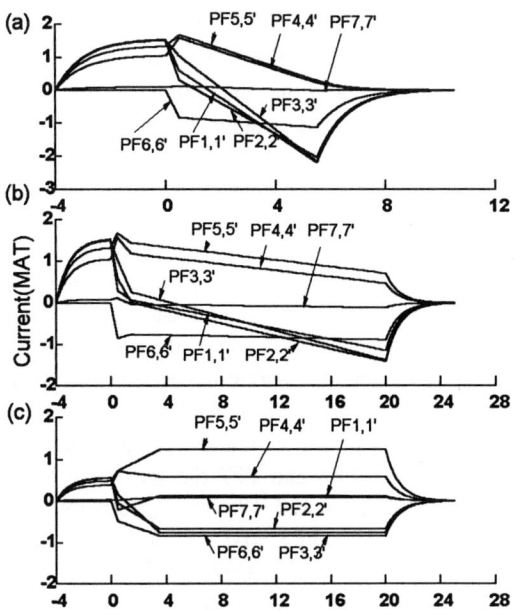

Figure 3. PF coil current snap-shot wave forms in (a) OH baseline, (b) Heating baseline, and (c) HiBS.

coils is also provided to cope the pulse repetition rate. The total required water flow rate of the TF & PF coil system is 194tons/h.

Under the temperature rise limit of 45℃ for PF coils and 50℃ for TF magnet, KT-2 has the maximum pulse length of 17s for heating baseline mode and 20s for HiBS mode

The stepwise maximum average power required to energize KT-2 magnets(PF+TF) is estimated to be about 69MW and peak power of 98MW in heating baseline mode.

3.2 Mechanical

1/8-th of the whole system including TF coils, PF coils, TF coil case and the support structures was modeled and analyzed. TF current is adjusted to produce 3T and the PF coil currents are taken from the OH baseline at the worst case, EOM. The maximum equivalent stress in the TF coil is about 45Mpa, much less than the yield strength of the conductor, which is about 300Mpa. The maximum equivalent stress in the TF coil case and antitorque beam is about 107Mpa, which is also less than the yield strength of SUS(about 500Mpa). And at the top and bottom of the straight part of the TF coil, the tensile stress maximum is 8.15 Mpa. It is possible for the shear stress inside the TF coils to destroy the conductor-epoxy adhesion. τ_{LR} was the dominant component, and its maximum is 6.22Mpa. So, the adhesion should sustain larger tensile and shear stresses than above numbers. Using a DZ80 primer coating, more than 30MPa in shear strength can be routinely achieved. The stress and displacements of adjacent TF coils and anti-torque beams for one TF coil failed mode are also analyzed. The maximum displacement of the coil case is 0.55mm, which is about two times larger than the normal conditions. However, the maximum equivalent stress is practically same as with normal discharge. The vertical forces acting on PF4 and PF5 are summed to be 640tons. For PF6 and PF7 the maximum vertical forces are 103tons and 18tons respectively. The worst case occurs for PF3, and the maximum equivalent stress is 119Mpa. However, this stress level is still within tolerable limits. The maximum stress and displacement on the structure are less than 39 MPa and 0.21mm respectively near the top cover plate.

3.3 VDE Controllability

The vertical motion in highly elongated plasmas as in KT-2 is primarily controlled by a set of passive stabilizer incorporating vacuum vessel, as well as the control coil set QQ' which takes care of the slower components left behind. Design and successful stabilizing action of the passive bars was checked with TSC[6]. The result of examination on vertical controllability is shown in figure 4. The plasmas are controlled effectively, with initial displacements of ~5mm controllable using a power supply with maximum output voltages and currents of ±105V and 20±7.7kA, respectively.

4. SUMMARY

Results from detailed thermal analysis of the PF magnets during the HiBS operation mode, where the bootstrap current fraction reaches as high as over 70% and the discharge lasts for over 20 s at reduced parameters of 2T and 300kA. In this mode, with a double-null configuration, the plasma cross section is highly shaped with $\kappa=1.8$, $\delta=0.6$, and the discharge is expected to maintain stable, hollow current profile with reversed magnetic shear. This 20s pulse can repeat every 10 minutes at maxi-

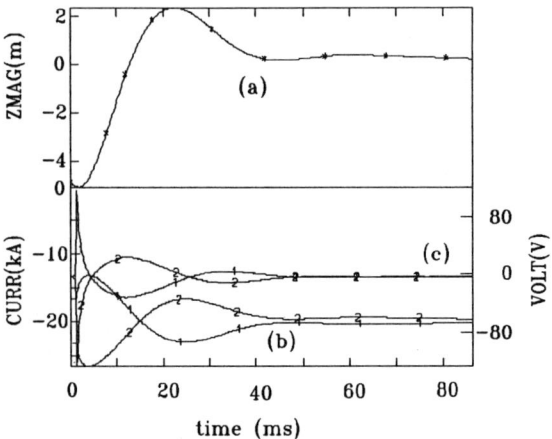

Figure 4. Vertical controllability for the typical KT-2 plasma with initial jump of 5mm. (a) Vertical position (m), (b) Currents(kA) and (c) Voltage(V) of Q and Q' coil.

mum. Despite the burden on the magnet system from such an operation, the analysis results indicate that reasonably long-pulse (about 7 times the current diffusion time) tokamak discharges are possible with ordinary copper magnets on a medium-sized tokamak such as KT-2, through careful design of the magnets and the cooling system as well as careful optimization of the operation scenarios.

ACKNOWLEDGMENT

This work has been supported by Korea Ministry of Science and Technology(MOST) and Korea Electric Power Co.(KEPCO).

REFERENCES

[1] I.S. Chang, et. al., "Concept Definition of KT-2 : a large aspect ratio divertor tokamak with FWCD", KAERI /TR-472/94, 1994
[2] B.G. Hong et. al., "Development of operation scenarios with high-bootstrap, negative-shear configuration for large-aspect-ratio tokamak KT-2", 16th SOFE, Oct. 1995
[3] S. Poedts et. al., "MHD Stability Analysis of the KT-2 Tokamak Plasma", 23rd EPS on Controlled Fusion and Plasma Physics, June 1996 and Rijnhuizen report 96-227, 1996
[4] P.J. Knight, "Design of a Plasma control System for the KT-2 Tokamak", UKAEA F/CO/KAERI/KT-2/2, 1996
[5] J.M. Han and K.W. Lee, "KT-2 Poloidal Field System Design", KAERI Report No. TR-584/95, 1995
[6] S.C. Jardin, N. Pomphrey, and J. Delucia, J. of Computational Phys. 66, pp.481-507, 1986.

AUTHOR INDEX

AUTHOR INDEX

Abdirashid, M.O. 1375
Abdrashitov, G.F. 885
Abramenkovs, A. 1463, 1507
Acero, J. 1119
Adachi, J. 1859
Adami, H.-D. 1189
Agarici, G. 203, 593, 713
Ageladarakis, P. 431, 743
Ahlfeld, C.E. 837
Aihara, Y. 363
Aikio, M. 1637
Ailisto, H. 1637
Akiba, M. 197, 279, 307, 359
Akino, N. 693
Albanese, R. 731, 735, 809
Alberdi, B. 981, 1119
Alberti, S. 533, 565
Albrecht, H. 1181
Aldrighi, C. 1051
Alekseev, A.B. 829, 1075, 1079
Alessandrini, C. 327
Alex, J. 569
Alexander, N.B. 1289
Alexandrov, E. 697
Ali, S.M. 1871
Almoguera, L. 981
Alonso, J. 985
Altmann, H. 483, 989
Alvarez-Armas, I. 1359
Amano, T. 1851
Amaro, P. 881
Ambrosino, G. 731, 735, 809
Amorim, P. 881
Anderson, P.M. 759
Ando, T. 849, 1083
Andreani, R. 813
Andres, J. 1125
Andrew, P. 751
Andritsos, F. 303, 1787
Ané, J.-M. 905, 1867
Angelini, B.M. 973
Angelone, M. 233, 1571
Anghel, A. 185
Angius, S. 179
Angoletta, M.E. 949
Antidormi, R. 1299, 1411, 1471

Antipenkov, A. 279, 1125, 1641
Antoniazzi, A.B. 1723
Antonov, N.V. 243
Aoki, I. 1791, 1859
Aoyagi, T. 693
Arai, T. 763, 1099
Araki, M. 307, 359, 719
Arhipov, V.V. 1145
Arkhipov, N.I. 507
Armas, A.F. 1359
Arneman, A.F. 829
Arnould, F. 1727
Artaud, J.F. 909
Arutunova, G. 1237
Aryev, N. 1285
Asano, K. 1019, 1027
Attura, F. 1387
Auerkari, P. 787
Augenstein, A. 921
Avalos, M. 1359
Avenhaus, R. 1257
Averin, Yu.P. 1233
Aymar, R. 9
Azizov, E.A. 833

Baba, T. 1133
Bacal, M. 585, 589, 1157
Bae, Y.D. 993
Baetens, I. 1621
Bainbridge, N. 227
Baity, F.W. 625, 629
Bakhtin, V.P. 507
Baldarelli, M. 291
Balden, M. 327
Ballinger, R.G. 1115
Balme, S. 909
Banks, D.M. 1689
Barabaschi, P. 719, 837, 1071, 1075, 1839
Barabash, V. 347
Barberi, D. 399
Barbier, F. 1475
Bareyt, B. 1059
Barinov, M. 697
Barratt, E. 1661
Barsuk, V.A. 383
Bartels, H.-W. 1755, 1851

Barthel, H. 921
Bartsoen, L. 1715
Baruah, U.K. 657
Batistone, P. 1571
Batistoni, P. 1555, 1563
Batistoni, R. 233
Baturo, I. 209
Baugh, W.A. 1289
Baxi, C.B. 391, 475
Bayetti, P. 673
Baylor, L.R. 1153
Bazilevski, V.P. 1149
Bazylev, B. 191
Beal, C.T. 1289
Beaumont, B. 593, 713
Becker, W. 597
Bell, A. 227
Bellina, F. 771
Belot, Y. 1751
Belyakov, V.A. 821, 829
Benamati, G. 1307, 1435, 1439, 1447, 1527, 1551
Benfatto, I. 1055, 1059
Benmansour, L. 1555
Berardinucci, L. 1427
Berezhko, P.G. 1281
Berger, H. 1253
Berger-By, G. 529
Bério, S. 203, 633, 637
Berkhov, N. 209
Bernard, C. 1419
Bernard, M. 1747
Bernier, F. 267
Berrebi, S. 443
Bertacci, G.C. 1335
Bertolini, E. 119, 1091, 1095
Bertrand, B. 435
Bertrand, C. 1539
Besenbruch, G.E. 1289
Besseghini, S. 1645
Besserer, U. 1189, 1261
Bessette, D. 1063, 1067
Bevilacqua, G. 179, 1047, 1051
Beyer, J. 1479
Beznosov, A.V. 247
Bhandal, I. 1637

Bhatnagar, V. 621
Bhise, M.K. 805
Bialek, J.B. 719
Bianchi, A. 813
Bibet, Ph. 203, 633, 637
Bielak, B. 259, 1299, 1303, 1471
Bijker, A. 557
Billone, M. 1315
Bisai, N. 657
Bisio, E. 1051
Bitter, H. 463
Blanquet, E. 1419
Blau, B. 185
Blaumoser, M. 981, 985
Blevins, J. 1709
Blomquist, R. 1819
Boccaccini, L.V. 1519, 1523
Bohn, H. 209
Boisset, L. 1693, 1697, 1727
Boissin, J.C. 1169
Boline, K.K. 1289
Bolt, H. 459, 779
Bonal, P. 327
Bondarchuk, E. 209
Bongers, W.A. 557, 561
Bonicelli, T. 1059, 1087
Borie, E. 545
Borsati, L. 1335
Boscary, J. 263
Bosch, H.-S. 101
Bosia, G. 609, 787
Botija, J. 985
Bottelier, P. 1395
Bottereau, J.M. 1059
Botti, S. 1379
Bottiglioni, F. 673
Boubée de Gramont, T. 1759
Boucher, D. 1851
Boulgakov, S.A. 829
Bourque, R. 1079, 1661
Boyer, D. 1751
Bozek, A.S. 475, 759
Braun, F. 597, 601
Braunsberger, U. 653, 801, 1011
Braz, O. 533, 545
Breitbach, G. 271, 499
Brelén, H.E.O. 705, 953
Brémond, S. 593, 905, 1867

Brennan, D. 227
Brinkschulte, H. 537
Brodén, K. 49, 1701
Broesch, J.D. 957
Brolatti, G. 291
Broocks, W. 1007
Brooks, N. 383
Brossaud, J. 637
Brown, L.C. 1289
Brown, P. 1673
Browne, A. 705, 893
Browne, M.L. 705
Brunet, F. 901
Brunnader, H. 1185
Bruno, L. 1331
Bruzzone, P. 1063
Bucalossi, J. 685
Buceti, G. 973
Bucké, S. 1479
Budd, T. 953
Bulgakov, S.A. 1059, 821
Bulkin, V. 1237
Buravand, Y. 909
Burgess, T. 1633, 1657
Busch, M. 645
Businaro, T. 747, 1637, 1645
Buzhinskij, O.I. 383
Buzio, M. 747, 751, 755, 989, 1095
Bykov, V.A. 825, 829

Cabral, J.A.C. 869, 977
Cai, G. 283
Callis, R.W. 629
Cambi, G. 299, 1783
Camous, B. 1555
Campbell, D.J. 949, 989
Campi, F. 1265, 1273, 1705
Capaldi, M.J. 1475
Caporali, R. 1779
Capriccioli, A. 723
Cardella, A. 197, 1315, 1327, 1331
Carrasco, R. 981
Carter, P. 1669
Carvalho, B.B. 167
Casagrande, A. 1435
Casagrande, E. 1387
Caso, A. 327
Cassarini, D. 1709

Castelli, S. 1205
Castro, R. 197
Ceccotti, G. 423
Cenacchi, G. 813
Cepraga, D.G. 299, 1783
Cerdan, G. 1633, 1709
Ceresara, S. 1645
Cerullo, N. 1863
Cesario, R. 613
Cétier, Ph. 1751
Chabrol, C. 1419
Chakraborty, A.K. 657
Challis, C.D. 661
Chang, H.Y. 861
Chankin, A. 387
Chappuis, Ph. 109, 259, 275, 439, 443, 447, 451, 491
Chatelier, E. 913
Chatelier, M. 447
Chaudhuri, P. 805
Chaudron, G.A. 525
Chaudron, V. 1727
Chaufour, Y. 1253
Chaumat, G. 439, 1295
Chavan, R. 1107
Chebotarev, V.V. 351
Chenna Reddy, D. 805
Cheron, C. 259
Chikaraishi, H. 1023
Chin, E. 391, 475
Chiocchio, S. 197, 279, 719, 723, 1641
Chitarin, G. 771, 775
Choh, K.K. 861
Choi, D.I. 861
Choi, J.H. 861
Choi, J.W. 861
Chung, Y.S. 861
Chuyanov, V. 837, 1839
Ciattaglia, S. 973
Ciazynski, D. 1039
Cierpka, P. 1129
Cirant, S. 941
Ciric, D. 661
Ciscato, D. 731
Claesen, R. 665, 989
Clair, C. 1831
Coates, K. 1843
Cocat, J.P. 435, 447
Coccorese, E. 731, 735, 809

Coenen, S. 1621
Cole, F.R. 1847
Collén, J. 411, 1819
Colombini, A. 1205, 1209
Colombo, P. 1375
Combs, S.K. 1153
Connor, K. 889
Conrad, R. 1511, 1611
Consano, L. 1637
Conti, S. 1047
Cook, I. 1831
Cool, R. 525
Coon, R.M. 957
Cooper, D. 893
Coppi, B. 813
Cordier, J.J. 435, 841, 1161
Corot, J.-P. 1249
Correia, C. 167
Costley, A. 849
Couturier, B. 909
Cozzani, F. 1615
Crenn, J.P. 529
Crisanti, F. 973
Crowley, T. 889
Crutzen, Y. 727
Cucchiaro, A. 613, 1111
Cumyn, Ph. 525
Cupido, L. 865, 873
Cusack, R. 1673

Dadonov, B.F. 1281
Daenner, W. 233, 1311
Dairaku, M. 307, 511
D'Alessandro, G. 1379
Dall'ava, D. 1253
Dalle Donne, M. 1427, 1483, 1487
Damiani, C. 1633, 1709
Dammertz, G. 545
Danger, K. 1253
Danilov, I. 1315
Dänner, W. 1315, 1327, 1331
D'Antona, G. 941, 989
Darses, Ph. 443
Darweschsad, M. 999, 1015
Datemichi, J. 857
Daverio, L. 1787
Davies, J.F. 893
Day, C. 1165
De Angelis, U. 1387

De Benedetti, M. 989
Deck, C. 203
Decool, P. 1039, 1043
Décoste, R. 525
Decréton, M. 1621, 1625
De Esch, H.P.L. 661
Deffain, J.-P. 1303
De Francesco, M. 1205
Degasperi, F. 581
De Gentile, B. 969
deGrassie, J.S. 625, 629
Degtyarev, L.M. 821
Dehne, J. 1185, 1261
deKock, L. 849
Deksnis, E.B. 387, 483
De la Peña, A. 981
del Río, J.M. 981, 1119
del Rió Bocio, C. 709
Dell'Orco, G. 1335
della Corte, A. 1047
Dellis, C. 1295
De Marchi, E. 747
Dementjev, S. 1531
Demers, Y. 525
Demidov, V. 1811
Deneuville, J.L. 1311
Denisov, G.G. 557
Denny, B.J. 1153
De Peña Hempel, S. 101
Derz, H. 331
De Santis, G. 1387
Deschamps, P. 435
Deschka, S. 427, 455
Desgranges, C. 685
Devillard, D. 1249, 1253
De Villers, P. 901
Devkin, B.V. 1575
Dhawan, S. 961
Diegele, E. 339
Dietz, J. 347
Dietz, K.J. 1125
Dietz, W. 57
Dilling, D. 837, 1641, 1661, 1689, 1839
DiMartino, M. 573
Diotalevi, M. 491
Di Pace, L. 299, 1763, 1871
Di Pietro, E. 355, 697
Dittrich, G. 1015
Divavin, V.A. 311, 825, 829

Dlougach, E. 697
Dmitrievsky, E. 1237
Doane, J.L. 573
Dobbing, T. 705
Doceul, L. 435, 447
Doerr, L. 1261
Doinikov, N. 209
Dolensky, B. 1043
Dolgetta, N. 747, 1091
Donato, A. 1375, 1379, 1607
Doncel, J. 985
Donne, M.D. 57
Donner, I. 921
Dougnac, H. 673
Doyle, P. 1087
Drews, L. 1011
Driemeyer, D. 279
Dubé, A. 525
Ducobu, L. 909
Dufayet, A. 443
Duglué, D. 1633, 1709
Dupin, L. 1751
Durodié, F. 577
Duval, B. 881
Duwe, R. 271, 275, 343
Dux, R. 101
Dvorkin, N.Ya. 833
Dworschak, H. 1213

Eagle, R.J. 897
Ebert, E. 1775, 1819
Ebisawa, K. 849
Ebisawa, N. 693
Edlund, O. 1771
Edwards, R.A.H. 1273, 1395, 1705
Egli, W. 1289
Egorov, K.E. 829
Egorov, S.M. 1063, 1145
Ehrenberg, J. 953
Eid, M. 1471, 1547
El Balghiti-Sube, F. 585, 589, 1157
Elbèze, D. 909
Elio, F. 609
Eliseev, L.G. 889
Elizarov, L.I. 589
Elkjaer, A. 1107
Ellis, J.J. 949
Elzendoorn, B.S.Q. 561

Empacher, L. 541
Engelmann, U. 1269
Englert, M. 1625
Enoeda, M. 1221, 1277
Erckmann, V. 541
Escourbiac, F. 109, 263
Evtikhin, V.A. 243

Falter, H.-D. 661
Fantechi, S. 727
Fasel, D. 569, 1107
Fasoli, A. 989
Faugel, H. 597
Favre, A. 569, 1107
Fazio, C. 1435
Febvre, M. 491, 1311
Federici, G. 197, 1125
Feist, J.-H. 645, 649, 697
Felten, F. 1419
Fenici, P. 419, 1391, 1395, 1443
Fermani, G. 1709
Fernandes, H. 977
Fernandes, J. 873
Fernández, A. 709
Ferrara, D. 1387
Ferrari, M. 723, 1315
Ferro, C. 813
Ferron, J.R. 965
Feuerstein, H. 1479
Figueiredo, H. 977
Filacchioni, G. 1379, 1387, 1607
Filatov, O.G. 833
Filatov, V. 209
Filin, V. 1237
Fink, S. 1015
Finken, K.H. 779
Fiorella, O. 1583
Fiorentin, P. 771
Fiorini, G.L. 1735
Fischer, U. 233, 1563, 1571, 1575, 1595
Fisher, P.W. 1125, 1153
Fisher, U. 1599
Fix, A. 537
Flerov, A. 1531
Floglietta, S. 1607
Floricourt, J.Y. 1249
Follin, J.F. 1289

Förster, W. 537, 541
Forty, C.B.A. 49, 1823, 1831, 1835, 1871
Foster, C.A. 1153
Foust, C.R. 1153
Frankrone, H. 921
Franzen, P. 925
Franzoni, G. 1731, 1735, 1739
Frattolillo, A. 1153
Freiesleben, H. 1567, 1571
Freitas, C.J. 977
Fresa, R. 809
Friedrich, B.-C. 1871
Friesinger, G. 1015
Fritsch, R. 597
Froger, C. 483
Froissard, P. 203, 633
Fuchinoue, K. 1407
Fuchs, A. 185
Fujisaki, H. 185
Fujiwara, Y. 681, 693, 697, 701
Fukatsu, S. 1653, 1657
Fukaya, K. 511
Fumelli, M. 685
Fursov, B.I. 1575
Furuya, K. 1339, 1343
Fütterer, M.A. 1295, 1303, 1307, 1539

Gabel, K. 1543, 1767
Gadeberg, M. 953
Gagey, B. 909
Gaggini, P.A. 1709
Gagliardi, P. 179, 1051
Gaio, E. 771, 1055
Galasso, G. 813, 1111
Galbiati, L. 1669
Galkin, S.A. 821
Gall, L. 1285
Gallix, R. 1071, 1075, 1839
Galvão, R. 581
Gangradey, R. 805
Gantenbein, G. 537, 541
Garampon, L. 203, 435, 443, 447
Garbet, X. 1867
Garcia-Cortes, I. 869
Garcia-Rosales, C. 455
Gardymov, G.P. 833
Gargiulo, L. 435

Garin, P. 447, 529, 533
Garkusha, I.E. 351
Garkusha, V.V. 351
Garnov, V.N. 515
Garré, R. 179, 1047
Garribba, M. 949
Garypov, I.S. 833
Gasparotto, M. 723, 813
Gay, J.-M. 1775
Geiler, V. 1483
Geist, T. 541
Gernhard, J. 933
Gerstenberg, H. 1003
Gervash, A. 271, 499
Ghendrih, Ph. 435
Giancarli, L. 57, 259, 1295, 1299, 1303, 1307, 1471, 1539
Gibson, C.R. 1289
Giegerich, M. 1315
Giese, H. 1603
Giesen, B. 209, 801, 845, 1011
Giguet, E. 533
Gilroy, J. 1661
Giorgi, R. 1375
Girard, C. 1743
Giroux, P. 1693, 1697
Glasbrenner, H. 1423, 1435
Glugla, M. 1193, 1197, 1261
Glukhikh, V.A. 833
Godden, D. 661
Gohar, Y. 1315
Golant, V.E. 833
Golikov, Y. 1237
Golubchikov, L.G. 243
Gondi, P. 1607
Goniche, M. 203, 633, 637
Goodin, D.T. 1289
Goodman, T.P. 565, 569
Gormezano, C. 621, 705
Götz, A. 1015
Gouge, M.J. 1125, 1153
Goulding, R.H. 487, 625, 629
Gowers, C.W. 897
Grattarola, M. 283
Gravanti, F. 973
Grebenshikov, Ju. 1803
Greuner, H. 463, 467
Gribov, Y. 731
Grieveson, B. 227

Grigoriev, S.A. 311, 825, 829
Grisham, L.R. 669
Grishmanov, V. 1451, 1507
Grisolia, C. 969
Grosman, A. 435
Grote, H. 327
Gruber, O. 427, 933
Gruber, P. 921
Grünhagen, A. 1015
Guay, J.M. 525
Guccini, M. 1551
Guilhem, D. 203, 439, 443, 451
Guillerminet, B. 909
Gulden, W. 1755
Günther, K. 1197
Gusev, V.K. 829

Haange, R. 75, 1649, 1653
Hackett, L. 893
Haferkamp, B. 1625
Hager, E.R. 1637
Hagrman, D.L. 1843
Haist, B. 1669, 1673
Halme, A. 1637
Hamada, K. 185
Hamada, Y. 889
Hamilton, D. 1677
Han, J.H. 861
Han, J.M. 993
Han, W.E. 1827, 1831
Hanada, M. 697
Hannula, H. 1637
Hansen, W. 1567, 1571
Hansink, M.J. 1289
Harmeyer, E. 1035
Harris, J.J. 957
Harrison, M.R. 415
Hasegawa, Y. 1499
Hashimoto, T. 1339
Hass, H. 1165
Hassanein, A. 379, 779, 1803
Hatamoto, K. 371
Hatano, T. 511, 1339, 1343
Hayashi, H. 1019
Hayashi, K. 549
Hayashi, T. 1277
Heer, B. 185
Heider, T. 1323
Heikinheimo, L. 787
Heikkinen, J.A. 787

Heikkinen, V. 1637
Heimsch, J. 1637
Heinemann, B. 697, 701
Heinmann, B. 645
Heller, R. 1015
Hellmann, G. 921
Hellriegel, W. 1261
Hemmerich, J.L. 227, 739
Hemsworth, R. 697
Henderson, M.A. 565, 569
Hennion, F. 913
Henry, D. 1161
Hering, C.-C. 653
Hernandez, M. 913
Herrmann, K.-D. 1007
Herrmann, P. 1193
Hertout, P. 203
Herz, W. 1015
Hidalgo, C. 869
Hino, T. 255, 363
Hinssen, H.-K. 335
Hirata, Y. 549
Hiratsuka, H. 763
Hirohata, Y. 255, 363
Hirst, P. 1723
Hiue, H. 1023
Hoang, G.T. 633
Hoang, T. 203
Hobbs, F.D. 1455
Hodgson, E. 787
Hoffman, D.J. 625
Hoffmann, E.H. 1289
Hofmann, A. 1015
Hofmann, F. 1107
Hofmann, G. 339, 455
Hofmann, M. 335
Hofmeister, F. 597, 605
Hohenauer, W. 459
Holland, D. 1755, 1795
Hollenberg, G.W. 1455
Hollerbach, M.A. 475, 759
Holloway, C. 697
Holmes, A.J.T. 661
Holzthüm, R. 463
Honda, A. 693
Honda, M. 1099
Honda, T. 1851
Hong, B.G. 817, 993
Hong, J. 861
Hopkins, M.B. 677

Horn, S. 1479
Hörner, L. 1479
Hosogane, N. 471
Housiadas, C. 1217
How, J.A. 705
Howie, J. 747
Hrabal, D. 1059
Huart, M. 1087, 1091
Hubbard, C. 961
Hubberstey, P. 1475
Huber, T. 467
Huguet, M. 23
Humphreys, D.A. 731, 759
Hurd, F. 387
Hurzlmeier, H.S. 1125, 1173
Hüttemann, P. 845
Hutter, E. 1181, 1185, 1189
Hwang, S.M. 861

Iagnov, V.A. 833
Iatrou, C.T. 545
Ibarreche, I. 1661
Icaran, J. 1661
Idei, H. 553
Ido, T. 617
Iguchi, H. 889
Iida, H. 1579, 1587
Iida, T. 857
Ikeda, Y. 857
Ikezi, H. 629
Illy, S. 545
Imagawa, S. 1019, 1027
Imai, T. 549, 637
Imaizumi, H. 1225
Imbeaux, F. 673
Inoue, T. 693, 697, 701, 1799
Inzaghi, A. 303
Iogansen, V.I. 833
Ioki, K. 197, 347, 719, 1315, 1327, 1331, 1347
Irving, M. 1677
Iseli, M. 1193
Ishimoto, K. 837
Ishitsuka, E. 1503
Ishitsuka, T. 1351
Isoz, P.-F. 569
Isozaki, N. 693
Israel, G. 989
Ito, T. 1083
Itoh, A. 1649

Itoh, H. 853
Itoh, I. 1023
Itoh, T. 693
Itou, Y. 1339, 1347, 1403
Ivanov, A.A. 585, 821, 885
Ivanov, D.P. 1807
Iwai, Y. 1277

Jacob, S. 805
Jacobson, D.M. 415
Jacquot, C. 677
Jager, B. 1039
Jakeman, R. 275, 279
Jaksic, N. 1031, 1035, 1371
Jakubik, P. 1637
Jana, M.R. 657
Jandl, O. 463
Janeschitz, G. 279, 347, 849, 1125, 1173, 1633, 1641, 1689
Jardin, S. 731
Jauregi, E. 981
Jayakumar, P.K. 657
Jayakumar, R. 1063
Jensen, F.A. 665, 705
Jeong, S.H. 817
Jernigan, T.C. 1153
Johner, J. 1867
Johnson, G. 1071
Johnson, L. 849
Johnson, W.R. 391, 479
Jones, G. 227
Jones, L.P.D.F. 1677
Jones, T.T.C. 705
Jong, C. 1075
Jonhson, G. 1347, 1839
Jordan, T. 783
Journeaux, J.Y. 203, 673, 913, 969
Joyer, P. 969
Junge, R.J. 391
Jüttner, B. 427

Kabanovsky, S.V. 515
Kabutomori, T. 1225
Kadi-Ogly, I.A. 833
Kaiser, E. 1487
Kakudate, S. 1649, 1653, 1657
Kalinin, G. 347
Kallenbach, A. 101

Kalyanam, K.M. 1719
Kaminaga, A. 763
Kamlah, M. 339
Kamperschroer, J. 669
Kanamori, N. 1347
Kaneko, O. 689
Kanno, M. 1133
Kappler, F. 191
Kapralov, V.G. 1145
Karditsas, P.J. 1823, 1835
Kareev, Yu.A. 1149
Kariya, T. 549
Karjalainen-Roikonen, P. 395
Karpov, D. 197
Kartikeyan, M.V. 545
Kasahara, F. 1795
Kasparek, W. 537, 541
Kasugai, A. 549
Katheder, H. 1003
Kato, Y. 403
Kaufmann, M. 101
Kavin, A.A. 731, 821, 829
Kawaguchi, I. 1339, 1859
Kawai, M. 693
Kawamata, Y. 795
Kawamura, H. 407, 1225, 1351, 1399, 1407, 1499, 1503, 1591
Kawamura, Y. 1277
Kazarin-Vibert, F. 203, 633
Kazawa, M. 693
Kemppainen, M. 787
Kerl, F. 463
Kern, S. 545
Kessel, C. 731
Kettyle, E. 1723
Khayrutdinov, R.R. 833
Kherani, N.P. 1245
Khirwadkar, S. 805
Khlopenkov, K.V. 853, 1145
Khrabrov, V.A. 515
Khrebtov, S.M. 869
Khripunov, B.I. 243
Khripunov, V. 1579
Khvostenko, P.P. 515
Kim, B.C. 861
Kim, D.E. 993
Kim, J. 861
Kim, S.K. 817
Kim, W.C. 861

Kim, Y.J. 861
Kimball, F. 1661
Kimura, S. 367
King, Q.A. 705
Kirillov, I.R. 251
Kirpitchev, L. 981
Kishiya, K. 471
Kissel, H. 1185
Kisslinger, J. 463, 467
Kitaev, B. 209
Kitagawa, S. 1023
Kitamura, K. 1339, 1347, 1403
Kizane, G. 1463, 1507
Kleefeldt, K. 1543, 1767
Klevtsov, V.G. 1233
Klingenstein, P. 921
Klischenko, A.V. 247
Knaster, J.R. 985
Knipe, S. 227
Kobayashi, K. 1277
Koblents, P.Yu. 1137, 1141, 1145
Kobozev, M.G. 1575
Kochin, V.A. 515
Kodama, K. 471, 763
Kogawa, H. 367
Koike, T. 1099
Koira, E.L. 1233
Koizumi, J. 693
Koizumi, K. 1347, 1403
Koizumi, N. 1083
Kokotvov, V. 719
Kolbasov, B.N. 1755, 1807
Kolbe, H. 1391, 1443
Kolchenko, A.I. 1149
Komarek, P. 999, 1015
Komarov, V.N. 825
Komen, E. 1715
Kondo, T. 1615
Königs, C. 577
Konishi, S. 1221, 1277
Konkashbaev, I. 379, 1803
Konobeyev, A.Yu. 1595
Konys, J. 1423
Kooijman, W. 557
Koppitz, T. 459
Kopytin, V.P. 1815
Korol'kov, M. 209
Korotkov, V.A. 829
Korovin, Yu.A. 1595

Korshakov, V.V. 833
Koskinen, K.T. 1629
Koslowski, H.R. 845
Kostsov, Y.A. 821
Kotikangas, K. 787
Kozhukhovskaja, N. 209
Kraemer, R. 1197
Krähling, E. 1003
Kraus, W. 645, 649
Krauth, H. 179
Krivchenkov, Y.M. 829
Kronhardt, H. 1007
Kruijt, O.G. 557
Krupnik, L.I. 869, 889
Krüssenberg, A.-K. 335
Krylov, A. 697
Krylov, V.K. 833
Kubo, S. 553
Kubota, Y. 403
Kuchinski, V. 1059
Kude, Y. 359
Kuitunen, S. 1637
Kulygin, V. 697
Kumazawa, R. 617
Kumrić, H. 541
Kuntze, M. 545
Kunugi, T. 1859
Kupschus, P. 1129
Kurasawa, T. 1339, 1343
Kurbatov, D.K. 1807
Kurihara, K. 795
Kurihara, R. 1791, 1859
Kuriyama, M. 693
Kurkin, S.M. 507
Kuroda, T. 143, 511, 617, 689, 1315, 1339, 1343, 1859
Kussel, E. 697
Kuteev, B.V. 495, 1137, 1141, 1145
Kuus, H. 593, 713
Kuyvenhoven, S. 557
Kuzmin, E.G. 825, 829

Labrador, I. 981
Ladd, P. 1125, 1173
Ladurelle, L. 593, 713
Laidani, N. 1447
Laillé, A. 399
Lamalle, P.U. 621
Lambertz, H.T. 845
Land, G. 557
Landman, I. 191
Lang, P.T. 427, 1129
Langhans, O. 1015
Langlais, G. 1743
Lanza, S. 1863
Lanzavecchia, L. 813, 1111
Lapayese, F. 981
Lappalainen, V.-P. 1637
Laqua, H. 541
Larose, D. 525
Lässer, R. 227
Last, J.R. 747, 1095
Latge, C. 1727
Lattaud, C. 1697
Laughon, G.J. 475
Laurens, J.M. 969
Laurent, A. 1727
Laurenti, A. 179, 1051
Lawson, S. 483
Laycock, B. 391
Lazzaro, E. 941
Le, T.L. 1197, 1261
Lebourg, P. 913
Lee, G.S. 861
Lee, H.G. 861
Lee, K.W. 993
Lee, S.G. 861
Lee, W. 1289
Lehmann, W. 1015
Leichtle, D. 1355
Leloup, C. 1867
Leluyer, M. 909
Le Marois, G. 197, 259, 267, 399, 1295
Lennnholm, M. 949
Lerche, E. 581
Leuterer, F. 537
Levy, R. 529
Leykin, I.N. 833
Libera, S. 355
Libeyre, P. 1043
Likin, K. 709
Lind, A. 411
Lindau, R. 1363
Lindberg, M. 49, 1701
Lindemuth, J.E. 487
Linden, J. 787
Lindholm, M. 1637
Lindmayer, M. 801

Linke, J. 271, 455, 459, 499, 779
Lipa, M. 109, 435, 439, 443
Lipko, A.V. 311, 825
Lison, R. 343, 459
Lister, J.B. 731, 735, 1107
Litaudon, X. 203, 633
Litunovsky, R. 751
Livshits, A.I. 1157
Loarer, T. 451
Lobanov, V.N. 1233
Lodato, A. 197, 1331
Lombard, G. 593, 713
Lopez, M. 1637
Lorenzetto, P. 1311, 1315
Lotte, Ph. 673, 697
Loureiro, C. 167
Lousteau, D. 1315, 1331
Loving, A.B. 1673, 1681
Lowry, C. 483
Lublinsky, I.E. 243
Lucca, F. 723
Lucia, C. 1119
Lüddecke, K. 929
Lukash, V.E. 833
Lukin, A.Ya. 1137
Lunev, V.P. 1595
Luodemaki, E. 1629
Lupo, J. 227
Luxon, J.L. 221, 759

Macaulay-Newcombe, R.G. 323
Mack, A. 1165, 1169
Maday, M.F. 1383
Maebara, S. 549, 637
Maekawa, F. 857
Maekawa, R. 1023
Magaud, P. 1867
Magne, R. 525
Maisonnier, D. 1633, 1637, 1641, 1657, 1661, 1709
Maix, R. 1003, 1047
Majumdar, S. 1315
Makhankov, A.N. 499
Makhlaj, V.A. 351
Maki, K. 1587, 1799
Mäkinen, E. 1629
Makowski, M. 609
Makris, Th.D. 1375

Malaquias, A. 869, 889
Malaquias, J.L. 167
Malara, C. 1201, 1213
Malkov, A.A. 829, 1075
Malléner, W. 459
Mamyrin, B. 1285
Mangano, R.A. 1289
Manintveld, P. 561
Manso, M.E. 865, 873, 877
Marbach, G. 1539, 1693, 1743, 1775
Marchese, V. 747
Marchiori, G. 771, 775
Marcus, F.B. 705
Marinucci, C. 185
Mariotti, F. 1335
Marra, A. 613
Marrs, R.A. 1125, 1173
Mart, J. 227
Martin, E. 1633, 1641, 1657, 1709
Martin, G. 203, 917, 969
Martin, R. 709
Martin, V. 443
Martinez, A. 1039
Martone, R. 809
Masaki, K. 471, 763
Mascherpa, C. 1265
Maschio, A. 1059
Masci, A. 1379
Masiello, A. 767, 775
Masset, R. 909
Massmann, P. 685, 697
Masuda, S. 617
Masuzaki, S. 1027
Matera, R. 197, 347
Mathis, V. 1471
Matsuhira, N. 1653
Mattas, R. 1315, 1327, 1331
Mattes, K. 427
Mattioli, M. 203
Mattoo, S.K. 657
Matveev, N. 709
Maurer, W. 1015
Maury, F. 1419
Maximenkova, N. 209
Mayaux, G. 673, 841
Mayor, J.-M. 1107
Mazul, I. 279
Mazza, M. 1335

Mazzitelli, G. 973
McBryan, H. 665
McCarthy, K.A. 1843
McCarthy, P. 933
Melkus, W. 641
Melnikov, A.V. 869, 889
Mencarelli, T. 1201
Mendez, P. 981
Meo, F. 901
Mercurio, G. 423
Merla, K. 1567
Merola, M. 275, 375, 419
Mertens, V. 101, 925
Meyer, I. 1015
Meyer, R. 1043
Michel, G. 545
Migliori, S. 1153
Mikhailov, N. 1059
Miki, N. 363
Milani, F. 949
Milburn, J.E. 993
Milechkine, I. 1285
Millard, J. 1709
Miller, A. 739
Miller, J.M. 1177, 1193
Mills, J. 665
Mills, S.F. 1681
Milora, S.L. 1153
Mineev, A.B. 833
Mingalev, B. 209
Miotello, A. 1447
Mironov, I.A. 825
Mishagin, V.V. 885
Mitchell, N.A. 1063, 1075, 1115
Mitin, D. 1125
Mitina, N.I. 351
Mito, T. 1023
Mitrishkin, Y. 731
Mitsunaka, Y. 549
Mitteau, R. 109, 435, 439, 443, 447
Mituyama, T. 1535
Miura, H. 1339, 1343, 1859
Miura, Y. 629
Miya, K. 791
Miyamoto, K. 677, 681, 693, 697, 701
Miyamoto, N. 693, 697, 701
Moal, A. 447
Möbius, A. 545

Moeller, C.P. 573, 629
Mogaki, K. 693
Mohri, K. 347, 1315, 1331
Moilanen, P. 395
Mollard, P. 713
Möllendorff, U.v. 1595
Monaco, F. 537
Mondino, P.L. 731, 1059
Montanari, R. 1607
Montvai, A. 557
Moons, F. 1467
Moormann, R. 327, 335
Morabito, C. 809
Morabito, F.C. 937
Moreau, R. 1527
Moreschi, L.F. 197, 295
Moret, F. 267
Moret, J.-M. 1107
Mori, K. 255
Mori, S. 1579
Morisaki, T. 403
Morley, N.B. 1303, 1539
Morra, M.M. 1115
Moshkin, S.A. 247
Motojima, O. 403, 945, 1019, 1023, 1027, 1133
Moulin, D. 909, 917
Mousdell, A. 1661
Mukhovatov, V. 849
Müller, G.A. 541
Müller, S. 339
Münich, M. 537
Muralidaran, C. 805
Murase, Y. 503
Muraviev, E.V. 251
Murdoch, D.K. 75, 1169
Murphy, G. 1091
Mustoe, J. 1871
Mutoh, T. 617

Na, H.K. 861
Nadkarni, A.V. 519
Nagakawa, J. 503
Naito, O. 215
Nakagawa, S. 363
Nakahira, M. 1083, 1347, 1403, 1649, 1653
Nakajima, H. 1083, 1115
Nakamichi, M. 1351, 1591
Nakamura, H. 1125, 1153,

1277
Nakamura, K. 307, 359
Nakanishi, K. 1027
Nakashima, Y. 1661, 1795
Nakazawa, N. 1591
Nannetti, C.A. 1379
Nardi, C. 1307, 1547
Nardoni, G. 1051
Nascimento, I.C. 581
Natalizio, A. 1719, 1819
Naumov, V. 697
Nedzelskij, I.S. 869, 889
Neffe, G. 1185
Neu, G. 925, 929
Neubauer, O. 209, 801
Neuhauser, J. 101
Neyatani, Y. 763
Nickel, H. 459
Nielsen, P. 897
Niiho, T. 1225, 1399, 1407
Nikonorov, A.I. 515
Nisan, S. 49, 1701, 1759, 1763
Nishida, K. 1499
Nishimura, K. 945, 1023
Nishio, S. 1859
Noda, K. 1455
Noda, N. 403
Noll, P. 751, 755
Nomura, G. 617
Norajitra, P. 1319
North, M.T. 487
Noterdaeme, J.-M. 597
Nöther, G. 999, 1015
Notkin, G.E. 833
Notkin, M.E. 1157
Novikov, V.P. 1149
Nowak, S. 941
Nunes, F.D. 877
Nunoya, Y. 1083
Nuutinen, S. 787

Obidenko, T. 209
O'Brien, D. 949
O'Connor, T. 669
Oda, Y. 1791
Ogawa, H. 945
Ogawa, T. 629
Oh, S.G. 861
Oh, Y.K. 861
Ohara, Y. 307, 693, 697, 701

Ohga, T. 693
O'Hira, S. 1277
Ohkubo, K. 553
Ohshima, K. 693
Ohyabu, N. 403
Oka, K. 1649, 1653
Oka, Y. 689
Okamoto, T. 553
Oksanen, M. 1391
Okumura, Y. 677, 693, 697, 701
Okuno, K. 1063, 1075, 1221, 1277
Oldaker, M.E. 669
Olstad, R.A. 573
O'Neill, R.C. 573
Onozuka, M. 1791
Oohara, H. 693
Oomens, A.A.M. 557
Opimach, I.V. 383
Orivuori, S. 787
Orlov, P.N. 515
Orsini, A. 355
Ortona, A. 1379
Osakabe, M. 689
Osaki, T. 1339
Ossiri, A. 1273
Ostrom, R. 989
Öström, R. 665
Ott, W. 641
Ottonello, G.B. 283
Ovchinnikov, I. 499
Oyama, Y. 857
Ozono, E. 581

Pacenti, P. 1265, 1273, 1705
Pacher, H.D. 275
Pacios, L. 981
Padligur, U. 921
Pain, M. 529, 533, 1867
Palmer, J. 1677
Palmieri, R.A. 287
Paméla, J. 685, 697
Panasenkov, A. 697
Panella, M. 973
Panin, A.G. 209, 829
Panteleev, L. 1237
Papastergiou, S. 431, 743
Papitto, P. 613
Park, H.K. 861

Parker, R.R. 33, 1315, 1331
Parlange, F. 969
Parshin, M. 1141
Parsons, W. 387
Parteder, E. 467
Pascual, C. 1661
Pascual, M. 1689
Pashkov, A.Yu. 1807
Patel, P.J. 657
Pater, S.L. 1665
Pathak, H.A. 805
Pautasso, G. 427
Peacock, A.T. 415, 483
Pearlstein, D. 731
Peeters, A. 537
Pégourié, B. 451, 1867
Pelzer, K.P. 845
Penaflor, B.G. 957, 965
Penco, R. 1335
Penningsfeld, F.-P. 641
Penzhorn, R.-D. 75, 1189, 1193, 1197, 1261
Peräniitty, M. 787
Perchermeier, J. 427
Pereslavtsev, P.E. 1595
Perevezentsev, A. 227, 1237
Perez, A. 569, 981, 1107, 1119
Périn, J.P. 1161
Perinic, G. 1015
Perujo, A. 327, 1415, 1439, 1443
Peruzzo, S. 767, 989
Pestchanyi, S.E. 191, 351, 507
Peterson, C. 1359
Petrizzi, L. 233, 1579
Petrov, V. 697
Petrov, V.B. 243
Petrov, V.S. 247
Petti, D. 1755
Petty, C.C. 625, 629
Peysson, Y. 203
Phelps, D.A. 629
Philibert, H. 1555
Philipps, V. 459
Piazza, G. 191, 1483
Pick, M. 415, 483
Piec, Z. 1063, 1661
Piel, D. 1319
Piet, S. 1755, 1795
Pillon, M. 233, 1555, 1563,

1571
Pilloni, L. 1387
Pillsticker, M. 1007
Pimanikhin, S.A. 1233
Pinna, T. 1779
Pinsker, R.I. 625, 629
Piosczyk, B. 545
Piovan, R. 1055
Pironti, A. 731, 735, 809
Pistunovich, V.I. 243
Pizzuto, A. 287, 291, 723, 813, 1111
Platnieks, I. 1531
Pleshakov, A.S. 243
Plöchl, L. 275, 439, 447, 467
Pochelon, A. 565
Podyminogin, A.A. 885
Poier, M. 1011
Poitevin, Y. 259, 1299
Polevoy, À.R. 833
Polman, R.W. 557
Polosukhin, B.G. 1415
Pomaro, N. 767, 775
Porfiri, M.T. 1759, 1779, 1783
Portafaix, C. 203
Portone, A. 731
Posadsky, I.A. 515
Poshekhonov, Y.Y. 821
Post, D. 1851
Potapenko, M.M. 1575
Pott, G. 331
Poucet, A. 1755
Pourrahimi, S. 185
Presle, P. 1095
Primakov, N.G. 1415
Prins, P.R. 561
Probst, F. 641, 645
Puppin, S. 949
Putvinski, S. 1851
Puzzolante, J.L. 1467

Quintric-Bossy, J. 259

Radway, T. 961
Raeder, J. 1755
Raff, S. 1043
Raffray, R. 197, 1315, 1331
Rahn, A. 1467, 1625
Raimondi, R. 1095
Raimondi, T. 747, 751, 755, 989
Rainer, F. 275
Rajasekar, K. 657
Rajendra Kumar, E. 805
Ramponi, G. 941
Ranga Nath, T. 805
Rapezzi, L. 1551
Raskob, W. 1771
Ratajczak, W. 1015
Ratel, G. 901
Rauch, J. 1111
Raupp, G. 925, 929, 933
Ravera, G.L. 613
Ravipragash, N. 805
Razuvanov, N.G. 1515
Rebelo, A.J.F. 1391
Reheis, N. 275
Reimann, G. 1319, 1323, 1423
Reimann, J. 1307, 1527, 1531
Reis, E.E. 319, 391, 475, 479, 759
Reitano, G. 809
Renner, H. 463, 467
Rey, G. 203, 633, 637
Reznichenko, P.V. 495
Riazantseva, N. 1285
Ricapito, I. 1201, 1213
Riccardi, B. 287, 291, 423
Riccardo, V. 751, 755, 1095
Ricci, M.V. 179, 1047
Richter, D. 1567, 1571
Riedl, R. 645
Riesch-Oppermann, H. 1323
Rietzschel, K. 921
Rimini, F.G. 705
Rita, C. 813
Rivkis, L. 1237
Riz, D. 677
Rizzello, C. 1209
Robert, A. 525
Robin, J.-C. 1763
Roccella, M. 291, 719, 723, 813
Rocco, P. 49
Roche, H. 443, 451
Rödig, M. 271, 275, 331
Rodrigo, L. 1177
Rodrigues, A.P. 167
Rogozin, A.I. 885
Röhrig, D. 1169
Rolfe, A.C. 91, 1685
Romanov, P.V. 247
Romer, O. 1487
Romero, J.A. 705
Rosenfeld, J.H. 487
Roshal, A. 1059
Rossi, L. 989
Rossi, S. 179, 1047
Rothan, B. 909
Rouleau, M. 1087
Roux, D. 529
Roux, N. 1411, 1471
Ruatto, P. 1519
Rubinacci, G. 809
Ruchko, L. 581
Rumyntsev, E.N. 821
Ruprecht, R. 1319
Rupyshev, A.S. 243
Ryder, R.H. 319
Ryle, C. 953
Ryter, F. 537

Saarelma, S. 787
Sabbioni, A. 1447
Sacchetti, M. 295
Sadakov, S. 1839
Safronov, V.M. 507
Sagara, A. 403
Sagawa, H. 1351, 1591
Saidoh, M. 471, 763, 1099
Saint-Antonin, F. 197, 267, 399
Saito, F. 553
Saito, M. 367, 371
Salto, Y. 553
Sakaki, K. 1023
Sakakibara, S. 945
Sakamoto, K. 549
Sakamoto, N. 407, 1499
Sakharov, N.V. 829
Saksagansky, G. 1141, 1169, 1233, 1285
Saksagansky, S. 1141
Sakurai, S. 471
Salavy, J.-F. 259, 1303, 1471
Salpietro, E. 179, 1047, 1051, 1063
Salvi, F. 295
Samaille, F. 1161
Sample, T. 1443
Sánchez, A. 709

Sandri, S. 1871
San Filippo, D. 267
Sangha, S.P.S. 415
Sannazzaro, G. 1347, 1839
Santagiustina, A. 989
Santamarina, A. 233, 1555
Santoro, R.T. 1579, 1583, 1587
Santos, J. 877
Sapper, J. 1031, 1371
Sardain, P. 1731, 1739
Sarigiannis, D.A. 1213, 1787
Sartori, F. 949, 989
Sartori, R. 949
Sasajima, T. 471, 763
Sasaki, S. 553
Sato, F. 857
Sato, K. 359
Sato, M. 553
Sato, S. 1339, 1339, 1343, 1587
Satoh, F. 693
Satoh, S. 307
Satow, T. 1019, 1023, 1027
Sauer, M. 459, 653
Sayer, R.S. 719
Sborchia, C. 1071, 1075, 1839
Scaffidi-Argentina, F. 1431, 1487, 1491
Scarcella, P. 1709
Scarinci, G. 1375
Schäfer, L. 1363, 1367
Schaffer, M.J. 759
Schalt, W. 653, 845
Schaubel, K.M. 1125, 1173
Schippl, K. 427
Schirra, M. 1363
Schittenhelm, H.P. 1015
Schleicher, S. 1641
Schleinkofer, G. 1015
Schleisiek, K. 1307, 1323
Schlosser, J. 109, 263, 447
Schmid, M. 621
Schmitt, F. 1261
Schmitt, R. 1359
Schnauder, H. 1547
Schneider, D. 783
Schneider, H. 427
Scholz, R. 419
Scholz, T. 779

Schrader, K.H. 1217
Schüller, F.C. 557
Schüller, P.G. 537, 541
Schultz, K.R. 1289
Schumacher, U. 541
Schuster, F. 1419
Schweikert, K. 1015
Schweizer, S. 455
Schwörer, K. 541
Scibile, L. 747
Scontrino, N. 1051
Scott, E. 1685
Scoville, J.T. 957
Sedano, L.A. 1415
Seidel, K. 233, 1567, 1571
Seidel, U. 427, 933
Seki, H. 693
Seki, M. 637
Seki, T. 617
Seki, Y. 1755, 1791, 1795, 1859
Senda, I. 731
Senohrabek, J. 1177
Serra, E. 1439
Serra, F. 865
Severi, Y. 259, 1539
Sevier, D.L. 475
Sevini, F. 375
Shakhovets, K.G. 833
Shannon, T. 1615
Shapkin, V.V. 243
Sharp, R.E. 1665
Shen, K. 1819
Sheppard, J. 1709
Shibanuma, K. 1633, 1649, 1653, 1657
Shibata, K. 697, 1799
Shikhovtsev, I.V. 885
Shimizu, K. 1347
Shimizu, M. 1099
Shimozuma, T. 553
Shin, N.S. 993
Shinbo, F. 617
Shiraishi, H. 503
Shlyahtenko, A. 1141
Shmayda, W.T. 1245, 1723
Shoji, M. 945
Shpanskij, Yu.S 247, 1515
Sibley, A.G.H. 621, 705
Sigalaev, V. 537

Signoret, J. 909
Sihler, Ch. 999, 1015
Sili, A. 1607
Silva, A. 865, 873
Simakov, A. 209
Simakov, S.P. 1575
Simon, K.H. 1197, 1261
Simon, M. 621
Simon-Weidner, J. 1031, 1035, 1371
Simonin, A. 681, 685
Sinha, P. 805
Sionov, A.B. 585
Sirén, M. 787
Sironi, M. 1633
Siuko, M. 1629
Skinner, N. 227
Skoblikov, S.V. 1137, 1141, 1145
Skripunov, V. 1141
Slagle, O.D. 1455
Smid, I. 263, 275, 291
Smith, D. 1685
Smith, J.P. 391, 475, 479
Smith, S. 185
Smolik, G.R. 1843
Sohda, Y. 359
Soikin, V.F. 829
Sokolov, Y.A. 243
Soksic-Kostic, M. 1595
Söldner, F. 705
Söll, M. 1003
Solomon, R.R. 407, 519
Soloviev, M.N. 1157
Solyakov, D.G. 351
Sonato, P. 767, 771
Song, W.S. 993
Sonnerup, L. 755
Sood, S.K. 1819
Sorokin, V.P. 1281
Sorolla, M. 709
Sousa, J. 167, 881, 977
Soussan, D. 1747
Spadoni, M. 179, 1047
Spannagel, G. 1257
Specht, E. 1015
Specht, T. 921
Speit, B. 1483
Sperger, Th. 597, 601
Speth, E. 641, 645, 649

Spiegel, H.J. 1015
Sprenger, D. 1483
Stagg, R. 227
Stankovsky, A.Yu. 1595
Stein, K. 1423
Stemke, R. 1289
Stepanov, B. 1067, 1075
Sterk, A.B. 557, 561
Sternini, E. 973
Stevenson, T.N. 669
Stigell, P. 1637
Stijkel, M. 1511
Stodilka, D. 1245
Stokes, R. 1673
Stoner, S. 697
Storai, S. 295
Stork, D. 431, 661
Stott, P.E. 157
Stöver, D. 343
Strebkov, Y. 1315, 1327, 1331
Streibl, B. 427, 455
Stubbe, E. 1715
Styhler, D. 801
Sublet, J.-Ch. 315
Sudo, S. 853, 1133, 1145
Suganuma, K. 637
Sugihara, M. 1125
Sugimoto, M. 1083
Sumner, R. 961
Suomela, J. 1637
Supe, A. 1507
Suppan, A. 1625
Surle, F. 203, 841
Süßer, M. 1015
Süss, R. 641
Suttrop, W. 101, 427
Suzuki, A. 1229, 1459
Suzuki, S. 197, 307, 359, 1019
Suzuki, Y. 553
Svensson, L. 893
Sviridov, V.G. 1515
Swain, D.W. 625
Szcepaniak, W. 645
Szczepanski, J. 259, 1303, 1471
Szulczyk, A. 179

Tabellini, M. 989
Tachikawa, N. 1331
Tada, E. 1347, 1403, 1633, 1649, 1653, 1657, 1661, 1709
Taguchi, K. 1649, 1653
Tähtinen, S. 395, 419, 787
Tai, E. 537
Takahashi, A. 857
Takahashi, K. 549, 1347
Takahashi, S. 471
Takahashi, T. 1019
Takahashi, Y. 185
Takanashi, T. 689
Takatsu, H. 511, 1315, 1327, 1331, 1339, 1343, 1403, 1587
Takeda, N. 1657
Takeiri, Y. 689
Takenouchi, T. 693
Takita, Y. 553
Talalaev, V.A. 1575
Tamura, H. 1019
Tanahashi, S. 1023
Tanaka, E. 837
Tanaka, S. 1229, 1451, 1459, 1463, 1495, 1503, 1507, 1535
Tanga, A. 989
Tanifuji, T. 1455
Taniguchi, M. 1495
Taniguchi, Y. 945
Tanii, M. 697, 701
Tarantini, M. 1709
Tarasyan, K.N. 889
Tareb, M. 203, 633
Tartaglia, G.P. 1395
Tatenuma, K. 1499
Taylor, N.P. 49, 1831
Tazhibaeva, I.L. 1415
Tcherednichenko-Alchevskiy, M.V. 1815
Tebus, V. 1237, 1811
Teller, G. 427
Tellini, B. 1523
Tenyaev, B.N. 1233
Terai, T. 1229, 1459, 1503, 1535
Teramoto, T. 367, 371
Tereshin, V.I. 351
Terlain, A. 1419
Terrani, S. 1265, 1273
Tesini, A. 1633, 1649, 1661, 1709
Thoener, M. 179
Thome, R.J. 1063, 1075
Thompson, D.A. 323
Thompson, E. 661
Thompson, H.M. 1871
Thouvenin, P. 533
Thumm, M. 533, 545
Tiliks, J. 1451, 1463, 1507
Tillack, M.S. 1855
Timms, M. 621
Timperi, A. 1629, 1637
Titus, P. 1075
Tivey, R. 279, 1633, 1641
Tobler, R.L. 1115
Toci, F. 1201
Toigo, V. 771, 1055
Tokami, I. 1339, 1343
Tonon, G. 203, 447, 633, 637, 1867
Tontegode, A.J. 589
Topilski, L. 1795
Toporkov, D.A. 507
Torres, T.A. 1289
Toschi, R. 1
Tosti, S. 1205, 1209
Toumi, I. 1759
Toyokawa, Y. 693
Trainham, R. 677
Tran, C. 533
Tran, M.Q. 533, 565, 569
Treutterer, W. 925, 929, 933
Trofimenko, Yu.V. 869, 889
Troxell, J.D. 519
Trubchaninov, S.A. 351
Trudel, C. 525
Tsarenko, A.V. 351
Tsaun, S.V. 833
Tsige-Tamirat, H. 1559, 1603
Tsitrone, E. 447, 451
Tsuchiya, K. 1225, 1399, 1407
Tsuji, H. 1063, 1083
Tsukamoto, H. 1083
Tsumori, K. 689
Tsunematsu, T. 1403
Tsuneoka, M. 549
Tsurumi, S. 471
Tubbing, B.J.D. 387, 705
Tuccillo, A. 973
Tuissi, A. 1645

Turck, B. 1043, 1867
Turner, A. 1641

Uchida, K. 1027
Uchimoto, T. 791
Uckan, N.A. 1851
Ueda, S. 1791, 1859
Ueda, Y. 1791
Uede, T. 1023
Uhlemann, R. 459
Ulbricht, A. 999, 1015
Ulrickson, M. 279
Umov, A.P. 1145
Unholzer, S. 1567, 1571
Urbanus, W.H. 561
Ustinov, A.V. 1515
Usui, K. 693
Utin, Y. 1347
Utsumi, T. 1587

Vaccari, M. 299
Vachon, L. 525
Vadgama, A.P. 665
Valente, P.L. 1335
Valenza, D. 1579, 1583
Vallcorba, R. 1161
van Amerongen, F.J. 557
van der Beken, H. 431
van der Laan, J. 1611
van der Laan, J.G. 1511
Van de Velde, J. 1467
van Houtte, D. 917, 969
Van Hove, W. 1715
van Veen, J. 747
Varandas, C.A.F. 167, 869, 881, 889, 977
Varela, P. 873
Vaßen, R. 343
Vasenin, S.G. 507
Vasiliev, V. 1067
Vassallo, G. 1269
Vayakis, G. 849
Vecsey, G. 185
Vedeneev, A.I. 1233, 1281
Velikhov, E.P. 833
Vella, G. 1583
Verdini, L. 355
Verhoeven, A.G.A. 557, 561
Verrecchia, M. 751
Versaci, M. 937

Vertiporokh, A.N. 515
Vertkov, A.V. 243
Vervier, M. 577
Vezzani, M. 1863
Viallet, H. 435
Vieider, G. 263, 275, 279, 291, 327, 355, 415
Vieira, R. 1063
Vilenius, M.J. 1629
Villar Colomé, J. 1739, 1867
Viniar, I.V. 495, 1137, 1141, 1145
Vinogradov, D. 557
Vinogradov, N.I. 833
Violante, V. 1205, 1209
Visca, E. 295, 355, 423
Vivaldi, F. 1661
Vivarlo, T. 1629
Völker, A. 1015
Vollmer, T. 1261
von Halle, A. 669
von Möllendorff, U. 1575, 1603
Vorobiev, G.M. 833
Voss, G.M. 1103

Wade, T. 621
Waganer, L.M. 1847
Wagner, D. 541
Wakisaka, Y. 1225
Walker, C.I. 849
Walker, K. 227
Walker, M.L. 731, 965
Wallura, E. 499
Walters, R.T. 1241
Ward, D.J. 731, 735
Warren, R. 227
Wasastjerna, F. 787
Watanabe, K.Y. 693, 697, 701, 945
Watari, T. 553, 617
Watarumi, K. 1407
Watson, M. 697
Watson, R.D. 487
Weckermann, B. 419
Wedemeyer, O. 1423
Wegrowe, J.G. 203, 633, 1867
Weigand, W. 999
Weimaar, P. 1487
Weisenburger, A. 1483
Weisse, J. 1867

Weissgerber, M. 427
Wendel, J. 1197, 1261
Werle, H. 327, 1431, 1463, 1483, 1487
Werner, B. 331
Wesley, J. 719, 731, 1851
Wesner, F. 597
West, P.W. 383
Whyte, D. 383
Wiechers, B. 271
Wiffen, F.W. 1615
Wijnands, T. 917, 969
Williams, M.D. 135
Williamson, D. 719, 1331
Willms, R.S. 1153
Wilson, D.J. 897
Wilson, P.P.H. 1595, 1599
Woll, D. 1595
Wong, F.M.G. 1075, 1115
Wu, C.H. 327, 335, 1751
Wu, Y. 1571, 1575
Wüchner, F. 999, 1015
Würz, G. 921
Würz, H. 191, 351, 507
Wykes, M.E.P. 1079, 1661, 1795

Yagnov, V.A. 821
Yagyu, J. 763
Yamada, H. 945, 1133
Yamada, M. 849, 1331
Yamada, S. 1023
Yamagiwa, T. 1027
Yamaguchi, K. 1351
Yamaguchi, S. 945, 1027
Yamaki, D. 1455
Yamamoto, J. 1019, 1023, 1027
Yamamoto, M. 693
Yamamoto, N. 503, 945
Yamamoto, S. 849
Yamamoto, T. 1027
Yamamura, C. 1591
Yamanishi, T. 1221, 1277
Yamashita, Y. 1799
Yamauchi, M. 1795
Yamauchi, Y. 255
Yamazaki, K. 945
Yamazaki, S. 1859
Yamazaki, T. 693

Yanagi, N. 1027
Yang, H.L. 861
Yang, J.G. 861
Yasukawa, Y. 1023
Yasutomi, S. 553
Yehia, A. 343
Ylikorpi, T. 1637
Yokoyama, K. 307
Yoneoka, T. 1535
Yoon, B.J. 993
Yoon, N.S. 861
Yorkshades, J. 227
Yoshida, K. 1075, 1661
Yoshimuta, S. 1407

You, J.H. 459
Youchison, D.L. 487
Young, D. 893

Zaccaria, P. 771
Zacchia, F. 283
Zahn, G. 1015
Zakhartsev, V. 1811
Zaluzhnyi, A.G. 1815
Zampelli, P. 613
Zanza, V. 973
Zasche, D. 925, 929
Zavialsky, L. 1615
Zehetbauer, T. 925, 929

Zhelamskij, M. 185
Zhitlukhin, A.M. 507
Zimeleva, L.G. 889
Zimmermann, E. 845
Zohm, H. 101
Zollino, G. 767, 771, 775
Zolti, E. 1315
Zornig, N.H. 705
Zucchetti, M. 49
Zullo, G. 989, 1087
Zwecker, V. 1015
Zyrionov, A. 1811